Human Anatomy and Physiology

Donna Van Wynsberghe

Professor and Chair of the Department of Biological Sciences
University of Wisconsin–Milwaukee

Charles R. Noback

Professor Emeritus of Anatomy and Cell Biology
College of Physicians and Surgeons, Columbia University

Robert Carola

McGraw-Hill, Inc.

New York   St. Louis   San Francisco   Auckland   Bogotá   Caracas   Lisbon   London   Madrid
Mexico City   Milan   Montreal   New Delhi   San Juan   Singapore   Sydney   Tokyo   Toronto

ABOUT THE AUTHORS

Donna Van Wynsberghe is Professor and Chair of the Department of Biological Sciences at the University of Wisconsin–Milwaukee. She is the recipient of two Distinguished Teaching Awards. Her articles have appeared in various physiology journals, and she has previously authored case history and study manuals for undergraduate students. Dr. Van Wynsberghe is a member of several professional societies for physiology research and teaching, including the American Physiological Society (APS) and the Human Anatomy and Physiology Society (HAPS), and she holds editorial positions in both these societies.

Charles R. Noback has been Professor of Anatomy and Cell Biology at the College of Physicians and Surgeons of Columbia University for 40 years. Dr. Noback is the coauthor of several textbooks, including *The Human Nervous System*, Third Edition, published by McGraw-Hill, and *The Nervous System: Introduction and Review*, Fourth Edition, published by Lea & Febiger. He is also the author of numerous publications of original researches in anatomy and neurobiology.

Robert Carola is a science writer who has written six textbooks as well as a variety of materials for many corporate clients. He is a member of the National Association of Science Writers.

HUMAN ANATOMY AND PHYSIOLOGY

Copyright ©1995, 1992, 1990 by McGraw-Hill, Inc. All rights reserved. Printed in the United States of America. Except as permitted under the United States Copyright Act of 1976, no part of this publication may be reproduced or distributed in any form or by any means, or stored in a database or retrieval system, without the prior written permission of the publisher.

This book is printed on acid-free paper.

1 2 3 4 5 6 7 8 9 0   VNH   VNH   9 0 9 8 7 6 5 4

ISBN 0-07-011171-5

This book was set in Janson by York Graphic Services, Inc. The editors were Kathi M. Prancan, Deena Cloud, and Holly Gordon; the designer was Gayle Jaeger; the cover illustration was done by Cynthia Turner; the production supervisor was Janelle S. Travers. The photo editor was Kathy Bendo; the photo researcher was Barbara Salz. Von Hoffmann Press, Inc., was printer and binder.

INTERNATIONAL EDITION

Copyright 1995. Exclusive rights by McGraw-Hill, Inc. for manufacture and export. This book cannot be re-exported from the country to which it is consigned by McGraw-Hill. The International Edition is not available in North America.

When ordering this title, use ISBN 0-07-113540-5.

Library of Congress Cataloging-in-Publication Data

Van Wynsberghe, Donna.
  Human anatomy and physiology.—3rd ed. / Donna Van Wynsberghe, Charles R. Noback, Robert Carola.
    p.   cm.
  Rev. ed. of: Human anatomy and physiology / Robert Carola, John P. Harley, Charles R. Noback. 2nd ed. c1992.
  Includes bibliographical references and index.
  ISBN 0-07-011171-5
  1. Human physiology.   2. Human anatomy.   I. Noback, Charles Robert, (date).   II. Carola, Robert.   Human anatomy and physiology.   III. Title.
  [DNLM: 1. Anatomy.   2. Physiology.   QS 4 V2258h 1995]
QP36.C28   1995
612—dc20
DNLM/DLC
for Library of Congress   94-25554

This book is dedicated to

Mary Jane Werderman, Eva Werderman, and Jeanne Seagard
Peter, Norma, and Teddy Noback
Ann Carola

Chapter-Opening Photographs

1 chromosomes in dividing cell, **2** crystals of the amino acid tyrosine, **3** human cheek cells, **4** human elastic cartilage, **5** basal cells of epidermis, **6-9** bone tissue, **10-11** skeletal muscle fibers, **12-16** neurons in the cerebral cortex, **17** rods and cones in human retina, **18** human thyroid tissue, **19-21** red blood cells, **22-23** lymphocytes, **24** capillary networks around alveoli of lung, **25** stomach tissue, **26** citric acid crystals, **27** urea crystals, **28** sodium chloride crystals, **29** ciliated epithelium of uterine tube, **30** computer-generated model of DNA molecule

Photo credits begin on page C.1.

oo- egg (*oocyte:* egg cell).

orth(o)- straight, normal, correct (*orthodontic:* pertaining to the straightening of teeth).

oste(o)- bone (*osteitis:* inflammation of the bone).

ot(o)- ear (*otology:* study of the ear).

ovi-, ovo- egg (*ovum:* egg cell).

para- beside, beyond (*paraspinal:* beside the spine).

path(o)- disease, suffering (*pathogenic:* capable of producing disease).

ped(ia)- child (*pediatrician:* physician who specializes in childhood diseases).

pend- hang down (*appendicular skeleton:* portion of the skeleton that hangs from the pectoral and pelvic girdles).

per- through (*permeate:* to pass through).

peri- around (*pericardium:* membrane surrounding the heart).

phag(o)- eat, consume (*phagocyte:* cell that ingests other cells or particles).

phleb(o)- vein (*phlebitis:* inflammation of a vein).

pleuro-, pleura- rib, side (*pleurisy:* inflammation of pleura, the membrane covering the lungs and lining the thoracic sac).

plur(i)- many, more (*pluriglandular:* several glands).

pneumo-, pneumono- lung, respiratory organs (*pneumonia:* chronic lung infection).

pod- foot (*podiatrist:* a specialist in care and treatment of the feet).

poly- many (*polycystic:* having many cysts).

pro- for, in front of, before (*prophylaxis:* preventive treatment, to keep guard before).

proct- rectum, anus (*proctoscope:* instrument for examining the rectum).

pseudo- false (*pseudocyst:* false cyst).

pulmo- lung (*pulmonary:* pertaining to lungs).

py(o)- pus (*pyocyst:* pus-filled cyst).

quadr(i)- four (*quadriplegia:* paralysis of all four limbs).

re- back, again (*reflex:* bend back).

ren- kidney (*adrenal:* relating to the position of the adrenal gland atop the kidney).

retro- behind, backward (*retrograde:* going backward).

rhin(o)- nose (*rhinitis:* inflammation of nasal mucous membranes).

sclero- hard (*sclerosis:* hardening of tissue).

steno- contracted, narrow (*stenosis:* condition of being narrowed).

super- above (*superior:* toward the head).

supra- above, over (*suprapubic:* above the pubic bone).

sym-, syn- together, union (*synapse:* meeting between two or more nerve cells).

tachy- fast (*tachycardia:* abnormally fast heart rate).

tel(o)- end (*telephase:* final phase of mitosis).

tetr(a)- four (*tetrad:* group of four chromosomes formed during meiosis).

therap- treatment (*therapeutic:* having a healing effect).

therm(o)- heat, warmth (*thermometer:* device to measure heat).

thorac(o)- chest (*thoracic cavity:* chest cavity).

thromb(o)- lump, clot (*thrombosis:* formation of a blood clot in a vessel or heart cavity).

tox- poison (*toxemia:* blood poisoning).

trans- across, through (*transepidermal:* occurring through or across the skin).

ultra- excessive, extreme (*ultrasonic:* sound waves beyond the audio-frequency range).

un(i)- one (*unilateral:* one side).

uria-, uro- urine, urinary tract (*polyuria:* production of excess urine).

vas(o)- vessel (*cardiovascular:* vessel related to the heart).

viscera-, viscero- organ (*visceral:* pertaining to a body organ).

SUFFIXES (with examples)

-ac pertaining to (*celiac:* pertaining to the abdominal region).

-ad toward, in the direction of (*cephalad:* toward the head).

-agra severe pain, seizure (*podagra:* severe pain in the foot).

-al pertaining to (*digital:* pertaining to the finger or toe).

-ar of or relating to, being, resembling (*valvular:* relating to a valve).

-ase enzyme (*lactase:* enzyme that catalyzes conversion of lactose into glucose and galactose).

-ate that which is acted upon, marked by having, to act on (*substrate:* substance acted on by enzyme; *lobate:* having lobes; *separate:* to keep apart).

-atresia abnormal closure (*proctatresia:* closed anus).

-blast sprout, growth (*osteoblast:* cell from which bone develops).

-cele swelling, tumor (*cystocele:* hernia of urinary bladder).

-centesis puncture of a cavity (*paracentesis:* puncture of a space around an organ or within a cavity to remove fluid).

-cide kill (*germicide:* killer of germs).

-cis cut (*excise:* cut out).

-clasis, -clasia breaking up (*bacterioclasis:* breaking up of bacteria).

-dynia pain (*pleurodynia:* pain on the side of the chest).

-ectasia, -ectasis dilation, expansion (*telangiectasis:* dilation of capillaries).

-ectomy excision, cutting out (*laryngectomy:* removal of larynx).

-emia condition of the blood (*leukemia:* blood condition characterized by excess leukocytes).

-ferent carrying (*efferent:* carrying away from).

-form structure, shape (*ossiform:* resembling the form of bone).

-fugal driving or traveling away from (*centrifugal:* moving away from the center).

-gen, -gene producer (*mutagen:* substance that increases the frequency of mutation).

-genesis production of, origin of (*glycogenesis:* production of glycogen).

-ia condition (*anuria:* condition of lack of urine).

-iasis a diseased condition (*cholelithiasis:* condition of bile stones).

-ic pertaining to (*colonic:* pertaining to the colon).

-ile having the qualities of or capability for (*febrile:* feverish).

-ism condition, act of, process of (*dwarfism:* condition of being a dwarf).

-itis inflammation (*phlebitis:* inflammation of a vein).

-logy study of (*cardiology:* study of the heart).

-lysis breaking down (*glycolysis:* breakdown of glucose).

-malacia softening (*osteomalacia:* softening of bone).

-meter instrument or means of measuring (*thermometer:* instrument to measure heat).

-ness quality of, state of (*illness:* state of being ill).

-oma tumor (*carcinoma:* malignant tumor).

-opia eye disorder (*myopia:* nearsightedness).

-ory process of, pertaining to, function of (*sensory:* pertaining to the senses).

-ose full of, having the qualities of (*comatose:* having the qualities of a coma).

-osis action, state, process, condition (*halitosis:* condition of having bad breath).

-ostomy creation of an opening (*colostomy:* creation of a new opening between the bowel and abdominal wall).

-otomy cutting into (*tracheotomy:* cut into the trachea).

-ous, -ious having the qualities of, capable of, full of (*infectious:* capable of being transmitted).

-pathy disease (*cardiopathy:* heart disease).

-penia deficiency, lack (*leukopenia:* lack of white blood cells).

-pexy fixation (*sigmoidopexy:* fixation of large intestine).

-philia love of, tendency toward (*hemophilia:* "love of blood," a blood disease).

-phobia abnormal fear (*claustrophobia:* fear of confinement).

-plasty molding or shaping (*osteoplasty:* plastic surgery on bone).

-plegia paralysis (*paraplegia:* paralysis of lower half of body).

-poiesis production, formation of (*hemopoiesis:* production of blood cells).

-rrhagia bursting forth, excessive discharge (*hemorrhage:* escape of blood from vessels).

-rrhaphy closure of by suturing, repair (*cystorrhaphy:* suture of bladder).

-rrhea flow, discharge (*galactorrhea:* excessive or spontaneous flow of milk).

-sect cut (*dissect:* cut into parts).

-sis process, action (*dialysis:* separation of particles through a semipermeable membrane).

-stasis state of being at a standstill (*hemostasis:* stopping of blood flow).

-sthenia strength (*asthenia:* loss of strength).

-stomy surgical opening (*colostomy:* surgical opening in the colon).

-tomy to cut (*lobotomy:* incision into a lobe).

-tonia stretching, putting under tension (*hypertonia:* excessive tension).

-tripsy rubbing, crushing (*lithotripsy:* surgical crushing of kidney stones).

-trophic related to nutrition, growth, development (*dystrophic:* faulty nutrition).

-tropic turning toward, changing (*hydrotropic:* turning toward water).

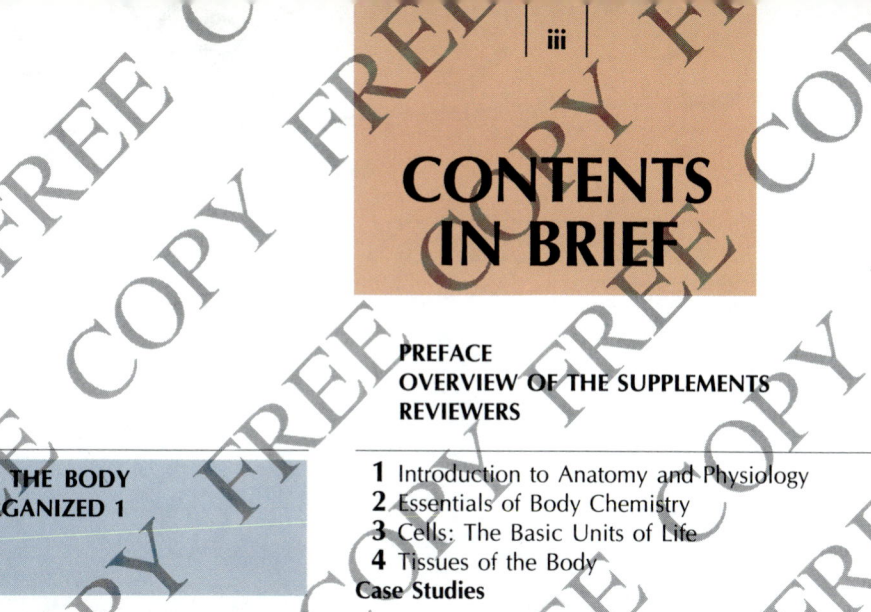

CONTENTS IN BRIEF

HOW THE BODY IS ORGANIZED 1

I

PROTECTION, SUPPORT, AND MOVEMENT 131

II

CONTROL, COMMUNICATION, AND COORDINATION 351

III

TRANSPORT AND MAINTENANCE 583

IV

REPRODUCTION AND DEVELOPMENT 939

V

APPENDIXES

| iv |

CONTENTS

PART II PROTECTION, SUPPORT, AND MOVEMENT 131

11 THE MUSCULAR SYSTEM 299

CASE STUDIES FOR PART II 350

PART III CONTROL, COMMUNICATION, AND COORDINATION 351

12 NERVOUS TISSUE: AN INTRODUCTION TO THE NERVOUS SYSTEM 352

13 THE CENTRAL NERVOUS SYSTEM I: THE BRAIN 384

14 THE CENTRAL NERVOUS SYSTEM II: THE SPINAL CORD 428

18 THE ENDOCRINE SYSTEM 547

CASE STUDIES FOR PART III 582

PART IV TRANSPORT AND MAINTENANCE 583

19 THE CARDIOVASCULAR SYSTEM: BLOOD 584

26 | **METABOLISM, NUTRITION, AND THE REGULATION OF BODY HEAT 850**

27 | **THE URINARY SYSTEM 887**

28 REGULATION OF BODY FLUIDS, ELECTROLYTES, AND ACID-BASE BALANCE 919

CASE STUDIES FOR PART IV 938

PART V REPRODUCTION AND DEVELOPMENT 939

29 THE REPRODUCTIVE SYSTEMS 940

PREFACE

Ordinarily, the topic of evolution is not emphasized in a textbook about anatomy and physiology. In this case, however, we would like to talk about the evolution of this book, from its first edition (1990) to this third edition (1995). Like evolution, textbook writing is a process that takes time; we believe that, for *Human Anatomy and Physiology*, each succeeding edition has been a refinement of the previous that better suits it to the learning environment.

A New Author, A New Order

In very simplistic terms, evolution can be described as "change over time." A significant change since the last edition of *Human Anatomy and Physiology* is the addition of a new coauthor to our team, Donna Van Wynsberghe. Dr. Van Wynsberghe, a physiologist, is a professor at the University of Wisconsin — Milwaukee and the chair of the Department of Biological Sciences. During her 20-year teaching career she has taught anatomy and physiology to literally thousands of students. By enhancing the coverage of physiological concepts, incorporating contemporary topics and examples of significance to health professionals, and ensuring the accuracy of text and art, Donna's contribution to the third edition has been invaluable. Her listing as first author on the cover and title page underscore that contribution.

As in previous editions, Dr. Charles R. Noback has made significant contributions to this text. Dr. Noback's skill and experience in teaching and research in anatomy and neurobiology are evident in many places throughout the book. A sample of his special approach to using graphics to elucidate complex anatomical structures can be seen on pages 406–407, where he and noted medical illustrator Robert J. Demarest have combined their talents to conceive five original drawings that illustrate the interior structure of the brain. Illustrations that place an emphasis not just on anatomical detail but on structural relationships can be found routinely throughout this edition. For the student encountering the vast amount of detail inherent in this course, this approach becomes an effective learning device.

Interesting and clear exposition has always been a strength of *Human Anatomy and Physiology*, and it remains so in this edition. Robert Carola has applied his skill and experience as a science writer and his innate interest in the human body and in health-related matters to each edition of this text. Writing each chapter with the goal of motivating the student to read further and integrating features such as word derivations, question-and-answer boxes, and When Things Go Wrong sections to pique a student's interest, he has enabled thousands of students to learn anatomy and physiology efficiently, effectively, and even enjoyably.

New to This Edition

As a result of feedback from users of previous editions, reviewers, and focus-group participants, we have made several changes and added a number of features in this edition.

A Reorganization

We have reorganized the chapters on the nervous system to reflect the way most instructors teach this difficult area of anatomy and physiology: The chapter on the nervous tissue is followed by one on the central nervous system, which is now covered in two chapters (the brain and the spinal cord). These chapters are followed by a new chapter on the peripheral nervous system, and the section ends with chapters on the autonomic nervous system and the senses.

Expanded Coverage of Physiology

A more balanced presentation of physiology enhances the outstanding coverage of basic anatomy this text has always been noted for. To support the textual information, this edition includes new physiological flowcharts that incorporate icons designed to help the student associate structures with the functions being described.

Updated Basic Science and Clinical Applications

Many sections of the text are either newly written or extensively updated. See, for example, Physiology of Muscle Contraction (page 278), Physiology of Neurons (page 363), Electrical Conducting System of the Heart (page 629), and Cardiac Cycle (page 633).

Several new boxes cover timely topics of clinical relevance to this audience. See, for example, "Replant Surgery of a Severed Body Part" (page 294) and "Excess Iron and Heart Disease" (page 646). Other boxes deal with up-to-date developments in sports physiology: See "Two Million Replaceable Joints" (page 259), "Pitching off the Cuff" (page 326), "New Hope for Spinal Cord Injuries" (page 446), and "Athletes and Anabolic Steroids" (page 564).

Pedagogical Adaptations

A unique presentation of the menstrual cycle places the ovarian and uterine events occurring simultaneously in a side-by-side format, making the correlation of events easier for the student to understand (page 958).

Practical case studies relating to real-life situations and requiring critical thinking on the part of the health professional have been added as a new feature at the end of each major section of the text. See page 130, for example.

Focus on Careers

New part openers describe a variety of career opportunities available for the future health professionals taking this course.

Features Retained and Enhanced

Enhanced Illustration Program

The illustrated summaries that so many of our previous adopters found to be effective learning tools for their students have been retained and increased in number. Careful and consistent placement of relevant text, art, and tables on the same page or on a two-page spread organizes the information for the student, providing clear explanations both textually and visually. See pages 668–669 and 678–679 in the chapters on the cardiovascular system for examples of this.

Photos of surface anatomy have been moved from the appendix into their relevant chapters for immediate student reference and reinforcement of the information being presented on the internal anatomy.

As before, all illustrative elements are color-coded so that a given structure or system is the same color throughout (related structures are different shades of the same color); for easy reference, table heads and various text elements also carry out this color-coding. As with previous editions, each new body system is introduced by an overview of that system, illustrating its major organs and their functions. System overviews are denoted by an icon: ★

Word Origins and Pronunciations
Because they provide simple and interesting links between words and their meanings, the inclusion of word origins has always been a popular feature of this text. In response to users and reviewers, the authors have expanded their incorporation of word origins and pronunciations.

Boxed Essays
In the last edition, three different types of boxed essays—Medical Update, To Your Health, and The Dynamic Body—were used to present interesting and relevant information in a meaningful way. In this edition we have replaced some of those boxes and updated others; we have also added a new type of box, Anatomy in Action, where the focus is on anatomical aspects of various human physical activities.

Questions/Answers
Interspersed throughout all chapters are question-and-answer boxes that present nontechnical, commonly asked questions about simple events of the body. This feature has been revised and expanded in this edition.

When Things Go Wrong
An abridged description and explanation of the many diseases and disorders that affect each system of the body appear in this section at the end of each chapter. Many conditions that have recently become problematic to human health have been added in this edition. See, for example, the discussion of computer keyboard stress syndrome on page 345.

Study Aids for Review
The extensive, end-of-chapter material has been rewritten and includes more "critical-thinking" questions in Understanding the Concepts. Answers to all questions have been carefully keyed to the text and appear in an appendix at the end of the text.

Other Features
This edition incorporates all the pedagogical features that proved so popular in the earlier editions:

Chapter-opening pages provide a chapter outline, which gives an overview of chapter topics, and a list of key concepts on which the student should focus while reading the material that follows.

Where clinical applications are integrated into the text, the caduceus symbol in the adjacent margin calls the reader's attention to these interesting observations: ⚕

The Ask Yourself boxes that occur at intervals throughout the text enable students to stop and review pertinent material before proceeding.

Developmental Anatomy sections at the end of selected chapters incorporate this aspect of various body systems for instructors who wish to integrate such material into their courses.

Chapter summaries are presented at the end of each chapter in an itemized format that includes page numbers for each main section for easy reference.

To the Student

These final years of the twentieth century have been designated as "the decade of the brain" in recognition of the major advances recently made in the field of neuroscience.

It's a good time to *think about thinking*. As you do, reflect on these words of Nobel laureate George Beadle:

> *However much we may want someone to make up our minds for us, there is no escape from doing it ourselves—if we want to preserve [and improve] the structure of our society. . . . Happily, we have available to us a splendid tool for this purpose: the human brain. It enables us not only to accumulate knowledge but also to modify behavior in accordance with what we have learned. It took us a million years of trial and error to forge it, but the finished product is a beauty. All we have to do now is use it.* (George Beadle and Muriel Beadle, The Language of Life. *New York: Anchor Books, 1967, p. 247)*

Acknowledgments

We feel that the artwork in *Human Anatomy and Physiology* is extraordinary, and we would like to thank the following artists for achieving such a high level of excellence: Marsha J. Dohrmann, Carol Donner, John V. Hagen, Neil O. Hardy, Steven T. Harrison, Jane Hurd, Joel Ito, and George Schwenk.

The authors would also like to thank the host of people who assisted them in making this edition even better than before. Gail Patt of Boston University wrote all the part-opening text, revised the Study Aids for Review sections at the end of each chapter, and authored several of the supplements. Her efforts in improving the correlation of study material to text material are greatly appreciated. Thanks go also to Holly Gordon, Kathy Bendo, Denise Schanck, Gayle Jaeger, Deena Cloud, and Kathi Prancan of McGraw-Hill.

Donna Van Wynsberghe
Charles R. Noback
Robert Carola

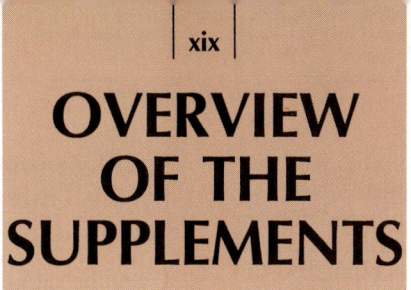

OVERVIEW OF THE SUPPLEMENTS

The following ancillary materials are available for instructors upon request, as are copies of the student materials. Ancillaries designed specifically for students are available for their purchase.

FOR THE INSTRUCTOR

Instructor's Resource Materials
by Gail R. Patt

Newly developed for this edition, the *Instructor's Resource Materials* offers the instructor a unique package of teaching aids. This supplement is designed around an extensive Lecture Outline that begins with a detailed introduction and overview of each chapter. The outline then walks the lecturer through the topics covered in the chapter, building on knowledge and synthesizing information. Integrated throughout the outline are various marginalia on additional resources that could be used to supplement the successful lecture. The instructor will find references to other supplements available — *Overhead Transparencies* and *Selected Line Art.* There are also special notes within the Outline featuring interesting teaching hints. Answers to Understanding the Facts and Understanding the Concepts, found in the text, are also presented.

Available with the *Instructor's Resource Materials* are tabbed file folders for each of the 30 text chapters. Instructors may order separately the individually packaged *Test Bank Manual, Selected Line Art,* and *Overhead Transparencies* and incorporate all materials into the folders for a handy chapter-by-chapter resource package.

Test Bank Manual
by Gail R. Patt

This manual has been completely rewritten for this edition and contains more than 75 questions per chapter for each of the 30 text chapters. Multiple-choice, matching, and fill-in questions are designated as either memory or reasoning exercises, depending on the particular student skills being tested. Clinical questions are also presented in a separate section for each chapter to test students' critical-thinking abilities. Answers are provided for all questions. The test bank is also available on diskette for IBM PC and Macintosh computers.

Selected Line Art

Black line drawings of major illustrations in the text are provided as part of a separate package. Labels and leaders have been removed from all pieces so that the instructor can make use of the figures in a variety of ways. Selected line art masters can be copied for handouts or for quizzes and can be made into overheads for lecture.

Guide for the *Slice of Life VI* Videodisk
by Barbara Cocanour and Patrick Scollin

The authors have provided a handy index to aid in accessing the more than 38,000 images available on the *Slice of Life VI* videodisk. Arranged by chapter headings that correspond to the text, barcodes and frame numbers are provided to call up images from various health-related disciplines: anatomy, hematology, histology, embryology, microbiology, pathology, radiology, and other clinical areas.

Instructor's Manual for the Laboratory Manuals
by Ted Namm, Alease S. Bruce, Barbara Cocanour, and Joseph P. Farina

This manual provides the laboratory instructor with valuable suggestions for the most efficient utilization of exercises within the time frame of the laboratory. Each chapter lists supplies and equipment, includes comments on teaching the exercises, and gives answers to the Study Questions.

Overhead Transparencies

Three hundred full-color transparencies of important illustrations, photographs, and electron micrographs from the text are available. Lettered titles and labels are consistently large and bold so that they can be viewed easily, even from the back of a large lecture hall. For many figures, the number of labels has been selectively limited so students are presented with simple and concise images.

Electronic Overhead Transparencies

Approximately 500 images of text art are available on CD-ROM for Macintosh and IBM. These images can be projected via computer to supplement your lectures or lab course, or students can view them directly on their computers.

Interactive Videodisk Software
by John Dustman

This software has been designed to quickly and easily access videodisk images and correlate them with specific chapters in the text. The *Slice of Life VI, Small-Animal Dissection* videodisk, and *Histology: A Photographic Atlas* can be easily manipulated to call up their contents for use in lecture or lab. In addition, the software offers a tutorial feature. For each chapter, various activities reinforce concepts and visual elements of the text. Included in the program are a glossary of terms and a testing component. If students get stuck at any point, they can access textual information that will help them proceed. The software is available for Macintosh and can be used with or without a videodisk.

Human Anatomy and Physiology Image Library

Available on Macintosh CD-ROM and on Macintosh 3.5-in. floppy disk, this supplement offers a database of over 600 images that allows instructors to create their own tutorials, transparencies, tests, and classroom handouts.

Small-Animal Dissection Videodisk
by the University of Delaware, Instructional Technology Group

Using the cat as a dissection animal, this interactive videodisk program is designed as a visual database for students to learn dissection techniques and the components of various body systems. Nearly 60 minutes of video footage show detailed dissections, still images, illustrations, and animations.

Voluntary Movement Videocassette

A videotape that includes basic concepts of bone and muscle anatomy and physiology is available to adopters.

Metabolism and Homeostasis Videocassette

A videotape that includes basic concepts of metabolism and homeostasis is available to adopters.

FOR THE STUDENT

Student's Study Manual
by Gail R. Patt

Newly developed for this edition, the *Student's Study Manual* features, in each chapter, an Active Reading section, which walks the student through the chapter, using a variety of question formats, key terms to test student knowledge of important terms and concepts, a Post-Test section that uses miscellaneous multiple-choice and matching questions to help students assess their proficiency with chapter material, integrative thinking questions that ask the student to pull together various information from the chapter to answer real-life clinical and application questions, and Your Turn, where students are encouraged to work in study groups to carry out increasingly difficult tasks, from writing their own test questions, to organizing a Jeopardy competition, to discussing issues such as the effects of smoking.

Laboratory Manuals
by Ted Namm, Alease S. Bruce, Barbara Cocanour, and Joseph P. Farina

The lab manuals present student-tested laboratory exercises designed to accommodate laboratory sessions of various focuses. The chapters are organized into five major categories: levels of organization; protection, movement, and support; control and integration; homeostatic systems; and continuity of life. Each chapter begins with a complete list of objectives followed by a general introduction. The material in each chapter, designed for laboratories of 3-hour length, is subdivided into exercises that may be adapted to shorter laboratory sessions. Each chapter ends with a selection of comprehensive Study Questions.

There are three versions of the lab manual. One (cat version) includes dissection exercises that use the cat as the principal specimen, one (fetal pig version) uses the fetal pig as the dissection specimen, and one (brief version) does not include dissection exercises at all. This latter version is intended for courses that focus on the human as a specimen and that use models, charts, and even prosected cadavers for demonstrating structures and systems.

Flash Cards
by Barbara Cocanour

Two-hundred flash cards are packaged in a separate box for students to self-test their understanding of muscles, skeletal components, nerves, and blood vessels.

Activities Manual
by Barbara Cocanour and William Farrar

This manual is a self-study tool that offers a coloring book, crossword puzzles, quotation puzzles, and anatomical flash cards. These are designed to help students achieve mastery of anatomical and physiological information. The coloring book includes modified illustrations from the text to be colored and labeled. Brief definitions and descriptions as well as a self-test are included for each illustration. The crossword puzzles appear within each section and provide an entertaining way for students to recall important definitions. A quotation puzzle appears at the end of each unit and combines clues from all the crossword puzzles in that unit. The flash cards, also available separately, are designed to review muscles, skeletal components, nerves, and blood vessels.

Diagnostic Reasoning Software
by Hurley Myers, Eldon Benz, and Kevin Dorsey, M.D.

This software provides five cases that simulate five separate patient-student encounters. It is designed to teach and evaluate critical-thinking skills to undergraduate allied health students. Students can apply their basic knowledge of anatomy and physiology to patient-related situations. Students will have an opportunity to interview the patient, perform certain aspects of a physiological examination, and receive results of tests that help to give an understanding of the causes of the problem and its treatment.

Radiographic Anatomy: A Working Atlas
by Harry W. Fischer, M.D.

This is a complete atlas of human anatomy as seen through today's imaging technologies. Approximately 200 beautifully reproduced radiographs, late-generation CT scans, and high-resolution NMR images are accompanied by clearly labeled drawings that highlight the anatomical structures that beginning anatomy and physiology students must know. This book was especially designed to help students relate their knowledge of gross anatomy to the clinical images they will see in the hospital or clinic environment. The emphasis throughout is on frequently used images and imaging modalities, making *Radiographic Anatomy* an indispensable aid for all students of gross human anatomy.

Radiographic Neuroanatomy: A Working Atlas
by Harry W. Fischer, M.D., and Leena Ketonen

This complete atlas of human neuroanatomy teaches students to correlate their knowledge of gross anatomy with its representation by modern imaging technology. The study of the central nervous system is founded on mastery of its anatomy. Plain radiographic images of the skull and spine are included, but the majority of images come from angiography and magnetic resonance imaging. Over 200 images are included with accompanying, clearly labeled drawings.

Atlas of Human Anatomy
by Frank H. Netter, M.D.

This atlas includes more than 500 beautifully rendered, full-color illustrations from the medical illustrator Frank Netter. Gross anatomy illustrations are classified and organized by system and region. They are clearly labeled with the most current anatomical terminology and nomenclature; a suitable level of labeling has been achieved to make these accessible to students and members of the medical and allied health professions alike.

REVIEWERS

Reviewers of the third edition

D. Andy Anderson, *Utah State University*; Karen M. Apel, *University of Wisconsin—Milwaukee*; Robert W. Bauman, Jr., *Amarillo College*; Leon Benefield, *Abraham Baldwin Agricultural College*; Edward J. Bradel, *University of Cincinatti—Raymond Walters College*; R. O. Bronander, *Yavapai College*; Doug Carmichael, *Pittsburg State University*; William H. Cliff, *Niagara University*; Peggy Rae Dorris, *Henderson State University*; John Dustman, *Indiana University Northwest*; Scott Fairgrieve, *Laurentian University*; James D. Fawcett, *University of Nebraska at Omaha*; Richard A. Fletcher, *University of Wisconsin—La Crosse*; Larry Ganion, *Ball State University*; Kathryn Hedges, *Indiana University Northwest*; Rita A. Hoots, *Yuba College*; Waiston Lee, *Wayne Community College*; Scott Ligman, *Walla Walla College*; Linda L. MacGregor, *Bucks County Community College*; Ted Markus, *Kingsborough Community College*; Aubrey Morris, *Pensacola Junior College*; Jesse J. Nance, *Volunteer State Community College*; Thomas E. Oldfield, *Ferris State University*; Michael J. Postula, *Parkland College*; Donald W. Puder, *College of Southern Idaho*; Warren Rosenberg, *Iona College*; Allan P. Russell, *Mount Wachusett Community College*; E. Edward Sheeley, *Valdosta State University*; Dale Smoak, *Piedmont Technical College*; Cynthia V. Sommer, *University of Wisconsin—Milwaukee*; Patti Thomas, *Delgado Community College*; Eugene Walton, *Tallahassee Community College*.

Focus-group participants for the third edition

Robert Brozanski, *Community College of Allegheny County*; William Michael Clark, *North Harris County College at Kingwood*; Annette Dawson, *Stephen F. Austin State University*; Eugene Fenster, *Longview Community College*; Susan K. Gibson, *Montana State University*; Herbert House, *Elon College*; Don Huff, *Odessa College*; Sarah C. Jackson, *Florida Community College at Jacksonville*; Helena T. Mobley, *Kilgore College*; Nancy L. Pencoe, *Dalton College*; Virginia Rivers, *Truckee Meadows Community College*; Jack Surridge, *Concordia University (Wisconsin)*; Susan O. Van Loon, *Our Lady of Holy Cross College*; Jan Whitley, *Columbia State Community College*.

Reviewers of the second edition

Edward Bradel, *University of Cincinnati—Raymond Walters College*; Jennifer Breckler, *San Francisco State University*; Craig Clifford, *Northeastern Oklahoma State University*; Charles F. Dock, *Marymount College*; Gerald Dotson, *Front Range Community College*; John Dustman, *Indiana University Northwest*; James Fawcett, *University of Nebraska at Omaha*; Larry Ganion, *Ball State University*; Stephen Gehnrich, *Salisbury State University*; Robert Given, *Marymount College*; Donald S. Kisiel, *Suffolk County Community College*; Richard Mostardi, *University of Akron*; William F. Nicholson, *University of Arkansas*; Michael J. Postula, *Parkland College*; Donald Puder, *College of Southern Idaho*; Marilyn M. Shannon, *Indiana University—Purdue University*; Stan Smith, *Bowling Green State University*; Bonnie G. Wood, *University of Maine*; Arthur Zeitlin, *Kingsborough Community College*.

Special thanks go to Clinton Benjamin of *Lower Columbia College* and Mary D. Healey of *Springfield College* who reviewed the entire manuscript for the second edition. Their suggestions were valuable in ensuring continuity in depth, level, and tone from chapter to chapter.

Reviewers of the first edition

Robert A. Altbaum, M.D., *Westport, CT*; Dean A. Beckwith, *Illinois Central College*; Jeffrey H. Black, *East Central University*; Robert J. Boettcher, *Lane Community College*; William Bonaudi, *Truckee Meadows Community College*; Clifton F. Bond, *Highland Park Community College*; James Bridger, *Prince George's Community College*; Jerry Button, *Portland Community College*; Kenneth H. Bynum, *University of North Carolina at Chapel Hill*; Lu Anne Clark, *Lansing Community College*; Glenna M. Cooper, *South Plains College*; Irene M. Cotton, *Lorain County Community College*; Darrell T. Davies, *Kalamazoo Valley Community College*; Clementine A. De'Angelis, *Tarrant County Junior College*; Edward A. DeSchuytner, *North Essex Community College*; William E. Dunscombe, Jr., *Union County College*; William W. Farrar, *Eastern Kentucky University*; Julian Wade Farrior, Jr., *Gwynedd-Mercy College*; Douglas Fonner, *Ferris State University*; Rose B. Galiger, *University of Bridgeport*; Gregory Gillis, *Bunker Hill Community College*; Harold E. Heidtke, *Andrews University*; William J. Higgins, *University of Maryland*; H. Kendrick Holden, Jr., *Northern Essex County Community College*; Gayle D. Insler, *Adelphi University*; David A. Kaufmann, *University of Florida*; Donald S. Kisiel, *Suffolk County Community College*; William G. Klopfenstein, *Sinclair Community College*; Thomas E. Kober, *Cincinnati Technical College*; Joseph R. Koke, *Southwest Texas State University*; Gordon Locklear, *Chabot College*; James A. Long, *Boise State University*; Dorothy R. Martin, *Black Hawk College*; Elden W. Martin, *Bowling Green State University*; Johnny L. Mattox, *Northeast Mississippi Community College*; R. J. McCloskey, *Boise State University*; Daniel McEuen, *Mesa Community College*; Robert C. McReynolds, *San Jacinto College*; Anne M. Miller, *Middlesex Community College*; William W. Miller III, *Grambling State University*; Gordon L. Novinger, *San Bernardino Valley College*; Dennis E. Peterson, *DeAnza College*; Joseph S. Rechtschaffen, M.D., *New York, NY*; Ralph E. Reiner, *College of the Redwoods*; Jackie S. Reynolds, *Richland College*; Donald D. Ritchie (deceased), *Barnard College*; Robert L. Ross, Optometrist, *Westport, CT*; Louis C. Renaud, *Prince George's Community College*; David Saltzman, *Santa Fe Community College*; David S. Smith, *San Antonio College*; Stan L. Smith, *Bowling Green State University*; Tracy L. Smith, *Housatonic Community College*; Alexander A. Turko, *Southern Connecticut State University*; Donald H. Whitmore, *University of Texas at Arlington*; Barry James Wicklow, *St. Anselm College*; Leonard B. Zaslow, D.D.S., *Westport, CT*; Stephen W. Ziser, *Austin Community College*.

How the Body Is Organized

CAREER IN FOCUS: MEDICAL ILLUSTRATION

A budding medical illustrator fascinated with the human form and with nature may have a conflict in choice of college major because talent in art and aptitude in biology seem to pull in opposite directions. Fortunately, there is an exciting career that resolves these conflicts. It is called biomedical visualization or, more simply, medical illustration, and its history dates back to Leonardo da Vinci.

In the fifteenth century, when Leonardo drew his anatomical figures, dissection of the human cadaver was illegal. While Leonardo's drawings took the viewer into a world of wondrous secrets, all the more powerful because they conveyed forbidden knowledge, the modern medical illustrator has other tasks at hand.

Today's medical artists are trained to communicate complex scientific ideas graphically and accurately in esthetically pleasing ways. They work at large research hospitals, medical centers, pharmaceutical companies, or prosthesis firms; or at publishing, software, or insurance companies; or in business for themselves. Some do general medical illustration for books, magazines, television, and film, while others specialize in medicolegal work, portraying anatomical aspects of crime scenes or victims for forensic documentation. Others are sculptors, working with patients to fashion reconstructions of limbs, often solving tragic problems with art when medicine or surgery do not suffice. Finally, a growing number of biomedical illustrators provide the art to support a growing range of computer and multimedia programs that aim to create simulated patients in a virtual-reality world.

To become a biomedical illustrator, one must have talent in art and aptitude in science. All the accredited programs are at the master's level, requiring applicants to have a bachelor's degree and to present a portfolio documenting their artistic ability and accomplishments. Each program specifies prerequisite courses in both art and biology. While the biology requirements vary, the usual prerequisites include some combination of undergraduate coursework in anatomy, embryology, histology, cell biology, and physiology. More advanced courses in human gross anatomy, neuroanatomy, pathology, microanatomy, physiology, and embryology, as well as courses in art, computer graphics, and design, constitute the master's degree curriculum.

In the near future, doctors will compare the actual patient morphology with a computer model of the average patient, and in the not-too-distant future, the artist will build the entire range of normal deviations into the program so that computer simulations will assist in diagnosis and treatment as well as training. For those whose talents bridge the gap between the world of art and the world of science, the future looks bright indeed.

In this section we look at a subject that has long fascinated human beings—the organization of the human body. Long before artists like Leonardo da Vinci and Michelangelo made realistic studies of human anatomy, other artists were engaged in their own, more primitive studies. While artists of all kinds continue to study the body, medical illustrators have a special interest in its overall structure and functions. In a sense, we look over the shoulder of the medical illustrator as we begin our study of the organization of the human body.

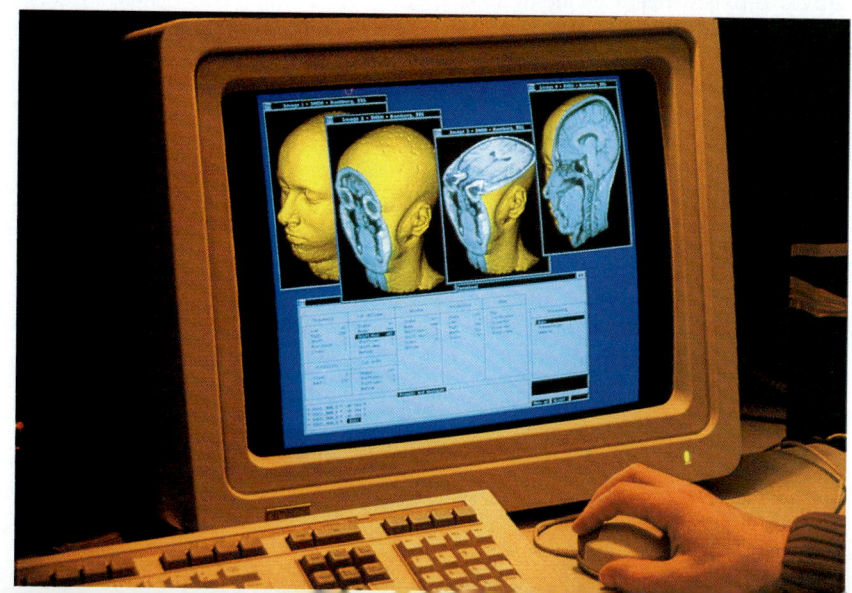

Introduction to Anatomy and Physiology

1

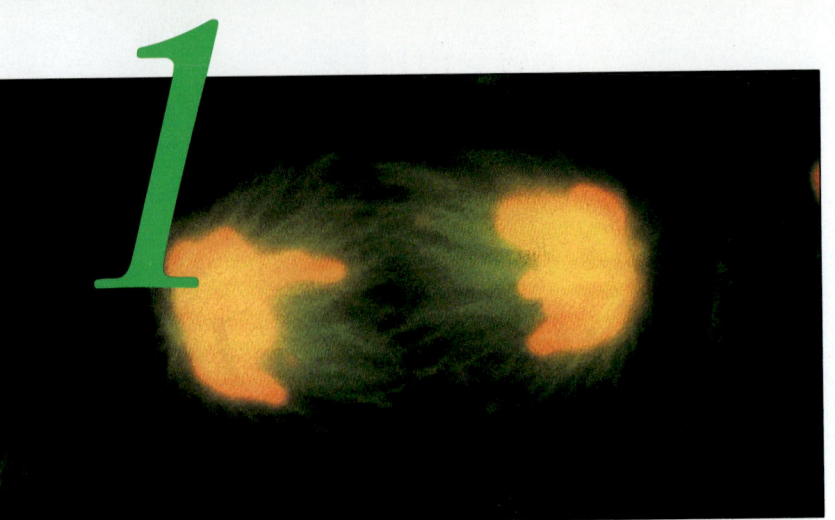

KEY CONCEPTS

1 Anatomy and physiology are discussed as a unit to help the reader appreciate how structure and function work together.

2 Homeostasis is the body's state of relative stability. All parts of the body contribute to the maintenance of homeostasis.

3 Stress can disrupt the body's normal state of equilibrium.

4 A progression of structural levels—including atoms, molecules, compounds, cells, tissues, organs, and systems—contributes to the body's order and stability.

5 Anatomical terminology is usually precise and logical, using descriptive root words, prefixes, and suffixes to name body parts and directional terms.

How does body temperature remain the same when the external temperature changes? Why does skin wrinkle? Why does a healing wound itch? Why does a foot "fall asleep"? Why can't we taste food when we have a cold? If millions of sperm are released, why does only one fertilize an egg?

These are just a few of the many questions that are answered in this book, questions that many of us have wondered about. Although we have learned a great deal about the human body and what makes it function, there are still many unanswered questions. And so we continue studying, hoping that someday we will be able to understand what sleep is, what causes cancer, and what makes one cell develop into a pancreas while a seemingly identical cell becomes a liver. Although the questions are many, the search is tireless, because nothing fascinates the human mind more than the human body.

WHAT ARE ANATOMY AND PHYSIOLOGY?

Anatomy (from the Greek word meaning *to cut up*) is the study of the many *structures* that make up the body and how those structures relate to each other. *Physiology* (from the Greek word for the *study of nature*) is the study of the *functions* of those structures. Anatomy and physiology are studied together to give students a full appreciation and understanding of the human body.

The study of anatomy includes many subspecialties, from gross anatomy and dissection to the most recent forms of microscopic and ultramicroscopic anatomy and noninvasive diagnostic techniques. (See "The Kindest Cuts of All" on page 9.)

Regional anatomy is the study of specific regions of the body, such as the head and neck, while *systemic anatomy* is the study of different systems, such as the reproductive and digestive systems. Regional anatomy and systemic anatomy are both branches of **gross anatomy,** a type of anatomy that can be undertaken without a microscope. In contrast, **microscopic** and **ultramicroscopic anatomy** require the use of a microscope. Some other subdivisions of anatomy are *embryological anatomy,* or *embryology,* the study of prenatal development; *developmental anatomy,* the study of human growth and development from fertilized egg to mature adult; and *radiographic anatomy,* or *radiology,* the study of the structures of the body using x-rays and other noninvasive imaging techniques.

As you begin to study anatomy and physiology, you will see that form and function go together. Parts of the body have specific shapes that make them suitable to perform their specific functions. Think of teeth. Three different kinds of teeth do three very different jobs. Flat molars grind food, sharp-pointed canines tear, and incisors cut. As you progress through this book, try to keep in mind the crucial connection between anatomy and physiology — form and function — and find as many examples of this connection as you can.

HOMEOSTASIS: COORDINATION CREATES STABILITY

Homeostasis (ho-mee-oh-STAY-siss; Gr. "staying the same") is the state of relative stability of the body's internal environment. Although the outside environment changes constantly, the internal environment of a healthy body remains the same, within normal limits. Under normal conditions, homeostasis is maintained by adaptive mechanisms in the body, ranging from control centers in the brain to chemical substances called hormones that are secreted by various organs directly into the bloodstream. Some functions controlled by homeostatic mechanisms are blood pressure, body temperature, breathing, and heart rate.

For a body to remain healthy, a chemical balance must be carefully maintained inside and outside its cells. To achieve this balance, the composition of the *extracellular fluid* — the fluid outside the cells that surrounds and bathes them — must remain fairly constant. Extracellular fluid circulates throughout the body, and many materials are passed into and out of cells by way of the extracellular fluid. Thus, the fluid is instrumental in helping the body's systems maintain optimal temperature and pressure levels, as well as the proper balance of acids and bases, oxygen and carbon dioxide, and the concentrations of water, nutrients, and the many chemicals that are found in the blood.

Practically everything that goes on in the body helps maintain homeostasis. For instance, the kidneys filter blood and remove a carefully regulated amount of water and wastes. The lungs work together with the heart, blood vessels, and blood to distribute oxygen throughout the body and remove carbon dioxide. From the digestive system, nutrients pass through the walls of the small intestine into the bloodstream, which carries them to parts of the body where they are needed. In general, all the systems of the body work together to contribute to the well-being that comes with inner stability.

Homeostasis and Negative Feedback

The whole regulation process of homeostasis is made possible by the coordinated feedback action of many organs and tissues under the direct or indirect control of sensitive networks in the nervous and hormonal systems. If all the systems are operating properly, we feel fine—homeostasis is taken for granted. If the coordination among the systems breaks down, we may begin to feel uncomfortable. The less coordination there is, the worse we feel. If homeostasis is not restored by the body itself or with the help of outside intervention, we die.

Feedback occurs whenever the body receives input from its sensors about a change in its internal condition, and as a result makes a positive or negative adjustment. For example, when we are too hot, our sweat glands are activated, and we perspire. The sweat evaporates and cools us. In contrast, when we are too cold, the muscles under the skin receive a message from the nervous system to contract and relax, contract and relax—we shiver—and the action gives off heat. Sweating and shivering are both involuntary responses initiated by the part of the nervous system that is linked to a temperature-regulating area in the brain. Although external temperatures may vary, our bodies must remain within a few degrees of the normal 37°C (98.6°F) to remain alive, let alone healthy [FIGURE 1.1].

Feedback activity such as sweating and shivering, which produces a response *opposite* to the initiating stim-

ulus, is called a ***negative feedback system*** [FIGURE 1.2]. In another example, if blood pressure decreases below normal, the response of a negative feedback system is to raise it back to the normal level; if blood pressure increases above normal, the response is to lower it. Each of these responses is a *negative* action, or *not the same as* the initial stimulus of abnormally low or high blood pressure.

In contrast, a ***positive feedback system*** operates when the initial stimulus is *reinforced* rather than opposed. For example, if the blood glucose level decreases, the response of a positive feedback system is to lower it further. If continued without restraint, such a response will lower the blood glucose level until the person eventually dies. Positive feedback is relatively rare in our bodies, mainly because it disrupts homeostasis.

Stress and Homeostasis

To maintain homeostasis, the mind and body must adjust to the various imbalances that arise in the internal and external environments. Factors that cause imbalances in the body are collectively called *stressors*, and the overall disruption that forces the body to make adaptive changes is called *stress*. (We usually use the word *stress* when we really mean "excessive stress" or "distress.") Hans Selye, a pioneer in stress research, has said that "the great capacity for *adaptation* is what makes life possible on all levels of complexity. It is the basis for homeostasis and of resistance to stress. . . . Adaptability is probably the most distinctive characteristic of life."* Selye defines stress as "the nonspecific response of the body to any demand made upon it. . . . No matter what kind of [specific action] is produced, all [stressors] have one thing in common; they also increase the demand for readjustment."†

Stressors may be physical (heat, noise), chemical (food, hormones), microbiological (viruses, bacteria), physiological (tumors, abnormal functions), developmental (old age, genetic changes), or psychological (emotional and mental disturbances). Some amount of stress is normal and may actually be beneficial. Walking, for instance, places some stress on bones, muscles, and joints, helping to strengthen them. Too little exercise (or stress) causes bones and muscles to weaken, while too much physical stress breaks bones and tears muscles.

FIGURE 1.1 SOME TYPICAL BODY TEMPERATURE RANGES

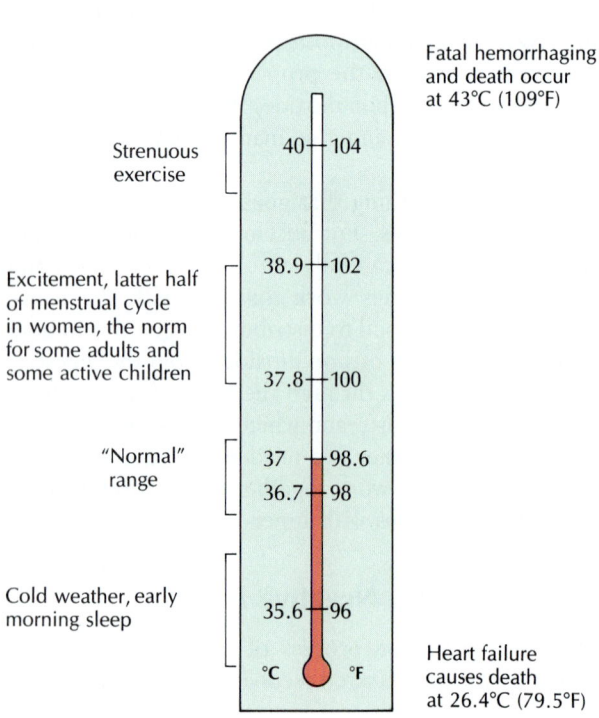

Fatal hemorrhaging and death occur at 43°C (109°F)

Strenuous exercise — 40 / 104

Excitement, latter half of menstrual cycle in women, the norm for some adults and some active children — 38.9 / 102 — 37.8 / 100

"Normal" range — 37 / 98.6 — 36.7 / 98

Cold weather, early morning sleep — 35.6 / 96

°C / °F

Heart failure causes death at 26.4°C (79.5°F)

*H. Selye, *The Stress of Life*, McGraw-Hill, New York, 1976, p. 74.

†H. Selye, *Stress without Distress*, Lippincott, Philadelphia, 1974, pp. 27–28.

FIGURE 1.2 NEGATIVE FEEDBACK

[A] A generalized negative feedback system. [B] A specific negative feedback system set in motion to balance the effects of a mild hemorrhage.

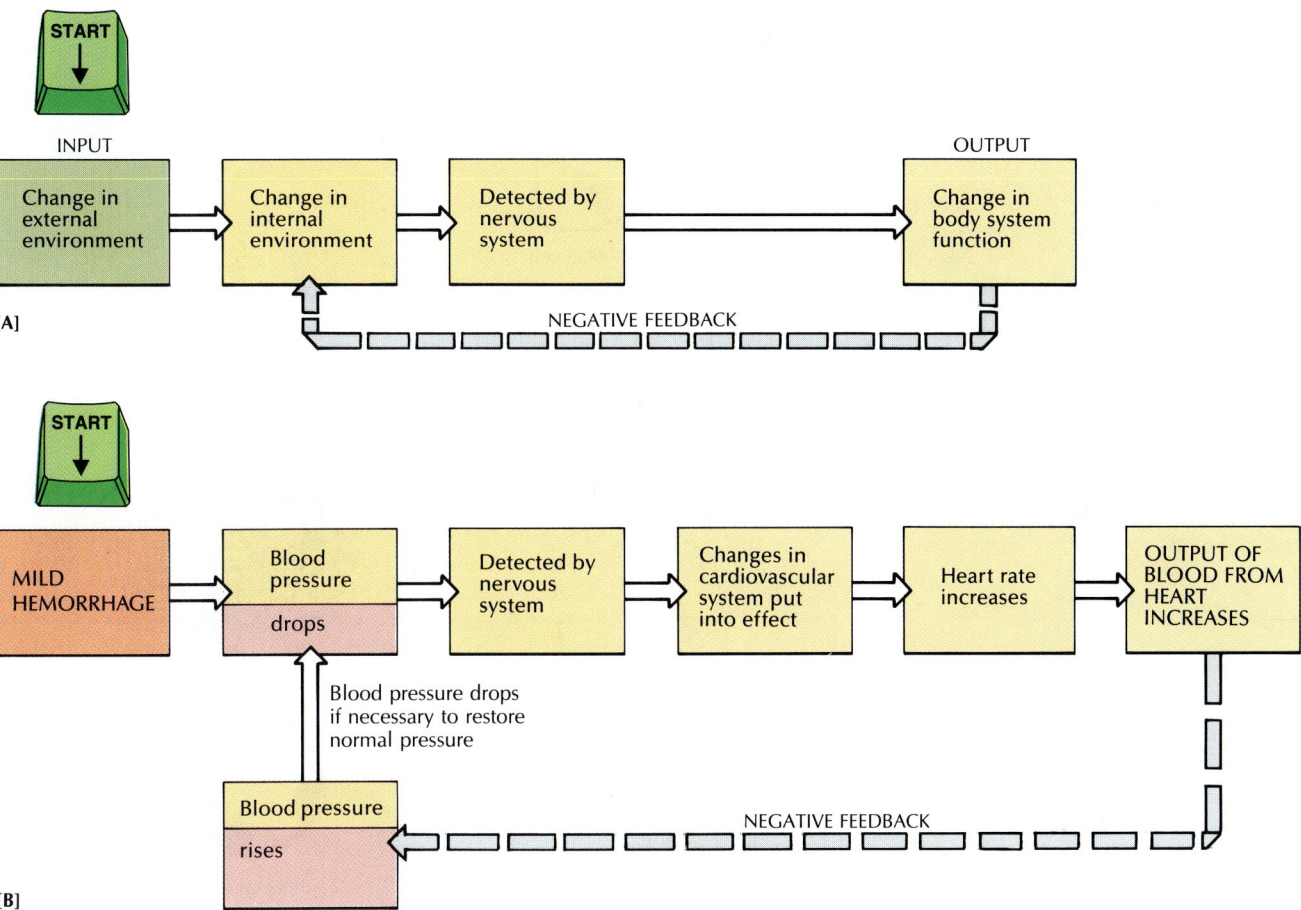

People who cannot control the mood or pace of their lives increase their susceptibility to disease. People who have little control over their jobs probably run the same risk of heart disease as do people who are heavy smokers or have high blood levels of cholesterol. Studies have shown that long-term stress decreases the effectiveness of the immune system, and some experts feel that stress is a key factor in most illnesses.

When stress is greater than the corrective feedback systems, it is considered to be an injury that may lead to disease, disability, or even death.

Stress is implicated either directly or indirectly in cancer, coronary heart disease, lung disorders, accidental injuries, and suicide. Not coincidentally, these are five of the leading causes of death in the United States. Two of the five best-selling prescription drugs in the United States are also stress-related: Zantac, an antiulcer medication, and Xanax, a tranquilizer.

ASK YOURSELF

1 What is homeostasis, and why is it important to the body?

2 What is the difference between negative feedback and positive feedback?

3 What is stress, and how can it affect homeostasis?

FIGURE 1.3 STRUCTURAL LEVELS OF THE BODY

Specialized cells and tissues of the respiratory system help the lungs and other parts of the system to perform their functions of breathing and gas exchange at an optimum level of efficiency. The same levels of organization are found in all body systems.

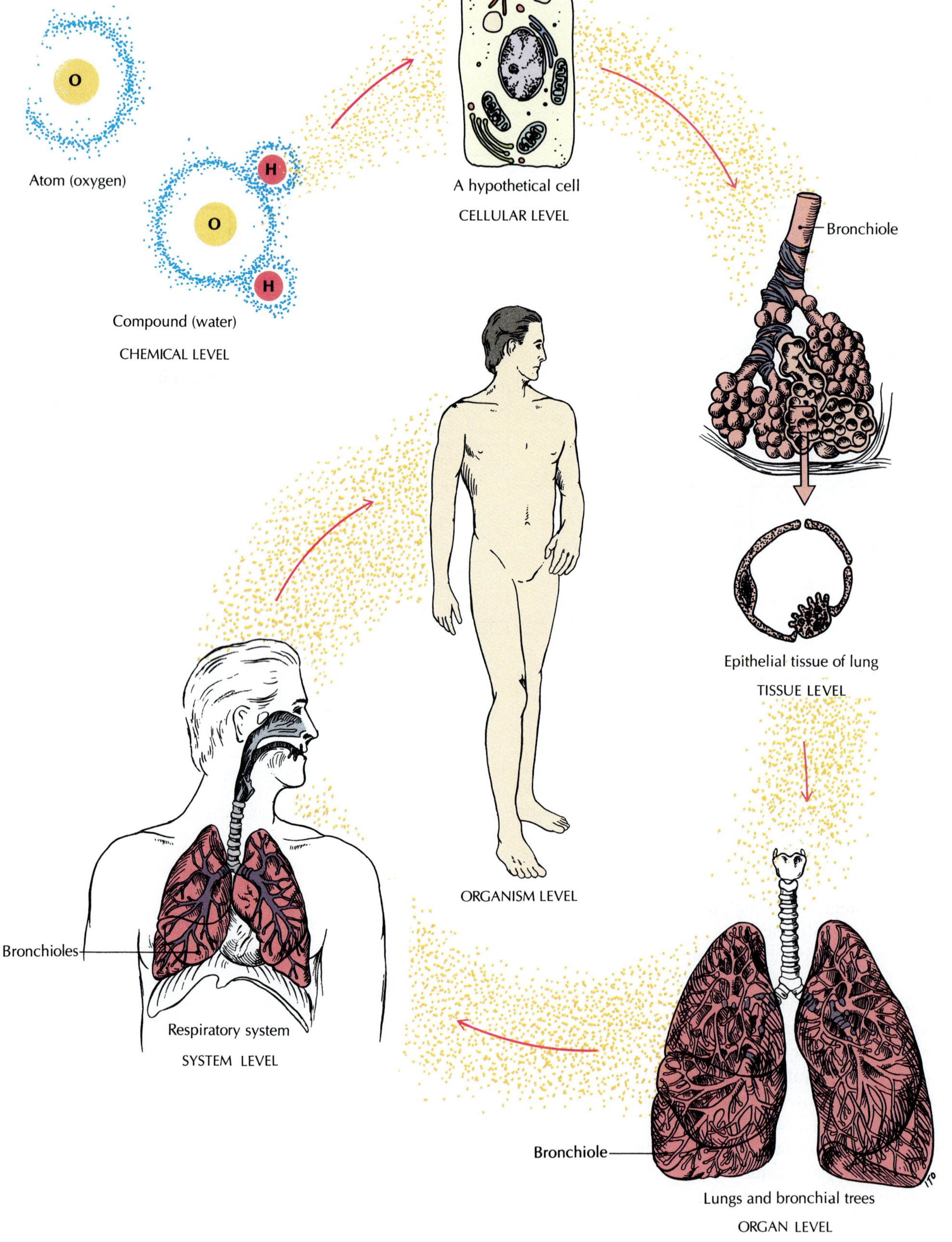

Atom (oxygen)

Compound (water)

CHEMICAL LEVEL

A hypothetical cell

CELLULAR LEVEL

Bronchiole

Epithelial tissue of lung

TISSUE LEVEL

ORGANISM LEVEL

Bronchioles

Respiratory system

SYSTEM LEVEL

Bronchiole

Lungs and bronchial trees

ORGAN LEVEL

FROM ATOM TO ORGANISM: STRUCTURAL LEVELS OF THE BODY

The human body has different structural levels of organization, starting with atoms, molecules, and compounds and increasing in size and complexity to cells, tissues, organs, and the systems that make up the complete organism [FIGURES 1.3, 1.4].

Atoms, Molecules, and Compounds

At its simplest level, the body is composed of *atoms* (Gr. *atomos*, indivisible), the basic units of all matter. Scientists have identified about 100 different kinds of atoms. Each kind is an *element*. The most common elements in living organisms are carbon, hydrogen, oxygen, nitrogen, phosphorus, and sulfur.

When two or more atoms combine, they form a *molecule.* For example, when two oxygen atoms combine, they form an oxygen molecule, O_2. A molecule containing atoms of more than one element is called a *compound.* Water (H_2O) and carbon dioxide (CO_2) are compounds, as are the carbohydrates, proteins, and lipids (fats) that are so important to our bodies.

Cells

Cells are the smallest independent units of life, and all life as we know it depends on the many chemical activities of cells. Some of the basic functions of cells are growth, metabolism, irritability, and reproduction. Cells vary in size from a sperm, which is about 5 micrometers (five-millionths of a meter) long, to a nerve cell with thin fibers that may be more than a meter long.

Tissues

Tissues are made up of many similar cells that perform a specific function. They are divided into four groups: epithelial, connective, muscle, and nervous [FIGURE 1.5].

FIGURE 1.4 SOME BASIC TISSUES AND STRUCTURES

This multilevel "step dissection" through the right lower leg (midcalf) shows typical tissue layers, bones, and a bundle of nerves and blood vessels.

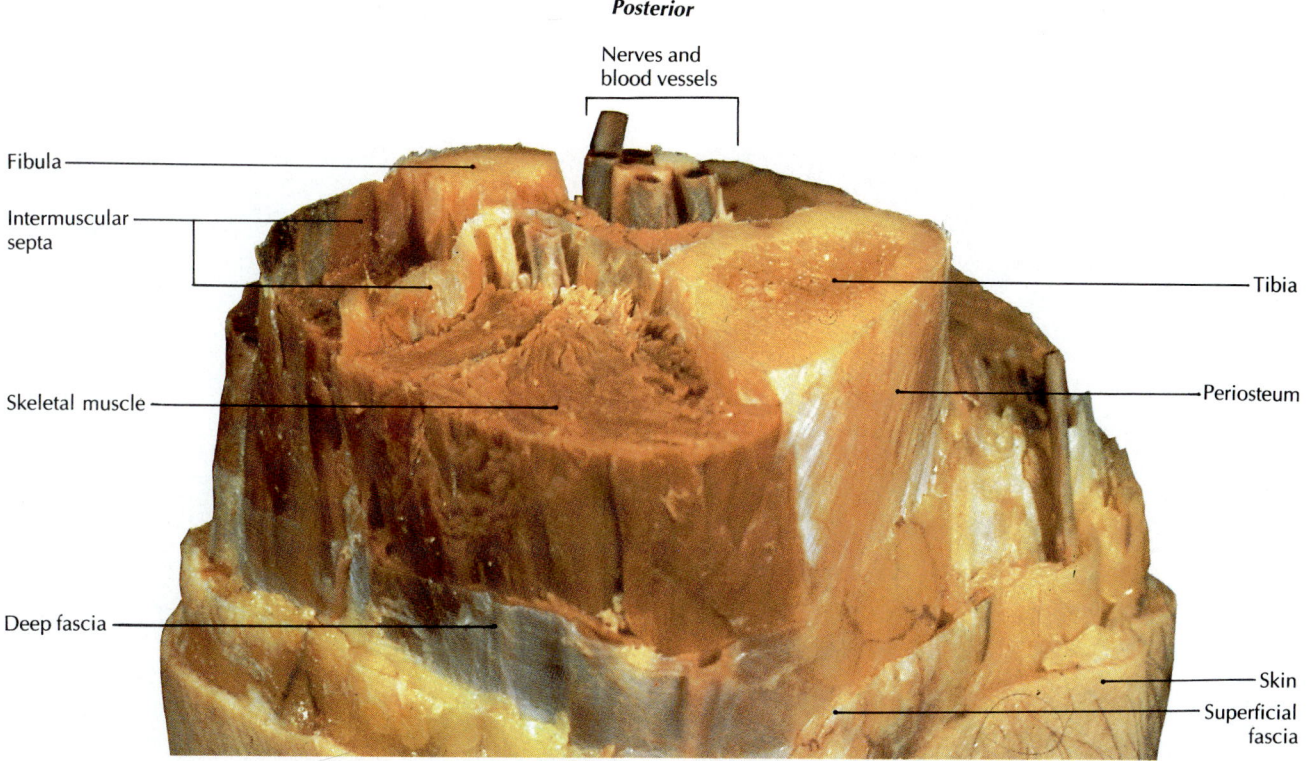

Posterior

Nerves and blood vessels

Fibula

Intermuscular septa

Skeletal muscle

Deep fascia

Tibia

Periosteum

Skin

Superficial fascia

Anterior

FIGURE 1.5 REPRESENTATIVE TYPES OF BODY TISSUES

[A] A highly magnified photograph of one type of epithelial tissue. The arrangement of flat, overlapping sheets is well suited for the shingles of a roof or the skin of your body, both of which protect the inside from the outside environment. One hair is visible. ×150. [B] Meshed fibers of connective tissue are seen in this photograph of interwoven collagen fibers. Connective tissue is abundant in the body and can resist stretching in several directions. ×3100. [C] Cardiac muscle tissue is found only in the heart. It consists of separate, but interconnected, cells, each with a centrally located nucleus. ×200. Other types of muscle tissue are found throughout the body. [D] Nervous tissue consists of branched nerve cells with slender processes that span relatively long distances in the body. ×200.

[A] EPITHELIAL TISSUE

Hair

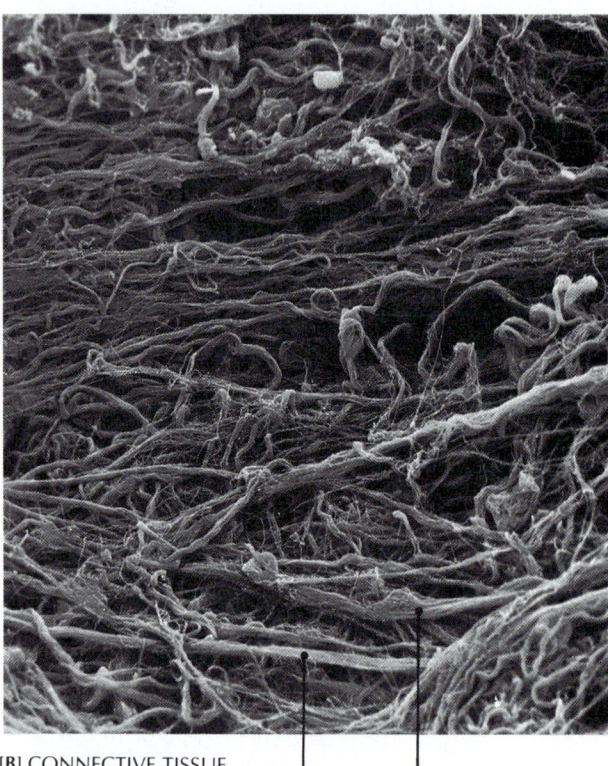

[B] CONNECTIVE TISSUE

Collagen fibers

[C] CARDIAC MUSCLE TISSUE

Branched processes of nerve cells (neurons)

[D] NERVOUS TISSUE

The Kindest Cuts of All

Like a kid intent on a Nintendo game, Dr. David Sugarbaker looks not at the patient lying senseless on the operating table but at the TV positioned by her side. "I think we're right on target," he exults. Displayed on the screen is a larger-than-life section of the woman's right lung, a rosy mass marred by a couple of suspicious lumps. "Fire away," Sugarbaker directs the assisting surgeon. On the screen a tiny pincer appears. Grabbing hold of the lung just above the lesion, the pincer makes a clean slice through the quivering tissue, simultaneously sealing the wound by laying down a triple row of surgical staples. A few more snips and the task is complete. Sugarbaker, chief of thoracic surgery at Boston's Brigham and Women's Hospital, draws a 10-cm-long sliver of lung through a finger-size hole in the patient's side and sends it for biopsy.

This nearly bloodless procedure, which Sugarbaker began performing [in 1991], is one of the most recent applications of a new approach to surgery that is rapidly displacing the dreaded knife and scalpel. "We are witnessing the greatest surgical revolution in the past 50 years," exclaims Dr. William Schuessler, a urological surgeon from San Antonio. The instrument sparking such enthusiasm is variously known as a laparoscope (when used in the abdomen), an arthroscope (when applied to the joints), a thoracoscope (when the chest is involved) and an angioscope (when the target lies inside blood vessel walls). But apart from differences in length and thickness, all these scopes are fundamentally alike: slender fiber-optic tubes that can be inserted deep inside the body through minute (1-cm-long or less) incisions. With the addition of a tiny telescopic lens, a miniature light source and a palm-size video camera, these tubes are transformed into videoscopes that project images of the patient's internal organs and, even more important, of the snippers, staplers and graspers that the surgeons manipulate.

The reason for the surging popularity of videoscope surgery is simple: correctly performed, it can dramatically reduce surgical trauma. Since 1987, when the first diseased gall bladder was removed in this fashion, rave reviews from patients have made it almost rare for a gall bladder to be removed the old-fashioned way. And for good reason. "Before," says Dr. Eddie Joe Reddick, a retired Nashville surgeon credited with popularizing the technique, "we were committing assault and battery on our patients. It wasn't what we did to their insides, but what we did in order to get there that was the problem." Now, instead of an 8-cm to 15-cm slash down their abdomens, patients wake up with four small incisions that not only heal more quickly but also are far less painful. In fact, most patients whose gall bladders are removed laparoscopically leave the hospital the next day and return to work within a week.

As their skills improve, videoscope surgeons are attempting more daring feats. In 1990, for example, a surgical team led by Dr. Ralph Clayman of Washington University in St. Louis devised a clever technique for removing problem-plagued kidneys laparoscopically. Because the kidney is a solid organ about the size of a fist, it has to be reduced in size before it can be drawn through a 2½-cm incision concealed in the patient's belly button. So after cutting the kidney free of connective tissue and sealing off the big artery that supplies it with blood, the surgeons move the organ into an impermeable sack and, while it is still inside the patient, chop it up with a tiny rotating blade. The sack and its pulverized contents can then be safely drawn out. . . .

Nowhere is videoscope surgery likely to have a greater impact than in the field of thoracic (chest) surgery. Only a [few years] ago, patients requiring a lung biopsy would inevitably be subjected to a muscle-slicing, rib-bruising operation that typically involves two or three days in intensive care followed by weeks of painful recovery. For elderly and frail patients, this often meant that a biopsy, and hence a firm diagnosis, was out of the question. Now a few pioneering surgeons are developing less traumatic ways of gaining access to the chest cavity. Sugarbaker, for example, makes a slash through the skin of his patient's side that looks no more serious than an accidental nick from a razor. Then he pushes a series of blunt-tipped probes through the bundles of muscle that lie between the ribs. Rather than tearing, the muscle fibers stretch to accommodate the probes, providing the surgical team with a temporary passageway about as thick as a man's finger. At the end of the operation, a couple of stitches and a Band-Aid suffice to close the patient up. (Unfortunately, if a biopsy reveals a malignancy, the patient will probably undergo an open-chest operation. At present there is no other way to remove a whole lung.)

For the surgeon, operating by videoscope means mastering a totally new set of skills. . . . But there are serious drawbacks. In open-lung surgery, for instance, when Sugarbaker can't see the lesion to be biopsied, he simply uses a gloved finger to locate it by feel. He can still do this, of course—provided the lesion is no more than a finger's length away. Even more challenging is the fact that the image displayed on operating-room TV screens is only two-dimensional. This makes it easy to misjudge the distance to a blood vessel or organ, which is a major hazard of videosurgery. A tiny nick to the lung, for instance, could unleash a bloody torrent that even the best surgeon would be pressed to stanch in time. . . .

Videoscope surgery will never completely replace open surgery, but it may come closer than anyone a [few years] ago might have imagined. Already, of nearly 600,000 gall bladders that are removed in the U.S. annually, an estimated three-quarters are removed laparoscopically. Other common operations, from hysterectomies to hernias, seem likely to follow suit. At Loyola University Medical Center near Chicago, a trauma team has begun using the technology to diagnose injuries from knife wounds and automobile crashes. Soon the team expects to move from diagnosis to laparoscopic repair of tears to the diaphragm and abdominal wall. Eventually, if doctors become convinced that operations performed in this manner do not inadvertently spread malignant cells, this kinder, gentler surgery will touch the lives of an even larger group of people: cancer patients.

J. Madeleine Nash, "The Kindest Cuts of All" (abridged), *Time*, March 23, 1992, pp. 52–53. Copyright 1992 Time Inc. Reprinted by permission.

Epithelial tissue is found in the outer layer of the skin and in the linings of organs, blood and lymph vessels, and body cavities. This type of tissue is well suited to its protective function because the cells are closely packed and arranged in sheets [FIGURE 1.5A] and because it can also add new cells when the old ones become worn or damaged. Some epithelial tissues are specialized for secretion.

Connective tissue connects and supports most parts of the body. It is found under the epithelial tissue in skin and also constitutes the major portion of bones and tendons. Connective tissue is the most widely distributed body tissue. Some connective tissues contain fibers that form a strong mesh [FIGURE 1.5B].

Muscle tissue produces movement through its ability to contract. *Skeletal muscle tissue* is found in the limbs, trunk, and face; *smooth muscle tissue* is found in the digestive tract, eyes, blood vessels, and ducts; and *cardiac muscle tissue* is found only in the heart [FIGURE 1.5C].

Nervous tissue is found in the brain, spinal cord, and nerves. It responds to various types of stimuli and transmits nerve impulses (messages) from one area of the body to another. Its long nerve fibers are well adapted to these functions [FIGURE 1.5D].

Organs

An *organ* is an integrated collection of two or more kinds of tissues that work together to perform a specific function. The stomach is an excellent example: Epithelial tissue lines the stomach and helps protect it; smooth muscle tissue churns food, breaking it down into smaller pieces and mixing it with digestive chemicals; nervous tissue transmits nerve impulses that initiate and coordinate muscle contractions; and connective tissue helps hold all the other tissues together.

Systems

A *system* is a group of organs that work together to perform a major function. The respiratory system, for example, contains several organs that provide a mechanism for the exchange of oxygen and carbon dioxide between the air outside the body and the blood inside. All the body systems are specialized, and their functions are coordinated to produce a dynamic and efficient *organism* — the human body.

A S K   Y O U R S E L F

1 What are the four types of body tissues?

2 What is the difference between organs and tissues?

BODY SYSTEMS

The structures of each system are closely related to their functions. The systems are illustrated and described briefly in FIGURE 1.6A–K and in the table below.

INTEGUMENTARY (skin, hair, nails, sweat and oil glands)

Covers body and protects internal organs; helps regulate body temperature; contains sensory receptors.

SKELETAL (bones, cartilage)

Supports body, protects organs; provides lever mechanism for movement; manufactures blood cells; contains reserve calcium, phosphorus.

MUSCULAR (skeletal, smooth, cardiac muscle)

Allows for body movement; moves materials through body parts; produces body heat.

NERVOUS (brain, spinal cord; peripheral nerves; sensory organs)

Regulates most body activities; receives and interprets information from sensory organs; initiates actions by muscles, glands.

ENDOCRINE (ductless glands)

Secretes hormones, which regulate many chemical actions within cells, including growth.

CARDIOVASCULAR (heart, blood, blood vessels)

Heart pumps blood through vessels; blood carries nutrients, other materials to tissues; transports tissue wastes for excretion.

LYMPHATIC (glands, lymph nodes, lymph, lymphatic vessels)

Returns excess fluid, protein to blood; part of immune system.

RESPIRATORY (airways, lungs)

Provides mechanism for breathing, exchange of gases between air and blood.

DIGESTIVE (stomach, intestines, other digestive structures)

Breaks down large food molecules into small molecules that can be absorbed into bloodstream; removes solid wastes.

URINARY (kidneys, ureters, urinary bladder, urethra)

Eliminates metabolic wastes; helps regulate blood pressure, acid-base and water-salt balance.

REPRODUCTIVE (ovaries, testes, reproductive cells, accessory glands and ducts)

Provides mechanism for reproduction, heredity.

FIGURE 1.6 SYSTEMS OF THE BODY

[A] The *integumentary system* consists of the skin and all the structures derived from it, including hair, nails, sweat glands, and oil glands.

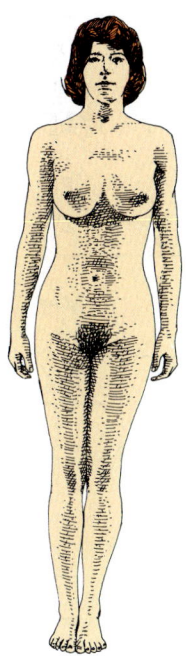

[B] The *skeletal system* consists of bones and cartilage. This illustration shows a typical bone, cartilage, and joint.

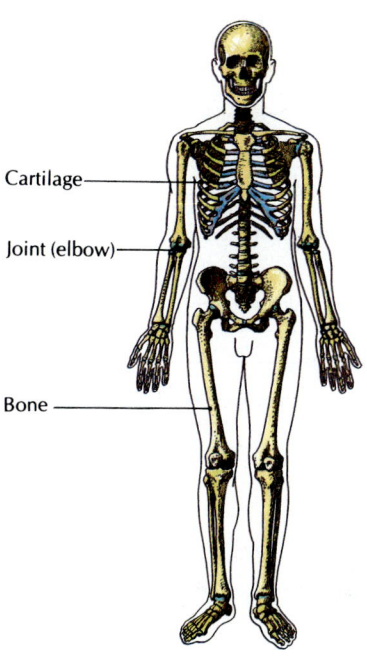

Cartilage

Joint (elbow)

Bone

[C] The *muscular system* consists of three different types of muscles: skeletal (some of the major ones are shown in the drawing), smooth, and cardiac (heart). It also includes tendons (the fibrous cords of connective tissue that attach muscles to bones).

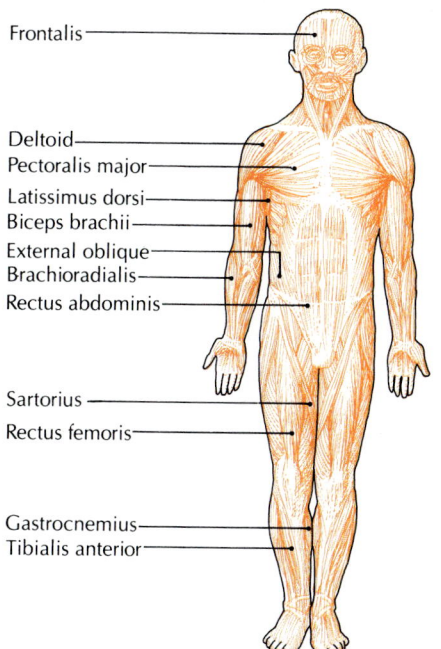

Frontalis

Deltoid
Pectoralis major
Latissimus dorsi
Biceps brachii
External oblique
Brachioradialis
Rectus abdominis

Sartorius
Rectus femoris

Gastrocnemius
Tibialis anterior

[D] The *nervous system* consists of the *central nervous system* (the brain and spinal cord) and the *peripheral nervous system* (the nerves that connect the brain and spinal cord with the rest of the body). It also includes the sensory organs.

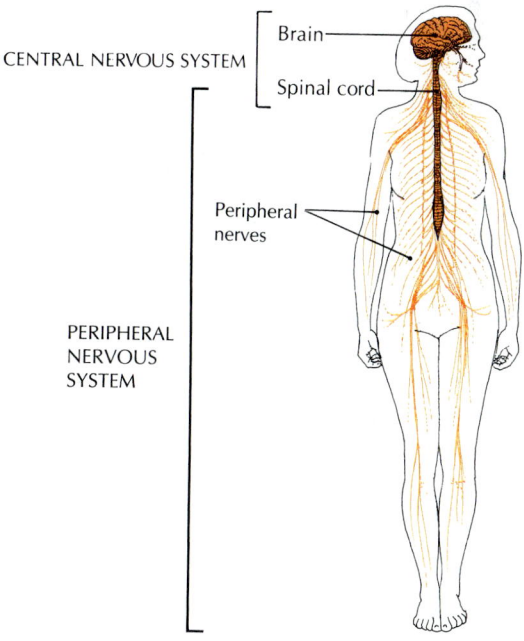

CENTRAL NERVOUS SYSTEM

Brain

Spinal cord

Peripheral nerves

PERIPHERAL NERVOUS SYSTEM

(Figure 1.6 continues on following pages)

FIGURE 1.6 SYSTEMS OF THE BODY (Continued)

[E] The **endocrine system** is the second major regulating system in the body, working closely with the nervous system. It is composed of ductless, or endocrine, glands. The locations of the major endocrine glands are shown in the drawing. (The parathyroids are not shown, since they are on the posterior side of the thyroid gland.)

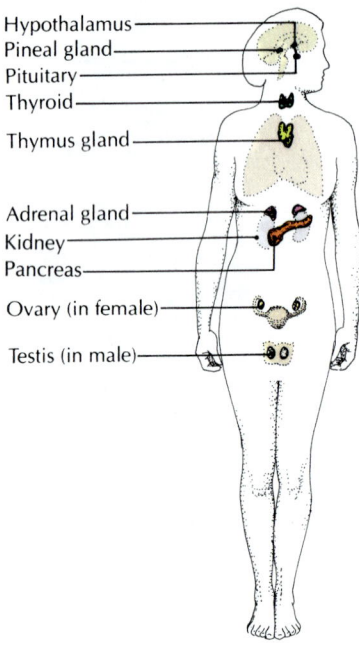

Hypothalamus
Pineal gland
Pituitary
Thyroid
Thymus gland
Adrenal gland
Kidney
Pancreas
Ovary (in female)
Testis (in male)

[F] The **cardiovascular system** consists of the heart, blood, and blood vessels. The illustration shows the heart, the aorta, an artery (red), and a vein (blue); capillaries, which connect arteries and veins, are not shown.

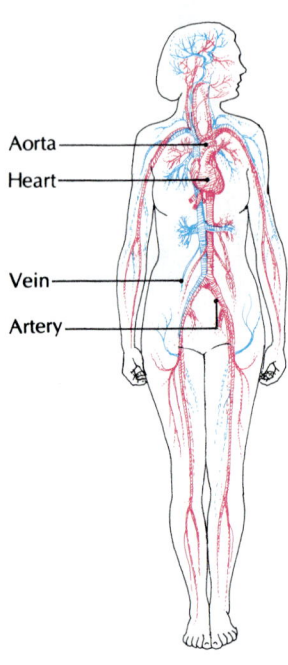

Aorta
Heart
Vein
Artery

[G] The **lymphatic system** is made up of glands (including the spleen, tonsils, and thymus gland), lymph nodes, and a network of thin-walled vessels that carry a clear fluid called lymph.

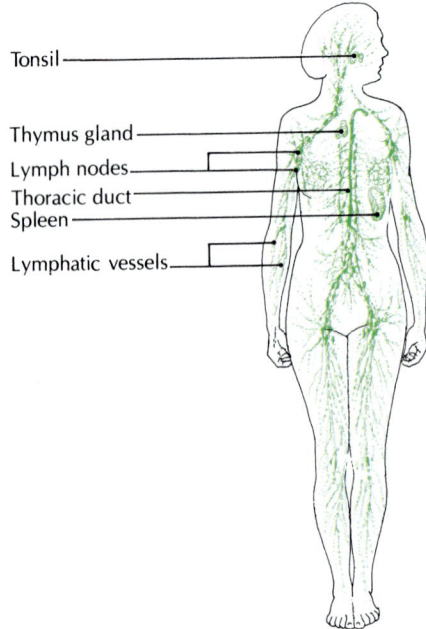

Tonsil
Thymus gland
Lymph nodes
Thoracic duct
Spleen
Lymphatic vessels

[H] The **respiratory system** is composed of the nose; a system of airways that includes the nasal cavity, pharynx, larynx, and trachea; the lungs; and the muscles that help move air into and out of the body, the most important of which is the diaphragm (dashed line).

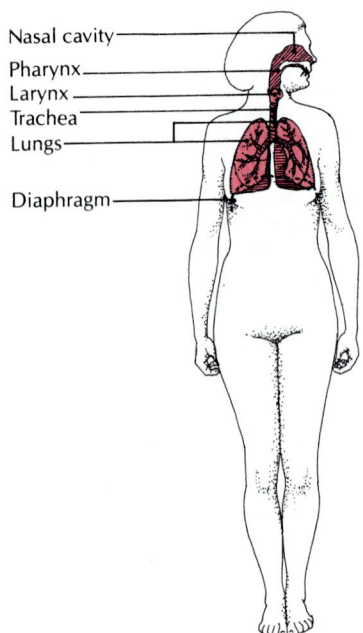

Nasal cavity
Pharynx
Larynx
Trachea
Lungs
Diaphragm

[I] The **digestive system** includes the teeth, tongue, salivary glands, esophagus (a tube leading from the mouth to the stomach), stomach, small and large intestines, rectum, anus, liver, gallbladder, and pancreas. It is compartmentalized, with each part adapted to a specific function.

[J] The **urinary system** consists of the kidneys, which produce urine; the ureters, which carry urine to the urinary bladder, where it is stored; and the urethra, which conveys urine to the outside.

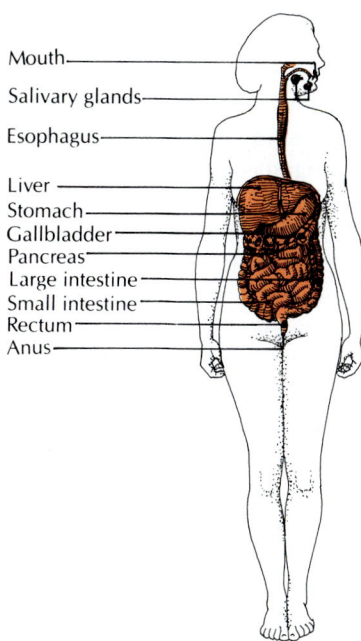

Mouth
Salivary glands
Esophagus
Liver
Stomach
Gallbladder
Pancreas
Large intestine
Small intestine
Rectum
Anus

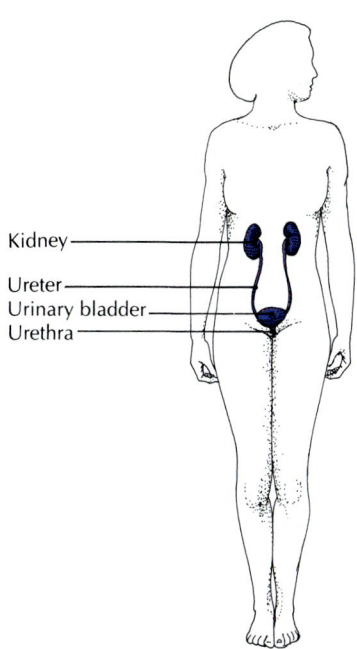

Kidney
Ureter
Urinary bladder
Urethra

[K] The **reproductive systems** are enlarged in the drawings for clarity. Each sex has reproductive organs (ovaries or testes) that secrete sex hormones and produce reproductive cells (eggs or sperm) and a set of ducts and accessory glands and organs such as the uterus, vagina, prostate gland, and penis.

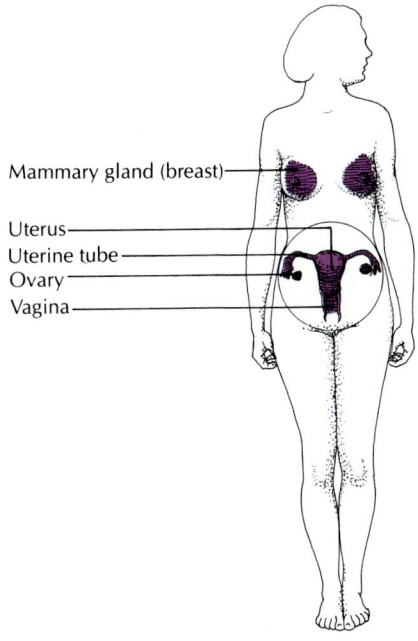

Mammary gland (breast)
Uterus
Uterine tube
Ovary
Vagina

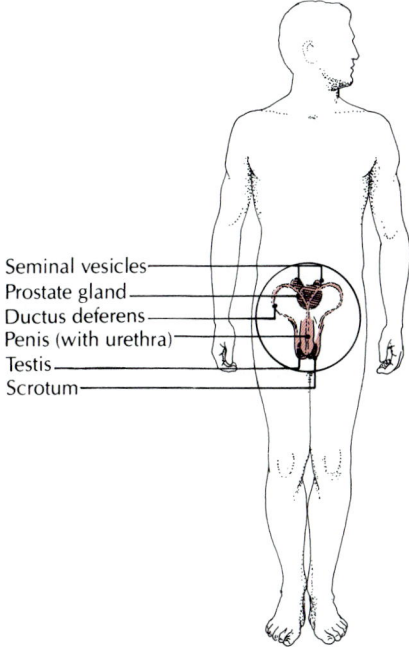

Seminal vesicles
Prostate gland
Ductus deferens
Penis (with urethra)
Testis
Scrotum

ANATOMICAL TERMINOLOGY

The language of anatomy will probably be unfamiliar to you at first. But once you have an understanding of the basic word roots, combining word forms, prefixes, and suffixes, you will find that anatomical terminology is not as difficult as you first imagined. For instance, if you know that *cardio* refers to the heart and that *myo* means muscle, you can figure out that *myocardium* refers to the muscle of the heart. See the front endpaper for a detailed list of prefixes, suffixes, and combining word forms. The Glossary at the back of this book can also help you learn the meanings of anatomical terms.

Anatomical Position

To describe the location of particular parts of the body, anatomists have defined the *anatomical position* [FIGURE 1.7A], which is universally accepted as the starting point for positional references to the body. In the anatomical position, the body is standing erect and facing forward, the feet are together, and the arms are hanging at the sides with the palms facing forward.

Relative Directional Terms of the Body

Standardized terms of reference are used when anatomists describe the location of a body part. Note that in the heading of this section we used the word *relative*, which means that the location of one part of the body is always described in relation to another part of the body. For instance, when you use standard anatomical terminology to locate the head, you say, "The head is *superior* to the neck," instead of saying, "The head is *above* the neck." When one is using directional terms, it is *always* assumed that the body is in the anatomical position.

Like so much else in anatomy, directional terms are used in pairs. If there is a term that means "above," there is also a term that means "below" [FIGURE 1.7]. If the thigh is *superior* to (above) the knee, the knee is *inferior* to (below) the thigh. The term *anterior* (or *ventral*) means toward the front of the body, and *posterior* (or *dorsal*) means toward the back of the body [FIGURE 1.7B]. The toes are anterior to the heel, and the heel is posterior to the toes.

Medial means nearer to the imaginary midline of the body or a body part [FIGURE 1.7A], and *lateral* means farther from the midline. The nose is medial to the eyes, and the eyes are lateral to the nose.

The terms *proximal* and *distal* are used mostly for body extremities, such as the arms, legs, and fingers. *Proximal* means nearer the trunk of the body (toward the attached end of a limb), and *distal* means farther from the trunk of the body (away from the attached end of a limb). The

shoulder is proximal to the wrist, and the wrist is distal to the forearm.

Superficial means nearer the surface of the body, and *deep* means farther from the surface. *External* means outside, and *internal* means inside; these terms are not the same as superficial and deep.

Peripheral is used at times to describe structures other than internal organs that are located or directed away from the central axis of the body. Peripheral nerves, for instance, radiate away from the brain and spinal cord.

The sole of the foot is called the *plantar* surface, and the upper surface of the foot is called the *dorsal* surface. The palm of the hand is the *palmar* surface, and the back of the hand is referred to as the dorsal surface.

The term *parietal* (puh-RYE-uh-tuhl) refers to the walls of a body cavity or the membrane lining the walls of a body cavity; *visceral* (VIHSS-er-uhl) refers to an internal organ contained in a body cavity (such as the liver in the abdominal cavity) or describes a membrane that covers an internal organ.

Body Regions

With the body in the anatomical position, the regional approach can be used to describe general areas of the body. The main divisions of the body are the *axial* part, consisting of the head, neck, thorax (chest), abdomen, and pelvis, and the *appendicular* part, which includes the *upper extremities* (shoulders, upper arms, forearms, wrists, and hands) and the *lower extremities* (hips, thighs, lower legs, ankles, and feet) [FIGURE 1.8A]. (The extremities are also called *limbs*.) FIGURE 1.8B and C presents additional anatomical terms for the body regions.

It is customary to subdivide the abdominal region with two vertical lines and two horizontal lines into nine regions, as shown in FIGURE 1.9A. (The two vertical lines are drawn downward from the centers of the collarbones; one horizontal line is drawn at the lower edge of the rib cage, and another horizontal line is placed at the upper edges of the hipbones.) The abdominal region may also be divided into four quadrants, as shown in FIGURE 1.9B. To help you remember the divisions of the abdominal region, study the following list:

1 *Upper abdomen:* right hypochondriac region, epigastric region, left hypochondriac region; roughly the upper third of the abdomen.

2 *Middle abdomen:* right lumbar (lateral) region, umbilical region, left lumbar (lateral) region; roughly the middle third of the abdomen.

3 *Lower abdomen:* right iliac (inguinal) region, hypogastric (pubic) region, left iliac (inguinal) region; roughly the lower third of the abdomen.

FIGURE 1.7 RELATIVE DIRECTIONAL TERMS OF THE BODY

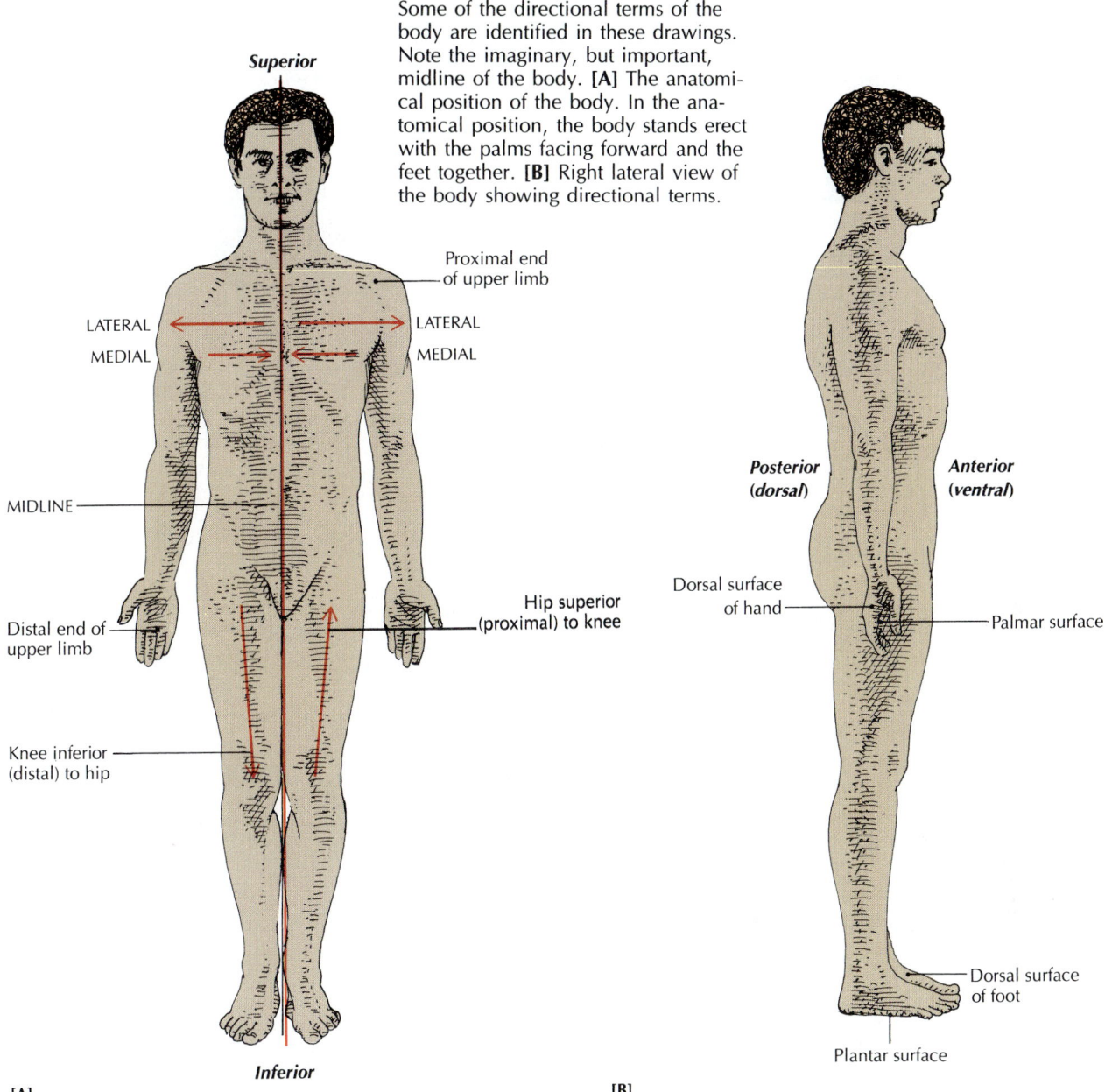

Some of the directional terms of the body are identified in these drawings. Note the imaginary, but important, midline of the body. **[A]** The anatomical position of the body. In the anatomical position, the body stands erect with the palms facing forward and the feet together. **[B]** Right lateral view of the body showing directional terms.

[A]

[B]

Term	Definition and example
Superior (cranial)	Toward the head: The leg is superior to the foot.
Inferior (caudal)	Toward the feet or tail region: The foot is inferior to the leg.
Anterior (ventral)	Toward the front of the body: The nose is anterior to the ears.
Posterior (dorsal)	Toward the back of the body: The ears are posterior to the nose.
Medial	Toward the midline of the body: The nose is medial to the eyes.
Lateral	Away from the midline of the body: The eyes are lateral to the nose.
Proximal	Toward (nearer) the trunk of the body or the attached end of a limb: The shoulder is proximal to the wrist.
Distal	Away (farther) from the trunk of the body or the attached end of a limb: The wrist is distal to the forearm.
Superficial	Nearer the surface of the body: The ribs are more superficial than the heart.
Deep	Farther from the surface of the body: The heart is deeper than the ribs.
Peripheral	Away from the central axis of the body: Peripheral nerves radiate away from the brain and spinal cord.

FIGURE 1.8 BODY REGIONS

[A] The basic portions of the body, as located within the axial and appendicular parts. **[B]** Ventral and **[C]** dorsal views of the body regions present a more detailed list of the many terms used to locate specific parts of the body.

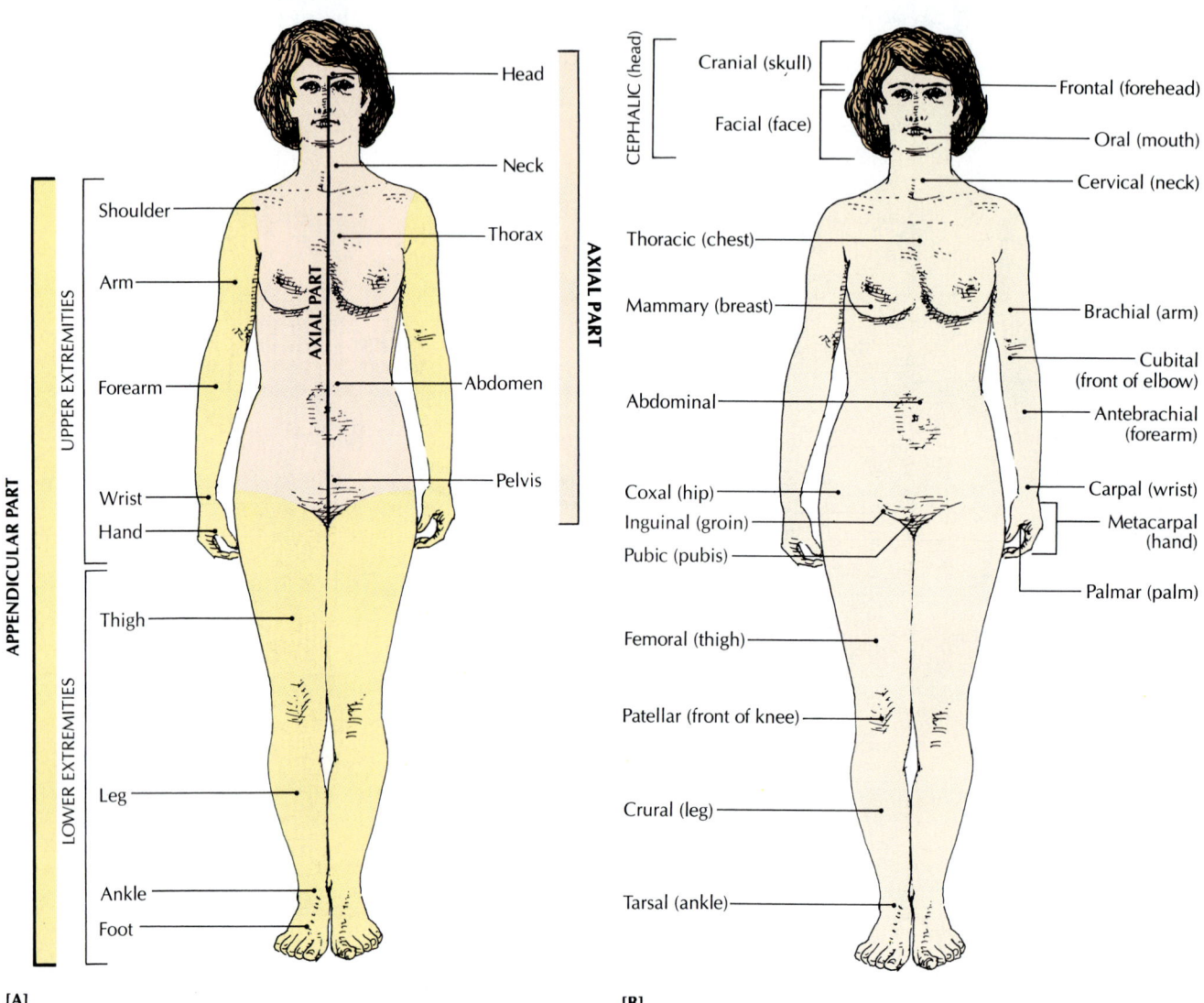

[A]

[B]

Body Planes

For further identification of specific areas, the body can be divided by imaginary flat surfaces, or *planes* [FIGURE 1.10]. The ***midsagittal plane*** divides the left and right sides of the body lengthwise along the midline into externally symmetrical sections. If a longitudinal plane is placed off-center and separates the body into asymmetrical left and right sections, it is called a ***sagittal plane.*** If you were to face the side of the body and make a lengthwise cut at right angles to the midsagittal plane, you would make a ***frontal*** (or *coronal*) ***plane,*** which divides the body into asymmetrical anterior and posterior sec-

tions [FIGURE 1.10B]. A ***transverse plane*** divides the body horizontally into upper (superior) and lower (inferior) sections. A transverse (or *horizontal*) plane is at right angles to the sagittal and frontal planes [FIGURE 1.10A]. Transverse planes do not produce symmetrical halves.

The system of planes is also used with *parts* of the body, including internal parts. FIGURE 1.11A shows how cross sections, oblique sections, and longitudinal sections of internal parts are made.

If your laboratory manual or any other book refers to a drawing of a *sagittal section*, a *frontal section*, or a *transverse section*, you should be aware of what is actually being

FIGURE 1.9 ABDOMINAL SUBDIVISIONS

[A] The nine subdivisions of the abdominal region. [B] The four quadrants of the abdominal region.

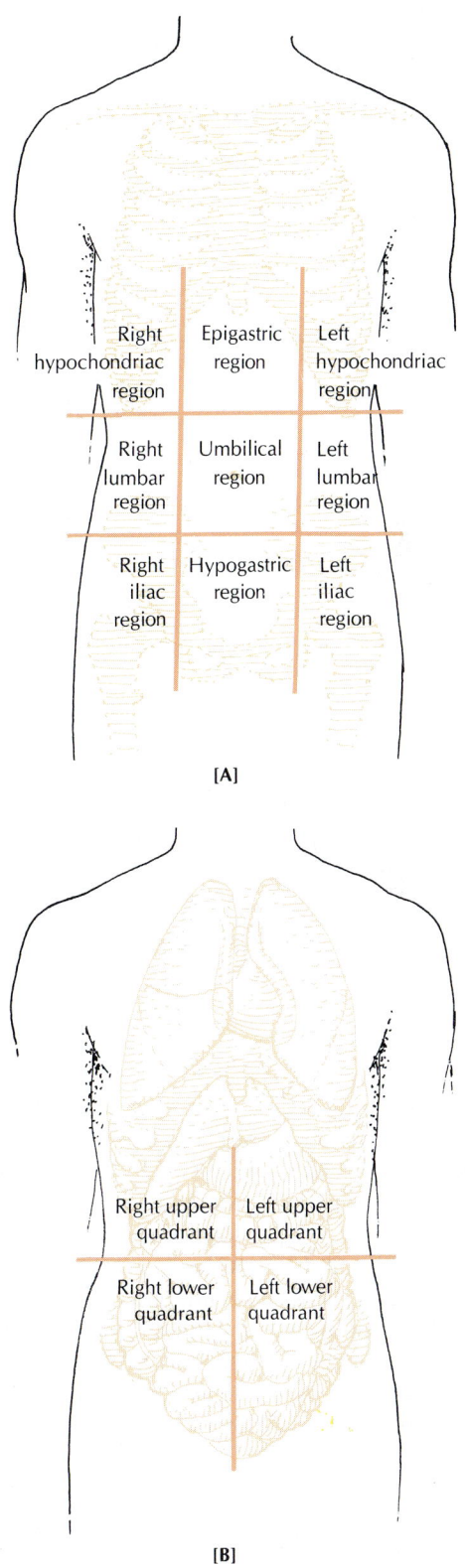

[A]

[B]

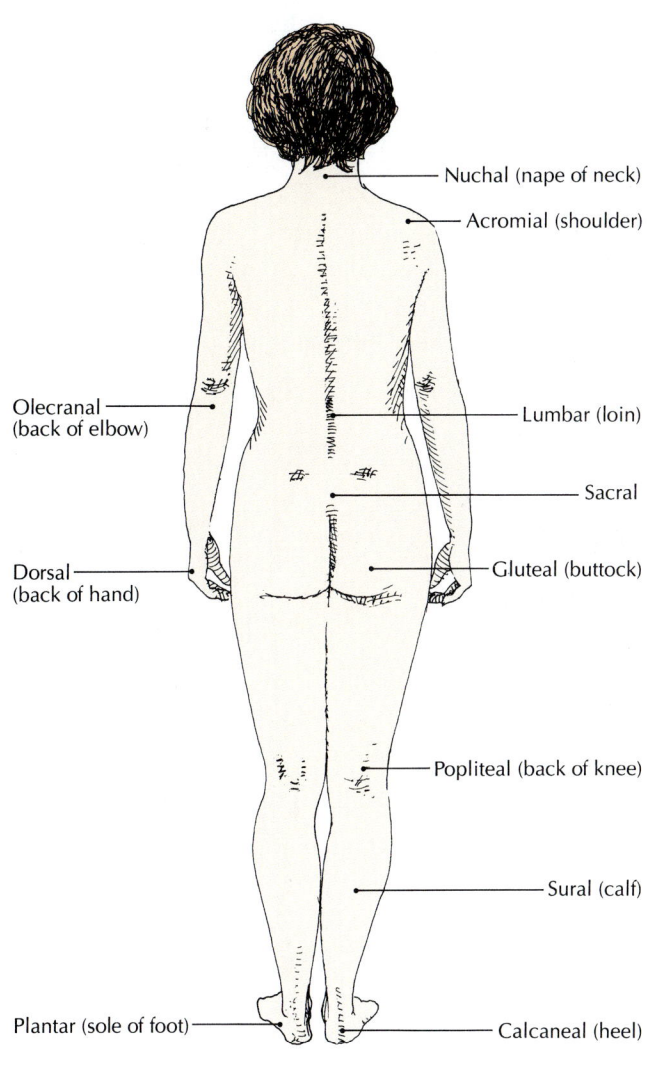

Nuchal (nape of neck)

Acromial (shoulder)

Olecranal (back of elbow)

Lumbar (loin)

Sacral

Dorsal (back of hand)

Gluteal (buttock)

Popliteal (back of knee)

Sural (calf)

Plantar (sole of foot)

Calcaneal (heel)

[C]

shown and how it relates to its corresponding plane. FIG-URE 1.11B shows a cut along the midsagittal plane of the head. Such a cut produces an exposed surface of the head called a *midsagittal section.* A cut along a frontal plane produces a *frontal section*; FIGURE 1.11C shows a frontal section of the brain. A cut along the transverse plane — in the case of FIGURE 1.11D, across the abdomen — produces a *transverse section.*

Until just a few years ago, the only way to obtain sections such as those shown in FIGURE 1.11 was to actually cut the organs or parts of the body. Now, techniques such as magnetic resonance imaging (MRI), computer-

FIGURE 1.10 BODY PLANES

The imaginary body planes are an additional source of identification and location. **[A]** Representations of the midsagittal, sagittal, and transverse planes. **[B]** Representation of the frontal plane.

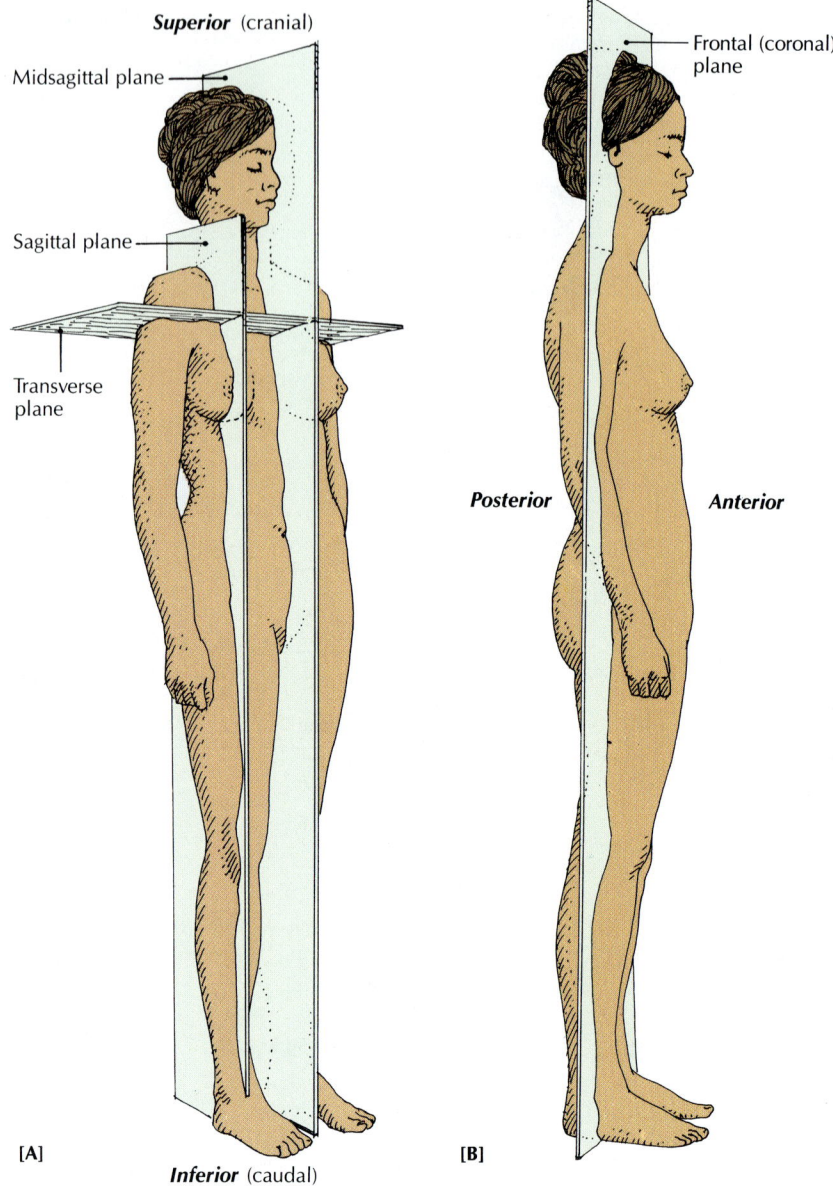

assisted tomography (CT) scanning,* and ultrasound make it possible to obtain clear pictures of planes of the body from living subjects [FIGURE 1.12]. Such pictures, which can be viewed on a computer monitor or video screen or printed on paper, can often show differences between healthy and diseased tissue or even indicate the metabolic state of the tissue being observed. In contrast, x-rays can only show a flat picture without indicating depth or isolating the different layers of an organ, and they do not always show differences between healthy and diseased tissue. A new development, called ultrafast CT,

operates in the same way as does a CT scan, but much faster. Such speed is important, for example, in recording clear images of the beating heart. Also, the ultrafast CT exposes the patient to less radiation.

Anatomical Variations

We take it for granted when we recognize people instantly. But why is it so easy to tell Aunt Rose from Aunt Marie? Because the shapes of heads, noses, ears, mouths, nostrils, eyes, arms, legs, and torsos are always different enough (except perhaps with identical twins) for us to be able to distinguish one person from another. Human structure is not rigidly consistent. Instead, some ***anatomical variations*** are normal and even expected.

*The word *tomography* means "a technique for making a picture of a section or slice of an object" and comes from the Greek *tomos*, a cut or section + *graphein*, to write or draw.

FIGURE 1.11 BODY SECTIONS

[A] Three different ways to cut sections for microscopic examination. [B] A cut along the midsagittal plane of the head (smaller drawing) produces a midsagittal section of the head. [C] A cut along the frontal plane of the brain (smaller drawing) produces a frontal section (right). [D] A cut along the transverse plane of the abdomen (smaller drawing) results in a transverse section, in this case looking down into the lower portion of the cut body.

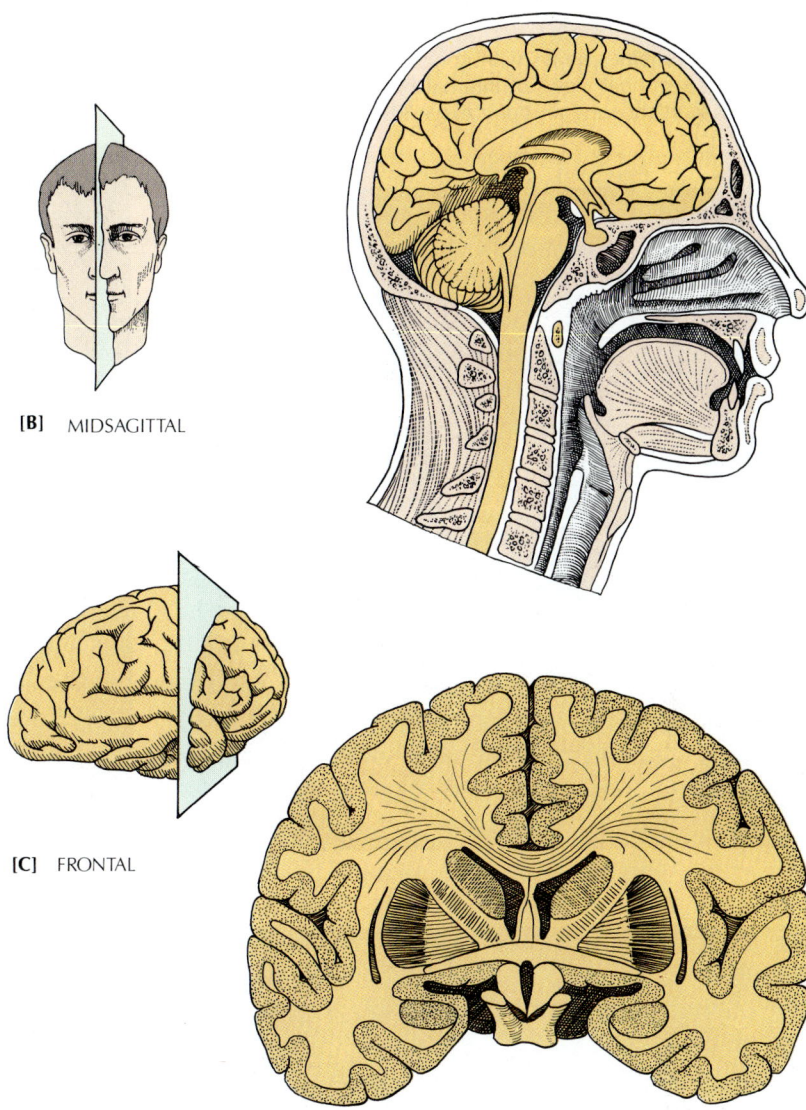

[B] MIDSAGITTAL

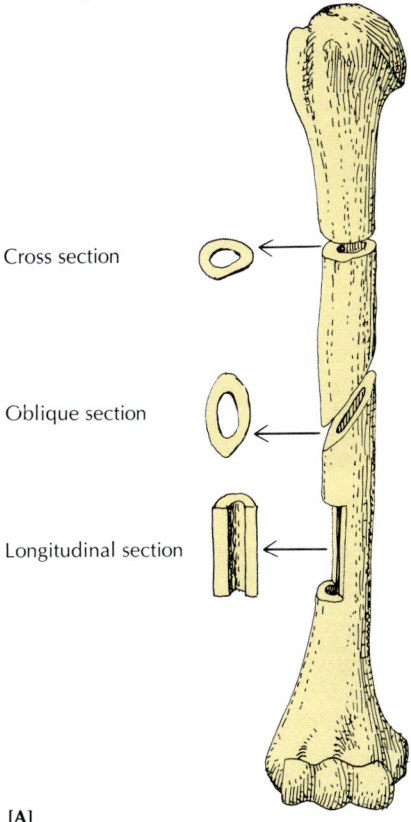

Cross section

Oblique section

Longitudinal section

[C] FRONTAL

[A]

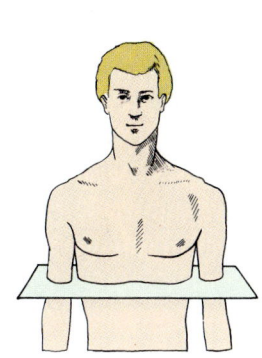

[D] TRANSVERSE

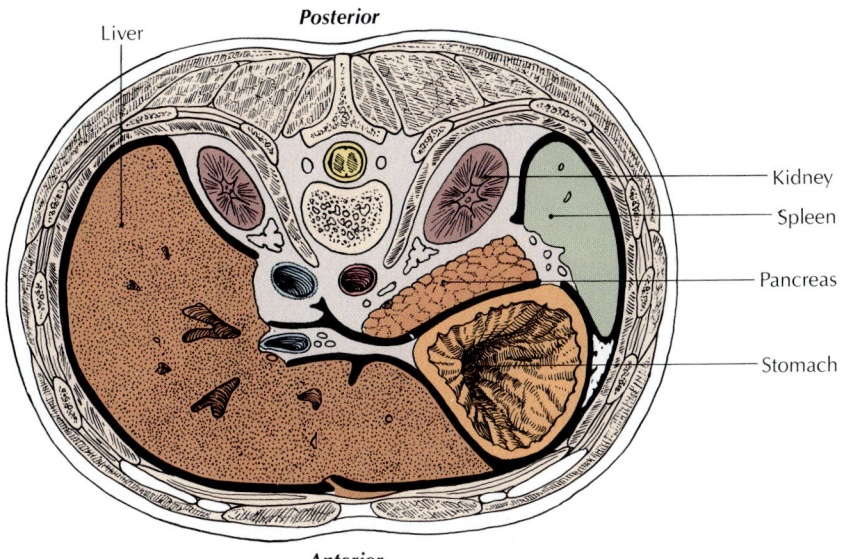

Liver

Posterior

Kidney

Spleen

Pancreas

Stomach

Anterior

FIGURE 1.12 NEW WAYS OF EXPLORING THE BODY

A **CT scan** combines x-rays with computer technology to show cross-sectional views of internal body structures. **[A]** A CT scan showing an optic nerve lesion.

[B] A three-dimensional CT scan image of a twisted and compressed spinal bone (vertebra) (center) after a motorcycle accident. Note the ruptured disks between the vertebrae.

Optic nerve lesion

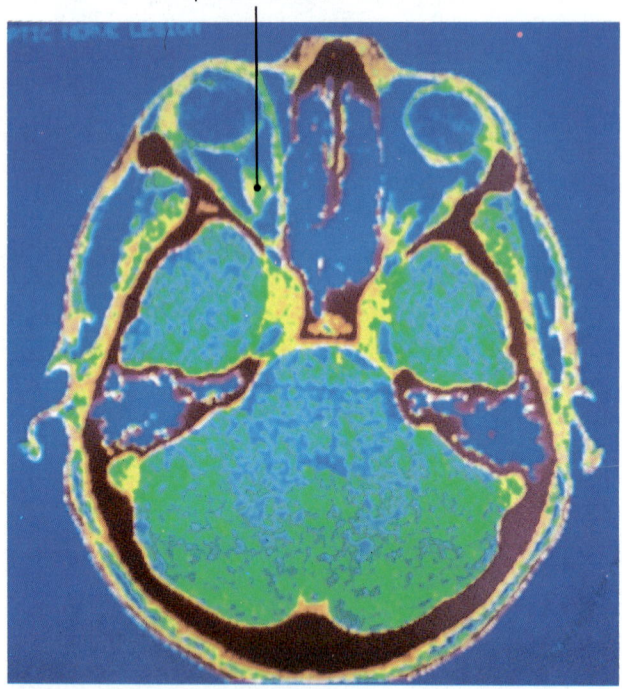

[A]

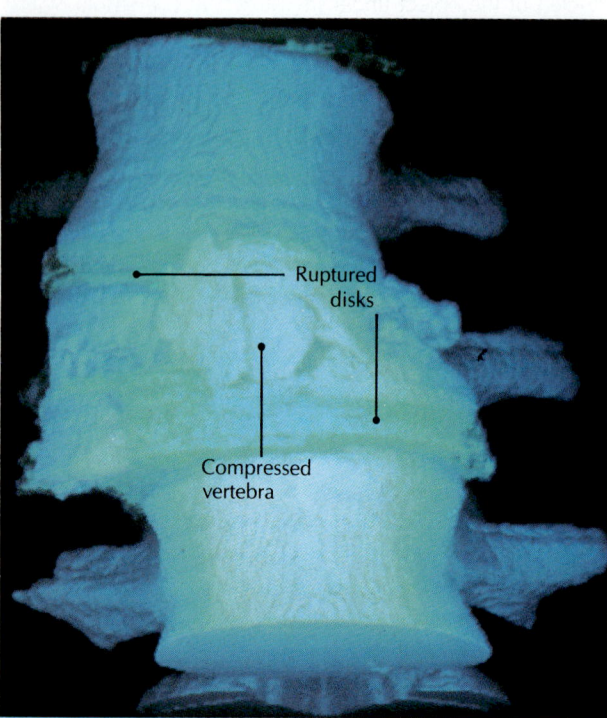

Ruptured disks

Compressed vertebra

[B]

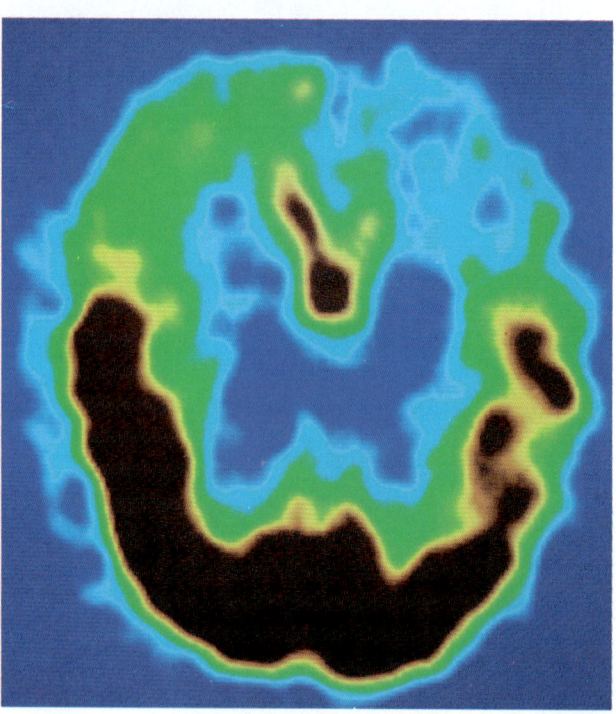

[C]

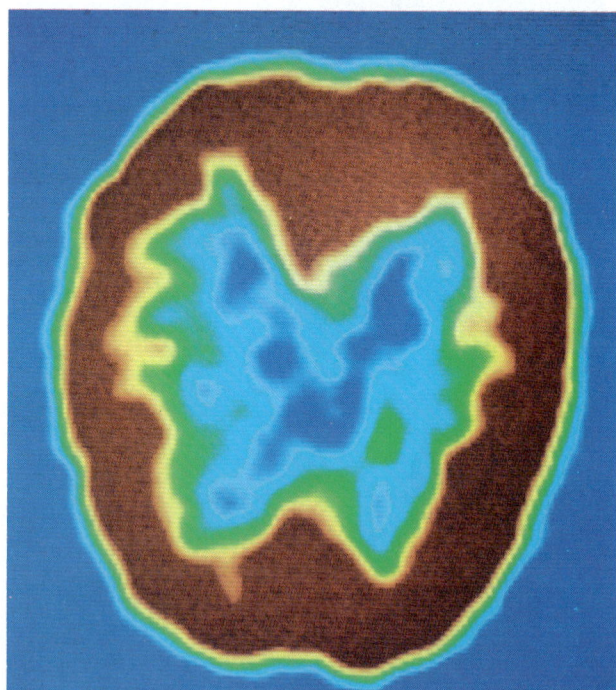

[D]

A **PET scan** (positron emission tomography) reveals the metabolic state (level of chemical activity) of an organ by indicating the rate at which its tissues consume injected biochemicals such as glucose. **[C]** A PET scan of a patient with Alzheimer's disease shows a decreased rate of metabolism (blue and green on the color scale) in the brain. **[D]** A PET scan of the brain of a normal individual.

Ultrasound (also called *sonography*) sends pulses of ultra-high-frequency sound waves into designated body cavities. From the echoes that result, video images can be constructed of the object under investigation. **[E]** An ultrasound image of the head and neck of a 19-week-old fetus inside its mother's uterus. Such pictures can assess fetal growth and development and assist a physician during fetal surgery, for example. Ultrasound images have also been used to detect cancer, arterial plaque, and other disorders.

Digital subtraction angiography (Gr. *angeion,* a vessel), or **DSA,** produces three-dimensional pictures of blood vessels. DSA has made it possible to predict and prevent imminent heart attacks. **[F]** A picture of the heart and the left coronary artery made by DSA reveals a constriction (arrow). The constriction has blocked about 60 percent of the blood flow to the lower heart (blue). The appearance of the heart was enhanced by an opaque dye.

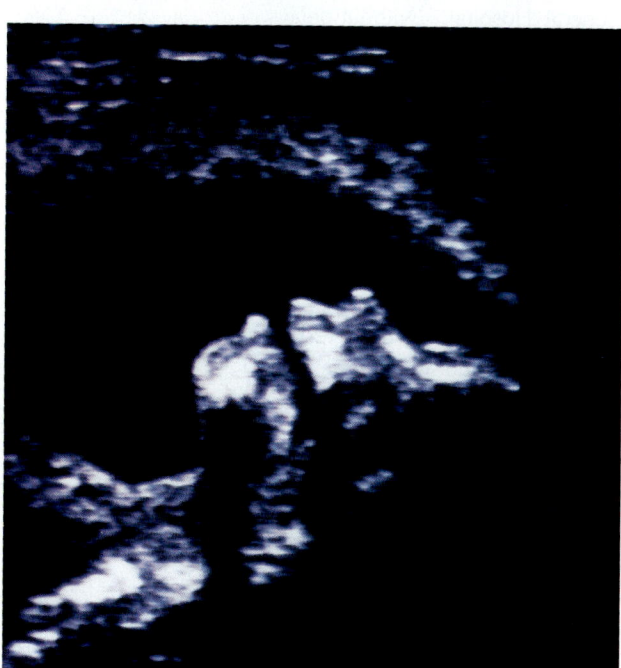

[E]

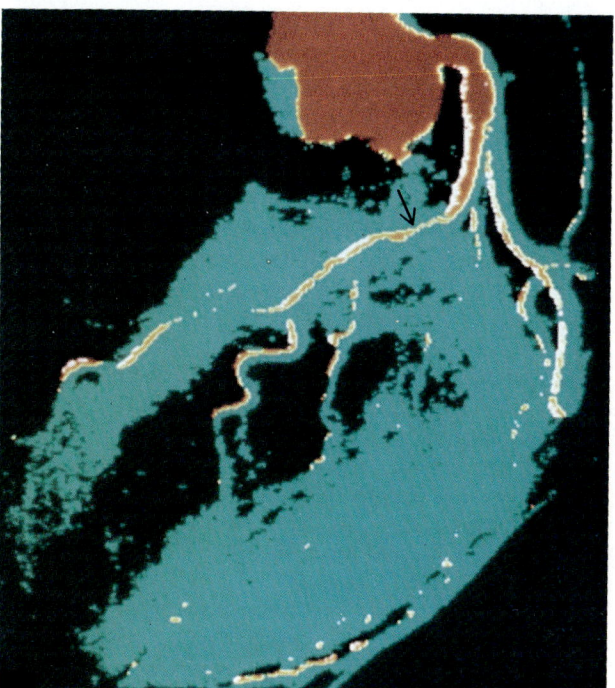

[F]

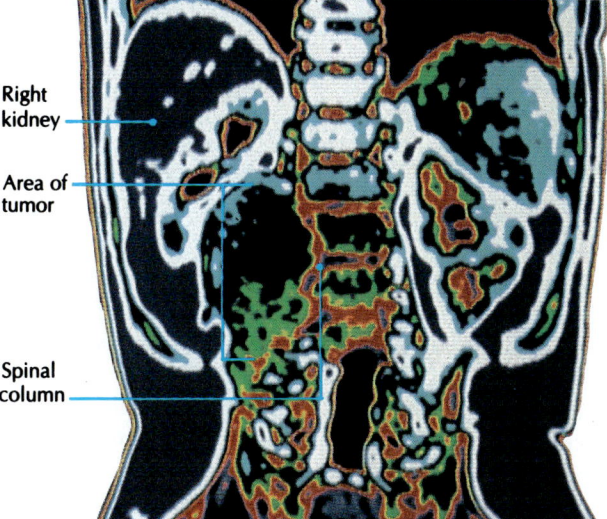

Right kidney

Area of tumor

Spinal column

[G]

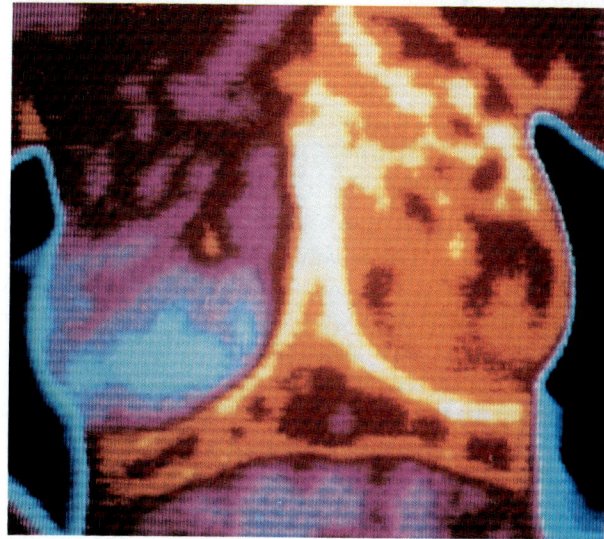

[H]

Magnetic resonance imaging (MRI) produces exceptionally clear images that show the difference between healthy and diseased tissue. It has been used to make an early diagnosis of multiple sclerosis, atherosclerosis, and brain tumors, among other disorders. **[G]** A color-enhanced MRI scan of a 7-month-old child shows a malignant tumor between the right kidney and the spinal column, pressing on the spinal cord.

Thermography ("heat writing") reveals chemical reactions taking place within the body, based on heat changes in the skin. In use since 1965, it has been employed to detect breast cancer, arthritis, circulatory problems, and other disorders. **[H]** This thermogram reveals cancer in the left breast of a woman (red areas) by indicating a marked temperature increase in the cancerous areas.

"Telemedicine"

On the television screen in the small room at the Medical College of Georgia [in Augusta] an image jumped and jerked as, 150 miles away, a doctor brought an instrument into position in the ear of a patient.

An operator [in Augusta] manipulated dials, zooming in, increasing the lighting and sharpening the focus of the image of the patient's inner ear and the scarring that was the focus of the examination. She and the specialist beside her had at least as good a view of what their colleague on the scene was seeing. The distance between them seemed to shrink to nothing as they examined the patient together.

Designers hope the two-way, interactive system will be to medical providers what teleconferencing is to business. The examination on this day saved the patient the cost of a trip to the Augusta hospital to see the same specialist. . . .

"The one thing we can't do with this apparatus is to palpate a patient's abdomen," said Dr. Frank Tedesco, president of the Medical College of Georgia. "But we're working on that." As an addition to the screens, cameras, fax machines and videocassette players, he said, a scientist at the Georgia Institute of Technology is working on a glovelike device that would transmit and receive something akin to tactile sensations.

The basics of telemedicine have been available for about 35 years. . . . It began for the most mundane of reasons: in one of its more notable early applications, in 1968, Massachusetts General Hospital used an interactive telemedicine hookup as a way of staffing the hospital's airport clinic and relieving the doctor on call of a daily three-hour drive.

The early efforts involved bulky and expensive equipment that required two flatbed trucks to transport. It depended on costly and slow analog-type transmission by microwave or satellite, and in many ways it was little more than teleconferencing with fixed cameras and very few hookups for other diagnostic tools. . . .

But recent advances have cut costs and led to less bulky equipment. There was the development of an attaché-case-sized coder-decoder, or "codec," which compresses and digitalizes signals used in telemedicine for sending and decompresses them at the receiving end. The older analog transmission used electrical wave patterns, which were subject to interference and decay over distances. But the newer digital transmission converted the audio and video signals into numerical representations, which are immune to those problems.

The speed, precision and versatility of digitalized transmission has sharpened the picture and sound and made interaction in telemedicine possible without delays.

Such transmission is also possible over relatively inexpensive fiber optic telephone lines. The Medical College of Georgia's telemedicine experiment was possible largely because in 1987 Southern Bell laid an extensive fiber optic network in the state in anticipation of future needs.

Other advances included the ability to hook stethoscopes, endoscopes, otoscopes, microscopes, electro- and echocardiograms and sonograms and other diagnostics devices directly into the system to send images and sounds instantaneously. The experiment in Georgia has all these features and is considered by experts in the field to be the most sophisticated of all the telemedicine experiments now operating.

There are many such variations besides the typical facial differences. For example, one person in 20 has an extra pair of ribs. We say that the body contains 206 bones, but that is only the most common number, not the constant one. The same muscle may have a different nerve or blood supply in two different people, or a small muscle may be lacking altogether. The shape of a particular bone may vary from one person to another, and connections of muscles to bones may be different. The size and placement of blood vessels vary relatively often, while nerves are remarkably constant. Variations are found most often from one side of a person's body to the other. But as long as the body functions properly, such variations are to be expected, and are not even noted as being variations. In contrast, major anatomical variations, such as extra fingers or toes, are considered abnormalities rather than variations.

BODY CAVITIES AND MEMBRANES

The *cavities of the body* house the internal organs, commonly referred to as the *viscera*. The two main body cavities are the dorsal and ventral cavities, each of which contains smaller cavities [FIGURE 1.13].

Body membranes are thin layers of epithelial and connective tissue that line the body cavities and cover or separate certain regions, structures, and organs. The three main types of membranes are mucous (nasal and oral cavities), serous (body cavities), and synovial (joint cavities); they are described in detail in Chapter 4.

Double-walled sacs of serous membranes line the body cavities and surround the organs. The two walls that make up these sacs are separated by a thin film of serous fluid, which acts as a lubricant to facilitate the nearly frictionless movements of the organs.

A simple analogy can be used to illustrate the concept of the body cavities and their surrounding membranes. Imagine a balloon that has been partially blown up so that it is still soft. Press your fist into the balloon [FIGURE 1.13C]. No matter how hard you press, your fist cannot get into the cavity of the balloon as long as the surface of the balloon is intact. The space within the balloon is equivalent to the body cavity, and your fist is equivalent to the organ. The portion of the balloon in contact with your fist is the *visceral* ("organ") portion of the membrane, and the outer wall of the balloon is the *parietal* ("outer wall") portion of the membrane [FIGURE 1.13C].

Dorsal Body Cavity and Membranes

The *dorsal body cavity* is located toward the posterior portion of the body. It contains the smaller cranial and spinal cavities. The *cranial cavity* contains the brain, and the *spinal* (or *vertebral*) *cavity* contains the spinal cord.

In addition to the protection given by the bony cranium and spinal column, the dorsal cavity is lined, and the brain and spinal cord are covered, by thin membranes called *meninges* (muh-NIHN-jeez).

Ventral Body Cavity and Membranes

The *ventral body cavity* is located toward the anterior portion of the body. It contains two large subdivisions, the thoracic and abdominopelvic cavities.

Thoracic cavity and membranes The *thoracic cavity* is separated from the inferior abdominopelvic cavity by the dome-shaped muscle called the diaphragm. The thoracic cavity contains the lungs and the *mediastinum* (mee-dee-as-TIE-nuhm; L. *medius*, middle), the partitioned area between the lungs. The mediastinum contains the heart and its attached blood vessels, the trachea, the esophagus, and all the other contents of the thoracic cavity except the lungs. The organs within the mediastinum are held together by dense connective tissue.

The thoracic cavity is protected by the rib cage and its associated musculature, as well as by the sternum (breastbone) anteriorly and the 12 thoracic vertebrae posteriorly.

Each lung is surrounded by a serous membrane sac called the *pleura* (pl. *pleurae*). The part of the sac that covers the lungs is called the *visceral pleura*, while the part that lines the thoracic body wall is called the *parietal pleura*. The very slight space between the two layers of the pleura is the *pleural cavity*. The cavity contains a small amount of serous fluid that serves as a lubricant, preventing friction between the membranes during breathing. An inflammation of the pleurae is called *pleurisy*.

The heart is surrounded by a double-walled serous membrane called the *pericardium* [FIGURE 1.13C]. The space between the two membrane layers is the *pericardial cavity*. It also contains a small amount of serous fluid that acts as a lubricant, allowing the heart to beat and move about slightly without friction.

Abdominopelvic cavity and membranes The *abdominopelvic cavity* extends from the diaphragm inferiorly to the floor of the pelvis [FIGURE 1.13B, C]. At the superior margin of the pelvis it is divided by an imaginary line into the superior *abdominal cavity* and the inferior *pelvic cavity*.

The *abdominal cavity*, the largest cavity in the body, contains the stomach, small and large intestines, liver, spleen, pancreas, kidneys, and gallbladder. The *pelvic cavity* contains the urinary bladder, the rectum, and the internal portions of the reproductive systems. The pelvic cavity projects backward from the abdominal cavity at about a 45-degree angle.

The abdominal and pelvic cavities are lined with, and the organs within them are covered with, a serous membrane called the *peritoneum* (per-uh-tuh-NEE-uhm). The layer covering the organs is the *visceral peritoneum*, while that lining the thoracic wall is the *parietal peritoneum*. An inflammation of the peritoneum is called *peritonitis*. The kidneys, duodenum, pancreas, and adrenal glands are located *behind* the peritoneum (adjacent to the posterior body wall) and are, therefore, referred to as *retroperitoneal* ("behind the peritoneum").

The *peritoneal cavity*, the space between the two layers of the peritoneum, contains only a thin film of lubricating fluid. The fluid reduces friction during movements of the intestines and contractions of the uterus during labor and delivery.

The Smaller Body Cavities

Some other smaller body cavities include the nasal, oral, orbital (eye), tympanic (middle ear), and synovial (movable joint) cavities. Each of these will be described further in their relevant chapters.

A S K   Y O U R S E L F

1 What is the anatomical position of the body?

2 What is the difference between superior and anterior?

3 How does a sagittal plane differ from a midsagittal plane?

4 What organs are within the two main body cavities?

5 How is the thoracic cavity separated from the abdominopelvic cavity?

6 What do you call the membrane that covers a body organ? That lines a body cavity?

FIGURE 1.13 MAIN BODY CAVITIES

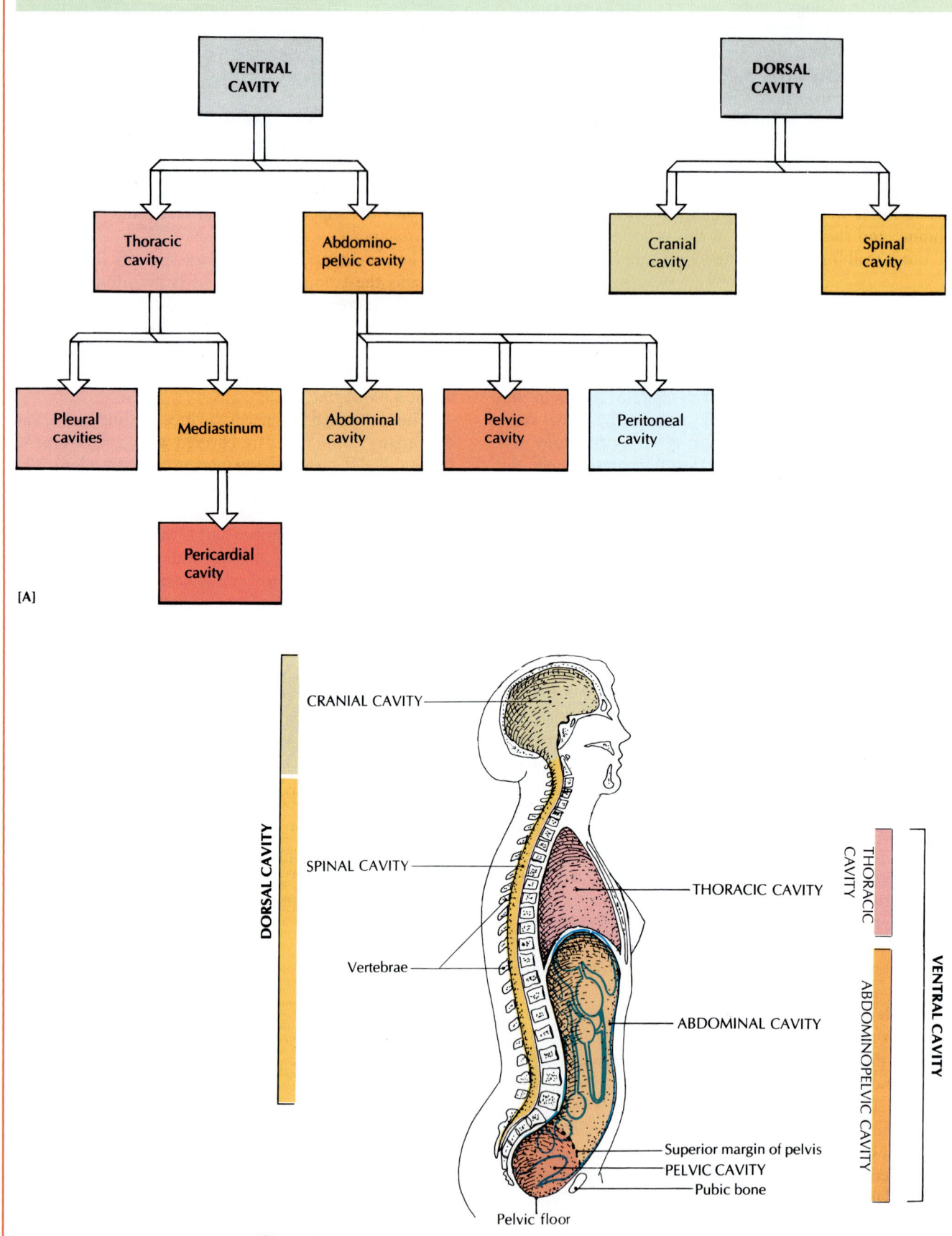

[A]

[B] MIDSAGITTAL SECTION

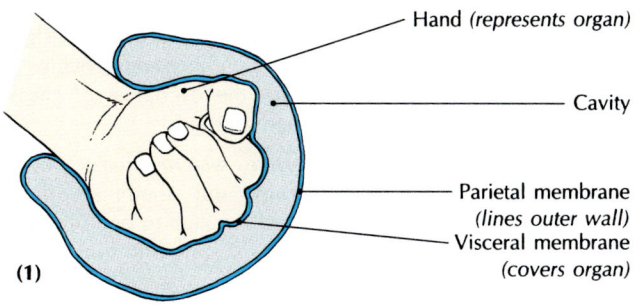

(1)

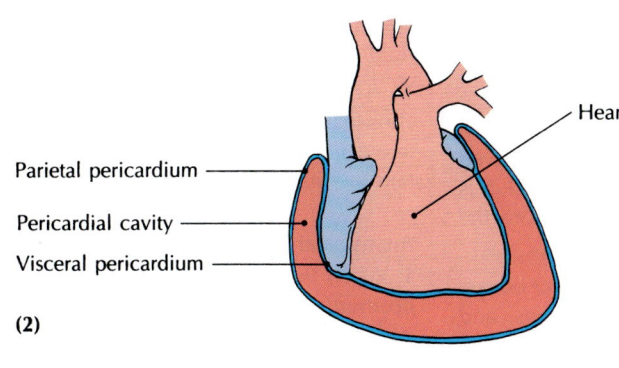

Parietal pericardium

Pericardial cavity

Visceral pericardium

Heart

(2)

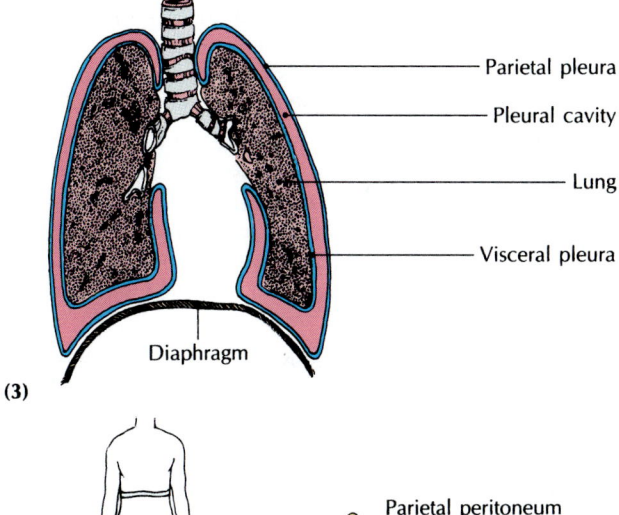

Parietal pleura

Pleural cavity

Lung

Visceral pleura

Diaphragm

(3)

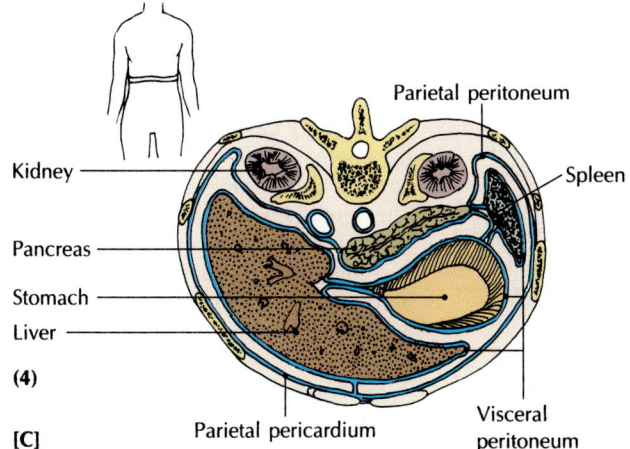

Kidney

Pancreas

Stomach

Liver

(4)

Parietal peritoneum

Parietal pericardium

Spleen

Visceral peritoneum

[C]

[A] A flowchart showing the ventral and dorsal cavities and their subdivisions. [B] The dorsal and ventral cavities and the smaller cavities within them; midsagittal section. [C] Schematic diagrams illustrating (1) the concept of body membranes and cavities, using a fist pushed into a partially blown-up balloon. The hand represents an organ; the balloon, the membrane. The outer membrane is the parietal membrane, and the inner membrane is the visceral membrane, covering the organ itself. (2) The heart enclosed by the pericardium. The pericardium covering the heart is the visceral pericardium, and the membrane lining the thoracic wall is the parietal pericardium. (3) The relationship of the lungs, pleurae, and pleural cavities. The visceral pleura covers the lung, and the parietal pleura lines the thoracic wall. (4) Transverse section through the upper abdomen, showing some organs and their relation to the peritoneum and peritoneal cavity. The visceral peritoneum covers the viscera (the organs), and the parietal peritoneum lines the abdominal wall. The mesentery is a fold of peritoneum from the parietal peritoneum to the visceral peritoneum; it attaches an abdominal organ to the abdominal wall.

MAIN BODY CAVITIES AND THEIR ORGAN CONTENTS*

Cavity	Contents
DORSAL (posterior)	Cranial cavity, spinal (vertebral) cavity
Cranial	Brain
Spinal (vertebral)	Spinal cord
VENTRAL (anterior)	Thoracic cavity, abdominopelvic cavity
Thoracic	Mediastinum (central region of thorax containing heart, great blood vessels, esophagus, and trachea) Lungs Pericardial and pleural cavities
Abdominopelvic	Abdominal and pelvic cavities
Abdominal cavity	Stomach, intestines, liver, spleen, pancreas, kidneys, gallbladder; peritoneal cavity
Pelvic cavity	Urinary bladder, rectum, internal portions of reproductive systems; peritoneal cavity

*The *pericardial cavity* is the space between the two membrane layers of the pericardium; the *peritoneal cavity* is the space between the two membrane layers of the peritoneum; the *pleural cavity* is the space between the two membrane layers of the pleura.

CHAPTER SUMMARY

What Are Anatomy and Physiology? (p. 3)

1 The study of **anatomy** deals with the *structure* of the body; **physiology** explains the *functions* of the parts of the body.

2 Some subdivisions of anatomy are *regional, embryological, systemic, developmental,* and *radiographic anatomy.* Any branch of anatomy that can be studied without a microscope is called *gross anatomy; microscopic* anatomy requires the use of a microscope.

3 In the human body, structure (anatomy) and function (physiology) work together to make the parts of the body operate at peak efficiency.

Homeostasis: Coordination Creates Stability (p. 3)

1 *Homeostasis* is an inner stability of the body that exists even if the environment outside the body changes. Homeostasis is achieved when structure and function are properly coordinated and all the body systems work together.

2 Practically everything that goes on in the body helps maintain homeostasis, and the entire process is made possible by the coordinated action of many organs and tissues under the control of the nervous and endocrine systems.

3 The maintenance of homeostasis often involves **negative feedback** in which the body's response is opposite to the initiating stimulus.

4 When homeostasis breaks down, we become sick or even die. One way to unbalance homeostasis is to introduce abnormally high levels of **stress.**

From Atom to Organism: Structural Levels of the Body (p. 7)

1 At its simplest level, the body is composed of **atoms,** the basic units of all matter. When two or more atoms combine, they form a **molecule.** If a molecule is composed of more than one element, it is a **compound.**

2 *Cells* are the smallest independent units of life. Some of the basic functions of cells are metabolism, irritability, growth, and reproduction.

3 *Tissues* are composed of many similar cells that perform a specific function.

Tissues are classified into four types: *epithelial, connective, muscle,* and *nervous.*

4 An **organ** is an integrated collection of two or more kinds of tissues that combine to perform a specific function.

5 A **system** is a group of organs that work together to perform a major body function. All the body systems are specialized within themselves and coordinated with each other to produce an **organism.**

Body Systems (p. 10)

1 The **integumentary system** consists of the skin and all the structures derived from it. The main function of the skin is to protect the internal organs from the external environment.

2 The **skeletal system** consists of bones, certain cartilages and membranes, and joints. It supports the body, protects the organs, enables the body to move, manufactures blood cells in the bone marrow, and stores usable calcium and phosphorus.

3 The **muscular system** consists of muscles and tendons. It allows for movement and generates body heat.

4 The **nervous system** consists of the *central nervous system* and the *peripheral nervous system;* it also includes special sensory organs. The nervous system and the endocrine system are the body's main control and regulatory systems.

5 The **endocrine system** consists of a group of ductless glands that secrete *hormones.* Hormones regulate metabolic reactions within cells, growth and development, stress and injury responses, reproduction, and many other critical functions.

6 The **cardiovascular system** consists of the heart, blood, and blood vessels. An important function of the cardiovascular system is to transport oxygen and other necessary substances throughout the body and to transport wastes to the lungs and kidneys for removal.

7 The **lymphatic system** consists of lymph nodes, lymph, and lymphatic vessels. It returns excess fluid and proteins to the blood and helps protect the body from foreign substances, organisms, and cancer cells.

8 The **respiratory system** accomplishes the process of breathing and provides a mechanism for the exchange of gases between blood and air.

9 The **digestive system** breaks down food physically and chemically into molecules small enough to be absorbed from the small intestine into the bloodstream or lymphatic system; it also removes solid, undigested wastes.

10 The **urinary system** eliminates wastes produced by cells and regulates fluid balance in the body.

11 The **reproductive systems** have organs that produce specialized reproductive cells that make it possible to maintain the human species.

Anatomical Terminology (p. 14)

1 When the body is in the **anatomical position,** it is erect and facing forward, with the feet together, arms hanging at the sides, and palms facing forward.

2 Some basic pairs of anatomical terms used to describe the relative position of parts of the body include **superior/inferior, anterior/posterior, medial/lateral, proximal/distal, external/internal,** and **superficial/deep.**

3 The main **regions** of the body are the **axial** part, consisting of the head, neck, thorax, abdomen, and pelvis, and the **appendicular** part, which includes the upper and lower extremities. The abdominal region is divided into nine subregions.

4 The body and the parts of the body may be divided by imaginary **planes,** including the **midsagittal, sagittal, frontal,** and **transverse planes.**

Body Cavities and Membranes (p. 22)

1 The **body cavities** house and protect the internal organs. The two main body cavities are the **dorsal** and the **ventral.** The ventral body cavity is separated by the diaphragm into the superior *thoracic* and inferior *abdominopelvic* cavities, and the dorsal cavity contains the *cranial* and *spinal* cavities.

2 **Membranes** line body cavities and cover or separate certain regions, structures, and organs. The main types of membranes are *mucous, serous,* and *synovial.*

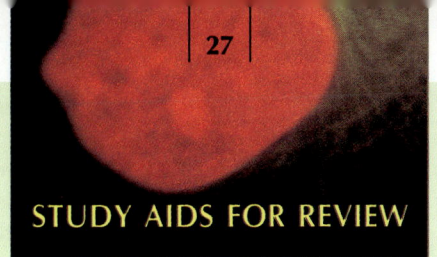

STUDY AIDS FOR REVIEW

KEY TERMS

UNDERSTANDING THE FACTS

1 What are anatomy and physiology?
2 Define homeostasis.
3 Contrast negative and positive feedback.
4 What is the difference between a tissue and an organ?
5 Which body systems are involved in waste disposal?
6 What is the anatomical position?
7 What are the meanings of superior, inferior, anterior, and posterior?
8 Define parietal and visceral.
9 What are the two main regions of the body?
10 What is a transverse plane? A frontal plane?
11 Name the two main subdivisions of the ventral cavity.

UNDERSTANDING THE CONCEPTS

1 Explain why negative feedback maintains homeostasis but positive feedback does not.
2 Which body systems act primarily to supply cells with raw materials for metabolism and which body systems act primarily as regulators and integrators?

SELF-QUIZ

Multiple-Choice Questions

1.1 The dorsal cavity contains the
   a heart
   b brain
   c spinal cord
   d a and b
   e b and c

1.2 The abdominal cavity contains the
   a abdominopelvic cavity
   b reproductive organs and urinary bladder
   c liver, spleen, and stomach
   d urinary bladder and lungs
   e entire peritoneal cavity

1.3 Which of the following does *not* correctly describe the anatomical position?
   a body facing forward
   b head turned to side
   c palms facing forward
   d body standing
   e feet together

1.4 The anatomical term most nearly opposite to lateral is
   a proximal
   b distal
   c medial
   d superficial
   e caudal

1.5 A plane that divides the body into anterior and posterior parts is a(n)
a medial plane
b coronal, or frontal, plane
c sagittal plane
d transverse plane
e oblique plane

1.6 Which of the following is a diagnostic technique that utilizes x-rays?
a CT scan
b PET scan
c sonography
d thermography
e laser surgery

1.7 Which of the following statements about stress or stressors is *false*?
a Stressors act to disrupt homeostasis.
b Stress can lead to adaptive changes.
c A life without stress or stressors would be the most healthful.
d Homeostatic mechanisms within the body counteract stress.
e Long-term stress may lead to disease, disability, or death.

1.8 Which of the following statements is correct for all of the body's organ systems?
a The organs in any system are composed of one kind of tissue.
b The organs in any system are linearly joined or linked.
c The organs in any system are discrete and isolated from those of all other systems.
d The organs in any system cooperate to perform one or more functions.
e The organs in any system may be composed either of tissues or of a group of diverse cells gathered in one location.

1.9 Which of the following structures includes all of the others?
a mediastinum
b pericardial cavity
c thoracic cavity
d pleural cavity
e ventral body cavity

1.10 Which of the following activities would be difficult or impossible if the visceral peritoneum and the parietal peritoneum were not serous membranes?
a sitting at a desk and writing
b thinking
c bending and twisting
d reading
e turning one's head

1.11 Under normal circumstances the level of sugar in the blood is controlled by a negative feedback response to the hormone insulin. What would be the effect of a rise in insulin levels?
a increased blood sugar, then increased insulin
b decreased blood sugar, then decreased insulin
c increased blood sugar, then decreased insulin
d decreased blood sugar, then increased insulin
e no change in blood sugar

1.12 For a cat standing on all fours, how would you describe the front appendages relative to the hind appendages?
a ventral
b anterior
c posterior
d superior
e inferior

1.13 Which of the following is *not* within the axial division of the body?
a chest
b abdomen
c head
d shoulder
e neck

1.14 Which of the following planes divides the body longitudinally?
a frontal
b coronal
c transverse
d sagittal
e horizontal

1.15 Which plane could result in a section through the lungs and the mediastinum?
a frontal
b coronal
c transverse
d horizontal
e all of the above

Matching

1.16	_____ proximal	a	the soles of the feet
1.17	_____ lateral	b	nearer the outside of the body
1.18	_____ superior	c	toward the head
1.19	_____ ventral	d	toward the back
1.20	_____ medial	e	toward the side of the body
1.21	_____ distal	f	toward the midline of the body
1.22	_____ dorsal	g	nearer the trunk of the body
1.23	_____ inferior	h	farther from the point of origin
1.24	_____ plantar	i	toward the front of the body
1.25	_____ superficial	j	toward the feet

A SECOND LOOK

1 In the following drawing, label the transverse, midsagittal, and sagittal planes.

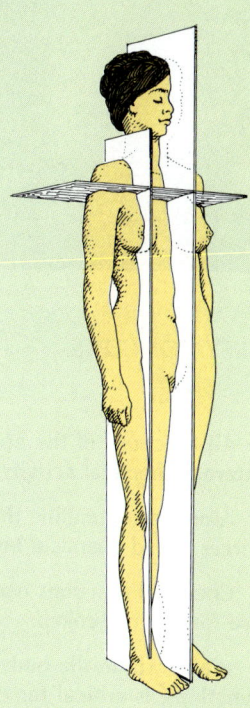

2 In the following drawing, fill in the empty boxes.

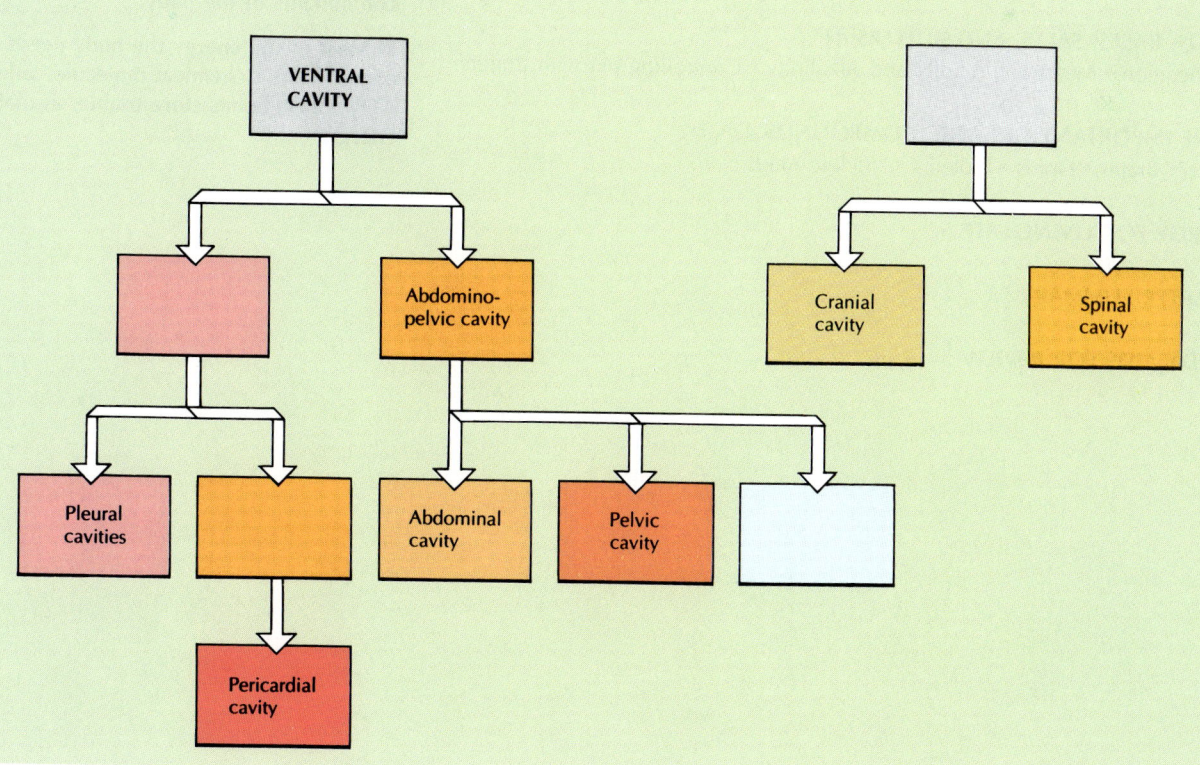

Essentials of Body Chemistry

CHAPTER OUTLINE

KEY CONCEPTS

1 All functions of the body depend on internal chemical activity.

2 Atoms can combine through electrical forces called chemical bonds.

3 Chemical reactions occur when bonds are formed or broken.

4 Water within the body, with its many functions, is critical for the maintenance of homeostasis.

5 Organic compounds—proteins, carbohydrates, lipids, and nucleic acids—along with water, are the main structural components of the body.

6 Most of the energy the body needs comes from a chemical molecule called ATP, whose bonds store usable, available energy.

Everything that goes on in the human body depends on some kind of chemical activity. Without chemical activity your muscles would be useless, your brain and nerves would be as ineffective as an unplugged television set, and the food you eat could not be chewed, swallowed, or digested. This chapter presents the chemistry you will need to understand the basic chemical processes that enable the human body to function.

MATTER, ELEMENTS, AND ATOMS

Everything in the universe is composed of **matter**, which is anything that occupies space and has mass. **Mass** refers to the amount of matter in an object. (Mass is not the same as weight. **Weight** is the measurable gravitational attraction of the earth for an object. Mass refers only to the *amount* of matter, not its weight.)

Matter is composed of **elements**, which are chemical substances that cannot be broken down into simpler substances by ordinary chemical means. Elements are represented by either one- or two-letter symbols. For instance, H is the symbol for the element hydrogen, O stands for oxygen, and Na stands for sodium. Currently 106 elements are known, 92 of which occur naturally. New elements are still being created in laboratories. About 24 elements are found in living organisms, and 4 of them — oxygen, carbon, hydrogen, and nitrogen — account for 96 percent of the body's weight: oxygen makes up about 65 percent, carbon about 18 percent, hydrogen about 10 percent, and nitrogen about 3 percent. Several of the elements most important to human beings are listed in TABLE 2.1, along with the roles they play in various vital chemical processes.

Each element is composed entirely of its own chemically distinct kind of atom. An **atom** is the smallest unit of an element that retains the chemical characteristics of that element.

Q: *How big is an atom?*

A: Atoms are so small that the only way we can visualize their size is to relate it to objects in our everyday lives. For instance, a child's balloon filled with hydrogen contains about 100 million million billion hydrogen atoms. A drop of water contains more than 100 billion billion hydrogen and oxygen atoms. If an atom were as large as the head of a pin, all the atoms in a grain of sand would fill a cube 1 mile (1.6 km) on a side.

TABLE 2.1 SOME CHEMICAL ELEMENTS ESSENTIAL TO PROPER BODY FUNCTION

Element	Symbol	Approximate percentage of human body (by weight)	Significance to human body
Oxygen	O	65.0	Part of water and organic compounds that are essential to many life processes.
Carbon	C	18.5	Basic component of all organic compounds.
Hydrogen	H	9.5	Part of water and organic compounds.
Nitrogen	N	3.2	Part of all protein and nucleic acid molecules.
Calcium	Ca	1.5	Necessary for healthy bones and teeth, muscle contraction, blood clotting, hormone production.
Phosphorus	P	1.0	Component of nucleic acids and ATP; especially important in nervous tissue, bones, and teeth.
Potassium	K	0.4	Essential for body growth; important for nerve conduction, muscle contraction, water-ion balance in body fluids.
Sulfur	S	0.3	Component of many proteins.
Chlorine	Cl	0.2	Important for water movement between cells.
Sodium	Na	0.2	Component of extracellular fluid; critical to nerve and muscle response; maintains proper water balance in blood.
Magnesium	Mg	0.1	Aids in muscle contraction and nerve transmission.
Iodine*	I	<0.1	Necessary for production of thyroid hormone by thyroid gland.
Iron*	Fe	<0.1	Basic oxygen-binding component of hemoglobin.

*Elements that make up less that (<) 0.1 percent of the body are known as *trace elements*. Some of the more common ones are cobalt, copper, fluorine, iodine, iron, manganese, silicon, and zinc (see Chapter 26).

Structure of Atoms

Atoms are not solid bits of matter. In fact, even atoms of the densest substances consist mostly of space. Atoms have two main parts — a central *nucleus* and the surrounding *electron field* [FIGURE 2.1]. The **nucleus** contains two kinds of particles: the positively charged **protons (p^+)** and the uncharged **neutrons (n^0).** (The hydrogen atom is the only one without any neutrons in the nucleus.) Moving around the nucleus are negatively charged particles called **electrons (e^-).** Modern atomic theory describes these electrons as existing in an electron field, or "cloud," which is a way of saying that any one electron cannot definitely be found at any given point at any particular moment in time.

Atomic number and atomic weight If all atoms (except hydrogen) are composed of protons, neutrons, and electrons, how do we account for the differences in the elements? The chemical and physical properties of an atom are determined by the number of protons and neutrons in its nucleus and by the number and arrangement of electrons surrounding the nucleus. The number of protons is critical. The **atomic number** of an element is the *number of protons in the nucleus of one of its atoms.* The atomic number alone can identify an element. For instance, if an atom has one proton, it is hydrogen; if an atom has eight protons, it is oxygen. FIGURE 2.2 shows how protons, neutrons, and electrons are distributed in the atoms of some common elements.

FIGURE 2.1 STRUCTURE OF AN ATOM

Negatively charged electrons moving around the nucleus form an electron field (or "cloud"). The nucleus is composed of electrically neutral neutrons (gray) and positively charged protons (red).

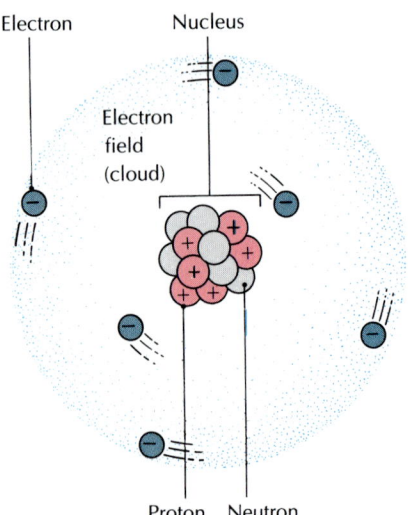

Electron   Nucleus

Electron
field
(cloud)

Proton   Neutron

Note that in FIGURE 2.2 the number of protons in any atom of a given element is always the same as the number of electrons. In other words, the number of *positive charges* (protons) and the number of *negative charges* (electrons) are equal. As a result, the negative and positive charges cancel each other and the atom is neutral — that is, it has no overall electrical charge.

Another measure of an atom is its **atomic weight,** which expresses the relative weight of an element compared with carbon. The atomic weight of carbon has been designated as 12. Thus, an atom of hydrogen (with an atomic weight of 1) is about one-twelfth as "heavy" as an atom of carbon, and an atom of calcium (with an atomic weight of 40) is about 3.3 times as heavy as a carbon atom. The approximate atomic weights of a few elements are given in the last column of the table in FIGURE 2.2; note that the *atomic weight equals the number of protons plus the number of neutrons.*

Isotopes All atoms of an element have the same number of protons in the nucleus, but some have *different numbers of neutrons* and thus different atomic weights. But because all forms of an element have the same number of electrons, they *all have the same chemical properties.* The different atomic forms of an element are called **isotopes.** Some elements have as many as 20 naturally occurring isotopes, while others have as few as 2. For example, about 99 percent of all carbon atoms have 6 protons and 6 neutrons, but other carbon atoms have 7 or 8 neutrons, giving them atomic weights of 13 and 14, respectively.* *Most* of the carbon in nature has 6 neutrons, giving that isotope an atomic weight of 12. For convenience, the atomic weight of this carbon isotope is used as the basis for determining the atomic weights of other elements.

Some isotopes are naturally radioactive, and others are made radioactive in the laboratory by bombarding their nuclei with subatomic particles. An element is considered radioactive when its nucleus undergoes spontaneous changes called *disintegration,* or "decay." In this process, nuclear particles and radiation are emitted. Many *radioisotopes* (short for radioactive isotopes) are used in biology and medicine. As their nuclei break down, or disintegrate, they emit radiation.

Experiments have shown that radioisotopes emit one or more of three major types of radioactivity: *alpha particles* (α), *beta particles* (β), and *gamma rays* (γ). An alpha particle consists of two neutrons (2n) and two protons (2p). These particles do not travel very far and are not very penetrating. By contrast, beta particles are fast-moving electrons that are capable of producing dense

---

*In a chemical atomic symbol, the atomic number is placed in an inferior position, and the atomic weight (sometimes called the *mass number*) in a superior position. For example, the isotopes of an atom of carbon would be written $^{12}_{6}C$, $^{13}_{6}C$, and $^{14}_{6}C$. Carbon-14 ($^{14}_{6}C$) is radioactive.

FIGURE 2.2 ATOMIC STRUCTURE OF COMMON ELEMENTS

The simplified diagrams of hydrogen, carbon, and oxygen atoms show the number of electrons (e^-) in orbit and the number of protons (p^+) and neutrons (n^0) in the nucleus. Note that the number of protons equals the atomic number and that the number of protons plus neutrons equals the atomic weight. The number of electrons equals the number of protons in any given atom.

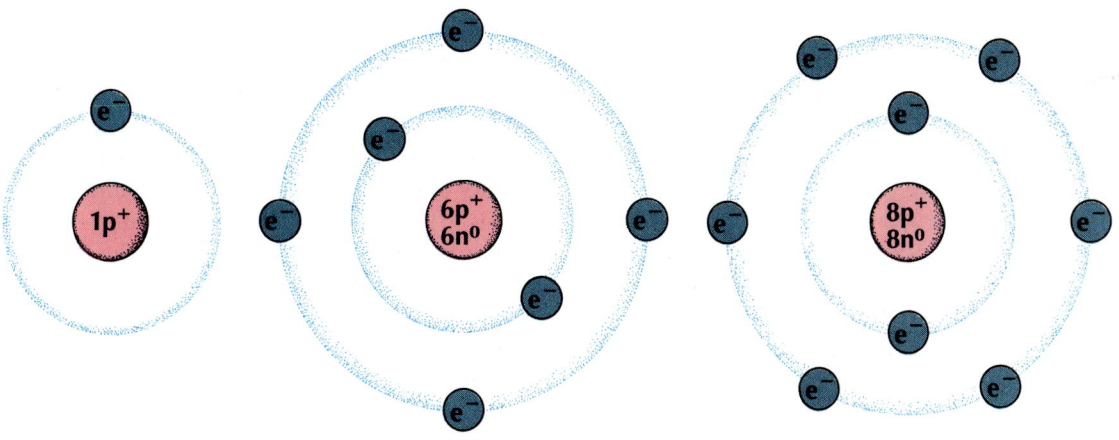

HYDROGEN
1 proton
1 electron
Atomic number: 1
Atomic weight: 1

CARBON
6 protons
6 neutrons
6 electrons
Atomic number: 6
Atomic weight: 12

OXYGEN
8 protons
8 neutrons
8 electrons
Atomic number: 8
Atomic weight: 16

| Element | Atomic number | Number of | | | Approximate atomic weight |
		Protons	Neutrons	Electrons	
Hydrogen (H)	1	1	0	1	1
Carbon (C)	6	6	6	6	12
Nitrogen (N)	7	7	7	7	14
Oxygen (O)	8	8	8	8	16
Sodium (Na)	11	11	12	11	23
Phosphorus (P)	15	15	16	15	31
Sulfur (S)	16	16	16	16	32
Chlorine (Cl)	17	17	18	17	35
Potassium (K)	19	19	20	19	39
Calcium (Ca)	20	20	20	20	40

ionization tracts. They have about 100 times more penetrating power than alpha particles do. Finally, gamma rays are not particles but a form of high-energy electromagnetic radiation. They travel at the speed of light and are so penetrating that they can be stopped only by lead or concrete, but they may also be captured eventually by nuclear collision.

Radioactive isotopes can be introduced into the body and then traced with detectors to help diagnose the presence of certain disorders and to treat specific diseases. For example, radioactive iodine is used to determine how the thyroid gland is functioning, to treat thyroid cancer, and to detect brain tumors. Radioactive cobalt is used to treat cancers; radioactive phosphorus is used in treating leukemia and bone cancer and in localizing breast cancers; radioactive sodium is used to study body fluids and circulation rates; radioactive carbon traces the absorption and cellular use of food and drugs; and radioactive iron and chromium are helpful in studying red blood cells and hemoglobin. Technetium is an isotope that is little known to the general public, but it is used extensively to help locate brain tumors before surgery is performed.

ASK YOURSELF

1 What are the basic components of an atom?

2 What are the electrical charges, respectively, of neutrons, protons, and electrons?

3 How is the number of protons in the nucleus of an atom related to that atom's atomic number?

4 What is an isotope?

Electron Shells

The electrons of an atom are distributed around the nucleus in layers called **electron shells.** Seven of these shells are known to exist. Each shell can hold only a certain number of electrons. There are never more than two electrons in the shell nearest the nucleus; as many as eight electrons can be in the second, and larger numbers fill the more distant shells.

Q: *How does a radioactive isotope such as cobalt help destroy cancer cells?*

A: The radiation emitted from radioactive cobalt can change the chemical nature of cancer cells enough to kill them. Cells that reproduce at a fast rate, such as cancer cells, are most susceptible to radiation, since radiation interferes with cellular division; this is why active cancer cells are supposed to die before healthy cells do. But radiation of any kind can be dangerous to all cells, especially those in bone marrow and those that line the intestines, which reproduce at a fairly fast rate.

The number of electrons and protons in an atom determines its chemical character, but the way the electrons are arranged in the shells, especially the outer shell (the one farthest from the nucleus), is most important. Only the electrons in the *outer shell* take part in a chemical reaction. When an atom has a complete outer shell—that is, when the shell holds the maximum number of electrons possible—the shell is stable. An atom with an incomplete outer shell tends to gain or lose electrons or to share electrons with another atom.

For instance, an oxygen atom has two electrons in its first shell but only six in its second shell. As a result, an oxygen atom tends to *gain* two more electrons in order to fill its outer shell with the stable number of eight electrons [FIGURE 2.3A]. A sodium atom—which has two electrons in its first shell, eight in the second, and only one in the third—tends to *lose* the one electron in its outer (third) shell, leaving the second shell with its stable number of eight electrons [FIGURE 2.3B]. A final example is carbon, which has four electrons in its second shell. In order for the shell to be stable, it must have eight electrons, which occurs when the atom *shares* electrons with other atoms [FIGURE 2.3C]. The ability to share electrons allows carbon to enter into many complex reactions.

ASK YOURSELF

1 What are the layers of electrons surrounding an atom called?

2 Which electron shell is critical for chemical reactions?

3 How do the shells of atoms become stable?

HOW ATOMS COMBINE

Atoms of elements rarely exist in nature by themselves. Instead, they combine chemically to form **molecules,** such as oxygen, O_2. (The subscript 2 indicates that the oxygen molecule contains two oxygen atoms.) **Compounds** are molecules that are made up of atoms of two or more elements (such as carbon dioxide, or CO_2; sodium chloride, or $NaCl$; and glucose, or $C_6H_{12}O_6$).

When atoms interact chemically to form molecules, the atoms are held together by electrical forces called **chemical bonds.** The kinds of bonds most important to the chemistry of the human body are the ionic, covalent, and hydrogen bonds.

Ionic Bonds

As we saw earlier, electrons have a negative electrical charge and protons have a positive electrical charge.

FIGURE 2.3 ELECTRON SHELLS

[A] How atoms gain electrons. (1) An oxygen atom with six electrons in its outermost (second) shell has two "available spaces" for electrons (dashed circles) to fill the shell. (2) The shell becomes stable with the addition of two electrons. **[B]** How atoms lose electrons. (1) A sodium atom with only one electron in its third shell tends to lose it so that the stable, complete second shell becomes the outermost shell. In this case, it is easier to lose one electron to achieve a stable second shell than it is to gain seven electrons to complete the third shell. (2) The shell becomes stable after it loses the single electron. **[C]** How carbon shares electrons. (1) An atom of carbon has four electrons in its second shell and needs four more to complete the shell. (2) Carbon forms a stable compound by sharing four electrons with other atoms. The partial shells indicate that the electrons are being shared with other atoms.

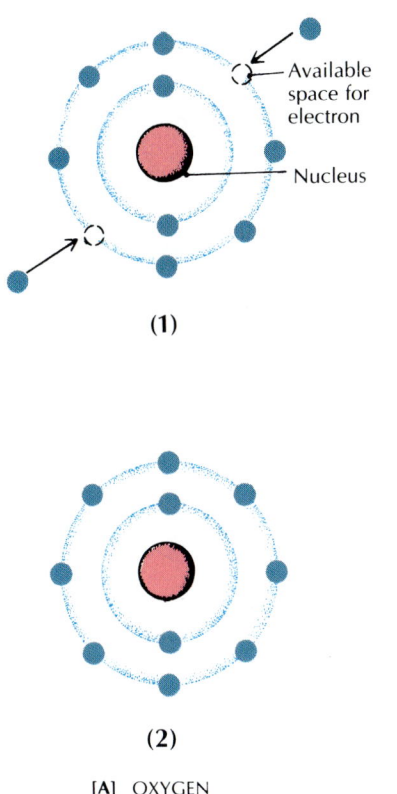

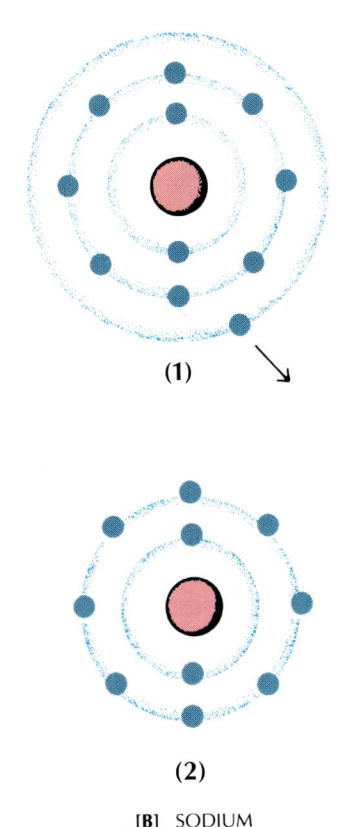

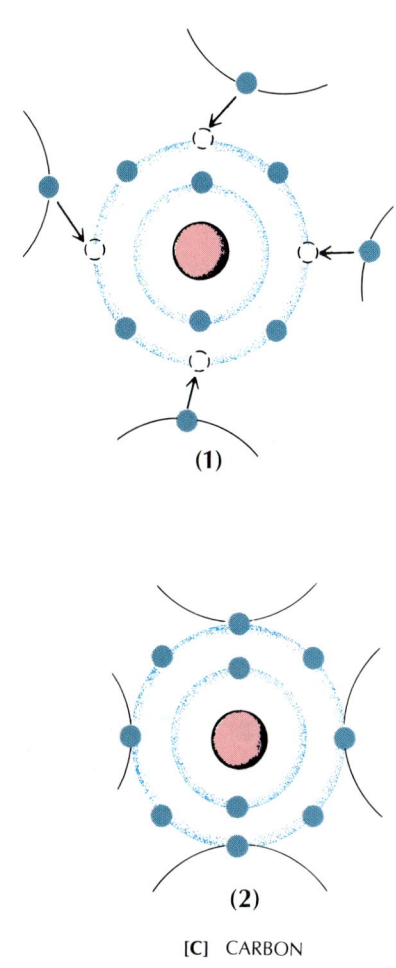

(1) (1) (1)

(2) (2) (2)

[A] OXYGEN   [B] SODIUM   [C] CARBON

When an atom gains or loses electrons and acquires an electrical charge, it is called an ***ion.*** If an atom *loses* one or more electrons, it becomes a *positively* charged *cation** because there are now more positively charged protons in the nucleus than negatively charged electrons surrounding the nucleus. On the other hand, if an atom *gains* one or more electrons, it becomes a *negatively* charged *anion†* because there are now more negative electrons than positive protons. The electrical charge of an ion is expressed as a superscript number and a plus or minus sign after the symbol for the element (for example, Ca^{2+}). A single charge is shown only with a plus or minus sign.

Ionic bonds are formed when an ion or group of ions becomes attracted to an oppositely charged ion or group of ions. FIGURE 2.4 shows how an ionic bond is formed between sodium and chlorine to produce sodium chloride, an ionic compound.

The ions in compounds behave differently from the atoms of the element from which they are formed. For example, sodium (Na) in its atomic form is a metal that reacts violently with water; chlorine (Cl) is a highly reactive gas. As ions, both are stable and do not react with

---

*The most important cations in the body are sodium (Na^+), potassium (K^+), hydrogen (H^+), magnesium (Mg^{2+}), calcium (Ca^{2+}), and iron (Fe^{3+}).

†The most important anions are chloride (Cl^-), hydroxyl (OH^-), bicarbonate (HCO_3^-), sulfate (SO_4^{2-}), phosphate (PO_4^{3-}), and carboxyl (COO^-).

FIGURE 2.4 IONIC BONDING

[A] A sodium atom has an available electron to give up to a chlorine atom with an incomplete outer shell. **[B]** After losing its outermost electron, the sodium atom acquires a positive charge and is now a sodium *ion,* represented by Na^+; the chlorine atom, having received an electron to stabilize its outer shell, acquires a negative charge and becomes a chloride *ion* (Cl^-). **[C]** The oppositely charged sodium and chloride ions are attracted to each other and form an ionic bond. The electrical attraction is indicated by a fuzzy electron field.

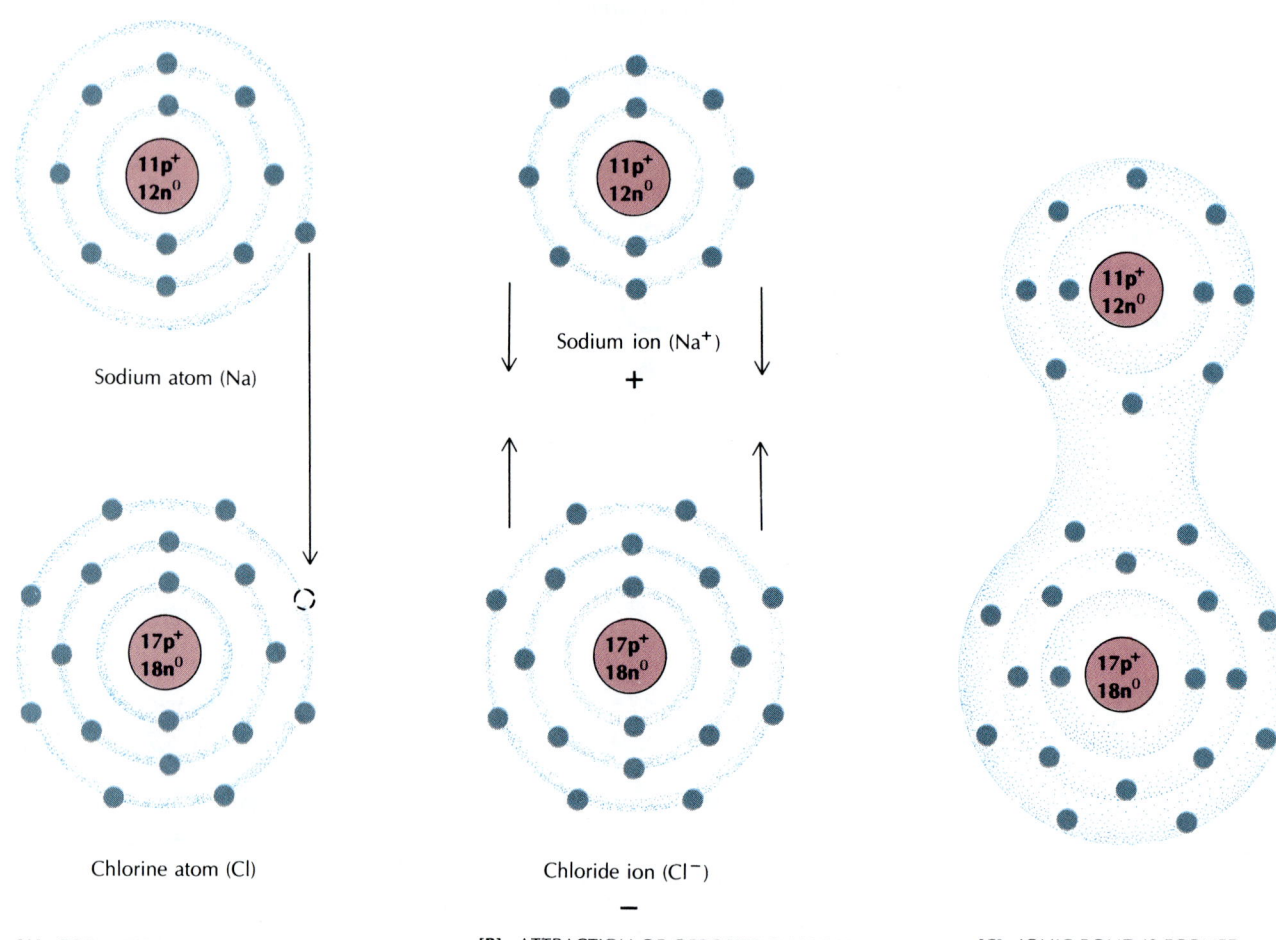

Sodium atom (Na)

Sodium ion (Na^+)
+

Chlorine atom (Cl)

Chloride ion (Cl^-)
−

[A] FORMATION OF IONS   **[B]** ATTRACTION OF OPPOSITE CHARGES   **[C]** IONIC BOND IS FORMED

water, although they dissolve in water. Together they form common table salt, NaCl. The human body is composed primarily of water, which has special properties (discussed later) that promote the breaking of ionic bonds. The breaking of ionic bonds results in the release of such important ions as potassium (K^+), which is required for the activation of many biological processes, and calcium (Ca^{2+}), which is necessary for the contraction of muscles.

Covalent Bonds

When atoms form ionic bonds, electrons are lost or gained. When atoms *share* electrons with other atoms,

the chemical bond that is formed is called a ***covalent bond*** [FIGURE 2.5]. In covalent bonding, electrons are always shared in pairs. When one pair of electrons is shared, a *single bond* is formed; when two pairs are shared, a *double bond* is formed.

Because covalent bonds are very strong, molecules and compounds with covalent bonds are very stable. There are two basic types of covalent bonds. In a *nonpolar covalent bond*, one or more electron pairs are distributed, or shared, equally between atoms. In a *polar covalent bond*, one or more electron pairs are distributed unequally between two atoms. In other words, the shared electrons are drawn more to one atom than to the other, creating a slight charge at each end of the molecule: positive for

FIGURE 2.5 COVALENT BONDING

[A] An oxygen atom needs two electrons to complete its outer shell, and a hydrogen atom needs one electron to complete its outer shell. **[B]** A water molecule is formed when an oxygen atom and two hydrogen atoms share their electrons. Two single bonds, each with a pair of electrons, are formed.

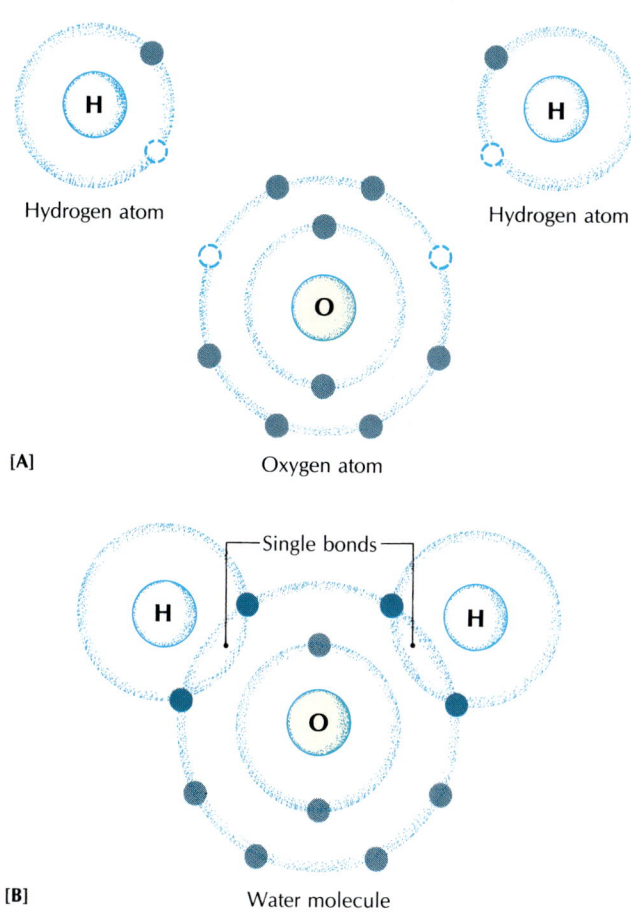

[A] Oxygen atom

[B] Water molecule

FIGURE 2.6 HYDROGEN BONDING

The formation of hydrogen bonds (dashed lines) between polar covalently bonded water molecules. A single hydrogen bond is relatively weak, but several bonds work together to maintain the molecular shape of water.

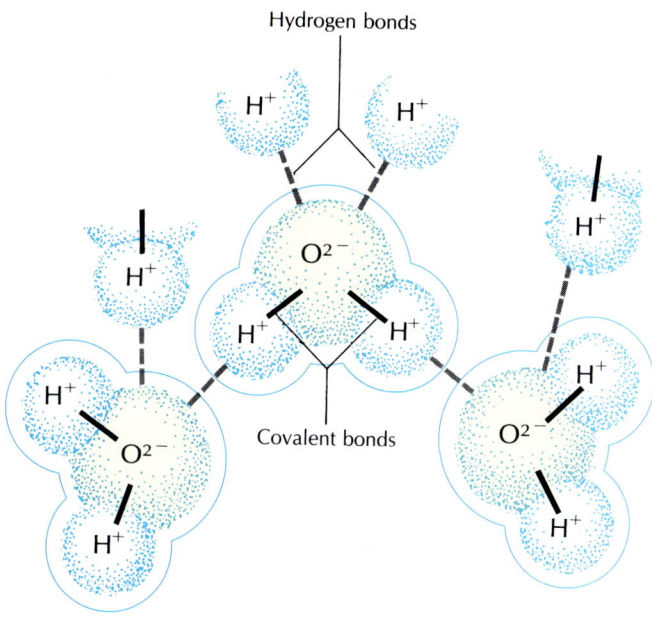

that portion in which the electrons spend less time and negative where they spend more time.

Hydrogen Bonds

In compounds containing covalently bonded hydrogen, the single hydrogen electron is attracted to the other atom whose nucleus has a greater positive charge. This leaves a single proton behind in the hydrogen nucleus. As a result, the hydrogen atom gains a slight positive charge. The remaining proton is attracted to negatively charged atoms in nearby molecules. This weak attractive force is a ***hydrogen bond.*** (No electrons are exchanged or shared in hydrogen bonding.) The hydrogen atom in one water

molecule, for example, forms a hydrogen bond with the oxygen atom in another water molecule, and so forth, until many water molecules are bonded together [FIGURE 2.6].

Hydrogen bonding is the intermolecular attractive force that gives water its special properties: a high melting point, a high boiling point, an ability to retain heat better than most liquids do, and an exceptional ability to dissolve ionic substances. Although hydrogen bonds are weaker than ionic or nonpolar covalent bonds, the many hydrogen bonds found in such substances as proteins, nucleic acids, enzymes, and hemoglobin encourage the formation of stable three-dimensional molecules.

A S K   Y O U R S E L F

1 What is an ion? A cation? An anion?

2 What is a covalent bond?

3 What is an ionic bond?

4 How is it possible for an atom that is already covalently bonded to form a hydrogen bond at the same time?

CHEMICAL REACTIONS

As you just saw, atoms combine by forming bonds. The combining of two or more atoms causes a chemical change called a chemical reaction. Because chemical reactions also occur when bonds are broken, we can say that in a **chemical reaction,** bonds between atoms are broken or joined and different combinations of atoms or molecules are formed.

Another way of talking about chemical reactions is to consider them as part of the overall process of **metabolism,** which includes all the chemical activities that go on in the body (see Chapter 26). Metabolism either builds up substances (synthesis) or breaks them down (decomposition). The synthesis process is called **anabolism,** and the decomposition process is called **catabolism.**

Combination and Decomposition Reactions

When two or more atoms, ions, or molecules combine to form a more complex substance, the process is called a **combination reaction:** $A + B \rightarrow AB$. (Another useful word for "combination" is *synthesis* [Gr. "to put together"].) The substances that are combined are the *reactants*, and the new molecule they form is the *product:*

C	+	O$_2$	$\longrightarrow$	CO$_2$
Carbon	plus	oxygen	produces	carbon dioxide
(reactant)		(reactant)		(product)

The arrow means "produces" or "yields." It shows the direction in which the reaction is moving.

Some chemical reactions are *reversible*. That is, after the reactants combine, the product can be decomposed in a later reaction to produce the original reactants. Such a reversible reaction is shown by special arrows going in both directions:

$$A + B \rightleftharpoons AB$$

Decomposition reactions are the opposite of combination reactions. They result in the breakage of chemical bonds to form two or more products (atoms, ions, or molecules): $AB \rightarrow A + B$.

2HgO	$\longrightarrow$	2Hg	+	O$_2$
Mercuric oxide	is decomposed to produce	mercury	and	oxygen

The digestion of food—which is simply the process of breaking down large, complex food molecules into smaller, simpler molecules—is an example of a decomposition reaction.

All chemical reactions either release or absorb energy. Combination (anabolic) reactions require energy, and chemical reactions that break the chemical bonds of large molecules release energy.

Oxidation-Reduction Reactions

Oxidation occurs when an atom or molecule *loses* electrons or hydrogen atoms:

Na	–	e$^-$	$\longrightarrow$	Na$^+$
When a sodium atom	loses	a negatively charged electron	it is oxidized and produces	a positively charged sodium ion

When an atom or molecule *gains* electrons or hydrogen atoms, the process is called **reduction:**

Cl$_2$	+	2e$^-$	$\longrightarrow$	2Cl$^-$
When a molecule of chlorine	gains	two negatively charged electrons	it is reduced and produces	two negatively charged chloride ions

When an *oxidation-reduction reaction* takes place, the oxidizing agent is reduced (gains electrons) and the reducing agent is oxidized (loses electrons). So *oxidation and reduction always occur together:*

O$_2$	+	2H$_2$	$\longrightarrow$	2H$_2$O
One oxygen molecule (the oxidizing agent)	reacts with	two hydrogen molecules (the reducing agent)	to produce	two water molecules (oxygen has been reduced and hydrogen has been oxidized)

Oxidation-reduction reactions are especially important to the body because they provide energy that is used in cellular activities such as muscle contractions.

Hydrolysis and Dehydration Synthesis

When **hydrolysis** (Gr. "water loosening") takes place, a molecule of water interacts with a reactant to break up the reactant's bonds. The original reactant is then rearranged, together with the water, into different molecules:

C$_{12}$H$_{22}$O$_{11}$	+	H$_2$O	$\longrightarrow$	C$_6$H$_{12}$O$_6$	+	C$_6$H$_{12}$O$_6$
Sucrose	plus	water	produces	glucose	and	fructose

The bonds of sucrose (table sugar) are broken (hydrolyzed) by water to form two simple sugars, glucose and fructose. Glucose and fructose have the same numbers and kinds of atoms, but in different arrangements (isomers).

Hydrolysis is important to the body. By means of hydrolysis, large molecules of proteins, nucleic acids, and fats are broken down into simpler, smaller, more usable molecules. Generally speaking, almost all digestive and degradative processes in the body occur by hydrolysis.

Dehydration synthesis reactions are essentially hydrolytic reactions in reverse: Small molecules are united into larger molecules, and one or more molecules of water are eliminated. For example, when two glucose molecules are chemically combined, they form maltose plus water:

$$C_6H_{12}O_6 \;+\; C_6H_{12}O_6 \;\longrightarrow\; C_{12}H_{22}O_{11} \;+\; H_2O$$

Glucose   plus   glucose   produces   maltose   plus   water

WATER

Chemical compounds are generally divided into two categories: inorganic and organic. ***Inorganic compounds*** are composed of relatively small molecules that are usually bonded ionically. ***Organic compounds*** contain carbon and hydrogen and are bonded covalently. Certain salts—such as sodium chloride (NaCl) and potassium chloride (KCl)—are examples of inorganic compounds, but the major example is ***water*** (H_2O), without which the other inorganic compounds would be useless, since many of their reactions occur only in the presence of water.

Water is one of the most plentiful compounds on earth. It is the most common component of the body, making up about 62 percent of the body weight of an adult [TABLE 2.2]. The percentage is usually higher in young children and lower in the elderly. Every day we take in about 2700 mL* (2¾ quarts) of water from various sources, and varying amounts of water are constantly leaving our cells and our bodies [TABLE 2.3]. Water is one of the chief regulators of homeostasis, and the body cannot function without it.

Chemical Properties of Water

The special properties of water depend on its bonding structure: two hydrogen atoms covalently bonded to one oxygen atom [FIGURE 2.7A]. The oxygen atom attracts electrons to itself and away from the hydrogen atoms. As a result, the oxygen atom picks up a slight negative charge, while the two hydrogen atoms bear slight positive charges. Because the two ends of a water molecule have different charges, it is said to be a ***polar molecule*** [FIGURE 2.7B]. Water can form hydrogen bonds with other water molecules and with a variety of other compounds.

*The abbreviation for milliliters is mL.

TABLE 2.2 APPROXIMATE WATER CONTENT OF THE ADULT HUMAN BODY

Tissues	Body weight, %	Water, %	Water in tissues, % body weight	Water, L
Muscle	41.7	75.6	31.53	22.10
Skin	18.0	72.0	12.96	9.07
Blood	8.0	83.0	6.64	4.65
Skeletal	15.9	22.0	3.50	2.45
Brain	2.0	74.8	1.50	1.05
Liver	2.3	68.3	1.57	1.10
Intestines	1.8	74.5	1.34	0.94
Fat	8.5	10.0	0.01	0.70
Lungs	0.7	79.0	0.55	0.39
Heart	0.5	79.2	0.40	0.28
Kidneys	0.4	82.7	0.33	0.23
Spleen	0.2	75.8	0.15	0.11
Total body	100.0%	62.0%	60.48%	43.07 L

Source: Robert L. Vick, *Contemporary Medical Physiology,* Addison-Wesley, Reading, MA, 1984, p. 540. Used with permission.

TABLE 2.3 COMPONENTS OF WATER BALANCE IN THE HUMAN BODY

Intake, mL/24 hr		Output, mL/24 hr	
Source	Amount	Route	Amount
Preformed water in food	750	Skin and lungs	840
Water of metabolism	320	Urine	1760
Drinking	1630	Feces	100
Total intake	2700 mL	Total output	2700 mL

Source: Robert L. Vick, *Contemporary Medical Physiology,* Addison-Wesley, Reading, MA, 1984, p. 621. Used with permission.

Water is able to dissolve other polar substances because of its own polarity. A general rule for predicting solubility is "Likes dissolve one another; unlikes do not"—where the terms *like* and *unlike* refer to the polarity of the substances. For example, nearly all large molecules, such as carbohydrates and proteins, have areas that are slightly charged. Because water molecules are attracted to those charged areas, they help keep these other molecules dissolved. Also, many particles within cells are charged; thus, water molecules are attracted to the surfaces of those particles [FIGURE 2.7C]. The neatly packed water molecules add stability to the surfaces of cell membranes and proteins.

FIGURE 2.7 MOLECULAR STRUCTURE OF WATER

[A] Two atoms of hydrogen are bonded to one atom of oxygen to form a molecule of water. [B] A simple model of a water molecule, showing the different electrical charges at either end. [C] The positive ends of water molecules are attracted to a small particle with a negatively charged surface, forming an orderly and stable condition.

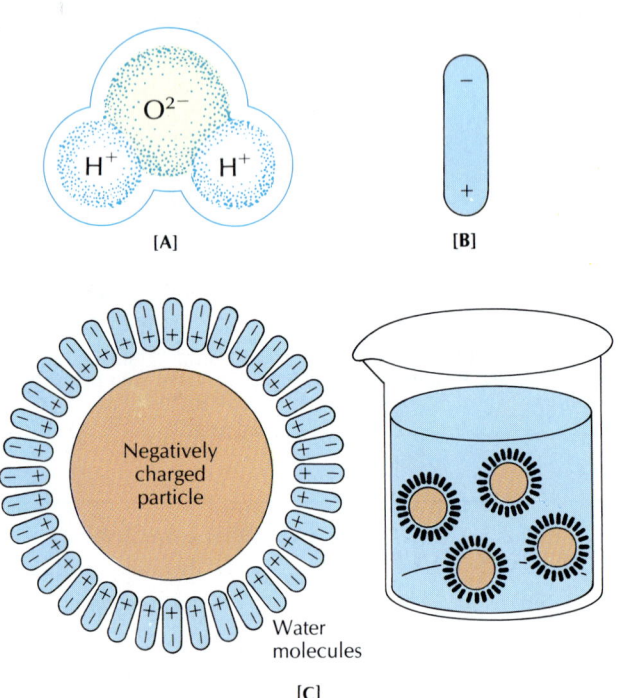

bloodstream. Inhaled oxygen can be used only when it is in solution with the water in blood, and waste carbon dioxide from cells is carried to the lungs in solution before it is exhaled.

Water as a Transporter

Water in the blood carries nutrients from the foods we eat to tissues throughout the body. It also transports waste products from the cells to the lungs, kidneys, and skin to be discarded as carbon dioxide, urine, or sweat. Besides nutrients and wastes, the liquid portion of blood (plasma) carries hormones to their specific target sites throughout the body. And, as you just saw, water is essential for blood circulation in general. Water also transports materials within cells.

Water as a Temperature Regulator

Because water retains heat better than most liquids do, it takes a great deal of heat to turn liquid water into water vapor (gas). For this reason, a lot of heat is lost when we perspire, and the body is cooled by losing only a little water. Even when temperatures are not very hot, water is constantly evaporating from the skin and lungs. This water loss, which is called *insensible perspiration* because we don't feel or see it, balances the heat produced through metabolic processes in the cells so that our body temperature remains constant.

Water as a Lubricant

The lubricating fluid (called *synovial fluid*) that helps joints such as knees and elbows move easily contains a great deal of water. The thin spaces around internal organs are also lubricated by fluids that contain water. In the chest cavity, for instance, lubrication is important because the rib cage slides over several internal organs during breathing. Water moistens food as it passes along the digestive tract, starting with saliva in the mouth and throat that makes swallowing easy, and ending with the passage of water-softened feces through the anus.

Water as a Solvent

A *solvent* is a liquid or gas capable of dissolving another substance. A *solute* is the substance being dissolved. A *solution* is the homogeneous mixture of a solvent and the dissolved solute. All our body fluids (including blood, urine, lymph, sweat, and digestive juices) are mostly water, and practically all the chemical reactions in the body take place in water. In fact, water is considered the "universal solvent" because it dissolves so many "like" polar substances. Water is used during digestion in a series of hydrolysis reactions that break down large compounds into smaller, more easily assimilated ones. Once the dissolved nutrients are in solution, they can be absorbed through the walls of the small intestine into the

---

Q: *How long can a person live without water?*

A: We can live without food for weeks, but without water we would be dead in about 3 days. An infant might not live that long. Death usually occurs when water loss is over 10 percent of a person's body weight.

A S K   Y O U R S E L F

1 How do organic compounds differ from inorganic compounds?

2 What is a polar molecule?

3 What is a solvent? A solution?

4 Why is water called the "universal solvent"?

5 How does water act as a transporter? As a regulator of body temperature? As a lubricant?

ACIDS, BASES, SALTS, AND BUFFERS

As you have seen, most of the chemicals in the body are not "dry" but are dissolved in a water solution. The idea of substances being chemically active only when they are dissolved in water, or when they are "in solution," certainly applies to acids and bases, two biologically important substances.

Any substance such as table salt (NaCl), which ionizes in water and whose solution conducts electricity, is called an **electrolyte**. Many fluids in our bodies contain strong electrolytes that break down into ions. Most water-soluble ionic compounds and most acids and bases are electrolytes.

An **acid** is a substance that releases hydrogen ions (H^+) when dissolved in water. For example, when hydrochloric acid (HCl) is placed in water, the following reaction takes place:

$$HCl \xrightarrow{\text{water}} H^+ + Cl^-$$

| One molecule of hydrochloric acid | dissolved in water produces | one hydrogen ion | and | one chloride ion |

In contrast, a **base** is a substance that releases hydroxyl ions (OH^-) when dissolved in water. The following reaction is typical:

$$NaOH \xrightarrow{\text{water}} Na^+ + OH^-$$

| One molecule of sodium hydroxide | dissolved in water produces | one sodium ion | and | one hydroxyl ion |

A **salt** is an ionic substance that contains an anion other than OH^- or O^{2-}. An acid reacts with a base to produce a salt. The reaction is called *neutralization*:

$$HCl + NaOH \longrightarrow NaCl + H_2O$$

| An acid | plus | a base | produce | a salt | and | water |

Acids have a sour taste and react with certain metals to release hydrogen gas; they can neutralize bases. Bases have a bitter taste and feel slimy; they can neutralize acids.

Dissociation

The tendency of some molecules to break up into charged ions in water is called **dissociation**. (When ions recombine to form a stable molecule, the process is called *association*.) Not all acids dissociate completely. Hydrochloric acid is considered a *strong acid* because it dissociates entirely in water. But acetic acid, the familiar ingredient of vinegar, is a *weak acid* because it does not completely dissociate in water. Combinations of weak acids are important in maintaining the concentration of hydrogen ions in cells.

Measuring Acidity and Alkalinity

The higher the concentration of hydrogen ions (H^+), the more acidic a solution is. Also, the higher the concentration of H^+ ions, the lower the concentration of hydroxyl ions (OH^-). TABLE 2.4 shows the relationship between

---

TABLE 2.4 RELATIONSHIP BETWEEN HYDROGEN ION (H^+) CONCENTRATION AND HYDROXYL ION (OH^-) CONCENTRATION

H^+ (hydrogen ion)		pH		OH^- (hydroxyl ion)
$10^0 = 1$		0		$10^{-14} = 0.00000000000001$
$10^{-1} = 0.1$		1		$10^{-13} = 0.0000000000001$
$10^{-2} = 0.01$		2		$10^{-12} = 0.000000000001$
$10^{-3} = 0.001$		3		$10^{-11} = 0.00000000001$
$10^{-4} = 0.0001$		4		$10^{-10} = 0.0000000001$
$10^{-5} = 0.00001$		5		$10^{-9} = 0.000000001$
$10^{-6} = 0.000001$		6		$10^{-8} = 0.00000001$
$10^{-7} = 0.0000001$	(Neutrality)	7	(Neutrality	$10^{-7} = 0.0000001$
$10^{-8} = 0.00000001$		8		$10^{-6} = 0.000001$
$10^{-9} = 0.000000001$		9		$10^{-5} = 0.00001$
$10^{-10} = 0.0000000001$		10		$10^{-4} = 0.0001$
$10^{-11} = 0.00000000001$		11		$10^{-3} = 0.001$
$10^{-12} = 0.000000000001$		12		$10^{-2} = 0.01$
$10^{-13} = 0.0000000000001$		13		$10^{-1} = 0.1$
$10^{-14} = 0.00000000000001$		14		$10^0 = 1$

Note: The superscripts are called *exponents;* they indicate a 10-fold difference in concentration.

hydrogen ion (H⁺) concentration and hydroxyl ion (OH⁻) concentration. A solution is considered neutral when the number of H^+ ions equals the number of OH^- ions.

The numerical scale that measures acidity and alkalinity is called the ***pH scale.*** It runs from 0 to 14, with neutrality at 7, and indicates the concentration of free hydrogen ions (H^+) in water [FIGURE 2.8]. Each whole number on the pH scale represents a 10-fold change in acidity. Therefore, a cup of black coffee at pH 5 has 10 times the concentration of H^+ ions that peas, at pH 6, have.

Controlling pH with Buffers

Homeostasis can be maintained only if there is a relatively constant pH of the blood and other body fluids. Too much of a strong acid or base can destroy the stability of cells. Also, too sudden a change in the pH balance may be destructive. Human blood, for example, cannot tolerate drastic changes from its normal pH of 7.4. In fact, the pH of blood rarely goes below 7.35 or above 7.45 because blood contains chemical substances that regulate the acid-base balance. These substances, called ***buffers,*** are combinations of weak acids or weak bases and their respective salts in solution. Buffer solutions help blood and other body fluids resist changes in pH when small amounts of strong acids or bases are added. So, even though black coffee has a pH of 5, the 7.4 pH of your blood does not change when you drink coffee. This is because of the stabilizing action of buffers.

The buffer control systems are one of the body's major ways of maintaining pH homeostasis. Some of these systems are the fastest of the body's regulators of acid-base balance, but other mechanisms provide more effective long-term control. For example, buffer systems are also present in the respiratory system, where the acid-base balance is controlled by regulating the exhalation rate of carbon dioxide. In the urinary system, excess acid is neutralized by adding bicarbonate ions (HCO_3^-) to the blood. Buffering systems in the lungs and kidneys are explained in Chapter 28.

Q: *What do the letters* pH *stand for?*

A: The letter *p* in pH stands for *potentia,* the Latin word for "power," and the letter *H* stands for "hydrogen." So pH means "the power (or concentration) of hydrogen ions in solution."

A S K   Y O U R S E L F

1 What is an acid? A base? A salt?

2 What does the pH scale indicate?

FIGURE 2.8 pH SCALE

Note that pure distilled water is neutral rather than being either acidic or basic (alkaline); this condition occurs because distilled water has equal numbers of hydrogen ions and hydroxyl ions. Each number on the pH scale represents a 10-fold change in acidity: black coffee at pH 5 has 10 times the acidity (hydrogen ions) of peas at pH 6.

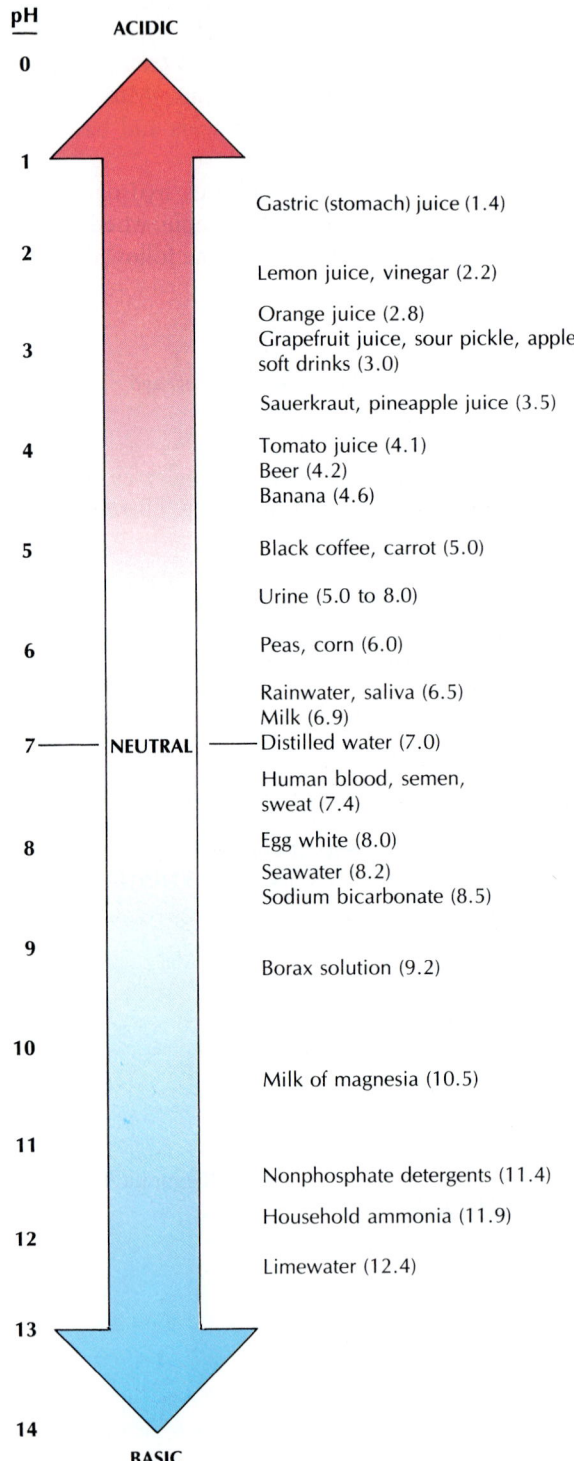

pH   ACIDIC

0

1

Gastric (stomach) juice (1.4)

2

Lemon juice, vinegar (2.2)

Orange juice (2.8)
Grapefruit juice, sour pickle, apple, soft drinks (3.0)

3

Sauerkraut, pineapple juice (3.5)

4

Tomato juice (4.1)
Beer (4.2)
Banana (4.6)

5

Black coffee, carrot (5.0)

Urine (5.0 to 8.0)

6

Peas, corn (6.0)

Rainwater, saliva (6.5)
Milk (6.9)

7 —— NEUTRAL —— Distilled water (7.0)

Human blood, semen, sweat (7.4)

8

Egg white (8.0)
Seawater (8.2)
Sodium bicarbonate (8.5)

9

Borax solution (9.2)

10

Milk of magnesia (10.5)

11

Nonphosphate detergents (11.4)

Household ammonia (11.9)

12

Limewater (12.4)

13

14

BASIC

SOME IMPORTANT ORGANIC COMPOUNDS

When we speak of the chemistry of the living body, we usually think of the essential body nutrients: carbohydrates, lipids (fats), proteins, and nucleic acids (the hereditary material). These substances are the organic chemicals of the body. Organic compounds always contain carbon and hydrogen. They are typically composed of large molecules that are held together by covalent bonds. Organic compounds, along with water, are the main materials that form the human body.

Carbohydrates

Carbohydrates are the major source of energy for most cells, since cells have the chemical machinery to break down these compounds more easily than others. Carbohydrates are made of carbon, hydrogen, and oxygen, with twice as much hydrogen as oxygen. This ratio is shown in the general formula for a carbohydrate, CH_2O.

Kinds of carbohydrates Carbohydrates are generally classified according to their molecular structure. The simplest types are called *monosaccharides* (Gr. *monos*, single + *sakkharon*, sugar). They are the building blocks of more complex sugars. Monosaccharides cannot be broken down by hydrolysis. Three common monosaccharides are glucose, fructose, and ribose* [FIGURE 2.9].

Two monosaccharides can be combined to form a *disaccharide* (*di*, two). Disaccharides all have the molecular formula $C_{12}H_{22}O_{11}$. Many compounds that have the same molecular formula have different molecular *structures*. Such compounds are called *isomers*. FIGURE 2.9 shows the structural differences between glucose and fructose, both of which have the formula $C_6H_{12}O_6$. Disaccharide isomers are sucrose, maltose, and lactose.

The kind of chemical reaction in which a molecule of water is lost when large organic molecules are synthesized from simpler organic molecules is called *dehydration synthesis*. In FIGURE 2.10 you can see how two monosaccharide glucose molecules unite to form the disaccharide maltose (malt sugar). Other typical reactions in which monosaccharides combine to form disaccharides are as follows:

Glucose + fructose $\longrightarrow$ sucrose + water

Glucose + galactose $\longrightarrow$ lactose + water

Note that in each case a molecule of water is formed in the dehydration synthesis. The water is given up when the two monosaccharides combine, as shown in FIGURE

*The suffix *-ose* usually stands for sugar. Prefixes such as *tri-* (three), *pent-* (five), and *hex-* (six) indicate the number of carbon atoms in the molecule. Pentoses, for example, are monosaccharides containing five carbons.

FIGURE 2.9 MOLECULAR STRUCTURES OF GLUCOSE, FRUCTOSE, AND RIBOSE

Glucose and fructose both have the same molecular formula, $C_6H_{12}O_6$, but different molecular *structures*. Each sugar can be represented in several different ways. Here, each is shown in open-chain and ring forms and by its molecular formula.

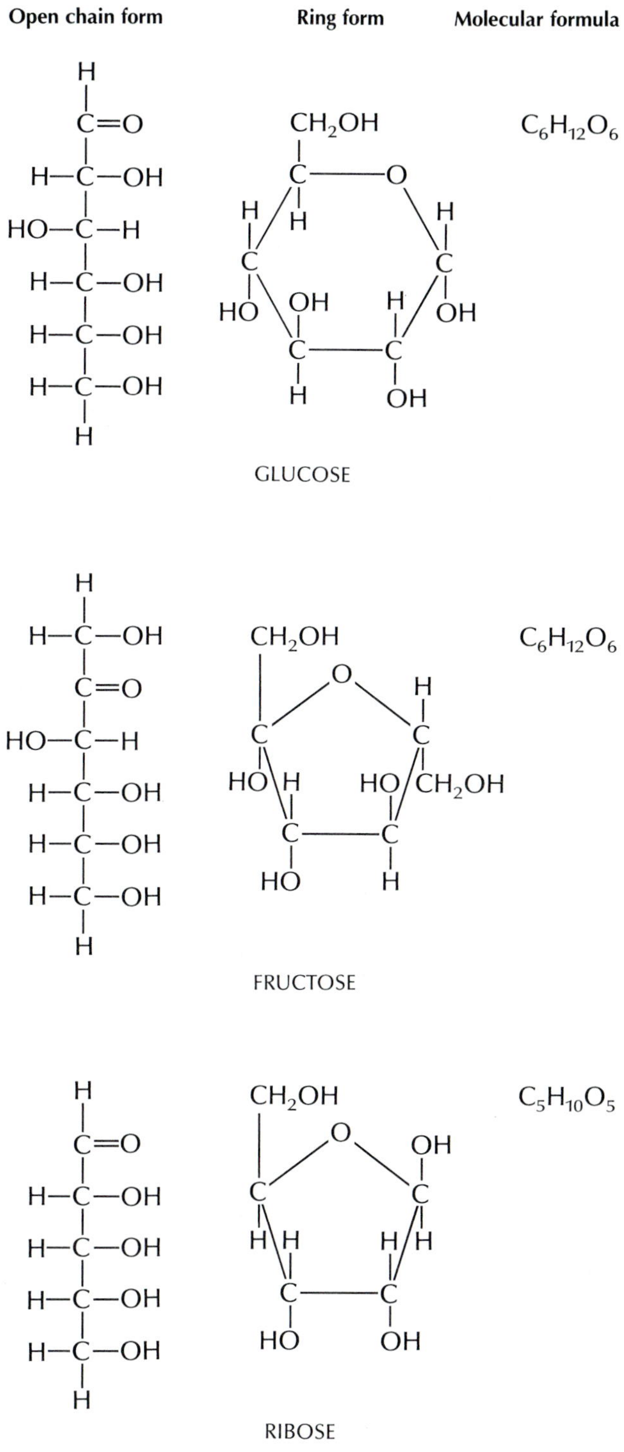

GLUCOSE

FRUCTOSE

RIBOSE

FIGURE 2.10 DEHYDRATION SYNTHESIS OF A DISACCHARIDE

When a molecule of water (shown in blue on the left) is removed during the reaction, a bond (shown in red) is formed between the two monosaccharide molecules, pro-ducing a single molecule of the disaccharide maltose and a molecule of water (shown in blue on the right).

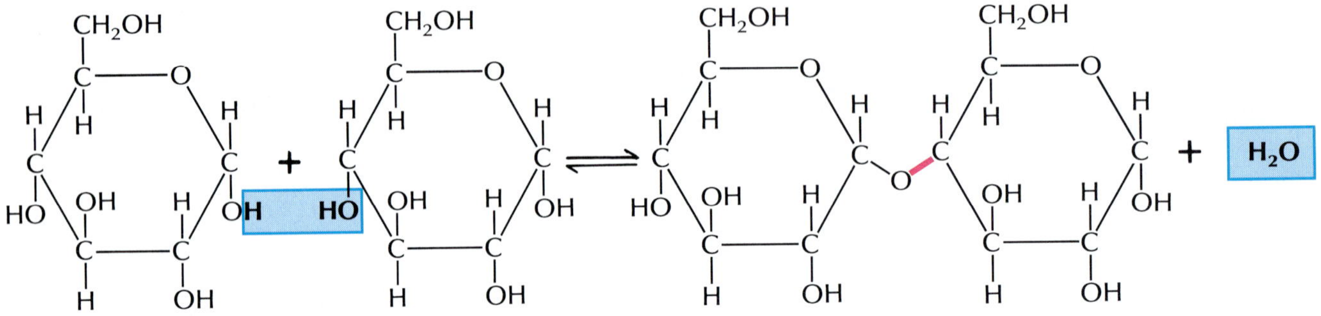

Glucose	plus	Glucose	yields	Maltose	plus	Water

Dehydration synthesis

$$C_6H_{12}O_6 \quad + \quad C_6H_{12}O_6 \; \underset{\text{Hydrolysis}}{\overset{\text{Dehydration synthesis}}{\rightleftharpoons}} \; C_{12}H_{22}O_{11} \quad + \quad H_2O$$

2.10. The reverse chemical reaction—hydrolysis—takes place when water is added. Disaccharides can thus be broken down into simpler monosaccharides by adding water:

Maltose + water $\longrightarrow$ glucose + glucose

The entire reversible processes of dehydration, the loss of a water molecule, and hydrolysis, the addition of a water molecule, may be written in a simple form as follows:

Monosaccharide + monosaccharide

$$\underset{\text{hydrolysis}}{\overset{\text{dehydration synthesis}}{\rightleftharpoons}} \text{disaccharide + water}$$

Carbohydrate molecules containing more than two monosaccharides are called **polysaccharides** (*poly*, many). Polysaccharides are formed by dehydration synthesis and can be broken down into their many monosaccharide components by hydrolysis. Polysaccharide chains may contain hundreds or thousands of monosaccharide molecules. These long chains may be branched or unbranched. Such long chains of repeating units are called *polymers*. An important polysaccharide is *glycogen*, which is the main storage form of carbohydrates in humans.

Uses of carbohydrates Carbohydrates are important in the body mainly as energy sources for cellular activities. When the body needs energy, it breaks down and uses carbohydrates first, fats second, and proteins last (see Chapter 26). Some of the more important carbohydrates and their uses are described in TABLE 2.5.

Lipids

Body fats belong to a diverse group of organic compounds called **lipids**. Lipids are insoluble in water but can be dissolved in organic solvents such as ether, alcohol, and chloroform. Like carbohydrates, they are composed mainly of carbon, hydrogen, and oxygen, but they may contain other elements, especially phosphorus and nitrogen. Also, lipids usually have more than twice as many hydrogen atoms as oxygen atoms.

Familiar lipids are fats and oils. Less well known but equally important are phospholipids, steroids, and prostaglandins.

Fats *Fats* are important mainly because they are energy-rich molecules (yielding about twice as much energy as carbohydrates) and are therefore a suitable source of reserve food. They are stored in the body in the form of *triglycerides*, also called neutral fats (because they are not electrically charged). Fats also provide the body with insulation, protection, and cushioning.

Each triglyceride molecule is composed of **glycerol** and three **fatty acids** [FIGURE 2.11]. The glycerol molecule contains three hydroxyl (OH) groups. Fatty acids are organic molecules with two regions: (1) a long hydrocarbon chain, which is not very reactive chemically, and (2) a single carboxyl group (COOH), which is highly reactive chemically. An example of a fatty acid, palmitic acid, is represented as

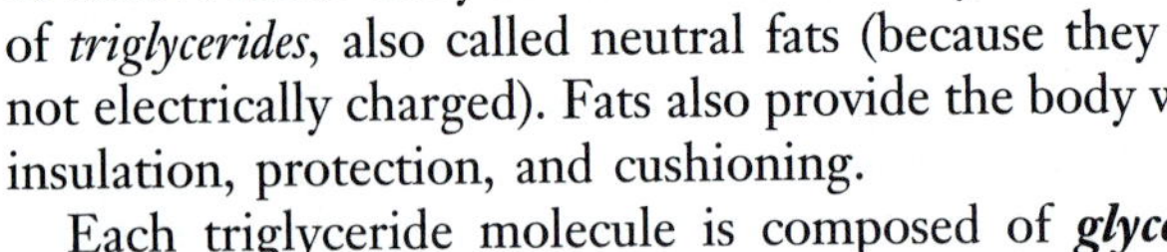

TABLE 2.5 SOME CARBOHYDRATES AND THEIR USES BY THE BODY

Name	Type	Description
Deoxyribose	Monosaccharide	Constituent of hereditary material, DNA.
Fructose	Monosaccharide	Important in cellular metabolism of carbohydrates. Found in fruits, it is the sweetest sugar.
Galactose	Monosaccharide	Present in brain and nerve tissue.
Glucose (dextrose)	Monosaccharide	Main energy source for the body. Breakdown of glucose produces ATP, which is used in almost all energy-requiring cellular activities. Used in intravenous feeding; does not have to be digested to be used by the body. The brain requires a constant supply.
Ribose	Monosaccharide	Constituent of RNA.
Lactose	Disaccharide	Milk sugar. Aids absorption of calcium. Yields glucose and galactose upon hydrolysis.
Sucrose	Disaccharide	Common cane or beet sugar. Yields glucose and fructose upon hydrolysis.
Cellulose	Polysaccharide	Undigestible by body but serves as important fiber that provides bulk for the proper movement of food through the intestines.
Glycogen	Polysaccharide	Main form of carbohydrate storage; stored in the liver and muscles until needed as energy source and converted to glucose.
Heparin	Polysaccharide	Prevents excessive blood clotting.
Starch	Polysaccharide	Chief food carbohydrate in human nutrition.

FIGURE 2.11 DEHYDRATION SYNTHESIS OF A FAT

Three water molecules (shown in blue on the left) are removed from a glycerol molecule and three fatty acid molecules. The $(CH_2)_{14}$ represents the 14 CH_2 units in each fatty acid molecule. The bonding (red) that results from this dehydration synthesis yields a fat (triglyceride) and three molecules of water (shown in blue on the right). The water molecules are produced by the removal of hydrogen atoms (H^+) from the glycerol molecule, and the hydroxyl (OH^-) groups from the fatty acid molecules.

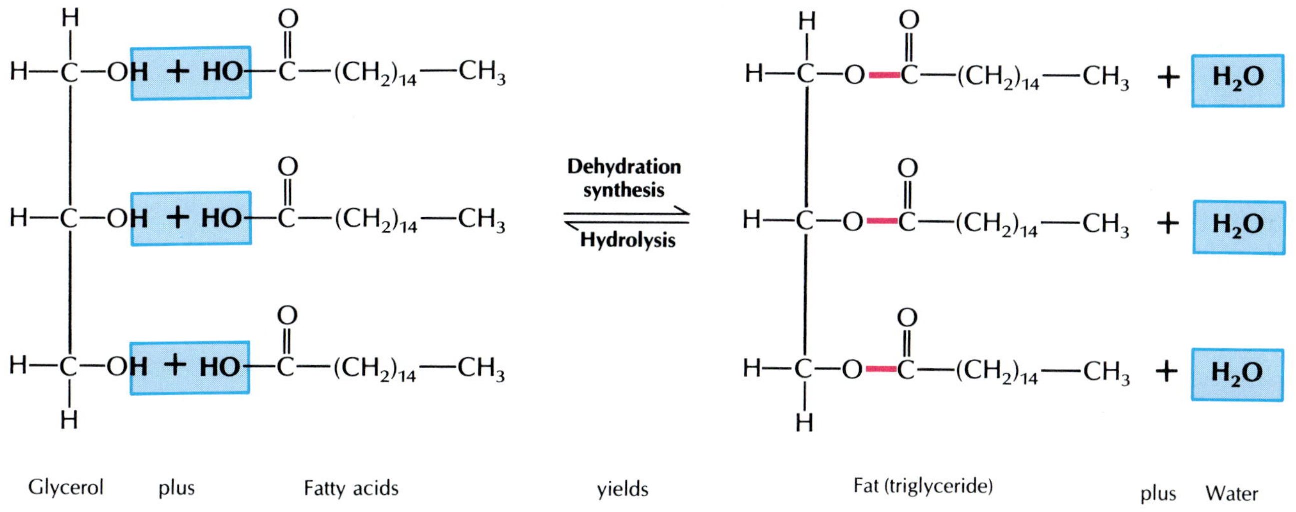

Glycerol   plus   Fatty acids   yields   Fat (triglyceride)   plus   Water

with the carboxyl group shown in color and the long hydrocarbon chain shown in black. Most fatty acid molecules are covalently linked to other molecules by their carboxyl group. The carboxyl groups are shown in FIGURE 2.11 as

$$HO-\overset{\overset{\textstyle O}{\|}}{C}-$$

with the double covalent bond (in green) representing two pairs of electrons being shared by two atoms.

Hydroxyl and carboxyl groups make the formation of fat possible. Each hydroxyl group may bind a fatty acid carboxyl group. A molecule of fat is formed when the three hydroxyl (OH) groups in a molecule of glycerol combine with the carboxyl (COOH) portion of three fatty acid molecules [FIGURE 2.11]. This reaction is a dehydration synthesis, similar to the dehydration reaction that produces disaccharides from the simpler monosaccharides. The reaction is reversible, and during digestion fats are broken down into their glycerol/fatty acid building blocks by hydrolysis, the addition of water.

Apparently, there is a connection between dietary fats and diseases of the heart and circulatory system. Dietary fats include both saturated and unsaturated fatty acids. *Saturated fatty acids* are solid at room temperature. They have no double bonds between their carbon atoms. Since *unsaturated fats* are liquid at room temperature, they are considered to be *oils*. Unsaturated fats have one or more double bonds between their carbon atoms. As the number of double bonds increases, the fats become more unsaturated, or oily. For this reason, unsaturated vegetable oils, which have many double bonds, are referred to as *poly*unsaturated fats. Animal fats generally contain more saturated fatty acids than vegetable fats do. Both saturated fats and cholesterol (an ingredient of fats) are thought to be involved in various cardiovascular diseases.

Phospholipids *Phospholipids* are fatty compounds that contain glycerol, two fatty acids, and a phosphate group (PO_4^{3-}) that is often linked with a nitrogen-containing group. The phosphate group is bonded to the glycerol at the point where a third fatty acid would be in a neutral fat:

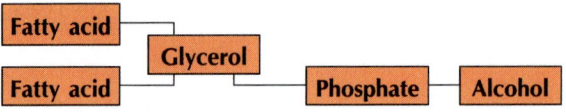

The polar end of a phospholipid molecule attracts water and thus is soluble in water, while the nonpolar end repels water and thus is insoluble [FIGURE 2.12]. This tendency to be soluble at one end and insoluble at the other is important in cell membranes, which are built from phospholipids, cholesterol, and proteins. In Chapter 3 you will see further how phospholipids give membranes

FIGURE 2.12 SIMPLIFIED DRAWING OF A PHOSPHOLIPID

The water-attracting polar end is shown by a sphere, and the water-repelling nonpolar end, in the form of hydrocarbon "tails," is shown by wavy lines.

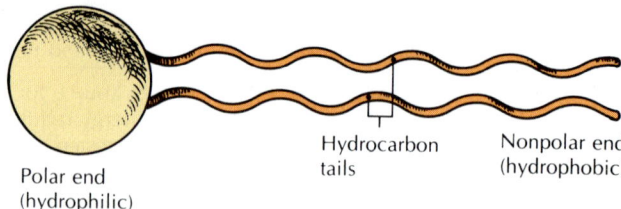

Polar end (hydrophilic)   Hydrocarbon tails   Nonpolar end (hydrophobic)

their distinct sheetlike structure and many of their functional properties.

Steroids *Steroids* are naturally occurring fat-soluble compounds composed of four bonded carbon rings [FIGURE 2.13]. Three of the rings are six-sided, and the fourth is five-sided. There are a total of 17 carbons in the four rings. Among the important steroids in the human body are cholesterol, bile salts, male and female sex hormones, and some of the hormones secreted by the adrenal glands. Steroid hormones help regulate certain phases of metabolism in the body.

Steroids that have a hydroxyl group and a chain of eight or more carbons on their five-sided ring are referred to as *sterols*. The sterol cholesterol is an important component of cell membranes.

Prostaglandins *Prostaglandins* are modified fatty acids found in many kinds of tissues throughout the body. Their varied functions are still being discovered, but among other things they can raise or lower blood pressure and cause the contraction of smooth muscle in the uterus.

Proteins

Proteins are extremely important in the structure and function of the body. They are the major components of most of the tissues of the body. Like carbohydrates and lipids, proteins are constructed from relatively simple building blocks, but their structural possibilities are almost limitless. Proteins always contain carbon, hydrogen, oxygen, and nitrogen; and sometimes proteins contain sulfur, phosphorus, and iron. The major classes of proteins in the body include enzymes, which control chemical activity; antibodies, which protect us against disease; hemoglobin, which carries oxygen in the blood; and numerous hormones, which are important in the control of cellular functions. Proteins also act as buffers and help blood to clot properly.

FIGURE 2.13 STRUCTURE OF A STEROID

Cholesterol, a steroid, is formed with four interlocking carbon rings. Note that one of the rings is five-sided, while the other three are six-sided.

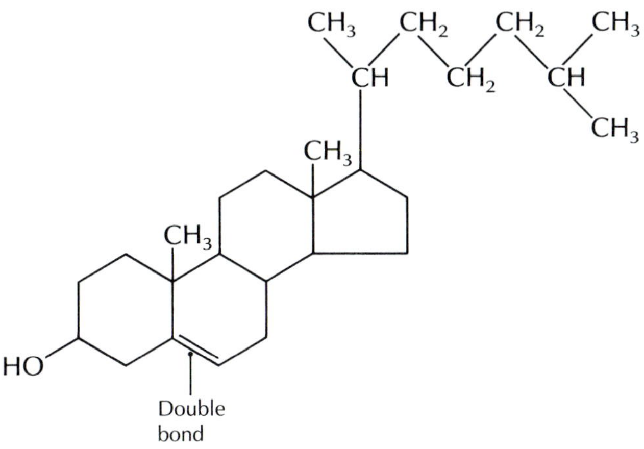

FIGURE 2.14 STRUCTURE OF AN AMINO ACID

[A] The general structure of an amino acid, with R representing a variable group of atoms. [B] The same general amino acid structure, showing the amino group, variable group, and carboxyl group linked to a central α-carbon.

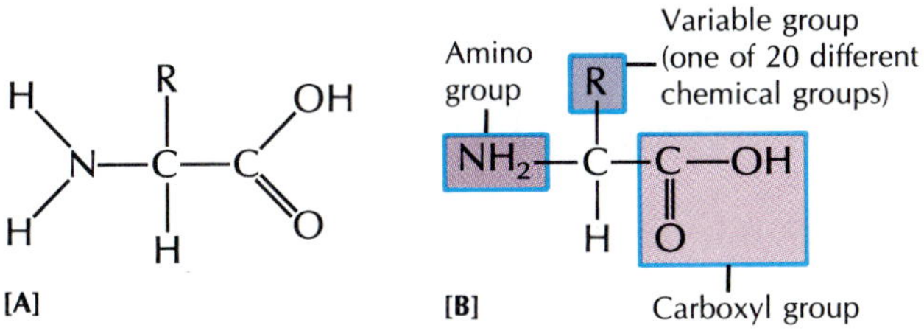

Amino acids: Building blocks of proteins

Proteins are large, complex molecules composed of smaller structural subunits called **amino acids**. Amino acids always contain an amino group (NH_2) at one end of the molecule, a carboxyl group (COOH) at the other end, and one variable component, shown as R in FIGURE 2.14. All these components are bonded to a central carbon atom, designated as the α-carbon. The variable R group contains an atom or group of atoms in different chemical combinations. The simplest amino acid, glycine, has a single hydrogen atom in the R position.

Only about 20 amino acids are known in living systems, but they can combine in many different ways to produce all the 50,000 or so different proteins in the body.

Protein formation When two amino acids are bound together, they form a protein unit called a *dipeptide*; three amino acids bonded together make a *tripeptide*. Many amino acids bonded together form a chain called a *polypeptide*. A protein usually consists of one or more polypeptide chains folded into a complex three-dimensional shape.

FIGURE 2.15 shows the dehydration synthesis of two amino acids into a single dipeptide molecule. This reaction takes place when the C in the carboxyl (COOH) group of one amino acid bonds to the N in the amino (NH_2) group of another amino acid. In this familiar dehydration synthesis, a water molecule splits out and a dipeptide is formed. Such a union between amino acids results in the formation of a strong covalent bond called a **peptide bond**.

Levels of protein structure Within a protein molecule, several different levels of structure can be distinguished. The *primary structure* is the linear sequence of amino acids in the polypeptide chains that constitute the protein molecule. The sequence of amino acids determines the overall three-dimensional shape of the molecule, which in turn determines how the protein func-

FIGURE 2.15 DEHYDRATION SYNTHESIS OF A PROTEIN

A peptide bond (red) is formed at the point where the OH⁻ and H⁺ (blue, left) are removed, and a molecule of water (blue, right) is produced along with the dipeptide molecule. Note that there is still a carboxyl group on the right-hand end of the new dipeptide molecule and an amino group on the left end.

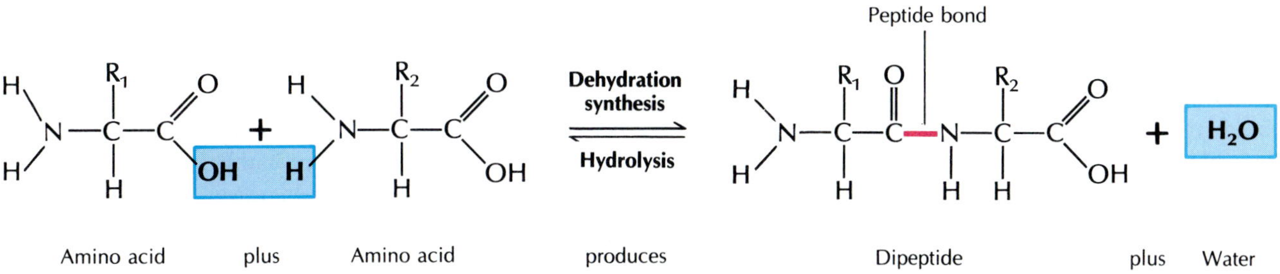

FIGURE 2.16 ORGANIZATIONAL LEVELS OF PROTEINS

[A] The *primary structure* of the two polypeptide chains (linked by disulfide bonds, —S—S—) that make up the human insulin molecule. The abbreviations (for example, Gly for glycine) refer to individual amino acids. **[B]** The *secondary (alpha-helical) structure* of a portion of a protein, showing the hydrogen bonds between amino acids. **[C]** The coiled and folded *tertiary structure* of a globin molecule, made up of a chain of amino acids. **[D]** The *quaternary structure* of human hemoglobin, made up of four tertiary structure molecules.

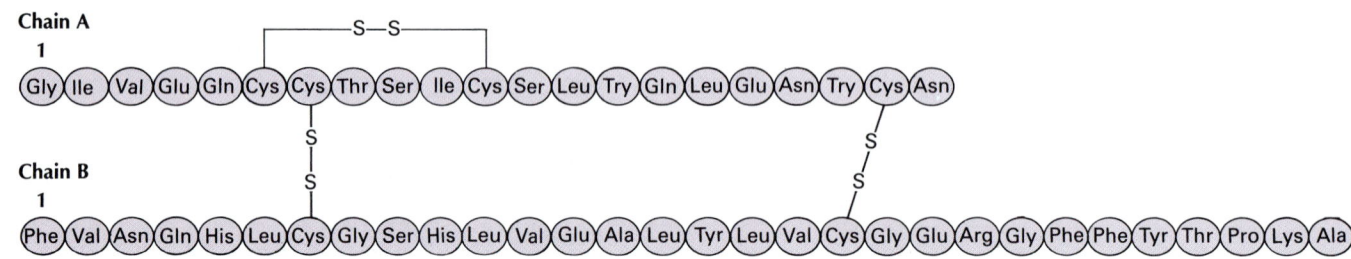

Chain A
1
Gly–Ile–Val–Glu–Gln–Cys–Cys–Thr–Ser–Ile–Cys–Ser–Leu–Try–Gln–Leu–Glu–Asn–Try–Cys–Asn

Chain B
1
Phe–Val–Asn–Gln–His–Leu–Cys–Gly–Ser–His–Leu–Val–Glu–Ala–Leu–Tyr–Leu–Val–Cys–Gly–Glu–Arg–Gly–Phe–Phe–Tyr–Thr–Pro–Lys–Ala

[A] PRIMARY STRUCTURE

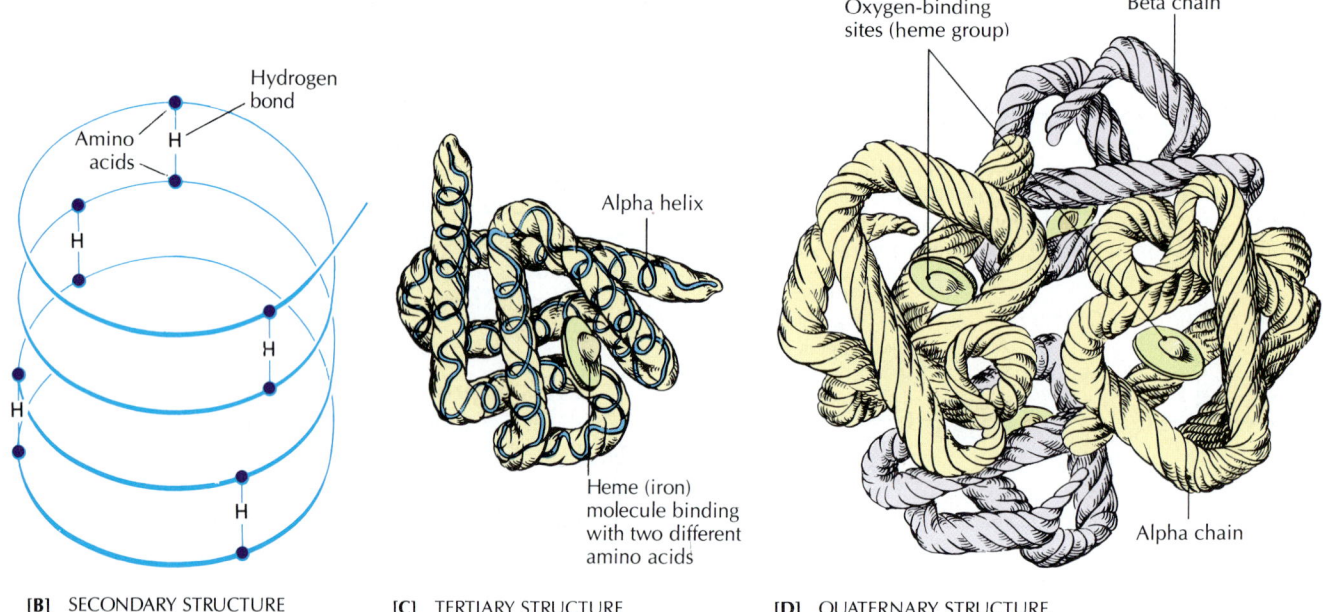

[B] SECONDARY STRUCTURE   **[C]** TERTIARY STRUCTURE   **[D]** QUATERNARY STRUCTURE

tions. FIGURE 2.16A illustrates the primary structure of the two polypeptide chains that make up the human insulin molecule. The way in which the chains are arranged in space results from their rotation around the hydrogen bonds within each chain of amino acids. This *secondary structure* is commonly a helix [FIGURE 2.16B]. The *tertiary structure* results from the folding of the helix into an overall fibrous or globular shape [FIGURE 2.16C]. Those proteins that consist of more than one polypeptide subunit form a *quaternary structure* [FIGURE 2.16D].

Enzymes The energy required to weaken chemical bonds enough to start a chemical reaction is called the *activation energy*. **Enzymes** are proteins that lower the amount of activation energy needed, increasing the rate of a chemical reaction without permanently entering into the reaction themselves. Enzymes are *catalysts*. The same enzyme can be used over and over to activate the same reaction. The enzyme-catalyzed reactions would take place without enzymes, but only at a uselessly slow pace.

The shape—the molecular structure—of an enzyme permits it to join specific reactants (or *substrates*) together, atom to atom, and then withdraw after the chemical union has been completed [FIGURE 2.17]. The action of some enzymes is reversible; they are thus able to drive a reaction in either direction. Under proper conditions,

FIGURE 2.17 ENZYMATIC ACTION

The catalytic action of an enzyme on two substrate molecules, X and Y, to make a new compound, molecule Z. Enzymes may also be used to split compounds apart. In this *induced-fit model*, the shape of the enzyme changes slightly when it is in contact with the substrates, but it springs back to its original form after the action is complete. In a different model, known as the *lock-and-key model*, the enzyme is unchanged throughout the process.

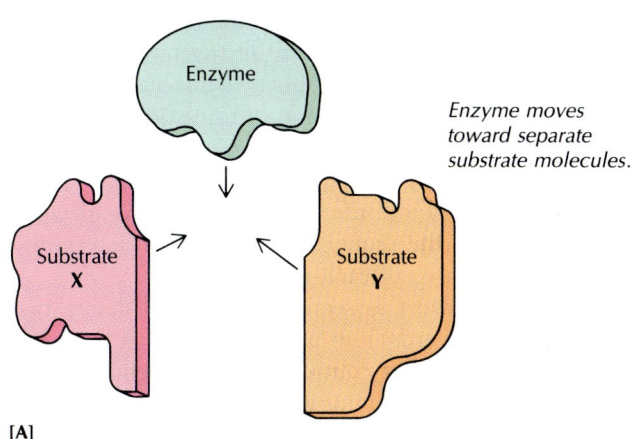

Enzyme moves toward separate substrate molecules.

[A]

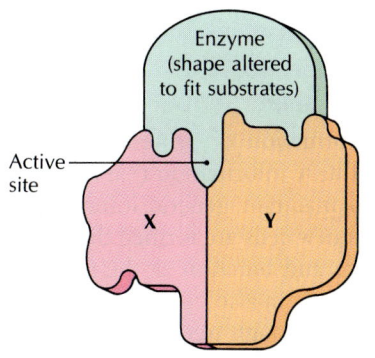

Enzyme unites molecules through a chemical reaction, temporarily forming an enzyme-substrate complex.

[B]

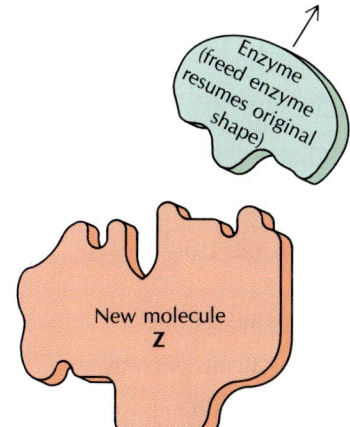

New molecule is formed, enzyme is free to be used in another reaction.

[C]

the same enzyme is capable of breaking a compound into its component parts:

$$\text{Substrate X + substrate Y} \xrightleftharpoons{\text{enzyme}} \text{product Z}$$

Besides being capable of reversible action, enzymes are *specific*; that is, there is one specific enzyme for any given chemical reaction. In the small intestine, for example, the enzyme *lactase** speeds the breakdown of lactose (which has 12 carbon atoms) into two 6-carbon molecules, glucose and galactose.

An enzyme functions by physically binding at a specific site on the enzyme (the *active site*) to the reactant molecules, actually altering their chemical bonds. Sometimes this results in breaking one molecule into two, as when lactose is converted into two monosaccharides. It may also result in bringing two compounds together, as when two amino acids are combined. *Enzyme inhibitors*, such as certain poisons, may prevent normal enzyme action by clogging the space usually occupied by the reactant or reactants.

Even though all enzymes are proteins, many of them are conjugated proteins; that is, they contain nonprotein components called *cofactors* that are bound tightly to enzymes and allow enzymes to function properly. Cofactors may be inorganic (metals), organic *(coenzymes)*, or both. Many vitamins in the diet are converted into coenzymes in the body; this explains why vitamins are so important. Cofactors, like enzymes, are not changed or used up by a chemical reaction, so they can be reused.

Several environmental requirements must be met if enzymes are to operate properly:

1 The temperature must remain within certain limits. Human enzymes generally function at optimum efficiency when the body temperature is 37°C (98.6°F). Cold slows down the reaction rate. When temperatures are too high, the hydrogen bonds of the enzyme are broken, and the enzyme becomes useless (denatured).

2 The pH of the cell environment affects an enzyme's rate of action. Each enzyme has an *optimal pH* at which it works best. Most enzymes operate effectively at pH values between 5 and 9. The pH has an effect on the hydrogen bonding, and therefore the shape, of an enzyme.

3 A sufficient amount of substrate must be present.

4 A certain amount of energy is required. Enzymes tend to lower the level of activation energy needed and allow energy-rich molecules, such as glucose, to supply usable energy slowly. Thus, all the available energy does not dissipate as heat.

---

*The *-ase* suffix identifies an enzyme. The prefix, *lact-* in this case, describes the substrate being acted upon. *Lact-* means milk, and *-ose* means sugar, so *lactose* (the substrate) is milk sugar.

Nucleic Acids

Like proteins, *nucleic acids* (noo-KLAY-ihk) are very large, complex molecules. The Swiss physician Friedrich Miescher first isolated them in 1869 from the nuclei of human pus cells. He called them nucleic acids because of their acidity and their location in the cell's nucleus. There are two types of nucleic acids: *deoxyribonucleic acid (DNA)* and *ribonucleic acid (RNA)*. DNA makes up the chromosomes within the cell's nucleus and is the main repository for the genetic information of the cell. RNA is present both in the nucleus and in the cytoplasm, the part of the cell outside the nucleus.

The DNA in a cell determines which proteins will be synthesized. The proteins (all enzymes are proteins) determine the shape and structure of the cell as well as what functions the cell carries on. Thus, DNA controls the cell's activities by determining which proteins are synthesized.

Structure of nucleic acids Nucleic acids are the largest organic molecules in the body. Their size results from the bonding together of small units called *nucleotides* (NOO-klee-uh-tides). Each nucleotide has three components: a phosphate group, a five-carbon sugar, and a *nitrogenous base* [FIGURE 2.18]. The nitrogenous base is bonded to the sugar by a covalent bond. DNA and RNA each have their own specific sugar. RNA contains ribose, a sugar that has five atoms of oxygen. Because DNA contains a sugar with one fewer oxygen atom, it is called *deoxy*ribose.

DNA contains four kinds of nitrogenous bases: the double-ring *purines* (adenine and guanine) and the single-ring *pyrimidines* (thymine and cytosine). Adenine, guanine, thymine, and cytosine are usually referred to as A, G, T, and C [FIGURE 2.19A]. Another nitrogenous base, uracil (U), is also shown in FIGURE 2.19A; RNA contains uracil instead of thymine. In double-stranded DNA the nucleotides of the two strands are connected by hydrogen bonds between the nitrogenous bases [FIGURE 2.19B].

The components of DNA have been known for almost 100 years, but the *structure* of a DNA molecule was not discovered until 1953, when it was proposed by James D. Watson and Francis H. C. Crick. DNA has the shape of a twisted ladder, or *double helix* [FIGURE 2.19C, D].

A double strand of DNA may be compared to a ladder. The "uprights" of the ladder are made of chains of alternating phosphate groups and deoxyribose sugars. The ladder also has "rungs" made of pairs of nitrogenous bases connected to each other by hydrogen bonds; the bases are bonded to the sugars on the uprights by covalent bonds [FIGURE 2.19B]. The nitrogenous bases can join only in specific relationships: adenine and thymine bond together, and guanine and cytosine bond together. In RNA, uracil, instead of thymine, always bonds with adenine. In both DNA and RNA, a purine always binds to a pyrimidine. Thus, the "rungs" are always of equal length. Overall, the sequence of base pairs determines the genetic makeup of each person.

Before we leave this section, we should define two very important words: *gene* and *chromosome*. A *gene* is a segment of DNA that controls a specific cellular function, either by determining which proteins will be synthesized or by regulating the action of other genes. Genes are the hereditary units that carry hereditary traits. A *chromosome* is a nucleoprotein (nucleic acid + protein) structure in the nucleus of a cell that contains the genes.

Functions of nucleic acids Nucleic acids determine and regulate the formation of proteins in cells. They also carry the hereditary information that makes it possible to transmit and maintain genetic information within the body so that as new cells are formed they have exactly the same structure and function as their parent cells. Hereditary information can also be passed on from one generation to the next through nucleic acids.

Q: *How much DNA does a person have?*

A: The average person has approximately 2.9×10^9 (2.9 billion) base pairs of DNA. The total length of all an adult's DNA strands is approximately 990 mm (about 1 m).

FIGURE 2.18 STRUCTURE OF A NUCLEOTIDE

Each nucleotide contains a phosphate group, a sugar, and a nitrogenous base; the base is covalently bonded to the sugar. Nucleotides bond together to form the nucleic acids RNA and DNA.

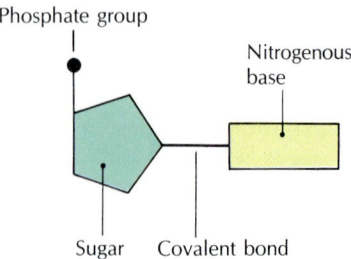

Phosphate group

Nitrogenous base

Sugar   Covalent bond

A S K   Y O U R S E L F

1 What are the components of a carbohydrate?

2 How does a monosaccharide differ from a polysaccharide?

3 What are some familiar kinds of lipids?

4 What are the structural subunits of proteins?

5 What are enzymes, and how do they enhance chemical activities in the body?

6 What are the two types of nucleic acids?

FIGURE 2.19 STRUCTURE OF NUCLEIC ACIDS

[A] The nitrogenous bases of nucleic acids with their molecular structures and symbols. The double-ring purines are adenine and guanine; the single-ring pyrimidines are thymine and cytosine in DNA and uracil and cytosine in RNA. [B] A small portion of a DNA strand showing 12 nucleotides. The nucleotides are connected by hydrogen bonds between their bases. Note that adenine can combine only with thymine and cytosine can combine only with guanine. The repeating sugars and phosphate groups can be thought of as the "uprights" of the ladder, and the nitrogenous bases A-T and C-G as the "rungs." [C] A computer-generated image of DNA. [D] A segment of a twisted, double-stranded DNA molecule. A complete DNA molecule has millions of cross-connecting nitrogenous base pairs.

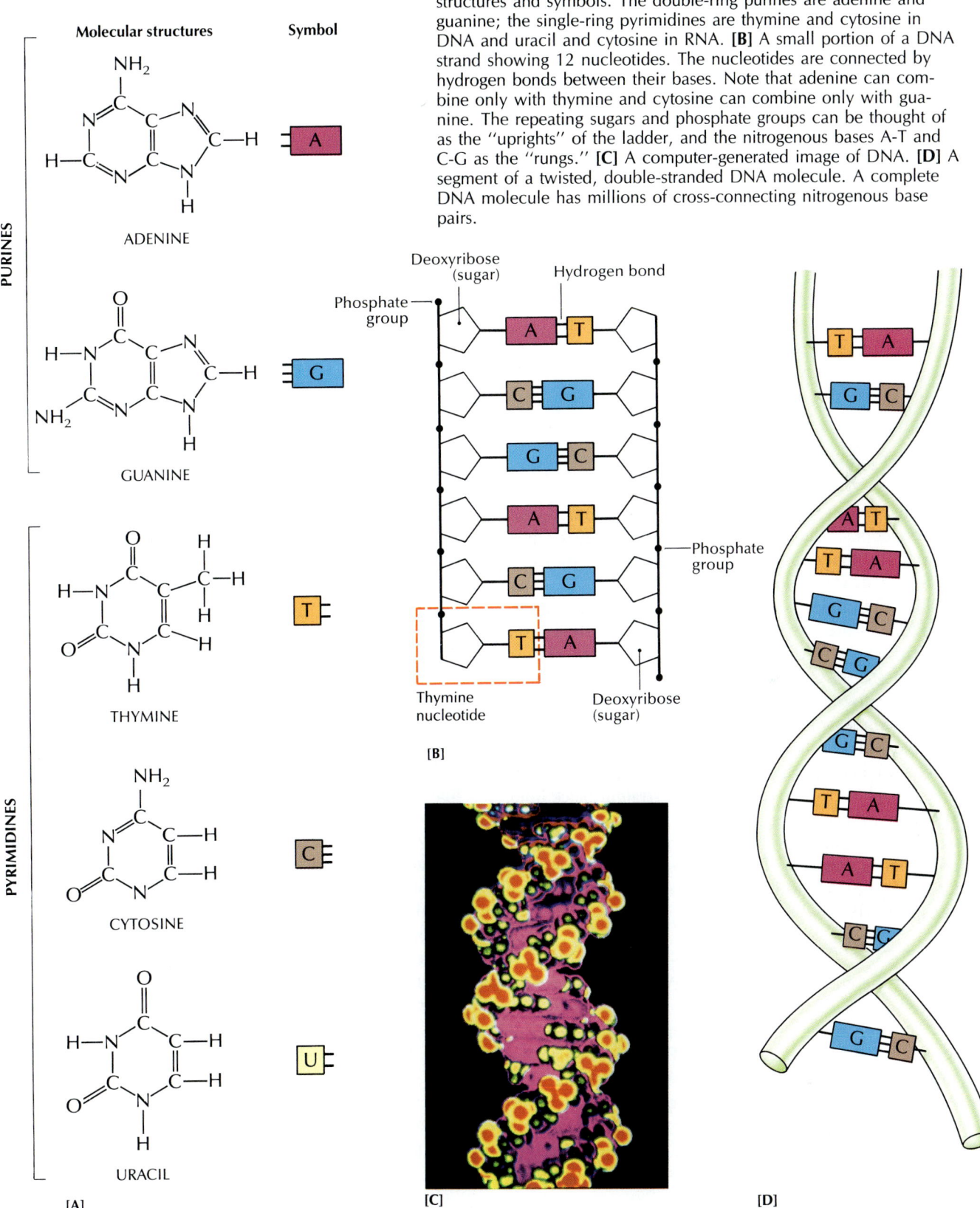

FIGURE 2.20 STRUCTURAL DIAGRAM OF ATP MOLECULE

Note the high-energy bonds (shown as wavy red lines) between the phosphate groups.

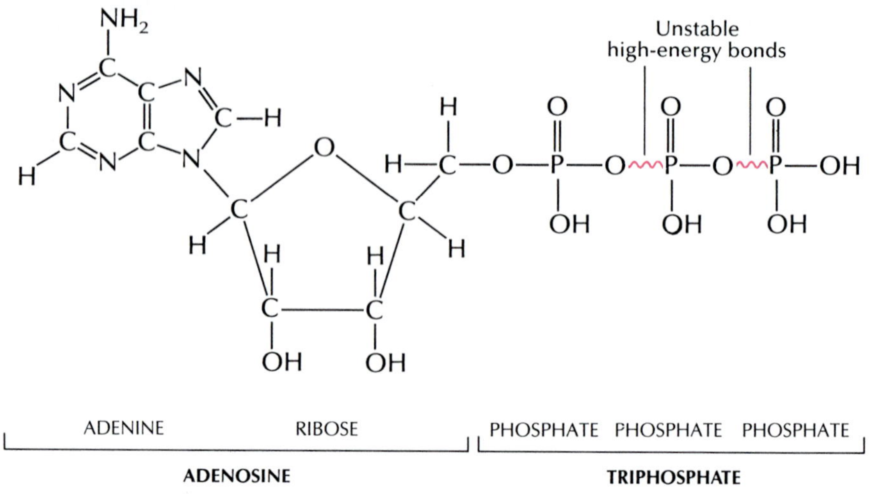

ADENINE   RIBOSE   PHOSPHATE   PHOSPHATE   PHOSPHATE

ADENOSINE   TRIPHOSPHATE

ENERGY FOR LIVING: ATP

The energy you need to live comes from chemicals and their reactions. The immediate source of energy for most biological activities in the cell is a small organic molecule known as **ATP** (adenosine triphosphate).

The ATP molecule is built from smaller units of adenine (one of the nitrogenous bases in nucleic acids), the five-carbon sugar ribose, and three phosphate groups. Note in FIGURE 2.20 that two of the three phosphates that make up the triphosphate portion of the ATP molecule are connected by high-energy bonds (shown as wavy red lines). These bonds are capable of releasing much more stored energy than does the bond that connects the first phosphate to ribose (shown as a straight black line). When energy is needed for cellular activity, ATP reacts with water (hydrolysis), and the last of the three phosphate bonds is broken. The last phosphate group is separated, yielding adenosine *di*phosphate (ADP), inorganic phosphate (P_i), and energy:

$$\text{ATP} \rightleftharpoons \text{ADP} + P_i + \text{energy}$$

Most of the energy released from ATP can be used for the cell's immediate needs.

Note that the chemical reaction described above is reversible. Converting ADP into ATP is a way of storing energy until it is needed. Although ATP stores energy, it is not produced and stored in large quantities. It is replaced as it is needed by the energy from decomposition reactions that take place within cells. This energy is combined with ADP and P_i to synthesize additional ATP.

The synthesis of ATP is an example of an *endergonic* ("energy in") reaction because energy is *required* for the reaction to take place. Activities that *liberate* energy, such as the breaking of ATP's high-energy bonds, are called *exergonic* ("energy out").

It should be understood that even though the bonds between the last two phosphate groups of an ATP molecule are usually referred to as being "energy-rich," the ATP molecule actually contains no more energy than many other organic molecules. ATP is a ready supplier of energy because it is a relatively unstable compound. What is most important about ATP is the ready *availability* and *transferability* of the energy stored in its bonds. Indeed, a molecule of sucrose (ordinary table sugar) has much more potential energy than an ATP molecule does, but cells cannot use the energy from sucrose directly. They can, however, readily convert the energy in ATP bonds into such useful forms as movement, heat, and chemical reactions.

Body cells need a slow, steady energy source that can be controlled and regulated, since the chemical reactions in body cells can use energy only in small amounts. The readily available energy in ATP provides cells with an effective way of regulating energy release.

ASK YOURSELF

1 What is the main function of ATP?

2 How does ATP store and release energy?

CHAPTER SUMMARY

Matter, Elements, and Atoms (p. 31)

1 *Matter* is anything that occupies space and has mass. *Mass* refers to the amount of matter in an object. Matter is composed of *elements,* basic substances that cannot be broken down into simpler substances by ordinary chemical means. An *atom* is the smallest unit of an element that retains the chemical characteristics of that element.

2 Some of the elements critical to proper body function are oxygen, carbon, hydrogen, nitrogen, calcium, and phosphorus.

3 The *nucleus* of an atom contains positively charged *protons* and uncharged *neutrons.* Orbiting the nucleus are negatively charged *electrons.*

4 An atom's *atomic number* indicates the number of protons in its nucleus. *Atomic weight* is calculated by adding the number of protons and the number of neutrons in the nucleus.

5 *Isotopes* are atoms of an element that have the typical number of protons and electrons but different numbers of neutrons. Radioactive isotopes have nuclei that are undergoing changes called "decay."

6 The electrons of atoms are distributed around the nucleus in layers called *electron shells.*

7 The number of electrons in the outermost shell determines how one atom will react with another atom. Atoms that have a filled outer shell are stable.

How Atoms Combine (p. 34)

1 Atoms of elements combine chemically to form *molecules; compounds* are composed of molecules made up of more than one element. The atoms that form molecules or compounds are held together by electrical forces called *chemical bonds.* When an atom or molecule gains or loses electrons, it acquires an electrical charge and is called an *ion.*

2 *Ionic bonds* are formed when an ion or group of ions is attracted to an ion or group of ions with an opposite electrical charge.

3 When atoms share one or more pairs of electrons, the resultant chemical bond is called a *covalent bond.*

4 *Hydrogen bonds* are formed when a covalently bonded hydrogen atom in a compound acquires a slight positive charge and becomes attracted to negatively charged atoms nearby.

Chemical Reactions (p. 38)

1 The combining of atoms or molecules or the breaking apart of molecules brings about a chemical change called a *chemical reaction.*

2 In *combination reactions,* two or more atoms or molecules combine to form a more complex substance.

3 *Decomposition reactions* break chemical bonds in molecules to form two or more products.

4 In *oxidation,* an atom or molecule *loses* electrons or hydrogen atoms. In *reduction,* an atom or molecule *gains* electrons or hydrogen atoms. Oxidation and reduction always occur together.

5 In *hydrolysis,* a molecule of water interacts with the reactant to break the reactant's bonds. The original reactant is then rearranged, together with the water, into different molecules. *Dehydration synthesis reactions* are chemically the reverse of hydrolysis reactions.

Water (p. 39)

1 *Inorganic compounds* are composed of relatively small molecules that are usually ionically bonded. Water is an inorganic compound.

2 *Water* is a *polar molecule;* each end of the molecule bears a different electrical charge. Water is a solvent, transporter, temperature regulator, and lubricant.

Acids, Bases, Salts, and Buffers (p. 41)

1 An *acid* is a substance that releases hydrogen ions when dissolved in water. A *base* is a substance that releases hydroxyl ions when dissolved in water.

2 A *salt* is a compound (other than water) formed during a neutralization reaction between acids and bases.

3 The breakdown of molecules into ions in water is called *dissociation.*

4 The degree of acidity or alkalinity of a solution is measured with the *pH scale.* This scale indicates the concentration of free hydrogen ions in water.

5 *Buffers* are chemical substances that regulate the body's acid-base balance.

Some Important Organic Compounds (p. 43)

1 *Organic* compounds always contain carbon and hydrogen. They are typically large and complex and are held together by covalent bonds.

2 *Carbohydrates* are made of carbon, hydrogen, and oxygen, with hydrogen and oxygen in a 2:1 ratio; they are an important source of energy. Carbohydrates are classified as *monosaccharides, disaccharides,* or *polysaccharides,* depending on how many monosaccharides they contain.

3 *Lipids* are a diverse group of organic compounds that include body fats and other fats, oils, *phospholipids, steroids,* and *prostaglandins.*

4 *Proteins* are large, complex molecules composed of *amino acids.* The amino acids are linked together in chains called *polypeptides.* A protein consists of one or more polypeptide chains.

5 *Enzymes* are proteins that increase the rate of a chemical reaction but are not permanently changed by the reaction.

6 *Nucleic acids* are very large molecules composed of chains of subunits called *nucleotides.* The two nucleic acids are *deoxyribonucleic acid (DNA)* and *ribonucleic acid (RNA).* Nucleic acids carry the body's hereditary information and regulate protein synthesis.

Energy for Living: ATP (p. 52)

1 The energy for most biological activities comes from *ATP* (adenosine triphosphate), which is composed of adenine, ribose, and three phosphate groups.

2 When the high-energy bonds in ATP are broken, they release stored energy that is usable for a cell's immediate needs.

STUDY AIDS FOR REVIEW

KEY TERMS

acid 41
amino acid 47
atom 31
atomic number 32
atomic weight 32
ATP 52
base 41
bond 34
buffer 42
carbohydrate 43
chemical reaction 38
covalent bond 36
dehydration synthesis 38, 43
deoxyribonucleic acid (DNA) 50
disaccharide 43
dissociation 41
electrolyte 41
electron 32
electron shell 34
element 31
enzyme 48
fats 44
fatty acids 44
hydrogen bond 37
hydrolysis 38
ion 35
ionic bond 35
isotope 32
lipid 44
monosaccharide 43
neutron 32
nitrogenous base 50
nucleic acid 50
nucleotide 50
nucleus 32
oxidation 38
peptide bond 47
pH scale 42
phospholipid 46
polar molecule 39
polysaccharide 44
protein 46
proton 32
reduction 38
ribonucleic acid (RNA) 50
salt 41
solute 40
solvent 40
steroid 46
water 39

UNDERSTANDING THE FACTS

1 Define atom and molecule.
2 Distinguish between ionic, covalent, and hydrogen bonds.
3 What characteristic of water accounts for its properties as a solvent?
4 Define pH, acid, base, salt.
5 Distinguish between ionization and oxidation.
6 Explain the difference between hydrolysis and dehydration synthesis.
7 List the molecular subunits of polysaccharides, lipids, proteins, and nucleic acids.
8 What effect do enzymes have on chemical reactions?
9 What is the function of DNA in the cell?
10 What is the role of ATP in metabolism?

UNDERSTANDING THE CONCEPTS

1 Describe why covalent bonds are the primary bonds that hold atoms together in molecules.
2 Describe the relation of hydrolysis and dehydration synthesis in metabolism.
3 Explain the role of water in the phenomenon of pH.
4 Explain why enzymes are a basic requirement for life.
5 Explain why cells require a molecule such as ATP.

SELF-QUIZ

Multiple-Choice Questions

2.1 The atomic number of an atom is determined by
   a the number of neutrons in its nucleus
   b its atomic weight
   c the number of filled electron shells
   d the number of protons in its nucleus
   e a and d

2.2 All atoms of a given element have
   a the same number of protons in their nuclei
   b filled outer electron shells
   c the same atomic weight
   d the same atomic number
   e a and d

2.3 The structural subunits that make up all matter are
   a monosaccharides   d compounds
   b molecules   e isotopes
   c atoms

2.4 A positively charged atom or group of atoms is a(n)

   a isotope
   b cation
   c molecule
   d anion
   e enzyme

2.5 A bond in which atoms share electrons is a(n)

   a covalent bond
   b hydrogen bond
   c ionic bond
   d peptide bond
   e none of the above

Matching

   a hydrogen bond
   b hydrolysis
   c dissociation
   d ionic bond
   e oxidation

2.6 _____ the removal of one or more atoms of hydrogen from a molecule

2.7 _____ the removal of a hydrogen ion from a molecule

2.8 _____ the electrostatic attraction between a hydrogen atom in one polar molecule and an oxygen or nitrogen atom in another polar molecule

2.9 _____ the electrostatic attraction between a cation and an anion within a crystalline salt

2.10 _____ a catabolic reaction that uses water as a substrate

Completion Exercises

2.11 Cholesterol is an example of a _____.

2.12 In a(n) _____, one end is polar; the other end is nonpolar.

2.13 The most common _____ is glucose.

2.14 Vegetable oils that contain many double covalent bonds are said to be _____.

2.15 The formation of a peptide bond between one _____ and another results in a dipeptide.

A SECOND LOOK

1 In the following drawing, label all covalent and hydrogen bonds.

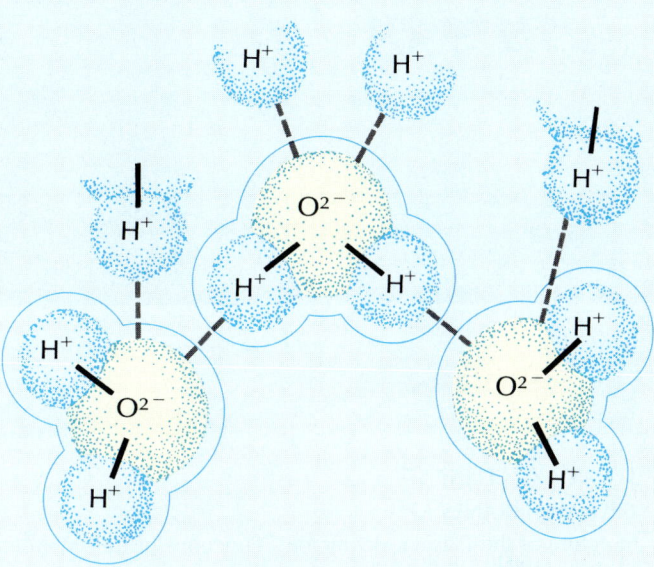

2 In the following drawing, label a phosphate group, a hydrogen bond, deoxyribose, and paired nitrogenous bases. What molecule is represented here?

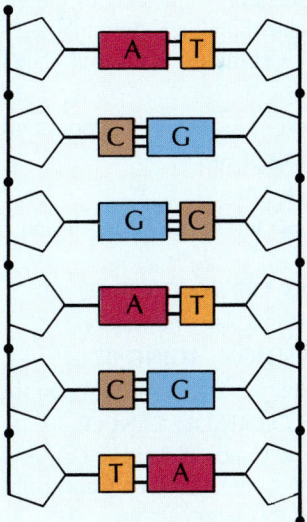

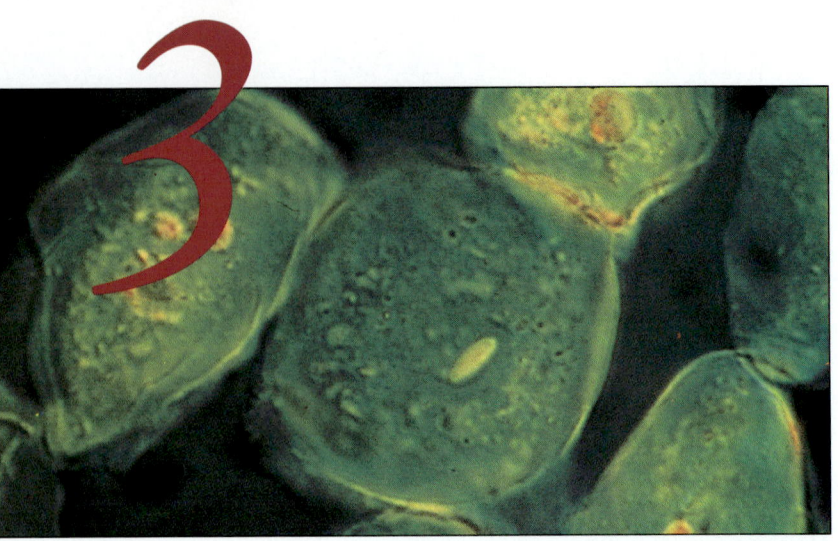

Cells: The Basic Units of Life

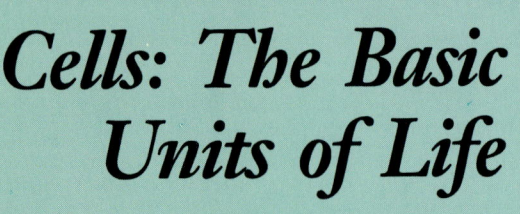

KEY CONCEPTS

1 Cells are the basic living structures of the body. They contain substructures that are specialized to perform specific functions that maintain homeostasis within the cell.

2 Although cell membranes are in a constant state of flux, they retain their fundamental structure and other properties, allowing some materials to pass through while blocking others.

3 The nucleus is the control center of a cell. It contains hereditary information in the form of DNA.

4 Cells divide in such a way that each new cell receives the same genetic information contained in the parent cell.

5 Aging is a natural cellular process, whereas cancer is an abnormal reproduction and distribution of cells.

We all begin life when a single sperm unites with a single egg — a fusion of two highly specialized cells. The resulting fertilized egg is the forerunner of all future cells in the body. The adult human body consists of more than 50 trillion (50 million million) cells, and most of these cells are specialized in structure and function. Whatever their specific functions, however, most cells are capable of carrying on such life-sustaining activities as breaking down food molecules for energy and generating energy-rich ATP, reproducing, synthesizing chains of polypeptides, engulfing foreign materials, and creating new cell structures and getting rid of old ones. Each cell works together with other cells to provide an environment that is compatible with all the processes of life.

WHAT ARE CELLS?

As you saw in Chapter 1, cells are the smallest independent units of life. The basic principles of the **cell theory** were formulated by Theodor Schwann and Matthias Jakob Schleiden about 155 years ago and were formalized later by Rudolf Virchow in 1855. Although these principles have been revised and updated since then, the four most important points remain the same:

1 All living organisms are composed of cells and cell products.

2 The cell is the basic unit of structure and function of all living organisms.

3 All cells come from the division of preexisting cells.

4 An organism as a whole can be understood through the collective activities and interactions of its cells.

Body cells vary enormously. They have different sizes, shapes, and colors related to their functions, but cells such as those that make up the liver are fairly typical. Several hundred liver cells would fit into a cube smaller than the small letters in this sentence.

Early *cytologists* (scientists who study cells) thought that the interior of cells consisted of a homogeneous fluid, which they called *protoplasm*. Today the word *protoplasm* is used only in a very general way, and scientists divide cells into four basic parts:

1 The **plasma membrane** is the outer boundary of the cell. It selectively allows substances to pass into and out of the cell.

2 **Cytoplasm** is the portion of the cell outside the nucleus and within the plasma membrane. Metabolic reactions take place here with the aid of specialized structures called organelles. The fluid portion of the cytoplasm is called **cytosol.**

3 The **nucleus** is the control center of the cell. Within the nucleus are the chromosomes, which contain the genes that direct reproduction, information flow, and the heredity of cells. The nucleus is a clearly defined body that is separated from the surrounding cytoplasm by a double nuclear envelope.

4 *Nucleoplasm* is the material within the nucleus.

Since cells vary so widely in form and function, no "typical" cell exists. To help you learn as much as possible about cells, FIGURE 3.1 shows a generalized version of a cell and its parts. TABLE 3.1 summarizes cellular components and their functions.

A S K   Y O U R S E L F

1 What are the four main points of the cell theory?

2 What are the four basic parts of a cell?

3 How do nucleoplasm and cytoplasm differ?

4 What is cytosol?

CELL MEMBRANES

The thin membrane that forms the *outermost* layer of a cell is called the **plasma membrane.** It maintains the boundary and integrity of the cell by keeping the cell and its contents separate and distinct from the surrounding environment. The plasma membrane is a complex, selective structure that allows only certain substances to enter and leave the cell. Because of the presence of specific receptor molecules on its surface, the plasma membrane is able to interact with other cells and with certain substances in the external environment.

Membranes that enclose or actually make up some organelles *inside* the cell are similar to the plasma membrane. They also help regulate cellular activities.

Structure of the Plasma Membrane

In 1972 the **fluid-mosaic model** of membrane structure was formulated. According to this model, the membrane is a fluid double layer, or bilayer, composed mainly of proteins and phospholipids [FIGURE 3.2A].

The phospholipid bilayer forms a fluid "sea," similar in consistency to vegetable oil, in which specific proteins float like icebergs. Because it is fluid, the membrane is in a constant state of flux, shifting and changing, yet at the same time retaining its basic structure and properties. The word *mosaic* refers to the many different proteins embedded on or within the phospholipid bilayer.

TABLE 3.1 CELLULAR COMPONENTS AND THEIR FUNCTIONS

Component	Description	Main functions
Centrioles [FIGURE 3.12]	Located within centrosome; contain nine triple microtubules.	Assist in cell reproduction; form basal body of cilia and flagella.
Cilia, flagella [FIGURE 3.13]	Cilia are short and threadlike; flagella are much longer.	Cilia move fluids or particles past fixed cells. Flagella provide means of movement for sperm.
Cytoplasm [FIGURE 3.1]	Semifluid enclosed within plasma membrane; consists of fluid cytosol and intracellular structures such as organelles.	Dissolves soluble proteins and substances necessary for cell's metabolic activities; houses organelles, vesicles, inclusions, and lipid droplets.
Cytoplasmic inclusions	Substances temporarily in cytoplasm.	Include food material or stored products of cell metabolism such as pigments, fats and carbohydrates, and secretions; also foreign substances.
Cytoplasmic vesicles (endosomes) [FIGURES 3.1]	Membrane-bound sacs.	Store and transport cellular materials.
Cytoskeleton [FIGURE 3.10]	Flexible cellular framework of interconnecting microfilaments, intermediate filaments, microtubules, other organelles.	Provides support; assists in cell movement; site for binding of specific enzymes.
Cytosol [FIGURE 3.1]	Fluid portion of cytoplasm; enclosed within plasma membrane; surrounds nucleus.	Houses organelles; serves as transporting medium for secretions; site of many metabolic activities.
Deoxyribonucleic acid (DNA) [FIGURE 3.17]	Nucleic acid that makes up the chromosomes.	Controls heredity and cellular activities.
Endoplasmic reticulum (ER) [FIGURE 3.6]	Membrane system extending throughout cytoplasm.	Provides for internal transport and storage; rough ER serves as point of attachment for ribosomes; smooth ER produces steroids.
Golgi apparatus [FIGURE 3.8]	Flattened stacks of disklike membranes.	Packages proteins for secretion.
Intermediate filaments [FIGURE 3.10]	Elongated, fibrillar structures composed of protein subunits.	Serve as key elements of cytoskeleton in most cells.
Lysosomes [FIGURE 3.1]	Small, membrane-bound spheres.	Digest materials; decompose harmful substances; play a role in cell death.
Microfilaments [FIGURE 3.10]	Solid, rodlike structures containing the protein actin.	Provide structural support and assist cell movement.
Microtubules [FIGURES 3.10–3.12]	Hollow, slender, cylindrical structures.	Support; assist movement of cilia and flagella and chromosomes; transport system.
Mitochondria [FIGURE 3.9]	Sacs with inner-folded, double membranes.	Produce most of the energy (ATP) required for cell; site of cellular respiration.
Nucleolus [FIGURE 3.14]	Rounded mass within nucleus; contains RNA and protein.	Serves as preassembly site for ribosomes.
Nucleus [FIGURE 3.14]	Large spherical structure surrounded by a double membrane; contains nucleolus and DNA.	Contains DNA that makes up the cell's genetic program and controls cellular activities.
Peroxisomes	Membranous sacs containing oxidative enzymes.	Carry out metabolic reactions and destroy hydrogen peroxide, which is toxic to cell.
Plasma membrane [FIGURES 3.1–3.5]	Outer bilayered boundary of cell; composed of lipids and proteins.	Regulates passage of substances into and out of cell; cell-to-cell recognition.
Ribonucleic acid (RNA) [FIGURE 3.20]	Three types of nucleic acid involved in transcription and translation of genetic code.	Messenger RNA (mRNA) carries genetic information from DNA; transfer RNA (tRNA) is involved in amino acid activation during protein synthesis; ribosomal RNA (rRNA) is involved in ribosome structure.
Ribosomes [FIGURES 3.7, 3.20]	Small structures containing RNA and protein; some attached to rough ER.	Serve as sites of protein synthesis.

FIGURE 3.1 A GENERALIZED CELL

A highly simplified version of a cell is sectioned (cut open) to show the various structures and organelles (several of which are also cut) within the cytoplasm. Organelles include the endoplasmic reticulum, ribosomes, mitochondria, and nucleus, among other structures. The nucleus is sliced open to show a nucleolus inside.

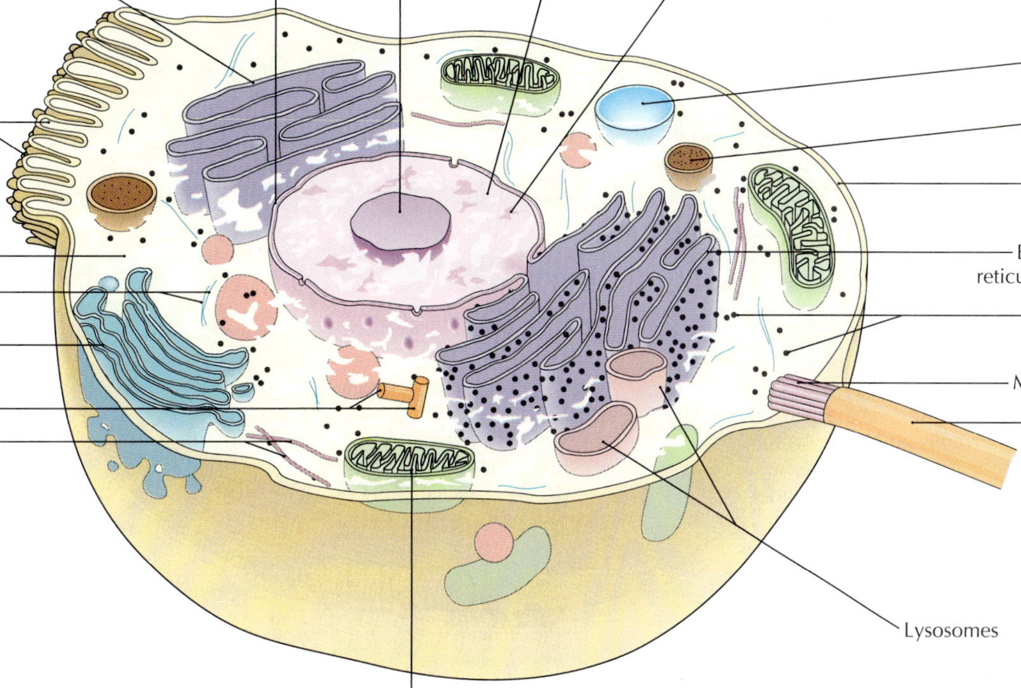

Nucleus

Endoplasmic reticulum (smooth)

Nuclear membrane

Nucleolus

Nucleoplasm

Chromatin

Vacuole

Vesicle (endosome)

Plasma membrane

Endoplasmic reticulum (rough)

Free ribosomes

Microvilli

Cytoplasm

Microfilaments

Golgi apparatus

Centrioles

Microtubules

Microtubules

Cilium

Lysosomes

Mitochondrion

The following are the most important points of the fluid-mosaic model:

1 *Phospholipids,* shown like balloons, each on two strings, in FIGURE 3.2B, are arranged in a fatty double layer consisting of a sheet two molecules thick. This arrangement is called a *bimolecular layer,* or *bilayer.*

2 Phospholipids have one electrically charged end and one uncharged end [FIGURE 3.2B]. The charged ends can stick into the watery cytosol inside the cell and into the similarly watery external environment; the uncharged ends face each other in the middle of the double layer.

3 The "tails" (the *lipid* portion) of the phospholipid molecules are attracted to each other and are repelled by water; they are *hydrophobic,* or "water-fearing." As a result, the polar, or charged, spherical "heads" (the *phosphate* portion) of the phospholipid molecules line up over the outer and inner cell surfaces. The heads are *hydrophilic,* or "water-attracting." The lipid portions constitute the center of the membrane between the two phosphate layers.

4 FIGURE 3.2A shows how *proteins* are embedded in the cell membranes like tiles in a mosaic or like icebergs floating in the fluid phospholipids. Some proteins penetrate only partially into the phospholipid layer (*surface,* or *peripheral, proteins*), some extend all the way through the membrane into the cytoplasm (*transmembrane integral proteins*), some are partially sunken in the outer layer, and some are partially sunken in the inner layer.

5 Protruding from some of the proteins and plasma membrane are antennalike *surface carbohydrates,* most of which are small, complex, nonrepeating sequences of monosaccharides. Proteins with surface carbohydrates extending from them are called *glycoproteins,* and lipids with attached carbohydrates are called *glycolipids* [FIGURE 3.2A]. Overall, these carbohydrates and the portions of the proteins on the surface of the plasma membrane make up the *glycocalyx,* the carbohydrate-rich peripheral

FIGURE 3.2 THE FLUID-MOSAIC MODEL

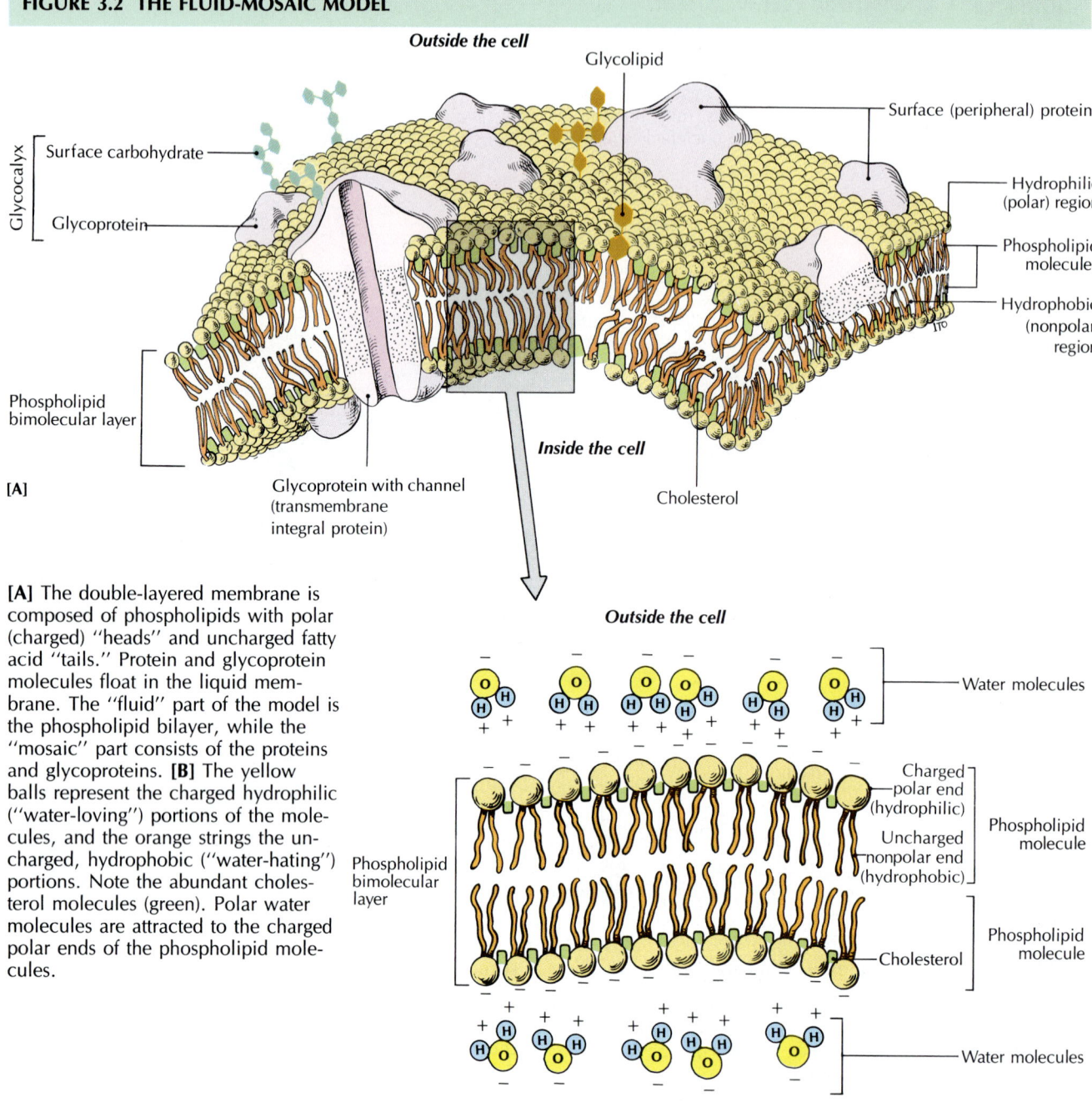

[A] The double-layered membrane is composed of phospholipids with polar (charged) "heads" and uncharged fatty acid "tails." Protein and glycoprotein molecules float in the liquid membrane. The "fluid" part of the model is the phospholipid bilayer, while the "mosaic" part consists of the proteins and glycoproteins. [B] The yellow balls represent the charged hydrophilic ("water-loving") portions of the molecules, and the orange strings the uncharged, hydrophobic ("water-hating") portions. Note the abundant cholesterol molecules (green). Polar water molecules are attracted to the charged polar ends of the phospholipid molecules.

zone of the cell surface. The glycoproteins and glycolipids enable a cell to recognize or reject another cell, as in immune reactions, or act as receptor sites for hormones such as insulin. Apparently, glycoproteins allow similar cells to "fit together" chemically, whereas a foreign cell is not "recognized" and is repelled or even destroyed.

Bacteria have surface hairlike appendages called fimbriae, which have receptors that bind only to specific surface carbohydrates [FIGURE 3.3A]. Researchers have noted that some surface carbohydrates actually inhibit binding with harmful bacteria. These surface carbohydrates could be used as "decoys" to intercept and bind with or destroy the bacteria before they can reach their target cells, thus preventing infection. Such a technique could be particularly useful in preventing common respiratory and urinary tract infections [FIGURE 3.3B].

FIGURE 3.3 SURFACE CARBOHYDRATES AND BACTERIAL INFECTION

[A] Some surface carbohydrates on a cell serve as attachment sites for viruses and bacteria. **[B]** Scanning electron micrograph of rod-shaped *Escherichia coli* bacteria bound to receptor sites of cells in urinary tract tissue. *E. coli* bacteria are the most common cause of urinary tract infections.

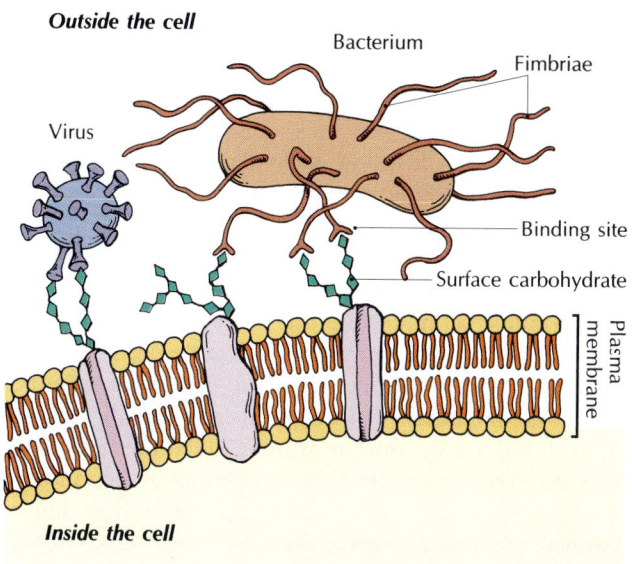

[A]

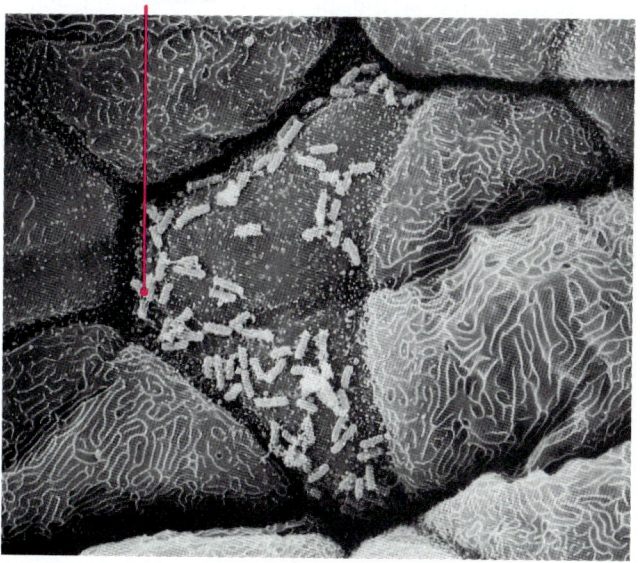

[B]

6 The plasma membrane of human cells contains *cholesterol.* The flexibility (fluidity) of the membrane is determined by the ratio of cholesterol to phospholipids. The plasma membranes of human cells contain relatively large amounts of cholesterol (as much as one molecule for every phospholipid molecule), which enhances the flexibility of the bilayer.

Functions of Cell Membranes

Plasma membranes serve several important functions. They (1) separate the cytoplasm inside a cell from the extracellular fluid outside, (2) separate cells from one another, (3) provide an abundant surface on which chemical reactions can occur, and (4) regulate the passage of materials into and out of cells. Other cell membranes (1) separate the cell into various compartments and (2) regulate the passage of materials from one part of a cell to another.

Some of the proteins embedded in a cell membrane provide structural support. Others are *enzymes* that enhance the chemical reactions that take place on the membrane. Other membrane proteins regulate the movement of water-soluble substances through channels; some are special receptors for specific chemical signals (hormones) and specific neurotransmitter molecules. Still other proteins, especially glycoproteins, play an important role in cell-to-cell recognition. The amount of a protein in a cell membrane may be a measure of that membrane's metabolic activity. The inner membranes of mitochondria, for example, contain more proteins than do the membranes of less metabolically active organelles.

Cells let some things in and keep others out, a quality called *semipermeability* (L. *permeare*, to pass through). This is essential for maintaining cellular homeostasis. Some substances, such as water and lipids, pass through plasma membranes readily. Others, such as potassium and calcium ions, pass only with the expenditure of energy by the cell, while still others cannot pass through at all.

In some cells, the surface area of the plasma membrane is greatly increased by the presence of *microvilli* [FIGURE 3.1]. These slender extensions of the plasma membrane function in either the absorption or secretion of molecules. The increased surface area allows many more molecules to be transported into or out of the cell.

Plasma Membranes and Cell Adhesion

Protruding from the plasma membranes of cells are cell surface glycoprotein receptors called *cell adhesion molecules,* or *CAMs.* The body contains four types of CAMs, which differ slightly in structure: (1) immunoglobulinlike adhesion receptors, (2) selectins, (3) integrins, and (4) cadherins.

CAMs have many different functions, from holding cells of the same kind together to assisting in the learning process. For example, an *immunoglobulinlike adhesion receptor* called apCAM—a type of neural cell adhesion molecule (see "The Guiding 'Glue' of the Nervous System" on page 363 for a description of CAMs of the nervous system)—may help reinforce the learning process, at least in a sea snail called *Aplysia californica* (hence the name, apCAM), where a learning neuron detaches itself from its neighboring cell by consuming the apCAMs that hold the cells together. Once detached, the learning neuron is free to form new connections (synapses) with other neurons.

Selectins protruding from cells that line blood vessels have been shown to slow down white blood cells as they pass through the vessels. The delayed white blood cells, which are part of the immune system, have more time to recognize and respond to distress signals from nearby damaged or infected tissues. *Integrins* have several functions, including assisting white blood cells to squeeze through gaps in the walls of blood vessels on their way to nearby tissues, possibly acting as tumor suppressors, helping the body resist harmful viruses, and shaping the human placenta. Researchers hope that integrins may someday be used to combat cancer, rheumatoid arthritis, and other diseases.

Of the four types of CAMs, *cadherins* are probably the least understood. They are cell surface glycoproteins that specialize in holding cells together and are probably also involved with tissue specialization during embryonic development.

A S K   Y O U R S E L F

1 To what do the terms *fluid* and *mosaic* refer in the fluid-mosaic model of membrane structure?

2 What is the function of cholesterol in cell membranes?

3 What are some of the functions of membrane proteins?

4 What is a semipermeable membrane?

PASSIVE MOVEMENT ACROSS MEMBRANES

When molecules pass through a cell membrane *without the use of cellular energy*, the movement is called *passive movement.* Passive movement processes include simple diffusion, facilitated diffusion, osmosis, and filtration [TABLE 3.2, page 68].

Simple Diffusion

Unless a substance is at absolute zero, $-273°C$, its molecules are in constant motion, especially those in liquids and gases. Molecules tend to move randomly from areas where they are heavily concentrated to areas of lesser concentration until they are evenly distributed in a state of *equilibrium.* The process in which molecules spread out randomly until they are evenly distributed throughout an enclosed space is called **simple diffusion** (L. *diffundere*, to spread). For example, FIGURE 3.4A shows how evaporating perfume molecules diffuse through a room full of air molecules until the perfume molecules are evenly distributed.

The rate of diffusion depends on several factors: (1) Gases diffuse rapidly, and liquids diffuse more slowly. (2) The higher the temperature, the faster the rate of diffusion. (3) Small molecules (glycerol, for example), diffuse faster than the large molecules of a fatty acid.

Certain small molecules can move across the lipid bilayer of plasma membranes simply in response to a difference in the concentration of the molecules on either side of the membrane (that is, inside and outside the cell). This difference in concentration is called the **concentration gradient.** A molecule moving from an area of high concentration to one of lower concentration moves "down" the concentration gradient.

Small nonpolar molecules can diffuse directly through the plasma membrane. In the human body, such diffusion occurs when oxygen in inhaled air passes by diffusion through the cells lining the air sacs of the lungs into the bloodstream and is transported throughout the body. Oxygen from the bloodstream diffuses into the body cells, and carbon dioxide diffuses out of the body cells and into the bloodstream.

For other small molecules—polar molecules *not* soluble in lipids—diffusion occurs not through the plasma membrane but through an *integral channel protein* embedded in the lipid bilayer. It is generally accepted that the integral channel protein provides a continuous pathway for specific molecules to get across the plasma membrane so that they never come in contact with the hydrophobic layer of the membrane or its polar surface.

Integral channel proteins form continuous pores through the plasma membrane to act as passageways, or *channels*, which are semipermeable to certain molecules, such as sodium, potassium, or calcium ions. Some channels are open permanently, while others open in response to chemical stimuli such as neurotransmitters and hormones, and still others open in response to changes in the membrane potential (voltage). Note that there are two types of channels that open and close: (1) *receptor-operated channels* and (2) *voltage-operated channels*. Operational defects in the channels can cause disorders such as diabetes, epilepsy, cystic fibrosis, and some forms of heart disease.

FIGURE 3.4 PASSIVE MOVEMENT

[A] Simple diffusion. Perfume molecules starting from an area of high concentration **(1)** move randomly among the air molecules in an area **(2)**. Eventually, a state of equilibrium is reached, and the perfume molecules are evenly distributed throughout the area **(3)**. Small polar molecules that cannot cross the hydrophobic interior of the plasma membrane diffuse through a channel protein **(4)**.

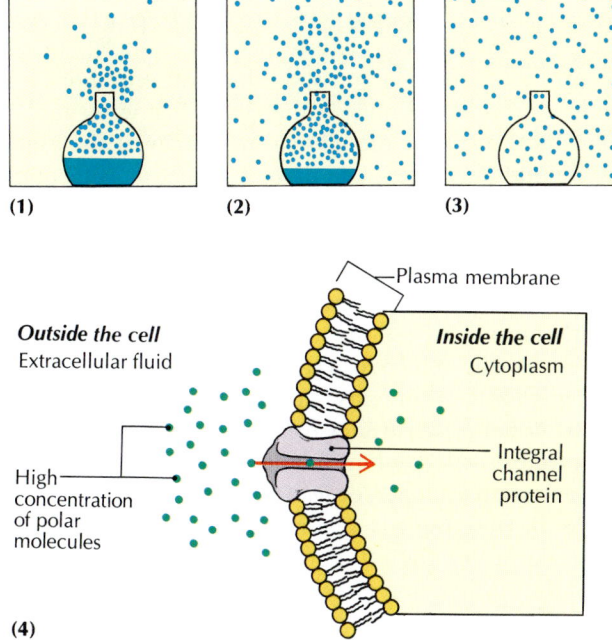

(1)　　　**(2)**　　　**(3)**

Plasma membrane

Outside the cell
Extracellular fluid

Inside the cell
Cytoplasm

Integral channel protein

High concentration of polar molecules

(4)

[B] Facilitated diffusion. A carrier protein in the plasma membrane binds temporarily with amino acids outside the cell and accelerates its movement across the membrane.

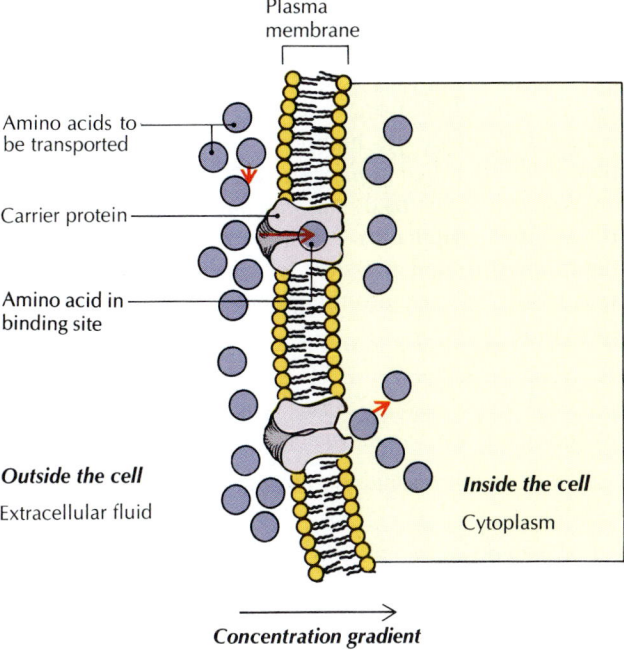

Plasma membrane

Amino acids to be transported

Carrier protein

Amino acid in binding site

Outside the cell
Extracellular fluid

Inside the cell
Cytoplasm

Concentration gradient

[C] Osmosis. A container of water is divided by a semipermeable membrane (permeable to water but not to sugar). **(1)** Sugar molecules are dissolved in the left compartment; thus the water concentration is higher in the right one. **(2)** Sugar molecules cannot cross the membrane, but water molecules move from the area of higher water concentration into the area of lower water concentration. **(3)** The water level rises on the left until the pressure from the greater volume of solution equals the tendency of the water on the right to cross the membrane, and equilibrium is reached.

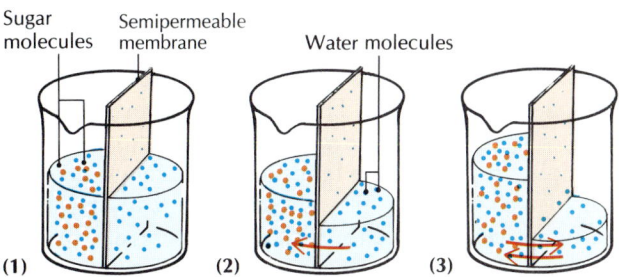

Sugar molecules　Semipermeable membrane　Water molecules

(1)　　　**(2)**　　　**(3)**

[D] Effects of solute concentration on human red blood cells. (1) In an *isotonic solution,* the pressure is the same inside and outside the cells. **(2)** In a *hypotonic solution,* water enters the cells, which eventually burst, leaving only the membranes. **(3)** In a *hypertonic solution,* water moves out of the cells, which shrivel. (Although the cells look like they are puffing up, they are actually collapsing.)

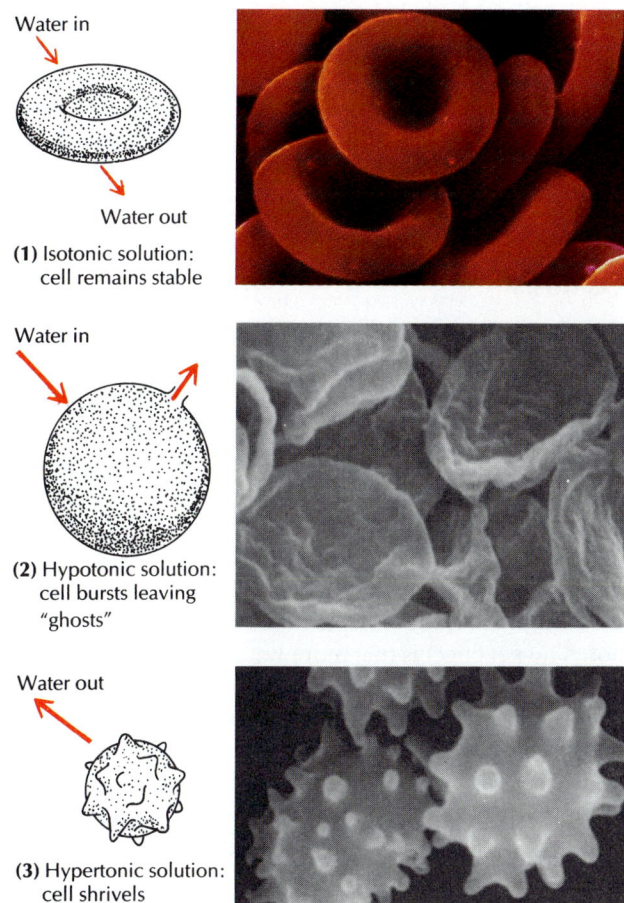

Water in

Water out

(1) Isotonic solution: cell remains stable

Water in

(2) Hypotonic solution: cell bursts leaving "ghosts"

Water out

(3) Hypertonic solution: cell shrivels

Other channels are specialized as *pumps* that facilitate the passage of ions (sodium, for instance) *against* the gradient from regions of lower concentration to regions of higher concentration. Some proteins on the surface of plasma membranes act as *receptors*. These receptor sites respond to special stimuli such as neurotransmitters or hormones to open channels, while others are involved in phagocytosis by macrophages. (See *receptor-mediated endocytosis* on page 65.)

Facilitated Diffusion

Larger molecules, especially those not soluble in lipids, require assistance to pass through the protein channels of the plasma membrane. The process used by such substances is called **facilitated diffusion.** Amino acids, for example, are insoluble in lipids. But since they are essential for cell function, they must have a way of getting inside the cell. To pass through the membrane, a specific amino acid temporarily binds with a *carrier protein* in the plasma membrane to form a new compound that is soluble in lipids [FIGURE 3.4B]. The amino acid–carrier protein complex moves through the plasma membrane, and the amino acid is released into the cytoplasm inside. As needed, the carrier protein then repeats the process for another amino acid. No direct cell energy in the form of ATP is required. Instead, only the internal structure of the protein changes.

The absorption of glucose into a cell also uses a carrier protein, or transporter, which apparently is a folded chain of 492 amino acids.

Like simple diffusion, facilitated diffusion transports substances only *from areas where they are in higher concentration to areas where they are in a lower concentration*, or "down" the concentration gradient.

Osmosis

The passage of water through a semipermeable membrane from an area of higher concentration (of water) to an area of lower concentration is called **osmosis** (Gr. "pushing"). The relative concentrations of water are determined by the amount of solute dissolved in the water on either side of the membrane. A higher concentration of solute (for example, sugar) on one side of the membrane means that less space is available there for the water molecules [FIGURE 3.4C(1)]. Water passes through the membrane from one side to the other; the solute cannot. The net effect is that more water moves into the area of lower water concentration than leaves, until the pressure of the increasing volume of solution counterbalances this tendency [FIGURE 3.4C(3)]. The **osmotic pressure** of the solution is the force (hydrostatic pressure) that is required to stop the net flow of water across a semipermeable membrane when the membrane separates solutions of different compositions.

The osmotic pressure on plasma membranes can be great enough to burst cells, so it is important for cells to have relatively constant internal and external pressures to maintain homeostasis. This principle can be seen, for example, in red blood cells, which exist in the body in an osmotically stable environment. The water and solute concentrations within the cells are the same as the water and solute concentrations in the surrounding fluid (the blood plasma). Thus, the cells are in an *isotonic* (Gr. *isos*, equal + *tonos*, tension) *solution* [FIGURE 3.4D(1)]. In an isotonic solution, the water and solute concentrations are the *same* inside and outside the cell. Thus water molecules move through the plasma membrane at the same rate in both directions, and there is no net movement of water in either direction.

In a *hypotonic* solution (Gr. *hypo*, under), the solute concentration is lower, or *less*, outside the cell than it is inside. Thus, the concentration of water molecules is greater outside the cell than it is inside. As a result, water moves *into* the cell by osmosis until the cell bursts (or *lyses*) and loses its contents [FIGURE 3.4D(2)]. This condition is called *hemolysis* in a red blood cell.

In a *hypertonic* solution (Gr. *hyper*, above), the solute concentration is higher, or *more*, outside the cell than it is inside. Because there is a greater concentration of water molecules inside the cell than outside, the water moves *out* of the cell by osmosis, causing the cell to shrivel like a raisin [FIGURE 3.4D(3)]. This condition is called *crenation* (L. *crena*, notch) in a red blood cell.

Filtration

Filtration is a process that *forces* small molecules through semipermeable membranes with the aid of hydrostatic (water) pressure or another externally applied force. As is usual with passive transport processes, movement is from areas of higher pressure or concentration to areas of lower pressure or concentration. In the body, filtration is evident when blood pressure forces water and dissolved particles through the highly permeable walls of small blood vessels called *capillaries*. In filtration, large molecules such as proteins do not pass through the walls; they remain in the capillaries. Filtration also occurs in the kidneys, where blood pressure forces water and dissolved wastes such as urea into the kidney tubules during urine formation.

ASK YOURSELF

1 How does passive movement work?

2 How do simple diffusion and facilitated diffusion differ?

3 Where do carrier proteins function?

4 What is osmosis?

5 Where in the body does filtration occur?

ACTIVE MOVEMENT ACROSS MEMBRANES

The basic difference between passive and active processes of movement is that ***active processes*** require energy to move substances through a semipermeable membrane [TABLE 3.2, page 68]. Passive movement across membranes does not require energy. Among the active processes are active transport, endocytosis, and exocytosis. Endocytosis and exocytosis are included in a general category called ***bulk transport,*** the movement of large molecules or particles across membranes.

Active Transport

Passive movement is satisfactory when substances enter a cell, because the concentration of molecules outside the cell is greater than the concentration inside. Then the small molecules diffuse naturally through the plasma membrane into the cell. But when a state of equilibrium is reached and still more molecules are needed within the cell, they must be pumped through the membrane *against the concentration gradient*. This movement of small molecules or ions is accomplished by ***active transport,*** which requires energy from the cell in the form of ATP.

In the human body, four specific active-transport systems have been studied extensively: (1) The *sodium-potassium pump* is common to many cells, especially nerve cells, where sodium and potassium concentration gradients are generated in order to produce an electrical charge; (2) the *calcium pump* is vital in muscle relaxation, where it transports calcium ions needed to terminate the contractile mechanism; (3) *sodium-linked cotransport* actively transports sugars and amino acids, while sodium ions tag along passively; and (4) in *hydrogen-linked cotransport*, hydrogen ions tag along while sugars are being actively transported.

These four active-transport mechanisms all use a similar process [FIGURE 3.5A]:

1 A molecule outside the cell binds with a carrier protein at the membrane boundary.

2 The molecule/carrier protein complex moves through the membrane.

3 Assisted by at least one enzyme and energy available from ATP, the molecule and carrier protein separate, and the molecule is released.

4 The carrier protein returns to its original shape and can repeat the process with another molecule.

Endocytosis

So far we have seen how small molecules can pass through plasma membranes. How do large molecules or particles move through? Large molecules such as proteins, as well as small amounts of fluids, can enter cells when the plasma membrane forms a pocket around the material, enclosing it and drawing it into the cytoplasm. This process of active movement is called ***endocytosis*** (Gr. *endon,* within). During endocytosis, only a small region of the plasma membrane folds inward, or *invaginates,* until it has formed a new intracellular vesicle that contains large molecules from outside the cell. Three types of endocytosis are known:

1 ***Pinocytosis*** ("cell drinking," from Gr. *pinein,* to drink + *cyto,* cell) is the nonspecific uptake of small droplets of extracellular fluid [FIGURE 3.5C(1)]. Any material dissolved in the fluid is also taken into the cell, including low-molecular-weight nutrients, amino acids, glucose, vitamins, and other substances. Kidney cells are an example of the many types of cells that use pinocytosis to help regulate their fluid environment.

2 The active movement that allows large particles and other cells to pass through the plasma membrane into the cytoplasm is called ***phagocytosis*** ("cell eating," from Gr. *phagein,* to eat + *cyto,* cell). The particle brought into a cell by phagocytosis is not always a nutrient. Probably the most familiar example of phagocytosis is that of a white blood cell engulfing a harmful bacterium before digesting it. FIGURE 3.5C(2) shows what happens during phagocytosis. When a bacterium or other particle makes contact with the plasma membrane of a *phagocyte* (Gr. "a cell that eats"), the membrane begins to form a pocket around the particle by sending out pseudopods (Gr. "false feet"). When the pocket is closed, it is called a *phagosome.* The particle and phagosome enter the cytoplasm. At least one lysosome attaches itself to the phagosome and releases its digestive enzymes inside the phagosome to digest the bacterium.

3 ***Receptor-mediated endocytosis*** involves a specific receptor on the plasma membrane that "recognizes" an extracellular macromolecule and binds with it. The substance bound to the receptor is called the *ligand* (L. *ligare,* to bind). The region of the plasma membrane that contains the receptor-ligand complex undergoes endocytosis. FIGURE 3.5C(3) show the specific example of LDL (low-density lipoprotein) as the ligand. Most cholesterol, which is needed to make new membranes, is transported into the cell by LDL.

Exocytosis

Exocytosis (Gr. *exo,* outside) is a process basically opposite to endocytosis [FIGURE 3.5B]. The cytoplasmic vesicle, with its undigested particles, fuses with the plasma

FIGURE 3.5 ACTIVE MOVEMENT

[A] Active transport uses energy from the breakdown of ATP to move substances across membranes against a concentration gradient. **[B]** Exocytosis uses energy from ATP to eject materials from the cell when a vesicle merges with the plasma membrane. **[C]** In endocytosis, energy is used to take materials into the cell by the invagination of the plasma membrane and the formation of vesicles.

[C] (1) Shows pinocytosis, in which fluids are taken into the cell. **[C] (2)** In phagocytosis, materials are surrounded by pseudopods and taken into the cell within a vesicle, which may fuse with a lysosome. Enzymes from the lysosome digest the contents of the vesicle. **[C] (3)** Shows receptor-mediated endocytosis. The electron micrographs show the same series of events as the drawing.

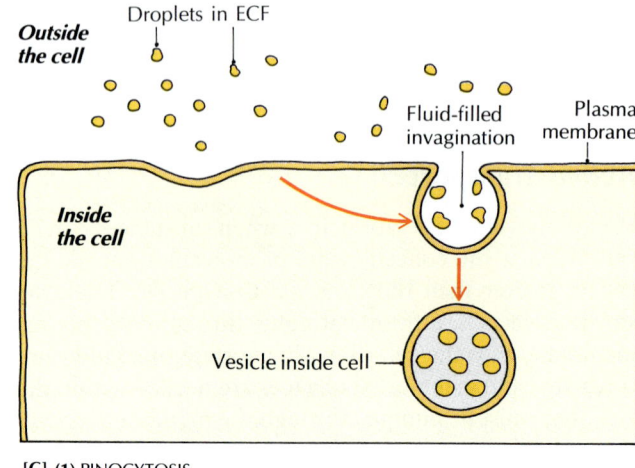

[C] (1) PINOCYTOSIS

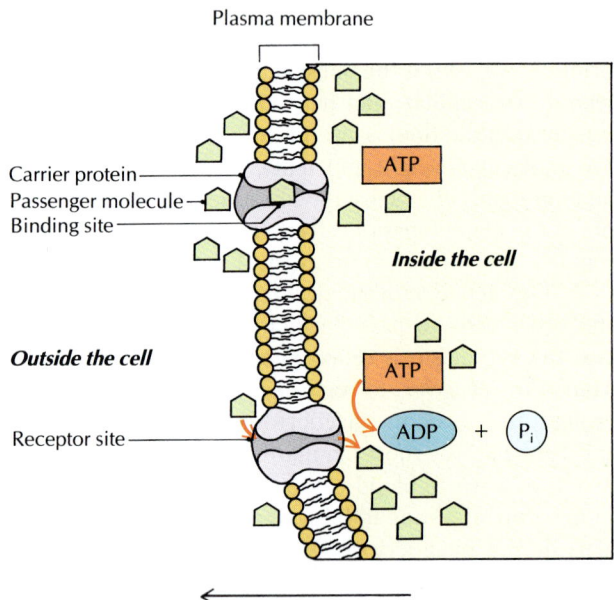

[A] ACTIVE TRANSPORT

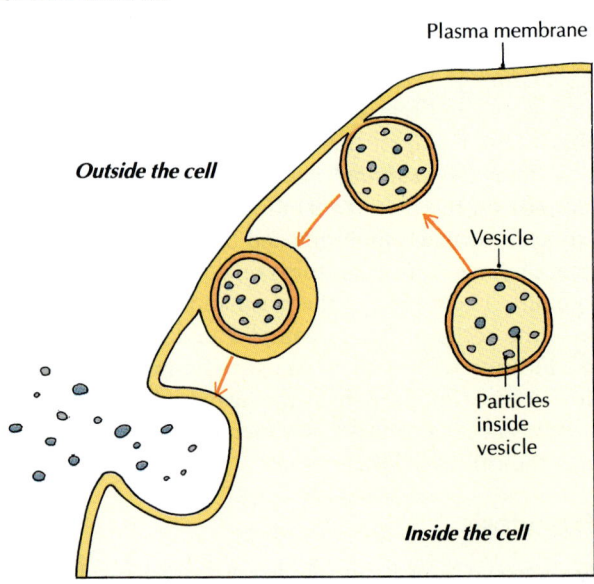

[B] EXOCYTOSIS

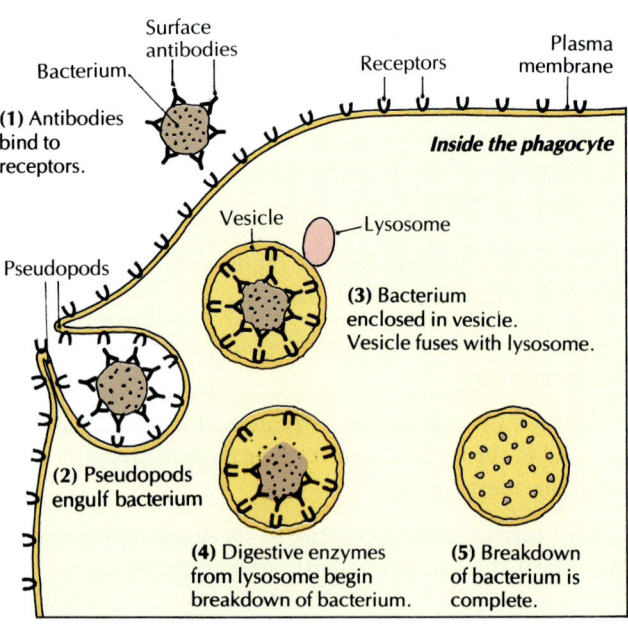

[C] (2) PHAGOCYTOSIS

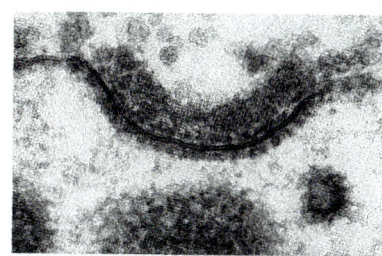

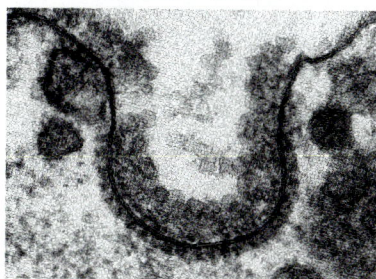

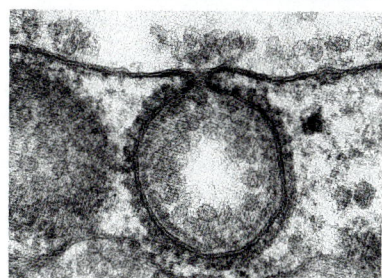

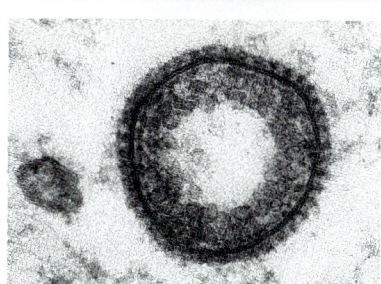

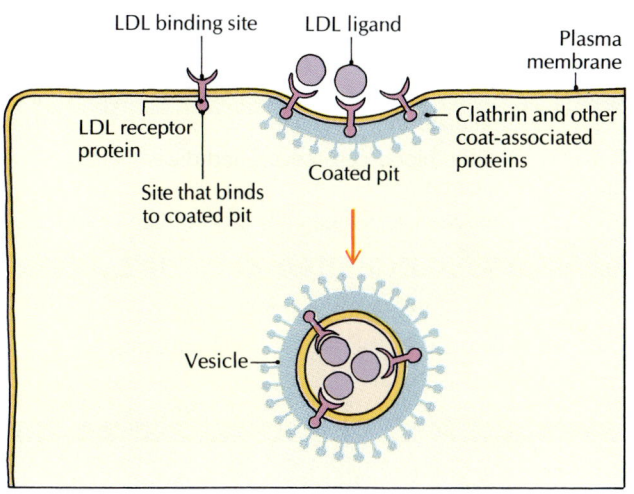

LDL binding site   LDL ligand

Plasma membrane

LDL receptor protein

Clathrin and other coat-associated proteins

Coated pit

Site that binds to coated pit

Vesicle

[C] (3) RECEPTOR–MEDIATED ENDOCYTOSIS

membrane and releases the unwanted particles from the cell. Exocytosis also allows cells to release useful substances to the rest of the body. For example, nerve cells release their chemical messengers, and various gland cells secrete proteins by exocytosis. Exocytosis and endocytosis are also used to remove receptors, channels, or markers from cell membranes or add them to the membranes.

CYTOPLASM

The *cytoplasm* of a cell is the portion outside the nucleus and within the plasma membrane [FIGURE 3.1]. It is composed of two distinct parts, or phases: (1) The *particulate phase* consists of well-defined structures, including membrane-bound vesicles (small fluid-filled structures) and organelles ("little organs"), lipid droplets, and *inclusions* (solid particles such as glycogen granules). (2) The *aqueous phase* consists of the fluid cytosol. Organelles and vesicles are suspended in the cytosol, and various proteins and substances required for some of the cell's metabolic activities are dissolved in it.

Like most other body materials, *cytosol* is mostly water (70 to 85 percent), but its other typical components are proteins (10 to 20 percent), lipids (about 2 percent), complex carbohydrates, nucleic acids, amino acids, vitamins, and electrolytes (substances, such as potassium and chloride, that dissolve in water and provide inorganic chemicals for cellular reactions).

Cytoplasm also contains a *cytoskeleton* [see FIGURE 3.10], a flexible lattice arrangement of microtubules, intermediate filaments, and microfilaments that supports the organelles and provides the machinery for the movement of cells and their organelles.

ORGANELLES

As you have seen, suspended within the cytosol and nucleoplasm are various membrane-bound structures called *organelles* [FIGURE 3.1]. Each organelle has its own specific structure and function, as described in the following sections.

TABLE 3.2 COMPARISON OF PASSIVE AND ACTIVE MOVEMENT ACROSS CELL MEMBRANES

Type of movement	Description	Example in body
PASSIVE MOVEMENT		
	Molecules move "down" the concentration gradient. No cell energy required.	
Simple diffusion [FIGURE 3.4A]	Molecules spread out randomly from areas of higher concentration to areas of lower concentration until they are distributed evenly. Movement is directly through membrane or integral channel protein; however, a membrane need not be present.	Inhaled oxygen is transported into lungs and diffuses through lung cells into bloodstream.
Facilitated diffusion [FIGURE 3.4B]	Carrier proteins in plasma membrane temporarily bind with molecules, allowing them to pass through membrane, via protein channels, from areas of higher concentration to areas of lower concentration.	Specific amino acids combine with carrier proteins to pass through plasma membrane into cell.
Osmosis [FIGURE 3.4C]	Water molecules move through semipermeable membrane from areas of higher water concentration to areas of lower water concentration.	Water moves into a red blood cell when concentration of water molecules outside cell is greater than it is inside cell.
Filtration	Hydrostatic pressure forces small molecules through semipermeable membrane from areas of higher pressure to areas of lower pressure.	Blood pressure inside capillaries is higher than in surrounding tissue fluid. This pressure forces water and dissolved small particles through capillary walls and into surrounding tissues. Blood pressure forces water and dissolved wastes into kidney tubules during formation of urine.
ACTIVE MOVEMENT		
	Cell energy allows substances to move through semipermeable membrane from areas of lower concentration to areas of higher concentration. Requires a living cell and energy in the form of ATP.	
Active transport [FIGURE 3.5A]	Carrier proteins in plasma membrane bind with ions or molecules to assist them across membrane "against" the concentration gradient.	Sodium ions are pumped out of resting nerve cells, although their concentration is much higher outside the cells.
Endocytosis [FIGURE 3.5C]	Membrane-bound vesicles enclose large molecules, draw them into cytoplasm, and release them.	
Pinocytosis	Plasma membrane encloses small amounts of fluid droplets and takes them into cell.	Kidney cells take in tissue fluids in order to maintain fluid balance.
Phagocytosis	Plasma membrane forms a pocket around a solid particle or cell and draws it into cell.	White blood cells engulf and digest harmful bacteria.
Receptor-mediated endocytosis	Extracellular molecules bind with specific receptor on plasma membrane, causing the membrane to invaginate and draw molecules into cell.	Intestinal epithelial cells take up large molecules.
Exocytosis [FIGURE 3.5B]	Vesicle (with undigested particles) fuses with plasma membrane and expels particles from cell through plasma membrane.	Nerve cells release chemical messengers.

Endoplasmic Reticulum

Inward from the plasma membrane is the ***endoplasmic reticulum (ER).*** The ER is a labyrinth of interconnected membrane-bound flattened sacs and tubules that branch and spread throughout the cytoplasm. In some places the membranes of the ER are connected with the nuclear envelope. FIGURE 3.6A shows the position of the endoplasmic reticulum in a simplified cell.

The ER functions as a series of channels that help circulate various materials throughout the cytoplasm. It also serves as a storage unit for enzymes and other proteins and as a point of attachment for ribosomes, which play an important role in forming proteins [FIGURE 3.6B, C].

The space within the channels and enclosed by the ER membranes is called the *ER lumen* [FIGURE 3.6B]. It is thought to be a single, continuous space. When endoplasmic reticulum does not have attached ribosomes, it is called *smooth ER.* Smooth ER serves as a site for the production of steroids, detoxifies a wide variety of organic molecules, and stores calcium in skeletal muscle cells. *Rough ER* refers to endoplasmic reticulum that does have ribosomes attached to its outer face, as shown in FIGURE 3.6B.

Ribosomes

Ribosomes (RYE-boh-sohmz) are nonmembrane-bound macromolecules that function in *protein synthesis,* the manufacture of proteins from amino acids. Ribosomes contain almost equal amounts of protein and a special kind of ribonucleic acid, *ribosomal RNA (rRNA).* Ribosomes are actually composed of two subunits with variable shapes [FIGURE 3.7A] but are usually pictured as slightly flattened spheres with "tucked-in" sections around the middle [FIGURE 3.7B].

Not all ribosomes are attached to the ER. Some float freely within the cytoplasm [FIGURE 3.1]. Whether ribosomes are free or attached, they are usually grouped in clusters called *polyribosomes* ("many ribosomes") [FIGURE 3.7C]. These clusters are connected by a strand of a single molecule of specialized nucleic acid called *messenger ribonucleic acid (messenger RNA, or mRNA).*

Golgi Apparatus

The ***Golgi apparatus*** (GOAL-jee), or ***Golgi complex*** (named for Camillo Golgi, who described it in 1898), is a collection of membranes associated physically and functionally with the ER in the cytoplasm. Golgi apparatuses are usually located near the nucleus [FIGURE 3.8A]. They are composed of flattened stacks of membrane-bound *cisternae* (sis-TUR-nee; Gr. *kiste,* basket), which are closed

FIGURE 3.6 ENDOPLASMIC RETICULUM

[A] Note that the endoplasmic reticulum (blue) actually touches the outer nuclear envelope, where a transfer of materials takes place. Rough ER and smooth ER (without attached ribosomes) are shown. Ribosomes can be seen in the drawing **[B]** and in the electron micrograph **[C]**. ×50,000.

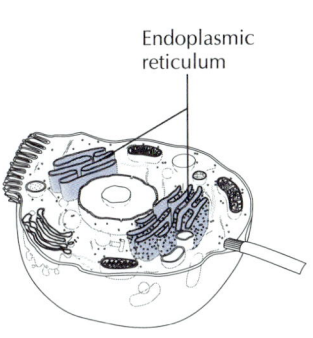

[A]

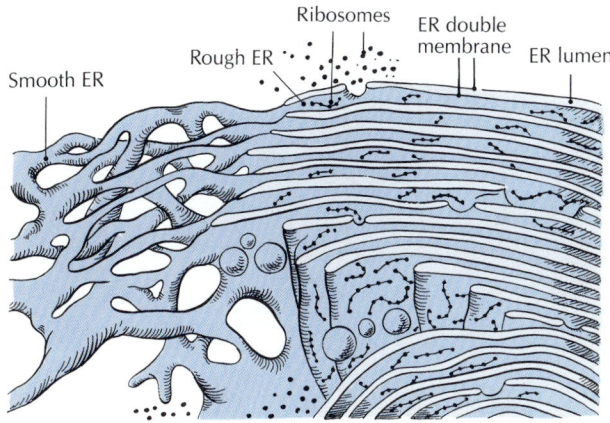

[B]

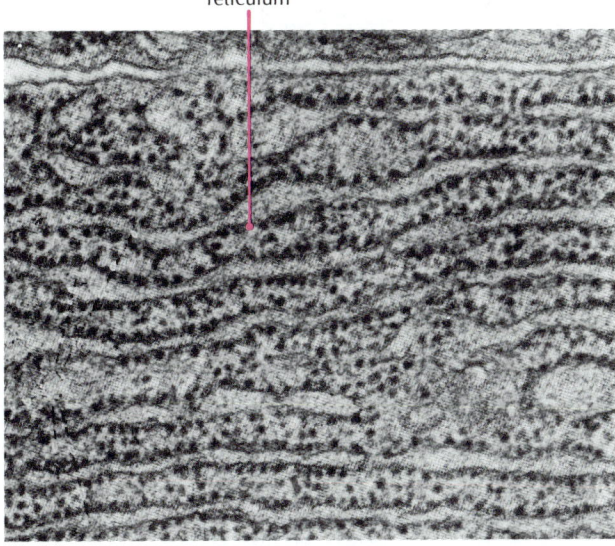

[C]

FIGURE 3.7 RIBOSOMES

[A] A current model of a ribosome with its two subunits. Apparently, the mRNA molecule is held in a channel between the large and small subunits. This representation is based on electron micrographs. **[B]** Simplified version of a ribosome, showing sites on the large subunit (A-site and P-site) for the binding of tRNA molecules, and the binding site for mRNA on the small subunit. **[C]** Electron micrograph of a string of ribosomes (a polyribosome) connected by a strand of mRNA. ×125,000.

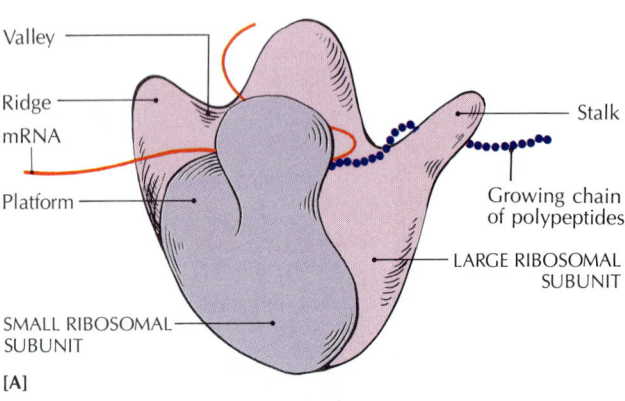

[A]

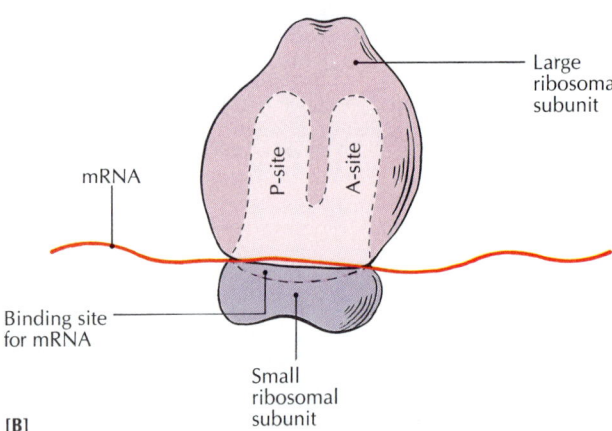

[B]

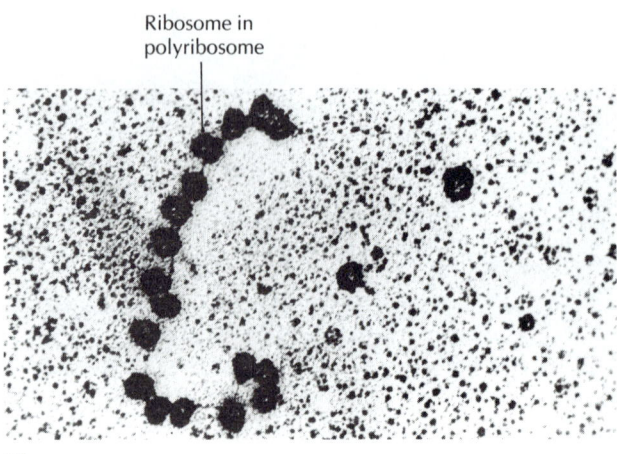

[C]

spaces serving as fluid reservoirs [FIGURE 3.8B]. The Golgi apparatus functions in the "packaging" and secretion of glycoproteins. Many of these glycoproteins are important components of the glycocalyx, constituting part of a system that allows a cell to recognize another cell of the same kind.

As proteins are synthesized by ribosomes attached to the rough ER, some of the proteins are sealed off in little packets called *transfer vesicles*. These vesicles bud off and fuse with the forming face of the Golgi apparatus close to the nucleus [FIGURE 3.8B]. Within the Golgi apparatus, the proteins and/or carbohydrates attached to them can be concentrated and modified. Eventually the proteins are packaged into relatively large *secretory vesicles* [FIGURE 3.8C], which are released from the cisterna of the mature face closest to the plasma membrane. When these vesicles reach the plasma membrane, they fuse with it and release some of their contents as secretory proteins to the outside of the cell by exocytosis. Other vesicles contain membrane proteins of the cell and proteins that remain within the cell.

Golgi apparatuses are most apparent and abundant in cells that are active in secretion, such as the pancreatic cells that secrete digestive enzymes and the nerve cells that secrete transmitter substances.

Lysosomes and Microbodies

Lysosomes (Gr. "dissolving bodies") are small membrane-bound organelles that contain digestive enzymes called *acid hydrolases*, which are capable of digesting proteins, nucleic acids, lipids, and carbohydrates. The component molecules of the hydrolyzed substances are discharged into the cytoplasm, where they are used in the synthesis of new materials. The acid hydrolases are synthesized in the ER, transported to the Golgi apparatus for processing, and then transported to the lysosomes via secretory vesicles.

Lysosomes are found in almost all cells, especially those in tissues that experience frequent cell divisions, such as the lungs, liver, and spleen, and in white blood cells. Lysosomes are considered to be the digestive system of the cell, because their membranes can engulf—and their acid hydrolases can digest—complex molecules, microorganisms, and "worn-out" or damaged cell parts. If only a single acid hydrolase is absent from the lysosomes, serious genetic diseases can occur because the foreign material that enters a cell cannot be digested.

Lysosomes are often confused with *microbodies*, since both appear as dense, granular structures. There are two types of microbodies, but only one type, the ***peroxisome***, is found in human cells. Peroxisomes are membrane-bound organelles that contain enzymes that use oxygen to carry out their metabolic reactions. Peroxisomes are so

FIGURE 3.8 GOLGI APPARATUS

[A] A simplified diagram of a cell, cut to show the position of a Golgi apparatus. **[B]** A three-dimensional drawing of a Golgi apparatus, showing how the transfer vesicles from the ER merge with the Golgi apparatus, and how secretory vesicles bud off from the opposite side of the Golgi apparatus on their way to delivering secretions. **[C]** An electron micrograph of a Golgi apparatus showing secretory vesicles. ×50,000.

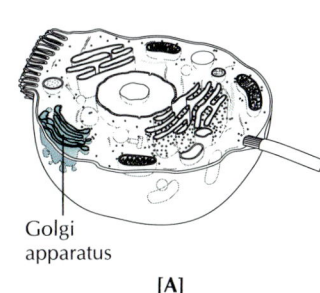

Golgi apparatus

[A]

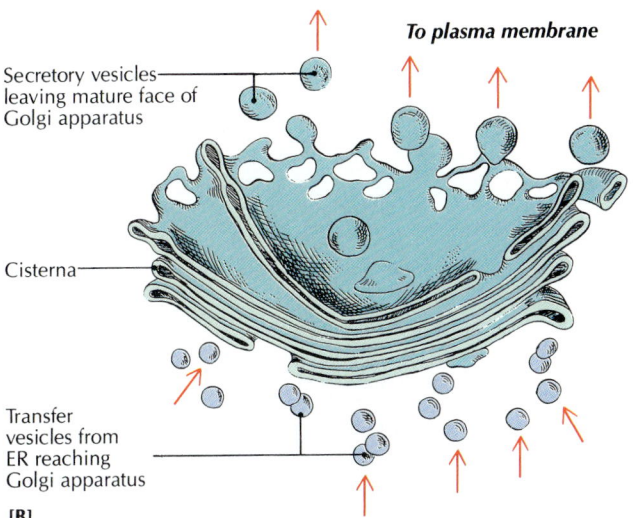

To plasma membrane

Secretory vesicles leaving mature face of Golgi apparatus

Cisterna

Transfer vesicles from ER reaching Golgi apparatus

[B]

Golgi apparatus   Secretory vesicles

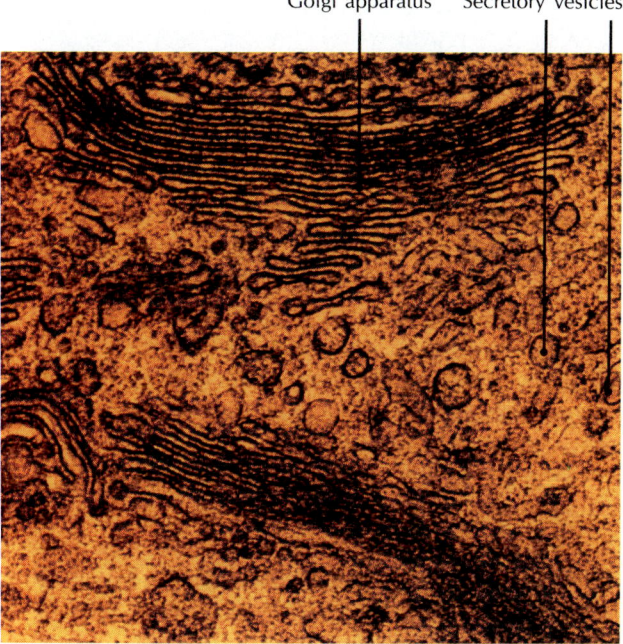

[C]

named because they can form hydrogen peroxide as they oxidize various substances and then destroy the peroxide with the enzymes they contain. The destruction of peroxides is crucial to the cell because they are toxic products of many metabolic processes, especially fatty acid oxidation. Peroxisomes are numerous in liver and kidney cells, where they are important in detoxifying certain compounds, such as alcohol.

Much of the damage that occurs to cells after injury results from the rupture of peroxisomes containing lytic enzymes. Anti-inflammatory steroids can stabilize cell membranes, thus limiting the inflammatory response after an injury.

Mitochondria

Mitochondria (Gr. *mitos*, a thread; sing. *mitochondrion*) are double-membraned organelles found throughout the cytoplasm [FIGURE 3.9A]. Their main function is the conversion of energy stored in carbon-containing molecules, especially glucose, into the high-energy bonds of ATP. (ATP is the only usable energy source for many cellular activities.) In fact, mitochondria are sometimes called the "powerhouses of the cell" because they provide about 95 percent of the cell's energy supply.

The internal structure of a mitochondrion is compatible with its function. A complex inner membrane folds and doubles in on itself to form incomplete partitions called *cristae* (KRIS-tee; L. crests) [FIGURE 3.9B, C]. The space between the cristae is called the *matrix*. The cristae greatly increase the available surface area for the chemical reactions that produce ATP. Enzymes for these reactions are located on the cristae and in the matrix.

The number of mitochondria per cell varies from as few as a hundred to as many as several thousand, depending on the energy needs of the cell. Muscle cells, which convert chemical energy into mechanical energy, have more mitochondria than do relatively inactive cells. Mitochondria contain their own DNA and ribosomes and are able to reproduce. They usually multiply when a cell needs increased amounts of ATP.

Mitochondrial DNA (mtDNA) contains genetic programs for the synthesis of some mitochondrial proteins. Most mitochondrial proteins, however, are made in the cytosol from genetic programs initiated by nuclear DNA and transported to the mitochondria. Because some mitochondria are present in egg cells but not in sperm, maternal mitochondria exert some genetic influence. Interestingly, the ancestors of mitochondria are thought to be aerobic bacteria, which originally gained access to eukaryotic cells (cells with a nucleus) as parasites or undigested prey. From this initial relationship developed a connection that was mutually beneficial to both the host cell and the mitochondria.

FIGURE 3.9 MITOCHONDRIA

[A] A drawing of mitochondria (cut) in a simplified cell. [B] A three-dimensional cutaway drawing of a mitochondrion. [C] An electron micrograph of a sectioned mitochondrion showing the outer membrane and the folds, or cristae, of the inner membrane. ×95,000.

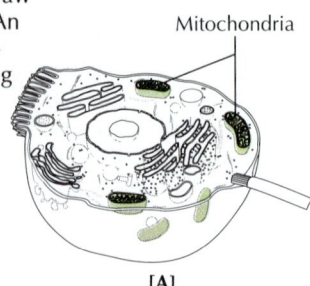

Mitochondria

[A]

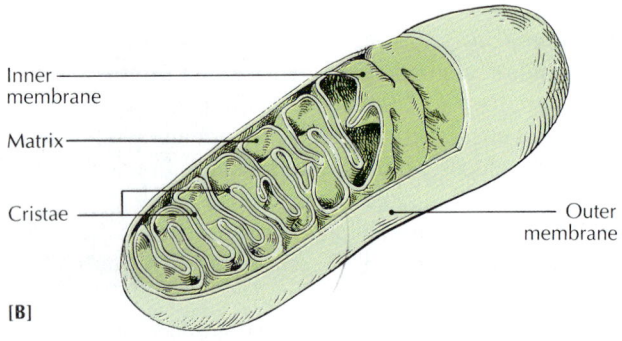

Inner membrane

Matrix

Cristae

Outer membrane

[B]

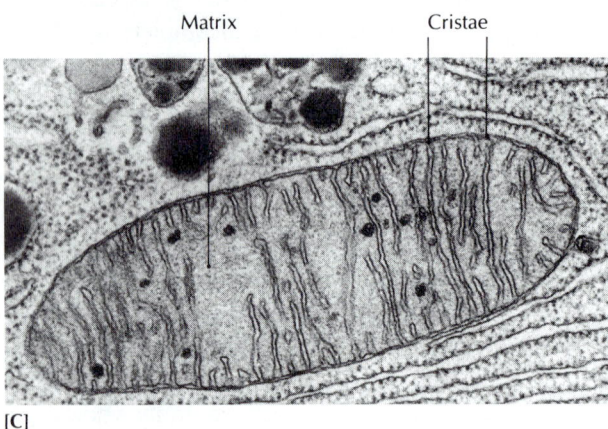

Matrix   Cristae

[C]

FIGURE 3.10 CYTOSKELETON

[A] This diagram of the cell cytoskeleton shows how fine microtrabecular strands and other structural units form a three-dimensional network that supports the principal structures of the cytoskeleton and other cellular structures.
[B] High-voltage electron micrograph showing three-dimensional microtrabecular lattice of cell. ×100,000.

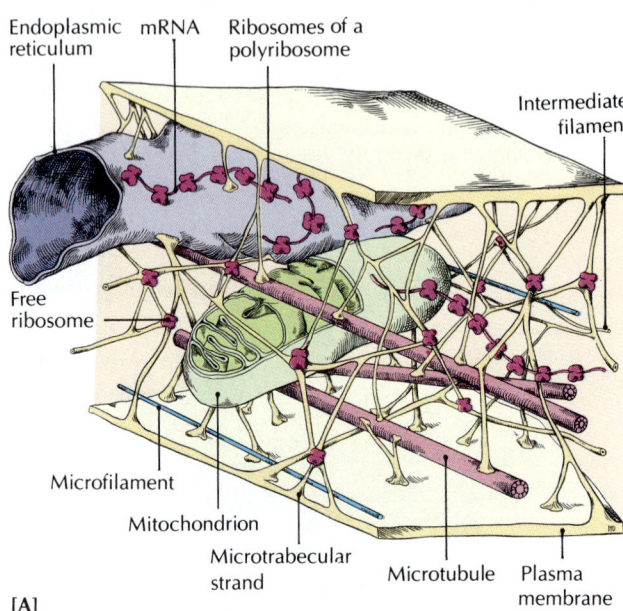

Endoplasmic reticulum   mRNA   Ribosomes of a polyribosome

Intermediate filament

Free ribosome

Microfilament

Mitochondrion

Microtrabecular strand

Microtubule   Plasma membrane

[A]

Microtrabecular strand

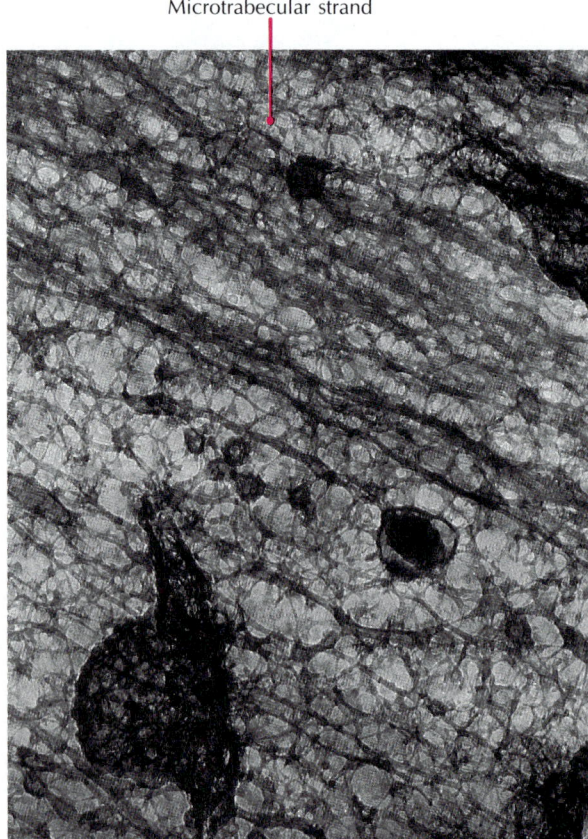

[B]

Microtubules, Intermediate Filaments, and Microfilaments

In most cells, a flexible latticed framework called the *cytoskeleton* is formed by microfilaments, intermediate filaments, and microtubules [FIGURE 3.10]. This framework extends throughout the cytoplasm, connecting its components with organelles and holding them more or less in place. The cytoskeleton also plays a role in cell movement and is a site for the binding of specific enzymes.

The most conspicuous elements of the cytoskeleton, *microtubules,* are hollow, slender cylindrical structures

FIGURE 3.11 MICROTUBULES

[A] Microtubules in a simplified cell. [B] An electron micrograph showing microtubules from the brain. ×100,000. [C] A drawing of a portion of a microtubule showing the arrangement of its spiraling protein.

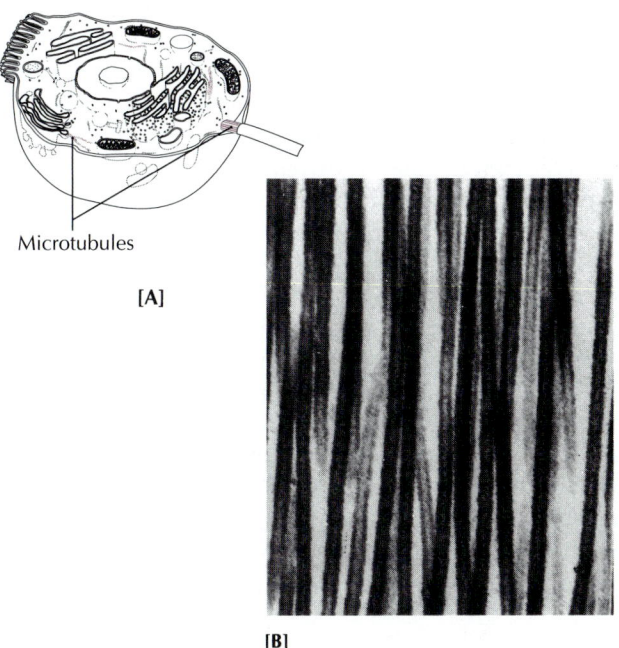

Microtubules

[A]

[B]

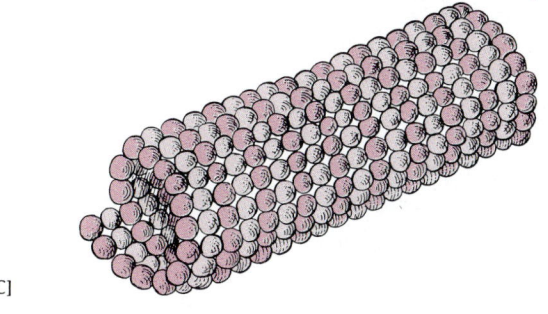

[C]

Intermediate filaments help maintain the shape of cells and the spatial organization of organelles, and promote mechanical activities in the cytoplasm. They are found in epithelial cells, glial cells in the brain, nerve cells, muscle cells, and almost all other types of cells. Intermediate filaments received their name because they are "intermediate" in diameter between the thin and thick filaments in muscle cells, where they were first described.

Solid cytoplasmic *microfilaments* are also found in most cells. They are composed of strings of protein molecules, including the protein actin. In nonmuscle cells, actin microfilaments provide mechanical support for various cellular structures and are part of the contractile systems responsible for many intracellular movements.

Centrioles

Near the nucleus of a cell lies a specialized region of cytoplasm called the *centrosome*, which contains two small structures called *centrioles* [FIGURE 3.12A]. Each centriole is a bundle of cylinders made up of microtubules, which resemble drinking straws. A centriole is composed of nine triplet microtubules that radiate from the center like the spokes of a wheel [FIGURE 3.12B, C]. The two centrioles lie at right angles to each other [FIGURE 3.12A, B] and are duplicated when the cell undergoes cell division. Centrioles are involved in the movement of chromosomes during cell division.

Cilia and Flagella

Cilia (SILL-ee-uh; L. originally "eyelids," later "eyelashes"; sing. *cilium*) and *flagella* (fluh-JELL-uh; L. "small whips"; sing. *flagellum*) are threadlike appendages of some cells. The shafts of cilia and flagella are composed of nine pairs of outer microtubules and two single microtubules in the center [FIGURE 3.13A]. Both types of shafts are anchored by a *basal body*, a specialized structure that acts as the template for the nine-plus-two arrangement of microtubules in the shaft. The basal body is structurally identical to a centriole, with a nine-plus-three arrangement; it protrudes into the cytoplasm of the cell to which it is attached.

Groups of cilia beat in unison, creating a rhythmic, wavelike movement in only one direction [FIGURE 3.13B]. Ciliated cells are found in the upper portions of the respiratory tract, where they sweep along mucus and foreign substances. They also appear in portions of the female reproductive system [FIGURE 3.13C], where they help move mucus and eggs along the uterine tube (Fallopian tube).

Flagella have the same internal structure as cilia but are much longer. In human beings, the only cells with flagella are sperm, which normally have only one flagel-

that are found in most cells [FIGURE 3.11A, B]. Each microtubule is made up of spiraling subunits of a protein called *tubulin* [FIGURE 3.11C]. Microtubules have several functions within a cell:

1 They are associated with movement, especially in cilia and flagella (whiplike appendages), and with chromosome movement when the cell nucleus is dividing.

2 They are part of a transport system within the cell; in nerve cells, for example, they help move materials along the outer surface of each tubule through the cytoplasm.

3 They form a supportive cytoskeleton within the cytoplasm.

4 They are involved in the overall changes in shape that cells go through during periods of cell specialization.

FIGURE 3.12 CENTRIOLES

[A] The position of centrioles near the cell nucleus. **[B]** A drawing of a pair of centrioles. Each has nine triplets of microtubules and lies at a right angle to the other. **[C]** An electron micrograph of a cross section of a centriole. ×400,000.

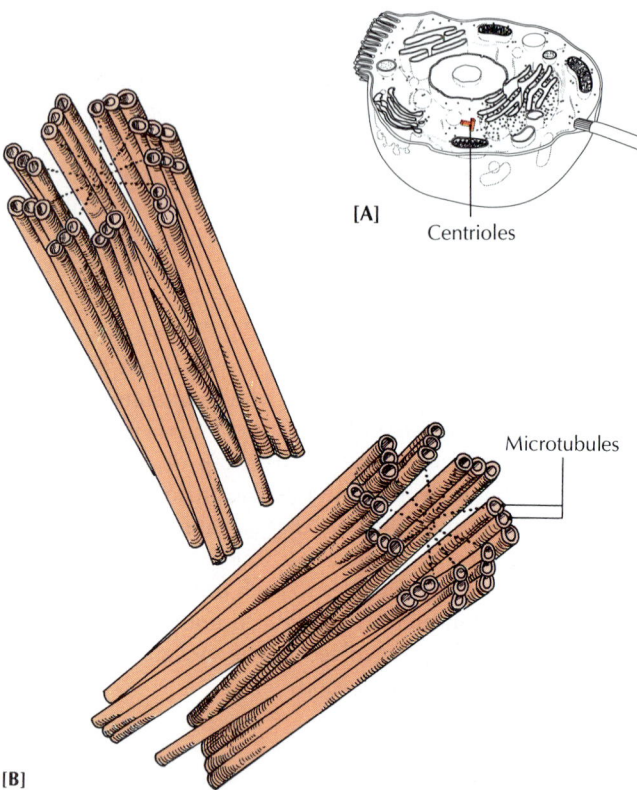

[A]
Centrioles

Microtubules

[B]

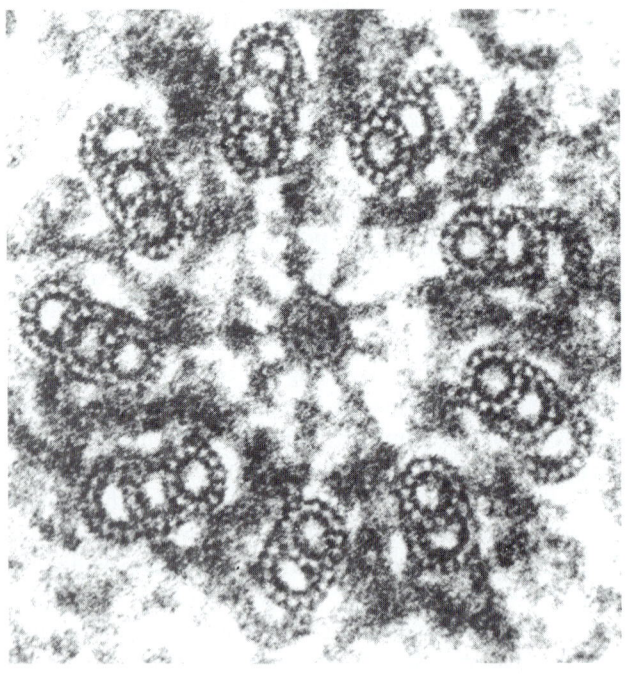

[C]

lum each. The tail of a sperm enables the free-swimming cell to move relatively great distances to reach an egg deep within the female reproductive system.

Cytoplasmic Inclusions

In addition to organelles and nonmembrane-bound macromolecules, cytoplasm contains *cytoplasmic inclusions,* chemical substances that are usually either food material or the stored products of cellular metabolism. Generally, inclusions are not permanent components of a cell, but are constantly being destroyed and replaced. Carbohydrates, for example, are stored mainly in liver and skeletal muscle cells in the form of *glycogen inclusions.* When the body needs energy, the glycogen is converted into glucose, the body's main source of energy. Lipids are also stored in cells as a reserve source of energy.

Also present in the cytoplasm of skin cells are several *pigments* (substances that produce color), probably the best known being *melanin,* the pigment that protects the skin by causing it to tan when exposed to excessive ultraviolet radiation. Melanin is also found in the hair and eyes. *Hemoglobin* is another well-known pigment found in mature red blood cells. Its major function is to carry oxygen or carbon dioxide in red blood cells.

A S K   Y O U R S E L F

1 What is the function of ribosomes?

2 How does a Golgi apparatus receive protein from the ER?

3 What do lysosomes do?

4 How is the internal structure of a mitochondrion compatible with its function?

5 What is the function of cilia and flagella, and how do they accomplish it?

6 What is a cytoplasmic inclusion?

THE NUCLEUS

The *nucleus* (NOO-klee-uhss; L. nut, kernel; pl. *nuclei,* NOO-klee-eye) contains the hereditary material DNA and is the control center of the cell. There are two important aspects of the nucleus:

1 The nucleus is the site of transcription, the process in which the hereditary information in DNA is copied (in code) into molecules of messenger RNA. The mRNA, in turn, determines the structure of the cell's proteins, which are synthesized in the cytoplasm.

FIGURE 3.13 CILIA AND FLAGELLA

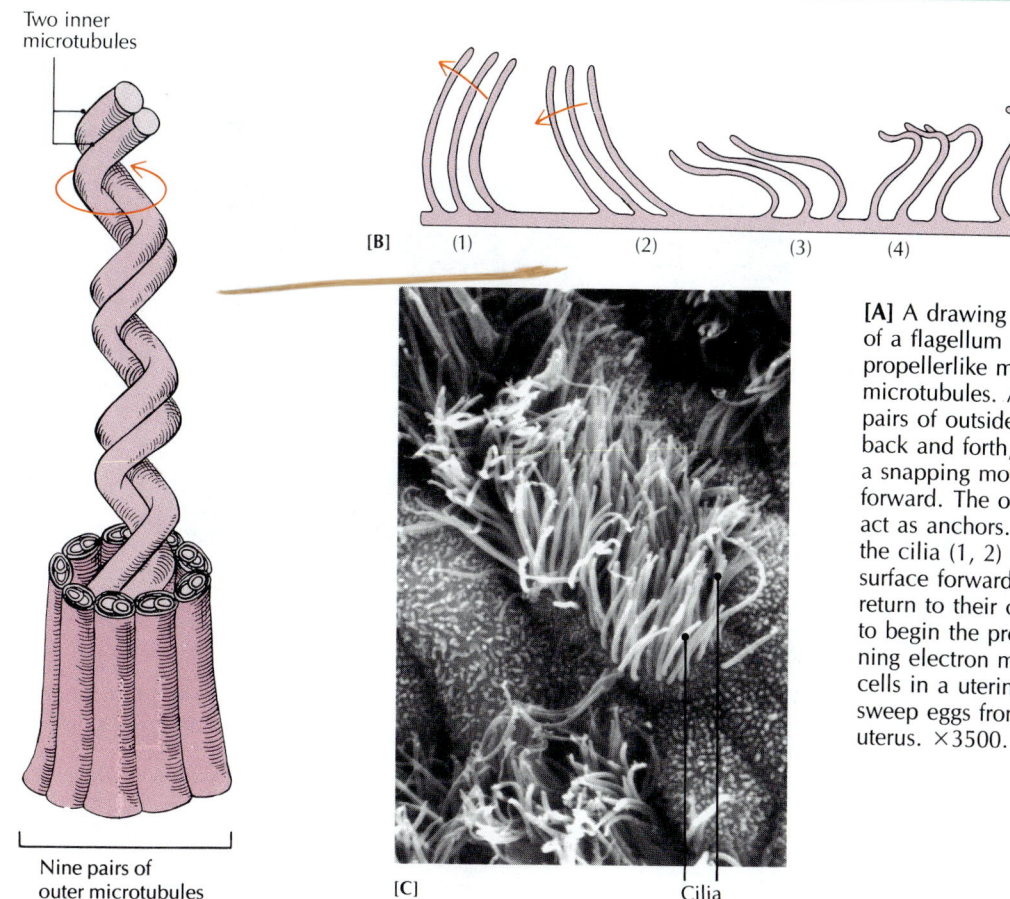

Two inner microtubules

Nine pairs of outer microtubules

[A]

[B] (1) (2) (3) (4) (5)

[C]

Cilia

[A] A drawing of a portion of the shaft of a flagellum shows the rotating, propellerlike motion of the two inner microtubules. Apparently, the nine pairs of outside microtubules slide back and forth, one at a time, creating a snapping motion that moves the cell forward. The outer microtubules also act as anchors. [B] The active stroke of the cilia (1, 2) moves fluid at the cell surface forward. The cilia relax and return to their original position (3–5) to begin the process again. [C] A scanning electron micrograph of ciliated cells in a uterine tube. The cilia help sweep eggs from the ovary to the uterus. ×3500.

2 The nucleus contains hereditary information that is transferred during cell division from one generation of cells to the next, and eventually from one generation of organisms to the next.

Almost all cells have nuclei. Mature red blood cells, which do not have nuclei or DNA, are incapable of producing mRNA. Thus, they cannot synthesize new proteins, duplicate themselves, or perform typical cellular activities. Red blood cells have a relatively short life span.

Depending on the cell type, nuclei vary in shape, location, and number. Nuclei are generally somewhat spherical and are located near the center of the cell, firmly embedded in the cytoplasm [FIGURE 3.14A]. They can also be elongated and found at the base of tall cells, or they can be flattened against the plasma membrane of compact cells.

Nuclear Envelope

The membrane of the nucleus looks somewhat like the plasma membrane of the cell, but with some important differences. It is not a single bilayered membrane, but two distinct membranes with a discernible "space" between them, leading cytologists to call the membrane a *nuclear envelope.* It contains many openings called *nuclear pores* [FIGURE 3.14B–D], giving the nucleus something of the appearance of a whiffle ball. The pores make it possible for larger materials to enter and leave the nucleus and for the nucleus to be in direct contact with the membranes of the endoplasmic reticulum [FIGURE 3.14E]. The movement of materials through the nuclear pores is controlled by charged molecules at the opening of the pores.

Nucleolus

Cell nuclei usually contain at least one dark, somewhat spherical nonmembrane-bound mass called a *nucleolus* (new-KLEE-oh-luhss; "little nucleus"; pl. *nucleoli,* noo-KLEE-oh-lye) [FIGURE 3.14A, B, D]. Besides containing DNA and RNA, the nucleolus contains protein, which assists in the construction of ribosomes. In fact, the nucleolus can be thought of as a preassembly site for ribosomes. Ribosomes are partially synthesized in the nucle-

FIGURE 3.14 NUCLEUS AND NUCLEAR ENVELOPE

[A] A simplified drawing of a cell showing the nucleus and nucleoli. [B] A drawing that shows a nucleolus floating in the nucleoplasm. [C] A scanning electron micrograph of a nucleus showing the nuclear envelope and its pores. ×15,000. [D] An electron micrograph of the nucleus and its constituents. Some of the nuclear pores are marked with arrows. ×21,000. [E] Simplified drawing of a two-dimensional thin section of a cell.

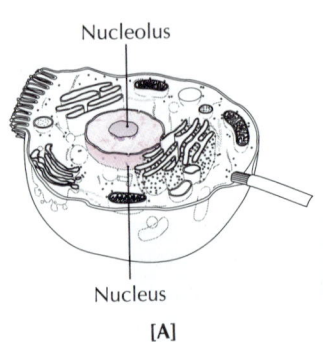

Nucleolus

Nucleus

[A]

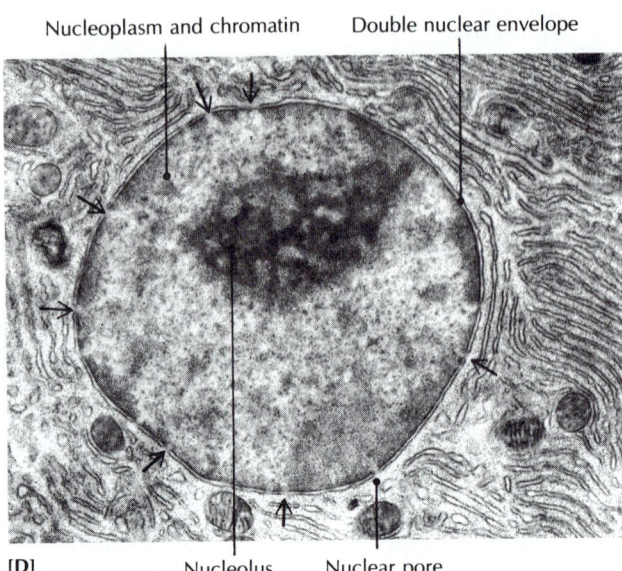

Nucleoplasm and chromatin   Double nuclear envelope

[D]   Nucleolus   Nuclear pore

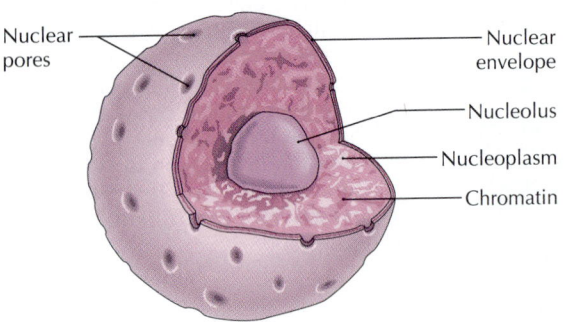

Nuclear pores

Nuclear envelope

Nucleolus

Nucleoplasm

Chromatin

[B]

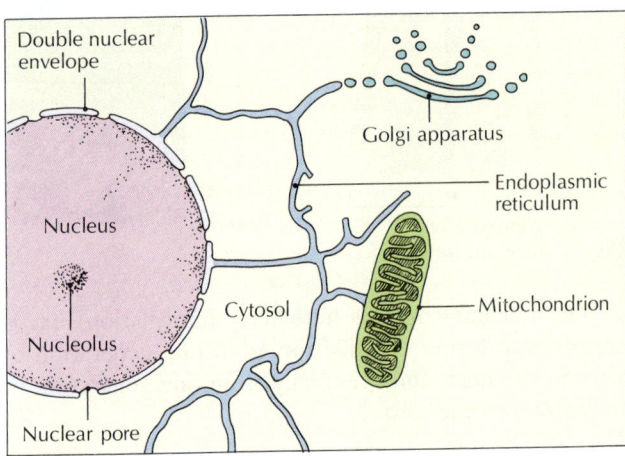

Double nuclear envelope

Golgi apparatus

Endoplasmic reticulum

Nucleus

Cytosol

Mitochondrion

Nucleolus

Nuclear pore

[E]

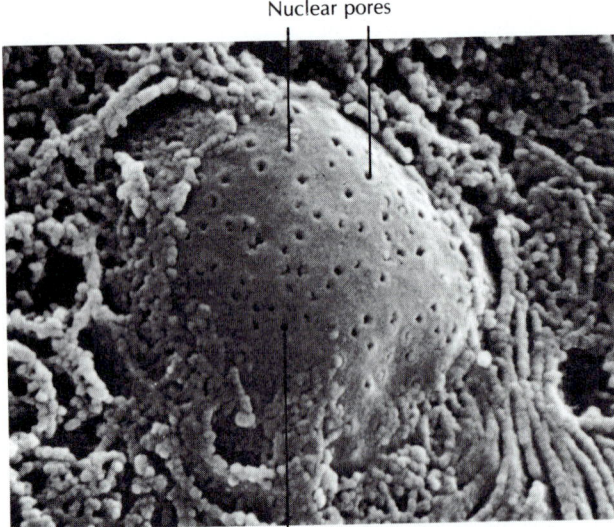

Nuclear pores

[C]   Inner nuclear membrane

olus before they pass into the cytoplasm through the pores of the nuclear envelope. Some of the ribosomes become attached to the endoplasmic reticulum, where they assist in the synthesis of proteins; others are unattached free ribosomes in the cytoplasm.

Chromosomes

The interior mass of the nucleus is the *nucleoplasm*, which contains genetic material in a form called *chromatin* [FIGURE 3.14B, D]. Chromatin consists of a combination of DNA and protein. During a type of cell division called mitosis (described later in this chapter), strands of protein-rich chromatin become arranged in coiled threads called **chromosomes**, which store the hereditary material in segments of DNA called *genes*.* Genes are located along the chromosomes in a specific sequence and position. In fact, the sequence and position of nucleotides in

*According to researchers at the California Institute of Technology, each cell in the body contains genetic information that, if translated into words, would fill 3000 books of 1000 pages each, with 1000 words on each page.

the DNA making up each gene create the *genetic code* that determines heredity and is responsible for protein structure.

CELLULAR METABOLISM

The original energy source for a cell is the food we eat, which eventually is used to generate ATP in the cell after a series of chemical reactions. The chemical processes that transform food into living tissue and energy are part of the body's *metabolism* (all the chemical reactions that take place in a living organism).

FIGURE 3.15 shows how metabolism works: (1) Food is eaten and moved into the digestive system. (2) In the digestive system, food is broken down into smaller molecules of simple sugars such as glucose and fatty acids and amino acids. (3) These small molecules are absorbed from the small intestine into the bloodstream and are carried into individual cells. (4) The simple sugars, fatty acids, and amino acids enter the cytoplasm of a cell. (5) The molecules are converted into smaller molecules of pyruvic acid, and a small amount of ATP is produced. (6) The pyruvic acid is converted into a compound called acetyl-CoA, which enters the mitochondria. (7) In the mitochondria, in the presence of oxygen, the molecules are broken down (reduced) further in a complicated series of chemical reactions (described in detail in Chapter 26). (8) The reactions produce enough ATP to provide the energy a cell needs to function properly. At the same time molecules of waste carbon dioxide and water are produced.

THE CELL CYCLE

In the next minute about 3 billion cells in your body will die. If all is well, 3 billion new cells will be created during that same minute. The correct number and kind of new cells are produced when healthy existing cells divide. Each original (parent) cell becomes two daughter* cells, both genetically identical to the parent.

This reproduction through duplication and division is the key to development and growth — it is the formula for

*The use of the word *daughter* has nothing to do with gender. A "daughter" cell is not necessarily a *female* cell but merely an *offspring*, a new cell formed from a parent cell or cells.

FIGURE 3.15 THE CELLULAR EVENTS OF METABOLISM

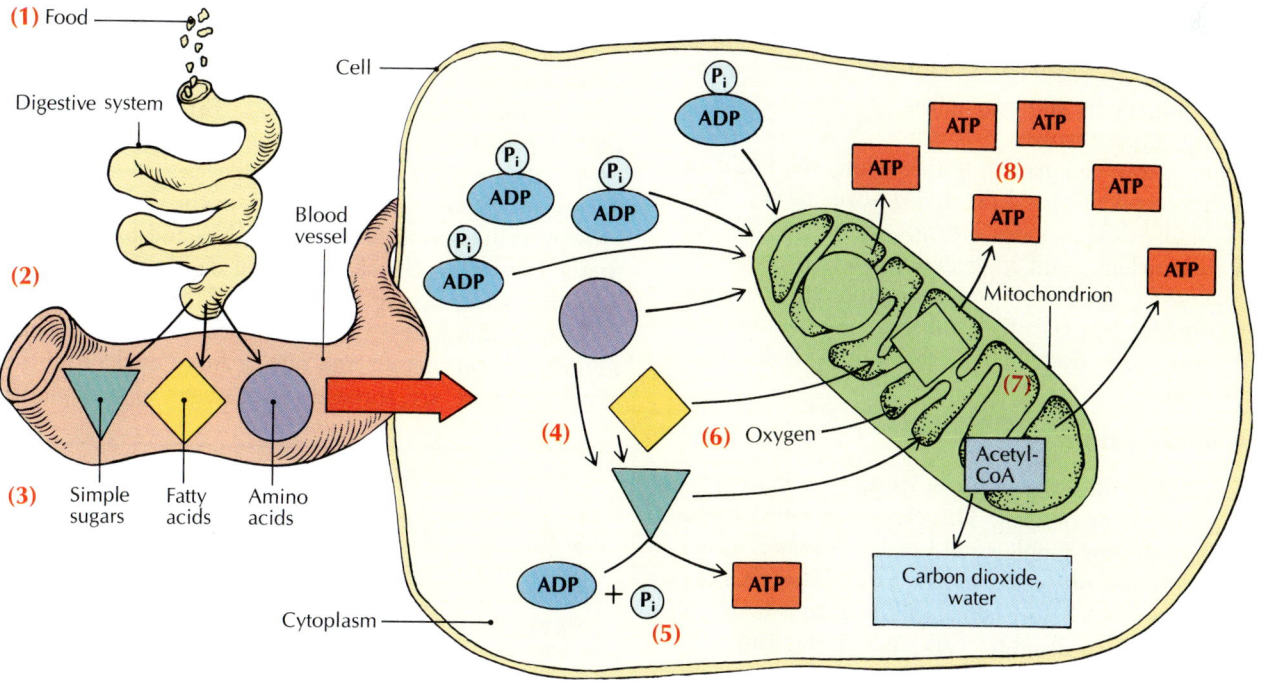

TABLE 3.3 LIFE SPANS OF SPECIALIZED CELLS

Type of cell	Approximate life span
Bone marrow cells (blood-forming)	10 hours
Stomach cells	2 days
Egg cells, sperm	2–3 days
Large intestine cells	3–4 days
White blood cells	13 days
Skin cells	19–34 days
Red blood cells	120 days
Liver cells	18 months
Nerve cells*	Lifetime of body

*Nerve cells, except for olfactory (smell) neurons in the nose, do not reproduce after early childhood.

life. The single fertilized egg that marks the beginning of your life doubles, then each of its daughter cells doubles, and so on until a predetermined number is reached. During the earlier stages of human growth, the doubling process is accompanied by cell specialization. The process in which cells become specialized in structure and function is called *differentiation.* Once cells specialize, their functions seldom change. For example, when a cell differentiates into an epithelial cell, it will never be any other kind of cell, and neither will its daughter cells.

The process of *cell division* marks the beginning and end of the *cell cycle,* or reproductive cycle, of a single cell. The cell cycle is that period from the beginning of one cell division to the beginning of the next cell division [TABLE 3.3]. Before cell division can begin, the parent cell must double its mass and contents. For cell division to be complete, both the nucleus and the cytoplasm must divide in such a way that each daughter cell gets one complete set of genetic material and all the necessary cytoplasmic constituents and organelles.

The cell cycle involves two major phases: (1) *interphase,* a period of cell growth during which the DNA in the nucleus replicates, and (2) the *M phase,* or cell division. Cell division includes *mitosis*—the period during which the nucleus (with its genetic material) divides—and *cytokinesis*—the division of the cytoplasm into two genetically identical cells [FIGURE 3.16]. All processes of the cell cycle are described in the following sections.

DNA Replication

Before a cell divides, the DNA in its nucleus must produce a perfect copy of itself. This process is called *replication,* because the double strand of DNA makes a *replica,* or duplicate, of itself [FIGURE 3.17]. The accurate replication of DNA before actual cell division is essential to ensure that each daughter cell receives the same genetic information as the parent cell.

FIGURE 3.16 THE CELL CYCLE

Simplified drawing of the cell cycle, which results in the reproduction of a cell. The cycle can be divided into two main phases: interphase and the M phase. The cell grows, and DNA is synthesized during interphase; the cell divides during the M phase, which includes the internal events of mitosis and cytokinesis, the actual division of the parent cell into two daughter cells. The full cycle is completed in about 16 hours; approximate times are given for each of the phases.

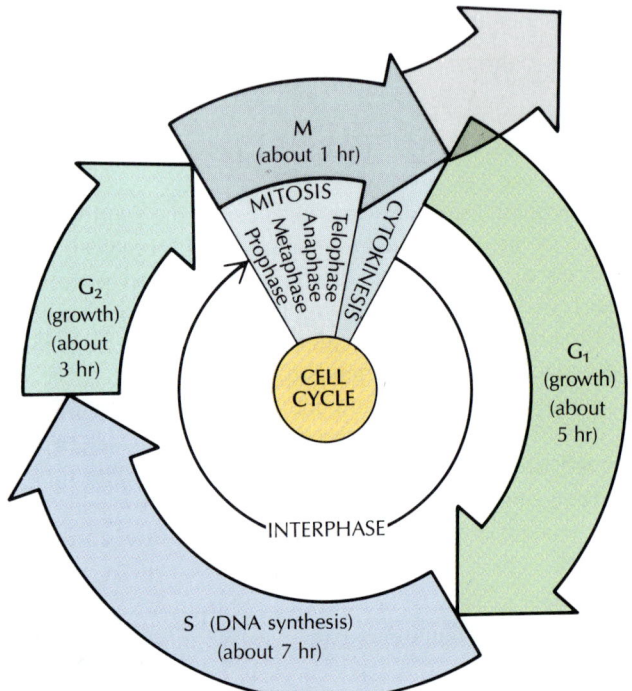

Interphase

Often erroneously described as a resting period between cell divisions, *interphase* [FIGURE 3.18] is actually a period of great metabolic activity, occupying about 90 percent of the total duration of the cell cycle. It is the period during which the normal activities of the cell take place: growth, cellular respiration, and RNA and protein synthesis. It also sets the stage for cell division, since *DNA replication occurs during interphase.*

Visually, interphase is distinguished by the presence of thin chromatin threads and one or more nucleoli in the nucleus. There are three distinct stages of interphase:

Q: *How often do cells divide?*

A: Most cells, such as skin cells and cells in the lining of the digestive system, divide continually. The new cells replace old or damaged ones. Some cells, however, rarely divide. Muscle cells seldom do, and except for the nerve cells involved with the sense of smell, after birth nerve cells never do. Apparently, the more specialized a cell is, the less frequently it divides.

FIGURE 3.17 THE REPLICATION OF DNA

The replication of a twisted, double-stranded DNA molecule begins with the untwisting and separation of its two complementary strands. The "unzipping" of the double strands takes place when enzymes break the hydrogen bonds connecting the bases of the two strands. Once the bases on each strand are exposed, the enzyme *DNA polymerase* attaches to each of the separated strands and moves along it one base at a time. The enzyme selects free nucleotides from the nucleoplasm and forms new hydrogen bonds with the exposed complementary bases on the unzipped strands. It is this principle of *complementary bonding* that ensures exact replication. In this way, each separated strand of the original twisted DNA molecule acts as a mold, or *template,* to make an exact copy, not of itself, but of the *other* strand of the unzipped DNA molecule. As soon as the new bonds are formed, the original strand and the newly formed complementary strand begin to twist together. So the complicated activities of untwisting, unzipping, rezipping, and retwisting all occur simultaneously along different portions of the DNA molecule. In this drawing, the untwisted ends of the original double DNA strand are serving as templates for the formation of new strands, which form from free nucleotides in the nucleoplasm. When replication is finished, each of the two newly formed DNA molecules will contain one strand from the original molecule and one new strand.

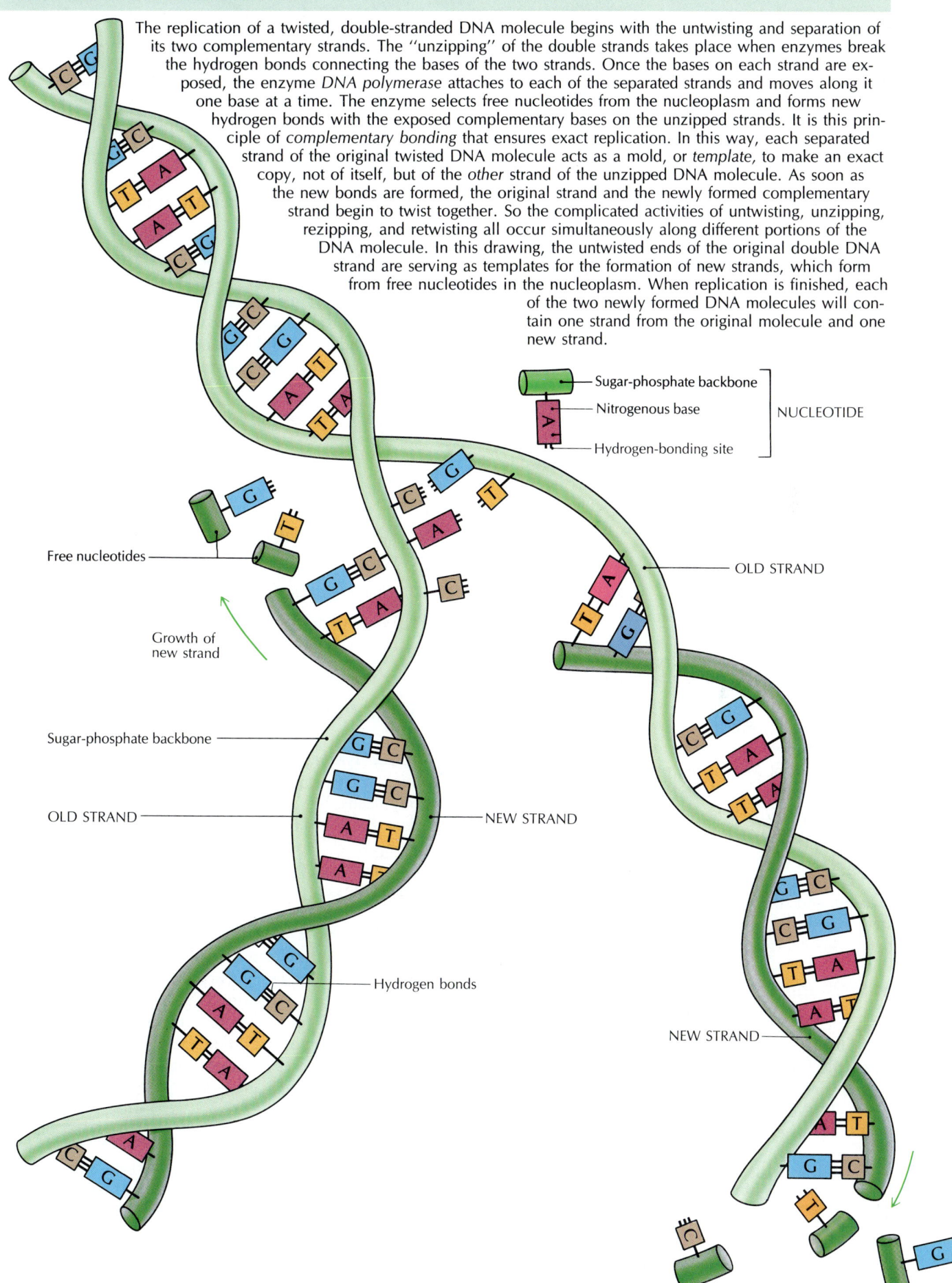

Sugar-phosphate backbone
Nitrogenous base } NUCLEOTIDE
Hydrogen-bonding site

Free nucleotides

Growth of new strand

Sugar-phosphate backbone

OLD STRAND

Hydrogen bonds

OLD STRAND

NEW STRAND

OLD STRAND

NEW STRAND

FIGURE 3.18 INTERPHASE

During interphase, the material of the chromosomes is in the form of chromatin threads.

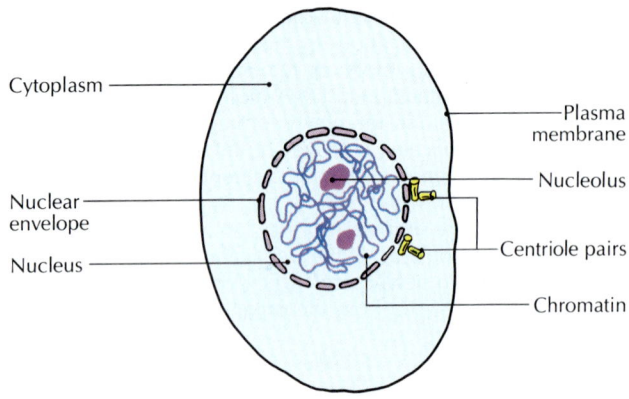

Cytoplasm

Nuclear envelope

Nucleus

Plasma membrane

Nucleolus

Centriole pairs

Chromatin

1 G_1: Immediately after new daughter cells are produced, they enter a period of growth also known as the "first gap" because it represents the "gap" between cell division and DNA replication. After cell division, one centriole pair separates and begins to replicate, as do other organelles.

2 S: After G_1 the cell enters the S phase (for *synthesis,* because DNA is synthesized then), and the DNA of the chromosome doubles. Important here is the exact replication of DNA, which is not just a doubling in quantity but a doubling in which every chromosome duplicates itself. Two pairs of centrioles appear during this phase. The onset of the S phase is controlled by an enzyme called *cdc2 kinase,* which is one subunit of a protein complex called *maturation promoting factor,* or *MPF.* The other subunit of *MPF* is the protein *cyclin,* which probably activates the cdc2 kinase.

3 G_2: With the S phase past and the chromosomes replicated, the cell goes into a "second gap" phase, when the final preparations for cell division occur. The centriole pairs start to move apart as a prelude to the next cell division, and structures directly associated with the next step, mitosis, begin to be assembled. ATP is accumulated before the energy-consuming process of mitosis begins.

Mitosis and Cytokinesis

Once the DNA of the parent cell has been replicated, the actual division of the cell can begin. This division has two parts; the first is nuclear division, or **mitosis,*** and the second is **cytokinesis,** the division of the cytoplasm [FIGURE 3.19]. Mitosis and cytokinesis together are referred to as the *M phase* [see FIGURE 3.16]. The *M phase* accom-

---

*The words *mitosis* and *mitochondria* are both derived from the same Greek word, *mitos,* which means "thread." This root word refers to the threadlike appearance of chromosomes during mitosis and to the threadlike shapes of mitochondria as viewed by early microscopists.

plishes two things: (1) the equal distribution of all cellular material between the daughter cells and (2) the actual division of the nuclear material (DNA) and the *equal distribution of DNA to each new cell.* (In the process, the number of chromosomes in each daughter cell is kept the same as in the parent cell.) The initiation of the M phase, like the S phase earlier, is triggered by cdc2 kinase.

The Cell Cycle as a Continuum

Although we have described the cell cycle in neatly defined stages, the events of the cycle actually occur in a dynamic environment. Under laboratory conditions, some events of metaphase and anaphase may seem to occur almost simultaneously. Cytokinesis may begin during telophase or anaphase, and the doubling of the centrioles may occur before cytokinesis is complete.

ASK YOURSELF

1 What is cellular differentiation?

2 Why is it necessary for DNA to replicate before mitosis begins?

3 What occurs during interphase?

4 What are the stages of mitosis?

5 What does mitosis accomplish?

6 What is cytokinesis?

HOW PROTEINS ARE SYNTHESIZED

Protein synthesis, the manufacture of proteins from amino acids, provides not only the structural proteins the body needs, but also the enzymes (which are also proteins) that control a cell's metabolism. By regulating protein synthesis, a cell regulates its own metabolism.

Proteins are synthesized from the amino acids available in the cell pool. The amino acids are assembled in specific sequences determined by each individual's genetic blueprint. This genetic program for protein synthesis is found in the chromosomal DNA in the cell nucleus. The DNA itself does not assemble the amino acids in the correct sequence. Instead, DNA works together with RNA, which is present in the cytoplasm as well as in the nucleus — an important difference. *RNA is not restricted to the nucleus* but is free to move about the cytoplasm, where protein synthesis actually occurs [FIGURE 3.20].

Q: *How often do mistakes occur during DNA replication?*

A: Less than one mistake is made for every billion nucleotides added. Genetic mistakes, or *mutations,* can have far-reaching effects, but most have little or no effect on the body.

FIGURE 3.19 MITOSIS AND CYTOKINESIS

MITOSIS

Although mitosis is a single, continuous process, it is usually described in four sequential stages known as *prophase, metaphase, anaphase,* and *telophase*. For the sake of clarity, the drawings that accompany the text show only four chromosomes and no organelles except the centrioles. The light micrographs show human white blood cells in various stages of mitosis. Human cells contain 46 chromosomes, but the principles of mitosis are the same no matter how many chromosomes are involved.

Prophase

After interphase, the cell enters the first stage of mitosis, **prophase.** In *early prophase,* the chromatin threads begin to coil, becoming shorter and thicker. (Before coiling begins, the total length of the DNA strands in a single human nucleus is greater than the length of the entire human body, but the strands are too thin to be seen with an ordinary light microscope.) The nucleoli and nuclear envelope begin to break up, and bursts of microtubules, called *asters* (L. stars), begin to radiate from the centrioles.

By *late prophase,* the chromatin threads form clearly defined chromosomes. Each chromosome has two strands, or **chromatids,** with a full complement of the replicated DNA formed during the S stage of interphase. Each pair of chromatids is joined somewhere along its length by a small spherical structure called a *centromere*. The fragments of the nuclear envelope and nucleoli disperse in the endoplasmic reticulum, and newly formed microtubules move in among the chromatid pairs. The two centriole pairs move apart.

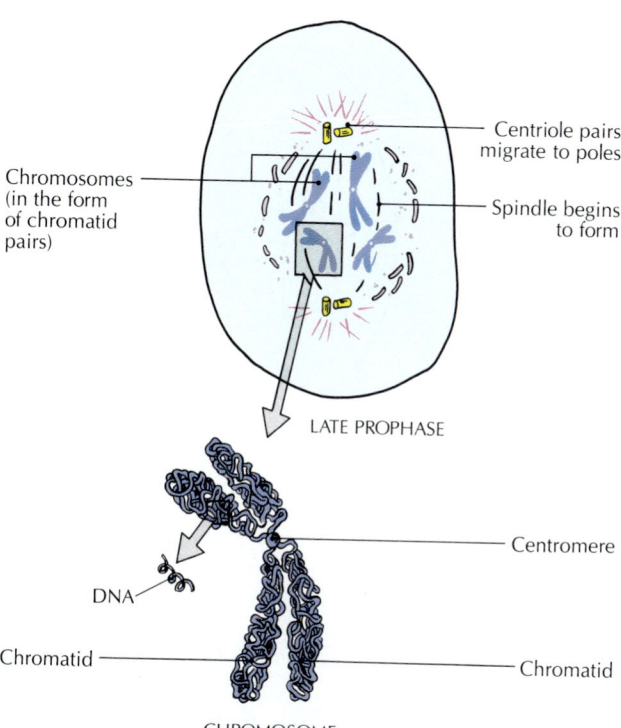

Chromosomes (in the form of chromatid pairs)

Centriole pairs migrate to poles

Spindle begins to form

LATE PROPHASE

Centromere

DNA

Chromatid

Chromatid

CHROMOSOME

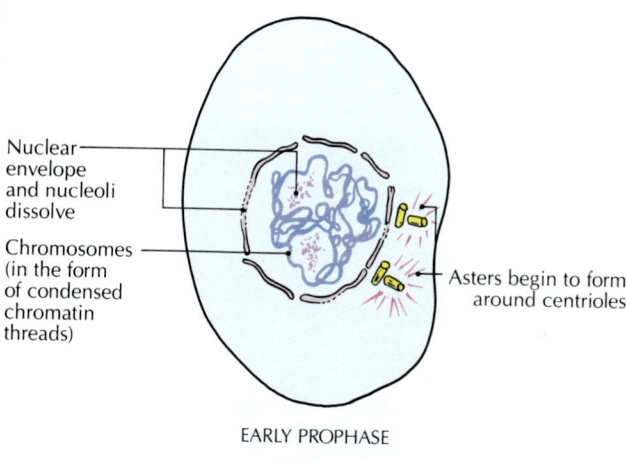

Nuclear envelope and nucleoli dissolve

Chromosomes (in the form of condensed chromatin threads)

Asters begin to form around centrioles

EARLY PROPHASE

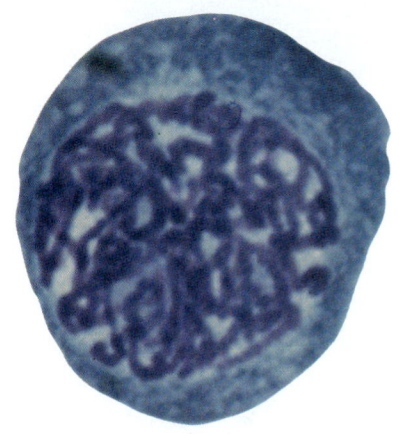

By the end of prophase, the centriole pairs have moved to opposite ends, or *poles,* of the nucleoplasm. The position of the centrioles at the poles determines the direction in which the cell divides. Between the centrioles, the microtubules form a *spindle* that extends from pole to pole. The chromatid pairs move to the center of the spindle. The asters, spindle, centrioles, and microtubules are called the *mitotic spindle*.

THE MITOTIC SPINDLE

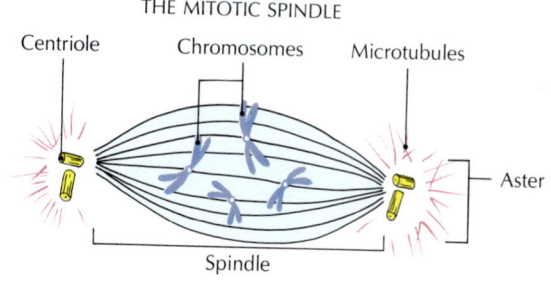

Centriole

Chromosomes

Microtubules

Aster

Spindle

(Figure 3.19 continues on following pages)

FIGURE 3.19 MITOSIS AND CYTOKINESIS (Continued)

Metaphase

During *metaphase,* the centromere of each chromatid pair attaches to one of the microtubules of the spindle. The centrioles are pushed apart as the spindle lengthens, and the double-armed chromosomes are pulled to the center of the nucleoplasm, lining up across the center of the spindle. Toward the end of metaphase, the centromeres double, so that each chromatid has its own centromere. At this point, each chromatid may be considered a complete chromosome, each with its double-stranded DNA molecule.

Anaphase

The shortest stage of mitosis is *anaphase.* In *early anaphase,* the chromosome pairs separate, and the members of each pair begin to move toward opposite poles of the cell. The microtubules are instrumental in moving the chromosomes toward the poles. By *late anaphase,* the poles themselves have moved farther apart. At this stage in human mitosis, 46 chromosomes are near one pole, and 46 are near the opposite pole.

Double-stranded chromosome attaches to spindle

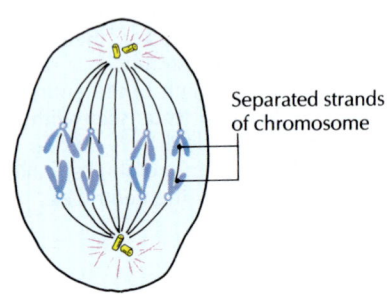

Separated strands of chromosome

EARLY ANAPHASE

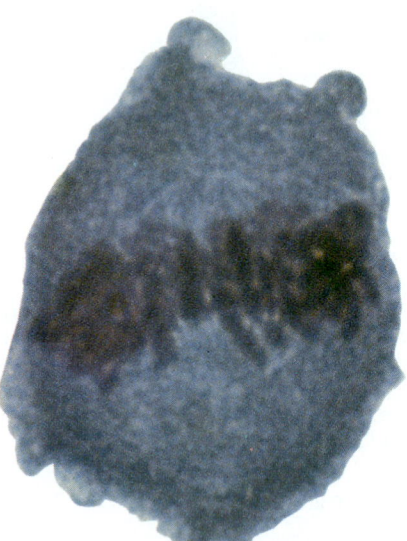

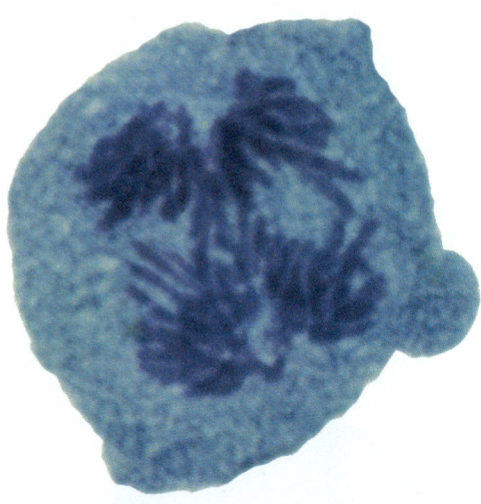

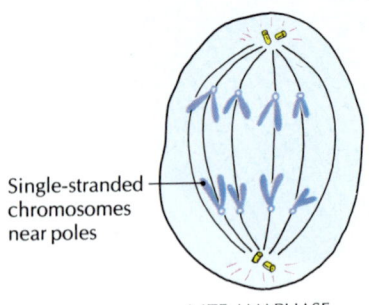

Single-stranded chromosomes near poles

LATE ANAPHASE

MITOSIS (Continued)

Telophase

In **telophase,** the chromosomes arrive at the poles. In *early telophase,* the chromosomes lose their distinctive rodlike form and begin to uncoil. They appear as they did during interphase. The spindle and asters dissolve. By *late telophase,* fragments from the endoplasmic reticulum spread out around each set of chromosomes, forming a new nuclear envelope. The cell begins to pinch in at the middle. New nucleoli form from the nucleolar regions of the chromosomes. *Mitosis is over, but cell division is not.*

CYTOKINESIS

The third major phase of the cell cycle, and the final stage of cell division, is **cytokinesis** (Gr. *cyto*, cell + *kinesis*, movement), the separation of the cytoplasm into two parts. Before cytokinesis, the two newly formed nuclei still share the same cytoplasmic compartment. The separation is accomplished by *cleavage*, a pinching of the plasma membrane. The *cleavage furrow*, where the pinching occurs, looks as though someone tied a cord around the middle of the cell and pulled it tight. Cytokinesis may begin during telophase or anaphase, and the doubling of the centrioles may begin before cytokinesis is complete. The two new, genetically identical daughter cells formed are each about half the size of the original parent cell.

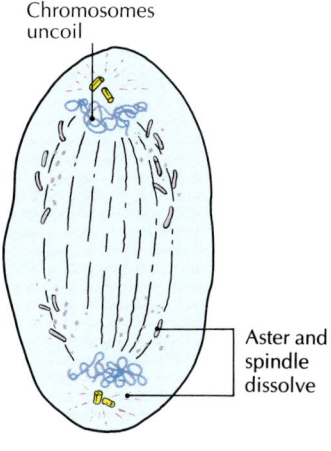

Chromosomes uncoil

Aster and spindle dissolve

EARLY TELOPHASE

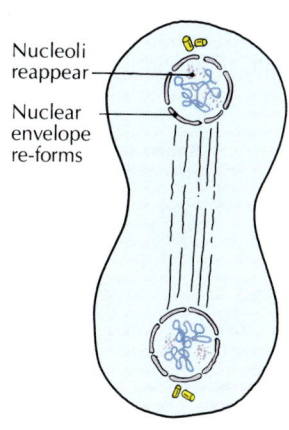

Nucleoli reappear

Nuclear envelope re-forms

LATE TELOPHASE

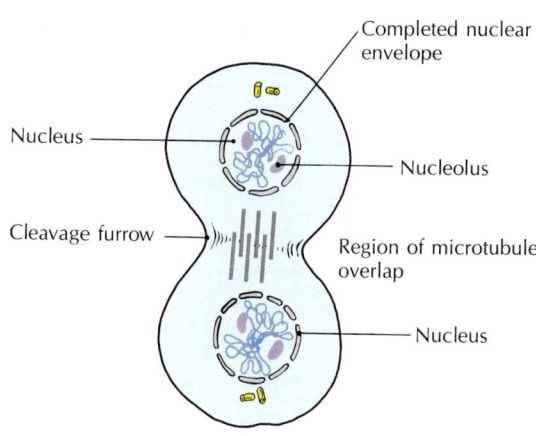

Completed nuclear envelope

Nucleus

Nucleolus

Cleavage furrow

Region of microtubule overlap

Nucleus

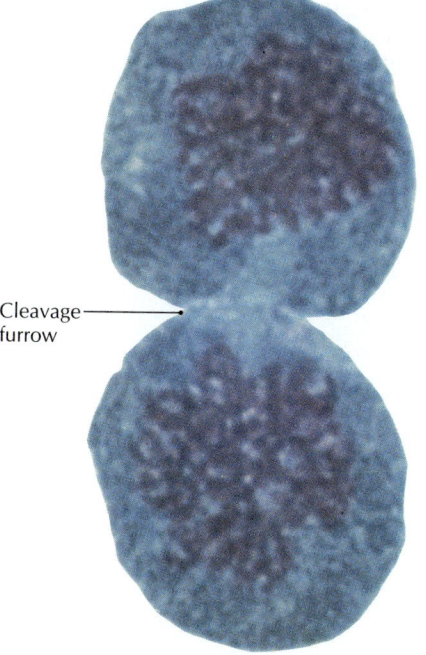

Cleavage furrow

FIGURE 3.20 TRANSCRIPTION AND PROTEIN SYNTHESIS

[A] Transcription

In transcription, portions of the DNA double helix untwist and separate. One of the two separated DNA strands serves as a template for the synthesis of a single-stranded messenger RNA (mRNA) molecule. The sequence of nucleotides in the forming mRNA is determined by the nucleotide sequence of the DNA strand on which it forms. The nucleotide sequence in the mRNA is complementary to that of the DNA template. For example, where there is a cytosine in the DNA, there is a guanine in the RNA. In DNA, adenine pairs with thymine. However, in the forming RNA, uracil pairs with adenines in the DNA. When transcription is complete, the mRNA molecule migrates out of the nucleus into the cytoplasm. Encoded in the mRNA molecule is the hereditary information needed for the synthesis of a particular protein or polypeptide.

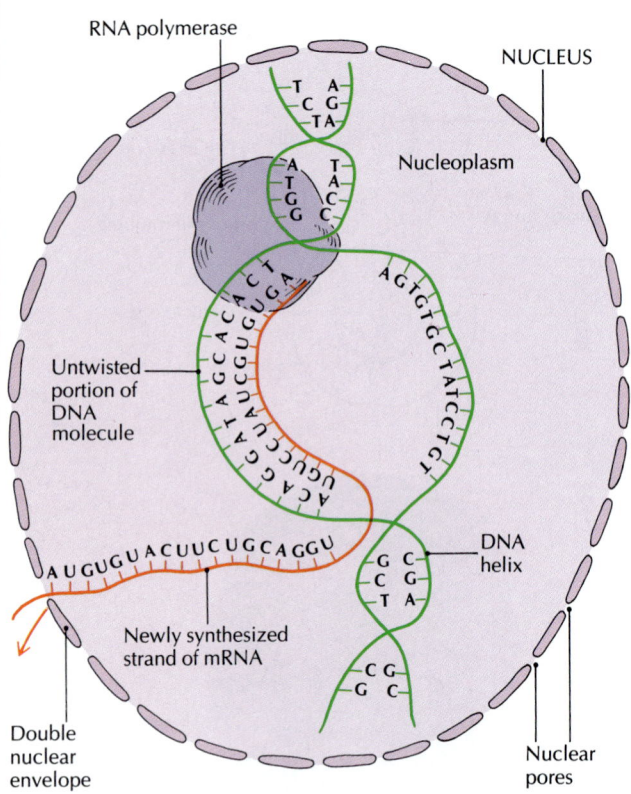

[B] Initiation

(1) A strand of mRNA moves out of the nucleus through a nuclear pore. An initiator tRNA molecule bonds with its amino acid, methionine, and the tRNA anticodon (UAC) binds to the small ribosome subunit.

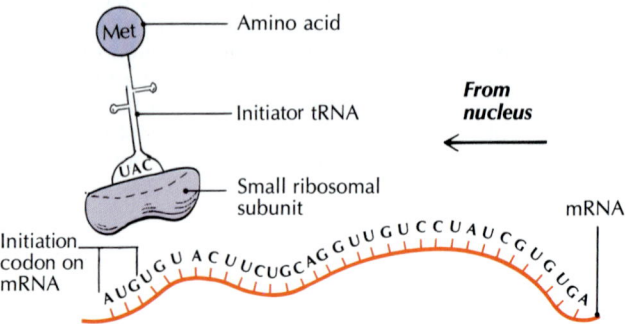

(2) The initiator tRNA–amino acid complex bonds to the complementary mRNA initiation ("start") codon on the mRNA strand. The small ribosomal subunit binds to mRNA.

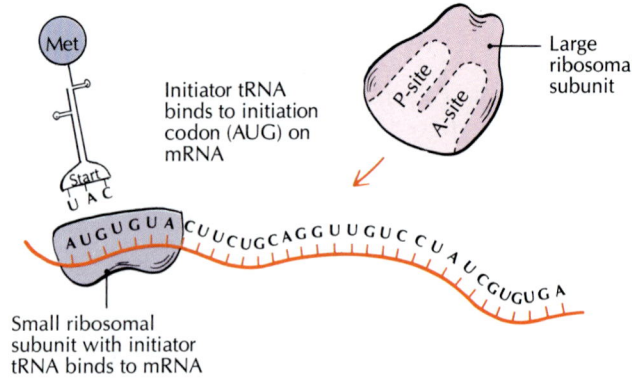

(3) The large ribosomal subunit binds to the small subunit, and the initiator tRNA and its amino acid move into the P-site on the large subunit. Initiation is complete.

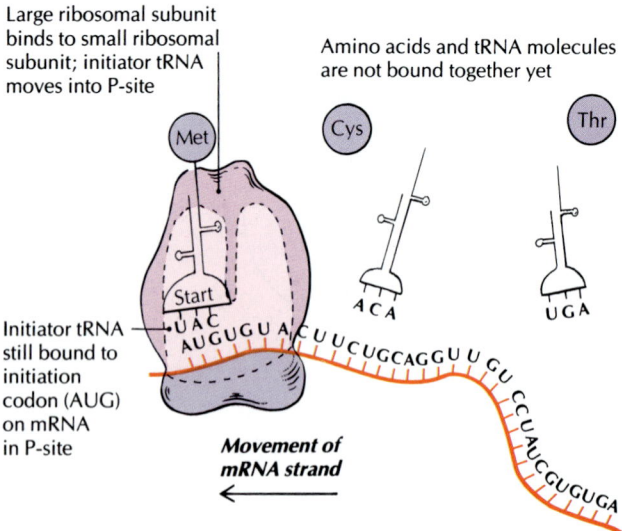

[C] Elongation

(1) Synthesis of a polypeptide begins when an amino acid carried to the ribosome by a tRNA molecule forms a peptide bond with the methionine carried by the initiator tRNA.

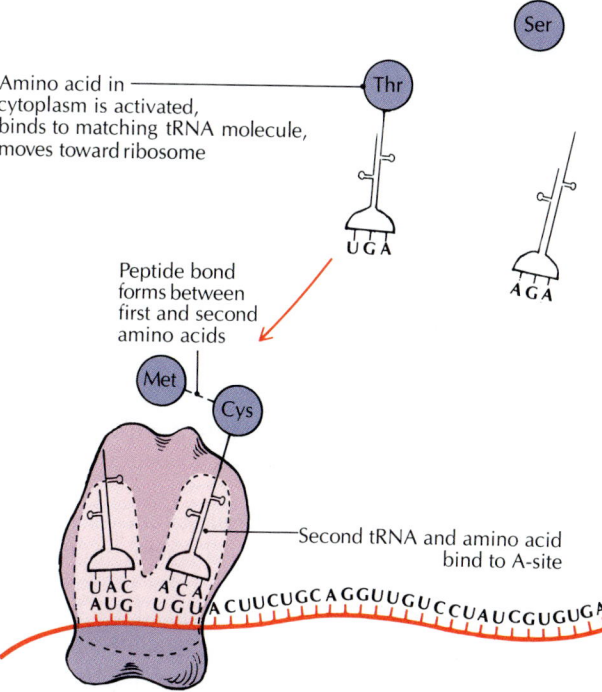

Amino acid in cytoplasm is activated, binds to matching tRNA molecule, moves toward ribosome

Peptide bond forms between first and second amino acids

Second tRNA and amino acid bind to A-site

(2) Amino acids bond together in a sequence determined by the nucleotide sequence of the mRNA.

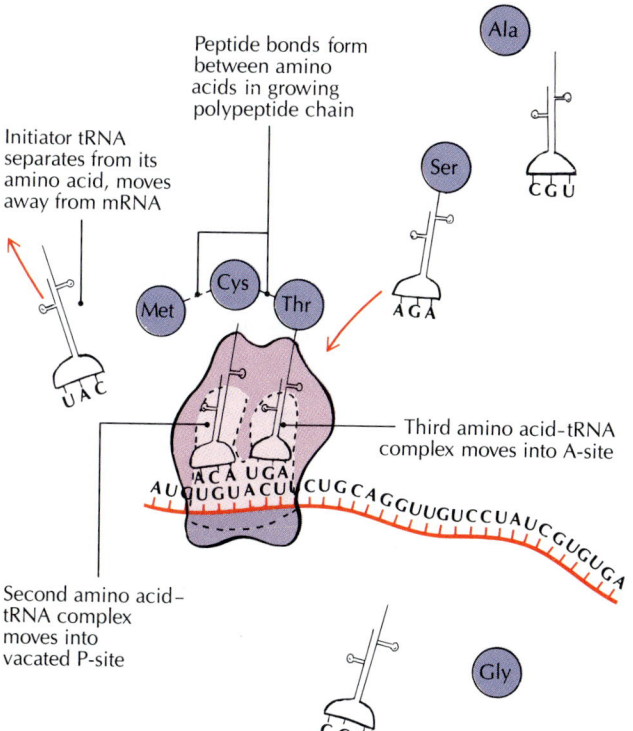

Peptide bonds form between amino acids in growing polypeptide chain

Initiator tRNA separates from its amino acid, moves away from mRNA

Third amino acid–tRNA complex moves into A-site

Second amino acid–tRNA complex moves into vacated P-site

[D] Termination

(1) Amino acids are added to the polypeptide until a "stop" codon reaches the A-site on the ribosome. At that point, a protein called a *release factor* bonds to the stop codon and breaks the bond between the polypeptide chain to the tRNA.

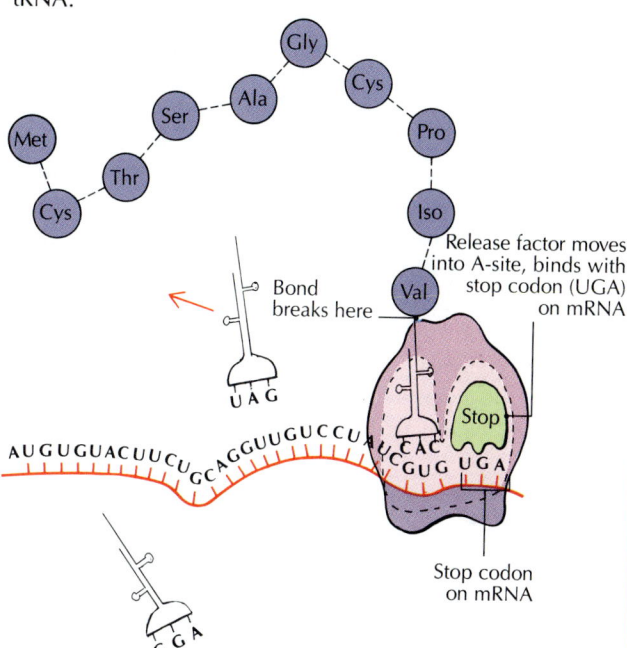

Bond breaks here

Release factor moves into A-site, binds with stop codon (UGA) on mRNA

Stop codon on mRNA

(2) The completed polypeptide is released into the cytoplasm along with the tRNA, the release factor, and the subunits of the ribosome.

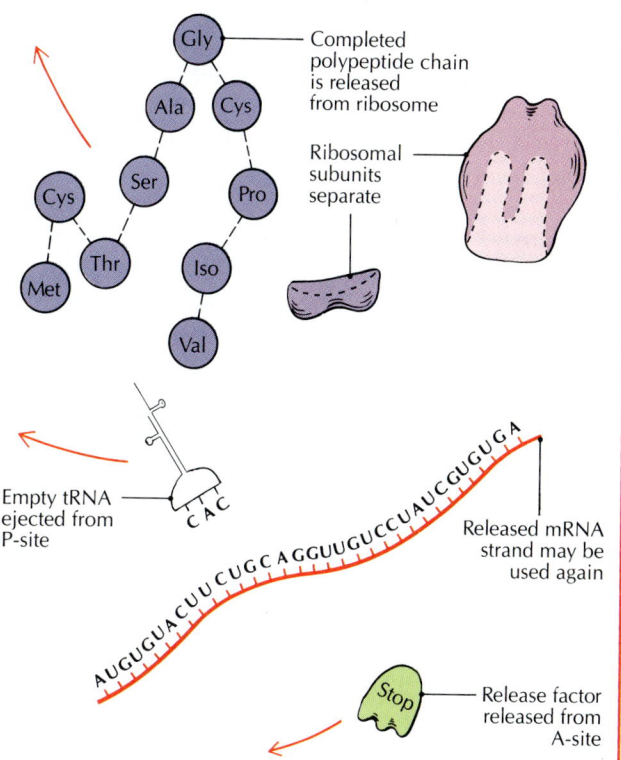

Completed polypeptide chain is released from ribosome

Ribosomal subunits separate

Empty tRNA ejected from P-site

Released mRNA strand may be used again

Release factor released from A-site

Transcription and mRNA

The first step in protein synthesis takes place in the nucleus and is called transcription. **Transcription** is the process by which the hereditary information encoded in the nucleotide sequence of DNA is encoded into the nucleotide sequence of RNA [FIGURE 3.20A]. During transcription, single-stranded RNA is synthesized under the direction of DNA with the aid of the enzyme *RNA polymerase*. The process begins when the two strands in a portion of the DNA double helix uncouple and the helix untwists. The sequence of nucleotide bases from one of the DNA strands serves as a template for the formation of a single-stranded RNA molecule. The RNA forms by complementary base pairing, similar to the base pairing in DNA replication [FIGURE 3.17]. However, in RNA molecules, the base uracil (U) is substituted for thymine (T). The RNA molecule formed from the DNA template is appropriately called **messenger RNA (mRNA)** because it carries the genetic information from the nucleus to the ribosomes, where it directs the synthesis of a protein.

Translation

The transcription of RNA from DNA is the first step in the process of protein synthesis. After a complete mRNA molecule is synthesized, the DNA and mRNA uncouple, and the DNA again assumes its original double helix form. The mRNA strand diffuses out of the nucleus and into the cytoplasm, where the genetic message must be read by a ribosome to produce a polypeptide. (One or more polypeptides make up a protein.) This part of the overall process of protein synthesis—the formation of a polypeptide under the control of mRNA in the cytoplasm—is called **translation**. The process of translation may be divided into three stages: *initiation, elongation,* and *termination.*

Transfer RNA As translation begins, the front end of the mRNA strand attaches to a ribosome and begins to move across it. The ribosome decodes the message by reading the sequence of nucleotide bases as a series of triplets, or **codons,** each coding for a specific amino acid. In addition to mRNA, translation requires another type of RNA found in cytoplasm, **transfer RNA (tRNA)**. Molecules of tRNA pick up amino acids in the cytoplasm and put them in position opposite the appropriate codons of mRNA. Because the nucleotide triplet on the tRNA is complementary to the mRNA codon, it is called an **anticodon**.

The molecules of tRNA are specialized; they provide the link between a specific codon on mRNA and a specific amino acid. There are 20 different kinds of tRNA molecules, one for each kind of amino acid. Activated by an enzyme and energy from ATP, a specific tRNA molecule bonds to one of the 20 amino acids found in cytoplasm. For example, the tRNA specific for the amino acid proline bonds only to that amino acid, the tRNA specific for cysteine bonds only to cysteine, and so on. The mRNA codon for proline (CCU) pairs with the tRNA anticodon for proline (GGA); the codon for cysteine (UGU) pairs with the anticodon for cysteine (ACA).

Initiation The first codon of mRNA is always AUG and is called the *initiation codon*. It codes for the amino acid methionine (Met), which functions as a "start" signal. **Initiation** of protein synthesis actually begins when a special *initiator tRNA* molecule carrying the amino acid methionine binds to the small ribosomal subunit [FIGURE 3.20B(1)]. The initiator tRNA anticodon (UAC), still attached to the small ribosomal subunit, recognizes the AUG initiation ("start") codon on the mRNA molecule and binds to it [FIGURE 3.20B(2)]. Next, the large ribosomal subunit binds to the small subunit, and the initiator tRNA moves into the P-site [FIGURE 3.20B(3); see also FIGURE 3.7B]. Initiation is now complete, and the elongation of a polypeptide chain is about to begin.

Elongation During *elongation,* amino acids are linked together to form an amino acid chain, or polypeptide. This process begins when molecules of tRNA and their specific amino acids become linked by covalent bonds. Each new amino acid–tRNA complex binds to the A-site on the large ribosomal subunit. The incoming amino acid forms a peptide bond with the first amino acid (methionine) as the polypeptide chain is begun [FIGURE 3.20C(1)].

The initiator tRNA moves out of the P-site and separates from its methionine. At the same time, the second tRNA–amino acid complex moves into the P-site, exposing the A-site to the next incoming tRNA and its amino acid. As the mRNA strand continues to move across the ribosome one codon at a time, another tRNA, with its attached amino acid, binds to the A-site [FIGURE 3.20C(2)]. The incoming amino acid links up with the previous one, and the polypeptide lengthens. After each tRNA molecule has released its amino acid, it returns to the cytoplasm where it can pick up another amino acid.

The mRNA strand moves steadily onward, and more links are made between mRNA codons and tRNA anticodons. One after another, amino acids are bound together in an ever-lengthening chain until a complete protein molecule is built up according to the codon order on the mRNA [FIGURE 3.20C(2)].

After the AUG "start" codon on the mRNA has moved well past the ribosome, another ribosome may bond to the strand and initiate the synthesis of a second, similar polypeptide chain. This process may occur repeatedly so that a string of five or more ribosomes is involved, forming a polyribosome [see FIGURE 3.7C]. The ribosomes in a polyribosome operate independently, each synthesizing a polypeptide chain.

Termination The *termination* of the translation process occurs when one of the three mRNA "stop" codons (UAA, UAG, or UGA) reaches the A-site on the ribosome. A protein called a *release factor, not* a tRNA, binds directly to the stop codon, causing hydrolysis of the bond linking the polypeptide chain to the tRNA at the P-site [FIGURE 3.20D(1)].

The completed polypeptide chain is released from the ribosome (the chain shown here is greatly shortened for simplicity), the empty tRNA at the P-site is ejected, the release factor is released from the A-site, and the ribosomal subunits separate and move into the cytoplasmic pool, where they will be available for another cycle of protein synthesis [FIGURE 3.20D(2)].

The polypeptide chain may combine with other polypeptide chains to form more complex protein molecules. The completed protein then performs its job. Some proteins remain inside the cell, and others, such as hormones, leave the cell and exert their effects elsewhere in the body.

A S K   Y O U R S E L F

1 Where in a cell is the genetic information located?

2 Why is it important for RNA to be able to leave a cell's nucleus?

3 What roles in protein synthesis are played by mRNA and tRNA?

4 What is a codon?

CELLS IN TRANSITION: AGING

Much evidence indicates that aging is a normal, genetically programmed process. It is usually not a simple, single process but a series of interrelated cellular events that accumulate until a change becomes noticeable and permanent.

Scientists have speculated that if the three main causes of death in old age—heart disease, cancer, and stroke—were eliminated, the life expectancy of human beings would be extended only 5 to 10 years beyond the present 76. If this is true, there must be a cause of aging and death besides disease. As more people reach old age and as the world's population continues to increase, the study of aging **(gerontology)** becomes more important.

Part of the aging process includes the death of cells, or **apoptosis** (app-uh-TOE-sihss; Gr. "a falling away"), formerly called *necrobiosis* (Gr. *nekros*, corpse, death + *biosis*, way of life). So-called "cell-death genes" control the death of cells under certain conditions. The study of apoptosis may lead to a better understanding of cancer, autoimmune diseases, and AIDS. Apoptosis is a natural death, as opposed to **necrosis**, which is the death of a cell or tissue resulting from irreversible damage caused by disease or accident [FIGURE 3.21].

When the natural hereditary changes of apoptosis occur in the structure and composition of cells and, of course, in the body, the process is called *primary aging, biological aging,* or **senescence** (L. *senex*, old). Changes that take place because of disease or accidental injury (necro-

---

FIGURE 3.21 THREE TYPES OF NECROSIS

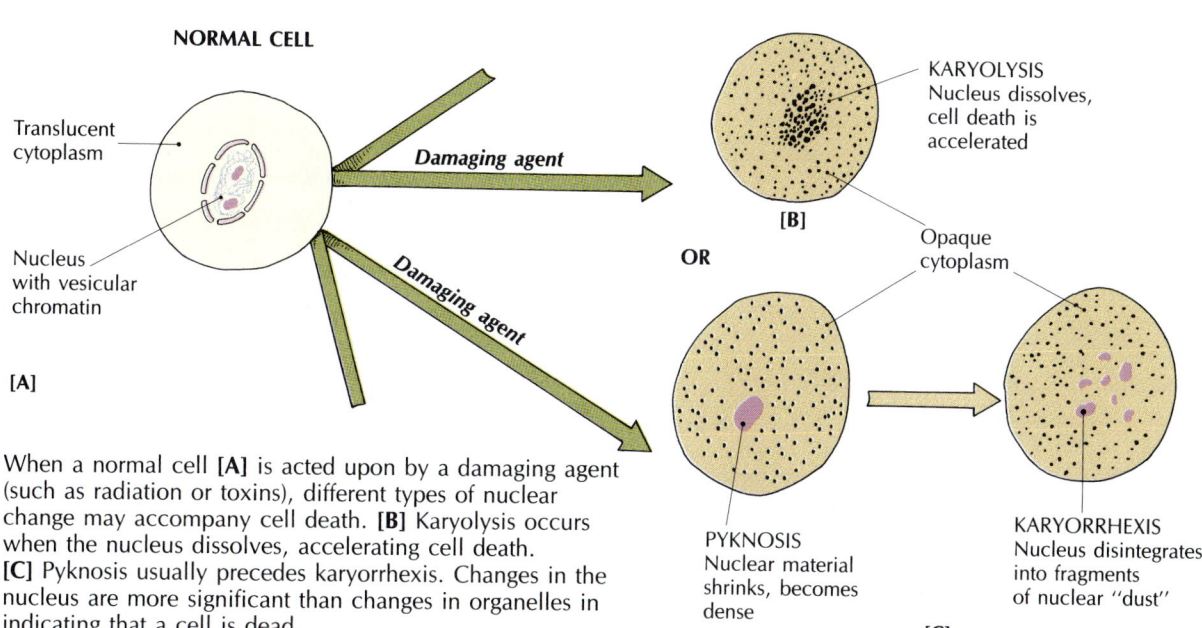

When a normal cell **[A]** is acted upon by a damaging agent (such as radiation or toxins), different types of nuclear change may accompany cell death. **[B]** Karyolysis occurs when the nucleus dissolves, accelerating cell death. **[C]** Pyknosis usually precedes karyorrhexis. Changes in the nucleus are more significant than changes in organelles in indicating that a cell is dead.

NORMAL CELL

Translucent cytoplasm

Nucleus with vesicular chromatin

[A]

Damaging agent

Damaging agent

NECROSIS

KARYOLYSIS
Nucleus dissolves, cell death is accelerated

[B]

OR

Opaque cytoplasm

PYKNOSIS
Nuclear material shrinks, becomes dense

KARYORRHEXIS
Nucleus disintegrates into fragments of nuclear "dust"

[C]

sis) are part of *secondary aging,* or **senility.** Aging is definitely *not* always accompanied by a loss of mental capabilities. In fact, there is usually little if any decline in intellectual abilities and skills unless the brain has been damaged by a stroke or a serious disease.

Hypotheses of Aging

There are several hypotheses that attempt to explain the basic mechanisms of aging. At least some of these mechanisms probably occur simultaneously to produce what we call aging. In other words, there is probably no *single* hypothesis that can explain the aging process.

One of the most important features of the aging process is the increasing **free-radical damage** that occurs within cells [TABLE 3.4]. When cells use oxygen to "burn" their fuel, the by-products include *free radicals* (or *oxidants*). Free radicals carry an unpaired electron, allowing them to oxidize molecules throughout the body, forming potentially harmful chemicals. Metabolic activity in mitochondria produces *superoxide radicals,* which are negatively charged oxygen molecules that normally are converted into hydrogen peroxide by a specific enzyme, superoxide dismutase. Normally, the potentially dangerous hydrogen peroxide is deactivated by another enzyme in the peroxisomes of the cell and converted into

harmless oxygen and water. As cells age, however, the necessary conversion enzymes are not produced fast enough and superoxide radicals and hydrogen peroxide begin to accumulate.

The radicals and hydrogen peroxide are highly reactive and can cause harmful free-radical chain reactions to occur within a cell, producing cross-linking of molecules that impair the normal functions of phospholipids, proteins, nucleic acids, and enzymes.

Another hypothesis of aging concerns the **limits on cell division.** Normal cells of the body are not programmed to grow and reproduce forever, and certain cells are capable of only a predetermined number of divisions. Connective tissue cells from human embryos divide about 50 times before they slow down and stop. Similar cells from a 40-year-old person stop dividing after about 40 divisions, while connective tissue cells from an 80-year-old stop dividing after about 30 doublings [FIGURE 3.22]. The permanent cessation of normal cell division is called **cell senescence.**

Cell growth and division are stimulated by varying combinations of specialized proteins called **growth factors,** such as platelet-derived growth factor (PDGF), fibroblast growth factor (FGF), and nerve growth factor (NGF). Cells that receive insufficient growth factors become inactive after mitosis, causing premature aging.

Occasional **mutations** (changes in DNA structure) occur as the genetic message in DNA is passed along to mRNA. Many of these mutations are corrected during RNA processing, but some are not, and they are expressed in the production of a defective enzyme or other protein. An enzyme defective at its active site may cause cells to operate inefficiently and eventually die.* Fortunately, DNA has a built-in repair mechanism that corrects most of its mutations before they are transferred to mRNA. However, the "repair DNA" itself does not remain totally effective for an indefinite period. Eventually the repair mechanism breaks down, allowing the mutations to accumulate. Cells become inefficient and eventually die, and "aging" occurs.

The *amount* of DNA may also play a role in aging. **Redundant,** or **"extra," DNA** exists in the nucleus of a cell. If a gene is damaged, it can be repaired immediately by an identical sequence of redundant DNA, and no permanent damage is done. After a period of time, however, the redundant DNA is used up and harmful mutations are free to function. Eventually, changes associated with aging appear in the body.

Aging is part of the normal **genetic message.** Normal events such as menopause and graying hair, which are harmless in themselves, indicate that cells are not operating as efficiently as they once were. Decreased cell effi-

---

TABLE 3.4 ANTIOXIDANTS AND THEIR FUNCTIONS

Compounds	Activity
ENZYMES	
Superoxide dismutases	Convert negatively charged superoxide radical (O_2^-) into hydrogen peroxide (H_2O_2).
Glutathione, peroxidases, catalases	Convert hydrogen peroxide into harmless oxygen and water.
OTHER ANTIOXIDANTS	
Vitamin E, beta carotene	React with free radicals, preventing them from attacking cells; because they are fat-soluble they can protect cell membranes.
Uric acid, vitamin C	React with free radicals in cytoplasm.
Metal chelators	Prevent iron, copper, other transition metals from catalyzing oxidation reactions.

Source: Kelvin J. A. Davies, Albany Medical College.

---

*Most mutations are at other than the active site and seldom result in more than a change in one amino acid. Such changes are rarely noticed.

FIGURE 3.22 LIMITS ON CELL DIVISION

[A] Young connective tissue cells are well organized, with distinctive nuclei (dark ovals). ×500. **[B]** Aging connective tissue cells after 50 cell divisions show degenerative changes. ×500.

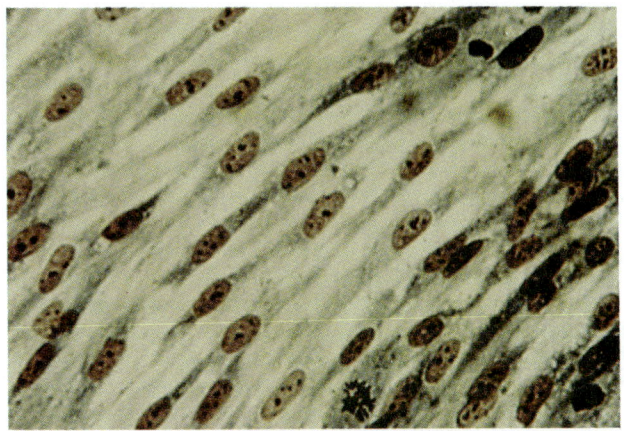

[A]

[B]

ciency causes an increased susceptibility to certain diseases. Aging occurs as the "genetic clock" begins to slow down, and when the clock finally stops, first the cells die and then the body does.

Other factors, such as the accumulation of cellular wastes, hormone imbalances, and cell starvation, probably contribute to the aging process. Also, the immune system may begin to destroy normal cells instead of foreign ones, producing cardiovascular disorders, diabetes, rheumatoid arthritis, or myasthenia gravis (a disease that produces muscular weakness).

The ***thickening of collagen*** seems to be an important factor in aging. The connective tissue collagen (KAHL-uh-juhn; Gr. *kolla*, glue + *genes*, born, produced) is the most common protein in the body. As the body ages, the protein *elastin* in elastic fibers decreases and collagen

thickens, resulting in hardening of the arteries, stiffening of the joints, sagging of the muscles, wrinkling of the skin, and other changes that decrease the body's efficiency. The thickening of collagen may be caused by *cross-linking*, in which large molecules both inside and outside the cell begin to stick together. (Some cross-linking is caused by free radicals.) Tissues and cells eventually become clogged, and normal functions are impaired.

Another process related to cross-linking is ***glycosylation,*** the chemical attachment of glucose to proteins and DNA without the aid of enzymes. It has been suggested that glycosylation adds glucose randomly to growing polypeptide chains, resulting in the formation of cross-links between protein molecules. Glycosylation could explain the cross-linking of proteins that contributes to the nonresilience of aging tissues, including heart muscle.

Effects of Aging on the Body

Elderly people have a lowered resistance to disease, and about 75 percent of people over age 65 have at least one chronic disease. The most common diseases of old age are cardiovascular diseases, cancer, arthritis and similar joint diseases, metabolic disorders, autoimmune diseases, diabetes, and diseases of the nervous system that affect the brain, such as Alzheimer's disease. Many diseases are directly related to arteriosclerosis (hardening of the arteries). Cancer is not unusual among the elderly, and people who live beyond 85 often have one or more types of cancer.

The most lethal diseases of old age are heart disorders, cancer, and cerebral hemorrhage (stroke). Elderly people die as a result of a disease or an accident; "old age" by itself is never a cause of death. TABLE 3.5 illustrates how some body functions are affected by age, and specific age-related disorders of the various systems are described in the relevant chapters.

TABLE 3.5 SOME EFFECTS OF AGING ON BODY FUNCTIONS

Body function	Percentage of function remaining at selected ages			
	Age 25	45	65	85
Maximum heart efficiency	100	94	87	81
Lung capacity	100	82	62	50
Muscle strength	100	90	75	55
Kidney function	100	88	78	69

FIGURE 3.23 DIFFERENCE BETWEEN A BENIGN AND MALIGNANT NEOPLASM: METASTASIS

[A] A benign neoplasm is made up of cells that grow abnormally, but remain enclosed within a sturdy capsule of connective tissue. [B] The cells of a malignant neoplasm burst out of the capsule and can spread to other areas of the body. [C] Cells that break away from a malignant neoplasm may enter the lymphatic system and eventually the bloodstream through small openings in lymphatic and blood vessels. Once in the bloodstream, cancer cells may establish secondary neoplasms far from the primary neoplasm. The success of metastasis depends on characteristics of the host tissue at the new site, as well as the characteristics of the cancer cells.

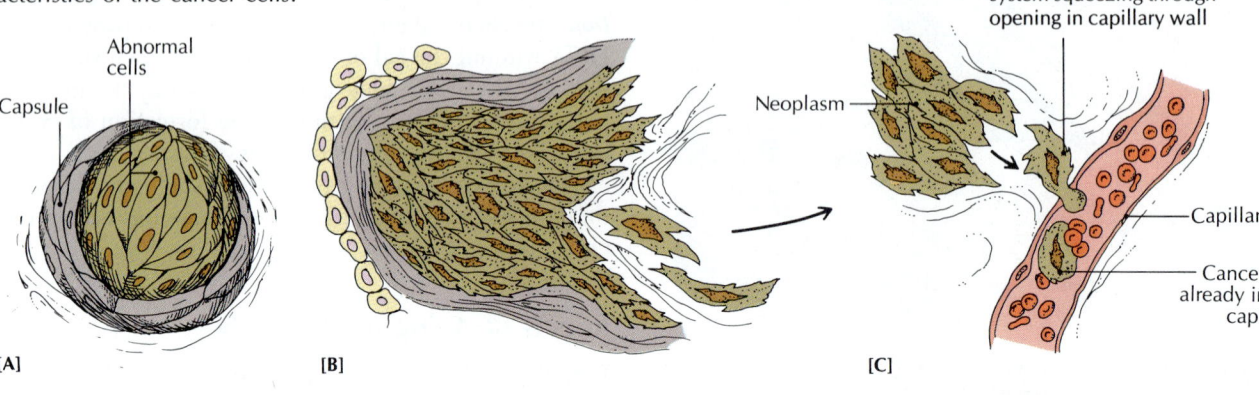

A Final Word about Growing Old

Many people in our society have been conditioned to regard old age as a time when many functions, such as sexual desire, memory, and learning ability, are inevitably decreased or even lost. However, such drastic changes in mental and physical prowess usually result from a serious disease or psychological problems, and they are not necessarily part of the normal aging process. Proper diet, physical activity, and positive mental attitude will overcome many of the routine problems of old age. Sometimes we give up before our bodies do.

A S K   Y O U R S E L F

1 What is the difference between apoptosis and necrosis? Between senescence and senility?

2 What are some of the current hypotheses of aging?

3 What are some of the effects of aging on the body?

CELLS OUT OF CONTROL: CANCER

Normal human body cells usually divide at a rate required to replace the dying ones. Cancer cells are different: They lack the controlling mechanisms that "tell" them when to stop reproducing. *Cancer* (L. crab, because to early anatomists the swollen veins surrounding the af-

fected body part looked like the limbs of a crab) occurs when cells grow and divide at abnormal rates and then spread beyond the original site.

Neoplasms

When cells do not stop reproducing after the normal number of cell divisions, they form an abnormal growth of new tissue called a tumor or *neoplasm* (Gr. *neos*, new + L. *plasma*, form). Neoplasms are either benign or malignant.

The cells of a *benign neoplasm* (L. *bene*, well) do not differ greatly in structure from those in normal tissue. Although these cells grow abnormally, their nuclei divide almost like those of normal cells, and there are usually few chromosomal changes. A benign neoplasm remains safely enclosed within a capsule of thick connective tissue and does not spread beyond its original site [FIGURE 3.23A]. A benign neoplasm generally can be located rather easily and removed surgically or treated with another appropriate procedure without further problems. However, if they are not removed in time, benign neoplasms may become large enough to press on nerves or blood vessels, and they occasionally become malignant.

The most important characteristic of a typical *malignant neoplasm* (L. *malus*, bad) is that its cells tend to break out of their connective tissue capsule and invade neighboring tissue [FIGURE 3.23B, C]. Malignant neoplasms, or cancers, grow rapidly in an uncontrollable and disorderly pattern that can be recognized by obvious

FIGURE 3.24 CHANGES IN CELL AND TISSUE STRUCTURE IN MALIGNANT NEOPLASMS

[A] Normal cells. ×500. [B] Cancerous cells. ×500. [C] Normal bronchial tissue. ×4000.
[D] Cancerous bronchial tissue. ×3000.

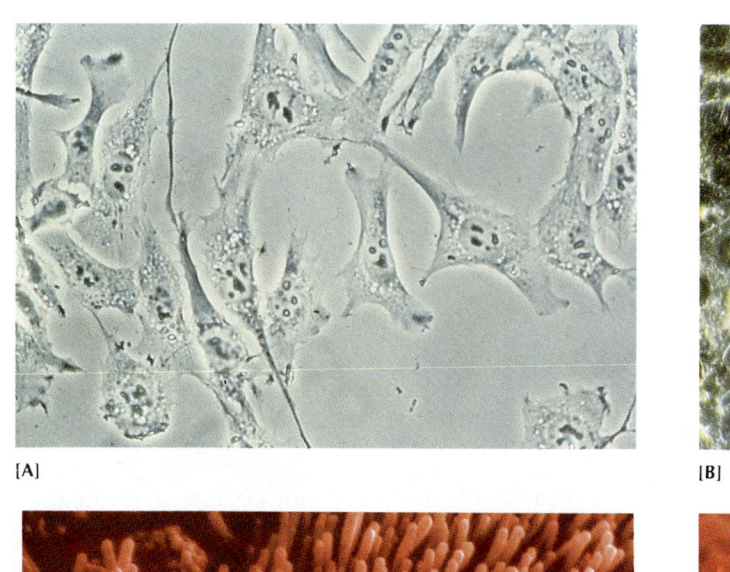

[A]

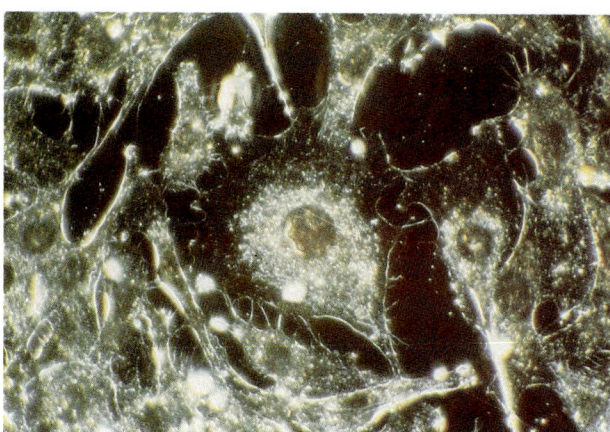

[B]

Cancer cells

[C]

[D]   Normal
ciliated epithelium

changes in cell and tissue structures [FIGURE 3.24] and in the chromosomes.

Malignant neoplasms are usually classified according to the type of body tissue in which they originate:

1 A *carcinoma* (L. cancerous ulcer; *oma*, tumor) originates in *epithelial tissues*. Carcinomas are the most common type of cancer, occurring most often in the skin, lungs, breasts, intestines, stomach, mouth, and uterus. They are usually spread by way of the lymphatic system.

2 A *sarcoma* (Gr. *sarkoun*, to make fleshy) originates in *muscle cells* or *connective tissue*. Sarcomas, which may form anywhere in the body, are usually spread to other tissues through the bloodstream.

3 A *mixed-tissue neoplasm* derives from tissue that is capable of differentiating into either epithelial or connective tissue; as a result, it is composed of several types of cells.

FIGURE 3.25 THE USUAL PROGRESSION OF CANCER CELLS

Cancer of the uterine cervix, showing the progression from dysplasia to a full-blown malignant carcinoma, where cancerous epithelial cells invade the underlying connective tissue and begin to spread.

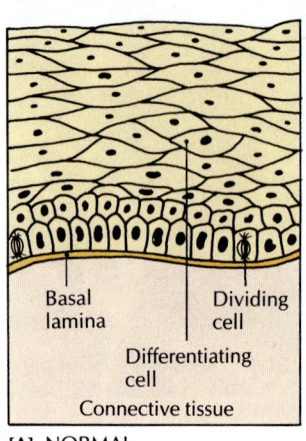
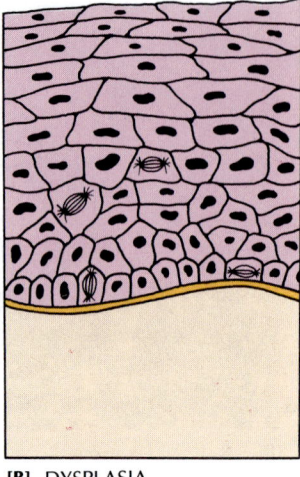
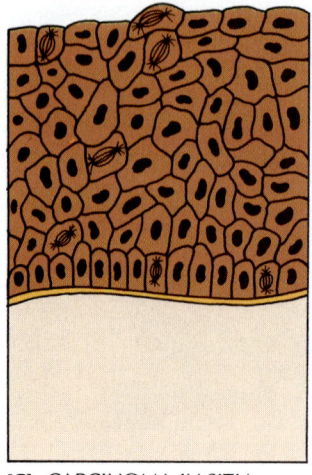
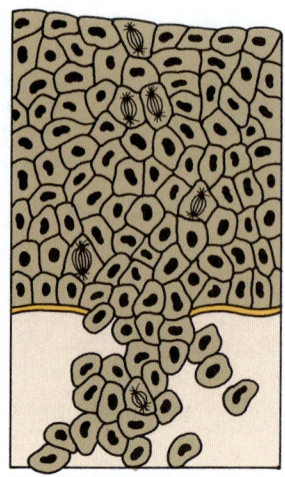

[A] NORMAL   [B] DYSPLASIA   [C] CARCINOMA *IN SITU*   [D] MALIGNANT CARCINOMA

4 Cancers that do not fit into these three categories include the various *leukemias* derived from tissues that produce certain abnormal blood cells and cancers that arise from cells of the nervous system.

How Cancer Spreads

The spread of a malignant neoplasm is called *metastasis* (muh-TASS-tuh-sihss; Gr. *meta,* involving change + *stasis,* state of standing) [FIGURE 3.23]. Metastasis begins when cancer cells break away from the original site, or *primary neoplasm,* and extend into normal tissues. Once cancer cells move from the original site, they may enter a body cavity, such as the thoracic or abdominal cavity, or may enter the bloodstream or the lymphatic system and be transported away from the primary neoplasm. They can then establish new malignant *secondary neoplasms* elsewhere in the body. Most deaths caused by cancer result from secondary neoplasms, not the original primary neoplasm.

As cancer cells spread, or *metastasize,* they encounter defensive cells of the body's immune system and are sometimes killed. Cancer is lethal when the body can no longer fight off the ever-increasing number of invading malignant cells that disrupt and starve healthy cells.

Cancers usually progress slowly. In epithelial tissues [FIGURE 3.25], for example, the first stage is usually *dysplasia,* a condition in which dividing epithelial cells are not restricted to the basal layers and show some abnormal differentiation [FIGURE 3.25B]. Dysplasia in cells of the uterine cervix can usually be detected with a "Pap smear." *Carcinoma in situ* (L. *"cancer in place"*) is an abnormal condition in which apparently undifferentiated dysplastic epithelial cells proliferate in all tissue layers [FIGURE 3.25C]. The final stage in the progression of epithelial cancer reveals a *malignant carcinoma* [FIGURE 3.25D], where cancerous epithelial cells break through the basal lamina to invade the underlying connective tissue.

Causes of Cancer

Although research scientists cannot yet say they have found *the* cause of cancer, they can say with certainty that there is no *single* cause of cancer. Some of the most evident causes are described in the following paragraphs.

In some cases the *tendency* to develop cancer may be inherited, but cancer researchers believe that most cancers are caused by repeated exposures to cancer-causing agents called *carcinogens* (kar-SIHN-uh-jehnz). The main categories of carcinogens are *chemicals, radiation,* and *viruses.* Carcinogens disrupt the normal metabolic functions of cells and may eventually cause them to become cancerous. Most cancers probably result from prolonged exposure to a *combination* of several carcinogens.

Several hypotheses concerning the development of cancer have been proposed. One hypothesis suggests that carcinogens cause genetic changes, or *mutations,* in DNA. The current thinking is that certain genes help prevent cancer until they are made ineffective by a mutation, or that an extensive rearrangement of genetic material caused by radiation, chemicals, or viruses produces effective mutations. A single gene, designated as *p53,* may be a common factor in initiating mutations of other

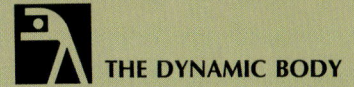

"Mick" and "Max": The Keys to Cellular Growth and Death?

Two of the most important questions in biology are these: How do cells start and stop dividing? and What causes some cells to divide uncontrollably? Cell biologists are tantalizingly close to finding the answers.

A gene named **myc*** (pronounced MICK) and its enzyme partner, **max,** are excessively abundant in cancerous tissue. The myc gene has been known for at least 20 years, but now research is accelerating rapidly, enabling cell biologists to come closer to answering the fundamental questions of how cellular division starts and stops and how it sometimes gets out of control.

If myc genes can be isolated and destroyed at an early stage—and there has already been some success in that area—there is a good chance that some cancers can be arrested. But the myc gene is

* myc got its name when it was first identified as a promoter of myeloma (bone marrow cancer) in chickens.

not just a probable cancer-causing factor; it also appears to be a necessary factor in normal cell division. myc seems to be the source of the signal for a cell to start dividing, and its signals to the dividing cell must be halted before a cell can mature. Also, when an abnormal division is taking place, myc can cause the cell to burst and die before the division proceeds to a stage where a defective cell might be produced.

But before the myc gene can be effective in any of these functions, it must bind in a zipper arrangement with max. (myc and max probably bind in a "parallel zippering," rather than in an alternating arrangement as in a clothing zipper.) Once the gene and enzyme are linked, the gene can direct a cell to divide, mature, or self-destruct.

A multifunctional oncogene (a gene that promotes cancer) called **bcl-2** is activated by a genetic mutation, counteracting myc's signal for cell death, keep-

ing the cells alive and allowing them to develop into tumors. (The bcl-2 gene is named after a cancer cell known as B-cell lymphoma, where it was first found.) The bcl-2 gene also increases the resistance of cancer cells to chemotherapy.

Apparently, among its other functions, bcl-2 creates a "memory" among immune-system cells that allows them to "remember" harmful agents that have attacked the body at an earlier time, destroying the agents and preventing infection. However, if the bcl-2 gene is switched on in error, it may direct cells to attack the body's own tissues, creating an autoimmune disease.

If a drug can be found that reduces the action of bcl-2, cancer cells may become more sensitive to chemotherapy, but researchers also hope to someday use bcl-2 to prevent nerve cells from dying in degenerative diseases such as Alzheimer's.

genes. New research, mostly with mice, has shown that $p53$ is a tumor-suppressor gene. Apparently, it acts to stop cell growth whenever DNA is damaged and its synthesis becomes abnormal. Researchers speculate that $p53$ responds, not to genetic damage, but to shortages of certain chemicals necessary for gene function and replication. If $p53$ is inoperative or absent for some reason, tumor cells may divide unchecked. Although a disabled $p53$ gene cannot be repaired in the laboratory as yet, it can be detected, allowing physicians to diagnose and treat a previously hidden malignancy.

Although $p53$ does not seem essential during early embryonic development, another tumor-suppressor gene, Rb, does appear to encourage normal early development. Without Rb, it is likely that proper cell differentiation will not occur, and fatal brain tumors may develop shortly after birth. People with only one of the usual pair of Rb genes often develop retinoblastoma, a tumor of the retina in the eye, while people who inherit only one of the usual two $p53$ genes have a predisposition toward cancer of the breast and other tissues.

Another hypothesis suggests that the forerunner of a cancer-causing gene—a *proto-oncogene* (Gr. *onkos,* tumor)—is a gene that helps regulate normal cell division. But when a proto-oncogene undergoes mutation, it can be transformed into an **oncogene,** which can transform a normal cell into a cancer cell. More than 50 proto-oncogenes have been identified.

High-energy radiation, especially gamma rays or x-rays, can cause severe cellular and tissue damage, or even necrosis. A large dose of high-energy radiation hydrolyzes the water in cytoplasm, producing harmful hydroxyl (OH) radicals and acute cell death. Lower doses of radiation can damage the DNA in the cell nucleus, producing cell death or abnormalities when the DNA fails to replicate or genetic mutations occur.

Considerable evidence supports the idea that at least some types of cancer are caused by **viruses.** Viruses are involved in Burkitt's lymphoma (Epstein-Barr virus), warts (benign papillomas of the skin), cervical cancer (herpes type II and certain papilloma viruses), and liver cancer (hepatitis B virus).

Testing for Cancer

Several new blood tests, called *tumor marker tests* (see table), are available to detect tumors while they are still very small and relatively harmless. Almost all these tests have a tendency to be overly sensitive, and alternative methods of testing, after a discussion with the attending physician, are usually recommended before surgery or expensive biopsies are performed. Although these tests may be flawed, they can be valuable early indicators of cancer if used and interpreted correctly.

TUMOR MARKER TESTS

Test and type of cancer	Screens for
AFP (testicular)	Alpha-fetoprotein, which occurs in elevated blood levels in 89% of testicular cancer patients.
CA 15-3 (breast)	Elevated blood levels of the protein CA 15-3, which are found in about 20% of patients with early breast cancer.
CA 19-9 (pancreatic)	Elevated blood levels of the protein CA 19-9, which are found in pancreatic cancer patients and are also a sign of pancreatic inflammation and liver disease.
CA 125 (ovarian)	Elevated blood levels of the protein CA 125, which are found in about 50% of women with ovarian cancer.
CEA (colon, possibly breast and lung)	Elevated blood levels of the protein CEA, which are found in about 28% of patients with early colon/rectal cancer.

Test and type of cancer	Screens for
HCG (testicular)	Decreasing blood levels of the hormone human chorionic gonadotropin, which indicate that chemotherapy is effective.
PSA (prostate)	Blood levels of a protein secreted in large amounts by enlarged and cancerous prostate glands.

GENETIC TESTS UNDER DEVELOPMENT

Test and type of cancer	Screens
Chromosomal abnormalities 3, 3p, 17, 13, 5q (lungs)	Sputum for early genetic mutations that precede lung cancer.
Her-2/neu (breast)	Blood for excessive copies of a gene found in 25–30% of breast cancer patients.
p53 (urinary bladder)	Urine for mutated p53 genes found in shed urinary bladder cells, a precursor condition of cancer.
Ras (colon)	Stool for mutant ras oncogenes linked to colon cancer and found in shed colon cells.

Cancer and the Immune System

Recent evidence suggests that cancer cells may release into the bloodstream a powerful chemical factor that suppresses defensive cells of the body's immune system. Without an efficient immune system to counteract them, cancer cells are then free to multiply and metastasize. Researchers are trying to isolate the factor released by tumors and to find out exactly how it works.

Some Common Cancer Sites

Cancer can strike anywhere in the body, but it usually occurs most often in sites where cell division is rapid and frequent, such as the lining of the digestive and respiratory tracts. The most common cancer sites in men are the lungs, the prostate gland, and the large intestine and rectum; the most common sites in women are the breasts, the large intestine and rectum, the uterus, and the lungs.

Treatment of Cancer

Most of the traditional methods used to treat cancer are not new. *Surgery* was first used in the second century A.D. to treat breast cancer; drug therapy *(chemotherapy)* was first used in 1941 to treat prostate cancer; and *radiation,* using cobalt-60, was introduced in 1952. *Combination therapy* uses more than one type of therapy to treat cancer. Surgery and radiation therapy have been used together for some time, and physicians are now including follow-up chemotherapy, especially for cancer of the breast, bones, large intestine, lungs, and stomach.

TO YOUR HEALTH

Cancer Treatment from Trees?

Will a new cancer treatment cause the extinction of the Pacific yew tree?

On December 29, 1992, the Food and Drug Administration (FDA) approved the use of Taxol, a new drug for the treatment of advanced ovarian cancer—and perhaps other types of cancer. What has made this approval noteworthy is that Taxol is derived from the bark of the rare Pacific yew tree. Because it takes about 60 pounds of the tree's bark to make enough Taxol to treat one patient, environmentalists are concerned for the future of the tree. (A similar European drug, Taxotere, is derived from the European yew tree, which is not threatened with extinction.)

Apparently, this is the first time that the FDA approval of a drug has had environmental ramifications—and the first time that an "environmental impact statement" was required for a new drug. Approval by the U.S. Forest Service and the Bureau of Land Management accompanied the FDA approval. The Environmental Defense Fund also supported the approval of Taxol but urged the timber industry to stop other logging of the yew tree.

The company that manufactures Taxol is attempting to accommodate the concerns. It has developed and begun the use of alternative sources to supplement the use of the yew bark. By 1995 the company expects to be using these alternative sources exclusively and has agreed to a reforestation program for the Pacific yew tree.

Now the growing knowledge of molecular biology is offering the possibility of new techniques that are much more specific and sophisticated than before. For example, experimental chemotherapy uses at least five different ways of combatting cancer cells: (1) It *blocks receptor sites* on the surface of cancer cells, stopping the flow of chemicals from the blood that are needed to stimulate the growth of cancer cells. (2) It stops the spread of tumors by *destroying their enzymatic ability* to penetrate body tissues. (3) It isolates cancer cells and *immobilizes them chemically.* (4) It stops the growth of a tumor by *preventing the growth of blood vessels* (angiogenesis) that branch into a tumor and carry nutrients to it. (5) It uses *monoclonal antibodies* (chemicals produced to combat the introduction of foreign protein into the body) against specific cancer cells. Potentially lethal cancer drugs or toxic molecules can be attached to these antibodies, which are then injected into the patient. The antibodies, with their "passenger" drugs, recognize only cancer cells, destroying them while doing little or no harm to normal body cells.

Although new forms of treatment are constantly being tested, nearly 75 million Americans will get cancer in their lifetimes, and about 500,000 will die of cancer this year (one death every minute). Next to cardiovascular diseases, cancer is the leading killer of adults.

Q: *How soon after exposure to a carcinogen does cancer appear?*

A: If cancer does occur (not all exposures to a carcinogen produce cancer), it usually develops slowly. Cancers of the liver, lungs, and urinary bladder may not appear until 30 years after exposure to vinyl chloride, asbestos, or benzidine. Cigarette smokers usually develop lung cancer about 20 years after they start smoking.

A S K   Y O U R S E L F

1 What is the difference between a benign neoplasm and a malignant neoplasm?

2 What is a carcinoma? A sarcoma?

3 What is the difference between a primary neoplasm and a secondary neoplasm?

4 What are carcinogens, and what are their main categories?

5 What is the "oncogene" hypothesis?

6 What are the main types of cancer treatment?

WHEN THINGS GO WRONG

Progeria

Adult progeria (L. *pro*, before; Gr. *geras*, old age) (Werner's syndrome) is a relatively rare disease in which a person grows old prematurely due to an abnormally small number of cell divisions. Fibroblasts taken from patients are unresponsive to PDGF and FGF, although they do respond to other growth factors. Adult progeria usually begins during early adulthood, progresses rapidly, and causes death before the age of 50. Its cause is unknown. Victims of progeria have abnormally high lev-els of carbonyl groups, which are by-products of protein oxidation. *Juvenile progeria* (Hutchinson-Gilford syndrome) is a form of premature old age that afflicts young children. It usually begins about age 4, and old-age symptoms (including gray hair and baldness; loss of fat, causing thin limbs and sagging skin; atherosclerosis — fatty deposits on the inner walls of arteries; and other internal degenerative changes) are fully developed by the age of 10 or 12. Death usually occurs around puberty.

CHAPTER SUMMARY

What Are Cells? (p. 57)

1 The *cell theory* states that (a) all living organisms are composed of cells and cell products, (b) the cell is the basic unit of structure and function in all living organisms, (c) all cells come from the division of preexisting cells, and (d) an organism as a whole can be understood through the collective activities and interactions of its cellular units.

2 Cells can range widely in size, color, shape, and function.

3 All cells have four basic parts: the *nucleus, nucleoplasm, cytoplasm,* and *plasma membrane.*

4 The part of a cell that is not cytoplasm is the *nucleus,* the control center of the cell. The nucleus contains the chromosomes that direct the reproduction of, and contain the genetic blueprint for, new cells. The nucleus is surrounded by a nuclear envelope and contains *nucleoplasm.*

5 The portion of the cell surrounding the nucleus is *cytoplasm.* The fluid portion of the cytoplasm is called *cytosol;* it contains various subcellular structures called *organelles* and *inclusions.*

6 The *plasma membrane* forms the outer boundary of the cell.

Cell Membranes (p. 57)

1 Cell membranes are selective screens that allow certain substances to get into and out of cells. That quality is called *semipermeability.*

2 Cell membranes are composed mainly of phospholipids and proteins and form a dynamic *fluid-mosaic model.*

Passive Movement across Membranes (p. 62)

1 *Passive movement* occurs when molecules use their own energy to pass through a cell membrane from areas of higher concentration to areas of lower concentration.

2 Examples of passive movement are *simple diffusion, facilitated diffusion, osmosis,* and *filtration.*

Active Movement across Membranes (p. 65)

1 Active processes of movement across membranes require energy from the cell and can move substances through a semipermeable membrane from areas of lower concentration to areas of higher concentration.

2 Active processes include *active transport, endocytosis,* and *exocytosis.* Endocytosis and exocytosis are examples of *bulk transport,* the movement of large molecules or particles across membranes.

Cytoplasm (p. 67)

The cytoplasm of a cell is composed of two parts, or phases. The *particulate phase* consists of vesicles and organelles, lipid droplets, and inclusions. The *aqueous phase* consists of fluid cytosol. The typical components of cytosol are water, nucleic acids, proteins, carbohydrates, lipids, and inorganic substances.

Organelles (p. 67)

1 The *endoplasmic reticulum (ER)* creates a series of channels for transport, stores enzymes and other proteins, and provides a point of attachment for ribosomes. Smooth ER does not contain ribosomes; rough ER does.

2 *Ribosomes* are the sites of protein synthesis.

3 The *Golgi apparatus* aids in the synthesis of glycoproteins. It also aids in the secretion of these products.

4 *Lysosomes* digest nutrients and clean away dead or damaged cell parts.

5 *Mitochondria* produce most of the energy (in the form of ATP) required by a cell for metabolic activities.

6 *Microtubules, intermediate filaments,* and *microfilaments* (the *cytoskeleton*) provide a transport system and a supportive framework and assist with organelle and chromosome movement.

7 *Centrioles* assist in cell reproduction and are involved with the movement of chromosomes during cell division.

8 *Cilia* are appendages that help move solids and liquids past fixed cells, and *flagella* help sperm to move.

9 *Cytoplasmic inclusions* are chemical substances that usually consist of either basic food material or the stored products of the cell's metabolic activities.

The Nucleus (p. 74)

1 The cell *nucleus* contains DNA, which controls the cell's heredity, protein structure, and other metabolic activities.

2 The *nuclear envelope* contains many *nuclear pores* that allow material to enter and leave the nucleus.

3 Within the nucleus are **chromosomes** composed of DNA and protein. *Genes* are segments of DNA that contain the hereditary information for protein synthesis.

4 The *nucleolus* is a preassembly point for ribosomes.

Cellular Metabolism (p. 77)

1 The chemical processes that transform food into living tissue and energy are part of the body's *metabolism,* which includes all the chemical reactions in a living organism.

2 Metabolism produces enough ATP to provide the energy needs of a cell.

The Cell Cycle (p. 77)

1 The *cell cycle* results in the reproduction of a cell. Its two main phases are interphase and the M phase, which includes mitosis and cytokinesis.

2 The process in which cells become specialized in structure and function is called **differentiation.**

3 *Replication* is the process by which DNA makes an exact copy of itself.

4 *Mitosis* distributes all nuclear material from the parent cell to the daughter cells equally. It is divided into four sequential stages: *prophase, metaphase, anaphase,* and *telophase.*

5 *Cytokinesis* is the division of the cytoplasm that produces two daughter cells after the nucleus has divided by mitosis.

How Proteins Are Synthesized (p. 80)

1 The many proteins that the body needs are produced from amino acids, which are assembled according to each person's genetic program contained in the DNA.

2 *Messenger RNA (mRNA)* carries the genetic message of DNA out of the nucleus to the ribosomes in the cytoplasm. Molecules of *transfer RNA (tRNA)* carry specific amino acids to ribosomes, where protein synthesis occurs.

3 Protein synthesis occurs in two major steps: transcription and translation.

4 The chromosomes in the nucleus contain DNA, whose sequence of nucleotides codes for the synthesis of specific proteins. The message is transcribed from one of the strands of the DNA to a single-stranded mRNA molecule. This is *transcription.* The DNA molecule serves as a template, or mold, so that the mRNA strand is complementary to the DNA strand.

5 The mRNA carries the message out of the nucleus through nuclear pores to the ribosomes in the cytoplasm. The formation of a polypeptide from mRNA in the cytoplasm is *translation.* The mRNA attaches itself to a ribosome. *Initiation* begins when the *anticodon* (a specific tRNA triplet of nucleotide bases on an *initiator tRNA* carrying the amino acid methionine) binds to the *codon* (the mRNA triplet that codes for methionine) and thereby signals "start." During *elongation,* the polypeptide chain lengthens as amino acids are carried to the ribosome by their matching tRNA, and each tRNA binds to its appropriate codon triplet on the mRNA strand. Amino acids are bonded together in a growing chain until a protein molecule is built up according to the codon order on the mRNA molecule. The appearance of a "stop" codon signals the end of the polypeptide chain and releases the chain from the ribosome. This is *termination.*

Cells in Transition: Aging (p. 87)

1 Part of the aging process includes the natural death of cells, or *apoptosis.*

2 *Senescence,* or *primary aging,* involves the natural hereditary changes in the structure and composition of cells that occur with age. *Senility,* or *secondary aging,* involves changes caused by disease or accidental injury.

3 Aging is a series of interrelated events that accumulate until a change becomes noticeable and permanent.

4 No single theory can explain aging. Among the mechanisms that produce aging are free-radical damage, genetically determined limit on cell division, DNA mutations, loss of redundant DNA, genetically determined decreased cell efficiency, thickening of collagen, and glycosylation.

5 The most lethal diseases of old age are cardiovascular diseases, cancer, and stroke.

6 Every body system is affected adversely in some way by the aging process. Among the most common age-related ailments are cardiovascular diseases, cancer, arthritis and other joint diseases, metabolic disorders, and diseases of the central nervous system.

Cells Out of Control: Cancer (p. 90)

1 When cells divide more frequently than normal and do not die after the typical number of divisions, they form an abnormal growth of tissue called a **neoplasm.**

2 The cells of a **benign neoplasm** grow abnormally but do not differ significantly in structure from those in normal tissue and do not spread beyond their original site. The cells in a **malignant neoplasm,** or **cancer,** have the ability to invade surrounding tissues.

3 Malignant neoplasms are usually classified according to the type of body tissue in which they originate. **Carcinomas** originate in epithelial tissue, **sarcomas** in connective tissue, and **mixed-tissue neoplasms** in tissue that is capable of differentiating into either epithelial or connective tissue.

4 The spread of cancer cells beyond the primary neoplasm and the establishment of secondary neoplasms in other parts of the body is called **metastasis.**

5 The main categories of **carcinogens** (cancer-causing agents) are *chemicals, radiation,* and *viruses.* It is thought that most cancers are caused by repeated and prolonged exposures to a *combination* of several carcinogens.

6 Hypotheses that explain the basic cause of cancer involve DNA mutations, oncogenes (cancer genes), and viruses.

7 The most common cancer sites are lungs, colon and rectum, breasts, prostate gland, uterus, and urinary bladder.

8 The main treatments for cancer are **surgery, radiation therapy,** and **chemotherapy. Combination therapy** uses more than one type of therapy to treat cancer.

9 Future cancer treatment may involve the use of **monoclonal antibodies.**

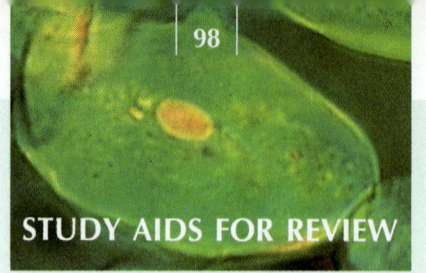

STUDY AIDS FOR REVIEW

KEY TERMS

MEDICAL TERMS FOR REFERENCE

ADENOMA (Gr. *aden,* gland + *-oma,* tumor) An epithelial tumor of glandular origin, usually benign or of low-grade malignancy.

ANAPLASIA (Gr. *ana,* reversion + *plasia,* growth) Reversion of cells to a less differentiated form.

BIOPSY (Gr. *bios,* life + *-opsy,* examination) The microscopic examination of tissue removed from the body.

GERIATRICS (Gr. *geras,* old age + *iatrikos,* physician, pertaining to a specific kind of medical treatment) The study of the physiology and ailments of old age.

HYPERPLASIA (Gr. *hyper,* above + *plasia,* growth) Enlargement of an organ or tissue as a result of an increase in the number of cells.

HYPERTROPHY (Gr. *hyper,* above + *trophe,* nourishment) Enlargement of an organ or tissue as a result of an increase in the size of cells.

MELANOMA (Gr. *melas,* black + *-oma,* tumor) A neoplasm made up of cells containing a dark pigment, usually melanin.

METAPLASIA (Gr. *meta,* involving change + *plasia,* growth) Change of cells from a normal to an abnormal state.

ONCOLOGY (Gr. *onkos,* mass, tumor + *logy,* the study of) The study of neoplasms; the study of cancer.

UNDERSTANDING THE FACTS

1 Describe the structure of the plasma membrane.

2 Distinguish between passive and active transport.

3 Which organelles function in endocytosis and exocytosis?

4 How does the endoplasmic reticulum differ from the Golgi apparatus?

5 Why are lysosomes and peroxisomes vital to the health of a cell?

6 Describe the structure and function of mitochondria.

7 What are the components of the cytoskeleton?

8 Which stages comprise the cell cycle, and what occurs during each stage?

9 Outline the major theories that attempt to explain the causes of aging.

10 What are malignant neoplasms, and how are they classified?

UNDERSTANDING THE CONCEPTS

1 What characteristics make the plasma membrane a semipermeable structure?

2 What roles are played by the proteins in the various cell membranes?

3 Contrast the functions of microtubules in the dividing cell and the interphase cell.

4 Explain the link between the DNA of a cell's genes and the physiological functioning of that cell.

SELF-QUIZ
Multiple-Choice Questions

3.1 Which of the following is *not* made of microtubules?

  a centrioles   d cilia

  b cytoskeleton   e flagella

  c microvilli

3.2 What distinguishes cell inclusions from organelles?

  a Inclusions are membrane-bound; organelles are not.

  b Organelles are membrane-bound; inclusions are not.

  c Inclusions contain proteins; organelles do not.

  d Organelles are generally more permanent structures within the cell than are inclusions.

  e Inclusions always contain pigment granules; only some organelles do.

3.3 The diffusion of water through a semipermeable membrane is called

  a osmosis   d endocytosis

  b facilitated transport   e receptor-mediated

  c facilitated diffusion   transport

3.4 Which of the following statements about the nuclear membrane is correct?

a The nuclear membrane is identical to the plasma membrane.

b Small molecules that can pass through the channels in the plasma membrane are too large to pass through the nuclear pores.

c Larger molecules can pass between the nucleus and the cytosol than can traverse the plasma membrane.

d The nuclear membrane is composed of two layers of phospholipid and protein.

e During interphase the nuclear membrane consists of four layers of phospholipids.

3.5 Within a cell, the flexible lattice that supports organelles and is involved in movement is called the

a endoplasmic reticulum   **d** aster
b cytoskeleton   **e** spindle
c Golgi apparatus

3.6 Cell division in body cells may be thought of as occurring in two stages:

a the G_1 and the G_2 phases
b anaphase and telophase
c the S and G_1 phases
d nuclear division and chromosome division
e mitosis and cytokinesis

3.7 Major components of the plasma membrane are

a phospholipids and carbohydrates
b proteins and carbohydrates
c phospholipids, proteins, and cholesterol
d carbohydrates and cholesterol
e carbohydrates and glycoproteins

3.8 The nucleolus

a is the site of ribosomal RNA synthesis
b has a surrounding membrane
c contains ATP used in chromosome replication
d is smaller in secretory than in nonsecretory cells
e specifies the chemical structure of enzymes

3.9 The most conspicuous structural elements of the cytoskeleton are

a microfilaments   **d** asters
b vacuoles   **e** microtubules
c ribosomes

3.10 Chromosomes that consist of two chromatids can only be found during

a early prophase
b early prophase and late prophase
c prophase and metaphase
d prophase, metaphase, and anaphase
e prophase, metaphase, anaphase, and telophase

Completion Exercises

3.11 In a cell that is not dividing, the genetic material is in the form of _____.

3.12 Division of the cell nucleus is called _____.

3.13 Within a cell, food materials and products of metabolism are generally found in _____.

3.14 The carbohydrate-rich zone of a plasma membrane is called the _____.

3.15 The characteristic of a membrane that allows some substances to enter the cell and blocks others is called _____.

3.16 The _____ is a complex labyrinth of flattened sheets and double membranes that spreads through the cytoplasm.

3.17 The small membrane-bound organelles that contain acid hydrolases are _____.

3.18 The spread of a neoplasm is called _____.

3.19 Any agent that causes cancer is a(n) _____.

3.20 A change in the genetic material is a(n) _____.

3.21 The process in which mRNA is synthesized from a DNA template is called _____.

3.22 The kind of nucleic acid that is found in the cytoplasm and carries amino acids is _____.

3.23 Codons are found on _____; they bind to anticodons on the _____.

3.24 The end product of the process of translation is a(n) _____.

3.25 Transcription occurs in the _____, while translation occurs in the _____.

Matching

3.26 _____ Golgi apparatus
3.27 _____ ATP
3.28 _____ DNA
3.29 _____ ribose
3.30 _____ mitochondrion
3.31 _____ microtubules
3.32 _____ lysosomes
3.33 _____ microbodies
3.34 _____ nucleotides
3.35 _____ mitosis
3.36 _____ chromosome replication

a nuclear division
b energy for cellular activity
c cilia and flagella
d a secretory organelle
e nucleic acid subunits
f makes up a gene
g peroxisomes
h site of ATP synthesis
i hydrolytic enzymes
j an RNA sugar
k interphase

A SECOND LOOK

In the following drawing of an idealized cell, label the Golgi apparatus, centrioles, nucleus, lysosome, microtubules, mitochondrion, and smooth ER.

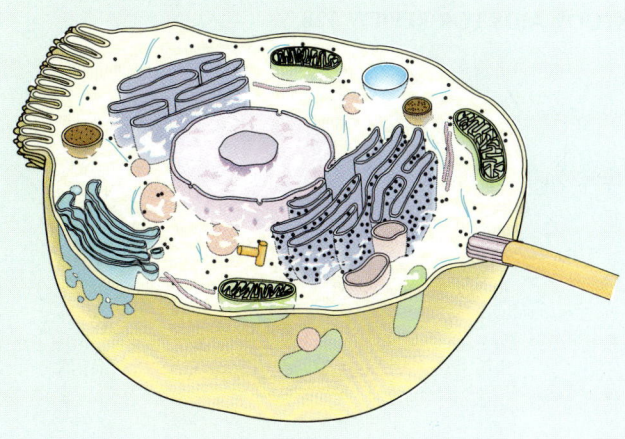

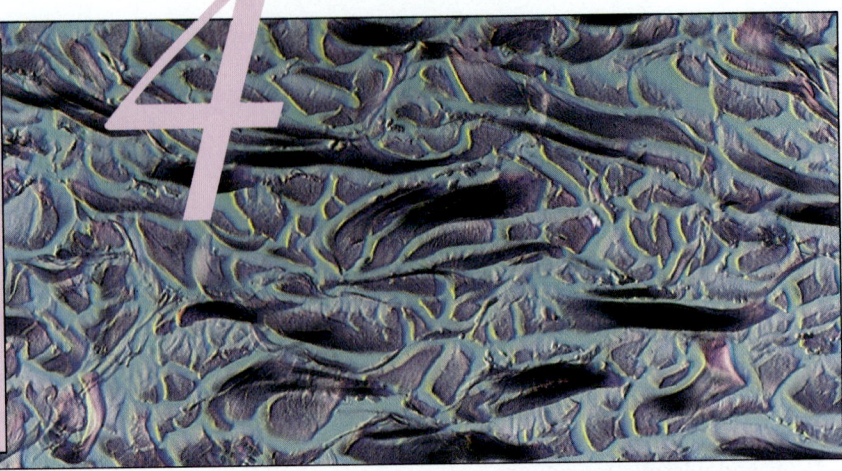

Tissues of the Body

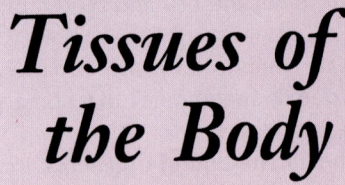

KEY CONCEPTS

1 Tissues are formed masses of specialized cells. Different types of cells form different types of tissues, with each tissue type performing a different function.

2 Epithelial tissues cover body surfaces or line body cavities or ducts.

3 Connective tissues function mainly to support and protect. Bone is considered a connective tissue.

4 Muscle tissue helps the body and its parts move.

5 Nervous tissue contains specialized cells that allow the entire nervous system—including the brain, spinal cord, and nerves—to function.

In a multicellular organism such as the human body, individual cells differentiate during development to perform special functions. These specialized cells are grouped together to carry out their functions as multicellular masses called *tissues.* The study of tissues, *histology* (Gr. *histos*, web), provides many excellent examples of how structures in the body have cellular arrangements that are closely related to their functions. The four major tissue types are epithelial, connective (including bone tissue), muscle, and nervous. In this chapter we look primarily at epithelial and connective tissues. Osseous (bone) tissue is described in detail in Chapter 6, muscle tissue (including heart muscle) in Chapter 10, and nervous tissue in Chapter 12.*

DEVELOPMENT OF PRIMARY GERM LAYERS

Repeated cell divisions in a newly fertilized egg create hundreds of identical daughter cells. However, once the resulting cell mass is implanted in the uterus, cell division stops temporarily. When cell division resumes, *the cells produced are not all alike.* **Differentiation,** the process by which cells develop into specialized tissues and organs, has begun. Soon the inner cell mass, which is now called an *embryo,* will be rearranged into three distinct **primary germ layers:** endoderm, ectoderm, and mesoderm [FIGURE 4.1; TABLE 4.1].

The exact sequence of the formation of germ layers is difficult to ascertain, especially since the layers of ectoderm and endoderm form almost simultaneously [FIGURE 4.1A, B]. During the second week of embryonic development the outer embryonic layer is formed. This is the **ectoderm** ("outside skin"), which will give rise to the outer layers of skin, hair, fingernails, tooth enamel, the nervous system, and other structures.

Also during the second week, a flattened layer of cells appears on the inner cell mass facing the inner cavity, or *blastocoel.* This layer of cells is the embryonic **endoderm** ("inside skin"). From this inner endodermal layer will develop the innermost organs, including the lining of the digestive tract and the glands and organs associated with it, the lining of the respiratory tract, and the lining of the urinary bladder, urethra, and vagina.

The third week is a period of rapid development for the embryo and is also the time when the prospective mother will miss her menstrual period and suspect that she is pregnant. At about the fifteenth day, the inner cell

mass develops into the three-layered **embryonic disk** [FIGURE 4.1C]. The cells of the embryonic disk fill the area between the endoderm and ectoderm and eventually form the bulk of the embryo. This final layer is the **mesoderm** ("middle skin") [FIGURE 4.1D], the forerunner of connective tissue in lower skin layers, bone, muscle, blood, and the inner lining of some internal organs.

EPITHELIAL TISSUES: FORM AND FUNCTION

Epithelial tissue (epp-uh-THEE-lee-uhl; L. *epi-*, on, over, upon + Gr. *thele*, nipple), or **epithelium,** exists in many structural forms, but in general it either *covers* or *lines* something. Epithelial tissues can be arranged in two different ways, corresponding to their two different functions:

1 Most epithelial tissues are composed of cells arranged in sheets one or more layers thick. The function of these sheets is to cover surfaces or line body cavities or *ducts* (passages) that often connect with the surface of the body. Some tissues line the surfaces of internal organs and are shaped to carry out the functions of absorption and protection.

2 Other epithelial tissues are organized into glands adapted for secretion. These modified epithelia are classified as *glandular epithelia.*

Functions of Epithelial Tissues

The typical functions of epithelial tissues are *absorption* (by the lining of the small intestine, for example), *secretion* (by glands), *transport* (by kidney tubules), *excretion* (by sweat glands), *protection* (by the skin), and *sensory reception* (by the taste buds). The size, shape, and arrangement of epithelial cells are directly related to these specific functions. Throughout this chapter you will see how these varied functions are enhanced by the different structures of epithelial tissues.

General Characteristics of Epithelial Tissues

A covering or lining epithelial tissue may be single-layered or multilayered, and its cells may have different shapes. However, all epithelial tissues except glandular tissues share certain characteristics.

Cell shapes and junctions The cells that make up epithelial tissues are relatively regular in shape and are closely packed in continuous sheets. There is little or none of the extracellular material known as *matrix* between epithelial cells. The matrix of other types of cells is

---

*Several other kinds of tissues are described with their relevant systems in separate chapters rather than being placed in one chapter. The authors have chosen this organization because they believe that it gives the reader a better understanding of how anatomy and physiology work together and how the parts of the body make up a cohesive whole.

FIGURE 4.1 FORMATION OF PRIMARY GERM LAYERS

The primary germ layers are the endoderm, mesoderm, and ectoderm. These layers give rise to all body tissues.

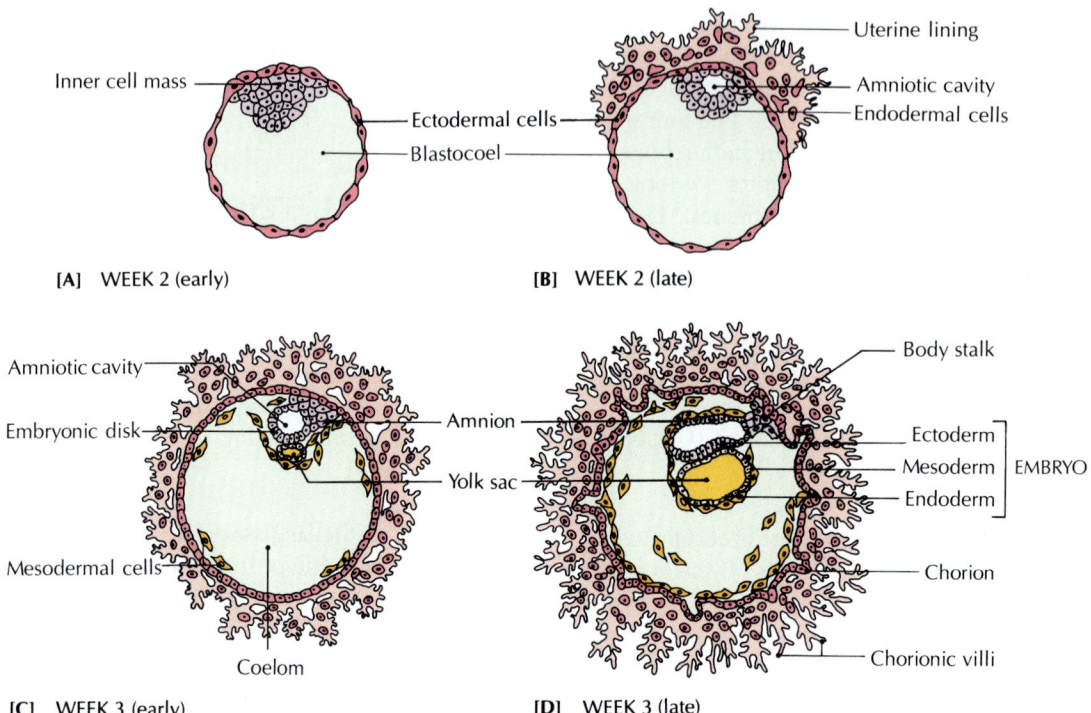

[A] WEEK 2 (early)

[B] WEEK 2 (late)

[C] WEEK 3 (early)

[D] WEEK 3 (late)

TABLE 4.1 MAJOR STRUCTURES DERIVED FROM PRIMARY GERM LAYERS

Endoderm	Mesoderm	Ectoderm
Epithelium of:	All muscle tissue (cardiac, smooth, skeletal), except in iris and sweat glands	All nervous tissue
Pharynx, larynx, trachea, lungs		Epidermis of skin; hair follicles, nails
Tonsils, adenoids	All connective tissue (fiborus, adipose, cartilage, bone, bone marrow)	Epithelium and myoepithelial cells of sweat glands, oil glands, mammary glands
Thyroid, thymus, parathyroids	Synovial and serous membranes	Lens and cornea of the eye
Esophagus, stomach, intestines, liver, pancreas, gallbladder; glands of alimentary canal (except salivary)	Lymphoid tissue: tonsils, lymph nodes	Receptor cells of sense organs
	Spleen	Enamel of teeth
Urinary bladder, urethra (except terminal male portion), prostate, bulbourethral glands	Blood cells	Adrenal medulla
	Reticuloendothelial system	Anterior pituitary gland
Vagina (partial)	Dermis of skin	Epithelium of: salivary glands, lips, cheeks, gums, hard palate, nasal cavity, sinuses
Auditory tubes	Teeth (except enamel)	Lower third of anal canal
	Endothelium of heart, blood vessels, lymphatics	Terminal portion of male urethra
	Epithelium of: gonads, reproductive ducts; adrenal cortex; kidneys, ureters; coelom, joint cavities	Vestibule of vagina and vestibular glands
		Muscle of iris
		Inner ear

FIGURE 4.2 JUNCTIONAL COMPLEXES OF EPITHELIAL CELLS

A unique feature of epithelial cells is the presence of various functionally and structurally distinct intercellular junctions between their plasma membranes. The three main kinds of junctions are shown here in simplified form. **[A]** A tight junction. **[B]** An electron micrograph of a tight junction between two heart muscle cells. ×190,920. **[C]** A spot desmosome. **[D]** A gap junction.

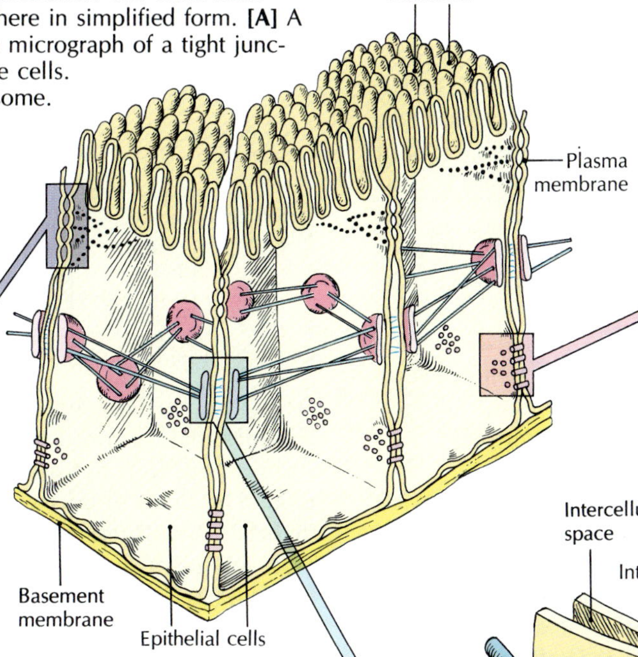

Microvilli

Plasma membrane

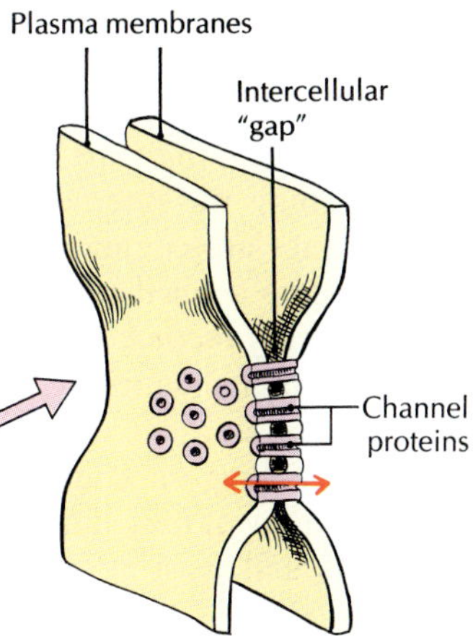

Plasma membranes

Intercellular "gap"

Channel proteins

[D] GAP JUNCTION

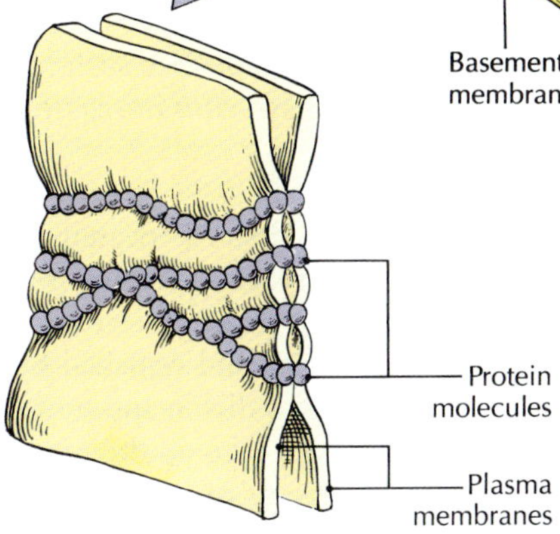

Basement membrane

Epithelial cells

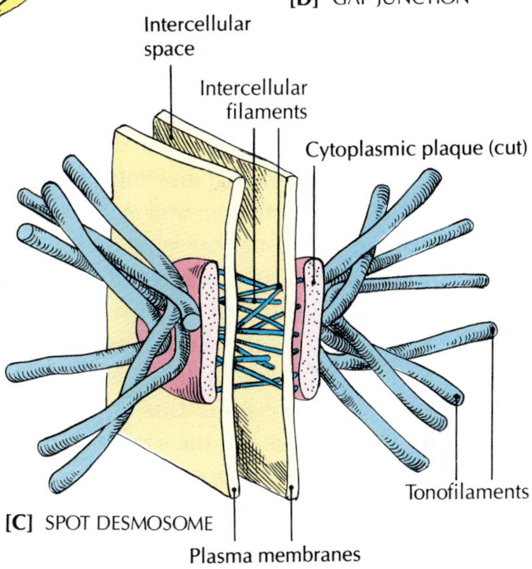

Intercellular space

Intercellular filaments

Cytoplasmic plaque (cut)

Tonofilaments

[C] SPOT DESMOSOME

Plasma membranes

Protein molecules

Plasma membranes

[A] TIGHT JUNCTION

Cell 1

Tight junction

Plasma membranes

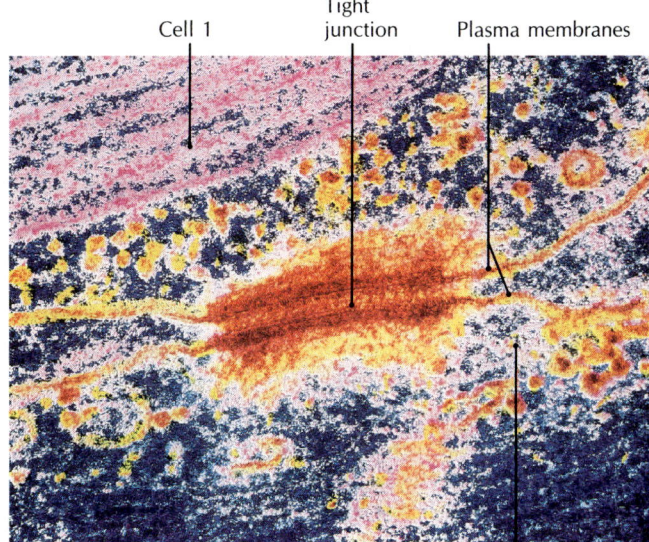

[B] TIGHT JUNCTION

Cell 2

composed of extracellular fibers and a homogeneous extracellular material called *ground substance*, which is composed mainly of glycoproteins. The framework of the matrix that does exist for epithelial tissue is composed of this ground substance. Instead of depending on a matrix for support, the tight-fitting epithelial cells are held in place by strong adhesions formed between the plasma membranes of adjacent cells. The specialized parts that hold the cells together are known as **junctional complexes;** they enable groups of cells to function as a unit. There are three main kinds of junctional complexes: tight junctions, spot desmosomes, and gap junctions:

1 A **tight junction** is the site of close connection between two plasma membranes, with little or no extracellular space between the cells [FIGURE 4.2A, B]. This "non-leaky" junction creates a permeability barrier that stretches across a continuous layer of epithelial cells, keeping material either in or out. Tight junctions are found in the epithelial tissues of the urinary bladder, for example, where they hold urine within the bladder.

2 A *spot desmosome* (Gr. *desmos*, binding) is a junction between adjacent plasma membranes with no direct contact [FIGURE 4.2C]. The plasma membranes are joined by a crisscrossed network of *intercellular filaments*. Thicker intracellular *tonofilaments* run parallel to the cell surface, strengthening the internal framework of the cell. Spot desmosomes are common in skin, where stress is constantly being applied to the junctions.

3 A *gap junction* is formed from several links of channel protein connecting two plasma membranes [FIGURE 4.2D]. Gap junctions are found in numerous locations, including intestinal epithelial cells, where they allow the flow of ions and small molecules between adjacent cells, and assist in spreading the rhythmic muscular contractions that move food along the intestines.

Basement membranes Most epithelial cells are anchored to the underlying connective tissue by a basement membrane. The **basement membrane** consists of (1) a *basal lamina* (a homogeneous layer of peptides and glycoproteins lacking fibers) and (2) a deeper layer of glycoproteins containing reticular and collagenous fibers [FIGURE 4.3]. *Basement membrane* and *basal lamina* are often used synonymously, but that is incorrect.

By firmly attaching the basal surface of the epithelial cells to the underlying connective tissue, the basement membrane assures that the epithelium is held in position, reducing the possibility of tearing. Basement membranes vary in thickness. They are quite thin in the skin and intestines, and rather thick in the trachea. Basement membranes provide elastic support and act as a partial barrier for diffusion and filtration. The basal lamina also provides a surface along which epithelial cells may migrate during cell renewal and wound healing.

Lack of blood vessels Epithelial tissues do not contain blood vessels, but they may contain nerves. Oxygen and other nutrients diffuse through their semipermeable basement membranes from capillaries in the underlying connective tissue. Wastes from the epithelial tissues diffuse into the connective tissue capillaries.

Surface specializations Epithelial tissues usually have several types of surface specializations. Some epithelial cells have smooth surfaces, but most epithelial cells have irregular surfaces that result from the complex folding of their outer plasma membranes. This extended folding produces *microvilli,* fingerlike projections that greatly increase the absorptive area of the cell [FIGURE 4.4A, B]. Microvilli usually appear on cells that are involved in absorption or secretion. The increased surface

FIGURE 4.3 COMPONENTS OF THE BASEMENT MEMBRANE

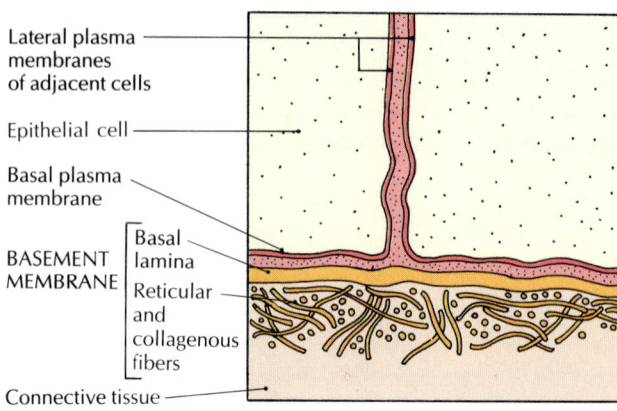

area of microvilli, coupled with the presence of surface enzymes that can break down large molecules, aids the absorptive process necessary for the efficient digestion of food. The shape of microvilli on the free surface of epithelial cells gives rise to the term **brush border.** The elaborate folding of the plasma membrane may occur on lateral and basal surfaces of cells, as well as on free surfaces [FIGURE 4.4A]. Some epithelial cells have **cilia** projecting from their free surfaces.

Regeneration Because epithelial tissue is located on surfaces, it is subject to constant injury. Fortunately, it is also capable of constant regeneration through cell division. Damaged epithelial cells are shed and replaced by new cells. These cells are specialized to divide and move into the damaged area, where they can take on the function of the cells they replace. The rate of renewal depends on the type of epithelium. For example, the outer layer of the skin and the lining of the intestinal villi are replaced entirely every few days. In contrast, the cells in glands and the lining of the respiratory tract usually are replaced every 5 to 6 weeks.

General Classification of Epithelial Tissues

Epithelial tissues are generally classified according to the *arrangement of cells* and the *number of cell layers;* they are further subdivided by the *shape of the cells* in the superficial (top or outer) layer. In this section we discuss the two main groupings of epithelial tissues: *simple* and *stratified.*

If an epithelial tissue is made up of a single layer of cells, it is called **simple;** if its cells are arranged in two or more layers, it is **stratified.** The basic shapes of the superficial cells are (1) **squamous** (SKWAY-muhss) (flat), (2) **cuboidal** (like a cube), and (3) **columnar** (elongated).

These general types of epithelia, as well as some exceptions to this classification, are summarized and illustrated in FIGURES 4.5 through 4.8.

FIGURE 4.4 MICROVILLI

[A] Elaborate folding of the plasma membrane is evident not only in the microvilli (the outer surface) but also on the lateral and basal surfaces. The clusters of mitochondria at the base indicate an area of cellular activity. Notice the tight junction between cells. **[B]** A scanning electron micrograph of the small intestine where the extra surface area created by the microvilli aids the absorption of digested food into the bloodstream. The microvilli resemble tufts of a shag rug, an appropriate image, since villi is Latin for "shaggy hairs." ×30,000.

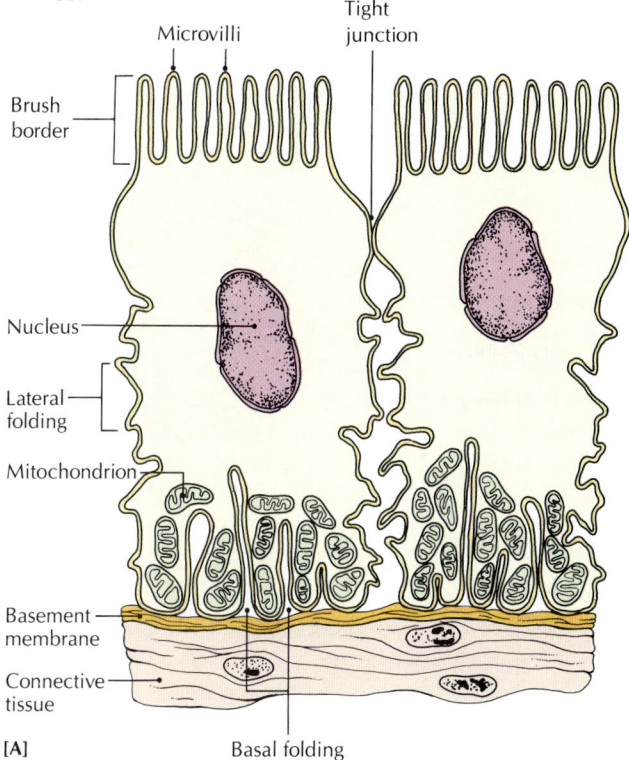

Tight
junction

Microvilli

Brush
border

Nucleus

Lateral
folding

Mitochondrion

Basement
membrane

Connective
tissue

[A]

Basal folding

Microvilli forming brush border

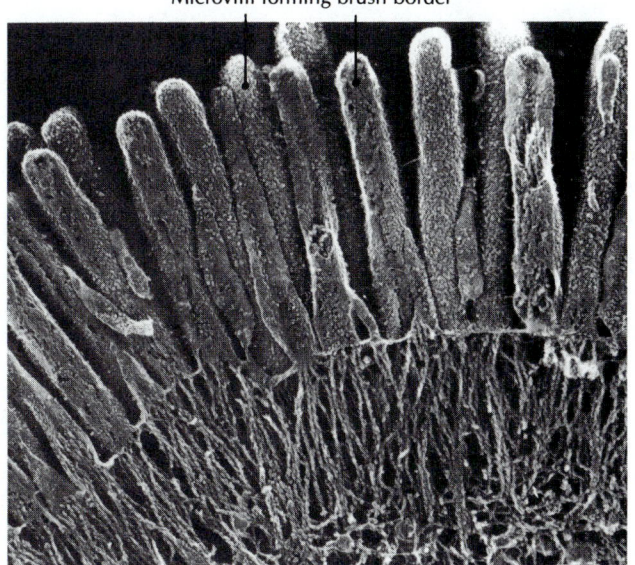

[B]

Exocrine Glands

Exocrine glands have ducts that carry their secretions to openings on the body surfaces. Exocrine glands may be classified in several ways. The most typical ways are described in the sections below.

Unicellular and multicellular glands The only example of a *unicellular* (one-celled) *gland* in the human body is the *goblet cell* [FIGURE 4.5, page 107], found in the lining of the intestines, other parts of the digestive system, the respiratory tract, and the conjunctiva of the eye. A goblet cell produces a carbohydrate-rich glycoprotein called *mucin* (MYOO-sihn), which later is secreted in the form of *mucus*, a thick lubricating fluid. (It should be noted, however, that goblet cells *do not* have ducts.) A *multicellular gland* contains many cells [TABLE 4.2, page 111].

Simple and compound glands A gland with only one *unbranched* duct is a *simple gland*. A gland with a *branched* duct system, resembling a tree trunk and its branches (although upside down), is a *compound gland*.

Tubular, alveolar, and tubuloalveolar glands All three of these categories are concerned with the shapes of the secretory portion of the glands. If the secretory portion of a gland is *tubular*, the gland is called a *tubular gland*. If the secretory portion is *rounded*, the gland is an *alveolar* (L. *alveolus*, hollow cavity), or *acinar* (ASS-ih-nur; L. *acinus*, grape, berry), *gland*. Glands whose secretory portions have both tubular and alveolar shapes are called *tubuloalveolar glands*. TABLE 4.2 lists the types of glands developed from combinations of tubular, alveolar, simple, and compound forms and shows their shapes.

Mucous, serous, and mixed glands Glands may be classified according to their secretions as well as their structures. *Mucous glands* secrete thick mucus; *serous glands* secrete a thinner, watery substance. The secretions of serous glands generally contain enzymes. *Mixed glands* contain both mucous and serous cells, and produce mucus and serous secretions.

Merocrine and holocrine glands Glands may also be classified according to how they release their secretions. *Merocrine glands* release their secretions without breaking the plasma membrane, using the process of *exocytosis*. An example of a merocrine gland is the pancreas, which secretes digestive enzymes. When *holocrine glands* release their secretions, whole cells become detached, die, and actually become the secretion. The sebaceous (oil) glands in the skin are the only holocrine glands in the body.

FIGURE 4.5 SIMPLE EPITHELIUM

SIMPLE SQUAMOUS EPITHELIUM (L. *squama,* scale)

Description
A single layer of flat cells that are wider than they are thick; nuclei are flattened and are parallel to the surface.

Main locations
Lining of lymph vessels, blood vessels, and heart (lining called *endothelium*); serous membranes lining peritoneal, pleural, pericardial, and scrotal body cavities (this epithelium, called *mesothelium*, is derived from mesoderm and has microvilli); glomerular (Bowman's) capsule in kidneys; lung air sacs (alveoli); small excretory ducts of many glands; membranes of inner ear.

Major functions
Permits diffusion through semipermeable surface; for example, blood is filtered in kidneys to form urine; oxygen in lung alveoli diffuses into blood, and waste carbon dioxide from blood diffuses into alveoli.

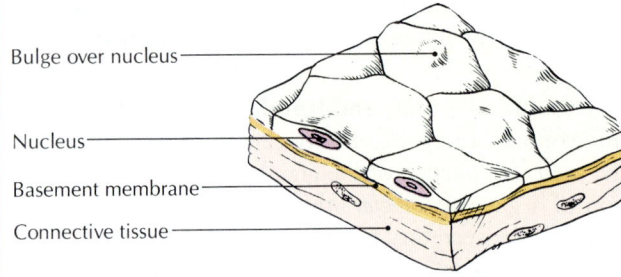

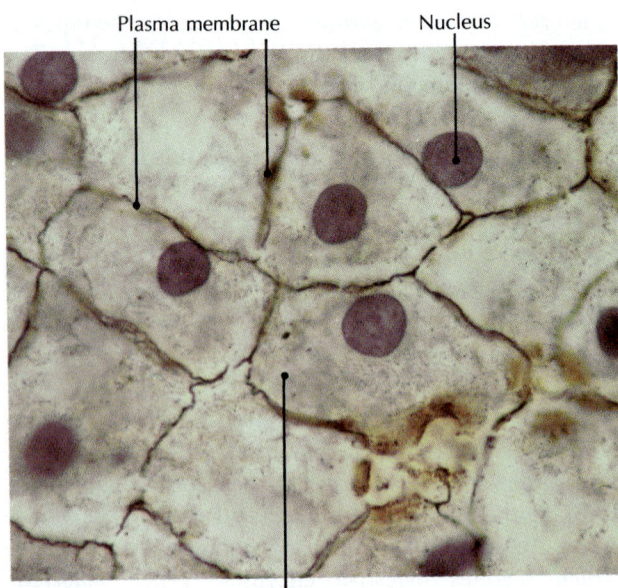

LINING OF ABDOMINAL CAVITY; light micrograph, ×1000

SIMPLE CUBOIDAL EPITHELIUM

Description
A single layer of approximately cube-shaped cells; nucleus large and centrally located.

Main locations
Lining of many glands and their ducts; surface of ovaries; inner surface of eye lens; pigmented epithelium of eye retina; some kidney tubules (with microvilli on cell surface).

Major functions
Secretion of mucus, sweat, enzymes, other substances; absorption of fluids, other substances.

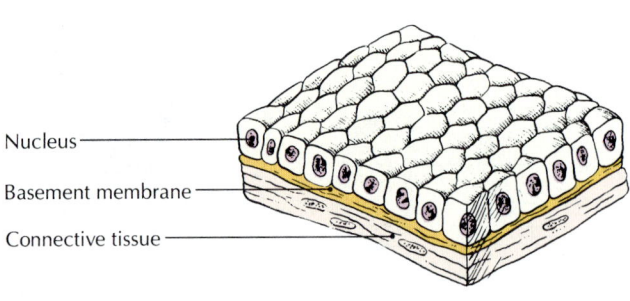

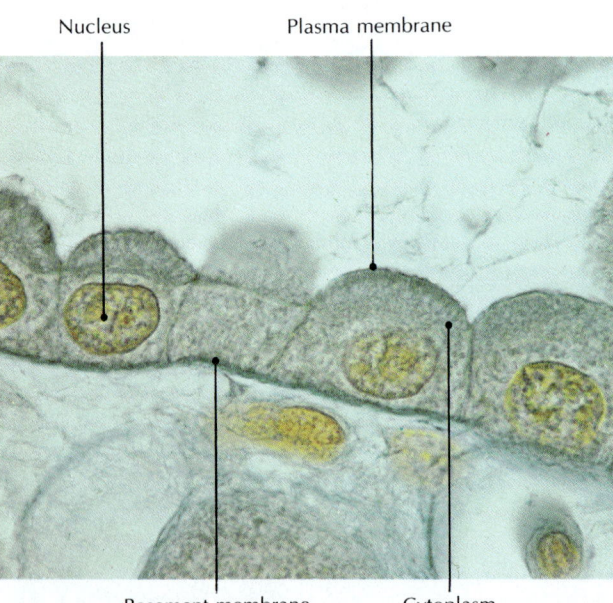

KIDNEY; light micrograph, ×900

SIMPLE COLUMNAR EPITHELIUM

Description
Single layer of cells that are taller than they are wide; looks like simple cuboidal epithelium with elongated cells; large, oval-shaped nucleus usually located at base of cell; may secrete mucus (goblet cells); may have cilia or microvilli on free surface of cell.

Main locations
Lining of stomach and intestines (with microvilli), digestive glands, and gallbladder.

Major functions
Secretion, absorption, protection, lubrication; mucus and cilia combine to trap and sweep away foreign substances; cilia may also help to move objects through a duct, such as an egg in the uterine tube.

GOBLET CELLS (in Simple Columnar Epithelium)

Description
Mucus produced by goblet cells accumulates near the top of cell, causing the cell to bulge in shape of a goblet; goblet cells are interspersed in layer of columnar cells.

Main locations
Lining of digestive tract, upper respiratory tract, uterine tube, uterus.

Major functions
Modified to secrete mucus from spaces at top of cell; mucus protects and lubricates walls of digestive tract.

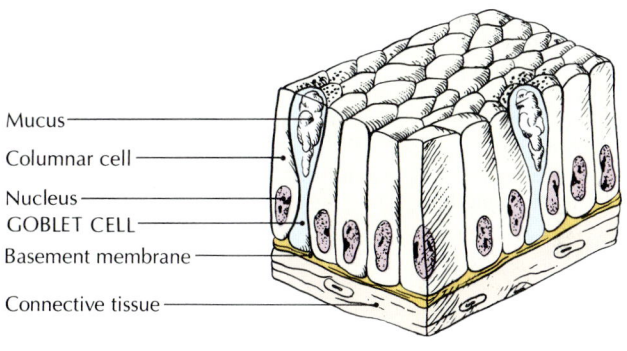

Mucus —
Columnar cell —
Nucleus —
GOBLET CELL —
Basement membrane —
Connective tissue —

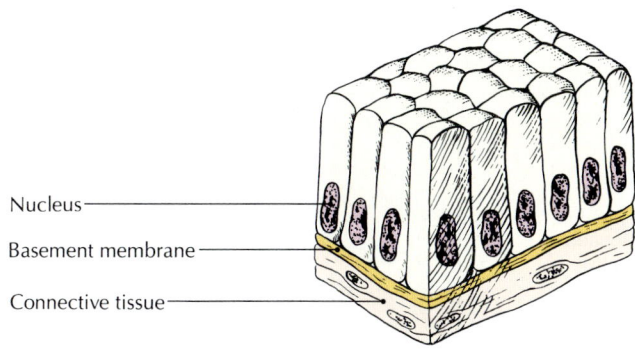

Nucleus —
Basement membrane —
Connective tissue —

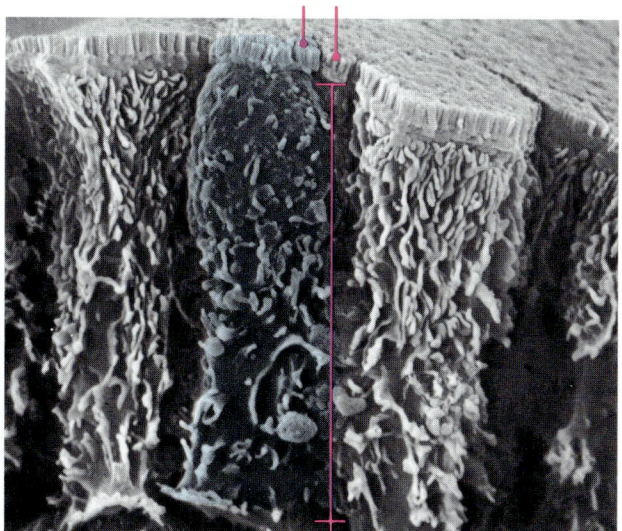

Microvilli

Goblet cell

EPITHELIAL LINING OF RAT SMALL INTESTINE; scanning electron micrograph, ×3000

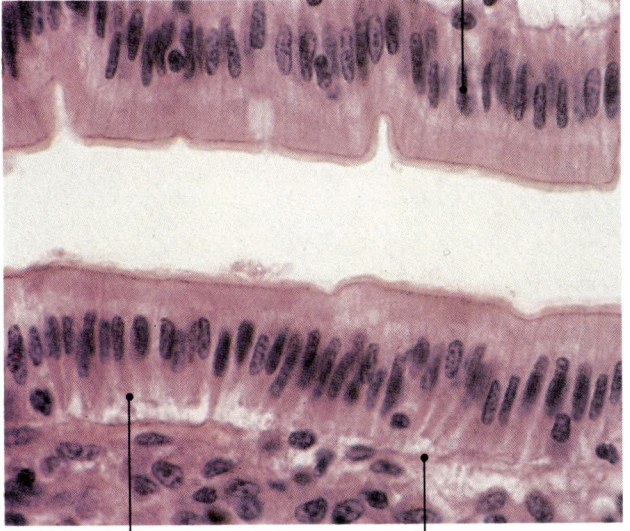

Nucleus

Columnar cell   Basement membrane

BILE DUCT; light micrograph, ×400

FIGURE 4.6 STRATIFIED EPITHELIUM

STRATIFIED SQUAMOUS EPITHELIUM

Description
Several layers of cells; only superficial layer composed of flat squamous cells; underlying basal cells are modified cuboidal or columnar cells that are pushed upward from near basement membrane to replace dying superficial cells; shedding process in outer layer is *desquamation*.

Main locations
Areas where friction or possibility of cellular injury or drying occurs, such as epidermis, vagina, mouth and esophagus, anal canal, distal end of urethra.

Major functions
Protection. In the skin, replacement cells rising to the surface from underlying layers produce **keratin** (KER-uh-tihn; Gr. *keras*, horn), a tough protein; their nuclei and organelles disappear, and by the time they reach the surface, they consist mainly of keratin. These *keratinized* cells flake off as the next batch of cells reaches the surface. Cells in a moist environment are *nonkeratinized;* they retain their nuclei and organelles.

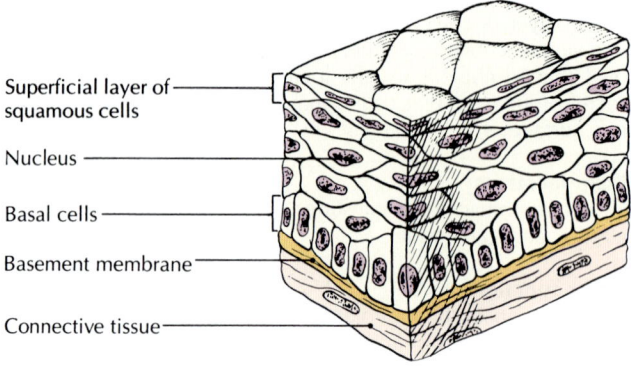

Superficial layer of squamous cells

Nucleus

Basal cells

Basement membrane

Connective tissue

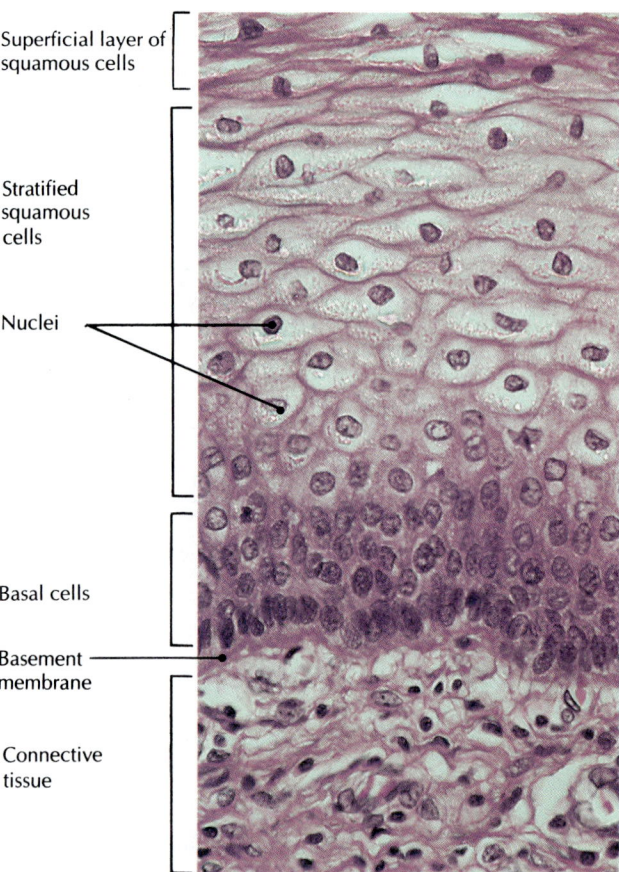

Superficial layer of squamous cells

Stratified squamous cells

Nuclei

Basal cells

Basement membrane

Connective tissue

VAGINA; light micrograph, ×400

STRATIFIED CUBOIDAL EPITHELIUM

Description
Multilayered arrangement with superficial layer composed of cuboidal cells; superficial cells are larger than basal cells; distinct basement membrane.

Main locations
Ducts of sudoriferous (sweat) glands, sebaceous (oil) glands, developing epithelium in ovaries and testes.

Major function
Secretion.

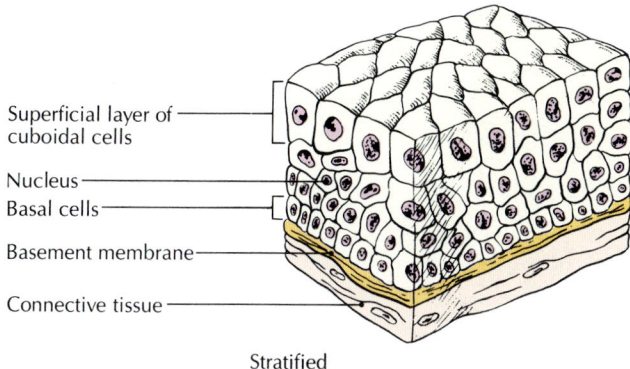

Superficial layer of cuboidal cells
Nucleus
Basal cells
Basement membrane
Connective tissue

Stratified cuboidal epithelium

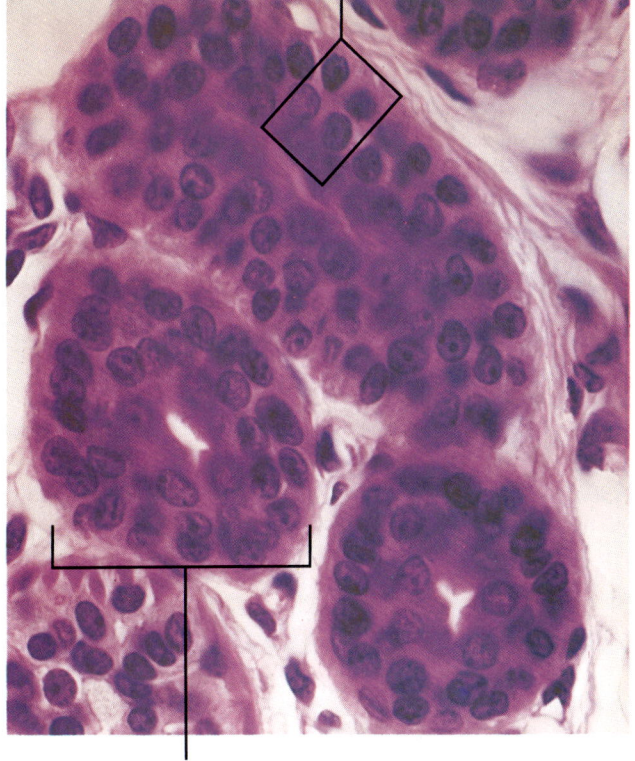

Duct of sweat gland

DUCTS OF SWEAT GLANDS; light micrograph, ×400

STRATIFIED COLUMNAR EPITHELIUM

Description
Multilayered arrangement with superficial layer composed of tall, thin columnar cells; underlying cuboidal-type cells in contact with basement membrane; sometimes ciliated, as in larynx and nasal surface of soft palate.

Main locations
Moist surfaces, such as larynx, nasal surface of soft palate, parts of pharynx, urethra, and excretory ducts of salivary glands and mammary glands.

Major functions
Secretion and movement of materials over cell surface.

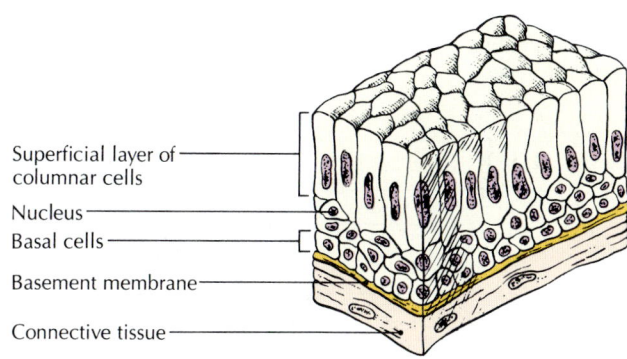

Superficial layer of columnar cells
Nucleus
Basal cells
Basement membrane
Connective tissue

Superficial layer of columnar cells

Basal cells

Basement membrane

MALE URETHRA; light micrograph, ×400

FIGURE 4.7 ATYPICAL EPITHELIUM

PSEUDOSTRATIFIED COLUMNAR EPITHELIUM (*pseudo* = false)

Description
Single layer of cells of varying height and shape; nuclei at different levels give false impression of multilayered structure; all cells in contact with basement membrane, but not all reach superficial layer; when ciliated, called *pseudostratified ciliated columnar epithelium.*

Main locations
Large excretory ducts, most of male reproductive tract, nasal cavity and other respiratory passages, and part of ear cavity.

Major functions
Protection, secretion, and movement of substances across surfaces; mucus from cells in respiratory tract traps foreign substances; mucus moved to throat by sweeping action of cilia and either coughed out or swallowed and later eliminated in feces.

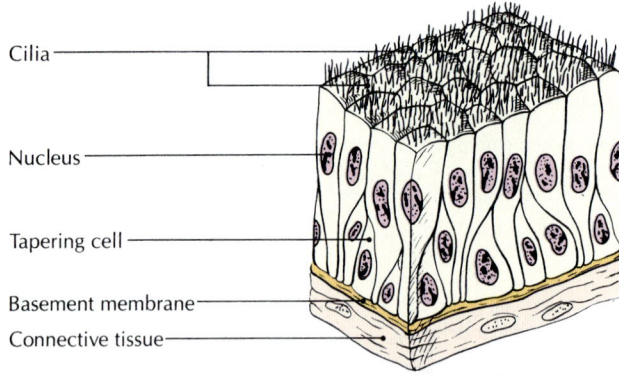

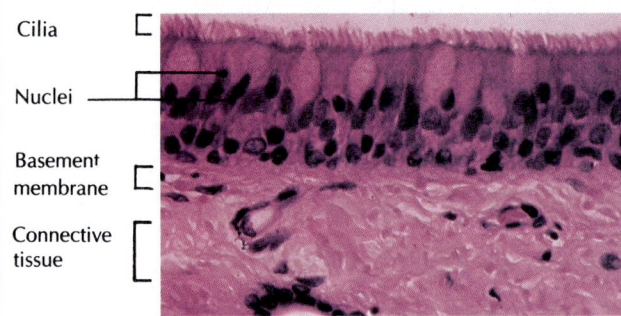

RESPIRATORY SYSTEM; light micrograph, ×600

TRANSITIONAL EPITHELIUM

Description
Stratified epithelium; shape of surface cells changes; they appear round when unstretched, flat when stretched; relaxed basal cells are cuboidal or slightly columnar, with layer above basal cells made up of irregularly shaped elongated cells; accordion-pleated membranes allow stretching and contracting.

Main locations
Urinary tract, where it lines bladder, ureters, urethra, and parts of kidney.

Major function
Allows for changes in shape.

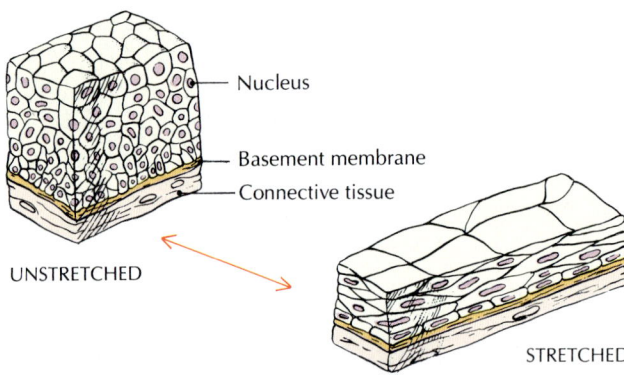

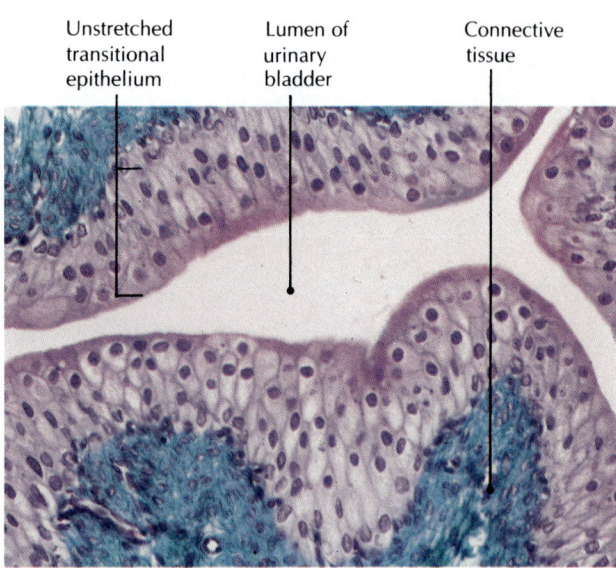

UNSTRETCHED URINARY BLADDER; light micrograph, ×400

TABLE 4.2 CLASSIFICATION OF MULTICELLULAR EXOCRINE GLANDS

Type of gland	General shape	Shape of secretory portion	Location
MULTICELLULAR SIMPLE GLANDS			
Simple tubular		Straight and tubular	Intestinal glands (crypts of Lieberkühn)
Simple coiled tubular		Coiled	Sudoriferous (sweat) glands
Simple branched tubular		Branched and tubular	Mouth, tongue, and esophagus
Simple alveolar (acinar)		Rounded	Seminal vesicle glands (in male reproductive system)
Simple branched alveolar		Branched and rounded	Sebaceous (oil) glands
MULTICELLULAR COMPOUND GLANDS			
Compound tubular		Tubular	Mucous glands of mouth, bulbourethral glands (in male reproductive system), kidney tubules, and testes
Compound alveolar (acinar)		Rounded	Mammary glands
Compound tubuloalveolar		Tubular and rounded	Salivary glands, glands of respiratory passages, and pancreas

---

FIGURE 4.8 GLANDULAR EPITHELIUM

DESCRIPTION
Epithelial cells modified to perform secretion. Main types are exocrine and endocrine; exocrine glands have ducts, endocrine glands secrete hormones directly into bloodstream where they travel to target cells. All glandular tissue is derived embryologically from epithelium.

MAIN LOCATIONS
Various sites throughout body, including skin glands such as sweat and mammary glands; digestive glands such as salivary glands; endocrine glands such as thyroid.

MAJOR FUNCTIONS
Synthesis, storage, and secretion of chemical products.

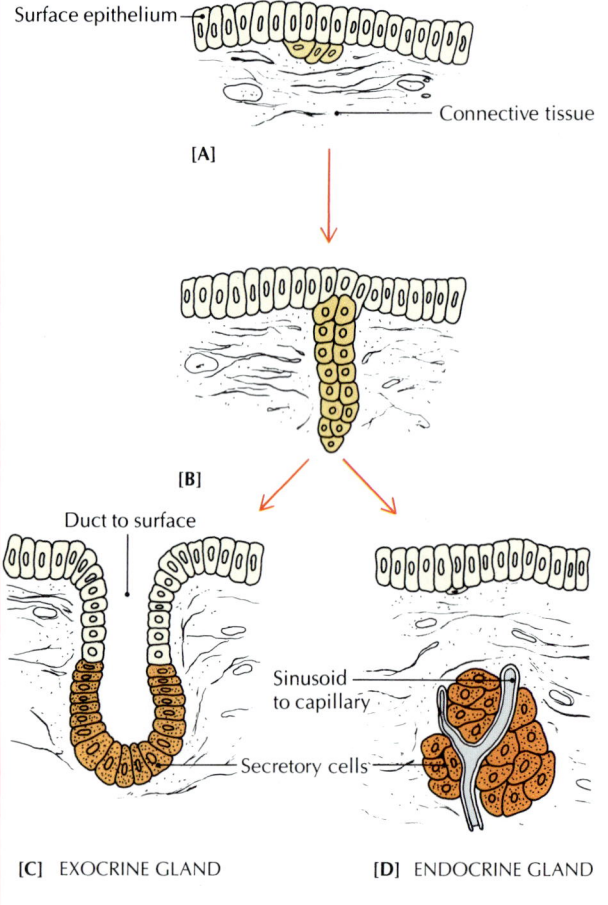

Surface epithelium

Connective tissue

[A]

[B]

Duct to surface

Sinusoid to capillary

Secretory cells

[C] EXOCRINE GLAND   [D] ENDOCRINE GLAND

CONNECTIVE TISSUES

Connective tissues are aptly named because they connect other tissues together. Unlike epithelial tissues, connective tissues have much extracellular fibrous material that helps support the cells of other tissues. In fact, when cartilage and bone are included in the overall classification of connective tissues, the whole group is often referred to as *supporting tissues*. Besides their supportive function, connective tissues form protective sheaths around hollow organs and are involved in storage, transport, and repair.

Connective tissues vary greatly in their structure and function. For this reason, we have grouped together the more generalized types—such as loose, dense, and elastic—under the section Connective Tissue Proper on page 116. The more specialized forms of connective tissue—cartilage, bone (osseous), and blood—are treated separately. In this chapter we discuss cartilage in some detail but give only a brief overview of bone tissue and blood. Bones and bone tissue are described in Chapter 6, the skeletal system in Chapters 7 and 8, and blood in Chapter 19.

All connective tissues consist of fibers, ground substance, cells, and some extracellular fluid [FIGURE 4.9 and TABLE 4.3]. The extracellular fibers and ground substance are known as the ***matrix*** (MAY-trihks). The arrangement, function, and composition of elements in the matrix vary in different kinds of connective tissues, as you will see in this and the following sections.

Fibers of Connective Tissues

Connective tissues are usually classified according to the arrangement and density of their extracellular fibers and the nature and consistency of the ground substance in which the fibers are embedded. Fibers of connective tissue may be collagenous, reticular, or elastic.

TABLE 4.3 MAJOR COMPONENTS OF CONNECTIVE TISSUES

Type of fiber or cell	Description	Major functions
EXTRACELLULAR COMPONENTS (MATRIX)		
Collagenous fibers [FIGURES 4.9, 4.11]	Thick bundles of whitish fibrils composed of collagen. Most common connective tissue fibers. Arranged randomly, ranging from loose and pliable to tightly packed and resistant.	Support and protect organs; connect muscles to bones and bones to bones.
Reticular fibers [FIGURES 4.9, 4.11]	Delicately branched networks of inelastic fibrils; have same chemical composition and molecular structure as collagenous fibers but are thinner.	Support fat cells, capillaries, nerves, muscle fibers, and secretory liver cells; form reticular framework of spleen, lymph nodes, and bone marrow.
Elastic fibers [FIGURES 4.9, 4.11]	Yellow fibers that appear only singly; branch to form networks. Less coarse than collagenous fibers; contain protein elastin.	Allow hollow organs and other structures to stretch and recoil; support; suspension.
Ground substance [FIGURE 4.9]	Homogeneous, extracellular material; main components are glycoproteins and proteoglycans.	Provides medium for passage of nutrients and wastes between cells and bloodstream; lubricant; shock absorber.
FIXED CELLS		
Fibroblasts [FIGURES 4.9, 4.11]	Large, long, flat branching cells, with large, pale nuclei; most common cells in connective tissue.	Synthesize matrix materials (fibers and ground substance); assist wound healing.
Adipose cells (fat cells) [FIGURES 4.9, 4.11]	May appear singly or in clusters; single cells are bloated spheres containing fat; nuclei and cytoplasm are flattened against side of cell.	Synthesize and store lipids.
Macrophages* [FIGURES 4.9, 4.10]	Irregularly shaped cells with short processes. Cells are normally fixed along bundles of collagenous fibers but become free-moving scavengers in cases of inflammation. Free macrophages are part of reticuloendothelial system.	Engulf and destroy foreign bodies in bloodstream and tissues.
Reticular cells [FIGURES 4.9, 4.11]	Flat, star-shaped cells resembling fibroblasts; associated with reticular fibers and such lymphoid organs as lymph nodes, bone marrow, and spleen.	Involved with immune response and probably with formation of reticular fibers, macrophages, blood cells, and fat cells.
WANDERING CELLS		
Plasma cells [FIGURES 4.9, 4.10]	Oval cells with large, dark nuclei off-center; found where bacteria enter through breaks in epithelium and in serous membranes, lymphoid tissue, and areas of chronic infection.	Main producers of antibodies, which help protect body against microbial infection.
Mast cells [FIGURE 4.9]	Relatively large cells with irregular shapes and small, pale nuclei; secretory granules abundant in cytoplasm.	Produce heparin, which reduces blood clotting, and histamine, which increases vascular permeability and contracts smooth muscle.
Leukocytes (white blood cells) [FIGURE 4.9]	Roundish cells with various-shaped nuclei. Five kinds of leukocytes exist, including lymphocytes.	Protect against infection by engulfing and destroying harmful microorganisms; also destroy tissue debris and play a role in inflammatory response.

*Macrophage cells can also be wandering cells. See page 115.

FIGURE 4.9 CONNECTIVE TISSUE

A diagrammatic rendering of connective tissue, showing a representative sample of cells and fibers bathed in tissue fluid.

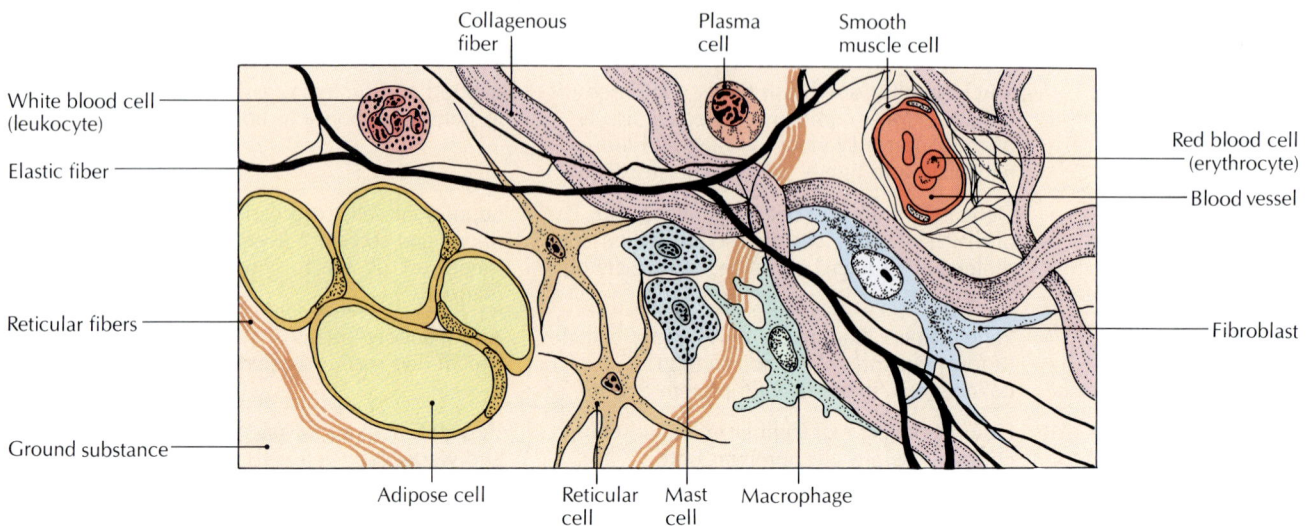

Collagenous fibers *Collagenous fibers* are whitish fibers composed mainly of the protein **collagen** (G. *kolla*, glue + *genes*, born, produced). Sturdy, flexible, and practically unstretchable collagenous fibers are the most common and are found in all kinds of connective tissue [FIGURE 4.12, page 120]. The arrangement of collagenous fibers varies from loose and pliable, as in the loosely woven connective tissue that supports most of the organs, to tightly packed and stretch-resistant, as in tendons. Generally, these fibers look wavy when they are not under pressure. No matter how collagenous fibers are arranged, they are extremely well suited for support. A collagenous fiber is actually a bundle of parallel *fibrils*, slender fibers composed of even smaller microfibrils. The number of fibrils in a bundle varies.*

Reticular fibers *Reticular fibers* form delicately branched networks (*rete* is Latin for net), compared to the thicker bundles of collagenous fibers [FIGURE 4.11, page 119]. Reticular fibers have the same molecular structure as collagenous fibers and practically the same chemical composition, and they also contain fibrils and microfibrils. Reticular fibers help support other cells.

Elastic fibers Yellowish *elastic fibers* appear singly, never in bundles, though they do branch and form net-

works. Elastic fibers contain microfibrils but do not have the coarseness of collagenous fibers. Like rubber bands, elastic fibers stretch easily when they are pulled and return to their original shape when the force is removed. The protein *elastin* in elastic fibers gives them the resilience that organs such as the skin need to move, stretch, and contract.

Ground Substance of Connective Tissues

Ground substance of connective tissue is a homogeneous extracellular material that ranges in consistency from a semifluid to a thick gel. Whereas fibers give connective tissue its strength and elasticity, ground substance provides a suitable medium for the passage of nutrients and wastes between the cells and the bloodstream. The main ingredients of ground substance are glycoproteins and proteoglycans,† including hyaluronic acid, chondroitin sulfate, and other specific sulfates. These compounds not only bind water and other tissue fluids that are needed for the exchange of nutrients and wastes, but also act as a lubricant and shock absorber. Hyaluronic acid forms a tough protective mesh that helps prevent invasion by microorganisms and other foreign substances. Interestingly, some bacteria can produce an enzyme, called *hyaluronidase*, that breaks down the mesh of hyaluronic acid and allows the bacteria to enter connective tissue.

---

*A high number of collagenous fibers can make meat tough. But when collagen is boiled in water, it turns into a soft gelatin. This is why tough meat simmered in a soup or stew becomes more tender. In contrast to this softening process, leather is toughened, or *tanned*, by the addition of tannic acid, which converts collagen into a firm, insoluble material.

†Proteoglycans have a protein core with covalently bonded sulfate side chains.

Cells of Connective Tissues

The cells of connective tissues can be classified as either *fixed cells*, such as fibroblasts and adipose cells, or *wandering cells*, such as plasma cells, mast cells, and granular leukocytes (white blood cells). Macrophage cells may be either fixed or wandering.

Fixed cells *Fixed cells* have a permanent site and are usually concerned with long-term functions such as synthesis, maintenance, and storage:

1 *Fibroblasts* (L. *fibra*, fiber; Gr. *blastos*, growth) are large, long, flat, branching cells that appear spindle-shaped [FIGURES 4.9, 4.11]. Their nuclei are light-colored and larger than in any other connective tissue cell. Fibroblasts are the most common cells in connective tissue and the only cells found in tendons. They synthesize and secrete the *matrix* materials (fibers and ground substance) and are considered to be secretory cells. In this way, fibroblasts assist wound healing.

2 *Adipose (fat) cells* synthesize and store lipids. A mature adipose cell accumulates so much fat that the nucleus and cytoplasm are flattened against the sides of the cell, forming a thin film around the rim. Adipose cells may appear singly or in clusters. Single cells are so bloated with one large droplet of stored fat they they look like spherical drops of oil [FIGURES 4.9 and 4.11].*

3 *Macrophage cells (macrophages)* are active phagocytes ("cell eaters"). The name *macrophage* is descriptive: *macro* means large, and *phage*, as you have seen, comes from the Greek word *phagein*, to eat; so a macrophage is literally a "big eater." Macrophages have an irregular shape and short cytoplasmic extensions (processes). They are normally fixed along the bundles of collagenous fibers and are then referred to as *fixed macrophages*. But in cases of inflammation, they become detached from the fibers, change their shape to resemble amoebas, and begin to move about actively through the body as *free macrophages*. Free macrophages are scavengers that engulf and destroy foreign material in the bloodstream and tissues.

4 *Reticular cells* are flat, star-shaped cells with long processes that enable the cells to come in contact with each other [FIGURES 4.9, 4.11]. The cells form the cellular framework in such netlike structures as bone marrow, lymph nodes, the spleen, and other lymphoid tissues. Reticular cells play a role in the immune response.

Wandering cells *Wandering cells* in connective tissues are usually involved with short-term activities such as protection and repair:

1 *Leukocytes (white blood cells)* are roundish cells with nuclei that have various shapes [FIGURE 4.9]. The cytoplasm may or may not contain secretory granules. Leukocytes multiply drastically in times of infection. Although leukocytes circulate in the bloodstream, they perform their protective function by moving through the walls of tiny blood vessels into the connective tissue by a process called *diapedesis*. Most leukocytes are phagocytes; they help protect the body against infection by engulfing and destroying harmful microorganisms.

2 *Plasma cells* are a specific type of leukocyte. They are derived from activated B lymphocytes. Plasma cells are oval-shaped, with large, dark nuclei located off-center [FIGURES 4.9, 4.10A]. They are the main producers of the antibodies that help defend the body against microbial infection and cancer. They are found in connective tissue under the moist epithelial linings of the respiratory and intestinal tracts, where microorganisms may enter through breaks in the epithelial membrane. Plasma cells are also present in serous membranes, lymphoid tissue, and areas of chronic infection.

3 *Mast cells* are relatively large cells with irregular shapes and small, pale nuclei [FIGURE 4.9]. They are often found near blood vessels. Their cytoplasm is crowded with secretory granules, which are bound by membranes. These granules contain *heparin*, a polysaccharide that prevents blood from clotting as it circulates throughout the body, and *histamine*, a protein that increases vascular permeability and causes smooth muscle to contract.

4 *Macrophages* become mobile when stimulated by inflammation; they are listed here as a reminder of their frequent wanderings as phagocytes. Free macrophages are part of the extensive **reticuloendothelial system,** which is made up of an army of specialized cells concerned with producing antibodies and removing dead cells, tissue debris, microorganisms, and foreign particles from the fluids and matrix of body tissues [FIGURE 4.10B].

---

*Single adipose cells are called *unilocular fat cells*; an accumulation of these cells is called *white fat*. A second type of adipose cell, called a *multilocular fat cell*, has many small fat droplets in its cytoplasm. An accumulation of multilocular fat is called *brown fat*. Brown fat is found only in fetuses and newborn infants, where it can serve as a source of heat energy.

ASK YOURSELF

1 What are the major components of connective tissues?

2 What are the three kinds of connective tissue fibers?

3 What types of cells are found in connective tissue?

FIGURE 4.10 PLASMA CELLS AND MACROPHAGE

[A] Plasma cells. [B] A scanning electron micrograph of a macrophage ingesting red blood cells. The macrophage is covered with microvilli. ×2500.

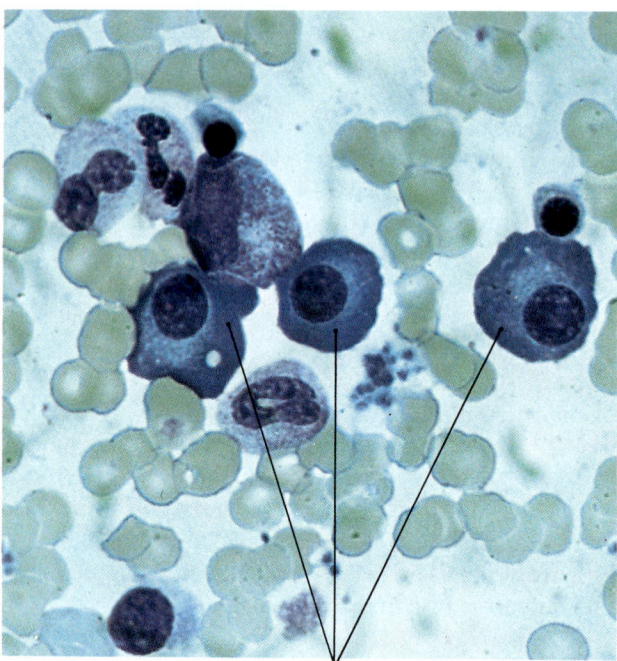

Plasma cells

[A]

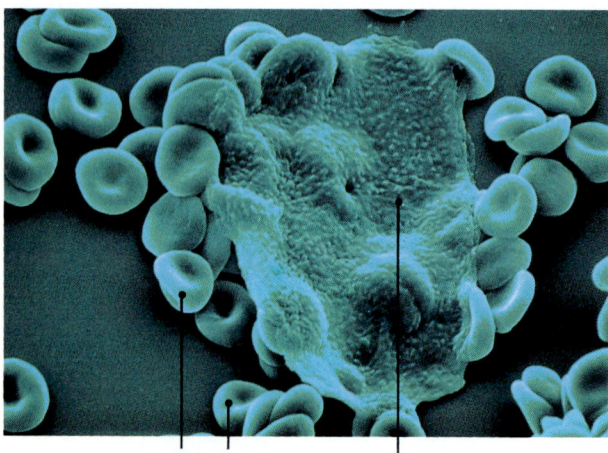

Red blood cells   Macrophage

[B]

Connective Tissue Proper

Connective tissues are usually classified according to the arrangement and density of their extracellular fibers [FIGURE 4.11]. *Loose (areolar) connective tissue* has irregularly arranged fibers and more tissue-fluid cells than fibers. *Dense (collagenous) connective tissue* has more fibers, and their arrangement is more regular. The following types of connective tissue are usually grouped under the heading connective tissue proper: embryonic, loose, dense, elastic, adipose, and reticular. Except for *embryonic* connective tissue, all the connective tissues we consider here are known as *adult connective tissues.*

Cartilage

Cartilage is a specialized connective tissue that provides support and aids the movement of joints. Like other connective tissue, it consists of cells, fibers, and ground substance. Cartilage cells are called *chondrocytes* (KON-droh-sites; Gr. *khondros*, cartilage); they are embedded within the matrix in small cavities called *lacunae* (luh-KYOO-nee; L. cavities). Cartilage does not contain blood vessels. Instead, oxygen, nutrients, and cellular wastes diffuse through the matrix. The three types of cartilage — hyaline, fibrocartilage, and elastic — are described and illustrated in FIGURE 4.12.

Bone as Connective Tissue

Bone is also a type of connective tissue. It is very similar to cartilage in that it consists mostly of matrix material that contains lacunae and specific cell types. However, unlike cartilage, bone is a highly vascular tissue. Bone histology is treated in detail in Chapter 6.

Blood as Connective Tissue

Blood cells and blood-forming tissues are mentioned in this section not because they connect anything — they do not — but because they have the same embryonic origin (mesenchyme) as the more typical connective tissues, and because blood has the three components of any connective tissue (cells, fibers, and ground substance). Blood is treated extensively in Chapter 19.

ASK YOURSELF

1 What types of connective tissue are classified as connective tissue proper?

2 What is the function of cartilage?

3 What are the three types of cartilage?

FIGURE 4.11 TYPES OF CONNECTIVE TISSUE: CONNECTIVE TISSUE PROPER

EMBRYONIC (Mesenchymal, Mucous Connective)

Description
Unspecialized packing material; cells vary in shape from stars to spindles; very little mesenchymal tissue remains after cellular differentiation in embryo; called *mesenchyme* during first two months of prenatal development, *mucous connective tissue* from two months to birth.

Main locations
Mesenchyme: around developing bones and under skin of embryo; mucous connective tissue: in umbilical cord of fetus.

Major functions
Mesenchyme differentiates into supporting tissues, tissues of blood vessels, blood, and smooth muscle. Mucous connective tissue forms padding for blood vessels in umbilical cord (where it is sometimes called *Wharton's jelly*) and prevents cord from snarling when fetus turns within uterus.

LOOSE (Areolar) (L. *areola*, "open place")

Description
Irregular, loosely woven fibers in a semifluid base; contains extracellular spaces filled with ground substance and tissue fluid and many types of cells, especially fibroblasts and macrophages; most common connective tissue.

Main locations
Most parts of body, especially around and between organs; wraps around nerves and blood vessels.

Major functions
Supports tissues, organs, blood vessels, nerves; forms subcutaneous layer that connects muscles to skin; allows for movement; acts as protective packing material between organs. Tissue fluid is medium for exchange of substances between capillaries and tissue cells; ground substance acts as barrier against harmful microorganisms.

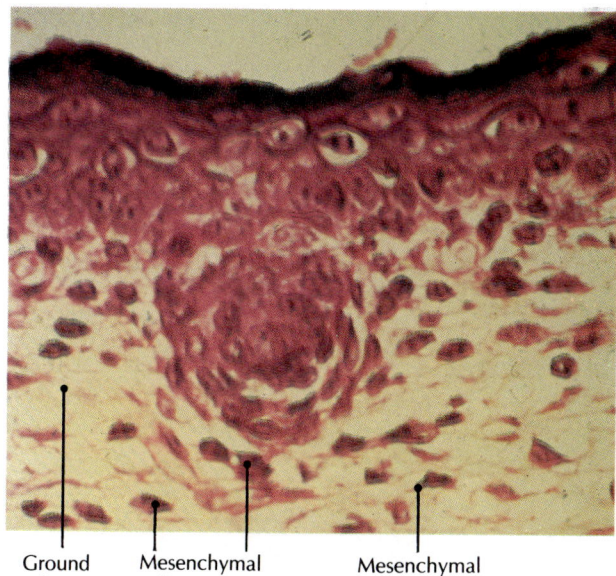

Ground substance   Mesenchymal cells   Mesenchymal process

MESENCHYMAL TISSUE IN EMBRYO; light micrograph, ×400

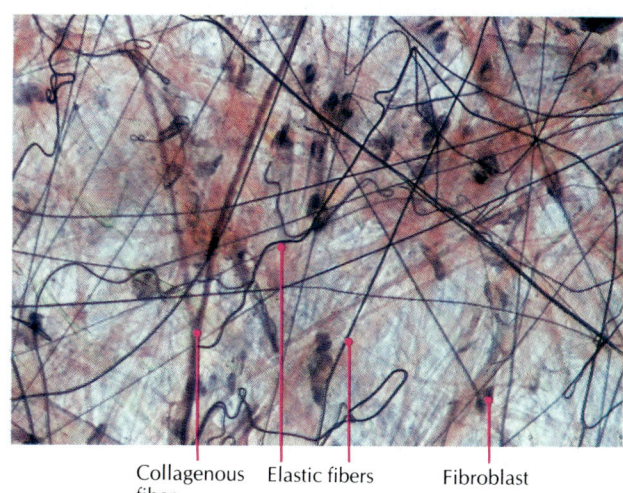

Collagenous fiber   Elastic fibers   Fibroblast

LOOSE AREOLAR TISSUE; light micrograph, ×64

(*Figure 4.11 continues on following pages*)

FIGURE 4.11 TYPES OF CONNECTIVE TISSUE: CONNECTIVE TISSUE PROPER *(Continued)*

DENSE (Collagenous)

Description
Tightly packed with coarse, collagenous fibers; distinctly white; classified as *regular* or *irregular*, depending on arrangement of fibers. Dense regular tissue has collagenous fibers arranged in parallel rows of thick, strong bundles; irregular tissue has thick bundle fibers arranged randomly in tough, resilient meshwork that can be stretched in more than one direction, as in the middle layer (dermis) of skin.

Main location
Regular: tendons, ligaments, aponeuroses.
Irregular: dermis of skin, capsules of many organs; covering sheaths of nerves, tendons, brain, spinal cord; deep fibrous coverings of muscles.

Major functions
Provides support and protection; connects muscles to bones (tendons) and bones to bones (ligaments).

ELASTIC

Description
Yellow elastic fibers branch freely, resembling a fishing net; spaces between are filled by fibroblasts and some collagenous fibers.

Main locations
Walls of hollow organs such as stomach; walls of largest arteries; some parts of heart; trachea, bronchi; ligaments between neural arches of spinal vertebrae; vocal cords; other parts of body where stretching, support, and suspension are required.

Major functions
Allows stretching, and provides support and suspension.

Collagenous fibers

DENSE REGULAR CONNECTIVE TISSUE; light micrograph, ×100

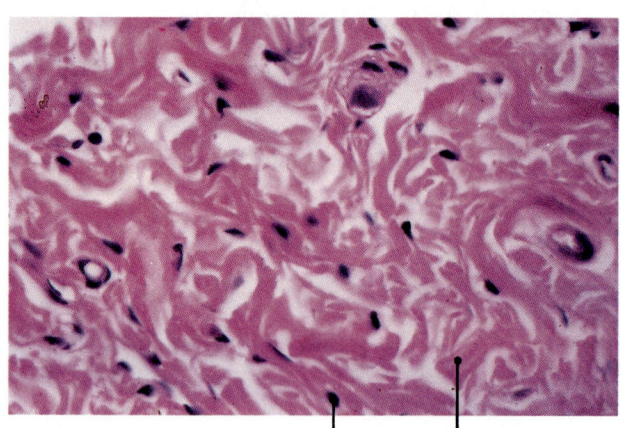

Fibroblast   Collagenous fiber

DENSE IRREGULAR CONNECTIVE TISSUE; light micrograph, ×310

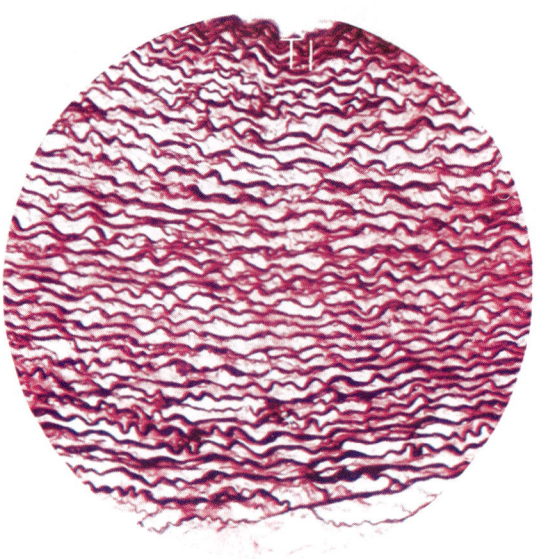

ELASTIC CONNECTIVE TISSUE FROM TUNICA MEDIA OF AN ELASTIC ARTERY; light micrograph, ×400

ADIPOSE (Fat)

Description
Consists of clustered adipocytes (cells specialized for fat storage); contains little or no extracellular substance; clusters of spherical, fat-bloated cells supported by collagenous and reticular fibers. Makes up about 10 percent of adult body weight; distributed differently in males and females.

Main locations
Beneath epidermis (outer layer of skin), around organs, many other sites.

Major functions
Provides reserve food supply by synthesizing and storing fat; cushions and protects organs; provides insulation against loss of body heat.

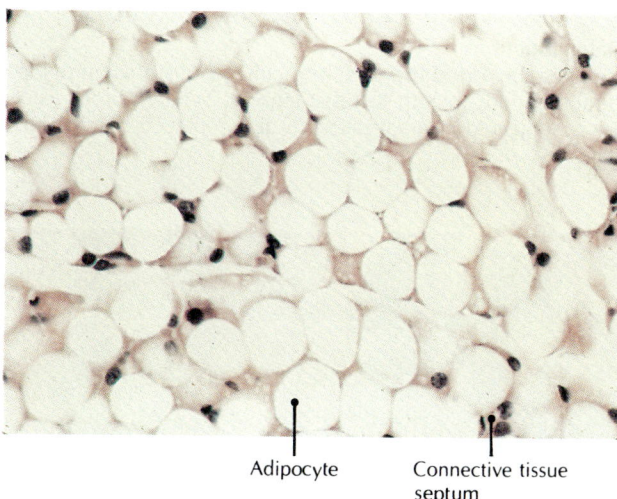

Adipocyte   Connective tissue
                         septum

ADIPOSE TISSUE; light micrograph, ×64

RETICULAR

Description
Lattice of fine, interwoven threads that branch freely, forming connecting and supporting framework; spaces between fibers are filled by reticular cells.

Main locations
Interior of liver, spleen, lymph nodes, tonsils, stomach, intestines, trachea, bronchi; supporting adipose cells.

Major functions
Forms connecting and supporting framework of reticular fibers for lymph nodes, bone marrow, spleen, thymus gland.

Lymphocytes   Reticular cell

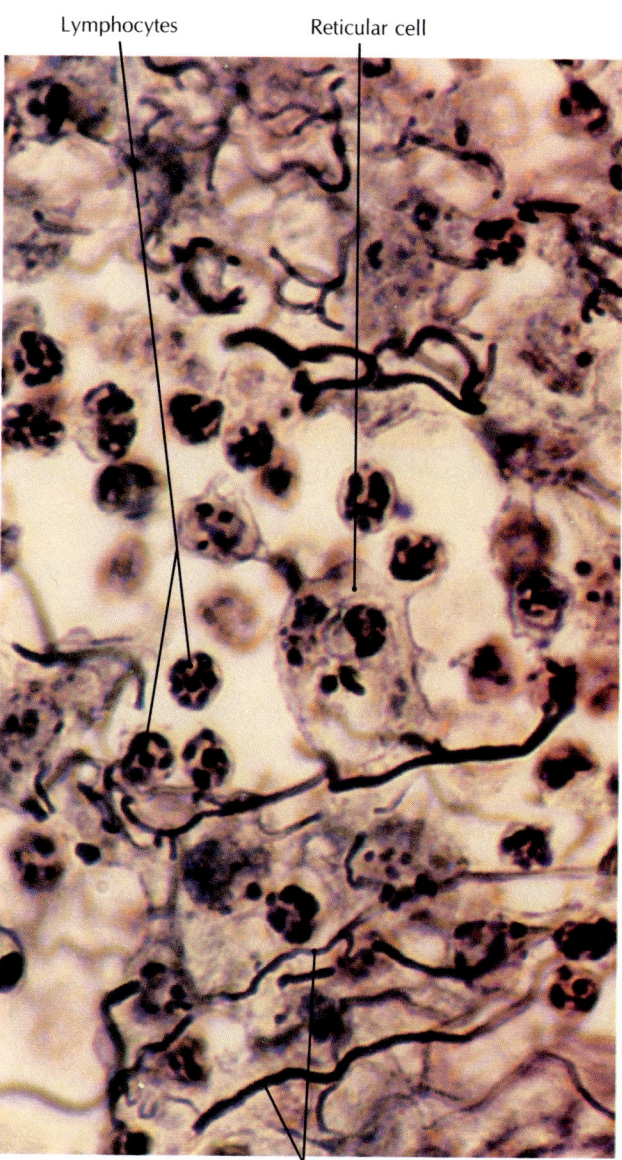

Reticular fibers

RETICULAR CONNECTIVE TISSUE; light micrograph, ×250

FIGURE 4.12 TYPES OF CONNECTIVE TISSUE: CARTILAGE

HYALINE CARTILAGE (Gr. *hyalos*, "glassy")

Description
Network of collagenous fibers with many spaces filled with ground substance; translucent, pearly blue-white appearance; strong, able to support weight; most common type of cartilage. Usually enclosed within fibrous **perichondrium,** which is composed of outer connective tissue layer and inner layer of cells capable of differentiating into chondrocytes.

Main locations
Trachea, larynx, bronchi, nose, costal (rib) cartilages, ends of long bones.

Major functions
Forms major part of embryonic skeleton; reinforces respiratory passages; aids free movement of joints; assists growth of long bones; allows rib cage to move during breathing.

FIBROCARTILAGE

Description
Consists of bundles of resilient, pliable collagenous fibers that leave little room for ground substance; usually merges with hyaline cartilage or fibrous connective tissue; no sheath; resembles dense connective tissue.

Main locations
Intervertebral disks, fleshy pad between pubic bones (symphysis pubis); lining of tendon grooves, rims of some sockets (for example, hip and shoulder joints), some areas where tendons and ligaments insert into ends of long bones; fibrocartilage disks located at temporomandibular (jaw) joint, knee joint (where they are called *menisci*), wrist joint, and sternoclavicular joint between clavicle and sternum; type of cartilage formed after injury.

Major functions
Provides support and protection; collagenous fibers provide durability and strength to resist tension. Where fibrous tissue is subjected to great pressure, it is replaced by fibrocartilage.

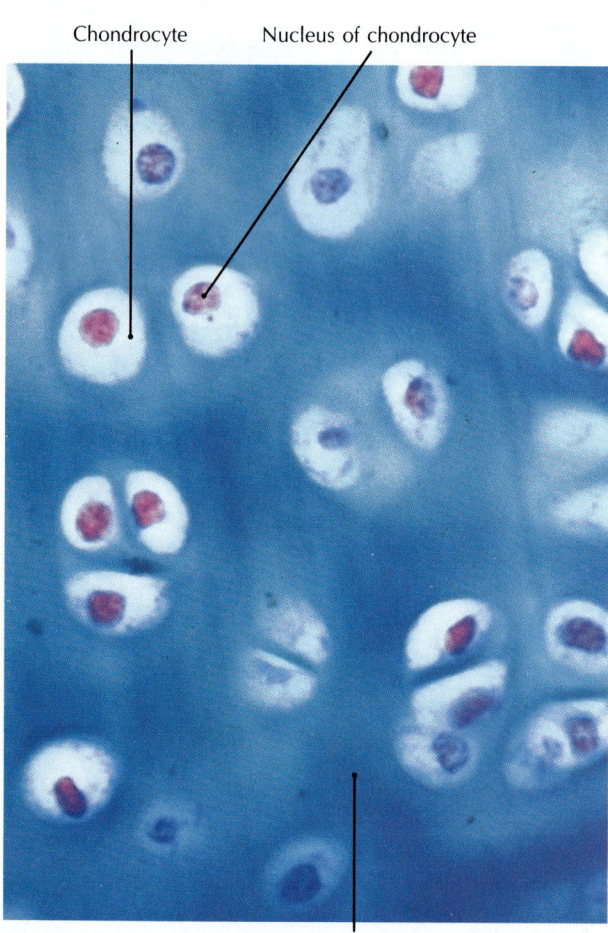

Chondrocyte   Nucleus of chondrocyte

Hyaline cartilage matrix

HYALINE CARTILAGE; light micrograph, ×125

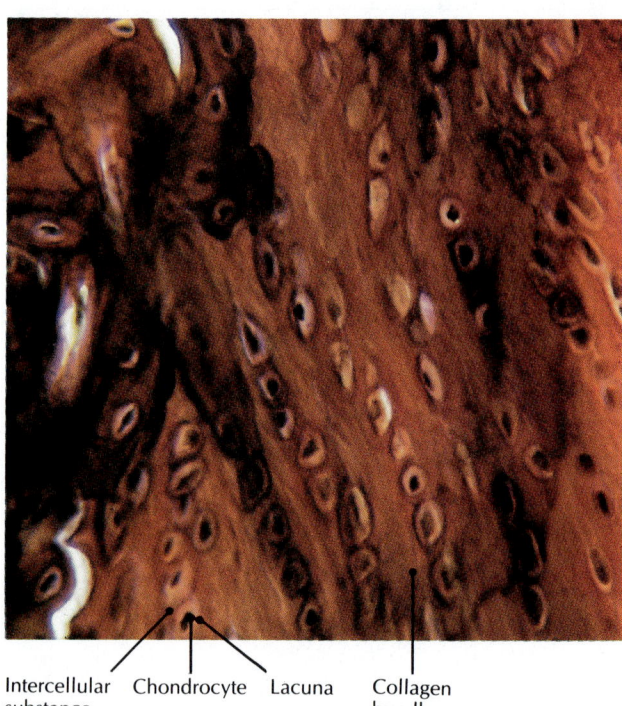

Intercellular   Chondrocyte   Lacuna   Collagen
substance   bundles

FIBROCARTILAGE; light micrograph, ×450

ELASTIC CARTILAGE

Description
Appears dark when stained because of dense elastic fibers, which are scattered in ground substance; cells can produce elastic and collagenous fibers; enclosed within perichondrial sheath; more flexible and elastic than hyaline cartilage.

Main locations
Areas that require lightweight support and flexibility, such as external ear, epiglottis in throat, auditory tube connecting middle ear to upper throat, larynx, nasopharynx.

Major functions
Provides flexibility and lightweight support.

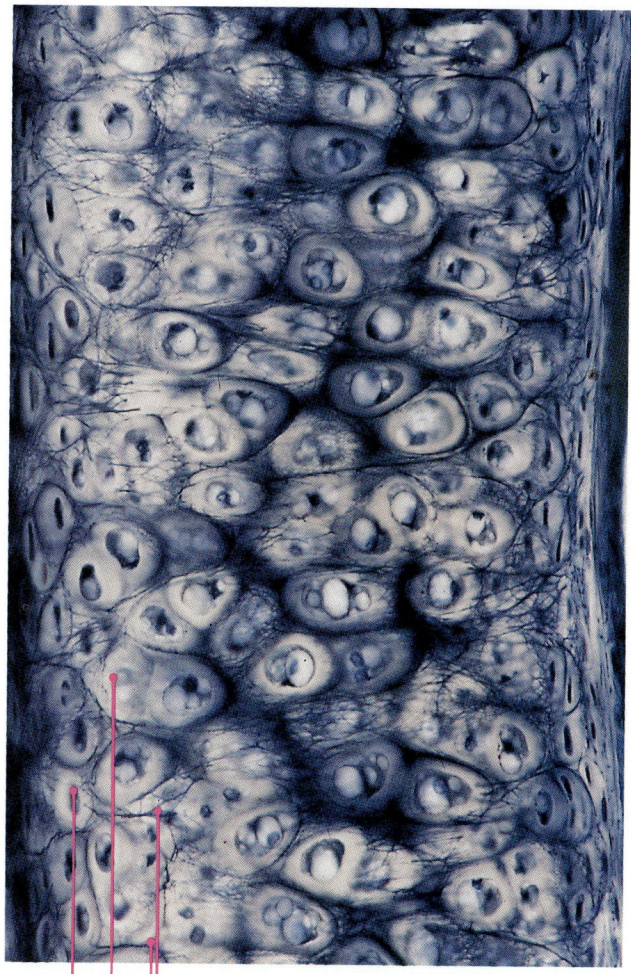

Chondrocytes   Elastic fibers

ELASTIC CARTILAGE; light micrograph, ×400

MUSCLE TISSUE: AN INTRODUCTION

Muscle tissue [FIGURE 4.13] helps the body and parts of the body move. For example, *skeletal muscle* is attached to bones and makes body movement possible; the rhythmic contractions of *smooth muscle* create a churning action (as in the stomach), help propel material through a tubular structure (as in the intestines and ureters), and control size changes in hollow organs such as the urinary bladder and uterus; and the contractions of *cardiac muscle* result in the heart beating. Muscle tissue is discussed more fully in Chapter 10.

NERVOUS TISSUE: AN INTRODUCTION

The *nervous tissue* of the nervous system is composed of several different types of cells, including impulse-conducting cells called *neurons* (the fundamental units of the nervous system) [FIGURE 4.14], nonconducting *neuroglia* (protective, supportive, and nourishing cells), and *peripheral glial cells*, which form various types of sheaths and help protect, nourish, and maintain cells of the peripheral nervous system. Nervous tissue is described more fully in Chapter 12.

MEMBRANES

Membranes are thin, pliable layers of epithelial and/or connective tissue. They line body cavities, cover surfaces, or separate or connect certain regions, structures, and organs of the body. The three kinds of membranes are mucous, serous, and synovial.

Mucous Membranes

Mucous membranes line body passageways that open to the outside of the body, such as the nasal and oral cavities, and tubes of the respiratory, digestive, urinary, and reproductive systems. These membranes are made up of varying kinds of epithelial tissue over a layer of loose connective tissue. For example, the small intestine is lined with simple columnar epithelium, and the oral cavity is lined with stratified squamous epithelium. Glands in the mucous membranes secrete the protective, lubricating mucus, which consists of water, salts, and a sticky protein called mucin. Besides providing lubrication, mucus helps trap invading microorganisms.

FIGURE 4.13 TYPES OF MUSCLE TISSUE

Light micrographs of **[A]** skeletal muscle (striated), ×106; **[B]** smooth muscle (unstriated), ×132; and **[C]** cardiac muscle, ×1000. Nuclei are elongated and prominent dark areas. Note intercalated disks and branches of cardiac muscle fibers.

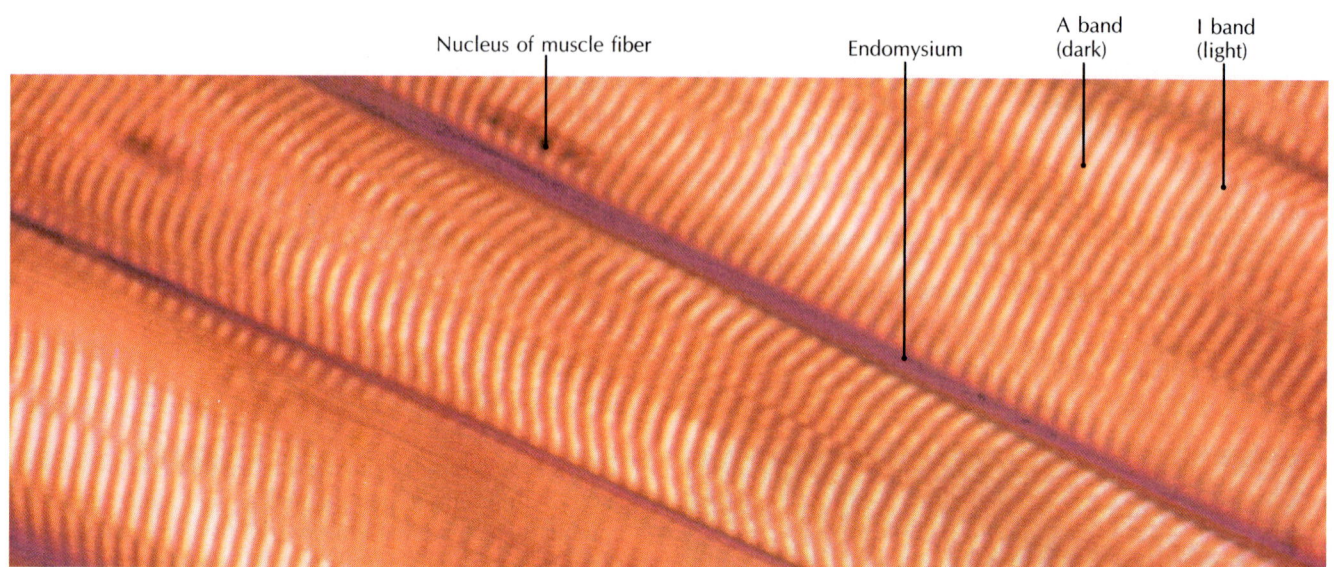

Nucleus of muscle fiber   Endomysium   A band (dark)   I band (light)

[A]

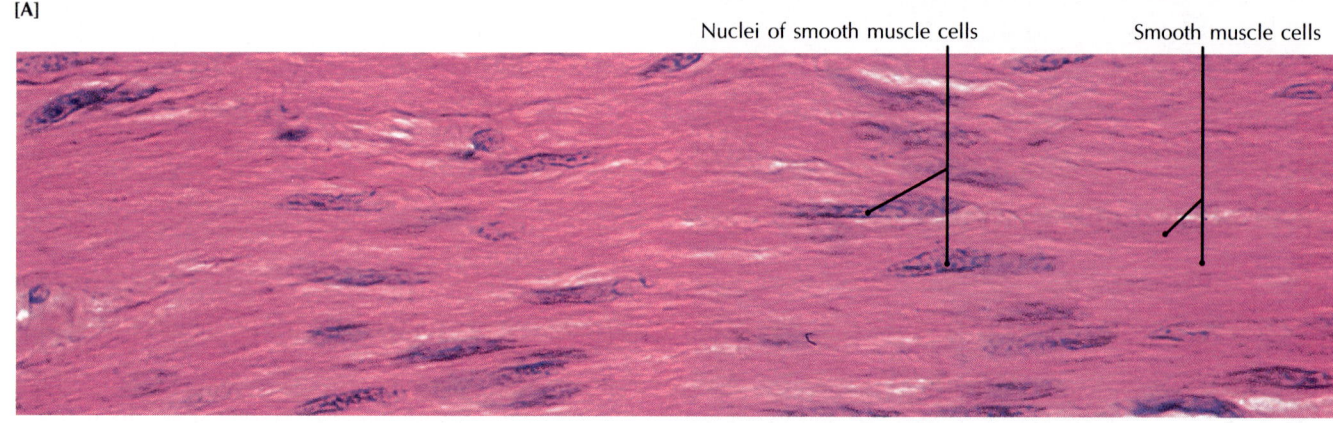

Nuclei of smooth muscle cells   Smooth muscle cells

[B]

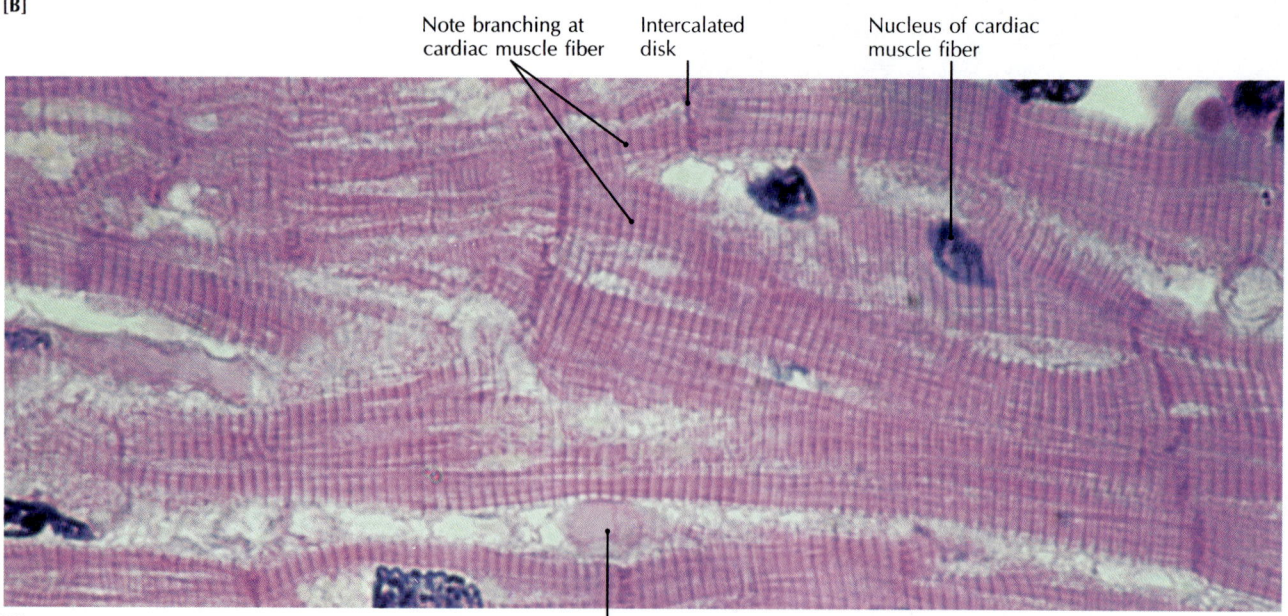

Note branching at cardiac muscle fiber   Intercalated disk   Nucleus of cardiac muscle fiber

[C]

Endomysium

FIGURE 4.14 NERVOUS TISSUE

Light micrograph of nervous tissue, showing its many *neurons,* or nerve cells.

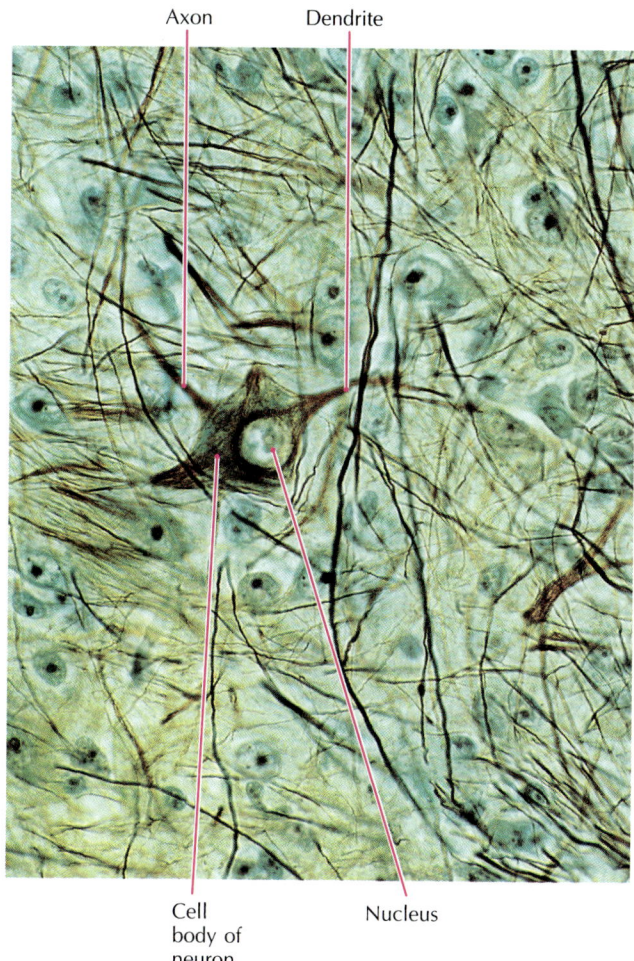

Axon   Dendrite

Cell body of neuron   Nucleus

Serous Membranes

Serous membranes are double membranes of loose connective tissue covered by a layer of simple squamous epithelium called *mesothelium.* They line some of the walls of the thoracic and abdominopelvic cavities and cover the organs that lie within these cavities [FIGURE 4.15; see also FIGURE 1.13B]. The part of a serous membrane that lines a body wall (such as the thoracic wall) is the *parietal layer* (*parietal* means "wall"), and the part that covers organs within the cavity is the *visceral layer* (*viscera* means "body organ").

Serous membranes include the *peritoneum, pericardium,* and *pleura:*

1 The ***peritoneum*** lines the peritoneal cavity and covers the organs inside the cavity. It also lines the abdominal and pelvic walls.

FIGURE 4.15 SEROUS MEMBRANES

Serous membranes **[A]** covering the heart and **[B]** lining the abdominopelvic cavity; sagittal section female.

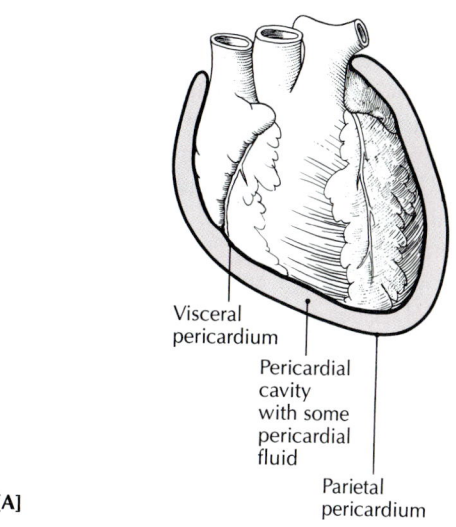

Visceral pericardium

Pericardial cavity with some pericardial fluid

Parietal pericardium

[A]

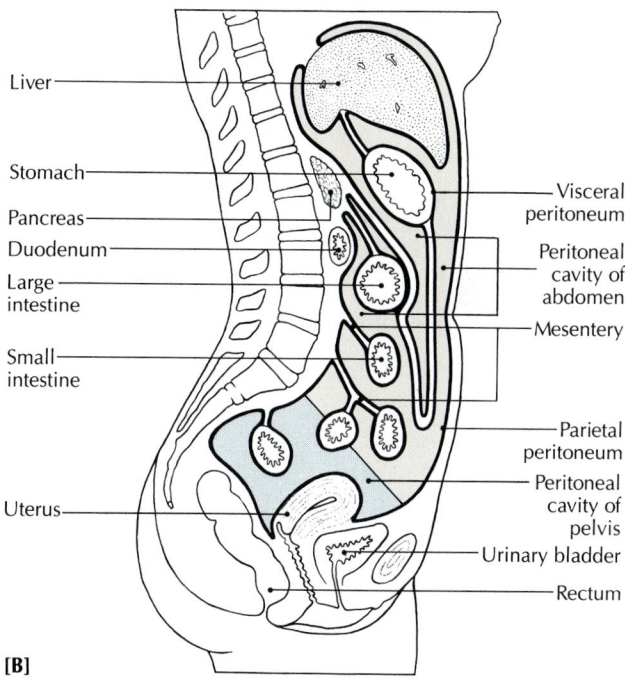

Liver

Stomach

Pancreas

Duodenum

Large intestine

Small intestine

Uterus

Visceral peritoneum

Peritoneal cavity of abdomen

Mesentery

Parietal peritoneum

Peritoneal cavity of pelvis

Urinary bladder

Rectum

[B]

2 The ***pericardium*** lines the pericardial cavity with parietal pericardium and covers the heart with a sac of visceral pericardium.

3 The ***pleura*** (pl. *pleurae*) lines the pleural cavity and covers the lungs. It also lines the wall of the thorax. (*Pleura* means "side" or "rib" but is normally used in connection with the lungs.)

The peritoneum, pericardium, and pleura are all double-layered membranes with a thin space between the layers. These spaces are the peritoneal, pericardial, and pleural *cavities*, respectively. This space between serous membranes receives secretions of *serous fluids* that act as protective lubricants around organs and help remove harmful substances through the lymphatic system. The serous fluid within the spaces of these cavities acts as a lubricating fluid that allows the heart, lungs, and abdominopelvic organs to shift in a virtually frictionless environment. *Pleurisy* is a disease that occurs when the pleural membranes become inflamed and stick together, making breathing painful.

Organs in the abdominopelvic cavity are suspended from the cavity wall by fused layers of visceral peritoneum. These fused tissues are called **mesenteries.** Besides providing points of attachment for organs, mesenteries permit vessels and nerves to connect with their organs through the otherwise impenetrable lining of the abdominopelvic cavity. Many abdominal and pelvic organs, such as the stomach, spleen, jejunum and ileum portions of the small intestine, uterus, and ovaries, are connected to the abdominopelvic wall by a mesentery. (In the case of the stomach, this mesentery is called an omentum; in the uterus, it is known as the broad ligament.) But the kidneys, pancreas, and some other structures are attached to the posterior wall of the abdominopelvic cavity by a mes-

entery *behind* the peritoneal cavity. For this reason, the kidneys and pancreas are considered to be *retroperitoneal* organs, or located behind the peritoneum (*retro* means "behind").

Synovial Membranes

Synovial membranes line the cavities of movable joints, such as the knee and elbow, and other similar areas where friction needs to be reduced. They are composed of loose connective and adipose tissues covered by fibrous connective tissue. No epithelial tissue is present. The inner surface of synovial membranes is generally smooth and shiny, and the synovial fluid is a thick liquid with the consistency of egg white. It helps reduce friction at the movable joints, and in fact, the synovial fluid, together with the covering of the bones, produces a mechanism that is practically frictionless.

A S K   Y O U R S E L F

1 Where are mucous membranes found?

2 What are the three kinds of serous membranes?

3 What are mesenteries?

4 How do synovial membranes aid movement?

WHEN THINGS GO WRONG

The prevention and treatment of certain diseases require an understanding of the structure and function of *collagen*, the important protein that forms the fibrous network in almost all tissues of the body.

The *pathogenesis* (development of disease conditions) of certain inherited and acquired diseases of connective tissues is now understood with respect to collagen synthesis and metabolic dysfunction caused by a lack of vitamin C (ascorbic acid) in the diet. Vitamin C is essential to collagen formation in connective tissues. In **scurvy**, collagen fibers are not formed. As a result, bone growth is abnormal, capillaries rupture easily, and wounds and fractures do not heal. In **rheumatoid arthritis,** collagen in articular cartilage (cartilage within a joint) is destroyed by specific enzymes from the inflamed synovial membranes. In **arteriosclerosis,** some of the damage to blood vessels is caused by the secretion of large amounts of collagen by the smooth muscle cells.

Biopsy

A **biopsy** (Gr. *bios*, life + *-opsy*, examination) is the microscopic examination of tissue removed from the body for the purpose of diagnosing a disease. It is especially useful in diagnosing cancer, since the difference between benign and malignant tissue is usually dramatic. Several types of biopsies can be performed, depending on the specific diagnostic requirement. In a *skin* or *muscle biopsy*, a small piece of skin or muscle tissue is removed under a local anesthetic, and the small wound is then sutured. In a *needle biopsy*, a biopsy needle is inserted through a small incision into the skin or organ to be examined, and a tissue sample is then removed. Another type of needle biopsy is an *aspiration biopsy*, in which cells are sucked from a tumorous tissue through a special needle. Needle biopsies usually require only a local anesthetic and are guided by ultrasound or CT scanning when the surgeon is unable to visually locate the tissue in question. During an *endoscopic biopsy*, an endoscope is passed into an accessible hollow organ such as the stomach or urinary bladder. A forceps attached to the penetrating end of the endoscope snips a tissue sample from the lining of the organ. Endoscopy usually requires sedation.

In an *open biopsy*, the surgeon actually opens a body cavity to obtain a sample of diseased tissue. This method is necessary when guided and endoscopic biopsies are impractical, and the symptoms indicate that the biopsy will be positive and the diseased organ will have to be removed. An *excisional biopsy* is used to remove a lump in a tissue or organ. After removal, the tissue is examined microscopically according to standard procedures.

Edema

Edema (ih-DEE-muh; Gr. swelling) is the condition in which excess interstitial fluid accumulates in the tissues. Such an increase of fluid also causes excessive pressure, and the tissues swell, creating a puffiness on the skin surface. (See Chapter 28 for a more detailed discussion of edema and its causes.)

Gangrene

Gangrene (Gr. *grainein*, to gnaw) is the necrosis (death) of the soft tissues of a body part (often a limb, fingers, or toes) when the arterial blood supply is obstructed as a result of bacterial infection or direct mechanical injury. In *gas gangrene*, foul-smelling gases are produced as bacteria digest the dead tissue, and the tissue becomes dark green or black as hemoglobin in the blood is broken down. *Intestinal gangrene* usually follows an obstruction of the blood supply to the intestines.

Marfan's Syndrome

Marfan's syndrome, or arachnodactyly (Gr. "spider fingers"), is a rare genetic disease of connective tissue that causes skeletal and cardiovascular defects and eye problems such as nearsightedness (caused by an abnormal position of the eye lens), detached retina, and glaucoma. People with this disease grow very tall and thin. Other characteristic skeletal deformities include abnormally long arms (arm span usually exceeds height), legs, fingers, and toes; concave or convex chest and spine caused by excessive growth of the ribs; and weak tendons, ligaments, and joint capsules, resulting in easily dislocated joints ("double-jointedness"). In almost all cases the heart valves or aorta (the major blood vessel leading from the heart) are abnormal. People with Marfan's syndrome are usually cautioned not to take part in strenuous physical activities that might cause a weak-walled aorta to burst, causing sudden death. Marfan's syndrome usually becomes apparent by about age 10, and those affected often die before they reach 50 from cardiovascular complications, especially those related to the expansion or constriction of the aorta. The defective gene that produces Marfan's syndrome has been isolated, and the disease can now be diagnosed during prenatal testing. The gene codes for *fibrillin*, a protein that helps support blood vessels and other body structures. Some of the effects of Marfan's syndrome probably result from abnormalities in elastin (the essential component of elastic connective tissue) and collagen. Symptomatic defects may be treated surgically. It has been suggested that Abraham Lincoln suffered from Marfan's syndrome, but there is little confirming evidence.

Systemic Lupus Erythematosus

Systemic lupus erythematosus (SLE) is an autoimmune disease that affects collagen in the lining of joints and other connective tissue. Women are affected about nine times more often than men. The first signs of the disease, which can be fatal in extreme cases, are general malaise, fever, a so-called butterfly rash on the face, appetite loss, sensitivity to sunlight, and pain in the joints. SLE causes cells in the immune system (B cells) to produce excess antibodies, called autoantibodies, which attack healthy cells. The basal layer of skin begins to deteriorate, and any organ can be affected. Because the body is unable to remove or destroy the autoantibodies, they may settle in such vital organs as the brain, heart, kidneys, and lungs, ultimately causing serious tissue damage.

Tissue Transplantations

The process of surgically *transplanting* healthy tissue to replace diseased or defective tissue is also called *grafting*. There are four basic types of tissue transplantations:

1 An *allograft* (Gr. *allos*, another) is a tissue transplant from one person (often recently deceased) to another person who is *not genetically identical.* The recipient usually begins to reject an allograft after 15 to 25 days, when the transplantation site becomes infiltrated with *graft-rejection cells* that recognize and destroy the foreign cells of the newly grafted tissue.

2 A *heterograft* (Gr. *heteros*, other) is a tissue transplant from an animal into a human being. Although re-jection may occur after a week or so with skin allografts and heterografts, they provide protection until the patient's own skin is ready to be transplanted.

3 An *autograft* (Gr. *autos*, self) is a tissue transplant from one site on a person to a different site on the *same person*. When performing a skin autograft operation, the surgeon does not have to connect the blood vessels of the graft to those of the host skin because the graft skin is thin enough to be nourished by blood seeping into it from the host skin. If the graft is not rejected, blood vessels from the host grow into the grafted skin.

4 A *syngeneic graft* is a tissue transplant from one person to a *genetically identical* person (in other words, from one identical twin to another).

Part of the body's rejection response to tissue grafts consists of the proliferation of activated "killer" T cells. These are specialized white blood cells that normally help destroy microorganisms that enter body tissues. The most successful tissue transplant involves the cornea of the eye precisely because the cornea, having no blood supply to provide entry to T cells, is not affected by the rejection response of the immune system. *Immunosuppressive drugs* such as cyclosporine can be used to suppress tissue rejection by reducing the body's supply of killer T cells, but most immunosuppressive drugs also reduce the patient's ability to fight infection.

A process called *tissue typing* is used to match donor and recipient tissues as closely as possible in an attempt to minimize the rejection response.

CHAPTER SUMMARY

Development of Primary Germ Layers (p. 101)

1 During week 2, the blastocyst develops into the *bilaminar embryonic disk* composed of the endoderm and ectoderm. The zygote is now an *embryo.*

2 The *differentiation* of cells into specialized tissues and organs begins shortly after the start of implantation. The inner cell mass of the blastocyst is rearranged into the *primary germ layers:* endoderm, ectoderm, and mesoderm.

3 *Endoderm* develops mainly into the innermost organs, *ectoderm* into the outer layers of skin and the nervous system, and *mesoderm* into connective tissue, bone, muscle, blood, and some inner epithelium.

Epithelial Tissues: Form and Function (p. 101)

1 *Epithelial tissues* may be in the form of sheets that *cover surfaces* or *line body cavities or ducts,* or they may be modified to become *glandular epithelia* that are organized into glands.

2 The typical functions of epithelial tissues are *absorption, secretion, transport, excretion,* and *protection.*

3 All types of epithelial tissues except glandular tissues share certain characteristics: (1) the cells are *regular in shape* and *packed in continuous sheets,* (2) the tissue is usually anchored to the underlying connective tissue by a *basement membrane,* (3) the tissue *has no blood* vessels, (4) there are several types of *surface specializations* (microvilli and cilia), and (5) the cells are capable of *regeneration.*

4 Epithelial tissues are classified according to the *arrangement of cells,* the *number of cell layers,* and the *shape of the cells* in the superficial layer.

5 Epithelial tissues one cell layer thick are *simple;* tissues with two or more cell layers are *stratified.* The basic shapes of the superficial cells are *squamous, cuboidal,* and *columnar.*

6 *Simple squamous epithelium* consists of flat cells in a single layer, suited for diffusion or filtration.

7 *Simple cuboidal epithelium* has a single layer of approximately cube-shaped cells. It functions in secretion and absorption.

8 *Simple columnar epithelium* is a single layer of elongated cells, with some cells modified to secrete mucus and others modified for absorption, protection, and cellular movement.

9 *Stratified squamous epithelium* usually has at least three layers of cells; only the superficial layer contains flat squamous cells. Underlying *basal cells* can replace damaged superficial squamous .cells. Its function is primarily protective.

10 *Stratified cuboidal epithelium* is a multilayered arrangement of somewhat cube-shaped secretory cells.

11 *Stratified columnar epithelium* has a superficial layer with a regular arrangement of columnar cells; the smaller cuboidal cells underneath are in contact with the basement membrane. Its main function is secretion.

12 Exceptions to the basic system of classification are *pseudostratified columnar epithelium, transitional epithelium,* and *glandular epithelium.*

13 *Pseudostratified columnar epithelium* has all its cells in contact with the basement membrane, but not all reach the surface. Functions include protection, secretion, and cellular movement.

14 *Transitional epithelium* is composed of cells capable of stretching as needed, allowing for changes in shape.

15 *Glandular epithelium* has cells specialized for the synthesis, storage, and secretion of their products. The two gland types are *exocrine* and *endocrine.*

16 *Endocrine glands* do not have ducts. They release their secretions into the bloodstream. *Exocrine glands* release their secretions into ducts.

17 Exocrine glands are classified as *unicellular* or *multicellular; simple* or *compound; tubular, alveolar,* or *tubuloalveolar; mucous, serous,* or *mixed;* and *merocrine* or *holocrine.*

Connective Tissues (p. 112)

1 Besides their supportive and protective functions, connective tissues form a sheath around hollow organs; they are involved in storage, transport, and repair.

2 Connective tissues consist of fibers, ground substance, some extracellular fluid, and cells. Fibers and ground substance make up the *matrix.*

3 Types of connective tissue fibers are *collagenous, reticular,* and *elastic.*

4 *Ground substance* is a homogeneous extracellular material that provides a medium for the passage of nutrients and wastes between cells and the blood.

5 The cells of connective tissues are usually classified as either *fixed cells (fibroblasts, adipose cells,* and *reticular cells)* or *wandering cells (plasma cells, mast cells,* and *leukocytes). Macrophage cells* may be fixed or wandering cells. Fixed cells are usually concerned with *long-term functions,* and wandering cells with *short-term activities.*

6 The *reticuloendothelial system* contains specialized cells that are active in defending the body against microorganisms and are involved with other cells in producing antibodies. These cells remove undesirable matter from the fluids and matrix of body tissues.

7 Connective tissues are classified according to the arrangement and density of their fibers. Typical connective tissues are *loose* (areolar), *dense* (collagenous), *elastic adipose,* and *reticular.*

8 *Embryonic connective tissue* is called *mesenchyme* in the embryo and *mucous connective tissue* in the fetus.

9 *Cartilage* is a specialized connective tissue that provides support and facilitates movement at the joints.

10 *Chondrocytes* are cartilage cells housed in *lacunae,* cavities within the matrix.

11 The three types of cartilage are *hyaline cartilage, fibrocartilage,* and *elastic cartilage.*

12 Hyaline cartilage and elastic cartilage are usually enclosed within a fibrous sheath called a *perichondrium.*

Muscle Tissue: An Introduction (p. 121)

1 *Muscle tissue* helps the body and parts of the body move.

2 The types of muscle tissue are *skeletal, smooth,* and *cardiac* (heart).

Nervous Tissue: An Introduction (p. 121)

1 *Nervous tissue* is composed of several different types of cells, including *neurons* (impulse-conducting cells), nonconducting *neuroglia,* and *peripheral glial cells.*

Membranes (p. 121)

1 *Membranes* are thin, pliable layers of epithelial or connective tissue that line body cavities, cover surfaces, or separate or connect certain regions, structures, or organs of the body.

2 *Mucous membranes* line body passageways that open to the outside.

3 *Serous membranes* line the pericardial, pleural, and peritoneal cavities within the closed thoracic and abdominopelvic cavities and cover the organs that lie within these cavities. Serous membranes include the *peritoneum, pericardium,* and *pleura.*

4 *Synovial membranes* line the cavities of joints and other similar areas.

STUDY AIDS FOR REVIEW

KEY TERMS

MEDICAL TERMS FOR REFERENCE

ATROPHY (AT-roe-fee; Gr. *atrophos*, ill-nourished; *a-*, without + *trophe*, nourishment) A wasting of body tissue, usually caused by a lack of use, such as when a broken arm is healing within an immobilizing cast, or when faulty blood circulation produces insufficient cell nutrition.

BIOPSY (BI-op-see; Gr. *bios*, life; *-opsy*, examination) The removal of tissue for microscopic examination as part of the procedure for diagnosing specific diseases, including cancer.

UNDERSTANDING THE FACTS

1 How are epithelial tissues classified?
2 Where are basement membranes found and what are their functions?
3 Contrast simple and compound glands. Are they exocrine or endocrine? Explain.
4 How are connective tissues classified?
5 What are the components of the connective tissue matrix?
6 Contrast mucous and serous membranes with respect to general location, type of secretion, and function.

UNDERSTANDING THE CONCEPTS

1 Why would you expect to find junctional complexes in epithelial tissues but not in connective tissue?
2 Contrast the role of the matrix in epithelial and connective tissues.
3 What is the source of nourishment for cells in epithelial and connective tissues, and how do the nutrients reach those cells?
4 Contrast the role of epithelial and connective tissues in mucous and serous membranes.
5 Contrast the general role of the cellular components of epithelial and connective tissues.

SELF-QUIZ
Multiple-Choice Questions

4.1 Which of the following is *not* a general characteristic of epithelial tissues?
 a regular shaped cells
 b cells at rest on a basement membrane
 c cells tightly joined to each other
 d both fixed and wandering cells
 e junctional complexes present

4.2 The principal role(s) of the basement membrane is/are
 a to provide elastic support
 b to act as a barrier for diffusion and filtration
 c to separate epithelium from underlying connective tissue
 d a and b   **e** a, b, and c

4.3 Which of the following structures is *not* found on the apical surface of simple epithelium?
 a cilia   **d** spot desmosome
 b microvilli   **e** mucus
 c brush border

4.4 Transitional epithelium is
 a found in the urinary bladder
 b stratified with surface cells that are larger and more rounded than those of underlying layers
 c able to change in thickness
 d a and b   **e** a, b, and c

4.5 Which of the following lines the abdominopelvic cavity and covers the organs inside the cavity?
 a peritoneum   **d** synovial membrane
 b pericardium   **e** mucous membrane
 c pleura

Matching

a simple squamous epithelium
b simple cuboidal epithelium
c stratified squamous epithelium
d merocrine glandular epithelium
e holocrine glandular epithelium

4.6 _____ functions in secretion and absorption
4.7 _____ usually located in areas subject to abrasion
4.8 _____ secretion composed of debris of whole cells
4.9 _____ lines hollow tubes and cavities; may act as a semipermeable surface
4.10 _____ secretes by exocytosis

a reticular cell
b macrophage
c mast cell
d plasma cell
e fibroblast

4.11 _____ synthesizes collagen
4.12 _____ synthesizes heparin and histamine
4.13 _____ secretes antibodies
4.14 _____ phagocytic; may be scavenger cell
4.15 _____ star-shaped fixed cell involved in immune response

a ground substance
b elastic fibers
c reticular fibers
d collagenous fibers

4.16 _____ thick bundles, whitish, unbranched fibrils
4.17 _____ composed mostly of glycoproteins
4.18 _____ allows hollow organs to stretch
4.19 _____ delicately formed networks of inelastic fibrils

Completion Exercises

4.20 Epithelium that makes up the serous membranes of various body cavities is called _____.

4.21 Overall, simple cuboidal cells are active in absorption and _____.
4.22 _____ epithelia have nuclei situated at different levels, giving a multilayered impression.
4.23 Fibers of connective tissue may be collagenous, reticular, or _____.
4.24 The most common cells in connective tissue, and the only cells found in tendons, are _____.
4.25 Connective tissues are usually classified according to the arrangement and density of their _____.
4.26 When connective tissue with wandering cells is found in the embryo it is called _____.
4.27 Cartilage cells are called _____.
4.28 _____ cartilage is the most prevalent type of cartilage in the human body.
4.29 The important protein that forms the fibrous network in almost all tissues of the body is _____.
4.30 Tendons and ligaments are classified as _____, _____ connective tissue.
4.31 _____, areolar connective tissue is found in the lower layers of the skin.
4.32 The only connective tissue with tightly clustered cells is called _____ tissue.
4.33 _____ connective tissue contains fibroblasts, some collagenous fibers, and many yellow branched fibers.

Problem-Solving/Critical Thinking Questions

4.34 A surgeon performing abdominal surgery first cuts through the skin, then loose _____ tissue, then _____, to reach the _____ lining of the inside wall of the abdomen.
a connective, muscle, parietal peritoneum
b adipose, connective, visceral peritoneum
c muscle, fat, visceral pleura
d areolar, smooth muscle, parietal pericardium
e connective, epithelium, elastic cartilage

A SECOND LOOK

Identify the type of tissue illustrated in each of the drawings.

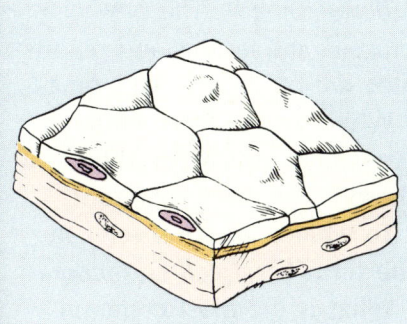
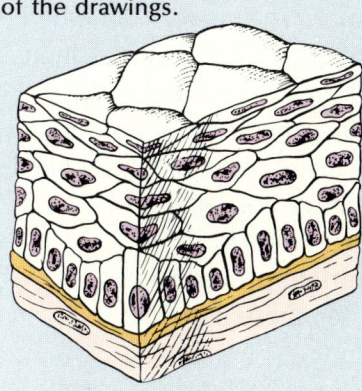
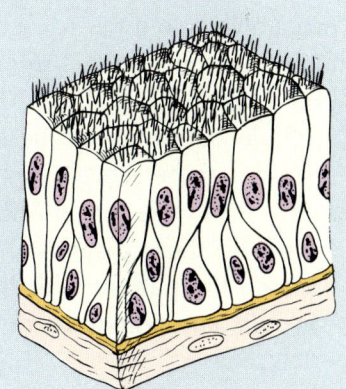

i case studies

1

A 23-year-old male received a gunshot wound. He was transported to the local emergency room where it was determined that the bullet had entered the thoracic cavity, passed through the middle lobe of the right lung, and lodged in the wall of the right ventricle of the heart.

- **Name the body cavities that the bullet entered or passed through.**
- **Through which internal membranes did the bullet pass?**
- **What conventional medical imaging device was probably used to locate the exact site where the bullet had lodged?**

2

While studying cell structure, each lab student received a vial containing an "unknown" cell suspension (5 mL of red blood cells in an isotonic solution). The students were to examine the suspension under the microscope and submit a description of what they had seen. When Jake, one of the students, dropped his vial, he decided that, rather than bothering the instructor, he would pour half of his lab partner's sample into an empty vial and then fill the vial to volume with water. After doing the lab exercise, Jake discovered that his description was very different from that of his partner, despite the fact that both had used the same cell suspension.

- **What would the cells in Jake's partner's vial have looked like?**
- **How would the cells in Jake's vial have differed from those in his partner's vial? Why?**
- **How does this incident illustrate the principle of osmosis?**

3

Three patients are in the waiting room that services Radiography and Nuclear Medicine. One, an elderly woman, has fallen while leaving her house. She is in great pain, is unable to walk, and has probably broken the neck of her femur. The second, an elderly man, has the slurred speech and drooping facial features indicative of a stroke. The third, a 20-year-old male, is known to have multiple sclerosis, but the extent of degeneration in his nervous system is to be reevaluated. One of these patients is scheduled for a PET scan, one for an MRI, and one for an x-ray.

- **How do these diagnostic procedures differ?**
- **Match the patient with the appropriate procedure.**
- **Would sonography be of any use for these patients?**

4

For many years, a fair-skinned male, now 40, enjoyed playing frequent games of volleyball on the beach. Though he often used a tanning lotion, he never used a brand with a sunblock ingredient. Now he has noticed several small, slow-growing, shiny, pinkish nodules on his forehead and his nose. At the urging of his friends, he goes to a dermatologist. The physician biopsies the lesions and reports they are basal cell carcinomas.

- **What are the three major types of skin cancer? Which of these is the most common?**
- **To what does *basal* refer in the term *basal cell carcinoma*?**
- **What is the most common form of treatment for this type of skin cancer?**

ii

Protection, Support, and Movement

CAREER IN FOCUS: PHYSICAL THERAPY

1984, Los Angeles, California: The first women's marathon ever held as part of the Olympic Games was won by the odds-on favorite, Joan Benoit Samuelson of the United States. For Samuelson, the drama had already peaked at the National Selection Trials for the Games. One month before the Trials, she had undergone arthroscopic knee surgery. The surgery was successful, but now there was less than a month to see whether intensive physical therapy—and a champion's determination—could pull it off.

1994, Detroit, Michigan, the National Selection Trials for the Olympic figure skating team: Nancy Kerrigan, the favorite for one of two places on the team, had just left a practice session on the ice when an assailant wielding a metal bar brutally clubbed her right leg just above the knee. Magnetic resonance imaging showed no significant damage, but now another determined champion had less than a month of intensive physical therapy to achieve the dream of a lifetime.

Anytime, anywhere: An elderly woman has fallen. Weakened by osteoporosis, her hip has fractured. The bones have been set and pinned, but without physical therapy it is clear that limited range of motion in the affected joint will prevent the resumption of an inde-

pendent life. Dressing, bathing, shopping for food—all require more effort than her musculoskeletal system can deliver.

A population of athletes whose prowess seemingly knows no limits and the increasing numbers of elderly in this nation have fueled a growing need for physical therapists. The U.S. Bureau of Labor Statistics estimates that by the year 2005, the number of jobs for physical therapists will have increased by 76 percent.

Physical therapists evaluate and implement rehabilitative programs for recovering medical and surgical patients; treat injured athletes and condition them for peak performances; and screen, evaluate, and develop exercise programs for

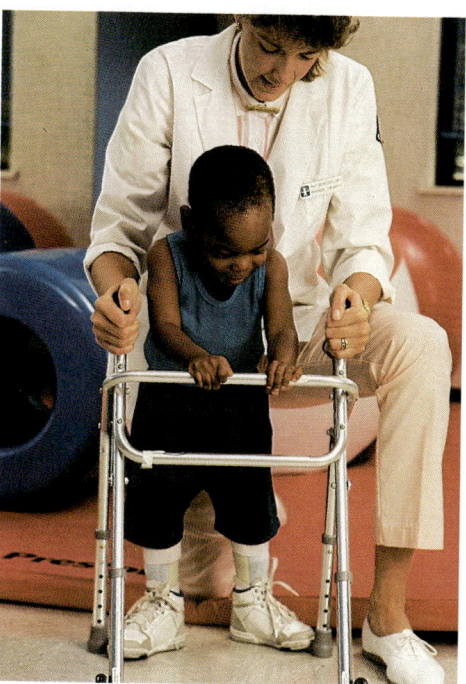

schoolchildren, pregnant women, and employees in physically demanding professions. While many physical therapists work in hospitals and rehabilitation facilities, today even more are in private practice. Many attain certification in one of seven specialty areas—cardiopulmonary, electrophysiological, geriatric, neurological, orthopedic, pediatric, and sports physical therapy.

The route into the profession is usually a postbaccalaureate or master's degree program. Prerequisites include good writing skills and a strong liberal arts background, with specific preparation in anatomy and physiology, chemistry, and psychology. More advanced work in anatomy, physiology, and kinesiology, capped by a clinical internship, is included in most 2- or 3-year graduate programs. Upon certification, a world of opportunity awaits these hands-on professionals.

This section explores the "nuts and bolts" of the body—skin, bones, muscles, and joints. We have learned an enormous amount about body conditioning in the last few years, largely because of the growing interest in athletics—both amateur and professional. In fact, physical therapists and other people who work in this area have often led the way in increasing our knowledge of the systems that protect, support, and move the human body.

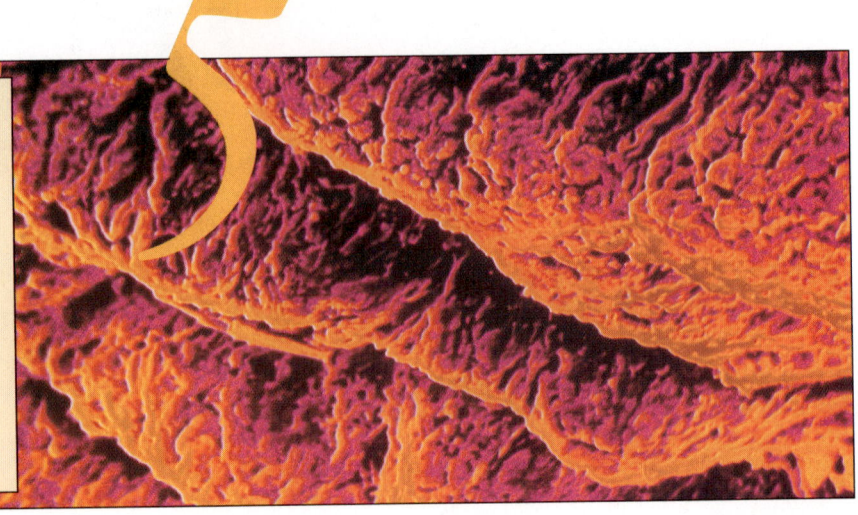

The Integumentary System

KEY CONCEPTS

1 The integumentary system consists mainly of skin but also includes hair, nails, and skin glands, which are all derived from the same embryonic tissue.

2 Skin serves many functions, including protection, regulation, and excretion. It is active in the healing process of wounds.

3 Skin glands include oil and sweat glands, which are important in the skin's overall functions.

4 To a certain extent, skin, hair, and nails are replaced as used-up cells are shed.

The ***integumentary system*** (L. *integumentum,* cover) consists of the skin and its derivatives, which include hair, nails, and several types of glands. The system functions in protection and in the regulation of body temperature, in the excretion of waste materials, in the synthesis of vitamin D_3 from the sun's rays, and in the reception of various stimuli perceived as pain, pressure, and temperature.

SKIN

The ***skin*** is the largest organ in the body, occupying almost 2 m² (21.5 ft²) of surface area. It varies in thickness on different parts of the body from less than 0.5 mm on the eyelids to more than 5 mm on the middle of the upper back. A typical thickness is 1 to 2 mm.

Skin has three main parts: The *epidermis* is the outermost layer of epithelial tissue, the *dermis* is a thicker layer of connective tissue beneath the epidermis, and the *hypodermis* is a layer of loose, fibrous connective tissue beneath the dermis [FIGURE 5.1A].

Epidermis

The outer layer of the skin, called the ***epidermis*** (Gr. *epi,* over + *derma,* skin), is stratified squamous epithelium. The epidermis contains no blood vessels, but most of it is so thin that even minor cuts reach the dermis and draw blood. In the thicker epidermis of the palms of the hands and the soles of the feet, there are five typical layers (strata). Starting with the outermost layer, they are *stratum corneum, stratum lucidum, stratum granulosum, stratum spinosum,* and *stratum basale* [FIGURE 5.1B]. The stratum spinosum and stratum basale together are known as the *stratum germinativum* (jer-mih-nuh-TEE-vuhm) because they generate new cells. In parts of the body other than the palms and soles, only the stratum corneum and stratum germinativum are regularly present.

Stratum corneum The ***stratum corneum*** (L. *corneus,* cornified tissue) is a flat, relatively thick layer of dead cells arranged in parallel rows. This keratinized* layer of the epidermis consists of soft keratin (compared with the hard keratin in fingernails and toenails), which

*__Keratin__ is a fibrous protein produced by epidermal cells called *keratinocytes.* It is insoluble in protein solvents and has a high content of sulfur and certain amino acids.

Q: *Why do the lips of some people appear reddish in color?*

A: The redness of lips is not just due to a rich blood supply, as is sometimes thought. They appear red because the stratified squamous epithelium of the lips is noncornified and relatively translucent. This translucency allows us to see the reddish color of the capillaries in the numerous connective tissue papillae (fingerlike projections) that extend into the epithelium.

helps keep the skin elastic. The soft keratin also protects living cells underneath from being exposed to the air and drying out. The cells in the stratum corneum are constantly being shed through normal abrasion. They are replaced by new cells that are formed by cell division and pushed up from the germinative layers below [FIGURE 5.2]. For example, cells on the soles of the feet are completely replaced every month or so.

Stratum lucidum The ***stratum lucidum*** (L. *lucidus,* bright, clear) consists of flat, translucent layers of dead cells that contain a protein called *eleidin* (ell-EE-ih-dihn). This protein is probably a transitional substance between the soft keratin of the stratum corneum and the precursor of soft keratin, *keratohyaline,* of the layer below. The stratum lucidum appears only in the palms of the hands and soles of the feet, acting as a protective shield against the ultraviolet (UV) rays of the sun and preventing sunburn of the palms and soles.

Stratum granulosum The ***stratum granulosum*** (L. *granum,* grain) lies just below the stratum lucidum. It is usually two to four cells thick. The cells contain granules of keratohyaline. The process of keratinization, which is associated with the dying of cells, begins in this layer.

Stratum spinosum The ***stratum spinosum*** (L. *spinosus,* spiny) is composed of several layers of polyhedral (many-sided) cells that have delicate "spines" protruding from their surface. The interlocking spinelike projections help support this binding layer and make it more difficult for surface bacteria to enter. Active protein synthesis takes place in the cells of the stratum spinosum, indicating that cell division and growth are occurring. Some new cells are formed here and pushed to the surface to replace keratinized cells of the stratum corneum.

Stratum basale The ***stratum basale*** (buh-SAY-lee; L. *basis,* base) rests on the basement membrane next to the dermis. It consists of a single layer of columnar or cuboidal cells. The stratum basale undergoes continuous cell division to produce new cells to replace those being shed in the exposed superficial layer. It is also the site of basal cell carcinoma.

Dermis

Most of the skin is composed of ***dermis*** ("true skin"), a strong, flexible connective tissue meshwork of collagenous, reticular, and elastic fibers. *Collagenous* fibers, which are formed from the protein collagen, are very

Q: *How much skin does a person shed in a lifetime?*

A: We shed about 1.5 lb of skin a year, so a 70-year-old person would have lost about 105 lb of skin.

FIGURE 5.1 THE SKIN

[A] A "textbook" drawing of the skin's structure, showing its many components in an ideal arrangement. The stratum lucidum would not appear on hairy skin. An enlarged portion of the epidermis is shown in the small drawing (top). **[B]** A vertical section of the thick epidermis of the fingertip. ×100.

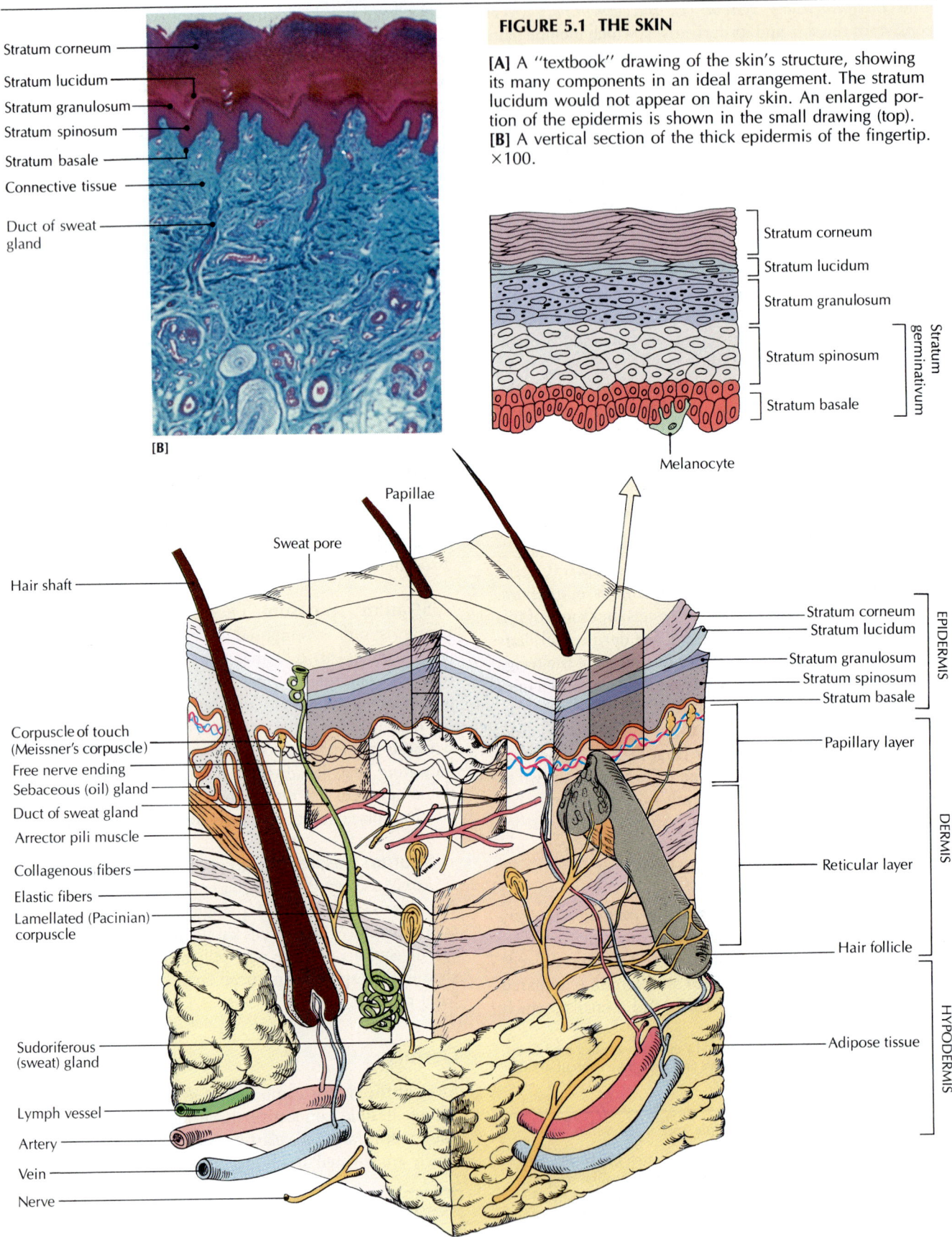

Stratum corneum
Stratum lucidum
Stratum granulosum
Stratum spinosum
Stratum basale
Connective tissue
Duct of sweat gland

[B]

Stratum corneum
Stratum lucidum
Stratum granulosum
Stratum spinosum
Stratum basale
Stratum germinativum
Melanocyte

Papillae
Sweat pore
Hair shaft
Stratum corneum
Stratum lucidum
Stratum granulosum
Stratum spinosum
Stratum basale
EPIDERMIS

Corpuscle of touch (Meissner's corpuscle)
Free nerve ending
Sebaceous (oil) gland
Duct of sweat gland
Arrector pili muscle
Collagenous fibers
Elastic fibers
Lamellated (Pacinian) corpuscle

Papillary layer
DERMIS
Reticular layer
Hair follicle

Sudoriferous (sweat) gland
HYPODERMIS
Adipose tissue

Lymph vessel
Artery
Vein
Nerve

[A]

FIGURE 5.2 THE KERATINIZING SYSTEM OF THE EPIDERMIS

The epidermal layer of the skin contains a large number of keratinocytes, which take part in the continuous renewal of the skin's surface through the four overlapping processes of (1) cell renewal, (2) differentiation (keratinization), (3) cell death, and (4) sloughing of dead cells (exfoliation). Each wave of activity pushes the new cells toward the surface of the skin. The entire process takes 15 to 30 days.

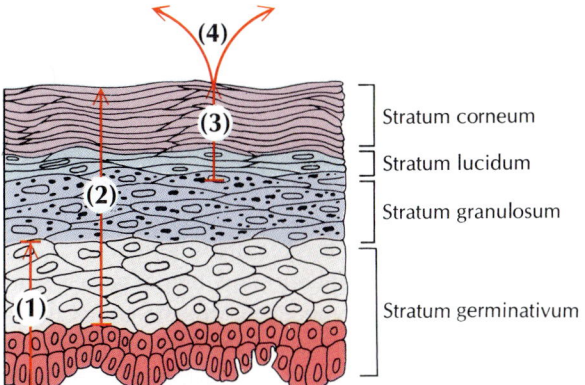

Stratum corneum
Stratum lucidum
Stratum granulosum
Stratum germinativum

FIGURE 5.3 DERMAL PAPILLAE

The boundary between the epidermis and dermis is marked by dermal papillae, which extend into the epidermis.

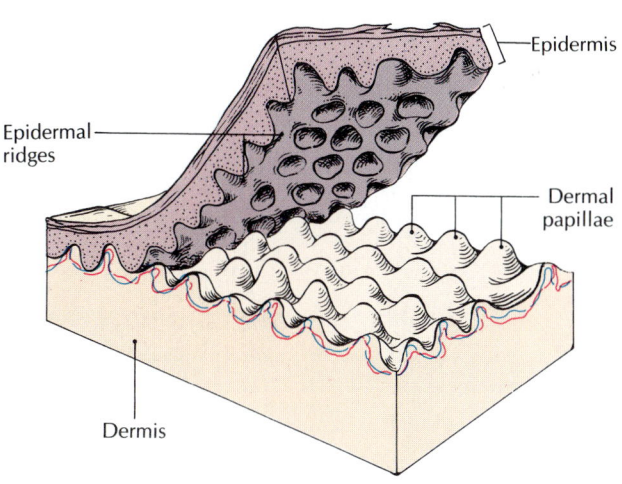

Epidermis
Epidermal ridges
Dermal papillae
Dermis

FIGURE 5.4 COLLAGENOUS FIBERS

Scanning electron micrograph of the meshed collagenous fibers of the skin. Collagen makes up about a third of the protein in the body. It is a major component of skin, tendons, ligaments, cartilage, and bone. ×38,000.

thick and give the skin much of its toughness. (The thickest dermis is located on the back, thighs, and abdomen, in that order.) Although *reticular* fibers are thinner, they provide a supporting network. *Elastic* fibers give the skin flexibility. The cells of the dermis are mostly *fibroblasts*, *fat cells*, and *macrophages*; macrophages digest foreign substances. Blood vessels, lymphatic vessels, nerve endings, hair follicles, and glands are also present. The dermis is composed of two layers that are not clearly separated. The thin *papillary layer* is directly beneath the epidermis; the deeper, thicker layer is called the *reticular layer*.

Papillary layer The *papillary* (L. dim. *papula*, pimple) *layer* of the dermis consists of loose connective tissue with thin bundles of collagenous fibers. The papillary layer is so named because of the papillae (tiny, fingerlike projections) that join it to the ridges of the epidermis [FIGURE 5.3]. Most of the papillae contain loops of capillaries that nourish the epidermis; others have special nerve endings, called *corpuscles of touch* (Meissner's corpuscles), that serve as sensitive touch receptors. A double row of papillae in the palms and soles produces ridges that help keep the skin from tearing, improve the gripping surfaces, and produce distinctive fingerprint patterns in the finger pads. Fingerprint patterns in the epi-

dermis follow the corrugated contours of the dermis underneath. (No two people, not even identical twins, have exactly the same fingerprints.)

Reticular layer The *reticular* ("netlike") *layer* of the dermis is made up of dense connective tissue with coarse collagenous fibers and fiber bundles that crisscross to form a strong and elastic network [FIGURE 5.4]. You

Q: *What is a blister?*

A: When the skin is burned or irritated severely, tiny blood vessels in the dermis dilate and a small amount of clear plasma leaks out. The plasma accumulates between the dermis and epidermis in a fluid-filled pocket, or blister.

FIGURE 5.5 CLEAVAGE (LANGER'S) LINES

Anterior [A] and posterior [B] views.

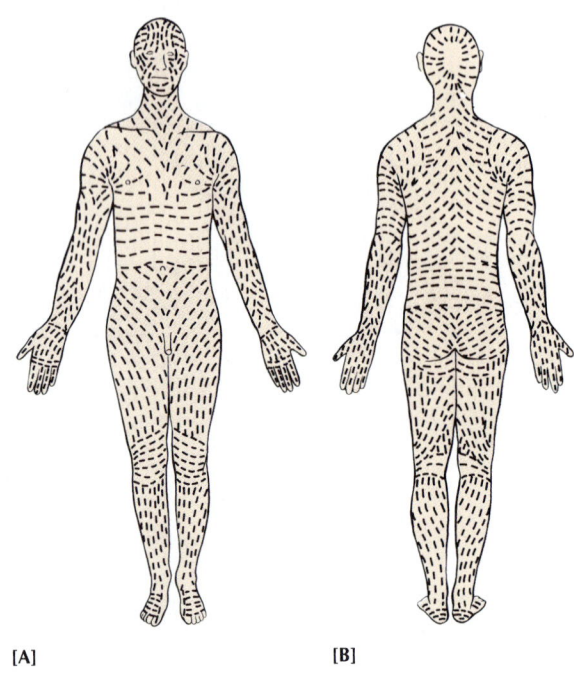

[A]   [B]

can understand how tough the reticular layer is when you realize that it is the part of an animal's skin that is processed commercially to make leather.

Although the collagenous fibers of the dermis appear to be arranged randomly, there is actually a dominant pattern. In fact, different directional patterns are found in each area of the body [FIGURE 5.5]. The resulting lines of tension over the skin are known as *cleavage (Langer's) lines.* They are very important during surgery because an incision made *across* the lines causes a gaping wound that bleeds more profusely, heals more slowly, and leaves a deeper scar than does an incision made *parallel* to the directional lines. An incision made parallel to the cleavage lines heals faster because it disrupts the collagenous fibers only slightly, and the wound needs only a small amount of scar tissue to heal.

Embedded in the reticular layer are many blood and lymphatic vessels, nerves, free nerve endings, fat cells, sebaceous (oil) glands, and hair roots. Receptors for deep

pressure (Pacinian corpuscles) are distributed throughout the dermis and hypodermis (subcutaneous layer). The deepest region of the reticular layer contains smooth muscle fibers, especially in the genital and nipple areas and at the base of hair follicles.

Although the dermis is highly flexible and resilient, it can be stretched beyond its limits. During pregnancy, for example, collagenous and elastic fibers may be torn. Characteristic abdominal "stretch marks" result from the repairing scar tissue. Weight lifters sometimes have stretch marks on their shoulders.

Hypodermis

Beneath the dermis lies the *hypodermis* (Gr. *hypo*, under), or *subcutaneous layer* (L. *sub*, under + *cutis*, skin). It is composed of loose, fibrous connective tissue. The hypodermis is generally much thicker than the dermis and is richly supplied with lymphatic and blood vessels and nerves. Also within the hypodermis are the coiled ducts of sudoriferous (sweat) glands and the bases of hair follicles. The boundary between the epidermis and dermis is distinct; that between the dermis and the hypodermis is not.

Where the skin is freely movable, hypodermis tissue fibers are scarce. Where it is attached to underlying bone or muscle, the hypodermis contains tightly woven fibers. In some areas of the hypodermis where extra padding is desirable, as in the breasts or the heels, thick sheets of fat cells are present. The distribution of fat in this layer is largely responsible for the characteristic body contours of the female.

Functions of Skin

An obvious function of the skin is to cover and protect the inner organs. However, this is only one of its many functions.

Protection The skin acts as a stretchable protective shield that prevents harmful microorganisms and foreign material from entering the body and also prevents the loss of body fluids. These functions are made possible by two features: (1) layered sheets of flat epithelial tissue that act like shingles on a roof [see FIGURE 4.5] and (2) a nearly waterproof layer of keratin in the outer layer of the skin. Skin also plays an active role in defending the body against various diseases.

Q: *Why is a tattoo so difficult to remove?*

A: Tattoo dyes are injected below the stratum germinativum, into the dermis, where the dyed cells do not move outward toward the surface as the skin is shed and replaced. In the same way, scars and stretch marks do not disappear easily because they involve the dermis.

Q: *Why does ringworm form a circle?*

A: Ringworm is caused by fungi, not worms, which spread outward from a central point. The fungi in the center die before the outer fungi, forming a circular shape.

We All Get Wrinkles

Everybody's skin gets wrinkled. The length and depth of a wrinkle may differ from person to person, but no one is exempt. In fact, wrinkling is so predictable that we can expect to wrinkle in the same place, and at the same time, as the generation ahead of us. It is relatively easy to estimate a person's age merely by studying the wrinkles on the face and neck.

Wrinkling usually starts in the mid-twenties in the areas of the greatest facial expression: around the eyes, mouth, and brow. The familiar "crow's feet," lines radiating from the corner of each eye,

usually appear by the age of 30. During the thirties and forties, new lines appear between the eyebrows and in front of the ears, followed in the fifties by wrinkles on the chin and bridge of the nose. New wrinkles appear on the upper lip in the sixties and seventies, and all of the established wrinkles become more pronounced. In the eighties, the ears become elongated, and we can see the "long ears" of old age.

Wrinkling happens when the protein *elastin* in the elastic tissue of the dermis loses its resiliency and degenerates into *elacin*. This causes the dermis to become

more closely bound to the underlying tissue. Also, the layer of fat beneath the skin diminishes, causing the skin to sag into wrinkles. The wrinkling process is reinforced because an adult's aging epidermis does not produce new cells as readily as it did during childhood and adolescence. The epidermal cells are simply not as vital as they used to be.

Exposure to the ultraviolet rays in sunlight speeds up the wrinkling process, causing the collagen and elastin in the dermis to degenerate. As a result, the skin becomes slack and prematurely aged.

Temperature regulation Although the skin is almost waterproof and is solid enough to keep out the water when you take a bath, it is also porous enough to allow some chemicals in* and to allow sweat to escape from sweat glands through ducts that end at the skin surface as tiny pores.

When sweat is excreted through the pores and then evaporates from the surface of the skin, a cooling effect occurs. On very humid days you are not cooled easily because the outside air is almost saturated with water, and sweat merely builds up on your skin instead of evaporating. On cold days your skin acts as a sheet of insulation that helps retain body heat and keep your body warm. In addition, the dermis contains a dense bed of blood vessels. On warm days the vessels dilate (enlarge), and heat radiation from the blood is increased; in this way, body heat is lost.

Excretion Through perspiration, small amounts of waste materials such as urea are excreted through the

skin. Up to 1 g of waste nitrogen may also be eliminated through the skin every hour.

Synthesis The skin helps screen out harmful excessive UV rays from the sun, but it also lets in some necessary UV rays, which convert a chemical in the skin called 7-dehydrocholesterol into vitamin D_3 (*cholecalciferol*).† Vitamin D in the diet is vital to the normal growth of bones and teeth. A lack of ultraviolet light and vitamin D impairs the absorption of calcium from the intestine into the bloodstream. When children are deprived of sunshine, they generally become deficient in vitamin D. Unless they receive cholecalciferol from another source, they develop rickets, a disease that may deform the bones.

Sensory reception The skin is an important sensory organ containing sensory receptors that respond to heat, cold, touch, pressure, and pain. (Sensory reception in the skin is described in detail in Chapter 17.) Skin helps protect us through its many nerve endings, which keep us responsive to things in the environment that might harm us: a hot stove, a sharp blade, a heavy weight. The nerve endings also help us sense the outside world so that adjustments can be made to maintain homeostasis.

*Unfortunately, several harmful substances can be absorbed through the skin and introduced into the bloodstream. Among the most common intruders are metallic nickel, mercury, many pesticides and herbicides, and a skin irritant called urushiol that is found in poison ivy, poison oak, and poison sumac. Even the smoke from burning these plants may produce skin eruptions.

†The general term "vitamin D" actually refers to a group of steroid vitamins, including vitamin D_3 (*cholecalciferol*) and vitamin D_2 (*calciferol*, or *ergocalciferol*).

Q: *Why do mosquito bites itch?*

A: We itch after being stung by a mosquito because the insect injects saliva that contains a substance that keeps the blood from clotting before the mosquito has drained the blood it needs. The anticlotting substance in the saliva seeps into surrounding tissue, causing an allergic reaction that produces an itch.

Q: *Where does the expression "getting cold feet," meaning being frightened, come from?*

A: When you become frightened, the blood vessels in your skin constrict. As a result of the decreased blood flow, especially to your extremities, you get "cold feet."

Diagnostic Information from Skin

Any physical examination includes a close look at the skin, because the skin provides many clues to the general health of the body. Skin color, especially in lighter-skinned individuals, can reveal several abnormal conditions. For example, yellow-orange skin indicates jaundice (which means *yellow*), a condition arising when bile pigments enter the blood. A person who has been poisoned by silver salts has bluish-gray skin; cyanide poisoning turns the skin "blue," indicating a lack of oxygen in the blood. The skin is a bronze tone when the cortex (outer layer) of the adrenal gland is underactive, and a person with unusually white skin may be anemic.

Different types of rashes are identifiable in diseases such as measles, chickenpox, scarlet fever, ringworm, and syphilis. An allergic reaction is often identified by the eruption of a specific skin rash. Flushed skin means that the underlying blood vessels are dilated; cold, clammy skin is a sign that the blood vessels have been constricted.

The skin is an especially good barometer of nutritional deficiencies. For example, dry skin and loss of skin color may indicate a protein deficiency; deficient amounts of essential fats may produce patchy baldness and eczema, and a shortage of zinc may cause hair loss or sores on the arms or legs; deficiencies in vitamin A produce dry, thickened skin; vitamin B_6, mouth sores, cracks at the corners of the mouth, and flaky skin; vitamin B_{12}, darkening of the skin and loss of hair pigmentation; vitamin C, bleeding gums and poor wound healing; riboflavin, oily skin, mouth sores, and sores at the corners of the mouth; and niacin, dark splotches in sun-exposed areas and mouth and rectal sores.

An itch or a tickle is simply a feeling of low-level pain that is not associated with the unpleasantness of pain. All pain sensations travel in identical nerve pathways to the brain.

Color of Skin

Skin color is determined by three factors: (1) the presence of **melanin** (Gr. *melas*, black), a dark pigment produced by specialized cells called *melanocytes*; (2) the accumulation of the yellow pigment *carotene*; and (3) the color of the blood reflected through the epidermis. Melanocytes are usually located in the deepest part of the stratum basale, where their long processes extend under and around the neighboring cells. Melanin is found in all areas of the skin, but the skin is darker in the external genitals, the nipples and the dark area around them, the anal region, and the armpits. In contrast, there is very little pigment in the palms and soles. Melanin is present not only in skin but also in hair and in the iris and retina of the eyes.

All people have some melanin in their skins. Although darker-skinned people have slightly more melanocytes than lighter-skinned people do, the main reason for their darkness is the wider distribution of melanin in the skin beyond the deep stratum basale into higher levels of the epidermis. Also, the melanocytes of a dark-skinned person produce more melanin than do the melanocytes of a light-skinned person. When present in sufficient amounts, melanin overshadows the yellow pigment carotene, but since Asian people have relatively little melanin, carotene is their dominant skin pigment.

A major function of melanin is to screen out excessive ultraviolet rays, especially protecting a cell's nucleus and genetic material. Extra protection is provided when melanin is darkened by the sun and transferred to the outer skin layers, producing a "suntan" that is less sensitive to the sun's rays than is previously unexposed skin. Increased exposure to the sun not only darkens existing melanin, but also speeds the production of the enzyme *tyrosinase*, which increases the rate and amount of melanin production.* In a sense, "freckles" can be thought of as a permanent tan produced by sun-darkened spots of melanin.

Blood Supply to the Skin

Although the epidermis is avascular, the skin receives an extensive blood supply through blood vessels in the reticular layer of the dermis. Dermal blood vessels hold about 4.5 percent of the body's total blood volume. Blood is supplied to the skin through four dermal sources: (1) arterial plexuses (networks of arteries), (2) papillary capil-

---

Albinism (L. *albus*, white; *albino*, a person afflicted with albinism) is a condition where no melanin is produced. It is most often caused by a genetic defect that eliminates the synthesis of tyrosinase.

Q: *How do transdermal "patches" work?*

A: A transdermal patch is a small, adhesive-backed patch that is attached to the skin. It contains a specific time-released medicine in a waterproof gel that passes through the epidermis into the blood vessels in the dermis, where it is carried through the bloodstream. Although the best-known transdermal patches are those that transmit small amounts of nicotine in an effort to alleviate a nicotine addiction, transdermal patches are used for many other purposes, including administration of nitroglycerine (for chest pain in heart patients), estrogen (during menopause), and scopolamine (for motion sickness).

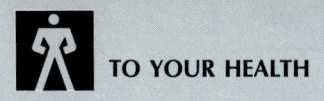

No Fun in the Sun

Why is sunbathing potentially dangerous? The process of tanning is activated by the ultraviolet rays in sunlight. These rays can kill some skin cells, damage others so that normal secretions are stopped temporarily, increase the risk of skin cancer and mutations by affecting the DNA in the nuclei of cells, and cause defects in the immune response. Ultraviolet rays can also damage enzymes and cell membranes and interfere with cellular metabolism. If tissue destruction is extensive, toxic waste products and other cellular debris enter the bloodstream and produce the fever associated with sun poisoning.

Ultraviolet rays cause tiny blood vessels below the epidermis to widen, allowing an increased flow of blood. This increased flow "colors" the skin of light-skinned persons red. (The skin of dark-skinned people usually takes on a darker, purplish hue.) Ultraviolet rays also stimulate melanocytes to produce melanin. The cells begin to divide more rapidly than usual, creating new cells that travel toward the surface in an attempt to repair the damaged skin. Ordinarily, these new cells take about four weeks to reach the surface, where they die and are shed unnoticed in the course of a normal day's activities. New cells produced in response to skin damage, however, are so numerous that they reach the surface in four or five days. At this point, sheets of old, sunburned skin begin to peel off, and the pigment-rich melanocytes, which have moved to just under the surface of the skin, produce the first signs of a tan. The tanned skin helps to prevent further ultraviolet damage by absorbing and scattering the harmful rays.

Unfortunately, many people think that they have to burn before they can tan, but dermatologists (physicians who specialize in treating skin problems) disagree. Instead, they suggest short periods of sun exposure (about 10 min) until the skin has a protective tan.

---

laries (small blood vessels connecting arteries, which carry blood away from the heart, and veins, which carry blood toward the heart), (3) venous plexuses (networks of veins), and (4) arteriovenous anastomoses (shunts).

A *plexus* is a network of interlacing blood vessels. The two **arterial plexuses** lie at the border between the papillary and reticular layers of the dermis and at the border between the dermis and hypodermis. Both plexuses branch out to arterioles (small arteries) that feed the papillary capillaries. The papillary layer of the dermis contains an abundant **capillary network** that supplies oxygen and nutrients to the epidermis. Each papillary capillary bed drains through a single venule (small vein) into one of the three **venous plexuses.**

The dermal plexuses contain many direct connections between arteries and veins. These connections are called **anastomoses** (uh-nas-tuh-MOH-seez). The nerves of these anastomoses control their constriction and dilation, which helps to regulate blood pressure and body temperature by controlling the amount of blood in the papillary capillaries. Dilation of the anastomoses directs blood *away from* the capillaries and into the larger blood vessels, which increases blood pressure and reduces heat loss through the skin; as a result, the skin is pale. Constriction of the anastomoses directs blood *into* the capillaries of the skin, which decreases blood pressure and increases heat loss; as a result, the skin is flushed.

How a Wound Heals

When the skin is cut, the healing process begins immediately. This process, which can take from a week to a month (depending on the severity of the cut), proceeds step by step.

1 Blood vessels are severed along with the dermis and epidermis. Blood cells leak from the severed vessels into the wound. Cell fragments called *platelets* (thrombocytes) and a blood-clotting protein called *fibrinogen* help start a blood clot. A network of fibers containing trapped cells forms, and the edges of the wound begin to join together again. Tissue-forming cells called *fibroblasts* begin to approach the wound [FIGURE 5.6A].

2 Less than 24 hours later, the clotted area becomes dehydrated and a scab forms. Phagocytes called *neutrophils* are attracted from blood vessels into the wound area, ingesting microorganisms, cellular debris, and other foreign material. Epidermal cells at the edge of the wound begin to divide and start to build a bridge across the wound. Other white blood cells called *monocytes* (precursors of macrophages) migrate toward the wound from surrounding tissue [FIGURE 5.6B].

3 Two or three days after the wounding, monocytes enter the wound and ingest the remaining foreign material. Epidermal cells complete the bridge of new skin under the scab [FIGURE 5.6C]. When a totally new epidermal surface has been formed, the protective scab is sloughed off. Fibroblasts build scar tissue with collagen.

4 About 10 days after the wounding, the epidermis has been restored and the scab is gone [FIGURE 5.6D]. Some monocytes and neutrophils remain in the area, but their work is usually completed by this time. Tough scar tissue continues to form, and bundles of collagen build up along the stress lines of the original cut.

FIGURE 5.6 HOW A WOUND HEALS

[A] At the site of a wound, the epidermis, dermis, and blood vessels are severed. Fibroblasts move toward the wound area. **[B]** A scab forms. Epidermal cells at the edge of the wound begin to divide. Neutrophils begin to ingest cellular debris; monocytes move toward the wound area.

[C] New skin is formed under the scab. Monocytes ingest the remaining foreign material in the wound area. Fibroblasts begin to build scar tissue. **[D]** The scab is gone. Scar tissue continues to form.

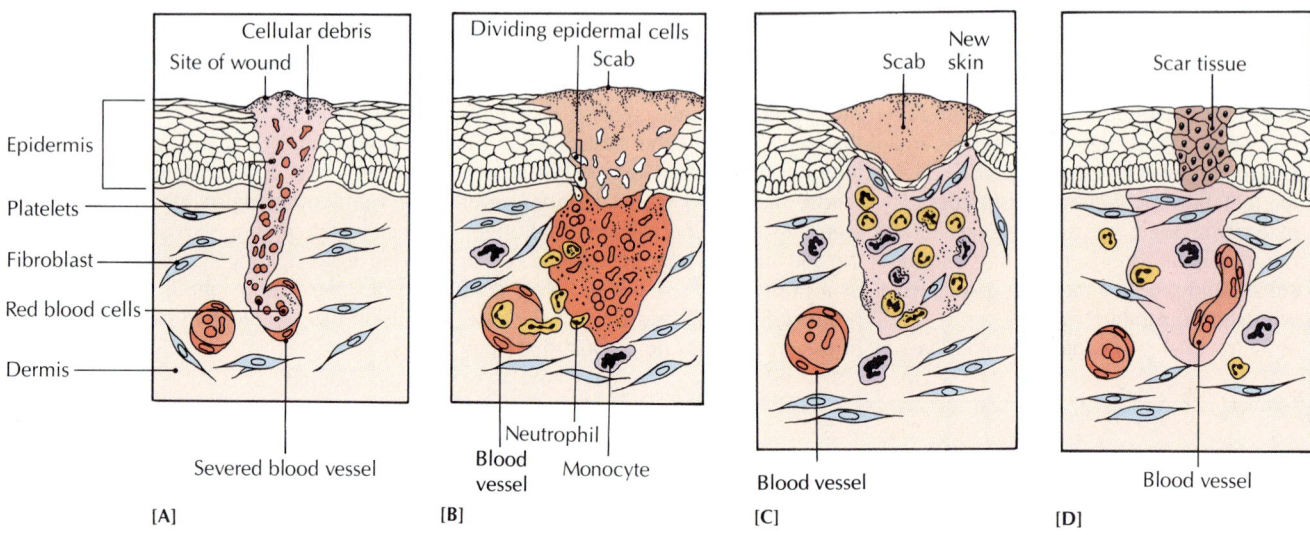

It is now known that wound healing is enhanced by a protein called *platelet-derived growth factor* (*PDGF*), which is released at the site of a wound by platelets and macrophages that are attracted to the site of blood vessel damage. PDGF is also released by endothelial cells that line blood vessels and by smooth muscle cells near the wound. Other growth factors that assist the process of wound healing are (1) *fibroblast growth factor* (*FGF*), which promotes the growth of new blood vessels that can supply local cells with oxygen and nutrients, (2) *transforming growth factor-beta* (*TGFb*), which helps to form a new matrix that works synergistically with (3) *epidermal growth factor* (*EGF*) to stimulate cells to grow and attach at the site of the wound.

As part of the overall healing process, an ***inflammatory response*** occurs, which includes redness, pain, swelling, scavenging by neutrophils and macrophages, and tissue repair by fibroblasts. (See the detailed discussion of the inflammatory response as part of the nonspecific defenses of the body in Chapter 23.) The following events occur during the inflammatory response:

1 Blood vessels dilate and blood flow increases, probably because *histamine* is liberated. (Histamine, a powerful dilator of blood vessels, may also be responsible for activating phagocytes and increasing the flow of extracellular fluid to damaged tissues.)

2 Cell death occurs as lysosome enzymes are released by injured cells in the immediate area.

3 Blood vessels become more permeable to protein, white blood cells, and the blood-clotting agent fibrinogen.

4 Swelling (*edema*) is caused by the increased movement of fluids to the wound site, which dilute the irritant.

5 White blood cells (neutrophils and monocytes) enter the area from nearby blood vessels and begin to remove cell debris and phagocytize microorganisms.

6 Dead white blood cells and tissue debris form *pus*.

7 Tissue is repaired with the aid of regenerative fibroblasts.

Q: *Why does a healing wound itch?*

A: When the skin is injured, there is usually some damage to superficial nerves. After about a week or so, the nerves begin to fire impulses as new nerve endings begin to grow. These impulses are interpreted as an itch on the skin, which usually subsides after a few days.

GLANDS OF THE SKIN

The glands of the skin are the sudoriferous and sebaceous glands.

Sudoriferous (Sweat) Glands

Sudoriferous glands (L. *sudor*, sweat) are also known as *sweat glands.* Two types of sudoriferous glands exist: eccrine (EKK-rihn) and apocrine (APP-uh-krihn). *Eccrine glands* (Gr. *ekkrinein*, to exude, secrete) are small sweat glands [FIGURE 5.7A]. They are distributed over nearly the entire body surface. (There are no sweat glands on the nail beds, margins of the lips, eardrums, inner lips of the vulva, or tip of the penis.) These glands are generally of the simple, coiled tubular type [see TABLE 4.2], with the secretory portion embedded in the hypodermis and a hollow, corkscrewlike duct leading up through the dermis to the surface of the epidermis [FIGURE 5.1A]. The duct generally straightens out somewhat as it reaches the surface. All these ducts combined would be about 9.5 km (6 mi) long.

Most eccrine glands secrete sweat by a physiological process called *perspiration** when the external temperature rises. Eccrine glands that respond to psychological stress are most numerous on the palms, fingers, and soles. Sweat from eccrine glands consists of a colorless aqueous fluid that holds in solution neutral fats, albumin and urea, lactic acid, sodium and potassium chloride, and traces of sugar and ascorbic acid. Its excretion helps in the regulation of body temperature, largely through the cooling effect of evaporation. Sweat glands are efficient. They secrete sweat even when the skin appears to be dry, and the constant combination of perspiration and evaporation keeps the body from overheating.†

Apocrine, or *odiferous,* *glands* are found in the armpits, the dark region around the nipples, the outer lips of the vulva, and the anal and genital regions. These are larger and more deeply situated than the eccrine glands [FIGURE 5.7B]. The female breasts contain apocrine glands that have become adapted to secrete and release milk instead of sweat. Apocrine glands become active at puberty and enlarge just before menstruation. Eccrine glands respond to heat, but apocrine glands in the skin respond to stress (including sexual activity) by secreting sweat with a characteristic odor. Secretions of apocrine glands contain specific human pheromones. (A *pheromone* is any substance secreted by an animal that provides communication with other members of the same species to elicit certain behavioral responses.) A smell we commonly call "body odor" results when bacteria on the skin decompose the secretions as they feed on them.

The *ceruminous glands* in the outer ear canals are also apocrine skin glands. They secrete the watery component of *cerumen* (ear wax), which helps trap foreign substances before they can enter the deeper portions of the ear.

Sebaceous (Oil) Glands

Sebaceous, or *oil, glands* (L. *sebum*, tallow, fat) are simple, branched alveolar glands found in the dermis. Their main functions are lubrication and protection. They are connected to hair follicles [FIGURE 5.1A]. The secretory portion is made up of a cluster of cells that are polyhedral in the center of the cluster and cuboidal on the edges. Secretions are produced by the breaking down of the interior cells, which become the oily secretion called *sebum* found at the base of hair follicles. Sebum is a semifluid substance composed almost entirely of lipids. About 60 percent of sebum consists of triglycerides and free fatty acids, but some waxes are also present. Destroyed cells are replaced from the outer layer of cuboidal cells.

---

†An inactive person produces about 70 mL (2.1 fl oz) of sweat a day; an athlete or manual laborer can produce a liter or more during an hour of vigorous physical activity. The inactive person loses sweat by simple diffusion, through a process called *insensible perspiration;* the active person perspires in a more noticeable way.

Q: *Do antiperspirants work by clogging the sweat pores?*

A: Antiperspirants work in part by clogging sweat pores, but they also use the principle of the negatively charged sweat being extracted from the pores by the positively charged skin surface. One of the ingredients of antiperspirants is the negatively charged aluminum. It neutralizes the electrical charges of sweat and skin, thereby stopping perspiration.

---

*The word *perspiration* literally means "to breathe through" (L. *per*, through + *spirare*, to breathe), in reference to the ancient belief that the skin breathes.

FIGURE 5.7 THE ANATOMY OF ECCRINE AND APOCRINE SWEAT GLANDS

The enlarged circular cross-sectional views show interior structure.

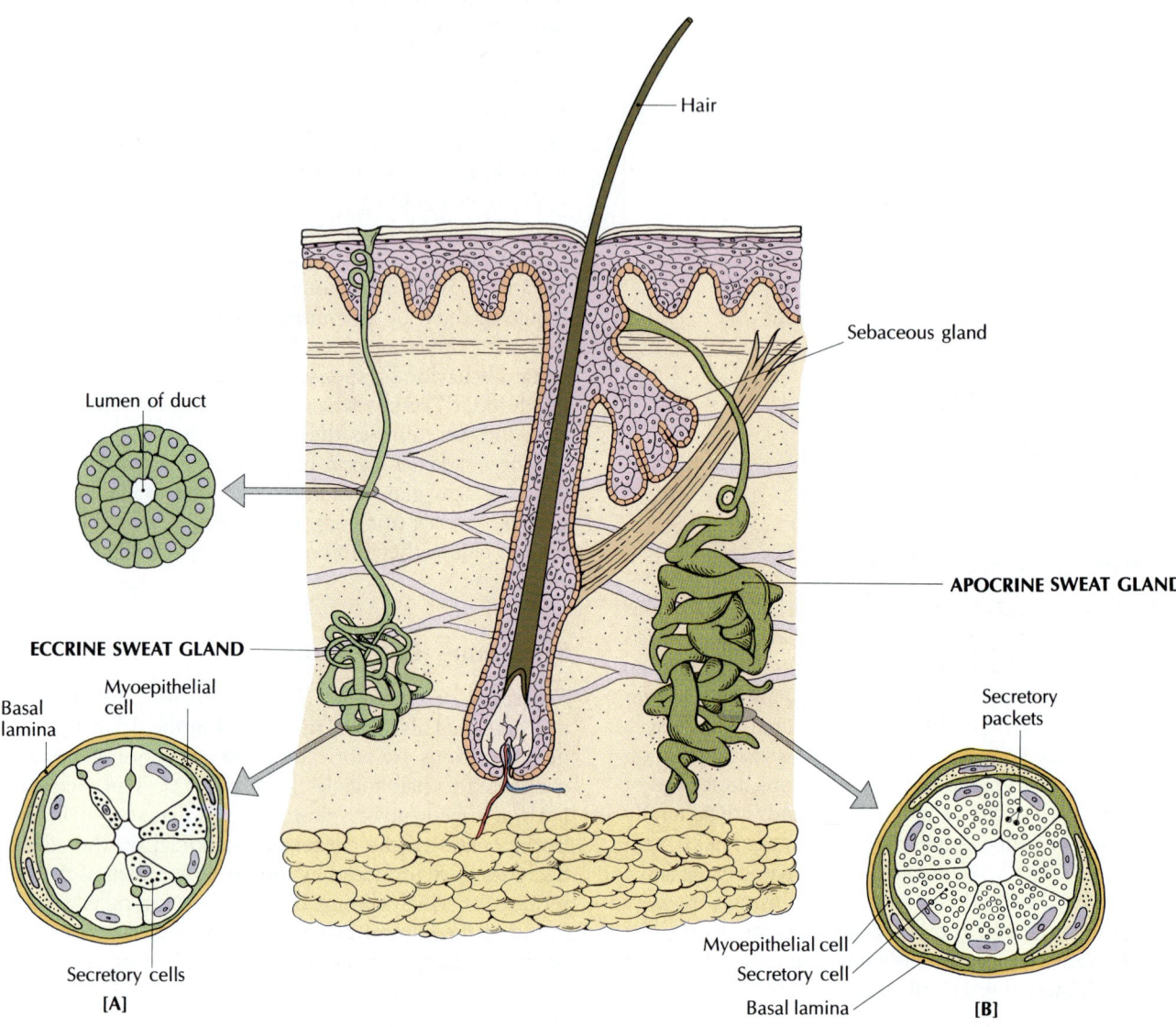

Hair

Sebaceous gland

Lumen of duct

APOCRINE SWEAT GLAND

ECCRINE SWEAT GLAND

Myoepithelial cell

Basal lamina

Secretory packets

Secretory cells

Myoepithelial cell

Secretory cell

Basal lamina

[A]

[B]

Sebum functions as a permeability barrier, a skin-softening agent, and a protective agent against bacteria and fungi. Sebum can also act as a pheromone.

When sebaceous glands become inflamed and accumulate sebum, the gland opening becomes plugged and a **blackhead** results. If the plugging is not relieved, a pimple or even a sebaceous cyst may develop. The "blackness" of blackheads is caused by air-exposed sebum, not dirt. Blackheads and pimples often accompany hormonal changes during puberty, when an oversecretion of sebum may enlarge the gland and plug the pore. The resulting skin disease is called *acne vulgaris* (see When Things Go Wrong on page 150).

Openings from sebaceous glands are found over the entire surface of the skin except in the palms and soles, where they would be a nuisance.

ASK YOURSELF

1 What are the two types of skin glands?

2 What are the functions of sweat glands?

3 What are the functions of oil glands?

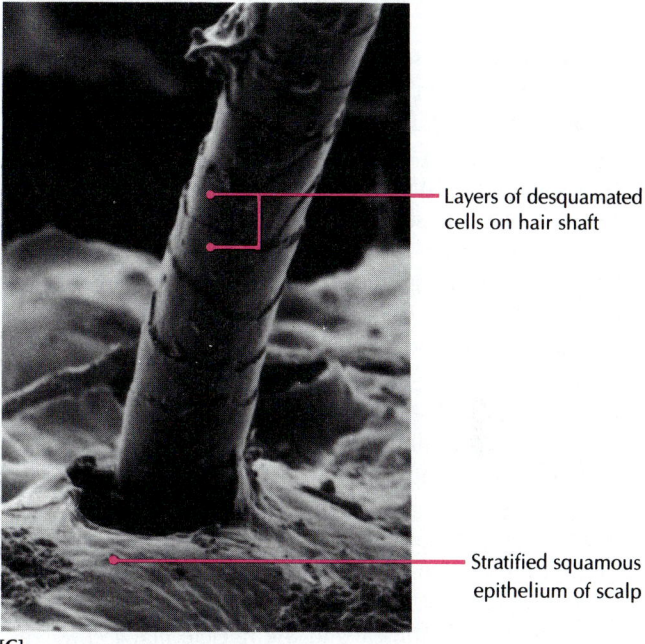

FIGURE 5.8 HAIR

[A] Longitudinal section of human hair, showing the hair root, follicle, and bulb. **[B]** The microscopic structure of hair. Notice how the scales of the cuticle all face upward. **[C]** A scanning electron micrograph of a hair shaft on the scalp. ×4375.

Labels in figure [A]:
Cortex, Medulla, Cuticle, Hair shaft, [B], EPIDERMIS, DERMIS, HYPODERMIS, Sebaceous (oil) gland, Arrector pili muscle, Hair root, Inner epithelial root sheath, Outer epithelial root sheath, Connective tissue sheath, HAIR FOLLICLE, Hair matrix, Hair papilla, HAIR BULB, [A]

Labels in figure [C]:
Layers of desquamated cells on hair shaft, Stratified squamous epithelium of scalp, [C]

HAIR

Hair is composed of keratinized threads of cells, a specialization that develops from the epidermis. Because hair arises from the skin, it is considered an appendage of the skin. It covers the entire body except for the palms, soles, lips, tip of the penis, inner lips of the vulva, and nipples.

Functions of Hair

Obviously, hair does not function as an insulative covering in humans as it does in many other mammals, but human hair does serve some protective functions. Scalp hair provides some insulation against cold air and the heat of the sun and also protects us from bumps. Eyebrows act as cushions in protecting the eyes and also help reduce glare and prevent sweat from running into the eyes. Eyelashes act as screens against foreign particles. Tiny hairs in the nostrils trap dust particles in inhaled air. Other openings in the body, such as the ears, anus, and vagina, are also protected by hair.

Hair is often used for diagnostic testing. For example, dry, brittle hair may indicate an underactive thyroid gland, since a healthy thyroid secretes enough hormones to produce healthy hair. Hair samples can also be used to detect metallic poisons such as arsenic, and drugs such as heroin and cocaine.

Structure of Hair

The two parts of a hair are the *shaft,* which is the portion that protrudes from the skin, and the *root,* the portion embedded in the skin [FIGURE 5.8A]. Hair consists of epithelial cells arranged in three layers. From the inside out of a hair shaft, these are the *medulla, cortex,* and *cuticle* [FIGURE 5.8B]. The medulla is composed of soft keratin, and the cuticle and cortex are composed of hard keratin. A strand of hair is stronger than an equally thick strand of nylon or copper.

The *medulla* forms the central core of the hair, and it usually contains loosely arranged cells separated by air "cells" or liquid in its extracellular spaces. The *cortex* is the thickest layer of hair, consisting of several layers of cells. Pigment in the cortex gives hair its color. When hair pigment fades from the cortex and the medulla becomes completely filled with air, the hair appears to be gray to white, with whiteness representing a total loss of pigment. The ***cuticle*** is made up of thin squamous cells that overlap to create a scalelike appearance [FIGURE 5.8B, C]. The cuticle can be softened or even dissolved by chlorine in pool water.

The lower portion of the root, located in the hypodermis, enlarges to form the ***bulb.*** The bulb is composed of a ***matrix*** of epithelial cells. The bulb pushes inward along its bottom surface to form a papilla of blood-rich connective tissue [FIGURE 5.8A]. The entire bulb is enclosed within a tubular ***follicle*** (L. *folliculus,* little bag). The hair follicle consists of three sheaths: (1) an inner epithelial root sheath, (2) an outer epithelial root sheath, and (3) a connective tissue sheath.

Since the hair follicle arises from the hypodermis, projecting at an angle, the hair shaft usually points away from the bulb. For this reason, hair covers the scalp more efficiently than it would if it grew at a right angle to the scalp. The shape of the follicle openings determines whether hair is straight, wavy, or curly. A round opening (in cross section) produces straight hair, an oval opening produces wavy hair, and a spiral-shaped opening produces curly hair.

An interesting part of the hair follicle is the bundle of smooth muscle attached about halfway down·the follicle. As you saw in FIGURES 5.1A and 5.8, an oil gland nestles between the hair follicle and the muscle like a child sitting in the crook of a tree. This muscle is the ***arrector pili*** muscle. When it contracts, it pulls the follicle and its hair to an erect position, elevates the skin above, and produces a "goose bump" on the skin. The contracting muscle also forces sebum from the oil gland.

Growth, Loss, and Replacement of Hair

We are constantly shedding and (we hope) replacing our hair. A hair is shed when its growth is complete. Just before a hair is to be shed, the matrix cells gradually become inactive and eventually die. The root of the hair becomes completely keratinized and detaches from the matrix. This keratinized root bulb begins to move along the follicle until it stops near the level of the oil gland. The papilla atrophies, and the outer root sheath collapses.

After a short period of rest, a new hair begins to grow when new hair cells from the outer root sheath start to develop near the old papilla. A new matrix develops, and a new hair starts to grow up the follicle, pushing the old hair out of the way if it hasn't been shed already.

Each type of hair has its own life cycle. Even the same type of hair (hair on the scalp, for example) grows at staggered rates, so no area of the body sheds all its old hair at the same time. Hair grows faster at night than during the day and faster in warm weather than in cold. The rate of hair growth also varies with age, usually slowing down as we get older. The fastest growth rate occurs in women between ages 16 and 24. Hair textures and locations also affect growth rates. Coarse black hair grows faster than fine blond hair. Scalp hairs last much longer (3 to 5 years) than eyebrow and eyelash hairs (about 10 weeks).

Human hair follicles alternate between growing and resting phases. Scalp hairs, which grow about 0.4 mm a day, usually last 3 to 5 years before they are shed and the follicles go into a resting phase of 3 to 4 months. Plenty of scalp hair is visible at any given time, however, because 80 to 90 percent of the follicles in the scalp are in the growing phase at the same time. (The average scalp contains about 125,000 hairs.) Because healthy, active follicles keep producing new hair, it is not unusual to lose 50 to 100 hairs a day without becoming bald. Some hairs have a longer resting phase than growing phase. Eyebrow hairs, for instance, grow 1 to 2 months before the follicles rest for the next 3 to 4 months. Eyelashes have an even shorter growing phase.

Although it seems that men have more body hair than women do, they actually have about the same number of hair follicles, about 21 million. Male hair is coarser and therefore more obvious. Baldness in men seems to occur when there is a genetic predisposition toward baldness. For *male pattern baldness* to be expressed, androgens (male sex hormones) are required. Baldness can also be caused by disease, stress, malnutrition, and many other external factors. Such baldness, called *alopecia* (al-oh-PEE-shuh; Gr. baldness) may be reversible (see page 154).

Q: *How long would scalp hair grow if left uncut?*

A: If scalp hair were left uncut, it would grow to a length of about 1 m (3.25 ft) in a lifetime. (Many exceptions of longer hair have been recorded.) If it did not grow in cycles that include inactivity and shedding, scalp hair would reach about 7.5 m (24.5 ft). Contrary to what some people believe, hair does not grow after a person dies. Instead, the skin shrinks, making the hair look longer.

ASK YOURSELF

1 What are some functions of hair?

2 What are the three layers of a hair called?

3 What is a hair follicle?

FIGURE 5.9 FINGERNAIL STRUCTURE

[A] The basic components of a fingernail.
[B] A drawing of a saggital section of a fingernail.

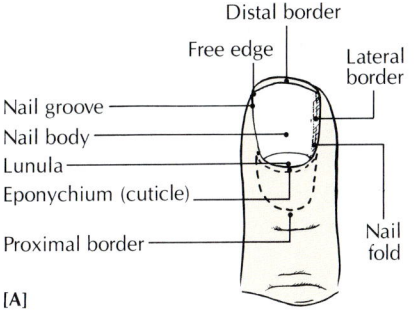

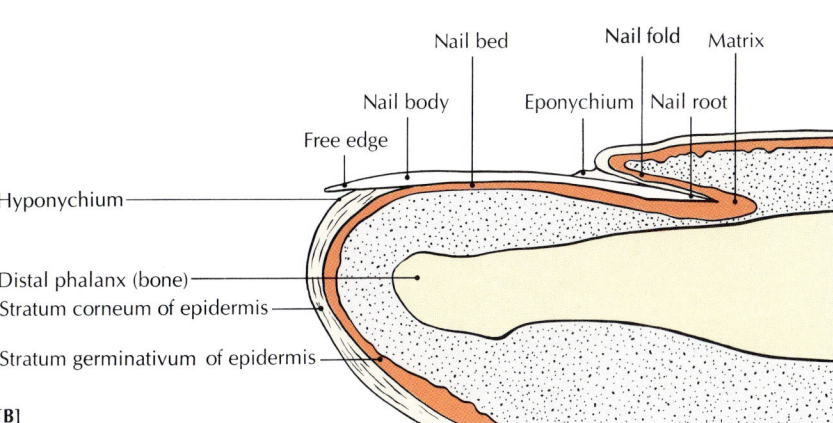

NAILS

Nails, like hair, are modifications of the epidermis. Also, like the cuticle and cortex of a hair, they are made of hard *keratin.* Nails are composed of flat, keratinized plates on the dorsal surface of the distal segment of the fingers and toes [FIGURE 5.9]. They appear pink because the nail is translucent, allowing the red color of the vascular tissue underneath to show through. White nails may be a sign of liver disease. The proximal part of the nail, the *lunula* (LOON-yuh-luh; L. "little moon"; commonly called "half-moon"), is white because the "red" capillaries in the underlying dermis do not show through.* The lunula usually does not appear on the little finger.

The body of the nail consists of keratinized dead cells containing shrunken and degenerated nuclei. It is the part that shows; the root is the part hidden under the skin folds of the nail groove [FIGURE 5.9A]. The nail ends with a *free edge* that overhangs the tip of the finger or toe. The nail rests on an epithelial layer of skin called the *nail bed,* and the thicker layer of skin beneath the nail root is the *matrix,* the area where new cells are generated for nail growth and repair [FIGURE 5.9B]. If the nail is injured, a new one will grow as long as the living matrix is intact. After birth, fingernails grow faster than toenails, with an average growth rate of 0.5 mm a week.

*Some anatomists contend that the lunula is white because it is the portion of the nail formed first. Relatively young nail substance such as the lunula is generally whiter, thicker, and more opaque than the more mature nail that extends over the nail bed.

Q: *What is a hangnail?*

A: A hangnail has nothing to do with a hanging nail. It is actually a torn cuticle, and its name is derived from the old Anglo-Saxon word for "painful nail" (*ang,* pain; *naegl,* nail).

The developing nail is originally covered by thin layers of epidermis (stratum corneum) called the *eponychium* (epp-oh-NICK-ee-uhm; Gr. *epi,* upon + *onyx,* nail). The eponychium remains at the base of the mature nail and is commonly called the *cuticle.* The stratum corneum of the skin beneath the free edge of the nail is the *hyponychium.*

Our nails protect our fingers and toes and allow us to pick up and grasp objects. And we use them to scratch.

EFFECTS OF AGING ON THE INTEGUMENTARY SYSTEM

With old age, the skin usually wrinkles as the elastic tissue becomes less resilient and the fatty layer and supportive tissue beneath the skin decrease in thickness. As oil glands and sweat glands decrease their activity, the skin becomes dehydrated. The inactivity of infection-fighting dermal cells and dry, brittle skin lead to itchiness and more frequent skin infections. Also, the dermis receives a reduced blood supply, causing wounds to heal more slowly. Sensitivity to changes in temperature increases as the thinning skin becomes less able to maintain a constant body temperature. Blotching of skin color often occurs because of the irregular growth of pigment cells, probably as a result of exposure to sunlight and other UV radiation. Smooth, flat, brown areas called *lentigines* (lehn-TIHJ-ih-neez; sing. *lentigo,* L. lentil) (commonly, and incorrectly, called "liver spots") may form on the face or back of the hands,† and damaged capillaries may give rise to small, bright red bumps known as *cherry angi-*

†The active ingredient in over-the-counter medications designed to minimize these "age spots" on the skin is *hydroquinone,* which inhibits melanin synthesis.

FIGURE 5.10 DEVELOPMENTAL ANATOMY OF SKIN

[A] Four weeks, [B] seven weeks, [C] 11 weeks, [D] newborn.

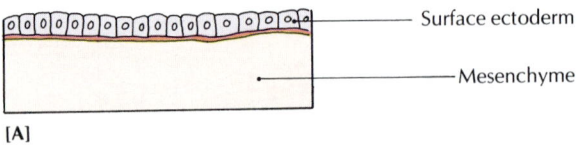

Surface ectoderm — Mesenchyme

[A]

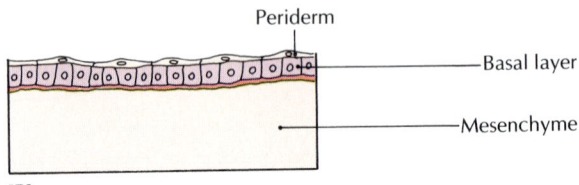

Periderm — Basal layer — Mesenchyme

[B]

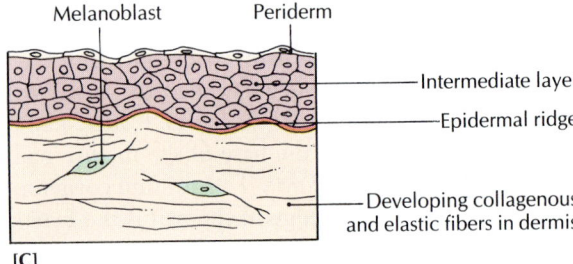

Melanoblast — Periderm — Intermediate layer — Epidermal ridge — Developing collagenous and elastic fibers in dermis

[C]

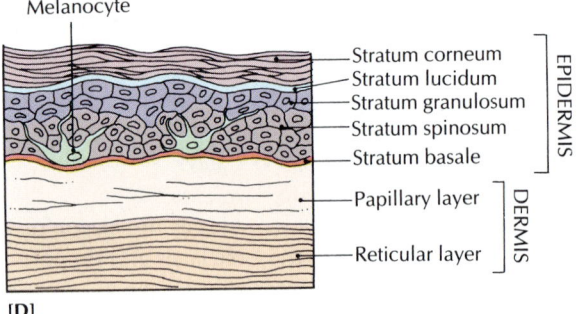

Melanocyte — Stratum corneum — Stratum lucidum — Stratum granulosum — Stratum spinosum — Stratum basale (EPIDERMIS) — Papillary layer — Reticular layer (DERMIS)

[D]

DEVELOPMENTAL ANATOMY OF THE INTEGUMENTARY SYSTEM

The *epidermis* and *dermis* are derived from different embryological tissues: The outer epidermis is derived from surface ectoderm (epithelium), and the inner dermis is derived from mesenchyme (connective tissue) beneath the surface ectoderm [FIGURE 5.10A]. At about 7 weeks after conception, the surface ectodermal cells form a superficial layer of simple squamous epithelium called the *periderm* [FIGURE 5.10B]. As the cells of the periderm undergo keratinization and begin to be shed, they are replaced by cells from the basal layer, which later forms the stratum germinativum.

At about week 11 the mesenchymal cells begin to produce collagenous and elastic fibers in the dermis [FIGURE 5.10C]. As these fibers migrate upward, they form the *dermal papillae* that project into the epidermis [see FIGURE 5.3]. The five distinct layers of the epidermis are recognizable by month 4.

FIGURE 5.11 DEVELOPMENTAL ANATOMY OF HAIR

The distinctive whorled pattern of the lanugo; fifth fetal month.

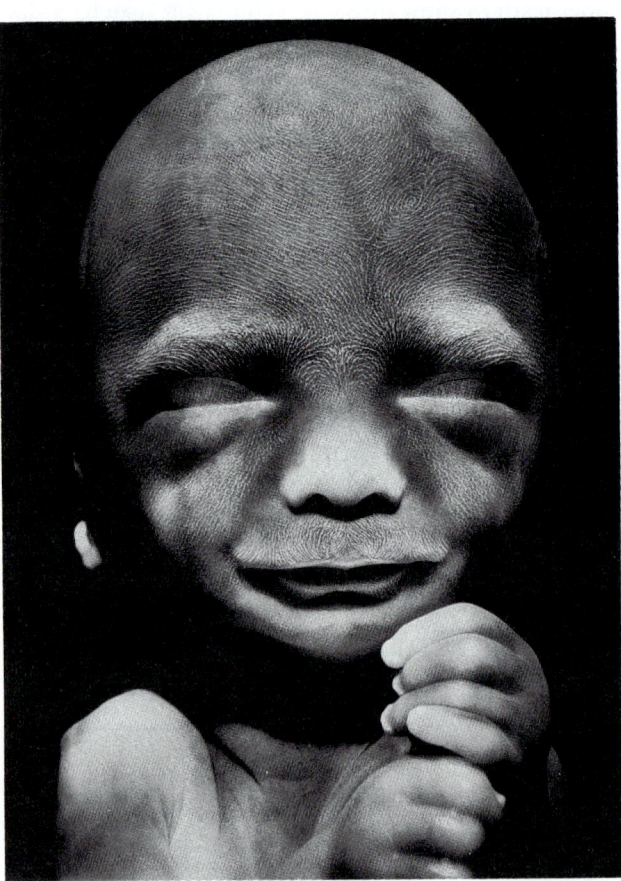

omas and also to larger purple blotches. Raised brown or black dime-sized growths called *seborrheic keratoses* may appear on the epidermal layer of the skin.

In addition to the typical inherited baldness in some men, there is a thinning of the hair in both sexes because of the atrophy of hair follicles. Growth cycles become shorter, especially in men, until follicles spend more time in resting periods than in growing periods. Hair usually becomes gray and brittle as melanin production slows and stops, and more hair appears in the nostrils, ears, and eyebrows. Postmenopausal women may grow longer hair on the upper lip and chin because of an increase in androgens and a decrease in estrogens. Most elderly women also experience some hair loss.

During the fifth fetal month, a waxy protective coating begins to cover the skin. This coating, called the *vernix caseosa*, is formed from sebum and desquamated peridermal skin cells. It protects the fetus from the fluids in its surrounding sac (the *amnion*). Some of the vernix caseosa is loosened by the amniotic fluid during childbirth, but enough remains to provide protection for the thin skin against infection.

Melanocytes are formed from melanoblasts that migrate upward toward the junction of the dermis and epidermis [FIGURE 5.10C]. Melanocytes begin producing melanin shortly before birth. The replacement of peridermal cells continues until about week 21, after which the periderm disappears. At birth [FIGURE 5.10D], all mature skin layers are present. After birth the baby's skin thickens, and underlying fat deposits continue to develop.

The *nails* begin to develop at about week 10, with the toenails starting before the fingernails. As the nail develops it is covered by superficial layers of epidermis, which later degenerate, except at the base of the nail, where they form the *cuticle.* Fingernails reach the fingertips by about week 32, and toenails reach the tips of the toes by about week 36.

The first *hairs* begin to develop during the third fetal month, when the epidermis begins to project downward into the dermis. Eventually these downgrowths form hair follicles that produce hairs. By the fifth month the fetus is covered with fine, downy hair called *lanugo* (luh-NOO-go; L. *lana*, fine wool) [FIGURE 5.11]. The lanugo helps the vernix caseosa adhere to the skin. Later in the fifth month, eyebrows and head hairs become apparent. Lanugo hair is shed before birth, except on the eyebrows and scalp, where it becomes thicker. After the baby is 5 or 6 months old, this hair is replaced by coarser hair called *terminal hair,* and the rest of the body becomes covered with a film of delicate short hair called *vellus* (L. "fleece" or "coarse wool"). Although the vellus and terminal hairs seem to be "new," no new hair follicles are developed after birth.

Coarse terminal hair appears at puberty in the genital area and armpits, and young men also develop chest and facial hair. Body hair usually thickens during puberty.

ASK YOURSELF

1 From which embryological tissues do the epidermis and dermis derive?

2 What is the difference between lanugo, vellus, and terminal hair?

WHEN THINGS GO WRONG

Burns

Burns occur when skin tissues are damaged by heat, electricity, radioactivity, or chemicals. The seriousness of burns can be classified according to (1) *extent* (how big an area of the body is involved) and (2) *depth* (how many layers of tissue are injured) [TABLE 5.1].

A *first-degree burn* (such as a sunburn) may be red and painful, but it is not serious. Generally, it damages only the epidermis and does not destroy it. Such a burn responds to simple first-aid treatment, including cold water and sterile bandages. Butter and other greasy substances should never be used on *any* burn, because they may actually damage the skin even further. Instead, the burn should be flushed or immersed in cold water (not ice water), or cold compresses (not ice) should be applied. Cold water helps reduce pain, swelling, fluid loss, and infection and also limits the extent of the damage.

A *second-degree burn* destroys the epidermis and also causes some cell destruction in the dermis. Oozing blisters form, and scarring usually results. After a second-degree burn, the body may be able to regenerate new skin. Second-degree burns require prompt medical attention. If left untreated, a second-degree burn can progress to a third-degree burn.

A *third-degree burn* involves the epidermis, dermis, and underlying tissue. Because the skin cannot be regenerated, this kind of burn must be treated with surgery and skin grafting (see illustrations on page 148). Ordinarily, the victim is in shock and feels no pain because nerve endings in the burned area have been destroyed. The damaged area is charred or pearly white, and fluid loss is severe.

One method of estimating the extent of burns on the skin is the *Lund-Browder method* (see drawings A and B, page 149). This method assigns percentages of surface area to certain parts of the body. For example, if both the anterior and posterior of one foot are burned, the burn is said to extend over 3.5 percent of the body. Another method is known as the *rule of nines* (see drawing C, page 149). It is generally less accurate than the Lund-Browder method. Physicians and therapists use these methods in conjunction with the specific area burned, the depth of the burn, and the age of the patient, especially in trauma and emergency centers.

TABLE 5.1 CLASSIFICATION OF BURNS

MINOR (FIRST-DEGREE) BURN

Surface area affected

Less than 10% of body surface.

Depth of tissue damage

Epidermis damaged but not destroyed.

Major effects

Mild swelling, reddening, pain; injured cells peel off and skin heals without scarring, usually within 2 weeks.

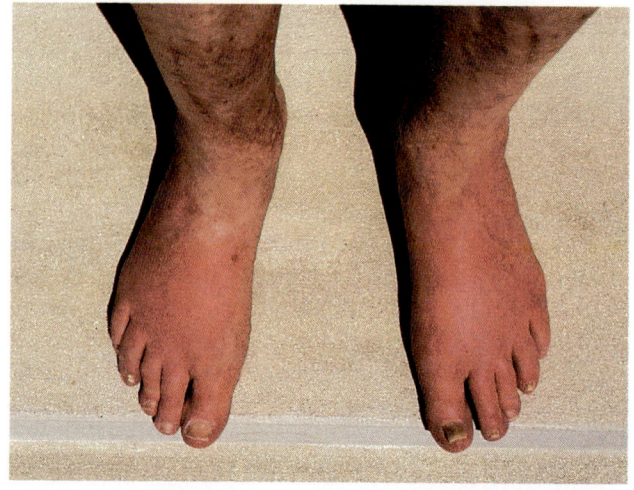

SERIOUS (SECOND-DEGREE) BURN

Surface area affected

Less than 15% of body surface for an adult, 10% for a child.

Depth of tissue damage

Epidermis and part of dermis destroyed. New skin may regenerate.

Major effects

Red or mottled appearance, blisters, swelling, wet surface due to plasma loss. Greater pain than third-degree burn (which destroys sensitive nerve endings).

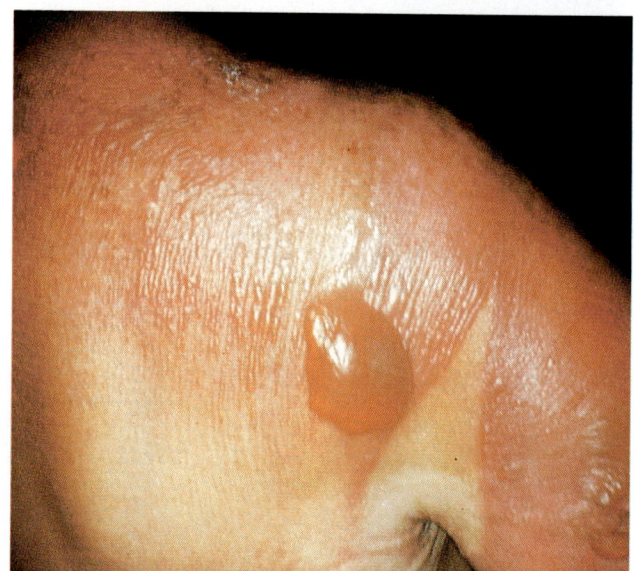

SEVERE (THIRD-DEGREE) BURN

Surface area affected

Includes burns of face, eyes, hands, feet, genitals, and more than 20% of body surface. Prompt medical attention required.

Depth of tissue damage

All skin layers destroyed; deep tissue destruction. Nerve endings in skin destroyed. Skin cannot be regenerated. Surgery and skin grafts necessary.

Major effects

White or charred appearance; severe loss of body fluids.

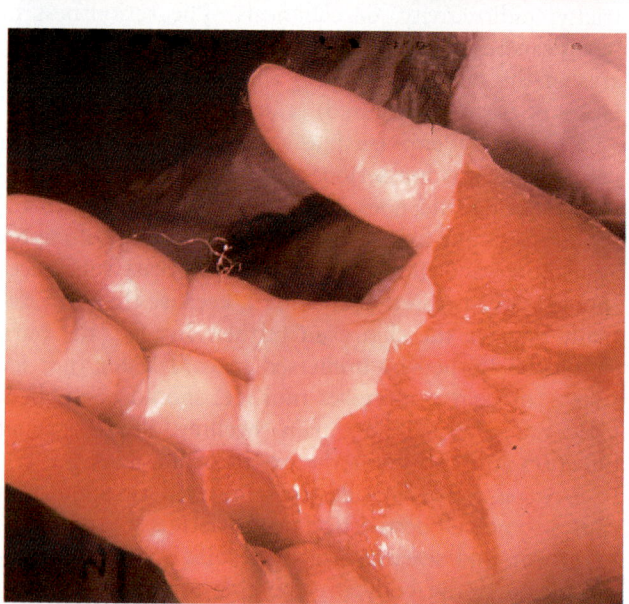

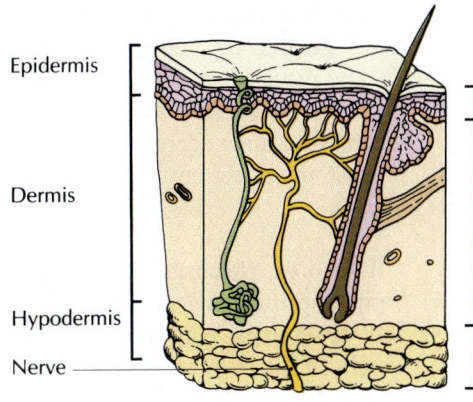

Epidermis

Dermis

Hypodermis

Nerve

FIRST-DEGREE BURN
(epidermis damaged)

SECOND-DEGREE BURN
(epidermis and part of dermis destroyed)

THIRD-DEGREE BURN
(all skin layers destroyed)

Estimating the extent of skin burns.
The *Lund-Browder method* for estimating
the extent of skin burns for **[A]** a 5-year-old child
and **[B]** an adult. Except for the perineum and
buttocks, percentages shown include both
anterior and posterior portions of the body.
[C] The *"rule of nines"* is also used for
estimating the extent of skin burns.
Except for the perineum, percentages
shown include both the anterior and
posterior portions of the body; the
buttocks are included in the
posterior trunk.

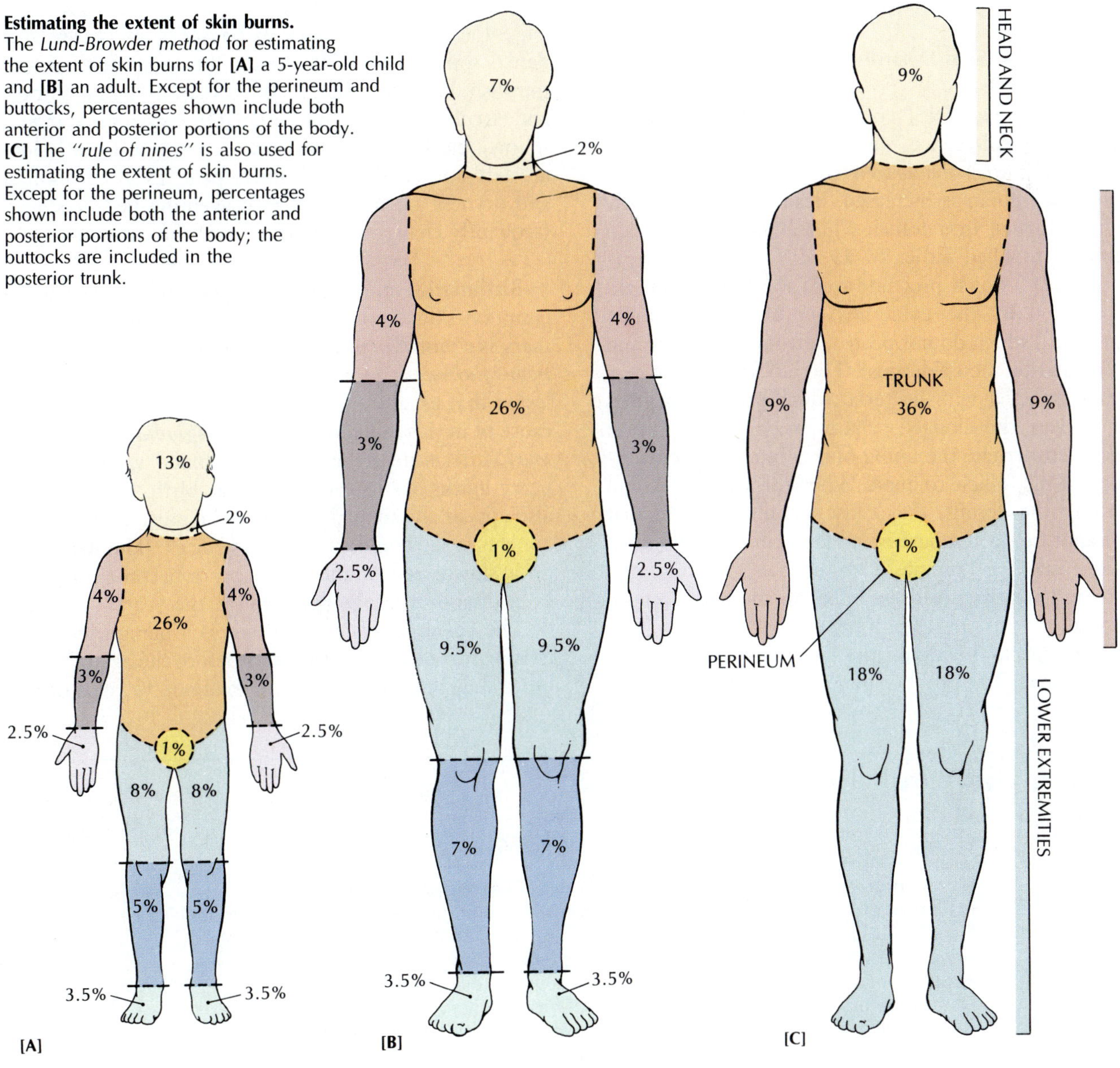

[A]

13%
2%
4%   4%
26%
3%   3%
2.5%   2.5%
1%
8%   8%
5%   5%
3.5%   3.5%

[B]

7%
2%
4%   4%
26%
3%   3%
2.5%   1%   2.5%
9.5%   9.5%
7%   7%
3.5%   3.5%

[C]

9%
TRUNK
9%   36%   9%
1%
PERINEUM
18%   18%

HEAD AND NECK

UPPER EXTREMITIES

LOWER EXTREMITIES

A burn is one of the most traumatic injuries the body can receive. Besides causing obvious tissue damage, serious burns expose the body to microorganisms, hamper blood circulation and urine production, and create a severe loss of body water, plasma, and plasma proteins that can produce shock. In fact, a major burn causes homeostatic imbalances in every system of the body. A severe burn leaves the skin more vulnerable to microbial infection than other types of wounds do. This happens because neutrophils, the skin's specialized infection-fighting cells, are practically immobilized by the burn. Instead of rushing to the infection site and releasing disease-fighting chemicals, the traumatized neutrophils release their chemicals prematurely. Because the chemicals interfere with the chemical signal from the infection site, the neutrophils do not know where to go, and only a few make it to the infected area.

Some Common Skin Disorders

Acne *Acne vulgaris* ("common acne") is most common during adolescence, when increased hormonal activity causes the oil glands to overproduce sebum. When the flow of sebum is increased, dead keratin cells may become clogged in a follicle. These plugs at the skin opening are called either *blackheads* (open *comedones* — sing. *comedo* — which protrude from the follicle and are not covered by the epidermis) or *whiteheads* (closed comedones, which do not protrude from the follicle and are covered by the epidermis). The blocked follicle may become infected with bacteria, which secrete enzymes that convert the clogged sebum into free fatty acids. These acids irritate the lining of the follicle and eventually cause the follicle to burst. When the acid and sebum seep into the dermis, they cause an inflammation that soon appears on the surface of the skin as a pus-filled papule called a "pimple" [TABLE 5.2]. Picking and scratching merely spread the infection and may produce scarring.

Acne appears mostly on the face, chest, upper back, and shoulders. The problem generally affects young men more severely than young women, probably because the causative hormones are *androgens*, which are found in much greater abundance in males than in females.

The most advanced form of acne is *cystic acne*, which produces deep skin lesions called *cysts*. It is produced when oil glands secrete excessive amounts of oil that nourish the infectious bacteria that cause acne in the first place.

Bedsores *Bedsores* (*decubitus ulcers*) are produced in bony, unprotected areas where the skin is close to a bone and undergoes constant pressure, usually from the weight of the body itself. The pressure causes blood vessels to be compressed, depriving the affected tissue of oxygen and nutrition and often leading to cell death. The most typical problem areas are the hips, elbows, tailbone, knees, heels, ankles, and shoulder blades (see drawing A, page 151). The first signs of bedsores are warm, reddened spots on the skin. Later, the spots may become purplish, indicating that blood vessels are being blocked and circulation is impaired. Actual breaks in the skin may follow, and bacterial infection is common if the lesions are left untreated (see photo B, page 151). Cleanliness and dryness are important in preventing bedsores, as is frequently changing the position of the patient.

Birthmarks and moles *Birthmarks* and *moles* are common skin lesions. The technical name for a birthmark is a *vascular nevus* (L. "birthmark"; plural *nevi*). A **nevus flammeus,** or *port-wine stain*, is a pink to bluish-red lesion that usually appears on the back of the neck. The cause of nevi is not known. A **hemangioma,** or *strawberry mark*, affects only the superficial blood vessels. Strawberry marks are usually present at birth, but they may also appear any time after birth. The most common sites are the face, shoulders, scalp, and neck. The mark may grow slowly, remain the same size, or become smaller or even disappear altogether through the years.

The **common mole,** or **nevus,** is a benign lesion (in most cases) that usually appears before the age of 5 or 6, but it may appear any time up to about 30 years of age.

TABLE 5.2 TYPES OF SKIN LESIONS (TISSUE ALTERATIONS)

Name	Example	Description
Bleb, bulla	Second-degree burn	A fluid-filled elevation of the skin.
Cyst	Epidermal cyst	A mass of fluid-filled tissue extending to dermis or hypodermis.
Macule	Freckle; flat, pigmented mole	A discolored spot, not elevated or depressed.
Papule	Acne, measles	Raised red area resembling small pimples.
Pustule	Acne vulgaris	Raised pus-filled pimple.
Tumor	Dermatofibroma	A fibrous tumor of the dermis.
Vesicle	Blister, chickenpox, herpes simplex	A small sac filled with serous fluid.
Wheal	Mosquito bite, hives	Local swelling, itching.

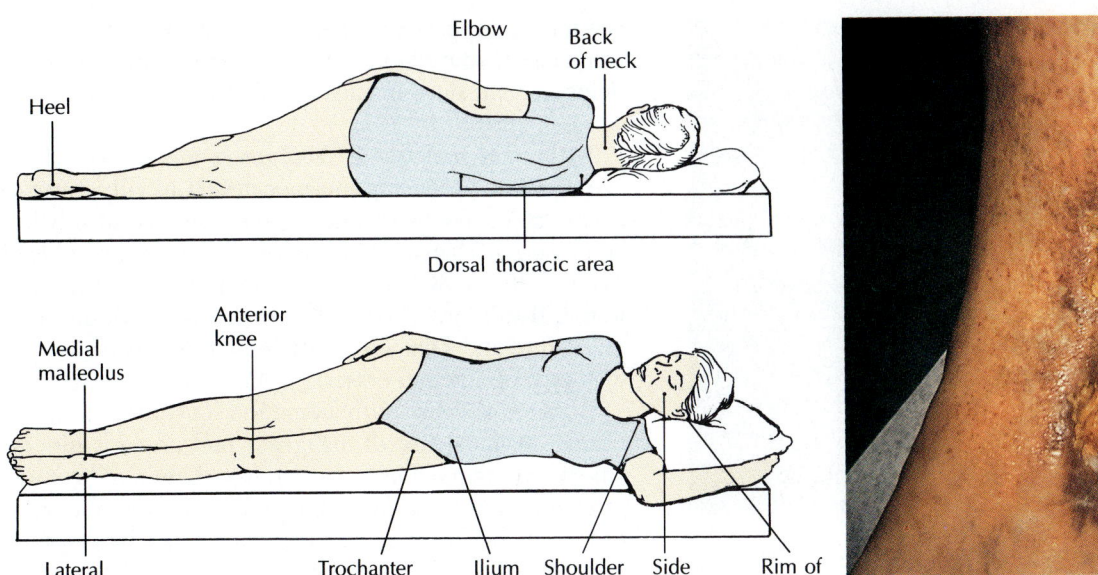

[A] Common sites of decubitus ulcers (bedsores).

[B] A decubitus ulcer on the right medial ankle.

Moles that darken, enlarge, bleed, or appear after a person is 30 should be checked by a physician, since an occasional mole may be transformed into a cancerous growth. Moles start out as flat brown or black spots composed of clusters of melanocytes and typically enlarge and become raised later, especially during adolescence and pregnancy. The tendency to have moles is thought to be an inherited characteristic. It is not unusual for a mole to contain hair.

Psoriasis *Psoriasis* (suh-RYE-uh-sihss; Gr. itch) is a chronic condition marked by lesions that are red, dry, and elevated and covered with silvery, scaly patches. The most usual sites of psoriasis are the elbows and knees, scalp, face, and lower back. Psoriasis is most common in adults, but may occur at any age. The cause of psoriasis is unknown, but there is general agreement that heredity plays a role. Attacks of psoriasis can be brought on by trauma, cold weather, pregnancy, hormonal changes, and emotional stress. The condition occurs when skin cells move from the basal layer to the stratum corneum in only 4 days instead of the usual 28. As a result, the cells do not mature and the stratum corneum becomes flaky.

Allergic responses *Poison ivy, poison oak,* and *poison sumac* all cause skin irritations when contact is made with those plants, which contain *urushiol,* a powerful skin irritant. (It is interesting that these plants have no effect the first time a person is exposed to them.) Exposed parts of the body usually begin to redden several hours (or even several days) after exposure. Redness, itching, and swelling generally progress to vesicles (raised red sacs), blisters, and finally a dry crust after serous fluid oozes from the blisters. Touching urushiol can cause the usual skin irritations, but touching the rash or blisters of an affected person cannot. Some people are so allergic to urushiol that they become affected by the smoke of the burning plants or by touching tools or pets that have touched the plants.

Warts *Warts* (*verrucae*) are benign epithelial tumors caused by at least 60 types of human papillovirus (HPV). A wart appears as a raised area of the skin that has a pitted surface. It is usually no darker than the skin color, except on the soles of the feet (plantar warts), where it is often yellowish. (Warts may appear darker than the skin because dirt becomes lodged in the tiny crevices between the fibers that make up the wart.) Although warts may appear anywhere on the body and on people of all ages, they are most common on the hands of children. This is probably so because the skin of the hands is likely to be irritated often, and a child's immune system is not yet effective against the virus. Warts are transmitted by direct contact and may be spread to other parts of the body by scratching and picking. They usually disappear after a year or so but may be removed by a physician if no surgical complications are expected.

Skin Cancer

The two most common forms of skin cancer are *basal cell epithelioma* (-*oma* means tumor), also called basal cell carcinoma, and *squamous cell carcinoma.* The most serious type of skin cancer is *malignant melanoma.* All three forms can be prevented to a great degree by avoiding overexposure to the ultraviolet rays in sunlight. Other causes of skin cancer include arsenic poisoning, radiation, and burns. It is believed that people who have moles may have an increased risk of developing melanomas.

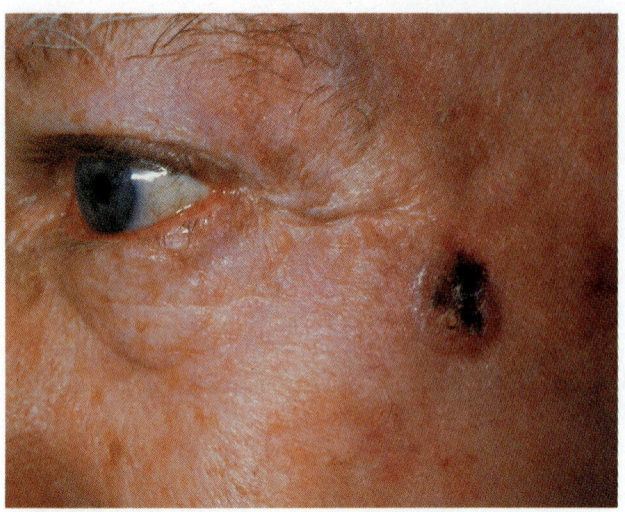

Basal cell epithelioma. The most common form of skin cancer.

Basal cell epithelioma *Basal cell epithelioma* generally appears on the face, where sweat glands, oil glands, and hair follicles are abundant (see photo). It occurs most frequently in fair-skinned males over 40. Three types of lesions are typical: (1) *Nodulo-ulcerative lesions* are small and pinkish during the early stage; eventually they enlarge and become ulcerated and scaly. (2) *Superficial basal cell epitheliomas* frequently erupt on the back and chest. These lightly pigmented areas are sharply defined and slightly elevated. They are associated with exposure to substances that contain arsenic. (3) *Sclerosing basal cell epitheliomas* are waxy, yellowish-white patches that appear on the head and neck.

Squamous cell carcinoma *Squamous cell carcinoma* usually appears as premalignant lesions, typically in the keratinizing epidermal cells of the lips, mouth, face, and ears. Unlike basal cell epithelioma, squamous cell carcinoma may metastasize actively, especially when the lesions occur on the ears and lower lip. Squamous cell carcinoma is most common in fair-skinned males over 60.

According to a recent study at Dartmouth Medical School, people with a history of basal cell and squamous cell cancers usually have about a 35-percent chance of developing another tumor within three years, and a 50-percent risk within five years. If treated promptly, basal cell and squamous cell skin cancers are rarely fatal. The study also revealed a link between cigarette smoking, excessive exposure to midday sunlight, and squamous cell carcinoma. Apparently, the ultraviolet rays in sunlight can cause a mutation in the *p53* gene, which normally helps prevent uncontrolled cell division.

Malignant melanoma *Malignant melanoma* affects the pigment-producing melanocytes. It usually starts as small, dark growths resembling moles that gradually become larger, change color, become ulcerated, and bleed easily. The "ABCDs" of melanoma are **A**symmetrical lesion, **B**order irregularity, **C**olor variations within a single lesion, and **D**iameter greater than 6 mm (about the diameter of a pencil eraser).

As with basal cell epithelioma and squamous cell carcinoma, the incidence of malignant melanoma is highest among fair-skinned persons, and it is slightly more common among women than among men. Besides the usual causes, malignant melanomas seem to be stimulated by hormonal changes during pregnancy. It is estimated that only 30 to 40 percent of all melanoma patients receive any benefit from chemotherapy. Most of these patients eventually have fatal recurrences.

Bruises

Bruises (Middle Eng. *brusen*, to crush) appear "black and blue" because a hard blow to the surface of the skin breaks capillaries and releases blood into the dermis. Although the blood is red, it creates a black-and-blue mark on the surface because the skin filters out all but the blue light that reflects off the bruise and makes it appear dark blue or purplish. Ordinarily, the darker the bruise, the deeper the blood has penetrated. Bruises sometimes turn yellow or green after several days. This is usually an indication that the leaked red blood cells have begun to break down into their components. Iron in the blood often gives the bruise a greenish color, as the decaying red pigment hemoglobin is transformed into a yellowish substance called *hemosiderin*. These color changes indicate that the bruise is in its final stages. Phagocytic white blood cells move into the affected area and ingest the hemosiderin and other debris, and the tissue returns to its normal color.

Q: *Is winter sun less dangerous than summer sun as a cause of skin cancer?*

A: No. Interestingly, the Skin Cancer Foundation points out that the level of ultraviolet rays in sunlight is about the same all year long. The application of a sunscreen of SPF 15 or higher is recommended about 30 minutes before outdoor activities.

CHAPTER SUMMARY

Skin (p. 133)

1 Skin is the major part of the *integumentary system,* which also includes the hair, nails, and glands of the skin.

2 The **epidermis** of the skin is the outermost layer of epithelial tissue, and the **dermis** is the thicker layer of connective tissue beneath the epidermis. The **hypodermis,** or **subcutaneous layer,** lies beneath the dermis.

3 A typical arrangement of strata in the epidermis, from the outermost stratum to the deepest one, is **stratum corneum, stratum lucidum, stratum granulosum, stratum spinosum,** and **stratum basale.** The stratum spinosum and stratum basale are collectively known as the **stratum germinativum.**

4 The **dermis** is composed of the thin **papillary layer** and the deeper, thicker **reticular layer.**

5 Skin serves as a stretchable protective shield and is also involved in the regulation of body temperature, the excretion of waste materials, and the synthesis of vitamin D_3. It screens out harmful ultraviolet rays and contains sensory receptors.

6 Skin color results from the presence of the pigments **melanin** and *carotene* and the color of the underlying blood.

7 Blood is supplied to the skin from blood vessels in the reticular layer of the dermis.

8 The wound-healing process involves the formation of blood clots by platelets and fibrinogen, phagocytosis by neutro-phils and monocytes, and the growth of new epidermal and collagenous tissue. The **inflammatory response** is also part of the overall healing process.

Glands of the Skin (p. 141)

1 The glands of the skin are the **sudoriferous (sweat) glands** and the **sebaceous (oil) glands.** Sudoriferous glands are either **eccrine** glands, small sudoriferous glands that secrete sweat, or **apocrine** glands, which become active at puberty and produce thicker secretions.

2 The major role of sweat glands is *temperature regulation,* while the main function of oil glands is *lubrication.*

Hair (p. 143)

1 *Hair* is composed of keratinized threads of cells that develop from the epidermis. Because of its embryonic origin, hair is considered to be an appendage of the skin.

2 Hair consists of epithelial cells arranged in three layers. From the inside out of a hair shaft, these are the **medulla, cortex,** and **cuticle.**

3 The hair **shaft** protrudes from the skin, the **root** is embedded beneath the skin, and the entire structure is enclosed within a **follicle.**

4 Human hair is constantly being shed and replaced; hair follicles alternate between growing and resting phases.

Nails (p. 145)

1 *Nails,* like hair, are modifications of the epidermis.

2 Composed of hard keratin, nails are flat, keratinized plates on the dorsal surface of the distal segment of the fingers and toes.

Effects of Aging on the Integumentary System (p. 145)

1 During old age the skin usually becomes less elastic and more wrinkled, brittle, and infection-prone.

2 Uneven pigmentation may lead to blotched skin color.

3 Baldness in men and facial hair growth in women is common.

Developmental Anatomy of the Integumentary System (p. 146)

1 The epidermis is developed from epithelium, and the dermis is developed from connective tissue.

2 Melanocytes begin producing melanin shortly after birth.

3 All mature skin layers are present at birth.

4 The first hairs begin to develop during the third fetal month. Terminal hairs appear at puberty in the genital area and armpits, and young men also develop chest and facial hair.

5 Toenails and fingernails begin to develop at about the tenth week of fetal development.

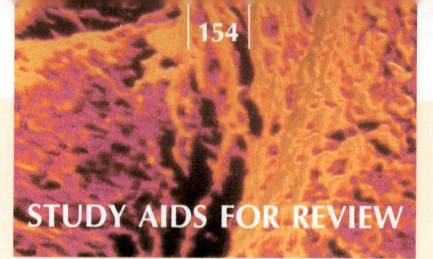

STUDY AIDS FOR REVIEW

KEY TERMS

apocrine gland 141

cleavage lines 136

dermis 133

eccrine gland 141

epidermis 133

follicle 144

hair 143

hypodermis 136

keratin 133f.

nails 145

sebaceous gland 141

sudoriferous gland 141

MEDICAL TERMS FOR REFERENCE

ALOPECIA AREATA (Gr. baldness, L. area, patchy) Sudden onset of patchy scalp and beard hair loss, usually without inflammation; usually reversible.

ALOPECIA TOTALIS (Gr. baldness, L. total) Sudden loss of all scalp hair.

ALOPECIA UNIVERSALIS (Gr. baldness, L. whole) Total loss of all body hair, including scalp and face.

ATHLETE'S FOOT (*tinea pedis*) A fungal infection that produces lesions of the foot, often accompanied by itching and pain.

CAFÉ AU LAIT SPOTS Light brown, flat spots on the skin; also known as von Recklinghausen's disease.

CALLUS (L. "hard") An area of the skin that has become hardened by repeated external pressure or friction.

CELLULITIS Inflammation of the skin and subcutaneous tissue, with or without the formation of pus.

CORN (L. *cornu*, horn) An area of the skin, usually on the feet, where the horny cells of the epidermis have become hardened by external pressure or friction.

DANDRUFF A scaly collection of dried sebum and flakes of dead skin on the scalp.

DERMABRASION A surgical technique to scrape away areas of the epidermis, usually to remove acne scars.

DERMATOLOGIST A physician who specializes in the treatment of skin disorders.

DERMATOLOGY The medical study of the physiology and pathology of the skin.

ECZEMA (Gr. "eruption") An inflammatory condition of the skin, producing red, papular, and vesicular lesions, crusts, and scales; characterized by itching and scaling.

HIVES A skin eruption of wheals, usually associated with intense itching.

IMPETIGO (*impetigo contagiosa*) (ihm-puh-TIE-go; L. "an attack") A contagious infection of the skin, caused by staphylococci or streptococci bacteria and characterized by small red macules that become pus-filled.

KELOID (Gr. "clawlike") An overgrowth of scar tissue.

KERATOSIS (Gr. *keras*, horn) A thickening and overactivity of the cornified cells of the epidermis.

MELASMA (Gr. *melas*, black) A condition characterized by dark patches on the skin, usually brought on by stress.

RINGWORM (*tinea*) Fungal infection, also known as *dermatophytosis*, characterized by red papules that spread in a circular pattern; affects tissues of the skin, nails, hair, and scalp.

SCABIES Contagious skin disease caused by the mite *Sarcoptes scabiei*, a parasite that lays eggs under the skin, causing itching and the eruption of vesicles.

VITILIGO (viht-ihl-EYE-go) An inherited condition characterized by splotchy white patches on the skin due to the destruction of melanocytes.

UNDERSTANDING THE FACTS

1 What layers are present in the epidermis of the thick skin on palms and soles? What layers are present elsewhere?

2 What are the layers of the dermis, and which structures are found in each layer?

3 List the five functions of the skin and the structures that are responsible for each of the functions.

4 What factors determine the color of skin?

5 Identify the glands of the skin by type, structure, secretion, and method of control.

6 What is the structure of a hair bulb, and how are the cells nourished?

7 What are the three most common forms of skin cancer, and which of the three is the most life-threatening?

UNDERSTANDING THE CONCEPTS

1 Outline the life cycle of a cell formed in the stratum basale.

2 Could any epidermal structure exist without an underlying dermis? Explain.

3 What does melanin protect the skin from, and how does it do so?

4 What roles do growth factors and hormones play in the functioning of the skin?

5 How are hair and nails similar, and how are they different?

6 Explain why third-degree burns are life-threatening.

SELF-QUIZ

Matching

a eponychium

b hair matrix

c lanugo

d arrector pili

e eccrine

f apocrine

g sebaceous

5.1 _____ smooth muscle attached to the hair follicle

5.2 _____ region of dividing cells that enables hair to grow

5.3 _____ a type of gland seen in modified form as a mammary gland

5.4 _____ a type of gland that plays a vital role in temperature regulation

5.5 _____ covers entire nail when first developed in the embryo

5.6 _____ fine hairs of the embryo that are mostly shed before birth

5.7 _____ a type of gland that produces an antibacterial waterproofing layer

Multiple-Choice Questions

5.8 Which of the following epidermal layers undergoes the most active cell division?
a stratum germinativum   d stratum lucidum
b stratum corneum   e a and b
c stratum granulosum

5.9 Elevated environmental temperature stimulates which of the following structures?
a apocrine glands and eccrine glands
b eccrine glands and sebaceous glands
c sebaceous glands and hair bulbs
d eccrine glands and hair bulbs
e apocrine glands and sebaceous glands

5.10 The bases of the sebaceous glands lie in the _____; the bases of the sweat glands lie in the _____.
a epidermis; dermis   d hypodermis; epidermis
b dermis; hypodermis   e dermis; epidermis
c dermis; epidermis

5.11 Capillary loops are found in the _____ layer, whereas cleavage lines result from patterns of collagenous fibers in the _____.
a reticular; papillary layer
b papillary; epidermis
c papillary; reticular layer
d reticular; hypodermis
e reticular; epidermis

5.12 Within the skin, melanin is found in the _____, but within hair, pigment is present in the _____.
a epidermis; cortex   d dermis; cortex
b epidermis; cuticle   e dermis; medulla
c epidermis; medulla

5.13 Nerve endings in the skin are stimulated after _____ burn(s).
a first-degree only
b second-degree only
c third-degree only
d first- and second-degree
e second- and third-degree

Completion Exercises

5.14 The outer layer of the skin is the _____.

5.15 In the skin, the stratum spinosum and stratum basale together are known as the _____.

5.16 The outermost layer of the epidermis is the _____.

5.17 A burn involving more than 20 percent of the body surface is a _____-degree burn.

5.18 The lower layer of the dermis is called the _____ layer.

5.19 Another name for the subcutaneous layer of the skin is the _____.

5.20 Melanin is produced by specialized cells called _____.

Problem-Solving/Critical Thinking Questions

5.21 *Pathology report:* A triangular segment of skin presenting an irregular pigmented lesion with irregular margins. Histological sections show a nest of bizarre and pleomorphic (variations in shape) malignant cells in the epidermis. Malignant cells are invading the papillary dermis but not into the reticular dermis. Final diagnosis:
a mole   d malignant melanoma
b hemangioma   e squamous cell
c basal cell carcinoma   carcinoma

A SECOND LOOK

In the following drawing, identify the arrector pili muscle, dermis, sudoriferous gland, and the stratum spinosum.

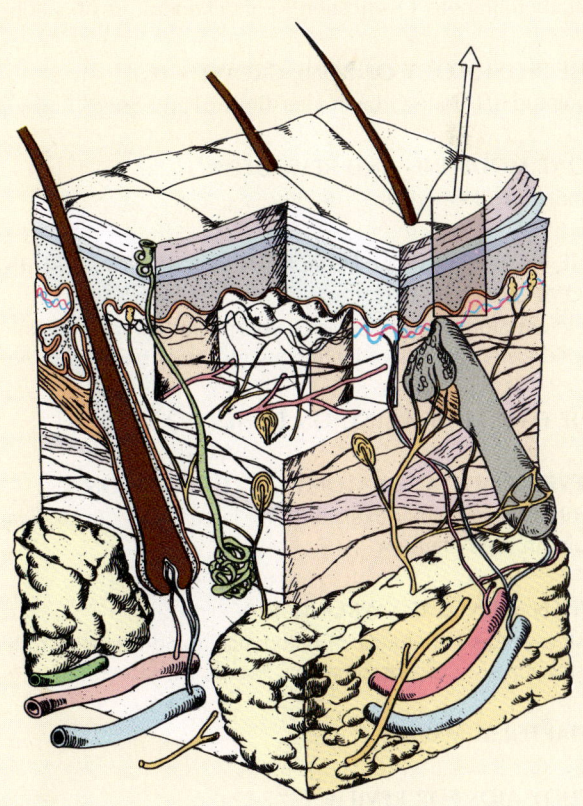

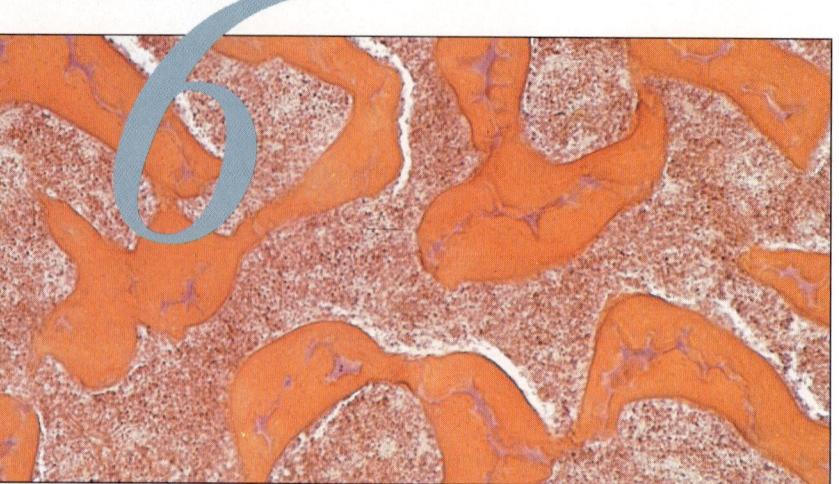

Bones and Bone Tissue

6

KEY CONCEPTS

1 Not only do bones support the body, protect the inner organs, and allow movement, but bone tissue stores and supplies calcium and phosphate and manufactures blood cells.

2 Bone cells can change their roles as the needs of the body change.

3 Although the mature skeleton may develop in different sequences, it always starts with a preexisting connective tissue skeleton, and mature bone is the same no matter how it develops.

4 Bones continue to grow in length and diameter after birth, and they are constantly changing shape slightly to accommodate shifting tensions.

From a structural point of view, the human skeletal system consists of two main types of supportive connective tissue: cartilage and bone. The histology of cartilage was described in Chapter 4. This chapter deals specifically with the histology, gross anatomy, and mechanical and physiological functions of bone.

Bone (osseous) tissue is specialized connective tissue that has the strength of iron and the lightness of pine wood. We usually think of bone tissue as constituting the skeleton that supports and protects our internal organs, but its functions go far beyond that. Bone tissue is the storehouse and main supply of reserve calcium and phosphate, and bone marrow serves as the site for the manufacture of red blood cells and some white blood cells. Bones aid movement by providing points of attachment for muscles and by transmitting the force of muscular contraction from one part of the body to another during movement. Living bone is not dry, brittle, or dead. It is a moist, changing, productive tissue that is continually resorbed (dissolved and assimilated), re-formed, and remodeled (replaced or renewed).

TYPES OF BONES AND THEIR MECHANICAL FUNCTIONS

The varied functions of bones may be classified as either mechanical or physiological. Bones make up the skeleton and provide the rigid framework that *supports* the body. They *protect* vulnerable internal organs such as the brain, heart, lungs, and organs of the pelvis by forming the sturdy walls of body cavities. They also make body *movement* possible by providing anchoring points for muscles and by acting as levers at the joints. These are the *mechanical aspects* of bone function. (The physiological aspects of bone function will be discussed in a later section of this chapter.)

Bones are usually classified by shape as long, short, flat, irregular, or sesamoid [FIGURE 6.1]. Accessory bones are included here as a separate group. The shapes of bones are generally related to their mechanical functions, those involving the support or movement of body parts. For example, a long bone acts as a lever, a short bone is usually a connecting bridge between other bones, and a flat bone is a protective shell.

Long Bones

A bone is classified as a *long bone* when its length is greater than its width. The most obvious long bones are in the arms and legs (the longest is the femur, or thighbone [FIGURE 6.1C]), but some are relatively short, as in the fingers and toes. Long bones act as levers that are pulled by contracting muscles. This lever action makes it possible for the body to move.

Short Bones

Short bones are about equal in length, width, and thickness, but they are shaped irregularly. They occur in the wrists (carpal bones) and ankles (tarsal bones [FIGURE 6.1E]), where only limited movement is required. Short bones are almost completely covered with articular surfaces, where one bone moves against another in a joint. However, there are some nonarticular areas on short bones where nutrient blood vessels enter, tendons attach muscles to bones, and ligaments connect bones.

Flat Bones

These bones are actually thin or curved more often than they are flat. *Flat bones* include the ribs, scapulae (shoulder blades), sternum (breastbone), and bones of the cranium (skull [FIGURE 6.1A]). The shape of flat bones is usually related to the expanse of the muscle attachment (origin or insertion), and the gently curved bones of the skull form a protective enclosure for the brain.

Irregular Bones

Irregular bones do not fit neatly into any other category. Examples are the vertebrae [FIGURE 6.1B], many facial bones, and the hipbones. The vertebrae have extensions that protrude from their main bony elements and serve as sites for muscle attachment. These bones support the spinal cord and protect it against compression forces.

Sesamoid Bones

Sesamoid bones are small bones embedded within certain tendons, cords that connect muscles to bones. Sesamoid bones usually occur where tendons pass over the joint of a long bone, as in the wrist or knee. Typical sesamoid bones include the patella, or kneecap [FIGURE 6.1D], which is within the tendon of a thigh muscle (quadriceps femoris), and the pisiform carpal bone, which is within the tendon of a wrist muscle (flexor carpi ulnaris). Be-

---

Q: *What color are living bones?*

A: The white bones of laboratory skeletons have actually been lightened in color. Living bones vary from pink or beige to light brown.

Q: *What are the largest and smallest bones in the body?*

A: The largest bone is the femur (thighbone), and the smallest is the stapes, a bone in the middle ear.

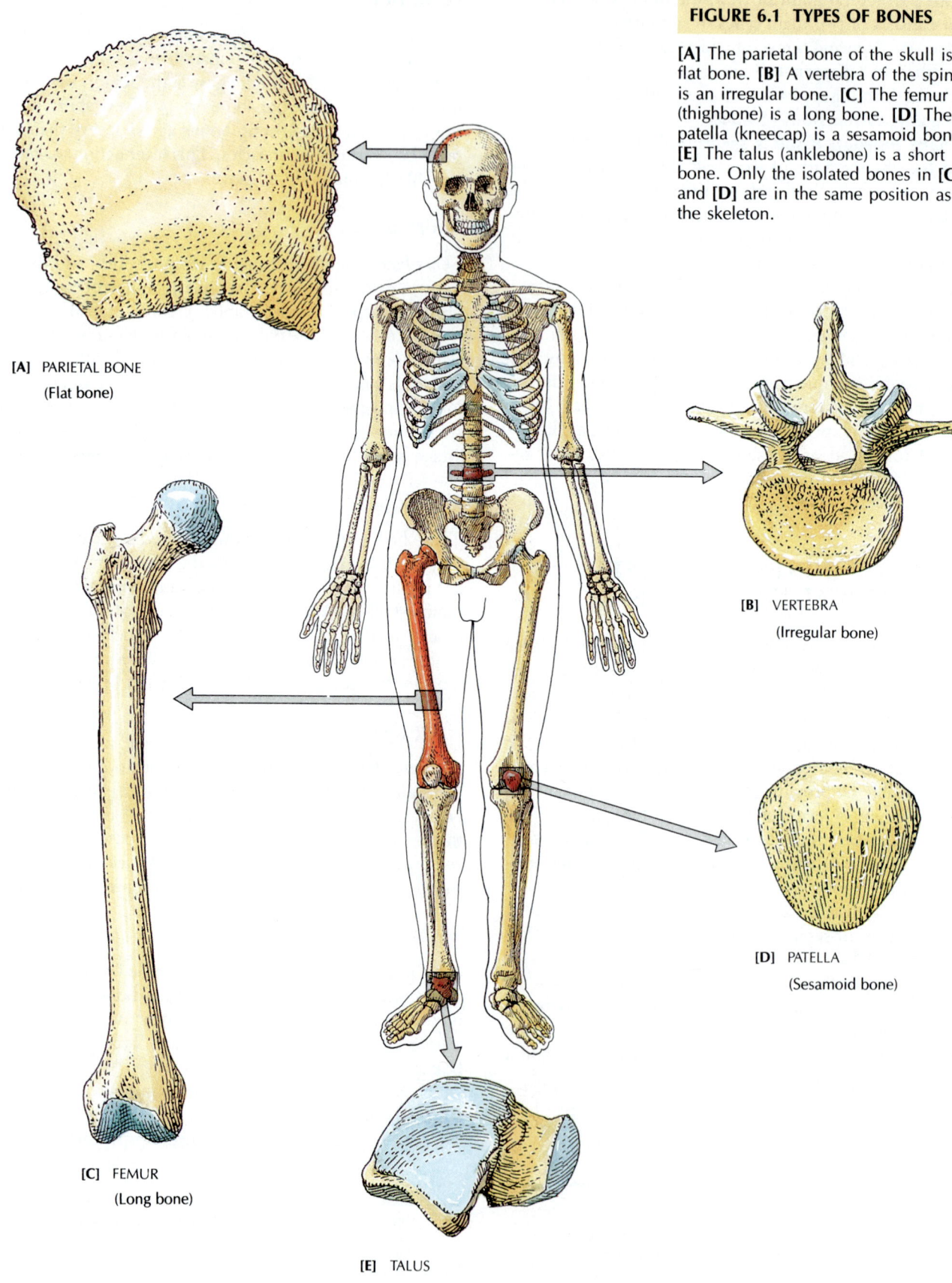

FIGURE 6.1 TYPES OF BONES

[A] The parietal bone of the skull is a flat bone. **[B]** A vertebra of the spine is an irregular bone. **[C]** The femur (thighbone) is a long bone. **[D]** The patella (kneecap) is a sesamoid bone. **[E]** The talus (anklebone) is a short bone. Only the isolated bones in **[C]** and **[D]** are in the same position as in the skeleton.

[A] PARIETAL BONE
   (Flat bone)

[B] VERTEBRA
   (Irregular bone)

[C] FEMUR
   (Long bone)

[D] PATELLA
   (Sesamoid bone)

[E] TALUS
   (Short bone)

METHODS OF CLASSIFYING BONES
SHAPE
Long: found in arms, forearms, thighs, legs, fingers, and toes
Short: found in wrists and ankles
Flat: found in skull and girdles
Irregular: found in axial skeleton (including vertebrae) and girdles
Sesamoid: found embedded within certain tendons, including knee and wrist
Accessory
Unfused bones: found mostly in feet
Sutural (Wormian) bones: small clusters of accessory bones between flat bones of skull
REGION
Axial skeleton: skull (cranium, face, ossicles of ear, hyoid); unfused vertebrae, coccyx, sacrum; sternum and ribs
Appendicular skeleton
Upper extremities: pectoral girdle; bones of arm, forearm, wrist, and hand
Lower extremities: pelvic girdle; bones of thigh, leg, ankle, and foot
EMBRYONIC DEVELOPMENT
Intramembranous ossification: bone (e.g., skull) develops directly from embryonic connective (mesenchymal) tissue
Endochondral ossification: bone (e.g., arms and legs) develops by replacing hyaline cartilage

sides helping protect the tendon, sesamoid bones help the tendon overcome compression forces, increasing the mechanical efficiency of joints. Sesamoid bones are so called because they sometimes resemble sesame seeds in shape. Their number varies from person to person.

Accessory Bones

Accessory bones are most commonly found in the feet. They usually occur when developing bones do not fuse completely. Unfused accessory bones may look like extra bones or broken bones in x-rays, and it is important not to confuse them with actual fractures. *Sutural (Wormian*) bones* are accessory bones that occur as small bone clusters between the joints of the flat bones of the skull [see FIGURE 7.5B]. The number of sutural bones in an individual varies. In general, accessory bones add some slight support and protection to the area of the skeleton where they are found.

**Wormian bones have nothing to do with worms. They are named after Olaus Worm, the seventeenth-century Danish anatomist.

A S K   Y O U R S E L F

1 What are some of the mechanical functions of bones?

2 What are the basic shapes of bones? How do their shapes relate to their functions?

3 What are accessory bones?

4 Where are sutural bones found?

GROSS ANATOMY OF A TYPICAL LONG BONE

The long bones of the body (for example, the humerus, tibia, and radius) provide an excellent descriptive model for the gross anatomy of a typical bone. Most adult long bones have a tubular shaft called the *diaphysis* (dye-AHF-uh-siss; Gr. "to grow between") [FIGURE 6.2]. The diaphysis is a hollow cylinder with walls of compact bone tissue. The center of the cylinder is the *medullary cavity*, which is filled with marrow. At each end of the bone is a roughly spherical *epiphysis* (ih-PIHF-uh-siss; Gr. "to grow upon") of spongy bone tissue. The epiphysis is usually wider than the shaft. The flat bones and irregular bones of the trunk and limbs have many epiphyses. The long bones of the fingers and toes have only one epiphysis.

Separating these two main sections at either end of the bone is the *metaphysis.* It is made up of the *epiphyseal (growth) plate* and the adjacent bony trabeculae (latticework) of spongy bone tissue on the diaphyseal side of the long bone. The epiphyseal plate is a thick plate of hyaline cartilage that provides the framework for the synthesis of the cancellous bone tissue within the metaphysis. The epiphyseal plates are the only places where long bones continue to grow in length after birth.

The medullary cavity running through the length of the diaphysis contains *yellow marrow,* which is mostly fat (adipose tissue). The porous latticework of the spongy epiphyses is filled with *red bone marrow.* The red bone marrow, also known as *myeloid tissue* (Gr. *myelos,* marrow), manufactures primarily red blood cells, which give the marrow its color. The thin membrane lining the medullary cavity of compact bone tissue and covering the trabeculae of spongy bone tissue is the *endosteum.*

Covering the outer surface of the bone (except in the joint) is the *periosteum,* a fibrous membrane that has the potential to form bone during growth periods and in fracture healing. The periosteum is often attached to the underlying bone by collagenous fibers called *periosteal perforating (Sharpey's) fibers* [see FIGURE 6.4A]. These fibers penetrate the inner layer of the periosteum and become embedded in the matrix of the bone. The periosteum contains nerves, lymphatic vessels, and many capillaries that provide nutrients to the bone and give

FIGURE 6.2 GROSS ANATOMY OF A TYPICAL LONG BONE

A typical long bone showing key anatomical features; the interior is partially exposed.

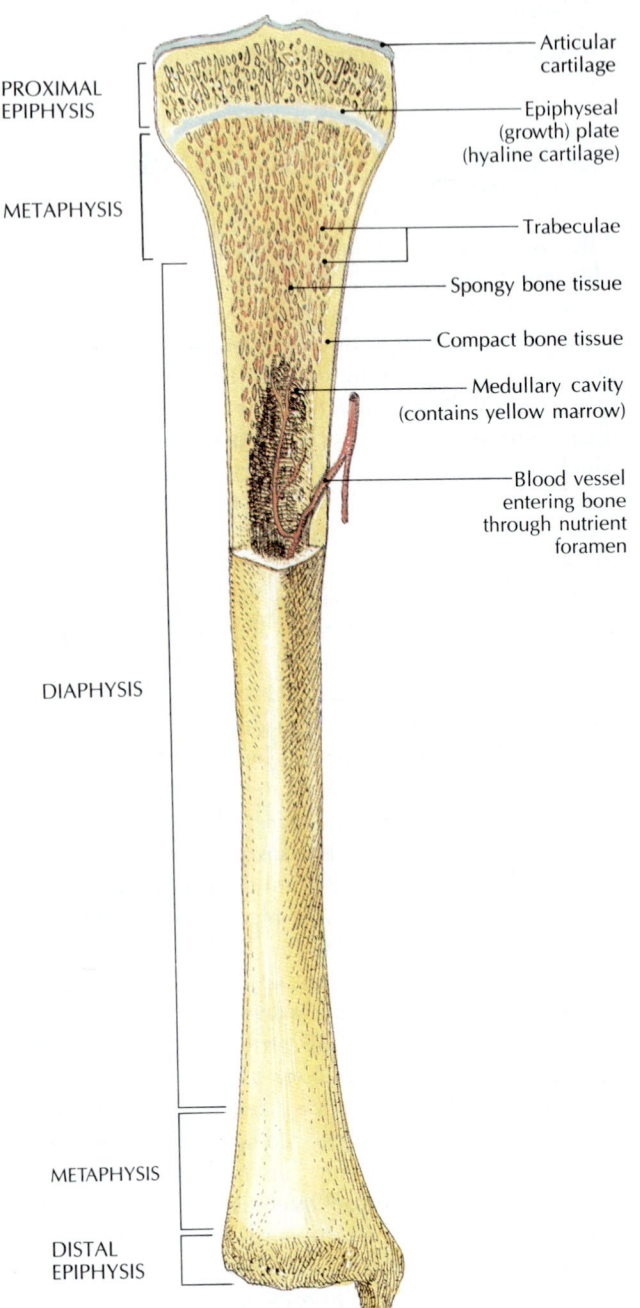

PROXIMAL EPIPHYSIS

METAPHYSIS

DIAPHYSIS

METAPHYSIS

DISTAL EPIPHYSIS

Articular cartilage

Epiphyseal (growth) plate (hyaline cartilage)

Trabeculae

Spongy bone tissue

Compact bone tissue

Medullary cavity (contains yellow marrow)

Blood vessel entering bone through nutrient foramen

living bone its distinctive pinkish color. Nutrients reach the marrow and spongy bone tissue by means of an artery that penetrates the compact bone tissue through a small opening called the **nutrient foramen.**

One of the few places where the periosteum is absent is at the joint surface of the epiphyses. There it is re-placed by articular cartilage, which provides a slick surface that reduces friction and allows the joint to work smoothly.

┌───┐
│ **A S K   Y O U R S E L F**

1 What is the diaphysis? The epiphysis? The metaphysis?

2 What is the specific function of red bone marrow?

3 What is the endosteum? The periosteum?
└───┘

THE HISTOLOGY OF BONE (OSSEOUS) TISSUE

The human body is about 62 percent water, but bone (osseous) tissue contains only about 20 percent water. As a result, your bones are stronger and more durable than your skin or your eyeballs, for instance. But even though most bones are as strong as iron, they are lighter and more flexible.

Like other types of connective tissue, bone tissue is composed of cells embedded in a *matrix* of ground sub-stance and fibers. However, bone tissue is more rigid than other tissues because its homogeneous organic ground substance also contains inorganic salts, mainly calcium phosphate and calcium carbonate. In the bones, these compounds plus others form hydroxyapatite crys-tals. When the body needs the calcium or phosphate that is stored within the bones, the hydroxyapatite crystals ionize and release the required amounts.

A network of collagenous fibers in the matrix gives bone tissue its strength and flexibility. Although the hardness of bone comes from the inorganic salts, its structure depends equally on the fibrous framework. When water and organic substances are removed from bone, it can crumble into a powdery chalk. However, if the inorganic salts are removed by a process called *decal-cification*, the bone becomes so flexible that it can be tied into a knot.

The older we get, the less organic matter and the more inorganic salts we have in our bones. Because of this shifting proportion of matrix to salts, the bones of older people are less flexible and more brittle than the bones of children.

Most bones have an outer shell of compact bone tissue enclosing an interior of spongy bone tissue, except where the spongy tissue is replaced by a marrow cavity or by air spaces called *sinuses*. For example, the leg bones contain marrow cavities, and some of the irregular skull bones contain sinuses that make the bones light. Irregular bones vary in the amount of spongy and compact tissue present. Short bones consist of spongy bone and marrow cavities.

The flat bones of the cranium (the part of the skull enclosing the brain) consist of two thin plates, called *tables*, of compact bone tissue with a layer of spongy bone tissue sandwiched between them [FIGURE 6.3]. The layer of spongy tissue contains marrow and veins (the diploic veins) and is called the *diploë* (DIHP-low-ee; Gr. "double"). Because of this protective arrangement, the outer table of the cranium can be fractured without harming the inner table and brain.

Compact Bone Tissue

The compact bone tissue that forms the outer shell of a bone is very hard and dense (like ivory) and appears to the naked eye to be solid, although it is not. It contains cylinders of calcified bone known as *osteons* (Gr. *osteon*, bone), or *Haversian systems.* These cylinders are made up of concentric layers, or *lamellae* (luh-MELL-ee; sing. *lamella*, luh-MELL-uh), of bone [FIGURE 6.4]. The term *lamellae* is derived from the Latin word for "thin plates." These lamellae are arranged like wider and wider drinking straws, each one nestled inside the next wider one. The structure of osteons provides the great strength needed to resist typical, everyday compressive forces on long bones.

In the center of the osteons are *central canals (Haversian canals*)*, longitudinal channels that contain blood vessels, nerves, and lymphatic vessels. Central canals usually have branches called *perforating canals (Volkmann's canals)* that run at right angles to the central canals and extend the system of nerves and vessels outward to the *periosteum* (outer covering) and inward to the *endosteum* (inner lining) of the bony marrow cavity. Unlike central canals, perforating canals are not enclosed by concentric lamellae.

Lamellae contain *lacunae* (luh-KYOO-nee; sing. *lacuna*, luh-KYOO-nuh; L. cavities, pods), or little spaces, which house the *osteocytes*, or bone cells [FIGURE 6.4]. Radiating like spokes from each lacuna are tiny *canaliculi* (KAN-uh-lick-yuh-lie; L. dim. *canalis*, channel) that contain the slender extensions of the osteocytes. Nutrients and waste products can pass to and from the blood vessels in the central canals (1) by normal processes of intracellular transport within each osteocyte, (2) over gap junctions from one osteocyte to another, or (3) possibly via tissue fluid in the tiny spaces between the osteocytes and their surrounding lacunae.

Spongy (Cancellous) Bone Tissue

Spongy, or cancellous (latticelike), bone tissue is in the form of an open interlaced pattern that withstands maxi-

*Note that a *Haversian system* and an osteon are the same structure; the *central (Haversian) canal* is just the longitudinal channel.

FIGURE 6.3 FLAT BONES OF THE CRANIUM

[A] In the flat bones of the cranium, two tables (plates) of compact bone tissue surround a center of spongy bone tissue and marrow, the diploë. **[B]** An enlarged view showing more detail.

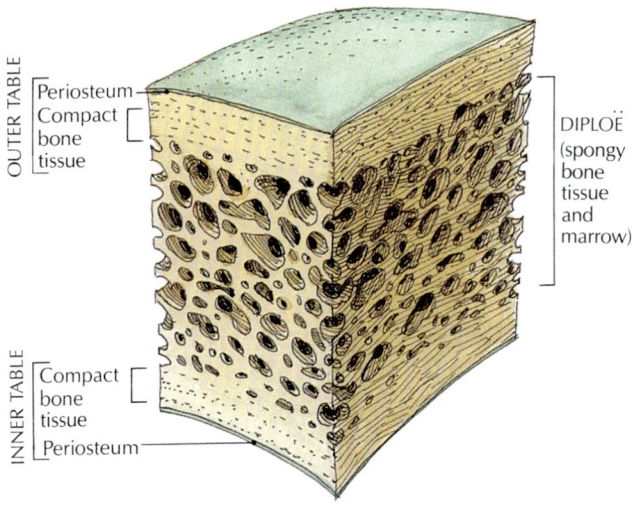

[A]

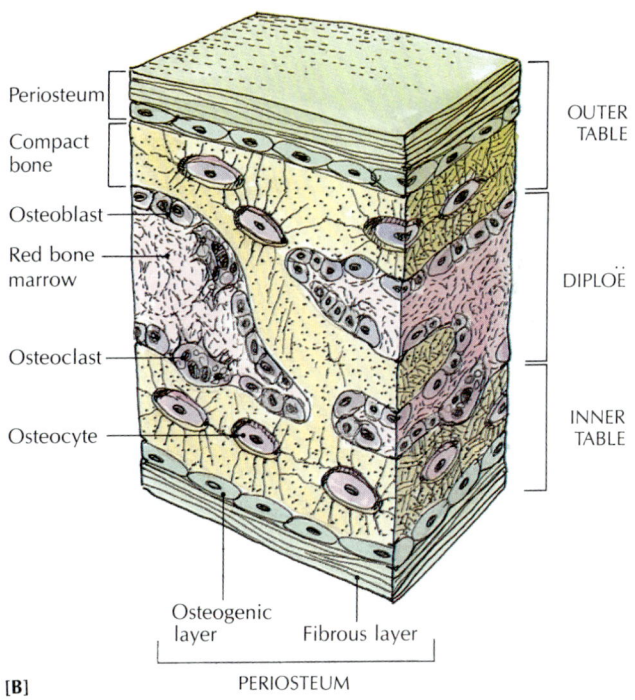

[B]

mum stress and supports shifts in weight distribution [FIGURE 6.5A]. Prominent in the interior structure of spongy bone tissue are *trabeculae* (truh-BECK-yuh-lee; L. dim. *trabs*, beam), tiny spikes of bone tissue [FIGURE 6.5B] surrounded by bone matrix that has *calcified*, or become

FIGURE 6.4 COMPACT BONE TISSUE

[A] An enlarged longitudinal section of compact bone tissue showing blood vessels, canals, and other internal structures. [B] An enlargement of a single osteon with lacunae, canaliculi, and a central (Haversian) canal visible. [C] An enlarged osteocyte (bone cell) inside a lacuna. [D] A scanning electron micrograph of compact bone tissue showing an osteon with its concentric lamellae housing lacunae and canaliculi. ×1000. (Richard G. Kessel and Randy H. Kardon, *Tissues and Organs: A Text-Atlas of Scanning Electron Microscopy*, San Francisco, W. H. Freeman, 1979.)

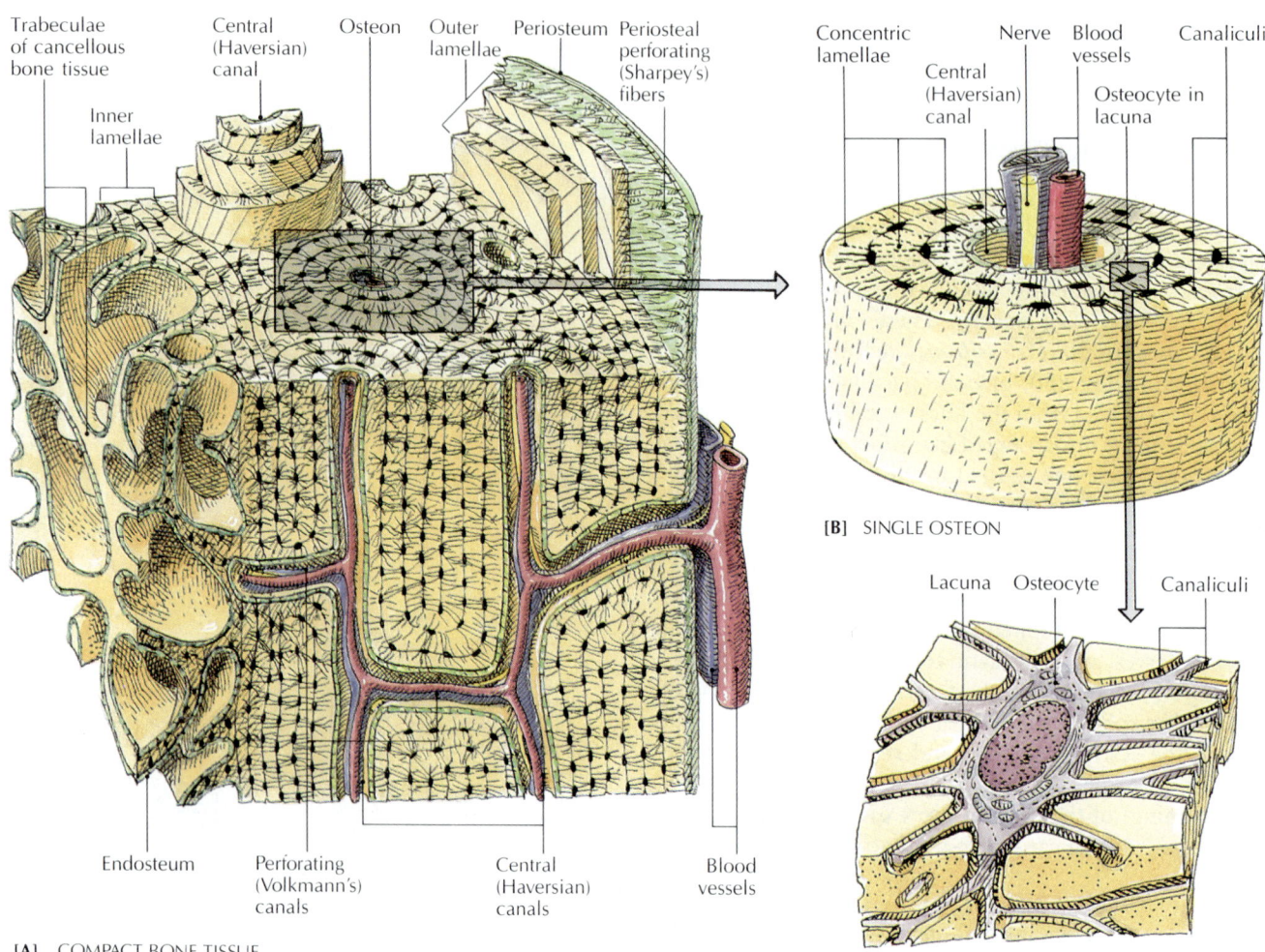

Trabeculae of cancellous bone tissue
Inner lamellae
Central (Haversian) canal
Osteon
Outer lamellae
Periosteum
Periosteal perforating (Sharpey's) fibers

Endosteum
Perforating (Volkmann's) canals
Central (Haversian) canals
Blood vessels

[A]   COMPACT BONE TISSUE

Concentric lamellae
Central (Haversian) canal
Nerve
Blood vessels
Osteocyte in lacuna
Canaliculi

[B]   SINGLE OSTEON

Lacuna   Osteocyte   Canaliculi

[C]   SINGLE BONE CELL

Canaliculi   Central (Haversian) canal   Lacunae

[D]

FIGURE 6.5 A SECTION THROUGH A HUMAN HIP JOINT

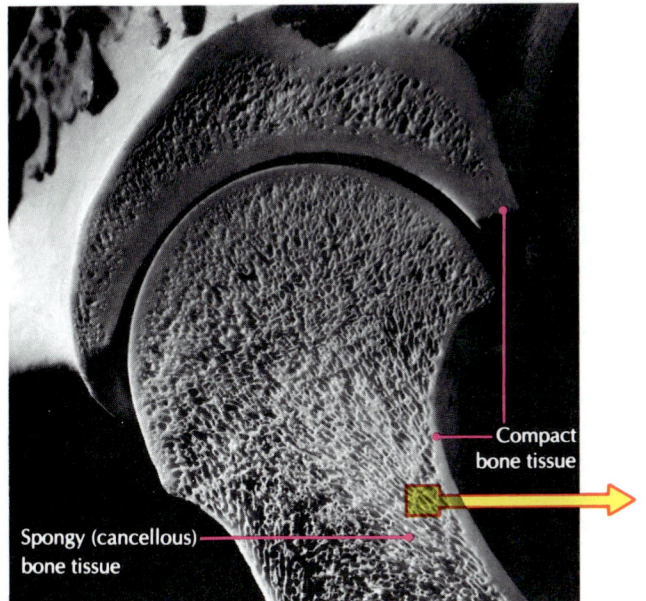

[A]

[A] The porous, spongy (cancellous) portion of the bone has a streaked appearance, which indicates how the bone is built up in the direction of the greatest stress. **[B]** A scanning electron micrograph of trabeculae in cancellous bone. ×4.

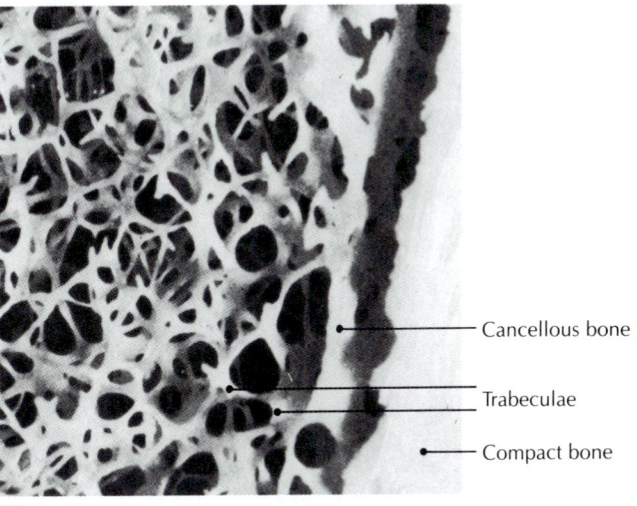

[B]

Labels in [A]: Compact bone tissue; Spongy (cancellous) bone tissue

Labels in [B]: Cancellous bone; Trabeculae; Compact bone

---

hardened by the deposition of calcium salts. The trabeculae form along the lines of greatest pressure or stress. This arrangement provides the greatest strength with the least weight. Spongy bone tissue is found inside most bones.

A S K   Y O U R S E L F
1 What does the bone matrix contain?
2 What is the structure of compact bone tissue? Of spongy bone tissue?
3 What is an osteon? A central canal?

BONE CELLS

Bones contain five types of cells that are capable of changing their roles as the needs of the body change in the growing and adult skeletons:

1 *Osteogenic (osteoprogenitor) cells* are small, spindle-shaped cells found mostly in the deepest layer of the periosteum and in the endosteum. These cells have a high mitotic potential and can be transformed into bone-forming cells *(osteoblasts)* during healing.

2 *Osteoblasts* (Gr. *osteon*, bone + *blastos*, bud or growth) synthesize and secrete unmineralized ground substance called *osteoid* (Gr. *osteon*, bone + *eidos*, form). When calcium salts are deposited in the fibrous osteoid, the osteoid calcifies into bone matrix. Osteoblasts act as pump cells to move calcium and phosphate into and out of bone tissue, thereby, respectively, calcifying or decalcifying it. Osteoblasts are usually found in the growing portions of bones, including the periosteum.

3 *Osteocytes* are the main cells of fully developed bones. Each osteocyte has a cell body that occupies a lacuna within the bone matrix and long cytoplasmic processes that extend through the matrix via canaliculi. These processes interconnect to form gap junctions between neighboring osteocytes. Osteocytes are derived from osteoblasts that have secreted bone tissue around themselves. They, along with osteoclasts, play an active role in homeostasis by helping to release calcium from bone tissue into the blood, thereby regulating the concentration of calcium in the body fluids. Osteocytes also keep the matrix in a stable and healthy state by secreting enzymes and maintaining its mineral content.

4 *Osteoclasts* (Gr. *klastes*, breaker) are multinuclear giant cells that move about on bone surfaces, resorbing (dissolving and assimilating) bone matrix from sites where it is either deteriorating or not needed. They are usually found where bone is resorbed during its normal growth.* Osteoclasts are derived from white blood cells called *monocytes.*

---

*To remember the difference between osteoblast and osteoclast, remember the *b* in osteoblast as standing for "building."

5 *Bone-lining cells* are found on the surface of most bones in the adult skeleton. These cells are believed to be derived from osteoblasts that cease their physiological activity and flatten out on the bone surface. These cells may have several functions. They may serve as osteogenic cells that can divide and differentiate into osteoblasts. Most probably they serve as an ion barrier around bone tissue. This barrier contributes to mineral homeostasis by regulating the movement of calcium and phosphate into and out of the bone matrix, and this in turn helps control the deposition of hydroxyapatite in the bone tissue.

A S K   Y O U R S E L F

1 What are the five kinds of bone cells?

2 What are the functions of osteoblasts? Osteocytes? Osteoclasts?

THE PHYSIOLOGY OF BONE FORMATION: OSSIFICATION

Bones develop through a process known as *ossification* (osteogenesis). Since the primitive skeleton of the human embryo is composed of hyaline cartilage and fibrous membrane, bones can develop in the embryo in two ways: *intramembranous ossification* or *endochondral ossification*. However, in both cases bones are formed from a preexisting connective tissue skeleton. Bone is the same no matter how it develops. Only the bone-making sequence is different.

Bones that develop by intramembranous ossification include the frontal and parietal bones, and bones of the face. Bones that develop by endochondral ossification include the temporal and occipital bones of the skull, and the bones of the vertebral column and extremities. FIGURE 6.6 compares the fetal and adult skeletons. The ages when some of the bones appear in fetal life and when they fuse with their epiphyses are shown on the skeletons. (See Developmental Anatomy of the Skeleton on page 173 for a more complete description of prenatal skeleton formation.)

Intramembranous Ossification

If bone tissue (spongy or compact) develops directly from mesenchymal (embryonic connective) tissue, the process is called *intramembranous ossification.** The vault

---

* Intramembranous means "within the membrane," and *ossification* means "bone formation." The term *intramembranous ossification* was originally used because this layer of mesenchyme was thought to be a sheet of membrane.

(arched part) of the skull, the flat bones of the face (including those lining the oral and nasal cavities), and part of the clavicle (collarbone) are formed this way. The skull is formed relatively early in the embryo to protect the developing brain. Some flexible dense connective tissue still exists between the flat skull bones at birth, so when the baby is born, its skull is flexible enough to pass unharmed through the mother's birth canal.

From the time of initial bone development, intramembranous ossification spreads rapidly from its center until large areas of the skull are covered with protecting and supporting bone. The first rapid phase begins when an *ossification center* (item 3 in the following list) first appears from the eighth through the twelfth fetal week and lasts through the end of the fifteenth fetal week, when the area is entirely covered [FIGURE 6.6A].

During the subsequent growth of the skull, the bones develop at a slower rate. The second phase continues into adolescence. Because the brain reaches full size by about the tenth year, the development of the protective skull vault is completed early in adolescence. In contrast, the bones of the face do not reach their adult size until the end of the adolescent growth spurt, anywhere from age 14 to age 17.

Because intramembranous ossification is studied most easily in the skull, we describe the process there:

1 A layer of embryonic connective tissue called *mesenchyme* forms between the developing scalp and brain.

2 A plentiful supply of blood vessels arises in the mesenchyme, where some mesenchymal cells are already connected to neighboring cells by long, thin fibers, or processes [FIGURE 6.7A].

3 Bone tissue development begins when thin strands that will eventually become branching trabeculae appear in the matrix. At about the same time, the mesenchymal cells become larger and more numerous, and their processes thicken and connect with other embryonic connective tissue cells, forming a ring of cells around a blood vessel [FIGURE 6.7B]. The site of this ring formation is a *center of osteogenesis* (also known as an *ossification center*). It begins about the second month of prenatal life.

4 The mesenchymal cells differentiate from osteogenic cells into osteoid-secreting osteoblasts [FIGURE 6.7C]. Then the osteoblasts begin to cause calcium salt deposits that form the spongy, latticelike bone matrix. (The trabeculae of cancellous bone tissue will develop later from the spongy matrix.)

5 In this process of calcification, some osteoblasts become trapped within lacunae in the developing matrix. The entrapped osteoblasts, now called *osteocytes* [FIGURE 6.7D], preserve the integrity of the matrix and also release calcium ions as they are needed. These osteocytes remain in contact through gap junctions with other osteocytes and osteoblasts through *canaliculi*, tiny channels formed

FIGURE 6.6 SOME DIFFERENCES BETWEEN THE FETAL AND ADULT SKELETONS

[A] Epiphyses begin to appear in the fetal skeleton about the fifth fetal week and continue to form until the sixteenth fetal week. [B] In the adult skeleton, zones of growth disappear, and epiphyses fuse with diaphyses between the ages of about 13 and 25.

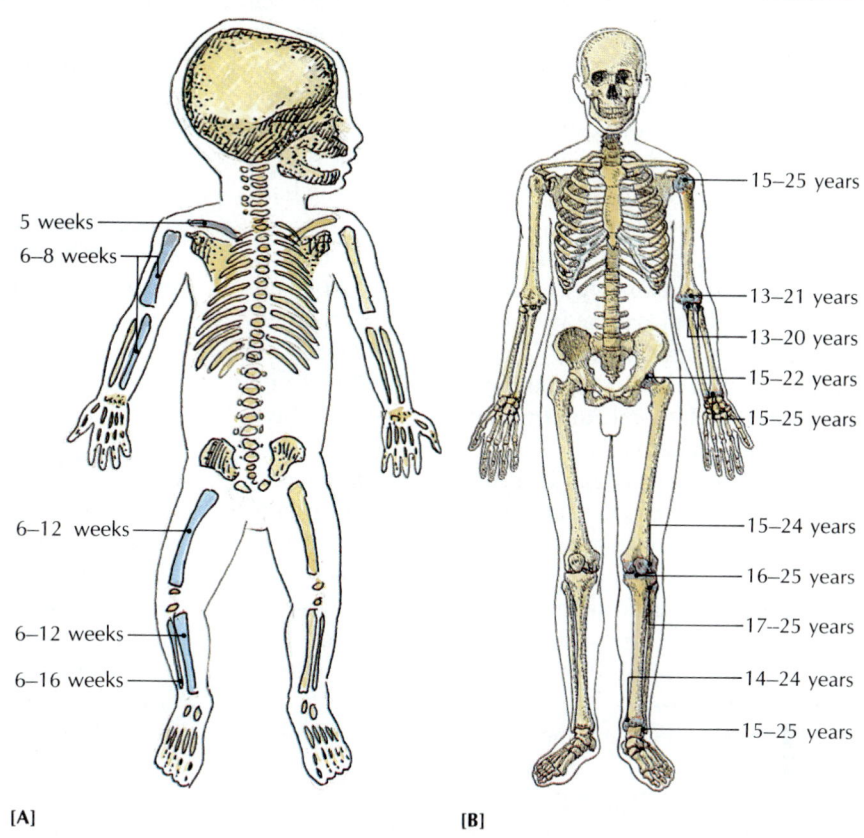

5 weeks

6–8 weeks

6–12 weeks

6–12 weeks

6–16 weeks

15–25 years

13–21 years

13–20 years

15–22 years

15–25 years

15–24 years

16–25 years

17–25 years

14–24 years

15–25 years

[A]   [B]

FIGURE 6.7 INTRAMEMBRANOUS OSSIFICATION

[A] The mesenchyme forms.
[B] The mesenchymal cells enlarge and form a ring around a blood vessel. [C] The mesenchymal cells differentiate into osteoblasts that secrete osteoid.
[D] As the osteoid calcifies into bone matrix, it entraps osteocytes in lacunae. [E] The osteoclasts remove small areas of bone (calcium) from the walls of the lacunae.

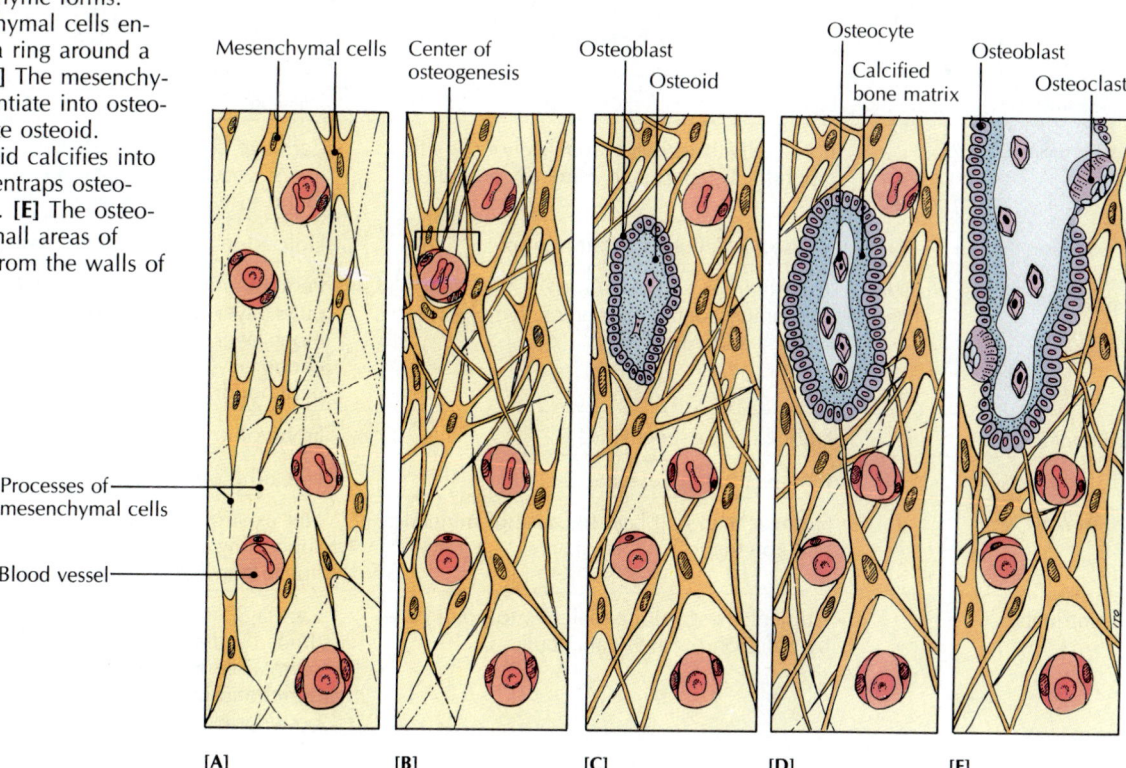

Mesenchymal cells   Center of osteogenesis   Osteoblast   Osteocyte   Osteoblast

Osteoid   Calcified bone matrix   Osteoclast

Processes of mesenchymal cells

Blood vessel

[A]   [B]   [C]   [D]   [E]

when the bone matrix is deposited around the long cell processes of the osteocytes. (In a real sense, bone may be considered to be a network of osteocytes and their processes surrounded by a bony matrix.) As bone-forming osteoblasts change into osteocytes, they are replaced by new osteoblasts that develop from the osteogenic cells in the surrounding connective tissue. In this way, the bone continues to grow.

The gap junctions facilitate the exchange of small molecules and presumably expedite the transport of parathyroid hormone from the Haversian canals to the osteocytes in the distant lamellae. The osteocytes respond to the parathyroid hormones and thereby contribute to the regulation of blood calcium by tapping the stores of calcium in the bone matrix.

6 While the osteoblasts are synthesizing and mineralizing the matrix, osteoclasts are playing a role in bone resorption by removing small areas of bone (calcium) from the walls surrounding the lacunae [FIGURE 6.7E]. Osteoclasts secrete a combination of an acid medium and acid hydrolase enzymes (derived from lysosomes). The acid medium creates a low pH in which the enzymes disintegrate bone tissue.

7 The trabeculae continue to thicken into the dense network that is typical of cancellous bone tissue [see FIGURE 6.5B]. The collagenous fibrils deposited on the trabeculae crowd nearby blood vessels, which eventually condense into blood-forming bone marrow.

8 The osteoblasts on the surface of the spongy bone tissue form the *periosteum* — the membrane that covers the outer surface of the bone. It is made up of an inner osteogenic layer (with osteoblasts) and a thick, fibrous outer layer. The inner layer eventually creates a protective layer of compact bone tissue over the interior cancellous tissue. Once intramembranous ossification has stopped, the osteogenic layer becomes inactive, at least temporarily. It becomes active again when necessary — to repair a bone fracture, for example.

Endochondral Ossification

When bone tissue develops by replacing hyaline cartilage, the process is known as ***endochondral ossification.*** The term *endochondral* (Gr. *endo,* within + *khondros,* cartilage) means "inside the cartilage" and is used to describe this type of bone formation because it takes place where the cartilage model is eroded. Endochondral ossification produces long bones and all other bones not formed by intramembranous ossification. Remember, however, that the *cartilage itself is not converted into bone.* The cartilage model of the skeleton is completely destroyed and replaced by newly formed bone. The compositions of cartilage and bone are compared in TABLE 6.1.

TABLE 6.1 COMPARISON OF BONE AND CARTILAGE

Feature	Bone	Cartilage
Components	Bone cells (osteocytes), collagenous fibers, ground substance, and mineral components.	Cartilage cells (chondrocytes), collagenous fibers, ground substance; elastic fibers are found in elastic cartilage.
Location of cells	Housed in lacunae within matrix.	Housed in lacunae within matrix.
Outer covering of tissue	Periosteum (except at joints), a fibrous membrane containing nerves, lymphatic vessels, and capillaries.	Perichondrium (except at joints); composed of outer layer of dense connective tissue and inner layer of cells that differentiate into chondrocytes; fibrocartilage has no perichondrium.
Derivation	Embryonic mesenchyme.	Embryonic mesenchyme.
Blood vessels	Contains blood vessels.	Has no blood vessels.
Strength	Stronger than cartilage because of minerals and abundant fibers in matrix.	Not as strong as bone but capable of supporting weight; matrix contains fewer fibers than bone matrix.
Nutrients	Nutrients pass from capillaries to bone cells by diffusion through canaliculi.	Cartilage cells receive nutrients from tissue fluids by diffusion through selectively permeable intercellular matrix (and perichondrium); some blood reaches outer boundaries.

Endochondral ossification at the cellular level

FIGURE 6.8 shows the stages of endochondral bone formation at the *cellular level:*

1 The ossification center is a cartilaginous matrix that includes *chondrocytes* (cartilage cells) in lacunae [FIGURE 6.8A].

2 The chondrocytes secrete alkaline phosphatase, which triggers a chemical reaction in the matrix that causes mineralization [FIGURE 6.8B]. As the ossification process begins, the chondrocytes and lacunae enlarge, while the cartilage between the chondrocytes becomes mineralized (calcified) into cartilaginous spicules. These spicules act as scaffolds upon which the osteoblasts lay down the osteoid matrix prior to ossification.

3 The calcified matrix blocks the diffusion of nutrients to the chondrocytes, which begin to die. Eventually they are resorbed, leaving irregular cavities in the ossifying matrix of trabeculae [FIGURE 6.8C]. Once the bone is formed, these cavities contain the bone marrow.

4 *Pluripotent cells* (which have the potential to divide into several distinct cell types) lining the cavities begin to differentiate into osteoblasts and osteocytes. The osteoblasts deposit osteoid on the mineralized cartilage cores, forming a thin layer of spongy bone tissue [FIGURE 6.8D]. Osteoclasts are involved in bone resorption during this stage of endochondral ossification.

Endochondral ossification at the tissue level

Endochondral ossification is slower than intramembranous ossification. FIGURE 6.9 depicts endochondral ossification at the *tissue level.* The most typical sequence takes place in long bones and begins in the diaphyseal area:

1 Bone tissue begins to develop in a limb bud in the embryo. About 6 to 8 weeks after conception, mesenchymal cells multiply rapidly and bunch together in a dense central core of precartilage tissue, which eventually forms the cartilage model [FIGURE 6.9A].

2 Soon after, an outline of a primitive *perichondrium* (the membrane that covers the surface of cartilage) appears [FIGURE 6.9B].

3 As it grows, the cartilage model is invaded by capillaries, triggering the transformation of the perichondrium into the bone-producing *periosteum*, the fibrous membrane that covers the outer surfaces of bones [FIGURE 6.9C].

4 As intramembranous ossification occurs in the periosteum, a hollow cylinder of trabecular bone called the *bone collar* forms around the cartilage of the diaphysis [FIGURE 6.9C].

FIGURE 6.8 EARLY STAGES OF ENDOCHONDRAL OSSIFICATION

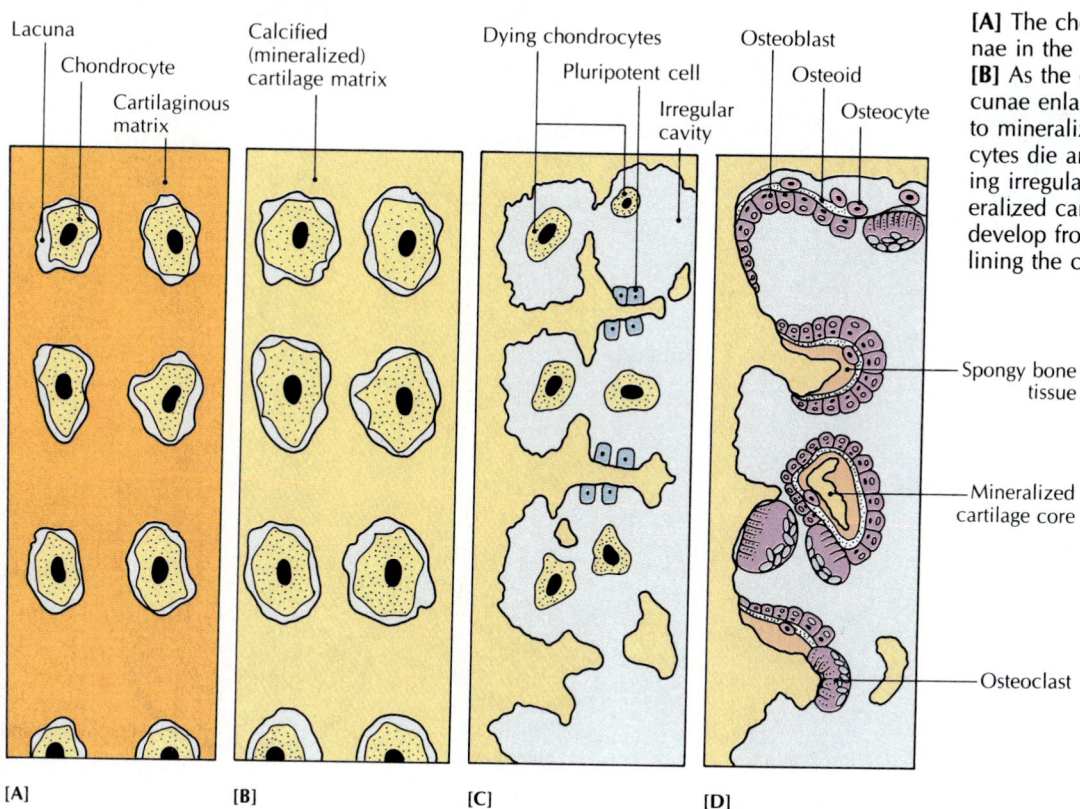

Lacuna
Chondrocyte
Cartilaginous matrix

Calcified (mineralized) cartilage matrix

Dying chondrocytes
Pluripotent cell
Irregular cavity

Osteoblast
Osteoid
Osteocyte

[A] The chondrocytes and lacunae in the cartilaginous matrix. **[B]** As the chondrocytes and lacunae enlarge, the matrix begins to mineralize. **[C]** The chondrocytes die and are resorbed, leaving irregular cavities in the mineralized cartilage. **[D]** Bone cells develop from pluripotent cells lining the cavities.

Spongy bone tissue

Mineralized cartilage core

Osteoclast

[A]   [B]   [C]   [D]

FIGURE 6.9 ENDOCHONDRAL OSSIFICATION AND LONGITUDINAL GROWTH IN A LONG BONE

Cartilage is shown in light blue, mineralized cartilage in darker blue. **[A]** Cartilage model. **[B]** The perichondrium forms. **[C]** The perichondrium is replaced by the periosteum. The bone collar forms. **[D]** Blood vessels enter the matrix. The primary center of ossification is established. **[E]** Blood vessels spread through the matrix. **[F]** The marrow cavity forms. **[G]** After birth, blood vessels enter the proximal epiphyseal cartilage, creating a secondary center of ossification. **[H]** Another secondary center of ossification forms in the distal epiphyseal cartilage. **[I]** Some cartilage remains on the joint edge of each epiphysis to form articular cartilage. **[J]** As growth in length of the bone slows, the distal epiphyseal plate disappears, and the epiphyseal line forms on the interior of the epiphysis. **[K]** The proximal plate disappears somewhat later.

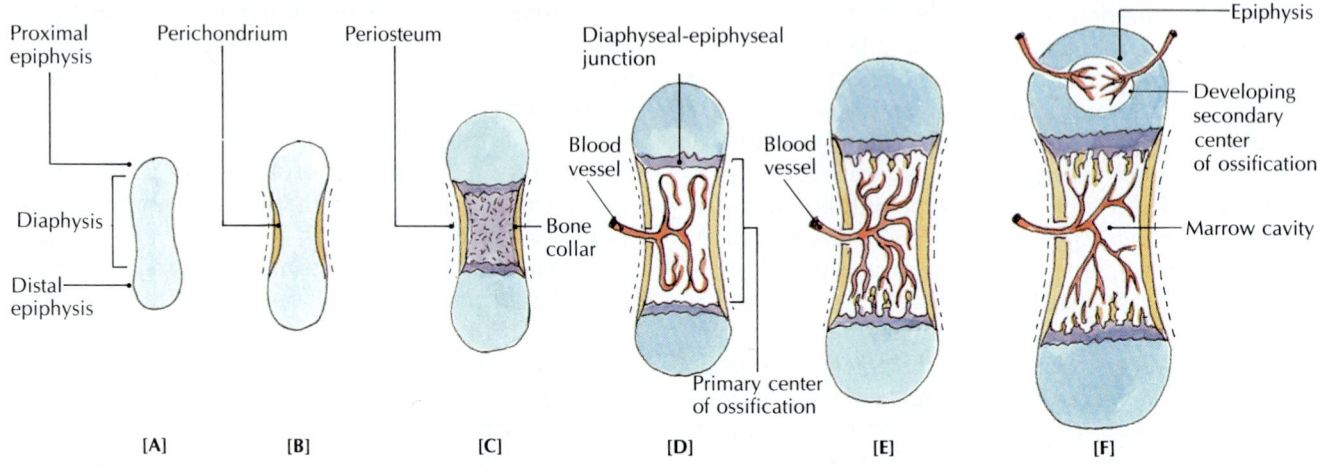

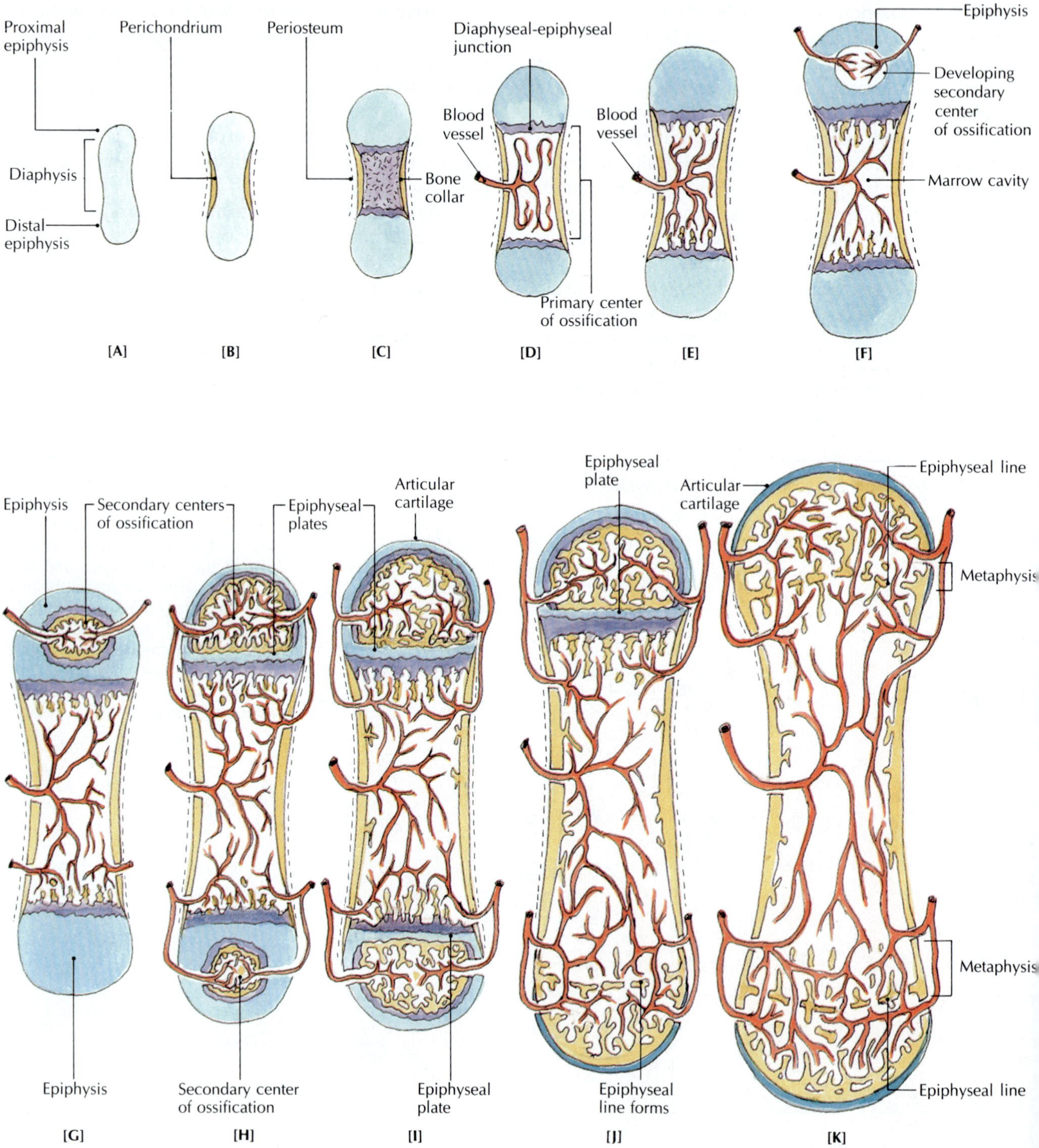

5 By the end of the second or third month of prenatal development, the ***primary center of ossification*** is established near the middle of what will become the diaphysis [FIGURE 6.9D]. The cartilaginous matrix mineralizes and begins to disintegrate. A developing *periosteal bud* derived from the periosteum penetrates the area left by the disintegrating cartilage. The periosteal bud consists of osteogenic cells that form the ossification center, blood vessels that form the vascular network in the future marrow, and pluripotent cells that form the progenitors of red blood cells, certain white blood cells, and other cells. At the same time, blood vessels develop in the periosteum and branch into the diaphysis [FIGURE 6.9D, E]. The diaphysis now has a well-developed bone collar under the periosteum and an increasing amount of marrow in the center of the shaft [FIGURE 6.9F]. The epiphyses, however, still consist of cartilage.

A S K   Y O U R S E L F

1 What is ossification?

2 What is the basic difference between intramembranous and endochondral ossification?

3 Can cartilage be converted into bone? Explain.

4 What is the primary center of ossification?

THE PHYSIOLOGY OF BONE GROWTH

The process of ossification allows embryonic bones to develop, but bones continue to lengthen and thicken *after* birth. The following sections describe the lengthening and thickening processes.

Longitudinal Bone Growth after Birth

After birth, chondrocytes in the epiphyses begin to mature and enlarge in the same way earlier chondrocytes did in the diaphysis. Here, too, the matrix mineralizes and the chondrocytes die. Blood vessels and osteogenic cells from the periosteum enter the epiphyses, where the cells develop into osteoblasts. The centers of this activity in the epiphyses are called ***secondary*** (or ***epiphyseal***) ***centers of ossification*** [FIGURE 6.9G, H]. Although a few secondary centers may be present before birth, most do not appear until childhood or even adolescence.

Some cartilage remains on the outer (joint) edge of each epiphysis to form the *articular cartilage* necessary for the smooth operation of the joints [FIGURE 6.9I]. The epiphyseal plate also remains throughout the growth period. It provides the framework for development of the cancellous bone tissue inside the metaphysis.

The longitudinal growth of bone after birth takes place in small spaces in the cartilaginous epiphyseal plates. During growth, new cartilage cells are generated in the epiphyseal plate, thickening and moving the superior portion of the epiphysis upward and thus producing lengthwise growth. The epiphyseal plates generate new cartilage cells until the age of about 17. Depending on the bone, the growing period stops altogether any time from adolescence until the early twenties [FIGURE 6.10].

As the growth in length of the long bone slows, the distal epiphyseal plate disappears. At this point — the *fu-*

FIGURE 6.10 LONGITUDINAL BONE GROWTH

Photomicrograph of the four zones of the epiphyseal plate. The photo shows an enlargement of the epiphysis and metaphysis in FIGURE 6.9K.

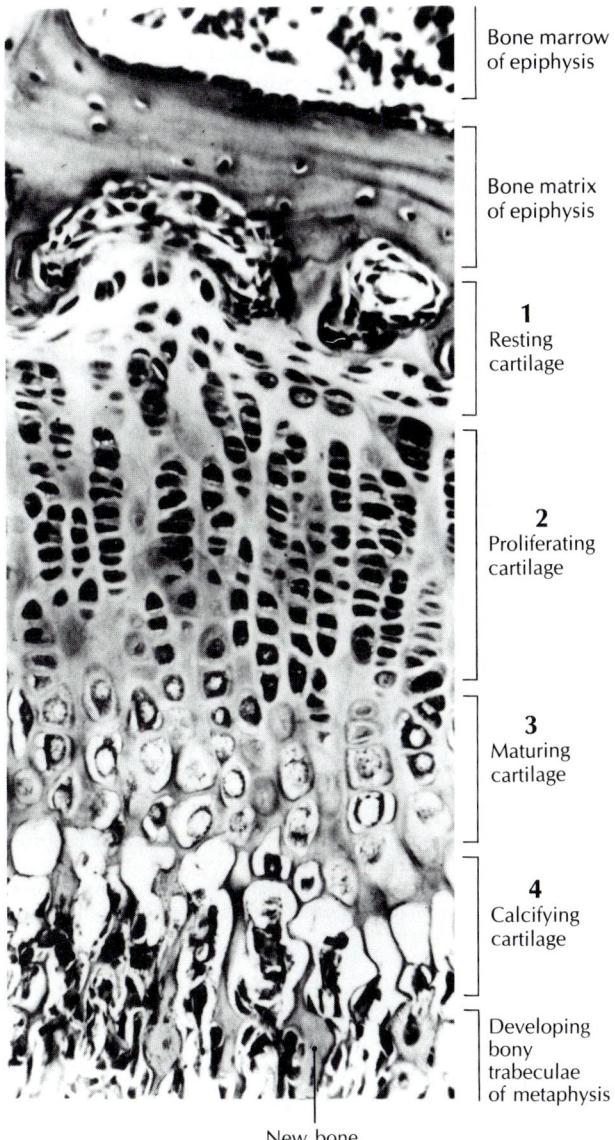

Bone marrow of epiphysis

Bone matrix of epiphysis

1 Resting cartilage

2 Proliferating cartilage

3 Maturing cartilage

4 Calcifying cartilage

Developing bony trabeculae of metaphysis

New bone

sion, or *closure,* of the epiphysis—all that remains of the epiphyseal plate is a thin epiphyseal line [FIGURE 6.9J]. The proximal epiphyseal plate disappears somewhat later [FIGURE 6.9K]. By the time longitudinal growth ends, the epiphyseal cartilage is completely replaced by bone tissue.

How Bones Grow in Diameter

Although bones do not grow in length after the age of 25 or so, the *diameter* of bones may continue to increase throughout most of our lives. The growth in diameter and circumference of a long bone results from the deposition of new intramembranous bone (bone that develops in the absence of a cartilage model) by the osteoblasts of the periosteum. This new bone is laid down on the surface of the compact bone as lamellae. In contrast, the trabeculae of spongy bone in the inner (marrow) surface of the bone are reorganized by the resorption of bone by osteoclasts and the addition of new bone. The new bone is laid down by osteoblasts as intramembranous bone on the surface of the bony trabeculae.

Bones can thicken, or become denser, to keep up with any physical changes in the body that may increase the stress, or load, the bones have to bear. For example, in people who greatly increase the size of their muscles, the bones can also be strengthened at the same time. If the bones are not made stronger, they may fracture because they are unable to cope with the increased pull of the stronger muscles. In the same way, the bones in people who are overweight can thicken enough to offset the additional stress caused by the extra pounds. This thickening occurs as a result of the mechanical tension that stimulates osteoblastic activity and, in turn, more bone (calcium) deposition. Such thickening is often seen in the ribs, femurs, tibiae, and radii.

A person who has one leg in a cast but continues to put weight on the other leg finds that the inactive leg may become thinner within a month or so. That occurs because the muscles in the injured leg atrophy, and the bones may decalcify by about 30 percent, while the active leg is not affected. Bone loss occurs because of inactivity, immobility, or anything that takes the load off the skeleton.*

*When astronauts are subjected to prolonged weightlessness in space, their bones begin to degenerate and lose calcium and other minerals. Dietary supplements of calcium and even attempts at strenuous exercise do not offset the degenerative effects of a lack of normal, everyday loads (stress) on bones.

ASK YOURSELF
1 What is a secondary center of ossification?
2 How do bones grow in diameter?

BONE MODELING AND REMODELING

Bone *modeling* is the alteration of bone size and shape during a bone's developmental growth. It alters the amount and distribution of bone tissue in the skeleton and determines the form of bones. Modeling occurs on different bone surfaces and at different rates during growth. Although bone is dense and hard, it is able to replace or renew itself through a process called *remodeling.* Remodeling occurs throughout adult life as a response to a variety of stresses.

Modeling

A dramatic example of bone modeling is the shaping of the skull bones. Bone yields to such soft structures as blood vessels and nerves by forming grooves, holes, and notches not only in the fetus, when bone forms around blood vessels and nerves, but even later on, especially when vessels or nerves are rerouted because of injuries or other traumatic changes.

As the body grows to adulthood, the brain and cranium continue to grow. As the cranium enlarges, the curvature of its four major bones must decrease. The changing of the curvature is accomplished by the growth of successive layers of new bone on the outer surface of the skull bones and, at the same time, the *resorption* (dissolution and assimilation) of old bone at the inner surface of the skull bones. When the cranium begins to shrink in old age (the brain mass also diminishes), bone is resorbed without being replaced. Most growth patterns are under hormonal control.

Remodeling

Just as for many other tissues of the body, the structural units of bone tissue must be replaced or renewed in order to maintain homeostasis. This remodeling occurs through the selective resorption of old bone and the simultaneous production of new bone [FIGURE 6.11]. For example, new bone may grow, and its internal patterns may change to accommodate added body weight or other stresses. In contrast, bone that is not used tends to lighten by losing bone cells. Remodeling occurs at specific locations on the bone called *foci* (FOH-sye; sing. *focus*). Within each focus are specialized cells that make up a bone-remodeling unit that is responsible for the erosion and refilling of the bone focus with new bone. Approximately 5 to 10 percent of the skeleton is remodeled each year. Recent studies suggest that the network of osteocytes throughout bone tissue may function as a mechanosensor that directs bone responses during remodeling when faced with the stimulation of a mechanical load.

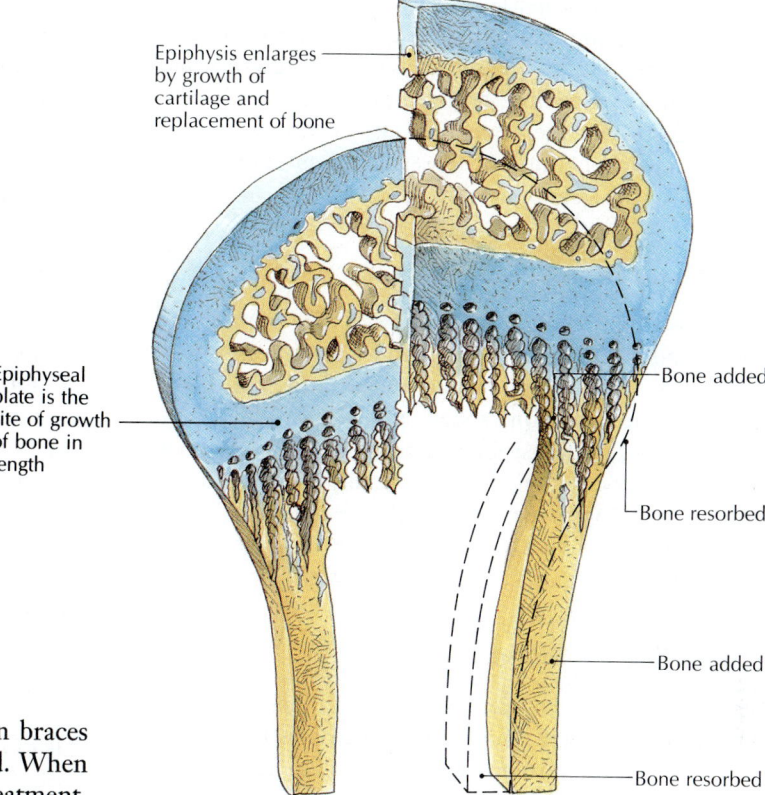

FIGURE 6.11 BONE REMODELING

The process of bone deposition and resorption during the remodeling of a typical long bone.

Epiphysis enlarges by growth of cartilage and replacement of bone

Epiphyseal plate is the site of growth of bone in length

Bone added

Bone resorbed

Bone added

Bone resorbed

An example of remodeling can be seen when braces are placed on teeth or when a tooth is extracted. When the teeth are repositioned during orthodontic treatment, force is placed on the bony tooth sockets. The slow shift of the teeth during their realignment takes place when some bone is resorbed on one side of the socket and added onto the other side. When a tooth is extracted, pressure on the socket is reduced, and some bone in the socket is resorbed. This reduction of pressure and bone mass allows the neighboring teeth to shift slightly.

A S K   Y O U R S E L F

1 What are the major differences between modeling and remodeling?

2 What are foci?

THE PHYSIOLOGICAL FUNCTIONS OF BONES IN MAINTAINING HOMEOSTASIS

Less obvious than the mechanical functions of bones are the *physiological* ones. Red blood cells, some white blood cells, and blood platelets are made in the red bone marrow in the ends of some long bones, ribs, and vertebrae, and in the diploë of the skull bones. Bone is also the body's main supplier and storehouse of calcium, phosphate, and magnesium salts. In fact, about 99 percent of all the calcium in the body is found in bones, along with 86 percent of the phosphate and 54 percent of the magnesium.

Calcium Storage and Release

Without calcium, some enzymes could not function, cells would come apart, the permeability of cell membranes would be affected, muscles (including the heart) could not contract, nerve function would be impaired, and blood would not clot. Of course, without calcium there would be no bones in the first place, since bones consist mainly of calcium.

If the diet does not provide enough calcium, the bones release it, and if there is too much calcium in the body, the bones store it. The actual storage and release of calcium take place in relatively young osteons. As osteons become more mature, they lose the ability to store and release calcium and become more involved with structural functions instead.

To maintain homeostasis, bones help regulate the amount and consistency of extracellular fluid by adding calcium to it or taking calcium out of it. Small decreases of calcium in plasma and extracellular fluid (*hypocalcemia*) can cause the nervous system to become more excitable because of increased neuronal membrane permeability, with resultant muscular spasms. Too much calcium (*hypercalcemia*) in body fluids depresses the nervous system and causes muscles to become sluggish and weak because of the effects of calcium on the muscles' plasma membranes. Also, increased calcium causes constipation and lack of appetite because it depresses the muscular contractility of the gastrointestinal tract.

Phosphate Storage and Release

Bones also help regulate the amount of phosphate in the body. Under hormonal control, bones can release phosphate salts when the salts are needed by the body. Changing the level of phosphate in the extracellular fluid or blood does not cause significant immediate effects on the body. However, the proper amount of phosphate is vital to the body's acid-base balance. Obviously, the physiological functions of bones are a vital factor in maintaining the body's overall homeostasis.

Production of Blood Cells

Besides helping to regulate the levels of minerals in the blood, the marrow within bones contributes to homeostasis by manufacturing some blood cells. Specifically, adult red bone marrow is the body's main blood-making, or **hemopoietic tissue** (Gr. *haima*, blood + *poiein*, to make). It produces *all* the red blood cells (erythrocytes), platelets (cellular fragments involved in blood clotting), and certain white blood cells (granular leukocytes and immature lymphocytes and monocytes). At birth all the bone marrow in the body is red marrow, but by adolescence most of it is replaced by yellow fat cells, which form the *yellow marrow*. Although yellow marrow is not active hemopoietic tissue, it has the potential to become active under stress and produce red blood cells.

In the adult, red marrow is found only in the proximal epiphyses of long bones such as the femur and humerus, in some short bones, and in the vertebrae, sternum, ribs, and cranium. Besides producing red blood cells, the red marrow contains macrophages and manufactures white blood cells, some of which help protect the body from disease. The continuing production of red blood cells is an important function because these cells live for approximately 120 days and need to be replaced as they die. About 2.5 million red blood cells are produced per *second*, and about 200 billion are produced daily. Likewise, about 200 billion red blood cells are destroyed daily by the liver and spleen, creating a balance.

Effects of Hormones on Bones

Several hormones, primarily *parathyroid hormone (PTH)* and *calcitonin (CT)*, have direct effects on bones. PTH is released from the parathyroid glands in response to *low* calcium levels in the blood and stimulates the uptake of calcium from bone, kidneys, and intestinal tract to return calcium levels to normal. PTH increases the number and activity of osteoclasts in bone to increase calcium (and phosphate) reabsorption from bone and stimulate release of the mineral(s) into the blood. This effect of PTH on bone is augmented by the effects of PTH on the kidneys, where it promotes the recovery of calcium from the urine and the elimination of phosphate into the urine; in the intestine, PTH increases the reabsorption of calcium into the blood.

Calcitonin (CT) is released from specific thyroid cells (parafollicular cells) in response to *elevated* calcium levels in the blood. CT causes calcium levels to be lowered by inhibiting osteoclastic activity in bone; it also favors calcium uptake by bone, promoting bone formation and decreasing blood calcium levels. Thus PTH increases blood calcium, while CT lowers it.

In addition to PTH and CT, human growth hormone (hGH), thyroid hormones, sex hormones, adrenal cortical hormones, and vitamins A, C, and D are important in bone function.

Effects of Nutrition on Bones

Proper nutrition is essential for normal bone development and maintenance. The most obvious nutritional needs are for calcium and phosphorus, since they constitute almost half the content of bone. If the body is deficient in either mineral, the bones become brittle and break easily.

If a diet is too low in vitamin D, the normal ossification process at the epiphyseal growth plates is upset and the bones are easily deformed. A deficiency in vitamin A may cause an imbalance in the ratio of osteoblasts and osteoclasts, thereby slowing the growth rate. Abnormally low amounts of vitamin C also inhibit growth by causing an insufficient production of collagen and bone matrix, a condition that also delays the healing of broken bones. Smoking may contribute to vitamin C deficiency, hindering the bone-healing process.

A S K   Y O U R S E L F

1 What are some of the physiological functions of bone?

2 How is calcium stored and released in bone?

3 How does calcium in bones contribute to homeostasis?

4 In the adult, where are most red blood cells produced?

5 How do hormones and nutrition affect bones?

A Bare-Bones History Lesson

Dead men tell no lies, they only leave clues in their bones. Deciphering them has become a science—forensic anthropology—that both the police and the historians now use to solve mysteries that are buried but still alive.

Soon after the British raised Henry VIII's flagship, the *Mary Rose,* from the ocean bottom off the Isle of Wight in 1982, scientists found one whole group of human remains that had extra deposits on a small bone at the tip of their shoulders, which had also become forked. The anomaly indicated that the men had been long-bow archers, their bones deformed by habitually drawing back and holding the taut bowstring, and showed how heavily defended the king's conveyance was. When human-rights campaigners in Argentina challenged revisionists who had denied that the Army had "disappeared" thousands of citizens between 1976 and 1983, forensic anthropologist Clyde Snow of Oklahoma confirmed, from the edges of bullet holes in bones, that the victims had been executed. "History," says Dr. Marc Micozzi, director of the National Museum of Health and Science in Washington, "is being rewritten by forensic scientists."

Even in burials short of mummification, bones, hair and nails can endure for decades. To the 100 or so forensic anthropologists in this country, who handled 2000 police cases last year, bones are biographies that reveal not only such basics as sex, race, age, and height but even such details as where the deceased grew up. Pelvic bones distinguish a male from a female with 95 percent accuracy, says Snow: women's pelvic openings are low and broad, to permit childbirth, while men's are high and narrow. Men's skulls are also thicker than women's and have a prominent brow ridge. Bones can also reveal whether a woman was a mother. A groove on a bone in front of the lower spine and hipbone widens and deepens when a baby passes through—which is how Snow determined that about 400 of Argentina's *desaparecidos* had given birth shortly before being executed.

Races differ from the neck up. Narrow nasal passages and a short distance between eye sockets mark a Caucasian, distinct cheekbones identify a Mongoloid and nasal openings shaped like an upside-down heart typify a Negroid. Scientists can deduce age at death by which teeth have erupted and by the size of gaps between still-fusing bones of the skull, pelvis and limbs. Another clue, says Judy Suchey of CalState Fullerton, is that the front part of the hipbone has furrows at the age of 16 that become flat and then concave by 35.

Now forensic anthropologists are putting flesh on the bones. Kenneth A. R. Kennedy of Cornell University has documented more than 140 correlations between bone marks and activity, including a forward-thrusting jaw in woodwind players and an extra bone in the joints of violinists' left hands. An enlarged forearm bone indicates that massive muscles were attached, in the service of, say, pitching a baseball. Finger bones retain the marks of ligaments exercised by habitual grasp of a pen and have identified Egyptian mummies as court scribes. Smithsonian anthropologist Douglas Owsley concluded that the skeleton of a woman from Colonial Virginia was that of a seamstress: she had notches in her teeth from holding pins or needles.

Meat Eater

Anthropologists find that you are what you eat: bones of carnivores contain more copper and zinc, vegetarians more magnesium and manganese. Different soils typically contain different isotopes of strontium, which find their way into the food chain and ultimately into teeth. Whatever was deposited over the first 12 to 15 years of a person's life is covered over by dental enamel and preserved. By matching strontium in teeth to strontium in soils, anthropologists can tell the source of the food the person ate— provided, of course, that he died before transcontinental food shipping began.

THE EFFECTS OF AGING ON BONES

As people age, their bones may undergo a loss of calcium, a decrease in calcium utilization, and a decreased ability to produce materials for the bone matrix. These metabolic changes make bones brittle so that they fracture more easily than the bones of a young person do. The most common fractures in older people occur in the clavicle, femur, wrist bones, and humerus. (Fractures in the wrist and arm often occur when a person attempts to break a fall.) Bone marrow decreases, and *osteoporosis*, a metabolic disorder, leads to a loss of skeletal mass and density, especially in postmenopausal women (see When Things Go Wrong at the end of this chapter). Adverse changes in bones usually begin earlier in women (usually in the early forties) than in men (late fifties or early sixties) because of a decrease in estrogen. Cartilage continues to grow, however, lengthening and widening the nose, lengthening the ears, and causing the earlobes to droop.

DEVELOPMENTAL ANATOMY OF THE SKELETON

This section on the gross developmental anatomy of the skeleton focuses on three separate regions: the skull, appendicular skeleton, and axial skeleton.

Skull

The skull is divisible into (1) the *neurocranium*, a protective wall surrounding the brain, and (2) the *viscerocranium*, the skeleton of the face.

The **neurocranium** is comprised of a *vault* and a *base*. The **vault** consists of *membrane bones*, which include the frontal and parietal bones and portions of the temporal and occipital bones. The ossification centers for these bones appear during the second and third fetal months. During development these membrane bones are literally floating within the membranous capsule surrounding the brain. Hence, as the membranous capsule expands in response to the rapidly growing brain, the bones of the vault are carried along until their growth ceases after birth. The **base** is initially a cartilage "model" formed by the chondrocranium as the floor of the neurocranium. It is transformed into bone by endochondral ossification. These bones include major portions of the ethmoid, sphenoid, temporal, and occipital bones, each of which has several primary ossification centers. They appear during the second through fifth fetal months. Other bones associated with cartilage models are the three ear bones (malleus, incus, and stapes).

The **viscerocranium** (bones of the face and pharynx) include the maxilla, zygomatic, palatine, vomer, inferior concha, nasal, lacrimal, and mandible, all of which are membrane bones. Their ossification centers appear during the second and third fetal months. Bones associated with cartilage models include the hyoid bone and the styloid process of the sphenoid bone.

At birth the viscerocranium is relatively small compared with the neurocranium. This is related largely to the virtual absence of the paranasal air sinuses in the viscerocranium and the small size of the face bones, especially the upper and lower jaws in the infant. With the development of air sinuses and the appearance and growth of teeth during infancy and childhood, the face assumes the features and proportions of the mature head. Some segments of the chondrocranium persist in the adult as cartilage; these include the nasal cartilages and the thyroid and cricoid cartilages of the larynx.

Appendicular Skeleton: The Extremities

The limb buds become visible as paddle-shaped growths by the fifth fetal week, and develop into recognizable limbs in a few weeks. The bones of the extremities (including their girdles) are all endochondral bones, except the clavicle, which is a membrane bone. By the end of the sixth week of development hyaline cartilage "models" of the future bones appear in the limb buds. With a few exceptions, the primary ossification centers of the bone extremities appear during prenatal life. The centers of some of the carpal and tarsal bones appear after birth. In general, the centers of all epiphyses appear after birth; the exceptions include the epiphyses of the femur and the tibia at the knee joint. These secondary ossification centers in the extremities appear at various times after birth. Radiologists use the timing of the appearance of these centers to ascertain whether a child has reached its proper maturation age. The time of closure of the epiphyses is variable and occurs normally, depending upon the specific bone, anywhere from 13 to 25 years of age.

Axial Skeleton: Ribs, Vertebral Column, and Sternum

The bones of the axial skeleton are all endochondral bones. The **ribs** commence to ossify during the second and third fetal months. Epiphyses on the processes of the ribs appear during puberty and fuse to the shaft of the ribs by early adulthood.

Each typical **vertebra** has three primary ossification centers (one for the body and a pair for the neural arches). Most appear during the second and third fetal months and some during late fetal life. Their epiphyseal centers appear during adolescence and fuse variably during early adulthood.

The **sternum** (breastbone) has several primary ossification centers. They appear during the fifth fetal month and fuse at variable times during adolescence and early adulthood.

WHEN THINGS GO WRONG

Osteogenesis Imperfecta

Osteogenesis imperfecta is an inherited condition in which the bones are abnormally brittle because of improper mineralization. The basic cause of this disorder is a decrease in the activity of the osteoblasts during bone formation. In some cases, fractures occur during prenatal life, and so the child is born with the deformities. In other cases, fractures occur as the child begins to walk. The tendency to fracture bones is reduced after puberty.

Osteomalacia and Rickets

Osteomalacia (ahss-teh-oh-muh-LAY-shee-uh; Gr. *osteon*, bone + *malakia*, soft) and *rickets* (variant of Gr. *rhakhitis*, disease of the spine) are skeletal defects resulting from a deficiency of vitamin D, which leads to a widening of the epiphyseal plates, an increased number of cartilage cells, wide osteoid seams, and a decrease in linear growth. A deficiency of vitamin D may result from an inadequate diet, an inability to absorb vitamin D, or too little exposure to sunlight.

Rickets is a childhood disease caused by insufficient mineralization. It occurs less frequently than it used to, primarily because of improved dietary habits. It is most common in black children, not necessarily because of inadequate diets but because highly pigmented skin absorbs fewer of the ultraviolet rays in sunlight. These rays are needed to convert 7-dehydrocholesterol in the skin into vitamin D (cholecalciferol). Skeletal deformities such as bowed legs, knock-knees, and a bulging forehead are typical in a young child with rickets.

Osteomalacia is the adult form of rickets and is sometimes referred to as "adult rickets." It leads to *demineralization*, an excessive loss of calcium and phosphorus. Although the skeletal deformities of rickets may be permanent, the similar skeletal abnormalities of osteomalacia may disappear with large doses of vitamin D.

Osteomyelitis

Osteomyelitis (Gr. *osteon*, bone + *myelos*, marrow) is an inflammation of bone and/or an infection of the bone marrow that can be either chronic or acute. It is frequently caused by *Staphylococcus aureus* and other bacteria, which can invade the bones or other sites in the body. Bacteria may reach the bone through the bloodstream or through a break in the skin caused by an injury. Although the disease often remains localized, it can spread to the marrow, cancellous tissue, and periosteum. *Acute osteomyelitis* is usually a blood-carried disease that most often affects rapidly growing children. *Chronic osteomyelitis*, which is more prevalent in adults, is characterized by draining sinuses and spreading lesions. The prompt use of antibiotics such as vancomycin is effective in treating the disease.

Osteoporosis

Osteoporosis (Gr. *osteon*, bone + *poros*, passage) is a loss of bone mass that can make the bones so porous that they crumble under the ordinary stress of moving about. (People with osteoporosis don't "fall and break a hip." They usually break a hip—or, more often, the neck of the femur—while walking and then fall when the hip support is gone.) Most victims of osteoporosis are postmenopausal women, usually over age 60. After menopause, the ovaries produce little if any estrogen, one of the female hormones. Without estrogen, old bone is destroyed faster than new bone can be remodeled, and so the bone becomes porous and brittle. Postmenopausal estrogen therapy has proved to be relatively successful.

Besides a lack of estrogen, osteoporosis can be hastened by smoking, a diet lacking in calcium or vitamin D, and lack of exercise. Older women who lose weight tend to lose bone mass also, even if their weight-loss program is complemented by regular exercise and adequate calcium. (A study of postmenopausal women at Tufts University revealed an interesting finding: calcium and exercise strengthen different bones. Exercise increased trabecular bone mass in the spine, while calcium had no effect. But calcium did increase bone density at the neck of the femur. Physicians at Tufts suggest an increase of calcium *and* physical activity for healthy women of all ages.)

Osteoporosis can cause vertebrae to crumble, producing a "dowager's hump" in the upper back. Most people over 70 are markedly shorter than they once were. This height loss occurs when aging vertebrae and other small bones lose mass, pressing close together as intervertebral disks become thinner.

A recent, interesting study at the Medical College of Georgia indicates that osteoporosis may be linked to an autoimmune reaction that leads to the resorption of bone.

Osteosarcomas

Osteosarcomas (Gr. *sark*, flesh + *oma*, tumor), or *osteogenic sarcomas,* are forms of bone cancer. Such malignant bone tumors are rare. Because the incidence of osteosarcomas is higher in growing adolescents than in children or adults, and because the adolescents affected are often taller than average, there is some speculation that areas of rapid growth are most vulnerable. No definite cause is known. Localized pain and tumors are common signs of malignancy.

The most common form of bone cancer is a *myeloma*, in which malignant tumors in the bone marrow interfere with the normal production of red blood cells. Anemia, osteoporosis, and fractures may occur. Myeloma occurs more frequently in women than in men.

Paget's Disease

Paget's disease (*osteitis deformans*) is a progressive bone disease in which a pattern of excessive bone destruction followed by bone formation contributes to thickening of bones. This deformity usually involves the skull, pelvis, and lower extremities. The cause is unknown. Paget's disease occurs after the age of 40 and most typically in the sixties.

CHAPTER SUMMARY

Types of Bones and Their Mechanical Functions (p. 157)

1 Bones may be classified according to their shape as *long, short, flat, irregular,* or *sesamoid. Accessory bones* are a minor category.

2 The shapes of bones are related to their functions. The mechanical functions of bones include support, protection, and movement.

Gross Anatomy of a Typical Long Bone (p. 159)

1 Most long bones consist of a tubular shaft called a *diaphysis,* with an *epiphysis* at either end of the bone.

2 Separating the diaphysis and epiphysis at each end of the bone is the *metaphysis.* It is made up of the *epiphyseal (growth) plate* and adjacent bony trabeculae of spongy bone tissue.

3 The epiphyseal plates and metaphyses are the only places long bones continue to grow in length after birth.

The Histology of Bone (Osseous) Tissue (p. 160)

1 *Bone (osseous) tissue* is composed of cells embedded in a matrix of ground substance, inorganic salts, and collagenous fibers. The inorganic salts give bone its hardness, and the organic fibers and ground substance give it strength and flexibility.

2 Most bones have an outer shell of compact bone tissue surrounding spongy bone tissue.

3 *Compact bone tissue* is made up of *osteons,* concentric cylinders of calcified bone. Within the osteons are *central (Haversian) canals* that carry nerves, lymphatic vessels, and blood vessels. These canals are connected with the outer surfaces of bones through *perforating (Volkmann's) canals.*

4 The *periosteum* is a fibrous membrane that covers the outer surfaces of bones, except in joints. It contains bone-forming cells, nerves, and vessels.

5 *Spongy* (or *cancellous) bone tissue*

forms a lacy pattern designed to withstand stress and support shifts in weight. Tiny spikes of bone tissue called *trabeculae,* surrounded by calcified matrix, give spongy bone its latticelike appearance.

Bone Cells (p. 163)

1 Bones contain five types of cells: *osteogenic cells,* which can be transformed into osteoblasts or osteoclasts; *osteoblasts,* which synthesize and secrete new bone matrix as needed; *osteocytes,* which help maintain homeostasis; *osteoclasts,* giant cells responsible for the resorption of bone; and *bone-lining cells,* which have several diverse functions.

2 Bone cells change their roles as the needs of the body change in growing and adult skeletons.

The Physiology of Bone Formation: Ossification (p. 164)

1 Bones develop through *ossification* (osteogenesis).

2 If bone is formed from mesenchymal tissue (embryonic connective tissue), the process is called *intramembranous ossification.*

3 If bone develops by replacing a cartilage model, the process is called *endochondral ossification.*

The Physiology of Bone Growth (p. 169)

1 The centers of growth activity after birth are the *secondary (epiphyseal) centers of ossification.*

2 Bones grow in diameter as osteogenic cells deposit new bone tissue beneath the periosteum and old bone tissue erodes.

3 The growth of the diameter of bones continues through most of life.

Bone Modeling and Remodeling (p. 170)

1 The alteration of bone size and shape during a bone's developmental growth is called *modeling.*

2 Bone replaces itself through the process called *remodeling.*

The Physiological Functions of Bones in Maintaining Homeostasis (p. 171)

1 Bones help maintain homeostasis by storing calcium and other minerals and releasing them as needed to maintain proper levels of those minerals in the blood and other tissues.

2 *Red bone marrow* produces red blood cells, contains macrophages, and manufactures some white blood cells that help fight disease.

3 Several hormones have a direct effect on bones, and bones have an effect on hormone secretion.

4 Calcium and phosphorus make up about half of the content of bone, and they must be supplied in a well-balanced diet. Adequate levels of vitamins A, C, and D are essential for the proper growth, mending, and strength of bones.

The Effects of Aging on Bones (p. 173)

1 Aging bones become brittle from loss of calcium, decreased calcium utilization, and decreased ability to produce matrix material.

2 Skeletal problems affect women more than men, usually because of a decrease in estrogen.

Developmental Anatomy of the Skeleton (p. 173)

1 The *vault* of the *neurocranium* consists of membrane bones, while the *base* is initially a cartilage model that is replaced by endochondral ossification.

2 The bones of the *viscerocranium* are all membrane bones.

3 The limb buds of the *appendicular skeleton* become visible by the fifth fetal week and develop into recognizable bones a few weeks later. The bones of the extremities are all endochondral bones except the clavicle, which is a membrane bone. Secondary ossification centers appear after birth.

4 The bones of the *axial skeleton* are endochondral bones.

STUDY AIDS FOR REVIEW

KEY TERMS

UNDERSTANDING THE FACTS

1 What are the functions of bone?
2 What are the components of osseous tissue?
3 What role does each component of the matrix play?
4 What kind of cells contribute to bone growth?
5 Distinguish between compact and spongy bone.
6 Contrast osteons with trabeculae.
7 Compare intramembranous and endochondral ossification.
8 How do bones grow in diameter?

UNDERSTANDING THE CONCEPTS

1 Why is endochondral ossification found in long bones and intramembranous ossification in flat bones?
2 Explain why bone is essential to the maintenance of homeostasis.
3 Why must osteoclasts and osteoblasts work together?
4 What functions of the periosteum are essential in the maintenance of a healthy, efficiently functioning bone?

SELF-QUIZ
Completion Exercises
6.1 The bony spicules in spongy bones are called _____.
6.2 _____ bones are small bones embedded in a tendon.
6.3 The central cavity within the diaphysis of a long bone is called the _____.
6.4 The _____ is the central cavity within an osteon.
6.5 A membrane called the _____ lines the internal cavities of bone.
6.6 Nutrients pass between osteocytes through _____.
6.7 Blood vessels enter a bone through an opening called a(n) _____.
6.8 Initially, the predominant cells in an ossification center of a long bone are _____.
6.9 The organic portion of the matrix of bone contains ground substance and _____ fibers.
6.10 The organic matrix of bone is synthesized and secreted by _____.
6.11 The only cartilage that remains after a long bone has completed its longitudinal growth is the _____ cartilage.
6.12 When bones grow in diameter, the cells that produce the new bone originate either in the _____ or the _____.
6.13 The structural and functional unit of compact bone is the _____.
6.14 Yellow marrow is composed mainly of _____.
6.15 Myeloid tissue is another name for red _____.
6.16 Whereas the function of red marrow is blood cell production, the function of yellow marrow is _____.
6.17 Calcitonin acts to decrease the level of calcium in the _____.
6.18 Parathyroid hormone increases the activity of bone cells known as _____.

6.19 The periosteum is attached to the underlying bone by collagenous fibers called _____ fibers.

6.20 If osteogenic cells, osteoblasts, osteocytes, and osteoclasts were compared for size, the largest would be the _____.

6.21 If bone develops without an intervening cartilaginous stage, the process is called _____.

6.22 An osteon is composed of concentric _____ of bone surrounding a central cavity.

6.23 When bones form by endochondral ossification, the perichondrium gives rise to the _____ of the bone.

Matching

a trabeculae	**g** short bones
b diaphysis	**h** accessory bones
c epiphysis	**i** lamellae
d metaphysis	**j** perforating canal
e long bones	**k** hydroxyapatite
f flat bones	

6.24 _____ contains yellow bone marrow

6.25 _____ contains the epiphyseal plate

6.26 _____ tiny spikes of bone surrounded by matrix

6.27 _____ a tubular shaft

6.28 _____ the end of a bone

6.29 _____ found in the feet

6.30 _____ found in the wrists

6.31 _____ contains diploë

6.32 _____ a branch off of a central canal

6.33 _____ inorganic salt of bone

6.34 _____ concentric layers

A SECOND LOOK

1 In the following drawing, label the lacuna and canaliculi.

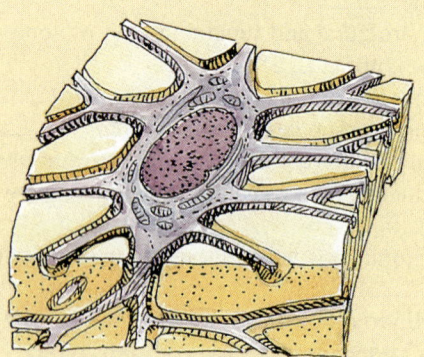

2 In the following drawing, label the proximal epiphysis, trabeculae, and metaphysis.

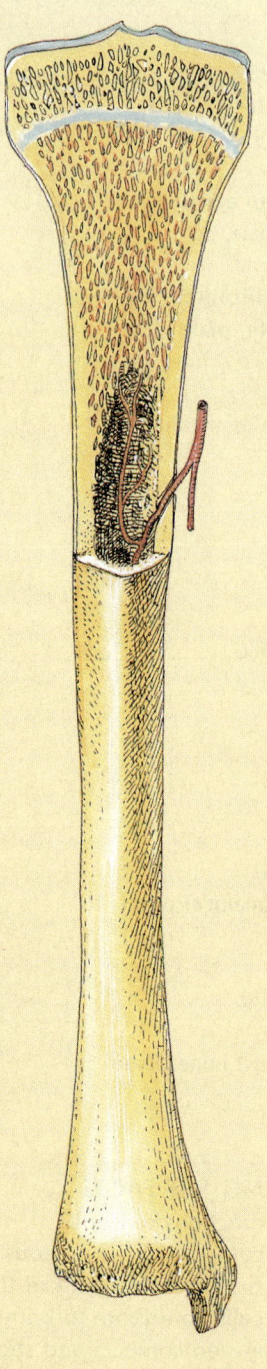

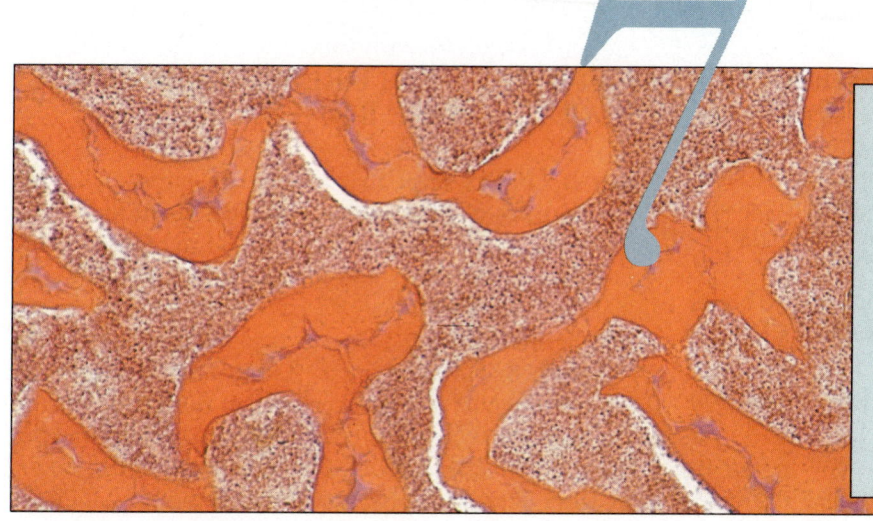

7

The Axial Skeleton

KEY CONCEPTS

1 A bone's surface features may provide clues about its specific function.

2 The skeleton is divided into the axial and appendicular portions. The axial skeleton forms the longitudinal axis of the body, and the appendicular skeleton is composed of the upper and lower extremities.

3 The vertebral column not only protects the spinal cord, it also provides the flexibility necessary for easy body movement.

We are born with as many as 300 bones, but many of them fuse during childhood to form the adult skeleton of 206 named bones. There are exceptions. For example, 1 person in 20 has an extra rib, and the number of small sutural bones in the skull varies from person to person. But for the most part, the bone count of 206 is the accepted one.

Other than the bones that come in pairs, no two bones are alike. Bones differ in size, shape, weight, and even composition. This diversity of form is directly related to many structural or mechanical functions of the skeleton itself.

The most obvious function of the skeleton is to provide support for the body. Without a bony skeleton to support our bag of skin and inner organs, we would collapse into a formless heap.

Besides supporting the inner organs, the skeleton surrounds and protects many of these organs within bony body cavities. For instance, the heart and lungs are safely enclosed within a roomy rib cage that offers protection and freedom of movement at the same time, and the brain is cushioned within the cranium, a shock-absorbing bone case that is designed to protect the brain from the many bumps of everyday life.

In addition to protecting the internal organs, the skeleton also protects many passageways. For instance, the bony scaffolding of the nasal region supports and protects the airway for the passage of air during breathing.

The bones of the skeleton also act as a system of levers for the pulley action of muscles. This lever-pulley arrangement provides attachment sites on bones for muscles, tendons, and ligaments. It allows us to move the entire body or just one finger or toe.

This chapter and the next present an introduction to the skeletal system itself, concentrating on its two main parts: the central "anchor" of the *axial skeleton* and the peripheral limbs of the *appendicular skeleton*. (Throughout the chapters each specific bone is the same color in each illustration. For example, the temporal bone is always dark blue.)

GENERAL FEATURES AND SURFACE MARKINGS OF BONES

Just as a bone's shape may indicate its purpose, the surface features of a bone often give clues about the bone's function. TABLE 7.1 describes some of the more important features of bones and their functions.

Some outgrowths or *processes* (such as tuberosities) may be attachment sites for the tendons of muscles or the ligaments that connect bones. A long, narrow ridge (linea aspera of femur) or a more prominent ridge (iliac crest of the pelvis) may be the site where broad sheets of muscles are attached. Large, rounded ends (condyle of the femur) or large depressions (glenoid fossa of the scapula) indicate adaptations of joints between bones. Grooves, foramina (holes), or notches are usually formed on bones to accommodate such structures as blood vessels, nerves, or tendons.

For example, a muscle is attached to a trochanter, tuberosity, tubercle, or ridge. Processes and other features form as a result of the continuous force exerted by muscles, nerves, or blood vessels on a bone. The bone responds by not invading the territory of the muscle, nerve, or blood vessel (for example, the foramen ovale for the mandibular nerve). The radial groove on the humerus is present because the humerus yields as the radial nerve presses on it slightly.

ASK YOURSELF

1 What is a process on a bone? Give examples of processes, openings, and depressions.

2 Give an example of how a particular feature or marking on a bone is related to the bone's function.

DIVISIONS OF THE SKELETON

The skeleton has two major divisions: the axial skeleton and the appendicular skeleton [FIGURE 7.1]. The *axial*

TABLE 7.1 SOME GENERAL FEATURES OR MARKINGS OF BONES

Feature or marking	Description and example
OPENINGS (HOLES) TO BONES	
Canal or meatus (L. channel) [FIGURE 7.5F]	Relatively narrow tubular channel, opening to a passageway. Carotid canal, external auditory meatus.
Fissure [FIGURE 7.5A]	Groove or cleft. Superior orbital fissure.
Foramen (L. opening) [FIGURE 7.5F]	Natural opening into or through a bone. Foramen magnum in occipital bone.

Feature or marking	Description and example
DEPRESSIONS ON BONES	
Fossa (L. trench) [FIGURE 7.5F]	Shallow depressed area. Mandibular fossa.
Groove or sulcus (L. groove) [FIGURE 8.5]	Deep furrow on the surface of a bone or other structure. Intertubercular and radial grooves of humerus.
Notch [FIGURE 8.6A]	Deep indentation, especially on the border of a bone. Radial notch of ulna.
Paranasal sinus (L. hollow) [FIGURE 7.9]	Air cavity within a bone in direct communication with nasal cavity. Maxillary sinus.
PROCESSES WHERE A BONE FORMS A JOINT WITH AN ADJACENT BONE	
Condyle (L. knuckle) [FIGURE 8.13]	Rounded, knuckle-shaped projection; concave or convex. Condyles of femur.
Facet (Fr. little face) [FIGURE 7.20B]	Small, flat surface. Head and tubercle of ribs.
Head (caput) (L. head) [FIGURES 8.5, 8.6, 8.13, 8.15]	Expanded, rounded surface at proximal end of a bone; often joined to shaft by a narrowed neck, and bearing the ball of a ball-and-socket joint. Head of humerus, radius, femur, fibula.
Trochlea (L. pulley) [FIGURE 8.5]	Grooved surface serving as a pulley. Trochlea of humerus.
PROCESSES OF CONSIDERABLE SIZE ATTACHED TO CONNECTIVE TISSUE	
Cornu (L. horn) [FIGURE 7.11]	Curved, hornlike protuberance. Cornu of hyoid bone.
Crest or lip ridge [FIGURE 8.11A]	Wide, prominent ridge, often on the long border of a bone. Iliac crest.
Line or linea	Narrow, low ridge. Linea aspera of femur.
Eminence (L. to stand out) [FIGURE 8.15]	Projecting part of bone, especially a projection from the surface of a bone. Intercondylar eminence of tibia.
Epicondyle [FIGURE 8.5]	Eminence upon a bone above its condyle. Epicondyles of humerus.
Malleolus (L. hammer) [FIGURE 8.15]	Hammer-shaped, rounded process. Malleoli of tibia and fibula.
Spine (spinous process) [FIGURES 7.13, 8.4C]	Sharp, elongated process. Spine of a vertebra and of scapula.
Sustentaculum [FIGURE 8.16C]	Process that supports. Sustentaculum tali of calcaneus bone of the foot.
Trochanter [FIGURE 8.13]	Either of the two large, roughly rounded processes found near the neck of the femur. Trochanters of femur.
Tubercle (L. small lump) [FIGURE 8.5]	Small, roughly rounded process. Tubercles of humerus.
Tuberosity (L. lump) [FIGURE 8.12]	Medium-sized, roughly rounded, elevated process. Ischial tuberosity.

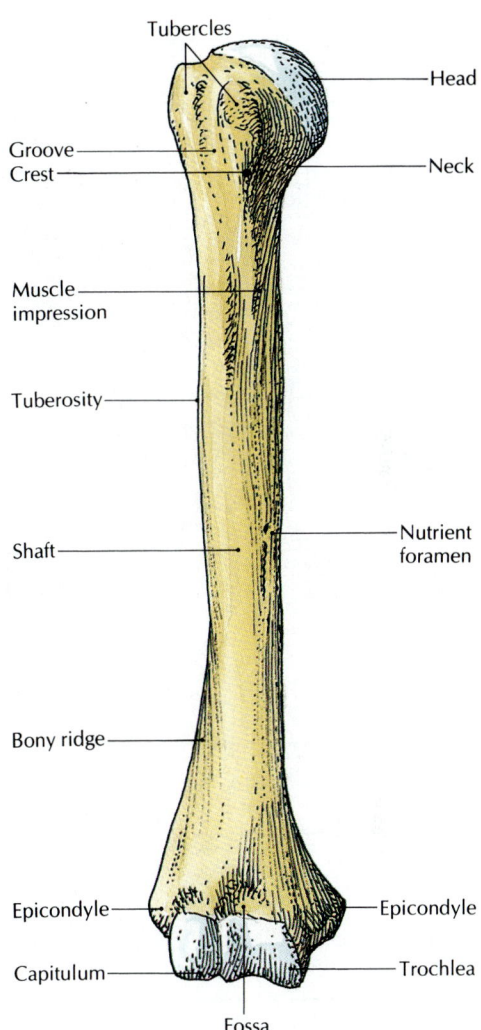

Some anatomical markings of a typical long bone.

FIGURE 7.1 AXIAL AND APPENDICULAR SKELETONS

[A] Anterior view of the skeleton. The axial skeleton is shown in orange. **[B]** Posterior view of the skeleton. The appendicular skeleton is shown in green.

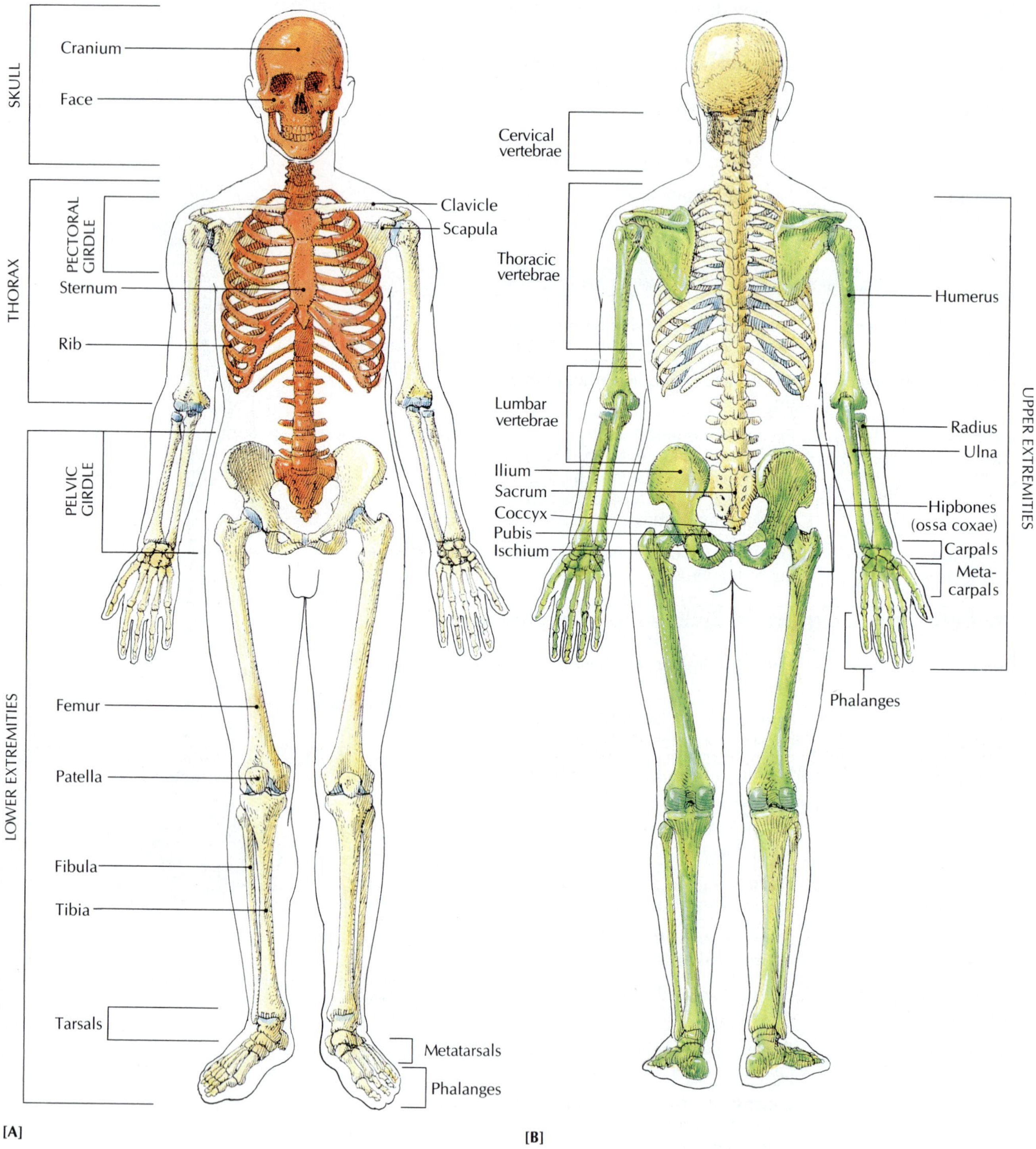

SKULL

- Cranium
- Face

THORAX

PECTORAL GIRDLE

- Sternum
- Rib

PELVIC GIRDLE

LOWER EXTREMITIES

- Femur
- Patella
- Fibula
- Tibia
- Tarsals

Metatarsals
Phalanges

Cervical vertebrae

Clavicle
Scapula

Thoracic vertebrae

Lumbar vertebrae

Ilium
Sacrum
Coccyx
Pubis
Ischium

Humerus

Radius
Ulna

Hipbones (ossa coxae)
Carpals
Meta-carpals

Phalanges

UPPER EXTREMITIES

[A]   [B]

DIVISIONS OF THE ADULT SKELETON (206 BONES)

AXIAL SKELETON (80 BONES)

Skull (29 bones)[*]
Cranium	8
Parietal (2)	
Temporal (2)	
Frontal (1)	
Ethmoid (1)	
Sphenoid (1)	
Occipital (1)	
Face	14
Maxillary (2)	
Zygomatic (malar) (2)	
Lacrimal (2)	
Nasal (2)	
Inferior nasal concha (2)	
Palatine (2)	
Mandible (1)	
Vomer (1)	
Ossicles of ear	6
Malleus (hammer) (2)	
Incus (anvil) (2)	
Stapes (stirrup) (2)	
Hyoid	1

Vertebral column (26 bones)
Cervical vertebrae	7
Thoracic vertebrae	12
Lumbar vertebrae	5
Sacrum (5 fused bones)	1
Coccyx (3 to 5 fused bones)	1

Thorax (25 bones)[†]
Ribs	24
Sternum	1
	80

APPENDICULAR SKELETON (126 BONES)

Upper extremities (64 bones)
Pectoral (shoulder) girdle	4
Clavicle (2)	
Scapula (2)	
Arm[‡]	2
Humerus (2)	
Forearm	4
Ulna (2)	
Radius (2)	
Wrist	16
Carpals (16)	
Hand and fingers	38
Metacarpals (10)	
Phalanges (28)	

Lower extremities (62 bones)
Pelvic girdle	2
Fused ilium, ischium, pubis[§]	
Thigh	4
Femur (2)	
Patella (2)	
Leg[‡]	4
Tibia (2)	
Fibula (2)	
Ankle	14
Tarsals (14)	
Foot and toes	38
Metatarsals (10)	
Phalanges (28)	
	126

Total (Axial and Appendicular)	**206**

[*]The number of skull bones is sometimes listed as 22, when the ossicles of the ears (6 bones) and the single hyoid bone are counted separately. Technically, the hyoid bone is not part of the skull; it belongs to the viscerocranium.
[†]The thoracic vertebrae are sometimes included in this category.

[‡]Technically, the term *arm* refers to the upper extremity between the shoulder and elbow; the *forearm* is between the elbow and wrist. The upper part of the lower extremity, between the pelvis and knee, is the *thigh*; the *leg* is between the knee and ankle.
[§]The sacrum (a component of the vertebral column) is included in the pelvic girdle.

skeleton is so named because it forms the longitudinal *axis* of the body. It is made up of the skull, vertebral column, sternum, and ribs. The *appendicular skeleton* is composed of the upper and lower extremities, which include the pectoral and pelvic girdles that attach the upper and lower appendages to the axial skeleton (see Chapter 8).

A S K   Y O U R S E L F

1 What are the two main divisions of the skeleton, and where are they joined together?

2 What bones make up the axial skeleton?

THE SKULL

At the top of the axial skeleton is the *skull,* which is usually defined as the skeleton of the head, with or without the mandible (lower jaw) [TABLE 7.2]. Other bones closely associated with the skull are the mandible, the auditory bones (ear ossicles), and the hyoid bone. (In this book we will treat the mandible, ear ossicles, and hyoid as associated bones of the skull.)

The bones of the skull form a supporting framework that combines a minimum of bony substance and weight with a maximum of strength and support. Bony capsules surround and protect the brain, eyes, inner ear, and nasal passages [FIGURE 7.2].

The skull can be divided into (1) the *cranial skull,* which supports and protects the brain, its surrounding membranes (meninges), and the cerebrospinal fluid, and (2) the *facial skull,* which forms the framework for the nasal cavities and the oral (mouth) cavity [FIGURE 7.3, TABLE 7.2].

The skull is lightened by small cavities called *paranasal sinuses.* Bony buttresses (like the delicate but strong buttresses that support the walls of many Gothic cathedrals) within the skull can sustain the enormous pressures exerted by the teeth during biting and chewing.

Besides containing buttresses, the skull has numerous *foramina* (L. openings; sing. *foramen,* fuh-RAY-muhn), openings of various sizes through which nerves and blood vessels pass. The main foramina of the skull are shown in FIGURES 7.5, 7.7, and 7.10.

Sutures and Fontanels

The skull contains 29 bones, 11 of which are paired (see table accompanying FIGURE 7.1). Except for the mandible, ear ossicles, and hyoid, they are joined by *sutures* (L. *sutura,* seam), wriggly, seamlike joints that make the skull bones of an adult immovable [FIGURE 7.4]. Four major cranial skull sutures are:

FIGURE 7.2 SKULL CAVITIES

Coronal (frontal) section through the skull, illustrating the bony capsules surrounding the brain (cranial cavity), eye (orbit), maxillary sinus, and nose (nasal skeleton surrounding nasal cavity).

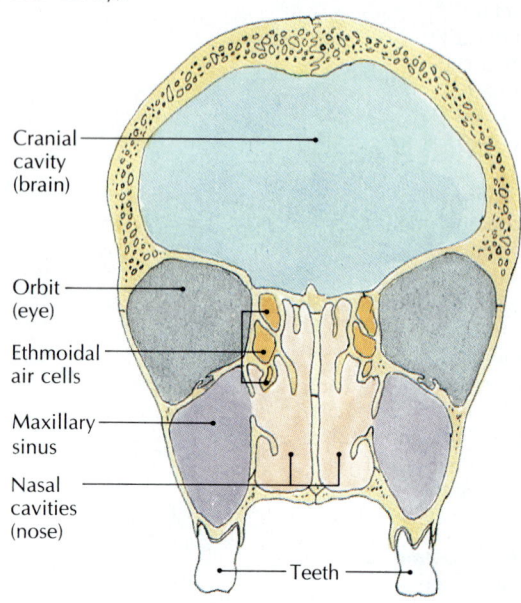

Cranial cavity (brain)

Orbit (eye)

Ethmoidal air cells

Maxillary sinus

Nasal cavities (nose)

Teeth

1 *Coronal,* between the frontal and parietal bones

2 *Lambdoidal,* between the parietal and occipital bones

3 *Sagittal,* between the right and left parietal bones

4 *Squamous,* between the temporal and parietal bones

Because the skull bones are joined by pliable membranes rather than tight-fitting sutures during fetal life and early childhood, it is relatively easy for the skull bones to move and overlap as the baby passes through the mother's narrow birth canal. Some of the larger membranous areas between such incompletely ossified bones are called *fontanels* [FIGURE 7.4A]. They allow the skull to expand as the child's brain completes its growth and development during the first few years of postnatal life. The membrane-filled fontanels (often called "soft spots") include the frontal, occipital, sphenoidal, and mastoid fontanels.

1 The *frontal (anterior) fontanel,* located between the angles of the two parietal bones and the two sections of the frontal bones. This diamond-shaped fontanel is the largest and usually does not close until 18 to 24 months after the child is born. The pulsating blood flow in the arteries of the brain can be felt at the frontal fontanel. If childbirth is difficult, a pulse monitor may be placed at this site to monitor the baby's heartbeat.

2 The *occipital (posterior) fontanel,* located between the occipital bone and the two parietal bones. This smaller,

FIGURE 7.3 SURFACE ANATOMY OF THE HEAD

The **head,** or **cephalic region,** can be divided anatomically into the *cranial* and *facial skulls* **[A].** The **cranial skull** is divided into the *frontal region* (tan), *parietal region* (orange), *temporal region* (gray), and *occipital region* (red). The **facial skull** is composed of four main regions: the *auricular region* (purple), *ocular region* (blue), *nasal region* (green), and *oral region* (pink). Surface features of the skull are shown in **[B]** and **[C].**

[A]

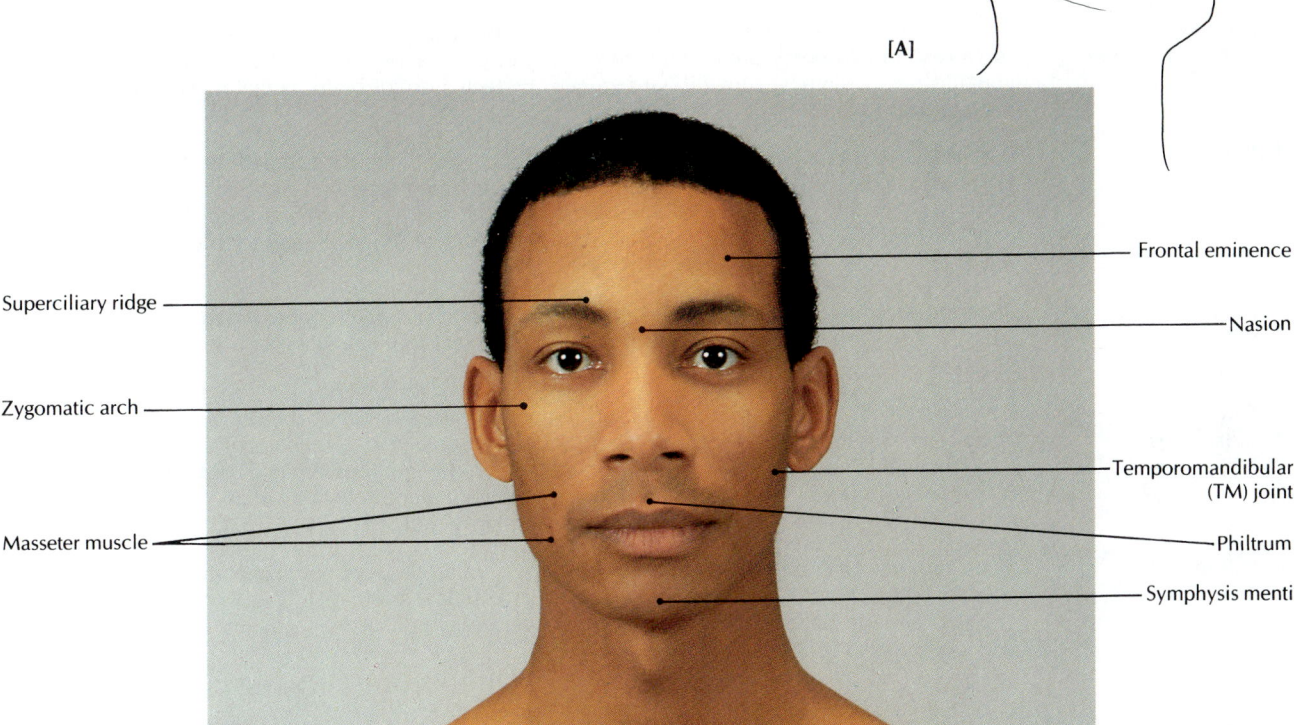

Superciliary ridge

Zygomatic arch

Masseter muscle

Frontal eminence

Nasion

Temporomandibular (TM) joint

Philtrum

Symphysis menti

[B]

Vertex

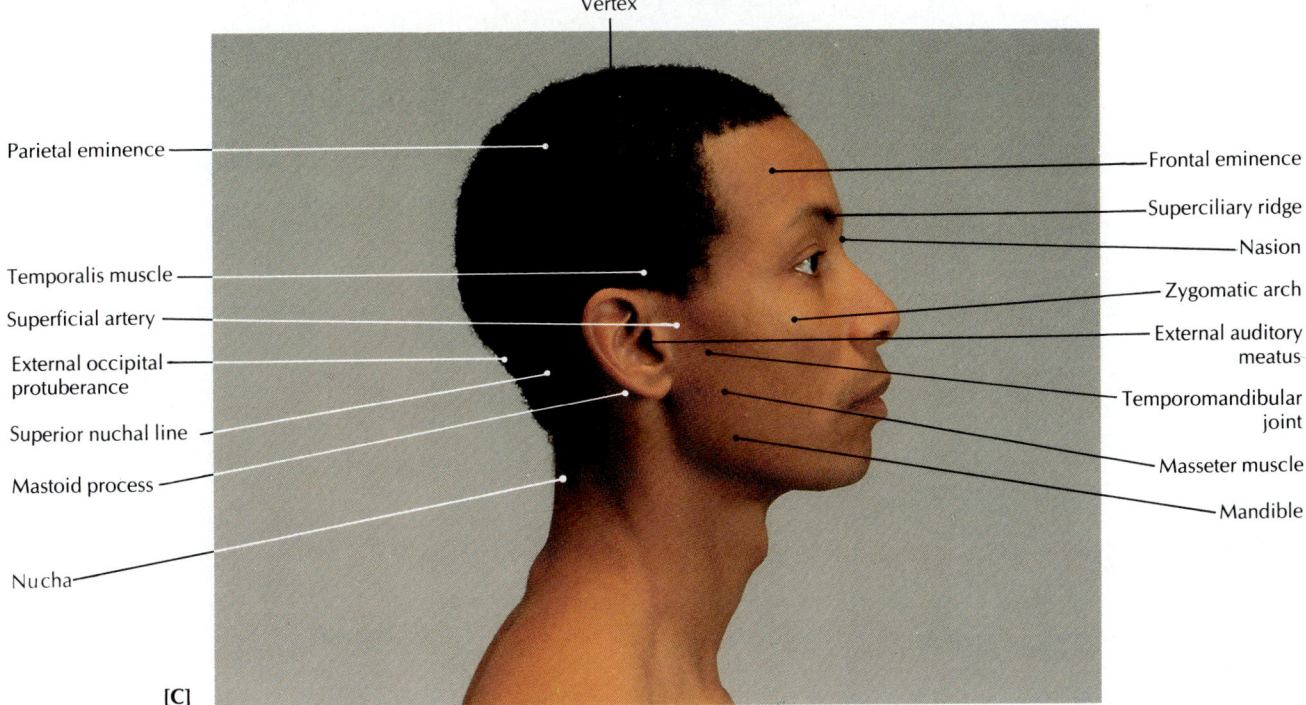

Parietal eminence

Temporalis muscle

Superficial artery

External occipital protuberance

Superior nuchal line

Mastoid process

Nucha

Frontal eminence

Superciliary ridge

Nasion

Zygomatic arch

External auditory meatus

Temporomandibular joint

Masseter muscle

Mandible

[C]

TABLE 7.2 BONES OF THE SKULL (29 BONES)

Bone	Description and function

CRANIAL SKULL (8 BONES)

Ethmoid (1)
[FIGURES 7.5D,E, 7.8]

Base of cranium, anterior to body of sphenoid. Made up of horizontal cribriform plate, median perpendicular plate, paired lateral masses; contains ethmoidal sinuses, crista galli, superior and middle conchae. Forms roof of nasal cavity and septum, part of cranium floor; site of attachment for membranes covering brain.

Frontal (1)
[FIGURE 7.5A,C–E,G]

Anterior and superior parts of cranium, forehead, brow areas. Shaped like large scoop; frontal squama forms forehead; orbital plate forms roof of orbit; supraorbital ridge forms brow ridge; contains frontal sinuses, supraorbital foramen. Protects front of brain; contains passageway for nerves, blood vessels.

Occipital (1)
[FIGURE 7.5B–G]

Posterior part of cranium, including base. Slightly curved plate, with turned-up edges; made up of squamous, base, and two lateral parts; contains foramen magnum, occipital condyles, hypoglossal canals, atlantooccipital joint, external occipital crests and protuberance. Protects posterior part of brain; has foramina for spinal cord and nerves; site of attachment for muscles, ligaments.

Parietal (2)
[FIGURE 7.5A–G]

Superior sides and roof of cranium, between frontal and occipital bones. Broad, slighly convex plates; smooth exteriors and internal depressions. Protect top, sides of brain; foramina for blood vessels.

Sphenoid (1)
[FIGURES 7.5A,C–G, 7.7]

Base of cranium, anterior to occipital and temporal bones. Wedge-shaped; made up of body, greater and lesser lateral wings, pterygoid processes; contains sphenoidal sinuses, sella turcica, optic foramen, superior orbital fissure, foramen ovale, foramen rotundum, foramen spinosum. Forms anterior part of base of cranium; houses pituitary gland; contains foramina for cranial nerves, meningeal artery supplying meninges and bony skull.

Temporal (2)
[FIGURE 7.5A–E,G]

Sides and base of cranium at temples. Made up of squamous, petrous, tympanic, mastoid areas; contain zygomatic process, mandibular fossa, ear ossicles, mastoid sinuses. Form temples, part of cheekbones; articulate with lower jaw; protect ear ossicles; site of attachment for neck muscles.

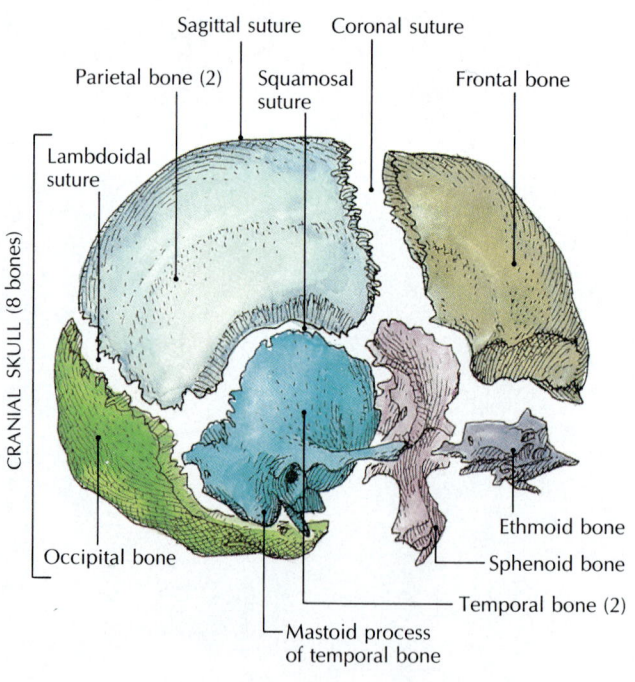

CRANIAL SKULL (8 bones)

Sagittal suture · Coronal suture · Parietal bone (2) · Squamosal suture · Frontal bone · Lambdoidal suture · Ethmoid bone · Sphenoid bone · Temporal bone (2) · Occipital bone · Mastoid process of temporal bone

[A] Exploded view of cranial skull. Eight bones; right lateral view.

Bone	Description and function
FACIAL SKULL (14 BONES)	
Inferior nasal conchae (2) [FIGURES 7.5A,E, 7.8A]	Lateral walls of nasal cavities, below superior and middle conchae of ethmoid bone. Thin, cancellous, shaped like curved leaves.
Lacrimal (2) [FIGURE 7.5A,C,D,F]	Medial wall of orbit, behind frontal process of maxilla. Small, thin, rectangular; contains depression for lacrimal sacs, nasolacrimal tear duct.
Mandible (1) [FIGURES 7.5A–E, 7.10]	Lower jaw, extending from chin to mandibular fossa of temporal bone. Largest, strongest facial bone; horseshoe-shaped horizontal body with two perpendicular rami; contains tooth sockets, coronoid, condylar, alveolar processes, mental foramina. Forms lower jaw, part of temporomandibular joint; site of attachment for muscles.
Maxillae (2) [FIGURE 7.5A–F]	Upper jaw and anterior part of hard palate. Made up of zygomatic, frontal, palatine, alveolar processes; contain infraorbital foramina, maxillary sinuses, tooth sockets. Form upper jaw, front of hard palate, part of eye sockets.
Nasal (2) [FIGURE 7.5A,C,D,E]	Upper bridge of nose between frontal processes of maxillae. Small, oblong; attached to a nasal cartilage. Form supports for bridge of upper nose.
Palatine (2) [FIGURE 7.5B,E]	Posterior part of hard palate, floor of nasal cavity and orbit; posterior to maxillae. L-shaped, with horizontal and vertical plates; contain greater and lesser palatine foramina. Horizontal plate forms posterior part of hard palate; vertical plate forms part of wall of nasal cavity, floor of orbit.
Vomer (1) [FIGURE 7.5A,B,E,F]	Posterior and inferior part of nasal septum. Thin, shaped like plowshare. Forms posterior and inferior nasal septum dividing nasal cavities.
Zygomatic (malar) (2) [FIGURE 7.5A,C,D,F]	Cheekbones below and lateral to orbit. Curved lateral part of cheekbones; made up of temporal process, zygomatic arch; contain zygomaticofacial and zygomaticotemporal foramina. Form cheekbones, outer part of eye sockets.
OTHER SKULL BONES (7)	
Hyoid (1) [FIGURE 7.11]	Below root of tongue, above larynx. U-shaped, suspended from styloid process of temporal bone; site of attachment for some muscles used in speaking, swallowing.
Ossicles of ear Incus (2) Malleus (2) Stapes (2) [FIGURE 7.6]	Inside cavity of petrous portion of temporal bone. Tiny bones shaped like anvil (incus), hammer (malleus), stirrup (stapes), articulating with one another and attached to tympanic membrane. Convey sound vibrations from eardrum to oval window (see Chapter 17).

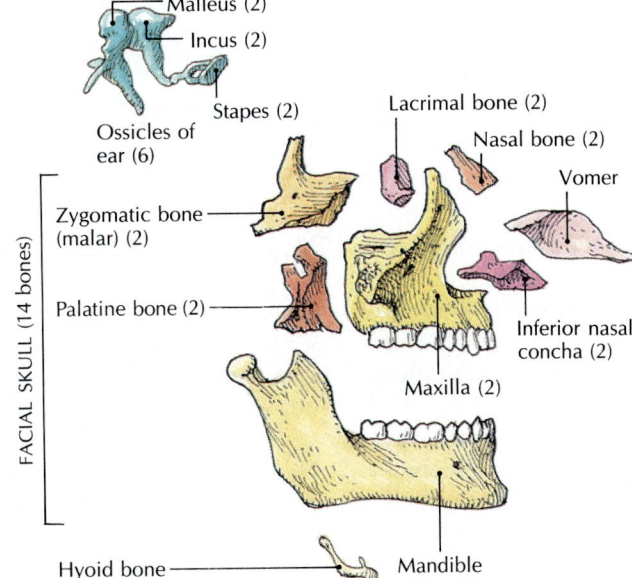

Malleus (2)
Incus (2)
Stapes (2)
Ossicles of ear (6)
Zygomatic bone (malar) (2)
Palatine bone (2)
Lacrimal bone (2)
Nasal bone (2)
Vomer
Inferior nasal concha (2)
Maxilla (2)
FACIAL SKULL (14 bones)
Hyoid bone
Mandible

[B] Exploded view of facial skull, ear ossicles, and hyoid bone. The facial skull has 14 bones; the 6 ossicles are enlarged for clarity. Right lateral view.

FIGURE 7.4 SUTURES AND FONTANELS

[A] Skull of a newborn infant illustrating the location of the fontanels; superior and right lateral views. **[B]** Sutures in an adult skull; superior and right lateral views. Note the absence of fontanels, which have closed.

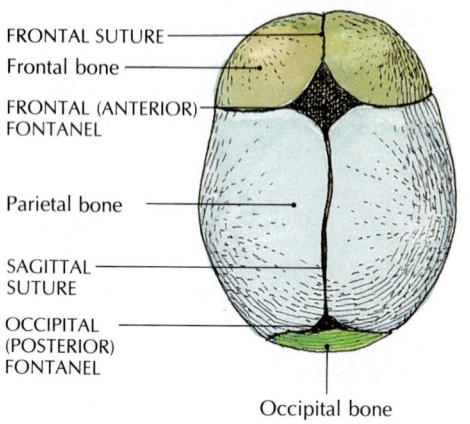

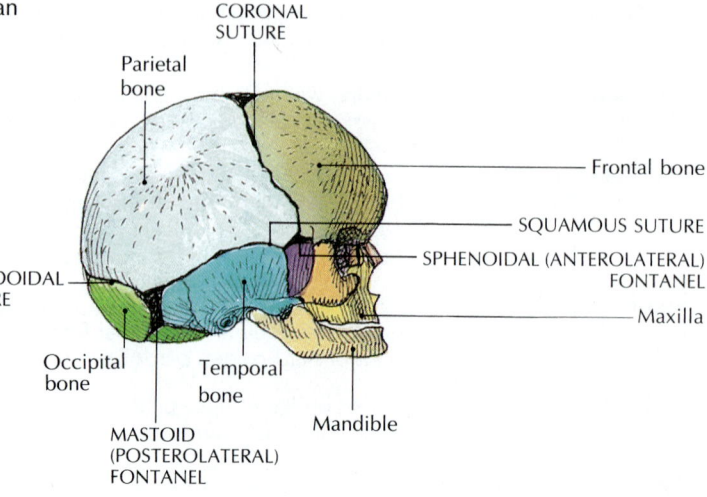

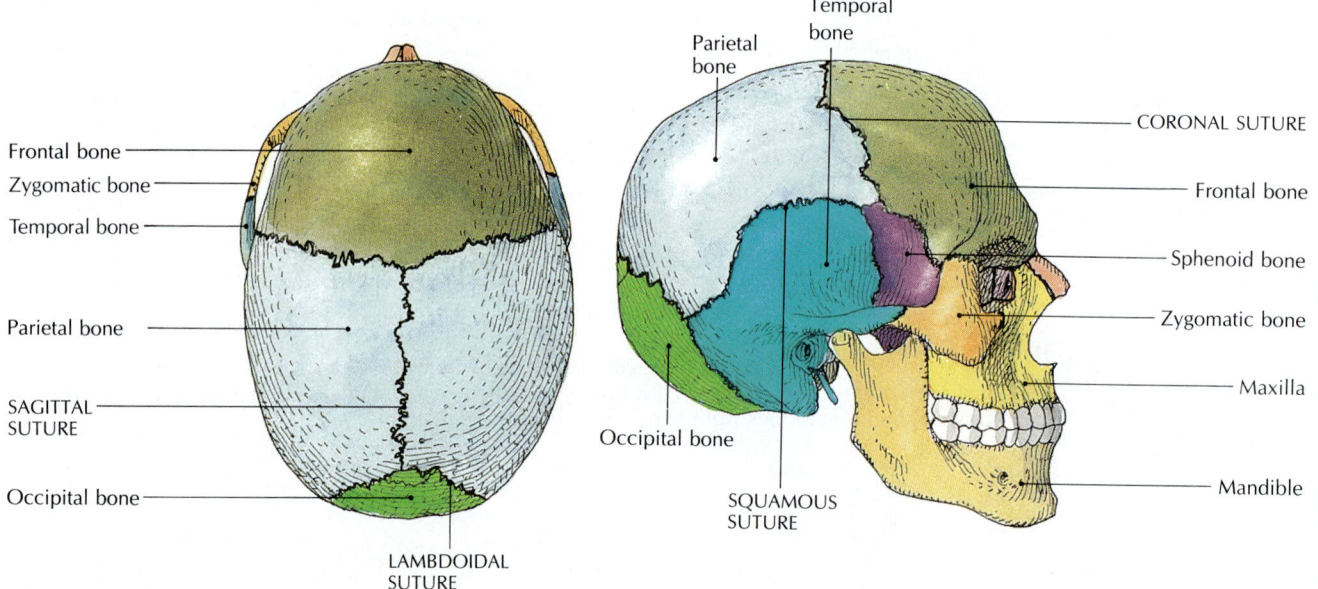

[A]

[B]

diamond-shaped fontanel closes about 2 months after birth.

3 The two *sphenoidal (anterolateral) fontanels*, situated at the junction of the frontal, parietal, temporal, and sphenoid bones. They usually close about 3 months after birth.

4 The two *mastoid (posterolateral) fontanels*, situated at the junction of the parietal, occipital, and temporal bones. They begin to close about 2 months after birth but do not close completely until about 12 months after birth.

Bones of the Cranium

The eight bones of the cranium are the frontal bone, the two parietal bones, the occipital bone, the two temporal bones, the sphenoid bone, and the ethmoid bone. The sutural bones are also considered part of the cranium.

The cranium consists of a roof and a base. The roof, called the *calvaria* (not calvarium), is made up of the squamous (L. *squama*, scale, which indicates that the bone is thin and relatively flat) part of the frontal bone (frontal squama), the parietal bones, and the occipital

FIGURE 7.5A SKULL, ANTERIOR VIEW

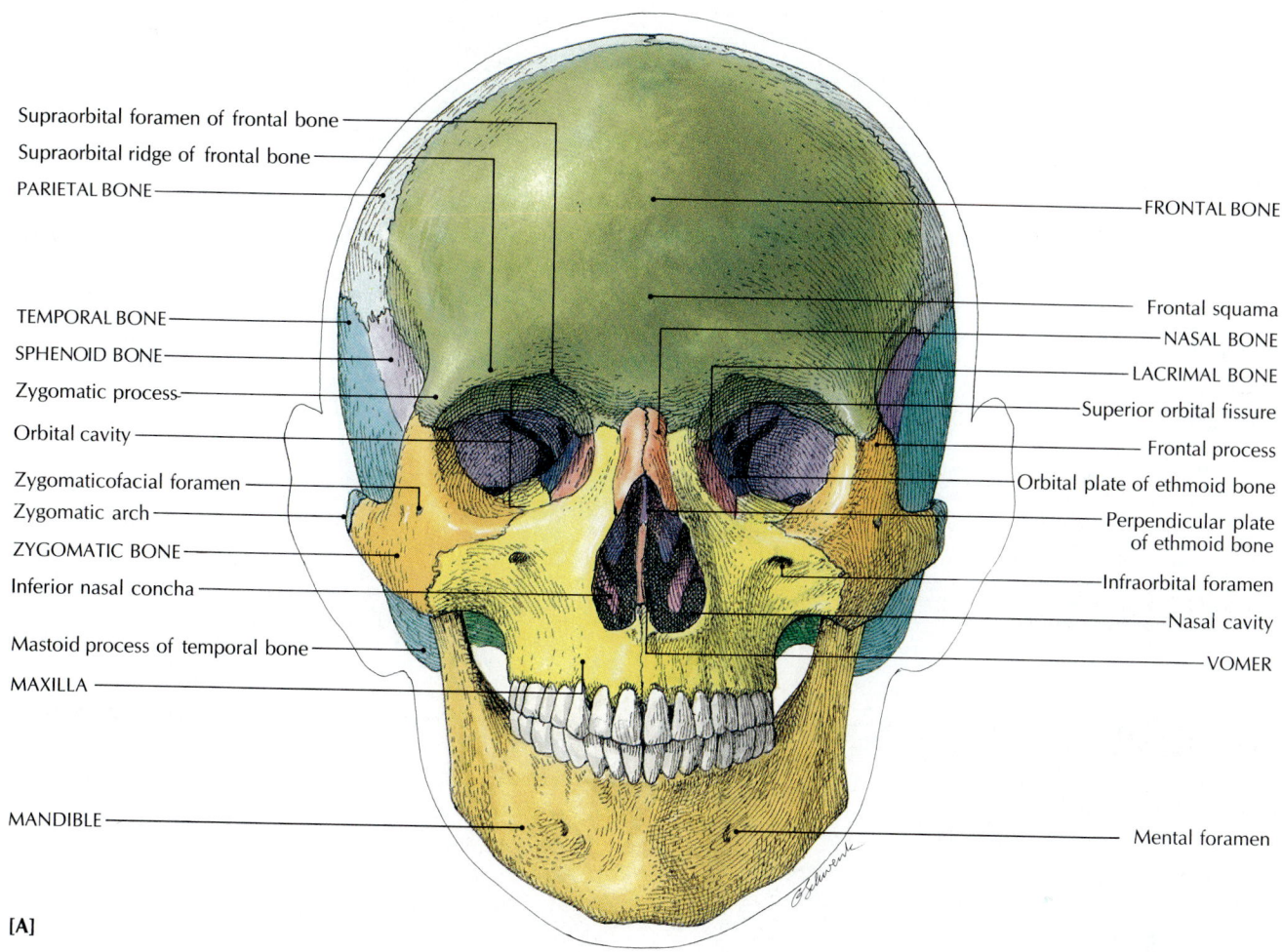

Supraorbital foramen of frontal bone
Supraorbital ridge of frontal bone
PARIETAL BONE

TEMPORAL BONE
SPHENOID BONE
Zygomatic process
Orbital cavity
Zygomaticofacial foramen
Zygomatic arch
ZYGOMATIC BONE
Inferior nasal concha
Mastoid process of temporal bone
MAXILLA

MANDIBLE

FRONTAL BONE
Frontal squama
NASAL BONE
LACRIMAL BONE
Superior orbital fissure
Frontal process
Orbital plate of ethmoid bone
Perpendicular plate of ethmoid bone
Infraorbital foramen
Nasal cavity
VOMER

Mental foramen

[A]

bone above the occipital protuberance [FIGURE 7.5A–E]. The *base* of the cranium is made up of portions of the ethmoid, sphenoid, and occipital bones. As viewed internally from above [FIGURE 7.5G], the base has three depressions, or *fossae*: the anterior, middle, and posterior fossae [see FIGURE 7.7A].

Frontal bone The large, scoop-shaped *frontal bone* forms the forehead and the upper part of the orbits (eye sockets) [FIGURE 7.5A, C, D, E, and G]. It forms the shell of the forehead as the *frontal squama* and the roof of each orbit as the *orbital plate*, and it protrudes over each eye as the *supraorbital ridge* to form the eyebrow ridge where the squamous and orbital portions meet.

The supraorbital nerve and blood vessels pass through the small supraorbital foramen (often only a notch) in the supraorbital ridge. Lining the inner surface of the frontal bone are many depressions, which follow the convolutions of the brain. At birth, the frontal bone has two parts

separated by a *frontal suture*. This suture generally disappears by the age of 6, when the frontal bones become united [see FIGURE 7.4].

Parietal bones FIGURE 7.5A–E, G shows the two *parietal bones* (L. *paries*, wall), which form the superior sides, the roof, and part of the back of the skull. These broad, slightly convex bones have smooth exteriors, but like the frontal bone, they have internal depressions, which accommodate the convolutions of the brain, and blood vessels, which supply the thin meninges that cover the brain. It is seldom mentioned that the parietal bones also form part of the back of the calvaria. In each parietal bone is a parietal foramen. Passing through the foramen is an emissary vein that connects the superior sagittal venous dural sinus inside the skull with scalp veins outside the skull. Emissary veins act as safety valves by which blood may flow in either direction to and from the cranial skull.

FIGURE 7.5B SKULL, POSTERIOR VIEW

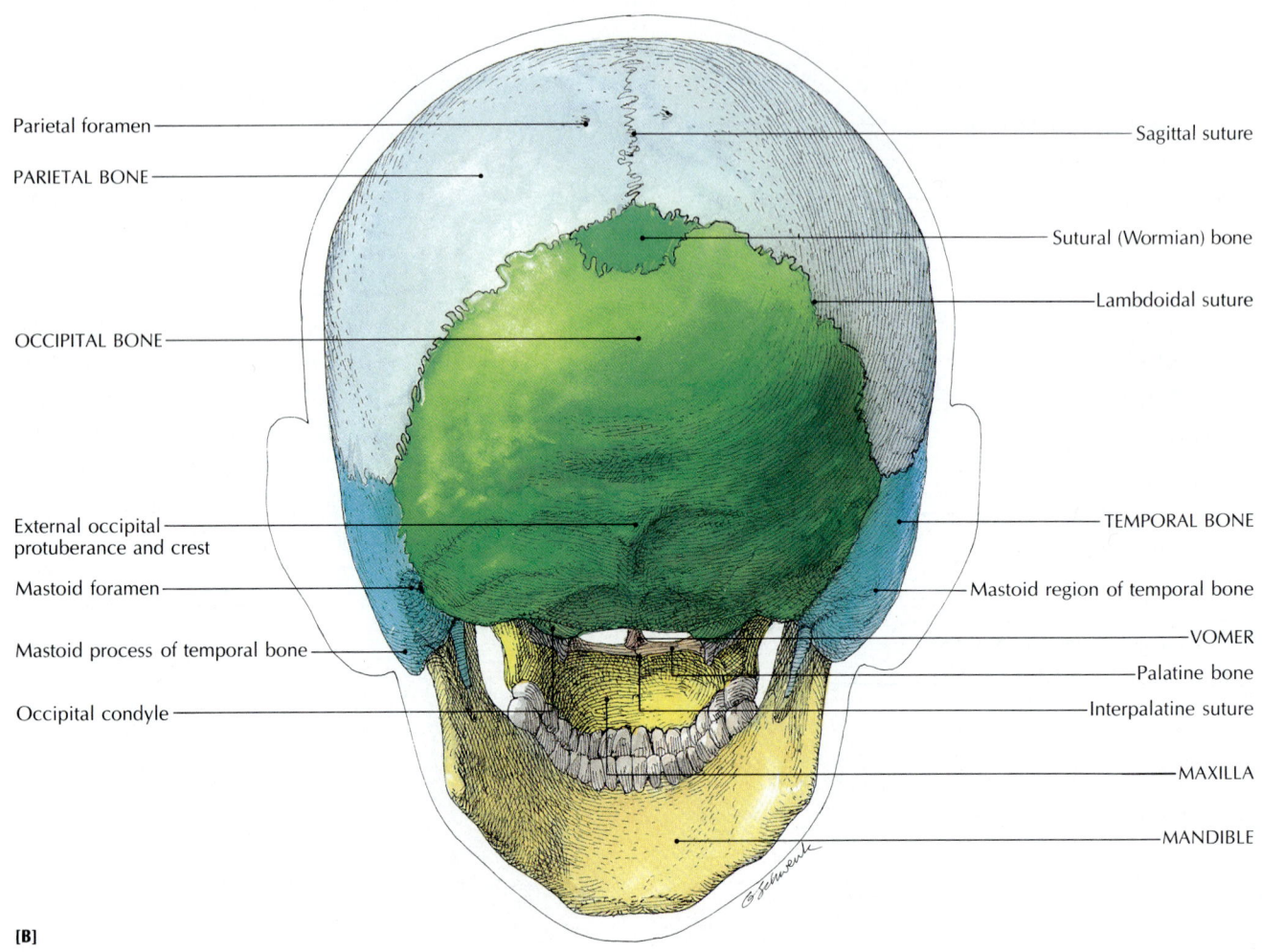

Parietal foramen

PARIETAL BONE

OCCIPITAL BONE

External occipital protuberance and crest

Mastoid foramen

Mastoid process of temporal bone

Occipital condyle

Sagittal suture

Sutural (Wormian) bone

Lambdoidal suture

TEMPORAL BONE

Mastoid region of temporal bone

VOMER

Palatine bone

Interpalatine suture

MAXILLA

MANDIBLE

[B]

Occipital bone FIGURE 7.5B–G shows the location of the *occipital bone* (L. *occiput*, in back of the head, *ob*, in back of + *caput*, head). This bone forms the posterior part of the cranial skull.

The occipital bone is a slightly curved plate that consists of a squamous part, a base part, and two lateral parts. At its base is a large oval opening called the *foramen magnum*, where the spinal cord passes through to join the medulla oblongata of the brain. On either side of the foramen magnum are the *occipital condyles* and the paired *hypoglossal canals*, through which the hypoglossal nerves pass. These nerves stimulate the muscles of the tongue. The *atlantooccipital joint* between the occipital condyles and the first cervical vertebra (the atlas) permits the head to nod up and down, such as in the "yes" movement.

The *external occipital crest* and the *external occipital protuberance* can be felt easily at the base of the occipital

bone. Both landmarks are sites of attachment for muscles and ligaments.

Temporal bones FIGURE 7.5A–E, G shows the paired *temporal bones*. With portions of the sphenoid bone, they form part of the sides and base of the cranium.

Each temporal bone has four parts: the squamous, petrous, and tympanic portions and the mastoid region:

1 The largest, and the superior part of the temporal bone, is the *squamous portion*. The *zygomatic process* of the temporal bone joins the temporal process of the zygomatic bone to form the slender *zygomatic arch* in the lateral part of the cheekbone [FIGURE 7.5A, C, D, and F]. The *mandibular fossa* is located posterior to the zygomatic arch. The fossa forms the socket that, together with the condyle of the mandible, forms the temporomandibular (TM, jaw) joint.

FIGURE 7.5C PHOTO OF SKULL, RIGHT LATERAL VIEW

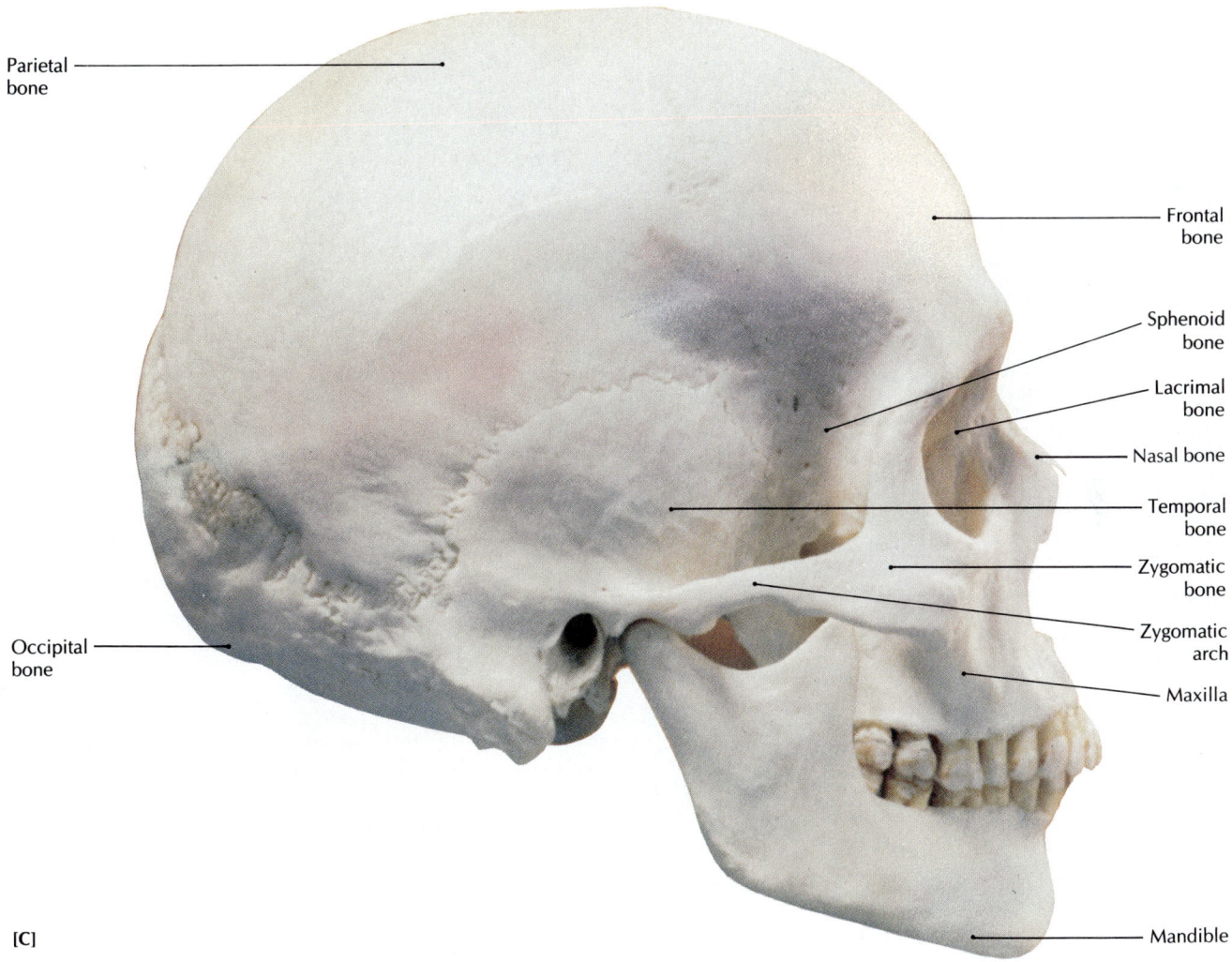

Parietal
bone

Frontal
bone

Sphenoid
bone

Lacrimal
bone

Nasal bone

Temporal
bone

Zygomatic
bone

Zygomatic
arch

Maxilla

Occipital
bone

Mandible

[C]

FIGURE 7.5D SKULL, RIGHT LATERAL VIEW

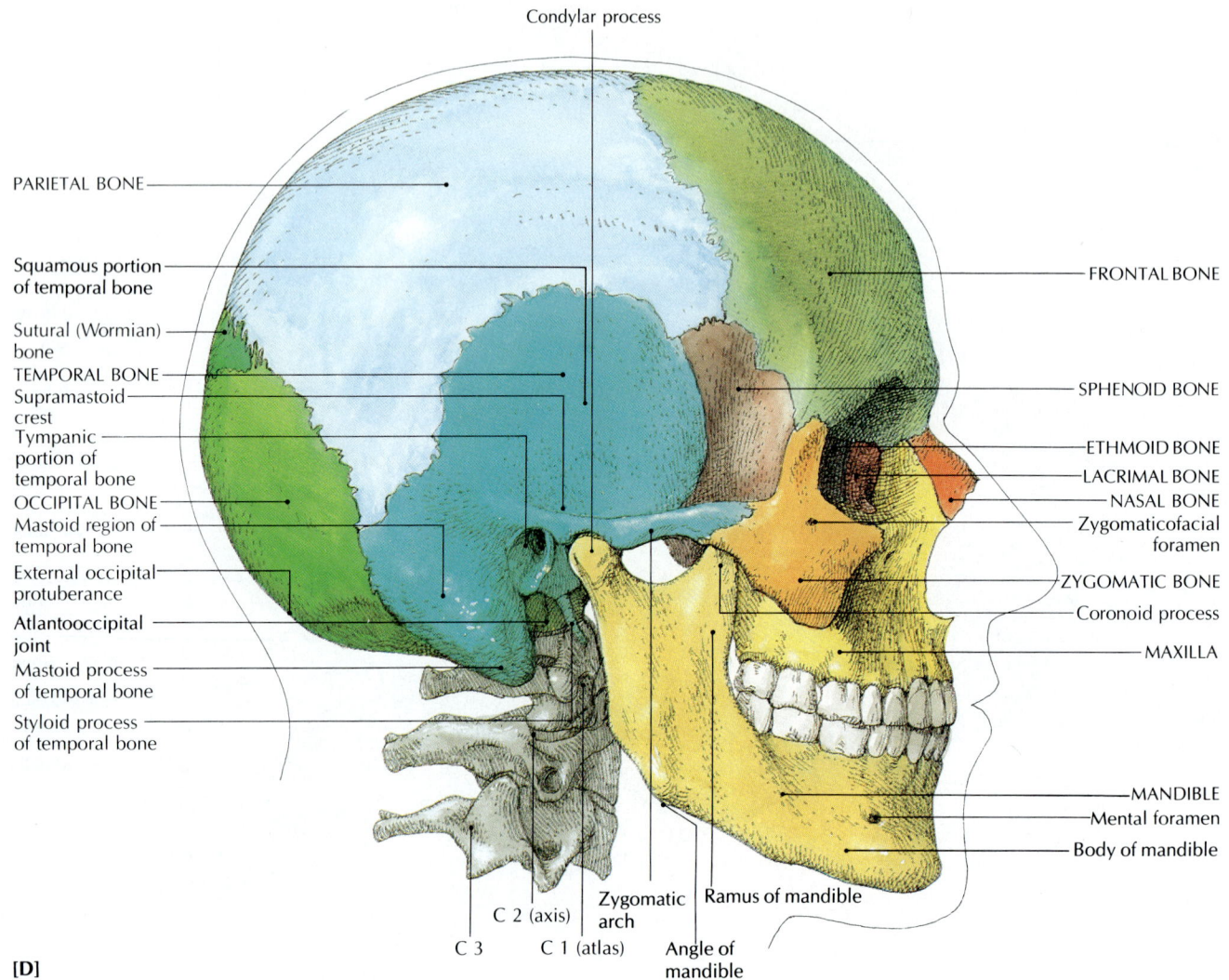

Condylar process

PARIETAL BONE

Squamous portion
of temporal bone

Sutural (Wormian)
bone

TEMPORAL BONE

Supramastoid
crest

Tympanic
portion of
temporal bone

OCCIPITAL BONE

Mastoid region of
temporal bone

External occipital
protuberance

Atlantooccipital
joint

Mastoid process
of temporal bone

Styloid process
of temporal bone

FRONTAL BONE

SPHENOID BONE

ETHMOID BONE

LACRIMAL BONE

NASAL BONE

Zygomaticofacial
foramen

ZYGOMATIC BONE

Coronoid process

MAXILLA

MANDIBLE

Mental foramen

Body of mandible

C 3   C 2 (axis)   C 1 (atlas)

Zygomatic
arch

Angle of
mandible

Ramus of mandible

[D]

FIGURE 7.5E SKULL, RIGHT MEDIAL VIEW

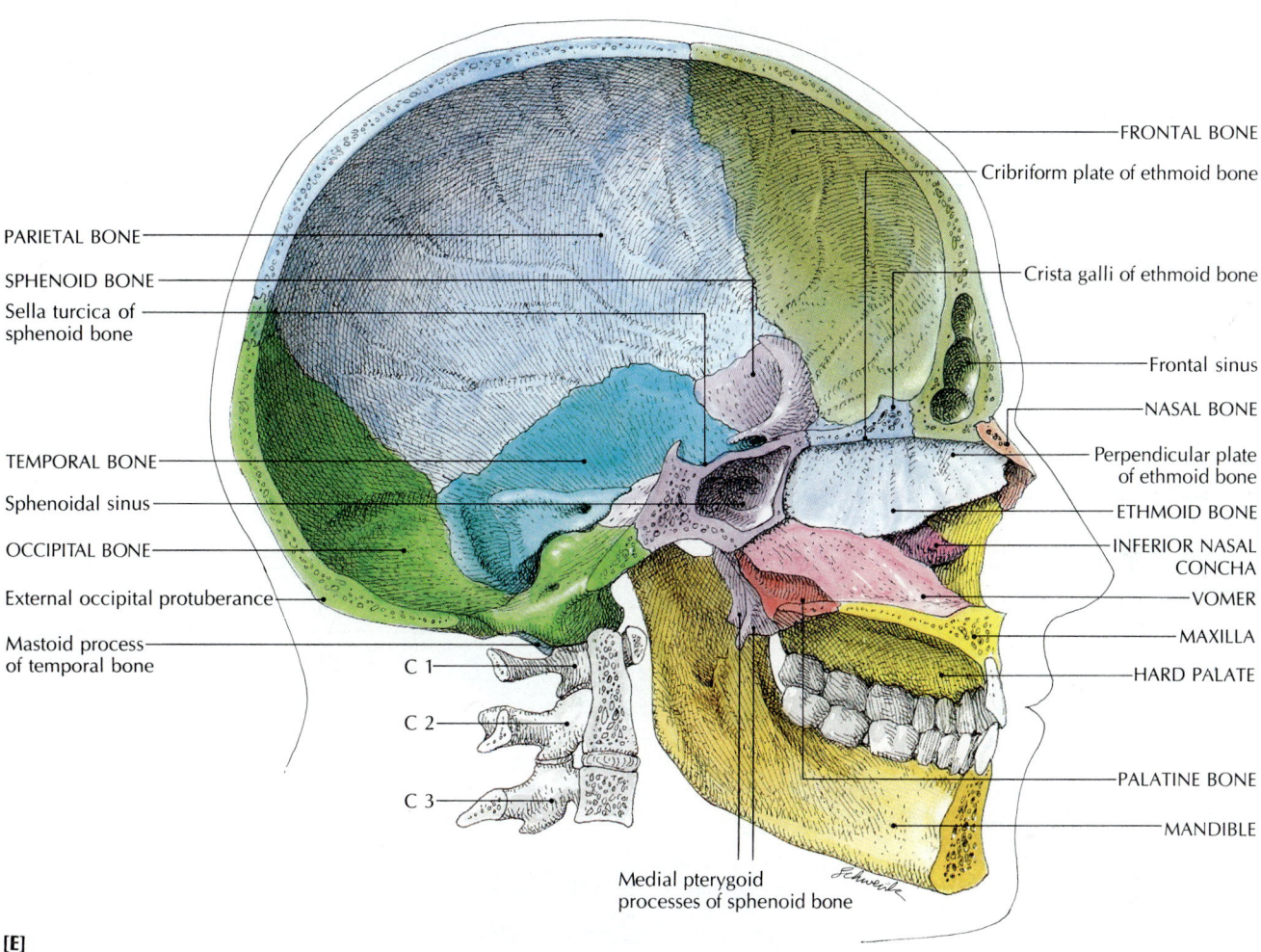

PARIETAL BONE

SPHENOID BONE

Sella turcica of
sphenoid bone

TEMPORAL BONE

Sphenoidal sinus

OCCIPITAL BONE

External occipital protuberance

Mastoid process
of temporal bone

C 1

C 2

C 3

Medial pterygoid
processes of sphenoid bone

FRONTAL BONE

Cribriform plate of ethmoid bone

Crista galli of ethmoid bone

Frontal sinus

NASAL BONE

Perpendicular plate
of ethmoid bone

ETHMOID BONE

INFERIOR NASAL
CONCHA

VOMER

MAXILLA

HARD PALATE

PALATINE BONE

MANDIBLE

[E]

FIGURE 7.5F SKULL, INFERIOR VIEW

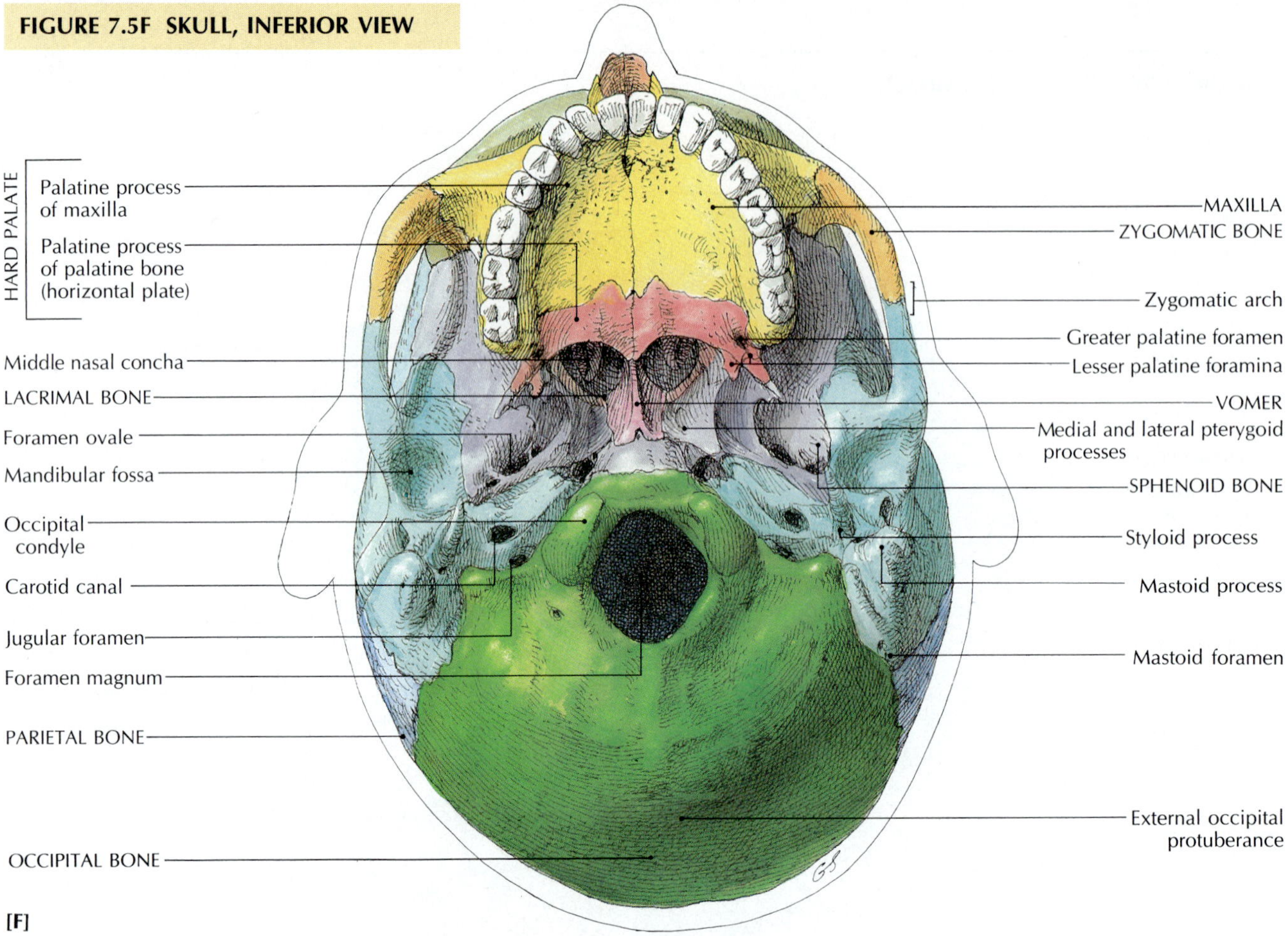

HARD PALATE

Palatine process of maxilla

Palatine process of palatine bone (horizontal plate)

Middle nasal concha

LACRIMAL BONE

Foramen ovale

Mandibular fossa

Occipital condyle

Carotid canal

Jugular foramen

Foramen magnum

PARIETAL BONE

OCCIPITAL BONE

MAXILLA

ZYGOMATIC BONE

Zygomatic arch

Greater palatine foramen

Lesser palatine foramina

VOMER

Medial and lateral pterygoid processes

SPHENOID BONE

Styloid process

Mastoid process

Mastoid foramen

External occipital protuberance

[F]

FIGURE 7.5G SKULL, SUPERIOR VIEW

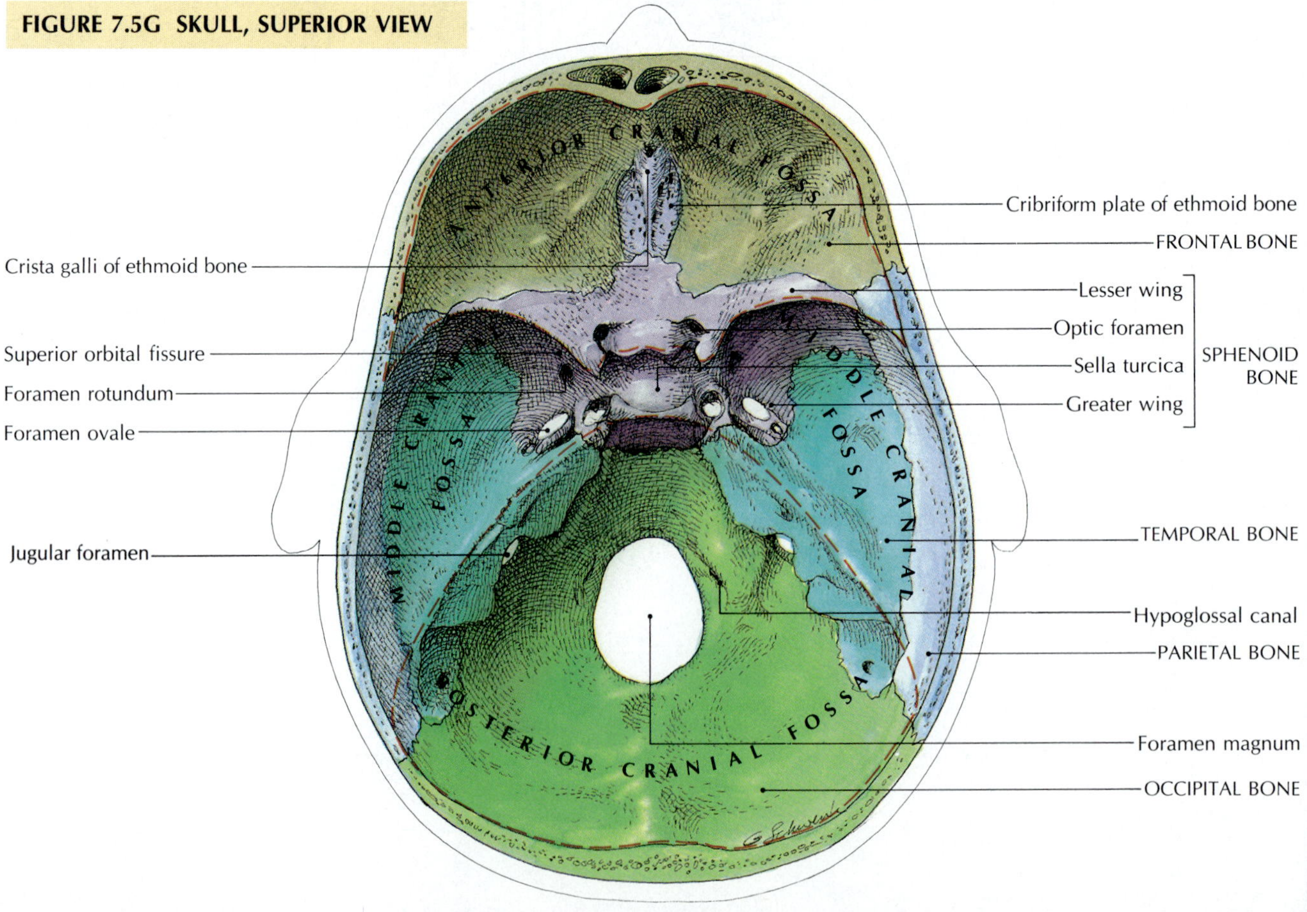

Crista galli of ethmoid bone

Superior orbital fissure

Foramen rotundum

Foramen ovale

Jugular foramen

ANTERIOR CRANIAL FOSSA

MIDDLE CRANIAL FOSSA

MIDDLE CRANIAL FOSSA

POSTERIOR CRANIAL FOSSA

Cribriform plate of ethmoid bone

FRONTAL BONE

Lesser wing

Optic foramen

Sella turcica

Greater wing

SPHENOID BONE

TEMPORAL BONE

Hypoglossal canal

PARIETAL BONE

Foramen magnum

OCCIPITAL BONE

[G]

FIGURE 7.6 MIDDLE EAR CAVITY

The middle ear cavity is within the petrous portion of the temporal bone. The drawing shows the ear bones (ossicles): malleus, incus, and stapes. (See drawing B figure in TABLE 7.2 and Chapter 17 for other illustrations of the ear bones.)

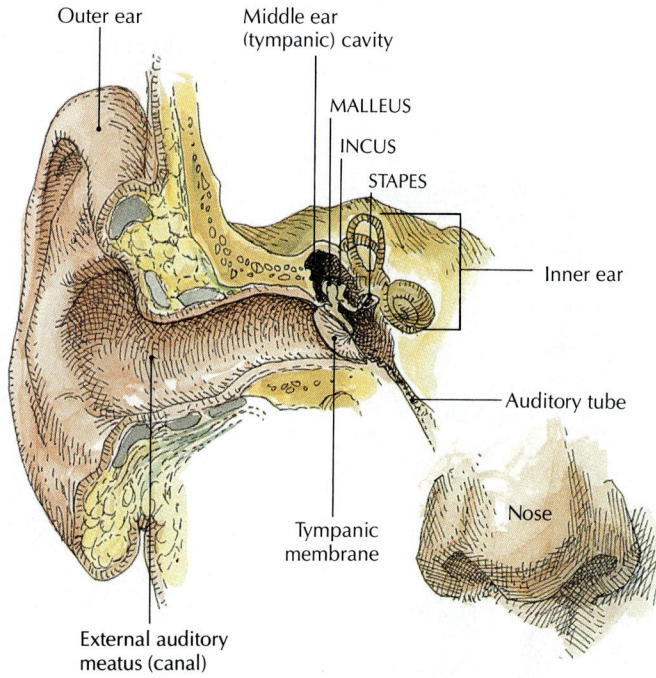

Outer ear
Middle ear (tympanic) cavity
MALLEUS
INCUS
STAPES
Inner ear
Auditory tube
Nose
Tympanic membrane
External auditory meatus (canal)

2 The *petrous portion* of the temporal bone is in the floor of the *middle cranial fossa*, wedged between the occipital and sphenoid bones [FIGURES 7.5G and 7.7A]. Projecting downward from the inferior surface of the petrous portion is the *styloid process.* The elongated styloid process is the site for the attachment of muscles and ligaments involved with some movements of the hyoid bone, tongue, and pharynx (throat) during speaking, swallowing, and chewing. Encased within the temporal bone are the three tiny auditory (hearing) bones of the middle ear, the *malleus, incus,* and *stapes* [FIGURE 7.6], and the sensory receptors for hearing (cochlea) and balance (semicircular ducts, utricle, and saccule).

3 The *tympanic portion* of the temporal bone forms part of the wall of the *external auditory canal* (meatus, the external opening of the ear) and part of the wall of the tympanic cavity, both of which are involved with the transmission and resonance of sound waves. The tympanic portion is the site for the attachment of the tympanic membrane (eardrum).

4 The *mastoid region* is actually the posterior portion of the temporal bone. Each prominent *mastoid process* in the posterior part of the temporal bone is located behind the ear; it provides the point of attachment for a neck muscle. These processes contain air spaces called *mastoid air cells* or *sinuses*, which connect with the middle ear. Because of this connection, infections within the middle ear can spread to the sinuses and cause an inflammation called *mastoiditis.*

Sphenoid bone As shown in FIGURES 7.7A and 7.5A, C–G, the *sphenoid bone* (SFEE-noid; Gr. *sphen*, wedge) forms the anterior part of the base of the cranium. It is generally described as being "wedge-shaped" or "wing-shaped" because it looks somewhat like a bat or a butterfly with two pairs of outstretched wings. It has a cube-shaped central *body,* two *lesser lateral wings,* two *greater lateral wings,* and *pterygoid processes* (Gr. *pteron,* wing), which project downward as medial and lateral pterygoid plates [FIGURE 7.7B, C].

The pterygoid processes form part of the walls of the nasal cavity, and the undersurface of the body forms part of the roof of the nasal cavity. A key feature of the sphenoid is a deep depression within its body that houses and protects the pituitary gland. This depression is the *sella turcica* (SEH-luh TUR-sihk-uh), so called because it is said to resemble a Turkish saddle (L. *sella,* saddle).

The lesser wings form a part of the floor of the *anterior cranial fossa.* At the base of each lesser wing is the *optic foramen (canal),* through which the optic nerve and ophthalmic artery pass into the orbit.

The greater wings of the sphenoid bone form a major part of the *middle cranial fossa* [FIGURE 7.7A]. Between the greater and lesser wings is the *superior orbital fissure* [FIGURE 7.7B, C], located between the middle cranial fossa and the orbit. Through the fissure pass three cranial nerves for the eye muscles and the sensory nerve to the orbital and forehead regions. Three important foramina in the greater wing are the *foramen rotundum* and the

FIGURE 7.7 SPHENOID BONE

[A] In the floor of the cranium, superior view. [B] Enlarged drawing, superior view. [C] Posterior view.

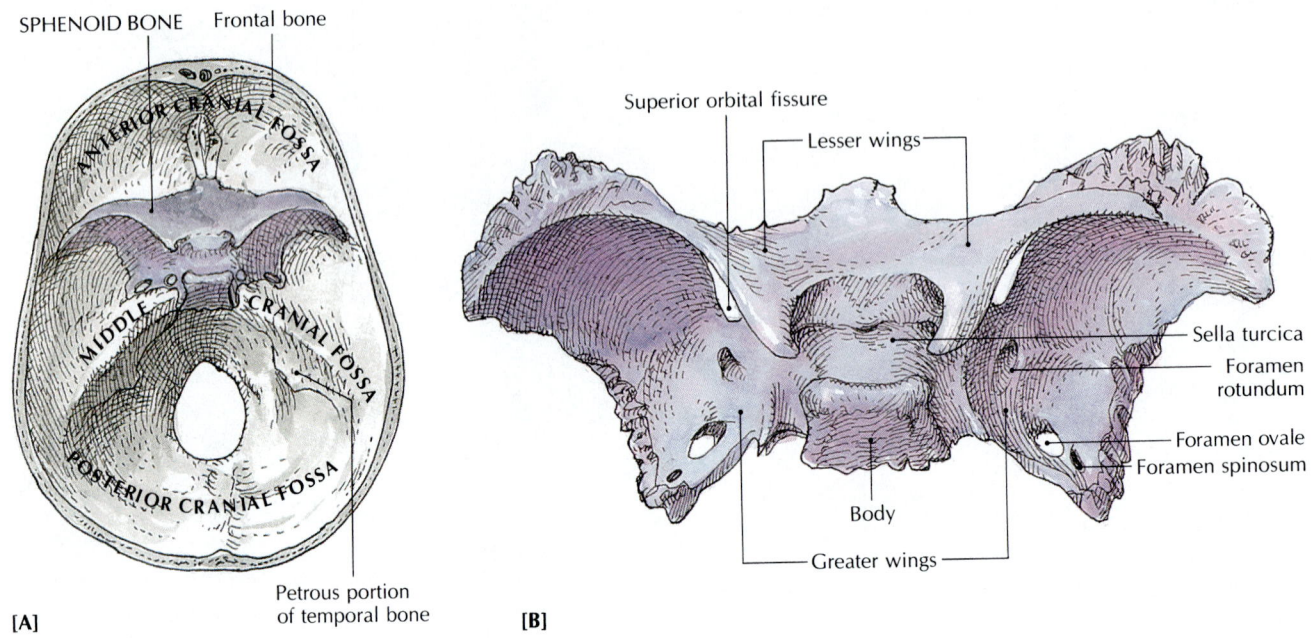

[A]

[B]

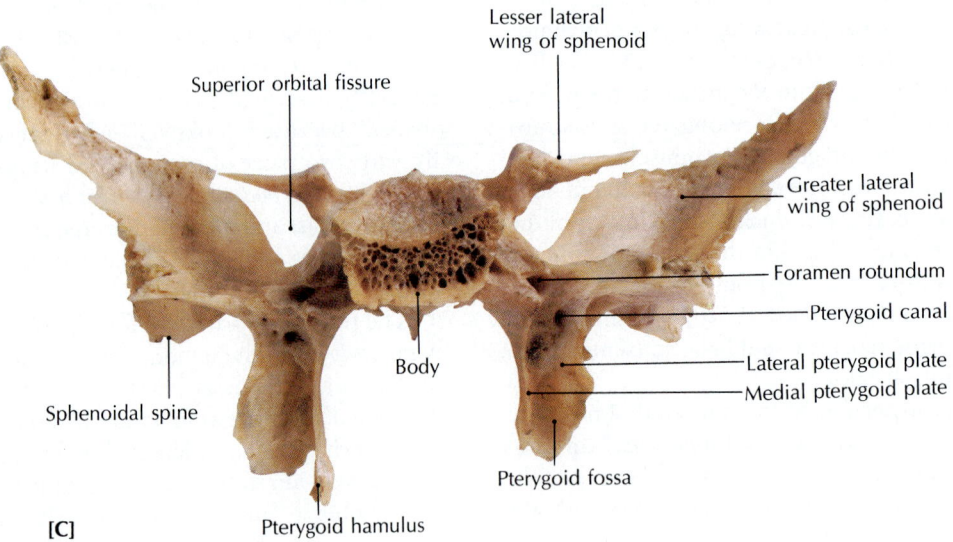

[C]

FIGURE 7.8 ETHMOID BONE

[A] In the cranium, right medial view.
[B] Right lateral view. [C] Superior view.

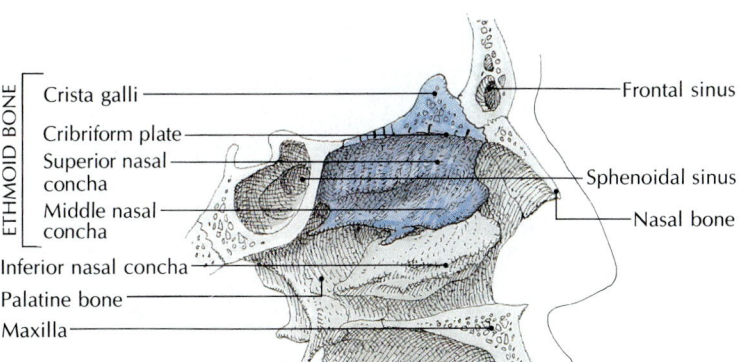

[A]

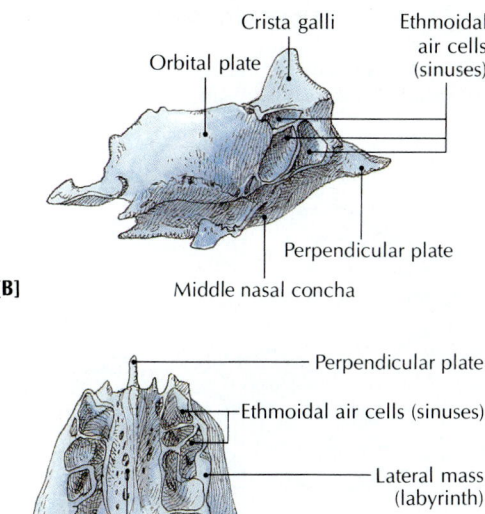

[B]

[C]

foramen ovale for different divisions of the trigeminal nerve, and the *foramen spinosum* for the middle meningeal artery to the side and roof of the skull [FIGURE 7.7B].

Ethmoid bone The *ethmoid bone* forms the roof of the nasal cavity and medial part of the floor of the anterior cranial fossa between the orbits [FIGURE 7.8]. This light and delicate bone consists of a horizontal cribriform plate (L. *cribrum*, strainer), the median perpendicular plate, and paired lateral masses, or labyrinths.

The inferior surface of the *cribriform plate* forms the roof of the nasal cavity, and the superior surface is part of the floor of the cranial cavity. The cribriform plate may have as many as 20 foramina. Through these tiny openings the branches of the olfactory (smell) nerve pass from the mucous membrane of the nose to the brain.

The *median perpendicular plate* forms a large part of the vertical nasal septum between the two nasal cavities. The *crista galli,* or cockscomb, is a vertical protuberance in the center of the cribriform plate. It provides a site of attachment for a membrane (the falx cerebri of the dura mater) that covers the brain.

Each lateral mass has a plate that forms the lateral wall of a nasal cavity and the orbital plate of an orbit. It also contains the superior and middle *conchae* (KONG-kee; L. "conch shell"), curved scrolls of bone that extend into the nasal cavity and are part of the lateral wall. In each lateral mass are *ethmoidal air cells* that open and drain into the nasal cavity.

Sutural bones *Sutural bones* (also called *Wormian* bones) are separate small bones found in the sutures of the calvaria [FIGURE 7.5B]. They are found most often in the lambdoidal suture. Their number may vary.

Paranasal Sinuses

In the interior of the ethmoid, maxillary, sphenoid, and frontal bones are four pairs of air cavities called *paranasal sinuses* [FIGURE 7.9]. They are named after the bones where they are located. The sinuses have two main functions. They act as resonating chambers for the voice and, being air-filled cavities, lighten the weight of the skull bones.

The paranasal sinuses are formed after birth as outgrowths from the nasal cavity, when the spongy part of the bone is resorbed, leaving air-filled cavities. The *maxillary* and *sphenoidal sinuses* are not fully formed until adolescence.

The walls of the paranasal sinuses are lined with mucous membranes, which communicate with the mucous membranes of the nasal cavity through small openings. An inflammation of the mucosa of the nasal cavity is called rhinitis, or a "head cold." When the inflammation occurs in the mucous membranes of the sinuses, or when the sinuses react adversely to foreign substances (allergy), the result is *sinusitis.* The *ethmoidal air cells* (one of the paranasal sinuses) are made up of many small cavities in the ethmoid bone between the orbit and the nasal cavity.

The *maxillary sinuses* are the largest of the paranasal sinuses. Mucus within the sinus drains into the nasal cavity through a single opening, which is high enough so that drainage is difficult when the nasal mucosa is swollen. Thus, fluid may accumulate and cause maxillary sinusitis.

The *sphenoidal sinuses* are contained within the body of the sphenoid bone. The size of the sphenoidal sinuses varies greatly, and they may extend into the occipital bone.

FIGURE 7.9 PARANASAL SINUSES

The paranasal sinuses include the frontal, ethmoidal, sphenoidal, and maxillary sinuses. The ethmoidal air cells, referred to as sinuses, look like honeycombs. **[A]** Anterior view. **[B]** Right medial view. **[C]** Photo of cleared specimen, anterior view.

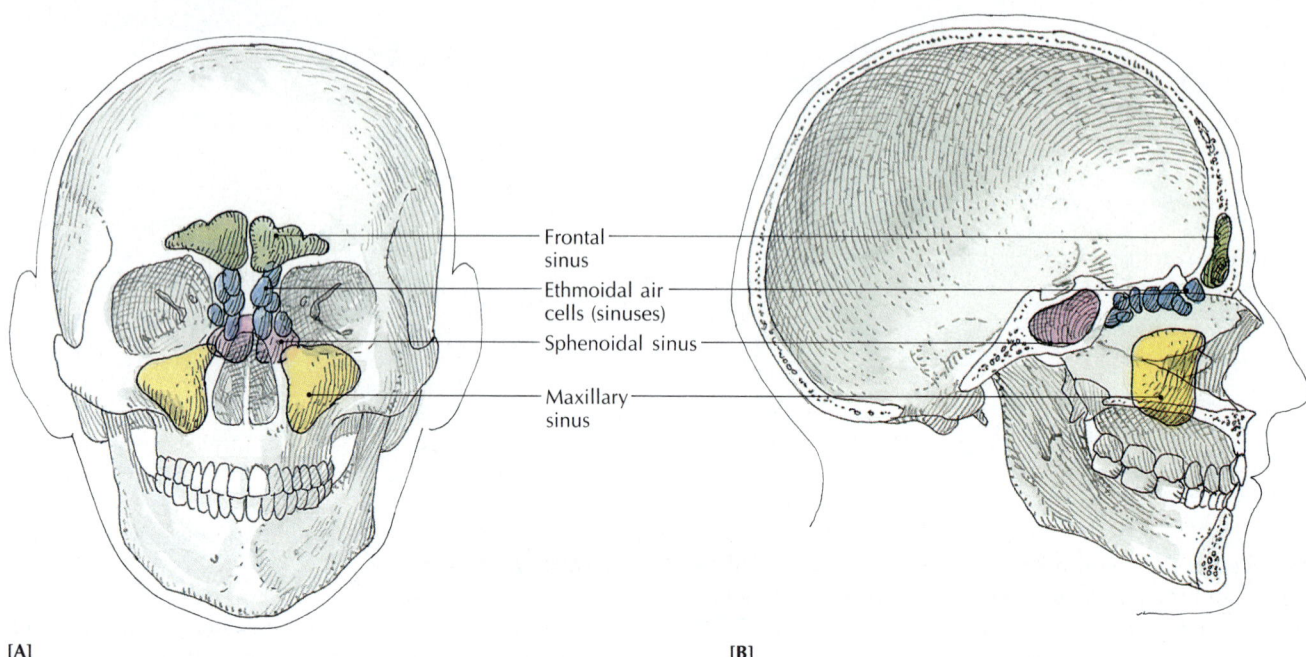

Frontal
sinus

Ethmoidal air
cells (sinuses)

Sphenoidal sinus

Maxillary
sinus

[A]   [B]

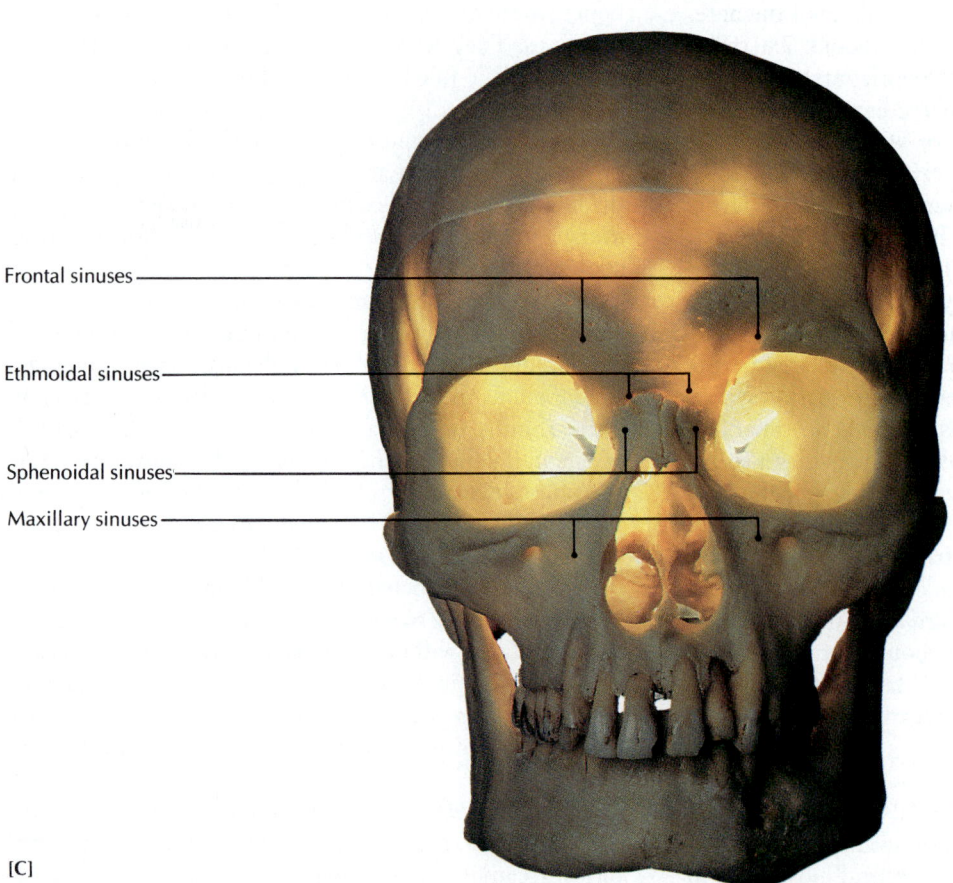

Frontal sinuses

Ethmoidal sinuses

Sphenoidal sinuses

Maxillary sinuses

[C]

The *frontal sinuses* are separated by a bony septum that is often bent to one side. Each sinus extends above the medial end of the eyebrow and anteriorly into the medial portion of the roof of the orbit [FIGURE 7.9].

Bones of the Face

The facial skull is anterior and inferior to most of the cranial skull, to which it is attached. Its main functions are to support and protect the structures associated with the nasal, oral, orbital, and pharyngeal cavities. It also provides a pair of joints between the condyles of the mandible and temporal bones (the temporomandibular, or TM, joints), which permit the lower jaw to open and close. Except for the mandible, all the bones of the facial skull are united by sutures and are therefore immovable.

The facial skull is composed of 14 irregularly shaped bones [FIGURE 7.3A, TABLE 7.2B]. They include the two inferior nasal conchae, the vomer, two palatine bones, two maxillae, two zygomatic bones, two nasal bones, two lacrimal bones, and the mandible. The single hyoid bone and the six ossicles of the ear are considered separately from cranial or facial bones.

Inferior nasal conchae FIGURES 7.5E and 7.8A show the location of the *inferior nasal conchae* in the lateral wall of each nasal cavity, below the superior and middle conchae of the ethmoid bone. These conchae, along with the conchae of the ethmoid bones, are thin, bony plates shaped somewhat like the curved leaves of a scroll. They increase the surface area of the nasal mucosa, a membrane containing ciliated epithelial cells and many blood vessels. The fluid portion of the blood leaks out of these blood vessels, continuously moistening (humidifying) the nasal cavity. Cilia in the nasal cavities cleanse the air, and the blood vessels warm it.

Vomer The *vomer* (L. plowshare), which is shown in FIGURE 7.5A, B, E, and F, forms part of the nasal septum. The single vomer, as its name suggests, is shaped like a plowshare. It is very thin and is often bent to one side. If the septum is pushed too far to one side and one nasal cavity is much smaller than the other, the condition is called a *deviated septum*. With such a condition, an allergy attack or severe head cold may swell the nasal mucosa enough to close the smaller cavity or even both nasal cavities.

Palatine bones FIGURE 7.5B and E shows the paired *palatine bones* (referring to the *palate*, or roof of the mouth). These bones lie behind the maxillae (upper jaw) and form the posterior part of the hard palate, parts of the floor and walls of the nasal cavity, and the floor of the orbit. The hard palate is composed of the palatine processes of the palatine and maxillary bones.

Each palatine bone is L-shaped and has a horizontal plate and a vertical plate. The horizontal plate (palatine process) forms the posterior part of the hard palate. The vertical plate forms the posterior lateral wall of the nasal cavity and part of the floor of the orbit. Nerves that serve the palate pass through the *greater* and *lesser palatine foramina* in the posterior lateral corner of the horizontal plate. In some dental procedures, dentists infiltrate these foramina with anesthetics.

Maxillae FIGURE 7.5A–F shows the two *maxillae* (L. upper jaw), which join to form the upper jaw. Each maxilla has a hollow body containing a large maxillary sinus and four processes: (1) The *zygomatic process* extends along the lateral orbital border and articulates with the zygomatic bone to form the *zygomatic arch*. (2) The *frontal process* extends along the medial wall of the orbit, where it articulates with the frontal bone. (3) The *palatine process* extends horizontally to meet the palatine process of the other maxilla and form the anterior part of the hard palate. (4) The *alveolar process* joins the alveolar process of the other maxilla to form the *alveolar arch*, which contains the bony sockets for the upper teeth. Beneath the orbital margin is the *infraorbital foramen*, through which pass the sensory infraorbital nerve and blood vessels. The lengthening of the face just before adolescence is caused by the growth of the maxillae.

Zygomatic bones The *zygomatic bones* are the cheekbones [FIGURE 7.5A, C, D, and F]. They lie below and lateral to the orbit. Each of the two zygomatic bones acts as a tie by connecting the maxilla with the frontal bone above and the temporal bone behind. Through the *zygomaticofacial* and *zygomaticotemporal foramina* in this bone pass the nerves bearing the same names (facial and temporal).

Nasal bones As shown in FIGURE 7.5A, C, D, and E, the two *nasal bones* lie side by side between the frontal processes of the maxillae. These two small, oblong bones unite to form the supportive bridge of the upper nose. In addition, they articulate with the frontal, ethmoid (perpendicular plate), and maxillary bones (frontal process) and are attached to the lower nasal cartilage.

Lacrimal bones Each *lacrimal bone* (L. *lacrima*, tear) is a thin bone located in the medial wall of the orbit behind the frontal process of the maxilla [FIGURE 7.5A, C, D, and F]. The rectangular lacrimal bones are the smallest facial bones. In a depression of each bone is a *lacrimal sac*, which collects excess tears from the surface of the eye. Tears from the lacrimal sacs drain through the nasolacrimal ducts and foramen into the nasal cavity, sometimes causing a runny nose.

FIGURE 7.10 MANDIBLE

[A] Right lateral view. [B] The mandible in the infant at birth (top), in the adult (center), and in the aged (bottom). Note how the alveolar process disappears in old age after the teeth have been lost.

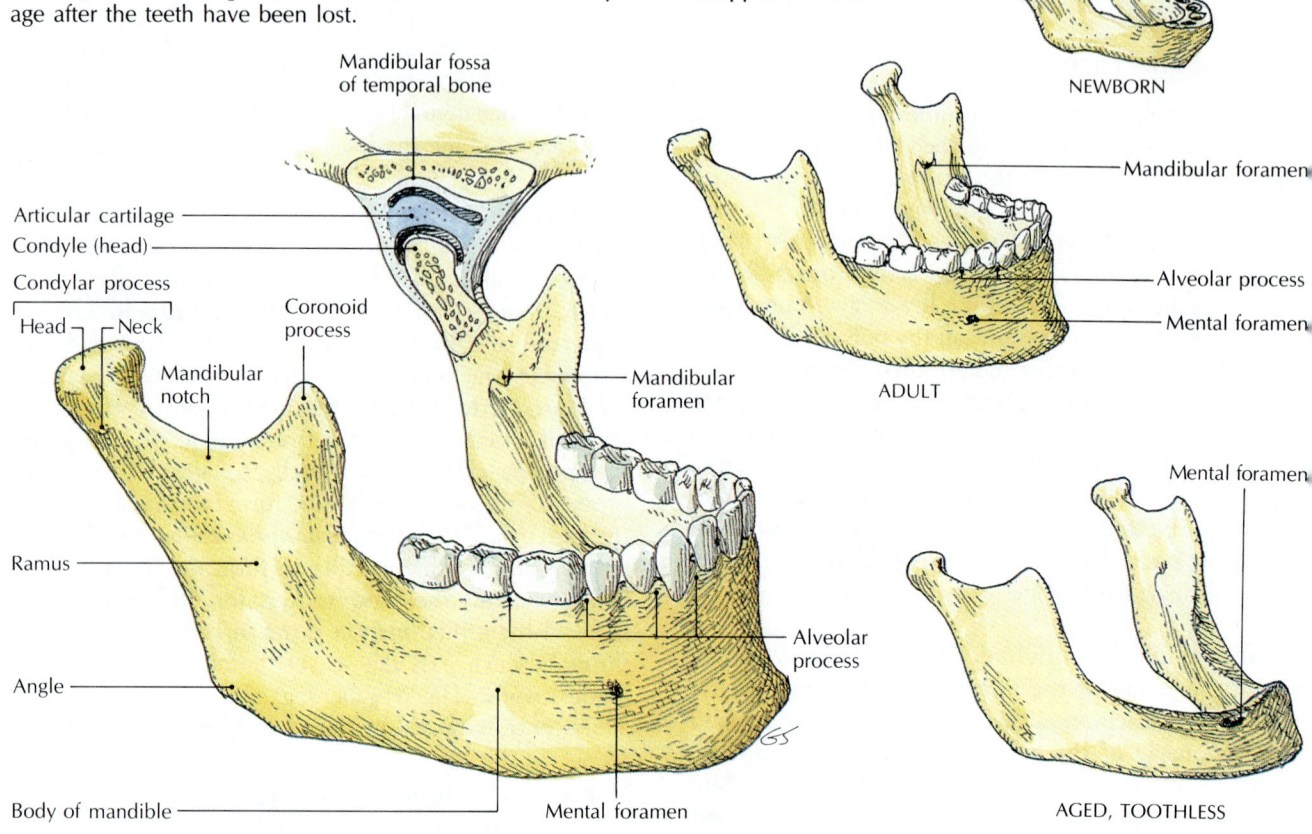

[A]

[B]

Mandible As shown in FIGURE 7.10A, the *mandible* (L. *mandere*, to chew) is the bone of the lower jaw. It is a single bone that extends from the mandibular fossa of one temporal bone to the mandibular fossa of the other, forming the chin. It is the largest and strongest facial bone. The right and left halves of the mandible fuse together in the center at the *symphysis menti* during the first or second year of life.

Except for the ossicles of the ear and the hyoid bone, the mandible is the only movable bone in the skull. It can be raised, lowered, drawn back, pushed forward, and moved from side to side. Try it the next time you are chewing food.*

The mandible consists of a horseshoe-shaped horizontal *body* joined to two perpendicular upright portions called *rami* (RAY-mye; L. branches; sing. *ramus*). The site where the body joins each ramus is known as the *angle* of the mandible. At the superior end of each ramus are two

processes separated by a deep depression called the *mandibular notch:* (1) The *coronoid process* is the attachment site for the temporalis muscle, and (2) the head of the *condylar process* articulates with the mandibular fossa of the temporal bone to form the temporomandibular joint. The inferior alveolar nerve enters the bone through the *mandibular foramen* and passes through the mandible, and a branch emerges as the mental nerve through the *mental foramen,* below the first molar tooth. (The mental foramen is anterior, whereas the mandibular foramen is located on the posterior surface of the bone.) Branches of the inferior alveolar nerve supply the teeth of the lower jaw. Blood vessels also pass through the mental foramen. The superior edge of the body of the mandible is the *alveolar process,* which contains the sockets for the lower teeth.

Hyoid Bone

FIGURE 7.11 shows the location of the U-shaped *hyoid bone* located inferior to the root of the tongue and superior to the larynx. When the chin is held up, the bone can be felt above the "Adam's apple," or thyroid cartilage.

*Because the joint between the mandible and the temporal bone is slightly loose, the head of the mandible may be dislocated forward during a wide yawn, locking the jaw in an open position. The condylar process can be realigned with the mandibular fossa of the temporal bone by depressing the jaw.

FIGURE 7.11 HYOID BONE

[A] Anterior view, in position in the neck, superior to the larynx. Note how the hyoid is suspended from the stylohyoid ligaments. **[B]** Anterior view. **[C]** Right lateral view.

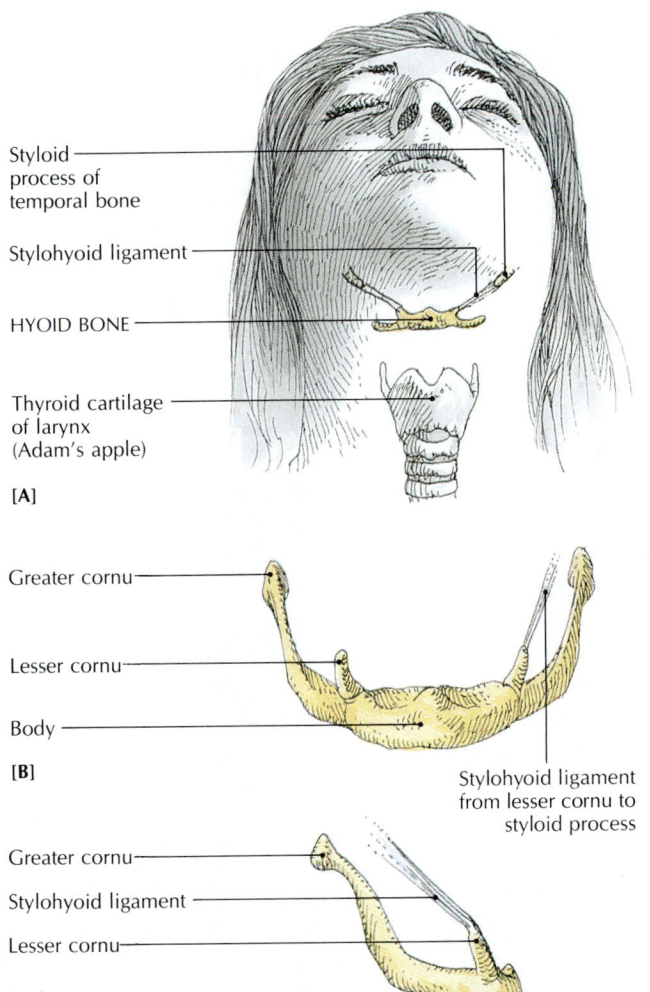

Styloid process of temporal bone

Stylohyoid ligament

HYOID BONE

Thyroid cartilage of larynx (Adam's apple)

[A]

Greater cornu

Lesser cornu

Body

[B]

Stylohyoid ligament from lesser cornu to styloid process

Greater cornu
Stylohyoid ligament
Lesser cornu
Body

[C]

The hyoid bone does not articulate directly with any other bone. Instead, it is held in position by muscles and the stylohyoid ligaments, which extend from the styloid process of each temporal bone to the hyoid.

In the center of the hyoid bone is a **body,** and projecting backward and upward are a pair of **lesser** and **greater horns,** or **cornua** (KOR-nyoo-uh; L. horns; sing. *cornu*). The hyoid supports the tongue and provides attachment sites for the muscles used in speaking and swallowing. In fact, the bone moves in a rotary motion (forward, up, back, and down) during the swallowing sequence. Because it is somewhat freely suspended, the hyoid bone can be held between the index finger and thumb and moved gently from side to side.

Ossicles of the Ear

Within the petrous portion of the temporal bone is the middle ear cavity, containing three pairs of tiny auditory bones, or **ossicles** [see FIGURE 7.6]. The ossicles are connected and transmit sound waves from the tympanic membrane (eardrum) to the inner ear. These bones are named according to their shapes. The outermost and largest bone is the **malleus** (L. hammer), or hammer. It is attached to the tympanic membrane. The middle bone is the **incus** (L. anvil), or anvil. The innermost bone, the **stapes** (STAY-peez; L. stirrup), is shaped like a stirrup. The stapes is the smallest bone in the body. It fits into a tiny membranous oval window, which separates the middle and inner ears.

A S K   Y O U R S E L F

1 How is the skull usually defined? What are the two typical divisions of the skull?

2 What are sutures? Name the four major skull sutures.

3 What are fontanels? Give some examples.

4 What are the eight bones of the cranium?

THE VERTEBRAL COLUMN

The skeleton of the trunk of the body consists of the vertebral column (commonly called the spine or the spinal column), the ribs, and the sternum. These bones, along with the skull, make up the axial skeleton.

The **vertebral column** is actually more like an S-shaped spring than a column [FIGURE 7.12B]. It extends from the base of the skull through the entire length of the trunk. The spine is composed of 26 separate bones called **vertebrae** (VER-tuh-bree; L. something to turn on, from *vertere*, to turn; sing. *vertebra*), which are united by a sequence of fibrocartilaginous **intervertebral disks** to form a strong but flexible support for the neck and trunk. The vertebral column is stabilized by ligaments and muscles that permit twisting and bending movements; however, they limit other movements that might be harmful to either the vertebral column or the spinal cord.

In addition to protecting the spinal cord and spinal nerve roots and providing support for the weight of the body, the vertebral column helps the body keep an erect posture. The resilient intervertebral disks act as shock absorbers when the load on the vertebral column is increased; they also allow the vertebrae to move without damaging each other. The disks account for about one-quarter of the length of the vertebral column and are thickest in the cervical and lumbar areas, where movement is greatest. The vertebral column is also the point of attachment for the muscles of the back.

FIGURE 7.12 THE VERTEBRAL COLUMN

[A] Anterior view. Note that the vertebrae are numbered in sequence within each region, starting from the head. [B] Right lateral view. Note the four spinal curves.

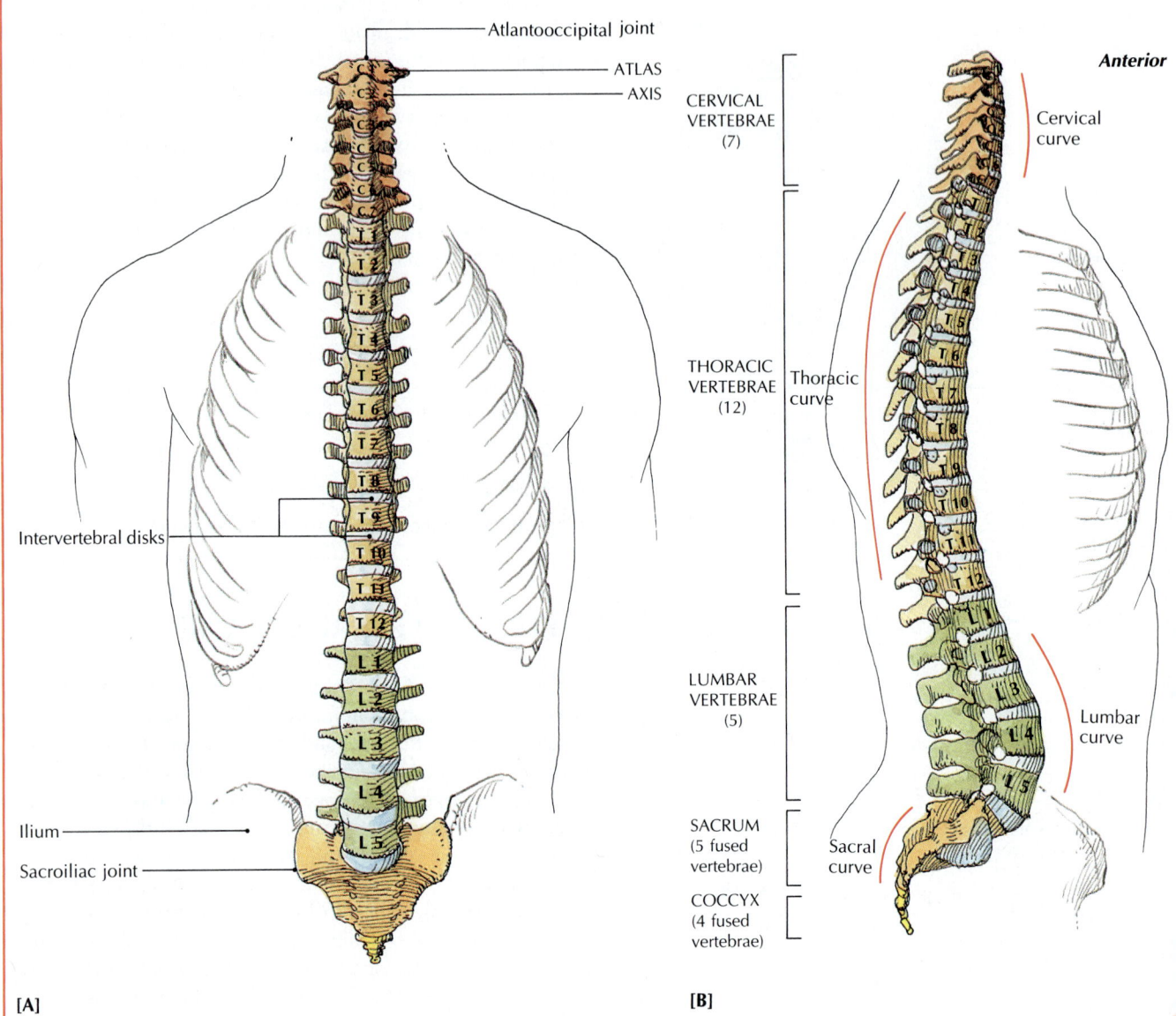

[A]

[B]

The adult vertebral column has 24 movable vertebrae, plus the fused vertebrae of the sacrum and coccyx:

Cervical (neck) vertebrae	7
Thoracic (chest) vertebrae	12
Lumbar (back) vertebrae	5
Sacral (5 fused vertebrae)	1
Coccygeal (3 to 5 fused vertebrae)	1
Total	26

In a child there are 33 separate vertebrae; the 9 in the sacrum and coccyx are not yet fused.

Curvatures of the Vertebral Column

When viewed from the side, the adult vertebral column exhibits four curves: (1) a forward cervical curve, (2) a backward thoracic curve, (3) a forward lumbar curve, and (4) a backward sacral curve [FIGURE 7.12B]. The thoracic and sacral curves are known as *primary curves* because they are present in the fetus. The cervical and lumbar curves are *secondary* because they do not appear until

BONES OF THE VERTEBRAL COLUMN (26 BONES)

Bones	Number of bones*	Description and function
Cervical vertebrae C1–C7 [FIGURE 7.14]	7	First (atlas), second (axis), and seventh vertebrae are modified; third through sixth are typical; all contain transverse foramina. Atlas supports head, permits "yes" motion of head at joint between skull and atlas; axis permits "no" motion at joint between axis and atlas.
Thoracic vertebrae T1–T12 [FIGURE 7.15]	12	Bodies and transverse processes have facets that articulate with ribs; laminae are short, thick, and broad. Articulate with ribs; allow some movement of spine in thoracic area.
Lumbar vertebrae L1–L5 [FIGURE 7.16]	5	Largest, strongest vertebrae; adapted for attachment of back muscles. Support back muscles; allow forward and backward bending of spine.
Sacrum (5 fused bones) [FIGURE 7.17]	1	Wedge-shaped, made up of five fused bodies united by four intervertebral disks. Support vertebral column; give strength and stability to pelvis.
Coccyx (3 to 5 fused bones) [FIGURE 7.17]	1	Triangular tailbone, united with sacrum by intervertebral disk. Vestige of an embryonic tail.
	26	

*In a child there are 33 separate vertebrae, the 9 in the sacrum and coccyx not yet being fused.

after birth. The cervical curve appears about 3 months after birth, when the infant begins to hold up its head, and continues to increase until about 9 months, when the child can sit upright. The lumbar curve appears at 12 to 18 months, when the child begins to walk.

The curves of the vertebral column provide the spring and resiliency necessary to cushion such ordinary actions as walking, and they are critical for maintaining a balanced center of gravity in the body.

The lumbar curve tends to be more pronounced in women than in men. During the later stages of pregnancy some women tend to increase the degree of the lumbar curve in an effort to maintain a balanced center of gravity. Backaches sometimes occur because of this exaggerated curve and the increased pressure on the posterior lumbar region. (See When Things Go Wrong at the end of this chapter for a further discussion of abnormal curvatures of the spine.)

A Typical Vertebra

Vertebrae are irregularly shaped bones, but the bones of each region are similar in shape. Except for the first and second cervical vertebrae, the sacrum, and the coccyx, they all have a similar structure. A "typical" vertebra consists of a *body*, a *vertebral (neural) arch*, and several *processes*, or projections [FIGURE 7.13]. The body supports weight, the arch supplies protection for the spinal cord, and the processes allow movement of the vertebral column.

Vertebral body The disk-shaped *vertebral body* is located anteriorly in each vertebra. The bodies of vertebrae from the third cervical to the first sacral become progressively larger in response to bearing increasing weight [FIGURE 7.12]. The upper and lower ends of each body are slightly larger than the middle. These roughened ends articulate with the fibrocartilaginous intervertebral disk between the bodies of the vertebrae above the sacrum. The anterior edge of the body contains tiny holes through which blood vessels pass.

Vertebral (neural) arch The *vertebral arch* is posterior to the vertebral body. Each arch has two thick *pedicles* (PED-uh-kuhlz; L. little feet), which form the lateral walls, and two *laminae* (LAMM-uh-nee; sing. *lamina*, LAMM-uh-nuh; L. thin plates), which form the posterior walls of the arch [FIGURE 7.13A]. The arch and the body together create the *vertebral foramen.* The sequence of all the vertebral foramina forms the *vertebral* (spinal) *canal,* which encloses the spinal cord and its surrounding meninges, nerve roots, and blood vessels [FIGURE 7.13]. The vertebral arch protects the spinal cord in the same way that the cranium protects the brain.

Each pedicle contains one vertebral notch on its inferior border and one vertebral notch on its superior border [FIGURE 7.13C]. These notches are arranged so that together they form an *intervertebral foramen,* through which the spinal nerves and their accompanying blood vessels pass.

Vertebral processes Seven *vertebral processes* extend from the lamina of a vertebra [FIGURE 7.13]. Three of these processes—the *spinous process* and the paired *transverse processes*—are the attachment sites for vertebral muscles. These processes act like levers, helping the attached muscles and ligaments move the vertebrae.

The other four processes are two superior and two inferior *articular processes.* All four processes join directly to other bones. The superior articular processes of one vertebra articulate with the inferior articular process of the vertebra above [FIGURE 7.13C]. The articular processes prevent the vertebrae from slipping forward and restrict movement between two vertebrae. Movement between two vertebrae occurs at the intervertebral disk and at the paired joints between the articular processes.

Cervical Vertebrae

The *cervical vertebrae* (L. *cervix*, neck) are the seven (C1 to C7) small neck vertebrae between the skull and the thoracic vertebrae [FIGURE 7.14; see also FIGURE 7.12B]. They support the head and enable it to move up, down, and sideways. Each of these vertebrae has a pair of openings called the *transverse foramina,* which are found only in these vertebrae. In all but the seventh, each foramen is large enough for the passage of a vertebral artery. The third to sixth cervical vertebrae are similar in shape, with a small, broad body and short, forked (bifid) spines. Their articular facets are always positioned in the same way. However, the first, second, and seventh cervical vertebrae are not typical.

First cervical vertebra (atlas) The first cervical vertebra is called the *atlas* because it supports the head, as the Greek god Atlas supported the heavens (not the world) on his shoulders. It is ringlike, with a short anterior arch and a long posterior arch, with large lateral masses on each side [FIGURE 7.14A]. Although the atlas is the widest of the cervical vertebrae, it has no spine and no body; its body is actually the dens of the axis. It articulates above with the occipital condyles and below with the second cervical vertebra (axis). This *atlantooccipital joint* makes it possible to make the nodding "yes" motion with the head.

Second cervical vertebra (axis) The second cervical vertebra is known as the *axis* because it forms a pivot point for the atlas to move the skull in a twisting "no" motion. The pivot is formed by a peglike protrusion from the body called the *dens* (Gr. tooth), which extends through the opening in the atlas; the dens is actually the body of the atlas [FIGURE 7.14B and D]. The dens is held in position against the anterior arch and displaced away from the spinal cord by a strong transverse ligament of the atlas.

Quick death from hanging occurs when the transverse ligament snaps and the dens crushes the lower medulla oblongata and the adjacent spinal cord. A small tear of this ligament may produce *whiplash* symptoms when the head is snapped backward during a violent automobile accident.

Seventh cervical vertebra The seventh cervical vertebra [FIGURE 7.14D] has an exceptionally long unforked spinous process with a tubercle at its tip that can be felt, and usually seen, through the skin. This vertebra is known as the *vertebra prominens.*

Thoracic Vertebrae

All 12 *thoracic vertebrae* (Gr. *thorax*, breastplate) increase in size as they progress downward. The first four (T1 to T4) are similar to the cervical vertebrae, and the

FIGURE 7.13 TYPICAL VERTEBRAE

[A] Drawing of a typical vertebra, illustrating its parts. The vertebral arch (dashed line) is made up of two pedicles and two laminae. Superior view. [B] Photo of a typical vertebra (L5), superior view; actual size. [C] Right lateral view. The sequence of vertebral foramina forms the vertebral canal, through which the spinal cord passes. The cauda equina is located caudal to the spinal cord. All three figures are shown approximately actual size.

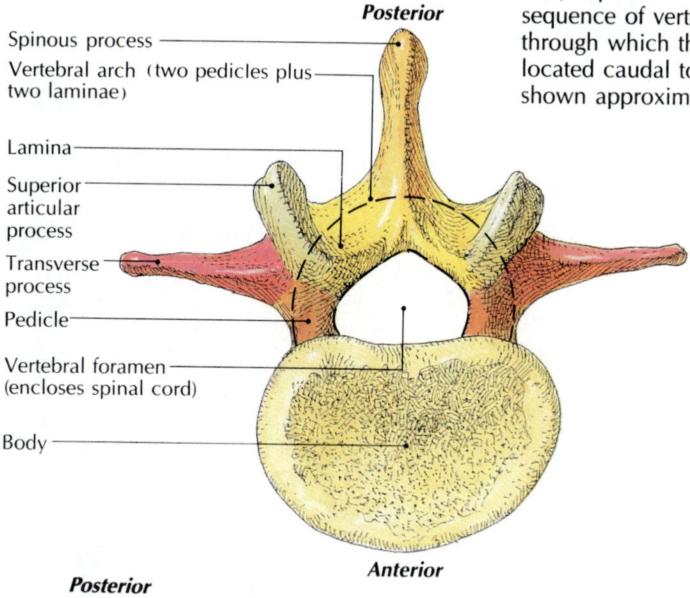

Posterior

Spinous process
Vertebral arch (two pedicles plus two laminae)
Lamina
Superior articular process
Transverse process
Pedicle
Vertebral foramen (encloses spinal cord)
Body

Anterior

[A]

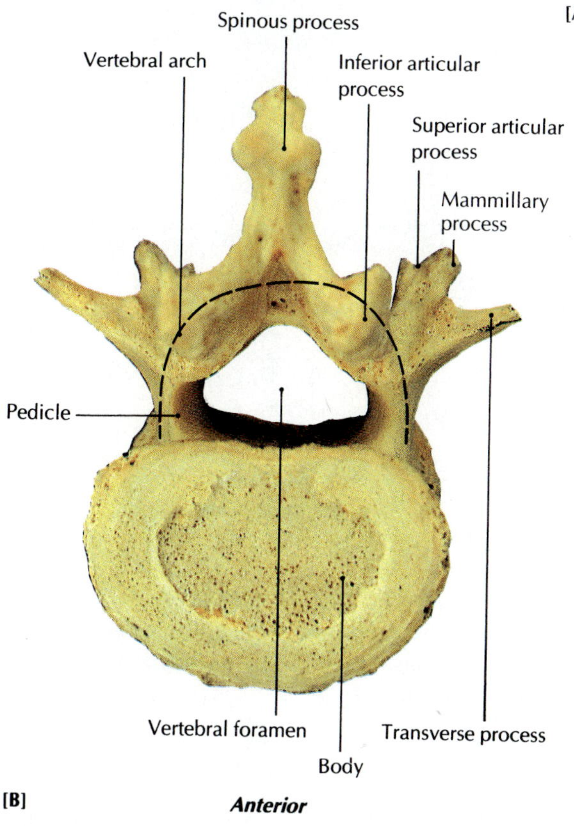

Posterior
Spinous process
Vertebral arch
Inferior articular process
Superior articular process
Mammillary process
Pedicle
Vertebral foramen
Transverse process
Body

Anterior

[B]

Cauda equina of spinal cord
Superior vertebral notch
Superior articular process
Pedicle
Transverse process
L 1
Body
Spinous process
Lamina
Intervertebral foramen
L 2

Posterior
Anterior

L 3

Inferior vertebral notch
Inferior articular process

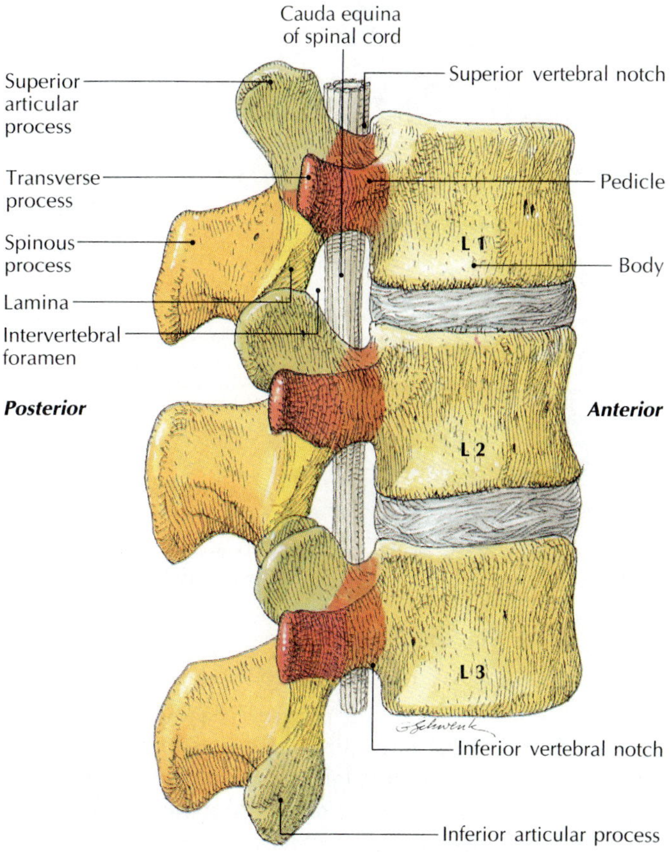

[C]

FIGURE 7.14 CERVICAL VERTEBRAE

[A] The atlas, or first cervical vertebra; superior view. [B] The axis, or second cervical vertebra; anterior view. [C] A typical cervical vertebra; superior view. Note that the vertebral artery passes through each transverse foramen, and the spinal cord through each vertebral foramen. [D] The articulation of the dens of the axis with the atlas, and their relationship to the other cervical vertebrae; posterior view.

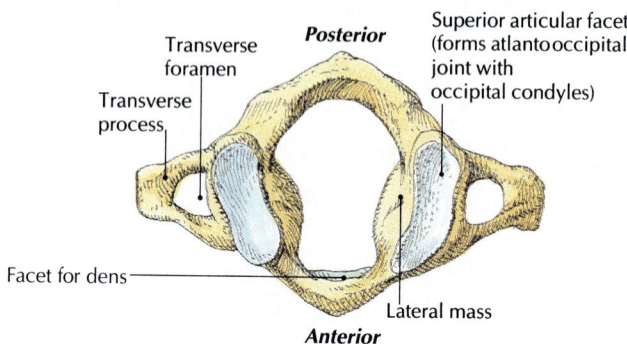

[A] ATLAS (FIRST CERVICAL VERTEBRA)

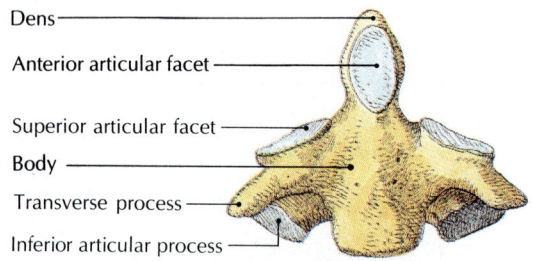

[B] AXIS (SECOND CERVICAL VERTEBRA)

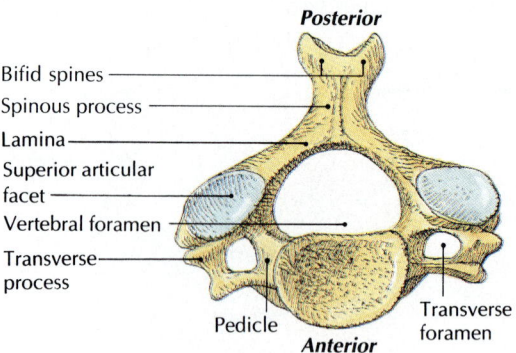

[C] TYPICAL CERVICAL VERTEBRA

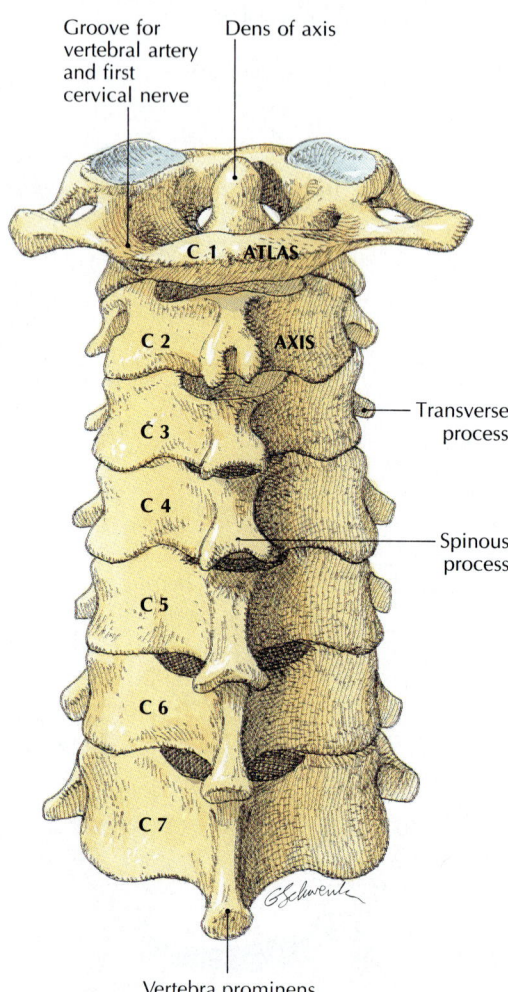

[D] CERVICAL VERTEBRAE

Because there are 12 thoracic vertebrae and 12 corresponding flexible intervertebral disks, the total mobility of this region is considerable, even with the presence of the ribs and sternum.

Lumbar Vertebrae

The five *lumbar vertebrae* (L1 to L5) (L. *lumbus*, loin) are the largest and strongest vertebrae. They are situated in the "small of the back," between the thorax and the pelvis [see FIGURES 7.1 and 7.12]. Their large kidney-shaped bodies have short, blunt, four-sided spinous processes, which are adapted for the attachment of the lower back muscles [FIGURE 7.16]. The arrangement of the fac-

last four (T9 to T12) have certain features in common with the lumbar vertebrae. Only the middle four (T5 to T8) are considered typical. Their bodies viewed from above are heart-shaped, and they have circular vertebral foramina [FIGURE 7.15B]. Each thoracic vertebra has three articular (*costal*) facets on each side that provide attachments for the ribs [see FIGURE 7.20].

FIGURE 7.15 THORACIC VERTEBRAE

[A] Right lateral view. **[B]** Superior view. The intervertebral notches between adjacent vertebrae form the intervertebral foramen, through which a spinal nerve passes.

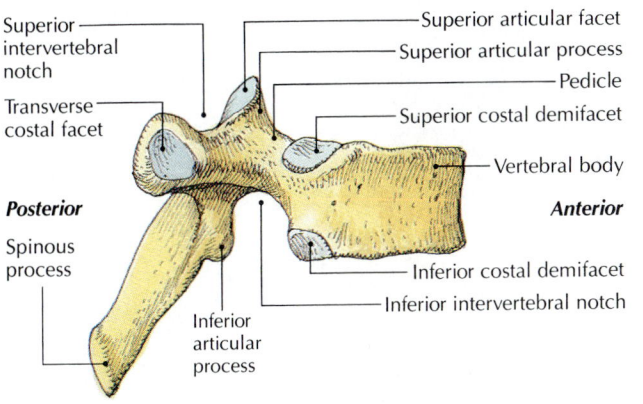

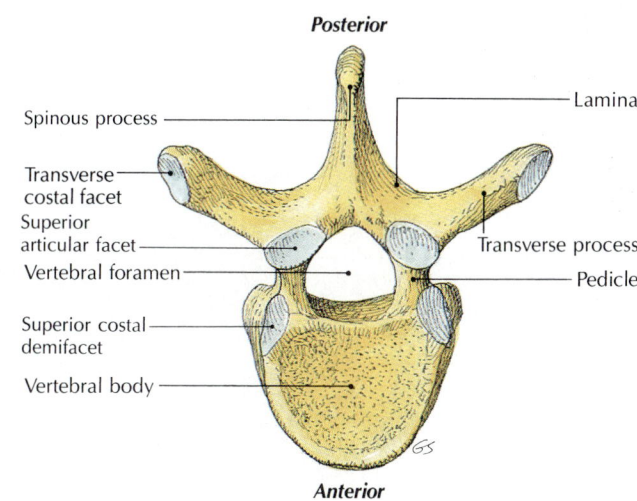

[A]

[B]

FIGURE 7.16 LUMBAR VERTEBRAE

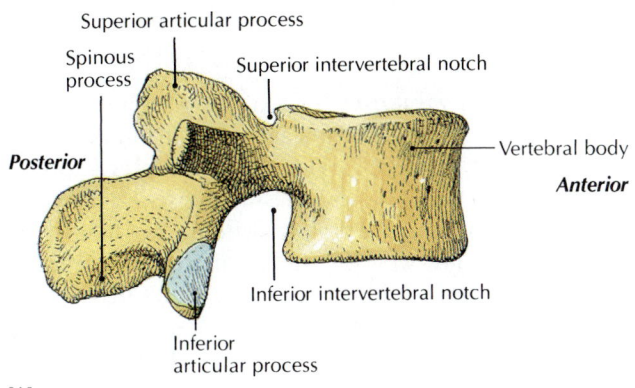

[A]

[A] Right lateral view. **[B]** Superior view. **[C]** X-ray, anterior view.

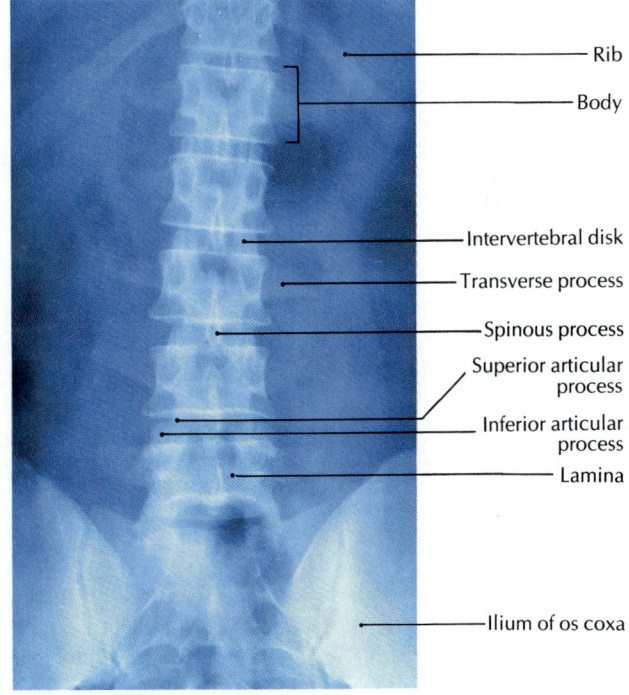

[C]

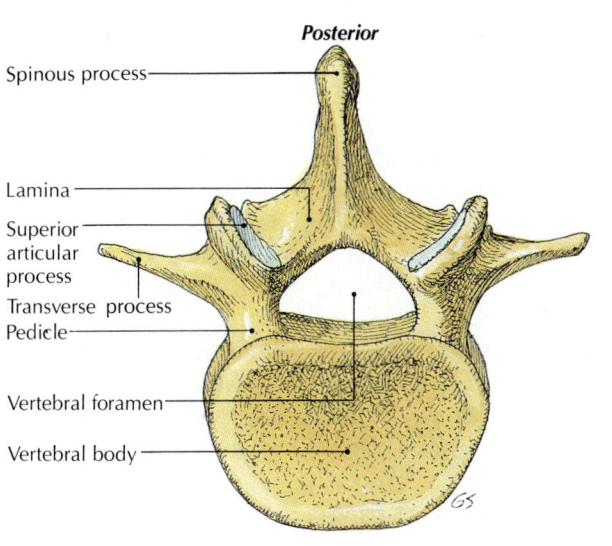

[B]

FIGURE 7.17 SACRUM AND COCCYX

[A] Ventral view. [B] Dorsal view.

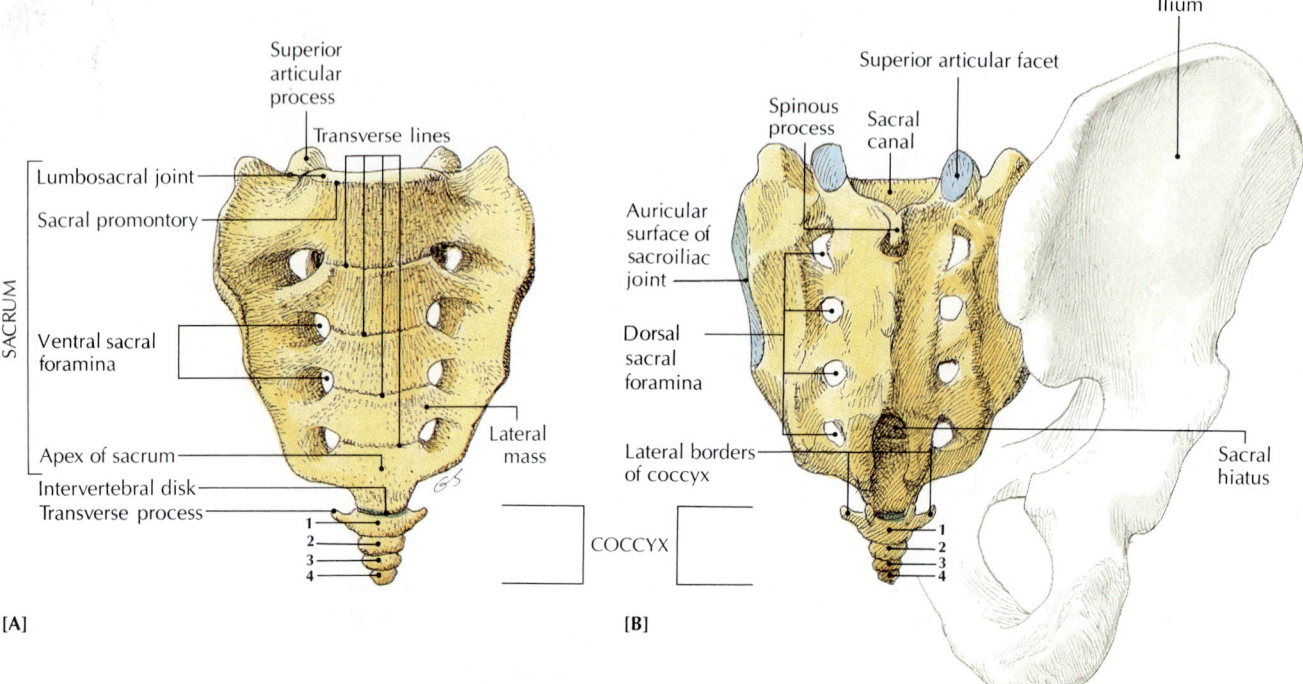

ets on the articular processes of each vertebra maximizes forward and backward bending, but lateral bending is limited, and rotation is practically eliminated. The lumbar vertebrae have no transverse foramina. The transverse processes are long and slender, and the vertebral foramina are oval or triangular.

In most people, the spinal cord ends between the first and second lumbar vertebrae. Therefore, a *lumbar puncture* (commonly called a *spinal tap*) can usually be made safely just below this point to obtain *cerebrospinal fluid*. (A lumbar puncture is usually made in the midline between the third and fourth or fourth and fifth vertebrae.) This clear fluid, which bathes the brain and spinal cord and acts as a shock absorber for the central nervous system, is useful in diagnosing certain diseases.

Sacrum and Coccyx

The wedge-shaped *sacrum* (SAY-kruhm) not only gives support to the vertebral column, but also provides strength and stability to the pelvis. It is composed of five vertebral bodies fused in an adult into one bone by four ossified intervertebral disks [FIGURE 7.17]. The sacrum is curved, forming a concave surface anteriorly and a convex surface posteriorly. The four *transverse lines* on the otherwise smooth concave anterior surface indicate the fusion sites of the originally separate vertebrae. At the lateral ends of these lines are two parallel rows of four

ventral sacral foramina [FIGURE 7.17A]. On the dorsal convex surface of the sacrum are four pairs of *dorsal sacral foramina* [FIGURE 7.17B]. The areas lateral to these foramina are known as the *lateral masses*.

The sacrum articulates above with the last lumbar vertebra (L5), below with the coccyx, and laterally from the *auricular surfaces* with the two iliac bones of the hip to form the *sacroiliac joints*.

The fused laminae and short spines form the roof of the *sacral canal*. Because the laminae of the lower sacral vertebrae (S4 and S5) are often absent or do not meet in the midline, a gap called the *sacral hiatus* [FIGURE 7.17B] is present at these levels. Caudal anesthesia, which spreads upward and acts directly on the spinal nerves, can be injected into the sacral hiatus.

The projecting anterior edge of the first sacral vertebra is called the *sacral promontory* [FIGURE 7.17A]. It is used as a landmark for making pelvic measurements. Just above the promontory is the *lumbosacral joint* between the fifth lumbar vertebra and the sacrum. The sacrum is tilted at this point, so that the articulation forms the *lumbosacral angle*.

The *coccyx* (KOCK-six), or tailbone, consists of three to five (usually four) fused vertebrae, the vestiges of an embryonic tail that usually disappears about the eighth fetal week.* Triangular in shape, it has an *apex, base, pelvic* and

* *Coccyx* comes from the Greek word for a cuckoo. The coccyx is so named because it was thought to resemble a cuckoo's beak.

dorsal surfaces, and two *lateral borders* [FIGURE 7.17B]. Its base articulates with the lower end of the sacrum by means of a fibrocartilaginous intervertebral disk.

THE THORAX

The *thorax* (Gr. breastplate) is the chest, which is part of the axial skeleton [FIGURES 7.18, 7.19]. The thoracic skeleton is formed posteriorly by the bodies and intervertebral disks of 12 thoracic vertebrae and anteriorly by 12 pairs of ribs, 12 costal cartilages,* and the sternum. The

*Although ribs 11 and 12 are not attached to the sternum, they are tipped anteriorly with costal cartilage.

thorax is fairly narrow at the top and broad below, and it is wider than it is deep.

The cagelike thoracic skeleton is a good example of a functional structure. It protects the heart, the lungs, and some abdominal organs. It supports the bones of the pectoral girdle and arms. The lever arms formed by the ribs and costal cartilages provide a flexible mechanism for breathing. Also, the sternum provides a point of attachment for the ribs.

Sternum

The *sternum* (Gr. *sternon,* breast), or breastbone, is the midline bony structure of the anterior chest wall [FIGURES 7.18A and 7.20A]. It resembles a dagger in the adult. The sternum consists of a manubrium, body, and xiphoid process.

The *manubrium* (L. handle) has a pair of *clavicular notches,* which articulate with the clavicle (collarbone). This *sternoclavicular joint* is the only site of direct attachment of the pectoral girdle to the axial skeleton [FIGURE 7.18A]. At the lateral *costal notches* the sternum articulates with the costal cartilages of the first ribs and part of the second ribs. On its upper border is the *jugular* (suprasternal) *notch.* The manubrium is united with the body of the sternum at the movable *manubriosternal joint,* which acts as a hinge to allow the sternum to move forward during inhalation [FIGURE 7.18A]. This joint may

FIGURE 7.18 SKELETON OF THE THORAX

[A] Anterior view. **[B]** Posterior view. Note the sternum in the anterior view. Ribs and vertebrae are numbered.

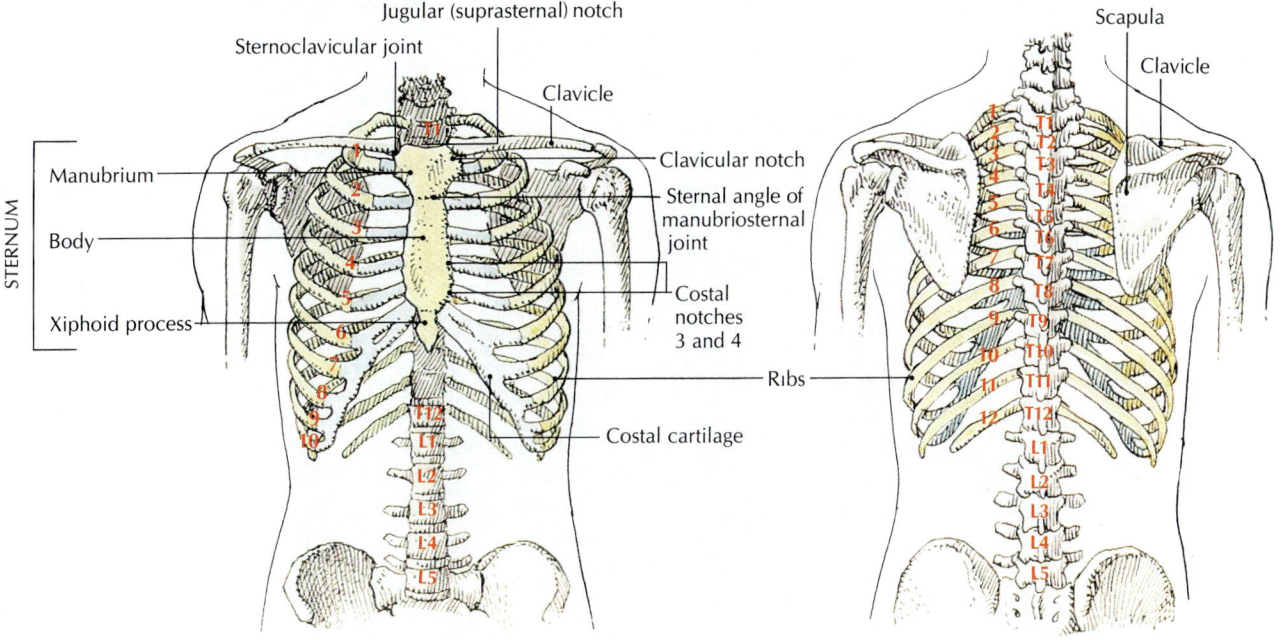

[A]   [B]

FIGURE 7.19 SURFACE ANATOMY OF THORAX AND BACK

[A] The bony *thorax* is composed of 12 thoracic vertebrae, 12 pairs of ribs, and the sternum (breastbone). A rib and its cartilage make up a *costa* (L. rib).

[B] The *back* is the posterior surface of the trunk. It extends from the top of the skull to the inferior tip of the coccyx (tailbone) and includes the posterior aspect of the neck.

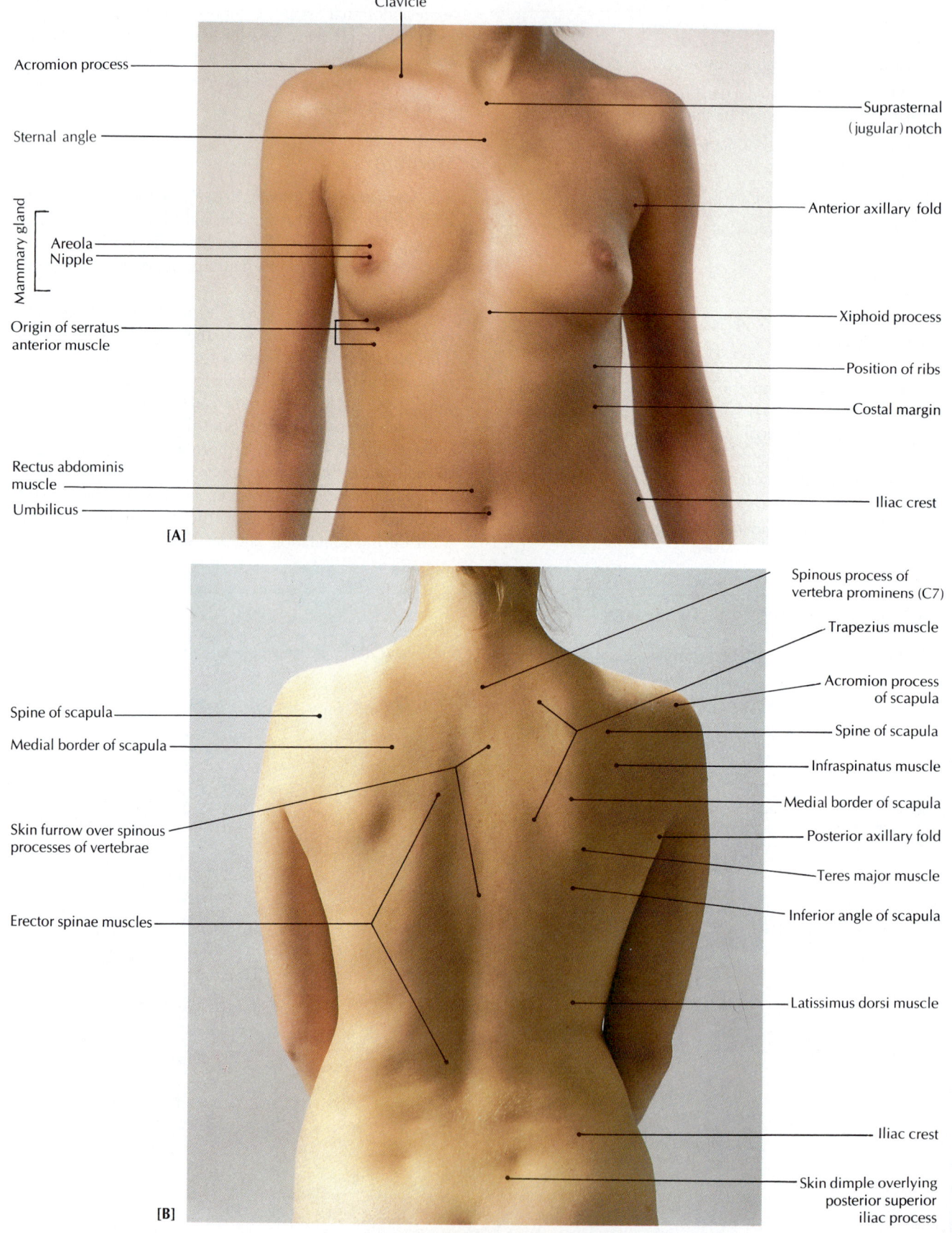

Clavicle

Acromion process

Sternal angle

Mammary gland
Areola
Nipple

Origin of serratus anterior muscle

Rectus abdominis muscle

Umbilicus

Suprasternal (jugular) notch

Anterior axillary fold

Xiphoid process

Position of ribs

Costal margin

Iliac crest

[A]

Spine of scapula

Medial border of scapula

Skin furrow over spinous processes of vertebrae

Erector spinae muscles

Spinous process of vertebra prominens (C7)

Trapezius muscle

Acromion process of scapula

Spine of scapula

Infraspinatus muscle

Medial border of scapula

Posterior axillary fold

Teres major muscle

Inferior angle of scapula

Latissimus dorsi muscle

Iliac crest

Skin dimple overlying posterior superior iliac process

[B]

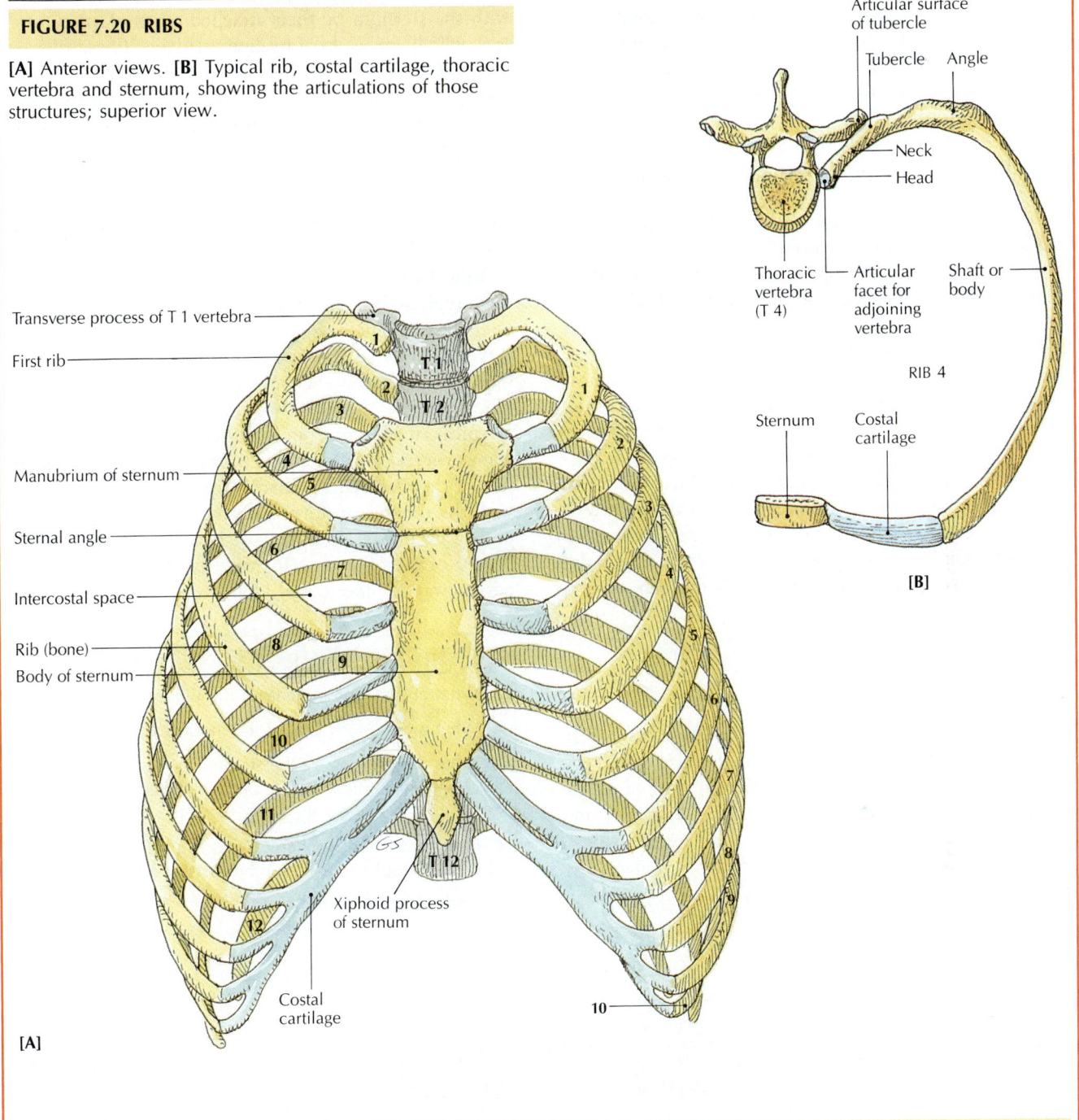

FIGURE 7.20 RIBS

[A] Anterior views. [B] Typical rib, costal cartilage, thoracic vertebra and sternum, showing the articulations of those structures; superior view.

Articular surface of tubercle

Tubercle   Angle

Neck

Head

Thoracic vertebra (T 4)

Articular facet for adjoining vertebra

Shaft or body

RIB 4

Sternum   Costal cartilage

[B]

Transverse process of T 1 vertebra

First rib

Manubrium of sternum

Sternal angle

Intercostal space

Rib (bone)

Body of sternum

Xiphoid process of sternum

Costal cartilage

[A]

Bone	Description and functions
Ribs (12 pairs)	Long, curved, varying in length and width; ribs 1–7 (true vertebrosternal ribs) attach directly to sternum; 8–12 (false ribs) do not attach directly to sternum; 8–10 (vertebrochondral ribs) attach to rib 7 cartilage; 11–12 attach only to vertebrae. With sternum and thoracic vertebrae, ribs form strong but lightweight cage to protect heart, lungs, other organs. With costal cartilages, ribs provide flexible mechanism for breathing and support bones of upper extremities.
Sternum	Dagger-shaped, about 15 cm long; made up of manubrium, body, xiphoid process; articulates with ribs 1–7 directly, 8–10 indirectly. Provides anterior attachment site for ribs; with ribs, forms protective cage for heart, lungs, other organs; supports bones of upper extremities.

become ossified in the elderly so that movements at the joint become restricted. The manubriosternal joint forms a slight angle, called the **sternal angle,** which can be felt through the skin [FIGURE 7.20A]. Because this angle is opposite the second rib, it is a reliable starting point for counting the ribs.

The **body** of the sternum is about twice as long as the manubrium. It articulates with the second through tenth pairs of ribs.

The **xiphoid process** (ZIFF-oid; Gr. sword-shaped) is the smallest, thinnest, and most variable part of the sternum. It is cartilaginous during infancy, but it usually is almost completely ossified and joined to the body of the sternum by the fortieth year. The xiphoid process does not articulate with any ribs or costal cartilages; however, several ligaments and muscles (including abdominal muscles) are attached to it. The joint between the body of the sternum and the xiphoid process is the *xiphisternal joint.* This joint is fragile, and if it is struck during cardiopulmonary resuscitation, the xiphoid process may break off.

Because the sternum is relatively accessible, a physician may insert a needle into its marrow cavity to obtain a specimen of the red blood cells that are developing in the bone marrow. Such a procedure is called a *sternal puncture.*

Ribs

The **ribs** are curved, slightly twisted strips of bone that form the widest and major part of the thoracic cage [FIGURE 7.20]. There are usually 12 pairs of ribs, all of which articulate posteriorly with the vertebral column. The ribs increase in length from rib 1 to rib 7 and decrease from 8 to 12. The space between the ribs is the *intercostal space,* which contains the intercostal muscles.

The upper seven pairs (ribs 1 to 7) connect directly with the sternum by their attached strips of *costal cartilage,* which are made of hyaline cartilage. These ribs are called **true ribs** because they attach to both the vertebrae and the sternum. The lower three pairs of ribs (8 to 10) are known as **false ribs** because they are attached to the sternum only indirectly. Of the false ribs, 8 to 10 have costal cartilages that connect with each other and also with the cartilage of rib 7. Ribs 11 and 12 are called **floating ribs** because they attach only to the vertebral column [FIGURE 7.20A].

Ribs 3 to 9 are known as **typical ribs** [FIGURE 7.20B]. Each typical rib has a wedge-shaped *head* on the end next to the spine. The head articulates with an intervertebral disk and body of an adjacent vertebra. The short, flattened *neck* is located between the head and the tubercle. The *tubercle* is located between the neck and the body of a typical rib. It forms an articular surface with a transverse process on a vertebra. The *shaft,* or *body,* is the main part of the rib. It curves sharply forward after its junction with the tubercle to form a distinct *angle;* it then arches downward until it joins the costal cartilage. On the lower border of the rib is a *costal groove* that forms a protective passageway for intercostal blood vessels and nerves.

ASK YOURSELF

1 What are the components of the thorax?

2 What are the functions of the thoracic cage?

3 What are the parts of the sternum? Where are the axial and appendicular skeletons connected to each other?

4 How many ribs are there? What is a true rib? A false rib? A floating rib? What are the parts of a typical rib?

WHEN THINGS GO WRONG

Herniated Disk

Herniated disk (also called ruptured or slipped disk) occurs when the soft, pulpy center (*nucleus pulposus*) of an intervertebral disk protrudes through a weakened or torn surrounding outer ring (*annulus fibrosus*) on the posteriolateral side of the disk (see drawings). The nucleus pulposus pushes against a spinal nerve or occasionally on the spinal cord itself. This produces a continuous pressure on the spinal cord, which may cause permanent injury. Actually, nothing "slips"; the nucleus pulposus "pushes out." Herniated disks occur most often in adult males. They may be caused by a straining injury or by degeneration of the intervertebral joint. The lumbar region is usually affected, but herniation may occur anywhere along the spine. Sharp pain usually accompanies a herniated disk, and because roots of spinal nerves may be compressed, the pain may radiate beyond the primary low back area to the buttocks, legs, and feet.

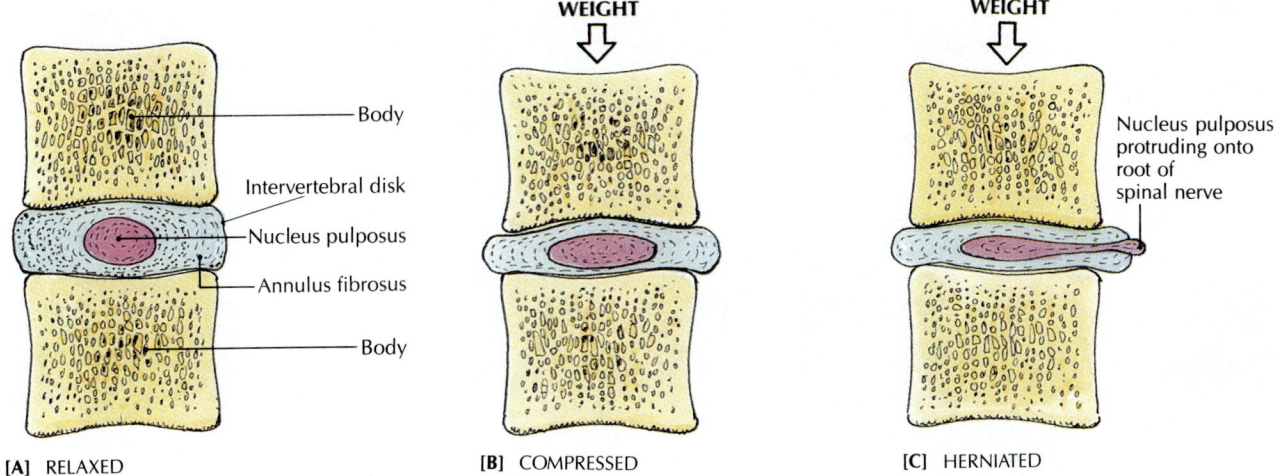

[A] RELAXED   **[B]** COMPRESSED   **[C]** HERNIATED

Labels: Body, Intervertebral disk, Nucleus pulposus, Annulus fibrosus, Body. WEIGHT. Nucleus pulposus protruding onto root of spinal nerve.

Because the posterior longitudinal ligament, which is on the posterior aspect of the vertebral bodies and intervertebral disks, is relatively narrow, herniated disks pass on only one side of the ligament. As a result, a herniated disk puts pressure on the nerve roots on one side only, producing sciatic nerve pain on only one side of the body, usually in a lower limb.

Treatment usually consists of bed rest, sometimes including traction. Heat applications, a regulated exercise program, and muscle-relaxing or pain-killing drugs may be prescribed. If such traditional treatment is ineffective, surgery is done to remove the protruding nucleus pulposus. To gain access to the nucleus pulposus, a portion of the vertebral arch (lamina) is removed. Such an operation is called a *laminectomy.*

If a spinal fusion is required, bone chips from the ilium are placed over the laminae. The bone that develops from the chips fuses to form a splint.

Cleft Palate
Cleft palate is the common name for a defect that occurs when the structures that form the palate do not fuse before birth. As a result, there is a gap in the midline of the roof of the mouth. Such a gap creates a continuity between the oral and nasal cavities. When a gap is forward, on the upper lip, the lip is separated and a "hare lip" ("cleft lip") results. A hare lip never appears at the midline and may be paired. An infant with a cleft palate may have difficulty suckling. In most cases, the defect can be repaired, at least partially, by surgery.

Microcephalus and Hydrocephalus
Sometimes the calvarial bones, along with the fontanels, close earlier or later than expected. If they fuse too early, brain growth may be retarded by the excessive pressure.

This condition is called *microcephalus* (Gr. *mikros,* small + *kephale,* head). It can be alleviated by removing the bone and widening the sutures. A contrasting problem is *hydrocephalus* (Gr. *hudor,* water + head), commonly called "water on the brain." It is usually a congenital condition in which an abnormal amount of cerebrospinal fluid accumulates around the brain and within the brain ventricles (cavities), causing an enlargement of the skull and putting pressure on the brain. Fusion of the skull sutures is delayed by the increased volume of the cranial cavity.

Physicians attend to the problem by inserting replaceable plastic tubes with pressure valves in a ventricle of the brain; the tubes drain excess fluid into a vein or body cavity. The valve prevents excessive rises in pressure and allows sufficient fluid to be retained. Even with drainage procedures, mental retardation and loss of vision may occur. Without drainage, the condition is usually fatal.

Spina Bifida
Spina bifida (SPY-nuh BIFF-ih-duh; L. *bifidus,* split into two parts), or cleft spine, affects approximately 1 of every 1000 children. In its severe form, it is the most common crippler of newborns. Spina bifida is a condition in which the neural arches of one or more vertebrae do not close completely during fetal development. In serious cases, the spinal cord and nerves in the area of the defective vertebrae form a fluid-filled sac, called a myelomeningocele, which protrudes through the skin. Because the myelomeningocele is covered by only a thin membrane, the protruding spinal cord and spinal nerves are easily damaged or infected.

Abnormal Spinal Curvatures
Three abnormal spinal curvatures are kyphosis, lordosis,

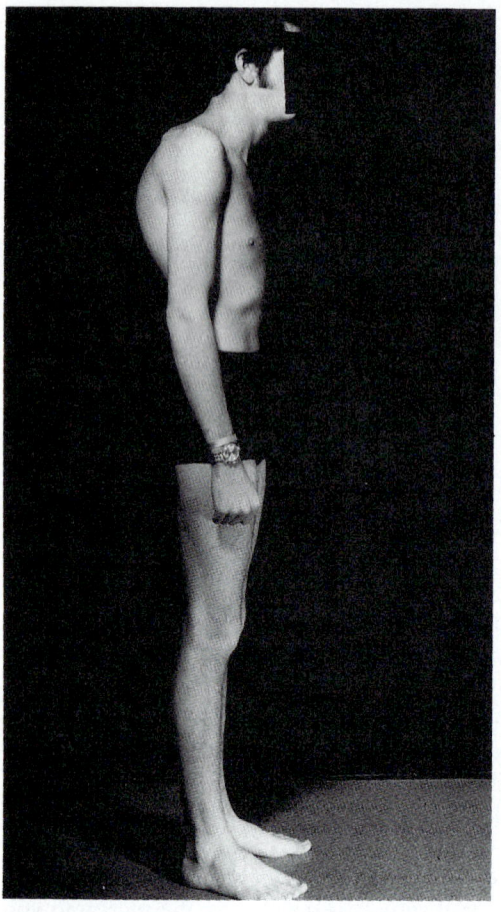

Kyphosis

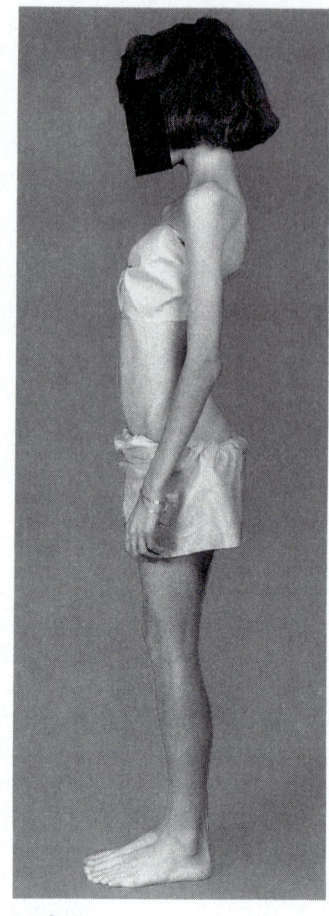

Lordosis

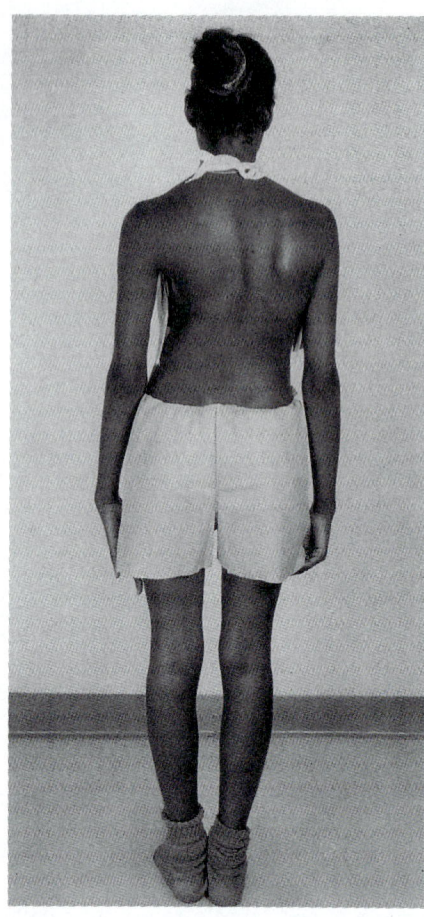

Scoliosis

and scoliosis. ***Kyphosis*** (Gr. hunchbacked) is a condition where the spine curves backward abnormally, usually at the thoracic level. A characteristic "hunchbacked" or "roundbacked" appearance results. Adolescent kyphosis is the most common form. It generally results from infection or other disturbances of the vertebral epiphysis during the active growth period. Adult kyphosis ("hunchback") is generally caused by a degeneration of the intervertebral disks, resulting in collapse of the vertebrae, but many other factors, such as poor posture and tuberculosis of the spine, may be responsible.

Lordosis (Gr. bent backward), also known as "sway-back," is an exaggerated forward curvature of the spine in the lumbar area. Among the causes of lordosis are the great muscular strain of advanced pregnancy, an extreme "pot-belly" or general obesity that places abnormal strain on the vertebral column, tuberculosis of the spine, rickets, and poor posture.

The most common spinal curvature is ***scoliosis*** (Gr. crookedness), an abnormal lateral curvature of the spine in the thoracic, lumbar, or thoracolumbar portion of the vertebral column. It is interesting to note that curves in the thoracic area are usually convex to the right and that lumbar deformities are usually convex to the left.

CHAPTER SUMMARY

The skeleton supports the body. It also protects the inner organs and passageways and acts as a system of levers that allows us to move.

General Features and Surface Markings of Bones (p. 180)

The markings on the surface of a bone are related to the bone's function. Some important features of bones are **processes,** or outgrowths on bones, such as a condyle, crest, or trochanter; **openings** to bones, such as a foramen or meatus; and **depressions** on bones, such as a fossa, groove, or notch.

Divisions of the Skeleton (p. 180)

1 The skeleton (206 bones) is divided into two major portions: the axial skeleton (80 bones) and the appendicular skeleton (126 bones). They are joined together at the pectoral girdle and pelvic girdle to form the overall skeleton.

2 The **axial skeleton** forms the longitudinal axis of the body. It is made up of the skull, vertebral column, sternum, and ribs.

3 The **appendicular skeleton** is composed of the upper and lower extremities, which include the pectoral and pelvic girdles.

4 Each **upper extremity** consists of the pectoral girdle, upper arm bone, two forearm bones, and the wrist and hand bones. Each **lower extremity** consists of the pelvic girdle, upper leg bone, two lower leg bones, and the ankle and foot bones.

The Skull (p. 184)

1 The **skull** is usually defined as the skeleton of the head, with or without the mandible. The skull can be divided into the **cranial skull** and the **facial skull.** The skull *protects* many structures, including the brain and eyes; *provides points of attachment* for muscles involved in eye

movements, chewing, swallowing, and other movements; and *supports* various structures such as the mouth, pharynx, and larynx.

2 Except for the mandible (and the ear ossicles and hyoid), the skull bones are joined together by **sutures,** seamlike joints that make the bones of an adult skull immovable. The four major skull sutures are the *coronal, lambdoidal, sagittal,* and *squamous* sutures.

3 The membrane-covered spaces between incompletely ossified sutures are the four **fontanels:** *anterior, posterior, anterolateral,* and *posteriolateral.*

4 The eight **cranial bones** are the *frontal,* two *parietal, occipital,* two *temporal, sphenoid,* and *ethmoid.* The sutural bones are also considered part of the cranium.

5 In the interior of the ethmoid, maxillary, sphenoid, and frontal bones are four pairs of air cavities called **paranasal sinuses.**

6 The 14 **facial bones** are 2 *inferior nasal conchae, vomer,* 2 *palatines,* 2 *maxillae,* 2 *zygomatic,* 2 *nasal,* 2 *lacrimal,* and *mandible.*

The Vertebral Column (p. 201)

1 The **vertebral column,** or spine, is the skeleton of the back. It is composed of 26 separate bones called **vertebrae.**

2 The vertebral column protects the spinal cord and nerves, supports the weight of the body, and keeps the body erect. **Intervertebral disks** act as shock absorbers and protect the **vertebrae.**

3 The adult vertebral column has four *curves,* which provide spring and resiliency.

4 A "typical" vertebra consists of a body, a vertebral (neural) arch, and several processes. The arch and the body meet to form an opening called the **vertebral foramen.** The sequence of foram-

ina forms the **vertebral canal,** which encloses the spinal cord. The processes are attachment sites for muscles and ligaments.

5 The **cervical vertebrae** are the seven between the skull and the thorax. The **atlas** supports the head and permits the "yes" motion; the **axis** permits the "no" motion. The 12 **thoracic vertebrae** articulate with the ribs. The five **lumbar vertebrae** are the largest and strongest vertebrae and provide attachments for lower back muscles. The adult **sacrum,** which is composed of five fused vertebral bodies, supports both the spinal column and the pelvis. The **coccyx** consists of three to five fused vertebrae.

The Thorax (p. 209)

1 The **thorax** is formed by the bodies and intervertebral disks of 12 thoracic vertebrae posteriorly, 12 pairs of ribs, 12 costal cartilages, and the sternum anteriorly.

2 The thoracic cage protects inner organs, provides a point of attachment for some bones and muscles of the upper extremities, and provides a flexible breathing mechanism.

3 The **sternum** consists of a manubrium, body, and xiphoid process. The articulation of the manubrium with the clavicle is the upper attachment of the axial skeleton to the appendicular skeleton.

4 There are usually 12 pairs of **ribs,** all of which articulate posteriorly with the vertebral column. The *true ribs* (1 to 7) attach to the vertebrae and sternum, but the *false ribs* (8 to 10) attach directly only to the vertebral column. Ribs 11 and 12 are called *floating ribs* because they are not even indirectly attached to the sternum or ribs above.

5 A *typical rib* is composed of a head, neck, and shaft.

STUDY AIDS FOR REVIEW

KEY TERMS

MEDICAL TERMS FOR REFERENCE

ABLATION (L. *ablatus*, removed) The surgical removal of part of a structure, such as part of a bone.

BONE MARROW TEST The withdrawal of bone marrow from the medullary cavity of a bone for microscopic examination.

CERVICAL RIB An overdevelopment of the costal projection of the seventh cervical vertebra. It resembles a rib and can be a separate bone.

CRANIOTOMY The surgical cutting or removal of part of the cranium.

CREPITATION (L. *crepitare*, to crackle) The grating sound caused by the movement of fractured bones or by other bones rubbing together.

KINESIOLOGY (Gr. *kinema*, motion) The study of movement and the active and passive structures involved.

OSTECTOMY The surgical excision of a bone.

OSTEOCLASIS The surgical refracture of an improperly healed broken bone.

OSTEOPLASTY The surgical reconstruction or repair of a bone.

POTT'S DISEASE Tuberculosis of the spine; may result in a partial destruction of vertebrae and a spinal curvature.

REDUCTION The nonsurgical manipulation of fractured bones to return (reduce) them to their normal positions.

REPLANTATION The reattachment of a severed limb.
SPINAL FUSION The fusion of two or more vertebrae.
SPONDYLITIS (Gr. *spondulos*, vertebra) Inflammation of one or more vertebrae.

UNDERSTANDING THE FACTS

1 What are the two main divisions of the skeleton, and what are their functions?
2 What bones form the cranial skull and facial skull?
3 As a group, how do the mandible, the hyoid, and the ossicles of the ear differ from the other bones of the skull?
4 Identify the general regions and curves of the vertebral column.
5 What are the functions of the body, vertebral arch, and processes of a typical vertebra?
6 How do true ribs, false ribs, and floating ribs differ?

UNDERSTANDING THE CONCEPTS

1 What are the functions of the depressions and processes found on bones, and how are they formed?
2 Other than increased size, how does the skull of an adult differ from that of a child?
3 Describe how the occipital bone and the first two cervical vertebrae produce the "yes" and "no" motions of the head.
4 Contrast the articulations between vertebral bodies with those of vertebral processes, and explain how both contribute to the physical protection and functional support of the spinal cord.

SELF-QUIZ
Multiple-Choice Questions
7.1 Unlike other vertebrae, thoracic vertebrae have
a facets for rib attachment
b pedicles
c laminae
d transverse processes
e a and c
7.2 A wide and prominent ridge or border often found on the surface of a long bone is a
a condyle c head e trochanter
b crest d line
7.3 A large, roughly rounded process found only on the femur is the
a crest c spine e ridge
b trochanter d tuberosity
7.4 The hard palate is made up of the
a lacrimal bone d maxillae and palatine bones
b ethmoid bone e sphenoid bones
c zygomatic bones

7.5 An infant's skull contains four _____, whereas an adult's skull exhibits _____ in the same regions.
  a Wormian bones, sutural bones
  b sutures, fontanels
  c air cells, conchae
  d fontanels, sutures
  e soft palatal bones, hard palatal bones
7.6 The major bone at the posterior part of the base of the skull is the
  a sphenoid   **d** lacrimal
  b occipital   **e** zygomatic
  c temporal
7.7 Which of the following is *not* a component of the axial skeleton?
  a sacrum   **d** vertebra
  b hipbone   **e** mandible
  c sternum
7.8 Which of the following does *not* articulate with one or more ribs?
  a manubrium   **d** body of thoracic vertebra
  b body of sternum   **e** transverse process of
  c xiphoid process   thoracic vertebra
7.9 Which of the following bones is least involved in protecting the brain?
  a frontal   **d** parietal
  b temporal   **e** occipital
  c mandible
7.10 Which of the following skull bones is *not* paired?
  a parietal   **d** zygomatic
  b nasal   **e** temporal
  c frontal

Matching
  a temporal bone   **c** maxilla   **e** ethmoid
  b occipital bone **d** sphenoid

7.11 _____ contains many tiny openings through which pass branches of the cranial nerve involved in olfaction (smell)
7.12 _____ forms the foramen magnum
7.13 _____ ear ossicles are contained within a portion of this bone

7.14 _____ a tooth-bearing bone that also contributes to the hard palate
7.15 _____ houses and protects the pituitary gland

Completion Exercises
7.16 Wormian bones are also called _____ bones.
7.17 In the interior of the ethmoid, maxillary, sphenoid, and frontal bones are four pairs of air cavities called _____.
7.18 The bone that is shaped like a plowshare and that forms part of the nasal septum is the _____.
7.19 The upper jaw consists of two _____.
7.20 The skeleton of the lower jaw is the _____.
7.21 The three ear ossicles are the _____, _____, and _____.
7.22 The first cervical vertebra is the _____.
7.23 The _____ consists of three to five fused bones at the end of the vertebral column.
7.24 A human has _____ pairs of ribs.

Problem-Solving/Critical Thinking Questions
7.25 Fill in the blanks: Mr. Ramey and Mr. Deaton were involved in a serious car accident on their way home from a fishing trip. Mr. Ramey had pronounced swelling of the upper right side of his head. X-rays showed a fracture of the _____ bone. X-rays also showed Mr. Deaton fractured his "tailbone." Correctly, this bone is called the _____.
7.26 The hyoid bone is _____ to the thyroid cartilage.
  a distal   **d** posterior
  b superior   **e** inferior
  c anterior
7.27 The most prominent bony landmark on the posterior aspect of the neck is the _____.
  a spinous process of the seventh cervical vertebra
  b transverse process of the sixth cervical vertebra
  c the hyoid bone
  d the transverse process of the atlas
  e the spinous process of the axis

A SECOND LOOK
Identify the bone in the following drawing.

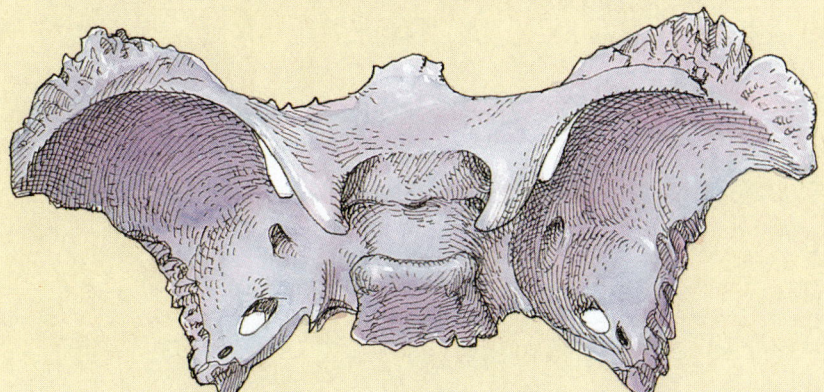

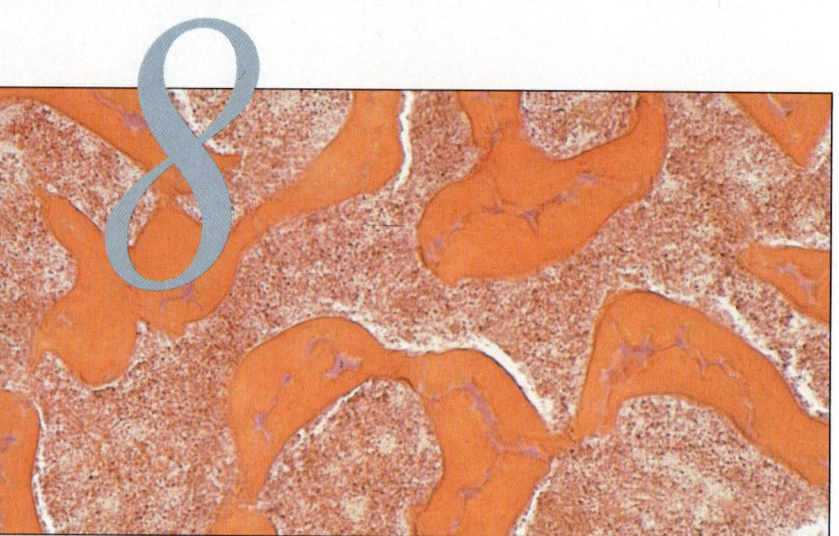

The Appendicular Skeleton

8

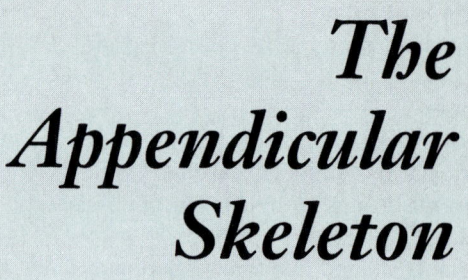

KEY CONCEPTS

1 The upper and lower extremities include the arms, legs, shoulder blades, collarbones, hipbones, hands, and feet.

2 Each shoulder girdle consists of a collarbone and shoulder blade. The pelvic girdle is formed by the paired hipbones.

The *appendicular skeleton,* although its name is derived from the Latin word *appendere* (to hang from), should not be thought of as consisting of only the hanging parts — the arms and legs.* In fact, the appendicular skeleton is composed of the *upper extremities,* which include the scapula (shoulder blade) and clavicle (collarbone) of the pectoral (shoulder) girdle in addition to the arms, forearms, and hands, and the *lower extremities,* which include the hipbones of the pelvic girdle as well as the thighs, legs, and feet.

THE UPPER EXTREMITIES (LIMBS)

The skeleton of the *upper extremity* or *limb* consists of 64 bones [FIGURES 8.1, 8.2]. These include the scapula and clavicle of the shoulder girdle, the humerus of the arm, the radius and ulna of the forearm, the carpals of the wrist, the metacarpals of the palm, and the phalanges of the fingers. The upper extremity is connected to and supported by the axial skeleton by only one joint and many muscles. The joint is the sternoclavicular joint between the manubrium of the sternum and the clavicle. The muscles form a complex of "suspension bands" from the vertebral column, ribs, and sternum to the shoulder girdle.

Pectoral Girdle

The upper limb girdle is known as the *pectoral,* or *shoulder, girdle* [FIGURE 8.2]. It consists of the clavicle and scapula. The clavicle is a long bone that extends from the sternum to the scapula in front of the thorax. The triangular scapula is located behind the thoracic cage.

The pectoral girdle is designed more for mobility than for stability. It is held in place and surrounded by muscles and ligaments. The stability of the shoulder is provided by these muscles and ligaments rather than by the shape of the bones and joints. Because of this arrangement, the shoulder is often the site of dislocation injuries such as shoulder separations.

Clavicle The paired *clavicles* (L. keys), or collarbones, are located anteriorly at the root of the neck. The clavicle is a horizontal double-curved long bone with a rounded medial end and a flattened lateral end [FIGURE 8.3]. The entire length of the clavicle can be felt through the skin. The medial part of the clavicle is curved anteriorly, and the lateral part is curved posteriorly. The medial or *sternal* end of the clavicle articulates with the manubrium just above the first rib at the *sternoclavicular*

joint, connecting the axial and appendicular skeletons. The lateral or *acromial* end articulates with the **acromion process.** This **acromioclavicular joint** involves the clavicle in all movements of the scapula.

The clavicle is held in place by strong ligaments at both ends. At the medial end is the **costal tuberosity,** where the costoclavicular ligament is attached. At the lateral end is the **conoid tubercle,** where the coracoclavicular ligament is attached. The clavicle is also a point of attachment for muscles of the shoulder girdle and neck. Large blood vessels and some nerves for the upper limb pass deep to the anterior curvature.

The main function of the clavicle is to act as a strut to hold the shoulder joint and arm away from the thoracic cage, allowing the upper limb much freedom of movement. Because of its vulnerable position and relative thinness, the clavicle is broken more often than any other bone in the body, and when it is, the whole shoulder is likely to collapse. For this reason, some self-defense strategies are designed to break an attacker's clavicle.

Scapula The *scapula* (L. shoulder), or shoulder blade (because of its resemblance to the blade of a shovel), is located on the posterior thoracic wall between ribs 2 and 7.

The flat, triangular *body* of the scapula has an obliquely oriented *spine* that can be felt on the posterior surface [FIGURE 8.4]. The prominent ridge of the spine separates the supraspinous fossa from the infraspinous fossa. The spine and fossae provide attachment sites for the muscles that move the arm. The spine ends in the large, flat **acromion,** which forms the point of the shoulder. It articulates with the clavicle and is an attachment site for chest and arm muscles. Below the acromion is the shallow **glenoid fossa,** which acts as a socket for the head of the humerus. This articulation forms the **glenohumeral** (shoulder) **joint.** Just medial to the upper end of the glenoid fossa is the anteriorly directed **coracoid process,** which serves as an attachment for the coracoclavicular ligament and several muscles (origins of pectoralis minor, short head of biceps brachii, and coracobrachialis muscles) of the arm and chest.

FIGURE 8.4 shows the location of the three borders and two angles of the scapula. The **medial border** runs parallel to the vertebral column. The **superior** and **inferior angles** are located at the ends of the medial border. Extending upward from the inferior angle to the glenoid fossa is the **lateral border.** Running horizontally from the glenoid fossa to the superior angle is the **superior border.**

Bones of the Arm, Forearm, and Hand

The bones of the arm, forearm, and hand are the humerus, ulna, radius, carpals (wrist), metacarpals, and the phalanges of the fingers.

---

*In terms of the appendicular skeleton, an *extremity* and a *limb* mean the same thing; a limb is not an arm or leg. Remember that the upper limb, or extremity, includes the scapula and clavicle *as well as* the arm, forearm, wrist, and hand.

FIGURE 8.1 SURFACE ANATOMY OF THE UPPER EXTREMITY

The *upper extremity* includes the *pectoral (shoulder) girdle,* axilla (armpit), arm, forearm, wrist, hands, and fingers.

[A] The bony framework of the shoulder girdle consists of the clavicle in front, the scapula in back, and the proximal portion of the humerus. The clavicle and scapula attach the upper extremity to the trunk. The *arm* (upper arm)

extends from the shoulder to the elbow. Its long bone, the *humerus,* extends from the shoulder joint to the *elbow,* the hinge-type joint between the humerus and forearm.

[B] The lower part of the upper extremity includes the forearm, wrist, and hand. The *forearm* is the region between the elbow and wrist. It contains two long bones: the *ulna* is the longer of the two, located on the side of the little finger; the *radius* is located lateral to the ulna, on the thumb side. The *wrist,* or *carpus,* is composed of eight short bones connected by ligaments.

[C] The *hand* is made up of the *metacarpal bones,* while the bones of the fingers and thumb are called *phalanges* (fuh-LAN-jeez).

Olecranon process of ulna

Triceps brachii muscle

Pectoralis major muscle

Bicipital aponeurosis

Biceps brachii muscle

Deltoid muscle

Clavicle

Medial epicondyle of humerus

Brachialis muscle

Posterior axillary fold

Anterior axillary fold

Axilla

Cubital fossa

Tendon of biceps brachii muscle

Site of palpation of brachial artery

Latissimus dorsi muscle

[A]

Tendon of extensor pollicis longus muscle

Distal interphalangeal joints

Proximal interphalangeal joints

Metacarpophalangeal joints (knuckles)

Anatomical snuffbox

Tendon of extensor pollicis brevis muscle

Tendon of extensor digiti minimi muscle

Tendons of extensor digitorum muscle

Base of first metacarpal bone

Styloid process of radius

Styloid process of ulna

[C]

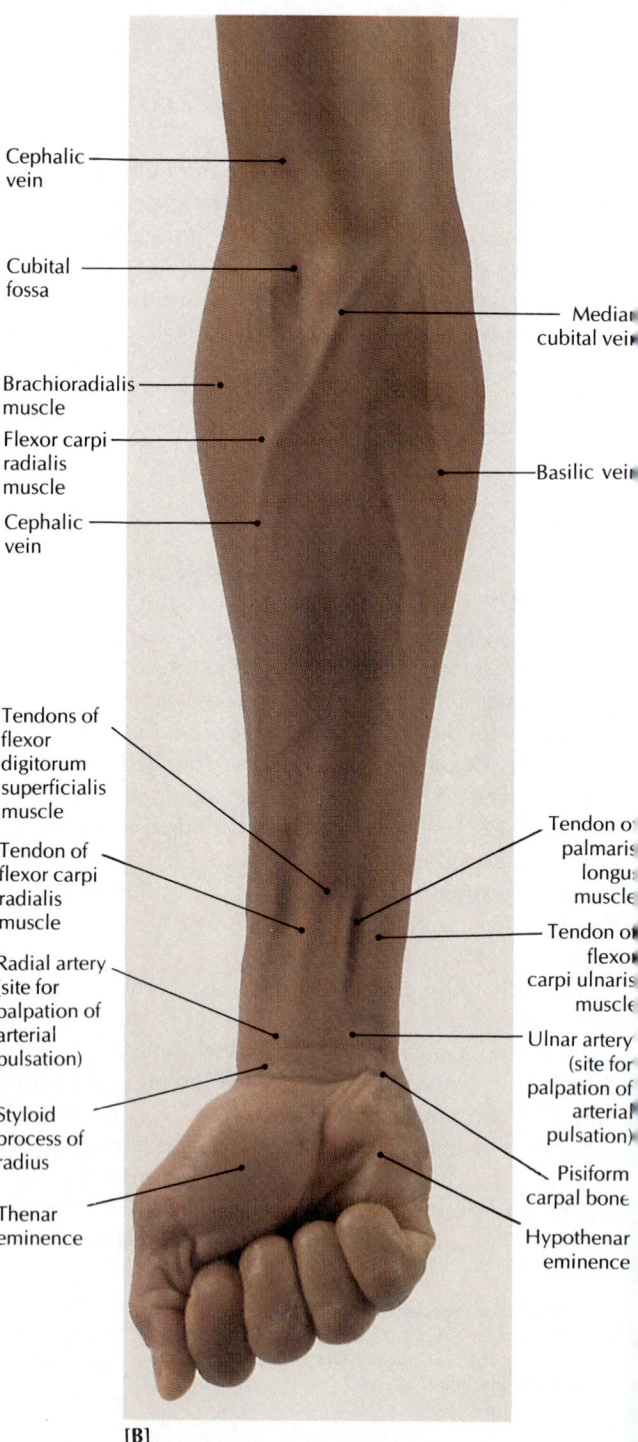

Cephalic vein

Cubital fossa

Median cubital vein

Brachioradialis muscle

Flexor carpi radialis muscle

Cephalic vein

Basilic vein

Tendons of flexor digitorum superficialis muscle

Tendon of flexor carpi radialis muscle

Radial artery (site for palpation of arterial pulsation)

Styloid process of radius

Thenar eminence

Tendon of palmaris longus muscle

Tendon of flexor carpi ulnaris muscle

Ulnar artery (site for palpation of arterial pulsation)

Pisiform carpal bone

Hypothenar eminence

[B]

FIGURE 8.2 THE UPPER EXTREMITY (64 BONES)

The upper extremity includes the pectoral (shoulder) girdle. **[A]** Anterior view. **[B]** Posterior view.

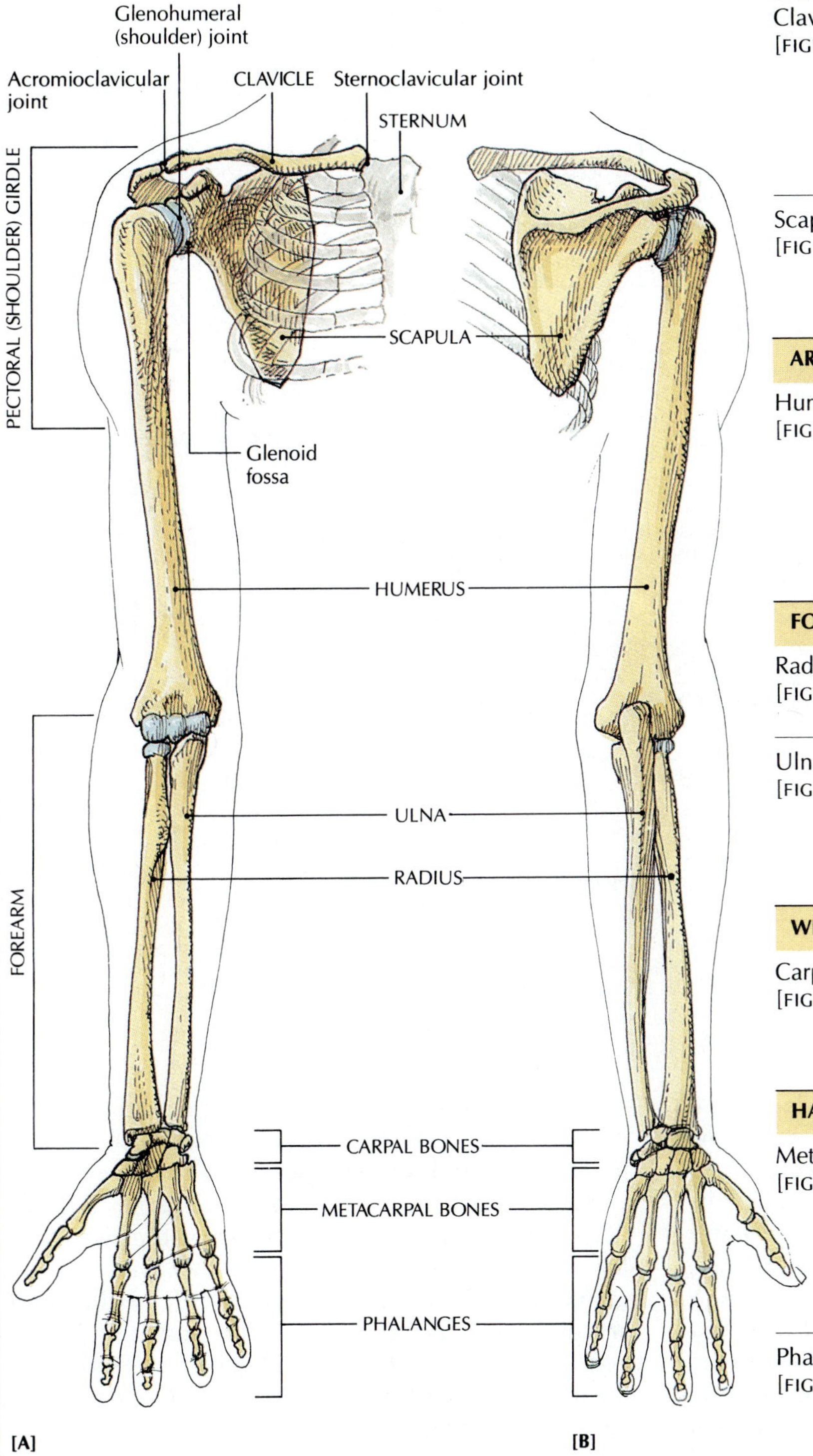

Bones	Description and functions
PECTORAL (SHOULDER) GIRDLE	
Clavicle (2) [FIGURE 8.3]	Collarbone; double-curved, long bone with rounded medial end and flattened lateral end; held in place by ligaments. Holds shoulder joint and arm away from thorax so upper limb can swing freely.
Scapula (2) [FIGURE 8.4]	Shoulder blade; flat, triangular bone with horizontal spine separating fossae. Site of attachment for muscles of arm and chest.
ARM	
Humerus (2) [FIGURE 8.5]	Longest, largest bone of upper limb; forms ball of ball-and-socket joint with glenoid fossa of scapula. Site of attachment for muscles of shoulder and arm, permitting arm to flex and extend at elbow.
FOREARM	
Radius (2) [FIGURE 8.6]	Shorter of two bones in forearm. Allows forearm to rotate in radial motion.
Ulna (2) [FIGURE 8.6]	Larger of two bones in forearm; large proximal end consists of olecranon process (prominence of elbow). Forms hinge joint at elbow.
WRIST	
Carpals (16) [FIGURE 8.7]	Small short bones; in each wrist, 8 carpals in 2 transverse rows of 4. With attached ligaments, allow slight gliding movement.
HANDS AND FINGERS	
Metacarpals (10) [FIGURE 8.7]	Five miniature long bones in each hand in fanlike arrangement; articulate with fingers at metacarpophalangeal joint (the knuckle). Aid opposition movement of thumb; enable cupping of hand.
Phalanges (28) [FIGURE 8.7]	Miniature long bones, 2 in each thumb, 3 in each finger; articulate with each other at interphalangeal joint. Allow fingers to participate in stable grips.

FIGURE 8.3 RIGHT CLAVICLE

Anterior view.

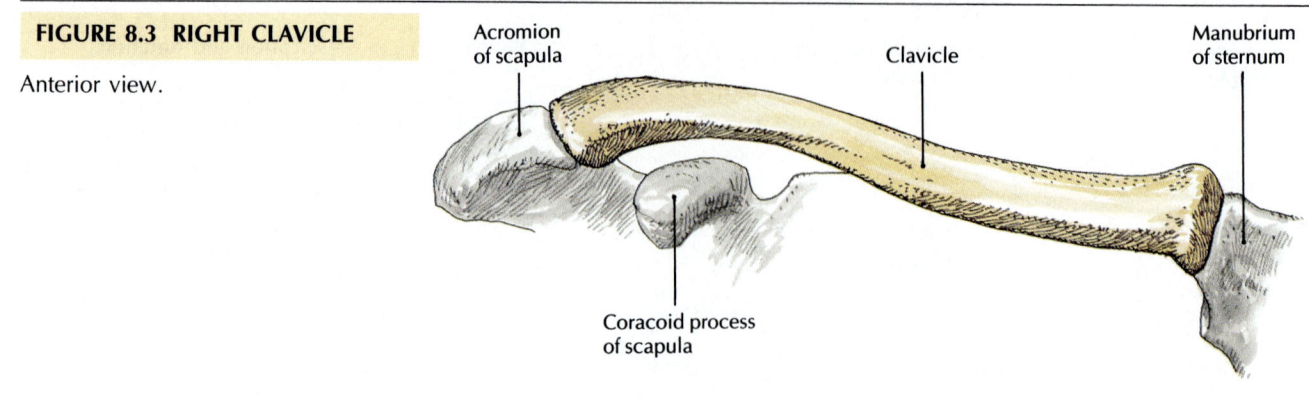

Acromion of scapula — Clavicle — Manubrium of sternum — Coracoid process of scapula

FIGURE 8.4 RIGHT SCAPULA

[A] Anterior view, costal surface. [B] Posterior view, dorsal surface. [C] Lateral view.

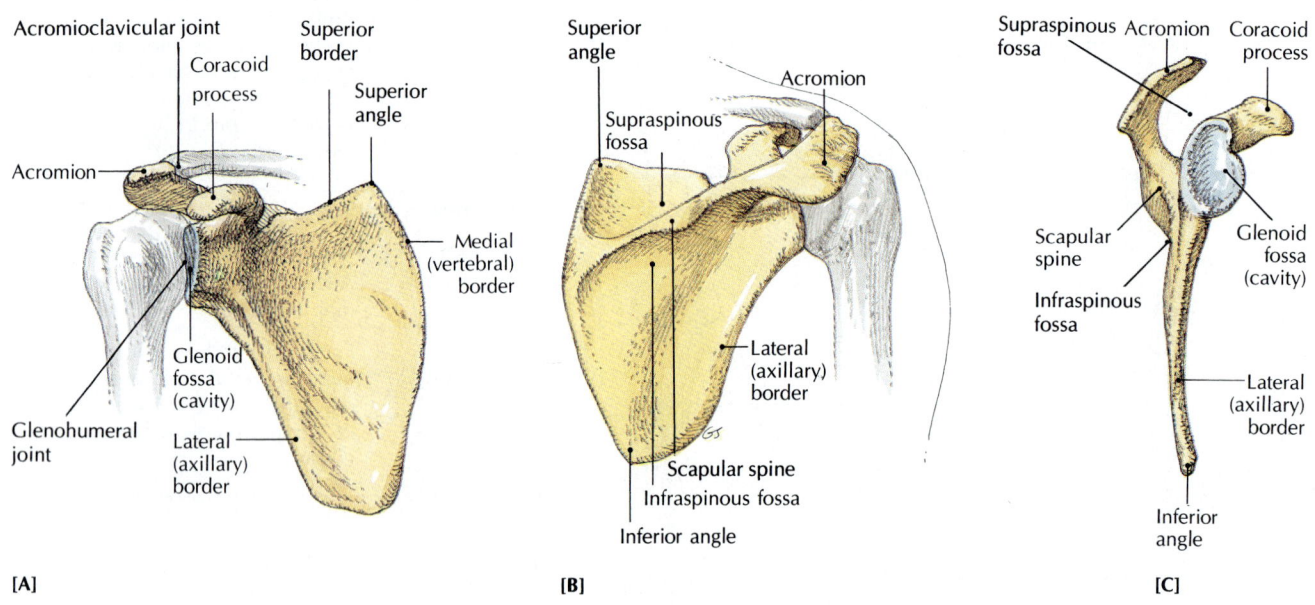

[A] [B] [C]

Humerus The *humerus* (L. upper arm) is the arm bone located between the shoulder and elbow. It is the longest and largest bone of the upper limb [FIGURE 8.5].

The *shaft* (or body) of the humerus is cylindrical in its upper half and flattened from front to back in its lower half. The *head* of the humerus is the ball of the ball-and-socket shoulder joint with the glenoid fossa of the scapula. Close to the head are the *greater* and *lesser tubercles,* which are attachment sites for muscles originating on the scapula. Passing along the *intertubercular groove* between the two tubercles is the tendon of the long head of the biceps brachii muscle for the forearm. The *anatomical neck* is located between the head and the tubercles. The *surgical neck,* so named because it is the site of frequent fractures of the upper end of the humerus, is located just below the tubercles. The *deltoid tuberosity,* halfway down the lateral side of the shaft, is the attachment site for the deltoid muscle. On the posterior surface is the *radial groove* spiraling from proximomedial to distolateral. Along this groove passes the radial nerve.

At the distal end of the humerus are the *trochlea* (TROHK-lee-uh; L. pulley), which is connected like a pulley with the olecranon process of the ulna, and the *capitulum* (kuh-PITCH-yoo-luhm; L. little head), which articulates with the head of the radius [FIGURES 8.5, 8.6]. Some muscles of the forearm and fingers are attached to the *lateral* and *medial epicondyles.* On the anterior surface are the *radial fossa* and *coronoid fossa,* which accommodate the head of the radius and the coronoid process of the ulna, respectively, when the arm is bent at the elbow. On the posterior surface is the *olecranon* (oh-LECK-ruh-non) *fossa,* which accommodates the olecranon process of the ulna when the arm is straightened.

FIGURE 8.5 RIGHT HUMERUS

[A] Anterior view. [B] Posterior view. [C] Photograph of upper and lower ends of right humerus, anterior view.

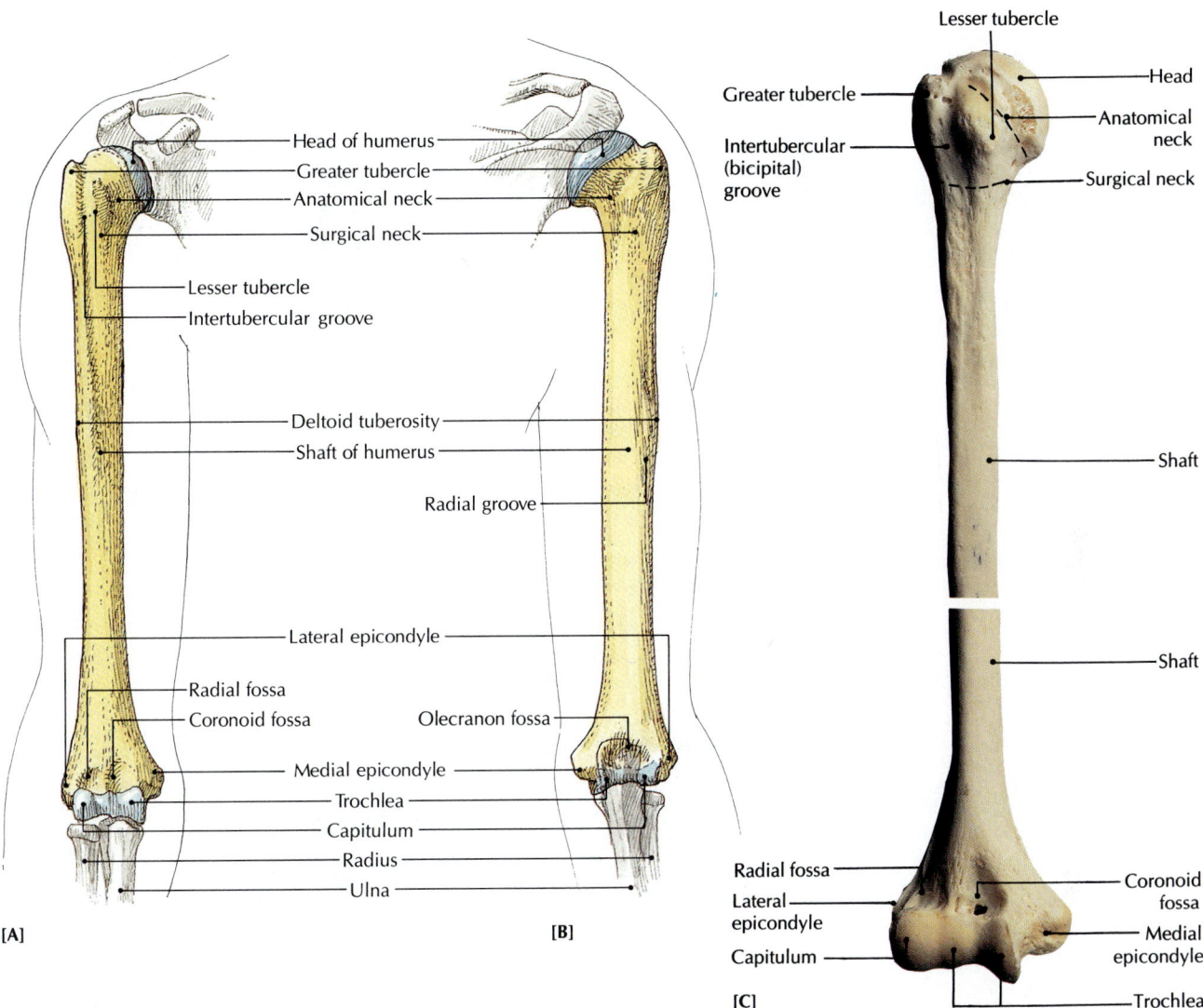

[A]   [B]

[C]

Ulna The *ulna* (L. elbow) is the longer of the two bones of the forearm between the elbow and the wrist. It is medially located (on the side of the little finger).

The large proximal end of the ulna consists of the *olecranon process,* which curves upward and forward to form the *semilunar* (half-moon), or *trochlear, notch* [FIGURE 8.6A]. This half-moon-shaped depression articulates with the trochlea of the humerus to form the hinged elbow joint. The *coronoid process* on the anterior surface of the ulna forms the lower border of the trochlear notch. The *radial notch* on the lateral surface of the coronoid process is the articulation site for the head of the radius. The ulna is narrow at its distal end, with a small head; a short, blunt peg called the *styloid process* is attached to a fibrocartilaginous disk that separates the ulna from the carpus.

Radius The *radius* (L. spoke of a wheel) is the long bone located lateral to the ulna (on the thumb side).

At the proximal end of the radius, its disk-shaped *head* articulates with the capitulum of the humerus [FIGURE 8.5A] and the radial notch of the ulna [FIGURE 8.6A]. The *radial tuberosity* on the medial side is the attachment site for the biceps brachii muscle. An interosseous membrane connects the shafts of the ulna and radius. The *shaft* becomes broader toward its large distal end, which articulates with two of the lunate carpal bones in the hand [FIGURE 8.7A]. The wrist joint is called the *radiocarpal joint* because it joins the forearm (radius) and wrist (carpus). At the lower end of the radius are the prominent *styloid process* (which may be felt on the outside of the wrist where it joins the hand) and the U-shaped *ulnar notch* into which fits the head of the ulna.

FIGURE 8.6 RIGHT RADIUS AND ULNA

[A] Anterior view. [B] Posterior view. [C] Right lateral view of articulations with the humerus at the elbow.

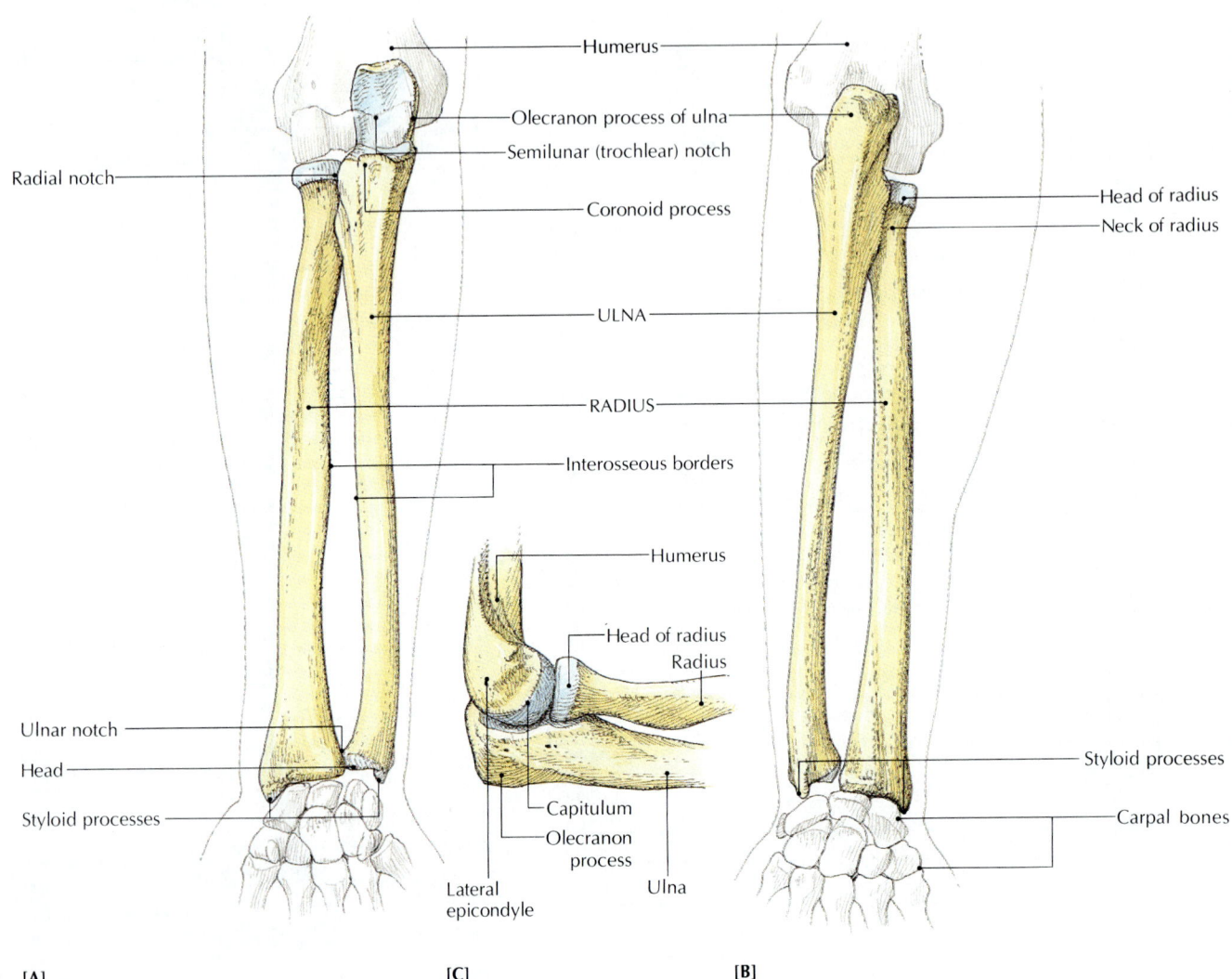

[A]   [C]   [B]

Carpus (wrist) The wrist or *carpus* (Gr. *karpos*, wrist) is composed of eight short bones connected to each other by ligaments that restrict their mobility primarily to gliding movements.*

As shown in FIGURE 8.7, the *carpal bones* are arranged in two transverse rows of four bones each. In the proximal row, from lateral to medial position, are the *scaphoid*, *lunate*, *triquetrum*, and *pisiform*. In the distal row are the *trapezium*, *trapezoid*, *capitate*, and *hamate*. The easily felt small pisiform is a clinically useful landmark. It is actually a sesamoid bone within the tendon of the flexor carpi

ulnaris muscle. Because of its shape and location, the scaphoid is the carpal bone most prone to fracture.

The carpal bones as a unit are shaped so that the back of the carpus is convex and the palmar side is concave. A connective tissue bridge called the flexor retinaculum stretches between the hamate and pisiform bones and the trapezium and scaphoid bones. This bridge converts the palmar concavity into a tunnel called the carpal tunnel. Nine long tendons and the median nerve pass through the carpal tunnel from the forearm to the hand. (See the discussion of *carpal tunnel syndrome* on page 345.)

Metacarpus and phalanges The 5 *metacarpal* bones make up the skeleton of the palm of the hand, or *metacarpus,* and the 14 *phalanges* are the finger bones [FIGURE 8.7].

*The wrist is the region between the forearm and the hand (i.e., distal to the forearm). A "wrist" watch is not usually worn around the wrist; instead, it encircles the distal end of the forearm, just proximal to the head of the ulna.

FIGURE 8.7 BONES OF THE RIGHT HAND

[A] Palmar (ventral, anterior) aspect. [B] Dorsal (posterior) aspect. [C] Photograph of right hand, dorsal aspect.

Key: **S** = Scaphoid **L** = Lunate **TRI** = Triquetral **P** = Pisiform **TRU** = Trapezium **TRO** = Trapezoid **C** = Capitate **H** = Hamate

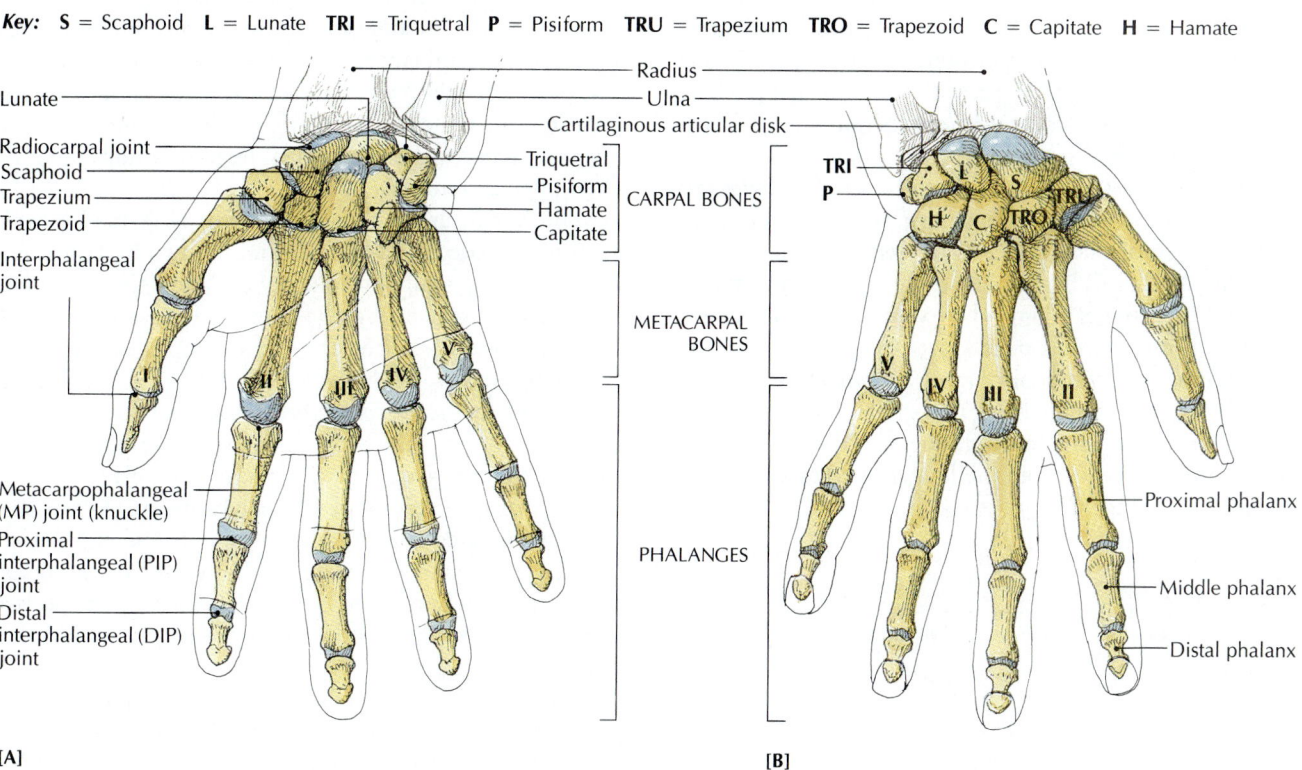

[A] [B]

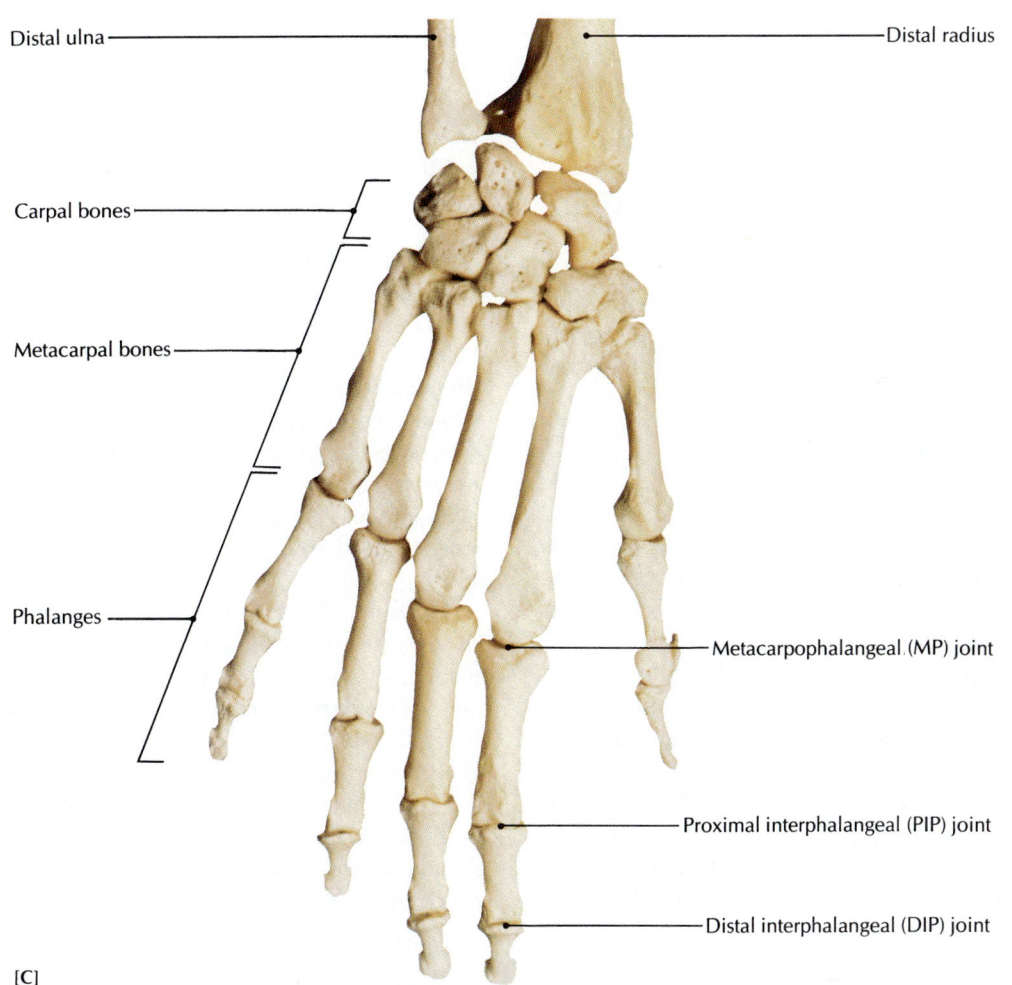

[C]

The *metacarpal* (L. behind the wrist) *bones* are miniature long bones. They are numbered from the lateral (thumb) side as metacarpals I to V. They are arranged as a fan from their proximal ends (bases), which articulate with the distal row of carpal bones, to their distal ends (heads). Each head articulates with the proximal phalanx of a digit. The *metacarpophalangeal joint* (MP joint) forms a "knuckle."

The bones of the digits (fingers) are the 14 *phalanges* (fuh-LAN-jeez; Gr. line of soldiers; sing. *phalanx*, FAY-langks). Each of these bones has a base, shaft, and head. The thumb (digit I) has two phalanges (proximal and distal), and each finger (digits II to V) has three phalanges (proximal, middle, and distal) [FIGURE 8.7]. Except for the thumb, which has only one *interphalangeal joint* (the IP joint), the digits have a proximal interphalangeal (PIP) joint and a distal interphalangeal (DIP) joint. (The word "pinkie" comes from the Dutch word *pinkje*, which means "little finger.")

A S K   Y O U R S E L F

1 What bones make up the upper extremities?

2 What bones make up the shoulder girdle?

3 What is the main function of the shoulder girdle?

4 What are the bones of the arm and hand, and how many of each type of bone are there?

5 What are the functions of the ulna and radius?

THE LOWER EXTREMITIES (LIMBS)

The *lower extremities* consist of 62 bones [FIGURES 8.8 and 8.9]. These include the hipbones of the pelvic girdle, the femur of the thigh, the tibia and fibula of the leg, the tarsal bones of the ankle, the metatarsals of the foot, and the phalanges of the toes.

Pelvic Girdle and Pelvis

The lower limb girdle, called the *pelvic girdle,* is formed by the right and left hipbones, which are also known as the *ossa coxae* (L. hipbones; sing. *os coxa*) [FIGURES 8.9, 8.10, 8.11]. The hipbones, sacrum, and coccyx form the bony pelvis. Thus, the pelvic girdle is made up of bones from both the appendicular and axial skeletons.

The paired hipbones are the broadest bones in the body. They are formed in the adult by the fusion of the ilium, ischium, and pubis. These three bones are separate in infants, children, and young adolescents, but they generally fuse between ages 15 and 17. On the lateral surface of each hipbone is a deep cup, called the *acetabulum* (ass-eh-TAB-yoo-luhm; L. vinegar cup), which is the

socket of the ball-and-socket joint with the head (ball) of the femur [FIGURES 8.11B and 8.12B]. The acetabulum is formed by parts of the ilium, ischium, and pubis.

Although the bones of the pelvic and shoulder girdles bear some resemblance, their functions are rather different. Because they have to hold the body in an upright position, the bones of the pelvic girdle are built more for support than for the exceptional degree of movement of the upper extremities.

The bowl-shaped *pelvis* (L. basin) is formed by the sacrum and coccyx posteriorly and the two hipbones anteriorly and laterally [FIGURE 8.11]. The pelvis is bound into a structural unit by ligaments at the lateral pairs of *sacroiliac joints* between the sacrum and the two ilia, at the *symphysis pubis* between the bodies of the pubic bones, and at the *sacrococcygeal joints* between the sacrum and coccyx. The symphysis pubis is especially important for the structural security of the pelvis.

The basic functions of the pelvis are to (1) provide attachment sites for muscles of the trunk and lower limbs, (2) transmit and transfer the weight of the body from the vertebral column to the femur of the lower limb, and (3) support and protect the organs within the pelvis.

The pelvis is usually divided into the *greater,* or *false, pelvis* and the *lesser,* or *true, pelvis,* as shown in FIGURE 8.11. Obstetricians often refer to the lesser pelvis in the female as the "obstetric pelvis" because it is the critical region during childbirth, providing the opening through which the baby must pass. The junction between the two pelves is the *pelvic brim,* which surrounds the superior pelvic aperture (pelvic inlet). This brim is the bony ring extending from the *sacral promontory* to the top of the symphysis pubis.

Ilium The *ilium* (L. flank) is the largest of the three fused hipbones [FIGURES 8.11, 8.12]. It forms the easily felt lateral prominence of the hip. (The *l* in ilium will help you remember that it occupies the *l*ateral position in the ossa coxae.) On its superior border is the *iliac crest,* which ends anteriorly as the *anterior superior iliac spine.* On the medial posterior part of the ilium is the *auricular surface* that articulates with the sacrum. Below this auricular surface is the *greater sciatic notch.* The internal surface of the ilium is the concave *iliac fossa.*

Ischium The *ischium* (IHSS-kee-uhm; L. hip joint) is the lowest and strongest bone of the ossa coxae. It is formed by the lower lateral portion of the acetabulum and the *ischial tuberosity,* the bony prominence that bears the weight of the body when we are seated [FIGURES 8.11, 8.12]. Extending from the body of the ischium are the slender superior and inferior *rami.* Above the body is the *ischial spine,* which is located between the greater sciatic notch of the ilium and the *lesser sciatic notch* of the ischium.

FIGURE 8.8 SURFACE ANATOMY OF THE LOWER EXTREMITY

The **lower extremity** is composed of the pelvic girdle, buttocks, thigh, leg, ankle, foot, and toes. The **pelvic girdle** is described in FIGURES 8.9 and 8.10.

[A] The **buttocks** (BUTT-uhks), or gluteal region, are bounded superiorly by the iliac crest and inferiorly by the fold of the buttock at the top of the thigh. The buttocks are composed largely of the gluteal muscles and fat.

[B] The **thigh,** or *upper leg,* extends between the hip and knee and contains the **femur,** the longest bone in the body. [C] The **knee** is the joint between the thigh and the lower leg.

[D] The **leg,** which extends between the knee and the ankle, contains two long bones: the *tibia* (shinbone) is located on the anterior and medial side of the leg; the *fibula* lies parallel and lateral to the tibia. The **ankle** is the joint between the leg and foot. The ankle is formed by 7 *tarsal bones,* which join the five *metatarsal bones* of the **foot;** the metatarsals connect to the 14 *phalanges,* or toe bones.

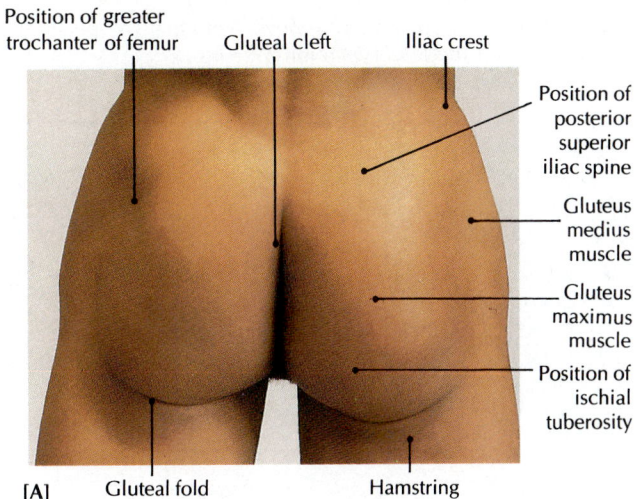

Position of greater trochanter of femur
Gluteal cleft
Iliac crest
Position of posterior superior iliac spine
Gluteus medius muscle
Gluteus maximus muscle
Position of ischial tuberosity
[A] Gluteal fold
Hamstring muscles

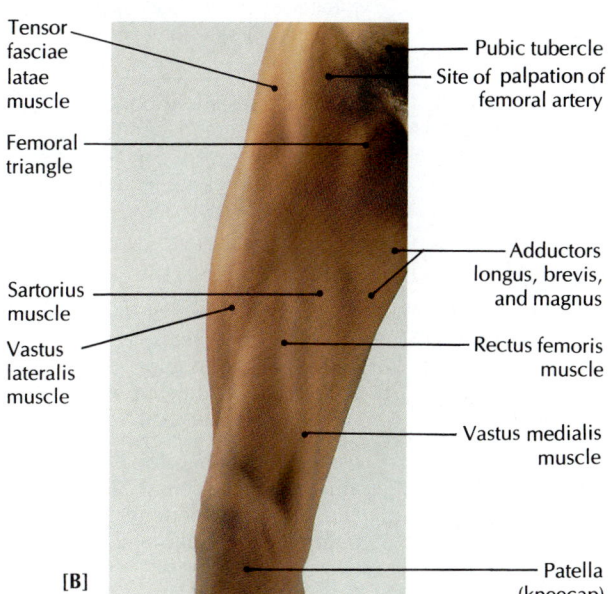

Tensor fasciae latae muscle
Femoral triangle
Sartorius muscle
Vastus lateralis muscle
Pubic tubercle
Site of palpation of femoral artery
Adductors longus, brevis, and magnus
Rectus femoris muscle
Vastus medialis muscle
Patella (kneecap)
[B]

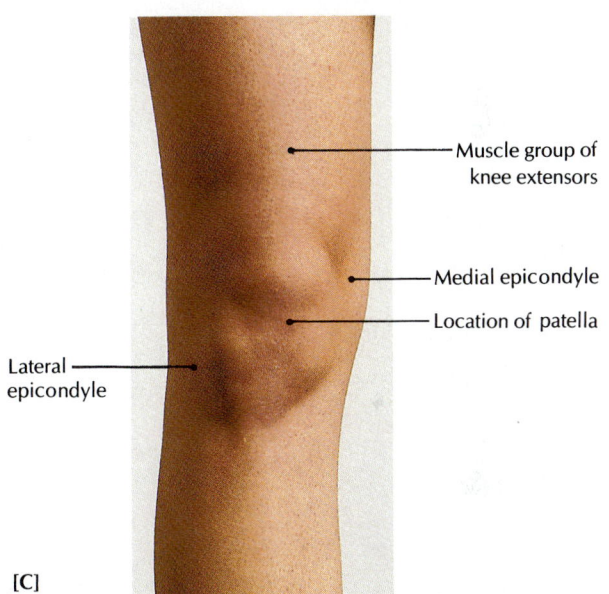

Muscle group of knee extensors
Medial epicondyle
Location of patella
Lateral epicondyle
[C]

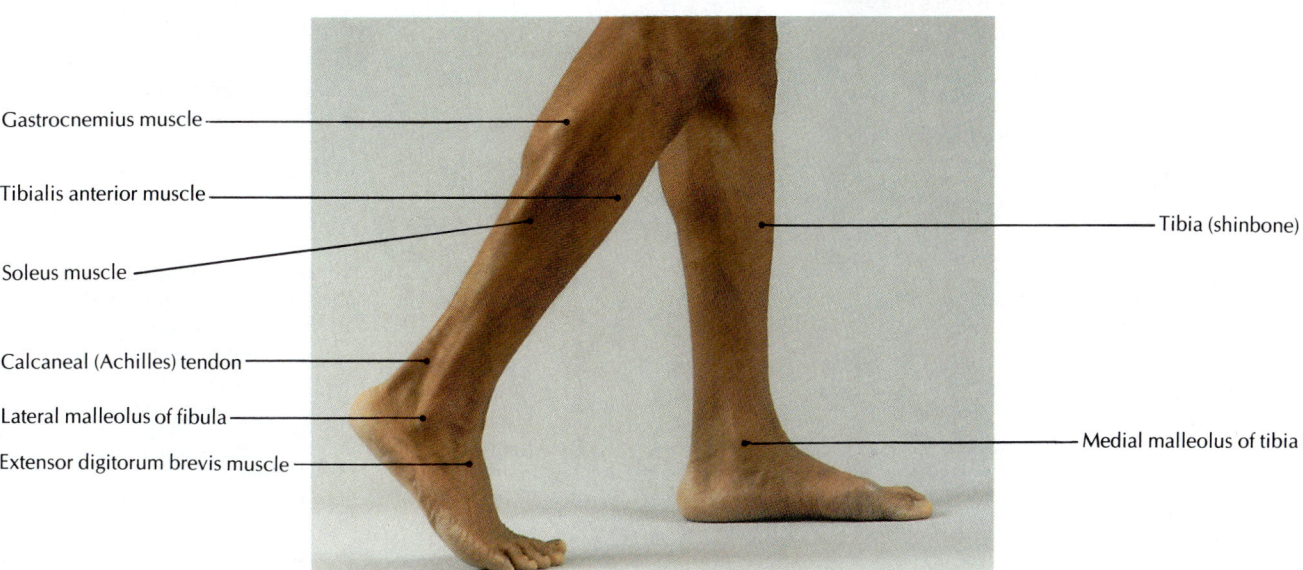

Gastrocnemius muscle
Tibialis anterior muscle
Soleus muscle
Calcaneal (Achilles) tendon
Lateral malleolus of fibula
Extensor digitorum brevis muscle
Tibia (shinbone)
Medial malleolus of tibia
[D]

FIGURE 8.9 THE LOWER EXTREMITY (62 BONES)

[A] Anterior view. [B] Posterior view. Note that the pelvic (hip) girdle is part of the lower extremity.

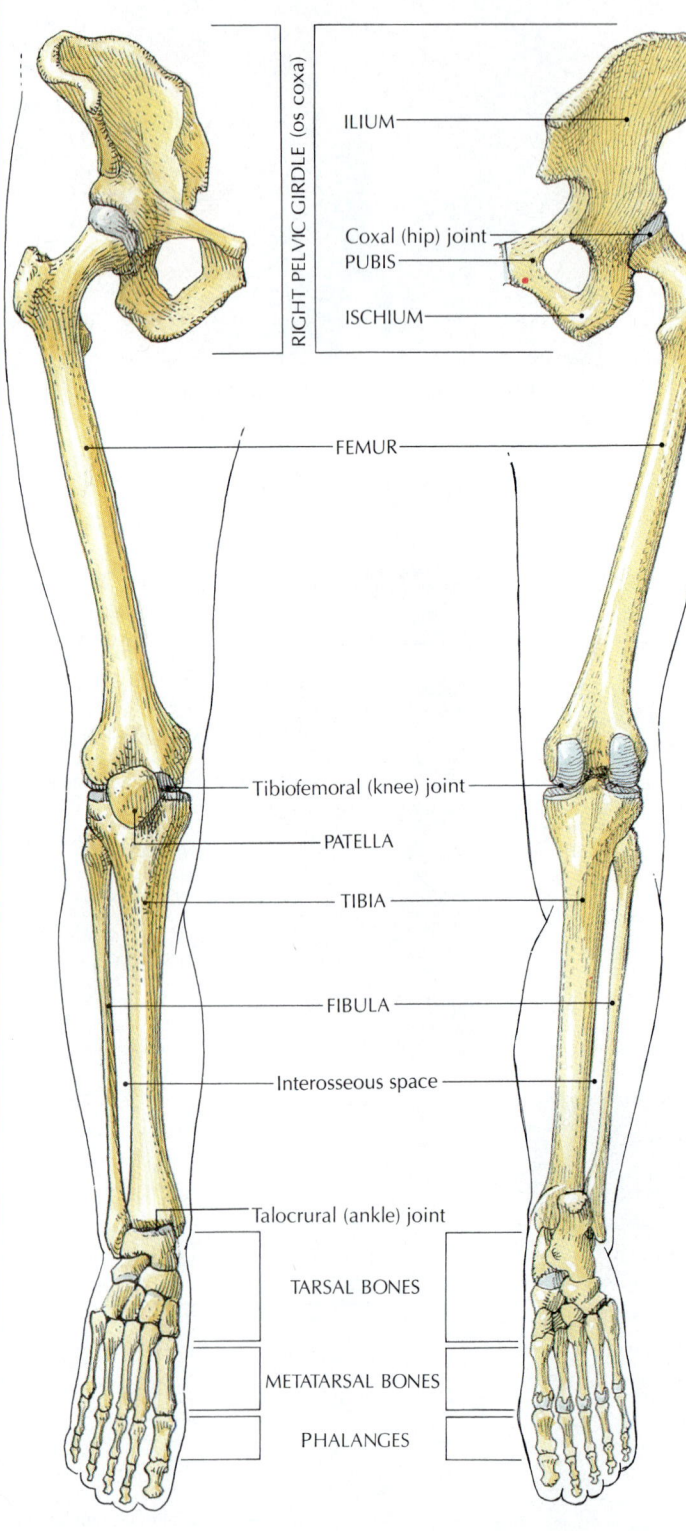

RIGHT PELVIC GIRDLE (os coxa)

ILIUM

Coxal (hip) joint
PUBIS

ISCHIUM

FEMUR

Tibiofemoral (knee) joint

PATELLA

TIBIA

FIBULA

Interosseous space

Talocrural (ankle) joint

TARSAL BONES

METATARSAL BONES

PHALANGES

[A] [B]

Bone	Description and functions
PELVIC GIRDLE	
Hipbone (os coxa) (2) [FIGURE 8.12]	Irregular bone formed by fusion of ilium, ischium, pubis; with sacrum and coccyx forms pelvis; forms socket of ball-and-socket joint with femur. Site of attachment for trunk and lower limb muscles; transmits body weight to femur.
THIGH	
Femur (2) [FIGURE 8.13]	Thighbone; typical long bone; longest, strongest, heaviest bone; forms ball of ball-and-socket joint with pelvic bones; provides articular surface for knee. Supports body.
Patella (2) [FIGURE 8.14]	Kneecap; sesamoid bone within quadriceps femoris tendon. Increases leverage for quadriceps muscle by keeping tendon away from axis of rotation.
LEG	
Fibula (2) [FIGURE 8.15]	Smaller long bone of lower leg; articulates proximally with tibia and distally with talus. Bears little body weight, but gives strength to ankle joint.
Tibia (2) [FIGURE 8.15]	Larger long bone of lower leg; articulates with femur, fibula, talus. Supports body weight, transmitting it from femur to talus.
ANKLE	
Tarsals (14) [FIGURE 8.16]	Ankle, heelbones; short bones; 7 in each ankle including talus, calcaneus, cuboid, navicular, 3 cuneiforms; with metatarsals, form arches of foot. Bear body weight; raise body and transmit thrust during running and walking.
FOOT AND TOES	
Metatarsals (10) [FIGURE 8.16]	Miniature long bones; 5 in each foot; form sole; with tarsals, form arches of feet. Improve stability while standing; absorb shocks; bear weight; aid in locomotion.
Phalanges (28) [FIGURE 8.16]	Toes; miniature long bones; 2 in each big toe, 3 in each other toe; arranged as in hand. Provide stability during locomotion.

FIGURE 8.10 SURFACE ANATOMY OF THE FEMALE PELVIS

The **pelvis** is the region of the trunk inferior to the abdomen. The **pelvic girdle** is formed by the fused hipbones (ilium, ischium, and pubis), sacrum, and coccyx. Anterior view.

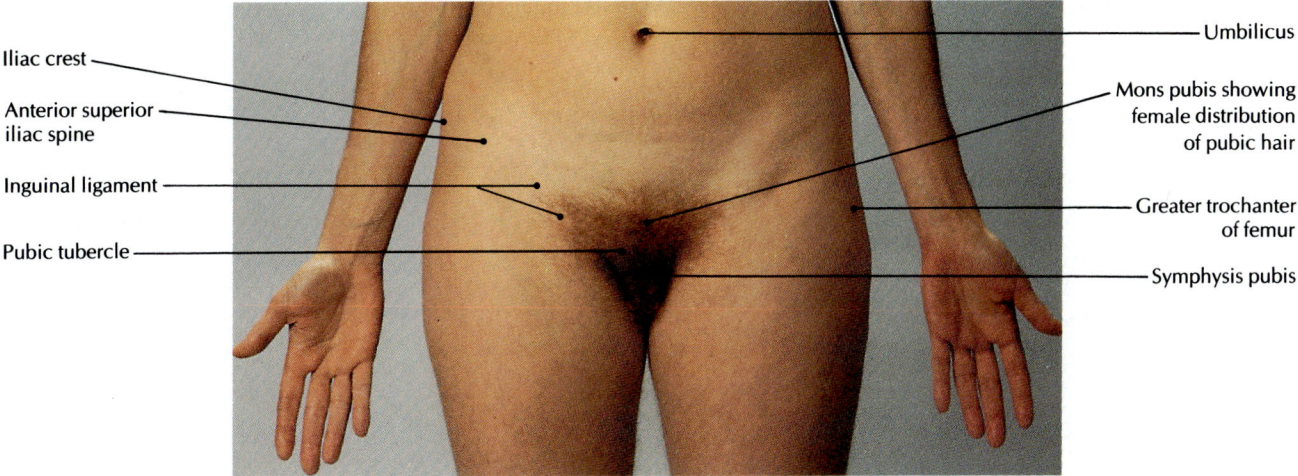

Iliac crest

Anterior superior iliac spine

Inguinal ligament

Pubic tubercle

Umbilicus

Mons pubis showing female distribution of pubic hair

Greater trochanter of femur

Symphysis pubis

FIGURE 8.11 MALE AND FEMALE PELVES

[A] Male pelvis; anterior view. [B] Female pelvis; anterior view. Note the wider pelvic aperture (inlet of true pelvis) in the female.

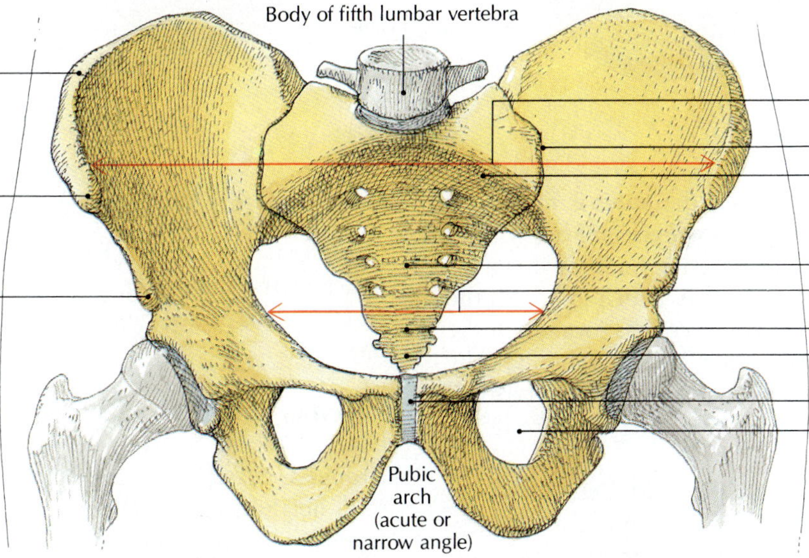

Body of fifth lumbar vertebra

Iliac crest

Anterior superior iliac spine

Anterior inferior iliac spine

Greater (false) pelvis

Sacroiliac joint

Sacral promontory

SACRUM

Inlet of lesser (true) pelvis

Sacrococcygeal joint

COCCYX

Symphysis pubis

Obturator foramen

Pubic arch (acute or narrow angle)

[A] MALE

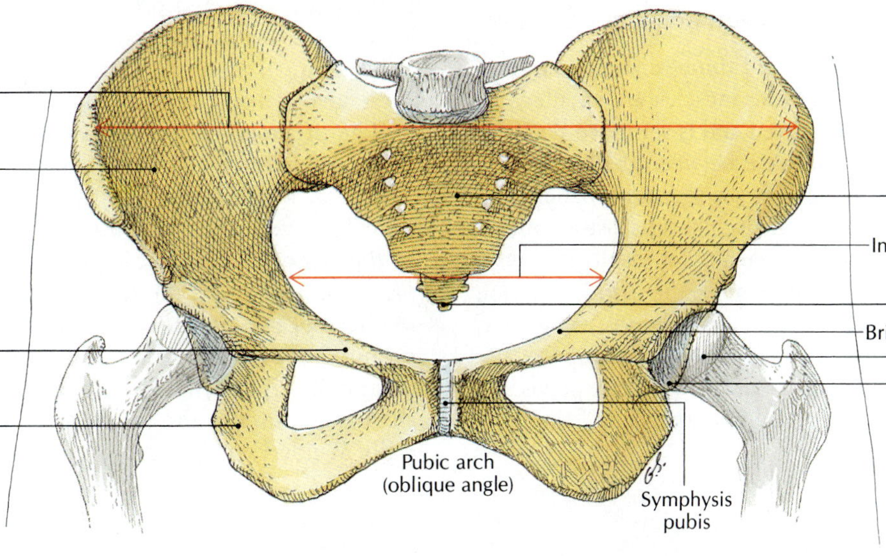

Greater (false) pelvis

RIGHT OS COXA

ILIUM

PUBIS

ISCHIUM

SACRUM

Inlet of lesser (true) pelvis

COCCYX

Brim of lesser (true) pelvis

Head of femur

Acetabulum

Pubic arch (oblique angle)

Symphysis pubis

[B] FEMALE

FIGURE 8.12 RIGHT OS COXA (HIPBONE)

[A] Right medial view. The borders of the different colored areas illustrate the lines of fusion of sutures at the junctions of the pubis, ischium, and ilium. **[B]** Right lateral view.

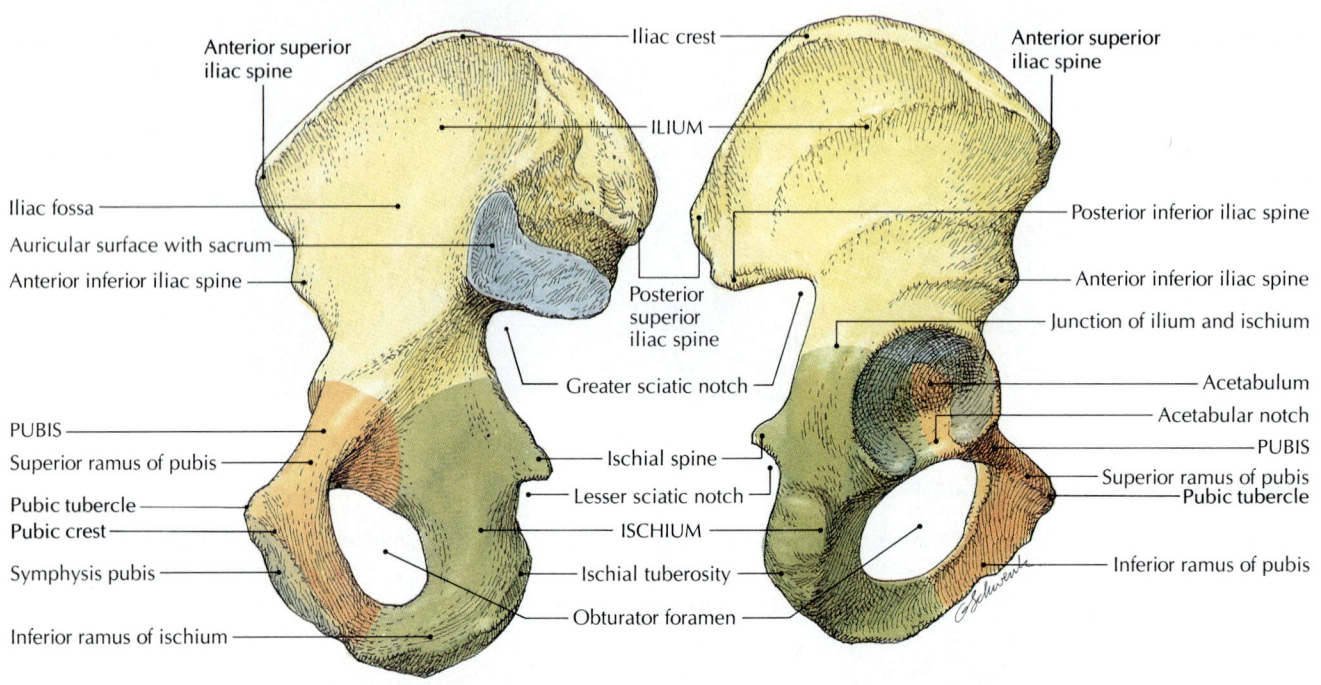

[A]

[B]

Pubis

Pubis The bilateral body of the *pubis* is joined together in front to form the *symphysis pubis* [FIGURES 8.11 and 8.12A]. Extending from the body of the pubis are the *superior* and *inferior rami,* which join with the ilium and ischium. The *pubic tubercle* is located on the body of the pubis. The *obturator foramen* is bounded by the rami and bodies of the pubis and ischium. Nerves and blood vessels pass through this opening into the thigh.

Pelves of the male and female Because of the structure of the female pelvis, a woman is able to carry and deliver a baby [FIGURE 8.11B]. The female pelvis usually shows the following differences from the male pelvis:

1 The bones are lighter and thinner. Bone markings are less prominent because the muscles are smaller.
2 The sacrum is less curved and is set more horizontally; this increases the distance between the coccyx and the symphysis pubis and makes the sacrum broader. The pubic rami are longer, and the ischial tuberosities are set farther apart and turned outward. As a result, the *pubic arch* has a wider angle. (This angle is the easiest criterion for distinguishing male from female skeletons.) Together, all these features create wider and shallower hips.
3 The pelvis has larger openings. The true pelvis and these openings surround and define the size of the birth canal.

Bones of the Thigh, Leg, and Foot

Completing our study of the lower extremities are the bones of the thigh, leg, and foot, including the femur, patella, tibia, fibula, tarsal bones, metatarsal bones, and phalanges.

Femur The *femur* (FEE-mur; L. thigh), or thighbone, is located between the hip and the knee. It is the strongest, heaviest, and longest bone in the body. (A person's height is usually about four times the length of the femur.) This strong bone plays an important role in supporting the body and provides mobility via the hip and knee joints. It can withstand a pressure of 3500 kg/cm^2 (1200 lb/in.^{2}), more than enough to cope with the pressures involved in normal walking, running, or jumping.

The proximal end of the femur consists of a head, a neck, and greater and lesser trochanters [FIGURE 8.13]. The *head,* which forms slightly more than half a sphere, articulates with the acetabulum (socket) of the hipbone. In the center of the head is a small depression called the *fovea capitis,* where the ligament and a blood vessel of the head of the femur are attached. If the blood vessel is ruptured as a result of trauma, the head of the femur may deteriorate. The thin *neck* connecting the head to the shaft is a common site for fractures of the femur, especially in elderly people. At the junction between the head

FIGURE 8.13 RIGHT FEMUR

[A] Anterior view. The fovea capitis is the site for the attachment of the ligament of the head of the femur [see FIGURE 9.5].
[B] Photograph of upper and lower ends of right femur, anterior view.

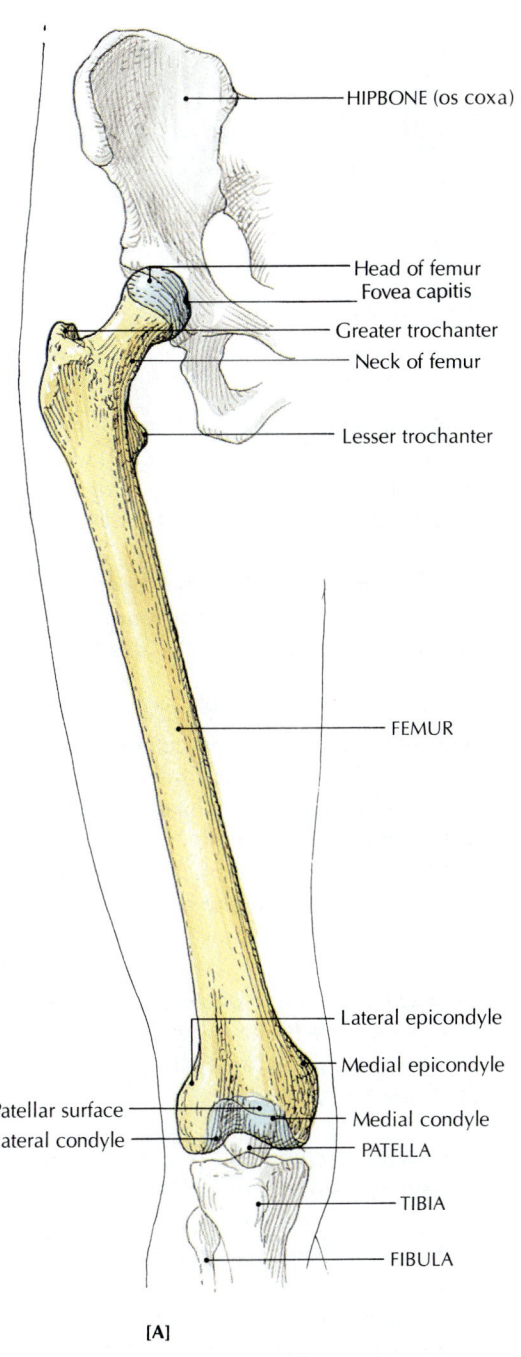

HIPBONE (os coxa)

Head of femur
Fovea capitis
Greater trochanter
Neck of femur

Lesser trochanter

FEMUR

Lateral epicondyle
Medial epicondyle
Patellar surface
Medial condyle
Lateral condyle
PATELLA
TIBIA
FIBULA

[A]

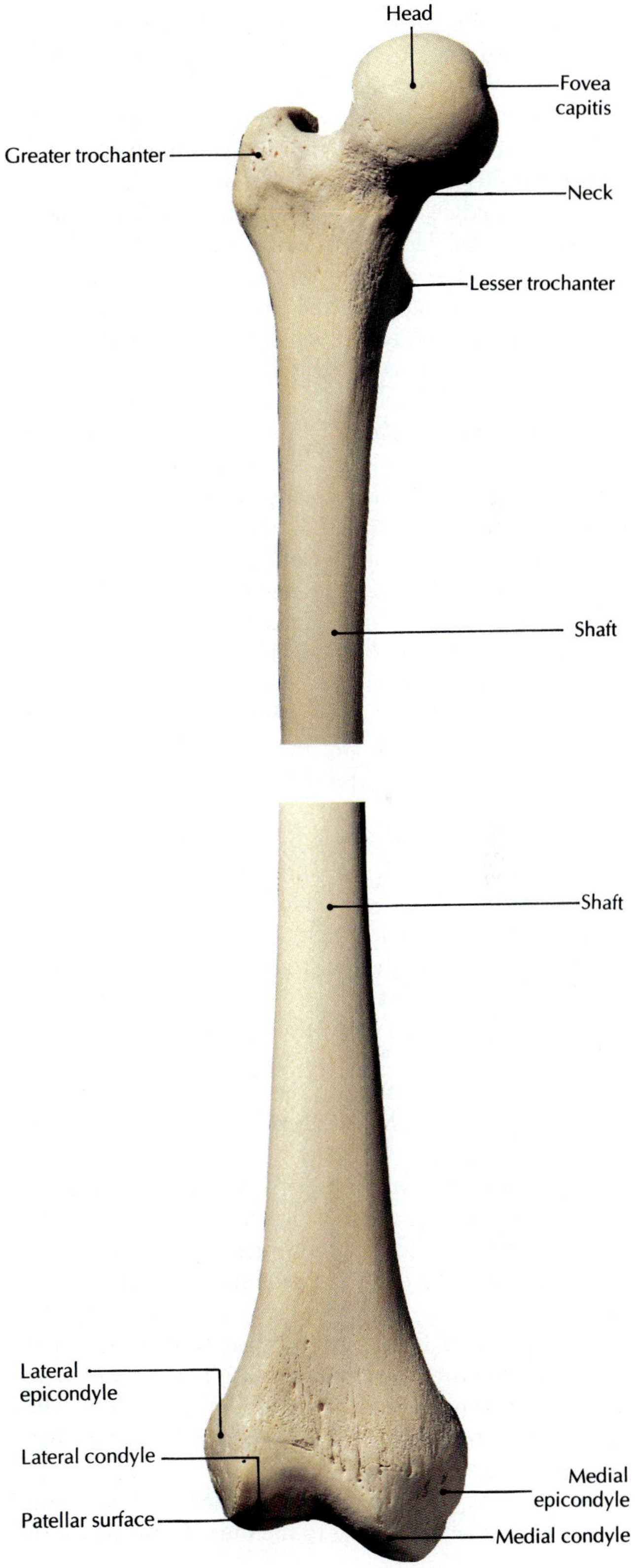

Head
Fovea capitis
Greater trochanter
Neck
Lesser trochanter

Shaft

Shaft

Lateral epicondyle
Lateral condyle
Patellar surface
Medial epicondyle
Medial condyle

[B]

and neck, laterally the **greater trochanter** (Gr. "to run") and medially the **lesser trochanter** are the sites of attachment for some large thigh and buttock muscles.

The *shaft*, or body, of the femur is slightly bowed anteriorly. It is fairly smooth except for a longitudinal posterior ridge called the **linea aspera** (L. rough line), which provides attachment sites for muscles. At the end of the femur toward the knee are **medial** and **lateral condyles,** which articulate with the tibia. Above the condyles are medial and lateral **epicondyles.** Epicondyles are the attachment sites for muscles, and condyles function in the movement of joints. The medial condyle is larger than the lateral condyle, so that when the knee is planted during walking, the femur rotates medially to "lock" the knee. Between the condyles on the anterior surface is a slight groove that separates the articular surface from the **patellar surface.** When the leg is bent or extended, the patella (kneecap) slides along this groove.

Patella The *patella* (L. little plate), or kneecap, is located within the quadriceps femoris tendon. This is the tendon of the muscle that extends the leg from the knee. Facets on the deep surface of the patella fit into the groove between the condyles of the femur.

The patella is the largest sesamoid bone in the body. It protects the knee, but more important, it increases the leverage for the action of the quadriceps femoris muscle by keeping its tendon farther away from the axis of rotation of the knee. The slightly pointed *apex* of the patella lies at the inferior end, and the rounded *base* is at the superior border [FIGURE 8.14].

Tibia The *tibia* (L. pipe), or shinbone, is located on the anterior and medial side of the leg, between the knee and the ankle. It is the second longest and heaviest bone in the body (the femur is first). This bone supports the body weight, transmitting it from the femur to the talus bone at the ankle joint.

At the proximal end of the tibia the **medial** and **lateral condyles** articulate with the condyles of the femur at the knee joint [FIGURE 8.15]. On the proximal anterior surface is the prominent **tibial tuberosity**, where the patellar tendon is attached.

At the distal end of the tibia is the **medial malleolus** (muh-LEE-oh-luhss), which articulates medially with the head of the talus. The junction of the talus and medial malleolus forms the easily felt prominence on the medial side of the ankle.

The tibia and fibula are attached throughout their lengths by an **interosseous membrane.** They articulate at both the proximal and the distal tibiofibular joints. The distal articulation site is the **fibular notch** on the tibia.

Fibula The *fibula* (L. pin or brooch) is a long, slender bone parallel and lateral to the tibia [FIGURE 8.15]. It

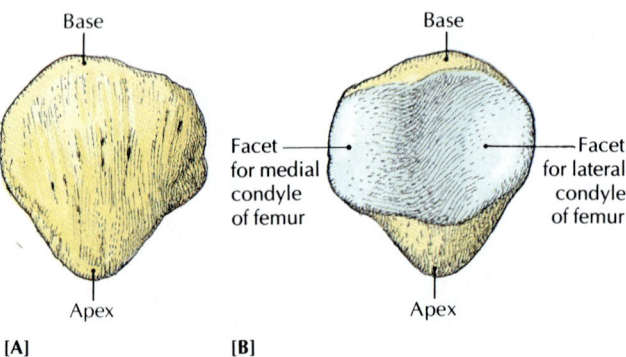

FIGURE 8.14 RIGHT PATELLA

[A] Anterior view. [B] Posterior view.

probably is so named because together with the tibia it somewhat resembles the clasp of a pin. The head at its proximal end articulates with the lateral condyle of the tibia but not with the femur. It articulates distally with the talus.

The slender shaft of the fibula bears little if any body weight, and it is not involved in the knee joint. However, the security of the ankle joint depends largely on the seemingly delicate fibula.

The medial malleolus of the tibia and the prominent **lateral malleolus** of the fibula both articulate tightly with the head of the talus. However, the fibula and talus are even more firmly bound together by ligaments to form a *mortise* (socket), which strengthens the ankle joint (*talocrural joint*) but limits the movement there to bending the foot up or down. Because the head of the talus is slightly wider anteriorly and the mortise is at the widest part of the talus, the ankle is most stable when the foot is dorsiflexed (as when a skier is crouching down). When the foot is extended (as when one is standing on tiptoe), the ankle joint is less stable because the mortise is at a narrower part of the talus.

Tarsus The *tarsus* is composed of the seven proximally located tarsal bones of the foot. They are classified as short bones.

The foot, unlike the hand, has relatively little free movement between its bones. The fingers, with their manipulative and gripping roles, are the functionally dominant structures in the hand. However, it is not the toes but the tarsal and metatarsal bones, with their weight-bearing and locomotive roles, that are the functionally significant structures of the foot. The bones work together as a lever, helping to raise the body and transmit the thrust during walking and running.

The *tarsal bones* are the talus, calcaneus, cuboid, navicular, and three cuneiforms [FIGURE 8.16]:

FIGURE 8.15 RIGHT TIBIA AND FIBULA

[A] Anterior view. [B] Photograph of upper and lower ends of right tibia and fibula, anterior view.

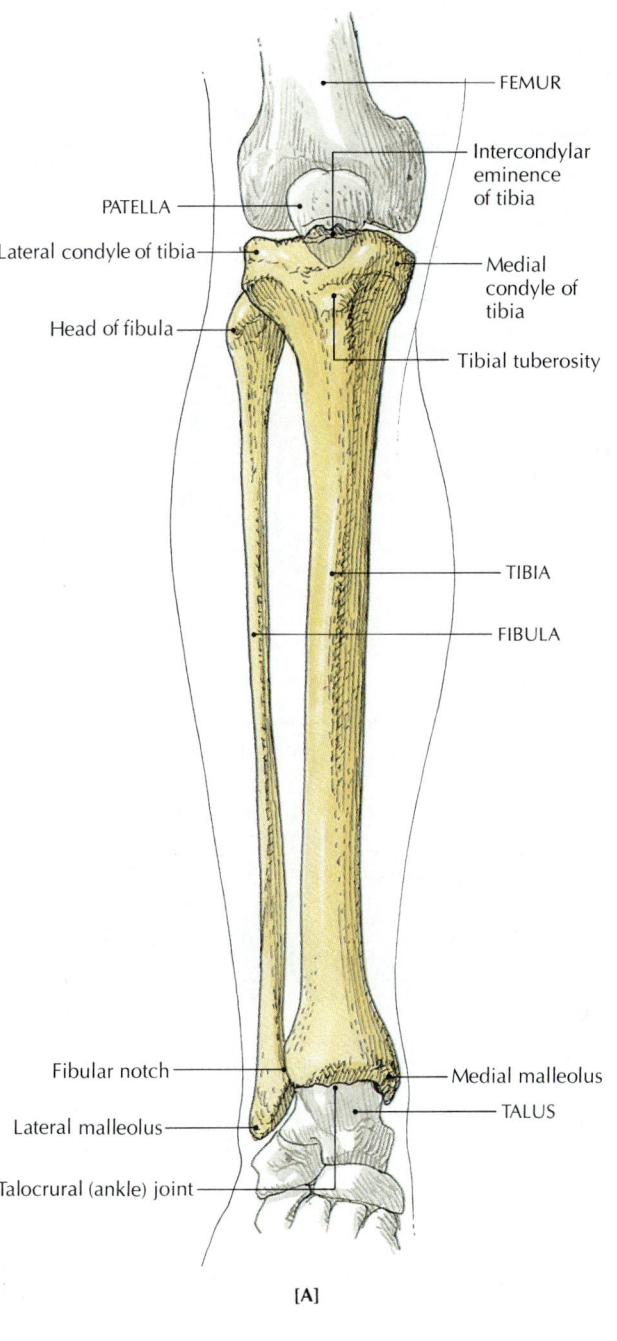

FEMUR

Intercondylar
eminence
of tibia

PATELLA

Lateral condyle of tibia

Medial
condyle of
tibia

Head of fibula

Tibial tuberosity

TIBIA

FIBULA

Fibular notch

Medial malleolus

Lateral malleolus

TALUS

Talocrural (ankle) joint

[A]

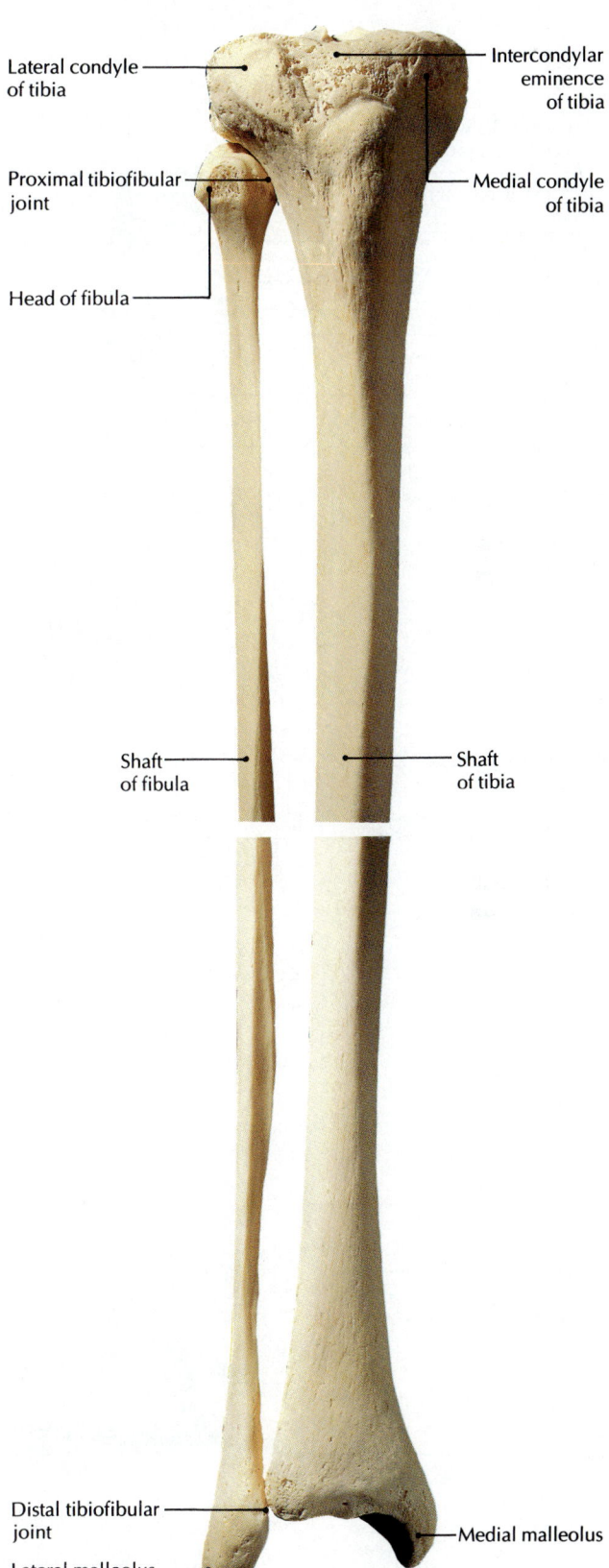

Lateral condyle
of tibia

Intercondylar
eminence
of tibia

Proximal tibiofibular
joint

Medial condyle
of tibia

Head of fibula

Shaft
of fibula

Shaft
of tibia

Distal tibiofibular
joint

Medial malleolus

Lateral malleolus

[B]

FIGURE 8.16 BONES OF THE RIGHT FOOT

[A] Superior view. [B] Photograph of bones of right foot, superior view. [C] Right medial view. [D] Right lateral view.

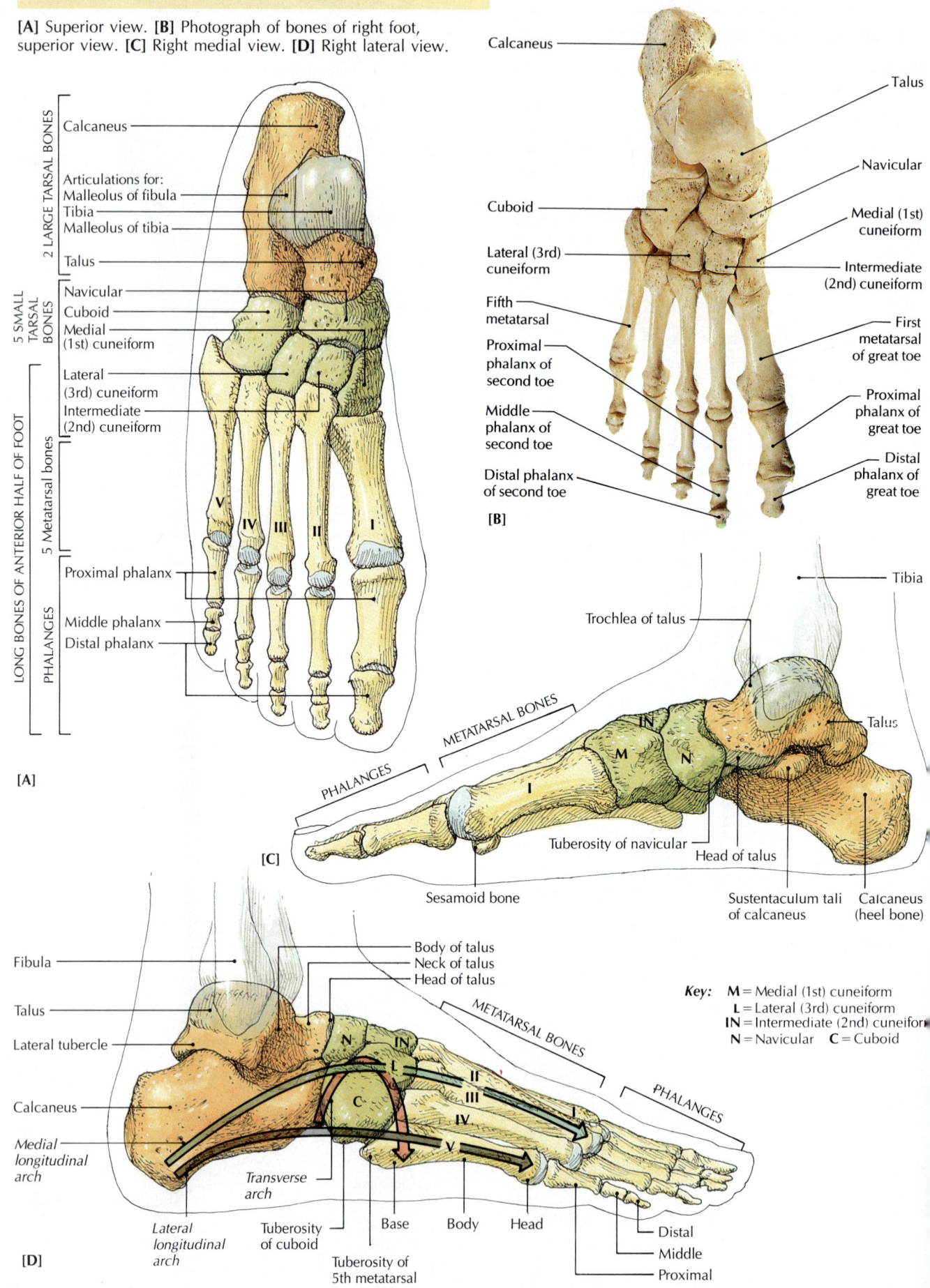

[A]

2 LARGE TARSAL BONES

Calcaneus

Articulations for:
Malleolus of fibula
Tibia
Malleolus of tibia

Talus

5 SMALL TARSAL BONES

Navicular
Cuboid
Medial (1st) cuneiform

Lateral (3rd) cuneiform
Intermediate (2nd) cuneiform

LONG BONES OF ANTERIOR HALF OF FOOT

5 Metatarsal bones

PHALANGES

Proximal phalanx

Middle phalanx
Distal phalanx

V IV III II I

[B]

Calcaneus
Talus
Navicular
Cuboid
Medial (1st) cuneiform
Lateral (3rd) cuneiform
Intermediate (2nd) cuneiform
Fifth metatarsal
First metatarsal of great toe
Proximal phalanx of second toe
Proximal phalanx of great toe
Middle phalanx of second toe
Distal phalanx of great toe
Distal phalanx of second toe

[C]

Trochlea of talus
Tibia
Talus
METATARSAL BONES
PHALANGES
Tuberosity of navicular
Head of talus
Sustentaculum tali of calcaneus
Calcaneus (heel bone)
Sesamoid bone

[D]

Body of talus
Neck of talus
Head of talus
METATARSAL BONES
PHALANGES
Fibula
Talus
Lateral tubercle
Calcaneus
Medial longitudinal arch
Lateral longitudinal arch
Transverse arch
Tuberosity of cuboid
Tuberosity of 5th metatarsal
Base
Body
Head
Distal
Middle
Proximal

Key: **M** = Medial (1st) cuneiform
L = Lateral (3rd) cuneiform
IN = Intermediate (2nd) cuneiform
N = Navicular **C** = Cuboid

1 The *talus* (TAY-luhss; L. ankle), or anklebone, is the central and highest foot bone. It articulates with the tibia and fibula to form the ankle joint. Together with the calcaneus it receives the weight of the body.

2 The *calcaneus* (kal-KAY-nee-us; L. heel) is commonly called the heelbone. It is the largest tarsal bone and is suited to supporting weight and adjusting to irregularities of the ground. It acts as a lever, providing a site of attachment for the large calf muscles through the *calcaneal* (Achilles) *tendon.* The sustentaculum tali is a process that projects medially from the calcaneus bone. It supports the talus bone.

3 The *cuboid* bone is usually, but not necessarily, shaped like a cube. It is the most lateral tarsal bone.

4 The *navicular* (L. little ship) is a flattened oval bone that some people think looks like a boat because of the depression on its posterior surface that houses the head of the talus. The three cuneiform bones line up on its anterior surface.

5 The three *cuneiform* bones (KYOO-nee-uh-form; L. wedge) are referred to as *medial* (first), *intermediate* (second), and *lateral* (third). They line up along the anterior surface of the navicular bone. Their wedgelike shapes contribute to the structure of the transverse arch of the foot.

Metatarsus and phalanges The five metatarsal (L. behind the ankle) bones form the skeleton of the sole of the foot, and the 14 phalanges are the toe bones.

The *metatarsus* consists of five *metatarsal bones,* numbered I to V from the medial side [FIGURE 8.16]. They are miniature long bones that consist of a proximal base, a body or shaft, and a distal head. The tarsus and metatarsus are arranged as arches, primarily to improve stability during standing, while the toes provide a stable support during locomotion.

The *phalanges* (fuh-LAN-jeez; Gr. line of soldiers), or toes, like the metatarsals, are miniature long bones with a base, shaft, and head. There are 14 phalanges in each foot, arranged similarly to the phalanges of the hand. The phalanges of the foot, however, are much shorter and are quite different functionally from those in the hand. They contribute to stability rather than to precise movements. The big toe has two phalanges (proximal and distal), and the other four toes each have three phalanges (proximal, middle, distal).

Arches of the Foot

The sole of your foot is arched for the same reason that your spine is curved: The elastic spring created by the arched bones of the foot absorbs enormous everyday shocks the way the spring-curved spine does. A running step may flatten the arch by as much as half an inch. When the pressure is released as the foot is raised, the arch springs back into shape, returning about 17 percent

of the energy of one step to the next. The arches combine with the calcaneal tendons to reduce by about half the amount of work the muscles need to expend when we run. Without arches, our feet would be unable to move properly or to cushion the normal pressure of several thousand pounds per square inch every time we take a step. In addition, the arches prevent nerves and blood vessels in the sole of the foot from being crushed.

Longitudinal and transverse arches There are two longitudinal arches of the foot: (1) the medial longitudinal arch and (2) the lateral longitudinal arch. The so-called *transverse arch* is located roughly along the distal tarsal bone and the tarsometatarsal joints. It runs across the foot between the heel and the ball of the foot. Technically, it is not an arch because only one side contacts the ground. The high *medial longitudinal arch* consists, in order, of the calcaneus, talus, navicular, three cuneiforms, and metatarsals I, II, and III [FIGURE 8.16D]. It runs from the heel to the ball of the foot on the inside. The low *lateral longitudinal arch* consists of the calcaneus, the cuboid, and metatarsals IV and V. It runs parallel to the medial longitudinal arch from the heel to the ball of the foot on the outside. The talus is the keystone, and the calcaneus and the distal ends of the metatarsal bones are the pillars that contact the ground.

The arches are held in place primarily by strong ligaments and in part by attached muscles and the bones themselves. The ligaments are largely responsible for the resiliency of the arches.

The *plantar* (underside of the foot) *calcaneonavicular ligament,* extending from the front of the calcaneus to the back of the navicular, is known as the "spring" ligament. It is important because it keeps the calcaneus and navicular bones together and holds the talus in its keystone position. The security of the high arch and, in a way, of the whole foot depends on this ligament. If it weakens and the calcaneus and navicular spread apart, the talus will sag and occupy the space between them; the result is a "flat foot" ("fallen arch"). The tendon of the powerful tibialis posterior muscle comes from the medial side of the foot and passes immediately below the "spring ligament" on its way to its lateral insertion. This tendon is presumed to act as an additional spring, helping to reinforce the ligament during excessive strain.

A S K   Y O U R S E L F

1 What bones constitute the lower extremities?

2 What bones constitute the pelvic girdle?

3 What are the differences between male and female pelves? Why are male and female pelves different?

4 What is the function of the arches of the foot? Name the major arches.

WHEN THINGS GO WRONG

Fractures

A *fracture* is a broken bone. Children have fractures more often than adults do because children have slender bones and are more active. Fortunately, the supple, healthy bones of children mend faster and better than the more brittle bones of older people. (A femur broken at birth is fully healed within 3 weeks, but a similar break in a person over 20 may take 4 or 5 months to heal completely.) Usually, broken bones that are reset soon after injury have an excellent chance of healing perfectly be-

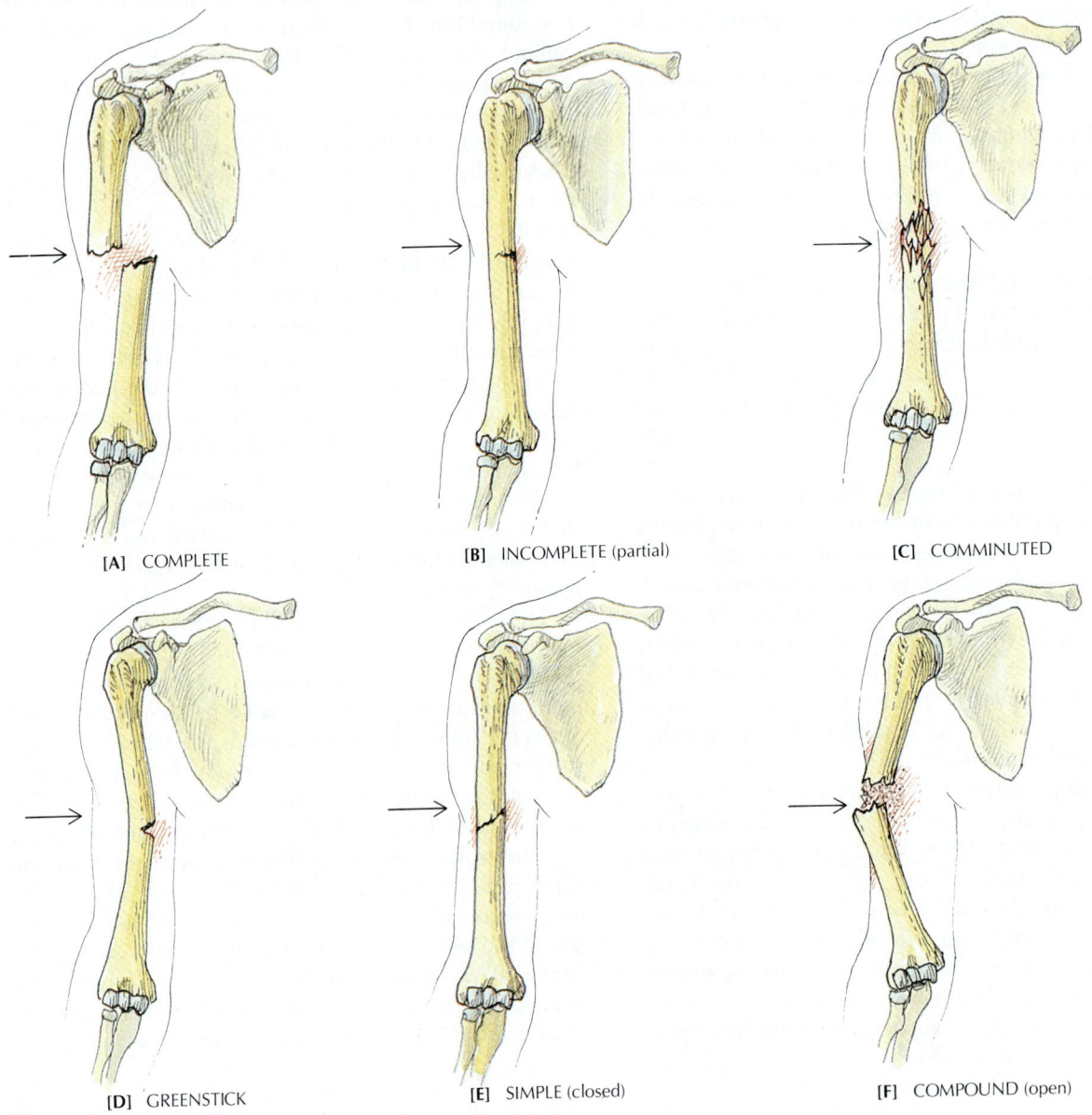

[A] COMPLETE   [B] INCOMPLETE (partial)   [C] COMMINUTED

[D] GREENSTICK   [E] SIMPLE (closed)   [F] COMPOUND (open)

Kinds of fractures. Fractures can be classified according to the type and complexity of the break, the location of the break, and certain other special features. The following commonly used types and classifications are shown in the respective illustrations.

[A] *Complete.* The bone breaks completely into two pieces.
[B] *Incomplete (partial).* The bone does not break completely into two or more pieces.

[C] *Comminuted.* The bone is splintered or crushed into small pieces.
[D] *Greenstick.* The bone is broken on one side and bent on the other; common in children.
[E] *Simple (closed).* The bone is broken but does not break through the skin.
[F] *Compound (open).* The bone is broken and cuts through the skin.

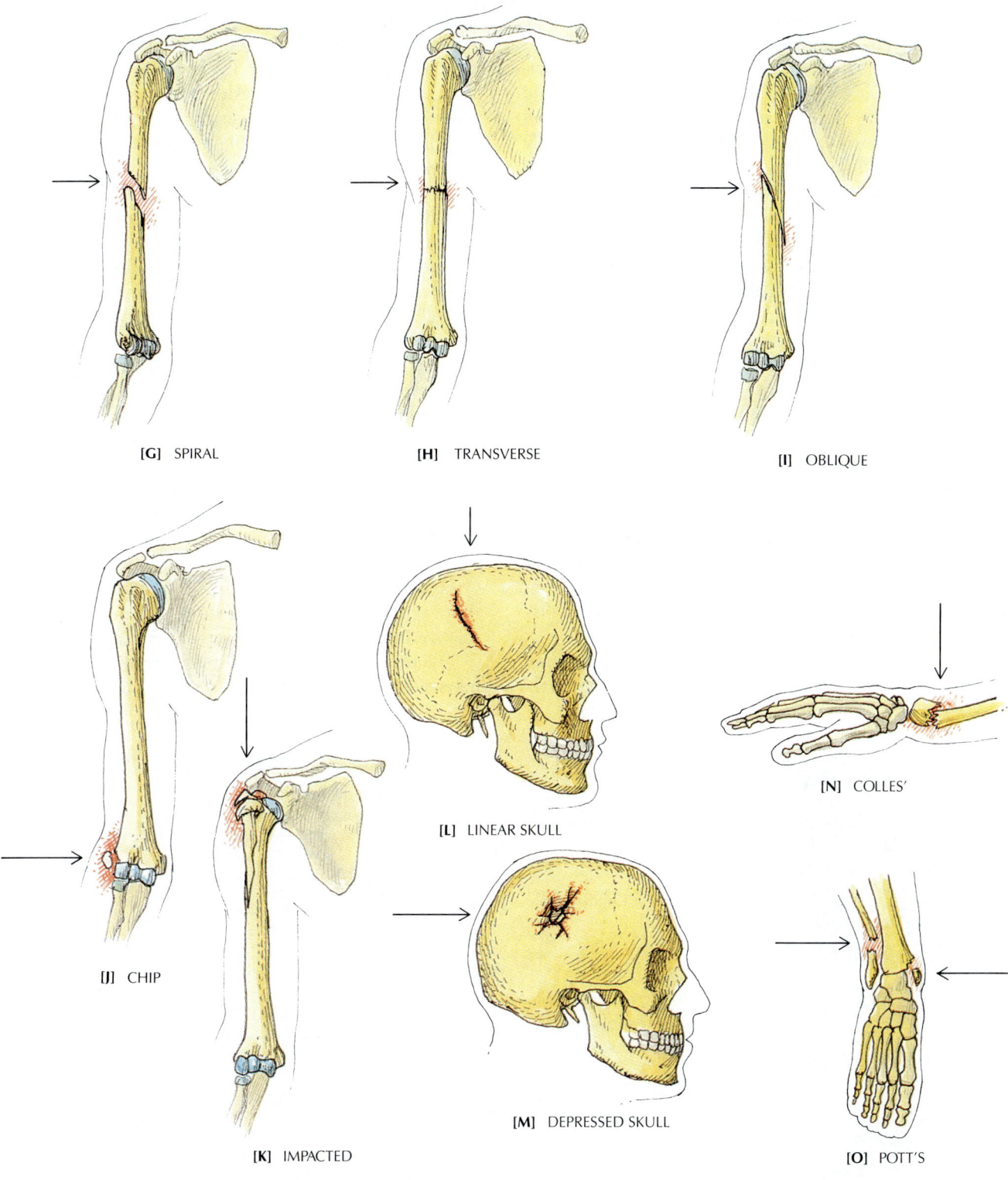

[G] SPIRAL

[H] TRANSVERSE

[I] OBLIQUE

[J] CHIP

[K] IMPACTED

[L] LINEAR SKULL

[M] DEPRESSED SKULL

[N] COLLES'

[O] POTT'S

[G] *Spiral*. The bone is broken by twisting.
[H] *Transverse*. The bone is broken directly across, at a right angle to the bone's long axis.
[I] *Oblique*. The bone is broken on a slant, at approximately a 45-degree angle to the bone's long axis.
[J] *Chip*. The bone is chipped where a protrusion is exposed.
[K] *Impacted*. The bone is broken when one part is forcefully driven into another, as at a shoulder or hip.

[L] *Linear skull*. The skull is broken in a line, lengthwise on the bone.
[M] *Depressed skull*. The skull is broken by a puncture, causing a depression below the surface.
[N] *Colles'*. The distal end of the radius is displaced posteriorly.
[O] *Pott's*. The distal part of the fibula and medial malleolus of the tibia are broken.

cause the living tissue and adequate blood supply at the fracture actually stimulate a natural repositioning.

In elderly people, bones contain relatively more calcified bone and less organic material. Consequently, old bones lose their elasticity and break more easily. A fall that a child hardly notices can be serious in an elderly person. "To fall and break a hip," a common disaster among the elderly, could frequently be better stated, "to break a hip and fall," because the fragile old bones may break merely under the strain of walking, making the legs give way. The hip (actually the neck of the femur, the most fragile part in elderly people) may be broken before the body hits the ground.

A fractured bone goes through several stages of healing. But even before healing can begin properly, the fragments of the broken bone must be manipulated, or *reduced*, back into their original positions by a physician. Usually the bone is immobilized by a cast, splint, or traction, and in severe cases surgery and a continuing program of physical therapy may be necessary.

Fractures of the vertebral column Many fractures of the vertebral column may be serious in themselves, but the real danger lies in injury to the spinal cord, which can result in paralysis or death. (Spinal-cord injuries are discussed in Chapter 14.)

The most common type of fracture is a *compression fracture*, which crushes the body of one or more verte-brae. Compressions often occur where there is the greatest spinal mobility: the middle or lower regions of the vertebral column and near the point where the lumbar and thoracic regions meet. Although a compression fracture crushes the body of a vertebra, the vertebral arches and the ligaments of the spine remain intact. As a result, the spinal cord is not injured.

In contrast to compression fractures, *extension fractures* and dislocations involve a pulling force, usually affecting the posterior portions of the vertebral column. When the neck is severely hyperextended (bent backward), as in a "whiplash" injury, the atlas may break at several points, and further extension may break off the arch of the axis at the isthmus. An even greater force may rupture one of the ligaments and the *annulus fibrosus* (outer ring) of the C2/C3 intervertebral disk. Such a great force separates the skull, atlas, and axis from the rest of the vertebral column, and the spinal cord is usually severed in the process.

The *interspinous ligaments* connect the spinous processes and restrict the range of movement of vertebrae. Some whiplash injuries may result when rapid stretching of these ligaments produces small tears in them.

Hyperextension injuries usually do not occur in the thoracic region because of the support of the ribs and the relative immobility of the thoracic vertebrae.

Fracture usually accompanies dislocations of vertebrae because the thoracic and lumbar articular processes in-

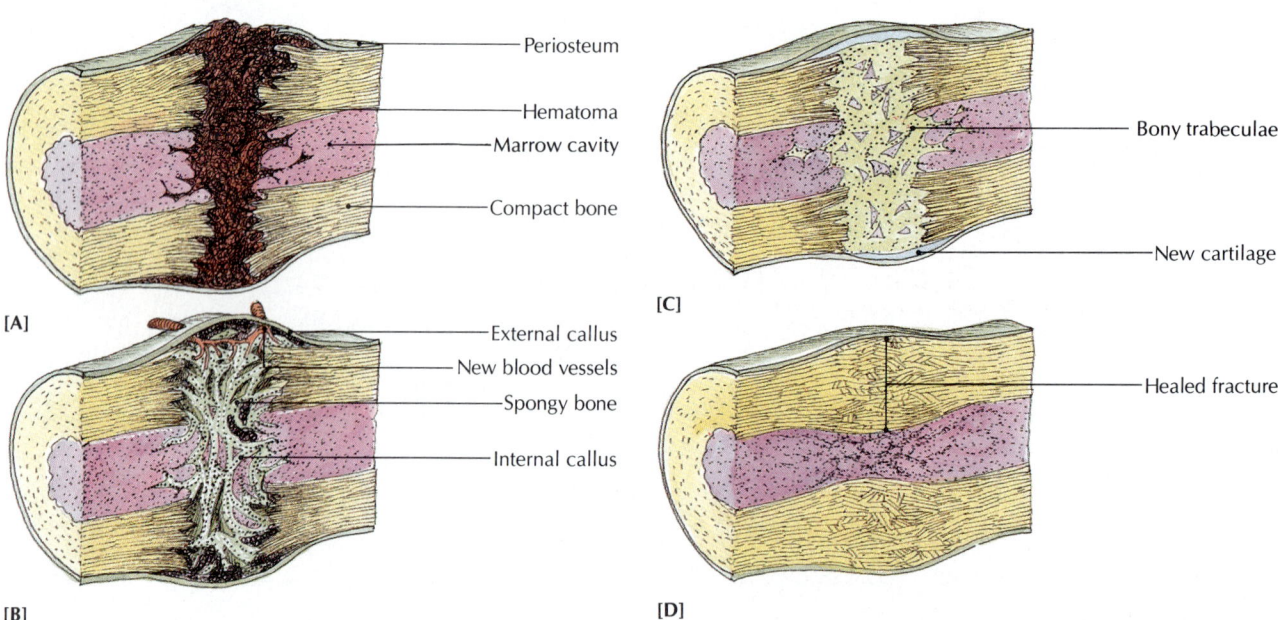

[A] Periosteum / Hematoma / Marrow cavity / Compact bone

[B] External callus / New blood vessels / Spongy bone / Internal callus

[C] Bony trabeculae / New cartilage

[D] Healed fracture

How a fracture heals. [A] Hemorrhaging occurs when tissue in the periosteum, osteon system, and marrow cavity are damaged. A hematoma (blood clot) forms several hours later. [B] Fibroblasts enter the damaged area, and a hard callus forms a few days later. [C] Osteogenic cells differentiate into new bony trabeculae, knitting the new fragments together. New cartilage forms on the outer collar of the fracture. [D] The cartilage is gradually replaced by spongy bone, and the fracture is repaired with compact bone.

terlock. The spinal cord is not necessarily severed when cervical vertebrae are dislocated because the vertebral canal in the cervical region is wide enough to allow some displacement without damaging the cord.

The primary treatment for injuries of the vertebral column is usually immobilization, which allows the bone to heal and prevents damage to the spinal cord. Surgery may be necessary to relieve pressure or repair severely damaged vertebrae or tissues. Exercises to strengthen back muscles are ordinarily prescribed after the fracture is healed.

The anterior longitudinal ligament interconnects the vertebral bodies and intervertebral disks along the anterior surface of the vertebral column. This broad ligament helps prevent intervertebral disk hernias from being directed anteriorly. The ligament can actually be used to *splint* a fractured vertebra when the trunk is cast in extension. When a vertebral fracture is suspected, the vertebral column should be kept in extension. Emergency paramedics are careful to keep the back in extension when removing a crash victim from a car, since flexion of the vertebral column can cause further injury to the spinal cord.

Bunion

A **bunion** (Old Fr. *buigne*, bump on the head) is a lateral deviation of the big toe toward the second toe, accompa-nied by the formation of a bursa and callus at the bony prominence of the first metatarsal. It may be caused by poorly fitted shoes that compress the toes. Bunions are most common among women.

Shin Splint

Shin splint is a common term for the pain, tenderness, and, at times, cramps caused by swelling of the muscles that dorsiflex the ankle. These muscles (tibialis anterior, extensor hallicus longus, extensor digitorum longus, and peroneus brevis) are surrounded and contained within the anterior compartment formed by the tibia and fibula, the interosseous membrane between these bones, and a heavy sheath of fascia. As a result of overuse, the muscles swell within the compartment and become painful and tender to the touch. A reduction of blood flow into the tight anterior compartment also produces painful muscle spasms (cramps). This condition typically afflicts poorly conditioned people who overuse their muscles, but it also occurs in trained athletes who do not warm up properly or who do not cool down sufficiently following strenuous physical activity. In some individuals shin splints may be caused by tendinitis of the tibialis posterior muscle. In these cases the resulting imbalance may be corrected with different running shoes or corrective shoe supports (orthotics).

CHAPTER SUMMARY

The Upper Extremities (Limbs) (p. 219)

1 Each **upper extremity** of the appendicular skeleton includes the scapula and clavicle of the pectoral girdle, the humerus of the arm, the radius and ulna of the forearm, the carpal bones of the wrist, the metacarpals of the palm, and the phalanges of the fingers. Some functions of the upper extremity include *balancing* while the body is moving, *grasping* of objects, and *manipulation* of objects.

2 The upper limb girdle is also known as the **pectoral girdle** or **shoulder girdle.** It consists of the clavicle and scapula.

3 The (upper) arm bone is the humerus, and the forearm bones are the ulna and radius. In each wrist are eight carpals, in each palm five metacarpals, in each thumb two phalanges, and in each finger three phalanges.

The Lower Extremities (Limbs) (p. 226)

1 The skeleton of the **lower extremity** or limb consists of the hipbones of the pelvic girdle, the femur of the thigh, the tibia and fibula of the leg, the tarsal bones of the ankle, the metatarsals of the foot, and the phalanges of the toes. Among the functions of the lower extremity are *movement,* such as in walking, and *balancing,* such as in standing.

2 The **pelvic girdle** is formed by the paired hipbones (ossa coxae), which help hold the body in an upright posi-tion. The *pelvis* is formed by the sacrum, coccyx, and hipbones.

3 The **hipbones** are formed in the adult by the fusion of the ilium, ischium, and pubis.

4 The female pelvis is lighter and wider than the male pelvis, enabling a woman to carry and deliver a baby.

5 The bones of the legs and feet are the femur (2), patella (2), tibia (2), fibula (2), tarsus (14), metatarsus (10), and phalanges (28).

6 The **arches** of the foot provide strength and resiliency. The two true arches are the *medial longitudinal arch* and the *lateral longitudinal arch.* The so-called *transverse arch* contacts the ground on only one side.

STUDY AIDS FOR REVIEW

KEY TERMS

MEDICAL TERMS FOR REFERENCE

COLLES' FRACTURE A type of fracture that occurs at the distal end of the radius. It often occurs when a person extends the hands to break a fall.

GENU VALGUM (L. *genu*, knee + *valgus*, bowlegged) Bowleggedness; a deformity typical of rickets.

GENU VARUM (L. *varus*, bent inward) Knock-kneed.

PELVIMETRY The measurement of the pelvic cavity and birth canal by a physician prior to the birth of a child. The procedure determines if the opening of the mother's lesser pelvis is large enough to allow the passage of the child's head and shoulders.

POTT'S FRACTURE A fracture and dislocation of the distal fibula. It is usually caused by the forceful turning outward of the foot, which destabilizes the ankle joint.

SHOULDER SEPARATION A separation of the acromioclavicular joint, not the shoulder. It is usually a serious injury only when the accompanying ligaments are torn.

SYNDACTYLISM (Gr. *syn*, together + *daktulos*, finger) The whole or partial fusion of two or more fingers or toes.

UNDERSTANDING THE FACTS

1 Name the bones in the upper extremity and the lower extremity.
2 Identify each of the following:
   a glenohumeral joint   d symphysis pubis
   b olecranon process   e ossa coxae
   c radiocarpal joint   f pelvis
3 Which bones form the knee and ankle joints?
4 How do the glenoid fossa and the acetabulum differ? What results from this difference?
5 How do male and female pelves differ?

UNDERSTANDING THE CONCEPTS

1 How do the pectoral and pelvic girdles differ structurally and functionally?
2 Describe how the weight of the body is transmitted to the ground when the body is in the standing position.
3 How do the arches of the foot contribute to the mobility of the body?

SELF-QUIZ

Multiple-Choice Questions

8.1 Which of the following is *not* part of the appendicular skeleton?
   a scapula   d ribs
   b clavicle   e tibia
   c radius

8.2 The deltoid tuberosity, radial groove, and intertubercular groove are structural features of the
   a ulna   d tibia
   b humerus   e scapula
   c femur

8.3 The nonweight-bearing bone of the lower extremity is the
   a tibia   d femur
   b fibula   e calcaneus
   c talus

8.4 The cuneiforms are
   a metatarsal bones   d carpal bones
   b phalanges   e tarsal bones
   c pelvic bones

8.5 The metacarpophalangeal joint forms the
   a shoulder   d wrist
   b knuckle   e symphysis pubis
   c ankle

8.6 Which of the following is *not* a tarsal bone?
   a navicular   d talus
   b calcaneus   e coccyx
   c cuboid

8.7 The fibula is lateral to the
   a tibia   c femur   e talus
   b ulna   d radius

8.8 The heelbone is the
   a fibula   c navicular   e lunate
   b talus   d calcaneus

Matching

  a tibia
  b ulna
  c scapula
  d ischium
  e talus
  f pisiform
  g ilium

8.9 _____ forms a joint with the axial skeleton

8.10 _____ glenoid fossa and acromion process are part of this bone

8.11 _____ a sesamoid bone in the wrist

8.12 _____ olecranon process that forms prominence of elbow

8.13 _____ one of several bones that form the knee joint

8.14 _____ strongest bone of the ossa coxae

8.15 _____ a weight-bearing anklebone that articulates with the heelbone

Completion Exercises

8.16 The bone extending between the shoulder and the elbow is the _____.

8.17 The shinbone is the _____.

8.18 The medial malleolus is part of the _____.

8.19 The bones of the wrist are the _____ bones.

8.20 The largest sesamoid bone in the body is the _____.

8.21 The tibia and fibula are attached along their lengths by a(n) _____.

8.22 The opening in the pelvis through which a baby passes during birth is called the _____.

8.23 The hipbones are formed by the fusion of the _____, _____, and _____.

8.24 The pelvis is formed by the _____, _____, and _____.

8.25 Muscles of the thigh and buttocks attach to the trochanters of the _____.

A SECOND LOOK

1 In the two drawings below, identify the radius, ulna, tibia, and fibula.

2 Identify the types of fractures in the two drawings below.

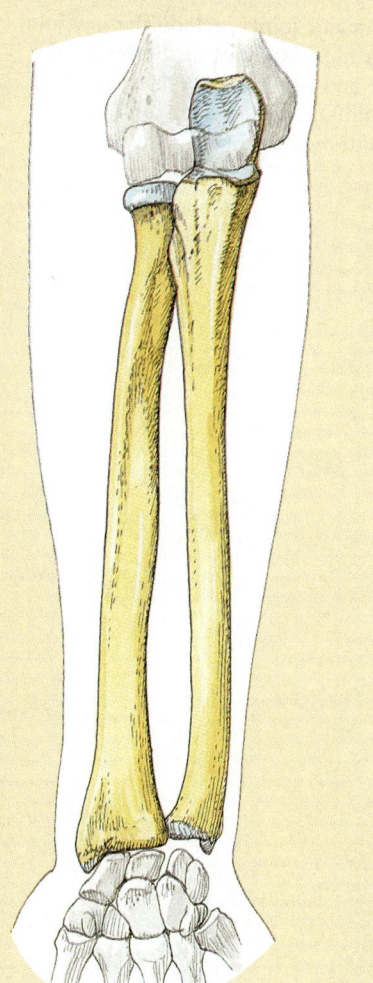
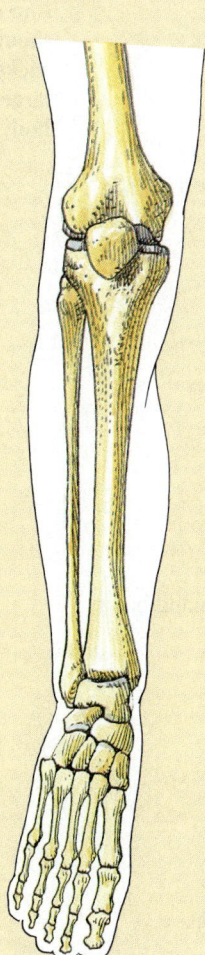
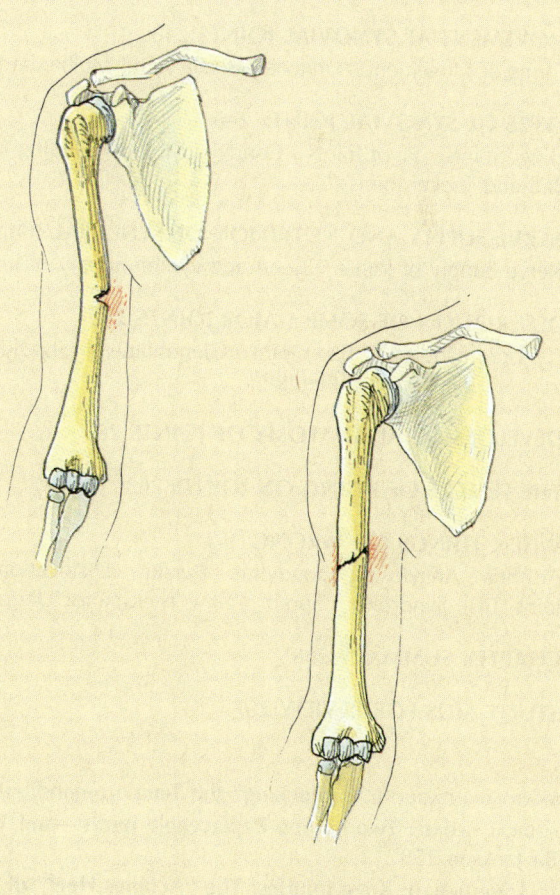

Articulations

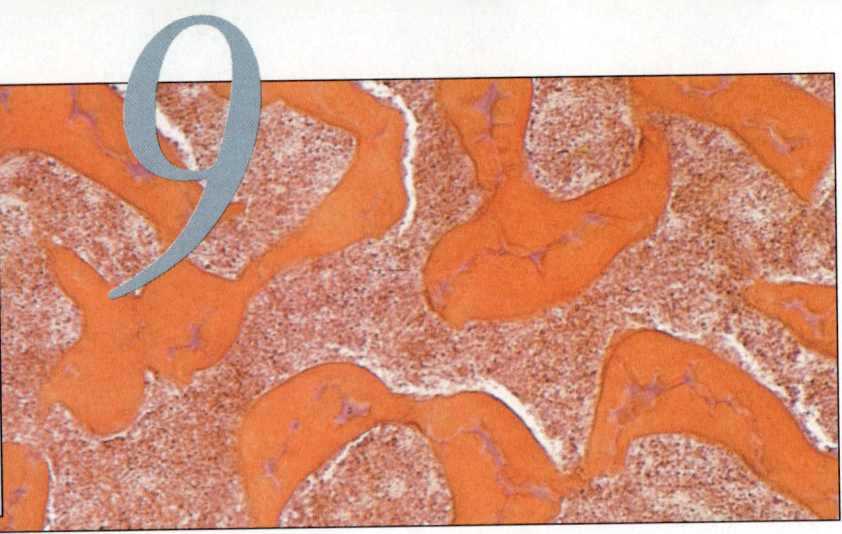

9

KEY CONCEPTS

1 Joints, which are formed where bones
meet, typically allow the body to move.
Although most joints are movable, some
are not.

2 The effectiveness of joints depends on
the coordination of the nervous, skeletal,
and muscular systems.

3 Some major joints include the jaw joint
(the only movable joint of the head), the
shoulder and hip joints (both ball-and-
socket joints), and the knee joint (the
largest and most complex joint in the
body).

Bones give the body its structural framework and muscles give it its power, but movable joints provide the mechanism that allows the body to move. A joint, or *articulation* (L. *articulus*, small joint), is the place where two adjacent bones or cartilages meet, or where adjacent cartilages or adjacent bones and cartilages are joined. Although most joints are movable, some are not. The effectiveness of the articular system involves the exquisite coordination of the nervous, muscular, and skeletal systems. To fully appreciate the usefulness of movable joints, try to walk or sit without bending your knees or eat dinner without bending your elbow or wrist.

CLASSIFICATION OF JOINTS

Joints are classified by two methods. One way to classify joints is by the extent of their *function*, that is, their *degree of movement*. According to this system, a joint may be immovable, slightly movable, or freely movable. An immovable joint, which is called a *synarthrosis* (Gr. *syn*, together + *arthrosis*, articulation), is an articulation in which the bones are rigidly joined together. A slightly movable joint, which is called an *amphiarthrosis* (Gr. *amphi*, on both sides), allows limited motion. A freely movable joint is a *diarthrosis* (Gr. *dia*, between).

Another way to classify joints is by their *structure*. This classification is based on the presence or absence of a joint cavity and on the kind of supporting tissue that binds the bones together. Based on structure, three types of joints are recognized: *fibrous, cartilaginous,* and *synovial.*

TABLE 9.1 sums up the classification of joints based on structure. As you can see, fibrous joints are generally synarthroses, cartilaginous joints are generally amphiarthroses, and synovial joints are generally diarthroses. However, the structural and functional categories are *not always* equivalent.

FIBROUS JOINTS

Fibrous joints lack a joint cavity, and fibrous connective tissue unites the bones. Because they are joined together tightly, fibrous joints are generally immovable in the adult, although some of them do allow slight movement. Three types of fibrous joints are generally recognized: sutures, syndesmoses, and gomphoses.

Sutures

A *suture* is usually such a tight union in an adult that movement rarely occurs between the two bones. Because sutures are found only in the skull, they are sometimes called "skull type" joints [see FIGURE 7.4]. Movement at

sutures can occur in fetuses and young children, since the joints have not yet grown together. In fact, the flexibility of the skull is necessary for a newborn baby to be able to pass through its mother's narrow birth canal. The flexibility of cranial sutures in fetuses and children also allows for the growth of the brain. In adults, the fibers of connective tissue between bones are replaced by bone, and the bones become permanently fused. This fusion provides complete protection for the brain from external factors. Such a sealed joint is called a *synostosis* (Gr. *syn*, together + *osteon*, bone).

Syndesmoses

When bones are close together but not touching, and are held together by collagenous fibers or interosseous ligaments, the joint is called a *syndesmosis* (sihn-dehz-MOH-sihss; Gr. to bond together; *syn* + *desmos*, bond). The amount of movement, if any, in a syndesmosis depends on the distance between the bones and the amount of flexibility of the fibrous connecting tissue. Ligaments (composed of collagenous connective tissue) make a firm articulation at the inferior tibiofibular joint so that very little movement occurs there. This limited movement adds strength to the joint. In contrast, the interosseous ligaments between the shafts of the radius and the ulna allow much more movement, including the twisting (pronation, supination) of the forearm [TABLE 9.1].

Gomphoses

A *gomphosis* (gahm-FOH-sihss; Gr. *gomphos*, bolt) is a fibrous joint made up of a peg and a socket. The root of each tooth (the peg) is anchored into its socket by the fibrous periodontal ("around the tooth") ligament, which extends from the root of the tooth to the alveolar processes of the maxillae or mandible.

CARTILAGINOUS JOINTS

In *cartilaginous joints* the bones are united by a plate of hyaline cartilage or a fibrocartilaginous disk [TABLE 9.1]. These joints lack a joint cavity and permit little or no movement. The two types of cartilaginous joints are the synchondrosis and the symphysis.

Synchondroses

A *synchondrosis* (sihn-kahn-DROH-sihss; Gr. *syn* + *chondros*, cartilage) is also called a *primary cartilaginous joint.* The chief function of a synchondrosis is to permit growth, not movement. It is a temporary joint composed of an epiphyseal plate of hyaline cartilage that joins the diaphysis and epiphysis of a growing long bone. A syn-

TABLE 9.1 CLASSIFICATION OF JOINTS

Structural and functional classification*	Type of structure and movement†	Examples

FIBROUS

Fibrous connective tissue unites articulating bones; no joint cavity. *Mostly immovable, some slightly movable; usually synarthroses.*

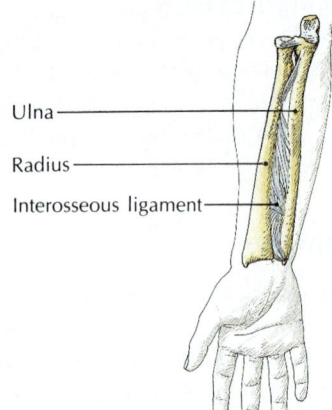

Ulna

Radius

Interosseous ligament

Suture: found only in skull; fibrous tissue between articulating bones in children, but permanently fused in adults. *Some movement in fetuses and young children; immovable in adults.*

Cranial sutures, such as coronal suture between frontal and parietal bones.

Syndesmosis: articulating bones held together (but not touching) by fibrous or interosseous ligaments. *Slight movement: twisting of forearm (pronation, supination).*

Inferior tibiofibular joint; interosseous ligament between shafts of radius and ulna.

Gomphosis: a peg fitting into a socket. *Mostly immovable; very slight movement of teeth in their sockets.*

Roots of teeth in alveolar processes of mandible and maxillae.

CARTILAGINOUS

Articulating bones united by plate of hyaline cartilage or fibrocartilaginous disk. *Mostly slightly movable, some immovable; usually amphiarthroses.*

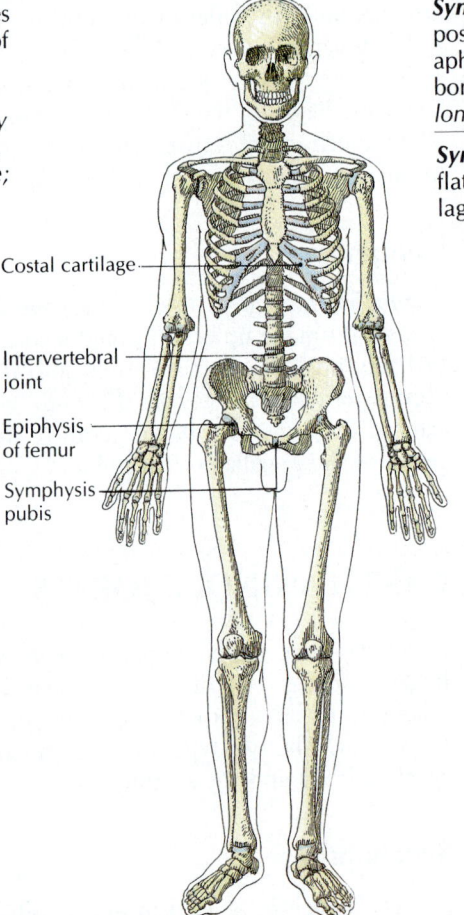

Costal cartilage

Intervertebral joint

Epiphysis of femur

Symphysis pubis

Synchondrosis: temporary joint composed of hyaline cartilage joining diaphysis and epiphysis of growing long bones. *Immovable; permits growth of long bones.*

Epiphyseal plate of femur; union of manubrium and body of sternum.

Symphysis: bony surfaces bridged by flattened plates or disks of fibrocartilage. *Slight movement.*

Symphysis pubis, manubriosternal joint, intervertebral joints between bodies of vertebrae.

---

*Note that synarthrosis and fibrous, amphiarthrosis and cartilaginous, and diarthrosis and synovial are *not always* synonymous.
†See TABLE 9.2 for an explanation of the terms for different types of movement.

Structural and functional classification*	Type of structure and movement†	Examples

SYNOVIAL

Articulating bones moving freely along smooth, lubricated articular cartilage; enclosed within flexible articular capsule. *Freely movable; usually diarthroses.*

UNIAXIAL: movement of bone about one axis of rotation.

Hinge: convex surface of one bone fitted into concave surface of other. *Flexion, extension.*

Elbow, interphalangeal joints, knee, ankle.

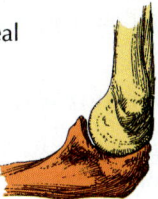

Pivot: central bony pivot surrounded by collar of bone and ligament. *Supination, pronation, rotation.*

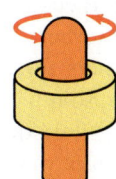

Proximal radioulnar joint, atlantoaxial joint.

BIAXIAL: movement of bone about two axes of rotation.

Condyloid (ellipsoidal): modified ball-and-socket. *Flexion, extension, abduction, adduction, circumduction.*

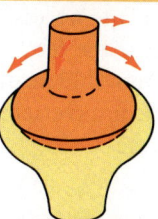

Metacarpophalangeal (knuckle) joints, except thumb.

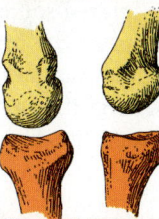

MULTIAXIAL: movement of bone about three axes.

Gliding: essentially flat articular surfaces. *Simple gliding movement within narrow limits.*

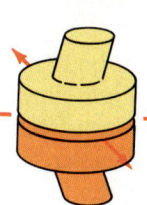

Between articular processes of vertebrae, acromioclavicular joint, some carpal and tarsal bones.

Saddle: opposing articular surfaces with both concave and convex surfaces that fit into one another. *Abduction, adduction, opposition, reposition.*

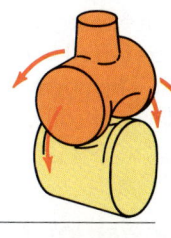

Carpometacarpal joint of thumb.

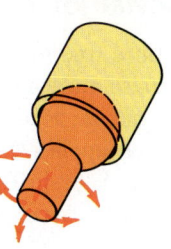

Ball-and-socket: globelike head of one bone fitted into cuplike concavity of another bone. *Flexion, extension, internal rotation, lateral rotation, abduction, adduction, circumduction.*

Shoulder joint, hip joint.

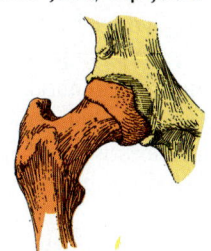

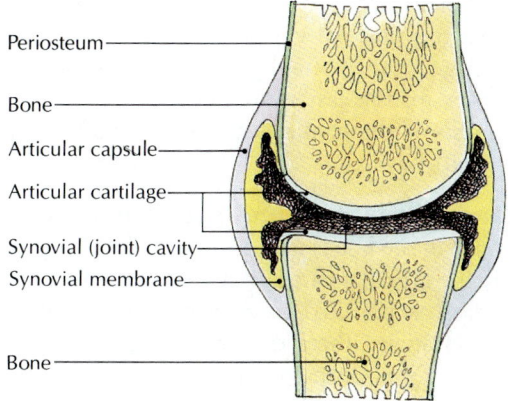

Periosteum

Bone

Articular capsule

Articular cartilage

Synovial (joint) cavity

Synovial membrane

Bone

SYNOVIAL JOINT WITHOUT ARTICULAR DISK

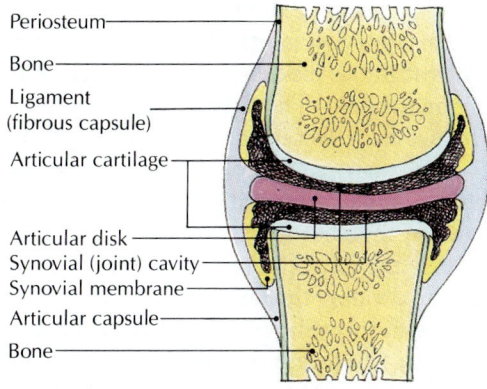

Periosteum

Bone

Ligament (fibrous capsule)

Articular cartilage

Articular disk

Synovial (joint) cavity

Synovial membrane

Articular capsule

Bone

SYNOVIAL JOINT WITH ARTICULAR DISK

chondrosis is eventually replaced by bone when the long bone stops growing, and it then becomes a synostosis. However, a few synchondroses are not replaced by synostoses and are still present in the adult. One such articulation is the synchondrosis of the sternoclavicular joint between the first pair of costal cartilages of the ribs and the manubrium. (The articulation between the sternum and clavicle is a diarthrodial joint.)

Symphyses

A *symphysis* (SIHM-fuh-sihss; Gr. growing together) is sometimes called a *secondary synchondrosis*. In such an articulation the two bony surfaces are covered by thin layers of hyaline cartilage. Between them are *fibrocartilaginous disks* that serve as shock absorbers. Fibrocartilage is a dense mass of collagenous fibers filled with cartilage cells and a scant cartilage matrix, so it has the firmness of cartilage and the toughness of a tendon. One of these slightly movable joints is the symphysis pubis, the midline joint between the bodies of the pubic portions of the paired hipbones. During pregnancy, the symphysis pubis is relaxed somewhat to allow for the necessary displacement of the mother's hipbones as the fetus grows. Two other examples are the slightly movable manubriosternal joint at the sternal angle and the intervertebral disks between the bodies of the vertebrae.

SYNOVIAL JOINTS

Most of the permanent joints in the body are synovial. Of all the types of joints, *synovial joints* allow the greatest range of movement [TABLE 9.1]. Such free movement is possible because the ends of the bones are covered with a smooth hyaline *articular cartilage*, the joint is lubricated and nourished by thick fluid called *synovial fluid* (Gr. *syn + ovum*, egg),* and the joint is enclosed by a flexible *articular capsule*. A synovial joint has a joint cavity.

Articular capsules that are reinforced with collagenous fibers are called *fibrous capsules*. The portion of a fibrous capsule reinforced by a thick layer of collagenous fiber is called a *ligament*.

Typical Structure of Synovial Joints

Because synovial joints allow more free movement than any other type of joint, they are also more complicated in structure. In general, *the more flexible a joint is, the less stable it is.* You will see how this rule applies when we look at some examples of major joints later in this chapter. The four essential structures of a synovial joint [TABLE

*Synovial fluid gets it name because its thick consistency resembles the white of an egg.

9.1] are the synovial cavity, the articular cartilage, the articular capsule, and ligaments.

Synovial cavity A *synovial cavity,* or joint cavity, is the space between the two articulating bones. It contains folds of *synovial membrane* that sometimes contain pads of fat. These fatty pads help fill spaces between articulating bones and also reduce friction. The synovial membrane contains two main types of cells: (1) phagocytic A cells and (2) B cells. The phagocytic A cells contain lysosomes that help to keep the articular cavity free of the debris formed as a result of friction between the articular cartilages. The B cells contain abundant rough endoplasmic reticulum and help produce the viscous *synovial fluid* that lubricates the synovial cavity. Combined with the articular cartilage, synovial fluid provides an almost friction-free surface for the easy movement of joints. Synovial fluid consists of plasma dialysate, protein, and mucin (hyaluronic acid).

Articular cartilage An *articular (hyaline) cartilage* caps the surface of the bones facing the synovial cavity. In a living body the articular cartilage has a silvery-blue luster and appears as polished as a pearl. Because of its thickness and elasticity, the articular cartilage acts like a shock absorber. If it is worn away (which occurs in some joint diseases), movement becomes restricted and painful. The cartilage itself is insensitive to feeling, since it has no nerve supply, but the other portions of the joint are supplied with pain-receptor nerve fibers.

Within several synovial joints there are *articular disks,* or fibrocartilaginous disks [TABLE 9.1]. Their roles vary depending on the joint. These disks may act as shock absorbers to reduce the effect of shearing (twisting) on a joint and to prevent jarring between bones. They also adjust the unequal articulating surfaces of the bones so that the surfaces fit together more evenly.

In some joints, the fibrocartilaginous disk forms a complete partition, dividing the joint into two cavities. In the knee joint, the fibrocartilages, called the medial and lateral *menisci* (muh-NISS-eye; sing. *meniscus;* Gr. *meniskos,* crescent), are crescent-shaped wedges that form incomplete partitions [FIGURES 9.1 and 9.6]. The menisci serve to cushion as well as guide the articulating bones. Many athletes, especially football players, tear these menisci, a condition commonly referred to as torn cartilages.

Articular capsule An *articular capsule* lines the synovial cavity in the noncartilaginous parts of the joint. This fibrous capsule is lax and pliable, permitting considerable movement. The inner lining of the capsule is the synovial membrane, which extends from the margins of the articular cartilages. The outer layer of the capsule is a fibrous membrane that extends from bone to bone across a joint and reinforces the capsule. So-called double-jointed people have loose articular capsules.

FIGURE 9.1 BURSAE AND TENDON SHEATHS

[A] Bursa in shoulder, positioned to reduce friction between articulating surfaces. [B] Bursae in knee, positioned to facilitate movement of tendons, muscles, and skin. [C] Diagram of a tendon sheath.

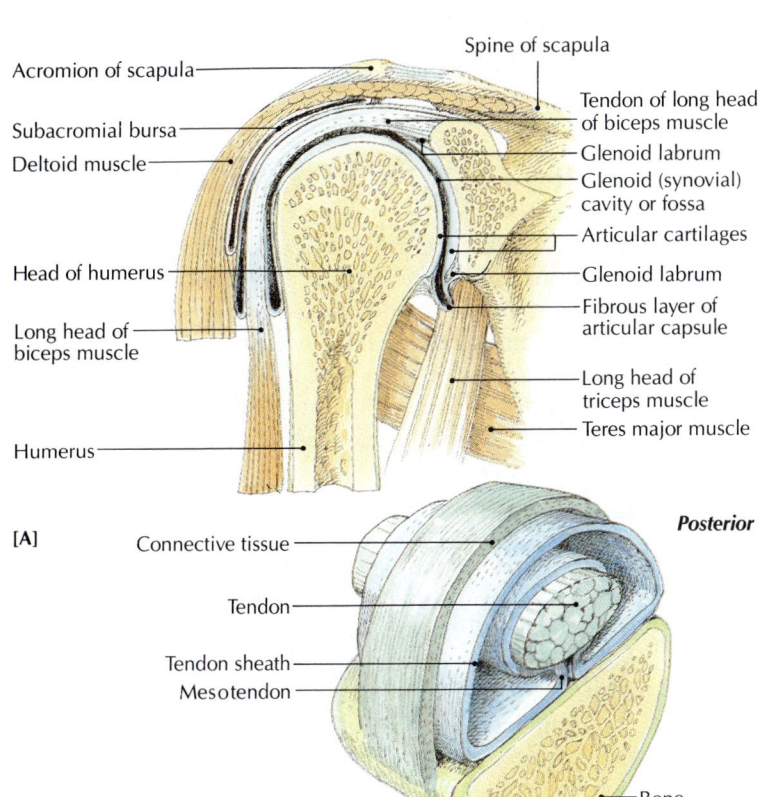

Acromion of scapula
Subacromial bursa
Deltoid muscle
Head of humerus
Long head of biceps muscle
Humerus

Spine of scapula
Tendon of long head of biceps muscle
Glenoid labrum
Glenoid (synovial) cavity or fossa
Articular cartilages
Glenoid labrum
Fibrous layer of articular capsule
Long head of triceps muscle
Teres major muscle

[A]

Connective tissue
Tendon
Tendon sheath
Mesotendon

Posterior

Bone
Periosteum

[C]

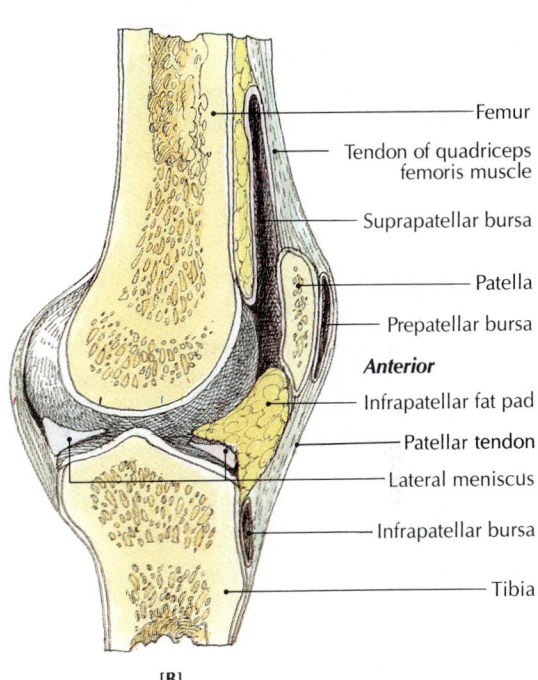

Femur
Tendon of quadriceps femoris muscle
Suprapatellar bursa
Patella
Prepatellar bursa
Infrapatellar fat pad
Patellar tendon
Lateral meniscus
Infrapatellar bursa
Tibia

Anterior

[B]

Ligaments *Ligaments* are fibrous thickenings of the articular capsule that join one bone to its articulating mate. They vary in shape and even in strength, depending on their specific roles. Most ligaments are considered inelastic, yet they are pliable enough to permit movement at the joints. However, ligaments will *tear rather than stretch* under excessive stress. For example, a sprained ankle results from excessive inversion of the foot, which causes a partial tearing of the anterior talofibular ligament and the calcaneofibular ligament. Torn ligaments are extremely painful and are accompanied by immediate local swelling. In general, ligaments are strong enough to prevent any excessive movement and strain, and a rich supply of sensory nerves helps prevent a person from overstretching ligaments.

Bursae and Tendon Sheaths

Two other structures associated with joints, but not part of them, are bursae and tendon sheaths. *Bursae* (BURR-see; sing. *bursa;* Gr. purse) resemble flattened sacs. They are filled with synovial fluid. Bursae are found wherever it is necessary to eliminate the friction that occurs when a

muscle or tendon rubs against another muscle, tendon, or bone [FIGURE 9.1A, B]. They also cushion certain muscles and facilitate the movement of muscles over bony surfaces. *Bursitis* results when bursae become inflamed.

A modification of a bursa is the **tendon** (synovial) **sheath** surrounding long tendons that are subjected to constant friction. Such sheaths surround the tendons of the wrist, palm, and finger muscles [FIGURE 9.1C]. Tendon sheaths are long cylindrical sacs filled with synovial fluid. Like bursae, tendon sheaths reduce friction and permit tendons to slide easily.

A S K   Y O U R S E L F

1 How do synarthroses, amphiarthroses, and diarthroses compare?

2 What are the three types of fibrous joints?

3 Where are the two kinds of cartilaginous joints found in the body?

4 What are the four main components of a synovial joint?

5 What are the functions of bursae and tendon sheaths?

MOVEMENT AT SYNOVIAL JOINTS

Any movement produced at a synovial joint is limited in some way. Among the limiting factors are the following:

1 *Interference by other structures.* For example, lowering the shoulder is limited by the presence of the thorax.

2 *Tension exerted by ligaments of the articular capsule.* For example, the thigh cannot be hyperextended at the hip joint because the iliofemoral ligament becomes taut as it passes in front of the hip joint.

3 *Muscle tension.* For example, when the knee is straight, it is more difficult to raise the thigh than it is when the knee is bent, because the stretched hamstring muscles exert tension on the back of the thigh.

Terms of Movement

Muscles and bones work together to allow different types of movement at different joints. Representative movements at joints and the terms used to describe those movements are shown in TABLE 9.2.

A S K   Y O U R S E L F

1 What three factors restrict movement at synovial joints?

2 How does circumduction form a "cone of movement"?

3 How does rotation differ from supination and pronation?

4 What movement is necessary to open the mouth?

Uniaxial, Biaxial, and Multiaxial Joints

In any movement about a joint, one member of a pair of articulating bones moves in relation to the other, with one bone maintaining a fixed position and the other bone moving about an axis (or axes) called an **axis of rotation**. For example, at the shoulder joint between the scapula and the humerus, the scapula is fixed in relation to the movement of the humerus.

When the movement of a bone at a joint is limited to rotation about one axis, as in the elbow joint, the joint is said to be **uniaxial** [FIGURE 9.2A]. In other words, the forearm may be flexed or extended from the elbow, moving in only one plane. When two movements can take place about two axes of rotation, as in the radiocarpal (wrist) joint, the joint is **biaxial** [FIGURE 9.2B]. The hand can be flexed or extended in one plane and moved from side to side (abduction and adduction) at the wrist in a second plane. When three independent rotations occur about three axes of rotation, as in the shoulder joint, the

joint is **multiaxial** or **triaxial** and has *three degrees of freedom* [FIGURE 9.2C]. The arm movements at the shoulder (glenohumeral) joint occur in three directions: flexion and extension, abduction and adduction, and medial and lateral rotation.

TYPES OF SYNOVIAL JOINTS

Synovial joints are freely movable joints that are classified according to the shape of their articulating surfaces and the types of joint movements those shapes permit. (Once again, structure influences function.) Six types of synovial joints are recognized: hinge, pivot, condyloid (ellipsoidal), gliding, saddle, and ball-and-socket. (TABLE 9.1 gives a complete classification of joints.)

Hinge Joints

Hinge joints roughly resemble the hinges on the lid of a box. The convex surface of one bone fits into the concave surface of another bone so that only a uniaxial, back-and-forth movement occurs around a single (transverse) axis [FIGURE 9.2A]. These uniaxial joints have strong collateral ligaments in the capsule around the joint. The rest of the capsule is thin and lax, permitting flexion and extension, the only movements possible in these joints. Hinge joints are found in the elbow, finger, knee, and ankle.

Pivot Joints

Another type of uniaxial joint is the **pivot joint,** which is only able to rotate around a central axis. Pivot joints are composed of a central bony pivot surrounded by a collar made partly of a bone and partly of a ligament. Pivot joints have a rotational movement around a long axis through the center of the pivot (like the hinges of a gate).

The atlantoaxial joint between the atlas and the axis is a pivot joint in which the collar (composed of the anterior bony arch of the atlas and the transverse ligament) rotates around the pivot, which is the dens of the axis. The "no" movement of the head occurs at the atlantoaxial pivot joint.

Condyloid Joints

Condyloid (Gr. *condylus,* knuckle) **joints** are modifications of the multiaxial ball-and-socket joint. However, because the ligaments and muscles around the joint limit the rotation to two axes of movement, the joint is classified as *biaxial.* Examples of condyloid joints are the metacarpophalangeal joints (knuckles) of the fingers (except the thumb). In condyloid joints the axes are at right angles to each other, permitting the usual flexion and extension

FIGURE 9.2 PLANES OF MOVEMENT AT SYNOVIAL JOINTS

[A] Uniaxial movement at elbow joint, illustrating axis for flexion and extension. [B] Biaxial movement at radiocarpal (wrist) joint, illustrating axes for flexion and extension and for abduction and adduction. [C] Multiaxial movement at glenohumeral (shoulder) joint, illustrating axes for flexion and extension, for abduction and adduction, and for medial rotation and lateral rotation.

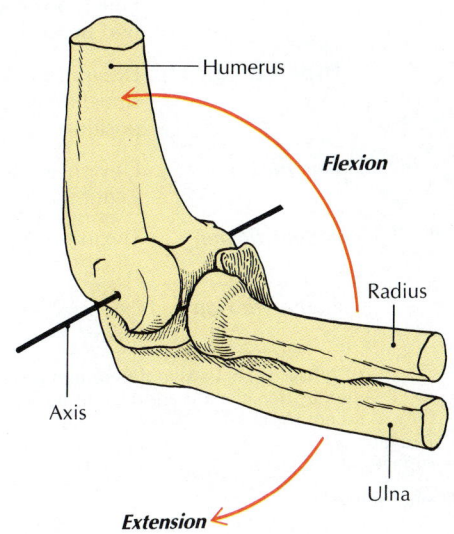

[A] UNIAXIAL (elbow)

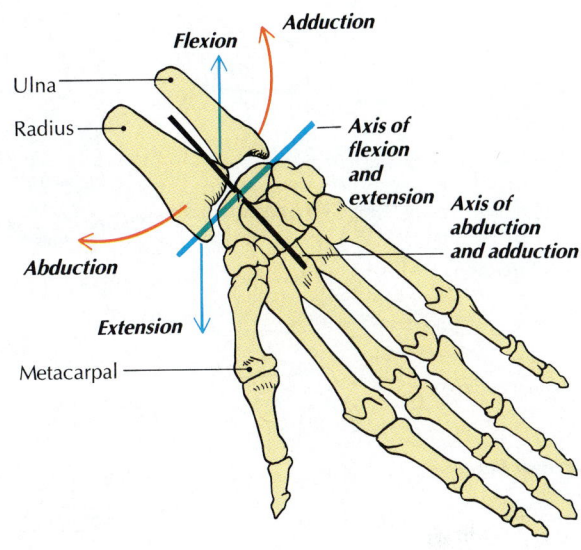

[B] BIAXIAL (wrist)

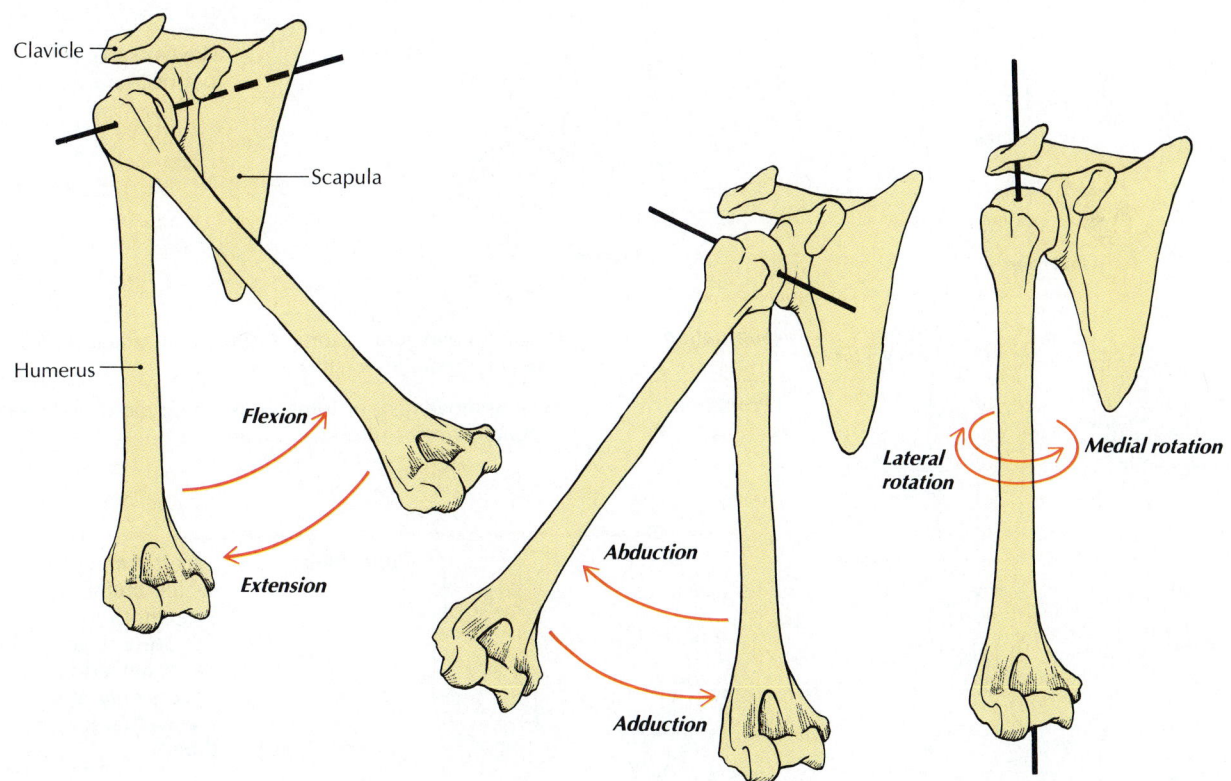

[C] MULTIAXIAL (shoulder)

TABLE 9.2 TYPES OF MOVEMENTS AT SYNOVIAL JOINTS

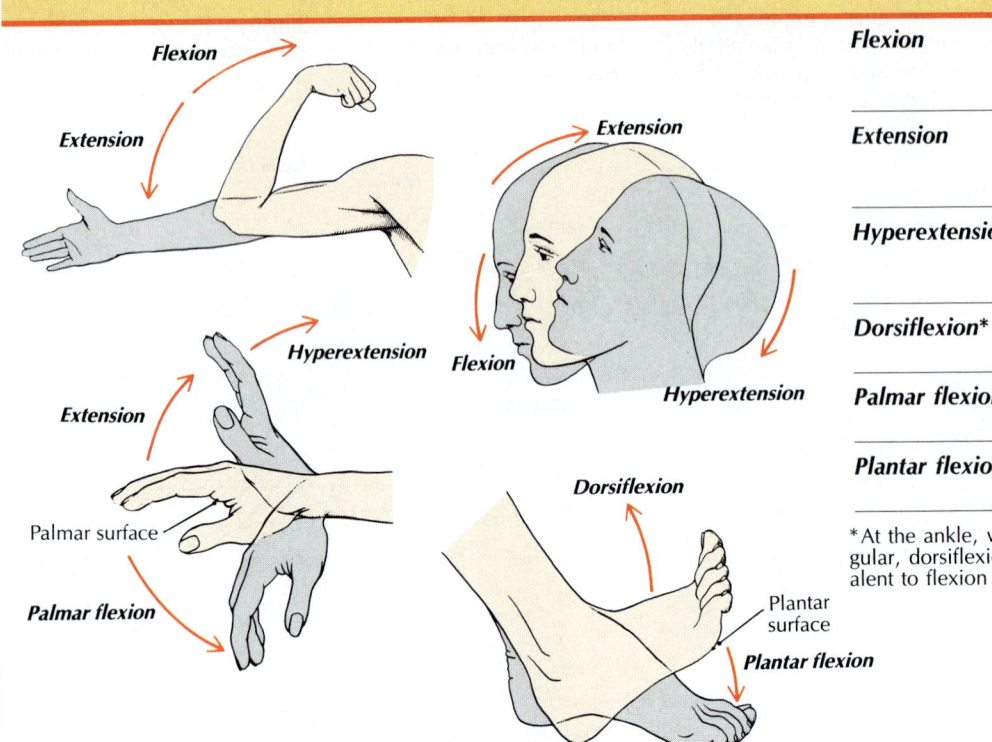

Flexion	Bending motion in which angle between two bones is decreased.
Extension	Straightening motion in which angle between two bones is increased.
Hyperextension	Extension beyond straight (anatomical) position.
*Dorsiflexion**	Flexion of foot at ankle joint.
Palmar flexion	Flexion of hand at wrist.
*Plantar flexion**	Extension of foot at ankle.

*At the ankle, where the normal position is angular, dorsiflexion and plantar flexion are equivalent to flexion and extension.

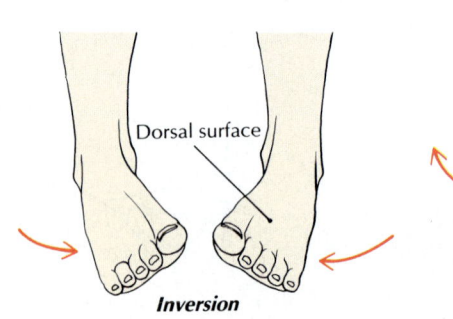

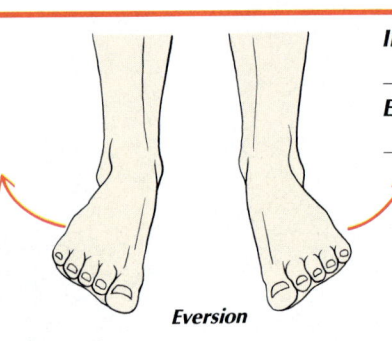

Inversion	Movement of sole of foot inward (medially).
Eversion	Movement of sole of foot outward (laterally).

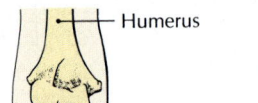

Supination	Pivoting movement of forearm in which radius is "rotated" to become parallel to ulna.
Pronation	Pivoting movement of forearm in which radius is "rotated" diagonally across ulna.

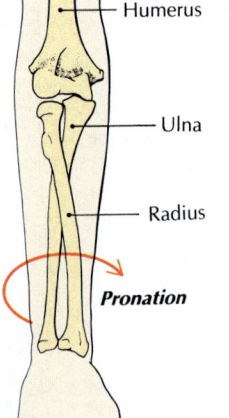

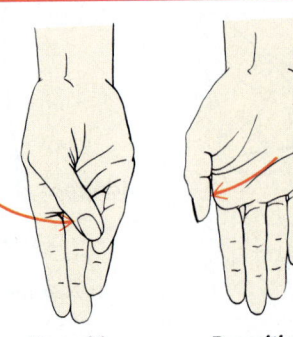

Opposition	Angular movement in which thumb pad is brought to touch and to oppose a finger pad of the extended fingers; occurs only at carpometacarpal joint of thumb.
Reposition	Movement that returns thumb to anatomical position; opposite of opposition.

TABLE 9.2 (CONTINUED)

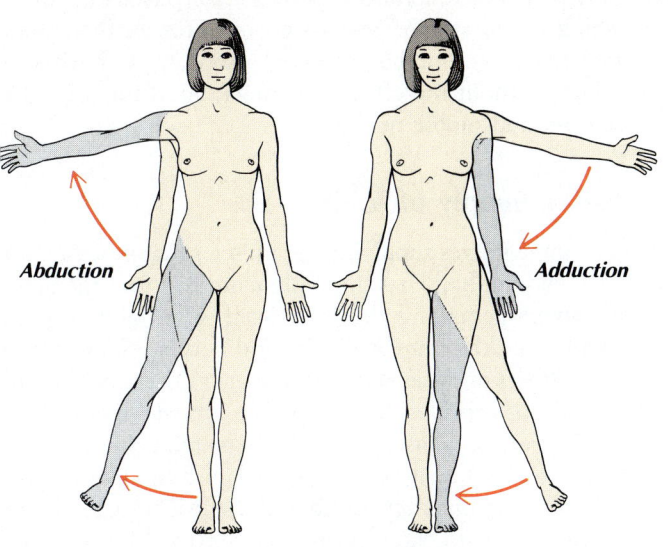

Abduction Adduction

Abduction	Movement of limb away from midline of body; movement of fingers or toes away from longitudinal axis of hand or foot.
Adduction	Movement of limb toward or beyond midline of body; movement of fingers or toes toward longitudinal axis of hand or foot.

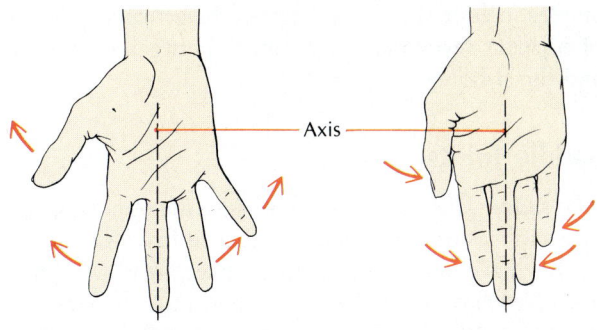

Axis

Abduction   Adduction

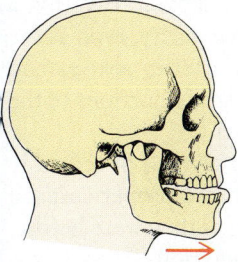

Circumduction	Movement in which distal end of bone moves in circular motion while proximal end remains stable; accomplished by successive flexion, abduction, extension, and adduction.

Circumduction

Imaginary cones of movement

Rotation	Movement of body part (usually entire extremity) around its own axis without any displacement of its axis.
Medial (internal)	Movement in which ventral surface of extremity rotates toward midline of body.
Lateral (external)	Movement in which ventral surface rotates outward away from midline.

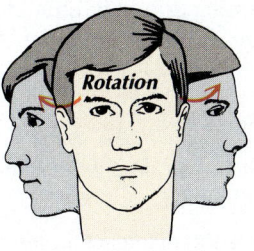

Rotation

Protraction
Forward movement.

Retraction
Backward movement.

Depression
Lowering a body part.

Elevation
Raising a body part.

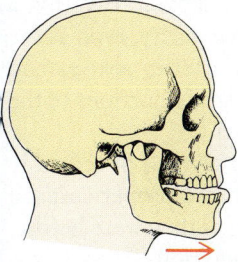 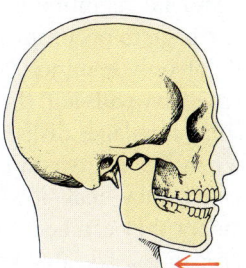 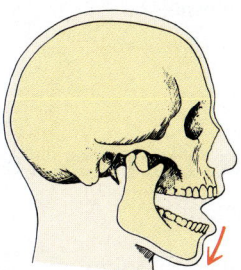 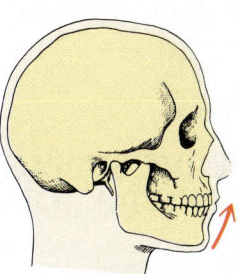

Protraction   Retraction   Depression   Elevation

of a hinge joint as well as abduction, adduction, and circumduction. However, rotational movement is not permitted.

Gliding Joints

Gliding joints are almost always small and are formed by essentially flat articular surfaces so that one bone *slides* on another bone with a minimal axis of rotation, if any. Examples include the joints between the articular processes of adjacent vertebrae and the joints between some carpal and tarsal bones.

Saddle Joints

The *saddle joint* is so named because the opposing articular surfaces of both bones are shaped like a saddle; that is, they have both concave and convex areas at right angles to each other. Movement is permitted in several directions, and the joint is considered *multiaxial.* Movements are abduction, adduction, opposition, and reposition. The carpometacarpal joint of the thumb is the best example of a saddle joint in the body.

Ball-and-Socket Joints

Ball-and-socket joints are composed of a globelike head of one bone that fits into a cuplike concavity of another bone. This is the most freely movable of all joints, permitting movement along three axes of rotation. Actually, an almost infinite number of axes are available, and the joint is classified as *multiaxial.* The movements of a ball-and-socket joint are flexion, extension, medial (internal) rotation, lateral (external) rotation, abduction, adduction, and circumduction. The shoulder and hip joints are ball-and-socket joints.

ASK YOURSELF

1 How can you describe uniaxial, biaxial, and multiaxial joints?

2 What is the difference in the direction of hinge motion in hinge joints and pivot joints?

3 What is an example of a condyloid joint?

4 Which type of joint is the most movable?

NERVE SUPPLY AND NUTRITION OF SYNOVIAL JOINTS

The nerve and blood supplies of synovial joints are discussed together because of the general rule "Where there is a nerve, there is also an artery." In other words, nerves cannot function without a readily available nourishing blood supply, and to go one step further, joint movement cannot be regulated properly, and without injury, without the effective stimulation of the joints by sensory and motor nerves.

Nerve Supply of Joints

The same nerves supplying the muscles that move a joint also send out branches that supply the skin over the muscle attachments and the joint itself. One joint may be supplied by the branches of several nerves. Many *sensory* nerve fibers of these nerves terminate as nerve endings in the fibrous capsules, ligaments, and synovial membranes of the joints. Sensory nerve fibers relay information about the joint activity to the spinal cord and brain. After this information is processed in the spinal cord and brain it contributes to the information conveyed by *motor* fibers of these nerves to the muscles controlling the movements of the joints.

Blood and Lymph Supply of Joints

All the tissues of synovial joints receive nutrients either directly or indirectly from blood vessels, except the articulating portions of the articular cartilages, disks, and menisci. (No articulating cartilage has a direct blood supply.) The articulating areas are nourished indirectly by the synovial fluid that is distributed over the surface of the articular cartilage. The flow of synovial fluid is stimulated by movement at the joints, and this is why physical activity is essential for the maintenance of healthy joints. Under normal conditions, only enough synovial fluid is secreted to produce a thin film over the joint surfaces. If a joint becomes inflamed, however, the secretion of synovial fluid may become overstimulated, causing the fluid to accumulate and producing swelling and discomfort at the joint.

Arteries near a synovial joint send out branches that join together freely on the outer surface of the joint. These branches penetrate the fibrous capsule and ligaments and form another branching network of capillaries that spreads throughout the synovial membrane. These capillaries are so numerous, and so close to the surface of the synovial membrane, that it is relatively easy for a hemorrhage to occur into the articular cavity, even as the result of only a minor injury. The arteries also extend into the fatty pads and the nonarticulating portions of the articular cartilage, disks, and menisci.

When ligaments and tendons are severed or seriously damaged, they heal slowly, but with the proper surgical treatment they can become as strong as they were before. The repair is associated with the generation of new fibroblasts and a good capillary blood supply, both of which grow into the damaged region. The fibroblasts become oriented parallel to the long axis of the ligament or tendon and form new collagenous fibers.

In addition to the blood vessels around a joint, there are many lymphatic vessels that form a network near the flexor part of the joint. These lymphatic vessels join adjacent lymphatic vessels of the body wall or relevant limb. They drain excessive tissue fluids from the region of the joint and return them to the bloodstream.

ASK YOURSELF

1 How is the nonarticulating portion of a joint nourished?

2 Why is a hemorrhage at a joint likely after an injury?

DESCRIPTION OF SOME MAJOR JOINTS

The jaw, shoulder, hip, and knee are major joints, each with its own distinctive anatomical characteristics. The following sections describe and illustrate these four joints in detail. In addition, TABLE 9.3 outlines some major articulations of the body, including the ones mentioned above.

Temporomandibular (Jaw) Joint

The *temporomandibular (TM) joint* (jaw) is the only movable joint of the head. It is a synovial joint with a combination of hinge and gliding structures [FIGURE 9.3].

The structure of the jaw is well fitted to the movements involved in chewing, biting, speaking, and so on. The articular surface of the temporal bone has a concave back part, called the *mandibular fossa,* and a convex part, called the *articular tubercle.* The joint cavity between the mandible and the temporal bone is divided into two compartments (actually two joints) by a biconcave fibrocartilaginous *articular disk,* which is fused to the *articular capsule.* The upper compartment is located between the articular disk and the mandibular fossa of the temporal bone. The lower compartment is located between the articular disk and the condyle of the mandible. The capsule is attached above to the temporal articular surface and below to the neck of the mandible.

When the mouth is closed, the convex head of the mandible fits into the mandibular fossa. Two simultaneous movements occur at the TM joint while the mouth is opening: (1) a forward gliding takes place in the upper compartment of the joint cavity as the disk and the head of the mandible ride forward in the mandibular fossa and move onto the articular tubercle, and (2) a hingelike rotation takes place in the lower compartment as the head of the mandible rotates on the articular disk. The reverse

---

FIGURES 9.3 THE TEMPOROMANDIBULAR JOINT

The temporomandibular (jaw or TM) joint is actually two joints. The upper joint is for gliding movements (protraction and retraction), and the lower joint is for hinge movements (elevation and depression). Note how the relation of the head of the mandible and articular disk to the articular tubercle changes when the mouth **[A]** closes and **[B]** opens. Right lateral views.

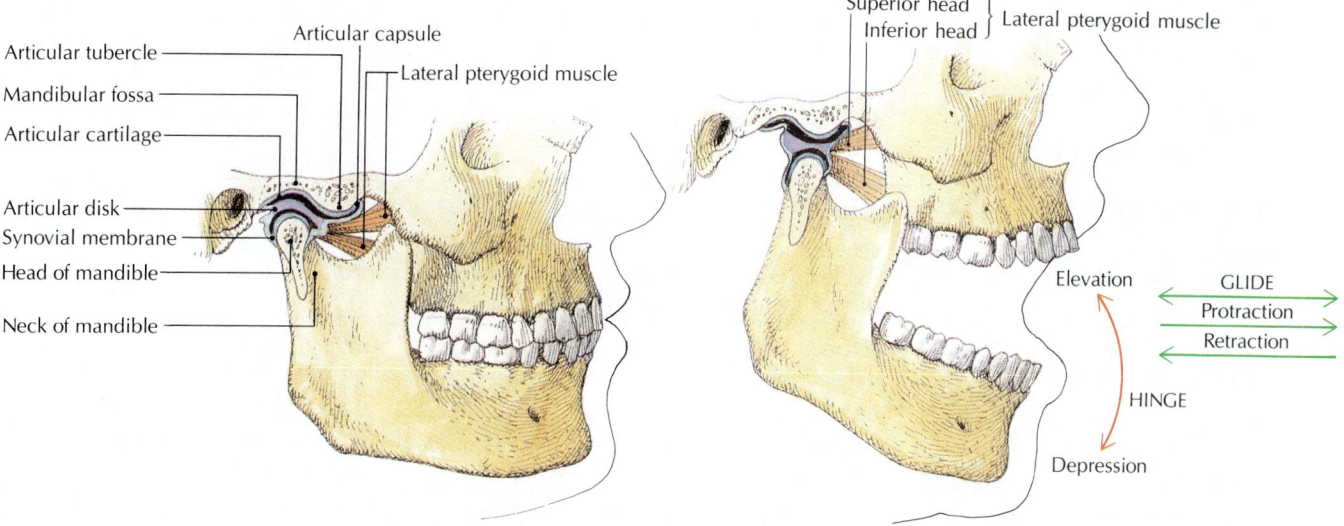

[A]   [B]

TABLE 9.3 SOME MAJOR ARTICULATIONS OF THE HUMAN BODY

Joint and classification	Articulation	Type of structure and movement
JOINTS ASSOCIATED WITH THE SKULL [FIGURES 7.5C, 7.5D, 7.10, 7.12, 9.3]		
Temporomandibular (TM, jaw) joint Diarthrosis (synovial)	Between head of mandible and mandibular fossa of temporal bones, divided into two compartments by articular disk.	Hinge (lower compartment); gliding (upper compartment). Simultaneous hinge and gliding movements open and close jaw. Protraction, retraction, grinding.
Atlantooccipital joint Diarthrosis (synovial)	Between atlas and occipital bone of skull.	Hinge. "Yes" movement of head.
JOINTS OF THE VERTEBRAL COLUMN [FIGURES 7.12, 7.14]		
Atlantoaxial joints (paired) Medial: Diarthrosis (synovial)	Between dens of axis and anterior arch of atlas and its transverse ligament.	Pivot. "No" movement of head.
Lateral: Diarthrosis (synovial)	Between articular processes of atlas and axis.	Gliding. "No" movement of head.
Intervertebral joints Diarthrosis (synovial)	Paired joints between articular processes of adjacent vertebrae.	Gliding. *Neck:* considerable variety and range of movement.
Amphiarthrosis (fibrocartilaginous)	Unpaired joints (intervertebral disks) between adjacent bodies of vertebrae.	Synchondrosis. *Thorax:* considerable variety, limited range. *Lumbar:* essentially flexion, extension, lateral bending.
JOINTS ASSOCIATED WITH THE RIBS AND STERNUM [FIGURES 7.18, 7.20]		
Costovertebral joints Diarthrosis (synovial)	Between articular facets on heads and tubercles of ribs and costal facets on transverse processes of thoracic vertebrae.	Gliding. Rotation.
Sternocostal joints Synarthrosis (fibrous)	Between rib 1 costal cartilage and sternum.	Syndesmosis (rib 1). No movement.
Diarthrosis (synovial)	Between ends of costal cartilages of ribs 2–7 and concavities on sides of sternum.	Gliding (true ribs). Rotation of true ribs (2–7).
Interchondral joints Diarthrosis (synovial)	Between successive costal cartilages of ribs 5–9.	Gliding. Adjustment during respiration.
Manubriosternal joint Amphiarthrosis (cartilaginous)	Between manubrium and body of sternum.	Symphysis. Slight movement at fibrocartilaginous disk between manubrium and sternal angle.
Xiphisternal joint Amphiarthrosis (cartilaginous)	Between xiphoid process and body of sternum.	Symphysis. Slight movement.
JOINTS OF THE CLAVICLE AND PECTORAL (SHOULDER) GIRDLE [FIGURES 8.2, 8.3, 8.4, 9.4]		
Sternoclavicular Diarthrosis (synovial)	Between medial end of clavicle, manubrium, and first costal cartilage; divided into two compartments by articular disk.	Gliding. Elevation, depression, protraction, retraction.
Acromioclavicular Diarthrosis (synovial)	Between lateral end of clavicle and medial surface of acromion.	Gliding. Essentially an accommodation between movement of clavicle and scapula.
Coracoclavicular Synarthrosis (fibrous)	Between clavicle and coracoid process of scapula; connected by coracoclavicular ligament.	Syndesmosis. Prevents separation of clavicle from scapula.
JOINTS OF THE ARM AND FOREARM [FIGURES 8.2, 8.5, 8.6, 9.4]		
Glenohumeral (shoulder) joint Diarthrosis (synovial)	Between head (ball) of humerus and glenoid cavity (socket) and scapula.	Ball-and-socket. Flexion, extension, abduction, adduction, medial and lateral rotation, circumduction.

Joint and classification	Articulation	Type of structure and movement
Elbow Diarthrosis (synovial)	Between trochlea and capitulum of humerus, trochlear notch of ulna, and head of radius.	Hinge. Flexion, extension.
Radioulnar articulation Proximal: Diarthrosis (synovial)	Between head of radius and radial notch of ulna.	Pivot. Pronation, supination.
Distal: Diarthrosis (synovial)	Between head of ulna and ulnar notch of radius.	Pivot. Pronation, supination.
Radiocarpal (wrist) joint Diarthrosis (synovial)	Between distal end of radius (and articular disk) and proximal row of carpal bones (scaphoid, lunate, triquetrum).	Ellipsoidal. Radial abduction, ulnar adduction, flexion, extension, hyperextension, circumduction.
JOINTS OF THE HAND [FIGURE 8.7]		
Midcarpal joints Diarthrosis (synovial)	Between proximal and distal rows.	Hinge. Slight flexion, extension.
Carpometacarpal joint of thumb Diarthrosis (synovial)	Between trapezium and proximal end of first metacarpal bone.	Saddle. Abduction, adduction, opposition, reposition.
Metacarpophalangeal (knuckle) joints Diarthrosis (synovial)	Between heads of metacarpal bones and bases of proximal phalanges.	Condyloid. Flexion, extension, abduction, adduction, circumduction.
Interphalangeal joints Diarthrosis (synovial)	Between heads of phalanges and concave base of adjacent phalanges.	Hinge. Flexion, extension.
JOINTS OF THE PELVIS [FIGURES 8.11, 9.5]		
Sacroiliac joint Anterior: Diarthrosis (synovial)	Between sacrum and ilium on anterior side.	Gliding. Slight gliding and rotary movement; gives resilience to joint.
Posterior: Synarthrosis (fibrous)	Between sacrum and ilium on posterior side.	Symphysis. Slight movement; gives security to joint.
Symphysis pubis Amphiarthrosis (fibrocartilaginous)	Between bodies of pubic bones.	Symphysis. Almost immovable, but accommodates during childbirth.
Coxal (hip) joint Diarthrosis (synovial)	Between head (ball) of femur and acetabulum (socket) of os coxa (hipbone).	Ball-and-socket. Flexion, extension, abduction, adduction, medial and lateral rotation, circumduction.
JOINTS OF THE LEG AND ANKLE [FIGURES 8.9, 8.15, 9.6]		
Tibiofemoral (knee) joint Diarthrosis (synovial)	Between medial and lateral condyles of distal femur and medial and lateral condyles of proximal femur.	Modified hinge. Flexion, extension, some rotation ("screw-home" action at end of extension).
Talocrural (ankle) joint Diarthrosis (synovial)	Between socket for talus (distal tibia flanked by medial and lateral malleolus of tibia) and upper surface of talus.	Hinge. Dorsiflexion, plantar flexion (hyperextension).
JOINTS OF THE FOOT [FIGURE 8.16]		
Subtalar Diarthrosis (synovial)	Posterior joint between talus and calcaneus.	Three joints articulate as a unit; axis of rotation forms a line called the subtalar axis. Eversion, inversion.
Talocalcaneonavicular Diarthrosis (synovial)	Combined anterior joint between talus and calcaneus and joint between talus and navicular.	
Transverse tarsal Diarthrosis (synovial)	Combined joint between calcaneus and cuboid and joint between talus and navicular.	
Tarsometatarsal Diarthrosis (synovial)	Between four anterior tarsal bones and bases of metatarsal bones.	Gliding. Slight movement.
Metatarsophalangeal Diarthrosis (synovial)	Between heads of metatarsal bones and bases of proximal phalanges.	Condyloid. Flexion, extension, abduction, adduction, circumduction.
Interphalangeal Diarthrosis (synovial)	Between heads of phalanges and concave bases of adjacent phalanges.	Hinge. Flexion, extension.

"Unlocking" the Temporomandibular Joint

Place your index fingers on your TM joints. Open your mouth, and note that the condyles of your mandible glide forward. If the condyles were to pass in front of the articular tubercle (as they occasionally do during a yawn), a spasm of the temporalis muscle could jam the mandible under the zygomatic arch, "locking" the jaw. To be "unlocked," the jaw must be pushed downward, overcoming the pull of the temporalis, masseter, and medial pterygoid muscles. As the condyles are forced back to their proper location (back and over the tubercle), the jaw snaps back into place. In performing such a maneuver, it is important *not* to force the jaw down by exerting pressure with the thumbs on the person's teeth because the sudden release of the jaw is accompanied by an opposition of the molars that is powerful enough to damage the thumbs seriously. In the correct procedure, downward force is generally exerted on the mandibular ridge just *below* the teeth.

---

action occurs as the mouth is closed. If the act of opening and closing the mouth were primarily a hinge movement rather than a movement that also incorporates gliding, the backward and forward motion would be strong enough to damage the inferior alveolar nerves and blood vessels.

In the acts of protraction (protrusion) and retraction of the mandible, the head of the mandible and the articular disk slide forward and then backward on the articular surface of the temporal bone. The grinding action of the teeth is produced by movements that combine the actions of the gliding joints and hinge joints of both left and right TM joints.

Glenohumeral (Shoulder) Joint

The *glenohumeral*, or *shoulder*, *joint* is a multiaxial ball-and-socket synovial joint (diarthrosis). The head of the humerus (the "ball") articulates with the shallow, concave *glenoid fossa* (the "socket") of the scapula [FIGURE 9.4 and TABLE 9.3; see also FIGURE 8.5A]. As is typical of a ball-and-socket joint, the movements of flexion, extension, abduction, adduction, medial rotation, lateral rotation, and circumduction occur at the shoulder joint.

The shoulder joint has a greater range of movement than any other joint in the body, but what it gains in mobility it loses in stability and security. As a result, the shoulder joint is dislocated more than any other joint. (See When Things Go Wrong at the end of this chapter.)

Q: *What causes the "popping" sound when you crack your knuckles?*

A: There are at least three likely possibilities: (1) When a joint is pulled tight, small ligaments or muscles tighten and snap across the bony protuberances of the joint. (2) When the joint is pulled apart, air can pop out from between the bones, creating a vacuum that produces a popping sound. (3) When the fluid pressure of the synovial fluid is reduced by the slow articulation of a joint, tiny gas bubbles in the fluid may burst, producing the sound.

The extensive range of movement results mainly from the laxity of the articular capsule and the shallowness of the glenoid fossa.

The shallow glenoid fossa is deepened slightly by a fibrocartilaginous rim called the *glenoid labrum.* The *articular capsule,* looser than in any other important joint, completely envelops the articulation. It is attached to the glenoid labrum of the scapula, to the articular margin of the head, and to part of the anatomical neck of the humerus. The articular capsule is under tension during adduction and lateral rotation, but the *coracohumeral ligament* between the scapula and humerus becomes taut during those movements, strengthening the joint.

The strength and stability of the shoulder joint also depend on four muscles, which nearly encircle the joint and hold the head of the humerus in the glenoid fossa. These joint-reinforcing muscles—the *supraspinatus, infraspinatus, teres minor,* and *subscapularis* muscles (sometimes referred to as SITS muscles for the first letter of each)—are collectively called the *rotator cuff muscles.* The tendons of the SITS muscles and an overall connective tissue sheath constitute the so-called *rotator cuff* itself. The rotator cuff acts as a dynamic "ligament," keeping the head of the humerus pressed into the glenoid fossa of the scapula, thus providing major stability in the shoulder joint. However, the rotator cuff offers little support to the joint inferiorly, and as a result, dislocations generally occur inferiorly. Rotator cuff injuries are common in baseball pitchers because pitching motions strain the rotator cuff muscles and their tendons.

Of the several bursa in the shoulder region, the large subacromial bursa is important because it is associated with subacromial bursitis. (See When Things Go Wrong at the end of this chapter.)

Coxal (Hip) Joint

The *coxal,* or *hip, joint* is a multiaxial ball-and-socket synovial joint (diarthrosis). The head of the femur articulates with the acetabulum of the os coxa [FIGURE 9.5].

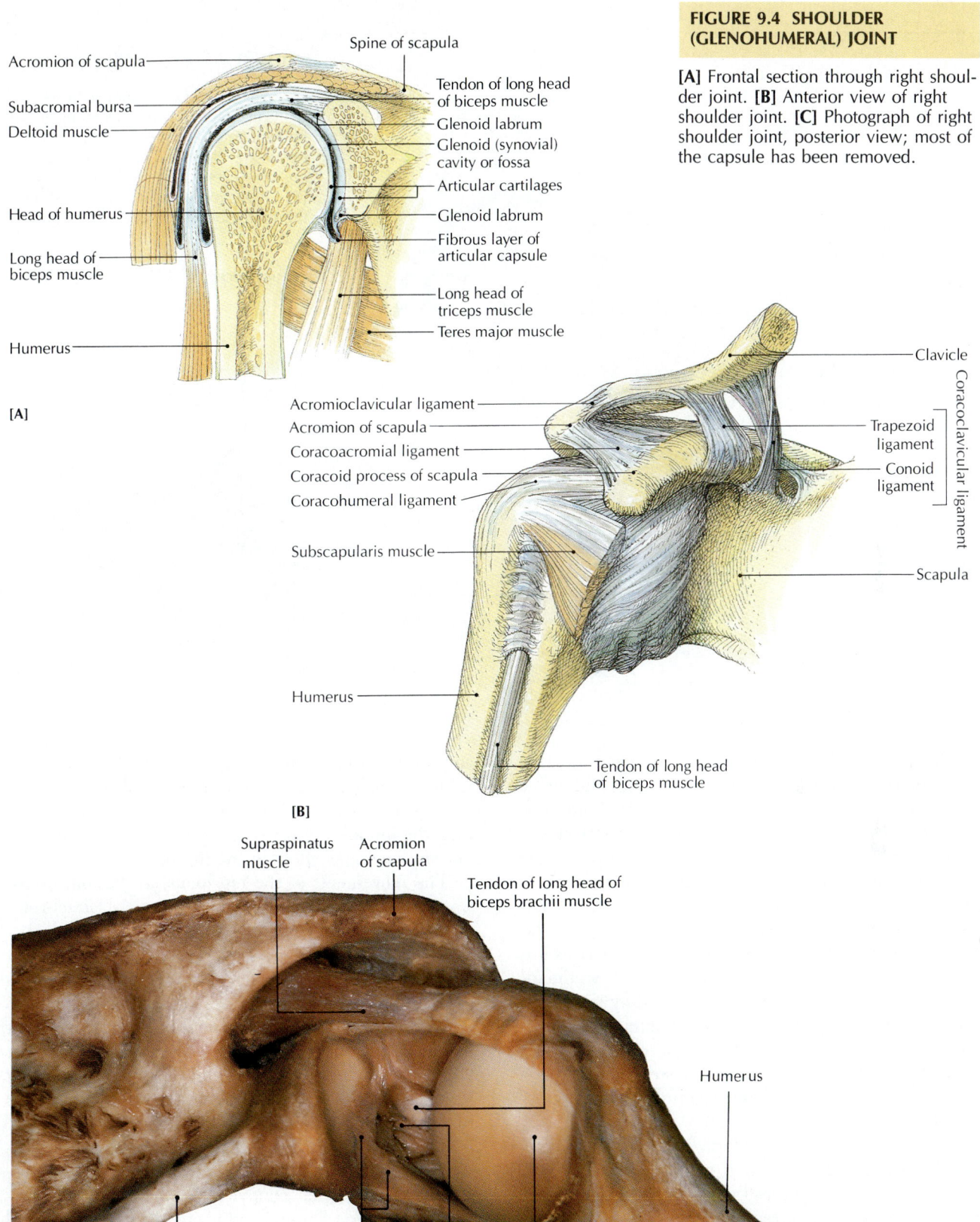

Acromion of scapula

Subacromial bursa

Deltoid muscle

Head of humerus

Long head of biceps muscle

Humerus

Spine of scapula

Tendon of long head of biceps muscle

Glenoid labrum

Glenoid (synovial) cavity or fossa

Articular cartilages

Glenoid labrum

Fibrous layer of articular capsule

Long head of triceps muscle

Teres major muscle

[A]

FIGURE 9.4 SHOULDER (GLENOHUMERAL) JOINT

[A] Frontal section through right shoulder joint. [B] Anterior view of right shoulder joint. [C] Photograph of right shoulder joint, posterior view; most of the capsule has been removed.

Clavicle

Acromioclavicular ligament

Acromion of scapula

Coracoacromial ligament

Coracoid process of scapula

Coracohumeral ligament

Subscapularis muscle

Humerus

Trapezoid ligament

Conoid ligament

Coracoclavicular ligament

Scapula

Tendon of long head of biceps muscle

[B]

Supraspinatus muscle

Acromion of scapula

Tendon of long head of biceps brachii muscle

Humerus

Long head of triceps brachii muscle

Glenoid labrum

Glenoid cavity

Head of humerus

[C]

FIGURE 9.5 HIP (COXAL) JOINT

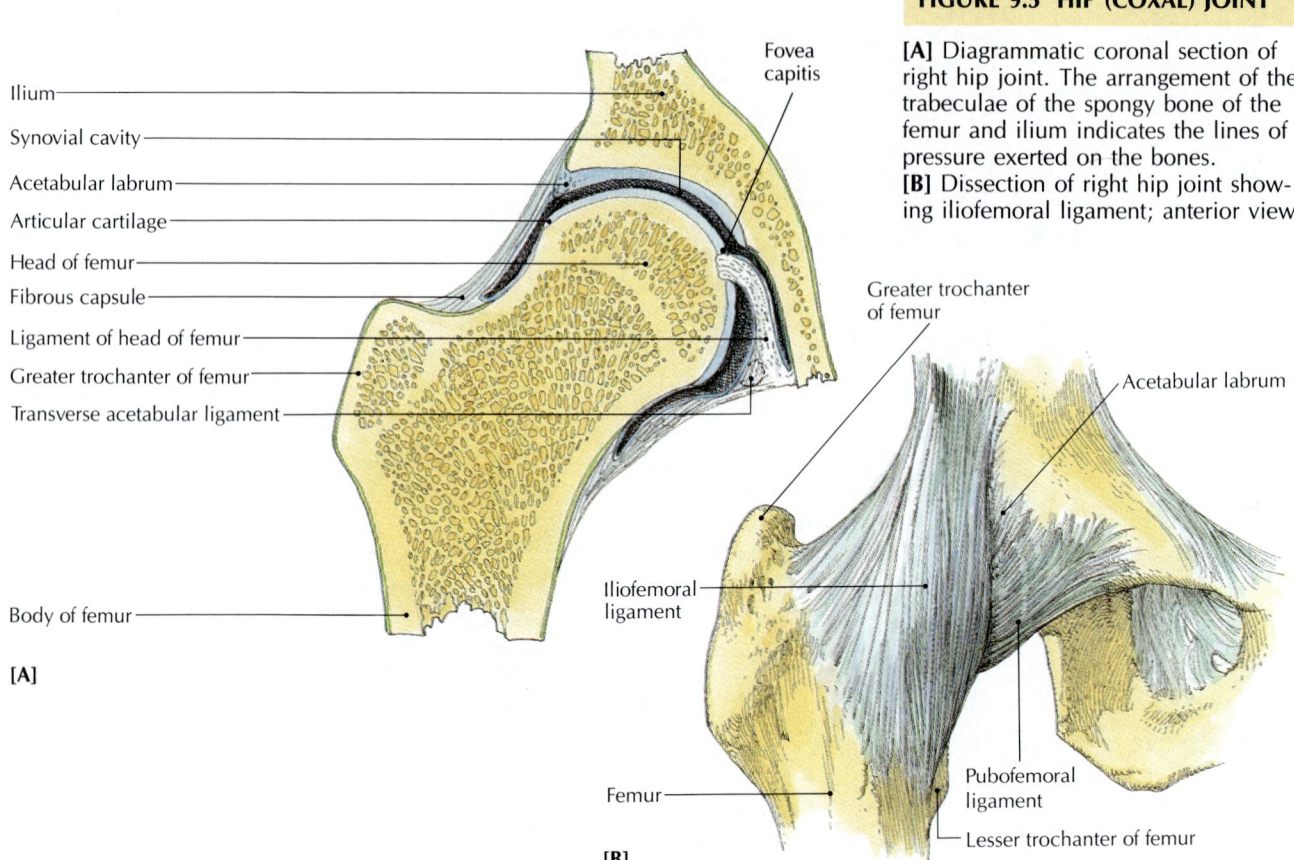

Ilium

Synovial cavity

Acetabular labrum

Articular cartilage

Head of femur

Fibrous capsule

Ligament of head of femur

Greater trochanter of femur

Transverse acetabular ligament

Body of femur

Fovea capitis

[A]

Greater trochanter of femur

Acetabular labrum

Iliofemoral ligament

Femur

Pubofemoral ligament

Lesser trochanter of femur

[B]

[A] Diagrammatic coronal section of right hip joint. The arrangement of the trabeculae of the spongy bone of the femur and ilium indicates the lines of pressure exerted on the bones. **[B]** Dissection of right hip joint showing iliofemoral ligament; anterior view.

The bones of the hip are arranged in such a way that the joint is one of the most secure, strong, and stable articulations in the body. The large, globular ***head*** (ball) of the femur fits snugly into the deep hemispherical ***acetabulum*** (socket). The acetabulum is made deeper by the fibrocartilaginous ***acetabular labrum,*** which forms a complete circle around the socket. The deep socket holds the head of the femur securely, but because of such stability, the movement at the hip joint is not as free as it is at the ball-and-socket shoulder joint. A notch on the inferior aspect of the acetabulum is bridged by the ***transverse acetabular ligament.*** The acetabulum faces downward, laterally, and forward. The head of the femur is mounted on the neck of the femur so that it forms a 125-degree angle with the long shaft of the femur [FIGURE 9.5].

The fibrous capsule in the hip is thick and tense compared with the relatively thin and lax capsule of the shoulder. The outer fibrous layer of the articular capsule is strengthened by three ligaments: the ***iliofemoral, pubofemoral,*** and ***ischiofemoral.*** The most important, the iliofemoral ligament, is shaped like an inverted Y and covers the anterior surface of the joint. The Y ligament is one of the strongest in the body. It prevents hyperextension (backward bending) of the hip joint, becoming taut and resisting the tensile stresses placed on it when the hip is fully extended. As a result, when the body is in a standing position, with the body weight centered slightly behind the hip joint, erect stance can be maintained without any muscular effort in the hip region.

The movements at the hip joints are flexion, extension, abduction, adduction, medial rotation, lateral rotation, and circumduction.

Tibiofemoral (Knee) Joint

The ***tibiofemoral,*** or ***knee, joint*** is the largest and most complex joint in the body, as well as one of the most vulnerable to injury [FIGURE 9.6]. It is a synovial joint (diarthrosis) with a modified hinge structure. It is capable of a small degree of rotational movement.

The knee joint is actually a composite of three synovial joints: (1) the articulation between the medial femoral and medial tibial condyles, (2) the articulation between the lateral femoral and lateral tibial condyles, and (3) the articulation between the patella and the femur (medial and lateral tibiofemoral joints and patellofemoral joint).

Flexion and extension are the primary movements at the knee joint, with a slight amount of rotation. With the

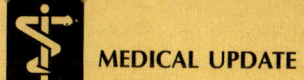

Two Million Replaceable Joints—and Then There's Bo Jackson

Physicians have been trying to create successful replaceable joints since the nineteenth century, when German surgeons used hip ball joints made of ivory. Infection and rejection by the body were the typical results. It wasn't until 1968 that the first successful total-hip replacement was performed in the United States, thanks mainly to the work of British orthopedic surgeon Sir John Charnley, who perfected the technique. He used a metal ball in a socket of high-density polyethylene and an acrylic cement to hold the ball and socket in place. By 1980 replaceable parts had been used successfully for practically every joint in the body.

Arthritis (inflammation of the joints) has been the major cause of damaged joints. About 40 million Americans suffer from arthritis, and more than 2 million of them have received artificial joints. The most successful joint replacement has been the hip; about 120,000 such operations are performed in the United States each year.

In total-hip replacement surgery, both the ball and the socket portions of the hip joint are replaced (see drawings). Following a 2- or 3-week hospitalization for the surgical procedure and several weeks of physical therapy, most hip replacement patients become reasonably independent in about 3 months. General physical fitness and normal daily activities are usually encouraged, but vigorous sports such as jogging, tennis, and baseball are not—unless your name is Bo Jackson.

Undoubtedly the most publicized hip replacement in recent years was the one performed on Bo Jackson, who was playing professional football with the Los Angeles Raiders in 1991 when he was tackled from behind and severely injured. After this hip injury ended his football career, Bo, who at the time was also playing major-league baseball, had to make some major decisions. One of the options offered him was to have total-hip replacement surgery, an option that would ostensibly allow him to continue his baseball career.

Bo chose to submit to the surgery, after which he completed a year of rigorous physical therapy and resumed his baseball career. Once an outfielder, Bo was converted into a designated hitter and first baseman, at which position he can more easily avoid crashing into a wall and dislocating his artificial hip or fracturing his femur. Also, he will be sliding headfirst from now on and will have to avoid collisions with burly catchers. A return to athletics on a professional level after total-hip replacement surgery was once unheard of. But this is Bo, and who knows?*

*In April 1993, during his first official time at bat since the injury, Bo hit a home run. His team went on to win the division championship.

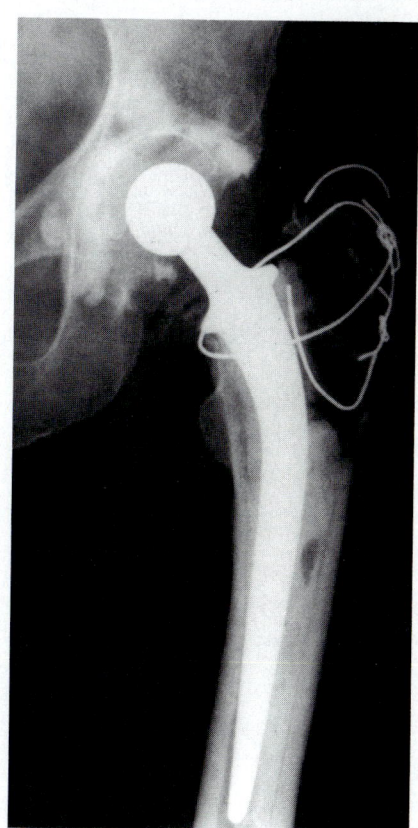

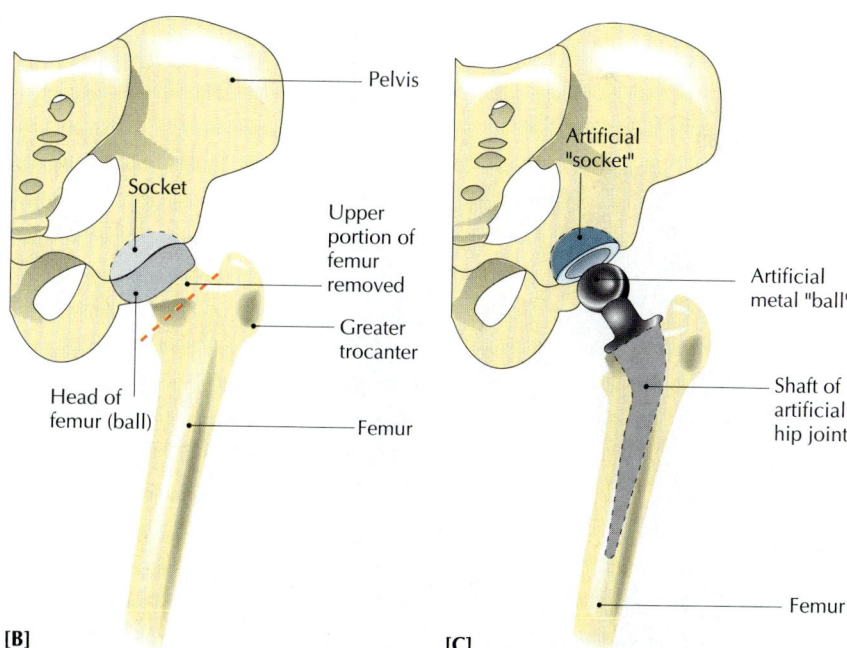

Total-hip replacement surgery. **[A]** An x-ray of an artificial hip joint. **[B]** The top of the femur, including the head (ball) and greater trochanter, is surgically detached, and the hip joint is dislocated to remove the old head of the femur. The damaged "socket" in the pelvis is reinforced and prepared to receive the new socket. **[C]** The shaft of the metal prothesis and some cement are inserted into a hole drilled into the cut surface of the femur. Then the metal "ball" of the prothesis is fitted within the new artificial socket. Finally, the greater trochanter is returned to its original site and secured by wires, as shown in **[A]**. In time, it will fuse to the shaft of the femur.

FIGURE 9.6 KNEE (TIBIOFEMORAL) JOINT

[A] Anterior view of right knee joint. The patella has been removed, and the patellar tendon has been cut. **[B]** Photograph of right knee joint, anterior view; the patellar tendon has been removed. **[C]** Sagittal section showing bursae of knee. **[D]** Cruciate ligaments of the knee joint in flexed knee, lateral view. The anterior cruciate ligament is slack during flexion.

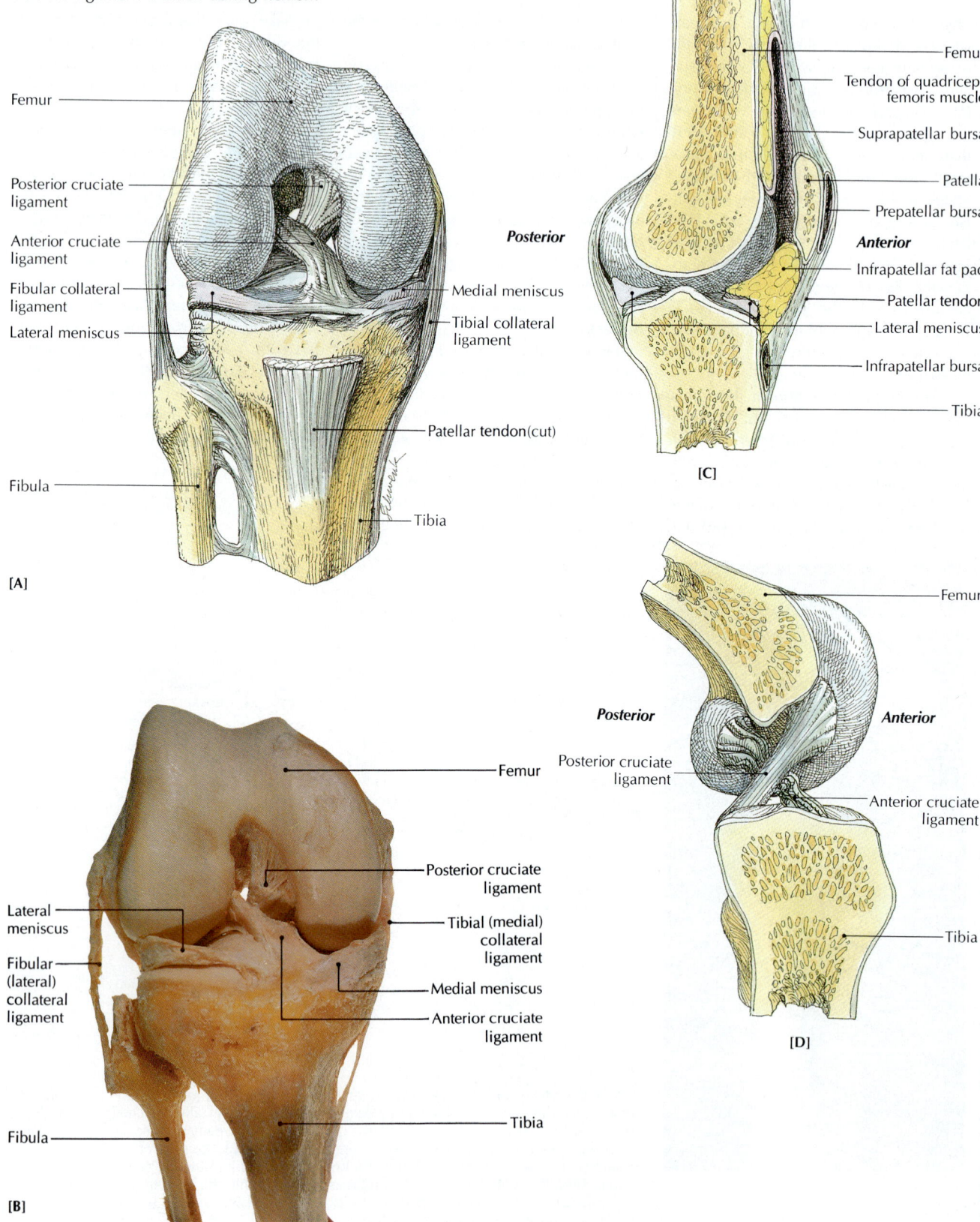

[A]

Femur

Posterior cruciate ligament

Anterior cruciate ligament

Fibular collateral ligament

Lateral meniscus

Fibula

Posterior

Medial meniscus

Tibial collateral ligament

Patellar tendon(cut)

Tibia

[C]

Femur

Tendon of quadriceps femoris muscle

Suprapatellar bursa

Patella

Prepatellar bursa

Anterior

Infrapatellar fat pad

Patellar tendon

Lateral meniscus

Infrapatellar bursa

Tibia

[D]

Posterior

Posterior cruciate ligament

Femur

Anterior

Anterior cruciate ligament

Tibia

[B]

Lateral meniscus

Fibular (lateral) collateral ligament

Fibula

Femur

Posterior cruciate ligament

Tibial (medial) collateral ligament

Medial meniscus

Anterior cruciate ligament

Tibia

Knee Injuries: The "Achilles Heel" of Athletes

The most common sports injury that requires surgery is commonly called "torn cartilage"; it usually occurs in the medial meniscus of the knee. When the cartilage is torn, it may become wedged between the articular surfaces of the femur and tibia, causing the joint to "lock." Until recently, such an injury might sideline an athlete for months, or even end his or her career altogether. Fifteen or twenty years ago, a football player recovering from surgery to repair knee cartilage that was torn on the opening day of the season would remain on crutches for about two months and would be unable to play for the remainder of the season. Now, a diagnostic and surgical technique called **arthroscopy** ("looking into a joint"), or arthroscopic surgery, makes it possible for an athlete with torn cartilage to return to the football field in about half the usual time.

Developed in Japan in 1970, arthroscopy permits a surgeon to place a lighted scope about the size of a pencil into the joint capsule to view the structural damage. If the damage is local and there do not seem to be complications, the surgeon makes another quarter-inch incision and inserts microsurgical instruments to clear away damaged cartilage.

Most incisions are closed with a single stitch, or even just a Band-Aid. The patient is usually encouraged to walk lightly on the day of the operation.

Such rapid recovery does not give leg muscles time to atrophy, so the patient does not have to worry about rehabilitating weakened muscles.

Surgical techniques are being constantly refined. For instance, damaged cartilage usually has frayed edges that tend to irritate the knee joint and restrict motion somewhat. To reduce this irritating friction, surgeons once trimmed away the damaged cartilage. But the trimmed cartilage tended to develop rough edges again, and still more had to be trimmed away. Eventually, the entire cartilage might be removed, leaving the knee without its natural shock absorber. The inability to withstand the shocks of everyday use sometimes led to arthritis, and in extreme cases, the entire knee joint had to be replaced with an artificial joint. Although a new joint usually relieves the pain, it cannot be expected to last more than 20 years or so. Now surgeons try to save the cartilage.

Contrary to what had been thought, it is now known that cartilage has enough blood supply to allow some healing, and

this new information has allowed surgeons to reattach tears in the outer boundary of the cartilage. New surgical techniques have even redirected blood from nearby blood vessels into the cartilage, increasing the possibility of successful repair. Finally, some surgeons place a clump of the patient's partially clotted blood over the sutured cartilage, not only to help bind the wound, but also to contribute growth factors that attract healing-inducing cells to the site. Because cartilage is resistant to rejection by the immune system, transplants may be the choice for the future.

Another common knee injury is a tear of the anterior and/or posterior cruciate ligaments. When the cruciate ligaments are torn, the knee joint becomes nonfunctional. To repair the cruciate ligaments, holes are drilled through the femur (see drawings). Sutures are stitched to the damaged ligaments, and the sutures are passed out through the holes. The ligaments are secured to the bone by knotting the sutures outside the femur. If the collateral ligament is separated from the bone, it is reattached by being sutured to the bone. Six to eight weeks of healing is usually followed by about a year of rehabilitative therapy.

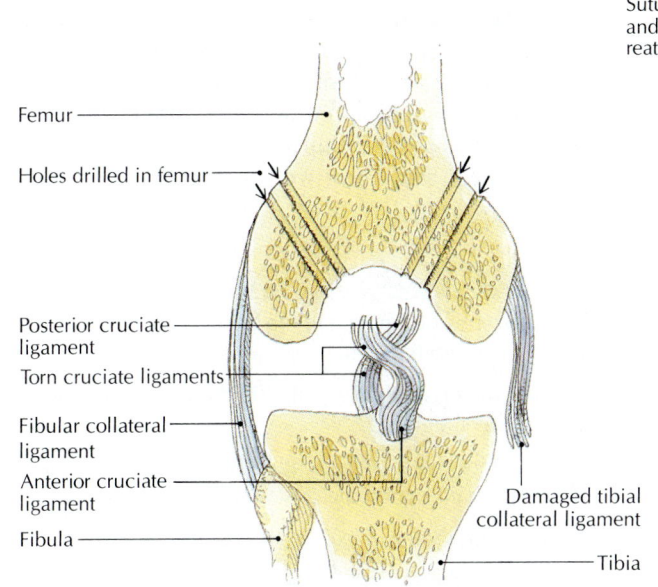

Femur
Holes drilled in femur
Posterior cruciate ligament
Torn cruciate ligaments
Fibular collateral ligament
Anterior cruciate ligament
Fibula
Damaged tibial collateral ligament
Tibia

INJURED KNEE

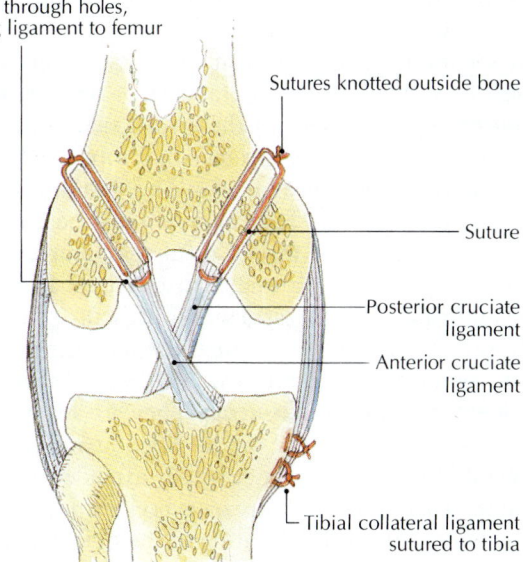

Suture stitched to ligaments and drawn through holes, reattaching ligament to femur
Sutures knotted outside bone
Suture
Posterior cruciate ligament
Anterior cruciate ligament
Tibial collateral ligament sutured to tibia

SURGICALLY REPAIRED KNEE

knee flexed, the leg can be rotated laterally and medially. A rotational "screw-home" movement* occurs just as the knee assumes full extension. If this screw-home action takes place when the foot is not free, as in the act of standing up, the femur is the bone that rotates medially in relation to the tibia until the knee is locked. If the action takes place when the foot is free, as when a punter kicks a football, the tibia rotates laterally in relation to the femur. The unlocking of the extended knee (the reverse of the screw-home action) is initiated by the *popliteus muscle.*

The screw-home phase that locks the knee occurs because the articular surface of the medial condyle is longer than that of the lateral condyle. As a result, the lateral condyle uses up its articular surface just before full extension is realized. The completion of extension occurs as the medial condyle continues to rotate on its longer articular surface, accompanied by the screw-home action and the locking of the knee. During this final phase, the lateral condyle acts as a pivot.

Several anatomical features are basic to knee movements. The curved condyles of the femur articulate with the flattened condyles of the tibia, allowing some rotation along with flexion and extension. Because the pelvis is wide, the shaft of the femur is medially directed from its proximal (hip) end to its distal (knee) end to assume an oblique set at the knee joint. Except in bowlegged people, the tibias of both legs are usually parallel.

In front, the synovial membrane is lax when the knee is extended. The membrane is extensive enough so that no strain is exerted on it when the knee is flexed. Thickenings within the articular capsule include the *tibial* (medial) *collateral ligament* and the *fibular* (lateral) *collateral ligament* extending from the sides of the femur. Both are strong ligaments that are slack during flexion and taut during extension.

A common knee injury among football players is caused by a blow on the lateral side of the leg, which produces excessive stress on the medial side of the knee that may cause a strained (stretched) or torn tibial collateral ligament. (See "Knee Injuries: The 'Achilles Heel' of Athletes" on page 261.) Such an injury may be accompanied by damage to the medial meniscus. Injuries to the knee usually involve the three Cs: collateral ligaments, cruciate ligaments, and cartilages (menisci).

In the center of the knee joint, between the condyles of the femur and tibia, are two ligaments: the *anterior* and *posterior cruciate ligaments.* They are strong cords that cross each other like an X. The anterior cruciate ligament is taut when the knee is fully extended and slack when the knee is flexed. It prevents backward dislocation of the femur, forward dislocation of the tibia, and exces-

sive extensor or rotational movements, especially hyperextension. The posterior cruciate ligament becomes tighter as flexion proceeds. It prevents forward dislocation of the femur, backward dislocation of the tibia, and hyperflexion of the knee.

Of the two ligaments, the anterior cruciate ligament is torn much more often. The damage may come from a blow driving the tibia forward relative to the femur, from a blow driving the femur back relative to the tibia, or from excessive hyperextension of the knee. The resulting instability of the knee is so serious that use of the limb may be impaired severely.

Within the knee joint are two crescent-shaped fibrocartilaginous plates called the *lateral* and *medial menisci.* They act as a cushion between the ends of bones that meet in a joint. The menisci at the knee joint deepen the articular surfaces between the femoral and tibial condyles and increase the security of the joint by adjusting the nonmatching surfaces of the tibial and femoral condyles. Injury to the medial meniscus occurs about 20 times more often than injury to the lateral meniscus because the medial meniscus is attached to the medial collateral ligament. A sudden twist of the flexed knee that is bearing weight can tear the medial meniscus.

In addition to the ligaments, the muscles and tendons surrounding the knee joint play a role in stabilizing it. Of the muscles, the most important is the quadriceps femoris.

Several bursae cushion the knee joint. They are (1) the *suprapatellar bursa* above the patella, between the quadriceps femoris muscle and the femur, (2) the *prepatellar bursa* between the skin and the patella, and (3) the *infrapatellar bursa* between the skin and the tibial tuberosity. The prepatellar bursa, which allows for the free movement of the skin over the patella, is subject to a friction bursitis known as "water on the knee." (Repeated blows or falls may also cause "water on the knee.") This condition occurs in miners and others who often work "on all fours." Bursitis of the infrapatellar bursa is found in roofers and others who kneel with their trunks upright.

---

*The "screw-home" movement occurs just as the knee straightens out and locks; at the same time, the foot turns out slightly.

A S K   Y O U R S E L F

1 What types of movement are possible at the temporomandibular joint?

2 What major structures make up the ball-and-socket joint of the shoulder?

3 How does the arrangement of the articulating bones of the hip joint contribute to its stability?

4 What three joints make up the composite knee joint?

5 What are the primary movements at the knee joint?

6 What are the functions of menisci?

DEVELOPMENTAL ANATOMY OF JOINTS

The mesenchyme (embryonic mesoderm that develops into connective tissue) between developing bones can differentiate in several ways, forming synovial, cartilaginous, or fibrous joints. A *synovial joint* (such as the knee or elbow) develops when the mesenchyme gives rise to the capsule and other ligaments peripherally, then disappears, leaving the joint cavity central. The mesenchymal cells that line the capsule and articular surfaces form the synovial membrane. A *cartilaginous joint* develops when the mesenchyme differentiates into hyaline cartilage, such as at embryonic vertebral joints that allow the vertebrae to grow as the spinal cord enlarges, or fibrocartilage, such as at the symphysis pubis. A *fibrous joint* develops when the mesenchyme differentiates into dense fibrous connective tissue, such as in the skull sutures.

THE EFFECTS OF AGING ON JOINTS

With age, there is usually a decrease in the synovial fluid in joints, and cartilage in joints becomes thinner. Arthritis, bursitis, and other joint diseases are prevalent; some of these are attributed to autoimmune reactions. Ligaments may become shorter and less flexible, bending the spinal column into a hunched-over position. Interestingly, a herniated disk is less likely after age 50 or 60 because the disk has become more fibrous, and therefore less susceptible to injury. See the discussion in When Things Go Wrong (below) of the three types of arthritis that are especially prevalent in the elderly: osteoarthritis, rheumatoid arthritis, and gouty arthritis.

WHEN THINGS GO WRONG

Arthritis

The word *arthritis* means "inflammation (*-itis*) of a joint (*arthro-*)." Arthritis may affect any joint in the body, but the most common sites are the shoulders, knees, neck, hands, low back, and hips. Arthritis may also affect different parts of a joint. For example, osteoarthritis is basically a degeneration of articular cartilage, while rheumatoid arthritis is basically an inflammation of the synovial membrane. Over 40 million Americans with arthritis require medical treatment. Arthritis is most common in men over 40, postmenopausal women, blacks, and people in low-income groups. Elderly people in general are most seriously affected. There are as many as 25 different specific forms of arthritis. The most common types are described in the following paragraphs.

Osteoarthritis *Osteoarthritis* (also called *degenerative arthritis* or *"wear-and-tear" arthritis*) is by far the most common form of arthritis, especially in the elderly. (The suffix *-itis* is misleading, since osteoarthritis is primarily degenerative rather than inflammatory.) Osteoarthritis is a chronic, progressive, degenerative disease that is usually produced by an infection or abnormality of chondrocyte metabolism that causes the breakdown of chondrocytes during the normal "wear and tear" of everyday living. The matrix becomes softer, and cartilage is lost until the bone eventually becomes exposed. Osteoarthritis occurs equally in men and women. It usually affects the weight-bearing joints such as the hips, knees, and lumbar region of the vertebral column. Typical

symptoms are pain in the joints, morning stiffness, grating of the joints, and restricted movement. Osteoarthritis of interphalangeal joints produces irreversible bony growths in the distal joints (Heberden's nodes) and proximal joints (Bouchard's nodes).

Rheumatoid arthritis *Rheumatoid arthritis,* the most debilitating form of chronic arthritis, involves inflammation of the synovial membrane. (Rheumatoid arthritis used to be called "rheumatism.") It is an autoimmune disease in which certain cells of the immune system (T cells) attack the joint cartilage. It usually involves matching joints on opposite sides of the body and their surrounding bursae, tendons, and tendon sheaths. Most movable joints can be affected, but the fingers, wrists, and knees are the most susceptible. Rheumatoid arthritis affects three times as many women as men, and it is most prevalent in women between 35 and 45. When it occurs in childhood it is usually severe. The disease generally starts with general symptoms such as fatigue, low-grade fever, and anemia before it begins to affect the joints. Eventually the tissue in the joint capsule becomes thickened (the tissue is then called a *pannus*) as a result of inflammation and the accumulation of synovial fluid. Soon the disease invades the interior of the joint and interferes with the nourishment of the articular cartilage. Slowly, the cartilage is destroyed, and crippling deformities are produced. Fibrous tissue then develops on the articulating surfaces of the joint and makes movement difficult (a condition called *fibrous ankylosis;* Gr. *ankulos,*

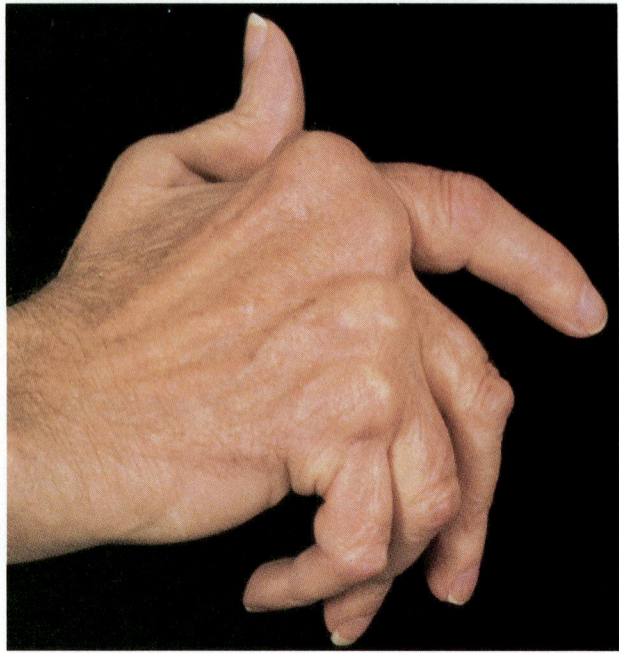

[A]

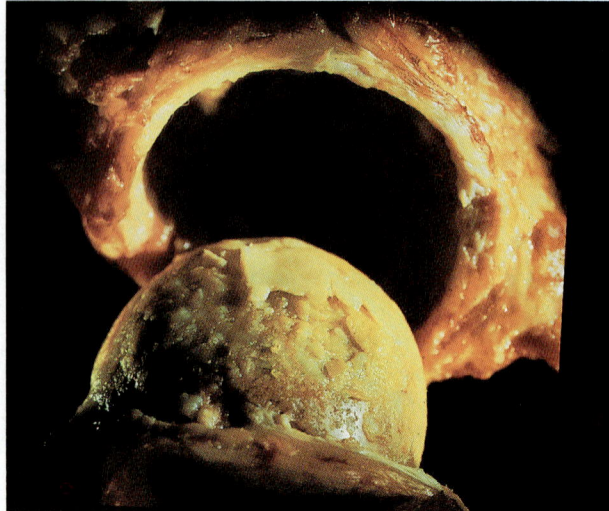

[C]

[A] Rheumatoid arthritis destroys articular cartilage and produces crippling deformities. Photographs of [B] the head of a healthy humerous and [C] the diseased joint of a person with rheumatoid arthritis. The articular cartilage has been removed.

Glenoid fossa of scapula

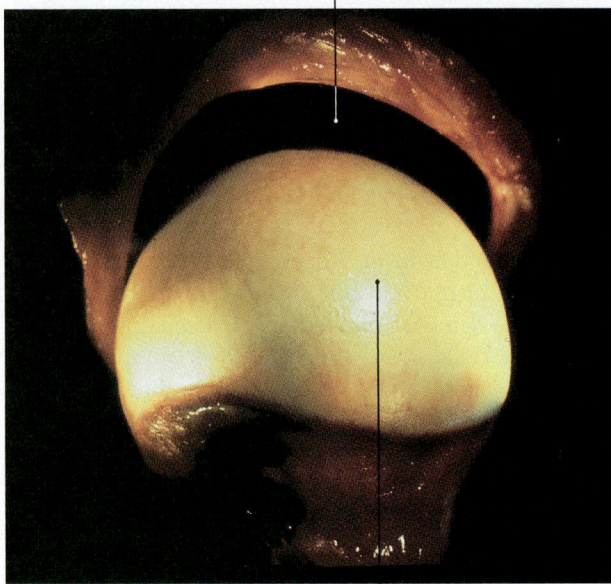

[B]   Head of humerus

the kidneys do not remove enough of it. The excess uric acid combines with sodium to form needle-sharp crystals of sodium urate salt that are deposited first in the articular cartilage and then settle in adjacent soft body tissues. The presence of the crystals produces an acute inflammatory reaction and severe pain. Any joint can be affected by gout, but the most common sites are the big toe and, oddly, since it is not a joint at all, the cartilage of the rim of the ear, just above the earlobe. The pain is sudden and intense, usually starting in the metatarsophalangeal (first) joint of the big toe (photo). The pain may progress to the instep, ankle, heel, knee, or wrist. Crystal deposits in the kidneys may lead to kidney stones or even kidney failure.

bent). In the final stage, the fibrous tissue becomes calcified and forms a solid fusion of bone so that the joint is completely nonfunctional *(bony ankylosis)*.

Gouty arthritis *Gouty arthritis,* or simply *gout,** is a metabolic disease resulting from chemical processes in the body. One product of the metabolism of purine is uric acid, which is normally excreted in urine. Sometimes, however, the body produces too much uric acid, or

*The word *gout* comes from the Latin word *gutta*, meaning "drop," because it was once thought that gout is caused by drops of unhealthy humors, or poisons, dripping into the joints.

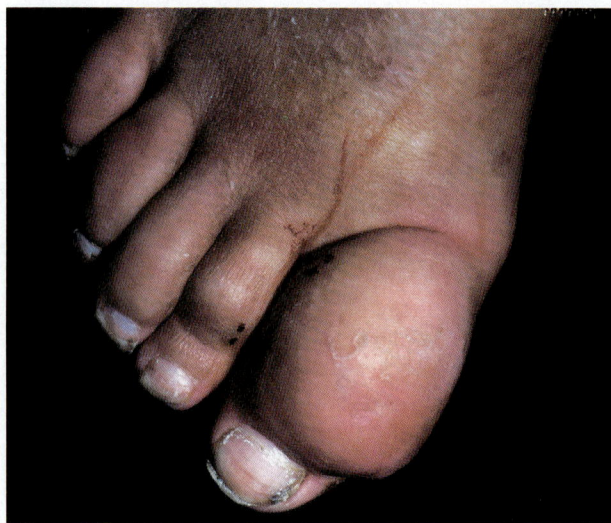

Gouty arthritis, one of the most painful forms of arthritis, often affects the big toe.

Ankylosing Spondylitis

Ankylosing spondylitis (Gr. *ankulos*, bent; *spondulos*, vertebra + *-itis*, inflammation) (sometimes called *rheumatoid spondylitis*) is a chronic, progressive disease that usually affects the sacroiliac, vertebral, and costovertebral joints and their adjacent soft tissue. It generally begins in the sacroiliac joints and gradually progresses upward along the spinal column. Ankylosing spondylitis is about three times more common in males (mostly young men) than in females, typically starting during young adulthood and progressing into middle age. The deterioration of bone and cartilage may eventually lead to the formation of fibrous tissue and the calcification and fusion of the vertebrae, making movement impossible.

The disease may be immunological in nature, since more than 90 percent of afflicted people show the presence of histocompatibility antigen HLA-B27 (a protein that causes an immune reaction) and other circulating immune complexes. The presence of HLA-B27 may be an inherited factor, although the disease itself is not. The typical symptoms — stiffness of the lumbar spine; inflammation and tenderness; peripheral arthritis in the knees, hips, and shoulders; and mild fatigue and fever — may progress irregularly, with sudden remission likely at any stage. No cure is known.

Bursitis

Bursitis is an inflammation of one or more *bursae*. Such an inflammation produces pain and swelling and restricts movement. It occurs most frequently in the subacromial bursa near the shoulder joint, the prepatellar bursa near the knee joint, and the olecranon bursa of the elbow. The pain associated with subacromial bursitis severely limits the mobility of the shoulder, especially during the initial stages of abduction or flexion of the shoulder joint.

Bursitis may be caused by an infection or by physical stress from repeated friction related to a person's activities (miner's knee, tennis elbow, house painter's shoulder) or from repeated blows or falls.

Dislocation

A *dislocation*, or *luxation* (L. *luxare*, to put out of joint), is a displacement movement of bones in a joint that causes two articulating surfaces to become separated. A partial dislocation is called a *subluxation*. A dislocation usually results from physical injury, but it can also be congenital (as in hip dysplasia) or the side effect of a disease such as Paget's disease.

The *shoulder joint* is dislocated more than any other joint, partly because of the shallowness of the glenoid cavity, which holds the head of the humerus, and the loose capsule. The *hip joint*, in contrast to the shoulder, is seldom dislocated because of the stability produced by the secure fit of the head of the femur inside the deep acetabulum, the strong articular capsule, and the strong ligaments and muscles that surround the joint.

The *TM joint* may also become dislocated. Following a vigorous laugh or yawn, the head of the mandible may move so far forward that it glides in front of the articular tubercle, with an accompanying spasm of the temporalis muscle that jams the mandible under the zygomatic bone and locks the jaw. As a result of such a dislocation, the mouth cannot be closed except by a knowledgeable second party who understands how to relocate the joint.

Immediate reduction (relocation) of the displaced joint prevents edema, muscle spasm, and further damage to tissues, nerves, and blood vessels. After reduction, the injured joint may need to be immobilized for 2 to 8 weeks.

Temporomandibular Joint (TMJ) Syndrome

The pain/dysfunction syndrome that places the chewing muscles and the TMJ muscles under painful stress is called *temporomandibular joint (TMJ) syndrome.* It may be caused by the clenching of teeth (bruxism), a misaligned bite (malocclusion), arthritis, or, more typically, psychologically related stress. Stress management and heat and muscle relaxation treatments may ease the condition. Dental braces or other such attempts to improve the bite should be temporary and reversible.

Sprain

A *sprain* is a tearing of ligaments that follows the sudden wrench of a joint. It is usually followed by pain, loss of mobility (which may not occur until several hours after the injury), and a "black-and-blue" discoloration of the skin caused by hemorrhaging into the tissue surrounding the joint. In a slight sprain the ligaments heal within a few days, but severe sprains (usually called "torn ligaments") may take weeks or even months to heal. In some cases the ligaments must be repaired surgically. A sprained ankle is the most common joint injury.

New Therapy for Damaged Joints

Ordinarily, damaged joints are immobilized in a cast for several weeks after surgery. According to recent reports, however, joints heal faster after surgery when they are kept in constant motion rather than being immobilized. When joints move, they stimulate the flow of synovial fluid, which prevents the synovial membrane from adhering to cartilage in the joint. Constant motion at the joint can be accomplished with a new technique called *continuous passive motion (CPM)*, in which a motorized apparatus moves the joint backward and forward gently. Usually, the apparatus does not disturb sleep after a short period of adjustment, and the muscles are not fatigued because the machine does all the work. CPM has proved to be more effective than standard therapy alone in improving the range of motion of the joint and decreasing postoperative swelling.

Generally, CPM patients are ready for physical therapy about 7 to 10 days after surgery. Immobilized patients usually begin therapy about 6 weeks after surgery, when the plaster cast is removed.

CHAPTER SUMMARY

Bones give the body its structural framework and muscles give it its power, but joints (articulations) provide the mechanism that allows the body to move.

Classification of Joints (p. 243)

1 Joints may be classified by function and degree of movement: an immovable joint is a *synarthrosis;* a slightly movable joint is an *amphiarthrosis;* a freely movable joint is a *diarthrosis.*

2 Based on structure, or the type of tissue that connects the bones, joints may be classified as *fibrous, cartilaginous,* or *synovial.*

Fibrous Joints (p. 243)

1 Fibrous connective tissue unites the bones in *fibrous joints.*

2 Three types of fibrous joints are *sutures, syndesmoses,* and *gomphoses.*

Cartilaginous Joints (p. 243)

1 In *cartilaginous joints,* bones are united by a plate of hyaline cartilage or a softer fibrocartilaginous disk.

2 The two types of cartilaginous joints are *synchondroses* and *symphyses.*

Synovial Joints (p. 246)

1 *Synovial joints* are the articulations where the bones move easily on each other. The ends of the bones are plated with a smooth *articular cartilage,* lubricated by *synovial fluid,* and bound together by an *articular capsule.*

2 The synovial joint is composed of the *synovial cavity, articular cartilage, articular capsule,* and *ligaments.*

3 *Bursae* are sacs filled with synovial fluid that helps eliminate friction when a muscle or tendon rubs against another muscle, tendon, or bone. A *tendon sheath* is a modification of a bursa that helps reduce friction around tendons.

Movement at Synovial Joints (p. 248)

1 Movement at synovial joints may be restricted by interference from other structures, tension exerted by ligaments of the articular capsule, and muscle tension.

2 In any movement about a joint, one member of a pair of articulating bones maintains a fixed position and the other bone moves in relation to it about an axis (or axes).

3 When movement is restricted to rotation about one axis, the joint is said to be *uniaxial.* When two movements can take place about two axes of rotation, the joint is *biaxial.* A *multiaxial* (triaxial) joint has three independent rotations about three axes.

Types of Synovial Joints (p. 248)

1 *Synovial joints* are classified on the basis of the shape of their articulating surfaces and the types of joint movements those shapes permit.

2 *Hinge joints* are uniaxial. The convex surface of one bone fits into the concave surface of another.

3 *Pivot joints* are uniaxial joints that have a rotational movement around a long axis through the center of the pivot.

4 *Condyloid joints* are modifications of ball-and-socket joints, with ligaments and muscles limiting movement to only two axes of rotation.

5 *Gliding* (plane) *joints* are small biaxial joints formed by essentially flat surfaces.

6 *Saddle joints* are multiaxial joints in which both articulating bones have saddle-shaped concave and convex areas.

7 *Ball-and-socket joints* are the most movable type. The globelike head (ball) of one bone fits into the cuplike concavity (socket) of another bone.

Nerve Supply and Nutrition of Synovial Joints (p. 252)

1 Joint movement requires stimulation by sensory and motor nerves, which must have a ready blood supply.

2 *Sensory nerve fibers* in joints relay information about joint activity to the spinal cord and brain, where it is processed and then conveyed by *motor nerve fibers* to muscles controlling joint movement.

3 Joints are nourished either directly or indirectly by blood vessels and are drained of excessive tissue fluids by lymphatic vessels.

Description of Some Major Joints (p. 253)

1 The *temporomandibular (jaw) joint,* the only movable joint of the head, is a multiaxial synovial joint. It includes the articular head of the mandible and the mandibular fossa and articular tubercle of the temporal bone.

2 The *glenohumeral (shoulder) joint* is a multiaxial ball-and-socket joint. The head of the humerus articulates with the shallow glenoid cavity of the scapula.

3 The *coxal (hip) joint* is a multiaxial ball-and-socket synovial joint. The head of the femur articulates with the acetabulum of the hipbone (os coxa).

4 The *tibiofemoral (knee) joint* is the largest, the most complex, and one of the weakest joints in the body. It is a synovial modified hinge joint. The knee joint is a composite of the articulation between (a) the medial femoral and medial tibial condyles, (b) the lateral femoral and lateral tibial condyles, and (c) the patella and the femur.

Developmental Anatomy of Joints (p. 263)

The mesenchyme between developing bones can differentiate into synovial, cartilaginous, or fibrous joints.

The Effects of Aging on Joints (p. 263)

1 With age, synovial fluid usually decreases, articular cartilage becomes thinner, and ligaments become shorter and less flexible.

2 Arthritis and other joint diseases become prevalent.

STUDY AIDS FOR REVIEW

KEY TERMS

abduction 251
adduction 251
amphiarthrosis 243
arthroscopy 261
articular capsule 246
articular cartilage 246
articulation 243
axis of rotation 251
ball-and-socket joint 252
biaxial joint 248
bursa 247
cartilaginous joint 243
circumduction 251
condyloid joint 248
coxal (hip) joint 256
depression 251
diarthrosis 243
dorsiflexion 250
elevation 251
eversion 250
extension 250
fibrocartilaginous disk 246
fibrous joint 243
flexion 250
glenohumeral (shoulder) joint 256
gliding joint 252
gomphosis 243
hinge joint 248
hyaline cartilage 246
hyperextension 250
inversion 250
ligament 247
meniscus 246
multiaxial joint 248
opposition 250
palmar flexion 250
pivot joint 248
plantar flexion 250
pronation 250
protraction 251
reposition 250
retraction 251
rotation 251
saddle joint 252
supination 250
suture 243
symphysis 246
synarthrosis 243
synchondrosis 243

syndesmosis 243
synovial cavity 246
synovial fluid 246
synovial joint 246, 248
temporomandibular (jaw) joint 253
tendon sheath 247
tibiofemoral (knee) joint 258
uniaxial joint 248

MEDICAL TERMS FOR REFERENCE

ANKYLOSIS (Gr. *ankulos*, bent + *-osis*, condition of) Stiffness or crookedness in joints.

ARTHRODESIS (Gr. *arthro*, joint + *desis*, binding) The surgical fusion of the bones of a joint.

ARTHROGRAM (joint + Gr. *grammos*, picture) An x-ray picture of a joint taken after the injection into the joint of a dye opaque to x-rays.

ARTHROPLASTY (joint + Gr. *plastos*, molded) The surgical repair of a joint, or the replacement of a deteriorated part of a joint with an artificial joint.

BURSECTOMY (bursa + *-ectomy*, removal of) The surgical removal of a bursa.

CAPSULORRHAPHY (capsule + Gr. *raphe*, suture) The surgical repair of a joint capsule to prevent recurrent dislocations.

CHRONDRITIS (Gr. *khondros*, cartilage + *-itis*, inflammation) Inflammation of a cartilage.

MENISCECTOMY (meniscus + *-ectomy*, removal of) The surgical removal of the menisci of the knee joint.

OSTEOPLASTY (Gr. *osteon*, bone + *plastos*, molded) The scraping away of deteriorated bone.

OSTEOTOMY (bone + *-tomy*, cutting of) The surgical cutting of bone, for example, the realignment of bone to relieve stress.

RHEUMATISM (Gr. *rheumatismos*, to suffer from a flux or stream) Any of several diseases of the muscles, tendons, joints, bones, or nerves. Generally replaced by the term *arthritis*.

RHEUMATOLOGY The study of joint diseases, especially arthritis.

SHOULDER SEPARATION A separation of the acromioclavicular joint, not the shoulder. It is usually a serious injury only when the accompanying ligaments are torn.

SUBLUXATION (L. *sub*, less than + *luxus*, dislocated) An incomplete dislocation.

SYNOVECTOMY (L. *synovia*, lubricating liquid + *-ectomy*, removal of) The surgical removal of the synovial membrane of a joint.

SYNOVITIS (*synovia* + *-itis*, inflammation) An inflammation of the synovial membrane.

UNDERSTANDING THE FACTS

1 What is an articulation?

2 What are the main functional classes of joints?

3 What are the main structural classes of joints?

4 Give an example of a suture, a syndesmosis, a synchondrosis, and a gomphosis.

5 What limits the motion of a synovial joint?

6 What features increase the stability of synovial joints?

7 What structures are included in a typical synovial joint?

8 Distinguish between articular cartilages and articular disks by composition and function.

9 What are the functions of bursae and tendon sheaths?

10 By shape, what kinds of joints produce uniaxial, biaxial, and multiaxial movements? Give an example of each.

11 What is an arthroscope? What is it used for?

12 Name the type of each of the following movements:

    a move thumb pad to touch an extended finger

    b bring wrist close to shoulder

    c move head to indicate "no"

    d extend foot and point toe

    e spread fingers

    f rotate forearm so that the outstretched hand can balance a ball

UNDERSTANDING THE CONCEPTS

1 How are the kind and the amount of connective tissue used to classify joints?

2 Are all fibrous joints synarthroses? Are all synarthroses fibrous? Explain.

3 By structure, what is the most common joint in a child? In a fully grown adult?

4 What is the role of the joint capsule in a synovial joint?

5 Contrast the relationship between the stability of a joint and its range of motion.

6 Explain why the articulating connective tissues in a healthy joint do not wear out.

7 Why do injuries to nonarticulating parts of joints produce intense pain, while those to articular cartilages do not?

8 Contrast the shoulder and hip joints with respect to stability and range of motion.

9 Contrast the effect of locking the temporomandibular joint and locking the knee joint.

10 Why are all types of arthritis painful? What occurs to make arthritis a crippling disorder?

SELF-QUIZ

Multiple-Choice Questions

9.1 Based on structure, which of the following types of joints contain a joint cavity?

    a fibrous   **d** suture

    b cartilaginous   **e** gomphosis

    c synovial

9.2 Which of the following is *not* a type of fibrous joint?

    a suture   **d** synchondrosis

    b syndesmosis   **e** a and b

    c gomphosis

9.3 A cartilaginous joint

    a contains a plate of hyaline cartilage

    b lacks a joint cavity   **d** a and b

    c is slightly movable   **e** a, b, and c

9.4 A synovial joint has

    a articular cartilage   **d** a joint cavity

    b synovial fluid   **e** all of the above

    c an articular capsule

9.5 Which of the following limits movement at synovial joints?

    a interference by other structures

    b tension exerted by ligaments of the articular capsule

    c muscle tension   **d** a and b   **e** a, b, and c

9.6 When the movement at a joint is limited to rotation about one axis, the joint is said to be

    a uniaxial   **d** triaxial

    b biaxial   **e** flexible

    c multiaxial

9.7 Which of the following is *not* a type of synovial joint?

    a hinge   **d** suture

    b pivot   **e** saddle

    c condyloid

9.8 Which of the following is a biaxial joint?

    a ball-and-socket   **d** pivot

    b condyloid   **e** gomphosis

    c hinge

9.9 The temporomandibular joint is a

    a fibrous joint   **d** hinge joint

    b cartilaginous joint   **e** c and d

    c gliding joint

9.10 The shoulder joint is

    a multiaxial   **d** the glenohumeral

    b a ball-and-socket joint   joint

    c a synovial joint   **e** all of the above

9.11 Which articulation is *not* part of the knee joint?

    a medial femoral to medial tibial condyles

    b medial femoral condyle to tibial fossa

    c lateral femoral to lateral tibial condyles

    d patella to femur

    e more than one of the above

9.12 The coxal (hip) joint is

    a multiaxial   **d** a fibrous joint

    b a hinge joint   **e** an unstable joint

    c a synchondrosis

Completion Exercises

9.13 The temporary joint that occurs between the epiphysis and the diaphysis of a growing bone is called a(n) _____.

9.14 Synovial fluid is produced by B cells in the _____.

9.15 Synovial fluid is found within joint cavities, bursae, and _____.

9.16 A cartilaginous pad that divides a synovial joint partially or completely is called a(n) _____.

9.17 The fibrous joint between teeth and their sockets is called a(n) _____.

9.18 Bones that are attached by an interosseous ligament form a joint called a(n) _____.

9.19 With respect to axes of rotation, condyloid joints are _____.

9.20 In the temporomandibular joint, a(n) _____ movement occurs as the mandibular head and articular disk ride forward in the mandibular fossa.

9.21 The slackness of the articular _____ contributes to the instability of the glenohumeral joint.

9.22 Cartilaginous joints called _____ are found between the bodies of successive vertebrae.

9.23 When a ballerina points her toe, she is _____ the foot.

9.24 The dens of the axis forms a(n) _____ joint with the atlas.

Matching

a extension
b flexion
c abduction
d adduction
e depression
f retraction
g eversion
h pronation
i none of the above

9.25 _____ increasing the angle between two bones
9.26 _____ movement of sole of foot outward laterally
9.27 _____ lowering a body part
9.28 _____ movement toward the midline
9.29 _____ bone moves in circular motion
9.30 _____ movement away from the midline
9.31 _____ backward pushing movement
9.32 _____ decreasing the angle between two bones
9.33 _____ pivoting movement of forearm

A SECOND LOOK

1 Identify the joints shown in the following drawings.

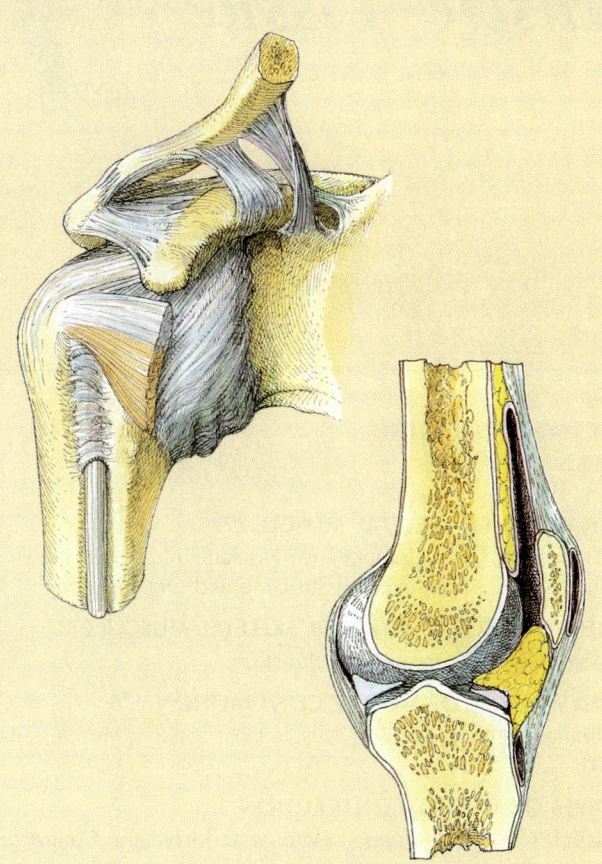

2 In the following drawing of the knee joint, label the anterior and posterior cruciate ligaments, lateral meniscus, and fibular collateral ligament.

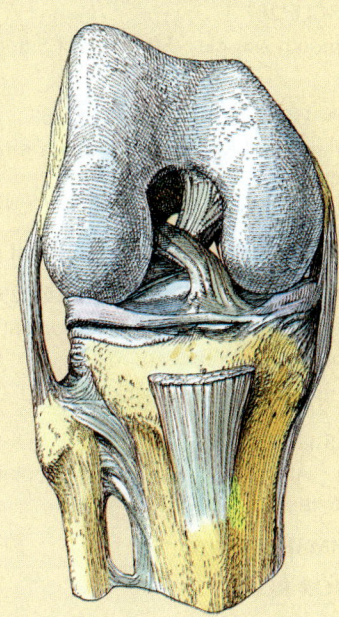

Muscle Tissue

10

KEY CONCEPTS

1 Muscles generally function to allow movement, adjust posture, and produce body heat.

2 The crucial properties of muscle tissue are its ability to contract, be excited by a stimulus, be stretched, and return to its original shape after stretching or contracting.

3 Skeletal muscles, which give the skeleton the power to move, are usually contracted voluntarily and consciously. Smooth muscle is typically involved with the involuntary rhythmic contraction of internal organs. Cardiac muscle is found only in the heart, and its specialized anatomy and physiology give it a unique method of contracting and relaxing.

Joints make a skeleton potentially movable, and bones provide a basic system of levers, but bones and joints cannot move by themselves. The driving force, the power behind movement, is muscle tissue. The three types of muscle tissue are skeletal, smooth, and cardiac [TABLE 10.1].

The basic physiological property of muscle tissue is *contractility,* the ability to *contract,* or shorten. In addition, muscle tissue has three other important physiological properties: *Excitability* (or irritability) is the capacity to receive and respond to a *stimulus,* *extensibility* is the ability to be stretched, and *elasticity* is the ability to return to its original shape after being stretched or contracted. These four properties of muscle tissue are related, and all involve movement.

In the process of contracting, work is done. For example, as certain *skeletal muscles* in your body contract, your lower limbs move at the ankle, knee, and hip, and you walk or run. Food is passed along the digestive tract by a series of rhythmic waves of *smooth muscle* contrac-

tions. The contractions of *cardiac muscle* pump blood with remarkable force and consistency from the heart to all parts of the body. Muscular contractions also help maintain body posture in a standing or sitting position, even when there is no obvious motion. Finally, the contractions of skeletal muscles produce much of the heat needed to maintain a stable body temperature. To sum up, the general functions of muscle tissue are *movement, posture,* and *heat production* (see page 284).

The muscular system follows the general rule that in the human body, structure and function are complementary. In the following sections you will see how the anatomy of each muscle type allows it to perform its function.

Q: *How much of a person's body weight is taken up by skeletal muscle?*

A: Skeletal muscle constitutes about 40 to 50 percent of an adult male's total body weight and about 30 to 40 percent of an adult female's total body weight.

TABLE 10.1 CHARACTERISTICS OF MUSCLE TISSUE

Features	Skeletal muscle	Smooth muscle	Cardiac muscle
Type of fibers	Thick, long, unbranched, cylindrical	Small, spindle-shaped	Short, branching, cylindrical
Nuclei per cell	Many, peripheral	One, central	One or two, central
Type of striation	Heavily striated	Nonstriated	Finely striated
Filament ratio	6 thin/1 thick	12 thin/1 thick	6 thin/1 thick
Sarcoplasmic reticulum and myofibrils	Highly organized sarcoplasmic reticulum surrounding myofibrils	Poorly organized sarcoplasmic reticulum; myofibrils present in properly done preparations	Poorly organized sarcoplasmic reticulum; myofibrils present in properly done preparations
Motor end plates	Present	Absent	Absent
Type of motor control	Voluntary	Involuntary	Involuntary
Location	Skeletal muscles	Walls of hollow organs of digestive, respiratory, reproductive, urinary systems; blood vessels	Heart
Function	Causes movement of skeleton, helps maintain posture, produces heat when contracted	Causes movement within internal organs and vessels	Causes heart to pump blood

Skeletal muscle, ×106

Smooth muscle, ×132

Cardiac muscle, ×1000

ANATOMY OF SKELETAL MUSCLE

Skeletal muscle tissue is so called because it makes up the muscles that are attached to the skeleton and make it move.* Skeletal muscle is also called *striated muscle* because its fibers (cells) are composed of alternating light and dark stripes, or striations. It is also called *voluntary muscle* because we can contract it when we want to. Although skeletal muscles can be contracted voluntarily, they are also capable of contraction *without* conscious control (involuntarily). Muscles are usually in a partially contracted state, which gives them *tonus*, or what we commonly refer to as "muscle tone." Tonus is necessary to keep a muscle ready to react to the stimulus preceding a complete contraction, hold parts of the body such as the head erect, and aid in the return of blood to the heart.

Connective Tissue Components: Fasciae

Skeletal muscles are covered and held together by fibrous connective tissue that is part of a network called *fascia* (FASH-ee-uh; pl. *fasciae*, FASH-ee-ee; L. band). Fascia appears in two major forms: superficial and deep.

The *superficial fascia* is located deep in the dermis of the skin and is found especially in the scalp, palms, and soles. Varying in thickness, the superficial fascia is generally composed of loose connective tissue containing blood vessels, nerves, lymphatic vessels, and many fat cells. The superficial fascia provides a protective layer of insulation and allows the skin to move freely over deeper structures.

Below the superficial fascia is the *deep fascia,* which is made up of several layers of dense connective tissue. Extensions of the deep fascia exist between muscles and groups of muscles. The deep fascia contains blood vessels, nerves, lymphatic vessels, and small amounts of fat.

The fascia that surrounds a muscle is a connective tissue sheath called the *epimysium* [FIGURE 10.1]. Extending inward from the epimysium is a layer of connective tissue, the *perimysium,* that encloses bundles of muscle fibers called *fascicles.* Further extensions of the connective tissue, called *endomysium,* wrap around each muscle fiber. These three sheaths of connective tissue contain many blood vessels, lymphatic vessels, and nerves.

Besides serving as packing material around muscles, protecting and separating them, the connective tissue sheaths provide a point of attachment to bones and other muscles. Extending from the sheaths that cover muscles or muscle fibers are tendons and aponeuroses. A *tendon* is a strong cord of fibrous connective tissue that extends

*The word *muscle* is based on the Latin *musculus*, which means "little mouse." Muscles were so named because the movement of muscles under the skin was thought to resemble a running mouse.

FIGURE 10.1 STRUCTURE OF A SKELETAL MUSCLE

[A] Cross section of a skeletal muscle showing how each muscle fiber is surrounded by the endomysium, each fascicle by the perimysium, and each muscle (group of fascicles) by the epimysium. [B] An enlargement of the endomysium, perimysium, and epimysium, showing blood vessels and nerves.

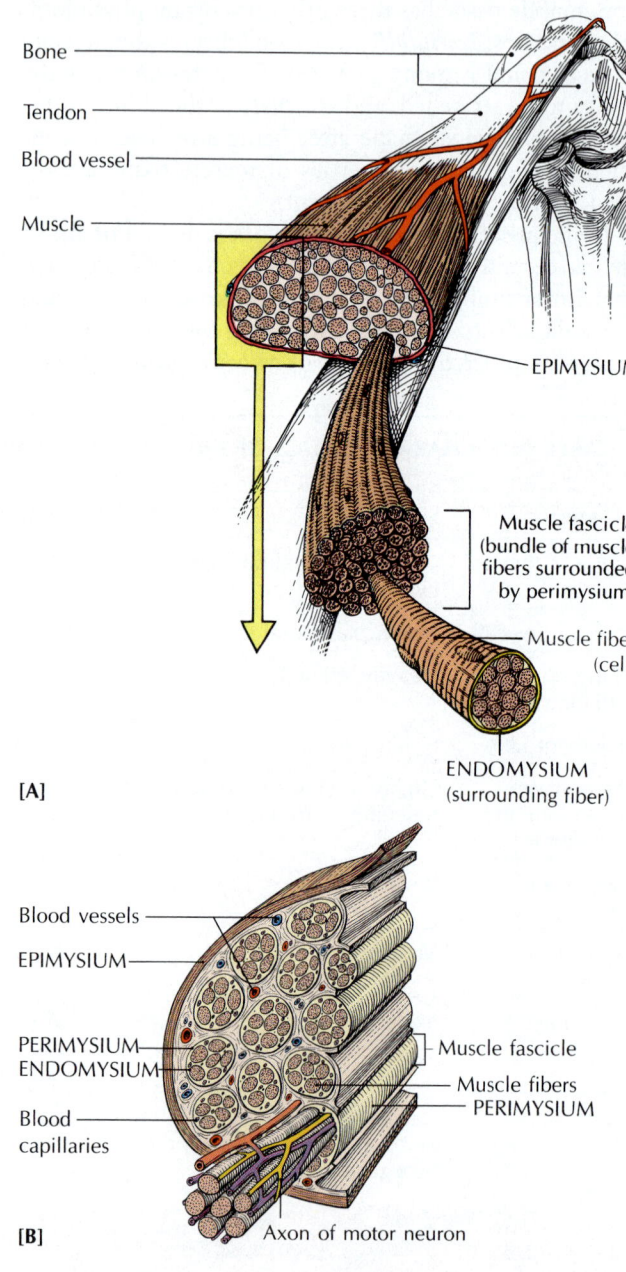

Bone

Tendon

Blood vessel

Muscle

EPIMYSIUM

Muscle fascicle (bundle of muscle fibers surrounded by perimysium)

Muscle fiber (cell)

ENDOMYSIUM (surrounding fiber)

[A]

Blood vessels

EPIMYSIUM

PERIMYSIUM
ENDOMYSIUM

Blood capillaries

Muscle fascicle

Muscle fibers
PERIMYSIUM

Axon of motor neuron

[B]

from the muscle to the periosteum of the bone, with many of its collagenous periosteal perforating fibers anchored into the bone. An *aponeurosis* is a broad, flat sheet of dense connective tissue that attaches to two or more muscles that work together, or to the coverings of a bone.

Blood and Nerve Supply

Skeletal muscles are well supplied with blood by arteries that penetrate the connective tissue coverings. These arteries branch out into tiny, thin-walled blood vessels called *capillaries*, which carry an abundant supply of oxygen-rich blood to the muscles. In fact, each muscle fiber is richly supplied with oxygen, glucose, and ATP by several capillaries that surround individual muscle cells. Without a steady and adequate oxygen and ATP supply, the muscles would be unable to contract properly, and if oxygen and ATP are cut off for too long, the muscles weaken and die. Accumulated wastes are removed from the muscles via capillaries and are carried in the blood toward the heart by veins.

Muscles are also well supplied (*innervated*) with nerves made up of *motor neurons*, which cause muscle cells to contract.

Microscopic Anatomy: Cell Structure and Organization

Skeletal muscle is composed of individual specialized cells called *muscle fibers* [FIGURE 10.2]. These cells are known as fibers because they have a long, cylindrical shape and numerous nuclei. Thus, they look more like fibers than like "typical cells." Skeletal muscle fibers average 3.0 cm (1.2 in.) in length, but some may be longer than 30 cm (12 in.) or as short as 0.1 cm (0.04 in.).* Diameters usually range from 0.01 cm (0.004 in.) to 0.001 cm (0.0004 in.).

Muscle fibers contain still smaller fibers called *myofibrils* (*myo*, muscle), which are made up of thick and thin threads called *myofilaments.* The thick myofilaments are composed of a fairly large protein called *myosin.* The thin myofilaments are composed mainly of a smaller protein, *actin.*

Each skeletal muscle fiber is enclosed by a plasma membrane called the *sarcolemma* (Gr. *sarkos*, flesh + *lemma*, husk). The fiber contains several nuclei, which are located near the plasma membrane, and a specialized type of cytoplasm called *sarcoplasm.* Within the sarcoplasm are many mitochondria and a large number of myofibrils, which run lengthwise and parallel to one another [FIGURE 10.3].

Around each myofibril and running parallel to it is the *sarcoplasmic reticulum.* This network of tubes and sacs contains calcium ions and is somewhat like the smooth endoplasmic reticulum found in other types of cells.

A SHORT GLOSSARY OF SKELETAL MUSCLE

SKELETAL MUSCLE

Contractile tissue composed of bundles (fascicles) of muscle fibers.

MUSCLE FASCICLE

A bundle of muscle fibers.

MUSCLE FIBER

A muscle cell; so named because it looks more like a long fiber than a "typical" cell.

MYOFIBRIL

Threadlike, parallel fibers that make up a muscle fiber.

MYOFILAMENTS

Thick and thin protein threads that make up a myofibril; thick myofilaments composed of myosin, thin myofilaments composed mainly of actin.

Crossing the sarcoplasmic reticulum at right angles are the *transverse tubules,* or *T tubules,* a series of tubular organelles that run across the fiber to the outside [FIGURE 10.3]. Note that the sarcolemma continues as the lining of the T tubule.

When viewed with a light microscope, skeletal muscle tissue shows a pattern of alternating light and dark bands. The bands are caused by the arrangement of the actin and myosin myofilaments. An overlapping of the thick myosin strands and the thin actin strands produces dark *A bands*; the thin actin strands appear alone as light *I bands* [FIGURE 10.2D, E, G]. Cutting across each I band is a dark *Z line.* Within the A band is a somewhat lighter *H zone,* which consists only of myosin strands. Extending across the H zone is a delicate *M line,* which connects adjacent myosin strands. The fundamental unit of muscle contraction is the *sarcomere* (Gr. *meros*, part), which is made up of a section of the muscle fiber that extends from one Z line to the next [FIGURES 10.2 and 10.3].

Q: *What makes one muscle larger than another?*

A: Ordinarily, one muscle is larger than another because it contains more bundles of fibers. The largest muscles are found where large, forceful movements are common, such as in the back and legs. When a muscle is enlarged by exercise, the fibers increase in diameter, but the *number* of fibers remains the same.

---

*The sartorius muscle of the anterior thigh is the longest muscle in the body. It may contain muscle fibers over 30 cm (12 in.) long, reaching from the hip to the knee. The shortest fibers are those of the stapedius muscle in the inner ear, which is about 0.1 cm (0.04 in.); these fibers attach to the tiniest bone, the stapes of the ear.

FIGURE 10.2 ANATOMY OF A SKELETAL MUSCLE

A progression from gross structure to molecular structure. **[A]** *Muscle in arm.* **[B]** *Muscle fascicle* (bundle of muscle fibers). **[C]** *Muscle fiber* (bundle of myofibrils). **[D]** *Myofibril.* Myofibrils, which are made up of myofilaments, show the banding that gives the designation *striated* to the muscle. **[E]** *Myofilaments.* The arrangement of myofilaments forms the sarcomere, the functional unit of muscle contraction, extending from Z line to Z line. **[F]** Light micrograph of skeletal muscle fibers. ×600. **[G]** Electron micrograph of skeletal muscle showing sarcomere. ×22,500.

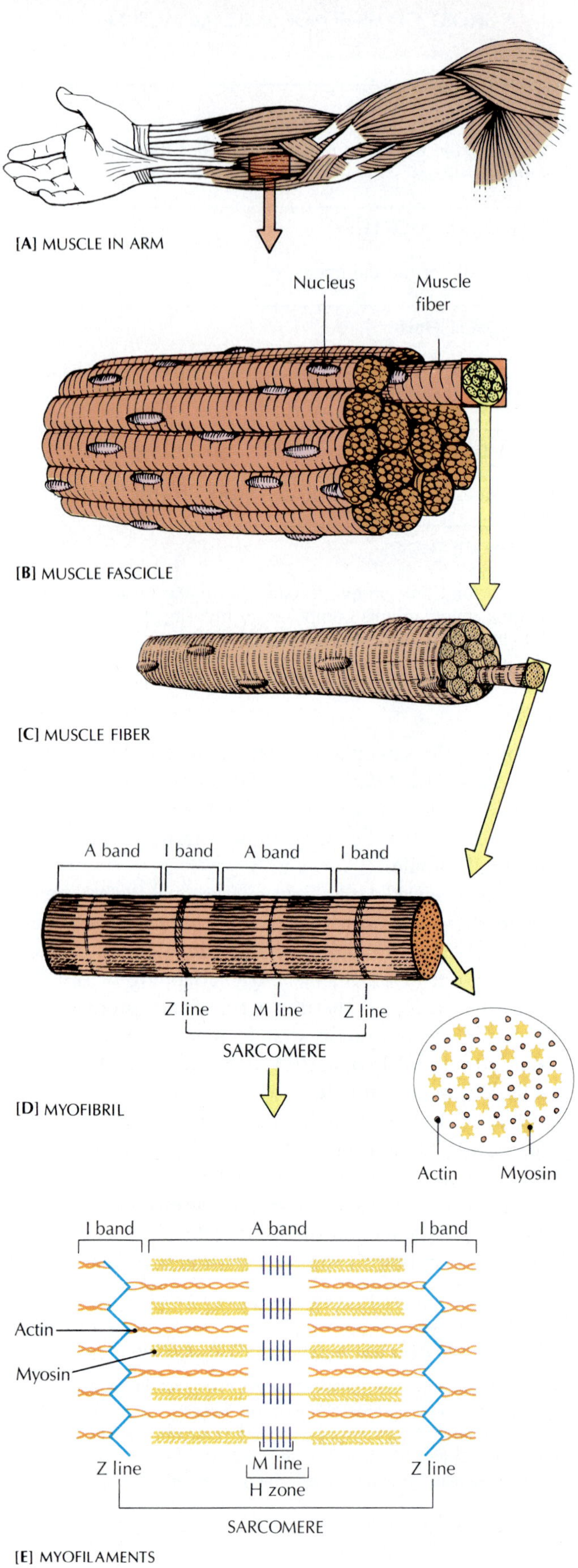

[A] MUSCLE IN ARM

Nucleus Muscle fiber

[B] MUSCLE FASCICLE

[C] MUSCLE FIBER

A band   I band   A band   I band

Z line   M line   Z line

SARCOMERE

[D] MYOFIBRIL

Actin   Myosin

I band   A band   I band

Actin

Myosin

Z line   M line   Z line

H zone

SARCOMERE

[E] MYOFILAMENTS

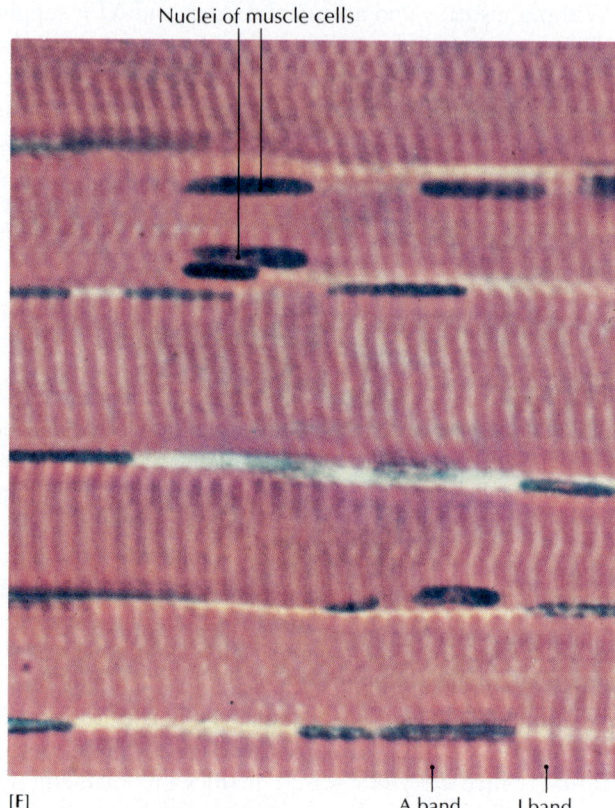

Nuclei of muscle cells

[F]   A band   I band

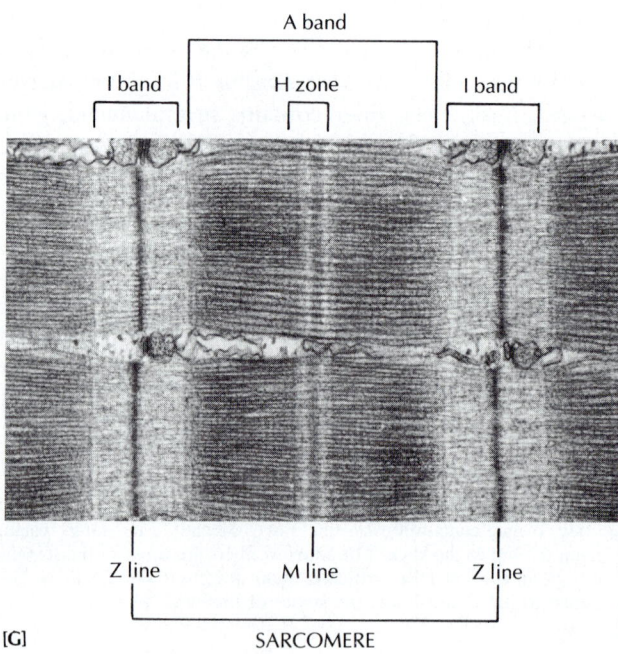

A band

I band   H zone   I band

Z line   M line   Z line

[G]   SARCOMERE

FIGURE 10.3 PORTION OF A MUSCLE FIBER

The myofibril on the right side of this bundle of myofibrils has the sarcoplasmic reticulum peeled away to show the A and I bands, H zone, and M and Z lines. The transverse (T) tubules conduct depolarization from the fiber surface into the muscle fiber.

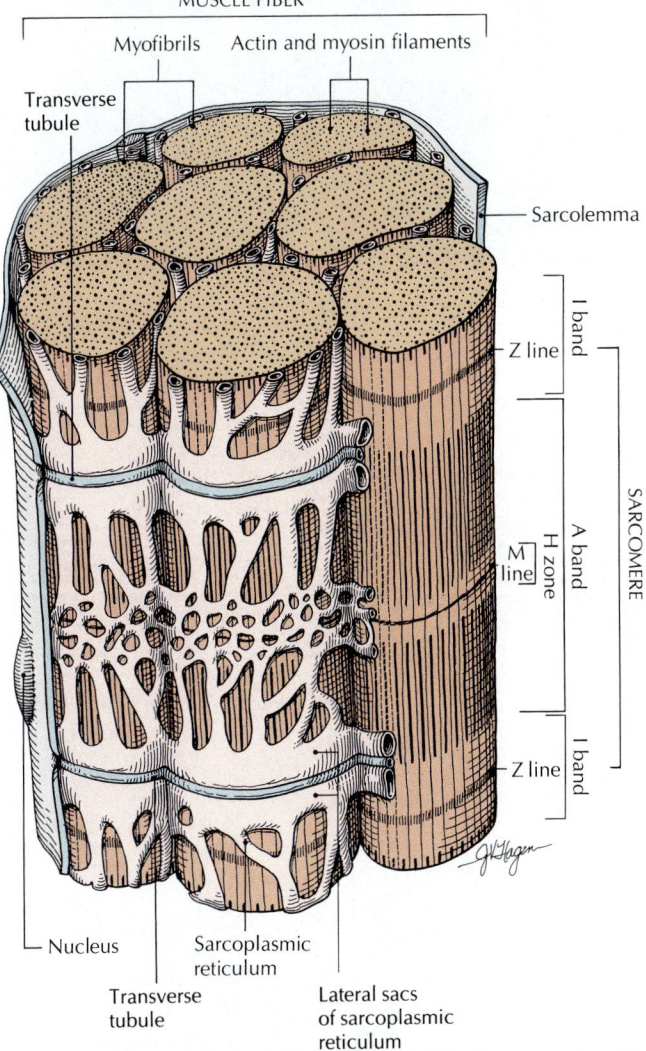

MUSCLE FIBER

Myofibrils   Actin and myosin filaments

Transverse tubule

Sarcolemma

I band

Z line

SARCOMERE

M line

H zone

A band

Z line

I band

Nucleus

Sarcoplasmic reticulum

Transverse tubule

Lateral sacs of sarcoplasmic reticulum

A S K   Y O U R S E L F

1 What are the important properties of muscle tissue?

2 What are the three types of muscle tissue?

3 Why is skeletal muscle also known as striated muscle and voluntary muscle?

4 What is the function of fasciae?

5 What is the difference between a fascicle and a fascia?

6 What are the main connective tissue sheaths associated with muscles?

7 How and why are muscles supplied with nerves and blood vessels?

NERVOUS STIMULATION OF SKELETAL MUSCLE

Contraction of skeletal muscle is under nervous control — that is, the muscle tissue alone cannot contract. Its contraction is initiated by a chemical released from a nerve cell, or neuron.

Motor Unit

A motor neuron, together with the muscle fibers it innervates, is called a *motor unit* [FIGURE 10.4]. In humans, a single motor unit causes the simultaneous contraction of from about 6 to 30 fibers (in some eye muscles) to over 1000 fibers (in powerful leg muscles).

Neuromuscular Junction

The junction between a motor neuron ending and a muscle fiber is called a *neuromuscular* (nerve + muscle) *junction (NMJ)* [FIGURE 10.5]. The end branches of the motor neuron, known as *axon terminals*, gain access to the muscle fiber through the endomysium. At the junction between the muscle fiber and the axon terminals, the muscle fiber membrane forms a *motor end plate.* The motor end plate is the specialized portion of the *sarcolemma* (plasma membrane) of a muscle fiber. It surrounds the synaptic end bulbs of the axon. As shown in FIGURE

FIGURE 10.4 A MOTOR UNIT

A motor unit is composed of a motor neuron and the muscle fibers it innervates. The junction of a motor unit and muscle fibers is a *neuromuscular junction.*

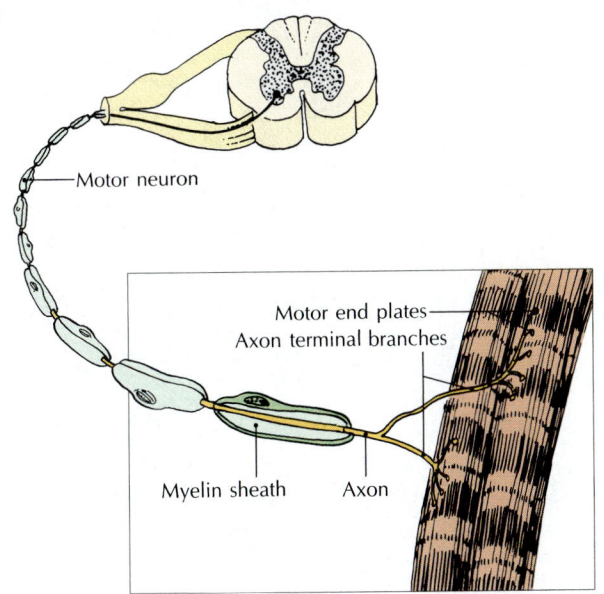

Motor neuron

Motor end plates

Axon terminal branches

Myelin sheath

Axon

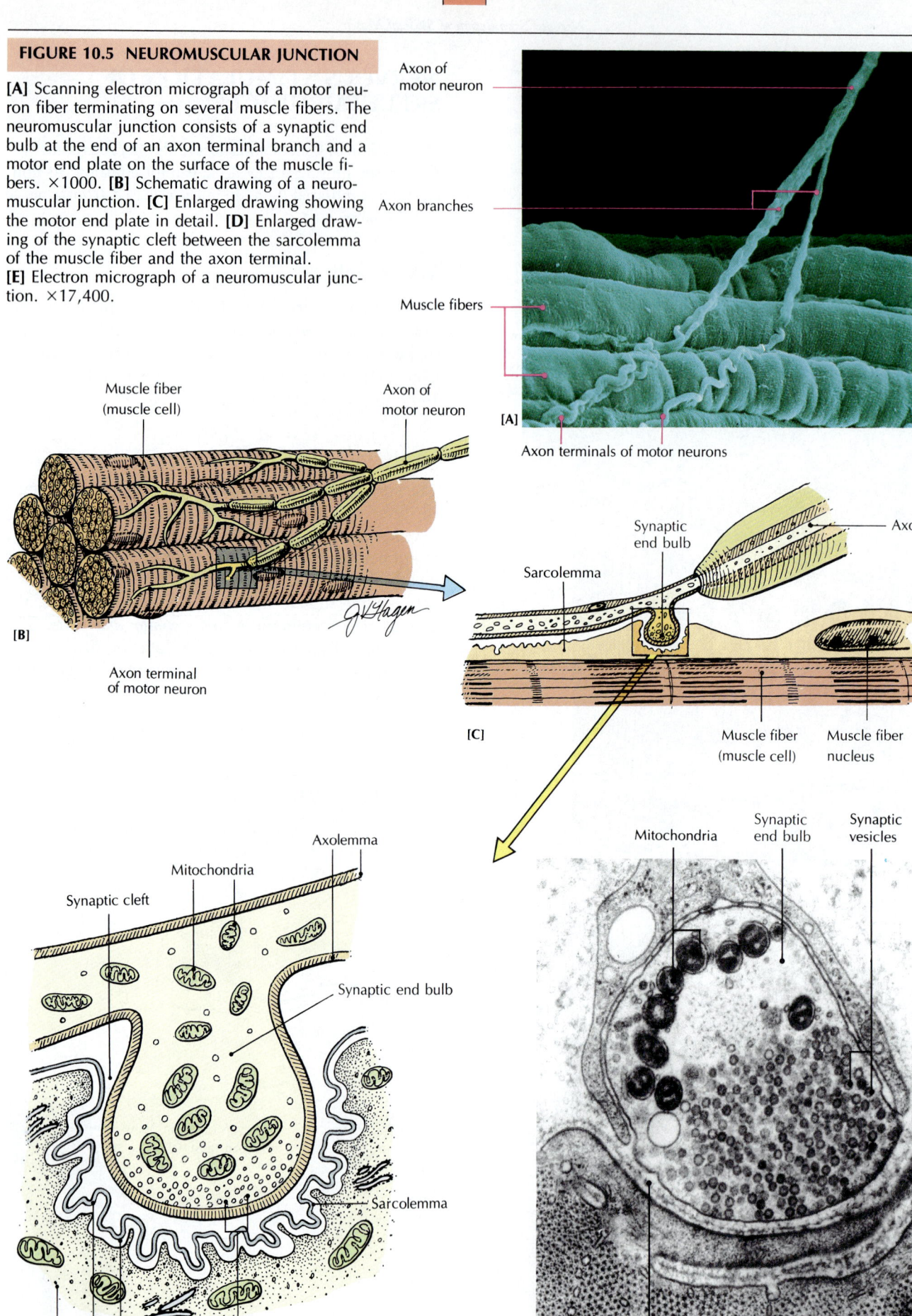

FIGURE 10.5 NEUROMUSCULAR JUNCTION

[A] Scanning electron micrograph of a motor neuron fiber terminating on several muscle fibers. The neuromuscular junction consists of a synaptic end bulb at the end of an axon terminal branch and a motor end plate on the surface of the muscle fibers. ×1000. **[B]** Schematic drawing of a neuromuscular junction. **[C]** Enlarged drawing showing the motor end plate in detail. **[D]** Enlarged drawing of the synaptic cleft between the sarcolemma of the muscle fiber and the axon terminal.
[E] Electron micrograph of a neuromuscular junction. ×17,400.

Axon of motor neuron

Axon branches

Muscle fibers

[A]

Axon terminals of motor neurons

Muscle fiber (muscle cell)

Axon of motor neuron

[B]

Axon terminal of motor neuron

Synaptic end bulb

Axon

Sarcolemma

[C]

Muscle fiber (muscle cell)

Muscle fiber nucleus

Axolemma

Mitochondria

Synaptic cleft

Synaptic end bulb

Sarcolemma

Sarcoplasm

[D]

Junctional folds in sarcolemma

Synaptic vesicles

Mitochondria

Synaptic end bulb

Synaptic vesicles

Synaptic cleft

Muscle fiber

[E]

10.5D, mitochondria are abundant near a motor end plate. In general, each muscle fiber is innervated by only one axon terminal.

The invaginated area of the sarcolemma under and around the axon terminal is called the *synaptic gutter,* and the clefts inside the folds along the sarcolemma are called *junctional folds* or *subneural clefts* [FIGURE 10.5D]. These folds greatly increase the surface area of the synaptic gutter.

At the motor end plate, nerve endings are separated from the sarcolemma of the muscle fiber by a tiny gap called a *synaptic cleft* [FIGURE 10.5D, E]. The chemical transmitter *acetylcholine* (uh-SEET-uhl-KOH-leen) is synthesized in the cytoplasm of the neuron and then packaged in the synaptic vesicles. The acetylcholine is released from the synaptic vesicles of the nerve endings, where it diffuses across the synaptic cleft and flows into the folds of the sarcolemma. Some acetylcholine then becomes attached to the receptor sites in the sarcolemma, initiating an electrochemical impulse across the sarcolemma of the muscle cells, so that sodium ions move *into* the sarcoplasm and potassium ions move *out.*

The result of this action of acetylcholine is a temporary disturbance in the permeability of the sarcolemma that leads to muscle contraction. The sequence of events at the neuromuscular junction is shown in FIGURE 10.6, which is numbered to correspond to the following account:

1 A nerve impulse (action potential) is propagated along the axon of a motor neuron, triggering the opening of voltage-gated calcium channels. Channel proteins in plasma membranes are usually involved with transporting ions across the membrane, and so they are called *ion channels.* These channels are selective, allowing some ions to pass but blocking others. The channels are not always open. Because the channels have "gates" that open and close with changes in voltage along the plasma membrane, they are called *voltage-gated channels.*

2 The opening of the calcium channels causes calcium ions (Ca^{2+}) to flow into the synaptic end bulb of the axon terminal.

3 The increase in Ca^{2+} concentration stimulates the exocytosis of the neurotransmitter acetylcholine from vesicles in the synaptic end bulb into the synaptic cleft.

4 Acetylcholine diffuses across the synaptic cleft at the neuromuscular junction and binds to specific receptor sites on the motor end plate of the muscle cell plasma membrane. (Each motor end plate may have 20 to 40 million acetylcholine receptor sites.)

5 When acetylcholine binds to the receptor sites, sodium and potassium channels in the plasma membrane of the motor end plate open. The flow of sodium ions (Na^+) into the muscle fiber is greater than the flow of potassium ions (K^+) out. The resulting change in the concentration of Na^+ and K^+ causes a *depolarization* (the reversal of the electrical charge across the plasma membrane) called an *end-plate potential* at the motor end plate.

6 The end-plate potential depolarizes the adjacent plasma membrane, which reaches its *threshold potential,* a degree of depolarization adequate to initiate an *action potential.* The action potential propagates over the surface of the entire muscle fiber and continues into the fiber along transverse tubules. In the transverse tubules it causes the release of Ca^{2+} from the sarcoplasmic reticu-

---

FIGURE 10.6 STIMULATION OF MUSCLE CONTRACTION BY A NERVE IMPULSE

An activated neuromuscular junction is shown in this schematic drawing. The numbered sequence follows the discussion in the text.

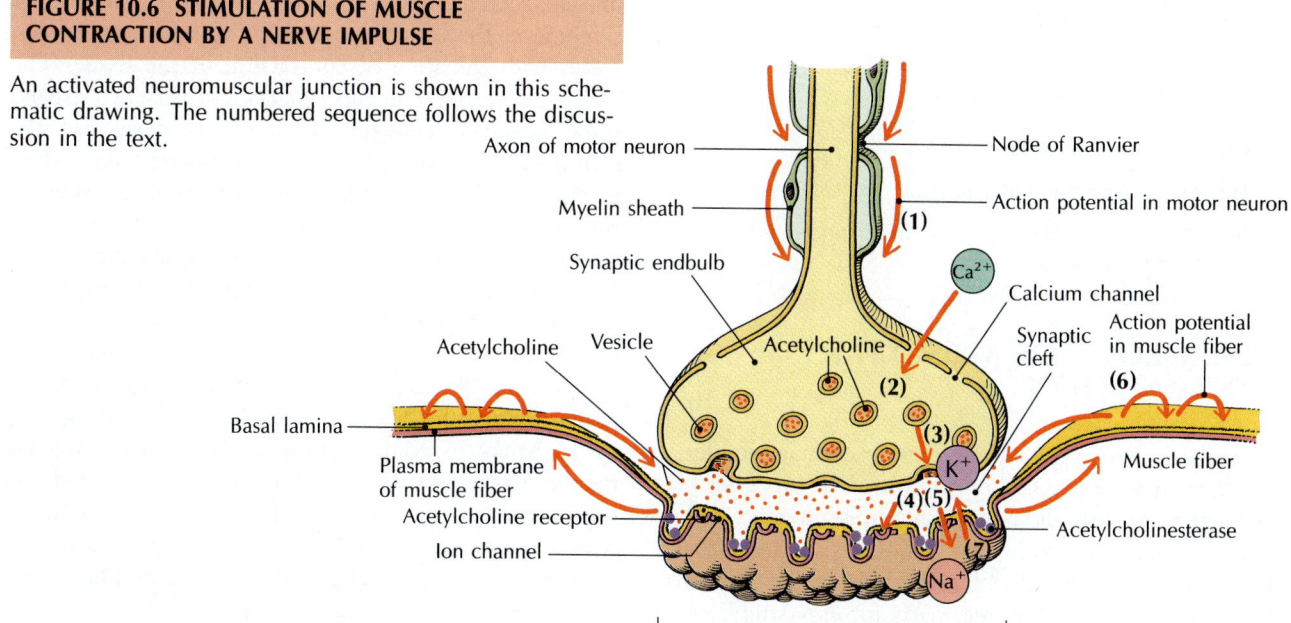

FIGURE 10.7   THE SLIDING-FILAMENT MODEL OF MUSCLE CONTRACTION AND RELAXATION

According to the sliding-filament model, in maximally contracted muscle **[A]**, the myosin filaments of a sarcomere are pulled inward toward the M line so that they actually overlap in the H zone. In partially contracted muscle **[B]**, the myosin filaments of a sarcomere are separated by a relatively small space in the H zone. In relaxed muscle **[C]**, the

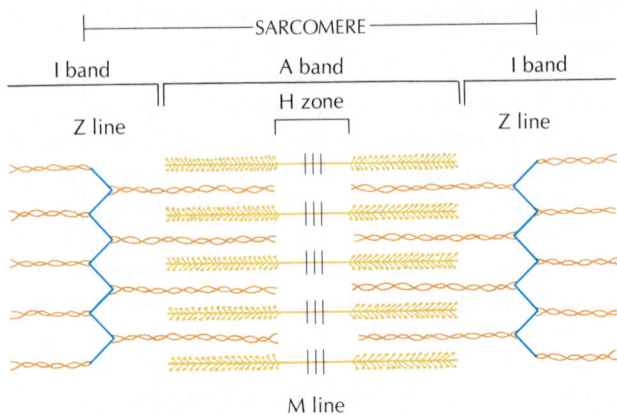

[A] MAXIMALLY CONTRACTED

[B] RESTING (PARTIALLY CONTRACTED)

lum; these ions stimulate all the myofibrils in the muscle cell to contract at the same time.

7 *Acetylcholinesterase*, an enzyme found on muscle fiber membranes, then breaks down acetylcholine into acetate and choline. When the receptor sites no longer contain bound acetylcholine, the ion channels in the motor end plate close; the plasma membrane of the motor end plate returns to its resting potential and can respond to another stimulus.

ASK YOURSELF

1 What structures make up a motor unit?

2 Where is a neuromuscular junction located?

3 What is the difference between ion channels and voltage-gated channels?

PHYSIOLOGY OF MUSCLE CONTRACTION

Because the physiology of muscle contraction is complex, the sequence of events that occur during the processes of contraction, resting, and relaxation is presented in a step-by-step format in the following sections. Also, the textual discussion is coordinated with several illustrations, which are broken down into discrete parts for easy study.

Sliding-Filament Model of Muscle Contraction

The arrangement of myosin and actin molecules in myofilaments is crucial to the mechanism of muscle contraction. As you can see in FIGURE 10.2D, E, and G, the actin and myosin myofilaments in myofibrils are parallel and partly overlapping. Electron micrographs of skeletal muscle, such as FIGURE 10.2G, show that the width of a sarcomere, the distance between two adjacent Z lines, varies with the state of contraction of the tissue. The widths of the H zone and I bands also vary. To explain these observed changes, physiologists developed the *sliding-filament model of muscle contraction.* According to this model, muscle tissue contracts, or shortens when the two kinds of myofilaments slide past each other, increasing the amount of overlap. The lengths of the myofilaments themselves do not change. In the course of muscle contraction, the heads of the myosin molecules move toward the actin myofilaments, forming *cross bridges* that act as hooks. The actin myofilaments are pulled toward the M line (the middle of the sarcomere) by a change in configuration of the myosin myofilaments. Movement of the actin myofilaments produces the changes shown in FIGURE 10.7A–C. In FIGURE 10.7A, which shows the maximally contracted condition, the ends of the actins actually overlap in the middle of the sarcomere.

distance between myosin filaments in the H zone is relatively large. Thus the length of the sarcomere can be seen to change with the degree of contraction.

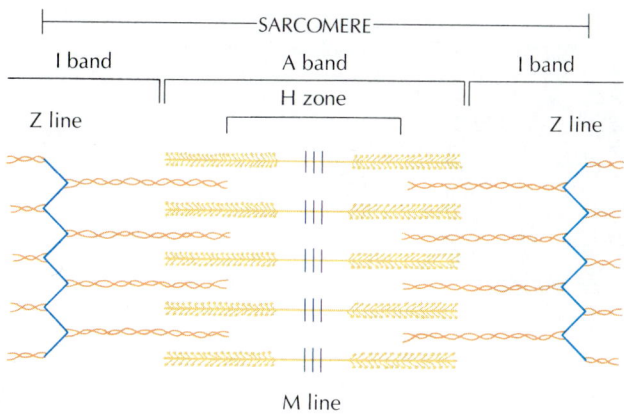

SARCOMERE

| I band | A band | I band |

H zone

Z line   Z line

M line

[C] RELAXED

FIGURE 10.8 ACTIN AND MYOSIN MYOFILAMENTS

[A] Thin actin myofilament in the resting state. The myofilament contains tropomyosin and troponin as well as actin. The troponin complex is represented by the divided blue sphere. [B] A thick myosin myofilament, showing the movable heads.

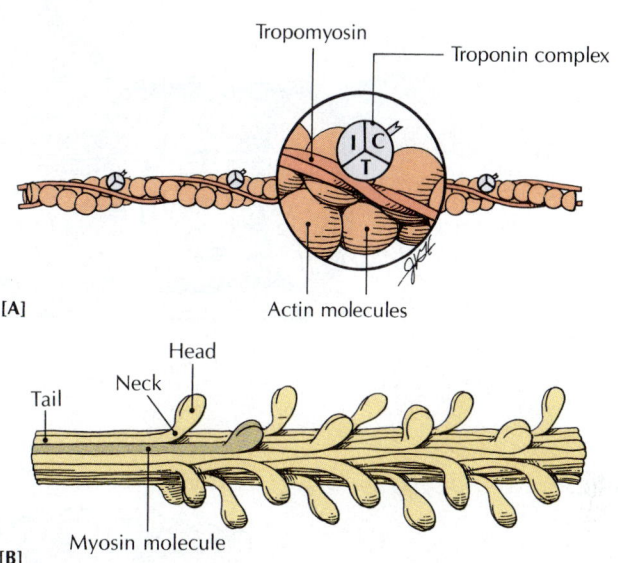

Tropomyosin

Troponin complex

I C T

[A]   Actin molecules

Head

Neck

Tail

Myosin molecule

[B]

Excitation-contraction coupling The sliding-filament model was first proposed in the 1950s. Since then, physiologists have discovered the molecular basis for muscle contraction. The process is called *excitation-contraction coupling* and it is powered by the hydrolysis of ATP. FIGURE 10.8 shows the structure of actin and myosin myofilaments. A complete actin myofilament is made up not only of actin, but also of the proteins *tropomyosin* and *troponin*. Troponin is actually composed of three subunits (I, T, and C) and is sometimes referred to as the troponin complex. The molecules of actin, tropomyosin, and troponin are arranged in thin, twisted strands. The thicker myosin myofilament is composed of myosin molecules, which have oval-shaped "heads" and long "tails."

A summary of excitation-contraction coupling and the mechanism of muscle contraction is given below and shown in FIGURE 10.9.

1 The myosin cross bridges are active only when they are attached to actin myofilaments. Before calcium ions are released from the sarcoplasmic reticulum in muscle fibers, tropomyosin blocks the binding sites on the actin from combining with myosin. The myosin heads are in a low-energy configuration when the myofibril is in this resting condition [FIGURE 10.9A].

2 When an action potential reaches the neuromuscular junction, the acetylcholine released from the axon terminal initiates a wave of electrical activity that spreads over the sarcolemma and T tubules within the muscle fiber, producing an action potential in the T tubules. The action potential triggers the release of calcium ions from the lateral sacs of the sarcoplasmic reticulum [FIGURE 10.9B, C].

3 Calcium ions bind to the troponin complex on the actin myofilament, causing the complex to shift and moving the tropomyosin strand away from its blocking position, leaving the binding site on the actin exposed. The myosin heads bind to actin [FIGURE 10.9C].

4 Cross-bridge binding allows myosin to act as an enzyme (myosin ATPase), splitting the ATP stored in myosin heads into ADP and inorganic phosphate (P_i) and releasing energy. The energy is used to move the myosin heads toward the actin myofilaments; the myosin head tilts, pulling the actin myofilament along so that the myosin and actin myofilaments slide past each other. The actin myofilaments from opposite ends of the sarcomere move toward each other, and the muscle *contracts* [FIGURE 10.9D]. The width of the A bands is unchanged, but the I bands shorten, and the Z lines move closer together. As the myofilaments slide, the cross bridges detach from one site and attach to the next site. Skeletal muscle contractions occur so rapidly that each myosin cross bridge may attach to and release actin binding sites as often as 100 times a second.

FIGURE 10.9 A SUMMARY OF EXCITATION-CONTRACTION COUPLING

(1) A muscle action potential is generated and propagated along the T tubules into the lateral sac of the sarcoplasmic reticulum. (2) Ca^{2+} is released from the lateral sac and binds to troponin, removing the blocking action of tropomyosin and exposing binding sites. ATP provides the energy for the cross bridge to move, causing contraction. (3) After the action potential, Ca^{2+} is taken up into the sarcoplasmic reticulum. Ca^{2+} removal from troponin restores the blocking action of tropomyosin; relaxation occurs.

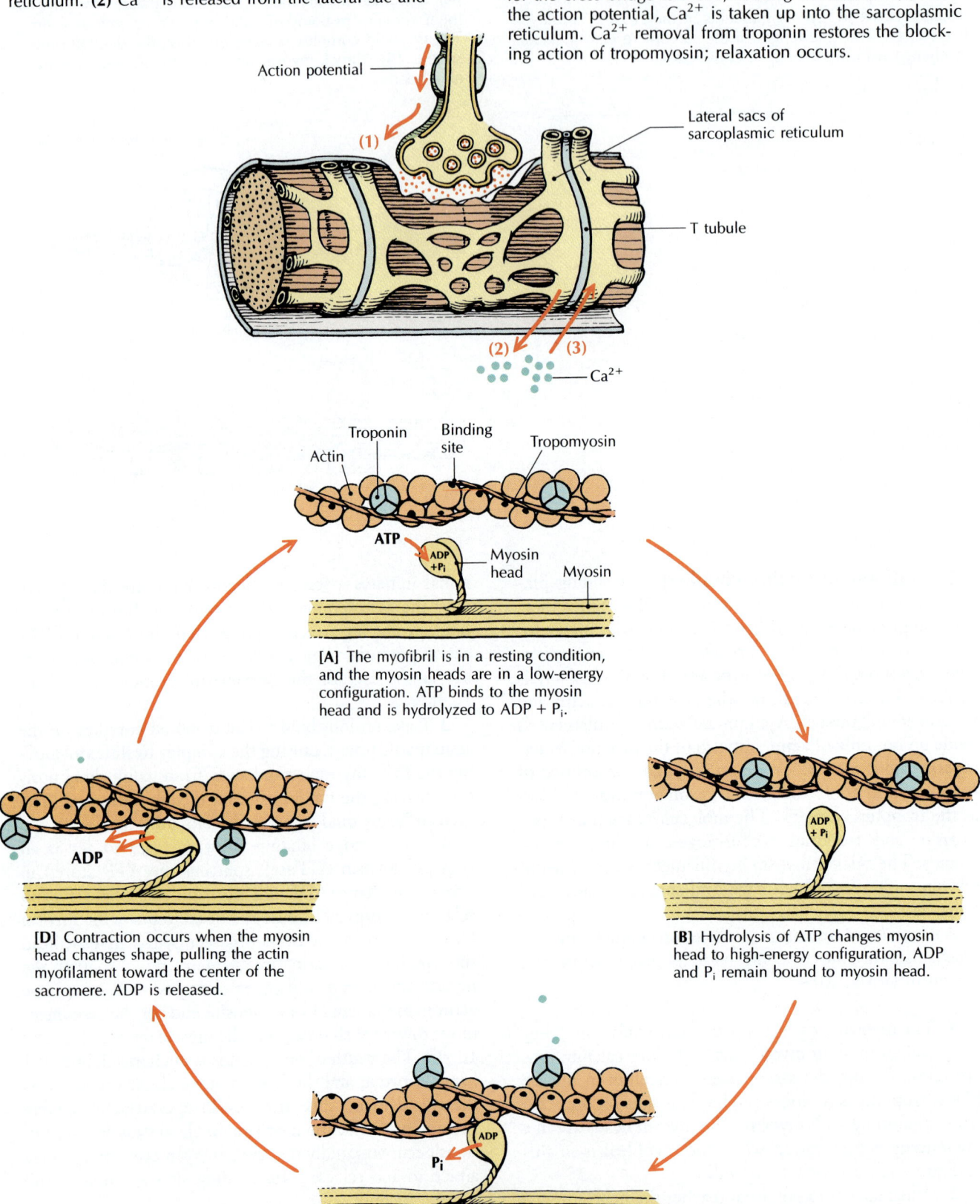

[A] The myofibril is in a resting condition, and the myosin heads are in a low-energy configuration. ATP binds to the myosin head and is hydrolyzed to ADP + P_i.

[B] Hydrolysis of ATP changes myosin head to high-energy configuration, ADP and P_i remain bound to myosin head.

[C] Calcium ions bind to troponin, and the complex of troponin and tropomyosin changes shape (moves), exposing the binding site on the actin molecule. The myosin head binds to the actin, and P_i is released.

[D] Contraction occurs when the myosin head changes shape, pulling the actin myofilament toward the center of the sacromere. ADP is released.

Muscle Relaxation

When a muscle *relaxes,* the following sequence occurs:

1 Acetylcholine is broken down by acetylcholinesterase, which is released from the plasma membrane of the muscle fiber. This breakdown prevents further stimulation of the muscle fiber by the motor neuron.

2 Without the stimulation of an action potential, the calcium ions move away from the myofilaments and are actively transported to the sarcoplasmic reticulum by Ca-ATPase.

3 Without calcium, troponin and tropomyosin once again block the binding sites on the actin myofilament, preventing myosin from forming cross bridges with actin.

4 As the myosin and actin myofilaments return to their original positions in the sarcomere, the I bands become broader and the Z lines move farther apart [FIGURE 10.7C]. The sarcomeres return to their original (resting) length, and the muscle fiber relaxes.

Sources of ATP

Enough ATP is present in a resting muscle fiber to allow it to contract for a few seconds. Since muscle fibers usually contract for more than a few seconds, where does the needed ATP come from? The primary source of extra ATP is the transfer of phosphate from *creatine phosphate* to ADP during a muscle contraction [FIGURE 10.10]:

$$\text{ADP} + \text{creatine phosphate} \xrightarrow{\underset{\text{phosphokinase}}{\text{creatine}}} \text{ATP} + \text{creatine}$$

Although the concentration of creatine phosphate in muscle fibers is about five times that of ATP, it still supplies only enough ATP for a few additional seconds of contraction. The ATP needed by muscle fibers for sustained contractions comes from the breakdown of stored muscle glycogen into glucose and the subsequent breakdown of glucose by cellular metabolism. Also, the additional breakdown of fats and amino acids provides extra ATP for sustained contractions. When the muscle fibers return to a resting state, they continue the breakdown of glucose to form ATP until the resting level of creatine

phosphate is restored. Glucose is then used to replenish the glycogen depleted during contractions.

TYPES OF MUSCLE CONTRACTION

Several types of muscle contraction have been identified, including twitch, tetanus, treppe, and isometric and isotonic contractions.

Twitch

A momentary contraction of a muscle in response to a single stimulus (such as an electric current or direct stimulation by a motor neuron) is a *twitch.* It is the simplest type of recordable muscle contraction. The short, jerky action of a twitch is usually produced artificially in order to record the response on a graph called a *myogram* [FIGURE 10.11].

Latent period The brief period between the time of the electrical stimulus and the beginning of contraction is the *latent period.* This is the time during which calcium is released from the sarcoplasmic reticulum to initiate shortening. This period lasts about 2 milliseconds (msec).

Contraction period The *contraction period* lasts about 10 to 80 msec; it is the upward part of the tracing.

Relaxation period The *relaxation period* is the downward part of the tracing, lasting also about 10 to 80 msec, and results from the active uptake of calcium into the sarcoplasmic reticulum, which causes relaxation.

Refractory period After depolarization, a skeletal muscle fiber cannot be depolarized again for about 0.005 sec. This recovery period is called the *refractory period.* (In contrast, the refractory period for cardiac muscle fibers is about 0.3 sec.) The refractory period is generally divided into the *absolute refractory period* (from the threshold stimulus until repolarization is about one-third complete) and the *relative refractory period* (from the end of the absolute period to the start of a new depolarization). During the absolute refractory period, no stimulus, no matter how long or strong, will cause a muscle

┌─────────────────────────────────────┐
Q: *What causes a hand or foot to "fall asleep"?*

A: Pressure on a limb (as when you cross your legs or fall asleep on your arm) may temporarily cut off its blood supply or the steady stream of nerve impulses that help maintain muscle tone. Sensation may be lost as long as pressure remains on the nerve or blood vessel, but when the blockage is removed, the nerve impulses are effective once more and the blood supply is returned. The combination of these renewed processes may cause an uncomfortable "pins and needles" feeling until normal functions resume completely.
└─────────────────────────────────────┘

FIGURE 10.10 SUMMARY OF EVENTS IN MUSCULAR CONTRACTION AND RELAXATION

Summary of events in muscular contraction and relaxation, according to the sliding-filament model.

CONTRACTION

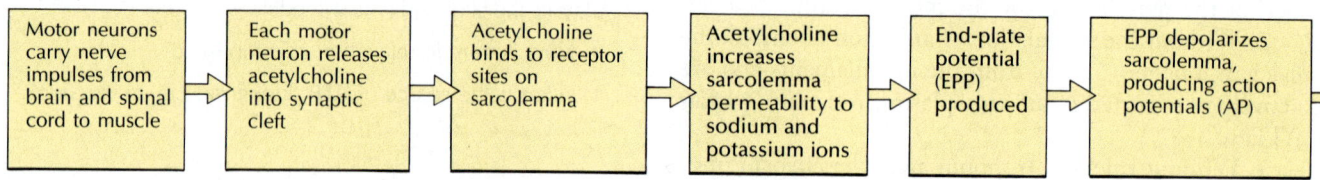

| Motor neurons carry nerve impulses from brain and spinal cord to muscle | → | Each motor neuron releases acetylcholine into synaptic cleft | → | Acetylcholine binds to receptor sites on sarcolemma | → | Acetylcholine increases sarcolemma permeability to sodium and potassium ions | → | End-plate potential (EPP) produced | → | EPP depolarizes sarcolemma, producing action potentials (AP) |

fiber to be stimulated. During the relative refractory period, stronger than normal stimuli are needed to stimulate a muscle fiber.

All-or-none principle The minimal stimulus needed to cause a muscle fiber to contract is called the *threshold stimulus*. If the intensity of the stimulation is greater than the threshold level, the magnitude of the contraction will *not* increase. If the stimulus is below the minimum required for contraction, it is considered to be a *subthreshold stimulus*, and the muscle fibers will not contract at all. The ***all-or-none principle*** refers to the fact that individual muscle fibers in a motor unit do not contract partially—they contract fully or not at all.

Although individual muscle fibers follow the all-or-none principle, whole muscles have graded contractions. (You need a stronger contraction to lift a suitcase than to lift a soupspoon, for example.) How is this possible if the all-or-none principle really works? The strength of a muscle contraction depends on the *frequency of stimulation* and on *how many motor units* are activated. All these processes involve *summation*, the contraction of varying numbers of muscle fibers all at once. For example, if the soupspoon is to be lifted, only a few motor units are contracted at the same time. Motor units do not all have the same threshold, and if only the motor units with low thresholds are stimulated, a relatively small number of motor units contract. At higher intensities of stimulation, other motor neurons respond, and this leads to the activation of more motor units.

Tetanus

If a muscle receives repeated stimuli at a rapid rate, it cannot relax completely between contractions. The tension achieved under such conditions is greater than the tension of a single muscle twitch and is called ***summation of twitches*** [FIGURE 10.11B]. A more or less continu-

FIGURE 10.11 MYOGRAMS OF SEVERAL TYPES OF CONTRACTIONS

[A] *Twitch.* The muscle is stimulated once at point A. There is a short latent period (A–B) of a few milliseconds before the contraction actually begins at point B and peaks at C. The contraction lasts about 0.04 sec. The following relaxation period (C–D) lasts about 0.05 sec. **[B] *Summation.*** Summation occurs when more than one stimulus is given before the muscle has relaxed completely after the first stimulus. **[C] *Incomplete tetanus.*** An incomplete tetanic contraction occurs when a muscle that is just beginning to relax receives another stimulus. **[D] *Complete tetanus.*** A complete tetanic contraction occurs when the rest periods between stimulations become so short that the muscle cannot relax at all.

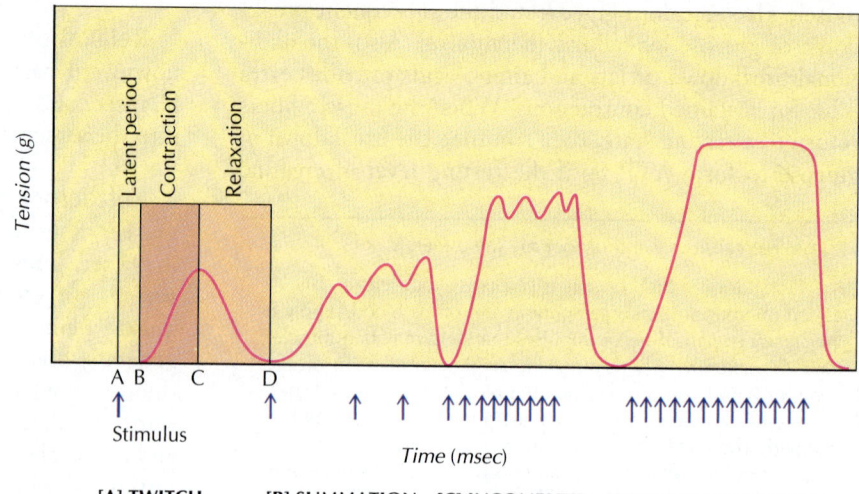

[A] TWITCH   [B] SUMMATION   [C] INCOMPLETE TETANUS   [D] COMPLETE TETANUS

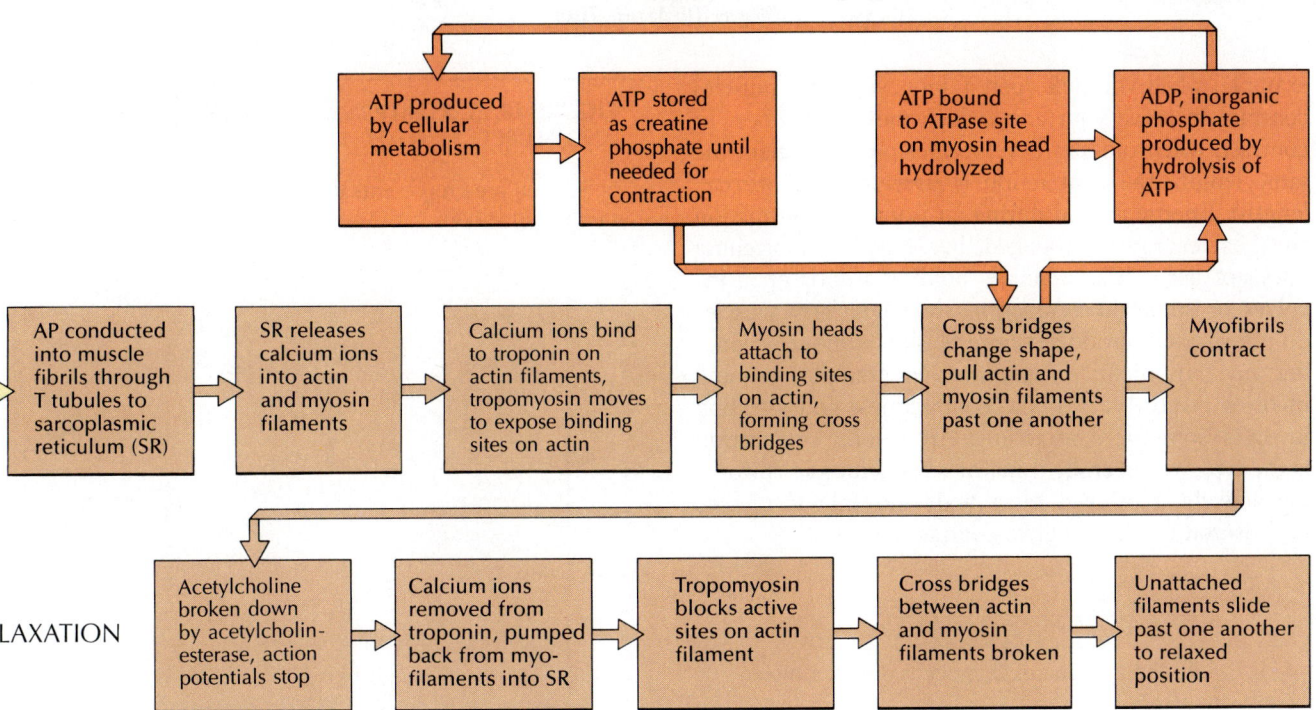

ous contraction of the muscle is called *tetanus.* When incomplete relaxations are still evident between contractions, the muscle is said to be in a state of *incomplete tetanus* [FIGURE 10.11C]. *Complete tetanus* occurs when the muscle is in a steady state of contraction with no relaxation at all between stimuli, as in "lockjaw" [FIGURE 10.11D; see When Things Go Wrong at the end of this chapter].

FIGURE 10.12 MYOGRAM OF A TREPPE CONTRACTION, OR "STAIRCASE" PHENOMENON

The stimulations occur repeatedly at about 0.5-sec intervals. Note the upward staircase effect during the first few seconds and then the leveling off of the strength of the contractions.

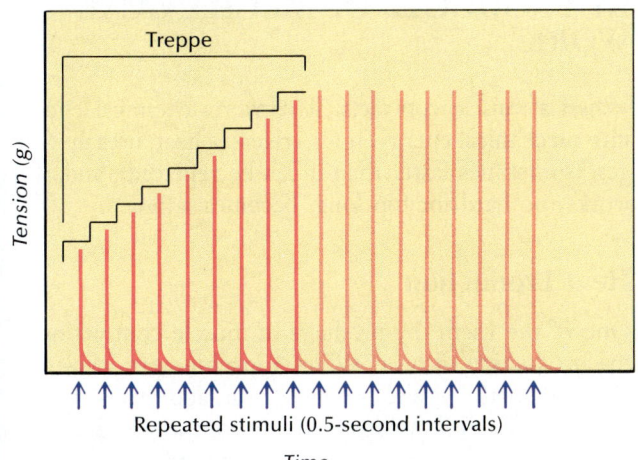

Treppe

When a rested muscle receives repeated stimuli over a prolonged period, the first few contractions increase in strength so that the myogram starts out looking like an upward staircase [FIGURE 10.12]. This type of contraction is known as *treppe* (TREHP-eh; Ger. staircase). After several contractions, a steady tension for each contraction is reached. This gradual increase in "peak performance" may be due to a progressive build-up of calcium available for muscle contraction and/or a change in the internal conditions in the muscle, such as in pH, temperature, or viscosity. "Warming up" may improve an athlete's performance by "getting past" this treppe phenomenon.

Isotonic and Isometric Contractions

When a muscle contracts by becoming shorter and thicker, the contraction is called *isotonic* (Gr. *isos*, equal + *tonos*, tension) because the amount of force, or tension, remains constant as movement takes place. For

Q: *What causes "rigor mortis"?*

A: Three or four hours after death, ATP in muscles breaks down and is not replaced. Thus, there is no ATP to release the cross bridges between the actin and myosin myofilaments. The myofilaments become locked in place, and the muscles become rigid. This condition is called rigor mortis ("death stiffening"). It is complete in about 12 hours. About 15 to 25 hours later, the muscle proteins are destroyed by enzymes in the cells, and rigor mortis disappears.

example, when you pull open a door, your arm muscles contract to move your arm, which moves the door. In contrast, when muscles develop tension but remain the same length, the contraction is *isometric* (Gr. *metron*, length). Energy is used during an isometric contraction and heat energy is produced, but a typical shortening does not occur. For example, holding a door open involves an isometric contraction, since your arm muscles do not shorten but do generate tension without producing any movement. To use another example, running involves noticeable isotonic contractions, and standing involves isometric contractions. An isotonic contraction is the type we generally think of when a muscle is working actively. However, most body movements involve both isotonic and isometric contractions.

A S K   Y O U R S E L F

1 What are the two phases that occur during a twitch contraction?

2 What happens during a refractory period?

3 How does the all-or-none principle operate?

4 What is the difference between incomplete and complete tetanus?

5 Why is treppe also called a "staircase" contraction?

6 How do isotonic and isometric contractions differ?

GRADING (VARYING STRENGTH) OF MUSCLE CONTRACTION

The main factors affecting *grading*, or the varying strength of muscle contraction, are (1) the number of muscle fibers contracting, (2) the relative size of the muscle, and (3) the degree of muscle stretch.

1 Number of fibers stimulated The greater the number of motor units recruited, the greater the strength of muscle contraction.

2 Muscle size The greater the muscle size (cross-sectional area), the greater the strength of contraction. Muscle size can increase (*hypertrophy*) through regular physical exercise, which can result in greater strength of contraction.

3 Length-tension relationship The ideal resting length of muscle fibers is the length at which the greatest tension can be developed. This is the length at which the greatest numbers of cross bridges between the actin and myosin molecules are formed, thus generating the greatest amount of tension. When the muscle fibers are stretched so that the actin and myosin filaments barely

FIGURE 10.13 RELATIONSHIP BETWEEN TENSION AND MUSCLE LENGTH

The shaded area represents the physiological range of tension/muscle length.

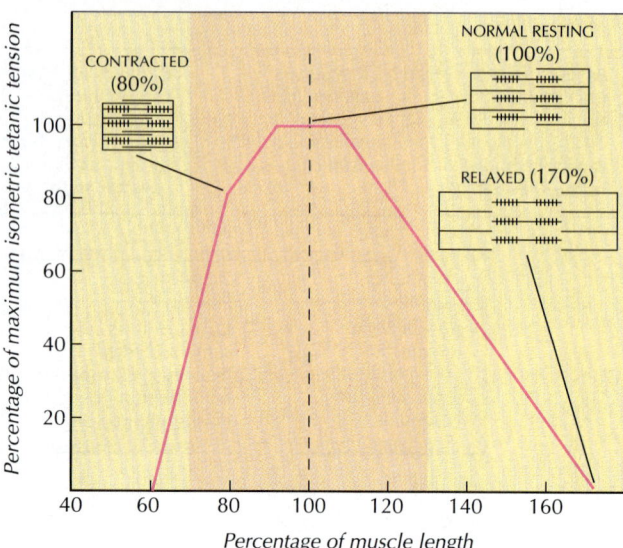

overlap, very little tension is generated. Or if the muscle fibers are compressed, so that the actin filaments overlap and interfere with cross-bridge formation, tension generation is minimal. The ideal operating length for whole muscle is about 80 to 120 percent of its normal resting length. If it is stretched severely beyond its optimal length (175 percent), it cannot develop significant tension; if it is contracted to 60 percent or less of its optimal length, further shortening is minimal [FIGURE 10.13].

BY-PRODUCTS OF MECHANICAL WORK

When a muscle contracts, it converts chemical energy into mechanical energy. It also releases heat, uses up oxygen (sometimes faster than it can be replaced), and, if it works too hard for too long, becomes fatigued.

Heat Production

One of the useful by-products of muscle contraction is the production of heat. When a machine produces mechanical energy, it also produces heat, and this is considered a mechanical disadvantage. By contrast, in the body, heat production is necessary to maintain a stable body

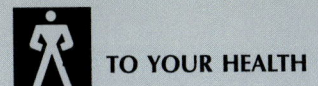

How Physical Activity Affects Skeletal Muscle

The number of skeletal muscle cells in the body does not increase as a result of strenuous physical activity, but the *diameter* of the individual muscle fibers increases, as do the number of myofibrils and mitochondria, the amount of sarcoplasmic reticulum, and the blood supply. This is called muscular **hypertrophy** (Gr. *hyper*, over + *trophe*, nourishment). A muscle fiber becomes larger and stronger through exercise, and its metabolic activity increases. In contrast, if a muscle is not used over a period of months, it will shrink and become weak, and its metabolic activity will decrease. This is muscular **atrophy** (Gr. *a*, without + *trophe*, nourishment). Astronauts show a decrease in muscle mass because their muscles do not need to work as hard in space, where gravity is reduced greatly. Apparently, about half the atrophied muscles also have damaged nerve endings, producing poor coordination as well as muscle weakness.

As an exercised muscle becomes larger, it also becomes stronger and uses oxygen and the available energy sources more efficiently. The efficiency of a trained athlete is usually about 20 to 25 percent, and some athletes have achieved 40-percent efficiency.

The amount of physical activity can change muscle fibers by altering blood flow to the muscle, by altering the amount of available ATP (by affecting the synthesis of enzymes involved in ATP production), and by increasing or decreasing the diameter of muscle fibers (by adding or subtracting myofibrils).

Different Types of Exercise Produce Different Effects

Low-intensity, long-duration exercise such as swimming has different effects than high-intensity, short-duration exercise such as weight lifting. *Low-intensity, long-duration exercise* increases the number of mitochondria in muscle fibers and the number of nearby capillaries and results in an overall increase in endurance and decrease in fatigue. Because low-intensity, long-duration activities also improve the efficiency of the respiratory and circulatory systems, the muscles receive more oxygen and other nutrients. However, fiber diameter decreases slightly, and there may be a corresponding decrease in strength.

High-intensity, short-duration activities mainly affect fast-twitch muscle fibers, which increase in diameter when more actin and myosin filaments are formed, producing additional myofibrils. Also, the number of metabolic enzymes increases. High-intensity, short-duration exercise increases the strength and bulk of muscles, but although such muscles may look impressive, they are generally low in endurance.

Cramps and Muscle Pain Due to Physical Activity

Cramps usually do not occur when a contracting muscle has an adequate blood supply, but an insufficient supply may produce a cramp. The pain of a cramp usually persists after the contraction is complete, lasting until the normal blood flow returns. Muscle pain can also occur with a normal blood supply if a muscle contracts for a prolonged period without relaxation because the contracted muscle eventually compresses the vessels that supply it with blood.

Cramps can also be caused by dehydration; low blood glucose; an imbalance of body salts such as sodium, potassium, and chloride; insufficient warm-up; overexercising; or poor conditioning. So-called heat cramps are usually caused by excessive sweating that leads to an extreme loss of salts and water.

Because different physical activities involve different sets of muscles, you may get muscle soreness when you perform a new activity and use muscles that are normally used only infrequently. Soreness usually is most evident about two days after the unusual activity and is probably due to tiny tears in the muscles. Such soreness is different from the muscle fatigue caused by a build-up of lactic acid. Excess lactic acid is usually removed from a fatigued muscle within 30 minutes after the exercise.

temperature. Muscular activity provides the major source of heat for temperature regulation. In response to cold, for example, skeletal muscles may begin to contract, causing shivering, muscular tremors that occur about 10 to 20 times a second. Heat is also produced during heavy physical activity and in a direct voluntary response to cold, as when we stamp our feet or rub our hands together.

Oxygen Debt and Muscle Fatigue

When you play tennis or jog at a leisurely pace, your body uses extra oxygen. It is made available when your heart rate increases, carrying oxygen to your muscles, and when you begin to breathe harder than usual. But sometimes, when a runner is sprinting, for instance, the skeletal muscles work so hard that they use up oxygen faster than it can be supplied. The body needs oxygen to carry out the chemical reactions that produce ATP from

Q: *Will drinking water, massaging the muscles, stretching, and eating bananas (as a source of potassium) help prevent muscle cramps while exercising vigorously on a very hot day?*

A: Everything but the bananas will help. The bananas won't be digested soon enough to supply the body with potassium *during* the activity.

the breakdown of glucose. ATP supplies the energy for most of the body's activities.

The production of ATP begins with the stepwise breakdown of a glucose molecule into two molecules of the compound pyruvic acid. The pyruvic acid is then gradually broken down to carbon dioxide and water. The breakdown of pyruvic acid requires oxygen:

$$Pyruvic\ acid + O_2 \longrightarrow CO_2 + H_2O + energy$$

Because oxygen is necessary for this process, it is called *aerobic* ("with oxygen") *respiration.*

During strenuous exercise the amount of oxygen supplied to the muscles may not be adequate to break down pyruvic acid completely into carbon dioxide and water. Instead, pyruvic acid is reduced to *lactic acid*, a process called *anaerobic* ("without oxygen") *respiration* because it does not require oxygen. Although anaerobic respiration allows muscles to contract, it is effective for only a short time. Most of the lactic acid is removed by the liver for conversion back to glucose. But excess lactic acid begins to accumulate in the skeletal muscles and makes contraction more and more difficult. Eventually, lactic acid builds up to the point where it causes the tired feeling we call **muscle fatigue.**

To overcome fatigue, lactic acid must be removed from the muscle fibers by being converted into carbon dioxide and water, a process that requires oxygen. But because the body has just completed a spurt of vigorous physical activity, oxygen is in short supply. Deep, rapid breaths provide oxygen for the fatigued muscle fibers. The amount of extra oxygen needed to metabolize the accumulated lactic acid and restore normal ATP levels represents the **oxygen debt.**

During a state of fatigue, there is no change in nervous impulses, neuromuscular junctions, or stimulation of muscle fibers. Contractions merely become weaker and weaker, and the muscle finally stops contracting altogether, because (1) ATP is not present in sufficient amounts, (2) toxic products (carbon dioxide and lactic acid) accumulate, and (3) circulatory disturbances to the muscle do not allow the delivery of needed substances or the removal of waste products.

When a muscle is completely fatigued, it does not contract, but it does not relax either. For this reason, complete fatigue may be confused with a muscle cramp. It is important not to try to use a fatigued muscle, because if a muscle's glycogen supply is depleted, it will begin to use the protein contained in its own fibers.

ASK YOURSELF

When does an oxygen debt occur, and how does it relate to muscle fatigue?

FAST-TWITCH AND SLOW-TWITCH MUSCLE FIBERS

Although there is a fairly continuous range in the speed of contraction of the body's muscles, muscle fibers generally are classified as fast-twitch (type II fibers) or slow-twitch (type I fibers). The differences between these types of fibers are due to different myosin ATPase enzymes that are also designated as "fast" or "slow."

The response of a muscle fiber or a whole muscle can be expressed by the duration of a muscle twitch (contraction-relaxation cycle). For example, the muscles that move the eyes — all fast muscles — have contraction-relaxation cycles lasting about 30 msec. A large leg muscle such as the soleus — a slow muscle — has a contraction-relaxation cycle lasting about 3000 msec. The contraction-relaxation cycles of other muscles lie between these two extremes.

Fast-twitch muscle fibers, such as those in the delicate eye and hand muscles, are well adapted to produce rapid contractions. Fast-twitch muscle fibers contain many energy-producing mitochondria (though fewer than slow-twitch muscle fibers do). They also have an extensive sarcoplasmic reticulum, providing a plentiful supply of calcium ions, which trigger a quick response to nerve impulses. However, these muscle fibers tire easily, since they contain relatively little *myoglobin*, an oxygen-binding protein that increases the rate of oxygen diffusion into a muscle fiber and provides a small store of oxygen within the fiber. Because fast-twitch muscle fibers also have fewer capillaries than slow-twitch muscles do, oxygen and other nutrients are supplied to them more slowly.

In contrast, **slow-twitch muscle fibers** are suited for prolonged, steady contractions. They have many mitochondria, large amounts of myoglobin, and a greater capillary blood supply. Slow-twitch muscle fibers do not tire as easily as fast-twitch muscle fibers do. The large back muscles that help us maintain an erect posture all through the day are an example of slow-twitch muscle fibers. Slow-twitch muscle fibers do not have, or need, as large a supply of calcium ions as fast-twitch muscle fibers do.

The color of fast-twitch and slow-twitch muscle fibers is different as well. Slow-twitch muscle fibers contain large amounts of myoglobin, which has a reddish color, and many capillaries with red blood cells. Thus slow-twitch muscle fibers are also called **red muscles.** Fast-twitch muscle fibers, with relatively little myoglobin and fewer capillaries with red blood cells, have a pale appearance and are referred to as **white muscles.**

Most people have approximately equal numbers of fast-twitch and slow-twitch fibers in running muscles such as the gastrocnemius. Sprinters, however, have more fast-twitch fibers, and distance runners have more slow-twitch fibers. Fast-twitch fibers are perfect for short, fast bursts of speed. They burn stored glycogen quickly, without using oxygen. In the process, however, lactic acid accumulates, and the muscles become fatigued when they run out of fuel.

The leg muscles of long-distance runners contain mostly slow-twitch fibers. (Tests of two world-class marathon runners showed that one had about 75 percent slow-twitch fibers, while the other had 93 percent.) Slow-twitch muscles take a longer time to reach a maximal contraction, but when they do, they can sustain it as long as oxygen is available.

A S K   Y O U R S E L F

What are some of the basic differences between fast-twitch and slow-twitch muscle fibers?

SMOOTH MUSCLE

Smooth muscle tissue is so called because it is not striated and therefore appears "smooth" when viewed with a microscope. It is also called *involuntary muscle* because it is controlled by the autonomic nervous system, the involuntary division of the nervous system.* Smooth muscle tissue is found most commonly in the circulatory, digestive, respiratory, and urogenital systems.

Unlike skeletal muscle, smooth muscle is not connected to bones. Instead, slender smooth muscle fibers [FIGURE 10.14A] generally form sheets in the walls of large, hollow organs such as the stomach and urinary bladder and in the walls of some blood vessels [FIGURE 10.14B]. Although smooth muscle fibers are often arranged in parallel layers, the exact arrangement of the fibers varies from one location to another. In the walls of the intestines, for example, smooth muscle fibers are arranged at right angles to each other, one layer running longitudinally and the next wrapped around the circumference of the tubular intestine. These layers work together in a coordinated action to supply the contractions that move the intestinal contents toward the anus prior to defecation. In the urinary bladder and uterus, the layers are poorly defined and are oriented in several different directions. Connective tissue outside the layers extends into the spaces between muscle fibers and binds them into bundles.

---

*Although smooth muscle is usually controlled involuntarily, techniques such as biofeedback may actually allow a person to regulate some functional activities of smooth muscle. (See "Biofeedback and the Autonomic Nervous System" on page 481.)

FIGURE 10.14 SMOOTH MUSCLE

[A] Light micrograph of longitudinal and cross section of smooth muscle tissue. ×760. **[B]** Scanning electron micrograph of smooth muscle of a vein. Nerve axons can be seen as thin white lines. ×1000.

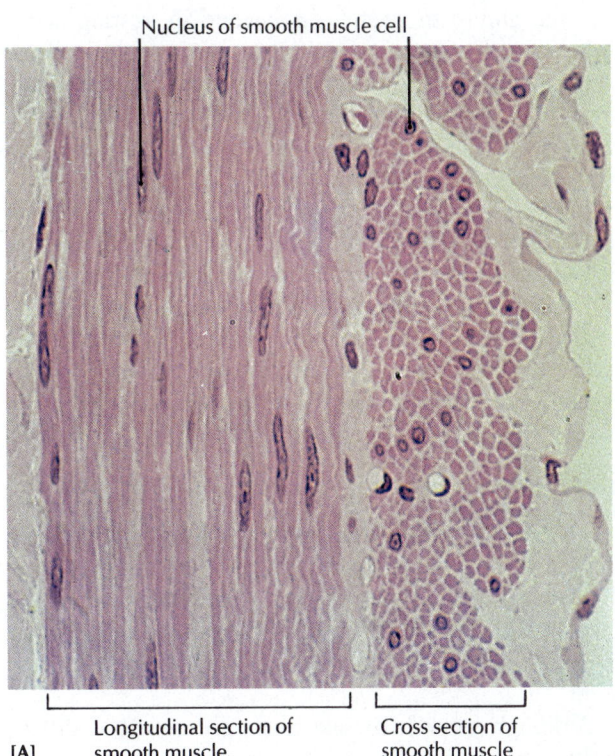

Nucleus of smooth muscle cell

[A] Longitudinal section of smooth muscle   Cross section of smooth muscle

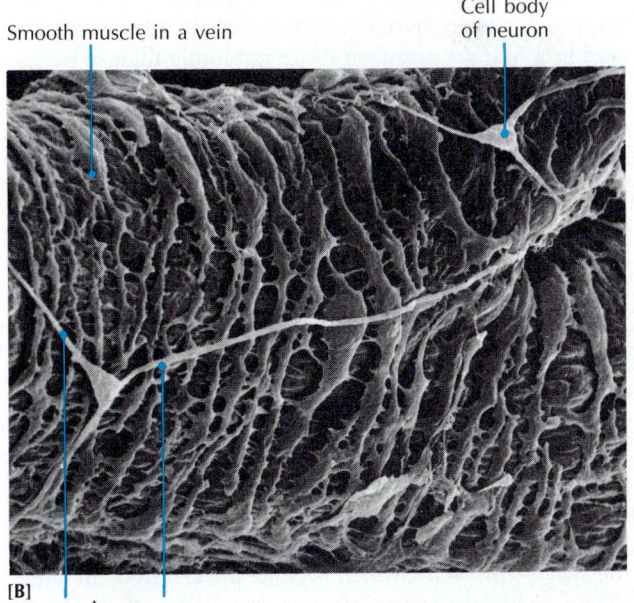

Smooth muscle in a vein   Cell body of neuron

[B] Axons

Properties of Smooth Muscle

The two main characteristics of smooth muscle are that (1) its contraction and relaxation periods are slower than those of any other type of muscle and (2) its action is rhythmical. Its contractions may last for 30 sec or more, but it does not tire easily. Such sustained contractions, plus the ability to stretch far beyond its resting state, make smooth muscle well suited to the muscular control of the stomach and intestines, the urinary bladder, and the uterus, especially during pregnancy and childbirth, when powerful, sustained contractions help move the baby out of the uterus.

More smooth muscle is found in the digestive system than in any other place in the body. Smooth muscle cells that line the walls of the stomach and intestines contract and relax rhythmically to help move food along the digestive tract. After we swallow food, all muscular contractions in the digestive system are involuntary until we consciously initiate the process of defecation.

Structure of Smooth Muscle

Like the cells of skeletal muscle tissue, the cells of smooth muscle tissue are called *fibers*. Each fiber is long (but not nearly as long as skeletal muscle fibers), spindle-shaped, and slender. It contains only one nucleus, which is usually located near the center of the fiber, at its widest point.

Although smooth muscle fibers are arranged differently from those in skeletal muscle, the basic contractile mechanism appears to be much the same. Actin, myosin, and tropomyosin are present, but troponin is absent. The actin and myosin myofilaments within the myofibrils are very thin and are arranged more randomly than in skeletal muscle cells. As a result, smooth muscle cells lack striations. Also, the myosin in smooth muscle is chemically different from that in skeletal muscle. The sarcoplasmic reticulum is poorly developed, and T tubules and Z lines are not present. The slowness of contractions is thought to be due to a limited amount of ATPase activity at the cross bridges.

Types of Smooth Muscle

Based on their arrangements as separate bodies or in bundles, smooth muscle fibers are usually classified as either the single-unit type or the multiunit type.

Single-unit smooth muscle Most smooth muscle is the ***single-unit*** type, which is generally arranged in large sheets of fibers [FIGURE 10.15A]. It is also called *visceral* (L. *viscera*, body organs) *smooth muscle* because it is present in the walls of hollow organs of the body, such as the uterus, stomach, intestines, and urinary bladder, as well as some blood vessels. The fibers are in contact with each

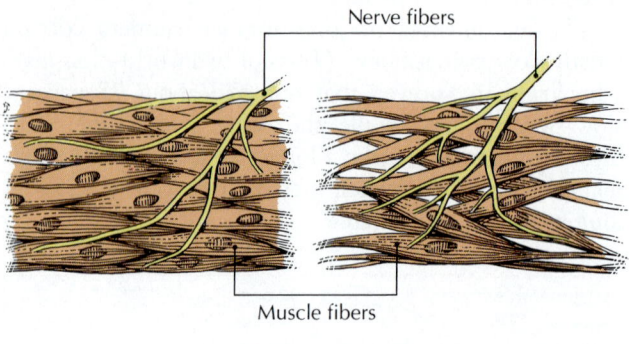

FIGURE 10.15 TYPES OF SMOOTH MUSCLE

[A] Single-unit (visceral), and [B] multiunit smooth muscle.

Nerve fibers

Muscle fibers

[A]   [B]

other through many gap junctions. This is critical because when a muscle cell is stimulated by an action potential, it contracts and spreads the action potential to adjacent cells. This method ensures a steady wave of contractions, such as those that push food through the intestines. The smooth muscle fiber that receives the stimulus from a motor neuron initially and passes it on to adjacent fibers is known as the ***pacemaker cell***. Pacemaker cells may receive impulses, but they do not require them for activity.

Two types of contractions take place in single-unit smooth muscle: tonic and rhythmic. *Tonic contractions* cause the muscle to remain in a constant state of partial contraction, or *tonus*. Tonus is necessary in the stomach and intestine, for example, where food is moved along the digestive tract. It is also found in ring-shaped muscles called *sphincters*, which regulate the openings from one part of the digestive tract to the next. Tonus also prevents stretchable organs such as the stomach and urinary bladder from becoming stretched out of shape permanently.

Smooth muscle helps retain the tension in the walls of expandable organs and tubes, such as the urinary bladder and blood vessels.

Rhythmic contractions are a pattern of repeated contractions produced by the presence of self-exciting muscle fibers from which spontaneous impulses travel. The digestive system provides excellent examples of the two types of rhythmic contractions: (1) *mixing movements*, which resemble the kneading of dough, blend the swallowed food with digestive juices, and (2) *propulsive movements*, or *peristalsis*, propel the swallowed food from the throat to the anus. FIGURE 10.16 shows how smooth muscle contracts like the folds of an accordion.

Multiunit smooth muscle *Multiunit smooth muscle* is so named because each of its individual fibers is stimulated by separate motor nerve endings. There are no gap-junction connections between the muscle cells, and each multiunit cell can function independently [FIG-

FIGURE 10.16 SMOOTH MUSCLE CONTRACTION

[A] Electron micrograph of partly contracted smooth muscle fibers. ×1000. **[B]** Enlarged drawing of partly contracted fiber. **[C]** Electron micrograph of fully contracted smooth muscle fibers. ×1000. **[D]** Enlarged drawing of fully contracted fiber showing how bundles of myofilaments in the fiber contract. The dense bodies anchor the myofilaments and have characteristics similar to the Z lines in skeletal muscle.

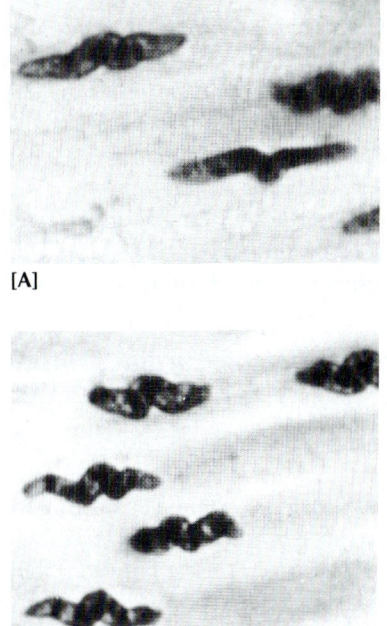

[A]

[C]

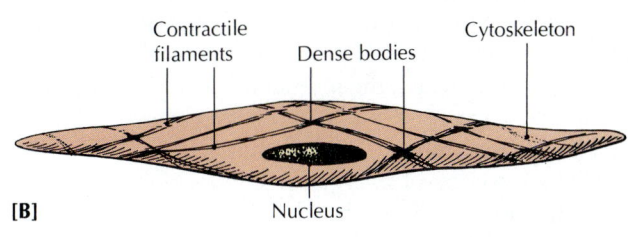

[B]

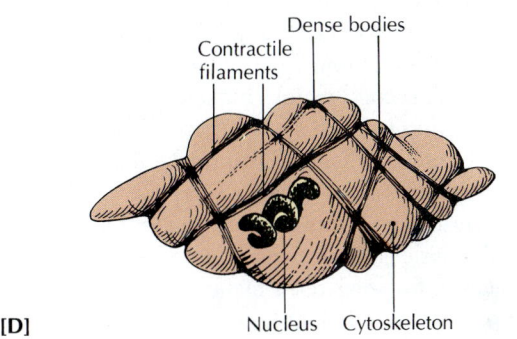

[D]

URE 10.15B]. Multiunit smooth muscle is found, for example, in the iris and ciliary muscles of the eye, where rapid muscular adjustments must be made in order for the eye to focus properly. The arrector pili muscles in the skin that cause the hair to stand on end and produce "goose bumps" are also of the multiunit type, as are the muscles in the wall of the ductus deferens, the tube that conveys sperm from the testes during ejaculation.

CONTRACTION OF SMOOTH MUSCLE

In smooth muscle cells, just as in skeletal muscle cells, changes in the intracellular calcium concentration indirectly control muscle contraction. However, smooth muscle cells do not have troponin, the calcium-binding protein that regulates contractile responses with respect to a change in calcium concentration within the cell.

Role of Calmodulin and Myosin Light-Chain Kinase

In smooth muscle, calcium can come from three sources: (1) intracellular stores within vesicular structures, (2) intracellular stores within mitochondria, and (3) extracellular stores. The effect of calcium ions (Ca^{2+}) is mediated by a calcium-binding protein called *calmodulin* [FIGURE 10.17].

FIGURE 10.17 CALCIUM-ACTIVATED SMOOTH MUSCLE CONTRACTION

The cascade of specific reactions by which calcium ions (Ca^{2+}) activate the contraction of smooth muscle.

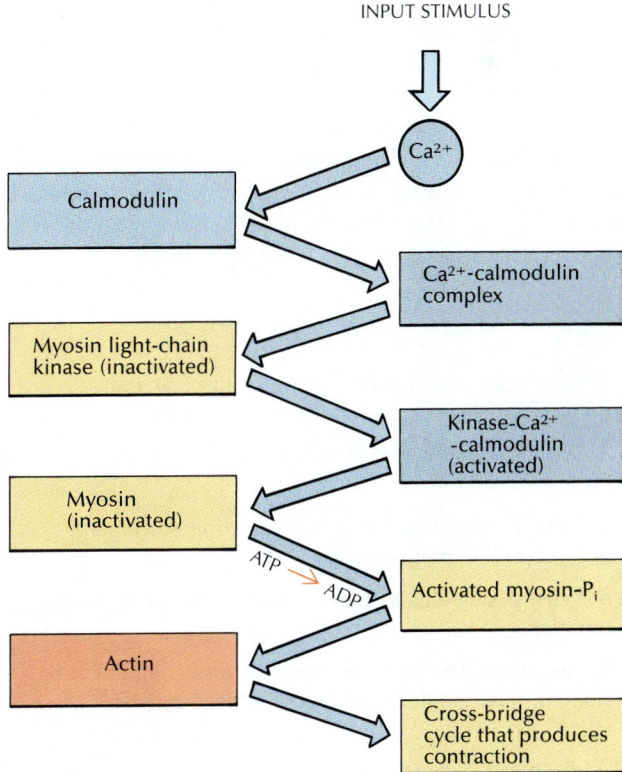

The complex of calmodulin with calcium activates the enzyme *myosin light-chain kinase* so that it adds a phosphate group from ATP to the myosin molecule. Once the phosphate is attached, the myosin binds to the actin, and contraction occurs.

Like skeletal muscle, smooth muscle uses ATP as an immediate source of energy for contraction. However, smooth muscle does not have the energy reserves, such as creatine phosphate, found in skeletal muscle. This means that in smooth muscle the short-term energy supply is very limited, and the muscle must rely on the synthesis of ATP from various substrates to meet its energy requirements for contraction. Thus, one major characteristic of smooth muscle is that it can use a wide variety of substrates for ATP production. Two examples are carbohydrates and fats. A second major characteristic is that the speed of shortening is much slower than in skeletal muscle. Thus, the rate at which ATP must be made for the contractile process is also lower.

Smooth muscle is a more efficient contractile unit than skeletal muscle is. As previously noted, smooth muscle is well suited for long-term maintenance of tension. The reason for this efficiency may involve a "latch" mechanism by which the cross bridges remain in the attached position for a long time, reducing the cycling rate and the rate of ATP consumption.

Regulation of Smooth Muscle Contraction

Smooth muscle contraction is controlled by a variety of inputs that act either on the sarcoplasmic reticulum or on specific calcium channels in the sarcolemma to increase calcium movement into the cell. These inputs include (1) stretching of the smooth muscle myofibrils, (2) spontaneous electrical activity (*pacemaker potential*) within the sarcolemma, (3) specific neurotransmitters released by autonomic neurons, (4) hormones and hormone modulators such as prostaglandins, and (5) locally induced changes in the extracellular fluid surrounding the smooth muscle cell (such as pH, O_2, and CO_2 levels).

A S K   Y O U R S E L F

1 How does the sheetlike arrangement of smooth muscle cells help the muscle perform its jobs?

2 Why is the slow and rhythmic action of smooth muscle important?

3 What is the difference between mixing movements and propulsive movements?

4 What is single-unit smooth muscle? Multiunit smooth muscle?

5 How does the chemical mechanism of contraction of smooth muscle differ from that of skeletal muscle?

CARDIAC MUSCLE

Cardiac muscle tissue (Gr. *kardia*, heart) is found only in the heart. It contains the same type of myofibrils and protein components found in skeletal muscle. Although the number of myofibrils varies in different parts of the heart, the contractile process is basically the same as that described for skeletal muscle. As expected in such hardworking tissue, cardiac muscle cells contain huge numbers of mitochondria.* A well-developed sarcoplasmic reticulum and T tubules are also present.

Structure of Cardiac Muscle

When viewed with a light microscope, cardiac muscle has striations similar to those of skeletal muscle. Although cardiac muscle cells are closely packed, they are *separate*, each with its own nucleus. The cells are joined end to end by specialized cell junctions, called *intercalated disks* (L. *intercalatus*, to insert between), between adjacent cardiac muscle cells [FIGURE 10.18A]. Each intercalated disk contains desmosomelike junctions and gap junctions [FIGURE 10.18B]. Each *desmosome junction* acts as a "spot weld" that strengthens the bond between cardiac cells [see FIGURE 4.2]. Each *gap junction* contains channel proteins that allow the almost instantaneous transmission of electrical impulses from one cardiac cell to the next, which helps synchronize cardiac muscle contractions.

In FIGURE 10.18 intercalated disks appear as dark bands that are wider than the Z lines in skeletal muscle. Intercalated disks always occur at the location of the Z lines.

Characteristics of Cardiac Muscle

Cardiac muscle depends on nervous impulses to some extent, but it is also able to contract rhythmically on its own. As you will see in Chapter 20, when the heart is discussed in detail, cardiac muscle is designed to conduct impulses that help the upper and lower chambers of the heart beat in a carefully synchronized way. Instead of contracting independently, the cells of the heart are all coordinated so that their rate and rhythm are appropriate for the job of pumping blood 24 hours a day.

---

*Mitochondria in cardiac muscle cells occupy about 35 percent of the cell, compared with 2 percent in skeletal muscle cells. This reflects the extreme dependence of cardiac muscle on aerobic metabolism (metabolism requiring oxygen).

Q: *Can cardiac muscle be regenerated?*

A: When cardiac muscle becomes injured, as in a heart attack, there is little chance that the muscle fibers will be able to regenerate. Any healing that does take place produces scar tissue, not muscle tissue.

FIGURE 10.18 CARDIAC MUSCLE

[A] Electron micrograph of cardiac muscle in longitudinal section, showing the intercalated disks between cardiac fibers. Notice the numerous mitochondria. ×12,000. [B] Detailed drawing of intercalated disks, showing two types of junctions involved.

Mitochondria

[A]   Intercalated disks

Transverse portion of
intercalated disk
(desmosome–like junction)

Longitudinal portion of
intercalated disk
(gap junction)

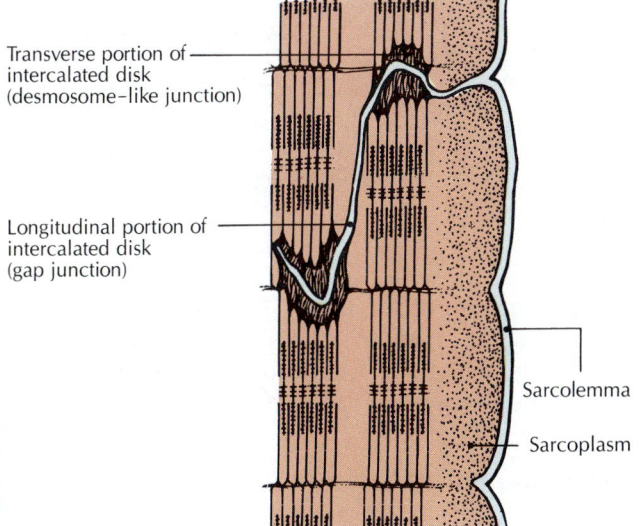

Sarcolemma

Sarcoplasm

[B]

Cardiac muscle and skeletal muscle differ in several ways. The sarcoplasmic reticulum is less extensive in cardiac muscle, but the calcium-ion sensitivity of intact cardiac muscle is much greater than that of skeletal muscle. Because a significant amount of calcium enters the cardiac muscle cell during contraction, the cell can actually contract for longer periods than a skeletal muscle cell can. However, since the excitation-contraction coupling is so calcium-dependent in cardiac tissue, cardiac muscle is affected by calcium imbalances sooner than any other excitable tissue.

One major difference between the electrical potential of cardiac muscle and that of the other types of muscle tissue is that cardiac muscle has a built-in safety feature that protects it from tetanic contraction. This protective device is crucial because a tetanic contraction in cardiac muscle would be fatal. Tetanization is avoided because of an extended depolarization (the *refractory period*) in cardiac muscle. The refractory period in skeletal muscles is about 1 to 2 msec, and the contraction itself takes 20 to 100 msec. In cardiac muscle, however, the refractory period lasts about 200 msec, almost as long as the contraction. As a result, a second contraction cannot be produced until the muscle relaxes (relaxation phase), not fast enough to cause a tetanic contraction.

Probably the most important characteristic of cardiac muscle is its ability to resist prolonged wear as it strenuously and continuously pumps blood from the heart to all parts of the body. Consider that the heart pumps without stopping for an entire lifetime, about 100,000 beats a day. If you live to be 70 years old, your heart will beat about 2.5 billion times.

Contractions of cardiac muscle last between 200 and 250 msec, longer than those of skeletal muscle. This sustained contraction allows the necessary time for blood to leave the chambers of the heart. Because the heart pumps all the time,* cardiac muscle needs a constant supply of oxygen. If oxygen deprivation lasts longer than about 30 sec, cardiac muscle cells may stop contracting, resulting in heart failure.

---

*The heart actually renews its resources during each relaxation phase. In this interval the heart is said to "rest," and the resistance to coronary blood flow is reduced, ensuring an adequate supply of oxygen and glucose.

DEVELOPMENTAL ANATOMY OF MUSCLES

The differentiation (specialization) of cells begins very early during embryonic development, forming three distinct layers of germ cells: the inner *endoderm*, the middle *mesoderm*, and the outer *ectoderm*. Endodermal cells develop into the epithelium of the digestive tract (including the liver and pancreas) and respiratory tract. The ectoderm develops into nervous tissue and outer body regions such as the epidermis of the skin and its derivatives, including hair, sweat glands, nails, and oil glands. Muscle (except in the iris of the eye) and connective tissue (including cartilage and bone) are derived from *mesodermal cells*. Primitive muscle cells are called *myoblasts*. The uninuclear myoblasts elongate, become arranged in parallel bundles, and then fuse to form multinucleated cells. Muscle development begins about the fifth week of embryonic development, and the muscles differentiate into their final shapes and relations throughout the body during the seventh and eighth weeks.

Development of Skeletal Muscle

Most *skeletal muscles* develop from mesodermal tissue arranged in paired, segmented cell masses called *somites* [FIGURE 10.19A]. The somites are located on both sides of the central neural tube of the primitive nervous system. The inner and outer walls of the somites differentiate into distinct layers that form different parts of the embryo [FIGURE 10.19B]: (1) The outer layer of a somite is the *dermatome* (Gr. skin slice). It develops into connective tissue, including the dermis of the skin. (2) The inner layer, called the *sclerotome* (Gr. hard slice), consists of rapidly dividing mesenchymal cells that migrate to the spaces surrounding the primitive spinal cord, where they eventually develop into connective tissue, cartilage, and bones associated with the vertebral column. (3) The middle layer of a somite is the *myotome* (Gr. muscle slice), one of the sources of myoblasts. Two other sources of myoblasts are the somitic mesoderm and the branchial (gill) arch mesenchyme. The mesenchymal cells of the myotomes eventually differentiate into the striated skeletal muscle cells of some head muscles and muscles of the neck, thorax, abdomen, and pelvis.

By the end of the third fetal month myofibrils having the characteristic cross striations of skeletal muscles appear. Myoblasts from myotomes differentiate into the extraocular muscles that move the eyes, tongue muscles, axial musculature, including the extensor and flexor muscles of the vertebral column, the musculature of the pelvic diaphragm, and the scalene and infrahyoid muscles. Somatic mesoderm gives rise to mesenchymal cells that surround the developing bones of the extremities. Many of these cells differentiate into myoblasts and then into

FIGURE 10.19 DEVELOPMENTAL ANATOMY OF SKELETAL MUSCLES

[A] Paired somites surround the neural tube of a 23-day-old embryo; dorsal view. [B] Transverse section of a somite.

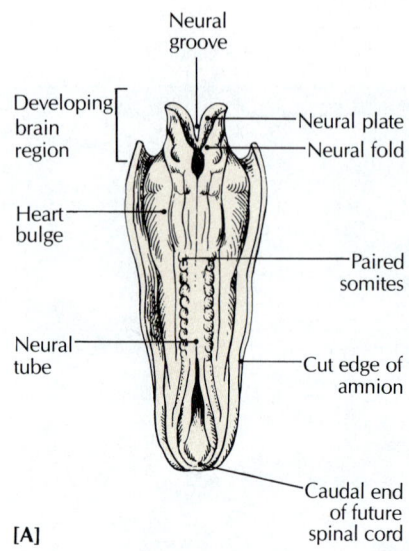

[A]

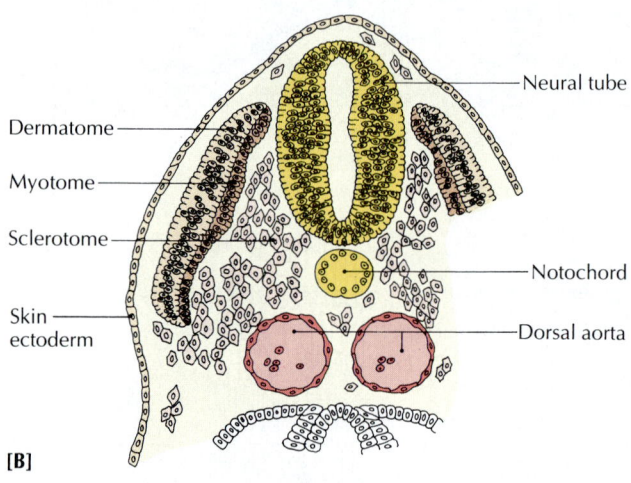

[B]

the musculature of the limbs. The migration of myoblasts from the branchial arches gives rise to musculature such as the muscles of facial expression, of mastication, and of the pharnyx and larynx.

Development of Smooth Muscle

Smooth muscle develops from primitive mesenchymal cells of mesodermal origin that migrate to the developing linings of hollow organs and vessels, including those in the digestive, circulatory, respiratory, urinary, and reproductive systems. This also includes the arrector pili muscles associated with hairs. During differentiation, mesenchymal cells elongate and accumulate myofilaments. Smooth muscles in the iris of the eye arise from ectoderm, and those of the ciliary body arise from mesoderm.

Development of Cardiac Muscle

Cardiac muscle develops from chains of elongated mesodermal cells that specialize early in embryonic life to eventually form the heart. The cells migrate to the developing heart while it is still a simple tube (see Developmental Anatomy of the Heart, Chapter 20). Cells in each chain develop connecting junctions, often branching and forming junctions with cells in adjacent chains. A branched network of myoblasts forms interwoven muscle fibers, but the myoblasts do not fuse.

THE EFFECTS OF AGING ON MUSCLES

Starting at about age 40, muscles begin to weaken and become smaller and dehydrated as the diameter of muscle fibers decreases. The force generated by a muscle decreases by 30 to 40 percent between the ages of 40 and 80, and by age 80 about half of the maximum muscle mass has been lost. Fibrous tissue appears, and muscles become increasingly infiltrated with fat. Muscle reflexes slow, and nocturnal cramps are common, especially among women.

For many reasons, elderly people tend to move about less than they should. As a result of their general reduced mobility, some of the following problems may arise: (1) cramps, loss of muscle tone, and osteoporosis; (2) blood clots (embolisms) within blood vessels; (3) digestive disturbances such as diarrhea, constipation, and general indigestion — some of these digestive problems may be related to decreased peristalsis (rhythmic muscular waves that help move food along the digestive tract); (4) accumulations of secretions within the lungs and the rest of the respiratory tract; (5) kidney stones and infections of the urinary tract; and (6) bedsores in bedridden patients whose movements are restrained.

Regular physical activity is one of the greatest aids to keeping muscles healthy. For example, astronauts in a zero-gravity spaceship for 90 days lose up to 20 percent of their leg strength through disuse.

WHEN THINGS GO WRONG

Muscle Injuries

Each type of muscle tissue responds to injury differently (although wound healing always involves the proliferation of fibroblasts and the synthesis of new connective tissue matrix materials). When *skeletal muscle* is injured, normally quiescent satellite cells under the endomysium of the muscle fibers begin to divide, differentiating into myoblasts and finally fusing to form new muscle fibers. An injury to *smooth muscle* causes the proliferation and differentiation of primitive smooth muscle cells into new muscle fibers. This same process occurs during pregnancy, as new muscle is added to the enlarging uterus. If *cardiac muscle* is injured during early childhood it can regenerate, but it loses this ability later in life. Injured adult cardiac tissue is replaced with connective tissue, which forms scars.

Atrophy, Myopathy, Dystrophy

The term *atrophy* (AT-ruh-fee; Gr. *a*, without + *trophe*, nourishment), when applied to muscles, refers (1) to the wasting of muscle tissue as a result of lack of use, where the nerve supply is intact *(disuse atrophy)*, or (2) to a disease typically associated with the nervous system, where motor neurons to a skeletal muscle are destroyed, the muscle fibers become smaller in diameter, and their amount of contractile proteins decreases *(denervation atrophy)*. As an example of disuse atrophy, consider the case of a broken leg immobilized in a cast for several weeks. When the cast is removed, the muscles appear shrunken. An example of denervation atrophy is spinal muscular atrophy, which is related to the lack of stimulation caused by damaged spinal nerves.

Myopathy (Gr. *mus, myo-*, muscle + *pathos*, suffering) refers to any disease of the muscles that is not associated with the nervous system or with an emotional disorder. The constant symptom of a myopathy is muscle weakness, but other symptoms may include impaired muscle relaxation *(myotonia)*; cramps; or the presence of free muscle hemoglobin (myoglobin) in urine *(myoglobinuria)*, which results in muscle necrosis.

A *dystrophy* (Gr. *dus, dys-*, faulty + *trophe*, nourishment) is a myopathy resulting from defective muscular nutrition. It usually shows four distinctive characteristics: (1) It is inherited, (2) its symptoms are due to a progressive weakness, (3) there are no tissue abnormalities other than degeneration and regeneration, and (4) there is no build-up of metabolic products. The term *dystrophy* is generally used in conjunction with *muscular dystrophy*.

Replant Surgery of a Severed Body Part

When a body part such as a finger or hand is severed—by a power saw, for example—the detached part may often be reunited at the original site by microsurgery. Successful surgery of this kind, called **replant surgery,** is usually followed by a significant return of normal function.

Assume that the index finger is severed between the proximal interphalangeal (IP) joint and the metacarpophalangeal (MP) joint. If possible, the severed segment should be wrapped in sterile gauze and placed in a sterile plastic bag filled with ice-cold saline solution containing an anticoagulant such as heparin. The bag should then be placed in an ice-cold bath, taking care to prevent frostbite by not allowing the severed segment to come in direct contact with the ice. (In a situation where such things as saline solution are not readily available, the severed segment should be wrapped in a towel or handkerchief and placed in a plastic bag filled with ice and water.)

The actual surgery to reattach a finger usually occurs in the following sequence: (1) The detached segment is cleaned carefully, (2) the bone of the phalanx is shaped and reattached, (3) the severed tendons are aligned and sutured together, (4) the cut ends of the severed nerves are aligned in order to facilitate nerve regeneration, (5) the severed veins and arteries are joined and sutured to those in the stump, using transplanted segments of veins to bridge any gaps, and (6) the wound is closed.

The functional recovery that occurs in a replanted finger, or even a replanted hand, is enhanced, because many of the muscles that manipulate the IP, MP, and wrist joints have their origins and bellies in the *forearm*. But the success rate depends as well on the amount of damage done to the nerves innervating these muscles.

Presently, functionally successful replants are achieved only with relatively small segments such as a thumb, finger, or hand. Attempts to replant arms severed between the elbow and shoulder have been unsuccessful, partly because the regeneration of the severed nerves takes so long that the muscles in the limb atrophy, with the result that by the time the atrophied muscles are reinnervated, they cannot be restored functionally.

Muscular Dystrophy

Muscular dystrophy (MD) is a general name for a group of inherited diseases that cause progressive weakness as a result of muscle degeneration. It is usually limited to the skeletal muscles, but cardiac muscle may also be affected. The characteristic symptoms include degeneration and reduction in the size of muscle fibers and an increase in connective tissue and fat deposits. Muscular dystrophy usually begins in childhood or early adolescence, but adults can also be afflicted.

Although the exact cause of muscular dystrophy is unknown, the defective gene that causes the *Duchenne* form of the disease has been identified.* In Duchenne muscular dystrophy (DMD) the calcium ion channels remain open longer than those of normal muscle cells, producing an accumulation of calcium ions in muscle cells. Apparently, the excess calcium leads to an accelerated degradation of protein. Progressive deterioration cannot be stopped, at least not permanently. Treatment usually includes exercise, physiotherapy, braces, and occasionally, surgery.

In the future, DMD may be treated with myoblast-transfer therapy, a treatment in which diseased muscles are injected with billions of healthy myoblasts (immature muscle cells). The healthy myoblasts carry the gene for dystrophin, an important muscle protein that is deficient in the muscles of DMD victims.

*The Duchenne form of muscular dystrophy affects only boys, generally of preschool age, and is usually fatal by age 20.

Myasthenia Gravis

Myasthenia gravis ("grave muscle weakness") is an autoimmune disease caused by antibodies (molecules that defend against a foreign substance) directed against acetylcholine receptors. The antibodies reduce the number of functional receptors or impede the interaction between acetylcholine and its receptors at the motor end plate. As a result, skeletal muscles become chronically weak, and even the slightest muscular exertion causes extreme fatigue. Although myasthenia may become progressively worse, it is usually not fatal unless the respiratory muscles fail and breathing becomes impossible. Although it can occur at any age, myasthenia gravis most commonly affects women between the ages of 20 and 40 and men over 40. It is about three times as common in women as in men.

Tetanus (Lockjaw)

Technically, *tetanus* (Gr. *tetanos*, stretched) is a disease that affects the nervous system, but it is discussed here because it produces spasms and painful convulsions of the skeletal muscles. Tetanus, commonly called *lockjaw* because of the characteristic spasms of the jaw muscles, is caused by the extracellular toxin produced by the bacterium *Clostridium tetani*. This bacterium produces an exotoxin 50 times stronger than poisonous cobra venom and 150 times stronger than strychnine. (An *exotoxin* is a toxin produced in the cell and released into the environment.) The exotoxin enhances the nervous activity that stimulates the contraction (spasms) of muscles. It accomplishes this by blocking the release of inhibitory neurotransmitters in the central nervous system.

Most often *C. tetani* is introduced into a puncture wound, cut, or burn by contaminated soil, especially soil that contains horse or cattle manure. These bacteria multiply only in anaerobic conditions, so deep puncture wounds are ideal for their growth. As the bacteria multiply, they produce the exotoxin that destroys surrounding tissue, entering the central nervous system by way of branching spinal nerves.

The early indication of tetanus is local pain and stiffening, but once the exotoxin begins to spread, painful spasms are felt in the muscles of the face and neck, chest, back, abdomen, arms, and legs. Prolonged convulsions may cause sudden death by asphyxiation.

Most children today receive a DPT vaccine, which permanently immunizes them against *d*iphtheria, *p*ertussis (whooping cough), and *t*etanus. If a child who has been immunized suffers an injury that might be conducive to a tetanus infection, a booster shot of tetanus toxoid is usually given.

CHAPTER SUMMARY

1 The basic properties of muscle tissue are **contractility, excitability, extensibility,** and **elasticity.**

2 The general functions of muscle tissue are movement, posture, and heat production. Muscles are specialized in their form to perform different functions. The three types of muscle tissue are **skeletal, smooth,** and **cardiac.**

Anatomy of Skeletal Muscle (p. 272)

1 Most **skeletal muscle tissue** is attached to the skeleton. It is also called **striated** muscle because it appears to be striped and **voluntary** muscle because it can be contracted voluntarily.

2 **Fascia** is a sheet of fibrous tissue enclosing muscles or groups of muscles. The three major types of fasciae are *superficial, deep,* and *subserous.*

3 Skeletal muscle tissue is composed of individual cells called **muscle fibers.** **Fascicles** are groups of fibers, and **muscles** are groups of fascicles.

4 A sheath of connective tissue called **endomysium** surrounds each muscle fiber, **perimysium** surrounds each fascicle, and **epimysium** encases muscles.

5 Muscles are supplied with blood by arteries. Each skeletal muscle fiber is contacted by at least one nerve ending.

Nervous Stimulation of Skeletal Muscle (p. 275)

1 A motor neuron, together with all the muscle fibers it stimulates, is a **motor unit.** The motor neuron endings make contact with muscle fibers at the **motor end plate,** and the actual electrical impulse is transmitted chemically to the fibers across a small space called a **synaptic cleft.** A nerve ending contacts a muscle fiber at a **neuromuscular junction.**

2 Skeletal muscle fibers contain several nuclei and a specialized type of cytoplasm called **sarcoplasm,** which contains many mitochondria and individual fibers called **myofibrils.** Each fiber is enclosed within a membrane, the **sarcolemma.**

3 Each myofibril is composed of myofilaments containing the proteins **myosin, actin, troponin,** and **tropomyosin.**

4 Impulses from the brain and spinal cord are carried to the muscle by **motor neurons,** which release *acetylcholine* at the neuromuscular junction. This produces an impulse in the muscle fiber membrane (sarcolemma). This impulse is conveyed to the **transverse tubules** and then to the **sarcoplasmic reticulum,** where it triggers the release of calcium. The release of calcium sets the sliding-filament mechanism in motion.

5 Myofibrils have alternating light and dark bands. The dark **A bands** contain myosin; the light **I bands** contain actin. Cutting across each I band is a **Z line.** Within the A band is a pale **H zone,** which contains the **M line.**

6 A section of myofibril from one Z line to the next makes up a **sarcomere,** the fundamental unit of muscle contraction.

Physiology of Muscle Contraction (p. 278)

1 The **sliding-filament model** describes the mechanism of muscle contraction. *Cross bridges* between myosin and actin filaments help slide the filaments past each other, producing contraction.

2 The main energy source for muscle contraction is **ATP.** Expended ATP is replenished through the breakdown of *creatine phosphate.*

Types of Muscle Contraction (p. 281)

1 The several types of muscle contraction include **twitch, tetanus, treppe, isotonic,** and **isometric.**

2 The recovery period after a contraction is called the **refractory period.**

3 The tendency of a muscle fiber to contract fully or not at all is called the **all-or-none principle.** The minimal nervous stimulation required to produce a muscle contraction is the **threshold.**

Grading (Varying Strength) of Muscle Contraction (p. 284)

The main factors affecting the strength of muscle contraction are number of muscle fibers contracting, relative size of the muscle, and degree of muscle stretch.

By-Products of Mechanical Work (p. 284)

1 Heat given off during muscle contraction helps maintain a stable body temperature.

2 *"Oxygen debt"* results during strenuous activity, when lactic acid is generated because the supply of oxygen to the muscles is inadequate for aerobic cellular respiration. This oxygen debt is repaid when sufficient oxygen is restored and the excess lactic acid is broken down to carbon dioxide and water. Heavy exercise can also produce *muscle fatigue,* when contractions grow weaker because not enough ATP is available.

Fast-Twitch and Slow-Twitch Muscle Fibers (p. 286)

Slow-twitch muscles and *fast-twitch muscles* respond to stimuli at different speeds. Slow-twitch muscles contain large amounts of *myoglobin* and are also called *red muscles;* fast-twitch muscles are also called *white muscles.*

Smooth Muscle (p. 287)

1 *Smooth muscle tissue* is not striated. It is also known as *involuntary muscle* because it is controlled by the involuntary division of the nervous system.

2 The slow and rhythmic contractions of smooth muscle make it suitable for the contractile control of internal organs (but not the heart), especially the stomach, intestines, urinary bladder, and uterus. It also lines hollow vessels, including blood vessels.

3 *Single-unit smooth muscle* fibers contract as a single unit in response to nervous stimulation transmitted across the junctions by calcium and potassium ions. *Multiunit smooth muscle* consists of individual fibers that are stimulated by separate motor nerve endings.

Contraction of Smooth Muscle (p. 289)

The contractile process of smooth muscle is essentially the same as that of skeletal muscle. However, biochemically, smooth muscle does not have troponin. Instead, the calcium-binding protein calmodulin mediates the phosphate group in myosin. Overall, smooth muscle is a more efficient contractile unit than is skeletal muscle.

Cardiac Muscle (p. 290)

1 *Cardiac muscle tissue* is found only in the heart. It is striated and involuntary.

2 Cardiac muscle cells are closely attached end to end but remain separate, each with its own nucleus. *Intercalated disks* strengthen the junction between cells and facilitate the passing of an impulse from one cell to the next.

3 Cardiac muscle depends on nervous impulses to some extent, but it is also able to contract rhythmically on its own.

4 Cardiac muscle can resist wear as it pumps blood continuously.

Developmental Anatomy of Muscles (p. 292)

1 The muscular system develops almost entirely from embryonic cells called *myoblasts,* which are derived from embryonic *mesodermal cells.*

2 Much skeletal muscle develops from mesodermal tissue arranged in paired cell masses called *somites.*

3 Myoblasts, which form the skeletal muscles, are derived from three sources: (1) the myotome of somites, (2) somatic mesoderm, and (3) branchial (gill) arch mesenchyme.

4 Myotomes develop into the axial muscles; somatic mesoderm develops into the musculature of the limbs; and the branchial arch mesenchyme develops into many muscles of the head and neck.

5 Most smooth and cardiac muscle derives from the mesoderm.

The Effects of Aging on Muscles (p. 293)

1 With aging, muscles become weaker, smaller, and dehydrated. Reflexes slow, and cramps may occur.

2 Because elderly people are usually less physically active than they used to be, many seemingly unrelated problems may begin to occur, including cardiovascular, digestive, and respiratory problems.

STUDY AIDS FOR REVIEW

KEY TERMS

acetylcholine 277
actin 273
all-or-none principle 282
cardiac muscle tissue 271, 290
endomysium 272
epimysium 272
excitation-contraction coupling 279
fascia 272
fast-twitch muscle 286
intercalated disk 290
isometric contraction 284
isotonic contraction 283
motor end plate 275
motor unit 275
multiunit smooth muscle 288
muscle fatigue 286
muscle fiber 273
myofibril 273
myofilament 273
myosin 273
myotome 292
neuromuscular junction 275

oxygen debt 286
pacemaker cell 288
perimysium 272
refractory period 281
sarcolemma 273
sarcomere 273
sarcoplasm 273
sarcoplasmic reticulum 273
single-unit smooth muscle 288
skeletal muscle tissue 271, 272
sliding-filament model 278
slow-twitch muscle 286
smooth muscle tissue 271, 287
somite 292
striated muscle tissue 271, 272
synaptic cleft 277
tetanus 283
transverse tubule 273
treppe 283
tropomyosin 279
troponin 279
twitch 281
voluntary muscle 272

MEDICAL TERMS FOR REFERENCE

ASTHENIA Loss or lack of bodily strength; weakness.

CHARLEY HORSE A cramp or stiffness of various muscles in the body, especially in the arm or leg, caused by injury or excessive exertion.

CONVULSION An intense involuntary tetanic contraction (or series of contractions) of a whole group of muscles.

CRAMP A sudden, involuntary, complete tetanic muscular contraction causing severe pain and temporary paralysis, often occurring in the leg or shoulder as the result of strain or chill.

FIBRILLATION Uncoordinated twitching of individual muscle fibers with little or no movement of the muscle as a whole.

MYOCARDITIS Inflammation of cardiac muscle tissue.

MYOMA Benign tumor of muscle tissue.

MYOPATHY A general term for any disease of a muscle.

MYOSARCOMA Malignant tumor of muscle tissue.

MYOSITIS Inflammation of a muscle, usually a skeletal muscle.

PHYSIOTHERAPY Treatment of muscle weakness by physical methods, such as heat, massage, and exercise.

SPASM A sudden, involuntary contraction of a muscle or group of muscles.

TIC A habitual, spasmodic muscular contraction, usually in the face, hands, or feet, and often of neurological origin.

TREMOR A trembling or quivering of muscles.

UNDERSTANDING THE FACTS

1 Compare and contrast the locations and general microscopic appearances of skeletal, smooth, and cardiac muscle tissue.
2 What is the molecular basis of the striations in skeletal muscle?
3 Contrast the location and structure of the sarcoplasmic reticulum with the T tubules.
4 Define sarcomere, motor unit, and motor end plate.
5 What is meant by the all-or-none principle?
6 Compare a muscle twitch with summation.
7 Contrast the microscopic appearances and functions of fast-twitch and slow-twitch fibers.
8 What is meant by oxygen debt?
9 Contrast single-unit and multiunit smooth muscle.
10 What is an intercalated disk? What is its function?

UNDERSTANDING THE CONCEPTS

1 Explain how the organization of the connective tissue wrappings assists contraction.
2 What roles do the sarcoplasmic reticulum and T tubules play in contraction?
3 Describe the function(s) of calcium ions in muscle contraction.
4 Reconcile the ability of muscles to exhibit graded contraction with the all-or-none principle.
5 Explain the functional difference between motor units with many skeletal muscle fibers and those with few.
6 Outline the sliding-filament model of muscle contraction, and explain the role of ATP in that process.
7 Contrast refractory period in skeletal muscle with muscle fatigue.
8 Compare the role of the nervous system in the contraction of skeletal, smooth, and cardiac muscles.

SELF-QUIZ
Multiple-Choice Questions

10.1 The all-or-none principle applies to
  a whole muscles only
  b muscle fibers only
  c motor units only
  d whole muscles and muscle fibers
  e muscle fibers and motor units

10.2 The period after a contraction when a muscle fiber cannot respond to a stimulus is called the
a depolarization period d relaxation phase
b isotonic period e tetanic period
c refractory period

10.3 Energy to move myosin heads during muscle contraction comes directly from
a ATP c glucose e pyruvic acid
b DNA d lactic acid

10.4 The ions that bind to the troponin complex in muscle contraction are
a sodium ions d ATPase
b potassum ions e calcium ions
c magnesium ions

10.5 During muscle contraction, the
a actin filaments shorten
b myosin filaments shorten
c troponin pulls on myosin
d Z lines move closer together
e actin pulls on tropomyosin

10.6 The type of muscular contraction in which the muscle develops tension but does not change length is
a treppe d tetanic contraction
b isometric contraction e a twitch
c isotonic contraction

Matching
a refractory period
b endomysium
c epimysium
d transverse tubules
e sarcoplasmic reticulum
f satellite cells
g sarcomere
h motor unit
i motor end plate
j calmodulin

10.7 _____ can differentiate into myoblasts
10.8 _____ also known as fascia
10.9 _____ extensions of sarcolemma that penetrate each fiber at right angles to the myofibrils
10.10 _____ the functional unit of a myofibril
10.11 _____ one motor neuron and all muscle fibers with which it makes synapses
10.12 _____ a network of tubes and calcium-containing sacs surrounding the myofibrils in a skeletal muscle fiber
10.13 _____ a specialized region of the sarcolemma that is part of the myoneural junction

10.14 _____ time during which a skeletal muscle fiber cannot be stimulated, irrespective of the strength of the stimulus
10.15 _____ a binding protein that helps to initiate contraction in smooth muscle
10.16 _____ connective tissue that surrounds each skeletal muscle fiber

Completion Exercises
10.17 An influx of _____ ions leads to the release of acetylcholine at the neuromuscular junction.
10.18 A motor neuron, together with the muscle fiber it innervates, is called a(n) _____.
10.19 The gap between a nerve ending and the sarcolemma of a muscle fiber is the _____.
10.20 The type of smooth muscle found in the walls of most hollow viscera is known as _____.
10.21 In place of Z lines, cardiac muscle fibers have _____.
10.22 Most muscle tissue develops from the embryonic _____ layer.
10.23 Skeletal muscle fibers develop from embryonic cells called _____.
10.24 Tendons connect _____ to _____.

A SECOND LOOK
In the following drawing, label the perimysium, epimysium, endomysium, muscle fiber, fascicle, and muscle.

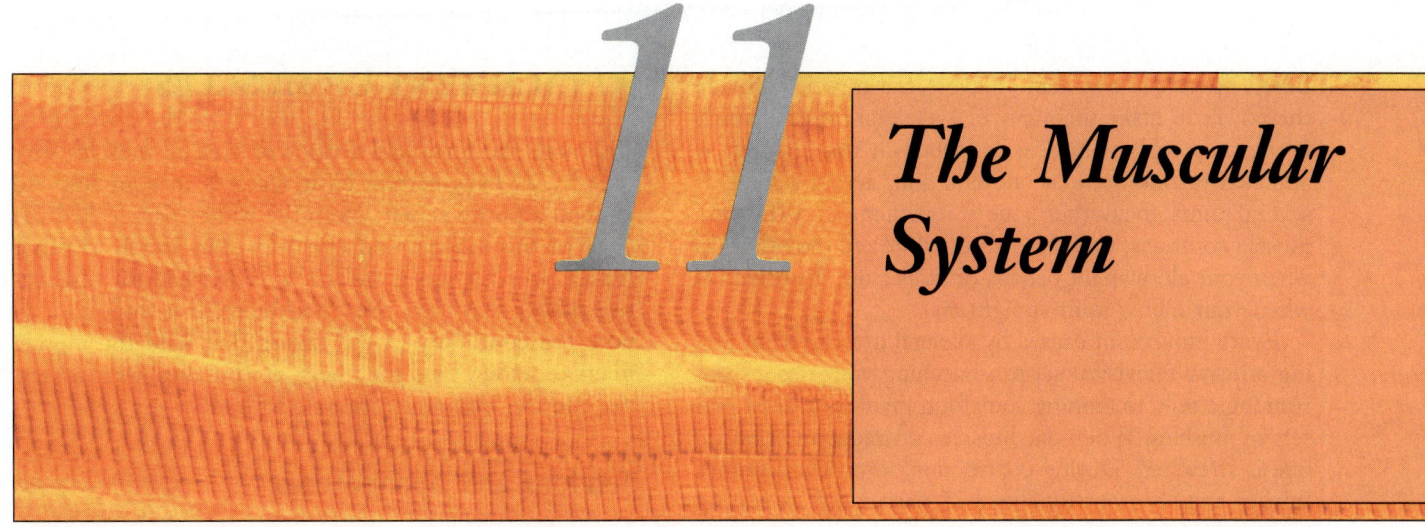

11

The Muscular System

KEY CONCEPTS

1 A muscle is usually attached to a bone or cartilage by a tendon.

2 Skeletal muscle fibers are grouped in bundles called fascicles, with the fibers running parallel to each other. The pattern of fascicles determines the range of movement and power of a muscle.

3 Most movements involve the actions of several muscles. Movements are usually accomplished through a system of levers.

4 The name of a muscle usually indicates its location, action, structure, size, or other physical characteristics.

Bones operate as a system of levers, but muscles provide the power to make them move. A skeletal muscle (which is the only type of muscle described in this chapter) is attached to one bone at one end and, across a joint or several joints, to another bone at its other end. When the muscle contracts, one bone is pulled *toward* the other (as when your elbow joint bends), or *away* from the other (as when your elbow joint straightens).

Every movement caused by skeletal muscle, from lifting a heavy television set to scratching your nose, from running a mile to tapping your foot, involves at least two sets of muscles. When one muscle contracts, an opposite one is stretched. During contraction, a muscle becomes thicker and shortens by about 15 percent of its resting length. A contracted muscle is brought back to its original condition by contraction of its opposing muscle. It can be stretched to about 120 percent of its resting length.

There are about 600 *skeletal muscles* in your body, which make up the **muscular system.*** In this chapter we examine the most important of them and find out how they operate on a gross level.

HOW MUSCLES ARE NAMED

The name of a muscle generally tells something about its structure, location, or function. **Shape** is described in such muscles as the trapezius (shaped like a trapezoid), deltoid (triangular or delta-shaped), or gracilis (slender). **Size** is clearly indicated in the gluteus *maximus* (largest) and gluteus *minimus* (smallest) muscles. We know the **location** of the supraspinatus and infraspinatus muscles (above and below, *supra* and *infra*, the spine of the scapula) and the tibialis anterior (in front of the tibia). Some muscles are named to indicate their **attachment sites**. For example, the sternohyoid muscle is attached to the sternum and hyoid bones. The number of **heads of origin** can be determined in muscles such as the biceps (*bi*, two) and triceps (*tri*, three). The **action** or function of a muscle is plain in the extensor digitorum, which extends the fingers (digits), and the levator scapulae, which raises (elevates) the scapula. The names of some muscles indicate the **direction of their fibers** with respect to the structures to which they are attached, for example, *transversus* (across) and *obliquus* (slanted or oblique).

Most muscles are named by a combination of the above methods. An example is the flexor digitorum profundus, which means the deep (profundus) flexor of the fingers (digitorum). The name of this muscle thus tells its depth, location, and action.

*We say "about" 600 skeletal muscles rather than giving a definite number because anatomists disagree about whether to count some muscles as separate or as pairs. It is generally agreed, however, that we have more than 600 skeletal muscles.

ATTACHMENT OF MUSCLES

Although the form and actions of skeletal muscles vary greatly, all have certain basic features in common. All contain a fleshy center, often the widest part, called the **belly** of the muscle, and two ends that attach to other tissues. In Chapters 4 and 10, we described the connective tissues that form this attachment. It is important to note the continuity of the tissues that hold the muscle fibers together, bind the fibers into bundles, wrap an entire muscle, and attach muscle to bone.

Tendons and Aponeuroses

A muscle is usually attached to a bone (or a cartilage) by a **tendon,** a tough cord of connective tissue composed of closely packed collagen fibers.† A tendon is an extension of the deep fascia and/or the epimysium surrounding the muscle. It also extends into the periosteum that covers the bone, and into the bone as the periosteal perforating fibers. The thickest tendon in the body is the calcaneal (kal-KAY-nee-uhl) tendon, commonly called the Achilles tendon [FIGURE 11.1A]. It attaches the calf muscles (gastrocnemius; gas-trahk-NEE-mee-uhss; Gr. "belly of the leg"; the soleus and plantaris are also calf muscles) to the heelbone (calcaneus). Tendons add length and thickness to muscles and are especially important in reducing strain on muscles.

In some parts of the body, such as the abdominal wall, the tendon spreads out in a broad, flat sheet called an **aponeurosis.** This sheetlike attachment is directly or indirectly connected with the various muscle sheaths, and in some cases with the periosteum covering a bone. Another example of an aponeurosis is the fibrous sheath beneath the skin of the palm, called the *palmar aponeurosis*.

Origin and Insertion of Muscles

When a muscle contracts, one of the bones attached to it remains stationary while the other bone moves along with the contraction. In reality, the mechanics are hardly ever that simple, but for convenience we say that the end of a muscle attached to the bone that *does not move* is the **origin,** and the point of attachment of the muscle on the bone that *moves* is the **insertion** [FIGURES 11.2 and 11.25]. Generally, the origin is the more proximal attachment (closer to the axial skeleton) and the insertion is the more distal attachment.

However, *origin* and *insertion* are only relative terms, and they can be reversed with the same muscle, depending upon the action. For example, muscles from the chest to the upper extremity generally move the extremity with

†Although most muscles are attached to bones, not all are. Some, such as the muscles of facial expression, are attached to skin, and the lumbrical muscles of the hand are attached to tendons of other muscles.

FIGURE 11.1 THE MUSCLE-TENDON-BONE RELATIONSHIP

[A] The calcaneal (Achilles) tendon connecting the gastrocnemius (calf) muscle and the soleus muscle to the calcaneus bone. **[B]** Detail showing collagenous periosteal perforating (Sharpey's) fibers passing into the connective tissue, periosteum, and bone.

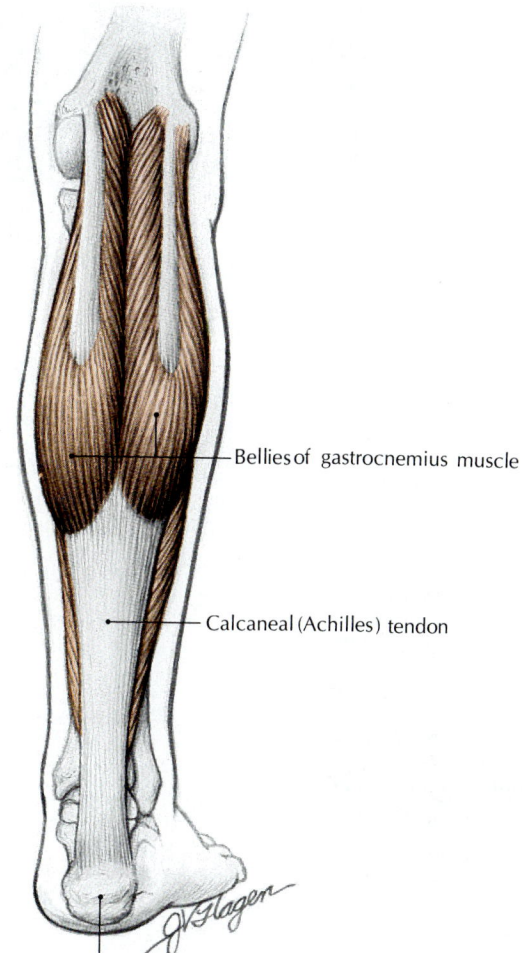

— Bellies of gastrocnemius muscle

— Calcaneal (Achilles) tendon

[A] Calcaneus (heel bone)

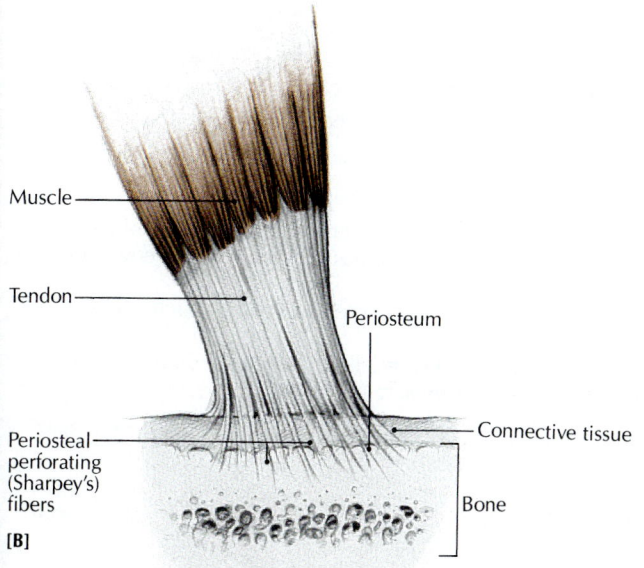

Muscle —

Tendon —

Periosteum

Periosteal perforating (Sharpey's) fibers

— Connective tissue

Bone

[B]

FIGURE 11.2 ORIGIN AND INSERTION OF ARM MUSCLES

The origins of the two heads of the biceps brachii muscle are on the scapula; the insertion is on the radius. Contraction of the biceps brachii flexes the elbow joint (arrow up).

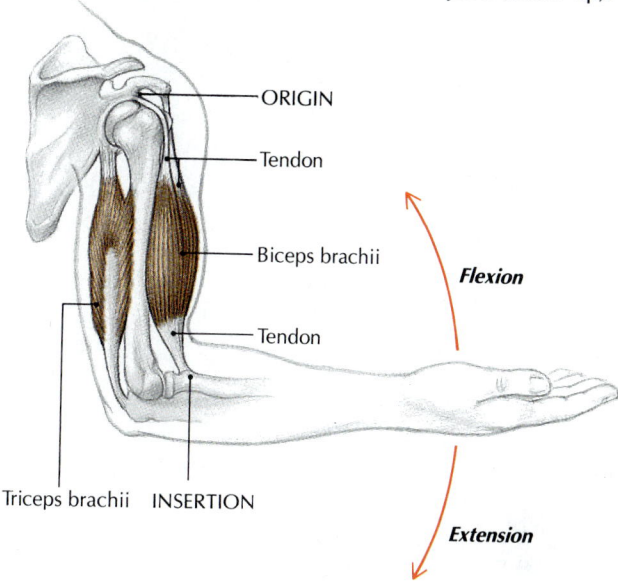

—ORIGIN

—Tendon

—Biceps brachii

Flexion

—Tendon

Triceps brachii   INSERTION

Extension

the origin on the thoracic skeleton and the insertion on the extremity. But when a person climbs a rope, the same muscles pull the body up, so the extremity becomes the origin and the insertion is on the thoracic skeleton.

A S K   Y O U R S E L F

1 What is the fleshy center of a muscle called?

2 What is the difference between a tendon and an aponeurosis?

3 What is the function of a tendon?

4 What is the difference between the origin and insertion of a muscle?

ARCHITECTURE OF MUSCLES

The fibers of skeletal muscles are grouped in small bundles called fascicles, with the fibers within each running parallel to one another [see FIGURE 10.5]. The fascicles, however, are organized in various architectural patterns in different muscles. The specific pattern determines the *range of movement* and the *power* of the muscle. The greater the length of the belly, the greater the range of movement. Also, the greater the number of fibers, the greater the total force generated by the muscle. The arrangement of fascicles and their tendons of attachment creates patterns such as strap muscles, fusiform muscles, pennate muscles, and circular muscles [FIGURE 11.3].

FIGURE 11.3 ARCHITECTURE OF MUSCLE

[A] A parallel strap muscle, sternohyoid. [B] A fusiform muscle, biceps brachii. [C] A unipennate muscle, flexor pollicis longus. [D] A bipennate muscle, rectus femoris. [E] A circular or sphincter muscle, orbicularis oris. [F] A multipennate muscle, deltoid.

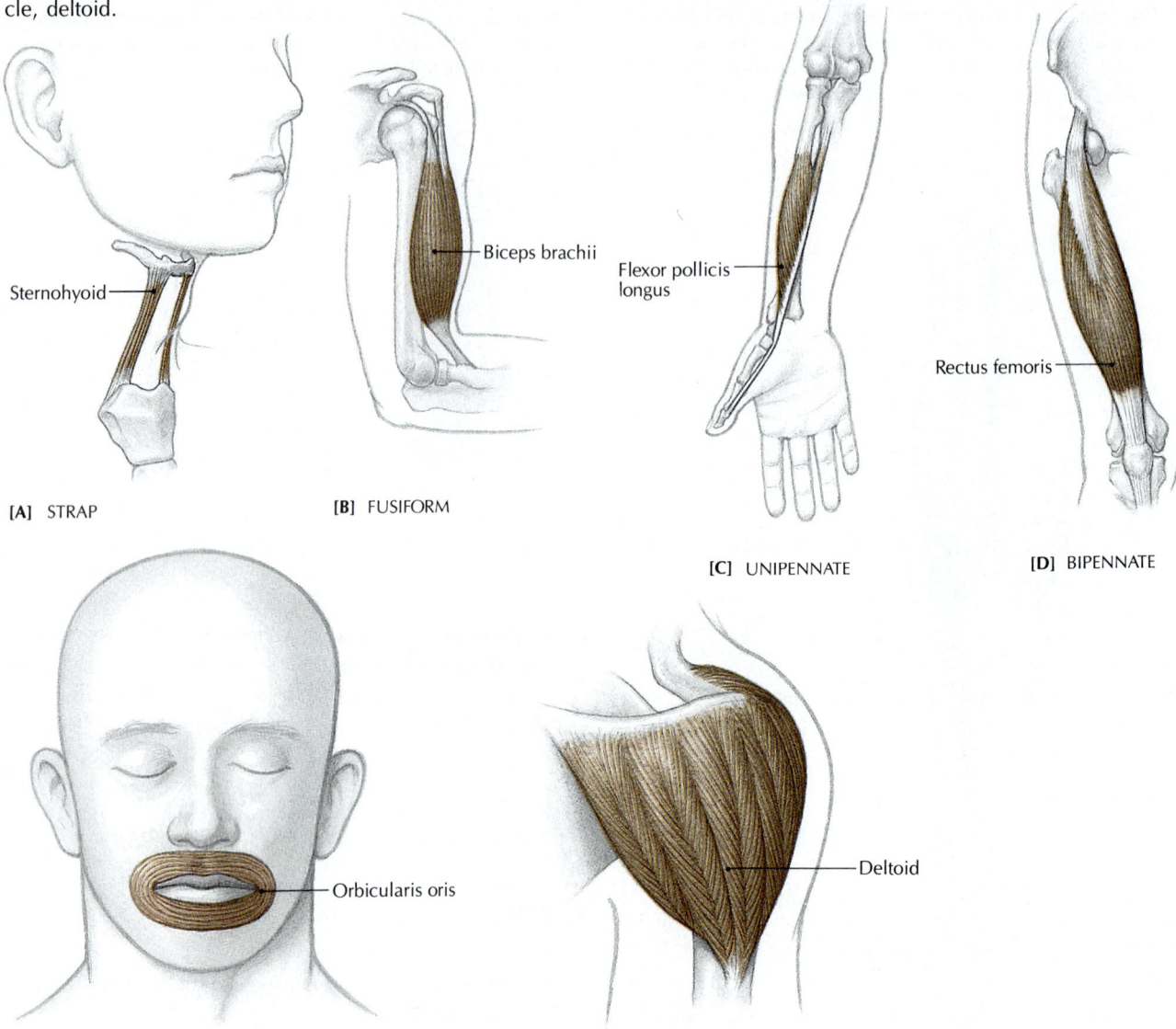

Sternohyoid

[A] STRAP

Biceps brachii

[B] FUSIFORM

Flexor pollicis longus

[C] UNIPENNATE

Rectus femoris

[D] BIPENNATE

Orbicularis oris

[E] CIRCULAR (SPHINCTER)

Deltoid

[F] MULTIPENNATE

Muscle type	Description and examples
Strap	Fascicles parallel to long axis; wide range of movement, but not very powerful. Sternohyoid (of neck), rectus abdominis (of abdominal wall).
Fusiform	Spindle-shaped with thick belly. Biceps brachii (arm).
Pennate	Short fascicles at angle (oblique) to a long tendon or tendons running length of muscle; resembles a feather. More fascicles attached directly to tendons than other muscle types; generally more powerful than other types.
Unipennate	Oblique fascicles on one side of tendon; direction of pull toward side of tendon with fascicles. Flexor pollicis longus (of thumb).
Bipennate	Oblique fascicles on both sides of tendon; equal pull on both sides of tendon. Rectus femoris (of leg).
Multipennate	Many oblique fascicles arranged along several tendons. Deltoid muscle (of shoulder).
Circular	Fascicles arranged in circular pattern around opening of structure. Orbicularis oris (of mouth), orbicularis oculi (of eye).

A S K   Y O U R S E L F

1 What are the four main patterns of fascicle arrangement?

2 What are the three types of pennate muscles?

3 Which arrangement of fascicles generally contributes to a greater range of movement?

INDIVIDUAL AND GROUP ACTIONS OF MUSCLES

Skeletal muscles can be classified according to the types of movement that they perform. TABLE 11.1 defines these actions and gives examples, and FIGURE 11.4 shows some of the major muscles.

Most movements, even those that seem simple, actually involve the complex interaction of several muscles or groups of muscles. Muscles produce or restrict movement by acting as agonists, antagonists, synergists, and fixators.

1 An *agonist* (Gr. *agonia*, contest, struggle), or *prime mover,* is the muscle that is primarily responsible for producing a movement.

2 An *antagonist* (Gr. "against the agonist") opposes the movement of a prime mover, but only in a subtle way. It does not oppose the agonist while it is contracting, but only at the end of a strong contraction to protect the joint. The antagonist helps produce a smooth movement by slowly relaxing as the agonist contracts, so it actually *cooperates* rather than "opposes."

3 A *synergist* (SIHN-uhr-jist; Gr. *syn,* together + *ergon,* work) works together with a prime mover by preventing movements at an "in-between" joint when a prime mover passes over more than one joint. In general, a synergist complements the action of a prime mover.

4 A *fixator,* or *postural muscle,* provides a stable base for the action of a prime mover. It usually steadies the proximal end (such as the arm), while the actual movement is taking place at the distal end (the hand).

The coordination of an action by a group of muscles can be demonstrated when a heavy object is held firmly in a clenched fist. The *prime movers* in this case are the long flexors [FIGURE 11.18A, D] of the fingers (flexor digitorum superficialis, flexor digitorum profundus, and flexor pollicis longus). When these muscles contract unopposed, the wrist also flexes. This undesired wrist flexion is eliminated, and the wrist is kept in a neutral or even hyperextended position by the contraction of the

TABLE 11.1 CLASSIFICATION OF MUSCLES BASED ON ACTION

Muscle type	Definition of action	Example
Flexor	Bending so angle between two bones decreases.	Flexor pollicis longus [FIGURE 11.18A, D]
Extensor	Bending so angle between two bones increases.	Extensor carpi ulnaris [FIGURE 11.18B]
Dorsiflexor	Bending of foot dorsally (toward back of foot).	Extensor digitorum longus [FIGURE 11.23A]
Palmar flexor	Bending (flexing) wrist ventrally (toward palm).	Flexor carpi ulnaris [FIGURE 11.18A, B]
Plantar flexor	Bending (extending) foot at ankle toward sole of foot.	Gastrocnemius [FIGURE 11.23A, B]
Abductor	Movement away from midline of body or structure.	Abductor pollicis brevis [FIGURE 11.18A, D]
Adductor	Movement toward midline of body or structure.	Adductor pollicis [FIGURE 11.18D]
Pronator	Turning of forearm so palm faces downward.	Pronator teres [FIGURE 11.18A]
Supinator	Turning of forearm so palm faces upward.	Supinator [FIGURE 11.18A]
Rotator	Turning movement around a longitudinal axis.	Sternocleidomastoid [FIGURES 11.9, 11.11]
Medial rotator	Turning movement so anterior surface faces median plane.	Subscapularis [FIGURE 11.17C]
Lateral rotator	Turning movement so anterior surface faces away from median plane.	Infraspinatus [FIGURES 11.16B, 11.17D]
Levator	Movement in an upward direction.	Levator scapulae [FIGURE 11.16A, B]
Depressor	Movement in a downward direction.	Depressor labii inferioris [FIGURE 11.6A, B]
Protractor	Movement in a forward direction.	Lateral pterygoid [FIGURE 11.8B]
Retractor	Movement in a backward direction.	Temporalis (horizontal fibers) [FIGURE 11.8A]
Tensor	Makes a body part more tense.	Tensor fasciae latae [FIGURE 11.21A, 11.22A]
Sphincter	Reduces size of an opening (orifice).	Orbicularis oris [FIGURE 11.6]
Evertor	Turning movement of foot so sole faces outward.	Peroneus longus [FIGURE 11.23A]
Invertor	Turning movement of foot so sole faces inward.	Tibialis anterior [FIGURE 11.23B]

FIGURE 11.4 THE MUSCULAR SYSTEM

[A] Anterior view. [B] Posterior view.

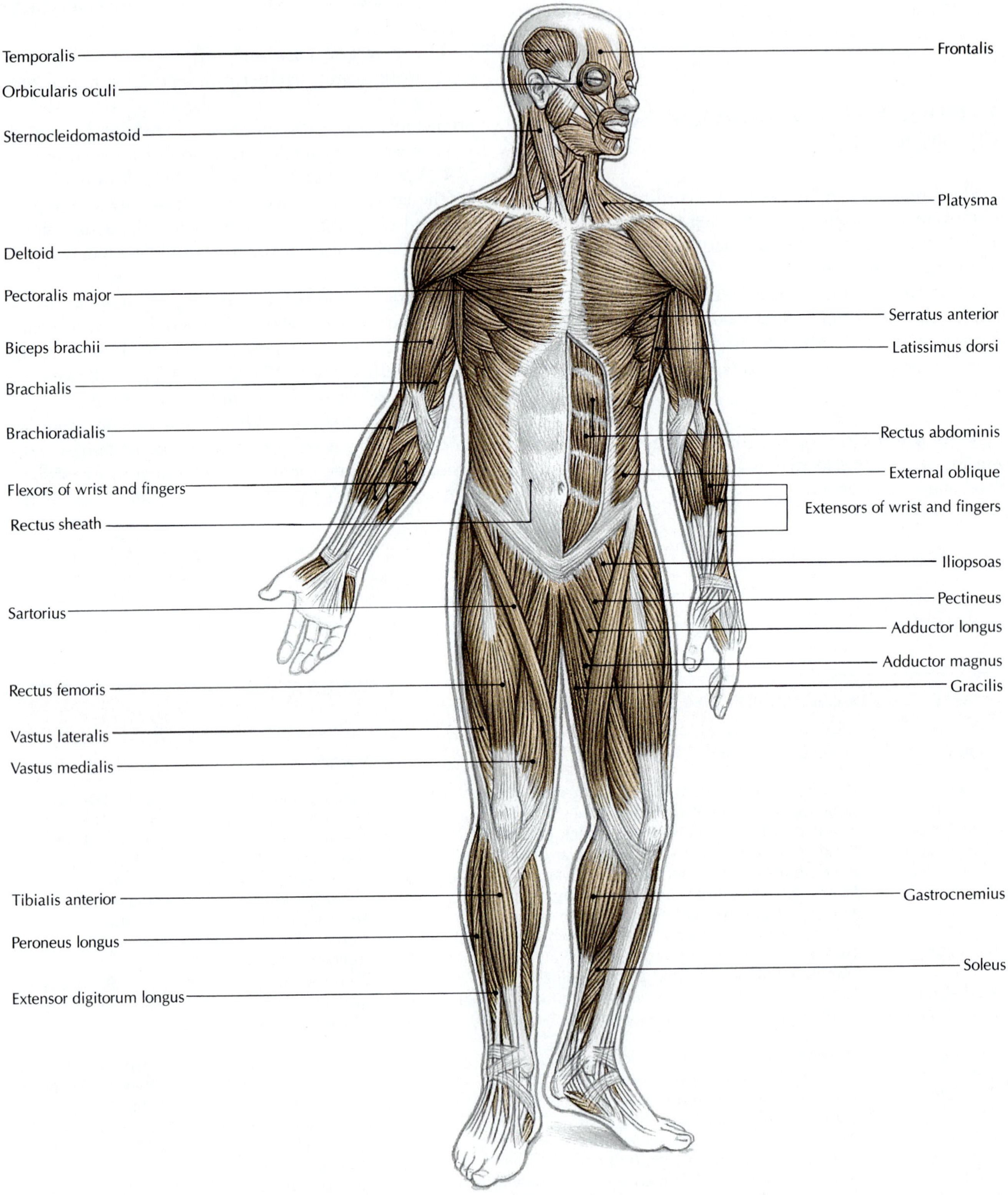

Temporalis

Orbicularis oculi

Sternocleidomastoid

Deltoid

Pectoralis major

Biceps brachii

Brachialis

Brachioradialis

Flexors of wrist and fingers

Rectus sheath

Sartorius

Rectus femoris

Vastus lateralis

Vastus medialis

Tibialis anterior

Peroneus longus

Extensor digitorum longus

Frontalis

Platysma

Serratus anterior

Latissimus dorsi

Rectus abdominis

External oblique

Extensors of wrist and fingers

Iliopsoas

Pectineus

Adductor longus

Adductor magnus

Gracilis

Gastrocnemius

Soleus

[A]

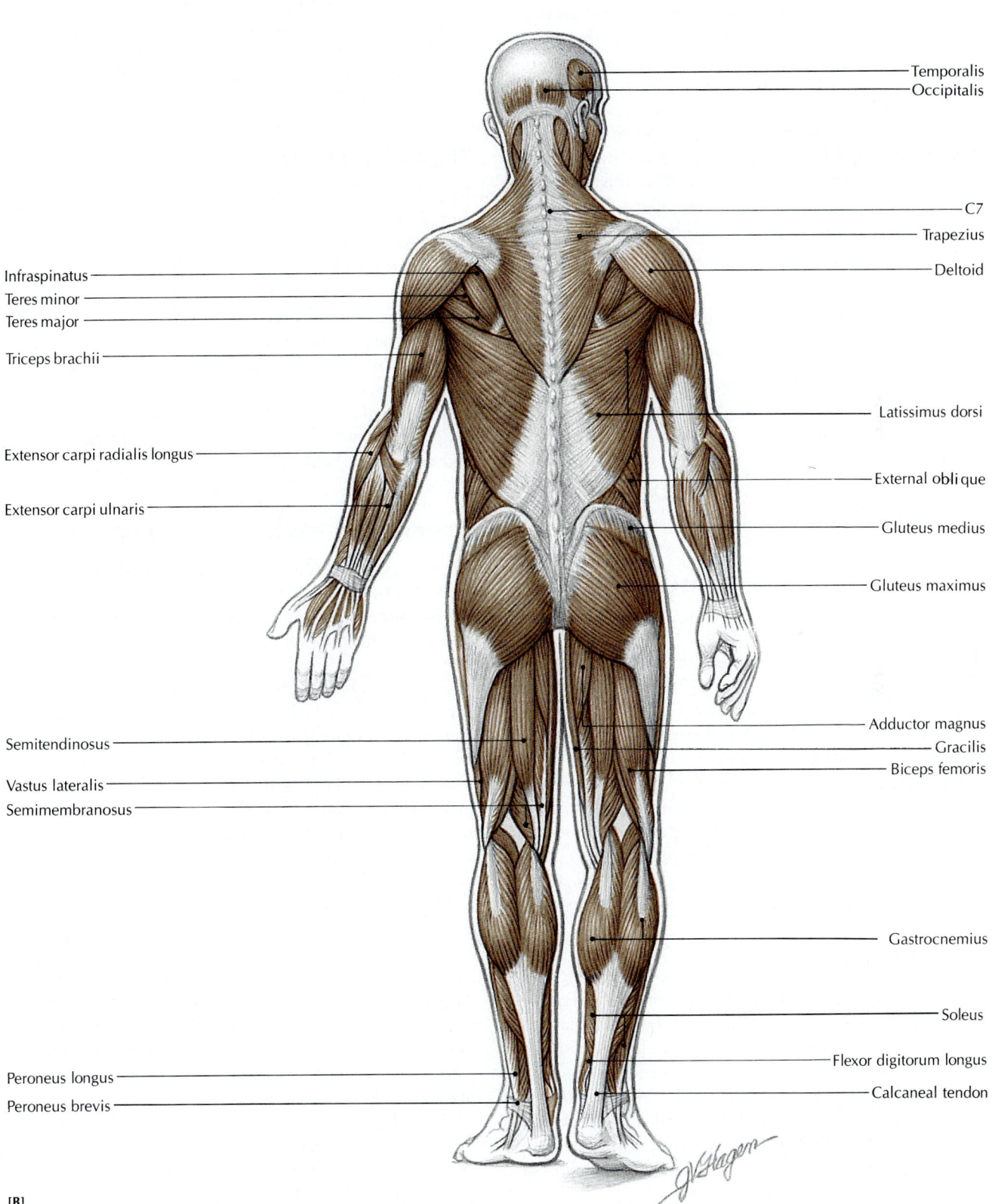

Temporalis
Occipitalis

C7
Trapezius
Deltoid

Infraspinatus
Teres minor
Teres major

Triceps brachii

Latissimus dorsi

Extensor carpi radialis longus

External oblique

Extensor carpi ulnaris

Gluteus medius

Gluteus maximus

Adductor magnus
Gracilis
Biceps femoris

Semitendinosus

Vastus lateralis
Semimembranosus

Gastrocnemius

Soleus

Flexor digitorum longus
Calcaneal tendon

Peroneus longus
Peroneus brevis

[B]

extensors [FIGURE 11.18B] of the wrist (extensor carpi radialis longus and brevis, and extensor carpi ulnaris muscles). These extensors of the wrist are the *synergists* in this action.

If the clenched fist is slowly flexed at the wrist, the extensors of the wrist can act as *antagonists* in the control of the activity by relaxing at the same time the prime movers are contracting.

As the hand clenches to hold the heavy object, the shoulder and elbow joints are stabilized to a greater or lesser degree, depending upon the weight of the object. The shoulder girdle, arm, and forearm are stabilized by the integrated actions of such muscles as the pectoralis major, deltoid, supraspinatus, subscapularis, biceps brachii, brachialis, and triceps brachii. These muscles are the *fixators.*

A S K   Y O U R S E L F

1 What is a pronator? Levator? Retractor? Sphincter?

2 What muscles are involved in pronation and supination?

3 What is the main job of an agonist muscle? Of a synergist?

LEVER SYSTEMS AND MUSCLE ACTIONS

The movements of most skeletal muscles are accomplished through a system of levers, with a rigid *lever arm* pivoting around a fixed point called a *fulcrum* (L. *fulcire,* to support). Also, acting on every lever are two different *forces:* (1) the weight to be moved (resistance to be overcome) and (2) the pull or effort applied (applied force). In the body, the bone acts as the lever arm, and the joint as the fulcrum. The weight of the body part to be moved is the resistance to be overcome. The applied force generated by the contraction of the muscle, or muscles, at the insertion is usually enough to produce a movement.

First-Class Levers

In a *first-class lever,* the force is applied at one end of the lever arm, the weight to be moved is at the other end, and the fulcrum is at a point between the two [FIGURE 11.5A]. A crowbar is an example. In the body, an example is raising the facial portion of the head where the atlantooccipital joint between the atlas and the occipital bone acts as a *fulcrum.* The vertebral muscles inserting at the back of the head generate the *applied force.* The facial portion of the head is the *weight* to be moved.

Second-Class Levers

In a *second-class lever,* the applied force is at one end of the lever arm, the fulcrum is at the other end, and the weight to be moved is at a point between the two. A wheelbarrow is a classic example [FIGURE 11.5B]. In the body, an example is raising the body by standing on "tiptoe." In this action, the ball of the foot is the *fulcrum,* the *applied force* is generated by the calf muscles on the back of the leg, and the *weight* to be moved is the entire body.

Third-Class Levers

In a *third-class lever,* the weight to be moved is at one end of the lever arm, the fulcrum is at the other end, and the applied force to move the weight is close to the fulcrum [FIGURE 11.5C]. A third-class lever is the most common lever in the body. Flexing the forearm at the elbow to lift a weight involves a third-class lever. The *applied force* is generated by the contraction of the biceps brachii muscle, which inserts at the proximal end of the radius, and the brachialis muscle, which inserts at the proximal end of the ulna. The *fulcrum* is the elbow joint, and the *weight* to be moved is the weight of the forearm, hand, and any object held in the hand.

Leverage and Mechanical Advantage

If you have ever used a screwdriver to pry open a can of paint, you know that the screwdriver (lever) made the work easier. When muscles use such a lever system, they also gain a mechanical advantage, which is called *leverage* (L. *levare,* to raise).

But most joints in the body are third-class levers, and they operate at a mechanical *disadvantage.* In fact, even some of the body's first-class levers operate at a mechanical disadvantage — for example, the action of the triceps muscle in extending the lower arm, as when one pounds a nail with a hammer. The only levers to operate exclusively at a mechanical *advantage* are second-class levers, and the only example in the body is rising up on the ball of the foot [FIGURE 11.5B].

Why do we have levers in our bodies that do not give us a mechanical advantage? Mostly because such a system allows for great *speed of motion* at the expense of great strength. After all, our bodies evolved to run away from dangerous animals, not to pick them up and throw them. One of the aims of the Industrial Revolution was to produce tools and machines that would enable our relatively weak bodies to gain a mechanical advantage when we did manual labor.

The brachialis muscle is a third-class lever, with its insertion to the ulna close to the elbow joint (fulcrum) and with the weight moved at the distal end of the hand (lever arm) [FIGURE 11.5C]. Like most skeletal muscles,

FIGURE 11.5 CLASSES OF LEVERS

[A] In a first-class lever, the fulcrum is between the force and the weight. **[B]** In a second-class lever, the weight is between the force and the fulcrum. **[C]** In a third-class lever, the force is between the weight and the fulcrum. The arrows indicate the applied force and the weight of the object to be moved.

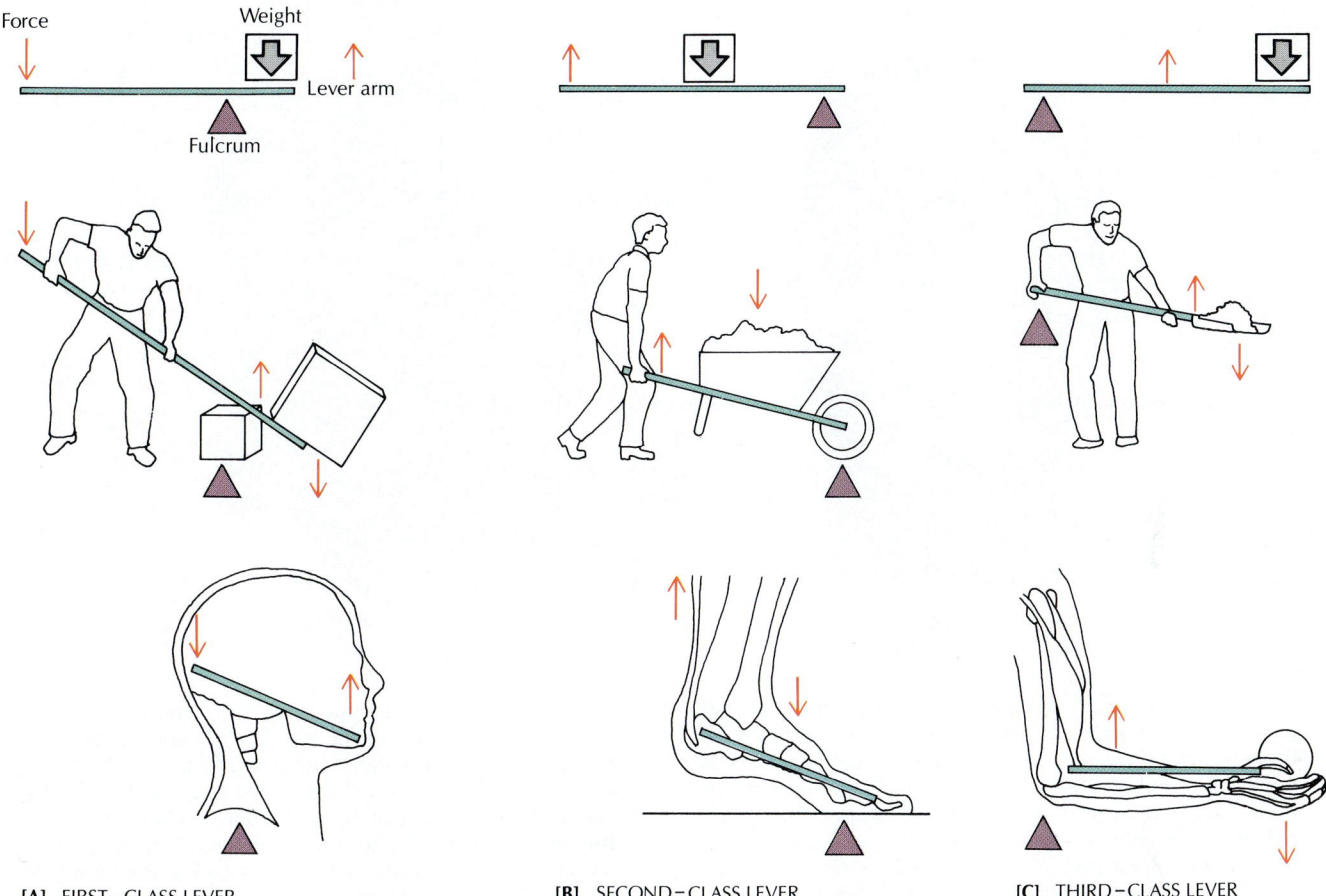

[A] FIRST–CLASS LEVER   **[B]** SECOND–CLASS LEVER   **[C]** THIRD–CLASS LEVER

the brachialis acts at a mechanical disadvantage because the length of the lever arm from the elbow (fulcrum) to the hand (weight) is greater than the length of the lever arm from the elbow to the insertion of the brachialis (the muscle producing the force to move the weight). A great deal of force is needed to move the weight.

The leverage of a muscle is improved as the distance from the insertion to the joint (fulcrum) is increased. In other words, a muscle with an insertion relatively far from a joint has *more force* than a muscle with an insertion closer to the joint because the longer lever arm between the joint and insertion produces more force. However, the longer lever arm must move a greater distance to produce this force, and the movement is slower. For example, the pectineus muscle, which is attached close to the axis of rotation at the hip joint, has less force than the adductor longus muscle, which is attached farther from the axis [FIGURE 11.22]. But the pectineus can produce a quicker movement than the adductor longus because it moves through a shorter distance.

ASK YOURSELF

1 What two forces are involved in a lever system?

2 What are the main differences in the three classes of levers? Give an example of each of the three types of levers in the body.

3 How does leverage operate in the human body?

SPECIFIC ACTIONS OF PRINCIPAL MUSCLES

In this section, which consists of FIGURES 11.6 through 11.25, the major muscles of the body are described and illustrated. The origins, insertions, innervation, and actions of these muscles are summarized in the accompanying tables.

FIGURE 11.6 MUSCLES OF FACIAL EXPRESSION

[A] Anterior view. Note that different layers of muscle are shown on each side of the face. [B] Right lateral view.

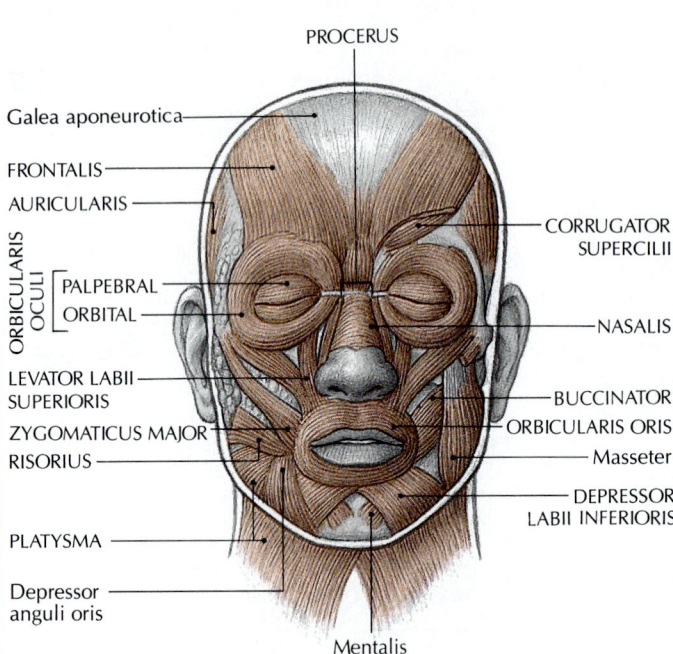

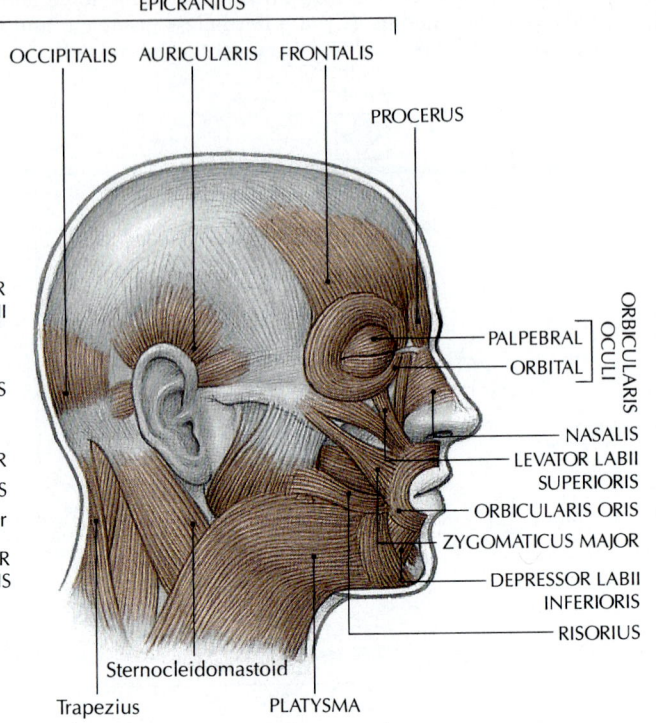

[A]

[B]

The ***muscles of facial expression*** are located not only in the face, but in the scalp and neck. Two features characterize these muscles—they are all innervated by the facial cranial nerve (VII), and they are used in the display of human emotions. They can move parts of the face because their principal insertions extend into the deep layers of the skin. Their origins may be located in a tendon, aponeurosis, bone, or the skin itself. Some of the facial muscles are arranged around the eyes, mouth, nose, and ears, and act as ***sphincters*** (ringlike muscles that normally constrict a bodily opening until relaxation is called for).

Facial muscles are used to express a wide variety of emotions. For example, when you smile, the muscles radiating from the angles of your mouth raise the corners of your upper lip. When you frown, certain muscles lower your eyebrows and depress the corners of your lower lip. (The cliché that it takes more effort to frown than to smile is true; you use 43 muscles to frown and only 17 to smile.)

Besides expressing emotion, the facial muscles have other uses. For example, the nasalis musculature dilates your nostrils as you inhale. The ***buccinator muscle*** (L. *bucca*, cheek) lining the cheek is essential for nudging food from between the cheek and teeth during chewing, and for blowing or sucking.* The scalp (epicranius) muscles originating from a broad, flat aponeurosis (galea aponeurotica) can wrinkle the forehead, producing a look of surprise or fright.

*When the buccinator muscles lose their tone and elasticity, the cheeks bulge uncontrollably when a person makes a blowing motion. This condition is known as "Gillespie's pouches," named for the jazz trumpeter Dizzy Gillespie, whose ineffective buccinator muscles caused his cheeks to bulge every time he blew into his trumpet. (However, the buccinator muscle was known as the trumpeter's muscle long before Dizzy Gillespie came along.)

Muscle	Origin	Insertion	Innervation	Action
MUSCLES OF SCALP				
Epicranius Frontalis	Galea aponeurotica	Fibers of orbicularis oculi	Temporal branches of facial nerve (VII)	Elevates eyebrows (surprised look); produces horizontal frown
Occipitalis	Occipital bone	Galea aponeurotica	Posterior auricular branch of facial nerve (VII)	Draws scalp backward
Auricularis	Galea aponeurotica	Vicinity of outer ear	Temporal branches of facial nerve (VII)	Wiggles ears in a few individuals

Muscle	Origin	Insertion	Innervation	Action
MUSCLES OF EYELID AND ORBIT				
Orbicularis oculi Palpebral (eyelid)	Medially from medial palpebral ligament	Sphincter fibers oriented in concentric circles around orbital margin and in eyelid	Temporal and zygomatic branches of facial nerve (VII)	Closes eyelid (as in winking)
Orbital	Medial orbital margin (maxilla), medial palpebral ligament	Sphincter muscle	Temporal and zygomatic branches of facial nerve (VII)	Closes eyelid tightly (against bright light)
Corrugator supercilii	Brow ridge of frontal bone	Skin of eyebrow	Temporal branch of facial nerve (VII)	Pulls eyebrows together
Levator palpebrae superioris	Central tendinous ring around optic foramen	Skin of upper eyelid	Oculomotor nerve (III)	Raises upper eyelid
MUSCLES OF NOSE				
Nasalis	Maxilla medial and inferior to orbit	Lower region of cartilage of nose	Facial nerve (VII)	Widens nasal aperture (as in deep breathing)
Procerus	Lower part of nasal bone, lateral nasal cartilage	Skin between eyebrows	Buccal branches of facial nerve (VII)	Pulls eyebrow downward and inward
MUSCLES OF MOUTH				
Orbicularis oris	Sphincter muscle within lips; encircles mouth and merges with other muscles of mouth		Buccal and mandibular branches of facial nerve (VII)	Closes mouth; purses lips; significant role in speech
Levator labii superioris	Above infraorbital foramen of maxilla	Skin of upper lip	Buccal branch of facial nerve (VII)	Raises lateral aspect of upper lip
Zygomaticus major	Zygomatic bone	Skin at angle of mouth	Zygomatic or buccal branch of facial nerve (VII)	Draws angle of mouth upward (as in smiling)
Risorius	Fascia over masseter muscle	Skin at angle of mouth	Buccal branch of facial nerve (VII)	Draws angle of mouth laterally (as in grinning)
Depressor labii inferioris	Mandible	Skin of lower lip	Mandibular branch of facial nerve (VII)	Lowers lateral aspect of lower lip
MUSCLE OF CHEEK				
Buccinator	Alveolar processes of mandible and maxilla, from pterygomandibular raphe (connective tissue) between sphenoid and mandible	Fibers of orbicularis oris	Buccal branch of facial nerve (VII)	Pulls cheek against teeth to move food during chewing and to make blowing or sucking motion
MUSCLE OF NECK				
Platysma	Fascia of upper thorax over pectoralis major, deltoid muscles	Mandible, skin of lower face, fibers of orbicularis oris at angle of mouth	Cervical branch of facial nerve (VII)	Draws outer part of lower lip down and back (as in frowning); tenses skin of neck

FIGURE 11.7 MUSCLES THAT MOVE THE EYEBALL

[A] Superior view, with horizontal and sagittal axes of the eyeball indicated. [B] Lateral view, with vertical axis of the eyeball indicated.

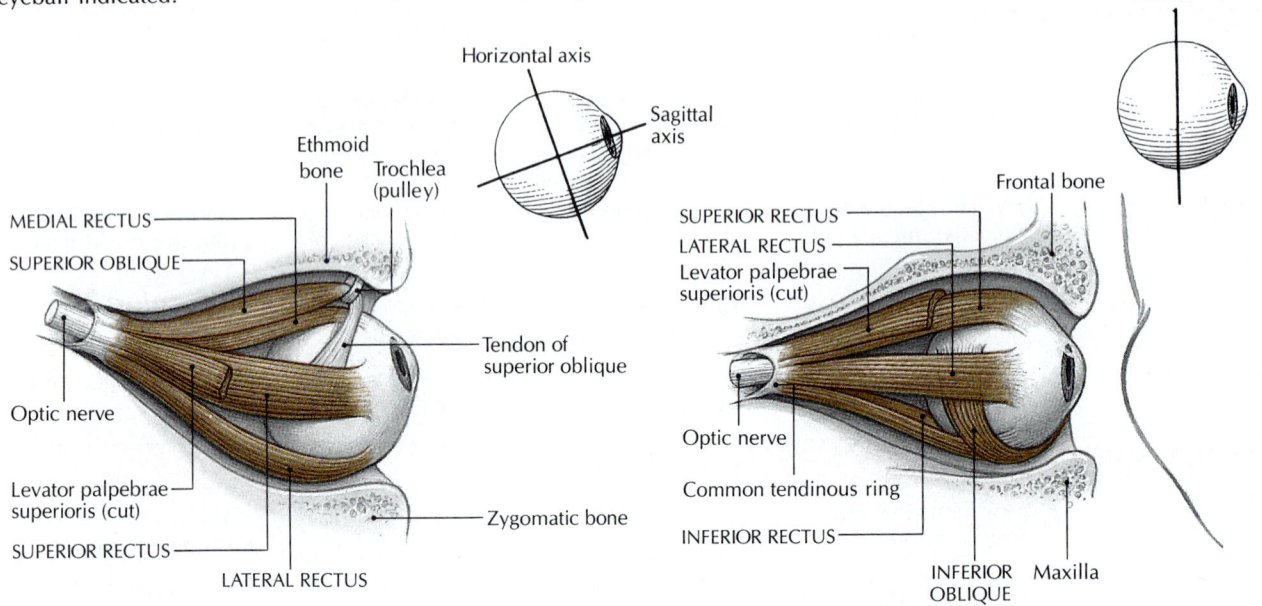

[A]

[B]

The movements of the eyeball are controlled by six rapidly responsive **extrinsic** (outside the eyeball) eye muscles. These muscles can rotate the eyeball around the horizontal, vertical, and sagittal axes passing through the center of the eyeball. The four **recti** ("upright") **muscles** insert in front of the horizontal axis, and the two *oblique muscles* insert behind the horizontal axis. The recti muscles pass slightly obliquely forward from their origin around the optic fora-

men. They form a cone as they spread out toward their insertions in the eyeball. The actions of the recti muscles can be understood by realizing that they pull on the eyeball in a plane slightly oblique to the visual (sagittal) axis. *Squint* (strabismus or cross-eye) is a disorder in which both eyes cannot be directed at the same point or object at the same time. It can result from an imbalance in the power and length of the recti muscles.

Muscle	Origin	Insertion	Innervation	Action
Superior rectus	Tendinous ring anchored to bony orbit around optic foramen	Superior and central part of eyeball	Oculomotor nerve (III)	Rolls eye upward; also adducts and rotates medially
Inferior rectus	Tendinous ring anchored to bony orbit around optic foramen	Inferior and central part of eyeball	Oculomotor nerve (III)	Rolls eye downward; also adducts and rotates laterally
Lateral rectus	Tendinous ring anchored to bony orbit around optic foramen	Lateral side of eyeball	Abducens nerve (VI)	Rolls (abducts) eye laterally (out)
Medial rectus	Tendinous ring anchored to bony orbit around optic foramen	Medial side of eyeball	Oculomotor nerve (III)	Rolls (adducts) eye medially (in)
Superior oblique	Sphenoid bone	Posterior and lateral to equator of eyeball under superior rectus*	Trochlear nerve (IV)	Depresses, abducts, and rotates eye medially
Inferior oblique	Lacrimal bone	Posterior and lateral to equator of eyeball under lateral rectus	Oculomotor nerve (III)	Elevates, abducts, and rotates eye laterally

*The tendon of the superior oblique changes its direction abruptly when it passes through the trochlea (a fibrocartilaginous pulley) located in the upper front of the bony orbit.

FIGURE 11.8 MUSCLES OF MASTICATION

[A] Temporal and masseter muscles; right lateral-superficial view. **[B]** Medial and lateral pterygoid muscles; right lateral view deep to the mandible.

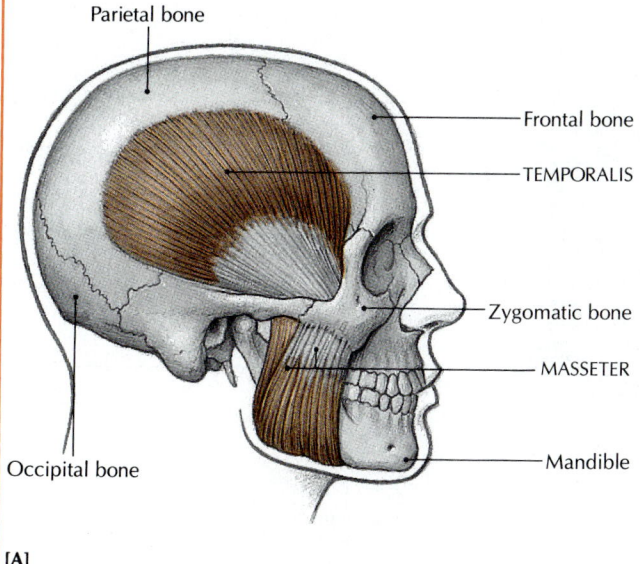

[A]

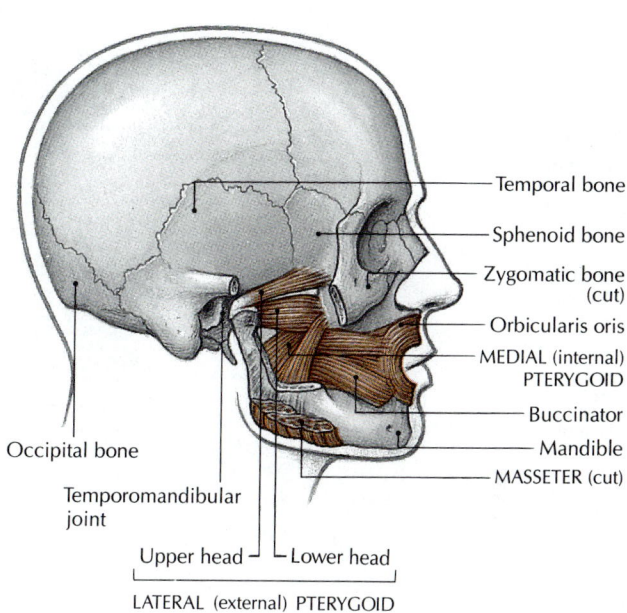

[B]

Four pairs of muscles produce biting and chewing movements: the ***masseter, temporalis, lateral pterygoid*** (Gr. "wing"; ter-uh-GOID), and ***medial pterygoid.*** The masseter (Gr. *maseter*, chewer) and the temporalis muscles lie close to the skin and can be felt easily when the teeth are clenched.

In chewing, the mandible may move from side to side in a grinding motion. This movement is controlled by the masseter and temporalis muscles on the same side as the direction of the movement, and by the lateral and medial pterygoid muscles on the opposite side.

Muscle	Origin	Insertion	Innervation	Action
Masseter	Zygomatic bone	Outer surface of angle and ramus of mandible	Mandibular division of trigeminal nerve (V)	Elevates mandible (closes mouth); slightly protrudes mandible
Temporalis	Temporal fossa of temporal bone	Coronoid process of mandible	Mandibular division of trigeminal nerve (V)	Elevates mandible; retracts mandible (pulls jaw back)
Medial pterygoid	Medial surface of lateral pterygoid plate of sphenoid bone, maxilla	Inner surface of angle and ramus of mandible	Mandibular division of trigeminal nerve (V)	Elevates mandible; slightly protrudes mandible; draws jaw toward opposite side in grinding movements
Lateral pterygoid	Lateral surface of lateral pterygoid plate of sphenoid bone	Condyle of mandible and fibrocartilage articular disk	Mandibular division of trigeminal nerve (V)	Protrudes mandible; depresses mandible (opens mouth); produces side-to-side movements during chewing and grinding

FIGURE 11.9 MUSCLES THAT MOVE THE HYOID BONE

[A] The sternocleidomastoid and sternohyoid muscles of the neck and the digastric muscle attached to the hyoid bone are shown on the right side of the diagram but not on the left; anterior view.

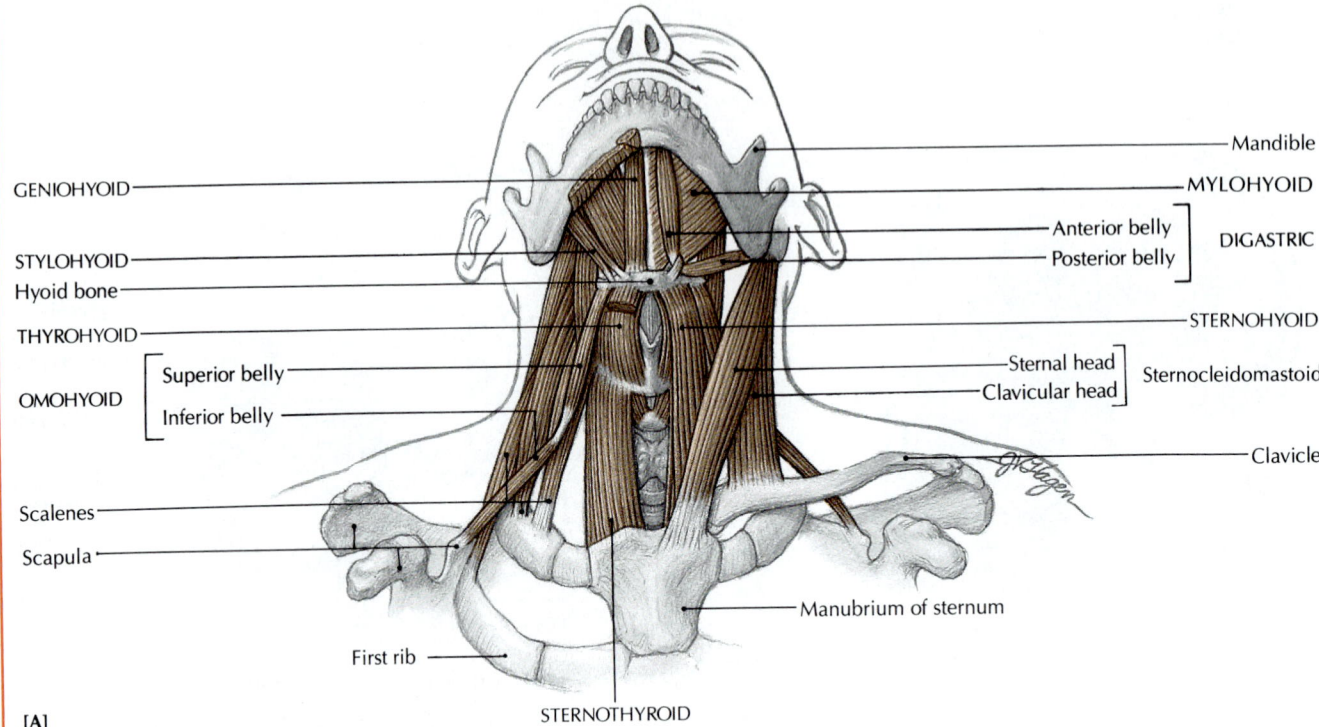

[A]

Many muscles associated with the mouth, throat (pharynx), and neck are attached to the hyoid bone. The *suprahyoid muscles* extend above the hyoid and attach it to the skull. The straplike *infrahyoid muscles* extend from the hyoid to skeletal structures below. The precise movements of the hyoid are controlled by coordinated muscular activity and are especially evident during swallowing, when the hyoid bone moves upward and forward and then downward and backward to its original position. The hyoid has attachments for some muscles of the tongue and larynx. When the mandible does not move, the hyoid is raised by the *mylohyoid, geniohyoid, digastric* (suprahyoid), and *stylohyoid muscles.* The infrahyoid muscles pull the hyoid downward. The diaphragmlike paired mylohyoid muscles form the floor of the mouth below the tongue. Both suprahyoid and infrahyoid muscles can aid in depressing the mandible.

INFRAHYOID MUSCLES (STRAP MUSCLES)

Sternohyoid	Manubrium of sternum	Body of hyoid	Cervical plexus*	Depresses hyoid
Sternothyroid	Manubrium of sternum	Thyroid cartilage of larynx	Cervical plexus	Depresses thyroid and hyoid
Thyrohyoid	Thyroid cartilage of larynx	Greater horn of hyoid	Cervical plexus	Depresses hyoid; raises thyroid cartilage
Omohyoid Inferior belly	Superior border of scapula (near coracoid process)	Intermediate tendon	Cervical plexus	Depresses hyoid
Superior belly	Intermediate tendon attached to medial end of clavicle	Body of hyoid	Cervical plexus	Depresses hyoid

*The cervical plexus innervating these muscles is formed from nerves derived from the first three cervical nerves.

Muscle	Origin	Insertion	Innervation	Action
SUPRAHYOID MUSCLES (MUSCLES OF FLOOR OF MOUTH)				
Digastric				
Posterior belly	Mastoid process of temporal bone	Common tendon attached to body of hyoid	Facial nerve (VII)	Raises hyoid
Anterior belly	Digastric fossa of mandible near symphysis		Mandibular division of trigeminal nerve (V)	Depresses mandible
Stylohyoid	Styloid process of temporal bone	Body of hyoid	Facial nerve (VII)	Raises hyoid; pulls hyoid backward
Mylohyoid	Mylohyoid line on internal surface of mandible	Hyoid bone, central raphe in floor of mouth	Mandibular division of trigeminal nerve (V)	Raises hyoid; forms and elevates floor of mouth
Geniohyoid	Adjacent to symphysis of mandible	Body of hyoid	First cervical nerve	Pulls hyoid upward and forward; helps to open jaw. Depresses mandible when hyoid is fixed

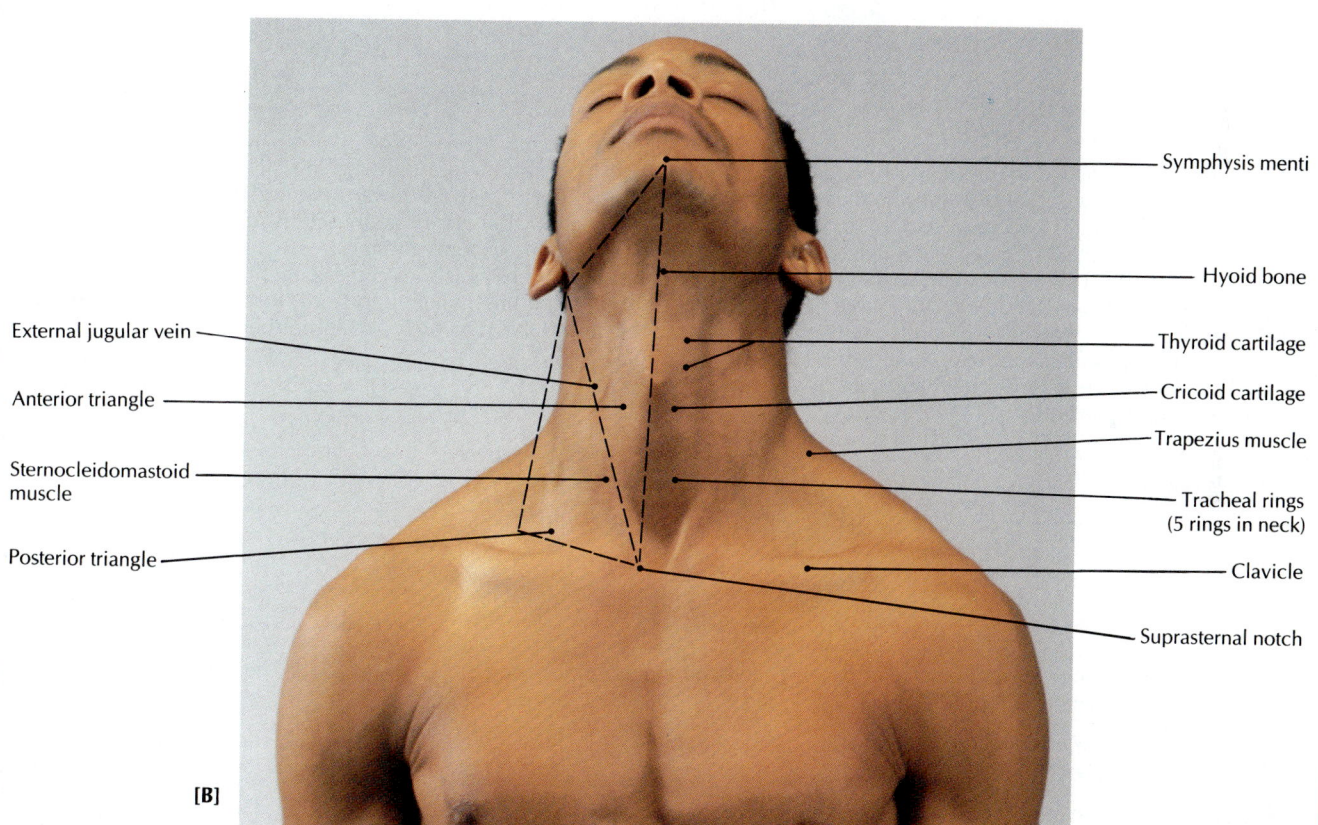

External jugular vein

Anterior triangle

Sternocleidomastoid muscle

Posterior triangle

Symphysis menti

Hyoid bone

Thyroid cartilage

Cricoid cartilage

Trapezius muscle

Tracheal rings (5 rings in neck)

Clavicle

Suprasternal notch

[B]

[B] Surface anatomy of the neck. The **neck** (*nucha*, NOO-kuh, or *collum*) is the region of the body between the lower margin of the mandible and the upper border of the clavicle. The neck region is divided into the anterior and posterior triangles by the sternocleidomastoid muscle, which is palpable over its entire length as it runs obliquely from the mastoid process under the ear to the front of the clavicle.

The **anterior triangle** is bounded by the body of the mandible, the sternocleidomastoid muscle, and the midline of the body. The **posterior triangle** is bounded in front by the sternocleidomastoid muscle, behind by the anterior border of the trapezius muscle, and inferiorly by the clavicle.

FIGURE 11.10 MUSCLES THAT MOVE THE TONGUE

Right lateral view.

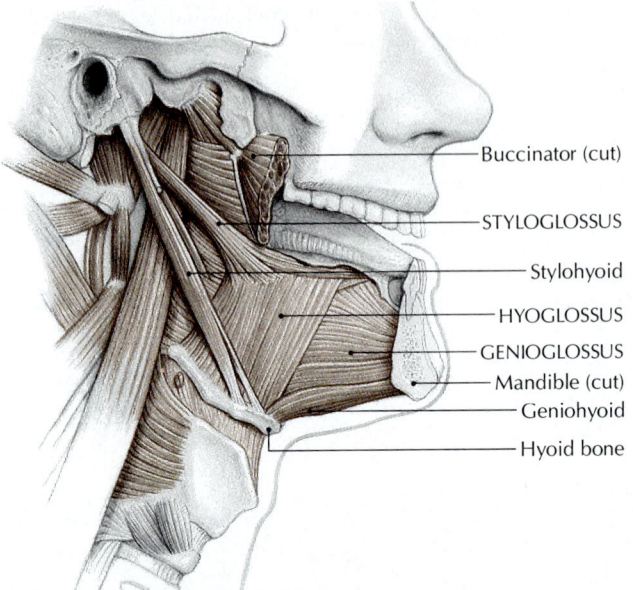

- Buccinator (cut)
- STYLOGLOSSUS
- Stylohyoid
- HYOGLOSSUS
- GENIOGLOSSUS
- Mandible (cut)
- Geniohyoid
- Hyoid bone

The muscles of the tongue are essential for normal speech and for the manipulation of food within the mouth. The *intrinsic muscles,* which are located within the tongue, are oriented in the horizontal, vertical, and longitudinal planes. They are able to fold, curve, and squeeze the tongue during speech and chewing. The action of the three *extrinsic muscles* of the tongue allow it to protrude **(genioglossus muscle),** retract **(styloglossus muscle),** and depress **(hyoglossus muscle).** All tongue muscles are innervated by the hypoglossal cranial nerve (XII).

Normally, the genioglossus muscle prevents the tongue from falling backward toward the throat, where it can cause suffocation. But when the muscle is paralyzed or totally relaxed during general anesthesia, it can block the respiratory passage. Anesthetists keep the tongue in a forward position by pushing the mandible forward, or by using a curved intubation tube. The tube should not be allowed to reach the larynx.

Muscle	Origin	Insertion	Innervation	Action
Genioglossus	Tubercle of anterior part of mandible beside midline	Fibers pass through tongue to insert in bottom of tongue		Protrudes and depresses tongue
Styloglossus	Tip of styloid process of temporal bone	Side of tongue	Hypoglossal nerve (XII)	Retracts and elevates tongue
Hyoglossus	Greater horn of hyoid bone	Side of tongue		Depresses tongue
Intrinsic muscles	Arranged in three planes within tongue in horizontal, vertical, and longitudinal planes			Alter shape of tongue

FIGURE 11.11 MUSCLES THAT MOVE THE HEAD

Posterior view.

In a sense, the head is balanced upon the vertebral column, with the atlas articulating with the occipital condyles. This *atlantooccipital* articulation is involved in flexing (bending forward) and extending (holding erect) the head, and is located about midway between the back of the head and the tips of the nose and chin. The head is normally bent slightly forward, with the neck muscles that insert in the occipital region (such as the splenius capitis and semispinalis capitis) partially contracted to prevent the head from falling forward. When the head flexes, the posterior neck muscles relax while the **sternocleidomastoid** and other anterior neck muscles contract. The *bilateral* (both sides) contraction of the **splenius capitis, longissimus capitis,** and **semispinalis capitis** extends the head. The *unilateral* (one side only) contraction of these muscles rotates the head, tilts the chin up, and turns the face toward the contracted side.

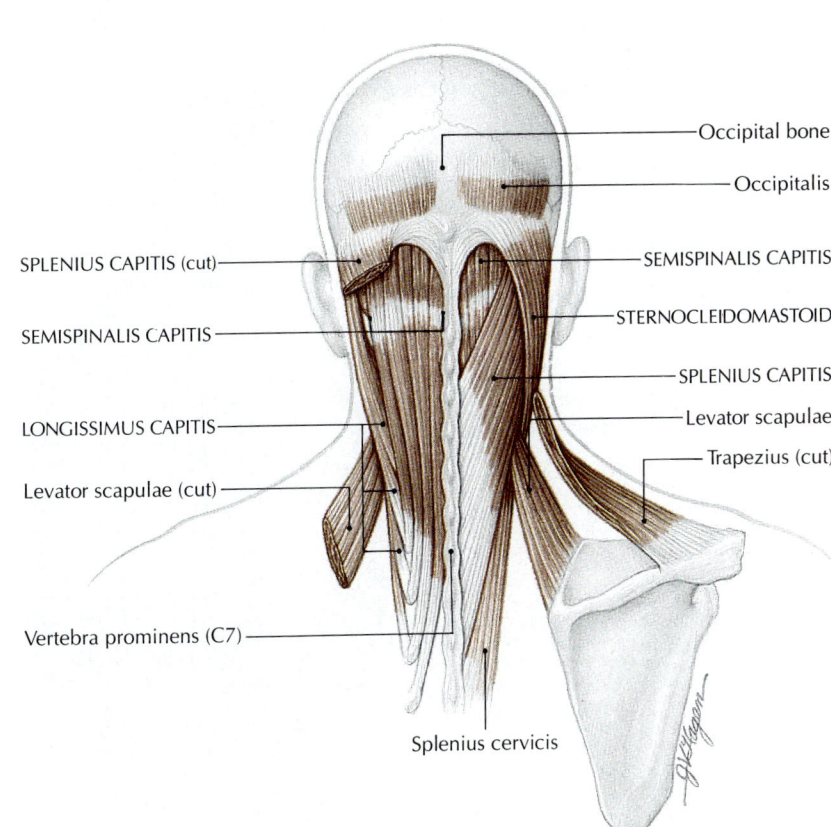

Occipital bone
Occipitalis
SEMISPINALIS CAPITIS
STERNOCLEIDOMASTOID
SPLENIUS CAPITIS
Levator scapulae
Trapezius (cut)

SPLENIUS CAPITIS (cut)
SEMISPINALIS CAPITIS
LONGISSIMUS CAPITIS
Levator scapulae (cut)
Vertebra prominens (C7)
Splenius cervicis

Muscle	Origin	Insertion	Innervation	Action
Sternocleido-mastoid	Sternum, clavicle	Mastoid process of temporal bone	Spinal accessory nerve (XI)	*Bilateral:* flex vertebral column, bringing head down *Unilateral:* bends vertebral column to same side, drawing head toward shoulder and rotating head so chin points to opposite side
Semispinalis capitis	Vertebral arches of C7 and T1 to T6 vertebrae	Occipital bone	Dorsal rami of cervical and upper thoracic nerves	*Bilateral:* extend head *Unilateral:* extends head; turns face to opposite side
Splenius capitis	Spines of C7 and T1 to T4 vertebrae, ligamentum nuchae (nape of neck)	Occipital bone, mastoid process of temporal bone	Dorsal rami of middle cervical nerves	*Bilateral:* extend head *Unilateral:* bends head to same side; rotates face to same side
Longissimus capitis	Transverse processes of T1 to T4 or T5 vertebrae	Mastoid process of temporal bone	Dorsal rami of middle and lower cervical nerves	*Bilateral:* extend head *Unilateral:* bends head to same side; rotates face to same side

FIGURE 11.12 INTRINSIC MUSCLES THAT MOVE THE VERTEBRAL COLUMN

Posterior view.

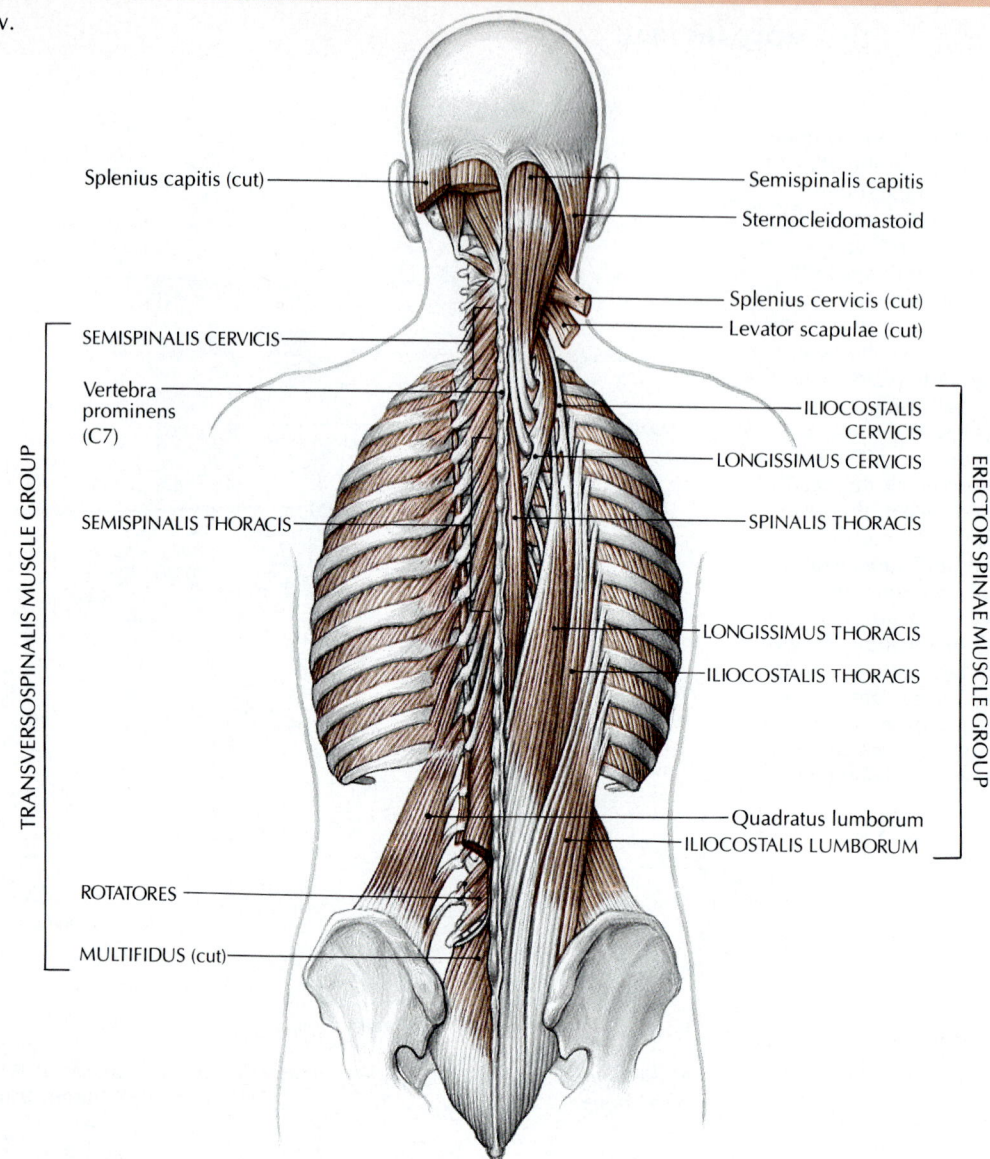

Labels (left side):
Splenius capitis (cut)
SEMISPINALIS CERVICIS
Vertebra prominens (C7)
SEMISPINALIS THORACIS
TRANSVERSOSPINALIS MUSCLE GROUP
ROTATORES
MULTIFIDUS (cut)

Labels (right side):
Semispinalis capitis
Sternocleidomastoid
Splenius cervicis (cut)
Levator scapulae (cut)
ILIOCOSTALIS CERVICIS
LONGISSIMUS CERVICIS
SPINALIS THORACIS
LONGISSIMUS THORACIS
ILIOCOSTALIS THORACIS
Quadratus lumborum
ILIOCOSTALIS LUMBORUM
ERECTOR SPINAE MUSCLE GROUP

The vast numbers of small muscles and muscle bundles that make up the intrinsic back muscles are located in a pair of broad, longitudinally oriented gutters on either side of the spines of the vertebrae. They are usually organized into two major groups: the *superficial group,* called the **erector spinae,** or **sacrospinalis, muscles,** and the *deep group,* called the **transversospinalis muscles.**

The superficial group consists of overlapping muscle fascicles. They mount "up" the vertebral column, and each bundle spans from origin to insertion over five or so vertebrae. This muscle group is subdivided into three columns: (1) the lateral column, the **iliocostalis,** originates on the iliac crest and ribs, and then after spanning five levels inserts on a rib or transverse process of a cervical vertebra; (2) the middle column, the **longissimus,** originates and inserts on the transverse processes of the thoracic and cervical vertebrae and mastoid process of the skull; and (3) the medial column, the **spinalis,** extends from spinous process to spinous process. Muscles of the superficial group are named **lumborum, thoracis,** and **cervicis,** depending on the region they are associated with.

The deep group consists of small bundles of shorter muscle fascicles called the **transversospinalis muscles.** Each

bundle originates on a tranverse process, extends upward and obliquely, and inserts on a spinous process. Some bundles **(rotatores)** run up one or two levels, others **(multifidus)** span two or three vertebrae, and still others **(semispinalis)** span four to five vertebrae. The semispinalis muscles are called **thoracis** or **cervicis,** depending on their location.

The intrinsic back muscles act to extend, bend laterally, and rotate the vertebral column and head. They also counteract the force of gravity, which tends to flex the vertebral column. The posture of the vertebral column is regulated by the vertebral muscles. Contrary to a common belief, these muscles are not active constantly. They are fairly relaxed during normal standing, and they are almost completely relaxed when the vertebral column is bent back (extension), because the ligaments of the vertebral column assume the load. The axiom to remember is: When ligaments suffice, the muscles will yield. In contrast, the vertebral muscles are most active when the vertebral column is bent forward. Extension and hyperextension of the column are used in forced inspiration (as in gasping for air). Gravity and the contraction of the rectus abdominus muscles flex the vertebral column.

FIGURE 11.13 MUSCLES USED IN QUIET BREATHING

[A] Note that the direction of the fibers of the internal and external intercostal muscles are at approximate right angles to each other. [B] Scalene muscles, which extend from the cervical vertebrae to the first two ribs. Right lateral views.

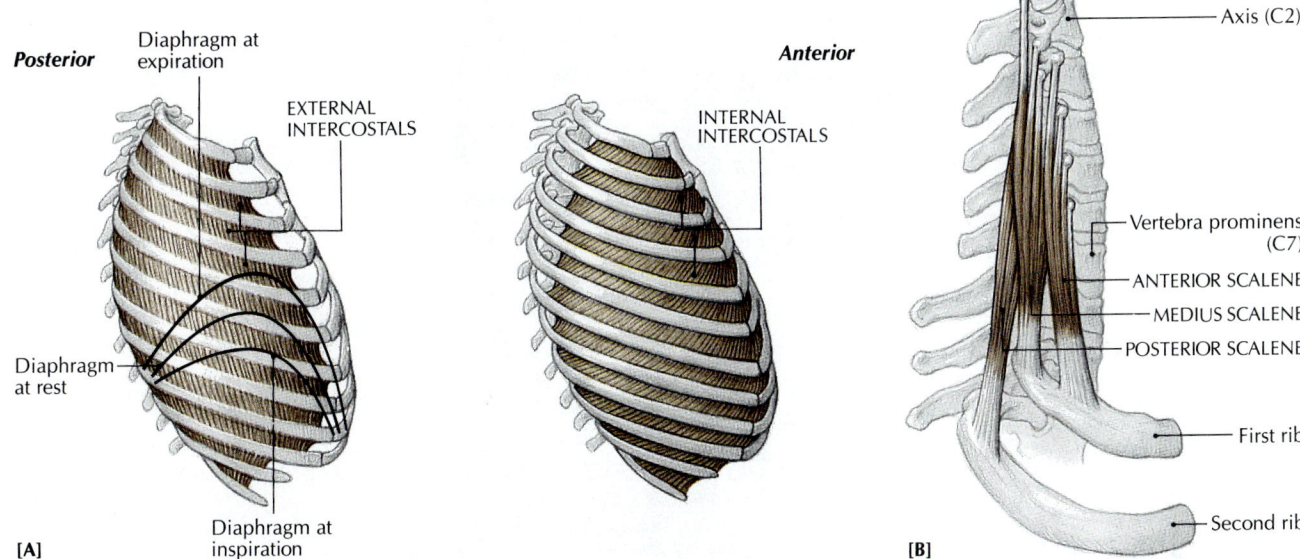

Quiet, or normal breathing, involves the coordinated activity of several muscles or groups of muscles. Inspiration (breathing in) requires the enlargement of the rib cage, so that air pressure inside the thorax is less than the atmospheric pressure. Because of this imbalance, air is sucked into the lungs as the dome-shaped **diaphragm** contracts and moves down. This movement is accompanied by the relaxation of the muscles of the abdominal wall. The contraction of the **external intercostal muscles** and **interchondral muscles** raises the ribs, in what is called the "bucket-handle movement," which increases the transverse diameter of the rib cage.

In expiration (breathing out), the diaphragm relaxes and is raised to its higher resting position, arching up into the thorax, by the pressure generated in the abdominal cavity by the contractile tone of the muscles in the abdominal wall. The rib cage becomes smaller when the ribs are lowered, largely by elastic recoil, and expiration may also be aided by the contraction of the **internal intercostal muslces.** The **scalene muscles** also assist in respiration by steadying the first two ribs and possibly elevating them.

Muscle	Origin	Insertion	Innervation	Action
Diaphragm	Lower six costal cartilages, xiphoid cartilage	Central tendon of diaphragm	Phrenic nerve (from C3 to C5 nerves) innervates portion of diaphragm attached to lumbar vertebrae	Contraction pulls central tendon down to flatten dome and increase vertical length and volume of thorax, resulting in inspiration
External intercostals	Each from lower border of each bony rib (except 12)	Upper border of next rib below	Intercostal nerves	Elevates ribs to increase all diameters of thorax, resulting in inspiration
Internal intercostals	Each from upper border of each bony rib (except 1)	Lower border of each bony rib above	Intercostal nerves	Depress (lower) ribs for forced inspiration
Internal interchondrals (portion of internal intercostals attached to costal cartilages)	Costal (cartilagenous) portion of ribs 2 to 11	Costal portion of ribs 1 to 10	Intercostal nerves	Elevates ribs during inspiration
Scalenes Anterior	Fronts of transverse processes in cervical vertebrae	Rib 1	Branches of cervical nerves	Steady ribs 1 and 2 during breathing; may assist in elevating them
Medius and posterior	Backs of transverse processes in cervical vertebrae	Medius to rib 1, posterior to rib 2	Branches of cervical nerves	Steady ribs 1 and 2 during breathing; may assist in elevating them

FIGURE 11.14 MUSCLES THAT SUPPORT THE ABDOMINAL WALL

[A] In this anterior view, some of the muscles have been cut to show the different layers of muscles, from superficial to deep.

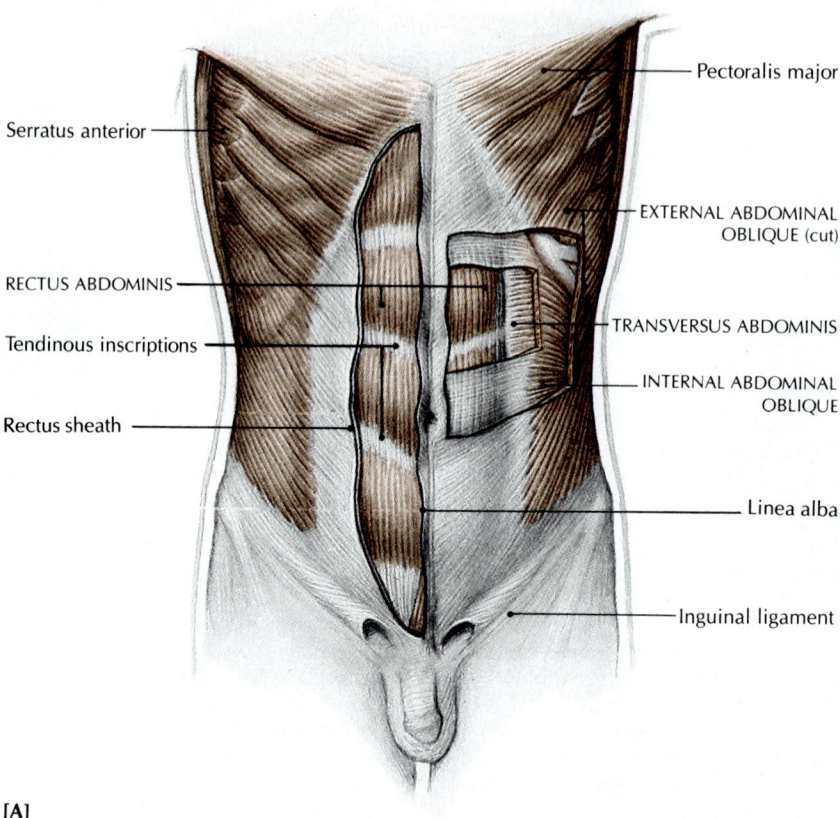

Pectoralis major

Serratus anterior

EXTERNAL ABDOMINAL OBLIQUE (cut)

RECTUS ABDOMINIS

TRANSVERSUS ABDOMINIS

Tendinous inscriptions

INTERNAL ABDOMINAL OBLIQUE

Rectus sheath

Linea alba

Inguinal ligament

[A]

The muscles of the abdominal wall act to support and protect the internal organs. Several layers of muscles running in different directions add strength to the abdominal wall:

1 As the outside layer of the anterior and lateral walls, the ***external abdominal oblique muscle*** extends forward and down from the ribs until it becomes an aponeurosis, which attaches to the linea alba at the midline. The ***linea alba*** (L. white line) is a tendon running from the xiphoid process of the sternum to the pubic symphysis.

2 The next layer inward is the ***internal abdominal oblique muscle,*** which runs forward and upward from the iliac crest and inguinal ligament, until it too becomes an aponeurosis that connects with the linea alba. The inguinal ligament extends from the anterior superior iliac spine to the pubis bone.

3 The innermost layer is the ***transversus abdominis muscle,*** which extends horizontally from the dorsolumbar fascia (the fascia of the lumbar region, attached to the lumbar vertebrae) to the inguinal ligament and ribs. After becoming an aponeurosis (passing deep to the rectus abdominis), it continues to meet the linea alba.

4 Running lengthwise on either side of the linea alba is the ***rectus abdominis*** muscle, which extends from the rib cage and sternum to the pubic crest. It is enveloped by the rectus sheath, which is actually made up of the aponeuroses of the other three layers of muscles. The transverse bands crossing the rectus abdominis muscle are called *tendinous inscriptions.*

Together these four abdominal muscles act as a dynamic corset around the abdomen, supplying the necessary pressure for respiration, urination, defecation, and parturition (childbirth). In addition, they also play roles in flexion and lateral bending of the vertebral column.

On the posterior wall of the abdomen, the ***quadratus lumborum muscle*** extends from the twelfth rib to the posterior iliac crest. Besides allowing the trunk to bend toward one side, this muscle gives the vertebral column stability and plays a role in respiration.

Muscle	Origin	Insertion	Innervation	Action
MUSCLES OF ANTERIOR AND LATERAL WALLS				
Rectus abdominis	Pubic crest, symphysis pubis	Costal cartilages of ribs 5 to 7, xiphoid process of sternum	Branches of intercostal nerves (T7 to T12)	Powerful flexor of lumbar vertebral column; depresses rib cage, plays role in stabilizing pelvis against leverage of thigh muscles during walking and running
External abdominal oblique	Ribs 5 to 12	Iliac crest, linea alba	Branches of intercostal (T8 to T12), iliohypogastric, and ilioinguinal nerves	Along with rectus abdominis, these muscles hold in and compress the abdominal contents, aiding in defecation, urination, and childbirth. Contract to protect abdominal contents against external blows; flex trunk, rotate vertebral column
Internal abdominal oblique	Iliac crest, inguinal ligament, lumbodorsal fascia	Costal cartilages and ribs 8 to 12, linea alba, xiphoid process of sternum	Branches of intercostal (T8 to T12), iliohypogastric, and ilioinguinal nerves	
Transversus abdominis	Iliac crest, inguinal ligament, lumbar fascia, cartilages of ribs 7 to 12	Xiphoid process of sternum, linea alba, pubis	Branches of intercostal (T8 to T12), iliohypogastric, and ilioinguinal nerves	
MUSCLE OF POSTERIOR WALL				
Quadratus lumborum [FIGURE 11.12]	Iliac crest, iliolumbar ligament, transverse processes of lower lumbar vertebrae	Rib 12, transverse processes of upper lumbar vertebrae	Ventral rami of T12 to L4 nerves	*Unilateral:* flexes (bends) trunk toward same side *Bilateral:* extends and stabilizes lumbar vertebral column; aids inspiration by increasing length of thorax

Tendinous inscription of rectus abdominis muscle

Linea semilunaris

External oblique muscle

McBurney's point

Anterior superior iliac spine

Inguinal ligament

Serratus anterior muscle

Linea alba

Umbilicus

Iliac crest

[B]

Pubic tubercle

[B] Surface anatomy of the abdomen and pelvis. The ***abdomen*** is the part of the body that lies below the diaphragm in the anterior thorax and above the superior margin of the pelvis. The shape and contours of the abdom[inal] wall result largely from the musculature, but the[y can] be modified by accumulations of subcutaneo[us...]

FIGURE 11.15 MUSCLES THAT FORM THE PELVIC OUTLET

Inferior views: **[A]** male, **[B]** female.

The muscles of the pelvic outlet are often divided into two groups: those of the pelvic diaphragm and those of the perineum. The **pelvic diaphragm** (pelvic floor) is the funnel-shaped muscular floor of the pelvic cavity. Its role is to support the pelvic organs and thus prevent organs such as the urinary bladder, uterus, and rectum from falling down (undergoing prolapse) and moving through the diaphragm (as in a hernia). The **levator ani muscle** (consisting of the **pubococcygeus** and **iliococcygeus muscles**), the **coccygeus muscle,** and fasciae compose the pelvic diaphragm. The *urogenital (UG) diaphragm* is composed of fasciae and the **deep transverse perinei muscles.** The UG diaphragm also functions as part of the **external sphincter of the urethra** and provides added reinforcement.

The urethra of both sexes and the vagina of the female pass through both the pelvic and UG diaphragms. The rectum passes through the pelvic diaphragm to become the anus. The tonus of the medial portion of the pubococcygeus muscle, called the *puborectalis,* acts as a major deterrent to fecal incontinence (the involuntary passage of feces). Certain muscles of these diaphragms participate in maintaining urinary continence (the ability to retain urine within the bladder until conditions are proper for urination).

The **perineum** (per-uh-NEE-uhm) is the diamond-shaped region extending from the symphysis pubis to the coccyx. The anterior half of the perineum, shaped like a triangle, is the urogenital diaphragm (or triangle). The posterior half, also triangular, is the anal region. The muscles of the urogenital triangle provide voluntary control of the urethra and the release of urine.

The perineum is of special importance to proctologists (physicians who treat disorders of the rectum and anus) and urologists (physicians who treat disorders of the urogenital organs). The female perineum is also of special importance to obstetricians and gynecologists, especially for the management of childbirth and disorders involving the ovaries, uterine tubes, uterus, and vagina.

Occasionally, labor and delivery during childbirth weaken the pelvic floor, and the pelvic diaphragm may be stretched or torn. Such a condition may be accompanied by a hernia or prolapse of the uterus or rectum. The damage can usually be repaired by surgery.

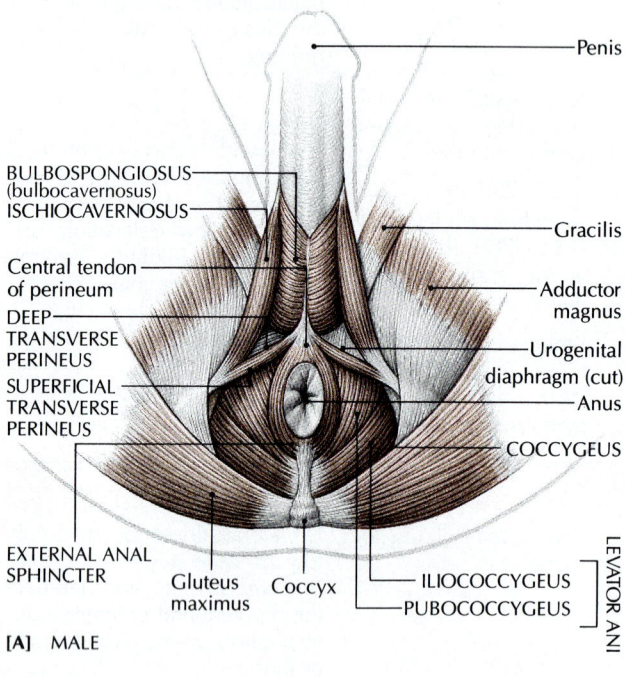

[A] MALE

Penis
BULBOSPONGIOSUS (bulbocavernosus)
ISCHIOCAVERNOSUS
Gracilis
Central tendon of perineum
Adductor magnus
DEEP TRANSVERSE PERINEUS
Urogenital diaphragm (cut)
SUPERFICIAL TRANSVERSE PERINEUS
Anus
COCCYGEUS
EXTERNAL ANAL SPHINCTER
Gluteus maximus
Coccyx
ILIOCOCCYGEUS
PUBOCOCCYGEUS
LEVATOR ANI

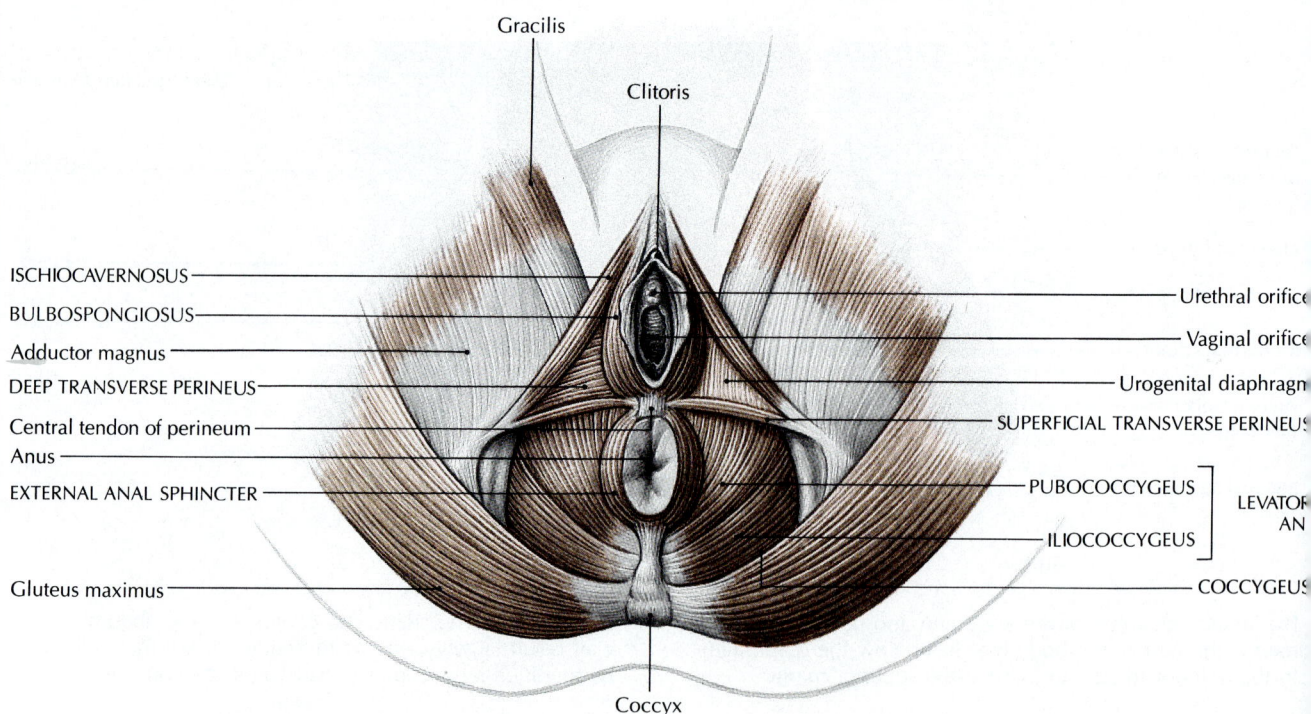

[B] FEMALE

Gracilis
Clitoris
ISCHIOCAVERNOSUS
Urethral orifice
BULBOSPONGIOSUS
Vaginal orifice
Adductor magnus
DEEP TRANSVERSE PERINEUS
Urogenital diaphragm
Central tendon of perineum
SUPERFICIAL TRANSVERSE PERINEUS
Anus
EXTERNAL ANAL SPHINCTER
PUBOCOCCYGEUS
LEVATOR ANI
ILIOCOCCYGEUS
Gluteus maximus
COCCYGEUS
Coccyx

Muscle	Origin	Insertion	Innervation	Action
MUSCLES OF PELVIC DIAPHRAGM (PELVIC FLOOR)				
Levator ani Pubococcygeus	Pubis	Coccyx, anal canal, central tendon of perineum and urethra	Branches of third and fourth sacral nerves and pudendal nerve (second, third, fourth sacral)	Supports pelvic viscera; helps maintain intra-abdominal pressure; acts as sphincter to constrict anus
Iliococcygeus	Ischial spine	Coccyx	Branches of third and fourth sacral nerves and pudendal nerve (second, third, fourth sacral)	Same as pubococcygeus
Puborectalis (pubic sling; most medial portion of levator ani; not illustrated)	Pubic arch	Pubic arch (loops around rectum to form sling)	Branches of fourth sacral nerve and pudendal nerve	Draws the rectum anteriorly (forward); acts as principal component for fecal continence
Coccygeus	Ischial spine	Fifth sacral vertebra, coccyx	Branches of fourth and fifth sacral nerves and pudendal nerve (second, third, fourth sacral)	Same as pubococcygeus
MUSCLE OF PERINEUM (UROGENITAL DIAPHRAGM)				
Superficial transverse perineus	Ischial tuberosity	Central tendon of perineum	Pudendal nerve (second, third, fourth sacral)	Probably acts to stabilize perineum
External sphincter of urethra	Central tendon of perineum	Midline in male; vaginal wall in female (not illustrated)	Pudendal nerve (second, third, fourth sacral)	Under voluntary control, prevents urination when bladder wall contracts; relaxes during urination; can contract to cut off stream of urine
Deep transverse perineus	Ischial rami	Central tendon	Pudendal nerve (second, third, fourth sacral)	Helps support perineum by steadying central perineal tendon
Ischiocavernosus	Ischial tuberosity and rami of pubis and ischium	Corpus cavernosum of penis in males; clitoris in females	Pudendal nerve (second, third, fourth sacral)	Contributes to maintaining erection of penis and clitoris
Bulbospongiosus (bulbocavernosus)	Central tendon of perineum	Fascia of urogenital triangle and penis in males; pubic arch, clitoris in females	Pudendal nerve (second, third, fourth sacral)	In male, contracts to expel last drops of urine, contracts rhythmically during ejaculation to propel semen along urethra; in female, acts as sphincter at vaginal orifice
MUSCLE OF PERINEUM (ANAL REGION)				
External anal sphincter	Anococcygeal raphe	Central tendon of perineum	Pudendal nerve (second, third, fourth sacral)	Constant state of tonic contraction to keep anal orifice closed; closes anus during efforts not associated with defecation

FIGURE 11.16 MUSCLES THAT MOVE THE SHOULDER GIRDLE

[A] Superficial muscles are shown on the right side of the body and deep muscles on the left; anterior view. **[B]** Superficial muscles are shown on the left side of body and deep on right; posterior view.

Three joints contribute to the movements of the shoulder girdle. They are (1) the sternoclavicular joint between the clavicle and the manubrium of the sternum, (2) the acromioclavicular joint between the acromion of the scapula and the clavicle, and (3) the scapulothoracic joint between the subscapularis and serratus anterior muscles attached to the scapula and the thoracic wall. (The scapulothoracic joint is not a true anatomical joint.) The acromioclavicular joint is a sliding joint of accommodation. The movements of the two axes of the sternoclavicular joint are (1) elevation and depression, and (2) protraction and retraction.

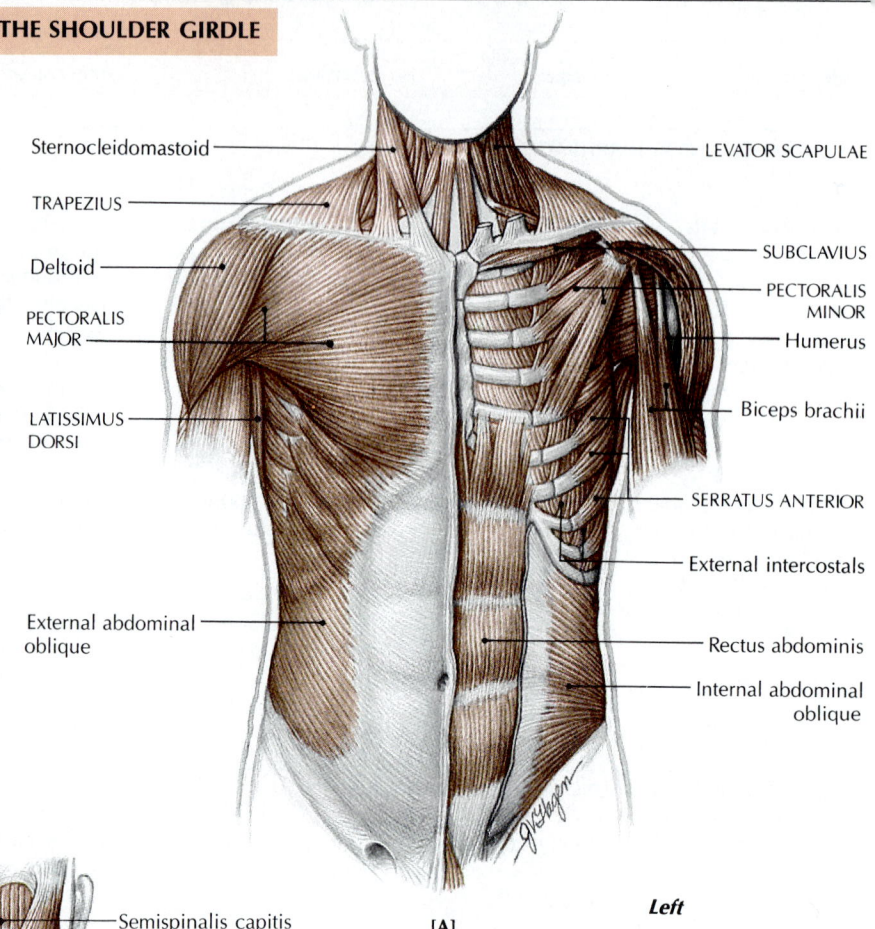

Sternocleidomastoid

TRAPEZIUS

Deltoid

PECTORALIS MAJOR

LATISSIMUS DORSI

External abdominal oblique

LEVATOR SCAPULAE

SUBCLAVIUS

PECTORALIS MINOR

Humerus

Biceps brachii

SERRATUS ANTERIOR

External intercostals

Rectus abdominis

Internal abdominal oblique

Left

[A]

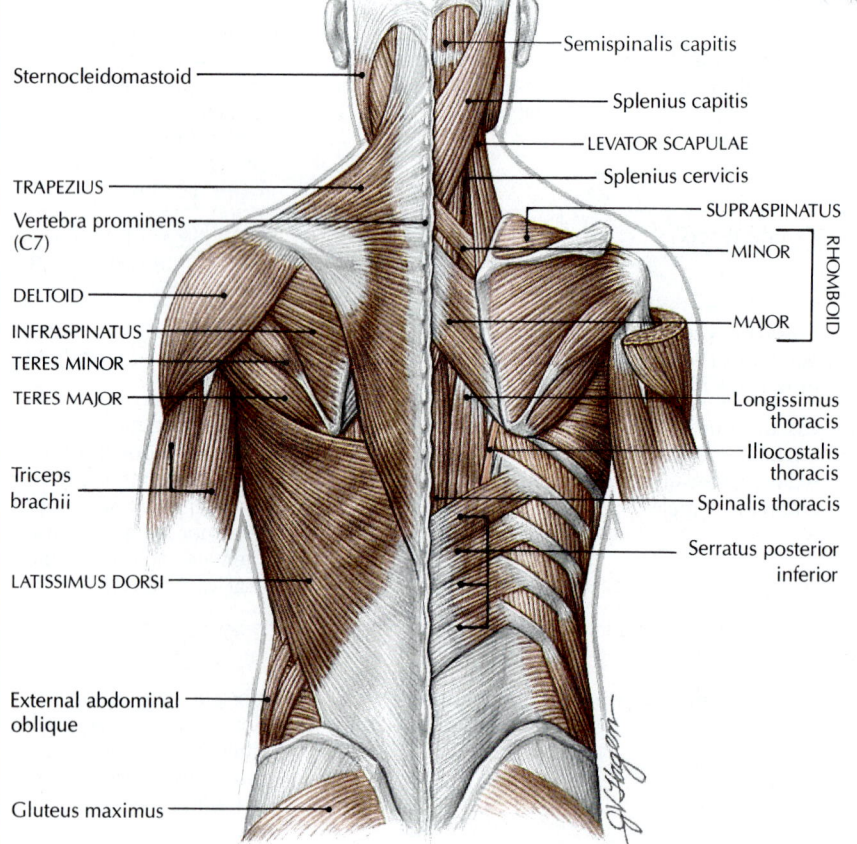

Sternocleidomastoid

TRAPEZIUS

Vertebra prominens (C7)

DELTOID

INFRASPINATUS

TERES MINOR

TERES MAJOR

Triceps brachii

LATISSIMUS DORSI

External abdominal oblique

Gluteus maximus

Semispinalis capitis

Splenius capitis

LEVATOR SCAPULAE

Splenius cervicis

SUPRASPINATUS

MINOR

RHOMBOID

MAJOR

Longissimus thoracis

Iliocostalis thoracis

Spinalis thoracis

Serratus posterior inferior

Left

[B]

The muscles of the shoulder girdle contract together to stabilize the girdle for certain movements, such as lifting a heavy object held in the hand. Muscles active during *elevation* include the upper fibers of the **trapezius, levator scapulae,** and **rhomboids.** Those active during depression include the **pectoralis major, pectoralis minor,** and **latissimus dorsi.** *Upward rotation* of the girdle (as in reaching upward) uses both the upper and lower fibers of the **trapezius** and lower fibers of the **serratus anterior.** *Downward rotation* uses the **levator scapulae, rhomboids major** and **minor, pectoralis major** and **minor,** and the **latissimus dorsi.** *Upward* and *downward rotary movements* of the scapula rotate around an axis located roughly in the center of the scapula just below the middle of the spine. The function of the **subclavius** is to steady the clavicle during shoulder movements. Because of its location, the subclavius can act as a protective cushion between a fractured clavicle and subclavian vein and artery.

Muscle	Origin	Insertion	Innervation	Action
Trapezius	Occipital bone, ligamentum nuchae, spines of C7 to T12 vertebrae	Lateral third of clavicle; acromion and spine of scapula	Spinal accessory motor nerve, C3 to C4 sensory nerves	Steadies, elevates, retracts, and rotates* scapula
Levator scapulae	Transverse processes of vertebrae C1 to C4	Upper vertebral border of scapula	Dorsal scapular nerve	Elevates and rotates* scapula downward
Rhomboids: major and minor	Ligamentum nuchae, spines of C7 to T5 vertebrae	Medial border of scapula	Dorsal scapular nerve	Retract and elevate scapula
Serratus anterior	Ribs 1 to 8, midway between angles and costal cartilages	Entire medial border of scapula	Long thoracic nerve	Steadies, rotates* upward, holds against chest wall, and protracts scapula
Pectoralis minor	Ribs 3, 4, and 5	Coracoid processes of scapula	Medial pectoral nerve	Steadies, depresses, rotates* downward, and protracts scapula
Pectoralis major	Medial half of clavicle (clavicular head) and sternum, ribs 1 to 6 (sternal head)	Lateral crest of intertubercular groove of humerus	Medial and lateral pectoral nerves	Adducts and rotates* arm medially. Clavicular head flexes shoulder joint (raises arm forward), sternal head extends shoulder joint (carries arm backward).
Latissimus dorsi	Spines of T7 to sacral vertebrae, iliac crest, ribs 7 to 12	Intertubercular groove of humerus	Thoracodorsal nerve (middle subscapular nerve)	Depresses, rotates* downward, and retracts scapula; also extends, adducts, and medially rotates humerus at shoulder (glenohumeral joint)
Subclavius	Rib 1 (median)	Clavicle	Nerve to subclavius	Draws clavicle toward sternoclavicular joints; also steadies clavicle

*The axis of rotation of the scapula is located in the center of the scapula. When the glenoid fossa (at shoulder joint) is directed downward, the scapula is said to rotate downward. When the glenoid fossa is directed upward, the scapula is said to rotate upward.

FIGURE 11.17 MUSCLES THAT MOVE THE HUMERUS AT THE SHOULDER JOINT

[A] Superficial muscles; anterior view; right shoulder. **[B]** Superficial muscles; posterior view. **[C]** Deep muscles; anterior view. **[D]** Deep muscles; posterior view. All views show the right upper limb.

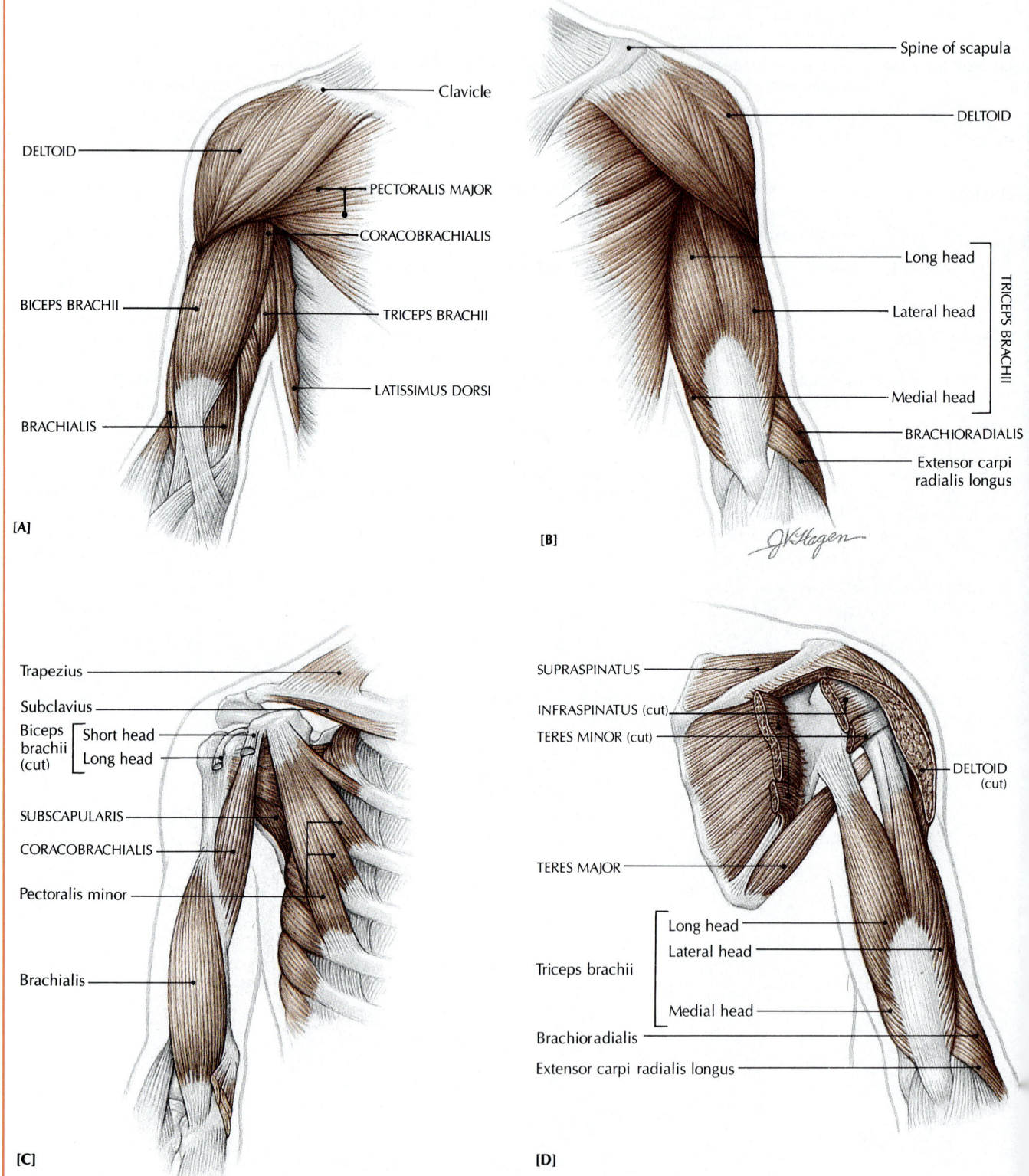

[A]

Clavicle
DELTOID
PECTORALIS MAJOR
CORACOBRACHIALIS
BICEPS BRACHII
TRICEPS BRACHII
LATISSIMUS DORSI
BRACHIALIS

[B]

Spine of scapula
DELTOID
Long head
Lateral head
Medial head
TRICEPS BRACHII
BRACHIORADIALIS
Extensor carpi radialis longus

[C]

Trapezius
Subclavius
Biceps brachii (cut)
Short head
Long head
SUBSCAPULARIS
CORACOBRACHIALIS
Pectoralis minor
Brachialis

[D]

SUPRASPINATUS
INFRASPINATUS (cut)
TERES MINOR (cut)
DELTOID (cut)
TERES MAJOR
Long head
Lateral head
Triceps brachii
Medial head
Brachioradialis
Extensor carpi radialis longus

Many muscles contribute to the movements of the shoulder joint. The four SITS muscles—**supraspinatus, infraspinatus, teres minor,** and **subscapularis**—are the *rotator cuff muscles*. They are located adjacent to the articular capsule of the joint. The tendons of the SITS muscles (also called the *rotator cuff*) prevent instability and dislocations in an otherwise potentially unstable joint. The deltoid muscle overrides the SIT muscles (supraspinatus, intraspinatus, teres minor) and helps form the roundness of the shoulder. Other contributing muscles are noted below.

The shoulder joint has three axes of rotation, resulting in (1) flexion and extension, (2) abduction and adduction, and (3) medial and lateral rotation. A combination of all these movements in succession is *circumduction*.

The *flexors* of the shoulder joint include the **pectoralis major** (clavicular head), **deltoid** (clavicular part), **coracobrachialis,** and **biceps brachii.** The extensors include the **latissimus dorsi, teres major, pectoralis major** (sternal head), **deltoid** (spinous part), and **triceps brachii** (long head). The abductors are the **supraspinatus** and **deltoid** (acromial part), and the adductors are the **pectoralis major** (both heads), **latissimus dorsi, teres major, coracobrachialis,** and **triceps brachii** (long head). The *lateral rotators* include the **infraspinatus, teres minor,** and **deltoid** (spinous part). The *medial rotators* include the **pectoralis major** (both clavicular and sternal heads), **teres major, latissimus dorsi, subscapularis,** and **deltoid** (clavicular part).

Muscle	Origin	Insertion	Innervation	Action
ROTATOR CUFF MUSCLES				
Supraspinatus	Supraspinatus fossa of scapula	Greater tubercle of humerus	Suprascapular nerve	Stabilizes joint; abducts arm
Infraspinatus	Infraspinatus fossa of scapula	Greater tubercle of humerus	Suprascapular nerve	Stabilizes joint; laterally rotates arm
Teres minor	Lower border of lateral scapula	Greater tubercle of humerus	Axillary nerve	Stabilizes joint; laterally rotates arm
Subscapularis	Subscapular fossa of scapula	Lesser tubercle of humerus	Upper and lower subscapular nerves	Stabilizes joint; medially rotates arm
OTHER CONTRIBUTING MUSCLES				
Teres major	Inferior angle of scapula	Medial crest of intertubercular groove of humerus	Lower subscapular nerve	Stabilizes upper arm during adduction; adducts, extends, and medially rotates humerus
Deltoid	Clavicle (lateral third), acromion process and spine of scapula	Deltoid tuberosity of humerus	Axillary nerve	*Clavicular part:* flexes and medially rotates arm; *acromial part:* abducts arm; *spinous part:* extends and laterally rotates arm
Pectoralis major	Medial half of clavicle (clavicular head) and sternum, and ribs 1 to 6 (sternal head)	Lateral aspect of intertubercular groove of humerus	Medial and lateral pectoral nerves	*Clavicular head:* flexes, adducts, and medially rotates arm; *sternal head:* extends, adducts, and medially rotates arm
Latissimus dorsi	Spines of lower thoracic, lumbar, and sacral vertebrae, iliac crest, and ribs 7 to 12	Intertubercular groove of humerus	Thoracodorsal (middle subscapular) nerve	Extends, adducts, and medially rotates arm; also depresses, retracts, and rotates scapula downward
Coracobrachialis	Coracoid process of scapula	Middle third of humerus	Musculocutaneous nerve	Flexes and adducts arm

Pitching Off the Cuff

Baseball pitchers are very familiar with the anatomy of the shoulder, especially the rotator cuff muscles, which are often damaged, along with associated cartilage, during strenuous pitching. The rotator cuff muscles enable a pitcher to throw fast and accurate pitches, help stabilize the upper arm in the shoulder socket, and also help fine-tune the other shoulder muscles involved in pitching (see drawings). In fact, when the rotator cuff muscles tear or become frayed, it is this fine tuning that is most affected.

The *subscapularis muscle* keeps the pitcher's arm from extending too far back at the start of the pitch (see photo) and begins the forward acceleration of the arm. Surgery has been fairly successful in repairing injured rotator cuff muscles—successful enough to allow a professional pitcher to return to action. The *supraspinatus muscle* helps bring the arm up, assists it through the pitching motion, and keeps the arm from coming out of the shoulder socket (see photo). The greatest strain is placed on the *infraspinatus* and *teres minor muscles,* since

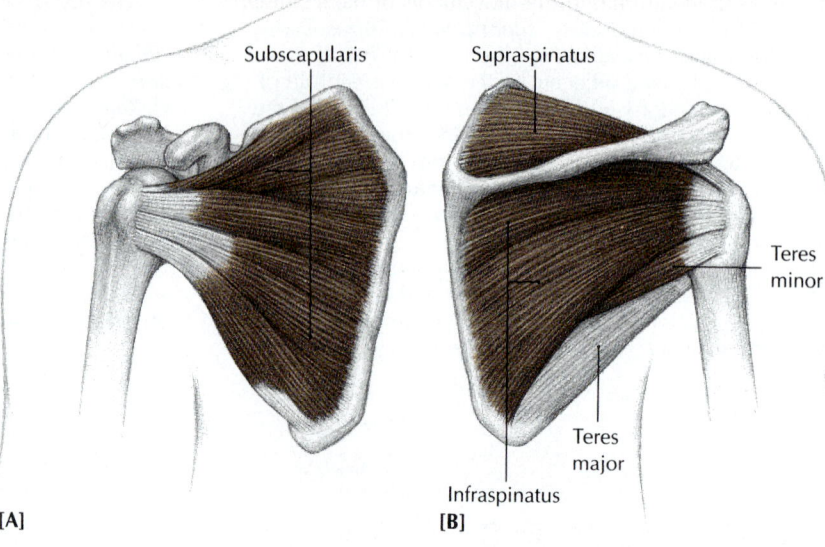

The rotator cuff muscles. [A] Anterior view; [B] posterior view.

they have to slow down the pitching arm at the end of a very forceful motion (see photo). As a result, the infraspinatus and teres minor are frequently damaged.

These two muscles also help cock the arm in the early phase of the pitching motion and are active when the ball is released.

Subscapularis

Supraspinatus

Infraspinatus, teres minor

FIGURE 11.18 MUSCLES THAT MOVE THE FOREARM AND WRIST

[A] Superficial muscles of right hand; ventral view. **[B]** Superficial muscles; dorsal view.

At the one-axis elbow joint, flexion and extension are the only movements. The *flexors* of the joint include the ***biceps brachii, brachialis,*** and ***brachioradialis.*** The *extensors* are the ***triceps brachii*** and the tiny ***anconeus.***

At the proximal and distal radioulnar joints, pronation and supination are the only movements. The *pronators* include the ***pronator teres*** and the ***pronator quadratus.*** The *supinators* are the ***supinator*** and the ***biceps brachii.*** Supination is more powerful than pronation. That is why screws, caps on jars, and door handles are designed the way they are. (For left-handed people, the motion involved in moving these objects is *pronation*.)

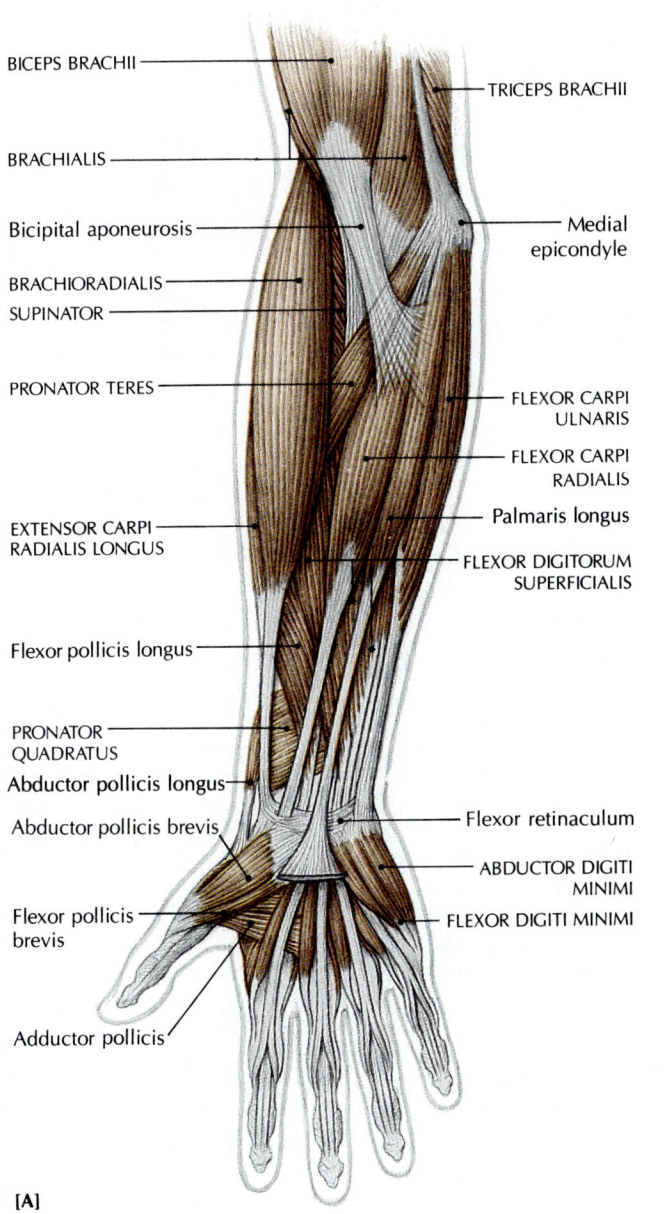

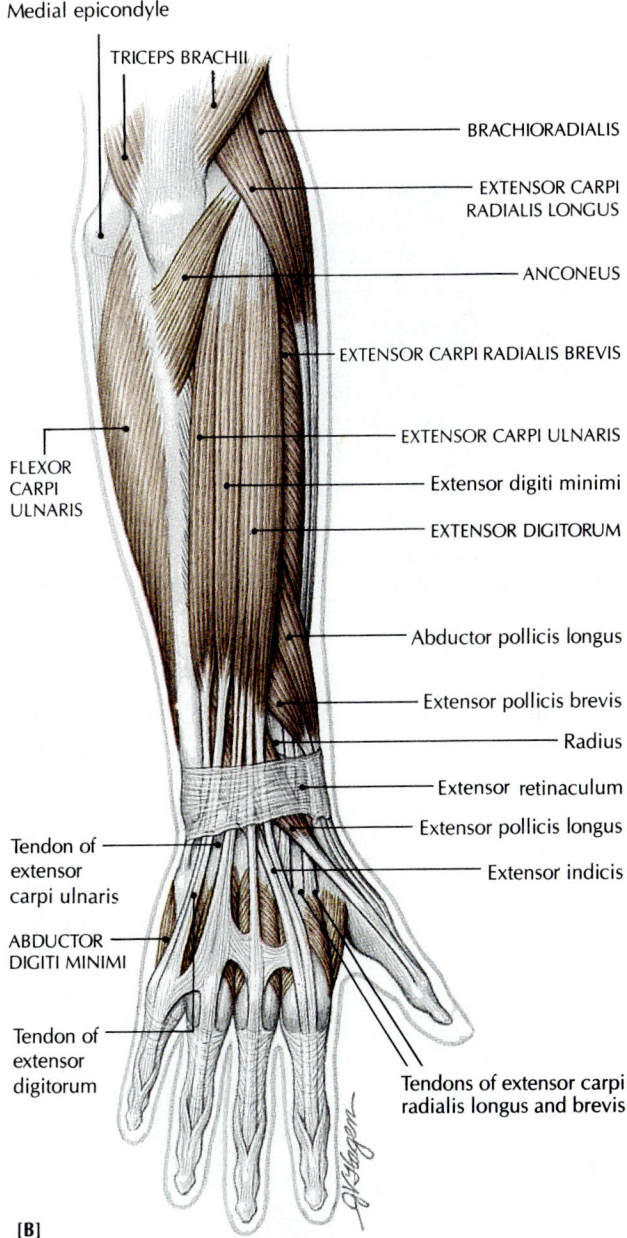

[A]

[B]

(Figure 11.18 continues on following pages)

FIGURE 11.18 MUSCLES THAT MOVE THE FOREARM AND WRIST *(Continued)*

[C] Superficial dissection of right forearm and hand; dorsal view. **[D]** Deep muscles; ventral view.

The wrist joints include the (1) radiocarpal joint (between the radius and proximal row of carpal bones), where most of the movements occur, and (2) the midcarpal joint (between the proximal and distal rows of carpal bones), where a slight amount of flexion and extension occurs. There is no movement between the distal carpal bones and the metacarpals. The radiocarpal joint is a two-axis condyloid joint where (1) flexion and extension (and hyperextension) and (2) abduction (radial deviation, toward the radius) and adduction (ulnar deviation, toward the ulna) occur.

The *flexor* muscles of the wrist joint include the **flexor carpi ulnaris, flexor carpi radialis,** and the **palmaris longus.** The *extensor* muscles are the **extensor carpi radialis longus, extensor carpi radialis brevis,** and the **extensor carpi ulnaris.** The *abductor* muscles include the **flexor carpi radialis, extensor carpus radialis longus** and **brevis,** and **abductor pollicis longus.** The *adductor* muscles are the **extensor carpi ulnaris** and the **flexor carpi ulnaris.** The contraction of any of these muscles in a sequential order can produce circumduction. Contracting in concert, these muscles act to stabilize the wrist for the effective use of the fingers.

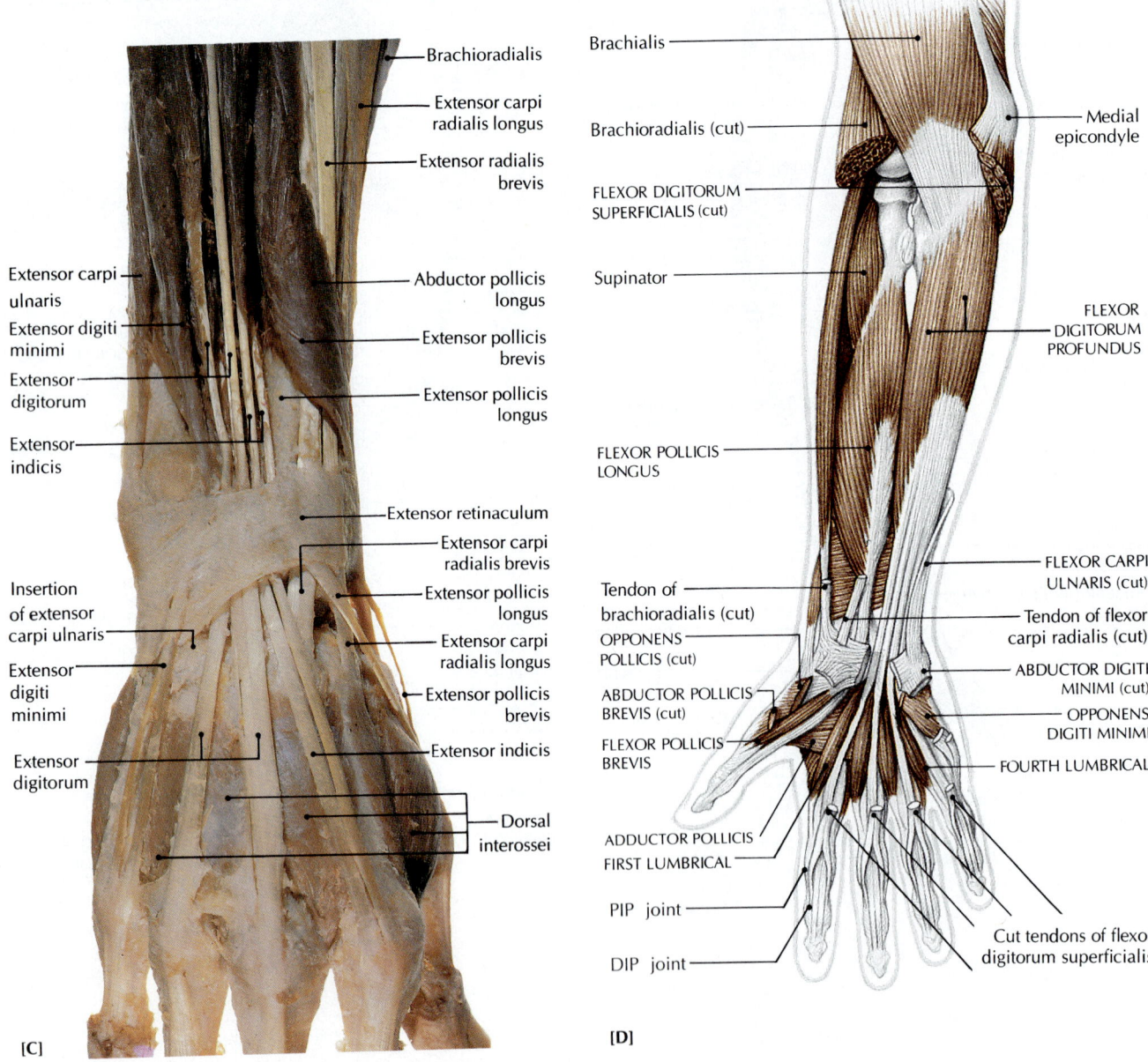

[C]

[D]

Muscle	Origin	Insertion	Innervation	Action
Biceps brachii	*Long head:* supraglenoid tubercle of scapula *Short head:* coracoid process of scapula	Tuberosity of radius, bicipital aponeurosis	Musculocutaneous nerve	Flexes elbow; supinates forearm
Brachialis	Anterior surface of humerus	Tuberosity and coronoid process of ulna	Musculocutaneous nerve	Flexes elbow
Triceps brachii	*Long head:* infraglenoid tubercle of scapula *Lateral head:* posterior and lateral humerus above radial groove *Medial head:* posterior humerus below radial groove	Olecranon process of ulna	Radial nerve	Extends elbow
Supinator	Lateral epicondyle of humerus, supinator crest of ulna	Lateral surface of radius (distal to head)	Radial nerve	Supinates forearm (rotates palm of hand anteriorly)
Pronator teres	Medial epicondyle of humerus, coronoid process of ulna	Midlateral surface of radius	Median nerve	Pronates forearm; flexes elbow
Pronator quadratus	Anterior distal end of ulna	Anterior distal end of radius	Median nerve	Pronates forearm
Brachioradialis	Lateral supracondylar ridge of humerus	Lateral surface of radius near base of styloid process	Radial nerve	Flexes elbow
Anconeus	Posterior lateral epicondyle of humerus	Lateral surface of olecranon process of ulna	Radial nerve	Extends elbow
Flexor carpi radialis	Medial epicondyle of humerus	Base of second and third metacarpals	Median nerve	Flexes and abducts wrist
Flexor carpi ulnaris	Medial epicondyle of humerus, olecranon process of ulna	Pisiform, hamate, base of fifth metacarpal	Ulnar nerve	Flexes and adducts wrist
Extensor carpi radialis longus	Lateral epicondyle of humerus	Base of second metacarpal	Radial nerve	Extends and abducts wrist
Extensor carpi radialis brevis	Lateral epicondyle of humerus	Base of third metacarpal	Radial nerve	Extends and abducts wrist
Extensor carpi ulnaris	Lateral epicondyle of humerus	Base of fifth metacarpal	Radial nerve	Extends and adducts wrist

FIGURE 11.19 MUSCLES THAT MOVE THE THUMB

Muscles of the back of the thumb; medial side of right hand.

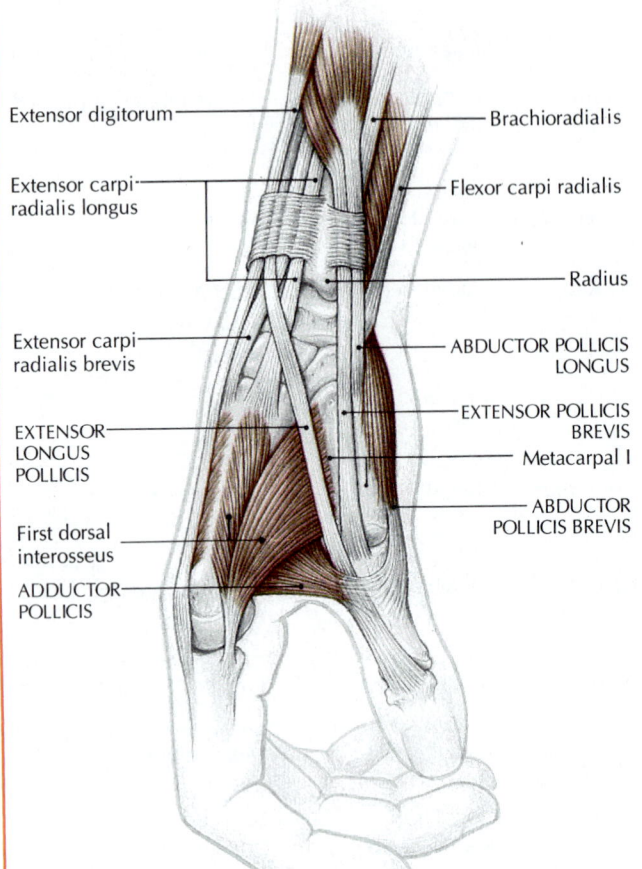

Extensor digitorum
Brachioradialis
Extensor carpi radialis longus
Flexor carpi radialis
Radius
Extensor carpi radialis brevis
ABDUCTOR POLLICIS LONGUS
EXTENSOR LONGUS POLLICIS
EXTENSOR POLLICIS BREVIS
Metacarpal I
First dorsal interosseus
ABDUCTOR POLLICIS BREVIS
ADDUCTOR POLLICIS

The movements of the thumb take place at (1) the carpometacarpal joint, which is a saddle joint with three axes of rotations, and (2) two hinge joints involving the phalanx. The movements at the saddle joint include (1) flexion and extension, (2) abduction and adduction, and (3) opposition and reposition. Crucial to the opposability of the thumb are opposition and reposition, and abduction and adduction. In opposition the metacarpal bone rolls toward the midline of the hand so that the ball of the thumb's distal phalanx can touch the balls of the distal phalanges of the other fingers. The muscular mound at the base of the thumb, the *thenar eminence*, is formed by the three thenar muscles: **abductor pollicis brevis, flexor pollicis brevis,** and **opponens pollicis.** At the carpometacarpal joint, the *flexor* is the flexor pollicis brevis. The *extensors* are the **extensor pollicis longus, extensor pollicis brevis,** and **abductor pollicis longus.** The *adductor* is the **adductor pollicis.** The *opposition* movement is by the **opponens pollicis** and the *reposition* movements by the **abductor pollicis longus** and **extensor pollicis longus** and **brevis.** At the phalangeal hinge joints, the *flexors* are the **flexor pollicis longus** and **flexor pollicis brevis,** and the *extensors* are the **extensor pollicis longus** and **brevis.**

Several features of the thumb contribute immensely to the versatility of the hand. The length of the thumb and the presence of the flexor pollicis longus and the opponens muscles permit the opposition movement and the great flexibility in using the thumb along with the other fingers.

The little finger is associated with the *hypothenar eminence*, formed by the flexor digiti minimi, abductor digiti minimi, and opponens muscles. They have roles similar to their counterparts of the thenar eminence.

Muscle	Origin	Insertion	Innervation	Action
Flexor pollicis longus	Anterior surface of radius; interosseous membrane	Base of distal phalanx of thumb	Median nerve	Flexes thumb
Flexor pollicis brevis	Trapezium and adjacent region	Base of proximal phalanx of thumb	Median nerve	Flexes thumb; helps in opposition and reposition
Opponens pollicis	Trapezium and adjacent region	First metacarpal of thumb	Median nerve	Rolls thumb toward midline of palm (opposition)
Abductor pollicis brevis	Scaphoid and trapezium bones and adjacent region	Base of proximal phalanx of thumb	Median nerve	Abducts thumb; helps in opposition
Abductor pollicis	Second and third metacarpals and capitate bones	Base of proximal phalanx of thumb	Ulnar nerve	Adducts, flexes thumb
Extensor pollicis longus	Dorsal surface of ulna; interosseous membrane	Base of distal phalanx of thumb	Radial nerve	Extends thumb; helps in reposition
Extensor pollicis brevis	Dorsal surface of radius; interosseous membrane	Base of proximal phalanx of thumb	Radial nerve	Extends thumb; helps in reposition, adduction
Abductor pollicis longus	Dorsal surfaces of ulna and radius	Base of first metacarpal bone	Radial nerve	Abducts and extends thumb; helps in reposition

The Incredible Hand

The human hand is a wonderful but paradoxical piece of anatomy. It can be clenched into a fist to bloody an opponent's nose, or it can be a sensitive and precise instrument that guides a surgeon's scalpel or a poet's pen. The hand combines mobility, muscle strength, dexterity, and sensitivity. The hand is so important that the primary role of the upper extremity is to place the hand in a position where it can perform its specific function. The sensitivity of the hand is noted when sensory information is used to perceive form, shape, and texture through the fingertips. The sensitive fingertips allow us to appreciate the smooth touch of silk. This same sensitivity enables a blind person to read Braille.

Anatomically, the hand is a relatively generalized and primitive structure. It is capable of only a limited range of movements, although there is much versatility within that range. The many purposeful actions of our hands are controlled by the brain and accomplished through the hand's various anatomical features.

The Hand Is Capable of Several Grips

The act of seizing or grasping an object is expressed by several different *grips*. All grips are variants of the power grip and the precision grip, and the two are combined in some activities. A major variant is the hook grip.

In the ***power grip***—the grip used in holding a hammer or climbing a rope, for example—the fingers are flexed around the object, with counterpressure from the thumb. Once the grip is made, the hand itself cannot move, and any movement is due to other joints of the upper limb. The power grip is strongest when the hand is held dorsiflexed. This is true because the tendons of the long flexor muscles of the fingers, such as the flexor digitorum profundus, have their origins in the forearm, and in the dorsiflexed position all the contractile energy of these muscles is exerted in flexing the fingers. In contrast, the power grip is weakest when the hand is palmarflexed, since in that position the tendons of the long flexor muscles are slack, and much of the contractile energy goes toward taking up that slack, with relatively little energy going into the actual flexing of the fingers. This explains why martial arts athletes flex the adversary's hand to loosen his or her power grip.

In the ***precision grip***—the grip used in grasping a pen or the bow of a violin, for example—the object is gripped between the tips of the index finger and thumb and is usually supported by the first digit of the middle finger. The small movements of the digits are usually, but not always, the main source of the delicate movements executed. The positioning of the object is carried out by movements of the wrist, forearm, and other parts of the upper limb. The everyday activities of picking up a small object for inspection or manipulation may actually combine the power and precision grips. For example, when you manipulate thread through the eye of a needle, you hold the needle still with a power grip and ease the thread through with a precision grip.

The ***hook grip*** is used to suspend or pull on objects—as in "chinning" on a horizontal bar or lifting a briefcase by its handle. The hook grip is not used for skillful manipulations. The fingers are looped around the object and flexed toward the palm; the thumb may or may not be involved.

FIGURE 11.20 MUSCLES THAT MOVE THE FINGERS (EXCEPT THE THUMB)

Deep intrinsic muscles of the right hand. **[A]** Palmar interossei; palmar view. **[B]** Dorsal interossei; dorsal view. **[C]** Lumbricals; palmar view.

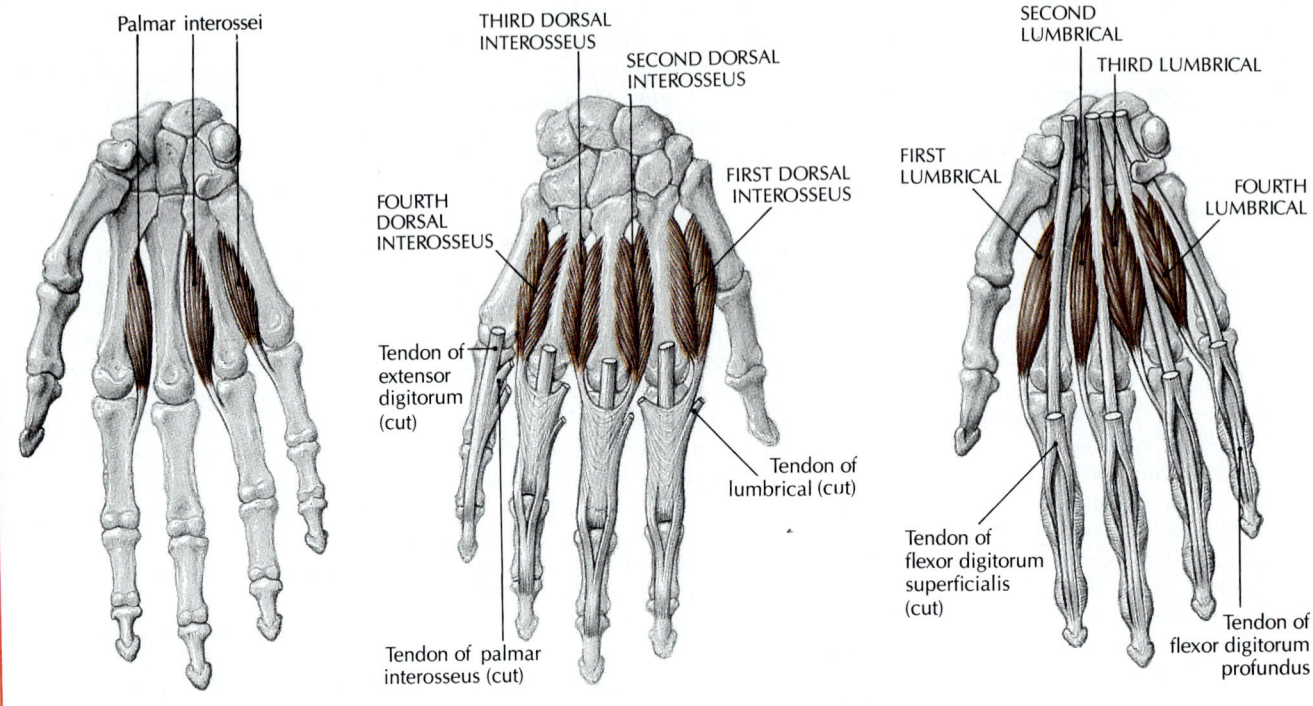

[A] PALMAR INTEROSSEI   **[B]** DORSAL INTEROSSEI   **[C]** LUMBRICALS

The movements of the fingers (except the thumb) are based on a remarkable interplay of several muscle groups, and especially on the role of the lumbrical muscles. To fully appreciate this interplay we must understand the movements at each joint and the anatomical relationships and actions of the tendons and muscles acting on those joints.

Each finger has three joints: (1) the *metacarpophalangeal joint* (MP or knuckle joint) between the metacarpal and proximal phalanges, (2) the *proximal interphalangeal joint* (PIP joint) between the proximal and middle phalanges, and (3) the *distal interphalangeal joint* (DIP joint) between the middle and distal phalanges. The metacarpophalangeal joint is a condyloid joint with two axes of rotation, permitting flexion and extension, abduction and adduction, and circumduction. The proximal and distal interphalangeal joints are hinge joints with one axis of rotation, permitting only flexion and extension.

The *flexors* of the metacarpophalangeal joints are the seven **interosseous muscles** and four **lumbrical muscles,** and the *extensor* is the **extensor digitorum** (including the extensor digitorum indicus and digitorum minimus). The *abductors* and *adductors* are the interosseous muscles and the abductor digiti minimi. The four **dorsal interossei** abduct (spread) the fingers (except the little finger*), and the three

palmar interossei adduct the fingers (bring them back together). The movements of abduction and adduction of the fingers are defined in relation to the third (middle) finger, which acts as the axis of the hand. Thus, spreading of the fingers away from the third finger is abduction, and movement toward the middle finger is adduction. The **abductor digiti minimi** abducts (spreads) the little finger away from the fourth finger. Circumduction combines all of these actions in sequence. Because the **flexor digitorum profundus** and **flexor digitorum superficialis** pass on the palmar side of the MP joint, they contribute to *flexion* at the joint. The *extension* of the MP joint is carried out by the **extensor digitorum muscles,** whose tendons attach to the base of each proximal phalanx on the dorsal side of the axis. The *extensors* for the index finger and little finger are the **extensor digitorum indicus** and the **extensor digitorum minimus,** respectively.

The flexor of the proximal interphalangeal joint is the **flexor digitorum superficialis,** with assistance from the

---

*Each finger has its own anatomical name. The thumb is the *pollex* (L. thumb), next to it is the *index* finger (L. pointer), then the *medius* or middle finger, *annularis* (L. ring) or fourth finger, and *minimus* (L. least) or little finger.

flexor digitorum profundus. The *flexor* of the distal interphalangeal joint is the **flexor digitorum profundus.** The *extensors* of the PIP and DIP joints are the interossei and lumbricals, with minor assistance from the extensor digitorum.

The "paradox" of the lumbrical and interosseous muscles acting to flex the MP joint and to extend the PIP and DIP joints is explained by the relationship of the tendons of these muscles to the joint. The tendons of these muscles (1) pass on the palmar side of the MP joint (flexing the joint) and (2) continue to join the extensor tendon on the dorsal side of each of the proximal phalanges II to IV (extending the PIP and DIP joints).

The roles of the lumbrical and interosseous muscles are critical in the extension of the PIP and DIP joints, and in the coordination of the finger movements. Although the tendons of the extensor digitorum muscles continue beyond the MP joint and attach to the base of the middle and distal phalanges on the dorsal sides, they contribute minimally to the extension of the PIP and DIP joints. This occurs because the force generated by the contraction of the attached muscles is almost completely expended at the MP joint. This complex arrangement between the flexors and extensors of the interphalangeal joints is all-important in producing delicate finger movements. The extensor activities of the lumbricals (largely) and interossei allow those muscles to act as effective antagonists in adjusting the activity of the flexor muscles during writing and other fine finger movements.

Muscle	Origin	Insertion	Innervation	Action
Flexor digitorum superficialis	Medial epicondyle of humerus, coronoid process of ulna, anterior border of radius	Both sides of middle phalanges of fingers 2 to 5	Median nerve	Flexes PIP and MP joints*
Flexor digitorum profundus	Anterior and medial surface of ulna	Palmar surfaces of distal phalanges of fingers 2 to 5	Median nerve, ulnar nerve	Flexes PIP, DIP, and MP joints*
Interossei (4 dorsal interossei and 3 palmar interossei)	Metacarpal bones	Bases of proximal phalanges and extensor expansions of fingers	Ulnar nerve	*Dorsal interossei:* abduct fingers; *palmar interossei:* adduct fingers, extend PIP and DIP joints, flex MP joints†
Lumbricals (4 muscles)	Tendons of flexor digitorum profundus in palm of hand	Extensor tendon (expansions) of digits distal to knuckles	Lateral two by median nerve; medial two by ulnar nerve	Flex MP joint (knuckles); extend PIP and DIP joints†
Extensor digitorum	Lateral epicondyle of humerus	Dorsal surface of phalanges in fingers 2 to 5	Radial nerve	Extends MP joint (and PIP slightly)
Flexor digiti minimi	Hook of hamate bone and adjacent region	Medial side of proximal phalanx of little finger	Ulnar nerve	Flexes proximal phalanx of little finger (MP joint)
Opponens digiti minimi	Hook of hamate bone and adjacent region	Medial half of palmar surface of little finger	Ulnar nerve	Rotates little finger toward midline of hand (opposition)
Abductor digiti minimi	Pisiform bone and adjacent region	Medial side of proximal phalanx of little finger	Ulnar nerve	Abducts little finger away from fourth finger

*Because of the pull they can generate, the long flexor muscles can help flex joints more proximal than that of their primary action.

†The extensor digitorum is not the primary extensor of the PIP and DIP joints because it expends most of its force at the MP joint. The lumbricals and interossei extend the PIP and DIP joints through their attachments to the extensor tendons.

FIGURE 11.21 MUSCLES THAT MOVE THE FEMUR AT THE HIP JOINT

[A] Right anterior view. [B] Right posterior view.

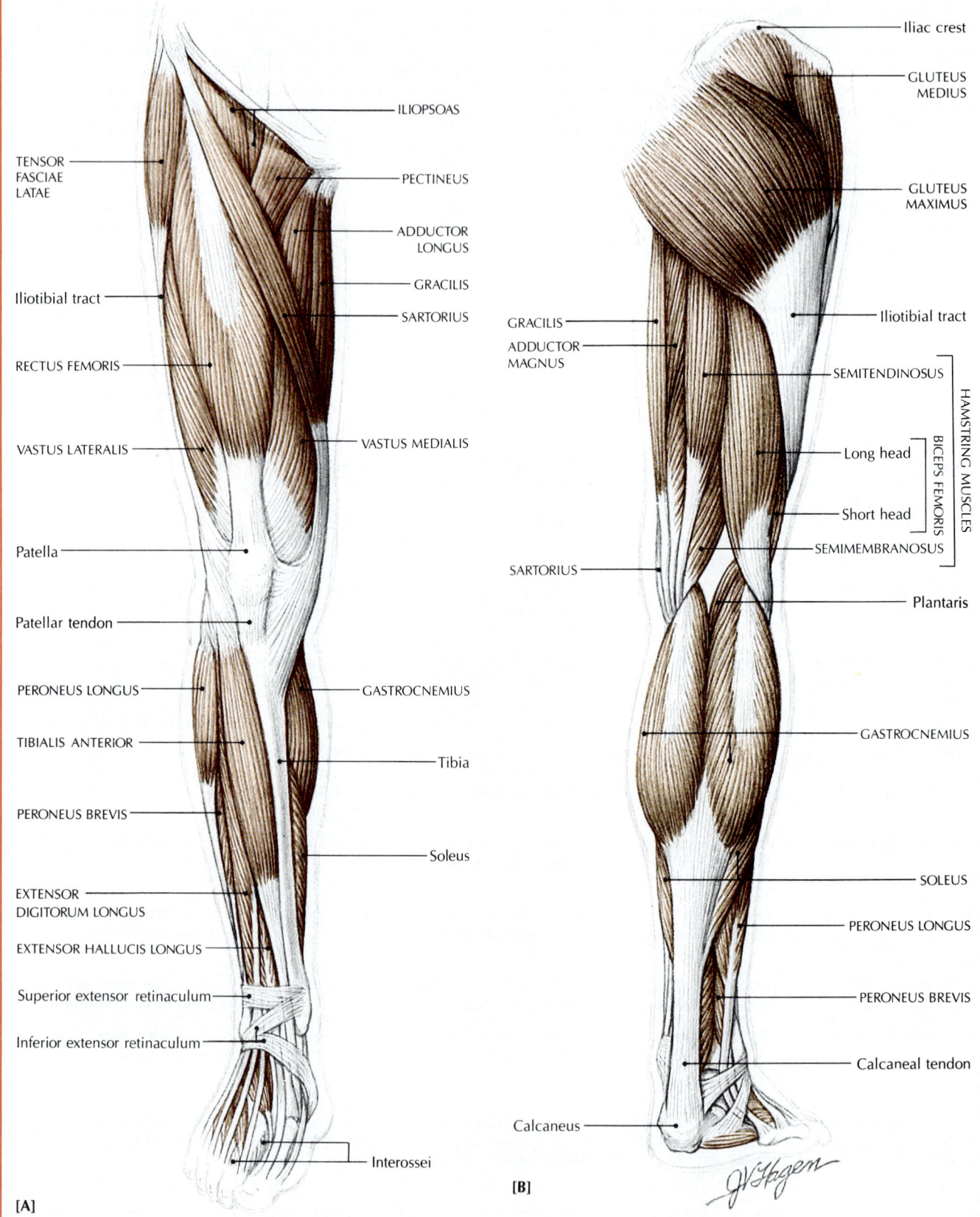

[A] (anterior view labels)
ILIOPSOAS
TENSOR FASCIAE LATAE
PECTINEUS
ADDUCTOR LONGUS
GRACILIS
Iliotibial tract
SARTORIUS
RECTUS FEMORIS
VASTUS LATERALIS
VASTUS MEDIALIS
Patella
Patellar tendon
PERONEUS LONGUS
GASTROCNEMIUS
TIBIALIS ANTERIOR
Tibia
PERONEUS BREVIS
Soleus
EXTENSOR DIGITORUM LONGUS
EXTENSOR HALLUCIS LONGUS
Superior extensor retinaculum
Inferior extensor retinaculum
Interossei

[B] (posterior view labels)
Iliac crest
GLUTEUS MEDIUS
GLUTEUS MAXIMUS
Iliotibial tract
GRACILIS
ADDUCTOR MAGNUS
SEMITENDINOSUS
Long head
Short head
BICEPS FEMORIS
HAMSTRING MUSCLES
SEMIMEMBRANOSUS
SARTORIUS
Plantaris
GASTROCNEMIUS
SOLEUS
PERONEUS LONGUS
PERONEUS BREVIS
Calcaneal tendon
Calcaneus

The ball-and-socket hip joint, with its three axes of rotation, is capable of (1) flexion and extension, (2) abduction and adduction, (3) medial and lateral rotation, and (4) circumduction. The powerful muscles surrounding the hip joint are active during walking, running, and climbing. They also act to stabilize the joint.

The **hamstring muscles (semitendinosus, semimembranosus,** and **biceps femoris)** act as extensors of the hip and flexors of the knee during walking on the level; these muscles relax as the hip flexors contract. The **gluteus maximus** acts as an extensor of the hip only when power is required, as in moving against gravity. For example, it is used in rising from a seated position, in climbing upstairs or up a steep hill, and in running. The **psoas major** contributes to the stability of the hip and is a powerful flexor. (The psoas major and the **iliacus** are sometimes considered together as

the **iliopsoas** muscle.) The **tensor fasciae latae** and gluteus maximus pull on the iliotibial tract (a modified fascial sheath located on the lateral thigh) extending beyond the knee joint. This muscle functions to brace the knee so the joint doesn't buckle while the other foot is off the ground during walking. The **adductor magnus** and **adductor longus** are powerful muscles used in kicking a soccer ball, for instance.

During locomotion the hip remains parallel to the ground through the action of the **gluteus medius** muscle. If this muscle is paralyzed on one side, a lurching gait results. If you place your hand on the gluteus muscle while walking, you will note that it contracts while the limb is grounded and relaxes when the limb is free and in motion. The contraction pulls down on the pelvis and prevents the opposite side from dropping, thereby preventing a lurch.

Muscle	Origin*	Insertion*	Innervation	Action
Iliopsoas†	*Iliacus:* iliac fossa of false pelvis *Psoas:* lumbar vertebrae	Lesser trochanter of femur	*Iliacus:* femoral nerve *Psoas:* directly by L2 and L3	Flexes hip joint; can flex and rotate vertebral column
Pectineus	Superior ramus of pubis	Pectineal line below lesser trochanter on posterior femur	Femoral and obturator nerves	Flexes and adducts hip joint
Adductor longus	Body of pubis	Middle third of linea aspera of femur	Obturator nerve	Adducts and flexes hip joint
Adductor brevis	Body and inferior ramus of pubis	Proximal part of linea aspera of femur	Obturator nerve	Adducts and helps flex hip joint
Adductor magnus	Inferior ramus of pubis to ischial tuberosity	Middle of linea aspera, adductor tubercle of femur	Obturator nerve and tibial portion of sciatic nerve	Adducts and helps flex and extend hip joint
Gluteus maximus	Iliac crest, sacrum, coccyx	Iliotibial tract of tensor fasciae latae and gluteal tuberosity of femur	Inferior gluteal nerve	Extends and laterally rotates hip joint
Gluteus medius	Ilium	Greater trochanter of femur	Superior gluteal nerve	Abducts and medially rotates hip joint; steadies pelvis
Gluteus minimus (not illustrated)	Ilium	Greater trochanter of femur	Superior gluteal nerve	Abducts and medially rotates hip joint; steadies pelvis
Tensor fasciae latae	Iliac crest, anterior superior iliac spine	Lateral condyle of tibia through iliotibial tract	Superior gluteal nerve	Abducts, flexes, and medially rotates hip joint; steadies trunk; extends knee joint
Small lateral rotators (piriformis, obturator internus, quadratus femoris)	Sacrum, obturator foramen, ischial tuberosity	Greater trochanter of femur	Branch from sacral plexus for each muscle	Laterally rotates and adducts hip joint

*The origin and insertion of a muscle acting on the hip and knee joints are reversible, depending on the movement. The insertion is on the femur when the thigh is moved and the pelvis is fixed in position, whereas the insertion is on the pelvis when the pelvis is moved and the thigh is fixed.

†The iliopsoas muscle is composed of an iliacus muscle and a psoas muscle.

FIGURE 11.22 MUSCLES THAT ACT AT THE KNEE JOINT

[A] Right anterior view. [B] Right posterior view.

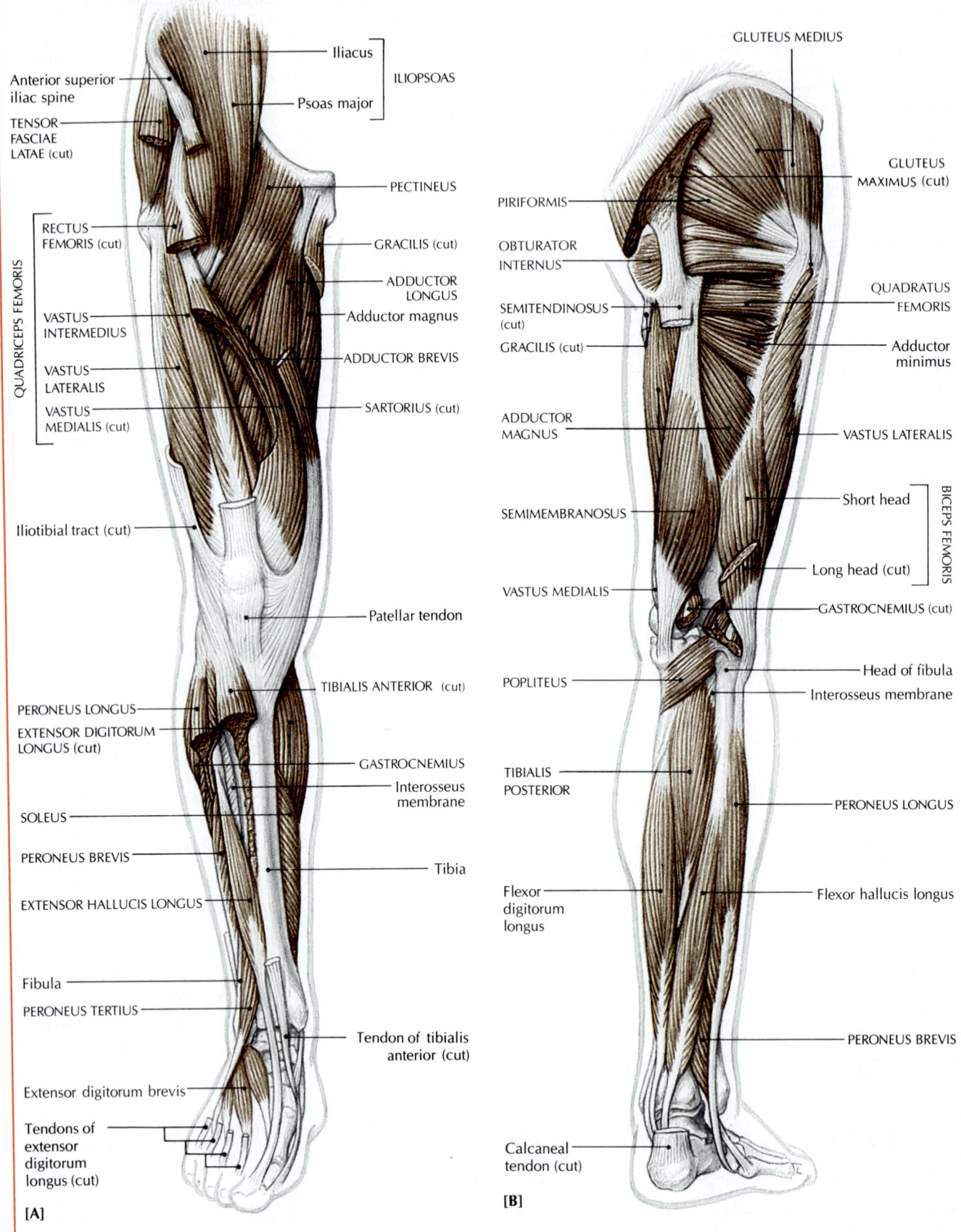

[A]

[B]

The knee is a hinge joint capable of flexion and extension, with a slight amount of medial and lateral rotation when not fully extended. During the final stage of full extension during walking, the femur rotates medially in relation to the tibia when the foot is planted on the ground. Try it. This is known as "locking" the knee to make it rigid. If the knee is extended while the limb is free, as when punting a football, the tibia rotates laterally in relation to the femur to lock the knee. Unlocking of the knee by the **popliteus muscle** is the first stage of flexion of the knee joint. The **hamstring muscles** are flexors **(biceps femoris, semimembranosus, semitendinosus),** lateral rotators **(biceps femoris),** and medial rotators **semitendinosus** and **semimembranosus).** The **sartorius, gracilis,** and **gastrocnemius** contribute to flexion.

The **quadriceps femoris muscle** is made up of four parts. The patella, the bone to which the quadriceps femoris is attached, acts to place the attachment of the patella and quadriceps femoris further away from the flexion-extension axis of rotation. Such an adjustment increases the length of the "lever arm" of the joint by increasing the leverage and power of the strong quadriceps femoris.

Four bursae are located in the vicinity of the patella, and are important to knee movement. The **suprapatellar bursa** is located just above the patella, between the femur and tendon of the quadriceps femoris muscle. It permits free movement of the quadriceps tendon over the distal end of the femur and facilitates the full range of extension and flexion of the knee joint. The **infrapatellar bursa** is located just below the patella, between the tibia and the patellar tendon. The **subcutaneous prepatellar bursa** lies between the skin and the anterior surface of the patella. It allows free movement of the skin over the underlying patella during flexion of the knee joint, for example. The subcutaneous prepatellar bursa can become the victim of *friction bursitis,* as a result of rubbing between the skin and patella. When inflammation is chronic, the bursa can become distended with fluid in individuals who frequently work on all fours. The **subcutaneous infrapetellar bursa** lies between the skin and the fascia anterior to the tibial tuberosity. It allows the skin to glide over the tibial tuberosity and to withstand pressure when kneeling with the trunk upright (when praying, for example).

Muscle	Origin	Insertion	Innervation	Action
QUADRICEPS FEMORIS				
Rectus femoris	Anterior inferior iliac spine	Patella, through patellar tendon to tibial tuberosity	Femoral nerve	Extends leg at knee joint; flexes thigh at hip joint
Vastus lateralis	Greater trochanter and linea aspera of femur	Patella, through patellar tendon to tibial tuberosity	Femoral nerve	Extends leg at knee joint
Vastus intermedius	Anterior femur	Patella, through patellar tendon to tibial tuberosity	Femoral nerve	Extends leg at knee joint
Vastus medialis	Linea aspera of femur	Patella, through patellar tendon to tibial tuberosity	Femoral nerve	Extends leg at knee joint
HAMSTRING MUSCLES				
Biceps femoris	*Long head:* ischial tuberosity *Short head:* linea aspera of femur	Head of fibula and lateral condyle of tibia	*Long head:* tibial nerve *Short head:* common peroneal nerve	Flexes and laterally rotates knee joint; extends thigh at hip joint
Semitendinosus	Ischial tuberosity	Medial surface of proximal tibia	Tibial nerve	Flexes and medially rotates knee joint; extends thigh at hip joint
Semimembranosus	Ischial tuberosity	Medial condyle of tibia	Tibial nerve	Flexes and medially rotates knee joint; extends thigh at hip joint
OTHER MUSCLES				
Gracilis	Body and inferior ramus of pubis	Medial surface of proximal tibia	Obturator nerve	Flexes and medially rotates knee joint; adducts thigh at hip joint
Sartorius	Anterior superior iliac spine	Medial surface of proximal tibia	Femoral nerve	Flexes leg at knee joint and thigh at hip joint
Popliteus	Lateral condyle of femur	Posterior surface of tibia	Tibial nerve	Flexes, rotates leg at knee joint; unlocks knee joint

*In the knee joint, the origin and insertion can be reversed depending upon whether the leg and foot are moved in relation to the thigh, as when one is seated and the leg freely moves, or whether the foot is planted and the thigh does much of the moving, as in squatting exercises or in rising from the seated position.

FIGURE 11.23 MUSCLES THAT MOVE THE FOOT

[A] Right lateral view. [B] Right medial view.

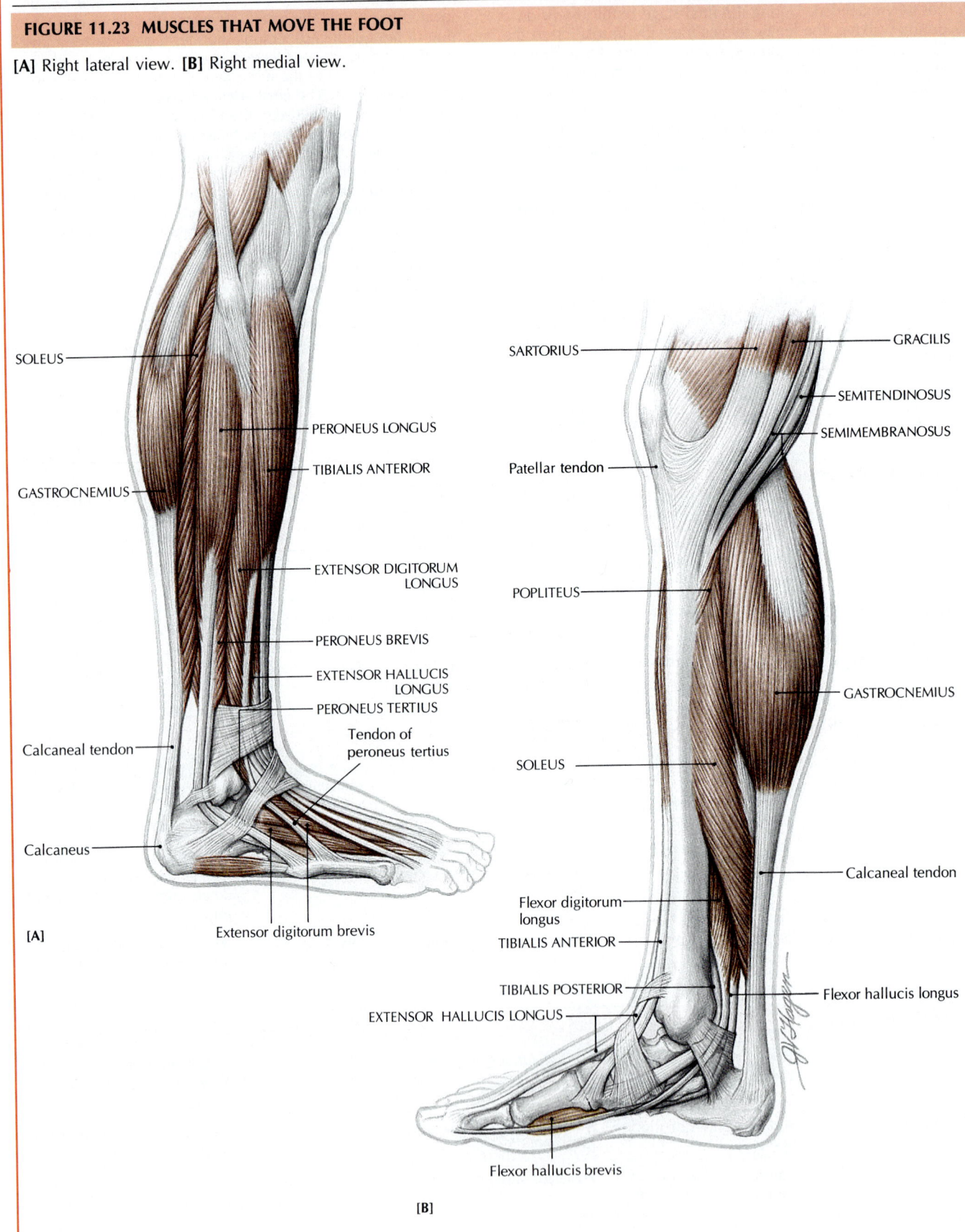

SOLEUS

PERONEUS LONGUS

TIBIALIS ANTERIOR

GASTROCNEMIUS

EXTENSOR DIGITORUM LONGUS

PERONEUS BREVIS

EXTENSOR HALLUCIS LONGUS

PERONEUS TERTIUS

Tendon of peroneus tertius

Calcaneal tendon

Calcaneus

Extensor digitorum brevis

[A]

SARTORIUS

GRACILIS

SEMITENDINOSUS

SEMIMEMBRANOSUS

Patellar tendon

POPLITEUS

GASTROCNEMIUS

SOLEUS

Calcaneal tendon

Flexor digitorum longus

TIBIALIS ANTERIOR

TIBIALIS POSTERIOR

Flexor hallucis longus

EXTENSOR HALLUCIS LONGUS

Flexor hallucis brevis

[B]

The ankle (talocrural) joint is a hinge joint capable of plantar flexion (moves sole of foot down) and dorsiflexion. The *soleus* and *gastrocnemius* muscles, both powerful plantar flexors, are active during the take-off phase of the foot in walking and running. Because of the shortness of its muscle belly, the gastrocnemius cannot flex the knee and plantar-flex the ankle joint at the same time, but it can do each independently. With the knee flexed, the soleus is the great plantar flexor of the ankle. In many individuals, these muscles are not active during ordinary standing.

The movements of eversion of the foot (turning so the sole faces outward) and inversion (turning so the sole faces inward)—the movements used, for instance, when we walk over rough ground—use the subtalar axis, or axis of Henke (the joint between the talus and calcaneus). When the foot is planted on the ground, it assumes an awkward position somewhere between eversion and inversion. The muscles responsible for these movements are the invertors and evertors. Should these muscles be "caught off guard," a sprained ligament might result. The *tibialis anterior* and *posterior muscles,* whose tendons pass on the great-toe side of the subtalar axis, are invertors. The *peroneus longus, brevis,* and **tertius,** whose tendons pass on the little-toe side of the axis, are evertors.

Muscle	Origin	Insertion	Innervation	Action
Soleus	Head of fibula, medial border of tibia	Posterior surface of calcaneus	Tibial nerve	Plantar flexes ankle joint; steadies leg during standing
Gastrocnemius	Posterior aspects of condyles of femur	Posterior surface of calcaneus	Tibial nerve	Plantar flexes ankle joint; flexes knee joint
Peroneus longus	Head and lateral surface of fibula	Lateral side and plantar surface of first metatarsal and medial cuneiform	Superficial peroneal nerve	Plantar flexes ankle joint; everts foot
Peroneus brevis	Lateral surface of fibula	Dorsal surface of fifth metatarsal	Superficial peroneal nerve	Plantar flexes ankle joint; everts foot
Tibialis posterior	Interosseous membrane, fibula, and tibia	Bases of second, third, and fourth metatarsals, navicular, cuneiform, cuboid, calcaneus	Tibial nerve	Plantar flexes ankle joint; inverts foot
Tibialis anterior	Lateral condyle and surface of tibia	First metatarsal, medial cuneiform	Deep peroneal nerve	Dorsiflexes ankle joint; inverts foot
Extensor digitorum longus	Lateral condyle of tibia, anterior surface of fibula	Middle and distal phalanges of four outer toes	Deep peroneal nerve	Dorsiflexes ankle joint; extends toes at metatarsophalangeal and interphalangeal joints
Extensor hallucis longus	Middle of fibula, interosseous membrane	Dorsal surface of distal phalanx of great toe	Deep peroneal nerve	Dorsiflexes ankle joint; extends great toe
Peroneus tertius	Distal third of fibula, interosseous membrane	Dorsal surface of fifth metatarsal	Deep peroneal nerve	Dorsiflexes ankle joint; everts foot

How Do We Stand?

We all know that walking and running involve the fine coordination of muscles, bones, and nerves, but how complicated is it to stand upright? The simplest answer is: It's more complicated than you think.

When you are standing in a relaxed position, several ligaments and muscle groups play a major role in counterbalancing the pulling forces of gravity. The key muscle groups are the extensor muscles of the back of the neck that hold the head upright, the lumbar erector spinal muscles that sustain the trunk, and the gastrocnemius and soleus leg muscles that maintain the position of the foot at the ankle joint. Essentially, all other muscles are relaxed.

When the body is in a relaxed, up-

right position, the gravitational force line through the lower extremity passes *behind* the hip joint and *in front of* the knee and ankle joints (see drawing). In effect, the iliofemoral ligament anterior to the hip joint balances the posterior force of gravity, the ligaments of the knee restrain the gravitational forces anterior to the knee joint, and no active muscle contractions are required. In addition, the contractions of the gastrocnemius and soleus muscles—acting through the calcaneal tendon—counterbalance the gravitational forces anterior to the ankle joint.

Proof that standing is not strictly a passive activity is the fact that when you stand in a relaxed position for one hour you use up about 140 calories.

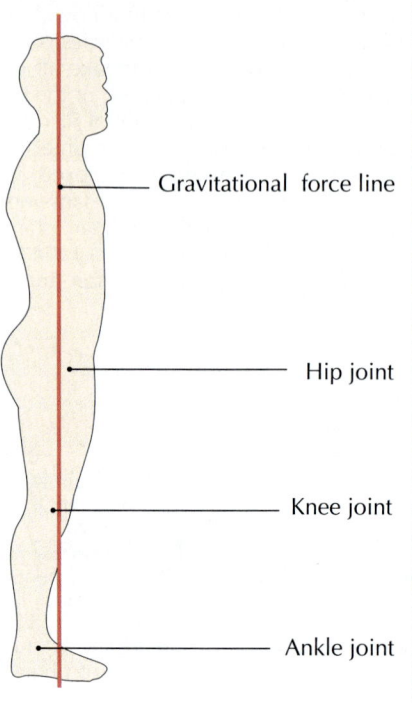

— Gravitational force line

— Hip joint

— Knee joint

— Ankle joint

FIGURE 11.24 MUSCLES OF THE TOES

The drawings show **[A]** superficial (first), **[B]** second, **[C]** third, and **[D]** deepest layers. Right plantar views.

The names and attachments of the small intrinsic muscles of the foot to the toes are somewhat similar to those in the hand. However, many of these muscles are ineffective in carrying out the presumed movements suggested by their names. Their primary functional role is to help maintain the stability of the resilient foot as it adjusts to the forces placed upon it.

The role of the short and long extensors of the toes is *dorsiflexion* (extension) at the metatarsophalangeal (MP) joints. The **lumbricals** are potential *extensors* of the interphalangeal (IP) joints. *Flexion* of the IP joints is accomplished by the **flexor digitorum longus** and the **quadratus plantae.** *Flexion* of the MP joints is the role of the **interossei, lumbricals,** and short flexors and abductors of the great and little toes. *Abduction* and *adduction* are accomplished by the **interossei, abductor** and **adductor hallucis,** and **abductor digiti minimi.** Apparently, the actions and functions of the lumbricals and interossei are not as important as their counterparts in the hand.

In walking, we plant the heel on the ground and then pass the weight forward on the lateral small-toe side of the foot, and then across the heads of the metatarsals to the metatarsal of the great toe and the pair of sesamoid bones of the two tendons of the flexor hallucis brevis for the takeoff. Try it. During the takeoff, the interossei muscles, by simultaneously flexing the MP joints and extending the IP

joints, permit transfer of some weight from the metatarsal heads to the toes. Note that the tendon of the flexor hallucis longus does not bear the weight at takeoff, because it is protected by passing between the sesamoids of the flexor hallucis brevis.

Specifically, there are four basic muscle layers of the foot. From the outside (superficial) to the inside (deep), the first layer contains the **abductor hallucis,** which abducts the great toe (hallux); the **flexor digitorum brevis,** which flexes the second phalanges of the four small toes; and the **abductor digiti minimi,** which abducts the small toe. In the second layer are the **quadratus plantae,** which flex the terminal phalanges of the four small toes, and the **lumbricals,** which flex the proximal phalanges and also extend the distal phalanges of the four little toes. The third layer consists of the **flexor hallucis brevis,** which flexes the proximal phalanx of the great toe; the **adductor hallucis,** which abducts the great toe; and the **flexor digiti minimi brevis,** which flexes the proximal phalanx of the little toe. The fourth or innermost muscle layer contains (1) the four **dorsal interossei,** which abduct the second, third, and fourth toes, and flex the proximal and extend the distal phalanges, and (2) the three **plantar interossei,** which adduct the third, fourth, and fifth toes. Remember that the axis of abduction and adduction of the foot is along the second toe. (The second toe is abducted by two dorsal interossei.)

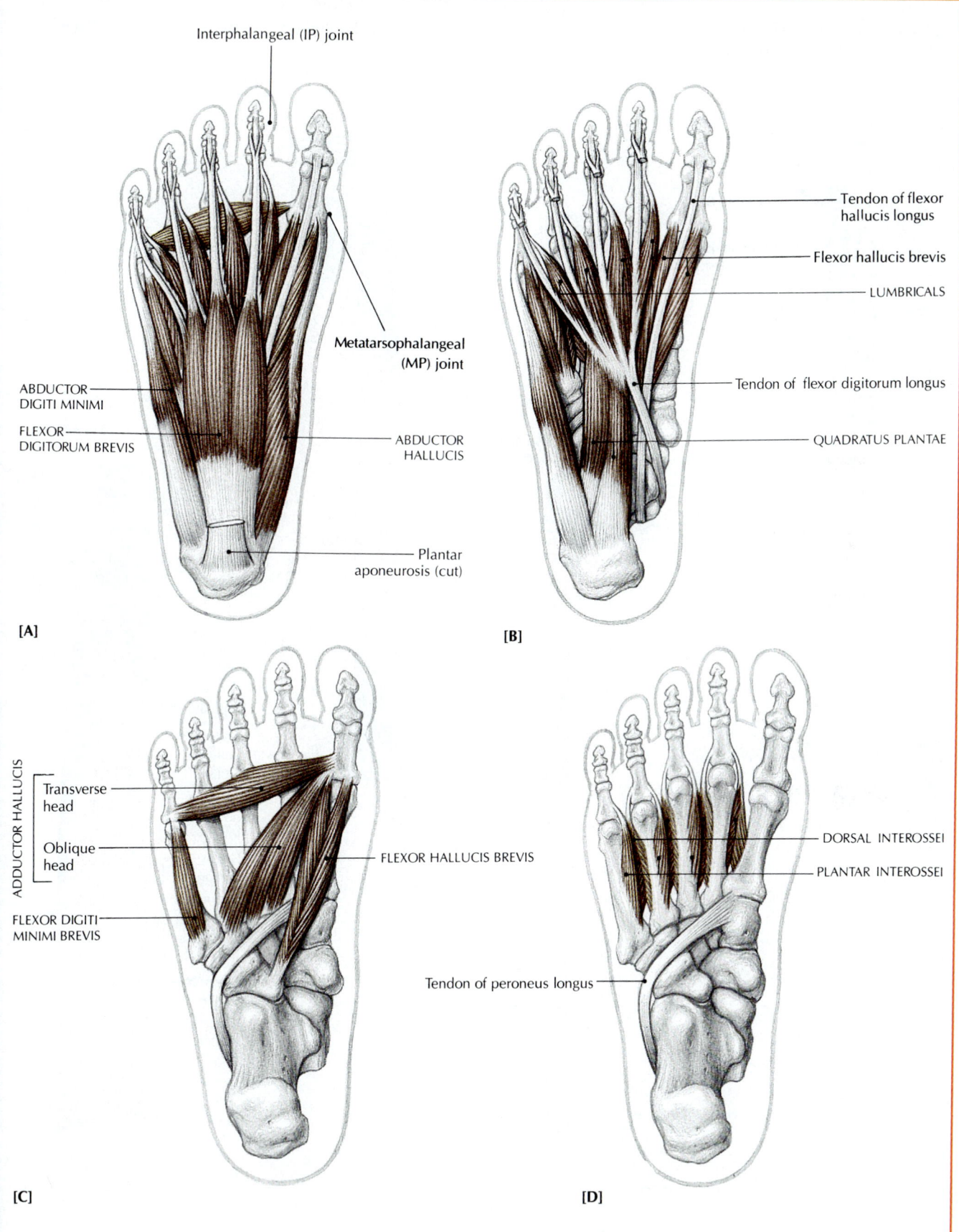

[A]

Interphalangeal (IP) joint

Metatarsophalangeal
(MP) joint

ABDUCTOR
DIGITI MINIMI

FLEXOR
DIGITORUM BREVIS

ABDUCTOR
HALLUCIS

Plantar
aponeurosis (cut)

[B]

Tendon of flexor
hallucis longus

Flexor hallucis brevis

LUMBRICALS

Tendon of flexor digitorum longus

QUADRATUS PLANTAE

[C]

ADDUCTOR HALLUCIS

Transverse
head

Oblique
head

FLEXOR HALLUCIS BREVIS

FLEXOR DIGITI
MINIMI BREVIS

[D]

DORSAL INTEROSSEI

PLANTAR INTEROSSEI

Tendon of peroneus longus

FIGURE 11.25 ORIGINS AND INSERTIONS OF SOME MAJOR MUSCLES

Pink areas indicate origins, and blue areas indicate insertions. Right humerus: [A] anterior view, [B] posterior view. Right radius and ulna: [C] anterior view, [D] posterior view.

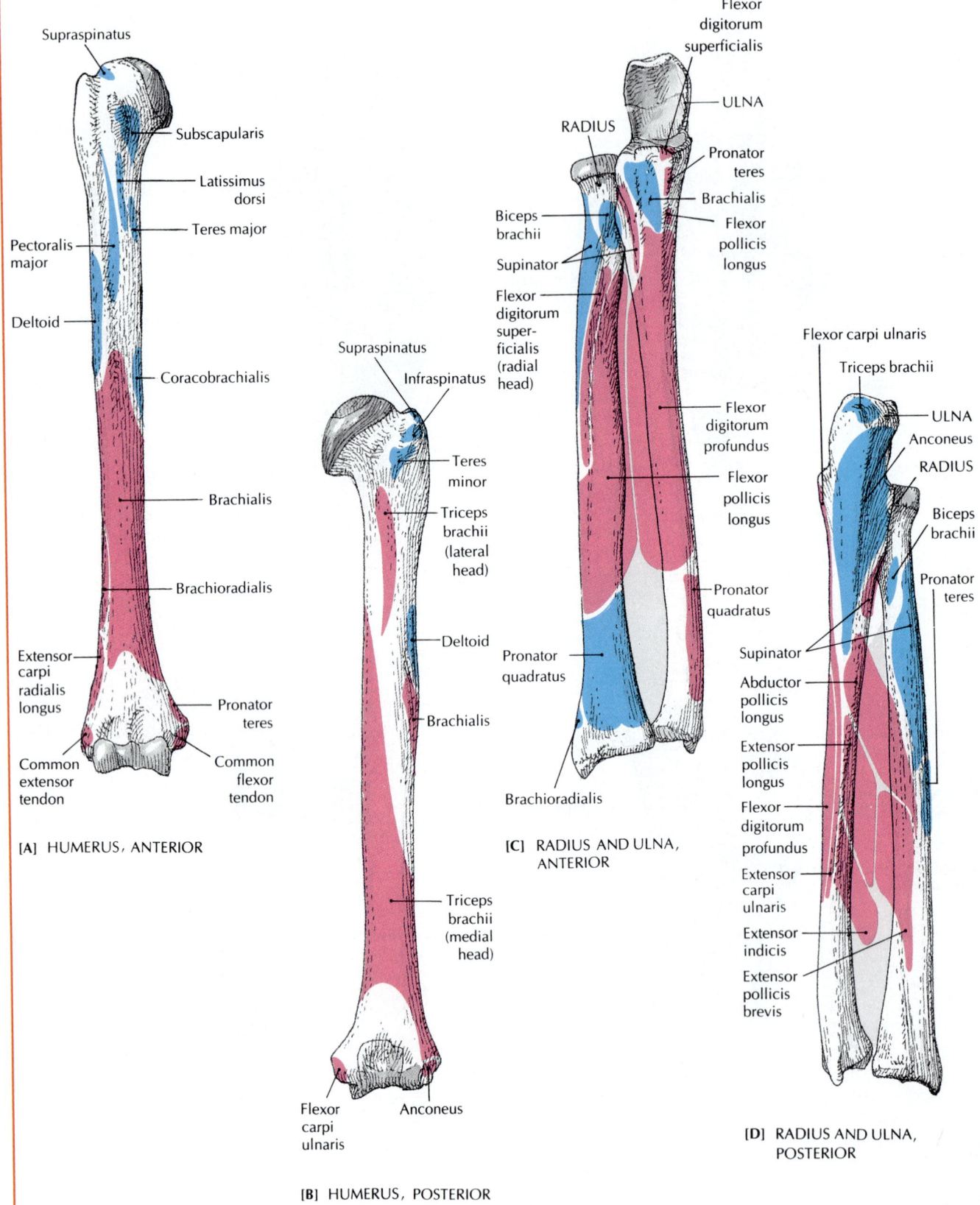

[A] HUMERUS, ANTERIOR

[B] HUMERUS, POSTERIOR

[C] RADIUS AND ULNA, ANTERIOR

[D] RADIUS AND ULNA, POSTERIOR

Right femur: **[E]** anterior view, **[F]** posterior view. Right tibia and fibula: **[G]** anterior view, **[H]** posterior view.

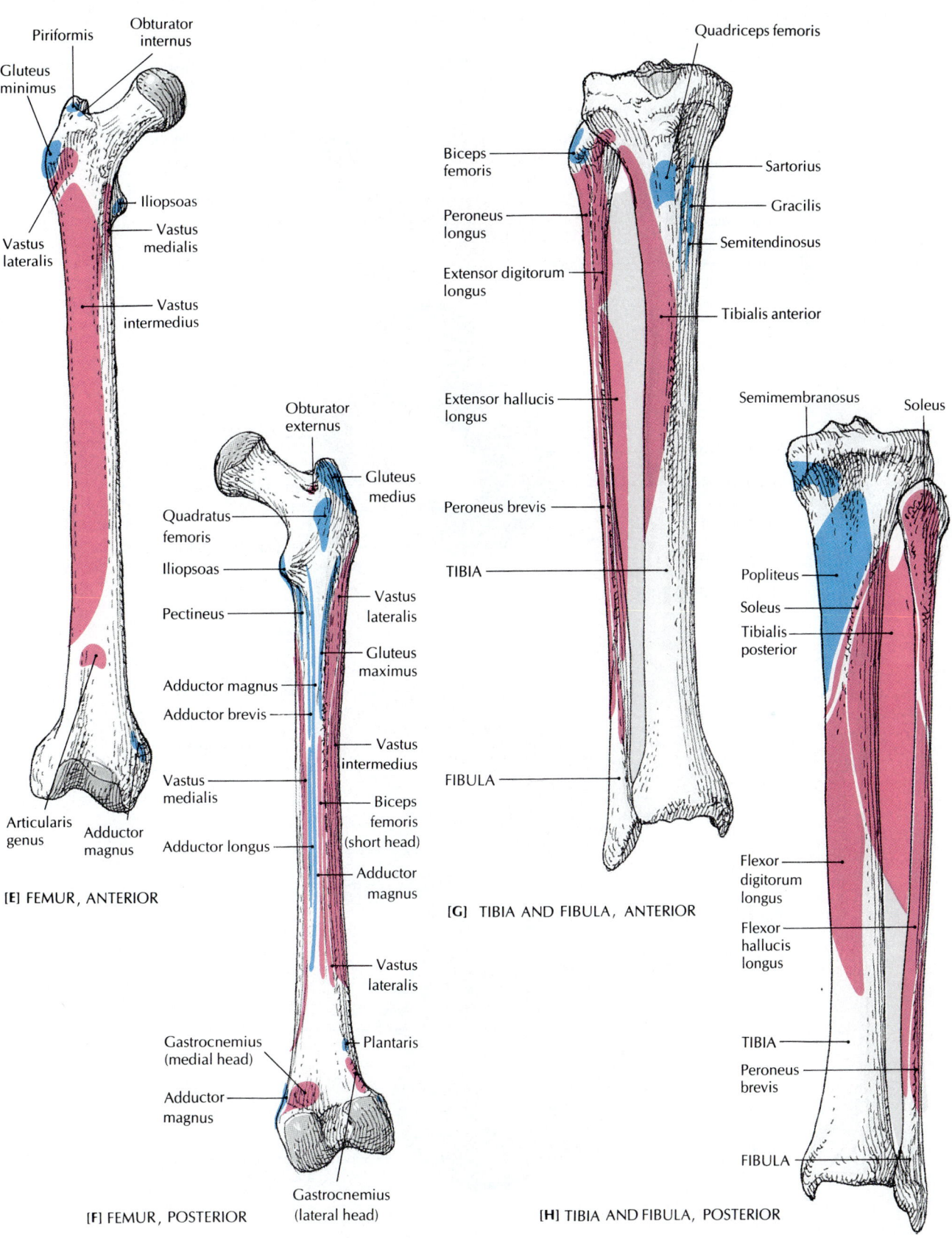

Piriformis

Obturator internus

Gluteus minimus

Iliopsoas

Vastus medialis

Vastus lateralis

Vastus intermedius

Articularis genus

Adductor magnus

[E] FEMUR, ANTERIOR

Obturator externus

Gluteus medius

Quadratus femoris

Iliopsoas

Pectineus

Vastus lateralis

Gluteus maximus

Adductor magnus

Adductor brevis

Vastus intermedius

Vastus medialis

Biceps femoris (short head)

Adductor longus

Adductor magnus

Vastus lateralis

Gastrocnemius (medial head)

Plantaris

Adductor magnus

Gastrocnemius (lateral head)

[F] FEMUR, POSTERIOR

Quadriceps femoris

Biceps femoris

Sartorius

Peroneus longus

Gracilis

Extensor digitorum longus

Semitendinosus

Tibialis anterior

Extensor hallucis longus

Peroneus brevis

TIBIA

FIBULA

[G] TIBIA AND FIBULA, ANTERIOR

Semimembranosus

Soleus

Popliteus

Soleus

Tibialis posterior

Flexor digitorum longus

Flexor hallucis longus

TIBIA

Peroneus brevis

FIBULA

[H] TIBIA AND FIBULA, POSTERIOR

(Figure 11.25 continues on following page)

FIGURE 11.25 ORIGINS AND INSERTIONS OF SOME MAJOR MUSCLES *(Continued)*

Right hipbone: **[I]** external surface, **[J]** internal surface.

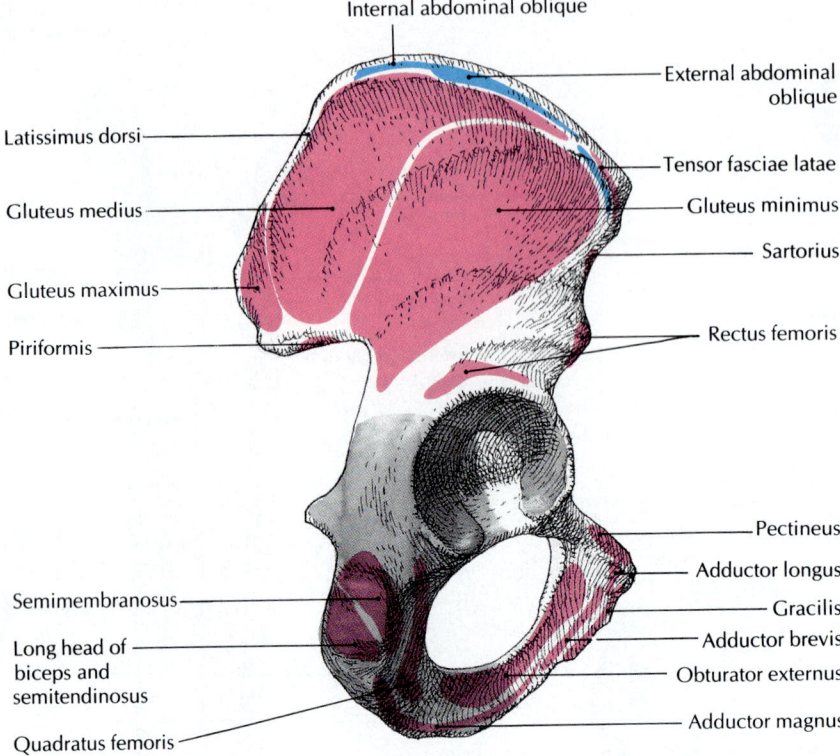

[I] HIPBONE, EXTERNAL SURFACE

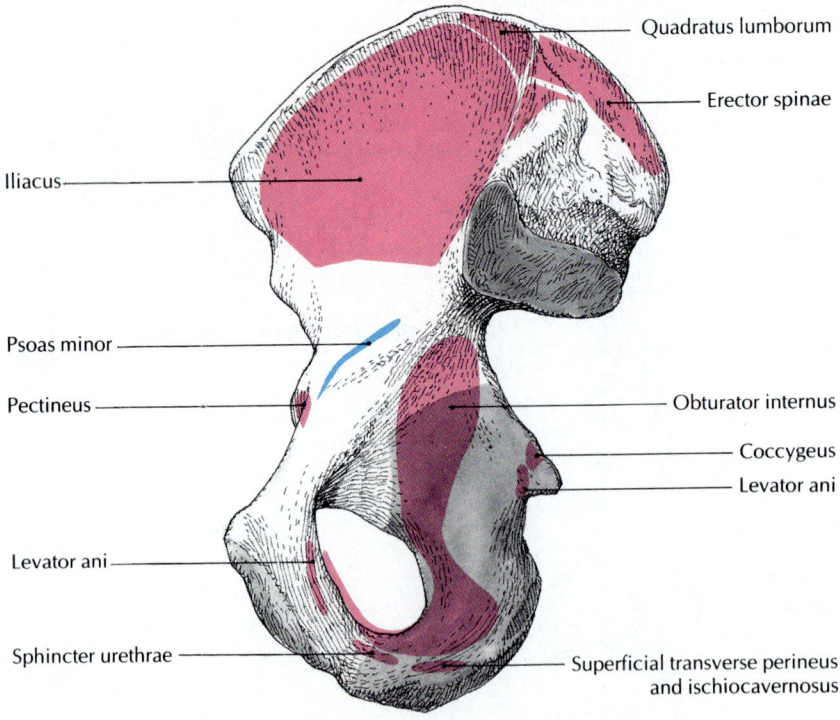

[J] HIPBONE, INTERNAL SURFACE

WHEN THINGS GO WRONG

Computer Keyboard Stress Syndrome

The recognition of this disorder is so new* it does not yet have an official name, being variously called cumulative trauma disorder, repetitive stress injury, keyboard overuse syndrome, work-related musculoskeletal disorder, repetitive motion injury, or the one devised here: *computer keyboard stress syndrome*. No matter what you call it, it means that your hands and forearms are undergoing excessive physical stress while typing at a computer keyboard.

Some of the specific injuries that can occur as a result of excessive (and sometimes improper) use of a computer keyboard are *carpal tunnel syndrome*, in which the median nerve in the wrist may be compressed by swelling tissues (see When Things Go Wrong at the end of Chapter 15, page 472); *flexor tendinitis*, in which friction within the sheath around tendons may cause pain and hamper movement of the fingers; *de Quervain's disease*, in which the tendon sheath over the thumb becomes inflamed by constant friction; and *overloaded forearm muscles and fasciae*, in which inflammation may lead to shortening of the muscles and the formation of scar tissue.

Intramuscular Injections

Because deep muscle tissue contains relatively few nerve endings for pain, irritating drugs are injected into a muscle instead of being taken orally or being injected under the skin. Also, when medication is injected intramuscularly, it is absorbed into the body rapidly because of the rich capillary beds in muscle tissue.

Intramuscular injections are usually given in the buttock (gluteus medius and gluteus maximus), the lateral part of the thigh, or the deltoid muscle of the arm. The most usual site for intramuscular injections is the gluteus maximus or gluteus medius muscle, about two or three inches below the iliac crest, in the upper outer quadrant of the buttock. When the gluteus maximus is used it is important to avoid injuring the nearby sciatic nerve. Also, drugs should not be injected directly into large blood vessels, no matter where they are located. A standing position tenses this muscle, and an injection into a tense muscle causes considerable pain. Therefore, the patient is usually placed in a prone (face down) position to relax the muscle before the injection.

*However, an Italian physician named Bernadino Ramazinni described cumulative microtrauma as a main cause of occupational disease in 1717. And musicians, assembly-line workers, and meatpackers have long been painfully aware of the effects of physical occupational stress on the hands and forearms.

Hernias

A *hernia* (L. protruded organ), or rupture, is the protrusion of any organ or body part through the muscular wall that usually contains it. *Inguinal* (IHNG-gwuh-null; L. groin) *hernias* are the most common type, occurring most often in males. Most inguinal hernias are caused by a weakness in the fascial wall that allows a loop of the intestines or fat to protrude in the area of the groin after passing superficially to the inguinal ligament and medial to the pubic tubercle, to which the inguinal ligament is attached. *Hiatal* (hye-A-tuhl; L. *hiatus*, gap) *hernias* develop when a defective diaphragm allows a portion of the stomach to pass through the opening for the esophagus in the diaphragm into the thoracic cavity. The stomach herniates into the thoracic cavity through the hiatus in the diaphragm as a consequence of the negative pressure within the thoracic cavity during the respiratory cycle, as compared to the positive pressure within the abdominal cavity. *Femoral hernias* are most common in women. They protrude deep to the inguinal ligament and extend into the groin, where the femoral artery passes into the thigh. Usually a fatty deposit within the femoral canal creates a gap large enough for a loop of intestine or fat to bulge through. *Umbilical hernias* protrude at the navel. They are most common in newborns, obese women, and women who have had several pregnancies. *Incisional hernias* result from the weakening around a surgical wound that does not heal properly.

Tendinitis

Tendinitis (which you might think should be spelled tendonitis) is a painful inflammation of the tendon and tendon sheath, typically resulting from a sports injury or similar strain on the tendon. It may also be associated with musculoskeletal diseases such as rheumatism or with abnormal posture. Tendinitis occurs most often in the shoulder area, calcaneal tendon, or hamstring muscles.

Tennis Elbow and Golf Elbow

Tennis elbow is often thought of as a form of tendinitis, but it is not. Its medical name is *lateral epicondylitis*; it results from the strain and subsequent inflammation of either the forearm extensor and supinator muscles of the fingers and wrist and/or their tendinous attachments to the lateral epicondyle of the humerus, or of the lateral (ulnar) collateral ligament of the elbow joint. These conditions are caused by the premature degeneration of the common tendon (origin) of the extensor muscles and produce tenderness over the anterior aspect of the lateral epicondyle. Tennis elbow afflicts people who supinate their forearms repetitively against resistance, as when driving a screw, or by the violent extension of the wrist with the hand pronated, as in the tennis backstroke. *Golf elbow* (*medial epicondylitis*) is a similar condition; it is associated with the flexor and pronator muscles and their tendinous attachments on the medial aspect of the elbow.

Tension Headache

Tension headaches, the most common type of headache,

are often caused by emotional stress, fatigue, or other factors that produce painful muscular contractions in the scalp and back of the neck. Muscle spasms may constrict blood vessels, increasing general discomfort. Pain usually spreads from the back of the head to the area above the eyes and may cause pain in the eyes themselves. The removal of toxic muscle wastes may be hampered by constricted blood vessels in the scalp, causing even further tenderness.

CHAPTER SUMMARY

How Muscles Are Named (p. 300)

The name of a muscle generally tells something about its location, action (or function), or structure, such as its size, shape, attachment sites, number of heads of origin, or direction of fibers.

Attachment of Muscles (p. 300)

1 A muscle is usually attached to a bone or cartilage by a **tendon.** Some tendons are expanded into a broad, flat sheet called an **aponeurosis.**

2 In addition to acting as attachments, tendons add useful length and thickness to muscles, reduce muscle strain, and add strength to muscle action.

3 The **origin** is the place on the bone that does not move when the muscle contracts, and the **insertion** is the place on the bone that does move when the muscle contracts. Generally, the origin is the attachment closer to the axial skeleton. Some muscles have more than one origin or insertion.

4 The origin and insertion of some muscles can be reversed, depending on which bone moves.

Architecture of Muscles (p. 301)

1 The fibers of skeletal muscles are grouped into small bundles called fascicles, with the fibers within each bundle running parallel to each other.

2 The specific architectural pattern of fascicles within a muscle determines the range of movement and power of the muscle.

3 Patterns of arrangement of fascicles and tendons of attachment include **strap, fusiform, pennate,** and **circular muscles.**

Individual and Group Actions of Muscles (p. 303)

1 Most movements involve the complex interactions of several muscles or even groups of muscles.

2 An **agonist,** or **prime mover,** is the muscle that is primarily responsible for producing a movement. An **antagonist** muscle helps produce a smooth movement by slowly relaxing as the agonist contracts. A **synergist** works with a prime mover by preventing movements at an "in-between" joint when the prime

mover passes over more than one joint. A **fixator** provides a stable base for the action of the prime mover.

3 In different situations, the same muscle can act as a prime mover, antagonist, synergist, or fixator.

Lever Systems and Muscle Actions (p. 306)

1 Most skeletal muscle movements are accomplished through a system of levers. A lever system includes a rigid **lever arm** that pivots around a fixed point called a **fulcrum,** with the bone acting as lever arm and the joint as fulcrum. Every lever system includes two different **forces:** the resistance to be overcome and the effort applied.

2 There are three types of levers, referred to as first-, second-, and third-class levers.

3 The mechanical advantage that muscles gain by using a lever system is called **leverage.**

Specific Actions of Principal Muscles (p. 307)

See FIGURES 11.6 through 11.25.

STUDY AIDS FOR REVIEW

KEY TERMS

agonist 303
antagonist 303
aponeurosis 300
belly 300
biceps brachii 325, 327
biceps femoris 335, 337
circular muscle 302
deltoid 325
erector spinae group 316
extrinsic muscle 310
fixator 303
force 306
fulcrum 306
fusiform muscle 302
gastrocnemius 337, 339
gluteus maximus 335
hamstring muscles 335
infraspinatus 325, 326
insertion 300
intercostal muscles 317
interossei 332, 340
intrinsic muscle 314
latissimus dorsi 322, 325
leverage 306
lever arm 306
linea alba 318
origin 300
pectoralis major 325
pennate muscle 302
perineum 320
postural muscle 303
prime mover 303
quadriceps femoris 337
rectus abdominis 318
rotator cuff muscles 325, 326
sartorius 337
soleus 339
sphincter 308
sternocleidomastoid 315
strap muscle 302
subscapularis 325, 326
supraspinatus 325, 326
synergist 303
tendon 300
teres major 325
teres minor 325, 326
transversospinalis group 316
trapezius 322
triceps brachii 325, 327

UNDERSTANDING THE FACTS

1 What can you deduce about the abductor pollicis longus just from its name?
2 Define the following terms: origin, insertion, agonist, antagonist, synergist, fixator.
3 Must a muscle have one fixed origin and one fixed insertion?
4 Distinguish between a tendon and an aponeurosis.
5 Describe the arrangement of fascicles in strap, fusiform, pennate, and circular muscles.
6 For each of the three classes of levers, which element of the system (force, weight, or fulcrum) is in the middle?
7 Identify the muscles that make up each of the following groups and indicate their primary function.
  a rotator cuff
  b hamstrings
  c quadriceps femoris (quads)
8 Are the muscles that control the movement of the fingers located in the phalanges? Explain.
9 Describe the differences between the muscles that move the thumb and those that move the other fingers.
10 List the muscles that
  a move the eyeball in the socket
  b are the primary muscles of mastication
  c extrinsically move the tongue
  d are the primary postural muscles of the spine
  e are active during quiet breathing
  f support the ventral abdominal wall
11 Distinguish between the function of the pelvic diaphragm and that of the perineum.

UNDERSTANDING THE CONCEPTS

1 Are all muscles attached to bones? Explain.
2 If a strap muscle and a pennate muscle are of equal overall size, why is the pennate muscle the more powerful?
3 Using the arm or the leg as an example, explain how antagonistic muscles are structurally organized.
4 Where is the insertion of the quadriceps femoris when a person is
  a rising from a seated position
  b swinging his or her leg while seated
5 What class of lever is exemplified by each of the following muscular actions? Explain.
  a flexing the forearm
  b abducting the thigh
  c hyperextending the head (touching the back with the back of the head)

6 What factors are important in the maintenance of an erect posture? Are flexors or extensors the major postural muscles of the spine? Explain.

SELF-QUIZ
Multiple-Choice Questions

11.1 Which of the following is a muscle of the thorax?
    a scalene   **d** gracilis
    b piriformis   **e** zygomaticus
    c transversus

11.2 Which of the following is *not* a muscle of mastication?
    a masseter   **d** lateral pterygoid
    b temporalis   **e** sternohyoid
    c medial pterygoid

11.3 All of the following are extrinsic muscles of the eye *except* the
    a superior rectus   **d** inferior oblique
    b styloglossus   **e** medial rectus
    c superior oblique

11.4 Which of the following is *not* an extrinsic muscle of the tongue?
    a genioglossus   **d** buccinator
    b styloglossus   **e** a, b, and c
    c hyoglossus

11.5 Movement of the shoulder girdle is accomplished by the
    a levator scapulae   **d** trapezius
    b rhomboid minor   **e** all of the above
    c pectoralis minor

11.6 All of the following muscles move the humerus *except* the
    a coracobrachialis   **d** deltoid
    b brachialis   **e** supraspinatus
    c teres major

11.7 The linea alba is a
    a region of the perineum
    b part of the calcaneal tendon
    c section of the tensor fasciae latae
    d muscle of the ventral abdominal wall
    e tendon of the ventral abdominal wall

11.8 Which of the following muscles move the forearm?
    a supinator   **d** a and b
    b pronator teres   **e** a, b, and c
    c pronator quadratus

11.9 Which of the following is *not* a muscle of the abdominal wall?
    a external oblique
    b internal oblique
    c transversus abdominis
    d rectus abdominis
    e quadratus lumborum

11.10 All of the following muscles insert on the femur *except* the

    a iliopsoas   **d** sartorius
    b gluteus maximus   **e** adductor brevis
    c gluteus medius

11.11 All of the following muscles are involved in moving the foot *except* the
    a sartorius   **d** peroneus
    b gastrocnemius   **e** tibialis
    c soleus

11.12 Which of the following muscles is *not* an antagonistic pair?
    a biceps brachii/triceps brachii
    b gluteus medius/adductor longus
    c rectus femoris/biceps femoris
    d lateral rectus/medial rectus
    e semimembranosus/semitendinosus

11.13 The adductor muscles of the thigh are the
    a piriformis and obturator
    b adductor brevis and obturator
    c adductor longus, brevis, and magnus
    d semitendinosis and deltoid
    e sternohyoid and transversus

11.14 Which of the following muscles has a belly that spans the distal interphalangeal joint?
    a abductor pollicis brevis   **d** lumbricals
    b palmar interossei   **e** none of the above
    c dorsal interossei

Matching
Choose the correct prime mover for each of the actions listed below.

11.15 _____ winking   **a** buccinator
11.16 _____ kissing   **b** zygomaticus major
11.17 _____ smiling   **c** orbicularis oris
11.18 _____ blowing out a   **d** orbicularis oculi palpe-
            candle   bral
11.19 _____ frowning   **e** orbicularis oculi orbital
11.20 _____ tightly closing   **f** platysma
            eyelid

Completion Exercises
11.21 The widest part of a muscle is called its _____.
11.22 A muscle is usually attached to a bone (or a cartilage) by a(n) _____.
11.23 Tendon sheaths that are broad and flat are called _____.
11.24 The end of a muscle attached to the bone that does *not* move is the _____.
11.25 The point of attachment of the muscle to the bone that moves is the _____.
11.26 The fibers of skeletal muscles are grouped into small bundles called _____.
11.27 A muscle action in which the angle between two bones is decreased is called _____.

11.28 The type of muscle that turns the forearm so that the palm faces downward is a(n) _____.

11.29 The type of muscle that turns the foot inward is a(n) _____.

11.30 In a _____-class lever, the force is applied at one end of the lever arm, the resistance at the other end, and the fulcrum is in between.

Matching

11.31 _____ strap muscles	**a** assists prime mover by stabilizing a joint
11.32 _____ fusiform muscle	**b** transverse fascicles
11.33 _____ pennate muscle	**c** movement toward the midline
11.34 _____ sphincter muscle	**d** opposes the movement of a prime mover
11.35 _____ agonist	**e** tapers at both ends
11.36 _____ antagonist	**f** circular fascicles
11.37 _____ synergist	**g** complements action of a prime mover
11.38 _____ fixator	**h** movement away from the midline
11.39 _____ abductor	**i** aneural muscles
11.40 _____ adductor	**j** parallel fascicles
	k prime mover
	l oblique fascicles

A SECOND LOOK

1 In the following drawing, label the deltoid, latissimus dorsi, pectoralis minor, and trapezius muscles.

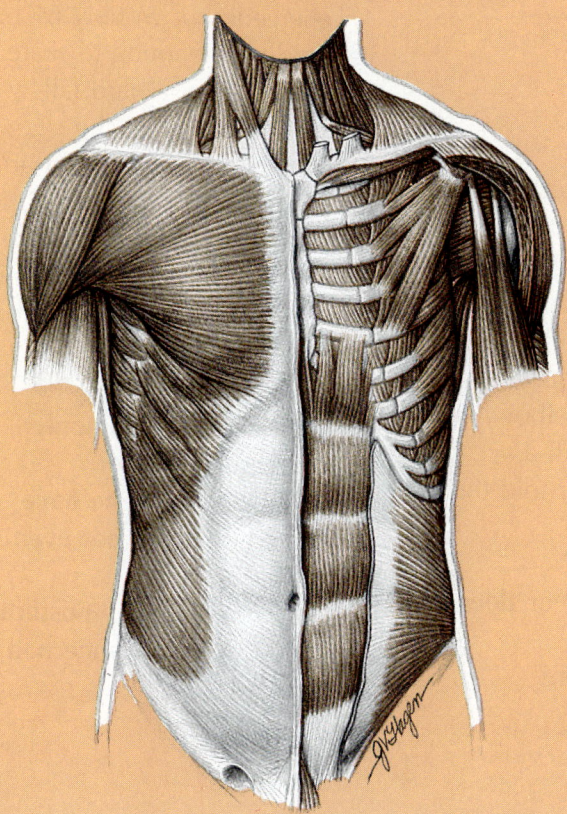

2 In the drawing below, label the following muscles: latissimus dorsi, triceps brachii, gluteus maximus, biceps femoris, and gastrocnemius.

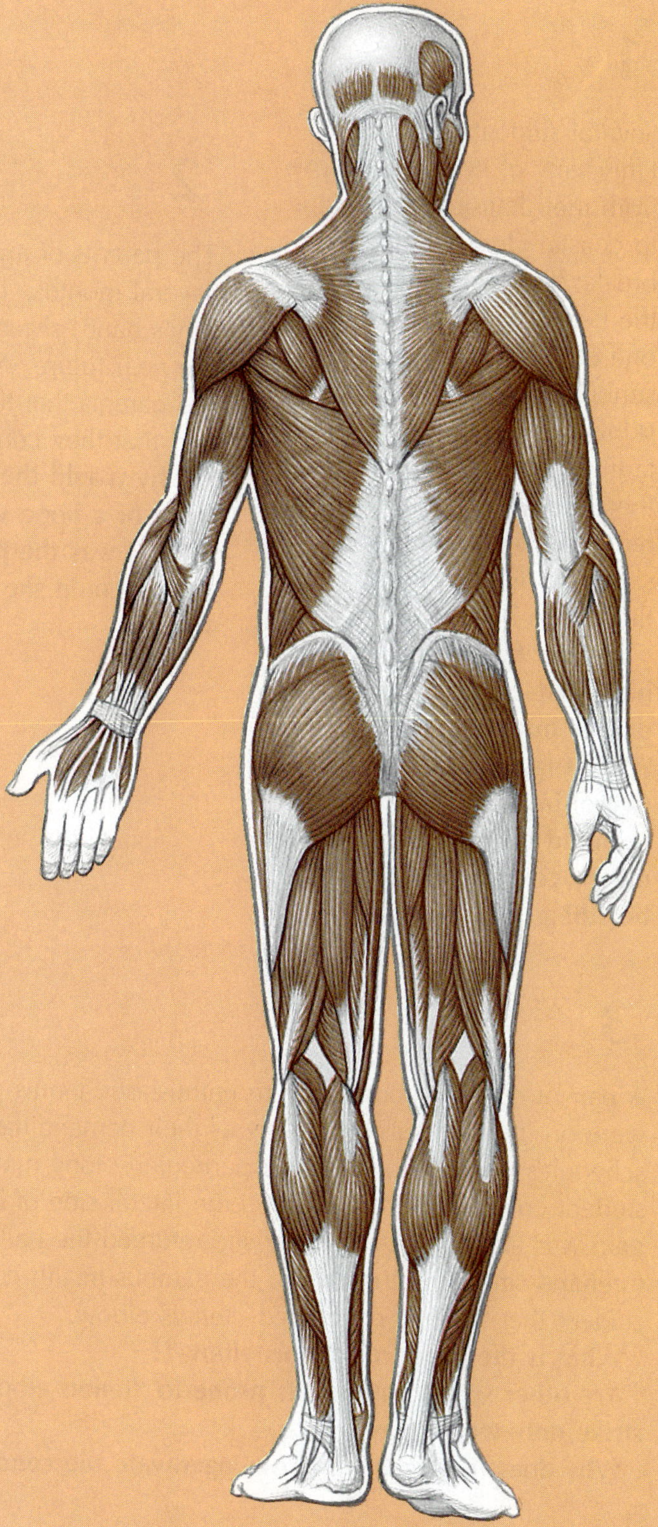

ii case studies

1

Several students from the University of Wisconsin drove from their half-frozen campus to coastal Florida for spring break. After only one day on the beach without sunscreen, one of the group was painfully sunburned, while his less-fair-skinned friends were nicely tanned. By the next morning, the sunburned student's dark-red and exquisitely tender shoulders were covered with fluid-filled blisters.

- **Did our fair-skinned student have first-, second-, or third-degree burns?**
- **What layer(s) of the skin was (were) involved?**
- **Should his severe sunburn be of any concern for his future health?**

2

The parents of an 11-year-old noticed that he had not grown in several months. The pediatrician told them that by x-raying the boy's hand, she could determine if the child would grow taller in the near future. After examining the x-rays, the physician informed the parents that their child's period of growth was not yet complete and that they could expect to see future growth spurts.

- **Why would the physician x-ray the child's fingers and not a skull bone or a bone such as the clavicle?**
- **What was the physician looking for in the x-rays?**
- **How could she predict that the child would grow taller in the future?**

3

A pair of college roomates were enthusiastic tennis players who were on the courts as frequently as their demanding academic schedules would allow. After a particularly long match, one student complained of a pain on the lateral side of her elbow. The pain was aggravated whenever she returned the ball with a forehand stroke. The doctor in the campus health center told the student that she had developed "tennis elbow."

- **What is the cause of "tennis elbow"?**
- **Are other sports enthusiasts prone to "tennis elbow," or does it strike only tennis players?**
- **Why does the forehand stroke aggravate the condition?**

4

An energetic but wobbly 6-year-old was learning to skate. When he began a hard fall forward onto the ice, he instinctively thrust his right arm outward to break the fall. Unfortunately, the child broke his collarbone. After having his arm immobilized in a sling for several weeks, he recovered and resumed skating.

- **Which bone is commonly called the collarbone?**
- **How could this bone have broken when it did not even touch the ice?**
- **Would there be any postural signs that the collarbone had broken?**

iii

Control, Communication, and Coordination

CAREER IN FOCUS: NURSING

1994 will be remembered as the year in which a national consensus crystallized as politicians and private citizens alike acknowledged the crisis in our health care system. While few of us can predict the exact course of future change, it is clear that a basic core of services will be available to more of our citizens. But as more patients need care, as ever-more expensive methods of managing diseases for an aging population, and as organized medicine becomes increasingly specialized, a radical restructuring must occur to meet the primary health needs of the nation.

Among the groups leading the way with positive suggestions for innovative change are the nation's 2.1 million nurses. They see first-hand the overcrowded emergency rooms. They are poignantly aware of infant mortality rates higher than in many other developed countries. They feel keenly the widespread public unhappiness with the present system. In response, they advocate more emphasis on screening and prevention, greater dependence on highly trained non-physician providers for broader-based primary care, and the transfer of some routine, preventive care to sites that are more accessible and more cost-effective than hospitals are—schools, workplaces, and community-based clinics, for instance.

In order to realize that vision, the field of nursing itself is in transition, even as it is growing. Applicants can still become licensed R.N.s with an associate's degree, but an increasing number of positions require R.N.s with a B.S.N. degree. The former are known as technical nurses, the latter as professional nurses. The professional nurse is empowered to manage the full scope of nursing care responsibilities as well as the supervision of other nursing personnel. The technical nurse acts within a narrower scope.

Beyond the R.N.s are four general categories of advanced practice nurses (A.P.N.s) who have met educational and clinical requirements substantially above those necessary for the R.N. Most have master's degrees and some have doctorates. All can deliver quality primary health care within their specialty. Nurse practitioners (N.P.s) conduct physical exams

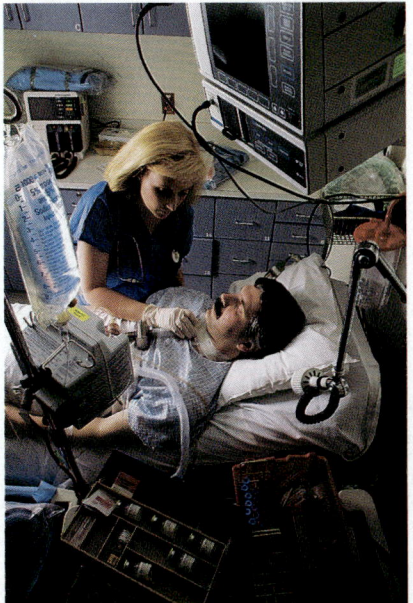

and diagnose and treat minor illnesses or injuries. They function as public health nurses and they counsel patients. Some work independently, others provide care at HMOs and hospitals or at industrial facilities. Certified nurse midwives (C.N.M.s) deliver babies and provide prenatal, postnatal, and gynecological care in the home, hospital, or birthing center. Clinical nurse specialists (C.N.S.s) provide primary care or function as consultants or administrators in specialized fields such as psychotherapy, cancer, and cardiac care. The certified registered nurse anesthetist (C.R.N.A.) provides more than half of the anesthesia services nationally in operating rooms and ambulatory surgical settings. They are the sole providers of anesthesia in 85 percent of rural hospitals.

In the next few years, as we debate our health care alternatives, nurses will already be implementing their vision of the future. Opportunities in this profession can only multiply as nursing grows to fill ever-expanding needs.

Two of the most rapidly developing areas in human physiology are the nervous and endocrine systems. New advances have brightened the future for patients who suffer from conditions or diseases that impair these systems. Nurses involved in the care of such patients benefit from new knowledge in the areas we study in this section: control, communication, and coordination systems of the human body.

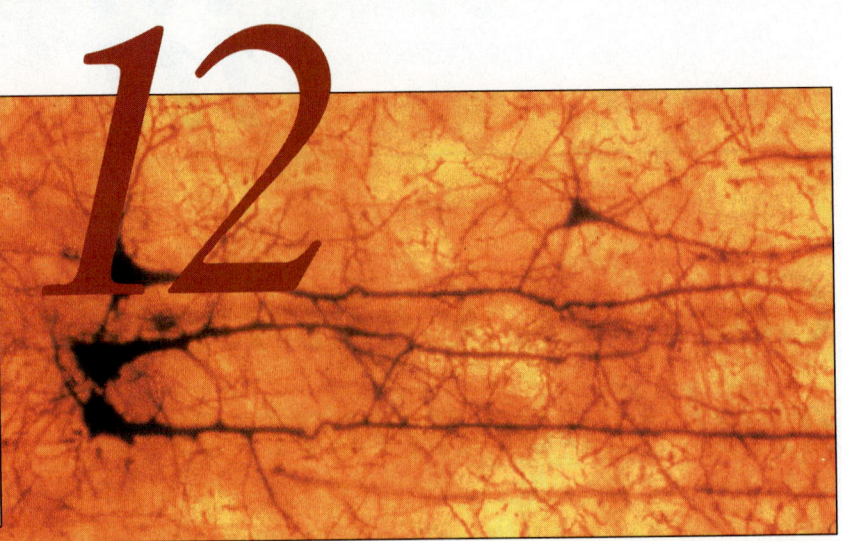

12

Nervous Tissue: An Introduction to the Nervous System

KEY CONCEPTS

1 Changes in the body's internal and external environments are called stimuli. They are detected by specialized receptors that initiate impulses in neurons. The impulses are relayed via neurons to the spinal cord and brain, where they are interpreted and acted upon.

2 The nervous system and the endocrine system are the body's main regulatory systems.

3 The central nervous system consists of the brain and spinal cord; the peripheral nervous system consists of nerve cells and fibers going to and emerging from the brain and spinal cord.

4 Nerve cells, or neurons, have the ability to respond to certain stimuli and conduct impulses.

5 Nerve impulses, which travel along nerve fibers, affect the activities of other neurons, muscle cells, or gland cells. The junction between nerve cells is a synapse.

In order to maintain homeostasis, the body is constantly reacting and adjusting to changes in the outside environment and in the internal environment (within the body itself). Such environmental changes, or *stimuli*, are detected and conveyed via nerves to the spinal cord and brain, where the messages (input) are analyzed, combined, compared, and coordinated by a process called *integration.*

After being sorted out, messages are conveyed by nerves to the muscles and glands of the body. The nervous system expresses itself through muscles and glands, causing muscles to contract or relax and glands to secrete or not secrete their products.

The nervous system and the endocrine system are the two major regulatory systems of the body, and both are specialized to make the proper responses to stimuli. The nervous system responds faster than the endocrine system. Stimuli received by the nervous system are processed rapidly (within a fraction of a second to seconds) through a combination of *electrochemical impulses* and chemical substances called *hormones* (Chapter 18). The endocrine system responds within minutes, hours, or days and typically regulates such processes as ion levels in the body, metabolic processes, and long-term processes such as growth and reproductive ability.

ORGANIZATION OF THE NERVOUS SYSTEM

Although the nervous system is a single, unified communications network, it can be divided on a gross anatomical basis into the central nervous system (Chapters 13 and 14) and the peripheral nervous system (Chapter 15).

Central Nervous System

The *central nervous system (CNS)* consists of the *brain* and *spinal cord* [FIGURE 12.1A], which are surrounded and protected by the skull and vertebral column, respectively. The central nervous system may be thought of as the body's central control center, receiving and interpreting or integrating all stimuli and relaying nerve impulses to muscles and glands, where the designated actions actually take place.

Peripheral Nervous System

The *peripheral nervous system (PNS)* consists of the nerve cells *(neurons)* and their fibers that emerge from and go to the brain (cranial nerves) and spinal cord (spinal nerves) [FIGURE 12.1A, B]. The peripheral nervous system is composed of an afferent (sensory) division and an efferent (motor) division.

Somatic nervous system The *somatic nervous system* is composed of (1) the *somatic afferent (sensory) division*, which receives sensory information and conveys it to the spinal cord and brain via nerves, and (2) the *somatic efferent (motor) division*, which regulates the contractions of skeletal muscles via neuronal pathways that descend from the brain and spinal cord to lower motor neurons.

Visceral nervous system The *visceral nervous system* consists of (1) the *afferent (sensory) division*, which conveys sensory information from sensory receptors in visceral* organs, and (2) the *efferent (motor) division*, more commonly called the *autonomic nervous system* (Chapter 16), which is involved in the motor activities that influence smooth muscle, cardiac muscle, and glands of the skin and viscera.

CELLS OF THE NERVOUS SYSTEM

The nervous system contains over 100 billion nerve cells, or *neurons* (Gr. nerve), which are specialized to transmit impulses. Neurons have two important properties: (1) *excitability,* or the ability to respond to stimuli, and (2) *conductivity,* or the ability to transmit a signal.

Supporting cells are specialized cells, other than neurons, in the central and peripheral nervous systems. For the most part, they support the neurons in some way. Although the neuron is considered the *functional unit* of the nervous system, most of the cells in the system are not neurons, but supporting cells called *neuroglia.* Depending on the region, neuroglial cells outnumber neurons 10 to 50 times.

Neurons

Neurons are among the most specialized types of cells. Although they vary greatly in shape and size, almost all neurons contain three principal parts, each associated with a specific function: (1) the cell body, (2) dendrites, and (3) an axon [FIGURE 12.2A].

Cell body The *cell body* of a neuron may be star-shaped, roundish, oval, or even pyramid-shaped, but its distinguishing structural features are its spreading processes (branches or fibers) that send or receive impulses to or from other cells. Besides varying in shape, cell bodies may vary in size.

A cell body has a large nucleus, which contains a prominent nucleolus, as well as several structures that are responsible for the metabolism, growth, and repair of the neuron. These structures include chromatophilic sub-

*The word *visceral* pertains to a *viscus* (L. body organ), one of several large body organs in any of the great body cavities; note that *viscera* is the plural of *viscus.* Do not confuse *viscus* with *viscous*, which means "flowing with relative difficulty."

FIGURE 12.1 THE HUMAN NERVOUS SYSTEM

[A] The brain and spinal cord make up the central nervous system (beige). The peripheral nervous system (yellow) is composed of nerves whose fibers convey nerve impulses between the brain and spinal cord and the rest of the body. The plexuses shown are networks of nerves, blood vessels, and lymphatic vessels. Only some nerves are shown. **[B]** A schematic drawing of the interrelationships between the various parts of the peripheral nervous system.

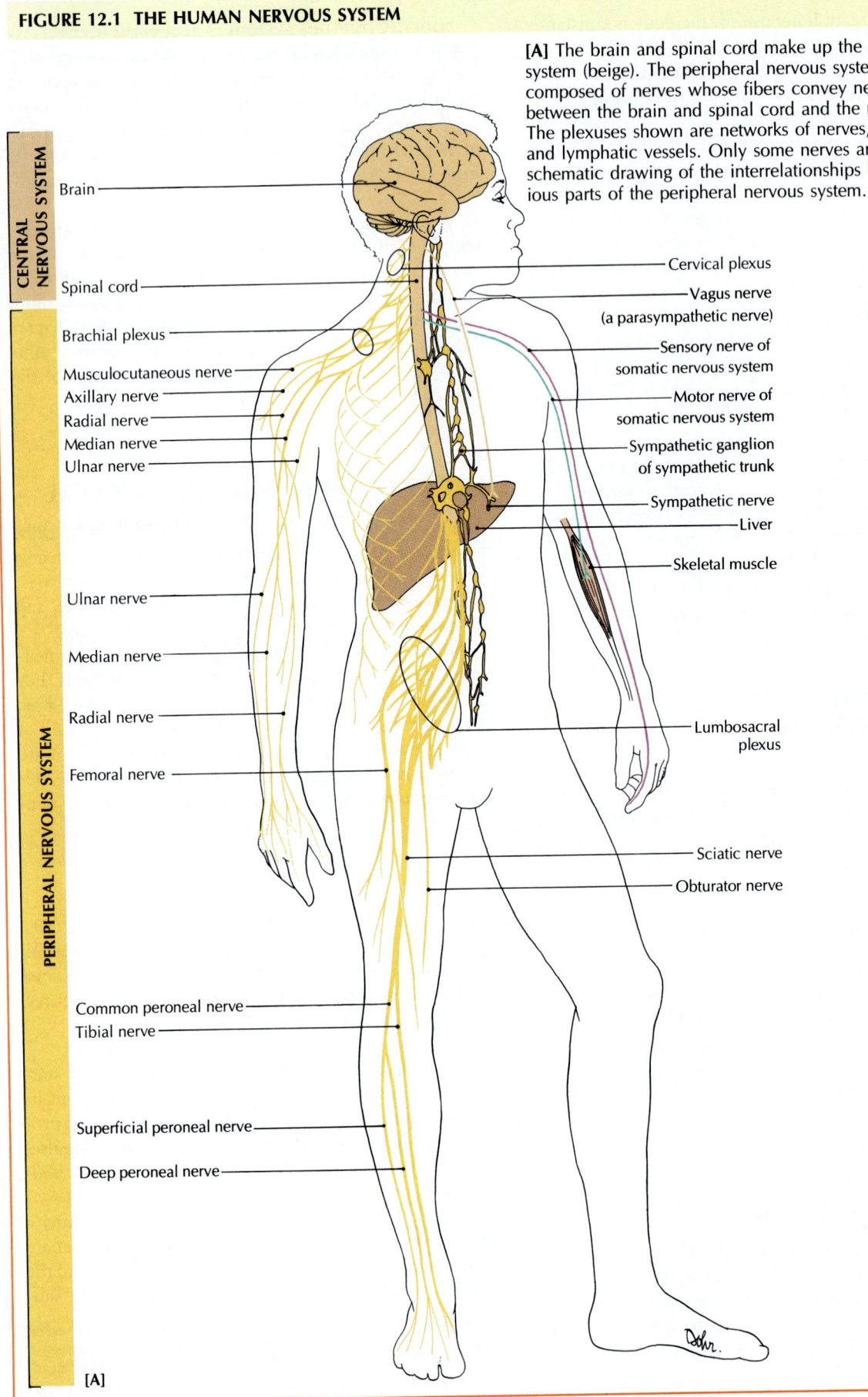

CENTRAL NERVOUS SYSTEM

Brain

Spinal cord

PERIPHERAL NERVOUS SYSTEM

Brachial plexus

Musculocutaneous nerve

Axillary nerve

Radial nerve

Median nerve

Ulnar nerve

Ulnar nerve

Median nerve

Radial nerve

Femoral nerve

Common peroneal nerve

Tibial nerve

Superficial peroneal nerve

Deep peroneal nerve

Cervical plexus

Vagus nerve
(a parasympathetic nerve)

Sensory nerve of
somatic nervous system

Motor nerve of
somatic nervous system

Sympathetic ganglion
of sympathetic trunk

Sympathetic nerve

Liver

Skeletal muscle

Lumbosacral
plexus

Sciatic nerve

Obturator nerve

[A]

Divisions	Functions

CENTRAL NERVOUS SYSTEM (CNS)

Brain and spinal cord.	Body's central control system. Receives impulses from sensory receptors, relays impulses for action to muscles and glands. Interpretive functions involved in thinking, learning, memory, etc.

PERIPHERAL NERVOUS SYSTEM (PNS)

Cranial and spinal nerves, with afferent (sensory) and efferent (motor) nerve cells.	Enables brain and spinal cord to communicate with entire body. *Afferent (sensory) cells:* Carry impulses from receptors to CNS. *Efferent (motor) cells:* Carry impulses from CNS to effectors.

Somatic nervous system

1 Composed of nerve fibers (axons) conveying information from receptors for pain, temperature, and mechanical stimuli in somatic structures such as skin, muscles, joints. 2 Characterized by nerve fibers of lower motor neurons that go directly from CNS to synapses with skeletal muscle.	*Afferent (sensory) division:* Receives and processes sensory input from skin, skeletal muscles, tendons, joints, eyes, ears. *Efferent (motor) division:* Excites skeletal muscles.

Visceral nervous system

1 Composed of nerve fibers conveying information from receptors in visceral organs (structures) such as the heart and digestive system. 2 Characterized by nerve fibers of motor neurons that go from CNS to interact with other nerve cells within a ganglion located outside CNS; nerve fibers of second nerve cells that innervate the effectors.	*Afferent (sensory) division*:* Receives and processes input from internal organs of cardiovascular, respiratory, digestive, urinary, and reproductive systems.
3 Includes the autonomic nervous system	*Efferent (motor) division* or *autonomic nervous system:* May inhibit or excite smooth muscle, cardiac muscle, glands.
Sympathetic nervous system	Relaxes intestinal wall muscles; increases sweating, heart rate, blood flow to skeletal muscles.
Parasympathetic nervous system	Contracts intestinal wall muscles; decreases sweating, heart rate, blood flow to skeletal muscles.
Enteric nervous system	Neural network within the gastrointestinal tract; integrates and coordinates activities such as peristalsis.

*Some textbooks include the general visceral afferent (sensory) division in the autonomic nervous system.

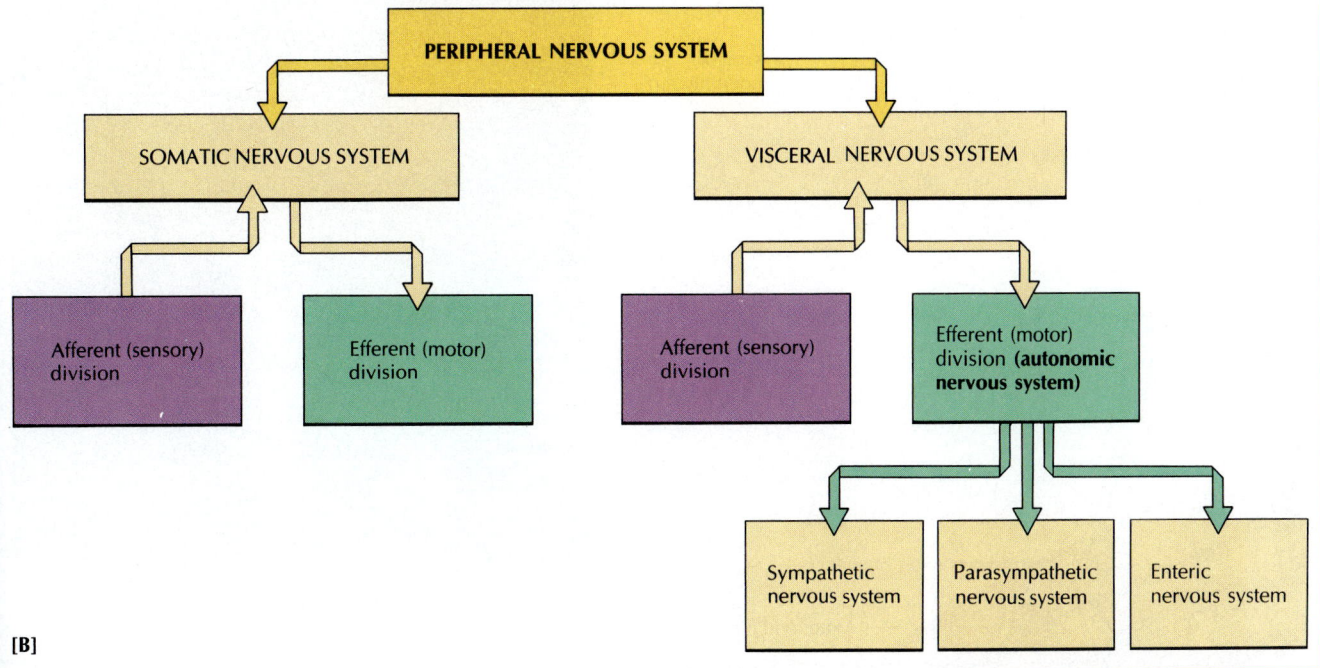

[B]

FIGURE 12.2 A NEURON

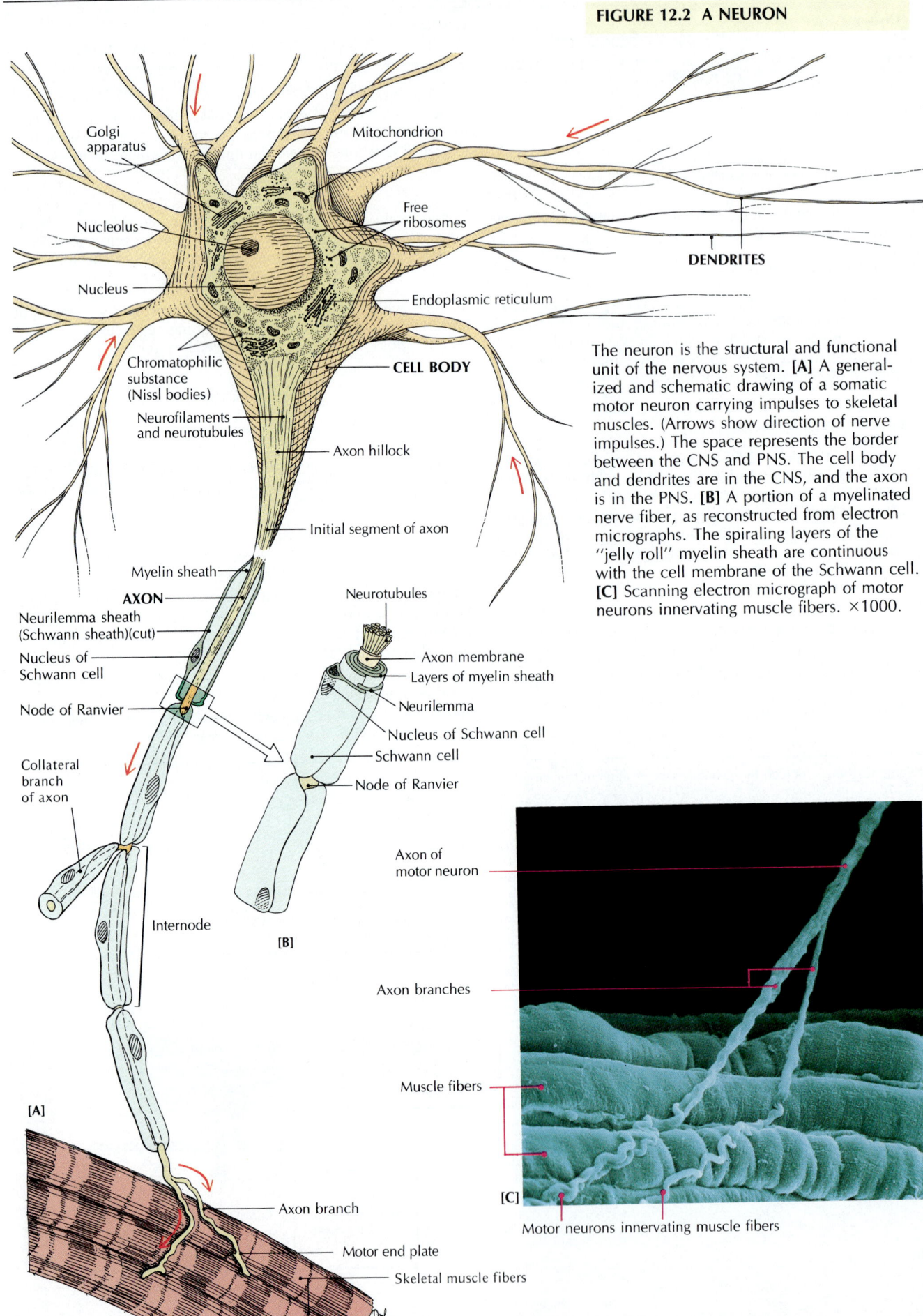

Golgi apparatus

Mitochondrion

Nucleolus

Free ribosomes

Nucleus

DENDRITES

Endoplasmic reticulum

Chromatophilic substance (Nissl bodies)

CELL BODY

Neurofilaments and neurotubules

Axon hillock

Initial segment of axon

Myelin sheath

AXON

Neurotubules

Neurilemma sheath (Schwann sheath)(cut)

Nucleus of Schwann cell

Axon membrane

Layers of myelin sheath

Node of Ranvier

Neurilemma

Nucleus of Schwann cell

Schwann cell

Collateral branch of axon

Node of Ranvier

Internode

[B]

[A]

Axon branch

Motor end plate

Skeletal muscle fibers

Axon of motor neuron

Axon branches

Muscle fibers

[C]

Motor neurons innervating muscle fibers

The neuron is the structural and functional unit of the nervous system. [A] A generalized and schematic drawing of a somatic motor neuron carrying impulses to skeletal muscles. (Arrows show direction of nerve impulses.) The space represents the border between the CNS and PNS. The cell body and dendrites are in the CNS, and the axon is in the PNS. [B] A portion of a myelinated nerve fiber, as reconstructed from electron micrographs. The spiraling layers of the "jelly roll" myelin sheath are continuous with the cell membrane of the Schwann cell. [C] Scanning electron micrograph of motor neurons innervating muscle fibers. ×1000.

stance, or Nissl bodies (made up of rough endoplasmic reticulum and free ribosomes); endoplasmic reticulum; lysosomes; mitochondria; neurofilaments (intermediate filaments); neurotubules (microtubules); and Golgi apparatus.

The neurotubules and neurofilaments of a cell body are tubelike protein structures; they actually extend into and throughout the cell body and processes of the neuron and run parallel to the long axis of each process. *Neurotubules* function in the intracellular transport of proteins and other substances, in both directions between the cell body and the ends of the processes. *Neurofilaments* are semirigid structures that provide a skeletal framework for the axon. In Alzheimer's disease, many neuron cell bodies in the cerebral cortex of the brain contain "tangles" of neurotubules and neurofilaments (see page 420).

Dendrites and axons Neurons have two types of processes: dendrites and axons. (An axon or dendrite of a neuron is also called a *nerve fiber.*) As shown in FIGURE 12.2A, a typical neuron has many short, threadlike processes called *dendrites* (Gr. *dendron*, tree), which are actually extensions of the cell body. Dendrites conduct nerve impulses *toward* the cell body. A neuron may have as many as 200 dendrites.

A neuron generally has just one *axon,* a slender process that extends from the cell body for from less than a millimeter (as in the brain) to more than a meter (as in the axons of the sciatic nerve, which extends from the spinal cord to innervate muscle cells in the leg). An axon carries nerve impulses *away from* the cell body to the next neuron, muscle cell, or gland cell.

In most neurons, the axon originates from a cone-shaped elevation of the cell body called the *axon hillock* [FIGURE 12.2A]. The thin part of the axon immediately after the axon hillock is the *initial segment* [see FIGURE 12.7], the site where a nerve impulse is initiated. The initial segment is important in generating a nerve impulse because the membrane of this region has the lowest threshold of excitation, making it the most excitable portion of the axon.

An axon may have some side processes called *collateral branches* [FIGURE 12.2A]. An axon and its collateral branches end in a spray of small axon branches, or *telodendria* [FIGURE 12.2A, C], whose even smaller branches usually have tiny swellings called **end bulbs.** A *synapse* (Gr. *synapsis,* a connection) is the junction between the end bulb of one axon and the cell body, dendrite, or axon of another neuron [see FIGURE 12.13].

Myelination of axons Some axons are covered with layers of a lipid sheath called *myelin* (MY-ih-lihn; Gr. *myelos,* marrow). The myelin sheath is formed by specialized nonneural cells called **Schwann cells,** in the peripheral nervous system, and by *oligodendrocytes* in the central nervous system [FIGURE 12.3]. The outer layer, or sheath, of these cells is the **neurilemma sheath** (Gr.

nerve + rind or husk). FIGURE 12.3A shows how a Schwann cell produces large amounts of myelin, which is wrapped around the axon in tight, spiraled layers. A myelin sheath may consist of as many as 100 layers. A myelinating Schwann cell covers only one axon, whereas an oligodendrocyte, with its many processes, may surround several axons [FIGURE 12.3D].

A myelin sheath is segmented, interrupted at regular intervals by gaps called *nodes of Ranvier* [FIGURES 12.2A, B and 12.4A]. The distance from one node to the next is called an *internode* [FIGURE 12.2A]. The thicker the diameter of the sheath, the longer the internode. In the peripheral nervous system, the myelin sheath of each internode is formed by one Schwann cell [FIGURE 12.4]. Myelin is absent at each node. In the central nervous system, each of the processes of an oligodendrocyte forms an internode of the myelin sheath.

An axon that has a myelin sheath is called a *myelinated axon,* while an axon that does not have a myelin sheath is called an *unmyelinated axon.* A myelinated nerve fiber transmits a nerve impulse faster than an unmyelinated fiber does.

Axoplasmic flow and axonal transport Many substances are transported from the cell body throughout the network of axons and dendrites, and from the axons and dendrites to the cell body. In this way, the cell is kept informed of the metabolic needs and conditions of its most distant parts. Such intracellular transport is routed primarily by neurotubules, which contain several different nerve tracts. Two types of transport mechanisms are known to occur within neurons: axoplasmic flow and axonal transport. Both mechanisms are active within all axons and dendrites.

Axoplasmic flow is a one-way transport *away from* the cell body. It produces a relatively slow movement of about 1 to 5 mm/day, conveying proteins and other substances to renew and maintain the cytosol of mature neurons. Axoplasmic flow supplies the materials necessary for the growth of axons and dendrites, as well as for their development and regeneration.

A two-way system of transport within neurons, called *axonal transport* [FIGURE 12.5], provides relatively fast movement of materials at about 400 to 2000 mm/day. Fast transport is made possible by neurotubules; when they are disrupted, fast transport ceases. Axonal transport conveys worn-out materials to lysosomes in the cell body for degradation, disposal, and recycling. It also conveys organelles such as mitochondria,

---

Q: *What are the relative sizes of the parts of a neuron?*

A: If the cell body of a motor neuron were enlarged to the size of a baseball, the axon would extend about a mile, and the dendrites would fill a large athletic field house or convention center.

FIGURE 12.3 MYELIN SHEATHS

[A] Schematic drawing of a single axon surrounded by spiraling layers of a myelin sheath in the peripheral nervous system. On the right is a schematic drawing of the early development of a myelin sheath around an axon. Note that the direction of growth (arrow) occurs at the inner part of the sheath. [B] Electron micrograph of a rat's sciatic nerve in cross section, showing the layers of myelin sheath around the axon (×29,000); the boxed area is enlarged (×129,000) in the inset (top). [C] Nine unmyelinated nerve fibers enclosed within individual troughs (invaginations) of a Schwann cell. [D] A schematic drawing of an oligodendrocyte, showing how it forms myelin sheaths around more than one axon.

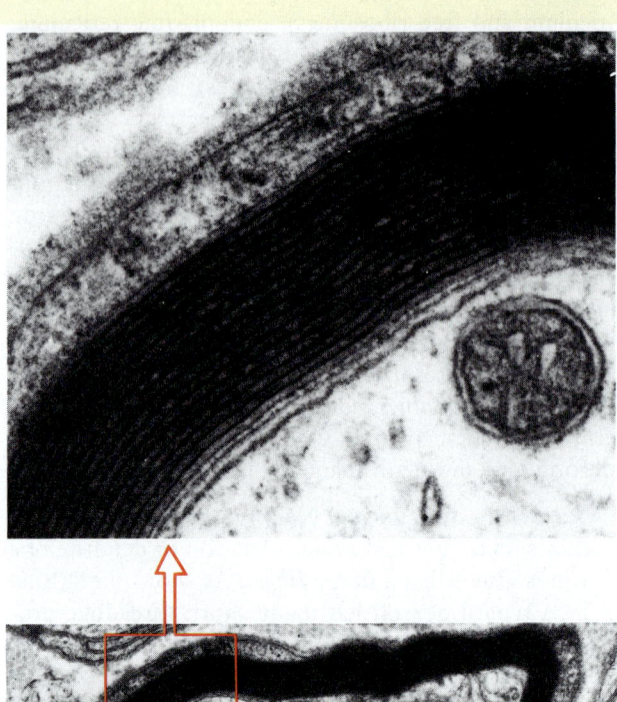

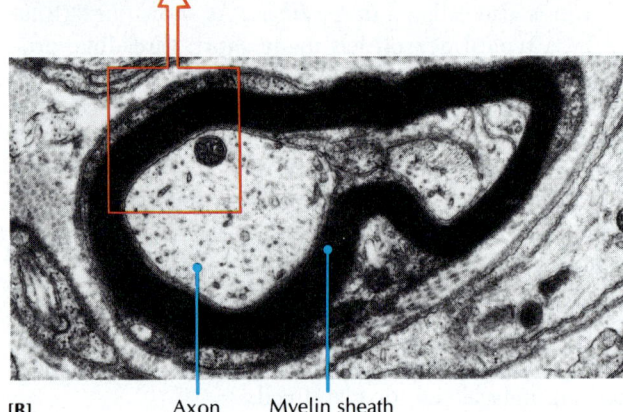

[B]   Axon   Myelin sheath

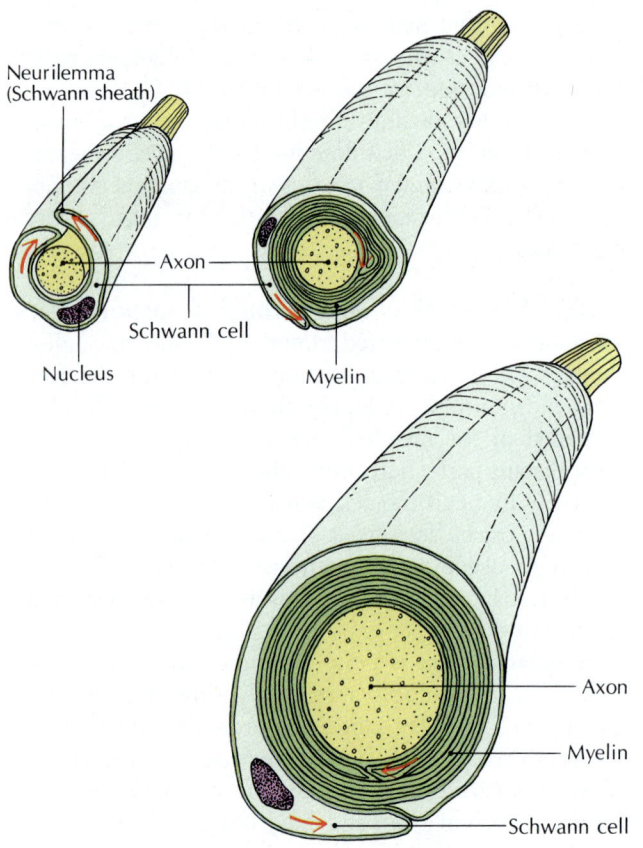

Neurilemma (Schwann sheath)

Axon

Schwann cell

Nucleus

Myelin

Axon

Myelin

Schwann cell

[A]

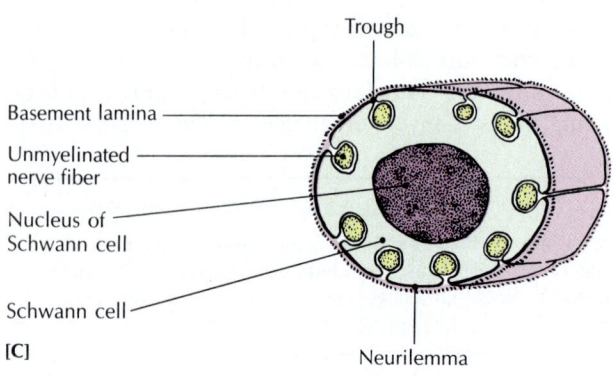

Trough

Basement lamina

Unmyelinated nerve fiber

Nucleus of Schwann cell

Schwann cell

Neurilemma

[C]

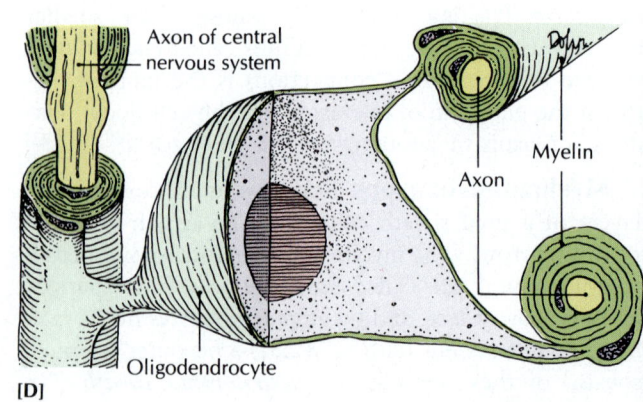

Axon of central nervous system

Myelin

Axon

Oligodendrocyte

[D]

FIGURE 12.4 NODES OF RANVIER

[A] The node of Ranvier is the unmyelinated gap between two myelinated segments (internodes) of an axon in the CNS. **[B]** Scanning electron micrograph of myelinated axons in the peripheral nervous system. The nodes of Ranvier are shown clearly, as are the many collagen fibers that bind the axons together. ×1885. (Richard G. Kessel and Randy H. Kardon, *Tissues and Organs: A Text-Atlas of Scanning Electron Microscopy,* San Francisco, W. H. Freeman, 1979.)

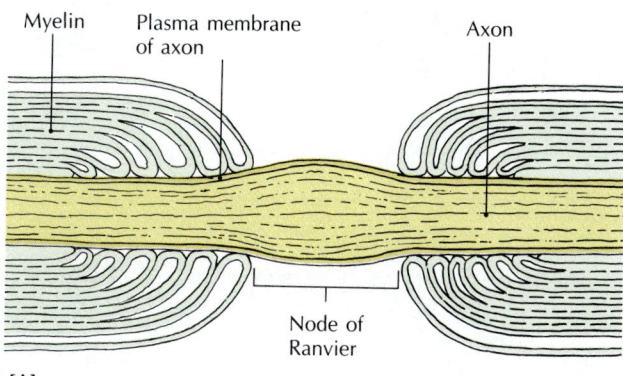

[A]

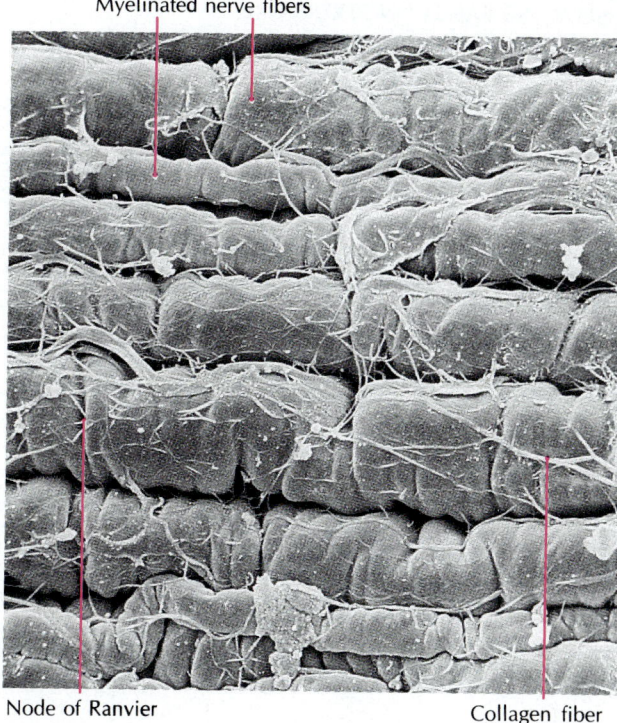

Myelinated nerve fibers

Node of Ranvier

Collagen fiber

[B]

FIGURE 12.5 AXONAL TRANSPORT

Neurotransmitters and other substances to be transported along an axon between the cell body and synaptic terminal of a neuron are contained within vesicles that are moved the length of the axon along the outer surface of neurotubules that act like conveyor belts (arrows). Precursors of neurotransmitters and plasma membrane proteins are produced by the Golgi apparatus in the cell body. Surplus substances are packaged and transported back to the cell body, where they are degraded by lysosomes. Mitochondria are transported in either direction in response to the energy needs of the cell.

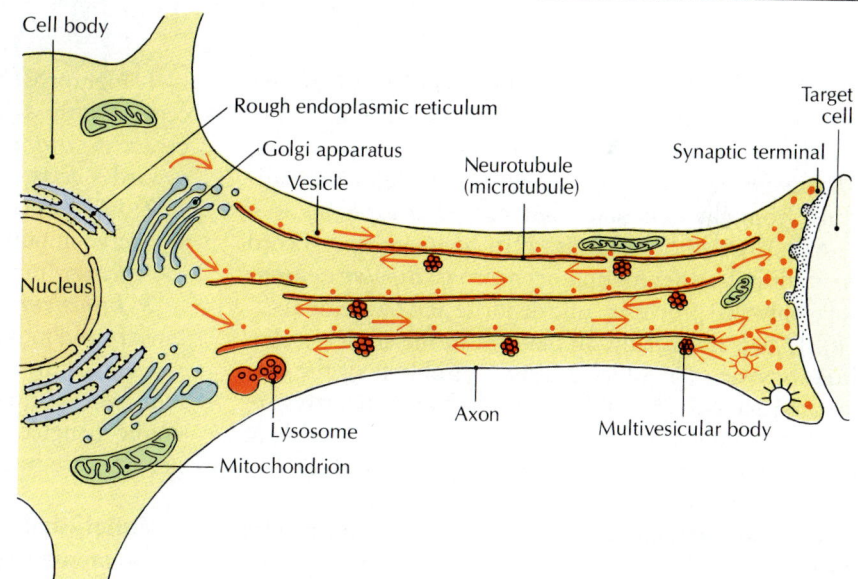

vesicles containing peptides and precursors of neurotransmitters, and components for cell membranes away from the cell body.

On a single neurotubule, a vesicle can pass another vesicle moving in the same direction on a separate tract, or two vesicles can move in opposite directions simultaneously on separate tracts of the same neurotubule. Vesicles may even shift from one neurotubule to another.

Q: *What is a nerve?*

A: In common usage, a nerve is simply a bundle of nerve fibers enclosed in a connective tissue sheath, like many telephone wires in a cable. The anatomical term for bundles of nerve fibers and their myelin sheaths in the central nervous system is *tracts;* the term for those in the peripheral nervous system is *nerves.*

FIGURE 12.6 CLASSIFICATION OF NEURONS BASED ON STRUCTURE

[A] Multipolar neuron. [B] Bipolar neuron. [C] Unipolar neuron. Arrows indicate the direction of nerve impulses.

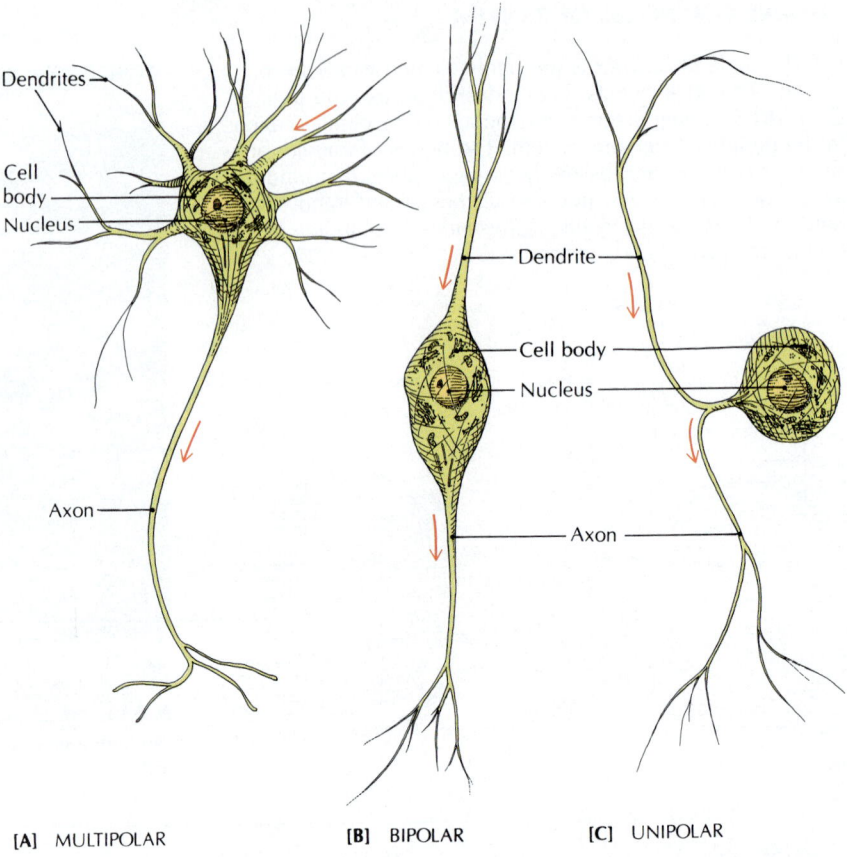

[A] MULTIPOLAR   [B] BIPOLAR   [C] UNIPOLAR

Types of neurons: Based on structure Neurons may be classified according to their structure or, more specifically, according to the number of their processes. *Multipolar neurons* generally have many dendrites radiating from the cell body, but only one axon [FIGURE 12.6A]. Most of the neurons of the brain and spinal cord are multipolar efferent (motor) neurons. *Bipolar neurons* have only two processes: one dendrite and one axon [FIGURE 12.6B]. Essentially, all neurons in the adult develop from bipolar cells, which are the neuroblasts of the embryonic nervous system. In the adult, bipolar neurons are located in only a few structures, including the retina of the eye, the cochlear and vestibular nerves of the ear, and the olfactory nerve in the upper nasal cavity.

Unipolar neurons have a single process that divides into two branches [FIGURE 12.6C]. One branch extends into the brain or spinal cord, conducting nerve impulses away from the cell body, and the other branch extends to a peripheral sensory receptor in a distal part of the body. Unipolar neurons are the most common afferent (sensory) neurons in the peripheral nervous system.

Types of neurons: Based on function The neurons of the peripheral nervous system may be classified as either afferent or efferent, according to the *direction* in which they transmit nerve impulses:

1 Neurons conveying information from sensory receptors in the skin, sense organs, muscles, joints, and viscera *to the central nervous system* are called *afferent* (L. *ad,* toward + *ferre,* to bring), or *sensory, neurons.*

2 *Efferent* (L. *ex,* away from), or *motor, neurons* convey nerve impulses *away from the central nervous system,* to the *effectors* (muscles and glands).

3 *Interneurons,* or *association neurons,* lie (1) between sensory and motor neurons in neural pathways and transmit signals through pathways of the central nervous system, where integration often occurs, and (2) in autonomic ganglia. Most neurons of the body (about 99 percent) are interneurons.

Functional segments of a neuron Most neurons are composed of four distinct functional segments, as shown in FIGURE 12.7:

1 The *receptive segment* of a multipolar neuron is composed of the cell body and its dendrites. It is the segment that receives synaptic inputs from numerous other neurons. These complex inputs are processed in the receptive segment, and if that segment is sufficiently stimulated, the resolution of this processing is conveyed to the initial segment at the junction of the cell body and axon hillock.

FIGURE 12.7 FUNCTIONAL ORGANIZATION OF A TYPICAL NEURON

[A] Most neurons contain a receptive segment, an initial segment, a conductive segment, and a transmissive segment. The cell body may be located within the receptive segment, as in the lower motor neuron shown in **[B]**, or within the conductive segment, as in the sensory neuron shown in **[C]**.

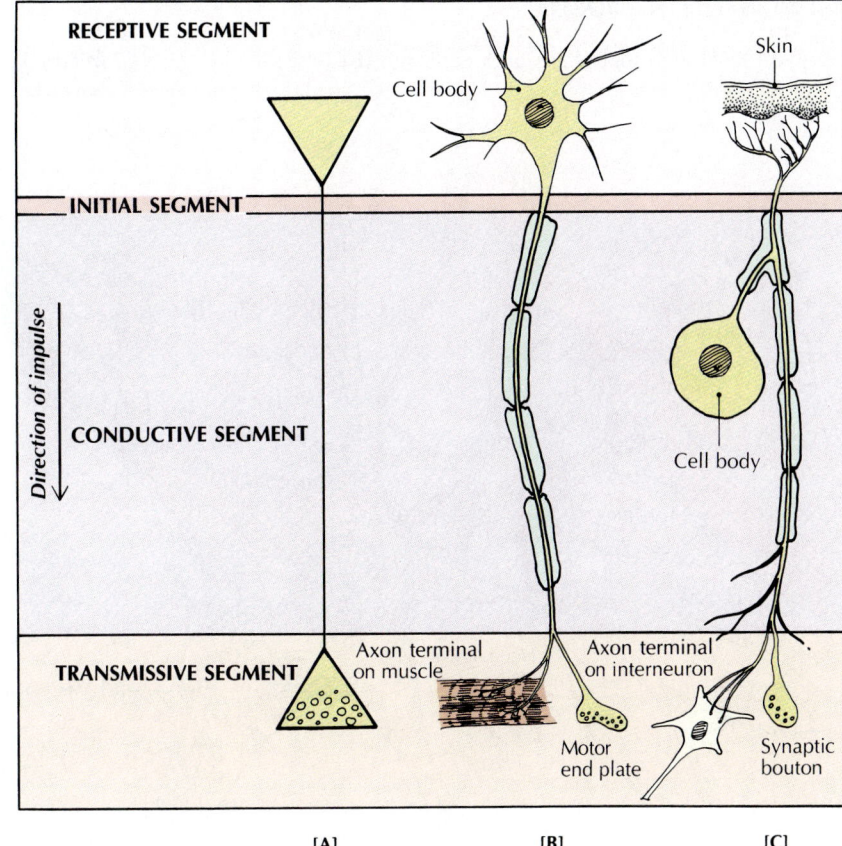

2 The *initial segment* is the trigger zone of the neuron, where the processed neural information of the receptive segment is converted into a nerve impulse. This initial segment, with its low threshold and sodium-ion channels, is critical for the generation of a nerve impulse.

3 The *conductive segment* (the axon) is specialized for the conduction of nerve impulses. It conveys the results of the neural processing of the receptive segment via nerve impulses to the terminal segment.

4 The *transmissive* (effector) *segment* contains the axon terminals that convert the stimulation of the nerve impulse to release chemical neurotransmitters at its synapses. These chemicals exert influences upon the receptor sites on an effector cell (another neuron, a muscle cell, or a gland cell).

Supporting Cells of the Nervous System

Certain cells in the nervous system function mainly to support neurons. The following sections describe supporting cells of the central nervous system (neuroglia) and peripheral nervous system (satellite cells and Schwann cells).

Supporting cells of the central nervous system: Neuroglia
In place of connective tissue, which is rela-
tively sparse in the central nervous system, are *neuroglia* ("nerve glue"; Gr. *glia*, glue), nonconducting cells that protect and nurture, as well as support the nervous system. There are four types of neuroglia: *astrocytes, oligodendrocytes, microglia,* and *ependyma*. They are shown and described in FIGURE 12.8.

Astrocytes and ependymal cells, along with the capillary bed of the central nervous system, form what is known as the *blood-brain barrier*. This barrier permits certain chemical substances to gain access to the neurons, and it also slows down or even prevents other substances, such as penicillin, from reaching the neurons. In this way, the nonneural cells and blood vessels of the brain and spinal cord act to maintain the homeostasis of the environment around each neuron and its processes. (See "The Blood-Brain Barrier" on page 396.)

Supporting cells of the peripheral nervous system: Satellite cells and Schwann cells
The supporting cells of the peripheral nervous system are the *satellite cells* and *Schwann cells*. A sequence of these supporting cells forms a continuous layer that ensheaths each neuron. *Satellite cells* surround each cell body, and *Schwann cells* ensheath the axon and dendrites. Satellite cells and Schwann cells, both of which nurture the neurons, are similar in structure and function. Schwann cells form the

FIGURE 12.8 NEUROGLIA

[A] Astrocyte with foot plates on a capillary. [B] Oligodendrocyte. [C] Microglial cell. [D] Ependyma.

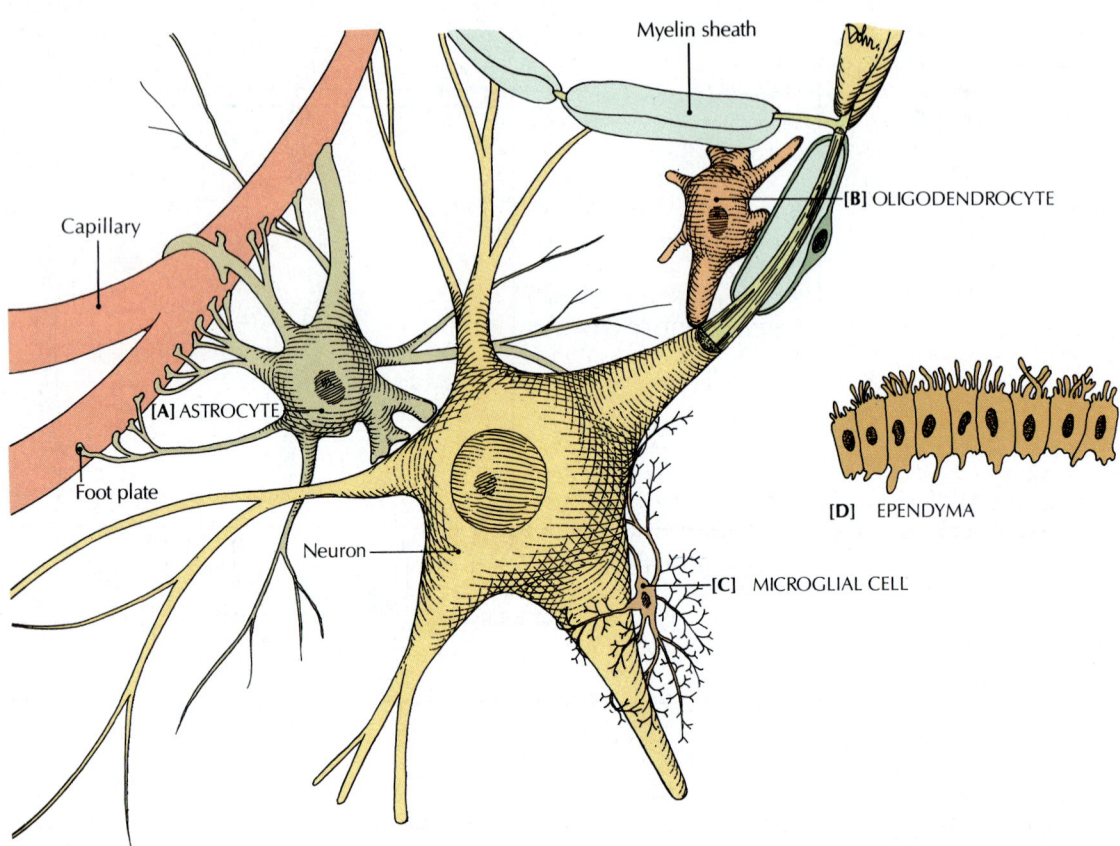

Type of cell	Description and functions
NEUROGLIA OF THE CNS	Associated cells of CNS. Nonconducting. Protect, nourish, support cells of CNS.
Astrocyte	Largest, most numerous glial cell, with long, starlike processes. Sustains neurons nutritionally, helps maintain concentration of chemicals in extracellular fluid, provides packing material and structural support, helps regulate transfer of substances from capillaries to nervous tissue.
Oligodendrocyte	Relatively small, with several branching processes. Found in gray and white matter of CNS. Similar to a Schwann cell. Produces and nurtures myelin sheath segments (internodes) of many nerve fibers (each process forming one internodal segment). Provides supportive framework, supplies nutrition for neurons.
Microglial cell	Small glial cell. Usually found between neurons and along blood vessels. Macrophage, not a true glial cell. Removes disintegration products of damaged neurons.
Ependymal cell	Elongated cell. Arranged in single layer to line central canal of spinal cord and ventricles of brain. Helps to form part of inner membrane of neural tube during embryonic growth, secretes cerebrospinal fluid.
PERIPHERAL GLIAL CELLS	Associated cells of PNS. Provide various types of sheaths. Help to protect, nourish, maintain cells of PNS.
Satellite cell (capsule cell)	Essentially the same structure as a Schwann cell. Forms capsule around cell bodies of neurons of peripheral ganglia; separates cell bodies from connective tissue framework of ganglion.
Schwann cell	Flattened cell. Arranged in single layer along axons and dendrites. Forms myelin sheaths of peripheral axons. Associated with both myelinated and unmyelinated fibers. Involved in regeneration of damaged peripheral nerve fibers.

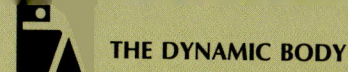

The Guiding "Glue" of the Nervous System

In 1982 a group of Rockefeller University researchers, led by Nobel laureate Gerald M. Edelman, discovered a cellular "glue" that may be involved in holding cells together. The glue in this "glue control model" is a glycoprotein of the plasma membrane of a cell called a *cell adhesion molecule (CAM)*. The CAM found in neurons is called *N-CAM*, and is present in the plasma membranes of embryonic and adult neurons and their processes.

N-CAMs are basically recognition molecules with the ability to recognize each other in adjacent neurons. During early development, neurons aggregate in prescribed groupings when the gene for N-CAM is turned on, and may migrate and have their processes grow when the gene that produces N-CAM is turned off. This enables the neurons to assemble into their final organization, and the processes make synaptic contact with other neurons and muscles when the N-CAM is turned on again. Developmental errors usually result when the genetic mechanism that turns the N-CAM on and off is faulty.

Subsequent research has indicated that N-CAM may be the substance that guides the specific embryonic development and growth of neurons by providing binding sites between neurons and other structures such as muscles. Experiments with animals suggest that a pathway containing N-CAM (among other guiding chemicals) guides the growth of axons from the eye to the brain. In contrast, a region of the brain where axons never grow (between the areas receiving input related to sight and smell) does not contain any N-CAM. N-CAM also seems to guide nerve growth at neuromuscular junctions.

---

myelin sheath of myelinated nerve fibers [FIGURE 12.2]. They also ensheath the axons and dendrites of the unmyelinated fibers but do not form myelin. (Cell bodies are usually unmyelinated.) Schwann cells enable damaged peripheral nerve fibers to regenerate.

ASK YOURSELF

1 What are the basic properties of neurons?

2 What are the main parts of a neuron?

3 What is myelin?

4 What is the difference between afferent and efferent neurons?

5 What is the main function of an interneuron?

6 What are the supporting cells of the central nervous system? What are their functions?

7 What are the supporting cells of the peripheral nervous system?

PHYSIOLOGY OF NEURONS

Neurons are specialized to respond to stimulation. They do this by generating several types of electrochemical impulses. These impulses are expressed as changes in the electrical potentials conducted along the plasma membranes of the dendrites, cell body, and axon of each neuron. The difference in potential across the plasma membrane of a neuron results from differences in the concentration of certain ions on either side of the membrane and differences in the permeability of the membrane to different ions.

In the following sections we will explain how impulses are generated; conducted along a nerve fiber; and transmitted to an adjacent neuron, muscle cell, or gland cell.

Resting Membrane Potential

A "resting" neuron is one that is not conducting a nerve impulse but *is* electrically charged. (It may be compared with a charged battery that does not generate an electrical current until it is switched on.) The plasma membrane of such a resting neuron is said to be **polarized**, meaning that there is a difference in electrical charge between the intracellular fluid (ICF) on the inner side of the membrane and the extracellular fluid (ECF) outside the membrane. The intracellular fluid is negatively charged *with respect to* the extracellular fluid, which is positively charged [FIGURE 12.9].

The difference in the electrical charge between the inside and outside of the plasma membrane at any given point is called the **potential difference**. It creates a potential for electrical activity, called the **resting membrane potential**, along the plasma membrane of a resting neuron.

The two major factors that contribute to the resting membrane potential are (1) the distribution of ions across the membrane and (2) the relative permeability of the membrane to Na^+ and K^+. The extracellular fluid is high in Na^+ and Cl^-, while the intracellular fluid is high in K^+, organic phosphates, and proteins. The extracellular fluid usually contains about 10 Na^+ ions for every K^+ ion. The ratio inside is reversed, with K^+ ions outnumbering Na^+ ions by at least 10 to 1. In addition, in a resting neuron, membrane permeability to K^+ is about 50 to 100 times greater than the permeability to Na^+.

FIGURE 12.9 RESTING MEMBRANE POTENTIAL

Sodium ions (red) and chloride ions (not shown) are concentrated on the outside of the plasma membrane of an unmyelinated nerve fiber; potassium ions (purple), negatively charged protein molecules (not shown), and some chloride ions (not shown) are concentrated on the inside. When the neuron is in this resting condition, it is said to be polarized. The movement of ions through selective channels across the plasma membrane is strictly regulated to maintain the proper concentration and the resultant electrical charge at any given point along the axon.

Key:

● = Sodium ion (Na +)

● = Potassium ion (K +)

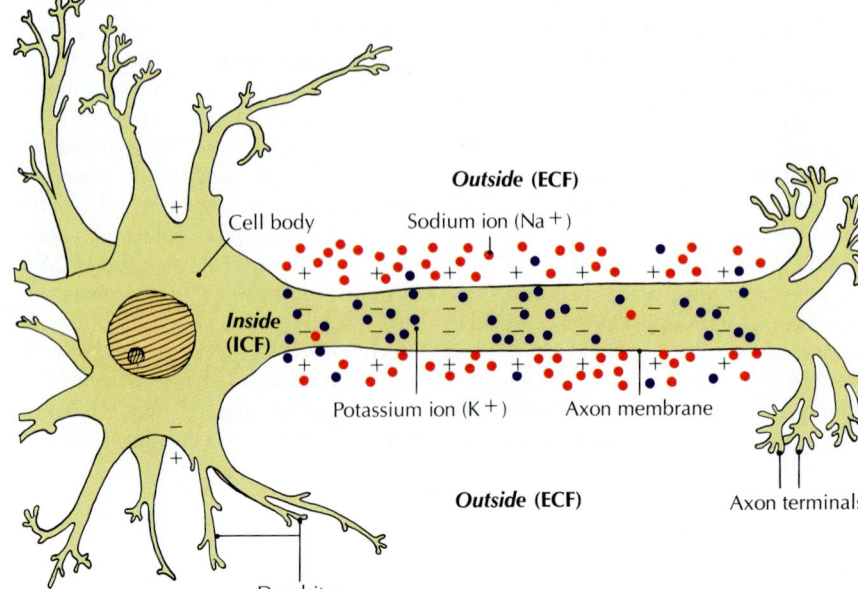

Thus, there is a slow, constant inflow of Na⁺ and an outflow of K⁺. The inward leakage of Cl⁻ down its concentration gradient, along with Na⁺, cannot balance the electrical effects of the outward flow of K⁺. Also, the negatively charged anions, like proteins, are not able to leave the cell with the outflow of K⁺. In summary, these electrical and ionic concentration gradients and permeabilities establish and maintain a resting membrane potential of about −70 mV.*

The sodium-potassium pump A very small part (about 4 percent) of the resting membrane potential is maintained by the ***sodium-potassium pump*** [FIGURE 12.10B], which is powered by the breakdown of ATP to ADP. The more sodium that leaks into the neuron through the plasma membrane, the more active the pump becomes, in order to restore the ionic concentrations that maintain the resting membrane potential. The sodium-potassium pump transports three sodium ions out of the cell for every two potassium ions it brings into the cell. So even though both ions are positive, more positive ions are moving out through the membrane than are moving in. It is partially for this reason that the inner surface of the plasma membrane is more negative than the outer surface.

*The resting membrane potential of −70 mV means that the electrical charge on the inside of the plasma membrane measures 70/1000 of a volt *more negative* with respect to the outside. (By convention, a minus sign denotes a negative charge inside the membrane, with respect to the outside.)

Membrane Channels

In addition to the sodium-potassium pump, there are two types of channels extending through the plasma membranes of neurons: open channels and gated channels.

Open channels *Open channels*, also called *passive channels*, are impermeable to large protein molecules but are always in a state that permits Na⁺, K⁺, Cl⁻, and Ca²⁺ ions to flow passively (diffuse) through the membrane at rates as high as 10⁷ ions per second per channel. Open channels are involved in maintaining resting membrane potentials.

Gated channels Other channels, which are not open all the time, are called ***gated channels***. Gated channels have a molecular "gate" that can open to permit ions to pass through a channel and then close so that the ions cannot pass. There are (1) *transmitter-gated* (chemically gated) channels [FIGURE 12.10A], (2) *voltage-gated channels* [FIGURE 12.10B], and (3) *modality-gated channels*.

Transmitter-gated (or *chemically gated*) *channels* have gates that respond to chemical agents. (Channels on the postsynaptic membrane at a synapse are one example.) A transmitter-gated acetylcholine channel opens for about 1 msec and then closes. During that time, each channel allows about 20,000 Na⁺ ions to flow into the neuron and somewhat fewer K⁺ ions to flow out.

Voltage-gated channels have gates that open or close when the electric potential across the plasma membrane changes. (Channels in the plasma membrane of an axon are one example.)

FIGURE 12.10 TRANSMITTER-GATED CHANNELS AND VOLTAGE-GATED ION CHANNELS

[A] Transmitter-gated (chemically gated) channels open and close their gates in sequence when a nerve impulse is being transmitted. The walls of the channel are thought to be lined with hydrophilic amino acid side chains; the gate is open when an aqueous surface is formed across the lipid bilayer of the plasma membrane. Hydrophobic side chains interact with the lipid bilayer to close the channel. An ion-selective filter determines which ions pass through. **[B]** The sodium-potassium pump is an integral protein. Its energy for the exchange of sodium and potassium ions comes from the hydrolysis of ATP. To attain the resting potential, the *active exchange* of these ions is driven by the pump (transport of two potassium ions into the axon for three sodium ions out of the axon), and the *passive exchange* of these same ions is due to ion diffusion. Thus, the net exchange of these ions (both by pump and by diffusion) across the plasma membrane is essentially balanced to attain a resting potential.

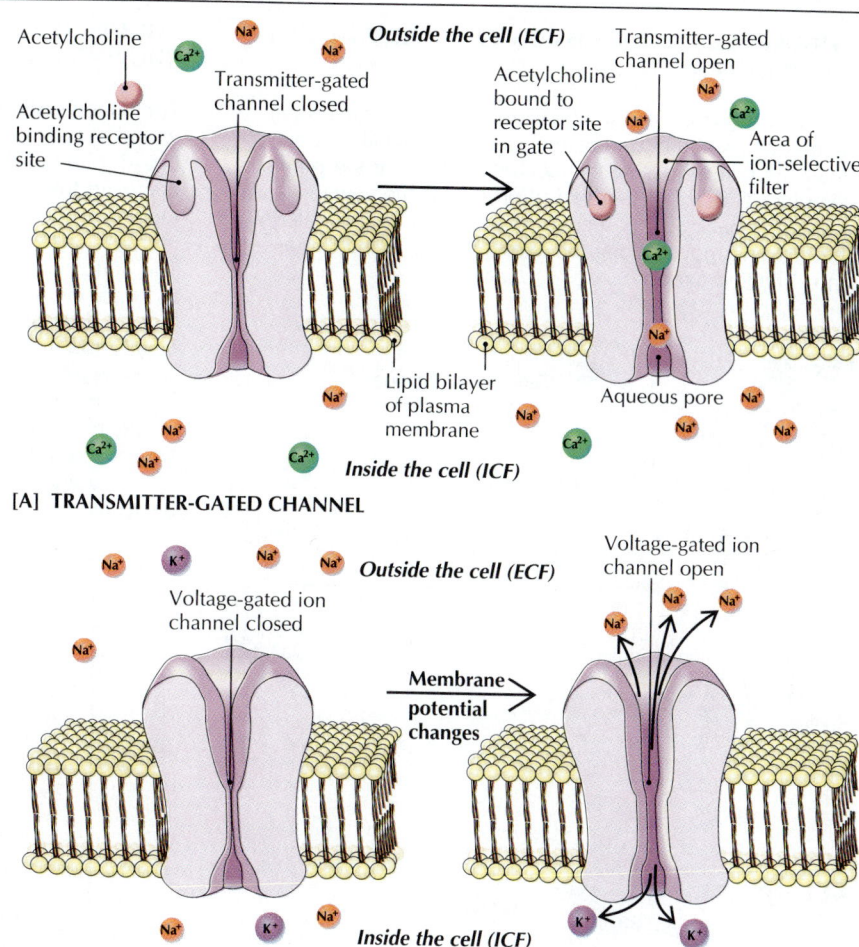

[A] TRANSMITTER-GATED CHANNEL

[B] VOLTAGE-GATED CHANNEL: Sodium -Potassium Pump

(1) Voltage = -70 mV (2) Voltage = -50 mV

Modality-gated channels respond by opening in response to a specific type of stimulus, such as touch or temperature change. (The nerve terminal of a sensory neuron is one example.)

Mechanism of Nerve Impulses (Action Potentials)

The *change* in the resting membrane potential across a plasma membrane is the key factor in the initiation and conduction of a nerve impulse. Conduction differs only slightly between unmyelinated and myelinated fibers, except for its greater velocity along myelinated fibers. The following steps describe conduction in unmyelinated fibers:

1 A stimulus that is strong enough to initiate an impulse in a neuron is called a *threshold stimulus*. When a threshold stimulus is applied to a resting axon membrane, voltage-gated sodium-ion channels open, and so-dium ions rush into the cell, reversing the electrical charge at the point of stimulus. This reversal of charge, giving the *inner* surface of the membrane a positive charge of about 30 mV relative to the *outer* surface [FIGURE 12.11], is called *depolarization*. When a stimulus is strong enough to depolarize the membrane to threshold level and generate a nerve impulse, the neuron is said to *fire*.

2 Once a small area on the axon is depolarized, it stimulates the adjacent area, which contains voltage-gated ion channels, and an *action potential* is initiated [FIGURE 12.12]. An action potential is the electrical current generated in a neuron or muscle cell during its activity, and a train of action potentials is called a *nerve impulse*.

3 Shortly after depolarization, there is (1) an opening of voltage-gated K^+ channels, which accelerates the *outflow* of K^+ ions, and (2) a closing of Na^+ channels, causing the membrane potential to change from +30 mV to the resting membrane potential of −70 mV. This is known as *repolarization*.

FIGURE 12.11 TRANSMISSION OF A NERVE IMPULSE ALONG AN AXON

[A] After a threshold stimulus is applied to a resting axon membrane, sodium-ion channels open and sodium ions enter the cell, reversing the electrical charge at the point of stimulus. This act of *depolarization* begins the generation of a nerve impulse **[B]**. A second impulse can be generated as the first one moves along the axon. (This drawing shows an unmyelinated axon.)

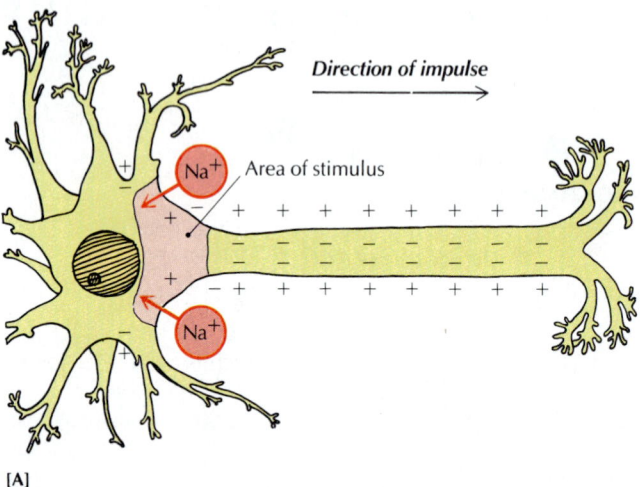

[A]

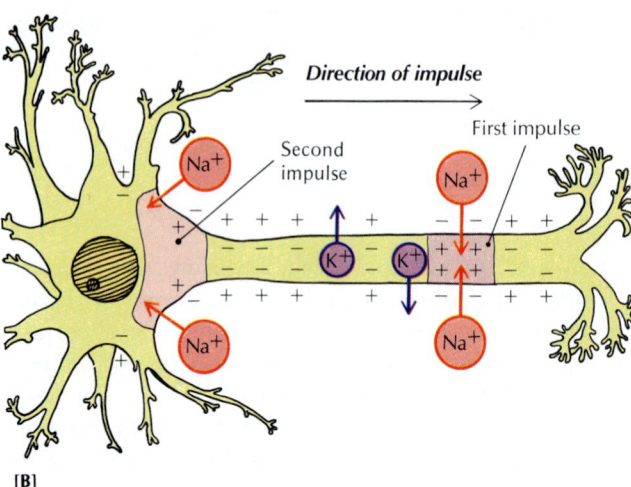

[B]

4 The outflow of K^+ ions during the time the voltage-gated K^+ channels are open may result in an "after-polarization," or ***hyperpolarization,*** a polarization even more negative than the resting membrane potential. (In hyperpolarization, the potential difference across the plasma membrane is greater than the resting membrane potential; that is, the inner surface is more negative with respect to the outer surface than it is during the resting membrane potential.) When the voltage-gated K^+ channels close, the membrane potential returns to its resting level of about -70 mV. The original balance of Na^+ and K^+ ions is restored by the increased permeability to K^+

FIGURE 12.12 VOLTAGE AND IONIC CHANGES DURING AN ACTION POTENTIAL

The changes during an action potential are shown as they would appear on an oscilloscope. During the depolarization phase (pink) of the action potential, there is a rapid increase in the flow of sodium ions (Na^+) into the axon. The repolarization phase (yellow) is characterized by a rapid increase in the flow of potassium ions (K^+) out of the axon. The action potential is sometimes called a spike because of its shape on a recorder. Note that this recording illustrates that (1) the upstroke resulting in an action potential is due largely to the flow of Na^+ into the axon and (2) the downstroke resulting in the resting membrane potential is due largely to the flow of K^+ out of the axon. Hyperpolarization results from the excess outflow of K^+ ions, corrected by the Na^+-K^+ pump. The symbols $[\uparrow P_{Na^+}]$ and $[\uparrow P_{K^+}]$ mean "permeability to Na^+ rises" and "permeability to K^+ rises," respectively.

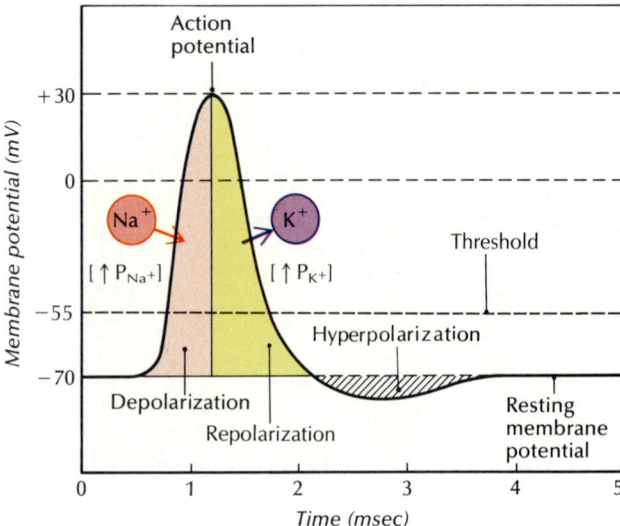

and the action of the sodium-potassium pump, which restore the negative charge inside the cell. Thus, the transmission of a nerve impulse along the plasma membrane may be visualized as a *wave* of depolarization and repolarization.

5 After each firing, there is a time interval of from 0.5 to 1 msec before an adequate stimulus can generate another action potential. This period of time when an excitable cell cannot generate another action potential is called the ***refractory period.*** Despite the refractory period, most nerve fibers are capable of generating about 300 impulses per second.

All-or-None Principle

A minimum stimulus is necessary to initiate an action potential, but an increase in the *intensity* of the stimulus does not increase the strength of the impulse. The principle is the same as when a gun is fired: If the trigger is

Principles of Electricity

Electricity is a phenomenon associated with the movement of charged particles, most commonly electrons. Although some molecules have no net electric charge because they contain equal numbers of electrons and protons, many molecules are charged; electrically charged particles are called *ions*.

Some important ions in the body are sodium (Na^+), potassium (K^+), and chloride (Cl^-). As you can see, some particles are positively charged and some are negative. Particles with like charges repel each other, while particles with opposite charges attract each other. The force that draws oppositely charged particles together and pushes particles with like charges apart has the potential to do work. Because this potential is determined by the *difference* in charge between two points, it is called the potential difference, or simply, the **potential.** The measure of this electrical potential is called **voltage** and is calculated in volts (V) or millivolts (mV). Because the charges in the human body are small, the preferred unit is millivolts (1 mV = 0.001 V).

The flow of an electrical charge between two points is called a **current.** The amount of current depends on (1) the potential difference (the difference in charge) between the two points and

(2) the nature of the material through which the current moves. The hindrance of current flow caused by the nature of the material is called **resistance.** The relationship between current (I), voltage (V), and resistance (R) is expressed as **Ohm's law:**

$$Current = \frac{voltage}{resistance},$$

or

$$I = \frac{V}{R}$$

Materials that have a high resistance (such as myelin sheaths around nerve fibers) are known as *insulators;* insulators do not conduct current well, if at all. Materials with a low resistance are called *conductors.* Although water is a poor conductor, it becomes a good conductor when sodium chloride (table salt) is added because the sodium and chloride ions carry the current. Extracellular fluid is a good conductor because it contains many ions, but because lipids have few charged groups, their resistance is too high for them to be able to carry a current. With this in mind, we can see how important plasma membranes are to the conduction of a current, with outer lipid layers of high resistance and inner water layers of low resistance (see table).

DISTRIBUTION OF MAJOR IONS ACROSS THE PLASMA MEMBRANE OF A TYPICAL NERVE CELL

Ion	Concentration, mmol/L	
	Extracellular	Intracellular
Na^+	150	15
Cl^-	110	10
K^+	5	150

Source: Arthur J. Vander, James H. Sherman, and Dorothy S. Luciano, *Human Physiology,* 6th ed. New York: McGraw-Hill, 1994, p. 188.

The flow of an ion through a membrane is determined by the *electrochemical gradient,* which is a combination of the voltage gradient and concentration gradient for that particular ion across the membrane. When both gradients are balanced, the electrochemical gradient is zero, and there is no net flow of ions through the membrane. The magnitude of this *equilibrium potential* is determined by the difference in ion concentrations between the two sides of the membrane. The greater the difference, the greater the tendency for the ions to diffuse in one direction (toward the lower concentration).

---

not pulled hard enough, nothing happens, but the minimum required pull on the trigger will fire the gun. However, pulling the trigger harder will not make the gun fire harder. The principle that states that a neuron will fire at full power or not at all is known as the **all-or-none principle.** It applies primarily to action potentials in axons.

But what about the difference we feel between a light touch and a strong shove, or between a soft sound and a loud one? Don't those differences contradict the all-or-none principle? No, they do not. Such differences are perceived when (1) the *frequency* of the impulses, not their strength, is changed. Some nerve fibers can conduct at different frequencies per second. The more frequent the impulses, the higher the level of excitation. Also, (2) the *number of neurons* involved makes a difference in how strong the stimulus is perceived to be. You feel the difference between a light push and a strong shove, for example, because the shove affects more of your neurons.

But when it comes to a single axon in an experimental situation, the all-or-none principle still applies.

Speed of Impulse Conduction

The speed of impulse conduction is determined primarily by two factors: (1) the diameter of the axon and (2) the presence or absence of a myelin sheath. Large-diameter fibers (5 to 20 μm) conduct impulses at speeds greater than smaller ones; the presence of myelin also increases conduction velocity. These large-diameter fibers are called *A fibers;* they are myelinated and conduct impulses at rates of from 12 to 120 m/sec. *A* fibers are present in nerves to skeletal muscles, as well as in large sensory nerves associated with heat and cold, touch, pressure, and position of joints. These are fast-conducting fibers that can alter the situation or environment.

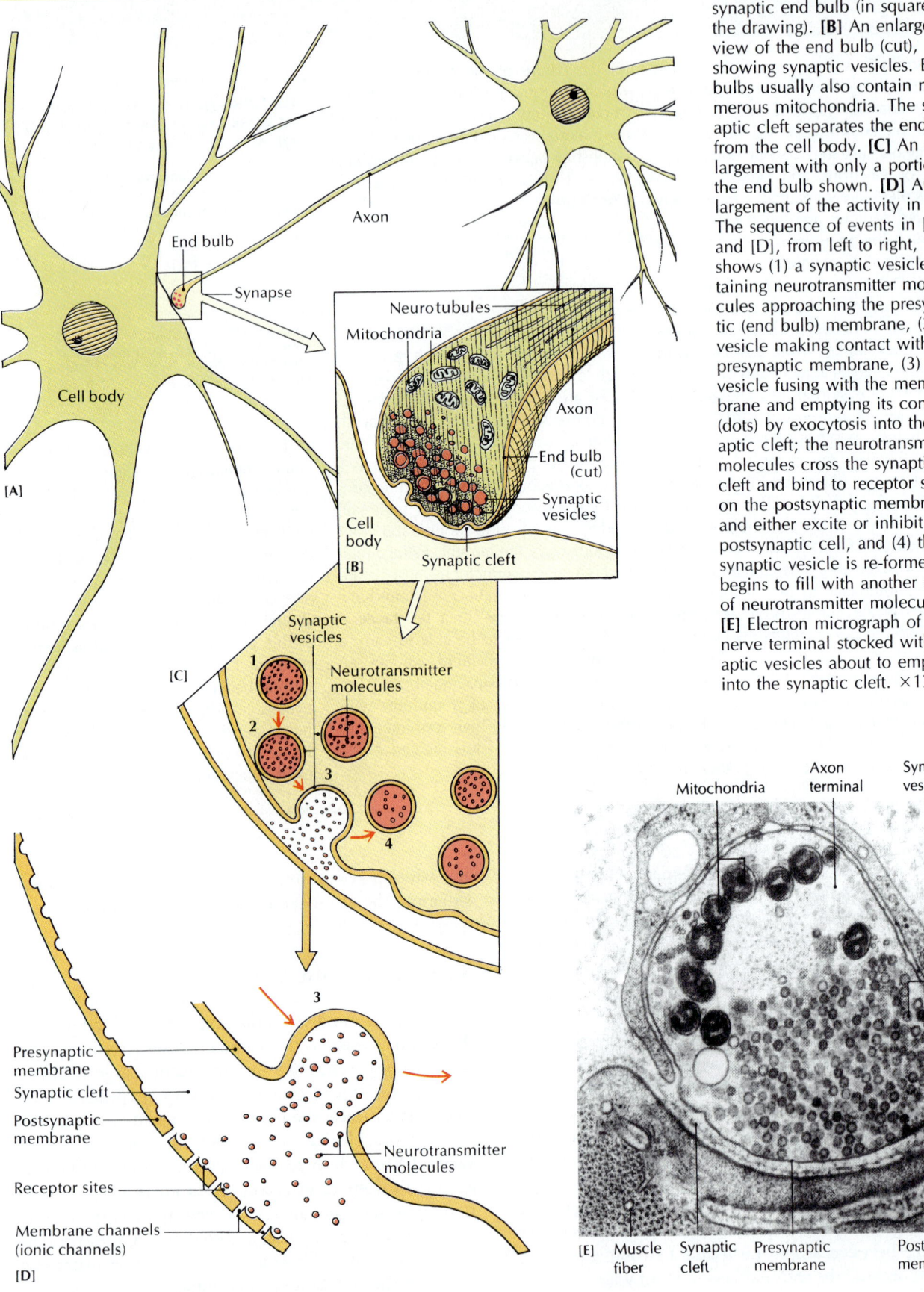

FIGURE 12.13 TRANSMISSION OF A NERVE IMPULSE ACROSS A SYNAPSE

[A] A low-magnification view of a neuron, adjacent to an axon synaptic end bulb (in square on the drawing). **[B]** An enlarged view of the end bulb (cut), showing synaptic vesicles. End bulbs usually also contain numerous mitochondria. The synaptic cleft separates the end bulb from the cell body. **[C]** An enlargement with only a portion of the end bulb shown. **[D]** An enlargement of the activity in [C]. The sequence of events in [C] and [D], from left to right, shows (1) a synaptic vesicle containing neurotransmitter molecules approaching the presynaptic (end bulb) membrane, (2) the vesicle making contact with the presynaptic membrane, (3) the vesicle fusing with the membrane and emptying its contents (dots) by exocytosis into the synaptic cleft; the neurotransmitter molecules cross the synaptic cleft and bind to receptor sites on the postsynaptic membrane and either excite or inhibit the postsynaptic cell, and (4) the synaptic vesicle is re-formed and begins to fill with another supply of neurotransmitter molecules. **[E]** Electron micrograph of a nerve terminal stocked with synaptic vesicles about to empty into the synaptic cleft. ×17,400.

Labels in figure:

[A] Axon, End bulb, Synapse, Cell body

[B] Neurotubules, Mitochondria, Axon, End bulb (cut), Synaptic vesicles, Cell body, Synaptic cleft

[C] Synaptic vesicles, Neurotransmitter molecules, 1, 2, 3, 4

[D] Presynaptic membrane, Synaptic cleft, Postsynaptic membrane, Receptor sites, Membrane channels (ionic channels), Neurotransmitter molecules, 3

[E] Mitochondria, Axon terminal, Synaptic vesicles, Muscle fiber, Synaptic cleft, Presynaptic membrane, Postsynaptic membrane

FIGURE 12.14 PRESYNAPTIC AND POSTSYNAPTIC NEURONS

Highly simplified representation of the relationship of presynaptic and post-synaptic neurons and the direction of nerve impulse transmission.

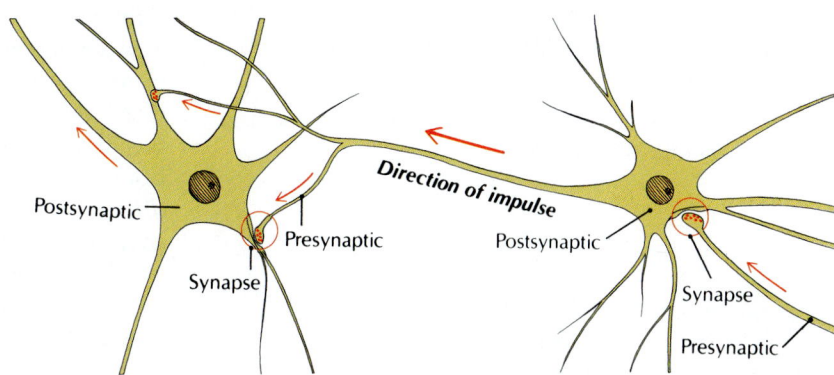

B fibers are smaller-diameter (1.5 to 5 μm), myelinated neurons that have slower conduction velocities (15 m/sec) than A fibers. B fibers are found in nerves that conduct impulses from the viscera to the brain and spinal cord and in the autonomic nervous system. Since B fibers conduct impulses more slowly than A fibers, they are found where immediate responses are not demanded.

C fibers are the smallest-diameter (0.5 to 1.5 μm), unmyelinated neurons and have the slowest conduction velocities (0.5 to 2 m/sec). Unmyelinated C fibers conduct some heat, cold, pain, touch, and pressure impulses from the skin and pain impulses from the viscera. C fibers are also found in the autonomic nervous system.

The diameter of an axon is roughly proportional to the length of the internode; thus, the longer the internode, the faster the speed of impulse conduction.

Saltatory conduction   Nerve impulses travel along unmyelinated fibers relatively slowly and steadily. In *myelinated* neurons, however, the passage of impulses is speeded up considerably. This occurs because the myelin sheath around a myelinated fiber acts as insulation, and for the impulse to move along the neuron, it must "jump" from one node of Ranvier to the voltage-gated channels of the next node. For this reason, conduction along myelinated neurons is known as *saltatory conduction* (L. *saltare,* to jump).

Voltage-gated channels are highly concentrated in the tiny portion of the axon exposed to the extracellular fluid at the node of Ranvier. An inactive node is hypersensitive to changes in voltage and is easily depolarized by an adjacent activated node. The inactive node is stimulated to fire, producing an action potential that jumps to the next node, and so on, all along the neuron. In essence, saltatory conduction occurs because the changes associated with the nerve impulse occur only at the exposed nodes of Ranvier, not along the insulated (myelinated) areas of the axons.

Saltatory conduction is significant because it provides a mechanism for achieving different speeds of conduction in different nerve fibers. It also requires less energy than continuous conduction because it depolarizes only the nodes of Ranvier, not the entire axon length.

Transmission across Synapses

After a nerve impulse travels along an axon, it reaches the branching axon terminals in the transmissive segment of the neuron. If the impulse is to be effective, it must be conveyed to another neuron, muscle cell, or gland cell. In this discussion we will concentrate on the junction between the axon terminal of one neuron and the dendrite, cell body, or axon of the next neuron. This junction between neurons is called a *synapse* (SIN-apps; Gr. connection) [FIGURE 12.13]. Functional contact between a neuron and a muscle cell or gland cell is also a synapse.

Presynaptic and postsynaptic neurons   The neuron carrying the impulse *toward* a synapse is called the *presynaptic* ("before the synapse") *neuron.* It initiates a response in the *postsynaptic* ("after the synapse") *neuron* leading *away from* the synapse [FIGURE 12.14]. The presynaptic cell is almost always a neuron, but the postsynaptic cell can be a neuron, muscle cell, or gland cell.

Synapses based on structure   There are several structural differences between types of synapses [FIGURE 12.15]: (1) A synapse between an axon and a dendrite is an *axodendritic synapse,* (2) a synapse between an axon and a cell body is *axosomatic,* and (3) a synapse between one axon and another axon is *axoaxonic.* The latter is associated with presynaptic inhibition and excitation. An axoaxonic synapse, like all synapses, can be formed only where the postsynaptic axon is not insulated with myelin, such as at the axon hillock, the nodes of Ranvier, and the end-bulb regions.

Electrical synapses   The transmission of a nerve impulse at a synapse can be either chemical or electrical. An *electrical synapse* is formed by a *gap junction* in which the two communicating cells are electrically coupled by tiny intercellular protein channels of adjacent plasma membranes called *connexons.* The free movement of ions occurs through the connexons between the presynaptic and postsynaptic cells. Because of the extremely low resistance associated with connexons, the nerve impulse virtually travels directly from the presynaptic to the postsynaptic cell. Also, since it involves no chemical neuro-

FIGURE 12.15 CLASSIFICATION OF SYNAPSES BASED ON STRUCTURE

Note the axosomatic, axoaxonic, and axodendritic synapses associated with the three neurons shown.

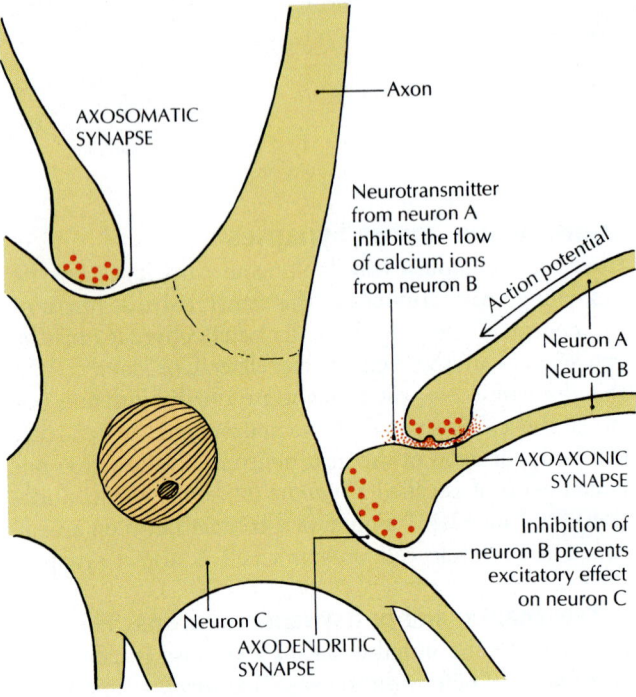

transmitters (which require time to interact), transmission at an electrical synapse is almost instantaneous.

In human beings and other mammals, electrical synapses (gap junctions) are found between cardiac cells, between smooth muscle cells of the intestinal tract, and between a few neurons — in the retina of the eye, for example.

Chemical synapses Chemical synapses are far more common than electrical synapses. In a **chemical synapse,** two cells are close, but they do not touch. They communicate by means of a chemical agent called a *neurotransmitter,* which is released by the presynaptic neuron. A neurotransmitter is capable of changing the resting membrane potential of the postsynaptic cell. (Neurotransmitters will be described in more detail in a later section of this chapter.)

When the nerve impulse of a presynaptic neuron reaches the end bulb of the axon, it depolarizes the presynaptic plasma membrane, causing voltage-gated cal-

cium channels to open. As a result, calcium ions (Ca^{2+}) flow down their concentration gradients from the synaptic area into the presynaptic terminal. The calcium ions cause the neurotransmitter storage vesicles to fuse with the plasma membrane, releasing the neurotransmitter from the vesicles by exocytosis into the narrow *synaptic cleft** [FIGURE 12.13B, D]. One such neurotransmitter is *acetylcholine;* another is *norepinephrine.*

The synapse is like a one-way valve, with the precursor of the neurotransmitter stored in the axons of the presynaptic neuron and all the chemically gated receptor sites for that neurotransmitter in the membrane of the postsynaptic neuron. The neurotransmitters can pass only from axon terminals across the synaptic cleft to the receptor site, not the other way. This directional control at the synapse prevents impulses from traveling in all directions at once. If they did, neural messages would be garbled.

Diffusion of a neurotransmitter across a synaptic cleft takes less than a millisecond; even so, that interval results in a *synaptic delay.* Some of the released neurotransmitter binds with chemically gated *receptor protein sites* in the chemically gated channels of the postsynaptic membrane, causing changes in the membrane's permeability to certain ions. The excess neurotransmitter in the synaptic cleft may diffuse away and be deactivated by enzymes. Its products may be recycled by being taken up by the presynaptic ending. As a result of the permeability change in the postsynaptic membrane, sodium and potassium ions rush through chemically gated channels and permit the same kind of depolarization that the presynaptic cell experienced. Waves of excitation pass along to the postsynaptic neuron, and the resulting nerve impulse is able to continue its path to an eventual effector cell.

The neurotransmitter acetylcholine is quickly broken down in the synaptic cleft into acetate and choline by the enzyme *acetylcholinesterase* and recycled for future use. Some chemical substances, such as certain insecticides and nerve gases, work by inhibiting the chemical action of acetylcholinesterase. Since acetylcholine is not destroyed, the stimulation of the postsynaptic muscle cells continues, and the cells remain in a state of contraction, paralyzing the breathing muscles. Death by asphyxiation results.

Excitatory and Inhibitory Presynaptic Potentials

One neuron (A) can exert influence through an axoaxonic synapse with another neuron (B), which in turn synapses through an axodendritic or axosomatic synapse with another neuron (C) [FIGURE 12.15]. The interactions

---

Q: *How fast can people react to stimuli?*

A: Olympic male sprinters react to the starter's gun in about 12/100 of a second.

---

*This narrow gap is usually less than 1/500 the width of a human hair (about 25 nm).

FIGURE 12.16 CHANGES IN ELECTRICAL POTENTIAL ACROSS THE PLASMA MEMBRANE OF A MOTOR NEURON

A cathode-ray oscilloscope records the potential difference across the plasma membrane with one electrode outside the neuron and the other inside the neuron. The resting membrane potential ranges between -60 and -70 mV, with the inside negative relative to the outside. When the membrane potential moves toward zero, the membrane becomes depolarized; when the membrane potential becomes more negative (such as -70 to -90 mV), the membrane becomes hyperpolarized. On the surface of the dendrites and cell body are excitatory and inhibitory synapses, which, when stimulated, produce local, graded, nonpropagating potentials. These are exhibited as an excitatory or depolarizing postsynaptic potential (EPSP) and an inhibitory or hyperpolarizing postsynaptic potential (IPSP). These local potentials are summated at the axon hillock and, if adequate, may trigger an integrated potential at the initial segment and an all-or-none action potential that is conducted along the axon to the motor end plate.

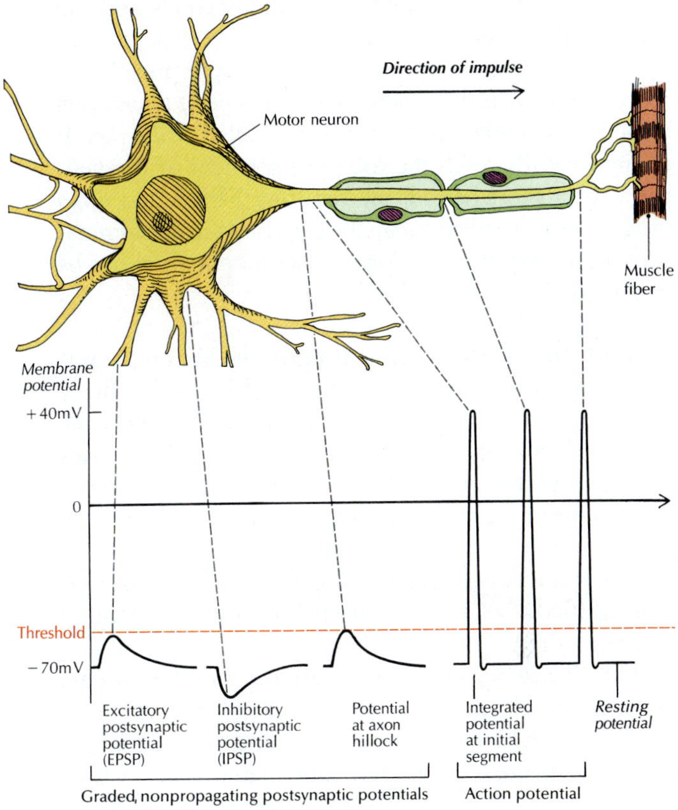

among these synaptic connections may result in either a presynaptic inhibition or a presynaptic excitation.

Presynaptic inhibition, which occurs before an action potential reaches the synapse, takes place in this way: An action potential of neuron A stimulates the release of a neurotransmitter that affects the axon of neuron B by reducing the flow of calcium ions through the channels into neuron B at the synapse of neurons B and C. As a result, an action potential of neuron B is inhibited from having an excitatory effect at its axodendritic (or axosomatic) synapse with neuron C.

In *presynaptic excitation,* neuron A releases a neurotransmitter that increases the flow of calcium ions through the channels into neuron B. The resulting action potential of neuron B exerts a greater excitatory activity with neuron C (postsynaptic excitation). Presynaptic inhibition and excitation act on the presynaptic neuron and "fine-tune" the postsynaptic activity.

Excitatory and Inhibitory Postsynaptic Potentials (Graded Local Potentials)

Neurotransmitters interact with the receptor sites of chemically gated channels on the postsynaptic (receptor) membranes of neurons, muscle cells, and gland cells,

producing a *postsynaptic potential (PSP)* [FIGURE 12.16]. Postsynaptic potentials are not typical all-or-none potentials, but are **graded local potentials,** which have no threshold or refractory period and fade out a short distance from the site of stimulation. Each postsynaptic potential propagates along the cell membrane for short distances and lasts for 1 to a few milliseconds. The two basic types of responses that can be produced at different synaptic sites on the postsynaptic membrane are (1) *excitatory postsynaptic potentials* and (2) *inhibitory postsynaptic potentials,* both of which can vary in intensity.

Excitatory postsynaptic potentials The excitatory response is associated with the opening of chemically gated cation channels (Na^+, K^+, and Ca^{2+}). The inflow of Na^+ is greater than either the inflow of Ca^{2+} or the outflow of K^+. This combined ion flow leads to an *excitatory postsynaptic potential (EPSP),* a partial depo-

Q: *How many synapses does a neuron have?*

A: Some neurons have only a few synapses, but a motor neuron may have as many as 2000, and a conducting fiber of the cerebellum may have as many as 200,000 on its dendrites alone.

larizing effect that lowers the membrane potential, but not to the threshold required to generate an action potential. An EPSP lasts only a few milliseconds and can spread only a short distance.

Because an EPSP makes the postsynaptic membrane more excitable, it does increase the chance of generating an action potential. Such an effect is called *facilitation*. If enough EPSPs reach the receptive segment of the neuron over a short period, the resulting EPSPs add up, a process known as *summation*. If there is sufficient summation (postsynaptic excitation), an action potential can be triggered.

Inhibitory postsynaptic potentials Whereas excitation brings a neuron to a state where it is more likely that an action potential will be generated, *inhibition* makes it less likely that an action potential will be triggered in the neuron. Inhibition occurs when the neurotransmitter interacts with a postsynaptic receptor site, *hyperpolarizing* the membrane, that is, making the inside more negative with respect to the outside, thus making it more difficult to generate an action potential [FIGURE 12.16].

This *inhibitory postsynaptic potential (IPSP)* occurs when chemically gated potassium and chloride channels are opened. Chloride ions move in, and potassium ions move out through the postsynaptic membrane. Such an inhibitory graded potential tends to prevent the postsynaptic potential from reaching a threshold level.

Spatial and temporal summation In a sense, the membrane of each neuron is a miniature integration center where the EPSPs and IPSPs summate either *spatially* (in space) or *temporally* (over time). The summing of the synaptic inputs from different neurons on the dendrites and cell body of one neuron is called *spatial summation*, indicating that each of these many synaptic inputs occupies a slightly different site on the postsynaptic membrane. *Temporal summation* occurs when one presynaptic neuron can increase its effect on one postsynaptic neuron by firing repeatedly. These two types of summation are important because synaptic input from presynaptic neurons activates the postsynaptic neuron.

If the resulting activity from spatial and temporal summations is sufficient to reach and excite the initial segment, then an action potential can be triggered in the axon. In effect, the action potential is an expression of part of a code message that conveys information from one neuron to another neuron.

Comparison of Action Potentials and Postsynaptic Potentials

There are three major differences between action potentials and the postsynaptic potentials described above:

1 Conduction velocity: Action potentials are propagated along an axon with no decrease in velocity, whereas postsynaptic potentials "die out" within a millimeter or so of the synapse.

2 Amplitude: Action potentials are all-or-none (same-size) responses, whereas postsynaptic potentials are graded in amplitude (size) as a function of the amount of neurotransmitter released at the synapse.

3 Refractory period: Action potentials have a refractory period, whereas postsynaptic potentials do not.

Importance of Inhibition

Inhibition is a neural activity that is just as important as excitation. Inhibitory neurons act like governors on an engine that control neural circuits and prevent them from firing unnecessarily. Imagine the consequences of uncontrolled excitatory influences that had no built-in inhibitions. (Imagine a car with an accelerator but no brakes.)

Inhibitory influences channel and modulate the effects of excitation and also direct activity to attain a desired end. In learning the intricate movements of penmanship, a child initially has control difficulties because, in part, too many unnecessary movements are made. During the learning process these movements are gradually reduced by inhibition, and the desired excitatory pathways are maintained to produce the focal movements. In this case, inhibition of nonessential movements is significant to learning. In a sense, the ultimate in inhibition may be the hallmark expressed by great musicians, singers, and athletes, who are constantly restraining themselves from giving everything they have.

The abnormal movements associated with afflictions such as cerebral palsy and Parkinson's disease result from the loss of certain inhibitory activities.

A S K   Y O U R S E L F

1 What is polarization? Depolarization?

2 What is a resting membrane potential?

3 What is a threshold stimulus?

4 What is the difference between an action potential and a nerve impulse?

5 What is the all-or-none principle?

6 How does the sodium-potassium pump function?

7 What is a synapse?

8 How do presynaptic and postsynaptic neurons differ?

NEUROTRANSMITTERS

Neurotransmitters are chemical messengers synthesized in neurons, stored in secretory vesicles in presynaptic axon terminals, and released into the synaptic cleft when a nerve impulse reaches the end bulb. Neurotransmitters bind to a receptor site on the postsynaptic membrane of another neuron or effector, where they affect ion channels. Later, neurotransmitters are removed from their site of action by a specific mechanism. A neurotransmitter may produce an excitatory or inhibitory response in the postsynaptic membrane. Either response is a result of the properties of the receptor. For example, acetylcholine evokes an excitatory response at the motor end plate of skeletal muscle, but it evokes an inhibitory response at a synapse with cardiac muscle pacemaker cells.

Kinds of Neurotransmitters

Many possible neurotransmitters are being investigated. At least 50 are present in the nervous system, and there may be over 100. Some substances already established as neurotransmitters are listed in TABLE 12.1. Known neurotransmitters are classified into four groups: acetylcholine, amino acid neurotransmitters, biogenic amines, and neuroactive peptides.

Acetylcholine *Acetylcholine* (ACh) is the main neurotransmitter released by neurons of the peripheral nervous system. (Only a restricted number of neurons in the central nervous system release acetylcholine.) The acetylcholine released by lower motor neurons* stimulates the contraction of skeletal muscle. Acetylcholine is synthesized by the enzyme *choline acetyltransferase* and is degraded by *acetylcholinesterase* within the synaptic cleft.

Amino acid neurotransmitters The *amino acid neurotransmitters* include gamma-aminobutyric acid (GABA), glycine, and glutamate. Although GABA is not an amino acid, it is classified with the amino acids because it is derived from glutamate. *GABA* is the major inhibitory neurotransmitter of the small local-circuit neurons in such structures as the cerebral cortex, cerebellum, and upper brainstem. *Glutamate* and *aspartate* are excitatory neurotransmitters found in many locations within the central nervous system. *Glycine*, the simplest amino acid, is the major inhibitory neurotransmitter of local-circuit neurons in the lower brainstem and spinal cord, selectively inhibiting certain neurons to refine and focus brain activity into meaningful patterns by eliminating and suppressing nonessential activity.

Biogenic amines *Biogenic amines,* or *monoamines,* include *catecholamines,* such as dopamine and norepinephrine, and *indoleamines,* such as serotonin. *Dopamine* is a neurotransmitter found in an area of the brain where certain neural circuits are involved with voluntary motor integration. It exerts inhibitory influences on postsynaptic neurons. Deficits of dopamine involving the basal ganglia of the brain are associated with Parkinson's disease.

Norepinephrine (NE, noradrenaline) is a neurotransmitter in both the central and peripheral nervous systems. It is found in neurons with cell bodies in the brainstem and in postganglionic neurons of the sympathetic nervous system. The fibers that release norepinephrine are distributed widely in such structures as the cerebral cortex, cerebellum, and spinal cord. Fibers that terminate in the cerebral cortex are involved with various levels of consciousness.

"Pep pills" ("speed"), such as those containing the drug amphetamine, increase the level of norepinephrine in the brain by blocking its reuptake in the neurons and by inhibiting the action of the enzyme monoamine oxidase (MAO). The euphoria and hallucinations produced by amphetamines are thought to result from the inhibition of MAO, which leads to a high level of norepinephrine. An "amphetamine psychosis," quite similar to schizophrenia, can result from the habitual use of amphetamines. Chlorpromazine is a drug used in the treatment of schizophrenia and similar disturbances. It also acts by blocking the reuptake of norepinephrine.

Serotonin is associated with various mood swings, including depression, elation, insomnia, and hallucinations. The relatively few serotonin-releasing neurons have their cell bodies in the brainstem. However, their fibers are widely distributed throughout the brain and spinal cord.

Neuroactive peptides *Neuroactive peptides* (chains of neuroactive amino acids) found in the central nervous system are somatostatin, the endorphins, the enkephalins, and substance P. *Somatostatin* inhibits the release of growth hormone from the pituitary gland. Apparently, it acts by altering the excitability of the postsynaptic membrane of target cells.

Endorphins (for "endogenous morphinelike substances") and *enkephalins* (for "substances in the head") are naturally occurring peptides found in several regions of the central nervous system. They function at postsynaptic receptor sites in the pain pathways to suppress synaptic activity leading to pain sensation. These opiates are not used therapeutically because their effectiveness is relatively short-lived and because they are addictive.

Substance P is generally an excitatory neurotransmitter located in the central nervous system and the intestinal tract. It is associated with the pain pathways, and its effects are long-lasting.

---

*Lower motor neurons have their cell bodies in the central nervous system, with their axons extending into the peripheral nervous system in order to synapse via the motor end plates on skeletal muscle cells.

TABLE 12.1 SOME MAJOR NEUROTRANSMITTERS

Transmitter group	Transmitter compound	Probable functions in the nervous system	Comments
Acetylcholine	Acetylcholine (ACh)	Excitatory and inhibitory. Neuromuscular, neuroglandular transmission. Also involved in memory.	Major transmitter of skeletal muscle, preganglionic sympathetic neurons; less prominent in CNS. Used by all motor axons arising from spinal cord. In Alzheimer's disease, cholinergic neurons do not function normally.
Amino acids	Gamma-aminobutyric acid (GABA)	Inhibitory. Evokes IPSPs in brain neurons.	Major inhibitory brain transmitter; most common transmitter in brain.
	Glutamate	Excitatory. Evokes EPSPs in brain neurons.	Major excitatory transmitter of cortical areas; together with aspartate, the most prevalent excitatory transmitter in brain.
	Glycine	Inhibitory. Excitatory in brain. Important in spinal cord.	Simplest amino acid.
Biogenic amines (monoamines)	Dopamine	Modulates activity of adrenergic neurons. Involved in arousal, motor activity. Evokes EPSPs in brain neurons.	Degeneration of dopaminergic synapses occurs in Parkinson's disease.
	Norepinephrine (NE, noradrenaline)	Excitatory and inhibitory. Involved in arousal, motor activity, visceral functions (such as heat regulation, reproduction). Acts as hormone when secreted into the bloodstream.	Cocaine blocks its receptor sites; release enhanced by amphetamines.
	Serotonin (5-hydroxytryptamine, or 5-HT)	Involved in sleep, mood, appetite, pain. Evokes EPSPs in brain neurons.	Produced from the amino acid tryptophan. Serotonergic neurons most active during waking hours, when motor awareness is enhanced and distracting stimuli are suppressed. LSD inhibits serotonergic neurons.
Neuroactive peptides	Somatostatin	Inhibits secretion of growth hormone.	The growth-hormone-release-inhibiting hormone of the hypothalamus, but its precise role in brain is unknown. In early onset of Alzheimer's disease the somatostatin in the cerebral cortex is depleted.
	Endorphins, enkephalins	Has analgesic properties.	Have opiatelike properties; their receptors are also the sites of action of morphine, codeine, and other opiate drugs.
	Substance P	Involved in mediating pain.	The first of many so-called gut-brain neuroactive peptides discovered. Localized in gastrointestinal tract and brain.

Many neuroactive peptides are classified as *neuromodulators*, chemical agents that are capable of altering (modulating) the responsiveness of neurons to a neurotransmitter. Some substances thought to function as neuromodulators include neuropeptides, histamine, prostaglandins, and the hormones cortisol and estrogen.

Some Neurotransmitter Candidates

Three well-known substances have recently been suggested as possible neurotransmitters: ATP, nitric oxide,* and carbon monoxide gas. *ATP* seems to work as a neurotransmitter in the peripheral nervous system and in isolated areas of the brain. The speed at which it operates is comparable to the fastest-known neurotransmitters, acetylcholine and glutamate, both of which produce a response within milliseconds.

Nitric oxide and **carbon monoxide** are the first gases suspected of being neurotransmitters, but researchers expect to find more. Nitric oxide in low concentrations acts as a neurotransmitter, but in high concentrations it is implicated in neurodegenerative diseases such as Alzheimer's disease and Huntington's disease. Nitric oxide may cause damage during a stroke, when neurons are overstimulated and fire repeatedly. Carbon monoxide may suppress such excessive activity, as well as counterbalance other adverse effects of nitric oxide.

It is believed that carbon monoxide and nitric oxide are capable of sending messages from the postsynaptic neuron back to the presynaptic neuron. Such a system of two-way communication may enhance the establishment of long-term memory.

Recycling of Neurotransmitters and Synaptic Vesicles

After neurotransmitter molecules are released from a vesicle by exocytosis, some molecules cross the synaptic cleft to activate receptor sites on the plasma membrane of the postsynaptic membrane. Other molecules are rapidly removed into the axon terminal by a sodium-dependent co-transport *reuptake system*. Others (acetylcholine, for example) have a breakdown product (choline) from enzymatic activity (cholinesterase) taken up into the axon terminal. Still other transmitter molecules diffuse away from the synaptic cleft.

Neuroactive transmitters such as serotonin, glutamate, and dopamine are taken up into the axon terminals by transmembrane proteins called *transporters*. Each transmitter is co-transported with sodium by the reuptake system and then incorporated into vesicles within the axon terminal. In the example of acetylcholine given

above, the choline molecules produced by the enzymatic action of cholinesterase in the synaptic cleft are taken up in the axon terminals. Acetyl coenzyme A (CoA, an activated form of acetate) combines with the choline to become the transmitter acetycholine, which is then concentrated in the vesicles in the axon terminal.

After the vesicle containing neurotransmitter molecules fuses with the synaptic membrane, it releases its contents into the synaptic cleft. The membrane of the vesicle is recycled in the following way. The vesicle membrane becomes incorporated into the presynaptic membrane and is then retrieved by the process of endocytosis, the opposite of exocytosis. The budding off of the vesicle membrane, which requires much energy, is aided by a coat of the protein *clathrin* on the membrane. Within a few seconds after the membrane is incorporated into the axon terminal, the clathrin is removed to be used again. The recycled membranes form new vesicles, which incorporate a new supply of reuptake transmitters.

Recycling is essential because of the rapid and almost constant release of neurotransmitters. If there were no recycling mechanism, the nucleus in the cell body would have to direct a continual production of the necessary neurotransmitter molecules and the proteins of the vesicular membrane through protein synthesis. However, the cell body on its own cannot direct the synthesis of these products fast enough to maintain sufficient quantities in the axon terminals.

Clinically useful results can be obtained by inhibiting the reuptake of specific neurotransmitters at synapses. For example, the popular antidepressant drug *fluoxetine* (Prozac) accomplishes its effects by blocking the reuptake transporter of the neurotransmitter serotonin. Serotonin is involved in the control of mood, as well as sensory reception, the onset of sleep, and temperature regulation. In essence, Prozac keeps serotonin in the circulation longer than if the neurotransmitter were taken up by the neuron immediately. In another example, certain drugs such as *prostigmine* inhibit the activity of acetylcholinesterase at neuromuscular junctions. This allows acetylcholine to remain in the synaptic cleft and influence receptor sites on the postsynaptic membrane of a muscle fiber for a longer time than usual, thus relieving some of the muscle weakness in patients with *myasthenia gravis*, a disorder caused by a shortage of acetylcholine receptors.

A S K   Y O U R S E L F
1 How does a neurotransmitter work?
2 What are the four kinds of neurotransmitters?
3 What is a neuromodulator?
4 Why is the recycling of neurotransmitters beneficial?

---

*Do not confuse nitric oxide (NO) with nitrous oxide (N_2O), also known as laughing gas.

FIGURE 12.17 NEURONAL CIRCUITS

Arrows indicate direction in which nerve impulses are propagated. **[A]** Principle of divergence. One presynaptic neuron branches and synapses with several postsynaptic neurons. **[B]** Principle of convergence. Several presynaptic neurons synapse with one postsynaptic neuron. **[C]** Simple feedback circuit in which the axon collateral branch of neuron A synapses with interneuron B, which, in turn, synapses with neuron A. **[D]** Parallel circuits in which neuron A synapses in neuron pools B–C and D–E, with each in turn projecting as parallel circuits. **[E]** Two-neuron sequence of afferent neuron synapsing with efferent neuron, found in a two-neuron reflex arc. **[F]** Three-neuron sequence of afferent neuron, interneuron, and efferent neuron, found in a three-neuron reflex arc.

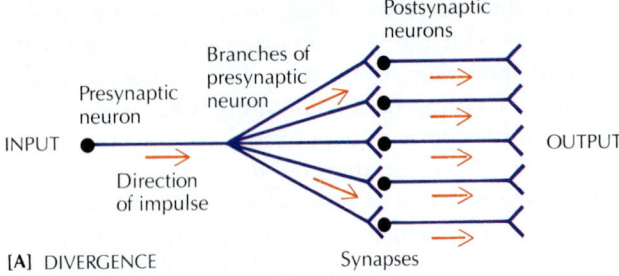

[A] DIVERGENCE

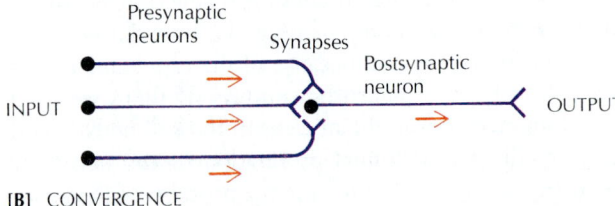

[B] CONVERGENCE

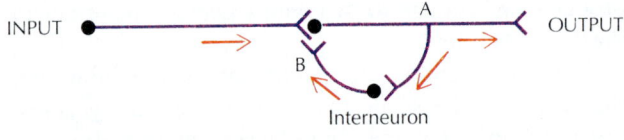

[C] FEEDBACK CIRCUIT

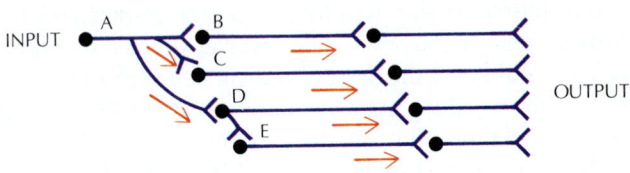

[D] PARALLEL CIRCUITS

[E] TWO-NEURON CIRCUIT

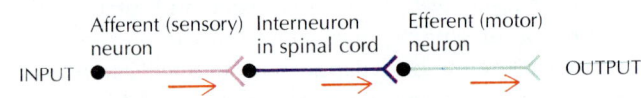

[F] THREE-NEURON CIRCUIT

NEURONAL CIRCUITS

The nervous system is an exquisitely structured network of neurons arranged in synaptically connected sequences called *neuronal circuits* [FIGURE 12.17]. The circuits are formed by input, intrinsic, and relay neurons. Several patterns and types of neuronal circuits exist. The circuits described in the following sections are the ones most commonly identified.

Divergence and Convergence

Some neurons in the central nervous system may synapse with as many as 25,000 other neurons. When the transmissive segment of a presynaptic neuron branches out to have many synaptic connections with the receptive segments of many other neurons, it is an example of *divergence* (L. *divergere*, to bend) [FIGURE 12.17A]. In diverging synapses, one neuron may excite or inhibit many others. The principle of *convergence* (L. *convergere*, to come together, to merge) is illustrated when the postsynaptic neuron is excited or inhibited by the axon terminals of many presynaptic neurons [FIGURE 12.17B].

Axons from as many as several thousand presynaptic neurons may converge on one postsynaptic neuron. If the excitatory stimuli are sufficient, an action potential can be generated.

Feedback Circuits

A *feedback circuit* is a mechanism for returning some of the output of a neuron (or neurons) for the purpose of modifying the output of a prior neuron (or neurons). Feedback in the nervous system is called *negative feedback*, because the feedback modulates an effect by *inhibiting* the prior output. The negative feedback shown in FIGURE 12.17C occurs when the lower motor neuron, which innervates a skeletal muscle, is excited. Its collateral axonal branch stimulates an interneuron through excitatory synapses. This interneuron then inhibits the lower motor neuron, readying it for restimulation by its presynaptic neurons.

Parallel circuits As a result of the neural processing of convergence and divergence, information is often conveyed by relay neurons to other neural levels through *parallel circuits* [FIGURE 12.17D]. Through these circuits, different forms of neural information can be relayed and ultimately recombined at the same time at other levels. For example, the sensory pathways for pain ascend to the brain from the spinal cord via two parallel systems: the spinothalamic pathway and the spinoreticulothalamic pathway. Much of the information conveyed via these pathways is recombined for processing at higher cerebral levels.

Two-neuron circuits The simplest neuronal circuit is the *two-neuron* (monosynaptic) *circuit.* It consists of a sequence of an afferent (sensory) neuron, one (*mono-*) synapse, and an efferent (motor) neuron [FIGURE 12.17E]. A familiar two-neuron sequence is the "knee-jerk" (patellar) extension reflex, where a tap on the patellar tendon of the flexed knee produces an extension of the knee.

Three-neuron circuits A *three-neuron circuit* is a sequence of an afferent (sensory) neuron, an interneuron, and an efferent (motor) neuron [FIGURE 12.17F]. Such a chain is used in flexor reflexes, such as the flexion of the forearm at the elbow when you touch a hot stove and pull your hand away. The circuit involved in the elbow flexion consists of (1) sensory neurons from the pain receptors in the hand to the spinal cord, (2) interneurons, located entirely within the spinal cord, which connect the sensory neurons with (3) the lower motor neurons to the biceps brachii muscle, which is stimulated to contract. Synapses are located between sensory neurons and the interneurons, and between the interneurons and the lower motor neurons. This type of circuit, where each neuron in the sequence is connected to many different neurons, is known as an *open circuit.*

A S K   Y O U R S E L F

1 How do diverging and converging circuits differ?

2 How do parallel circuits operate?

3 Besides the number of neurons involved, what is the difference between a two-neuron circuit and a three-neuron circuit?

THE EFFECTS OF AGING ON THE NERVOUS SYSTEM

The number of neurons in the brain usually decreases with age; at age 75 the loss is estimated to be about 10 percent. This moderate loss does not usually produce noticeable consequences because the remaining neurons may compensate by sprouting new collateral branches that form functional synaptic connections. This probably explains why a large number of elderly people retain their intellectual capacities. The brains of such people typically show a loss in weight and a minimal atrophy of the cerebral cortex but no demonstrable damage. Although the number of brain cells is reduced, there is no appreciable loss of brain function unless the blood supply is cut off temporarily by a stroke or other disorder. A stroke may lead to the progressive loss of intellectual abilities, personality, and memory.

Short-term memory may be impaired somewhat in old age, especially if the blood supply to the brain is decreased by diseased arteries. There is usually little or no change in *learning ability*, although many elderly people have been conditioned to think that their mental faculties are seriously diminished by the aging process.

Several age-related diseases affect brain functions. In *Alzheimer's disease*, which often appears between ages 40 and 60, some spaces between parts of the brain are enlarged, degenerative changes occur, and memory loss is common. *Parkinson's disease*, a motor disability, is commonly called shaking palsy because it is characterized by tremors of the head and hands, slow movements, rigid joints, and sagging facial muscles. It is an ailment that usually appears between ages 55 and 70, and it occurs more often in men than in women.

DEVELOPMENTAL PATTERNS OF NEURAL TISSUE

In the following chapters, the developmental anatomy of the brain, spinal cord, and spinal nerves will each be dealt with specifically. Here we consider a phenomenon exhibited by all developing neural tissue: plasticity.

In prenatal life, each developing neuron possesses more potential for change than it will ultimately express. Thus an immature neuron begins with an ability to generate more branches, synapses, and receptor sites than are actually present in the mature neuron. This capacity for change is a function of *plasticity*. Plasticity is probably retained during the entire life span of a neuron, enabling neurons to sustain and, in the case of injury, to repair the nervous system. This may be one means by which the organism compensates for its inability, with a few exceptions, to generate new neurons during postnatal life.

Plasticity is possible as long as the genetic machinery within the neuron is intact and able to produce the chemical substrates essential for change. When an axon is severed, the injured neuron is activated to synthesize those substrates essential in regenerating new branches and synapses to reinnervate the denervated target area. Even neurons of the central nervous system, following axonal injury, can regenerate collateral branches from the axon and attempt to regenerate branches at their severed ends.

Some neurons possess the genetic potential to produce several transmitters, yet end up releasing only one or two. For example, the postganglionic neurons of the sympathetic nervous system that innervate sweat glands, release the transmitter epinephrine during prenatal life in human fetuses and change postnatally (exhibit plasticity) by releasing acetylcholine.

In both the developing and mature neuron, there are means by which expression of the genetic potential is sustained and modified to generate and maintain the mature neuron. This potential is encouraged or inhibited by chemical factors in the neuron's environment. (See "Specialization of Brain Neurons: Nature and Nurture"

on page 419.) Following the generation of numerous branches, synapses, and receptor sites, the excess connections are pruned to the essential number compatible with the target region. The stability of neurons and their connections are dependent upon an adequate subsequent environment.

To sustain the optimal efficiency of the neural circuitry, stimulation and activity must be maintained. Evidence demonstrates that neural connection patterns within the mature cerebral cortex are not static but are maintained by usage through voluminous sensory inputs and by active employment of the motor systems. In brief, maximize neural activity to sustain the functioning nervous system, which is another way of saying, "Use it or lose it."

WHEN THINGS GO WRONG

Multiple Sclerosis

Multiple sclerosis (MS) is a progressive demyelination of neurons in the central nervous system (accompanied by the destruction of oligodendrocytes) that interferes with the conduction of nerve impulses and results in impaired sensory perceptions and motor coordination. Because almost any myelinated site in the brain or spinal cord may be involved, the symptoms of the disease may be diverse. With repeated attacks of inflammation at myelinated sites, scarring (sclerosis) takes place and some permanent loss of function occurs. The disease usually affects young adults between ages 18 and 40 and is five times more prevalent in whites than in blacks.

Although MS is very disabling, it progresses slowly, and most patients lead productive lives, especially during the recurring periods of remission. Among the typical symptoms are problems with vision, muscle weakness and spasms, urinary infections and bladder incontinence, and drastic mood changes. Recent experiments have shown that beta interferon lowers the activity of immune cells, reducing the number and severity of attacks.

The specific cause of MS is not known, but two theories are currently held: (1) a slow-acting virus infects the central nervous system and (2) the body's immune system attacks its own central nervous system (an autoimmune response).

Amyotrophic Lateral Sclerosis

Amyotrophic lateral sclerosis (ALS; commonly known as Lou Gehrig's disease; Gehrig was a professional baseball player with the New York Yankees who died of ALS) is a well-known motor neuron disease. Both upper motor neurons and lower motor neurons are affected. ALS is confined to the voluntary motor system, with progressive degeneration of corticospinal tracts (upper motor neu-

rons) and alpha motor neurons (lower motor neurons). The disease causes skeletal muscles to atrophy when excess levels of neurofilaments accumulate, blocking the passage of nutrients to axons.

In about 10 percent of cases, ALS is inherited, affecting men and women almost equally. (In the inherited, or familial, form of ALS the gene that causes the disease when mutated has been localized on the long arm of chromosome 21.) In other cases, ALS generally affects men four times more often than women, and is more common among whites than blacks.

Victims of ALS generally have weakened and atrophied muscles, especially in the hands and forearms. They may also exhibit impaired speech and have difficulty breathing, chewing, and swallowing, resulting in choking or excessive drooling. No effective treatment or cure exists, and the disease is invariably fatal.

Soon after the inherited form of ALS was traced to a genetic defect on chromosome 21, it was discovered that the normal gene codes for the enzyme superoxide dismutase (SOD), which helps inactivate harmful free radicals (highly reactive oxygen molecules) that may destroy motor neurons. Current genetic research is intense, and a study of the causative gene and its function may lead to a therapy for both the inherited and noninherited (sporadic) forms of ALS.

Parkinson's Disease

Parkinson's disease (Parkinsonism or "shaky palsy") is a motor disability characterized by symptoms such as stiff posture, tremors, and reduced spontaneity of facial expressions. It results from a deficiency of the neurotransmitter dopamine in certain brain neurons involved with motor activity. Parkinson's disease is discussed in greater detail in Chapter 13 on page 422.

MEDICAL UPDATE

Medicines for Mental Disorders

Drugs may act at several points in the synapse. Antidepressants that affect the presynaptic cell include (1) those that block the cell's reuptake of mono-amines. These drugs include tricyclic antidepressants, such as imipramine, which block the reuptake of several monoamines, and more specific block-ers such as fluoxetine, for serotonin, and buproprion, for dopamine. (2) Other antidepressants known as monoamine oxidase inhibitors prevent the presynap-tic cell from metabolizing monoamines. Drugs that affect the postsynaptic cell include agents that either block mono-amine receptors or stimulate their ability to respond. (3) Haloperidol, an antipsy-chotic, is a dopamine receptor blocker. (4) Finally, some drugs affect the second messenger that is normally produced after a receptor has been activated. For example, lithium carbonate, an antide-pressant and antimanic agent, works by inhibiting the synthesis of phosphatidyl inositol. Here a postsynaptic receptor is shown coupled to a stimulatory G pro-tein; this is its activated state, which causes more second-messenger chemi-cals to be synthesized, triggering molec-ular cascades that determine how the postsynaptic cell will respond.

From Elliot S. Gershon and Ronald O. Rieder, "Medicines for Mental Disorders," illustrated by Ian Worpol. Copyright © 1992 by Scientific American, Inc. All rights reserved.

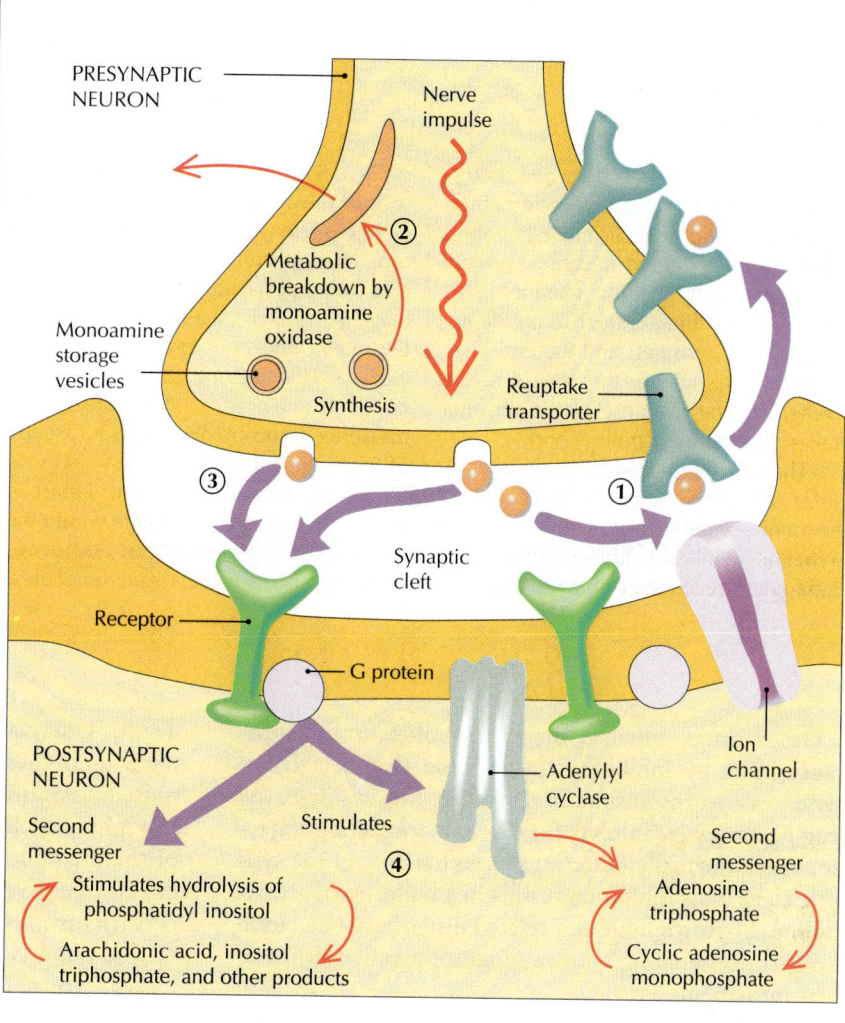

Huntington's Disease

Huntington's disease is a fatal hereditary brain disease that has been associated with insufficient amounts of the neurotransmitter GABA (see description on page 373).

Therapeutic Use of Drugs for Nervous System Disorders

The therapeutic use of drugs to treat various diseases of the nervous system is based on the effects of these drugs at synapses. (See "Medicines for Mental Disorders" above.) Drugs can alter the events that take place at syn-apses in several ways, including (1) increasing the release of a neurotransmitter from its vesicles into the cytoplasm of the axon, which exposes the neurotransmitter to enzy-matic breakdown; (2) increasing the release of a neuro-transmitter from the axon terminal; (3) blocking the re-lease of a neurotransmitter; (4) inhibiting the synthesis of a neurotransmitter; (5) blocking the reuptake of a neuro-transmitter; (6) blocking enzymes that metabolize the neurotransmitter; (7) binding to a receptor site to block or mimic the action of a neurotransmitter; and (8) inhib-iting or enhancing the activity of a "second messenger" such as cyclic AMP or calcium.

Some drugs are effective because their actions are sim-ilar to those of the natural neurotransmitters. Drugs that mimic acetylcholine are called *parasympathomimetic* ("have parasympathetic effects"), or *cholinergic drugs*. Drugs that mimic norepinephrine are *sympathomimetic*, or *adrenergic*, *drugs*. Epinephrine and amphetamines are sympathomimetic drugs. Because they constrict blood vessels and inhibit mucus secretion, they are effective as nasal decongestants.

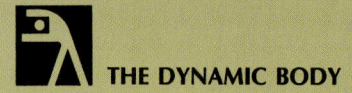
Chemical Messengers

There are two transmitter-mediated systems in the human nervous system: (1) The *first-messenger system* uses a "first messenger" and (2) the *second-messenger-system* uses a "second messenger."

The **first messenger** is a *neurotransmitter* released by a presynaptic neuron in the synaptic cleft. It interacts with a specific postsynaptic receptor protein associated with the ion channel. This protein has a direct influence on the activity of the channel, and the entire sequence is fast: From the release of the protein to the response of the channel takes only about a millisecond.

The **second messenger** is an *intracellular signaling molecule.* Following a sequence of interactions within the postsynaptic membrane, the second messenger triggers reactions within the cell. The entire sequence, referred to as the **second-messenger system,** is relatively slow, lasting from a few milliseconds to minutes.

In the well-known second-messenger system, in which cyclic AMP (cAMP, cyclic adenosine monophosphate) is the second messenger, a transmitter (or hormone) is released by a presynaptic neuron. The transmitter (or hormone) interacts with a postsynaptic receptor protein that is linked sequentially with a G-protein and the enzyme adenylate cyclase. This enzyme converts ATP into cAMP, which acts as a diffusible second messenger. The cAMP triggers a cascade of biochemical reactions that activate other enzymes (the protein kinases), which alter the chemical state within the neuron. Protein kinases can add or remove phosphate groups from molecules, activating or inactivating them, through a process called *phosphorylation,* a common method used by cells to activate or inactivate biochemical reactions.

Receptors, G-proteins, second messengers, and enzyme kinases interact in subtle and highly integrated biochemical reactions that operate throughout the neuron: in the nucleus (where gene expression can be altered), in the cytoplasm (where protein synthesis can be modified), and in the plasma membrane (where channel activity can be modulated).

Many other second-messenger systems are known to have significant roles in the chemical reactions within cells. Undoubtedly, many new seond-messenger systems are still to be discovered.

Other drugs enhance the effects of natural neurotransmitters by inhibiting the enzyme that deactivates any excess neurotransmitter substance. For example, the drug *physostigmine* inhibits the enzyme acetylcholinesterase, which normally deactivates acetylcholine in the synaptic cleft. This inhibition results in a build-up of acetylcholine. Physostigmine and related drugs have been used therapeutically along with immunosuppressive drugs to treat myasthenia gravis.

Some drugs, known as **blocking agents,** act by occupying some receptor sites on the postsynaptic membrane and thus preventing (blocking) the normal neurotransmitter from exerting its full influence. *Curare*, a poison used on arrow tips by certain South American Indians to kill wild animals, is such a drug. Curare acts by blocking the acetylcholine receptors at the motor end plate. The subsequent paralysis of the respiratory muscles causes death. A derivative of curare called d-tubocurarine was once used therapeutically in small doses as a skeletal muscle relaxant.

Ophthalmologists use *homatropine* to dilate the pupils of the eyes during examinations. It acts by blocking the effects of acetylcholine at the smooth muscle cholinergic constrictor muscle of the pupil. As a result, the receptor on the pupil is dilated.

Like drugs, diseases can affect synapses. For example, the toxin of the tetanus bacterium works by interfering with the inhibition of the motor neurons that innervate skeletal muscles. As a result, the excitatory input to these muscles is unimpeded, and the muscles contract continuously. The condition is called *tetanus*, or more commonly, *lockjaw* (because spasms of the jaw muscles occur in the early stages of the disease).

CHAPTER SUMMARY

The body uses a combination of electrical impulses and chemical messengers to react and adjust to stimuli in order to maintain homeostasis.

Organization of the Nervous System (p. 353)

The nervous system may be divided into the **central nervous system** (the brain and spinal cord) and the **peripheral nervous system** (peripheral nerves and their ganglia located outside the central nervous system).

Cells of the Nervous System (p. 353)

1 *Neurons* are nerve cells specialized to transmit nerve impulses throughout the body. Their basic properties are *excitability* and *conductivity.*

2 A neuron is composed of a *cell body,* branching *dendrites,* and an *axon.* Dendrites conduct action potentials toward the cell body, and the axon usually carries impulses away from the cell body. Axons may be coated with a sheath of *myelin,* which enhances the speed of conduction.

3 *Afferent* (sensory) *neurons* of the peripheral nervous system carry information from sensory receptor cells to the CNS. *Efferent* (motor) *neurons* of the PNS carry neural information away from the CNS to muscles and glands. *Interneurons* are connecting neurons that carry impulses from one neuron to another.

4 Neurons and their peripheral fibers may be classified into *functional components* as *somatic, visceral, general, special, afferent,* and *efferent.*

5 Neurons may be classified according to their structure as *multipolar, bipolar,* and *unipolar.*

6 Most neurons are composed of specific functional segments known as *receptive, initial, conductive,* and *transmissive.*

7 *Supporting cells* are specialized cells other than neurons in the central and peripheral nervous systems. They are involved in nourishing, protecting, and supporting neurons. Supporting cells of the PNS, *satellite cells* and *Schwann cells,* provide various types of sheaths.

8 *Neuroglia* are nonconducting cells that protect, nourish, and support neurons within the central nervous system. The types of neuroglia are astrocytes, oligodendrocytes, microglia, and ependymal cells.

Physiology of Neurons (p. 363)

1 *Nerve impulses,* or *action potentials,* are conducted along a nerve fiber when specific changes occur in the electrical charges of the fiber membrane.

2 A resting cell that is not conducting an impulse is *polarized,* with the outside of its plasma membrane positively charged with respect to its negatively charged interior. The imbalance of electrical charges creates a potential for electrical activity *(resting membrane potential).*

3 The resting potential of a neuron is maintained by a *sodium-potassium pump* that regulates the concentration of sodium and potassium ions inside and outside the plasma membrane. The pumps are complemented by selective *ion channels.*

4 A *threshold stimulus* is the minimal stimulus required to initiate a nerve impulse.

5 *Depolarization* is the decrease of the plasma membrane potential that can make the outside of the cell negative relative to the inside. When a stimulus is strong enough to cause sufficient depolarization, the nerve cell *fires.* Depolarization can produce an action potential.

6 When the original electrical charges are restored, the membrane is *repolarized.* The *refractory period* is the time during which a neuron will not respond to a second stimulus.

7 *Saltatory conduction* occurs when a nerve impulse jumps from node to node on a myelinated fiber.

8 The phenomenon of a neuron firing at full power or not at all is the *all-or-none principle.*

9 A *synapse* is the junction between the axon terminal of one neuron and the dendrite, the cell body, or specific parts of the axon of the next neuron. The functional contact between a neuron and a muscle cell or gland cell is also a synapse. The cell carrying the impulse toward a synapse is a *presynaptic neuron;* it initiates a response in the receptive segment of a *postsynaptic neuron* leading away from the synapse.

10 Neurotransmitters interact with the receptor sites of the postsynaptic membrane, producing a *postsynaptic poten-* *tial (PSP),* or graded potential. Postsynaptic potentials may be *excitatory (EPSP)* or *inhibitory (IPSP).* A *receptor potential* is the graded potential exhibited by the terminal segment of a nerve fiber located within a sensory receptor.

Neurotransmitters (p. 373)

1 Synapses can be electrical or chemical. In a chemical synapse, two cells communicate by means of a chemical agent called a *neurotransmitter,* such as acetylcholine or norepinephrine. In an electrical synapse, the electrical activity of one neuron spreads readily to the next neuron.

2 Neurotransmitters are classified in four groups: acetylcholine, amino acids, biogenic amines, and neuroactive peptides.

Neuronal Circuits (p. 376)

1 Neurons are organized in networks arranged in synaptically connected sequences called *neuronal circuits.*

2 The principle of *divergence* is shown when the transmissive segment of a neuron branches to have many synaptic connections with the receptive segments of many other neurons. When the axon terminals from many presynaptic neurons synapse with the receptive segment of only one postsynaptic neuron, it is an example of *convergence.*

3 Neuronal circuits may be *divergent, convergent, feedback circuits, parallel circuits, two-neuron circuits,* or *three-neuron circuits.*

The Effects of Aging on the Nervous System (p. 377)

1 Although some neurons continue to die during old age, there is usually no shortage. Moreover, there is only about a 10 percent decrease in the *velocity of nerve impulses.*

2 *Short-term memory* may be impaired, especially if the blood supply to the brain is decreased.

3 Several *age-related diseases* (such as Alzheimer's disease and Parkinson's disease) affect brain functions adversely.

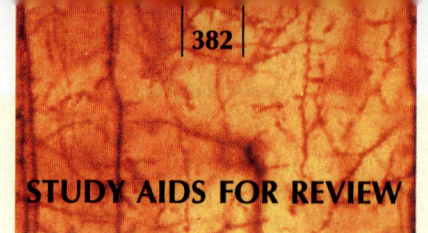
STUDY AIDS FOR REVIEW

KEY TERMS

acetylcholine 373
action potential 365
afferent (sensory)
   neurons 360
all-or-none principle
   367
axon 357
axonal transport 357
axoplasmic flow 357
cell body 353
central nervous
   system (CNS) 353
chemically gated
   channels 364
conductivity 353
dendrite 357
depolarization 365
efferent (motor)
   neurons 360
end bulb 357
excitability 353
gated channel 364
interneuron 360
modality-gated
   channels 365
myelin 357
nerve 359
nerve impulse 365

neuroglia 361
neuron 353
neuronal circuit 376
neurotransmitter 373
node of Ranvier 357
open ion channels 364
peripheral nervous system
   (PNS) 353
polarization 363
postsynaptic neuron 369
potential difference 363
presynaptic neuron 369
refractory period 366
repolarization 365
resting membrane
   potential 363
saltatory conduction 369
satellite cell 361
Schwann cell 357, 361
sodium-potassium pump 364
somatic nervous system 353
stimuli 353
supporting cells 353
synapse 357, 369
threshold stimulus 365
transmitter-gated channels 364
visceral nervous system 353
voltage-gated channels 364

UNDERSTANDING THE FACTS

1 What are two primary functions of nervous tissue?
2 What are two major divisions of the nervous system?
3 Name the three principal parts of a typical neuron and identify the function of each part.
4 Name the three structural and the three functional types of neurons.
5 What is meant by the "resting potential"? How is it maintained?
6 Distinguish between chemically gated, voltage-gated, and modality-gated channels.
7 How does synaptic transmission assure the unidirectional transmission of nerve impulses?
8 Besides neurons, what other kinds of cells compose nervous tissue? What is the function of these cells?
9 Where are myelin sheaths found and what is their function?
10 Name the four different chemical groups of neurotransmitters and indicate whether each includes excitatory compounds, inhibitory compounds, or both.

11 What is meant by each of the following terms?
   a all-or-none
   b postsynaptic potential
   c saltatory conduction
   d divergent and convergent circuits
   e electrical synapses

UNDERSTANDING THE CONCEPTS

1 What is the role of the sodium-potassium pump in maintaining the resting potential?
2 Distinguish between depolarization and an action potential.
3 How do we perceive differences in the intensity of stimuli?
4 How are neurotransmitters processed after they have been liberated into the synaptic cleft?
5 What happens to the polarity of the postsynaptic membrane as a result of an IPSP and an EPSP?
6 What is the physiological importance of inhibition?

SELF-QUIZ

Multiple-Choice Questions

12.1 Which of the following is/are not part of the peripheral nervous system?
   a cranial nerves
   b spinal nerves
   c ganglia
   d spinal cord
   e c and d

12.2 Which of the following is not or does not include a neurotransmitter?
   a acetylcholine
   b neuroactive peptides
   c N-CAM
   d amino acids
   e biogenic amines

12.3 An example of a disease that shows progressive degeneration of myelin sheaths in neurons of the CNS is
   a myasthenia gravis
   b multiple sclerosis
   c Parkinson's disease
   d Huntington's disease
   e senility

12.4 A stimulus that is just strong enough to initiate an impulse in a neuron is called a _____ stimulus
   a base line
   b synaptic
   c threshold
   d subliminal
   e differential

12.5 Acetylcholine is secreted by
   a Schwann cells
   b the receptive segment of motor neurons
   c electrical synapses
   d presynaptic motor neurons
   e all of the neurons in the brain

12.6 The myelin sheaths of axons in the PNS are formed by
   a astrocytes
   b microglial cells
   c oligodendrocytes
   d Schwann cells
   e motor neurons

12.7 The intervals between myelin sheath segments are called

a sheath segments   d internodal segments
b nodes of Ranvier   e nodes of Schwann
c dendritic segments

12.8 IPSPs are characterized by

a potassium ions rushing out of the neuron
b chloride ions rushing into the neuron
c depolarization of the membrane
d a and b
e a, b, and c

12.9 A drug that acts as a blocking agent by selectively occupying postsynaptic receptor sites is

a physostigmine   d CAM
b sympathomimetic   e N-CAM
c curare

12.10 Which kind of neuronal circuit is necessarily characterized by a greater number of presynaptic neurons than postsynaptic neurons?

a divergent   d three-neuron
b convergent   e parallel
c two-neuron

12.11 In which region of a neuron is its neurotransmitter synthesized?

a dendrites   d axon terminals
b cell body   e nodes of Ranvier
c axon hillock

12.12 Which of the following would be most important in the formation of patterns of synapses in the embryonic nervous system?

a beta receptor blockers
b alpha receptor blockers
c N-CAM
d acetylcholine receptor blockers
e acetylcholinesterase inhibitors

Completion Exercises

12.13 The arrival of an action potential at a synaptic junction causes voltage-gated channels in the presynaptic membrane to open, which allows an influx of _____ ions.

12.14 This influx of ions causes the release of molecules of the neurotransmitter into the _____.

12.15 Gap junctions and connexons are characteristic of _____ synapses.

12.16 During depolarization of a region of an axon, sodium and potassium ions flow through _____-gated channels.

12.17 A neurotransmitter such as acetylcholine results in the opening of _____-gated channels.

12.18 The _____ principle states that a neuron will fire at full power or not fire at all.

12.19 In a myelinated neuron, the "jumping" of an action potential from one node of Ranvier to the next is called _____.

12.20 A collection of cell bodies of neurons of peripheral nerves is called a _____.

12.21 The central nervous system consists of the _____ and the _____.

12.22 The peripheral nervous system can be divided on a functional basis into the _____ and the _____ nervous systems.

Matching

a serotonin   e ganglion
b nitric oxide   f GABA
c neuroglia   g neurilemma
d nerve   h acetylcholinesterase

12.23 _____ an organized collection of axons and dendrites in the peripheral nervous system

12.24 _____ improper function of this substance causes postsynaptic muscle cells to remain in a state of constant contraction

12.25 _____ imbalances of this neurotransmitter prompt depression, insomnia, and hallucinations

12.26 _____ acts as neurotransmitter in low concentrations, but is implicated in destruction of neural tissue in high concentrations

12.27 _____ outer layer of the myelin sheath

12.28 _____ most common inhibitory neurotransmitter in the brain

12.29 _____ a collection of cell bodies of neurons in the peripheral nervous system

12.30 _____ a group of cells that perform various functions, including formation of the myelin sheath and of the blood-brain barrier

A SECOND LOOK

In the micrograph below, label the synaptic cleft, synaptic vesicles, axon terminal, and muscle cell.

13

The Central Nervous System I: The Brain

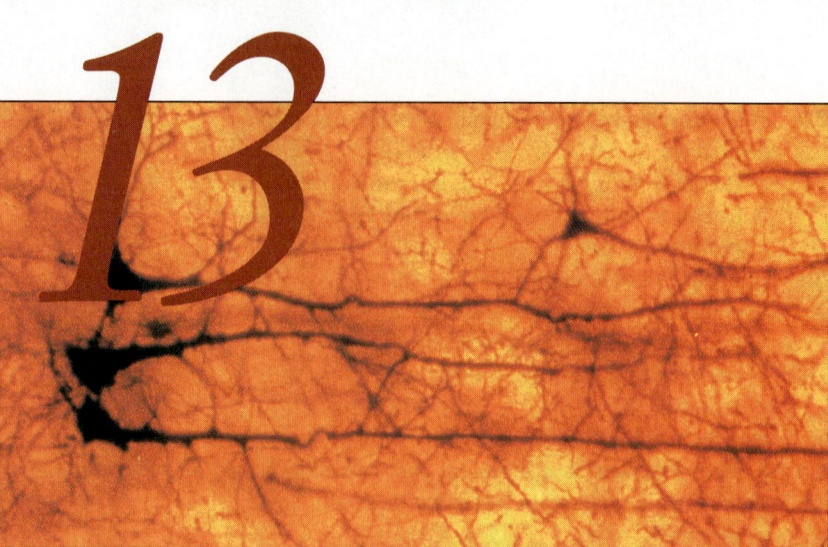

KEY CONCEPTS

1 The major parts of the brain are the brainstem, cerebellum, cerebrum, and diencephalon.

2 The brain is protected by the meninges, cerebrospinal fluid, and cranium.

3 The blood-brain barrier is a network of membranes that allows only some substances to enter the brain.

4 The brainstem relays messages between the spinal cord and cerebrum.

5 The cerebellum is involved primarily with balance and muscular coordination.

6 The cerebrum is made up of two convoluted hemispheres and a connecting bridge of nerve fibers. It is the center of conscious thought.

7 The diencephalon connects the midbrain to the cerebrum. It has many regulatory functions.

8 No single area of the brain controls learning.

The human brain weighs only about 1400 g (3 lb), yet it contains more than 100 billion neurons, about the same number as stars in the Milky Way.* In addition, each neuron may have from 1000 to 10,000 synaptic connections with other nerve cells. There may be as many as 100 *trillion* synapses in the brain. It is tempting to compare the human brain with a computer, but there is really no comparison. Nothing we know can match the exquisite complexity of the brain.

Your brain does much more than help you think and make decisions. It is your body's main key to homeostasis, regulating body processes from cellular metabolism to the overall functioning of organs and systems. Your nerves and their specialized receptors may *receive* the stimuli of sound, touch, vision, smell, and taste, but you actually *experience* the sensation in your brain. Your brain is working even when you are sleeping to activate, coordinate, and regulate the body's many functions and their relationships to the outside world.

GENERAL STRUCTURE OF THE BRAIN

The brain is technically called the *encephalon* (en-SEFF-uh-lon; Gr. *en*, in + *kephale*, head). It has four major divisions: brainstem, cerebellum, cerebrum, and diencephalon [FIGURE 13.1]. The major divisions and structures of the brain are summarized in TABLE 13.1.

The **brainstem** is composed of the midbrain, pons, and medulla oblongata. The **cerebellum** is a coordinating center for skeletal muscle movement. The **diencephalon** is composed of the thalamus, hypothalamus, epithalamus, and ventral thalamus.

Probably the most obvious physical feature of the brain is the large pair of hemispheres that make up about 85 percent of the brain tissue. These are the two hemispheres of the **cerebrum.** Another obvious feature is the outer portion of the cerebrum, called the *cerebral cortex*, with its many folds or *convolutions*. Lying below the gray matter of the cerebral cortex are both white matter and deep, large masses of gray matter called the *basal ganglia*. Connecting the two cerebral hemispheres is a bundle of nerve fibers called the *corpus callosum*, which relays nerve impulses between the hemispheres. Emerging from the cerebrum are 2 of the 12 cranial nerves (the olfactory and optic nerves, I and II). The other 10 arise from the brainstem. The brain is surrounded by cranial bones and three membranes called cranial meninges: the outer *dura mater*, middle *arachnoid*, and inner *pia mater*, all of which will be discussed later.

*To get some idea of the size of 100 billion, think of this: you would have to spend almost *$5.5 million a day* to spend $100 billion in 50 years. If you wanted to spend $100 billion in *one* year, you would have to spend $274 million *a day*. One hundred billion is a lot of neurons.

MENINGES, VENTRICLES, AND CEREBROSPINAL FLUID

The human brain is mostly water (about 75 percent in an adult). It has the consistency of gelatin, and if it were not supported, the brain would slump and sag. Fortunately, it has ample support. The brain is protected by the scalp, with its hair, skin, fat, and other tissues, and by the cranium, one of the strongest structures in the body. It also floats shockproof in cerebrospinal fluid and is encased by layers of cranial meninges.

Cranial Meninges

The brain is surrounded by three layers of protective membranes called *meninges* (muh-NIHN-jezz; Gr. plural of *meninx*; MEE-ningks; membrane): the dura mater, arachnoid, and pia mater.

Dura mater The outermost cranial meninx is the *dura mater* (DURE-uh MAY-ter; L. hard mother). It consists of two fused layers: (1) an inner dura mater that is continuous with the spinal dura mater and (2) an outer dura mater, which is actually the periosteal layer of the skull bones. The outer cranial dura mater is a tough, fibrous layer containing veins and arteries that nourish the bones. The inner dura mater extends into the fissure (the *falx cerebri*) that divides the left and right hemispheres of the cerebrum [FIGURE 13.2A] and reaches into the fissure between the cerebrum and cerebellum (*tentorium cerebelli*). By dividing the cranial cavity into three distinct compartments, the dura mater adds considerable support to the brain.

The fused inner and outer cranial dura mater is closely attached to the bones of the cranial cavity. As a result, there is no epidural space between the membrane and the bones, as there is surrounding the spinal cord. Between the inner dura mater and the arachnoid is a potential space called the **subdural space.** It does not contain cerebrospinal fluid.

In certain locations the two layers of dura mater separate to form channels [FIGURE 13.2A] that are lined with endothelium and contain venous blood. These spaces are the *dural sinuses of the dura mater*, which drain venous blood from the brain.

Arachnoid The middle layer of the meninges, between the dura mater and pia mater, is the **arachnoid** (Gr. cobweblike), a delicate connective tissue. Between the arachnoid and the pia mater is a membranous network [FIGURE 13.2] and the *subarachnoid space*, which contains cerebrospinal fluid. The arachnoid contains no blood vessels of its own, but blood vessels are present in the subarachnoid space.

FIGURE 13.1 MAJOR STRUCTURES OF THE BRAIN

[A] Right lateral view of the external surface of the brain, showing the cerebellum and four of the six cerebral lobes. **[B]** Right sagittal section. **[C]** Right sagittal section showing the major subdivisions of the brain and its connection to the spinal cord.

THE FOUR MAJOR DIVISIONS OF THE BRAIN

BRAINSTEM
{ Midbrain
  Pons
  Medulla oblongata

CEREBELLUM

CEREBRUM

DIENCEPHALON
{ Thalamus
  Hypothalamus
  Epithalamus
  Ventral thalamus

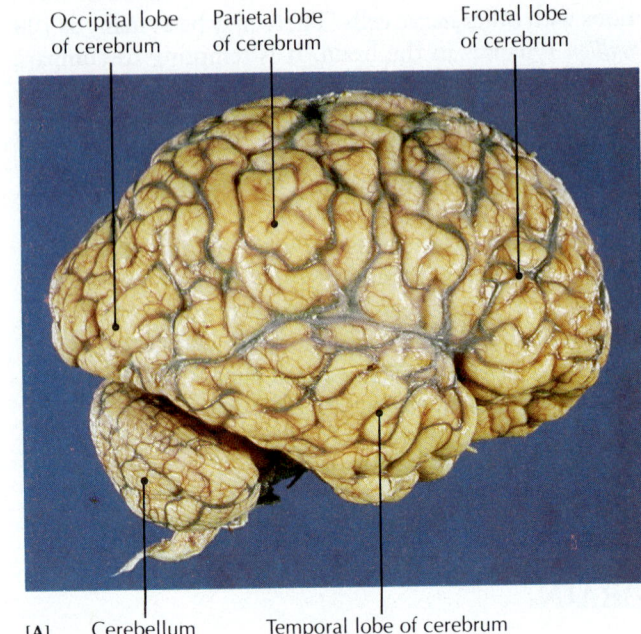

Occipital lobe of cerebrum Parietal lobe of cerebrum Frontal lobe of cerebrum

[A] Cerebellum Temporal lobe of cerebrum

DIENCEPHALON
Thalamus Hypothalamus Corpus callosum

CEREBRUM

CEREBELLUM Medulla oblongata Pons Midbrain

BRAINSTEM

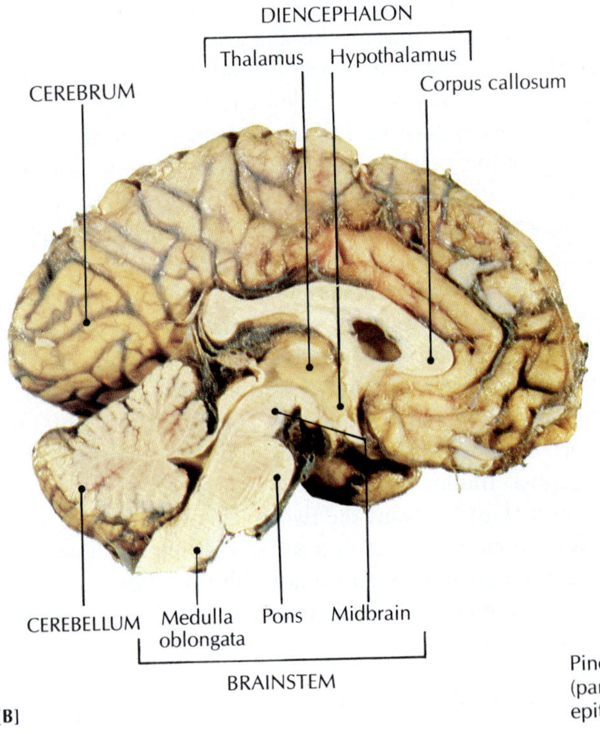

[B]

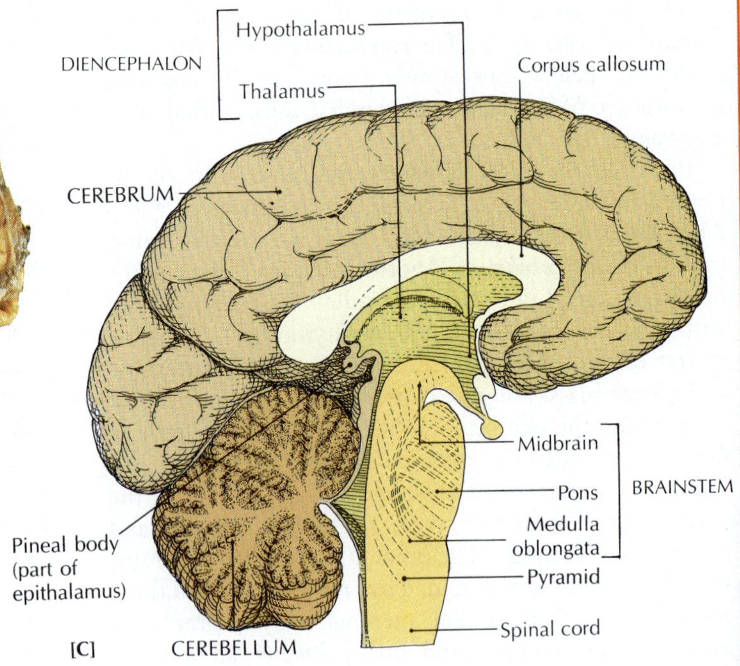

DIENCEPHALON
Hypothalamus Corpus callosum
Thalamus

CEREBRUM

Pineal body (part of epithalamus)

Midbrain
Pons BRAINSTEM
Medulla oblongata
Pyramid
Spinal cord

[C] CEREBELLUM

FIGURE 13.2 CRANIAL MENINGES

[A] Frontal section through the superior sagittal (dural) sinus. Note the supportive trabeculae in the subarachnoid space, the falx cerebri separating the cerebral hemispheres, and the triangular superior sagittal (dural) sinus. Most of the cerebrospinal fluid returns into the blood through the arachnoid villi. Note the detailed drawing. **[B]** Electron micrograph of cranial meninges. The space between the dura mater and arachnoid occurred during the processing of the specimen; it is normally only a potential space.

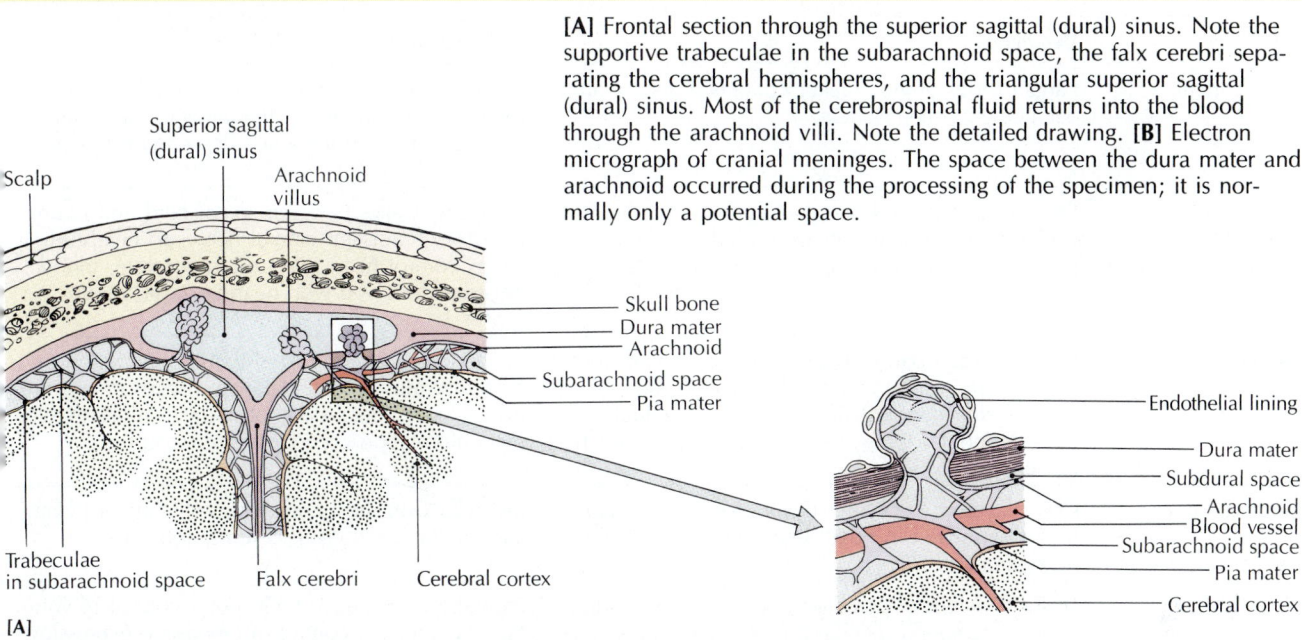

[A]

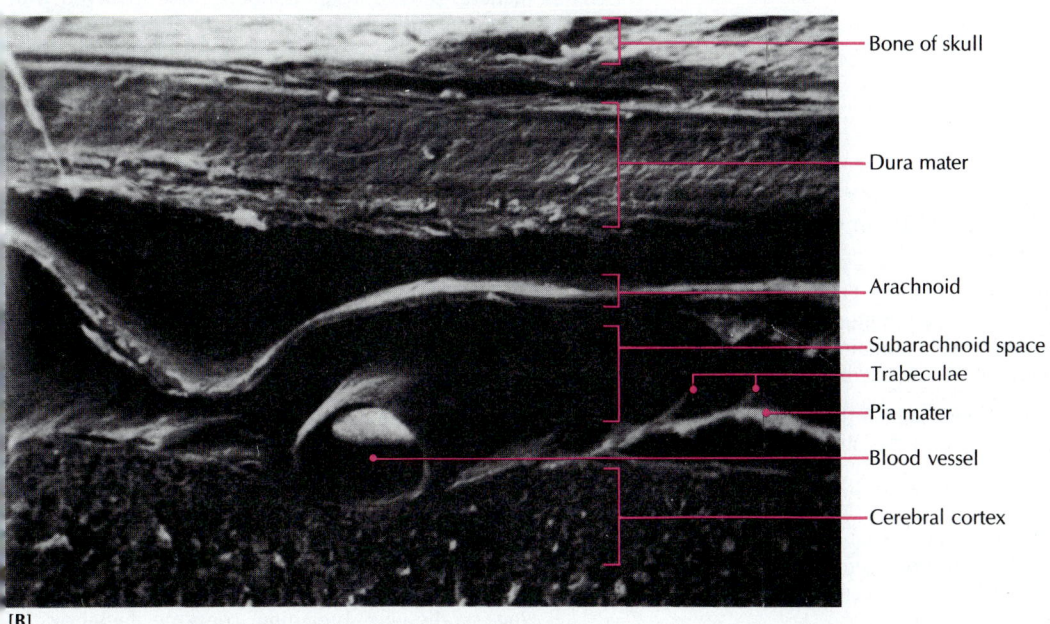

[B]

Pia mater   The *pia mater* (PEE-uh MAY-ter; L. tender mother) is the delicate innermost meningeal layer. It directly covers and is attached to the surface of the brain and dips down into the fissures between the raised ridges of the brain. Most of the blood to the brain is supplied by the large number of small blood vessels in the pia mater.

In head injuries, blood may flow from a severed blood vessel into the potential space between the skull and cranial dura mater (an extradural or epidural hemorrhage), into the potential subdural space (a subdural hemorrhage), into the subarachnoid space (a subarachnoid hemorrhage), or into the brain itself (an intracerebral hemorrhage).

Ventricles of the Brain

Within the brain is a series of connected cavities called *ventricles* (L. little bellies). Each cranial ventricle is filled

TABLE 13.1 MAJOR STRUCTURES OF THE BRAIN

Structure	Description	Major functions
Basal ganglia	Large masses of gray matter contained deep within each cerebral hemisphere.	Help to coordinate skeletal muscle movements by relaying information via thalamus to motor area of cerebral cortex to influence descending motor tracts.
Brainstem	Stemlike portion of brain continuous with diencephalon above and spinal cord below. Composed of midbrain, pons, medulla oblongata.	Relays messages between spinal cord and brain, and from brainstem cranial nerves to cerebrum. Helps control heart rate, respiratory rate, blood pressure. Involved with hearing, taste, other senses.
Cerebellum	Second largest part of brain. Located behind pons, in posterior section of cranial cavity. Composed of cerebellar cortex, two lateral lobes, central flocculonodular lobes, medial vermis, some deep nuclei.	Processing center involved with coordination of muscular movements, balance, precision and timing of movements, body positions. Processes sensory information used by motor systems.
Cerebral cortex	Outer layer of cerebrum. Composed of gray matter and arranged in ridges (gyri), grooves (sulci), depressions (fissures).	Involved with most conscious activities of living. (See major functions of cerebral lobes.)
Cerebral lobes	Major divisions of cerebrum, consisting of frontal, parietal, temporal, occipital lobes (named for bones under which they lie), insula, and limbic lobe.	Frontal lobe involved with motor control of voluntary movements, control of emotional expression and moral behavior. Parietal lobe involved with general senses, taste. Temporal lobe involved with hearing, equilibrium, emotion, memory. Occipital lobe organized for vision and associated forms of expression. Insula may be involved with gastrointestinal and other visceral activities. Limbic lobe (along with the limbic system) is involved with emotions, behavioral expressions, recent memory, smell.
Cerebrospinal fluid	Fluid that circulates in ventricles and subarachnoid space.	Supports and cushions brain. Helps control chemical environment of central nervous system.
Cerebrum	Largest part of brain. Divided into left and right hemispheres by longitudinal fissure and divided into lobes. Also contains cerebral cortex (gray matter), white matter, basal ganglia.	Controls voluntary movements, coordinates mental activity. Center for all conscious living.
Corpus callosum	Bridge of nerve fibers that connects one cerebral hemisphere with the other.	Connects cerebral hemispheres, relaying sensory information between them. Allows left and right hemispheres to share information, helps to unify attention.
Cranial nerves	Twelve pairs of cranial nerves. Olfactory (I) and optic (II) arise from cerebrum; others (III through XII) arise from brainstem. Sensory, motor, or mixed.	Concerned with senses of smell, taste, vision hearing, balance. Also involved with specialized motor activities, including eye movement, chewing, swallowing, breathing, speaking, facial expression.
Diencephalon	Deep portion of brain. Composed of thalamus, hypothalamus, epithalamus, ventral thalamus.	Connects midbrain with cerebral hemispheres. (See major functions of thalamus and hypothalamus.)

Structure	Description	Major functions
Hypothalamus	Small mass below the thalamus; forms floor and part of lateral walls of third ventricle.	Highest integrating center for autonomic nervous system. Controls most of endocrine system through its relationship with the pituitary gland. Regulates body temperature, water balance, sleep-wake patterns, food intake, behavioral responses associated with emotion.
Medulla oblongata	Lowermost portion of brainstem. Connects pons and spinal cord. Site of decussation of descending corticospinal (motor) tract and an ascending sensory pathway from spinal cord to thalamus; emergence of cranial nerves VI through XII; movement of cerebrospinal fluid from ventricle to subarachnoid space.	Contains vital centers that regulate heart rate, respiratory rate, constriction and dilation of blood vessels, blood pressure, swallowing, vomiting, sneezing, coughing.
Meningeal spaces	Spaces associated with meninges. Potential epidural space between skull and dura mater; potential subdural space between dura mater and arachnoid; subarachnoid space between arachnoid and pia mater (contains cerebrospinal fluid).	Provide subarachnoid circulatory paths for cerebrospinal fluid, protective cushion.
Meninges	Three layers of membranes covering brain. Outer tough dura mater; arachnoid; inner delicate pia mater, which adheres closely to brain.	Dura mater adds support and protection. Arachnoid provides space between the arachnoid and pia mater for circulation of cerebrospinal fluid. Choroid plexuses, at places where pia mater and ependymal cells meet, are site of formation of much cerebrospinal fluid. Pia mater contains blood vessels that supply blood to brain.
Midbrain	Located at upper end of brainstem. Connects pons and cerebellum with cerebrum. Site of emergence of cranial nerves III, IV.	Involved with visual reflexes, movement of eyes, focusing of lens, dilation of pupils.
Pons	Short, bridgelike structure composed mainly of fibers that connect midbrain and medulla oblongata, cerebellar hemispheres, and cerebellum and cerebrum. Lies anterior to cerebellum and between midbrain and medulla oblongata. Site of emergence of cranial nerve V.	Controls certain respiratory functions. Serves as relay station from medulla oblongata to higher structures in brain.
Reticular formation	Complex network of nerve cells organized into ascending (sensory) and descending (motor) pathways. Located throughout core of entire brainstem.	Specific functions for different neurons, including involvement with respiratory and cardiovascular centers, regulation of individual's level of awareness.
Thalamus	Composed of two separate bilateral masses of gray matter. Located in center of cerebrum.	Intermediate relay structure and processing center for all sensory information (except smell) going to cerebrum.
Ventricles	Cavities within brain that are filled with cerebrospinal fluid. Left and right lateral ventricles in cerebral hemispheres, third ventricle in diencephalon, fourth ventricle in pons and medulla oblongata.	Provide circulatory paths for cerebrospinal fluid. Choroid plexuses associated with ventricles are site of formation of most cerebrospinal fluid.

FIGURE 13.3 VENTRICLES OF THE BRAIN AND CIRCULATION OF CEREBROSPINAL FLUID

[A] Frontal view of ventricles of the brain. Note that the anterior horn and body of lateral ventricles are oriented in a plane that is medial to that of the inferior horn. [B] Right lateral view showing the ventricles of the brain and the flow of cerebrospinal fluid. The arrows indicate the direction of flow.

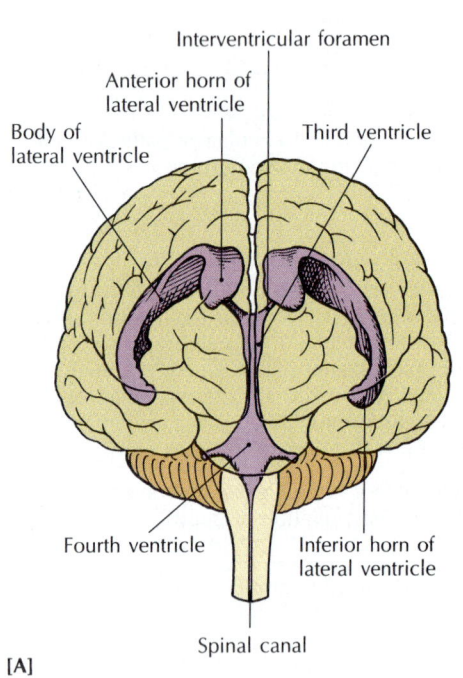

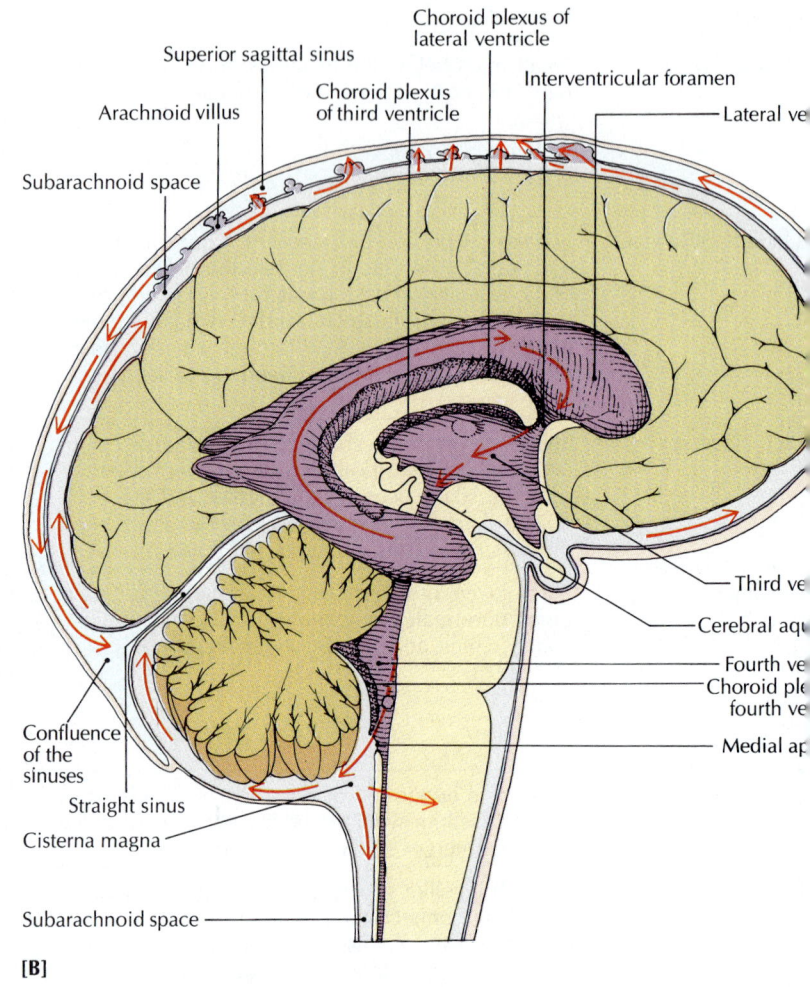

[A]

[B]

with cerebrospinal fluid and is lined by cuboidal epithelial cells known as *ependymal cells*. A network of blood vessels called a *choroid plexus* is formed in several places where the ependyma contacts the pia mater. The four ventricles are numbered from the top of the brain downward. They are the *left* and *right lateral ventricles* of the cerebral hemispheres, the *third ventricle* of the diencephalon, and the *fourth ventricle* of the pons and medulla oblongata [FIGURE 13.3]. Each lateral ventricle is connected to the third ventricle of the diencephalon through a small *interventricular foramen*. The third ventricle is continuous with the fourth ventricle through a narrow channel called the *cerebral aqueduct* of the midbrain.

Cerebrospinal Fluid in the Brain

Cerebrospinal fluid (CSF) is a clear, colorless liquid that is similar to blood plasma. It consists of water; a small amount of protein; oxygen and carbon dioxide in solu-

tion; sodium, potassium, calcium, magnesium, and chloride ions; glucose; a few white blood cells; and many other organic compounds. The subarachnoid space contains about 75 mL (60 percent) of the total amount (about 125 mL) of the cerebrospinal fluid in the central nervous system. This fluid in the subarachnoid space provides a special environment in which the brain floats, cushioning it against hard blows and sudden movements. Floating in cerebrospinal fluid, the brain weighs only about 14 percent of its actual weight.

Q: *Does the weight of the brain increase after childhood?*

A: Although the number of brain cells does not increase after infancy, the cells do grow in size and degree of myelination as the body grows. In addition, the number of glial cells increases after birth. These changes account in large measure for the fact that the adult brain is about three times as heavy as it was at birth. After the age of about 20, the brain begins to lose about 1 g a year as neurons die and are not replaced.

Besides providing a protective buoyancy for the brain, the cerebrospinal fluid aids in maintaining the chemical environment of the central nervous system. It conveys excess components and unwanted substances away from the extracellular fluid and into the venous portion of the blood circulatory system.

Formation of cerebrospinal fluid The ependymal cells that line the ventricles and the pia mater, with its rich blood supply, form the *choroid plexuses,* with their intricate networks of capillaries. The choroid plexuses are considered components of the blood-brain barrier (blood-cerebrospinal fluid barrier). Most of the cerebrospinal fluid is formed continuously at the choroid plexuses of the two lateral, third, and fourth ventricles by a combination of diffusion and active transport.

The choroid plexuses are able to produce cerebrospinal fluid primarily because of the semipermeability of their blood vessels. The choroid plexuses do not allow blood cells or the largest protein molecules to pass into the ventricles. However, they do permit the passage of traces of protein; oxygen and carbon dioxide in solution; sodium, potassium, calcium, magnesium, and chloride ions; glucose; and a few white blood cells.

Circulation of cerebrospinal fluid Cerebrospinal fluid moves from the ventricles inside the brain to the subarachnoid space outside the brain. The fluid flows slowly from the two lateral ventricles of the brain, where much of the fluid is formed, through the paired interventricular foramina to the third ventricle [FIGURE 13.3]. From there, it passes through the cerebral aqueduct into the fourth ventricle. The fluid leaves the ventricular system through three apertures (two lateral and one median) in the roof of the fourth ventricle. From there, according to a new concept, the fluid flows into the subarachnoid space (the *cisterna magna*) around the posterior surface of the posterior surface of the medulla oblongata. From the cisterna magna, cerebrospinal fluid flows slowly up in the subarachnoid space past the brainstem and cerebrum to the arachnoid villi, where it passes into the venous blood of the superior sagittal sinus. Some cerebrospinal fluid flows down from the cisterna magna within the subarach-

noid space surrounding the spinal cord and cauda equina. Most of this fluid is now presumed to return to the venous blood by passing through microarachnoid villi associated with the veins of spinal nerve roots located within the spinal canal.

According to one concept, the spongelike arachnoid villi (fingerlike extensions of the arachnoid that project into the dural venous sinuses) contain pressure-sensitive valves (a series of small tubules) that permit a one-way flow of cerebrospinal fluid from the subarachnoid space into the superior sagittal sinus, which drains venous blood from the brain [FIGURES 13.2A and 13.3]. When the pressure of the fluid in the subarachnoid space exceeds that of the blood in the superior sagittal sinus, the valves open. When the pressures in these two channels are equal or reversed, the valves close.

According to another concept, the arachnoid cells of the villi imbibe ("drink up") cerebrospinal fluid, form vacuoles of the fluid, transport the vacuoles, and discharge them (vesicular transport and discharge) into the venous blood of the superior sagittal sinus. The mechanisms described in both concepts may be operative.

Cerebrospinal fluid is continuously formed and reabsorbed at a rate of about 20 mL/hour (480 mL/day).

A S K   Y O U R S E L F

1 What are the major structures of the brain?

2 What is contained in the subarachnoid space?

3 Which meningeal layer envelops the brain directly?

4 What are cranial ventricles?

5 What are the functions of cerebrospinal fluid?

6 Where is cerebrospinal fluid formed?

7 Describe the circulation of cerebrospinal fluid in the brain.

NUTRITION OF THE BRAIN

The nutrients needed by the brain can reach it only through the blood, and about 750 mL (3 pints) of blood circulates through the brain every minute. Blood reaching the brain contains glucose as well as oxygen. A

Q: *What is the difference between a concussion and a contusion?*

A: A *concussion* (L. shake violently) is an abrupt and momentary loss of consciousness after a violent blow to the head. It results from the sudden movement of the brain within the skull. The loss of consciousness may be due to the sudden pressure on neurons essential to the conscious state or to sudden changes in the polarization of certain neurons. A *contusion* (L. a bruising) is a cerebral injury that produces a bruising of the brain due to blood leakage. Unconsciousness follows a contusion and may last minutes or hours. A concussion is rarely serious, but a contusion can be.

Q: *How long can the brain be deprived of oxygen before it becomes damaged?*

A: If the brain is deprived of oxygen for more than 5 sec, we lose consciousness. After 15 to 20 sec, the muscles begin to twitch in convulsions, and after 9 min, brain cells are damaged permanently. (There is a big difference between holding your breath for 15 sec and depriving the brain of oxygen for the same amount of time, because the oxygen in the air within the lungs is available for exchange with the circulating blood.)

steady supply of glucose is necessary not only because it is the body's chief source of usable energy, but also because the brain cannot store it. The brain also requires about 20 percent of all the oxygen used by the body, and the need for oxygen is high even when the brain is "at rest."

Effects of Deprivation

A lack of either oxygen (*hypoxemia*) or glucose (*hypoglycemia*) in the blood will damage brain tissue faster than any other tissue. Deprivation for even a few minutes may produce permanent brain damage.

Proper and continuous nourishment is so important to the brain that it has built-in homeostatic mechanisms that make it almost impossible to constrict blood vessels that would reduce the incoming blood supply. It is likely that the brain's blood vessels are prevented from constricting by the products of cellular metabolism, which are formed in the brain itself or carried to the brain by the blood. An increase in carbon dioxide or a decrease in oxygen dilates the blood vessels leading to the brain. The presence of too much oxygen and too little carbon dioxide is one of the few conditions that allow *constriction* of cerebral blood vessels. A hyperoxic person (too much oxygen) may suffer dizziness, mental confusion, convulsions, or even unconsciousness until the brain's regulatory system balances the supply of oxygen and carbon dioxide.

ASK YOURSELF

1 What two substances does the brain need continuously?

2 Why is it important that blood vessels to the brain not constrict easily?

BRAINSTEM

The *brainstem* is the stalk of the brain, and it relays messages between the spinal cord and the cerebrum. Its three segments are the *midbrain*, *pons*, and *medulla oblongata* [FIGURE 13.4; see also FIGURE 13.1]. The brainstem is continuous with the diencephalon above and the spinal cord below. It narrows slightly as it leaves the skull, passing through the foramen magnum to merge with the spinal cord.

FIGURE 13.4 THE BRAINSTEM

All cranial nerves except the olfactory and optic nerves emerge from the brainstem. Only the cranial nerves on the right side of the body are shown in this dorsal view. The oculomotor, abducens, and hypoglossal nerves emerge on the ventral side.

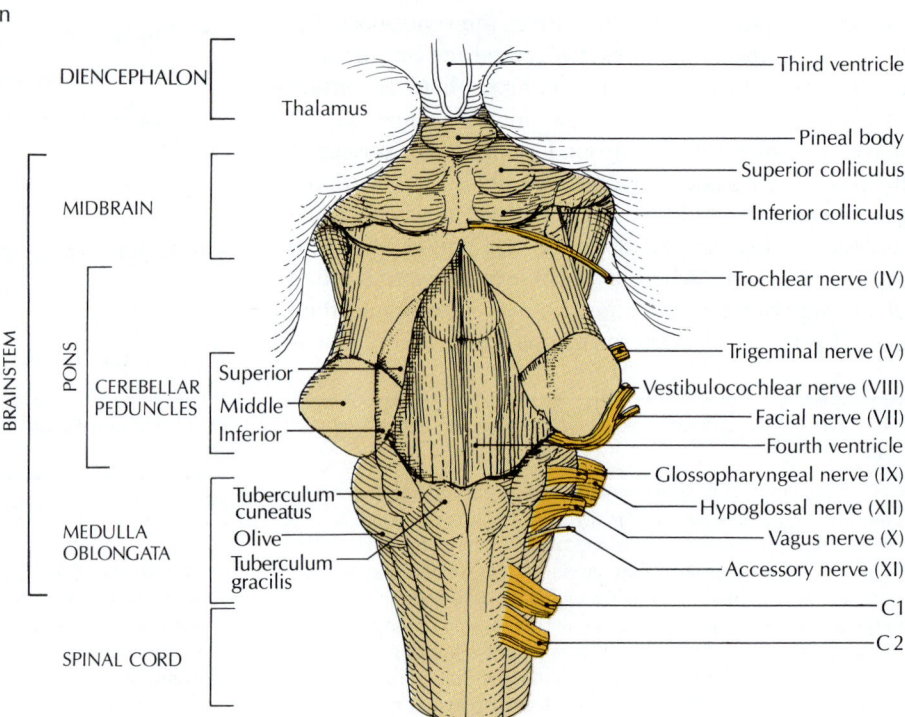

Other structures of the brainstem that are very important functionally include the *long tracts of ascending and descending pathways* conducting information between the spinal cord and parts of the cerebrum, which all pass through the brainstem. A network of nerve cell bodies and fibers, which make up the *reticular formation*, is also located throughout the core of the entire brainstem. It plays a vital role in maintaining life, being involved with the respiratory (breathing) center and the cardiovascular center, among others. Finally, all *cranial nerves* except I (olfactory) and II (optic) emerge from the brainstem. The sensory and motor components of the cranial nerves are associated with the so-called *cranial nerve nuclei* within the brainstem. The cell bodies of sensory neurons conveying sensory (afferent) input to the brainstem terminate in *sensory nuclei.* The cell bodies of motor neurons conveying motor (efferent) output from the brainstem originate in *motor nuclei.*

Sensory and Motor Pathways through the Brainstem

In the nervous system, the word *nucleus* means a collection of nerve cell bodies located *inside* the central nervous system. A similar collection of nerve cell bodies found *outside* the central nervous system is a *ganglion.*

Sensory (ascending) pathways composed of interneurons extend upward from the spinal cord through the brainstem to the cerebrum and cerebellum. Such ascending pathways include those for pain, touch, and vision.

Adequate stimulation of the sensory receptors for pain, touch, and vision results in action potentials. The action potentials are conveyed to the spinal cord and brain by afferent fibers. This afferent input is processed within the central nervous system and is then conveyed through bundles of axons called *tracts* to other nuclei for further processing.

Motor (descending) pathways extend from the cerebral cortex and cerebellum to the brainstem, and from the cerebral cortex and brainstem to the spinal cord. (TABLES 14.1 and 14.2 on pages 434 and 436 summarize some of the major pathways.)

Reticular Formation

Deep within the brainstem is a slender network of neurons and fibers called the *reticular formation* (L. *reticulum*, netlike) [FIGURE 13.5]. It runs through the entire length of the brainstem, with axons extending into the spinal cord and diencephalon.

Neurons within the reticular formation are organized into several groups, each having a specific and life-sustaining function. The reticular formation contains the respiratory and cardiovascular centers, which help regulate such functions as breathing, heart rate, and the diameter of blood vessels. It also helps regulate our level of awareness. (A lesion in the reticular formation of the upper brainstem may result in loss of consciousness, or *coma*, which may last for months or even years.) When the effects of sensory stimuli pass through the brainstem on their way to the highest centers of the brain, they stimulate the reticular formation, which in turn leads to the increased activity of the cerebral cortex.

The reticular formation also contains the spinoreticulothalamic pain pathway, which is associated with the dull and diffuse qualities of pain. Descending motor pathways include the reticulospinal pathways, which convey motor impulses to the spinal cord.

Reticular Activating System

The reticular formation of the brainstem also receives a great variety of sensory information as input from our internal and external environments, alerting the cortex to incoming sensory signals. This input enters an area of the reticular formation known as the *reticular activating system (RAS),* which has roles associated with many behavioral activities, including adjusting certain aspects of the sleep-wake cycle, awareness, alertness, levels of sensory perception, emotions, and motivation. The reticular activating system also helps the cerebellum modulate selected motor units to produce smooth, coordinated contractions of skeletal muscles, as well as maintaining muscle tone.

Medulla Oblongata

The inferior portion of the brainstem is the *medulla oblongata.* It is situated in the inferior part of the cranial cavity, and is about the same length (3 cm) as the midbrain and pons.

The medulla oblongata is continuous with the spinal cord and extends from the foramen magnum to the pons [FIGURES 13.1 and 13.4]. Within the reticular formation of the medulla oblongata are the nuclei that make up the vital cardiac, vasomotor (constriction and dilation of blood vessels), and respiratory centers. The medulla oblongata monitors the level of carbon dioxide through the hydrogen-ion concentration so closely that it will cause the respiration rate to double if the carbon dioxide concentration in the blood rises by as little as 0.03 percent. The medulla oblongata also regulates vomiting, sneezing, coughing, and swallowing.

> **Q:** *Why do we eventually stop hearing a clock ticking?*
>
> **A:** The ticking of a clock usually ceases to arouse neurons in the reticular formation. If the same clock stops, however, we usually notice the silence. People who live near a highway rarely notice the repetitive traffic noise. This phenomenon is called *habituation.*

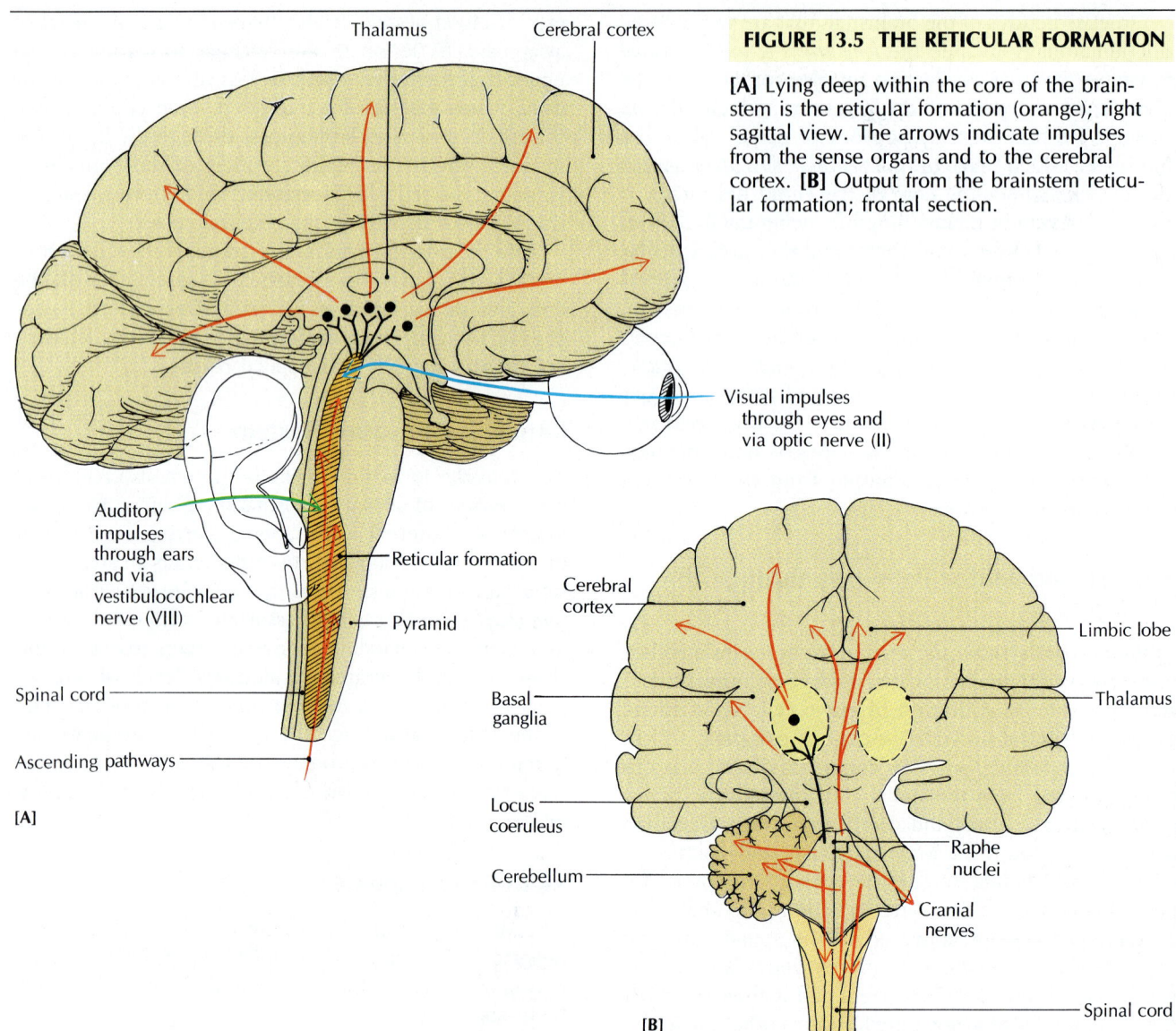

FIGURE 13.5 THE RETICULAR FORMATION

[A] Lying deep within the core of the brainstem is the reticular formation (orange); right sagittal view. The arrows indicate impulses from the sense organs and to the cerebral cortex. [B] Output from the brainstem reticular formation; frontal section.

The medulla oblongata is connected to the pons by longitudinal bundles of nerve fibers. It is joined to the cerebellum by the paired bundles of fibers called the *inferior cerebellar peduncles* (peh-DUNG-kuhlz; L. little feet).

The ventral surface of the medulla oblongata contains bilateral elevated ridges called the **pyramids** [FIGURES 13.1 and 13.9]. The pyramids are composed of the fibers of motor tracts from the motor cerebral cortex to the spinal cord. These **corticospinal** (pyramidal) **tracts** cross over, or *decussate*, in the lower part of the medulla oblongata to the opposite side of the spinal cord, forming an X.* These motor nerve fibers that cross over in the me-

dulla oblongata [FIGURE 13.6] form the lateral corticospinal tracts. The few pyramidal fibers that do not decussate become the uncrossed fibers of the anterior corticospinal tract [see TABLE 14.2 on page 436], which finally cross over within the spinal cord before terminating on the opposite side.

Because almost all motor and sensory pathways cross over, *each side of the brain controls the opposite side of the body.* Hence the left cerebral hemisphere activates the skeletal muscles on the right side of the body, and the right cerebral hemisphere is similarly linked to the left side of the body.

On the dorsal surface of the medulla oblongata are two pairs of bumps: the *tuberculum gracilis* (L. lump + slender) and the *tuberculum cuneatus* (L. wedge) [FIGURE 13.4]. These tubercula are formed by the relay nuclei

---

*Decussate comes from the Latin word *decussare*, which in turn is derived from *dec*, meaning "10." The Latin symbol for 10 is X, which represents the crossing over of the tracts.

FIGURE 13.6 DECUSSATION OF NERVE FIBERS IN THE MEDULLA OBLONGATA

[A] Schematic drawing of ascending touch pathway and descending motor (corticospinal) tract; anterior view.
[B] The ascending touch pathway is composed of a sequence of three neurons. The second neuron in the sequence decussates in the medulla oblongata; anterior view.
[C] The following pathways decussate: (1) corticoponto cerebellar pathway from cerebral cortex to pontine nuclei to cerebellum; (2) cerebellothalamic pathway from cerebellum to thalamus to motor cerebral cortex; (3) corticospinal (py-

ramidal) tract from motor cortex to spinal cord; anterior view. Ninety percent of the fibers cross over (pyramidal decussation) in the lower medulla oblongata, descend as the lateral corticospinal tract, and terminate in all levels of the spinal cord. Ten percent of the fibers descend as the uncrossed fibers of the anterior corticospinal tract, which cross over within the spinal cord before terminating in the cervical and upper thoracic levels of the spinal cord.

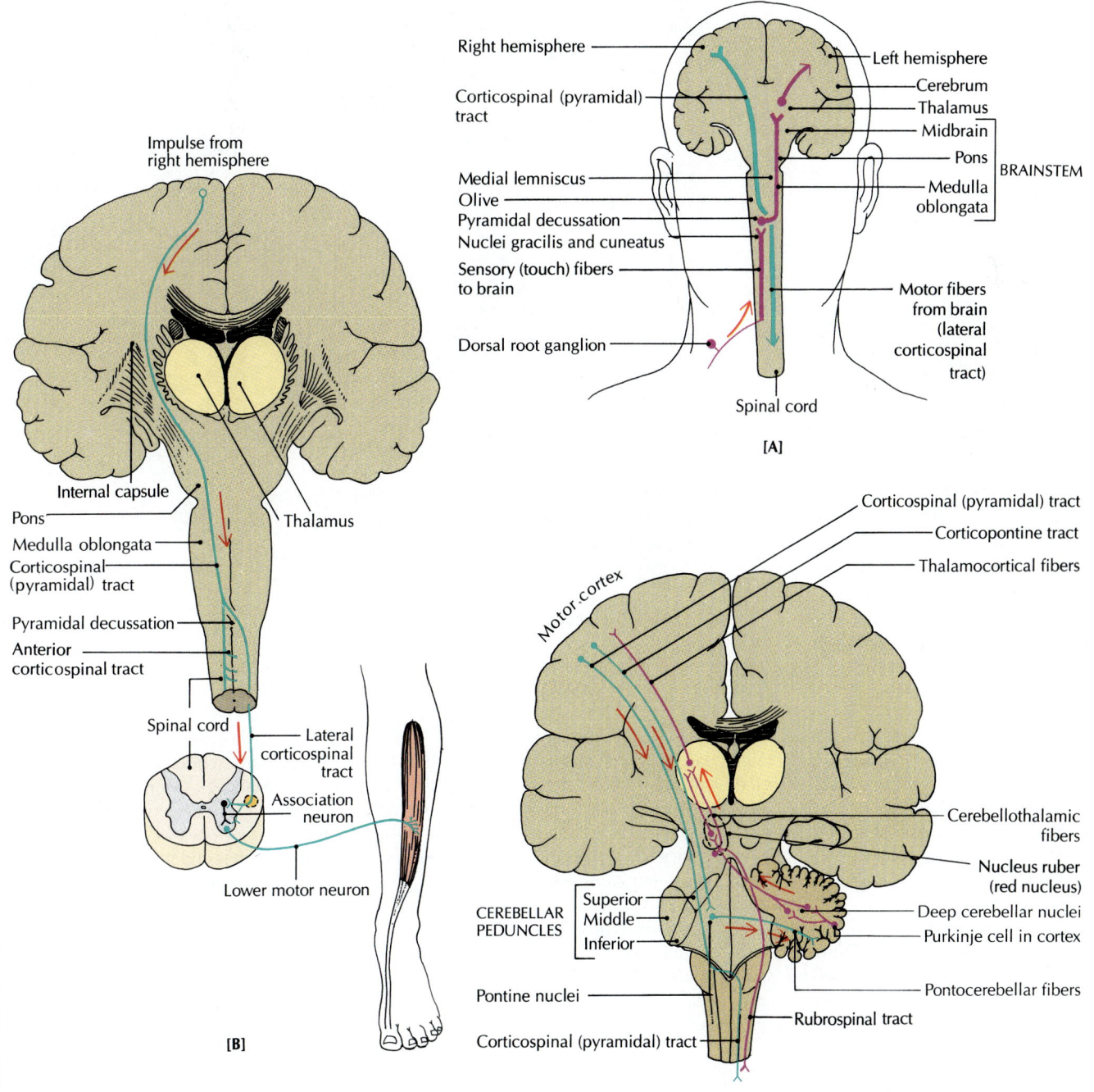

The Blood-Brain Barrier

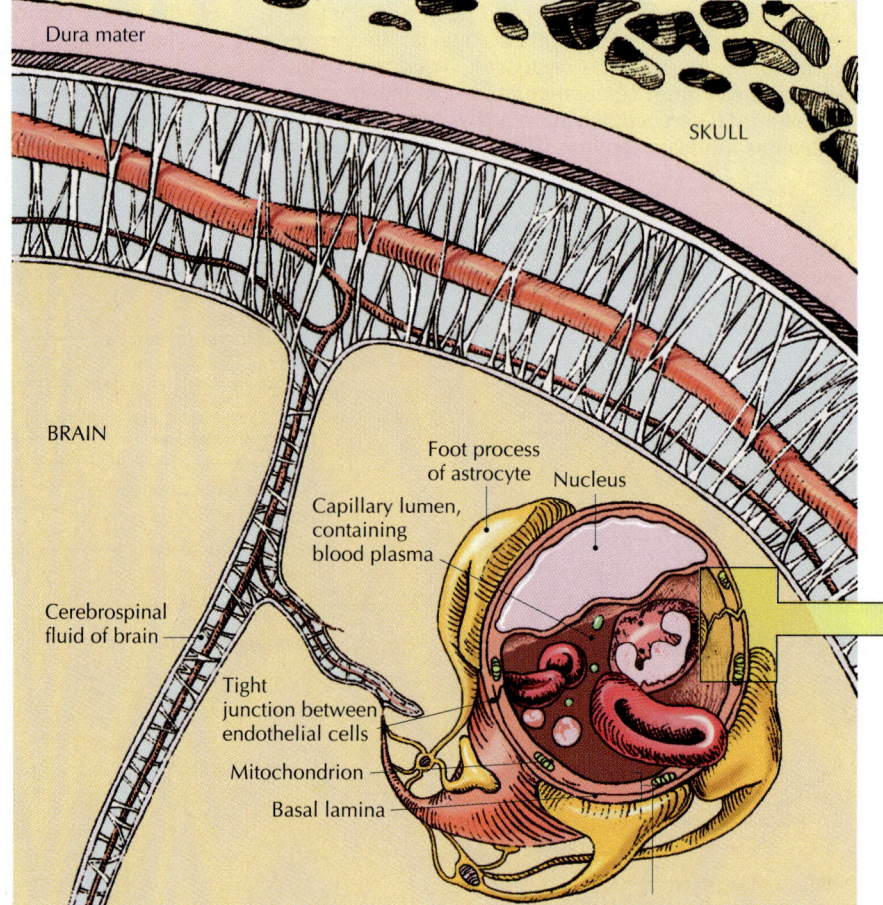

Dura mater

SKULL

BRAIN

Foot process
of astrocyte

Nucleus

Capillary lumen,
containing
blood plasma

Cerebrospinal
fluid of brain

Tight
junction between
endothelial cells

Mitochondrion

Basal lamina

BLOOD-BRAIN BARRIER

The blood-brain barrier.

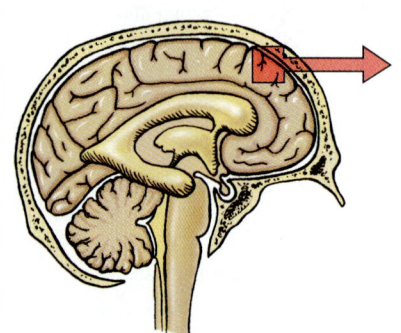

Nowhere in the body is the need for homeostasis greater than in the brain, and no other organ is so well protected against chemical imbalances. The mechanism for maintaining homeostasis in the brain is a special one, focusing on the capillary network that supplies blood to the brain.

The brain is able to absorb some substances, but not others. The barrier that selects which substances reach the brain is called the **blood-brain barrier** (see drawing). The blood-brain barrier is both a physical barrier and a system of cellular transport mechanisms. It helps to maintain the delicate homeostasis of neurons in the brain by restricting the entrance of potentially harmful substances from the blood, and by allowing the entrance of essential nutrients.

Specialized Structure of Brain Capillaries

Brain capillaries have certain unique features: (1) The walls of brain capillaries are formed by endothelial cells joined by *tight junctions*. (The walls of other capillaries contain gaps that allow most substances to pass through.) (2) Brain capillaries are contacted by many *foot processes of astrocytes* that store metabolites and transfer them from capillaries to neurons, take up excess potassium ions from the extracellular fluid during intense neuronal activity, and take up excess neurotransmitters. The astrocytes separate the capillaries from the neurons. (3) The endothelial cells that make up the walls of the brain

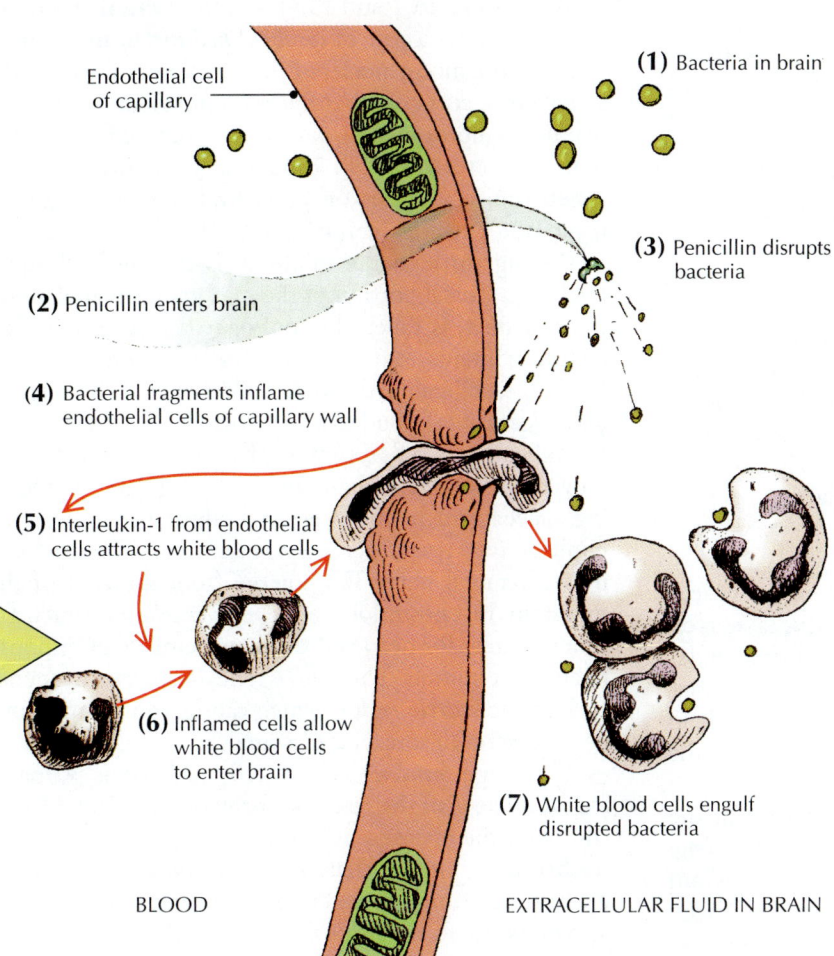

Endothelial cell
of capillary

(1) Bacteria in brain

(3) Penicillin disrupts
bacteria

(2) Penicillin enters brain

(4) Bacterial fragments inflame
endothelial cells of capillary wall

(5) Interleukin-1 from endothelial
cells attracts white blood cells

(6) Inflamed cells allow
white blood cells
to enter brain

(7) White blood cells engulf
disrupted bacteria

BLOOD

EXTRACELLULAR FLUID IN BRAIN

How penicillin enters the brain and encounters meningitis bacteria.

material. In this way, the transport systems contribute directly to the homeostatic regulation of the brain.

Some drugs pass through the blood-brain barrier easily. They include barbiturates; anesthetics such as thiopental sodium, ether, and nitrous oxide (laughing gas); carbon monoxide; cyanide; strychnine; hallucinogenic drugs such as LSD and mescaline; and alcohol in large quantities.

Some other substances, such as large protein molecules and most antibiotics, cannot enter at all. For example, tetracycline crosses the barrier easily, but penicillin is admitted only in trace amounts. The drawing shows what happens when penicillin does enter the brain and encounters the bacteria that cause meningitis.

Modifying the Barriers

How can scientists modify the barriers to allow antibiotics and other useful drugs to reach the brain in order to treat disorders such as Parkinson's disease and cancer? Most anticancer drugs fail to destroy brain tumors because insufficient amounts of the drugs are able to reach the tumors. One experimental approach aims to target specific brain tumors by linking anticancer drugs with lipid-soluble antibodies, immunologic molecules that can "recognize" and bond selectively to cancer cells. When an antibody-drug complex attacks the tumor, the anticancer drug is also delivered at the precise tumor site.

Another promising idea is to bond water-soluble drugs to lipid-soluble carrier molecules that can penetrate the blood-brain barrier. Once inside the brain tissue, the entire complex is modified enzymatically to become water-soluble, so that it is unable to escape from the brain tissue. At this point, the drug would be activated by enzymes in the brain by separating it from its carrier, allowing the drug to provide a sustained release.

capillaries have a large number of *mitochondria,* thus allowing for a high level of oxidative metabolism. The endothelial cells also function much like a semi-permeable membrane.

Passage of Lipid-Soluble Molecules

Lipid-soluble substances pass through the blood-brain barrier rather easily because the plasma membranes of endothelial cells are composed primarily of lipid molecules. Such lipid-soluble substances include nicotine, caffeine, etha-

nol, and heroin. In contrast, water-soluble substances such as sodium, potassium, and chloride ions are unable to cross the barrier without the assistance of carrier-mediated transport mechanisms.

Interestingly, the brain's main energy source—glucose—and those amino acids that the brain cannot synthesize have to be transported across the barrier by carrier proteins. Transport systems not only carry such essential substances into the brain, they also remove surplus

(nuclei gracilis and cuneatus) in the posterior column-medial lemniscus pathway that conveys touch and related sensations to higher brain regions.

Several cranial nerves emerge from the medulla oblongata. The rootlets of cranial nerve XII emerge from the groove located anterior to the olive, a prominent oval mass on each lateral surface of the superior part of the medulla oblongata [FIGURE 13.4]. The rootlets of nerves IX, X, and XI emerge from the groove located posterior to the olive. Cranial nerves VI, VII, and VIII emerge from the junction of the medulla oblongata and pons.

Pons

Just superior to the medulla oblongata is the *pons* (L. bridge), so named because it forms a connecting bridge between the medulla oblongata and the midbrain, the uppermost portion of the brainstem [FIGURES 13.1 and 13.4]. The posterior portion of the pons is called the *dorsal pons*, and the anterior portion is the *ventral pons*. Cranial nerve V emerges from the pons.

Dorsal pons The *dorsal pons* consists of the reticular formation, some nuclei associated with cranial nerves, ascending (sensory) pathways, and some fibers of descending (motor) pathways. Within the reticular formation are the *pneumotaxic* and *apneustic centers*, which help regulate breathing. These centers are integrated with the respiratory centers of the medulla oblongata.

The ascending pathways include the neurons of the ascending reticular system and such sensory tracts as the medial lemniscus (touch-pressure and proprioception) and spinothalamic and trigeminothalamic tracts (pain, temperature, and light touch). The descending pathways are made up of the corticobulbar* fibers, corticoreticular fibers, and rubrospinal tract. The corticobulbar fibers are pathways influencing the motor nuclei of the cranial nerves. The corticoreticular fibers regulate the activity of the reticulospinal tracts that project to the spinal cord. The rubrospinal tracts convey motor impulses from one side of the midbrain to the opposite side of the body.

Ventral pons The *ventral pons* contains the pontine nuclei, which are the relay nuclei of the corticopontocerebellar pathway. The pontocerebellar fibers of this pathway cross over and convey excitatory influences to the cerebellum via the middle cerebellar peduncle [FIGURE 13.6C]. The corticopontocerebellar pathway is the means by which the cerebral cortex communicates with the cerebellum of the opposite side. The corticobulbar and corticospinal tracts pass through the ventral pons.

*The "bulb" refers to the medulla oblongata (sometimes the pons and medulla oblongata). "Corticobulbar fibers" refer to descending motor fibers from the cerebral cortex that influence the nuclei of the motor neurons, which in turn innervate the voluntary muscles of the head.

Midbrain

The *midbrain,* or mesencephalon, is the segment of the brainstem located between the diencephalon and the pons [FIGURES 13.1 and 13.4]. On the ventral surface of the midbrain is a pair of *cerebral peduncles,* made up of fibers to the motor nuclei of the spinal nerves within the spinal cord, corticobulbar fibers (motor fibers to the cranial nerve motor nuclei), and corticopontine fibers to the pons [see FIGURE 13.9]. Emerging from the fossa between the penduncles on its ventral side is the pair of oculomotor nerves (cranial nerve III).

Passing through the midbrain is the cerebral aqueduct. The dorsal portion of the midbrain, situated above the aqueduct, is called the roof or *tectum* (L. roof). The tectum contains four elevations called *colliculi* (L. little hills); the colliculi are known collectively as the *corpora quadrigemina* (L. bodies of four twins). The *superior* pair of colliculi are reflex centers that help coordinate the movements of the eyeballs and head, regulate the focusing mechanism in the eyes, and adjust the size of the pupils in response to certain visual stimuli. The trochlear nerves (cranial nerve IV) emerge from the roof of the midbrain. Just posterior are the *inferior colliculi*, which are reflex centers for head and trunk movements in response to auditory stimuli. They allow reflex responses to sound, such as the *startle reflex,* which causes you to turn your head toward an unexpected sound.

The large *nucleus ruber* (red nucleus) is a major motor nucleus of the reticular formation [FIGURE 13.6C]. (It is so named because it has a reddish appearance in the fresh state.) This nucleus is the termination point for nerve fibers from the cerebral cortex and cerebellum. It also gives rise to the rubrospinal tract, which crosses over in the midbrain before descending to the spinal cord. The corticorubrospinal pathway, along with the corticospinal tract, is involved in somatic motor activities.

Also in the midbrain are heavily black-pigmented paired nuclei called the *substantia nigra* (L. *nigra,* black) [see FIGURE 13.10C], which play a role in controlling subconscious muscle activity.

A S K   Y O U R S E L F

1 What are the three major components of the brainstem?

2 What is the reticular formation?

3 What are cerebral peduncles?

4 How do ganglia and nuclei differ?

5 What is decussation?

CEREBELLUM

The main role of the *cerebellum* (L. little brain) is to regulate balance, timing and precision of body movements, and body positions. It processes input from sensory receptors in the head, body, and limbs. Through connections with the cerebral cortex, vestibular system, and reticular formation, the cerebellum refines balance and coordinates muscular movements. It does not initiate any movements and is not involved in the conscious perception of sensations.

Anatomy of the Cerebellum

The cerebellum is located posterior to the pons in the posterior cranial fossa [FIGURE 13.7A]. It is the second largest part of the brain, the cerebrum being the largest. The cerebellum is separated from the occipital lobes of the cerebrum by a fold of dura mater called the *tentorium cerebelli* and by the transverse cerebral fissure.

The cerebellum may be divided into three parts: (1) a midline portion called the *vermis* (L. wormlike), (2) two small flocculonodular lobes (vestibular cerebellum), and (3) two large lateral lobes. The *flocculonodular lobes* (L. *flocculus*, little tuft of wool) [FIGURE 13.7D], together with the centrally placed wormlike *vermis,* play a role in maintaining skeletal muscle tone, equilibrium, and posture through their influence on the motor pathways that regulate the activity of the back and trunk (axial) muscles.

The much larger *lateral lobes,* or *hemispheres,* of the cerebellum help smooth out muscle movement. They synchronize the delicate and precise timing of the many skeletal muscles involved with any complex activity. Such synchronization is especially apparent in the movements of the upper and lower extremities.

The lobes of the cerebellum are covered by a surface layer of gray matter called the *cerebellar cortex* (L. bark or shell [FIGURE 13.7A]), which is composed of a network of billions of neurons. The cerebellar cortex is corrugated, with long, parallel ridges called *folia cerebelli,* which are more regular than the gyri of the cerebral cortex [FIGURE 13.7B]. The folia are separated by deep folds, or fissures. Under the cortex is a mass of white matter composed of nerve fibers. Lying deep within the white matter are the deep cerebellar nuclei, from which axons project out of the cerebellum to the cerebral cortex. A sagittal section of the vermis reveals a branched arrangement of white matter called the *arbor vitae* (VYE-tee; L. tree of life) [FIGURE 13.7A, B].

Functions of the Cerebellum

The cerebellum integrates the contractions of skeletal muscles in relation to each other as they participate in a movement or series of movements. It is especially involved with coordinating agonists and antagonists in a cooperative way. The cerebellum smoothes out the action of each muscle group by regulating and grading muscle tension and tone in a precise and delicate way. Although the cerebellum *does not initiate any movements*, it participates in each movement through connections to and from the cerebral cortex.

The cerebellum continuously monitors sensory input from muscles, tendons, joints, and vestibular (balance) organs. These sensory *proprioceptive inputs* (the sense of the relative position of one body part to another) are derived on an unconscious level from neuromuscular spindles, tendon (Golgi) organs, the vestibular system, and other sensory endings (see Chapter 17).

One way to understand the function of the cerebellum is to observe the results of a cerebellar lesion, such as might be caused by a stroke or a tumor on one side of the cerebellum. In such a lesion, the reflexes are diminished, and the absence of perfect coordination is shown through tremors and jerky, puppetlike movements. The condition is called *ataxia* (Gr. lack of order). The patient has the symptoms on the same side of the body as the lesion because the fibers cross at two sites, making a "double cross" instead of the usual single crossing over.

The double-cross expression of cerebellar activity is first conveyed (1) from the cerebellum via a crossing in the midbrain to the thalamus, and then relayed to the cerebral cortex, where it is finally processed; then (2) the control of the movement is conveyed by way of the corticobulbar and corticospinal tracts via a second crossing in the lower medulla oblongata to the opposite side of the spinal cord, where it influences the movements on the same side as the cerebellar lesion.

In addition to the cerebellum's involvement with ongoing motor activities, it is involved with the neural processing *just prior to* the actual voluntary movement.

ASK YOURSELF

1 What are the parts of the cerebellum?

2 What is the cerebellar cortex?

3 What are the main functions of the cerebellum?

CEREBRUM

The largest and most complex structure of the nervous system is the *cerebrum* (suh-REE-bruhm; L. brain). It consists of two cerebral hemispheres [FIGURE 13.1]. Each hemisphere is composed of a cortex (gray matter), white matter, and basal ganglia. The cortex on each side is further divided into six lobes: the frontal, parietal, temporal,

FIGURE 13.7 CEREBELLUM

[A] Right sagittal view. [B] Sagittal section, showing the arbor vitae. [C] Superior surface, showing the lobes and vermis. [D] Inferior surface showing the flocculonodular lobe.

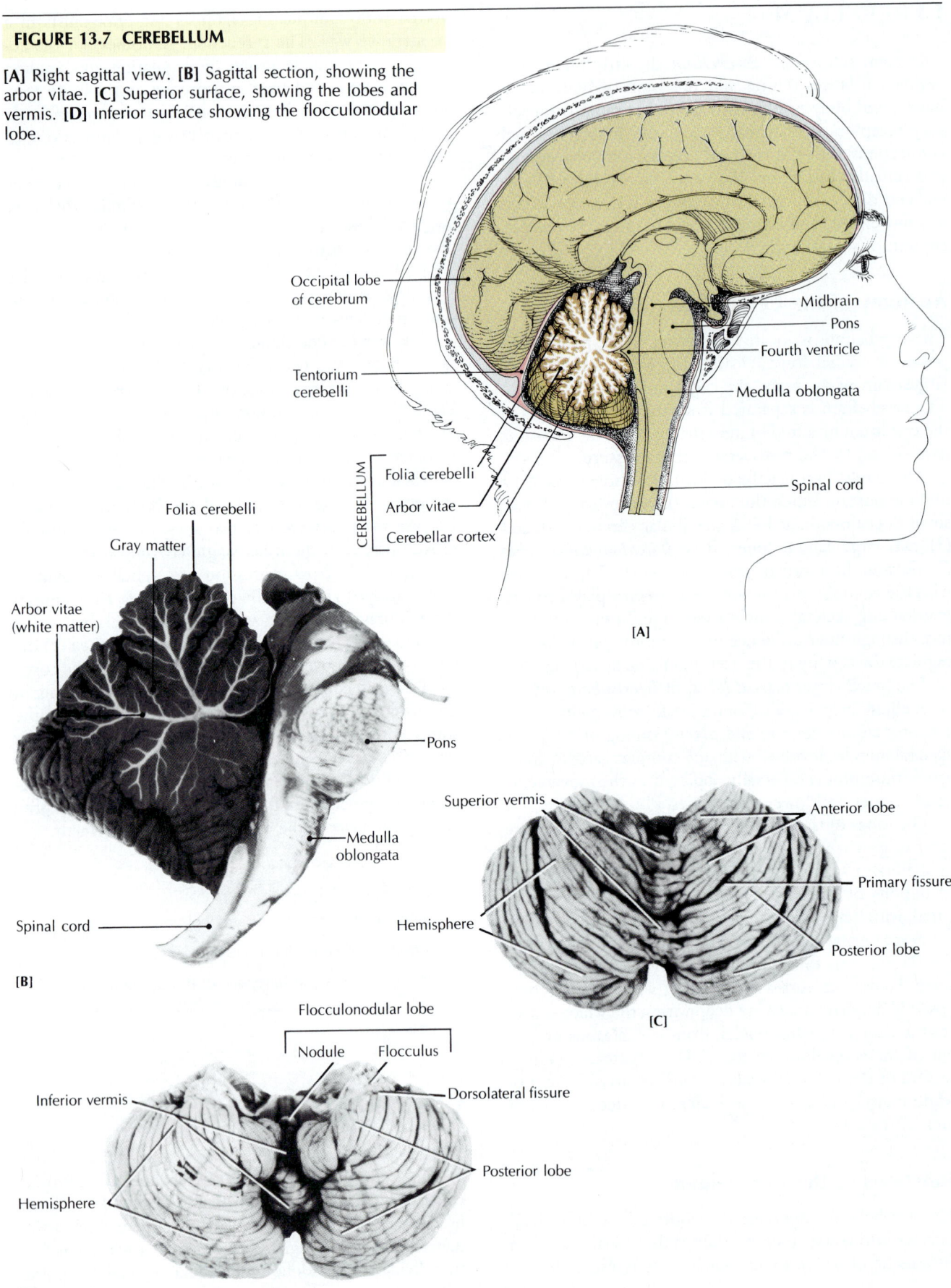

Occipital lobe of cerebrum

Tentorium cerebelli

Midbrain

Pons

Fourth ventricle

Medulla oblongata

Spinal cord

CEREBELLUM

Folia cerebelli

Arbor vitae

Cerebellar cortex

[A]

Folia cerebelli

Gray matter

Arbor vitae (white matter)

Pons

Medulla oblongata

Spinal cord

[B]

Superior vermis

Anterior lobe

Primary fissure

Hemisphere

Posterior lobe

[C]

Flocculonodular lobe

Nodule   Flocculus

Inferior vermis

Dorsolateral fissure

Hemisphere

Posterior lobe

[D]

FIGURE 13.8 GYRI, SULCI, AND FISSURES OF THE CEREBRAL HEMISPHERES

[A] Superior view. [B] Right lateral view.

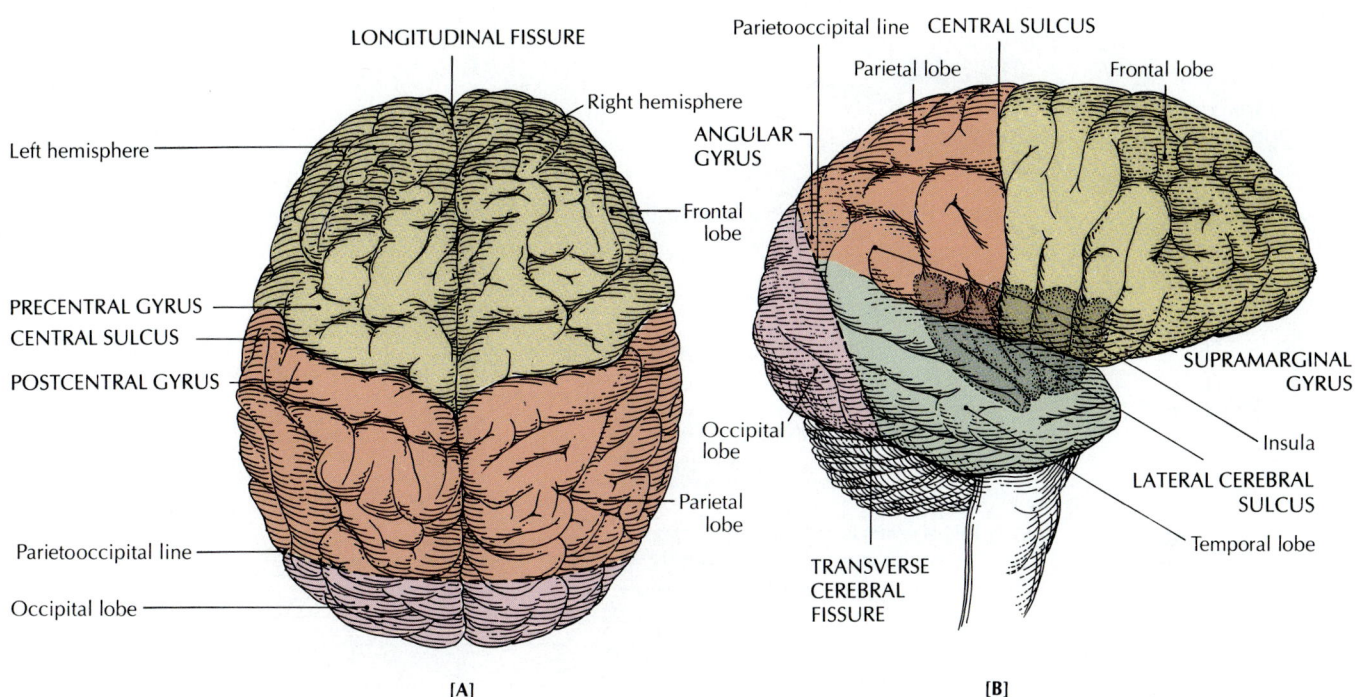

occipital, limbic, and insula (central lobe) [FIGURE 13.8]. Each of the first four lobes contains special functional areas, including speech, hearing, vision, movement, and the appreciation of general sensations. The olfactory nerve and bulb are located beneath the frontal lobe [see FIGURE 13.12].

All our conscious living depends on the cerebrum. Popularly, it is considered the region where thinking is done, but no specific parts have been identified as the exact sites of consciousness or learning.

Anatomy of the Cerebrum

The cerebrum has a surface covering of gray matter called the *cerebral cortex.* The cortex is a thin (about 4.5 mm) convoluted covering containing over 50 billion neurons and 250 billion glial cells (estimated to be 70 percent of the brain cells). The raised ridges of the cortex are called convolutions or *gyri* (JYE-rye; sing. *gyrus*), which are separated by slitlike grooves called *sulci* (SUHL-kye; sing. *sulcus*). The gyri and sulci increase the surface area of the cerebral cortex, resulting in a 3:1 proportion of cortical gray matter to the underlying white matter. Although each person has a specific pattern of gyri and sulci, the overall developmental design of the cerebrum produces a fairly consistent pattern.

Extremely deep cerebral grooves or depressions are called *fissures* (L. *fissus,* crack). The cerebral hemispheres are separated by the *longitudinal fissure,* and the cerebrum is separated from the cerebellum by the *transverse cerebral fissure* [FIGURE 13.8B].

Beneath the cortex lies a thick layer of white matter. The white matter consists of interconnecting groups of axons projecting in two basic directions. One group projects *from the cortex* to other cortical areas of the same and opposite hemisphere, the thalamus, the basal ganglia, the brainstem, or the spinal cord. The other group projects *from the thalamus* to the cortex. The thalamus is functionally integrated with the cerebral cortex in the highest sensory and motor functions of the nervous system.

Thus, the white matter consists of three types of fibers [FIGURE 13.9]:

1 *Association fibers* are axons that link one area of the cortex to another area of the cortex of the *same hemisphere.*

2 *Commissural fibers* are the axons that project from a cortical area of one hemisphere to a corresponding cortical area of the *opposite hemisphere.* The two major cerebral commissures are the *anterior commissure* and the massive bundle of axons called the *corpus callosum* (L. hard body). Both connect the two cerebral hemi-

FIGURE 13.9 FIBER TRACTS IN THE WHITE MATTER OF THE CEREBRUM

[A] Right lateral view. [B] Anterior view.

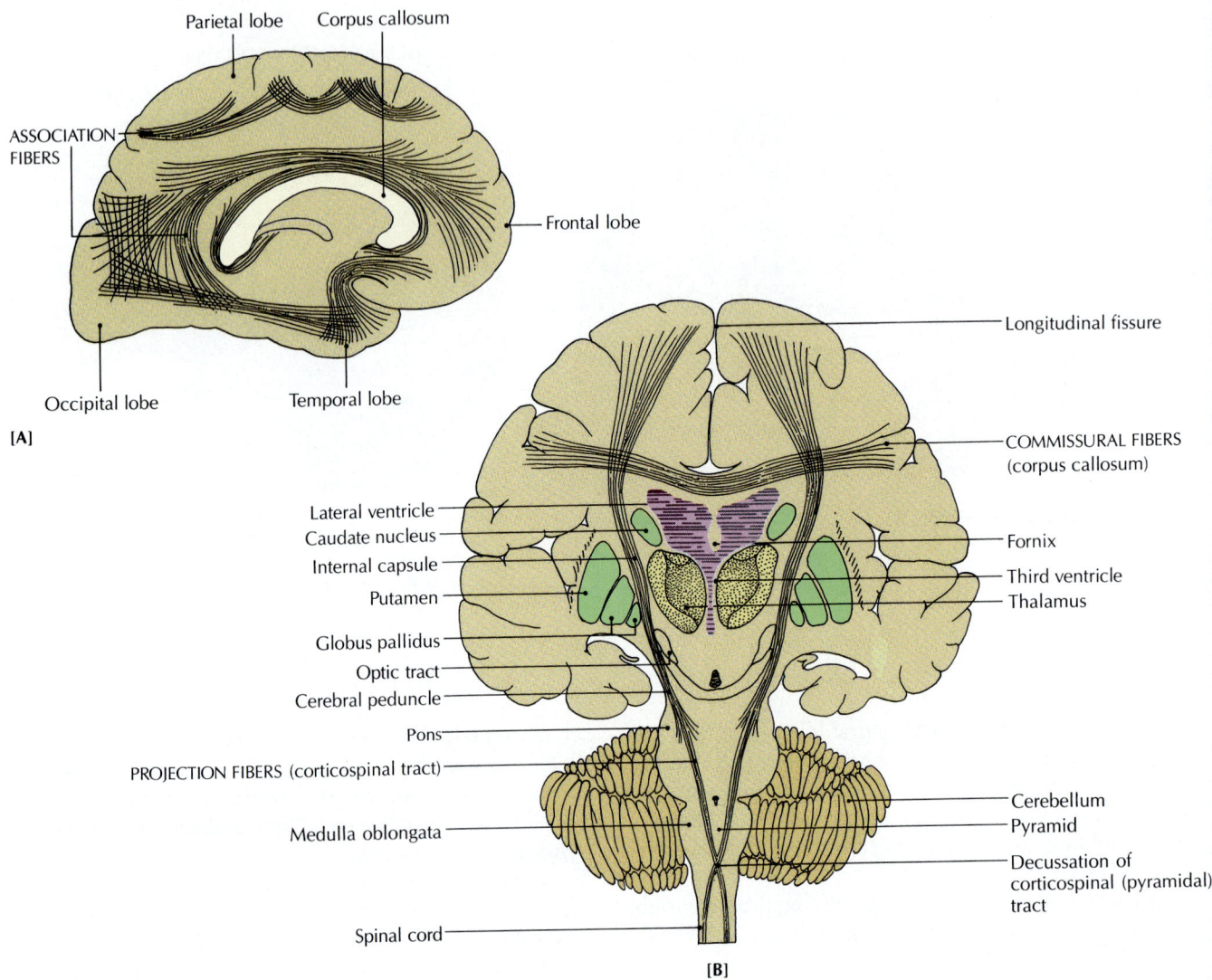

[A]

[B]

spheres and relay nerve impulses between them [FIGURE 13.1B, C]. The corpus callosum contains about 200 million axons.

3 Projection fibers include the axons that project from the cerebral cortex *to other structures* of the brain, such as the basal ganglia, thalamus, and brainstem, and to the spinal cord. The corticospinal tract from the motor cortex to the spinal cord is composed of projection fibers.

Basal ganglia Deep within each cerebral hemisphere is a group of five subcortical nuclei called the *basal ganglia* (see "The Anatomical Organization of the Brain" on page 407). It is important to note that the use of the word *ganglia* when referring to the basal ganglia is actually inconsistent with modern terminology. Such

usage goes back to the nineteenth century, when any collection of neurons was called a ganglion. However, the use of the term *basal ganglia* persists when applied to these five subcortical nuclei, because it is still used by neuroscientists, neurologists, neurosurgeons, physiological psychologists, and other professionals.

The basal ganglia include the *caudate* ("tail-shaped") *nucleus, putamen, globus pallidus* ("pale ball"), *subthalamic nucleus,* and *substantia nigra* [FIGURES 13.9 and 13.10]. The **caudate nucleus** and the *putamen* (pyoo-TAY-muhn; L. prunings) are collectively called the **striatum.** The putamen and globus pallidus make up the **lentiform** ("lens-shaped") **nucleus.** The basal ganglia are organized as an intricate network of neural circuits and processing centers. Along with the cerebellum, the basal ganglia act

FIGURE 13.10 BASAL GANGLIA

[A] Section through the brain showing the parts of the basal ganglia (green). **[B]** Structures that make up the basal ganglia. **[C]** Frontal view. The structures of the basal ganglia shown in these drawings include the head and tail of the caudate nucleus, globus pallidus, putamen, lentiform nucleus, subthalamic nucleus, and substantia nigra. The other structures are not part of the basal ganglia.

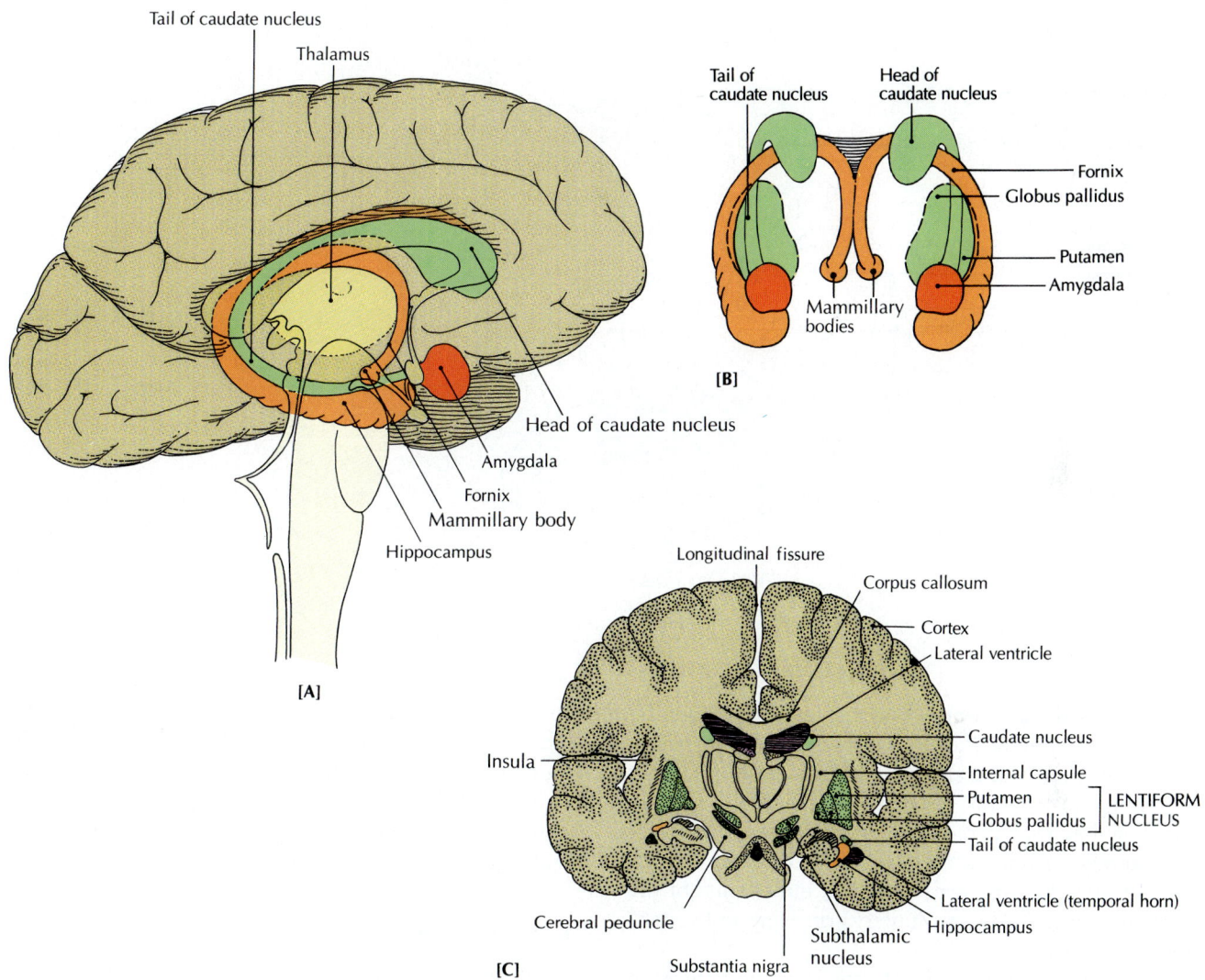

at the interface between the sensory systems and many motor responses, affecting motor, emotional, and cognitive behaviors. Clinically, the terms *movement disorders* and *malfunctioning of the basal ganglia* are essentially synonymous. Such disorders as Parkinson's disease and Huntington's disease are basal ganglia disturbances.

The basic circuitry associated with the basal nuclei may be summarized as follows: The basal ganglia receive information from many areas of the cerebral cortex, process and integrate these inputs, and then relay them to the thalamus, which in turn projects its output to the motor, premotor, and limbic cortical areas. These areas then exert their influences on the upper motor neuron pathways and other systems that affect motor, emotional,

and cognitive behaviors. The abnormal involuntary movements associated with malfunctioning of the basal ganglia are called *dyskinesias*. These movements are expressions of *release phenomena*, in which certain inhibitory influences are reduced or lost. The consequence of this is that certain neuronal centers are deprived of regulatory and modulatory influences. An analogy is the brakes (inhibitors) that act to control a car. Malfunctioning of the brakes results in uninhibited movement. In Parkinson's disease there is a reduction in the neurotransmitter dopamine in the neurons with cell bodies in the substantia nigra and axons that terminate in the striatum (caudate nucleus and putamen). The resulting reduction in the "inhibitory" activity of dopamine elicits the symptoms of

FIGURE 13.11 CEREBRUM

A partial Brodmann's numbered map of the right lateral cerebral hemisphere. Brodmann's maps are useful mainly to show regions of different neural structure, but some areas also have a functional designation. In keeping with the other illustrations in the book, this one shows the *right* side of the body. Note, however, that Broca's (anterior) speech area (44, 45) and Wernicke's (posterior) speech area (39, 40, 22) are almost always in the *left* cerebral hemisphere.

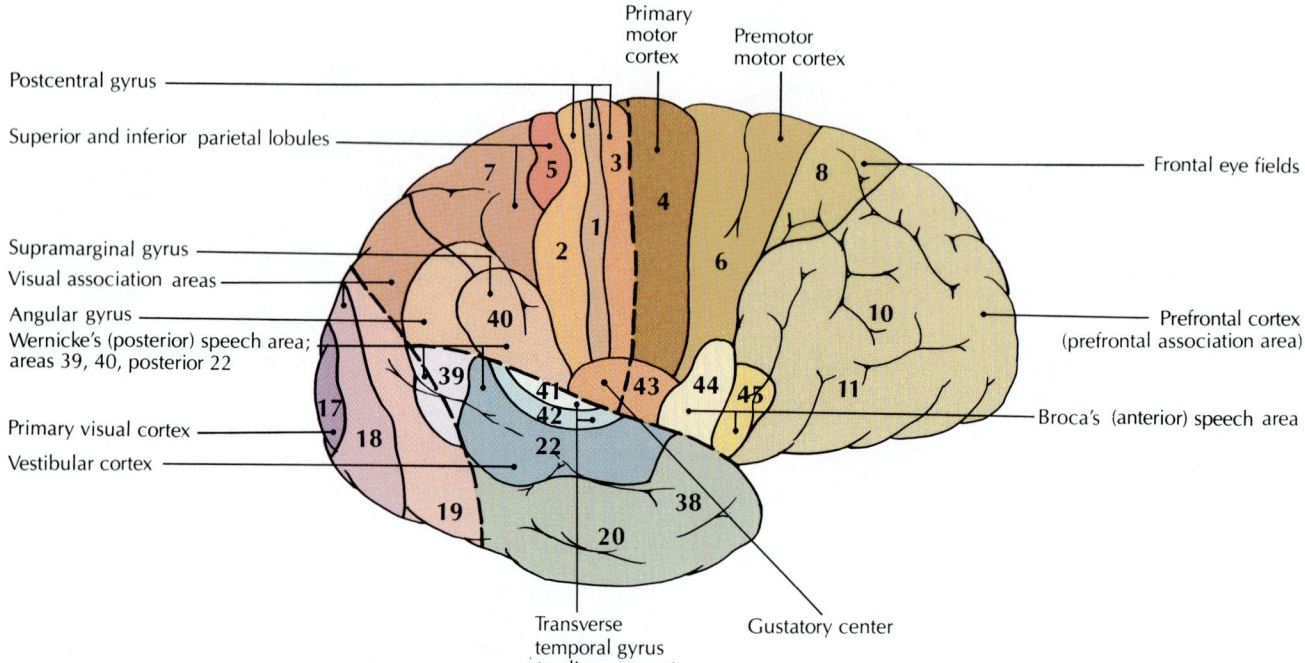

Primary motor cortex
Premotor motor cortex
Postcentral gyrus
Superior and inferior parietal lobules
Frontal eye fields
Supramarginal gyrus
Visual association areas
Angular gyrus
Wernicke's (posterior) speech area; areas 39, 40, posterior 22
Prefrontal cortex (prefrontal association area)
Primary visual cortex
Vestibular cortex
Broca's (anterior) speech area
Transverse temporal gyrus (auditory cortex)
Gustatory center

Parkinson's disease, which is characterized by rigidity, tremor, and akinesia. *Rigidity* is increased tonus (hypertonus) in the skeletal muscles; *tremor* is rhythmic oscillatory trembling movements, especially of the forearm and hand; and *akinesia* is the tendency to be immobile.

Huntington's disease is a hereditary disorder characterized by jerky, irregular, and brisk movements of the limbs accompanied by involuntary grimacing and twitching of the face.

The *precise* role of the basal ganglia in the regulation of normal movements is still under investigation.

Functions of the Cerebrum

The two cerebral hemispheres have the same general appearance, but each has different functions. One of the hemispheres, usually the left, is active in speech, writing, calculation, language comprehension, and analytic thought processes [FIGURE 13.11]. The other hemisphere, usually the right, is more specialized for the appreciation of spatial relationships, conceptual nonverbal ideas, simple language comprehension, and general thought processes. The left hemisphere sorts out the *parts* of things, while the right hemisphere concentrates

on the *whole*. In a manner of speaking, the left hemisphere sees the trees but not much of the forest, while the right hemisphere sees the forest but not many of the trees.

The specific functions of the cerebrum and their localized areas are discussed further in the following sections on the cerebral lobes.

Cerebral Lobes

Each cerebral hemisphere is subdivided into six lobes: the frontal, parietal, temporal, occipital, and limbic lobes, and the insula (central lobe) [FIGURE 13.8]. The first four lobes are named for the skull bones covering them. The frontal lobe is separated from the parietal lobe by the **central sulcus**. The **lateral cerebral sulcus** divides the frontal and parietal lobes from the temporal lobe. The parietal and temporal lobes are separated from the occipital lobe by the arbitrary *parietooccipital line*. Buried deep in the lateral sulcus is the small central lobe, or *insula*.

Another subdivision of each hemisphere, the **limbic lobe**, is on the medial surface of the cerebral hemisphere [FIGURE 13.12B]. It consists largely of the **cingulate gyrus** ("girdling convolution") and the **parahippocampal gyrus**. The **parietooccipital sulcus** on the medial surfaces separates the occipital lobe from the parietal lobe.

FIGURE 13.12 LIMBIC SYSTEM

[A] Major components of the limbic system. **[B]** Median surface of the cerebral hemisphere; the limbic lobe is gray. The structures of the limbic system illustrated in these drawings include the amygdala, hippocampus, limbic lobe, prefrontal cortex, mammillary body, and fornix.

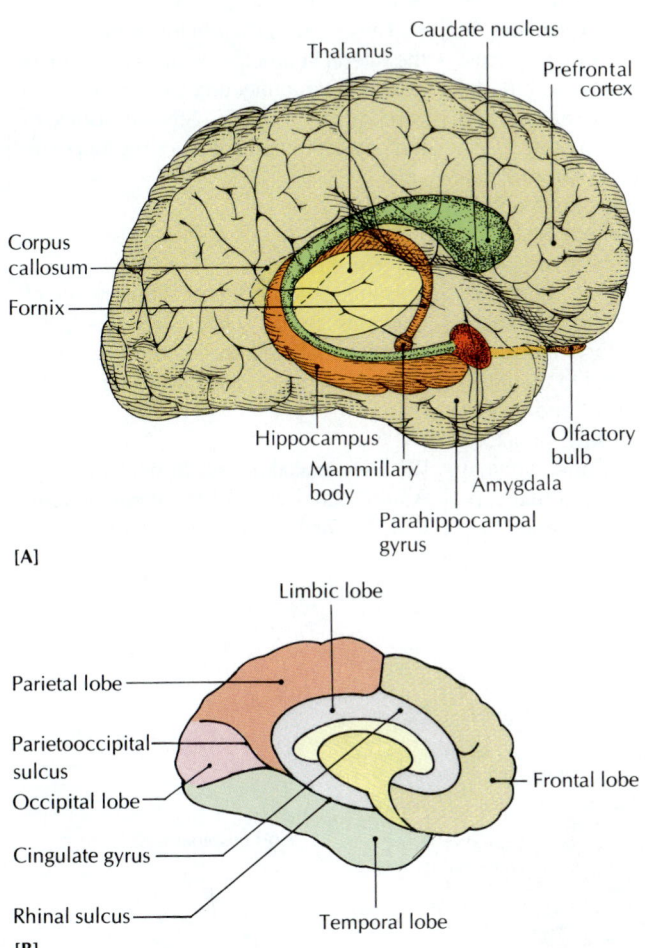

[A]

[B]

FIGURE 13.13 CORTICAL SPACE ALLOTTED TO SENSATIONS

The human body is distorted in this model, so that the body parts are proportional to the space allotted in the brain to sensations from different parts of the body. If the eyes were drawn to scale, they would be larger than the entire body.

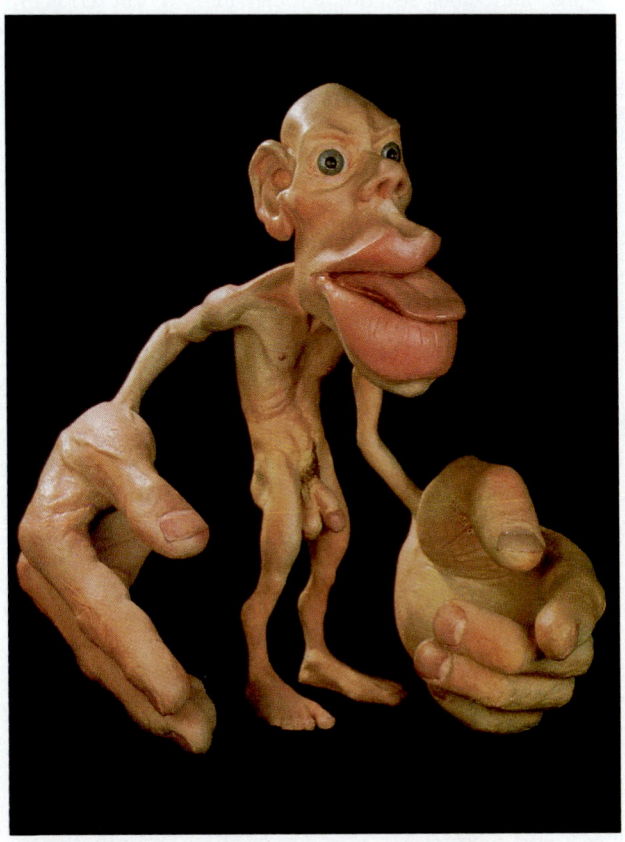

Frontal lobe The *frontal lobe* (also called the "motor lobe") is involved with two basic cerebral functions: (1) the motor control of voluntary movements, including those associated with speech, and (2) the control of emotional expressions and moral and ethical behavior. The motor activity is expressed through the *primary motor cortex* (Brodmann area 4), *supplementary* and *premotor motor cortex* (area 6), the *frontal eye field* (area 8), and *Broca's speech area** (areas 44, 45) [FIGURE 13.11].

From the *primary,* **supplementary,** and **premotor motor cortices,** nerve impulses are conveyed through the motor pathways to the brainstem and spinal cord to processing centers that stimulate the motor nerves of the

---

*This area was named for Pierre Paul Broca, who first described it in the late nineteenth century.

skeletal muscles. FIGURE 13.13 shows the body drawn in proportion to the cortical space allotted to sensations and movement patterns of different parts of the body. Relatively large cortical areas are devoted to the face, larynx, tongue, lips, and fingers, especially the thumb. The thumb is so important for our dexterity that more of the brain's gray matter is devoted to manipulating the thumb than to controlling the thorax and abdomen. The disproportionate allotment of the cortical areas to these body parts reflects the delicacy with which facial expressions, vocalizations, and manual manipulations can be controlled.

The *frontal eye field* is a cortical area that regulates the scanning movements of the eyes, such as searching the sky to locate an airplane. *Broca's speech area (anterior speech area)* is critically involved with the formulation of words.

The cortex in front of the motor areas is called the *prefrontal cortex.* It has a role in various forms of emo-

The Anatomical Organization of the Brain

These illustrations are designed to help you visualize the internal three-dimensional structure of the brain. An understanding of these relationships is useful in identifying structures in CT scans, MRI scans, and other sections through the brain.

1 The blue tubelike lines represent the nerve fibers that convey sensory information to the cerebral cortex and motor information from the cerebral cortex. The concentration of these fibers in the region of the ventricles [C] and basal ganglia [D] is called the *internal capsule*. As the fibers continue toward the cortex, they spread out, forming the *corona radiata*. The internal capsule is lateral to the diencephalon [E]; it is also lateral to the head of the caudate núcleus and medial to the lentiform nucleus [D].
2 The *anterior horn* and the *body* of the *lateral ventricle* are in a plane medial to the internal capsule; the *inferior horn* of the lateral ventricle is in a plane lateral to the internal capsule. The third ventricle is in the midsagittal plane between the bilateral diencephalon.
3 The *corticospinal tract* originates in the motor cortex of the precentral gyrus. Its axons pass successively within the internal capsule and the basilar portion of the brainstem before crossing to the opposite side as the decussation of the corticospinal tract at the medulla oblongata–spinal cord junction into the lateral funiculus of the spinal cord. (In drawings [A–E], the dashed black line represents the main axis of the corticospinal tract.)
4 The *amygdala* and *hippocampus* of the *limbic system* are lateral to the internal capsule [drawings [A, C] and FIGURE 13.12]. The *fornix* (pathway of axons from the hippocampus) forms an arc that finally passes medial to the internal capsule to terminate in the mammillary body of the diencephalon (hypothalamus).
5 The centrally located *diencephalon* is flanked laterally by the internal capsule of each hemisphere.
6 The *corpus callosum* forms the roof of the lateral ventricles. It consists of fibers (axons) interconnecting the cortex of one cerebral hemisphere with the cortex of the cerebral hemisphere of the opposite side.

[A] The Brain
This illustration is a partial composite of diagrams [B–D]. Unless specified, all structures labeled are either in the midsagittal plane or the right side of the brain.

[B] Diencephalon and Brainstem
A midsagittal view of the **diencephalon** and **brainstem.** The diencephalon is

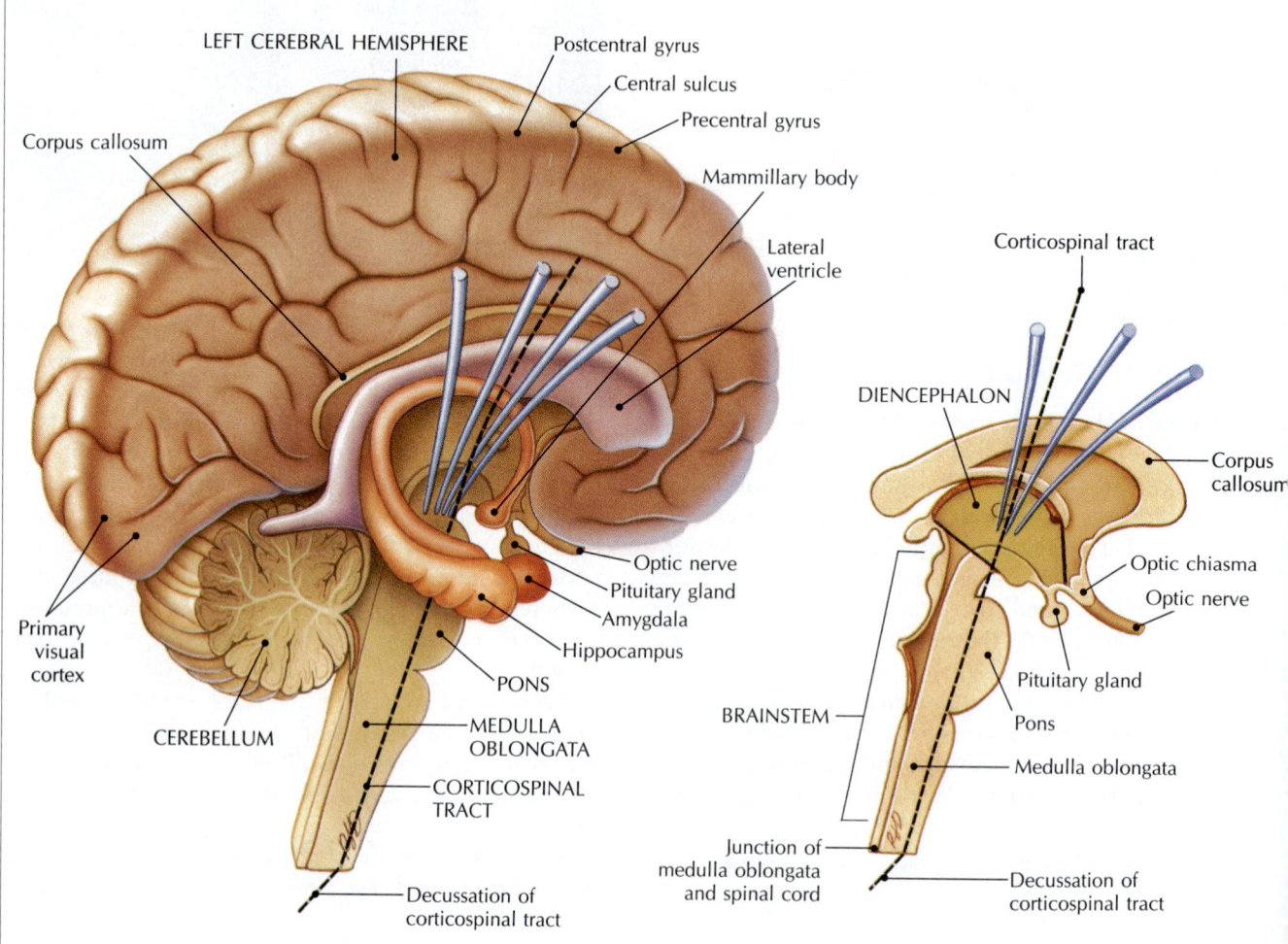

[A] Internal organization of the brain

[B] Diencephalon and brainstem

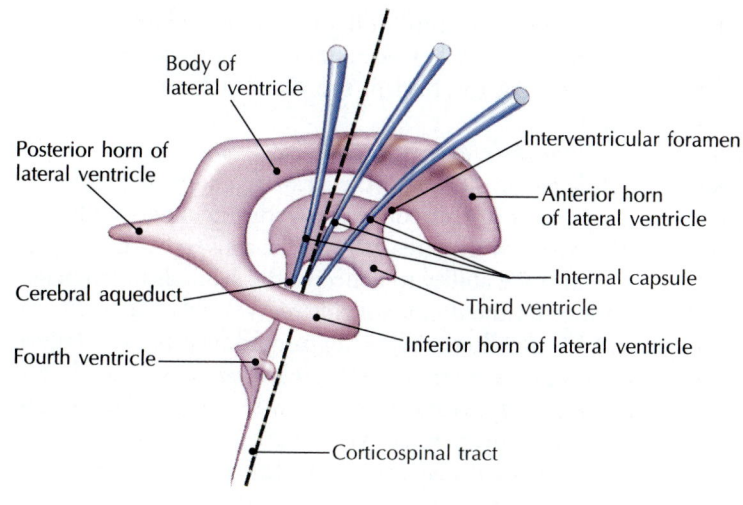

Body of
lateral ventricle

Interventricular foramen

Posterior horn of
lateral ventricle

Anterior horn
of lateral ventricle

Internal capsule

Cerebral aqueduct

Third ventricle

Inferior horn of lateral ventricle

Fourth ventricle

Corticospinal tract

[C] Ventricles

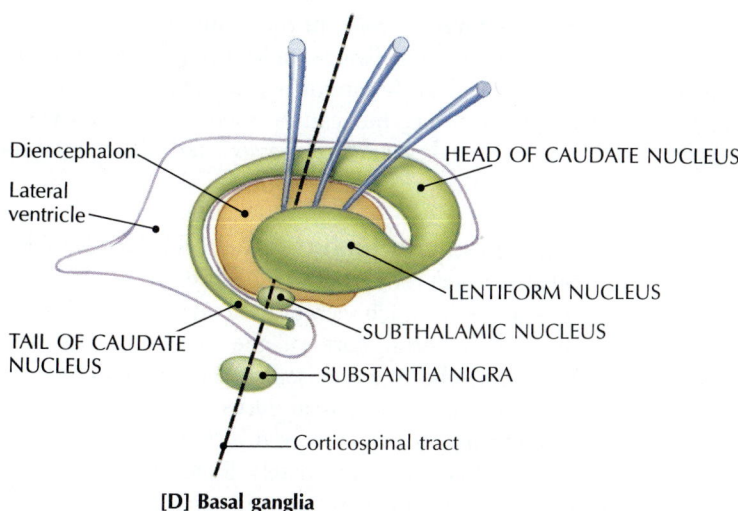

Diencephalon

HEAD OF CAUDATE NUCLEUS

Lateral
ventricle

LENTIFORM NUCLEUS

SUBTHALAMIC NUCLEUS

TAIL OF CAUDATE
NUCLEUS

SUBSTANTIA NIGRA

Corticospinal tract

[D] Basal ganglia

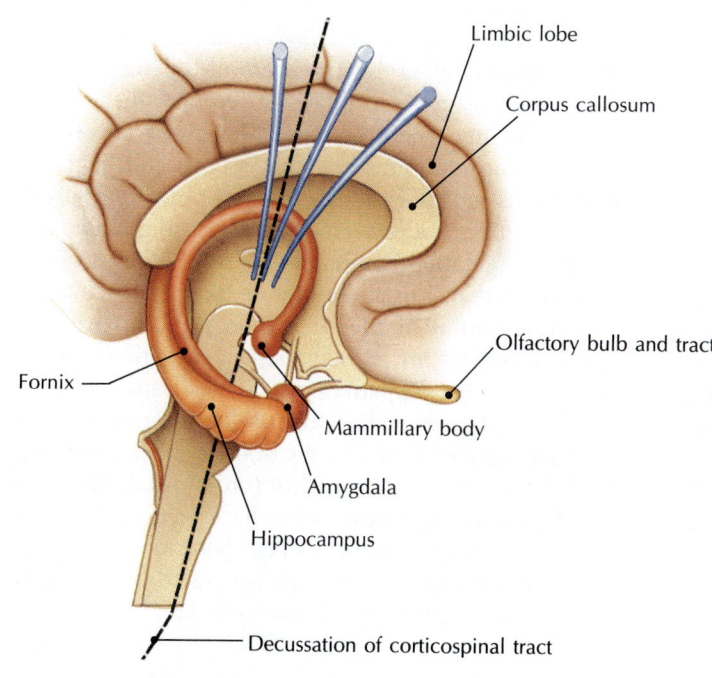

Limbic lobe

Corpus callosum

Olfactory bulb and tract

Fornix

Mammillary body

Amygdala

Hippocampus

Decussation of corticospinal tract

[E] Limbic system

flanked by the cerebral hemispheres. The third ventricle, in the midsagittal plane, extends caudally from the diencephalon into the cerebral aqueduct of the midbrain and the fourth ventricle of the pons and medulla oblongata. The medulla oblongata is continuous with the spinal cord.

[C] Ventricles

The **ventricles** are fluid-filled cavities within the brain. Each *lateral ventricle* is located in a cerebral hemisphere, and its location is an important anatomical reference site. It outlines an *arc* composed of the *anterior horn* (in the frontal lobe), *body* (in the parietal lobe), and *inferior horn* (in the temporal lobe). The *posterior horn* extends into the occipital lobe. The *anterior horn* and *body* are located in a medial plane, while the *inferior horn* is in a more lateral plane of each cerebral hemisphere. The *interventricular foramen* is the opening between a lateral ventricle and the third ventricle. The *third ventricle* of the diencephalon, the *cerebral aqueduct* of the midbrain, and the *fourth ventricle* of the pons and medulla oblongata are located in the midsagittal region of the diencephalon and brainstem [B].

[D] Major Components of the Basal Ganglia

The **basal ganglia,** or basal nuclei, include the *caudate nucleus, lentiform nucleus, subthalamic nucleus,* and *substantia nigra*. Note the anatomical relations of these structures to the lateral ventricle and to the internal capsule. The caudate nucleus borders the lateral ventricle, with its head adjacent to the anterior horn and body and tail following the arc of the inferior horn. The head of the caudate nucleus is in a plane medial to, and the tail in a plane lateral to, the internal capsule. The lentiform nucleus is located lateral to the internal capsule. The subthalamic nucleus is in the base of the diencephalon, and the substantia nigra is in the basal aspect of the midbrain.

[E] Major Structures of the Limbic System

The major structures of the **limbic system** are the *amygdala, hippocampus, fornix,* and *mammillary body*. They form an arc that roughly parallels the arc formed by each lateral ventricle. Note that the amygdala and hippocampus are located close to the inferior horn and that the fornix (composed of axons from the hippocampus) gradually shifts to a medial position before reaching the mammillary body of the diencephalon. The hippocampus and amygdala are lateral to the internal capsule. The limbic lobe is a lobe of each cerebral hemisphere.

tional expression. The functional role of the prefrontal cortex has been deduced largely from patients who have had the white matter of various areas of the prefrontal lobe surgically removed or cut, a procedure called a *prefrontal lobotomy*. Following the operation, the patients are usually less excitable and less creative than before, but they vent their feelings frankly, without the typical restraint. Physical drive is lowered, but intelligence is not. An awareness of pain remains, but normal feelings associated with the bothersome aspects of pain are lost.

Parietal lobe The *parietal lobe* is concerned with the evaluation of the general senses. It integrates the general information that is necessary to create an awareness of the body and its relation to its external environment. The parietal lobe is composed of (1) the *postcentral gyrus* (areas 3, 1, 2), which receives and integrates general sensory input from the ascending tracts, (2) the *superior* and *inferior parietal lobules* (areas 5, 7), and (3) the *supramarginal gyrus* (area 40), which further processes this neural information toward a higher level, leading to the perception of general sensations. The postcentral gyrus is called the ***primary somesthetic association area*** (general sensory area) because it receives information about the general senses from receptors in the skin, joints, muscles, and body organs [FIGURE 13.11].

Temporal lobe The *temporal lobe* is the lobe located closest to the ears. It has critical functional roles in hearing, equilibrium, and, to a certain degree, emotion and memory. The ***auditory cortex*** is located in the *transverse temporal gyrus* (areas 41, 42), just anterior to the vestibular (equilibrium) area [FIGURE 13.11]. Electrical stimulation of the auditory cortex elicits such elementary sounds as clicks and roaring, while the same stimulation of the vestibular area may produce feelings of dizziness and vertigo (a sense of rotation). The understanding of the spoken word involves the participation of the *superior temporal gyrus* (area 22).

Occipital lobe Although the *occipital lobe* is relatively small, it is important because it contains the visual cortex. It is made up of several areas organized for vision and its associated forms of expression.

Visual images from the retina of the eye are transmitted and projected to the ***primary visual cortex*** (area 17) [FIGURE 13.11]. Each specific site in the retina is represented by a specific site in area 17. Visual information from this area is conveyed for further processing and elaboration to the visual association area (areas 18 and 19) and then to other areas such as the *angular gyrus* (area 39). Areas 18 and 19 are involved with assembling the features of a visual image and making it meaningful.

The *occipital eye field* of the occipital lobe is the cortical area involved in *visual* pursuit. For example, after a person has located a bird in flight with the aid of the frontal eye fields, his or her occipital eye fields take over as the eyes lock on the bird in the visual field and track it in flight.

Wernicke's area (*posterior speech area*) (areas 39, 40, 22), located inferior to the primary auditory area in the temporal cortex, is concerned with the understanding of the written word (area 39) and the spoken word (area 22) and with the ability to conceive the symbols of language. Lesions on the dominant side of Wernicke's area 39, 40, and 22 (usually the left side) can produce *alexia*, a failure to recognize written words. (*Dyslexia* is associated with reading and writing disabilities; see page 423.) At the same time, the primary visual perception remains intact, and the patient may accurately describe the shape, color, and size of an object.

Insula Little is known about the role of the ***insula*** (L. island), or ***central lobe***, of the brain, which lies deep in the lateral cerebral fissure under the parietal, frontal, and temporal lobes. It cannot be seen from the exterior surface of the brain, but can be seen in a frontal view [FIGURE 13.10C]. It seems to be associated with gastrointestinal and other visceral activities.

Limbic lobe and limbic system The *limbic lobe* (L. *limbus*, border) is the ring of the cortex, located on the medial surface of each cerebral hemisphere and surrounding the central core of the cerebrum [FIGURE 13.12B]. This lobe is composed of the cingulate gyrus, isthmus, and parahippocampal gyrus.

The ***olfactory cortex*** is involved with the perception of odors, and olfaction is intimately linked to the limbic system. Irritation of this region may produce hallucinations of an odor (usually foul), along with fear and feelings of unreality.

The ***limbic system*** (L. *limbus*, border) is a ring of structures encircling the brainstem. It is defined in functional terms as an assemblage of cerebral, diencephalic, and midbrain structures that are actively involved in memory and emotions and the visceral and behavioral responses associated with them. This physiologically defined system includes such structures as the limbic lobe, amygdala, and hippocampus and parts of other structures such as the thalamus and midbrain. (See "The Anatomical Organization of the Brain" on page 407.) Two of these structures are closely associated with the limbic lobe. One is the ***hippocampus*** (Gr. sea horse; so named because it is S-shaped), which is located medial to the parahippocampal gyrus, in the floor of the lateral ventricle. The other is the ***amygdala*** (uh-MIG-duh-luh; L. almond, so named because of its shape), a complex of nuclei located deep within the uncus [FIGURE 13.12A]. The amygdala and hippocampus are integrated with the functional expressions of the limbic system.

The hippocampus is involved with memory traces for recent events. Patients with Korsakoff's syndrome, in which lesions are present in the hippocampus, have a loss of short-term memory and of a sense of time; they tend to become confused easily. They forget questions that were just asked and may reply to questions with irrelevant answers. All such expressions of the amygdala and hippocampus are associated with an interaction with other centers. For example, the impact of inputs of the sensory systems from the sensory cortical areas upon these structures can trigger a variety of emotional and behavioral responses.

DIENCEPHALON

The *diencephalon* (L. "between brain") is the deep part of the brain, connecting the midbrain with the cerebral hemispheres. It houses the third ventricle and is composed of the thalamus, hypothalamus, epithalamus, and ventral thalamus. The pituitary gland is connected to the hypothalamus.

Flanking the diencephalon laterally is the *internal capsule* [FIGURE 13.9B], a massive structure of white matter made up of ascending fibers that convey information to the cerebral cortex and descending fibers *from* the cerebral cortex. Many fibers of the internal capsule continue caudally into the ventral aspect of the brainstem.

Thalamus

The *thalamus* (Gr. inner chamber, so named because early anatomists thought it was hollow, resembling a room) is composed of two egg-shaped masses of gray matter covered by a thin layer of white matter. It is located in the center of the cranial cavity, directly beneath the cerebrum and above the hypothalamus. It forms the lateral walls of the third ventricle [FIGURE 13.3].

The thalamus is the intermediate relay point and processing center for all sensory impulses (except the sense of smell) ascending to the cerebral cortex from the spinal cord, brainstem, cerebellum, basal ganglia, and other sources. After processing the input, the thalamus relays its output to the cerebral cortex. The thalamus is involved with four major areas of activity:

1 *Sensory systems.* Nerve fibers from the thalamic nuclei project into the sensory areas of the cerebral cortex, where the sensory input is "decoded" and translated into the appropriate sensory reaction. For example, light is "seen" and sound is "heard." Crude sensations and some aspects of the general senses may also be brought to consciousness in the thalamus.

2 *Motor systems.* The thalamus has a critical role in influencing the motor cortex. Some thalamic nuclei receive neural input from the cerebellum and basal nuclei and then project into the motor cortex. The motor pathways that regulate the skeletal muscles innervated by the cranial and spinal nerves originate in the motor cortex.

3 *General neural background activity.* Background neurophysiological activities of the brain, such as the sleep-wake cycles and the electrical brain waves, are expressed in the cerebral cortex. These cortical rhythms are generated and monitored by thalamic nuclei, which receive much of their input from the ascending reticular systems in the brainstem.

4 *Expression of the cerebral cortex.* The thalamus, through its connections with the limbic system, helps regulate many expressions of emotion and uniquely human behaviors. In fact, it is linked with the highest expressions of the nervous system, such as thought, creativity, interpretation and understanding of the written and spoken word, and the identification of objects sensed by touch. Such accomplishments are possible because of the two-way communication between the thalamus and the association areas of the cortex.

Hypothalamus

The *hypothalamus* ("under the thalamus") lies directly under the thalamus [FIGURE 13.14; see also FIGURE 13.1]. It is a small region (only 1/300 of the brain's total volume) located in the floor of the diencephalon, forming the floor and part of the lateral walls of the third ventricle. Extending from the hypothalamus is the *hypophysis* (*pituitary gland*), which is neatly housed within the sella turcica in the body of the sphenoid bone.

The most important functions of the hypothalamus are the following:

1 *Integration within the autonomic nervous system.* The hypothalamus is considered to be the highest integrative center associated with the autonomic nervous system. Through the autonomic nervous system, it adjusts the activities of other regulatory centers, such as the cardiovascular centers in the brainstem. The hypothalamus modifies blood pressure, peristalsis (muscular movements that push food along the digestive tract) and glandular secretions in the digestive system, secretion of sweat glands and salivary glands, control of the urinary bladder, and rate and force of the heartbeat.

2 *Temperature regulation.* The hypothalamus plays a vital role in the regulation and maintenance of body temperature. Specialized nuclei within the hypothalamus monitor the temperature of the body. A decrease in temperature activates a neuronal center in the posterior hypothalamus. This triggers mechanisms for heat production and heat conservation, such as increased metabolic

FIGURE 13.14 HYPOTHALAMUS

Hypothalamic nuclei are shown, as are the relationships between the hypothalamus and hypophysis, and between the hypothalamus, and brainstem.

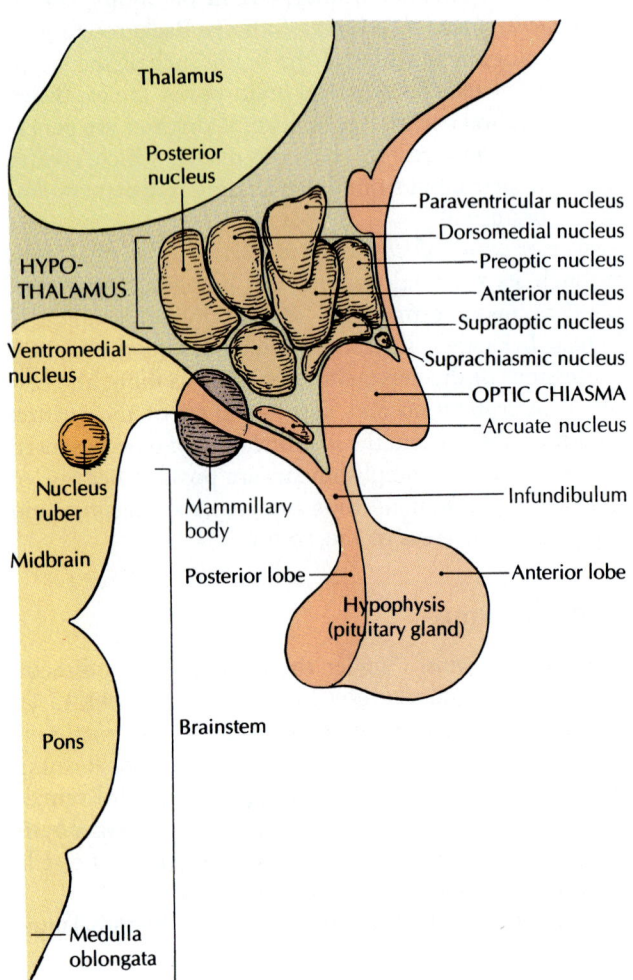

centration of water in the blood is reduced, causing the electrolyte concentration to rise, a hormone called *antidiuretic hormone* (ADH) is produced by the hypothalamus and released into the bloodstream by the posterior lobe of the pituitary gland. ADH stimulates the kidneys to absorb water from newly formed urine and return the water to the blood.

4 *Sleep-wake patterns.* The hypothalamus is integrated with the neural circuitry that regulates sleep-wake patterns and the state of awareness.

5 *Food intake.* The hypothalamus has a crucial role in food consumption. When neural centers collectively called the "appestat" or "hunger or feeding centers" are stimulated by low levels of glucose, fatty acids, amino acids in the blood, and possibly other factors, they activate appropriate body responses to satisfy the deficiency: you experience hunger. After food has been eaten, the "satiety (satisfaction) center" is stimulated to inhibit the feeding center.

6 *Behavioral responses associated with emotion.* The subjective feelings of emotion (pleasure, pain, anger, fear, love) are expressed as visible physiological and physical changes when the cerebral cortex activates the autonomic nervous system by way of the hypothalamus. In turn, the autonomic nervous system is responsible for changes in the heart rate and blood pressure, blushing, dryness of the mouth, clammy hands, crying, gastrointestinal activity, and many other emotional expressions.

7 *Endocrine control.* The hypothalamus, the central control unit of the endocrine system, produces the hormones oxytocin and antidiuretic hormone, which are stored in and released by the posterior lobe of the pituitary gland, as well as certain other hormones that control the release of hormones by the anterior lobe of the pituitary gland.

8 *Sexual responses.* The dorsal region of the hypothalamus contains specialized nuclei that respond to sexual stimulation of the genital organs. The sensation of an orgasm involves neural processing within this center.

activity, shivering, constriction of blood vessels in the skin, and decreased sweating. A rise in blood temperature activates a neuronal center in the anterior hypothalamus that triggers activities that prevent a rise in the body temperature, such as the dilation of blood vessels of the skin and increased sweating.

3 *Water and electrolyte balance.* The hypothalamus has a "thirst center" and a "thirst-satiety (satisfaction) center" that help produce a balance of fluids and electrolytes in the body. These centers regulate the intake of water (through drinking) and its output (through the kidneys and sweat glands). Certain neurons in the hypothalamus, known as *osmoreceptors*, maintain homeostasis by sensing and monitoring the osmotic concentration of the blood. As the blood passes through the nuclei, the osmoreceptor neurons trigger activities that help maintain a normal extracellular fluid volume. For example, when the con-

Epithalamus

The *epithalamus* is the dorsal portion of the diencephalon, located near the third ventricle. It contains the pineal body, a glandlike structure that extends outward from the posterior end of the epithalamus by way of a stalk.

Ventral Thalamus

The *ventral thalamus,* or *subthalamus,* located ventrally in the diencephalon, contains the subthalamic nucleus, which is one of the basal ganglia. According to recent evidence, the subthalamic nucleus is the driving force that regulates and modulates the output of the basal ganglia and their influences on motor activity.

ASK YOURSELF

1 What are the basal ganglia?

2 What is the cerebral cortex?

3 What is the benefit of cerebral gyri and sulci?

4 What is the function of the corpus callosum?

5 What are the six lobes of the cerebrum? Where is each located?

6 What are the roles of the limbic system and the limbic lobe?

7 What are the main parts of the diencephalon?

8 What are the chief functions of the thalamus? The hypothalamus? The epithalamus?

REGENERATION IN THE CENTRAL NERVOUS SYSTEM

When axons of the mature central nervous system are severed, they do not regenerate to a functional state, except in a few exceptional cases. Damaged neurons make an attempt to regenerate axonal sprouts at the severed end, but they are unsuccessful in reaching their original terminals, probably for four reasons: (1) The sprouts cannot penetrate the glial scar tissue formed at the site of injury; (2) there is no basal lamina associated with the oligodendroglia in the central nervous system to guide the regeneration of fibers; (3) oligodendrocytes do not form continuous cords (as do Schwann cells), and so cannot guide the fibers in an organized regeneration; and (4) the oligodendrocytes, through chemical substances in the plasma membrane, inhibit axonal growth and regeneration in the central nervous system. This lack of functional regeneration can have serious effects, such as those that follow some strokes or spinal cord injuries.

LEARNING AND MEMORY

It now seems certain that *no single area or structure of the brain controls learning*. In fact, long-term studies of children with learning disabilities due to brain damage have shown that alternative areas of the brain can be trained to "relearn" what had once been learned by the damaged portion. Apparently, the brain can make such learning adjustments as long as there is a memory trace of what had been learned in the first place. More and more scientists believe that learning involves a variety of mechanisms, not just one.

Young children recover from an injury to the cerebral cortex more completely than teenagers or adults do be-cause the cerebral hemispheres of a child have not yet become specialized. As a result, an undamaged area of the cortex is able to learn to some degree what had been lost in the damaged portion. Apparently, the dominance of one cerebral hemisphere over the other develops gradually during childhood and does not become relatively well fixed until about age 10. Before that age, it appears that language and speech capabilities are found relatively equally in both hemispheres and that either hemisphere can assume the functional role in language or speech. The natural "twin-brain" of a child explains why a right-handed child can be taught to write with the left hand and why some athletes who have been trained since childhood to use both hands remain ambidextrous as adults. In contrast, adult patients with cerebral damage usually find it difficult or impossible to switch from using one hand to another or to relearn lost language and speech abilities.

The ability of the brain to "relearn" involves the concept of "redundant neurons and circuitry." Many studies have confirmed that extra (redundant) neurons are produced during embryonic development and that some neurons eventually degenerate and die through the years. Some redundant neurons and circuits may be "trained" to take over certain functions of neurons that are injured or die. This system is analogous to the "extra circuitry" used in many computers.

Memory is related to learning because in order to remember something, we have to "learn" it first. Apparently, *memory traces* are formed during learning and are imprinted on the brain when neurons record and store information. *Short-term memory* is the process by which we remember recent events that have no permanent importance (the score of yesterday's football game or a telephone number that can be forgotten as soon as the call is made). *Long-term memory* is the process by which we remember information that for some reason is interpreted as being important enough to store for a lifetime.

Current evidence suggests that the gases nitrous oxide (NO) and carbon monoxide (CO) have a major physiological role, acting like neurotransmitters, in the functioning of the nervous system. These gases are unlike the known neurotransmitters in that they are *not* stored in synaptic vesicles, are *not* released by exocytosis, and do *not* act on conventional plasma-membrane-associated receptors. Apparently, NO and CO are formed when needed and diffuse to the neurons they act on. Both gases are thought to be involved in the complex circuitry of the hippocampus by which short-term memory is converted into long-term memory.

The ability to convert short-term memory into long-term memory is lost with the removal of the medial portions of both temporal lobes (which includes the hippocampus and amygdala) and/or the nearby diencephalon [FIGURE 13.15]. Patients who lose these portions of the

FIGURE 13.15 MEMORY CENTERS

Shown are sites of brain damage that can lead to memory loss, indicating that these sites may be important for memory formation.

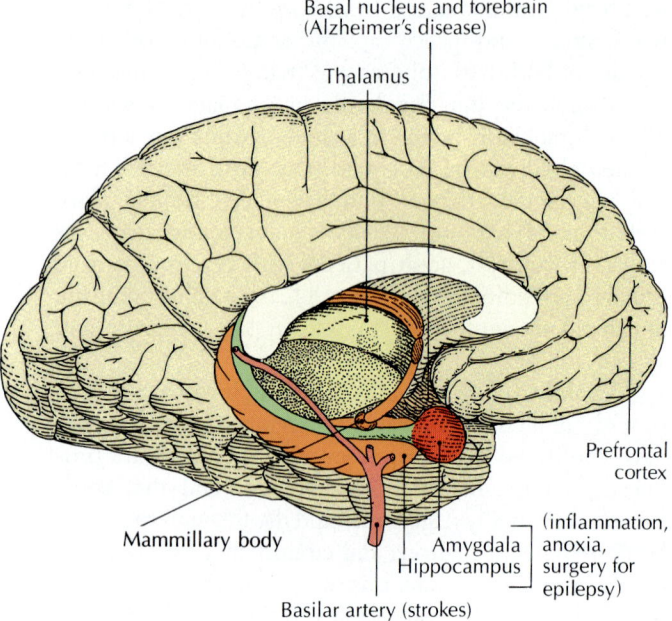

Basal nucleus and forebrain (Alzheimer's disease)

Thalamus

Prefrontal cortex

Mammillary body

Amygdala
Hippocampus

(inflammation, anoxia, surgery for epilepsy)

Basilar artery (strokes)

brain retain their long-term memory but are unable to form and store *new* long-term memories. Neither short-term nor long-term memory is impaired when only one temporal lobe is removed.

Considerable experimental evidence supports the theory that the neural connections of the hippocampus and amygdala with the diencephalon, medial temporal lobe, and prefrontal cortex have a crucial role in the storage of recently acquired memory. After the perception and awareness of short-term memory sensations are formed in the highest levels of the neocortex, including the temporal lobe, they are conveyed from the temporal lobe neocortex via neural circuits to the hippocampus and amygdala. These two structures are considered to be the neural processing centers through which the sensations of short-term memories are processed and conveyed into the long-term memory bank.

ASK YOURSELF

1 Does one area of the brain control learning?

2 What are redundant neurons?

3 What is the difference between short-term and long-term memory?

SLEEP AND DREAMS

The brain is always active; in fact, some cortical areas of the human brain are more active and require a greater blood supply when the body is asleep than when it is awake. Brain activity in the form of waves may be recorded with an instrument called an *electroencephalograph*. The tracing this instrument produces, the *electroencephalogram* ("electric writing in the head"), is referred to by the initials *EEG*. The EEG shows the changing levels of electrical activity in the brain during various levels of consciousness, and it is frequently used to detect brain damage by locating areas of altered wave patterns. The four distinct wave patterns in the brain are known as alpha, beta, delta, and theta waves [FIGURE 13.16]. *Alpha waves* (8 to 12 cycles per second, or cps), which are evident in relaxed adults whose eyes are closed, usually indicate a state of well-being. *Beta waves* (5 to 10 cps) are typical of an alert, stimulated brain. *Delta*

FIGURE 13.16 BRAIN WAVES

Alpha, beta, delta, and theta brain waves recorded by an electroencephalograph (EEG).

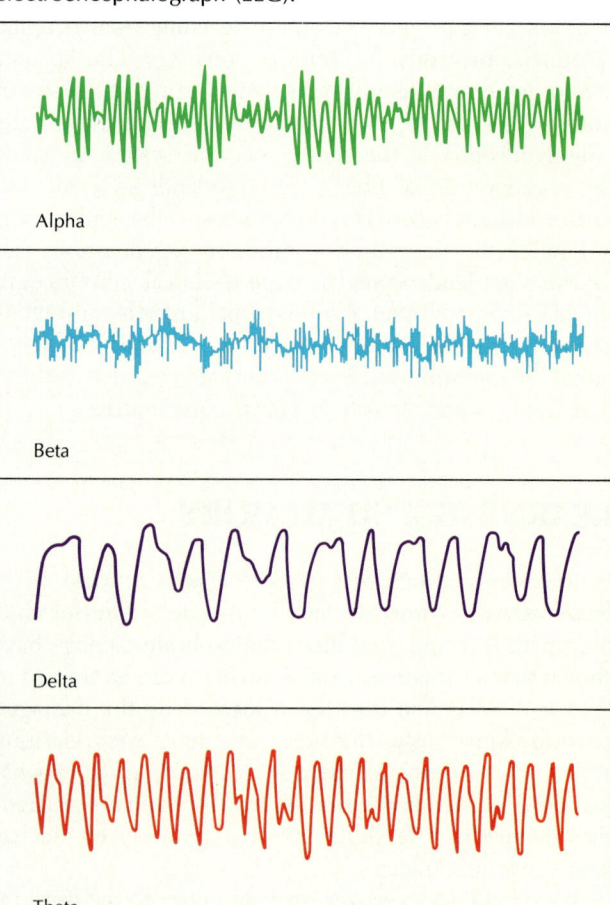

Alpha

Beta

Delta

Theta

FIGURE 13.17 SLEEP PATTERNS

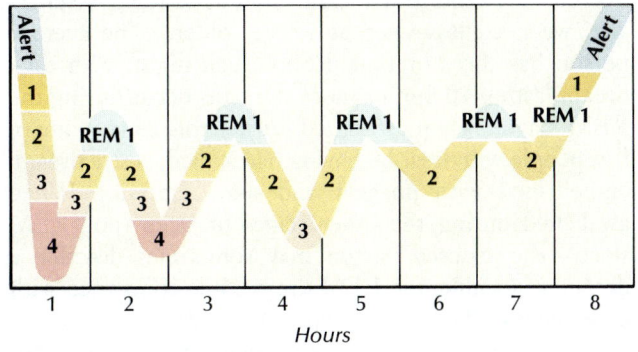

[A]

[A] Sleep cycles vary during the night, ending with lengthening REM periods and less deep sleep than in the first hours of sleep. [B] Brain-wave patterns during the various stages of sleep, as shown in EEGs.

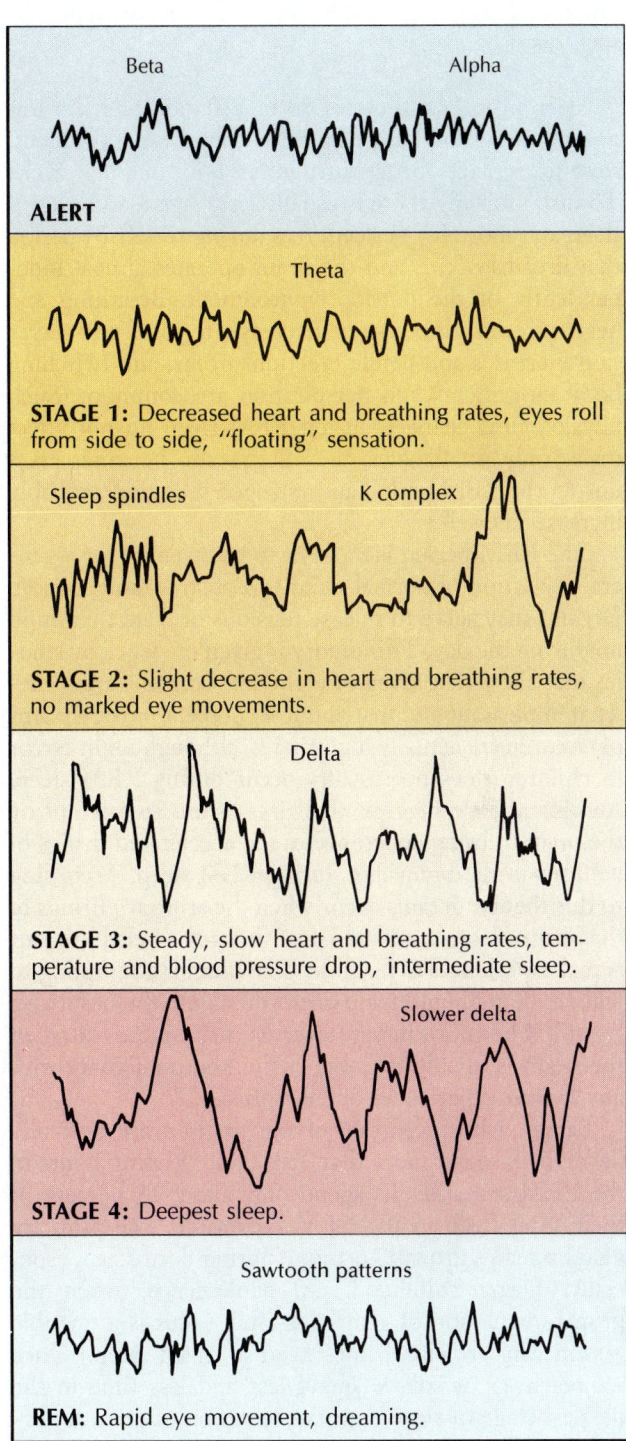

[B]

waves (0.5 to 2 cps) are seen during deep sleep, in damaged brains, and in infants. *Theta waves* (3 to 7 cps) usually occur during sleep.

It is generally agreed that sleep is one of the more important activities of human beings (lack of sleep will cause death faster than lack of food will), but the exact benefits of sleep are not known, and the causes of sleep are equally unknown. A group of scientists at Harvard University have isolated an extremely potent sleep-inducing hormone, appropriately named *factor S*, for sleep. Factor S is a small peptide that appears to build up gradually while we are awake, finally reaching the point where the amount is great enough to cause sleep. This hypothesis has yet to be confirmed.

It is particularly ironic that so little is known about sleep, because if you are 20 years old, you have already spent about 8 years of your life asleep. By the time you are 60, you will have slept about 20 years.

The EEG has made possible the detection of at least four separate stages of sleep:

1 During *stage 1 sleep* the rhythm of alpha waves slows down, and slower theta waves appear [FIGURE 13.17]. Heart and breathing rates decrease slightly, the eyes roll slowly from side to side, and the individual experiences a floating sensation. This stage is usually not classified as true sleep, and individuals awakened from stage 1 are quick to agree, insisting that they were merely "resting their eyes." Stage 1 sleep usually lasts less than 5 min.

2 *Stage 2 sleep* is indicated by the appearance on the EEG of short bursts of waves known as "sleep spindles" (12 to 14 cps), along with "K complexes," which are high-voltage bursts that occur before and after a sleep spindle. Eyes are generally still, and heart and breathing rates decrease only slightly. Sleep is not deep.

3 Slow delta waves appear during *stage 3 sleep.* This stage of intermediate sleep is characterized by steady, slow breathing, slow pulse rate (about 60), and a decrease in temperature and blood pressure.

4 *Stage 4 sleep,* also known as oblivious sleep, is the deepest stage. It usually begins about an hour after a person has fallen asleep. Delta waves become even slower, and heart and breathing rates drop to 20 or 30 percent below those in the waking state. Although the EEG indicates that the brain acknowledges outside noises and other external stimuli, the sleeper is not awakened by such disturbances.

Sleep proceeds in cycles 80 to 120 min long. An important condition known as ***REM*** (rapid eye movement) ***sleep*** takes place during entry into stage 1 of a new cycle. Toward morning there is usually less stage 3 and stage 4 sleep and more REM sleep. It is during the REM period that dreams occur, and the brain operates almost independently of the outside environment. Breathing and heart rate may increase or decrease, testosterone secretion increases and penile erections occur, and twitching body movements and perspiration are common. (Such physiological changes during REM sleep do not affect males only. In females, the clitoris also becomes erect during the REM period and estrogen secretion probably increases as well.)

The REM period is thought to be essential for a general relaxation of normal build-ups of stress and tension. Dreams may serve to release tensions or fears that build up during the day. This theory is given credence by studies that show that seven of eight adult dreams are somewhat unpleasant and that about 40 percent of the dreams of children are actually nightmares, although night terror in children does not usually occur during REM sleep. Another theory suggests that dreams are an attempt by the brain to bring coherence to the accelerated firings of neurons in the brainstem during REM sleep. According to this theory, dreams occur when the arbitrary firings of the neurons are translated into images as close to our actual experiences as possible. The brain is attempting to take random impulses and direct them into themes with a "plot." The more bizarre dreams may be caused when the brain is unable to connect the neuron activity with any logical experiences or memories.

Babies, whose central nervous systems are not fully developed, spend more than half their sleeping hours in the REM state. Adults spend only about 15 percent of their total sleep in the REM condition. The pituitary gland releases growth hormone during deep sleep, especially during childhood and adolescence, when the proper regulation of growth is vital. (This is a possible reason why growing babies need so much sleep.) Once we reach 45 or so, we spend less and less time in the deepest stage of sleep, and the secretion of growth hormone slows down until it practically stops.

Sleep is a highly individual process. No two people have precisely the same pattern of the sleep-wake cycle, nor do they require the same amount of sleep. (Generally, we need less sleep as we get older.) The average person has three or four dreams each night, with each dream lasting 10 min or more. Dreams occur during the REM stage. When awakened during this stage, four of five people will describe having had a vivid, active dream, embellished with imagery and some fantasies. When awakened during the other stages of sleep (non-REM sleep), the aroused person may sometimes describe a dream, but unlike the REM dream, it is concerned with experiences related to thought processes.

It appears that learning and memory are reinforced during periods of REM sleep. Apparently, new stored information is consolidated and differentiated more during REM sleep than during non-REM sleep.

Most people will experience extreme psychological discomfort after a few days of sleep (and dream) deprivation, although occasional rare individuals can live for years with only 2 or 3 hours of sleep per day. Without sufficient sleep, the ATP reserve in the brain declines precariously, adrenal stress hormones are secreted into the blood steadily, and mental and physical levels of performance falter markedly. Large amounts of a chemical similar to LSD appear in the blood and may cause hallucinations (such as those that happen to sleepy drivers) and psychotic behavior. Sleep is the only relief.

Sleep Patterns and Age

A normal 20-year-old usually falls asleep in 8 to 10 min, spends about 95 percent of the night asleep, is in deep REM sleep for 30 min or more, dreams about 2 hours a night, and sleeps for 7.5 to 8 hours. In contrast, a typical 80-year-old takes 18 to 20 min to fall asleep (unless he or she is watching television, when sleep seems to come instantaneously), spends about 80 percent of the night asleep, is in deep REM sleep for a few minutes, dreams for about an hour, and sleeps for about 6 hours. Lack of short-term memory retention in the elderly may be partially due to the reduced amount of REM sleep, when new memories and learned tasks are usually reinforced.

A S K   Y O U R S E L F

1 What is an EEG, and what is it used for?

2 What are the four distinct brain-wave patterns?

3 What is REM sleep, and why is it important?

4 Do we dream every night?

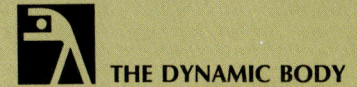
The Body's Own "Tranquilizers"

Scientists have long searched for alternatives to such pain-killing drugs as morphine and opium, which are highly addictive. The answer may come from the body itself.

In the 1970s, American researchers discovered how opiates work. Opiate molecules lock onto special receptor sites of certain neurons in the central nervous system, slowing down the firing rate of those neurons. Apparently, the decreased firing rate decreases the sensation of pain. Many opiate receptors are located in the spinal cord, where the pain impulse is first introduced into the central nervous system. It was also found that opiates have an especially strong effect on the thalamus, where pain is eventually processed into an actual sensation.

Within a short time after the discovery of the mechanism of opiates, teams of scientists around the world isolated natural short-chain neuropeptides in the brain and pituitary gland, which they called the **endorphins** (endogenous + morphine). Also discovered were the breakdown products of endorphins, smaller peptides that were named the **enkephalins** (en-KEFF-uh-lihnz; "in the head").

Endorphins and enkephalins are morphinelike substances that occur naturally in the nervous system. They are the brain's own opiates, having the pain-killing effects of opiates such as morphine. It is likely that endorphins and enkephalins work by binding to the same neuronal receptors that bind opiate drugs. These receptor sites in the central nervous system are associated with the intrinsic pain pathways. In addition, the endorphins and enkephalins apparently act as neurotransmitters in the pain-inhibiting pathways. In addition to moderating pain, enkephalins in the limbic system seem to counteract psychological depression by producing a state of euphoria similar to the feelings produced by opiate drugs.

Some researchers believe that the "natural high" experienced by serious joggers is caused by the release of endorphins. Most scientists, while not disagreeing with the existence and function of endorphins as pain-moderating substances, feel that only a few athletes are consistently vigorous enough to secrete sufficient endorphins to produce a natural euphoria.

It is thought that other neurotransmitters are associated with the pain pathways. Some other neurotransmitters that play important roles in pain are substance P, serotonin, dopamine, norepinephrine, and dynorphin.

CHEMICALS AND THE NERVOUS SYSTEM

Certain chemicals are essential for the proper functioning of the nervous system. Among these are the **neurotransmitters** and **neuromodulators,** which were discussed in Chapter 12. More than 50 neurotransmitters or transmitterlike substances have been identified in the brain, and many other transmitter substances are under investigation.

Chemicals that are not natural constituents of the nervous system may be used to alter the normal functions of the nervous system or to correct abnormal functions. These chemicals or drugs fall into several major categories [TABLE 13.2]:

1 **Stimulants** increase the activity of the central nervous system, especially the brain. Among these drugs are amphetamines, caffeine, cocaine, and nicotine. Depending on the type and strength of the particular drug taken, their effects may range from mild (for caffeine) to powerful (for amphetamine, or "speed").

2 **Depressants** are drugs that inhibit the activity of the central nervous system, especially the brain. Among the depressants are anesthetics, barbiturates, alcohol, opiates, and tranquilizers. In small amounts, barbiturates induce sleep, alcohol reduces tension, opiates relieve pain, and tranquilizers reduce anxiety. But in larger doses, all depressants may cause damage and abnormal reactions. Certain stimulants and depressants may be addictive.

3 **Antidepressants** are drugs that relieve the symptoms of psychological depression. The tricyclic antidepressants, such as Tofranil and Elavil, block the uptake of norepinephrine and serotonin at nerve endings. Chemically unrelated to tricyclic, tetracyclic, or other available antidepressants is fluoxetine hydrochloride (Prozac), which inhibits the uptake of serotonin in the central nervous system.

4 **Psychedelic** and **hallucinogenic drugs** alter perception and mood. Extreme effects include *hallucinations,* or distortions of sense perceptions, and even *psychoses,* or severe mental illnesses.

5 **Analgesics** are pain-relieving drugs, which may be classified as *nonopioid* analgesics, or *opiates* and *opioid* analgesics. Opiates are derived from natural substances, while opioids are synthetic drugs. They both have similar pharmacological properties, including the ability to relieve pain and the likelihood of dependence (addiction) with prolonged use. The **nonopioids** include *aspirin,* which is used to relieve mild pain, reduce fever and inflammation, and prevent blood clots; *acetaminophen* (Tylenol), which is used to relieve mild pain and reduce fever; and *ibuprofen* (Nuprin and Advil), which is used to relieve pain. **Opiates** and **opioids** include *codeine, meperidine hydrochlo-*

TABLE 13.2 EFFECTS OF DRUGS ON THE NERVOUS SYSTEM

Type of drug	Major effects	Mechanism of action
STIMULANTS		
Amphetamines ("uppers"): Benzedrine, Dexedrine, amphetamine ("speed"), phenmetrazine (Preludin)	Elevate mood, produce sense of increased energy, decrease appetite, increase anxiety and irritability. *Powerful reaction:* accelerate heart rate, dilate pupils, increase blood glucose.	Stimulate sympathetic nervous system.
Caffeine: coffee, tea, cocoa, cola drinks	Relieves drowsiness, muscle fatigue; prolongs physical and intellectual activity. *Mild reaction:* temporarily accelerates heart rate, dilates pupils, increases blood glucose.	Stimulates sympathetic nervous system. Facilitates synaptic transmission.
Cocaine	More powerful than amphetamines, but similar reactions. Temporary sense of well-being and alertness.	Stimulates central nervous system. Inhibits uptake of norepinephrine.
Nicotine: tobacco	Similar to caffeine. *Medium reaction:* accelerates heart rate, dilates pupils, increases blood glucose.	Stimulates sympathetic nervous system. Facilitates synaptic transmission.
DEPRESSANTS		
Anesthetics: ether, chloroform, benzene, toluene, carbon tetrachloride	Induce unconsciousness. Similar to alcohol intoxication.	Depress central nervous system. Block transmission of electrical impulse that triggers contraction of ventricles.
Barbiturates: Nembutal, Seconal, Amytal, Tuinal, phenobarbital	Sedative action, induce sleep. Addictive. High doses may induce respiratory failure. Combination of barbiturates and alcohol may produce extreme depression of central nervous system and may cause death.	Depress reticular formation and, in large doses, medulla oblongata. Interfere with synthesis and secretion of norepinephrine and serotonin.
Ethyl alcohol: whiskey, gin, beer, wine	Small amounts reduce tension, produce feeling of well-being, stimulation. Additional amounts depress the lower brain centers, impair motor coordination, produce insensitivity to touch, distorted vision, difficulty with speech. Prolonged use may lead to cirrhosis of liver, unconsciousness, coma, brain damage.	Exact mechanism unknown. May inhibit activity of thalamus and interfere with action of acetylcholine, norepinephrine, serotonin. Reduces neuron function in brain. Inhibits motor and sensory regions of cortex. Depresses visual, auditory, speech centers of cortex. Inhibits cerebellum. Depresses reticular formation.
Opiates and opioids: heroin, morphine, codeine, methadone, opium	Reduce pain, tension, anxiety; induce muscle relaxation, lethargy. May produce loss of appetite, constipation, fatal depression of respiratory center. Highly addictive.	Depress thalamus.
Tranquilizers: meprobamate (Miltown, Equanil), chlorpromazine (Thorazine), chlordiazepoxide (Librium)	Reduce anxiety, tensions. Antipsychotic.	Block receptors of epinephrine and acetylcholine. Depress activity of reticular formation.
ANTIDEPRESSANTS		
Dibenzapines: Tofranil, Elavil	Reverse effects of psychological depression.	Increase levels of norepinephrine in brain.
MAOIs (monoamine oxidase inhibitors): Nardil, Parnate	Reverse effects of psychological depression.	Increase levels of norepinephrine in brain.
Fluoxetine hydrochloride (Prozac)	Reverses effects of psychological depression.	Increases levels of serotonin in brain.

Type of drug	Major effects	Mechanism of action
PSYCHEDELICS, HALLUCINOGENS		
Cannabis (marijuana), hashish	Alter perception of sensory phenomena. Usually produce a sense of well-being. May produce anxiety.	Unknown.
LSD (lysergic acid diethylamide)	Distorts visual and auditory imagery. Produces sense of increased sensory awareness.	Inhibits brain serotonin, perhaps by decreasing activity of serotonin-producing cells in reticular system.
Mescaline, psilocybin, DMT (dimethyltryptamine), DET (diethyltryptamine), DOM or STP (dimethoxymethylamphetamine), Ditran, phencyclidine (Sernyl)	Produce hallucinations, increased sensory awareness, psychoses.	Mimic molecular structure of serotonin. Duplicate effects of nervous system activity. May enhance effects of biogenic amines (naturally occurring amines), which they resemble.

ride (Demerol), and *methadone hydrochloride* (Dolophine). *Morphine* is an opiate used to relieve severe pain and also to allay the fear and anxiety of pain. Opiates and opioids tend to produce *tolerance*, the necessity of increasing the dosage in order to maintain the effect of the drug.

6 *Antianxiety drugs* are used to suppress anxiety (tranquilizers) or to induce sleep (sleeping pills). They are among the most commonly used drugs. If antianxiety drugs are taken alone, they are relatively safe, although they have the potential to be addictive when abused. When taken in combination with central nervous system depressants or alcohol, however, they can be dangerous. Antianxiety drugs include the groups called benzodiazepines, three of which are *diazepam* (Valium), *alprazolam* (Xanax), and *chlordiazepoxide hydrochloride* (Librium).

Drugs and the Elderly

The fastest-growing segment of our population is the group over age 75. By the turn of the century, 13 percent of the U.S. population will be over 65, and this group will probably live about 20 years longer than their grandparents did. Elderly Americans are already taking about twice as much medication as other age groups; many of them take 10 or more different medications every day. Interactions among these drugs may cause additional problems that are difficult to treat, and to make it worse, some elderly people see so poorly that they frequently take the wrong pills.

Elderly people often take longer than usual to absorb drugs, and a drug that is expected to be completely cleared from the body in 24 hours may remain in the bloodstream for several days. This problem may be aggravated by decreased blood flow through the intestines, liver, and kidneys.

DEVELOPMENTAL ANATOMY OF THE BRAIN

FIGURE 13.18 shows the gross development anatomy of the brain, and the following paragraphs also describe the cellular development.

The brain develops from undifferentiated ectodermal cells that receive a chemical message from the mesoderm of the embryo about 18 days after fertilization to form a flat *neural plate* along the dorsal midline of the future body. At about day 22, the neural plate differentiates into the *neural tube* [FIGURE 13.18B]. The cephalic (head) portion of the tube differentiates into the brain, and the caudal (tail) portion develops into the spinal cord. The hollow part of the tube will become the ventricular system of the brain and the central canal of the spinal cord.

The neuroepithelial cells that form the neural tube begin to divide, differentiate, and develop into immature neurons and glial cells (potential astrocytes and oligodendrocytes). The immature neurons and glial cells migrate from the matrix layer of the neural tube to their final prescribed location in the brain. Each neuron forms dendrites and an axon. The axon elongates and forms branches that terminate in the vicinity of their target areas, and the axon terminals then form synaptic connections with a select subset of target neurons, muscle fibers, or gland cells.

The events of this early period include (1) an orderly migration of hundreds of billions of neurons; (2) the growth of neuronal axons, many of which extend widely throughout the brain and spinal cord; and (3) the formation of thousands of synapses between axons and their target neurons.

By about the fourth fetal week, the neural tube enlarges to form three cavities, or vesicles, called the

FIGURE 13.18 DEVELOPMENTAL ANATOMY OF THE BRAIN

[A, B] Dorsal views. [C–I] Right lateral views.

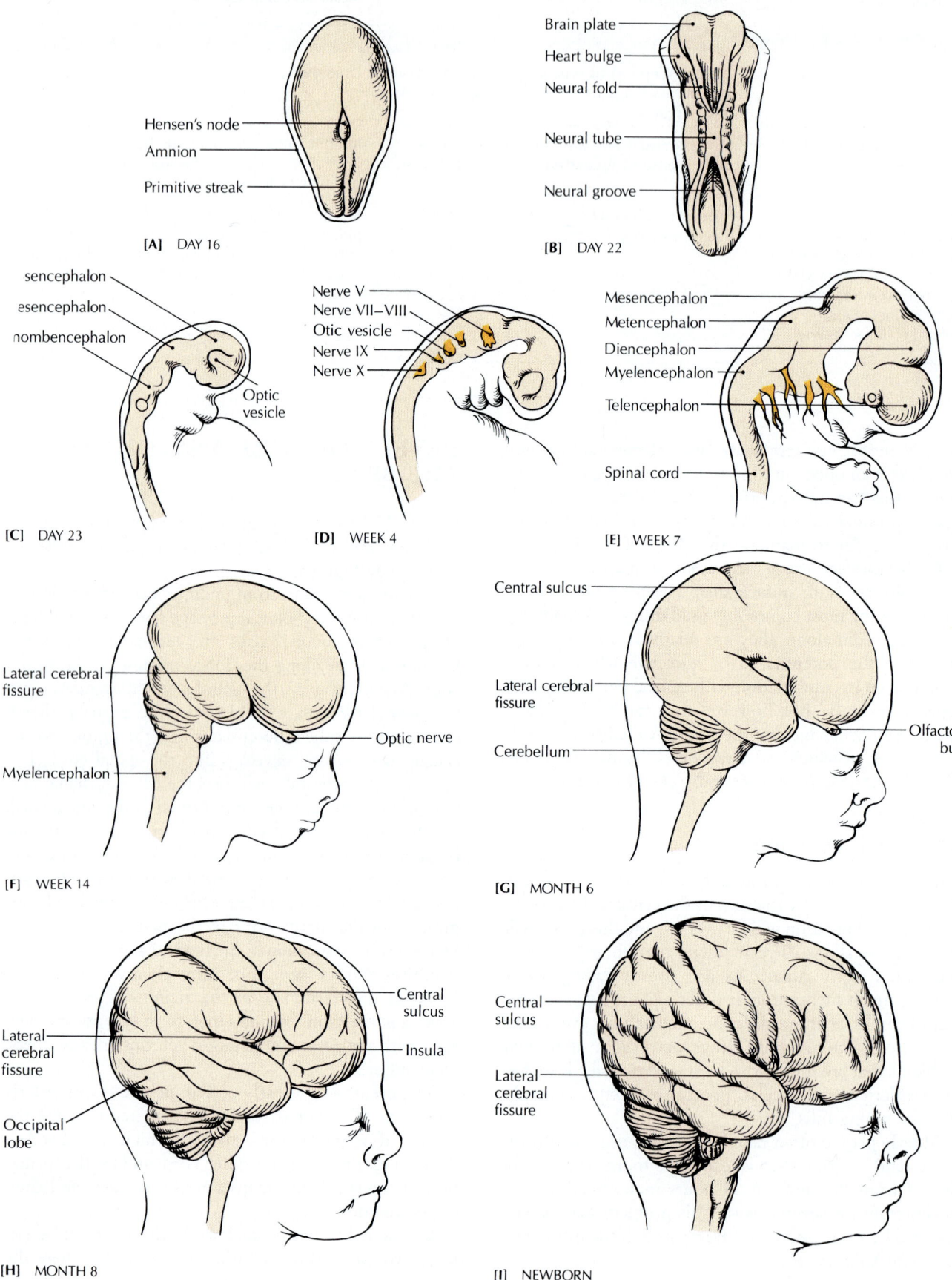

Hensen's node
Amnion
Primitive streak

[A] DAY 16

Brain plate
Heart bulge
Neural fold
Neural tube
Neural groove

[B] DAY 22

...sencephalon
...esencephalon
...hombencephalon
Optic vesicle

[C] DAY 23

Nerve V
Nerve VII–VIII
Otic vesicle
Nerve IX
Nerve X

[D] WEEK 4

Mesencephalon
Metencephalon
Diencephalon
Myelencephalon
Telencephalon
Spinal cord

[E] WEEK 7

Lateral cerebral fissure
Myelencephalon
Optic nerve

[F] WEEK 14

Central sulcus
Lateral cerebral fissure
Cerebellum
Olfactory bulb

[G] MONTH 6

Central sulcus
Insula
Lateral cerebral fissure
Occipital lobe

[H] MONTH 8

Central sulcus
Lateral cerebral fissure

[I] NEWBORN

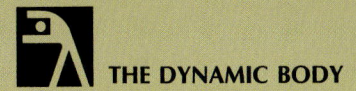

Specialization of Brain Neurons: Nature and Nurture

Not enough genes are present in the human genome to specify the total number of neurons and all their synaptic connections. Other influences have a role in determining the final structural and functional features of each neuron. This is epitomized in the central thesis of development—the roles of "nature" and "nurture." *Nature* refers to the intrinsic genetic influences within the neural cell that are involved in each neuron's development. For example, it is clear that timing of neuronal development is controlled by innate genetic instructions. *Nurture* refers to the extrinsic forces that shape each neuron's mature state. Recent evidence indicates that developing neurons are more flexible than they would be if they were locked into a firm genetic message.

It can be seen that development follows *nature* in that once the immature neuron leaves the marginal layer of the neural tube it is fated, and marked, to become a neuron. This innate capacity finds expression as the immature neuron migrates in an ordered and precisely timed sequence to its proper place. Following the neuron's arrival at its destination, its axon begins to grow, and guided by cues, extends itself to its destination. These organized growth patterns are critical to the establishment of the basic structural and functional organization of the nervous system. In their migration, the immature neurons and, subsequently, their axons are guided along their routes by (1) structural features such as the fibers of glial cells that act as scaffolds, and (2) environmental signals from other cells—such as the extracellular, diffusible molecules that act as guideposts, even at times preventing growth along inappropriate routes.

Development also follows *nurture* in that local conditions—such as nutrition, sensory experiences, social interaction, and learning—can determine how an immature neuron eventually differentiates. For example, the development of a progenitor cell into any one of several types of retinal cells is determined by the influence of signals received by neighboring cells.

prosencephalon (forebrain), *mesencephalon* (midbrain), and *rhombencephalon* (hindbrain), which grow and differentiate according to their own patterns of development [FIGURE 13.18C]. By the fifth week, cranial nerves III through XII of the embryo have already begun to develop as offshoots from the brainstem. In the 7-week embryo, the prosencephalon has divided into two secondary vesicles known as the *telencephalon* and the *diencephalon;* the rhombencephalon has also divided into two vesicles, called the *metencephalon* (cerebellum and pons) and the *myelencephalon* (medulla oblongata) [FIGURE 13.18E].

By the end of the third fetal month, the overall form of the brain is recognizable, but the surface is still smooth [FIGURE 13.18F]. *Sulci* begin to develop in the cerebrum during the fourth month, and the folds of the cerebellum are clearly formed in a 6-month fetus [FIGURE 13.18G]. By the eighth month, the major sulci are present, and the occipital lobe dominates the underlying cerebellum [FIGURE 13.18H].

The brain weighs about 350 g at birth, which occurs at a developmental stage when the infant is not so helpless that it would be unable to survive, and yet is small enough to be delivered through the maternal pelvis and birth canal. If the brain at birth were large enough to support intelligent behavior, a normal delivery through the female bony pelvis would be impossible.

After a year the brain weighs about 1000 g. Brain growth slows down thereafter, until full growth is attained during puberty, when the brain weighs about 1300 g. The adult brain weighs about 1400 to 1500 g, with the weight increase due mostly to the growth of preexisting neurons, new glial cells, and the myelination of axons.

Q. *How fast are neurons produced during prenatal life?*

A. An average of 1.25 million neurons are produced each minute during prenatal life, providing the mature brain with its estimated 500 billion neurons.

WHEN THINGS GO WRONG

The nervous system receives information about the environment from outside the body and from within the body through the sensory system, and it expresses itself through the motor system. Malfunctioning of any of these systems as a result of injury or disease can be noted through clinical signs and symptoms. The signs may be an increase, decrease, or distortion of a sensation, as, for example, greater pain, less or absent pain, or a distortion of pain, as in a phantom limb. (Following the amputation of a foot, for example, the patient can often feel excruciating pain where the foot used to be; this is the so-called phantom pain.) A malfunction in the motor system may be indicated by greater, less, or distorted activity by the muscular and glandular systems. The enhancement of sensations and motor activities after a lesion can be explained in the following way: the injured structure may contain centers that regulate other centers by inhibiting them. Following the injury, the inhibition is decreased, and the ordinarily inhibited centers express themselves without the usual controls. The result may be a more intense feeling of sensations and greater or unwanted motor activity.

Because disturbances of the brain are so numerous, we will give only a brief description of some of the major ones here.

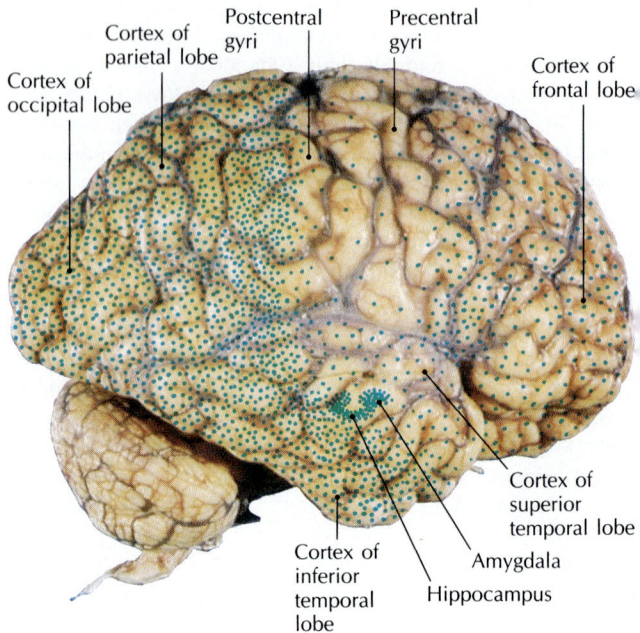

Brain of a patient with Alzheimer's disease shows specific areas (stippled) where senile plaques of dying neurons and amyloid beta-protein accumulate. Many of these plaques are located in memory centers.

Senile Dementia

Some brain neurons, which are irreplaceable, begin to degenerate at a relatively early age. Because of a surplus of brain neurons, however, a slowing down of mental processes usually does not occur until after the age of 70; sometimes it does not occur at all. *Severe* atrophy of the brain is commonly called "senility." The more technical name is *senile dementia* (L. *senex*, old man + *demens*, out of mind). It is characterized by progressive mental deterioration, including anxiety, irritability, difficulty with speech, and irrationality. Only 10 to 20 percent of the U.S. population over age 65 shows signs of mild to severe mental impairment. It should be emphasized that senility is a disease and definitely *not* a normal condition of aging.

Alzheimer's Disease

In a few cases, brain atrophy may occur much earlier than usual during the normal course of aging, even as early as 30 or 40 years of age, with the symptoms being identical to those of senile dementia. Such a condition is known as *Alzheimer's disease,* a neurodegenerative disease characterized by a progressive loss of memory, intellectual functions, and speech. Like senility, Alzheimer's disease is not a part of the normal aging process. Alzheimer's

disease afflicts more than 3 million Americans; and about 120,000 of its victims die each year, making it the fourth leading cause of death among the elderly (after cardiovascular disease, cancer, and stroke). Death usually occurs less than 10 years after the first symptoms appear.

The death of certain brain cells appears to be the basis for many of the symptoms associated with Alzheimer's disease. Autopsies of victims invariably reveal significant changes in the brain, including a reduced number of neurons, especially in areas related to thought processes; certain neurons have an accumulation of *neuritic* (*senile*) *plaques* located outside the neurons. The senile plaques are composed of dying neurons enmeshed in fibers called *neurofibrillary tangles.* These tangled neurons surround a core of *amyloid beta-protein.* The excessive accumulation of amyloid beta-protein is a leading suspect in the development of Alzheimer's disease (see photo). Victims of Alzheimer's disease also have a reduced amount of *protein kinase C,* a phosphorylating enzyme. It is possible that increased amounts of calcium in the elderly brain help inhibit protein kinase C. Finally, there is a marked reduction in the release of acetylcholine. This reduction occurs when certain cholinergic neurons which have their cell bodies in the basal nucleus (of Meynert, located at the base of the cerebrum) and whose fibers terminate in wide areas of the cerebral cortex are functionally impaired. Presumably, the loss of acetylcholine inhibits memory formation.

Apparently, Alzheimer's disease is caused by a defective gene on chromosome 19. The three known forms of the gene are called E2, E3, and E4. (The most harmful form seems to be E4.) The gene codes for a blood protein called ApoE (apolipoprotein E), whose main function seems to be transporting cholesterol throughout the body. However, another function of ApoE may be to carry amyloid beta-protein to brain cells. Although Alzheimer's disease has a genetic basis, the disease is not always inherited, and some patients do not even have the E4 gene.

There already is a simple blood test to detect Alzheimer's disease, but now researchers are seeking a clue as to how ApoE actually functions so that its harmful effects can be halted, probably by drug treatment.

Cerebral Palsy (CP)

*Cerebral palsy** actually comprises a group of neuromuscular disorders that usually result from damage to a child before it is born, during childbirth, or shortly after birth. Actually, the brain can be affected at any time from fertilization through infancy. The major types of cerebral palsy are *spastic* (increased muscle tone and stiffness of a muscle), *athetoid* (a condition marked by slow, involuntary, writhing hand movements), and *ataxic* (lack of muscular coordination). All three forms involve an impairment of skeletal motor activity to some degree, ranging from muscular weakness to complete paralysis. Related disorders such as mental retardation and speech difficulties may accompany the disease. If the functional handicap affects motor performance primarily, it is assumed that the basal nuclei of the brain are predominantly involved, and the patient is said to have *cerebral* palsy specifically. Causes are varied, from infection and malnutrition of the mother to prolonged labor, brain infection, or circulatory problems. There is no known cure.

Cerebrovascular Accident (CVA)

A *cerebrovascular accident (CVA),* commonly called a *stroke,* is a sudden withdrawal of sufficient blood supply to the brain caused by the impairment of incoming blood vessels. The resulting oxygen deficiency causes brain tissue to be damaged or even destroyed (*infarction*). Unconsciousness results if the blood and oxygen supply to the brain is insufficient for as little as 10 sec. Irreversible brain damage can occur if the brain is deprived of oxygen for 5 min or more. CVA kills about 30 percent of the people it strikes, and about a third of those who survive are permanently disabled. Cerebrovascular accident is

*"Palsy" is a distorted form of *paralysis,* which comes from the Greek word meaning "to loosen," because the muscles seem to "fall loose."

the most common brain disorder, the third most common cause of death in the United States (after cardiovascular disease and cancer), and the most common cause of adult disability.

The major causes of CVA are *thrombosis* (Gr. a clotting), the clotting of blood in a blood vessel; *embolism* (Gr. "to throw in"), a blockage in a blood vessel; and *hemorrhage,* a rupture of a cerebral artery as the result of an *aneurysm,* local dilation or ballooning of an artery. As a result, the blood supply to the brain is reduced, and blood leaks through the dura mater and puts harmful pressure on brain tissue, sometimes destroying it. Researchers are using a compound called *tissue plasminogen activator* (tPA) to break up arterial blood clots before they lead to a stroke. A new brain-scanning technique can show a stroke in progress or give a clear, noninvasive picture of damaged areas of the brain within an hour or so after a stroke. The new technique is made possible by refinements in magnetic resonance imaging (MRI).

The risk of thrombosis, embolism, or hemorrhage is increased by atherosclerosis (artery disease), hypertension (high blood pressure), diabetes mellitus, a high-fat diet, cigarette smoking, and lack of exercise.

Epilepsy

Epilepsy (Gr. *epilambanein,* seize upon) is a nervous disorder characterized by recurring attacks of motor, sensory, or psychological malfunction, with or without unconsciousness or convulsive movements. The term *epilepsy* refers to a group of symptoms with many causes rather than to a specific disease. In *symptomatic* epilepsy, seizures can be traced to one of several known causes, including a brain tumor or abscess, diseases that affect central blood vessels, and poisons. Epilepsy is often caused by brain damage before, during, or shortly after birth. The more common occurrence is *idiopathic* (Gr. a disease of unknown cause) epilepsy, in which brain cells act abnormally for no apparent reason. The disease is known as *grand mal* (Fr. great illness) when the motor areas of the brain are affected and severe spasms and loss of consciousness are involved. It is called *petit mal* (small illness) when the sensory areas are affected, without convulsions and prolonged unconsciousness. During an epileptic seizure, neurons in the brain fire at unpredictable times, even without a stimulus.

About half of all epileptics are able to control their seizures with prescription drugs, and another 25 to 30 percent obtain partial relief from medication. In cases that do not respond to medication, radical brain surgery may be performed. Surgeons have achieved success in curing epilepsy in young children (most notably under 2) with a radical technique that removes the abnormal portion of the cortex or severs connections between one cortical hemisphere and the rest of the brain. The brains of infants seem able to make extensive reconnections after surgery, but the brains of older children do not reorganize as much, although the epileptic condition does improve. Adult brains have not shown an ability to reorganize after radical brain surgery.

Headache

Headaches related to muscle tension were discussed in Chapter 11. Here we consider headaches that are connected with cranial nerves, blood vessels, and meninges.

The brain itself is not sensitive to pain, but the cerebral veins on the surface of the brain, the cerebral arteries, the cranial nerves, and parts of the dura mater are. If one of these areas is disturbed, a **headache** may result. Pressure on cranial veins, arteries, or meninges is sometimes produced by tumors, hemorrhage, meningitis, or an inflamed trigeminal nerve root. However, some of the more typical causes are emotional stress, increased blood pressure, and food allergies that make blood vessels dilate or constrict, stimulating the pain-sensitive nerve endings in the vessels. The resulting headache may be accompanied by dizziness or vertigo.

Migraine headaches are severe, recurring headaches that usually affect only one side of the head. They are often preceded by fatigue, nausea, vomiting, and the visual sensation of zigzag lines or brightness; they may be accompanied by intense pain, nausea, vomiting, and sensitivity to light and noise. Among the causes are emotional stress, hypertension, menstruation, and certain foods and beverages, such as chocolate, animal fats, and alcohol. New research suggests that migraine headaches may be associated with an excited state of glial cells. Migraine headaches frequently occur within families (suggesting a genetic basis in some cases); in compulsive, tense people, and, interestingly, on weekends and holidays, when the normal rhythm of life is disrupted.

Cranial arteritis is marked by intense pain and tenderness in the temples when the arteries in that region become inflamed. Untreated cranial arteritis may block the artery leading to the retina (central retinal artery) and cause blindness, especially in the elderly.

Huntington's Disease

Huntington's disease is a hereditary disease involving the malfunctioning of certain circuits of the basal ganglia, a region of the brain that controls movement. The disease causes rapid, jerky, involuntary movements that usually start unilaterally in the face and arms and then progress to both arms and legs. Also evident is progressive intellectual deterioration that is associated with personality changes, memory loss, and irritability. The symptoms become progressively worse.

In 1993, the specific gene that causes Huntington's disease was isolated on chromosome 4. The defective gene contains numerous copies of a repetitive DNA triplet. Apparently, the longer the repetitive DNA sequence,

the sooner the disease appears. Each child of a parent with this disease has a 50-percent chance of inheriting it. A child of afflicted parents who does not inherit the disease cannot pass it on to his or her children.

Huntington's disease strikes most commonly around age 35, often after the victim has married and had children and hence too late to avoid passing the disease on to children. Now that the causative gene has been located, researchers are hoping to be able to devise more accurate diagnostic tests. Huntington's disease has no present cure, although current research using transplants of fetal brain tissue shows potential for lessening symptoms. (See Chapter 30 for a description of fetal-tissue transplants.) The disease is usually fatal about 10 to 20 years after its appearance.

Parkinson's Disease

Parkinson's disease (also called Parkinsonism, paralysis agitans, and shaking palsy) is a progressive neurological disease characterized by stiff posture, an expressionless face, slowness in voluntary movements, a resting tremor, and a shuffling gait. The involuntary tremor, especially of the hands, often disappears when the upper limb moves. The tremor is accompanied by "pill-rolling" actions of the thumb and fingers. Parkinson's disease results from a deficiency of the neurotransmitter *dopamine*, which is released by neurons with their cell bodies in the substantia nigra and their axons terminating in the striatum (caudate nucleus and putamen). Such a shortage of dopamine usually occurs when the neurons projecting from the substantia nigra to the striatum of the basal ganglia degenerate. The deficiency prevents brain cells from performing their usual inhibitory functions within the intricate circuitry of the basal ganglia in the central nervous system.

Parkinson's disease affects more men than women, typically those over 60. It usually progresses for about 10 years, at which time death generally occurs from other causes, such as pneumonia or another infection.

At present, there is no known cure. Treatment consists of drugs and physical therapy that keep the patient functional as long as possible. The patient with established Parkinsonian symptoms is best treated with individualized dosages of *levodopa* (L-dopa) combined with carbidopa (Sinemet). Both drugs are dopamine substitutes that may cause unpleasant side effects, including nausea and vomiting. (Dopamine itself cannot be administered as a drug because it cannot penetrate the blood-brain barrier.) This treatment is optimal during the first three years of use; later the effectiveness of the drugs declines, and the incidence of side effects increases. No completely satisfactory means of treatment exists for this phase of the disease.

A surgical approach, still in the developmental stage, utilizes neonatal substantia nigral cells transplanted to the striatum (caudate nucleus and putamen) of the basal ganglia. To date, only a limited number of patients have undergone the procedure, with reported results varying from dramatic reversal of the Parkinsonian state to only modest improvement. The feasibility of transplanting cells into the brain appears to have been established, but considerable investigation is still necessary to determine the long-term survival rate of such grafts. At this time, this operative procedure is considered experimental and is not an accepted treatment for Parkinson's disease.

Dyslexia

Dyslexia (L. *dys*, faulty + Gr. *lexis*, speech) is an extreme difficulty in learning to identify printed words. It is most commonly seen as a reading and writing disability, and it is usually identified in children when they are learning to read and write. Typical problems include letters that appear to be backward, words that seem to move on the page, and transposed letters. The disorder is not related to intelligence.

Reading and writing are usually dominated by the left cerebral hemisphere, but dyslexics appear to have an overdeveloped right hemisphere that competes with the left hemisphere for the control of language skills. Apparently, the abnormal distribution of nerve cells takes place during the middle third of pregnancy, when the outer cortex of the fetal brain is formed. Prenatal disturbances such as maternal stress or a viral infection may underlie the abnormal fetal development. An early discovery of reading and writing problems and improved remedial teaching techniques appear to be the main factors in treating dyslexics.

Encephalitis

Encephalitis is an acute inflammatory disease of the brain, due to a direct viral invasion or to hypersensitivity initiated by a virus or foreign protein. It often involves a virus transmitted by a mosquito or tick. Several other viral causes are known. Brain tissue is infiltrated by an increased number of infection-fighting white blood cells called lymphocytes, cerebral edema occurs, ganglion cells in the brain degenerate, and neuron destruction may be widespread. Symptoms generally include the sudden onset of fever, headache and neck pain (indicating meningeal inflammation), drowsiness, coma, and paralysis. Extreme cases may be fatal.

Lead Poisoning

Lead poisoning, the accumulation of high blood levels of lead that can lead to permanent brain damage, is considered the main environmental problem for children in the United States. Children can receive unwanted lead from drinking water that comes through old plumbing (that is, through lead pipes, which are now banned); from ingesting chips of peeling, lead-based paint (lead-based paint was banned in 1977 but is still found in millions of older homes) or food grown in soil containing lead, stored in lead-glazed pottery or lead crystal, or packaged in lead-soldered cans; from inhaling industrially polluted air, exhaust emissions from automobiles that use leaded gasoline (now banned), tobacco smoke, or smoke from burned solid wastes.

Lead levels that were once considered safe are now known to lower intelligence and cause abnormal aggressiveness and other behavioral problems. Brain damage can even begin during fetal life if the pregnant woman has high levels of lead in her blood. Children are considered to have lead poisoning if they have 10 or more micrograms of lead per deciliter of blood. Such an amount can be found in 10 to 15 percent of preschool children in the United States. Public health experts have concluded that there is no "safe" level of lead in the blood of young children.

CHAPTER SUMMARY

General Structure of the Brain (p. 385)

1 The major parts of the brain are the **cerebrum, diencephalon, cerebellum,** and **brainstem.** The cerebrum consists of the two *cerebral hemispheres.* The diencephalon is composed primarily of the thalamus, hypothalamus, epithalamus, and ventral thalamus (subthalamus). The brainstem is composed of the midbrain, pons, and medulla oblongata.

2 Each hemisphere of the cerebrum consists of outer gray matter, deep white matter, and deep gray matter *(basal ganglia).* The outer gray matter, called the **cerebral cortex,** has many folds and convolutions. The **corpus callosum,** a bundle of nerve fibers, connects the hemispheres and relays messages between them.

Meninges, Ventricles, and Cerebrospinal Fluid (p. 385)

1 The brain is covered by three **meninges.** The outermost **dura mater** consists of an inner dura mater that is continuous with the spinal dura mater and an outer dura mater, which is actually the periosteum of the skull bones. The cranial dura mater contains venous blood vessels called *dura sinuses.* The potential space between the dural mater and the arachnoid is the subdural space.

2 The middle **arachnoid** is a thin layer between the dura mater and pia mater. Between the arachnoid and pia mater is the *subarachnoid space,* which contains cerebrospinal fluid.

3 The innermost **pia mater** directly covers the surface of the brain. In certain locations, the pia mater joins together with modified ependymal cells to form the **choroid plexuses,** networks of small blood vessels.

4 *Cerebrospinal fluid* supports and cushions the brain and helps control the chemical environment of the central nervous system.

5 Cerebrospinal fluid circulates through the *ventricular system* of the brain, which consists of a series of connected cavities called **ventricles.**

6 Cerebrospinal fluid is essentially an ultrafiltrate of blood. Most of it is formed continuously at the choroid plexuses of the lateral, third, and fourth ventricles.

7 Cerebrospinal fluid circulates from the choroid plexuses through the ventricles and then the subarachnoid space in the vicinity of the medulla oblongata until it reaches the subarachnoid villi, through which it passes back into the venous blood of the superior sagittal sinus.

Nutrition of the Brain (p. 391)

1 A lack of oxygen or glucose will damage brain tissue faster than any other tissue. Glucose, the body's chief source of usable energy, cannot be stored in the brain.

2 The brain has built-in regulating devices that make it almost impossible to constrict the blood vessels that carry the incoming blood supply.

3 The **blood-brain barrier** is an anatomical and physiological network of semipermeable membranes that prevents some substances from entering the brain while allowing relatively free passage to others.

Brainstem (p. 392)

1 The **brainstem** is the stalk of the brain that relays messages between the spinal cord and the cerebrum.

2 Within the brainstem are ascending and descending pathways between the spinal cord and parts of the brain.

3 Contained within the core of the brainstem is the **reticular formation,** with its pathways and integrative functions.

4 Of the 12 cranial nerves, all but the olfactory and optic nerves emerge from the brainstem.

5 The lowermost portion of the brainstem is the **medulla oblongata.** It contains vital cardiac, vasomotor, and respiratory centers; and it also regulates vomiting, sneezing, coughing, and swallowing.

6 The **pons** forms a connecting bridge between the medulla oblongata and the midbrain. It also contains fibers that project to the hemispheres of the cerebellum and link the cerebellum with the cerebrum. Within the reticular formation of the pons are respiratory centers that are integrated with those in the medulla oblongata.

7 The **midbrain** connects the pons and cerebellum with the cerebrum. It contains all the ascending fibers projecting to the cerebrum. It also contains oculomotor nerves, which are involved with movements of the eyes, focusing, and constriction of the pupils of the eyes.

Cerebellum (p. 399)

1 The **cerebellum** is located behind the pons in the posterior cranial fossa. It is mainly a coordinating center for muscular movement, involved with balance, precision, timing, and body positions. It has no role in perception, conscious sensation, or intelligence.

2 The cerebellum is divided into the unpaired midline **vermis,** the small bilateral **flocculonodular lobes,** and the two large **lateral lobes** or hemispheres.

3 The cerebellum is covered by a surface layer of gray matter called the **cerebellar cortex,** which is composed of a neuronal network of circuits.

4 The cerebellum is connected to the brainstem by bundles of nerve fibers called **cerebellar peduncles.**

Cerebrum (p. 399)

1 The **cerebrum** is the largest and most complex structure of the nervous system. It is composed of the cerebral *hemispheres, white matter,* and *basal ganglia.* Each hemisphere is further divided into six lobes—the frontal, parietal, occipital, temporal, limbic, and insula.

2 The surface of the cerebrum is a mantle of gray matter called the **cerebral cortex.** The cortex contains **gyri, sulci,** and deep **fissures,** all of which increase the surface area of the cerebrum.

3 A deep groove called the *longitudinal fissure* runs between the cerebral hemispheres. The **corpus callosum,** a bridge of nerve fibers between the hemispheres, relays nerve impulses between them.

4 Beneath the cortex is a thick layer of white matter consisting of association, commissural, and projection fibers.

5 Each cerebral hemisphere contains a large core of *gray matter* called the **basal ganglia,** which help coordinate muscle movements by relaying neural inputs from the cerebral cortex to the thalamus and finally back to the motor cortex of the cerebrum.

6 The **frontal lobe** is involved with the motor control of voluntary movements and the control of a variety of emotional expressions and moral behavior. The **parietal lobe** is concerned with the evaluation of the general senses and of taste. The **temporal lobe** has critical roles in hearing, equilibrium, and, to a certain degree, emotion and memory. The **occipital lobe** is involved with vision and its associated forms of expression. The **insula** (central lobe) appears to be associated with gastrointestinal and other visceral activities.

7 The limbic lobe is an integral part of the **limbic system,** which is involved with emotions, behavioral expressions, recent memory, and smell.

Diencephalon (p. 409)

1 The **diencephalon** is located deep to the cerebrum; it connects the midbrain with the cerebral hemispheres. It is composed of the thalamus, hypothalamus, epithalamus, and ventral thalamus.

2 The **thalamus** is the intermediate relay point and processing center for all sensory impulses (except smell) going to the cerebral cortex. The projections from the thalamus to the cerebral cortex mediate the sleep-wake cycle, alerting responses and consciousness in general. In addition, the thalamus integrates neural information from the basal nuclei and cerebellum that, after being processed, is conveyed to the motor cortex.

3 The **hypothalamus** is the highest integrating center for the autonomic nervous system. It regulates many physiological and endocrine activities through its relationship with the *pituitary gland* (hypophysis). It is involved with the regulation of body temperature, water balance, sleep-wake patterns, food intake, behavioral responses associated with emotion, endocrine control, and sexual responses.

4 The **epithalamus** acts as a relay station to the midbrain. It contains the *pineal body.*

5 The **ventral thalamus,** or **subthalamus,** is functionally integrated with the basal ganglia.

Regeneration in the Central Nervous System (p. 411)

Except in very few cases, axons of the mature central nervous system that have been severed lack the ability to regenerate themselves to a functional state.

Learning and Memory (p. 411)

1 Little is known about the neuronal processes of human learning, but it has been shown that specific areas of the brain are involved with different types of learning. No single area of the brain controls learning, and scientists believe that learning involves a variety of mechanisms, not just one.

2 **Short-term memory** refers to the remembrance of recent events, and **long-term memory** refers to the remembrance of information that may last a lifetime. It is not known if both types of memory use the same basic mechanism, but all memory is related to electrical and chemical activity in neuronal circuits.

Sleep and Dreams (p. 412)

1 Brain activity may be recorded on an **electroencephalograph (EEG).** The four distinct brain-wave patterns are **alpha, beta, delta,** and **theta waves.**

2 The EEG has made it possible to detect four separate stages of sleep and an important condition known as **REM** (rapid eye movement) **sleep,** which occurs during reentry into a new stage 1 phase. Dreams occur during REM sleep. It is thought that REM sleep is essential for relaxation of normal stress and for brain maturation. Learning and memory are probably reinforced during REM sleep.

3 The factor that produces sleep is unknown.

Chemicals and the Nervous System (p. 415)

1 Neurotransmitters and other chemicals that are natural constituents of the body are essential for proper functioning of the nervous system. Drugs may alter the normal workings of the nervous system. Most drugs operate at the synaptic level.

2 Some major categories of drugs are **stimulants, depressants, antidepressants, psychedelics and hallucinogens, analgesics,** and **antianxiety drugs.**

Developmental Anatomy of the Brain (p. 417)

1 The CNS develops from the embryonic **neural plate,** which is soon transformed into the **neural tube.**

2 By about week 4, the neural tube develops into cavities called the **prosencephalon** (forebrain), **mesencephalon** (midbrain), and **rhombencephalon** (hindbrain). By about week 7, the prosencephalon divides into the **telencephalon** (which develops into the cerebrum) and **diencephalon,** and the rhombencephalon divides into the **metencephalon** (cerebellum and pons) and **myelencephalon** (medulla oblongata).

3 By the end of month 3, the overall form of the brain is recognizable. *Sulci* begin to develop during month 4, and major sulci are present by month 8.

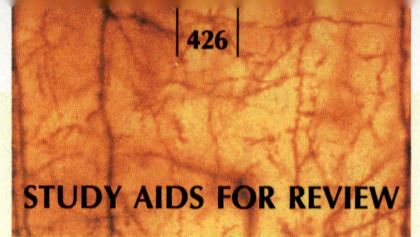

STUDY AIDS FOR REVIEW

KEY TERMS

MEDICAL TERMS FOR REFERENCE

AGNOSIA (Gr. *a*, without + *gnosis*, knowledge) Inability to recognize objects.

ANENCEPHALY (Gr. *an*, without + *enkephalos*, brain) A condition in which a child is born without most of its brain.

APHASIA (Gr. *a*, without + *phasis*, speech) Partial or total loss of the ability to speak and understand words, resulting from brain damage.

APOPLEXY (Gr. *apoplessein*, to cripple by a stroke) Sudden loss of muscular control, with partial or total loss of sensation and consciousness, resulting from rupture or blocking of a blood vessel in the brain.

ATAXIA (Gr. *a*, without + *taktos*, order) Lack of coordination of voluntary muscles.

BRADYKINESIA (Gr. *bradus*, slow + *kinesis*, movement) A condition characterized by abnormally slow movements.

CEREBRAL ANEURYSM (Gr. *aneurusma*, dilation) A bubble-like sac formed by the enlargement of a cerebral blood vessel.

DYSARTHRIA (L. *dys*, faulty + Gr. *arthron*, joint) Lack of coordination of the muscles that control speech.

DYSKINESIA (L. *dys*, faulty + Gr. *kinesis*, movement) Difficulty in performing voluntary muscular movements as a result of a brain lesion.

ENCEPHALOPATHY A general term for any brain disease.

MENINGIOMA A meningeal tumor.

MENINGOCELE A protrusion of the meninges through an opening in the skull or spinal column.

NARCOLEPSY (Gr. *narke*, numbness + *lepsia*, to seize) A condition of uncontrollable sleepiness and sleep.

SUBDURAL HEMATOMA (Gr. *haimato*, blood + *oma*, tumor) A localized pool of venous blood in the subdural space.

UNDERSTANDING THE FACTS

1 Name the four major parts of the brain and their main subdivisions.

2 What parts of the brain comprise the brainstem?

3 Trace the route followed by cerebrospinal fluid from its synthesis to its reabsorption back into the circulatory system.

4 What are the names of the meninges of the brain? How do the meninges support and/or protect the brain?

5 What is the function of the blood-brain barrier? How is the barrier formed?

6 Name the lobes of the cerebrum and state the primary function(s) of each.

7 What is the function of each of the following?

  a corpus callosum   **f** thalamus
  b cerebral aqueduct   **g** reticular activating system
  c pons   **h** basal ganglia
  d corpora quadrigemina **i** medulla oblongata
  e hypothalamus   **j** limbic system

8 Which areas of the brain are associated with the conversion of short-term memory to long-term memory?

9 Which brain waves are characteristic of each of the four stages of sleep?

10 In what ways are the actions of stimulants and antidepressants similar? In what ways are they different?

UNDERSTANDING THE CONCEPTS

1 What functions of the brain are enhanced by the presence of a blood-brain barrier? What is the influence of the blood-brain barrier on medical treatments for diseases of the brain?

2 Which regions of the brain monitor the input to the conscious regions of the cerebral cortex? Why do you think that such monitoring is necessary?

3 In what functional ways are the hypothalamus and the medulla oblongata related? How do they differ in function?

4 What are the functional implications of decussation?

5 In light of decussation, how can you explain the fact that a lesion in the right cerebellum will be expressed by the right side of the body?

6 Contrast the location of white matter and gray matter in the brainstem, cerebellum, and cerebrum.

7 What is the functional significance of the gyri and sulci of the cerebral cortex?

8 What activities are believed to occur during REM sleep?

9 Why is the hypothalamus called the crossroad between the nervous system and the endocrine system?

SELF-QUIZ
Multiple-Choice Questions

13.1 Of the following, which is *not* associated with the medulla oblongata?
- **a** pyramids
- **b** vasomotor center
- **c** decussation
- **d** apneustic center
- **e** olive

13.2 Which is *not* a function of the cerebellum?
- **a** initiation of body movements
- **b** balance
- **c** timing of body movements
- **d** body positions
- **e** processing input from receptors in the head

13.3 Which are parts of the cerebellum?
- **a** vermis
- **b** flocculonodular lobes
- **c** lateral lobes
- **d** a and b
- **e** a, b, and c

13.4 The thalamus and hypothalamus are parts of the
- **a** cerebellum
- **b** cerebrum
- **c** brainstem
- **d** medulla oblongata
- **e** diencephalon

13.5 Which of the following is *not* an important function of the hypothalamus?
- **a** integration with the autonomic nervous system
- **b** temperature regulation
- **c** water and electrolyte balance
- **d** regulation of heart rate
- **e** food intake

13.6 The reticular formation can best be described as
- **a** a collection of neurons deep in the cerebrum
- **b** a series of nuclei within the diencephalon
- **c** a region of white matter within the cerebellum
- **d** cross-connecting axons in the cerebral cortex
- **e** a collection of neurons within the brainstem

13.7 The most inferior part of the brainstem is the
- **a** pons
- **b** medulla oblongata
- **c** midbrain
- **d** reticular formation
- **e** diencephalon

13.8 The complex network of tiny islands of gray matter within the brain that acts as a filter for incoming sensory impulses is the
- **a** denticulate nucleus
- **b** reticular formation
- **c** limbic system
- **d** corpora quadrigemina
- **e** none of the above

13.9 If you were inside the limbic system, you could be in contact with all the following *except* the
- **a** putamen
- **b** amygdala
- **c** hippocampus
- **d** a and c
- **e** a and b

Completion Exercises

13.10 Neuroglial cells called _____ help to form the blood-brain barrier.

13.11 Cerebrospinal fluid flows through the _____ into the superior sagittal sinus.

13.12 The space external to the pia mater that contains cerebrospinal fluid is the _____.

13.13 The outer portion of the cerebrum is called the _____.

13.14 The bundle of nerve fibers that connects the two cerebral hemispheres is the _____.

13.15 A network of blood vessels called the _____ is formed where the ependyma contacts the pia mater.

13.16 The portion of the brain that is located between the diencephalon and pons is the _____.

13.17 The cerebral peduncles and the nucleus ruber are stuctures within the _____.

13.18 The thalamus processes all sensory impulses ascending from the spinal cord to the _____.

Matching

- **a** gyri
- **b** sulci
- **c** fissures
- **d** basal ganglia
- **e** frontal lobe
- **f** parietal lobe
- **g** temporal lobe
- **h** occipital lobe
- **i** hippocampus
- **j** limbic lobe
- **k** diencephalon
- **l** thalamus

13.19 _____ involved with memory, part of limbic system

13.20 _____ gray matter deep within the cerebrum

13.21 _____ extremely deep depressions

13.22 _____ houses third ventricle

13.23 _____ motor control over voluntary movements

13.24 _____ raised ridges of the cerebrum cortex

13.25 _____ involved with hearing

13.26 _____ part of cerebrum and also limbic system

13.27 _____ contains visual cortex

13.28 _____ evaluation of the general senses

13.29 _____ slitlike grooves

13.30 _____ a diencephalic relay center

A SECOND LOOK

1 Label the cerebellum, midbrain, pons, medulla oblongata, spinal cord, and cerebellar cortex.

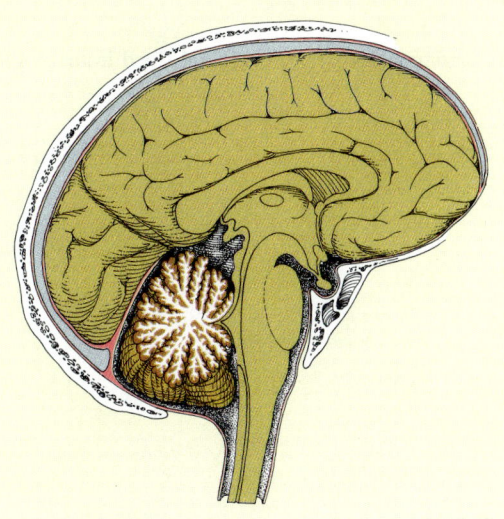

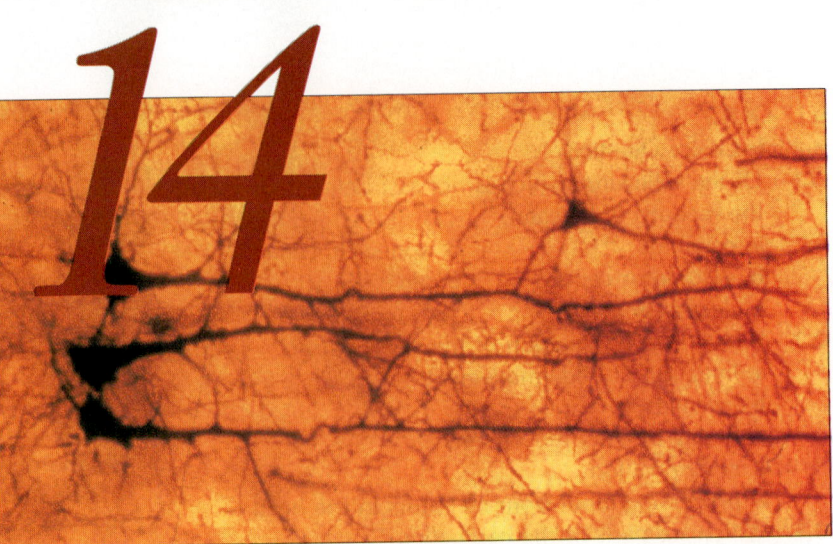

The Central Nervous System II: The Spinal Cord

14

KEY CONCEPTS

1 The upper end of the spinal cord is continuous with the lowermost part of the brain. Thirty-one pairs of spinal nerves connect the spinal cord with the rest of the body.

2 The spinal cord is protected by three layers of membranes, a surrounding fluid, and the bony spinal column.

3 Within the spinal cord, ascending sensory tracts carry impulses to the brain, and descending motor tracts carry impulses from the brain to effectors, such as those in muscles and glands.

The spinal cord, with its 31 pairs of spinal nerves, serves two important functions: (1) It is the connecting link between the brain and most of the body, and (2) it is involved in spinal reflex actions, both somatic and visceral. *Somatic spinal reflexes* involve a series of responses to the stimulation of sensory receptors and skeletal muscles. These reflexes bring about movement and maintain posture. *Visceral spinal reflexes* occur in visceral organs, for example, when the urinary bladder becomes distended and evokes urination. Visceral spinal reflexes also help to regulate blood pressure, and they influence the action of glands. Thus, both voluntary and involuntary movements of the limbs, as well as many visceral processes, depend on the spinal cord, a vital link between the brain and the rest of the body.

BASIC ANATOMY OF THE SPINAL CORD

The *spinal cord* is the part of the central nervous system that extends from the foramen magnum of the skull caudally for about 45 cm (18 in.) to the level of the first lumbar vertebra (L1) in adults. Its superior end is continuous with the lowermost part of the brain (the medulla oblongata). Its inferior end tapers off as the cone-shaped *conus terminalis*, located in the vicinity of the first lumbar vertebra [FIGURE 14.1A]. Extending caudally from the conus is a nonneural fiber called the *filum terminale*, which attaches to the coccyx. The filum consists mainly of fibrous connective tissue.

Surrounding and protecting the spinal cord is the vertebral column. The vertebral canal of the vertebral column (the opening through which the spinal cord passes) has a diameter slightly larger than your index finger (about 1 cm). Inside the column, the cylindrical cord itself is slightly thicker than a pencil. The spinal cord is slightly flattened dorsally and ventrally, with two prominent enlargements known as the *cervical* and *lumbosacral enlargements* [FIGURE 14.1B]. Emerging from these enlargements are the spinal nerves that innervate the upper and lower limbs.

Of the 31 pairs of spinal nerves, there are 8 cervical (C) nerve pairs, 12 thoracic (T), 5 lumbar (L), 5 sacral (S), and 1 coccygeal (Co). Each pair of spinal nerves typically passes through a pair of intervertebral foramina located between two successive vertebrae; the nerves are then distributed to a specific pair of segments of the body [FIGURE 14.1A]. Note that there is a pair of spinal nerves for each vertebra.

The roots of all nerves passing caudally below the conus terminalis (below L1 vertebral level) resemble flowing, coarse strands of hair. For this reason, the lumbar and sacral roots are collectively called the *cauda equina* (KAW-duh ee-KWY-nuh), which means "horse's tail" in Latin.

The spinal cord and the roots of its nerves are protected not only by the flexible vertebral column and its ligaments but also by the spinal meninges and cerebrospinal fluid.

Spinal Meninges

The spinal *meninges* [FIGURE 14.2] are continuous with the same membrane layers that cover the brain. The outer layer, called the *dura mater,* is a tough, fibrous membrane that merges with the filum terminale. The middle layer, the *arachnoid,* runs caudally to the S2 vertebral level, where it joins the filum terminale. The arachnoid is so named because it is delicate and transparent and has some connective strands running across the space that separates it from the innermost layer, the *pia mater.* The thin, highly vascular pia mater is tightly attached to the spinal cord and its roots. It contains blood vessels that nourish the spinal cord. The fibrous bands of pia mater that extend from each side of the entire length of the spinal cord are the paired *denticulate ligaments* [FIGURE 14.2A]. Each ligament is attached to the dura mater at regular intervals.

Between the dura mater and the periosteum of the vertebrae is the *epidural space* [FIGURE 14.2B], containing many blood vessels and some fat. Anesthetics can be injected into the epidural space below the L3 vertebral level, from which it ascends to act upon sensory neurons that help dull pain. This procedure, known as a *caudal block,* permits patients to be conscious during operations on the pelvic region and also during childbirth.

Between the dura mater and the arachnoid is the *subdural space,* which is merely a slit and contains no cerebrospinal fluid. The *subarachnoid space,* which separates the arachnoid and the pia mater, contains cerebrospinal fluid, blood vessels, and spinal roots.

Cerebrospinal Fluid

Cerebrospinal fluid (CSF) is a clear, watery ultrafiltrate solution formed primarily from blood in the choroid plexus of the brain. The basic mechanism of its circulation involves both an active transport system and passive diffusion into the four ventricles. Then the fluid passes through small openings in the roof of the fourth ventricle into the subarachnoid spaces, through which it circulates around the brain and spinal cord. Under normal conditions, it returns to the blood at about the same rate at which it is formed (about 300 mL/day) [see FIGURE 13.3].

The cerebrospinal fluid provides a cushion that, together with the vertebral column and meninges, protects the delicate tissues of the spinal cord. It is also involved in

FIGURE 14.1 SPINAL CORD, SPINAL NERVES, AND VERTEBRAL COLUMN

[A[Right lateral view of the spinal cord and vertebral column, showing the various spinal segments and the emergence of spinal nerves from the vertebral column. [B] Posterior view.

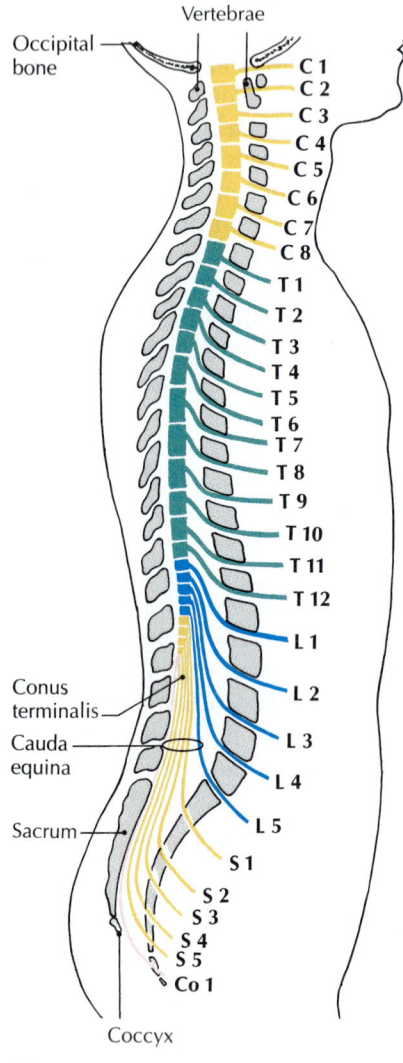

Vertebrae
Occipital bone
C 1
C 2
C 3
C 4
C 5
C 6
C 7
C 8
T 1
T 2
T 3
T 4
T 5
T 6
T 7
T 8
T 9
T 10
T 11
T 12
L 1
L 2
L 3
L 4
L 5
Conus terminalis
Cauda equina
Sacrum
S 1
S 2
S 3
S 4
S 5
Co 1
Coccyx

[A]

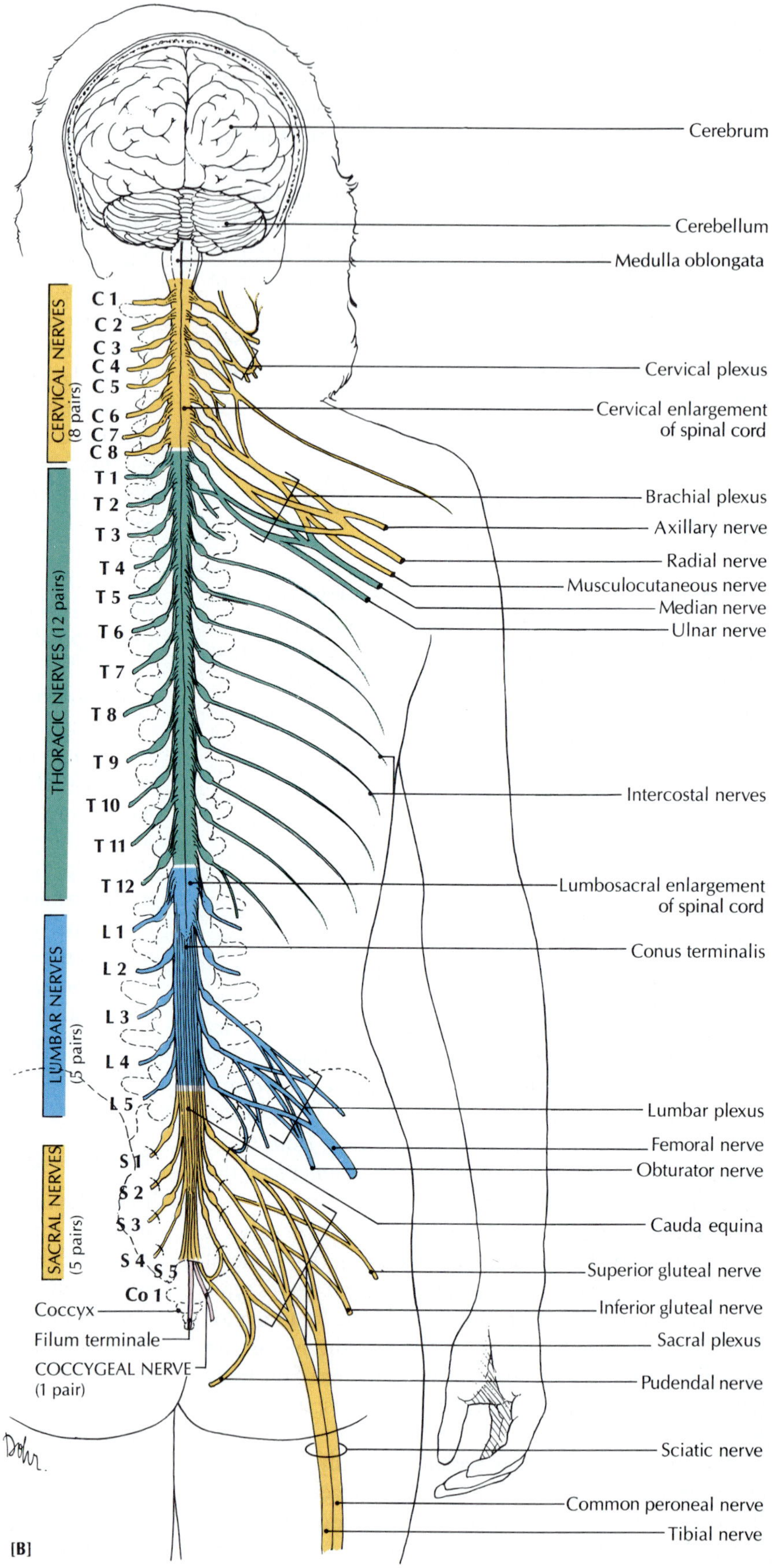

Cerebrum
Cerebellum
Medulla oblongata
CERVICAL NERVES (8 pairs)
C 1
C 2
C 3
C 4
C 5
C 6
C 7
C 8
Cervical plexus
Cervical enlargement of spinal cord
Brachial plexus
Axillary nerve
Radial nerve
Musculocutaneous nerve
Median nerve
Ulnar nerve
THORACIC NERVES (12 pairs)
T 1
T 2
T 3
T 4
T 5
T 6
T 7
T 8
T 9
T 10
T 11
T 12
Intercostal nerves
Lumbosacral enlargement of spinal cord
Conus terminalis
LUMBAR NERVES (5 pairs)
L 1
L 2
L 3
L 4
L 5
Lumbar plexus
Femoral nerve
Obturator nerve
SACRAL NERVES (5 pairs)
S 1
S 2
S 3
S 4
S 5
Co 1
Cauda equina
Superior gluteal nerve
Inferior gluteal nerve
Sacral plexus
Coccyx
Filum terminale
COCCYGEAL NERVE (1 pair)
Pudendal nerve
Sciatic nerve
Common peroneal nerve
Tibial nerve
Dohn.

[B]

FIGURE 14.2 THE SPINAL MENINGES

[A] A drawing of the spinal meninges. [B] The top layers are peeled away to show the triple-layered meninges and the spaces between them; right lateral view.

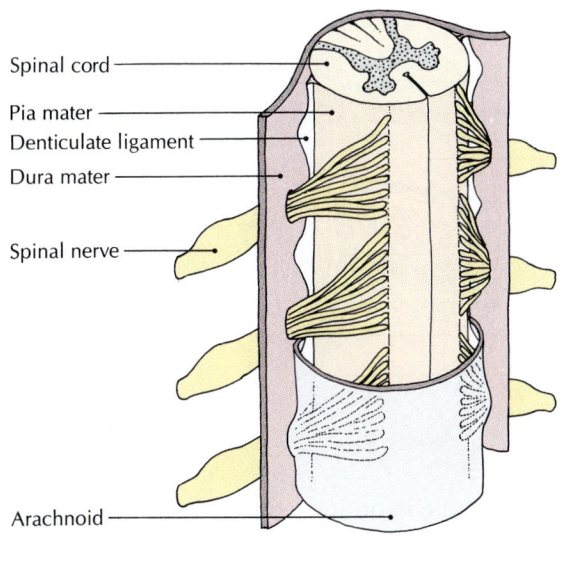

[A]

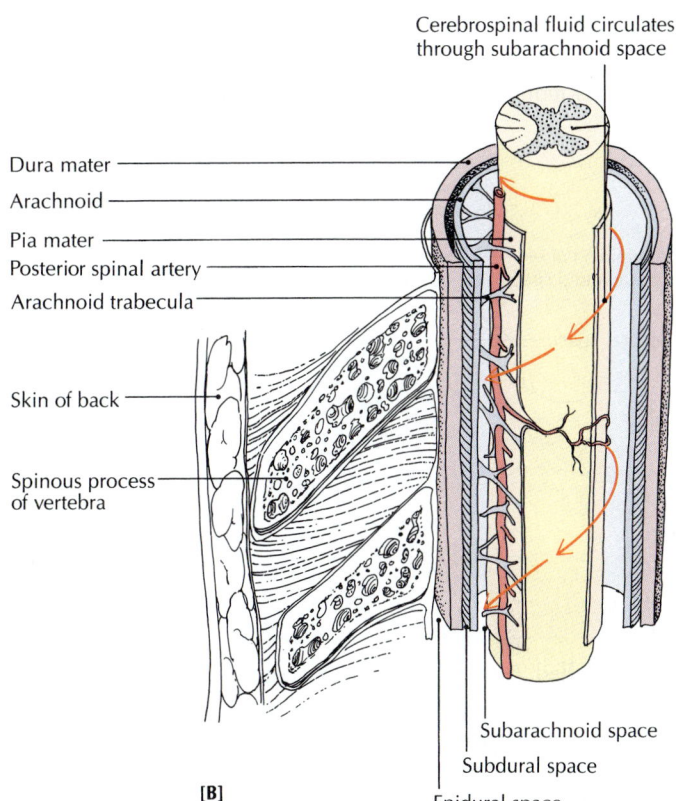

[B]

the exchange of nutrients and wastes between the blood and the neurons of the brain and spinal cord.

Cerebrospinal fluid has many important clinical implications. In certain neurological disorders, alterations can occur in the cellular and chemical content as well as its pressure. The fluid can be removed for analysis by inserting a needle into the subarachnoid space between L3 and L4, with the patient usually lying curled up on one side [FIGURE 14.3]. Such a *lumbar puncture*, or "spinal tap," is performed on this lower lumbar region to avoid injury to the spinal cord, which ends between L1 and L2. In special situations, a *cisternal puncture* can be performed by a neurologist by inserting a needle between the occipital bone and atlas (C1), entering the cisterna cerebellomedullaris, and withdrawing spinal fluid. (A puncture of the medulla oblongata can be fatal.)

A marked increase in white blood cells in the cerebrospinal fluid occurs in acute bacterial meningitis, and a moderate increase may indicate the presence of a viral infection or cerebral tissue damage. The protein (gamma globulin) content is increased in multiple sclerosis, and glucose levels are reduced during active bacterial infections.

Internal Structure

If you were to cut the spinal cord in cross section, you would see a tiny central canal, which contains cerebrospinal fluid, and a dark portion of H-shaped or butterfly-shaped "gray matter" surrounded by a larger area of "white matter" [FIGURE 14.4]. The spinal cord is divided into more or less symmetrical left and right halves by a deep groove, called the *anterior (ventral) median fissure*, and a median septum, called the *posterior (dorsal) median sulcus*. Extending out from the spinal cord are the ventral and dorsal roots of the spinal nerves. (The spinal cord and nerves cut vertically are shown in FIGURE 14.5; the roots are shown in FIGURE 14.6.)

Gray matter The *gray matter* of the spinal cord consists of nerve cell bodies, dendrites and axon terminals or bundles of unmyelinated axons, and neuroglia. The gray matter forms an H shape and is composed of three columns of neurons (posterior, anterior, and lateral horns) running up and down the length of the spinal cord from the upper cervical level to the sacral level [FIGURE 14.4]. The columns that form the two vertical bars of the

FIGURE 14.3 LUMBAR PUNCTURE, OR SPINAL TAP

The spine is in a flexed position **[A]**, which separates the spinous processes of the vertebrae and allows the lumbar puncture needle to enter the subarachnoid space well below the termination of the spinal cord **[B]**.

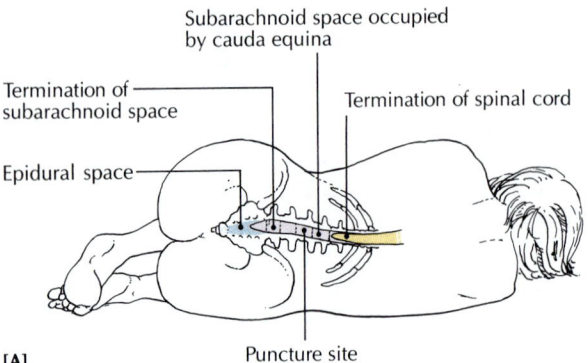

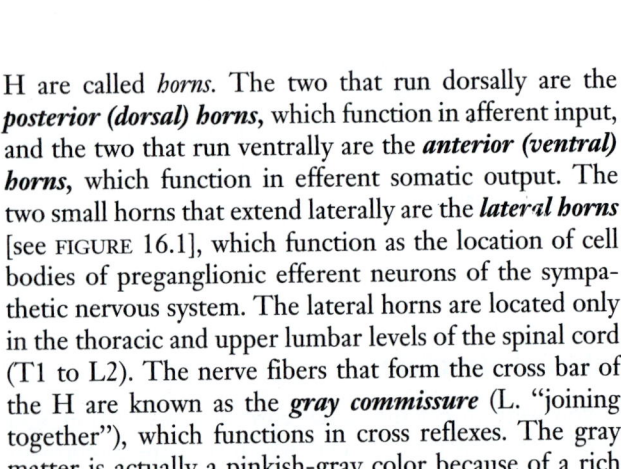

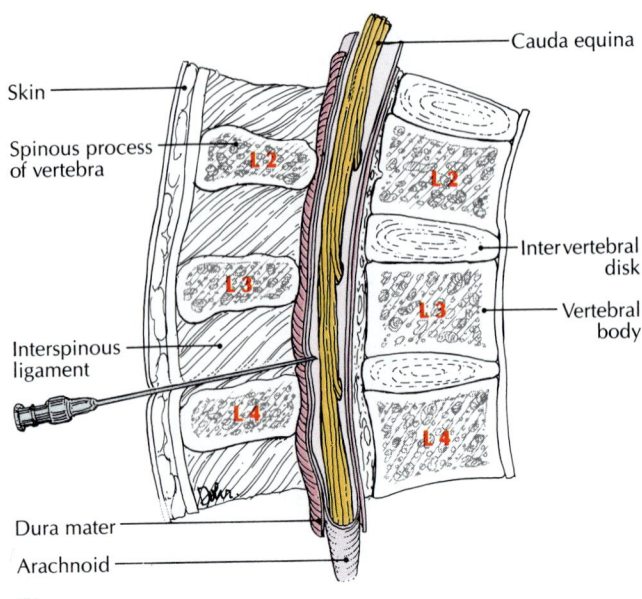

H are called *horns*. The two that run dorsally are the ***posterior (dorsal) horns,*** which function in afferent input, and the two that run ventrally are the ***anterior (ventral) horns,*** which function in efferent somatic output. The two small horns that extend laterally are the ***lateral horns*** [see FIGURE 16.1], which function as the location of cell bodies of preganglionic efferent neurons of the sympathetic nervous system. The lateral horns are located only in the thoracic and upper lumbar levels of the spinal cord (T1 to L2). The nerve fibers that form the cross bar of the H are known as the ***gray commissure*** (L. "joining together"), which functions in cross reflexes. The gray matter is actually a pinkish-gray color because of a rich network of blood vessels.

White matter The *white matter* of the spinal cord gets its name because it is composed mainly of myelinated nerve fibers, and myelin has a whitish color. The white matter is divided into three pairs of columns, or ***funiculi*** (fyoo-NICK-yoo-lie; L. little ropes), of myelinated fibers that run the entire length of the cord. The funiculi consist of the anterior (ventral) column, the posterior (dorsal) column, the lateral column, and a commissure area [FIGURE 14.4B].

The bundles of fibers within each funiculus are subdivided into ***tracts*** called *fasciculi* (fah-SICK-yoo-lie; L. little bundles). *Ascending tracts* are made up of sensory fibers that carry impulses *up* the spinal cord to the brain [TABLE 14.1]; *descending tracts* of motor fibers transmit impulses from the brain *down* the spinal cord to the efferent neurons [TABLE 14.2]. The longer tracts carry nerve im-

pulses up to the brain or down through the cord to neurons that innervate the muscles or glands. The shorter tracts convey impulses from one level of the cord to another. The major ascending and descending tracts are described in TABLES 14.1 and 14.2. Their functional roles are discussed in the next section.

Ventral and dorsal roots In the vicinity of the cord, each spinal nerve divides into a ventral (anterior, motor) root and a dorsal (posterior, sensory) root [FIGURE 14.6]. *Ventral roots* contain efferent nerve fibers* and convey motor information. *Dorsal roots* contain afferent nerve fibers, which enter the cord with sensory information. The axons of *motor neurons* whose cell bodies are located within the CNS in the anterior horn (gray matter) emerge from the spinal cord to form the ventral (motor) roots. In contrast, groups of cell bodies of *sensory neurons,* whose axons make up the dorsal roots, lie *outside* the cord in the ***dorsal root ganglia,*** or *spinal ganglia,* of the PNS.

*Some fibers of ventral roots are now known to be afferent.

A S K   Y O U R S E L F

1 What is the cauda equina? Where is it?

2 What are the three spinal meninges called?

3 What are some functions of cerebrospinal fluid?

4 What does the gray matter of the spinal cord consist of?

FIGURE 14.4 INTERNAL STRUCTURE OF THE SPINAL CORD

[A] Cross section of the spinal cord with the bones that protect it. Spinal nerves connect with the spinal cord in *pairs* at the ventral and dorsal roots. **[B]** Cross section of the spinal cord showing some prominent internal features, including columns, or funiculi, of myelinated nerve fibers. The insets show the composition of white matter (right) and gray matter (left).

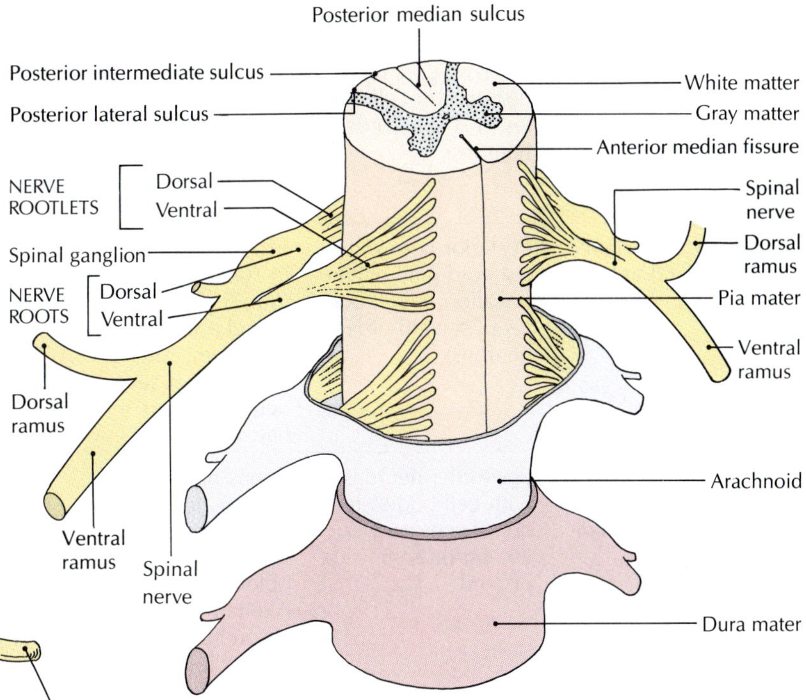

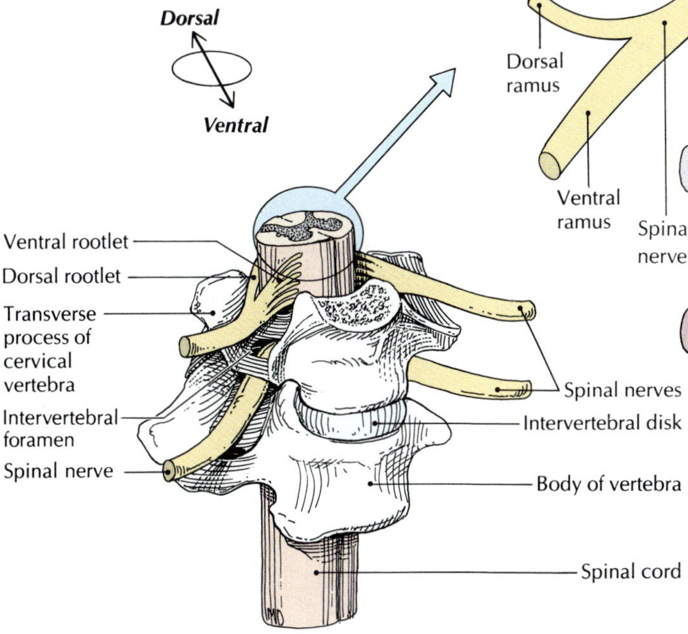

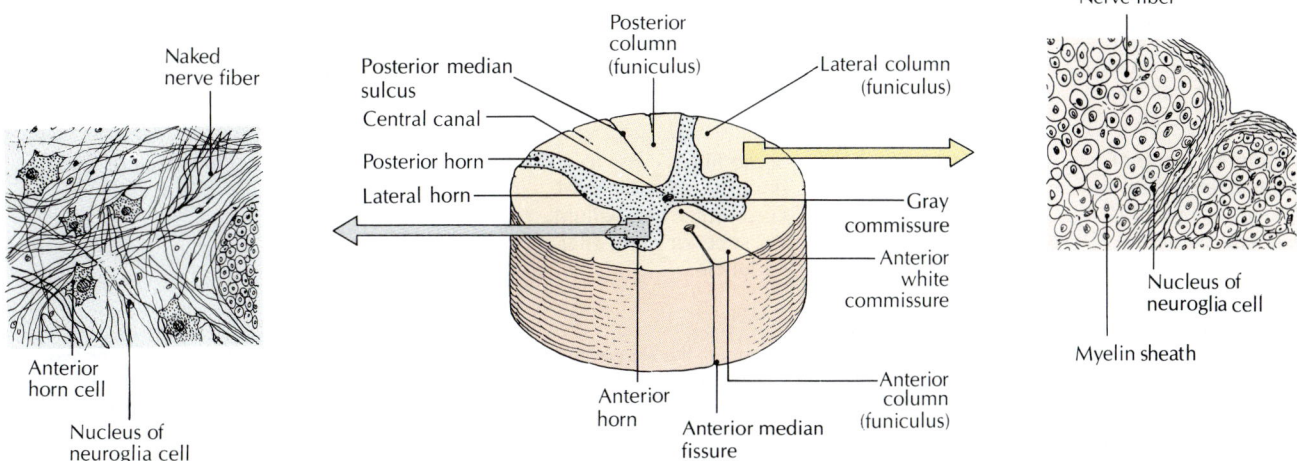

Dorsal

Ventral

Ventral rootlet
Dorsal rootlet
Transverse process of cervical vertebra
Intervertebral foramen
Spinal nerve

Spinal nerves
Intervertebral disk
Body of vertebra
Spinal cord

[A]

Posterior median sulcus
Posterior intermediate sulcus
Posterior lateral sulcus
White matter
Gray matter
Anterior median fissure

NERVE ROOTLETS — Dorsal / Ventral
Spinal ganglion
NERVE ROOTS — Dorsal / Ventral

Spinal nerve
Dorsal ramus
Pia mater
Ventral ramus

Dorsal ramus
Ventral ramus
Spinal nerve

Arachnoid
Dura mater

Naked nerve fiber
Anterior horn cell
Nucleus of neuroglia cell

Posterior column (funiculus)
Posterior median sulcus
Central canal
Posterior horn
Lateral horn
Lateral column (funiculus)
Gray commissure
Anterior white commissure
Anterior horn
Anterior median fissure
Anterior column (funiculus)

Nerve fiber
Nucleus of neuroglia cell
Myelin sheath

[B]

TABLE 14.1 MAJOR ASCENDING (SENSORY) TRACTS FROM SPINAL CORD TO BRAIN

Tract and location in spinal cord	Origin	Description and course to thalamus of brain	Course from thalamus to termination in cerebral cortex	Sensations conveyed
Lateral spinothalamic, in anterior half of lateral column*	Posterior horn of spinal cord gray matter (location of cell bodies of second-order neurons)	Fibers of second-order neurons cross to the opposite side at each spinal cord level, ascend and terminate in thalamus.	Axonal fibers of third-order neurons with cell bodies in thalamus terminate in postcentral gyrus and association cortex of cerebral cortex.	Pain and temperature
Spinoreticulothalamic, in lateral and anterior columns*	Posterior horn of spinal cord gray matter (location of cell bodies of second-order neurons)	Crosses to opposite side in spinal cord, ascends to and terminates in reticular formation of brainstem. After a sequence of several neurons, fibers terminate in thalamus.	Fibers of neurons of cell bodies in thalamus terminate in many areas of cerebral cortex (postcentral gyrus and association cortex).	Pain
Fasciculus gracilis and fasciculus cuneatus, in posterior column (posterior column-medial lemniscus pathway)	First-order neurons with cell bodies in dorsal root ganglia; ascend on same side of cord	Axons of first-order neurons terminate in nuclei gracilis and cuneatus in medulla oblongata of same side. Axonal fibers of second-order neurons with cell bodies in nuclei gracilis and cuneatus cross to opposite side to form *medial lemniscus,* which ascends to, and terminates in, thalamus.	Axonal fibers of third-order neurons with cell bodies in thalamus terminate in postcentral gyrus of cerebral cortex.	Touch-pressure two-point discrimination, vibratory sense, position sense (proprioception);[†] also some light (coarse) touch
Anterior spinothalamic, in anterior column*	Posterior horn of spinal cord gray matter (location of cell bodies of second-order neurons)	Crosses to opposite side at each spinal cord level. Joins medial lemniscus of brainstem, which terminates in thalamus.	Axonal fibers of third-order neurons with cell bodies in thalamus terminate in postcentral gyrus of cerebral cortex.	Light (coarse) touch[‡]

Tract and location in cerebellum of brain	Origin	Description and course in cerebellum	Termination	Modality conveyed
Posterior (dorsal) spinocerebellar, in posterior half of lateral column	Posterior horn of spinal cord gray matter	Uncrossed tract. Enters cerebellum via inferior cerebellar peduncle.	Cerebellum	Unconscious proprioceptions[§]
Anterior (ventral) spinocerebellar, in anterior half of lateral column	Posterior horn of spinal cord gray matter	Some fibers cross to opposite side in spinal cord. Enter cerebellum via superior cerebellar peduncle.	Cerebellum	Unconscious proprioception

*These three tracts are collectively called the *anterolateral system.*
[†]*Touch-pressure* (deep touch) sensation is obtained by deforming the skin with pressure. The ability to discriminate between one or two points applied to the skin is called *two-point discrimination;* it is acute on the ball of a finger, but poor on the back of the body. *Vibratory sense* is the sensation of feeling vibrations when the stem of a vibrating tuning fork is placed on the bones of a joint. *Position sense* enables us to know where the various body parts are with our eyes closed.
[‡]*Light touch* is the sensation obtained by touching, but not deforming, the skin or by moving a hair.
[§]*Unconscious proprioception* (*proprio* = self) is the sensory information from the body, which does not result in any sensation. Such information about movement and limb position is continuously streaming in from the body and is utilized by the nervous system in the coordination of voluntary muscles during movements.

FIGURE 14.5 SPINAL CORD AND SPINAL NERVES

Photographs of posterior views of **[A]** the spinal cord and spinal nerves in the cervical region, and **[B]** the cauda equina and spinal nerves in the lumbosacral region.

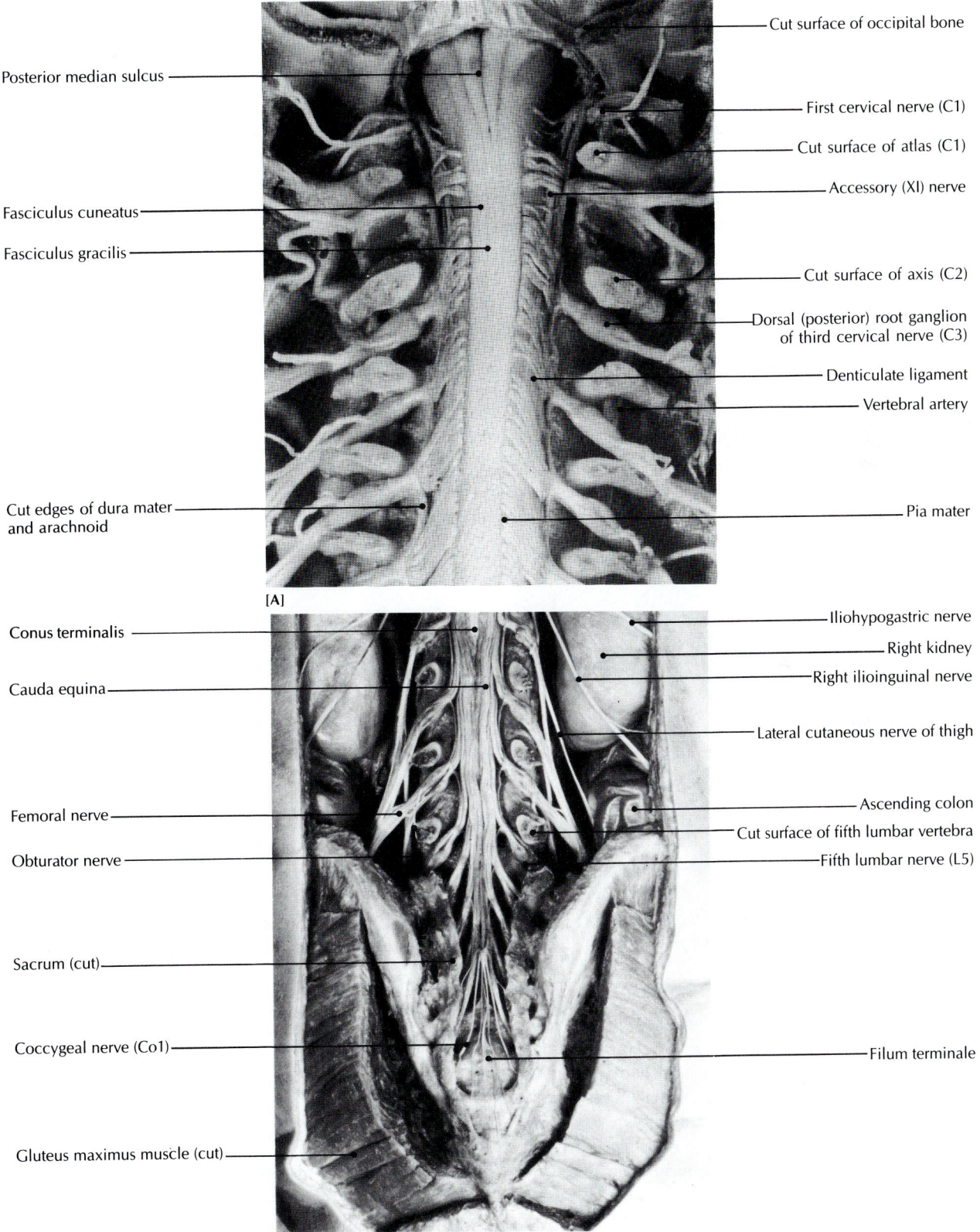

[A]

Cut surface of occipital bone

Posterior median sulcus

First cervical nerve (C1)

Cut surface of atlas (C1)

Accessory (XI) nerve

Fasciculus cuneatus

Fasciculus gracilis

Cut surface of axis (C2)

Dorsal (posterior) root ganglion of third cervical nerve (C3)

Denticulate ligament

Vertebral artery

Cut edges of dura mater and arachnoid

Pia mater

[B]

Conus terminalis

Iliohypogastric nerve

Right kidney

Cauda equina

Right ilioinguinal nerve

Lateral cutaneous nerve of thigh

Femoral nerve

Ascending colon

Cut surface of fifth lumbar vertebra

Obturator nerve

Fifth lumbar nerve (L5)

Sacrum (cut)

Coccygeal nerve (Co1)

Filum terminale

Gluteus maximus muscle (cut)

TABLE 14.2 MAJOR DESCENDING (MOTOR) TRACTS FROM BRAIN TO SPINAL CORD

Tract and location in spinal cord	Origin	Description	Termination	Motor impulse conveyed
Lateral (pyramidal) corticospinal, in lateral column	Cerebral cortex	Crosses to opposite side between medulla oblongata and spinal cord.	Gray matter, but primarily anterior horn	Manipulative movements of extremities, especially delicate finger movements.
Anterior (pyramidal) corticospinal	Cerebral cortex	Mainly uncrossed tracts; crosses near termination.	Anterior horn of spinal cord	Voluntary movements of axial musculature.
Rubrospinal, in lateral column	Nucleus ruber of midbrain	Crosses in midbrain to opposite side in midbrain.	Gray matter, but primarily anterior horn	Manipulative movements of extremities.
Medullary and pontine reticulospinal, in anterior column and anterior half of lateral column	Reticular formation of medulla oblongata and pons	Mainly uncrossed tracts.	Anterior horn	Maintenance of erect posture. Integrate movements of axial body and girdles of limbs. Involved with autonomic nervous system activity.
Vestibulospinal, in anterior column	Vestibular nuclei of medulla oblongata	Uncrossed tract.	Anterior horn	Maintenance of erect posture. Integrates movements of body and limbs.

DESCENDING (MOTOR) TRACTS OF THE SPINAL CORD   ASCENDING (SENSORY) TRACTS OF THE SPINAL CORD

FUNCTIONAL ROLES OF PATHWAYS OF THE CENTRAL NERVOUS SYSTEM

Each pathway of the central nervous system is basically composed of organized sequences of neurons. Upper motor neurons in the brain influence the activity of lower motor neurons in the cranial and spinal nerves. Some neurons have long axons that terminate in *processing centers,* where the processing of neural information takes place. These processing centers may be considered the "computers" of the pathways. They are called by one of several names: nucleus, ganglion, gray matter of spinal cord, or cortex of the brain.

Structural and Functional Aspects of the General Somatic Efferent (Motor) System

The brain exerts active and subtle influences on the activity of skeletal muscles through descending motor pathways that make up the *upper motor neurons.* Originating from cell bodies in the cerebral cortex and brainstem, these upper motor neuron pathways act by regulating and modulating the activity of the *lower motor neurons* of the cranial and spinal nerves.

Lower motor neurons The *lower motor neurons* include alpha and gamma motor neurons. *Alpha motor neurons* have their cell bodies in the central nervous system. Their axons course through cranial and spinal

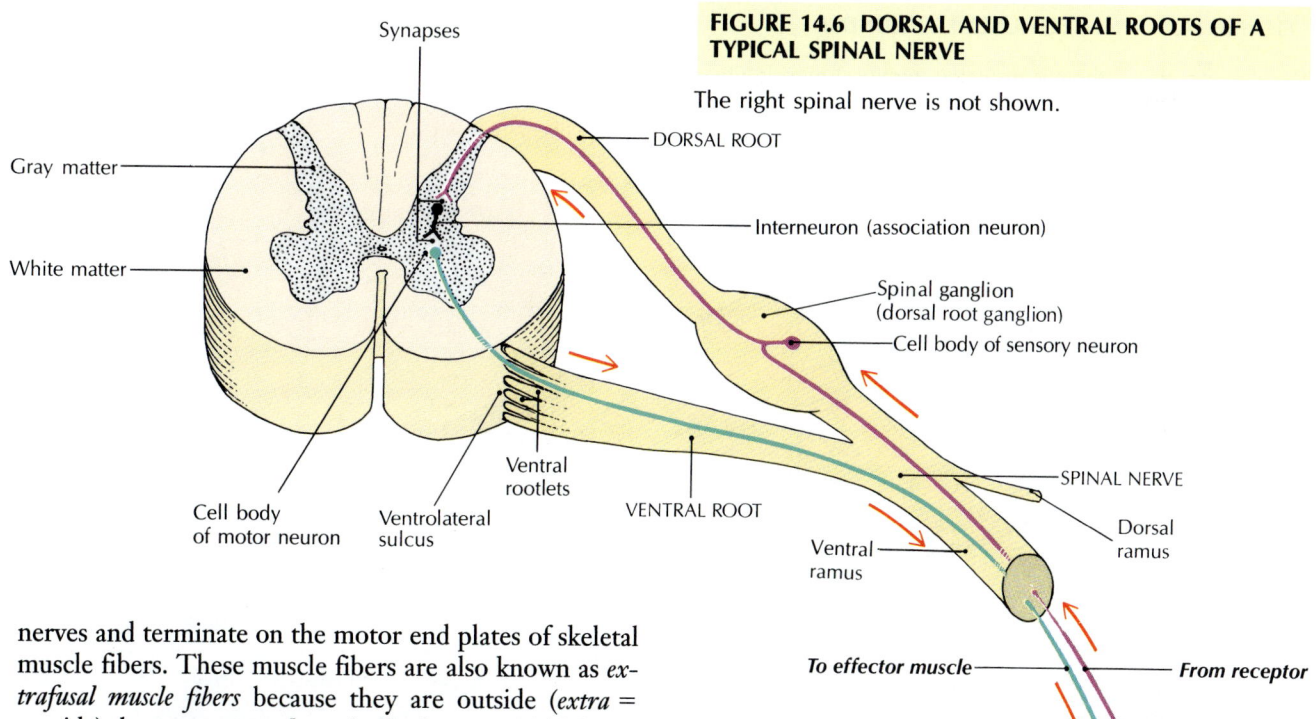

FIGURE 14.6 DORSAL AND VENTRAL ROOTS OF A TYPICAL SPINAL NERVE

The right spinal nerve is not shown.

Synapses

Gray matter

White matter

Cell body of motor neuron

Ventrolateral sulcus

Ventral rootlets

VENTRAL ROOT

DORSAL ROOT

Interneuron (association neuron)

Spinal ganglion (dorsal root ganglion)

Cell body of sensory neuron

SPINAL NERVE

Dorsal ramus

Ventral ramus

To effector muscle

From receptor

nerves and terminate on the motor end plates of skeletal muscle fibers. These muscle fibers are also known as *ex-trafusal muscle fibers* because they are outside (*extra* = outside) the **neuromuscular spindles** [FIGURE 14.7]. Neuromuscular spindles are sensory receptors that monitor the extent and rate of muscle lengthening, and thus are called *stretch receptors. Gamma motor neurons* also have their cell bodies within the central nervous system. Their axons pass through the cranial and spinal nerves to innervate the *intrafusal muscle fibers* inside the neuromuscular spindles. Alpha motor neurons are involved in the stretch reflex (see page 440), and gamma motor neurons are involved in the gamma motor neuron reflex (see page 441).

Because the lower motor neurons are the only neurons that innervate the skeletal muscle fibers, they function as the *final common pathway,* the final link between the central nervous system and skeletal muscles. Axons from lower motor neurons are located in both cranial and spinal nerves. Those in cranial nerves innervate the skeletal muscles associated with the movements of the eyes and tongue, the muscles used in chewing, and the muscles of facial expression, swallowing, and vocalizing.

Upper motor neurons Lower motor neurons are influenced by two sources: (1) sensory receptors in the body that are integrated into reflexes and (2) **upper motor neurons** from the brain forming the "voluntary descending pathways." The *upper motor neuron pathways* include the (1) corticospinal (pyramidal) and corticobulbar tracts originating from neurons in the motor areas of the cerebral cortex, (2) the rubrospinal tract originating from neurons in the red nucleus in the midbrain, (3) the reticulospinal tracts originating from reticular nuclei of the lower brainstem (pons and medulla oblongata), and (4) the vestibulospinal tracts originating from vestibular

nuclei in the lower brainstem. The nucleus ruber and reticular nuclei receive input from the motor areas of the cerebral cortex and other sources. The vestibular nuclei receive their primary inputs from the vestibular sensors and the cerebellum.

In contrast to a lower motor neuron, which has parts located in both the central and peripheral nervous systems, an upper motor neuron is located wholly within the central nervous system.

Roles of Nuclei and Processing Centers

Each nucleus or processing center consists of (1) the terminal branches of sensory (afferent) axons entering the center, (2) cell bodies and dendrites of neurons whose axons form the tract, and (3) intrinsic neurons whose dendrites, cell bodies, and axons are located wholly within the center. Interactions among neurons within each nucleus cause input information to be altered (*processed*) in some way before it is conveyed to the next center. Many centers receive input from neurons originating in more than one location.

Some pathways contribute to the complex processing within the brain itself. These are neuronal circuits of interconnected sequences of a number of processing centers (some are called *feedback circuits*). They integrate and process information at the higher levels of brain functions, including behavioral activities, complex voluntary movements, and thinking.

FIGURE 14.7 A NEUROMUSCULAR SPINDLE

The drawing shows the encapsulated intrafusal muscle fibers and the extrafusal muscle fibers outside the spindle.

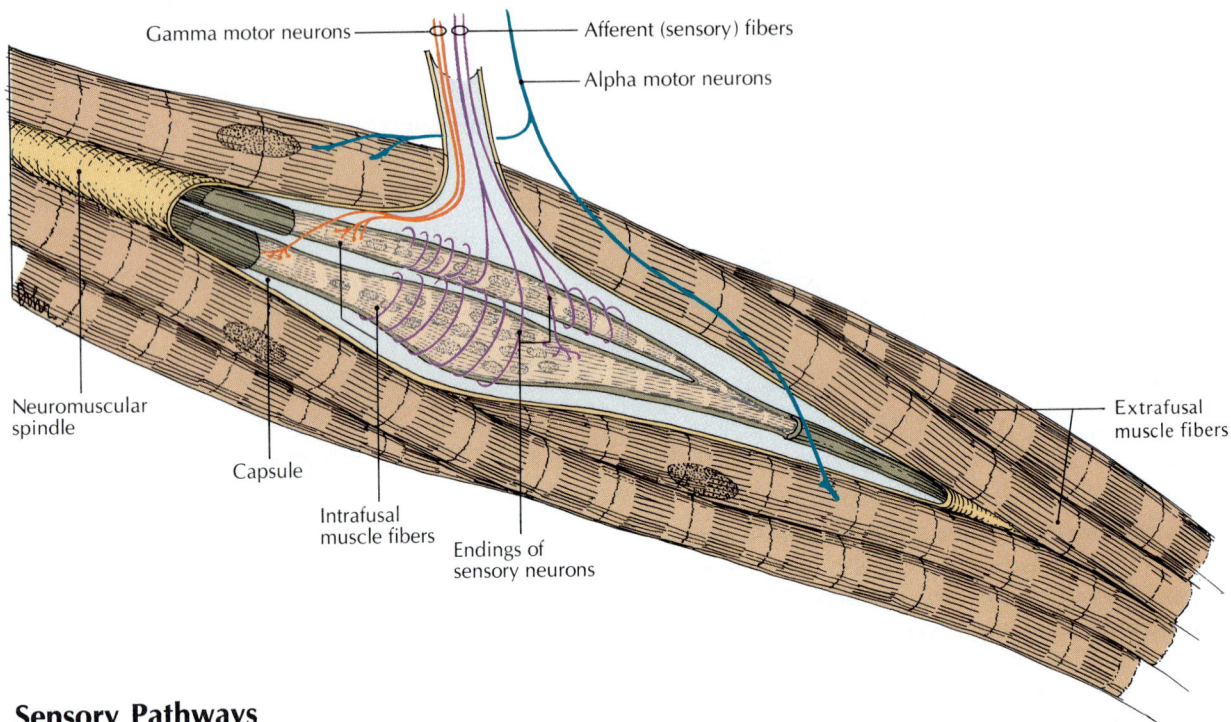

Gamma motor neurons

Afferent (sensory) fibers

Alpha motor neurons

Neuromuscular spindle

Capsule

Intrafusal muscle fibers

Endings of sensory neurons

Extrafusal muscle fibers

Sensory Pathways

Some sensory pathways have sequences that are made up of three neurons. In some pathways, the neurons in the sequence are called first-, second-, and third-order neurons. A *first-order neuron* extends from the sensory receptor to the central nervous system, a *second-order neuron* extends from the spinal cord or brainstem to a nucleus in the thalamus, and a *third-order neuron* extends from the thalamus to a sensory area of the cerebral cortex [FIGURE 14.8].

A critical feature of many pathways is that the axons of a tract (for example, the second-order neuron of many sensory pathways) *cross over (decussate)* from one side of the spinal cord or brainstem to the other side [FIGURE 14.8A, B and TABLES 14.1 and 14.2]. By knowing where a pathway crosses over, a physician can use this information to help locate the site of an injury (lesion) in the central nervous system. For example, the touch-pressure pathway decussates in the medulla oblongata. A lesion in this pathway on one side of the spinal cord results in the loss of sensation on the *same* side (ipsilateral) of the body as the lesion, whereas a lesion above the decussation in the brain results in the loss of sensation on the *opposite* side (contralateral) of the body.

Many tracts are named after their nuclei of origin and termination, as well as their location in the spinal cord. For example, the lateral spinothalamic tract (a tract in a pain pathway) originates in the gray matter of the spinal cord (*spino-*), terminates in the thalamus of the cerebrum

(thalamic), and is located laterally in the spinal cord (*lateral*).

Anterolateral system The *anterolateral system* consists of the lateral spinothalamic tract [FIGURE 14.8A], spinoreticulothalamic pathway [FIGURE 14.8A], and anterior spinothalamic tract [TABLE 14.1]. This system involves the general somatic sensory sensations of pain, temperature, and light touch that are conveyed from the body via the spinal nerves to the spinal cord.

Posterior column-medial lemniscus pathway The *posterior column-medial lemniscus pathway* [FIGURE 14.8B and TABLE 14.1] conveys impulses for touch-pressure, vibratory sense, two-point discrimination, position sense (proprioception), and some light touch. It is composed of (1) first-order neurons whose axons ascend within the posterior columns on the same side (do not decussate) and terminate in the nucleus gracilis and nucleus cuneatus of the lower medulla oblongata and (2) second-order neurons with cell bodies in the nucleus gracilis and nucleus cuneatus and axons that decussate in the lower medulla oblongata and then ascend as the medial lemniscus to the thalamus.

A note about pyramidal and extrapyramidal systems The motor pathways and associated neural circuits

FIGURE 14.8 SENSORY PATHWAYS

[A] The lateral spinothalamic tract and spinoreticulo-thalamic pathway of the anterolateral system. [B] The posterior column-medial lemniscus pathway.

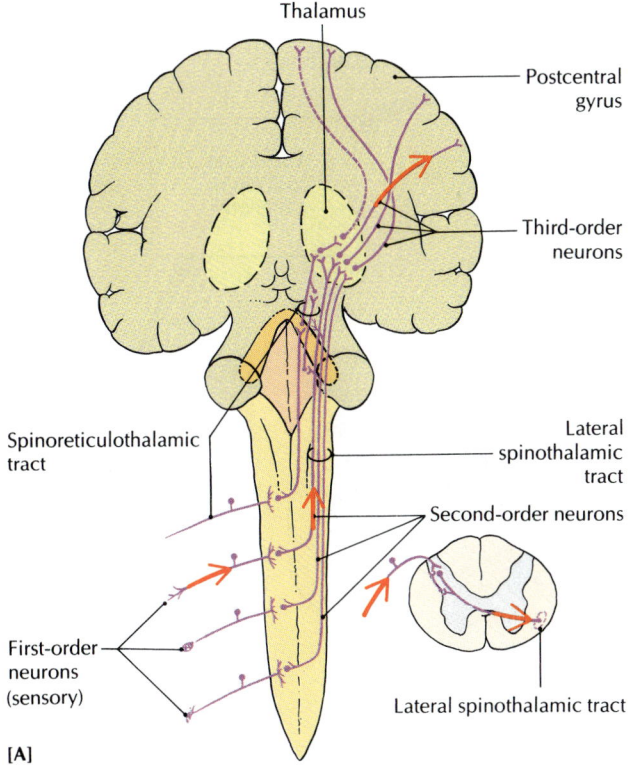

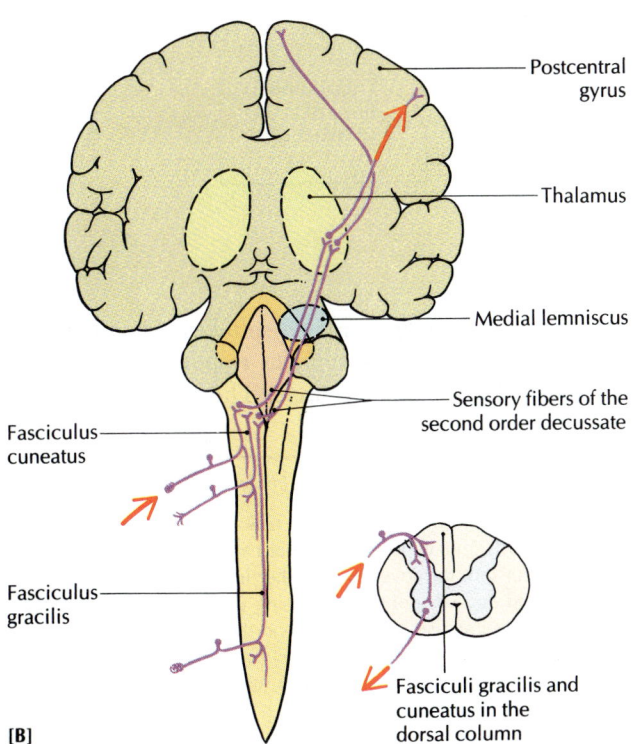

were formerly called the pyramidal and extrapyramidal systems [TABLE 14.2]. The *pyramidal system* (so called because its fibers pass through the pyramids of the medulla oblongata) referred to the corticobulbar and corticospinal tracts, and the *extrapyramidal system* consisted of the other upper motor neurons: the rubrospinal, reticulospinal, and vestibulospinal tracts. Because the pyramidal-extrapyramidal classification has little significance in modern neuroscience and neurology, its use is being discontinued.

SPINAL REFLEXES

In addition to linking the brain and most of the body, the spinal cord coordinates reflex action. A *reflex* (L. to bend back) is a predictable involuntary response to a stimulus, such as quickly pulling your hand away from a hot stove. A reflex involving the skeletal muscles is called a *somatic reflex*. A reflex involving responses of smooth muscle, cardiac muscle, or a gland is a *visceral reflex*. Visceral reflexes influence the heart rate, respiratory rate, digestion, and many other body functions, as described in Chapter 16. Both types of reflex action allow the body to respond quickly to internal and external changes in order to maintain homeostasis.

All *spinal reflexes*, that is, reflexes carried out by neurons in the spinal cord alone and not immediately involving the brain, are based on the sequence shown in FIGURE 14.9. Such a system is called a *reflex arc.* In a *monosynaptic* (one-synapse, two-neuron) *reflex arc*, the sensory and motor neurons synapse directly. More often, however, one or more interneurons (association neurons) synapse with the sensory and motor neurons in a *polysynaptic reflex arc.* An example of a simple polysynaptic reflex is pulling your foot away when you step barefoot on a tack.

Most reflexes in the human body are more complex than the one that occurs when you step on a tack. The more usual reflexes are actually *chains* of reflexes, with the possibility of several skeletal muscles being activated almost simultaneously.

Reflex actions save time because the "message" being transmitted by the impulse does not have to travel from the stimulated receptor all the way to the brain. Instead, most reflex actions never travel any higher than the spinal cord, though some extend to the brainstem.

FIGURE 14.9 PATHWAY OF A SIMPLE THREE-NEURON REFLEX ARC

A reflex arc always starts with a sensory neuron and ends with a motor neuron. The arc pictured here begins with a tack pricking the skin surface of a big toe. The impulse (arrows) travels from the toe to the spinal cord and back to a muscle in the foot, which jerks away from the tack—a flexor, or withdrawal, reflex.

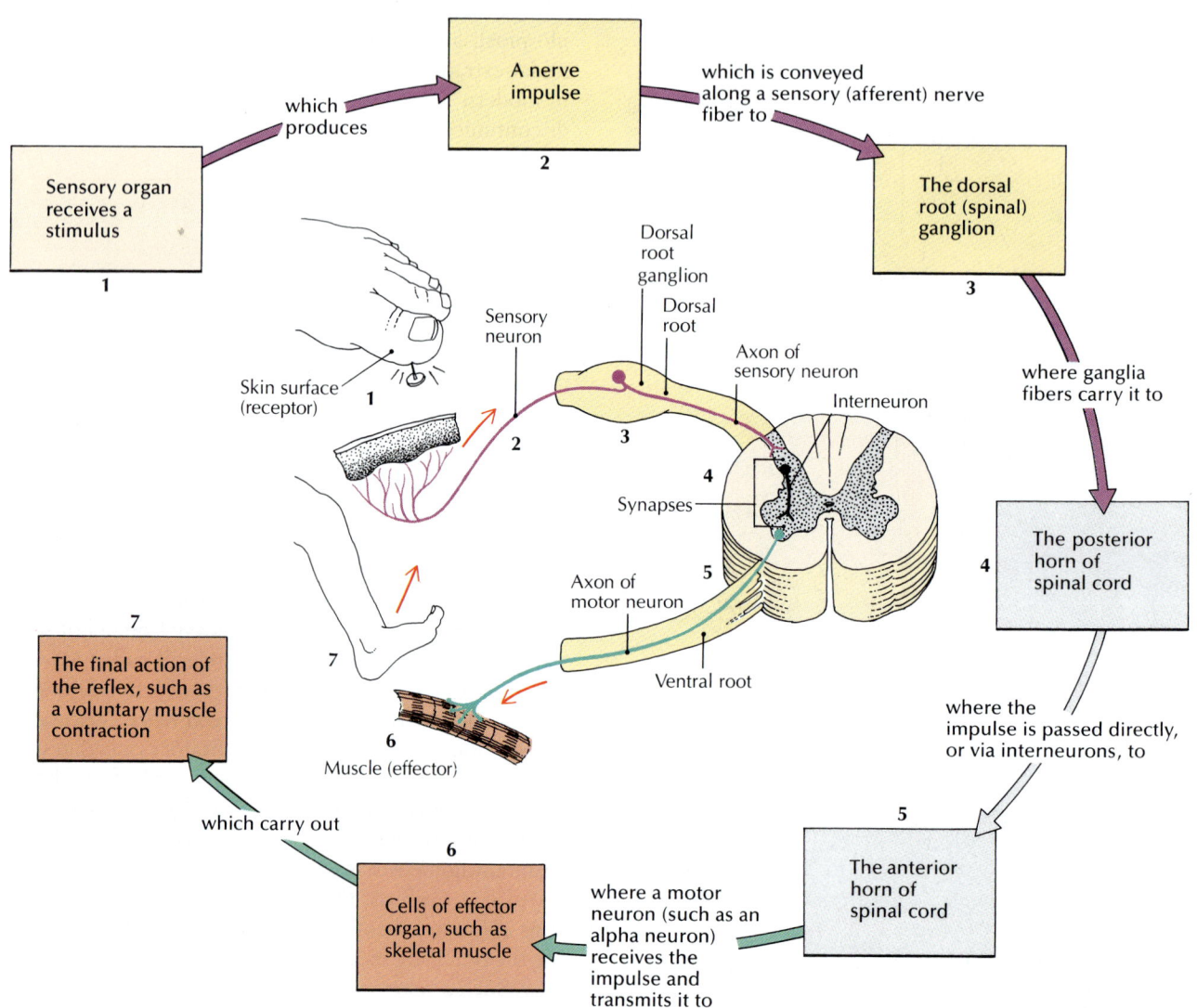

Stretch (Myotatic) Reflex

The **stretch (myotatic) reflex** is a two-neuron (monosynaptic) reflex arc. It acts to maintain erect upright posture and stance by exciting the extensor muscles of the lower limbs, back, neck, and head. A well-known example of this reflex is the *knee jerk,* or *patellar reflex,* which is produced by tapping the patellar tendon of the relaxed quadriceps femoris muscle [FIGURE 14.10]. Such a reflex is described as *ipsilateral (ipsi,* same + *lateral,* side) because the response occurs on the same side of the body and spinal cord where the stimulus is received.

In the knee jerk, a tap on the patellar tendon suddenly stretches the quadriceps femoris tendon and its muscles

and some of the neuromuscular spindles within the muscle. The neuromuscular spindles respond to changes in the length (stretch) of a muscle. The stretched spindles

---

Q: *How "simple" is a so-called simple reflex?*

A: The puncture of your foot by a tack stimulates sensory cells, which stimulate other cells that carry the impulse that causes your foot to move away. Other impulses enable you to hop around on one foot without losing your balance and falling down. Other impulses also go beyond the spinal reflex arc and reach your brain, causing you to feel pain. The reflex response is "simple" only in comparison with other, more complicated responses.

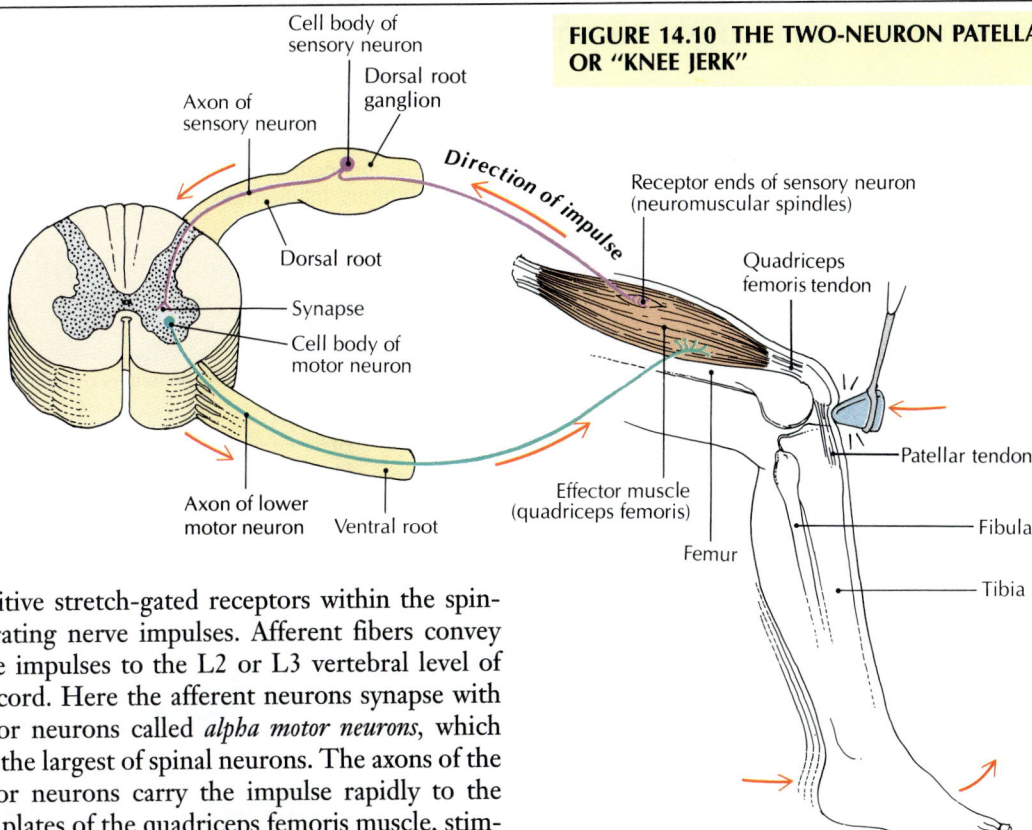

FIGURE 14.10 THE TWO-NEURON PATELLAR REFLEX, OR "KNEE JERK"

excite sensitive stretch-gated receptors within the spindles, generating nerve impulses. Afferent fibers convey these nerve impulses to the L2 or L3 vertebral level of the spinal cord. Here the afferent neurons synapse with lower motor neurons called *alpha motor neurons*, which are among the largest of spinal neurons. The axons of the alpha motor neurons carry the impulse rapidly to the motor end plates of the quadriceps femoris muscle, stimulating it to contract, thus causing the lower leg to swing forward (extend) [FIGURE 14.10].

Gamma Motor Neuron Reflex Arc

The *gamma motor neuron reflex arc* (named after the gamma lower motor neurons that innervate the intrafusal muscle fibers of the neuromuscular spindles) acts to smooth out the movements of muscle contractions or to sustain the contraction of a muscle, as in holding a heavy object, by regulating the state of contraction of the intrafusal fibers of the muscle spindles [FIGURE 14.7]. Stretching of the intrafusal fibers maintains the tension on the muscle spindles during contraction. So the brain always "knows" the state of muscle contraction.

Plantar Reflex

The *plantar reflex* clinically tests the integrity of the spinal cord from L4 to S2, as well as the proper functioning of the corticospinal tracts. The reflex is tested by drawing a blunt instrument down along the lateral aspect of the sole (plantar surface) of the foot. A normal response is a curling or downward flexion of the toes. If there is corticospinal damage, there will be an abnormal reflex response called *Babinski's sign,* in which the great toe flexes upward and the other toes fan laterally. Infants normally have a Babinski sign while their nervous systems are still developing, generally until about 1 year of age.

Withdrawal Reflex Arc

The *withdrawal reflexes* are also known as protective or escape reflexes [FIGURE 14.9]. They are elicited primarily by stimuli for pain and heat great enough to be painful. Withdrawal reflexes are a combination of ipsilateral reflexes [FIGURE 14.11A] and crossed extensor reflexes [FIGURE 14.11B]. For example, when you touch a hot stove or step on a tack, you immediately pull away the injured hand or foot. These reflexes are polysynaptic, and the circuit includes, in order, (1) sensory receptors, (2) afferent neurons, (3) spinal interneurons, (4) alpha motor neurons, and (5) skeletal muscles.

Withdrawal reflexes, which are basically flexor responses, are initiated by a wide variety of receptors, all of which are associated with afferent fibers. Such receptors, located in the skin, muscles, and joints, are called *flexor reflex afferents.*

The flexor reflex afferents can activate circuits within the spinal cord that have synaptic connections with both sides and at different levels of the spinal cord. When the interneurons within the circuit are excited by an intense stimulus, the flexor reflex activity can spread to (1) the opposite side of the spinal cord to influence circuits that evoke responses in the *opposite extremity* and (2) other

FIGURE 14.11 WITHDRAWAL AND CROSSED EXTENSOR REFLEXES

In this highly schematic drawing, the pin-prick stimulus results in a withdrawal (ipsilateral) reflex **[A]** and a crossed extensor (contralateral) reflex **[B]**. In both reflexes there is reciprocal innervation of the agonist and antagonist muscles. Inhibitory interneurons are shown in red.

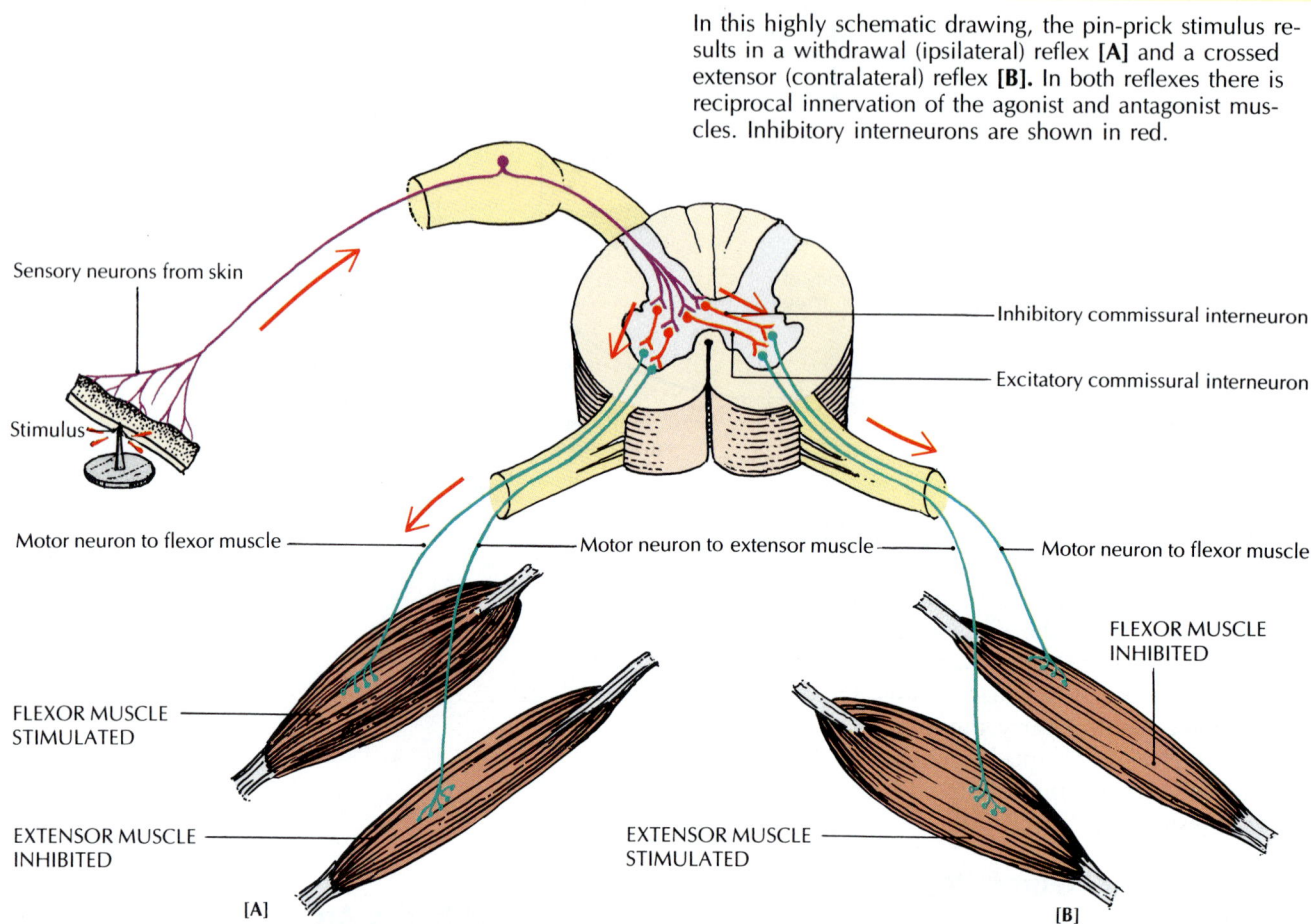

Sensory neurons from skin

Stimulus

Inhibitory commissural interneuron

Excitatory commissural interneuron

Motor neuron to flexor muscle — — Motor neuron to extensor muscle — — Motor neuron to flexor muscle

FLEXOR MUSCLE
INHIBITED

FLEXOR MUSCLE
STIMULATED

EXTENSOR MUSCLE
INHIBITED

EXTENSOR MUSCLE
STIMULATED

[A]

[B]

spinal segments that produce movements of the *other extremity on the same side*. The latter movements involve both flexor and extensor reflexes.

For example, if you pick up a hot frying pan with your right hand, a painful stimulus that reaches the cerebral cortex, you will most likely drop the pan, pull your injured right hand away, and make compensating movements with both your legs and your left arm as you move away from the stove. Such a complex of movements illustrates both the principle of reciprocal innervation and the crossed extensor reflex.

The levels of the spinal cord are in communication through spinal interneurons. Some interneurons, known as *commissural interneurons*, convey influences from one side of the spinal cord to the other side at each spinal level. Within the lumbosacral enlargement levels, these commissural interneurons are involved in the *crossed extensor reflex* [FIGURE 14.11B]. The interneuron circuitry of this reflex is basic in coordinating the activity of the lower limbs during walking and running.

The crossed extensor reflex is important because it

helps the lower limbs support the body. When the right leg moves up from the ground by flexion, the left leg is activated to extend so that it supports the body. Such a reciprocal arrangement occurs when we walk or run. The spinal reflex circuits are also involved in the maintenance of body position and posture by stimulating motor nerves to regulate and sustain tonus of the body musculature.

Functional and Clinical Aspects of Reflex Responses

In a resting skeletal muscle, some of the fibers are always partially contracted because of the continual stimulation to receptors of certain reflex arcs, especially the stretch reflex. This continuous state of contraction is known as *muscle tone* (tonus). Muscle tone can also be described as the minimal degree of contraction exhibited by a normal muscle at "rest." When an examiner passively manipulates a limb in a relaxed patient (flexion and extension), the muscle tone is expressed as the *amount of resistance* perceived by the examiner.

When some of the lower motor neurons or afferent fibers innervating neuromuscular spindles are injured so that there is a reduction in the stimulation of a muscle, the muscle loses some of its tonus, a condition called *hypotonia*. During hypotonia a muscle has a decreased resistance to passive movement. Poliomyelitis (polio) is a viral infection of lower motor neurons. When this condition is followed by an absence of stimuli to the muscles, they lose much or all of their tonus. If a muscle is completely deprived of its motor innervation, it will lose all its tonus, a condition called *atony*. *Hypertonia* is a condition in which a muscle has increased tonus. During hypertonia a muscle has an increased resistance to passive movement, as in the later stages following some strokes.

In addition to changes in tonus, lesions in the nervous system may result in changes in reflex activity (reflexia). *Hyporeflexia* is a condition in which a reflex is less responsive than normal, and *hyperreflexia* is a condition in which a reflex is more responsive than normal. Both conditions are associated with changes in the excitability of the stretch reflex. Hyporeflexia, like hypotonia, is caused by lesions of afferent fibers from the spindles or of the lower motor neurons to a muscle. Hyporeflexia accompanies poliomyelitis.

Hyperreflexia and hypertonia are symptoms of lesions in certain descending upper motor pathways from the brain. It is believed that these lesions remove more inhibitory influences than excitatory ones. Many victims of a stroke exhibit both hypertonia and hyperreflexia as well as Babinski's sign, a reflex that is absent in the normal adult [TABLE 14.3].

Any abnormal reflex response may indicate a disorder of the nervous system. Some of the major diagnostic reflexes are described in TABLE 14.3.

TABLE 14.3 SOME DIAGNOSTIC REFLEXES OF THE CENTRAL NERVOUS SYSTEM

Reflex	Description	Indication
Abdominal reflex	Anterior stroking of the sides of lower torso causes contraction of abdominal muscles.	Absence of reflex indicates lesions of peripheral nerves or in reflex centers in lower thoracic segments of spinal cord; may also indicate multiple sclerosis.
Achilles reflex (ankle jerk)	Tapping of calcaneal (Achilles) tendon of soleus and gastrocnemius muscles causes both muscles to contract, producing plantar flexion of foot.	Absence of reflex may indicate damage to nerves innervating posterior leg muscles or to lumbosacral neurons; may also indicate chronic diabetes, alcoholism, syphilis, subarachnoid hemorrhage.
Biceps reflex	Tapping of biceps tendon in elbow produces contraction of brachialis and biceps muscles, producing flexion at elbow.	Absence of reflex may indicate damage at the C5 or C6 vertebral level.
Brudzinski's reflex	Forceful flexion of neck produces flexion of legs, thighs.	Reflex indicates irritation of meninges.
Hoffmann's reflex	Flicking of index finger produces flexion in all fingers and thumb.	Reflex indicates damage to upper motor neurons.
Kernig's reflex	Flexion of hip, with knee straight and patient lying on back, produces flexion of knee.	Reflex indicates irritation of meninges or herniated intervertebral disk.
Patellar reflex (knee jerk)	Tapping of patellar tendon causes contraction of quadriceps femoris muscle, producing upward jerk of leg.	Absence of reflex may indicate damage at the L2, L3, or L4 vertebral level; may also indicate chronic diabetes, syphilis.
Plantar reflex	Stroking of the lateral part of sole causes toes to curl down. If corticospinal damage, great toes flexes upward and other toes fan out **(Babinski's sign).**	Reflex indicates damage to upper motor neurons. Normal in children less than 1 year old.
Romberg's sign	Inability to maintain balance when standing with eyes closed.	Reflex indicates dorsal column injury.
Triceps reflex	Tapping of triceps tendon at elbow causes contraction of triceps muscle, producing extension at elbow.	Absence of reflex may indicate damage at C6, C7, or C8 vertebral level.

FIGURE 14.12 DEVELOPMENT OF THE SPINAL CORD

Transverse sections at [A] 19 days after fertilization, [B] 21 days, [C] 26 days, [D] 30 days.

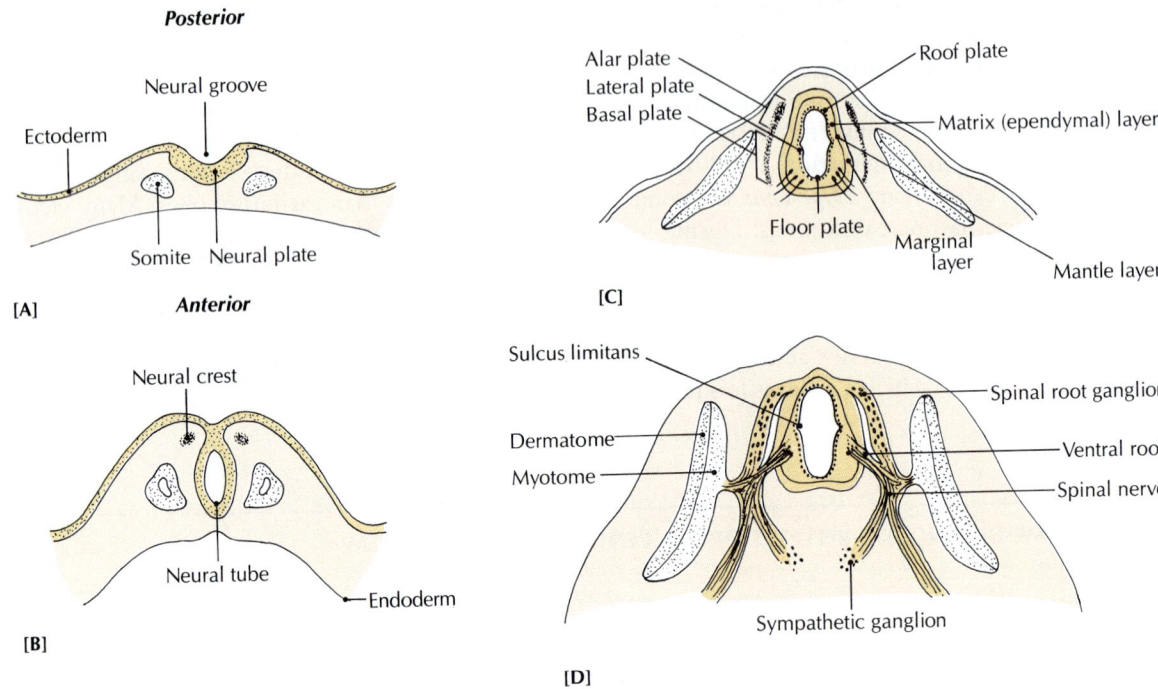

DEVELOPMENTAL ANATOMY OF THE SPINAL CORD

The embryonic development of the nervous system is initiated by the *notochord*, a rod that defines the longitudinal axis of the embryo. The notochord releases a *chemical inducing factor* that induces the embryonic ectoderm to develop into the *neural plate* and *neural crest*. On about day 19, the neural plate folds in on itself to form a *neural groove* with *neural folds* on each side [FIGURE 14.12A]. By day 21, the neural folds have fused, forming the **neural tube** [FIGURE 14.12B]. The cephalic end of the tube develops into the brain, and the caudal portion becomes the spinal cord. At the site of the closure forming the neural tube, there develops a paired segmented series of outgrowths called the **neural crest** [FIGURE 14.12B]. It contains the cells that form all the future sensory neurons, satellite cells, Schwann cells, meninges, some cartilage cells, and postganglionic neurons of the autonomic nervous system [FIGURE 14.12D].

The spinal cord and brain portions of the neural tube differentiate into three layers [FIGURE 14.12C]: (1) The inner *matrix (ependymal) layer*, which lines the central canal of the cord, contains cells called *neuroblasts* that develop into all neurons, and cells called *glioblasts* that develop into the astrocytes and oligodendrocytes of the central nervous system. Most divisions of these cells occur within the matrix layer. Once a potential neuron or macroglial cell leaves the matrix layer, it migrates to its final destination and does not divide. (2) The middle *mantle layer* develops into the butterfly-shaped gray matter, which contains the cell bodies of neurons, including their processes, and macroglia. (3) The outer *marginal layer* develops into the white matter of the spinal cord, which is composed of supporting cells and myelinated and unmyelinated axons. The axons of the white matter ascend to other spinal levels and to the brain, and descend from the brain and to other levels of the spinal cord.

At the outset, the development of the spinal cord and vertebrae proceeds at an even pace, and the spinal cord extends throughout the entire length of the bony vertebral column. But after the third fetal month the vertebral column continues to elongate faster than the spinal cord, and thus the spinal cord becomes relatively shorter than the vertebral column. The spinal cord of a newborn child ends at about the third lumbar vertebra; it extends to between the first and second lumbar vertebrae during adolescence and adulthood. As a result of the differing developmental rates of the spinal cord and vertebral column, the spinal nerves do not align with the intervertebral spaces they pass through, with several of the spinal nerves projecting downward.

WHEN THINGS GO WRONG

Injury and disease can severely impair the functioning of the spinal cord and spinal nerves. Only a few examples are given here.

Spinal Cord Injury

Spinal cord injury is any lesion of the spinal cord that bruises, cuts, or otherwise damages the neurons of the cord. Each year in the United States there are over 10,000 such injuries, most from motor vehicle accidents.

Transection, or severing, of the spinal cord may be complete or incomplete. Complete transections immediately produce total paralysis of all skeletal muscles, loss of bowel or bladder control, loss of reflex activity *(areflexia)*, and a total loss of sensation below the level of the injury. Incomplete transections produce partial loss of voluntary movements and sensations below the level of the injury.

Immediately after a spinal cord injury, *spinal shock* occurs, lasting from several hours to several weeks. Spinal shock involves paralysis, areflexia, and loss of sensation above the level of the injury. In the case of partial transection, spinal shock is followed by a period of spasticity, exaggerated spinal reflexes, and a decreased sensitivity to pain and temperature. Babinski's sign may also appear.

People with incomplete lesions may recover partially, but until recently there was no hope of any restoration of motor and sensory functions with complete transections. Researchers are now experimenting with nerve grafts, nerve regeneration, enzymes, hormones, steroids, growth-associated proteins, and various drugs. (See "New Hope for Spinal Cord Injuries" on page 446.)

Paraplegia is the loss of motor and sensory function in both lower extremities (see drawing). It results from transection of the spinal cord in the thoracic and upper lumbar regions. When the spinal cord is severed completely at a spinal level below the cervical enlargement

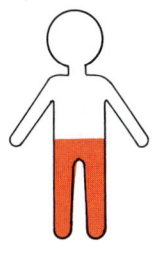  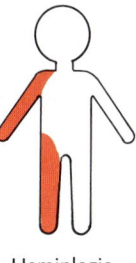

Paraplegia   Quadriplegia   Hemiplegia

The red areas indicate paralysis

and above the upper lumbosacral enlargement, paralysis below the lesion generally occurs immediately afterward, impairing excretory and sexual functions. If damage to the spinal cord is incomplete, however, some sensory and motor capability will remain below the lesion.

Quadriplegia is a paralysis of all four extremities, as well as any body part below the level of injury to the cord. It usually results from injury at the C8 to T1 level. Quadriplegia is more complicated than paraplegia because it affects other body systems. For example, the cardiovascular and respiratory systems may not function properly because of insufficient respiratory muscle action.

Hemiplegia is a paralysis of the upper and lower limbs on one side of the body. It usually results from a stroke following the rupture of a branch of the middle cerebral artery. The resulting lesion interrupts the fibers of the corticospinal tract as they pass through the internal capsule of the cerebrum. Paralysis occurs in the limbs opposite the lesion site because the corticospinal tract decussates in the lower medulla oblongata. Symptoms include spasticity and exaggerated spinal reflexes.

New Hope for Spinal Cord Injuries

In December 1992, New York Jets defensive end Dennis Byrd attempted to tackle the opposing quarterback. Instead, he jammed his head into his own player. Because Byrd's head was down, he compressed his spinal column, fracturing his C5 vertebra. The crushed vertebra traumatized the spinal cord, but did not sever it. Nevertheless, when Byrd regained consciousness he was paralyzed from the chest down.

The typical prognosis for such an injury is quadriplegia; the patient may never walk again. But in April 1993, just five months after the injury, Byrd was able to walk to the pitcher's mound at Shea Stadium and throw out the first ball of the season for the New York Mets baseball team.

This amazing recovery, although still only partial, was made possible by daily sessions of intensive physical therapy, and by use of several new treatments:

1 Within hours after the injury, Byrd received large doses of an anti-inflammatory steroid, methylprednisolone (prednisone). The drug relieved the swelling, reducing pressure on the spinal cord that could have blocked blood vessels and caused further damage. Apparently, prednisone attaches itself to free radicals that are released by injured neurons, preventing the radicals from destroying neurons, first in the central gray matter of the spinal cord and then in the surrounding white matter. The key here is to administer the drug as soon as possible, no later than eight hours after the accident. Proposals are being considered to allow paramedics to inject the drug at the scene of an accident.

2 Soon after Byrd reached the hospital, he underwent seven hours of surgery, during which three titanium plates were implanted to support the spinal column and to prevent further damage to the spinal cord. Although it used to be common to wait a week or so until the swelling subsided before performing surgery, it is now believed that prompt surgery produces a quicker, more thorough recovery and fewer complications.

3 In addition to prednisone, Byrd was given GM-1, a ganglioside produced by the plasma membranes of cells. Apparently, GM-1 helps cells to communicate, prevents further damage to the white matter in the spinal cord, and stimulates nerve repair.

4 In Byrd's case, physicians also prescribed an experimental drug called Sygen to help regenerate damaged neurons. Research is in progress to find ways to clone chemicals that regulate nerve growth, and to use grafts of specialized Schwann cells that manufacture nerve growth factors. Other researchers are experimenting with fetal tissue for the same purpose.

Poliomyelitis

Poliomyelitis is a contagious viral infection that affects both the brain and spinal cord and sometimes causes the destruction of neurons. The poliomyelitis virus shows a preference for infiltrating the lower motor neurons of the spinal cord and brainstem. The initial symptoms may be sore throat and fever, diarrhea, or pain in the back and limbs. In cases of *nonparalytic polio*, these symptoms disappear in less than a week. When motor neurons in the spinal cord are damaged, there is obvious lower motor paralysis of muscles within a few days. Paralysis may be limited to the limbs (especially the lower limbs), or it may also affect the muscles used for breathing and swallowing.

Treatment is supportive rather than curative. Medication is ineffective against the polio virus except as a preventive measure. The Salk vaccine, which became available in 1955, virtually eliminated the disease among those immunized, and the Sabin oral vaccine has been shown to be even more effective and easier to use.

Spinal Meningitis

Spinal meningitis is an inflammation of the spinal meninges, especially the arachnoid and pia mater, which increases the amount of cerebrospinal fluid and alters its composition. The cause may be either viral or bacterial. Meningitis may also follow a penetrating wound or an infection in another area. Meninges infected by bacteria produce large amounts of pus, which infiltrates the cerebrospinal fluid in the subarachnoid space between the arachnoid and pia mater. Pus is not usually formed when the infection is viral.

The first sign of meningitis is usually a headache accompanied by fever, chills, and vomiting. Also evident may be rigidity in the neck region, exaggerated deep tendon reflexes, and back spasms that cause the body to arch upward. Coma may develop. Diagnosis is usually made by analyzing cerebrospinal fluid withdrawn by a lumbar puncture. The fluid is examined for its content of protein, carbohydrates, and bacteria. The chances of recovery are excellent if the disease is diagnosed early and if the infecting microorganisms respond to antibiotics.

CHAPTER SUMMARY

The spinal cord is the connecting link between the brain and most of the body. It also controls many reflex actions. Thus, many voluntary and involuntary actions depend on it.

Basic Anatomy of the Spinal Cord (p. 429)

1 The *spinal cord* extends caudally from the brain for about 45 cm. Its upper end is continuous with the brain.

2 There are 31 pairs of spinal nerves: 8 cervical, 12 thoracic, 5 lumbar, 5 sacral, and 1 coccygeal. The roots of the lumbar and sacral spinal nerves are called the *cauda equina.*

3 The spinal nerves emerging from the *cervical enlargement* of the cord innervate the upper limbs, and the nerves emerging from the *lumbosacral enlargement* of the cord innervate the lower limbs.

4 The spinal cord and spinal nerve roots are protected by the bony vertebral column and its ligaments, the triple-layered *spinal meninges* (inner pia mater, arachnoid, and outer dura mater), and the cerebrospinal fluid.

5 The *gray matter* of the spinal cord consists primarily of nerve cells. Three pairs of nerve cell columns in the gray matter are the *posterior (dorsal) horns, anterior (ventral) horns,* and a median connecting column that forms the anterior and posterior *gray commissure.*

6 Each spinal nerve emerges from the cord as ventral rootlets that form a *ventral root* and as dorsal rootlets that form a *dorsal root.* The cell bodies of neurons whose axons constitute the dorsal roots are located in *dorsal root ganglia.*

7 The *white matter* of the cord is composed mainly of bundles of myelinated nerve fibers. These bundles are *ascending* (sensory) *tracts* and *descending* (motor) *tracts.* The white matter is divided into longitudinal columns called *funiculi.*

Functional Roles of Pathways of the Central Nervous System (p. 436)

1 The brain influences the activity of skeletal muscles through descending *upper motor neurons* from the brain that regulate the activity of *lower motor neurons* of the PNS.

2 Each pathway of the central nervous system can be viewed as consisting of sequences of *processing centers,* with each center having terminal branches of axons entering the center, cell bodies and dendrites of neurons whose axons form the tract, and intrinsic neurons whose dendrites, cell bodies, and axons are located within the center.

3 Sensory pathways may contain *first-order, second-order,* and/or *third-order neurons.*

4 The *anterolateral system* consists of the lateral and anterior spinothalamic tracts and the spinoreticulothalamic pathway. The *posterior column-medial lemniscus pathway* is involved with touch-pressure, vibratory sense, two-point discrimination, position sense, and some light touch.

Spinal Reflexes (p. 439)

1 The spinal cord is involved with spinal reflex actions. A *reflex* is a predictable involuntary response to a stimulus that enables the body to adapt quickly to environmental changes. The system of a sensory cell, an effector cell, and usually one or more connecting nerve cells is a *reflex arc.*

2 The *stretch (myotatic) reflex* is a two-neuron ipsilateral reflex. An example is the knee-jerk reflex. It consists of a neuromuscular spindle, an afferent neuron, an alpha motor neuron, and a skeletal muscle. It plays a role in maintaining body position and is integrated with other spinal reflexes in normal voluntary motor activities.

3 A *gamma motor neuron reflex arc* consists of gamma motor neurons, a neuromuscular spindle, afferent neurons, an alpha motor neuron, and a skeletal muscle. The gamma motor neuron arc acts to smooth the movement of muscle contractions or to maintain the contraction when an object is being held or lifted.

4 The *withdrawal flexor reflexes* are protective escape reflexes. The flexor reflex circuit is composed of sensory receptors, afferent neurons, spinal interneurons, alpha motor neurons, and skeletal muscles.

5 Reflexes used for diagnostic purposes include the patellar reflex (knee jerk), the Achilles reflex, and Babinski's sign.

Developmental Anatomy of the Spinal Cord (p. 444)

1 The embryonic development of the nervous system begins with the formation of the *notochord.* By day 21, the *neural tube* forms, with its cephalic and caudal ends developing into the brain and spinal cord, respectively.

2 The *neural crest* contains the cells that form the future neurons of all spinal ganglia, satellite and Schwann cells of the peripheral nerves, and neurons of autonomic ganglia.

3 The spinal cord portion of the neural tube differentiates into the inner *matrix* (ependymal) *layer,* the middle *mantle layer,* and the outer *marginal layer.*

4 At about month 3, the vertebral column becomes longer than the spinal cord, causing the spinal nerves to be slightly higher than the intervertebral spaces through which they pass.

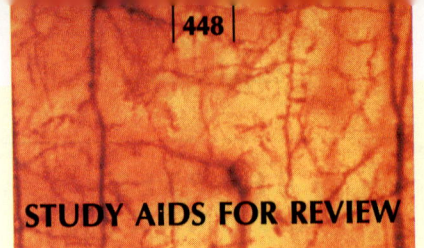

STUDY AIDS FOR REVIEW

KEY TERMS

MEDICAL TERMS FOR REFERENCE

ATAXIA (Gr. "not ordered") Lack of motor coordination as a result of disease of the nervous system or certain genetic disorders.

EPIDURAL ANESTHESIA (caudal block) An injection of anesthesia into the epidural space outside the dura mater, which anesthetizes the nerves of the cauda equina.

HEMATOMYELIA Hemorrhaging within or on the spinal cord.

MYELITIS (Gr. myelo, marrow, spinal cord) A general term for inflammation of the spinal cord.

MYELOGRAM X-ray of the spinal cord obtained by injecting contrast fluid into the subarachnoid space surrounding the cord, a technique called myelography.

SPINA BIFIDA (L. bi, two + fidus, split) Certain congenital defects producing the absence of a neural arch, with varying degrees of protrusion of a portion of the spinal cord or meninges.

SPINAL ANESTHESIA The injection of anesthesia into the epidural space to block the nerves below and numb the lower part of the body.

UNDERSTANDING THE FACTS

1 What are the major functions of the spinal cord?
2 Describe the relationship between the spinal cord and the vertebral column.
3 What are the meninges of the spinal cord? What is the function of each?
4 Describe the relationship between gray matter and white matter in the spinal cord.
5 What is a lumbar puncture, what structures are penetrated, and at what level is it generally performed?
6 Contrast the dorsal and ventral roots with respect to location and function.
7 How do spinal "roots" differ from "horns"?

8 What are the composition and function of dorsal root ganglia?
9 Name the major ascending and descending spinal tracts, indicating point of origin, point of termination, and location within the spinal cord.
10 List in order the components of a simple reflex arc.

UNDERSTANDING THE CONCEPTS

1 What is the role of a processing center in the spinal cord?
2 Why are lower motor neurons called the "final common pathway"?
3 What are the structural and functional relationships between first-, second-, and third-order sensory neurons?
4 Explain the relationship between alpha and gamma motor neurons.
5 Contrast the level of complexity of stretch reflexes and withdrawal reflexes.

SELF-QUIZ
Multiple-Choice Questions

14.1 The major functions of the spinal cord include
  a linking the brain with most of the body
  b its involvement in spinal reflex actions
  c integration of sensory and motor functions in the trunk and limbs
  d a and b
  e a, b, and c

14.2 The spinal cord and the roots of its nerves are supported and protected by
  a the flexible bony vertebral column
  b the filum terminale
  c the dura mater
  d cerebrospinal fluid
  e all of the above

14.3 The gray matter of the spinal cord consists of
  a cell bodies and dendrites of motor neurons
  b unmyelinated axons of spinal neurons
  c interneurons
  d axon terminals of sensory neurons
  e all of the above

14.4 The spinal cord funiculi consist of the
  a anterior column   d commissure area
  b posterior column   e all of the above
  c lateral column

14.5 Processing centers of the CNS consist of
  a terminal branches of axons entering the center
  b cell bodies and dendrites of neurons whose axons form the tract

c intrinsic neurons whose dendrites, cell bodies, and axons are within the center

d a and b

e a, b, and c

14.6 The lateral spinothalamic tract

a originates in the gray matter of the spinal cord

b terminates in the thalamus

c is located laterally in the spinal cord

d is a tract in the pain pathway

e all of the above

14.7 The reflex that helps support the body against gravity is the

a flexor reflex

b extensor reflex

c bireflex

d monosynaptic reflex

e crossed extensor reflex

14.8 The cauda equina consists of

a strands of connective tissue

b nerve roots

c dorsal root ganglia

d strands of gray matter

e denticulate ligaments

14.9 The lateral and anterior corticospinal tracts

a originate in the cerebral cortex

b terminate in the spinal cord

c carry motor impulses

d a and b

e a, b, and c

14.10 The fasciculus gracilis and fasciculus cuneatus

a carry sensory information from the spinal cord to the brain

b are located in the posterior funiculus

c contain cell bodies of sensory neurons

d a and b

e a, b, and c

14.11 Which pairs are both motor tracts?

a spinoreticular and spinocerebellar

b rubrospinal and reticulospinal

c spinoreticular and spinothalamic

d spinothalamic and vestibulospinal

e corticospinal and spinothalamic

14.12 Where does a first-order sensory neuron have its cell body?

a dorsal horn

b ventral horn

c spinal ganglion

d ventral root

e central canal

14.13 What is the minimum number of neurons in a reflex arc?

a none

b one

c two

d three

e four

14.14 Which of the following statements about the patellar reflex is *not correct*?

a It is an ipsilateral reflex.

b It is a stretch reflex.

c The pathway includes no interneurons.

d Alpha motor neurons are stimulated.

e The reflex is initiated when the motor neurons to the quadriceps are stretched.

14.15 Which of the following occurs in a crossed extensor reflex but *not* in a stretch reflex?

a Sense organs are stimulated.

b Sensory neurons are stimulated.

c Commissural interneurons are stimulated.

d Motor neurons are stimulated.

e Muscles are stimulated.

14.16 Lesions in the upper motor neurons of some descending motor pathways are associated with

a hypotonia and hyporeflexia

b hypertonia and hyperreflexia

c hypotonia and hyperreflexia

d hypertonia and hyporeflexia

e atony

Completion Exercises

14.17 The neural terminal end of the spinal cord is called the _____.

14.18 The nonneural fibers at the end of the spinal cord are called the _____.

14.19 Groups of cell bodies whose axons make up the dorsal root lie outside the spinal cord and are called _____.

14.20 In the spinal cord, the white matter is divided into three pairs of columns, or _____.

14.21 In the central nervous system, the crossing over of tracts from one side of the spinal cord or brainstem to the other side is called _____.

14.22 A _____ is a predictable involuntary response to a stimulus.

14.23 The structure that anchors the pia mater to the dura mater is called the _____.

14.24 Cerebrospinal fluid is found in the central canal and the _____.

A SECOND LOOK

Label the following: dorsal root, ventral root, gray matter, white matter, spinal ganglion, spinal nerve, sensory neuron, motor neuron, and interneuron.

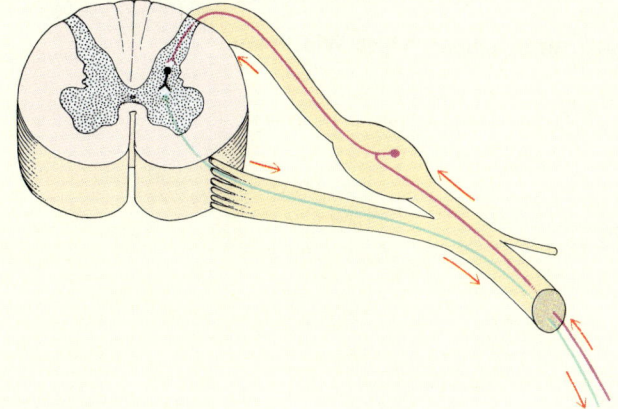

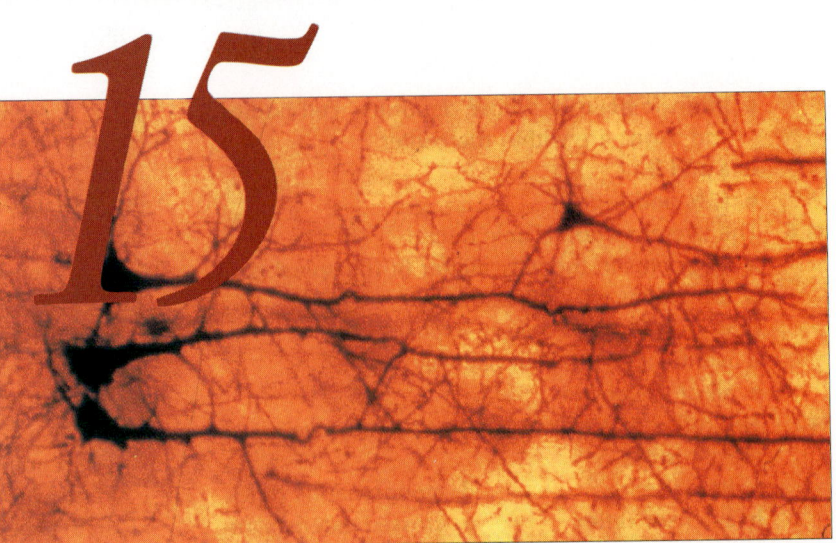

The Peripheral Nervous System

15

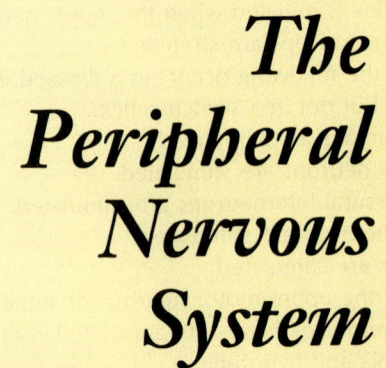

KEY CONCEPTS

1 The peripheral nervous system consists of nerve cells and fibers emerging from and going to the brain and spinal cord.

2 Afferent (sensory) neurons carry impulses from receptors to the central nervous system. Efferent (motor) neurons carry impulses from the brain and spinal cord to muscles and glands.

3 The peripheral nervous system may be divided on a functional basis into the somatic nervous system and the visceral nervous system.

4 Twelve pairs of cranial nerves emerge from the brain.

5 Thirty-one pairs of spinal nerves emerge from the spinal cord. One pair exits at each segment of the spinal cord.

The *peripheral nervous system (PNS)* consists of sensory and motor neurons that emerge from and go to the brain (*cranial nerves*) and those that emerge from and go to the spinal cord (*spinal nerves*). This system allows the brain and spinal cord to communicate with the rest of the body.

CLASSIFICATION OF FUNCTIONAL COMPONENTS OF NERVES

In terms of function, two types of neurons are present in the peripheral nervous system: (1) *afferent* (L. *ad*, toward + *ferre*, to bring), or *sensory, neurons* carry nerve impulses from sensory receptors in the body *to* the central nervous system, where the information is processed, and (2) *efferent* (L. *ex*, away from), or *motor, neurons* convey information *away from* the central nervous system to the effectors (muscles and glands).

The functional components of the neurons in the peripheral nervous system are as follows:

1 *General somatic afferent* neurons carry sensory information from the skin, skeletal muscles, joints, and connective tissues to the central nervous system.

2 *General visceral afferent* neurons carry information from the visceral organs to the central nervous system.

3 *General somatic efferent* neurons carry nerve impulses from the central nervous system to most of the skeletal muscles. These impulses result in the contraction of skeletal muscles.

4 *General visceral efferent* neurons carry impulses from the central nervous system that modify the activities of the heart, smooth muscles, and glands. These are the neurons of the autonomic nervous system, as discussed in Chapter 16.

5 *Special visceral efferent* neurons carry impulses from the brain to the other skeletal muscles. This "visceral" musculature is found in the jaw, muscles of facial expression, pharynx (throat), and larynx.

6 *Special afferent* neurons carry neural information from the receptors of the olfactory (smell), optic (sight), auditory (hearing), vestibular (balance), and gustatory (taste) systems to the central nervous system.

The peripheral nervous system may be divided on a purely functional basis into the *somatic nervous system* and the *visceral nervous system* [FIGURE 15.1]. Each of these systems is composed of an afferent (sensory) division and an efferent (motor) division.

Somatic Nervous System

The *somatic nervous system* is composed of afferent and efferent divisions. The *somatic afferent (sensory) division* consists of nerve cells that receive and process sensory input from the skin, skeletal muscles, tendons, joints, eyes, tongue, nose, and ears. This input is conveyed to the spinal cord and brain via the spinal and some cranial nerves and is utilized by the nervous system at an unconscious level. On a conscious level, the sensory input is perceived as sensations such as touch, pain, heat, cold, balance, sight, taste, smell, and sound.

The *somatic efferent (motor) division* is composed of lower motor neurons that conduct impulses from the CNS to skeletal muscle. When these lower motor neurons are stimulated, they always excite (never inhibit) the skeletal muscles to contract. This system regulates the "voluntary" contraction of skeletal muscles. (As you saw in Chapter 10, not all such activity is actually voluntary or under our conscious control.)

Visceral Nervous System

The *visceral nervous system* is also composed of afferent and efferent divisions. The *afferent (sensory) division* includes the neural structures involved in conveying sensory information from sensory receptors in the visceral organs of the cardiovascular, respiratory, digestive, urinary, and reproductive systems. Some of the input from these sensory receptors may be utilized on a conscious level and perceived as sensations such as pain, intestinal discomfort, urinary bladder fullness, taste, and smell. The *efferent (motor) division*, more commonly known as the *autonomic nervous system*, includes the neural structures involved in the motor activities that influence smooth muscle, cardiac muscle, and glands of the skin and viscera. The autonomic nervous system is dealt with in greater detail in Chapter 16.

A S K   Y O U R S E L F

1 What are the main components of the peripheral nervous system?

2 What are the major differences between the somatic and visceral nervous systems?

FIGURE 15.1 ORGANIZATION OF THE PERIPHERAL NERVOUS SYSTEM

A schematic drawing of the interrelationships between the various parts of the peripheral nervous system.

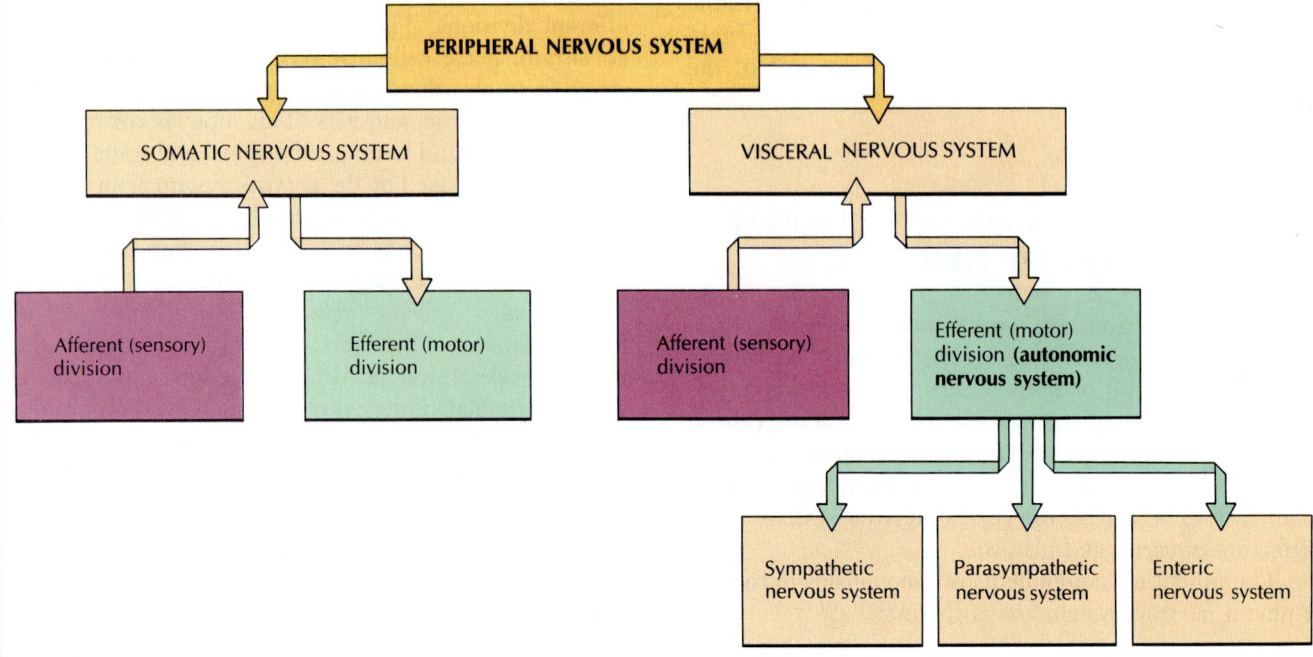

Division	Functions
PERIPHERAL NERVOUS SYSTEM (PNS)	
Cranial and spinal nerves, with afferent (sensory) and efferent (motor) nerve cells.	Enables brain and spinal cord to communicate with entire body. *Afferent (sensory) cells:* Carry impulses from receptors to CNS. *Efferent (motor) cells:* Carry impulses from CNS to effectors (muscles and glands).
Somatic nervous system	
1 Composed of nerve fibers (axons) conveying information from receptors for pain, temperature, and mechanical stimuli in somatic structures such as skin, muscles, and joints. 2 Characterized by nerve fibers of lower motor neurons that go directly from CNS to synapses with skeletal muscle.	*Afferent (sensory) division:* Receives and processes sensory input from skin, skeletal muscles, tendons, joints, eyes, ears. *Efferent (motor) division:* Excites skeletal muscles.
Visceral nervous system	
1 Composed of nerve fibers conveying information from receptors in visceral organs (structures) such as the heart and digestive system. 2 Characterized by nerve fibers of motor neurons that go from CNS to interact with other nerve cells within a ganglion located outside CNS; nerve fibers of second nerve cells that innervate the effectors. 3 Includes the autonomic nervous system.	*Afferent (sensory) division:** Receives and processes input from internal organs of cardiovascular, respiratory, digestive, urinary, and reproductive systems. *Efferent (motor) division, or autonomic nervous system:* May inhibit or excite smooth muscle, cardiac muscle, glands.
Sympathetic nervous system	Relaxes gastric and intestinal wall muscles; increases sweating, heart rate, blood flow to skeletal muscles.
Parasympathetic nervous system	Contracts gastric and intestinal wall muscles; decreases sweating, heart rate, blood flow to skeletal muscles.
Enteric nervous system	Coordinates activity of muscles and glands of gastric and intestinal walls.

* Some textbooks include the general visceral afferent (sensory) division in the autonomic nervous system.

CRANIAL NERVES

The 12 pairs of *cranial nerves* are the peripheral nerves of the brain. Their names are an indication of some of their anatomical or functional features, and their numbers (in Roman numerals) indicate the sequential order in which they emerge from the brain.

Cranial nerves I and II are nerves of the cerebrum that originate from the forebrain, and nerves III through XII are nerves of the brainstem. (Part of nerve XI emerges from the cervical spinal cord.) Of the ten brainstem nerves, one (VIII) is a purely sensory nerve, four are primarily motor nerves with some proprioceptive fibers (III, IV, VI, and XII), and five are mixed nerves containing both sensory and motor fibers (V, VII, IX, X, and XI) [FIGURE 15.2].

The *motor (efferent) axons* of the cranial nerves emerge from the brainstem. They arise from groups called *motor nuclei,* which are stimulated by nerve impulses from many outside sources, including the cerebral cortex and the sense organs. These axons have two roles. They either (1) stimulate skeletal muscles or (2) synapse with ganglia of the autonomic nervous system. These ganglia in turn relay nerve impulses to cardiac muscle, smooth muscle, or glands.

The *sensory (afferent) axons* of cranial nerves emerge from neurons with cell bodies outside the brain. These cell bodies are found in *sensory ganglia,* which are groups of cell bodies situated on the trunks of cranial nerves, or are located in peripheral sense organs, such as the inner ear. Axons of sensory neurons enter the brain, synapse with interneurons, and produce the appropriate sensation (hearing, for example, in the case of the ear).

The cranial nerves are concerned with the senses of smell, taste, vision, hearing, and balance. They are also involved with the motor activities of eye movement, chewing, swallowing, breathing, speaking, and facial expression.

Cranial Nerve I: Olfactory

The *olfactory nerve* (L. *olfacere,* to smell) is strictly a sensory nerve. The 10 million to 20 million bipolar neurons (*olfactory cells*) of the olfactory nerve are located high in the nasal cavities within nasal epithelium. They act both as chemoreceptors that sense odors and as conductors of impulses that ultimately result in the perception called smell. The unmyelinated axons of olfactory cells pass through the foramina of the cribriform plate of the ethmoid bone and synapse with axons of neurons in the *olfactory bulb,* which is actually an "appendage" of the brain [FIGURE 15.3]. The axons of these olfactory bulb neurons form the *olfactory tract,* which conveys impulses to the primary olfactory cortex of the cerebrum.

Olfactory neurons are the only neurons that are replaced by new neurons throughout life. Each olfactory neuron has a life span of only about 30 days and is replaced by "basal cells" in the olfactory mucosa. These basal cells are continuously differentiating into neurons that form new synaptic connections in the olfactory bulb.

Cranial Nerve II: Optic

The *optic nerve* is a special sensory nerve. It conveys impulses that result in vision and in reflexes associated with vision. Each optic nerve is actually a tract composed of about a million axons that arise from the ganglion cells of each retina. In the retina, rods and cones (the photoreceptive cells), interneurons, and bipolar neurons form a circuitry that interacts with ganglion neurons.

The retina and the optic nerve are unusual in that the retina is the only mobile portion of the brain and the optic nerve is not a true peripheral nerve. The retina is actually a portion of the brain. Both are derived from a common embryologic origin, the neural tube. The cells of the neuroretina of the optic cup [see FIGURE 17.24] differentiate into the cells of the retina, including its glial cells (called Muller's cells).

The axons of the ganglion cells of the retina are the myelinated nerve fibers that compose the optic nerve, chiasma, and tract. The myelin of the optic nerve axons is formed by oligodendrocytes (as are all myelinated fibers of the tracts of the central nervous system), not by Schwann cells (as are all myelinated fibers of nerves of the peripheral nervous system). Hence the optic nerve is a *tract* and not a true peripheral nerve.

Each optic nerve passes out of the orbit through the optic foramen into the cranial cavity [FIGURE 15.4]. While passing through the X-shaped *optic chiasma* (kye-AZ-muh; so named because its X shape reminded early anatomists of the Greek letter *chi* [KYE], which also has an X shape), the axons of the medial half of the retinas of both eyes decussate across the midline. The axons from the lateral half of the retinas of both eyes do not cross the midline. One-half of the fibers of each optic nerve of each eye then cross over to the other side to form the optic chiasma. After passing through the optic chiasma, the axons continue, forming the two *optic tracts.*

The axons of each optic tract terminate mainly in the lateral geniculate body of the thalamus. They synapse with neurons that project as the optic radiations to the primary visual cortex in the occipital lobe of the cerebrum. Other fibers from the retina terminate in the superior colliculus of the midbrain. The midbrain, through its connections with cranial nerves III, IV, and VI, is involved with subconscious visual reflexes (eye movements and pupillary responses).

FIGURE 15.2 ORIGINS OF THE CRANIAL NERVES

Basal view of the brain (center), showing the distribution of cranial nerves to relevant body parts.

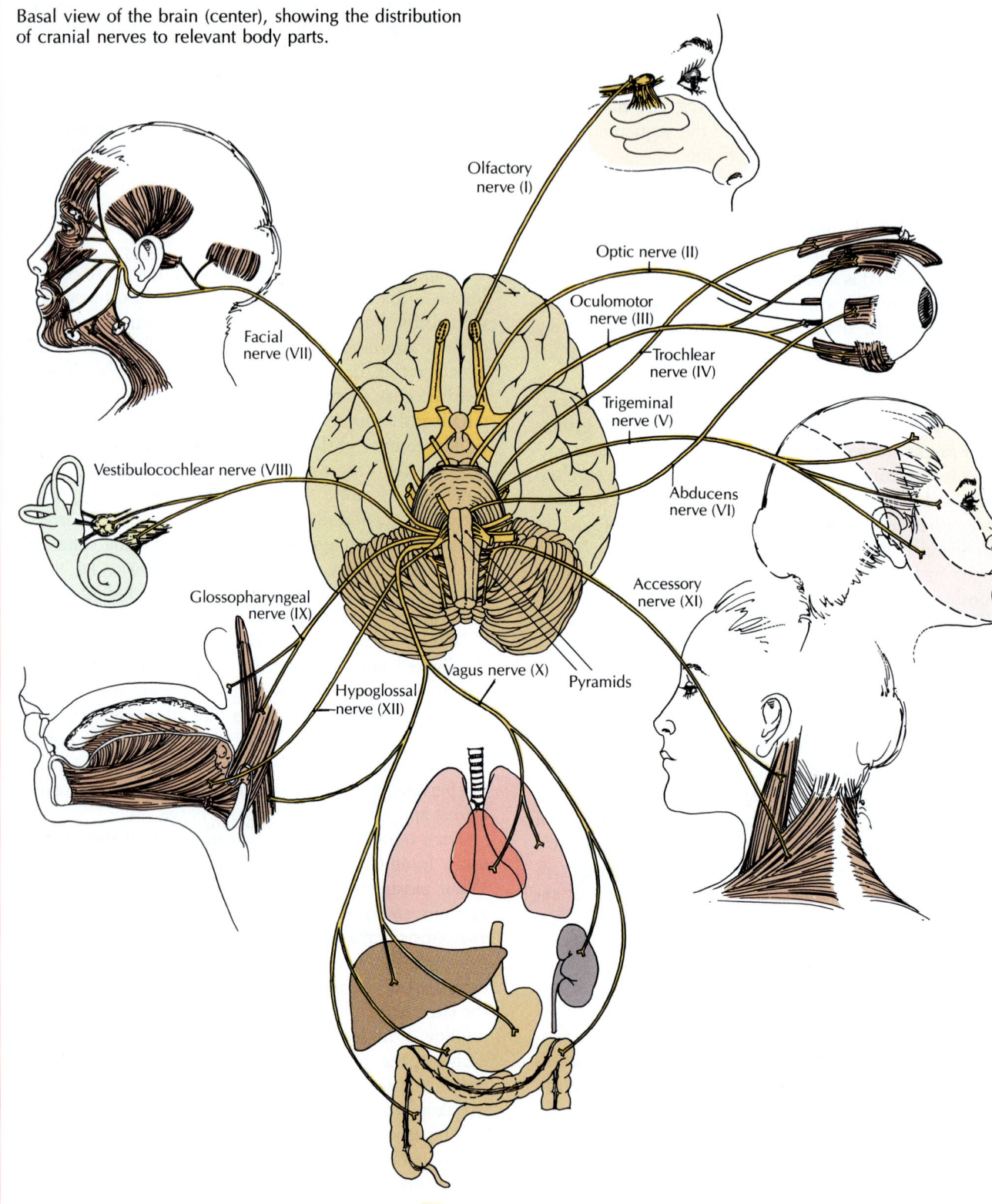

Olfactory nerve (I)

Optic nerve (II)

Oculomotor nerve (III)

Trochlear nerve (IV)

Trigeminal nerve (V)

Facial nerve (VII)

Abducens nerve (VI)

Vestibulocochlear nerve (VIII)

Accessory nerve (XI)

Glossopharyngeal nerve (IX)

Vagus nerve (X)

Pyramids

Hypoglossal nerve (XII)

Nerve/type	Origin	Distribution	Function
I Olfactory (sensory) [FIGURE 15.3]	Nasal mucous membrane high in nasal cavities	Terminates in olfactory bulb of cerebrum.	Smell (olfaction).
II Optic (sensory) [FIGURE 15.4]	Retina of eye	Terminates in lateral geniculate body of thalamus and superior colliculus of midbrain.	Vision. Afferent limb of reflex of focusing by adjusting lens and constricting pupil.

Nerve/type	Origin	Distribution	Function
III Oculomotor (motor) [FIGURE 15.4]	Midbrain	To all extrinsic muscles of eyeball except superior oblique and lateral rectus; also autonomic fibers to ciliary muscles of lens and constrictor muscles of iris.	Movements of eyeball, elevation of upper eyelid, constriction of pupil, focusing by lens (accommodation).
IV Trochlear (motor) [FIGURE 15.4]	Caudal midbrain	Innervates superior oblique muscle of eye.	Eye movements (down and out).
V Trigeminal [FIGURE 15.5]			
Ophthalmic nerve (V^1) sensory)	General area of forehead, eyes	General area of forehead, eyes.	Conveys general senses from cornea of eyeball, upper nasal cavity, front of scalp, forehead, upper eyelid, conjunctive, lacrimal (tear) gland.
Maxillary nerve (V^2) (sensory)	General area of maxillary region	General area of maxillary region.	Conveys general senses from cheek, upper lip, upper teeth, mucosa of nasal cavity, palate, parts of pharynx.
Mandibular nerve (V^3) (mixed)	*Sensory:* General area of mandibular region	*Sensory:* general area of mandibular region.	*Sensory:* conveys general senses from tongue (not taste), lower teeth, skin of lower jaw.
	Motor: pons	*Motor:* innervates muscles of mastication.	*Motor:* chewing.
VI Abducens (motor) [FIGURE 15.4]	Caudal pons	Innervates lateral rectus muscle of eye.	Abduction of eye (lateral movement).
VII Facial (mixed) [FIGURE 15.6]	Pons	*Sensory:* innervates taste buds of anterior two-thirds of tongue.	*Sensory:* taste.
		Motor: innervates muscles of facial expression; autonomic fibers to salivary glands, lacrimal glands.	*Motor:* salivation, lacrimation, movement of muscles of facial expression.
VIII Vestibulocochlear [FIGURE 15.7]			
Cochlear (sensory)	Cochlea of inner ear	Cochlea of inner ear.	Hearing.
Vestibular (sensory)	Semicircular ducts, utricle, and saccule of inner ear	Semicircular ducts, utricle and saccule of inner ear.	Equilibrium.
IX Glossopharyngeal (mixed) [FIGURE 15.8]	Medulla oblongata	*Sensory:* conveys taste from posterior third of tongue, general senses from upper pharynx.	Taste, other sensations of tongue.
		Motor: innervates stylopharyngeus muscle; autonomic fibers stimulate parotid gland.	Secretion of saliva; swallowing.
X Vagus* (mixed) [FIGURE 15.9]	Medulla oblongata	Voluntary muscles of soft palate, cardiac muscle, smooth muscle in respiratory, cardiovascular, digestive systems.	Swallowing. Monitors oxygen and carbon dioxide concentrations in blood, senses blood pressure, other visceral activities of affected systems.
XI Accessory* (spinal accessory) (motor) [FIGURE 15.10]	Medulla oblongata, cervical spinal cord	Muscles of larynx, sternocleidomastoid, trapezius.	Voice production (larynx); muscle sense; movement of head, shoulders.
XII Hypoglossal† (motor) [FIGURE 15.11]	Medulla oblongata	Tongue muscles.	Movements of tongue during speech, swallowing; muscle sense.

*Some fibers of spinal accessory nerve innervating muscles of the larynx join the vagus nerve; hence, these muscles are usually said to be innervated by the vagus nerve.

†Contains some sensory proprioceptive fibers (from extraocular muscles of eyes and tongue muscles), which leave nerve and join trigeminal nerve (see illustration).

FIGURE 15.3 DISTRIBUTION OF THE OLFACTORY NERVE (I)

Right lateral view.

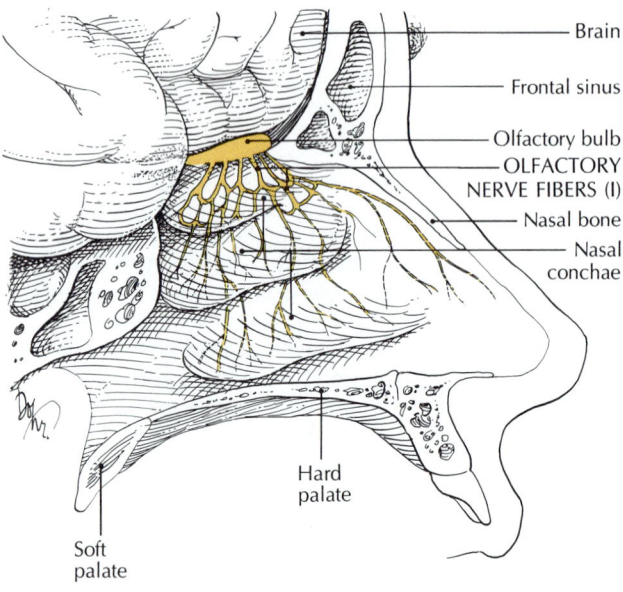

Cranial Nerve III: Oculomotor

Cranial nerves III (oculomotor), IV (trochlear), and VI (abducens) are classified as the *extraocular motor nerves* because they innervate the extraocular (extrinsic) muscles that move the eyeball. The *oculomotor nerve* (L. *oculus*, eye + motor, or "eye movement") innervates four of the muscles that move the eye. In addition, it innervates the levator palpebrae superioris muscles, which elevate the upper eyelids. The parasympathetic fibers of the oculomotor nerves innervate the smooth muscles within the eyeball that constrict the pupil (usually in response to bright light) and also the ciliary muscles, which regulate the tension on the lens for focusing (accommodation).

The brainstem motor nucleus of the oculomotor nerve is located in the midbrain [FIGURE 15.4]. The oculomotor nerve emerges from the ventral surface of the midbrain, passes through the superior orbital fissure, and enters the orbit. The sensory fibers from the proprioceptors (neuromuscular spindles) in the muscles innervated by cranial nerves III, IV, and VI join the ophthalmic nerve and have their cell bodies in the trigeminal ganglion [FIGURE 15.5].

Cranial Nerve IV: Trochlear

The *trochlear nerve* (L. pulley) is so named because it innervates the superior oblique muscle, whose tendon passes through a cartilaginous pulleylike sling on its way to its insertion into the sclera (outer covering) of the eyeball. The nerve arises from its nucleus in the caudal midbrain, emerges from the dorsal surface of the midbrain, and passes forward through the superior orbital fissure into the orbit, where it innervates the superior oblique muscle. The trochlear nerve is long and slender and follows a winding course [FIGURE 15.4].

When the trochlear nerve is injured, double vision may result from lack of motor control, especially when looking downward. As a result, a person with an injured trochlear nerve might have trouble walking downstairs or stepping off a curb into the street. (See TABLE 15.1 for an outline of some major disorders of the cranial nerves.)

Cranial Nerve V: Trigeminal

The *trigeminal nerve* (L. *tri*, three + *geminus*, twin, referring to its three major branches and two roots) is the largest, but not the longest, cranial nerve [FIGURE 15.5]. It is a mixed nerve, being the chief sensory nerve of the face and oral and nasal regions and the motor nerve of the chewing muscles. The sensory nerves convey impulses of the general senses, similar to those found in the spinal nerves. All the sensory fibers terminate in trigeminal nuclei in the brainstem. The motor fibers originate from the trigeminal motor nucleus in the pons.

FIGURE 15.4 DISTRIBUTION OF THE OPTIC, OCULOMOTOR, TROCHLEAR, AND ABDUCENS NERVES (II, III, IV, VI)

Right lateral view.

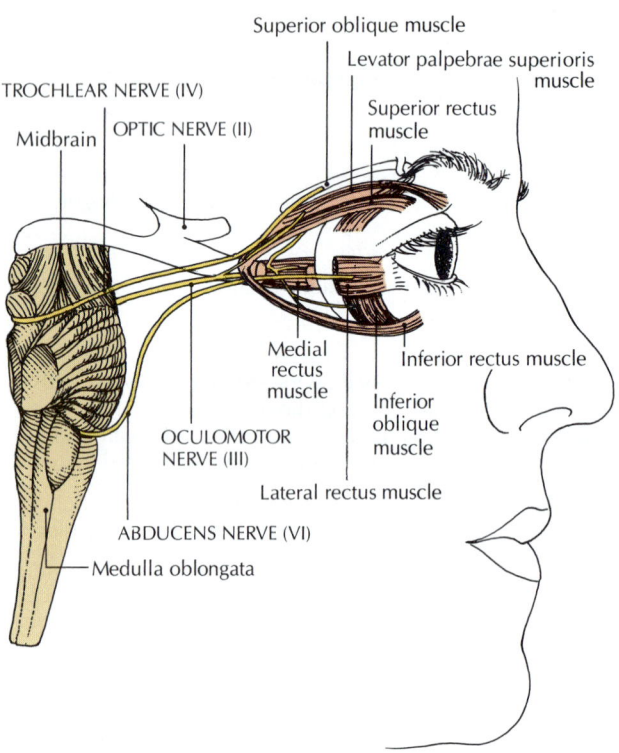

FIGURE 15.5 DISTRIBUTION OF THE TRIGEMINAL NERVE (V)

[A] Motor nerves. [B] Sensory nerves. Right lateral views.

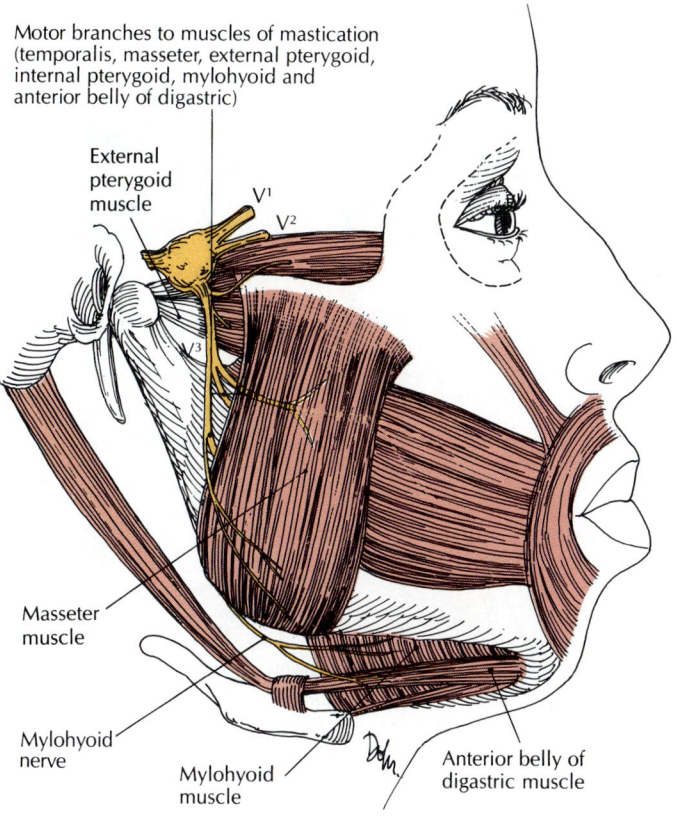

Motor branches to muscles of mastication (temporalis, masseter, external pterygoid, internal pterygoid, mylohyoid and anterior belly of digastric)

External pterygoid muscle

V¹

V²

V³

Masseter muscle

Mylohyoid nerve

Mylohyoid muscle

Anterior belly of digastric muscle

[A]

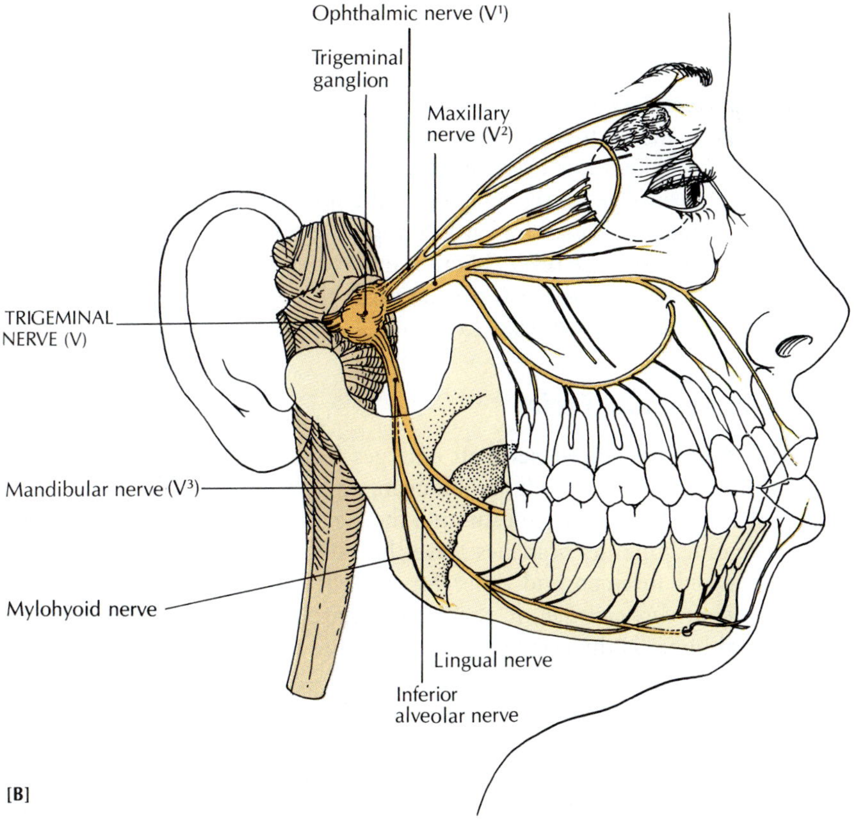

Ophthalmic nerve (V¹)

Trigeminal ganglion

Maxillary nerve (V²)

TRIGEMINAL NERVE (V)

Mandibular nerve (V³)

Mylohyoid nerve

Lingual nerve

Inferior alveolar nerve

[B]

The sensory root has a sensory ganglion located close to where the nerve emerges from the pons within the middle cranial fossa. The cell bodies of the sensory nerves are located in this *trigeminal ganglion* (the equivalent of the dorsal root ganglion of a spinal nerve).

After the trigeminal nerve emerges from the pons, it divides into three branches, the ophthalmic, maxillary, and mandibular nerves [FIGURE 15.2]:

1 The *ophthalmic nerve* (V^1),* a sensory nerve, passes through the superior orbital fissure, through the orbit, and into the upper head. Its branches convey the general senses from the front of the scalp, forehead, upper eyelid, conjunctiva (inner membrane of the eyelid), cornea of the eyeball, and upper nasal cavity.

2 The *maxillary nerve* (V^2), a sensory nerve, passes through the foramen rotundum. Its branches convey general sensations from the skin of the cheek, upper lip, upper teeth, mucosa of the nasal cavity, palate, and parts of the pharynx. It covers the general area of the maxillary bone.

3 The *mandibular nerve* (V^3), a mixed nerve, passes from the skull through the foramen ovale. Its branches convey general senses from the mucosa of the mouth (including the tongue but not sensations of taste), lower teeth, and skin around the lower jaw. The motor fibers innervate the chewing muscles. The mandibular nerve is distributed to the general region of the mandible. (See "Open Wide" on page 459.)

Cranial Nerve VI: Abducens

The *abducens nerve* (L. *abducere*, to lead away) is an extraocular motor nerve, together with cranial nerves III and IV. The brainstem nucleus is located in the caudal pons. The nerve emerges at the junction of the pons and medulla oblongata and passes forward through the superior orbital fissure into the orbit, where it innervates the lateral rectus muscle to *abduct* (turn outward) the eye, hence the name *abducens* [FIGURE 15.4].

Cranial Nerve VII: Facial

The *facial nerve* is a mixed nerve. It emerges from the junction of the pons and medulla oblongata and passes through the internal auditory meatus and the facial canal of the temporal bone; its branches leave the skull from the stylomastoid foramen and other foramina [FIGURE 15.6]. The *geniculate ganglion*, in the petrous portion of the temporal bone, is the sensory ganglion of the facial nerve.

The sensory fibers of the chorda tympani nerve innervate the taste buds of the anterior two-thirds of the

*Branches of cranial nerves are designated by superscript numerals in the order of their branching from the main cranial nerve.

FIGURE 15.6 DISTRIBUTION OF THE FACIAL NERVE (VII)

Right lateral view.

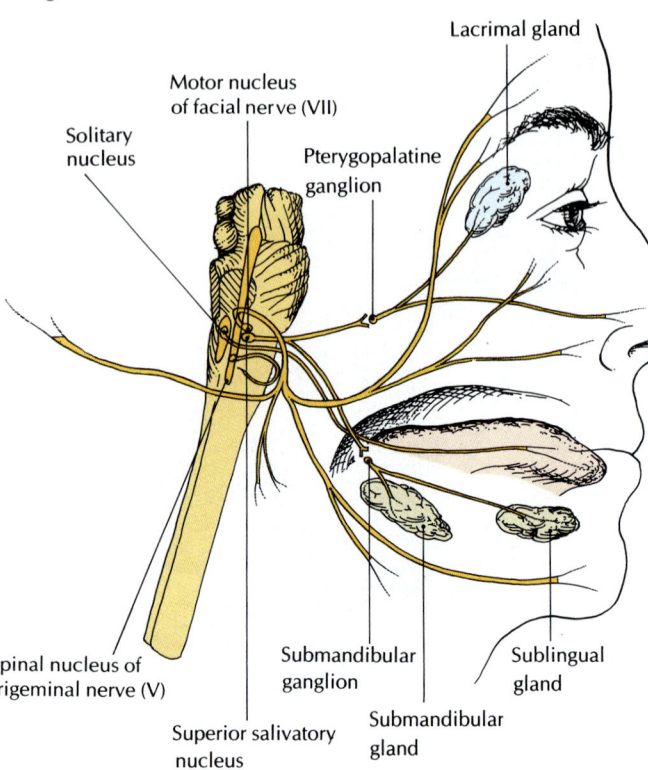

Motor nucleus of facial nerve (VII)

Solitary nucleus

Lacrimal gland

Pterygopalatine ganglion

Spinal nucleus of trigeminal nerve (V)

Superior salivatory nucleus

Submandibular ganglion

Submandibular gland

Sublingual gland

tongue. The fibers terminate in the solitary tract nucleus in the brainstem. Fibers project from this nucleus into the thalamus where they synapse with neurons that terminate in the gustatory (taste) area of the cerebral cortex.

The motor fibers originate from the motor nucleus of the facial nerve in the pons and innervate all the muscles of facial expression. Parasympathetic (autonomic nervous system) efferent fibers originating in the pterygopalatine and submandibular ganglia innervate the lacrimal (tear) glands, the submandibular and sublingual salivary glands, and glands of the nasal and oral mucosa.

Cranial Nerve VIII: Vestibulocochlear

The *vestibulocochlear nerve* (L. entranceway + snail shell) is a sensory nerve composed of two branches: the cochlear nerve and the vestibular nerve. The cochlear nerve conveys impulses concerned with hearing, and the vestibular nerve conveys information about equilibrium or balance and the position and movements of the head.

The fibers of the *cochlear nerve* arise at their synapses with the hair cells of the spiral organ in the snail-shaped cochlea of the inner ear [FIGURE 15.7]. The cell bodies of the cochlear nerve are located in the spiral ganglion adja-

"Open Wide"

A toothache on the lower left-hand side of your mouth brings you to your dentist. After being x-rayed, you learn that the filling in your tooth is cracked and that you may have an abscess (a pus-filled localized area of infection).

Before repairing the tooth, your dentist gives you some anesthetic. The dentist first applies a topical anesthetic on your gums, where a needle will be inserted to administer a major anesthetic. The anesthetic desensitizes the *inferior alveolar branch of the trigeminal nerve* [FIGURE 15.5B]. The dentist takes care not to actually touch the nerve with the needle. Touching the nerve would give

you an "electric shock," because the nerve would be depolarized instantly. The dentist also avoids the blood vessels adjacent to the nerve trunk, pulling back once or twice on the syringe to make sure no blood is drawn out.

Now the anesthetic diffuses through the soft tissue along the forward path of the inferior alveolar nerve. Branches from the nerve radiate to each tooth on the lower left-hand side, the chin, and the lower lip [FIGURE 15.5]. The anesthetic will also reach the nearby *lingual nerve*, anesthetizing the left lateral edge of your tongue. You can feel the numb-

ness move from the back of your mouth toward the front, following the course of the nerve to the midline of your lower lip.

In a few minutes, your tooth is completely insensitive to pain, and your dentist takes care of the problem without causing you any discomfort. Your only problem now is being able to feel the normal sensations again. As you get up to leave, your dentist reminds you not to test the numbness by biting your tongue or grinding your teeth before the feeling returns, since either action could cause a serious injury in the absence of pain.

---

cent to the cochlea. The axons of the cochlear neurons terminate in the cochlear nuclei of the upper medulla oblongata. Auditory pathways from the cochlear nuclei terminate in the medial geniculate body of the thalamus. From the thalamus, impulses are relayed via fibers of the

FIGURE 15.7 DISTRIBUTION OF THE VESTIBULOCOCHLEAR NERVE (VIII)

Right lateral view.

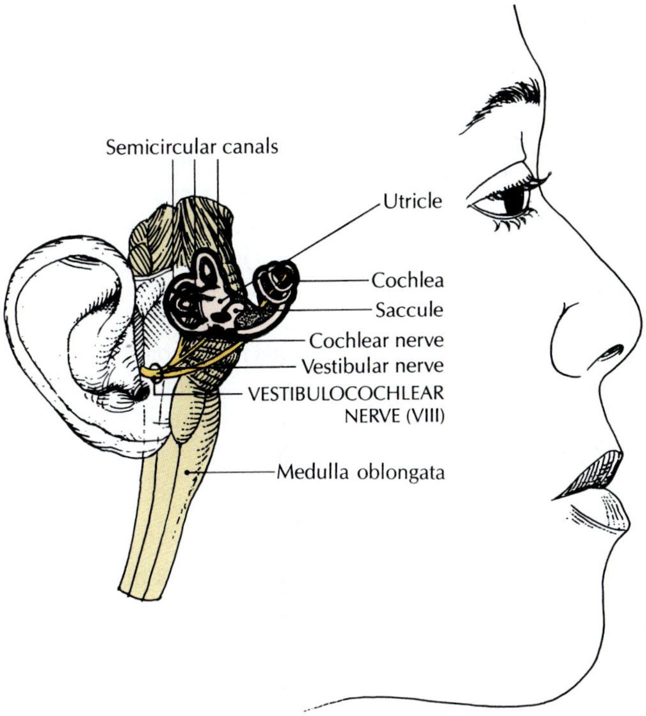

Semicircular canals
Utricle
Cochlea
Saccule
Cochlear nerve
Vestibular nerve
VESTIBULOCOCHLEAR NERVE (VIII)
Medulla oblongata

auditory radiations to the auditory cortex in the temporal lobe of the cerebrum.

The *vestibular nerve* arises from receptors in the hair cells in the semicircular canals, utricle, and saccule of the inner ear. The cell bodies of the vestibular nerve are located in the vestibular ganglion within the petrous part of the temporal bone. Its axons terminate in the vestibular nuclei in the upper medulla oblongata.

Neural information from the vestibular nuclei is conveyed via neural connections to the spinal cord, brainstem, cerebellum, and cerebral cortex. These neural inputs (1) convey information concerning postural movements of the body and head for maintaining balance and muscular coordination; (2) are involved with involuntary conjugate movements of the eyes (see *nystagmus*, TABLE 15.1); and (3) via a pathway to the thalamus and cerebral cortex, carry the conscious sensations of vertigo (whirling) and dizziness.

Cranial Nerve IX: Glossopharyngeal

The *glossopharyngeal nerve* (Gr. tongue + throat) is a small mixed nerve. After the nerve emerges from the upper medulla oblongata, it leaves the posterior cranial fossa through the jugular foramen [FIGURE 15.8]. Its branches are distributed to the region of the posterior third of the tongue and the upper pharynx. The sensory fibers terminate in several sensory nuclei in the medulla oblongata, and its motor fibers originate from motor nuclei in the medulla oblongata.

Sensory fibers convey taste and general senses from the posterior third of the tongue, and general senses from adjacent structures of the upper pharynx. When the sen-

FIGURE 15.8 DISTRIBUTION OF THE GLOSSOPHARYNGEAL NERVE (IX)

Right lateral view.

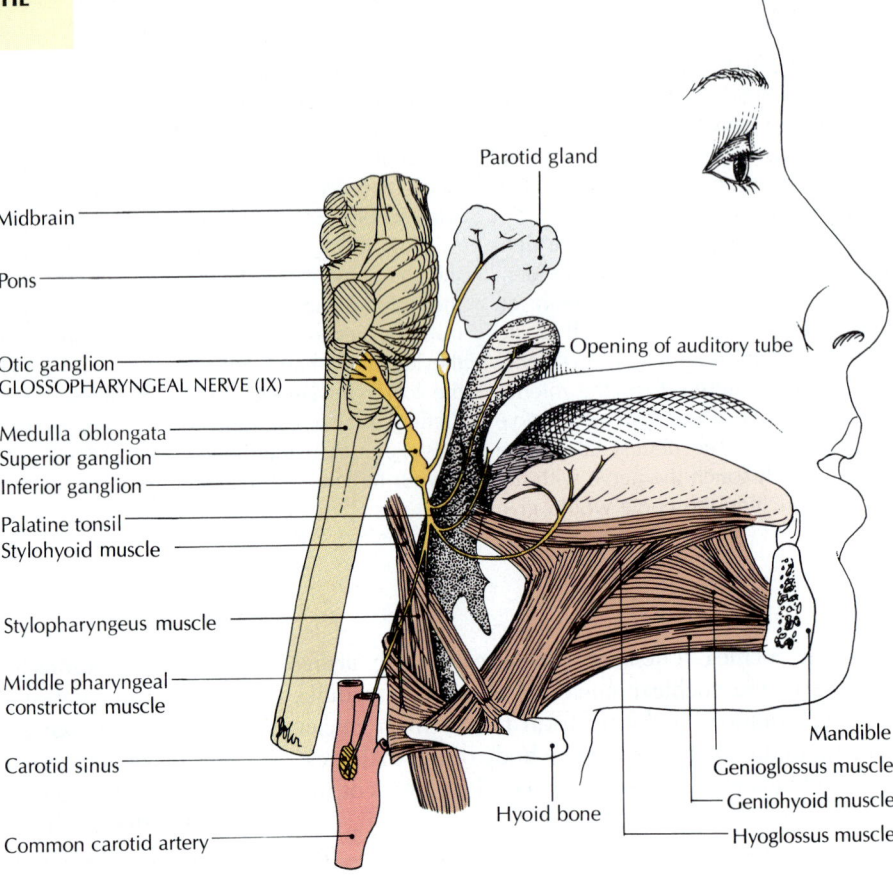

Parotid gland

Midbrain

Pons

Opening of auditory tube

Otic ganglion
GLOSSOPHARYNGEAL NERVE (IX)

Medulla oblongata
Superior ganglion
Inferior ganglion

Palatine tonsil
Stylohyoid muscle

Stylopharyngeus muscle

Middle pharyngeal
constrictor muscle

Carotid sinus

Mandible
Genioglossus muscle
Geniohyoid muscle
Hyoglossus muscle

Hyoid bone

Common carotid artery

sory nerve in the back of the mouth is stimulated, the act of swallowing and the gag reflex are triggered. A small branch of the nerve terminating in the carotid sinus (in the neck) monitors blood pressure. Motor fibers (from the nucleus ambiguus) innervate the stylopharyngeus muscle, which helps elevate the pharynx during swallowing. Autonomic (parasympathetic) fibers in the glossopharyngeal nerve stimulate the parotid gland (anterior to the ear) to secrete saliva.

The carotid branch of the glossopharyngeal nerve innervates the carotid sinus and carotid body. Its afferent fibers respond to the blood concentrations of oxygen and carbon dioxide and to blood-pressure changes in the carotid arteries, conveying the information to the solitary nucleus of the brainstem.

Cranial Nerve X: Vagus

The *vagus nerve* (L. wandering) is a mixed nerve. It is so named because it has the most extensive distribution of any cranial nerve [FIGURE 15.9]. The vagus nerve is the longest cranial nerve, innervating structures in the head, neck, thorax, and abdomen. Two different types of motor fibers are present:

1 The fibers that innervate the skeletal muscles of the soft palate, pharynx, and larynx originate in a motor nucleus (nucleus ambiguus) in the medulla oblongata. These muscles are involved in swallowing and also in speaking.

2 Originating from another nucleus (dorsal vagal nucleus) in the medulla oblongata are fibers of the parasympathetic system, which convey impulses involved with the activity of cardiac muscle, smooth muscle, and exocrine glands (glands with ducts) of the cardiovascular, respiratory, and digestive systems.

The sensory fibers of the vagus nerve convey impulses from all the structures of the systems innervated by the motor fibers. The sensory fibers terminate in the solitary nucleus in the medulla oblongata.

Cranial Nerve XI: Accessory

The *accessory* (or *spinal accessory*) *nerve* is a mixed nerve, originating from both the medulla oblongata and the cervical spinal cord [FIGURE 15.10]. Both the bulbar (medullary) and spinal roots are composed primarily of motor fibers.

FIGURE 15.9 DISTRIBUTION OF THE VAGUS NERVE (X)

Right lateral view.

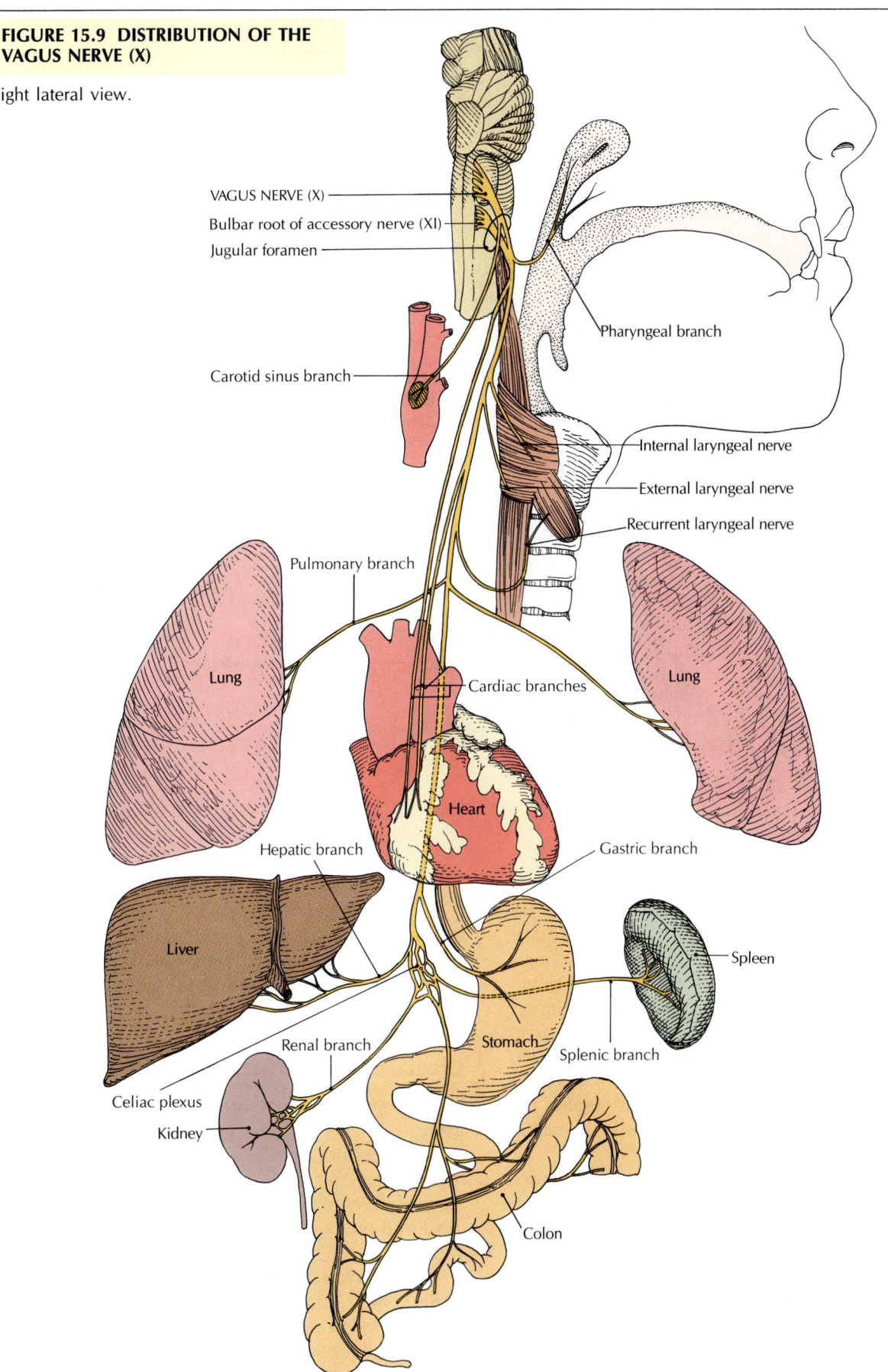

VAGUS NERVE (X)

Bulbar root of accessory nerve (XI)

Jugular foramen

Carotid sinus branch

Pharyngeal branch

Internal laryngeal nerve

External laryngeal nerve

Recurrent laryngeal nerve

Pulmonary branch

Lung

Lung

Cardiac branches

Heart

Hepatic branch

Gastric branch

Liver

Spleen

Renal branch

Stomach

Splenic branch

Celiac plexus

Kidney

Colon

FIGURE 15.10 DISTRIBUTION OF THE ACCESSORY NERVE (XI)

Right lateral view.

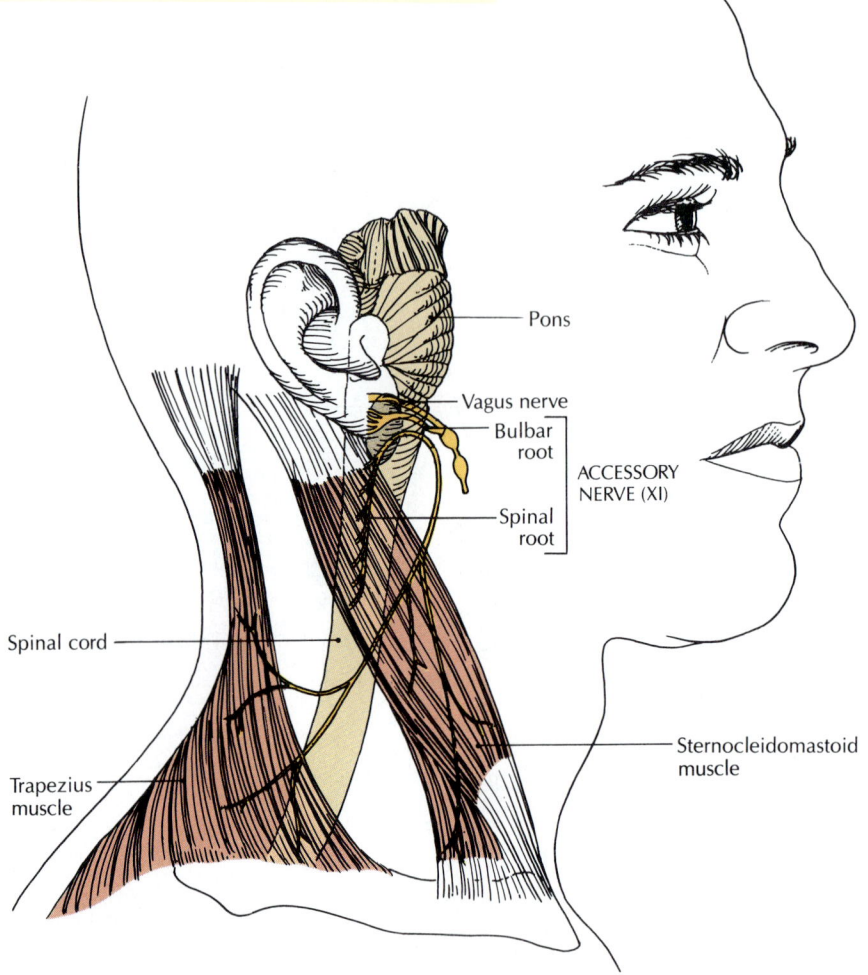

The *bulbar root* arises from a motor nucleus (nucleus ambiguus) in the medulla oblongata, joins the spinal root, and passes through the jugular foramen. The fibers of the bulbar root leave the fibers of the spinal root and join the vagus nerve. Thus the larynx is actually innervated by the accessory nerve, which joins the vagus nerve. The fibers eventually form the recurrent laryngeal nerve, which innervates the muscles of the larynx that control the vocal cords. The laryngeal nerve also supplies the sensory innervation of the upper trachea and the region around the vocal cords.

The *spinal root* originates from neurons in the anterior gray horn of the first five cervical spinal levels. The fibers of the spinal root emerge from the spinal cord, ascend within the vertebral canal, pass through the foramen magnum, join the bulbar root, and pass through the jugular foramen. The spinal fibers leave the bulbar root fibers and pass into the neck to innervate the sternocleidomastoid and trapezius muscles. Both muscles are involved with movements of the head.

The proprioceptive fibers from these muscles pass through the cervical nerves to the spinal cord. Other afferent fibers, for example, those from the region of the vocal cords, terminate in trigeminal nuclei of the medulla oblongata.

Cranial Nerve XII: Hypoglossal

The *hypoglossal nerve* ("under the tongue") is a motor nerve that innervates the muscles of the tongue. The motor fibers originate from the motor nucleus of the hypoglossal nerve in the medulla oblongata, emerge from the medulla oblongata, and pass through the hypoglossal canal into the region of the floor of the mouth [FIGURE 15.11].

The branches of the hypoglossal nerve innervate the intrinsic muscles within the tongue and three extrinsic muscles: hyoglossus, genioglossus, and styloglossus. These muscles are important in manipulating food within the mouth, in speaking, and in swallowing. The

FIGURE 15.11 DISTRIBUTION OF THE HYPOGLOSSAL NERVE (XII)

Right lateral view.

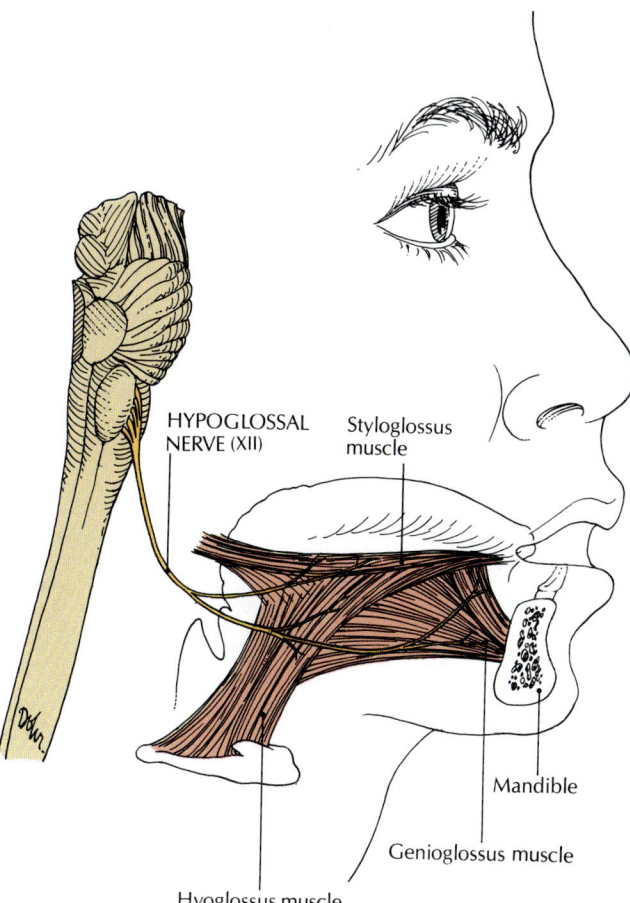

HYPOGLOSSAL NERVE (XII)

Styloglossus muscle

Mandible

Genioglossus muscle

Hyoglossus muscle

proprioceptive fibers from the tongue join the mandibular nerve and have their cell bodies in the trigeminal ganglion of cranial nerve V.

ASK YOURSELF

1 How many cranial nerves are there?

2 Which cranial nerve is actually a tract?

3 Which are the extraocular motor nerves?

4 What are the three main branches of the trigeminal nerve?

5 What are the two main functions of the vestibulocochlear nerve?

6 Which cranial nerve has the widest distribution?

7 Which cranial nerves are mixed nerves? Motor nerves? Sensory nerves?

STRUCTURE AND DISTRIBUTION OF SPINAL NERVES

At each segment of the spinal cord, a pair of nerves exits the H-shaped gray matter. One nerve of the pair exits to the left, entering the left side of the body. The other nerve exits to the right, entering the right side of the body. Each nerve has a *ventral* (anterior) *root* and a *dorsal* (posterior) *root* [see FIGURE 14.6], which meet shortly after leaving the spinal cord to form a single *mixed nerve*. All spinal nerves are mixed, containing both sensory and motor fibers that, together with cranial nerves, form part of the peripheral nervous system.

How Spinal Nerves Are Named

The 31 pairs of spinal nerves are named and numbered according to the region and level of the spinal cord from which they emerge (cervical, thoracic, lumbar, sacral, or coccygeal) [see FIGURE 14.1A]. The first cervical nerve passes between the occipital bone and the first cervical vertebra. All other spinal nerves pass through the intervertebral foramina between adjoining vertebrae. The numbering of each cervical nerve corresponds to the vertebra *inferior* to its exit. For example, the third cervical (C3) nerve emerges above the third cervical vertebra. But because there are only seven cervical vertebrae, the eighth cervical (C8) nerve passes through the intervertebral foramen between the seventh cervical vertebra and the first thoracic vertebra. The numbering of each spinal nerve *other than the cervical nerves* corresponds to the vertebra *superior* to its exit from the vertebral column.

Structure of Spinal Nerves

Nerves of the peripheral nervous system are less fragile and more cordlike than the tissue of the central nervous system because of three sheaths of connective tissue around the nerve fibers [FIGURE 15.12].

Each nerve is made up of nerve fibers enclosed in distinct bundles called *fascicles* [FIGURE 15.12B]. Surrounding the entire peripheral nerve to bind together the many fascicles is a connective tissue sheath called the *epineurium* ("upon the nerve"). Also within the epineurium are blood vessels and lymph vessels [FIGURE 15.12]. A thinner sheath of connective tissue, the *perineurium* ("around the nerve"), encases each fascicle of nerve fibers. Each of these fibers is also covered by the *endoneurium* ("within the nerve"), the delicate interstitial connective tissue that separates individual nerve fibers.

Branches of Spinal Nerves

A short distance after the dorsal and ventral roots join together to form a spinal nerve proper, the nerve divides

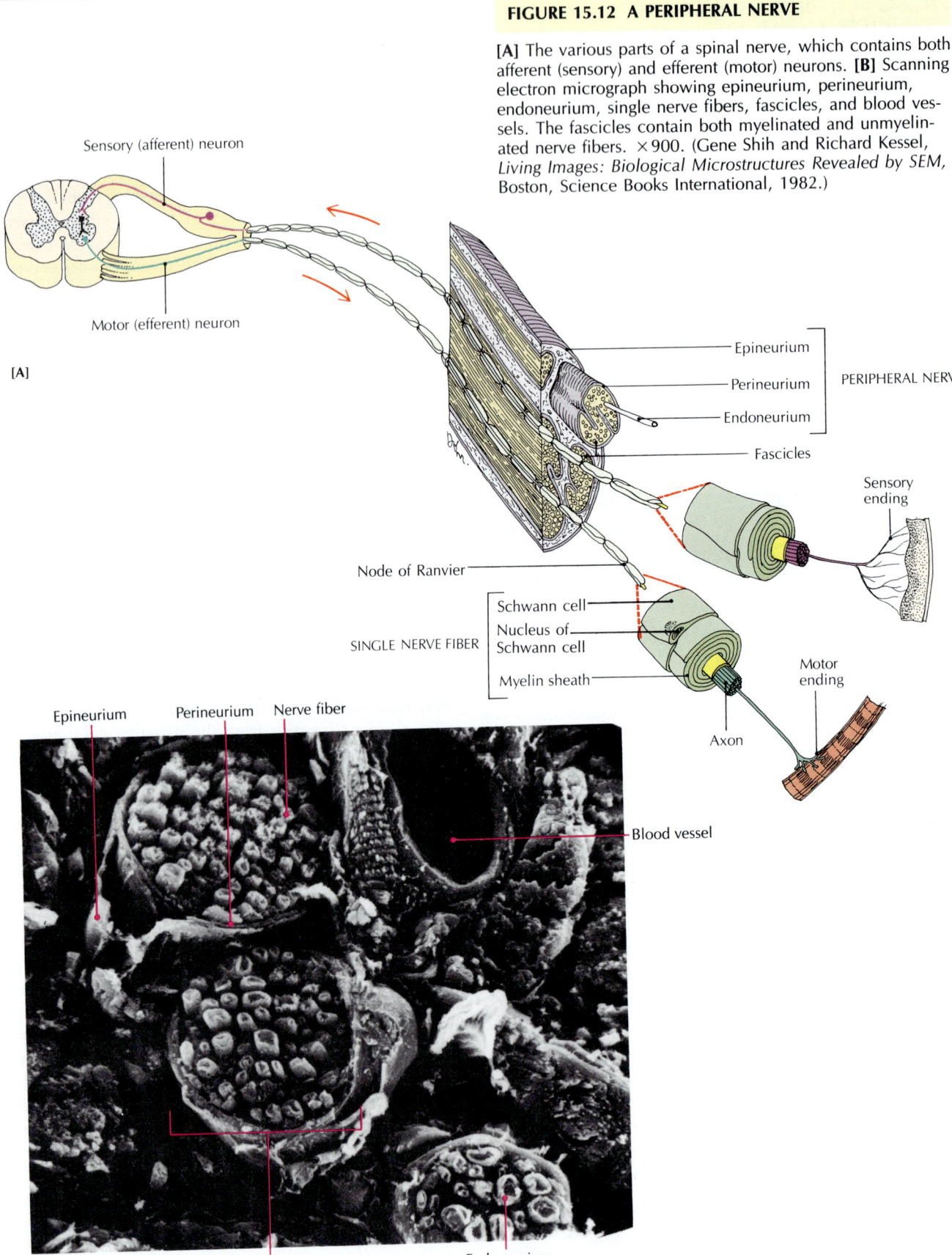

FIGURE 15.12 A PERIPHERAL NERVE

[A] The various parts of a spinal nerve, which contains both afferent (sensory) and efferent (motor) neurons. **[B]** Scanning electron micrograph showing epineurium, perineurium, endoneurium, single nerve fibers, fascicles, and blood vessels. The fascicles contain both myelinated and unmyelinated nerve fibers. ×900. (Gene Shih and Richard Kessel, *Living Images: Biological Microstructures Revealed by SEM*, Boston, Science Books International, 1982.)

Sensory (afferent) neuron

Motor (efferent) neuron

[A]

Epineurium

Perineurium

Endoneurium

PERIPHERAL NERVE

Fascicles

Sensory ending

Node of Ranvier

Schwann cell

Nucleus of Schwann cell

SINGLE NERVE FIBER

Myelin sheath

Motor ending

Axon

Epineurium

Perineurium

Nerve fiber

Blood vessel

Fascicle

Endoneurium

[B]

FIGURE 15.13 RAMI AND OTHER PARTS OF A SPINAL NERVE

Dorsal and ventral roots join to form a spinal nerve, in this case, a thoracic nerve. The nerve divides into a dorsal (posterior) ramus, a ventral (anterior) ramus, a meningeal ramus, and white and gray rami communicantes. Each sympathetic ganglion of the autonomic nervous system near the spinal cord is a paravertebral ganglion (adjacent to the spinal column), and each one near the abdominal viscera is a prevertebral ganglion.

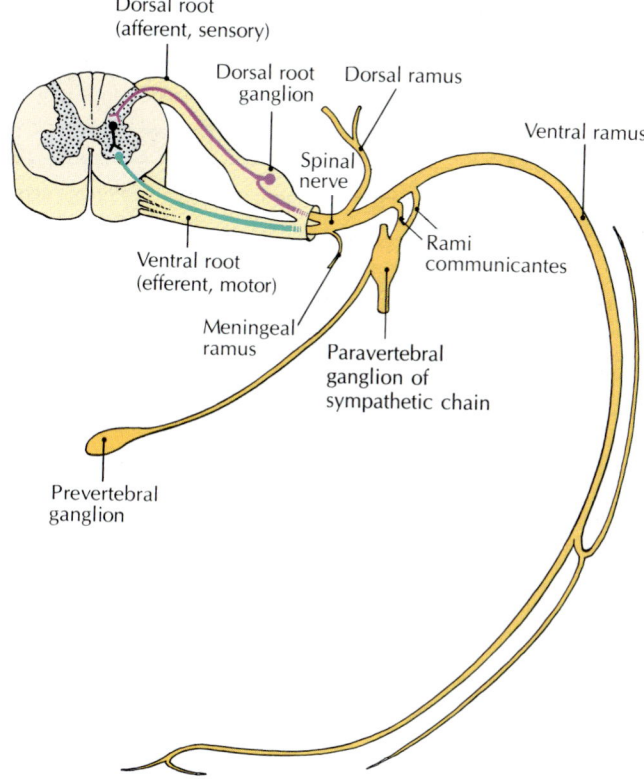

into several branches called **rami** (RAY-mye; sing. *ramus*, RAY-muhss). These branches are the dorsal (posterior) ramus, the ventral (anterior) ramus, the meningeal ramus, and the rami communicantes [FIGURE 15.13].

The branches of the **dorsal ramus** (a mixed nerve) innervate the skin of the back, the skin on the back of the head, and the tissues and intrinsic (deep) muscles of the back. Branches of the **ventral ramus** (a mixed nerve) innervate the skin, tissues, and muscles of the neck, chest, abdominal wall, both pairs of limbs, and pelvic area. The **meningeal ramus** innervates the vertebrae, spinal meninges, and spinal blood vessels. The **rami communicantes** (sing. *ramus communicans*) are composed of sensory fibers associated with the general visceral afferent system and motor nerve fibers associated with the autonomic nervous system (general visceral efferent system) innervating the visceral structures.

Plexuses

The ventral rami of the spinal nerves (except T2 through T12) are arranged to form several networks of nerves called **plexuses** (sing. *plexus*; L. braided) [FIGURE 15.14]. In a plexus, the nerve fibers of the different spinal nerves are sorted and recombined so that fibers associated with a particular peripheral nerve are composed of the fibers from several different rami. Plexuses include (1) the *cervical plexus*, (2) the *brachial plexus*, (3) the *lumbar plexus*, (4) the *sacral plexus* (sometimes the lumbar and sacral plexuses are referred to collectively as the *lumbosacral plexus*), and (5) the *coccygeal plexus*. Note that the ventral rami of T2 through T12 do not form plexuses. Instead, each ramus innervates a segment of the thoracic and abdominal walls.

Cervical plexus   The **cervical plexus** is composed of the ventral rami of spinal nerves C1 through C4 [FIGURE 15.14]. Its branches can be placed into the following groups:

1 The *cutaneous* sensory nerves innervate the skin and scalp behind the ears, the skin of the sides and front of the neck, and the upper portions of the thorax and shoulder.

2 The branches of the *ansa* (looplike) *cervicalis* and its superior and inferior roots provide motor innervation to the strap muscles attached to the hyoid bone in the neck.

3 The *phrenic* nerves from C3, C4, and C5 descend into the thorax to innervate the diaphragm.

Lesions of both phrenic nerves cause paralysis of the diaphragm. The major symptoms are shortness of breath and difficulty in coughing or sneezing. A severe case of hiccups (spasmodic, sharp contractions of the diaphragm) can be relieved by surgically crushing a phrenic nerve in the lower neck, which temporarily paralyzes the diaphragm. (The nerve fibers regenerate over time.) A less drastic method is the injection of an anesthetic solution around the phrenic nerve. The anesthesia concentrates on the anterior surface of the middle third of the anterior scalene muscle, producing a temporary paralysis of half of the diaphragm.

Brachial plexus   The **brachial plexus** is made up primarily of the ventral rami of C5 to C8 and T1 spinal nerves [FIGURES 15.14, 15.15]. The plexus extends downward and laterally, passing behind the clavicle and into the armpit. In its course downward, the brachial plexus consists of branches (nerves) and recombinations that are called, in order, roots, trunks, divisions, cords, and main branches.

The brachial plexus gives rise to a number of nerves to the upper limb. Five major nerves constitute the terminal branches of the cords. These nerves — the axillary, ulnar,

FIGURE 15.14 SPINAL PLEXUSES AND SOME MAJOR SPINAL NERVES

Anterior view.

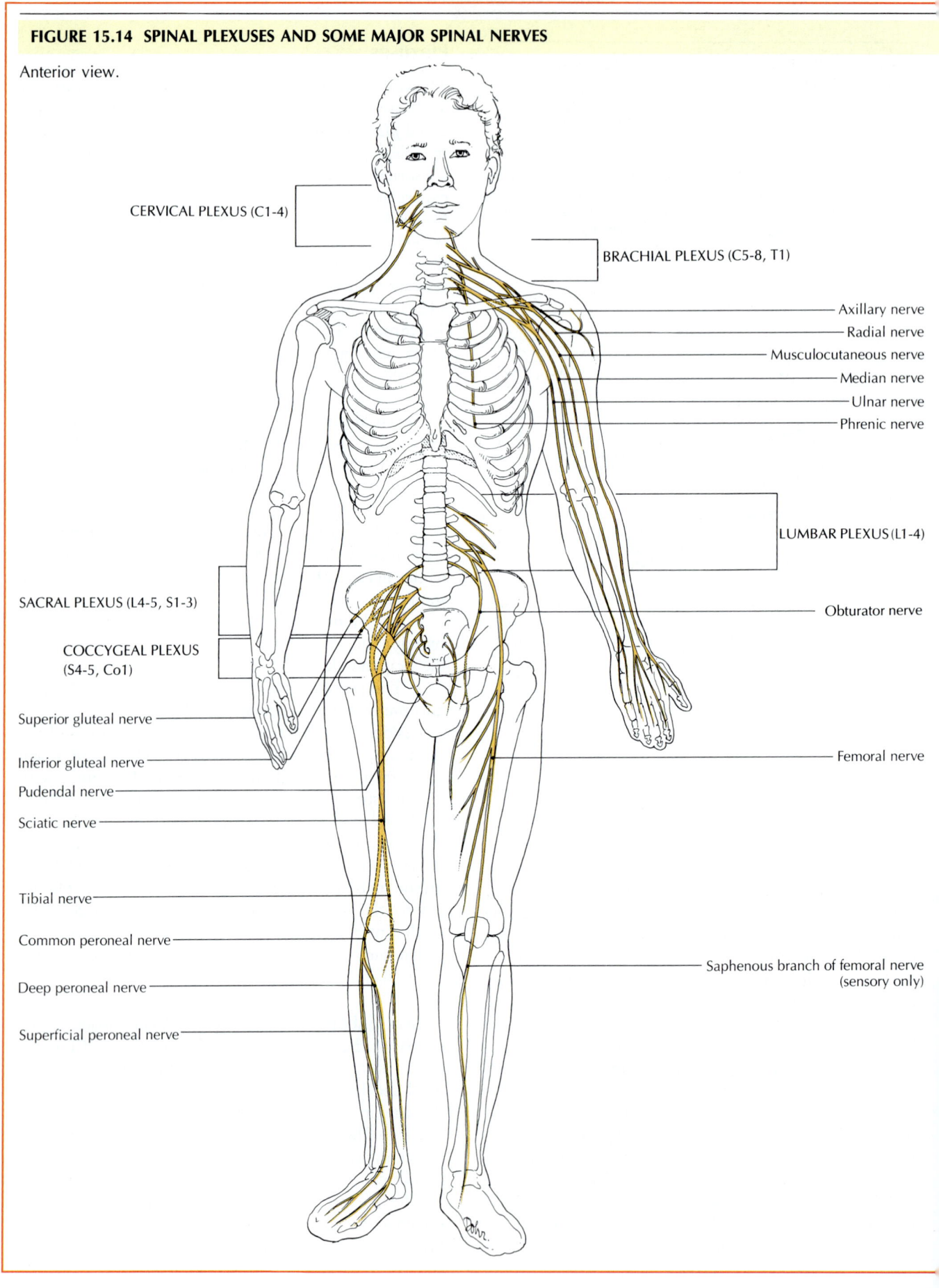

CERVICAL PLEXUS (C1-4)

BRACHIAL PLEXUS (C5-8, T1)

Axillary nerve

Radial nerve

Musculocutaneous nerve

Median nerve

Ulnar nerve

Phrenic nerve

LUMBAR PLEXUS (L1-4)

SACRAL PLEXUS (L4-5, S1-3)

Obturator nerve

COCCYGEAL PLEXUS (S4-5, Co1)

Superior gluteal nerve

Inferior gluteal nerve

Femoral nerve

Pudendal nerve

Sciatic nerve

Tibial nerve

Common peroneal nerve

Deep peroneal nerve

Saphenous branch of femoral nerve (sensory only)

Superficial peroneal nerve

Plexus	Components	Location	Major nerve branches	Regions innervated	Result of damage to specific nerves
Cervical	Ventral rami of C1 to C4 nerves	Neck region; origin covered by sterno-cleidomastoid	Cutaneous, muscular, communicating, phrenic, ansa cervicalis	Skin and some muscles of back of head, neck; diaphragm.	*Phrenic:* paralysis of diaphragm (respiratory muscles).
Brachial	Ventral rami of C5 to C8 and T1 nerves	Lower neck, axilla	Axillary, ulnar, median, radial, musculocutaneous	Muscles and skin of neck, shoulder, arm, forearm, wrist, hand.	*Axillary:* impaired ability to abduct and rotate arm. *Ulnar:* "clawhand" (fingers partially flexed). *Median:* "ape hand" (thumb adducted against index finger); impaired ability to oppose thumb; carpal tunnel syndrome. *Radial:* "wristdrop" (inability to extend hand at wrist).
Lumbar*	Ventral rami of L1 to L4 nerves	Interior of posterior abdominal wall	Femoral, obturator	Muscles, skin of abdominal wall.	*Femoral:* inability to extend leg; marked weakness in flexing hip; inflammation leads to lumbago (back pain).
Sacral*	Ventral rami of L4, L5, and S1 to S3 nerves	Posterior pelvic wall	Superior gluteal, inferior gluteal; sciatic nerve branches:		*Superior gluteal:* walk with lurching gait. *Inferior gluteal:* difficulty walking up stairs. *Sciatic:* severely impaired ability to extend hip, flex knee; impaired ability to plantar-flex foot.
			Tibial	Buttocks, medial thigh, muscles and skin of posterior thigh, posterior leg, plantar (sole) of foot.	
			Common peroneal	Muscles and skin of lateral posterior thigh, anterior leg, dorsal foot.	*Common peroneal:* "footdrop" (inability to dorsiflex foot, toes).
			Pudendal	Voluntary sphincters of urethra, anus.	*Pudendal:* incontinence in urination, defecation.
Coccygeal	Co1 nerve, plus communications from S4 and S5 nerves	Coccyx region	A few fine filaments	Skin in coccyx region.	Possibly some loss of sensation in coccygeal region.

*Both are part of the lumbosacral plexus.

FIGURE 15.15 PHOTOGRAPH OF DISSECTED RIGHT BRACHIAL PLEXUS

All blood vessels have been removed, and some cords have been displaced to reveal the branches in this anterior view.

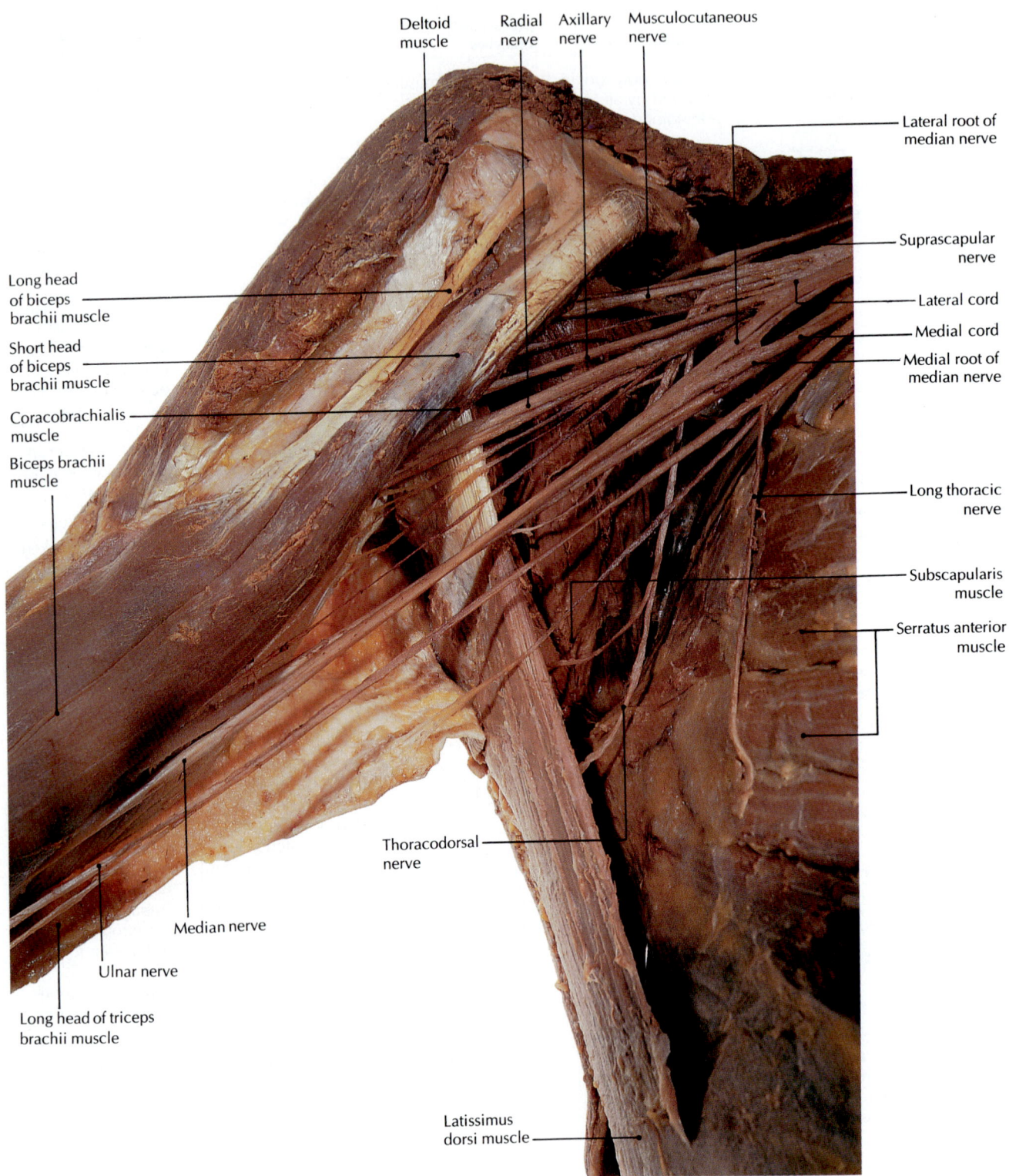

median, radial, and musculocutaneous—innervate the shoulder, arm, forearm, and hand.

Lumbar plexus The *lumbar plexus* is composed of fibers of the ventral rami of L1 to L4 nerves [FIGURE 15.14]. It supplies the anterior and lateral abdominal wall, the external genitals, and the thigh. Two major nerves, the femoral and obturator, and some lesser nerves (genitofemoral, ilioinguinal, iliohypogastric) are derived from the recombining branches.

The *femoral nerve* is composed of fibers from L2, L3, and L4. It innervates the muscles that flex the hip joint and extend the knee joint (knee jerk, or patellar tendon reflex). Inflammation of the femoral nerve leads to lumbago (pain in the lumbar region). The *obturator nerve*, also composed of fibers from L2, L3, and L4, innervates the adductor muscles of the thigh as well as the skin.

Sacral plexus The *sacral plexus* is composed of fibers of the ventral rami of spinal nerves L4, L5, and S1 through S3 [FIGURE 15.14]. It passes in front of the sacrum and into the regions of the buttocks. Several important nerves are derived from the plexus. The *sciatic nerve*, the thickest and longest nerve in the body, extends from the pelvic area to the foot, dividing above the knee into the *common peroneal* and *tibial nerves*. Together, those nerves innervate the thigh, leg, and foot muscles. Inflammation of the sciatic nerve, or *sciatica*, is described in When Things Go Wrong at the end of the chapter.

The *superior gluteal nerve* innervates the gluteus medius, gluteal minimus, and tensor fascia lata muscles, and the *inferior gluteal nerve* innervates the gluteus maximus muscle. The *pudendal nerve* innervates the voluntary muscles of the perineum, especially the voluntary sphincters of the urethra and anus. It is the pudendal nerve, among others, that is blocked by local anesthesia in the "saddle block" procedure that is sometimes used during childbirth.

Coccygeal plexus The *coccygeal plexus* is formed by the coccygeal nerve (Co1) and sacral nerves S4 and S5. A few fine nerve filaments supply the skin in the coccyx region.

Intercostal Nerves

The *intercostal nerves* (ventral rami of thoracic nerves) are the second through twelfth thoracic nerves (T2 to T12). The intercostal nerves innervate muscles and skin in the thoracic and abdominal walls. After these nerves leave the intervertebral foramina, they take a course parallel to the ribs. Intercostal nerves T2 through T6 innervate intercostal muscles, plus the skin on the lateral and anterior thoracic walls. Nerves T7 through T12 inner-

vate the intercostal muscles, plus the abdominal wall and its overlying skin.

Dermatomes

The dorsal root of each spinal nerve is distributed to a specific region or segment of the body and supplies the sensory innervation to a segment of the skin known as a *dermatome* (Gr. *derma*, skin + *tomos*, cutting). There are 30 dermatomes, one for each spinal nerve, except C1, which does not innervate the skin [FIGURE 15.16]. The face and scalp in front of the ears are innervated by the trigeminal cranial nerve, whose three divisions (ophthalmic, maxillary, and mandibular nerves) each innervate a separate region.

When the function of even a single dorsal nerve root is interrupted, there is a faint but definite decrease in sensitivity in the dermatome. The area of diminished sensitivity is detected by using a light pin scratch to stimulate a pain sensation. Using this method, a map of the dermatomes can be used to locate the sites of injury to dorsal roots of the spinal cord.

The sensory innervations of adjacent dermatomes overlap in such a way that if a spinal nerve is cut, the loss of sensation in its dermatome is minimal. However, if one dorsal root is irritated, as in shingles (described in When Things Go Wrong), the resultant pain does spread to adjacent, overlapping dermatomes.

ASK YOURSELF

1 How are spinal nerves named?

2 What are the epineurium, perineurium, and endoneurium?

3 What is a plexus? A ramus?

4 How do the thoracic nerves differ from the other spinal nerves?

5 What nerves make up the cervical plexus? The brachial plexus? The lumbar and sacral plexuses?

6 What is a dermatome?

Q: *Why does your foot sometimes "fall asleep"?*

A: When you cross your legs or otherwise put pressure on one of your legs, you may press down on a nerve and temporarily cut off the feeling in your foot. When the local pressure is removed, you may feel a numbness we often call "pins and needles" as the nerve endings become reactivated. The same thing may happen when a blood vessel that supplies a peripheral nerve is temporarily forced shut. In that case, the "pins and needles" feeling occurs when the many tiny capillaries in the foot restore the nerve's blood supply.

FIGURE 15.16 DERMATOMES, WITH THEIR SPINAL NERVES

[A] Anterior view.
[B] Posterior view.

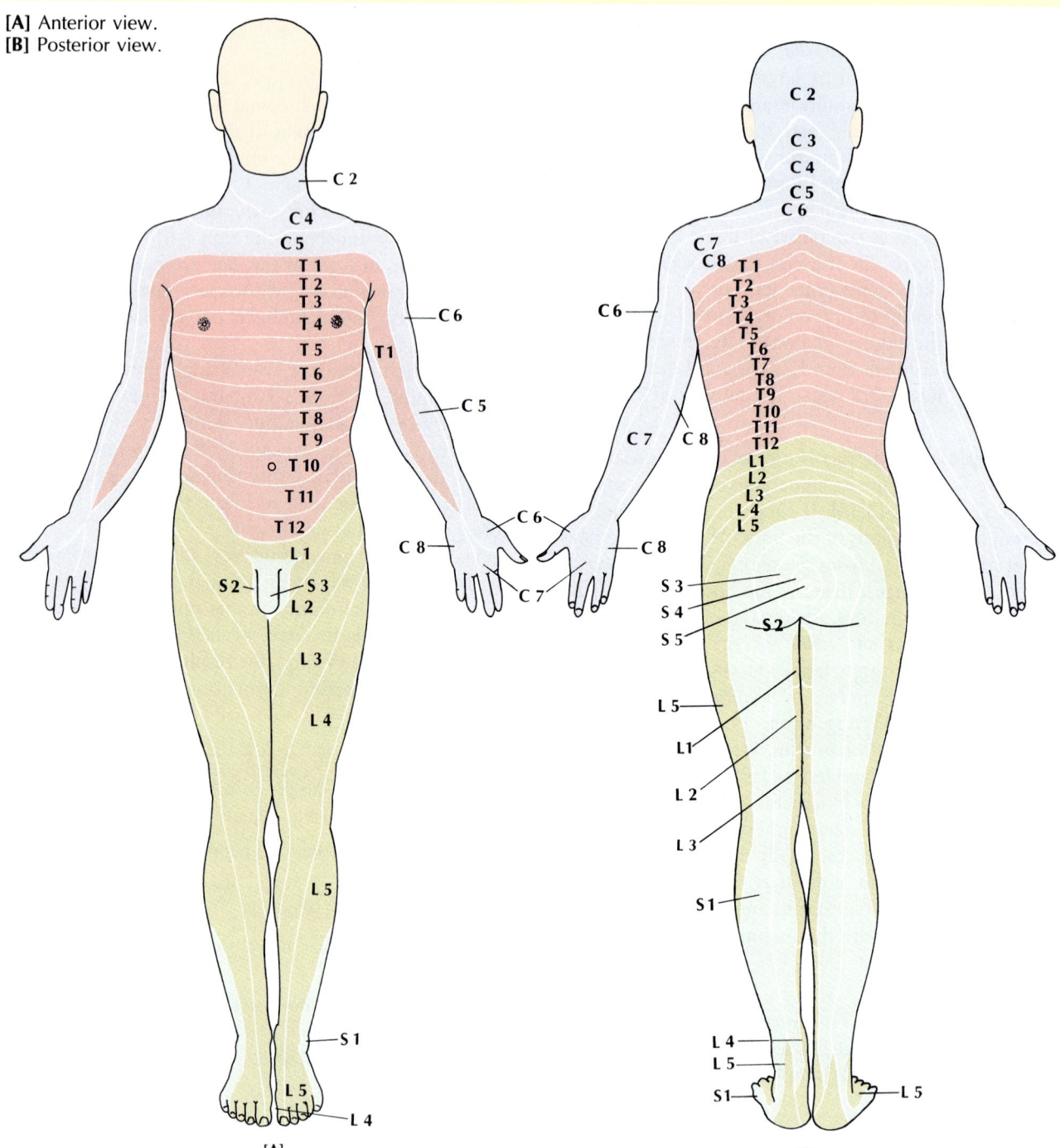

[A]

[B]

DEGENERATION AND REGENERATION OF PERIPHERAL NERVE FIBERS

The development of neurons begins before birth and is not completed until infancy. Before this completion, some neurons *degenerate* and die, and they continue to degenerate and die throughout our lives, never again equaling the number at infancy. (Such a loss of brain cells, for instance, is usually no problem in a healthy person, because we start out with many more neurons than we actually need.) But even though neurons are not replaced when they die, severed peripheral nerve *fibers* are sometimes able to regenerate if the cell body is undamaged and the myelin sheath is intact. The only neurons that are replaced continuously are the olfactory (smell) neurons of the olfactory nerve.

FIGURE 15.17 DEGENERATION AND REGENERATION OF SEVERED PERIPHERAL MOTOR FIBERS

The regenerated portions of the fibers conduct nerve impulses more slowly than before because the distance between nodes (internodes) is shorter in regenerated nerves.

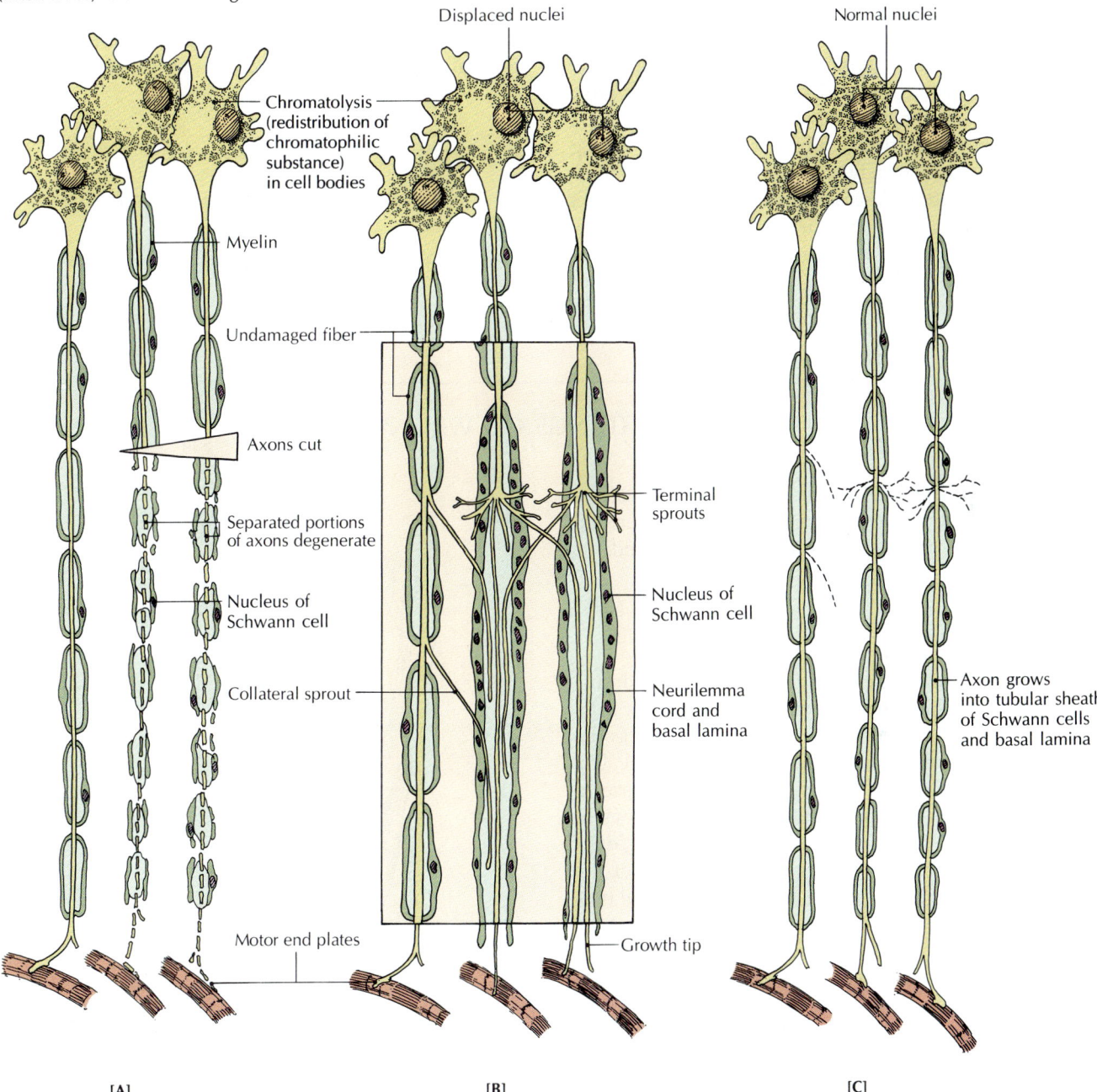

When a peripheral nerve fiber is cut, the motor neuron has the capacity to regenerate its axon, and a sensory neuron has the capacity to regenerate its dendrites. First, the part of the axon severed from its cell body begins to degenerate in the stump distal to the cut, and in a few weeks it disappears altogether [FIGURE 15.17A].

As part of the process of *regeneration,* the cell body enlarges. The chromatophilic substance increases its activity and produces extra protein, which is needed to support the growth of new branches called *terminal sprouts.*

These sprouts grow from the proximal portion of the axon still connected to the cell body. At the same time, Schwann cells in the distal stump divide profusely and arrange themselves into continuous cords. These cords extend from the cut end of the stump to the sensory and motor nerve endings at the distal end of the nerve [FIGURE 15.17B].

Each regenerating fiber can form 50 or more terminal sprouts. The sprouts grow along the cords of Schwann cells, which act as guides the way a trellis guides the path of a growing vine. The actual guide is the basal lamina of each Schwann cell. (The basal lamina contains a chemical substance involved with the guiding process. The proliferating Schwann cells secrete *laminin*, which promotes axon lengthening. Adhesion molecules, N-CAMs, found on the surface of Schwann cells may also contribute to axonal growth.) The fiber proceeds to grow into the newly formed cord [FIGURE 15.17B]. Most sprouts grow along cords and do not form viable functional contacts. A regenerating sprout that finally reaches a nerve ending compatible with its own functional role can make a physiological connection. A sensory fiber may connect with a similar sensory ending, or a fiber of a motor neuron may connect with a motor end plate or fibers of a skeletal muscle [FIGURE 15.17C]. Such "successful" sprouts become myelinated by the surrounding Schwann cells. Sev-

ered fibers may also receive support from adjacent undamaged fibers that send out *collateral sprouts* from nodes of Ranvier [FIGURE 15.17B].

Depending on the severity of the damage to the nerve and the distance of the injury from the nerve endings at the motor end plate, functional recovery of nerve fibers may take from about a month to more than a year.

ASK YOURSELF

1 Compare the functions of terminal and collateral sprouts.

2 Why are severed neurons in the central nervous system unable to regenerate?

WHEN THINGS GO WRONG

Disorders involving cranial nerves range from infections that may damage nerve fibers to serious injuries, cerebral strokes, or tumors that may produce paralysis and the loss of some sensations [TABLE 15.1].

Carpal Tunnel Syndrome

Carpal tunnel syndrome is the most common syndrome in which a peripheral nerve is entrapped. It results from the compression of the median nerve at the wrist as the nerve passes through the *carpal tunnel* (see drawing). The carpal tunnel also provides passage from the forearm to the hand for the eight flexor tendons (four tendons of flexor digitorum superficialis and four of flexor digitorum profunda) to the fingers and one tendon of the thumb (tendon of flexor pollicis longus). The carpal tunnel is formed by a concave arch on the palmar side of the carpal bones, which is covered by a connective tissue ligament called the *flexor retinaculum* (transverse carpal ligament). Carpal tunnel syndrome is usually associated with arthritis of the wrist, dislocation of the lunate carpal bone, or inflammation of the tendon sheaths of the flexor tendons. Compression on the median nerve produces tingling and the loss of some sensation in the palmar surfaces of the thumb and the index and middle fingers, weakness, and pain that may worsen at night. Fine coordinated movements of the hand (especially the thumb) may suffer, and the affected person is usually unable to make a fist. This condition is most common in people who use their fingers excessively in rigorous work and is often found among operators of computer terminals. Treatment ranges from rest and corticosteroid injections to severing the flexor retinaculum at the wrist.

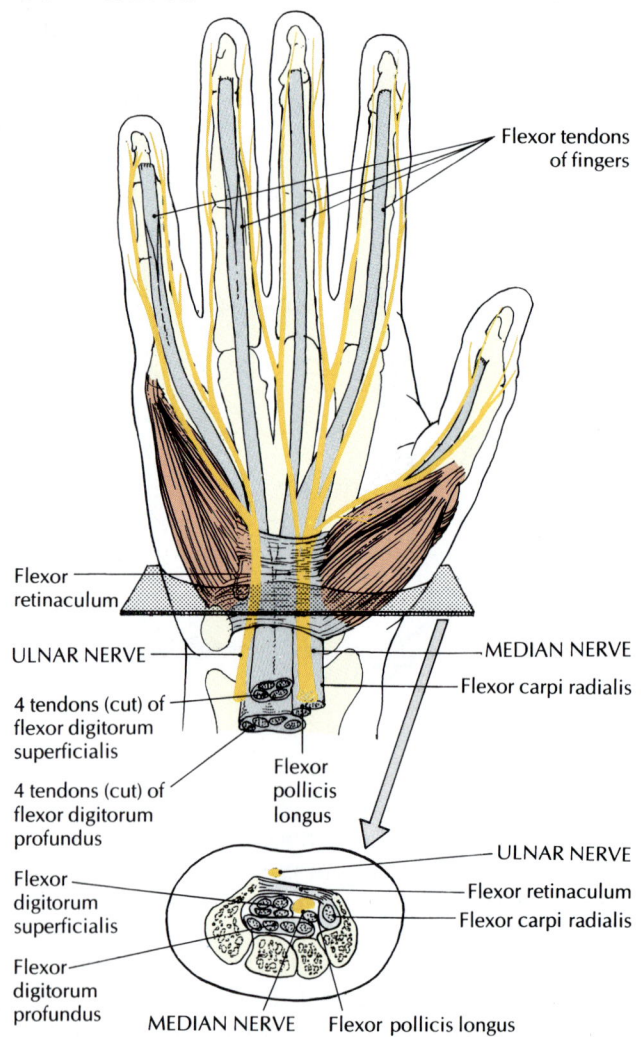

Carpal tunnel syndrome, as seen in this palmar view (top) and cross section (bottom) of the right hand.

TABLE 15.1 SOME DISORDERS OF CRANIAL NERVES

Nerve	Disorder
I Olfactory	Fracture of cribriform plate of ethmoid bone or lesions along olfactory pathway may produce total inability to smell (*anosmia*); as a result, food tastes flat.
II Optic	Trauma to orbit or eyeball, or fracture involving optic foramen produces inability of pupil to constrict. Certain poisons (such as wood alcohol) or infections (such as syphilis) may damage nerve fibers. Pressure on optic pathway may cause blindness in certain regions of visual fields.
III Oculomotor	Lesion or pressure on nerve produces dilated pupil, drooping upper eyelid (*ptosis*), absence of direct pupil reflex, squinting (*strabismus*), double vision (*diplopia*).
V Trigeminal	Injury to terminal branches or trigeminal ganglion causes loss of pain and touch sensation, abnormal tingling, itching, numbness (*paresthesias*), inability to contract chewing muscles, deviation of mandible to side of lesion when mouth is opened. Severe spasms in nerve branches cause nerve pain (*trigeminal neuralgia*, or *tic douloureux*).
VI Abducens	Nerve lesion causes inability of eye to move laterally, with eye cocked in; diplopia; strabismus.
VII Facial	Damage causes Bell's palsy, loss of taste on anterior two-thirds of tongue of side of lesion. Cerebral stroke may produce paralysis of lower muscles of facial expression (below eye) on opposite side of lesion. Sounds are louder on side of injury because stapes muscle is paralyzed.
VIII Vestibulocochlear	Tumor or other injury to nerve produces progressive hearing loss, noises in ear (*tinnitus*), involuntary rapid eye movement (*nystagmus*), whirling dizziness (*vertigo*).
IX Glossopharyngeal	Injury to nerve produces loss of gag reflex, difficulty in swallowing (*dysphagia*), loss of taste on posterior third of tongue (not noticed by patient unless tested), loss of sensation on affected side of soft palate, decrease in salivation.
X Vagus	Unilateral lesion of nerve produces sagging of soft palate, hoarseness due to paralysis of vocal fold, dysphagia.
XI Accessory	Laceration of neck produces inability to contract sternocleidomastoid and upper fibers of trapezius muscles, drooping shoulders, inability to rotate head (*wry neck*).
XII Hypoglossal	Damage to nerve produces protruded tongue deviated toward affected side, moderate difficulty in speaking (*dysarthria*), chewing, and dysphagia.

Peripheral Neuritis

Peripheral neuritis is a progressive degeneration of the axons and myelin sheaths of peripheral nerves, especially those that supply the distal ends of muscles of the limbs. It results in muscle atrophy and weakness, the loss of tendon reflexes, and some sensory loss. The disease is associated with infectious diseases such as syphilis and pneumonia, metabolic and inflammatory disorders such as gout and diabetes mellitus, nutritional deficiencies, and chronic intoxication, including lead and arsenic poisoning and chronic exposure to benzene, sulfonamides, and other chemicals. The prognosis is usually good if the underlying cause can be identified and removed.

Sciatica

Sciatica (sye-AT-ih-kuh) is a form of *neuritis* (nerve inflammation) characterized by sharp pains along the sciatic nerve and its branches. Pain usually radiates from the buttocks to the hip, back, posterior thigh, leg, ankle, and foot. One of the most common causes is pressure from a herniated intervertebral disk on one or more dorsal or ventral roots of the sciatic nerve.

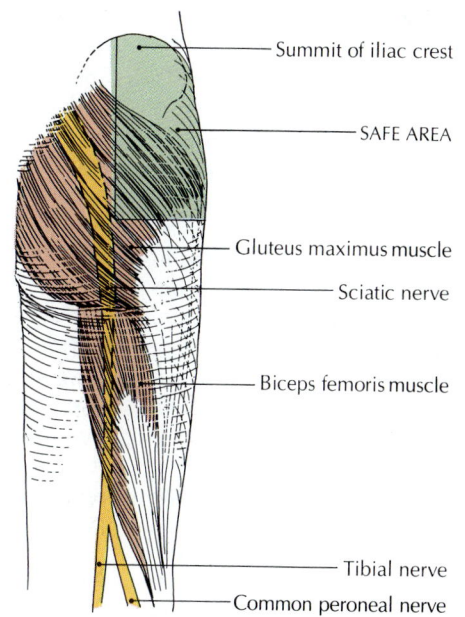

Summit of iliac crest
SAFE AREA
Gluteus maximus muscle
Sciatic nerve
Biceps femoris muscle
Tibial nerve
Common peroneal nerve

The safe area for intramuscular injection is indicated in green, and the area where the sciatic nerve is vulnerable is shown in red.

The sciatic nerve may be injured by an intramuscular injection into the buttocks. The safe area for injection is the upper, outside quadrant (see drawing on page 473).

Shingles

Shingles, or *herpes zoster,* is an acute inflammation of the dorsal root ganglia. Shingles occurs when the virus that caused childhood chickenpox lies dormant in the ganglia of cranial nerves or the ganglia of posterior nerve roots and then becomes reactivated and attacks the root ganglia.

Shingles begins with fever and a general feeling of illness, followed in 2 to 4 days by severe deep pain in the trunk and occasionally in the arms and legs. A rash typically appears in a dermatome around one side of the chest, trunk, or abdomen, and the pain and rash always progress along the course of one or more spinal nerves (usually intercostal nerves) beneath the skin. Sometimes the ganglion of the trigeminal cranial nerve is affected, producing pain in the eyeball and a rash from the eyelid to the hairline. The rash usually disappears within 2 or 3 weeks, but some pain may persist for months. Shingles seldom recurs.

If the disease is diagnosed within 48 hours after its onset, it can be treated with the antiviral drug acyclovir (Zovirax).

CHAPTER SUMMARY

The **peripheral nervous system** consists of the nerve cells and their fibers emerging from and going to the brain (cranial nerves) and spinal cord (spinal nerves). It allows the central nervous system to communicate with the rest of the body.

Classification of Functional Components of Nerves (p. 451)

1 The peripheral nervous system contains **afferent** (sensory) nerve cells, **efferent** (motor) nerve cells, and sensory and autonomic ganglia.

2 The peripheral nervous system may be divided on a functional basis into the **somatic nervous system** and the **visceral nervous system,** each composed of afferent and efferent neurons and their fibers. The *somatic afferent division* conveys sensory information from the skin, muscles, joints, ears, eyes, and associated structures; the *visceral afferent division* conveys sensory information from the viscera, nose (smell), and taste buds to the CNS. The *somatic efferent division* conveys influences from the CNS, regulating the activity of the skeletal muscles. The *visceral efferent division,* also known as the *autonomic nervous sys-têm,* conveys impulses from the CNS involved in the regulation of cardiac muscle, smooth muscle, and glands. Cranial nerves V, VII, IX, and X have motor nerve fibers that convey influences from the brainstem to skeletal muscles of the jaw, facial expression, pharynx, and larynx.

Cranial Nerves (p. 453)

1 The 12 pairs of **cranial nerves** are the peripheral nerves of the brain. Their numbers indicate the sequential order in which they emerge from the brain.

2 The **motor (efferent) fibers** of the cranial nerves emerge from the brainstem. They arise from groups of neurons called **motor nuclei.** The **sensory (afferent) fibers** of cranial nerves arise from cell bodies located in **sensory ganglia** or in peripheral sense organs such as the eye. Axons of sensory neurons synapse with **sensory nuclei.** Fibers transmit input to the cerebrum, where sensations are brought to a conscious level.

3 The cranial nerves, in ascending numerical order, are the **olfactory** (I), **optic** (II), **oculomotor** (III), **trochlear** (IV), **trigeminal** (V), **abducens** (VI), **facial** (VII), **vestibulocochlear** (VIII), **glosso- pharyngeal** (IX), **vagus** (X), **accessory** (XI), and **hypoglossal** (XII).

Structure and Distribution of Spinal Nerves (p. 463)

1 At each segment of the spinal cord a pair of nerves is distributed to each side of the body. Each nerve has a **ventral** (anterior) **root** and a **dorsal** (posterior) **root.** Motor fibers emerge from the ventral root, and the sensory fibers emerge as the dorsal root. Both roots meet to form a single **mixed nerve.**

2 Each nerve is made up of nerve fibers enclosed in bundles of connective tissue called **fascicles.** Several fascicles are held together by a sheath called the **epineurium,** each fascicle is encased by the **perineurium,** and each nerve fiber is covered by the **endoneurium.**

3 After a spinal nerve leaves the vertebral column, it divides into initial branches called **rami.** The ventral rami (except T2 to T12 nerves) are arranged to form networks of nerves called the cervical, brachial, lumbar and sacral (lumbosacral), and coccygeal **plexuses.**

4 The **intercostal nerves** are the T2 through T12 spinal nerves.

5 The dorsal root of each spinal nerve arises from a specific region or segment of the body and receives sensory innervation from a segment of the skin called a **dermatome.**

Degeneration and Regeneration of Peripheral Nerve Fibers (p. 470)

Regeneration of severed peripheral nerves is usually possible if the cell body is undamaged and the myelin sheath is intact.

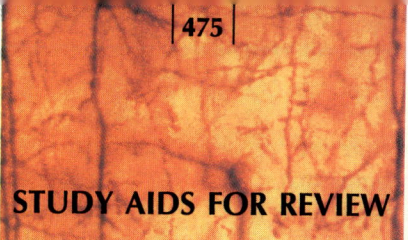

STUDY AIDS FOR REVIEW

KEY TERMS

MEDICAL TERMS FOR REFERENCE

NERVE BLOCK An injection of anesthesia near a nerve that supplies the area to be treated.

NEURALGIA (*algia* = pain) Pain along the length of a nerve.

NEURECTOMY The removal of part of a nerve.

NEURITIS Inflammation of a nerve.

NEUROLYSIS (Gr. *lys,* loosening) Relief of tension on a nerve caused by adhesions.

NEUROTRIPSY (Gr. *trips,* friction) The surgical crushing of a nerve.

UNDERSTANDING THE FACTS

1 Distinguish between the somatic nervous system and the visceral nervous system, indicating the functional kinds of neurons found in each.

2 List the cranial nerves and indicate whether they are sensory, motor, or mixed.

3 Which cranial nerves are the extraocular nerves? What is the function of each of them?

4 Which is the largest cranial nerve? Which is the longest cranial nerve?

5 Identify the following:

a fascicle   d endoneurium
b epineurium   e plexus
c perineurium   f dermatome

6 Distinguish between the roots and rami of spinal nerves, giving the functions of each of the roots and rami.

7 Which nerves are involved in Bell's palsy, shingles, vertigo, anosmia, and swallowing disorders?

UNDERSTANDING THE CONCEPTS

1 Cranial nerves may be sensory, motor, or mixed. Is the same true for spinal nerves? Explain.

2 In what ways are cranial nerves I and II different from the other cranial nerves?

3 Which cranial nerves carry axons of general visceral (autonomic) neurons?

4 Which cranial nerves carry impulses from the organs of the special senses (vision, hearing, smell, taste, and balance) in the head?

5 Which parts of the peripheral nervous system arise from the neural crest?

6 Explain the larger size and greater complexity of the ventral ramus compared with the dorsal ramus of a typical spinal nerve.

7 What general regions are innervated by the cervical, brachial, lumbar, and sacral plexuses?

8 When a spinal nerve is cut, why may the sensation in its dermatome still be felt?

9 What conditions appear to be necessary for successful regeneration of an axon in the peripheral nervous system?

SELF-QUIZ

Multiple-Choice Questions

15.1 The _____ cranial nerve primarily carries processes of special afferent neurons.
a optic   d vagus
b trochlear   e hypoglossal
c abducens

15.2 Which of the following is a mixed nerve?
a olfactory   d hypoglossal
b trochlear   e vestibulocochlear
c trigeminal

15.3 Special visceral efferent fibers are carried in cranial nerve _____.
a II   b III   c IV   d VI   e VII

15.4 The first cervical nerve exits below the
a rootlets of cranial nerve XI   d a and b
b occipital bone   e a and c
c first cervical vertebra

15.5 The local anesthesia given by a dentist to dull the pain of inserting a filling primarily affects the _____ nerve.
a trigeminal   d hypoglossal
b facial   e vagus
c glossopharyngeal

15.6 Which cranial nerve enables us to hear as well as to know whether we are in a balanced posture?
a III   b VI   c VII   d VIII   e X

15.7 Cranial nerves IX and XII both innervate the tongue. One difference between them is that nerve IX is a _____ nerve whereas nerve XII is a _____ nerve.

a sensory, motor   d mixed, motor
b motor, sensory   e cranial, spinal
c sensory, mixed

15.8 The only cranial nerve to innervate abdominal viscera is the

a accessory   d trigeminal
b glossopharyngeal   e vagus
c hypoglossal

15.9 Which cranial nerve(s) innervates muscles that are used for swallowing as well as for speaking?

a X   d a and b
b XI   e a and c
c XII

15.10 Neurons that comprise cranial nerves develop from the

a neural tube   d a and b
b neural crest   e a, b, and c
c vesicles and placodes

Completion Exercises

15.11 Efferent neurons conduct _____ impulses toward the central nervous system.

15.12 Neurons that carry sensory information from the skin, skeletal muscles, and joints to the central nervous system are called _____ _____ _____ neurons.

15.13 Neurons that carry motor impulses to the skeletal muscles of the limbs are called _____ neurons.

15.14 The autonomic nervous system is the efferent division of the _____ nervous system.

15.15 Motor neurons carried by cranial nerves have their cell bodies in the _____.

15.16 The unmyelinated axons that pass through the cribriform plate of the ethmoid bone belong to the _____ nerve.

15.17 The optic tract runs between the _____ and the lateral geniculate body of the thalamus.

15.18 The movement of the eyeball in its socket is controlled by eye muscles that are innervated by cranial nerves _____, _____, and _____.

15.19 Sensations of taste are conveyed to the brain along the _____ and the _____ nerves.

15.20 In the thoracic, lumbar, and sacral regions, spinal nerves are named for the vertebra _____ the point of exit from the spinal cord.

15.21 The bones that form the intervertebral foramen through which the eighth cervical nerve passes are the _____ and the _____ vertebrae.

Matching

a dorsal ramus   f cervical plexus
b ventral ramus   g brachial plexus
c rami communicantes   h lumbar plexus
d meningeal ramus   i sacral plexus
e dermatome   j coccygeal plexus

15.22 _____ supplies the skin in the region below the sacrum

15.23 _____ network of which the phrenic nerve that innervates the diaphragm is one branch

15.24 _____ a mixed nerve that innervates the skin and muscles of the back

15.25 _____ network of which the sciatic nerve is one branch

15.26 _____ a mixed nerve that innervates the vertebral column and meninges

15.27 _____ a mixed nerve that innervates the limbs

15.28 _____ a mixed nerve that carries autonomic motor fibers

15.29 _____ comprised of ventral rami of C5 to C8 and T1 spinal nerves

15.30 _____ a body segment that is innervated by a single dorsal root

15.31 _____ network of which the femoral nerve is a major branch

A SECOND LOOK

In the following drawing, label the dorsal root, ventral root, spinal nerve, dorsal ramus, ventral ramus, and rami communicantes.

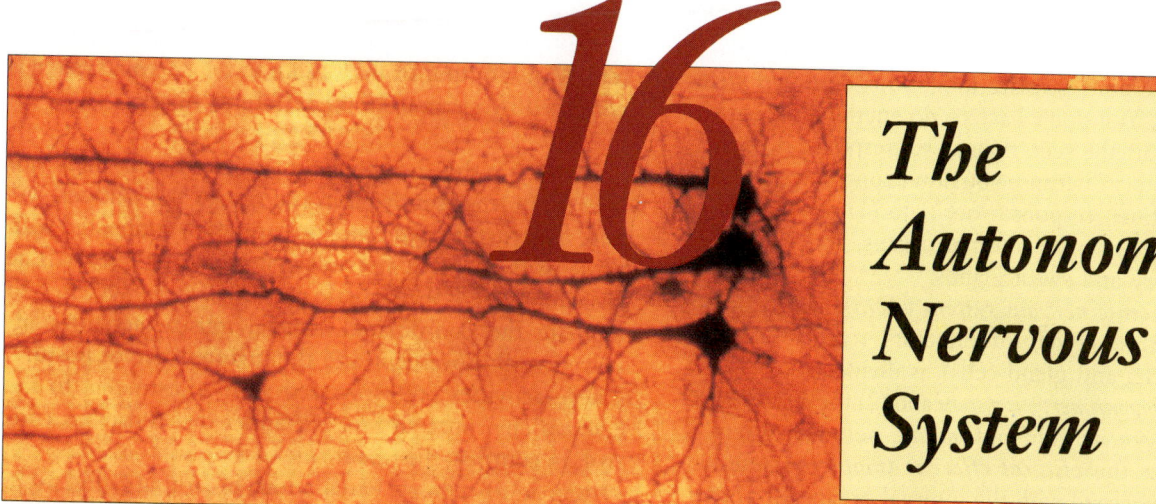

16

The Autonomic Nervous System

KEY CONCEPTS

1 The main function of the autonomic nervous system is to promote homeostasis. Most regulatory actions of the autonomic nervous system are not under conscious control.

2 The autonomic nervous system is divided into the sympathetic, parasympathetic, and enteric divisions.

3 Several centers in the brain and spinal cord are involved with autonomic control of internal organs.

4 Apart from the cerebrum, the hypothalamus is the main regulatory center of the autonomic nervous system.

5 Responses to impulses from nerves of the autonomic nervous system are due partly to the particular neurotransmitter released and partly to the nature of receptor sites on the effectors.

When you see your favorite food, your digestive juices begin to flow. You don't have to do anything except see the food to start the secretions. This response is considered "involuntary" because you don't control it on a conscious level. Such responses occur in (1) *cardiac muscle,* (2) *smooth muscle* of the internal organs, and (3) *glands,* such as salivary and sweat glands.

The **autonomic nervous system (ANS),** which innervates these three types of effectors (cardiac muscle, smooth muscle, and gland cells), is divided into (1) the *peripheral autonomic nervous system* and (2) the *central autonomic control centers.* The autonomic nervous system is also known as the **visceral efferent (motor) system** because it is concerned with the internal organs, or *viscera. Its primary function is to regulate visceral activities* to maintain homeostasis.

Two important points about the autonomic nervous system must be understood at the outset:

1 The autonomic nervous system is conventionally defined as a specific part of the peripheral nervous system that influences the activity of cardiac muscle, smooth muscle, and glands.

2 Sensory input to the autonomic nervous system is derived from sensory receptors and is conveyed by *both* the somatic afferent system and the visceral afferent system. Motor output from the autonomic nervous system is conveyed by visceral motor (efferent) fibers.

Although the autonomic nervous system does not operate independently from the rest of the nervous system, its structure and functions are often contrasted with those of the somatic nervous system [FIGURE 16.1]. The basic function of the somatic motor (efferent) system is to regulate the coordinated skeletal muscle activities involved with movement and the maintenance of posture. These activities are associated with adjustments to the *external* environment and are under our conscious control. In contrast, the major role of the autonomic nervous system is to regulate circulation, breathing,* digestion, and other internal functions *not* usually subject to our conscious control, in order to maintain a relatively stable internal environment. The two systems often work together. For example, when you are exposed to the cold, you shiver, and the blood vessels in your skin constrict. Shivering, a reaction of skeletal muscles that produces heat, is controlled by the somatic motor system. Constriction of blood vessels is an action of the autonomic nervous system that conserves heat.

*Inhalation and exhalation are controlled by the somatic motor system, while the smooth muscle of the trachea, bronchi, and bronchioles is controlled by the autonomic nervous system.

FIGURE 16.1 CROSS SECTION OF THE SPINAL CORD

The drawing shows the pathways of **[A]** somatic nerves and **[B]** visceral nerves. Afferent nerve fibers are purple, and efferent nerve fibers are blue. Dotted lines represent efferent postganglionic fibers.

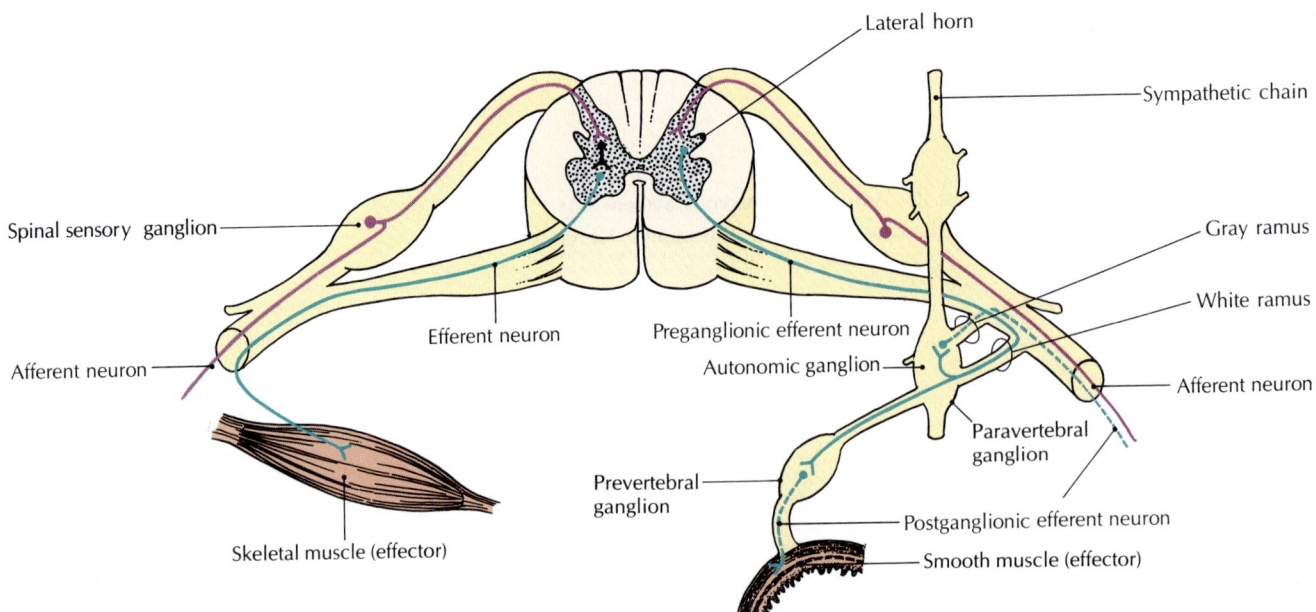

[A] SOMATIC REFLEX ARC

[B] AUTONOMIC (VISCERAL) REFLEX ARC

STRUCTURE OF THE PERIPHERAL AUTONOMIC NERVOUS SYSTEM

The *peripheral autonomic nervous system* is a motor system consisting of two divisions, the *sympathetic* and *parasympathetic*. Each division sends efferent nerve fibers to the muscle, organ, or gland it innervates. In general, but not always, the two divisions have opposite effects. In the broadest terms, we can say that the sympathetic division helps the body adjust to stressful situations and that the parasympathetic division is active when the body is operating under normal conditions.

Anatomical Divisions of the Peripheral Autonomic Nervous System

The efferent nerve fibers of the autonomic nervous system emerge from the central nervous system at several different levels. The visceral efferent fibers that emerge with the thoracic and lumbar spinal nerves constitute the *thoracolumbar* (thuh-RASS-oh-LUM-bar) outflow of the sympathetic division. At the *cranial* level, visceral efferent fibers leave the central nervous system by way of cranial nerves III, VII, IX, and X from the brainstem. At the *sacral* level, other visceral efferent fibers leave by way of sacral spinal nerves 2, 3, and 4. These two groups make up the *craniosacral* outflow of the parasympathetic division of the autonomic nervous system [FIGURE 16.2].

Preganglionic and Postganglionic Neurons

The number of neurons between the central nervous system and effectors (muscles and glands) differ in the somatic nervous system and the autonomic nervous system: The somatic system utilizes *one neuron*, while the autonomic system uses *two neurons*.

Each somatic lower motor neuron has an axon that courses from its cell body in the brainstem or spinal cord through a cranial or spinal nerve directly to a skeletal muscle. Thus one neuron links the central nervous system and the effector, and there are no synapses in the pathway.

In contrast, in the autonomic nervous system, the two neurons in the pathway synapse in *autonomic ganglia,* which are outside the central nervous system. (Autonomic ganglia are clusters of cell bodies, dendrites, and short interneurons that are contained wholly within the ganglion.) The first neuron in the autonomic two-neuron pathway is called a *preganglionic neuron* [FIGURE 16.3]. Its cell body is in the brainstem or spinal cord, and its *myelinated axon* courses through a cranial or spinal peripheral nerve. The axon terminates with the dendrites and cell body of one or more neurons in an autonomic ganglion. The second neuron in a two-neuron linkage is called a *postganglionic neuron.* Its cell body is in an autonomic ganglion, and its *unmyelinated axon* courses

FIGURE 16.2 ANATOMICAL DIVISIONS OF THE AUTONOMIC NERVOUS SYSTEM

Visceral efferent fibers that emerge from the CNS at the cranial and sacral levels make up the craniosacral, or parasympathetic, division of the ANS, while visceral efferent fibers that emerge from the thoracic and lumbar levels make up the thoracolumbar, or sympathetic, division of the ANS.

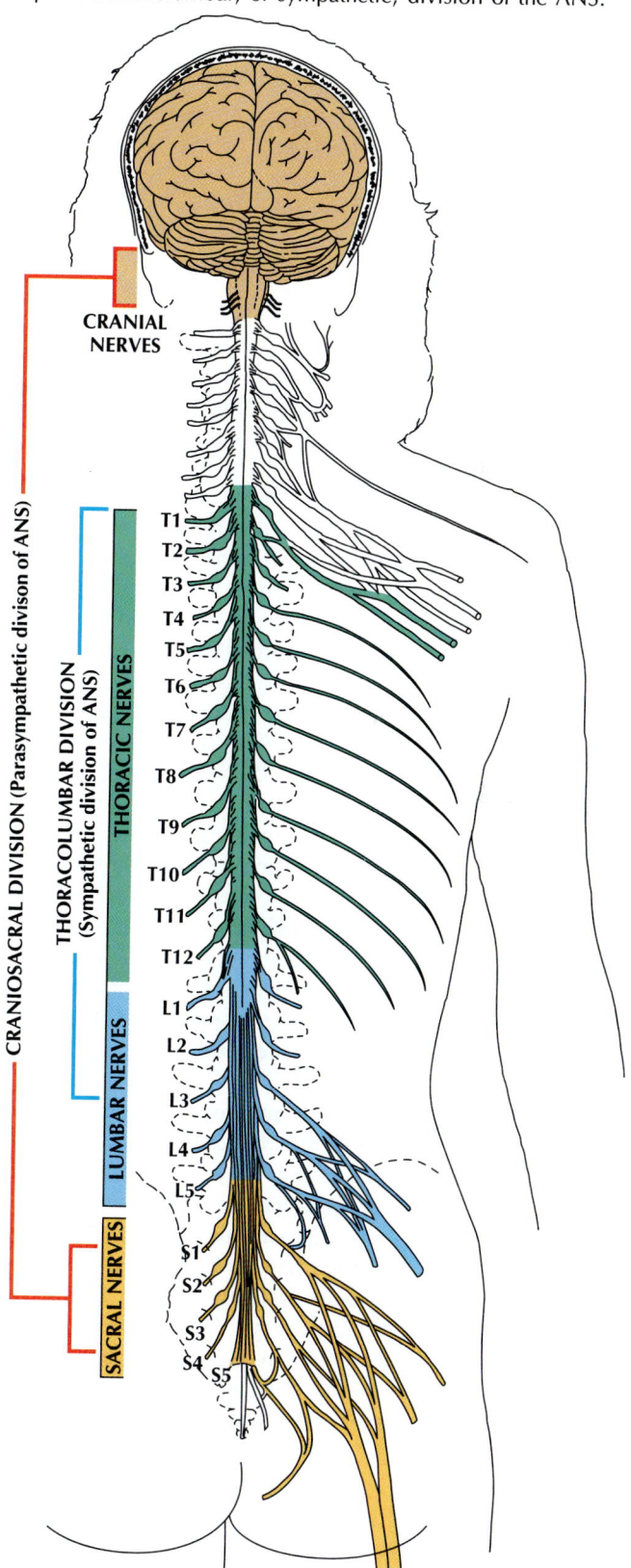

CRANIAL NERVES

CRANIOSACRAL DIVISION (Parasympathetic divison of ANS)

THORACOLUMBAR DIVISION (Sympathetic division of ANS)

THORACIC NERVES

LUMBAR NERVES

SACRAL NERVES

T1 T2 T3 T4 T5 T6 T7 T8 T9 T10 T11 T12
L1 L2 L3 L4 L5
S1 S2 S3 S4 S5

FIGURE 16.3 PREGANGLIONIC AND POSTGANGLIONIC NEURONS

This simplified representation shows preganglionic and postganglionic neurons in an efferent pathway of the autonomic nervous system.

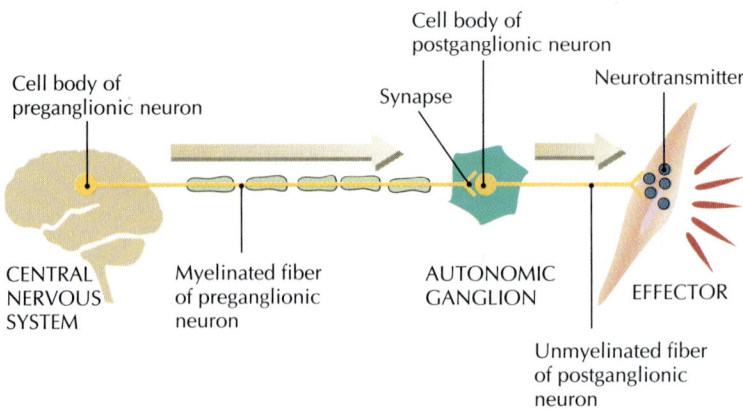

through nerves and plexuses to a motor ending on cardiac muscle, smooth muscle, or a gland.

Autonomic Ganglia

Autonomic ganglia are arranged in three groups: (1) paravertebral, or lateral, ganglia, (2) prevertebral, or collateral, ganglia, and (3) terminal, or peripheral, ganglia. The first two are part of the sympathetic division, and the third is part of the parasympathetic division:

Paravertebral ganglia The sympathetic *paravertebral* (*para*, beside) (lateral) *ganglia* form beadlike rows of 21 or 22 swellings (3 cervical, 10 or 11 thoracic, 4 lumbar, 4 sacral) that run down both sides of the vertebral column, *outside* the vertebral column. These beadlike ganglia are connected by intervening nerve fibers to form two bilateral vertical ganglionic chains, which are also known as the *sympathetic trunk ganglia* [FIGURE 16.4A]. These ganglia extend from the upper cervical vertebrae to the coccyx. Because these paravertebral (sympathetic trunk) ganglia are located near the spinal cord, the sympathetic preganglionic fibers are short.

Prevertebral ganglia Sometimes a preganglionic fiber passes through the sympathetic trunk ganglia without forming a synapse. Instead, it joins with sympathetic *prevertebral* (collateral) *ganglia,* which lie in front of the vertebrae near the large thoracic, abdominal, and pelvic arteries that give them their names. The largest prevertebral ganglia are the *celiac* (solar) ganglion just below the diaphragm (near the celiac artery), the *superior mesenteric* ganglion in the upper abdomen (near the superior mesenteric artery), and the *inferior mesenteric* ganglion near the lower abdomen (near the inferior mesenteric artery) [FIGURE 16.4A].

Terminal ganglia The preganglionic fibers of the parasympathetic division synapse in *terminal* (peripheral) *ganglia.* These ganglia are composed of small collections of ganglion cells, which are very close to or within the organs they innervate [FIGURES 16.4B and 16.5]. Thus, these parasympathetic preganglionic fibers, which extend from the central nervous system to the terminal ganglia in the innervated organ, are long. Terminal ganglia are especially common in the gastrointestinal tract and urinary bladder.

Plexuses

Some postganglionic nerve fibers are distributed like cables to branching, interlaced networks that follow along the blood vessels in the thoracic, abdominal, and pelvic cavities and are called *autonomic plexuses.* The autonomic plexuses include the following:

1 The *cardiac plexus* lies among the large blood vessels emerging from the base of the heart. It has a regulatory effect on the heart.

2 Most of the *pulmonary* (L. lung) *plexus* is located posterior to each lung. Sympathetic nerve fibers dilate the bronchi (delicate air passages leading from the trachea to the lungs), and parasympathetic fibers constrict them.

3 The *celiac* (Gr. abdomen) *plexus,* or *solar plexus,* is one of the largest masses of nerve cells outside the central nervous system. It lies on the aorta in the vicinity of the celiac artery, behind the stomach. A sharp blow to the solar plexus, just under the diaphragm below the sternum, may cause unconsciousness by slowing the heart rate and reducing the blood supply to the brain.

4 The *hypogastric* ("under the stomach") *plexus* connects the celiac plexus with the pelvic plexuses below. It

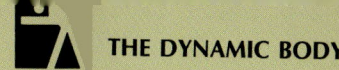
Biofeedback and the Autonomic Nervous System

The activity of the autonomic nervous system in response to an ever-changing internal environment goes on without our conscious awareness or control. Some of these automatic and involuntary functions involve heart rate, blood pressure, respiratory rate, blood-glucose concentration, and body temperature. We can have some *conscious* control over these activities through **biofeedback,** a technique that allows a person to monitor and control his or her own bodily functions.

In biofeedback, a mechanical device is used to register and display signs of physiological responses to make the person aware of them and to enable him or her to monitor changes in them. In general, any physiological function that can be recorded, amplified by electronic in-

struments, and fed back to the person through any of the five senses can be regulated to some extent by that person.

Biofeedback can work because every change in a physiological state is accompanied by a responsive change in a mental (or emotional) state. Conversely, every change in a mental state is accompanied by a change in the physiological state. The autonomic nervous system thus acts as the connecting link between the mental and physiological states.

Biofeedback is not new. Yoga and Zen masters of Eastern cultures have long demonstrated their ability to control bodily functions that were thought to be strictly involuntary. Now, Western culture has devised a technique through which, under the proper conditions,

anyone can learn to duplicate these once "impossible" physiological feats.

Biofeedback control holds considerable promise as a means of treating some psychosomatic problems (illnesses caused mainly by stress or psychological factors) such as ulcers, anxiety, and phobias, as well as many other disorders. Biofeedback can also relieve muscle-tension headaches, reduce the pain of migraine headaches, achieve relaxation during childbirth, lower blood pressure, alleviate irregular heart rhythms, and control epileptic seizures. The greatest value of these biofeedback techniques is that they demonstrate that the autonomic nervous system is not entirely autonomous and automatic; some visceral responses can be controlled.

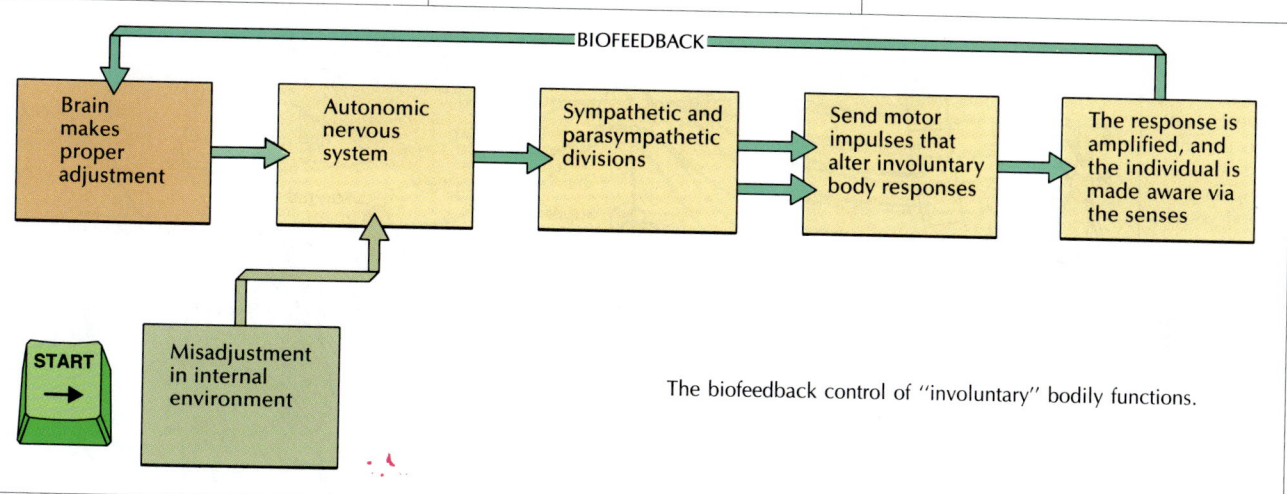

The biofeedback control of "involuntary" bodily functions.

innervates the organs and blood vessels of the pelvic region.

5 The *enteric* (Gr. intestine) *plexuses* contain both sympathetic and parasympathetic fibers, as well as fibers of the enteric division. Located between the longitudinal and circular muscles of the digestive system, they help regulate *peristalsis* (rhythmic contraction of the smooth muscles in the digestive system).

Certain nerves that innervate the viscera of the thorax, abdomen, and pelvis are called *splanchnic* (SPLANK-nick) (*visceral*) *nerves*. These nerves are composed of preganglionic or postganglionic fibers of either the sympathetic or the parasympathetic nervous system; thoracic splanchnic nerves innervate the heart and lungs, and pelvic splanchnic nerves innervate the pelvic viscera.

Sympathetic Division

The **sympathetic division** of the autonomic nervous system arises from cell bodies in the lateral gray horn of the spinal cord. The myelinated nerve fibers emerge from the spinal cord in the ventral nerve roots of the 12 thoracic and first 2 or 3 lumbar spinal nerves. This emergence of fibers is known as the *thoracolumbar outflow*. These preganglionic fibers form small nerve bundles called **white rami** (sing. *ramus*) [FIGURE 16.1]. They are white because the nerve fibers are myelinated. The fibers then pass to the paravertebral ganglia of the sympathetic chain (trunk) [FIGURE 16.5].

When a preganglionic neuron reaches the sympathetic chain, it may take one of several pathways:*

*Although we speak of several pathways, the impulses actually spread over all the pathways at once.

FIGURE 16.4 THE DIVISIONS OF THE AUTONOMIC NERVOUS SYSTEM

Nerve fibers actually emerge from both sides of the cord. **[A]** The sympathetic division. **[B]** The parasympathetic division.

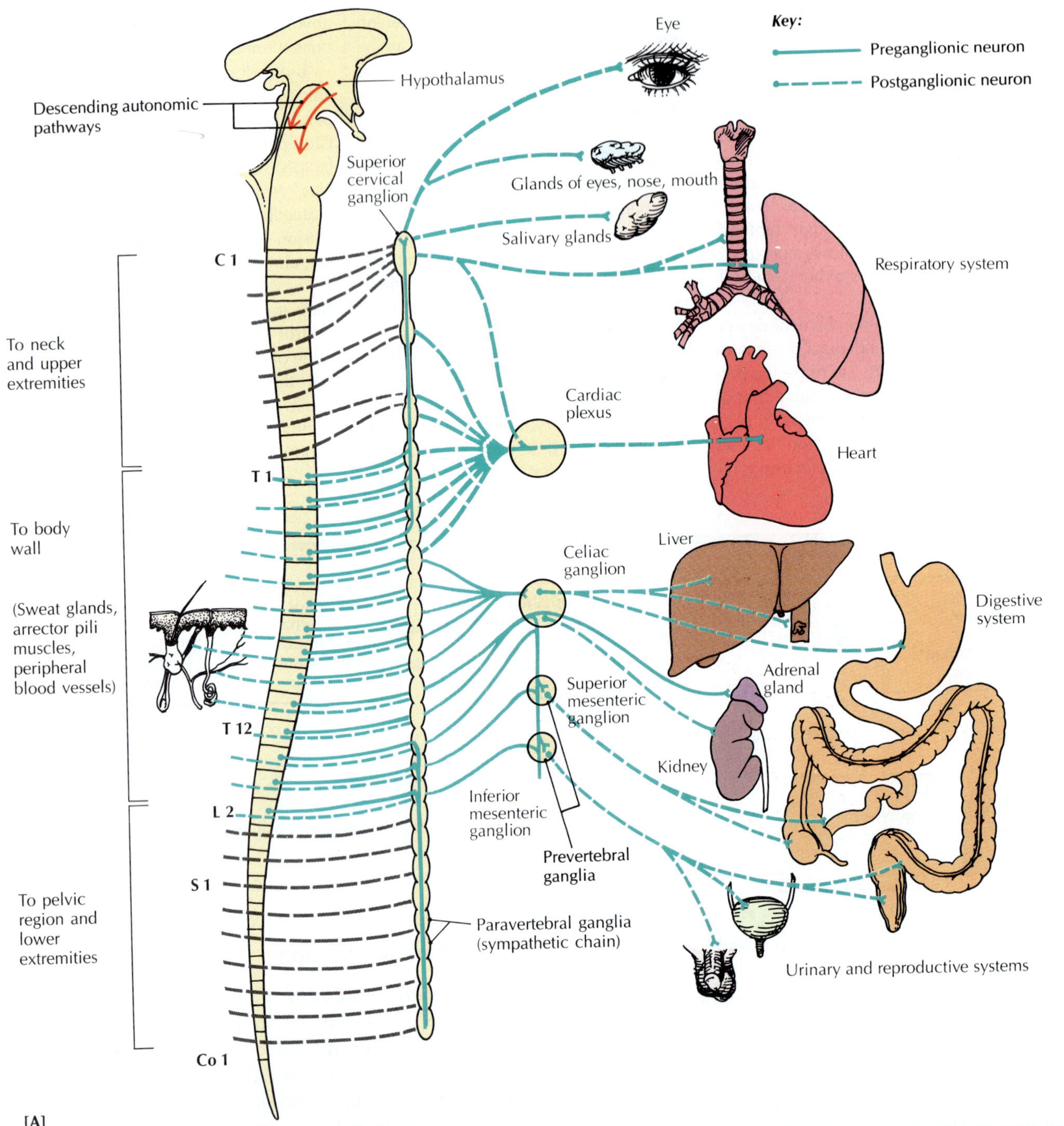

Key:
———— Preganglionic neuron
- - - - - Postganglionic neuron

Hypothalamus

Descending autonomic pathways

Superior cervical ganglion

Eye

Glands of eyes, nose, mouth

Salivary glands

Respiratory system

C 1

To neck and upper extremities

Cardiac plexus

Heart

T 1

To body wall

(Sweat glands, arrector pili muscles, peripheral blood vessels)

Celiac ganglion

Liver

Digestive system

Superior mesenteric ganglion

Adrenal gland

T 12

Kidney

Inferior mesenteric ganglion

L 2

Prevertebral ganglia

S 1

To pelvic region and lower extremities

Paravertebral ganglia (sympathetic chain)

Urinary and reproductive systems

Co 1

[A]

Part of body	Sympathetic effect
Eye (iris)	Dilates pupil
Salivary glands	Increases secretion of thick, viscous saliva
Lungs (bronchial tubes)	Causes bronchodilation
Heart	Increases heart rate and blood pressure, dilates coronary arteries, increases blood flow to voluntary muscles
Intestinal walls	Causes relaxation, decreases peristalsis, contracts sphincters
Urinary bladder	Inhibits constriction of bladder wall, contracts sphincters
Sweat glands	Increases secretion
Peripheral blood vessels	Constricts

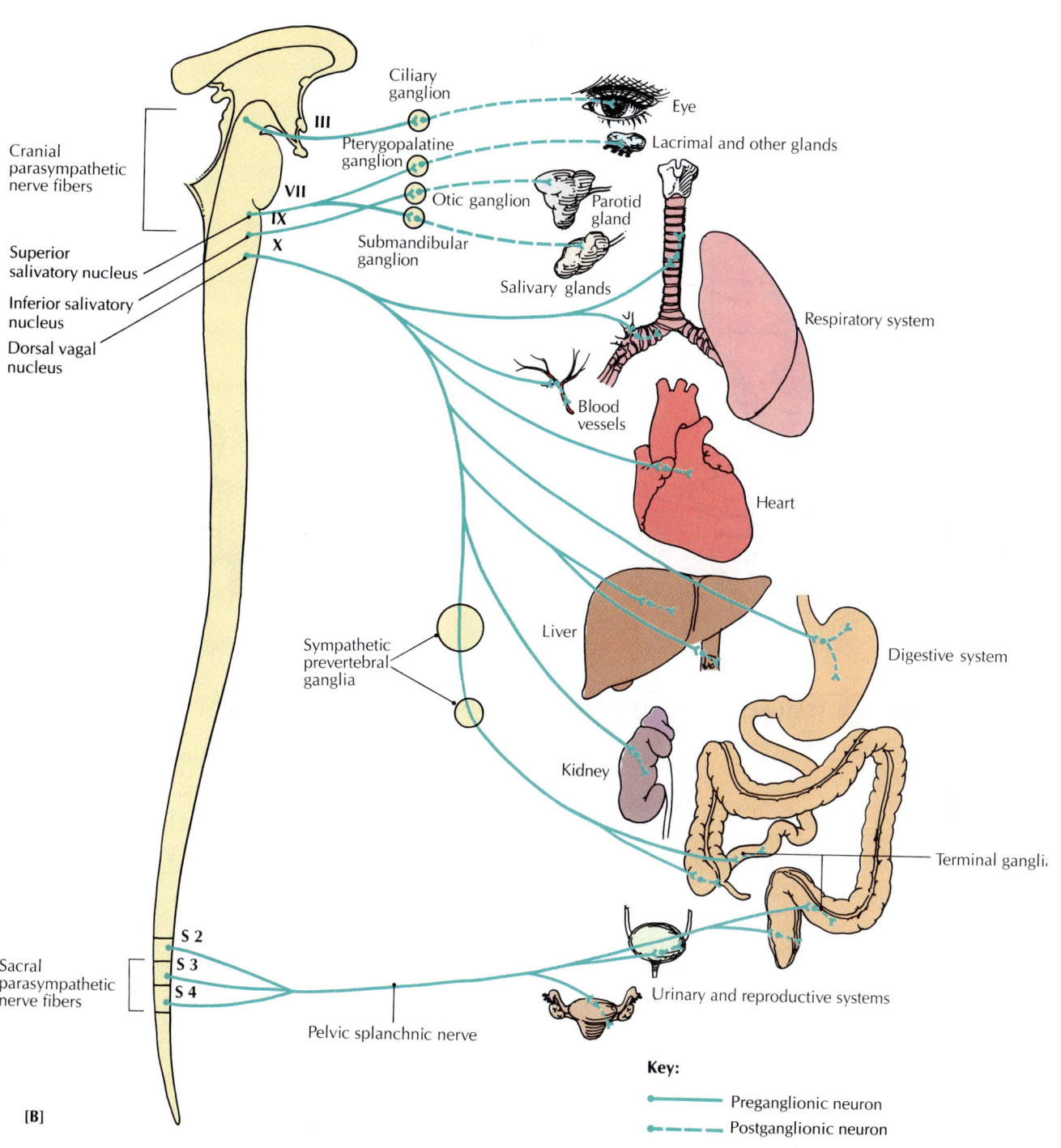

Cranial parasympathetic nerve fibers

III

Ciliary ganglion

Pterygopalatine ganglion

VII

Otic ganglion

IX

X

Submandibular ganglion

Superior salivatory nucleus

Inferior salivatory nucleus

Dorsal vagal nucleus

Eye

Lacrimal and other glands

Parotid gland

Salivary glands

Respiratory system

Blood vessels

Heart

Sympathetic prevertebral ganglia

Liver

Digestive system

Kidney

Terminal ganglia

S 2

S 3

S 4

Sacral parasympathetic nerve fibers

Pelvic splanchnic nerve

Urinary and reproductive systems

[B]

Key:

⎯⎯⎯ Preganglionic neuron

- - - Postganglionic neuron

Part of body	Parasympathetic effect
Eye (iris)	Constricts pupil
Salivary glands	Increases secretion of thin watery saliva
Lungs (bronchial tubes)	Causes bronchoconstriction.
Heart	Decreases heart rate, constricts coronary arteries, decreases blood flow to skeletal muscles
Intestinal walls	Causes contraction, increases peristalsis, relaxes sphincters
Urinary bladder	Stimulates contraction of bladder wall, relaxes sphincters
Sweat glands	No innervation
Peripheral blood vessels	No innervation for many vessels

FIGURE 16.5 SYMPATHETIC AND PARASYMPATHETIC PATHWAYS

Note that the parasympathetic synapse occurs close to the effector and that the neurotransmitters released onto the effector organ are different. The parasympathetic postganglionic fiber releases acetylcholine, and the sympathetic postganglionic fiber releases norepinephrine.

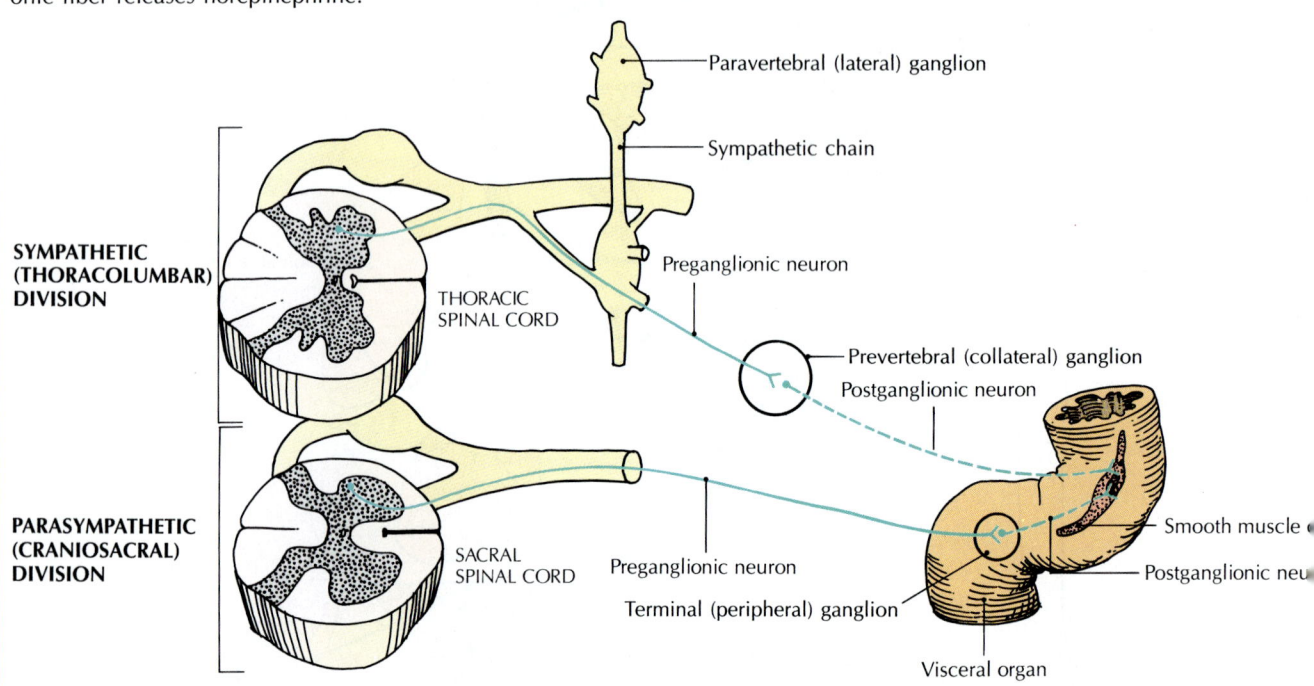

COMPARISON OF SOME FEATURES OF THE SYMPATHETIC AND PARASYMPATHETIC DIVISIONS

Feature	Sympathetic division	Parasympathetic division
Distribution	Throughout the body	Viscera of head, thorax, abdomen, pelvis
Location of ganglia	Paravertebral and prevertebral ganglia close to CNS	Terminal ganglia near effectors
Outflow from CNS	Thoracolumbar levels	Craniosacral levels
Ratio of preganglionic to postganglionic neurons	Each preganglionic neuron synapses with many postganglionic neurons	Each preganglionic neuron synapses with a few postganglionic neurons
Neurotransmitter at preganglionic nerve terminals	Acetylcholine	Acetylcholine
Neurotransmitter at neuroeffector junction	Usually norepinephrine	Acetylcholine
Deactivation of neurotransmitter	Slow, by way of monoamine oxidase or catechol-O-methyl transferase	Rapid, by way of acetylcholinesterase

SOME EFFECTS OF THE AUTONOMIC NERVOUS SYSTEM

Visceral effector	Sympathetic division	Parasympathetic division
Blood vessels	Generally constricted	Generally dilated slightly
External genitalia	Constricted	Dilated
Salivary glands	Constricted	Dilated
Skeletal muscles	Dilated (as a result of cholinergic neurons)	No innervation
Skin	Constricted	Dilated slightly
Viscera	Constricted	Dilated
Eye		
Muscles of iris	Radial (dilator) muscle contracted (pupil dilated)	Sphincter muscle contracted (pupil contracted)
Ciliary muscle	Relaxed for far vision	Contracted for near vision
Heart		
Rate	Increased	Decreased
Strength of contraction	Increased	Decreased
Gastrointestinal tract		
Motility and tone	Inhibited	Stimulated
Sphincters	Stimulated (contracted)	Inhibited (relaxed)
Secretion	Probably inhibited	Stimulated
Gallbladder	Inhibited	Stimulated
Liver	Increased glycogenolysis (resulting in increased blood glucose)	No innervation
Lungs		
Bronchial tubes	Lumen dilated	Lumen constricted
Bronchial glands	No innervation	Secretion stimulated
Bronchial muscle	Relaxed	Contracted
Glands of head		
Lacrimal glands	No innervation	Secretion stimulated
Salivary glands	Increased viscous secretion	Profuse water secretion
Urinary bladder	Relaxed	Contracted
Skin		
Sweat glands	Stimulated to secrete	No innervation
Blood vessels	Constricted	No innervation
Piloerector muscles	Contracted (hair stands erect)	No innervation
Sex organs	Vasoconstricted (orgasm)	Vasodilated (erection in male, lubrication in female)
Adrenal medulla	Epinephrine, norepinephrine secreted (innervated by preganglionic cholinergic sympathetic neurons)	No innervation

Source: Adapted from Charles R. Noback, Norman L. Strominger, and Robert J. Demarest, *The Human Nervous System,* 4th ed., Lea & Febiger, Philadelphia, 1991, pp. 330–331. Used with permission.

FIGURE 16.6 COMPARISON OF SYMPATHETIC AND PARASYMPATHETIC SYSTEMS AND THEIR NEUROTRANSMITTERS

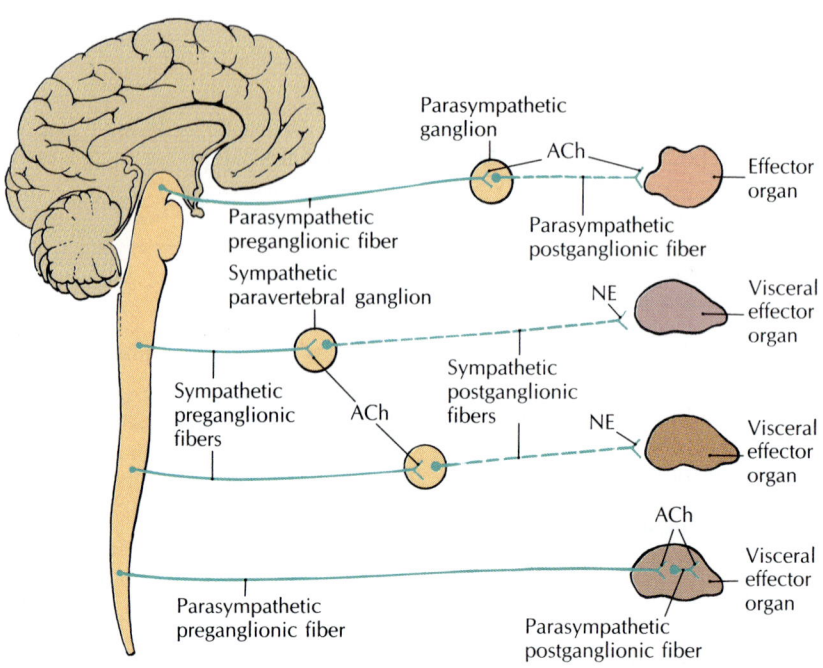

1 It may synapse with a postganglionic neuron, which then terminates on an effector. A preganglionic neuron may synapse with a postganglionic neuron at the same level in the sympathetic trunk. Then the axons of the postganglionic neuron joins the spinal nerve by way of the *gray rami* [FIGURE 16.1] before it terminates in an effector. These rami are gray because their fibers are unmyelinated.

2 In another pathway, the preganglionic neuron may course up or down to a different level of the sympathetic chain before synapsing with postganglionic neurons at many levels. The axons of the postganglionic neuron join the spinal nerve by way of the gray rami.

3 In still another pathway, the preganglionic neuron may pass directly through a ganglion (without synapsing) in the sympathetic trunk to synapse with postganglionic neurons within a prevertebral (collateral) ganglion [FIGURE 16.4A].

Each preganglionic fiber may synapse with 20 or more postganglionic neurons in a ganglion. Some axons of the postganglionic neurons may pass directly to an effector (such as the heart or lungs), but most postganglionic axons pass through the gray rami and travel back to the spinal nerve to be distributed for the innervation of sweat glands, as well as of the smooth muscles of blood vessels, the arrector pili muscles of the skin (which make the hair stand erect), the body wall, and the limbs.

Sympathetic neurotransmitters and receptors
The neurotransmitter released by the preganglionic nerve terminals is *acetylcholine (ACh)* [FIGURE 16.6]. It is inactivated to choline and acetate by the enzyme *acetylcholinesterase*. The neurotransmitter released by the postganglionic nerve terminal is *norepinephrine (NE)*. However, there are a few exceptions. Sympathetic postganglionic fibers to most sweat glands and to some blood vessels release acetylcholine instead of norepinephrine.

Only a small amount of the released norepinephrine stimulates the receptors of the postganglionic neuron or effector. The excess neurotransmitter is rapidly inactivated by the enzyme *catechol-O-methyl transferase* (located in the region of the synaptic cleft) or taken up by the nerve terminal for recycling. The norepinephrine that is taken up is inactivated within the terminal by the enzyme *monoamine oxidase*.

Based on the type of neurotransmitter released, neurons are classified as either *adrenergic* or *cholinergic*. *Adrenergic neurons* release norepinephrine or epinephrine, and *cholinergic neurons* release acetylcholine. Sympathetic postganglionic neurons are classified as adrenergic because most postganglionic neurons of this division release norepinephrine (also called *noradrenaline*). The *preganglionic* neurons of the sympathetic division release acetylcholine and thus are classified as cholinergic neurons.

Adrenergic receptors Located in or on the plasma membrane of the cells of effector organs are specific protein receptors that respond to norepinephrine and epinephrine. (These two neurotransmitters are called *catecholamines* because they both have a six-carbon catechol

FIGURE 16.7 TWO CATECHOLAMINES

Note that the only difference between norepinephrine and epinephrine is a methyl group (—CH₃) in epinephrine (red), which replaces hydrogen.

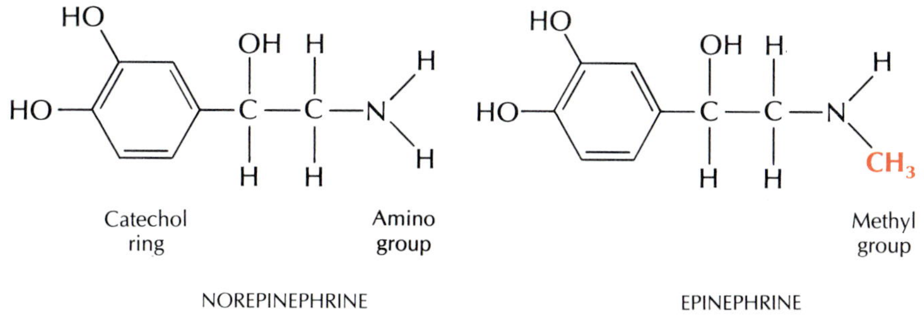

Catechol ring   Amino group   Methyl group

NOREPINEPHRINE   EPINEPHRINE

FIGURE 16.8 PLASMA MEMBRANE OF AN EFFECTOR ORGAN

Shown are alpha-receptor protein sites for norepinephrine and beta-receptor protein sites for epinephrine.

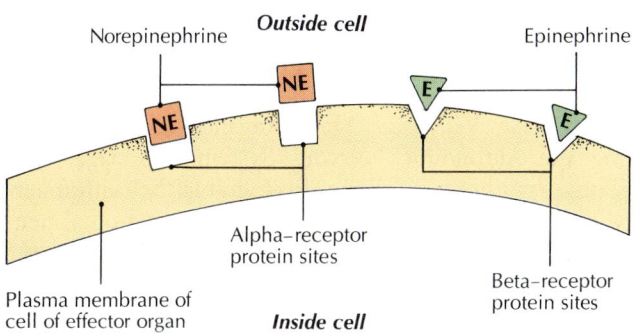

Norepinephrine   *Outside cell*   Epinephrine

Alpha–receptor protein sites

Plasma membrane of cell of effector organ

Inside cell

Beta–receptor protein sites

ring and an amine group [FIGURE 16.7].) There are two main types of postsynaptic adrenergic receptors: *alpha (α) receptors* and *beta (β) (heart) receptors* [FIGURE 16.8]. Each group has subtypes: α_1, α_2, β_1, β_2. The type of receptor determines the specific physiological response it produces and its high specificity for certain drugs that either excite or inhibit it. Norepinephrine interacts mainly with alpha receptors, and epinephrine interacts equally with α_2 and beta receptors. Alpha receptors usually elicit excitatory effects, whereas some beta receptors elicit excitatory effects and others elicit inhibitory effects.

A response by a particular effector is due partly to the specific neurotransmitter and partly to the nature of the receptor site on which the transmitter acts. The same neurotransmitter may have a different effect on different effectors. For example, norepinephrine stimulates the smooth muscle of an arteriole (a small, terminal branch of an artery) to contract, but it stimulates the smooth muscle of the bronchial tubes to dilate. Acetylcholine stimulates skeletal muscle to contract, but it may cause smooth muscle to relax.

Widespread effects of sympathetic stimulation

The sympathetic division of the autonomic nervous system is anatomically and physiologically organized to affect *widespread* regions of the body, or even the entire body, for sustained periods of time. For example, each short preganglionic neuron synapses with numerous long postganglionic neurons, which terminate in neuroeffector synapses over a large area. Long-lasting sympathetic effects are directly related to the slow inactivation of norepinephrine and to the extensive distribution of norepinephrine and release of epinephrine from the adrenal medulla into the bloodstream.

Parasympathetic Division

The *parasympathetic division* (Gr. *para*, beside, beyond; that is, located beside or around the sympathetic division) of the autonomic nervous system has long preganglionic neurons that originate in the brainstem and *lateral horns* of sacral levels of the spinal cord and short postganglionic neurons. The cranial portion supplies parasympathetic innervation to the muscles and glands in the head, neck, thorax, and most of the abdominal viscera [FIGURE 16.4B]. The sacral portion supplies parasympathetic innervation to the smooth muscle and glands of the viscera in the lower abdomen and pelvis.

The preganglionic fibers from the cranial portion of the parasympathetic division are known as the *cranial parasympathetic outflow*. These fibers emerge from the brainstem through cranial nerves III, VII, IX, and X. The preganglionic fibers from the sacral portion are the *sacral parasympathetic outflow*. They leave the spinal cord by way of the ventral roots of spinal nerves S2, S3, and S4.

Parasympathetic neurotransmitter and receptors The neurotransmitter of the parasympathetic division is acetylcholine [FIGURE 16.6]. Both the preganglionic and postganglionic neurons release acetycholine, and therefore are classified as *cholinergic neurons.* Two types of cholinergic receptors are (1) *nicotinic receptors* (receptors that respond to nicotine, the tobacco plant derivative) and (2) *muscarinic receptors* (receptors that respond to muscarine, an alkaloid produced by the poisonous mushroom *Amanita muscaria*).

Nicotinic receptors are found on sympathetic and parasympathetic postganglionic neurons of both the sympathetic and parasympathetic divisions, and on the postsynaptic membranes (sarcolemma) of skeletal muscle cells in the somatic nervous system. Thus, nicotinic receptors respond to acetylcholine released from both sympathetic and parasympathetic preganglionic fibers. Muscarinic receptors are found on effectors innervated by the parasympathetic postganglionic neurons of the autonomic nervous system (for example, the heart, salivary and sweat glands, and urinary bladder). These receptors bind with acetylcholine, which is released from parasympathetic postganglionic neurons and usually inhibits (relaxes) smooth muscle (in some organs) and cardiac pacemaker cells, but excites smooth muscle in other areas.

Two examples of useful anticholinergic drugs and their derivatives are *atropine* and *homatropine.* Atropine is an antimuscarinic alkaloid agent derived from the belladonna plant (also called "deadly nightshade"). Its inhibitory action on the activity of acetylcholine is utilized to decrease bronchial secretions and sweating or to increase heart rate. Homatropine, a semisynthetic drug used by ophthalmologists when examining the eyes, blocks the sphincter muscles of the iris that regulate the size of the pupil and the ciliary muscles that adjust the lens. As a result, the pupil is dilated and the ability to accommodate (focus) is lost, producing an extreme sensitivity to light and a lens that is fixed for distant vision only. (This is why your near vision becomes blurred after your ophthalmologist puts drops in your eyes.)

Discrete effects of parasympathetic stimulation In general, the parasympathetic division responds to a specific stimulus in a discrete region for a short time. This effect occurs because the postganglionic neurons have short axons that are distributed for short distances to specific areas [FIGURE 16.5]. Also, the rapid deactivation of acetylcholine by acetylcholinesterase results in a short-term effect by the neurotransmitter.

Enteric Division

The *enteric division* (ED, enteric nervous system, or "gut brain") is now considered to be a distinct division of the autonomic nervous system. Its neural networks and plexuses extend from the esophagus to the rectum (that is, through the gastrointestinal tract). Its neuronal population is about 100 million neurons—approximately equivalent to the number of neurons in the spinal cord. The enteric division can function independently even when completely deprived of innervation by the central nervous system. In essence, the sympathetic and parasympathetic divisions modulate and regulate the activity of the enteric circuitry.

It is important to draw a clear distinction between the autonomic nerves that are *intrinsic* to the gut (enteric division) and those that are *extrinsic* (sympathetic and parasympathetic divisions). The neurons of the enteric division do not have axons that synapse directly with neurons within the central nervous system.

The enteric division is organized as the gastrointestinal tract's own miniature nervous system; it consists of neural networks composed of sensory neurons, interneurons, and motor neurons. Its sensory neurons, whose dendrites orginate from the mucosal lining cells, act as pressure receptors responding to the distension forces exerted by the gut. More than a dozen neurotransmitters and neuromodulators are present in the gut neurons. They include acetylcholine, norepinephrine, adenosine triphosphate (ATP), serotonin, dopamine, and many peptides such as somatostatin and vasoactive intestine polypeptide (VIP) (see "Major Advances in Understanding the Autonomic Nervous System" on page 489). These "gut hormones" are also found in the central nervous system, where they are thought to function as neurotransmitters or neuromodulators just as they do in the enteric division.

ASK YOURSELF

1 Why is the sympathetic division called the thoracolumbar division and the parasympathetic division called the craniosacral division?

2 What is the main function of autonomic ganglia?

3 What is the difference between a preganglionic neuron and a postganglionic neuron?

4 What are the neurotransmitters released from the preganglionic and postganglionic neurons in the sympathetic and parasympathetic divisions of the ANS?

5 What is the sympathetic chain?

6 Where in the body is the parasympathetic division distributed?

7 What is the extent of the enteric division?

Major Advances in Understanding the Autonomic Nervous System

Some recent advances have drastically altered our understanding of the autonomic nervous system. These new findings, as discussed below, indicate that some hitherto unappreciated factors are at work in the system, including a multiplicity of transmitters, cotransmission by those transmitters, neuromodulation, and plasticity. In essence, these findings stress the greater complexity of the autonomic nervous system in modulating and controlling its effector organs.

Many neurons of sympathetic ganglia are now thought to release such acknowledged "autonomic transmitters" as serotonin, dopamine, and gamma aminobutyric acid (GABA). These neurons are said to be nonadrenergic, noncholinergic neurons (designated as NANC neurons). Even nitrous oxide

(NO) is said to have an important role in transmitting influences to certain smooth muscles. Thus, a variety of neurotransmitters are found in autonomic nerves; these neurotransmitters include such peptides as VIP (vasoactive intestinal polypeptide), enkephalin, cholecystokinin, somatostatin, and substance P. It is currently believed that some, if not all, autonomic neurons store and release more than one transmitter. For example, acetylcholine and VIP are cotransmitters of the parasympathetic neurons that innervate the salivary glands; acetylcholine increases salivary secretion, and VIP acts as a neuromodulator that enhances the action of acetycholine. (A *neuromodulator* is a substance that modifies the process of neurotransmission.)

Within sympathetic ganglia are microscopically small, intensely fluorescent (SIF) cells, which contain peptide transmitters that are thought to act as modulators at synaptic sites within each ganglion.

Many recent studies also demonstrate that autonomic nerves exhibit *plasticity*—the capacity for a neuron to modify its synaptic transmitter or receptor activity—during development and aging. For example, the postganglionic sympathetic nerves innervating the sweat glands are adrenergic before birth and become cholinergic after birth.

The significance of these advances is that they give evidence of the dynamic nature of the autonomic nervous system and its role in health and disease.

AUTONOMIC CONTROL CENTERS IN THE CENTRAL NERVOUS SYSTEM

Neural centers in many regions of the central nervous system are involved with producing specific sympathetic and parasympathetic responses in the visceral organs. These higher centers are located in the brainstem, reticular formation, hypothalamus, cerebral cortex, and structures of the limbic system. The hypothalamus is considered to be the highest and primary subcortical regulatory center of the autonomic nervous system.

The neural influences from the higher centers are exerted on the activities of the autonomic nervous system through visceral reflex arcs. Sensory inputs from receptors in the viscera are conveyed via visceral afferent fibers to the central nervous system. There, after being processed and modulated by visceral centers within the central nervous system, appropriate efferent outputs are conveyed via the preganglionic neurons of the sympathetic and parasympathetic outflow to the postganglionic neurons innervating the viscera.

Brainstem and Spinal Cord

Regulatory centers in the medulla oblongata of the brainstem are involved with visceral reflex arcs. For ex-

ample, the cardiovascular center monitors heart rate, blood pressure, and blood-vessel tone, and the respiratory center monitors arterial carbon dioxide and pH. After receiving input from sensory receptors, the cardiovascular center processes the neural information and elicits appropriate output through the autonomic preganglionic neurons in the brainstem (expressed through the cranial nerve parasympathetic outflow) and the spinal cord (thoracolumbar sympathetic outflow).

Respiratory centers in the brainstem are involved with reflex circuits that regulate the tone of the delicate bronchial tubes of the lungs, as well as of the somatic skeletal muscles such as the diaphragm and intercostal muscles involved in breathing.

Hypothalamus

The hypothalamus is the highest subcortical controlling neural *regulator* and is the *coordinating center* of the autonomic nervous system. In fact, virtually any autonomic response can be evoked by stimulating a site within the hypothalamus. Allowing for a considerable overlap, the stimulation of sites in the anterior and medial hypothalamus tends to trigger parasympathetic responses, and the stimulation of sites in the posterior and lateral hypothalamus triggers sympathetic responses. Certain somatic responses involving skeletal muscles, such as shivering, are also associated with the hypothalamus.

Within the hypothalamus are many neural circuits (nuclei), called *control centers*, which regulate such vital autonomic activities as body temperature, heart rate, blood pressure, blood osmolarity, and the desire for food and water. The hypothalamus is also involved with behavioral expressions associated with emotion, such as blushing.

Limbic System

Structures of the limbic system, such as the limbic lobe, amygdala, and hippocampus, are connected to the hypothalamus and use the hypothalamus to express their activities. These expressions include many visceral and behavioral responses associated with self-preservation (such as feeding and fighting) and preservation of the species (such as mating and care of the offspring). Electrical stimulation of the limbic lobe and hippocampus produces changes in the cardiovascular system, including alterations in the heart rate and the tone of blood vessels. Stimulation of the amygdala and limbic lobe may alter the secretory activity of digestive glands.

Visceral Reflex Arc

A *visceral reflex* is an action involving sequences of visceral afferent neurons, interneurons, and visceral efferent neurons. Visceral efferent neurons innervate cardiac muscle, smooth muscle, or glands. When stimulated by autonomic neurons, smooth muscle or cardiac muscle contracts and glands release their secretions. Such a reflex, like a somatic motor reflex, does not involve the cerebral cortex, and most visceral adjustments are made through regulatory centers, for example, in the medulla oblongata, or spinal cord, without our conscious control or knowledge.

Unlike the somatic reflex, which uses only one efferent neuron, a sequence of two efferent neurons is involved in an autonomic visceral reflex arc [see FIGURE 16.1]. A *visceral reflex arc* is made up of (1) a *receptor,* (2) an *afferent neuron* that conveys sensory influences to the central nervous system, (3) *interneurons* within the gray matter of the central nervous system that connect with preganglionic neurons in the sympathetic division, (4) *two efferent neurons* that are part of the sequence composed of preganglionic neurons, an autonomic ganglion, and a postganglionic neuron, and (5) a *visceral effector.*

Some examples of autonomic visceral reflex arcs that occur in the spinal cord are the contraction of a full urinary bladder, muscular contraction of the intestines, and constriction or dilation of blood vessels. Examples of reflex arcs integrated in the medulla oblongata include the regulation of blood pressure, heart rate, respiration, and the evocation of vomiting.

FUNCTIONS OF THE AUTONOMIC NERVOUS SYSTEM

In this section, we provide an overall picture of the autonomic nervous system as a two-part regulatory system by looking at the way the sympathetic and parasympathetic divisions balance their influences to help us react to changes and maintain homeostasis. As an example, we show how the system operates during a downhill ski race.

Example of the Operation of the Autonomic Nervous System: A Ski Race

An Olympic skier on a twisting downhill slope is concentrating every part of the body on negotiating the course faster than anyone else in the world. The skier's heart, beating up to three times faster than yours is right now, is also pumping more blood, faster, to the skeletal muscles than yours is now. The skier's pupils are dilated. The blood vessels to the skin, body organs, and salivary glands — all but those to the skeletal muscles — are constricted. The sweat glands are stimulated. Epinephrine and norepinephrine virtually pour out of the adrenal glands. Obviously ready for action, the skier shows the so-called fright, fight, or flight response, a state of heightened readiness.

At the same time, organ system functions not needed in the race are practically shut down. Digestion, urination, and defecation can wait until the race is over. Blood is directed toward the skeletal muscles. The *sympathetic division* is dominant.

When the race is over and it is time to relax and enjoy a leisurely meal, the emphasis is on the "maintenance" of normal body functions and a restoration of the body's energy resources. The *parasympathetic division* is dominant now. The heart rate and the force of its contractions are reduced, saliva and digestive juices flow freely, the pupils are constricted to protect the eyes from excessive brightness, skeletal muscles are relaxed, and the body is free once again to devote time to ridding itself of wastes.

Some of the glucose that was coming from the liver into the skeletal muscles during the race is now being diverted to other organs. So is blood. Blood vessels in the skin that were constricted to lessen the chance of serious bleeding in case of a wound are now back to normal diameter. The extra sweat that helped keep the skier cool during the race is not needed now, and the sweat glands secrete minimally, if at all. Blood vessels in the intestines dilate, while those in the skeletal muscles constrict to their normal diameter. The autonomic nervous system, together with the central nervous system, has kept the body in balance with the rigorous demands of its surroundings.

Coordination of the Two ANS Divisions

Many bodily activities involve either one autonomic division or the other. However, sexual activities require the coordinated, sequential involvement of both the sympathetic and parasympathetic divisions. Also, although the two divisions may be said to usually have opposite, or "antagonistic," effects on viscera and glands, not every structure innervated by the autonomic nervous system receives innervation from both divisions. For example, piloerector muscles in the skin may be innervated by only one (sympathetic) of the two divisions, because relaxation of the muscles is achieved simply by the termination of the sympathetic impulses that cause contraction.

Many organs receive a *dual innervation*, with apparently opposite responses to stimulation from the sympathetic and parasympathetic divisions. However, such responses do not mean that the two divisions are antagonistic in the sense that they work against each other. Rather, like antagonistic muscles, they are coordi-nated to achieve a single functional goal. The eye, for example, shows an interesting dual response to the degree of light intensity affecting it. The pupil dilates when the radial (dilator) smooth muscle cells of the iris are stimulated to contract by sympathetic fibers. When the circular sphincter (constrictor) muscle cells of the iris are stimulated to contract by parasympathetic fibers, the pupil constricts.

Responses of Specific Organs

The autonomic nervous system does not control the basic activity of the organs it innervates, but it does alter that activity. The organs innervated by the autonomic nervous system are not fully dependent on autonomic innervation. For example, if the heart is deprived of its autonomic innervation, it will still contract, but it will not respond to the changing demands of the body and will not increase its rate to as great a degree when physical activity is increased. (This ability of the heart to contract without innervation is advantageous in its transplantation from one person to another.) In contrast, a skeletal muscle deprived of its lower motor neuron innervation will not contract.

ASK YOURSELF

1 Which division of the autonomic nervous system prepares a person for intense muscular activity?

2 Do both divisions operate simultaneously? Explain.

3 What is dual innervation?

4 Does a neurotransmitter always have the same effect? Explain.

WHEN THINGS GO WRONG

Horner's Syndrome
In *Horner's syndrome,* either sympathetic preganglionics or postganglionics are interrupted. Lesions typically occur from the superior cervical ganglion to the head, or from the preganglionics from T1 to T4 to the superior cervical ganglion. The lesion blocks the flow of sympathetic activity to the head, resulting in a drooping eyelid (ptosis), constricted pupil, sunken eyeball, and flushed, dry skin. All these symptoms are ipsilateral to the lesion.

Achalasia
Achalasia (ACK-uh-lay-zhuh; Gr. failure to relax) can occur at any point along the gastrointestinal tract where the parasympathetics fail to relax the smooth muscle. In *achalasia of the esophagus,* the individual has difficulty swallowing because of a persistent contraction of the esophagus where it enters the stomach.

In *Hirschsprung's disease,* the large intestine becomes enlarged and distended because a short segment of it is continuously constricted, thereby obstructing the passage of feces. The constricted segment contains the flaw because the neurons normally present in the intestinal wall are lacking in this segment. These neurons, which are essential for normal peristalsis, are necessary for the expression of the influences of the autonomic nervous system. Treatment usually concentrates on resectioning the segment that does not have innervation.

CHAPTER SUMMARY

The *autonomic nervous system* is also called the *visceral efferent (motor) system.* Most of the functions of our internal organs (viscera) are not under our conscious control.

Structure of the Peripheral Autonomic Nervous System (p. 479)

1 The autonomic nervous system is divided into the *sympathetic, parasympathetic,* and *enteric divisions.* The sympathetic division is also called the *thoracolumbar division* because its nerve fibers emerge from the thoracic and upper lumbar spinal nerves. The parasympathetic division is called the *craniosacral division* because its visceral efferent fibers leave the CNS via cranial nerves in the brainstem and spinal nerves in the sacral region of the spinal cord. The *enteric division* is composed of neural networks and plexuses that are intrinsic to the gastrointestinal tract.

2 In the one-neuron linkage of the somatic motor system, each lower motor neuron connects directly with its effector, with no synapse outside the central nervous system. In the two-neuron linkage of the autonomic system, one neuron (*preganglionic*) synapses with a second neuron (*postganglionic*) in an *autonomic ganglion* before it synapses with an effector.

3 *Paravertebral* (lateral) and *prevertebral* (collateral) autonomic *ganglia* occur in the sympathetic division, and *terminal* (peripheral) *ganglia* occur in the parasympathetic division. Rows of paravertebral ganglia form the *sympathetic chains.*

4 Some postganglionic nerve fibers are distributed to branching networks in the thoracic, abdominal, and pelvic cavities as *autonomic plexuses.*

5 The sympathetic division arises from cell bodies in the lateral gray horn of the spinal cord. The preganglionic nerve fibers emerge from the spinal cord in the ventral roots of the 12 thoracic and first 2 or 3 lumbar spinal nerves. This emergence of fibers is the *thoracolumbar outflow;* they form the *white rami.*

6 The neurotransmitter released by sympathetic preganglionic nerve terminals is *acetylcholine.* With a few exceptions, the neurotransmitter released by the sympathetic postganglionic nerve terminals is *norepinephrine* (noradrenaline), and therefore the division is called the *adrenergic division.* The effects of norepinephrine are usually widespread and relatively long lasting.

7 *Adrenergic receptors* are specific protein-receptor sites in the plasma membrane of cells of effector organs. They respond to norepinephrine and epinephrine.

8 The preganglionic fibers from the cranial portion of the parasympathetic division are called the *cranial parasympathetic outflow.* The preganglionic fibers from the lateral gray horn of the sacral portion constitute the *sacral parasympathetic outflow.*

9 The neurotransmitter of the parasympathetic division is *acetylcholine,* and the division is classified as *cholinergic.* The effects of acetylcholine are usually short range and relatively short term.

10 *Cholinergic receptors* respond to acetylcholine at synapses. Preganglionic receptors are *nicotinic,* and postganglionic receptors are *muscarinic.*

Autonomic Control Centers in the Central Nervous System (p. 489)

1 *Neural centers* in the central nervous system, including the brainstem, reticular formation, spinal cord, hypothalamus, cerebral cortex, and structures of the limbic system, are involved with producing sympathetic and parasympathetic responses in the visceral organs.

2 Regulatory centers in the medulla oblongata of the *brainstem* are involved with influencing visceral reflex arcs.

3 The *hypothalamus* is the highest and main subcortical regulatory center of the autonomic nervous system. It modulates autonomic centers in the brainstem and spinal cord.

4 Structures of the *limbic system* utilize the hypothalamus to express their activities, which include many visceral and behavioral responses associated with self-preservation and preservation of the species.

5 The *cerebral cortex* utilizes the limbic system and hypothalamus to express some emotions.

6 A *visceral reflex arc* innervates cardiac muscle, smooth muscle, and glands. Its components are an afferent receptor, afferent neuron, interneurons, two efferent neurons, and visceral effector.

Functions of the Autonomic Nervous System (p. 490)

1 The main function of the autonomic nervous system is to promote homeostasis by regulating visceral activities, specifically activities of cardiac muscle, smooth muscle, and glands.

2 The sympathetic and parasympathetic divisions are coordinated into a balanced, complementary system that helps the body adjust to constantly changing environmental conditions.

3 Generally, the sympathetic division prepares the body for stressful situations, and the parasympathetic division is active when the body is at rest.

4 Many organs receive a *dual innervation,* with opposite responses to stimulation by the sympathetic and parasympathetic divisions.

5 A response is due partly to the particular neurotransmitter involved and partly to the nature of the receptor sites on the effectors.

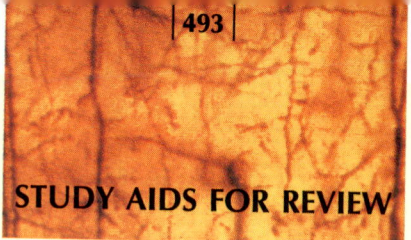

STUDY AIDS FOR REVIEW

KEY TERMS

UNDERSTANDING THE FACTS

1 What is the primary function of the autonomic nervous system?

2 Are the cell bodies of sympathetic and parasympathetic neurons in the central nervous system, in the peripheral nervous system, or in both? Explain.

3 Name three groups of autonomic ganglia and the five plexuses.

4 Name the major neurotransmitters usually released by the preganglionic and postganglionic neurons in the sympathetic and parasympathetic divisions.

5 Through which nerves do parasympathetic neurons emerge from the central nervous system?

6 Identify the location of alpha, beta, nicotinic, and muscarinic receptors.

7 What functional kinds of neurons constitute the enteric division of the autonomic nervous system?

8 What kinds of fibers constitute the gray and white rami? What causes the difference in color?

9 What are the components of a visceral reflex arc involving the sympathetic division of the autonomic nervous system?

10 What is meant by dual innervation?

UNDERSTANDING THE CONCEPTS

1 How do the general functions of the sympathetic and parasympathetic divisions differ?

2 Describe the pathways of sympathetic fibers from the spinal cord to effectors.

3 Describe the pathways of parasympathetic fibers from the central nervous system to effectors.

4 How do somatic and visceral reflex arcs differ?

5 Compare the neurotransmitters of the sympathetic, parasympathetic, and enteric divisions.

6 The sympathetic and parasympathetic divisions are said to be motor systems. Is the same true for the enteric division? Explain.

7 Explain the relationship of the brain to the sympathetic, parasympathetic, and enteric divisions of the autonomic system.

SELF-QUIZ
Multiple-Choice Questions

16.1 The sympathetic and parasympathetic divisions of the autonomic nervous system consist of
  a visceral afferent fibers
  b somatic afferent fibers
  c visceral efferent fibers
  d somatic sensory fibers
  e **a** and **b**

16.2 The sympathetic pathway between the central nervous system and the heart is via
  a a two-neuron linkage
  b an autonomic ganglion
  c pre- and postganglionic neurons
  d the sympathetic trunk ganglia
  e all of the above

16.3 In the autonomic nervous system, a preganglionic neuron
  a has its cell body in an autonomic ganglion
  b has its cell body in the brainstem or spinal cord
  c has a myelinated axon
  d has an unmyelinated axon
  e **b** and **c**

16.4 The sympathetic trunk ganglia are also known as the
  a prevertebral ganglia
  b paravertebral ganglia
  c terminal ganglia
  d collateral ganglia
  e peripheral ganglia

16.5 The sympathetic division of the autonomic nervous system is also known as the
  a thoracolumbar, or cholinergic, division
  b craniosacral, or adrenergic, division
  c splanchnic division
  d thoracolumbar, or adrenergic, division
  e craniosacral, or cholinergic, division

16.6 Which of the following must be included as an active component of every visceral reflex arc?
  a hypothalamic centers
  b one or more interneurons in the central nervous system

c somatic motor neurons
d centers in the limbic system
e nuclei in the cerebral cortex

16.7 A particular autonomic neuron may have different effects on different effectors because
a of the nature of the receptor
b of the presence of neuromodulators
c of the stage of development of the nervous system
d a single neuron may produce more than one neurotransmitter
e all of the above

Matching

a sympathetic division
b parasympathetic division
c enteric division
d sympathetic and parasympathetic divisions
e sympathetic and enteric divisions
f sympathetic, parasympathetic, and enteric divisions
g none of the above

16.8 _____ thoracolumbar
16.9 _____ sensory neurons, interneurons, and motor neurons
16.10 _____ acetylcholine
16.11 _____ noradrenaline
16.12 _____ form motor end plates on skeletal muscle fibers
16.13 _____ paravertebral ganglia
16.14 _____ postganglionic axons may be longer than preganglionic axon

16.15 _____ preganglionic cell bodies in brainstem
16.16 _____ preganglionic axons exit from cervical region of spinal cord
16.17 _____ synapse with smooth muscle fibers in stomach and intestines

Completion Exercises

16.18 Under conditions of stress, visceral functions are most strongly influenced by the _____ division of the autonomic nervous system.
16.19 A preganglionic autonomic neuron carried in the oculomotor nerve must be part of the _____ division.
16.20 The two-neuron linkage of the autonomic nervous system involves synapses within a(n) _____.
16.21 Alpha and beta receptors react with _____.
16.22 The types of tissue influenced by neurons of the autonomic nervous system are _____, _____, and _____.
16.23 A visceral reflex arc involves _____ efferent neuron(s) and _____ afferent neuron.
16.24 The major neurotransmitter of postganglionic parasympathetic fibers is _____.
16.25 Ganglia that are close to or within the organs they innervate are called _____ ganglia.

A SECOND LOOK

In the following drawing, label the central nervous system, autonomic ganglion, and effector. Also label the preganglionic and postganglionic neurons.

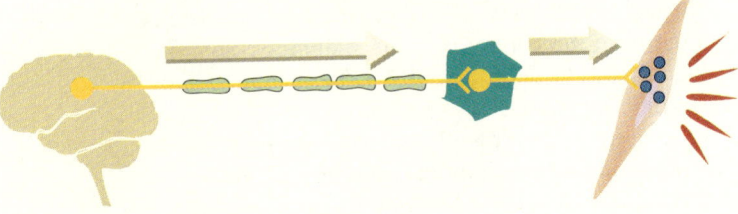

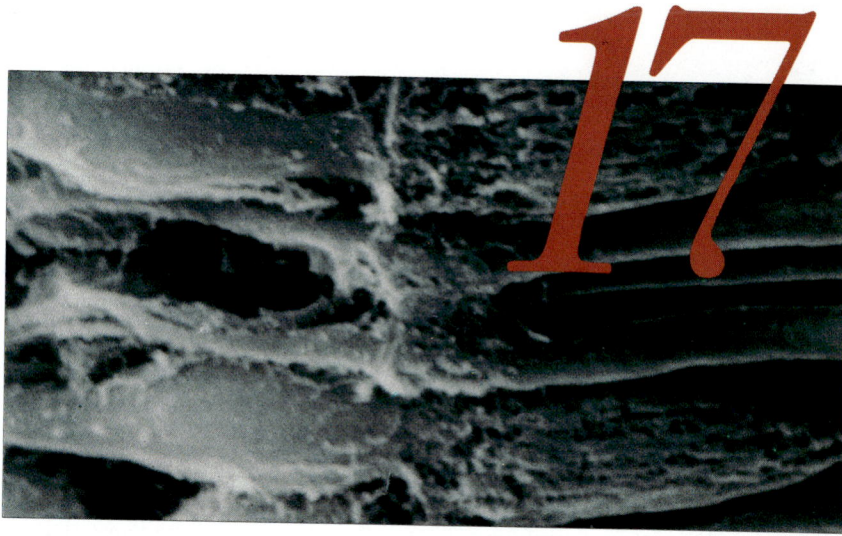

17

The Senses

KEY CONCEPTS

1 All our knowledge and awareness depends on the reception and decoding of stimuli from the outside world and from within our bodies.

2 All sensory receptors contain sensitive receptor cells, are designed to receive a specific kind of stimulus, and interact with nerve fibers that convey impulses to the central nervous system.

3 The type of sensation elicited by a stimulus depends on which part(s) of the brain is (are) stimulated.

4 The stimuli for hearing and balance are both received by the inner ear.

5 Sight is our dominant sense, with receptors in the eyes accounting for about 70 percent of the receptors in the entire body.

About 2000 years ago Aristotle identified five senses—touch, taste, smell, hearing, and sight—and it is still common to refer to the "five senses" of the body. But in fact, the skin alone is involved with the sensations of light touch, touch-pressure (deep pressure), heat, cold, and pain. Also included in a more complete list of the senses are a sense of balance, or equilibrium, and a sense of body movement. In addition, receptors in the circulatory system register changes in blood pressure and blood concentrations of carbon dioxide and hydrogen ions, and receptors in the digestive system are involved in the feelings of hunger and thirst.

Our impressions of the world are limited and defined by our senses. In fact, all knowledge and awareness depend on the reception and decoding of stimuli from the outside world and from within our bodies.

SENSORY RECEPTION

We are able to cope with change because part of our nervous system is specialized to make sure we have a suitable reaction to any stimulus. Structures that are capable of perceiving and processing such stimuli are called sense organs, or *receptors* [FIGURE 17.1]. These sensory receptors are the body's links to the outside world and the world within us.

In terms of the nervous system, a receptor is (or is associated with) the peripheral end of sensory afferent neurons. Sensory receptors are stimulated by specific stimuli. The eye (the receptor) is stimulated by visible light waves (the stimulus), and specialized receptors inside the ear are stimulated by audible sound waves. Sound waves do not stimulate the sensory receptors that are specialized to receive *light* waves, and light waves have no effect on our ears.

A favorite question is: Do sounds occur when there is no living thing to hear them? The answer is no. A "sound" is something that is received (*sensed*) by the ear and "heard" (*perceived*) by the brain. If there is no ear to receive sound waves and no brain to translate those sound waves into what we consciously recognize as the "sound" of thunder, for example, the thunder will send out *sound waves* but there will be no perceived "sound." The same is true of the other senses.

The different sensations are brought about when nerve fibers connect with specialized portions of the brain. We "see" a tree, for example, not when its image enters our eyes but when that coded image stimulates the vision centers of the brain.

All sensory receptors are structures that are capable of converting environmental information (stimuli) into nerve impulses. Thus, all receptors are *transducers;* that is, they convert one form of energy into another. *Since all nerve impulses are the same, different types of receptors convert different kinds of stimuli* (such as light or heat) *into the same kind of impulse.*

When a stimulus is strong enough, the stimulated terminal receptive segments of the sensory neurons generate *receptor,* or *generator, potentials.* These receptor potentials are *graded potentials.* They travel only short distances along the plasma membrane because, unlike action potentials, they are not self-propagating. When the receptor potentials of the receptor segment of the neuron reach the first node of Ranvier (the first node acts like an initial segment), they trigger the generation of *action potentials* (nerve impulses), which can travel long distances to a synapse.

Basic Characteristics of Sensory Receptors

Sensory receptors may be either (1) a nerve ending of a neuron (for example, a pain receptor) or (2) a specialized receptor cell that is not a neuron (for example, a hair cell in the cochlea of the auditory system). Although there are

---

FIGURE 17.1 BASIC ARCHITECTURE OF A SENSORY PATHWAY

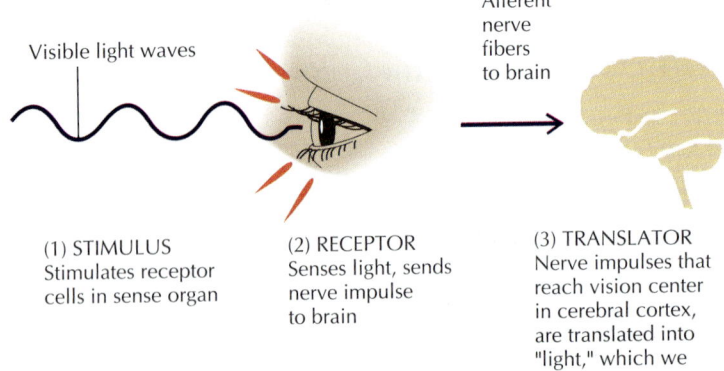

Visible light waves

Afferent nerve fibers to brain

(1) STIMULUS
Stimulates receptor cells in sense organ

(2) RECEPTOR
Senses light, sends nerve impulse to brain

(3) TRANSLATOR
Nerve impulses that reach vision center in cerebral cortex, are translated into "light," which we then "see."

several types of sensory receptors and many ways to classify them, certain features are basic to all sensory receptors:

1 They *contain sensitive receptor cells* that respond to certain minimum (threshold) levels of stimulus intensity. That is, the stimulus must be strong enough to generate a receptor potential in the nerve ending and then an action potential.

2 Their *structure is designed to receive a specific kind of stimulus.* For example, the eye contains an elastic adjustable lens, a nonadjustable lens (cornea), and other structures that direct light waves in the visible spectrum to the receptor cells (rods and cones), which contain light-absorbing pigments.

3 Their primary receptor cells *interact with afferent neurons* that convey impulses to the central nervous system along spinal or cranial nerves. Some receptors in the skin are connected to neurons that have axons that extend for about a meter before they reach the spinal cord and form a synapse. In contrast, the primary receptor cells in the eye have short axons that synapse with other cells in the retina before projecting impulses to the brain via the optic nerve.

After receptor cells stimulate afferent neurons, the nerve impulses are *conveyed along neural pathways* through the brainstem and diencephalon to the cerebral cortex of the brain. It is not until the nerve impulses reach the diencephalon and cerebral cortex, where neural processing occurs, that they are translated into a recognizable perception such as sight or sound.

Classification of Sensory Receptors

Sensory receptors may be classified according to their location, type of sensation, type of stimulus, or structure (the presence or absence of a sheath).

Location of receptor Four kinds of sensory receptors are recognized on the basis of their location:

1 *Exteroceptors* (L. "received from the outside") are skin receptors found in widespread regions of the body. They respond to external environmental stimuli that affect the skin directly. These stimuli result in the sensations of touch-pressure, pain, and temperature.

2 *Teleceptors* (Gr. "received from a distance") are the special exteroceptors located in restricted regions of the body, namely, the eyes, ears, and nose. They detect environmental changes (stimuli) that occur some distance away from the body. These stimuli are ultimately perceived as sight, sound, and smell.

3 *Interoceptors* (L. "received from the inside"), also called *visceroceptors* (L. "received from the viscera"), respond to stimuli from within the body, such as changes in blood pressure, carbon dioxide, oxygen, and hydrogen-ion concentrations in blood, and the stretching action of smooth muscle in organs and blood vessels. Interoceptors are located within organs that have motor innervation from the autonomic nervous system.

4 *Proprioceptors* (L. "received from one's own self") respond to stimuli in such deep body structures as joints, tendons, muscles, and the vestibular apparatus of the ear. They are involved with sensing where parts of the body are in relation to each other and the position of the body in space.

Sight, hearing, equilibrium, smell, and taste are known as the **special senses** because their receptors are found in restricted regions of the body. Their sensory receptors are also more specialized and complex than those of the **general senses** (also called the *somatic,* or *visceral, senses*), which include touch-pressure, heat, cold, pain, body position (proprioception), and light (crude) touch.

Type of sensation Sensory receptors can detect several types of sensations that are associated with the general senses. **Thermal** sensations include cold and warmth. **Pain** sensations are the feelings initiated by harmful stimuli. Both **light touch** and **touch-pressure** sensations are also produced by mechanical stimuli that come in contact with the body. Light touch involves a finer discrimination than touch-pressure does. **Position sense** is elicited by the movement of joints and muscles. It includes both the sense of position when the body is not moving and the sense of body movement, which is called *kinesthesia* (Gr. *kinesis*, motion + *esthesis*, perception).

Type of stimulus Another way to classify sensory receptors is by the stimuli to which they respond. **Thermoreceptors** (Gr. "heat receivers") respond to temperature changes. **Nociceptors** (NO-see; L. "injury receivers") respond to potentially harmful stimuli that produce pain. **Chemoreceptors** (L. "chemical receivers") respond to chemical stimuli that result in taste and smell; they also respond to changes in the concentrations of carbon dioxide, oxygen, and hydrogen ions in the blood, as well as other chemical changes. **Photoreceptors** (Gr. "light receivers") in the retina of the eye respond to the visual stimuli of visible light waves. **Mechanoreceptors** (Gr. "mechanical receivers") respond to and monitor such physical stimuli as touch-pressure, muscle tension, joint position changes, air vibrations in the cochlear system of the ear that produce hearing, and head movements detected by the vestibular system of the ear that result in sensing body equilibrium. Mechanoreceptors are the most widespread of all the sensory receptors; they are also the most varied in sensitivity and structure. *Baroreceptors* (Gr. "pressure receivers") are mechanoreceptors that respond to changes in blood pressure.

Structure of receptors A final method of classifying sensory receptors is by their structure. The distal terminals of spinal nerves are located in sensory receptors, or *nerve endings*. Two kinds of nerve endings are usually distinguished:

1 Terminals that lack Schwann cells, myelin, and other cellular coverings are called *free nerve endings.* Free nerve endings are the naked telodendria in the surface epithelium of the skin, connective tissues, blood vessels, and other tissues. They are the sensors for such perceived sensations as pain, light touch, and temperature.

2 Receptors that are covered with various types of capsules are known as *encapsulated endings.* Encapsulated endings are located in the skin, muscles, tendons, joints, and body organs. Two such endings are *lamellated (Pacinian) corpuscles*, which are involved with vibratory sense and touch-pressure on the skin, and *tactile (Meissner's) corpuscles*, which are skin sensors that detect light touch [FIGURE 17.2].

Corpuscles of Ruffini and *bulbous corpuscles (of Krause)*, previously thought to be the skin receptors for temperature, are now believed to be sensors for touch-pressure, position sense of a body part, and movement. They are probably variants of other encapsulated endings, such as lamellated corpuscles. Other important encapsulated receptors are the *neuromuscular spindles* (stretch receptors in muscles) and *tendon (Golgi) organs* (tension receptors in tendons), which monitor, respectively, the stretch and tension in muscles and the tension in tendons. These receptors are involved with skeletal muscle reflexes.

Sensory Receptors and the Brain

Our senses inform us about environmental conditions and changes that may be beneficial or harmful. The sensory system is an effective survival mechanism because it provides information from the environment that enables us to react appropriately.

Stimuli derived from both the internal and external environments activate the sensory receptors. The responses of the receptors are converted into action potentials in first-order neurons that convey impulses from the receptors to the central nervous system. In turn, these impulses are processed in the nuclei of the central nervous system and utilized in reflex activities, in decision making, and in effecting behavioral changes. This neural processing uses the motor neurons and the peripheral nerves to express its activity through the effectors: muscles contract or relax, glands secrete enzymes and hor-

mones, and blood vessels constrict or dilate. To accomplish these functions, the brain receives nerve impulses, suppresses what is irrelevant, compares it with information already stored, and coordinates the final impulses to the effectors. The brain's involvement between the stimulus and the response makes possible complex behavioral patterns.

Individual nerve impulses are essentially identical, but an impulse for sight is distinguished from an impulse for taste by *the relationship between where the impulse is coming from and the places in the brain where it is going*. For example, taste receptors in neuroepithelial cells of the taste buds on the tongue respond to chemical stimuli and stimulate the nerve endings of taste neurons to generate nerve impulses. Afferent fibers carry these impulses to a location in the brain that is specialized to receive only taste impulses. The type of sensation elicited by the brain depends on *where* the brain is stimulated, not *how* it is stimulated.

A S K Y O U R S E L F

1 What is a sensory receptor?

2 What is a transducer?

3 How does a generator potential differ from an action potential?

4 What are the basic characteristics of a sensory receptor?

5 What are the minimal requirements for the perception of a sensation?

6 What are the types of sensory receptors based on location?

7 What are the types of sensory receptors based on the types of stimuli to which they respond?

8 What is the role of the brain in the sensory system?

GENERAL SENSES

Sensory receptors (mechanoreceptors) in the skin detect tactile (touch) sensations that the brain interprets as light touch, touch-pressure, vibration, and itch and tickle. Several other general sensations, such as heat, cold, and pain, will also be described.

Light Touch

Light touch is perceived when the skin is touched but not deformed. Receptors for light touch are most numerous in the dermis, especially in the tips of the fingers and toes, the tip of the tongue, and the lips.

FIGURE 17.2 SENSORY RECEPTORS AND OTHER STRUCTURES IN THE SKIN

Nerve endings, sensory receptors, and root hair plexuses around hair follicles in the skin.

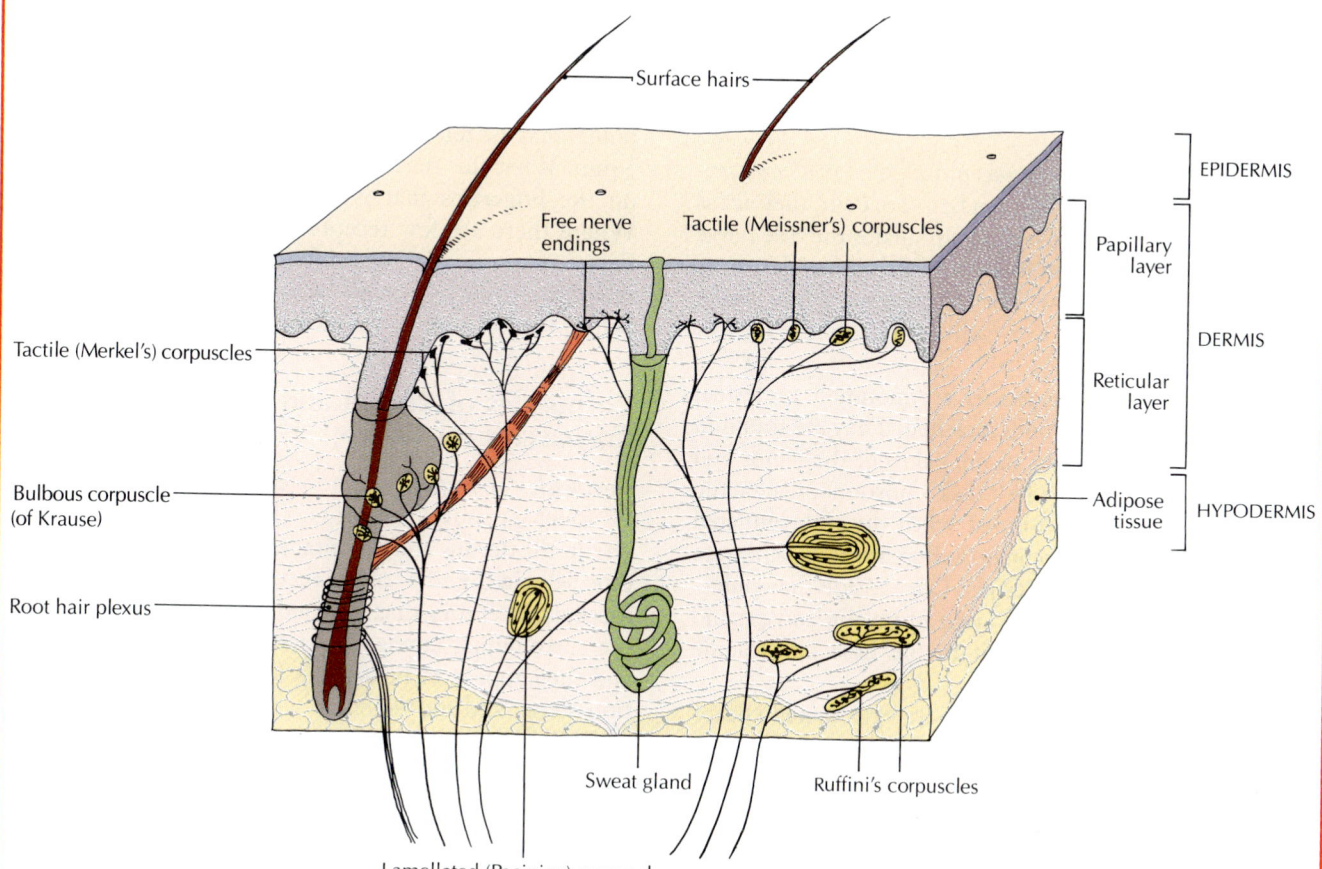

Cutaneous sensation	Receptors	Neural pathways
Light touch (skin not deformed)	Free nerve endings, tactile (Merkel's) corpuscles, tactile (Meissner's) corpuscles	Posterior column-medial lemniscal pathway, anterior spinothalamic tracts of anterolateral system
Touch-pressure (deep pressure), two-point discrimination, vibratory sense (tuning fork)	Hair plexuses, lamellated (Pacinian) corpuscles, corpuscles of Ruffini, bulbous corpuscles (of Krause)*	Posterior column-medial lemniscal pathway, spinocervicothalamic pathway (for touch)
Heat	Free nerve endings	Anterolateral system, including lateral spinothalamic tract
Cold	Free nerve endings	Anterolateral system, including lateral spinothalamic tract
Pain	Specialized free nerve endings	Lateral spinothalamic tract, indirect spinoreticulothalamic pathway
Proprioception	Specialized "spray" endings, lamellated corpuscles in synovia and ligaments	Posterior column-medial lemniscal system (conscious), spinocerebellar system (unconscious), spinocervicothalamic pathway

*Bulbous corpuscles (of Krause) (once thought to be cold detectors) and corpuscles of Ruffini (once thought to be heat receptors) are actually mechanoreceptors.

Receptors of light touch include free nerve endings and tactile (Merkel's) corpuscles within the epidermis; just below, in the uppermost (papillary) layers of the dermis, are tactile (Meissner's) corpuscles [FIGURE 17.2]. *Free nerve endings* are the most widely distributed receptors in the body and are involved with pain and thermal stimuli as well as light touch. The next most numerous cutaneous receptors are those for touch, cold, and heat.

There are many sensory receptors called *root hair plexuses* around hair follicles. When hairs are bent, they act as levers, and the slight movement stimulates the free nerve endings surrounding the follicles, which act as detectors of touch and movement. For this reason, a tiny insect crawling along a hairy arm will be felt even if its feet never touch the skin.

Tactile (Merkel's) corpuscles are modified epidermal cells with free nerve endings attached. They are found in the deep epidermal layers of the palms of the hands and the soles of the feet. The "disk" portion of a tactile (Merkel's) corpuscle is formed when the unmyelinated terminal branches of myelinated afferent axons penetrate the basal layer of the epidermis. Once inside the epidermis, the axons lose their covering of myelin and expand into a terminal disk attached to the base of a tactile (Merkel's) corpuscle.

Tactile (Meissner's) corpuscles are egg-shaped encapsulated nerve endings [FIGURE 17.2]. They are found in abundance on the palms of the hands, soles of the feet, lips, eyelids, external genitals, and nipples. These receptors are situated in the papillary layer of the dermis, just below the epidermis.

Touch-Pressure (Deep Pressure)

The difference between light touch and touch-pressure (deep pressure) on your skin can be shown by gently touching a pencil (light touch) and then squeezing it hard (touch-pressure). Touch-pressure results from a deformation of the skin, no matter how slight. Sensations of touch-pressure last longer than do sensations of light touch and are felt over a larger area. Receptors for touch-pressure are primarily **lamellated,** or **Pacinian** (pah-SIN-ee-an), **corpuscles** [FIGURE 17.2]. They are mechanoreceptors that actually measure *changes* in pressure rather than pressure itself. Lamellated corpuscles are distributed throughout the dermis and subcutaneous layer, especially in the fingers, external genitals, and breasts, but they are also found in muscles, joint capsules, the wall of the urinary bladder, and other areas that are regularly subjected to pressure. Other touch-pressure receptors include the corpuscles of Ruffini and the bulbous corpuscles (of Krause).

In contrast to such sensitive areas of the skin as the fingertips, the torso (especially the back) and back of the neck are relatively insensitive to light touch. Sensitivity can be measured with a test called *two-point discrimination*, which measures the minimal distance that two stimuli must be separated to be felt as two distinct stimuli. Usually, one or two points of a compass are applied to the skin without the subject seeing how many points are being used. In areas where sensory receptors are abundant, for example, the fingertips, two distinct compass points may be felt when they are separated by only 2 or 3 mm. What this means is that there is virtually no spot on the fingertips that is insensitive to tactile stimuli. Where there are few receptors and those receptors are far apart, as on the posterior (dorsal) torso, the points may have to be separated by as much as 60 or 70 mm before they can be felt as two points.

Corpuscles of Ruffini and Bulbous Corpuscles (of Krause)

Corpuscles of Ruffini are now considered to be variants of touch-pressure receptors. They are located deep within the dermis and subcutaneous tissue [FIGURE 17.2], especially in the soles of the feet. They are thought to be mechanoreceptors that respond to the displacement of the surrounding connective tissue within the corpuscle, and they appear to be sensors for touch-pressure, position sense, and movement.

Bulbous corpuscles (of Krause) are found in the dermis of the conjunctiva (the covering of the whites of the eyes and the lining of the eyelids), tongue, and external genitals. They too are thought to be mechanoreceptors.

Vibration

Most tactile receptors are involved to some degree in the detection of vibration. The term **vibration** refers to the continuing periodic change in a displacement with respect to a fixed reference. This change, per unit time, is termed the *frequency*. Different receptors detect different frequencies. For example, lamellated corpuscles can detect vibrations (frequencies) as high as 700 cycles per second (cps). Tactile (Meissner's) corpuscles and corpuscles of Ruffini, by contrast, respond to low-frequency vibrations up to 100 cps.

Thermal Sensations

Until recently it was believed that the cutaneous receptors for heat were the corpuscles of Ruffini, and the receptors for cold were the bulbous corpuscles (of Krause). Currently, the cutaneous receptors for heat and cold are considered to be naked nerve endings. Cold receptors respond to temperatures below skin temperature, and heat receptors respond to temperatures above skin temperature.

So-called *cold spots* and *warm spots* are found over the surface of the entire body, with cold spots being more numerous. A *spot* refers to a small area that, when stimulated, yields a temperature sensation of warmth or cold. A spot is associated with several nerve endings. The lips have both cold and warm spots, but the tongue is only slightly sensitive to warmth. Nerve endings that innervate teeth are usually sensitive to cold but much less sensitive to heat. The face is less sensitive to cold than are parts of the body that are usually covered by clothing.

Itch and Tickle

Itch is probably produced by the repetitive low-key stimulation of slow-conducting nerve fibers in the skin. *Tickle* is caused by a mild stimulation of the same type of fibers, especially when the stimulus moves across the skin. Receptors for both sensations are found almost exclusively in the superficial layers of the skin. It is thought that the sensations result from the activation of several sensory endings and that the information is conveyed via a combination of pathways. Like the areas most sensitive to pain, itch usually occurs where naked endings of unmyelinated fibers are abundant. Itch occurs on the skin, in the eyes, and in certain mucous membranes (such as in the nose and anus) but not in deep tissues or viscera.

Although itching may be produced by a repetitive, mechanical stimulation of the skin, it is also produced by chemical stimuli, including polypeptides called *kinins* and by histamine that mast cells release during an allergic or inflammatory response.

Pain Sensations

The subjective sensation we call *pain* is a warning signal that alerts the body to a harmful or unpleasant stimulus. The sensation may be initiated by *nociceptors* that are sensitive to mechanical, thermal, electrical, or chemical stimuli. These pain receptors are **specialized free nerve endings** [FIGURE 17.2] that are present in most parts of the body. (Brain tissue has no pain receptors.) Most nociceptors are probably chemoreceptors that respond to chemical substances such as histamines that accumulate in response to the local trauma. For example, potassium is released from damaged cells, and histamine is released from mast cells and other types of cells.

Types of pain include (1) fast-conducted, sharp, prickling pain, (2) slow-conducted, burning pain, and (3) deep, aching pain in joints, tendons, and viscera. Other distinctions are sometimes made: *Superficial somatic pain* originates from stimulation of skin receptors; *deep somatic pain* arises from stimulation of receptors in joints, tendons, and muscles; *visceral pain* originates from stimulation of receptors in body organs.

Some tissues are more sensitive to pain than others: A needle inserted into the skin produces great pain, but the same needle inserted into a muscle produces little pain. An arterial puncture is painful, but a venous puncture is almost painless. A kidney stone that distends a ureter (the tube leading from the kidney to the urinary bladder) produces excruciating pain. In contrast, the intestines are not sensitive to pain if they are cut, but *are* sensitive if they are distended or markedly contracted (cramps).

We adapt to most of our senses so that they don't become a bother. If we didn't, we would be continuously aware of the touch-pressure of clothing on our skin, for example. (This phenomenon of *adaptation* refers to the decline in the response of receptors to continuous, even stimulation.) However, it is to our benefit that we do not adapt completely to pain. If pain is to be useful as a warning signal that prevents serious tissue damage, it must be felt each time it occurs, even when we are asleep. For example, who has not turned over during the night to relieve pain caused by sleeping on a twisted arm?

Referred pain A visceral pain felt subjectively in a somatic area is known as **referred pain.** For example, the pain of myocardial ischemia (lack of blood flow to the heart) may be felt in the left shoulder, arm, and armpit; an irritation of the gallbladder may be felt under the shoulder blades.

One possible explanation for the brain's "misinterpretation" of most visceral pain is that certain neurons use a common dorsal root to innervate both the visceral and somatic locations involved in referred pain. It is thought that the visceral sensory fibers and somatic sensory fibers both discharge into a common pool of CNS neurons.

Another explanation of referred pain is that the area of the body to which the pain is referred usually is a part of the body that develops from the same embryological structure as the real source of the pain. These structures are supplied by branches of the same peripheral nerves. Thus, angina pectoris can be felt in the left shoulder and arm.

Phantom pain Another unusual phenomenon in the sensing of pain is the **phantom pain** that is felt in an amputated limb (*phantom limb*). Such pain may be intense, and it is actually felt. The sensations of pain, "pins and needles," and temperature change are often felt by amputees in their amputated limbs for several months. Ordinarily, the pain is felt more in the joints than in other regions of the phantom limb and more in the distal portion of the amputated segment than in the proximal portion. Phantom pain usually persists longest in regions that have the largest representation in the cerebral cortex: the thumb, hand, and foot.

The neural mechanism for phantom pain is not known

TABLE 17.1 BASIC ASCENDING SENSORY NEURAL PATHWAYS

System and pathway or tract	Final destination	Sensations involved
Anterolateral system*	Cerebral cortex	Pain, light touch, temperature
Lateral spinothalamic tract	Cerebral cortex	Pain, temperature
Anterior spinothalamic tract	Cerebral cortex	Light touch
Indirect spinoreticulothalamic pathway	Cerebral cortex	Pain
Posterior column-medial lemniscal system*	Cerebral cortex	Touch-pressure, proprioception, vibration
Dorsal column-medial lemniscal pathway	Cerebral cortex	Touch-pressure, light touch, stereognosis, associated tactile discrimination senses, vibration, two-point discrimination, conscious proprioception
Spinocervicothalamic (spinocervicolemniscal) pathway	Cerebral cortex	Touch-pressure, proprioception
Spinocerebellar system	Cerebellar cortex	Unconscious proprioception
Anterior spinocerebellar tract	Cerebellar cortex	Unconscious proprioception (from lower half of body)
Posterior spinocerebellar tract	Cerebellar cortex	Unconscious proprioception (from lower half of body)
Cuneocerebellar tract	Cerebellar cortex	Unconscious proprioception (from upper half of body)
Rostral spinocerebellar tract	Cerebellar cortex	Unconscious proprioception (from upper half of body)

*The trigeminothalamic tracts and pathways convey all the sensations in this system from the head to the thalamus and cerebral cortex.

completely. It appears that pools of neurons associated with the sensations of the missing limb persist in the brain, are somehow activated, and result in the perception. Impulses in the pools of neurons may be triggered by the irritation of peripheral nerves in the proximal stump.

Proprioception

Receptors in muscles, tendons, and joints transmit impulses about our position sense up the dorsal columns of the spinal cord. These impulses help us to be aware of the position of our body and its parts without actually seeing them. This sense of position is called *proprioception* (L. *proprius*, one's self + receptor). The receptors in or near joints that are responsible for proprioception are specialized "spray" endings. Lamellated corpuscles in the synovial membranes and ligaments may also be involved.

Stereognosis

The ability to identify unseen objects by handling them is *stereognosis* (STEHR-ee-og-NO-sis; Gr. *stereos*, solid, three-dimensional; *gnosis*, knowledge). This ability de-

pends on the sensations of touch and pressure, as well as on the participation of sensory areas in the parietal lobe of the cerebral cortex. Damage to certain areas of the cortex of the parietal lobe usually impairs stereognosis, even if the cutaneous sensations remain intact.

Receptor (Generator) Potentials

The sensory receptors of the body (such as those involved with sensing touch-pressure and monitoring blood pressure) are associated with the axon terminal of a sensory nerve fiber. When the axon terminal is stimulated, it can exhibit a graded local potential called a *receptor potential*. An example of a receptor potential is the touch-pressure receptor of a lamellated (Pacinian) corpuscle in the skin. This ending has sodium and potassium channels that are *modality gated*. (The word *modality* refers to a sensory component such as touch-pressure.) A stimulus at the nerve ending causes an inflow of sodium ions through Na^+ channels into the unmyelinated nerve terminal and an outflow of potassium ions through K^+ channels. This produces a graded receptor potential called a *generator potential*, which is similar to an EPSP.

Q: *Do painkillers work by desensitizing the receptors?*

A: No. Painkillers (analgesics) have no effect on sensory receptors. Instead, they modify the perception of pain or the emotional reaction to pain.

Q: *Why does scratching relieve the itching of insect bites?*

A: Scratching an insect bite temporarily stimulates and soothes the naked endings of unmyelinated fibers near the surface of the skin. However, scratching usually makes the inflammation worse by irritating the skin and releasing the chemical stimulants—kinins and histamine—that caused the itch in the first place.

The generator potential spreads along the plasma membrane to reach the first node of Ranvier (low-threshold site of the myelinated portion of the nerve fiber), where an action potential may be generated. The first node of Ranvier is the initial segment of a sensory neuron.

Neural Pathways for General Senses

The neural pathways involved in relaying influences from specific general sensory receptors to the cerebral cortex include the *dorsal column-medial lemniscus pathway,* the *spinothalamic tracts,* and the *trigeminothalamic tract.* The other ascending tracts and pathways for taste, smell, hearing, and vision are discussed in later sections of this chapter, and a summary of the basic ascending neural pathways is presented in TABLE 17.1.

In general, afferent nerves that convey highly localized and discriminative sensations are larger, are more heavily myelinated, and conduct faster than those that convey less-defined sensations. The information is conveyed in the dorsal column and in the medial lemniscus and trigeminothalamic pathways. The afferent nerves that convey the less-defined sensations travel in the spinothalamic tracts.

Neural pathways for light touch The sensory area of the brain specialized for touch is located in the general sensory region of the parietal lobe of the cerebral cortex. Light touch is mediated by at least three neural pathways from the spinal cord to the cerebral cortex, including (1) the *dorsal column-medial lemniscal pathway,* (2) the *spinocervicothalamic pathway* of the dorsal column-medial lemniscal system, and (3) the *anterior spinothalamic tracts* of the anterolateral system.

Neural pathways for touch-pressure Touch-pressure is mediated by the *dorsal column-medial lemniscal pathway* and probably by the *spinocervicothalamic pathway* of the posterior column-medial lemniscal system [see FIGURE 14.8].

Neural pathways for temperature The sensory area of the brain for temperature is the same as that for touch: the parietal lobe of the cerebral cortex. Crude sensations of temperature may be experienced in the thalamus. The sensation of temperature change is mediated by the *lateral spinothalamic tract* of the anterolateral system.

Neural pathways for pain It is thought that pain impulses are conveyed by two or more pathway systems, including (1) the *lateral spinothalamic tract,* which consists of a sequence of at least three neurons with long axonal processes that relay pain impulses from the spinal cord to the thalamus, and (2) the *indirect spinoreticulothalamic pathway,* which consists of a sequence of many neurons that relay pain impulses to the reticular

formation and thalamus. These pathways are part of the anterolateral system. Fibers from the spinotectal tract to the midbrain may also be involved in the transmission of pain. The perception of pain occurs in the thalamus, but the discrimination (judgment) of the type of pain and its intensity occurs in the parietal lobe of the cerebral cortex.

Neural pathways for proprioception Many proprioceptive impulses are relayed to the cerebellum, but some are conveyed to the cerebral cortex through the **medial lemniscal pathway** and from **thalamic projections.** The sensory area for conscious position sense is located in the parietal lobe of the cerebral cortex. The neural pathways are the spinocerebellar tracts (unconscious proprioception) and the dorsal column-medial lemniscal pathway (conscious perception). Degenerative diseases of the dorsal column of the spinal cord produce *ataxia* (lack of muscular coordination) because proprioceptive impulses are not conveyed to the cerebellum and to the thalamus and cerebral cortex.

Neural pathways for somesthetic sensations from the head Afferent nerves that convey highly localized discriminative sensations from the face, mouth, nasal cavities, and associated structures, as well as those conveying cruder sensations, form the **trigeminothalamic tracts.** These tracts originate from the spinal trigeminal nucleus (pain and temperature) and the principal sensory nuclei (touch-pressure), respectively, and they parallel the spinothalamic and dorsal column-medial lemniscus tracts, respectively. The sensations are felt primarily on the skin and the mucosal membranes of the nasal and oral cavities. They involve the sensory region of the parietal lobe of the cerebral cortex.

ASK YOURSELF

1 What are the receptors for light touch? Touch-pressure? Pain?

2 How do light touch and touch-pressure differ?

3 What types of stimuli can initiate pain?

4 What is the difference between somatic pain and visceral pain?

5 What is proprioception? Stereognosis?

6 What are the three ascending neural pathways involved in the general senses?

Q: *How much outside stimuli does the nervous system screen out as irrelevant or distracting?*

A: The brain suppresses about 99 percent of the input it receives from sensory receptors.

TASTE (GUSTATION)

The receptors for taste, or **gustation** (L. *gustus*, taste), and smell are both chemoreceptors, and the two sensations are clearly interrelated. (A person whose nasal passages are "blocked" by a cold cannot "taste" food as effectively.) But despite some similarities, taste and smell are separate and distinct senses and will be treated separately here.

Structure of Taste Receptors

The surface of the tongue is covered with many small protuberances called **papillae** (puh-PILL-ee; sing. *papilla*, puh-PILL-uh, L. diminutive of *papula*, nipple, pimple). Papillae give the tongue its bumpy appearance [FIGURE 17.3]. They are most numerous on the dorsal surface of the tongue and are also found on the palate (roof of the mouth), throat, and posterior surface of the epiglottis.

The three main types of papillae are the following: (1) *Fungiform* (L. mushroomlike) *papillae* are scattered singly, especially near the tip of the tongue. Each fungiform papilla contains 1 to 8 taste buds. (2) From 10 to 12 *circumvallate* (L. "wall around") *papillae* form two rows parallel to the V-shaped sulcus terminalis near the posterior third of the tongue. Each circumvallate papilla may contain 90 to 250 taste buds. (3) *Filiform* (L. threadlike) *papillae* are pointed structures near the anterior two-thirds of the tongue. Filiform papillae do not necessarily contain taste buds.

Located within the crevices of most papillae are approximately 10,000 taste receptors called **taste buds.** They are barrel-shaped clusters of *chemoreceptor* (taste or gustatory) *cells* and *sustentacular* (supporting) *cells* arranged like alternating segments of an orange [FIGURE 17.3B]. Each taste bud contains 25 or more receptor cells. Basal cells act as reserve cells, producing new receptor cells to replace those that die. Mature receptor cells have a life span of only about 10 days, and they usually can be replaced from reserve cells in about 10 hours. Taste receptor cells are replaced with decreasing frequency as we get older. This explains in part why our sense of taste may diminish with age and may also explain why babies dislike spicy foods and tend to favor relatively bland baby foods.

Q: *Why do some astronauts have trouble tasting food while traveling in space?*

A: Apparently, the loss of taste is caused by the lack of gravity. The heart, accustomed to pumping against gravity, forces more blood into the head than is necessary, and other body fluids accumulate there as well. The result is congestion, the same feeling we get when we have a head cold.

FIGURE 17.3 TASTE-SENSITIVE AREAS AND TASTE BUDS OF THE TONGUE

[A] The specific taste-sensitive areas are shown in color. Sour sensations (green) are perceived most acutely on the sides of the tongue, stimulated by hydrogen ions in acids. Saltiness (pink) is tasted mainly at the sides and tip of the tongue. Bitter sensations (yellow) are perceived mainly at the back of the tongue. Sweetness (purple) is tasted optimally at the tip of the tongue. The center of the tongue, with only a few taste buds, is relatively insensitive to taste. **[B]** Drawing of section through taste bud.

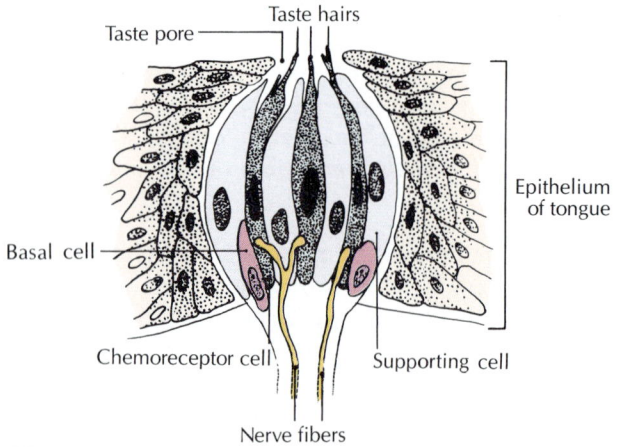

Extending from the free end of each receptor cell are short *taste hairs* (microvilli) that project through the tiny outer opening of the taste bud, called the *taste pore* [FIGURE 17.3B]. It is thought that gustatory sensations are initiated on the taste hairs, but before a substance can be tasted it must be in solution. Saliva, containing ions or dissolved molecules of the substance to be tasted, enters the taste pore and interacts with receptor sites on the taste hairs.

Basic Taste Sensations

Although all taste cells are structurally identical, each cell has many different types of receptor sites. Because the proportion of different types varies from cell to cell, each taste cell can respond to a variety of stimuli. The four generally recognized basic taste sensations are sweet, sour, bitter, and salty. We can taste many subtle flavors because of combinations of the four basic sensations, complemented by an overlay of odors. Taste perception is also aided by information about the texture, temperature, spiciness, and odor of food. The maximal areas of response to the four basic tastes are located on specific parts of the tongue [FIGURE 17.3A]. Salt and sweet are perceived most acutely at the tip of the tongue, but bitter and sour are perceived more acutely on the sides and back of the tongue.

The Taste Mechanism

Although the sensory systems of taste and smell are often considered to differ somewhat from those of the other senses, recent advances indicate that they do share certain fundamental features with the other systems. All senses apparently utilize the same basic principles of processing and have the same general organization of pathways: (1) In all systems, the membrane receptors are activated by sensory stimuli (transduction), (2) intracellular second messengers* play a role in all systems, and (3) similar ion channels are used in the systems. Chemical stimuli trigger the surface proteins of the receptor plasma membrane to alter the permeability of the membrane of the receptor cells. The initial signals generated by the stimulation are graded potentials. The presence of the G-protein and the second messengers *gustducin* and *cGMP* within the taste cells implies that they have roles

*Second messengers are intracellular signaling molecules that alter the biochemical state of neurons; they can trigger events that enhance synaptic transmission.

Q: *Is sugar the sweetest substance?*

A: No. Saccharine (a synthetic organic compound), for example, is about 600 times "sweeter" than is sugar.

---

similar to those of transducin and cGMP in the photoreceptor cells of the retina.

Taste cells contain voltage-gated Na^+, K^+, and Ca^{2+} channels similar to those in neurons. Thus, in response to chemical stimulation, the activities of the membrane channels, graded potentials of the taste cells, and second messengers lead to the release of a chemical neurotransmitter that generates action potentials in the afferent fibers of the nerves innervating the taste buds.

A given gustatory (taste) neuron may respond to some extent to many chemical taste stimuli, but the response of a given nerve fiber is optimal for one of the four taste qualities. The coding of the gustatory sensation is not simple. The actual taste sensation perceived depends upon a pattern of nerve fibers activated by specific chemical stimuli and the nuances of the taste flavors, which are, in part, related to the degree of intensity of the stimulation of different taste cells. In addition, the olfactory sense has a significant role in the appreciation of taste.

Neural Pathways for Gustation

Taste impulses are conveyed from the anterior two-thirds of the tongue to the brain by a branch of the facial nerve. Impulses from the posterior third of the tongue are carried to the brain by the glossopharyngeal (IX) nerve, and from the palate and pharynx by the vagus (X) nerve. The taste fibers of all three cranial nerves terminate in the *nucleus solitarius* in the medulla oblongata. From there, axons project to the thalamus and then to the "taste center" in the parietal lobe of the cerebral cortex.

In one respect, the taste pathway is unusual compared with other sensory pathways. Because the pathway is an ipsilateral system (does not decussate), the taste buds of one side communicate via the gustatory pathway of nuclei and tracts with the cerebral cortex of the *same* side.

A S K   Y O U R S E L F

1 What are papillae?

2 Describe the structure of a taste bud.

3 What is the sensory role of taste hairs?

4 Which areas of the tongue respond to each of the four basic taste sensations?

SMELL (OLFACTION)

Our sense of smell, or **olfaction** (L. *olere*, to smell + *facere*, to make), is perhaps as much as 20,000 times more sensitive than our sense of taste. We can taste quinine in

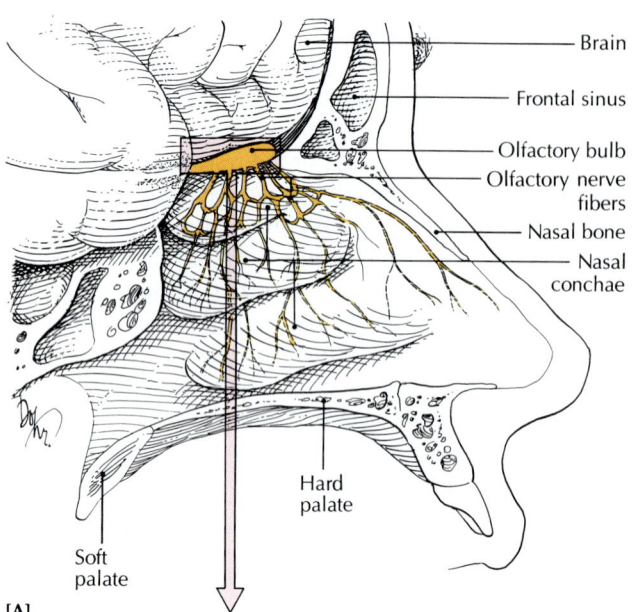

Brain
Frontal sinus
Olfactory bulb
Olfactory nerve fibers
Nasal bone
Nasal conchae
Hard palate
Soft palate

[A]

FIGURE 17.4 OLFACTORY RECEPTORS

[A] Olfactory receptive area in the roof of the nasal cavity; medial view. [B] Enlarged drawing of the olfactory epithelium, showing receptor cells, sustentacular cells, basal cells, and the formation of the olfactory tract inside the olfactory bulb. [C] Scanning electron micrograph of olfactory epithelial surface, showing olfactory vesicle, cilia extending into the nasal mucosa, and microvilli of sustentacular cells. ×9000. (Gene Shih and Richard Kessel, *Living Images: Biological Microstructures Revealed by SEM*, Boston, Science Books International, 1982.)

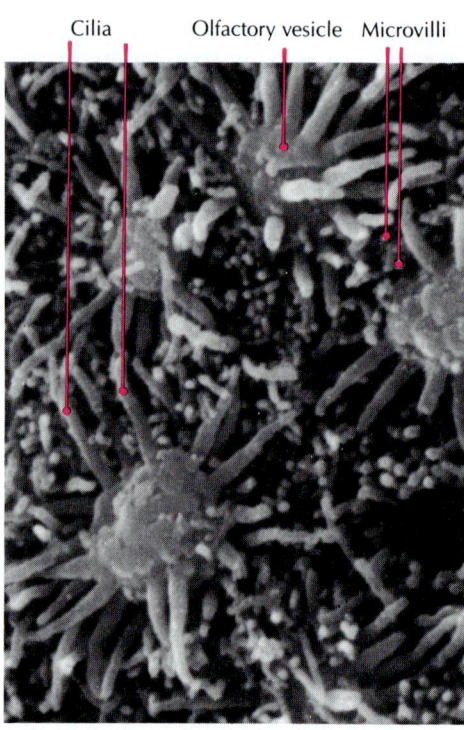

Cilia   Olfactory vesicle   Microvilli

[C]

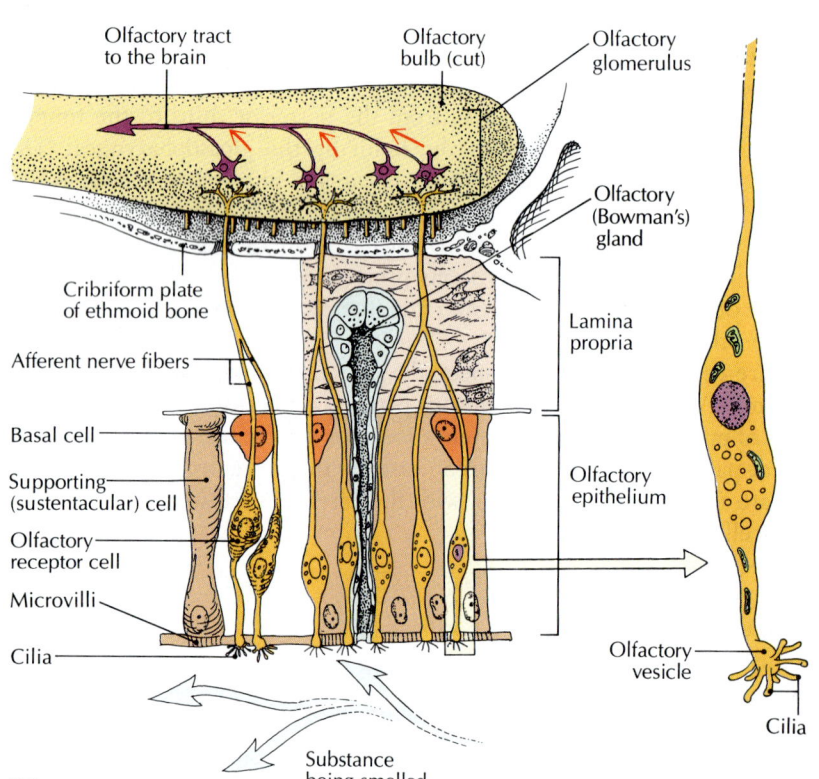

Olfactory tract to the brain
Olfactory bulb (cut)
Olfactory glomerulus
Olfactory (Bowman's) gland
Cribriform plate of ethmoid bone
Lamina propria
Afferent nerve fibers
Basal cell
Supporting (sustentacular) cell
Olfactory receptor cell
Microvilli
Cilia
Olfactory epithelium
Olfactory vesicle
Cilia
Substance being smelled

[B]

a concentration of 1 part in 2 million, but we can smell mercaptans (the type of chemical released by skunks) in a concentration of 1 part in 30 *billion*. Adults can usually sense up to 10,000 different odors, and children can do even better. Unfortunately, our sense of smell is not perfect. Several poisonous gases, including carbon monoxide, are not detectable by our olfactory receptors.

Structure of Olfactory Receptors

The *olfactory receptor cells* are located high in the roof of the nasal cavity, in specialized areas of the nasal mucosa called the *olfactory epithelium* [FIGURE 17.4]. Each nostril contains a small patch of pseudostratified, columnar olfactory epithelium about 2.5 cm^2 (about the size of a thumbnail). The epithelium consists of three types of cells: (1) *receptor cells*, which actually are the olfactory neurons, (2) *sustentacular* (supporting) *cells*, and (3) a thin layer of small *basal cells*. These basal cells are capable of undergoing cell division and replacing degenerating receptor and sustentacular cells. Each receptor cell has a lifetime of about 30 days. They are replaced by basal cells, which are continually differentiating into new olfactory neurons and forming new synaptic connections in the olfactory bulb.

We have more than 25 million bipolar receptor cells (a hunting dog has about 220 million), each of which is surrounded by sustentacular cells.* Each thin receptor cell has a short dendrite extending from its superficial end to the surface epithelium. The receptor cell ends in a bulbous *olfactory vesicle* [FIGURE 17.4B]. From this swelling, 6 to 20 long *cilia* project through the mucuslike fluid that covers the surface epithelium. The fluid is secreted by the sustentacular cells and *olfactory glands*. This secretion is important because odoriferous substances need to be dissolved before they can stimulate receptor sites.

The receptive sites of the cilia are exposed to the molecules responsible for odors. The axons of the bipolar receptive cells pass through the *basal lamina* of the olfactory mucosa and join other axons to form fascicles of the olfactory nerve (cranial nerve I). These unmyelinated olfactory nerve fibers are among the smallest and slowest-conducting fibers of the nervous system. They pass through the olfactory foramina in the cribriform plate of the ethmoid bone on their way to the olfactory bulbs. The *olfactory bulbs* are specialized structures of gray matter, stemlike extensions of the olfactory region of the brain.

Once inside the olfactory bulbs, the terminal axons of the receptor cells synapse with dendrites of *tufted cells*, *granule cells*, and *mitral cells*. These complex, ball-like synapses are called **olfactory glomeruli** (gluh-MARE-you-lie; sing. *glomerulus*; L. *glomus*, ball). Each glomerulus receives impulses from about 26,000 receptor cell axons. These impulses are conveyed along the axons of mitral and tufted cells, which form the *olfactory tract* running posteriorly to the olfactory cortex in the temporal lobe of the cerebrum. Olfaction is the only sense that does not

project fibers into the thalamus before reaching the cerebral cortex.

The Odor Mechanism

Many theories have been proposed to account for the mechanism that enables us to perceive such a vast variety of odors, but as yet, the enigma remains unresolved. The olfactory receptor cells (neurons) are unusual in that each acts as a chemoreceptor, a transducer, and a first-order neuron.

After the odorants are first absorbed into the mucus layer, they either diffuse directly to the protein receptor sites or, after being attached to a nasal-specific protein called an *olfactory binding protein*, are presented to the receptor sites on the cilia of the receptor cells. The resulting graded receptor potentials are thought to be due to the opening of specific ion channels for Na^+. This stimulates the activity of the intracellular second messenger cAMP. Summation of the graded potentials generates the nerve impulses that travel to the olfactory bulb. According to one current theory, the discrimination of different odors results, in part, from the combination of the simultaneous but subtle differences in the degree of stimulation among the receptor cells of a given population.

The unique capabilities of a rapid turnover and the brief life span of olfactory receptor cells, combined with the retention and stability of the odors perceived over time, continue to be intriguing phenomena.

Neural Pathways for Olfaction

Mitral cells in the olfactory bulbs project axonal branches through the olfactory tract to the primary olfactory cortex. The *primary olfactory cortex* is composed of the cortex of the uncus and adjacent areas, located in the temporal lobe of the cerebral cortex.

Q: *Why do we "sniff" when we want to detect an odor?*

A: Ordinarily, odors are carried to the nose by air currents. Sniffing creates eddying currents that greatly increase the amount of air that reaches the olfactory receptors high in the nasal cavities. Sniffing is a semireflexive response, usually occurring when an odor attracts our attention.

ASK YOURSELF

1 What are the three types of cells in the olfactory epithelium?

2 What are olfactory bulbs?

3 What happens in the olfactory receptor cells?

4 What is one current theory to explain the mechanism of olfaction?

HEARING AND EQUILIBRIUM

Hearing and equilibrium are considered in the same section because both sensations are received in the inner ear.

---

*These olfactory neurons (receptor cells) are among the most primitive neurons in the nervous system. They are chemoreceptor cells. Each olfactory receptor cell is a chemical detector, a transducer of a stimulus, and the transmitter of the nerve impulse to the olfactory bulb.

FIGURE 17.5 THE EAR

[A] Frontal cutaway diagram of the right ear, showing the external, middle, and inner ears. A section of the cochlear duct has been removed to show the spiral organ. [B] Structural components of the external ear; right lateral view. [C] An enlarged portion, emphasizing the auditory ossicles; right lateral view.

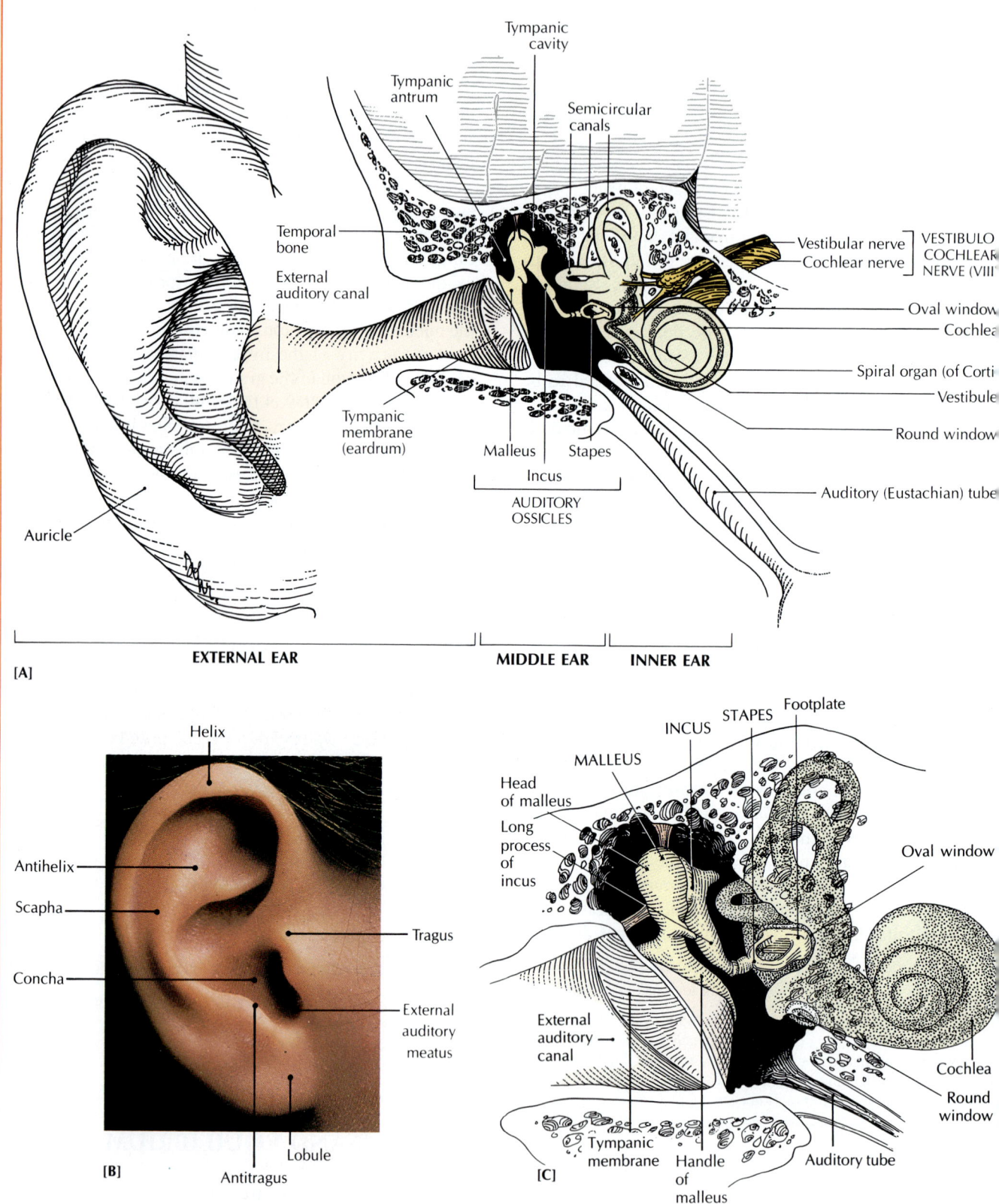

Structure	Description
EXTERNAL EAR	
Auricle	Cartilaginous, exterior "flap" of ear; conveys sound waves to middle ear.
External auditory canal (meatus)	Canal leading from floor of concha in outer ear to tympanic membrane.
MIDDLE EAR	
Tympanic membrane (eardrum)	Fibrous tissue extending across deep inner end of external auditory canal, forming partition between external and middle ears.
Tympanic cavity	Air-filled space in temporal bone; separated from external auditory canal by tympanic membrane, and from inner ear by bony wall with round and oval windows.
Auditory tube (Eustachian tube)	Tube leading downward and inward from tympanic cavity to nasopharynx.
Auditory ossicles (ear bones)	Malleus, incus, and stapes form chain from tympanic membrane to oval window of inner ear.
INNER EAR (LABYRINTH)	
Vestibule	Central chamber of inner ear; includes utricle and saccule filled with endolymph and surrounded by perilymph.
Semicircular ducts*	Three small ducts lying at right angles to each other; posterior to the vestibule; suspended in perilymph. Each duct has an expanded end, the ampulla, which contains a receptor structure, the crista ampularis.
Cochlea	Spiral structure containing perilymph-filled scala vestibuli and scala tympani, and endolymph-filled scala media (cochlear duct); anterior to the vestibule.
Spiral organ (of Corti)	Organ of hearing resting on the basilar membrane of the cochlea; a complex of supporting cells and hair cells.

*There are three semicircular canals surrounded by bone. Within each canal is a semicircular duct (about one-quarter the diameter of the canal), which contains endolymph. Surrounding each duct, and contained within each canal, is perilymph.

The ear actually has two functional units: (1) the **auditory apparatus,** concerned with hearing, and (2) the **vestibular apparatus,** concerned with posture and balance. The auditory apparatus is innervated by the *cochlear nerve,* and the vestibular apparatus is innervated by the *vestibular nerve.* The two nerves are collectively known as the *vestibulocochlear nerve* (cranial nerve VIII).

Anatomy of the Ear

The auditory system is organized to detect several aspects of sound, including pitch, loudness, and direction. The anatomical components of this system are the external ear, the middle ear, and the inner ear [FIGURE 17.5].

External ear The *external ear* is composed of the auricle (the part you can see) and the external auditory canal. The *auricle* (L. *auris,* ear) is composed of a thin plate of elastic cartilage covered by a close-fitting layer of skin. The funnel-like curves of the auricle are well designed to collect sound waves and direct them to the middle ear [FIGURE 17.5A]. The deepest depression, the *concha* (KONG-kuh; L. conch shell), leads directly to the external auditory canal (meatus) [FIGURE 17.5C]. The *helix* is the prominent ridge that forms the rim of the uppermost portion of the auricle [FIGURE 17.5B]. The *lobule,* or earlobe, is the fatty, lowermost portion of the auricle. It is the only part of the external ear without any cartilage.

The *external auditory canal* is a slightly curved canal extending about 2.5 cm (1 in.) from the floor of the concha to the tympanic membrane, which separates the external ear from the middle ear. The outer third of the wall of the external auditory canal is composed of elastic cartilage, and the inner two-thirds is carved out of the temporal bone [FIGURE 17.5C]. The canal and tympanic membrane are covered with skin. Fine hairs in the external ear are directed outward, and oil glands and modified sweat glands (ceruminous glands) secrete *cerumen* (suh-ROO-muhn), or earwax. The hairs and wax make it difficult for tiny insects and other foreign matter to enter the canal. Cerumen also prevents the skin of the external ear from drying out. The canal also acts as a buffer against humidity and temperature changes that can alter the elasticity of the eardrum.

Q: *Why does your voice sound different on a tape recording?*

A: When you hear yourself speak, you are hearing some extra resonance produced by the conduction of sound waves through the bones of your skull. Your voice as played by a tape recorder is the way it sounds to a listener, who receives the sound waves only through air conduction.

MEDICAL UPDATE

Cochlear Implants

Miranda is a typical 3-year-old whirlwind, dashing from one activity to another, shouting with sheer childish joy. It is not until she skids to a stop, turns to her mother, and begins using sign language that one realizes she is deaf.

Correction. Miranda *was* deaf—in both ears—but new technology is enabling her to hear voices and sounds for the first time in her short life. Miranda is being exposed to sounds because of the technological breakthrough called a **cochlear implant.** This is a surgically implanted prosthesis consisting of a receiver-stimulator and an electrode array activated by external components. The external components include a microphone that picks up sound, a speech processor that changes sound into an electrical signal, and a transmitter that sends the signal to the implanted receiver-stimulator.

An implant is not a high-tech hearing aid, but a system that transforms sounds into usable signals. Miranda's 22-electrode implant is reprogrammed every 3 months to her hearing level. She is rapidly approaching what is considered the normal hearing range. The implant is not brain surgery. The electrodes are threaded into the cochlea, and the implant receiver is placed securely on the skull behind the ear and just under the scalp. (The cochlea, which is the same size in children as it is in adults, is very small—approximately 1¼ in. in length and ¹⁄₁₀ in. in diameter.)

The basic assembly of all cochlear implants is the same. The microphone in Miranda's ear is connected to a sound processor, which she carries on her belt. The processor splits the sounds into bands of frequencies and sends them to a receiver implanted in Miranda's skull. From there the signals travel down an insulated cable, which ends as electrode wires snaked into the cochlea. Each electrode receives information about a single band of frequencies, and it stimulates the nerve endings around it in a matching pattern. The brain can then convert the pattern of stimulation into the range of frequencies that make up the sounds.

There have been over 7000 implant surgeries worldwide, 2000 of them on children, with no serious complications reported. This FDA-approved procedure costs approximately $45,000.

Middle ear The *middle ear* is a small chamber between the tympanic membrane and the inner ear. It consists of the tympanic cavity (Gr. *tumpanon*, drum) and contains the auditory ossicles (ear bones).

The *tympanic membrane,* popularly called the *eardrum,* forms a partition between the external ear and middle ear. It is a thin layer of fibrous tissue that is continuous externally with skin and internally with the mucous membrane that lines the middle ear. Between its concave external surface and convex internal surface is a layer of circular and radial fibers that give the membrane its firm elastic tension. The tympanic membrane is attached to a ring of bone (the *tympanic annulus*) and vibrates in response to sound waves entering the external auditory canal. The tympanic membrane is well endowed with blood vessels and nerve endings, so a "punctured eardrum" usually produces considerable bleeding and pain.

The *tympanic cavity* (middle ear cavity) is a narrow, irregular, air-filled space in the temporal bone. It is separated laterally from the external auditory canal by the tympanic membrane and medially from the inner ear by the bony wall, which has two openings: the *oval window* and the *round window* [FIGURE 17.5A]. An opening in the posterior wall of the cavity leads into the *tympanic antrum*, a chamber continuous with the small air cells in the mastoid process of the temporal bone. When an infection of the middle ear progresses through the tympanic antrum into the mastoid cells, it can cause *mastoiditis*.

In the anterior wall of the tympanic cavity is the *auditory tube,* commonly called the **Eustachian tube** (yoo-STAY-shun). It leads downward and inward from the tympanic cavity to the nasopharynx, the space above the soft palate that is continuous with the nasal passages [FIGURE 17.5A]. The mucous membrane lining of the nasopharynx is also continuous with the membrane of the tympanic cavity. As a result, an infection may spread from the nose or throat into the middle ear, producing a middle ear infection, or *otitis media*. (It is a good idea to keep your mouth slightly open when you blow your nose. Blowing the nose hard with the mouth closed creates a pressure that may force infectious microorganisms from the nose into the middle ear.)

The main purpose of the auditory tube is to maintain equal air pressure on both sides of the tympanic membrane by permitting air to pass from the nasal cavity into the middle ear. The pharyngeal opening of the tube remains closed when the external pressure is greater but opens during swallowing, yawning, and nose blowing so that minor differences in pressure are adjusted without conscious effort. The tube may remain closed when the pressure change is sudden, as when an airplane takes off

Q: *Why do you cough when your ear is probed?*

A: Stimulation of the external auditory canal can initiate a coughing reflex involving the auricular branch of the vagus nerve, which innervates the skin of the external auditory canal.

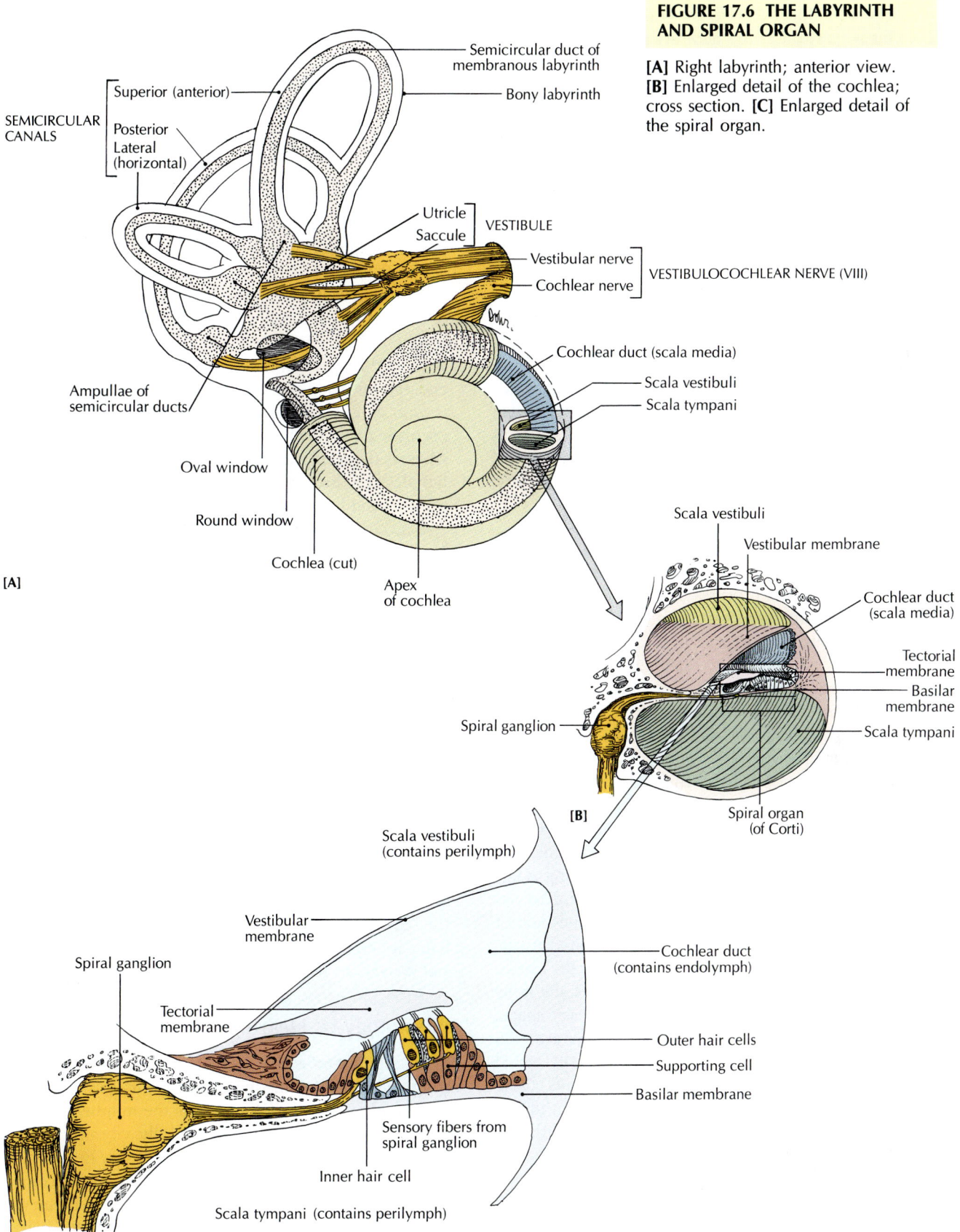

SEMICIRCULAR CANALS
- Superior (anterior)
- Posterior
- Lateral (horizontal)

Semicircular duct of membranous labyrinth

Bony labyrinth

Utricle
Saccule
VESTIBULE

Vestibular nerve
Cochlear nerve
VESTIBULOCOCHLEAR NERVE (VIII)

Cochlear duct (scala media)
Scala vestibuli
Scala tympani

Ampullae of semicircular ducts

Oval window

Round window

Cochlea (cut)

Apex of cochlea

[A]

FIGURE 17.6 THE LABYRINTH AND SPIRAL ORGAN

[A] Right labyrinth; anterior view. [B] Enlarged detail of the cochlea; cross section. [C] Enlarged detail of the spiral organ.

Scala vestibuli
Vestibular membrane
Cochlear duct (scala media)
Tectorial membrane
Basilar membrane
Scala tympani

Spiral ganglion

Spiral organ (of Corti)

[B]

Scala vestibuli (contains perilymph)

Vestibular membrane

Spiral ganglion

Tectorial membrane

Cochlear duct (contains endolymph)

Outer hair cells
Supporting cell
Basilar membrane

Sensory fibers from spiral ganglion

Inner hair cell

Scala tympani (contains perilymph)

[C]

or lands. However, the pressure can usually be equalized, and the discomfort relieved, by swallowing or yawning. This maneuver causes the auditory tube to open.

The three *auditory ossicles* of the middle ear form a chain of levers extending from the tympanic membrane to the inner ear. This lever system transmits sound waves from the external ear to the inner ear. From the outside in, the tiny, movable bones are the *malleus* (hammer), *incus* (anvil), and *stapes* (STAY-peez) (stirrup) [FIGURE 17.5C]. The ear bones are the smallest bones in the body, with the stapes being the smallest of all.

The auditory ossicles are held in place and attached to each other by ligaments. Two tiny muscles — the *tensor tympani* and the *stapedius* — are attached to the ear bones. The *tensor tympani* is attached to the handle of the malleus. When this muscle contracts, it pulls the malleus inward, increasing the tension on the tympanic membrane and reducing the amplitude of vibrations transmitted through the chain of auditory ossicles. The *stapedius* attaches to the neck of the stapes. Its contraction pulls the footplate of the stapes, decreasing the amplitude of vibrations at the oval window. This reduces the intensity, or loudness, of the sound.

Inner ear   The *inner ear* is also called the *labyrinth* (Gr. maze) because of its intricate structure of interconnecting chambers and passages [FIGURE 17.6]. It consists of two main structural parts, one inside the other: (1) The *bony labyrinth* is a series of channels hollowed out of the petrous portion of the temporal bone. It is filled with a fluid called *perilymph*. (2) The bony labyrinth surrounds the inner *membranous labyrinth*, which contains a fluid called *endolymph* and all the sensory receptors for hearing and equilibrium.

The membranous labyrinth consists of three semicircular ducts as well as the utricle, saccule, and cochlear duct, all of which are filled with endolymph and contain various sensory receptors (cristae ampullaris maculae for the vestibular senses and spiral organ for hearing). The semicircular ducts are located within the semicircular canals of the bony labyrinth. They are about one-quarter the diameter of the canals. Perilymph is located in the space between the ducts and the bony walls of the canals. Because the membranous labyrinth fits inside the bony labyrinth, these two channels have the same basic shape.

The bony labyrinth consists of (1) the vestibule, (2) three semicircular canals, and (3) the spirally coiled cochlea. The *vestibule* (L. entrance) is the central chamber of the labyrinth. Within the vestibule are the two endolymph-filled sacs of the membranous labyrinth: the *utricle* (YOO-trih-kuhl; L. little bottle) and the smaller *saccule* (SACK-yool; L. little sack). Each sac contains a sensory patch called a *macule*.

The three *semicircular ducts* and *canals* are perpendicular to each other, allowing each one to be oriented in one of the three planes of space [FIGURE 17.6A]. The ducts are lined by the membranous labyrinth, while the canals are surrounded and lined by bone. On the basis of their locations, the semicircular ducts are called *superior*, *lateral*, and *posterior*. Each duct has an expanded end called an *ampulla* (am-POOL-uh), which contains a receptor structure, the *crista ampullaris*. The utricle, saccule, and semicircular ducts are concerned with equilibrium, not hearing, and will be discussed in further detail later in the chapter.

Beyond the semicircular ducts is the spiral *cochlea* (KAHK-lee-uh; Gr. *kokhlos*, snail), so named because it resembles a snail's shell. It may be thought of as a bony tube wound 2¾ times in the form of a spiral. The cochlea is divided longitudinally into three spiral ducts: (1) the *scala* (L. staircase) *vestibuli*, which communicates with the vestibule and oval window, (2) the *scala tympani*, which ends at the round window, and (3) the *scala media*, or *cochlear duct*, which lies between the other two ducts. The cochlear duct contains endolymph and the spiral organ, whereas the scala vestibuli and scala tympani contain perilymph. The three ducts, arranged in parallel, ascend in a spiral around the bony core, or *modiolus* (L. hub).

The cochlear duct is separated from the scala vestibuli by the *vestibular membrane*, and from the scala tympani by the *basilar membrane*. Resting on the basilar membrane is the **spiral organ (of Corti)**, the organ of hearing [FIGURE 17.6B, C]. The spiral organ is an organized complex of supporting cells and hair cells. The hair cells are arranged in rows along the length of the coil. The outer hair cells are arranged in three rows, and the inner hair cells are in a single row along the inner edge of the basilar membrane. There are about 3500 inner hair cells and 20,000 outer hair cells.

Both the inner and outer hair cells have bristlelike *sensory hairs*, or *stereocilia* — which are specialized microvilli — and a basal body on one side [FIGURE 17.7]. Each outer hair cell has 80 to 100 sensory hairs, and each inner hair cell has 40 to 60 sensory hairs. In each hair cell, the hairs are arranged in rows that form the letter W or U, with the base of the letter directed laterally. The tips of many hairs are embedded within and firmly bound to the *tectorial membrane* above the spiral organ.

Physiology of Hearing

Physically, *sound* is the alternating compression and decompression of the medium (usually air) through which the sound is passing. *Frequency* is the number of sound waves per second. Very few naturally vibrating bodies produce simple vibrations or a "pure" tone. Instead they produce combinations of frequencies. When the combinations are subjectively interpreted by the nervous system, they give various sounds their *timbre* (TIM-ber or TAM-ber; Fr. a bell struck with a hammer), or "quality."

FIGURE 17.7 SENSORY HAIR CELLS IN THE SPIRAL ORGAN

Drawing of inner and outer hair cells showing the rows of stereocilia.

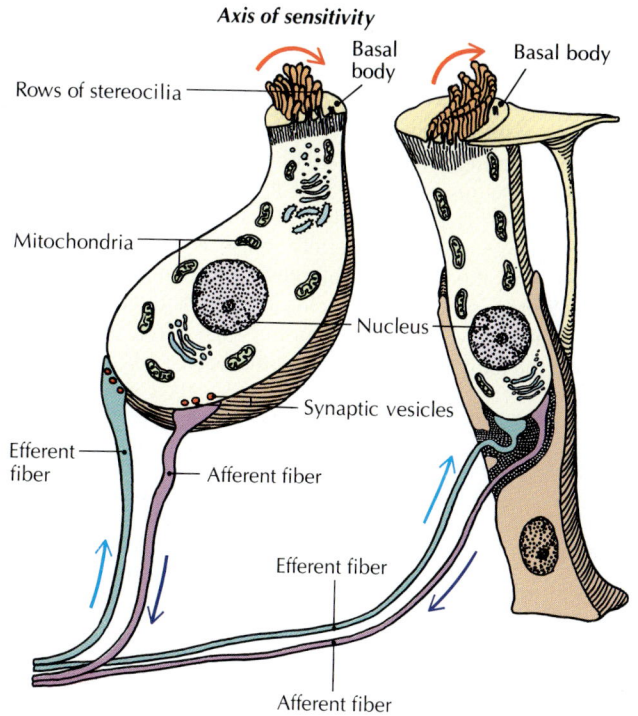

Axis of sensitivity

Basal body

Basal body

Rows of stereocilia

Mitochondria

Nucleus

Synaptic vesicles

Efferent fiber

Afferent fiber

Efferent fiber

Afferent fiber

INNER HAIR CELL

OUTER HAIR CELL

It is not known exactly how the brain interprets distinctive sounds, but we do know that *loudness* or intensity is determined by the size of the sound waves, *timbre* by their shape, and *pitch* by their frequency. We also understand that the difference between noise and music lies in the synchronous regularity of musical sound waves.

Sound waves are conducted through air in the external ear (air conduction), through solids in the middle ear (bone conduction), and through fluid in the inner ear (fluid conduction). The transition is important because sound waves in the air do not pass readily into a fluid medium, as you may know if you swim underwater. Although our three-phase conduction of sound waves (air-bone-fluid) works admirably, there is a loss of energy

Q: *How can you tell the direction of a sound?*

A: Depending on the position of the head, sound reaches the closer ear about 1/1500 of a second sooner than it reaches the other ear. Also, the sound is a little louder in the closer ear. These differences are recognized and analyzed by the brain to tell you the direction from which a sound is coming.

when sound waves pass from air to fluid, since fluid is more difficult to move than air is. This energy loss is just about balanced by the amplifying action of the ossicles in the middle ear, however, and there is no appreciable loss of energy.

The efficiency of the energy transfer from the tympanic membrane (which responds to air vibrations) to the oval window (which transfers the vibration to the endolymph in the cochlea) is enhanced because the surface area of the tympanic membrane is about 20 times greater than that of the foot of the stapes in the oval window. This contributes to overcoming much of the energy lost during the transfer from air to endolymph. The three-phase conduction system is estimated to work at 99.9 percent transmission efficiency. The ear is considered a more efficient energy converter than the eye.

The hair cells of the spiral organ are the mechanoreceptors in which the mechanical energy of sound is transduced into the receptor (generator) potentials of the cochlear nerve endings. Each hair cell is innervated by several neurons. There are about 23,500 hair cells in the spiral organ and about 30,000 neurons in the cochlear nerve. The conversion of sound waves to generator potentials is described in FIGURE 17.8.

Neural Pathways for Hearing

Axons of neurons with cell bodies in the spiral ganglion of the cochlear nerve extend from the spiral organ and terminate centrally in the dorsal and ventral cochlear nuclei. The cochlear nerve enters the brainstem at the junction of the pons and medulla oblongata and terminates in the cochlear nuclear complex. After entering, each fiber divides into two main branches.

Axons of neurons in the dorsal and ventral cochlear nuclei *ascend* as crossed and uncrossed fibers in the lateral lemnisci to the inferior colliculus. Thus, the auditory pathways ascend bilaterally. From cell bodies in the inferior colliculus, the pathway continues as the brachium of the inferior colliculus to the medial geniculate body of the thalamus. From cell bodies in the medial geniculate body, axons project via the auditory radiations to the primary auditory cortex (transverse temporal gyrus of Heschl, areas 41 and 42; see FIGURE 13.11).

A lesion of the auditory pathway on one side results in a decrease of hearing acuity in both ears, but the loss is more pronounced on the opposite (contralateral) side because the ascending auditory pathways are composed of mainly crossed and some uncrossed projections.

Projecting from the auditory cortex and other nuclei of the auditory pathways are *descending pathways*, which accompany the ascending pathways described above. The descending fibers have a role in processing ascending auditory impulses, enhancing the signals and suppressing the "noise."

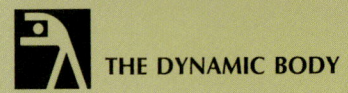
The Sounds We Hear

Human ears are generally responsive to frequencies from about 30 to 20,000 cycles per second (cps), but some people can hear from about 16 to 30,000 cps, especially during their early years. (Some animals, notably bats and dogs, can hear much higher frequencies.) Hearing ability declines steadily from early childhood, probably because the basilar membrane loses some of its elasticity. In addition, harmful calcium deposits may form, and hair cells begin to degenerate. The human ear is most acute to frequencies ranging from 300 to 3000 cps, the approximate range of the human voice. (*Cycles per second* may also be referred to as *hertz*, abbreviated *Hz*.)

Current estimates are that some environmental noises are twice as intense now as they were in the 1960s, and their intensity is expected to continue to double every 10 years. Studies have shown that factory workers have twice as much hearing loss as white-collar office workers. Approximately 10 million Americans use hearing aids, and many of these individuals may have suffered hearing impairment through prolonged exposure to sounds that were not thought of as excessively loud. The accompanying table shows the decibel scale and the range of some common noise sources. A *decibel* represents the relative intensity of a sound, zero decibels being the faintest sound the average person can hear. Each increment of 10 increases the intensity 10 times (exponentially). The ratio between 140 decibels and 0 decibels is about 100 trillion to 1. Ear damage depends on the length of exposure as well as the decibel level.

It is interesting that you are more susceptible to hearing damage when you find the sound unpleasant. This is probably why rock musicians suffer less hearing impairment than would be expected. But generally, noise is a stressor, constricting blood vessels, damaging ear tissue, reducing oxygen and nutrients to the ear, causing high blood pressure, and increasing the heart rate.

Exposure level	Decibels	Common noise sources	Permissible exposure
Harmful to hearing	140	Jet engine 25 m away; shotgun blast	None (any exposure dangerous)
	130	Threshold of pain	1 min
		Jet takeoff 100 m away	3 min
		Air raid siren	5 min
	120	Propeller aircraft; discotheque	10 min
Possible hearing loss	110	Live rock band; cassette player; jet takeoff 600 m away	26 min
	100	Power mower; jackhammer	2 hr
		Subway station with train coming; farm tractor	3½ hr
	90	Convertible ride on freeway; motorcycle 8 m away	8 hr
		Average street traffic; food blender; heavy traffic 5 m away	16 hr
Very noisy	80	Alarm clock	
		Garbage disposal	
Moderately noisy (urban area) (upper limit for hearing normal conversation)	70	Vacuum cleaner; private car; noisy business office	No limit
	60	Conversation; singing birds	
Quiet (suburban and small town)	50	Light traffic 30 m away; quiet business office	
	40	Soft radio music; faucet dripping; library	
Very quiet	30	Soft whisper at 5 m	
	20	Quiet suburban dwelling; broadcasting studio	
	10	Leaves rustling	
	0	Threshold of hearing	

FIGURE 17.8 THE PHYSIOLOGY OF HEARING

The drawing shows how sound waves are converted into mechanical vibrations. The cochlea has been partially un-rolled, and some anatomical details have been simplified for clarity. The numerals refer to the itemized description in the following text:

1 Sound waves are pressure waves that enter the external ear. After crossing the external auditory meatus, the waves reach the tympanic membrane.

2 Air molecules under pressure cause the tympanic membrane to vibrate. Low-frequency sound waves produce slow vibrations, and high-frequency waves produce rapid vibrations. The vibrations move the malleus, on the other side of the membrane.

3 The handle of the malleus articulates with the incus, causing it to vibrate.

4 The vibrating incus moves the stapes as it oscillates into and out of the cochlea at the oval window.

5 The sound waves that reach the inner ear through the oval window set up pressure changes that vibrate the peri-lymph in the scala vestibuli.

6 Vibrations in the perilymph are transmitted across the vestibular membrane to the endolymph of the cochlear duct and also up the scala vestibuli and down the scala tympani. The vibrations are transmitted to the basilar membrane, causing the membrane to ripple. The fundamental vibratory

ripples result in the perception of pure tones. Overtones such as musical sounds, chords, and harmonics result from secondary vibrations superimposed on the fundamental vibrations of the spiral organ.

The ripples in the long axis of the basilar membrane are concerned with the *frequency* and *intensity* of sound. The spiral organ is organized so that the high tones are encoded near the base of the cochlea, and the low tones are encoded near the apex. Loudness is associated with the amplitude of the vibrations (the amount of displacement of the basilar membrane).

7 Receptor hair cells of the spiral organ that are in contact with the overlying tectorial membrane are bent, causing them to generate graded receptor (generator) potentials. These generator potentials excite the cochlear nerve to generate action potentials. When the hairs are displaced toward the basal body (axis of sensitivity), the hair cells are excited; when the hairs are displaced away from the basal body, the hair cells are inhibited.

8 The nerve impulses are conveyed along the cochlear branch of the vestibulocochlear nerve. These fibers activate the auditory pathways in the central nervous system, which terminate in the auditory area of the temporal lobe of the cerebral cortex, where the appropriate sound is perceived.

9 Vibrations in the scala tympani are dissipated out of the cochlea through the round window into the middle ear.

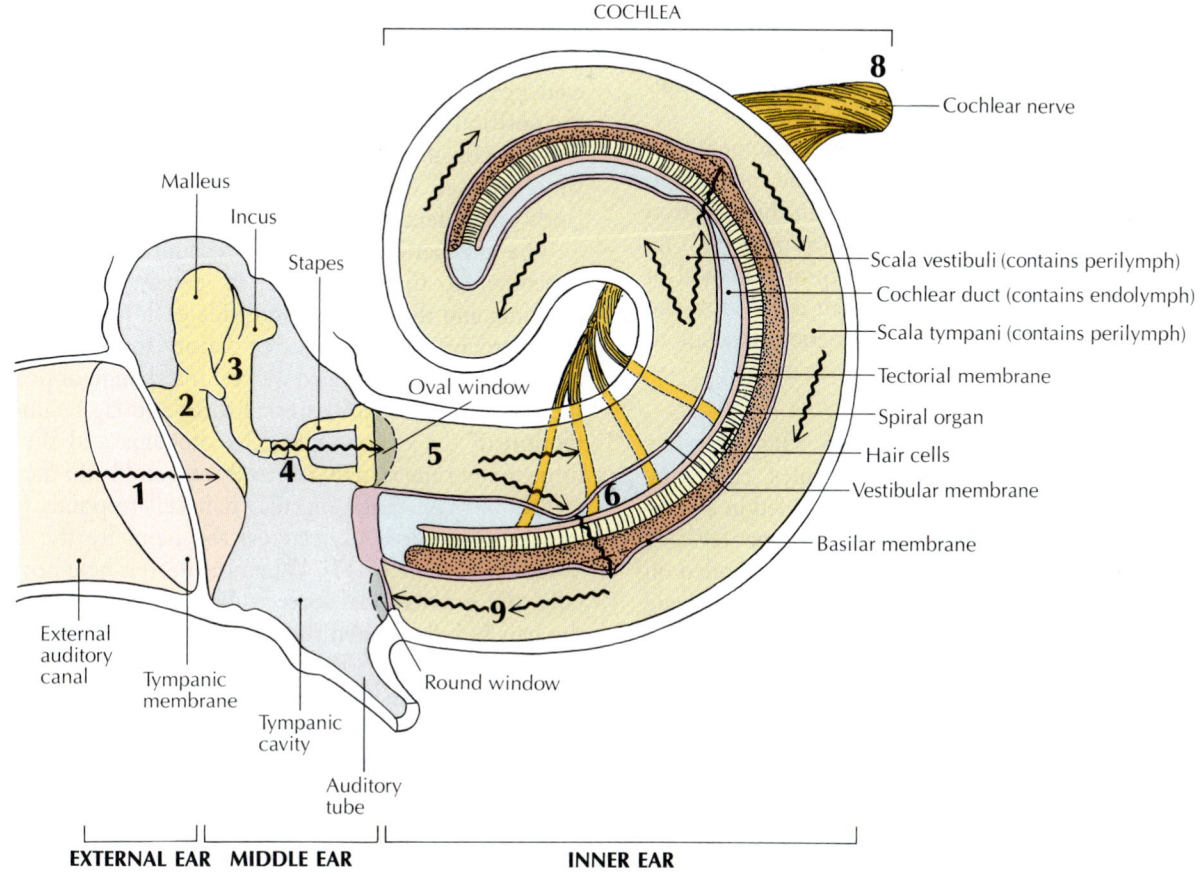

Vestibular Apparatus and Equilibrium

The inner ear helps the body cope with changes in position and acceleration and deceleration. This *vestibular apparatus* [FIGURE 17.9] signals changes in the *motion* of the head *(dynamic equilibrium)* and in the *position* of the head with respect to gravity *(static equilibrium).*

The main components of the vestibular apparatus are the utricle, the saccule, and the three fluid-filled semicircular ducts of the membranous labyrinth. The receptors in the utricle and saccule regulate static equilibrium, and the receptors in the ampullae of the semicircular ducts respond to acceleration of the head. The equilibrium system also receives input from the eyes and from some proprioceptors in the body, especially the joints. (Try standing on your toes with your eyes closed. Without your eyes to guide your body, you invariably begin to fall forward.)

Specialized proprioceptors of the vestibular apparatus, known as *hair cells,* are arranged in clusters of *hair bundles* [FIGURE 17.10]. Hair cells are extremely sensitive receptors that convert a mechanical force applied to a hair cell into an electrical signal that is relayed to the brain. An extremely slight movement of a hair bundle can cause the hair cells to respond.

Two types of sensory hairs are present in hair bundles: (1) *stereocilia,* which are actually modified microvilli, and (2) *kinocilia,* which are modified cilia. Each hair cell has about 100 stereocilia in a tuft and 1 kinocilium at the edge of the tuft [FIGURE 17.10]. Because of its asymmetry, a receptor hair cell is said to be polarized.

The electrical signal triggered by the movement of a hair bundle depends on the direction of the movement. When the hairs of the hair cell are bent in the direction of the kinocilium (known as the *axis of sensitivity;* FIGURE 17.7), the hair cell can generate a receptor potential in the nerve ending, synapsing with the hair cell. When the hairs of the hair cell are bent in the direction opposite to the axis of sensitivity, the hair cell is not excited.

Static equilibrium: utricles and saccules The receptor region of the utricles and saccules, called the *macula,* contains receptor hair cells embedded in a jelly-like *otolithic* (Gr. "ear stones") *membrane* [FIGURE 17.10A]. Loosely attached to the membrane and piled on top of it are hundreds of thousands of calcium carbonate crystals called *otoconia.* The utricles and saccules are both filled with endolymph. Hair cells in the utricle respond to the motion changes that occur during the back-and-forth movements of the head, while the saccules respond to up-and-down movements of the head. The hair cells also monitor the position of the head in space, controlling posture. For example, the next time you dive into a pool, notice how you turn and swim upward without having to decide consciously which way is up. This is

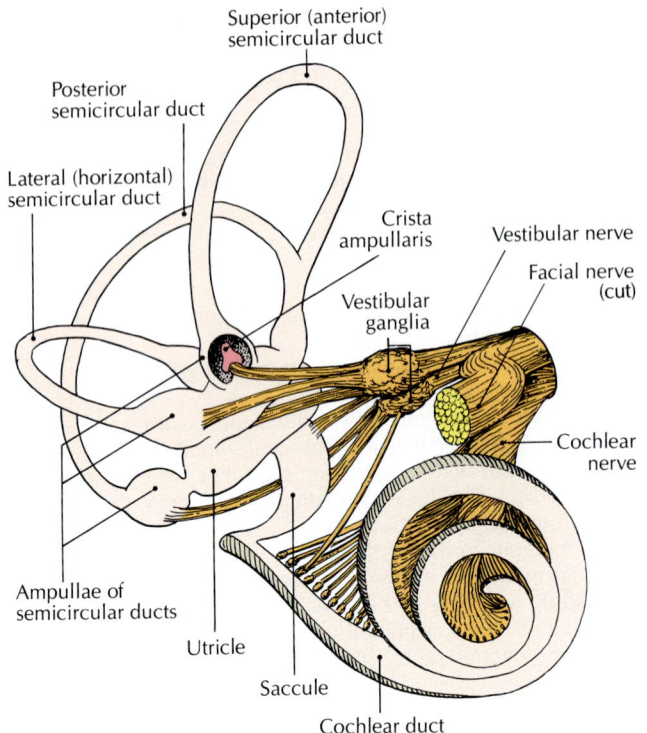

FIGURE 17.9 THE VESTIBULAR APPARATUS

Superior (anterior) semicircular duct
Posterior semicircular duct
Lateral (horizontal) semicircular duct
Crista ampullaris
Vestibular nerve
Facial nerve (cut)
Vestibular ganglia
Cochlear nerve
Ampullae of semicircular ducts
Utricle
Saccule
Cochlear duct

evidence of your utricles and saccules at work, telling you the position of your head in relation to gravity. They are also responsible for initiating the "righting reflex," which we see when a cat is dropped upside down and lands on its feet.

The mechanism of the static equilibrium response depends on the difference in density between the otoconia and the endolymph inside each utricle and saccule. Otoconia have a greater density than endolymph. As a result, when the head moves in a change of posture, the otoconia resist the external force and lag behind the motion of the endolymph. The otoconia and the otolithic membrane remain relatively still and bend the hairs of the hair cell. Each macular hair cell responds to the gravitational force exerted on the hairs by the dense otoconia [FIGURE 17.11]. When the head is held horizontally, the gravitational force is directed downward upon the hair bundle. When the head is tilted to the side, the hair bundle of each cell is displaced along the axis of sensitivity and can excite the afferent nerve fiber. A tilt of the head that bends the hair against the axis of sensitivity has the opposite effect, inhibiting the vestibular afferent neuron.

The bending of the hair cells alters the permeability of the cells to sodium and potassium ions. The resulting graded receptor potential stimulates nerve endings of the

FIGURE 17.10 MACULA AND HAIR CELLS

[A] Macula, receptor region of utricle and saccule. The stereocilia of hair cells extend into the adjacent gelatinous otolithic membrane, which contains embedded otoconia crystals. [B] Drawing of a hair cell in cross section.

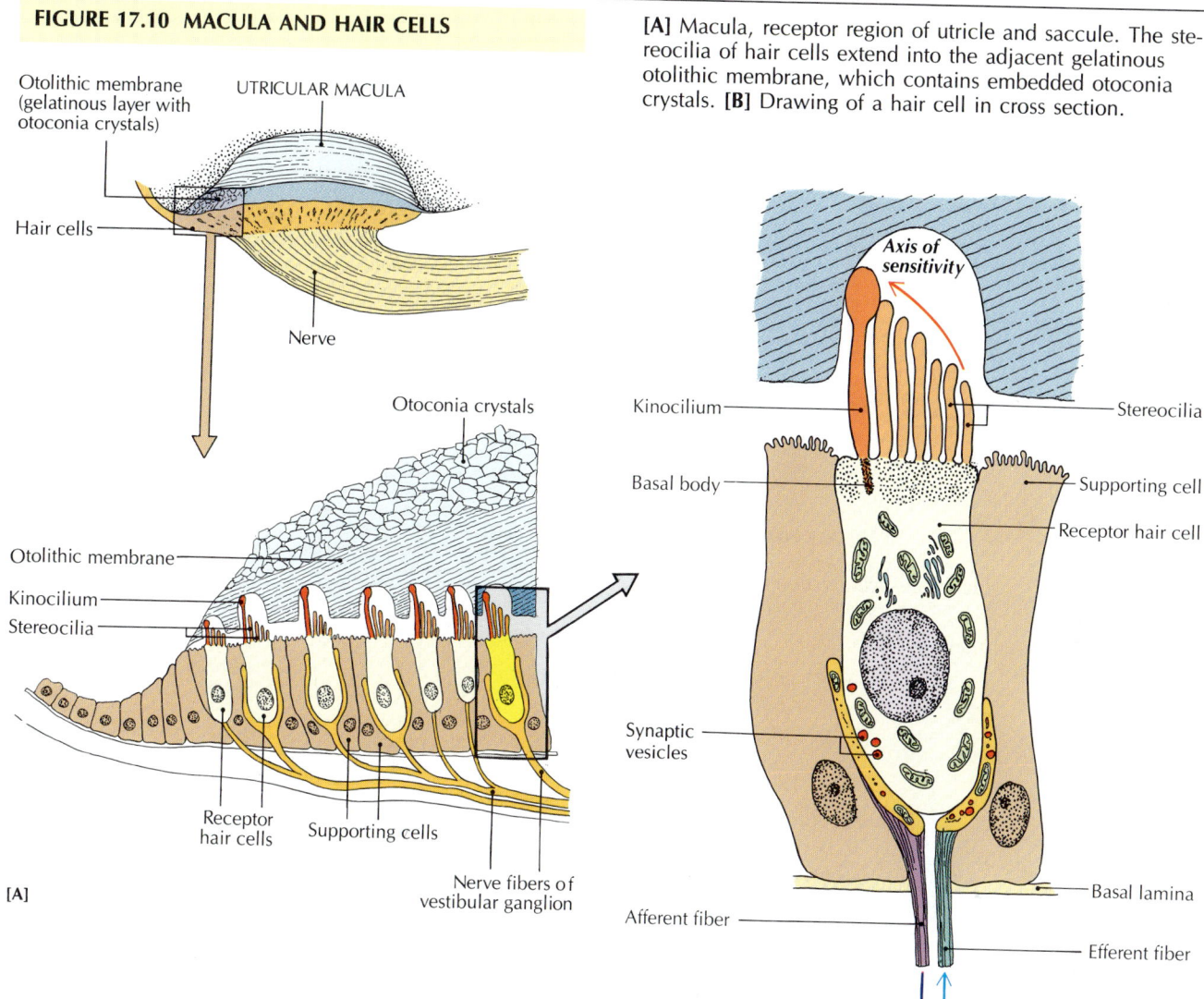

Otolithic membrane (gelatinous layer with otoconia crystals)

UTRICULAR MACULA

Hair cells

Nerve

Otoconia crystals

Otolithic membrane

Kinocilium

Stereocilia

Receptor hair cells

Supporting cells

Nerve fibers of vestibular ganglion

[A]

Axis of sensitivity

Kinocilium

Stereocilia

Basal body

Supporting cell

Receptor hair cell

Synaptic vesicles

Afferent fiber

Basal lamina

Efferent fiber

[B]

vestibular nerve fibers to generate another graded receptor potential, and subsequently an action potential in a vestibular nerve fiber, which is transmitted to the vestibular nuclei in the brainstem and cerebellum.

As efficient as human utricles and saccules are, they are certainly not perfect. This lack of efficiency is demonstrated when an airplane pilot does not realize that he or she is flying through clouds upside down. Also, the slowness of the utricle in registering deceleration can be seen in the delayed stumbling response of a standee in a bus when the bus stops suddenly.

Dynamic equilibrium: semicircular ducts

The utricles and saccules are *organs of gravitation,* responding to movements of the head in a straight line: forward, backward, up, or down. In contrast, the crista ampullaris of the semicircular ducts responds to changes in acceleration in the *direction of head movements,* specifically turning

and rotating. These movements are called angular movements, in contrast to straight-line movements.

Because each of the three ducts is situated in a different plane, at right angles to each other, at least one duct is affected by every head movement. Each duct has a bulge, the *ampulla,* that contains a patch of hair cells and supporting cells embedded in the *crista ampullaris* [FIGURE 17.12]. The hairs of the hair cells project into a gelatinous flap called the *cupula* (KYOO-pyuh-luh; L. little cask or tub). The cupula acts like a swinging door, with the crista as the hinge. The free edge of the cupula brushes against the curved wall of the ampulla. When the head rotates, the endolymph in the semicircular ducts lags behind because of the inertia, displacing the cupula and the hairs projecting into it in the opposite direction. (The semicircular ducts do not sense movement at slow, steady speeds because the head and the endolymph move at the same rate.) As a result of the slight displacement of

FIGURE 17.11 STATIC EQUILIBRIUM

[A] Displacement of the hair bundle stimulates the hair cells in the utricle and saccule.

[B] As a hair cell responds to the displacement of the hair bundle, nerve impulses are transmitted to the brain along an afferent nerve fiber at the base of the cell. (1) When the hair bundle is displaced, (2) depolarization occurs and spreads through the cell. (3) Depolarization causes semipermeable channels in the base of the cell to open. Vesicles release their neurotransmitter substances, which diffuse across the synaptic cleft between the hair cell and the adjacent neuron, transmitting the impulse from one cell to the next. (4) The message is sent to the brain along an afferent fiber of the vestibular branch of the vestibulocochlear nerve (VIII).

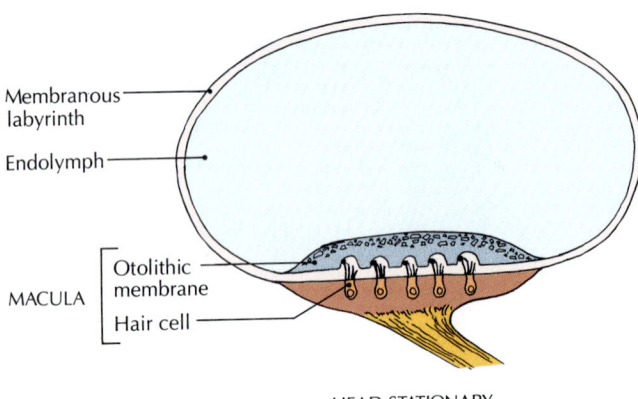

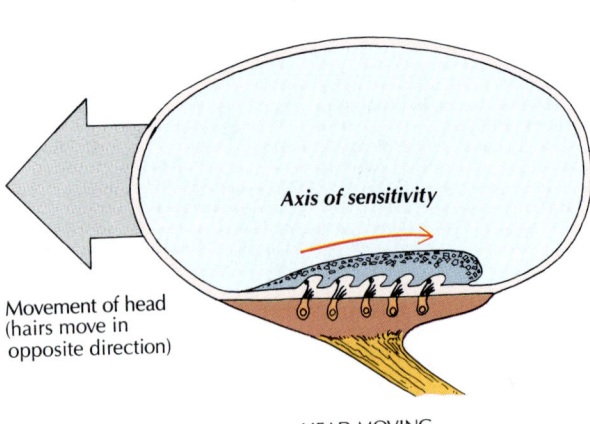

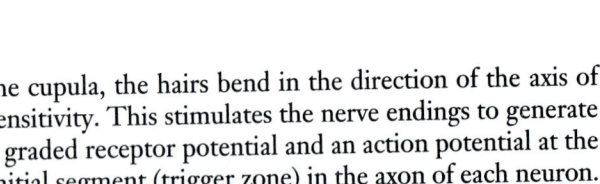

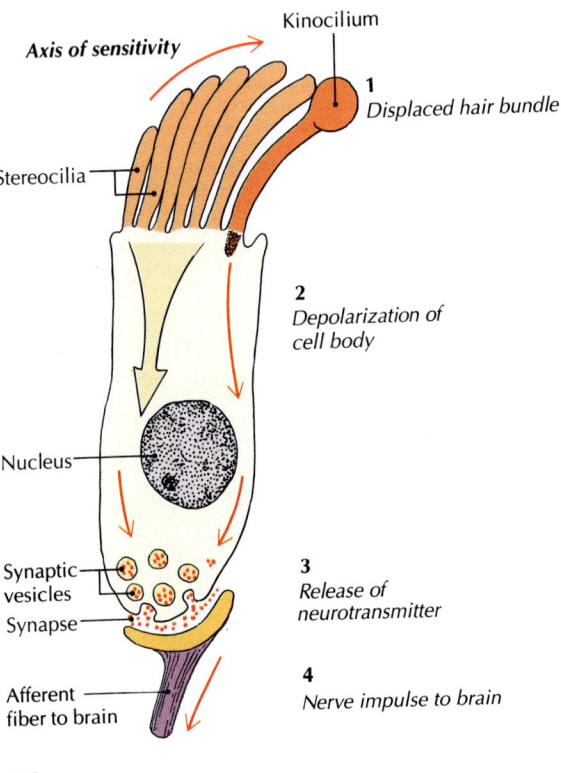

[A]

[B]

the cupula, the hairs bend in the direction of the axis of sensitivity. This stimulates the nerve endings to generate a graded receptor potential and an action potential at the initial segment (trigger zone) in the axon of each neuron. The brain receives input and then signals the appropriate muscles to contract or relax in order to maintain the body's equilibrium.

A feeling of dizziness occurs when you spin about or move violently and then stop suddenly. Because of inertia, the endolymph keeps moving (the way you keep moving forward when your car stops quickly), and it continues to stimulate the hair cells. Although you know you have stopped moving, the signals being sent from your inner ear to your brain make you feel that motion is still occurring, but in the reverse direction. In other words, while in motion, the inertia of the endolymph bends the hairs in the direction opposite to the direction of the movement. Immediately after stopping, the endolymph continues to move relative to the ducts. As a result, the hairs are now bent in the direction of the prior movement.

Deaf individuals, who do not have functional receptors in their vestibular labyrinths, are immune to dizziness and motion sickness and rely on visual cues for the maintenance of normal locomotion and posture. (Without visual cues, a swimmer who has lost the use of the labyrinth may navigate down instead of up in an attempt to reach the surface.)

FIGURE 17.12 DYNAMIC EQUILIBRIUM

[A] The ampulla in transparent full view. [B] The ampulla in cross section. Hairs anchored in the crista ampullaris project into the gelatinous cupula. [C] and [D] When the head rotates, displacement of the hair bundle stimulates the hair cells in the crista ampullaris of the semicircular ducts.

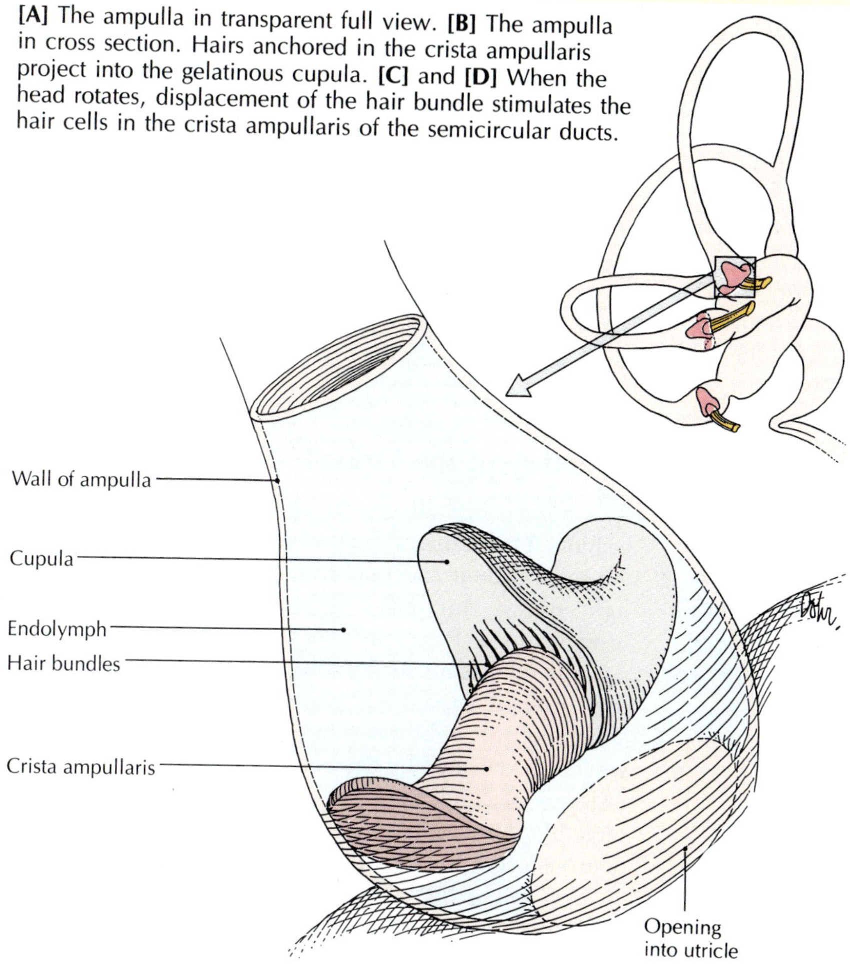

Wall of ampulla

Cupula

Endolymph

Hair bundles

Crista ampullaris

Opening into utricle

[A]

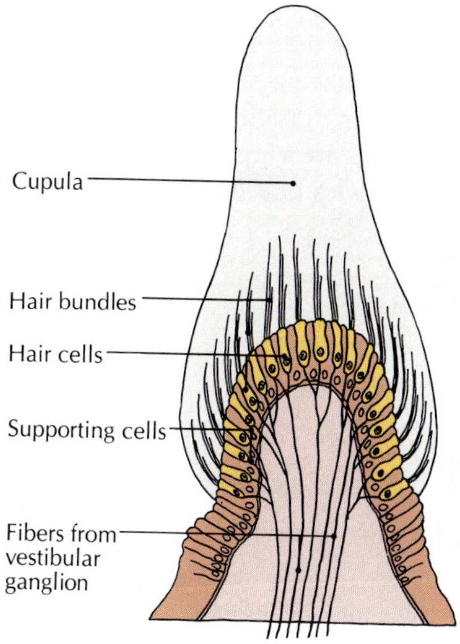

Cupula

Hair bundles

Hair cells

Supporting cells

Fibers from vestibular ganglion

[B]

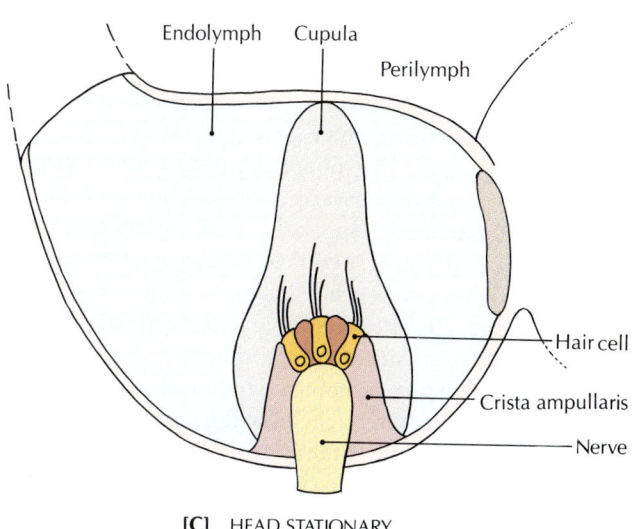

Endolymph   Cupula

Perilymph

Hair cell

Crista ampullaris

Nerve

[C]   HEAD STATIONARY

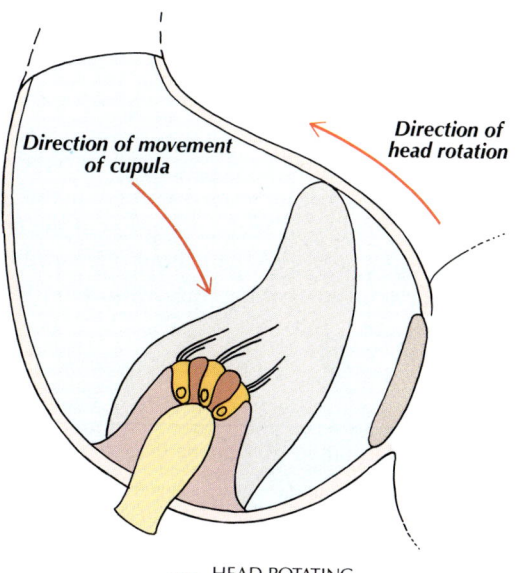

Direction of movement of cupula

Direction of head rotation

[D]   HEAD ROTATING

Neural Pathways for Equilibrium

The vestibular tracts consist of pathways to the brainstem, spinal cord, cerebellum, and cerebral cortex. The approximately 19,000 neurons of each vestibular nerve have their cell bodies in the vestibular ganglion near the membranous labyrinth. The primary vestibular fibers from the vestibular nerve pass into the upper medulla oblongata and terminate (1) in each of the four vestibular nuclei in the upper lateral medulla oblongata and (2) in specific regions of the cerebellum.

The sensory signals from the vestibular sensors of the labyrinth are indicators of the position and movements of the head. These inputs from the vestibular receptors are critical in (1) generating compensatory movements to maintain balance and an erect posture in response to gravity, (2) producing the conjugate (coupled) movements of the eyes that compensate for changes in the position of the constantly moving head, and (3) supplying information for the conscious awareness of position, acceleration, deceleration, and rotation. The vestibular functions are supplemented by proprioceptive input from the muscles and joints as well as the visual system.

A S K   Y O U R S E L F

1 What are the main parts of the middle ear?

2 What is the spiral organ?

3 How are sound waves converted into nerve impulses?

4 What is the function of otoconia?

5 What is the importance of hair cells?

6 What is the difference between static equilibrium and dynamic equilibrium?

VISION

We live primarily in a visual world, and sight is our dominant sense. The specialized exteroceptors in our eyes constitute about 70 percent of the receptors of the entire body, and the optic nerves contain about one-third of all the afferent nerve fibers carrying information to the central nervous system.

Q: *Do the vestibular organs work during space travel?*

A: The utricle and saccule do not work under zero-gravity conditions, but the semicircular canals do. Most astronauts experience motion sickness at some time during space travel.

Although we can rely on our eyes to bring us many of the sights of the external world, they are not able to reveal everything. We "see" only objects that emit or are illuminated by light waves in our receptive range, representing only 1/70 of the entire electromagnetic spectrum. Some organisms, such as insects, are sensitive to shorter wavelengths in the range of ultraviolet, and other organisms can "see" longer wavelengths in the infrared range of the spectrum.

Light reaches our light-sensitive "film," or *retina*, through a transparent window, the *cornea*. In addition to the basic and accessory structures of the eye, vision involves the brain and the optic nerve. We consider first the basic anatomy of the eye and then the physiology of vision.

Structure of the Eyeball

The human eyeball can be compared to a simple old-fashioned box camera. Instead of being a box, the eyeball is a sphere about 2.5 cm (1 in.) in diameter. In both cases, light passes through a lens. The external image is brought to a focus on the sensitive retina, which is roughly equivalent to the film in a camera. Over a hundred million specialized photoreceptor cells (rods and cones) convert light waves into electrochemical impulses, which are decoded by the brain. The retina is composed of layers of slender photoreceptors and a complex of interacting, processing neurons. Some of these sensory neurons send axons to the brain via the optic nerve.

The wall of the eyeball consists of three layers of tissue: the outer supporting layer, the vascular middle layer, and the inner retinal layer. The eyeball is divided into three cavities: the anterior chamber, the posterior chamber, and the vitreous chamber [FIGURE 17.13].

Supporting layer   The outer *supporting layer* of the eyeball consists mainly of a thick membrane of tough, fibrous connective tissue. The posterior segment, which makes up five-sixths of the tough outer layer, is the opaque white *sclera* (Gr. *skleros*, hard). The sclera forms the "white" of the eye, giving the eyeball its shape and protecting the delicate inner layers [FIGURE 17.13A]. The anterior segment of the supporting layer is the transparent *cornea* (L. *corneus*, horny tissue), which constitutes the modified anterior one-sixth of the outer layer. The cornea bulges slightly. If you close your eyes, place your finger lightly on your eyelid, and move your eye, you will feel the bulge. Light enters the eye through the cornea.

The sclera and cornea are continuous. The cornea of this supporting layer contains no blood vessels. The supporting layer completely encloses the eyeball, except for the posterior portion, where small perforations in the sclera allow the fibers of the optic nerve to leave the eyeball on their way to the brain.

Cornea transplants from one individual to another have been very successful; over 30,000 cornea transplants are performed in the United States each year, with a success rate of approximately 95 percent. The typical problem of tissue rejection is usually avoided because the cornea has no blood or lymphatic vessels. As a result, antibodies and lymphocytes that cause rejection cannot reach the cornea.

Vascular layer Because the middle layer of the eyeball contains many blood vessels, it is called the *vascular layer* (L. *vasculum*, diminutive of *vessel*). The dark color of the middle layer is produced by pigments that absorb stray light and help to reduce excessive reflection inside the eyeball. The posterior two-thirds of the vascular layer consists of a thin membrane called the *choroid*, which is essentially a layer of blood vessels and connective tissue sandwiched between the sclera and the retina [FIGURE 17.13B].

The vascular layer becomes thickened toward the anterior portion to form the *ciliary body* [FIGURE 17.13A]. Extending inward from the ciliary body are the fine *ciliary processes*. The smooth muscles in the ciliary body (*ciliary muscles*) contract to ease the tension on the suspensory ligaments of the lens, which consist of fibrils that extend from the ciliary processes to the lens. The *lens* of the eye is a flexible, transparent, colorless, avascular body of epithelial cells behind the iris, the colored part of the eye. The lens is held in place by the *suspensory ligaments of the lens* and by the ciliary processes. The shape of the lens can be adjusted so that objects at different distances can be brought into focus on the retina. This mechanism is called *accommodation* (see page 529). The lens loses much of its elasticity with aging, making it difficult to focus efficiently without corrective eyeglasses.

The anterior extension of the choroid is a thin muscular layer called the *iris* (Gr. rainbow). It is the colored part of the eyeball that can be seen through the cornea. In the center of the iris is an adjustable circular aperture, the *pupil* (L. doll), so called because when you look into someone else's eyes, you can see a reflected image of yourself that looks like a little doll. The pupil appears black because most of the light that enters the eye is absorbed, not reflected outward. The iris, acting as a diaphragm, is able to regulate the amount of light entering the eye because it contains smooth muscles that contract or relax in an involuntary reflex (called the *light reflex*) in response to the amount of light available, causing the pupil to become larger or smaller. The smaller the pupil, the less light entering the eye. This mechanism is called *adaptation* (see page 532). The pupil may shrink with age and some melanin may be lost, causing the iris to appear lighter. This is why the eyes of some elderly people appear to be lighter than they once were.

Retinal layer The innermost layer of the eyeball is the *retina* (REH-tin-uh; L. *rete*, net), an egg-shaped, multilayered, light-sensitive membrane containing a network of specialized nerve cells [FIGURE 17.14]. It is connected to the brain by a circuit of over a million axons in each optic nerve. The retina has a thick layer and a thin layer. The thick layer is nervous tissue, called the *neuroretina*, that connects with the optic nerve. Behind it is a thin layer of *pigmented epithelium* that prevents reflection from the back of the retina. The pigmented layer, along with the choroid, actually absorbs stray light (light that is not used by the photoreceptor cells) and prevents reflection back to the neuroretina. Stray light in the eye can restimulate the photoreceptors. Albinos, who have no eye pigment, are abnormally sensitive to light because the stray light is not absorbed by pigment.

The function of the neuroretina is to receive focused light waves and convert them into nerve impulses that can be conveyed to the brain and converted into visual perceptions. The neuroretina does not extend into the anterior portion of the eyeball, where light could not be focused on it.

The neuroretina consists of highly specialized photoreceptor nerve cells, the *rods* and *cones* [FIGURE 17.15]. The outer segment of a rod or cone contains most of the elements necessary (including light-sensitive photopigments) to absorb light and produce a receptor (generator) potential. The inner segment contains mitochondria, Golgi apparatuses, endoplasmic reticulum, the nucleus, and other structures necessary for generating energy and renewing molecules in the outer segment. The inner segment also contains a synaptic terminal, which allows the photoreceptor cells to communicate chemically with other retinal cells.

In addition to rods and cones, the neuroretina contains several other cells, which are actually neurons. These include bipolar cells, outer horizontal cells, amacrine cells (inner horizontal cells), and ganglion neurons. These neurons form the complex neuronal circuitry for the processing of light waves within the retina [FIGURE 17.14]. The ganglion neurons have axons that leave the eye and constitute the nerve fibers of the optic nerve, optic chiasma, and optic tracts.

Q: *Why do most newborn babies have blue eyes?*

A: Eye color depends on the number and placement of pigment cells (melanocytes) in the iris of the eye. Darker eye color is a result of a greater concentration of melanocytes. At birth, melanocytes are still being distributed in the eyes; they first appear at the back of the iris. However, the eyes appear brown only when melanocytes are deposited in the *anterior* part of the iris, in front of the muscles of the iris. This deposition does not occur until a few months after birth. However, babies of dark-skinned, dark-eyed parents are usually born with dark eyes.

FIGURE 17.13 THE HUMAN EYE

[A] Horizontal section through the eye. [B] Horizontal section through the anterior portion of the eye (enlarged drawing).

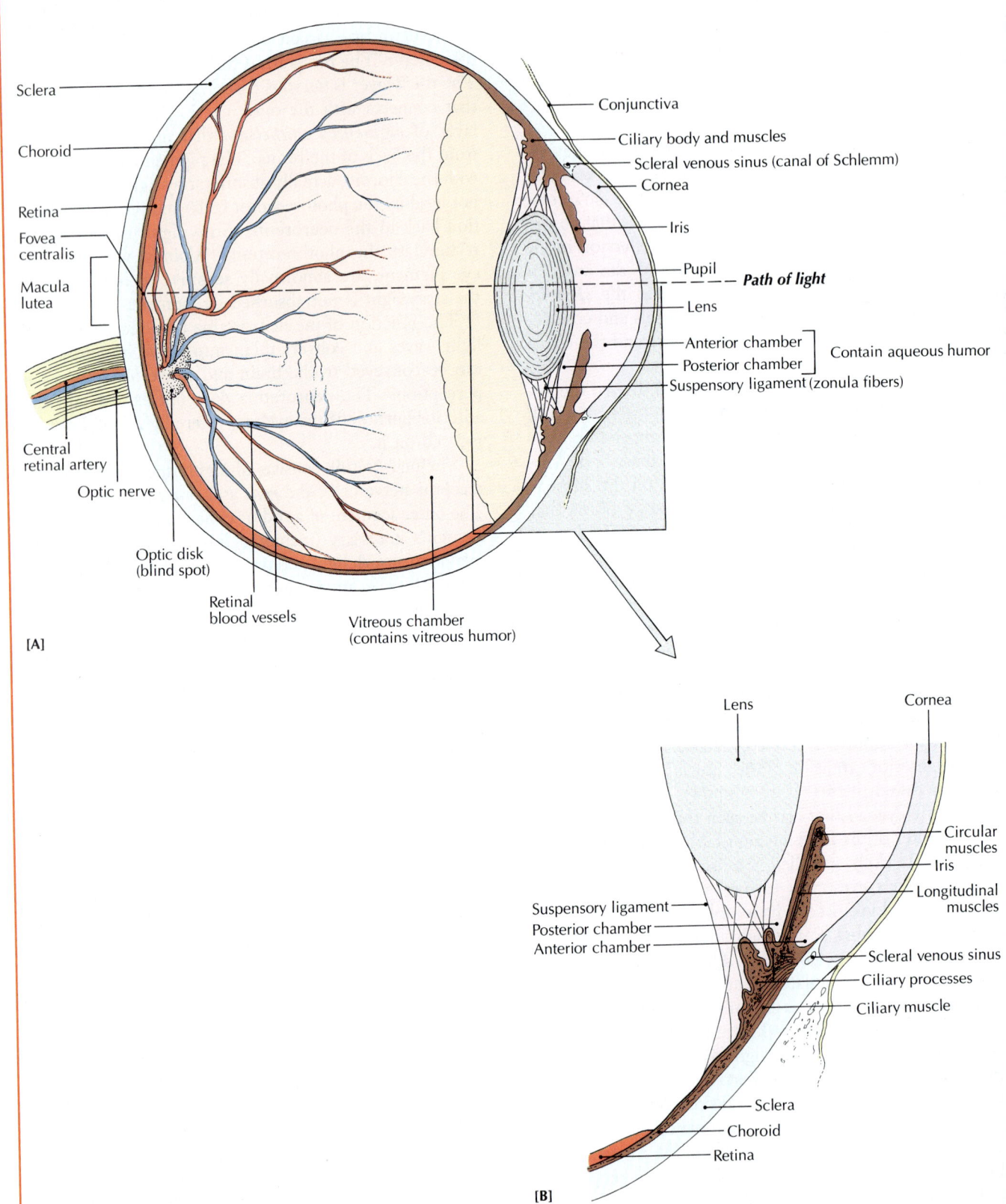

Structure	Description
SUPPORTING LAYER OF EYEBALL	
Sclera	Opaque layer of connective tissue over posterior five-sixths of outer layer of eyeball; "white" of the eye. Gives eyeball its shape, protects inner layers. Perforated to allow optic nerve fibers to exit.
Cornea	Transparent anterior portion of outer layer of eyeball. Light enters eye through cornea, a nonadjustable lens.
VASCULAR LAYER OF EYEBALL	
Choroid	Thin membrane of blood vessels and connective tissue between sclera and retina. Posterior two-thirds of vascular layer of eyeball.
Ciliary body	Thickened vascular layer in anterior portion of eyeball. Ciliary muscles help lens to focus by either increasing or decreasing tension on suspensory ligaments of lens. Produces aqueous humor and some elements of vitreous humor.
Ciliary processes	Inward extensions of ciliary body. Help hold lens in place.
Lens	Elastic, colorless, transparent biconvex body of epithelial cells posterior to iris. Shape modified to focus on subjects at different distances (accommodation) through action of ciliary muscles. The lens is adjustable.
Iris	Colored part of eye. Thin, muscular layer; anterior extension of choroid. Regulates size of pupil, and thus amount of light entering eye, by controlling degree of constriction and dilation of iris.
Pupil	Adjustable circular opening in iris. Opened and closed reflexively relative to amount of light available (adaptation).
RETINAL LAYER OF EYEBALL	
Retina	Multilayered, light-sensitive membrane; innermost layer of eyeball. Connected to brain by optic nerve. Consists of neural layer and pigmented layer, which prevents reflection from back of retina. Receives focused light waves, transduces them into nerve impulses that the brain converts into visible perceptions.
Fovea (fovea centralis)	Depressed area in center of retina containing only cones. Area of most acute image formation and color vision.
CAVITIES OF EYEBALL	
Aqueous chambers Anterior chamber Posterior chamber	Between cornea and iris. Contains aqueous humor. Between iris and lens. Contains aqueous humor.
Vitreous chamber	Largest cavity, fills entire space behind lens. Contains vitreous humor.

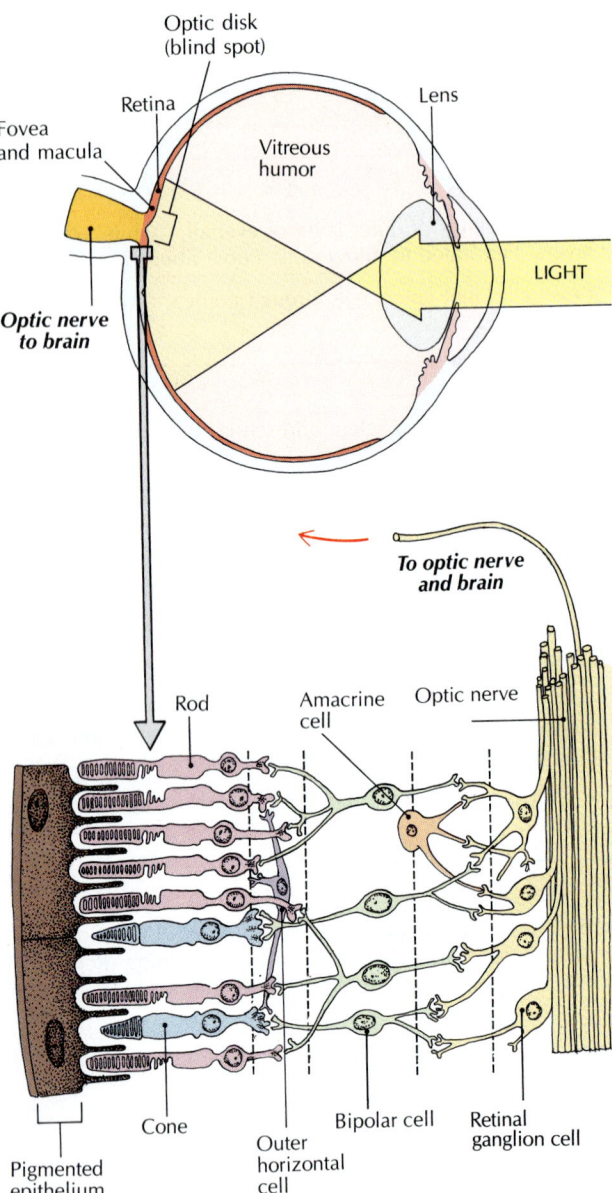

FIGURE 17.14 THE RETINA

[A] Diagram of a section through the retina. Light first passes through the rest of the eye and finally through several layers of retinal cells before reaching the light-sensitive rods and cones. Beyond the rods and cones is the pigmented epithelial layer, which absorbs stray light and prevents reflection from the back of the retina. When light energy stimulates a rod or cone, the resulting neuronal activity passes from the receptor cells through an intermediate set of bipolar cells and finally to ganglion cells, whose axons form the optic nerve. **[B]** Scanning electron micrograph of a retina. ×1090. (Gene Shih and Richard Kessel, *Living Images: Biological Microstructures Revealed by SEM,* Boston, Science Books International, 1982.)

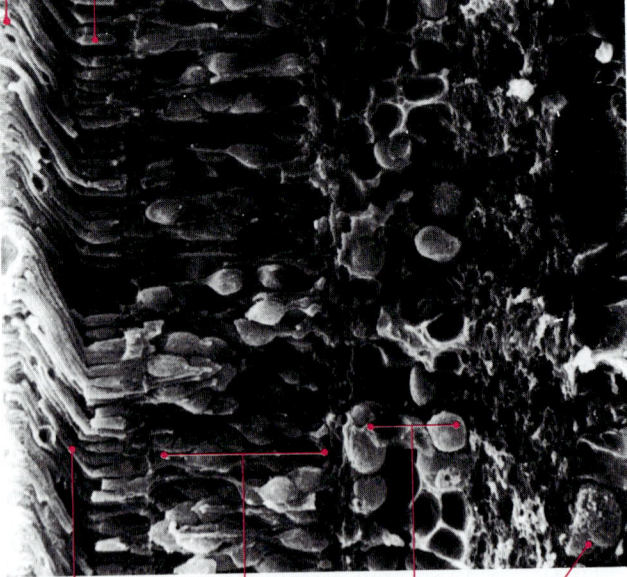

FIGURE 17.15 RODS AND CONES

[A] Detailed drawing of rod cell. The outer segment of each rod contains approximately 2000 disks stacked in an orderly pile. The disks contain most of the light-absorbing protein molecules that initiate the generation of a local potential. **[B]** Detailed drawing of a cone cell.

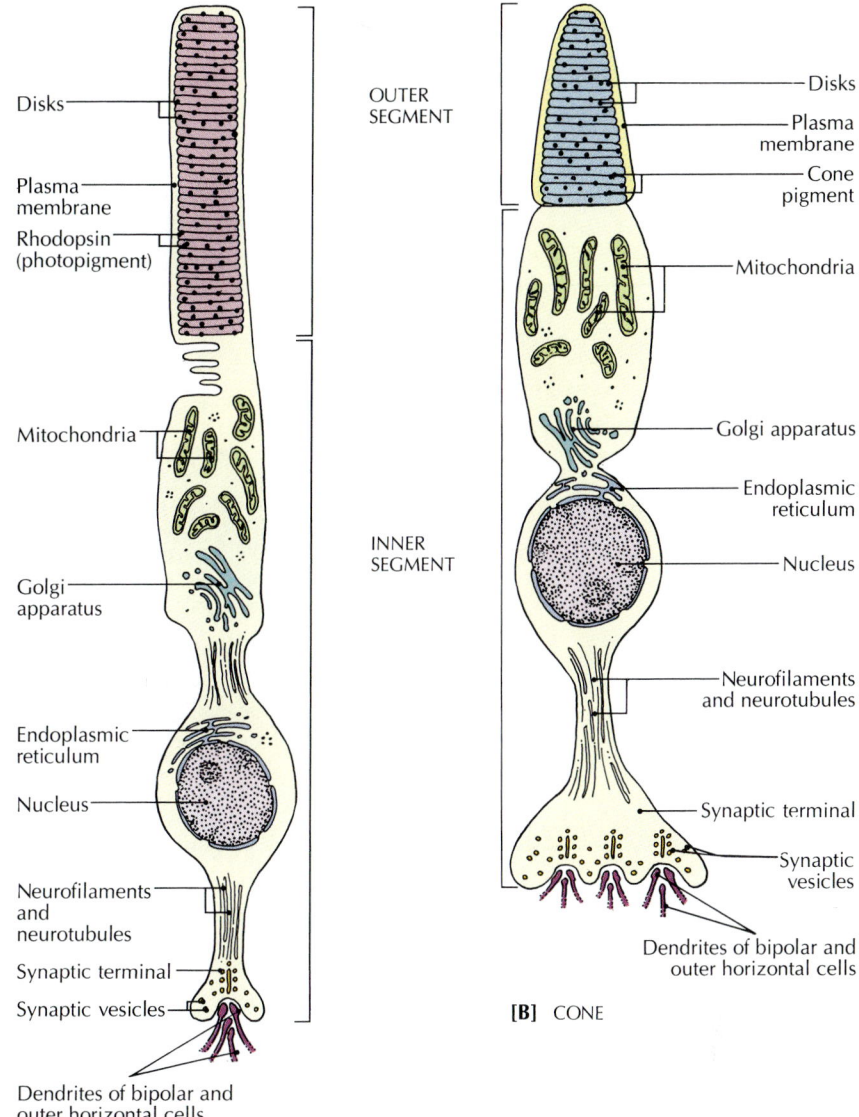

[A] ROD

[B] CONE

Each eye has about 125 million rods and 7 million cones. Most of the cones are concentrated in the center of the retina directly behind the lens in an area called the *macula lutea* (MACK-yool-uh LEW-tee-uh; L. "yellow spot"), especially in a small depressed rod-free area called the *fovea,* or *fovea centralis* (FOE-vee-uh; L. small pit) [FIGURE 17.13A]. The rods and some cones are located in the remainder of the retina, called the *peripheral retina.* The rods are the sensors for the perception of black-to-white shades, and the cones are the sensors for the perception of color. Night vision is almost totally rod vision, since the color-sensitive cones require 50 to 100 times more stimulation than rods do. (In order to distinguish the functions of rods and cones, remember that the *c* in *cone* can stand for *color.*)

Vision is sharpest, and color perception is optimal, on the fovea of the macula lutea. The nonmacular retina (for peripheral vision) is sensitive to weak light intensities and is associated with black-and-white vision. In normal light we can see best by looking directly at an object so that the image falls on the cones in the fovea. However, in poor light we can see best by *not* looking directly at an object (as one does when stargazing) so that the image falls on the light-sensitive rods located on the nonmacular portion of the retina.

The portion of the retina where the optic nerve exits from the eyeball contains neither rods nor cones; it is called the optic disk, or *blind spot,* because it is not sensitive to light [FIGURE 17.13A].

Chambers of the eyeball The eyeball contains three chambers: (1) anterior chamber, (2) posterior chamber, and (3) vitreous chamber. The region between the cornea and iris is the ***anterior chamber*** [FIGURE

FIGURE 17.16 THE EXTERNAL RIGHT EYE

The eye is seen from the front, showing some accessory structures.

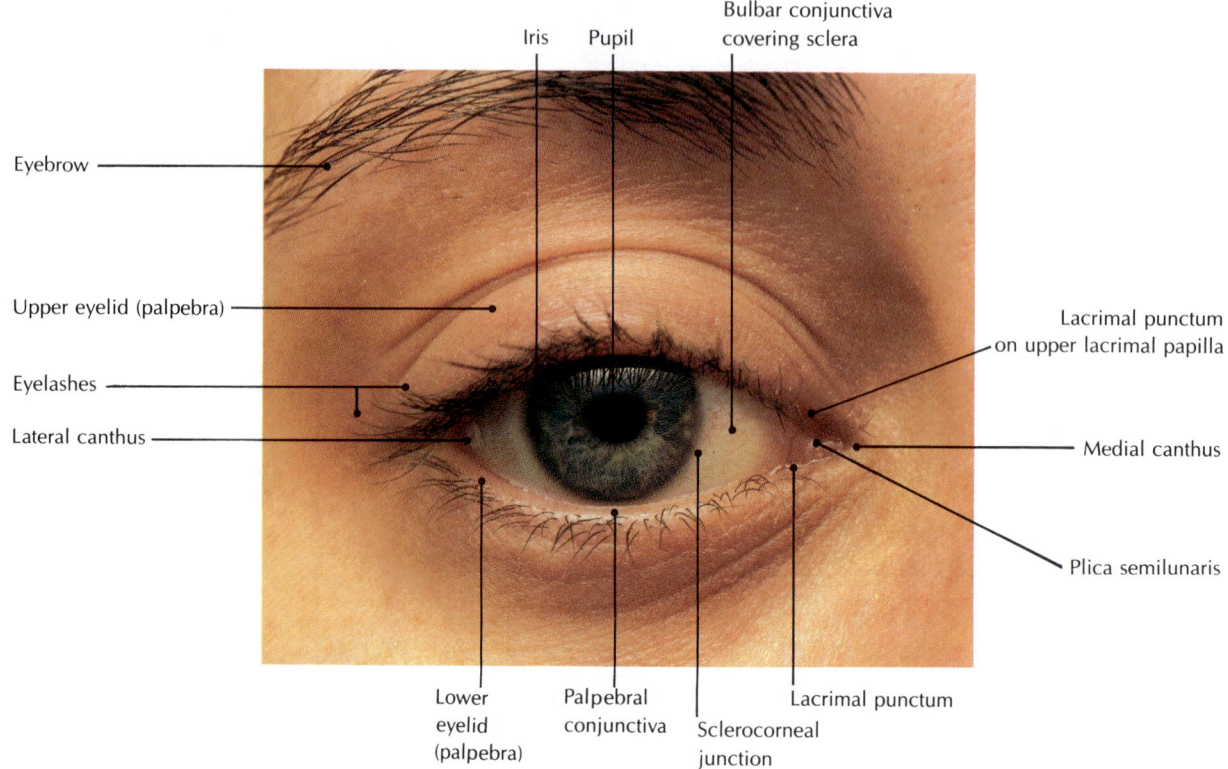

Iris Pupil Bulbar conjunctiva covering sclera

Eyebrow

Upper eyelid (palpebra)

Eyelashes

Lateral canthus

Lacrimal punctum on upper lacrimal papilla

Medial canthus

Plica semilunaris

Lower eyelid (palpebra) Palpebral conjunctiva Sclerocorneal junction Lacrimal punctum

---

17.13A]. The **posterior chamber** lies between the iris and lens. Both chambers are filled with *aqueous humor,* a thin, watery fluid that is essentially an ultrafiltrate of blood similar to cerebrospinal fluid. Aqueous humor is largely responsible for maintaining a constant pressure within the eyeball. It also provides such essential nourishment as oxygen, glucose, and amino acids for the lens and cornea, which do not have blood vessels to nourish them. Aqueous humor is produced by capillaries in the ciliary body. It passes through the posterior chamber and then the anterior chamber and diffuses into a drainage vein called the *scleral venous sinus* (canal of Schlemm) at the base of the cornea.

The third and largest chamber of the eyeball is the **vitreous chamber,** which occupies about 80 percent of the eyeball. It fills the entire space behind the lens. This chamber contains *vitreous humor,* a gelatinous substance. The humor is actually a modified connective tissue. Its function is to keep the eyeball from collapsing as a result of external pressure. Except for the addition of collagen and hyaluronic acid, the chemical composition is similar to that of aqueous humor. The vitreous humor also provides another source of nourishment for the lens and possibly the retina. The vitreous humor is formed by the ciliary body.

Accessory Structures of the Eye

Most of the accessory structures of the eye are either protective devices or muscles. They include the orbits of the skull, eyelids, eyelashes, eyebrows, conjunctiva, lacrimal (tear) apparatus, and muscles that move the eyeball and eyelid [FIGURE 17.16].

Orbit The eye is enclosed in an orbital cavity, or *orbit,* which protects it from external buffeting. The floor of the orbit is composed of parts of the maxilla, zygomatic, and palatine bones. The roof is composed of the orbital plate of the frontal bone and the lesser wing of the sphenoid. Several openings in the bones of the orbit allow the passage of nerves and blood vessels. Between

---

Q: *Why do we sometimes see "spots" in front of our eyes?*

A: Such "spots" are called "floaters" or *muscae volitantes,* which is Latin for "flying flies." They are actually the shadows of red blood cells that have escaped from the capillaries in the retina or of collagenous particles. These particles and cells move slowly through the vitreous humor of the eye. Their presence is usually normal and harmless.

the orbit and the eyeball is a layer of fatty tissue that cushions the eyeball and permits its smooth rotation.

Eyelids, eyelashes, eyebrows The *eyelids* (palpebrae) are folds of skin that create an almond-shaped opening around the eyeball when the eye is open [FIGURE 17.16]. The points of the almond, where the upper and lower eyelids meet, are called *canthi* (KAN-thigh). The medial (or inner) canthus is the one closest to the nose, and the lateral (or outer) canthus is the point closest to the ear. The eyelid may be divided into four layers: (1) The skin layer contains the eyelashes, (2) the muscular layer contains the orbicularis oculi muscle, which lowers the eyelid to close the eye, (3) the fibrous connective tissue layer contains many modified oil glands, whose secretions keep the eyelids from sticking together, and (4) the innermost layer is composed of a portion of the lining of the eyelid, the conjunctiva.

Eyelids protect the eyeball from dust and other harmful external objects. In addition, the periodic blinking of the eyelids sweeps glandular secretions (tears) over the eyeball, keeping the cornea moist. During sleep, the closed eyelids prevent evaporation of the secretions. The eyelids also protect the eye by closing reflexively when an external object threatens the eye, as when a piece of paper is suddenly blown toward your face.

The edges of the eyelids are lined with short, thick hairs, the *eyelashes,* which act as strainers to prevent foreign materials from entering the eye. The eyelids of each eye contain about 200 eyelashes. Each eyelash lasts 3 to 5 months before it is shed and replaced. *Eyebrows* are thickened ridges of skin over the protruding frontal bone, covered with short, flattened hairs. They protect the eye from sweat, excessive sunlight, and foreign materials and also help absorb the force of blows to the eye and forehead.

Conjunctiva The *conjunctiva* (L. connective) is a thin, transparent mucous membrane that lines the eyelids and bends back over the surface of the eyeball, terminating at the transparent cornea, which is uncovered. The portion that lines the eyelid is the *palpebral conjunctiva,* and the portion that covers the white of the eye is the *bulbar conjunctiva.* Between both portions of the conjunctiva are two recesses called the *conjunctival sacs.* The looseness of the sacs makes movement of the eyeball and eyelid possible. Your ophthalmologist usually pulls back your lower eyelid to place eyedrops in the inferior conjunctival sac.

Lacrimal apparatus The *lacrimal apparatus* (LACK-ruh-mull; L. "tear") is made up of the lacrimal gland, lacrimal sac, lacrimal ducts, and nasolacrimal duct [FIGURE 17.17]. The eyeball is kept moist by the secretions of the *lacrimal gland,* or tear gland, located under

FIGURE 17.17 THE LACRIMAL APPARATUS

Anterior view.

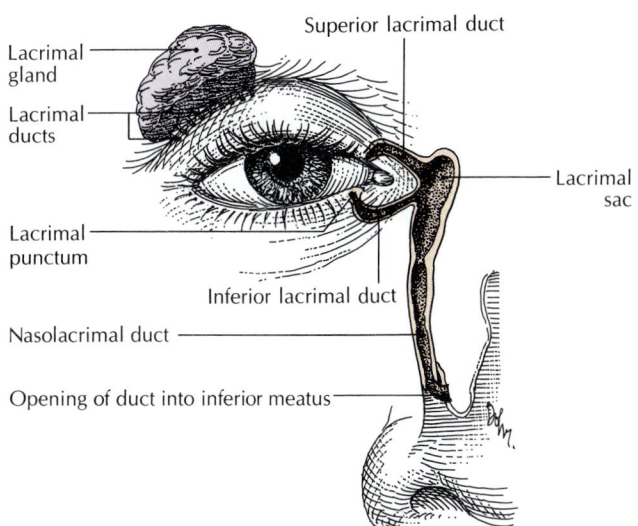

the upper lateral eyelid (palpebra) and extending inward from the outer canthus of each eye. The lacrimal glands of infants take about 4 months to develop fully. As a result, the eyes of a newborn baby should be protected from dust, bright light, and other irritants.

The eye usually blinks every 2 to 10 sec, with each blink lasting only 0.3 to 0.4 sec. Blinking stimulates the lacrimal gland to secrete a sterile fluid that serves at least four purposes: (1) It washes foreign particles off the eye. (2) It kills invading bacteria with a mild antibacterial enzyme, lysozyme. (3) It distributes water and nutrients to the cornea and lens. (4) It gives the eyeball a clear, moist, smooth surface.

Tears are composed of salts, water, the enzyme lysozyme, and mucin (organic compounds produced by mucous membranes). In addition to the steady secretion of tears, *reflex* tears are produced in emergencies, as when the fumes from a sliced onion irritate the eyes.

Approximately 3 to 12 *lacrimal ducts* lead from each gland onto the superior conjunctival sac in the upper eyelid. From there tears flow down across the eye into small openings near the inner canthus called *lacrimal puncta.* The puncta open into the *lacrimal ducts* (superior and inferior canaliculi), which drain excess tears from the area of the inner canthus to the *lacrimal sac,* the dilated upper end of the *nasolacrimal duct* [FIGURE 17.17]. The nasolacrimal duct is a longitudinal tube that delivers excess tears into the nasal cavity.

Ordinarily, tears are carried away by the nasolacrimal duct to the nasal cavity, but when a person is crying or has conjunctivitis or hay fever, the tears form faster than they can be removed. In such cases, tears run down the cheeks, and the nasal cavity becomes overloaded. This is why you have to blow your nose when you cry. Also, a watery fluid sometimes flows out of the nose after the nose has been blown. This fluid is tears flowing out of the "unplugged" nasolacrimal duct.

Muscles of the eye and eyelid   A set of six muscles moves the eyeball in its socket. (The action of these muscles is described in FIGURE 11.7, page 310.) The muscles are the four *rectus muscles* and the *superior* and *inferior oblique muscles*. They are called **extrinsic** or **extraocular muscles** because they are outside the eyeball (*extra* = outside). One end of each muscle is attached to a skull bone, and the other end is attached to the sclera of the eyeball. The extraocular muscles are coordinated and synchronized so that both eyes move together in order to center on a single image. These movements are called the *conjugate movements* of the eyes.

Other muscles move the eyelid. The *orbicularis oculi* lowers the eyelid to close the eye, and the *levator palpebrae superioris* raises the eyelid to open the eye. The *superior tarsal* muscle is a smooth muscle innervated by the sympathetic nervous system. It helps raise the upper eyelid, and when it is paralyzed (as in Horner's syndrome), it causes a slight drooping (ptosis) of the upper eyelid.

Inside the eyes are three smooth **intrinsic muscles.** The *ciliary muscle* eases tension on the suspensory ligaments of the lens and allows the lens to change its shape in order for the eye to focus (accommodate) properly. The *circular muscle* (pupillary sphincter) of the iris constricts the pupil, and the *radial muscle* (pupillary dilator) dilates it.

Physiology of Vision

The visual process can be subdivided into five phases:

1 **Refraction** of light rays entering the eye.
2 Focusing of images on the retina by **accommodation** of the lens.
3 **Convergence** of the images viewed by both eyes.
4 Conversion of light waves by **photochemical reactions** into neural impulses.
5 **Neural pathways** for vision.

Let us follow the process through each phase in detail.

Q: *What causes "bloodshot eyes"?*

A: Although the bulbar conjunctiva is normally colorless, its blood vessels can become dilated and congested because of infection or external irritants such as smoke. The result is "bloodshot eyes."

Refraction   Light waves travel parallel to each other, but they bend when they pass from one medium to another medium with a different density.* Such bending is called **refraction.** Light waves that enter the eye from the external air are refracted so that they converge at the retina at a sharp, focused point called the *focal point* [FIGURE 17.18A].

Before light reaches the retina, it passes through (1) the cornea, (2) the aqueous humor of the anterior chamber between the iris and lens, (3) the lens, and (4) the gelatinous vitreous humor in the vitreous chamber behind the lens. Refraction takes place as the light passes through both surfaces of the cornea (which is a convex, nonadjustable lens) and again as it passes through the anterior and posterior surfaces of the lens (a convex, adjustable lens).

A normal eye can bring distant objects more than 6 m (20 ft) away to a sharp focus on the retina. When parallel light rays are focused exactly on the retina and vision is perfect, the condition is called *emmetropia* (Gr. "in measure"). Nearsightedness, or *myopia* (Gr. "contracting the eyes"), occurs when light rays come to a focus *before* they reach the retina.† As a result, when the rays do reach the retina, they form an unfocused circle instead of a sharp point, and distant objects appear blurred [FIGURE 17.18B]. Farsightedness, or *hypermetropia* (Gr. "beyond measure"), occurs when light rays are focused *beyond* the retina, and as a result near objects appear blurred [FIGURE 17.18C].‡

---

*People who fish have learned that when they try to grab a fish swimming below the surface of the water, they must reach a little to the side of the image to compensate for the bending of light waves from air to water and vice versa.

†*Myopia* is so named because a nearsighted person often squints through narrowed eyelids in an effort to focus better. Although the resultant tiny opening requires little or no focusing, the amount of light entering the eye is decreased, and strain on the relevant eye muscles may cause headaches.

‡Some textbooks show a corrective lens for myopia as)(and a corrective lens for hypermetropia as (). Such lenses are capable of refracting light rays, according to the fundamental laws of physics, but are actually never used to correct vision defects. The corrective lenses shown in FIGURE 17.18B and C are meant to approximate the actual shapes of corrective *eyeglass* lenses. In any given lens, the relationship of one curve to another changes according to the specific prescription, but the basic shapes for correcting myopia and hypermetropia remain the same, as shown in FIGURE 17.18B and C, with the center of the lens being the thinnest point for myopia and the thickest point for hypermetropia.

Q: *What does a newborn baby see?*

A: A newborn baby can detect the shape of faces and is also aware of peripheral form and motion. At about 7 weeks a baby develops some reflexes, such as blinking its eyes when something is thrust at its face. By 3 months a baby can focus well on objects and can follow them, and by 6 months a child usually can see very well.

FIGURE 17.18 REFRACTION

[A] In a normal eye, light rays come to a focal point exactly on the retina. **[B]** In a myopic (nearsighted) eye, the focal point falls *in front* of the retina. This may be caused by an elongated eyeball (as shown here) or a thickened lens. Corrective lenses (eyeglasses) placed in front of the eye make the focal point fall on the retina. **[C]** In a farsighted eye, the focal point falls *behind* the retina. This may be caused by a shortened eyeball or a thinned lens. Corrective lenses redirect the light rays to produce a focal point on the retina.

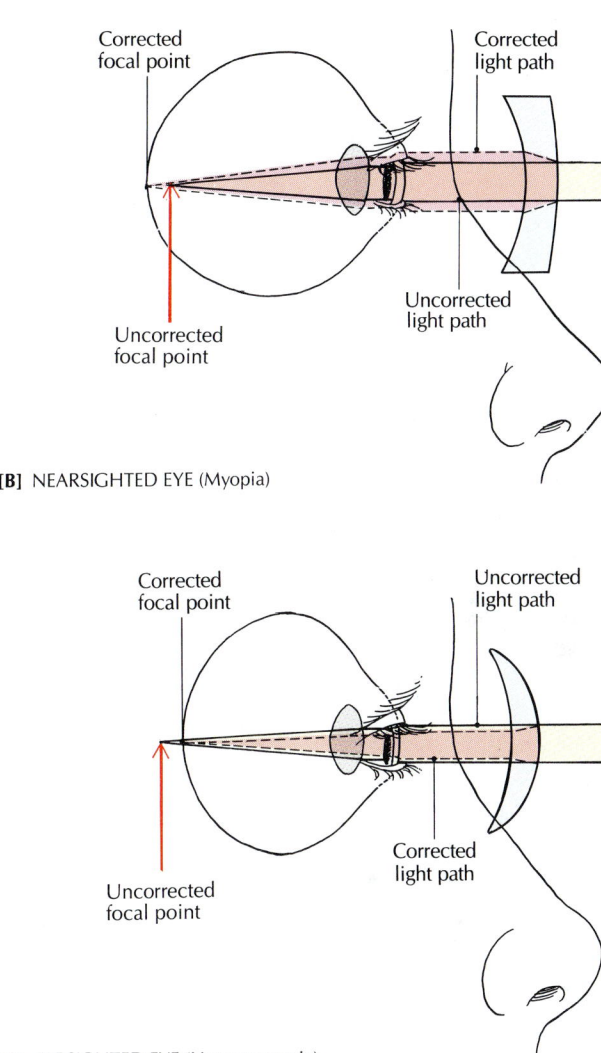

[B] NEARSIGHTED EYE (Myopia)

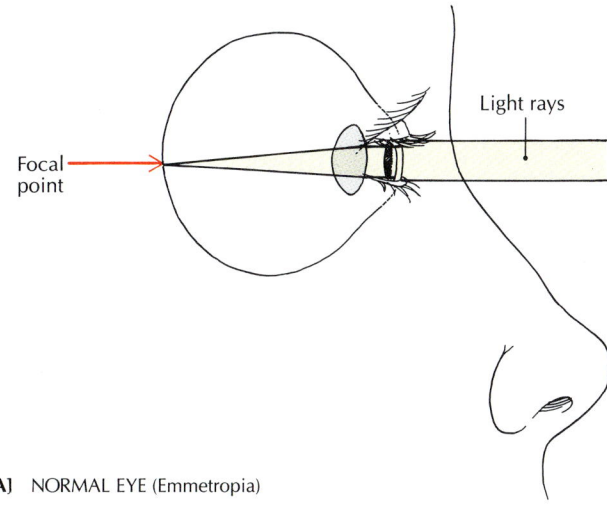

[A] NORMAL EYE (Emmetropia)

[C] FARSIGHTED EYE (Hypermetropia)

Both myopia and hypermetropia can be corrected by wearing prescription eyeglasses or contact lenses, which are specially ground lenses placed in front of the eye to change the angle of refraction.

Astigmatism (Gr. "without focus") occurs when the curvature of the cornea or lens is not uniform. As a result, part of the image formed on the retina is unfocused. This condition can usually be corrected with lenses that have greater bending power in one axis than in others.

Accommodation Images from an object less than 6 m (20 ft) away would normally be focused behind the retina instead of on it. To bring images into perfect focus on the retina is the role of the adjustable lens. This is accomplished by a reflex called *accommodation.* When you want to focus your eyes on an object close to you (near-sight vision), you involuntarily contract the ciliary muscles in your eye, which pull the ciliary body slightly forward and inward, reducing the tension on the suspensory ligaments attached to the lens capsule. When the tension is reduced, the elastic lens becomes thicker (rounds up). The rounder lens is able to focus on a close object [FIGURE 17.19A].

When distant objects are viewed, the ciliary muscles relax. As they relax, the tension on the suspensory liga-

ments becomes greater, so that the lens becomes thinner (flattened) [FIGURE 17.19B]. Looking at a distant building or tree is restful to your eyes when you are doing long-term close work because during far-sight vision your ciliary muscles are relaxed.

As the lens becomes harder and less elastic with age, it becomes more and more difficult to focus on near objects, a condition known as *presbyopia* (Gr. *presbus*, old man + eye). When you were a young child you probably held a book about 7 cm (3 in.) from your eyes.* By the

*Parents, who forget what it was like to read a book at 10 years old, are constantly complaining that their children are ruining their eyes by reading in poor light and by holding the book too close. In fact, because of the acute sensitivity of their rods and cones and the elasticity of their lenses, healthy children are not hurting their eyes at all.

FIGURE 17.19 ACCOMMODATION

How the eye focuses on a nearby object **[A]** and a distant one **[B]**.

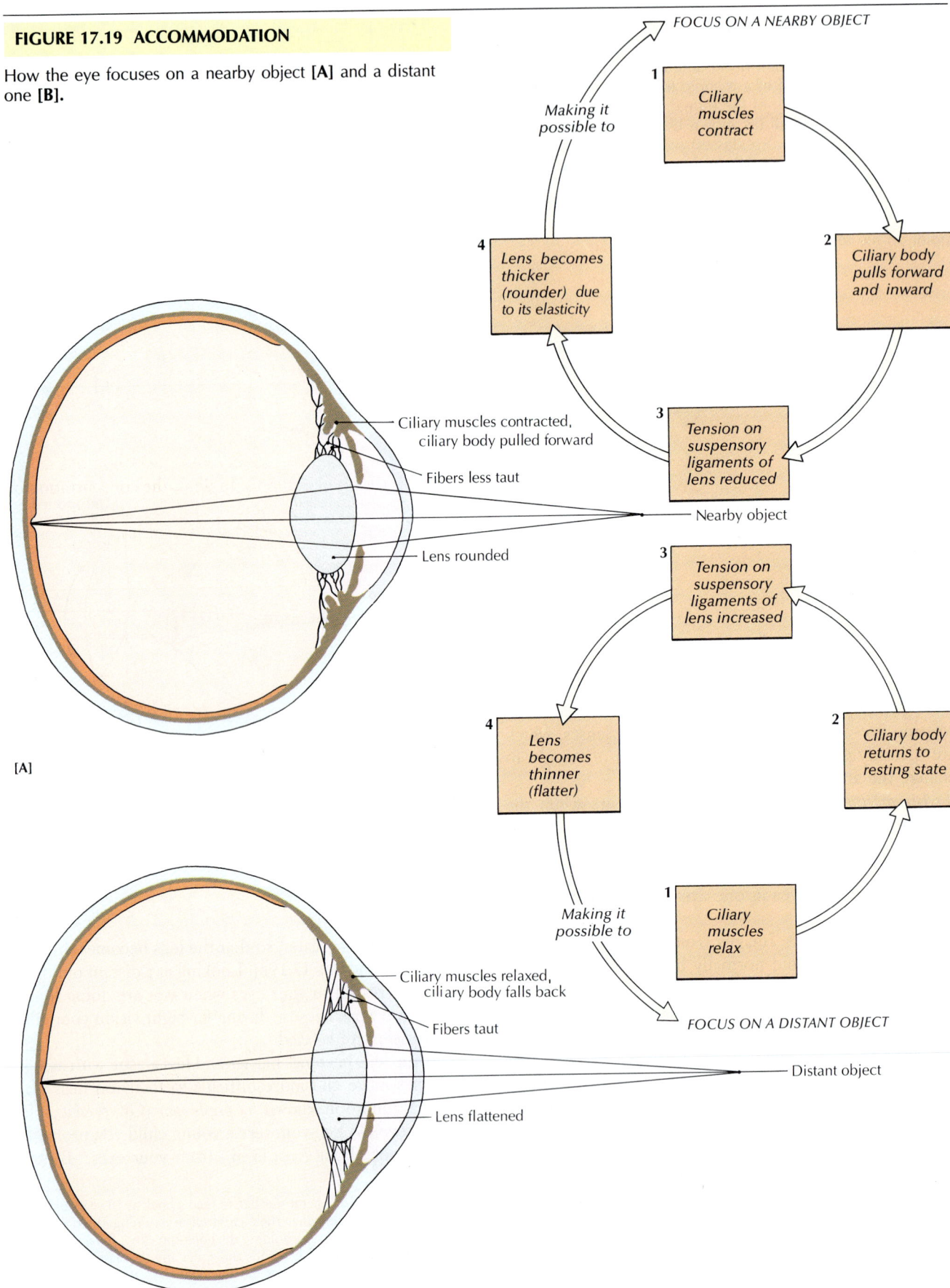

FOCUS ON A NEARBY OBJECT

Making it possible to

1 Ciliary muscles contract

2 Ciliary body pulls forward and inward

3 Tension on suspensory ligaments of lens reduced

4 Lens becomes thicker (rounder) due to its elasticity

Ciliary muscles contracted, ciliary body pulled forward

Fibers less taut

Nearby object

Lens rounded

[A]

3 Tension on suspensory ligaments of lens increased

2 Ciliary body returns to resting state

4 Lens becomes thinner (flatter)

Making it possible to

1 Ciliary muscles relax

FOCUS ON A DISTANT OBJECT

Ciliary muscles relaxed, ciliary body falls back

Fibers taut

Distant object

Lens flattened

[B]

time you are 40 years old you will probably be holding your book about 16 cm (6 in.) or more away, and by 60 the distance may increase to about 1 m (39 in.). At that age, if not sooner, corrective convex lenses will probably be necessary to focus on close objects. In old age, the lens also gradually acquires a cloudy, yellow tint.

Convergence Human beings have *binocular vision,* meaning that although we have two eyes, we perceive only one image. Each eye receives an image from a slightly different angle, and this creates the impression of distance, depth, and three-dimensionality. For this reason, binocular vision is sometimes called *stereoscopic vision* (Gr. *stereos,* solid + *skopein,* to see).

In binocular vision, the two eyeballs turn slightly inward to focus on a close object so that both images fall on the corresponding points of both retinas at the same time. This action is called **convergence** (L. *com,* together + *vergere,* to bend). In order to produce a single image, the six pairs of extraocular muscles must move together with perfect coordination. As a result of convergence, the simultaneous stimulation of both retinas produces the perception of a single image in the occipital lobe of the cerebral cortex.

Photochemical reactions in rods Vision begins when "packets" of electromagnetic energy called *photons* are converted into neural signals that the brain can decode and analyze. (A photon is the smallest possible quantity of light.) The translation of photons into neural signals is accomplished by rods and cones, the photoreceptor cells of the eye. Each eye contains about 125 million rods, which are located in most of the neuroretina but are absent in the fovea and blind spot. How is the absorption of light waves by rods translated into vision?

The extracellular fluid surrounding the rods contains a high concentration of positively charged sodium ions and a low concentration of positively charged potassium ions. The distribution of ions inside the cell is the opposite: abundant potassium ions and few sodium ions. These concentrations are maintained by a sodium-potassium pump. In a resting state, potassium ions tend to diffuse to the outside, creating a relatively negative charge inside the cell. Ordinarily, a photoreceptor cell in a dark environment has a permeability to sodium ions, which move from the extracellular fluid into the outer

segment of the cell. This inward flow of sodium ions into the outer segment is balanced by the outward flow of potassium ions from the rest of the cell.

When light is absorbed by a photoreceptor cell (rod or cone), however, sodium ions cease to flow into the cell. This causes the negative charge inside the cell to increase relative to the outside, producing a hyperpolarization at the outer segment that spreads to the synaptic ending. The hyperpolarization controls the flow of neural information across synapses to other retinal cells.

In rods, the photosensitive membrane is made up of an orderly pile of disks inside the surface membrane [FIGURE 17.15A]. When the disks absorb light, the photopigments of the rods respond, resulting in the release of the neurotransmitter *glutamate* to which the neurons of the retina respond. This neurotransmitter induces graded, local potentials. In darkness, rods contain large amounts of the second messenger *cyclic guanosine monophosphate* (cGMP), which binds to sodium channels in the surface membrane and holds them open. Sodium ions enter through the open channels. In light, the concentration of cGMP decreases, the sodium channels close, and a hyperpolarization is produced as sodium ions are blocked from entering the cell.

The hyperpolarization of photoreceptor cells occurs as follows. Each rod contains about 100 million molecules of a reddish, light-sensitive pigment called **rhodopsin** (roh-DOPP-sin). Rhodopsin contains a light-absorbing organic molecule called 11-*cis* retinal, which is an isomer of **retinal,** a derivative of vitamin A [FIGURE 17.20].* Retinal is coupled with a specific protein for each type of photopigment in receptor cells. The protein coupled with retinal in rods is called **scotopsin** (sko-TOPP-sin).

When 11-*cis* retinal absorbs a photon of light, it changes its molecular configuration to form all-*trans* retinal, another isomer of retinal, and activates scotopsin's ability to act as an enzyme. The reaction produces large amounts of the activated G-protein called *transducin.* (The role of transducin in seeing is analogous to the role of gustducin in tasting.) Transducin activates a specific enzyme called a *phosphodiesterase,* which hydrolyzes cGMP molecules to an inactive form. Sodium ions are prevented from entering the rods, hyperpolarization occurs, and the subsequent neural signals are processed by bipolar, amacrine, horizontal, and ganglion cells in the retina; then action potentials generated by the ganglion cells are conveyed to the brain via the optic nerve, and light is finally perceived. Afterward, retinal is enzymatically combined with scotopsin to synthesize rhodopsin again. FIGURE 17.21 illustrates the process in detail.

---

*A condition known as *night blindness* — the inability to see in the dark — can result from a deficiency of vitamin A, which is essential for the synthesis of retinal.

Q: *What does 20/20 vision mean?*

A: If you have "20/20 vision," your eyes can see at 20 ft what the normal eye can see at that distance. The larger the second number, the worse the *visual acuity* (sharpness); 20/60 means that you can see at 20 ft what a normal eye sees at 60 ft. (Legal blindness is 20/200 or worse in both eyes, with or without corrective eyeglasses.)

FIGURE 17.20 ISOMERS OF RETINAL

[A] The molecular structure of 11-*cis* retinal is converted by light into all-*trans* retinal. After the cycle of light transmission and perception is complete, all-*trans* retinal is converted back into 11-*cis* retinal by enzymatic action. **[B]** A three-dimensional representation of how 11-*cis* retinal changes its shape when it absorbs light. In the top drawing, the hydrogen atoms (blue) attached to carbon-11 and carbon-12 are on the same side of the carbon backbone, causing the backbone to bend. When light is absorbed, the backbone rotates between carbon-11 and carbon-12, straightening out the chain to form all-*trans* retinal (bottom).

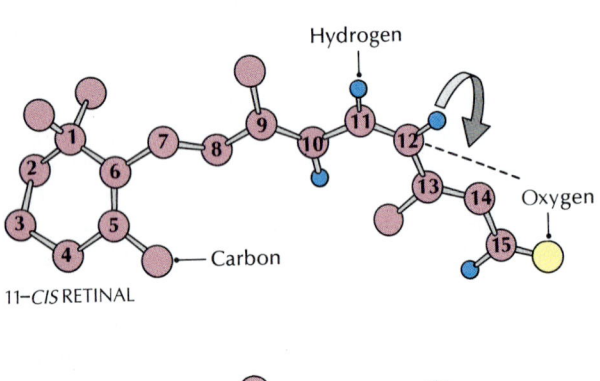

[A] 11-*CIS* RETINAL ALL-*TRANS* RETINAL

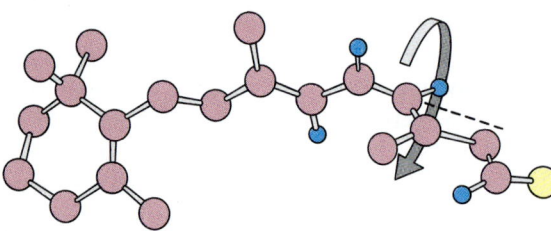

11-*CIS* RETINAL

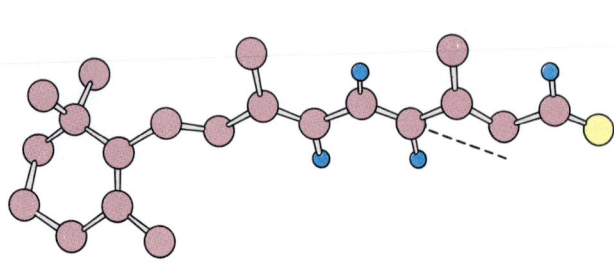

ALL-*TRANS* RETINAL

[B]

Photochemical reactions in cones Each eye contains about 7 million cones, located primarily in the macula of the fovea. The events of visual excitation are similar in rods and cones. The major difference is that there are three different types of cones, each one containing separate photopigments that respond to red, green, or blue light. These photopigments all contain 11-*cis* retinal and have the same basic molecular structure as rhodopsin, but they contain **photopsins** (foh-TOPP-sihnz), which are slightly different from the scotopsin of rods.

The perception of color depends on which cones are stimulated. The final perceived color, which can combine all three types of cones and is almost unlimited in its color possibilities, is determined by the combinations of different levels of excitation of each type of cone. The proper mix of all three basic colors* produces the perception of white, and the absence of all three colors produces the perception of black.

In "red," "green," and "blue" cones, the same retinal absorbs a different frequency of light. The red cones respond to the long waves of the visible spectrum; the green cones respond to the middle waves; and the blue cones to the short waves.

Adaptation to light and dark Rods function best in dim light. One reason is that bright light "bleaches" (breaks down) rhodopsin, which decreases the amount of photopigment available to rods. As a result, the light-sensitive rods become overloaded, even in ordinary day-

---

*Note that cone pigments respond to red, green, or blue light. These are not the same as the primary colors of paint pigments: red, yellow, and blue.

FIGURE 17.21 THE PHYSIOLOGY OF LIGHT ABSORPTION AND PERCEPTION

The process begins when a photon of light is absorbed by 11-*cis* retinal in a rhodopsin molecule and ends when impulses are sent to the brain and interpreted as visual images. After the process is complete, the split molecule of rhodopsin is resynthesized.

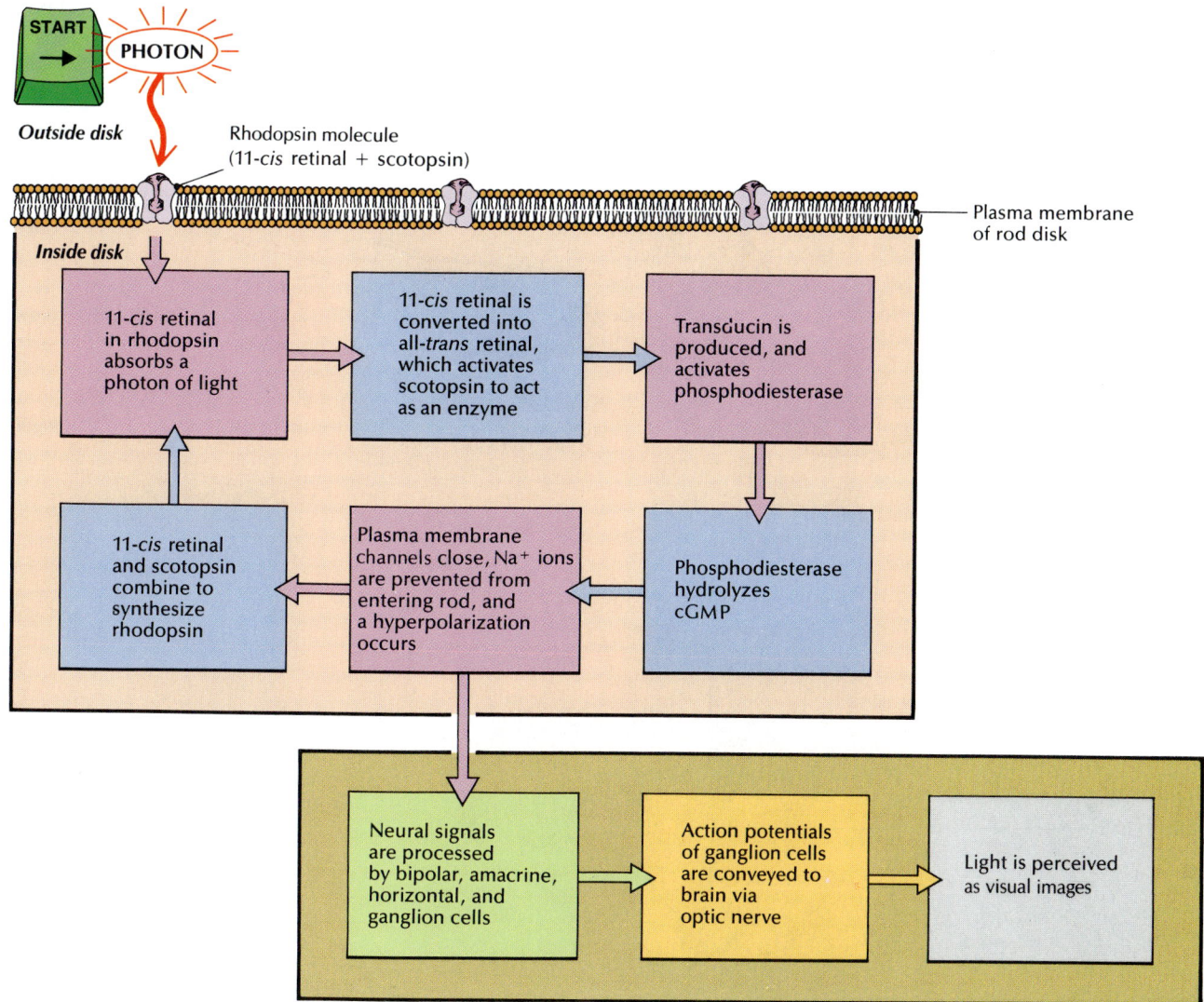

light, and stop transmitting neural signals. The capability of rods in bright light is also decreased because cones have neural connections that inhibit rods when the light conditions are more appropriate for cone function.

Rods begin to regain their functional levels of sensitivity when the light source is diminished. Some sensitivity returns after a few minutes, as when we enter a darkened movie theater from a sunlit street. However, a complete sensitivity in dim light, or *dark adaptation,* usually takes from 20 to 30 min while the "bleached" rhodopsin is resynthesized. Dark adaptation consists of a rapid phase and a slow phase. The *rapid phase* (neural adaptation) is usually complete in a few seconds. It is thought to take place at the neuronal level in the retina. The *slow phase* (photochemical adaptation) takes 20 to 30 min to complete as the "bleached" rhodopsin in the rods is resynthe-

sized. Adaptation to the dark is complemented by the dilation of the pupils, which is a separate process.

The adjustment to bright light, known as *light adaptation,* may take only 5 min to complete. It occurs dramatically during the first few minutes after we enter a brightly lighted area. The glare may be uncomfortable initially, but the discomfort decreases gradually as the eyes adapt.

Neural Pathways for Vision

All visual information originates with the stimulation of the rods and cones of the retina and is conveyed to the brain by way of the axons of the ganglion cells. These axons form a visual pathway that begins in the eyes and ends in the occipital lobes of the cerebral cortex.

17.22 THE AFFERENT VISUAL PATHWAY

[A] The output from the retina is conveyed to the lateral geniculate body by ganglion cell axons of the optic nerves. Half of the axons of each eye cross over at the optic chiasma, so that each half of the brain receives impulses from both eyes. The right half of the visual field is projected to the primary visual cortex of the left occipital lobe, and the left half of the visual field to the primary visual cortex of the right occipital lobe. [B] Photograph of the afferent visual pathway; ventral aspect of brain, inferior view.

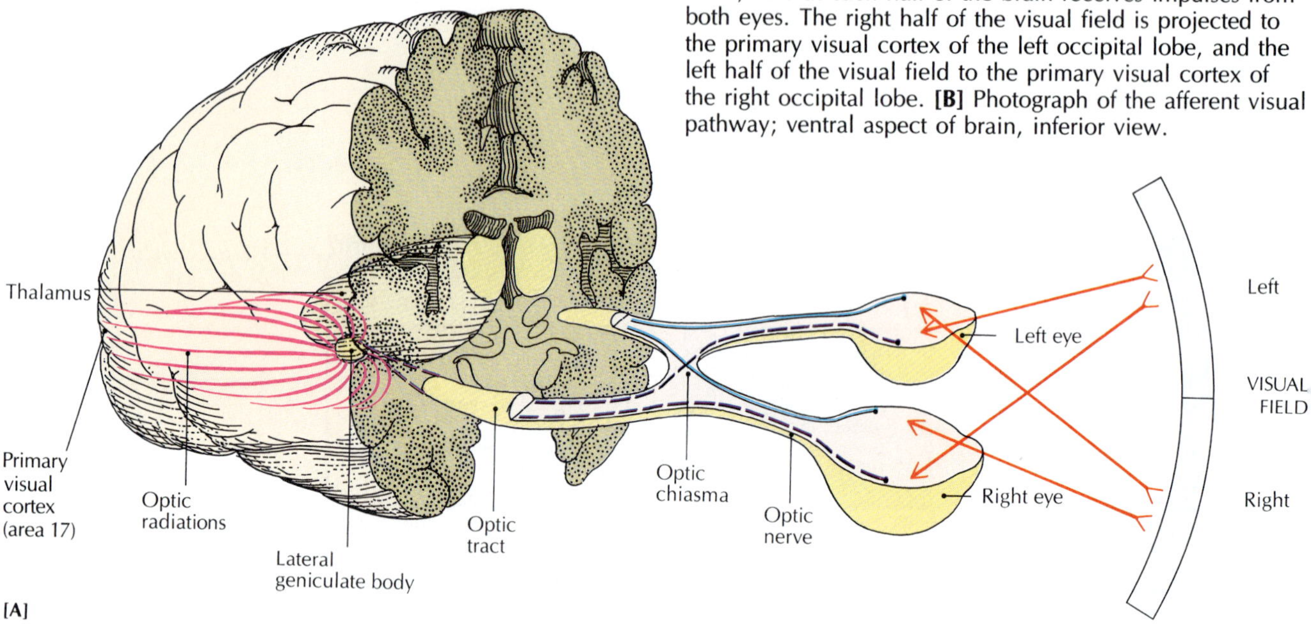

Thalamus

Primary visual cortex (area 17)

Optic radiations

Lateral geniculate body

Optic tract

Optic chiasma

Optic nerve

Left eye

Right eye

Left

VISUAL FIELD

Right

[A]

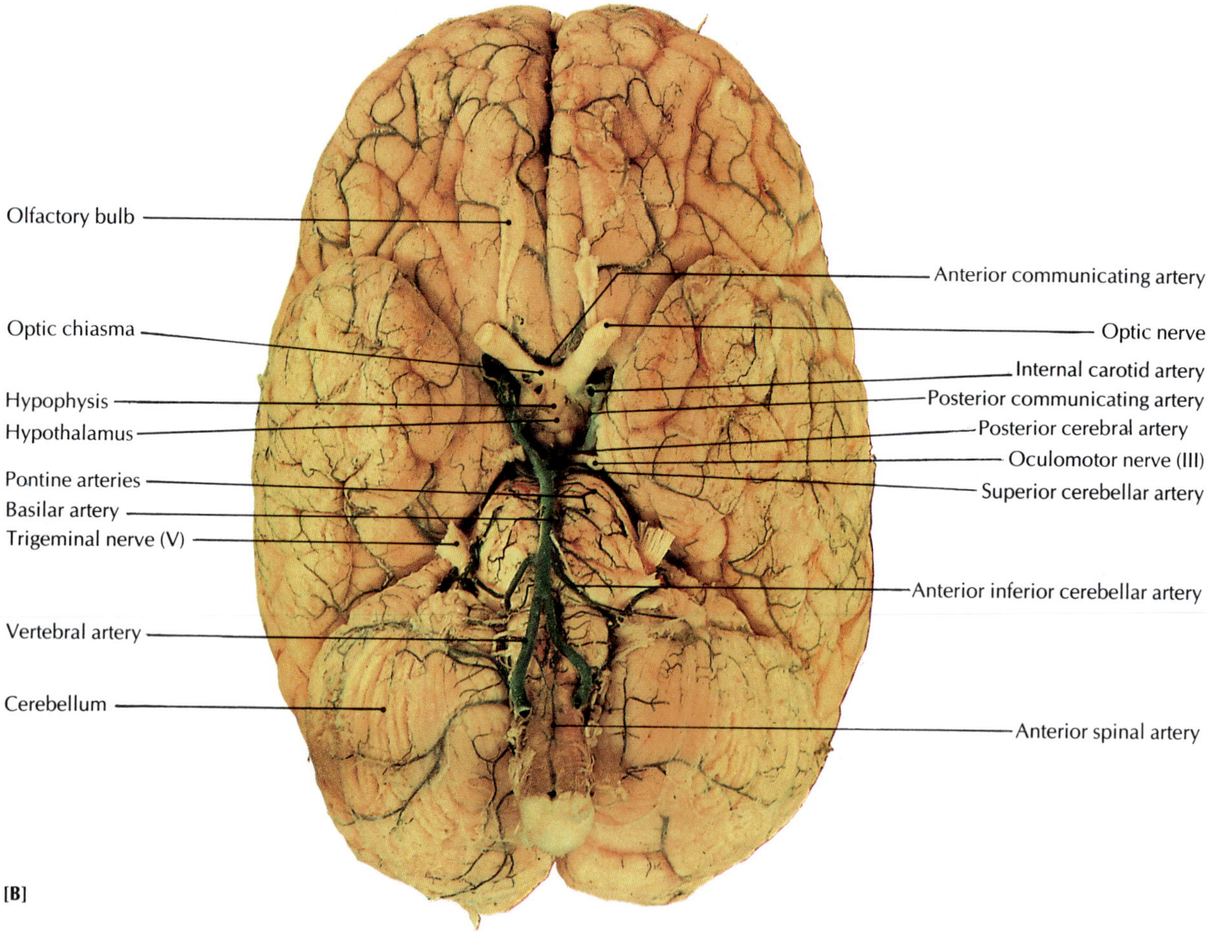

Olfactory bulb

Optic chiasma

Hypophysis

Hypothalamus

Pontine arteries

Basilar artery

Trigeminal nerve (V)

Vertebral artery

Cerebellum

Anterior communicating artery

Optic nerve

Internal carotid artery

Posterior communicating artery

Posterior cerebral artery

Oculomotor nerve (III)

Superior cerebellar artery

Anterior inferior cerebellar artery

Anterior spinal artery

[B]

The field of vision is the environment viewed by the eyes. For each eye it is divided into (1) an outer, lateral *(temporal)* half and an inner, medial *(nasal)* half and (2) an upper half and a lower half. Rays of light entering the eye move diagonally across the eyeball. Because the lens in the eye acts like the lens in a camera, the field of vision is reversed in both the vertical and horizontal planes.

Thus, in the horizontal plane, the light waves from the temporal visual field fall on the nasal half of each retina, and the light waves from the nasal field fall on the temporal half of the retina [FIGURE 17.22A]. In the vertical plane, the rays of light from the upper half of the visual field fall on the lower half of the retina, and rays from the lower half of the visual field fall on the upper half of the retina. Then nerve fibers of the ganglion cells from both eyes carry the impulses along two optic nerves. These two nerves meet at the *optic chiasma,* where the fibers from the nasal half of each retina cross over; the fibers from the temporal half of each retina do not cross over. Because half the nerve fibers from each eye cross over at the optic chiasma, each side of the brain receives visual messages from both eyes. After passing through the optic chiasma, the nerve fibers are called the *optic tracts.* Each optic tract contains nasal fibers from the opposite side and temporal fibers from the same side.

Each optic tract continues posteriorly until it synapses with neurons in the thalamic nucleus called the *lateral geniculate body.* From there, the axons of neurons in the lateral geniculate body project ("go to") to the primary visual cortex in the occipital lobe of the cerebral cortex.

A reflex pathway proceeds from the retinas directly to the superior colliculi in the midbrain. This reflex system is involved in unconscious movements of the eye, including contraction and dilation of the pupil, and coordinated movements of both eyes.

Neural cells of the retina The retina is composed of five main classes of neurons, which are interconnected by synapses. The three types of retinal neurons that form a *direct pathway* from the retina to the brain are *photoreceptor cells* (rods and cones), *bipolar cells,* and *ganglion cells* [see FIGURE 17.14]. The two remaining classes of retinal neurons—*horizontal cells* and *amacrine cells* (neurons without axons)—synapse with the bipolar and ganglion cells. These five classes of neurons have many subtypes of cells, bringing the total number of functional neural elements in the retina to about 60.

The neural elements in the retina interact with each other and begin processing the stimuli received by the photoreceptors. In the direct pathway, this evaluation is accomplished by the convergence of signals from a number of photoreceptors onto a single ganglion cell whose axon projects via the optic nerve to the brain. The horizontal and amacrine cells, acting laterally, connect adjacent neurons and allow them to modify the signals as they are conveyed along the direct pathway to the brain.

Direct pathway and lateral influences
Photoreceptors are stimulated by the light from the visual field in the environment. The resulting activity is conveyed to the ganglion cells via the bipolar cells. A single ganglion cell and the photoreceptors that converge to influence its activity describe the *receptive field* of that ganglion cell.

The receptive field of a ganglion cell is organized into two zones: (1) a small circular zone called the *center* and (2) a surrounding, concentric zone called the *surround* [FIGURE 17.23A]. The center may be likened to the hole of a doughnut, while the surround is the doughnut itself. Receptive fields may be classified as either (1) on-center, off-surround (excitatory center, inhibitory surround) [FIGURE 17.23B] or (2) off-center, on-surround (inhibitory center, excitatory surround) [FIGURE 17.23C]. Neurons with a center-surround organization include (1) ganglion cells of the retina, (2) neurons of the lateral geniculate body, and (3) some neurons in the primary visual cortex (area 17).

Stimulating the receptors in the center of an excitatory center, inhibitory-surround receptive field increases the firing rate of the ganglion cell. Stimulating the surround of the same receptive field decreases the firing rate of the ganglion cell. The reverse is true for the inhibitory-center, excitatory-surround cell. When the entire field is stimulated, the response of the ganglion cell results from the interactions among inhibitory and excitatory photoreceptor signals.

Horizontal and amacrine cells modify communication among the elements of the direct pathway. During *lateral inhibition,* these lateral elements sharpen the contrast between different colors and different intensities of light projected from the visual field onto the retina. By allowing inhibitory signals to pass from a strongly stimulated area to an adjacent weakly stimulated area, the elements further depress the response from the weakly stimulated area. As a result, there is a greater distinction between the strongly and weakly stimulated areas. This ensures that contrast from the visual field passes, via these modified signals, along the direct pathway to the optic nerve and the brain.

In summary, each of the *one million ganglion cells* of the retina of each eye receives stimulation from a spot in the environment. This spot is physiologically expressed by one ganglion cell as *center-surround response.* Thus the retina is a mosaic of ganglion cells capable of sending a continuous mosaic stream of *one million center-surround signals* to the lateral geniculate body and from there to the primary visual cortex (area 17). Each stream has a *visuotopic organization* throughout its course from the retina to the primary visual cortex. "Visuotopic" means that a small neighboring group of neurons in the retina project upon ("goes to") a neighboring group of neurons in the lateral geniculate body which, in turn, project to a neighboring group in the primary visual cortex. (The

FIGURE 17.23 ZONES OF RECEPTIVE FIELDS OF NEURONS OF THE VISUAL PATHWAY

[A] A receptive field has an interior central portion called the *center* and a peripheral concentric portion called the *surround*. [B] An on-center, off-surround receptive field. Photoreceptor cells, bipolar cells, and a ganglion cell form a direct pathway to the brain. Note that although light entering the retina passes through the ganglion cell and other neural cells, it stimulates only the photoreceptor cells, which then stimulate neural cells of the retina. The neural cells send impulses to the brain via the optic nerve. [C] The opposite type of receptive field, with an inhibitory center (off-center) and an excitatory surround (on-surround). Amacrine cells and horizontal cells with axons that project laterally in the retina are involved with neural processing within the retina. Horizontal cells may have an effect (small arrows) on the bipolar cells that is opposite to the direct effect of the rods and cones.

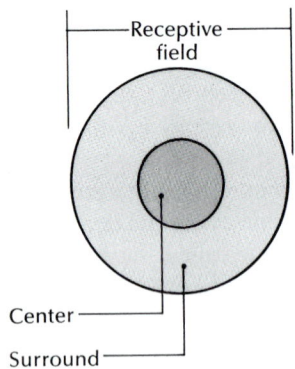

Receptive field
Center
Surround

[A]

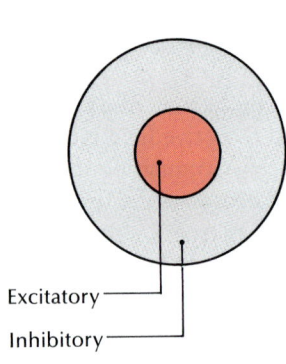

Excitatory
Inhibitory

[B] ON-CENTER, OFF-SURROUND

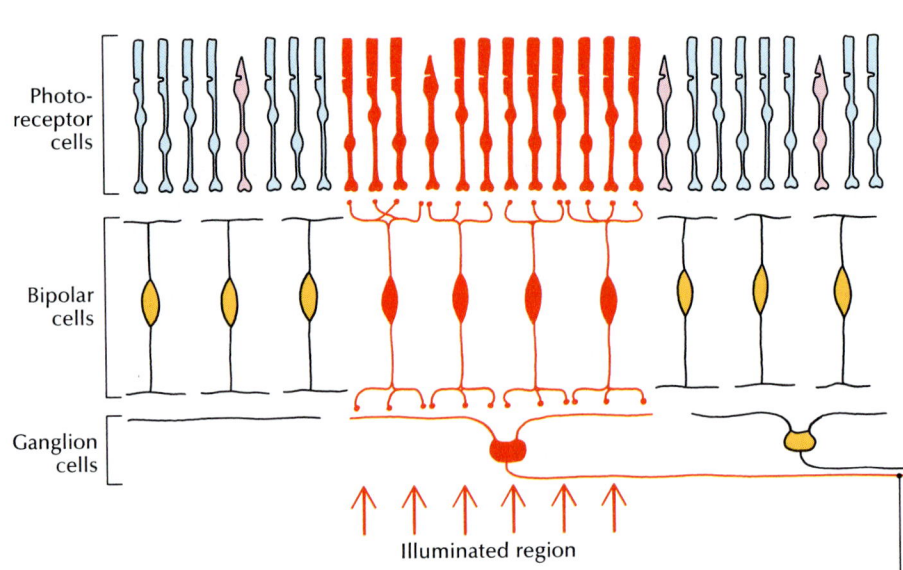

Photoreceptor cells
Bipolar cells
Ganglion cells
Illuminated region
Optic nerve to the brain

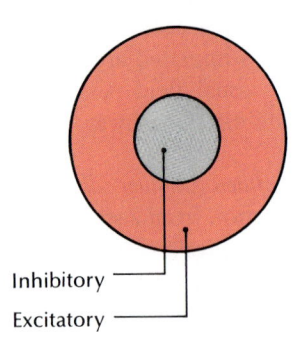

Inhibitory
Excitatory

[C] OFF-CENTER, ON-SURROUND

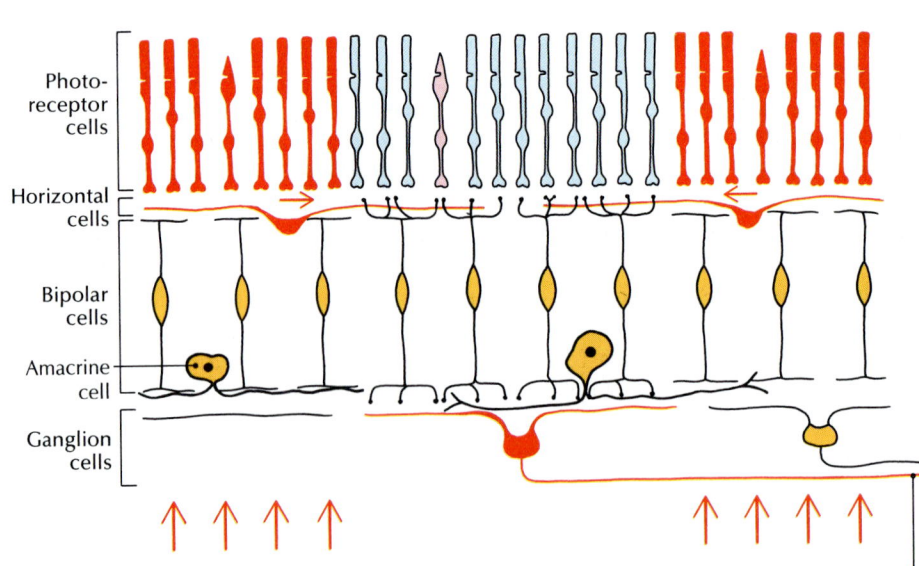

Photoreceptor cells
Horizontal cells
Bipolar cells
Amacrine cell
Ganglion cells
Optic nerve to the brain

word "project" is not meant to imply a projection such as that from a movie projector.)

Normally we perceive one image of an object we see in our field of vision even though two images are formed, one on each retina. The two images are processed into one by a phenomenon known as *fusion*. The sequences of transformations leading to (physiological) fusion of the two streams of center-surrounds from the two eyes begin in the primary visual cortex. This is the *first level* at which neurons are influenced by inputs from both eyes. The subsequent processing of fusion occurs in the other cortical areas associated with vision, primarily those of the occipital lobes. This fusion permits us to see one visual image at each moment in time. When the eyes are not perfectly aligned (as in cross-eyed individuals), two images will be perceived (called *diplopia*; dih-PLO-pee-uh; Gr. *di*, double + *ope*, vision). Normally, the brain eliminates diplopia by ultimately blocking out the input from one eye.

The primary visual cortex is considered to be the site of the analysis and sorting out of raw visual data, which are further processed in occipital visual association areas 18, 19, and others. All association areas have roles for the expression of our conscious visual experiences. The features of visual imaging include form and shape, color, depth and stereoscopic perception, visual acuity, object identification ("what"), spatial features, and movement detection and location ("where"). Different cortical regions, or a combination of them, are involved with processing each of these features. Some noncortical areas (such as the thalamus) may also contribute. How the various features are linked together to produce the total perception of a visual image is a subject of much inquiry and speculation. Added to this is the problem of the storage of images into the memory for future recall.

Functional Subdivisions of the Visual System

Several independent subdivisions of the visual system have been defined. (Such a distinction is not easily accomplished in other systems.) Each subdivision analyzes different aspects of neural information conveyed from the retina. Two subdivisions (the P and M pathways) are components of the direct pathway from the retina via the optic nerves, chiasma, and tracts to the lateral geniculate bodies, and then via the optic radiations to the primary visual cortex [FIGURE 17.22]. The impulses are then relayed for further processing in the higher visual cortical areas. These two pathway systems are functionally associated with visual perception. The systems — the *P* (parvocellular) *pathway* and the *M* (magnocellular) *pathway* — are named, respectively, for the four *parvocellular* (small-celled) layers and the two *magnocellular* (large-celled) layers of each lateral geniculate body.

Functionally, the M pathway processes neural information relatively rapidly, being associated with input bringing good depth perception (stereoscopic vision) and low resolution (low visual acuity) and lacking color. In contrast, the P pathway processes information relatively slowly, since it is associated with poor depth perception, high resolution (hyperacuity), and color-sensitivity.

Recent studies of these pathway systems in dyslexic and nondyslexic children have shown evidence that there is a problem in the magnocellular pathway of dyslexics. Dyslexia ("word blindness") is a selective impairment of reading skills despite normal intelligence, sensory acuity, motivation, and instruction. Even in normal children and adults, reading is difficult when the letters and background are different colors but have no luminance contrast, a condition under which the M pathway responds poorly. The incorrect timing in the integration of the neural information of the P and M pathways may contribute to the vision problems experienced by dyslexics.

The third subdivision — called the *tectal pathway* — projects via the optic nerves, chiasma, and tracts to the superior colliculi of the midbrain tectum. It is involved with directing visual attention to the target of interest, with detecting movement, and with the adjustments associated with the light reflex and accommodation.

The fourth subdivision is the pathway from the retina directly to the suprachiasmatic nucleus of the hypothalamus [see FIGURE 13.14], a nucleus integrated into the circadian rhythms of the body.

ASK YOURSELF

1 How much of the entire electromagnetic spectrum do we see?

2 What are the three layers of the eyeball?

3 What is the importance of the fovea?

4 What are the differences between aqueous humor and vitreous humor?

5 Why do we blink our eyes periodically?

6 What is refraction? Accommodation? Convergence?

7 What is the function of 11-*cis* retinal?

8 How is the sodium-potassium pump involved in the physiology of vision?

9 What is the neurotransmitter in rods and cones?

10 What is the significance of the differences in cone pigments?

11 How do eyes adapt to sudden darkness?

12 What is the function of the optic chiasma?

13 What are the five main classes of retinal neurons?

FIGURE 17.24 EARLY DEVELOPMENTAL ANATOMY OF THE EYE

The drawings **[A]**–**[E]** show the developmental progression of the optic nerve, cornea, lens, neuroretina, and pigmented layer of the retina. The scanning electron micrographs show **[F]** the optic cup and **[G]** the eyelids, optic cup, and lens.

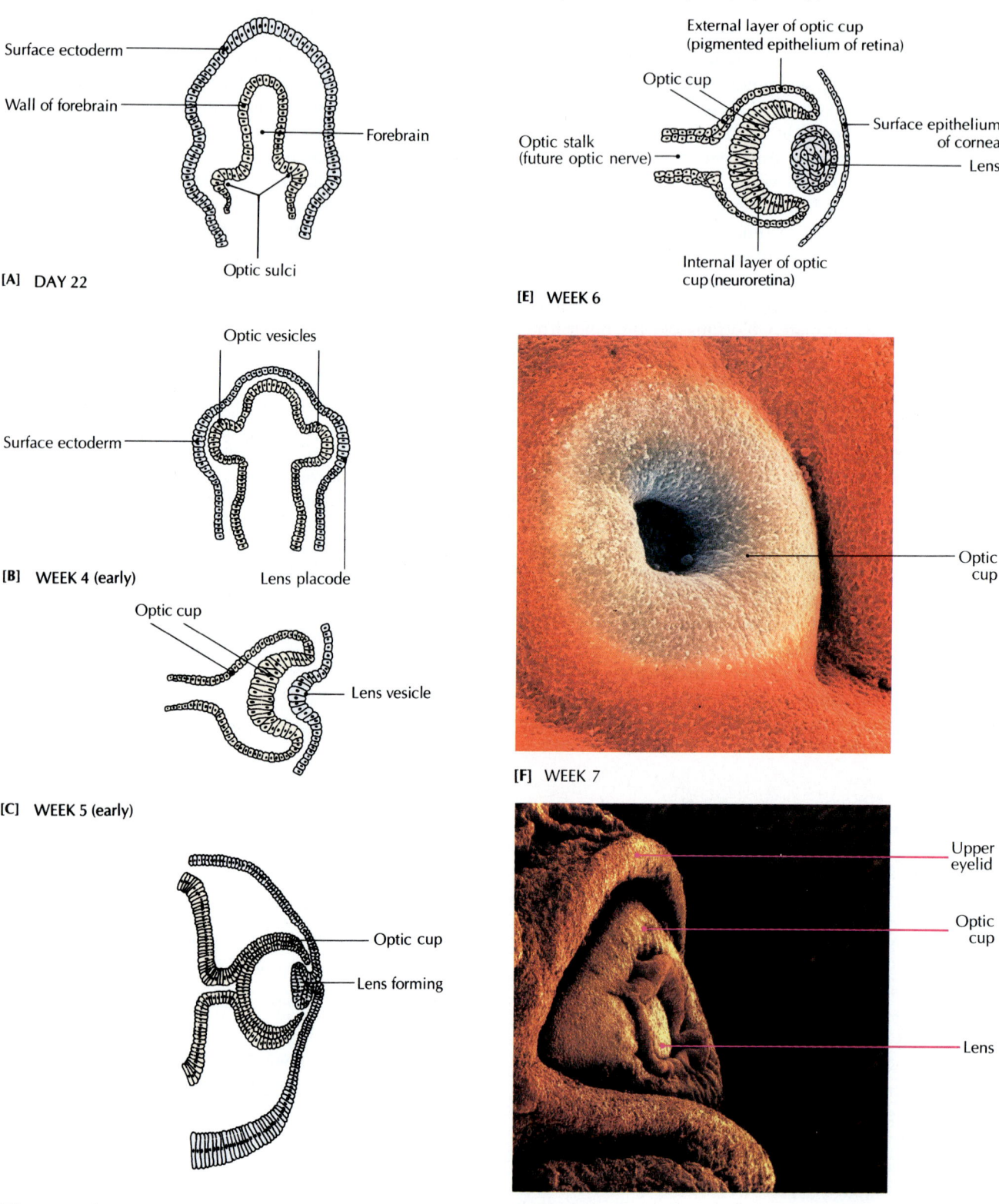

[A]   DAY 22

Surface ectoderm
Wall of forebrain
Forebrain
Optic sulci

[B]   WEEK 4 (early)

Optic vesicles
Surface ectoderm
Lens placode

[C]   WEEK 5 (early)

Optic cup
Lens vesicle

[D]   WEEK 5 (late)

Optic cup
Lens forming

[E]   WEEK 6

External layer of optic cup
(pigmented epithelium of retina)
Optic cup
Optic stalk
(future optic nerve)
Surface epithelium
of cornea
Lens
Internal layer of optic
cup (neuroretina)

[F]   WEEK 7

Optic cup

[G]   WEEK 8

Upper eyelid
Optic cup
Lens

Developmental Anatomy of the Eye

The eyes develop from three embryonic sources: neuroectoderm, surface ectoderm, and mesoderm. The first sign of eye development appears in the 22-day embryo when a pair of shallow *optic sulci* (grooves) form from the neuroectoderm in the forebrain at the cephalic end of the embryo [FIGURE 17.24A]. The sulci develop into a pair of lateral outpockets called *optic vesicles* [FIGURE 17.24B, F]. The distal portion of each optic vesicle dilates, while the proximal portion remains constricted to form the *optic stalk* [FIGURE 17.24B, E]. Eventually, axons of ganglion cells elongate and extend into the optic stalk to become the *optic nerve*. The optic vesicle soon reaches the surface ectoderm and induces* it to thicken into a *lens placode* [FIGURE 17.24B]. By the beginning of week 5, the lens placode induces each optic vesicle to turn in on itself (invaginate), forming a double-layered *optic cup* [FIGURE 17.24E, F, G]. The internal layer of the optic cup differentiates into the complex *neural layer* (neuroretina) of the retina, while the external layer develops into the *pigmented epithelium* of the retina. The retina is well developed at birth except for the central foveal region, which completes its development by the fourth month after birth.

The optic cup induces the lens placode to invaginate and to form a *lens vesicle* that is partially surrounded by the optic cup [FIGURE 17.24C, D]. The vesicle sinks below the outer surface and pinches off like a submerged bubble. This "bubble," now almost completely surrounded by the optic cup, forms the *lens,* while the surface ectoderm becomes the thin surface epithelial layer of the *cornea* [FIGURE 17.24E, F]. The *iris* eventually forms from the rim of the optic cup, which partially covers the lens.

*The principle of *induction* occurs when one tissue is caused, or *induced,* to change its developmental pattern as a result of the influence of a different tissue. These changes are probably produced by chemical stimuli.

Linear grooves called *optic fissures* develop on the inferior surface of the optic cups and optic stalks. Blood vessels develop within these fissures. The hyaloid artery supplies the optic cup and lens vesicle, and the hyaloid vein returns blood. The hyaloid vessels are eventually enclosed within the optic nerve when the edges of the optic fissure fuse. The open end of the optic cup now forms a circular opening, the future *pupil.* The distal portions of the hyaloid vessels eventually degenerate, but the proximal portions become the central artery and vein of the retina.

THE EFFECTS OF AGING ON THE SENSES

Eyesight is usually impaired because of the degeneration of fibers in the optic nerve and an accumulation of injuries to the eyes. (Only one of six 80-year-olds has 20/20 vision.) Common ailments are *presbyopia* (loss of the ability to focus on close objects), *cataract* (opaque density of the lens of the eye), and *glaucoma* (increased fluid pressure within the eyeball). *Senile macular degeneration* is the leading cause of blindness in people over 65. It is characterized by scarring of the central retinal zone and a loss of retinal cells, leading to impaired vision. The cause is unknown. Color perception is reduced as cones in the retina degenerate. People over 60 usually need twice as much light as a 40-year-old because some rods in the retina have degenerated.

Hearing, smell, and *taste* are all reduced. Most elderly people have adequate hearing, but high-pitched sounds become increasingly difficult to hear as the eardrum loses its elasticity, and fibers in the auditory nerve degenerate. Hearing may also be impaired by the accumulation of hardened wax in the outer ear. The abilities to smell and taste begin to deteriorate at about age 60, as the lining of the mucous membrane becomes thinner and less sensitive, and as the number of active taste buds is reduced.

WHEN THINGS GO WRONG

The Ear

More than 28 million Americans have some hearing loss, and 80 percent of them have irreversible hearing damage. About 2 million Americans are completely deaf, and more than one-third of the population over 65 years old has significant hearing impairment. More than half of all cases of severe childhood hearing impairment are caused by genetic factors.

Otosclerosis The stapes may become fused to the bone in the region of the oval window and become immobile, a condition called *otosclerosis* (Gr. *otos,* ear + *skleros,* hard). When this happens, the bones cannot vibrate properly, and the transmission of sound waves to the inner ear may be almost impossible. A hearing aid, which transmits sound waves to the inner ear by conduction through the bones of the skull rather than by air conduction along the auditory ossicles, can make hearing almost normal in people who have been deafened by otosclerosis. An operation that allows the stapes to move again has also had some success.

Labyrinthitis *Labyrinthitis,* or inflammation of the inner ear, can cause discharge from the ear, vomiting, hearing loss, and vertigo (the feeling of spinning) and other forms of dizziness. Vertigo upsets the balance, and the patient tends to fall in the direction of the affected ear. Within 3 to 6 weeks the symptoms usually subside. The causes of labyrinthitis include bacterial infections, allergies, toxic drugs, severe fatigue, and overindulgence of alcohol.

Ménière's disease Head noises, dizziness, and hearing loss are all characteristic of *Ménière's disease,* an inner-ear disorder. It is thought to be caused by an excess of endolymph and dilation of the labyrinth. The basilar membrane is distorted, the semicircular ducts are affected by pressure, and some cochlear hair cells degenerate. A patient may have residual *tinnitus* (ringing or whistling in the ears) (tih-NYE-tuhss; L. *tinnire,* to ring) and hearing loss after repeated attacks over many years.

Motion sickness Many people know the familiar queasy feeling of *motion sickness,* which results from the sensation of motion or from repeated rhythmic movements. Symptoms of nausea, vomiting, pallor, and cold sweats are common when people travel on boats, planes, cars, and trains. However, suffering from one type of motion sickness, such as seasickness, does not necessarily mean that a person is susceptible to other forms. Motion sickness arises from the excessive stimulation of the vestibular receptors of the inner ear. Tension, fear, offensive odors, and visual stimuli are also important factors.

Another theory of motion sickness suggests that it results from conflicting perceptions. For example, when a person is seated in the lounge of a ship, that person's vestibular system may detect the rocking motion of the ship, but the visual system may not. Alternatively, in a car, the vestibular system may not detect motion (since the forward motion of the car is virtually constant), while the visual system detects motion by observing the passing scene through the side windows. According to this theory, matching the two perceptions should help relieve the motion sickness. In other words, watch the horizon list with the ship or look out the front windshield of the car to perceive less motion.

Motion sickness can be prevented by drugs such as dimenhydrinate (Dramamine) that suppress vestibular function.

Otitis media *Otitis media,* or inflammation of the middle ear, causes the tympanic membrane to redden and bulge out. If untreated, the membrane may rupture. Otitis media is commonly seen in children and is most prevalent between the ages of 6 and 24 months. If the inflammation is severe enough or prolonged, scarring, structural damage, and hearing loss may result. This disorder seems to arise from a malfunction of the auditory tube, in which bacteria from the nasopharynx enters the middle ear. Otitis media may be acute or chronic. Both forms are bacterial infections, caused by organisms such as *Streptococcus pneumoniae* or *Haemophilus influenzae.*

The Eye

Detached retina Normally no space exists between the two retinal layers, but sometimes the layers become separated. Such a condition is commonly called a *detached retina.* The detachment takes place between the pigmented epithelium and neuroretina — the site where the two layers of the optic cup were separated in the embryo. The portion of the neuroretina that is detached usually stops functioning because its blood supply is impaired. New surgical procedures, including photocoagulation by laser beam, and cryosurgery, can successfully reattach the retinal layers in about 90 percent of cases, and thus arrest further detachment.

Cataract One of the most common causes of blindness is *cataract formation,* in which the lens becomes opaque. As the opaque areas increase, there is a progressive loss of vision, and if the cataract is severe enough, total blindness may result. The cause is thought to be the oxidative damage to the lens proteins, reducing solubility and eventually forming insoluble opacities in the otherwise transparent tissue. The result is a cataract. Apparently, regular doses of vitamins C and E may help to prevent cataracts.

Cataracts are generally associated with people over 70, but they can occur at any age. There are various types of cataracts, including senile, congenital, traumatic, and toxic. Cataracts are also common in people with diabetes, since the abnormal glucose metabolism of diabetes may affect the vitality of the lens. Treatment usually begins with frequent changes in eyeglasses to compensate for gradual vision loss. When new prescription eyeglasses are no longer helpful, cataract surgery is the preferred procedure. During surgery the lens is removed from its capsule, and in its place an artificial lens is implanted permanently. Soon after the operation, the patient can see with the new "fixed-focus" lens and does not have to wear a contact lens. Newly prescribed eyeglasses can be used to make adjustments.

Conjunctivitis *Conjunctivitis,* or inflammation of the conjunctiva, is a common eye disorder. It may have several causes. Bacteria, viruses, pollen, smoke, pollutants, and excessive glare all affect the conjunctiva, causing discharge, tearing, and pain. Vision, however, generally is not affected. "Pinkeye," a form of conjunctivitis caused by *Haemophilus* bacteria, is contagious. In such cases, affected people should avoid spreading the infection by not rubbing the infected eye and not sharing towels and pillows. Conjunctivitis can be acute or chronic, and treatment varies with the cause.

Glaucoma *Glaucoma* (glaw-KOH-muh) is a leading cause of blindness in the United States, affecting over 1 million people. The disease strikes people of all ages, but mainly those over 40; women are more susceptible than men. Glaucoma occurs when the aqueous humor does not drain properly. Since more fluid is formed than is drained, it builds up in the eyeball and increases the intraocular pressure. If the pressure continues, it destroys the neurons of the retina, and blindness usually results. Glaucoma may be chronic (90 percent of cases) or acute. Chronic forms result in progressively reduced vision. A common symptom is a visual defect in which lights appear to be surrounded by halos. Acute forms can occur suddenly at any age, causing pain, pressure, and blurring. Chronic glaucoma may be genetically linked. Close relatives of patients with glaucoma are five times more susceptible to developing the disease than are those with no glaucoma in their family history.

CHAPTER SUMMARY

Sensory Reception (p. 496)

1 Sense organs, or **receptors,** are the body's link to the outside world and to changes within the body.

2 All receptors are *transducers;* that is, they convert one form of energy into another.

3 A receptor cell that responds directly to stimuli may be either a specialized neuroepithelial cell or the ending of a sensory neuron.

4 The sensory neurons respond initially with *receptor (graded) potentials,* which are converted into *action potentials.* All stimulated receptors generate the same type of action potential. Different sensations occur when nerve fibers connect with specialized portions of the central nervous system.

5 Certain features are basic to all types of sensory receptors: (1) They contain *sensitive receptor cells* that respond to threshold levels of intensity, (2) they have a *structure* that is conducive to receiving a specific stimulus, (3) their primary receptor cells *synapse with afferent nerve fibers* that travel to the central nervous system along neural pathways.

6 The sensory neurons associated with specific perceived sensations, such as pain and sound, convey neural influences to specific **neural pathways** within the CNS. These pathways process the neural influences, which are finally perceived as the relevant sensations.

7 Sensory receptors may be classified according to their *location* (**exteroceptor, teleceptor, interoceptor, proprioceptor**); *type of sensation* (**thermal, pain, light touch, touch-pressure, position sense**); *type of stimulus* (**thermoceptor, nociceptor, chemoreceptor, photoreceptor, mechanoreceptor**); or *sensory ending* (**free nerve ending, encapsulated ending**).

8 The **general senses** include light touch, touch-pressure (deep pressure), two-point discrimination, vibratory sense, heat, cold, pain, and body position. Taste, smell, hearing, equilibrium, and vision are called the **special senses** because they originate from sensors in restricted (special) regions of the head.

9 Individual nerve impulses are essentially identical, but an impulse for sight is distinguished from an impulse for taste by the *relationship between where the impulse is coming from and the place in the brain where it is going.*

General Senses (p. 498)

1 Sensory receptors in the skin detect stimuli that the brain interprets as light touch, touch-pressure, heat, cold, pain, proprioception, and stereognosis. Other miscellaneous cutaneous sensations include itch and tickle.

2 Two receptors for light touch are **tactile corpuscles (of Merkel)** and **tactile corpuscles (of Meissner).** Receptors for touch-pressure are **lamellated (Pacinian) corpuscles. Corpuscles (of Ruffini)** and **bulbous corpuscles (of Krause)** are mechanoreceptors for touch.

3 The cutaneous receptors for heat and cold are probably free nerve endings.

4 *Itch* and *tickle* are probably produced by the activation of several sensory types of nerve endings. The information resulting in the sensation is probably conveyed by a combination of pathways.

5 Pain receptors are **specialized free nerve endings** that are present in most parts of the body.

6 *Referred pain* is a visceral pain that is felt in a somatic area. *Phantom pain* is felt in an amputated limb.

7 The sense of position of body parts in relation to each other is called *proprioception. Stereognosis* is the ability to identify unseen objects merely by handling them.

Taste (Gustation) (p. 504)

1 The receptors for taste and smell are both chemoreceptors. Both sensations are interrelated, but taste (*gustation*) and smell (*olfaction*) are separate.

2 The surface of the tongue is covered with many small protuberances called *papillae,* including three types: filiform, fungiform, and circumvallate. Located within the crevices of papillae are *taste buds,* the receptor organs for the sense of taste.

3 The four basic taste sensations are sweet, sour, bitter, and salty.

Smell (Olfaction) (p. 505)

1 The **olfactory receptor cells** are located high in the roof of the nasal cavity, in specialized areas of the nasal mucosa called the *olfactory epithelium.* The epithelium consists of *receptor cells, sustentacular (supporting) cells,* and *basal cells.*

2 The olfactory receptor cell (neuron) ends in a bulbous *olfactory vesicle,* from which extend *cilia* that project through the mucuslike fluid that covers the surface epithelium.

3 Olfactory nerve fibers extend through foramina in the cribriform plate of the ethmoid bone and terminate in the **olfactory bulbs,** from which axons of neurons project to the olfactory cortex of the cerebrum.

Hearing and Equilibrium (p. 507)

1 The ear consists of two functional units: the **auditory apparatus,** concerned with hearing, and the **vestibular apparatus,** concerned with posture and balance.

2 The anatomical components of the auditory apparatus are the external ear, the middle ear, and the inner ear. The *external ear* is composed of the **auricle** and **external auditory canal;** the *middle ear* is made up of the **tympanic membrane** (eardrum), **tympanic cavity, auditory** (Eustachian) **tube,** and the three **auditory ossicles** (ear bones); the *inner ear* is composed of the **vestibule** (which contains the **utricle** and **saccule**), **semicircular canals** and **ducts,** and **cochlea** (which contains the **spiral organ of Corti**).

3 *Sound* is the alternating compression and decompression of the medium through which the sound is passing.

4 Sound waves reach the spiral organ through a sequence of vibrations that start in the external ear and tympanic membrane and progress through the ear bones into the inner ear. The displacement of the hairs of the hair cells in the spiral organ generates receptor (generator) potentials and, subsequently, nerve impulses in the cochlear nerve. Their influences are conveyed to the auditory area of the temporal lobe via the auditory pathways.

5 The main receptors for **equilibrium** are hair cells in the utricle, saccule, and semicircular ducts in the inner ear. The equilibrium system also receives input from the eyes and from some proprioceptors in the skin and joints.

6 The purpose of the **vestibular system** is to signal changes in the motion of the head **(dynamic equilibrium)** and the position of the head with respect to gravity **(static equilibrium,** or posture).

7 Specialized receptor cells of the vestibular sense organs are **hair cells,** which are arranged in clusters called hair bundles. *Stereocilia* and a *kinocilium* are present in each hair bundle. When hairs are bent in the direction of the stereokinocilium, the hair cells convert a mechanical force into an electrical signal that is conveyed to the brain via the vestibular nerve.

8 When the head moves in a change of posture, calcium carbonate crystals **(otoconia)** in the inner ear respond to gravity, resulting in the bending of the hairs of hair cells. The bending stimulates nerve fibers to generate a generator potential and then an action potential, which is transmitted to the brain. The brain signals appropriate muscles to contract, and body posture is adjusted to follow the new head position.

9 The utricles and saccules are *organs of gravitation,* responding to movements of the head in a straight line: forward, backward, up, or down. In contrast, the **crista ampullaris** of the semicircular ducts responds to changes in the *direction of head movements,* including turning, rotating, and bending. Hair cells in the crista project into a gelatinous flap called the **cupula.** When the head rotates, the endolymph in the semicircular canals lags behind, displacing the cupula and the hairs projecting into it. The resulting action potentials are sent to the neural centers in the brain, which signals certain muscles to respond appropriately to maintain the body's equilibrium.

Vision (p. 522)

1 The human eye can "see" only objects that emit or are illuminated by light waves representing about 1/70 of the electromagnetic spectrum.

2 The wall of the eyeball consists of three layers of tissue: the outer **supporting layer,** the **vascular layer,** and the inner **retinal layer.**

3 The posterior five-sixths of the supporting layer is the **sclera** ("white" of the eye). It gives the eyeball its shape and protects the delicate inner layers. The anterior segment of the supporting layer is the transparent **cornea,** through which light enters the eye.

4 The vascular layer contains many blood vessels. The posterior two-thirds is a thin membrane, the **choroid.** The **ciliary body** is the thickened anterior portion of the choroid layer. The **lens** is an elastic body that changes shape to focus on objects at different distances.

5 The anterior extension of the choroid is a muscular layer, the **iris,** which is the colored part of the eye. An adjustable opening in the center of the iris is the **pupil.** It opens and closes in response to the amount of light available.

6 The innermost portion of the eyeball is the **retina.** It contains (1) a *layer of nervous tissue* **(neuroretina)** that receives focused light waves, transduces them, and processes their effects into neural impulses, which are converted into visual perceptions in the brain, and (2) a **pigmented layer** behind the neural layer that absorbs light not utilized by rods and cones and thus prevents reflection of light that could restimulate the rods and cones.

7 The retina contains highly specialized photoreceptor nerve cells, the **rods** and **cones,** as well as other types of nerve cells. Rod cells respond to the entire visual spectrum and produce black-to-white vision. They function in dim light and peripheral vision. The color-sensitive cone cells are mainly concentrated in the center of the retina in the **fovea centralis.**

8 The hollow eyeball is divided into three cavities. The **anterior chamber** lies between the iris and cornea. The **posterior chamber** lies between the iris, lens, and vitreous chamber. Both chambers are filled with *aqueous humor.* The largest cavity of the eyeball is the **vitreous chamber** (the space behind the lens), which contains *vitreous humor.*

9 The accessory structures of the eye are mainly protective and supportive devices. They include the **orbits, eyelids, eyelashes, eyebrows, conjunctiva,**

lacrimal apparatus, and *muscles* that act on the eyeball and eyelid.

10 *Refraction* is the bending of light waves as they pass from one medium to another medium with a different density. *Accommodation* is the mechanical process that involves the ciliary muscle and the lens to bring images to exact focus on the retina. *Convergence* is the movement of both eyes inward to focus on nearby objects in the visual field so that the objects fall on corresponding points of the retinas of both eyes simultaneously.

11 Vision begins when a *photon* is absorbed by a photoreceptor cell (rod or cone). This produces a hyperpolarization that spreads to the synaptic ending. When photoreceptor cells absorb light, they release the neurotransmitter *glutamate.*

12 The hyperpolarization in rods occurs as the light-absorbing 11-*cis* retinal (an isomer of *retinal*) in the photopigment *rhodopsin* absorbs light. 11-*cis* retinal is converted into all-*trans* retinal, which activates *scotopsin* (the protein in rhodopsin) to act as an enzyme. The activated protein *transducin* activates a phosphodiesterase enzyme, which hydrolyzes cGMP to an inactive form. Sodium ions are blocked from entering the rods, hyperpolarization occurs, neural signals are processed by bipolar, amacrine, horizontal, and ganglion cells in the retina, and action potentials of the ganglion cells are conveyed to the brain, where ''photon'' activity is perceived as light.

13 Three types of cones contain separate photopigments that respond to red, green, or blue light. Cone pigments contain 11-*cis* retinal and *photopsins.* The final color is determined in the cerebrum by the combinations of different levels of excitation of different types of cones.

14 *Dark adaptation* occurs as eyes adapt slowly to darkness, and *light adaptation* occurs when eyes adjust to bright light.

15 The field of vision for each eye is divided into (1) an outer, lateral *(temporal)* half and an inner, medial *(nasal)* half and (2) an upper half and a lower half. Nerve fibers of ganglion cells in the retina carry impulses from both eyes along two optic nerves. The nerves meet at the *optic chiasma,* where the fibers from the nasal half of each retina cross over; the temporal fibers do not cross over. As a result of the crossover, each side of the brain receives visual messages from both eyes.

16 After passing through the optic chiasma, the nerve fibers are called the *optic tracts.* Each optic tract continues until it synapses with neurons of the *lateral geniculate body* of the thalamus. From there, axons pass to the primary visual cortex in the cerebral cortex.

17 The retina is composed of five main classes of synaptically interconnected neurons. *Photoreceptors, bipolar cells,* and *ganglion cells* form a direct pathway from the retina to the brain; *horizontal cells* and *amacrine cells* help process visual information.

18 The visual system can be divided into four functional subdivisions: (1) the P (parvocellular) pathway, (2) the M (magnocellular) pathway, (3) the tectal pathway, and (4) the pathway from the retina to the suprachiasmatic nucleus of the hypothalamus.

19 The eyes develop from neuroectoderm, surface ectoderm, and mesoderm, starting at about day 22, when optic sulci form. The sulci develop into *optic vesicles.*

20 The optic vesicle induces the surface ectoderm to thicken into *lens placodes,* which then induce each optic vesicle to invaginate, forming *optic cups.*

21 The internal layer of the optic cups differentiates into the *neuroretina,* while the external layer develops into the *pigmented epithelium* of the retina.

22 The optic cups induce the lens placodes to invaginate, forming *lens vesicles* that eventually form the *lens.*

23 The *cornea* develops from the surface ectoderm.

24 Blood vessels develop within optic fissures.

The Effects of Aging on the Senses (p. 539)

Eyesight, hearing, smell, and *taste* are usually reduced. Presybopia, cataract, and glaucoma are relatively common eye disorders of old age; senile macular degeneration can lead to blindness. Otosclerosis is an ear disease in which the ear ossicles fuse.

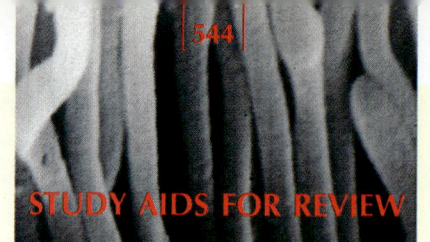

KEY TERMS

MEDICAL TERMS FOR REFERENCE

The Ear

EUSTACHITIS Inflammation of the auditory (Eustachian) tube.

IMPACTED CERUMEN An accumulation of cerumen, or earwax, that blocks the ear canal and prevents sound waves from reaching the tympanic membrane.

MYRINGITIS Inflammation of the tympanic membrane or eardrum. Also known as *tympanitis*.

NYSTAGMUS (nih-STAG-muhss; Gr. *nustagmos*, drowsiness) The involuntary, rhythmic movement of the eyeballs with a rapid movement in one direction followed by a slow movement in the opposite direction. These movements can give the illusion of revolving motion, either of oneself or of the surroundings *(vertigo)*. Nystagmus is related to an imbalance of synchronized impulses from vestibular sources and usually produces blurred vision and difficulty in focusing. By convention, nystagmus is classified according to the direction of the fast movement (the direction of the illusion of spinning). Horizontal nystagmus, or spinning in a horizontal plane, occurs more commonly than vertical, or rotatory, nystagmus.

OTALGIA (Gr. *ous*, ear + *algia*, pain) Earache.

OTOPLASTY (Gr. *ous*, ear + *plastos*, molded) Surgery of the outer ear.

OTORRHEA (Gr. *ous*, ear + *rrhea*, discharge) Fluid discharge from the ear.

OTOSCOPE (Gr. *ous*, ear + *skopein*, to see) Instrument used to look into the ear.

TINNITUS (tih-NYE-tuhss; L. *tinnire*, to ring) A temporary or permanent condition in which there is a ringing, buzzing, hissing, humming, whistling, or roaring in one or both ears; may be caused by Ménière's disease, Paget's disease (osteitis deformans), a perforated eardrum, loud noises, or infection.

The Eye

ACHROMATOPSIA (Gr. *a*, not + *kroma*, color + *ope*, vision) Complete color blindness.

AMBLYOPIA (Gr. *amblus*, dim + *ops*, eye) Reduced visual acuity from not using the eye. Also called "lazy eye."

AMETROPIA (Gr. *ametros*, without measure + *ops*, eye) Inability of the eye to focus images correctly on the retina, resulting from a refractive disorder.

ANOPIA (Gr. *a*, no + *ops*, eye) No vision, especially in one eye.

BLEPHARITIS (Gr. *blepharon*, eyelid + *itis*, inflammation) Inflammation of the eyelid.

CHALAZION (Gr. *khalaza*, hard lump) Swelling in the oil glands of the eyelid.

DIPLOPIA (Gr. *di*, double + *ope*, vision) Double vision.

ESOTROPIA Medial deviation (turning inward) of the eyeball, resulting in diplopia, caused by a muscular defect or weakness in coordination. Also called "crosseye" and *convergent strabismus;* see **STRABISMUS.**

EXOPHTHALMIA (Gr. *ex*, out + *ophalmos*, eye) Abnormal protrusion of the eyeball.

EXOTROPIA (Gr. *ex*, out + *tropos*, turn) Lateral deviation (turning) of the eyeball, resulting in diplopia. Also called "walleye" and *divergent strabismus.*

KERATITIS Inflammation of the cornea.

KERATOPLASTY Plastic surgery of the cornea, such as a corneal transplant.

MIOTIC Drug that causes the pupil to constrict.

MYDRIASIS Dilation of the pupil.

MYDRIATIC Drug that causes the pupil to dilate.

OPHTHALMOSCOPE Instrument used to visually examine the interior of the eyeball.

OPTIC NEURITIS Inflammation of the optic nerve.

PTOSIS Drooping (prolapse) of an organ or part. Specifically, drooping of the eyelid.

RETINITIS Inflammation of the retina.

RETINITIS PIGMENTOSA A group of inherited, progressively blinding diseases.

SCLERITIS Inflammation of the sclera.

SCOTOMA (Gr. *skotos*, darkness) A blind spot or dark spot seen as a result of vision loss in part of the visual field.

STRABISMUS (Gr. *strabos*, squinting) "Crossed eyes." An eye muscle defect in which the eyes are not aimed in the same direction, resulting from dysfunction of the extrinsic eye muscles.

TRACHOMA Disease caused by *Chlamydia trachomatis*, a bacterium spread by insects, body contact, poor hygiene, and contaminated water. It is a leading cause of blindness in many countries.

UNDERSTANDING THE FACTS

1 Which senses are included among the general senses?

2 How are receptors classified?

3 Which sensory receptors are activated by

   a light touch   **c** temperature

   b touch-pressure   **d** pain

4 What is the function of sustentacular cells, and where are they found?

5 What is the function of hair cells, and where are they found?

6 Distinguish between a taste bud and a taste papilla.

7 List the four qualities of taste that our taste buds distinguish.

8 List the characteristics that a substance must have in order to be smelled.

9 Distinguish between the bony labyrinth and the membranous labyrinth.

10 List, in order, the structures in the ear that are affected by incoming sound waves.

11 Which parts of the vestibular apparatus are associated with static equilibrium, and which parts are associated with dynamic equilibrium?

12 List, in order, the parts of the brain in the neural pathway for hearing.

13 List, in order, the parts of the brain in the neural pathway for balance.

14 What are the three tissue layers and the three chambers that comprise the eyeball?

15 What are the accessory structures of the eye and what are their functions?

16 Trace the route of light through the eyeball from the cornea to the retina.

17 Define the following processes, indicating which structures in the eyeball are responsible for each.

   a refraction   **c** dark adaptation

   b accommodation **d** convergence

18 Define the following terms:

   a presbyopia   **c** hypermetropia

   b myopia   **d** astigmatism

19 List, in order, the parts of the neural pathway for vision within the brain.

20 What are the names of the receptor cells that are involved in each of the special senses?

UNDERSTANDING THE CONCEPTS

1 What important features do all sensory receptors have in common?

2 What is meant by the statement "All receptors are transducers"?

3 In what ways are receptor potentials and action potentials alike? How do they differ?

4 Given the fact that sensory receptors either are neurons or activate neurons, and given that fact that action potentials in neurons are alike, how do we distinguish the various senses from each other?

5 Distinguish the role of the brain in referred pain from its role in phantom pain.

6 What are the roles of gustducin and transducin? Why are they called second messengers?

7 In what ways is the sense of smell different from any of the other special senses?

8 Although the senses of hearing and balance are separate, the sensory structures responsible for each develop from the same embryonic structures. Explain.

9 What are the structural and functional relationships between stereocilia and kinocilia of hair cells in the utricle?

10 Describe the structural relationship between rods and cones, bipolar cells, horizontal cells, amacrine cells, and ganglion cells in the retina.

SELF-QUIZ
Multiple-Choice Questions

17.1 Which of the following body parts have no pain receptors?

   a lungs   **d** skin

   b brain   **e** ureter

   c stomach

17.2 Light touch is mediated by which of the following neural pathways?

   a dorsal column-medial lemniscal pathway

   b spinocervicothalamic tract

   c anterior spinothalamic tract

   d trigeminothalamic tract

   e all but **d**

17.3 The neural pathway(s) for pain include the
a lateral spinothalamic tract
b indirect spinoreticulothalamic pathway
c dorsal column-medial lemniscal pathway
d a and b
e a, b, and c

17.4 Which of the following help the body deal with changes in position and acceleration?
a utricle   d proprioceptors
b saccule   e all of the above
c semicircular ducts

17.5 Which of the following is *not* a skeletal muscle of the eye?
a ciliary muscle   d superior oblique
b medial rectus   e levator palpebrae
c lateral rectus   superioris

17.6 Sensations of taste that are conveyed to the brain by the glossopharyngeal nerve (IX) arise in taste buds within ____ papillae.
a fungiform   d sustentacular
b circumvallate   e solitarius
c filiform

17.7 Which of the following cells are the receptor structures for olfaction?
a sustentacular cells   d mitral cells
b olfactory bulbs   e olfactory glomeruli
c bipolar cells

17.8 The tympanic cavity is located within the
a external ear   d cochlea
b middle ear   e semicircular canals
c inner ear

17.9 Which of the following structures does *not* contain endolymph?
a utricle   d scala tympani
b saccule   e semicircular canals
c cochlear duct

17.10 The crista ampullaris is found within the
a utricle   d cochlear duct
b saccule   e ampulla
c scala vestibuli

Completion Exercises

17.11 In bright light, the most active photoreceptor cells are the ____.

17.12 The lateral geniculate body is part of the ____ pathway, whereas the medial geniculate body is part of the ____ pathway.

17.13 All of the photoreceptor cells that influence the activity of one ganglion cell are known as the ____ of the ganglion cell.

17.14 The optic nerve is composed of axons of ____ cells.

17.15 Visceral pain felt subjectively in a somatic area is called ____.

17.16 The inner ear is also called the ____ because of its intricate structure.

17.17 Calcium carbonate crystals involved in static equilibrium are called ____.

17.18 The transparent membrane lining the inner eyelid is the ____.

17.19 When retinal detachment occurs, a space develops between the neuroretina and the ____.

17.20 The eyeball is kept moist by secretions of the ____.

Matching

17.21 ____ retina
17.22 ____ cochlea
17.23 ____ cornea
17.24 ____ fovea
17.25 ____ scala media
17.26 ____ auditory tube

a in the inner ear
b depressed area made up only of cones
c transparent anterior portion of eyeball
d light-sensitive membrane
e in the middle ear
f contains endolymph

A SECOND LOOK

1 In the following drawing, label the external auditory canal, malleus, oval window, spiral organ, middle ear, and inner ear.

2 In the following drawing of the eye, label the iris, pupil, retina, optic nerve, cornea, and vitreous chamber.

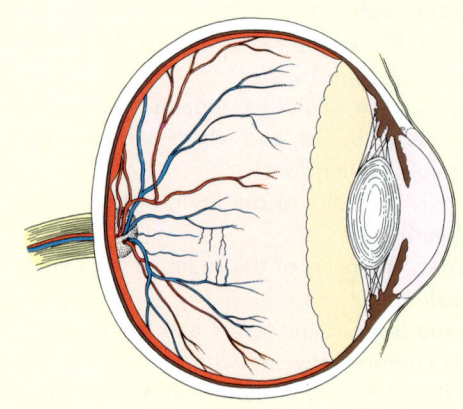

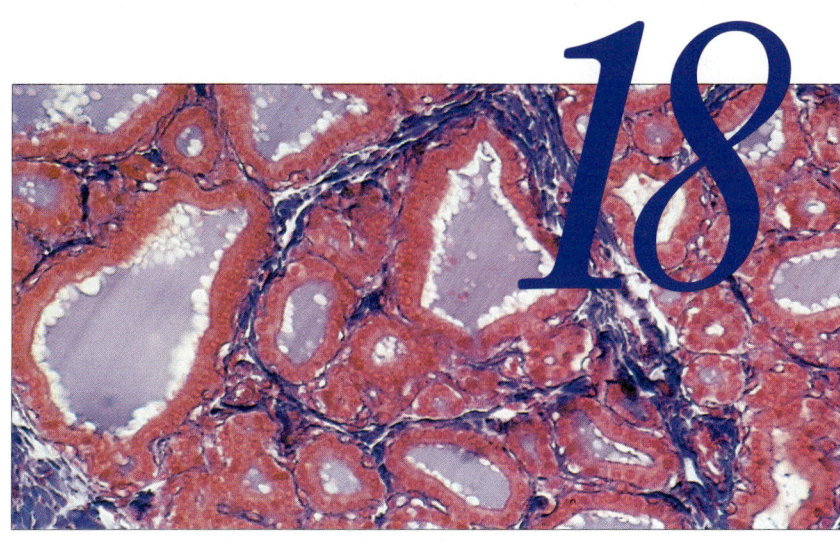

18

The Endocrine System

KEY CONCEPTS

1 The endocrine system helps regulate growth, development, and many metabolic activities.

2 A hormone is the secretion of an endocrine gland. Hormones usually work together with other hormones in a self-regulating feedback system.

3 Hormones released from the hypothalamus and pituitary gland regulate many of the activities of the endocrine system.

4 Some organs that are not classified as endocrine glands, for example, the heart, also produce hormones.

The endocrine system is a control system that is concerned mainly with three functions: (1) It helps maintain homeostasis by regulating activities such as the concentration of chemicals in body fluids and the metabolism of proteins, lipids, and carbohydrates. (2) Its secretions act in concert with the nervous system to help the body react to stress properly. (3) It is a major regulator of growth and development, including sexual development and reproduction.

The endocrine system is made up of discrete tissues or organs called *endocrine glands*, as well as scattered cells in other organs [FIGURE 18.1]. These glands and specific cells are *ductless*, secreting products called hormones into extracellular spaces from which they enter the bloodstream and circulate throughout the body to their target areas. In contrast, the secretions of *exocrine glands*, such as sweat and salivary glands, empty directly into *ducts* that transport them to specific locations.

HORMONES AND THEIR FEEDBACK SYSTEMS

A *hormone* is a specialized chemical substance produced and secreted by an endocrine cell or tissue. Hormones are effective only at specific *target cells* — cells with compatible receptors on the surface of the plasma membrane or within the cytoplasm or nucleus.

Hormones may be classified according to the cells they act upon: (1) Hormones that are released into the bloodstream and interact with distinct target cells are called *endocrines*. (2) *Paracrines* are local hormones that act on nearby cells. (3) *Autocrines* are local hormones that act on the cell that released them.

Hormones are effective in extremely small amounts. Only a few molecules of a hormone may be enough to produce a dramatic response in a target cell. Typically, one hormone influences, depends on, and balances another in a controlling feedback system.

Biochemistry of Hormones

Hormones are either (1) lipid-soluble steroids, (2) derivatives of amino acids (biogenic amines), (3) water-soluble proteins and peptides, or (4) eicosanoids (prostaglandins and leukotrienes) [TABLE 18.1]. The type of secretion depends on the embryonic origin of the gland. Glands derived from endoderm secrete proteins, glands derived from ectoderm secrete biogenic amines, and glands derived from mesoderm secrete steroids.

Hormones are synthesized from raw materials in the cell and secreted into the extracellular space, usually entering the bloodstream by way of capillaries that flow between the cells of the gland. Steroid hormones, being

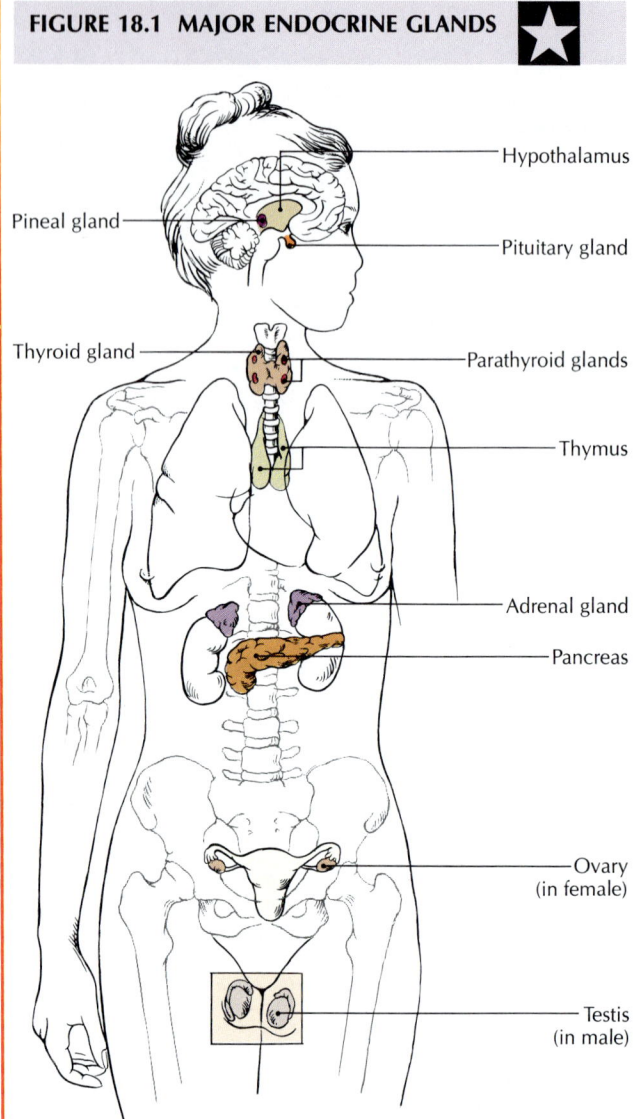

FIGURE 18.1 MAJOR ENDOCRINE GLANDS

Structure	Main function
Hypothalamus	Regulates anterior pituitary hormones
Pineal gland	May affect sleep
Pituitary gland	Regulates growth and various metabolic activities of other endocrine glands
Thyroid gland	Controls rate of metabolism
Parathyroid glands	Regulate levels of calcium and phosphate
Thymus	Processes developing T and B cells
Adrenal gland	Affects metabolism, blood pressure, sodium and potassium levels
Pancreas	Regulates blood glucose levels
Ovaries	Produce ova and female sex hormones
Testes	Produce sperm and male sex hormones

TABLE 18.1 CHEMICAL CLASSES OF HORMONES

Chemical class	Representative examples	Source
STEROIDS	Estrogens, progesterone	Ovaries
	Testosterone	Testes
	Aldosterone, cortisol	Adrenal cortex
BIOGENIC AMINES	Thyroxine, triiodothyronine	Thyroid gland
	Catecholamines (epinephrine, norepinephrine)	Adrenal medulla
PROTEINS AND PEPTIDES	Insulin	Pancreas
	Oxytocin	Posterior pituitary
	Growth hormone	Anterior pituitary
EICOSANOIDS	Prostaglandins, leukotrienes, thromboxanes	Most cells

A feedback system that produces a response that reduces the initiating stimulus is called a ***negative feedback system*** [FIGURE 18.2]. The system is called negative because its response to the initial stimulus is a *negative* action (or *opposite* to the initial stimulus). In contrast, a ***positive feedback system*** is one in which the initial stimulus is reinforced. That is, the response intensifies the initiating stimulus. Positive feedback systems are relatively rare in our bodies, mainly because they usually lead to instability. An example of positive feedback occurs during childbirth. The hormone oxytocin stimulates the contraction of the uterus, which stimulates a further release of oxytocin.

With some endocrine glands, negative feedback systems monitor the level of cellular activity and alter the amount of hormone secreted to maintain homeostasis. For example, suppose the rate of chemical activity in the cells (the *metabolic rate*) slows down [FIGURE 18.2]. The hypothalamus responds to this lowered rate by releasing *thyrotropin-releasing hormone* (TRH), which causes the anterior pituitary gland to secrete *thyroid-stimulating hormone* (TSH). This hormone causes the thyroid gland to secrete other hormones, *thyroxine* and *triiodothyronine*.

lipid-soluble, are poorly soluble in the water portion of plasma and are transported in the blood in combination with plasma proteins. Some peptide hormones (thyroid hormones) also require *carrier proteins* to transport them to their target cells. Because protein hormones are *water-soluble*, many are transported freely in the circulating blood.

Hormones help regulate the *rate* of biochemical reactions [TABLE 18.2]. Like enzymes, hormones are not changed by the reactions they regulate. The secretion rate of hormones varies from one gland to another and also for each gland, depending on physiological conditions. For example, the hormone epinephrine (commonly called adrenaline) is secreted by the adrenal glands in response to an emergency that calls for an immediate increase of blood flow to the muscles. In contrast, ovarian hormones are secreted cyclically over extended periods of time to regulate the ongoing menstrual cycle.

Feedback Control System of Hormone Secretion

Although hormones are always present in various amounts in endocrine glands, they are not secreted continuously. Instead, endocrine glands tend to secrete only the amount of hormones that the body needs to maintain homeostasis. This regulation of homeostasis occurs through a ***feedback control system*** in which changes in the body or the environment are detected by a central control unit (such as the brain), where the adjustments to maintain homeostasis are made.

FIGURE 18.2 NEGATIVE FEEDBACK AND METABOLIC RATE

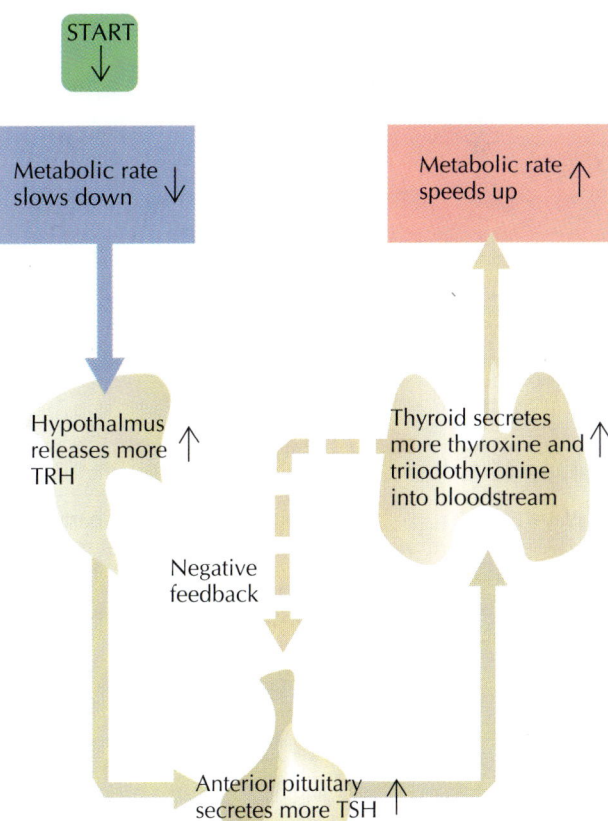

START

Metabolic rate slows down ↓

Metabolic rate speeds up ↑

Hypothalmus releases more TRH ↑

Thyroid secretes more thyroxine and triiodothyronine into bloodstream ↑

Negative feedback

Anterior pituitary secretes more TSH ↑

TABLE 18.2 MAJOR ENDOCRINE GLANDS: HORMONES AND FUNCTIONS

Gland	Hormone	Function of hormone	Means of control
HYPOTHALAMUS	Releasing and inhibiting hormones (CRH, TRH, GnRH, GHRH, GHIH, PRH, PIH)	Control release of anterior pituitary hormones.	Blood and local levels of specific substances
POSTERIOR PITUITARY (NEUROHYPOPHYSIS)	Antidiuretic hormone (ADH; vasopressin)	Increases water absorption from kidney tubules; raises blood pressure.	Synthesized in hypothalamus, released from neurohypophysis
	Oxytocin	Stimulates contractions of pregnant uterus, milk ejection from breasts after childbirth.	Synthesized in hypothalamus, released from neurohypophysis
ANTERIOR PITUITARY (ADENOHYPOPHYSIS)	Growth hormone (GH; somatotropic hormone, STH)	Stimulates growth of bone, muscle; promotes protein synthesis, fat mobilization; slows carbohydrate metabolism.	Hypothalamic growth-hormone releasing hormone (GHRH); growth-hormone inhibiting hormone (GHIH)
	Prolactin	Promotes breast development during pregnancy, milk production after childbirth.	Hypothalamic prolactin-inhibiting hormone (PIH); prolactin-releasing hormone (PRH)
	Thyroid-stimulating hormone (TSH)	Stimulates production and secretion of thyroid hormones.	Hypothalamic thyrotropin-releasing hormone (TRH)
	Adrenocorticotropic hormone (ACTH)	Stimulates production and secretion of adrenal cortex steroids (glucocorticoids).	Hypothalamic corticotropin-releasing hormone (CRH)
	Luteinizing hormone (LH)	*Female:* stimulates development of corpus luteum, release of oocyte, production of progesterone and estrogen. *Male:* stimulates secretion of testosterone, development of interstitial tissue of testes.	Hypothalamic gonadotropin-releasing hormone (GnRH)
	Follicle-stimulating hormone (FSH)	*Female:* stimulates growth of ovarian follicle, ovulation. *Male:* stimulates sperm production.	Hypothalamic gonadotropin-releasing hormone (GnRH)
	Melanocyte-stimulating hormone (MSH)	Apparently involved with skin color (melanocytes); role uncertain.	Uncertain
THYROID (FOLLICULAR CELLS)	Thyroid hormones: thyroxine (T_4), triiodothyronine (T_3)	Increase metabolic rate, sensitivity of cardiovascular system to sympathetic nervous activity; affect maturation, homeostasis of skeletal muscle.	Thyroid-stimulating hormone (TSH) from adenohypophysis; TSH regulated by thyrotropin-releasing hormone (TRH) from brain
THYROID (PARAFOLLICULAR	Calcitonin (CT)	Lowers blood calcium and phosphate concentrations.	Blood calcium concentration
PARATHYROID	Parathormone (PTH; parathyroid hormone)	Increases blood calcium concentration, decreases blood phosphate level.	Blood calcium concentration
ADRENAL MEDULLA	Epinephrine (adrenaline)	Increases heart rate; blood pressure; regulates diameter of arterioles; stimulates contraction of smooth muscle. Raises blood glucose levels.	Sympathetic nervous system
	Norepinephrine (noradrenaline)	Constricts arterioles; increases metabolic rate.	Sympathetic nervous system

Gland	Hormone	Function of hormone	Means of control
ADRENAL CORTEX	Glucocorticoids, mainly cortisol (hydrocortisone), corticosterone, 11-deoxycorticosterone	Affect metabolism of all nutrients; regulate blood glucose concentration; affect growth; decrease effects of stress, ACTH secretion, anti-inflammatories.	Corticotropin-releasing hormone (CRH) from hypothalamus, ACTH from adenohypophysis
	Mineralocorticoids, mainly aldosterone	Increase sodium retention and potassium loss in kidney tubules.	Angiotensin II, blood potassium concentration
	Gonadocorticoids (adrenal sex hormones)	Slight effect on ovaries and testes.	ACTH
PANCREAS (BETA CELLS IN PANCREATIC ISLETS)	Insulin	Lowers blood glucose by facilitating glucose transport across plasma membranes and increasing glycogen storage; affects muscle, liver, adipose tissue.	Blood glucose concentration
PANCREAS (ALPHA CELLS IN PANCREATIC ISLETS)	Glucagon	Increases blood glucose concentration.	Blood glucose concentration
OVARIES (FOLLICLE)	Estrogens	Affect development of sex organs and female characteristics.	Follicle-stimulating hormone (FSH)
OVARIES (CORPUS LUTEUM)	Progesterone, estrogens	Influence menstrual cycle; stimulate growth of uterine wall; maintain pregnancy.	Luteinizing hormone (LH)
PLACENTA	Estrogens, progesterone, human chorionic gonadotropin (hCG)	Maintain pregnancy.	Uncertain
TESTES	Androgens, mainly testosterone	Affect development of sex organs and male characteristics; aid sperm production.	Luteinizing hormone (LH)
THYMUS	Thymosin alpha, thymosin B_1 to B_5, thymopoietin I and II, thymic humoral factor (THF), thymostimulin, factor thymic serum (FTS)	Help develop T cells in thymus, maintain T cells in other lymphoid tissue; involved in development of some B cells into antibody-producing plasma cells.	Uncertain
DIGESTIVE SYSTEM	Secretin	Stimulates release of pancreatic juice to neutralize stomach acid.	Acid in small intestine
	Gastrin	Produces digestive enzymes and hydrochloric acid in stomach.	Food entering stomach
	Cholecystokinin (CCK)	Stimulates release of pancreatic enzymes, gallbladder contraction.	Food in duodenum
HEART	Atrial natriuretic peptide (ANP)	Helps maintain balance of fluids, electrolytes; decreases blood pressure and volume.	Salt concentration, blood pressure, blood volume
PINEAL GLAND	Melatonin	Antigonadotropic effect; exposure to light decreases release, darkness increases release.	Exposure to light

These hormones increase the metabolic rate, restoring homeostasis. Conversely, if the metabolic rate of the body increases, the hypothalamus secretes less TRH, thus reducing the anterior pituitary's secretion of TSH. As a result, the thyroid gland secretes less thyroxine and triiodothyronine, and the metabolic rate decreases.

MECHANISMS OF HORMONE CONTROL

The function of a hormone is to modify the physiological activity of a target cell or tissue. Because hormones are secreted from endocrine glands into the bloodstream and have their effects elsewhere, all hormones are considered to be chemical "messengers." (Chemical messengers such as hormones that function *outside* a target cell, in the extracellular fluid, are called "first messengers." Others that function *inside* a cell are called "second messengers.")

Much remains to be learned about how hormones work, but two mechanisms seem to operate for most hormones: (1) The first mechanism applies to hormones that are proteins, peptides, or catecholamines. Since they are water-soluble and cannot diffuse through the plasma membrane easily, they bind to specialized fixed receptors on the outer surface of the plasma membrane of the target cell. (2) The second mechanism applies mainly to steroid and thyroid hormones. Since they are lipid-soluble and diffuse freely into the cytoplasm, they initiate their physiological response by binding to cytoplasmic receptors located *inside* the target cell.

Fixed-Membrane-Receptor Mechanism for Protein and Peptide Hormones

Water-soluble protein and peptide hormones do not enter cells readily because of the resistance of the lipid bilayer of the plasma membrane. According to the *fixed-membrane-receptor mechanism* (also called the *second-messenger mechanism*), a water-soluble hormone is secreted by an endocrine gland and circulates through the bloodstream until it reaches its target organ [FIGURE 18.3]. At the cells of the target organ, the hormone, or "first messenger," binds to specific receptor sites on the plasma membrane [event **1** in FIGURE 18.3].

The hormone-receptor complex stimulates a nearby G protein (**2**), which links the receptors on the outer surface of the plasma membrane to the enzyme *adenylyl cyclase* on the inner surface (**3**). The activated adenylyl cyclase converts ATP into the "second messenger," *cyclic AMP* (cAMP, cyclic adenosine 3′,5′-monophosphate)*

---

*Cyclic AMP is called "cyclic" because of the ring structure formed from phosphate groups when ATP is converted into cAMP.

FIGURE 18.3 FIXED-MEMBRANE-RECEPTOR MECHANISM OF HORMONE ACTION

(1) The hormone, or "first messenger," circulates through the bloodstream until it reaches its target organ, where it binds to its specific receptor. **(2)** The hormone-receptor complex stimulates a G protein, which activates the enzyme adenylyl cyclase in the plasma membrane **(3)**. **(4)** The activated adenylyl cyclase converts ATP into the "second messenger," cyclic AMP. **(5)** Cyclic AMP diffuses throughout the cell and activates a cascade of enzymatic reactions that cause the cell to carry out its specific physiological function.

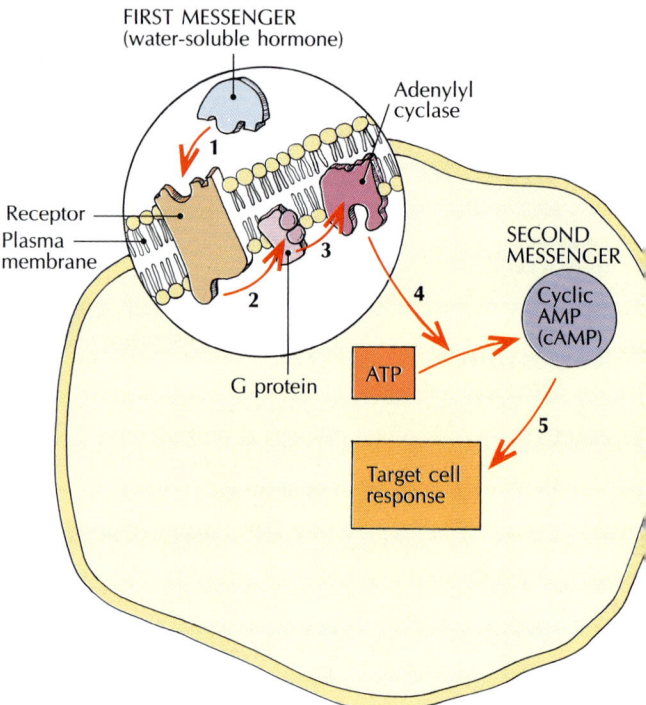

(4). A single molecule of a hormone may cause the production of thousands of molecules of cAMP. Cyclic AMP activates protein kinases that trigger a cascade of enzymatic reactions **(5)** that enzymatically induce the cell to perform its specific functions. An increase in intracellular cAMP may cause many different biochemical changes within the cell. Cyclic AMP amplifies the signal that causes the cell to perform its specific function in the production of energy in liver cells, hormone secretion by thyroid cells, or increased contraction of cardiac muscle.

After the cell responds with its specific physiological function, cAMP is inactivated by one or more enzymes called *phosphodiesterases*. The receptor sites on the plasma membrane of the target cell now become available for new reactions as the cAMP is inactivated and the hormone level declines.

Mobile-Receptor Mechanism of Steroid Hormone Action

Because steroid hormones are lipid-soluble and pass easily through the plasma membrane, their receptors are *inside* the target cells. The hormonal mechanism for steroid hormones, called the ***mobile-receptor mechanism,*** or the *direct-gene mechanism*, involves the stimulation of protein synthesis [FIGURE 18.4].

Cortisol, aldosterone, progesterone, estrogen, testosterone, and thyroid hormones are mobile-receptor hormones and have their receptors localized in the cytoplasm. All the other hormones usually use the fixed-membrane-receptor mechanism.

A S K   Y O U R S E L F

1 What is a hormone? A target cell?

2. What is the main functional difference between negative feedback and positive feedback?

3 If a hormone is the "first messenger," what is the "second messenger"?

4 What is the function of a receptor site on a plasma membrane?

5 What kind of hormones stimulate protein synthesis within a cell?

PITUITARY GLAND (HYPOPHYSIS)

The ***pituitary gland,**** also known as the ***hypophysis*** (hye-PAHF-uh-sihss; Gr. "undergrowth"), is located directly below the hypothalamus and rests in the *sella turcica*, a depression in the sphenoid bone [FIGURE 18.5]. This important gland is protected on three sides by bone and on top by a tough membrane. It is about 1.0 cm long, 1.0 to 1.5 cm wide, and 0.5 cm thick—about the size and shape of a plump lima bean. Because of the closeness of the pituitary to the optic chiasma, an enlarged pituitary may affect vision by impinging on the optic pathways.

The pituitary gland has two anatomically and functionally distinct lobes—the anterior lobe and the posterior lobe—and is actually two distinct glands.

The Anterior and Posterior Pituitary Lobes

The ***anterior pituitary*** is composed of glandular epithelial tissue derived embryonically from an outpouching that buds off from the roof of the mouth (Rathke's

---

*The pituitary gland received its name, which means "mucus" in Latin, because it was thought to transfer mucus from the brain into the nose through the cribriform plate of the ethmoid bone.

FIGURE 18.4 MOBILE-RECEPTOR MECHANISM OF STEROID HORMONE ACTION

(1) The steroid hormone diffuses from the blood into the target cell. **(2)** It binds to a specific steroid-receptor protein in the target-cell cytoplasm. **(3)** The hormone-receptor complex enters the nucleus and binds to DNA. **(4)** The binding of the hormone-receptor complex to DNA initiates the transcription of specific genes on the DNA and produces new mRNA. **(5)** The newly transcribed mRNA moves to the cytoplasm and initiates protein synthesis. **(6)** New proteins alter the cell's functions.

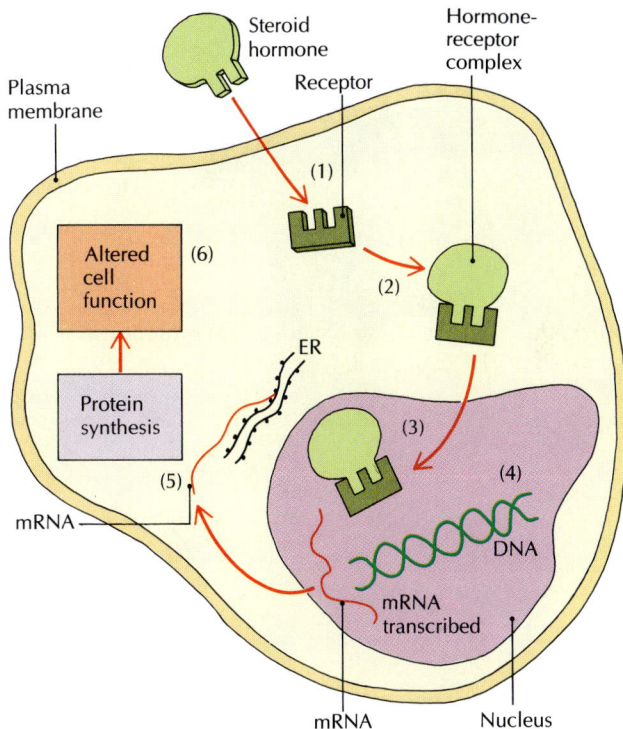

pouch). It is also known as the ***adenohypophysis*** (ADD-ee-noh-hye-POFF-ih-sihss; *adeno-* means "glandular") because it contains many glandular epithelial cells. The anterior pituitary is the larger of the two lobes, accounting for about 75 percent of the total weight of the pituitary gland.

The ***posterior pituitary*** is derived embryonically from an outgrowth of the hypothalamus, is composed of nervous tissue, and is appropriately termed the ***neurohypophysis***. It consists of the axons and axon terminals of thousands of neurons that have their cell bodies in the supraoptic and paraventricular nuclei of the hypothalamus.

The anterior and posterior pituitary lobes have only their location in common. The anterior pituitary is connected to the hypothalamus by a vascular link, while the posterior pituitary has a neural link through a stalk of nerve cells and blood vessels called the ***infundibular***

FIGURE 18.5 LOCATION OF THE PITUITARY GLAND

[A] Relationship of the pituitary gland to the brain; right medial view. [B] Enlarged drawing showing the relationship of the pituitary gland to the hypothalamus and sella turcica.

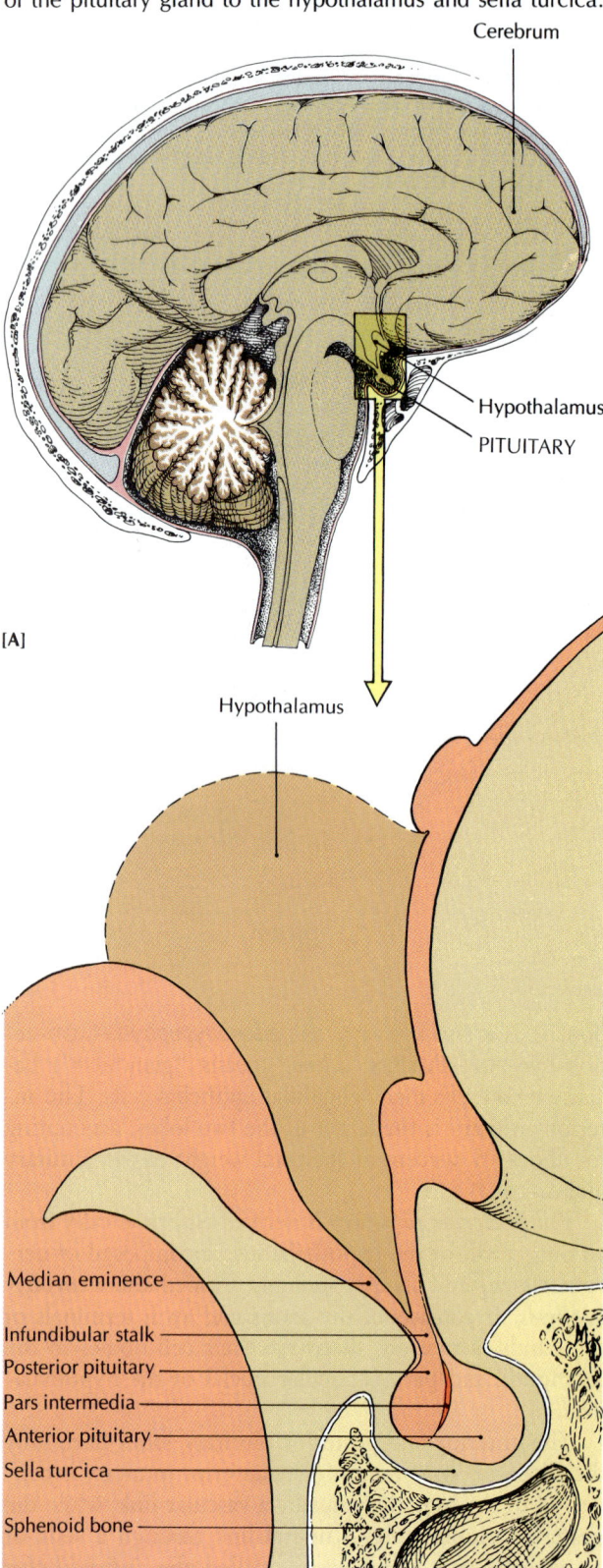

stalk, or *infundibulum* (in-fuhn-DIBB-yuh-luhm; L. funnel) [FIGURE 18.6]. Between the two lobes is a small zone called the *pars intermedia* (intermediate lobe), whose function in humans is unknown.

Relationship between the Pituitary and Hypothalamus

Although the daily secretions of the pituitary gland are less than one-millionth of a gram, the pituitary was once called the "master gland" because of its control over most of the other endocrine glands and body organs [FIGURE 18.7]. In truth, the **hypothalamus** might better deserve the title "master gland," since hormones released from the hypothalamus control the secretions of the adenohypophysis, and hormones secreted from the neurohypophysis are synthesized and regulated by neuronal centers in the hypothalamus.

The hypothalamus and the adenohypophysis are connected by an extensive system of blood vessels called the **hypothalamic-hypophyseal portal** (blood vessel network) **system** [see FIGURE 18.6]. Hormones produced in the hypothalamus are transported through the portal vessels to the adenohypophysis, where they either stimulate or inhibit the release of the appropriate pituitary hormones.

The link between the hypothalamus and the neurohypophysis relies on nerve impulses, hence the name *neuro*hypophysis for the posterior pituitary. The neurohypophysis is composed of unmyelinated axons of neurons whose cell bodies are in the hypothalamus, and pituicytes, which have a supporting rather than a secretory function.

Hormones of the Neurohypophysis

The neurohypophysis does not actually manufacture hormones. Instead, hormones synthesized in the cell bodies of the hypothalamus are transported down the nerve fibers as secretory vesicles and pass through the infundibular stalk to axon terminals lying on capillaries in the neurohypophysis. The secretory granules are stored in the terminals until nerve impulses from the hypothalamus stimulate the secretion of hormones. Hypothalamic nerve cells are called *neurosecretory cells* because they are neurons with a secretory ability.

The nerve endings from the hypothalamus secrete two hormones — *antidiuretic hormone* (ADH) and *oxytocin* — into the capillaries. ADH is produced by certain neurons and oxytocin by other neurons, in both the *supraoptic nucleus* and the *paraventricular nucleus* [FIGURE 18.7]. Thus the hormones of the neurohypophysis are actually neurosecretions from hypothalamic nerve cells.

Antidiuretic hormone A *diuretic* is a substance that stimulates the excretion of urine, whereas an *antidiuretic*

FIGURE 18.6 RELATIONSHIP OF HYPOTHALAMUS AND PITUITARY GLAND

The blood vessels that make up the hypothalamic-hypophyseal (pituitary) portal system pass through the infundibular stalk and thus provide the link between the hypothalamus and the adenohypophysis.

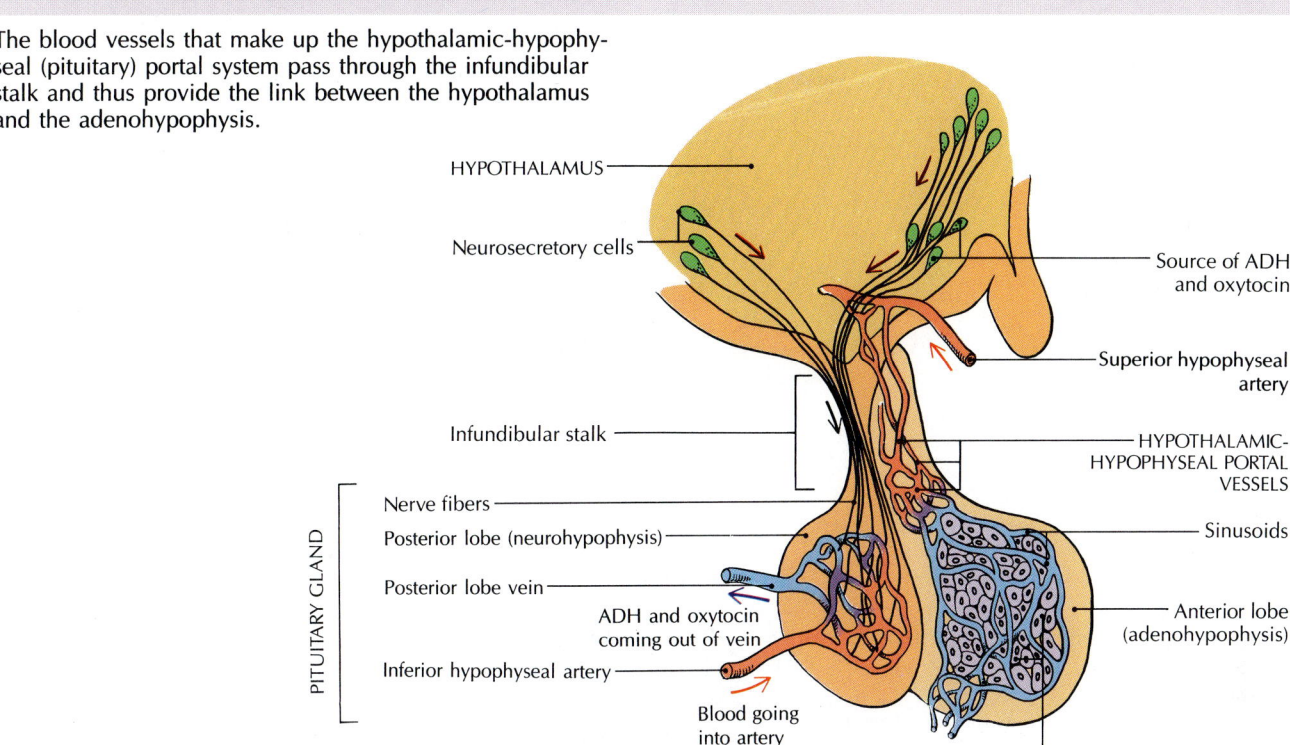

decreases urine excretion. ***Antidiuretic hormone (ADH),*** or *vasopressin,** is a peptide composed of nine amino acids. Its main role is to help regulate the body's fluid balance.

The main target organs of ADH are the kidneys. ADH exerts its effects by increasing the permeability of the kidney tubules to water, thereby allowing more water to be reabsorbed into the body rather than excreted in urine. Such a regulating mechanism is important if the body loses volume, as in blood loss (hemorrhaging). Hemorrhaging causes an increase in ADH secretion so that water is reabsorbed in an attempt to increase the body fluids back toward normal. Emotional or physical stress, increased plasma osmotic pressure,† decreased extracellular fluid volume, very strenuous exercise, and drugs such as nicotine and barbiturates all *increase* the secretion of ADH, thus increasing water reabsorption by the kidneys and decreasing urine formation.

The secretion of ADH is *decreased* by a drop in plasma osmotic pressure, by increased extracellular fluid volume,

or by increased alcohol levels. If you consume large quantities of beer, for example, you will increase urine output, not only because you increase the liquid content of your body, but also because the alcohol in the beer inhibits the secretion of ADH, resulting in an increase in urine excretion. In addition, another component of beer, lupulin from hops, is a diuretic.

A deficiency of ADH results in a disease known as *diabetes insipidus.*‡ Without any ADH, large volumes of dilute urine are excreted, usually about 23 L/day.

Oxytocin *Oxytocin* (AHK-suh-TOE-sihn; Gr. "sharp childbirth") is a hormone that stimulates uterine contractions during childbirth and milk ejection (for feeding the infant) after childbirth. Oxytocin works during childbirth by stimulating smooth muscle contractions in the uterus. The uterus becomes very sensitive to oxytocin during the late stages of pregnancy, and secretions increase during the actual process of childbirth. The attending physician may even give synthetic oxytocin (Pitocin) intravenously to stimulate labor.

The ejection of milk from the mammary glands occurs after childbirth, when oxytocin stimulates myoepithelial cells that surround the ducts of the breasts. The myoepi-

---

*ADH is also known as vasopressin because it also causes the constriction of arterioles (small arteries), thus raising blood pressure. This is beneficial during hemorrhaging when blood pressure falls below normal.

†Specific cells in the hypothalamus act as osmoreceptors that respond to changes in the osmotic pressure (caused by the presence of solutes such as electrolytes) of the interstitial fluid.

‡This form of diabetes is not to be confused with *diabetes mellitus*; see When Things Go Wrong, page 577.

FIGURE 18.7 HORMONES RELEASED BY THE HYPOTHALAMUS AND PITUITARY GLAND

Target areas for each hormone are shown under the relevant box.

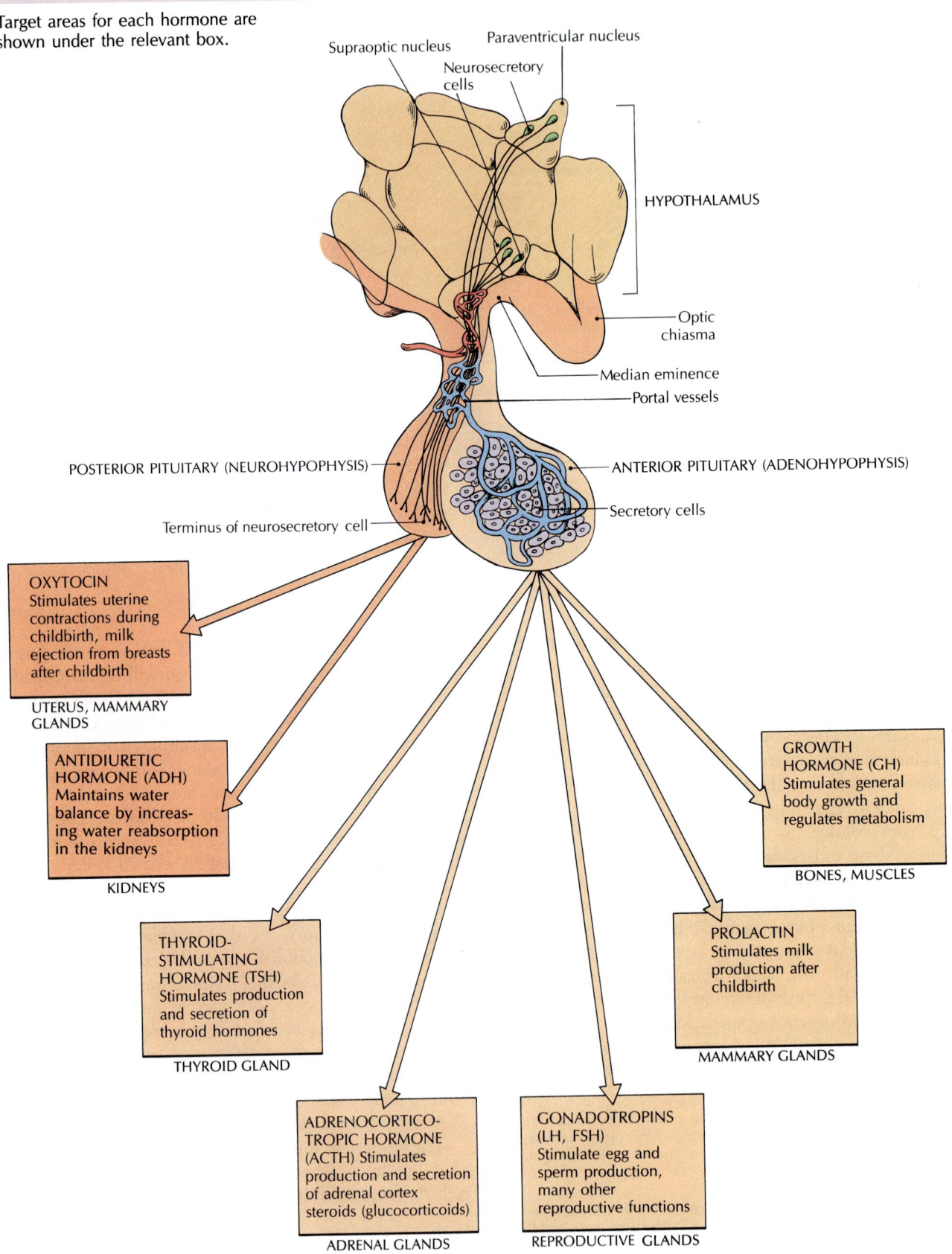

thelial cells contract, forcing milk out of the network of milk-containing sinuses (alveoli) in the breasts and through ducts in the nipples. Suckling by the newborn infant transmits impulses to the brain that stimulate nerve cells in the hypothalamus to cause the release of oxytocin. The infant does not receive any milk for about a minute after it starts suckling, but once oxytocin is secreted, the contraction of the myoepithelial cells begins, and milk flows from the nipple. Although oxytocin stimulates milk *ejection*, the actual *production* of milk is stimulated by *prolactin*, a pituitary hormone secreted by the adenohypophysis.

Like ADH, oxytocin is a peptide composed of nine amino acids. The structure of the two hormones is similar except for a difference in two of the amino acids.

Hormones of the Adenohypophysis

The true endocrine portion of the pituitary is the adenohypophysis, which synthesizes seven separate hormones. All hormones secreted by the adenohypophysis are polypeptides, and all but two are considered to be *tropic* (TROE-pihk) *hormones*, hormones whose primary target is another endocrine gland. (*Tropic* is an adjective derived from the Greek word *trophikos*, meaning "turning toward" or "to change.") The two nontropic hormones are growth hormone and prolactin, which are involved with growth and milk production, respectively.

Growth hormone *Growth hormone (GH)* is also called *somatotropin* or *somatotropic hormone* (STH). Rather than influencing a specific target organ, it affects all parts of the body that are associated with growth. Growth hormone has its most dramatic effect on the growth rate of children and adolescents, increasing tissue mass and stimulating cell division.

The release of growth hormone from the anterior pituitary (adenohypohysis) is controlled by two hormones produced by the hypothalamus: (1) *growth-hormone releasing hormone* (GHRH), which stimulates release, and (2) *growth-hormone inhibiting hormone* (GHIH), or *somatostatin*, which inhibits the release of growth hormone. Growth hormone is usually released from the anterior pituitary in a pulsatile rhythm.

The most direct effect of growth hormone is to maintain the epiphyseal disks of the long bones. If a young person is deficient in growth hormone, there is a premature closure of the epiphyseal disks, the body stops growing, and *dwarfism* results. In contrast, if the secretion of growth hormone does not decrease toward the end of adolescence, as it usually does, *giantism* occurs, and the person continues to grow to 7 or even 8 ft tall. When growth hormone is overproduced after normal growth has stopped, a condition called *acromegaly* occurs, with bones in the head, hands, and feet thickening rather than

lengthening (see When Things Go Wrong at the end of the chapter).

Growth hormone stimulates the growth rate of body cells by increasing the formation of RNA, which increases the rate of protein synthesis. At the same time, growth hormone decreases the breakdown of proteins. It also causes a shift from the use of carbohydrates for energy to the use of fats. By increasing the use of the body's adipose tissue to produce fatty acids and decreasing the body's use of glucose, growth hormone conserves carbohydrates. The overall effect is a change in the body's composition so that muscle mass is increased and deposits of fat are decreased. When blood glucose levels are increased by growth hormone, the result is *hyperglycemia*. A prolonged hyperglycemia may lead to excessive stimulation and early "burnout" (or failure to produce insulin) of the beta cells of the pancreas, which can cause a *diabetogenic* ("producing diabetes") *effect*, or the disease known as diabetes mellitus.

Recently, recombinant human growth hormone has become readily available (although it is expensive), and researchers are exploring many possible uses. They are testing its use in (1) boosting the responsiveness of the pituitary gland to GHRH during middle age, when responsiveness begins to slow down, (2) reversing some of the effects of aging, (3) improving healing in severely burned children, (4) inducing ovulation (in combination with sex-hormone treatment), (5) preventing osteoporosis, and (6) stimulating the development of some immune-system cells.

Prolactin The other hormone considered to be nontropic is **prolactin (PRL)**, or *lactogenic hormone*. Prolactin has two functions in females: (1) Together with the female sex hormone estrogen, it stimulates the development of the duct system in the mammary glands during pregnancy. (2) It stimulates milk production from the mammary tissue after childbirth. (Recall that oxytocin stimulates milk *ejection* from the mammary glands during breastfeeding, but not milk *production*.)

Because a woman is *not* pregnant most of the time, the secretion of prolactin is usually inhibited. Inhibition is accomplished by the secretion of *prolactin-inhibiting hormone* (PIH) (which is dopamine) by the hypothalamus. The inhibition of prolactin secretion diminishes during pregnancy, and by the time the baby is born and the placenta is expelled from the uterus, most of the inhibi-

---

Q: *Does milk production cease once the menstrual cycle resumes after pregnancy?*

A: The drive to produce food for offspring is so strong that prolactin levels remain high, and milk continues to be produced. This is true not only after the menstrual cycle starts again, but even if a woman becomes pregnant again, as long as she is still nursing her child.

tory effects of PIH are removed. In addition, the process of nursing a baby apparently causes the hypothalamus to secrete a *prolactin-releasing hormone* (PRH) which *stimulates* the secretion of prolactin.

Thyroid-stimulating hormone *Thyroid-stimulating hormone (TSH)* stimulates the synthesis and secretion of thyroid hormones in several ways. The main effect of TSH is to stimulate the secretion of the thyroid hormones, *thyroxine* (T$_4$) and *triiodothyronine* (T$_3$). An excessive amount of TSH may produce an enlarged thyroid gland, called a *goiter* (see When Things Go Wrong at the end of the chapter).

The secretion of TSH is controlled by *thyrotropin-releasing hormone* (TRH) produced by the hypothalamus. The secretion of TRH depends on the amount of thyroid hormones circulating in the blood. Low blood levels stimulate an increase in TRH. When a normal level of thyroid hormones is reached, the production of TRH returns to a rate that maintains a homeostatic condition. The production of TRH is also affected by blood levels of TSH, thyroid hormones, metabolic rate, and environmental temperature.

Adrenocorticotropic hormone *Adrenocorticotropic hormone (ACTH)* is also called *corticotropin*. It stimulates the adrenal cortex (the outer part of the adrenal glands) to produce and secrete steroid hormones called *glucocorticoids*. Secretions of ACTH are regulated by the liberation of corticotropin-releasing hormone (CRH) from the hypothalamus, which is in turn regulated by a feedback system influenced by such factors as stress, insulin, interleukin 1 (IL-1), and other hormones.

Luteinizing hormone *Luteinizing hormone (LH)* received its name from the *corpus luteum*, a temporary endocrine tissue in the ovaries that secretes the female sex hormones progesterone and estrogens. LH, a gonadotropic hormone, stimulates ovulation, the monthly release of a mature egg from an ovary. In the male, the same hormone used to be called *interstitial cell-stimulating hormone* (ICSH), but it now is called luteinizing hormone in both sexes. Its target cells in the male are interstitial ("between spaces") cells in the testes that secrete the male hormone testosterone. The mechanism for the control of LH depends on a specific gonadotropin-releasing hormone (GnRH) from the hypothalamus, which is regulated by a typical negative feedback

---

Q: *How can LH and FSH affect both male and female organs?*

A: The same hormone is able to stimulate similar kinds of actions in two very different glands (testes and ovaries) because male and female sex organs have the same embryonic origin. Thus FSH is associated with sex cell production in males and females, and LH is associated with sex hormone secretion in males and females.

---

system involving levels of progesterone, estrogens, and testosterone.

Follicle-stimulating hormone *Follicle-stimulating hormone (FSH)* is also a gonadotropic hormone of the adenohypophysis. In females, FSH stimulates the growth of follicle cells in the ovaries during each menstrual cycle; it also stimulates the follicle cells to secrete estrogen. In the male, FSH stimulates the cells of the testes that produce sperm. The regulatory mechanism for FSH is similar to the negative feedback systems of other hormones of the adenohypophysis. The specific regulating factor from the hypothalamus is gonadotropin-releasing hormone (GnRH).

Melanocyte-stimulating hormone In humans, the exact hormonal function of *melanocyte-stimulating hormone (MSH)* is uncertain. It may play a role in skin darkening, since the skin becomes very pale without MSH. Two hypothalamic hormones regulate the release of MSH: (1) MSH-releasing hormone (MRH) increases MSH release, and (2) MSH-inhibiting hormone (MIH) suppresses MSH release.

ASK YOURSELF

1 What are the two lobes of the pituitary gland called?

2 What is the function of the hypothalamic-hypophyseal portal system?

3 What hormones are secreted by the adenohypophysis?

THYROID GLAND

The *thyroid gland* is located in the neck, anterior to the trachea [FIGURE 18.8A]. It consists of two lobes, one on each side of the junction between the larynx and trachea. The lobes are connected across the second and third tracheal rings by a bridge of thyroid tissue called an *isthmus* (IHSS-muhss). In about half the cases there is a *pyramidal process* extending upward from the isthmus.

The thyroid gland has a well-developed circulatory system (80–100 mL of blood/min) through which amino

---

Q: *How can the tissues of the adenohypophysis be specialized to synthesize so many different hormones?*

A: There are three types of cells in the adenohypophysis: acidophils, basophils, and chromophobes. *Acidophils,* which stain red, secrete growth hormone and prolactin. *Basophils,* which stain blue, secrete TSH, FSH, LH, and MSH and may be involved in the secretion of ACTH. The function of *chromophobes* is not firmly established, but they may be acidophils and basophils that have lost their granules after secreting, or reserve cells capable of differentiating into acidophils or basophils; they may also take part in the secretion of ACTH. Chromophobes stain little, if at all, hence their name, which means "fear of color."

FIGURE 18.8 THYROID GLAND

[A] Anterior view. [B] Photomicrograph of thyroid follicles. ×160. [C] Cross section of a thyroid follicle. Some parafollicular cells lie outside the follicle.

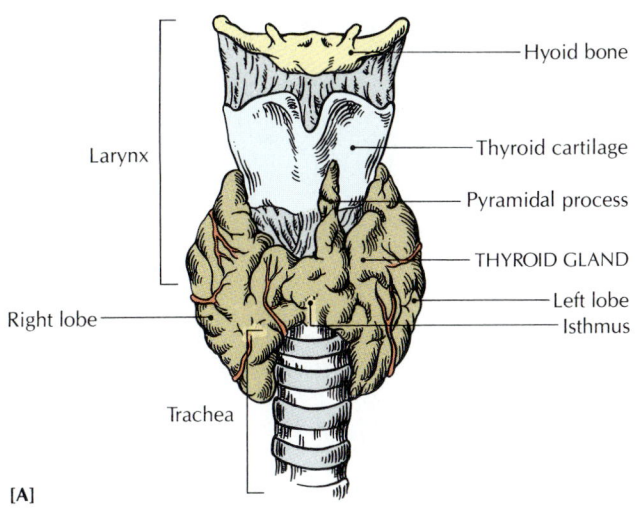

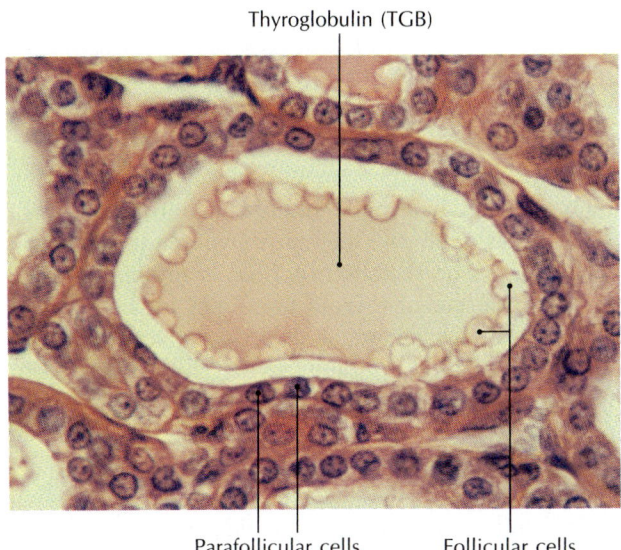

Thyroglobulin (TGB)

Parafollicular cells   Follicular cells

[B]

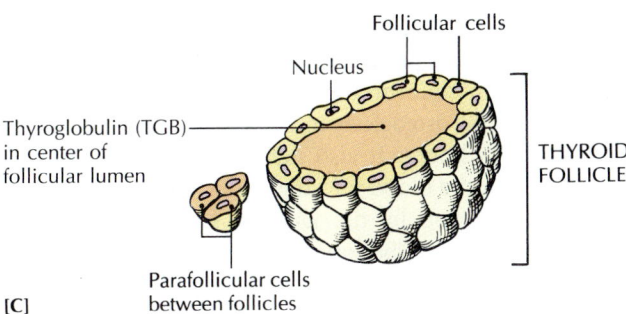

[C]

acids, iodine, gland secretions, and other substances are transported [FIGURE 18.8B]. The gland is composed of hundreds of thousands of spherical sacs, or *follicles*, which are filled with a gelatinous colloid in which the thyroid hormones are stored [FIGURE 18.8C]. The follicles are made up of a single layer of follicular cells.

Two types of cells make up the thyroid gland. The *follicular cells* are the most prevalent cells. They synthesize and secrete the thyroid hormones (thyroxine and triiodothyronine), which increase the rate of metabolism in most body cells. The *parafollicular*, or C (clear), *cells*, though usually larger than follicular cells, are not as plentiful. Clusters of parafollicular cells are found between follicles. Parafollicular cells synthesize and secrete the hormone *calcitonin*, which reduces the concentration of calcium in the blood.

Thyroxine and Triiodothyronine

Follicular cells secrete the most abundant thyroid hormone, *thyroxine* [FIGURE 18.9]. The thyroxine molecule, which contains four atoms of iodine, is often referred to as tetraiodothyronine, or T_4. Follicular cells also synthesize *triiodothyronine*, often called T_3 because its molecules contain only three atoms of iodine. Thyroxine and triiodothyronine are collectively called the *thyroid hormones*.

Although thyroxine accounts for about 90 percent of the thyroid's secretions, triiodothyronine is highly concentrated and is several times more potent. Both hormones consist mainly of iodine and the amino acid tyrosine. Iodine is obtained directly from dietary sources. People who live in inland areas where the iodine content of the water is low may become deficient in iodine if they do not have a supplementary source, such as iodized salt. Tyrosine is synthesized by the body from a wide variety of dietary sources, so there is no problem in obtaining it.

Both hormones appear to have the same endocrine functions. They accelerate cellular reactions in most body cells. They increase body metabolism (the rate at which cells use oxygen and organic molecules to produce energy and heat); cause the cardiovascular system to be more sensitive to sympathetic innervation, thus increasing cardiac output and heart rate; affect the maturation and homeostasis of the skeletal and central nervous systems; stimulate cellular differentiation and protein synthesis; and affect growth rate, water balance, and several other physiological processes.

In general, thyroid hormones increase the oxygen consumption of most body cells, but not in the brain, spleen, or testes. In so doing, energy is produced and body heat is given off in a temperature-raising process called a *calorigenic effect*. Thyroid hormones are essential for proper skeletal growth in children, and they operate together with growth hormone from the pituitary to reg-

FIGURE 18.9 THE SYNTHESIS OF THYROID HORMONES

Iodination of tyrosine takes place in the colloid of thyroid follicles, while the thyroid hormone molecules are bound in peptide linkage in thyroglobulin.

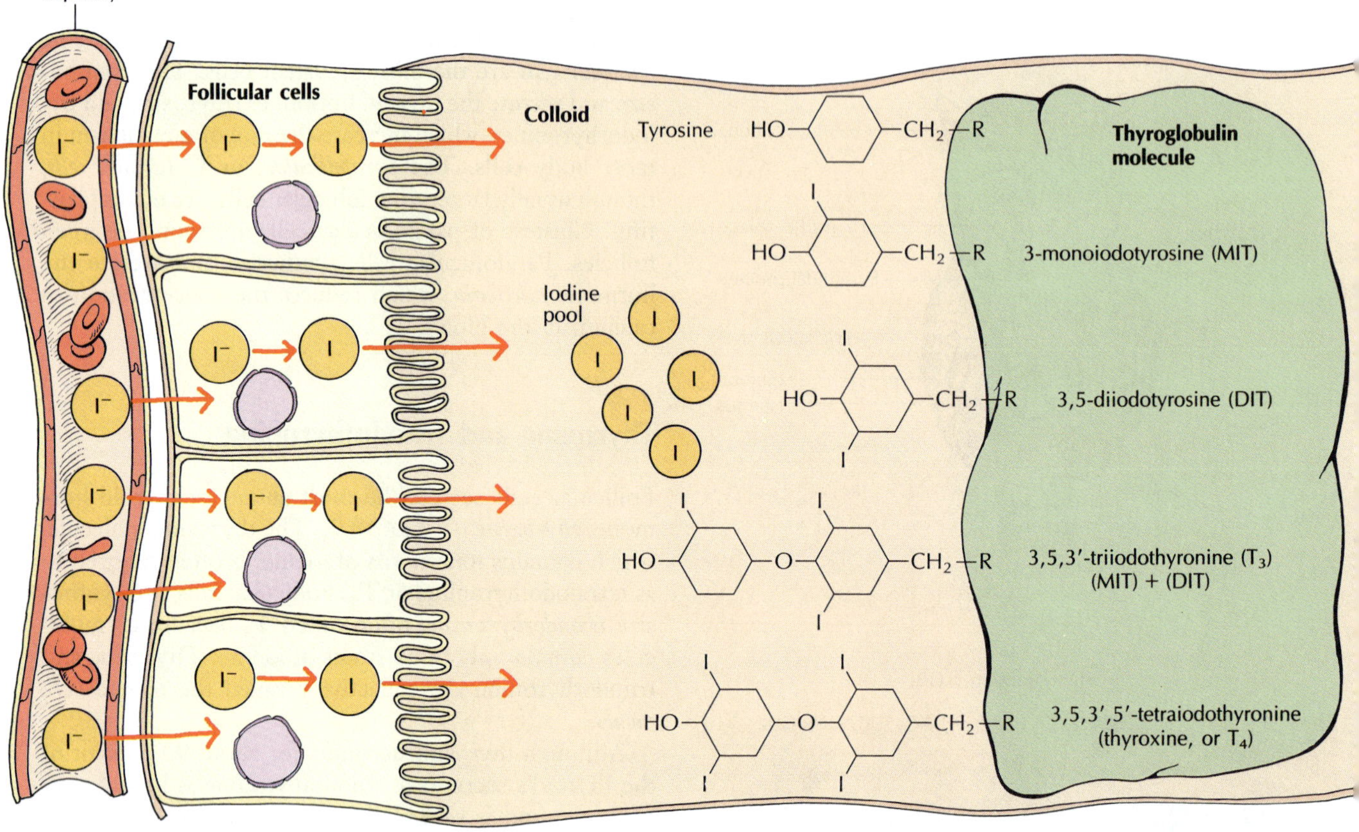

ulate overall growth and maturation, including the development of the nervous system.

Because the thyroid hormones affect most cells of the body rather than a specific target area, the possibilities of malfunctions are far-reaching and can be serious. Overactivity of the thyroid gland results in *hyperthyroidism*, which usually produces a goiter. Underactivity of the thyroid gland results in *hypothyroidism* and may also result in a goiter through an increased output of TSH. An underactive thyroid gland during prenatal development or infancy causes *cretinism*, which results in mental retardation and irregular development of bones and muscles. (Both conditions are discussed in When Things Go Wrong at the end of this chapter.)

The ionized form of iodine, *iodide* (I^-), is found in food. This trace element is taken up by the follicular cells of the thyroid gland as inorganic iodide. The iodide is bound and collected by a large glycoprotein called *thyroglobulin;* it is then used to iodinate tyrosine residues in the thyroglobulin to produce the thyroid hormones. When thyroid hormones are released, the iodinated tyrosine molecules are removed from the thyroglobulin and secreted into the bloodstream.

The entire process of thyroid production and secre-

tion is regulated by thyroid-stimulating hormone (TSH), which is secreted by the adenohypophysis [FIGURES 18.10, 18.2]. The secretion of TSH is stimulated when the concentration of thyroid hormones in the blood is low, and by such factors as cold, stress, or pregnancy. The output of TSH is inhibited when the blood levels of thyroid hormones are high.

Calcitonin

The parafollicular cells of the thyroid produce a polypeptide hormone called *calcitonin (CT)*. It lowers the calcium and phosphate concentrations in the blood, in direct contrast to the action of parathyroid hormone secreted by the parathyroid glands, which increases blood calcium. As the calcium concentration increases, so does the secretion of calcitonin, which then reduces the blood calcium concentration toward normal. Calcitonin acts directly on osteoclasts to reduce the reabsorption of calcium from bones. It also increases the movement of calcium from the blood into the bones and decreases calcium reabsorption by the kidneys. Calcitonin acts for short periods only; the long-term regulation of calcium in the blood is a function of parathyroid hormone. The

FIGURE 18.10 CONTROL OF THYROID HORMONE SECRETION

STRESS

COLD

Hypothalamus

↑ Thyrotropin-releasing hormone (TRH) released into hypophyseal portal veins

Low blood level of thyroxine, triiodothyronine

Anterior pituitary

↑ Thyroid-stimulating hormone (TSH) released into blood stimulates thyroid follicular cells

↑ Thyroxine, triiodothyronine released into blood by follicular cells

↑ Metabolic rate increases

secretion of calcitonin is directly regulated by the concentration of calcium in the blood rather than depending on the higher control of a hypothalamic-pituitary feedback system.

PARATHYROID GLANDS

The **parathyroid glands** are tiny lentil-sized glands embedded in the posterior of the thyroid lobes, usually two glands in each lobe [FIGURE 18.11A]. The parathyroid glands are so small that they were not discovered until 1850. Before then, the parathyroids were frequently removed unknowingly during goiter surgery, and the patients died "mysteriously." The adult parathyroids are composed mainly of small *principal cells*, which secrete most of the parathyroid hormone *(parathormone, PTH)*,

FIGURE 18.11 PARATHYROID GLANDS

[A] Posterior view. **[B]** Section of a human parathyroid gland showing the small principal cells and the larger oxyphilic cells. ×600.

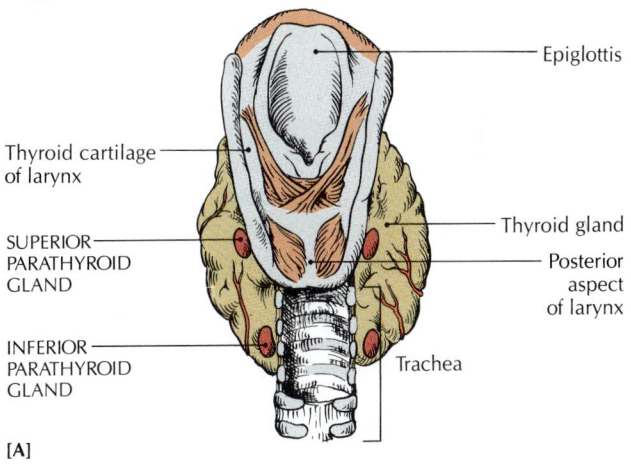

Epiglottis

Thyroid cartilage of larynx

SUPERIOR PARATHYROID GLAND

Thyroid gland

Posterior aspect of larynx

INFERIOR PARATHYROID GLAND

Trachea

[A]

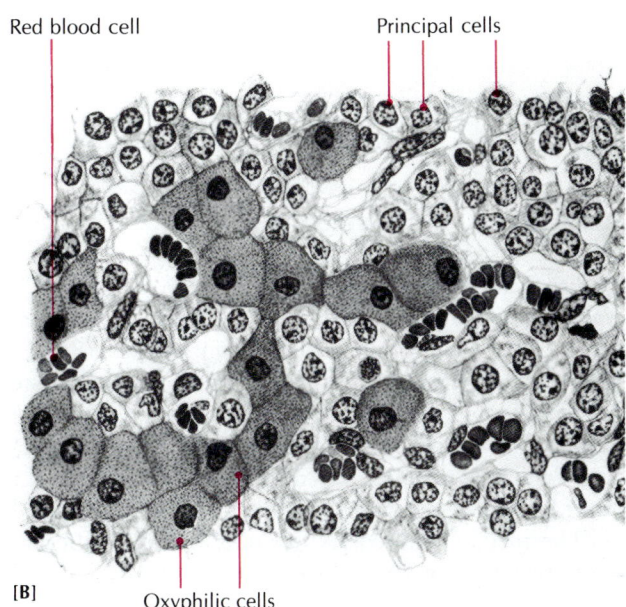

Red blood cell

Principal cells

Oxyphilic cells

[B]

and larger *oxyphilic cells* whose function is unknown [FIGURE 18.11B].

The main function of parathormone is to regulate the concentrations of calcium (Ca^{2+}) and phosphate (HPO_2^{4-}) in the blood. An imbalance of calcium and phosphate ions in the blood can cause altered transmission of nerve impulses, destruction of bone tissue, hampered bone growth, and muscle tetany.

When the concentration of calcium in the blood is low, parathormone *increases* the level of blood calcium in several ways: (1) It stimulates the activity of osteoclasts, which break down bone tissue (*osteolysis*), releasing calcium ions from bones into the blood. (2) It enhances the absorption of calcium and phosphate from the small intestine into the blood. Adequate dietary vitamin D and the hormone *1,25-dihydroxyvitamin D₃* (produced in the kidneys) are important for this process. (3) It promotes the reabsorption of calcium by the kidney tubules so that the amount of calcium excreted in urine is decreased.

Parathormone *decreases* the concentration of phosphate in the blood by inhibiting its reabsorption by the kidney tubules and thereby increasing its excretion in the urine. The decrease in blood phosphate tends to raise the concentration of calcium in the blood. The interaction between parathormone and calcitonin, in which the regulation of blood calcium levels is controlled by both PTH and calcitonin, is illustrated in FIGURE 18.12.

A S K   Y O U R S E L F

1 What effect does parathormone have on the levels of calcium and phosphate in the blood?

2 How do calcitonin, parathormone, and 1,25-dihydroxyvitamin D₃ work together in the body to regulate circulating calcium levels?

ADRENAL GLANDS

The two **adrenal** (uh-DREEN-uhl; L. "upon the kidneys") **glands** are triangular-shaped organs that rest like tilted berets on the superior tip of each kidney [FIGURE 18.13A]. Each adrenal gland is actually made up of two separate endocrine glands. The inner portion of each gland is the *medulla* (L. "marrow," meaning "inside"); the outer portion, which surrounds the medulla, is the *cortex* (L. "bark," as in the outer bark of a tree) [FIGURE 18.13B and C]. The medulla and cortex not only produce different hormones but also have separate target organs.

Adrenal Cortex

The **adrenal cortex** accounts for about 90 percent of the weight of the adrenal gland, which weighs from 5 to 7 g. Like other glands with a mesodermal embryonic origin, its hormones are steroids. The adrenal cortex secretes three classes of general steroid hormones: *glucocorticoids*, *mineralocorticoid*, and small quantities of sex hormones. The corticoids of the adrenal cortex primarily affect the

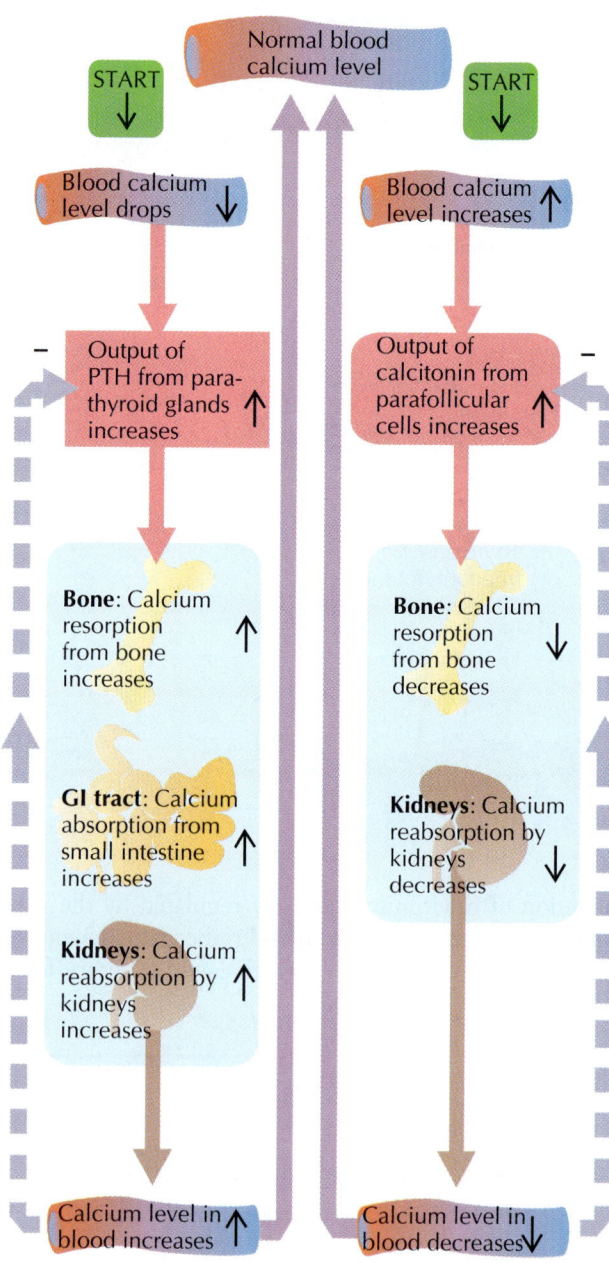

FIGURE 18.12 REGULATION OF CALCIUM LEVELS BY PARATHYROID HORMONE (PTH) AND CALCITONIN (CT)

FIGURE 18.13 ADRENAL GLANDS

[A] The adrenal glands rest on top of the kidneys.
[B] Cross section of the adrenal gland. [C] Enlarged section of the adrenal cortex of an adult male showing the layered regions from the outer capsule to the zona reticularis, which merges with the inner medulla.

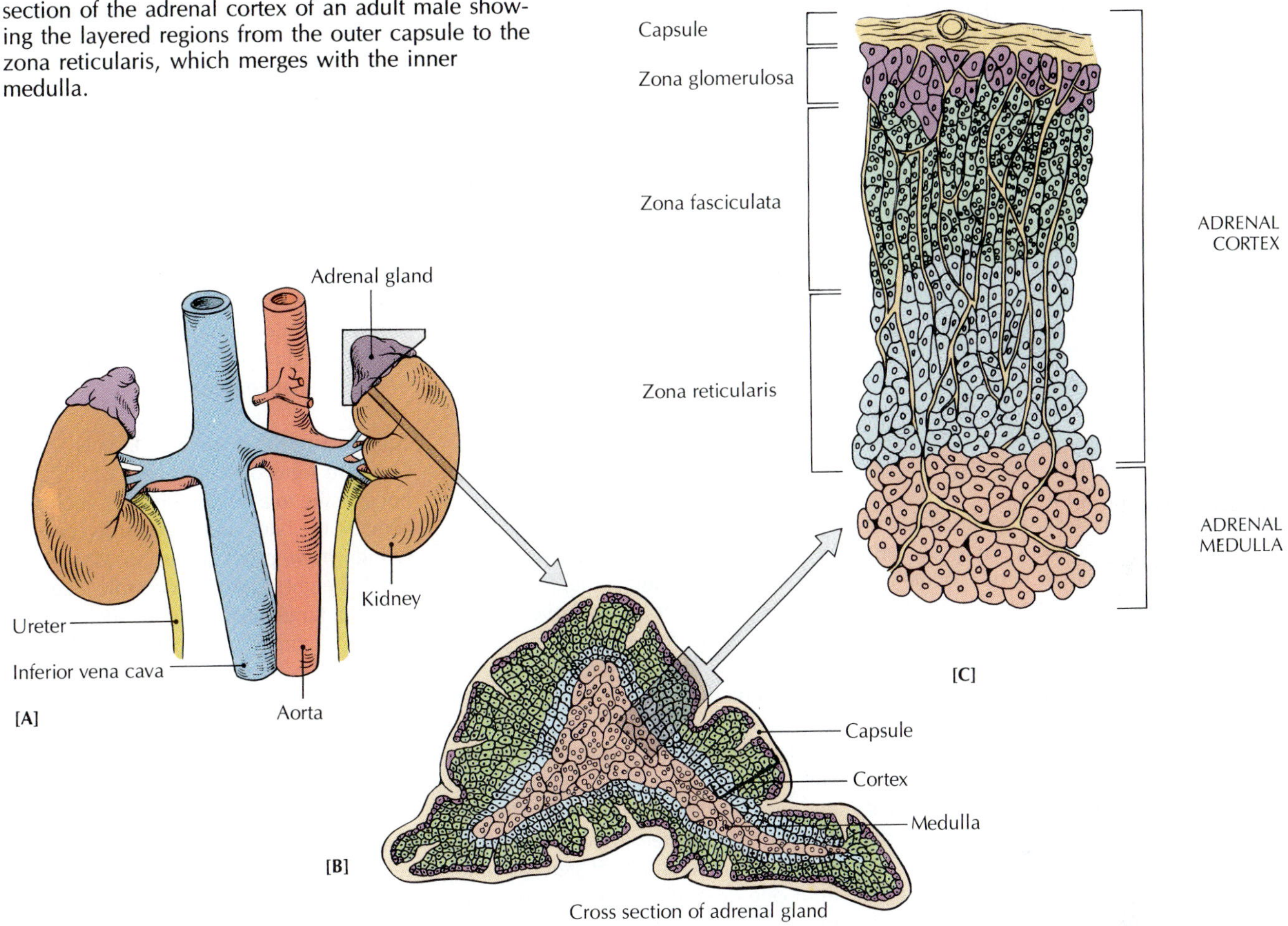

Capsule

Zona glomerulosa

Zona fasciculata

Zona reticularis

ADRENAL CORTEX

ADRENAL MEDULLA

[C]

Adrenal gland

Kidney

Ureter

Inferior vena cava

Aorta

[A]

Capsule

Cortex

Medulla

[B]

Cross section of adrenal gland

metabolism of glucose and other nutrients, as well as sodium and potassium.

The tissue of the cortex has three distinct zones lying beneath the outermost capsule [FIGURE 18.13C]: (1) Directly beneath the capsule is the thin *zona glomerulosa,* which secretes primarily mineralocorticoids (which affect the metabolism of minerals). (2) The second level down is the thick *zona fasciculata,* which makes up the bulk of the adrenal cortex and secretes primarily glucocorticoids (which affect glucose metabolism). Cholesterol, the precursor of steroid hormones, is more concentrated in the cells of the zona fasciculata than in any other part of the body. (3) The *zona reticularis* is the third and deepest layer of the adrenal cortex. It secretes small amounts of the sex steroids, mainly androgens.

Glucocorticoids The word *glucocorticoid* (gloo-koh-KORE-tih-koid) itself explains that the secretions come from the cortex (*-corticoid*) and that they help con-

trol the concentration of blood glucose (*gluco-* refers to glucose). There are three major types of glucocorticoids: *cortisol, corticosterone,* and *cortisone.* Cortisol accounts for about 95 percent of the activity of glucocorticoids. In addition to regulating the concentration of glucose, the glucocorticoids affect the metabolism of all types of foods, act as anti-inflammatory agents, affect growth, and decrease the effects of physical or emotional stress.

One of the most important effects of glucocorticoids is the stimulation of *gluconeogenesis,* the synthesis by the liver of glucose from noncarbohydrate sources such as amino acids and fatty acids.* (*Gluco,* glucose + *neo,*

---

*Another process promoted by glucocorticoids is *glycogenesis,* the production of glycogen from other carbohydrates. Glycogen stored in the liver becomes effective as an energy source only when enzymes in the liver break down the large organic molecules of glycogen into smaller molecules of glucose that can circulate in the blood and become available to cells when they need energy.

Athletes and Anabolic Steroids

Anabolic steroids (fully, *anabolic-androgenic steroids*) are synthetic derivatives of testosterone, the male hormone responsible for the *anabolic* (tissue-building) and *androgenic* (masculinizing) effects seen in normal postpubertal males.

Athletes use anabolic steroids mostly to build muscle mass, strength, and endurance. Unless prescribed by a physician, the distribution and use of anabolic steroids are illegal. Even with a prescription from a physician, steroids may not be used in conjunction with athletic competitions. College, professional, and Olympic athletics are given random urine tests after competition to determine if the athlete has used performance-enhancing drugs.

Some athletes take other drugs along with anabolic steroids to enhance their physical performance or to counteract the common side effects of steroids. Such drugs may include diuretics, stimulants, human growth hormone, insulin, human chorionic gonadotropin, gonadotropin-releasing hormone, and L-dopa. All these substances, as well as many others, are banned by the International Olympic Committee. When athletes take more than one steroid at a time the procedure is called *stacking*.

Probably the most publicized incident of illicit steroid use among athletes involved the Canadian sprinter Ben Johnson, who won the 100-m dash in the 1988 Olympics but was eventually disqualified when he was unable to pass postmeet drug tests. At that time, an anonymous Soviet coach said, "I feel sorry for Ben Johnson. All sportsmen— not all, but maybe 90 percent, including our own—use drugs."

Testing for Steroid Use

The use of drugs to enhance athletic performance is not new. But the testing of human athletes is a relatively recent happening. (Racehorses have been tested for some time.) The formal testing of Olympic athletes took place for the first time at the 1972 Olympic Games in Munich. These tests, however, were for nonsteroidal drug use. Official testing for anabolic steroids began during the 1976 Olympics in Montreal, but so-called modern drug testing did not begin until 1987.

In most cases, it is the presence of the metabolites of anabolic steroids, rather than the steroids themselves, that show up in the athlete's urine. If the metabolites are there at testing time, it is inferred that the athlete took anabolic steroids earlier. Testing for anabolic steroids is imperfect for several reasons. For one thing, athletes have become adept at disrupting the test results, using diet or other means to add misleading ingredients to their urine. But there are other reasons for inadequacies in testing: Some anabolic steroids are difficult to detect, the metabolism of steroids varies from one drug to another, some testing methods do not detect the specific metabolites of anabolic steroids, some of the metabolites of lesser known steroids are not yet familiar to testers, and information about the metabolites of some anabolic steroids obtained from other countries is not always available.

The original tests for anabolic steroids were based on radioimmunoassay (RIA), a procedure that relies on an antibody response to a specific steroid with a radioactive tracer. Currently, a technique known as gas chromatography-mass spectrometry (GC-MS) is used. In GC-MS analysis, a urine sample from an athlete is injected into a long tube. The components of the urine are identified by the time it takes them to pass through the tube, each component having a specific passage time. Because even GC-MS is not foolproof, laboratories use a high safety margin so that innocent athletes are not penalized unfairly.

Telltale Signs and Adverse Effects

Outward signs of steroid use, which are not specific to anabolic steroids, include acne on the back and upper arms, mood swings from pleasantness to violence, rapid weight gain, hypodermic needle marks in large muscle groups such as the gluteus maximus, male pattern baldness, decreased body fat, jaundice, excessive body hair in females (hirsutism), enlarged breasts in males (gynecomastia), reduced breasts and deepened voice in females, testicular atrophy and prostatic hypertrophy in males, clitoral hypertrophy in females, excessive muscular development of the upper torso, and edema in the extremities.

Laboratory examinations can reveal many signs of steroid use, including elevations in bilirubin, creatine phosphokinase, LDL cholesterol, and triglyceride levels; decreased levels of HDL cholesterol, luteinizing hormone, and follicle-stimulating hormone; and decreased sperm count and motility.

---

new + *genesis*, production or generation; so gluconeogenesis is simply "the production of new glucose.")

Besides facilitating the catabolism (metabolic breakdown) of proteins, glucocorticoids also promote the catabolism of fats, as well as removing fatty acids from adipose tissue and making them available as an energy source.

Another effect of glucocorticoids is to suppress allergic reactions and inflammatory responses. They may operate by decreasing the inflammatory effect of enzymes released by lysosomes. Glucocorticoids may also decrease the permeability of blood capillaries, thereby reducing the flow of plasma that normally causes swelling and irritation. By decreasing the ability of white blood cells to engulf foreign substances, glucocorticoids can slow down the inflammatory process. Also, by decreasing the efficiency of the thymus gland, spleen, and other lymphatic tissues related to the immune system, glucocorticoids can reduce the concentration of antibodies that may cause tissue irritation.

The secretion of glucocorticoids is controlled by a negative feedback system between the adrenal cortex and the hypothalamic-pituitary axis. When the blood concentration of glucocorticoids (mainly cortisol) drops, the hypothalamus releases corticotropin-releasing hormone (CRH), which promotes the release of ACTH from the adenohypophysis. In turn, ACTH stimulates the secretion of the glucocorticoid cortisol. As the concentration of cortisol in the blood goes up, the secretion of ACTH decreases until the concentration of cortisol returns to normal. Pain, anxiety, and other types of stress induce the release of CRH in a similar manner, and the same negative feedback system is put into effect to regulate the glucocorticoid levels in the blood.

Mineralocorticoids *Mineralocorticoids* (mihn-uh-ruhl-oh-KORE-tih-koidz), another type of adrenocortical hormone, are a group of steroid hormones produced in the cells of the zona glomerulosa. They regulate water and ion concentrations, primarily sodium and potassium. The main mineralocorticoid is **aldosterone,** which promotes the retention of sodium by the kidneys and the loss of potassium in urine. Mineralocorticoid activity also increases the reabsorption of sodium from sweat, saliva, and gastric juice. Mineralocorticoids are not as dependent on CRH and ACTH as the glucocorticoids are. Instead, their control is effected through a combination of the renin-angiotensin-aldosterone pathway and the level of potassium in the blood.

The main target area of aldosterone is the kidney, where the kidney tubules and collecting duct epithelia are stimulated to reabsorb sodium into the blood, which leads to chloride and bicarbonate reabsorption as well as the retention of water. At the same time, aldosterone stimulates the excretion of potassium ions from the kidney tubules into the urine. Aldosterone also promotes the excretion of hydrogen ions into the urine, thus regulating normal acid-base balance as well as a normal electrolyte balance in the body fluids.

Renin-angiotensin-aldosterone mechanism The release of aldosterone is controlled primarily by the **renin-angiotensin-aldosterone (RAA) mechanism** [FIGURE 18.14]. A drop in blood volume (from hemorrhage or dehydration, for instance) causes certain specialized kidney cells to release the enzyme *renin*. Renin converts the plasma protein *angiotensinogen* from the liver into *angiotensin I*. As angiotensin I passes through the lungs, it is converted into *angiotensin II* by the lung enzyme *angiotensin-converting enzyme (ACE)*. Angiotensin II, which is a peptide, stimulates the adrenal cortex to release aldosterone, which causes the kidneys to increase the reabsorption of sodium. Thus water follows, increasing the volume of the blood and returning blood pressure to normal. Angiotensin II also constricts arterioles, which aids in returning blood pressure to normal.

Aldosterone release is also controlled by extracellular

FIGURE 18.14 ALDOSTERONE AND REGULATION OF BLOOD PRESSURE

The flowchart shows the negative feedback system that regulates the secretion of aldosterone and controls blood pressure.

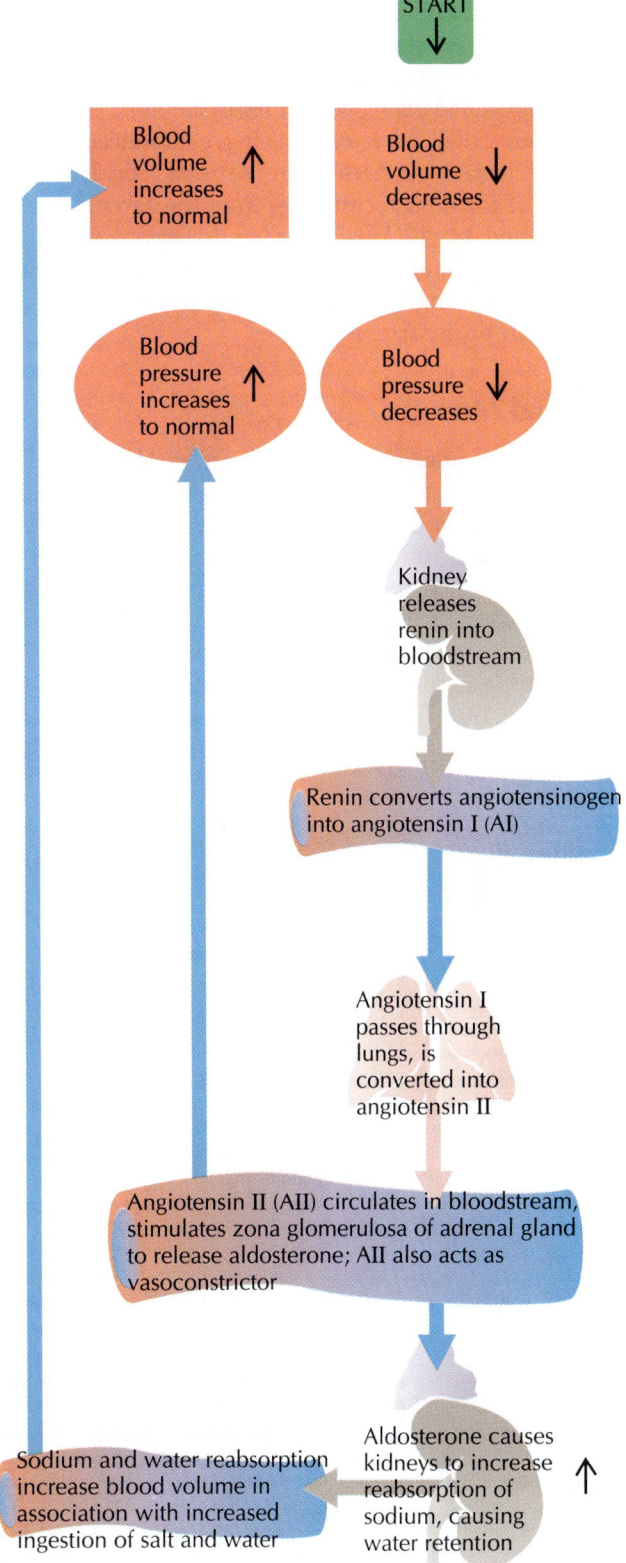

potassium concentrations, or the Na^+/K^+ ratio. Increased extracellular concentrations of K^+ increase the release of aldosterone, which causes the kidneys to eliminate the excess K^+. Decreased extracellular concentration of K^+ does just the opposite.

Gonadocorticoids The *gonadocorticoids,* or adrenal sex hormones, normally have only a slight effect on the sex organs, or gonads. They consist mainly of male hormones called *androgens* and lesser amounts of female hormones called *estrogens,* which are produced in the zona fasciculata and perhaps in the zona reticularis. Secretion of the gonadocorticoids from the adrenal cortex is regulated by ACTH. Disorders (overproduction) of the adrenal gland androgens can cause masculinizing characteristics, such as facial hair, deep voice, and a reduction in breast size, in females.

Adrenal Medulla

The *adrenal medulla,* the inner portion of the adrenal gland, should be thought of as an entirely separate endocrine gland from the adrenal cortex. It secretes different hormones, has a different tissue structure, is derived from different embryonic tissue, and, unlike the adrenal cortex, is not essential to life.

The cells of the adrenal medulla may extend somewhat into the innermost portion of the zona reticularis [FIGURE 18.13C]. They are usually grouped in clumps and are surrounded by blood vessels, especially capillaries.

Granules in the cells store high concentrations of two hormones, epinephrine and norepinephrine, both of which are catecholamines.

The secretory cells of the adrenal medulla (sometimes called *chromaffin cells* because of their tendency to stain a dark color) are derived from the same embryonic ectodermal tissue as the ganglia of the sympathetic nervous system. Both types of cells receive direct innervation from preganglionic neurons of the sympathetic division of the autonomic nervous system. Adrenal chromaffin cells synthesize, store, and secrete epinephrine and norepinephrine, which work in concert with glucocorticoids to prepare the body for "fright, fight, or flight."

Epinephrine and norepinephrine The main secretion (80 percent) of the adrenal medulla is *epinephrine* (ep-ih-NEFF-rihn), commonly known as adrenaline. The other hormone, *norepinephrine (NE;* noradrenaline), though closely related to epinephrine, is produced and released in much smaller amounts. The hormones produce effects similar to stimulation by the sympathetic nervous system (sympathomimetic effects). The major effects of epinephrine and norepinephrine are listed in TABLE 18.3.

The effectors on which epinephrine and norepinephrine act can be separated into two groups, based on their response to each hormone. Effector organs may contain alpha receptors, beta receptors, or both on their plasma membranes. Even though epinephrine and norepinephrine both bind to alpha and beta receptors, they do not

---

TABLE 18.3 MAJOR EFFECTS OF EPINEPHRINE AND NOREPINEPHRINE

Organ or function	Effects of epinephrine	Effects of norepinephrine
Heart and blood vessels	Dilates coronary vessels; dilates arterioles in skeletal muscles and viscera; increases heart rate and cardiac output.	Dilates coronary vessels; causes vasoconstriction in other organs; increases heart rate and cardiac output.
Blood pressure	Raises blood pressure because of increased cardiac output and peripheral vasoconstriction.	Raises blood pressure because of peripheral vasoconstriction.
Muscles	Inhibits contraction of smooth muscles of digestive system, producing relaxation; dilates respiratory passageways; decreases rate of fatigue in skeletal muscle; increases respiratory rate.	Relaxes smooth muscle of gastrointestinal tract.
Metabolism	Stimulates glycogenolysis (the breakdown of glycogen to glucose) in liver and muscles, which elevates concentrations of blood glucose and muscle lactic acid; increases oxygen consumption; enhances lipid metabolism; inhibits insulin release from pancreas, providing a ready supply of fatty acids for fuel for skeletal muscles and making glucose available, especially to the skeletal muscles and central nervous system during emergencies.	Enhances lipid metabolism and release of free fatty acids from adipose tissue.

necessarily evoke the same response. This is due, in part, to the existence of several subtypes of receptors. Beta-receptor activation is linked to *stimulating* the intracellular second messenger cyclic AMP, whereas alpha-receptor activation is linked to *inhibiting* cyclic AMP. Some alpha-receptor activation is linked to the second messenger calcium. Norepinephrine interacts predominantly with alpha receptors and less strongly with beta receptors. Epinephrine reacts well with beta receptors but its potency is less with alpha receptors (see page 566). Thus the response of a cell having alpha receptors may be completely different from the response of a gland, muscle, or other neuron with beta receptors for the same transmitters or hormones. This is consistent with the rule that the transmitter triggers the receptor, but it is the *receptor* that defines the nature of the response.

Stress and the Adrenal Cortex

A stressful condition is first perceived by the brain and other sensory receptors. The hypothalamus releases corticotropin-releasing hormone (CRH), which in turn stimulates the release of ACTH from the anterior pituitary, which causes the release of cortisol from the adrenal cortex [FIGURE 18.15].

Increased amounts of cortisol help the body deal with stress in the following ways:

1 The catabolism of protein releases amino acids as an energy source and also is a means of tissue repair in case of injury.

2 Amino acids are converted into glucose by the liver (gluconeogenesis). This provides a source of energy during stress, when eating and digestion are reduced.

3 The breakdown of triglycerides into fatty acids and glycerol (lipolysis) provides another energy source.

Stress and the Adrenal Medulla

In times of stress, epinephrine produces a condition sometimes called the "fright, fight, or flight" response, which permits the body to react quickly and strongly to emergencies. The secretion of epinephrine causes intense effects that last a very short time. In fact, enzymes in the liver inactivate epinephrine in about 3 min.

After epinephrine has been secreted, some blood vessels constrict and others dilate, redistributing blood to such organs as the brain and muscles, where it is most needed. Digestion, an energy-consuming function, is halted during the emergency by the diversion of blood from the stomach and intestines to the skeletal muscles, where extra energy may be needed. Blood pressure rises, heart rate increases, and the time needed for blood clotting is reduced. Respiratory rate increases and bronchioles dilate. Enzymes in the liver are activated to release

FIGURE 18.15 RESPONSE TO STRESS

Stress indirectly stimulates the adrenal cortex to secrete cortisol and directly stimulates the adrenal medulla to secrete epinephrine and norepinephrine.

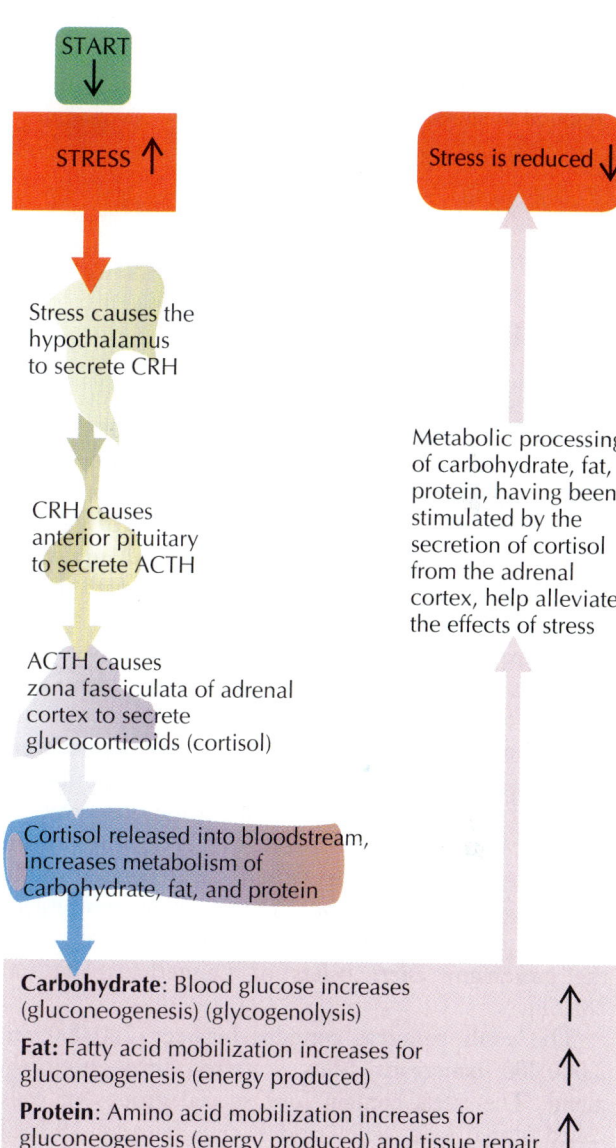

START

STRESS ↑

Stress causes the hypothalamus to secrete CRH

CRH causes anterior pituitary to secrete ACTH

ACTH causes zona fasciculata of adrenal cortex to secrete glucocorticoids (cortisol)

Cortisol released into bloodstream, increases metabolism of carbohydrate, fat, and protein

Carbohydrate: Blood glucose increases (gluconeogenesis) (glycogenolysis) ↑

Fat: Fatty acid mobilization increases for gluconeogenesis (energy produced) ↑

Protein: Amino acid mobilization increases for gluconeogenesis (energy produced) and tissue repair ↑

Stress is reduced ↓

Metabolic processing of carbohydrate, fat, protein, having been stimulated by the secretion of cortisol from the adrenal cortex, help alleviate the effects of stress

glucose from glycogen for increased energy. In each case, the secretion of epinephrine favors the body's survival in times of stress.

PANCREAS

The *pancreas* (Gr. "all flesh") is an elongated (12 to 15 cm), fleshy organ consisting of a head, body, and tail [FIGURE 18.16A]. The area where the head and body join is the neck. The pancreas is located posterior to the stomach, with the head tucked into the curve of the duodenum (where the stomach meets the small intestine). The body and tail extend laterally to the left, with the tail making contact with the spleen.

The pancreas is considered a *mixed gland* because it functions both with ducts, as an exocrine gland, and without ducts, as an endocrine gland. As an exocrine gland, it acts as a digestive organ, secreting digestive enzymes and alkaline materials into a duct that empties into the small intestine. As an endocrine gland, it secretes its hormones into the bloodstream. The endocrine portion of the pancreas makes up only about 1 percent of the total weight of the gland. This portion synthesizes, stores, and secretes hormones from clusters of cells called the *pancreatic islets* (islets of Langerhans) [FIGURE 18.16B].

The adult pancreas contains between 200,000 and 2,000,000 pancreatic islets scattered throughout the gland. The islets contain four special groups of cells, called alpha, beta, delta, and F cells [FIGURE 18.16C]: (1) *Alpha cells* produce glucagon, which raises the blood glucose level, and (2) *beta cells* produce insulin, which lowers the blood glucose level. Both hormones help regulate glucose metabolism. Beta cells are the most common type of islet cells. They are generally located near the center of the islet and are surrounded by the other cell types. (3) *Delta cells* secrete somatostatin, the hypothalamic growth-hormone inhibiting hormone that also inhibits the secretion of both glucagon and insulin. (4) *F cells* secrete pancreatic polypeptide, which is released into the bloodstream after a meal and regulates the release of pancreatic digestive enzymes.

Glucagon

Alpha cells in the pancreatic islets synthesize, store, and secrete the peptide hormone *glucagon* (GLOO-kuh-gon). When the concentration of blood glucose falls, (1) glucagon stimulates the liver to convert glycogen into glucose (*glycogenolysis*), which causes the blood glucose concentration to rise. (2) Glucagon also stimulates *gluconeogenesis*, the formation of glucose from noncarbohydrate sources such as amino acids and lactic acid. Of these two processes, glycogenolysis is the more important source of glucose. (3) Glucagon increases the concentration of cyclic AMP from ATP in liver cells, causing the enzyme *phosphorylase a* to be activated. This enzyme separates glucose units from the large glycogen molecule, and the freed glucose units enter the bloodstream rapidly. (4) Glucagon also stimulates the release of fatty acids and glycerol from adipose tissue. As the result of a negative feedback system, glucagon causes the level of blood glucose to increase during periods of fasting or any other time when the concentration of blood glucose drops below normal (70–110 mg/dL). For this reason, glucagon is considered to be a *hyperglycemic* (elevated blood glucose) *factor*.

Insulin

Like glucagon, *insulin* is a peptide hormone.* Beta cells secrete insulin when the blood glucose concentration rises, as it does immediately after a meal. The most important effects of insulin include the following:

1 It facilitates glucose transport across plasma membranes into cells and enhances the conversion of glucose to glycogen (*glycogenesis*), which is then stored in the liver as a ready source of blood glucose.

2 It increases glucose transport into cells, which increases the metabolism of carbohydrates and decreases blood glucose (*hypoglycemic effect*).

3 It increases amino acid transport into cells, which increases protein synthesis.

4 It increases the conversion of glucose into fatty acids (*lipogenesis*).

5 It inhibits glycogenolysis and gluconeogenesis.

---

*Insulin was the first protein to have its full structure determined. This was done in 1955 by Frederick Sanger, who received the Nobel Prize in 1958 in recognition of his work.

Q: *Why do diabetics have to inject insulin instead of taking it orally?*

A: Insulin is a peptide, which is a small protein molecule. It cannot be taken by mouth because it would be rapidly inactivated by protein-digesting enzymes in the digestive tract.

FIGURE 18.16 PANCREAS

[A] Location and anatomy. [B] Section of pancreatic tissue showing a pancreatic islet. ×600. [C] A highly magnified human pancreatic islet with alpha, beta, and delta cells clearly delineated. ×5000.

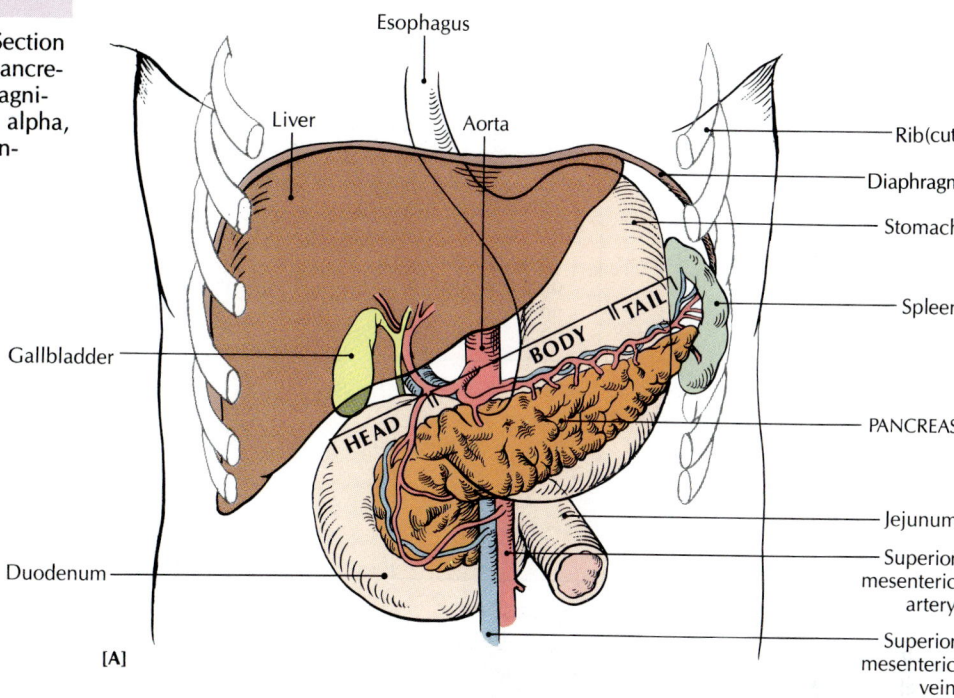

Esophagus
Liver
Aorta
Rib(cut)
Diaphragm
Stomach
Spleen
Gallbladder
BODY
TAIL
HEAD
PANCREAS
Jejunum
Superior mesenteric artery
Duodenum
Superior mesenteric vein

[A]

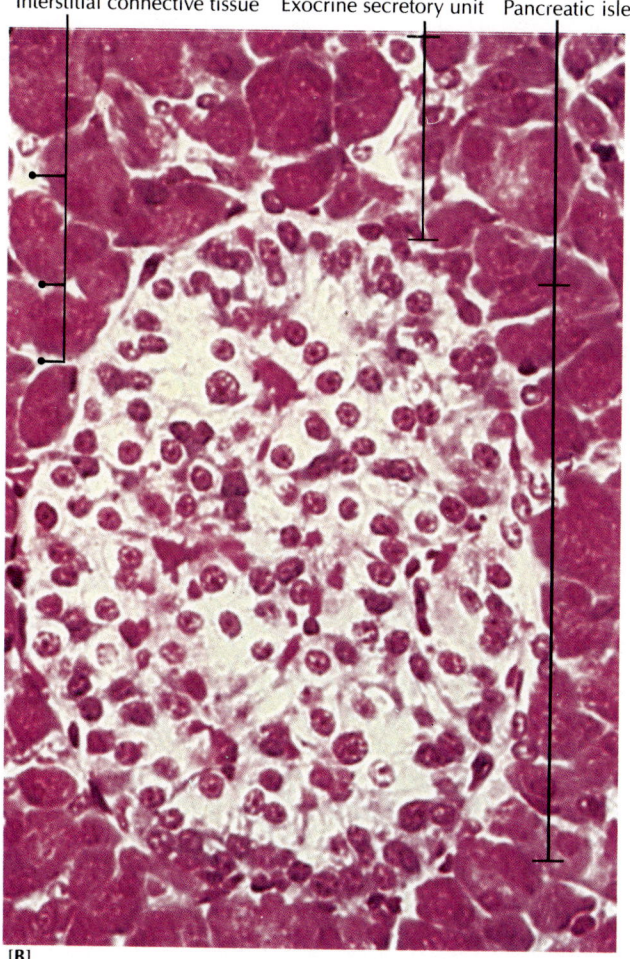

Interstitial connective tissue   Exocrine secretory unit   Pancreatic islet

[B]

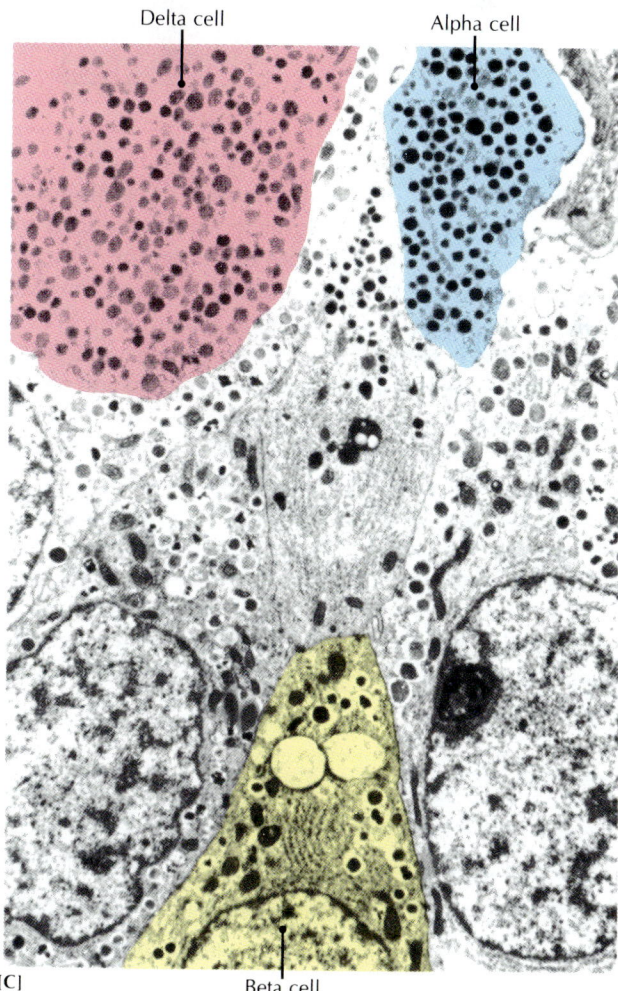

Delta cell   Alpha cell

Beta cell

[C]

FIGURE 18.17 GLUCAGON AND INSULIN REGULATION

Negative feedback mechanism regulating the secretion of glucagon and insulin. Note that both the pancreas and liver are involved.

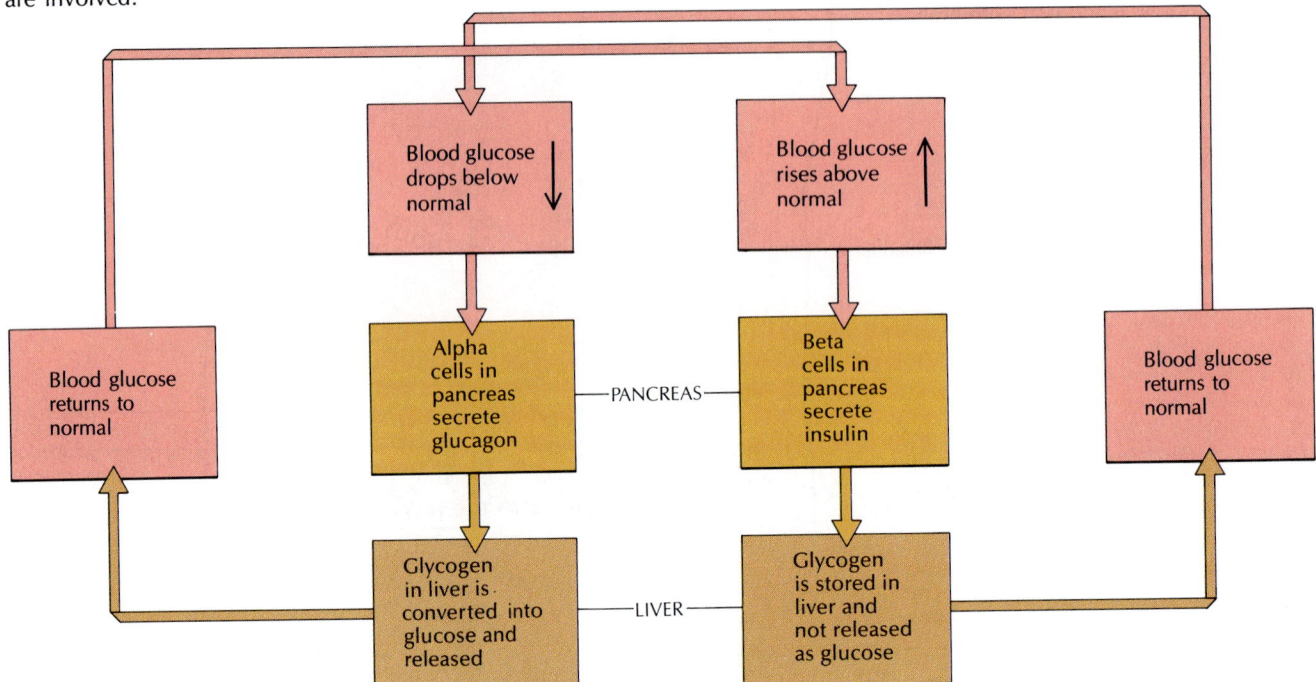

The functions of insulin are opposite those of glucagon, and the two hormones usually work in concert to maintain a normal blood glucose concentration. FIGURE 18.17 illustrates the negative feedback mechanisms that regulate the secretion of glucagon and insulin. When blood glucose is high, glucose moves into the liver cells easily. Insulin then stimulates liver enzymes to add phosphate groups to glucose molecules. The glucose 1-phosphate can now be added to the glycogen molecule for storage. Thus insulin indirectly inhibits the conversion of glycogen into glucose and its release into the bloodstream. As a result, the concentration of blood glucose drops toward a normal level.

When beta cells do not produce enough insulin, *diabetes mellitus* results. In contrast, excessive amounts of insulin usually produce *hypoglycemia,* or "low blood sugar."

A S K   Y O U R S E L F

1 What is the function of the pancreatic islet cells?

2 Why is glucagon called a hyperglycemic factor?

3 How are insulin and glucagon opposite in function?

4 How does insulin regulate blood glucose levels?

GONADS

The *gonads* (Gr. *gonos,* offspring), which are the *ovaries* in the female and the *testes* in the male, secrete hormones that help regulate reproductive functions. These sex hormones include the male *androgens* and the female *estrogens, progestins,* and *relaxin.* The gonads are discussed in detail in Chapter 29.

Male Sex Hormones

At least three hormones help regulate male reproductive functioning. The most important one is *testosterone,* which is produced by interstitial cells in the testes. The other major hormones are *luteinizing hormone (LH)* and *follicle-stimulating hormone (FSH),* both produced by the adenohypophysis.

Testosterone acts with LH and FSH to stimulate the production of sperm (*spermatogenesis*). It is necessary for the growth and maintenance of the male sex organs and promotes the development and maintenance of sexual behavior. It also stimulates the growth of facial and pubic hair, as well as the enlargement of the larynx, which causes the voice to deepen. The testes also produce the hormone *inhibin,* which inhibits the secretion of FSH.

Female Sex Hormones

Three major classes of ovarian hormones help regulate female reproductive functioning. *Estrogens* (estrin, estrone, and estradiol) help regulate the menstrual cycle and the development of the mammary glands and female secondary sex characteristics. The *progestins* (progesterone) also regulate the menstrual cycle and the development of the mammary glands, and aid in the formation of the placenta during pregnancy. *Relaxin,* which is produced in small quantities, is involved in childbirth, softening the cervix at the time of delivery and causing the relaxation of ligaments of the symphysis pubis to ease the baby's passage through the birth canal.

Overall, the secretion of hormones from the ovaries is regulated by two anterior pituitary hormones: follicle-stimulating hormone and luteinizing hormone. *Follicle-stimulating hormone* initiates the monthly development of a follicle within the ovary. As a result, the level of estrogen rises. Estradiol, which is secreted by the follicular cells of the ovary, inhibits the production of FSH. *Luteinizing hormone* initiates the production of progestin by the ovarian follicle.

The intricate interrelationships of hormones that regulate reproductive mechanisms are discussed fully in Chapters 29 and 30.

ASK YOURSELF

1 What are the male and female gonads called?

2 What are the functions of the major male and female hormones?

OTHER SOURCES OF HORMONES

In addition to the major endocrine glands just discussed, other glands and organs also have hormonal activity. In the following sections we briefly discuss the kidneys, pineal gland, thymus gland, heart, digestive system, placenta, and the hormones known as eicosanoids.

Kidneys

The paired *kidneys* are primarily organs for the excretion of wastes, but they also produce several hormones, including erythropoietin, 1,25-dihydroxyvitamin D_3, prekallikreins, and prostaglandins. In addition, the kidneys produce *renin* (*renes* is Latin for "kidneys"), an enzyme whose natural substrate is the plasma protein angiotensinogen, which helps form the renin-angiotensin-aldosterone system, which influences blood pressure and volume, the intake of salt, and salt-water balance [see FIGURE 18.14].

The function of *erythropoietin* (Gr. *eruthros*, red + *poiesis*, making) is to stimulate the production of erythrocytes (red blood cells) by facilitating the synthesis of hemoglobin and the release of erythrocytes from bone marrow.

Pineal Gland

The *pineal gland* (PIHN-ee-uhl; L. *pinea*, pine cone) is also known as the *pineal body*. It is a pea-sized body located in the roof of the diencephalon, deep within the cerebral hemispheres of the brain, at the posterior end of the third ventricle [see FIGURE 18.1]. The pineal gland has been called a "neuroendocrine transducer" — a system that converts a signal received through the nervous system (dark and light, for instance) into an endocrine signal (shifting concentrations of hormone secretion). Information about daily cycles of light and dark is detected by the eyes and conveyed via the optic nerve to the hypothalamus. From there, sympathetic nerves convey the signal to the pineal gland.

Several chemicals that have hormonal activity have been isolated from the pineal, but the functions of these substances have not been fully established. One of these is *melatonin,* which is derived from serotonin. The pineal gland produces steady secretions of melatonin throughout the night; light inhibits the production of melatonin. It has been observed that melatonin causes drowsiness, and the pineal gland may affect the sleep cycle. The typical lack of light during long winters may contribute to *seasonal affective disorder* (SAD), a condition characterized by lack of energy and mood swings that border on depression. Some researchers speculate that the pineal gland is involved in SAD, but no firm evidence is available.

Thymus Gland

The *thymus gland* is a double-lobed lymphoid organ located behind the sternum in the anterior mediastinum [see FIGURE 18.1]. It has an outer cortex containing many lymphocytes and an inner medulla containing fewer lymphocytes, as well as clusters of cells called thymic (Hassall's) corpuscles, whose function is unknown. The thymus gland is well supplied with blood vessels but has only a few nerve fibers.

The thymus gland reaches its maximum effectiveness during early adolescence. After that time, the gland begins to involute (turn in on itself) gradually. This process is accelerated by adrenal corticosteroids and sex hormones. The thymus gland also becomes smaller with advancing age, although it apparently continues to produce new lymphocytes. The current speculation that involution and shrinkage of the thymus gland lead to a total loss of function may be unjustified.

The main function of the thymus gland is the processing of T cells (T lymphocytes). These cells are responsible for one type of immunity, called *cellular-mediated immunity* (see Chapter 23). There is some evidence that the thymus gland may be a true endocrine gland, since it produces thymic hormones, or "factors," that play a role in the development of T cells in the thymus gland and their maintenance within other lymphoid tissue. Some of the hormones and factors include *thymosin alpha, thymosin B_1 to B_5, thymopoietin I and II, thymic humoral hormone (THH), thymostimulin,* and *factor thymic serum (FTS).* The thymic hormones also play a role in the development of some B cells into plasma cells, which produce antibodies.

It had been thought until recently that thymic hormones are produced exclusively in the thymus gland and that their main function is to assist in the processing of bone marrow cells into infection-fighting cells of the immune system (T cells). Recent discoveries, however, indicate that thymosin B_4 and B_5 influence hormones of the reproductive system and that thymosin B_4 is also synthesized by macrophages in the immune system.

Heart

Recent findings have revealed that in addition to being the pump that maintains circulation, the heart acts as an endocrine organ. Cardiac muscle cells in both atria (the upper chambers of the heart) contain secretory granules that produce, store, and secrete a peptide hormone called *atrial natriuretic peptide,* or *ANP.*

Atrial natriuretic peptide is secreted continuously in minute amounts into the general blood circulation. Secretion increases when excess salt accumulates in the body, when blood volume increases enough to stimulate stretch receptors in the atria, or when blood pressure rises significantly. Special target-cell receptors have been found in blood vessels, kidneys, and adrenal glands. ANP also affects neurons in the brain, especially the hypothalamus, where control and regulation occur for blood pressure and the excretion of sodium, potassium, and water by the kidneys.

Current evidence suggests that ANP helps maintain a proper balance of fluid and electrolytes by increasing the output of sodium in urine; relaxes blood vessels directly, thus lowering blood pressure by reducing resistance to blood flow; lowers blood pressure by blocking the actions of hormones such as aldosterone, which tends to raise blood pressure; and reduces blood volume by stimulating the kidneys to filter more blood and produce more urine. Thus ANP is known as a natriuretic, diuretic, and hypotensive hormone.

Digestive System

Among the major hormones of the digestive system are gastrin, secretin, and cholecystokinin. *Gastrin* is a polypeptide secreted by the mucosa (lining) of the stomach. Its function is to stimulate the production of hydrochloric acid and the digestive enzyme pepsin when food enters the stomach. Thus the stomach is both the producer and the target organ of gastrin.

Secretin is a polypeptide secreted by the mucosa of the duodenum. It stimulates a bicarbonate-rich secretion from the pancreas that neutralizes stomach acid as the acid contents pass to the small intestine. Secretin was the first hormone to be discovered (by the British scientists William M. Bayliss and Ernest H. Starling in 1902) and the first substance to actually be called a "hormone."

Cholecystokinin (CCK; koh-lee-SIS-toe-kine-in) is secreted from the wall of the duodenum. It stimulates the contraction of the gallbladder, which releases bile when food (particularly fats) enters the duodenum. It also stimulates the secretion of enzyme-rich digestive juices from the pancreas.

Placenta

The *placenta* is a specialized organ that develops in a pregnant female as a source of nourishment for the developing fetus (see Chapter 30). It secretes estrogens, progesterone, and human chorionic gonadotropin (hCG), which help maintain pregnancy. Target areas are the ovaries, mammary glands, and uterus.

Eicosanoids

One of the reasons that fatty acids are necessary for good health is that they are the source of prostaglandins (including thromboxanes and prostacyclin) and leukotrienes. The precursor substances for prostaglandins and leukotrienes are *eicosanoids* (eye-KOH-suh-noidz), substances that belong to the 20-carbon unsaturated fatty acid, *arachidonic acid,* and are 20-carbon derivatives of linoleic and linolenic acids. Eicosanoids are synthesized from the fatty acids that make up the structural parts of plasma membranes. (The specific type of prostaglandin or leukotriene produced depends on which fatty acids and enzymes are present in the membrane.)

Prostaglandins *Prostaglandins (PGs)* were originally thought to come from the prostate gland, hence their name. They are actually produced by all nucleated cells and are found in practically all tissues of the body. There are more than 16 different prostaglandins, belonging to 9 classes.

Prostaglandins are considered to be chemical modulators or messengers rather than true hormones, since they are not produced by specific endocrine glands and secreted into the bloodstream for transport to target cells. Instead, they are produced locally, close to their site of action. Apparently, disturbances near a cell cause the release of fatty acids from the cellular membrane, and enzymes called cyclooxygenases convert the fatty acids into prostaglandins. Prostaglandins, unlike true hormones, are not stored. Once they are synthesized in response to a stimulus, they are released immediately, act quickly and locally, and are then inactivated and degraded.

The actions of prostaglandins are as varied as their origins. At the cellular level, prostaglandins appear to work along with specific hormones to increase or decrease the effect of cyclic AMP on target cells. At the tissue level, prostaglandins can raise or lower blood pressure, regulate digestive secretions, inhibit progesterone secretion by the corpus luteum, reduce infection by stimulating microorganism-destroying blood cells, and regulate blood clotting. In addition, prostaglandins cause the contraction of smooth muscle in the uterus and can dilate air passages to the lungs.

Little is known about how prostaglandins work. These hormonal modulators may operate inside cells to turn chemical reactions on and off, or they may function by changing the composition of cellular membranes. *Thromboxanes* (TXs) are modified PGs that promote platelet aggregation and constrict blood vessels. They are secreted by all cells in the body except red blood cells and exert potent effects that are inactivated rapidly. Extensive research on eicosanoids is helping scientists to better understand their paracrine and autocrine functions.

Prostaglandins have many clinical uses. Prostaglandin drugs may be useful in inducing labor and in treating asthma, arthritis, ulcers, and hypertension (high blood pressure). One of the first clinical uses of prostaglandins was to induce labor in pregnant women and to induce abortions. There is also some evidence that prostaglandin drugs may inhibit the growth of viruses and may be useful in treating multiple sclerosis and other diseases. Clinical studies indicate that separate forms of prostaglandins may be useful in treating diseases of the blood vessels by increasing blood flow, promoting the healing of stomach and duodenal ulcers, and protecting the stomach and small intestine from irritating drugs such as aspirin. A specialized use of prostaglandin drugs is to delay the natural closure of a blood vessel in newborn babies who require corrective surgery for a congenital

heart disease. The prostaglandin called *prostacyclin* relaxes and dilates blood vessels and also suppresses the action of blood-clotting structures called platelets. Prostacyclin may be useful in preventing heart attacks and strokes and in preventing blood clots from forming during heart bypass operations, when the blood is circulated outside the body.

Drugs that inhibit the effects of prostaglandins are being tested, and some are already in use. Such a drug has been used to relieve menstrual cramps, excessive muscular contractions produced by high levels of uterine prostaglandins.

Leukotrienes *Leukotrienes* (LTs; loo-koh-TRY-enz) are a class of eicosanoids that function as mediators of inflammatory and allergic responses. They are synthesized by granulocytes through the action of cytoplasmic enzymes called lipoxygenases. Leukotrienes are released when specific allergens (foreign substances) combine with antibodies on the surface of mast cells. The effects of leukotrienes include the increased permeability of blood vessels, the attraction of scavenger cells — neutrophils and eosinophils — to the site of inflammation, and the constriction of lung bronchioles and small arteries called arterioles. Such bronchoconstriction is typical in cases of bronchial asthma and *anaphylaxis* (shock produced by a drop in blood pressure).

Like prostaglandins, leukotrienes are short-acting and are considered to be *local hormones*, acting mainly in the tissues in which they are produced.

ASK YOURSELF

1 What hormones do the kidneys secrete?

2 What is one possible function of the pineal gland?

3 Why do diseases become more prevalent as the thymus atrophies?

4 What appear to be the major functions of ANP?

5 What are the main endocrine secretions of the digestive system?

6 Why are prostaglandins considered to be hormonal modulators rather than true hormones?

7 What are some of the clinical uses of prostaglandin drugs?

Q: *What is the connection between prostaglandins and aspirin?*

A: Aspirin reduces pain, inflammation, and fever by inhibiting cyclooxygenase, the enzyme needed for the production of prostaglandins.

THE EFFECTS OF AGING ON THE ENDOCRINE SYSTEM

The aging process affects many functional activities of the endocrine system, including the production of blood

FIGURE 18.18 DEVELOPMENTAL ANATOMY OF THE PITUITARY GLAND

[A] Lateral view. [B–E] Midsagittal views.

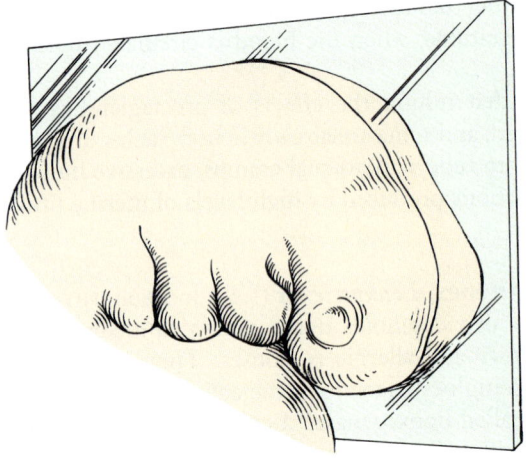

[A] WEEK 4

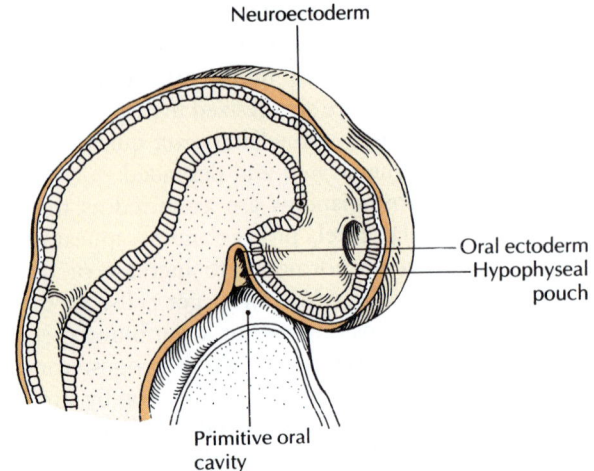

[B] WEEK 4

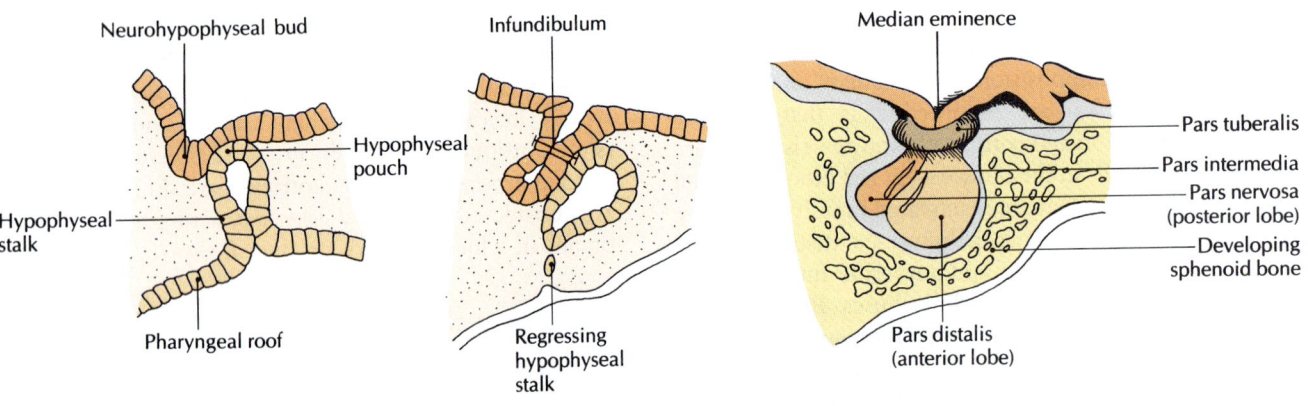

[C] WEEK 5   [D] WEEK 8   [E] FETAL STAGE—POST WEEK 8

glucose, regulation of blood pressure, reaction to stress, and production of urine. For example, as the thyroid gland becomes smaller with age and its output is reduced, the basal metabolic rate (BMR) of the body (the amount of energy used at rest) is lowered. In a similar way, the metabolism of protein and glucose is diminished as the aging pancreas secretes less insulin, or the cells become unresponsive to the insulin that *is* secreted. The most common disease attributable to aging and an altered hormone secretion or response is diabetes mellitus. In younger individuals diabetes is *usually* the result of insufficient insulin output from the pancreas, but in older people it is due to the inability to use insulin rather than to a lack of insulin.

Both men and women experience changes in sexual function beginning in the middle years, with the most obvious change being a decrease in the secretion of sex hormones. Starting at about 40 or 45, men usually experience a gradual reduction in the desire for sexual activity and some loss in sexual responsiveness. The physiological changes are more drastic for women, as they experience menopause (the cessation of menstrual periods) and a gradual decrease in estrogen secretion between 45 and 55 years of age. Hormone replacement therapy is often recommended.

DEVELOPMENTAL ANATOMY OF THE PITUITARY AND THYROID GLANDS

The *pituitary gland* is derived in part from *oral ectoderm* of the embryonic oral (mouth) cavity, which develops

TABLE 18.4 DERIVATIONS OF COMPONENTS OF THE PITUITARY GLAND

Components	Division*	Lobe of pituitary*	Derivation
Pars tuberalis Pars distalis Pars intermedia	Adenohypophysis	Anterior Anterior Posterior	Oral ectoderm
Pars nervosa (neural lobe) Infundibular stalk	Neurohypophysis	Posterior	Neuro-ectoderm

*We have adopted the frequently used terminology that equates the adenohypophysis with the anterior lobe and the neurohypophysis with the posterior lobe.

neurohypophyseal bud at about week 8 [FIGURE 18.18D]. The connection between the hypophyseal pouch and the oral cavity is called the **hypophyseal stalk.** It disappears between weeks 6 and 8 as the pouch closes completely [FIGURE 18.18D].

Fetal development continues after week 8 as the hypophyseal pouch differentiates into the components of the adenohypophysis [FIGURE 18.18E; see also TABLE 18.4]. The anterior wall of the pouch develops into the **pars distalis,** and the posterior wall develops into the **pars intermedia** [FIGURE 18.18E]. Later, a small extension from the pars distalis called the **pars tuberalis** surrounds the infundibular stalk. The neurohypophyseal bud and embryonic infundibulum differentiate into the components of the neurohypophysis [TABLE 18.4]. The extension of the neurohypophysis downward toward the developing pars distalis is the **pars nervosa** (neural lobe). The nerve fibers from the neurosecretory neurons of the supraoptic and paraventricular nuclei pass from the hypothalamus through the infundibular stalk, terminating in the pars nervosa.

The **thyroid gland** originates late in week 4 as a mass of endoderm of the pharyngeal pouches. The primitive thyroid migrates downward along the slender **thyroglossal duct,** which breaks down late in week 5. The developing thyroid reaches its final destination just inferior to the cricoid cartilage by week 7. The thyroid gland may begin to function as early as week 10.

into the adenohypophysis, and in part from *neuroectoderm* of the diencephalon of the brain, which develops into the neurohypophysis [TABLE 18.4]. During week 4, the **hypophyseal** (Rathke's) **pouch** differentiates from the oral ectoderm of the pharyngeal roof and extends toward the brain [FIGURE 18.18B]. By week 5, the pouch has elongated and makes contact with the **neurohypophyseal bud,** which develops from the hypothalamus of the diencephalon [FIGURE 18.18C].

The **infundibulum** is the stalk connecting the pituitary gland to the hypothalamus. It develops from the

---

WHEN THINGS GO WRONG

Endocrine disorders are usually caused by underfunctioning (*hypo-*) or overfunctioning (*hyper-*), inflammation, or tumors of the endocrine glands. Certain disorders arise in the hypothalamus or pituitary; others develop in the endocrine glands that come under the influence of these higher control centers. Some of the major disorders of the endocrine system are reviewed here briefly.

Pituitary Gland: Adenohypophysis

Among the disorders of the adenohypophysis is **giantism,** which is caused by the oversecretion of growth hormone during the period of skeletal development (see photo). The body of a person with this disorder grows much larger than normal, sometimes to more than $2\frac{1}{2}$ m (8 ft) and 180 kg (400 lb). In most cases, death occurs before the age of 30.*

A normal-sized person (center) with a "giant," whose pituitary gland was overactive, and a pituitary dwarf, whose pituitary gland was underactive.

*The tallest person for whom there are authenticated records was Robert Wadlow of Illinois. He was 8 ft 11 in. tall and still growing when he died at the age of 22.

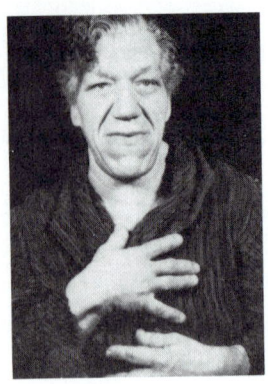

The effects of acromegaly, such as thickening of the jaw, nose, and hands, are shown here at three different stages of an afflicted woman's life.

Acromegaly (ack-roh-MEG-uh-lee; Gr. *akros*, extremity + *megas*, big) is a form of giantism that affects adults *after* the skeletal system is fully developed. It is caused by a pituitary tumor. After maturity, oversecretion of growth hormone does not lengthen bones of the skeleton but does cause some cartilage and bone to thicken. The jaw, hands, feet, and eyebrow ridges may widen noticeably (see photos), and enlargement of the heart, liver, and other internal organs is also possible. Blood pressure may increase, and subsequent congestive heart failure is likely. Muscles grow weak, and osteoporosis and painful enlargements at the joints may occur. Acromegaly may be treated with some success by the surgical removal of the adenohypophysis or by using radiation treatments.

Persons with an undersecretion of growth hormone are known as ***pituitary dwarfs.*** Although their intelligence and body proportions are normal, they do not grow any taller than a normal 6-year-old child. Some pituitary dwarfs become prematurely senile, and most die before the age of 50. Usually, their sex organs and reproductive ability are not fully developed, but some pituitary dwarfs are capable of producing normal-sized children. Some dwarfism (*cretinism*) is caused by a deficiency of the thyroid hormones rather than by a deficiency of pituitary growth hormone. Treatment by hormonal replacement may increase growth somewhat and help retain fertility in sexually active patients.

Pituitary Gland: Neurohypophysis

When ADH (vasopressin) is undersecreted, excessively large amounts of water are excreted in the urine (*polyuria*). This condition, known as ***diabetes insipidus,*** is accompanied by dehydration and unrelenting thirst (*polydipsia*). Diabetes insipidus can often be treated successfully by administering controlled doses of ADH.

Thyroid Gland

Overactivity of the thyroid gland, known as ***hyperthyroidism*** (Graves' disease), may be caused by long-acting thyroid stimulator (LATS). LATS is an antibody that acts on the thyroid gland in the same way that TSH does, but it is not regulated by the normal negative feedback control. Smoking also increases the possibility of developing Graves' disease. Graves' disease is inherited and is likely to appear in genetically predisposed people. Among the symptoms are nervousness, irritability, increased heart rate and blood pressure, weakness, weight loss, and high oxygen use, even at rest. The bulging eyes (*exophthalmos*) typical of this condition are due partly to increased fluid behind the eyes caused by an exophthalmos-producing substance (EPS). Drug therapy that inhibits thyroxine production and the administration of radioactive iodine to selectively destroy the overactive thyroid tissue have successfully replaced surgery as treatments for hyperthyroidism.

Underactivity of the thyroid, or ***hypothyroidism,*** is often associated with *goiter* (L. "throat"), an enlarged thyroid (see photo). Such a swelling in the neck is caused when insufficient iodine in the diet causes the thyroid to enlarge in an attempt to produce more thyroxine. The adaptive responses are triggered in large part by an increased secretion of TSH, which attempts to stimulate the iodine-trapping mechanism and the subsequent steps in the metabolism of iodine. Most cases of goiter used to be found in areas away from the ocean, where iodine content in the soil and water supply is low. With the addition of minute amounts of iodine to ordinary table salt and drinking water in recent years, this type of goiter has practically disappeared as a common ailment. Symptoms of hypothyroidism include decreased heart rate,

A woman with a massive goiter. Such a large goiter would not occur with proper medical treatment.

blood pressure, and body temperature; lowered basal metabolic rate; and underactivity of the nervous system.

Underactivity of the thyroid during the last three months of pregnancy and the first few months after birth causes *cretinism* (CREH-tin-ism), which is characterized by mental and skeletal retardation (dwarfism). The skin is jaundiced (yellow) and dry, the eyelids are puffy, the hair is brittle, and the shoulders sag. Early diagnosis and treatment of cretinism with L-thyroxine (a drug form of thyroxine) may arrest the disease before the nervous system is damaged.

If the thyroid becomes underactive during adulthood, *myxedema* (mick-suh-DEE-muh) may result, producing swollen facial features, dry skin, low basal metabolic rate, tiredness, possible mental retardation, and intolerance to cold in spite of increased body weight. Like cretinism, myxedema may be corrected with oral thyroxine if it is treated early.

An underactive thyroid gland in newborns interferes with the normal growth of cartilage and bone, resulting in an irreversible dwarfism. Such dwarfs are usually intelligent and active. They have stubby arms and legs, a relatively large chest and head, and flattened facial features.

Parathyroid Glands

The most common causes of underactivity of the parathyroid glands (*hypoparathyroidism*) are (1) damage to or removal of the parathyroids during surgery and (2) parathyroid adenoma. Hypoparathyroidism results in low concentrations of calcium and high levels of phosphorus in the blood. Such an imbalance may alter the transmission of nerve impulses and cause osteoporosis, impaired bone growth, and muscle tetany (paralysis). Controlled doses of vitamin D and calcium salts may restore normal calcium concentrations. Low-phosphate diets and drugs that increase the excretion of phosphorus in urine may also be successful.

Overactivity of the parathyroid glands (*hyperparathyroidism*) is usually caused by an adenoma. It results in excessive amounts of calcium and low levels of phosphorus in the blood. As in hypoparathyroidism, osteoporosis is evident, and many other general symptoms may occur, including loss of appetite, nausea, weight loss, personality changes, stupor, kidney stones, duodenal ulcers, kidney failure, increased blood pressure, and congestive heart failure. Treatment may include the surgical removal of excess parathyroid tissue, drugs that lower the calcium concentration, and in severe cases of kidney failure, an artificial kidney or kidney transplant.

Adrenal Cortex

Two diseases caused by overactivity of the adrenal cortex (*hyperadrenalism*) are Cushing's disease and adrenogenital syndrome. *Cushing's disease* is usually caused by a corti-cal tumor that overproduces glucocorticoids. Symptoms include a redistribution of fat to the face, chest, and abdomen (the limbs remain normal), accompanied by abdominal striations and a tendency toward diabetes mellitus caused by increased blood glucose levels. Protein is lost, and the muscles become weak. Surgical removal of the causative tumor usually brings about remission.

Adrenogenital syndrome is also caused by an overactive adrenocortical tumor, which stimulates excessive production of the cortical male sex hormones known as *androgens*. These androgens cause male characteristics to appear in a female and accelerate sexual development in a male. Hormonal disturbances during the fetal development of a female child may cause a distortion of the genitals, so that the clitoris and labia become enlarged and resemble a penis and scrotum. In a mature woman an extreme case of adrenogenital syndrome may produce a beard. (Such pronounced male characteristics in a female may be caused by other hormonal malfunctions besides defects in the adrenal cortex.)

Underactivity of the adrenal cortex (*hypoadrenalism*), resulting in a decreased output of glucocorticoids and aldosterone, produces *Addison's disease,* whose symptoms include anemia (deficiency of red blood cells), weakness and fatigue, increased blood potassium, and decreased blood sodium. Skin color becomes bronzed because excess ACTH, due to the loss of negative feedback inhibition by cortisol, mimics the skin-darkening effects of MSH (melanocyte-stimulating hormone).* (ACTH and MSH have the same first 13 amino acids.) Until recently Addison's disease was usually fatal, but now it can be controlled with regular doses of cortisone and aldosterone.

Pancreas

About 4 percent of the people in the United States will develop *diabetes mellitus* at some time in their lives. It can occur as either *Type I (insulin-dependent) diabetes,* which usually begins early in life, or *Type II (insulin-independent) diabetes*, which occurs later in life, mainly in overweight people. Heredity plays a major role in the development of both types. Type I diabetes results when beta cells do not produce enough insulin. When this happens, glucose accumulates in the blood and spills into the urine, but does not enter the cells. Excess glucose in the urine is a diuretic and causes excessive urine production (*polyuria*), excessive thirst (*polydipsia*), and dehydration. Because the cells are unable to use the accumulated blood glucose (the most readily available energy source in the body), the body must shift to fat metabolism — the second-preferred body fuel. Appetite may increase, leading to excessive eating (*polyphagia*).

---

*President Kennedy's permanent "suntan" was probably due to Addison's disease.

In Type II diabetics there is a nearly normal plasma concentration of insulin. The problem is hyporesponsiveness, or no response, to insulin, a condition termed *insulin resistance.* Insulin's target cells do not respond normally to the circulating insulin because of alteration in the insulin receptors. Also, an insufficient number of insulin receptors per target cell may be a cause of the insulin resistance, especially in those individuals who are overweight. The majority of diabetics (about 79 percent) are Type II.

Because the removal of glucose by the kidneys requires large amounts of water, a diabetic person produces excessive "sugary" urine; he or she may excrete as much as 20 L of sugary urine per day.* In response to the increased urine production, with the possibility of serious body dehydration, diabetics become extremely thirsty and drink large amounts of liquids.

---

*The word *diabetes* comes from the Greek word for "siphon" or "to pass through," referring to the seemingly instant elimination of liquids. The word was actually used by the Greeks as early as the first century. In the seventeenth century, the sweetness of diabetic urine was discovered, and the name of the disease was lengthened to diabetes mellitus. *Mellitus* comes from the Greek *meli,* honey.

The use of fats (to replace glucose) for energy production in the diabetic causes the accumulation of acetoacetic acid, β-hydroxybutyric acid, and keto acids in the blood and body fluids. This leads to acidosis, which can lead to coma and death.

Diabetes is incurable in any form, but mild diabetes can usually be controlled by strict dietary regulation, weight control, and exercise. More serious cases may require treatment with regular injections of insulin (Type I) or oral hypoglycemic agents (Type II). If the disease is untreated, almost every part of the diabetic's body will be affected, and gangrene, hardening of the arteries (arteriosclerosis), other circulatory problems, and further complications may occur.

Low blood glucose, **hypoglycemia,** may be caused by an excessive secretion of insulin. Glucose is not released from the liver, and the brain is deprived of its necessary glucose. Hypoglycemia can be controlled by regulating the diet, especially carbohydrate intake. On a short-term basis, the glucose in a glass of orange juice may restore the normal glucose balance in the blood, but a long-term treatment usually consists of a reduction of carbohydrates in the diet. Carbohydrates tend to stimulate large secretions of insulin, thereby removing glucose from the bloodstream too quickly.

CHAPTER SUMMARY

The endocrine system and the nervous system together constitute the two great regulatory systems of the body. The endocrine system is made up of tissues or organs called **endocrine glands,** which secrete chemicals directly into the bloodstream. These chemical secretions are called **hormones.**

Hormones and Their Feedback Systems (p. 548)

1 Most hormones are steroids, derivatives of amino acids, or proteins.

2 Hormones travel through the bloodstream to all parts of the body but affect only **target cells** that have compatible receptors.

3 Hormonal secretions are usually kept at a normal level by *negative feedback systems* involving other glands and hormones.

Mechanisms of Hormone Control (p. 552)

1 Water-soluble hormones (amine and protein hormones) regulate cellular responses through a mechanism called the **fixed-membrane-receptor mechanism.** The end result of the chemical activation is the diffusion of **cyclic AMP** throughout the cell, which causes the cell to perform its distinctive function.

2 Lipid-soluble hormones (steroids) regulate cellular activity through the **mobile-receptor mechanism,** involving synthesis of proteins that affect the cell's function.

Pituitary Gland (Hypophysis) (p. 553)

1 The **pituitary gland,** or **hypophysis,** consists of the anterior lobe **(adenohypophysis)** and the posterior lobe **(neurohypophysis).** The adenohypophysis has an abundance of secretory cells, while the neurohypophysis contains many nerve endings.

2 The pituitary is connected to the hypothalamus by a stalk of nerve cells and blood vessels called the *infundibular stalk,* which provides a direct link between the nervous system and the endocrine system.

3 The **hypothalamus** releases regulating substances called **releasing** and **inhibiting hormones** that control the secretions of the adenohypophysis. Nerve centers in the hypothalamus regulate secretions from the neurohypophysis. The connection between the hypothalamus and the adenohypophysis is facilitated by a system of blood vessels called the **hypothalamic-hypophyseal portal system,** whereas the hypothalamic link with the neurohypophysis relies on nerve impulses.

4 The hypothalamus synthesizes **antidiuretic hormone** (ADH) and **oxytocin,** which are stored in and secreted from the neurohypophysis.

5 The true endocrine portion of the pituitary is the adenohypophysis, which synthesizes and secretes seven separate hormones: **growth hormone (GH), prolactin (PRL), thyroid-stimulating hormone (TSH), adrenocorticotropic hormone (ACTH), luteinizing hormone (LH), follicle-stimulating hormone (FSH),** and **melanocyte-stimulating hormone (MSH).**

Thyroid Gland (p. 558)

1 The **thyroid gland** secretes the thyroid hormones **thyroxine** and **triiodothyronine** from its follicular cells and **calcitonin (CT)** from its parafollicular cells. Thyroid hormones increase basal metabolism, accelerate growth, and stimulate cellular differentiation and protein synthesis.

2 Calcitonin lowers calcium concentrations in the blood.

Parathyroid Glands (p. 561)

Parathormone (PTH), the hormone of the **parathyroid glands,** increases the concentration of calcium in the blood and decreases the concentration of phosphate.

Adrenal Glands (p. 562)

1 The **adrenal glands** are composed of an inner **medulla** and an outer **cortex.** The adrenal cortex secretes the three types of steroid hormones: glucocorticoids, mineralocorticoids, and gonadocorticoids.

2 **Glucocorticoids,** mainly **cortisol,** are essential to the proper metabolism of carbohydrates, proteins, and fats, most critically through the stimulation of *gluconeogenesis,* the synthesis by the liver of glucose from noncarbohydrate sources. They also act to suppress allergic reactions and inflammatory responses.

3 **Mineralocorticoids,** the most important of which is **aldosterone,** control mineral balance through the retention and loss of sodium and potassium.

4 **Gonadocorticoids,** or adrenal sex hormones, affect the sex organs, but only slightly.

5 The adrenal medulla secretes **epinephrine** and **norepinephrine (NE),** which cause sympathomimetic effects.

6 Stressful conditions stimulate increased production of ACTH by the pituitary gland, which stimulates the adrenal cortex and adrenal medulla to prepare the body's muscular, digestive, circulatory, and respiratory systems to cope with the stress.

Pancreas (p. 568)

1 The **pancreas** functions as an exocrine gland, secreting digestive enzymes into ducts, and as an endocrine organ, secreting hormones into the bloodstream. The endocrine portion synthesizes, stores, and secretes hormones from the **pancreatic islets.**

2 The pancreatic islets contain alpha cells that secrete **glucagon,** beta cells that secrete **insulin,** delta cells that secrete somatostatin, and F cells that secrete pancreatic polypeptide.

3 Glucagon raises the concentration of blood glucose and stimulates the release of fatty acids and glycerol from adipose tissue. The most important effect of insulin is to facilitate glucose transport across cellular membranes.

Gonads (p. 570)

1 The **gonads**—**ovaries** in a female and **testes** in a male—secrete hormones that control reproductive functions.

2 The major hormones are **testosterone** in males and **estrogens, progestins,** and **relaxin** in females.

Other Sources of Hormones (p. 571)

1 The **kidneys** secrete several hormones, including erythropoietin, 1,25-dihydroxyvitamin D_3, prekallikreins, and prostaglandins. In addition, the kidneys produce **renin,** an enzyme that causes blood pressure to increase.

2 The **pineal gland** contains several chemicals, whose functions have not been established. The secretion of **melatonin,** which may affect skin pigmentation, seems to be affected by light signals and may be involved with the sleep cycle, seasonal affective disorder, and puberty.

3 The **thymus gland** secretes several hormones related to immunity and the maturation of T cells.

4 The **heart** produces, stores, and secretes the peptide hormone called **atrial natriuretic peptide (ANP).** It is believed that atriopeptin helps maintain the proper balance of fluids and electrolytes and lowers excessively high blood pressure and volume.

5 The **digestive system** secretes several digestive hormones, especially **gastrin,** which aids digestion in the stomach; **secretin,** which helps neutralize stomach acid; and **cholecystokinin,** which stimulates the release of bile and enzymes from the gallbladder and pancreas.

6 The **placenta** produces hormones that help maintain pregnancy.

7 **Eicosanoids** are fatty acid hormonal mediators or messengers that are found in many parts of the body. They appear to be involved in regulating blood flow and pressure, blood clotting, digestive secretion, contraction of the uterus, and microorganism-destroying blood cells.

The Effects of Aging on the Endocrine System (p. 573)

1 The endocrine system is relatively unaffected by aging.

2 Some changes include a lowering of the basal metabolic rate as the thyroid becomes smaller and insulin secretion and/or response is altered.

Developmental Anatomy of the Pituitary and Thyroid Glands (p. 574)

1 The embryonic pituitary gland develops from **ectoderm** of the primitive mouth cavity and **neuroectoderm** of the diencephalon.

2 The ectoderm develops into the **adenohypophysis,** consisting of the **pars distalis, pars tuberalis,** and **pars intermedia.** The neuroectoderm develops into the **neurohypophysis,** which is made up of the **infundibulum** and **pars nervosa.**

3 The thyroid gland originates as a mass of endoderm of the pharyngeal pouches.

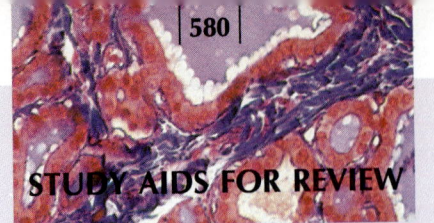

STUDY AIDS FOR REVIEW

KEY TERMS

adenohypophysis 553
adrenal cortex 562
adrenal glands 562
adrenal medulla 566
adrenocorticotropic
  hormone (ACTH) 558
aldosterone 565
anabolic steroid 564
antidiuretic hormone
  (ADH) 555
atrial natriuretic
  peptide (ANP) 572
calcitonin 560
cholecystokinin 572
cortisol 563
diabetes mellitus 570, 577
eicosanoids 572
endocrine gland 548
epinephrine 566
estrogens 571
fixed-membrane-receptor
  mechanism 552
follicle-stimulating hormone
  (FSH) 558, 570, 571
gastrin 572
glucagon 568
glucocorticoid 563
gonadocorticoid 566
growth hormone (GH) 557
hormone 548
hypoglycemia 570, 578
hypophysis 553

hypothalamic-hypophyseal
  portal system 554
hypothalamus 554
insulin 568
leukotrienes 573
luteinizing hormone
  (LH) 558, 570, 571
mineralocorticoid 565
mobile-receptor
  mechanism 553
neurohypophysis 553
norepinephrine 566
oxytocin 555
pancreatic islet 568
parathormone 561
parathyroid glands 561
pineal gland 571
pituitary gland 553
progestins 571
prolactin 557
prostaglandins 572
renin 565
secretin 572
target cell 548
testosterone 570
thymus gland 571
thyroid gland 558
thyroid-stimulating hormone
  (TSH) 558
thyroxine 559
triiodothyronine 559

MEDICAL TERMS FOR REFERENCE

HYPERPLASIA (Gr. *hyper*, over + *plasis*, change, growth) A nontumorous growth in a tissue or organ.

HYPOPLASIA Incomplete or arrested development of a tissue or organ.

POSTPRANDIAL (L. *post*, after + *prandium*, late breakfast, meal) Usually pertaining to an examination of blood glucose content after a meal.

RADIOIMMUNOASSAY (RIA) (RAY-dee-oh-im-myu-noh-ASS-ay) A technique that measures minute quantities of a substance, such as a hormone, in the blood.

REPLACEMENT THERAPY A method of treatment where insufficient secretions of hormones are replaced with natural or synthetic chemicals.

STEROID THERAPY Use of steroids to treat certain endocrine disorders.

VIRILISM (L. *vir*, man) Masculinization in women.

UNDERSTANDING THE FACTS

1 What is the difference between an endocrine gland and an exocrine gland?

2 Name the four chemical categories of hormones, give two examples of each kind, and state where each kind is synthesized.

3 What are the functions of adenylyl cyclase, cyclic AMP, and cyclic AMP phosphodiesterase in hormonal action?

4 Contrast the source of synthesis of the hormones of the anterior pituitary with the source of synthesis for those of the posterior pituitary.

5 List the pituitary hormones and state the principal function(s) of each.

6 What are the clinical uses of oxytocin, growth hormone, and prostaglandins?

7 List the hormones of the thyroid gland and state the primary function(s) of each.

8 In which region of the adrenal cortex are the three kinds of adrenal cortical hormones synthesized? What is the primary function of each kind?

9 Identify the hormones synthesized in the pancreas and identify the cell type that produces each one.

10 List the hormones that are produced by the kidney, pineal gland, heart, and digestive system and state the principal function(s) of each.

11 Identify the hormonal imbalance that is the cause of each of the following illnesses or conditions:

a acromegaly	f diabetes mellitus
b giantism	g Graves' disease
c goiter	h Cushing's disease
d myxedema	i Addison's disease
e diabetes insipidus	

12 Where is each of the following structures?

a neurosecretory cells	c parafollicular cells
b infundibulum	d pineal gland

13 When and where is melatonin produced?

UNDERSTANDING THE CONCEPTS

1 What is meant by a fixed-membrane receptor and a mobile receptor? Why are both kinds of receptors required in the endocrine system?

2 Contrast the influence of the hypothalamus on the adenohypophysis and the neurohypophysis.

3 What hormones act in the production and release of milk from the mammary glands?

4 Which hormones affect the growth of children and adolescents?

5 How do hormones produced by the thyroid and the parathyroid glands control calcium blood levels?

6 Which hormones control the level of each of the following?

a growth hormone	**d** TSH	**f** FSH
b prolactin	**e** ACTH	**g** LH
c thyroxine		

7 Explain how the levels of glucocorticoids and mineralocorticoids are regulated.

8 Why are chromaffin cells considered to be equivalent to sympathetic postganglionic neurons?

9 Contrast the stress response of the cortical and medullary cells of the adrenal gland.

10 How do glucagon and insulin work together to regulate the level of blood glucose?

11 What is the difference between gluconeogenesis and glycogenesis? What are the primary hormones that act in each process?

12 Why are prostaglandins not considered true hormones?

13 Contrast Type I diabetes with Type II diabetes.

SELF-QUIZ

Multiple-Choice Questions

18.1 Epinephrine and norepinephrine are secreted by the

a stomach mucosa	**d** adrenal medulla
b adrenal cortex	**e** anterior pituitary
c pancreatic islets	

18.2 The major hormone regulating the metabolic rate is

a thyroxine	**d** calcitonin
b parathormone	**e** TSH
c adrenaline	

18.3 The release of hormones from the adenohypophysis is controlled by secretions of the

a adrenal medulla	**d** posterior pituitary
b kidneys	**e** small intestine
c hypothalamus	

18.4 The part of the pituitary that contains secretory cells is the

a sella turcica	**d** posterior lobe
b anterior lobe	**e** hypothalamic-hypophyseal portal system
c infundibulum	

Matching

Identify the source of each of the hormones:

a hypothalamus	**d** gonads
b adenohypophysis	**e** many sites of production throughout the body
c adrenal cortex	

18.5 _____ thromboxanes
18.6 _____ somatostatin
18.7 _____ gonadotropins
18.8 _____ prolactin
18.9 _____ aldosterone

18.10 _____ prostaglandins
18.11 _____ cortisol
18.12 _____ thyroid-stimulating hormone
18.13 _____ thyrotropin-releasing hormone
18.14 _____ leuteinizing hormone

Completion Exercise

18.15 Allergic reactions and inflammatory responses are inhibited by _____.

18.16 The production of _____ by the pineal gland is inhibited by _____.

18.17 Inhibin is secreted by the _____, whereas relaxin is secreted by the _____.

18.18 The precursor molecule for all of the steroid hormones is _____.

18.19 The adenohypophysis is controlled by hypothalamic hormones that travel within a system of blood vessels called the _____.

18.20 Gluconeogenesis is stimulated by hormones produced in the _____ and _____, whereas glycogenesis is stimulated by the pancreatic hormone _____.

18.21 Hypothalamic neurons that produce hormones are called _____ cells.

18.22 The hormones that enter the bloodstream from the neurohypophysis are synthesized in the _____.

18.23 The parathyroid glands are embedded in the _____.

18.24 Chromaffin cells secrete the hormones _____ and _____.

A SECOND LOOK

Label the hypothalamus, neurohypophysis, adenohypophysis, secretory cells, and neurosecretory cells.

iii case studies

1

Recently, an experienced 16-year-old gymnast began having severe lower-back pain. The pain appears to radiate to her left buttock as well as down the posterior aspect of her left leg, and it intensifies during coughing and sneezing. In addition, there is a mild, diverse numbness and minor loss of motor functions in the left leg.
- **What neurological disorder probably exists?**
- **Which spinal nerve is involved?**
- **What are the neurological causes of the symptoms?**
- **Which imaging techniques would be used to confirm the diagnosis?**

2

One winter day, while riding his motorcycle to work, a young man began to experience an excruciating, stabbing pain and weakness on the left side of his face. Within a day or two, the facial muscles on the left side were paralyzed, leaving the patient unable to smile or even to close his left eye. Taste perception on the anterior portion of the tongue was also distorted. Fortunately, with time, both the sensory and motor symptoms disappeared completely.
- **What is the name of the condition described and what nerve is probably involved?**
- **What are the major branches of this nerve?**
- **What is the underlying cause of the condition?**

3

A 50-year-old professor has been having increasing difficulty, while giving his lectures, in keeping in focus both his notes on the lectern and the faces of his students in the audience. He wears corrective lenses for distance viewing and has no difficulty seeing his students' faces clearly. However, when he shifts his gaze to the lectern, the writing on the pages appears blurred. He is puzzled because he normally reads well without corrective lenses.
- **Is the professor nearsighted or farsighted?**
- **What term describes the reflex that enables us to change the focal depth of our gaze?**
- **How does the reflex work?**
- **What is the cause of the professor's problem?**

4

A research executive is in charge of developing drugs that block the effects of sympathetic and parasympathetic postganglionic neurotransmitters. To design such drugs, she must understand the neurochemistry of the autonomic system. She must be able to answer the following questions.
- **What neurotransmitters are released by sympathetic and parasympathetic postganglionic neurons?**
- **With what types of receptors do these neurotransmitters interact?**
- **What drugs can block the effects of sympathetic stimulation at autonomic effector organs? What about parasympathetic stimulation?**
- **What single type of drug can block both sympathetic and parasympathetic effects at the same time?**

Transport and Maintenance

CAREER IN FOCUS: EXERCISE PHYSIOLOGY

More than 30 years ago exercise physiologists established a connection between exercise and cardiovascular rehabilitation. Shortly thereafter, the medical community realized that exercise was more than a cure for numerous disorders—it could be a preventative. And there was more to rivet public attention as study after study showed that weight loss depended not only on reduction of caloric intake, but also on the physiological effects of exercise. Certainly everyone has come to realize how important it is to be physically active. Rather than just helping us fit into last year's clothes, physical activity is vitally important in keeping our body systems healthy. Exercise physiologists are now recognized for the major contributions they are making toward improving public health.

The breadth of the field of exercise physiology is apparent in the certification standards set by the American College of Sports Medicine. The ACSM offers certification at three levels of training and responsibility in its health and fitness track. These professionals work with an apparently healthy public in the design and implementation of competitive sports programs and in the training and rehabilitation of athletes after injury. The entry-level position is the exercise leader. Higher levels include the health/fitness instructor and health/fitness director. In addition to passing written and practical tests, instructors must have an

undergraduate degree and directors a graduate degree, both in an allied health field.

The ACSM's clinical track offers three levels of certification for exercise physiologists who work in a clinical setting with patients suffering primarily from cardiovascular, pulmonary, or metabolic diseases. Here the entry-level certification is for the exercise test technologist, an individual who is competent to administer safe and valid exercise-related tests. Higher certifications include those for the exercise specialist, who assesses the needs of individual patients, and the clinical program director, who oversees the design and implementation of effective clinical programs. Once again, the specialist must have an undergraduate degree and the program director a graduate degree in an allied health field.

The work of clinical and fitness professionals rests on the basic scientific research of still other exercise physiologists who work primarily in the academic settings of universities and medical schools. Active areas of research include the methods by which a muscle cell actually generates force, the ways in which exercise prevents osteoporosis, and the mechanisms by which exercise regulates the body's metabolic "set point." These researchers usually have doctoral degrees in one of the biomedical sciences, and they form a substantial portion of the faculty within the allied health and medical fields.

Many universities offer bachelor's and/or master's programs in

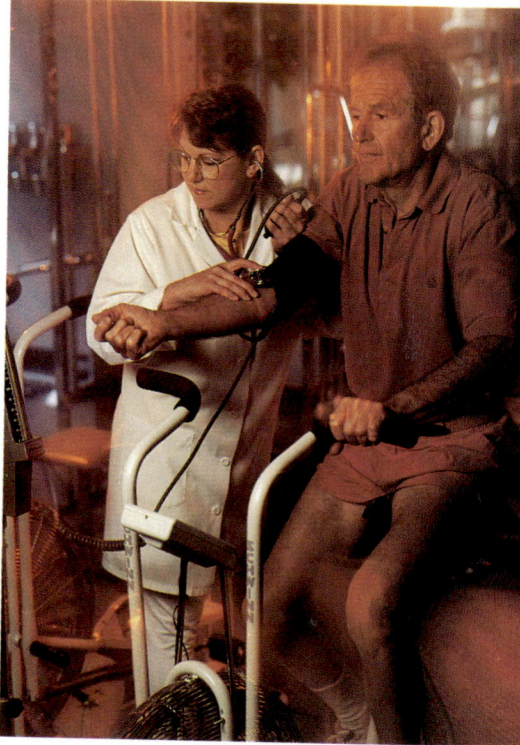

clinical exercise physiology. All of these programs include such courses as anatomy and physiology, kinesiology, biochemistry, biomechanics, electrocardiography, and advanced cardiopulmonary physiology; and they all culminate in a clinical internship.

The focus of exercise physiology is prevention of disease. As prevention is clearly seen to be both humane and cost-effective, careers that fall under its umbrella will become even more attractive.

The exercise physiologist needs to be thoroughly grounded in the areas of anatomy and physiology that we look at in this section—the transport and maintenance systems of the human body.

The Cardiovascular System: Blood

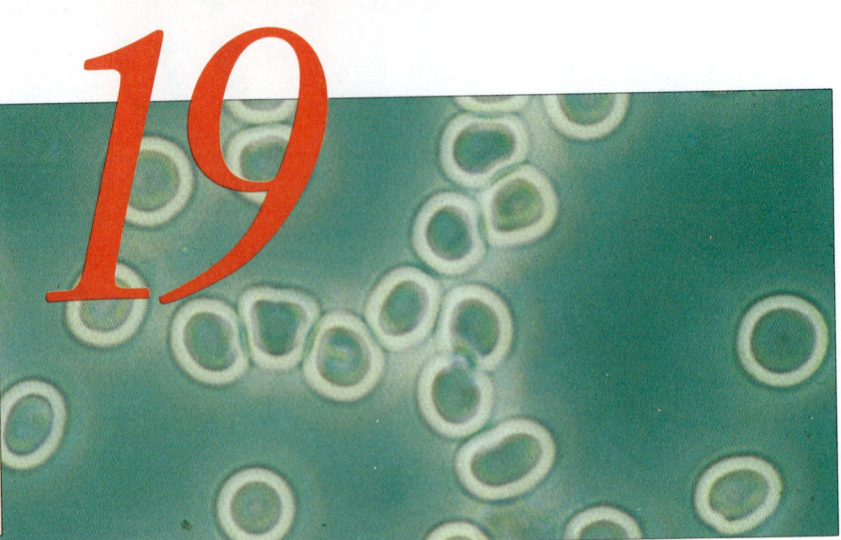

19

KEY CONCEPTS

1 As blood circulates throughout the body's extensive system of blood vessels, nutrients, hormones, and other substances diffuse out of the blood and into body tissues, while waste products diffuse from the tissues into the blood.

2 Red blood cells transport oxygen from the lungs to the body tissues and transport carbon dioxide from the tissues to the lungs. White blood cells are scavengers that serve a protective and housecleaning function, and platelets help blood to clot at appropriate times.

3 Hemostasis, or the prevention of blood loss, is a three-step process that involves blood vessels, platelets, and the overall process of blood clotting.

4 Because of differing surface proteins, not all blood types are compatible during blood transfusions.

The *cardiovascular system* is made up of the heart (a pump), blood vessels (a means of transport), and blood, the transport medium. Blood can also be classified as part of the *circulatory system*, which is a more general term that includes not only the blood, blood vessels, and heart, but also the lymph and lymph vessels. Blood is classified as a type of connective tissue because it shows the characteristics of connective tissue and because it is derived from mesodermal mesenchyme.

Blood consists of an extracellular liquid matrix, known as ***plasma*** (about 55 percent by volume), and cells, or the ***formed elements*** (about 45 percent by volume), which are mostly blood cells suspended in the plasma. The formed elements include red blood cells (erythrocytes), white blood cells (leukocytes), and fragmented cells called platelets (thrombocytes) [FIGURES 19.1, 19.2].

Both the formed elements and plasma play important roles in homeostasis. Throughout the chapter, emphasis is placed on showing the many ways in which blood helps maintain homeostasis.

FIGURE 19.1 HUMAN BLOOD CELLS

This scanning electron micrograph shows thrombocytes, erythrocytes, and leukocytes. ×4000.

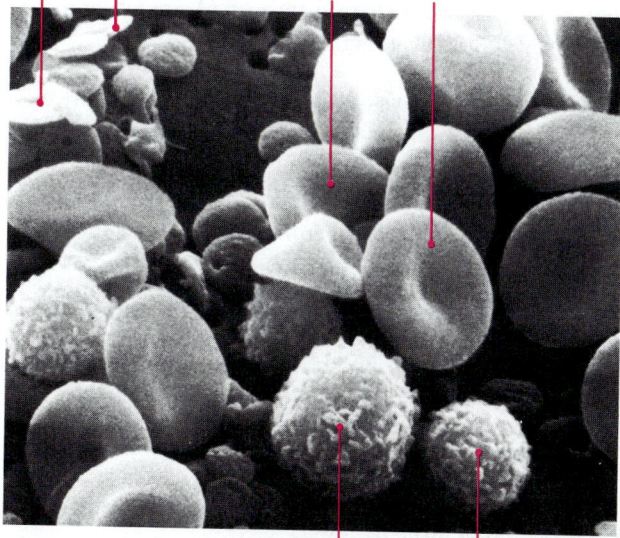

Thrombocytes (platelets) Erythrocytes (red blood cells)

Leukocytes (white blood cells)

FIGURE 19.2 COMPONENTS OF BLOOD (NORMAL ADULT)

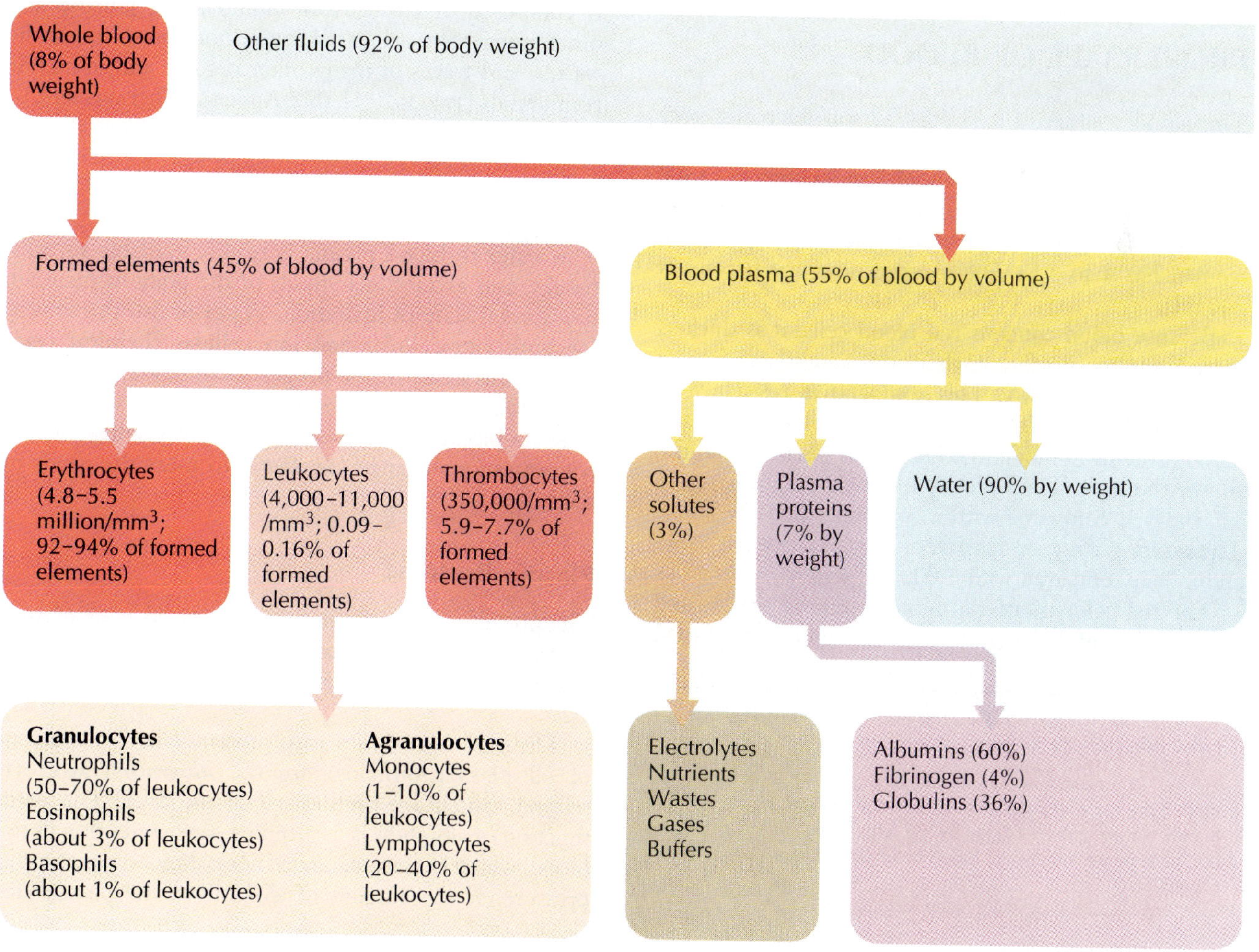

Whole blood (8% of body weight)

Other fluids (92% of body weight)

Formed elements (45% of blood by volume)

Blood plasma (55% of blood by volume)

Erythrocytes (4.8–5.5 million/mm^3; 92–94% of formed elements)

Leukocytes (4,000–11,000 /mm^3; 0.09–0.16% of formed elements)

Thrombocytes (350,000/mm^3; 5.9–7.7% of formed elements)

Other solutes (3%)

Plasma proteins (7% by weight)

Water (90% by weight)

Granulocytes
Neutrophils (50–70% of leukocytes)
Eosinophils (about 3% of leukocytes)
Basophils (about 1% of leukocytes)

Agranulocytes
Monocytes (1–10% of leukocytes)
Lymphocytes (20–40% of leukocytes)

Electrolytes
Nutrients
Wastes
Gases
Buffers

Albumins (60%)
Fibrinogen (4%)
Globulins (36%)

FUNCTIONS OF BLOOD

As blood circulates throughout the body, tissues are continuously adding to it their waste products, secretions, and metabolites, and taking from it vital nutrients, oxygen, hormones, and other substances. Overall, blood performs the following functions:

1 *Transports* oxygen from the lungs to body tissues and transports the waste products of cellular metabolism from body tissues to the kidneys, liver, lungs, and sweat glands for eventual removal from the body; it also transports nutrients, hormones, and enzymes throughout the body.

2 *Regulates* blood clotting to stop bleeding; body temperature, mainly by increasing or decreasing blood flow to the skin for heat exchange; acid-base balance (pH) through the distribution of buffers; and the renal (kidney) control of the amount of water and electrolytes in body fluids.

3 *Protects* against harmful microorganisms and other substances by contributing white blood cells, proteins, and antibodies to the inflammatory and immune responses.

PROPERTIES OF BLOOD

The blood volume of a healthy person fluctuates very little. Even when blood is lost, it is replaced rapidly. The blood of an average adult is about 7 to 9 percent of total body weight, or 79 mL/kg of body weight. An average-sized man has 5 to 6 L of blood, and an average-sized woman has 4 to 5 L. The average newborn has about 240 mL.

Because blood contains red blood cells, it is thicker, denser, and more adhesive than water and flows four to five times more slowly. This comparative resistance to flow is called *viscosity*. The more red blood cells and blood proteins in blood, the higher the viscosity and the slower the flow. The viscosity of blood ranges between 3.5 and 5.5, compared with a constant 1.000 for water. The *specific gravity*, or density, of blood is between 1.045 and 1.065, compared with 1.000 for water.

The red color of arterial blood is due to oxygenated *hemoglobin*, a globular protein carried by the red blood

---

Q: *Does the circulation of a pregnant woman contain more blood than that of a nonpregnant woman?*

A: Yes. A pregnant woman has about 0.5 L (1 pint) of blood more than a nonpregnant woman because the metabolic demands placed upon her body by the fetus stimulate the production of a greater number of blood cells and a larger volume of plasma.

---

cells. When oxygen is removed, the blood appears darker. White blood cells and platelets in the blood are clear, and the plasma is yellowish.

Blood is slightly alkaline, with a pH between 7.35 and 7.45. Arterial blood is more alkaline than venous blood because it has less carbon dioxide. As the concentration of carbon dioxide increases, it reacts with water to form carbonic acid, which then lowers the blood pH. The temperature of blood averages about 38°C (100.4°F).

A S K   Y O U R S E L F

1 What is the normal viscosity of blood? Normal pH? Normal temperature?

2 Why is blood usually red?

PLASMA

Plasma is the liquid part of blood. It is about 90 percent water and provides the solvent for dissolving and transporting nutrients. A group of proteins constitutes another 7 percent of the plasma. The remaining 3 percent is composed of electrolytes, amino acids, glucose and other nutrients, various enzymes, hormones, metabolic wastes, and traces of many other organic and inorganic compounds [TABLE 19.1]. (See Appendix D, TABLES D.1–D.4.)

Water in Plasma

The water in blood plasma is readily available to cells, tissues, and extracellular fluids of the body to maintain the normal state of hydration. Water is also the solvent for both extracellular and intracellular chemical reactions. Thus many of the properties of blood plasma can be correlated with the basic properties of water. The water in blood plasma also contains many solutes whose concentrations change constantly to meet the needs of the body.

Plasma Proteins

The proteins in blood plasma are referred to as *plasma proteins*. The total protein component of blood plasma can be divided into the *albumins*, *fibrinogen*, and the *globulins*.

The most abundant plasma proteins (about 60 percent of all plasma proteins) are the *albumins* (al-BYOO-mihnz), which are synthesized in the liver. The main function of albumins is to promote water retention in the blood, which in turn maintains normal blood volume and pressure. If the amount of albumin in the plasma de-

TABLE 19.1 COMPONENTS OF BLOOD PLASMA

Substance	Description
WATER	Constitutes about 90 percent of plasma; maintains normal body hydration and is a solvent for extracellular and intracellular chemical reactions; contains solutes whose concentrations change to meet body needs.
PLASMA PROTEINS	
Albumins	Constitute about 60 percent of all plasma proteins; are synthesized in liver; provide colloid osmotic or oncotic pressure that regulates passage of water and diffusible solids through capillaries; promote water retention in blood, which maintains normal blood volume and pressure; act as carrier molecules for hormones, other substances transported in plasma.
Fibrinogen	Constitutes about 4 percent of all plasma proteins; is synthesized in liver; is essential for blood clotting.
Globulins	Constitute about 36 percent of all plasma proteins; are synthesized in liver and plasma cells derived from B lymphocytes. Divided into alpha, beta, gamma globulins: alpha and beta globulins transport lipids and fat-soluble vitamins in blood; gamma globulins are antibodies that help prevent diseases such as measles, tetanus, polio.
PLASMA ELECTROLYTES	Inorganic molecules that separate into ions when dissolved in water. Na^+ has important effect on osmotic pressure, fluid movements, ECF volume. Other important ions include Cl^-, K^+, Ca^{2+}, PO_4^{2-}, I^-, Mg^{2+}.
NUTRIENTS	*Glucose* is body's most available source of usable energy. *Amino acids* are used rapidly by cells, provide building blocks for protein synthesis. *Lipids* are important components of nerve cells and steroid hormones; some serve as fuel source.
WASTE PRODUCTS	Composed of metabolic wastes, including lactic acid and nitrogenous wastes of protein metabolism; transported by plasma.
GASES	Principal gases dissolved in plasma are oxygen, nitrogen, carbon dioxide: oxygen transported by red blood cells and dissolved in plasma, nitrogen transported in dissolved plasma state, carbon dioxide transported by red blood cells in plasma both in dissolved state and in form of bicarbonate ion (HCO^{3-}).

creases, fluid leaves the bloodstream and accumulates in the surrounding tissue, causing a swelling known as *edema* (ih-DEE-muh). Albumins also act as carrier molecules by binding to molecules of other substances, such as hormones, that are transported in plasma.

Also produced by the liver is *fibrinogen* (fye-BRIHN-uh-jehn) (about 4 percent of all plasma proteins), a plasma protein essential for blood clotting. When fibrinogen and several other proteins involved in clotting are removed from plasma, the remaining liquid is called *serum.* The role of fibrinogen and related proteins in blood clotting will be described later in this chapter.

The *globulins* (GLAHB-yoo-lihnz) (about 36 percent of all plasma proteins) are divided into three classes based on their structure and function: alpha (α), beta (β), and gamma (γ). The *alpha* and *beta globulins* are produced by the liver. Their function is to transport lipids and fat-soluble vitamins in the blood. One form of these trans-

port molecules, *low-density lipoproteins* (LDL), transports cholesterol from its site of synthesis in the liver to various body cells. Another transporter, *high-density lipoproteins* (HDL), removes cholesterol and triglycerides from arteries, preventing their deposition there.

Gamma globulins are the immunoglobulins, antibodies that help prevent diseases such as measles, tetanus, and poliomyelitis. (An *antibody* is any one of millions of dis-

Q: *Does coffee help remove alcohol from the blood and thus increase the rate of "sobering up"?*

A: Time, rest, and the intake of fluids (including the fluid in a cup of coffee) help to remove alcohol from the blood, but not coffee itself. Alcohol is eliminated via the liver at a fairly steady, unchangeable rate. So, a person who is too drunk to drive will still be too drunk to drive after drinking a cup of coffee.

tinct proteins produced by the body that is capable of inactivating a specific bacterium, virus, protein, or cancer cell that it "recognizes" as foreign. The antibody combines with the foreign body, known as an *antigen*, forming an *antigen-antibody complex*.) The five classes of antibodies that make up the immunoglobulins are designated as IgG, IgA, IgM, IgD, and IgE (see Chapter 23).

Plasma Electrolytes

Electrolytes are inorganic molecules that separate into ions when they are dissolved in water. The ions are either positively charged (*cations*) or negatively charged (*anions*). The major cation of plasma is sodium (Na^+), which has an important effect on osmotic pressure and fluid movements and helps determine the total volume of extracellular fluid. The principal anion is chloride (Cl^-). The other major ions in plasma are potassium (K^+), calcium (Ca^{2+}), phosphate (PO_4^{2-}), iodide (I^-), and magnesium (Mg^{2+}).

Nutrients and Waste Products

Although *glucose* appears in plasma in low concentrations (70–110 mg/100 mL of blood), it is the body's most readily available source of usable energy.

Like glucose, *amino acids* are used rapidly by body cells and usually appear in the plasma in low concentrations. Amino acids are also important because they provide the building blocks for protein synthesis.

Lipids are found in plasma in the form of phospholipids, triglycerides, free fatty acids, and cholesterol. They are important components of nerve cells and steroid hormones, and some serve the body as an excellent source of fuel.

Several *metabolic wastes* — especially lactic acid, along with some nitrogenous waste products of protein metabolism — are transported by the plasma.

Gases

Oxygen, nitrogen, and carbon dioxide are the principal gases dissolved in plasma. Oxygen is transported by the red blood cells and is dissolved in the plasma. Nitrogen is transported in the dissolved state in plasma. Carbon dioxide is transported by red blood cells and in the plasma, both in the dissolved state and in the form of the bicarbonate ion (HCO_3^-).

ASK YOURSELF

1 What are the main functions of blood?

2 What are the major components of plasma?

3 What are the three classes of plasma proteins?

FORMED ELEMENTS

The solid formed elements of the blood consist mainly of red blood cells, white blood cells, and platelets.

Red Blood Cells (Erythrocytes)

Red blood cells, or *erythrocytes* (ih-RITH-roh-sites; Gr. *erythros*, red + cells), make up about half the volume of human blood. There are about 25 trillion erythrocytes in the body, and each cubic millimeter of blood contains 4 to 6 million erythrocytes. (The average-sized man has about 5.5 million erythrocytes per cubic millimeter of blood, and the average-sized woman has about 4.8 million.) Erythrocytes measure about 7 micrometers (μm) in diameter and are about 2 μm thick.* If five or six red blood cells were placed in a row, they would reach across the period at the end of this sentence.

The erythrocyte is shaped like a biconcave disk, slightly concave on top and bottom, like a doughnut without the hole poked completely through [FIGURE 19.3]. This shape provides a larger surface area for gas diffusion than a flat disk or a sphere and gives the erythrocyte greater flexibility to squeeze through narrow capillaries. When an erythrocyte is mature, it no longer has a nucleus or many organelles, such as mitochondria, so it can neither reproduce nor carry on metabolic activities. Thus it must rely on its store of already-produced proteins, enzymes, and RNA.

Hemoglobin Almost the entire weight of an erythrocyte consists of *hemoglobin*, Hb (HEE-moh-gloh-bihn;

---

*Hemoglobin got its name in a roundabout way. Early microscopes showed the shape of a red blood cell as a sphere, not the biconcave disk we can identify today. As a result, the cells were called "globules," the proteins they contained were called "globulins" or "globins," and the substance was named "hemoglobin" or "blood protein."

FIGURE 19.3 ERYTHROCYTE STRUCTURE

Drawing of an erythrocyte cut open to show its biconcave shape.

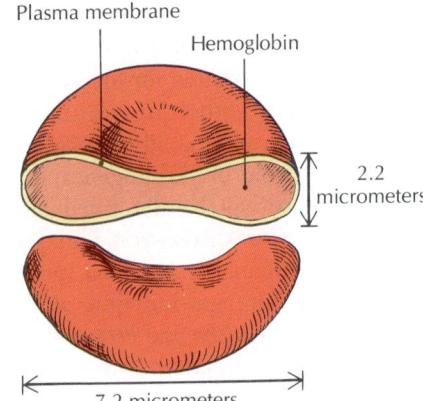

Plasma membrane

Hemoglobin

2.2 micrometers

7.2 micrometers

Gr. *haima*, blood + L. *globulus*, little globe), an oxygen-carrying globular protein.

Each adult hemoglobin molecule consists of 5 percent *heme*, an iron-containing pigment, and 95 percent *globin*, a polypeptide protein.* Attached to each of hemoglobin's four polypeptide chains (two α and two β) is a heme group, which gives blood its color. Attached to each heme group is an iron atom [FIGURE 19.4], which is the binding site for oxygen. The iron atom plays a key role in hemoglobin's function as an oxygen-carrying substance by binding oxygen and releasing it to tissues at the appropriate times.

The function of hemoglobin depends on its ability to combine with oxygen where the oxygen concentration is high (in the lungs) and to release oxygen where the concentration is low (in body tissues). Hemoglobin also carries waste carbon dioxide from the tissues to the lungs, where the carbon dioxide is exhaled. By removing carbon dioxide (an acid) from the plasma, hemoglobin helps maintain a stable acid-base balance in the blood.

The red color of arterial blood is due to oxygenated hemoglobin, called *oxyhemoglobin* (HbO₂). The oxyhemoglobin in the erythrocytes of subcutaneous capillaries gives cheeks and lips their pink color. The darker red (bluish) appearance of blood in the veins is due to the presence of *reduced hemoglobin*, hemoglobin that is *not* combined with oxygen.

The blood of a fetus contains *fetal hemoglobin* (HbF), which has a greater affinity for oxygen than adult hemoglobin does. As a result, the vital movement of oxygen from mother to fetus is enhanced. Fetal hemoglobin is usually replaced by adult hemoglobin within a few days after birth.

Males usually have more hemoglobin than females do—about 13.5 to 18 g/dL for males, about 12 to 16 g/dL for females, and about 14 to 20 g/dL for infants.

Transport of oxygen in blood The major role of erythrocytes is to transport oxygen from the lungs to the body tissues. The abundant oxygen in the alveoli (air sacs) of the lungs combines reversibly with the iron in hemoglobin to form oxyhemoglobin. The oxyhemoglobin is then transported in the red blood cells to the tissues. Because oxygen is constantly being used up by cellular oxidation, there is more oxygen in the blood coming from the lungs than there is in the tissues. Thus oxygen released from oxyhemoglobin tends to diffuse into the cells of the tissues.

Unfortunately, some toxic agents bind to hemoglobin even more readily than oxygen does. Air pollutants such as insecticides and sulfur dioxide bind to hemoglobin and prevent it from carrying oxygen effectively. Such poisons cause the numbness or dizziness that is characteristic of a

*A micrometer (μm) equals 1/1000 of a millimeter; it replaces the older term micron (μ).

FIGURE 19.4 HEMOGLOBIN

[A] Quaternary structure of a hemoglobin molecule. The folded chain represents globulin (protein), and the ovals represent the iron-containing heme groups. Note that the entire molecule consists of two alpha polypeptide chains and two beta chains. [B] Primary chemical structure of the heme group from a single hemoglobin molecule. Within each heme group is an iron atom (Fe²⁺), where a single oxygen molecule (O=O) may bind.

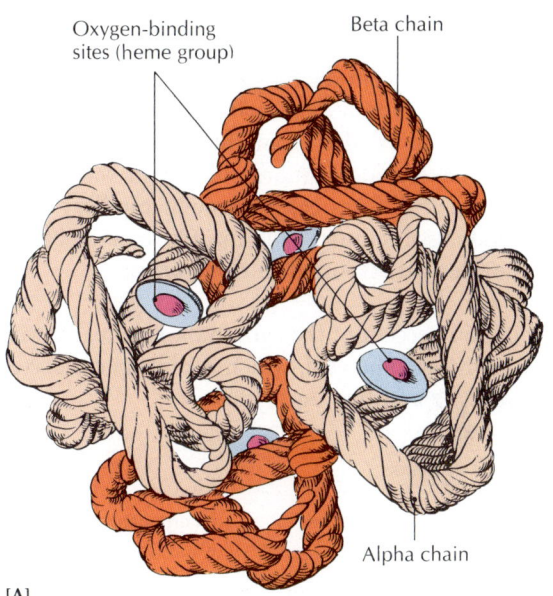

[A]

[B]

lack of oxygen. Probably the best-known hemoglobin poison is carbon monoxide, which is found in automobile exhaust fumes and cigarette smoke.* Carbon monoxide (CO) has an affinity for hemoglobin about 210 times that of oxygen, and it forms a stable compound. Even at concentrations of 0.1 or 0.2 percent in the air, carbon monoxide is dangerous, and increased amounts may cause death by blocking the uptake of oxygen by hemoglobin. When this happens, tissues die from lack of oxygen.†

Transport of carbon dioxide in blood In addition to carrying oxygen from the lungs to body tissues, erythrocytes transport carbon dioxide from the tissues to the lungs, where it is exhaled from the body as described earlier. Carbon dioxide is more soluble in water than oxygen is, and it diffuses easily through capillary walls from the body tissues into the plasma. Once in the blood, carbon dioxide is transported in three ways:

1 About 60 percent of the carbon dioxide reacts with water to form carbonic acid, which is quickly converted into bicarbonate and hydrogen ions:

$$CO_2 + H_2O \xrightleftharpoons{\text{carbonic anhydrase}} \underset{\substack{\text{carbonic} \\ \text{acid}}}{H_2CO_3} \rightleftharpoons \underset{\substack{\text{bicarbonate} \\ \text{ion}}}{HCO_3^-} + \underset{\substack{\text{hydrogen} \\ \text{ion}}}{H^+}$$

Thus the actual amount of carbonic acid in the blood at any time is very small. The two reactions above occur primarily in red blood cells, which contain large amounts of carbonic anhydrase (an enzyme that facilitates the reactions). Once the bicarbonate ion is formed, it moves out of the red blood cells and into the plasma, while chloride ions move into the red blood cells from the plasma. This ionic movement is carried out by an electrically neutral transport protein, which is driven by the difference in the concentrations of the bicarbonate and chloride ions. This exchange of chloride ions for bicarbonate ions is called the *chloride shift*. The process is reversed when blood reaches the lungs.

2 About 30 percent of the carbon dioxide reacts directly with hemoglobin to form the carbaminohemoglobin compound ($HbCO_2$), which is carried from the tissues to the lungs. When carbaminohemoglobin arrives in the lungs, the carbon dioxide is exchanged for oxygen.

3 The remaining 10 percent of the carbon dioxide is dissolved directly in the plasma and red blood cells as molecular carbon dioxide (CO_2).

---

*Twenty percent of a cigarette smoker's hemoglobin is nonfunctional because it has carbon monoxide bound to it instead of oxygen.

†Hemoglobin that binds to carbon monoxide instead of oxygen is an even brighter red than oxyhemoglobin. It is ironic that victims of carbon monoxide poisoning, whose tissues are fatally starved for oxygen, have bright-red, healthy-looking lips.

Erythrocyte plasma membranes and tonicity

Erythrocytes have a thin plasma membrane that is strong and flexible, allowing the erythrocytes to move easily through small blood vessels. The plasma membrane is permeable to water, oxygen, carbon dioxide, glucose, urea, and several other substances, but it is impermeable to hemoglobin and other large proteins.

Under normal conditions in the body, the total concentrations of solutes (osmotically active particles) on both sides of the erythrocyte membrane are the same. Thus the concentration of osmotically active particles in the plasma is *isotonic* to the fluid inside the erythrocyte. This relationship maintains a constant osmotic pressure on both sides of the plasma membrane, as well as a normal cell shape. If the plasma concentration of solutes increases above normal, the amount of water outside the erythrocyte *decreases* relative to the water concentration inside, and the plasma becomes *hypertonic*. As a result, water leaves the cell faster than it enters, and the cell shrinks, or *crenates* (L. notch or cleft). In contrast, if the plasma concentration of solutes decreases below normal, the amount of water outside the erythrocyte *increases* relative to the water inside, and the plasma becomes *hypotonic*. As a result, water enters the cell faster than it leaves, and the cell swells and eventually bursts. The rupturing of erythrocytes is called *hemolysis* (Gr. blood + loosening) [see FIGURE 3.4B]. Thus, the homeostasis of erythrocytes depends to a large extent on the concentration of solutes in the blood plasma.

Erythrocyte production (erythropoiesis)

Before birth, the fetus produces blood cells progressively in the yolk sac, liver, and spleen. During the fifth fetal month, blood cell production decreases in these sites and increases in the bone marrow cavities. After birth, erythrocytes are manufactured primarily, and continuously, in the red marrow of certain bones, especially in the vertebrae, ribs, sternum, pelvis, and upper ends of the femur and humerus. This process is called *erythropoiesis* (ih-RITH-roh-poy-EE-sis; Gr. red + to make).

Erythrocytes are derived from large embryonic cells in the bone marrow called *committed stem cells* or *hemocytoblasts,* which are destined to become different kinds of blood cells [FIGURE 19.5]. Some hemocytoblasts differentiate within the bone marrow into *common myeloid progenitor cells,* some of which then differentiate into *erythroblasts* and begin synthesizing hemoglobin. After several cell divisions, the erythroblast loses its nucleus and is now an immature red blood cell called a *reticulocyte* (Gr. "network cell") because it contains an intricate netlike pattern of endoplasmic reticulum. The reticulocyte leaves the bone marrow and enters the bloodstream, where it continues to synthesize hemoglobin. About 2 or 3 days later it loses its endoplasmic reticulum, mitochondria, and ribosomes as it matures into an *erythrocyte.*

The production and destruction (in the spleen) of erythrocytes are maintained at an equal rate. If red blood cells are lost from the circulatory system, the rate of erythropoiesis is increased until the normal erythrocyte number is regained.

Normal adult erythropoiesis produces about 10 billion cells an hour, enough in about a week for a pint of blood. (Blood donors are advised, however, to wait about 8 weeks before donating again.) Certain nutrients are necessary to maintain this pace, especially amino acids for the production of hemoglobin and other proteins and iron for the production of heme. Required in trace amounts are riboflavin; vitamin B_{12} and folic acid, which are necessary for the cell to mature; and vitamin B_6, which is required for the synthesis of hemoglobin.

The main controller of the rate of erythrocyte production is **erythropoietin** (ih-RITH-roh-poy-EE-tihn), a glycoprotein hormone produced mostly in the kidneys. It operates in a negative feedback system, as follows: The rate of erythropoiesis is sensitive to the pressure of oxygen in arterial blood, which is detected by the kidneys. Any process that decreases the delivery of oxygen to the tissues (such as hemorrhage) leads to *hypoxia* (low oxygen concentration) and a decreased oxygen pressure. This decrease in oxygen stimulates the production of erythropoietin, which then increases the number of hemocytoblasts committed to the production of erythrocytes. The increased number of erythrocytes increases the total amount of oxygen in the blood and the amount of oxygen that reaches the tissues. This relieves the hypoxia, which in turn inhibits the production of erythropoietin, and homeostasis is restored.

Destruction and removal of erythrocytes The life span of an erythrocyte is only about 80 to 120 days, primarily because it has no nucleus and is unable to replace the enzymes and other proteins that it needs for metabolism. Although erythrocytes are able to use some glucose as a source of energy, they cannot synthesize much protein. As the cells age, their protein goes through a normal process of degradation, which cannot be repaired by the cells. As a result, the plasma membrane begins to leak, and it leaks more and more as the integrity of the membrane protein continues to be lost.

As these aged and fragile erythrocytes pass through the narrow capillaries (sinusoids) of the spleen, liver, and other parts of the reticuloendothelial system (a system of macrophages throughout the body), their leaky membranes rupture, and the cellular remnants are engulfed by phagocytic cells called *macrophages*. The macrophages digest the hemoglobin into smaller amino acids, which are then returned to the body's amino acid pool for the future synthesis of new proteins. The heme portion of the hemoglobin molecule is converted first into *biliverdin* and then into the pigment *bilirubin*, which binds to plasma albumin and is transported to the liver. Within the liver, bilirubin is bound to glucuronic acid and is eventually secreted in bile. (If the liver is faulty, as in alcoholism or malaria, bilirubin may accumulate in abnormally high amounts and cause the skin to turn yellow—a condition known as *jaundice*.) The iron portion of the heme is bound to protein and stored in the bone marrow as *ferritin* (L. *ferrum*, iron). Iron is thus recycled and is available for the synthesis of new heme during the formation of new erythrocytes.

White Blood Cells (Leukocytes)

White blood cells, or **leukocytes** (LYOO-koh-sites; Gr. *leukos*, white + cells), serve as scavengers that destroy microorganisms at infection sites, help remove foreign molecules, and remove debris that results from dead or injured tissue cells. Leukocytes range from slightly larger to much larger than erythrocytes [FIGURE 19.5]. Unlike erythrocytes, leukocytes do have nuclei, and the cells are able to move about independently and pass through blood vessel walls into the tissues.

Leukocytes are able to produce a continuous supply of energy, and their anabolic and catabolic chemical processes are much more complex than those of erythrocytes. For example, they are able to produce mRNA in their nuclei and thus synthesize protein.

In adults, there are about 700 erythrocytes for every leukocyte. The normal adult leukocyte count is between 4,000 and 11,000 per cubic millimeter, which may increase as a result of infection to 25,000 per cubic millimeter, about the same number that newborn infants have. An increase in the number of white blood cells is called *leukocytosis*, and an abnormally low level of white blood cells is called *leukopenia*. Some white blood cells, especially lymphocytes, may live for months or years, but most live for just a few days. During periods of infection, phagocytic white blood cells may live for only a few hours.

Q: *If the blood in our veins is a very dark red (bluish), why does it appear lighter red when we cut a vein and begin to bleed?*

A: The deoxygenated venous blood turns bright red as soon as it is exposed to oxygen in the air.

Q: *Where is most of the body's blood located?*

A: Most of your blood—about 64 percent—is located in your veins, with about 20 percent in the arteries and capillaries. The lungs contain about 9 percent, and the brain contains about 7 percent.

FIGURE 19.5 DEVELOPMENT OF FORMED ELEMENTS

Dashed lines indicate the omission of some intermediate stages. The mature cells are shown enlarged.

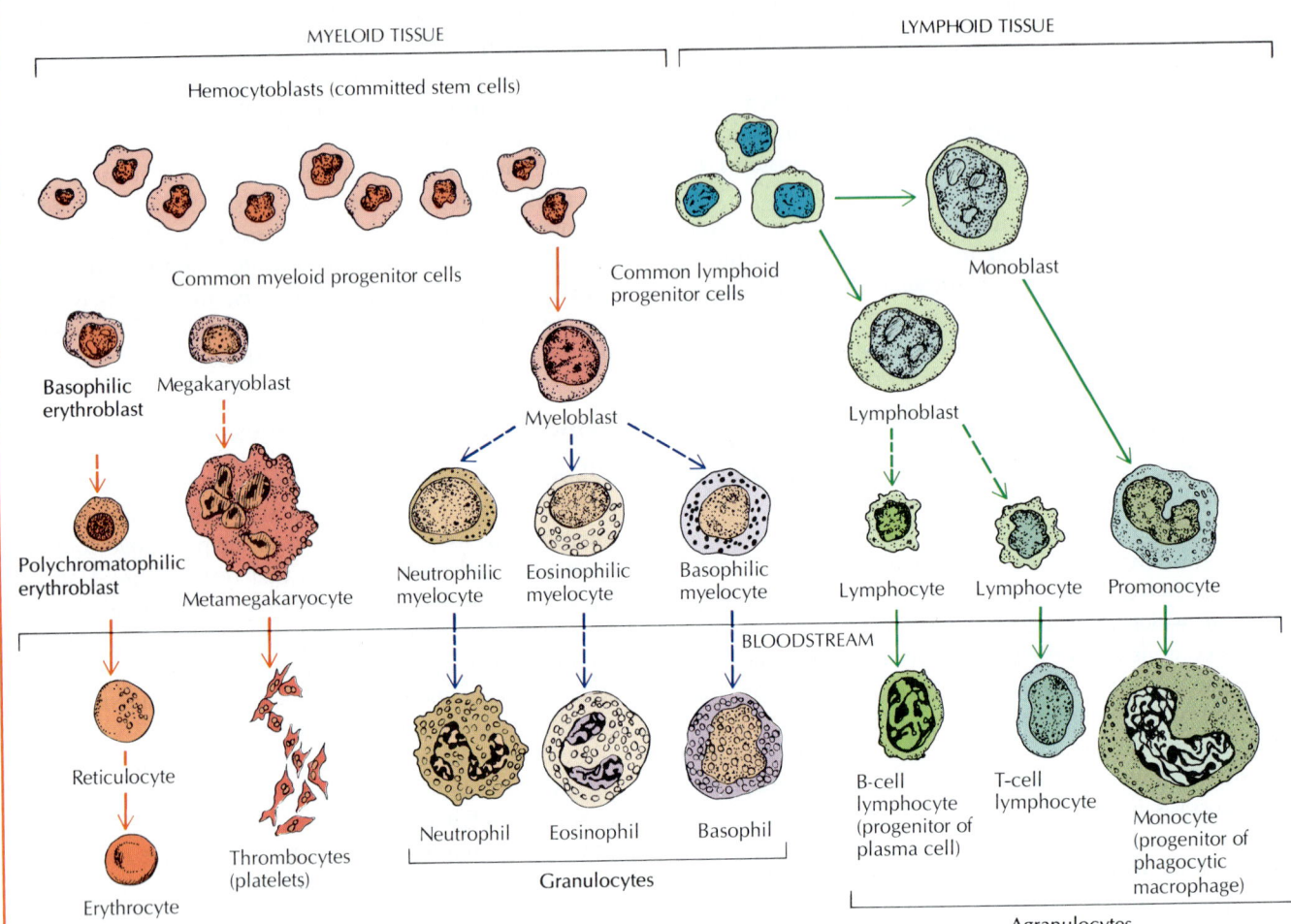

Type of cell	Description

ERYTHROCYTE (RED BLOOD CELL)

Biconcave disk; no nucleus, principal component hemoglobin; diameter about 7 μm. About 50% of total blood volume (4.8–5 million/mm^3). Transports oxygen and carbon dioxide. Life span 80–120 days.

LEUKOCYTE (WHITE BLOOD CELL)

Several forms of cells; capable of independent amoeboid movement. Less than 1% of total blood volume (4000–11,000/mm^3). Defends body against harmful microorganisms and foreign particles.

Granulocytes (polymorphonuclear):
  Neutrophil

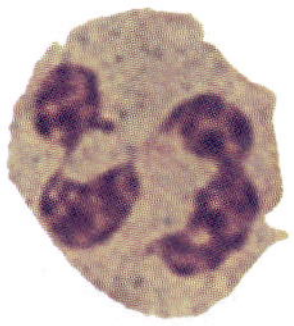

Larger than erythrocytes; contain many granules in cytoplasm. About 50–70% of total leukocytes. Multilobed nucleus; granules appear pink to blue-black in a neutral stain; diameter 10–12 μm. Destroys microorganisms and other foreign particles by phagocytosis and lytic enzymes. Life span 4–5 days.

  Eosinophil

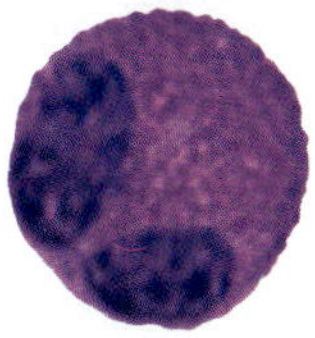

B-shaped nucleus; granules appear red to red-orange in an acid stain; diameter 10–12 μm. Modulates allergic and inflammatory reactions, destroys antigen-antibody complexes. Phagocytes containing lysosomal enzymes, peroxidase. Life span several weeks.

  Basophil

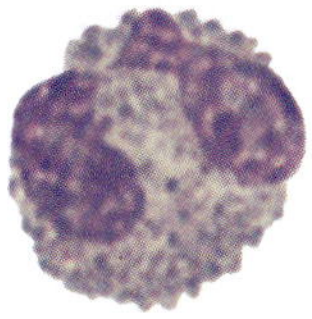

Lobed nucleus; granules appear blue-black to red-purple in basic stain; diameter 8–10 μm. About 1% of total leukocytes. Involved in allergic reactions and inflammation; contains histamine, heparin, and slow-reacting substance (SRS-A).

Agranulocyte (mononuclear):
  Monocyte

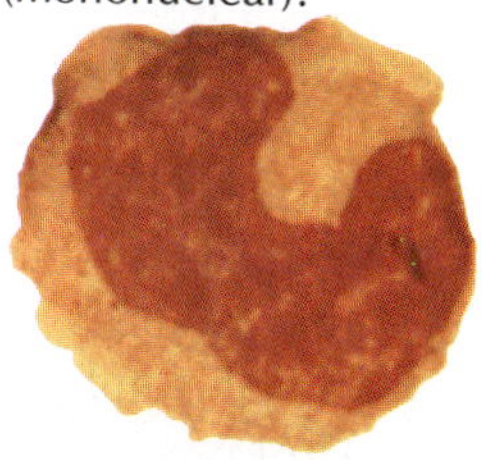

Largest blood cells, approximately two to three times larger than erythrocytes; diameter 15–20 μm. About 1–10% of total leukocytes. Under stress conditions, becomes a macrophage, a large mobile phagocyte that ingests and destroys harmful particles. Life span days to years.

  Lymphocyte

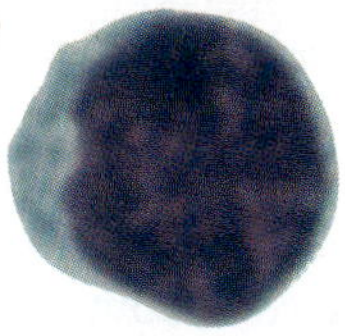

Large, round nucleus that nearly fills the cell; diameter 6–20 μm. About 20–40% of total leukocytes. Involved in immune response and synthesis of antibodies. Abundant in lymphoid tissue. Life span days to years.

THROMBOCYTE (PLATELET)

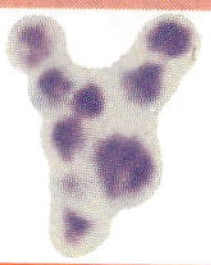

Fragments of megakaryocytes; no nucleus; diameter about 2–4 μm. Less than 1% of total blood volume (350,000/mm^3). Important in blood clotting, releases serotonin. Life span 7–8 days in bloodstream.

The production of white blood cells is called *leukopoiesis* (LYOO-koh-poy-EE-sihss). Some leukocytes (called granulocytes) originate from the same undifferentiated hemocytoblasts of blood-forming bone marrow that erythrocytes do. However, leukopoiesis also occurs throughout the body in lymphoid tissues such as lymph nodes, spleen, and tonsils.

The term *leukocyte* is used to cover a number of different cell types that circulate in the blood. The two basic classifications of leukocytes are granulocytes (polymorphonuclear leukocytes, or polymorphs) and agranulocytes (mononuclear leukocytes).

Granulocytes The most numerous of the white blood cells are *granulocytes,* so named because they contain large numbers of granules in the cytoplasm outside their multilobed nuclei [FIGURE 19.5]. The three types of granulocytes are *neutrophils, eosinophils,* and *basophils.* When stained with Wright's stain, the granules of neutrophils appear light pink to blue-black, eosinophil granules appear red to red-orange, and granules of basophils appear blue-black to red-purple.*

Neutrophils, which represent about 60 percent of the granulocytes, are phagocytes that engulf and destroy microorganisms and other foreign materials. The granules inside the cytoplasm are packets of enzymes (lysozymes) that digest the intruders, break them down, and eventually destroy them. In the course of phagocytosis, the neutrophils may also be destroyed as their granules are depleted. Dead microorganisms and neutrophils make up the thick, whitish fluid we call *pus.*

Specific chemicals released at the site of infection or tissue injury attract neutrophils and other types of leukocytes, such as monocytes and macrophages, and cause them to migrate quickly to the problem areas. This process of chemical attraction is called *chemotaxis.* Neutrophils (and leukocytes in general) are able to deform, elongate, and squeeze through the pores of capillary walls by a process called *diapedesis* (dye-uh-puh-DEE-sis; Gr. "a leaping through").

Eosinophils (EE-oh-SIHN-uh-fillz) have B-shaped nuclei. Like neutrophils, eosinophils are phagocytes that have an amoeboid movement (slow cytoplastic flow). Their granules contain lysosomal enzymes and peroxidase that, when released, destroy phagocytized material. For unknown reasons, the eosinophil count increases during allergy attacks, with certain parasitic infections, some autoimmune diseases (the production of antibodies that attack one's own tissues), and in certain types of can-

cer. In addition, eosinophils contain the protein plasminogen, which helps dissolve blood clots.

Basophils are granulocytes with elongated, indistinctly lobed nuclei. They are the least numerous of all the granulocytes. Their granules contain *heparin* (an anticoagulant), *histamine* (which dilates general body blood vessels and constricts blood vessels in the lungs), and *slow-reacting substance-A* (SRS-A) of allergies. SRS-A produces some of the allergic symptoms, such as bronchial constriction. Basophils play an important role in providing immunity against parasites.

Agranulocytes Despite their name, *agranulocytes* (or *nongranular leukocytes*) usually have a few nonspecific lysosome granules in their cytoplasm. Agranulocytes include *monocytes* and *lymphocytes,* which account for about 5 and 30 percent of the leukocytes, respectively.

Monocytes, which are mobile phagocytes, are the largest blood cells. They have large, folded nuclei and often have fine granules in the cytoplasm [FIGURE 19.5].

Monocytes develop in bone marrow from *monoblasts,* enter the bloodstream for about 30 to 70 hours, and then leave by diapedesis. Once in the tissue spaces, monocytes enlarge to 5 to 10 times their normal size and become phagocytic macrophages. These macrophages form a key portion of the *reticuloendothelial system,* which lines the vascular portions of the liver, lungs, lymph nodes, thymus gland, and bone marrow. In the connective tissue of these regions, macrophages phagocytize microorganisms and cellular debris. Macrophages also play a role in the immune system by processing specific antigens. Monocytes and macrophages also produce a group of substances collectively called a *colony-forming unit* (CFU), which stimulates the bone marrow to produce more monocytes and neutrophils.

Lymphocytes are small, mononuclear, agranular leukocytes with a large round nucleus that occupies most of the cell [FIGURE 19.5]. They get their name from lymph, the fluid that transports them. Lymphocytes move sluggishly and do not travel the same routes through the bloodstream as other leukocytes. They originate from the hemocytoblasts of bone marrow and then invade lymphoid tissues, where they establish colonies. These colonies then produce additional lymphocytes *without* involving the bone marrow. Most lymphocytes are found in the body's tissues, especially in lymph nodes, the spleen, the thymus gland, tonsils, adenoids, and the lymphoid tissue of the gastrointestinal tract. They can leave the blood more easily than other cells to enter lymphoid tissue, and they differ from other leukocytes in being able to reenter the circulatory system. Some lymphocytes live for years, recirculating between blood and lymphoid organs. The biggest difference between lymphocytes and other white blood cells is that lymphocytes are not phagocytes.

*The names of granulocytes are derived from the type of stain with which they are most easily stained for laboratory preparations. The suffix *phil* comes from the Greek *philos,* which means "loving" or "having a preference for." Neutrophils "prefer" a *neutral* dye, eosinophils an *eosin* (acid) dye, and basophils a *basic* dye.

Two distinct types of lymphocytes are recognized: B cells and T cells [FIGURE 19.5]. **B cells** originate in the bone marrow and colonize lymphoid tissue. In contrast, **T cells** are associated with, and influenced by, the thymus gland before they colonize lymphoid tissue. T cells regulate the cellular immune response that is a part of the body's own defense system, and they secrete chemicals that destroy bacteria, poisons, viruses, and tissue and chemical debris. When B cells are activated, they enlarge and become *plasma cells*. Plasma cells have much more cytoplasm than B cells do, and this enables them to accommodate the biochemical machinery necessary for the production of antibodies.

Platelets (Thrombocytes)

Platelets (so named because of their platelike flatness), or *thrombocytes* (Gr. *thrombus*, clot + cells), are fragments of cells and are about one-quarter the size of erythrocytes. Their main function is to start the intricate process of blood clotting. Platelets are much more numerous than leukocytes, averaging about 350,000 per cubic millimeter. (An average-sized adult has about a trillion platelets.) Platelets lack nuclei and are incapable of cell division, but they have a complex metabolism and internal structure [FIGURE 19.6]. Once in the bloodstream, platelets have a life span of 7 to 8 days.

About 200 billion platelets are produced every day. They originate from committed hemocytoblasts in the bone marrow. The hemocytoblasts involved in platelet formation develop into myeloid progenitor cells from which large cells called **megakaryoblasts** arise. The megakaryoblasts differentiate into **megakaryocytes**. Platelets break off from the pseudopods of these cells in the bone marrow and then enter the bloodstream. Thus the platelets that appear in the circulating blood are not actually blood cells but are cellular fragments of megakaryocytes. After entering the bloodstream, platelets begin to pick up and store chemical substances that can be released later to help seal vessel breaks.

Platelets adhere to each other and to the collagen in connective tissue but not to red or white blood cells, a property that is essential for blood clotting and the overall process of *hemostasis*, the prevention and control of bleeding. When a blood vessel is injured (capillaries may

---

Q: *What is platelet donation?*

A: Platelet donation is a form of blood donation in which only platelets are separated from the whole blood of the donor and transferred to the recipient, who has a shortage of platelets. The unused portions of blood are returned to the donor intravenously.

FIGURE 19.6 PLATELETS

[A] Flash-contact x-ray micrograph of a human platelet. This new technique shows *live* cells. Note the forming pseudopod, which will become part of the meshwork essential for blood clotting. ×30,000. **[B]** Electron micrograph of cross section of an injured capillary. Note the platelet plugging a tiny break in the capillary wall. Larger breaks in blood vessels attract many platelets to the injured site. Sections of four erythrocytes appear above the platelet. ×8000.

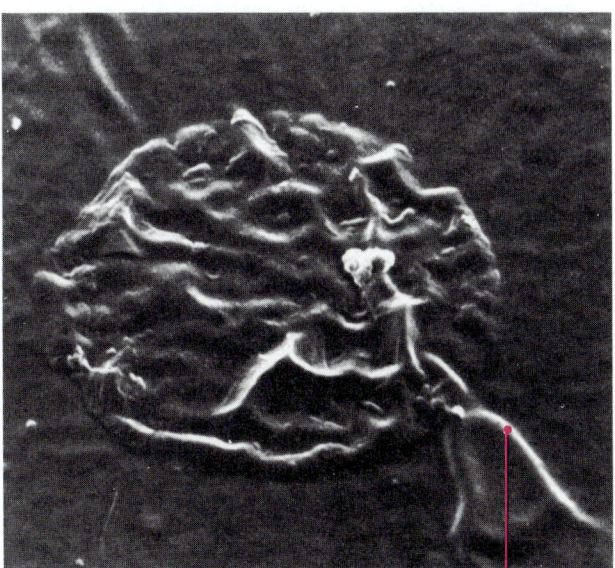

[A]   Pseudopod

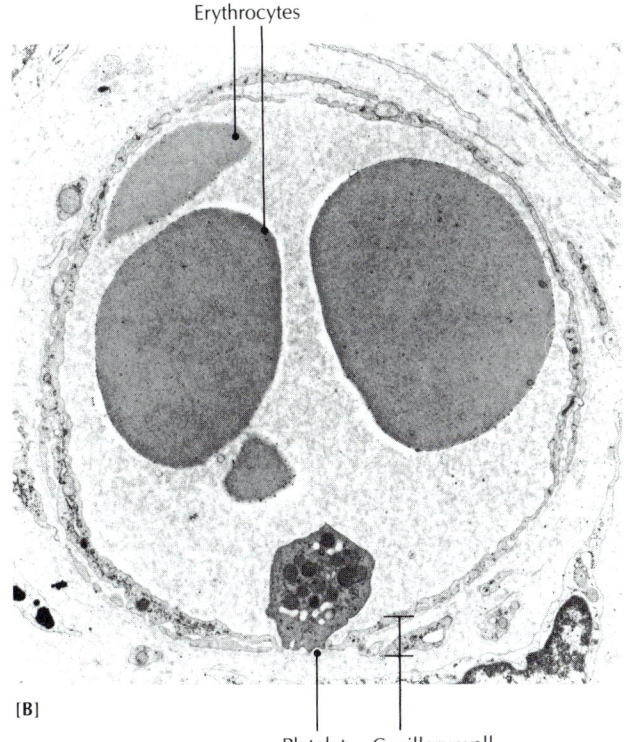

Erythrocytes

[B]

Platelet   Capillary wall

rupture many times a day), platelets immediately move to the site and begin to clump together, attaching themselves to the damaged area. The platelets release granules that contain *serotonin*, which constricts broken or injured vessels and retards bleeding, and *adenosine diphosphate* (ADP), which attracts more platelets to the damaged area. If the break is small enough, it is repaired by the platelet plug [FIGURE 19.6B]. However, if there are not enough platelets to make the repair, they begin the process of blood clotting, or **coagulation** (koh-AGG-yoo-LAY-shuhn; L. *coagulare*, to curdle), described in a following section (page 598).

Hemopoiesis

Blood cell formation is referred to as **hemopoiesis** or *hemotopoiesis* (Gr. *hemato*, blood; *poiein*, to make). Historically, two concepts have existed concerning the origin of blood cells: (1) The *monophyletic (unitarian) theory* states that all blood cells develop from one common stem cell. (2) The *polyphyletic theory* states that each type of blood cell develops from a unique type of stem cell.

Recent experiments support the monophyletic theory. The evidence strongly indicates that a lymphocytelike cell is the stem cell. They are thought to form groups, each called a *colony-forming unit (CFU)*. Each cell in a CFU in such blood-forming organs as bone marrow and the spleen has the potential to differentiate, under appropriate conditions, into each type of blood cell. The origin of the stem cell remains controversial. CFU cells are presumed to make their commitment to the formation of a specific cell line before the known characteristics of the differentiated cell line develop; for example, before any trace of hemoglobin is found in the erythrocyte line or surface immunoglobulins in the lymphocyte line. The CFU cells are presumed to form the distinct cell lines for the production of each of the following: erythrocyte; granulocyte (neutrophils, eosinophils, and basophils); agranuloctye (B-cell lymphocytes, T-cell lymphocytes and monocytes); and blood platelet lines [see FIGURE 19.5].

ASK YOURSELF

1 What are the formed elements of blood?

2 What is the main function of hemoglobin?

3 Can you describe the process of erythropoiesis?

4 What are the three types of granulocytes? The two types of agranulocytes?

5 What is hemopoiesis?

CLINICAL BLOOD TESTS

Several blood tests are usually performed as part of a routine physical examination. Three commonly performed tests are (1) red blood cell (erythrocyte) count and hematocrit, (2) hemoglobin content, and (3) white blood cell (leukocyte) count and differential white count.

Red Blood Cell Count and Hematocrit

The **red blood cell (RBC) count** may be determined by using a *hemocytometer*, a device that includes a ruled glass slide for counting the number of cells per square as they are viewed under a microscope. Or more commonly, highly specialized machines are able to compute the RBC count from injected samples. A normal RBC count for males is 4.6 to 6.0 million per cubic millimeter of blood and for females is 4.0 to 5.0 million per cubic millimeter of blood.

Hematocrit (Hct) (Gr. "to separate blood") is the volume percentage of red blood cells in whole blood obtained by centrifuging whole blood. A hematocrit of 46 percent [FIGURE 19.7] means that in every 100 mL of whole blood there are 46 mL of red blood cells, or that 46 percent of the volume of blood is composed of RBCs. A normal hematocrit range for males is 40 to 54 percent and for females is 36 to 46 percent. A hematocrit determination may be used to diagnose *anemia* (abnormally low percentage of RBC; Hct less than 30 percent) and *polycythemia* (abnormally high percentage of RBCs; Hct greater than 65 percent).

From the hematocrit and an estimate of the total blood volume in the body, the total volumes of plasma and erythrocytes can be determined as follows: total blood volume is approximately 8 percent of body weight. Thus, in a 70-kg person,

0.08×70 kg = 5.6 kg

Since 1 kg of blood occupies about 1 L,

Total blood volume = 5.6 L

Using our example of a 46-percent hematocrit, then

Total red blood cell volume = 0.46×5.6 L = **2.58** L

and

Plasma volume = 5.6 L − 2.58 L = **3.02** L

Hemoglobin Content

The **hemoglobin (Hb)** in the blood is measured by a *hemoglobinometer* (hemometer), a specially adapted photometer. A sample of whole blood is treated chemically to form a stable pigment. The density (concentration) of hemoglo-

FIGURE 19.7 HEMATOCRIT

[A] The test tube is filled with whole blood to the 100 mark and then centrifuged. The red blood cells become packed at the bottom. **[B]** The percentage of red blood cells (hematocrit) can then be determined; in this case, it is 46 percent.

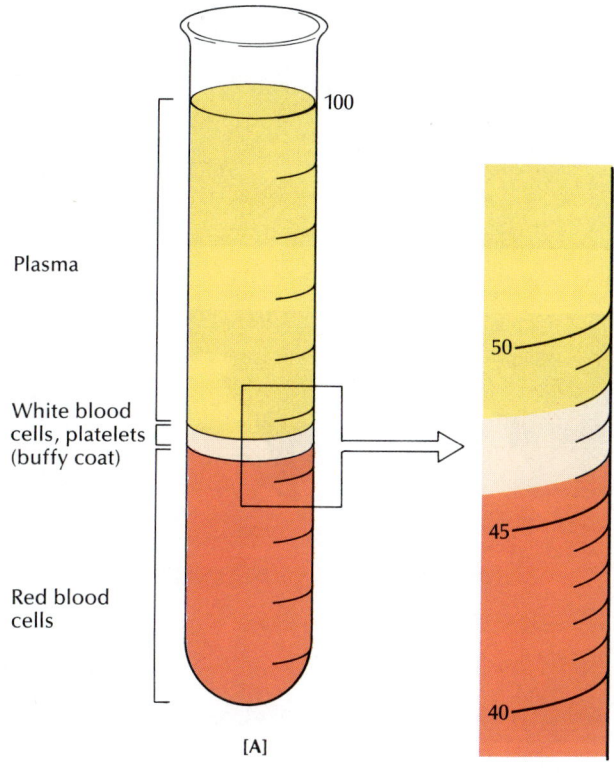

Plasma

White blood cells, platelets (buffy coat)

Red blood cells

[A]

100

50

45

40

[B]

TABLE 19.2 VALUES FOR A WHITE BLOOD CELL (LEUKOCYTE) DIFFERENTIAL WHITE COUNT

Type of leukocyte	Differential white count	
	Normal range, %	Actual count, %*
Granulocytes		
Neutrophils	50–70	65
Eosinophils	1–4	3
Basophils	0.5–1	1
Agranulocytes		
Lymphocytes	20–40	25
Monocytes	2–8	6
		Total = 100

*In a differential white blood cell count the sum of the different leukocytes must equal 100 percent. The values listed are for adults.

sure or certain drugs. (3) *Certain viral infections* may elevate lymphocytes, while immunosuppressive agents will depress them. (4) *Allergic reactions* may elevate eosinophils and/or basophils. (5) *Various types of leukemias* may elevate basophils, lymphocytes, or monocytes.

A S K   Y O U R S E L F

1 What is the difference between a hemocytometer and a hematocrit?

2 How is hemoglobin measured?

3 What is a differential white count?

---

bin is then measured and compared to a standard color scale. The normal range is 13 to 18 g Hb/100 mL of peripheral blood for men and 12 to 16 g Hb/100 mL for women.

White Blood Cell Count and Differential White Count

The ***white blood cell (WBC) count*** may also be determined by a special hemocytometer or a sorting machine. WBC counts normally are 5,000 to 10,000 per cubic millimeter of blood. (Red blood cells outnumber white blood cells by about 700 to 1.) In addition to a WBC count, a ***differential white count*** is also obtained, which indicates the percentage of each type of white blood cell in the blood sample [TABLE 19.2]. Many different infections, diseases, and drugs can differentially elevate or depress specific types of white blood cells. Some examples include the following: (1) *Elevated neutrophils* may result from stress, inflammation, or bacterial infections. (2) *Depressed neutrophils* may be caused by radiation expo-

HEMOSTASIS: THE PREVENTION OF BLOOD LOSS

One drawback of a circulatory system such as ours, in which the blood is under high pressure, is that serious bleeding can take place after even a slight injury. To prevent the possibility of uncontrolled bleeding, we have a three-part specific homeostatic mechanism called ***hemostasis*** (HEE-moh-STAY-sihss) consisting of (1) the constriction of blood vessels (*vasoconstrictive phase*), (2) the clumping together (aggregation) of platelets (*platelet phase*), and (3) blood clotting (*coagulation phase*). Each of these phases is described in detail below.

Vasoconstrictive Phase: 1

Normally, when a tissue is damaged and blood escapes from a blood vessel, the vessel wall constricts in order to narrow the opening of the vessel and slow the flow of blood. This vasoconstriction is due to contraction of the smooth muscle of the vessel wall as a direct result of the injury and the release of vasoconstrictor chemicals from platelets. Proper vasoconstriction is also enhanced by pain reflexes, producing constriction in proportion to the extent of the injury. Constriction of capillaries, which do not have muscular layers, is due to the vascular compression caused by the pressure of lost blood that accumulates in surrounding tissues. Injured blood vessels may continue to constrict for 20 min or more.

Platelet Phase: 2

The next event in hemostasis is the escape from blood vessels of platelets, which swell and adhere to the collagen in adjacent connective tissues. This attachment stimulates vasoconstriction. The platelets become very sticky, so that as more and more of them move into the injured area they stick together [FIGURE 19.8A]. In about a minute they can clog a small opening in the vessel with a *platelet plug* (also called a hemostatic plug or adherent platelet) [FIGURE 19.8B].

Circulating platelets do not adhere to normal endothelium or to each other until the vessel's endothelial lining is broken to expose the subendothelium. Adhesion requires participation of an endothelial cell secretion of a protein called *von Willebrand factor (VWF)*. Within the platelets, a mechanism activates the contraction of platelet actinomyosin. The platelet plug is compressed and consolidated, further securing it to the site of injury.

The process involving the formation of platelet plugs is called *platelet aggregation.* It is important partly because it successfully stops hundreds of small hemorrhages every day and partly because it triggers the blood-clotting mechanism.

Coagulation Phase: Basic Mechanism of Blood Clotting: 3

If the blood vessel damage is so extensive that the platelet plug cannot stop the bleeding, the complicated process of blood clotting—the *coagulation phase*—begins. The simplified basic clotting mechanism involves the following events:

1 Aided by a plasma globulin called *antihemophilic factor (AHF)*, blood platelets disintegrate and release the enzyme *thromboplastinogenase* and platelet factor 3 [TABLE 19.3].

2 Thromboplastinogenase combines with AHF to convert the plasma globulin *thromboplastinogen* into the enzyme *thromboplastin*.

FIGURE 19.8 STAGES OF PLATELET AGGREGATION

Scanning electron micrographs. **[A]** When exposed to ADP at the site of an injured blood vessel, platelets stick together. ×30,000. **[B]** Platelets begin to swell, become spiny, and adhere to even more platelets to form a plug to close the opening in the vessel. ×30,000.

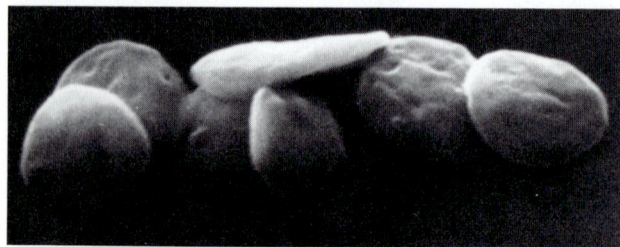

[A]

[B]

3 Thromboplastin combines with *calcium ions* to convert the inactive plasma protein *prothrombin* into the active enzyme *thrombin.*

4 Thrombin acts as a catalyst to convert the soluble plasma protein *fibrinogen* ("giving birth to fibrin") into the insoluble, stringy plasma protein *fibrin.*

5 The fibrin threads entangle the blood cells and create a clot [FIGURE 19.9].

This basic process may be summarized as follows:

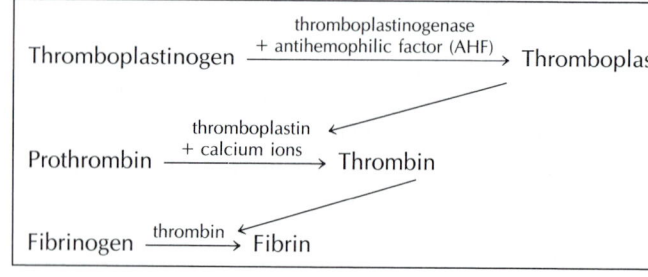

TABLE 19.3 BLOOD-CLOTTING FACTORS

Coagulation factor number and name*	Description and origin	Function
PLASMA COAGULATION FACTORS		
I Fibrinogen	Plasma protein synthesized in liver.	Precursor of fibrin; converted to fibrin in final stage of clotting. Serum is plasma minus fibrinogen.
II Prothrombin	Plasma protein synthesized in liver. Synthesis requires vitamin K.	Precursor of thrombin, the enzyme that converts fibrinogen into fibrin.
III Thromboplastin	Complex lipoprotein formed from disintegrating platelets or tissues.	Combines with calcium to convert prothrombin into active thrombin.
IV Calcium ions	Inorganic ion in plasma, acquired from bones and diet.	Necessary for formation of thrombin and for all stages of clotting.
V Proaccelerin, labile factor, or accelerator globulin	Plasma protein synthesized in liver.	Necessary for extrinsic and intrinsic pathways.
VI	No longer thought to be a separate entity; possibly a combination of activated factors V and X.	
VII Serum prothrombin conversion accelerator (SPCA), stable factor, or proconvertin	Plasma protein synthesized in liver. Synthesis requires vitamin K.	Necessary for first phase of extrinsic pathway.
VIII Antihemophilic factor (AHF), antihemophilic factor A, or antihemolytic globulin (AHG)	Plasma protein (globulin) synthesized in liver and other tissues.	Necessary for first phase of intrinsic pathway. Deficiency causes hemophilia A, a genetic disorder.
IX Plasma thromboplastin component (PTC), Christmas factor, or antihemophilic factor B	Plasma protein synthesized in liver. Synthesis requires vitamin K.	Necessary for first phase of intrinsic pathway. Deficiency causes hemophilia B.
X Stuart-Prower factor or Stuart factor	Plasma protein synthesized in liver. Synthesis requires vitamin K.	Necessary for early phases of extrinsic and intrinsic pathways.
XI Plasma thromboplastin antecedent (PTA) or antihemophilic factor C	Plasma protein synthesized in liver.	Necessary for first phase of intrinsic pathway. Deficiency causes hemophilia C.
XII Hageman factor or glass factor	Plasma protein; source unknown.	Necessary for first phase of intrinsic pathway; activates plasmin; activated by contact with glass, probably involved with clotting outside body.
XIII Fibrin-stabilizing factor (FSF) or Laki-Lorand factor	Protein present in plasma and platelets; source unknown.	Necessary for final phase of clotting.
PLATELET COAGULATION FACTORS		
Pf$_1$ Platelet factor 1 or platelet accelerator	Platelets.	Same as factor V; accelerates action of platelets.
Pf$_2$ Platelet factor 2 or thrombin accelerator	Platelets, phospholipid.	Accelerates thrombin formation at start of intrinsic pathway; accelerates conversion of fibrinogen into fibrin.
Pf$_3$ Platelet factor 3 or platelet thromboplastic factor	Platelets, phospholipid.	Necessary for first phase of intrinsic pathway.
Pf$_4$ Platelet factor 4	Platelets.	Binds the anticoagulant heparin during clotting.

*Coagulation (clotting) factors are substances in the plasma that are essential for the maintenance of normal hemostasis. Thirteen factors are recognized and identified by Roman numerals I to XIII. The platelet coagulation factors are identified as Pf$_1$ to Pf$_4$.

FIGURE 19.9 THE BEGINNING OF A BLOOD CLOT

Scanning electron micrograph showing tangled threads of fibrin, which bind the clot into an insoluble mass. The large circular cells are erythrocytes, and the smaller ones are platelets.

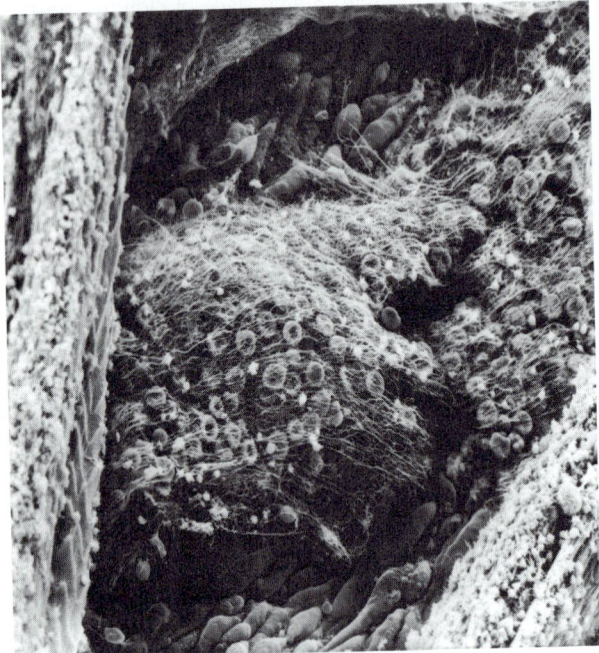

Extrinsic and Intrinsic Pathways

Intense research on blood coagulation took place in the 1960s. At that time, two partially independent pathways were identified for the triggering of a blood clot: (1) The *extrinsic pathway* is a rapid clotting system activated when blood vessels are ruptured and tissues are damaged. (2) The *intrinsic pathway* is activated when the inner walls of blood vessels become damaged. (Remember that *extrinsic* means "outside" and *intrinsic* means "inside.")

Extrinsic pathway Damaged tissue triggers the *extrinsic pathway,* which initiates blood clotting by the release of *thromboplastin,* also known in this form as *coagulation factor III.* (A somewhat different form of thromboplastin is at work at the site of ruptured vessels, triggered by the disintegration of platelets.) Thromboplastin combines with calcium, coagulation factor VIII, and a mixture of enzymes and the phospholipids from damaged cell membranes released by the injured tissue to produce a substance called *factor X* [FIGURE 19.10]. At this point, the extrinsic system merges with the intrinsic system to activate yet another mechanism (called the *common pathway*) that actually produces the clot. The common pathway includes steps 3, 4, and 5 described above in the coagulation phase.

Intrinsic pathway The *intrinsic pathway* for initiating blood clotting uses only substances found in the blood. These substances are called *clotting factors* and are described in TABLE 19.3. Injury to the inner wall of a blood vessel activates clotting factor XII, which triggers a series of rapid chemical reactions usually called the "cascade effect." Each step activates the next step in the sequence until factor X is activated. After factor X forms the prothrombin activator in the common pathway, the basic blood-clotting process proceeds. The activation of the extrinsic pathway usually produces a clot in as little as 15 sec, while the activation of the intrinsic pathway requires 2 to 6 min.

At this point, the fibrin threads form only a weak mesh, and the clot must be strengthened if it is to hold. Platelets and plasma globulins release a *fibrin-stabilizing factor* (see blood coagulation factor XIII in TABLE 19.3) that responds to thrombin to create an interlacing pattern of fibrin threads. Within a few minutes after the clot is formed, it begins to contract, squeezing out serum and helping the clot solidify. The power to contract comes from platelets, which contain actin and myosin, the same proteins that make muscle contraction possible. (Platelets contain more actin and myosin than any tissue in the body except muscle.) Most of the serum is drained within an hour, and the solid clot is finally complete. A "scab" forms, dries up, and in a few days falls off as the underlying tissue heals.

A well-known dietary substance involved with blood clotting is vitamin K.* Found in leafy green vegetables, tomatoes, vegetable oils, and also produced by intestinal bacteria, vitamin K is necessary for the production of prothrombin and other clotting factors by the liver.

Hemostasis and the Nervous System

The sympathetic nervous system helps compensate for massive blood loss when a blood clot is inadequate to stop the flow. When the body loses more than 10 percent of its blood, there is a sudden drop in blood pressure, and the body goes into *hypovolemic* (low blood volume) *shock.* The decreased blood pressure triggers reflexes in the sympathetic nervous system that constrict veins and the small terminal branches of arteries (arterioles) in an attempt to limit the decrease in blood pressure. Also, the heart rate may rise from a normal 72 to 84 beats per minute to as many as 200 beats per minute, increasing the blood flow to vital areas to counteract the reduced blood pressure. Blood flow to the brain and heart is especially increased. Without the functioning of the sympathetic reflexes, a person would probably die after losing 15 to 20 percent of the total blood volume. When the

*The K stands for *Koagulation,* the German word for "clotting." The discoverer of vitamin K originally named it the Koagulation-Vitamin.

FIGURE 19.10 EXTRINSIC AND INTRINSIC PATHWAYS OF BLOOD CLOTTING

Note how the two pathways combine to form the common pathway.
Factor VIII is often absent in hemophiliacs.

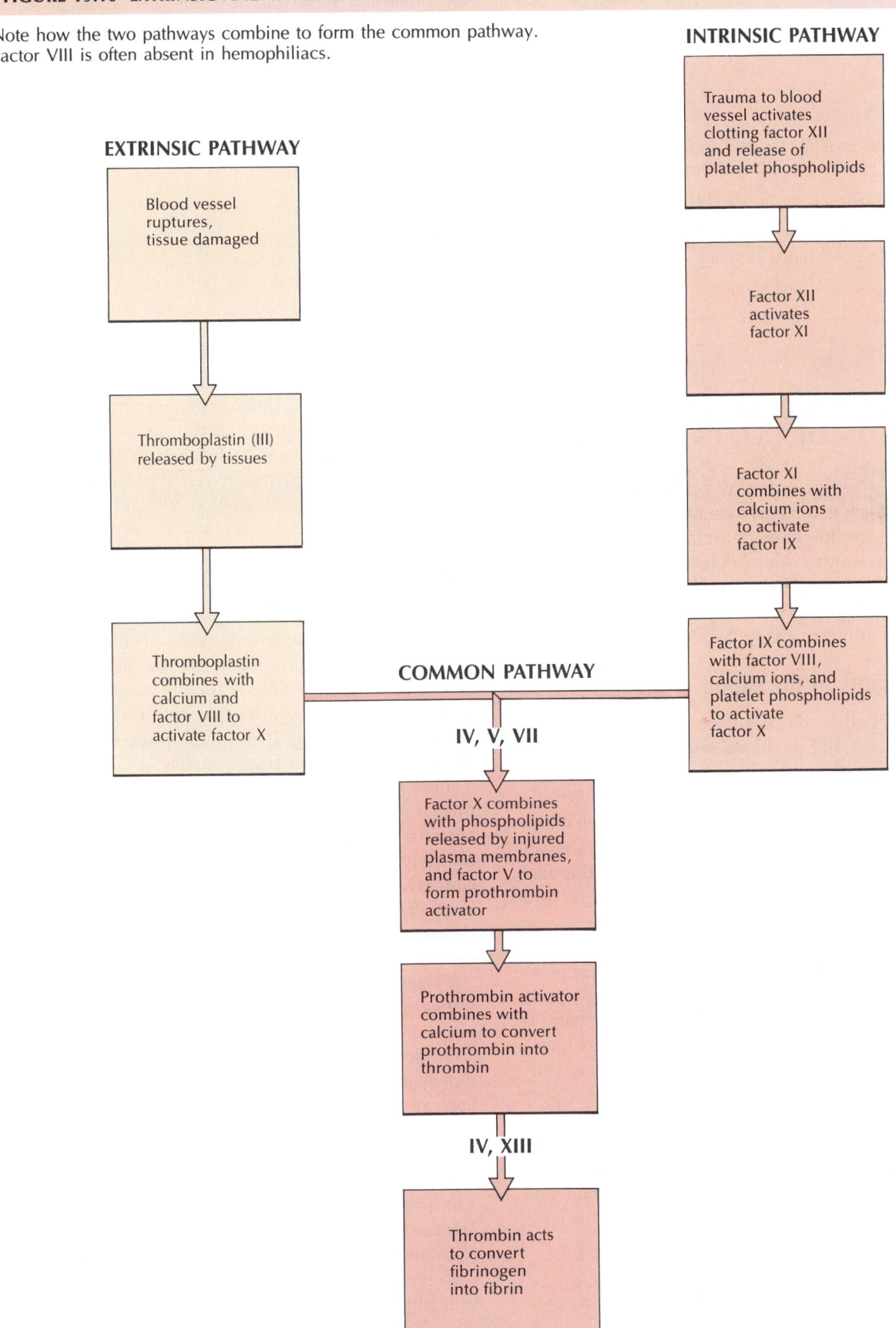

feedback system of blood-pressure control is operating properly, a person may still be alive after losing as much as 40 percent of the total blood volume. If the lost blood is not replaced quickly, however, death may follow.

ASK YOURSELF

1 Starting with prothrombin and ending with fibrin, what are the basic steps in blood clotting?

2 How do platelets assist in blood clotting?

3 What is the difference between the extrinsic and intrinsic pathways of blood clotting?

4 What are blood-clotting factors?

ANTICOAGULATION: THE INHIBITION OF CLOTTING

As many as 35 compounds may be required for blood coagulation. Such a complex system of checks and balances is necessary to prevent clotting when there is no bleeding. An unwanted clot in a blood vessel that cuts off the blood supply to a vital organ is one of the body's worst enemies. How are blood clots prevented, and how are they broken down if they do form?

Anticoagulants in the Blood

Most of the body's anticoagulant substances circulate within the blood, and the blood vessels themselves help prevent clotting. The blood vessels contribute in two ways. First, the smoothness of the inner walls normally prevents activation of the intrinsic clotting mechanism. Second, a thin layer of negatively charged protein molecules attached to the inner walls repels the clotting factors, preventing the initiation of clotting. Injury to a blood vessel removes both of these safeguards.

Heparin and antithrombin One of the most powerful anticoagulants in the blood is *heparin,* a polysaccharide produced by mast cells and basophils.* Heparin is concentrated mostly in the liver and lungs. Minute quantities of heparin in normal circulating blood prevent clotting by combining with the *antithrombin-heparin cofactor* (also called *antithrombin* or *antithrombin III*) to induce the cofactor to combine with thrombin 1000 times more rapidly than usual. Such a rapid binding to thrombin removes it almost instantly from the bloodstream and makes clotting almost impossible. Without heparin, anti-

*Mast cells arise from hemocytoblasts. They are not normally found in blood but instead lodge in connective tissues throughout the body.

thrombin-heparin cofactor binds to thrombin molecule for molecule, removing it from the blood in about 15 min.

The combination of heparin and antithrombin-heparin cofactor also reacts with several clotting factors in the extrinsic and intrinsic pathways, further inhibiting blood clotting. Thrombin itself acts as an anticoagulant. When its concentration becomes too high, it destroys factor VIII to prevent clotting.

Fibrinolysis

Clot prevention is important, but so is clot *destruction,* or *fibrinolysis* ("fibrin breaking"). Small blood clots form continually in blood vessels throughout the body. If they are not removed promptly, the blood vessels become clogged. In the process of fibrinolysis, a blood protein called *plasminogen* is activated into an enzyme called *plasmin.* The plasmin digests the threads of fibrin by first making them soluble and then breaking them into small fragments. The fragments are removed from the bloodstream by phagocytic white blood cells and macrophages.

Anticoagulant and Clot-Dissolving Drugs

When used under medical supervision, anticoagulant drugs can sometimes remove blood clots in the body or prevent them from forming in the first place. The best-known anticoagulant drug is *aspirin* (acetylsalicylic acid), which works by preventing platelets from sticking together to form a plug. It also inhibits the release of clot-promoting substances from platelets.

One drug that digests the fibrin threads of a clot is *streptokinase,* which is released by certain streptococcal bacteria. Streptokinase activates plasminogen to speed up fibrinolysis (clot destruction). It is used to dissolve blood clots (thrombi) in veins and arteries. Streptokinase also helps dissolve the fibrin threads in a blood clot by converting plasminogen into plasmin, the fibrin-destroying enzyme.

Genetically engineered *(recombinant) tissue-plasminogen activator (tPA)* is effective in dissolving intravascular blood clots when delivered directly to a clotted area through a catheter. For example, if tPA is used shortly after a blood clot forms in a coronary artery, the heart is often spared serious damage.

Dicumarol is a compound that resembles vitamin K to such an extent that the liver enzymes that form prothrombin will pick up dicumarol instead of vitamin K. The anticoagulatory effect of dicumarol is often used to prevent clotting after surgery.

In addition to being used to remove blood clots and keep blood from coagulating during surgery, anticoagulant drugs may be necessary to prevent clotting in blood that will be used later for blood transfusions. (Once blood clots, it cannot be used for transfusions.) To avoid such clotting, a dilute, sterile solution of a *citrate* or an *oxalate salt* is added to collected blood. Clotting does not occur because citrate ions or oxalate ions combine with the available calcium ions, making calcium unavailable for its usual blood-clotting functions.

Blood Coagulation Tests

Several tests are used to determine blood-clotting time. The most popular ones are platelet count, bleeding time, clotting time, and prothrombin time.

The **blood platelet count** must be greater than 150,000 per cubic millimeter in order for normal coagulation to take place. Also, if platelet *function* is not normal, normal coagulation may not occur.

A pierced fingertip or earlobe usually bleeds for 3 to 6 min. A longer **bleeding time** for this wound generally (but not always) indicates a platelet deficiency.

Clotting time is determined by placing blood in a test tube and tipping it back and forth every 30 sec or until it clots. This usually occurs in 5 to 8 min. Because the condition and size of test tubes vary, standardization is necessary to obtain accurate results.

The test for **prothrombin time (PT)** indicates the amount of prothrombin in the blood. Immediately after blood is removed, oxalate is added to prevent the prothrombin from being converted into thrombin. Then calcium ions and tissue extract containing thromboplastin are added to the blood sample. The calcium offsets the effect of the oxalate, and the tissue extract activates the conversion of prothrombin. The time usually required for blood to clot, referred to as the *prothrombin time*, is about 12 sec. (A longer prothrombin time may

also mean a decreased quantity of some factor other than prothrombin.) Similar tests are used to determine the relative quantities of other clotting factors.

A S K   Y O U R S E L F

1 How is heparin involved with anticoagulation?

2 What is fibrinolysis?

BLOOD TYPES

In 1900, a Viennese pathologist named Karl Landsteiner proved that there are individual differences in blood. He isolated two distinct glycoproteins (antigens) on the surface of red blood cells that could, in combination with certain samples of incompatible blood, cause the red blood cells to clump together. These two proteins were called A and B *antigens* and are now referred to as **agglutinogens** (uh-GLOOT-n-oh-jehnz). The clumping together of cells in general is called **agglutination,** and the clumping of red blood cells is called *hemagglutination*. Based on the possible combinations of two of the agglutinogens in blood, four types of blood were identified: A person with agglutinogen A is *blood type A*, someone with agglutinogen B is *type B*, someone with both agglutinogens is *type AB*, and someone with neither A nor B agglutinogen is *type O* [TABLE 19.4]. These blood types are inheritable characteristics, passed on from parents to their children.

ABO Blood Grouping and Transfusions

The hemagglutination Landsteiner witnessed, which was the cause of sometimes fatal blood transfusions, was the result of an antigen-antibody reaction produced when two incompatible blood types are combined. This incompatibility is due to the presence of one of two antibodies in the blood plasma. These *antibodies* are referred to as **agglutinins** (uh-GLOO-tih-nihnz) because they agglutinate the red blood cells from other individuals whose blood is incompatible. Thus the blood plasma of a

---

Q: *Why doesn't blood clot in the blood vessels?*

A: Blood doesn't clot in the blood vessels because the enzyme thrombin does not exist in the normal circulation but is generated from a precursor, prothrombin, only in the presence of clumped platelets.

TABLE 19.4 ABO BLOOD GROUPING

Blood type	Agglutinogens (antigens) on erythrocytes	Agglutinins (antibodies) in plasma	Can donate blood to*	Can receive blood from*
A	A	Anti-B	A, AB	A, O
B	B	Anti-A	B, AB	B, O
AB	A, B	None	AB	A, B, AB, O
O	None	Anti-A, Anti-B	A, B, AB, O	O

*In practice, only matched blood types are used for transfusions.

FIGURE 19.11 ABO BLOOD GROUPING

[A] Schematic representation of blood groups. **[B]** Schematic representation of antibody-antigen complex formed when type B red blood cells are transfused into a recipient with type A blood. The agglutinins of the donated blood have little or no effect on the red blood cells of the recipient.

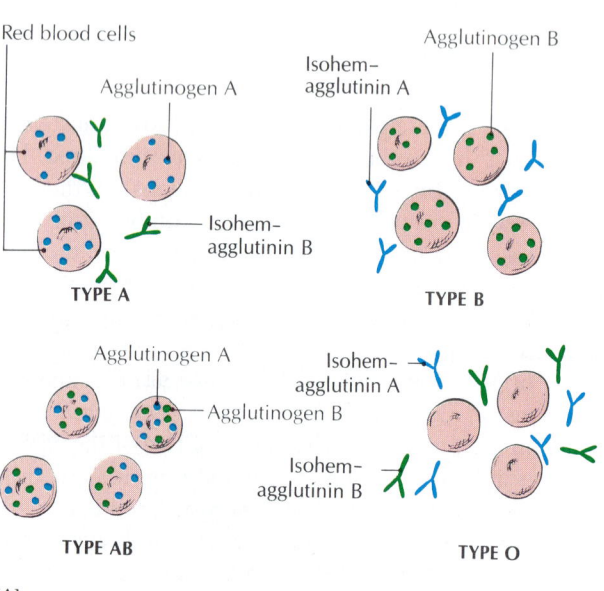

[A]

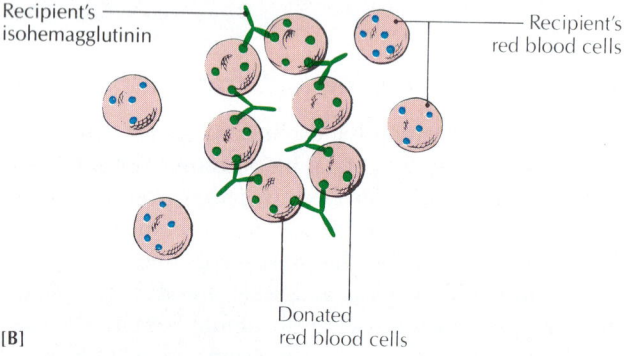

[B]

person with blood type A has the agglutinin against type B red blood cells. If blood type B or AB is introduced, the agglutinin reacts violently with the B agglutinogens present on the donated red blood cells, causing these "invaders" to hemagglutinate [FIGURE 19.11]. TABLE 19.4 shows the complementary distribution of agglutinogens and agglutinins for the four blood types.

The blood-matching system for blood types A, B, AB, and O is known as the ***ABO blood-grouping system***. It operates as follows during blood transfusions from one person to another: Type A blood is incompatible with B, and these two blood types will form clumps (agglutinate) if combined. Theoretically, a person with AB blood, con-

taining no agglutinins, can *receive* any type of blood (formerly called a universal recipient), and someone with type O blood, containing no agglutinogens, can safely *give* blood to any of the other types (formerly called a universal donor). However, in actual practice this is misleading and dangerous. There are other agglutinogens and agglutinins in blood, in addition to those associated with the ABO system, which can cause transfusion problems. Thus blood should be carefully typed and cross matched* before a transfusion, except in an extreme emergency.

In a transfusion, the donor's blood (usually 0.5 L per transfusion) is diluted substantially by the recipient's blood when transfused into the recipient's circulatory system. Because of this dilution, the agglutinins in the donor's blood cause little hemagglutination of the recipient's red blood cells. Thus agglutinins in the recipient's blood and the type of red blood cells of the donor are important in transfusions. In practice, the blood types of the donor and recipient are tested, or *cross matched*, to be sure that they are compatible for a transfusion.

Rh Factor

About 40 years after formulating the ABO blood-grouping system, Landsteiner and A. S. Wiener identified another factor in blood: the ***Rh factor***. (The Rh factor was so named because it was first discovered in *Rh*esus monkeys.) Of the several genes responsible for different blood agglutinogens, this one is of clinical interest because it sometimes causes Rh disease, or ***hemolytic disease of the newborn (HDN)*** (formerly called *erythroblastosis fetalis*).

About 85 percent of white Americans have the Rh factor (in addition to the ABO grouping) and are called "Rh-positive." The remaining 15 percent, without the Rh factor, are "Rh-negative." We now know that the Rh system contains six erythrocyte antigens (D, C, E, c, d, and e). Of these, the D Rh antigen is most important in causing the production of antibodies. If blood from an Rh-positive person is transfused into an Rh-negative person, the Rh-negative person will form antibodies against the Rh-positive red blood cells. During the next several months, antibodies build up in the Rh-negative blood plasma. In itself, this build-up is harmless. However, if the Rh-negative person receives additional Rh-positive blood, the newly formed antibodies will agglutinate the Rh-positive red blood cells.

The child of an Rh-positive father and an Rh-negative mother has an equal chance of inheriting either factor. A potential problem arises if an Rh-negative mother conceives an Rh-positive child [FIGURE 19.12]. Normally the

*In cross matching, the possible donor's red blood cells are mixed with the recipient's plasma to see if agglutination occurs.

FIGURE 19.12 DEVELOPMENT OF HEMOLYTIC DISEASE OF THE NEWBORN

[A] In an Rh-negative woman pregnant for the first time with an Rh-positive fetus, Rh-positive agglutinogens (antigens) (red) from the fetus's blood may diffuse through the placenta. (The fetus may be Rh-positive if its father is Rh-positive.) **[B]** Over time, the Rh-negative mother develops anti-Rh agglutinins (antibodies) (green). The first child will have been born before it could be affected by the antibodies. **[C]** A second Rh-positive child may receive some of its mother's anti-Rh antibodies through the placenta, and this may destroy that second child's red blood cells unless appropriate countermeasures are taken.

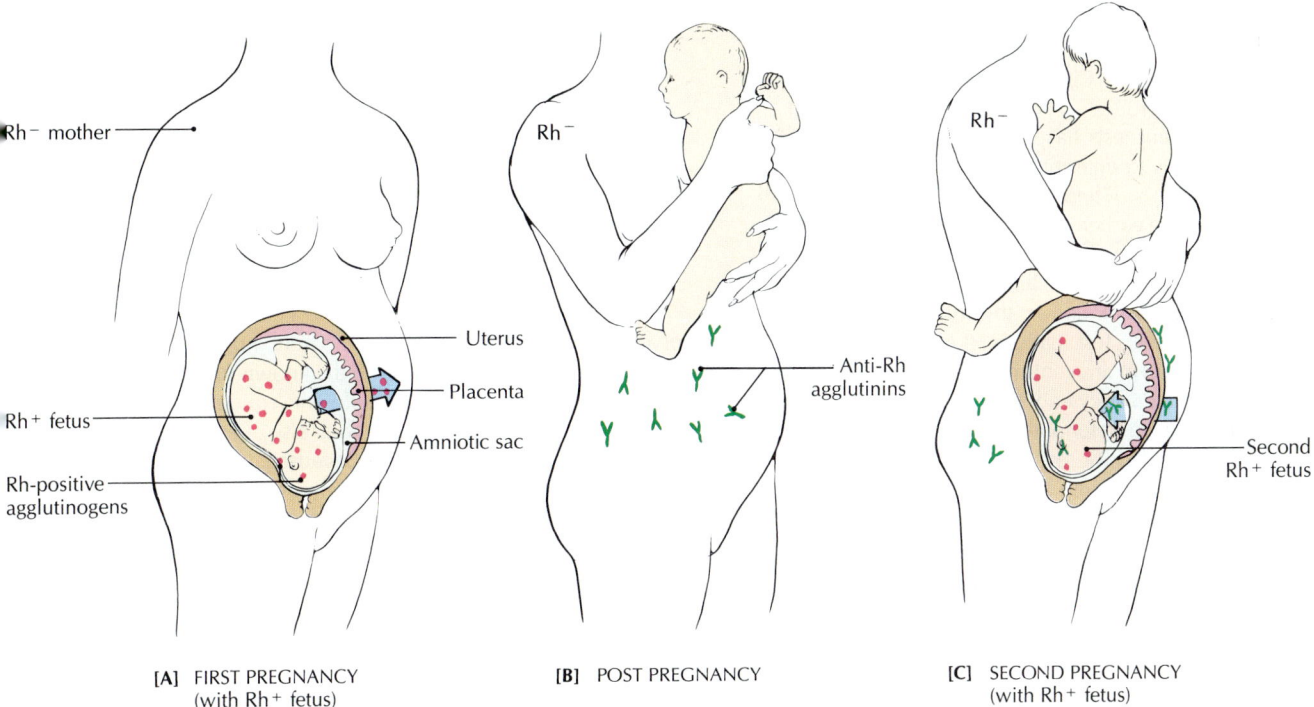

[A] FIRST PREGNANCY
(with Rh+ fetus)

[B] POST PREGNANCY

[C] SECOND PREGNANCY
(with Rh+ fetus)

circulation of a fetus is separate from the mother's circulation, but sometimes some fetal blood cells leak into the mother's blood during late pregnancy or at birth. If some of the antigens (Rh, C, D, E) in the fetus's blood leak through the placenta into the mother's bloodstream, the mother will produce antibodies in response to the fetal Rh-positive antigen. These antibodies are not yet developed enough to harm the first baby, but if a later fetus is Rh-positive, the mother's antibodies can cross the placenta, enter the fetus's blood, and endanger the fetus.

In hemolytic disease of the newborn, the fetus's agglutinated erythrocytes gradually undergo hemolysis and release hemoglobin into the blood of the fetus. Reticuloendothelial cells convert the hemoglobin into bilirubin, which causes jaundice of the fetus's skin. If born alive, the child must be given immediate blood transfusions of Rh-negative blood. The disease can now be prevented by giving all Rh-negative women who have given birth to Rh-positive infants an injection of a drug called Rho-GAM (an anti-Rh gamma globulin preparation) after delivery. The drug prevents her from making the anti-Rh antibodies and thus protects a future Rh-positive fetus.

Other Blood Antigen Systems

Although only the ABO and the Rh systems have been included here, there are about 100 described blood antigen systems, such as Kell, Lewis, M, N, P, and S. (Currently, these antigens are used mainly in legal decisions.) When one considers the number of combinations of blood antigens possible, it appears likely that there are enough different arrangements to provide every human being with an individual blood type. Since only a few types seem to be medically important, those few are well known, but the rest are present and functioning and are only important in isolated situations such as organ transplants.

A S K   Y O U R S E L F

1 What is an agglutinogen? An agglutinin?

2 What is the Rh factor?

3 What is hemolytic disease of the newborn?

Blood Tests and Paternity Suits

Blood types are sometimes used in paternity suits, even though most tests of blood types cannot show who *is* the child's father but can prove only who is *not*. A famous paternity case involved Charlie Chaplin. His blood type was O, the baby's was B, and the mother who made the claim had type A. Look at the table to see if Mr. Chaplin could have been the father.

More recent tests have been used in paternity suits to determine who the father actually is. These tests are more costly than the earlier ones, but they can prove paternity by showing that the child and alleged father both have five rare blood factors not found in the mother. An even newer test called HLA (for human leukocyte antigen) matches substances in the leukocytes of the child and alleged father to determine paternity.

Another paternity test employs new genetic techniques with a high degree of accuracy. Called "DNA fingerprinting," the technique involves short stretches of DNA called intervening sequences, which provide distinct markers of human differences. DNA is extracted from white blood cells, and enzymes are used to cleave specific locations near the intervening sequences. The resulting fragments of DNA vary in length from one person to another. The DNA of at least one parent must match those of the child. If the known mother and her child have different sequences and the alleged father does also, he cannot be the true father. Because DNA is more stable than blood, hair or skin samples from deceased people can also be tested for paternity.

INHERITANCE OF ABO BLOOD TYPE

Blood types of parents	Children's blood type possible	Children's blood type not possible
A + A	A, O	AB, B
A + B	A, B, AB, O	—
A + AB	A, B, AB	O
A + O	A, O	AB, B
B + B	B, O	A, AB
B + AB	A, B, AB	O
B + O	B, O	A, AB
AB + AB	A, B, AB	O
AB + O	A, B	AB, O
O + O	O	A, B, AB

WHEN THINGS GO WRONG

Anemias

Anemia is a condition in which the number of red blood cells, the normal concentration of hemoglobin, or the hematocrit is below normal. In general, anemia decreases the blood's oxygen-carrying capacity. When anemia is due to the excessive destruction of erythrocytes, there is usually an abnormally high amount of bilirubin in the plasma, producing a characteristic jaundice of the skin and a darkening of the feces. Anemia is usually a symptom of an underlying disease. Descriptions of several types of anemia follow.

Hemorrhagic anemia results from heavy blood loss and is sometimes seen in cases of heavy menstrual bleeding, severe wounds, or the sort of internal bleeding that accompanies a serious stomach ulcer.

Iron-deficiency anemia is the most common type of anemia. It can be caused by long-term blood loss, low intake of iron, or faulty iron absorption. If iron is not available to make erythrocytes, a *hypochromic* (low concentration in erythrocytes) *microcytic* (small erythrocytes) anemia develops.

Aplastic anemia is characterized by failure of the bone marrow to function normally. It is usually caused by poisons such as lead, benzene, or arsenic that hamper the production of red blood cells, or by radiation such as x-rays or atomic radiation that damages the bone marrow. With the marrow not functioning properly, the total count of erythrocytes and leukocytes decreases drastically.

Hemolytic anemia is produced when an infecting organism such as the malarial parasite enters red blood cells and reproduces until the cells actually burst. Such a rupture is called *hemolysis* ("blood destruction") and causes the anemia. (Other cases may be hereditary or due to sickle-cell anemia or an adverse reaction to a drug.) *Thalassemia* is the name of a group of hereditary hemolytic anemias characterized by impaired hemoglobin production. As a result, the synthesis of erythrocytes is also impaired. It is most common among people of Mediterranean descent. Sickle-cell anemia (discussed on page 607) is another hereditary hemolytic anemic condition.

Pernicious anemia is usually caused by improper absorption of dietary vitamin B_{12}, which is required for the complete maturation of red blood cells. The majority of cases are associated with antibodies that destroy the parietal cells that produce *intrinsic factor* in the stomach mucosa. Intrinsic factor is a protein that combines with vitamin B_{12} and facilitates its absorption by the small intestine. Without intrinsic factor, immature red blood cells accumulate in the bone marrow. The number of mature red blood cells circulating in the blood may drop as low as 20 percent of normal, and immature cells may enter the bloodstream. If diagnosed early, pernicious anemia is usually treated successfully with intramuscular injections of vitamin B_{12} and an improved diet. If left untreated, the disease may affect the nervous system, causing decreased mobility, general weakness, and damage to the brain and spinal cord.

Sickle-cell anemia is a genetic hemolytic anemia that occurs most often in blacks; it affects 1 in 600 American blacks. It results from the inheritance of the hemoglobin S-producing gene, which causes the substitution of the amino acid valine for glutamic acid in two of the four chains that compose the hemoglobin molecule. Sickle-cell anemia is so named because the abnormal red blood cells become rigid, rough, and crescent-shaped like a sickle (see photo). Sickled cells do not carry or release oxygen as well as normal erythrocytes do. Such cells make the blood more viscous by clogging capillaries and other small blood vessels, reducing blood supply to some tissues and producing swelling, pain, and tissue destruction. No effective permanent treatment has been discovered. Chronic fatigue and pain, increased susceptibility to infection, decreased bone marrow activity, and even death are common results. Geneticists believe that sickle-cell anemia is an adaptational response to the tropical disease *malaria*, because children with sickle-cell anemia cope better with malaria than those children without the sickle-cell trait.

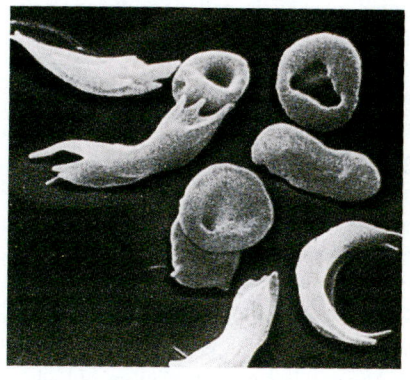

Scanning electron micrograph of erythrocytes from a patient suffering from sickle-cell anemia. Instead of being normal biconcave disks, these cells are variously distorted. ×2000.

Hemophilia

Hemophilia is a genetically inherited disease in which blood fails to clot or clots very slowly. The gene that causes the condition is carried by females, who are not affected themselves, and is expressed mostly in males. Females can be born with the disease only when the mother is a carrier and the father already has hemophilia.

The untreated hemophiliac bleeds very easily, especially into joints. Until recently, hemophilia created a severe disability, but excellent transfusion therapy to reduce fatal bleeding is now available. However, there is no known permanent cure.

Leukemia

Leukemia is a malignant (cancerous) disease characterized by uncontrolled leukocyte production. Leukocytes increase in number as much as 50- or 60-fold, and millions of abnormal immature leukocytes are released into the bloodstream and lymphatic system. The white blood cells tend to use the oxygen and nutrients that normally go to other body cells, causing the death of otherwise healthy cells. Infiltration of the bone marrow by abnormal leukocytes prevents the normal production of white and red blood cells, causing anemia. Because almost all the new white blood cells are immature and incapable of normal function, victims of leukemia have little resistance to infection. In most cases, the combination of starving cells and a deficiency in the immune system is enough to cause death. The exact cause of leukemia is unknown, although a viral infection is considered likely.

The two major forms of leukemia are lymphocytic and myelocytic. *Acute lymphocytic leukemia* (ALL) appears suddenly and progresses rapidly. It occurs most often in children and is much more serious than *chronic lymphocytic leukemia* (CLL), which occurs later in life. Both of these leukemias are caused by the cancerous production of lymphoid cells in lymph nodes and the spread of these cells to other parts of the body. In *myelocytic leukemia* (ML), the cancer begins in myelocytic cells such as early neutrophils and monocytes in the bone marrow and then spreads throughout the body, where white blood cells can be formed in extraneous organs. Acute myelocytic leukemia can occur at any age, whereas chronic myelocytic leukemia occurs during middle age.

Recent evidence shows that ALL occurs when pieces of two chromosomes break off and exchange places (translocation), resulting in the abnormal growth of both chromosomes. Researchers speculate that the new genes created by translocation create an abnormal "fusion" transcription factor that may cause leukemia by prevent-

Q: *Do people with hemophilia usually bleed to death?*

A: No. The blood of hemophiliacs usually has the ability to clot—it just takes longer. Bleeding from a minor cut would not be fatal, but massive bleeding, including internal hemorrhage, could be a serious problem. Most hemophiliacs are protected by modern drugs that enhance blood clotting.

ing the normal transcription factor from activating the correct genes, or by activating the production of proteins that activate oncogenes.

Chronic lymphocytic leukemia also appears to have a genetic cause, with mutations affecting a protein that regulates the production of interferon and helps maintain the homeostasis of cell growth. Scientists are experimenting with "antisense" RNA and other compounds that block the expression of leukemia-causing genes. It is interesting to note that a 1993 study in the United States revealed that cigarette smokers may increase their risk of contracting leukemia.

Bone Marrow Transplantation

Another possible method of combating disorders of blood-forming tissues such as aplastic anemia, leukemia, myeloma, and lymphoma, is *bone marrow transplantation.* The purpose of bone marrow transplantation is to introduce normal blood cells into the bloodstream of the patient in an attempt to replace the diseased cells. Progenitor cells are introduced to recolonize the recipient's marrow. To ensure a high degree of compatibility between donor and recipient so that the recipient will not reject the donated marrow, physicians try to match six tissue factors called *human leukocyte antigens* (HLA). In

bone marrow transplantation, about 500 to 700 mL of marrow from a compatible donor (preferably a twin or sibling) is aspirated (removed by suction) from the donor's pelvic bones (usually the iliac crest) with a large hypodermic needle attached to a syringe. Only a small incision is necessary. The removed marrow is mixed with an anticoagulant (heparin) and then filtered to separate the donor's T cells, specialized cells of the immune system that would otherwise encourage rejection of the marrow by the recipient. The recipient receives the donor's marrow through an intravenous infusion. Two units of the donor's blood, which were removed during two sessions about a week before the transplantation, are transfused back into the donor to replace the blood aspirated along with the marrow. Approximately 3000 to 5000 bone marrow transplants are performed annually in the United States.

Bone marrow transplantation is a hazardous procedure. The patient is treated before the transplantation with high doses of radiation and/or chemotherapy to remove all cells from the bone marrow. At this time, and for a period following the transplantation, the patient is vulnerable because the immune system is compromised. The success of the transplant is not assured fully until the reseeding of the marrow with healthy cells is established.

CHAPTER SUMMARY

Functions of Blood (p. 586)

1 *Blood* is the circulating fluid that transports nutrients, oxygen, carbon dioxide, gaseous waste products, and regulatory substances throughout the body.

2 Blood also defends against harmful microorganisms; is involved in inflammation, coagulation, and the immune response; and helps regulate the pH of body fluids. In general, blood helps maintain homeostasis by providing a constant cellular environment.

3 Blood consists basically of two parts: liquid *plasma* and solid *formed elements* (red blood cells, white blood cells, platelets) suspended in the plasma.

Properties of Blood (p. 586)

1 The average person has about 5 L of blood.

2 Blood *viscosity* and *specific gravity* are greater than those of water.

3 Blood *pH* ranges from 7.35 to 7.45.

4 Blood *temperature* averages about 38°C (100.4°F).

Plasma (p. 586)

Plasma provides the solvent for dissolved nutrients. It is about 90 percent water, 7 percent dissolved plasma proteins (albumin, globulins, fibrinogen), and 3 percent electrolytes, amino acids, glucose and other nutrients, enzymes, antibodies, hormones, metabolic waste, and traces of other materials.

Formed Elements (p. 588)

1 Red blood cells, or *erythrocytes,* make up about half the volume of blood. Their biconcave shape provides a large surface area for gas diffusion. They contain the globular protein **hemoglobin,** which transports oxygen from the lungs to all body cells and helps remove waste carbon dioxide. Hemoglobin contains a

small amount of *heme,* an iron-containing pigment that binds to oxygen and releases it at appropriate times.

2 The production of erythrocytes in bone marrow is called **erythropoiesis.** The rate of erythropoiesis is controlled mainly in a negative feedback system involving the glycoprotein hormone *erythropoietin.*

3 White blood cells, or **leukocytes,** destroy microorganisms at infection sites and remove foreign substances and body debris. Leukocytes use amoeboid movement to creep along the inner walls of blood vessels, and they can pass through blood vessel walls into tissue spaces.

4 The two basic classifications of leukocytes are **granulocytes** and **agranulocytes.** The three types of granulocytes are neutrophils, eosinophils, and basophils. **Neutrophils** destroy harmful microorganisms and other foreign particles, **eosinophils** help destroy parasites and antibody-antigen complexes, and **basophils** are involved in allergic reactions and inflammation, although their specific function is unknown.

5 The two types of agranulocytes are monocytes and lymphocytes. **Monocytes** can become macrophages that ingest and destroy harmful substances. **Lymphocytes** are involved in the immune response and the synthesis of antibodies.

6 Platelets, or **thrombocytes,** initiate the blood-clotting process when a blood vessel is injured and are important in the overall process of **hemostasis,** the prevention and control of bleeding. They are fragmented from **megakaryocytes.**

Clinical Blood Tests (p. 596)

Blood studies commonly performed for routine physical examinations include red blood cell count and hematocrit, hemoglobin content, and white blood cell and differential white count.

Hemostasis: The Prevention of Blood Loss (p. 597)

1 The body has a three-phase hemostatic mechanism consisting of constriction of blood vessels *(vasoconstriction)*, clumping *(aggregation)* of platelets, and blood clotting *(coagulation)*.

2 The final product of coagulation is the conversion of the soluble plasma protein **fibrinogen** into the insoluble, stringy plasma protein **fibrin.** This reaction requires the enzyme **thrombin.** Fibrin threads entangle the escaping blood cells and form a **clot.**

3 The **extrinsic pathway** of blood clotting is a rapid clotting system activated when blood vessels are ruptured and tissues are damaged. The **intrinsic pathway** is activated when the inner walls of blood vessels become damaged or irregular. Either pathway may precede the **common pathway**—the basic clotting mechanism of the conversion of fibrinogen to fibrin.

4 **Clotting factors** are substances found in the blood that are specialized to enhance blood clotting.

Anticoagulation: The Inhibition of Clotting (p. 602)

1 Substances in the blood, such as heparin and the antithrombin-heparin cofactor, prevent blood from clotting when there is no bleeding.

2 **Fibrinolysis,** a process in which fibrin is broken down, promptly removes naturally forming small clots from blood vessels.

3 Blood coagulation tests include platelet count, bleeding time, clotting time, and prothrombin time.

Blood Types (p. 603)

1 Blood can be classified into groups based on the presence of an A or B antigen *(agglutinogen)* on the surface of erythrocytes. Antibodies in blood plasma are called **agglutinins.** Agglutinins of one type of blood react violently with agglutinins of other types of blood, causing incompatible cells to clump together, a process called **agglutination.** The four blood types in the **ABO blood grouping** are A, B, AB, and O.

2 Another factor in blood is the **Rh factor.** It can cause **hemolytic disease of the newborn,** a disease that endangers a fetus if the fetus is Rh-positive, the mother is Rh-negative, and the mother has had a previous pregnancy in which the fetus was Rh-positive.

3 There are about 100 blood antigen systems in addition to the ABO and Rh systems.

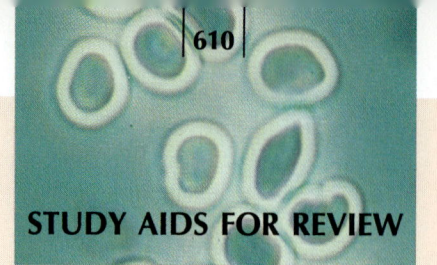

STUDY AIDS FOR REVIEW

KEY TERMS

MEDICAL TERMS FOR REFERENCE

ANOXIA (*an-*, without + oxygen) An oxygen deficiency in the blood.

APHERESIS (uh-FUR-ee-sis; Gr. *aphairein*, to take away from) A medical technique for cleansing the blood, in which a portion (such as plasma) suspected of containing harmful substances is removed and replaced with fresh ingredients.

BLOOD PLASMA SUBSTITUTE A chemical substance that imitates plasma characteristics. It is used to keep up the blood volume temporarily during emergencies until the blood is matched and to help replace fluids and electrolytes.

CITRATED WHOLE BLOOD Whole blood placed in a solution of acid citrate or a similar compound to prevent coagulation.

CYANOSIS (Gr. *kyanos*, blue) A bluish discoloration, especially of the skin and mucosa, as a result of an excessive amount of deoxyhemoglobin.

DIRECT (IMMEDIATE) TRANSFUSION Transfer of blood directly from one person to another without exposure of the blood to air.

DRIED PLASMA Normal plasma that has been vacuum-dried to prevent microorganisms from growing in it.

ERYTHROPENIA (Gr. red + poverty, lack) Decreased red blood cell count due to disease or hemorrhage.

EXCHANGE TRANSFUSION Direct transfer of blood from the donor to replace blood as it is removed from the recipient. This technique is used in poisonings and other conditions.

FRACTIONED BLOOD Blood separated into its components.

HEMORRHAGE (Gr. blood + bursting forth) An abundant discharge of blood from a broken vessel.

HEPARINIZED WHOLE BLOOD Whole blood placed in a heparin solution to prevent coagulation.

INDIRECT (MEDIATE) TRANSFUSION Transfer of blood in which whole or fractioned donor blood is stored for later delivery to a recipient. The blood can be separated into components, and patients receive only the needed portions.

LEUKOPENIA (Gr. white + poverty, lack) Decreased white blood cell count due to disease or hemorrhage.

NORMAL PLASMA Plasma that has had blood cells removed but retains the normal concentrations of solutes; employed to bring blood volume back to normal levels.

PACKED RED CELLS The concentrated solution of erythrocytes that remains when plasma is removed from whole blood.

PLATELET CONCENTRATE Platelets separated from fresh whole blood; used for platelet-deficiency blood disorders.

POLYCYTHEMIA (pahl-ee-sigh-THEE-mee-uh; "many cells in the blood") A condition in which the number of red blood cells is above normal.

PURPURA (PUR-pyoo-ruh; Gr. *porphura*, shellfish yielding a purple dye) Purple spots on the skin resulting from escaped erythrocytes from capillaries or larger hemorrhagic areas; also appear when platelets are drastically reduced in number.

SEPTICEMIA (sehp-tih-SEE-mee-uh; Gr. *septos*, rotten + blood) The presence of harmful substances, such as bacteria or toxins, in the blood. Also called *blood poisoning*.

THROMBOCYTOPENIA A decreased number of platelets, resulting from impaired production or increased destruction.

THROMBOCYTOSIS An increased number of platelets, usually due to increased production accompanying diseases such as leukemia and polycythemia or following the removal of the spleen. Also called *thrombocythemia*.

VENISECTION Opening a vein to withdraw blood. Also called *phlebotomy*.

VON WILLEBRAND'S DISEASE An inherited antihemophilic blood factor disorder and capillary defect, characterized by abnormal bleeding of the nose, gums, and skin.

WHOLE BLOOD Blood that has all its components (formed elements, plasma, and plasma solutes) in the natural concentration.

UNDERSTANDING THE FACTS

1 What are the major functions of blood?

2 What are the components of blood plasma?

3 What is the difference between serum and plasma?

4 What kinds of cells are characteristically found in blood? Which are the most numerous? The least numerous?

5 What constituent(s) of the blood is (are) responsible for the transport of oxygen? Of carbon dioxide?

6 What is the chloride shift and when does it occur?

7 Define

  a hemolysis   e diapedesis

  b hemostasis   f hematocrit

  c bilirubin   g agglutinin

  d leukopenia   h RhoGAM

8 Where and how are red blood cells destroyed?

9 List the types of granulocytes and agranulocytes, indicating the major functions of each type.

10 Which blood cells arise from hemocytoblasts? From megakaryoblasts? From myoblasts? From monoblasts?

11 List the three phases of hemostasis and describe the major events that occur in each stage.

12 What is the specific role of platelets in blood coagulation?

13 Distinguish between the extrinsic and intrinsic mechanisms of blood coagulation; outline the steps in each process.

14 What is the function of vitamin K with respect to blood coagulation?

15 Distinguish the causes of hemorrhagic, iron-deficiency, aplastic, hemolytic, and pernicious anemias.

UNDERSTANDING THE CONCEPTS

1 Compare blood to water with respect to viscosity, specific gravity, and pH. Explain which constituents in blood give rise to the differences.

2 Explain why arterial blood is more alkaline than venous blood.

3 Contrast the roles of each of the plasma proteins.

4 What is erythropoietin? Where is it synthesized and what is its role?

5 What roles do leukocytes play in the immune response?

6 Explain the role of the sympathetic nervous system in hemostasis.

7 How is blood normally prevented from clotting within blood vessels?

8 Explain what would occur if type A blood was transfused into a type B individual.

9 Explain why hemolytic disease of the newborn is unlikely in a firstborn child.

10 Are there treatments and/or cures for hemophilia or sickle-cell anemia? Explain.

SELF-QUIZ
Multiple-Choice Questions

19.1 The proteins of blood plasma that function as antibodies are

  a albumins   d a and b

  b globulins   e a, b, and c

  c fibrinogen

19.2 Lipids are found in plasma in the form of

  a phospholipids   d cholesterol

  b triglycerides   e all of the above

  c free fatty acids

19.3 Carbon dioxide is not transported in the blood in the form of

  a carbonic acid   d $CaCO_3$

  b bicarbonate ions e dissolved CO_2

  c $HbCO_2$

19.4 Which of the following is neither a granulocyte nor a progenitor of a granulocyte?

  a myeloblast   d eosinophil

  b neutrophil   e basophil

  c monocyte

19.5 Granulocytes develop in bone marrow from undifferentiated cells called

  a myeloblasts   d eosinophilic myelocytes

  b promyelocytes   e none of the above

  c metamyelocytes

19.6 The smallest blood cells are

  a basophils   d neutrophils

  b monocytes   e erythrocytes

  c eosinophils

19.7 Which is *not* a phagocyte?

  a neutrophil   d monocyte

  b eosinophil   e a and b

  c lymphocyte

19.8 Which of the following develops from megakaryocytes?

  a lymphocytes   d platelets

  b hemocytoblasts e erythrocytes

  c monocytes

19.9 All but one of the following might be diagnosed as a result of an abnormal differential white count. Which is the exception?
- **a** bacterial infection
- **d** viral infection
- **b** polycythemia
- **e** allergic reaction
- **c** leukemia

19.10 Which is *not* an enzyme?
- **a** fibrinogen
- **d** thromboplastinogenase
- **b** thromboplastin
- **e** a and b
- **c** thrombin

19.11 Which is neither an anticoagulant nor a fibrinolytic agent?
- **a** streptokinase
- **d** antithrombin
- **b** heparin
- **e** vitamin K
- **c** plasmin

19.12 Which combination is most likely to result in hemolytic disease of the newborn?
- **a** both parents are Rh-positive
- **b** both parents are Rh-negative
- **c** Rh-positive mother, Rh-negative father
- **d** Rh-negative mother, Rh-positive father
- **e** either c or d

19.13 A person with blood type O can safely receive
- **a** type B blood
- **d** type AB blood
- **b** type A blood
- **e** b and c
- **c** type O blood

19.14 The protein threads that trap blood cells and create a clot are
- **a** fibrinogen
- **d** thromboplastin
- **b** thrombin
- **e** prothrombin
- **c** fibrin

19.15 A hematocrit is a measure of
- **a** blood volume
- **b** hemoglobin concentration
- **c** volume percentage of red blood cells
- **d** total number of blood cells
- **e** number of white blood cells

Matching

Classify each of the following as a clotting compound or a clot-dissolving compound.

a clotting compound   **b** anticlotting compound

19.16 _____ streptokinase
19.17 _____ prothrombin
19.18 _____ von Willebrand factor
19.19 _____ thromboplastinogenase
19.20 _____ heparin
19.21 _____ (recombinant) tissue-plasminogen activator

Completion Exercises

19.22 The process by which some white blood cells move through vessel walls is _____.

19.23 The proteins in blood plasma are referred to as _____.

19.24 The synthesis of carbonic acid from CO_2 and H_2O occurs in the _____.

19.25 The production of red blood cells is called _____.

19.26 The rupturing of erythrocytes is called _____.

19.27 The production of white blood cells is termed _____.

A SECOND LOOK

In the following drawing, identify a monoblast, myeloblast, reticulocyte, lymphocyte, and monocyte; also identify myeloid and lymphoid cells.

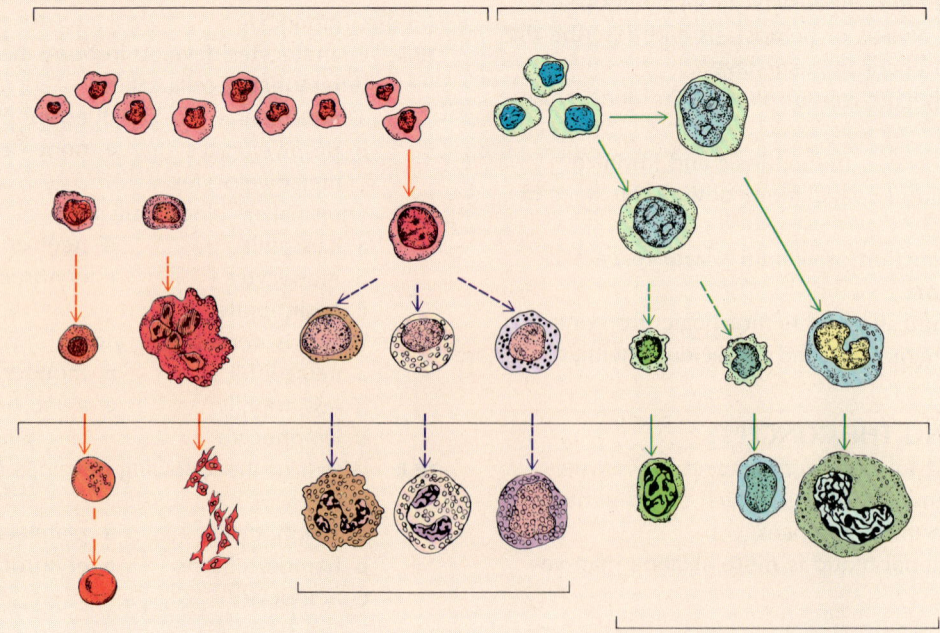

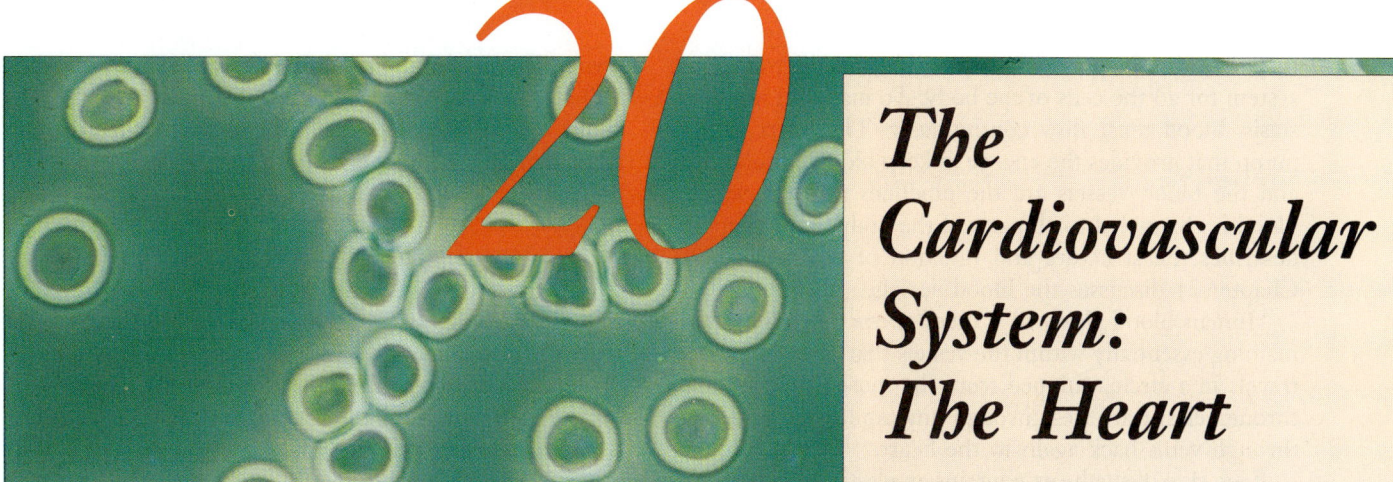

20

The Cardiovascular System: The Heart

KEY CONCEPTS

1 The heart is a double pump. The right ventricle pumps oxygen-poor blood to the lungs, where it picks up a fresh supply of oxygen; the left ventricle pumps the oxygen-rich blood to the rest of the body.

2 When the ventricles pump simultaneously, equal volumes of blood leave the heart.

3 One-way valves between the upper and lower heart chambers prevent the backflow of blood and ensure that blood passes through the heart in only one direction.

4 Unlike skeletal muscle, cardiac muscle has a built-in pacemaker that initiates the heartbeat independently of the nervous system.

As you saw in the previous chapter, blood is the transport system for all the cells of the body. To maintain homeostasis, blood must flow continuously. The heart is the pump that provides the energy to keep blood circulating, and the blood vessels are the pipelines through which blood is channeled to all parts of the body. This chapter describes the structure and function of the heart, and Chapter 21 discusses the blood vessels.

Human blood flows in a *closed system* of vessels, remaining essentially within the vessels that carry it. Blood travels in a circle, pumped from the muscular heart out through elastic arteries, through tiny capillaries, and then through veins back again to the heart.

Each side of the heart contains an elastic upper chamber called an *atrium*, where blood enters the heart, and a lower pumping chamber called a *ventricle*, where blood leaves the heart. Thus the heart is actually a *double pump*. The oxygen-poor blood returning to the right atrium from the body tissues is pumped by the right ventricle (the first pump) into the lungs, where carbon dioxide is exchanged for oxygen. The oxygenated blood moves from the lungs into the left atrium and then into the left ventricle (the second pump), which pumps blood to all parts of the body. The circulation between the heart and the lungs is called the **pulmonary circulation** (L. *pulmo*, lung); the circulation between the heart and the rest of the body is called the **systemic circulation** [FIGURE 20.1].

Developmentally, the heart, together with its vessels, takes shape and begins to function long before any other major organ. It begins beating in a human embryo during the fourth week of development and continues throughout the life of a person at a rate of about 70 beats per minute, 100,000 beats a day, or about 2.5 billion beats during a 70-year lifetime. The specialized cardiac muscle tissue that performs this extraordinary feat is found only in the wall of the heart.

ANATOMY OF THE HEART

The **heart** (Gr. *kardia*, as in *cardiac*; L. *cor*)* is shaped like a blunt cone. It is about the size of the clenched fist of its

*The word *coronary* is often used in reference to structures and events involving the heart. (A "coronary thrombosis" is a blockage of a coronary artery by a blood clot. The resultant heart attack is informally called a "coronary.") However, the term actually comes from the Latin word *corona*, which means "crown."

Q: *How much more work does the heart of a physically inactive person do than that of an active person?*

A: The heart of an inactive person beats about 36,000 times a day more than the heart of an active person. Such an added burden on the heart is almost like running a marathon every day.

owner. It averages about 12 cm long and about 9 cm wide (5.0 in. by 3.5 in.). The heart of an adult male weighs about 250 to 390 g (8.8 to 13.8 oz), and the heart of an adult female usually weighs between 200 and 275 g (7.0 to 9.7 oz).

Location of the Heart

The heart is located in the center of the thorax. It is slanted diagonally, with about two-thirds of its bulk to the left of the body's midline [FIGURE 20.2A]. The heart is turned on its longitudinal axis so that the right ventricle is partially in front of the left, directly behind the sternum. The left ventricle faces the left side and the back of the thorax. The heart lies closer to the front of the thorax than to the back.

The pointed end of the blunt cone is called the **apex** (L. point). It extends forward, downward, and to the left. Normally the apex is located between the fifth and sixth ribs (the fifth intercostal space) on the midclavicular line (a perpendicular line from the middle of the clavicle to the diaphragm). The uppermost part of the heart, the **base,** extends upward, backward, and to the right. Anteriorly, it lies just below the second rib.

The base is in a relatively fixed position because of its attachments to the great vessels, but the apex is able to move. When the ventricles contract, they change shape just enough so that the apex moves forward and strikes the left thoracic wall near the fifth intercostal space. This thrust of the apex is what we normally feel from the outside as a heartbeat.

Covering of the Heart: Pericardium

The heart does not hang freely in the thorax. It hangs by the great blood vessels inside a protective sac called the **pericardium** ("around the heart") [FIGURE 20.3]. This sac is composed of an outer, fibrous layer of connective tissue, the *fibrous pericardium*, and an inner layer of serous tissue, the *serous pericardium*. The serous pericardium is divided into an outer, or *parietal*, layer, which lines the inner surface of the fibrous pericardium, and an inner, or *visceral*, layer, which covers the outer surface of the heart and the adjoining portions of the large blood vessels. The serous pericardium surrounds the *pericardial cavity*, which contains a small amount of serous *pericardial fluid*. Because the visceral pericardium forms the outer layer of

Q: *During a physical examination, why does a physician tap the chest wall while listening with a stethoscope?*

A: A physician can estimate the size of the heart by tapping the chest wall progressively and listening for sound changes. If a problem appears to exist, the actual size can be determined by using x-ray and ultrasound.

FIGURE 20.1 SYSTEMIC AND PULMONARY CIRCULATIONS

The systemic circulation starts in the heart; flows through the muscles, organs, and tissues of the body; and then returns to the heart. The pulmonary circulation flows only from the heart to the lungs and back to the heart. Oxygenated blood is shown in red, deoxygenated blood in blue.

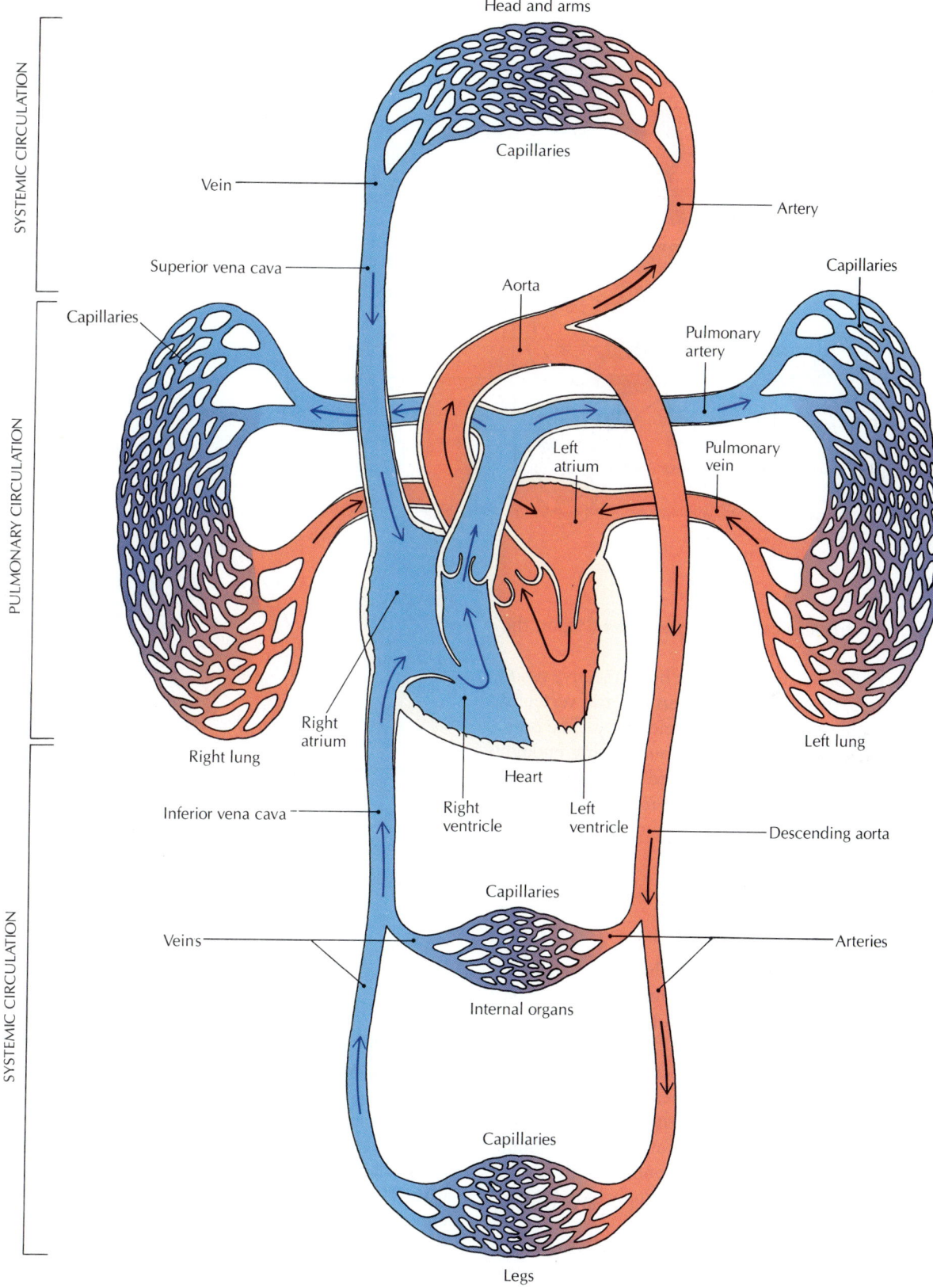

FIGURE 20.2 POSITION OF THE HEART

The position of the heart in the thorax is shown in relation to the ribs, sternum, and diaphragm. **[A]** Anterior view. **[B]** Left lateral view.

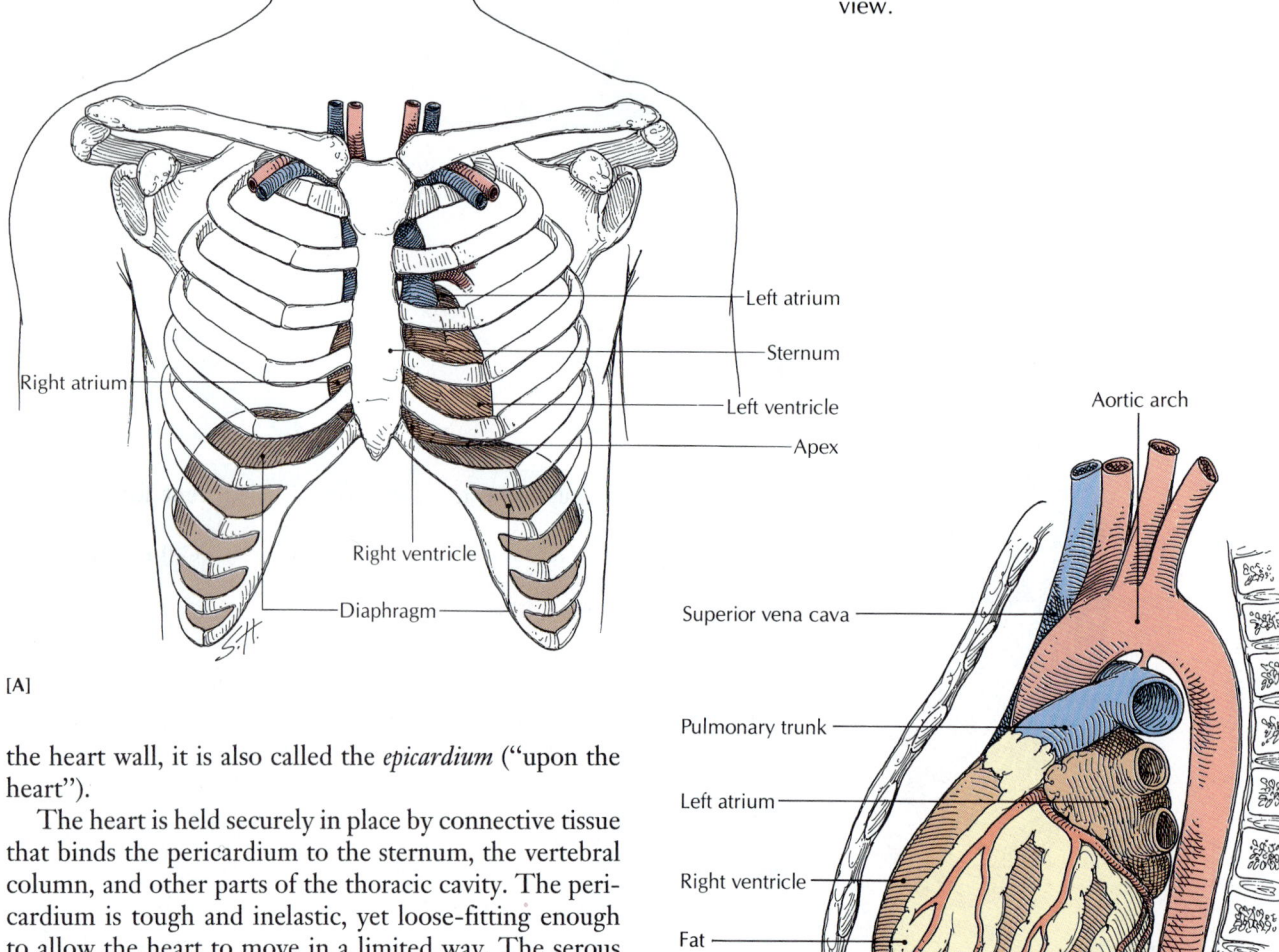

[A]

[B]

the heart wall, it is also called the *epicardium* ("upon the heart").

The heart is held securely in place by connective tissue that binds the pericardium to the sternum, the vertebral column, and other parts of the thoracic cavity. The pericardium is tough and inelastic, yet loose-fitting enough to allow the heart to move in a limited way. The serous pericardial fluid moistens the sac and minimizes friction between the membranes as the heart moves during its contraction-relaxation phases.

Wall of the Heart

The wall of the heart is made up of three layers: (1) the outer *epicardium* (*epi*, upon), (2) the middle *myocardium*, or muscular layer (*myo*, muscle), and (3) the inner *endocardium* (*endo*, inside) [FIGURE 20.3B].

If you were to cut away the parietal pericardium, you would see that the surface of the heart itself is reddish and shiny. This shiny membrane is the **epicardium**. The epicardium is continuous with the parietal pericardium. Inside the epicardium, and often surrounded with fat, are the main coronary blood vessels that supply blood to and drain blood from the heart.

Directly beneath the epicardium is the middle layer, the **myocardium** ("heart muscle"), which is a thick layer

of cardiac muscle that gives the heart its special pumping ability. Cardiac muscle cells in the atria and ventricles are involuntary, striated, and branched. They function as a *coordinated unit* in response to physiological stimulation rather than as a group of separate units as skeletal muscle does. These myocardial muscle cells act in this way because they are connected end to end by transverse thickenings of the sarcolemma called *intercalated disks*, which contain gap junctions and desmosomes. **Gap junctions** allow action potentials to be transmitted from one car-

FIGURE 20.3 COVERING AND WALL OF THE HEART

[A] Layers of the pericardial sac and the heart wall. [B] Enlarged view of the structure of the pericardium and the ventricular heart wall. [C] Schematic diagram of the spiral arrangement of ventricular muscles. The cardiac (fibrous) skeleton of the heart is shown above the dashed line. Also shown are the four rings of the cardiac skeleton (top).

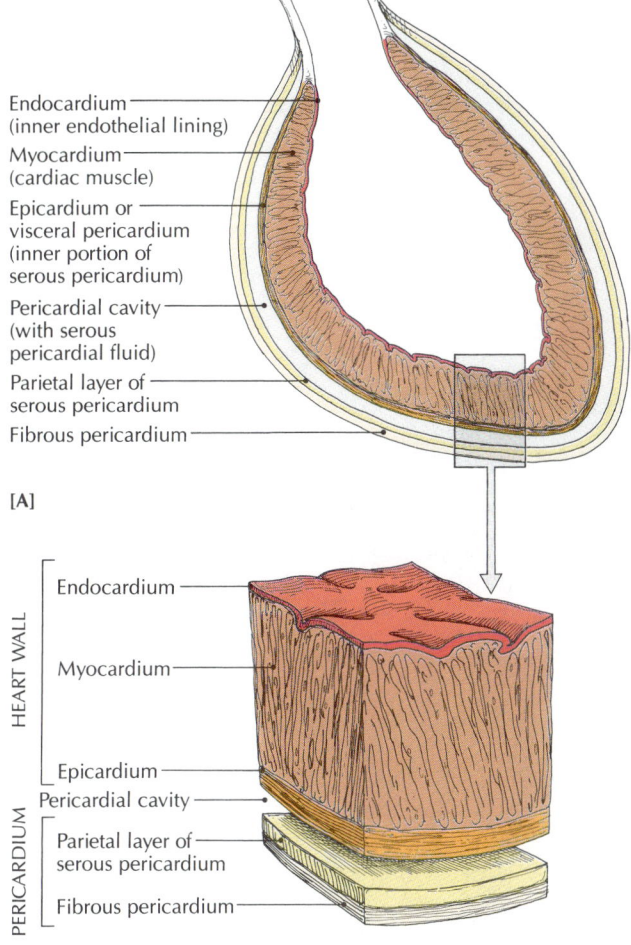

Endocardium
(inner endothelial lining)

Myocardium
(cardiac muscle)

Epicardium or
visceral pericardium
(inner portion of
serous pericardium)

Pericardial cavity
(with serous
pericardial fluid)

Parietal layer of
serous pericardium

Fibrous pericardium

[A]

HEART WALL

Endocardium

Myocardium

Epicardium

PERICARDIUM

Pericardial cavity

Parietal layer of
serous pericardium

Fibrous pericardium

[B]

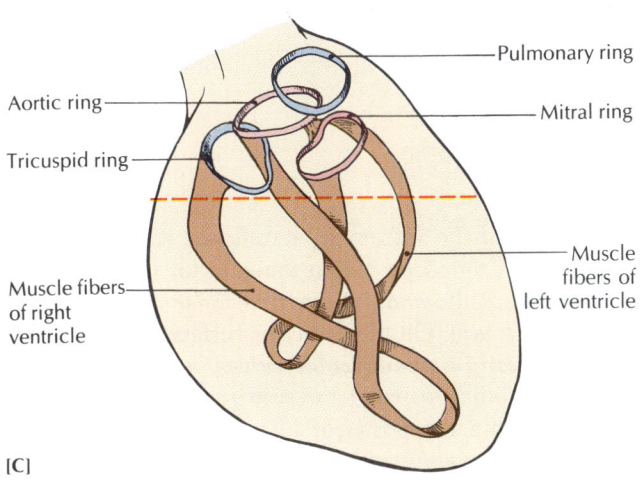

Pulmonary ring

Aortic ring

Mitral ring

Tricuspid ring

Muscle
fibers of
left ventricle

Muscle fibers
of right
ventricle

[C]

diac cell to another, and ***desmosomes*** hold the cells together and serve as the attachment sites for myofibrils. This connection maintains cell-to-cell cohesion so that the "pull" of one contractile unit is transmitted to the next one.

Overall, this series of interconnected cells forms a latticework called a *syncytium* (sihn-SIE-shum). The importance of a functional syncytium is that when the atrial or ventricular muscle mass is stimulated, the action potential spreads over the entire atrial or ventricular syncytium, causing the muscle cells in the entire chamber to contract almost in unison.

The myocardium contains a large number of mitochondria, which provide a constant source of ATP for the hardworking heart muscle. It also has an abundant blood supply and a high concentration of *myoglobin*, a muscle protein that stores oxygen.

The myocardium has two spiral layers of cardiac muscle, which are attached to a fibrous ring (fibrous trigone) that forms the cardiac (fibrous) skeleton of the heart. The spiral is the most effective arrangement for squeezing blood out of the heart's chambers [FIGURE 20.3C].

The inside cavities of the heart and all the associated valves and muscles are covered with the ***endocardium***. The endocardium is a thin, fibrous layer composed of simple squamous epithelial tissue (endothelium) and some connective tissue. The endothelium of the heart is continuous with the endothelium of the blood vessels that enter and leave the heart.

Cardiac (Fibrous) Skeleton

When blood is pumped to the lungs and the rest of the body, it is wrung out of the ventricles like water from a wet cloth. To accomplish this wringing motion, the heart has a fibrous ***cardiac skeleton*** of tough connective tissue that provides attachment sites for the valves and muscular fibers.

The cardiac skeleton is made up of a *fibrous trigone* (Gr. *trigonos*, triangular) and four rings (each ring is called an *annulus fibrosus*), one surrounding each of the heart's four major orifices [FIGURES 20.3C and 20.6C]. These heart openings are regulated by valves. Four rings of the cardiac skeleton provide sites of attachment (origin) for the muscle layers of the atria and ventricles [FIGURE 20.3C] and support the bases of the valves and prevent them from stretching. This anatomical arrangement of cardiac skeleton and valves prevents blood from flowing backward from the ventricles into the atria and from the arteries back into the ventricles.

Surface Markings of the Heart

On the surface of the heart are some *sulci* (depressions) that are helpful in locating certain features of the ventri-

FIGURE 20.4 EXTERNAL HEART

[A] Anterior external view of the heart, showing surface features and great vessels.

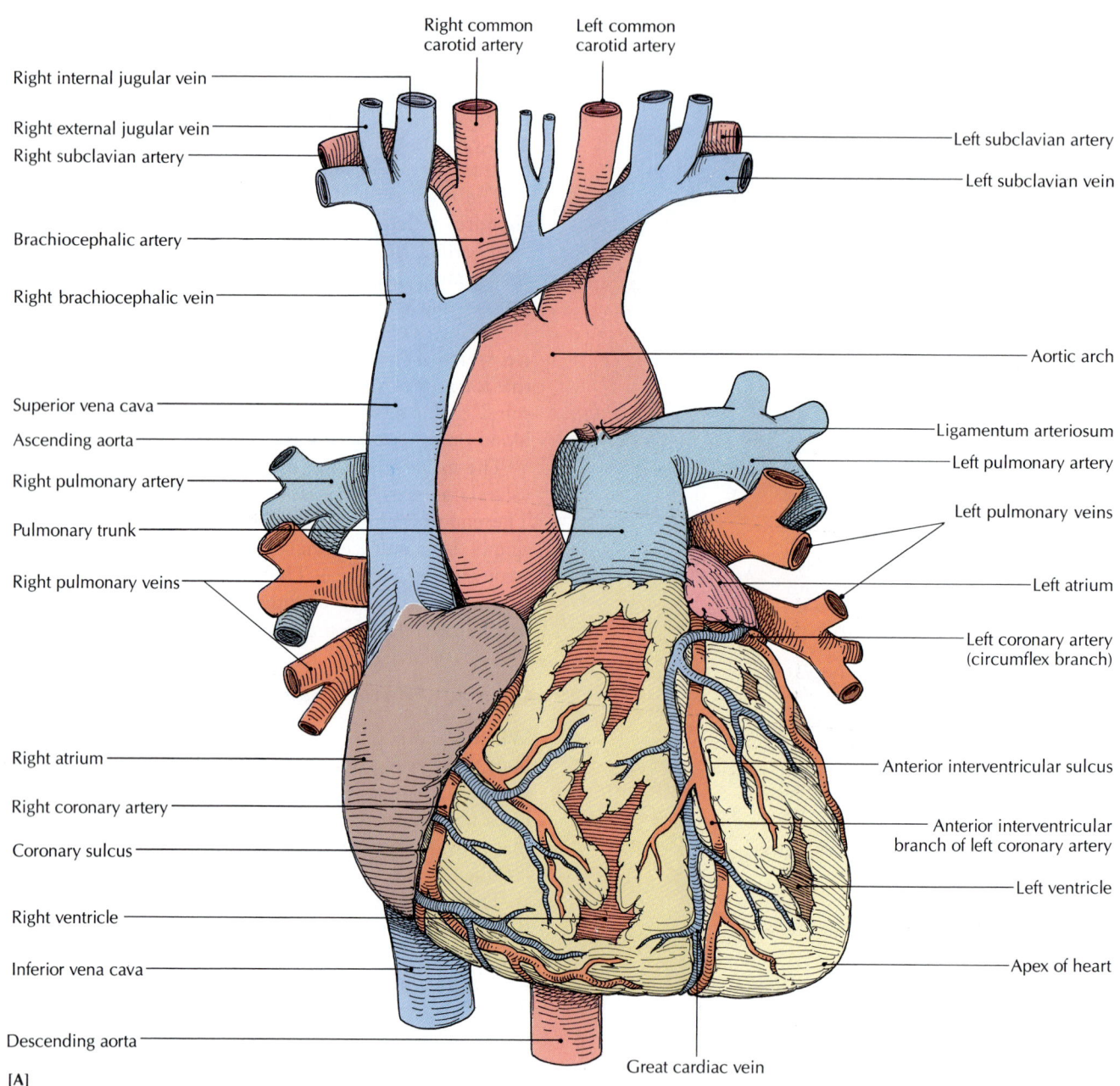

Right common carotid artery

Left common carotid artery

Right internal jugular vein

Right external jugular vein

Right subclavian artery

Brachiocephalic artery

Right brachiocephalic vein

Superior vena cava

Ascending aorta

Right pulmonary artery

Pulmonary trunk

Right pulmonary veins

Right atrium

Right coronary artery

Coronary sulcus

Right ventricle

Inferior vena cava

Descending aorta

Left subclavian artery

Left subclavian vein

Aortic arch

Ligamentum arteriosum

Left pulmonary artery

Left pulmonary veins

Left atrium

Left coronary artery (circumflex branch)

Anterior interventricular sulcus

Anterior interventricular branch of left coronary artery

Left ventricle

Apex of heart

Great cardiac vein

[A]

cles and atria as well as coronary vessels [FIGURE 20.4A]. The ***coronary sulcus,*** encircling the heart, indicates the border between the atria and ventricles. Embedded in fat within the coronary sulcus are the *right* and *left coronary arteries* and the coronary sinus (not shown in FIGURE 20.4; see FIGURE 20.7).

On the anterior surface of the heart is the ***anterior interventricular*** ("between the ventricles") ***sulcus,*** which marks the location of the *interventricular septum*, the an-

terior part of the septum separating the ventricles [FIGURE 20.4]. Embedded within the anterior interventricular sulcus are the *anterior interventricular artery* and the *great cardiac vein*. On the posterior surface of the heart is the ***posterior interventricular sulcus,*** which marks the location of the posterior interventricular septum. Embedded within the posterior interventricular sulcus are the *posterior interventricular descending artery* and the *middle cardiac vein*.

[B] Photograph of anterior external heart, life size.

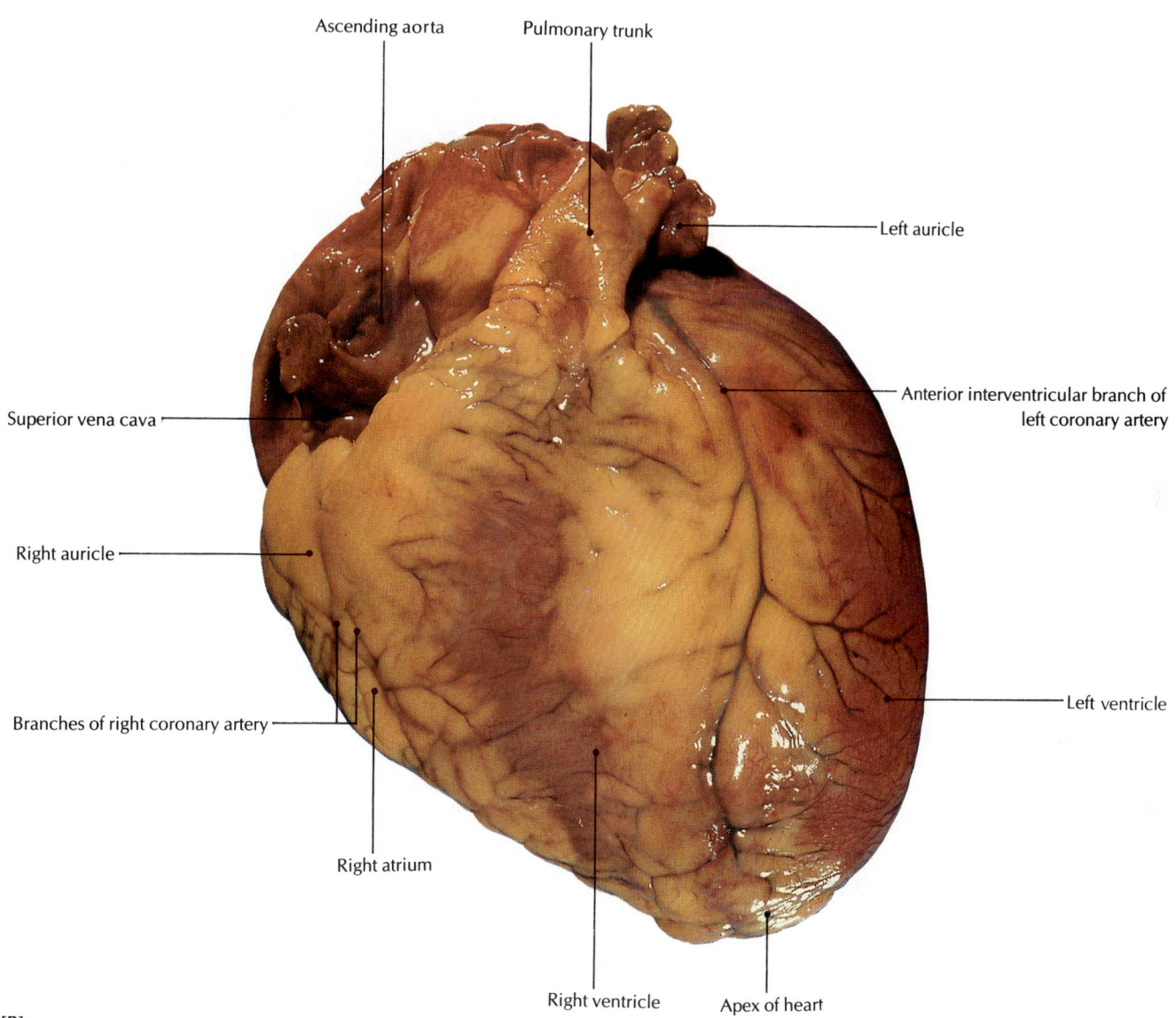

Ascending aorta

Pulmonary trunk

Left auricle

Superior vena cava

Anterior interventricular branch of left coronary artery

Right auricle

Branches of right coronary artery

Left ventricle

Right atrium

Right ventricle

Apex of heart

[B]

Chambers of the Heart

The heart is a hollow organ that contains four chambers [FIGURE 20.5A]. Dividing the heart vertically into a *right heart* and a *left heart* is a wall of muscle called the *septum* (L. *sepire*, to separate with a hedge). At the top of each half of the heart is a chamber called an *atrium* (AY-tree-uhm; L. porch, antechamber). Below it is another chamber, the *ventricle* ("little belly"). Each atrium has a flaplike appendage of unknown function called an auricle ("little ear") [FIGURE 20.4B]. The atria lead to the ventricles by way of openings called *atrioventricular orifices*. In a normal heart, blood flows from the atria to the ventricles, never the other way.

On the septal wall of the right atrium is an oval depression called the *fossa ovalis*, the site of the fetal foramen ovale before it closed at birth (see Adaptations in Fetal Circulation in Chapter 21). The lining of both atria is smooth except for some ridges called *musculi pectinati* (pectinate muscles) [FIGURE 20.5B; see also FIGURE 20.6],

FIGURE 20.5 CHAMBERS OF THE HEART

[A] Schematic drawing of interior of the heart with blood vessels removed to show the four chambers; anterior view. [B] Photograph of heart chambers; internal anterior view. [C] Cutaway drawing of the right and left ventricles, showing the thicker wall of the left ventricle.

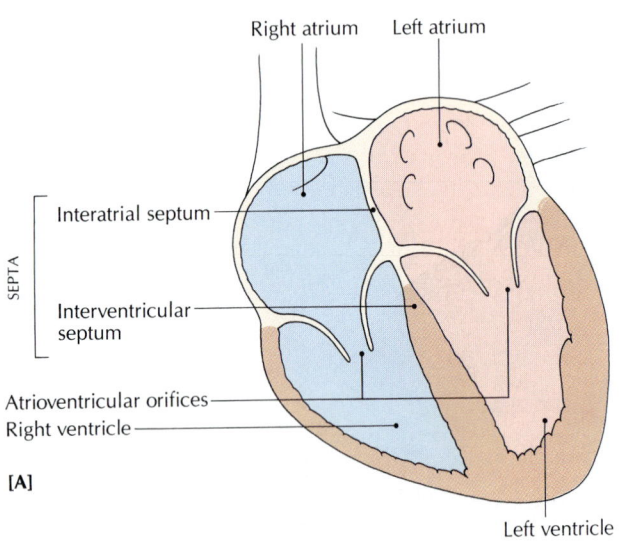

[A]

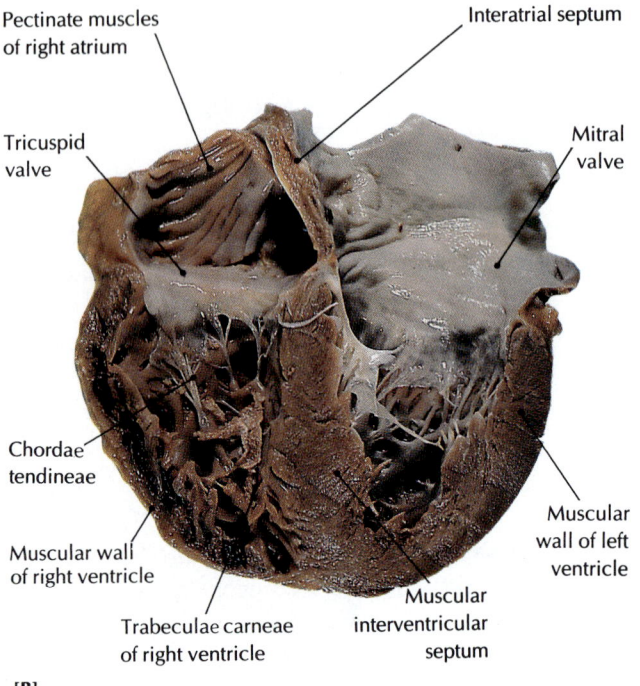

[B]

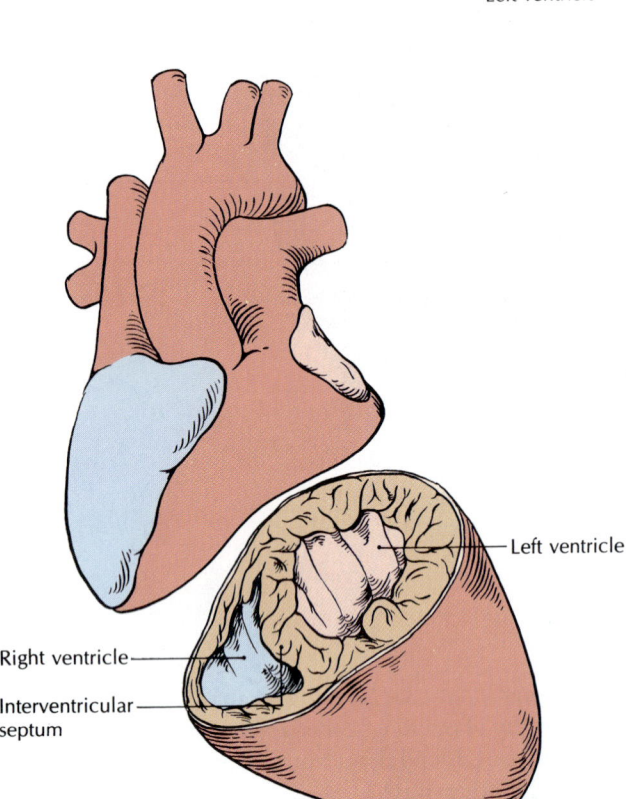

[C]

which are formed by parallel muscle bundles that look like a comb. The musculi pectinati are located on both right and left auricles and on the anterior wall of the right atrium.

The right and left ventricles have walls that contain ridges called *trabeculae carneae* (L. "little beams of flesh"), which are formed by coarse bundles of cardiac muscle fibers. Both the right and left ventricles contain papillary muscles and supportive cords called *chordae tendineae*, which are attached to the free edges of the atrioventricular valves [FIGURE 20.5B; see also FIGURE 20.6]. The upper part of the right ventricular wall, called the *conus* or *infundibulum*, is smooth and funnel-shaped. This area leads to the pulmonary artery [see FIGURE 20.6A]. The left ventricle does not have this conus.

Because the heart is a closed system of chambers, when the left and right ventricles pump simultaneously, equal volumes of blood leave the heart. The wall of the left ventricle is thicker, however, because it must be strong enough to supply blood to all parts of the body, whereas the right ventricle supplies blood only to the lungs. As a result, the left ventricular blood pressure is higher (120 mmHg) during contraction than that in the right ventricle (20 mmHg). The left ventricle is thicker to accommodate the higher pressure required to move blood a greater distance against a high resistance. The walls of the atria are thinner than the ventricular walls because less pressure is required to move blood only a short distance (and against less resistance) into the ventricles.

Valves of the Heart

The four heart valves allow blood to flow through the heart in only one direction, thus preventing the backflow of blood. The two *atrioventricular valves* allow blood to flow from the atria to the ventricles, and the two *semilunar valves* allow blood to flow from the right ventricle to the pulmonary artery (on the way to the lungs) and from the left ventricle to the aorta (on the way to the rest of the body) [FIGURE 20.6]. A heart valve consists of flaplike folds called *cusps* of the endocardium reinforced by dense connective tissue. The dense connective tissue of the heart valves is continuous with that of the fibrous rings of the skeleton of the heart. These rings completely surround the outlet through which blood flows, and they prevent an outlet from becoming dilated when the heart contracts and forces blood through it.

Atrioventricular valves The two *atrioventricular (AV) valves* differ in structure but operate in the same way. The right atrioventricular is known as the *tricuspid valve* of the right heart because it consists of three cusps (L. *tri*, three + *cuspis*, point) [FIGURE 20.6B]. The left atrioventricular, or *bicuspid valve* of the left heart, has two cusps (*bi* = two). Because the cusps resemble a bishop's miter (a tall, pointed hat), the valve is also called the *mitral* (MY-truhl) *valve*.

Each cusp is a thin, strong fibrous flap covered by endocardium. Its broad base is anchored into a ring of the cardiac skeleton [FIGURE 20.6C]. Attached to its free end are strong yet delicate tendinous cords called *chordae tendineae* [FIGURE 20.6E and F]. The chordae tendineae, which resemble the cords of a parachute, are continuous with the nipplelike papillary muscles (L. *papilla*, nipple) in the wall of the ventricle [FIGURE 20.6E and F].

When blood is flowing from the atria to the ventricles, the cusps of the AV valves are forced against the ventricular walls, thereby opening each atrioventricular orifice. When the ventricles contract, the cusps are brought together by the increasing ventricular blood pressure, and the atrioventricular orifices are closed [FIGURE 20.6D]. At the same time, the papillary muscles contract, putting tension on the chordae tendineae. The chordae tendineae pull on the cusps, preventing them from being forced into the atria. Otherwise, blood would flow backward from the ventricles into the atria.

Semilunar valves The *semilunar* ("half-moon") *valves* prevent blood in the pulmonary trunk and aorta from flowing back into the ventricles. The right semilunar valve is the heart valve in the opening between the right ventricle and the pulmonary trunk and is called the *pulmonary semilunar valve*. It allows oxygen-poor blood to enter the pulmonary trunk on its way to the lungs from the right ventricle. The left semilunar heart valve is the *aortic semilunar valve,* which allows freshly oxygenated blood to enter the aorta from the left ventricle. The left semilunar valve is slightly larger and stronger than the right.

Each semilunar valve contains three cusps. The base of each cusp is anchored to the fibrous ring of the cardiac skeleton. The free borders of the cusps curve outward and extend into the respective artery (aorta and pulmonary artery). These cusps do not have chordae tendineae attached to their free margins, as the atrioventricular valves do. During a ventricular contraction, the blood rushes out of the ventricles and pushes the cusps up and against the wall of each artery, allowing blood to flow into the artery. When the ventricles relax, some blood flows back, filling the space between the cusps and the walls of the aorta, forcing the edges of the cusps together. This motion closes off the artery completely and prevents blood from flowing back into the relaxed ventricles.

Great Vessels of the Heart

The largest arteries and veins of the heart form the beginnings of the pulmonary and systemic circulations. The vessels carrying oxygen-poor blood *from the heart to the lungs* are the *pulmonary arteries* [FIGURE 20.1]. Those returning oxygen-rich blood *from the lungs to the heart* are the *pulmonary veins*. Draining venous blood from the upper and lower parts of the body to the heart are the *superior vena cava* and the *inferior vena cava,* respectively.* The artery carrying highly oxygenated blood away from the heart is the *aorta*. It is called the *ascending aorta* as it extends upward from the left ventricle, the *aortic arch* where it bends (arches), and the *descending aorta* as it continues downward.

Blood Supply to the Heart

Heart muscle needs more oxygen than any organ except the brain. To obtain this oxygen, the heart must have a generous supply of blood. Like other organs, the heart receives its blood supply from arterial branches that arise from the aorta. The flow of blood that supplies the heart tissue itself is the *coronary circulation.* (The heart needs its own separate blood supply because blood in the pulmonary and systemic circulations cannot penetrate through the lining of the heart — endocardium — from the heart chambers to nourish cardiac tissue.) The heart pumps about 380 L (100 gal) to its own muscle tissue through the coronary circulation every day, or about 5 percent of all the blood pumped by the heart.

*The superior and inferior veins are called *cava* (L. hollow, cavern) because of their great size.

FIGURE 20.6 VALVES OF THE HEART

[A] The two atrioventricular valves (tricuspid and bicuspid) separate the atria and ventricles. (Note that one cusp of the tricuspid valve does not show in this section.) The two semilunar valves (pulmonary and aortic) permit the flow of blood out of the heart to the lungs and body; anterior view.

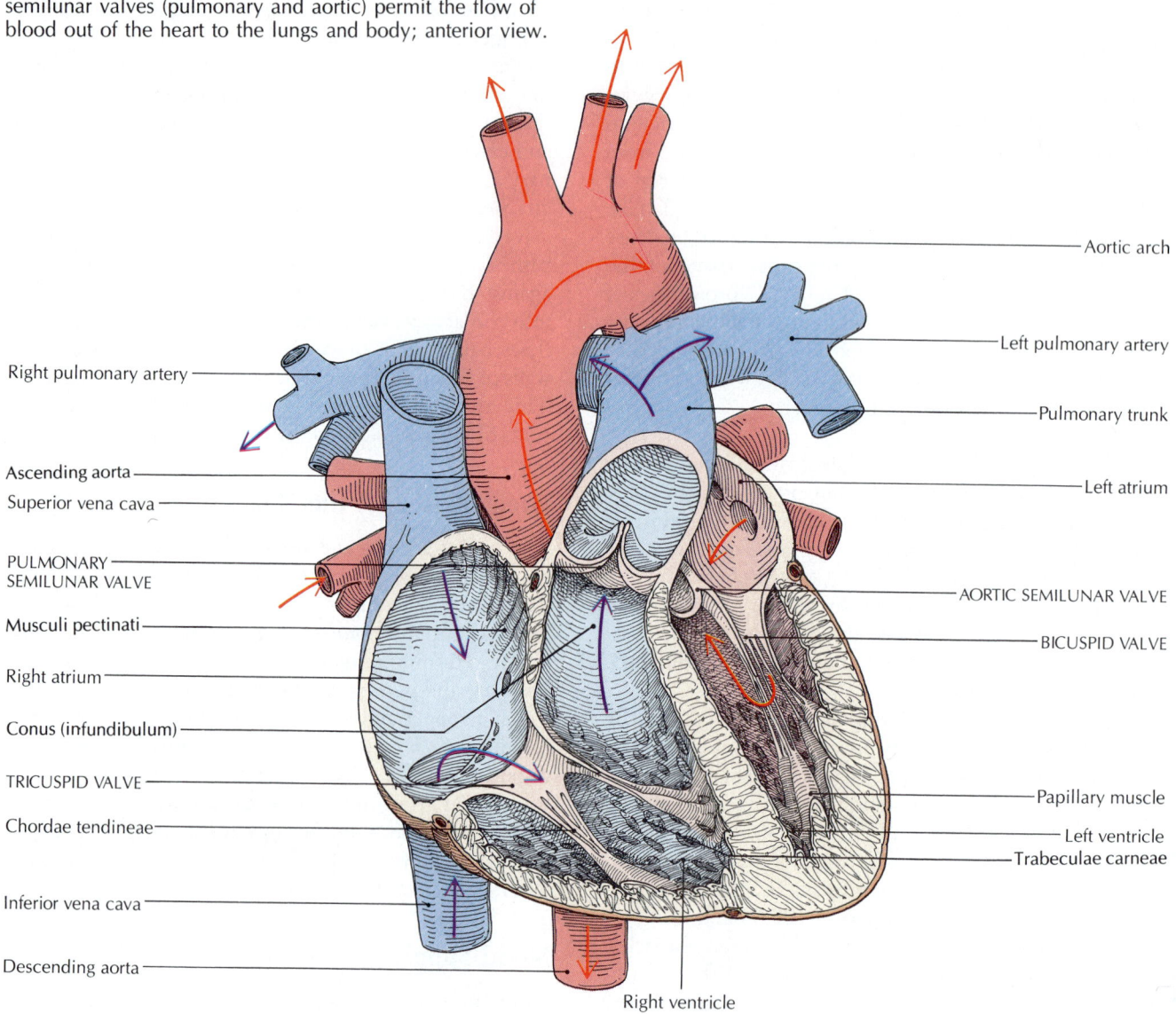

[A]

Coronary arteries Blood is supplied to the heart by the right and left *coronary arteries*, which are the first branches off the aorta, just beyond the aortic semilunar valve. This valve partially covers the openings of these arteries while blood is being pumped into the aorta from the heart. Blood can pass into the coronary arteries only when the left ventricle has relaxed and the cusps do not cover the openings.

The branching of the coronary arteries varies from person to person, but the following arrangement is the most common [FIGURE 20.7]:

1 The *right coronary artery*, which arises from the aorta near the right cusp of the aortic semilunar valve, courses between the right atrium and the right ventricle. Its major branches include (1) the *nodal branch* that supplies the sinoatrial (pacemaker) node, (2) the *right marginal branch (artery)*, which feeds both the ventral and dorsal surfaces of the right ventricle, (3) a *nodal branch to the atrioventricular node*, (4) *small branches* to the right branch of the AV bundle, and (5) *branches on the posterior aspect of the heart* that anastomose (join) with terminals of both the anterior descending branch and the circumflex artery of the left coronary artery.

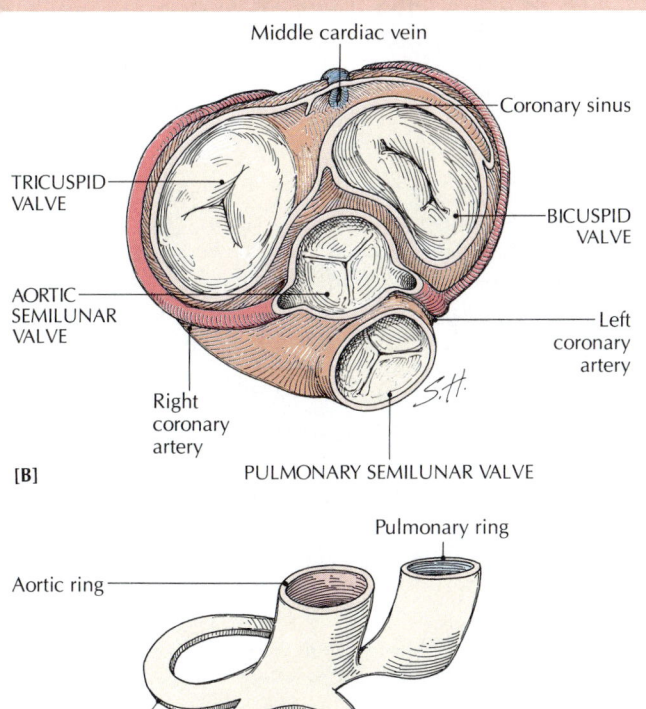

Middle cardiac vein

Coronary sinus

TRICUSPID VALVE

BICUSPID VALVE

AORTIC SEMILUNAR VALVE

Left coronary artery

Right coronary artery

PULMONARY SEMILUNAR VALVE

[B]

[B] The closed valves of the heart, viewed from above; the atria have been removed. [C] The fibrous skeleton of the heart. [D] The seemingly fragile pulmonary valve opens (left) and closes (right) about once a second. [E] Cusps of the tricuspid (right atrioventricular) valve remain open, allowing blood to flow from the right atrium into the right ventricle, when the chordae tendineae and papillary muscles are relaxed. The valve closes, blocking blood flow, when the chordae tendineae are taut and the papillary muscles are contracted. [F] A photo inside the right ventricle showing the branching chordae tendineae rising from strong papillary muscles.

Pulmonary ring

Aortic ring

Bicuspid (mitral) ring

Tricuspid ring

Membranous part of interventricular septum

[C]

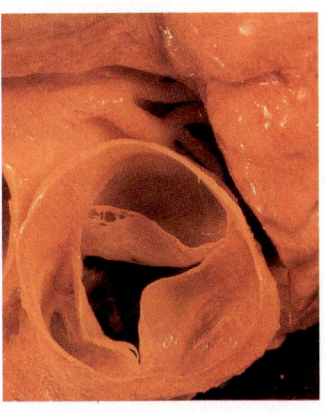
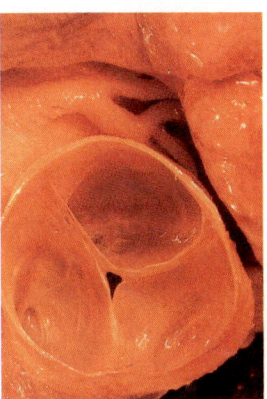

[D]   Valve open   Valve closed

Cusps of tricuspid valve

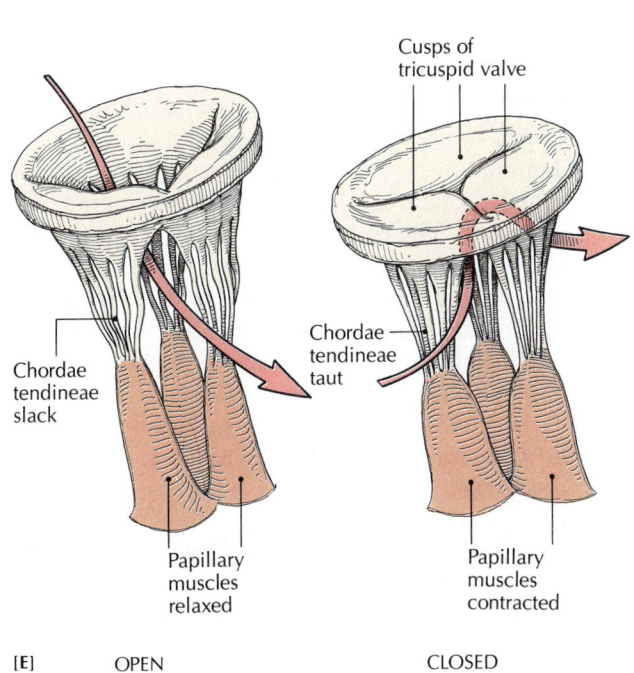

Chordae tendineae slack

Chordae tendineae taut

Papillary muscles relaxed

Papillary muscles contracted

[E]   OPEN   CLOSED

Chordae tendineae

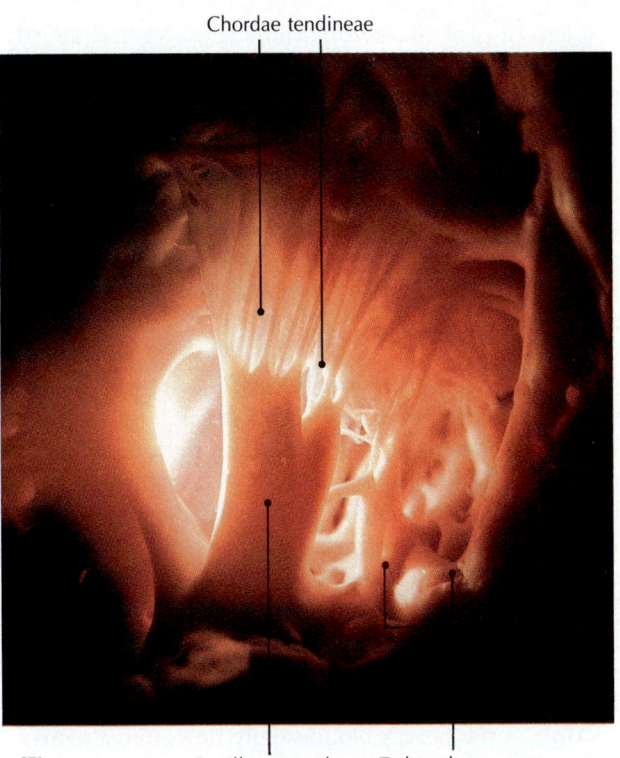

[F]   Papillary muscles   Trabeculae carneae

Anterior view. See also FIGURE 20.4.

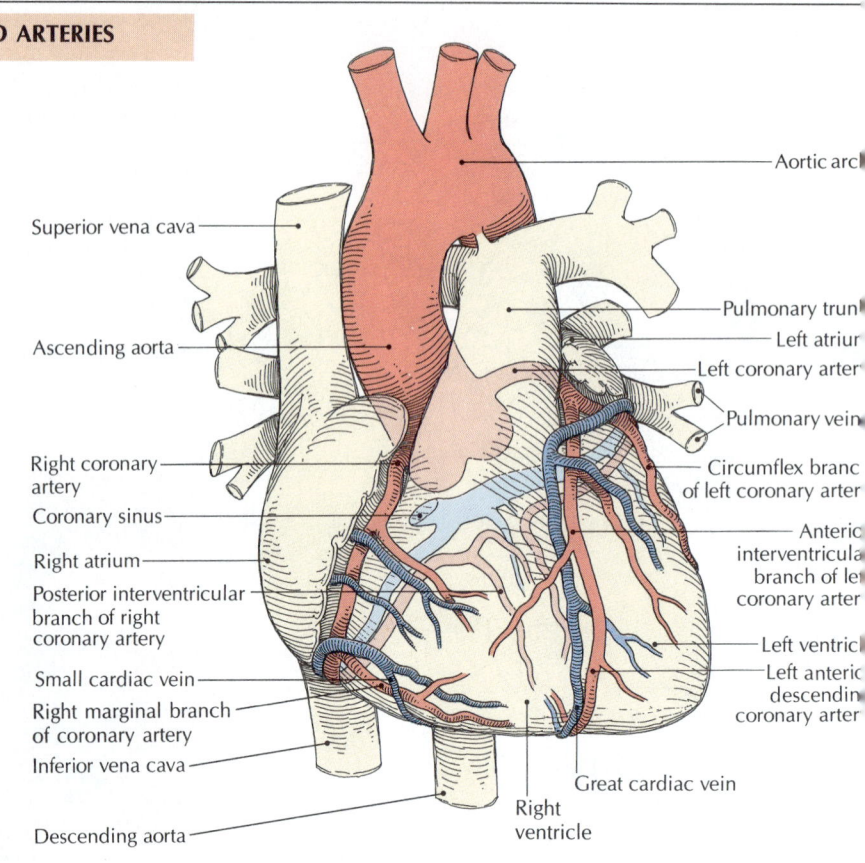

Superior vena cava

Ascending aorta

Right coronary artery

Coronary sinus

Right atrium

Posterior interventricular branch of right coronary artery

Small cardiac vein

Right marginal branch of coronary artery

Inferior vena cava

Descending aorta

Aortic arch

Pulmonary trunk

Left atrium

Left coronary artery

Pulmonary veins

Circumflex branch of left coronary artery

Anterior interventricular branch of left coronary artery

Left ventricle

Left anterior descending coronary artery

Great cardiac vein

Right ventricle

---

2 The *left coronary artery* arises from the aorta near the left cusp of the aortic semilunar valve and shortly divides into two branches: (1) The *anterior interventricular branch (artery)* courses along the anterior interventricular sulcus and anastomoses with the terminals of the right coronary artery in the vicinity of the apex. Small branches supply the left branches of the AV bundle. (2) The *circumflex branch (artery)* courses in the coronary sulcus to the lateral and posterior aspects of the heart. Its terminal branches anastomose with the right coronary artery.

Although some variation exists in coronary circulation, the largest volume of blood generally goes to the musculature of the left ventricle. The right coronary artery has branches that mainly supply the right side of the heart, but they also send some blood to the left ventricle. Branches of the left coronary artery send the main blood supply to the left side of the heart, but a small volume is also sent to the musculature of the right ventricle. The branches of the right and left coronary arteries anastomose on the surface of the heart and within the heart muscle, providing some collateral circulation.

Most of the heart's blood supply flows into the myocardium by way of the coronary arteries when the heart muscle is relaxed. This occurs because contraction of the heart muscle compresses the coronary vessels where they enter the muscular heart wall. This partially or totally blocks these vessels during contraction.

Coronary sinus and cardiac veins Most of the cardiac veins drain into the *coronary sinus,* a large vein located in the coronary sulcus on the posterior surface of the heart [FIGURE 20.7]. The coronary sinus empties into the right atrium. The coronary sinus receives blood from the *great cardiac vein,* which drains the anterior portion of the heart; the *middle cardiac vein* and *oblique vein,* which drain the posterior aspect of the heart; and the *small cardiac vein,* which drains the right side of the heart. Other cardiac veins include the *anterior cardiac vein,* which drains directly into the right atrium, and tiny veins called *thebesian veins,* which open directly into each of the four heart chambers.

Regulation of coronary blood flow Blood flow to the heart tissue is self-regulating, and two processes are thought to be involved. (1) In the *metabolic process of autoregulation,* cellular metabolites (waste products) tend to build up in active heart tissue. When they reach a certain concentration, local blood vessels dilate, blood

Cardiac Catheterization

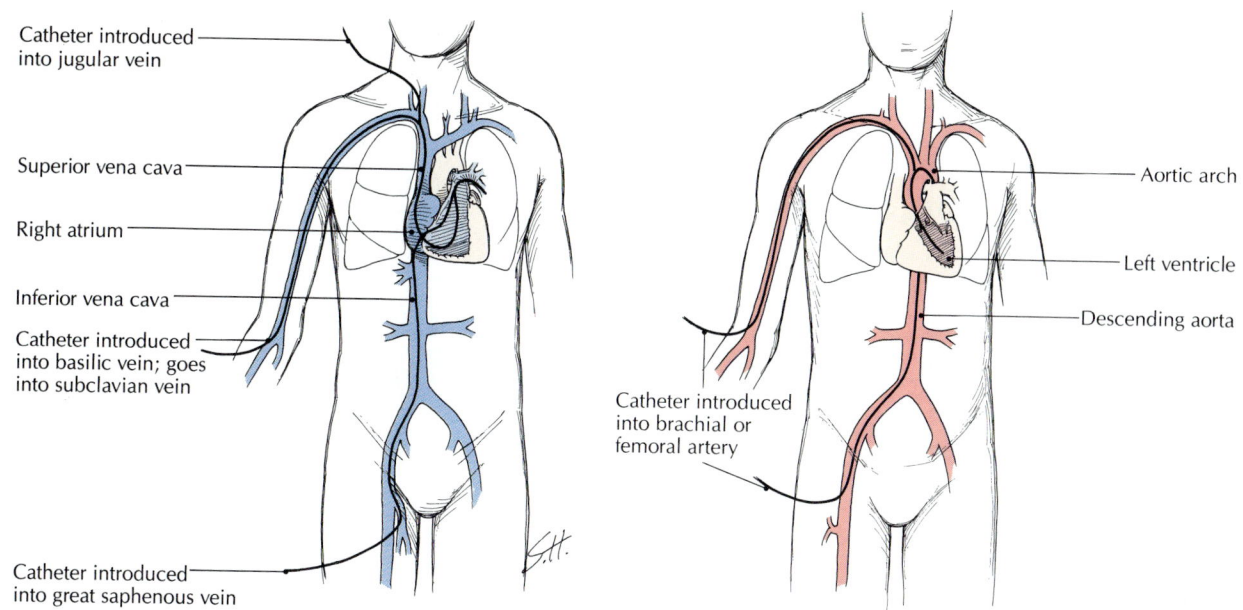

Catheter introduced into jugular vein

Superior vena cava

Right atrium

Inferior vena cava

Catheter introduced into basilic vein; goes into subclavian vein

Catheter introduced into great saphenous vein

Aortic arch

Left ventricle

Descending aorta

Catheter introduced into brachial or femoral artery

Cardiac catheterization through jugular, basilic, or saphenous vein (left), or through brachial or femoral artery (right).

A precise way of looking *inside* the heart is via **cardiac catheterization** (KATH-uh-tuh-rihz-AY-shun). A *catheter* (Gr. something inserted) is a hollow, flexible tube. It is inserted into a vein, usually in the arm, thigh, or neck, and then moved along slowly with the aid of a fluoroscope into the chambers of the heart (see drawings).

With cardiac catheterization, it is possible to measure bloodflow through the heart, the volume and pressure of blood in the heart, the pressure of blood passing through a heart valve, and the oxygen content of blood in the heart and its vessels. It is also possible to inject harmless radiopaque dyes through the catheter; the dye can be viewed on a television monitor to show septal openings and other congenital disorders in the aorta and pulmonary arteries. This technique is called *angiocardiography* (see photo). Currently a motion-picture x-ray machine is used with dye injected through cardiac catheterization. This technique, called *cineangiocardiography,* can indicate a heart abnormality by tracing the path of dye through the heart and its blood vessels.

An application of cardiac cathe-

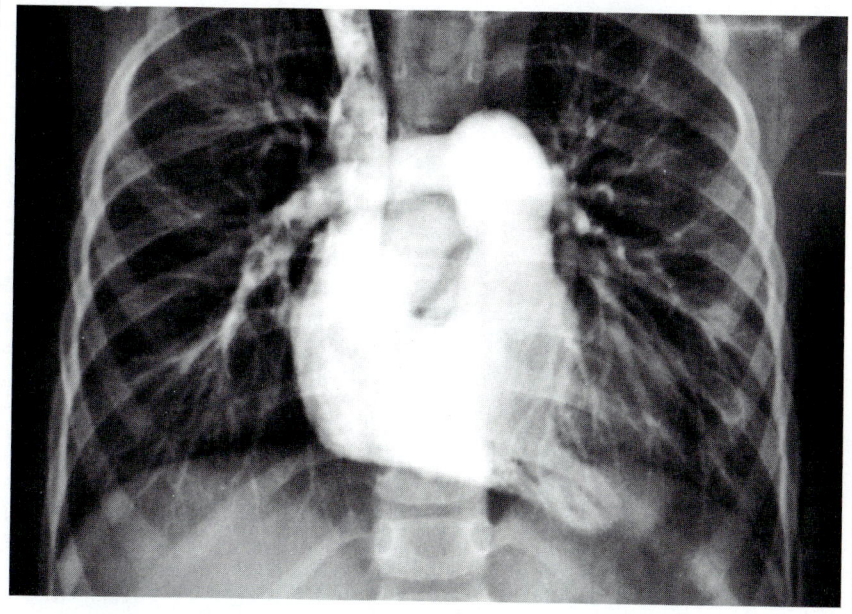

Angiocardiogram, anterior view.

terization is *balloon angioplasty*. In this procedure, a catheter with a small, uninflated balloon on its tip is inserted into an obstructed artery and inflated. The material causing the obstruction is pushed aside, the catheter is withdrawn, and blood flow is improved. Also, clot-

dissolving enzymes (such as streptokinase and tPA) can be injected directly into coronary arteries after a heart attack. Cardiac catheterization does not normally affect heart function and is usually not uncomfortable for the patient.

FIGURE 20.8 THE PATH OF BLOOD THROUGH THE HEART

[A] Blood enters the atria and ventricles. **[B]** Blood is pumped into the ventricles. **[C]** The ventricles relax. **[D]** The ventricles contract, pumping blood through the pulmonary trunk and aorta to the lungs and to the rest of the body.

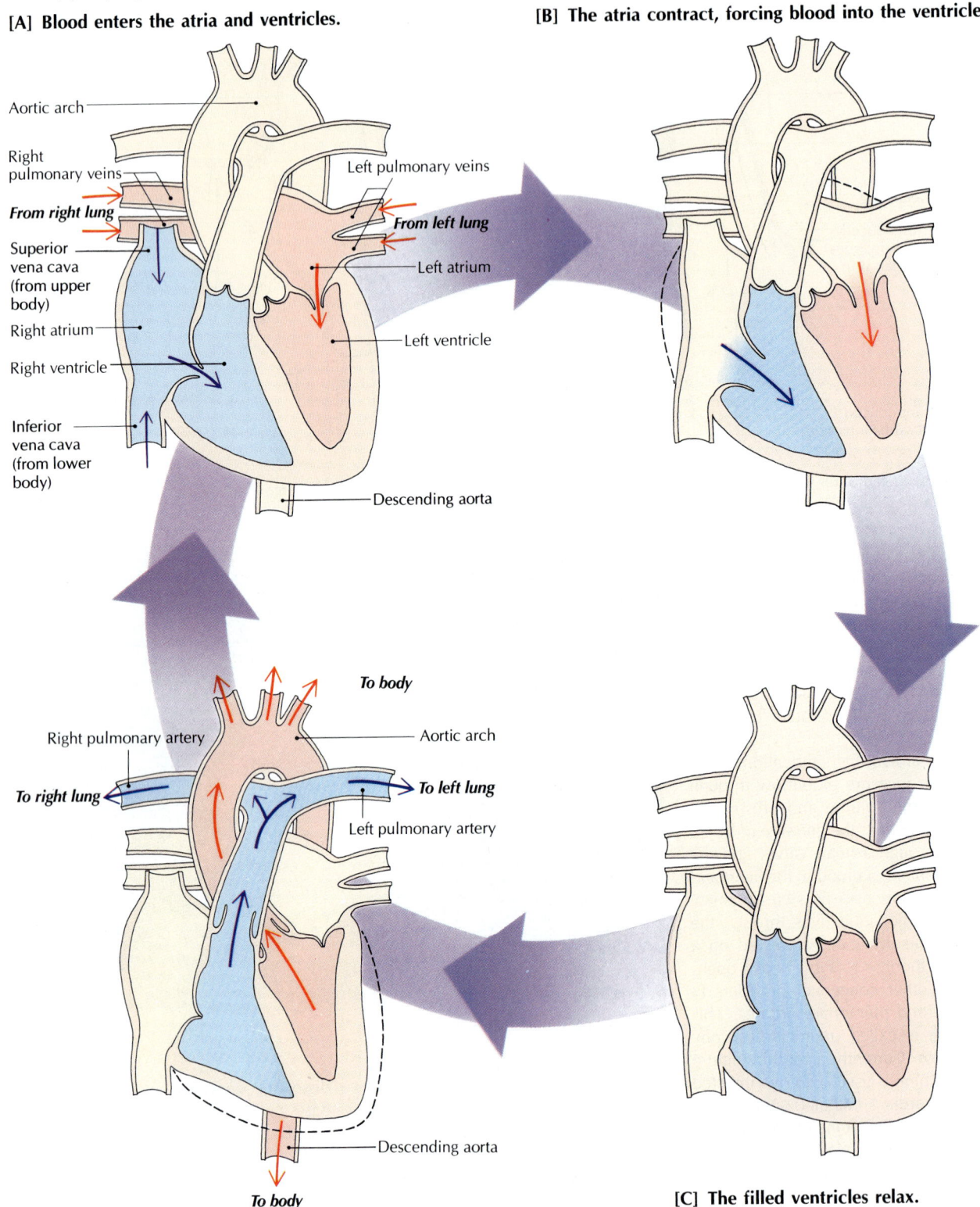

[A] Blood enters the atria and ventricles.

Aortic arch

Right pulmonary veins

From right lung

Superior vena cava (from upper body)

Right atrium

Right ventricle

Inferior vena cava (from lower body)

Left pulmonary veins

From left lung

Left atrium

Left ventricle

Descending aorta

[B] The atria contract, forcing blood into the ventricles.

To body

Right pulmonary artery

To right lung

Aortic arch

To left lung

Left pulmonary artery

Descending aorta

To body

[C] The filled ventricles relax.

[D] The ventricles contract, pumping blood into the pulmonary trunk and aorta.

flow increases, the excess metabolites are removed, and homeostasis is restored. (2) In the *myogenic process of autoregulation,* the stimulus arises in the smooth muscle within arterioles in the myocardium, which have a resting "tone," or tension. As blood pressure rises, blood flow increases and the vessels are distended. This causes the smooth muscle fibers in the vessel walls to contract, decreasing blood flow, lowering blood pressure, and restoring a homeostatic blood flow.

Path of Blood through the Heart

Before we present the physiological functions of the heart, we will describe the path of blood flow through the heart:

1 **Blood enters the atria** [FIGURE 20.8A]. Oxygen-poor blood from the body flows into the right atrium at about the same time as newly oxygenated blood from the lungs flows into the left atrium: (a) The *superior vena cava* returns blood from all body structures above the diaphragm (except the heart and lungs). (b) The *inferior vena cava* returns almost all blood to the right atrium from all regions below the diaphragm. (c) The *coronary sinus* returns about 85 percent of the blood from the coronary circulation to the right atrium. (d) The *pulmonary veins* carry oxygenated blood from the lungs into the left atrium. The blood entering the right atrium (blue in FIGURE 20.8A) is low in oxygen and high in carbon dioxide because it has just returned from supplying oxygen to the body tissues. The blood entering the left atrium (red in FIGURE 20.8A) is high in oxygen because it has just passed through the lungs, where it has picked up a new supply of oxygen and released its carbon dioxide. (This is the *only* time or place in the adult where oxygen-rich blood is carried by veins. All other times, oxygen-rich blood is carried by arteries.) Approximately 75 percent of this blood in the atria passes into the ventricles prior to atrial contraction.

2 **Blood is pumped into the ventricles** [FIGURE 20.8B]. The heart's natural pacemaker (the SA node) elicits an electrical impulse that coordinates the contractions of both atria (atrial systole). Blood remaining in the atria is forced through the one-way atrioventricular valves into the relaxed ventricles. This is called the "capping off" of ventricular filling.

3 **The ventricles, filled with blood, relax during the AV nodal delay** [FIGURE 20.8C].

4 **The ventricles contract, sending blood to the body (systemic circulation) and lungs (pulmonary circulation)** [FIGURE 20.8C]. The ventricular contraction generates pressure that closes the atrioventricular valves between the atria and ventricles while opening the two semilunar valves leading out of the ventricles. The right ventricle forces blood low in oxygen out through the right and left

pulmonary arteries to the lungs. The left ventricle pumps the newly oxygenated blood through the aortic semilunar valve into the aorta. The aorta branches into the ascending and descending arteries that carry oxygenated blood to all parts of the body [FIGURE 20.8D]. The left and right ventricles pump simultaneously, so that equal volumes of blood leave the heart. By this time, the atria have already started to refill, preparing for another cycle.

A S K   Y O U R S E L F

1 In what way is the heart a double pump?

2 What are the layers of the heart wall, and how do they differ from the pericardium?

3 What is the function of the cardiac skeleton?

4 What are the separate functions of the four heart chambers?

5 What are the functions of the atrioventricular and semilunar valves?

6 Why does the heart need a separate blood supply?

7 How does the heart regulate its own blood flow?

8 What is the path of blood through the heart?

PHYSIOLOGY OF THE HEART

The electrical and mechanical properties of the heart allow it to be a functional pump that beats reliably every day of our lives. The following sections explain how the heart maintains this steady pumping action.

Electrical Properties of Cardiac Muscle Cells

When an action potential travels through the heart, each cardiac muscle cell produces and conducts its own action potential. The resting membrane potential of individual cardiac muscle cells is about -90 mV (that is, the inside of the cell is 90 mV negative with respect to the outside). The cardiac muscle *action potential* is similar to that in nerve and skeletal muscle, although it lasts longer. As with skeletal muscle, stimulation produces a *propagated action potential* (one that travels in all directions) that initiates contraction. The cardiac muscle action potential can be divided into five phases: depolarization, early repolarization, plateau, repolarization, and resting potential [FIGURE 20.9]. These phases are described below for a ventricular muscle cell.

Depolarization in the cardiac muscle results from a large increase in the inward movement of sodium ions through fast voltage-gated sodium channels, which causes the membrane potential to reverse from its resting

FIGURE 20.9 CARDIAC ACTION POTENTIAL

[A] A cardiac action potential (red curve), its period of contraction (tension development), and absolute and relative refractory periods. [B] Relative permeability changes for ions during a cardiac action potential.

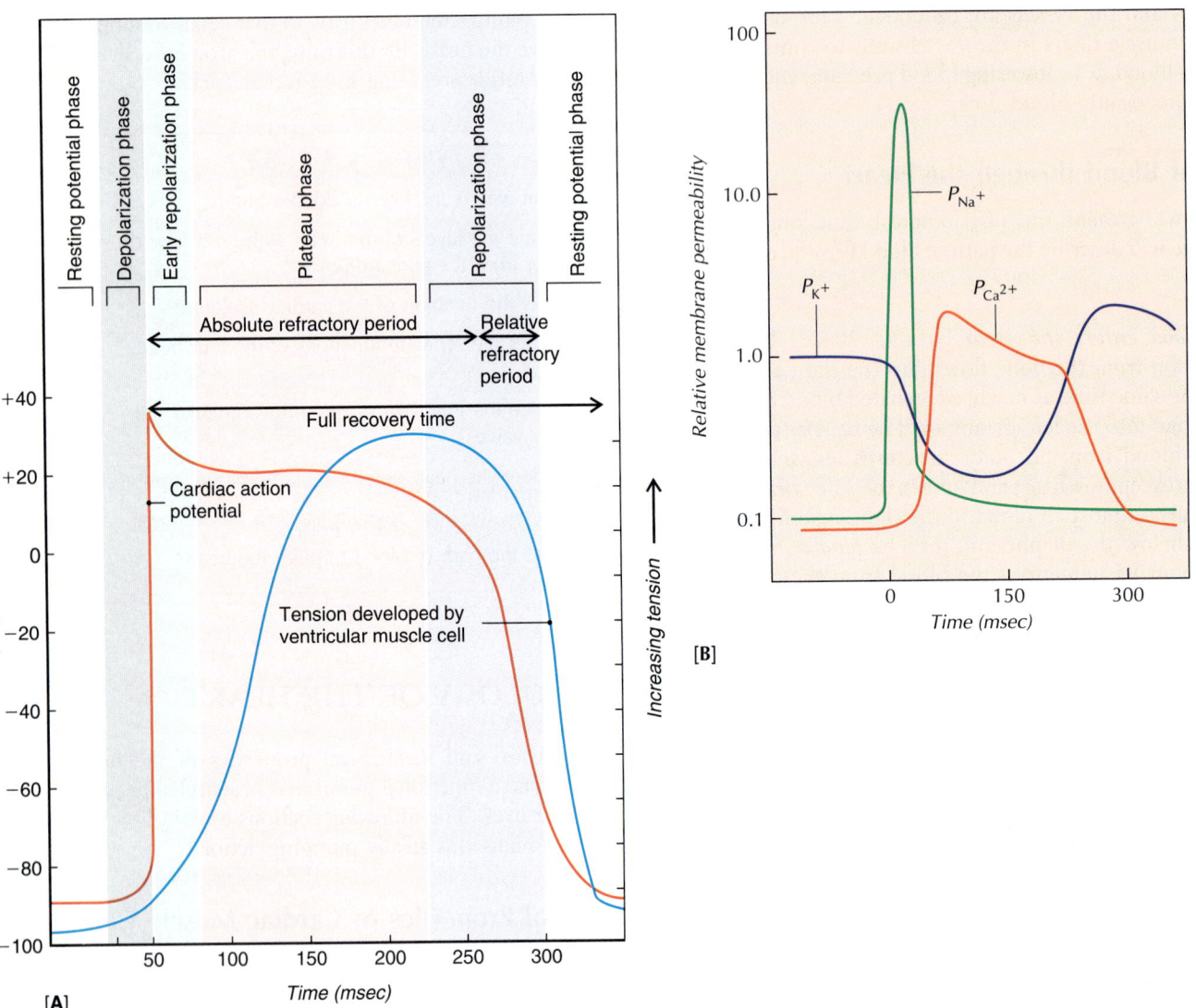

[A]

[B]

potential of −90 mV to a potential of about +30 mV (increase in sodium permeability). At almost the same time, membrane permeability to potassium decreases as the slow potassium channels close; this also contributes to membrane depolarization, since there is less potassium leaving the cell. Depolarization lasts about 2 msec.

During the ***early repolarization phase*** and ***plateau phase,*** positively charged calcium ions move into the cell through slow voltage-gated calcium channels (increase in calcium permeability); the calcium helps initiate muscle contraction. This movement prevents the membrane from returning to its normal electrical potential of −90 mV, and the repolarization levels off, or *plateaus.* The absolute refractory period, during which the cells cannot respond to a stimulus, is about nine times longer than in skeletal muscle. The plateau phase lasts about 200 msec.

During ***repolarization,*** potassium ion channels reopen and calcium ion channels close (increase in potassium permeability and decrease in calcium permeability), so that potassium ions move out of the heart cell, causing the inside of the cell to become more negative as the positive potassium ions move out. The increasing negativity inside the cell returns the membrane to its normal resting membrane potential of −90 mV. It is important to note that at this point the electrical potential across the plasma membrane is returning to normal, but the ion distributions are altered. More potassium ions are just outside the plasma membrane, while sodium ions are more concentrated just inside. Active transport mechanisms now pump sodium out and potassium in. The active transport of these ions continues to maintain a −90 mV potential in the resting state, since ion diffusion occurs to some degree at all times.

The phases of the cardiac action potential just described represent only the excitation portion of cardiac muscle function. If this electrical excitation is to cause a contraction, the two events must be "coupled." Calcium ions are involved in the coupling of excitation and contraction. The role of calcium is the same as in skeletal muscle. The sarcoplasmic reticulum supplies most of the calcium. However, a second important source is diffusion across the plasma membrane during the action potential. From both of these sources, the concentration of calcium in the cytoplasm increases. Calcium then combines with the regulator muscle protein, troponin. This combination removes tropomyosin's inhibition of cross-bridge formation between the two muscle proteins, actin and myosin. Cross-bridge formation occurs, and the cardiac muscle cells contract. During repolarization, the cytoplasmic calcium is restored to its low concentration by a gradient-mediated countertransport mechanism, which transports calcium ions back into the sarcoplasmic reticulum and out of the cell through the plasma membrane.

As you have seen, under normal circumstances both atrial and ventricular muscle cells have such a high *stable negative resting membrane potential* that they show no spontaneous electrical activity. (However, if there is sufficient time or under abnormal conditions, such as electrolyte imbalances, hypoxia, or caffeine or nicotine influence, any myocardial cell may depolarize.) These cells are thus excited only by electrical impulses from adjacent cells. However, the cells of the SA node in the heart do not have a *steady* resting potential. Instead, they show what is called *pacemaker activity* or a *pacemaker potential,* which is shown in FIGURE 20.10. Note that the unstable resting membrane continuously depolarizes toward a threshold, leading to an action potential, which establishes the heart rate.

Electrical Conducting System of the Heart

You saw in Chapter 10 that skeletal muscles cannot contract unless they receive an electrochemical stimulation from the nervous system. Although the central nervous system does exert some control over the heart, cardiac muscle has built-in autorhythmic *pacemaker cells,* which initiate a beat independently of the central nervous system, establishing the rhythm for the heartbeat. These pacemaker cells also form a part of the system for conducting impulses throughout the heart muscle. FIGURE 20.11 shows the components of the electrical conducting system:

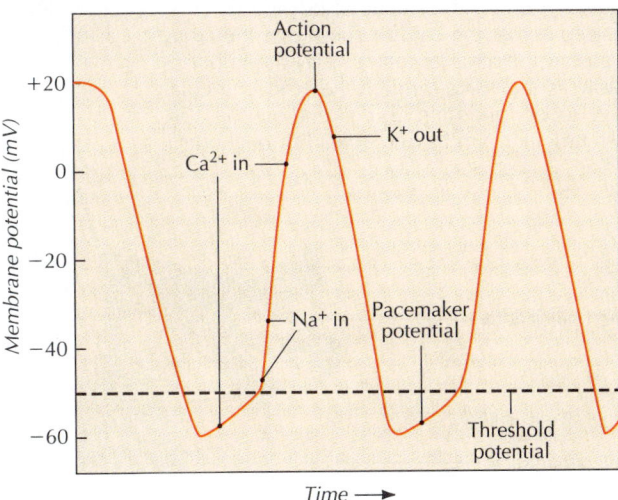

FIGURE 20.10 PACEMAKER POTENTIAL

1 The sinoatrial (SA) node
2 The atrioventricular (AV) node
3 The atrioventricular (AV) bundle (bundle of His)
4 The right and left bundle branches
5 The cardiac conducting myofibers (Purkinje fibers)

The electrical stimulation that starts the heartbeat and controls its rhythm originates in the superior wall of the right atrium, near the entry point of the superior vena cava, in a mass of specialized heart muscle tissue called the *sinoatrial node,* or *SA node* [FIGURE 20.11A].* The SA node is the normal pacemaker of the heart because its inherent rhythmicity is faster than any other area of the heart.

The pacemaker activity results from the spontaneous depolarization of the SA node at regular intervals, 70 to 80 times a minute. The SA node makes contact with adjacent atrial muscle cells and causes them to be depolarized by electrical conduction through the gap junctions of the intercalated disks. These atrial cells then cause their neighboring cells to initiate action potentials. In this way, a wave of electrical activity, assisted by interatrial tracts, spreads throughout the right atrium and then the left atrium, much like ripples spreading on a pond.

In addition, some specialized atrial conducting pathways of cardiac conducting myofibers (tracts between the SA and AV nodes) help carry the electrical activity through the atrium. Electrical activity is spread as one cell depolarizes the adjacent cell to threshold voltage. The electrical stimulation causes the atria to contract.

*The sinoatrial node is so named because it is located in the wall of the *sinus venosus* during early embryonic development and is absorbed into the right atrium along with the sinus venosus.

FIGURE 20.11 THE CONDUCTION SYSTEM OF THE HEART

[A] The cardiac conduction system synchronizes the events of the cardiac cycle. Impulses from (1) the sinoatrial (SA) node sweep across the cardiac muscle of the atria, causing the atria to contract; (2) the impulses slow as they pass through the atrioventricular (AV) node, and then (3) descend via the atrioventricular bundle and spread throughout

the myocardium of the ventricles. Ventricular contraction results. **[B]** The numbers show the time, in seconds, required for the impulse from the SA node to reach various parts of the heart. **[C]** An action potential from each part of the electrical conducting system.

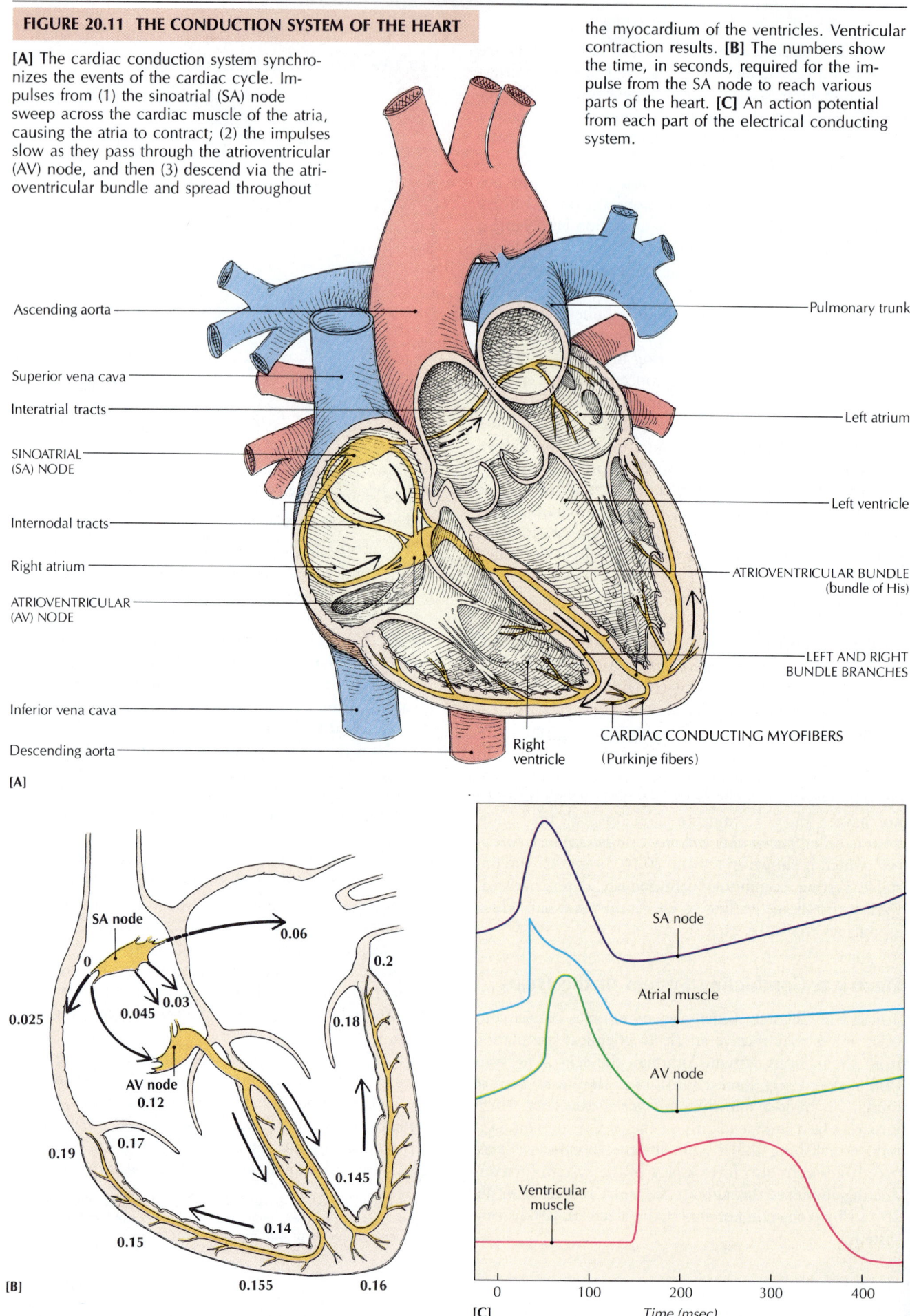

[A]

[B]

[C]

Time (msec)

A few hundredths of a second after leaving the SA node, the wave of electrical activity reaches the *atrioventricular node,* or *AV node,* which lies at the base of the right atrium, between the atrium and ventricle [FIGURE 20.11]. The AV node delays the conduction of electrical activity about a tenth of a second before allowing it to pass into the ventricles. The AV-nodal delay allows time for the atria to contract while the ventricles are relaxed, thus allowing blood to continue moving from the atria into the ventricles. If the SA node is damaged or diseased, the AV node may become the pacemaker, "pacing" between 40 and 50 beats per minute.

From the AV node, a group of conducting fibers in the interventricular septum called the *atrioventricular bundle (bundle of His)* runs for a short distance and then divides into two branches, right and left, which spread along the septum, one branch for each ventricle. Because a sheet of connective tissue separates the atria from the ventricles, the atrioventricular bundle is the only electrical link between the atria and the ventricles.

When the branches reach the apex of the ventricles, they divide into hundreds of tiny, large-diameter, specialized cardiac muscle fibers called *cardiac conducting myofibers (Purkinje fibers)* that follow along the muscular walls of the ventricles. Such an arrangement directs the electrical impulse in a definite pathway, so that it can make contact with all areas of the ventricular muscle. Because of the large diameter of cardiac conducting myofibers, an impulse traveling along them is conducted rapidly and directly into the cardiac muscle, so that all the cardiac muscle cells contract in unison, producing a coordinated pumping effort.

Cardiac conducting myofibers are actually modified cardiac muscle cells that conduct action potentials at higher velocities than normal cardiac muscle cells do. The myofibers are continuous with the myocardial muscle cells, but have only feeble contractile properties.

If the SA and AV nodes are ineffective, the AV bundle, bundle branches, or cardiac conducting myofibers may take over pacing the heart, but their inherent rhythmicity is very slow — 20 to 40 beats per minute. If this condition occurs, an artificial pacemaker may be required. (See "Artificial Pacemaker" on page 635.)

Electrocardiogram

The rhythm of the heart and the passage of an electrical current generated by an action potential from the SA node through the atria, down into the AV node, and through the atrioventricular bundle and cardiac conducting myofibers of the ventricles can be measured easily and accurately with an instrument called the *electrocardiograph,* which produces a recording of the electrical waves of the heart [FIGURE 20.12A]. The recording is called an *electrocardiogram,* and the printed record is abbreviated

as either *ECG* or *EKG* (C for cardio or K for the Greek *kardia*).

The electrocardiograph has electrodes, which when placed at certain points on the body can detect the electrical activity in the heart. (The body fluids are an excellent conductor of electricity.) To enhance electrical contact, a jelly containing an electrolyte is put on the skin where the electrodes (or leads) are to be attached. The three standard leads for an ECG are connected in the following way: Lead I to the right wrist and left wrist, lead II to the right wrist and left ankle, and Lead III to the left wrist and left ankle. In addition, six electrodes are placed on the chest in standard positions numbered V1 to V6 [FIGURE 20.12C].

Electrical activities recorded Different electrical impulses during the cardiac cycle are recorded in an ECG as distinct *deflection waves.* The first activity in the

> **Q:** *What does a patient's cardiac monitor show?*
>
> **A:** In the past, cardiac monitors showed only heart rate from ECG tracings. New techniques now allow monitors to show heart rate and rhythm, arterial pressure, pulmonary artery pressure, central venous pressure, and the S-T segment of the ECG, which can detect periods of ischemia. (See the photo below.)

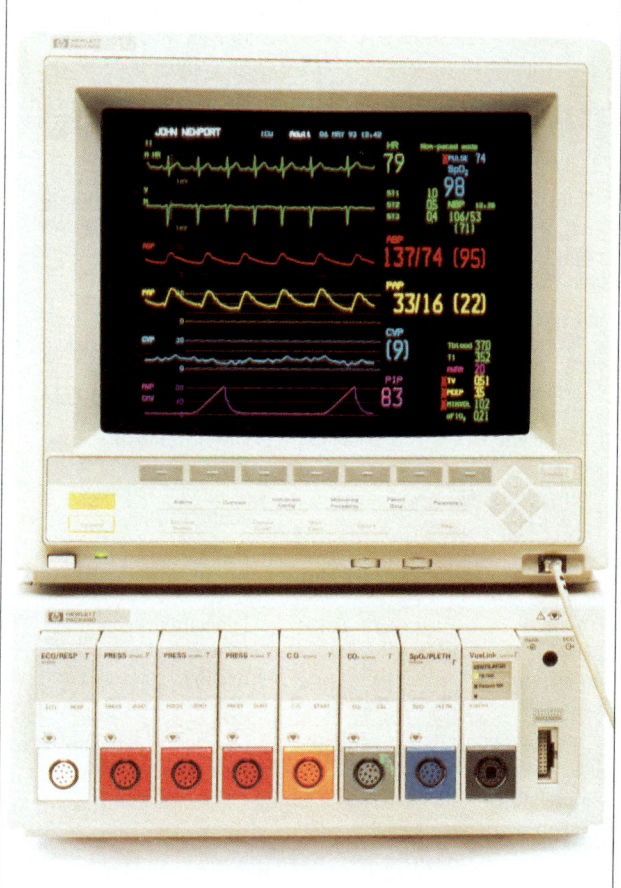

FIGURE 20.12 ELECTROCARDIOGRAM (ECG)

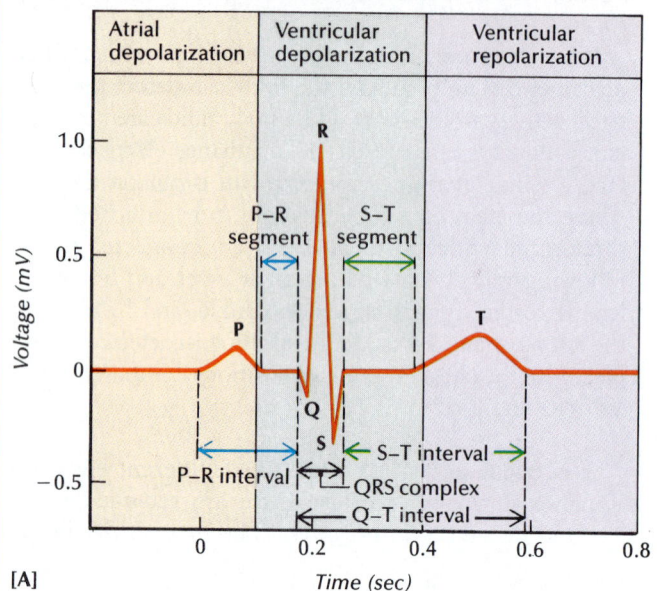

[A]

[A] Events of the cardiac cycle as they are recorded on an ECG. [B] Summary of ECG events. [C] The three standard leads and the six chest leads for an ECG.

ECG event	Range of duration, sec	Corresponding physiological events in heart
P wave	0.06–0.11	Depolarization (excitation) of atria prior to their contraction. Impulse begins in SA node, spreads through muscles of atria to AV node.
P–R segment	0.06–0.10	Atrial depolarization and conduction through AV node.
P–R interval (onset of P wave to onset of QRS complex)	0.12–0.21	Time between onset of atrial depolarization and onset of ventricular depolarization.
QRS complex	0.03–0.10	Depolarization of ventricles; repolarization of atria is masked on ECG by ventricular depolarization.
S–T segment (end of QRS complex to onset of T wave)	0.10–0.15	End of ventricular depolarization to beginning of repolarization of ventricles.
T wave	Varies	Repolarization of ventricles.
S–T interval (end of QRS complex to end of T wave)	0.23–0.39	Interval between completion of depolarization and end of repolarization.
Q–T interval (onset of QRS complex to end of T wave)	0.26–0.49	Ventricular depolarization plus ventricular repolarization.

[B]

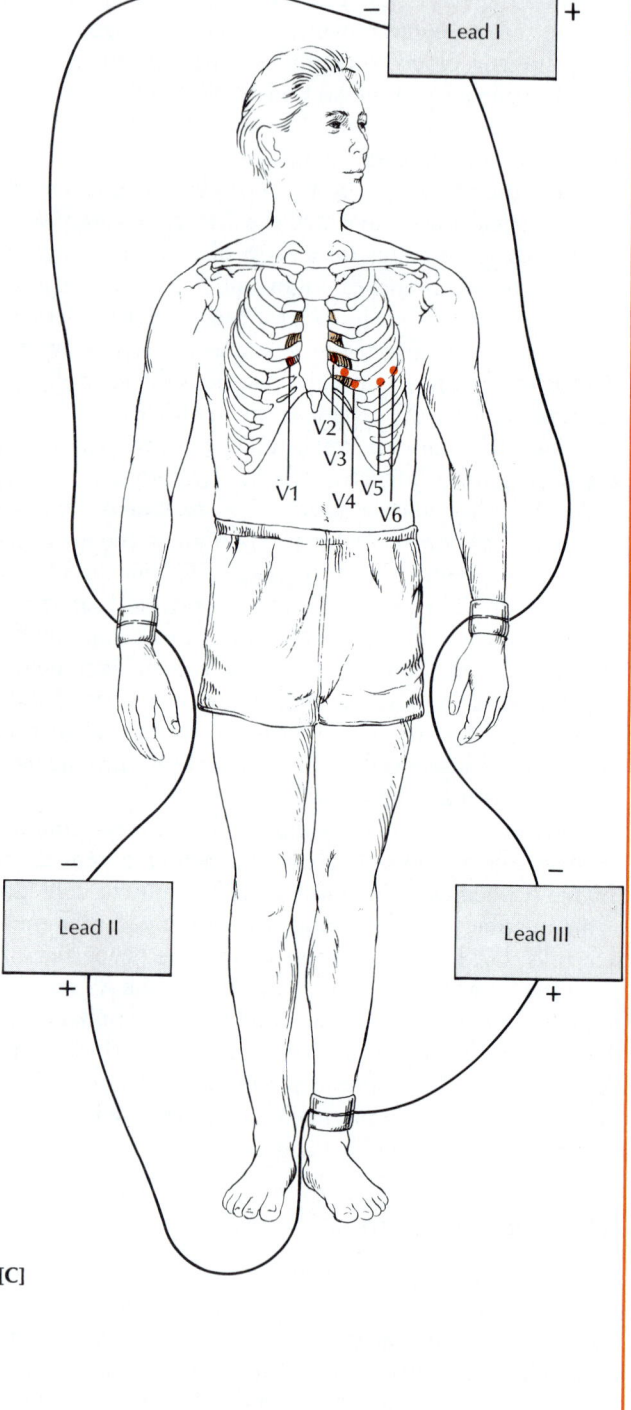

[C]

electrocardiogram is the *P wave* [FIGURE 20.12A, B]. It is caused by the electrical potential generated by the passage of the impulse from the SA node, through the muscle fibers of the atria, and reaching the AV node. The P wave represents the depolarization (excitation) of both atria.

The passage of the wave between the atria and ventricles is marked by a short horizontal segment immediately following the P wave. This is the *P-R segment*. Next, the depolarization of the ventricles produces a short dip (Q), a tall spiked peak (R), and a sharp dip (S). This triple-wave activity is the **QRS complex** (wave), recorded as the ventricles are depolarized. The repolarization of the atria also takes place at this time but is masked on the ECG by the stronger ventricular depolarization. After a short horizontal line called the *S-T segment*, a repolarization wave is shown by a rounded peak called the **T wave**. It represents the repolarization of the ventricles.

Electrocardiogram as a diagnostic tool Examining the frequency and duration of ECG deflection waves is helpful in evaluating heart function. For example, a heightened P wave indicates an enlarged atrium, a higher than normal Q wave may indicate a heart attack (myocardial infarction), and a heightened R wave usually indicates enlarged ventricles. An acute myocardial infarction raises the S-T segment above the horizontal, and an S-T segment below the horizontal may indicate an abnormally high concentration of blood potassium.

Any deviation from the normal rate or sequence of excitation is called *cardiac arrhythmia*. It is the result of structural or functional disorders such as abnormal heart rhythm or damage to the myocardium. Because the electrical activity of the heart is very sensitive to changes in ion concentration, especially potassium, the ECG may be altered by abnormal electrolyte levels in the blood. Some of the more easily detectable abnormalities are described in When Things Go Wrong at the end of the chapter.

Cardiac Cycle

The human body contains 4 to 6 L (about 1 to 1.5 gal) of blood, but the heart takes about a minute to pump this volume of blood throughout the body. In times of strenuous exercise, the heart can quintuple this output. On the average, the adult heart pumps about 7500 L (2000 gal) of blood throughout the body every day.

All the physiological events associated with a single heartbeat are known as the **cardiac cycle**. These events include the (1) *electrical events* (ECG), (2) *mechanical events: pressures* in the various chambers and vessels (atria and ventricles, aorta and pulmonary arteries) and *volumes* in these chambers, and (3) *heart sounds*. In each cardiac cycle, as the atria and ventricles alternately contract and relax, pressures in these chambers increase and decrease.

As the chambers contract, pressures increase and outflow valves open. Blood then flows from areas of higher blood pressure to areas of lower blood pressure. During the cardiac cycle, the two atria contract (this is atrial **systole;** SISS-toe-lee; Gr. from *systellein*, to contract), while the two ventricles relax (this is ventricular **diastole;** die-ASS-foe-lee; Gr. dilatation). Then the two ventricles contract during the time the two atria relax.

NOTE: Periods of **systole,** whether atrial or ventricular, consist of *contraction* and *emptying;* **diastole** consists of *relaxation* and *filling.* One complete cardiac cycle begins and ends with ventricular diastole.

The cardiac cycle proceeds in four stages:

1 During *atrial systole* (which lasts 0.1 sec), both atria contract, forcing the blood in the atria into the ventricles.

2 During *ventricular systole* (0.3 sec), both ventricles contract, forcing blood out through the pulmonary artery (to the lungs) and aorta (to the rest of the body).

3 During *atrial diastole* (0.7 sec), or relaxation of the atria, the atria refill with blood from the large veins (vena cavae) leading to the heart from the body.

4 *Ventricular diastole* (0.5 sec), or relaxation of the ventricles, begins before atrial systole, allowing the ventricles to fill *passively* with blood from the atria.

FIGURE 20.13 shows the changes that take place in the left side of the heart and aorta during a single cardiac cycle. These include the electrical events (ECG) and mechanical events with the resultant changes in aortic, left ventricular, and left atrial pressures as well as left ventricular volume. Since this is a drawing of the left side of the heart, the pressures are higher than they are in the right side of the heart. The right side of the heart is thinner-walled and pumps blood only to the lungs. However, the ventricular volumes for both sides of the heart are the same (about 70 mL per beat). Refer to FIGURE 20.13 as you read the following description of the electrical and mechanical events for one complete cardiac cycle, beginning and ending with ventricular diastole.

Mechanical events of the cardiac cycle The numbered events (1–25) of the cardiac cycle that follow correspond to the boldface numerals in FIGURE 20.13: As the cardiac cycle begins, blood flows into the atrium, and atrial pressure becomes slightly higher than ventricular pressure **(1).** Because of this difference in pressure, the AV (mitral) valve opens, and blood flows from the atrium into the ventricle **(2)** (see also heart drawing A). Because the AV valve is open, ventricular volume rises even before atrial contraction occurs. The SA nodal cells reach threshold, spreading action potentials throughout the atrium. The event is shown as the **P wave (3).** Atrial depolarization causes the atrial muscle to contract, and the atrial pressure curve rises **(4),** forcing additional

FIGURE 20.13 THE CARDIAC CYCLE

One full cardiac cycle proceeds from one ventricular diastole to another. (Events of the cycle are numbered according to the account in the text.) The graph shows simultaneous events of the cardiac cycle. Follow the columns downward to see what simultaneous events are occurring at each phase of the cycle. Changes in the electrocardiogram, in heart pressures, volume, and sounds can be followed from left to right. The drawings of the heart (bottom row) show the flow of oxygen-poor (blue) and oxygen-rich (red) blood in and out of the heart during the cardiac cycle.

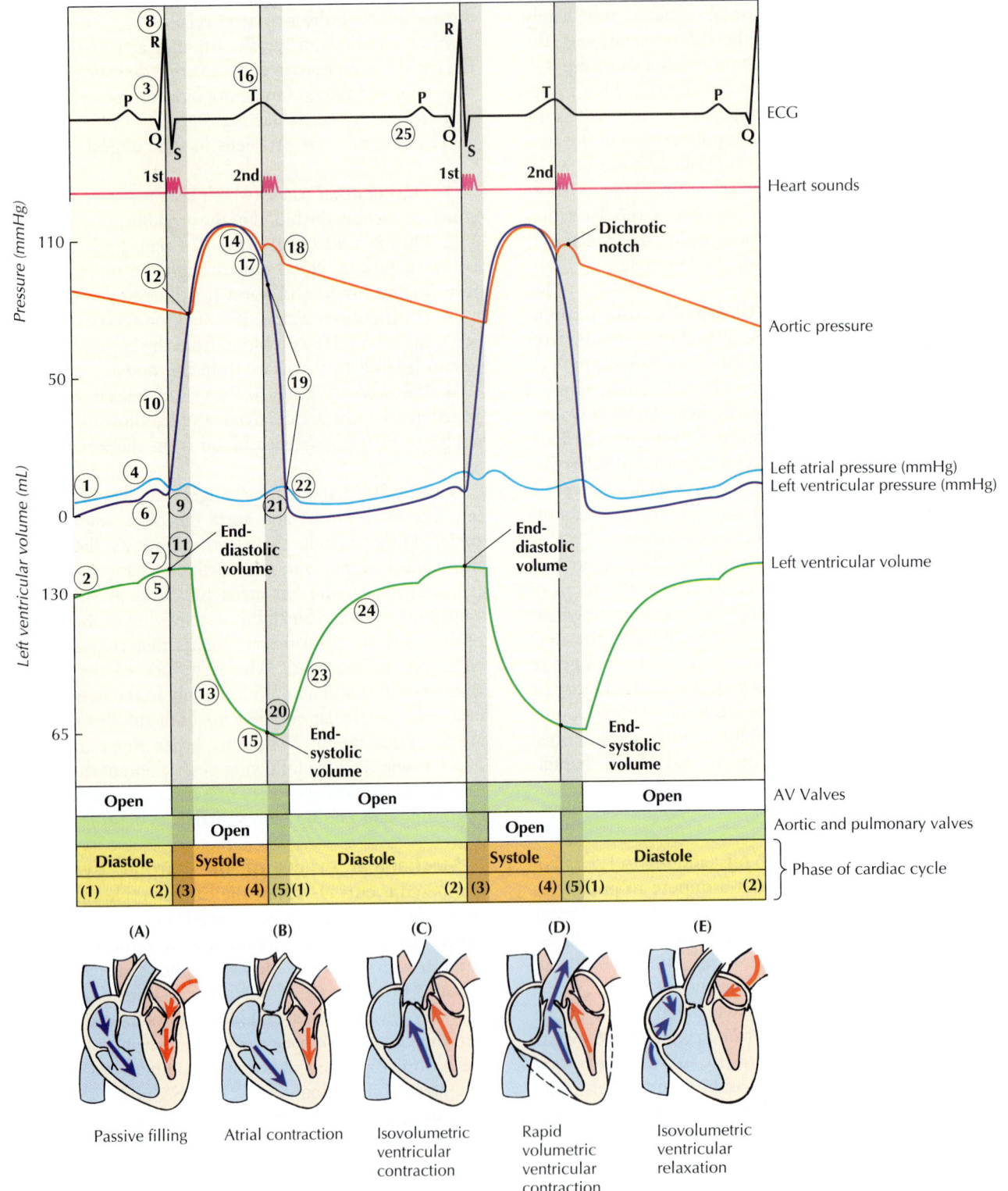

(A) Passive filling

(B) Atrial contraction

(C) Isovolumetric ventricular contraction

(D) Rapid volumetric ventricular contraction

(E) Isovolumetric ventricular relaxation

MEDICAL UPDATE

Artificial Pacemaker

An **artificial pacemaker** is a battery-operated electronic device that is implanted in the chest with electrical leads to the heart of a person whose natural pacemaker (the SA node) has become erratic. In a relatively simple operation, electrode leads (catheters) from the pacemaker are passed beneath the skin, through the external jugular vein (or other neck vein), into the superior vena cava, into the right atrium, through the tricuspid valve, and into the myocardium of the right ventricle (see drawing). If the patient's veins are damaged or too narrow to receive the typical chest implant of the pacemaker with its connecting wires, the pacemaker is implanted in the left abdominal area, with a connecting lead inserted into the epicardium.

Three basic types of artificial pacemakers are available. The first type delivers impulses when the patient's heart rate is slower than that set for the pacemaker, and shuts off when the natural pacemaker is working adequately. The second is a fixed-rate model that delivers constant electrical impulses at a preset rate. The third is a transistorized model that picks up impulses from the patient's SA node and operates at about 72 beats per minute when the natural pacemaker fails.

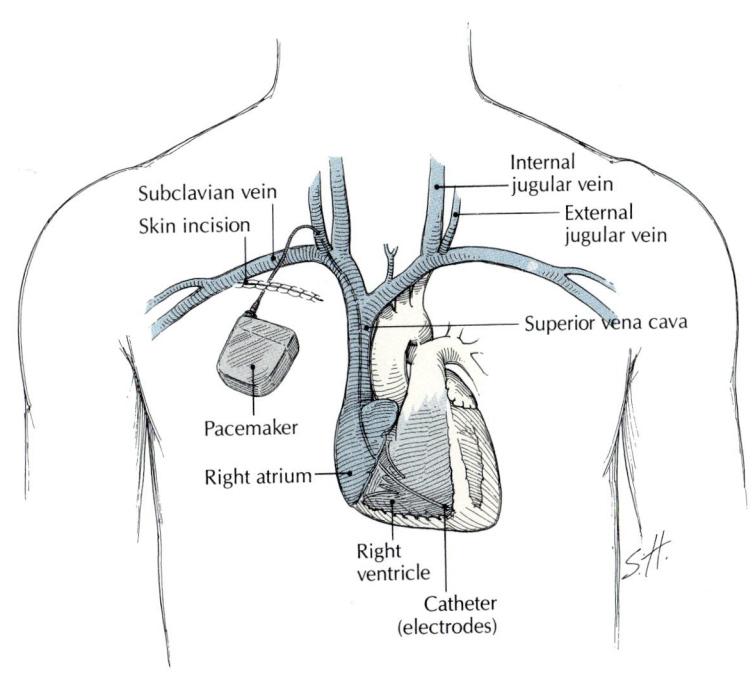

An implanted artificial pacemaker.

---

blood into the ventricle **(5)** (see also heart drawing **B**) and causing a simultaneous increase in ventricular pressure **(6)**. The AV valve remains open during atrial contraction because the atrial pressure is still slightly higher than the ventricular pressure. Atrial contraction ends, and ventricular filling is completed. The blood volume in the ventricle is called the **end-diastolic volume (7)**; it is the largest volume the ventricle will contain during the cardiac cycle.

Ventricular diastole ends as ventricular contraction begins. Ventricular muscle depolarization, which produces the **QRS complex (8)**, causes ventricular muscle contraction. Ventricular systole is indicated by a rise in the ventricular pressure curve accompanying the QRS complex. At the start of systole, ventricular pressure rises above atrial pressure, forcing the AV valve closed **(9)**. As ventricular pressure rises **(10)**, the ventricle remains sealed between the time the AV valve closes and the aortic valve opens. No blood goes into or out of the ventricle, and the ventricle remains at a constant volume **(11)**.

This is the period of **isovolumetric** (same volume) **ventricular contraction** (heart drawing **C**).

When ventricular pressure becomes greater than aortic pressure, the aortic valve opens **(12)**, and blood is forced out rapidly **(13)** (heart drawing **D**). Since blood is forced from the ventricle into the aorta faster than blood is drained off into smaller vessels, aortic pressure rises **(14)**. About half the blood in the ventricle is pumped out during ventricular ejection. The amount of blood remaining in the ventricle at the end of systole is called the **end-systolic volume,** the least amount of blood contained in the ventricle during the cardiac cycle **(15)**.

The amount of blood pumped out of a ventricle during systole (contraction) is the **stroke volume (SV):**

SV = end-diastolic volume (EDV) −

end-systolic volume (ESV)

Repolarization of the ventricular muscle, as shown by the **T wave (16)**, ends ventricular contraction and the muscle relaxes, causing ventricular pressure to fall below

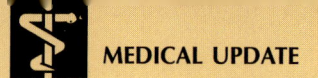

Noninvasive Cardiac Diagnosis

Several noninvasive techniques are available for exploring the heart without even the simplest form of surgery. Two of these are the ultrasound echocardiogram and the nuclear scanner. The **echocardiogram** passes high-frequency sound waves through the chest. After the waves bounce off the heart, they are converted into electric signals and displayed on a television monitor. Echoes obtained from the walls of the ventricles and the flaps of the mitral valve can be used to estimate end-diastolic and end-systolic ventricular volumes. With this information, stroke volume and cardiac output can be calculated. With two-dimensional echocardiography, physicians can see cross sections of all four chambers in action and can detect blood clots, tumors, defective valves, and other problems.

Nuclear scanning uses a specialized camera to pick up radioactive traces from isotopes that have been injected into the bloodstream. The scanner then creates a computer-assisted color-coded scan of the beating heart. When different isotopes are used, such a picture can reveal the amount of blood pumped by the left ventricle, the motion of blood through the heart, and dead or damaged tissue following a heart attack. It can even locate tissue that is not receiving sufficient blood and oxygen.

Another innovation in noninvasive diagnosis is the **dynamic spatial reconstructor (DSR)**. The DSR sends x-rays through the body and produces images of the internal organs on a television monitor. It produces a full-sized, three-dimensional image of the heart (or any other organ) in motion.

---

aortic pressure. When ventricular pressure becomes less than aortic pressure, the aortic valve closes **(17)**, and no more blood is able to leave the ventricle. The closing of the aortic valve causes a notch on the aortic pressure curve called the **dichrotic notch (18)**. Because the AV valve was previously closed, no blood can enter the ventricle from the atrium. This brief period, when ventricular pressure is decreasing **(19)** and all valves are closed, is called *isovolumetric ventricular relaxation* (heart drawing E). The ventricular volume remains constant, as no blood enters or leaves the relaxed ventricle **(20)**. However, the atrium is in diastole, and blood continues to flow into the left atrium from the pulmonary veins. Thus atrial pressure rises **(21)**.

When ventricular pressure drops below atrial pressure, the AV valve opens **(22)**, and rapid ventricular filling occurs **(23)**. This is the blood that has accumulated in the atrium during ventricular systole which can now flow into the ventricle. After this period of rapid filling, the rate of ventricular filling slows **(24)**, but blood continues to flow from the pulmonary veins into the left atrium and through the AV valve into the left ventricle. The SA node fires during late ventricular diastole, initiating atrial depolarization (P wave) **(25)**, and the cycle begins again.

Heart sounds Detectable **heart (valve) sounds** are produced with each heartbeat. These sounds represent the *auscultatory events* (L. *auscultare*, to listen to) of the cardiac cycle and can be heard best in the areas indicated in FIGURE 20.14. Heart sounds can be amplified and recorded by placing an electronically amplified microphone on the chest. The recording is called a *phonocardiogram*, which shows heart sounds as waves.

There are four heart sounds associated with the cardiac cycle, although only the first and second sounds (traditionally referred to as *lubb* and *dupp*, respectively) can be heard easily with a stethoscope:

1 The *first heart sound (S1)*, described as a *lubb*, is more complex, is louder, and lasts longer than the second sound. The first heart sound occurs as a result of the ventricles contracting, which forces blood against the AV valves and causes them to bulge backward toward the atria until the chordae tendineae abruptly stop the bulging. The elastic nature of the valves then causes the blood to bounce forward into each ventricle. This causes the blood, ventricular walls, and valves to vibrate. It is these vibrations that produce the first heart sound, which can be heard most clearly with a stethoscope in the area of the apex of the heart.

2 The *second heart sound (S2)*, described as a *dupp*, results from the sudden closing of the semilunar valves, which bulge backward toward the ventricles until their elastic stretch recoils the blood back into the arteries. This recoil produces vibrations that reverberate back and forth between the heart walls, arteries, and valves. When the vibrations contact the chest wall, they create what can be heard as the second heart sound. The second heart sound is heard best over the second intercostal space, where the aorta is closest to the surface. Immediately after the second sound there is a short interval of silence.

3 A low-pitched *third heart sound (S3)* is heard occasionally. It is caused by the vibration of the ventricular walls after the atrioventricular valves open and the blood gushes into the ventricles. This sound is heard best in the tricuspid area.

4 A *fourth heart sound (S4)* is usually not heard with an unamplified stethoscope in normal hearts because of its low sound. It is caused by blood rushing into the ventricles. It is best heard in the mitral area.

FIGURE 20.14 LOCATION OF HEART SOUNDS

The chest areas where adult heart sounds can be heard most clearly. The ribs are numbered for clarity.

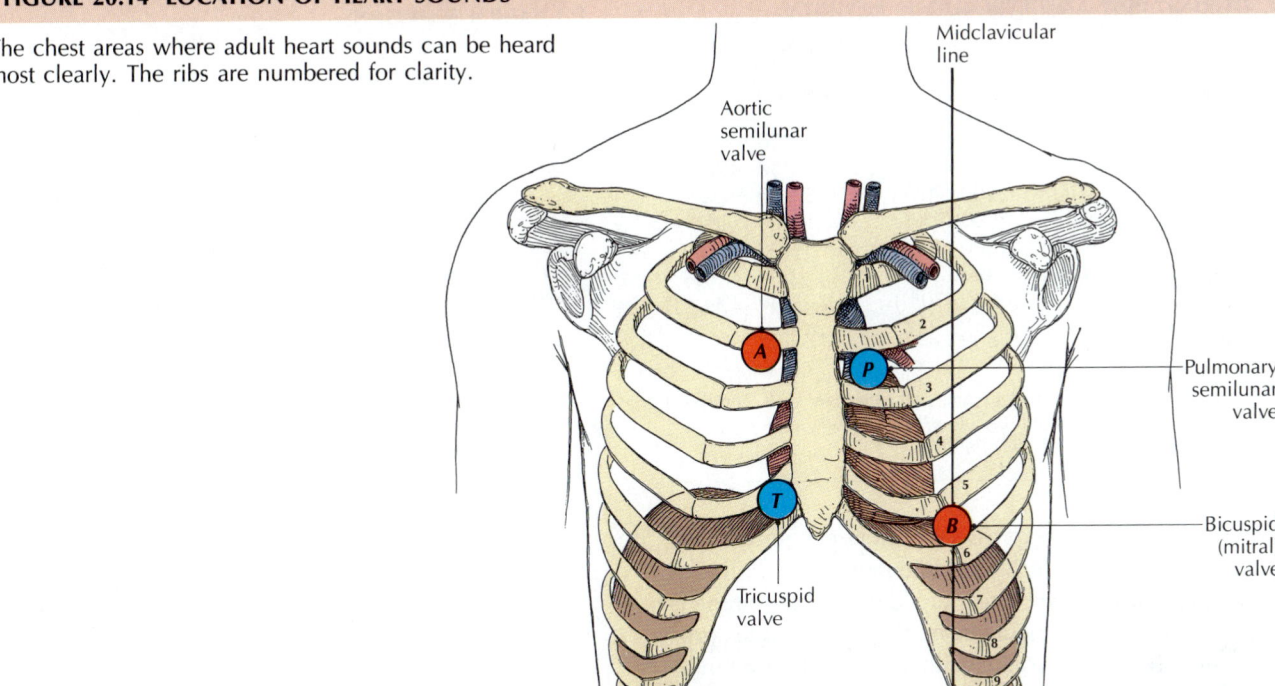

Heart sounds are an important tool in diagnosing valvular abnormalities. Any unusual sound is called a *murmur*, but not all murmurs indicate a valve problem, and many have no clinical significance. By listening carefully, a physician can detect *resting heart murmurs* that may be symptoms of valvular malfunctions, congenital heart disease, high blood pressure, and many other serious problems. If the atrioventricular valves are faulty, for instance, a gentle blowing or hissing sound can be heard between the first and second heart sounds: lubb-hiss-dupp — lubb-hiss-dupp.

Cardiac Output

The heart at rest pumps about 70 mL of blood with every beat. At an average rate of 72 beats per minute, the heart pumps more than 5 L a minute, 300 L an hour, 7200 L a day, and 2,628,000 L a year.

Cardiac output (CO) is the quantity of blood (in liters) pumped by either ventricle (not both) in 1 min. (The left ventricle is the one usually measured.)

Cardiac output is determined by multiplying the stroke volume by the heart rate. If the normal stroke volume is about 70 mL per beat, and the normal heart rate is 72 beats per minute, then

$$\text{Cardiac output} = \text{stroke volume} \times \text{heart rate}$$
$$\text{(mL/min)} \qquad \text{(mL/beat)} \qquad \text{(beats/min)}$$

$$= 70 \text{ mL/beat} \times 72 \text{ beats/min}$$

$$= 5040 \text{ mL/min} \quad \text{or} \quad 5.04 \text{ L/min}$$

TABLE 20.1 shows cardiac output for individuals at rest and during strenuous physical activity, and FIGURE 20.15 shows the distribution of blood to various organs. The *cardiac reserve* is the difference between the actual volume of blood pumped and the volume the heart is capable of pumping under stressful conditions. Cardiac reserve measures the potential blood-pumping ability of the heart, while cardiac output measures the actual work done. For example, in a normal young adult, the cardiac reserve is between 300 to 400 percent. In a well-trained athlete, it is occasionally as high as 500 to 600 percent, whereas in a physically inactive person, it may be as low as 200 percent.

The *cardiac index (CI)* is a measurement of the cardiac output of a resting subject as related to body surface area (BSA):

$$CI = \frac{CO}{BSA}$$

TABLE 20.1 DISTRIBUTION OF CARDIAC OUTPUT DURING REST AND STRENUOUS ACTIVITY

Organ	At rest, mL/min	During exercise, mL/min
Brain	750 (13%)	750 (4%)
Heart	250 (4%)	750 (4%)
Skeletal muscle	1,200 (20%)	12,500 (72%)
Skin	500 (9%)	1,900 (11%)
Kidneys	1,100 (20%)	600 (3.5%)
Abdominal organs	1,400 (24%)	600 (3.5%)
Other	600 (10%)	400 (2%)
Total	5,800	17,500

Source: From Arthur J. Vander, James H. Sherman, and Dorothy S. Luciano, *Human Physiology*, 6th ed. New York, McGraw-Hill, 1993, p. 406. Used with permission.

The cardiac index is expressed in liters per minute per square meter (L/min/m^2). Since the rate of metabolism is related to body surface area and the cardiac output is related to the rate of metabolism, the cardiac index provides an index for the comparison of different-sized individuals to normal values. A normal cardiac index is between 2.5 and 4.0 L/min/m^2.

Regulation of cardiac output: Heart rate We have shown that variations in cardiac output can be produced by changes in the heart rate, stroke volume, or both, depending on the physiological situation. The major controlling factor is heart rate. Heart rate can change by more than three times, but stroke volume can vary by only about half that amount. FIGURE 20.16 summarizes the major factors that influence heart rate and stroke volume and, thus, cardiac output.

The normal heart rate is determined primarily by the rhythmic pacemaker potentials of the SA node. It is also influenced by the autonomic nervous system and certain hormones. Both sympathetic and parasympathetic fibers innervate the SA node as well as other parts of the heart's conductive system.

Sympathetic stimulation increases the rate at which the pacemaker potential develops, causing the SA cells to reach their action potential threshold faster. As a result, the heart rate increases. Parasympathetic stimulation has the opposite effect, decreasing the pacemaker potential, causing the pacemaker cells to take longer to reach their action potential threshold, and decreasing the heart rate.

Q: *Do children and adults have the same heart rate and blood pressure?*

A: No. Children usually have *higher* heart rates and *lower* blood pressure.

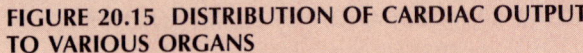

FIGURE 20.15 DISTRIBUTION OF CARDIAC OUTPUT TO VARIOUS ORGANS

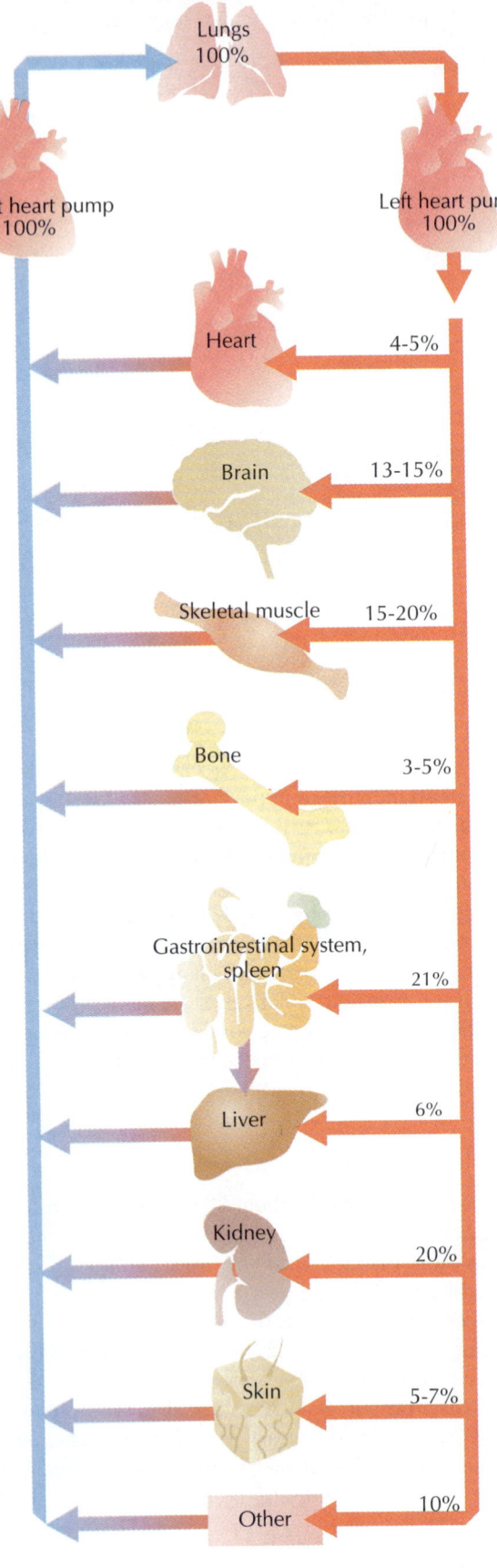

Lungs 100%

Right heart pump 100%

Left heart pump 100%

Heart 4-5%

Brain 13-15%

Skeletal muscle 15-20%

Bone 3-5%

Gastrointestinal system, spleen 21%

Liver 6%

Kidney 20%

Skin 5-7%

Other 10%

FIGURE 20.16 FACTORS AFFECTING CARDIAC OUTPUT

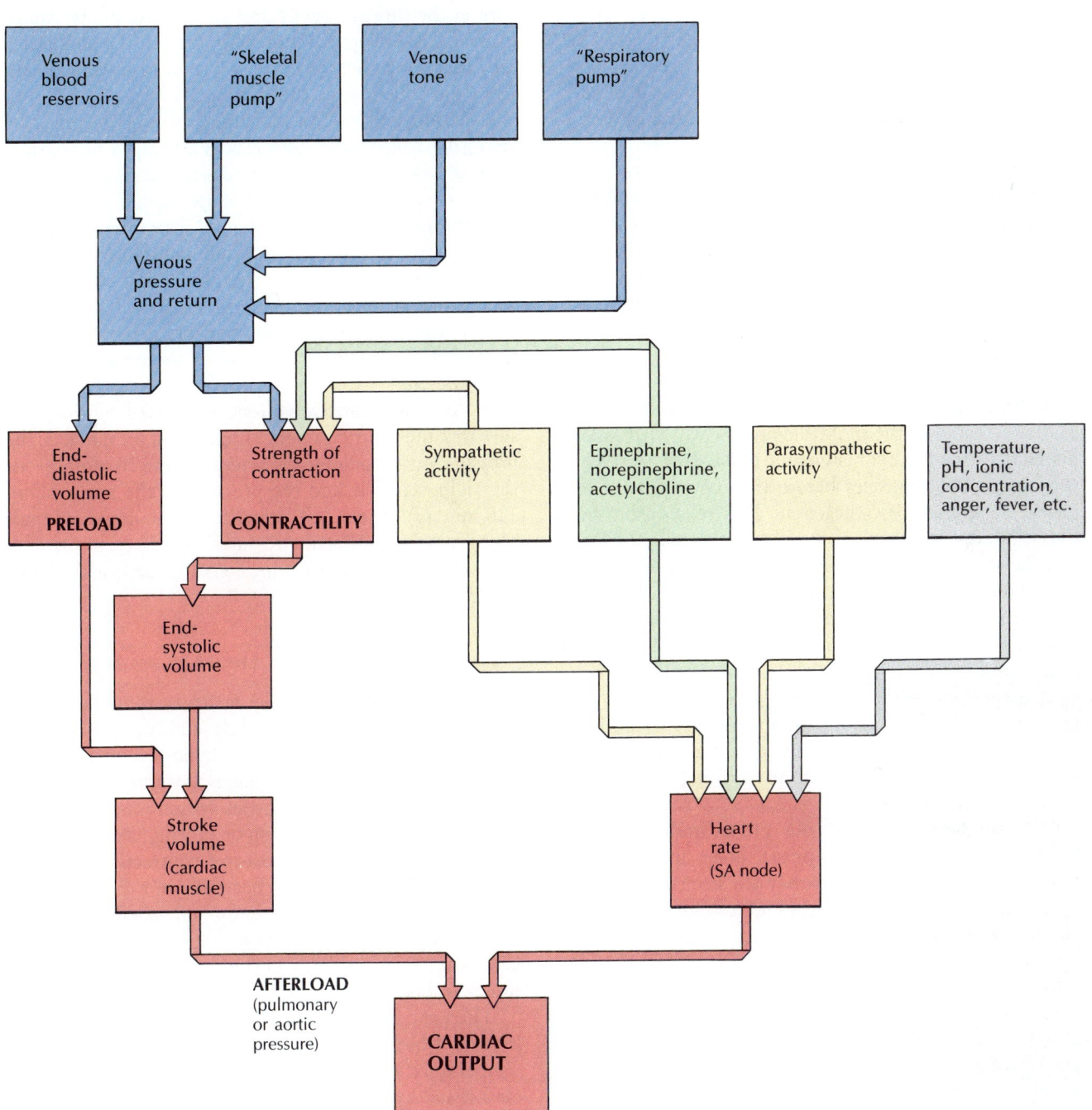

In the resting homeostatic state, the heart rate (*inherent rate*) is set by the tone of the parasympathetic system, in the range of 60 to 100 beats per minute. Because the autonomic neurons innervate other parts of the heart's conductive system, sympathetic stimulation also accelerates the spread of the electrical impulse through the AV node, atrioventricular bundle, and cardiac conducting myofibers. Parasympathetic stimulation slows the electrical impulse along this same conducting pathway.

The regulatory effects of the autonomic nervous system are produced by the release of acetylcholine from the parasympathetic neurons and norepinephrine from the sympathetic neurons. Acetylcholine affects the pacemaker potential by increasing the permeability of the SA node cells to potassium, which slows the rate of depolarization. Norepinephrine enhances the depolarization rate of SA node cells by increasing the flow of calcium ions into the cells.

The heart rate can be altered to a lesser degree by factors outside of autonomic control. For example, blood temperature, pH, ionic concentrations, hormones, anger, pain, exercise, fever, and grief all have an effect on heart rate.

Regulation of cardiac output: Stroke volume

Stroke volume is the difference between the blood volumes in the left ventricle at the beginning and end of each contraction. If either of these volumes is varied, the stroke volume and cardiac output are altered. Stroke volume is regulated by changing the (1) preload, (2) contractility, or (3) afterload.

1 Preload. The more ventricles are filled during diastole (relaxation), the more blood they will eject upon systole (contraction). This greater *preload,* or "stretch" (filling), of the ventricles before they contract will increase their force of contraction. This relationship between the resting length of the ventricular muscle fibers and the strength of heart contraction was first measured in 1895 by O. Z. Frank, and later by E. H. Starling, and resulted in the *Frank-Starling law of the heart,* which states that within physiological limits, the more the ventricles are filled during diastole, the more blood they will eject upon systole. The greater the preload (filling), the greater the force of contraction. The Frank-Starling mechanism equalizes the outputs of both ventricles, so that equal volumes of blood go to both pulmonary and systemic circulations.

The mechanism behind the Frank-Starling law is based on the fact that cardiac muscle fibers increase their strength of contraction when they are stretched. An increase in *venous return* to the heart increases end-diastolic volume, which distends the ventricles, stretches cardiac muscle fibers, increases stroke volume, and finally increases cardiac output.

The factors that influence venous return and pressure also have a direct effect on end-diastolic volume and stroke volume [FIGURE 20.16]. Since veins are more distensible than arteries, they can serve as *reservoirs* that store blood, which in turn can either increase or decrease the volume of blood returning to the heart. The contraction of skeletal muscles compresses veins during movement and squeezes blood toward the heart. This action is known as the *skeletal muscle pump.* Venous blood reserves can be actively mobilized by contracting the smooth muscle in the walls of the veins. Sympathetic stimulation causes this constriction (*venous tone*), which increases venous pressure and, in turn, end-diastolic volume.

Venous return is also influenced during the breathing process by a mechanism called the *respiratory pump.* During inspiration (taking air in), there is a decrease in pressure in the thoracic cavity and an increase in abdominal pressure. This forces venous blood from the abdominal area back to the heart by increasing the pressure gradient from the abdomen to the thorax.

2 Contractility. Stroke volume is also controlled by *changing ventricular contractility* (strength of contraction) at any given preload. Factors that increase contractility are known as *positive inotropic agents* and include such things as increases in calcium, sympathetic stimulation, norepinephrine, epinephrine, and glucagon. *Negative inotropic agents* decrease ventricular contractility, and thus stroke volume. Such agents include increased potassium in the extracellular fluid, acidosis, anoxia, or inhibition of the sympathetic division of the autonomic nervous system.

3 Afterload. Stroke volume is affected by the pressure that the ventricles pump their blood into, that is, the pressure in the pulmonary artery (about 20 mmHg) for the right ventricle and the pressure in the aorta (about 120 mmHg) for the left ventricle. If the *afterload* (pulmonary artery or aortic pressures) increases, stroke volume decreases, resulting in a greater ventricular end-systolic volume.

Neural Control of the Heart

The major function of the heart is to pump blood through a closed system of vessels. This function is regulated by means of the cerebrum, hypothalamus, medulla oblongata, and autonomic nerves [FIGURE 20.17]. The effects of the autonomic nervous system on the heart are strictly *regulatory*, speeding up or slowing down the heart rate and increasing or decreasing contractility, and are not essential for the heart to beat. (If you were to sever all nerve connections from the brain and spinal cord to the heart, the heart would still beat. This is why a heart that is removed from the body during transplant surgery continues to beat.)

The main control center is located in the medulla oblongata, which receives afferent information about body temperature, emotions, feelings, and stress from the cerebrum and hypothalamus. It also receives afferent information from receptors in the walls of the aortic arch and carotid artery sinuses. Information about the changing chemical composition of blood (O_2, CO_2, H^+) is transmitted from chemically sensitive receptors called *chemoreceptors*, and information about how arteries are stretched by changes in blood pressure is received from pressure-sensitive clusters of cells called *baroreceptors*. Both types of receptors help the body make the adjustments necessary to restore normal conditions.

FIGURE 20.17 NEUROREGULATION OF THE HEART

The vagus nerve usually sends steady (tonic) inhibitory impulses (green) to the heart, keeping the heart rate regular. Under stressful conditions, impulses from sympathetic nerves (black) from the cardioregulatory center in the medulla oblongata increase the heart rate and contractility.

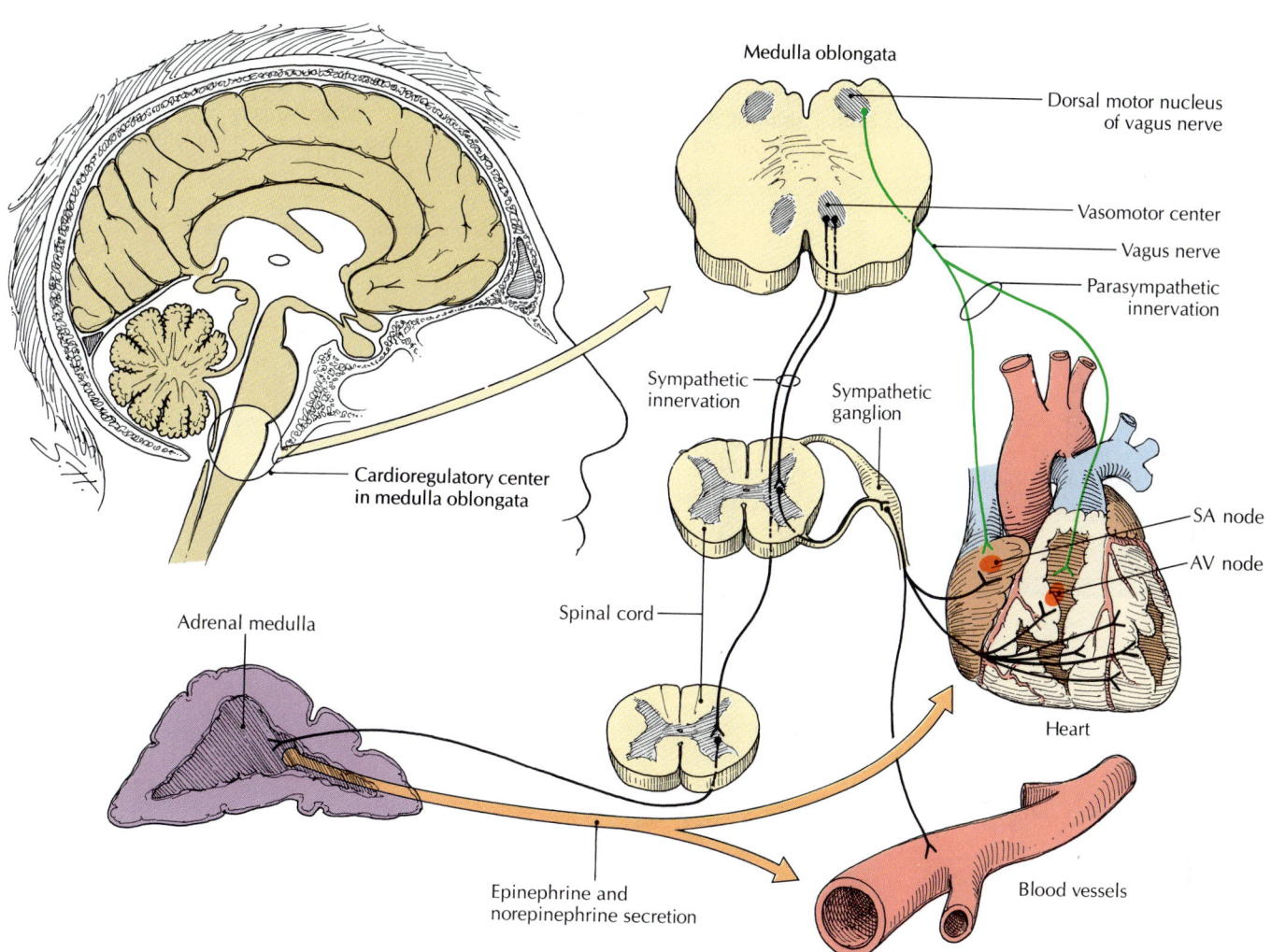

Medulla oblongata

Dorsal motor nucleus of vagus nerve

Vasomotor center

Vagus nerve

Parasympathetic innervation

Cardioregulatory center in medulla oblongata

Sympathetic innervation

Sympathetic ganglion

SA node

AV node

Heart

Adrenal medulla

Spinal cord

Epinephrine and norepinephrine secretion

Blood vessels

The upper part of the medulla oblongata contains an area called the *cardioacceleratory center (CAC)*, or *pressor center*; the lower part contains the *cardioinhibitory center (CIC)*, or *depressor center*. Because the neurons of both centers interact to regulate the heart rate, they are collectively called the **cardioregulatory center.**

Sympathetic nerve fibers arise from the cardioacceleratory center, travel down the spinal cord through specific tracts, emerge from the cord by way of the cardiac nerves, and innervate the heart, where neuronal branches release norepinephrine [FIGURE 20.17]. Norepinephrine accelerates the heart rate and strength of contraction. Parasympathetic (vagal) nerve fibers arising from the cardioinhibitory center go directly to the SA and AV nodes of the heart, where neuronal branches release acetylcholine, which decreases the heart rate.

Endocrine Control of the Heart

As you just saw, chemical transmitters are used by the nervous system to regulate the activity of the heart. Any generalized sympathetic activity within the autonomic nervous system affects the medullary region of the adrenal glands. When cells of the adrenal medulla are stimulated, they secrete epinephrine primarily and some norepinephrine into the bloodstream. Epinephrine and norepinephrine increase the heart rate and strength of contraction.

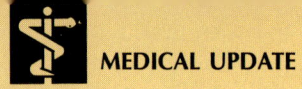

Heart Surgery

Surgery on blood vessels and the heart has been done since the early 1930s. However, it was not until the heart-lung machine was introduced in 1953 that true open-heart surgery was possible. In December of 1982, the first operation took place in which an artificial heart was placed into a person.

Heart-Lung Machine

Open-heart surgery is possible only if the heart is quiet and empty of blood. The **heart-lung machine** takes over the job of pumping *and* oxygenating blood while open-heart surgery is in progress (see drawing A). Tubes are inserted into the inferior and superior vena cavae to lead blood through a pump and oxygenator, where carbon dioxide is removed and oxygen is added, just as occurs in the lungs. A pump then returns the oxygenated blood into the arterial circulation by way of the aorta or one of its branches.

While surgery is taking place and blood is being circulated through the

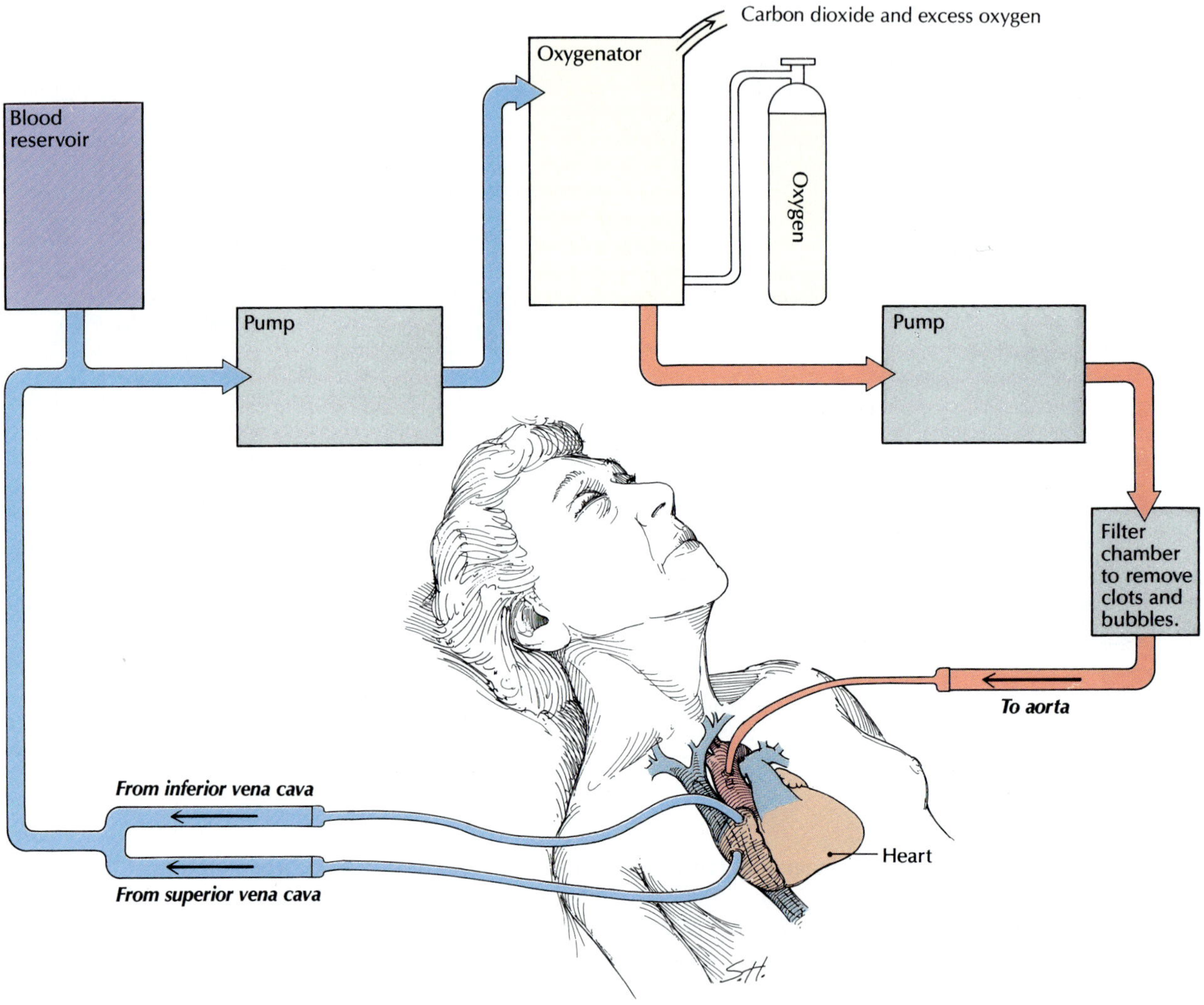

Carbon dioxide and excess oxygen

Oxygenator

Oxygen

Blood reservoir

Pump

Pump

Filter chamber to remove clots and bubbles.

To aorta

From inferior vena cava

From superior vena cava

Heart

[A] The heart-lung machine (simplified).

heart-lung machine, blood clotting is prevented by introducing the anticoagulant heparin into the circulation. When the natural circulation through the heart is restored, the effect of heparin is reversed by the introduction of protamine sulfate.

An important technique used with the heart-lung machine is *hypothermia,* in which the patient's body is cooled enough to induce ventricular fibrillation, thus providing a quiet state of the heart. Recent techniques allow cooling of only the heart instead of the whole body.

The newer techniques eliminate all heart activity during the operation, and reduce metabolism so that absolutely no strain is placed on the myocardium during surgery.

Coronary Bypass Surgery

The most common serious heart disease is the obstruction or narrowing of the coronary arteries by atherosclerosis. When coronary arteries become blocked, the flow of blood to the heart is reduced or cut off completely, and angina pectoris or myocardial infarction may result. In *coronary bypass surgery,* the surgeon removes the diseased portion of the coronary artery and replaces it with a segment of the internal thoracic artery or saphenous vein from the patient's own body (see drawing B). (Because the replacement vessel is taken from the patient's body, there is virtually no danger of tissue rejection.) One end of the vein is sutured to the aorta, and the other end is sutured to a coronary artery beyond the point of obstruction. Thus the diseased artery is "bypassed," and normal blood flow is reestablished.

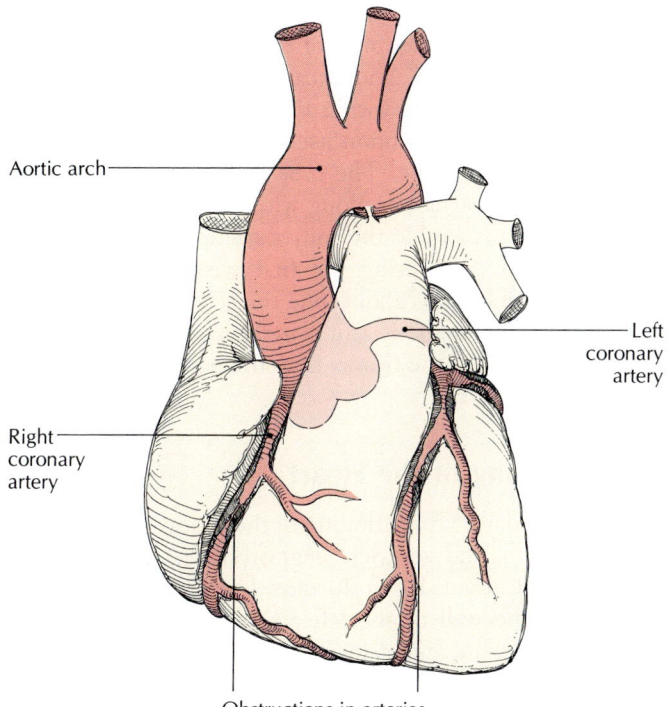
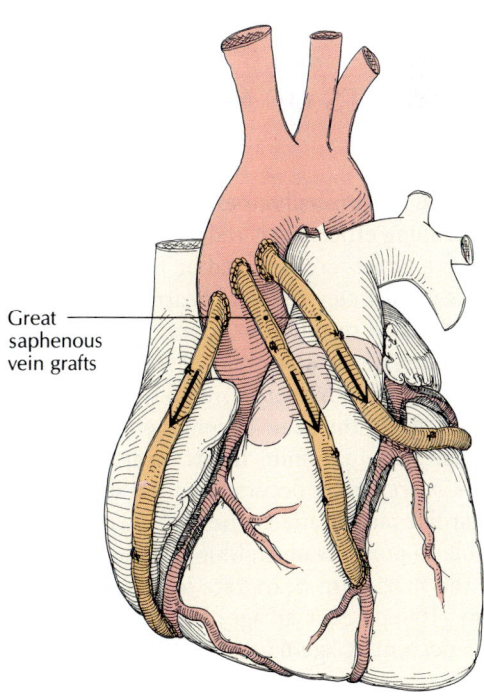

[B] Coronary arteries before (left) and after (right) coronary bypass surgery.

THE EFFECTS OF AGING ON THE HEART

Although the heart usually does not decrease in size with age, its pumping efficiency is reduced because some muscle and valve tissues are replaced by fibrous tissue. Blood pressure is usually elevated, and heart rate does not compensate as well in response to stress. If hardening of the arteries occurs, the arterial walls become less elastic, and the heart must work harder to pump the same amount of blood. As a result, the heart becomes enlarged, and if the heart reaches its maximum limits, heart failure occurs. *Congestive heart failure* occurs when weakening or scarring of cardiac muscle affects the heart to the point where it is unable to pump as much blood as the body needs. At age 45, the maximum heart rate is about 94 percent of optimum (100 percent at age 25), and it decreases to about 87 percent at age 65 and to 81 percent at age 85.

DEVELOPMENTAL ANATOMY OF THE HEART

The embryonic circulatory system forms and becomes functional before any other system, and by the end of the third week after fertilization the system is fulfilling the nutritional needs of the embryo.

Early Development of the Heart

The heart begins as a network of cells, from which two longitudinal **heart cords** (also called *cardiogenic cords*) are formed about 18 or 19 days after fertilization. On about day 20, the heart cords develop canals, forming **heart tubes** (also called *endocardial heart tubes*). On days 21 and 22 the heart tubes fuse, forming a single median **endocardial heart tube** [FIGURE 20.18A, B]. Immediately following, the endocardial heart tube elongates; dilations and constrictions develop, forming first the **bulbus cordis, ventricle,** and **atrium,** and then the **truncus arteriosus** and **sinus venosus** [FIGURE 20.18C]. At this time (day 22), the sinus venosus develops *left and right horns*, the venous end of the heart is established as the **pacemaker,** heart contractions resembling peristaltic waves begin in the sinus venosus, and on about day 28 coordinated contractions of the heart begin to pump blood in one direction only.

Each horn of the sinus venosus receives an *umbilical vein* from the chorion (primitive placenta), a *vitelline vein* from the yolk sac, and a *common cardinal vein* from the embryo itself [FIGURE 20.18D]. Eventually, the left horn forms the coronary sinus, and the right horn becomes incorporated into the wall of the right atrium.

Growth is rapid, and because the heart tube is anchored at both ends within the pericardium, by day 25 the tube bends back upon itself, forming first a U shape and then an S shape. The tube develops alternate dilations and constrictions along with further bending until it begins to resemble the fully developed heart [FIGURE 20.18C and D]. During the first weeks of embryonic growth, the heart is about nine times as large in proportion to the whole body as it is in the adult. Also, its position is higher in the thorax than the permanent position it will assume later.

Partitioning of the Heart

On about day 25, partitioning of the atrioventricular orifice, atrium, and ventricle begins [FIGURE 20.18E]; it is completed about 10 to 20 days later [FIGURE 20.18F]. Most of the wall of the left atrium develops from the pulmonary vein, and as mentioned above, the right horn of the sinus venosus becomes incorporated into the wall of the right atrium. By about day 28, the heart wall has formed its three layers (endocardium, pericardium, epicardium). At about day 32, the *interatrial septum* forms, dividing the single atrium into left and right atria. An opening in the septum, the *foramen ovale,* closes at birth. The partitioning of the single ventricle into left and right ventricles begins with the formation of an upward fold in the floor of the ventricle, called the *interventricular septum* [FIGURE 20.18F], and is completed by about day 48. The development of the atrioventricular valves, chordae tendineae, and papillary muscles proceeds from about the fifth week until the fifth month. The external form of the heart continues to develop from about day 28 to day 60.

FIGURE 20.18 EARLY EMBRYONIC DEVELOPMENT OF THE HEART

[A, B] The primitive heart tubes fuse together and form a single endocardial heart tube during days 21 and 22 after fertilization; ventral views. [C] On about day 22 or 23, the first major structures form, and the heart tube begins to bend and twist; ventral view. [D] On about day 25, the tube has formed an S shape. [E] A frontal view at about 32 days shows the remaining three pairs of aortic arches.

[F] At about 5 weeks, the interior partitions can be seen clearly in this frontal section. [G] A frontal section at 6 weeks shows the aorta and pulmonary trunk after the bulbus cordis has been incorporated into the ventricles to become the infundibulum. Black arrows indicate the movement of the heart, and red arrows indicate the flow of blood.

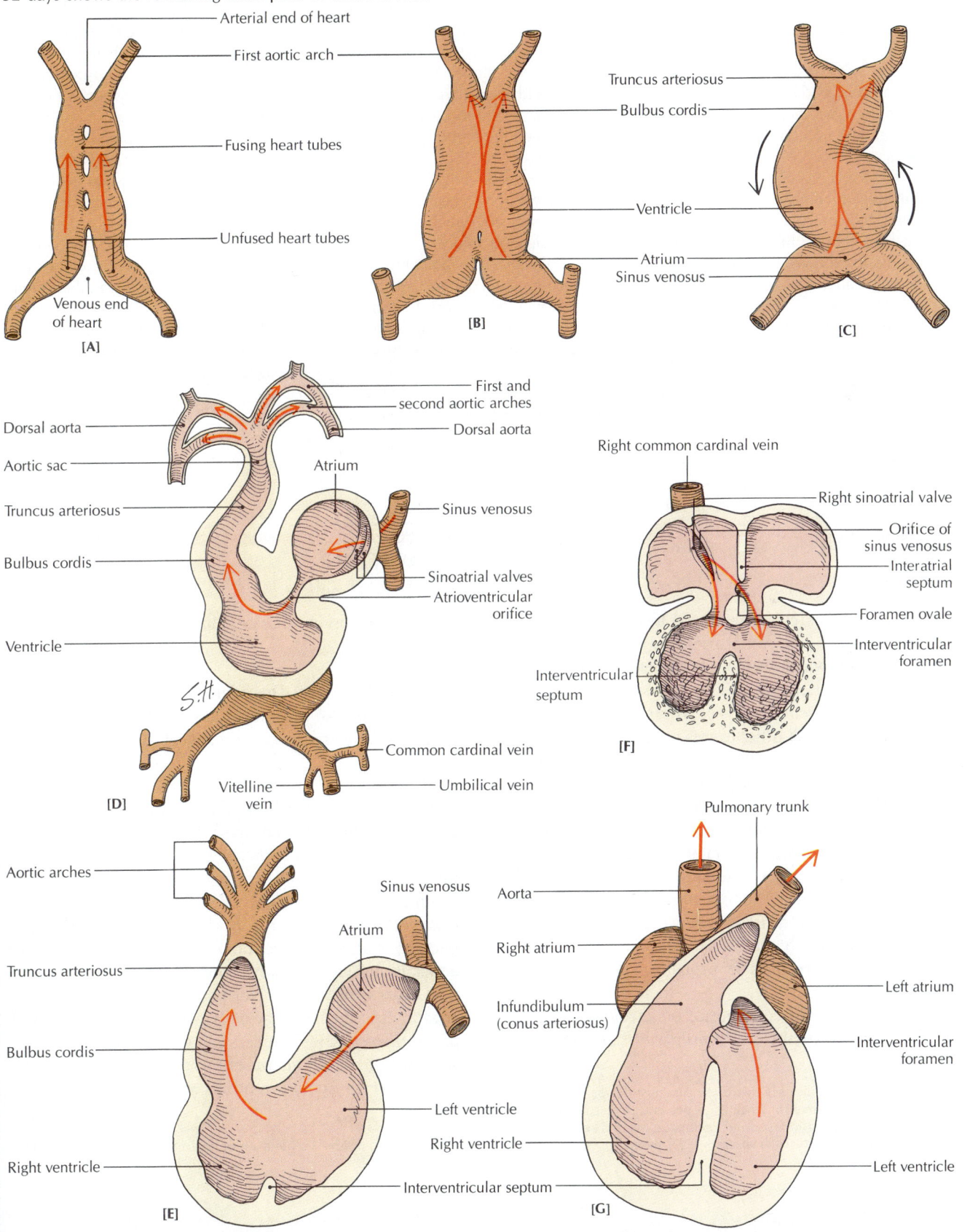

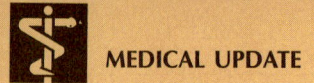

Excess Iron and Heart Disease

It has been known for many years that men have more heart attacks than women but that the risk of heart attacks in postmenopausal women begins to approach that of men. Until recently, the reason was a mystery (except for the decrease of estrogen after menopause), but plausible evidence reveals that the main culprit may be iron.

What does iron have to do with heart attacks?

An Intruiging Finnish Study

A 1992 study in Finland of 1931 middle-aged men with no history of cardiovascular disease showed that men with high blood concentrations of ferritin (a molecule that stores iron)—more than 200 μg/L—had twice the risk of a heart attack as men with lower ferritin levels. In fact, the study showed that high levels of iron stored in the body may outweigh blood cholesterol as a precursor to a heart attack. (Of the 19 risk factors monitored by the Finnish study, only cigarette smoking was a stronger predictor of heart attacks.) The combination of cholesterol and iron count was shown to be a more reliable predictor of heart disease than cholesterol alone. Men with both high iron and cholesterol counts were four times as likely to have a heart attack—double the rate of men with only high iron levels.

The body needs only trace amounts of iron. Excess iron can cause the formation of free radicals, which may damage heart muscle and injure the cells that line arterial walls. Free radicals may also cause the formation of the "bad" cholesterol—oxidized low-density lipoprotein (LDL). (Oxidized LDL adheres to arterial walls better and builds up fatty plaque faster than nonoxidized LDL does.)

Postmenopausal women begin accumulating iron because they retain the iron they once lost in their monthly menstrual flow. Postmenopausal iron builds up rapidly. The problem is complicated when men and women over 50 or 60 begin to take iron supplements as a matter of course, without a physician's recommendation. Ordinarily, iron supplements are needed only when a person has iron-deficiency anemia, a condition that should be diagnosed and treated by a physician. A further complication is the typical American diet that is high in red meat, which contains large amounts of iron.

The Finnish results are fascinating, and researchers all over the world are currently testing the results in their own laboratories. Perhaps the *dual* factors of loss of estrogen and increase in iron are the real causes of heart attacks in postmenopausal women.

WHEN THINGS GO WRONG

Heart disease is a commonly used term for any disease that affects the heart. A more appropriate term is *cardiovascular disease*, which includes both heart and blood vessel disorders. About 40 million Americans have some form of cardiovascular disease, and it is responsible for more deaths than all other causes of death *combined.** Although the American death rate from cardiovascular disease is decreasing, it is still one of the highest in the world. This year more than 1.5 million Americans will suffer heart attacks, and of the 550,000 who survive, 100,000 will have another, fatal, attack within a year.

In 1993, heart disease became the leading cause of death among women. At that time, almost half of the people who died of heart disease were women, and it is predicted that fatal heart disease will soon occur more in women than in men. The probable cause? Cigarette smoking.

This section will concentrate on cardiovascular diseases that involve the heart primarily. The following chapter will discuss diseases that originate in the blood vessels.

Degenerative Heart Disorders

The first group of heart disorders results from the deterioration of the tissues or organs of the cardiovascular system.

Myocardial infarction (heart attack) When the blood flow through a coronary artery is reduced for any reason (usually because of a clot or plaque† build-up), the myocardium is deprived of oxygen and begins to die. The result is a *myocardial infarction* (L. *infercire,* to stuff), or *heart attack.* An *infarct* is an area of tissue that has died because of an inadequate blood supply. Although a heart attack is always serious, it is not always diagnosed easily. The usual symptoms include pain in the midchest (*angina pectoris,* described on page 648), which travels up the neck or out through the shoulders and arms, especially the left shoulder or arm. Sometimes there may be only a shortness of breath or no symptoms at all.

*The cardiovascular diseases that cause the most deaths each year are heart attack (56.3 percent), stroke (17.4 percent), hypertensive disease (3.1 percent), rheumatic fever and rheumatic heart disease (0.8 percent), all other cardiovascular diseases (22.4 percent).

†Plaque in an artery is a build-up of cholesterol and other fatty acids.

Congestive heart failure *Congestive heart failure (CHF)* occurs when either ventricle fails to pump blood out of the heart as quickly as it enters the atria or when the ventricles do not pump equal amounts of blood. For example, left heart failure occurs if the left ventricle is weakened and pumps less blood than normal; the right ventricle will then pump more blood *into* the lungs than can be pumped by the left ventricle. As a result, the lungs become engorged with blood in a condition called *pulmonary edema*. The old-fashioned term *dropsy* refers to the accumulation of fluid in the abdomen and legs that also accompanies congestive heart failure.

Valvular Heart Diseases

As you saw earlier, the two major heart sounds provide information about the heart's valves. Abnormal heart sounds, or *murmurs*, may be indicative of a **valvular heart disease,** in which one or more cardiac valves operate improperly. Certain types of valvular malfunction may produce *regurgitation* (backflow of blood through an incompletely closed valve) or *stenosis* (an incompletely opened valve). Either type of malfunction can lead to congestive heart failure. Some forms of valvular heart disease are (1) *mitral insufficiency* (blood from the left ventricle flows back into the left atrium during systole), (2) *mitral stenosis* (narrowing of the valve reduces flow from the left atrium to the left ventricle), (3) *aortic insufficiency* (blood flows back into the left ventricle during diastole), (4) *aortic stenosis* (pressure in the left ventricle increases in response to a narrowed valve opening, and ischemia, a localized decrease in blood supply, results), (5) *pulmonary insufficiency* (blood flows back from the pulmonary artery into the right ventricle during diastole), (6) *pulmonary stenosis* (blood is prevented from leaving the right ventricle, causing ventricular hypertrophy and eventual failure), (7) *tricuspid insufficiency* (blood flows back into the right atrium during systole), and (8) *tricuspid stenosis* (blood is obstructed from flowing from the right atrium to the right ventricle). Valvular heart diseases may sometimes be corrected with the surgical substitution of an artificial valve.

Congenital Heart Diseases

A congenital disease is one that is present at birth but is not necessarily hereditary.

Ventricular septal defect *Ventricular septal defect* is the most common congenital heart disease. It is an opening in the ventricular septum that allows blood to move back and forth between the ventricles. Small openings usually close naturally, before permanent damage occurs, or are repaired surgically. Large openings are not always reparable, and patients often die in their first year from biventricular congestive heart failure or secondary complications.

Interatrial septal defect In an *interatrial septal defect,* the foramen ovale between the two atria fails to close at birth, and the child is born with an opening between the right and left atria that allows blood to flow through the foramen. After birth, some blood usually flows from the left atrium to the right atrium because atrial pressure is higher in the left atrium. Eventually, the entire right side of the heart becomes enlarged and heart failure occurs.

Tetralogy of Fallot *Tetralogy of Fallot* is a combination of four (Gr. *tetra,* four) congenital defects: (1) ventricular septal defect, (2) pulmonary stenosis, (3) enlargement of the right ventricle, and (4) emergence of the aorta from both ventricles. Blood usually flows from the right to the left ventricle, although it may flow from the left to the right when pulmonary stenosis is mild. The most common symptom of tetralogy of Fallot is *cyanosis* (Gr. dark blue), a bluish discoloration of the skin resulting from inadequate oxygenation of the blood. Ventricular septal defect allows oxygenated blood to mix with deoxygenated blood in the ventricles. Pulmonary stenosis obstructs blood from leaving the right ventricle, producing the third defect, an enlarged right ventricle. The emergence of the aorta from within the ventricles, coupled with stenosis of the pulmonary artery, allows insufficient blood to reach the lungs.

Infectious Heart Diseases

Severe damage to the heart valves or heart walls can result from certain infectious diseases.

Rheumatic fever and rheumatic heart disease *Rheumatic fever* is a severe infectious disease occurring mostly in children. It is characterized by fever and painful inflammation of the joints and frequently results in permanent damage to the heart valves. Rheumatic fever is a hypersensitive reaction to a specific streptococcal bacterial infection in which antibodies cause inflammatory reactions in the joints and heart. *Rheumatic heart disease* refers to the secondary complications that affect the heart. They include *pancarditis* (*pan* = all, because it encompasses pericarditis, myocarditis, and endocarditis) in the early acute phase and valvular heart disease later.

Pericarditis, myocarditis, endocarditis Inflammations of the three layers of the heart wall are called, from the outside in, pericarditis, myocarditis, and endocarditis.
Pericarditis is an inflammation of the parietal and/or visceral pericardium. When it is caused by a bacterial, fungal, or viral infection, it is known as *infectious pericarditis*. It may also be caused by uremia, high-dose radiation, rheumatic fever, cancer, drugs, and trauma such as myocardial infarction. Pain is typical, especially when the heart presses against its covering membranes.
Myocarditis is an inflammation of the myocardium. It

may result from viral, bacterial, parasitic, or other infections; hypersensitive immune reactions; large doses of radiation; and poisons such as alcohol. Myocarditis usually proceeds without complications, but occasionally it leads to heart failure, arrhythmias, and other complications.

Endocarditis is caused by a bacterial or fungal infection of the endocardium, the heart valves, and the endothelium of adjacent blood vessels. Untreated, endocarditis is usually fatal.

Cardiac Complications

Among the complications that may affect the heart are circulatory shock, rapid rise in pressure, and abnormal rhythms.

Circulatory shock *Circulatory shock* refers to a generalized inadequacy of blood flow throughout the body. It may be caused by *reduced blood volume* as a result of hemorrhage, severe diarrhea or vomiting, lack of water, or severe burns that involve serious fluid loss. Another possible cause is *increased capacity of veins* as a result of bacterial toxins. A third cause is *damage to the myocardium,* usually as result of a heart attack.

When blood volume drops below about four-fifths of normal, one of the results is **hypovolemic shock,** which impairs circulation and the flow of liquids to the tissues. Such a condition is usually caused by hemorrhage, severe burns, or any other occurrence that reduces the fluid content of the body. Hypovolemic shock is usually accompanied by low blood pressure, rapid heart rate, rapid and shallow breathing, and cold, clammy skin.

In contrast to hypovolemic shock, **cardiogenic shock** occurs when the cardiac output is decreased, resulting in too little fluid reaching the tissues. It is caused by severe failure of the left ventricle, usually as a complication of myocardial infarction, and most patients die within 24 hours.

Cardiac tamponade In *cardiac (pericardial)* **tamponade** (Fr. *tampon,* plug), a rapid rise in pressure occurs inside the pericardial sac, usually because of the accumulation of blood or other fluid as a result of an infection or severe hemorrhage in the chest area.

Cardiac arrhythmias Abnormal electrical conduction in the conductive tissues or changes in heart rate and rhythm are called **cardiac arrhythmias.** In *paroxysmal ventricular tachycardia* (Gr. *paroxunein,* to stimulate; *takhus,* swift), the heart suddenly starts to race at a steady rhythm of 200 to 300 beats per minute. In medical terms, *paroxysm* refers to something that starts and stops suddenly, and *tachycardia* (tack-ih-KAR-dee-uh) is a heart rate faster than 100 beats per minute.

Atrial fibrillation (L. *fibrilla,* fibril) occurs when the atria suddenly begin to beat with a fast but feeble twitching movement. The ventricles contract normally for a short time, but their beat soon becomes irregular also. Because of the irregular beat, the ventricles may contract when they are not full, reducing the cardiac output. Blood may also stagnate and clot in the atria. If these clots get into the circulation, they can clog arteries and cause heart attacks, stroke, or other serious problems. The leading cause of atrial fibrillation is mitral stenosis.

Atrial flutter occurs when the atria beat regularly, but at a very rapid, "flapping" pace of about 200 to 300 beats per minute. Ordinarily, every third or fourth atrial beat will stimulate a ventricular beat, but the relationship can vary from two to one, to as much as six to one. The causes of atrial flutter are usually similar to those of atrial fibrillation.

Atrioventricular block (or *AV block*) occurs when cells of the atrioventricular node become diseased and cannot transmit adequate electrical impulses. The condition may progress from a pattern where the rhythm of the impulse is almost imperceptibly slowed to an irregular, slow atrial beat that may be accompanied by dizziness. If the cells of the AV node totally lose their ability to conduct, the beat of the heart is taken over by one of the emergency pacemakers in the atria or ventricles. At this stage, called *complete heart block*, the atria and ventricles beat independently. If the ventricles stop beating for a few seconds, fainting and convulsions may occur.

Ventricular fibrillation is caused by a continuous recycling of electrical waves through the ventricular myocardium. As a result, abnormal contraction patterns with varied rates are set up. During ventricular fibrillation, the strong, steady contractions of the ventricles are replaced by a feeble twitching that pumps little or no blood. These effects can be produced by damage to the myocardium, which usually results from an inadequate blood supply. Without treatment, the victim usually dies in a matter of minutes.

Angina Pectoris (Myocardial Ischemia)

Angina pectoris (an-JYE-nuh PECK-tuh-rihss; L. strangling + chest) is an example of referred pain. It occurs when not enough blood gets to the heart muscle because of a damaged or blocked artery. Sometimes exercising or stress will cause angina (pain). Stopping the exercising or the stress may relieve the pain, drugs may be prescribed, or surgery may be necessary to replace the damaged artery.

Angina does not always lead to a heart attack. Sometimes collateral circulation develops, more blood reaches the heart muscle, and the pain decreases. Angina may even disappear altogether if the heart muscle is receiving enough blood.

CHAPTER SUMMARY

Human blood flows in a *closed system,* remaining essentially within the vessels that carry it. The heart is a *double pump,* pumping oxygen-poor blood to the lungs—the *pulmonary circulation*—and oxygen-rich blood to the rest of the body—the *systemic circulation.*

Anatomy of the Heart (p. 614)

1 The cone-shaped heart is about the size of its owner's fist. It is located in the center of the thorax. It is oriented obliquely, with about two-thirds of its bulk to the left of the body's midline.

2 The heart lies within a protective sac called the *pericardium (pericardial sac).* It is composed of an outer *fibrous pericardium* and an inner *serous pericardium.* Pericardial fluid between the sac and the heart helps minimize friction when the heart moves. The serous pericardium is divided into an outer *parietal layer* and an inner *visceral layer,* separated by the *pericardial cavity.* The visceral layer forms part of the heart wall.

3 The wall of the heart is composed of the *epicardium* or outer layer (the visceral pericardium), the *myocardium* or middle muscular layer, and the *endocardium* or inner layer.

4 Cardiac muscle cells function as a single unit in response to physiological stimulation because they are connected by *intercalated disks.* The interconnected cells form a latticework called a *syncytium.* Action potentials spread over a syncytium, causing all the cardiac muscle cells to contract in unison.

5 The *cardiac (fibrous) skeleton* is a structure of tough connective tissue inside the heart. It provides attachment sites and support for the valves and muscular fibers that allow the heart to wring blood out of the ventricles.

6 The heart is made up of two separate, parallel pumps, often called the *right heart* and *left heart.* Each of the two pumps has a receiving chamber on top called an *atrium* and a discharge pumping chamber below called a *ventricle.* Separating the left and right hearts is a thick wall of muscle called the *septum.*

7 Visible features on the surface of the heart include some sulci and coronary veins and arteries that carry blood to and from the heart.

8 The two *atrioventricular (AV) valves* permit blood to flow from the atria to the ventricles, and the two *semilunar valves* permit the flow from the ventricles to the pulmonary artery and aorta. The atrioventricular valves are the *tricuspid* valve of the right heart and the *bicuspid,* or *mitral,* valve of the left heart. The right semilunar valve is the *pulmonary semilunar valve,* and the left semilunar valve is the *aortic semilunar valve.*

9 The great vessels of the heart are the *superior vena cava, inferior vena cava, pulmonary artery, pulmonary veins,* and *aorta.*

10 Circulation to and from the tissues of the heart is the *coronary circulation.* Blood is supplied to the heart by the right and left *coronary arteries.* Most of the cardiac veins drain into the *coronary sinus.*

11 Blood proceeds through the heart as follows: (1) Oxygen-poor blood from the body enters the right atrium, and oxygen-rich blood from the lungs enters the left atrium. (2) Blood from the atria is forced into the ventricles. (3) Contraction of the right ventricle pumps oxygen-poor blood to the lungs, and the left ventricle pumps oxygen-rich blood (which just entered the heart from the lungs) through the aorta to the body.

Physiology of the Heart (p. 627)

1 The *cardiac action potential* can be divided into five phases: *depolarization, early repolarization, plateau, repolarization,* and *resting potential.*

2 The conduction system of the heart consists of the *sinoatrial (SA) node* (*pacemaker*) in the right atrium, the *atrioventricular (AV) node* between the atrium and ventricle, a tract of conducting fibers called the *atrioventricular bundle* that divides into a branch for each ventricle, and modified nerve fibrils called *cardiac conducting myofibers* in the walls of the ventricles.

3 An *electrocardiogram,* or *ECG,* is a recording of the electrical activity of the heart.

4 The *cardiac cycle* is the carefully regulated sequence of steps that constitutes a heartbeat. A complete cardiac cycle begins with a ventricular relaxation, or *diastole,* and ends with an atrial contraction, or *systole.*

5 The mechanical events of the cardiac cycle include: *ventricular filling, isovolumetric ventricular contraction, ventricular emptying,* and *isovolumetric ventricular relaxation.*

6 *Heart sounds* are caused by the closing of the heart valves and vibrations in the heart wall.

7 *Cardiac output* is the quantity of blood pumped by either ventricle in 1 min. The amount of blood expelled with each ventricular contraction is the *stroke volume. Cardiac reserve* is the difference between the actual volume of blood pumped and the volume the heart is capable of pumping under stressful conditions. *Cardiac index* is a measurement of cardiac output in relation to body surface area.

8 The central mechanism regulating the heartbeat, rate, and volume is the *cardioregulatory center* in the medulla oblongata. The autonomic nervous system increases or decreases the heartrate but does not initiate it. Endocrine control is involved also.

9 Nervous control operates through a negative feedback system involving *baroreceptors* and *chemoreceptors* in the carotid sinuses and aorta.

10 The *Frank-Starling law* of the heart states that the heart will pump all the blood it receives, within physiological limits.

The Effects of Aging on the Heart (p. 644)

The pumping efficiency of the heart is reduced with age, and congestive heart failure may occur.

Developmental Anatomy of the Heart (p. 644)

The embryonic circulatory system forms and becomes functional before any other system, fulfilling the nutritional needs of the embryo by the end of the third week after fertilization.

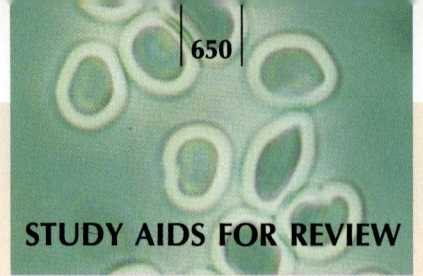
STUDY AIDS FOR REVIEW

KEY TERMS

aorta 621
atrioventricular (AV) bundle 631
atrioventricular (AV) node 631
atrioventricular (AV) valves 621
atrium 619
bicuspid valve 621
cardiac conducting myofibers 631
cardiac cycle 633
cardiac output 637
cardiac skeleton 617
cardioregulatory center 641
coronary circulation 621
coronary sinus 624
diastole 633
electrocardiogram 631
endocardium 617
epicardium 616
Frank-Starling law 640
inferior vena cava 621
mitral valve 621
myocardium 616
P wave 633
pacemaker cells 629
pericardium 614
pulmonary arteries 621
pulmonary circulation 614
pulmonary veins 621
QRS complex 633, 635
semilunar valves 621
septum 619
sinoatrial (SA) node 629
stroke volume 635, 640
superior vena cava 621
systemic circulation 614
systole 633
T wave 635
tricuspid valve 621
ventricle 619

MEDICAL TERMS FOR REFERENCE

ANEURYSM (AN-yoo-rihz-uhm) A bulging of a portion of the heart, aorta, or other artery.
ASYSTOLE Failure of the myocardium to contract.
AUSCULTATION Listening for heart sounds.
BRADYCARDIA A slower-than-normal heartbeat.
CARDIAC ARREST Cessation of normal, effective heart action, usually caused by asystole or ventricular fibrillation.

CARDIAC MASSAGE Manual stimulation of the heart when asystole occurs.
CARDIOMEGALY Enlargement of the heart.
CARDIOTONIC DRUG Drug that strengthens the heart.
CHRONOTROPIC DRUG Drug that changes the timing of the heart rhythm. A positive chronotropic drug increases the heart rate, and a negative drug decreases the heart rate.
COMMISSUROTOMY An operation to widen a heart valve that has been thickened by scar tissue.
CONSTRICTIVE PERICARDITIS A condition in which the heart muscle cannot expand and contract properly because the pericardium has shrunken and thickened.
CORONARY OCCLUSION Blockage in the circulation of the heart.
COR PULMONALE (KORE PULL-mo-nay-lee) Heart disease resulting from disease of the lungs in which the right ventricle hypertrophies and there is pulmonary hypertension.
DEFIBRILLATOR Instrument that corrects abnormal cardiac rhythms by applying an electric shock to the heart.
EMBOLISM (Gr. emballein, to throw in) Obstruction or occlusion of a blood vessel by an air bubble, blood clot, mass of bacteria, or other foreign material.
PALPITATION Skipping, pounding, or racing heartbeats.
SINUS RHYTHM Normal cardiac rhythm regulated by the SA node.
STETHOSCOPE Instrument used to listen to sounds in the chest.
STOKES-ADAMS SYNDROME Sudden seizures of unconsciousness which may accompany heart attacks.
TACHYCARDIA A faster-than-normal heartbeat.
THROMBUS (Gr. a clotting) A blood clot obstructing a blood vessel or heart cavity; the condition is thrombosis.

UNDERSTANDING THE FACTS

1 Which chambers of the heart are primarily involved with the pulmonary circulation? With systemic circulation?
2 What is the specific location of
  a myocardium   d mitral valve
  b chordae tendineae e SA node
  c fossa ovalis   f fibrous pericardium
3 Describe the size, shape, position, and function of the heart.
4 Describe the structure and function of the cardiac skeleton.
5 Where is the location of pericardial fluid, and what is its function?

6 Explain why the cells in a particular region of the myocardium receive nourishment from the coronary circulation only when that region is relaxed.

7 Name the great vessels of the heart.

8 List, in correct operating order, the structures that compose the conduction system of the heart.

9 What are the actual causes of the heart sounds?

10 In the electrocardiogram, what is indicated by the P wave? By the QRS complex? By the T wave?

11 List, in order, the mechanical events of the cardiac cycle.

12 Define cardiac output and explain how it is calculated.

13 Define cardiac reserve and cardiac index.

14 What factors regulate stroke volume?

15 Define myocardial infarction.

16 What is a congenital disease? What is the most common congenital disease of the heart?

17 Explain the difference between rheumatic fever and rheumatic heart disease.

18 Define shock. Distinguish between hypovolemic and cardiogenic shock.

UNDERSTANDING THE CONCEPTS

1 Trace the pathway of blood through the heart, starting with the right atrium. Be sure to include all the valves in the pathway.

2 Distinguish the epicardium from the fibrous pericardium and the serous pericardium.

3 Correlate the thickness of the wall of each chamber of the heart with its function.

4 Where are the papillary muscles, and what is their function?

5 What factors are important in regulating blood flow to the tissues of the heart?

6 Explain how the action potential generated in a cardiac muscle cell is coupled to contraction of that cell.

7 Explain how the heart can contract as a unit when each myocardial cell is not directly innervated by the nervous system.

8 Contrast the resting potential of a typical ventricular cell with that of the SA node.

9 Correlate atrial and ventricular contraction with the electrical conducting system of the heart.

10 What role does the endocrine system play in the regulation of the heart?

11 What role does the nervous system play in generating and/or regulating the rate at which the heart beats?

SELF-QUIZ

Multiple-Choice Questions

20.1 The part of the serous pericardium that lines the outer portion of the pericardial sac is called the
a visceral pericardium   d endocardium
b parietal pericardium   e myocardium
c epicardium

20.2 The outermost layer of the heart is called the
a myocardium   d myocardial cortex
b endocardium   e visceral pericardium
c epicardium

20.3 The oval depression on the septal wall of the right atrium is the
a musculi pectinati   d fossa ovalis
b trabeculae carneae   e infundibulum
c conus

20.4 Which of the following valves allows deoxygenated blood to enter the pulmonary artery?
a atrioventricular valve   d bicuspid valve
b aortic semilunar valve   e tricuspid valve
c pulmonary semilunar valve

20.5 Of the following which is *not* one of the great vessels of the heart?
a pulmonary arteries   d superior vena cava
b coronary arteries   e inferior vena cava
c pulmonary veins

20.6 The tiny veins that open directly into each of the four heart chambers are the
a great cardiac veins   d oblique veins
b small cardiac veins   e anterior cardiac veins
c thebesian veins

20.7 Which of the following is *not* part of the heart's conducting system?
a SA node
b AV node
c atrioventricular bundle
d cardiac conducting myofibers
e vagus nerve endings

20.8 Which heart sound usually requires amplification in order to be heard?
a first   d fourth
b second   e fifth
c third

20.9 The cardioregulatory center is located in the
a cerebrum   d cerebellum
b hypothalamus   e pons
c medulla oblongata

20.10 The skeleton of the heart consists of
a bone within the myocardium
b bone within the interatrial septum
c fibrous connective tissue in the pulmonary trunk
d fibrous connective tissue encircling the atrioventricular orifices
e mostly cartilage

20.11 Which endocrine gland produces a hormone that affects the heart rate?
a thymus gland   d pancreas
b pineal gland   e adrenal gland
c thyroid gland

Matching

  a chordae tendineae
  b trabeculae carneae
  c coronary sulcus
  d musculi pectinati
  e fibrous trigone

20.12 _____ part of the cardiac skeleton

20.13 _____ a depressed ring that encircles the heart between the atria and the ventricles

20.14 _____ muscular bundles in the wall of the atria

20.15 _____ muscular ridges in the ventricles

20.16 _____ structures that prevent backflow through the atrioventricular valves

  a bicuspid valve
  b tricuspid valve
  c pulmonary semilunar valve
  d aortic semilunar valve

20.17 _____ located between the left atrium and left ventricle

20.18 _____ blood passes through this valve upon leaving the right ventricle

20.19 _____ located between the right atrium and right ventricle

20.20 _____ located close to the origin of the coronary arteries

  a atrial systole
  b ventricular systole
  c atrial diastole
  d ventricular diastole

20.21 _____ ventricles fill completely

20.22 _____ blood forced into ventricles

20.23 _____ blood forced into systemic and pulmonary arteries

20.24 _____ blood from vena cavae enters heart

Completion Exercises

20.25 A continuously depolarizing, unstable resting potential is characteristic of the _____.

20.26 During an action potential, _____ ions diffuse into cardiac muscle cells where they subsequently combine with troponin.

20.27 The only electrical link between the atria and the ventricles is the _____.

20.28 If the SA node is damaged or ineffective, the pacemaking function may be taken over by the _____, which causes the heart to beat 50 to 60 times per minute.

20.29 Cardiac output is calculated by multiplying the _____ by the _____.

20.30 Sympathetic and parasympathetic innervation to the heart is primarily restricted to the _____.

A SECOND LOOK

1 In the following drawing, label the endocardium myocardium, epicardium, and pericardial cavity.

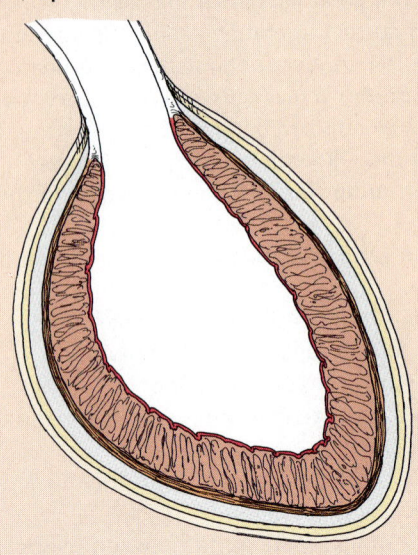

2 In the following drawing, label the ascending and descending aorta, aortic arch, and inferior vena cava.

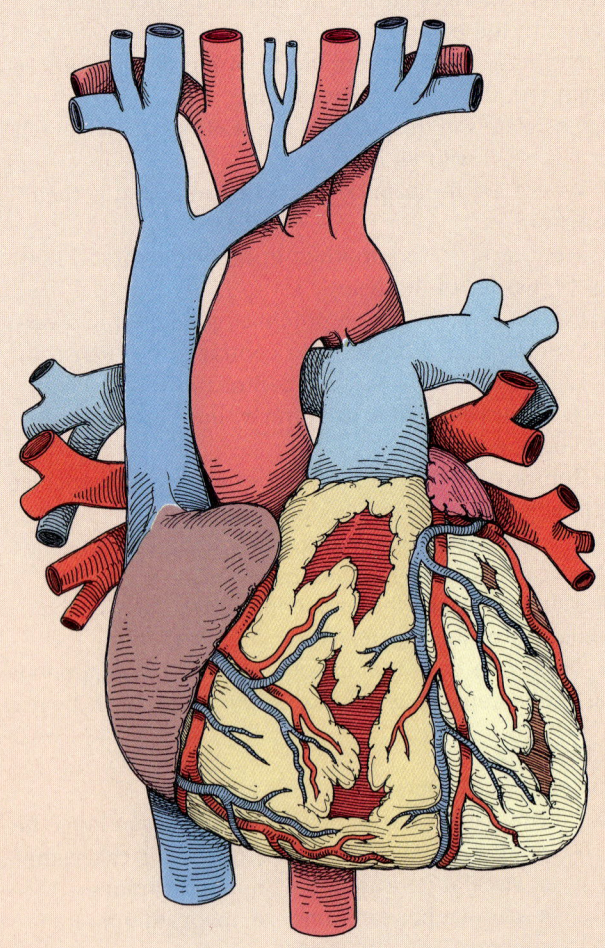

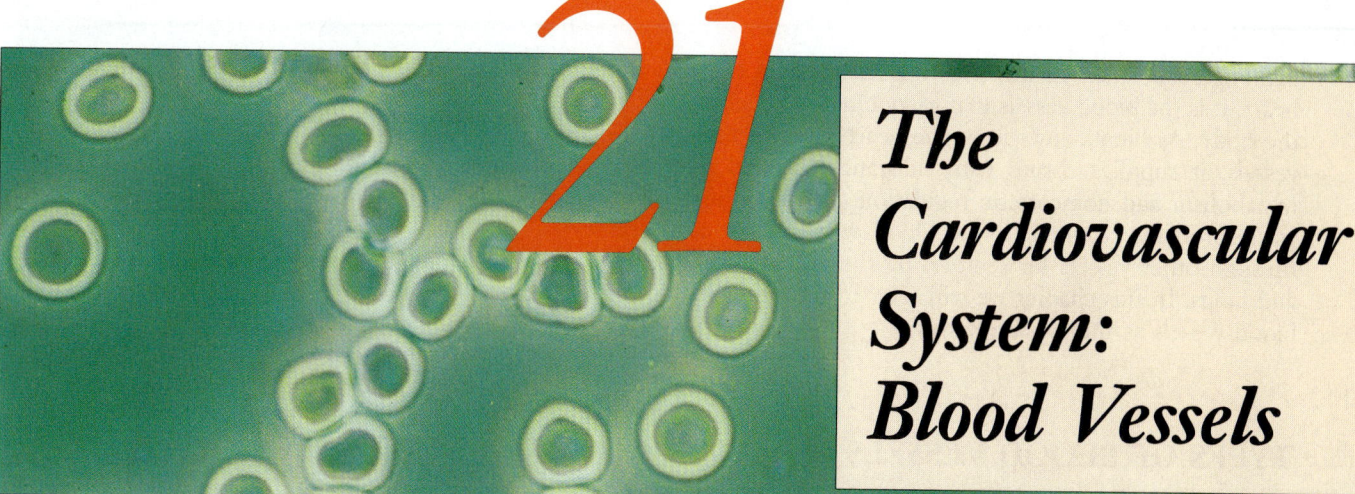

21

The Cardiovascular System: Blood Vessels

KEY CONCEPTS

1 Arteries carry blood away from the heart, and veins carry blood toward the heart.

2 Capillaries connect the arterial and venous systems. Their thin walls allow the exchange of gases, wastes, nutrients, and other substances between body tissues and the blood.

3 The pulmonary circulation carries deoxygenated blood to the lungs, where carbon dioxide is removed and oxygen is added. The oxygenated blood is returned to the heart for dispersal throughout the body through the systemic circulation.

4 Optimum blood pressure and flow are maintained by regulatory systems within the circulatory, endocrine, and nervous systems.

5 Major adjustments occur in the circulation of a newborn baby to enable it to breathe as soon as it is born.

Although the center of the cardiovascular system is the heart, it is the blood vessels that carry blood throughout the body. As blood flows from the heart through these vessels, it supplies tissues with substances needed for metabolism and homeostatic regulation. When cellular waste products accumulate, the blood carries them off through other vessels for removal by the kidneys, skin, and lungs. In this chapter we will refer to the system of blood vessels as the *circulatory system*.

TYPES OF BLOOD VESSELS

Arteries are blood vessels that carry blood *away from* the heart to the organs and tissues of the body. Because different arteries contain varying amounts of elastic fibers and smooth muscle tissue, some are called *elastic* (conducting) *arteries*, and others are called *muscular* (distributing) *arteries*. The walls of elastic arteries expand slightly with each heartbeat. The elastic arteries include *large* arteries such as the aorta, brachiocephalic, common carotid, vertebral, and common iliac arteries, which conduct blood from the heart to medium-sized muscular arteries. The muscular arteries distribute blood to various parts of the body. They include the axillary, brachial, radial, splenic, femoral, popliteal, and tibial arteries. The muscular arteries branch into smaller *arterioles*. These vessels play important roles in determining the amount of blood going to any organ or tissue and in regulating blood pressure. The muscular arteries and arterioles can be either dilated or constricted by autonomic nervous control. Arterioles branch into smaller vessels called terminal arterioles, or *metarterioles*, which carry blood into the smallest vessels in the body, the *capillaries* [FIGURE 21.1].

Capillaries are microscopic vessels with walls mostly one cell thick. The thin capillary wall is porous and allows the passage of water and small particles of dissolved materials. Capillaries are distributed throughout the body, except in the dead outer layers of skin and in special places such as the lenses of the eyes. Capillaries converge into larger vessels called *venules*, which merge to form even larger vessels called *veins*.

Veins carry blood *toward* the heart. They are generally more flexible than arteries, and they collapse if blood pressure is not maintained.

Arteries

Most *arteries** are efferent vessels that carry blood *away from* the heart to the capillary beds throughout the body.

---

*The word *artery* comes from the Greek word for "windpipe." When the ancient Greeks dissected corpses, they found blood in the veins but none in the arteries. As a result, they thought the arteries carried air.

FIGURE 21.1 TYPES OF BLOOD VESSELS

This simplified drawing shows the basic arrangement of blood vessels in relation to the heart.

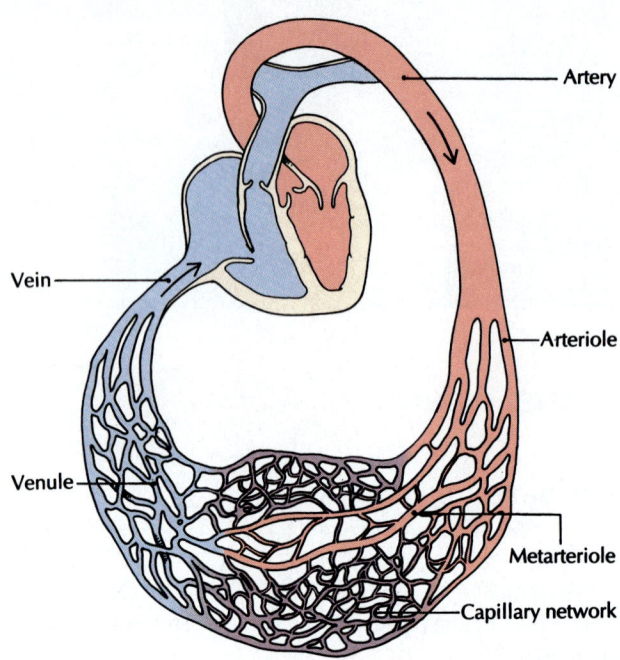

In the adult, all arteries except the pulmonary trunk and pulmonary arteries carry oxygenated blood. The left and right pulmonary arteries carry *deoxygenated* blood from the heart to the lungs. Thus the most reliable way to classify blood vessels is by the *direction* in which they carry blood, either toward the heart or away from it.

The great arteries that emerge from the heart are often called *trunks*. The major arterial trunks are the **aorta** from the left ventricle and the **pulmonary trunk** from the right ventricle.

The central canal of all blood vessels, including arteries, is called the *lumen*. Surrounding the lumen of a large elastic artery is a thick wall composed of three layers, or *tunicae* (TEW-nih-see; pl. of L. *tunica*, covering) [FIGURE 21.2].

The **tunica intima** ("innermost covering") has an inner lining of endothelial cells (simple squamous epithelium) attached to a basement membrane, a thin subendothelial layer of fine connective tissue (collagenous fibers), and an external elastic layer called the internal elastic lamina.

The **tunica media** ("middle covering") is the thickest layer of arterial wall in the large arteries. It is composed mainly of collagenous fibers, connective tissue, smooth muscle cells, and elastic fibers. The walls of the largest arteries (elastic arteries) have more elastic tissue and less smooth muscle. In the smaller arteries (muscular arteries), the elastic fibers in the tunica media contain more smooth muscle cells and less elastic tissue.

FIGURE 21.2 TYPICAL ELASTIC ARTERY

Cutaway drawing showing three layers of an elastic artery. Not shown are the vasa vasorum (blood vessels that supply nutrients to the large arteries and veins).

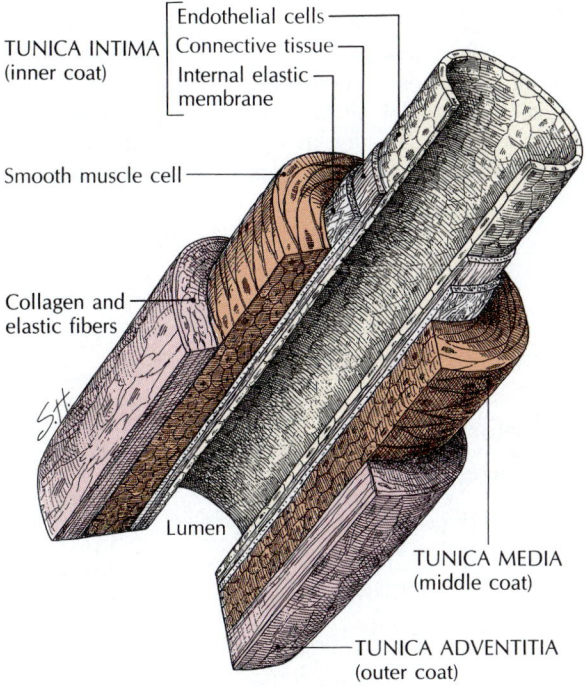

TUNICA INTIMA (inner coat)
— Endothelial cells
— Connective tissue
— Internal elastic membrane

Smooth muscle cell

Collagen and elastic fibers

Lumen

TUNICA MEDIA (middle coat)

TUNICA ADVENTITIA (outer coat)

The ***tunica adventitia*** (ad-vehn-TISH-uh; "outermost covering") is composed mainly of collagen fibers and elastic fibers. Nerves and lymphatic vessels are also found within this layer. The walls of the large arteries (thicker than 20 mm) are nourished by small blood vessels called the ***vasa vasorum*** ("vessels of the vessels"), which form capillary networks within the tunica adventitia and the outer part of the tunica media.

The strong elastic walls of the largest elastic (conducting) arteries allow these vessels to adjust to the great pressure created by the contraction of the ventricles during systole. The arteries stretch when blood is ejected, storing energy, and recoil when the heart relaxes, converting stored (potential) energy into kinetic energy of flow. (Blood is ejected into the aorta at a speed of about 30 to 40 cm/sec—almost a mile an hour.)

Arterioles

The arteries nearest the heart are the largest. As their distance from the heart increases, the arteries branch into smaller and smaller arteries and then into ***arterioles*** shortly before reaching the capillary networks. Arterioles are covered by the three tunicae [FIGURE 21.2]. Because their walls contain a large number of smooth muscle cells, arterioles can dilate or constrict more easily than arteries can, thus controlling the flow of blood from ar-

teries into capillaries and later into the organs. If necessary, an arteriole can dilate to increase blood flow to capillaries by as much as 400 percent.

Terminal arterioles (those closest to a capillary) are muscular and well supplied with nerves, but they do not have an internal elastic layer or their own blood vessels (vasa vasorum), as larger arteries do. The tunica media is particularly well supplied with sympathetic nerves, which cause the smooth muscle cells to contract and the lumen to constrict.

Capillaries

Terminal arterioles branch out to form ***capillaries*** (L. *capillus*, hair), which connect the arterial and venous systems [FIGURE 21.3]. Capillaries are generally com-

FIGURE 21.3 CAPILLARY

Electron micrograph of a cross section of a capillary. The capillary wall is composed of two endothelial cells, and the large nucleus of one of the cells is prominent. Note the clefts at the junctions between the endothelial cells. ×41,000.

Capillary wall   Capillary lumen   Nucleus of endothelial cell

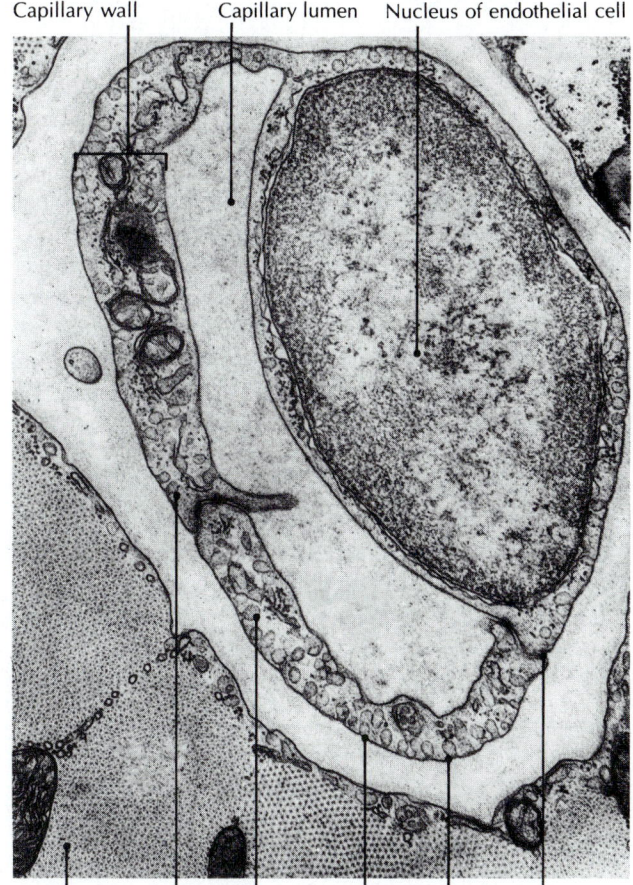

Muscle cell   Endothelial cells   Pinocytic vesicles   Cleft

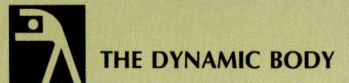

Tumors and the Growth of Blood Vessels

Malignant tumors require a generous blood supply. They manage to acquire that supply by encouraging the sprouting and extension of new capillary networks, a process called *angiogenesis*. The new capillaries connect the tumor to nearby veins and arteries, keeping it well nourished as it grows. The cellular mechanism for angiogenesis was unknown until 1985, when Bert L. Vallee and his coworkers at Harvard Medical School announced the isolation of a protein that stimulates angiogenesis. The researchers, together with scientists at the University of Washington in Seattle, also succeeded in cloning the protein.

The growth-stimulating protein has been named *angiogenin*. It is about a thousand times more potent than normal growth factors. If future research is successful, scientists will learn what turns angiogenin on and off, and the protein may be used to stimulate the growth of new blood vessels after a heart attack, burn, or other serious tissue damage. Also, tumors may be controlled by inhibiting the action of angiogenin. The detection of angiogenin in the blood or urine may also provide an early warning of cancer.

TABLE 21.1 APPROXIMATE AVERAGE PHYSICAL CHARACTERISTICS OF BLOOD VESSELS

Type of vessel	Diameter, mm	Wall thickness, mm	Length, cm	Internal pressure, mmHg	Total cross-sectional area, sq cm	Percentage of total body blood volume
Aorta	25.000	2.000	40.000	100	2.5	6
Medium-sized arteries	4.000	0.800	15.000	90	20.0	13
Arterioles	0.300	0.020	0.200	60	40.0	2
Capillaries	0.008	0.001	0.075	30	2500.0	5
Venules	0.020	0.002	0.200	20	250.0	5
Medium-sized veins	5.000	0.500	15.000	15	80.0	20
Large veins	15.000	0.800	20.000	10	20.0	39
Venae cavae	30.000	1.500	40.000	10	8.0	10

posed of only a single layer (tunica intima) of endothelial cells on a thin basement membrane of glycoprotein. Capillaries are the smallest and most numerous blood vessels in the body. If all the capillaries in an adult body were connected, they would stretch about 96,000 km (60,000 mi). This abundance of capillaries makes an enormous surface area available for the exchange of gases, fluids, nutrients, and wastes between the blood and nearby cells.

Capillaries are about 600 times narrower than a medium-sized vein, and about 500 times narrower than a medium-sized artery [TABLE 21.1]. The diameter of capillaries varies with the function of the tissue.

Types of capillaries At least three different types of capillaries are recognized, and each one performs a specific function: (1) **Continuous capillaries** are found in skeletal muscle tissue. Their walls are made up of one continuous endothelial cell, with the ends overlapping in a tight endothelial cell junction [FIGURE 21.4A]. Apparently, intracellular (pinocytic) vesicles help move fluids across the membrane of the cell by exocytosis and endocytosis. (2) The walls of **fenestrated capillaries** (L. *fenestra*, window) consist of two or more adjacent endothelial cells connected by thin endothelial membranes called *fenestrations* or *pores* [FIGURE 21.4B]. Fenestrated capillaries are usually found in the kidneys, endocrine glands, and intestines. (3) **Discontinuous capillaries,** also known as **sinusoids,** have fenestrations and a much wider lumen than the other types [FIGURE 21.4C]. Such an open, irregular structure is highly permeable. Sinusoid capillary systems are found in the liver, spleen, anterior pituitary, bone marrow, and parathyroid glands. Liver sinusoids also contain active phagocytes called *stellate reticuloendothelial cells* (*Kupffer cells;* KOOP-fur), which are part of the reticuloendothelial system.

Q: *Since blood is already within large arteries and veins, why do these vessels need their own blood supply?*

A: The walls of large arteries and veins are so thick that nutrients cannot diffuse far enough to reach all their cells.

FIGURE 21.4 TYPES OF CAPILLARIES

[A] Continuous capillary in muscle, showing the overlapping ends of the single endothelial cell. [B] Fenestrated capillary of an endocrine gland, showing two endothelial cells connected by thin stretches of membrane called fenestrations. [C] Discontinuous capillaries (sinusoids) in the liver, showing open, irregular structure and a phagocytic stellate reticuloendothelial (Kupffer) cell.

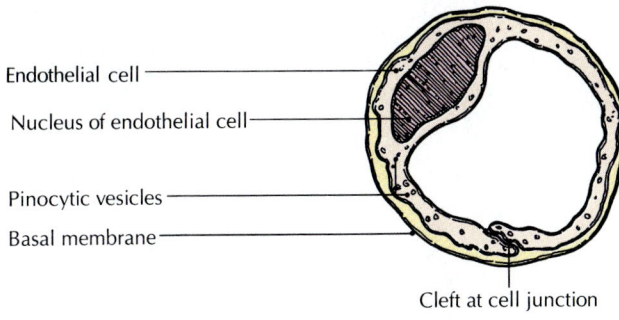

Endothelial cell

Nucleus of endothelial cell

Pinocytic vesicles

Basal membrane

Cleft at cell junction

[A]

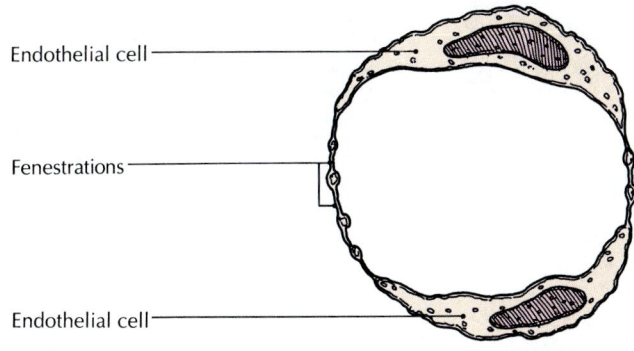

Endothelial cell

Fenestrations

Endothelial cell

[B]

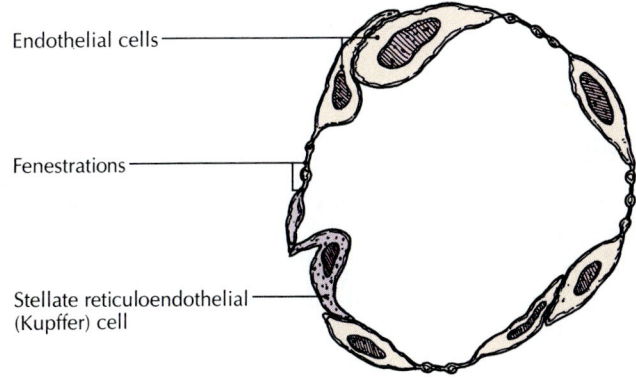

Endothelial cells

Fenestrations

Stellate reticuloendothelial (Kupffer) cell

[C]

Capillary blood flow Blood leaves the heart traveling about 30 to 40 cm/sec (12 in./sec), but it is slowed to only 2.5 cm/sec (1 in./sec) by the time it reaches the arterioles and to less than 1 mm/sec (0.04 in./sec) in the capillaries. Blood remains in the capillaries for only a second or two, but given the short length of capillaries (about 1 mm), that is long enough for the crucial exchanges of nutrients and wastes.

Because capillary walls are usually only one cell thick, certain materials pass through rather easily. Small molecules, including gases such as oxygen and carbon dioxide, certain waste products, salts, glucose, and amino acids, pass through freely, but the large molecules of plasma proteins pass through only with difficulty, if at all. Red blood cells cannot pass through the capillary walls.

Microcirculation of the blood The capillaries and their associated structures (including terminal arterioles, metarterioles, and venules) constitute the *microcirculation* of the blood [FIGURE 21.5]. This name reflects the extremely narrow diameters of the vessels [TABLE 21.1].

A *metarteriole* is a vessel that emerges from an arteriole, traverses the capillary network, and empties into a venule. At the junction of the metarteriole and capillary is the *precapillary sphincter*, a ringlike smooth muscle cell that regulates the flow of blood into the capillaries. The smooth muscle cells of the precapillary sphincters are not innervated but constrict or dilate in response to local changes in oxygen and carbon dioxide concentrations, pH, temperature, and circulating chemical agents. Such a response, which does not depend on hormonal or nervous stimulation, is called *autoregulation*.

Some metarterioles connect directly to venules by way of *thoroughfare channels*. Similar to thoroughfare channels are *collateral channels* (*capillary shunts* or *arteriovenous anastomoses*), which bypass the capillary beds and act as direct links between arterioles and venules. Collateral channels are numerous in the fingers, palms, and earlobes, where they control heat loss. Because these collateral channels have thick walls, they are not exchange vessels in the way that capillaries are. They are heavily innervated and are muscular at the arteriole end and elastic at the venule end.

Q: *Why do we sometimes have dark circles under our eyes?*

A: In the skin below your eyes are hundreds of tiny blood vessels, which help drain blood from the head. When you are ill or tired, these vessels dilate, concentrating blood in the vessels and causing them to swell. Because the skin under the eyes is very thin, the engorged vessels become visible. The dark circles are actually pools of blood.

FIGURE 21.5 MICROCIRCULATION OF THE BLOOD

Metarterioles provide the path of least resistance between the arterioles and venules. Note the absence of smooth muscle in the true capillaries. Arrows indicate direction of blood flow.

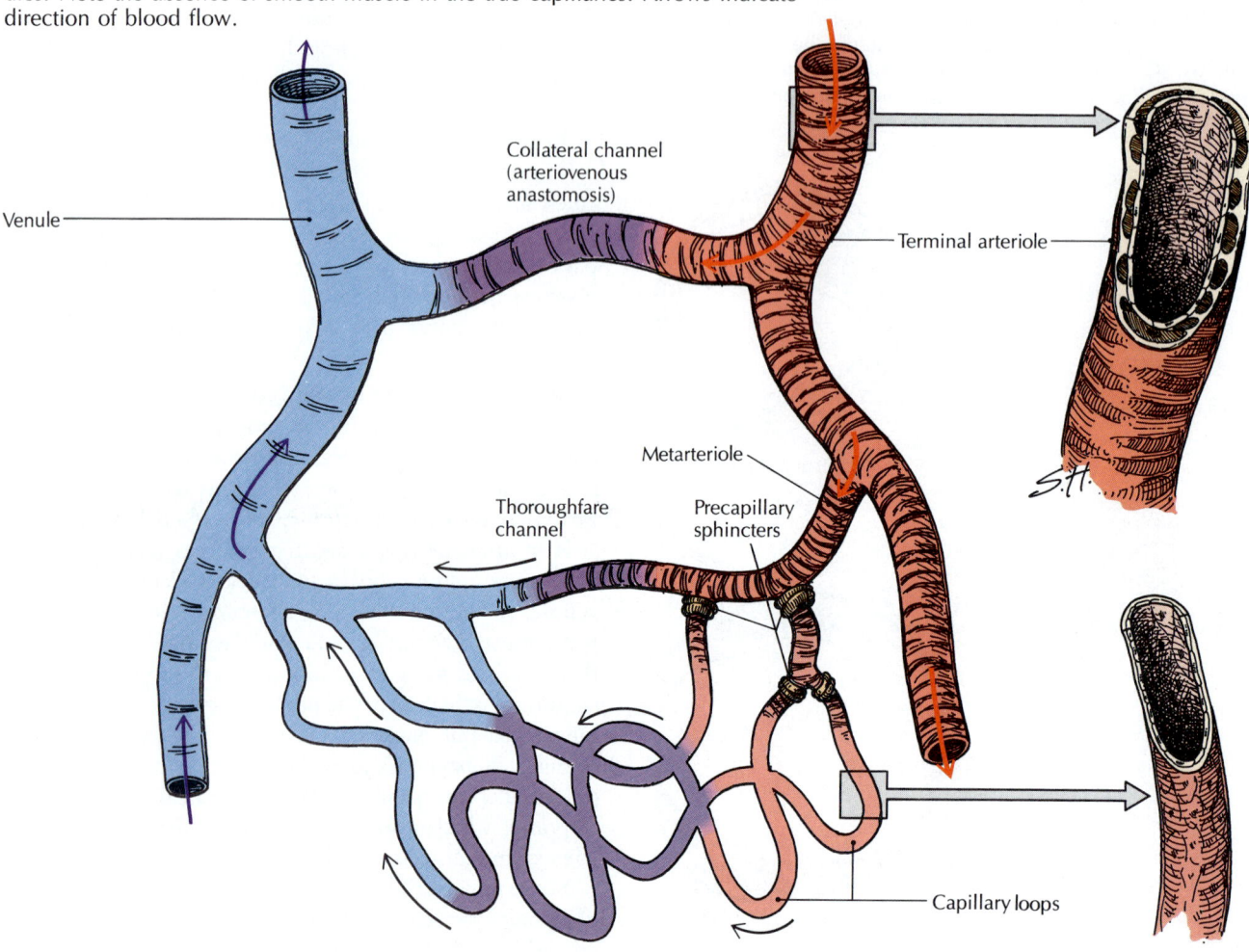

Collateral channel (arteriovenous anastomosis)

Venule

Terminal arteriole

Metarteriole

Thoroughfare channel

Precapillary sphincters

Capillary loops

Venules

Blood drains from the capillaries into **venules** (VEHN-yoolz), tiny veins that unite to form larger venules and veins. The transition from capillaries to venules occurs gradually. The immediate postcapillary venules consist mainly of endothelium and a thin tunica adventitia. Postcapillary venules play an important role in the exchange between blood and interstitial fluid. Unlike capillaries, these venules are easily affected by inflammation and allergic reactions. They respond to these conditions by opening their pores, allowing water, solutes, and white blood cells to move out into the extracellular space.

Veins

Venules join together to form **veins.** Superficial veins are found in areas where blood is collected near the surface of the body and are especially abundant in the limbs. Veins become larger and less branched as they move away from the capillaries and toward the heart.

The walls of veins contain the same three layers (tunicae) as arterial walls, but the tunica media is much thinner [FIGURE 21.6]. The venous walls contain less elastic tissue, collagenous tissue, and smooth muscle. As a result, veins are very distensible and compressible. Like arteries, veins are nourished by small vasa vasorum.

FIGURE 21.6 VEINS

[A] Cutaway drawing of a medium-sized vein, showing a valve and the three-layered wall. Compare the thickness of the layers with those of the artery shown in FIGURE 21.2. [B] Scanning electron micrograph of a medium-sized vein and artery, showing the larger lumen and thinner wall of the vein. Connective tissue surrounds the vessels. ×305. (Richard G. Kessel and Randy H. Kardon, *Tissues and Organs: A Text-Atlas of Scanning Electron Microscopy*, San Francisco, W. H. Freeman, 1979.)

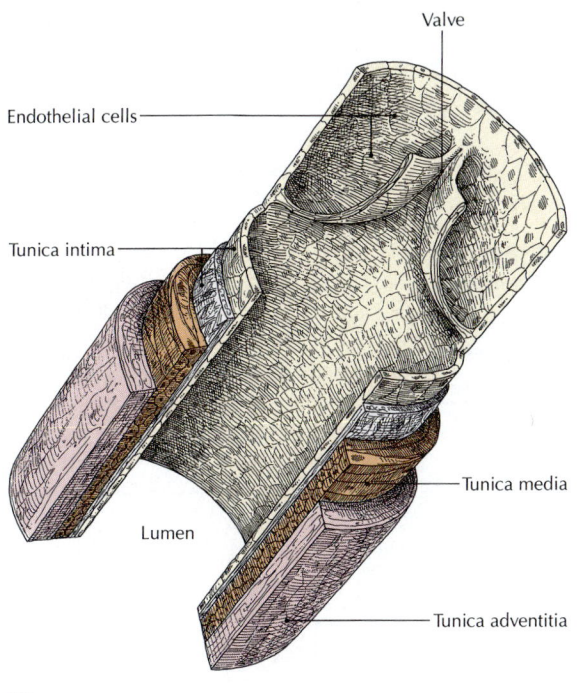

Valve
Endothelial cells
Tunica intima
Tunica media
Lumen
Tunica adventitia

[A]

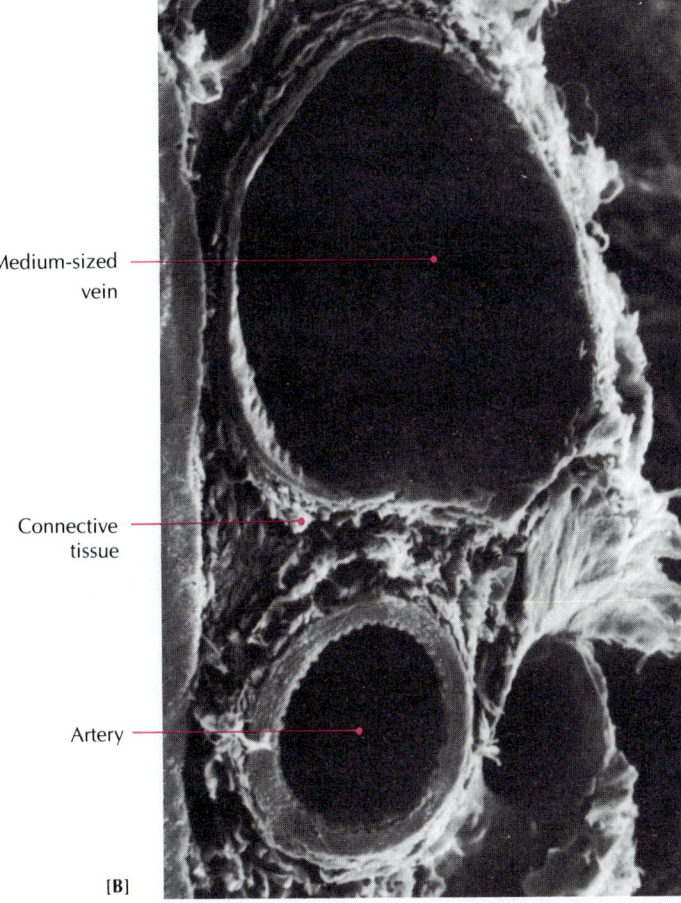

Medium-sized vein
Connective tissue
Artery

[B]

Veins usually contain paired semilunar bicuspid valves that permit blood to flow in only one direction, restricting any backflow [FIGURE 21.6A]. Blood pressure in the veins is low, and the venous blood is helped along by the skeletal muscle pump (the compression of the venous walls by the contraction of surrounding skeletal muscles). The venous valves, which are derived from folds of the tunica intima, are especially abundant in the legs, where gravity opposes the return of blood to the heart. There are no valves in veins narrower than 1 mm or in veins in the thoracic and abdominal cavities.

Blood Distribution

The total amount of blood in the circulatory system is known as the ***blood volume.*** It is about 5 L in the average adult. Almost 80 to 90 percent of the blood is in the systemic circulation at any given time, and the rest is in the pulmonary circulation. Of the blood in the systemic circulation, about 75 percent is in the veins, 20 percent in the arteries, and 5 percent in the capillaries.

ASK YOURSELF

1 What are the basic functions of arteries? Veins?

2 What are the three layers of arterial walls?

3 What are the functions of arterioles? Venules?

4 Why are capillaries sometimes called exchange vessels?

5 What are the three types of capillaries?

6 What is meant by the microcirculation of the blood?

7 How do venous valves function?

PULMONARY AND SYSTEMIC CIRCULATION OF THE BLOOD

Blood circulates throughout the body in two main circuits: the pulmonary and systemic circuits [FIGURE 21.7; see also FIGURE 20.1]. In the following sections we will describe these two circuits as well as some subdivisions or special areas of circulation. (Coronary circulation was described in Chapter 20.)

Pulmonary Circulation

The ***pulmonary circulation*** supplies blood only to the lungs. It carries deoxygenated blood from the heart to the lungs, where carbon dioxide is removed and oxygen is added. It then returns the newly oxygenated blood to the heart for distribution to the rest of the body [FIGURE 21.8]. Pulmonary circulation takes 4 to 8 sec.

The major blood vessels of the pulmonary circulation are the *pulmonary trunk* and two *pulmonary arteries*, which carry deoxygenated blood from the right ventricle of the heart to the lungs; *pulmonary capillaries*, which are the site of the exchange of oxygen and carbon dioxide in the lungs; and the four *pulmonary veins* (two from each lung), which carry oxygenated blood from the lungs to the left atrium of the heart. Each of the pulmonary veins enters the left atrium through a separate opening.

Systemic Circulation

The ***systemic circulation*** supplies all the cells, tissues, and organs of the body with oxygenated blood, and also returns deoxygenated blood to the heart [FIGURE 21.9]. The "systemic circuit" from the heart and back again takes about 25 to 30 sec. The systemic circulation is divided into the arterial and venous divisions. The main vessel of the ***arterial division*** is the *aorta*, which emerges from the left ventricle as the *ascending aorta*, curves backward over the top of the heart as the *aortic arch*, and continues down through the thorax and abdomen as the *descending aorta*. The ascending aorta and arch give off branches to the heart, head, neck, and upper extremities. The descending aorta gives off branches to internal organs in the thoracic and abdominal cavities and then terminates in the two *common iliac arteries*, which supply the lower extremities.

The ***venous division*** of the systemic circulation is linked to the arterial system by capillary beds. All the venous blood from the upper part of the body eventually drains into the large *superior vena cava*, and the venous blood from the lower extremities, pelvis, and abdomen drains into the *inferior vena cava*. Both venae cavae empty their deoxygenated blood into the right atrium of the heart. The *coronary sinus* drains blood from the walls of

FIGURE 21.7 CIRCULATORY SYSTEM

Diagrammatic representation of the circulatory system showing the systemic and pulmonary circulations. Oxygenated blood is indicated by pink arrows and boxes, and deoxygenated blood by blue arrows and boxes.

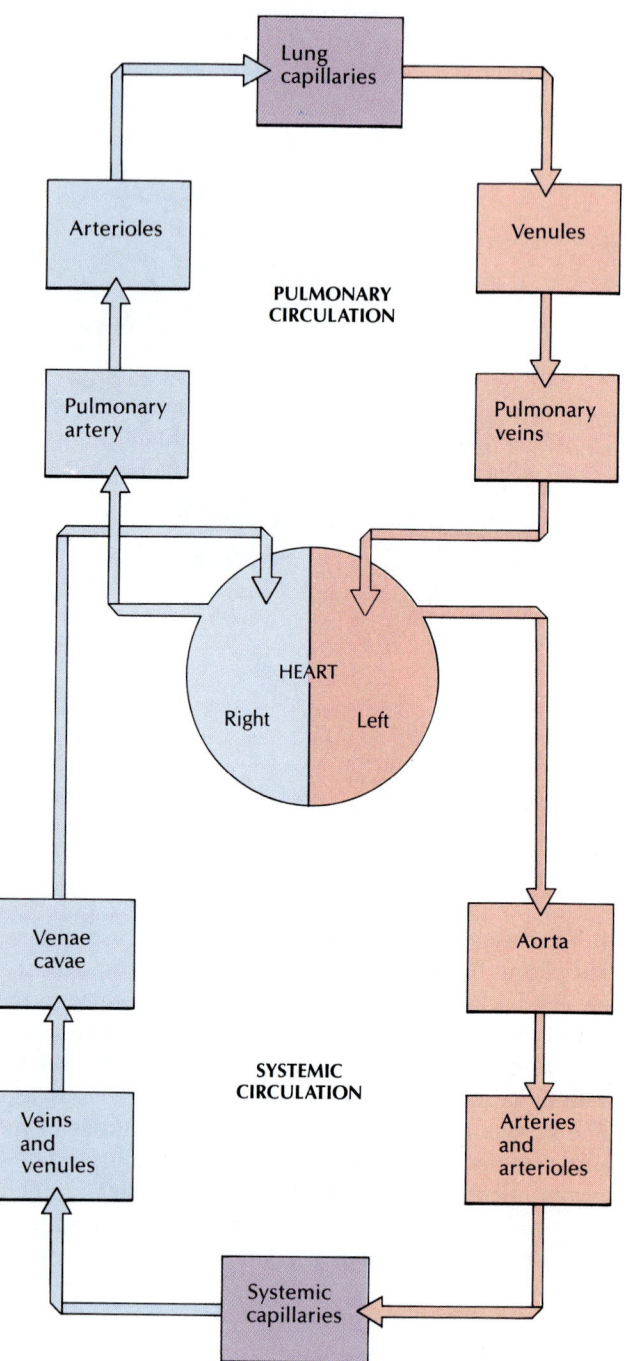

FIGURE 21.8 PULMONARY CIRCULATION

The freshest, most highly oxygenated blood in the body enters the left heart
from the lungs through the pulmonary *veins,* not the pulmonary arteries.
Arrows indicate direction of blood flow.

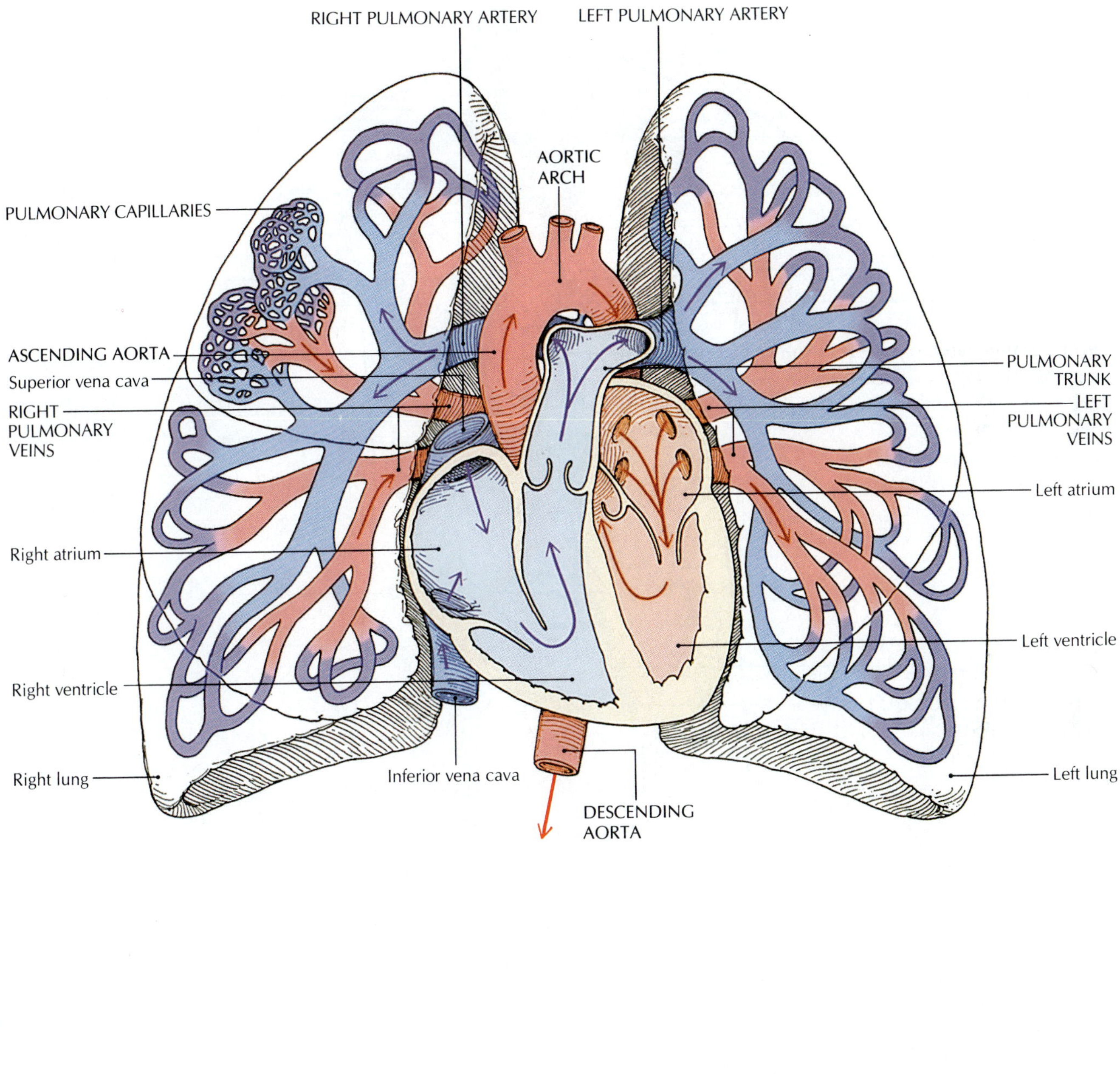

RIGHT PULMONARY ARTERY

LEFT PULMONARY ARTERY

AORTIC ARCH

PULMONARY CAPILLARIES

ASCENDING AORTA

Superior vena cava

RIGHT PULMONARY VEINS

Right atrium

Right ventricle

Right lung

Inferior vena cava

DESCENDING AORTA

PULMONARY TRUNK

LEFT PULMONARY VEINS

Left atrium

Left ventricle

Left lung

FIGURE 21.9 SYSTEMIC CIRCULATION

The drawing shows the major arteries and veins of the systemic circulation.
The pulmonary circulation (between the heart and lungs) is not shown.

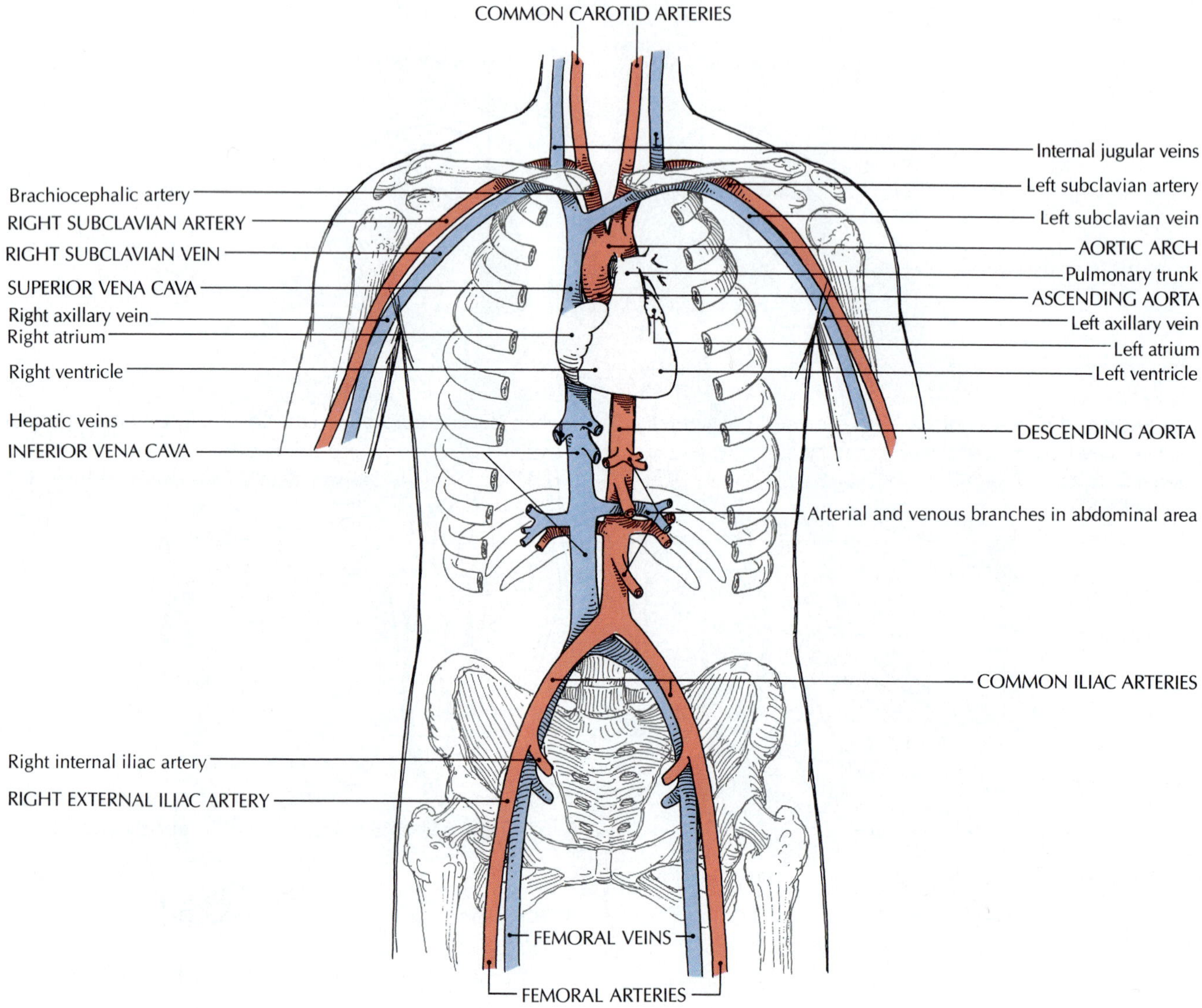

COMMON CAROTID ARTERIES

Internal jugular veins

Brachiocephalic artery

Left subclavian artery

RIGHT SUBCLAVIAN ARTERY

Left subclavian vein

RIGHT SUBCLAVIAN VEIN

AORTIC ARCH

Pulmonary trunk

SUPERIOR VENA CAVA

ASCENDING AORTA

Right axillary vein

Left axillary vein

Right atrium

Left atrium

Right ventricle

Left ventricle

Hepatic veins

INFERIOR VENA CAVA

DESCENDING AORTA

Arterial and venous branches in abdominal area

COMMON ILIAC ARTERIES

Right internal iliac artery

RIGHT EXTERNAL ILIAC ARTERY

FEMORAL VEINS

FEMORAL ARTERIES

the heart and is part of the venous division of the systemic circulation.

Subdivisions of the Systemic Circulation

Subdivisions of the systemic circulation include the portal, cerebral, cutaneous, and skeletal muscle circulations.

Portal circulations Most veins transport blood directly back to the heart from a capillary network, but in the case of a *portal system,* the blood passes through *two* sets of capillaries on its way to the venous system. The human body has two portal systems: (1) The **hepatic portal system** (Gr. *hepatikos,* liver) moves blood from the capillary beds of the intestines, spleen, and pancreas to the sinusoidal beds of the liver [FIGURE 21.10]. (2) The **hypothalamic-hypophyseal portal system** moves blood from the capillary bed of the hypothalamus directly, by way of veins, to the sinusoidal bed of the pituitary gland [see FIGURE 18.7].

Because the veins of the hepatic portal system lack valves, they are vulnerable to the excessive pressures of venous blood. This situation can produce *hemorrhoids* (dilated veins in the anal region) in the internal hemorrhoidal plexus [FIGURE 21.10].

Much of the venous blood returning to the heart from the capillaries of the spleen, stomach, pancreas, gallbladder, and intestines contains products of digestion. This nutrient-rich blood is carried by the *hepatic portal vein* to the *sinusoids* of the liver. After leaving the sinusoids of the liver, the blood is collected in the **hepatic veins** and drained into the **inferior vena cava,** which returns it to the right atrium of the heart.

Cerebral circulation The brain is supplied with blood by four major arteries: two **vertebral arteries** and two **internal carotid arteries** [see FIGURE 21.12A]. Blood from the vertebral arteries supplies the cerebellum, the brainstem, and the posterior (occipital) part of the cerebrum. The internal carotid arteries supply the rest of the cerebrum and both eyes.

The vertebral arteries merge on the ventral surface of the brainstem to form the *basilar artery,* which terminates by forming the left and right *posterior cerebral arteries.* The internal carotid arteries enter the cranial cavity and then branch into the *ophthalmic arteries,* which supply the eyes, and the *anterior* and *middle cerebral arteries,* which supply the medial and lateral parts of the cerebral hemispheres, respectively.

All the blood entering the cerebrum must first pass through the **cerebral arterial circle** (*circle of Willis*) [see FIGURE 21.12B]. This circular *anastomosis,* or shunt, at the base of the brain consists of the proximal portion of the posterior cerebral arteries, the posterior communicating

arteries, the internal carotid arteries, the anterior cerebral arteries, and the anterior communicating artery. The vascular connections among the major branches of the basilar and vertebral arteries tend to make adjustments between differences in the arterial pressures within these branches.

Blood flows from the brain capillaries into large *venous sinuses* located between the inner and outer dura mater [see FIGURE 21.16]. The sinuses then drain into the *internal jugular veins* on each side of the neck at the base of the skull. Blood returns to the heart by way of the *brachiocephalic veins.* The junction of the left and right brachiocephalic veins forms the *superior vena cava,* which conveys blood to the right atrium of the heart.

Cutaneous circulation The arrangement of blood vessels in the skin allows for the increase or decrease of heat radiation from the skin. When the body temperature increases, more blood flows to the superficial layers, from which heat radiates from the body. In contrast, when the body needs to conserve heat, blood is shunted away from the surface of the skin through deep arteriovenous anastomoses (vessels that bypass capillary beds; see FIGURE 21.5). In addition to these shunts, the skin has an extensive system of venous plexuses that can hold a great deal of blood. From these plexuses, heat is radiated to the surface of the skin. The amount of blood flowing through the plexuses can be controlled by the constriction or dilation of the appropriate vessels. The diameter of the vessels is regulated by the hypothalamic temperature control center by way of sympathetic nerves to the vessels.

Skeletal muscle circulation Blood nourishes skeletal muscles and also removes wastes during both physical activity and rest. Because the total body mass of skeletal muscle is so large, the blood vessels play an important role in homeostasis, especially during physical activity. The neural regulation of blood flow within skeletal muscles is controlled largely by complex interactions of neurotransmitters of the sympathetic (vasomotor) nervous system. Norepinephrine or large concentrations of epinephrine produce vasoconstriction, and low concentrations of epinephrine or acetylcholine produce vasodilation. The autoregulatory response of precapillary sphincters also dilates blood vessels in response to a decrease in the oxygen concentration in active muscles.

Q: *Why do your nose and cheeks turn red on a cold day?*

A: When the skin is very cold, oxygen in the capillaries is not needed for metabolism by the relatively inactive skin. Thus large numbers of red blood cells containing red hemoglobin accumulate in the capillaries, showing through the skin.

FIGURE 21.10 HEPATIC PORTAL CIRCULATION

The hepatic portal system transports venous blood containing absorbed nutrients from the gastrointestinal tract to the liver. The liver also receives oxygenated blood from the hepatic artery. Arrows indicate the direction of blood flow.

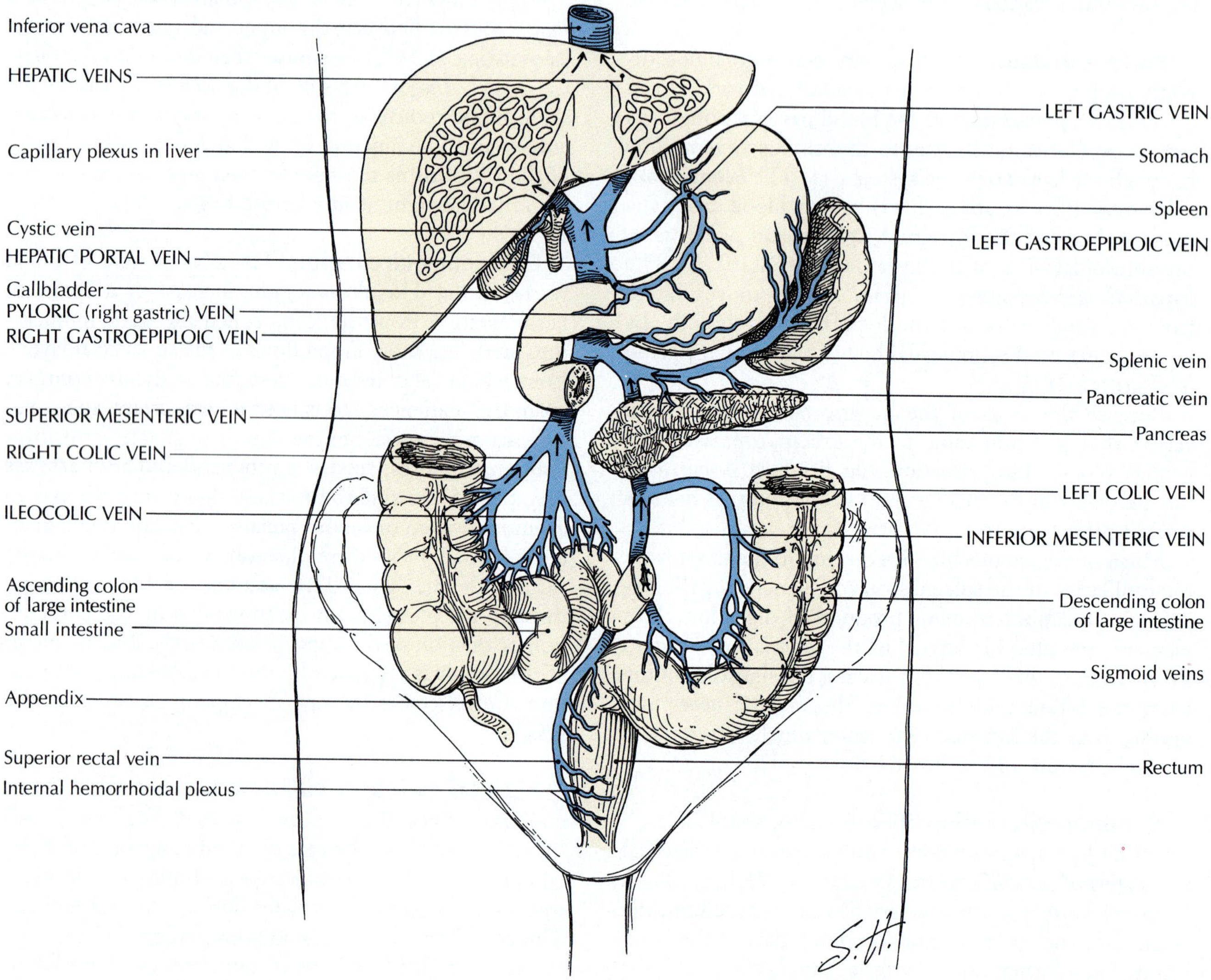

Inferior vena cava

HEPATIC VEINS

Capillary plexus in liver

Cystic vein
HEPATIC PORTAL VEIN

Gallbladder
PYLORIC (right gastric) VEIN
RIGHT GASTROEPIPLOIC VEIN

SUPERIOR MESENTERIC VEIN

RIGHT COLIC VEIN

ILEOCOLIC VEIN

Ascending colon
of large intestine
Small intestine

Appendix

Superior rectal vein
Internal hemorrhoidal plexus

LEFT GASTRIC VEIN

Stomach

Spleen
LEFT GASTROEPIPLOIC VEIN

Splenic vein
Pancreatic vein
Pancreas

LEFT COLIC VEIN

INFERIOR MESENTERIC VEIN

Descending colon
of large intestine

Sigmoid veins

Rectum

S.H.

A S K   Y O U R S E L F

1 How does the systemic circulation differ from the pulmonary circulation?

2 What is the hepatic portal system? The hypothalamic-hypophyseal portal system?

3 What are some special areas of circulation?

A S K   Y O U R S E L F

1 What are the major segments of the aorta?

2 Which arteries supply blood to the head and neck?

3 What region is supplied by the axillary artery?

4 What are the major arteries of the lower limb?

5 What are the major veins of the head and neck?

6 Which veins of the upper limb are deep veins? Superficial veins?

7 What region is drained by the azygos vein and its tributaries?

8 What are some of the criteria for naming blood vessels?

MAJOR ARTERIES AND VEINS: REGIONS SUPPLIED AND DRAINED

FIGURE 21.11A gives an overall view of the major arteries of the body. Anatomic representations of the major arteries of specific parts of the body are presented in FIGURES 21.12 through 21.15.

FIGURE 21.11B is an overall view of the major veins of the body. FIGURES 21.16 through 21.19 are anatomic representations of the major veins of specific parts of the body.

How Blood Vessels Are Named

The majority of blood vessels are named according to the major organ or anatomical site supplied (by arteries) or drained (by veins), or according to their location in the body. For example, the *renal artery* supplies the kidneys and the *renal vein* drains the kidneys. The depth of the vessel is also used. For example, the *internal jugular vein* lies deeper in the neck than the *external jugular vein* does. In some instances, the name of a blood vessel changes as it passes into a different part of the body. For example, the *subclavian artery* runs underneath the clavicle in the neck region. When it enters the region of the armpit (axilla), its name changes to the *axillary artery*, and it becomes the *brachial artery* when it enters the arm (brachium). In general, veins run parallel to most arteries and are given the same names, although there are exceptions.

Q: *Why do relatively minor bumps on the head produce such distinctive lumps?*

A: The many blood vessels in the scalp rupture and bleed fairly easily. But because the scalp is pulled so tightly over the skull, there is little room for drainage, so the blood accumulates as a prominent "goose egg."

FIGURE 21.11 MAJOR ARTERIES AND VEINS

Anterior view of [**A**] the aorta and its principal branches.

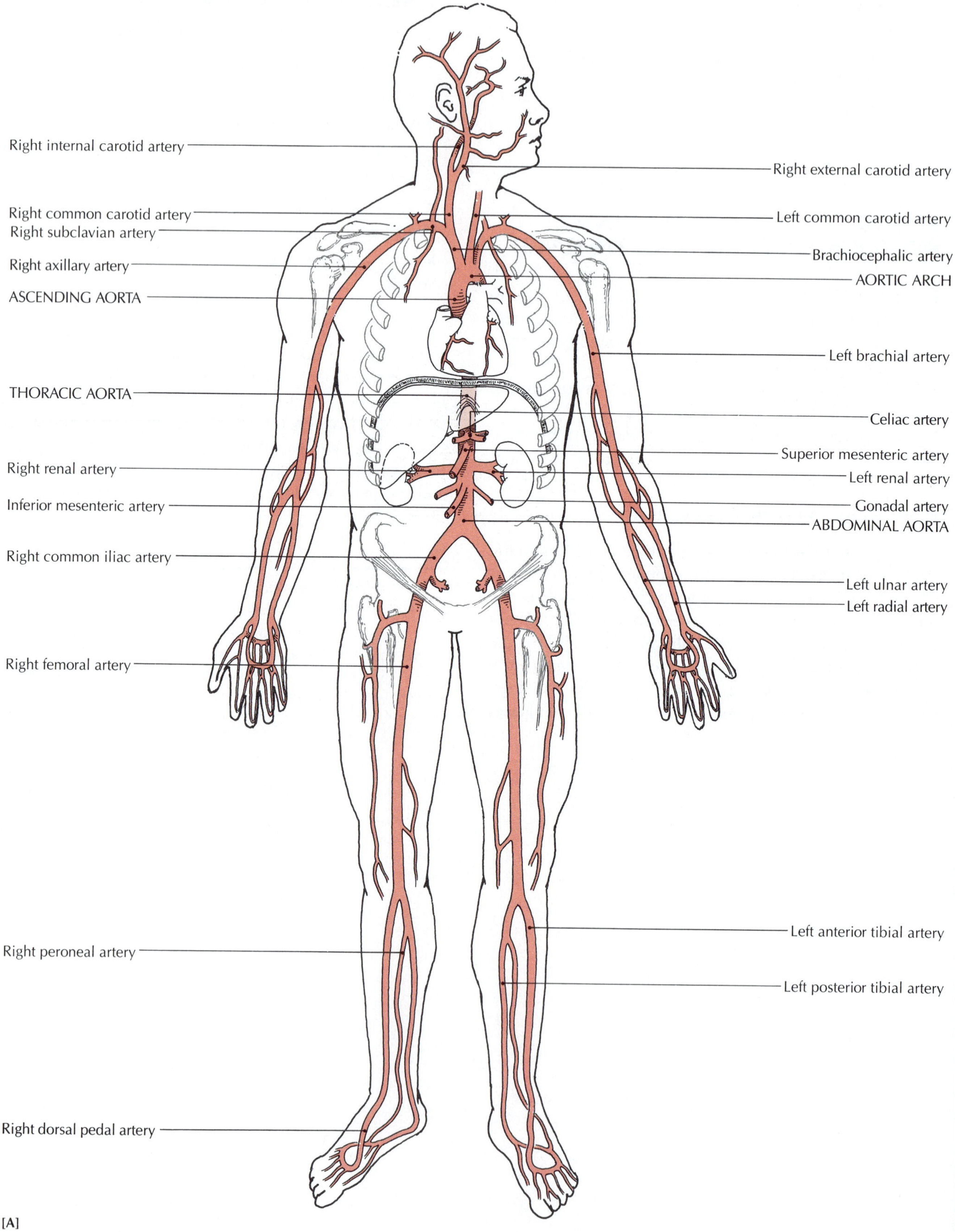

Right internal carotid artery

Right common carotid artery
Right subclavian artery

Right axillary artery

ASCENDING AORTA

THORACIC AORTA

Right renal artery

Inferior mesenteric artery

Right common iliac artery

Right femoral artery

Right peroneal artery

Right dorsal pedal artery

Right external carotid artery

Left common carotid artery

Brachiocephalic artery
AORTIC ARCH

Left brachial artery

Celiac artery

Superior mesenteric artery
Left renal artery

Gonadal artery
ABDOMINAL AORTA

Left ulnar artery
Left radial artery

Left anterior tibial artery

Left posterior tibial artery

[A]

[B] Principal veins.

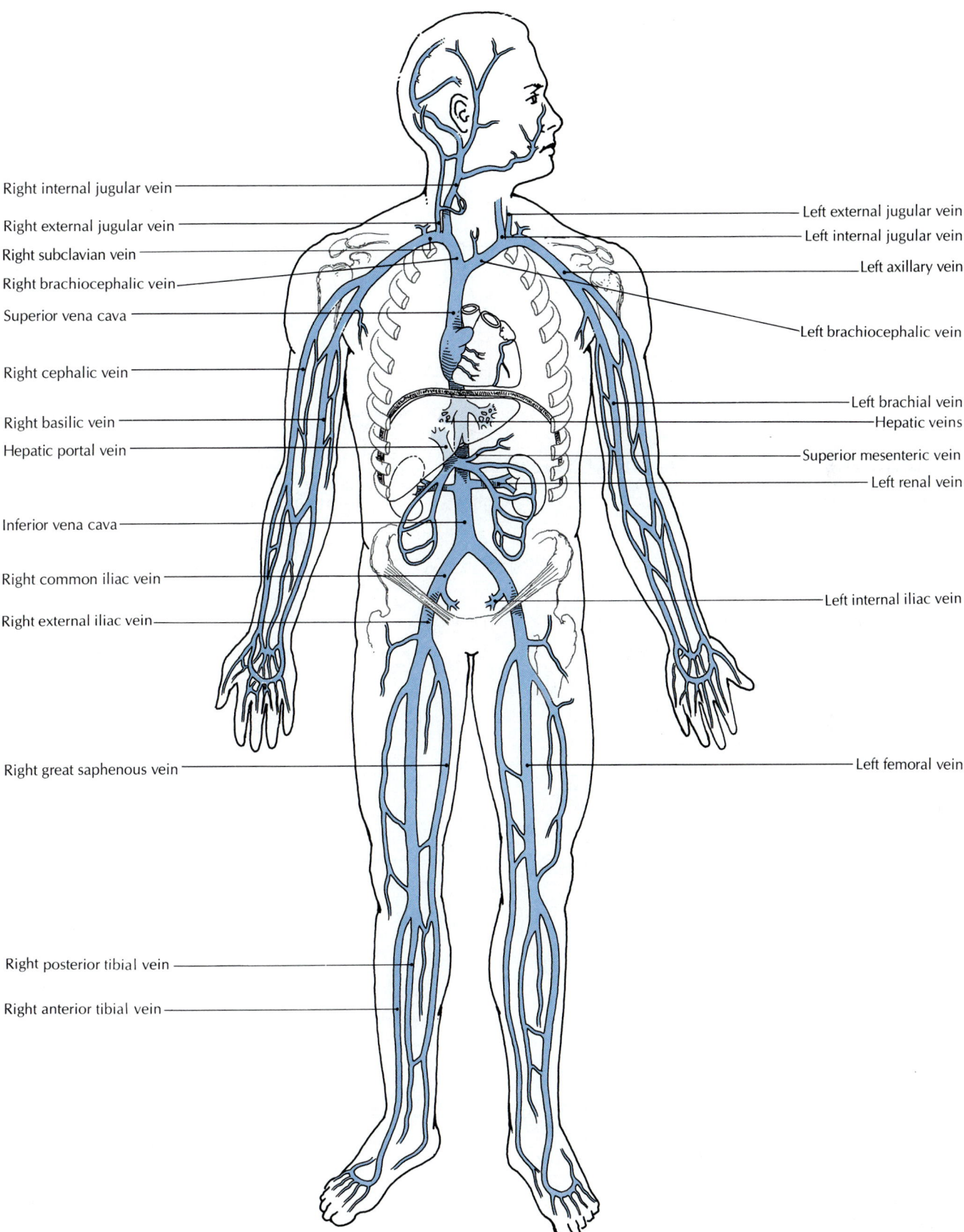

Right internal jugular vein

Right external jugular vein

Right subclavian vein

Right brachiocephalic vein

Superior vena cava

Right cephalic vein

Right basilic vein

Hepatic portal vein

Inferior vena cava

Right common iliac vein

Right external iliac vein

Right great saphenous vein

Right posterior tibial vein

Right anterior tibial vein

Left external jugular vein

Left internal jugular vein

Left axillary vein

Left brachiocephalic vein

Left brachial vein

Hepatic veins

Superior mesenteric vein

Left renal vein

Left internal iliac vein

Left femoral vein

[B]

FIGURE 21.12 MAJOR ARTERIES OF THE HEAD AND NECK

[A] Arteries of the head and neck, right lateral view.

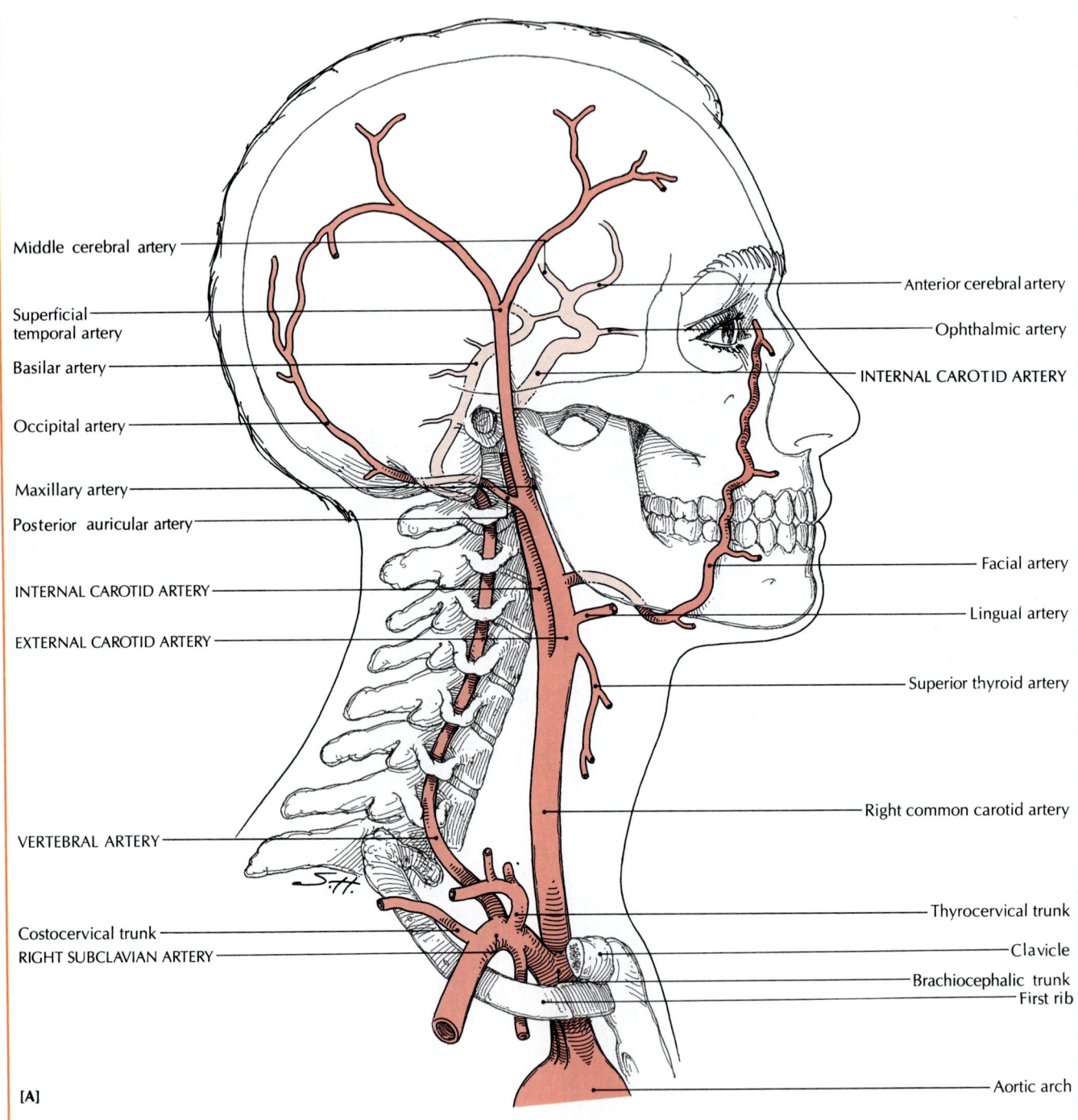

Middle cerebral artery

Superficial temporal artery

Basilar artery

Occipital artery

Maxillary artery

Posterior auricular artery

INTERNAL CAROTID ARTERY

EXTERNAL CAROTID ARTERY

VERTEBRAL ARTERY

Costocervical trunk

RIGHT SUBCLAVIAN ARTERY

Anterior cerebral artery

Ophthalmic artery

INTERNAL CAROTID ARTERY

Facial artery

Lingual artery

Superior thyroid artery

Right common carotid artery

Thyrocervical trunk

Clavicle

Brachiocephalic trunk

First rib

Aortic arch

[A]

Artery	Major branches and region supplied	Comments
Internal carotid artery*	Anterior cerebral artery; brain Middle cerebral artery; brain	Internal carotid has no neck branches; passes through carotid canal into cranial cavity, gives off ophthalmic artery, terminates by dividing into anterior and middle cerebral arteries.

[B] Ventral surface of the brain showing the cerebral arterial circle (circle of Willis).

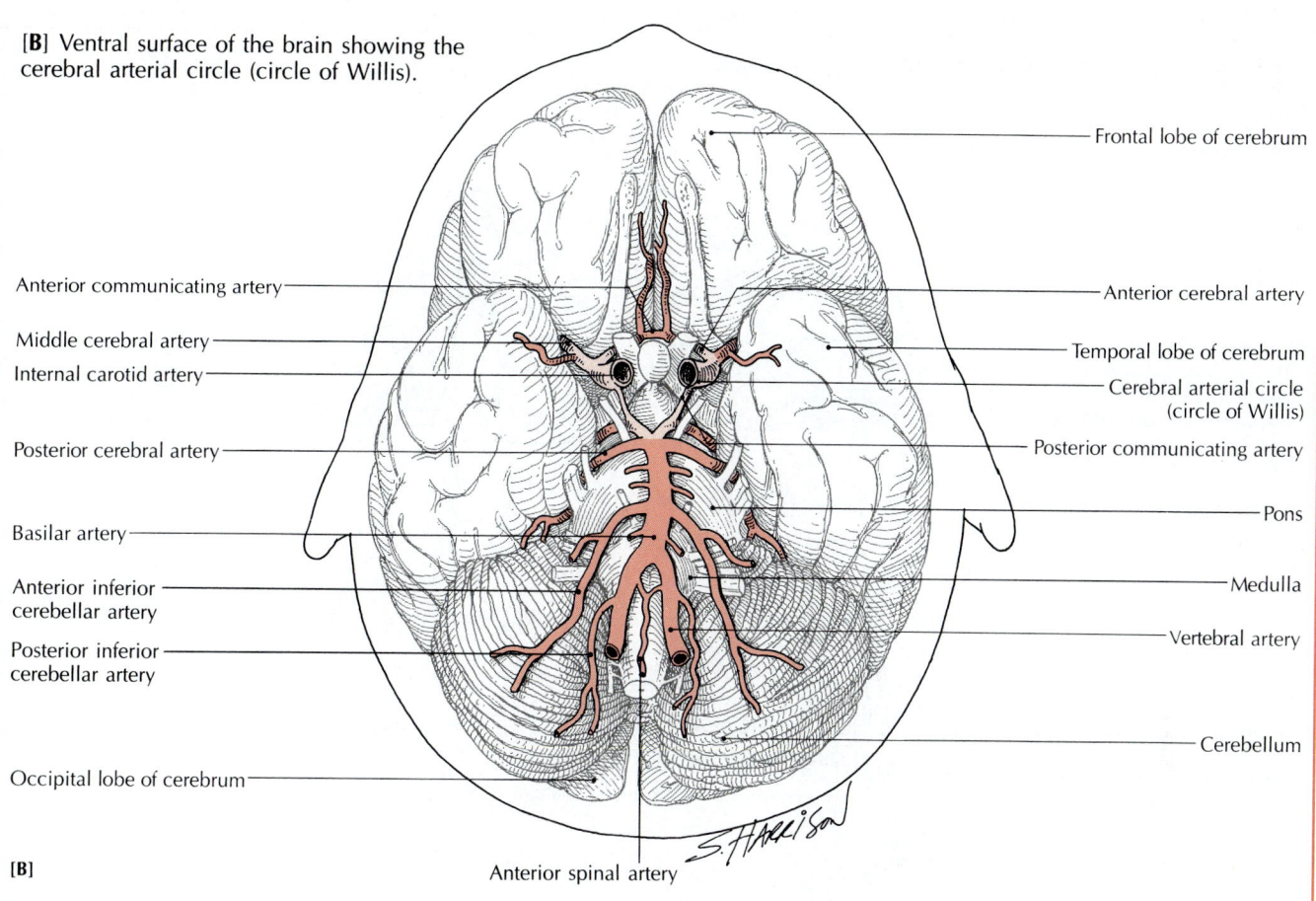

Anterior communicating artery

Middle cerebral artery

Internal carotid artery

Posterior cerebral artery

Basilar artery

Anterior inferior cerebellar artery

Posterior inferior cerebellar artery

Occipital lobe of cerebrum

Frontal lobe of cerebrum

Anterior cerebral artery

Temporal lobe of cerebrum

Cerebral arterial circle (circle of Willis)

Posterior communicating artery

Pons

Medulla

Vertebral artery

Cerebellum

Anterior spinal artery

[B]

Artery	Major branches and region supplied	Comments
Subclavian artery	Vertebral artery; brain, spinal cord (cervical region), vertebrae, other deep neck structures	Right and left vertebral arteries unite to form one basilar artery, which bifurcates and terminates as the paired posterior cerebral arteries.
	Thyrocervical trunk; trachea, esophagus, thyroid gland, neck muscles	
	Costocervical trunk; muscles of back of neck, neck vertebrae	
Internal thoracic artery	Anterior thoracic wall, muscles, skin, breasts	
External carotid artery*	Superior thyroid artery; thyroid gland, neighboring muscles, larynx	External carotid terminates in region of parotid gland by bifurcating into superficial temporal and maxillary arteries.
	Lingual artery; floor of mouth, tongue, neighboring muscles, salivary glands	
	Facial artery; face below eyes, including soft palate, tonsils, and submandibular gland	
	Occipital artery; skin, muscles, and associated tissues of back of head (scalp), brain meninges	
	Posterior auricular artery; skin, muscles, and associated tissues of and near ear and posterior scalp	
	Superficial temporal artery; outer ear, parotid gland, forehead, scalp and muscles in temporal region	
	Maxillary artery; face, including upper and lower jaws, teeth, palate, nose, muscles of mastication, and brain dura mater	

*The internal and external carotid arteries are branches of the common carotid artery.

FIGURE 21.13 MAJOR ARTERIES OF THE UPPER LIMB

[A] Drawing of arteries, right anterior view.
[B] Arteriogram of right wrist, anterior view.

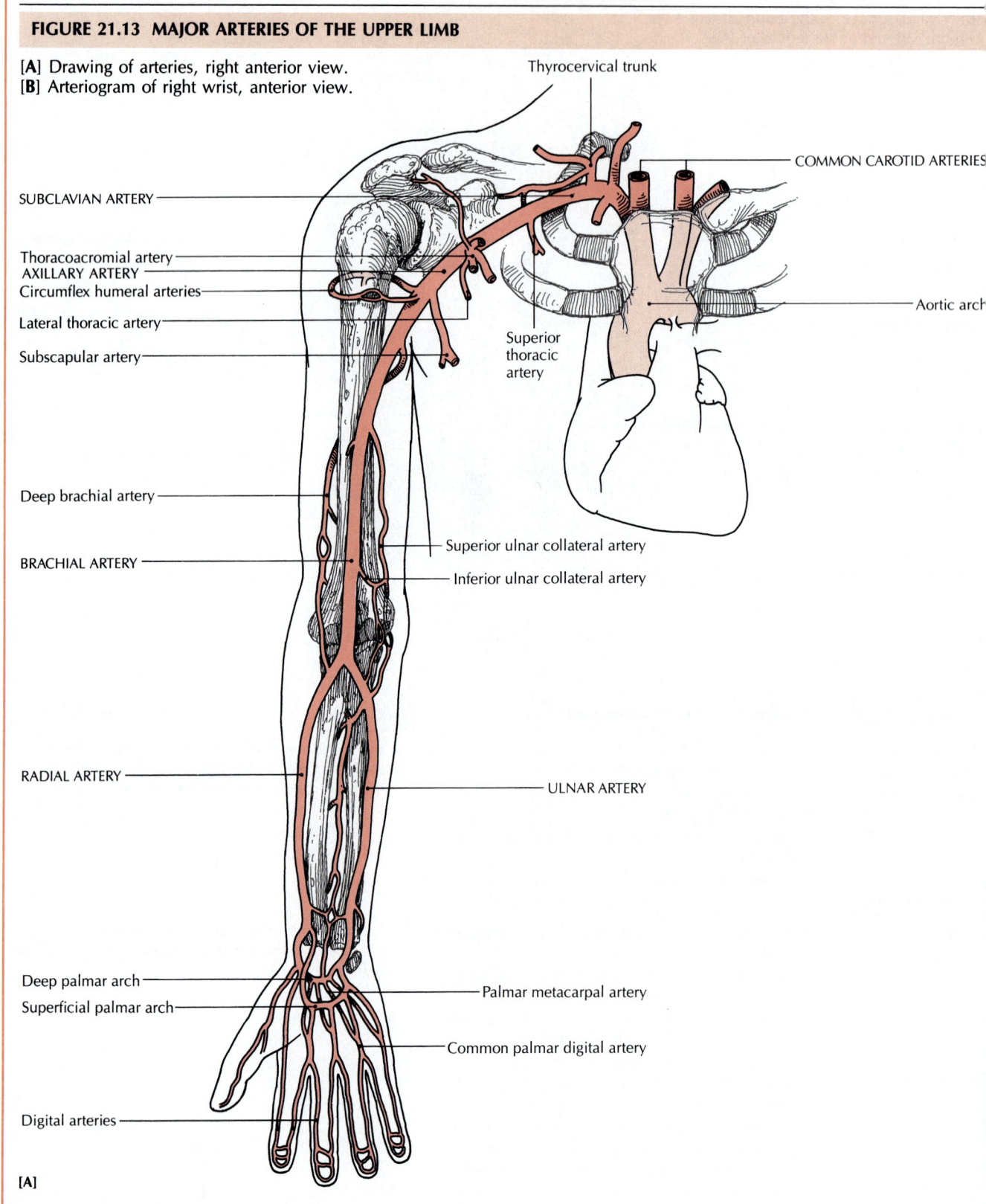

Thyrocervical trunk

COMMON CAROTID ARTERIES

SUBCLAVIAN ARTERY

Thoracoacromial artery
AXILLARY ARTERY
Circumflex humeral arteries
Lateral thoracic artery
Subscapular artery

Superior
thoracic
artery

Aortic arch

Deep brachial artery

BRACHIAL ARTERY

Superior ulnar collateral artery

Inferior ulnar collateral artery

RADIAL ARTERY

ULNAR ARTERY

Deep palmar arch
Superficial palmar arch

Palmar metacarpal artery

Common palmar digital artery

Digital arteries

[A]

Radial artery

Palmar carpal branch of radial artery

Palmar carpal arch

Princeps pollicis artery

Radialis indicis artery

Ulnar artery

Deep palmar branch of ulnar artery

Deep palmar arch

Common palmar digital arteries

Proper palmar digital arteries

[B]

Artery	Major branches and region supplied	Comments
Axillary artery	Superior thoracic artery, thoracoacromial artery, lateral thoracic artery, subscapular artery, circumflex humeral artery; chest wall, muscles of chest wall, shoulder girdle, shoulder joint	
Brachial artery	Deep (profunda) brachial artery, ulnar collateral arteries; muscles of arm, upper forearm, humerus, and elbow joint	Brachial artery in anterior elbow is site used for measurement of blood pressure.
Radial artery	Branches in forearm; elbow joint, radius, muscles of radial (thumb) side of forearm Continues as deep palmar arch and palmar branches; carpal and metacarpal bones and muscles in palm Arch (along with superior palmar arch) provides palmar membranes, which continue as digital arteries; fingers	Pulse of radial artery can be felt at base of thumb between the tendons of the brachioradialis and flexor carpi radialis muscles, and on back of wrist between tendons of thumb muscles (extensor pollicis longus and brevis muscles). Anastomoses among arteries of lower forearm and those of branches of arches of palm are substantial.
Ulnar artery	Branches in forearm; elbow joint, ulna, muscles of ulnar (small-finger) side of forearm Continues as superficial palmar arch and palmar branches; carpal and metacarpal bones and muscles in palm Arch (along with deep palmar arch) provides palmar and digital branches; structures of palm and fingers	Pulse of ulnar artery can be felt just to the radial (lateral) side of the pisiform bone on the ulnar side of the wrist.

FIGURE 21.14 MAJOR ARTERIES OF THE THORAX AND ABDOMEN

[A] Drawing, anterior view. [B] Photograph, abdominal aorta and branches, anterior view. Renal veins and part of the inferior vena cava have been removed.

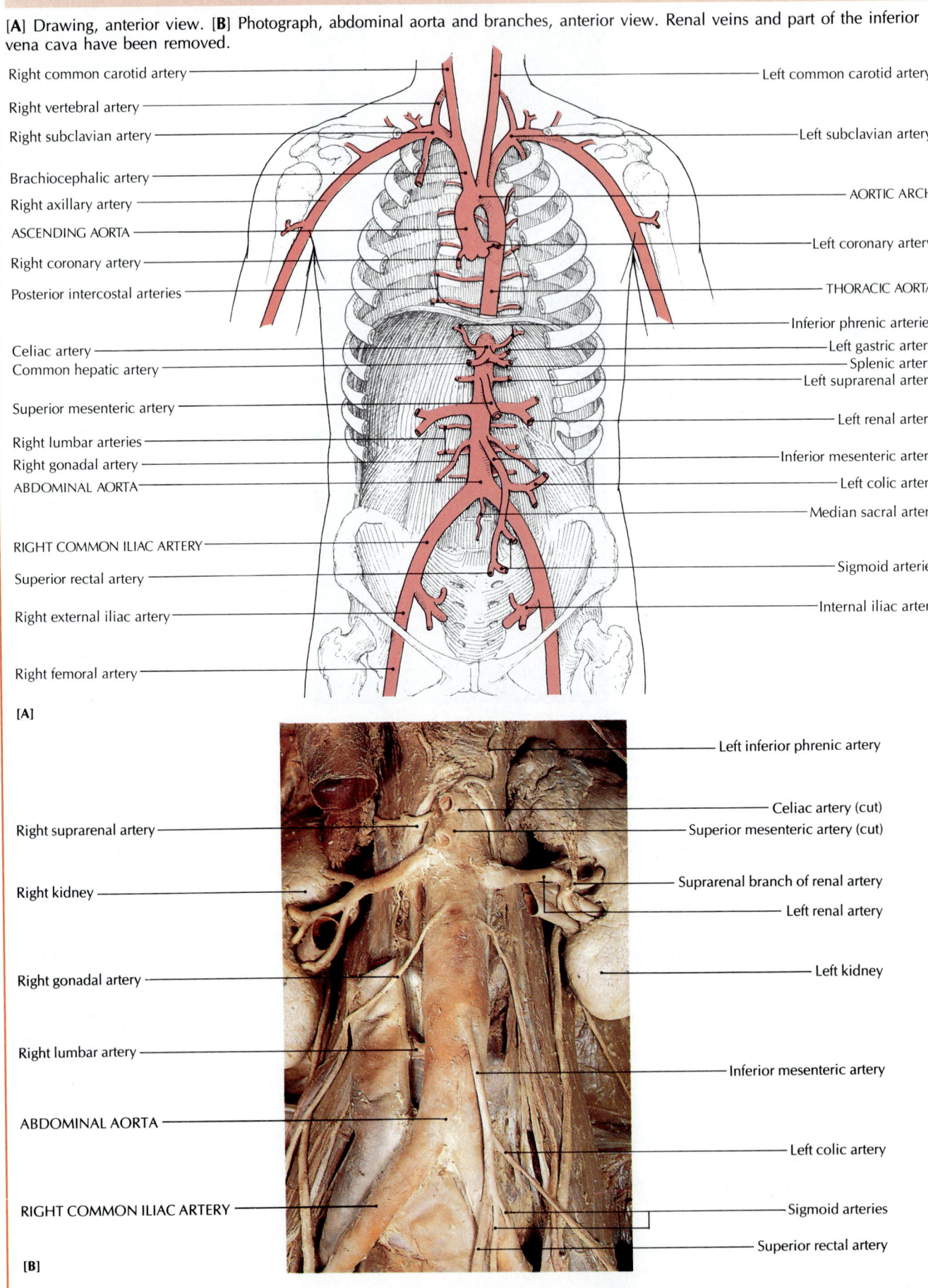

Right common carotid artery

Right vertebral artery

Right subclavian artery

Brachiocephalic artery

Right axillary artery

ASCENDING AORTA

Right coronary artery

Posterior intercostal arteries

Celiac artery

Common hepatic artery

Superior mesenteric artery

Right lumbar arteries

Right gonadal artery

ABDOMINAL AORTA

RIGHT COMMON ILIAC ARTERY

Superior rectal artery

Right external iliac artery

Right femoral artery

[A]

Left common carotid artery

Left subclavian artery

AORTIC ARCH

Left coronary artery

THORACIC AORTA

Inferior phrenic arteries

Left gastric artery

Splenic artery

Left suprarenal artery

Left renal artery

Inferior mesenteric artery

Left colic artery

Median sacral artery

Sigmoid arteries

Internal iliac artery

Right suprarenal artery

Right kidney

Right gonadal artery

Right lumbar artery

ABDOMINAL AORTA

RIGHT COMMON ILIAC ARTERY

[B]

Left inferior phrenic artery

Celiac artery (cut)

Superior mesenteric artery (cut)

Suprarenal branch of renal artery

Left renal artery

Left kidney

Inferior mesenteric artery

Left colic artery

Sigmoid arteries

Superior rectal artery

Artery	Major branches and region supplied
ASCENDING AORTA	
Right coronary artery	Marginal and posterior interventricular branches; nutrient arteries to heart wall
Left coronary artery	Circumflex and anterior interventricular branches; nutrient arteries to heart wall
AORTIC ARCH	
Brachiocephalic trunk (innominate artery)	Right common carotid artery*; head, neck Right subclavian artery; right upper limb
Left common carotid artery*	Head, neck
Left subclavian artery	Left upper limb
THORACIC AORTA (thoracic portion of descending aorta)	
Bronchial arteries (unpaired)	Bronchi, bronchioles
Esophageal arteries (unpaired)	Esophagus
Intercostal arteries (9 pairs)	Dorsal branch; spinal cord, muscles and skin of back
Posterior intercostal arteries (1 pair below rib 12)	Muscular branch; chest muscles Cutaneous branch; skin of thorax Mammary branch; breasts
Superior phrenic arteries	Diaphragm
ABDOMINAL AORTA (abdominal portion of descending aorta)	
Inferior phrenic arteries	Diaphragm, lower esophagus
Celiac artery (trunk) (unpaired)	Left gastric artery; stomach, esophagus Common hepatic artery; liver, stomach, pancreas, duodenum, gallbladder, bile duct Splenic artery; stomach, pancreas, spleen, omentum
Superior mesenteric artery (unpaired)	Inferior pancreaticoduodenal artery; head of pancreas, duodenum Jejuneal and ileal arteries; small intestine Ileocolic artery; lower ileum, appendix, ascending and part of transverse colon Right colic and middle colic arteries; transverse and ascending colon
Inferior mesenteric artery (unpaired)	Left colic artery; transverse and descending colon Sigmoid artery; descending and sigmoid colon Superior rectal artery; rectum, anal region
Renal arteries	Inferior suprarenal abdominal artery; kidneys, ureters, adrenal glands
Gonadal (testicular or ovarian) arteries	Testes or ovaries
Lumbar arteries (4 or 5 pairs)	Skin, muscle, and vertebrae of lumbar region of back, spinal cord and meninges of lower back, caudal equina
Median sacral (middle) artery	Sacrum, coccyx
Common iliac arteries	External iliac arteries; lower limb Internal iliac arteries; viscera, walls of pelvis, perineum, rectum, gluteal region

*Right and left common carotid arteries ascend into the neck and then bifurcate at the upper part of the larynx into right and left external and internal carotid arteries.

674

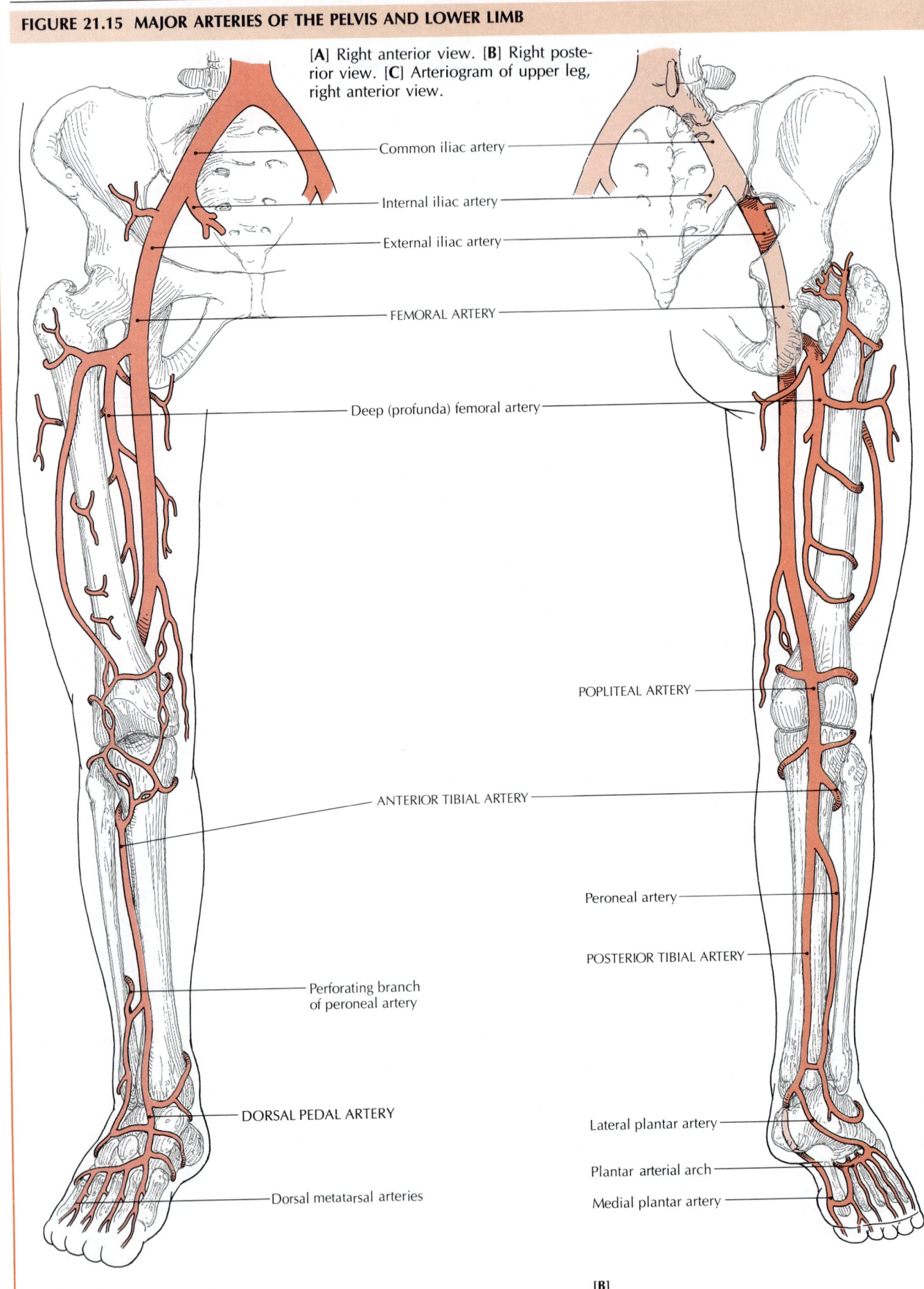

FIGURE 21.15 MAJOR ARTERIES OF THE PELVIS AND LOWER LIMB

[A] Right anterior view. [B] Right posterior view. [C] Arteriogram of upper leg, right anterior view.

Common iliac artery

Internal iliac artery

External iliac artery

FEMORAL ARTERY

Deep (profunda) femoral artery

POPLITEAL ARTERY

ANTERIOR TIBIAL ARTERY

Peroneal artery

POSTERIOR TIBIAL ARTERY

Perforating branch of peroneal artery

DORSAL PEDAL ARTERY

Lateral plantar artery

Plantar arterial arch

Medial plantar artery

Dorsal metatarsal arteries

[A]

[B]

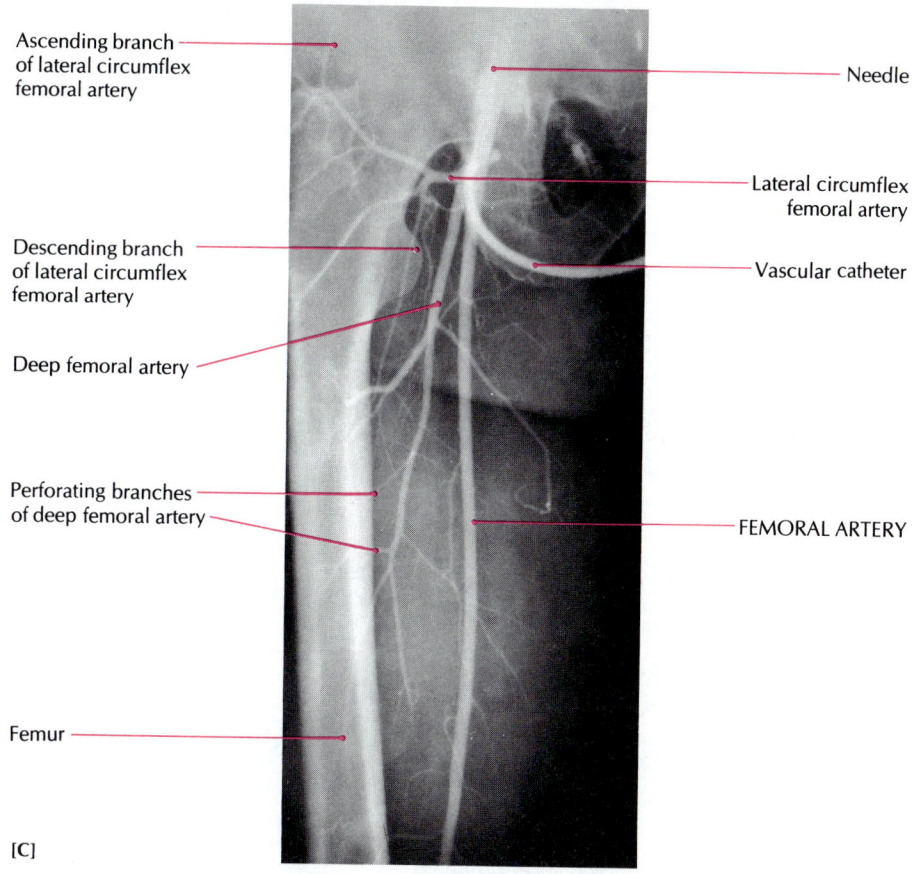

Ascending branch of lateral circumflex femoral artery

Needle

Lateral circumflex femoral artery

Descending branch of lateral circumflex femoral artery

Vascular catheter

Deep femoral artery

Perforating branches of deep femoral artery

FEMORAL ARTERY

Femur

[C]

Artery	Major branches and region supplied	Comments
Common iliac artery	Internal iliac artery; pelvic viscera, including urinary bladder, prostate gland, ducts deferens, uterus, vagina, rectum	
	Peroneal region, including anal canal, external genitalia; walls of pelvic cavity, including muscles. Gluteal region (buttocks), including muscles, hip joint	
	External iliac artery; lower limb	External iliac artery enters thigh behind inguinal ligament and becomes femoral artery.
Femoral artery	Muscular branches, including deep (profunda) femoral artery; skin and muscles of thigh, including anterior flexors, medial adductors, extensors (hamstrings)	Femoral artery, located on anterior thigh, passes backward through adductor magnus to become popliteal artery on back of knee (popliteal fossa).
Popliteal artery	Muscular branches; knee joint, gastrocnemius and soleus muscles	Popliteal artery terminates just beyond knee joint by dividing into anterior tibial artery and posterior tibial artery.
Anterior tibial artery	Muscular branches; knee joint, muscles, and skin in front of leg, ankle joint	Anterior tibial artery continues over ankle to become dorsal pedal artery.
Dorsal pedal artery	Muscles, skin, and joints (including ankle) on dorsal side of foot	
Posterior tibial artery	Muscular branches, including peroneal artery on lateral leg; knee joint, muscles and skin on back of leg, ankle joint	
	Peroneal, medial malleolar, calcaneal, and plantar arteries; fibula, muscles on fibula, ankle joint, heel, toes	

FIGURE 21.16 MAJOR VEINS OF THE HEAD AND NECK

[A] Drawing, right lateral view.

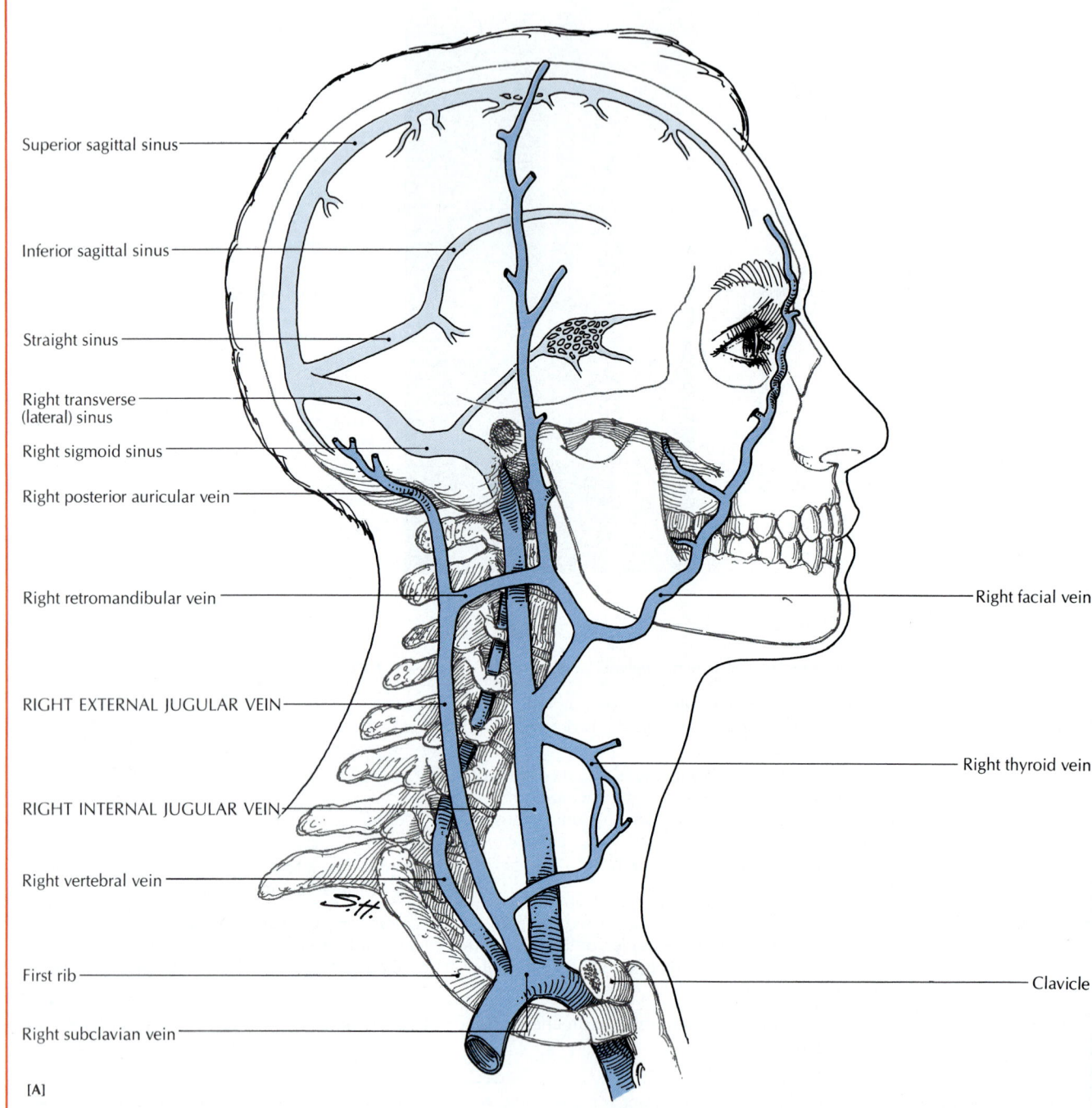

Superior sagittal sinus

Inferior sagittal sinus

Straight sinus

Right transverse (lateral) sinus

Right sigmoid sinus

Right posterior auricular vein

Right retromandibular vein

RIGHT EXTERNAL JUGULAR VEIN

RIGHT INTERNAL JUGULAR VEIN

Right vertebral vein

First rib

Right subclavian vein

Right facial vein

Right thyroid vein

Clavicle

[A]

[B] Cast of blood vessels of
the head and neck, anterior view.

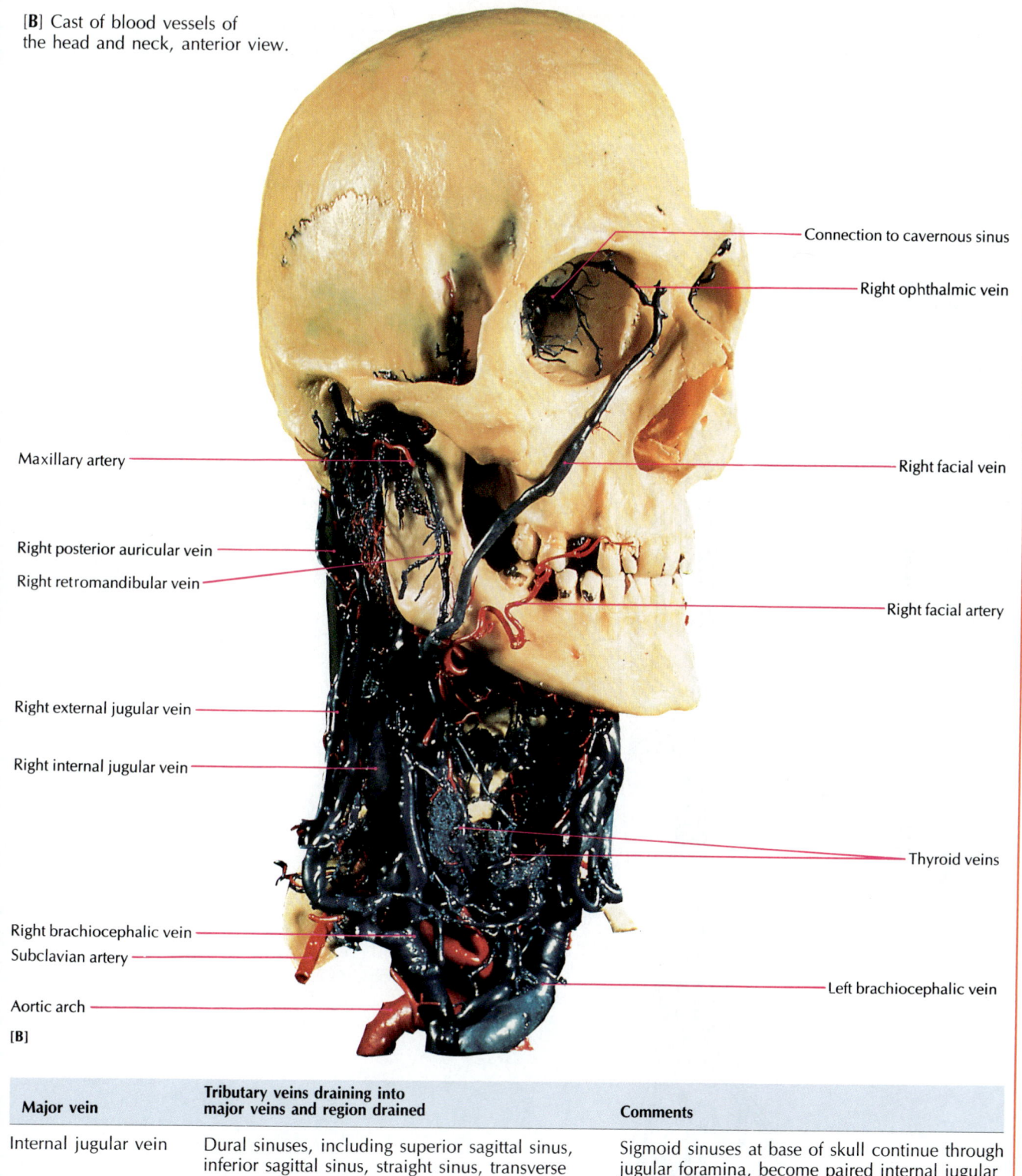

Connection to cavernous sinus

Right ophthalmic vein

Maxillary artery

Right facial vein

Right posterior auricular vein

Right retromandibular vein

Right facial artery

Right external jugular vein

Right internal jugular vein

Thyroid veins

Right brachiocephalic vein

Subclavian artery

Aortic arch

Left brachiocephalic vein

[B]

Major vein	Tributary veins draining into major veins and region drained	Comments
Internal jugular vein	Dural sinuses, including superior sagittal sinus, inferior sagittal sinus, straight sinus, transverse sinuses, sigmoid sinuses; brain, essentially all venous blood within cranial cavity Facial and thyroid veins; face, neck	Sigmoid sinuses at base of skull continue through jugular foramina, become paired internal jugular veins; venous dural sinuses are between layers of dura mater. Internal jugular veins join subclavian veins to form brachiocephalic veins, which drain into superior vena cava.
External jugular vein	Retromandibular vein; posterior face Posterior auricular vein; scalp behind ear Other veins; parotid glands, scalp, neck muscles	External jugular formed by retromandibular and posterior auricular veins and drains into subclavian vein.

FIGURE 21.17 MAJOR VEINS OF THE UPPER LIMB

[A] Right anterior view.

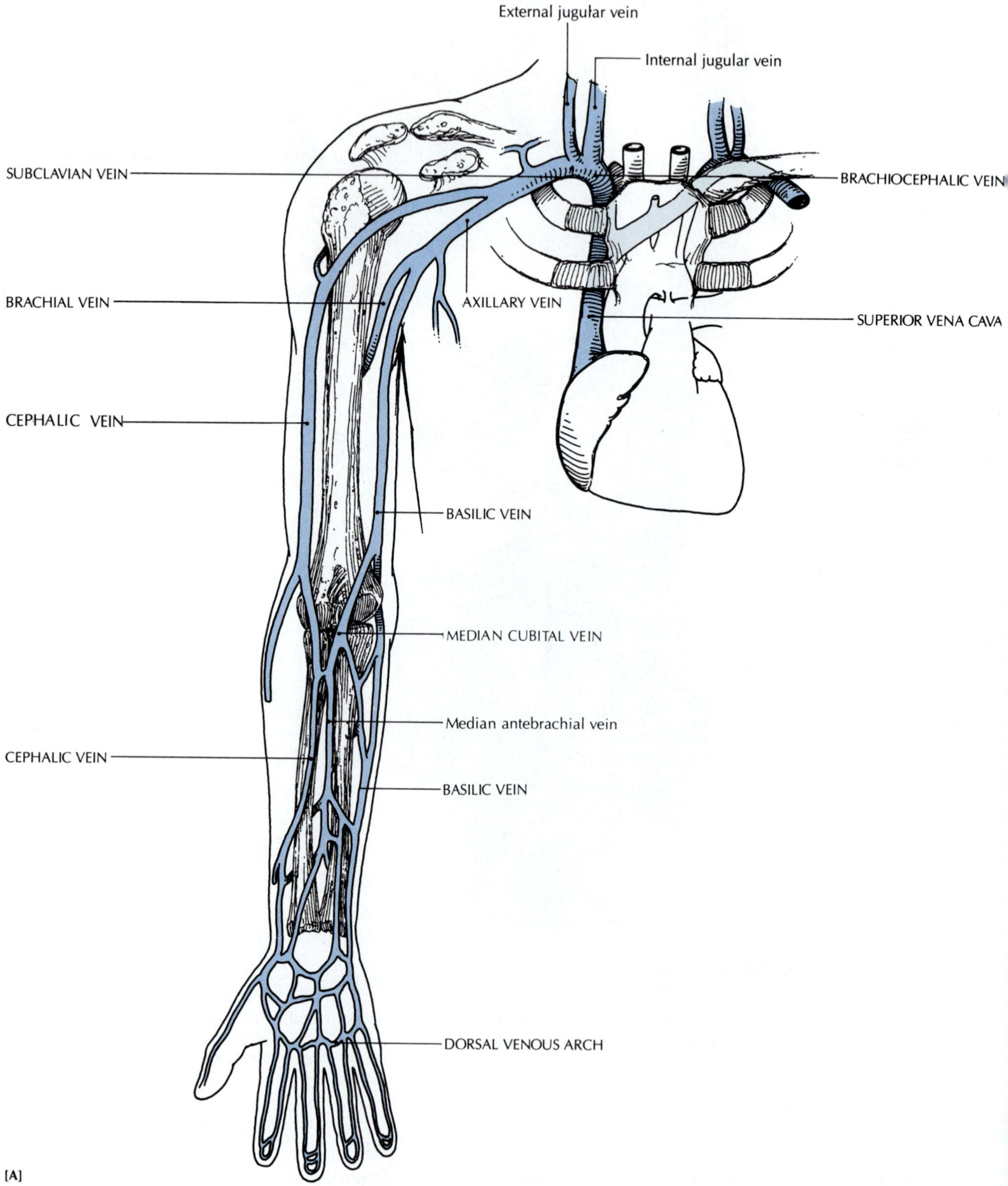

External jugular vein

Internal jugular vein

SUBCLAVIAN VEIN

BRACHIOCEPHALIC VEIN

BRACHIAL VEIN

AXILLARY VEIN

SUPERIOR VENA CAVA

CEPHALIC VEIN

BASILIC VEIN

MEDIAN CUBITAL VEIN

Median antebrachial vein

CEPHALIC VEIN

BASILIC VEIN

DORSAL VENOUS ARCH

[A]

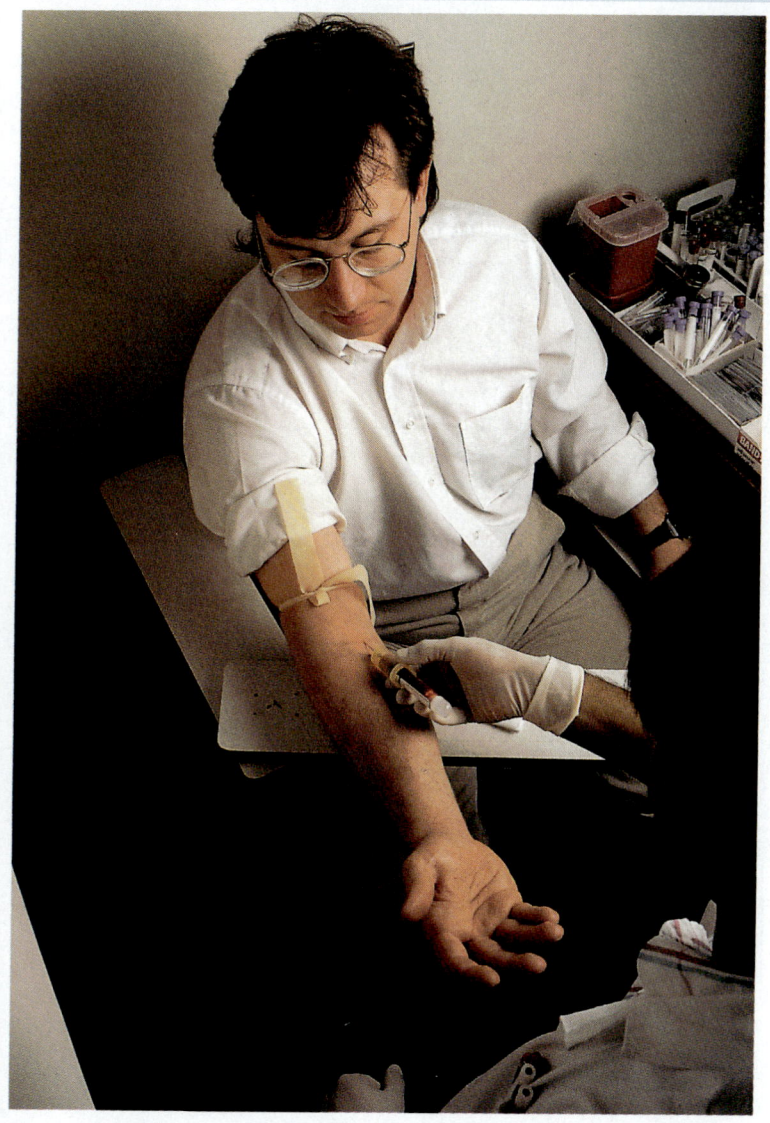

[B] The photograph shows a medical technician drawing blood from the patient's median cubital vein into a hypodermic needle. The tourniquet above the elbow should be snug, but not tight, to close off (occlude) the superficial venous return, distending the vein. If the tourniquet is too tight, the brachial artery will be occluded, obstructing the flow of arterial blood to the forearm and hand.

[B]

Major veins	Region drained	Comments
Deep veins		
Brachial vein Axillary vein Subclavian vein	Upper limb	Deep veins may accompany an artery and have transverse anastomoses (*venae comitantes*) with arteries. Axillary and subclavian veins accompany corresponding artery; right and left subclavian veins join with internal jugular veins to form brachiocephalic veins.
Superficial veins		
Dorsal venous arch Basilic vein	Back of hand Medial forearm and arm	Superficial veins are cutaneous and have anastomoses with deep veins. Basilic vein originates from dorsal venous arch, extends along medial side of limb, and terminates by joining brachial vein to form axillary vein.
Cephalic vein	Lateral forearm and arm	Cephalic vein originates from dorsal venous arch, extends along lateral side of limb, and terminates by emptying into axillary vein near lateral end of clavicle.
Median cubital vein	Anterior elbow	Median cubital vein communicates between cephalic and basilic veins, which form variable patterns; cubital vein is often used when vein must be punctured for injection, transfusion, or withdrawal of blood.

FIGURE 21.18 MAJOR VEINS OF THE THORAX, ABDOMEN, AND PELVIS

[A] Anterior view. [B] Photograph of interior vena cava, anterior view.

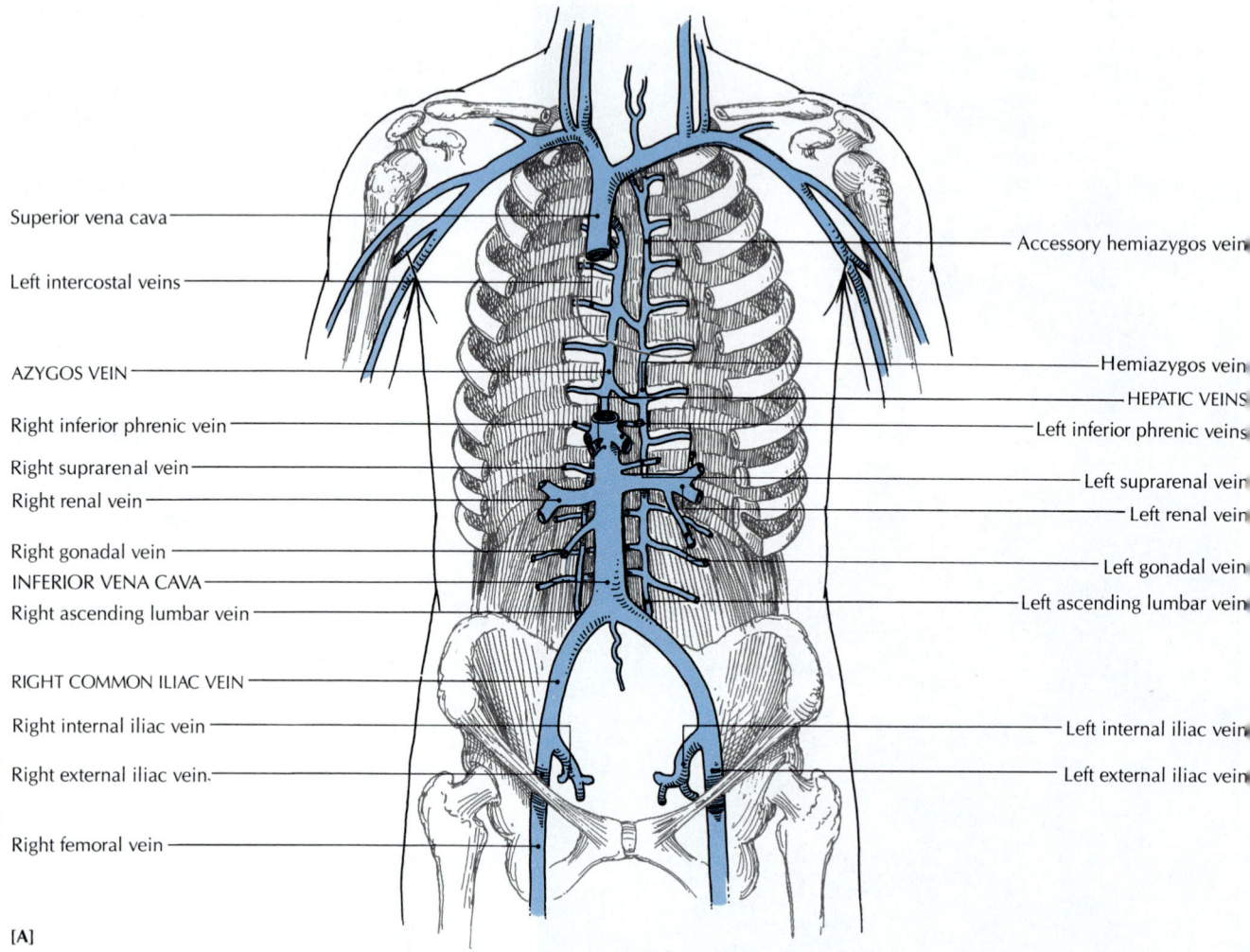

Superior vena cava

Left intercostal veins

AZYGOS VEIN

Right inferior phrenic vein

Right suprarenal vein

Right renal vein

Right gonadal vein

INFERIOR VENA CAVA

Right ascending lumbar vein

RIGHT COMMON ILIAC VEIN

Right internal iliac vein

Right external iliac vein

Right femoral vein

Accessory hemiazygos vein

Hemiazygos vein

HEPATIC VEINS

Left inferior phrenic veins

Left suprarenal vein

Left renal vein

Left gonadal vein

Left ascending lumbar vein

Left internal iliac vein

Left external iliac vein

[A]

Major vein	Tributary veins draining into major veins and region drained	Comments
Common iliac vein	External iliac vein; lower limb	Common iliac vein is continuation of femoral vein.
	Internal iliac vein; pelvis, buttocks, pelvic viscera, including gluteal muscles, rectum, urinary bladder, prostate, uterus, vagina, external genitalia	Common iliac veins unite to form inferior vena cava.
Inferior vena cava (abdominal area)	Ascending lumbar veins; body wall Renal veins; kidneys	Inferior vena cava is largest blood vessel in body (approximately 3.5 cm in diameter). It receives tributaries from lumbar, visceral, and hepatic areas, but not directly from digestive tract, pancreas, or spleen.
	Gonadal veins (ovarian and testicular veins)	Left gonadal vein drains directly into renal vein; right gonadal vein drains directly into inferior vena cava.
	Suprarenal veins Inferior phrenic veins; diaphragm Hepatic veins; liver	Left suprarenal vein drains into renal vein. Hepatic vein drains blood from liver to inferior vena cava.

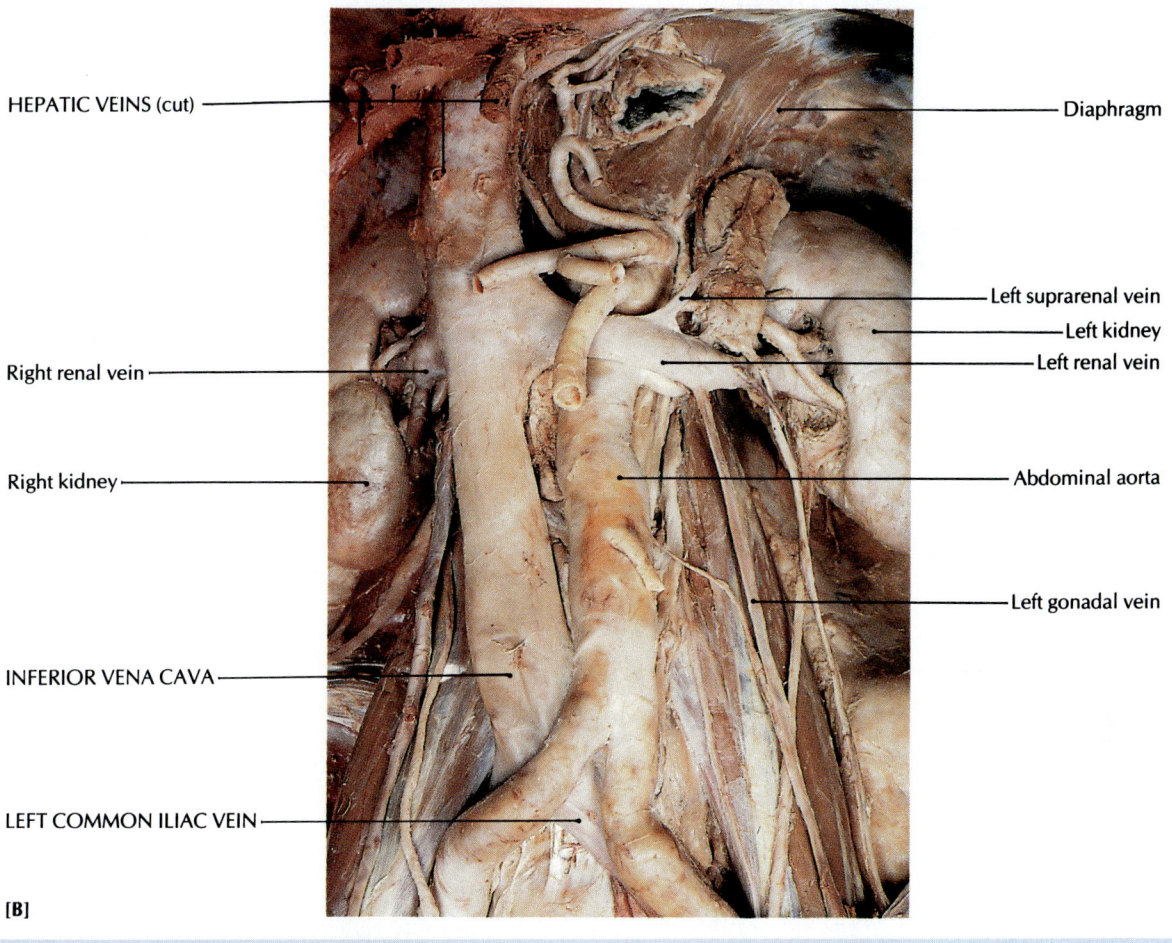

HEPATIC VEINS (cut) ———— Diaphragm

Left suprarenal vein

Left kidney

Right renal vein ———— Left renal vein

Right kidney ———— Abdominal aorta

Left gonadal vein

INFERIOR VENA CAVA ————

LEFT COMMON ILIAC VEIN ————

[B]

Major vein	Tributary veins draining into major veins and region drained	Comments
Portal system (hepatic portal vein)	Splenic vein; spleen Superior mesenteric vein; small intestine, part of large intestine Inferior mesenteric vein; large intestine	Portal system drains abdominal and pelvic viscera, which are supplied by celiac, superior mesenteric, and inferior mesenteric arteries. Venous blood of portal system drains into liver and its sinusoids and then into hepatic vein.

Azygos vein	Azygos vein
	  Intercostal veins; chest wall
	  Right ascending lumbar vein; abdominal body wall
	  Right bronchial vein; bronchi
		Azygos vein drains into superior vena cava.

	Hemiazygos vein
	  Intercostal veins (lower 4 or 5); chest wall
	  Left ascending lumbar vein; abdominal body wall
	  Left bronchial vein; bronchi
		Hemiazygos vein drains into azygos vein at level of ninth thoracic vein.

	Accessory hemiazygos vein
	  Intercostal veins (upper 3 or 4); chest wall
	  Right bronchial vein; bronchi
		Accessory hemiazygos vein joins azygos vein at level of eighth thoracic vein.
		Azygos system drains blood from thoracic wall and posterior wall of abdomen. It has venous connections with veins draining inferior vena cava, including those of esophagus, mediastinum, pericardium, and bronchi.
		When inferior vena cava of hepatic portal vein is obstructed, venous blood can be diverted to azygos system and superior vena cava.

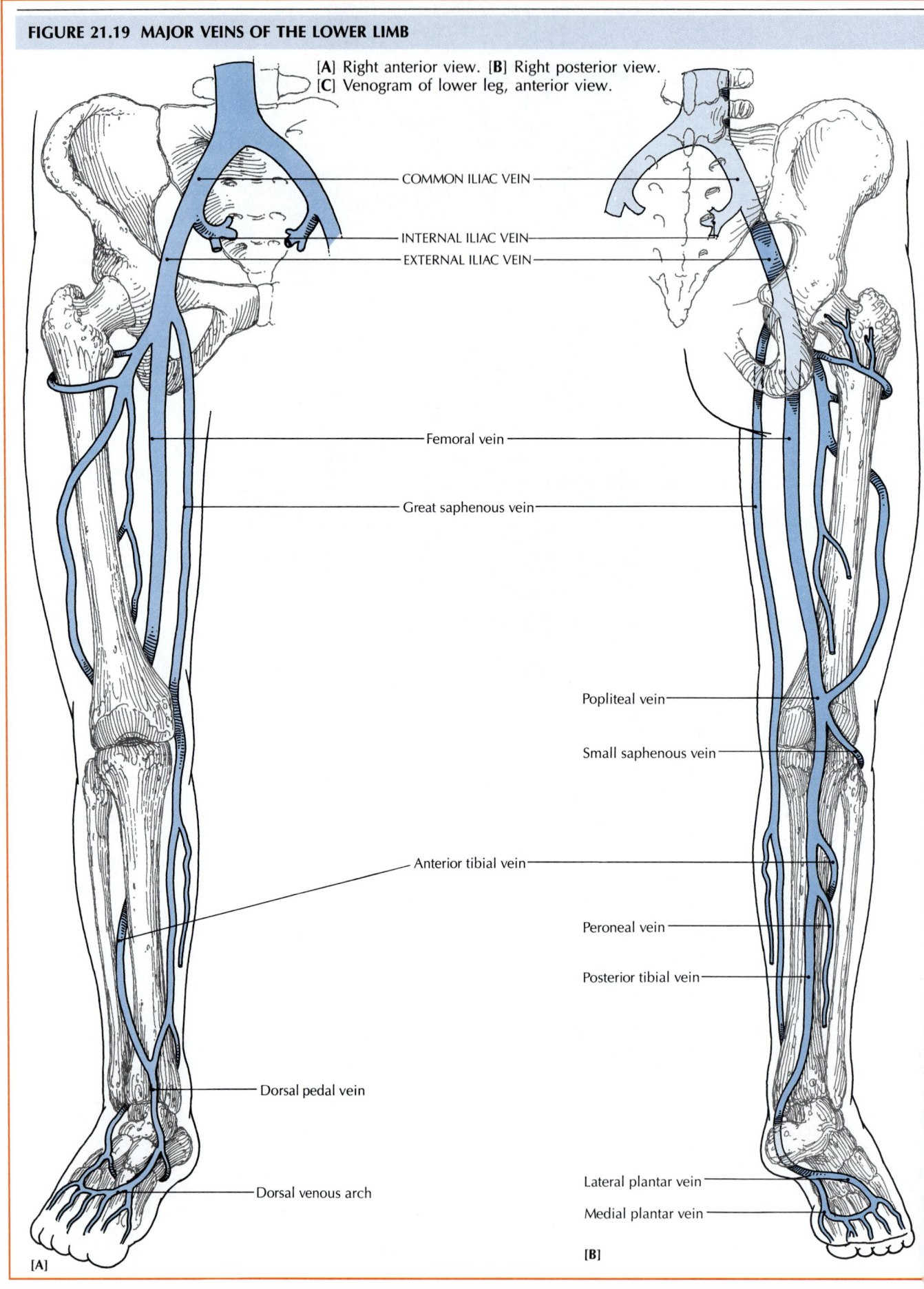

FIGURE 21.19 MAJOR VEINS OF THE LOWER LIMB

[A] Right anterior view. [B] Right posterior view.
[C] Venogram of lower leg, anterior view.

COMMON ILIAC VEIN

INTERNAL ILIAC VEIN

EXTERNAL ILIAC VEIN

Femoral vein

Great saphenous vein

Popliteal vein

Small saphenous vein

Anterior tibial vein

Peroneal vein

Posterior tibial vein

Dorsal pedal vein

Dorsal venous arch

Lateral plantar vein

Medial plantar vein

[A]

[B]

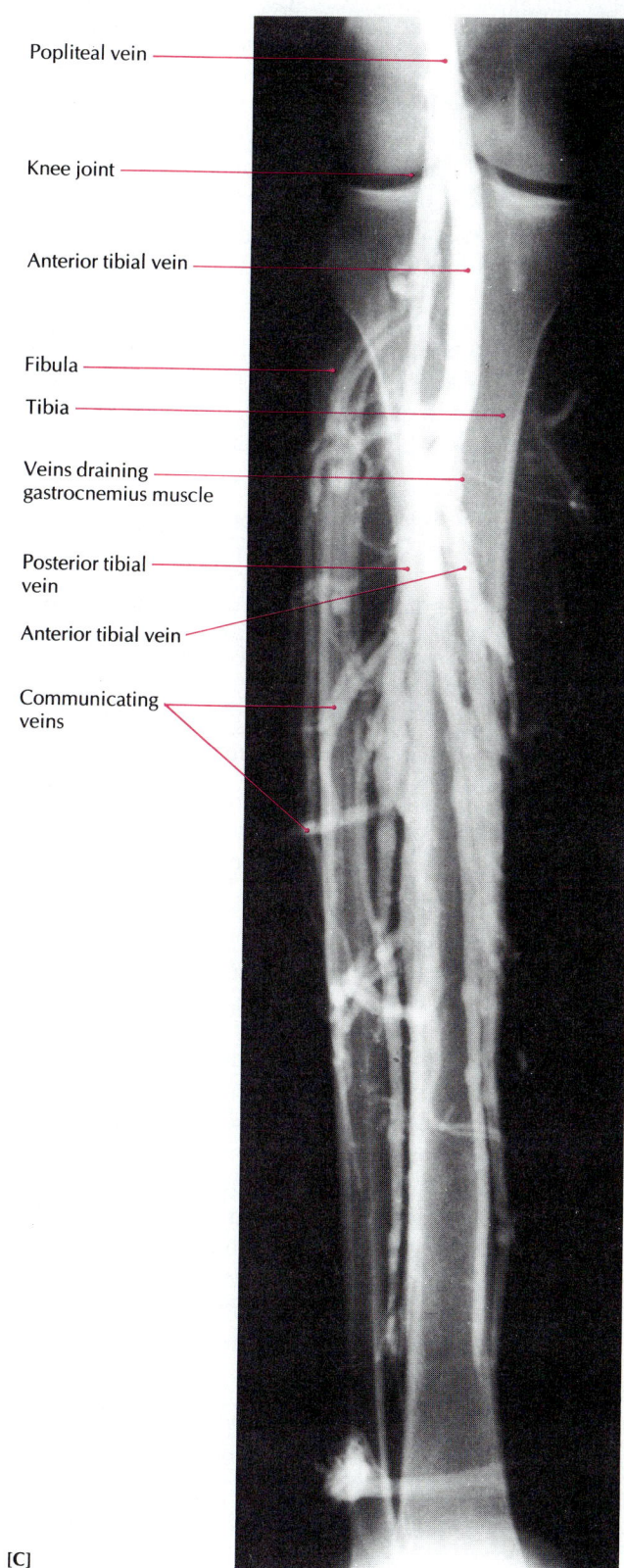

Popliteal vein

Knee joint

Anterior tibial vein

Fibula

Tibia

Veins draining
gastrocnemius muscle

Posterior tibial
vein

Anterior tibial vein

Communicating
veins

[C]

Major veins	Region drained	Comments
Deep veins		
Femoral vein	Lower limb	Deep veins are organized as *venae comitantes*. Femoral vein passes under inguinal ligament to become external iliac vein of pelvis.
Popliteal vein		
Anterior tibial vein		
Posterior tibial vein		
Peroneal vein		
Dorsal pedal vein		
Medial plantar vein		
Lateral plantar vein		
Superficial veins		
Small saphenous vein	Lateral side of leg	Superficial veins are cutaneous and have anastomoses with deep veins. The small saphenous vein arises at lateral side of dorsal venous arch of foot, ascends on posterolateral side of calf, and joins deep popliteal vein.
Great saphenous vein	Medial side of lower limb	Great saphenous vein arises from medial side of dorsal venous arch of the foot, ascends upward on medial aspect of leg and thigh, and joins deep femoral vein in groin.

PHYSIOLOGY OF THE CIRCULATORY SYSTEM

The overall function of the cardiovascular system is to transport life-sustaining blood from one part of the body to another. Although the heart and all the blood vessels play important roles in circulating blood, the capillaries are the focal point of the system for nutrient and waste exchange. In fact, all the physiological mechanisms that help regulate cardiac output and the diameter of blood vessels function to supply the capillaries with the precise volume of blood per unit time that will maintain homeostasis in the body's cells. These cardiovascular activities are integrated to maintain an optimum blood pressure and blood flow through the capillaries.

Hemodynamics: Principles of Blood Flow

Hemodynamics ("blood power") is the study of the physical principles that govern *blood flow* through the blood vessels and heart. Blood is forced out of the heart and through the blood vessels under great *pressure*. Blood flows through the narrow arterioles, capillaries, and venules with much difficulty. In other words, these vessels offer *resistance* to the flow of blood. The physiological regulation of blood flow, pressure, and resistance in maintaining homeostasis is based on some simple laws of hemodynamics.

Blood flow The term *blood flow* refers to the volume (quantity) of blood flowing through a vessel or group of vessels during a specific period of time. It is usually measured in milliliters per minute. The volume of blood that circulates through the systemic or pulmonary circulation each minute is the *cardiac output* (CO). In Chapter 20 it was noted that CO equals stroke volume (SV) × heart rate (HR). Two other hemodynamic factors that influence cardiac output are blood pressure and resistance.

Blood pressure *Blood pressure* is the force (energy) with which blood is pushed against the walls of blood vessels. It is usually measured in millimeters of mercury (mmHg). The rhythmic expansion and contraction of the vessels create a pulsating pressure wave. Blood flows down a *pressure gradient*, from one vessel or vessel bed where a higher pressure exists (P_1) to the other end of the vessel or vessel bed (P_2) where a lower pressure exists. It is the difference between these two points ($P_1 - P_2$) that determines blood flow down this pressure gradient. The greater the pressure gradient, the greater the blood flow through that vessel or vessel bed.

Peripheral resistance As circulating blood cells touch vessel walls and other blood cells, their flow is impeded by friction. The friction offered by the entire system of blood vessels is called *total peripheral resistance*. The amount of resistance to flow depends on (1) the *length* and (2) the *radius* of the blood vessels, and (3) the *viscosity* of the blood. Longer vessels and vessels of smaller radii, as well as higher blood viscosity, all result in a greater resistance to flow. However, the most important and most variable of these factors in influencing resistance to flow is the radius of the vessels: The greater the radius of the vessels, the greater the flow; the smaller the radius, the less the flow.

Interrelationship of pressure, flow, and resistance You saw above that blood flow (F) is caused by a blood-pressure difference ($P_1 - P_2$), and is opposed by resistance to flow (R). This relationship can be expressed as follows:

$$\text{Blood flow (F)} = \frac{\text{pressure difference } (P_1 - P_2)}{\text{resistance to flow (R)}}$$

You have also seen above that resistance to blood flow can be expressed as a ratio of the factors that govern blood flow:

$$\text{Resistance (R)} = \frac{\text{vessel length} \times \text{viscosity of blood}}{\text{vessel radius}^4}$$

The interrelationship of blood flow (also previously equated to cardiac output, CO), pressure difference, and resistance (also referred to as total peripheral resistance, TPR) is as follows:

$$\text{Cardiac output (CO)} = \frac{\text{pressure difference } (P_1 - P_2)}{\text{total peripheral resistance (TPR)}}$$

In this equation P_1 equals the pressure generated during the ejection of blood from the left ventricle into the aorta during systole. P_2 equals the pressure in the right atrium during diastole. Thus the total drop in pressure through the entire systemic circulation is equal to $P_1 - P_2$. Because the pressure in the right atrium during diastole (P_2) is approximately 0 mmHg, P_1, which represents the mean (the middle point between systolic and diastolic) arterial pressure, becomes the significant driving force at the beginning of the systemic circulation. The equation can finally be written as

$$\text{Cardiac output} = \frac{\text{mean arterial pressure}}{\text{total peripheral resistance}}$$

that is,

$$CO = \frac{MAP}{TPR}$$

or

Mean arterial pressure = cardiac output
$\qquad\qquad\qquad$ × total peripheral resistance

that is,

MAP = CO × TPR

As you will see in the following sections, the body has various mechanisms for regulating vessel diameter (and total peripheral resistance) and cardiac output, which in turn help maintain a homeostatic mean arterial pressure and a steady blood flow to the tissues.

Blood velocity *Blood velocity* is the distance blood flows along a vessel during a specific time period; it is measured in units of distance per unit time, such as centimeters per second. The velocity (v) of blood flow is inversely related to the total cross-sectional area (A) of blood vessels ($v = 1/A$). This means that the velocity of blood flow is slowest where the total cross-sectional area is the greatest. The velocity of blood flow in the aorta is approximately 30 cm/sec, in arterioles 1.5 cm/sec, in capillaries 0.04 cm/sec, in venules 0.5 cm/sec, and in the venae cavae 8 cm/sec [FIGURE 21.20]. Generally, as the vessels branch out from an artery like limbs of a tree, the diameter of a vessel gets smaller, but the total cross-sectional area increases and the blood velocity decreases. When venules merge into veins, the total cross-sectional area decreases and blood velocity increases [FIGURE 21.20].

Capillaries, the most numerous of the blood vessels, have the smallest diameter and the greatest total cross-sectional area. (See TABLE 21.1 on page 656.) As a result, the velocity of blood flow is slowest in capillaries. (Blood remains in each capillary about 1 to 2 sec.) This slow blood flow allows enough time for the exchange of gases, nutrients, and waste products between the blood and extracellular tissue fluid.

A S K Y O U R S E L F

1 What is hemodynamics?

2 How is blood pressure related to blood flow and resistance?

Factors Affecting Resistance

The most important factor in determining resistance to blood flow is the size of the lumen (radius or diameter) of a blood vessel. This opening is regulated primarily by (1) neural, (2) hormonal, and (3) autoregulatory mechanisms that affect the smooth muscle in the walls of arterioles and precapillary sphincters.

The processes of constricting blood vessels *(vasoconstriction)* and dilating blood vessels *(vasodilation)* are collectively called *vasomotion.* Vasomotion affects both

FIGURE 21.20 RELATIONSHIP BETWEEN BLOOD PRESSURE, BLOOD VELOCITY, AND CROSS-SECTIONAL AREA

[A] Blood pressure and relative cross-sectional area of vessels ranging from the aorta (largest artery) to the tiny capillaries to the venae cavae (largest veins). [B] Relationship between blood velocity (red curve) and total cross-sectional area of blood vessels (black curve).

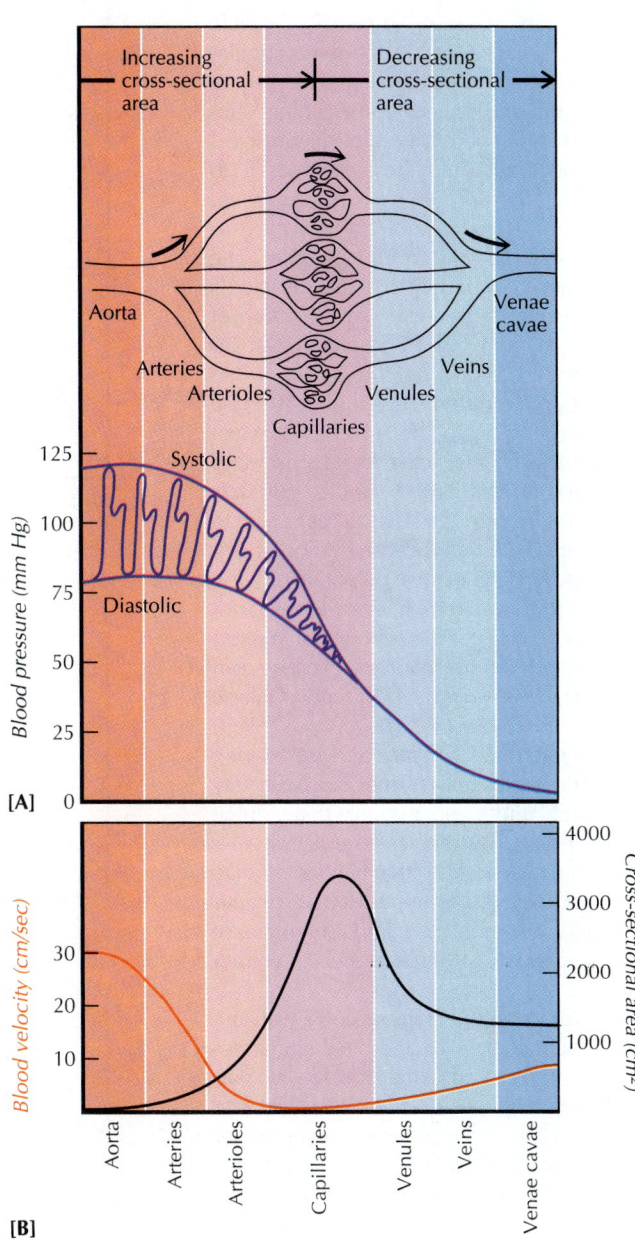

the arteries and arterioles (known as *resistance vessels*) and the venules and veins (known as *capacitance vessels*). Intense vasoconstriction of arterioles can completely stop the flow of blood to a capillary bed, whereas complete dilation can cause more than a 20-fold increase in the rate of blood flow through the same bed.

Effect of Exercise on the Cardiovascular System

The human body works best when it is physically active. Exercise is especially beneficial to the cardiovascular system. To obtain a maximum and permanent benefit from exercise, however, it must be done regularly.

During exercise (especially aerobic exercise), changes in the microcirculation of the body increase the delivery of oxygen and other life-sustaining substances to the active skeletal muscles. Blood flow during rest averages 6 to 10 mL/min per 100 g of muscle tissue, but during exercise blood flow may increase to 80 mL or more (graph A). The extra oxygen supply to active muscles is produced by an increased extraction of oxygen from the blood and by an increase in cardiac output.

Cardiac output rises during exercise in proportion to the increased oxygen demand by the active skeletal muscles (graph B). During rest, cardiac output is approximately 5 L/min. It can rise to about 35 L/min in well-conditioned athletes. A rise in cardiac output of 120 percent is quite common, even during moderate exercise. The amount of oxygen released during vigorous exercise can increase up to 15 mL/100 mL of blood (graph B).

The **stroke volume** increases during moderate exercise from approximately 70 mL/min to about double that volume (graph C). This increase is due primarily to an increased contractility of cardiac muscle caused by impulses from sympathetic nerves. These impulses increase the strength of the ventricular contraction.

Heart rate increases along with the amount of exercise and is the major factor in increasing the cardiac output. During vigorous exercise, the heart rate can reach approximately 200 beats per minute. As the physical conditioning of an individual improves, however, the amount of increase in heart rate becomes progressively less. This happens because the stroke volume is greater at the beginning of exercise, keeping the

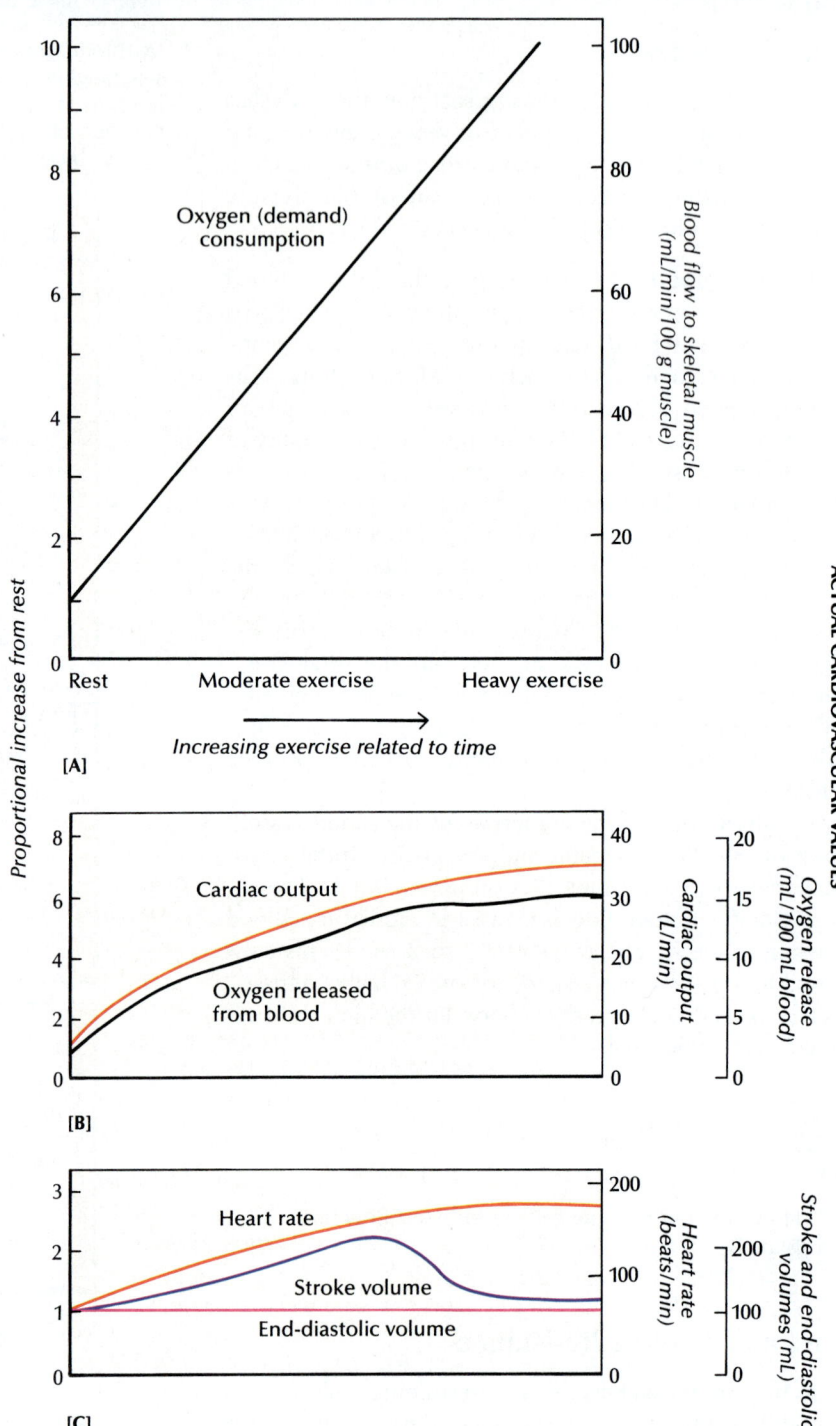

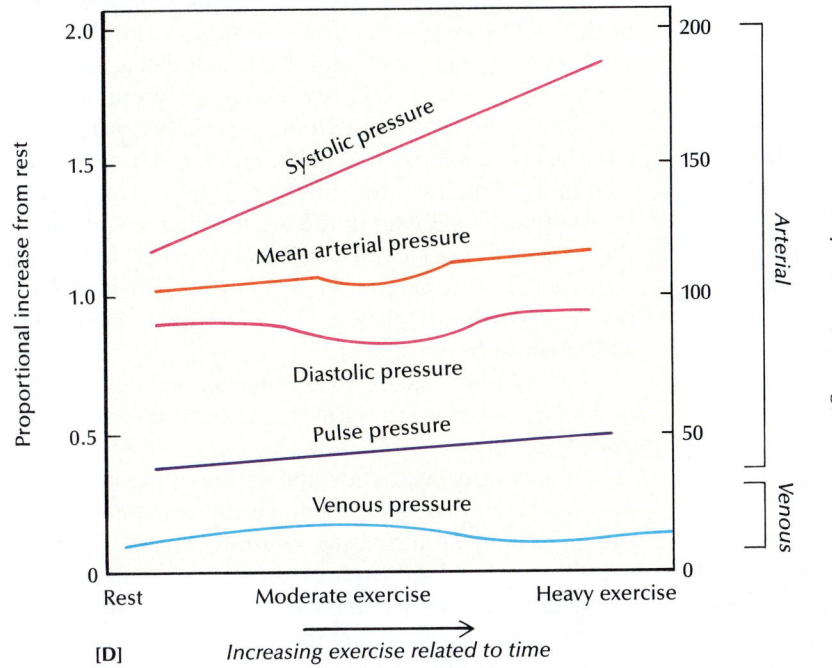

Effect of exercise on various cardiovascular functions. [A] An increase in exercise causes a proportional increase in oxygen consumption by skeletal muscles. [B] The increased demand for oxygen causes an increased cardiac output, which increases blood flow to skeletal muscles, and leads to a greater amount of oxygen extraction from the blood. [C] The increase in cardiac output is due primarily to an increase in heart rate and only secondarily to a small increase in stroke volume. Note that the stroke volume actually falls after prolonged exercise with an increased cardiac output due to less time during diastole for complete filling. [D] Arterial pressure changes more than venous blood pressure during exercise. The increase in pulse pressure (systolic — diastolic) is related primarily to the increase in systolic pressure. This is due to the quicker ejection of blood by the heart due to sympathetic nervous stimulation.

cardiac output the same. Because the cardiac output increases sevenfold while the heart rate increases threefold, the stroke volume must be more than double (graph C). Note also that there is no change in the end-diastolic ventricular volume.

Blood pressure in the systemic circulatory system increases temporarily as exercise increases (graph D). Systolic blood pressure increases more than diastolic pressure does, but it seldom rises above 180 mmHg during heavy exercise. In contrast, diastolic pressure does not increase significantly. In fact, it actually *decreases* during moderate exercise. Because the systolic pressure increases more than does the diastolic, the pulse pressure increases. The increase in mean arterial pressure is, by comparison, less. There is only a slight increase in venous pressure at the beginning of exercise, but it drops eventually to resting values (averaging 15 mmHg).

There is a marked dilation of blood vessels in the skeletal muscles during exercise. At the same time, vasoconstriction occurs in the skin and other organs.

The ***mass of the heart*** increases permanently after prolonged periods of exercise. This enlargement is caused by either the increased thickness of the ventricular walls (cardiac hypertrophy) or a lengthening of cardiac muscle fibers, depending on the exercise regimen (isometric or isotonic exercise, respectively). This increase in mass benefits the heart by strengthening it.

The density of capillaries in cardiac muscle increases after a period of regular exercise, and the number of capillaries in skeletal muscle tissue also increases, allowing more blood to flow to the muscle fibers. Both factors improve the transport of oxygen to active muscles.

Cardiovascular responses to exercise depend on the frequency, duration, and intensity of the exercise being done. Strenuous exercise is the most effective way to increase cardiac output.

Neural control of blood flow Sympathetic stimuli cause the blood vessels (specifically, arterioles, metarterioles, and venules) of the skin and abdominal organs to constrict moderately, producing a vasomotor tone in those areas and establishing a base, or resting, level of vascular flow. In contrast, sympathetic stimuli cause blood vessels in the heart and skeletal muscles to dilate, increasing blood flow. Parasympathetic stimuli cause the blood vessels of the reproductive organs and digestive tract to dilate by releasing acetylcholine, which inhibits smooth muscle contraction.

As presented in Chapter 16, most postganglionic sympathetic neurons release norepinephrine, which combines with alpha-adrenergic receptors on the smooth muscle in the blood vessel walls, causing vasoconstriction and thus raising blood pressure. The blood vessels of the heart and skeletal muscle have beta-adrenergic receptors, which respond to sympathetic stimulation, causing vasodilation rather than vasoconstriction. This ensures that they receive adequate blood flow during stressful situations.

Some sympathetic neurons to the blood vessels of skeletal muscle are also cholinergic, releasing acetylcholine and causing vasodilation. In summary, sympathetic innervation to vascular smooth muscle may cause vasoconstriction in some tissue and vasodilation in others.

During exercise, both vasoconstriction and vasodilation may occur, but in different tissues, thus changing the total peripheral resistance and blood pressure. Sympathetic stimulation to the veins primarily causes vasoconstriction (venoconstriction), increasing venous return to the heart from the venous reservoirs and increasing blood pressure.

Hormonal and chemical control of blood flow Blood flow is also regulated by hormonal and other chemical substances that can cause constriction and dilation. The *catecholamines* (KAT-ih-KOHL-uh-meenz; epinephrine and norepinephrine) that are released by the adrenal medulla vasoconstrict skin and abdominal arterioles and veins (alpha receptors), and vasodilate heart and skeletal muscle arterioles (beta receptors). The hypothalamic peptide *vasopressin (ADH)* and the blood-borne *angiotensin II* are powerful vasoconstrictors. Specific tissues also produce chemicals that have localized effects. For example, eosinophils and mast cells in most tissues can release *histamine,* which causes vessel dilation. The heart produces a hormone, *atrial natriuretic peptide (ANP),* that is a vasodilator. Finally, some tissues, such as blood, contain *bradykinin* (BRAD-ee-kye-nihn) which is a strong vasodilator.

Autoregulation of blood flow *Autoregulation* is the local adjustment of blood flow to a given tissue due primarily to chemical and physical factors in that area.

Chemical (metabolic) factors that cause *vasodilation* include a decrease in oxygen or nutrient levels, or an increase in the carbon dioxide level in that area; a decrease in pH; and increases in adenosine, lactic acid, and nitric oxide (NO). *Vasoconstrictor factors* include the local release of eicosanoids and endothelins from endothelial cells that line blood vessel walls. *Endothelins* are polypeptides that act as powerful vasoconstrictors, apparently assisting the entry of calcium into the smooth muscle of blood vessels. The body contains three different forms of endothelin: (1) *endothelin 1* is found in the brain, kidneys, and endothelial cells; (2) *endothelin 2* is found primarily in the intestine; and (3) *endothelin 3* is found primarily in the intestine and adrenal glands. The exact functions of the endothelins are not known.

Local physical factors also influence the autoregulation of blood flow to various areas. Heat application causes vasodilation and increases blood flow, whereas cold causes vasoconstriction and decreases blood flow in the area. Arteriolar smooth muscle also responds to passive stretching by increasing its tone (*myogenic* [myo = muscle; gen = origin] *response*), thereby resisting the passive stretch and causing vasoconstriction. Thus metabolic and myogenic responses are instrumental in the autoregulation of blood flow to various tissues and organs in the body.

A S K   Y O U R S E L F

1 What factors affect peripheral resistance?

2 How is the nervous system involved in the regulation of blood flow?

3 What are some of the hormonal and chemical controls of blood flow?

4 How are the arterioles involved in autoregulation?

Factors Affecting Blood Pressure

Several factors affect blood pressure in both the arteries and the veins. We consider these factors in the following sections before turning to a discussion of the regulation of blood pressure.

Hydrostatic pressure and gravity The pressure that results from gravity acting on the fluids of the body influences *hydrostatic pressure.* It determines the weight of the blood in the various vessels, and therefore affects venous return. For example, when a person is standing quietly, the pressure in the right atrium is approximately 0 mmHg because the heart pumps into the arteries any excess blood that accumulates. Because of gravity, however, blood pressure increases between the heart and feet. For example, blood pressure is approximately 8 mmHg in the arm veins, 22 mmHg in the inferior vena cava,

40 mmHg in the great saphenous vein, and 90 mmHg in the foot. This gravity-induced high venous pressure in the lower extremities must be overcome continuously in order to prevent blood from pooling in the extremities. When people must stand for long periods, the high venous pressure cannot be overcome. As a result, the veins in the leg may become permanently distended, a condition called *varicose veins* (see When Things Go Wrong, page 703).

Because the blood has volume, hydrostatic pressure results from the mere weight of the blood and the position of the body. When a person is standing or seated, gravity operates to speed the return of blood to the heart from the veins above the heart. However, the weight of the blood from the heart down requires considerable extra pressure in these veins to drive the blood "uphill." On the other hand, if a person lies down and elevates the legs above the level of the heart, venous blood pressure, following the effects of gravity, moves blood more easily from the extremities back to the heart.

Vessel elasticity It is primarily the *vessel elasticity* (flexibility) of the large arteries that determines the pressure within a vessel when the ventricles contract. The elasticity results from the resilient nature of the arterial wall. If a vessel becomes less elastic, as often happens with aging (arteriosclerosis), the blood pressure rises, and the recoil is lost.

Cardiac output As described in Chapter 20, the volume of blood ejected by the left ventricle per minute is the *cardiac output* (mL/min). It is equal to the stroke volume (mL/beat) × heart rate (beats/min). After leaving the left ventricle, blood enters a closed system of vessels, which has a given capacity at a given time. If the cardiac output increases and the volume in the vessels remains the same, blood pressure will increase momentarily. If the cardiac output decreases and blood volume remains the same, the blood pressure will decrease momentarily.

Regulation of Arterial Blood Pressure

The blood pressure within the large systemic arteries must be maintained precisely to ensure adequate blood flow to the tissues. The major factors that regulate arterial blood pressure are those that regulate cardiac output and total peripheral resistance. FIGURE 21.21 briefly summarizes these factors, which have been discussed previously. Take time to review them.

Baroreceptor reflex Arterial pressure is constantly monitored by *baroreceptors* (pressure sensors) located primarily in the carotid sinus and aortic arch, but to a lesser extent in the walls of arteries, veins, and right atrium. (The carotid sinus is a widening in the internal carotid artery where the external and internal carotid arteries branch from the common carotid artery.) These carotid and aortic baroreceptors sense changes in arterial pressure and continuously generate action potentials that correspond to blood pressure changes [FIGURE 21.22]. When arterial pressure increases, the action potential firing rate from the baroreceptors increases, and when blood pressure decreases, the firing rate from the baroreceptors decreases. The baroreceptors send their "information" (afferent impulses) regarding the status of blood pressure to the *cardioregulatory center* in the medulla oblongata. The outflow from the cardioregulatory center (efferent impulses) is by way of the autonomic nervous system (sympathetic and parasympathetic divisions) to the heart and blood vessels.

Chemoreceptor reflex *Chemoreceptors* (chemical sensors) also help regulate blood pressure by monitoring certain blood-borne substances (oxygen, carbon dioxide, hydrogen-ion concentrations). The chemoreceptors are located close to the aortic and carotid sinus baroreceptors in small structures called *aortic* and *carotid bodies.* Their main function is to increase respiration in order to increase oxygen intake and carbon dioxide output, but they also reflexly increase blood pressure by sending afferent impulses via the glossopharyngeal (IX) and vagus (X) nerves to the cardiovascular regulatory center in response to a low oxygen concentration or high carbon dioxide or hydrogen-ion concentrations.

Other factors influencing the control of blood pressure Several other factors help regulate blood pressure:

1 *Left atrial volume receptors and hypothalamic osmoreceptors* help regulate the body's salt and water balance. This helps control blood pressure through the regulation of *blood volume.*

2 The *higher brain centers* and *emotions* play important roles in affecting arterial pressure. For example, any excitatory emotion such as fear or rage that stimulates the sympathetic nervous system also stimulates the *vasomotor center,* located in the lower part of the pons and medulla oblongata, which causes constriction of arterioles and a subsequent rise in blood pressure. Opposite emotions such as grief, loneliness, and depression cause a decrease in sympathetic stimulation, dilation of arterioles, and a decrease in arterial blood pressure. Many other physiological, psychological, and pathological factors can affect arterial blood pressure.

3 The *viscosity of blood* determines its resistance to flow. A thin fluid flows more easily than a thicker one, and less force (blood pressure) is required to move it along. In contrast, a thick fluid has more resistance than a thin one and requires a greater blood pressure to make it flow.

FIGURE 21.21 REGULATION OF MEAN SYSTEMIC ARTERIAL BLOOD PRESSURE

A summary of the major factors involved in the regulation of mean systemic arterial blood pressure. This flowchart is a composite of the physiological factors that affect cardiac output and total peripheral resistance.

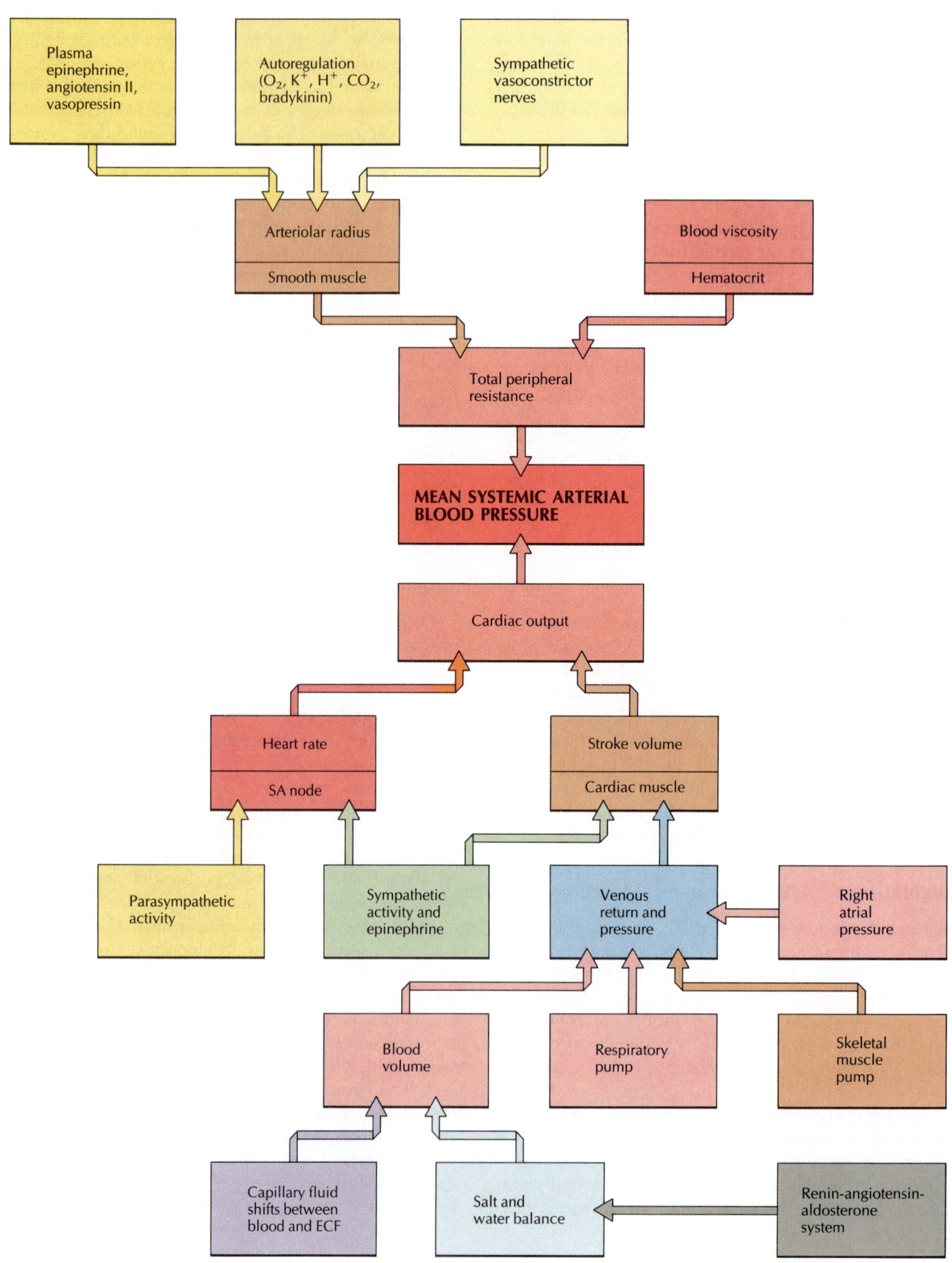

FIGURE 21.22 BARORECEPTOR REFLEXES TO CORRECT BLOOD PRESSURE CHANGES

[A] Baroreceptor reflex response to elevated blood pressure.
[B] Baroreceptor reflex response to decreased blood pressure.

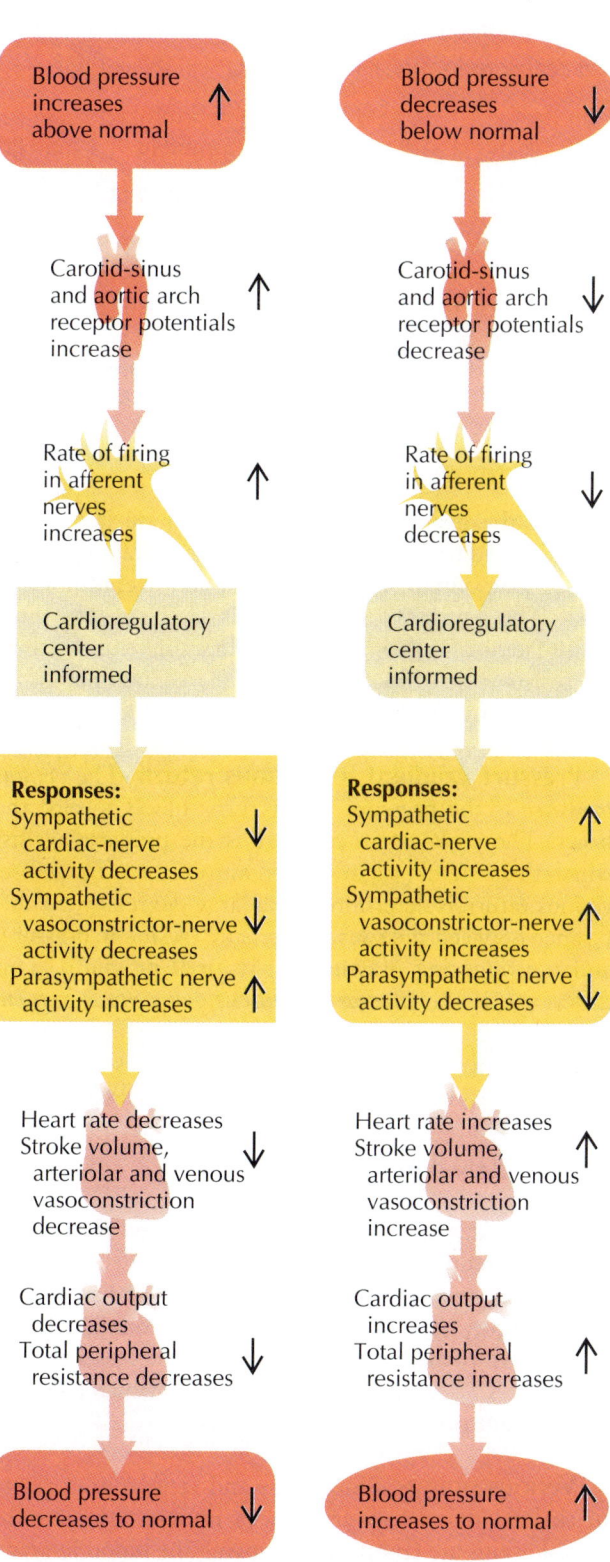

[A] RESPONSE TO ELEVATED BLOOD PRESSURE

[B] RESPONSE TO DECREASED BLOOD PRESSURE

4 Several *hormones* affect arterial pressure. For example, the ***renin-angiotensin system*** takes several hours to alter arterial pressure. When too little blood reaches the kidneys, they produce an excess of an enzyme called *renin* (L. *renes*, kidneys). Renin splits off the protein *angiotensin I* from the large plasma protein angiotensinogen. Angiotensin I then undergoes further cleavage in the lungs to form *angiotensin II*, a powerful but short-acting vasoconstrictor that raises arterial pressure. Angiotensin II also increases blood pressure by stimulating the release of aldosterone from the adrenal cortex, which increases water reabsorption and blood volume, and by stimulating neural centers in the hypothalamus that increase the intake of water to increase blood volume, blood flow, and blood pressure.

A constriction of vessels that leads to increased arterial pressure may also be caused by increased levels of other hormones, including vasopressin, epinephrine, and norepinephrine.

5 In addition to exerting a hormonal control, the ***kidneys*** help control arterial pressure by regulating blood volume. If the arterial blood pressure increases, the kidneys filter a greater amount of blood, which increases the quantity of fluids and solutes excreted in urine. As a result, blood volume is lowered and arterial pressure drops. In contrast, if the pressure decreases, fluid is retained, blood volume increases, and arterial pressure rises. Overall, the kidneys play a major role in the long-term regulation of arterial pressure.

6 A mechanism called the ***capillary fluid shift*** regulates arterial pressure by altering blood volume [FIGURE 21.23]. A rise in arterial pressure forces fluid out of the permeable capillaries and into the extracellular fluid, lowering blood volume and, consequently, arterial pressure. A fall in arterial pressure causes more fluid to enter the capillaries, which raises blood volume and arterial pressure. This same mechanism plays an important role in the second-by-second exchange of materials between blood, extracellular tissue fluids, and the tissue cells themselves.

The mechanism responsible for the capillary fluid shift is called the *Starling-Landis concept*, named after the individuals who first described it in 1896. Basically, there are four mechanisms that can move materials through capillary walls: *diffusion, filtration, osmosis,* and *pinocytosis.* Oxygen, for example, *diffuses* out of systemic capillaries from an area of high concentration to one of lower concentration. Many capillaries have fenestrations, and small molecules can also diffuse through the endothelial cell membranes. Diffusion, along with pinocytosis to a limited degree, controls most of the metabolic exchange of nutrients, gases, and wastes that occurs in capillaries.

The force behind *filtration*, which pushes fluid out of the capillary, is blood pressure, or ***hydrostatic pressure (HP).*** An opposing force called ***osmotic pressure (OP)***

FIGURE 21.23 CAPILLARY FLUID SHIFT

The drawing shows the forces involved in filtration and re-absorption in a capillary. The values are general and do not apply to all capillaries. Values are in mmHg, where HP_c = capillary hydrostatic pressure, OP_t = tissue osmotic pressure, OP_c = capillary osmotic pressure, and HP_t = tissue hydrostatic pressure. The thickness of the arrows indicates the relative amounts of fluid flowing in a particular direction in response to the specific pressure differences.

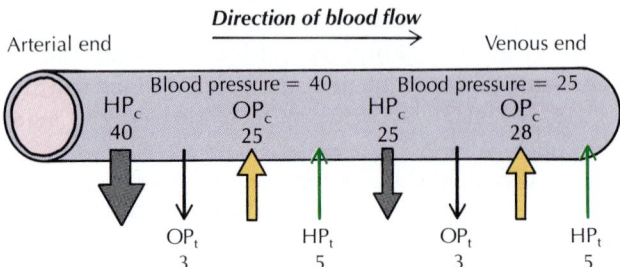

helps prevent fluid from accumulating by returning it to the blood. Osmotic pressure is produced by the plasma proteins in the blood serum. Recall that water moves across a semipermeable membrane from a region of low osmotic pressure (high water pressure) to a region of higher osmotic pressure (lower water pressure). This specific osmotic pressure promotes *reabsorption*, which draws fluid back into the capillaries. Extracellular fluid exerts a slight hydrostatic pressure that opposes the movement of fluid out of the capillary. It also has a slight osmotic pressure because of the proteins in the fluid.

The *net force* for fluid movement, either out of or into the capillary, is the difference between those forces moving fluid *out* of the capillary (capillary HP_c + tissue OP_t) and those moving fluid *into* the capillary (capillary OP_c + tissue HP_t).

Using typical values for the arterial end of a capillary (where t = tissue values and c = capillary values), we can calculate the net filtration (*fluid out*) at the arterial end of a capillary as follows:

Net filtration = fluid out − fluid in

or

Net filtration out = $(HP_c + OP_t) - (OP_c + HP_t)$

$$= (40 \text{ mmHg} + 3 \text{ mmHg})$$
$$- (25 \text{ mmHg} + 5 \text{ mmHg})$$
$$= +13 \text{ mmHg forcing fluid out}$$

The inward movement of fluid results from the marked drop in blood pressure (from 40 to 25 mmHg) as blood moves through the capillary and from a slight rise in blood osmotic pressure (from 25 to 28 mmHg). This rise in blood OP_c occurs when water filters out of the capil-

lary, leaving protein in. Thus, the net reabsorption (*fluid in*) at the venous end of a capillary can be calculated as follows:

Net reabsorption = fluid in − fluid out

or

Net reabsorption in = $(OP_c + HP_t) - (HP_c + OP_t)$

$$= (28 \text{ mmHg} + 5 \text{ mmHg})$$
$$- (25 \text{ mmHg} + 3 \text{ mmHg})$$
$$= +5 \text{ mmHg forcing fluid in}$$

Notice that the net pressure forcing fluid out (+13 mmHg) is higher than the pressure forcing fluid in (+5 mmHg). The fluid that is not reabsorbed, along with any protein that escapes from the plasma and cells, is returned to the venous circulation by the lymphatic vessels. This return maintains fluid homeostasis.

Regulation of Venous Pressure

Because the heart pumps out all the blood it receives, venous blood pressure is important in maintaining a normal cardiac output, blood pressure, and overall homeostasis. The return of venous blood is determined by three major factors: (1) the pressure gradient for venous return, (2) resistance to blood flow, and (3) venous pumps. A summary of these factors is shown in FIGURE 21.24.

Pressure gradient for venous return The *venous pressure* is the pressure that returns blood toward the heart. The *right atrial pressure* is the pressure against which the systemic venous blood must return. Therefore, venous return is determined to a large degree by the difference between the causing pressure (systemic pressure) and the opposing pressure (right atrial pressure). This difference is the *pressure gradient for venous return*.

Resistance to blood flow Since venous blood flow is determined by the same formula as arterial blood flow,

$$\text{Venous blood flow} = \frac{P_1 - P_2}{R},$$

any factor that increases *resistance to blood flow* in the veins also reduces the amount of blood returning to the right heart *(venous return)* and in turn, cardiac output. For example, because the walls of veins contain smooth muscle, sympathetic stimulation causes the vessel walls to constrict. This raises the pressure within, which increases venous return.

Venous pumps Since veins and venules are present in skeletal muscles, every time a muscle contracts the flexible veins are compressed, moving the blood forward. Since most veins have valves that prevent backflow, blood

FIGURE 21.24 FACTORS THAT INFLUENCE VENOUS RETURN

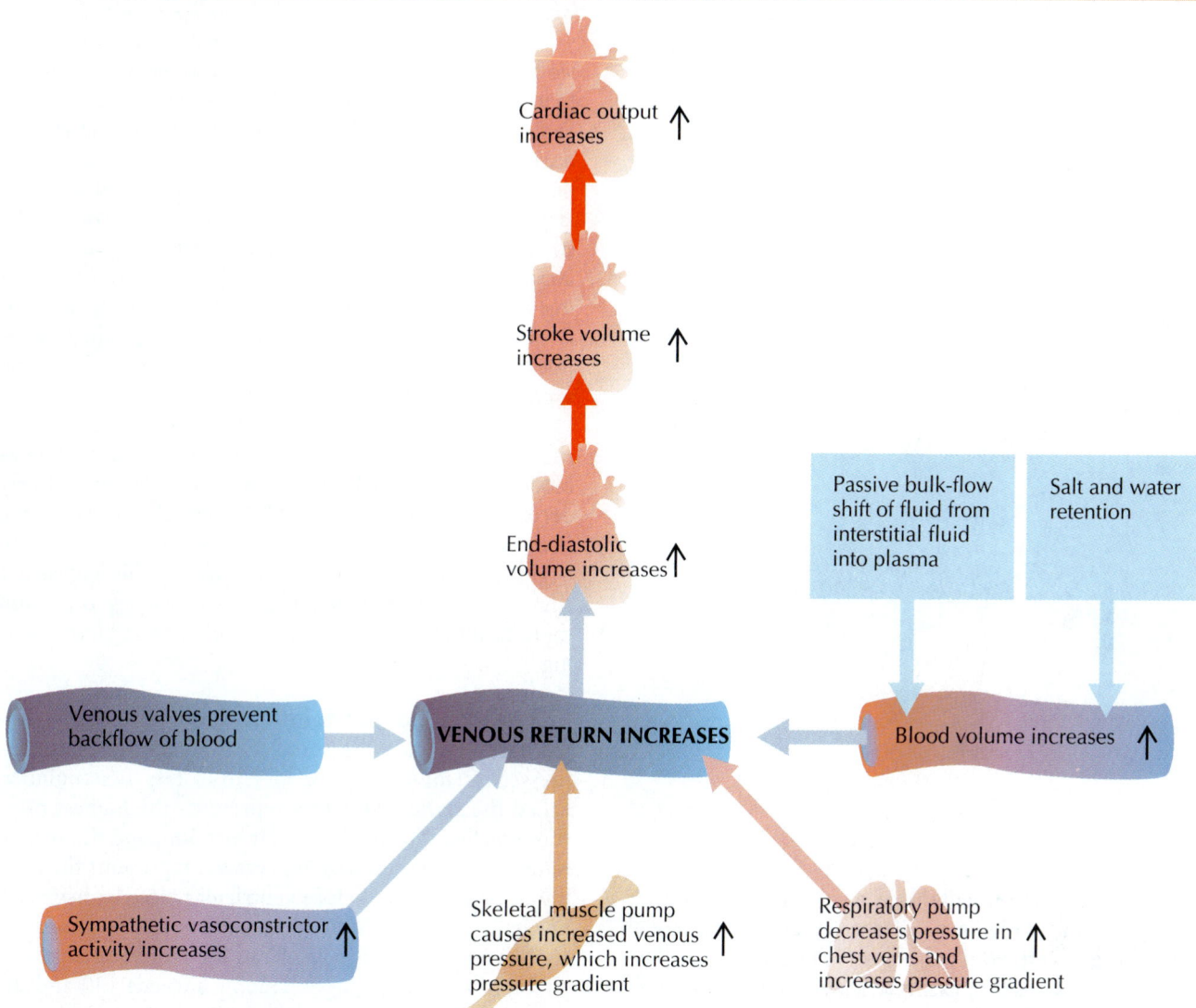

is allowed to flow only *toward* the heart. This one-way mechanism is called the ***skeletal muscle venous pump.*** During inspiration (breathing in), the diaphragm pushes on the abdominal viscera, increasing abdominal and decreasing thoracic pressures. As a result, the venous pressure gradient between the veins in the abdomen and those in the thorax is increased. The final effect is an auxiliary ***respiratory pump,*** which increases venous return during inspiration.

If a person stands still for a long time, the leg muscles cannot push the venous blood up rapidly enough, and the venous blood tends to accumulate in the leg veins. Reduced return of blood to the heart may result in poor blood flow to the rest of the body. Decreased blood flow to the brain can cause fainting. (Fainting usually results in a person lying down, helping blood flow to the brain without the resistance of gravity.)

Measuring Pulse Rate and Arterial Blood Pressure

Pulse   A *pulse* can be felt from an artery that is close to the surface of the skin. This ***arterial pulse*** corresponds to the beating of the heart and the alternating expansion and elastic recoil of the arterial wall. The pulse is produced when the left ventricle forces blood against the wall of the aorta, and the impact creates a continuing pressure wave along the branches of the aorta and the rest of the elastic arterial walls. A ***venous pulse*** occurs only in the largest veins. It is a reflected pulse produced by the changes in pressure that accompany atrial contractions. All arteries have a pulse, but it can be felt most easily at the points shown in FIGURE 21.25.

The most common site for checking the pulse rate is the radial artery on the underside of the wrist. The three

FIGURE 21.25 THE ARTERIAL PULSE

The pulse can be felt most easily at the points shown.

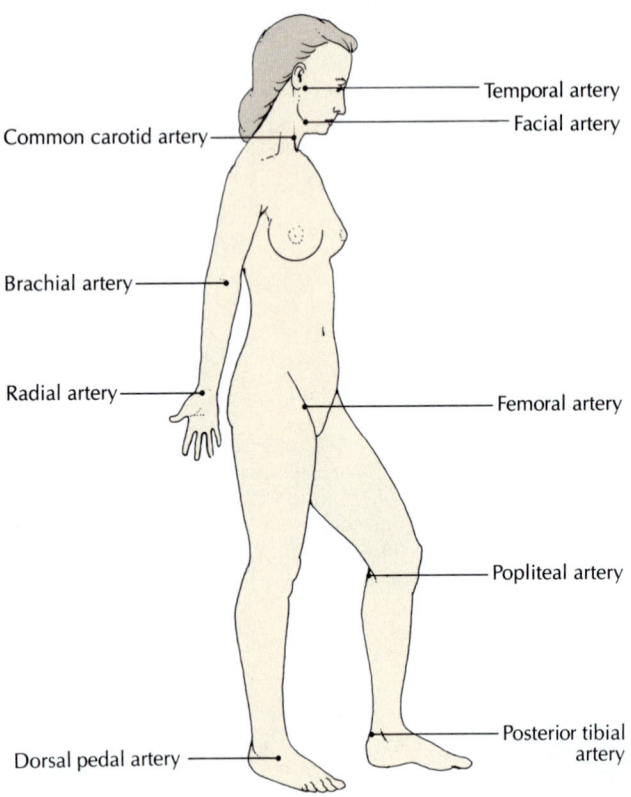

middle fingers are usually used to check the pulse (but not the thumb, which may be close enough to the radial artery to reflect its own pulse beat). The pulse is checked for several reasons. For example, a practitioner can detect the number of heartbeats per minute (heart rate), the strength of the beat, the tension of the artery, the rhythm of the beat, and several other diagnostic factors. The average resting pulse rate can range from 70 to 90 beats per minute in adults and from 80 to 140 in children. When the pulse rate exceeds 100 beats per minute, the condition is called *tachycardia* (Gr. *takhus*, swift); when the rate is below 60 beats per minute, the condition is called *bradycardia* (Gr. *bradus*, slow).

The pulse rate normally decreases during sleep and increases after eating or exercising. During a fever it may increase about five beats per minute for every degree Fahrenheit above normal body temperature. Pulse rates tend to increase significantly after severe blood loss and are usually high in cases of serious anemia.

Pulse-wave propulsion *Arterial blood pressure* depends on the volume of blood in the arteries and the elasticity of the arterial walls, as well as on the rate and force of ventricular contractions. Because the arteries are elastic, they can be stretched by large volumes of blood without an appreciable rise in blood pressure.

When a volume of blood (SV) is ejected into the arteries by the ventricles during systole, an equal amount of blood is *not* simultaneously released out of the arteries. In fact, only about one-third of the blood leaves the arteries during systole, and the excess volume raises the arterial pressure. For example, when blood is ejected into the large arteries during ventricular systole, most of the stroke volume is used to stretch the walls of the arteries— almost like blowing up a balloon [FIGURE 21.26A]. During diastole, the elastic recoil of the arteries is used to keep the blood moving forward, even though the heart is not contracting, just as the elasticity of a balloon can be used to expel air out of the balloon.

After systole, when the ventricular contraction is over, the arterial walls return to their unstretched condition as blood continues to leave the arteries. Pressure slowly decreases, but before all the blood has left the artery, the next ventricular contraction occurs, and the pressure begins to build up again. Because of this consistent rhythm, the arterial pressure never reaches zero, and there is always enough pressure to keep the blood flowing.

Systolic and diastolic pressure Arterial blood-pressure levels are measured by two numbers, both expressed as millimeters of mercury. The first number, called the *systolic pressure,* represents the highest pressure reached during ventricular ejection, and the second number, called the *diastolic pressure,* represents the minimum pressure just before ventricular ejection begins. A normal young adult's blood pressure is 120/80 mmHg or *less.* Blood pressure is considered high, or *hypertensive,* in an adult when the systolic reading exceeds 140 and the diastolic reading is higher than 95.

Blood pressure varies with age. The systolic pressure of a newborn baby may be only 40, rising to 80 after a month. During adolescence it may progress from 100 to 120, and it continues to rise slightly throughout adulthood. The normal pressure of a 60-year-old man is about 140/90. Most physicians agree that blood pressure need not rise above an acceptable middle-age level, even in old age, if good health is maintained. Blood pressure usually rises temporarily during exercise or stressful conditions, and a systolic reading of 200 would not be considered abnormal under those circumstances.

Pulse pressure The difference between systolic and diastolic pressure is called the *pulse pressure,* which results in a pulsation or throb that can be felt in certain arteries with each heartbeat. In the example above of a blood pressure of 120/80, the pulse pressure equals 40 mmHg, as calculated below:

FIGURE 21.26 PULSE-WAVE PROPULSION OF BLOOD DURING SYSTOLE AND DIASTOLE

The elastic arteries help push the blood forward by contracting behind the systolic rush of blood that stretches the arterial walls. The elastic recoil of arteries during cardiac diastole continues to drive the blood forward, even when the heart is not pumping.

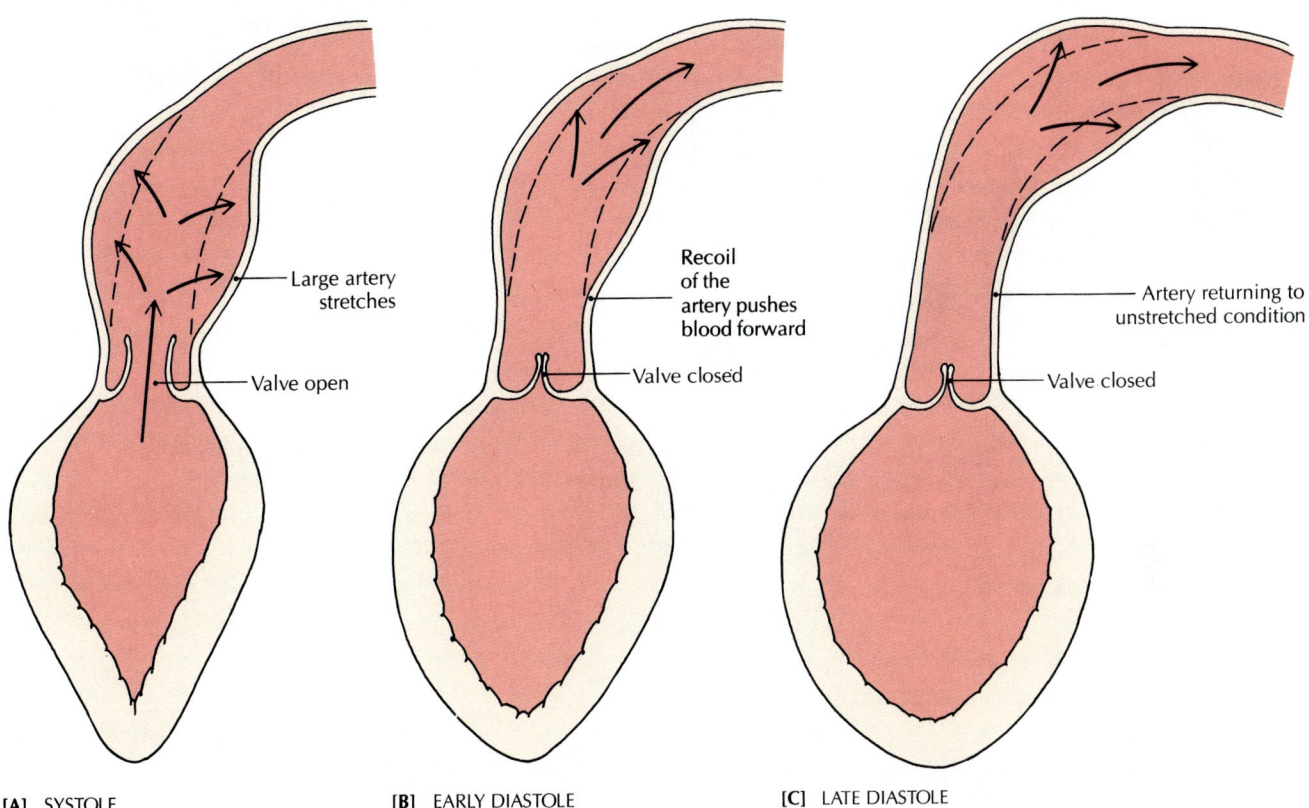

[A] SYSTOLE [B] EARLY DIASTOLE [C] LATE DIASTOLE

Pulse pressure = systolic pressure − diastolic pressure

= 120 mmHg − 80 mmHg

= 40 mmHg

Pulse pressure is determined by three factors: (1) arterial distensibility (the greater the distensibility, the higher the pulse pressure), (2) stroke volume (the greater the stroke volume, the higher the pulse pressure), and (3) the speed of the blood ejected from the left ventricle into the systemic circulation (the greater the speed, the higher the pulse pressure). Many cardiovascular diseases can be detected by the presence of abnormal pulse pressure. Some examples include atherosclerosis and patent ductus arteriosus (see When Things Go Wrong, pages 701 and 704).

Mean arterial blood pressure Arterial blood pressure changes constantly throughout the cardiac cycle. One of the most useful measurements indicating these pressure changes is the *mean arterial pressure (MAP)*. Mean arterial pressure is the average pressure that drives the blood through the systemic circulatory system. It is determined by two factors: (1) the cardiac output and (2) the rate at which blood drains from the arteries. Arterial drainage is controlled by the vascular resistance and blood flow. Therefore

MAP = cardiac output × total peripheral resistance

Since diastole usually lasts longer than systole, MAP is approximately equal to the diastolic pressure plus one-third of the pulse pressure (SP − DP), or

$MAP = DP + \frac{1}{3}(SP − DP)$

Using our previous example of 120/80, we have

$MAP = 80 + \frac{1}{3}(120 − 80) = 93.3$ mmHg

Under normal conditions, the mean arterial pressure is close to 90 mmHg. This pressure is crucial for the maintenance of a steady blood flow to capillaries during the entire cardiac cycle. Several regulatory mechanisms are available for achieving this stable blood pressure, including cardiac output, blood volume, and peripheral

FIGURE 21.27 MEASUREMENT OF ARTERIAL BLOOD PRESSURE

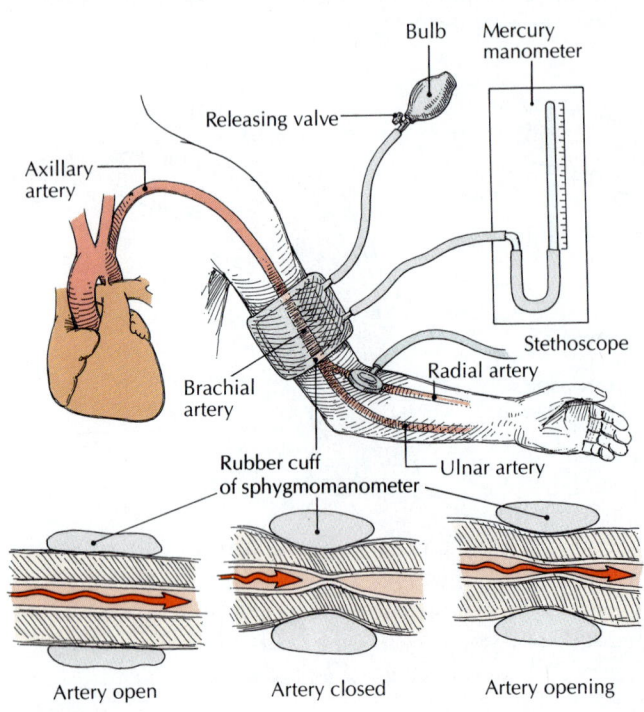

[A] The materials needed for measuring blood pressure. Arterial events are shown. [B] Sounds heard through a stethoscope when the cuff pressure of a sphygmomanometer is gradually reduced. The first sounds indicate systolic pressure, and the cessation of sound indicates diastolic pressure.

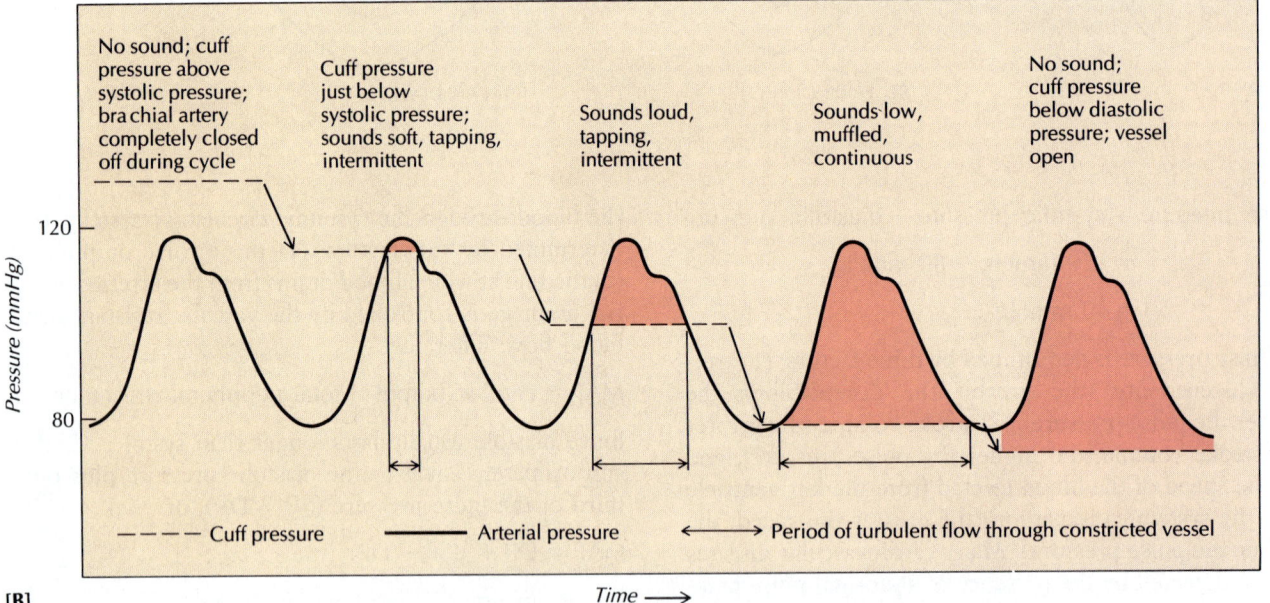

resistance. In Chapter 20 we examined specific neural, chemical, and hormonal factors that help regulate cardiac output by acting directly on the heart. The same factors also help regulate mean arterial blood pressure by acting directly on blood vessels. FIGURE 21.21 presents an overall view of this blood-pressure regulation.

Measuring arterial blood pressure The instrument used for measuring arterial blood pressure is the *sphygmomanometer* (sfig-moh-muh-NOM-ih-ter). It consists of an inflatable cuff, a rubber bulb, and either a column of mercury, an air gauge, or an electronic display [FIGURE 21.27]. The cuff is wrapped around the upper arm and inflated with the rubber bulb to a pressure above the systolic pressure. (This high cuff pressure compresses the brachial artery under the cuff, stopping the flow of blood.) The pressure within the cuff is shown on the scale of the sphygmomanometer. A stethoscope is placed over the brachial artery just below the cuff, and the pressure in the cuff is slowly released until the blood begins to flow. At that point, there is a faint tapping sound in the stethoscope. When the sound is first noted, the blood-pressure figure represents the *systolic* blood pressure. As the cuff pressure is lowered further, the sounds become progressively louder and then softer. When the sound is no longer heard, the *diastolic* blood pressure is noted. The absence of sound indicates that blood is flowing freely and smoothly through the brachial artery.

A S K   Y O U R S E L F

1 How is arterial blood pressure calculated?

2 What are the major factors that help maintain normal blood pressure?

3 What is the function of the vasomotor center?

4 How is venous pressure regulated?

5 What is a venous pump?

6 How does gravity affect venous blood pressure?

7 What is meant by capillary fluid shift?

8 What is the physiological difference between systolic and diastolic blood pressure?

9 What is the difference between pulse and pulse pressure?

10 How does a sphygmomanometer work?

THE EFFECTS OF AGING ON BLOOD VESSELS

One of the body systems most affected by aging is the circulatory system. Fortunately, many of the typical problems can be prevented by a low-fat, high-fiber diet and a regular program of physical activity.

Cerebral arteriosclerosis is a disease in which the arteries that supply blood to the brain harden and narrow. If blood vessels leading to the brain become severely narrowed or blocked or leak, so that the brain is denied blood for even a short time, serious brain damage and even death may occur. *Arteriosclerosis* and *hypertension* (high blood pressure) are the chief causes of *cerebral hemorrhage* (stroke). In the United States, about 98 percent of the 100,000 people who die of a stroke each year are over age 50.

Transient ischemic attacks (TIAs, or "little strokes") occur when momentary blood clots form in arteries leading to the brain, and *multiple infarct dementia* occurs when recurrent small clots in the brain destroy brain tissue, impairing thinking and memory. *Hypertension* usually accompanies *atherosclerosis*, a condition in which plaque builds up in the arteries, narrowing the lumen and reducing the elasticity of the vessels.

Other common effects of aging on blood vessels are *varicose veins*, *phlebitis*, *thrombophlebitis*, and *aneurysms*. (See When Things Go Wrong at the end of this chapter.)

DEVELOPMENTAL ANATOMY OF MAJOR BLOOD VESSELS

Blood leaves the embryonic heart through the arterial end (truncus arteriosus) and returns through the horns of the sinus venosus at the venous end, passing into the single atrium through the sinoatrial orifice. The orifice is equipped with two flaps of endocardium that function as primitive valves to prevent the backflow of blood. Blood from the atrium flows into the ventricle through the atrioventricular orifice. When the ventricle contracts, blood is pumped into the bulbus cordis. The bulbus cordis narrows to become the truncus arteriosus, which pierces the roof of the pericardium to open into the **aortic sac.** From the somewhat dilated aortic sac emerge the **aortic arches** [see FIGURE 20.18E].

Six pairs of aortic arches develop, but only the left arch of the fourth pair forms part of the arch of the mature aorta. The first, second, and fifth pairs disappear during embryonic life. The third pair forms the common carotid arteries and the internal carotid arteries; the right arch of the fourth pair becomes the right subclavian artery; the proximal part of the left sixth aortic arch develops into the proximal part of the left pulmonary artery, and the distal part forms the ductus arteriosus; and the proximal part of the right sixth aortic arch develops into the proximal part of the right pulmonary artery, while the distal part of the arch degenerates.

At about day 35 to 42, the truncus arteriosus is divided into the "great arteries," the **pulmonary trunk** and **aorta** [see FIGURE 20.18G]. The pulmonary trunk carries blood to the lungs, and the aorta carries blood to the blood vessels that supply the rest of the body.

Adaptations in Fetal Circulation

The circulatory system of a fetus differs from that of a child and adult for two main reasons: (1) The fetus gets oxygen and nutrients and eliminates carbon dioxide and waste products by way of its mother's blood. (2) The fetal lungs, kidneys, and digestive system (except for the liver) are not functional. At birth, however, the baby must make several rapid physiological and anatomical adjustments as it shifts to an essentially adult circulation.

Fetal circulation The *placenta,* a thick bed of tissues and blood vessels embedded in the wall of the mother's uterus, provides an indirect connection between the mother and the fetus. It contains arteries and veins of both the mother and the fetus. These vessels intermingle but do not join. As a result of this close intermingling, the fetus obtains nutrients and eliminates wastes through its mother's blood instead of its own organs. The *umbilical cord* connects the placenta to the fetus. The fully developed cord contains a single *umbilical vein,* which carries oxygenated, nutrient-rich blood from the placenta to the fetus [FIGURE 21.28A]. Coiled within the umbilical cord are two *umbilical arteries,* which carry both deoxygenated blood and waste material from the fetus to the placenta.

The umbilical vein, carrying fully oxygenated blood, enters the abdomen of the fetus, where it branches. One branch joins the fetal portal vein, which goes to the liver, while the other branch, called the *ductus venosus,* bypasses the liver and then joins the inferior vena cava [FIGURE 21.28A]. Before the oxygenated blood that enters the vena cava actually enters the heart, it becomes mixed with deoxygenated blood being returned from the lower extremities of the fetus.

The fetal inferior vena cava drains into the right atrium. A large opening with two flaps called the *foramen ovale* ("oval window") connects the two atria. Most of the blood passes through the foramen ovale into the left atrium. Flaps of tissue in the left atrium act as a valve, preventing the backflow of blood into the right atrium. From the left atrium the blood flows into the left ventricle, and from there it is pumped into the aorta for distribution primarily to the brain and upper extremities, with some going to the rest of the body.

Blood from the fetal head and upper extremities drains into the right atrium via the superior vena cava, where it is deflected mainly into the right ventricle. The blood then leaves the right ventricle through the pulmonary trunk. However, since the fetal lungs are still collapsed and nonfunctional, most of the blood passes into the *ductus arteriosus,* a shunt between the pulmonary trunk and aortic arch, where it mixes with blood coming from the left ventricle.

By connecting the pulmonary artery to the aorta, the ductus arteriosus allows fetal blood to bypass the lungs and flow from the aorta to the umbilical arteries to the placenta. Some blood, however, does go from the pulmonary arteries to the lungs, not only to supply cells with oxygen and other nutrients, but to ensure a circulation that will accept the blood flow when the newborn takes its first breaths. The blood is drained by the pulmonary veins.

Because of the mixing of oxygenated and deoxygenated blood, the fetal blood has a lower oxygen content than adult blood does. To help compensate for this deficiency, *fetal hemoglobin* can combine with oxygen more easily than adult hemoglobin can. This special property of fetal hemoglobin is lost within a few days after birth, when the newborn starts producing adult hemoglobin. The blood in the pulmonary artery is low in oxygen and nutrients and high in waste products. It flows through the ductus arteriosus into the aorta, down toward the lower part of the body, and eventually to the two umbilical arteries. The umbilical arteries terminate in the blood vessels of the placenta, where carbon dioxide is exchanged for oxygen and waste products are exchanged for nutrients.

Circulation in the newborn At birth, several important changes must take place in the cardiovascular system of the newborn infant [FIGURE 21.28B]. When an infant takes its first breath, the collapsed lungs inflate and the pulmonary vessels become functional. As a result of the increased pulmonary flow, the pressure in the left atrium is raised above that in the right atrium. The increased pressure closes the foramen ovale by pressing on the overlapping valves of the interatrial septum, and the heart begins to function like a double pump. It takes several months for the foramen ovale to fuse completely. Eventually, only a depression called the *fossa ovalis* remains at the former site of the foramen ovale [FIGURE 21.28B]. At the same time that the foramen ovale is closing, the ductus arteriosus is gradually constricting. In about 6 weeks it becomes the *ligamentum arteriosum.*

During the first few days after birth, the stump of the umbilical cord dries up and drops off. Inside the body, the umbilical vessels atrophy. The umbilical vein becomes the *round ligament* of the liver, the ductus venosus forms the *ligamentum venosum* (a fibrous cord embedded in the wall of the liver), and the umbilical arteries become the *lateral umbilical ligaments.*

If either the foramen ovale or the ductus arteriosus does not close off (a condition called *patent ductus arteriosus*), the oxygen content of the blood will be low and the baby's skin will appear slightly blue. A newborn baby with this condition is therefore called a "blue baby." Most often, a defective ductus arteriosus must be corrected surgically.

A S K   Y O U R S E L F

1 What are the two main differences between fetal circulation and a young child's circulation?

2 What is the ductus arteriosus?

3 What causes the condition that produces a "blue baby"?

FIGURE 21.28 CONVERSION OF CIRCULATORY SYSTEM: FETUS TO NEWBORN

[A] Before birth the lungs are essentially bypassed because the foramen ovale permits the passage of deoxygenated blood from the right atrium to the left atrium in the fetal heart, and the ductus arteriosus carries blood from the pul- monary artery (from the right ventricle) directly into the aorta. Note that much of the blood bypasses the liver through the ductus venosus.

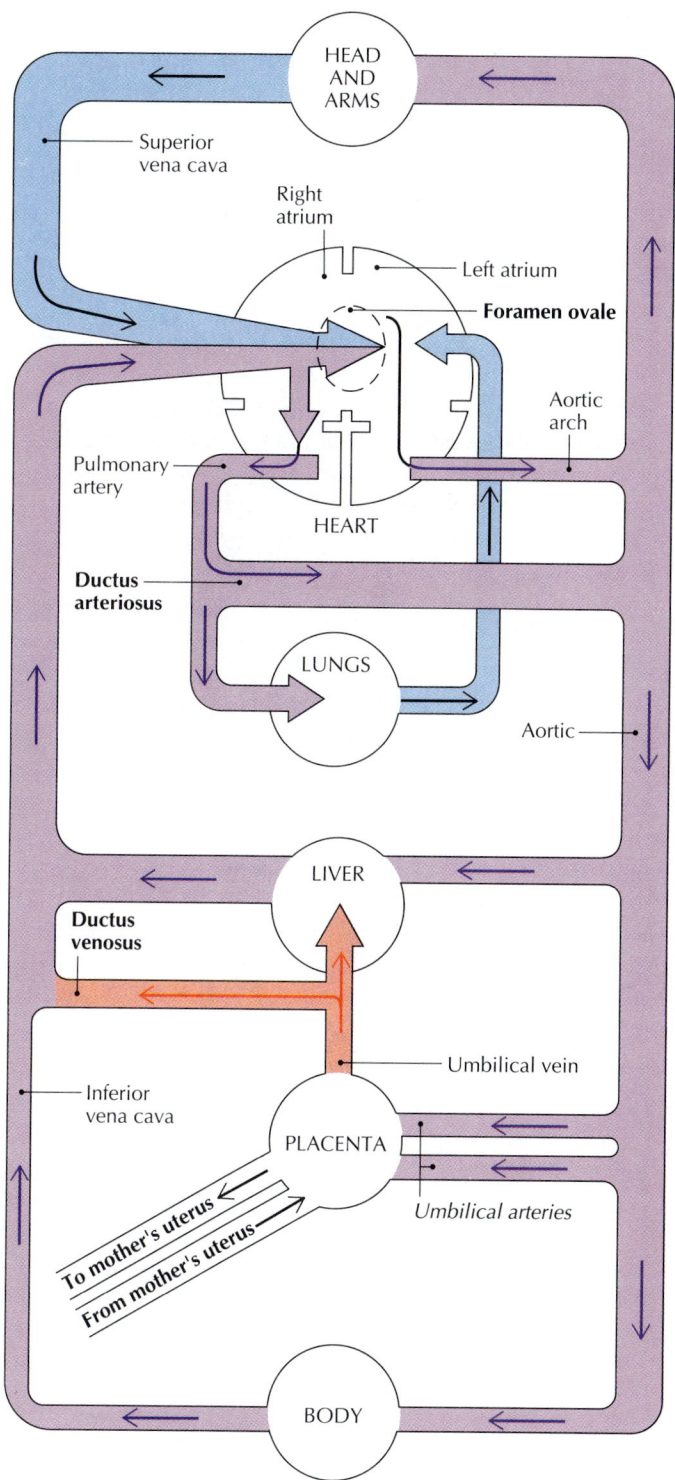

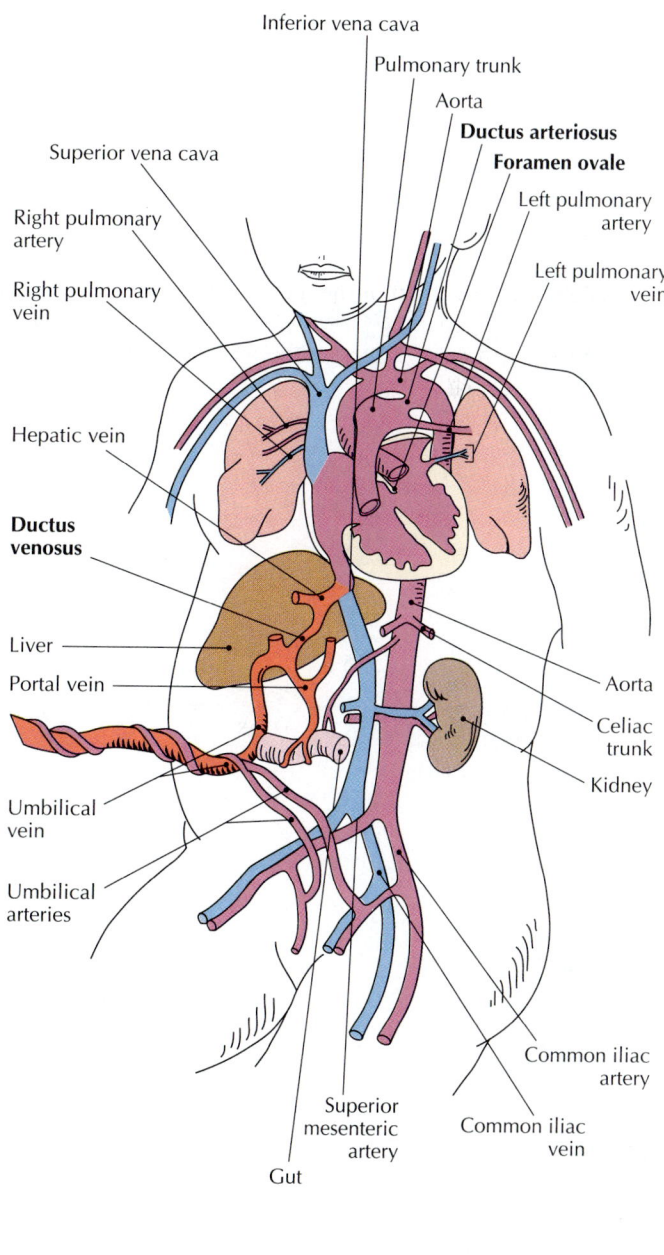

[A]

(Figure 21.28 continues on following page)

FIGURE 21.28 CONVERSION OF CIRCULATORY SYSTEM *(Continued)*

[**B**] After birth, the foramen ovale closes, and the ductus arteriosus is converted into the ligamentum arteriosum, resulting in an adult-type circulation. Also, the ductus veno-sus becomes the ligamentum venosum, the umbilical vein becomes the round ligament, and the umbilical arteries become the lateral umbilical ligaments.

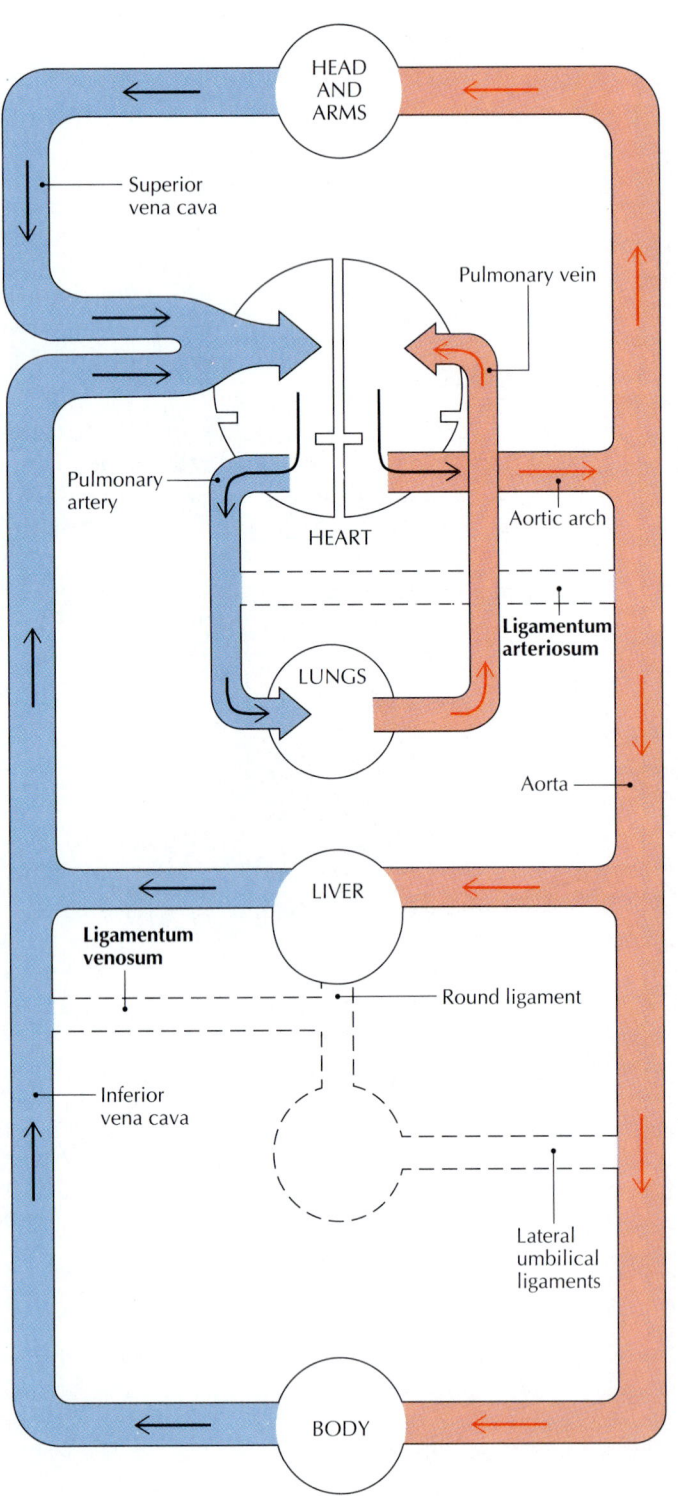

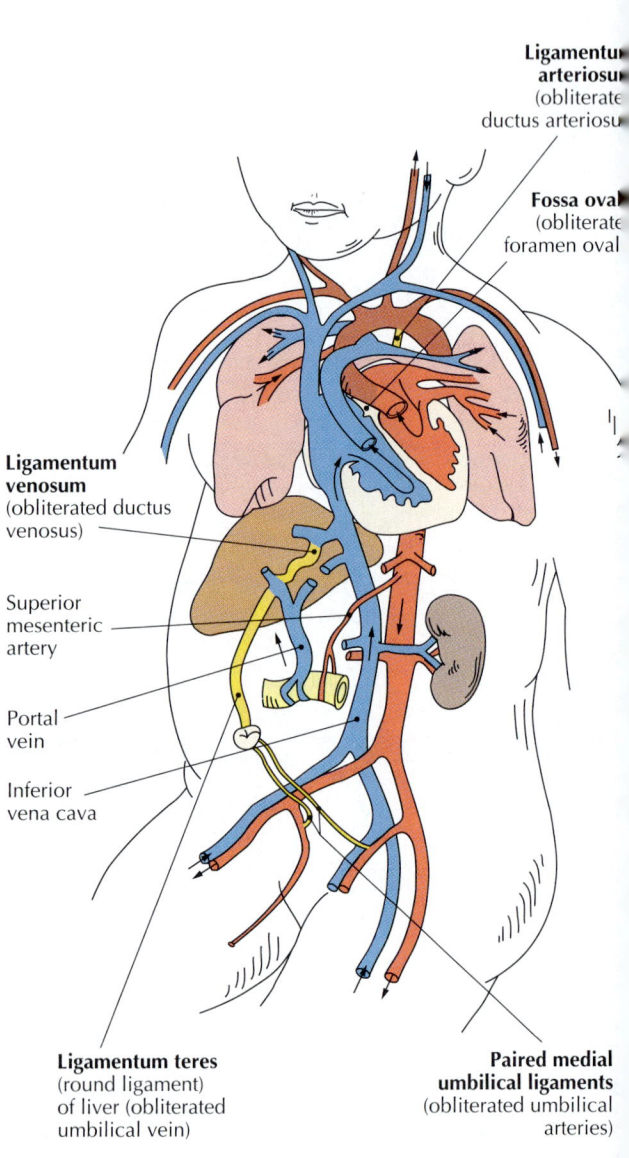

[B]

WHEN THINGS GO WRONG

Aneurysms

An *aneurysm* (AN-yuh-rihz-ihm; Gr. "to dilate") is a balloonlike, blood-filled dilation of a blood vessel. It occurs most often in the larger arteries, especially in the brachial artery, abdominal aorta, femoral artery, and popliteal artery. Cerebral aneurysms are also common. Aneurysms happen when the muscle cells of the tunica media become elongated and the wall weakens. The wall may also be weakened by degenerative changes related to such diseases as *arteriosclerosis* (hardening of the arteries). If an aneurysm is not treated, the vessel may eventually burst, and if a cerebral vessel is affected, a stroke may result.

Atherosclerosis

Atherosclerosis (Gr. "porridgelike hardening") is the leading cause of coronary artery disease. It is characterized by deposits of fat, fibrin, cellular debris, and calcium on the inside of arterial walls and in the underlying tissues (see photo). These built-up materials, called *atheromatous plaques*, adhere to the tunica intima, narrowing the lumen and reducing the elasticity of the vessel.

Although the basic cause of atherosclerosis is not known, the earliest stage in its development is believed to be damage to the endothelial cells and tunica intima of the vessel wall. Once the damage occurs, the endothelial cells proliferate and attract lipid substances. Several factors increase its progress. Among them are cigarette smoking and excessive animal fat and cholesterol in the diet. Other factors that may have an adverse effect are hypertension, diabetes, age, stress, heredity, and the male sex hormones.

There are three ways in which atherosclerosis can cause a heart attack: (1) It can completely clog a coronary artery. (2) It can provide a rough surface where a blood clot (*thrombus*) can form so that it closes off the artery and causes a *coronary thrombosis*. (3) It can partially block blood to the myocardium and regions of the impulse-conducting system of the heart; this may result in the heart beating arhythmically.

An effective method of treating atherosclerosis is *balloon angioplasty*, a surgical technique that uses catheters with inflatable balloons on their tips to open clogged vessels. When the balloons are inflated they press the narrowed areas of the vessels outward, creating a larger opening for blood to pass through.

Coronary Artery Disease

The typical effect of *coronary artery disease (CAD)* is a reduced supply of oxygen and other nutrients to the myocardium. The condition is usually brought on by atherosclerosis. As a result, less blood reaches the heart, and *myocardial ischemia* (ih-SKEE-mee-uh; Gr. *iskhein*, to keep back + *haima*, blood) or local ischemia in cardiac muscle occurs.

Lipoproteins play an important role in atherosclerosis and CAD. They are formed when lipids combine with certain blood proteins. In fact, nearly all the lipid in the blood is carried as lipoproteins. They are classified according to their densities and function: (1) *Chylomicrons* are aggregates of triglycerides and cholesterol and are derived from the digestive tract. Because they contain less than 2 percent protein, they are of low density. (2) *Very low density lipoproteins (VLDL)* are synthesized in the liver and transported by the blood to adipose tissue, where they are stored. (3) *Low-density lipoproteins (LDL)* are the primary source of cholesterol for all tissues. (4) *High-density lipoproteins (HDL)* are rich in phospholipids and cholesterol. They are produced in many tissues of the body and transported to the liver for storage.

The incidence of CAD is closely related to the concentration of these specific lipoproteins in the blood. On the one hand, there is a strong positive correlation between CAD and high blood levels of LDL as related to total cholesterol. On the other hand, there is a strong negative correlation between CAD and the ratio of HDL to the total cholesterol. That is, the higher the level of HDL, the lower the chance of developing CAD. Several ways to improve the HDL-LDL ratio are: exercising regularly, reducing animal fat in the diet, giving up cigarettes, and increasing the consumption of unsaturated omega-3 oils from coldwater fish.

In 1987, Joseph Goldstein and Michael Brown won the Nobel Prize in physiology or medicine for discovering an LDL-receptor on the surface of cells. These receptors control the amount of cholesterol in the blood.

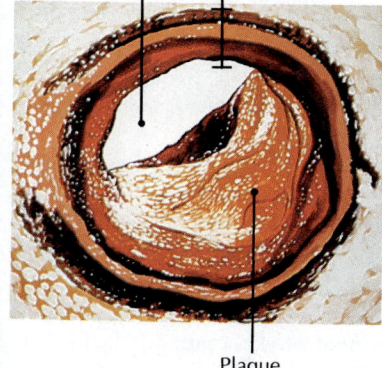

Existing lumen   Arterial wall

Atherosclerotic plaque can narrow the lumen of an artery substantially.

Plaque

Several new drugs have been used successfully to lower cholesterol levels and dissolve arterial blood clots. *Lovastatin* (Mevacor) inhibits an enzyme in the liver that helps make cholesterol. As a result, the liver produces more LDL-receptors that help remove cholesterol from the blood, reducing blood cholesterol levels. Lovastatin combined with other drugs can reduce the blood cholesterol level by about 50 percent. *Gemfibrozil* reduces the risk of coronary heart disease by slightly lowering the total cholesterol level while moderately increasing HDL and moderately reducing LDL.

Hypertension (High Blood Pressure)

Hypertension, commonly called *high blood pressure,* is characterized by systolic or diastolic pressure that is above normal all the time — not just as a result of specific activities or conditions. (See the chart on the right.) About 35 million Americans have hypertension, and more than a million die annually of related diseases. The two types of hypertension are (1) the more common *primary* or *essential hypertension,* an above-normal blood pressure that cannot be attributed to any particular cause *(idiopathic),* and (2) *secondary hypertension,* which results from a disorder such as kidney disease or arteriosclerosis.

Hypertensive arterial walls are hard and thick, and their elasticity is reduced, forcing the heart to work harder to pump enough blood. If hypertension persists, the heart may become enlarged, a condition called *hypertensive heart disease* [TABLE 21.2]. High blood pressure can also cause a stroke by rupturing an artery in the brain, resulting in a cerebral hemorrhage.

Although hypertension may be accompanied by headache, dizziness, or other symptoms, one of the problems in treating hypertension is that often there are no external signs that blood pressure is abnormally high. The prognosis is optimistic if the disease is diagnosed and treated early, before complications develop. If untreated, hypertension is accompanied by a high mortality rate.

Hypertension may be an inherited problem, but it is also related to stress, obesity, a diet that is high in sodium and saturated fats, aging, lack of physical activity, race (it is most common in blacks), and the use of tobacco and oral contraceptives (especially if used together). Hypertension may also occur when regulatory mechanisms in the central nervous system break down.

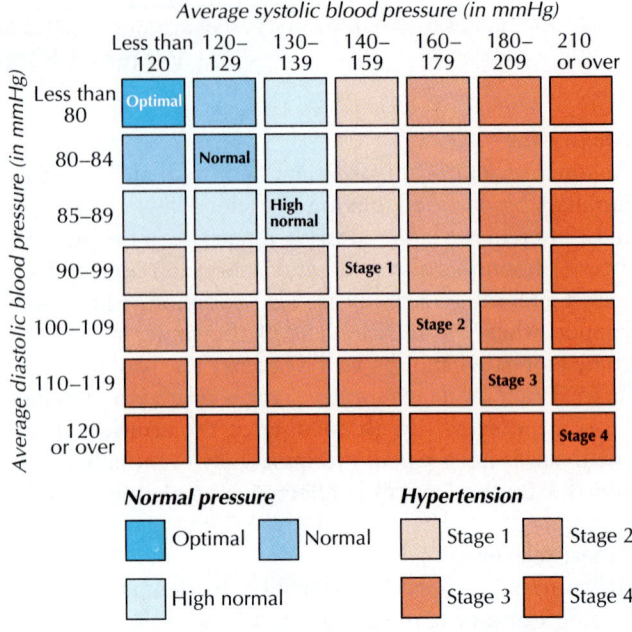

For the first time a blood-pressure classification system uses the systolic pressure (top number) as well as the diastolic pressure (bottom number) in assessing the severity of hypertension. The new guidelines also emphasize the belief that there is no precise distinction between normal and abnormal. The risk of death and disability from heart attack and stroke increases progressively with higher levels of pressure. Even people whose pressure is in the high normal range (systolic between 130 and 139 and diastolic between 85 and 89) are at risk of developing definite high blood pressure and therefore should attempt lifestyle modifications. (National High Blood Pressure Education Program)

Stroke

A *stroke* (cerebrovascular accident, or CVA) occurs when the blood supply to a portion of the brain is cut off. There are several ways a stroke may occur: (1) Atherosclerosis in the arteries of the brain or neck may block the flow of blood. (2) A thrombus may form in the atherosclerotic vessel, closing off the artery and causing a *cerebral thrombosis.* (3) A traveling blood clot *(embolus)* can become wedged in a small artery of the brain or neck; this kind of stroke is an *embolism.* (4) A weak spot in a

---

Q: *How does hypertension affect the risk of coronary heart disease?*

A: Women with hypertension increase their risk for coronary heart disease about 3 times; men with hypertension increase their risk about 2.5 times.

Q: *Is blood pressure higher at one time of the day than another?*

A: Yes. Blood pressure is at its lowest at about 3 A.M., but it rises rapidly to its highest point between 6 A.M. and noon, the six-hour period when most heart attacks occur.

Q: *Is cholesterol bad for the body?*

A: *Excessive levels* of cholesterol in the blood may be associated with cardiovascular disease, but cholesterol is an essential component of certain cells and is necessary for several physiological processes. For instance, it is found in steroid hormones such as cortisone and testosterone and is also a major component of brain tissue. The body actually produces low levels of essential cholesterol for itself.

TABLE 21.2 A SUMMARY OF MAJOR ANTIHYPERTENSIVE DRUGS

Category of drug	Examples in category	How they work	Comments
Angiotensin converting enzyme inhibitors (ACE)	Captopril, Enalapril maleate, Lisinopril	Dilate blood vessels by preventing constriction	Especially effective for people with heart problems, diabetes
Beta blockers	Acebutolol, Atenolol, Nadolol, Penbutolol sulfate, Propranolol hydrochloride, Timolol	Reduce heart output by blocking nerve signal that increases heart rate	Most often used with people under 60 who have had heart attacks; high success rate when used with diuretic
Calcium-channel blockers	Diltazem hydrochloride, Nicardipine, Nifedipine, Verapamil, Verapamil SR	Dilate vessel walls by blocking entry of calcium	Effective for patients over 60
Centrally acting alpha stimulants	Clonidine, Clonidine TTS, Guanabenz, Guanfacine hydrochloride, Methyldopa	Lower blood pressure by dilating blood vessels	Effective for patients who have had heart attacks, diabetes
Diuretics	*Loop diuretics:* Furosemide, Bumetanide, Ethacrynic acid *Potassium-sparing:* Amiloride, Spironolactone, Triamterene *Thiazide or thiazidelike:* Chlorthalidone, Indapamide, Metolazone, Quinethazone, drugs ending in "thiazide"	Reduce blood volume by increasing excretion of sodium and water from kidneys (in urine)	Effective for elderly people; enhance effectiveness of other drugs
Peripherally acting alpha blockers	Prazosin, Terazosin	Cause dilation of vessels by blocking nerve signal that constricts vessels	Used when other drugs fail; do not have sedating effects
Peripheral adrenergic antagonists	Guanadrel sulfate, Guanethidine monosulfate, Rauwolfia, Reserpine	Block adrenaline release during stress	Sedative effect; safe for pregnant women; high success rate when used with diuretics
Vasodilators	Hydralazine, Minoxidil	Dilate arteries	Effective for patients with heart, kidney problems; usually combined with beta blocker and diuretic

blood vessel may break; this a *cerebral hemorrhage.* (When the weak spot bulges, it is called a cerebral *aneurysm.*) (5) In rare cases, a brain tumor may press on a blood vessel and shut off the blood supply.

Because the brain controls the body's movements, any part of the body can be affected by a stroke. The damage may be temporary or permanent. If a brain artery is blocked in the area that controls speaking, speech will be affected. A stroke is often preceded by transient ischemic attacks, which should be taken as a warning and reported to a physician immediately. It is now known that smoking doubles the risk of stroke.

Thrombophlebitis
Thrombophlebitis (Gr. *thrombos,* clot + *phleps,* blood vessel + *itis,* inflammation) is an acute condition character-

ized by clot formation and inflammation of deep or superficial veins. Thrombophlebitis usually results from an alteration in the endothelial lining of the vein. Platelets begin to gather on the roughened surface, and the consequent formation of fibrin traps red blood cells, white blood cells, and more platelets. The result is a thrombus. If untreated, the thrombus may become detached from the vein and begin to move through the circulatory system. Such a mobile blood clot is called an **embolus.** The result may be pulmonary embolism, a blockage of the pulmonary veins that is a potentially fatal condition.

Varicose Veins
Varicose veins (L. *varix,* swollen veins) are abnormally dilated and twisted veins. The saphenous veins and their branches in the legs are often affected. They become per-

manently dilated and stretched when the one-way valves in the legs weaken. As a result, some blood flows backward and pools in the veins. Varicose veins can result from a hereditary weakness of the valves but can also be caused or aggravated by vein inflammation (*phlebitis*), blood clot formation (*thrombophlebitis*), pregnancy, lack of exercise, loss of elasticity in old age, smoking, low-fiber diet, and occupations that require long periods of standing. Untreated varicose veins may lead to edema in the ankles and lower legs, leg ulcers, dizziness, pain, fatigue, and nocturnal cramps.

Portal Cirrhosis of the Liver

Portal cirrhosis of the liver results from chronic alcoholism, the ingestion of poison, viral diseases (such as infectious hepatitis), or other infections in the bile ducts. It is a condition in which large amounts of fibrous tissue develop within the liver. This fibrous tissue destroys many of the functioning liver cells and eventually grows around the blood vessels, constricting them and greatly impeding the flow of portal blood through the liver. This impediment blocks the return of blood from the intestines and spleen. As a result, blood pressure increases so much that fluid moves out through the capillary fluid shift into the peritoneal cavity, leading to *ascites*, the accumulation of free fluid in the peritoneal cavity. Because this fluid is mostly plasma, which contains large quantities of protein, a high colloid osmotic pressure is created in the abdominal fluid, pulling more fluid by osmosis from the blood, liver, and gastrointestinal tract in a positive feedback cycle. The patient often dies because of excessive fluid loss.

Patent Ductus Arteriosus

Patent ductus arteriosus occurs when the ductus arteriosus of the fetus, which carries blood from the pulmonary artery to the aorta, does not shut at birth (see page 698). Patent ductus arteriosus creates a left-to-right movement of some arterial blood from the aorta to the pulmonary artery, which recirculates arterial blood through the lungs.

Transposition of the Great Vessels

In *transposition of the great vessels,* the arteries are reversed. The aorta emerges from the right ventricle, and the pulmonary artery emerges from the left ventricle. As a result, oxygenated blood returning to the left side of the heart is carried back to the lungs by the pulmonary artery, and unoxygenated blood traveling to the right side of the heart is carried into the systemic circulation by the transposed aorta.

CHAPTER SUMMARY

Types of Blood Vessels (p. 654)

1 *Arteries* carry blood away from the heart to capillary beds throughout the body. Arterial blood is oxygenated, with the exception of the blood in the *pulmonary arteries,* which carry deoxygenated blood from the heart to the lungs.

2 The major arterial trunks are the *aorta* from the left ventricle and the *pulmonary trunk* from the right ventricle.

3 Arterial walls are composed of three layers: the inner *tunica intima,* the middle *tunica media,* and the outer *tunica adventitia.*

4 Arteries branch into smaller arteries and then into small *arterioles* shortly before reaching the capillary networks. Terminal arterioles control the flow of blood from arteries into capillaries.

5 Arterioles enter the body tissues and branch out further to form *capillaries,* the bridge between the arterial and venous systems.

6 The *microcirculation* of the blood consists of the capillaries, terminal arterioles, metarterioles, and venules.

7 The three types of capillaries are con-*tinuous capillaries, fenestrated capillaries,* and *sinusoids,* each differentiated to perform a specific function.

8 Blood drains from capillaries into *venules,* tiny veins that unite to form larger venules and veins. *Veins* carry deoxygenated blood from the body tissues to the heart, with the exception of the *pulmonary veins,* the *hepatic portal system* that carries blood from the capillaries of the intestines to the capillaries of the liver, and the *hypothalamic-hypophyseal portal system* in which veins formed from the capillaries of the hypothalamus divide into the capillaries of the anterior pituitary gland.

9 Veins usually contain paired semilunar bicuspid valves that permit blood to flow only toward the heart. Their walls contain the same layers as arterial walls.

10 Venous blood pressure is low, and blood is assisted toward the heart by the skeletal muscle pump.

Pulmonary and Systemic Circulation of the Blood (p. 660)

1 The *pulmonary circulation* carries deoxygenated blood from the heart to the lungs, where carbon dioxide is removed and oxygen is added. It then carries oxygenated blood back to the heart.

2 The *systemic circulation* supplies the tissues of the body with blood high in oxygen concentration and also removes blood high in carbon dioxide. The main vessels of the *arterial division* are the aorta, ascending aorta, aortic arch, descending aorta, and common iliac arteries. The main vessels of the **venous division** are the superior vena cava, inferior vena cava, and coronary sinus.

3 Veins ordinarily transport blood directly back to the heart from a capillary network, but the two **portal systems** of the body (hepatic and hypothalamic-hypophyseal) transport the blood to a second set of capillaries on its way to the venous system.

4 The brain is supplied with blood by two **vertebral arteries** and two **internal carotid arteries.** All the blood entering the cerebrum must first pass through the **cerebral arterial circle.**

5 The arrangement of blood vessels in the skin allows for the increase or decrease of heat radiation from the integumentary system.

6 Blood nourishes skeletal muscles and also removes wastes. The main controllers of skeletal muscle blood flow are the sympathetic vasodilator fibers that cause the blood vessels to dilate.

Major Arteries and Veins: Regions Supplied and Drained (p. 665)

Figures 21.11 through 21.19 summarize the major arteries and veins of the body.

Physiology of the Circulatory System (p. 684)

1 **Hemodynamics** is the study of the principles that govern blood flow.

2 **Blood flow** refers to the volume (quantity) of blood flowing through a vessel during a specific period of time. **Blood pressure** is the force with which blood is pushed against the walls of blood vessels. **Blood velocity** is the distance blood moves along a vessel during a specific time period. **Peripheral resistance** is the impediment to blood flow by friction. It depends on the length, radius, and total cross-sectional area of the vessel and the viscosity of the blood.

3 Blood-flow homeostasis is maintained by a combination of **neural** control, **hormonal** and **chemical** control, and **autoregulation**.

4 *Factors affecting blood pressure* are gravity (hydrostatic pressure), vessel elasticity, and cardiac output.

5 Specific neural, chemical, and hormonal factors help regulate arterial blood pressure by acting directly on blood vessels. They include **baroreceptors** and **chemoreceptors, various hormones,** the **kidneys, capillary fluid shift,** and **higher brain centers.** *Viscosity* of the blood and the body's *salt and water balance* are also factors.

6 Because the heart pumps all the blood it receives, the pressure of the venous blood and its return to the heart are important in maintaining homeostasis. The regulation of venous blood return is determined by the **pressure gradient, resistance to blood flow** through vessels, and **venous pumps.**

7 A **pulse** is a beat felt on the surface of the skin over a nearby artery. It corresponds to the beat of the heart and the alternating expansion and recoil of the arterial wall.

8 Blood-pressure levels are expressed by two numbers. The first number is the **systolic pressure,** and the second is the **diastolic pressure.**

9 The difference between systolic and diastolic pressure is called the **pulse pressure.** It is determined by *arterial distensibility, stroke volume,* and the *speed of blood ejected* from the left ventricle into the systemic circulation.

10 Arterial blood pressure changes throughout the cardiac cycle. *Mean arterial pressure (MAP)* is the average pressure that drives the blood through the systemic circulatory system.

11 An instrument used for measuring blood pressure is the **sphygmomanometer.**

The Effects of Aging on Blood Vessels (p. 697)

1 The effects of aging generally cause blood vessels to become narrowed and less elastic.

2 Some major disorders associated with aging and blood vessels are cerebral arteriosclerosis, cerebral hemorrhage, hypertension, thrombophlebitis, and varicose veins.

Developmental Anatomy of Major Blood Vessels (p. 697)

1 Six pairs of *aortic arches* develop from the aortic sac, but only the left arch of the fourth pair forms part of the mature aortic arch.

2 At about day 35 to 42, the truncus arteriosus is divided into the *pulmonary trunk* and the *aorta.*

3 The *circulatory system of a fetus* differs from that of a child or adult in that the fetus gets nutrients and removes its wastes through the placenta, and its lungs, kidneys, and digestive system (except for the liver) do not function.

4 The fetus has an opening in the septum between the atria called the **foramen ovale** and a vessel called the **ductus arteriosus** that bypasses the lungs by carrying blood from the pulmonary artery to the aorta. In a normal child, both close at birth.

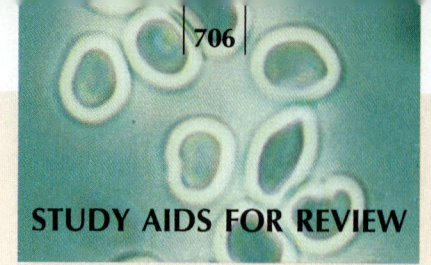

STUDY AIDS FOR REVIEW

KEY TERMS

aorta 654
arteriole 655
artery 654
baroreceptors 689
blood flow 684
blood pressure 684
blood velocity 685
capillary 654, 655
capillary fluid shift 691
cardioregulatory center 689
cerebral arterial circle 663
diastolic pressure 694
ductus arteriosus 698
hepatic portal system 663
hydrostatic pressure 688, 691
hypothalamic-hypophyseal portal system 663
mean arterial pressure 695
microcirculation 657
pulmonary circulation 660
pulmonary trunk 654
pulse 693
pulse pressure 694
resistance to blood flow 685, 692
systemic circulation 660
systolic pressure 694
total peripheral resistance 684
tunica adventitia 655
tunica intima 654
tunica media 654
vasomotor center 689
vein 654, 658
venous pump 692
venule 658

MEDICAL TERMINOLOGY

ANGIOGRAM An x-ray of blood vessels, taken after a radiopaque dye is injected into the vessels.

ARTERIOGRAPHY Technique of x-raying arteries, after a radiopaque dye is injected.

ARTERIOSCLEROSIS OBLITERANS A form of arteriosclerosis that affects the lower limbs.

AVASCULAR NECROSIS Condition in which tissue dies from the lack of blood (from the blood vessels).

CLAUDICATION Improper circulation of blood in vessels of a limb, which causes pain and lameness.

CYANOSIS A bluish discoloration of tissue due to an oxygen deficiency in the systemic blood.

EDEMA Swelling due to abnormal accumulation of fluid in intercellular tissue spaces.

ENDARTERECTOMY Removal of an obstructing region of the inner wall of an artery.

HEMATOMA Tumor or swelling in tissue due to an accumulation of blood from a break in a blood vessel.

HYPOTENSION Low blood pressure.

OCCLUSION Clot or closure in the lumen of a blood vessel or other structure.

PHLEBOSCLEROSIS Thickening or hardening of walls of veins.

SHUNT Connection between two blood vessels or between two sides of the heart.

THROMBECTOMY Removal of a blood clot from a blood vessel.

VALVOTOMY Cutting into a valve.

VASCULITIS A general term for inflammation of blood vessels.

VENIPUNCTURE Inserting a catheter or needle into a vein.

UNDERSTANDING THE FACTS

1 How do elastic arteries differ from muscular arteries in structure and location?

2 List the types of capillaries and distinguish between them structurally and functionally.

3 List the kinds of vessels that comprise the microcirculation, and briefly describe their arrangement.

4 Which arteries carry deoxygenated blood? Which veins carry oxygenated blood? Consider both the adult circulation and the fetal circulation in your answer.

5 Compare the structure of a typical arterial wall with that of a vein.

6 What tissue layer(s) comprise valves? Where are valves found?

7 What is the distinguishing feature of a portal system?

8 What is the role of each of the following in the fetal circulation?

  a ductus venosus

  b foramen ovale

  c ductus arteriosus

9 List the major arteries that branch from the aorta.

10 List the major veins that drain the head, arms, and legs.

11 What is a pulse?

12 Define

a mean arterial pressure	**d** baroreceptor
b stroke volume	**e** systolic pressure
c vasomotion	**f** sphygmomanometer

UNDERSTANDING THE CONCEPTS

1 How are the cells that make up the walls of arteries and veins nourished?

2 Compare the thickness of each of the layers that constitute the wall of arteries and veins, and correlate the difference with the function of each kind of vessel.

3 Explain how the structure of arterioles enables them to determine the pattern of blood flow in the body.

4 Explain the functional importance of capillary shunts and arteriovenous anastomoses.

5 Describe the modifications to the fetal circulation that occur after birth.

6 Define cardiac output, heart rate, and stroke volume, and explain the functional relationship between them.

7 What factors determine peripheral resistance, and how are they regulated?

8 Where are baroreceptors, and how do they act to maintain normal blood pressure?

9 What is meant by autoregulation of blood flow?

10 Contrast the roles of the sympathetic and parasympathetic divisions of the nervous system in the regulation of blood flow.

11 Describe the forces that enable fluid to move across capillary walls.

12 What forces act to move venous blood toward the heart?

13 Explain how atherosclerosis can cause a heart attack.

SELF-QUIZ

Multiple-Choice Questions

21.1 Most of the exchange between the blood and extracellular fluid occurs at the level of the
a heart   **d** capillaries
b arteries   **e** veins
c arterioles

21.2 Which of the following layers contains connective tissue and smooth muscle cells?
a tunica intima   **d** a and b
b tunica media   **e** a, b, and c
c tunica adventitia

21.3 Which is *not* a function or characteristic of arterioles?
a helping to regulate body temperature
b directing the flow of blood
c responding to emotional stimuli
d allowing gas exchange to occur
e acting as shunts

21.4 Which type of vessel has walls made up only of a single layer of continuous endothelial cells, each of which fully encircles the vessel?
a sinusoids   **d** terminal arterioles
b continuous capillaries   **e** venules
c fenestrated capillaries

21.5 Veins that carry deoxygenated blood are the
a pulmonary   **d** umbilical
b hepatic portal   **e** all of the above
c ductus venosus

21.6 Which of the following is *not* part of the microcirculation?
a terminal arterioles   **d** muscular arteries
b metarterioles   **e** c and d
c venules

21.7 Portal veins lead directly into which kind of vessel?
a arteries   **d** venules
b arterioles   **e** metarterioles
c capillaries

21.8 All the blood entering the cerebrum must first pass through the
a vertebral artery   **d** cerebral arterial circle
b carotid artery   **e** cerebral artery
c basilar artery

21.9 The umbilical artery is called an artery because it carries
a oxygenated blood away from the mother's heart
b oxygenated blood away from the fetal heart
c blood away from the fetal heart
d oxygenated blood away from the placenta
e blood from the mother to the placenta

21.10 Which of the following vessels have valves?
a arteries   **d** venules
b arterioles   **e** veins
c capillaries

21.11 Which factor(s) influence(s) the resistance of blood flow?
a lumen radius   **d** a and b
b blood viscosity   **e** a, b, and c
c vessel length

21.12 Where is the flow of blood most turbulent?
a elastic arteries   **d** venules
b muscular arteries   **e** veins
c arterioles

Matching
a fossa ovalis
b ligamentum venosum
c lateral umbilical ligaments
d ligamentum arteriosum
e round ligament of liver

21.13 _____ ductus arteriosis
21.14 _____ foramen ovale
21.15 _____ ductus venosus
21.16 _____ umbilical artery
21.17 _____ umbilical vein

a supplies the kidney
b supplies the arm
c supplies the stomach
d supplies the peroneal artery
e supplies the lower limbs
f supplies the muscles of the abdomen

21.18 _____ celiac artery
21.19 _____ iliac artery
21.20 _____ renal artery
21.21 _____ brachial artery
21.22 _____ femoral artery

a unite to form inferior vena cava
b drains diaphragm
c drains large intestine
d drains the liver
e drains into the superior vena cava
f drains into the jugular vein

21.23 _____ brachiocephalic vein
21.24 _____ external iliac veins
21.25 _____ inferior phrenic vein
21.26 _____ inferior mesenteric vein
21.27 _____ hepatic vein

Completion Exercises

21.28 The walls of the large arteries are nourished by small blood vessels called the _____.

21.29 _____ arterioles are those closest to a capillary.

21.30 The difference between systolic and diastolic pressure is called _____.

21.31 The outward force that causes capillary filtration is called _____.

21.32 At the arterial end of a capillary bed hydrostatic pressure is _____ than capillary osmotic pressure.

21.33 As the total cross-sectional area increases from arteries to capillaries, _____ and velocity of the blood decrease.

21.34 Resistance to blood flow is directly proportional to _____ and _____ and inversely proportional to _____.

21.35 Cardiac output equals mean arterial pressure divided by _____.

21.36 The aortic and carotid bodies contain _____ that reflexively increase blood pressure.

A SECOND LOOK

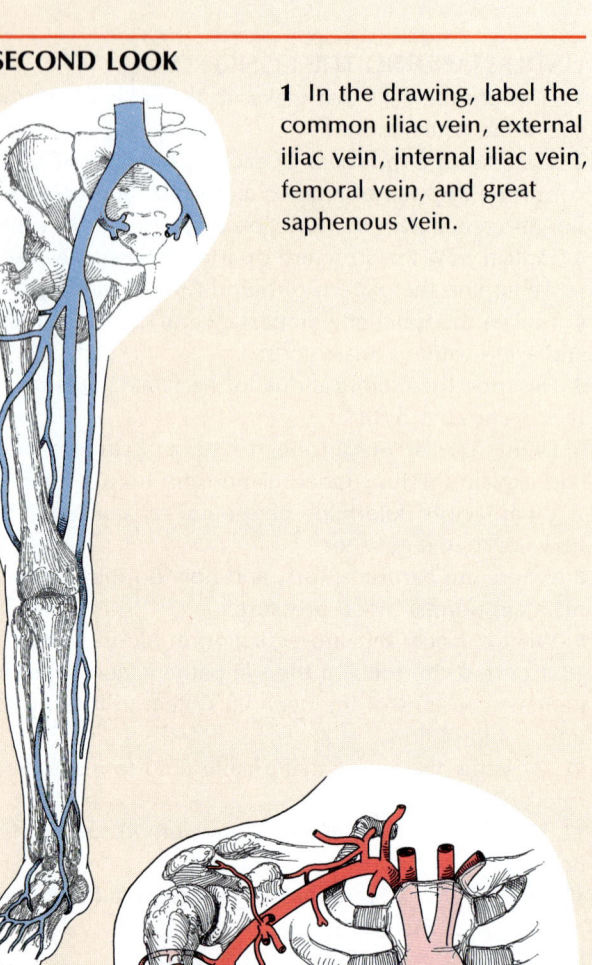

1 In the drawing, label the common iliac vein, external iliac vein, internal iliac vein, femoral vein, and great saphenous vein.

2 In the drawing, label the brachial, ulnar, and axillary arteries and the common carotid arteries.

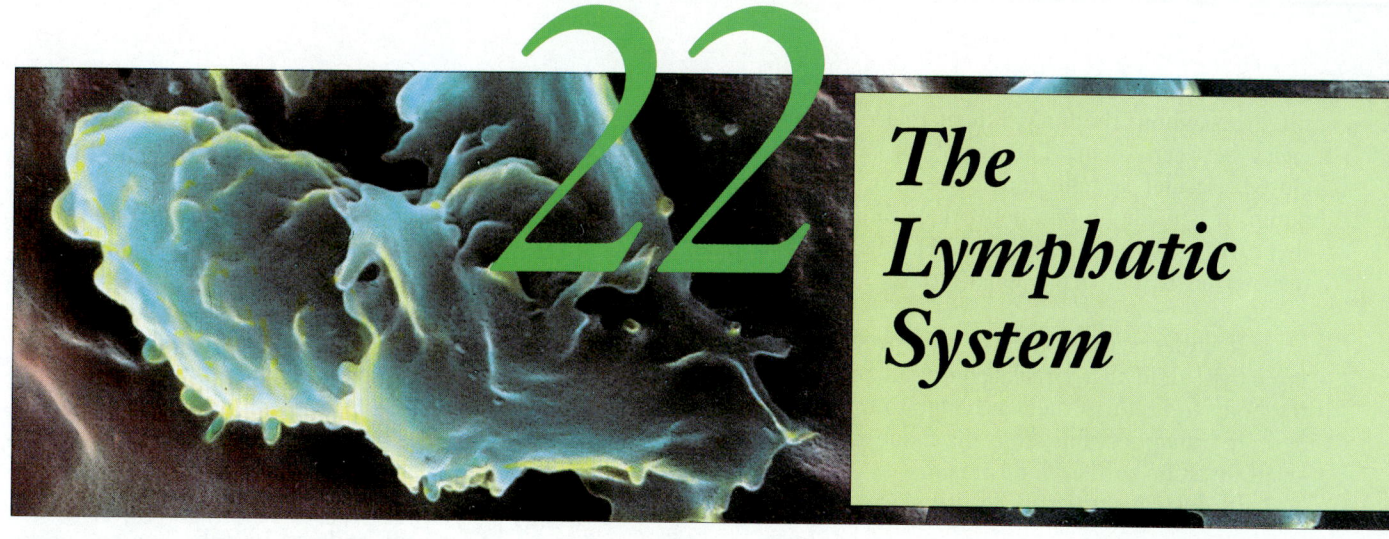

22

The Lymphatic System

KEY CONCEPTS

1 The lymphatic system helps protect the body from harmful substances and prevents excessive accumulation of interstitial fluids.

2 Most of the cells in the lymphatic system are leukocytes with protective roles similar to those of blood leukocytes.

3 The tonsils, spleen, and thymus gland are part of the lymphatic system, which also includes lymph nodes.

Both the cardiovascular and lymphatic systems move fluid throughout the body, but there are distinct differences between them. The lymphatic system is not a closed, circular system, and it does not have a central pump. It is made up of a network of thin-walled vessels that carry a clear fluid [FIGURE 22.1]. The lymphatic system performs four major functions:

1 It **collects excess water and proteins** from the interstitial fluid that circulates around cells throughout the body and returns them as lymph to the bloodstream.

2 It **transports fats** from the tissue surrounding the small intestine to the bloodstream.

3 It **destroys microorganisms** and other foreign substances.

4 It **provides long-term protection** against microorganisms and other foreign substances.

THE LYMPHATIC SYSTEM

The **lymphatic system** begins with very small, closed-ended vessels called *lymphatic capillaries*, which are in direct contact with interstitial fluid and surrounding tissues [FIGURE 22.1B and C]. The system collects and drains most of the fluid that is forced out of the bloodstream and accumulates in the spaces between cells. The small lymphatic capillaries merge to form larger *lymphatic vessels*, which pass through specialized structures called *lymph nodes*. These larger vessels converge to form two main drainage ducts that return the excess fluid to the blood circulation through the subclavian veins above the heart. All the tissues of the body except those of the central nervous system and the cornea are drained by the lymphatic system.

In addition to lymphatic capillaries, lymphatic vessels, and lymph nodes, the lymphatic system comprises aggregates of encapsulated lymphoid nodules called *tonsils*, the *spleen*, the *thymus gland*, and aggregated unencapsulated lymph nodules. The tonsils are actually lymphatic tissue of the throat region. Lymphatic tissue is a special type of tissue that is set apart from the fundamental tissue types already described. It is a variety of reticular connective tissue containing varying amounts of specialized leukocytes called lymphocytes. All lymphatic structures contain relatively large numbers of lymphocytes within a framework of reticular cells and fibers. Finally, aggregated unencapsulated lymph nodules, which are scattered throughout the small intestine, help initiate the secretion of antibodies.

Each of these components of the lymphatic system will be described in greater detail after a discussion of lymph fluid and its circulation.

Lymph

Blood pressure inside blood vessels, which is generated by the ventricular contraction of the heart, causes water, small proteins (albumin), and other materials to be forced out of capillaries into the spaces between cells. This *interstitial fluid* bathes and nourishes surrounding body tissues. However, if there were not some means of draining it from the tissue, excess interstitial fluid would cause the tissues to swell, producing *edema*. This excess fluid moves from the body tissues into the small lymphatic capillaries and through the lymphatic system until it returns to the bloodstream. Once inside the lymphatic capillaries, the fluid is called *lymph* (LIMF; L. "clear water").

Composition and cells of lymph The body contains about 1 to 2 L of lymph, or about 1 to 3 percent of body weight. Lymph is similar to blood, but it does not contain erythrocytes and most of the proteins found in blood. The cells in the lymph fluid are *leukocytes*, and they perform the same protective functions as leukocytes in the blood. Most of the leukocytes essential to the function of lymph are produced in the bone marrow and are found in the blood, as well as in the organs and lymph nodes of the lymphatic system.

Monocytes are one class of leukocytes capable of developing into phagocytic **macrophage cells,** which ingest and digest microorganisms, cell debris, cancer cells, and any other particulate matter in the lymph. Macrophages are collectively referred to as the *reticuloendothelial system* ("network of endothelial cells"). These cells are found in lymph, in lymph nodes, and adhering to the walls of blood vessels and lymph vessels. Macrophages also migrate into and become attached to many other tissues and are then referred to as *tissue macrophages* or **histocytes** ("tissue cells"). Histocytes are common in the walls of the lung (alveolar macrophages), the sinusoids of the liver (stellate reticuloendothelial cells or Kupffer cells), the kidney (mesangial phagocytes), the spleen (littoral cells), and the bone marrow. These cells group into large clusters that surround and isolate foreign particles that are too large to phagocytize. This "walling-off" process occurs in certain chronic infections, such as tuberculosis, and limits the spread of the infectious microorganism.

A second class of leukocytes are the *lymphocytes*. The two fundamental types of lymphocytes are **B cells** and **T cells** (also called **B lymphocytes** and **T lymphocytes**). The B cells generate daughter cells called *plasma cells*, which produce specific antibodies against a particular foreign substance, and the T cells attack specific foreign cells. The human body contains about 2 trillion lymphocytes. They are the backbone of the immune system and are the *basis of the immune response*. The specific functions of B cells and T cells are discussed in the next chapter.

FIGURE 22.1 THE LYMPHATIC SYSTEM ★

[A] Major components of the lymphatic system. The right lymphatic duct drains the upper right quadrant of the body (green), and the thoracic duct drains the rest of the body. **[B]** Relationship of the blood vessels and lymphatic vessels (green). Excess fluid and proteins in the tissue spaces enter lymphatic capillaries and return to the venous system by lymphatic vessels. Arrows indicate direction of fluid flow.

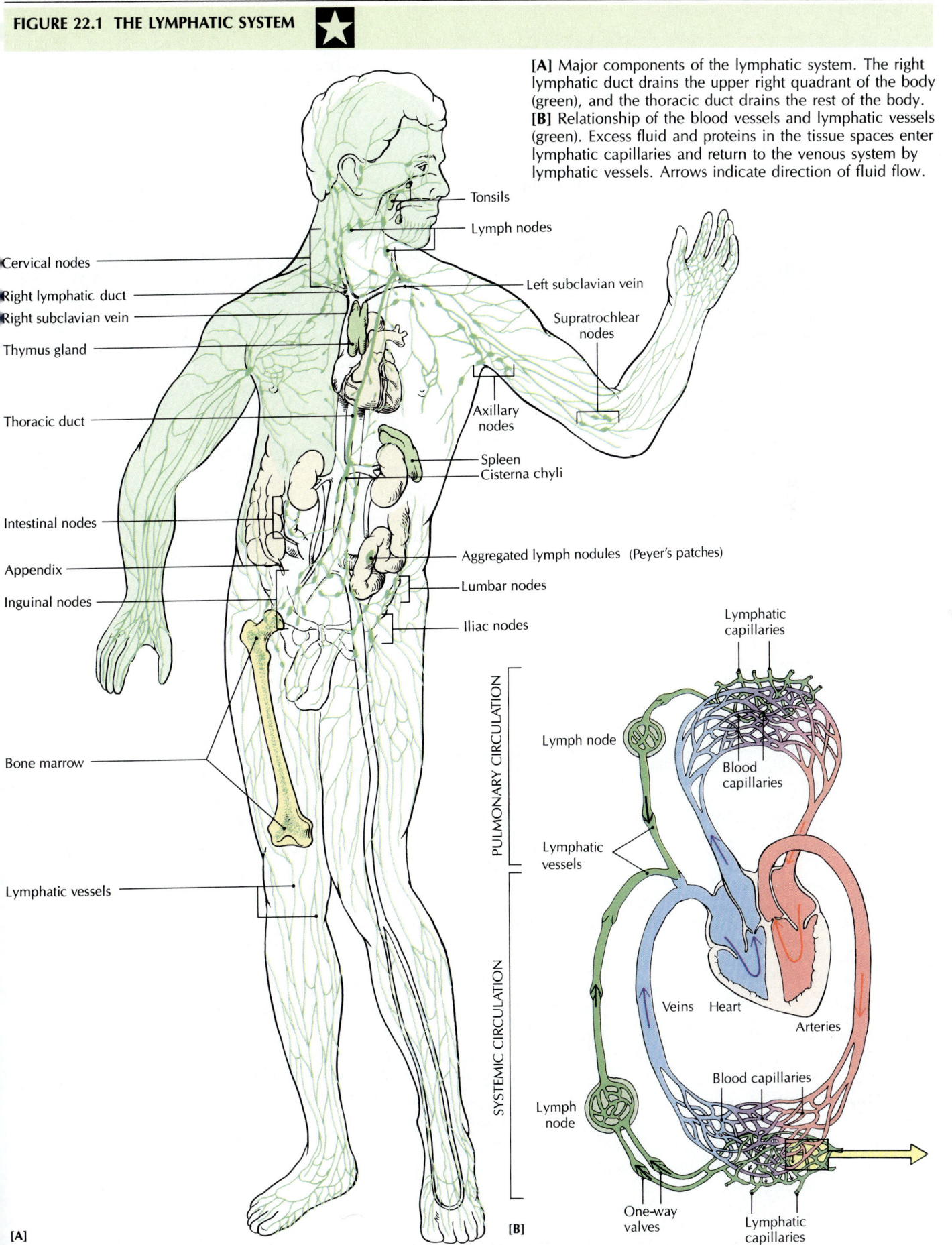

Tonsils
Lymph nodes
Cervical nodes
Right lymphatic duct
Right subclavian vein
Thymus gland
Thoracic duct
Left subclavian vein
Supratrochlear nodes
Axillary nodes
Spleen
Cisterna chyli
Intestinal nodes
Appendix
Inguinal nodes
Aggregated lymph nodules (Peyer's patches)
Lumbar nodes
Iliac nodes
Bone marrow
Lymphatic vessels

Lymphatic capillaries
Lymph node
Blood capillaries
PULMONARY CIRCULATION
Lymphatic vessels
SYSTEMIC CIRCULATION
Lymph node
Veins   Heart
Arteries
Blood capillaries
One-way valves
Lymphatic capillaries

[A]   [B]

(Figure 22.1 continues on following page)

[C] Enlargement showing the relationship of lymphatic and blood capillaries. [D] Basic structure of lymphatic tissue.

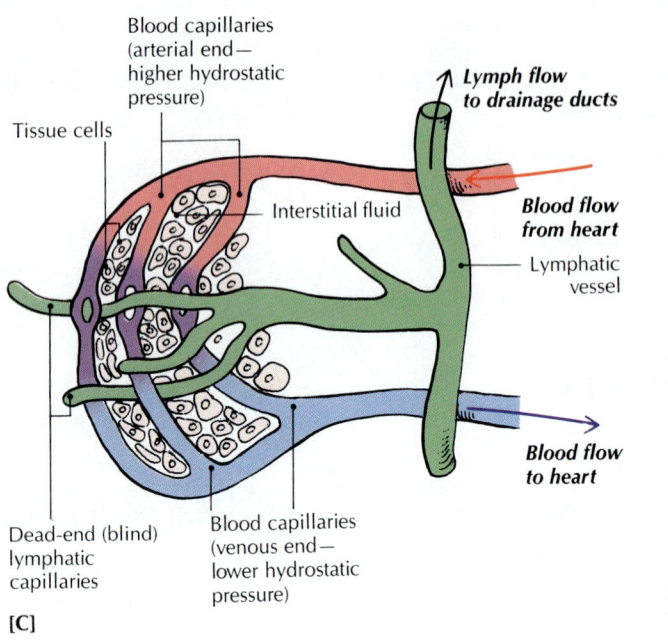

[C]

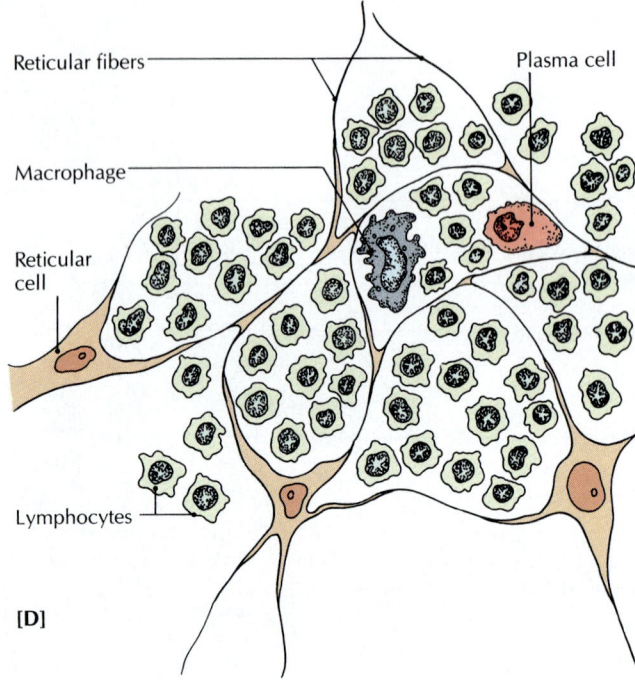

[D]

Structure	Major functions
Lymphatic capillaries	Collect excess fluid from tissues.
Lymphatic vessels (collecting vessels)	Carry lymph from lymphatic capillaries to veins in neck.
Lymph nodes	Situated along collecting lymphatic vessels; filter foreign material from lymph.
Axillary nodes	Drain arms, most of thoracic wall, breasts, upper abdominal wall.
Supratrochlear nodes	Drain hands, forearms.
Cervical nodes	Drain scalp, face, nasal cavity, pharynx.
Intestinal nodes	Drain abdominal viscera.
Inguinal nodes	Drain legs, external genitalia, lower abdominal wall.
Iliac nodes	Drain pelvic viscera.
Lumbar nodes	Drain pelvic viscera.
Tonsils	Destroy foreign substances at upper entrances of respiratory and digestive systems.
Spleen	Filters foreign substances from blood, manufactures phagocytic lymphocytes, stores red blood cells, releases blood to body in case of extreme blood loss.
Thymus gland	Forms antibodies in newborn; involved in initial development of immune system; site of differentiation of lymphocytes into T cells; produces thymosin.
Aggregated unencapsulated lymph nodules (Peyer's patches)	Respond to antigens in intestine by generating plasma cells that secrete antibodies.

Circulation of Lymph

Since there is no central pump to circulate the lymph, the actual movement is accomplished primarily by three forces: (1) The action of circular and some longitudinal smooth muscles in the lymphatic vessels other than capillaries moves lymph. (2) The squeezing action of skeletal muscles during normal body movement helps move lymph through the vessels. (Lymph flow may increase 5- to 15-fold during vigorous exercise.) (3) The lymphatic system runs parallel with the venous system in the thorax, where a subatmospheric pressure exists. This pressure gradient creates a "pull factor," called the auxiliary respiratory pump, that aids lymph flow.

All lymph vessels are directed toward the thoracic cavity. The upper right quadrant of the body contains lymphatics that drain their contents into the right subclavian vein through the *right lymphatic duct* [FIGURE 22.1A]. This drainage includes the right side of the head, right upper extremity, right thorax and lung, right side of the heart, and the upper portion of the liver. The remainder of the lymphatic system returns its fluid to the left subclavian vein through the *thoracic* (left lymphatic) *duct,* the largest of the lymphatics. The following description of the circulation of lymph through the lymphatic system [FIGURE 22.2] is not meant to indicate a specific *sequence* of events:

1 The upper right quadrant of the body drains into the right lymphatic duct [see FIGURE 22.1A].

2 The right lymphatic duct empties into the right subclavian vein.

3 The three-quarters of the body not drained by the right lymphatic duct is drained by the thoracic (left lymphatic) duct.

4 The thoracic duct begins as a dilated portion called the *cisterna chyli* (KYE-lee) within the abdomen, below the diaphragm. It extends upward through the diaphragm, along the posterior wall of the thorax, into the left side of the base of the neck.

5 At the base of the neck, the thoracic duct receives the left jugular lymphatic trunk from the head and neck, the left subclavian trunk from the left upper extremity, and other lymphatic vessels from the thorax and related parts.

6 The thoracic duct then opens into the left subclavian vein, returning the lymph to the blood.

Lymphatic Capillaries and Other Vessels

Lymphatic capillaries and blood capillaries are somewhat similar in structure. Both consist of a single layer of endothelial tissue that permits fluid absorption and *diapedesis,* the passage of white blood cells through capillary walls. The major structural difference is an important

one: lymphatic capillaries are one-way vessels that begin with a "dead-end" (blind) end. Lymphatic capillaries are not part of a circuit of vessels as blood capillaries are.

Lymphatic capillaries have a slightly larger internal diameter than blood capillaries. The endothelial cells that make up the walls of lymphatic capillaries regulate the passage of materials into and out of the lymph. The ends of adjacent endothelial cells overlap to form *flap valves* that open to permit fluid to enter the lymphatic capillary and close when the lymphatic capillary contracts [FIGURE 22.3].

Lymphatic capillaries are most abundant near the innermost and outermost surfaces of the body, for example, the dermis of the skin, and the mucosal and submucosal layers of the respiratory and digestive systems. Lymphatic capillaries are also numerous beneath the mucous membrane that lines the body cavities and covers the surface of organs. Very few lymphatic capillaries are found in muscles, bones, or connective tissue. There are none in the central nervous system or the cornea of the eyeball.

Specialized lymphatic capillaries called *lacteals* (L. *lacteus,* of milk) extend into the intestinal villi. Lacteals absorb fat from the small intestine and transport it into the blood for distribution throughout the body. The lymph in the lacteals takes on a milky appearance (hence the name *lacteal*) because of the presence of many small droplets of fat. At that point, the mixture of lymph and finely emulsified fat is known as *chyle* (KILE; L. *chylus,* juice).

The lymphatic capillaries join with other capillaries to become larger collecting *lymphatic vessels,* sometimes called *lymphatics.* Lymphatic vessels resemble veins, but their walls are thinner than venous walls, they contain more valves, and they pass through specialized masses of tissue (the lymph nodes). Lymphatic vessels are usually found in loose connective tissue, running parallel to blood vessels. Lymphatic vessels are arranged into a superficial set and a deep set, and they pass through various lymph nodes. The superficial lymphatic vessels in the skin and subcutaneous tissue tend to follow the course of superficial veins, and the deeper vessels follow the deep veins and arteries.

Lymphatic vessels join with one another to form two large ducts—the *right lymphatic duct* and the *thoracic duct*—that empty their contents into the subclavian veins above the heart, as described previously.

Lymph travels in only one direction (toward the subclavian veins) because of valves within the lymphatics that do not allow fluid to flow back. The lymphatic valves operate in the same way as the one-way valves in veins.

Lymph Nodes

Scattered along the lymphatic vessels like beads on a string are small (1 to 25 mm in length) bean-shaped masses of tissue called *lymph nodes* [FIGURE 22.4]. These

FIGURE 22.2 LYMPHATIC CIRCULATION

[A] Drawing showing how lymph drains from the lymphatic system (green) into the bloodstream; anterior view. Arrows indicate the direction of lymph flow into the subclavian veins. [B] Photograph of the thoracic duct and cisterna chyli; anterior view.

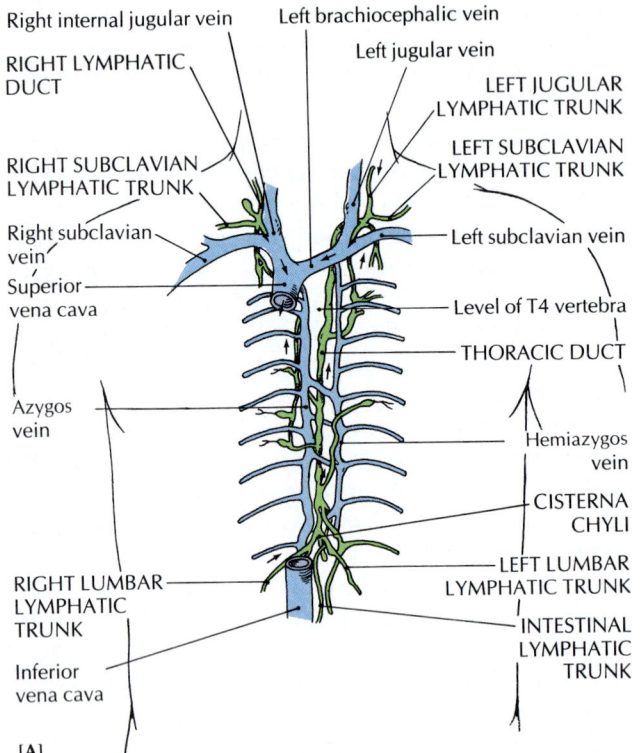

[A]

nodes are found in the largest concentrations at the neck, armpit, thorax, abdomen, and groin. Lesser concentrations are found behind the elbow and knee. The *superficial* lymph nodes are located near the body surface in the neck, armpit, and groin. The *deep* nodes are located deep within the groin area, near the lumbar vertebrae, at the base of the lungs, attached to the tissue surrounding the small intestines, and in the liver. Most lymph passes through at least one lymph node on its way back to the bloodstream.

Lymph nodes, which are sometimes incorrectly called glands,* filter out harmful microorganisms and other foreign substances from the lymph, trapping them in a mesh of reticular fibers. Lymph nodes are also the initiating sites for the specific defenses of the immune response. These functions are intimately related to the structure of the lymph node and the cells it contains.

Lymph nodes are covered by a *capsule* of fibrous connective tissue. Projections of connective tissue *trabeculae* extend inward from the capsule toward the center of the lymph node, dividing it into compartments. The outer

---

*Lymph nodes were originally called *lymph glands* (L. *glans*, acorn) because they seemed to resemble acorns. Soon, other small bits of tissue came to be called glands even if they did not look anything like acorns. When it was discovered that some of these bits of tissue secrete various fluids, it was decided that a gland was a structure that formed secretions. Ironically, the original "glands," the lymph glands, do not secrete fluids and were misnamed in the first place. The word *node*, derived from the Latin word for "knob," seems to describe the structure more appropriately.

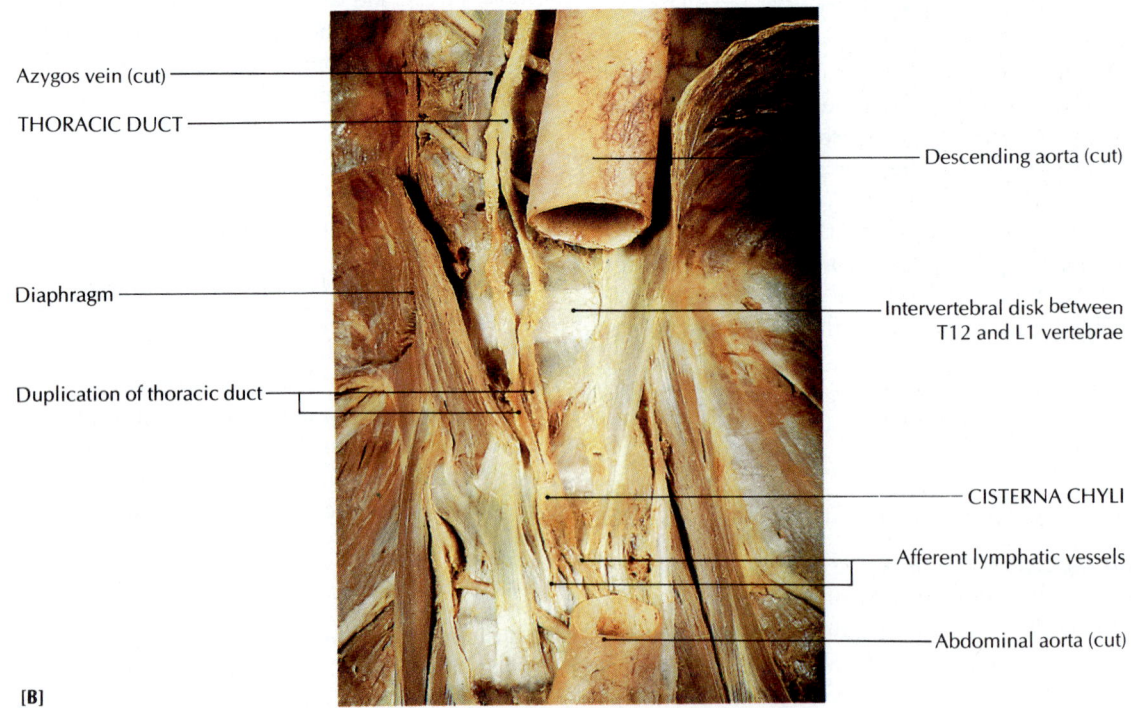

[B]

FIGURE 22.3 LYMPHATIC CAPILLARIES

[A] Flap valves between adjacent endothelial cells in lymphatic capillaries open to permit tissue fluid to enter and close to prevent leakage. The red arrow indicates the direction of fluid flow. [B] Electron micrograph of "loose junction" between the overlapping ends of two endothelial cells in the wall of a lymphatic capillary. ×64,000.

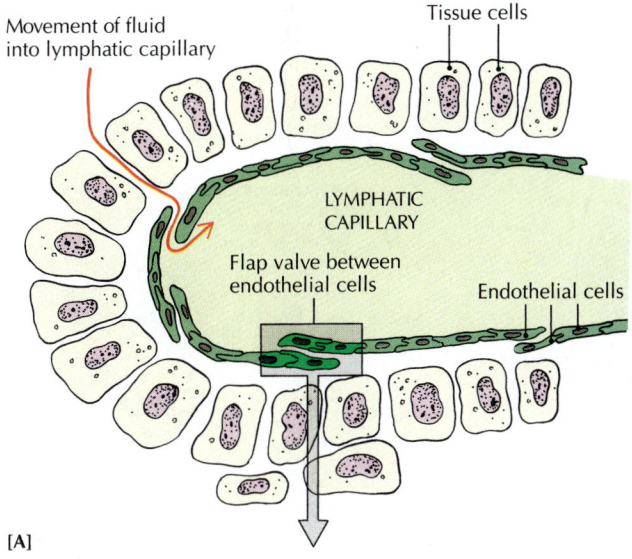

[A]

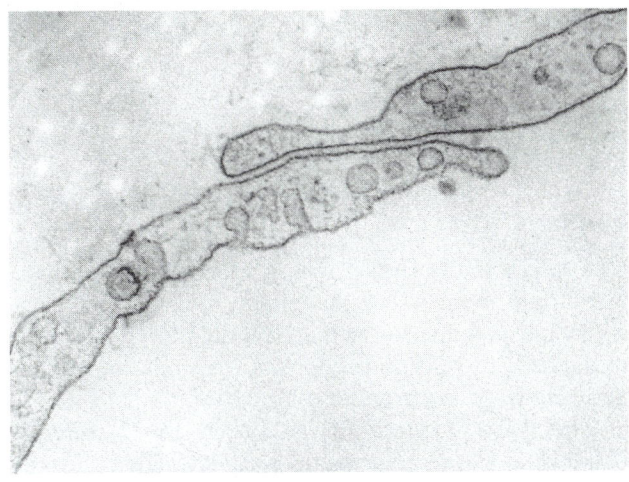

[B]

Running from the cortex to the medulla, and surrounding the nodules and medullary cords, are *medullary sinuses*, through which the lymph flows before it leaves the node. The lymph node effectively funnels foreign materials in the lymph through the sinuses so that the lymph comes in contact with lymphocytes and macrophages. Afferent (*to* the node) lymphatic vessels with lymph enter the node at various points along the outer capsule. Efferent (*from* the node) lymphatic vessels leave the node from a small depressed area called the *hilus*. Blood vessels enter, as well as leave, through the hilus.

Tonsils

The *tonsils* are aggregates of lymphatic nodules enclosed in a capsule of connective tissue. There are three types of tonsils: (1) the single *pharyngeal* tonsil in the upper posterior wall of the pharynx behind the nose, (2) the *palatine* tonsils on each side of the soft palate, and (3) the *lingual* tonsils at the base of the tongue.

The tonsils lack afferent lymphatic vessels, a feature that distinguishes them from lymph nodes. The efferent lymphatic vessels of the tonsils contribute many lymphocytes to the lymph. These cells are capable of leaving the tonsils and destroying invading microorganisms in other parts of the body. Together the tonsils form a band of lymphoid tissue that is strategically placed at the upper entrances to the digestive and respiratory systems, where foreign substances may enter easily. Most infectious microorganisms are killed by lymphocytes at the surface of the pharynx or are killed later, after the initial defenses are set in motion by the tonsils. The presence of plasma cells (antibody-producing cells that develop from B lymphocytes) within the tonsils indicates the formation of antibodies.

Although tonsils usually function to prevent infection, they may become infected repeatedly themselves. In such cases, some physicians recommend their removal. The palatine tonsils are the ones most frequently removed (the familiar operation is called a *tonsillectomy*). The lingual tonsils are rarely removed. The pharyngeal tonsil, popularly called the *adenoid*, may also become enlarged. If this occurs, obstruction of the nasal pharynx caused by swelling can interfere with breathing, necessitating the removal of the adenoid (*adenoidectomy*).

portion of each compartment is the *cortex* of the node. It contains lymphocytes in dense clusters called *lymph nodules*. In the middle of each nodule is the *germinal center*, where lymphocytes are produced by cell division. The lymph nodes produce about 10 billion lymphocytes every day. The inner part of a lymph node is the *medulla*. It contains strands of lymphocytes extending from the nodule. These strands are appropriately called the *medullary cords*.

Q: *Why are lymph nodes sometimes removed during cancer operations?*

A: Lymph nodes near a cancer site may contain viable cancer cells. On entering the lymph nodes, cancer cells can multiply and establish secondary cancers by dispersing throughout the body via the lymphatic system.

FIGURE 22.4 LYMPH NODE

[A] Drawing of a lymph node, showing how lymph flows (red arrows) through the sinuses surrounding the nodules. This arrangement brings potential antigens into contact with lymphocytes, monocytes, and macrophages, enhancing defensive reactions. The smaller drawing on the right shows an enlarged portion of the lymph node. **[B]** Scanning electron micrograph of the interior of a lymph node. ×1200.

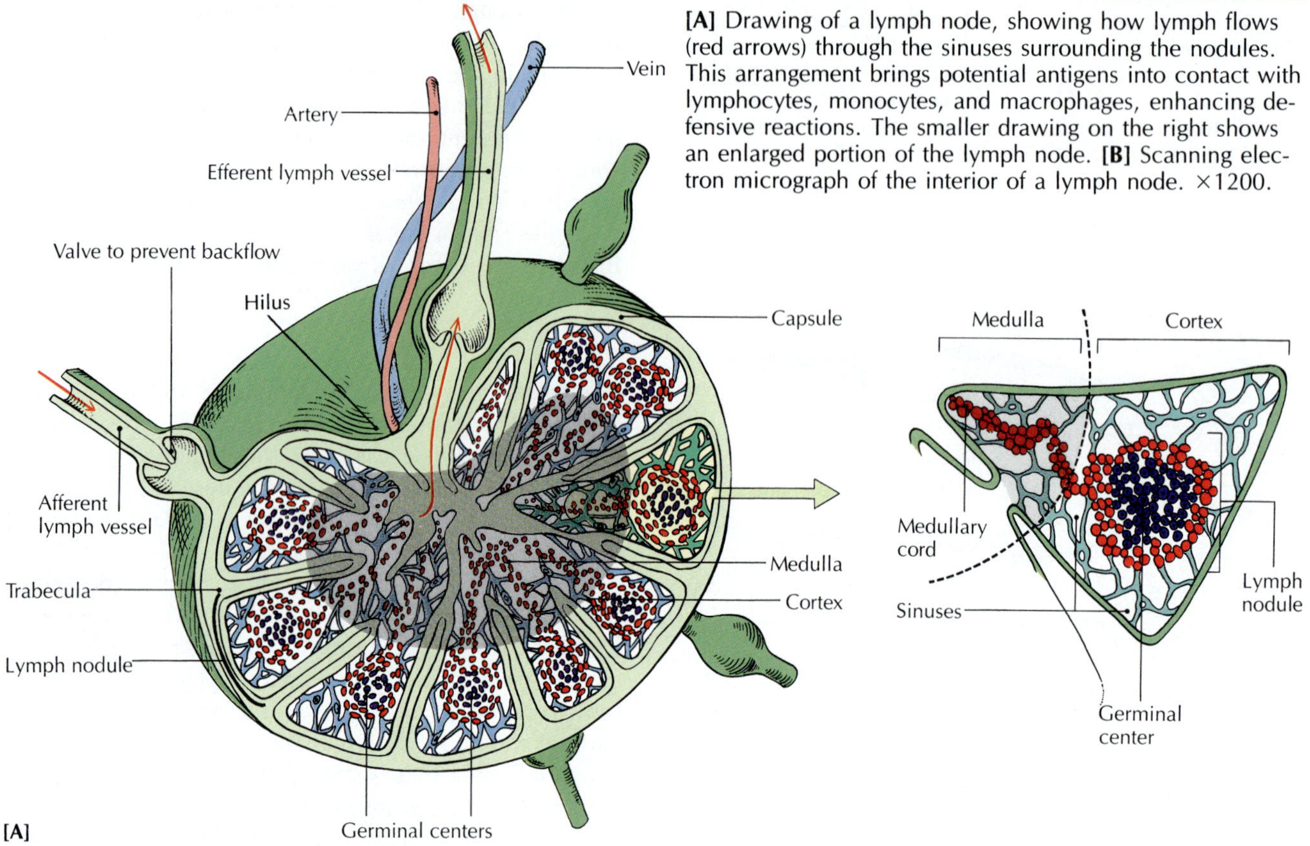

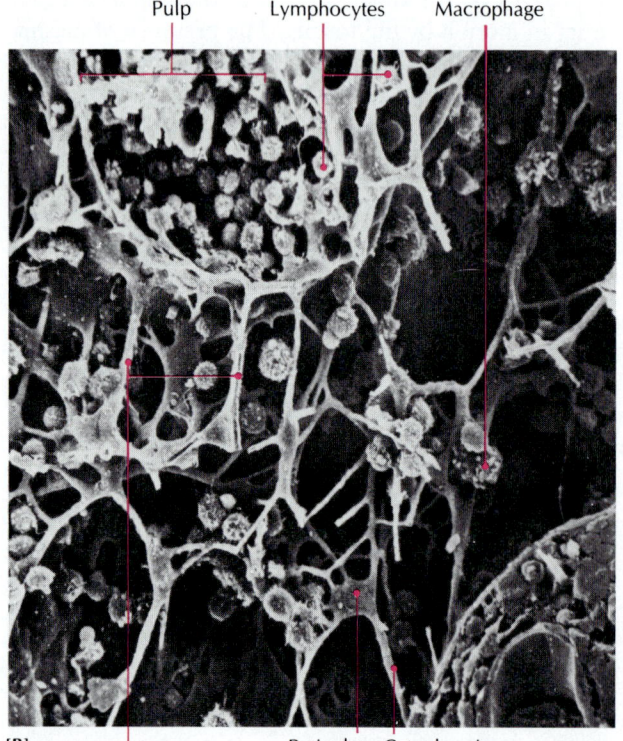

Spleen

The *spleen* is the largest lymphoid organ in the body, measuring about 12 cm in length, and is purplish in color. Located inferior to the diaphragm on the left side of the body, it rests on portions of the stomach, left kidney, and large intestine [FIGURE 22.1A].

The main functions of the spleen are (1) filtering blood and (2) manufacturing phagocytic lymphocytes and monocytes. It also contributes to the functioning of the cardiovascular and lymphatic systems as follows:

1 Macrophages, which are abundant in the spleen, help remove damaged or dead erythrocytes and platelets, microorganisms, and other debris from the blood as it circulates through the spleen. Macrophages also remove iron from the hemoglobin of old red blood cells and return it to the circulation for use by the bone marrow in producing new red blood cells.

2 Antigens in blood entering the spleen activate lymphocytes that develop into cells that produce antibodies or are otherwise involved in the immune reaction.

FIGURE 22.5 SPLEEN

[A] Drawing of the spleen showing blood flow (colored arrows) through the white pulp to the venous sinuses in the red pulp. Note that the white pulp consists of nodules and lymphocytes, while the red pulp is an open mesh with venous sinuses running through it. **[B]** Scanning electron micrograph of a section through the spleen. ×1100.

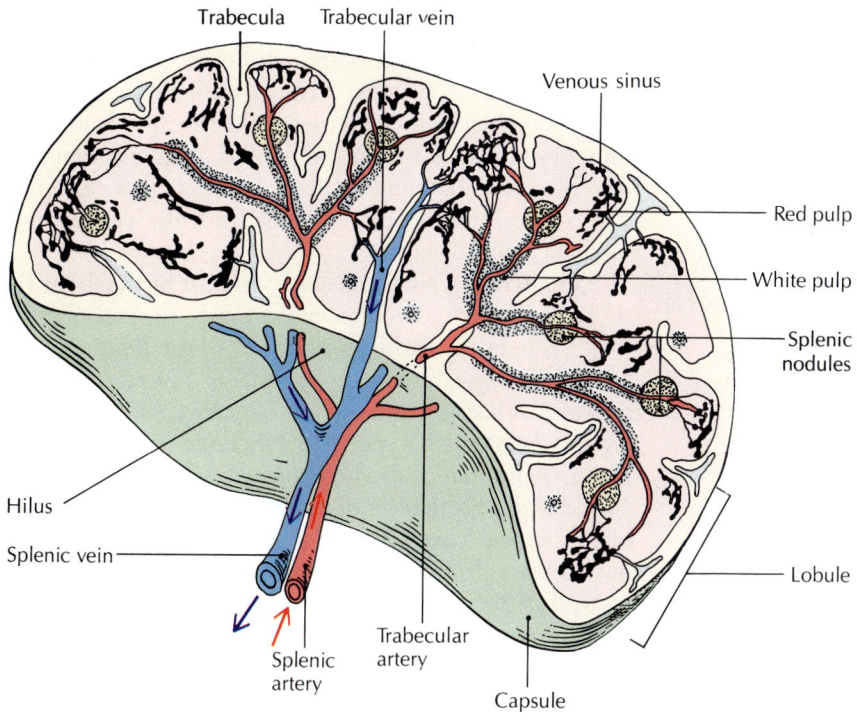

[A]

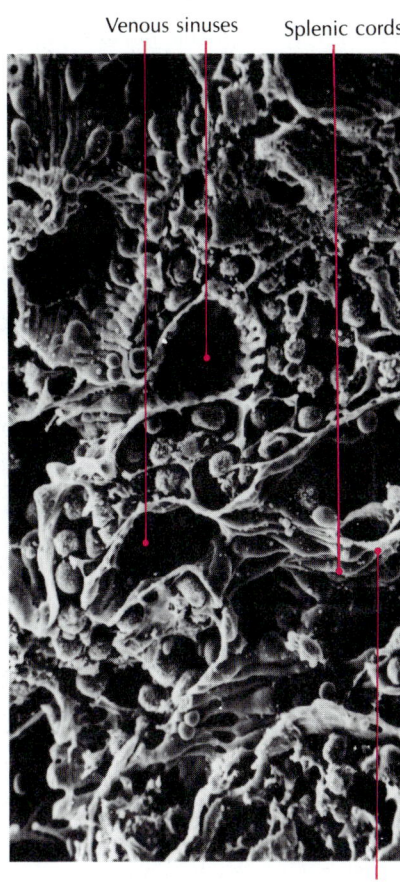

[B]

Reticular fiber of sinus wall

3 The spleen produces red blood cells during fetal life. In later life, it stores newly formed red blood cells and platelets and releases them into the bloodstream as they are needed.

4 Because the spleen contains a large volume of blood, it serves as a blood reservoir. If the body loses blood suddenly, the spleen contracts and adds blood to the general circulation. (The spleen is capable of releasing approximately 200 mL of blood into the general circulation in 1 min.)

The tissue structure of the spleen is similar to that of lymph nodes [FIGURE 22.5]. Surrounded by a *capsule* of connective tissue, the spleen is divided by *trabeculae* into compartments called *lobules*. The functional part of the medulla consists of *splenic pulp*, which contains small islands of white pulp scattered throughout red pulp.

The *white pulp* is made up of compact masses of lymphocytes surrounding small branches of the splenic artery. These masses, which occur in intervals, are called *splenic nodules (Malpighian corpuscles).* Within the *red pulp*

are *venous sinuses* filled with blood and lined with monocytes and macrophages. (The pulp is red because of the many erythrocytes in the blood.) This arrangement brings the blood and any foreign materials it may contain to the lymphocytes, monocytes, and macrophages for cleansing. Since the spleen does not receive lymphatic vessels and lymph, it cannot be considered a filter in the same way that a lymph node is a filter. Efferent lymphatic vessels, which leave the spleen, contribute lymphocytes, monocytes, and macrophages to the lymph.

Thymus Gland

The ***thymus gland*** is a ductless pinkish-gray mass of flattened lymphoid tissue located in the thorax. It overlies the heart and its major blood vessels. The thymus gland consists of two *lobes* joined by connective tissue and surrounded by a *capsule* of connective tissue. Each lobe is divided into *lobules* by coarse *trabeculae*, with each lobule having an outer *cortex* and an inner *medulla* [FIGURE 22.6]. The cortex contains many lymphocytes, most of which

FIGURE 22.6 INTERNAL ANATOMY OF THYMUS GLAND

The drawing of a lobule of the thymus gland shows mature lymphocytes in the medulla and many more immature lymphocytes in the cortex.

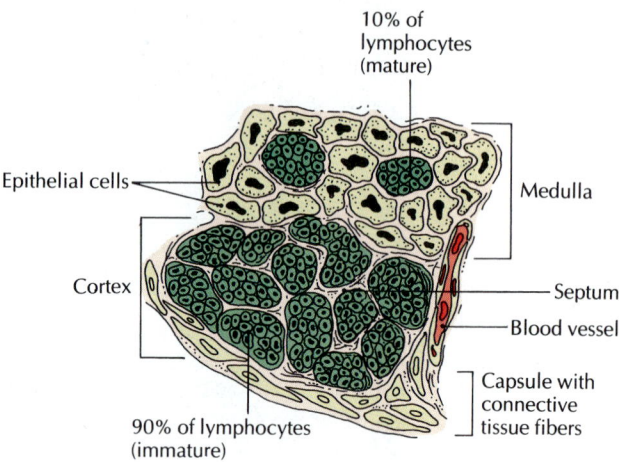

are immature. Undifferentiated lymphocytes migrate from the bone marrow to the thymus gland to become specialized, or immune-competent, T cells (T for "thymus-dependent") with defined roles in specific defenses against foreign cells.

T lymphocytes mature in the cortex and migrate to the medulla. The mature T lymphocytes soon leave the medulla and enter the peripheral blood circulation, which transports them to the spleen and lymph nodes, where they encounter foreign substances and respond to them [FIGURE 22.7]. Most thymic lymphocytes, however, degenerate before they ever leave the thymus gland. The only blood vessels within the thymus gland are capillaries. The medulla contains fewer lymphocytes than the cortex does, making the epithelial reticular cells in the medulla more obvious.

The thymus gland is relatively large at birth (about 12 to 15 g), forms antibodies in the newborn, and plays a major role in the early development of the immune system. It increases in size from birth to puberty, is most active in childhood and early adolescence, and atrophies after puberty. Other possible functions of the thymus gland are not understood completely, but it is clearly involved in the formation of a permanent immune system of antibodies. The thymus gland also secretes a group of hormones collectively called *thymosin*, which may be necessary for the differentiation of T cells from stem cells. Like the tonsils and spleen, the thymus gland has efferent lymph vessels but no afferent ones. As a result, no lymph drains into it.

FIGURE 22.7 THE MATURATION OF LYMPHOCYTES

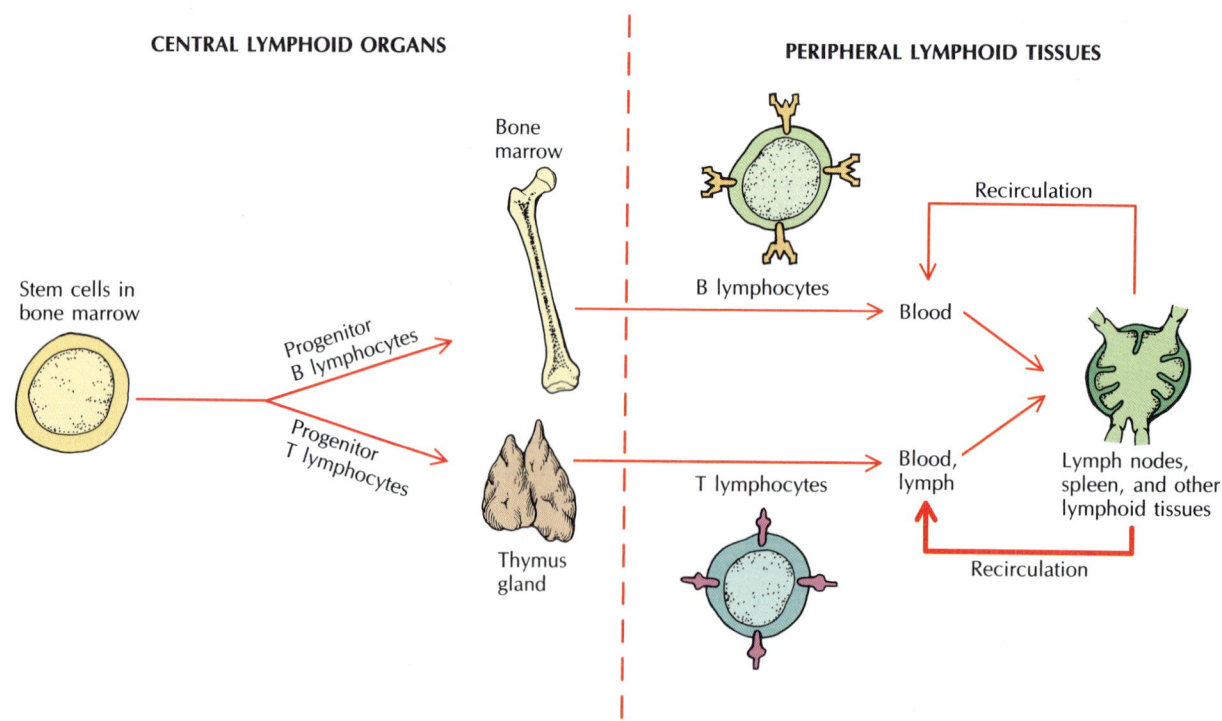

FIGURE 22.8 AGGREGATED UNENCAPSULATED LYMPH NODULES (PEYER'S PATCHES)

[A] Light micrograph of tissue from an aggregated unencapsulated lymph nodule. [B] Aggregated lymph nodules in the intestine expose lymphocytes (B cells) to antigens in the intestinal lumen. [C] Lymphocytes leave the aggregated lymph nodules and move through the lymphatic system and bloodstream to locate as mature plasma cells beneath the intestinal epithelium [D].

Aggregated Unencapsulated Lymph Nodules (Peyer's Patches)

Aggregated unencapsulated lymph nodules (Peyer's patches) are clusters of unencapsulated lymphoid tissue found in the tonsils, small intestine, and appendix. Because of their location, they are also called *gut-associated lymphoid tissue* (GALT). Also, similar clusters of lymphoid tissue are found throughout the body in mucosal tissues, such as in the lining of the respiratory and genitourinary tracts, where they are called *mucosa-associated lymphoid tissue* (MALT).

Aggregated lymph nodules are the sites where B cells mature into plasma cells that secrete antibodies in large quantities in response to antigens in the intestine [FIGURE 22.8]. Such plasma cells do not remain clustered in aggregated lymph nodules, but are distributed along the length of the intestine in the following way: Inactive B cells migrate from bone marrow to aggregated lymph nodules. B cells activated by exposure to antigens leave the aggregated lymph nodules through efferent lymphatics, migrate to lymph nodes, and then later enter the bloodstream through the thoracic duct. The B cells are carried by the bloodstream throughout the body and eventually come to reside as plasma cells beneath the mucosal surface of the intestine. Here they recognize an enormous variety of specific antigens (foreign substances) and produce corresponding antibodies, primarily against bacteria and some viruses.

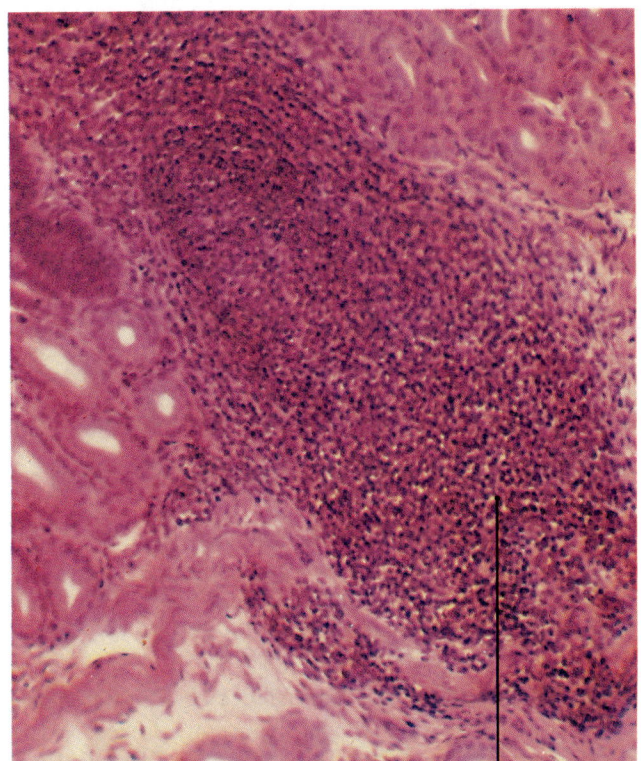

[A]   Aggregated lymph nodules

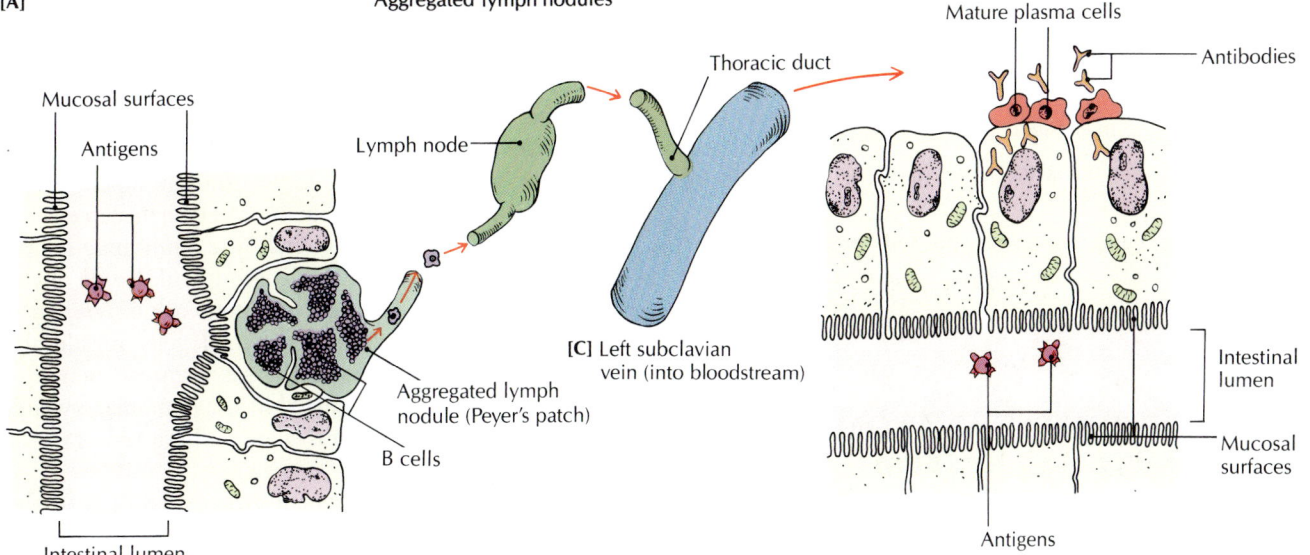

[B]

[C] Left subclavian vein (into bloodstream)

[D]

FIGURE 22.9 DEVELOPMENTAL ANATOMY OF THE LYMPHATIC SYSTEM

[A] Drawing of a 7-week-old embryo, left lateral view; note the five primary lymph sacs. **[B]** The paired thoracic ducts are developed at about week 9; ventral view. **[C]** Later stage of development showing the thoracic duct and right lymphatic duct; ventral view.

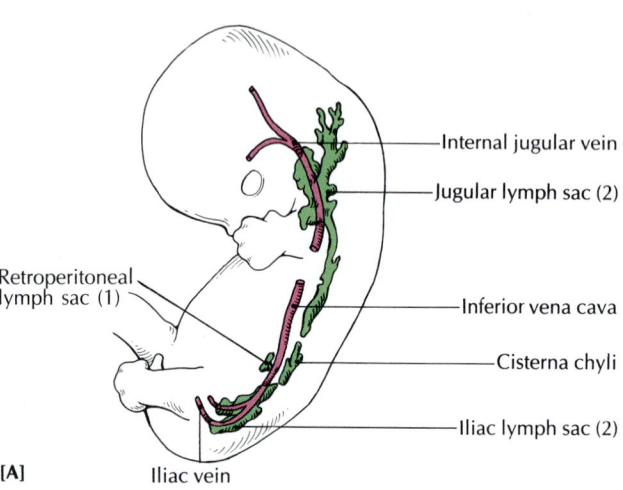

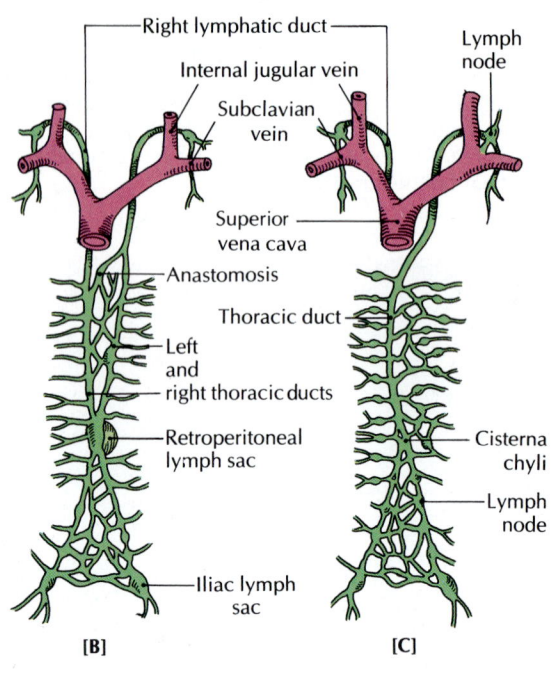

DEVELOPMENTAL ANATOMY OF THE LYMPHATIC SYSTEM

The development of the lymphatic system begins at the end of the fifth week after fertilization. Lymphatic vessels emerge from five primary lymph sacs [FIGURE 22.9A] and branch from the paired *jugular lymph sacs* to the head, neck, and arms; from the paired *iliac lymph sacs* to the lower trunk and lower limbs; and from the single *retroperitoneal lymph sac* and the *cisterna chyli* to the digestive system. The jugular lymph sacs are linked to the cisterna chyli by the left and right thoracic ducts; at about week 9, these ducts become connected by an anastomosis [FIGURE 22.9B].

The mature *thoracic duct* develops from the caudal portion of the right thoracic duct, the anastomosis, and the cranial portion of the left thoracic duct. The *right lymphatic duct* develops from the cranial portion of the right thoracic duct [FIGURE 22.9C], and the superior portion of the cisterna chyli remains after birth. Except for this portion of the cisterna chyli, the embryonic lymph sacs develop into recognizable groups of *lymph nodes* during the early fetal period. The network of *lymphatic sinuses* develops when mesenchymal cells invade the lymph sacs and break down the sac cavities. Immature fetal lymphocytes are produced in the bone marrow and later migrate to the thymus gland, where they mature into inactive lymphocytes (T cells).

The *spleen* develops from mesenchymal cells in the stomach mesentery (greater omentum), and the *tonsils* develop from various lymph nodules near the mouth and throat.

WHEN THINGS GO WRONG

Hodgkin's Disease

Hodgkin's disease is a form of cancer of the lymph nodes typified by the presence of large, multinucleate cells (Reed-Sternberg cells) in the affected lymphoid tissue. The first sign of Hodgkin's disease is usually a painless swelling of the lymph nodes, most commonly in the neck area but occasionally in the axilla, inguinal, or femoral region. Late symptoms of the disease include edema of the face and neck, anemia, jaundice, and increased susceptibility to infection.

Hodgkin's disease occurs most often in young adults between the ages of 15 to 38 and in people over 50. Interestingly, it occurs in Japan only in people over 50. Its cause is unknown, but the disease is potentially curable. It is usually treated with considerable success with a combination of radiation therapy and chemotherapy. If untreated, Hodgkin's disease is fatal.

CHAPTER SUMMARY

The **lymphatic system** returns to the blood excess fluid and proteins from the spaces around cells, plays a major role in the transport of fats from the tissue surrounding the small intestine to the blood, filters and destroys microorganisms and other foreign substances, and aids in providing long-term protection for the body.

The Lymphatic System (p. 710)

1 The lymphatic system consists of vessels, lymph nodes, lymph, leukocytes, lymphatic organs (spleen and thymus gland), and specialized lymphoid tissues.

2 Excess interstitial fluid is drained from tissues by the lymphatic system. Once inside the lymphatic capillaries, the fluid is called **lymph.**

3 The composition of lymph is similar to that of blood, except that lymph lacks red blood cells and most of the blood proteins.

4 Leukocytes in lymph include monocytes, which develop into macrophages, and two types of lymphocytes, **B cells** and **T cells.** Specific defenses are accomplished by **lymphocytes,** a type of leukocyte. Lymphocytes are the basis of the immune response.

5 **Lymphatic capillaries** are one-way vessels that begin with a closed end. They join together to form **lymphatic vessels,** which drain into the **right lymphatic duct** and the **thoracic duct.** The ducts return fluid to the blood through the right and left subclavian veins.

6 **Lymph nodes** are small masses of lymphoid tissue scattered along the lymphatic vessels. Lymphocytes in the nodes filter out harmful substances from the lymph and are the initiating sites for the specific defenses of the immune system.

7 The three **tonsils** (pharyngeal, palatine, and lingual) form a band of lymphoid tissue, which prevents foreign substances from entering the body through the throat.

8 The **spleen** is the largest lymphoid organ. Its major functions are filtering blood and manufacturing phagocytic lymphocytes.

9 The **thymus gland** forms antibodies in the newborn and plays a major role in the early development of the immune system.

10 **Aggregated unencapsulated lymph nodules** are clusters of lymphoid tissue in the tonsils, small intestine, and appendix. They respond to antigens from the intestine by generating plasma cells that secrete antibodies in large quantities. Lymph nodules are also in the bronchi of the respiratory tract.

Developmental Anatomy of the Lymphatic System (p. 720)

1 The development of the lymphatic system begins at the end of week 5.

2 **Lymphatic vessels** develop from **lymph sacs** and the **cisterna chyli.** The lymph sacs develop into **lymph nodes.**

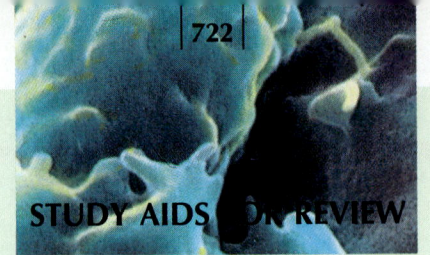

STUDY AIDS FOR REVIEW

KEY TERMS

MEDICAL TERMS FOR REFERENCE

ELEPHANTIASIS Obstruction (caused by a parasitic filarial worm) of the return of lymph to the lymphatic ducts that causes enlargement of a limb (usually a lower limb) or the genital area.

HYPERSPLENISM A condition in which the spleen is abnormally active and blood cell destruction is increased.

LYMPHADENECTOMY Removal of lymph nodes.

LYMPHADENOPATHY A disease of lymph nodes, resulting in enlarged nodes.

LYMPHANGIOMA Benign tumor of lymph tissue.

LYMPHANGITIS Inflammation of lymphatic vessels.

LYMPHATIC METASTASIS A condition in which a disease travels around the body via the lymphatic system.

LYMPHOMA Tumor of the lymph nodes.

LYMPHOSARCOMA Malignant tumor of lymph tissue.

SPLENECTOMY Total removal of the spleen.

SPLENOMEGALY Enlargement of the spleen, following infectious diseases such as scarlet fever, typhus, or syphilis.

UNDERSTANDING THE FACTS

1 List the four major functions of the lymphatic system.
2 How does lymph differ from blood?
3 Describe the structure of a lymph node.
4 How do lymphatic capillaries differ from blood capillaries?
5 What portion of the body is drained by the thoracic duct? The right lymphatic duct?
6 Into which blood vessels do the vessels of the lymphatic system drain?
7 In addition to lymph nodes, what other encapsulated structures are found in the lymphatic system? Unencapsulated structures?
8 What is the major function of each of the following?
  a macrophage
  b B cell
  c T cell
  d plasma cell
9 Identify
  a cisterna chyli
  b lacteal
  c chyle

UNDERSTANDING THE CONCEPTS

1 What forces propel lymph through the vessels of the lymphatic system?
2 Contrast the structure of a lymphatic vessel with that of a vein.
3 To what does the term *reticuloendothelial system* refer?
4 Which organs of the lymphatic system lack afferent vessels? How do the components of lymph enter these structures?

SELF-QUIZ
Multiple-Choice Questions

22.1 Excess interstitial fluid drains directly from body tissues into
  a the right lymphatic duct
  b the subclavian veins
  c the thoracic duct
  d lymphatic capillaries
  e aggregated unencapsulated lymph nodules

22.2 Which of the following cell types is *not* found in lymph?
  a monocytes
  b erythrocytes
  c leukocytes
  d macrophages
  e B cells

22.3 Flap valves are found in
  a lymphatic vessels
  b the thoracic duct
  c lymphatic capillaries
  d a and b
  e a, b, and c

22.4 Lacteals contain fluid drained from the
  a liver
  b stomach

c small intestine
d large intestine
e spleen

22.5 A lymphatic structure that serves as a blood reservoir is the
a spleen
b tonsil
c thymus gland
d aggregated unencapsulated lymph nodule
e adenoid

Matching

a right lymphatic duct
b right subclavian vein
c left jugular lymphatic trunk
d left subclavian vein
e thoracic duct

22.6 _____ right lymphatic duct drains into this vessel
22.7 _____ begins as cisterna chyli
22.8 _____ the lymphoid vessel that drains the upper right quadrant of the body
22.9 _____ the thoracic duct empties into this vessel
22.10 _____ drains the head and neck; empties into the thoracic duct

a germinal center
b trabeculae
c aggregated unencapsulated lymph nodule
d medullary cord
e medullary sinus

22.11 _____ connective tissue strands that divide a lymph node into compartments
22.12 _____ site of lymphocyte production within a lymph nodule
22.13 _____ region of lymph flow within a lymph node
22.14 _____ associated with the gut
22.15 _____ strands of lymphocytes within the center of a lymph node

Completion Exercises

22.16 The presence of excess interstitial fluid in the body tissues produces a condition called _____.
22.17 Lymphatic vessels pass through bean-shaped masses of tissue called _____.
22.18 Blood vessels enter and leave lymph nodes through an area called the _____.
22.19 The three types of tonsils are the _____, _____, and _____ tonsils.
22.20 "Adenoid" is the common name for the _____.
22.21 A lymphatic structure that filters blood as well as recycling iron from hemoglobin is the _____.

22.22 Materials forced out of the capillaries by the hydrostatic pressure of the blood initially enter the _____ that surrounds the body's cells.
22.23 The products of fat digestion in the intestine are absorbed into _____.
22.24 Tissue macrophages are also called _____.
22.25 Within lymphatic vessels, lymph can flow in only one direction because of the presence of _____.
22.26 The two basic types of lymphocytes are _____ and _____.
22.27 The lymphoid organ that produces red blood cells in the fetus is the _____.
22.28 Disease-causing microorganisms in the throat are attacked by lymphocytes produced by the _____.

A SECOND LOOK

In the following drawing, label the thymus gland, tonsils, iliac nodes, axillary nodes, lymphatic vessels, and spleen.

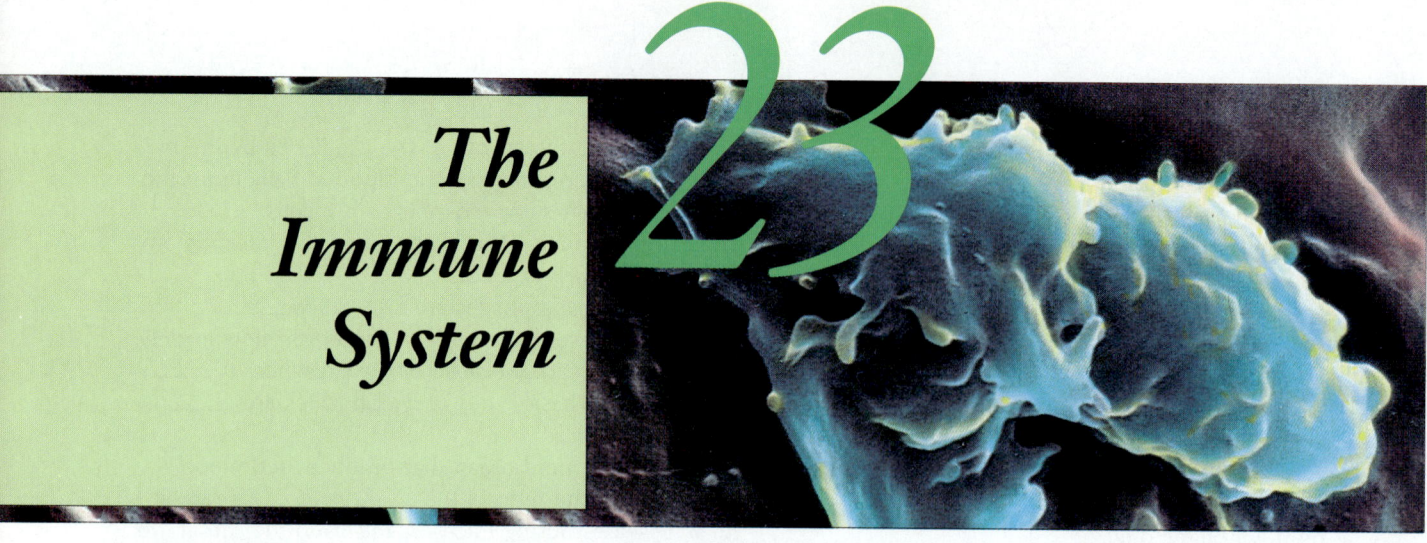

The Immune System

23

KEY CONCEPTS

1 The body's specific defenses involve the formation of antibodies, while the nonspecific defenses do not.

2 The functional cells of the immune system are lymphocytes, which produce antibodies that attack different specific invading microorganisms (antigens).

3 Several different types of lymphocytes perform different functions within the immune system.

4 One manifestation of the body's response to foreign substances is an allergy.

5 Autoimmune diseases are caused by the formation of antibodies against the body's own tissues. Conditions called immunodeficiencies decrease the effectiveness of the immune system.

Our immune system protects us from the everyday threats of infection from disease-causing agents, or **pathogens,** such as viruses, bacteria, and fungi. A healthy immune system is so effective that if it were not able to distinguish the body ("self") from invading foreign substances ("nonself"), it would destroy the body as ruthlessly as it destroys foreign substances. This ability to tell the difference between "self" and "nonself" is at the core of the immune system.

The body has several ways of protecting itself against invading pathogens: (1) Some *nonspecific defenses* protect us from a variety of pathogens, while (2) *specific defenses* are used against specific disease-causing agents. The **nonspecific defenses** provide a *general* response to destroy foreign substances, but do *not* involve the exact recognition of the pathogen or the production of antibodies. The **specific defenses,** usually referred to as the immune response, involve the formation of antibodies, which help recognize and destroy foreign substances.

NONSPECIFIC DEFENSES OF THE BODY

In this section we discuss some of the body's nonspecific defenses: skin and mucous membranes, nonspecific antiviral and antibacterial substances, the inflammatory response, and phagocytosis.

Skin and Mucous Membranes: First Line of Defense

The skin (primarily the epidermis) is an excellent *mechanical barrier* that prevents many microorganisms from entering the body. The acidic surface and high fat content of the skin also restrict the growth of microorganisms. Finally, microorganisms on the skin are removed as the outer layer of squamous cell epithelium is constantly shed and replaced.

Some areas of the body that have openings to the outside, such as the nasal passages, mouth, lungs, digestive system, and reproductive system, are lined with *mucous membranes*, which provide a mechanical barrier. The stickiness of the secreted mucus traps many microorganisms and prevents them from attaching to the epithelium or entering the tissues. Other nonspecific defensive factors include *chemical barriers,* such as *sweat,* which washes microorganisms from the pores and skin surface, and tears from the *lacrimal glands,* which wash away foreign substances from the eyes. Tears contain the enzyme *lysozyme,* which attacks bacterial cell walls to kill the invading organisms.

Many other general mechanical and chemical defenses exist. For example, *nasal hairs* filter the external air before it enters the upper respiratory tract, and *cilia* of the upper respiratory tract sweep bacteria and other particles trapped in mucus toward the digestive tract for eventual elimination. When microorganisms do enter the stomach, the extreme *acidity* of the stomach kills most of them. Finally, the normal microorganisms that live on the skin and mucous membranes provide a major barrier to potentially dangerous nonresident microorganisms by suppressing their growth. (See "The Skin as an Infection Fighter" on page 731.)

Nonspecific Antiviral and Antibacterial Substances

To protect itself from infection, the body is also able to produce several important molecular substances that attack bacteria and viruses, including *interferons* and *complement.* The actions of these substances are less specific than the antibody defenses that respond to an antigen, although complement does require a prior reaction of antibodies. This is an example of how nonspecific and specific defenses sometimes interact to maintain the body's homeostasis.

Interferons *Interferons (IFNs)* are small protein molecules produced by virus-infected cells. In humans there are three types of interferons: (1) *alpha interferon,* made by leukocytes; (2) *beta interferon,* made by fibroblasts; and (3) *gamma interferon,* made by T cells. These interferons have the ability to protect uninfected adjacent body cells by binding to their surface receptors and causing the cells to produce *antiviral proteins* that prevent viruses within the cells from multiplying [FIGURE 23.1].

Alpha interferon is used for the treatment of hepatitis C; the AIDS-related cancer, Kaposi's sarcoma; genital warts; and hairy-cell leukemia, a rare form of blood cancer. It has also been tried in an attempt to prevent nonsymptomatic patients infected with the HIV virus from developing the full-blown expression of AIDS and is used with the drug azidothymidine (AZT) in the treatment of AIDS. Interferons and their numerous subtypes, in combination with chemotherapeutic agents, continue to be studied for their potential anticancer effects.

Complement *Complement* is a group of inactive plasma proteins, designated C1 through C9 (with three forms of C1) and factors B, D, and P (properdin). These proteins are named for their ability to enhance (or complement) both the nonspecific and specific defense systems of the body. Each protein in the complement group has the capacity to interact with the preceding complement protein, forming a "cascade reaction," which amplifies the inflammatory process as well as bacterial cell lysis (disruption). Complement may be activated by ei-

FIGURE 23.1 THE MECHANISM OF INTERFERON ACTIVATION

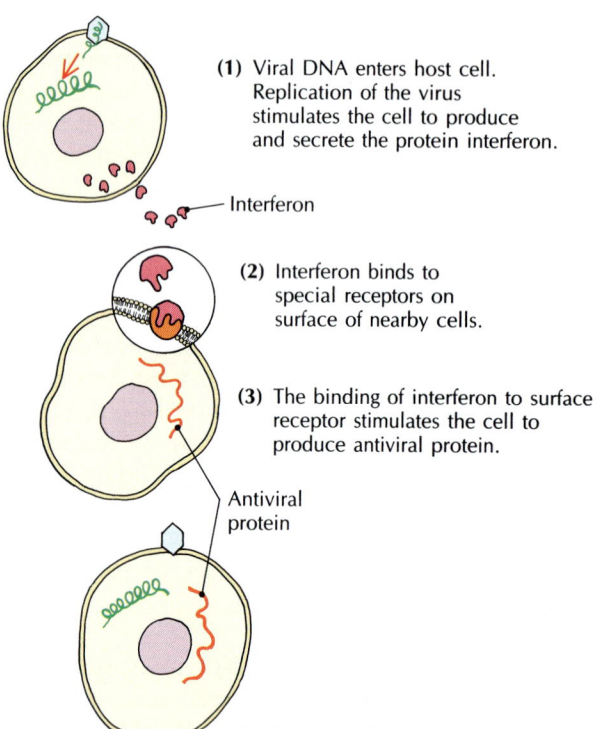

(1) Viral DNA enters host cell. Replication of the virus stimulates the cell to produce and secrete the protein interferon.

Interferon

(2) Interferon binds to special receptors on surface of nearby cells.

(3) The binding of interferon to surface receptor stimulates the cell to produce antiviral protein.

Antiviral protein

(4) Presence of antiviral protein blocks replication of the virus.

ther of two different ways: (1) the *classical pathway* or (2) the *alternate pathway* [FIGURE 23.2].

In the ***classical pathway,*** antibodies bind to invading microorganisms (*antigenic microbes*) and then C1 binds to the antigen-antibody complex (*complement fixation*). This binding activates C1 so that it can act on C2 and C4. C2 and C4 can then act on C3, a key molecule in the complement pathway. In the ***alternate pathway,*** certain cell-wall polysaccharides on microorganisms and factors B, D, and P activate C3. Each pathway, whether classical or alternate, through its series of "cascading" steps results in the activation of C3, a key protein, which immediately splits into C3a and C3b and forms a ***common pathway.*** This common pathway thus leads to eventual destruction of microorganisms through activation of phagocytosis, cell lysis, and the inflammatory process. C3b is primarily involved in opsonization and membrane attack, whereas C3a contributes to the inflammatory response (see the following section). During *opsonization,* or *immune adherence,* C3b binds to the surface of the microorganisms that have been coated with antibodies (*opsonins*) and is instrumental in "feeding" the microorganisms to phagocytes. During *membrane attack,* C3b also initiates the C5, C6, C7, C8, and C9 complex known as the *membrane attack complex (MAC),* which creates holes in the plasma membranes of invading microorganisms, causing *cell lysis* or

FIGURE 23.2 ACTIVATION OF COMPLEMENT

Complement may be activated by the classical pathway (left) or the alternate pathway (right), either of which eventually becomes the common pathway (bottom, center).

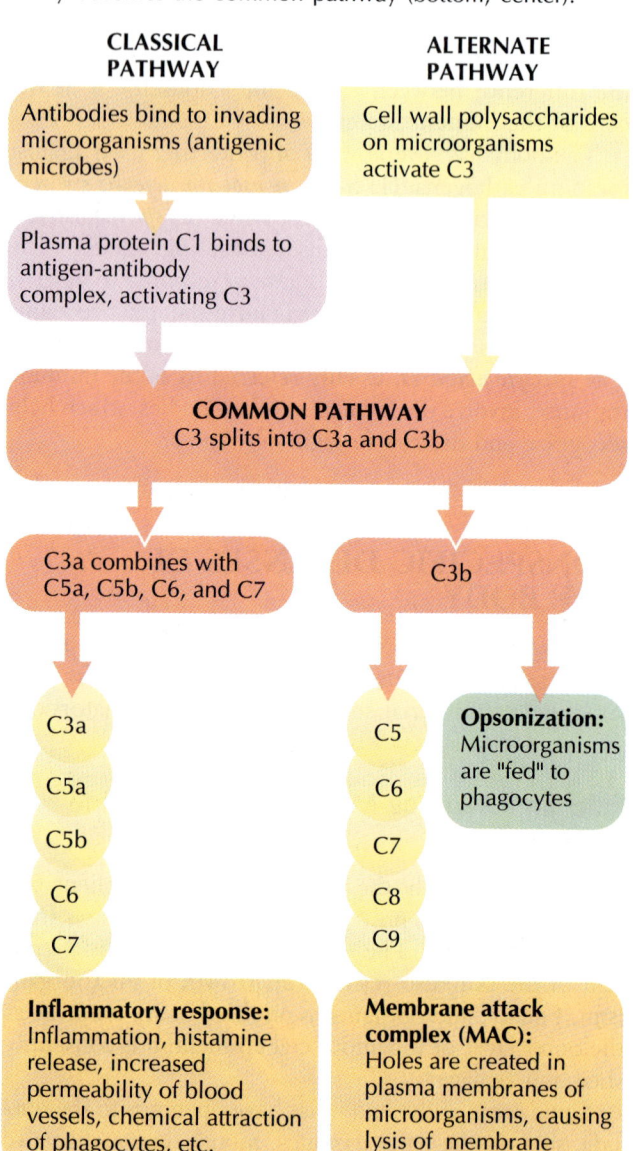

CLASSICAL PATHWAY

Antibodies bind to invading microorganisms (antigenic microbes)

Plasma protein C1 binds to antigen-antibody complex, activating C3

ALTERNATE PATHWAY

Cell wall polysaccharides on microorganisms activate C3

COMMON PATHWAY
C3 splits into C3a and C3b

C3a combines with C5a, C5b, C6, and C7

C3b

C3a
C5a
C5b
C6
C7

C5
C6
C7
C8
C9

Opsonization: Microorganisms are "fed" to phagocytes

Inflammatory response: Inflammation, histamine release, increased permeability of blood vessels, chemical attraction of phagocytes, etc.

Membrane attack complex (MAC): Holes are created in plasma membranes of microorganisms, causing lysis of membrane

destruction. This process is also known as *immune cytolysis.* During the process of *inflammation,* C3a, with C5a, C5b, C6, and C7, causes the *inflammatory response.*

Q: *Can exercise strengthen the immune system?*

A: Studies have shown that *moderate* exercise is beneficial to the immune system. Exercise helps combat minor infections; usually increases the blood levels of several immune-related cells and substances that may improve the resistance to infection; and increases the concentration of endogenous pyrogens, proteins that cause infection-fighting fever.

Inflammatory Response

When tissues are damaged by punctures, abrasions, burns, foreign objects, infections, or toxins, the normal result is *inflammation* of the tissue. The sequence of events that follows such injuries is the **inflammatory response**. Inflammation progresses through several phases. Localized effects of tissue damage and infection are *redness*, *heat*, *swelling*, and *pain*. The redness and heat are produced by vasodilation, which increases blood flow to the area. The localized edema (swelling) is caused by the leakiness of blood vessels, resulting in an increased amount and pressure of interstitial fluid. The pain is usually a result of the pressure of edema on receptors and nerves and of chemical by-products that irritate nerve endings.

If tissue damage is extensive or if the infection is widespread, inflammation will have *systemic effects*, such as changes in heart and respiratory rates, increased body temperature, and a general feeling of malaise. When large numbers of neutrophils are fighting such an infection, they release endogenous *pyrogens*, chemical substances that raise the body temperature by acting on the temperature-regulating center in the hypothalamus. The resultant fever may help fight infection by increasing the blood flow and leukocyte activity in the area while inhibiting the growth of some infectious organisms.

Sequence of events in the inflammatory response
When harmful bacteria enter the body (for example, after the skin has been broken by a small cut or puncture), the following sequence of anti-inflammatory events takes place:

1 Several *local inflammatory mediators* are released into the extracellular fluid: Eicosanoids and platelet-activating factor are released from injured cells; histamine is released from mast cells; kinins, complement, and products of blood clotting are released from plasma; monokines, including interleukin-1 and tumor necrosis factor, are released from monocytes and macrophages; lysosomal enzymes and oxygen-derived substances are released by neutrophils and macrophages; and antibodies and cytokines (cell signal molecules), including gamma interferon, are released by lymphocytes.

2 These chemical mediators cause *vasodilation*, which increases blood flow to the inflamed area. As a result, phagocytic leukocytes, antibodies, and clot-forming plasma proteins in the blood are transported to the injured area, enhancing the immune response. The mediators also cause the capillaries in the area to become more permeable to proteins [FIGURE 23.3A, B]. This increased permeability guarantees that the phagocytes, antibodies, and clot-forming plasma proteins will reach the inflamed site.

3 The increased blood flow produces a characteristic *redness* and *heat*, and the increased filtration of plasma into the interstitial fluid produces the characteristic *swelling* (edema) and *pain*.

4 Healthy tissue is somewhat isolated from damaged tissue by the *formation of fibrinogen clots* at the site of the injury. As a result, tissue damage and the spread of infectious microorganisms are normally restricted to the immediate area of the wound.

5 Within an hour after the inflammatory response begins, neutrophils begin to adhere to the inner walls of blood vessels near the infected area. This action is called *margination* [FIGURE 23.3C].

6 After margination, the neutrophils pass through tiny spaces between endothelial cells in the vessel walls (this action is called *diapedesis*) (DYE-uh-peh-dee-sihss), and the neutrophils move into the interstitial fluid, along with liquid from the capillaries (*exudation*). The neutrophils move toward the infectious microorganisms in the inflamed area. The entire process, beginning with the attachment of neutrophils to the vessel walls, is called *chemotaxis*. Several of the inflammatory mediators already mentioned are classified as *chemotaxins*, chemical mediators that initiate chemotaxis.

7 Monocytes, macrophages, and histocytes are also chemically attracted to the site, and begin to arrive shortly after the neutrophils do [FIGURE 23.3D].

8 The neutrophils are mature cells that are fully capable of *phagocytizing* bacteria and small particles of damaged tissue [FIGURE 23.3D]. Each neutrophil may consume up to 20 bacteria before dying.

9 When monocytes enter the tissues from the blood, they are still immature cells. After arriving in the tissues, however, they *differentiate* or mature into macrophages, which migrate toward the damage site along with activated macrophages already in the tissue. Because macrophages are larger and more active than neutrophils, they can phagocytize as many as 100 bacteria, larger cells such as protozoa, and cell debris from dead or damaged tissues [FIGURE 23.3D]. As macrophages become more numerous and active, neutrophils become fewer and less active. The combination of tissue fluid, dead microorganisms, and phagocytes is known as *pus*. The pus that is not released at the skin surface is eventually absorbed by the body.

10 Some chemical mediators, called *opsonins* (Gr. "to prepare for eating"), help bind the phagocytes to microorganisms. This process (*opsonization*) greatly enhances phagocytosis. Also, some phagocytes can kill microorganisms indirectly, by secreting hydrolytic enzymes and oxygen derivatives into the extracellular fluid.

11 Another indirect method of destroying microorganisms without prior phagocytosis is accomplished by using the large family of proteins known as *complement*. When five of the complement proteins (C5b, C6, C7, C8, and C9) are activated, they form a complex called the *membrane attack complex (MAC)*, as noted above, which

FIGURE 23.3 THE INFLAMMATORY RESPONSE

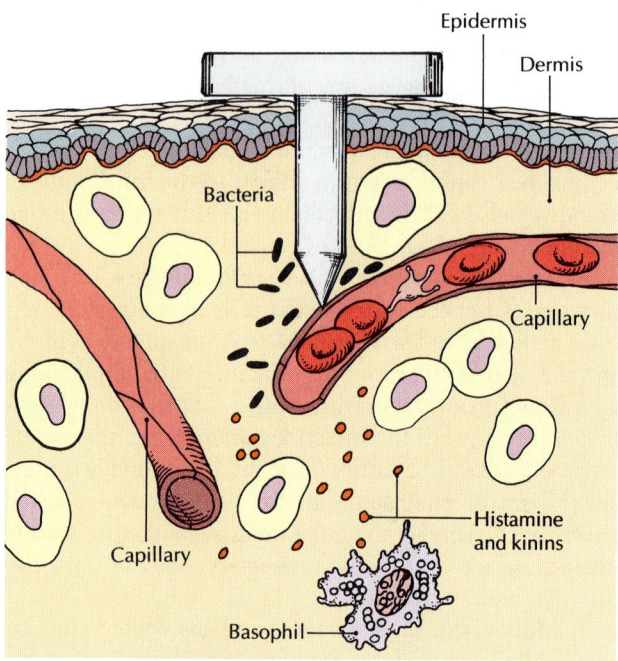

[A] A puncture damages tissue and introduces bacteria into the body. Injured cells, basophils, and mast cells release histamine and kinins.

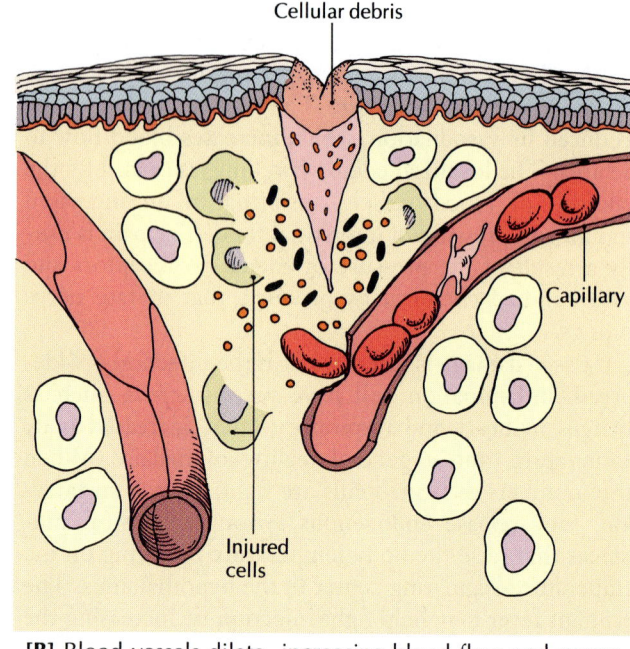

[B] Blood vessels dilate, increasing blood flow and permeability; this results in the accumulation of fluid in the interstitial space, which causes swelling.

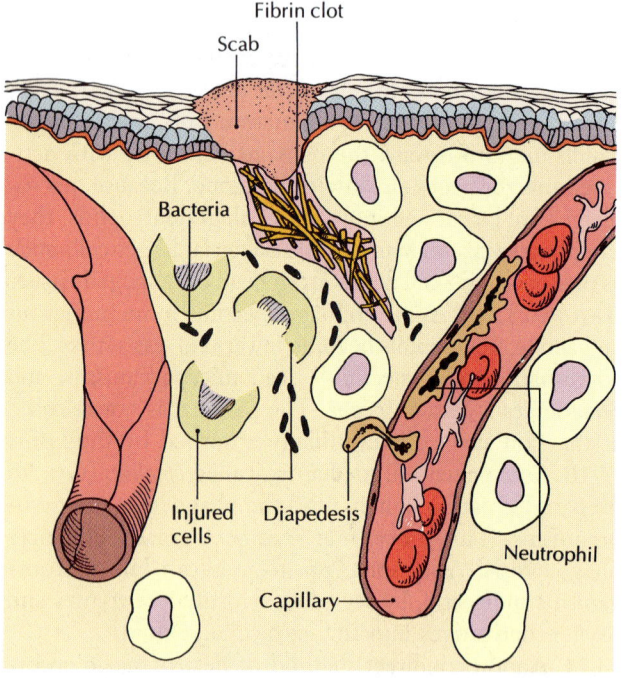

[C] A fibrin clot forms, separating healthy from injured cells. Neutrophils adhere to the capillary wall, then migrate through the capillary wall by diapedesis.

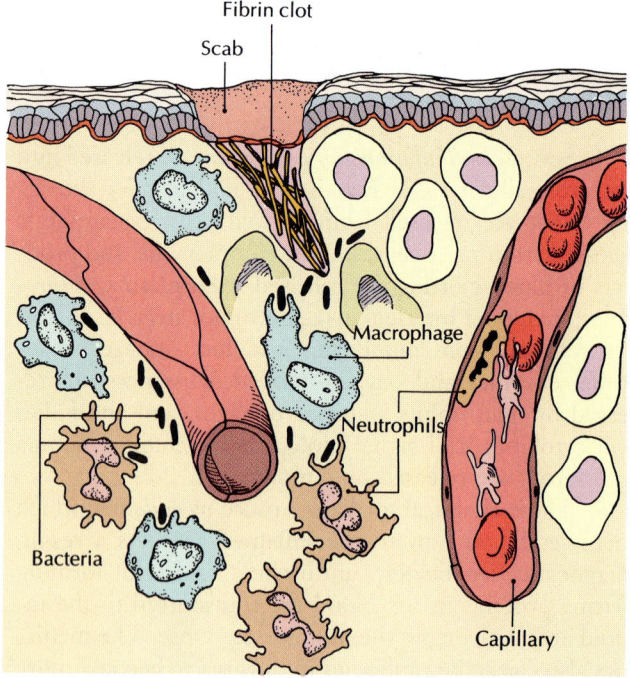

[D] Macrophages and neutrophils phagocytize bacteria and dead cells.

causes the plasma membranes of the infectious microorganisms to leak. Eventually, the internal homeostasis of the cells is disrupted, and the cells die. The complement system also causes vasodilation, increased protein permeability, and opsonization.

12 The final stage of the inflammatory response is the *repair of damaged tissue*, a process that is initiated and enhanced by inflammatory mediators.

In summary, the main events of the inflammatory response occur as follows:

1 Tissue damage and entry of bacteria into the body tissues

2 Vasodilation in the infected area, causing increased blood flow

3 Increased capillary permeability to proteins, accompanied by entry of fluids into tissues, causing edema

4 Entry of neutrophils, monocytes, and other cells into the extracellular fluid near the infected area

5 Destruction of bacteria, mostly by phagocytosis

6 Repair of damaged tissue

The details of phagocytosis are described in the following section.

Phagocytosis

Nonspecific defenses rely on **phagocytosis** ("the process of eating cells") to combat infection. This process normally involves leukocytes (monocytes, macrophages, and neutrophils) that consume foreign substances or dead tissue rather than healthy body cells. This distinction between "nonself" and "self" may be aided by the coating of foreign matter by a special group of globulin molecules, the *opsonins* (Gr. "to prepare for eating"), prior to phagocytosis. (Opsonins do not react with normal, healthy body cells.)

FIGURE 23.4A illustrates the following steps in the process of phagocytosis:

FIGURE 23.4 PHAGOCYTOSIS

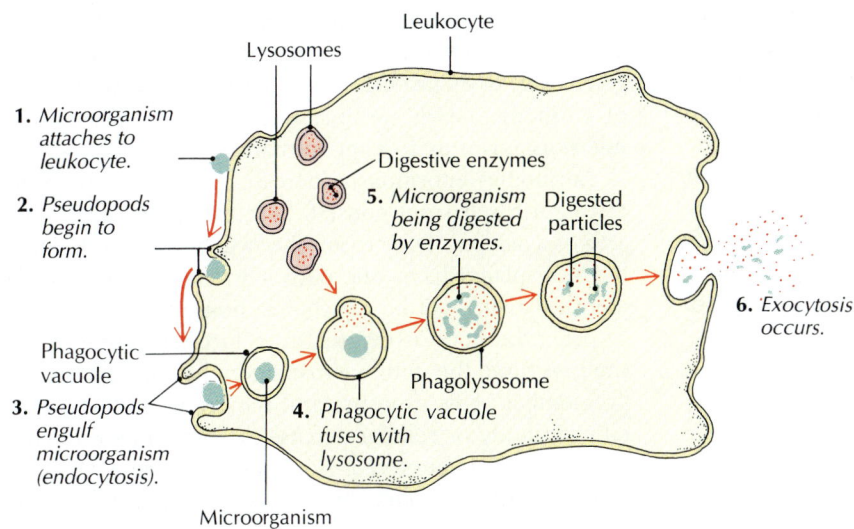

1. *Microorganism attaches to leukocyte.*

2. *Pseudopods begin to form.*

3. *Pseudopods engulf microorganism (endocytosis).*

4. *Phagocytic vacuole fuses with lysosome.*

5. *Microorganism being digested by enzymes.*

6. *Exocytosis occurs.*

Leukocyte
Lysosomes
Digestive enzymes
Digested particles
Phagocytic vacuole
Phagolysosome
Microorganism

[A]

[A] Diagram of the events of phagocytosis in neutrophils and macrophages.
[B] Scanning electron micrographs of a macrophage devouring a colony of bacteria. The phagocytic process usually takes less than a second. ×3000.

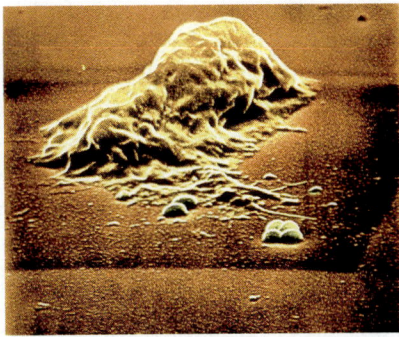

[B](1)

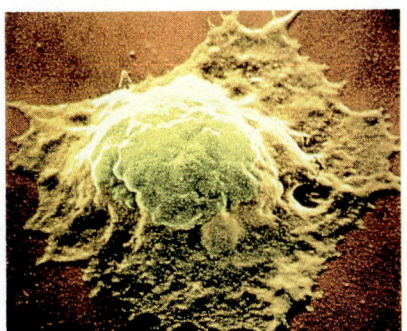

(2)

(3)

1 Foreign particles or microorganisms attach to a leukocyte.

2 Leukocytes form protrusions called pseudopods ("false feet") around the microorganism.

3 Pseudopods engulf (by endocytosis) the microorganism so it is completely surrounded by the leukocyte and enclosed inside a *phagocytic vacuole* within the leukocyte's cytoplasm.

4 The phagocytic vacuole fuses with a lysosome, which contains hydrolytic (digestive) enzymes capable of breaking down proteins, lipids, nucleic acids, and complex polysaccharides.

5 After the phagocytic vacuole and lysosome fuse, the foreign particles and enzymes are contained within a single membrane contributed by both the phagocytic vacuole and the lysosome membrane. The combination is now called a *phagolysosome.* The foreign particles are digested in the fused compartment.

6 Some of the digested contents may contribute nutrients to the leukocyte, while the rest are emptied outside the cell when the phagolysosome fuses with the leukocyte's surface membrane. The expulsion process is known as *exocytosis.*

Defensins It has long been known that neutrophils are potent infection fighters, but now an antibiotic *inside* neutrophils has been identified. This group of related proteins, called *defensins,* may constitute a useful step toward producing new antibiotics in the laboratory. One of the most exciting things about defensins compared with existing antibiotics is their ability to destroy a wider range of bacteria, fungi, and viruses, including the HIV-type "enveloped" virus. Also, because defensins are a natural part of cells, they may be safer than other antibiotics.

A S K   Y O U R S E L F

1 How are the nonspecific defenses of the body characterized?

2 What are some ways in which the body keeps out foreign substances?

3 What are interferons? How do they function?

4 What is the function of complement?

5 What are the events in the inflammatory response?

6 How is phagocytosis accomplished?

SPECIFIC DEFENSES: THE IMMUNE RESPONSE

The nonspecific defenses of the body do not discriminate among types of foreign materials. The *specific defenses,* which constitute the *immune response,* do discriminate among foreign substances (*antigens*) by forming specific proteins, called *antibodies,* and/or specific cells that react with the foreign substances and help destroy them. This overall protective mechanism is known as *immunity* (L. *immunis,* exempt).

The so-called *immune system* is made up of lymphatic tissues and organs located throughout the body. These separate tissues and organs communicate with each other via cells circulated by lymphatic and blood vessels. The functional units of the immune system are a specific class of leukocytes called *lymphocytes.* The two different types of lymphocytes directly involved in defending the body against infection are *B cells,* which develop into plasma cells and produce antibodies, and *T cells.* About 2 trillion B lymphocytes in the body produce about 100 million trillion antibodies during a lifetime. What is even more impressive than this number is the fact that *not all antibodies are alike.* For example, antibodies against the measles virus will not react with the polio virus, and those against the *Streptococcus* bacterium will not react against *Staphylococcus.* Millions of different molecular configurations exist to cope with the huge number of possible antigens and combinations of antigens. It is estimated that the human body can recognize approximately 1 million *antigenic determinants* (small, identifiable portions of antigens). The body produces a seemingly unlimited number of antibodies, each specifically targeted at an invading microorganism or foreign substance.

Antibodies enormously increase the body's ability to distinguish self from nonself. Skin transplants from one area of a person's body to another will be accepted as self, but transplants from one person to another (except an identical twin) will be rejected as nonself.

After the body is exposed to a foreign substance for the first time, the immune machinery has the capacity to "remember" this exposure and mount a rapid and specific antibody or cellular defense the next time the same substance is encountered. Both the killing and memory capacities of antibodies are derived from lymphocytes.

Antigens

An *antigen* is a protein or polysaccharide molecule that, when introduced into the body as a foreign substance, causes the production of specific antibodies, the proliferation of specific T cells, or both. This characteristic is called *antigen immunogenicity.* Antigens also react chemically (demonstrate *reactivity*) with the specific antibody to form a stable complex called the *antigen-antibody complex.* The antigen can also bind to specific receptor molecules on the T and B cells and cause them to produce their specific responses.

An antibody does not form against, or react with, the entire antigen. Instead, it combines with specific sites on

The Skin as an Infection Fighter

Only recently has the skin, with all its other protective functions, been recognized as an integral part of the immune system that protects the body from infection. This discovery was made after it was determined that the epithelial cells of the thymus gland and epidermis have genetic and structural similarities. (The thymus gland functions not only as part of the endocrine system but also as an important component of the immune system.)

One of the most startling findings in the search for *functional* similarities between cells of the skin and the immune system was that keratinocytes in the skin secrete interleukin-1, a chemical also secreted by macrophages throughout the immune system. Prior to this discovery, keratinocytes were known to generate the outer layer of keratin that protects hair and skin. Now it could be shown that these cells also secrete hormonelike chemicals that affect the maturation of skin T cells.

Mobilization of Defensive Cells

The secretion of interleukin-1 by keratinocytes is critical for the eventual production of T cells in the skin. When molecules of a foreign substance (antigen) penetrate the skin's outer layer of keratin, they bind to the surface of nonpigmented granular dendrocytes (Langerhans' cells) in the skin, which then present the antigen molecules to skin T cells that have surface receptors that

bind to the antigen (see drawing). This binding exposes a second receptor on the T cells, which receives molecules of interleukin-1 from nearby keratinocytes. The T cells are then stimulated to release interleukin-2, which binds to interleukin-2 receptors on other T cells that are also binding antigen. These T cells are stimulated to start the proliferation of many identical T cells, which respond to the antigens. Defensive lymphocytes can enter vessels of the lymphatic system, traveling throughout the body to attack antigens that may have spread beyond the skin.

Interaction in the Skin between the Nervous System and the Immune System

Recently, scientists reported a connection between the nervous system and the immune system. Apparently, nerve endings in the epidermis of the skin and nonpigmented granular dendrocytes of the immune system are in direct contact. The nerve endings secrete *calcitonin gene-related peptide* (CGRP), which normally helps transmit messages between neurons. When excess CGRP is released—usually in times of stress—it may form a coat on the nonpigmented granular dendrocytes, diminishing their ability to begin the attack on a foreign substance. This connection may explain why skin diseases such as psoriasis become aggravated under stressful conditions.

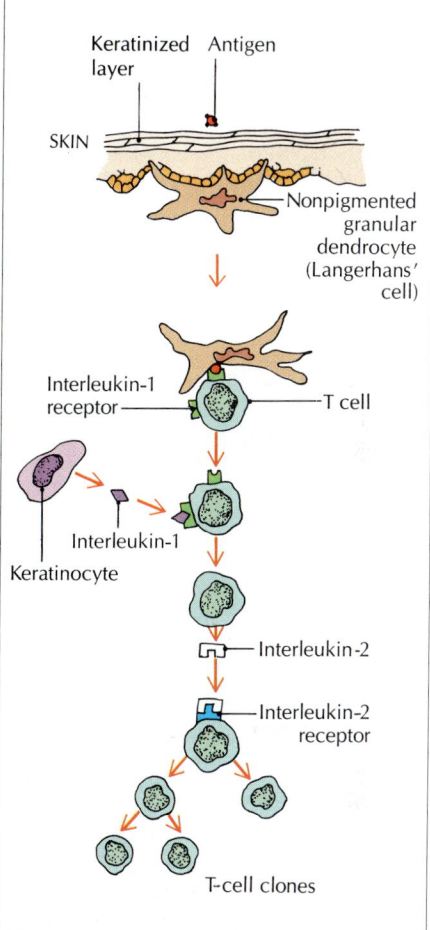

the surface of the antigen. These combining sites, called **antigenic determinants,** or **epitopes** [FIGURE 23.5A], are composed of parts of distinctive proteins or polysaccharides. B and T cells not only can recognize antigens from their surface markings, but also can distinguish one specific antigen from among the over 100,000 known ones and react accordingly.

Almost any protein is capable of acting as an antigen. Incompatible blood cells, proteins on the surfaces of transplanted tissues, pollen, dust, animal dander (flaking skin), animal hair, and some components of foods are common examples of potential antigens. Also, because only antigenic determinants, not the entire organism, are necessary to produce an immune reaction, bacterial components such as flagella, capsules, cell walls, and toxins can be antigenic.

Certain small substances with very low molecular weights can be antigenic. Such substances, called **haptens,** are reactive only when they are combined with a much larger *carrier molecule,* such as a protein. The antibiotic *penicillin* is an example of a hapten. In some patients who receive penicillin for the first time, it combines with a protein carrier to cause the formation of antibodies against penicillin. Subsequent doses of penicillin will cause the antibodies to react with the small penicillin molecule, producing a severe immune response (an allergy) to the penicillin.

Q: *Why don't we become immune to colds?*

A: The common cold is caused by many strains of viruses, and immunity to one strain does not convey immunity to another.

FIGURE 23.5 MOLECULAR STRUCTURE OF AN ANTIBODY

[A] Schematic drawing of an antibody molecule. The variable regions have a different amino acid structure for each specific antibody molecule. The function of the carbohydrate group is not known.
[B] Model of an IgG molecule. Each sphere represents an amino acid.

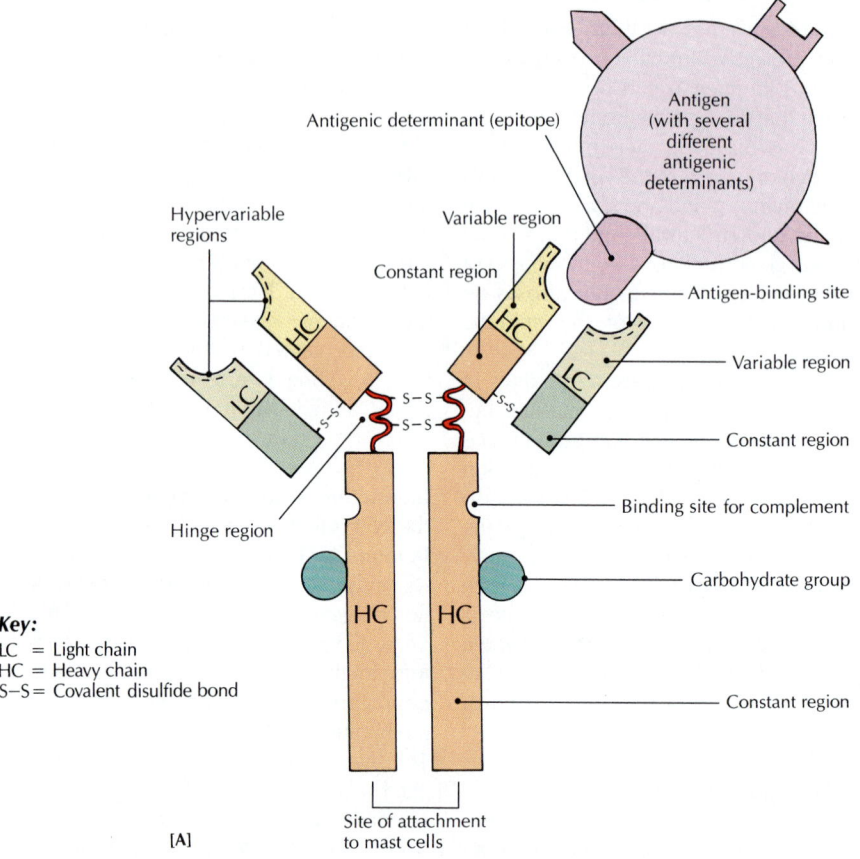

Antigenic determinant (epitope)

Antigen (with several different antigenic determinants)

Hypervariable regions

Variable region

Constant region

Antigen-binding site

Variable region

Constant region

Binding site for complement

Carbohydrate group

Constant region

Hinge region

Key:
LC = Light chain
HC = Heavy chain
S—S = Covalent disulfide bond

Site of attachment to mast cells

[A]

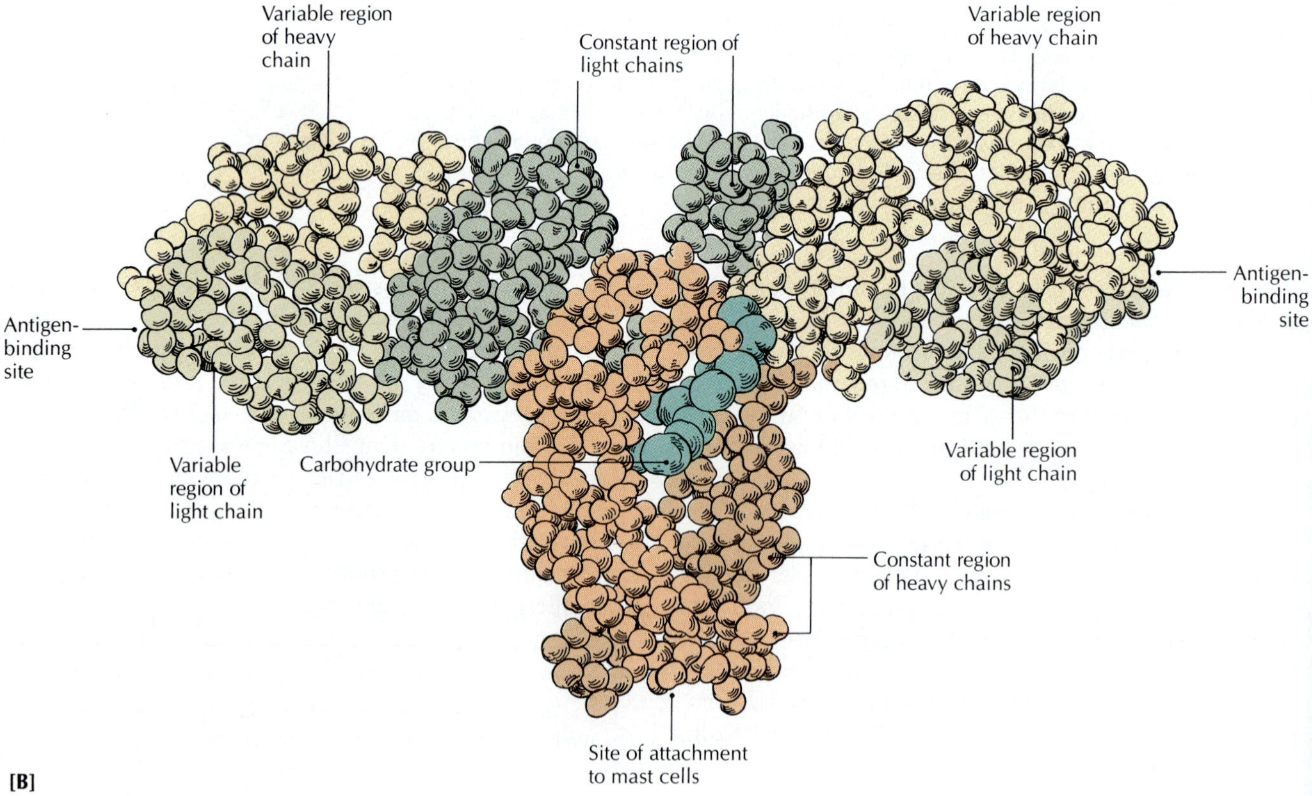

Variable region of heavy chain

Constant region of light chains

Variable region of heavy chain

Antigen-binding site

Antigen-binding site

Variable region of light chain

Variable region of light chain

Carbohydrate group

Constant region of heavy chains

Site of attachment to mast cells

[B]

CLASSES OF BLOOD ANTIBODIES, OR IMMUNOGLOBULINS (Ig)

Immunoglobulin name and structure	Characteristics and functions
IgA 	**IgA molecules** are found mainly in the mucous membranes of the nose and throat, where they help fight respiratory allergens. IgA is the major class of antibody in saliva, tears, secretions from the intestinal and respiratory tracts, and a mother's milk, constituting 5–15% of total blood antibodies. IgA molecules bind to antigens on microorganisms, activating complement. In general, IgA molecules act as localized protective barriers against microorganisms at potential points of entry.
IgD 	**IgD molecules** are present only in small numbers (0.2% of total blood antibodies). Although they are receptors for antigens, their specific function is still unknown. Current evidence suggests that IgD molecules play a role in antigen-triggered lymphocyte differentiation.
IgE 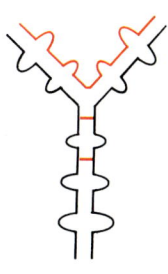	**IgE molecules** are responsible for immediate allergic reactions (such as asthma and hay fever). After the first exposure to an allergen, IgE molecules are synthesized and become tightly bound to the surface of mast cells. When these IgE molecules combine with their specific allergens, they trigger the release of histamine and other inflammatory substances from the mast cells, producing increases in capillary permeability and mucus. IgE (less than 0.01% of total blood antibodies) may also help leukocytes, antibodies, and complement components reach inflammation sites.
IgG 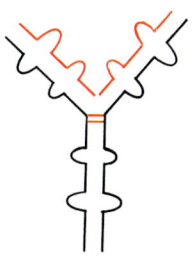	**IgG molecules** are the most common type of immunoglobulin (75–85% of total blood antibodies). They are produced in abundance the second and subsequent times the body is exposed to a specific antigen. IgG molecules are the only immunoglobulins that pass through the placenta during pregnancy from mother to fetus, providing the newborn with natural passive immunity. Such immunity is accomplished because placental cells have receptors that bind only to IgG molecules. Phagocytes that become coated with IgG molecules have an increased efficiency to ingest and destroy infectious microorganisms [FIGURE 23.6]. IgG also activates the first component of the complement system. IgG molecules move through the walls of blood vessels into interstitial spaces most efficiently, and it is no coincidence that half the body's IgG is found in the interstitial fluid and the other half is found in the blood. The four other classes of circulating antibodies are restricted mainly to the bloodstream and local sites of the immune response.
IgM 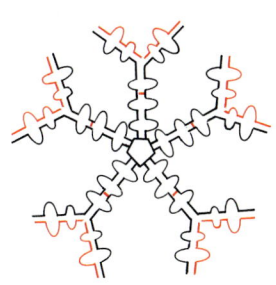	**IgM molecules** (5–10% of total blood antibodies) are the largest antibodies and are major antigen fighters. They are the first antibodies to arrive at an infection site, are the predominant antibodies secreted into the blood during the early stages of a first-time exposure, and can be made by both activated B cells and plasma cells. Because IgM is a pentamer, composed of 5 four-chain units with a total of 10 antigen-binding sites, it is even more efficient than IgG in activating the first component of the complement system. IgM molecules also stimulate the activity of macrophages. They are the first antibodies made by the body, since they are synthesized by the fetus.

The Fate of Antigens

Antigens that penetrate the skin or mucosal barriers of the body may follow one of three routes:

1 Antigens that enter the body through the **blood-stream** are carried to the spleen, where they interact with macrophages, B cells, and T cells in an antigen-specific immune response. The plasma cells in the spleen then release antibodies into the circulation.

2 Antigens that enter the body through the **skin** may produce an inflammatory response. The antigens either remain free or are trapped by macrophages and transported from the epidermal, dermal, or subcutaneous tissue of the skin through the afferent lymphatics into the draining lymph node, where the antigen, node, macrophages, T cells, and B cells interact to produce an immune response.

3 Antigens that enter the **respiratory** or **gastrointestinal tracts** become lodged in the mucosa-associated lymphoid tissue, where they interact with macrophages and lymphocytes.

Antibodies

Antibodies are proteins produced by B cells in response to an antigen. They are specialized to react with antigens, triggering a complex process, called *immunity*, that protects the body by destroying the invader. In contrast to phagocytes, which provide an immediate defense against infections, antibodies contribute an *active immunity* that provides relatively long-term protection against reinfection by the same microorganism or chronic infections. Antibodies cannot kill foreign organisms on their own, but they initiate the killing of such organisms by activating complement, phagocytes, and natural killer T cells. Antibodies can also combine with viruses or bacterial toxins to prevent them from binding to receptors on their target cells.

Structure and specificity of antibodies The molecular structure of antibodies is a crucial part of their ability to react with specific antigens. An antibody molecule is Y-shaped and symmetrical [FIGURE 23.5]. (Because the molecule is flexible, it can also assume a T shape.) The molecule is composed of four polypeptide chains: two identical heavy chains and two identical light chains. The light chains contain about 220 amino acids, and the heavy chains contain about 440 amino acids.

The single stem of the Y is formed by two parallel heavy chains joined together by covalent disulfide bonds $(-S-S-)$. The two "arms" of the Y are formed by a bending out of the heavy chains. Parallel to each bent portion of the heavy chain is a short light chain [FIGURE 23.5A]. The light chains are joined to the bent heavy chains by disulfide bonds.

The molecular properties of antibodies varies only in the regions where reactions with antigens can take place. This region, which is the tip of each arm of the Y, is therefore known as the *variable region*, and it contains a specific *antigen-binding site* (FIGURE 23.5A). It has the structural potential to bind to specific antigens like a lock (antibody) accepting a key (antigen). The remainder of the arms of the Y, as well as the entire stem, are always identical from one antibody class to another. This portion of the antibody is the *constant region*. Each heavy chain also contains a flexible *hinge region*, so that antigen-binding sites need not remain a fixed distance apart.

An antibody is very specific and will react only with the specific antigen that stimulated its synthesis in the first place. This *specificity* is an important property of antibodies. The exact fit of the "lock-and-key" model will not work if the antigen has a different molecular shape.

Classes of antibodies Because antibodies belong to the *globulin* group of proteins and are involved with the *immune* response, they are called **immunoglobulins** (abbreviated as **Ig**). Immunoglobulins are abundant in the body, making up about 20 percent of the total weight of plasma proteins. Five classes of immunoglobulins are known: IgA, IgD, IgE, IgG, and IgM. The constant region on the heavy chain of an antibody identifies its class. Immunoglobulins may have two or more identical reactive sites for an antigen. Those with two sites are called *bivalent*; those with more than two are called *multivalent*. Immunoglobulins D, E, and G are bivalent, and A and M are multivalent. All classes of immunoglobulins except IgM and IgA are composed of a single molecule with the basic Y structure. IgM immunoglobulins are composed of five identical molecules bonded together into a *pentamer* (a structure with five similar parts), and IgA is usually composed of two identical molecules bound together (a *dimer*).

Each of the five classes of immunoglobulins has a separate defensive role, as outlined in the tabular portion of FIGURE 23.5.

Genetic basis of immunoglobulin diversity The body is capable of producing extremely large numbers of antibodies, many of which are for antigens they will never be exposed to. The basis for this diversity lies in the unique genetic mechanism responsible for the production of many different configurations of immunoglobulins in the body's lymphocytes. This diversity is due in part to the fact that antibodies can have two different kinds of light chains and five different kinds of heavy chains.

The variable portion of an antibody's heavy and light chains is coded by several hundred coding regions. These antibody gene segments can be joined in different combinations to produce the millions of antibodies that our bodies need.

Activation of complement by antibodies

Complement is involved in several nonspecific defenses, but it also plays a crucial role, together with antibodies, in the defense against specific antigens [FIGURE 23.6]. The sites for complement binding are exposed only after the antibody has reacted with the antigen. When two IgG molecules are bound side by side on an antigen cell, they can activate the first protein component of complement, which then binds to both antibodies. A single IgM molecule can also activate complement. The first protein in the complement group becomes an active enzyme that activate the next protein in the group. The complement proteins continue to bind to one another in sequence. Each protein that binds to the antigen-antibody-complement complex changes it slightly so that the next protein can bind to it.

When all the proteins are attached to the antigen cell, they make a hole in its plasma membrane, which allows water and ions to flow freely into the cell, causing the cell to burst. As you read earlier, the complement protein complex is appropriately called the *membrane attack complex* (MAC). In a system that is meant to minimize the possibility of destroying the wrong cell, all but the last complement protein must be set in place on the plasma membrane before the activation of that last protein finally blows a hole in the membrane. (Part of the overall complement system is involved in allergic reactions, causing the exploded cell to release histamine.)

Formation and Maturation of Lymphocytes

The specific defenses that recognize and respond to antigenic determinants are found in the closely related (but distinct) B cells and T cells.* These two types of lymphocytes are fairly similar in origin. Both are formed from precursor hemopoietic stem cells in bone marrow. B cells pass through several stages of development in the bone marrow. The mature B cells can then migrate to peripheral lymphoid tissue, such as the spleen, or move to other areas of bone marrow. In response to antigen, these mature B cells can then become *plasma cells,* which secrete antibodies. Some stem cells leave the marrow and migrate to the thymus gland, where they mature into inactive T cells, becoming activated T cells in peripheral lymphoid tissue. Activated T cells *do not* secrete antibodies.

The tissues of the thymus gland and bone marrow, where stem cells develop into lymphocytes [FIGURE 23.7], are called *central lymphoid tissues*. The newly formed lymphocytes, still inactive, can eventually migrate via the blood to the *peripheral lymphoid tissues* — lymph nodes, spleen, the gut-associated lymphoid tissue (GALT) —

*The *B* in B cells stands for *bursa of Fabricius,* a small patch of lymphoid tissue in the intestine of birds, where B cells of birds are processed. T cells get their name from their association with the thymus gland.

FIGURE 23.6 IgG MOLECULES AND PHAGOCYTOSIS

A bacterium coated with IgG molecules is ingested by phagocytosis by a macrophage that has cell-surface receptors that bind to the IgG molecules. Phagocytosis is activated when the molecules on the bacterium bind to the receptors on the macrophage.

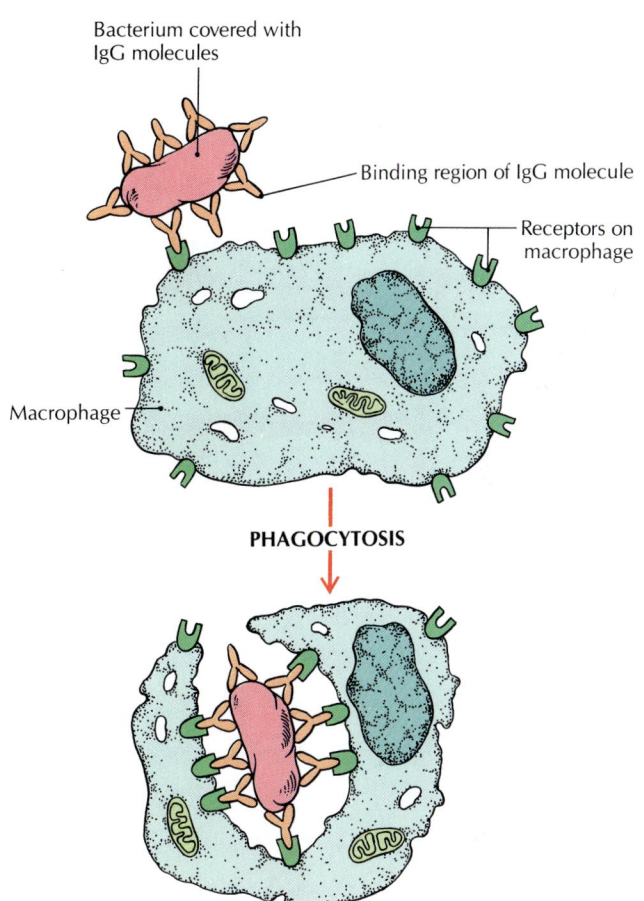

Bacterium covered with IgG molecules

Binding region of IgG molecule

Receptors on macrophage

Macrophage

PHAGOCYTOSIS

which include aggregated unencapsulated lymph nodules in the small intestine, appendix, tonsils, and adenoid. There they react with foreign antigens and become activated.

T and B cells differentiate and proliferate fully only after they have been *activated* by a specific antigen. When antigens enter the body, they are phagocytized by macrophages, displayed on the surface of the macrophages, and then presented to T cells and B cells, which "recognize" the antigen. Simultaneously, the macrophages are also secreting *interleukin-1,* a protein that stimulates the initial activation of T cells and B cells. After the T and B cells are activated by their recognition of the specific antigen, they cause macrophages in the vicinity of the infection to become more efficient at phagocytizing and digesting invading antigens. (TABLE 23.1 outlines some of the major differences and similarities between B cells and T cells.)

FIGURE 23.7 DEVELOPMENT OF T AND B CELLS

[A] The drawing shows the development of T and B cells from hemopoietic stem cells in the fetal liver and postnatal bone marrow. Some stem cells mature in the bone marrow into B cells, which become plasma cells. T cells mature only after the precursor stem cells migrate to the thymus gland via the bloodstream. T and B cells are activated when they come in contact with a specific antigen in peripheral lymphoid tissue. **[B]** The relationship of T and B cells.

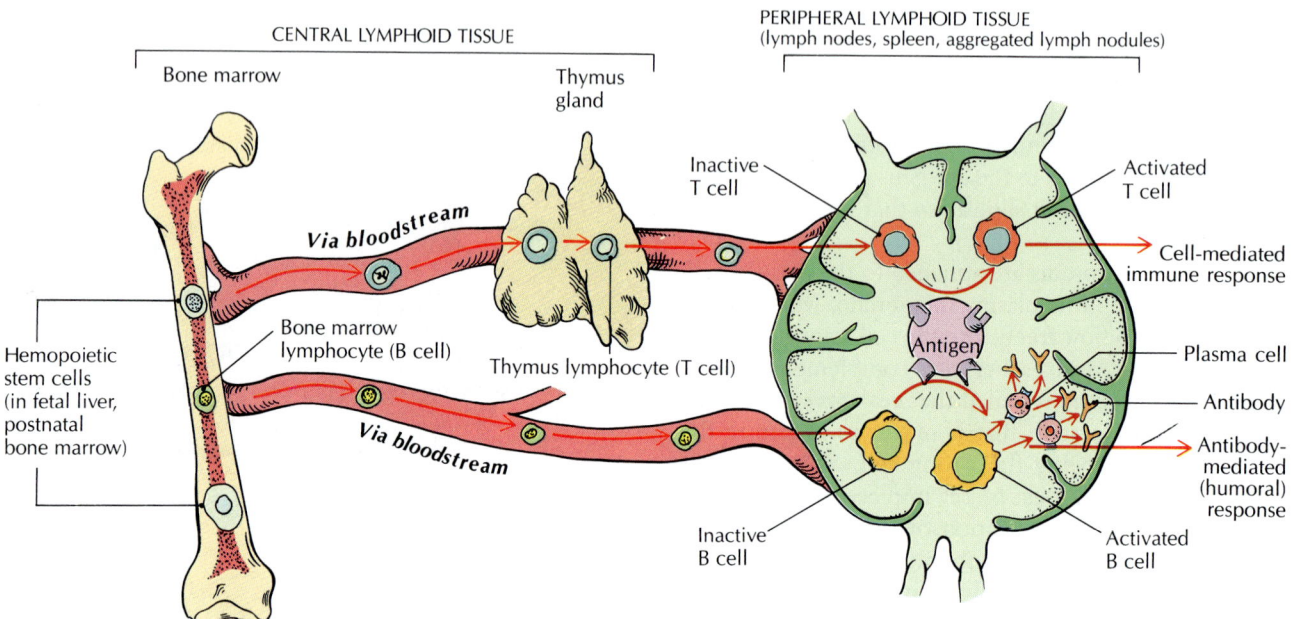

[A]

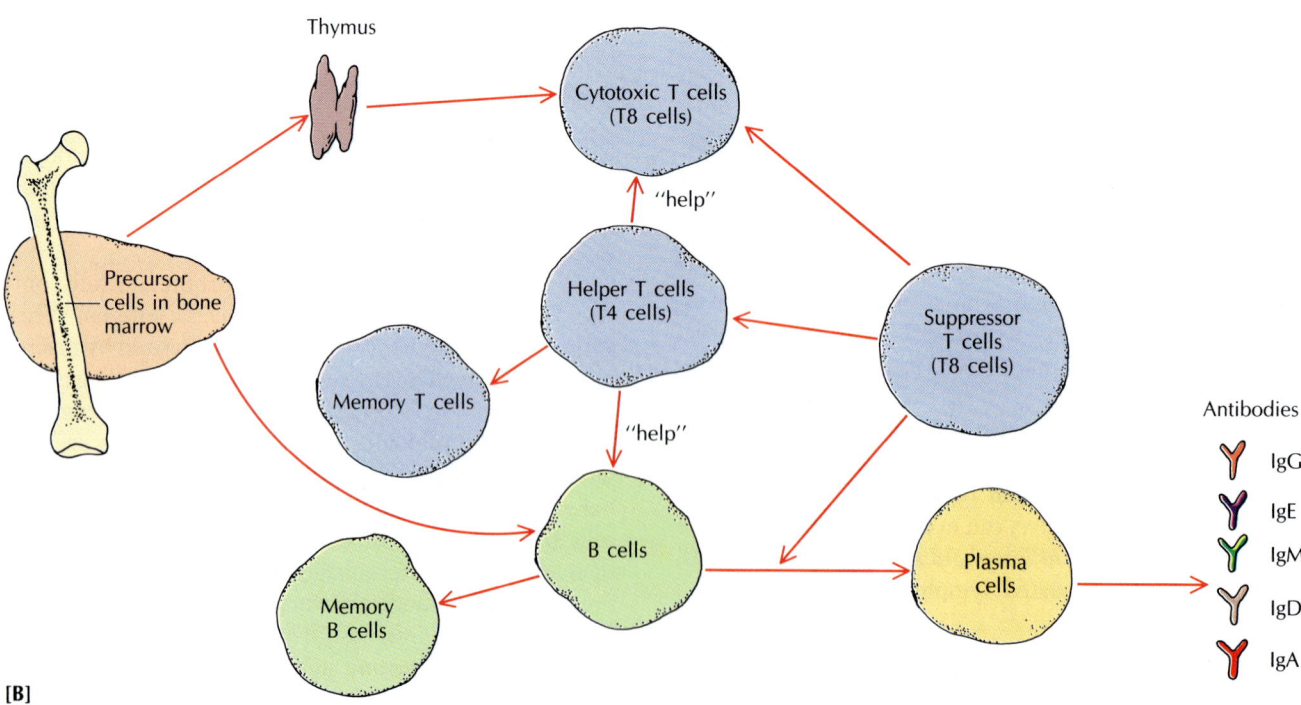

[B]

TABLE 23.1 COMPARISON OF B CELLS AND T CELLS

Property	B cells	T cells
Site of production of undifferentiated cell	Bone marrow	Bone marrow
Site of differentiation	Bone marrow	Thymus gland
Response after binding to antigen	Become enlarged, multiply repeatedly to produce plasma cells; plasma cells release specific antibodies	Become enlarged, multiply repeatedly, release cytokines
Antibody production	Synthesize and release specific antibodies	Stimulate B cells to produce specific antibodies
Type of immunity produced	Antibody-mediated (humoral) response	Cell-mediated response
Cytotoxic activity	None	Activated T cells kill specific antigen-bearing cells on contact
Factor influencing response to antigens	Macrophages	Macrophages
Effect on macrophages	None	Stimulate phagocytic activity
Basic functions	Release specific antibodies	Secrete specific toxins, stimulate production of specific antibodies by B cells, stimulate phagocytic activity of macrophages, produce cell-mediated immunity

Immune Responses

The two types of lymphocytes (B and T) display different responses to antigens: (1) the antibody-mediated (humoral) response and (2) the cell-mediated response.

Antibody-mediated (humoral) response When B cells come into contact with a specific antigen for the first time, they respond by first dividing and then developing into *plasma cells,* which produce specific soluble antibodies and secrete them into the blood and lymph. The reaction is called the *primary immune response* [see FIGURE 23.10].

Antibodies are not capable of independent movement. Instead they are carried by the blood and lymph to the site of an infection or injury, where they bind to the specific antigen that caused the antibodies' production. Because such body fluids were once referred to as *humors,* this type of immunity involving B cells in the production of antibodies is also called *humoral immunity.* **Antibody-mediated (humoral) immunity** is most active against extracellular pathogens such as bacteria, viruses, toxins, and other soluble foreign proteins.

B-cell activation and maturation into antibody-secreting plasma cells occur following the binding of antigen to the B cell's surface receptor. The helper T cells amplify this process by secreting chemical signals called *cytokines.* The types of cytokines important for the production of plasma cells are called *interleukins* because they transmit messages between ("inter") white blood cells (leukocytes).

Not all activated B cells develop immediately into antibody-secreting plasma cells. Some retain their previous appearance as smaller, inactivated cells, and instead of circulating for a short time in the blood or lymph, they remain in lymphoid tissue for a long time. Such activated but apparently inactive B cells are called **B memory cells** because they have the ability to "remember" the sensitizing antigen and react to it the next time it appears. Such a *secondary immune response* occurs at the second and subsequent exposures to the same antigen that produced the primary immune response.

A secondary response starts faster and releases many more antibodies than a primary response does [FIGURE 23.8]. For this reason, a second tissue or organ transplant from a single donor is usually rejected much faster than the first transplant. Memory cells are capable of multiplying during a secondary response, which makes them even more effective against a recurring antigen. If a totally new antigen is introduced, the primary immune response is elicited all over again, indicating that lymphocytes have an antigen-specific memory, reacting appropriately to each exposure to an antigen.

Cell-mediated response T cells also respond to specific antigens, but they do not produce antibodies as B cells do. T cells give protection by (1) producing chemicals that destroy antigens if the antigen is an infecting virus, (2) inducing macrophages or other host cells to destroy the antigen, (3) stimulating cytotoxic T cells to destroy infected host cells, or (4) regulating the immune

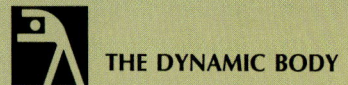

Macrophages and the Immune Response

We may tend to think of macrophages only as large cells that "eat" other cells and foreign particles [see FIGURE 23.4B], but actually macrophages (mononuclear phagocytes), which do not have the narrow antigenic specificity of lymphocytes, play several active roles in the immune response. They are involved both in the production of antibodies and in cell-mediated responses. They also

1 Process and present antigens to lymphocytes.

2 Produce immunoregulatory molecules, including the protein interleukin-1, which is required for the proliferation and survival of T and B lymphocytes.

3 Display antigens on their surface (macrophage-bound antigens), trigger-

ing the proliferation and activation of T and B lymphocytes.

4 Produce cytokines, which stimulate inflammatory cells such as neutrophils to produce fever and other systemic effects of the inflammatory response.

5 Promote the repair of injured tissues by producing growth factors for fibroblasts and vascular endothelium.

FIGURE 23.8 PRIMARY AND SECONDARY RESPONSES TO ANTIGEN-ANTIBODY CONTACT

Memory cells are responsible for the faster initiation of antibody synthesis and a higher rate of production after the second exposure to an antigen. In the secondary response the antibody reaches a greater concentration in the blood. Exposure to a new antigen (antigen B) initiates a new primary response, which is unrelated to the response to antigen A.

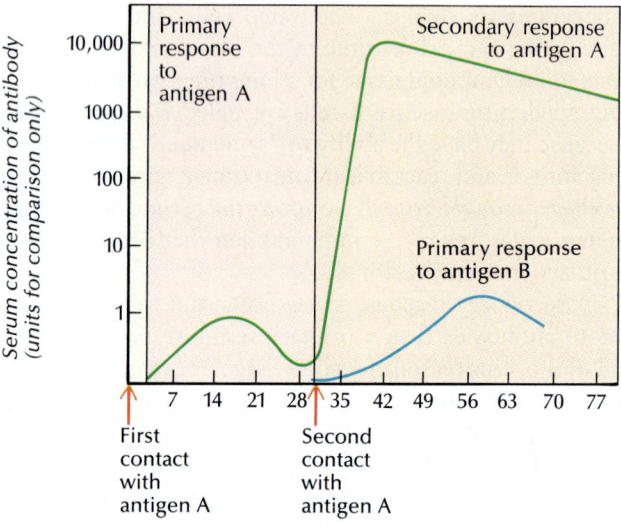

against any cells that contain an infecting bacterium or virus. Some intracellular parasites are not attacked by the soluble antibodies prominent in antibody-mediated immunity because the antibodies are unable to cross the host-cell membranes to reach the parasites.

Clonal Selection Theory of Antibody Formation

The unifying theory of antibody formation is called the ***clonal selection theory*** of antibody formation. Basically, it states that an antigen influences the *amount* of antibody produced but does not affect its three-dimensional structure (its combining sites). The theory also proposes that each B cell is genetically programmed to react *only to a particular antigen*. The clonal selection process works in the following ways:

1 Each B cell uses a unique DNA pattern that makes it genetically programmed to respond to a specific antigen even *before it meets that antigen* [FIGURE 23.10]. Each lymphocyte produces only one type of antibody in response to its specific corresponding antigen.

2 Small amounts of specific antibody structures become bound to the plasma membrane surface of each lymphocyte. These structures act as receptors for specific antigens.

3 Any lymphocytes that have antibodies corresponding to body cells or molecules are killed during fetal life, so that all normal lymphocytes become tolerant of, or nonreactive to, self.

4 When mature lymphocytes encounter their specific antigen, they are activated and begin to enlarge and divide.

5 Repeated cell divisions form a ***clone,*** a group of genetically identical cells descended from a single activated lymphocyte.

response to make certain that the system does not over-react to the point where it damages the body. The protective method of T cells is called ***cell-mediated immunity*** because it involves direct contact between T cells and antigens [FIGURE 23.9].

Cell-mediated immunity operates against intracellular pathogens such as multicellular parasites and fungi, and also against cancer cells and tissue transplants. It is active

FIGURE 23.9 CELL-MEDIATED IMMUNITY

[A] Clonal selection and proliferation of T cells in cell-mediated immunity. [B] Summary of cell-mediated immunity.

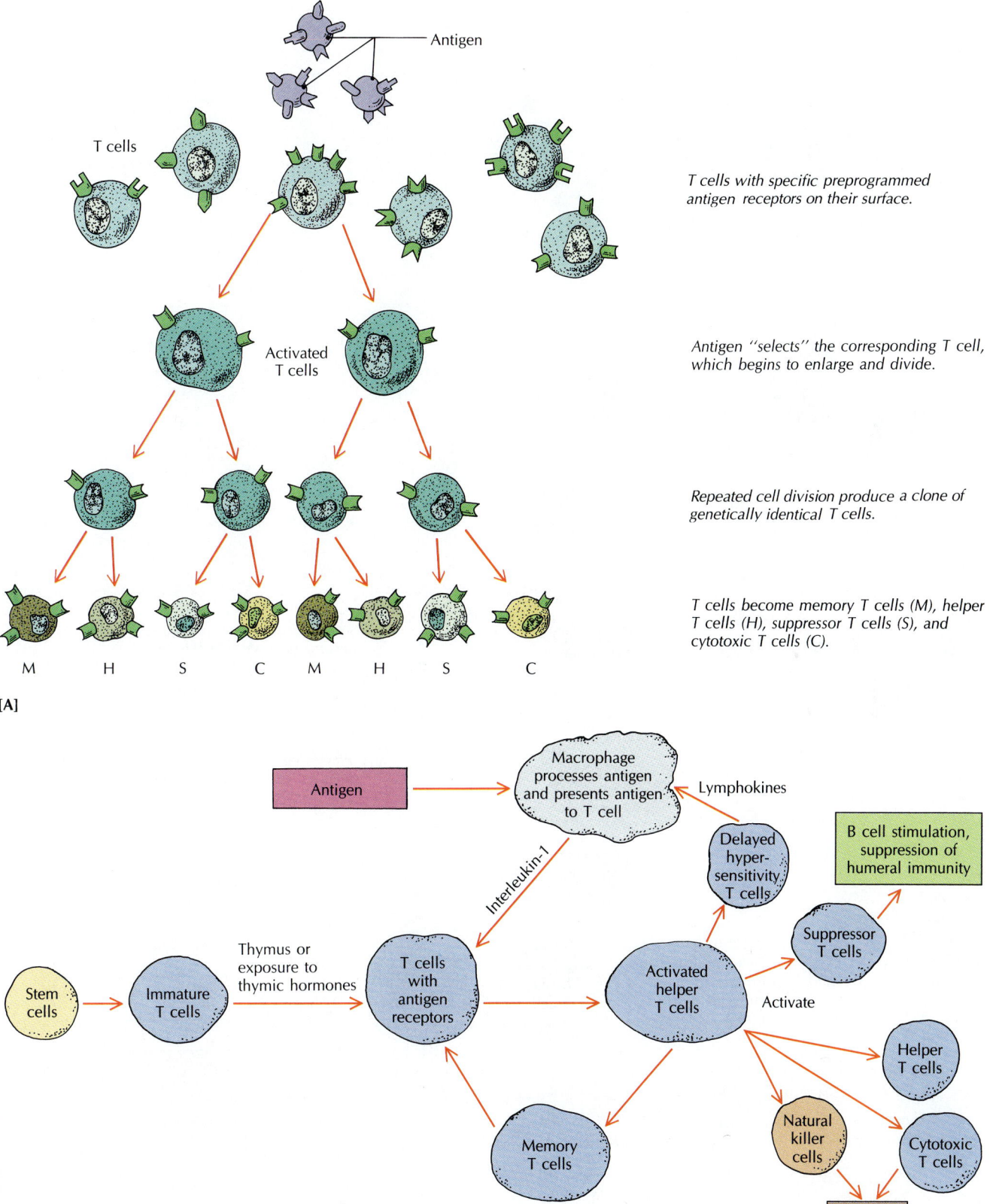

T cells with specific preprogrammed antigen receptors on their surface.

Antigen "selects" the corresponding T cell, which begins to enlarge and divide.

Repeated cell division produce a clone of genetically identical T cells.

T cells become memory T cells (M), helper T cells (H), suppressor T cells (S), and cytotoxic T cells (C).

FIGURE 23.10 ANTIBODY-MEDIATED (HUMORAL) IMMUNITY

[A] Clonal selection and proliferation of B cells in antibody-mediated (humoral) immunity. [B] Summary of antibody-mediated (humoral) immunity.

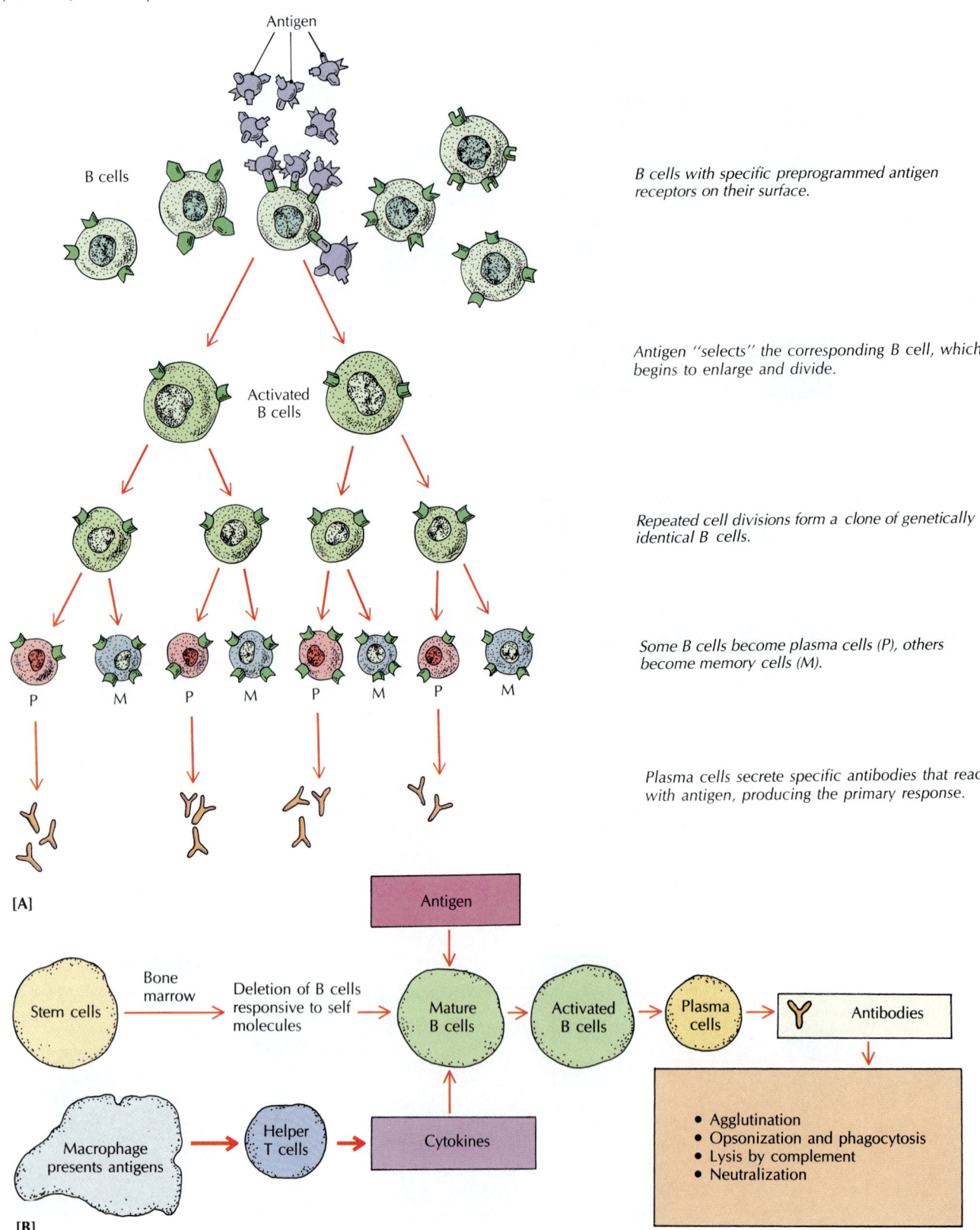

Antigen

B cells

B cells with specific preprogrammed antigen receptors on their surface.

Activated B cells

Antigen "selects" the corresponding B cell, which begins to enlarge and divide.

Repeated cell divisions form a clone of genetically identical B cells.

P   M   P   M   P   M   P   M

Some B cells become plasma cells (P), others become memory cells (M).

Plasma cells secrete specific antibodies that react with antigen, producing the primary response.

[A]

Antigen

Stem cells — Bone marrow → Deletion of B cells responsive to self molecules → Mature B cells → Activated B cells → Plasma cells → Antibodies

Macrophage presents antigens → Helper T cells → Cytokines

- Agglutination
- Opsonization and phagocytosis
- Lysis by complement
- Neutralization

[B]

As cell division continues, some lymphocytes become plasma cells and others become memory cells. The plasma cells produce the large amounts of antibody necessary to inactivate the antigen. All the newly formed plasma cells make the same specific antibody.

Memory cells persist in the body after the antigen disappears, and in fact, they may remain potentially active for a person's entire lifetime. Upon a second enounter with the antigen, memory cells proliferate, differentiate into plasma cells, and secrete antibody so rapidly that the symptoms of the disease may not even be observed. Because of this capability, memory cells are said to have an *immunological memory.*

The surface proteins that identify each cell as self or nonself are encoded by genes called ***major histocompatibility*** ("tissue compatibility") ***complex (MHC) genes.*** Because there are many variants of the MHC genes that encode each protein, it is highly unlikely that two people (other than identical twins) will have the same MHC-encoded proteins. The antigenic determinants and MHC-encoded proteins must form a complex before a T cell can bind to them.

Differentiation of T Cells

The clonal selection theory also applies in part to T cells, which provide cell-mediated immunity [FIGURE 23.9]. The major difference is that although some T cells that remain potentially active after an antigen is inactivated persist as *memory cells*, others differentiate further in lymphoid tissue into four types of cells with specialized functions: the helper T cell, the suppressor T cell, the cytotoxic T cell, and the delayed hypersensitivity T cell [TABLE 23.2]. Refer to FIGURE 23.9 as you read about the different types of T cells.

Helper T cells *Helper T cells* are essential to the differentiation of B cells into plasma cells and their subsequent secretion of antibodies. Each helper T cell is capable of activating hundreds of specific B cells. In order for a programmed B cell to become activated, two successive events must take place. First, the B cell must encounter its specific antigen [FIGURE 23.11]. Second, it must be assisted by helper T cells, which completes the activation of the B cell. Helper T cells are required for the appropriate response of cytotoxic T cells—by the release of gamma interferon—and suppressor T cells—by the release of interleukin-2—to an antigen, and they also activate macrophages by activating chemical factors called cytokines by means of the delayed hypersensitivity T cells.

Suppressor T cells *Suppressor T cells,* after being stimulated by helper T cells, suppress the response of B cells and other T cells to antigens. They inhibit the development of B cells into plasma cells, regulate the activity of cytotoxic T cells, and suppress the production of antibodies when they become excessive. Suppressor T

TABLE 23.2 PRINCIPAL CELLS OF THE IMMUNE SYSTEM	
Type of cell	**Major functions**
B cell	Differentiates into antibody-secreting plasma cells when stimulated by antigen
T cell	
Helper T cell	Activates B cells after B cells encounter specific antigens by releasing B-cell growth factor; necessary for appropriate responses of cytotoxic T cells and suppressor T cells to antigens; activates lymphokines that activate macrophages
Suppressor T cell	Suppresses autoimmune responses, excessive antibody production; regulates activities of cytotoxic T cells; inhibits development of B cells into plasma cells
Cytotoxic T cell	Attacks virus-infected cells and tumor cells; is antigen-specific
Delayed hypersensitivity T cell	Releases lymphokines, which assist in macrophage activity
Natural killer cell	Attacks any intracellular foreign microorganism; not antigen-specific
Plasma cell	Secretes antibodies that mark foreign substances for destruction
Macrophage	Ingests microorganisms, presents antigens to T cells to initiate specific immune response

FIGURE 23.11 HELPER T CELLS AND ANTIBODY PRODUCTION

(1) Helper T cells recognize and respond to antigens and stimulate B cells. Activated B cells make antibodies that destroy toxins and other foreign antigens in the blood. (2) Helper T cells multiply through cell division, (3) become activated, and (4) secrete interleukins that (5) stimulate B cells to multiply and (6) become plasma cells that release antigen-fighting antibodies.

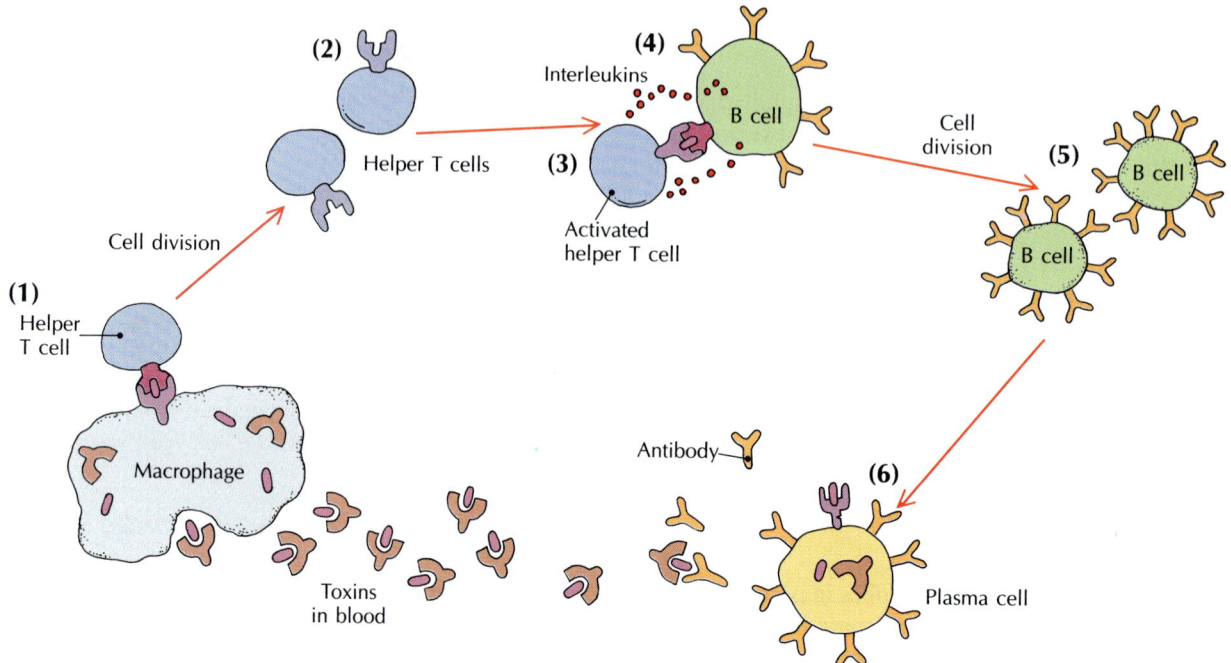

cells also suppress autoimmune responses, in which the body forms antibodies against its own antigens. Helper T cells that activate suppressor T cells are themselves inhibited in a feedback circuit that regulates the activity of both types of cells. Many such self-regulatory circuits are active within the complex interaction of lymphocytes.

Cytotoxic T cells *Cytotoxic T cells (Tc)* have specific receptors for antigenic determinants. Cytotoxic T cells migrate from lymphoid tissue to the site of foreign-cell invasion. There they make cell-to-cell contact with the foreign cells or virus. A cytotoxic T cell destroys a foreign cell by secreting a pore-forming protein called *perforin* into the plasma membrane of the foreign cell. The perforin molecules create transmembrane pores in the foreign cells, causing them to rupture and die within 2 hours [FIGURE 23.12]. This reaction is accomplished without the use of complement.

Apparently, a Tc cell can dissociate itself from one destroyed foreign cell to attach to another. This reaction is very specific, and only cells that have the specific surface antigen are attacked. Tc cells are the main cause of rejections of tissue or organ transplants from one person to another, and are also responsible for activity against tumors and cancer cells.

Delayed hypersensitivity T cells *Delayed hyper-*

sensitivity T cells release various lymphokines that assist in a variety of functions. They (1) help macrophages find microorganisms, (2) stimulate phagocytic activity, (3) prevent microorganisms from leaving infection sites, and (4) cause macrophages to congregate at infection sites. Delayed hypersensitivity T cells also participate in delayed hypersensitivity reactions, which are types of allergic responses (see below).

FIGURE 23.13 summarizes the activity of various types of T, B, and killer cells.

ASK YOURSELF

1 What is the basic difference between specific and nonspecific defenses?

2 What is an antigen? An antibody? An antigen-antibody complex?

3 What are the five classes of antibodies?

4 What is the difference between antibody-mediated (humoral) immunity and cell-mediated immunity?

5 How do antibodies activate complement?

6 What is the clonal theory of antibody formation?

7 What are some basic differences between B and T cells?

FIGURE 23.12 HOW CYTOTOXIC T CELLS FUNCTION

[A] **(1)** Cytotoxic T cell recognizes the foreign antigen displayed by an abnormal cell and binds with it. **(2)** Cell division produces many cytotoxic T cells. **(3)** Cytotoxic T cells bind with abnormal cells and destroy them by creating holes in their membranes. **[B]** Cytotoxic T cells have produced holes in the plasma membrane of the cancer cell, producing holes that cause the cell to leak its contents. The rough surface of the cancer cell indicates that the cell is dying. ×7000.

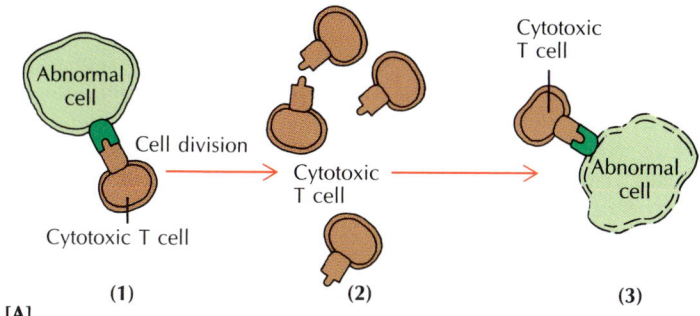

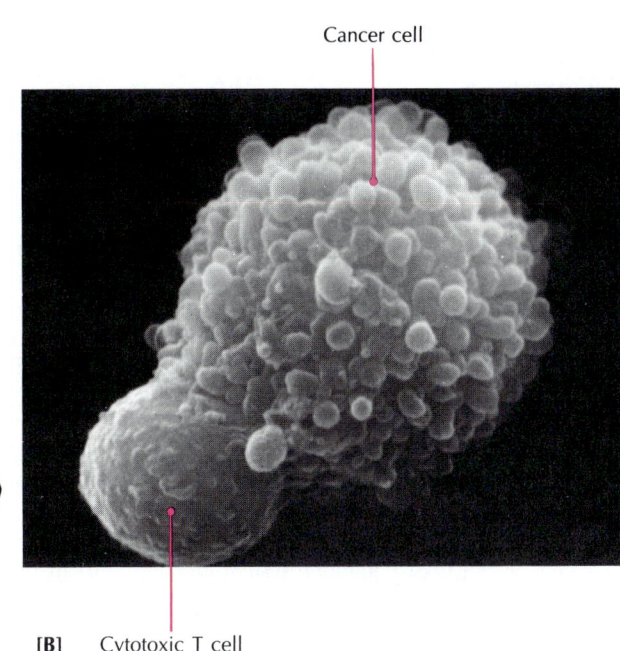

FIGURE 23.13 SUMMARY OF ACTIVATION OF VARIOUS TYPES OF T, B, AND KILLER CELLS

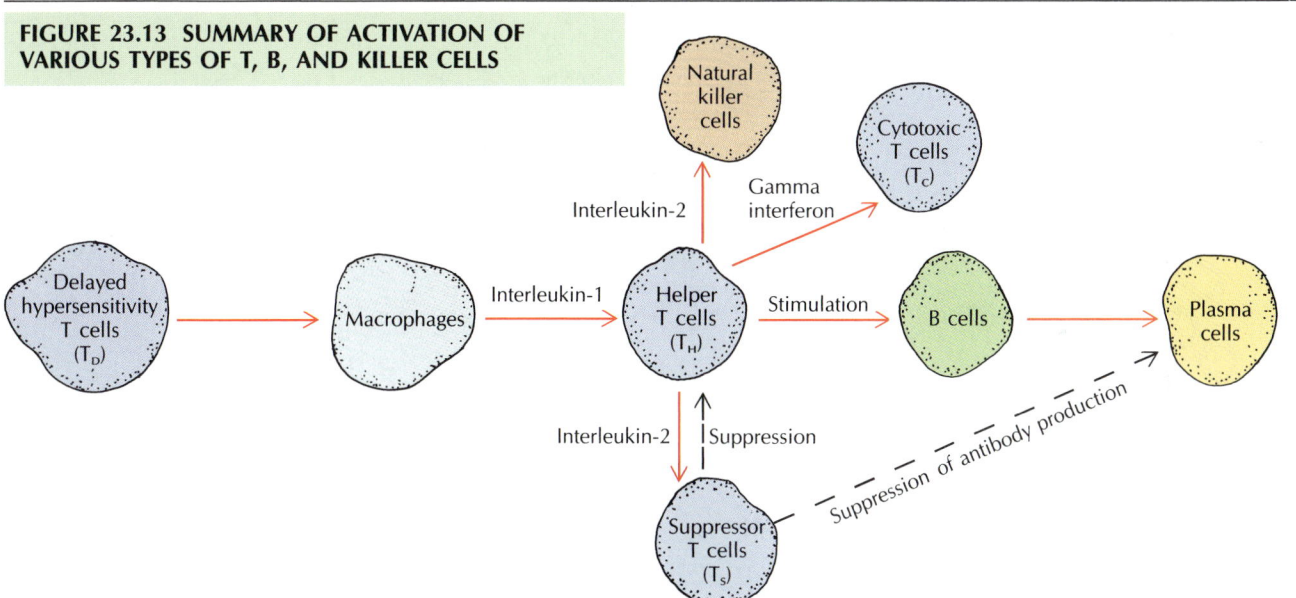

HYPERSENSITIVITY (ALLERGY)

About one of every five people is ***hypersensitive,*** or ***allergic,*** to antigens such as dust, pollen, or certain foods and chemicals in amounts that do not affect most people [TABLE 23.3]. In the case of allergies, the causative antigen is known as an *allergen.* Allergens elicit the production of high levels of IgE, which coats mast cells and basophils. The allergen becomes attached to the IgE, and the coated cells release amines such as *histamine.* Allergies develop when suppressor T cells, which normally suppress the production of IgE by plasma cells, are either absent or deficient.

Allergens may enter the body through the integumentary, respiratory, or digestive systems, where they trigger antigen-antibody reactions. The location of the meeting between allergen and antibody determines the severity

---

Q: *Can children outgrow food allergies?*

A: Children who have *mild* allergic reactions to food usually outgrow the allergy.

How T Lymphocytes Help Fight Cancer Cells

One of the problems in treating cancer is the fact that chemotherapy, while possibly destroying cancer cells and prolonging the life of the patient, usually damages the immune system, and a damaged immune system has little or no chance to destroy cancer cells effectively and permanently. Now researchers at the Ludwig Institute for Cancer Research in Brussels have proven that most tumor cells display antigens that can stimulate antigen receptors on cytotoxic T cells, causing the T cells to destroy the tumor cell.

Cancer researchers are looking for ways to use tumor-cell antigens to trigger T cells in the body's immune system in the most efficient way. The production of antigens within tumor cells takes place in several steps (see drawing): **(1–3)** A gene inside the nucleus directs protein synthesis. **(4)** Enzymes cut the synthesized proteins into smaller peptides. **(5)** Some peptides are transported into the endoplasmic reticulum, where the peptides combine with class 1 major histocompatibility (MHC) molecules. **(6)** The resulting peptide-MHC complexes are transported to the surface of the cell, where they are held within the plasma membrane as cytotoxic T cells examine them. T cells that have appropriate antigen receptors can then attack the cell. In this way, cancerous tumor cells can be destroyed by a cellular mechanism without injuring nearby tissue or harming the immune system.

In a related procedure, scientists have removed cancer cells from patients suffering from B-cell lymphoma and treated the cells with a vaccine that makes them more "stimulating" to the immune system. The vaccinated cells are then reinjected under the patient's skin, increasing the activity of the immune system. If the technique is successful with a large group of patients, researchers intend to use it for other diseases, including diabetes, multiple sclerosis, and rheumatoid arthritis.

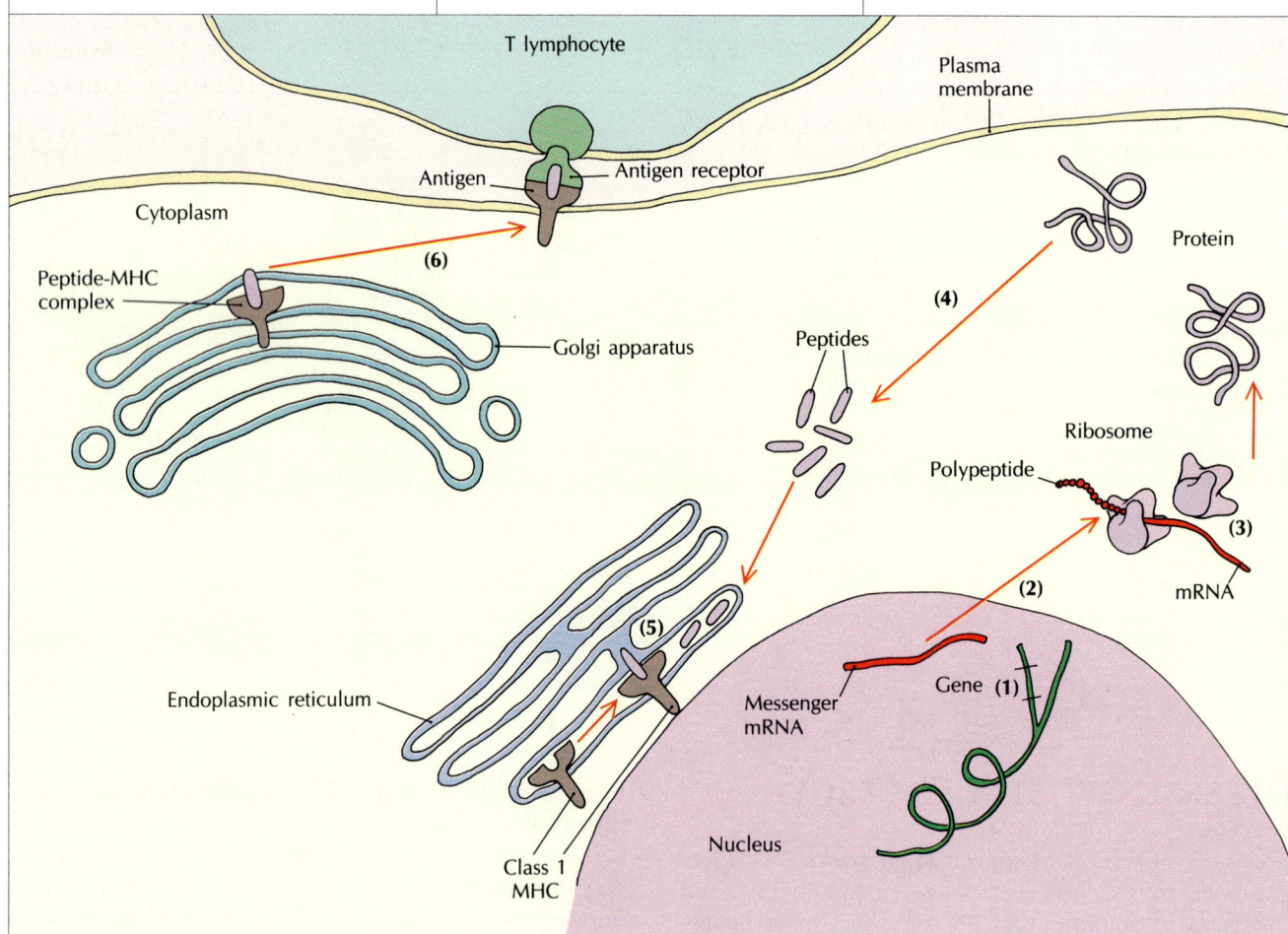

and types of symptoms of the allergic reaction. If the antibody meets the allergen in or on a cell that can react with the release of agents such as histamine, neutralization occurs, but it disturbs the cell. *It is this disturbance and the release of specific substances such as toxic agents that produce the specific allergic reaction.* The hypersensitivity response can be either immediate or delayed.

Immediate Hypersensitivity

When a person who has already been exposed to an allergen is exposed again, the allergic reaction sometimes occurs rapidly, within minutes or hours. Such a reaction to an allergen is called *immediate hypersensitivity*. The person's first exposure results in the production of IgE,

TABLE 23.3 LEADING CAUSES OF ALLERGIC REACTIONS

Seasonal causes	Perennial causes
Ragweed pollen	Pet dander
Tree pollen	Medicines (e.g., insulin, aspirin,
Grass pollen	cephalosporin, penicillin)
Stinging insects	Food *Adults:* peanuts, fish, shellfish
	Children: wheat, eggs, milk, nuts
	Dust-mite feces
	Cockroach feces
	House molds

which binds to the surface of mast cells (or circulating basophils). The IgE molecules have no effect on the mast cells until the IgE encounters the specific allergens that combine with adjacent IgE molecules [FIGURE 23.14]. When this combination occurs, the mast cells immediately release inflammatory substances such as histamine and leukotrienes into the extracellular spaces. The release of these substances is called *degranulation. Histamine* increases the permeability of blood vessels and causes smooth muscles to contract. *Leukotrienes* are synthesized by mast cells and basophils after their reaction with the allergen. Their effects are similar to those of histamine but are longer lasting.

The inflammation resulting from the release of chemicals from activated mast cells may be mild or severe. Inflammation associated with the contraction of bronchial smooth muscles may be serious enough to interfere with breathing, as occurs in *asthma.*

In most hypersensitivity reactions, the inflammation is *localized.* In allergies to pollen (popularly called *hay fever*), the inflammation is localized in the mucous membrane of the nose, where the allergens first make contact with the responsive tissue [FIGURE 23.14B].

If an allergen enters the circulation, the effects may be distributed throughout the body, affecting several organ systems. Such a *systemic reaction* may result from insect stings, some therapeutic drugs (such as penicillin), and components of certain foods. In systemic reactions, inflammation and edema are extensive, lowering blood pressure and possibly causing *anaphylactic shock.* Occasionally the reaction is severe enough to constrict air passages in the lungs. Suffocation may result unless the reaction is treated rapidly with an injection of epinephrine or an antihistamine.

Allergies may be treated with *antihistamine drugs,* which reduce the effects of histamine on the skin and blood vessels; *immunosuppressive drugs,* such as cortisone or cyclosporine, which suppress the formation of antibodies; or *desensitization,* a procedure in which first small doses and then progressively larger doses of allergens are injected into the body until the allergic response is minimized.

Delayed Hypersensitivity

When the effects of an allergic reaction take several days or longer to develop and be expressed, the reaction is called *delayed hypersensitivity.* T cells are responsible for delayed hypersensitivity, and the mechanism of sensitization and response is the same as described earlier for T cells. In fact, delayed hypersensitivity may be no different from the normal activity of T cells directed at intracellular infections, cancer cells, or tumors.

Delayed hypersensitivity is responsible in cases where the body rejects tissues and organs transplanted from another person. Rejection reactions are mediated mainly by T cells. The rejection of donor tissue is an immune response to foreign antigens on the surface of the cells in the transplanted tissue that are different from those of the host (except in the case of identical twins). The cornea is an exception. Because corneas do not contain blood vessels or lymphatic vessels (which carry T cells), they can usually be transplanted to another person without rejection.

After receiving nonself tissue from a donor, the host begins manufacturing T cells that are active against the foreign cells bearing surface antigens. The T cells migrate to the transplant site and attack the tissue.

Immunosuppressive Drugs

Drugs that help suppress a typical immune reaction are called, logically enough, *immunosuppressive drugs.* Such drugs can be very useful in helping the body accept a tissue transplant. The longer foreign tissue can remain in the host body without causing delayed hypersensitivity, the greater is the possibility that the tissue will be accepted as self, a phenomenon known as *tolerance.*

Q: *If you have an allergy, will your children inherit it?*

A: Possibly, but it is more likely that they will inherit the *susceptibility* to allergy rather than an allergy to a specific substance.

Q: *Why do some people get an allergic reaction when they rake leaves?*

A: If you get an allergic reaction when you rake leaves it probably means that you are allergic to molds. You will probably have a less severe reaction, or none at all, if you rake the leaves no later than a day or two after they fall, before they develop molds and mildew.

FIGURE 23.14 ALLERGIC RESPONSE

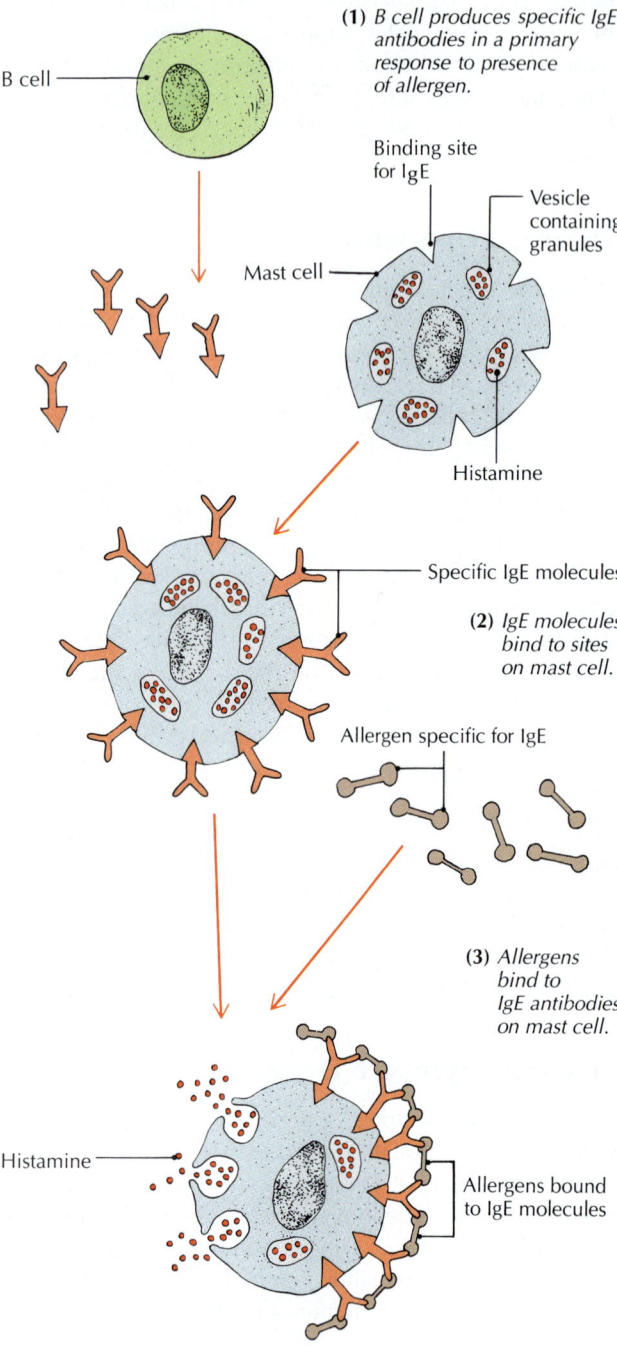

B cell

(1) *B cell produces specific IgE antibodies in a primary response to presence of allergen.*

Binding site for IgE

Vesicle containing granules

Mast cell

Histamine

Specific IgE molecules

(2) *IgE molecules bind to sites on mast cell.*

Allergen specific for IgE

(3) *Allergens bind to IgE antibodies on mast cell.*

Histamine

Allergens bound to IgE molecules

(4) *Binding of allergen to IgE antibodies on cell surface triggers the release of histamine, leukotriene, and other inflammatory substances. This process is called degranulation.*

(5) *Plasma escapes from capillaries when capillary permeability is increased in response to the release of inflammatory substances.*

[A]

[A] Role of IgE and mast cells in the allergic response (immediate hypersensitivity). **[B]** Scanning electron micrograph of a mast cell releasing histamine-containing granules that cause sneezing and watery eyes.

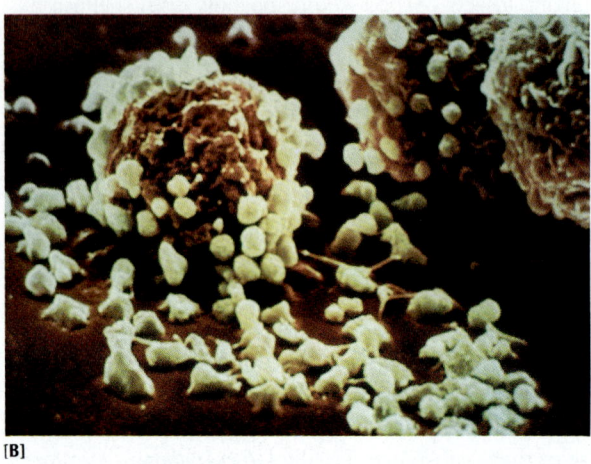

[B]

Several drugs have been used to suppress the immune response following transplant surgery. These drugs usually are given when the foreign tissue and its surface antigens are introduced. Improved immunosuppressive drugs have made it possible to transplant kidneys, livers, hearts, and lungs with some degree of success.

Some immunosuppressive drugs, such as *azathioprine* (Imuran) and *6-mercaptopurine*, work by inhibiting the synthesis of DNA, which consequently limits the proliferation of lymphocytes and their differentiation into Tc cells. Other drugs, such as *cortisone*, act to suppress inflammation. Another immunosuppressive drug is *cyclosporine*, which blocks either the production or the action of interleukin-2, the lymphokine produced by helper T cells.

A S K   Y O U R S E L F

1 What is an allergen?

2 How do the two types of hypersensitivity differ?

3 What is the purpose of immunosuppressive drugs?

AUTOIMMUNITY AND IMMUNODEFICIENCY

Early in embryonic development, lymphocytes with receptors for antigenic determinants on self molecules are either eliminated or suppressed. This allows the immune system to respond only to foreign (nonself) antigens and to show tolerance for its own (self) antigens. When tolerance to self antigens breaks down, however, the body forms antibodies to its own antigens, resulting in *autoimmunity* ("protection against one's self") and *autoimmune disease* [TABLE 23.4]. One such disease is *myasthenia gravis*, in which a person makes antibodies against the acetylcholine receptors on skeletal muscle cells. The receptors cease to function properly, causing, among other things, difficult breathing that may lead to death.

Immunodeficiencies

Conditions that decrease the effectiveness of the immune system, or destroy its ability to respond to antigens altogether, are called *immunodeficiencies.* Such conditions may result from a complete absence of B or T cells or a decrease in their number or activity. If B cells are few in number or absent, the person will be vulnerable to bacterial infections (but not viral infections if T cells are normal). If T cells are absent or low in number, the person is particularly vulnerable to viral infections, parasites, and tumor cells. *Acquired immune deficiency syndrome (AIDS)* is a disease in which the number of helper T cells is decreased, making the affected person vulnerable to infections such as *Pneumocystis carinii* pneumonia and cancers such as *Kaposi's sarcoma.* (See When Things Go Wrong on page 751.)

Some people are born without either B or T cells and have practically no protection against infection. This condition, known as *severe combined immunodeficiency (SCID)*, is usually fatal in the first few years of life. In some cases, this disease has been treated with limited success with bone marrow transplants. Experimental gene therapy is also being used, since it has been shown that some forms of SCID—an X-linked genetic disorder—is caused by a faulty gene, one that causes immune-system cells to produce defective copies of a receptor for interleukin-2, an important regulator of the immune system.

A S K   Y O U R S E L F

1 What causes autoimmunity?

2 What are immunodeficiencies?

TYPES OF ACQUIRED IMMUNITY

Resistance to disease may be innate, or it may be acquired in several different ways. Some individuals and species are not susceptible to diseases that infect other individuals or species. This resistance to disease, which is

TABLE 23.4 SOME AUTOIMMUNE DISEASES

Disease	Specificity of autoantibodies against	Result
Addison's disease	Adrenal gland	Weakness, skin pigmentation changes, weight loss, electrolyte imbalance
Autoimmune hemolytic anemia	Erythrocyte antigens	Hemolysis, anemia
Goodpasture's syndrome	Basement membrane (kidney, lung)	Pulmonary hemorrhage, kidney failure
Hashimoto's thyroiditis	Thyroid antigens	Goiter, abnormal changes in thyroid gland
Idiopathic thrombocytopenic purpura	Platelets	Hemorrhages in skin and mucous membranes
Diabetes (type I)	T cells	Beta cells in pancreas destroyed
Multiple sclerosis	Macrophages, T cells	Myelin progressively destroyed
Myasthenia gravis	Acetylcholine receptors on skeletal muscle cells	Progressive neuromuscular weakness, breathing difficulty
Pernicious anemia	Intrinsic factor	B_{12} absorption from intestine prevented, severe anemia
Rheumatoid arthritis	Immunoglobulin	Inflammation and deterioration of joints and connective tissue
Systemic lupus erythematosus	DNA, nuclear antigens	Facial rash, lesions of blood vessels, heart, and kidneys

present at birth, is called *innate resistance.* For example, human beings are not susceptible to canine distemper, and dogs are not susceptible to tetanus. Some individuals are not susceptible to some diseases that other members of their species are susceptible to. In a sense, innate resistance is not an immunity, since it does not involve the specific defenses of the immune response.

Acquired immunity is the resistance to infection by both antibody- and cell-mediated immunity. Such immunity can be *active* or *passive*. It may be acquired *naturally*, through an encounter with the infectious microorganism, or *artificially*, through the injection of a noninfectious antigenic determinant such as a vaccine [TABLE 23.5].

Active immunity results when the body produces its own antibodies or T cells in response to a foreign anti-

gen. *Passive immunity* results when antibodies, not antigens, are transferred from one person to another. This happens when maternal antibodies (IgM) cross the placenta and provide short-term (weeks to months) protection for the fetus. Maternal antibodies are also passed to the infant from the mother's milk during breastfeeding.

Artificial passive immunity occurs when specific antibodies are injected into the body. Such antibodies are taken from the blood of someone who has been infected or immunized, or from animals that have been deliberately immunized to provide antibodies. Artificial passive immunity is particularly useful when a person has been exposed to a dangerous disease and must be immunized as quickly as possible.

Passive immunity is used against botulism, diphtheria, pertussis (whooping cough), rabies, tetanus, and rattlesnake venom. The serum containing the antibodies is called the *immune serum*. Although the protection of artificial-passive immunity is immediate upon injection, it lasts only a few weeks.

If an antigen is artificially prepared and injected as a vaccine into a person, the person is said to have an *artificial acquired active immunity*. Such prepared antigens are infectious organisms that are either dead or severely weakened by special laboratory culture procedures or by using genetic engineering to change the properties of some of its molecules. Today, vaccination is available against many diseases, including hepatitis B, polio, tetanus, diphtheria, pertussis, German measles (rubella), mumps, and some forms of influenza [TABLE 23.6]. No vaccine is currently available for chickenpox (varicella), although a vaccine that has been tested and found to be safe and effective should be available in the near future.

TABLE 23.5 TYPES OF IMMUNITY

Type of immunity	Description
Innate (native) resistance (immunity)	Certain individuals or species are born with resistance to specific substances that affect other individuals or species. Permanent.
Acquired immunity	Resistance to infection, which may be acquired naturally or through the injection of a vaccine; can be active or passive.
Active immunity	Antibodies are formed in the body in response to exposure to an antigen.
Natural active immunity	Antibodies are formed during the course of a disease, sometimes providing permanent immunity (measles, chickenpox, yellow fever). Long-lasting (years to lifetime).
Artificial active immunity	Prepared dead or weakened antigens in vaccines are injected into individual to stimulate the production of specific antibodies. May require booster injections. Long-lasting (years to lifetime).
Passive immunity	Antibodies are acquired from a source outside the body.
Natural passive immunity	Antibodies from an immune pregnant woman pass to the fetus through the placenta or to the baby in milk during breastfeeding. The newborn baby receives temporary (several weeks or months) immunity against diseases to which the mother has active immunity. May be effective for several months.
Artificial passive immunity	Immune serum from immunized animals or human beings is injected into exposed individual, who receives specific antibodies (diphtheria, tetanus). Effective for weeks to months.

TABLE 23.6 A VACCINATION SCHEDULE*

Age	Diseases
2 months	Diphtheria, pertussis (whooping cough), tetanus—given in combination, called DPT—polio
4 months	Diphtheria, pertussis, tetanus, polio
6 months	Diphtheria, pertussis, tetanus
15 months	Measles, mumps, rubella (German measles), hemophilus influenza type B
15–18 months	Diphtheria, pertussis, tetanus, polio
4–6 years	Diphtheria, pertussis, tetanus, polio
11–12 years	Measles, mumps, rubella
14–16 years	Tetanus, diphtheria

*As recommended by the American Academy of Pediatrics.

MEDICAL UPDATE

Chronic Fatigue Syndrome

What used to be dismissed as "hypochondria for yuppies" is now being taken seriously as **chronic fatigue syndrome** (CFS), a disease that affects the immune system. Victims of CFS exhibit the same chronic fatigue and flulike symptoms that are found in Epstein-Barr virus syndrome and chronic mononucleosis, and in fact, CFS appears to be identical to those diseases. (In Japan the disease is called *low natural killer cell syndrome,* and in England it is known as *myalogic encephalomyelitis.* In the United States, it has been called the yuppie flu, among other names, including *chronic fatigue immune dysfunction syndrome,* or CFIDS.)

About 2 million Americans have CFS, and previously unreported cases are now being reported regularly. Unlike AIDS, which has somewhat similar symptoms, CFS is not fatal. Instead, it envelops the patient in a much broader spectrum of difficult-to-treat problems such as insomnia, hair loss, acne, mood swings, fatigue, fever, headaches, night sweats, diarrhea, joint and muscle pain, swollen lymph nodes, confusion, visual problems, and memory loss. Sometimes the symptoms disappear permanently after a few months or years, and sometimes they return.

The Centers for Disease Control (CDC) has been investigating CFS, but still has not determined its exact cause or discovered whether it is one disease or several working at the same time.

What *is* known about CFS is that initially the immune system is weakened, allowing normally manageable viruses to infect the body. Then helper T cells produce large amounts of cytokines (including interleukin-2), proteins that act as defensive agents against infectious microorganisms. Unfortunately, cytokines may be produced in excess, causing the same symptoms found in CFS. Also, NK cells become ineffective, certain cytotoxic T cells become either overactive or underactive, and blood flow decreases to the hippocampus and one or both temporal lobes in the brain, decreasing mental activity. (These sites are related to memory formation.)

For the time being, CFS seems to be linked to a viral infection, a possible genetic susceptibility, and excessive stress. Major studies are continuing.

Active immunity, whether it is acquired naturally or artificially, may last a lifetime for diseases such as measles and chickenpox. For other diseases, such as diphtheria and tetanus, immunity may last only a few years. In such cases, an additional ("booster") vaccination may be given to retain immunity. Certain viruses, such as those that cause the common cold and influenza, may occur in so many different forms that natural or artificial immunity to one form will not be effective against another form, in which case the disease may recur.

ASK YOURSELF

1 What is acquired immunity?

2 What is innate immunity?

3 How do active and passive immunity differ?

4 What is artificial immunity? Natural immunity?

MONOCLONAL ANTIBODIES

In 1975 British scientists Georges Kohler and Cesar Milstein fused an activated lymphocyte from the spleen of a mouse with a cancer (myeloma) cell from the bone marrow. The result was a hybrid cell called a **hybridoma** (HIGH-brihd-OH-muh) that divided over and over again, producing identical copies (clones) of itself [FIGURE 23.15]. The clones of the hybridoma produced a limitless supply of antibodies called **monoclonal antibodies,** all of which react with one specific antigen.

The technique of producing monoclonal antibodies is important because it provides the possibility of developing monoclonal antibodies from human cells that could be used to manufacture vaccines against cancer and other diseases. Also, when radioactive tracers are attached to circulating monoclonal antibodies, it is possible to locate cancer cells anywhere in the body, because the antibodies react with cancer cells as they are encountered. This is especially important in determining if a primary cancer has spread to other parts of the body.

Hybridoma technology also offers the possibility of attaching cancer-killing chemicals to monoclonal antibodies. This "targeted drug therapy" would permit chemotherapy to be directed *only* at cancer cells. Currently, cancer chemotherapy destroys some body cells as well as cancer cells because it cannot be focused precisely on the cancer cells alone.

The availability of monoclonal antibodies makes it relatively easy to identify the molecules on donor tissues or organs that would function as antigens in the recipient. Blood samples from the donor and recipient are analyzed for the types of tissue antigens (MHC molecules) before a tissue transplant. The closer the types can be matched, the better the chance of reducing tissue rejection.

About 9000 kidney transplants are performed each year in the United States, and about 60 percent result in acute rejection. (The *hyperactive* rejection stage takes place immediately after a transplant. *Acute rejection* occurs days or weeks following the transplant. The *chronic* stage of rejection follows, lasting as long as the transplanted organ does.) A 1991 study showed that 92 percent of those who receive a kidney transplant from a cadaver live at least one year after the transplant. In 1988 the 1-year survival rate was only 71 percent.

In 1986 the Food and Drug Administration (FDA) approved the first monoclonal antibody for use in acute rejection following human kidney transplants. This monoclonal antibody, which is called OKT3, attacks the T cells responsible for tissue rejection. In one trial, OKT3 reversed kidney-transplant rejections in 94 percent of transplant recipients. In comparable tests, treatment with cyclosporine produced a 75-percent success rate.

FIGURE 23.15 MONOCLONAL ANTIBODIES

[A] The mechanism for the production of monoclonal antibodies. **[B]** Scanning electron micrograph of a hybridoma in the process of cloning. ×10,000.

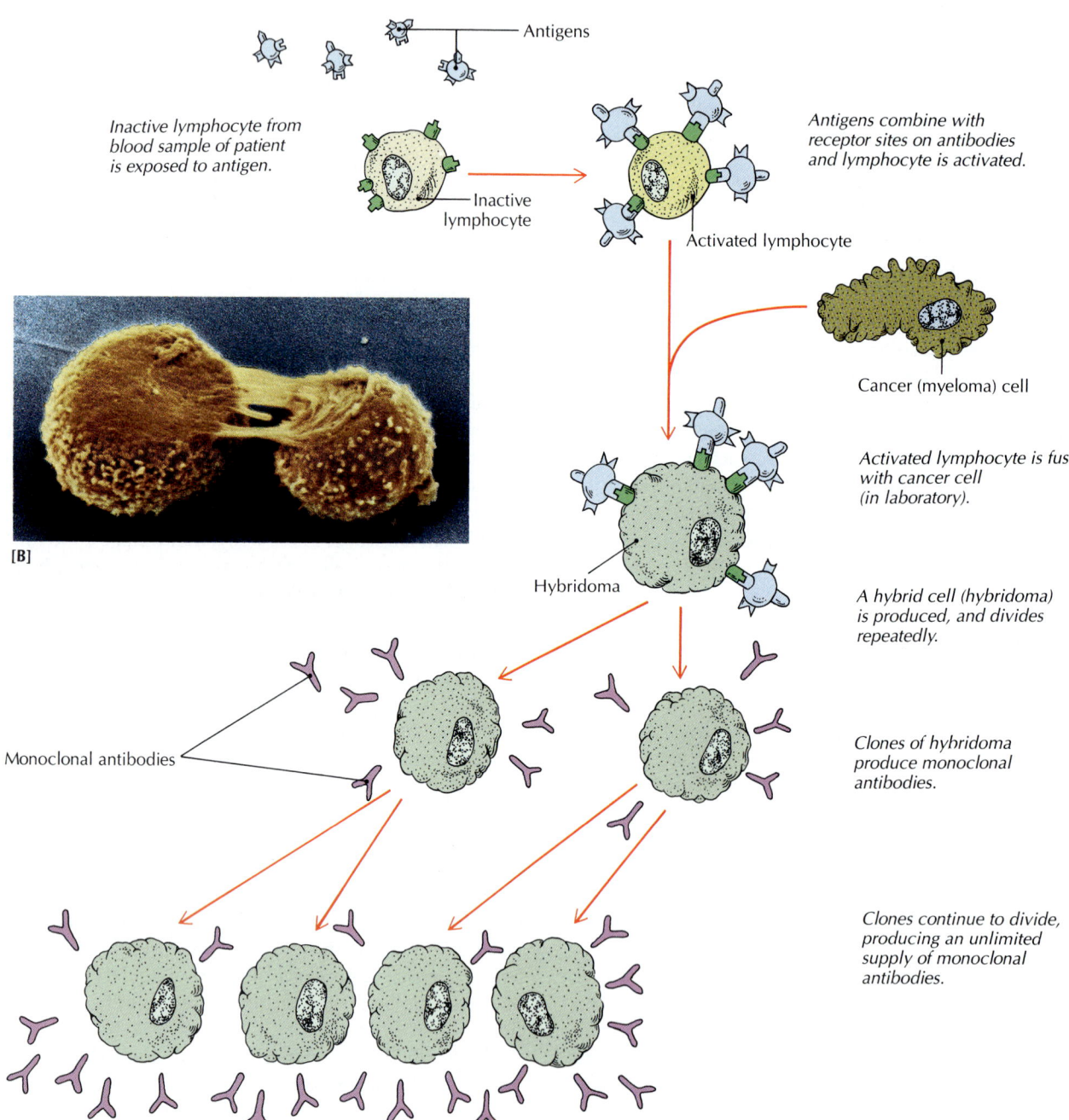

Antigens

Inactive lymphocyte from blood sample of patient is exposed to antigen.

Inactive lymphocyte

Antigens combine with receptor sites on antibodies and lymphocyte is activated.

Activated lymphocyte

Cancer (myeloma) cell

Activated lymphocyte is fused with cancer cell (in laboratory).

Hybridoma

A hybrid cell (hybridoma) is produced, and divides repeatedly.

Monoclonal antibodies

Clones of hybridoma produce monoclonal antibodies.

Clones continue to divide, producing an unlimited supply of monoclonal antibodies.

[B]

[A]

WHEN THINGS GO WRONG

Acquired Immune Deficiency Syndrome (AIDS)

As its name suggests, *acquired immune deficiency syndrome,* or *AIDS,** is a disease that cripples the body's immune system. It is caused by the *human immunodeficiency virus,* or *HIV,* which invades and kills huge numbers of *helper T cells,* also called *T4 cells,* whose plasma membranes display the CD4 receptor molecules.

There are two different types of HIV: *HIV-1* and *HIV-2.* Most AIDS cases in the United States are caused by HIV-1. HIV-2 is most common in Africa and may be a less virulent form of the virus. Blood tests for HIV test for HIV-1, but testing for HIV-2 may be necessary in the near future. The following discussion concerns HIV-1.

Because T4 cells are essential effectors of the immune response, the result of their depletion is a severe and (so far) irreversible immunodeficiency. With an insufficient number of T4 cells, the immune system becomes ineffective, and the continued destruction of T4 cells triggers the collapse of the entire immune system. As a result, a person infected with HIV becomes vulnerable to infections that rarely affect healthy people. (Such infections include pneumonia and tuberculosis and are referred to as *opportunistic infections.*)

Also, relatively uncommon forms of cancer of the blood and lymphatic system seem to be unusually prevalent among AIDS victims. In fact, the first sign that a new disease had emerged was the sudden appearance in 1981 of *Pneumocystis carinii* pneumonia (PCP) and Kaposi's sarcoma (KS; a formerly rare cancer of the lining of blood vessels that usually shows externally as purplish spots on the skin) among young, middle-class white males. Until then, KS had appeared mainly among older Italian and Jewish men and in some Africans. The young males turned out to be predominantly homosexual; PCP and KS were also spreading among drug users who used contaminated hypodermic needles and among patients who received frequent blood transfusions or blood products such as clotting factors.

In 1989, 70 percent of AIDS victims in the United States were active male homosexuals or bisexuals; only 4 percent were heterosexuals. But the trend is changing. In 1993 those percentages were 58 and 6, respectively [TABLE 23.7]. The incidence of AIDS in women and children is rising also, and intravenous drug users remain at

TABLE 23.7 REPORTED AIDS CASES IN ADULTS IN THE UNITED STATES*	
Type of transmission	Percentage of cases
Homosexual sex	58
Intravenous drug use	23
Homosexual sex and intravenous drug use	6
Heterosexual sex	6
Blood transfusions	2
Treatments for hemophilia and other blood-clotting disorders	1
Other	4

*By the end of 1993, almost 335,000 cases of AIDS had been diagnosed in the United States since the first known infection in 1981.
Source: Centers for Disease Control.

high risk. (Worldwide — as of 1992 — about one in three people infected with HIV is female.)

HIV is transmitted more often via unprotected receptive anal intercourse than through any other activity. Anal intercourse is an especially effective method of transmission because rectal capillaries are so close to the skin that infection may occur through microscopic breaks in the skin. However, researchers stress that AIDS is not exclusively a disease of homosexuals, bisexuals, or drug addicts. In Africa, for example, it is a disease of heterosexuals, and of the 13 million people infected with HIV there since 1981, most were exposed through heterosexual intercourse.

HIV is a special kind of virus called a *retrovirus.* Like ordinary viruses, retroviruses need host cells in order to reproduce. However, the genetic material of a retrovirus is RNA, not DNA, and RNA needs the help of an enzyme called *reverse transcriptase* to convert viral RNA into viral DNA (see drawing on page 752).

Both the RNA and reverse transcriptase from HIV are injected into a host cell, which is usually a T4 cell (see photos on page 752). Reverse transcriptase "fools" the T4 cell into accepting the viral RNA as its own, and the viral RNA is then converted into viral DNA. The viral DNA enters the nucleus of the T4 cell, where it inserts itself into the DNA of the cell and once again "fools" the cell into accepting it as its own DNA. The viral DNA proceeds to make viral protein that is used to assemble new HIVs. After producing new HIVs, the T4 cell eventually dies (probably because the budding of viruses from the cell produces tiny holes in its plasma membrane), and

*The acronym AIDS has come to stand for a slightly shortened phrase: *acquired immunodeficiency syndrome.* The original name was purposely descriptive. The condition was *acquired* rather than inherited, produced a *deficient immune system,* and was a collection of several diseases (a *syndrome*) rather than a single, easily classified disease.

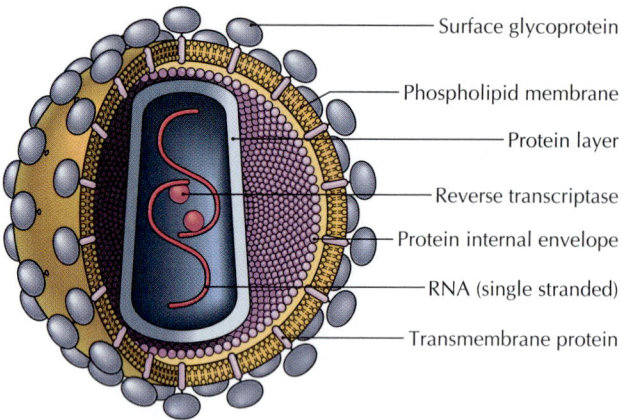

The structure of HIV.

Labels: Surface glycoprotein · Phospholipid membrane · Protein layer · Reverse transcriptase · Protein internal envelope · RNA (single stranded) · Transmembrane protein

inhibits immunological functions. During this period, diseases that are normally rejected by a healthy immune system become established.

The first symptoms of HIV infection often resemble the symptoms of the flu or mononucleosis; they include persistent or recurrent fever, severe dry cough and sore throat, blurred vision, enlarged lymph nodes, night sweats, chronic fatigue, decreased appetite and rapid weight loss, blood or mucus in the feces, and white blemishes inside the mouth. An upper-body rash may also be present. These initial symptoms usually last from a few days to a month or more. About 85 percent of all AIDS patients eventually show signs of damage to the nervous system, since HIV infects macrophages, altering them so they can penetrate the blood-brain barrier and enter the brain. The infected macrophages then destroy the myelin of myelinated nerves, causing behavioral dementia. This symptom may appear long before the other symptoms mentioned above.

It is believed that about 5 percent of the 1 million Americans infected with HIV may never have "full-blown AIDS." (It is important to remember that a person can be infected with HIV — that is, be HIV-positive — but still not have AIDS.)

It is likely that HIV may have been infecting human beings long before the recognized outbreak in 1981. HIV has been traced to south central Africa before 1970, but it may have originated elsewhere. It arrived in New York City around 1977 (probably from the Caribbean), and by

the newly created viruses spread to infect other T4 cells. This sequence is repeated over and over until the supply of T4 cells is seriously depleted.

Because the body does not produce an unlimited supply of T4 cells, they soon reach a low point, at which time the immune system is no longer effective. As the T4 (helper) cells are destroyed, the ratio of suppressor T cells to helper T cells increases, thus depressing other immune responses. Also, with the decrease in helper T cells, the B cells are not appropriately stimulated to produce antibodies to fight off infections. HIV-infected T4 cells also release a *soluble suppressor factor*, which further

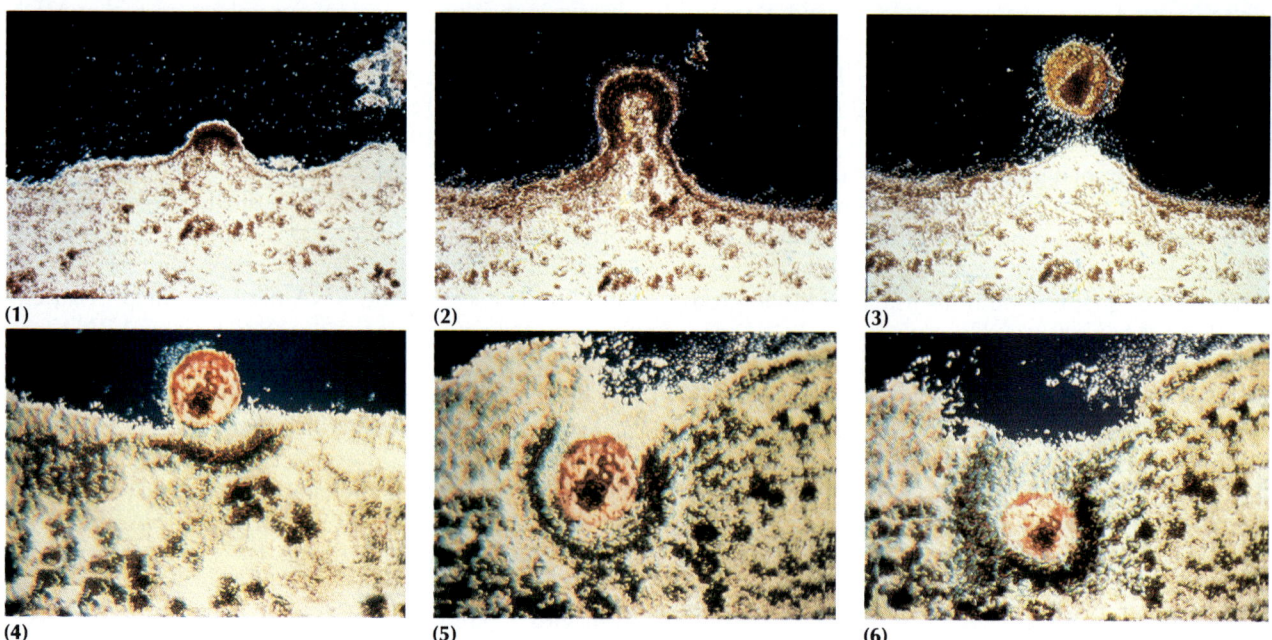

(1) (2) (3)

(4) (5) (6)

HIV in action. (1–3) A newly replicated HIV buds off from a T lymphocyte, wrapping itself in the T cell's plasma membrane. The coated virus travels to another T cell, **(4)** fuses with its plasma membrane, and **(5)** enters. **(6)** The second T cell is now infected.

TABLE 23.8 ESTIMATED NUMBER OF AIDS CASES (1992)*

Geographic location	Percentage of all cases	Number of cases
Africa	69	1,170,000
United States	16	272,000
Latin America, South America	9	153,000
Europe	6	102,000
Other	1	17,000
Total	101†	1,714,000

*Estimated number of adult HIV infections worldwide: 10 to 12 million.
†Because of rounding, numbers do not add up to 100 percent.

dideoxycytidine) and *interferon*. Also, the clinical production of genetically engineered T4 proteins is currently in progress. Finally, the National Institutes of Health (NIH) has approved preliminary trials involving gene therapy for AIDS patients.

Prevention may be the most effective approach to conquering AIDS, and prevention has been helped considerably by the development of blood tests that can detect the presence of antibodies to HIV in the blood before transfusions are given [TABLE 23.9].

The Centers for Disease Control (CDC) in Atlanta operates a 24-hour, toll-free telephone number that provides free and confidential information about AIDS. The hotline number is 1-800-342-AIDS.

Infectious Mononucleosis

Infectious mononucleosis is a viral disease caused by the Epstein-Barr virus, a member of the herpes group. It appears most often in 15- to 25-year-olds. It is usually accompanied by fever, increased lymphocyte production, and enlarged lymph nodes in the neck. Secondary symptoms include dysfunction of the liver and increased numbers of monocytes. Infectious mononucleosis is not contagious in adults. In adolescents, however, the Epstein-Barr virus may be spread during close contact or via the exchange of body fluids, as when drinking glasses are shared or through kissing. The virus may also be spread from infant to infant in saliva, from the mother's breast, or in food.

1978 it had reached Los Angeles, San Francisco, and Miami. By 1990 HIV was found in all 50 states and throughout the world. As many as 2 million Americans may be infected (twice the reported number), and the situation is much worse in Africa and the Caribbean [TABLE 23.8].

At present there is no known cure for AIDS. Antiviral agents are difficult to develop, especially for retroviruses. Drugs that have been used in the treatment of AIDS include *AZT* (3'-azido-2', 3'-dideoxythymidine), which interferes with the synthesis of viral DNA. AZT is best used with infected individuals who have not developed full-blown symptoms. Other drugs include *ddC* (2', 3'-

TABLE 23.9 A SUMMARY OF AIDS TESTS

Test	Benefits	Disadvantages
HIV-1 ELISA blood test (1985)	Highly sensitive; usually the first test to detect antibodies to HIV virus; preferred test for mass screenings.	May give false results, so results must be confirmed by second test; not effective before HIV-1 antibodies are present (2 weeks to 3 months after infection); results take 1 week.
Western blot blood test (1987)	Highly accurate; often used to confirm HIV-1 ELISA test.	Not effective before HIV-1 antibodies are present; results take 1 week.
HIV-1/HIV-2 blood test (1991)	Used in same way as HIV-1 ELISA, but also screens for HIV-2 antibodies.	Often registers false positives because of high sensitivity; must be confirmed for both HIV-1 and HIV-2.
10-minute blood test (1992)	Used in same way as HIV-1 ELISA, but requires smaller samples; can be processed immediately.	Not useful for mass screening; may register false positives; patient knows results before confirmation.
PCR (polymerase chain reaction) blood test (under review)	Highly accurate in detecting HIV virus itself early in infection; excellent confirming test.	Samples reviewed individually by technician; expensive.

CHAPTER SUMMARY

Nonspecific Defenses of the Body (p. 725)

1 The **nonspecific defenses** of the body are those that do not involve the production of antibodies. They include the skin, mucous membranes and mucus, lacrimal glands, nasal hairs, cilia in the respiratory tract, and the acidity of the stomach.

2 The body produces nonspecific chemical substances such as **complement** and **interferons** that attack bacteria and viruses.

3 When tissues are damaged, the normal result is the **inflammatory response**, which initiates healing and prevents further damage.

4 **Phagocytosis** is the destruction of foreign substances and dead tissue by leukocytes.

Specific Defenses: The Immune Response (p. 730)

1 The **specific defenses** of the body, which constitute the **immune response**, discriminate among foreign substances (*antigens*) by forming specific proteins (*antibodies*) to react with the foreign substances and destroy them.

2 An **antigen** causes the production of specific antibodies and reacts chemically with the antibody to form a stable interaction called the **antigen-antibody complex.**

3 An antibody reacts with specific sites on the surface of an antigen called **antigenic determinants,** which are formed by distinctive proteins or polysaccharides.

4 **Antibodies** are proteins produced by lymphocytes in response to an antigen. They contribute an **active immunity** that provides relatively long-term protection against chronic infections.

5 An **antibody molecule** is composed of four polypeptide chains. The molecular

structure of an antibody is crucial to its ability to react with specific antigens. The *variable region* of an antibody has the structural potential to bind to specific antigens like a lock (antibody) accepting a key (antigen).

6 Antibodies are **immunoglobulins.** Five known classes of immunoglobulins are IgA, IgD, IgE, IgG, and IgM.

7 The proteins that make up **complement** must bind in sequence to antibodies attached to antigens before the antigen cell membrane is ruptured and the antigen cell is destroyed.

8 B and T cells are formed from precursor **stem cells** in the fetal bone marrow. Some stem cells migrate to the thymus gland, where they mature into T cells. It is not certain where the remaining stem cells mature into B cells, but they may remain in the bone marrow.

9 **Antibody-mediated (humoral) immunity** involves B cells, which differentiate into **plasma cells** that secrete antibodies into the blood or lymph. Some B cells become **memory cells** that react to an antigen the second time it appears. The direct contact between T cells and antigens is **cell-mediated immunity.**

10 According to the **clonal selection theory** of antibody formation, B cells are genetically programmed, before meeting an antigen, to react only to a particular antigen.

11 T cells differentiate into *helper T cells, suppressor T cells, cytotoxic T cells,* and *delayed hypersensitivity T cells.*

Hypersensitivity (Allergy) (p. 743)

1 The two types of **hypersensitivity (allergy)** are **immediate** and **delayed.** Severe systemic reactions can produce *anaphylactic shock.* Delayed hypersensitivity is responsible for the body's rejection of tissues or organs transplanted from another person.

2 **Immunosuppressive drugs** help suppress a typical immune reaction. Such drugs usually cause the immune system to become less effective, lowering the body's resistance against infection.

Autoimmunity and Immunodeficiency (p. 746)

1 **Autoimmunity** is the formation of antibodies against the body's own antigens.

2 Conditions that decrease the effectiveness of the immune system or that destroy its ability to respond to antigens altogether are called **immunodeficiencies.**

Types of Acquired Immunity (p. 747)

1 Some individuals are not susceptible to diseases that infect other individuals. This inborn resistance to disease is called **innate resistance.**

2 **Acquired immunity** is the resistance to infection by antibody-mediated (humoral) immunity or cell-mediated immunity. Such immunity can be active or passive, acquired naturally or artificially.

3 **Active immunity** results when the body manufactures its own antibodies or T cells in response to a foreign antigen. **Passive immunity** results when antibodies, not antigens, are transferred from one person to another.

Monoclonal Antibodies (p. 749)

1 The laboratory fusion of an activated lymphocyte with a cancer cell produces a hybrid cell called a **hybridoma.** Clones of the hybridoma produce a limitless supply of antibodies called **monoclonal antibodies.**

2 Monoclonal antibodies may be useful in producing a vaccine against cancer, tracing the location of cancer cells in the body, and directing or transporting specific antibodies or cancer-killing chemicals to cancer cells.

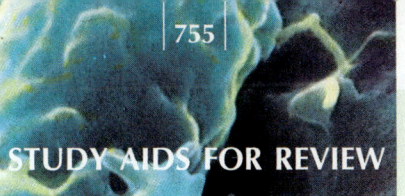

STUDY AIDS FOR REVIEW

KEY TERMS
acquired immune deficiency syndrome (AIDS) 747, 751
acquired immunity 747
active immunity 748
allergy 743
antibody 734
antibody-mediated (humoral) immunity 737
antigen 730
antigen-antibody complex 730
autoimmunity 746
cell-mediated immunity 737
clonal selection theory 738
complement 725
cytotoxic T cell 742
delayed hypersensitivity T cell 742, 745
helper T cell 741
human immunodeficiency virus (HIV) 751
hypersensitivity 743
immune response 730
immunodeficiencies 747
immunoglobulin 734
inflammatory response 727, 737
innate resistance 747
interferons 725
memory cell 737
monoclonal antibody 749
nonspecific defenses 725
passive immunity 748
phagocytosis 729
plasma cell 735, 737
specific defenses 725, 730
suppressor T cell 741
T4 cells 751

MEDICAL TERMS FOR REFERENCE
ANAPHYLAXIS (an-uh-fuh-LACK-sis; L. *ana*, intensification + Gr. *phulassein*, to guard) Hypersensitivity to a foreign substance.
VACCINE General term for the immunization preparation used against specific diseases.

UNDERSTANDING THE FACTS
1 What is the chemical nature of interferon, and what is its role?
2 What is the chemical nature of complement, and what is its role?
3 List in order the events that occur during an inflammatory response.
4 List in order the events that occur during phagocytosis.

5 Describe the basic molecular shape of an antibody molecule.
6 Name the five classes of antibodies, and indicate the function of each class.
7 Distinguish between the origin and role of B lymphocytes, T lymphocytes, and plasma cells.
8 List the different kinds of T cells, and identify the role of each.
9 What are the primary cells of the immune system that are defective in SCID? In AIDS?

UNDERSTANDING THE CONCEPTS
1 Which mechanisms for fighting disease are said to be nonspecific, and which are specific? Explain.
2 How does the skin defend against microorganisms?
3 What is the basis of immunologic memory?
4 Explain the nature of clones, and identify the agent of selection in the clonal selection theory.
5 What roles are played by macrophages in the immune response?
6 Explain the difference between sensitivity and hypersensitivity.
7 How do immunosuppressive drugs work?
8 How are monoclonal antibodies produced, and how are they used?

SELF-QUIZ
Matching

a IgA c IgE e IgM
b IgD d IgG f more than one of the above

23.1 _____ includes bivalent antibodies
23.2 _____ includes multivalent antibodies
23.3 _____ most common immunoglobulin; can pass through the placenta
23.4 _____ found in tears and saliva
23.5 _____ a pentameric molecule that is the first antibody to appear at the site of an infection
23.6 _____ immunoglobulin that is responsible for immediate allergic responses
23.7 _____ the least common of the antibody classes

a MHC proteins d hybridoma
b cytokine e leukotriene
c perforin

23.8 _____ a cell clone formed by fusing a human myeloma cell with a mouse lymphocyte
23.9 _____ a compound that attracts phagocytes
23.10 _____ a protein that results in leaky plasma membranes

23.11 _____ an inflammatory substance produced by mast cells

23.12 _____ cell surface proteins that are virtually unique in each individual

a helper T cells   d memory cells
b suppressor T cells   e plasma cells
c cytotoxic T cells

23.13 _____ primary role is secretion of blood-borne immunoglobulins

23.14 _____ inactive, immature B cells that respond during a secondary response

23.15 _____ essential for the maturation of B cells to plasma cells

23.16 _____ inhibit the maturation of B cells to plasma cells

23.17 _____ cytotoxic cells that secrete perforin

Multiple-Choice Questions

23.18 Which of the following apply primarily to the antibody-mediated response?
a B lymphocytes are active
b plasma cells are active
c effective against extracellular, unicellular pathogens
d a and b
e a, b, and c

23.19 T lymphocytes
a produce chemicals that destroy cells
b stimulate the maturation of B lymphocytes
c regulate the termination of an immune response
d stimulate macrophages
e all of the above

23.20 Which of the following are used to treat allergies?
a antihistamines   d a and b
b immunosuppressive drugs   e a, b, and c
c desensitization

23.21 Cell-mediated immunity operates against
a cancer cells   d multicellular
b tissue transplants   parasites
c intracellular pathogens   e all of the above

23.22 The combination of tissue fluid, dead microorganisms, and macrophages is called
a opsonin   d leukotriene
b pyrogen   e pus
c histamine

23.23 According to the clonal selection theory,
a selected antibodies are continuously circulating in the blood
b there exists at least one lymphocyte capable of synthesizing antibodies specific to each antigen prior to exposure to the antigen

c each antigen serves as a template for the synthesis of a corresponding antibody
d the immune system selects the proper antibody to use in each infection from a large number of existing clones
e clones of antibodies are produced by self-replication of immunoglobulins

23.24 If an immunological preventive to dental caries were to be developed, it would have to elicit the production of
a IgG b IgM c IgA d IgD e IgE

23.25 The secondary humoral response is generally stronger than the primary response because
a all classes of immunoglobulins are active
b both specific and nonspecific defenses are active
c the antigen is weakened by the primary response
d a larger population of B cells is active
e the ineffective B cells have been removed by clonal selection

23.26 Patients with severe combined immunodeficiency disease (SCID) lack the ability to
a reject grafts
b produce antibodies against viral pathogens
c produce antibodies against bacterial pathogens
d a and b
e a, b, and c

23.27 Which of the following statements is true?
a All classes of antibodies consist of two polypeptide chains.
b All classes of antibodies are multivalent.
c All immunoglobulins consist of constant regions and a variable region.
d All classes of antibodies are produced by T cells.
e All of the immunoglobulins are found primarily in the blood.

23.28 Which of the following statements is _not_ true?
a Delayed hypersensitivity involves the action of B memory cells.
b Immunosuppressive drugs work by blocking the action of T cells.
c Macrophages can display antigens on their surface.
d Although T cells do not produce antibodies, they are essential to the process.
e Antibodies alone cannot destroy pathogens.

23.29 When immune serum from an immunized animal is injected into a human patient, it produces
a innate immunity
b artificial, active immunity
c natural, passive immunity
d artificial, passive immunity
e a cell-mediated response

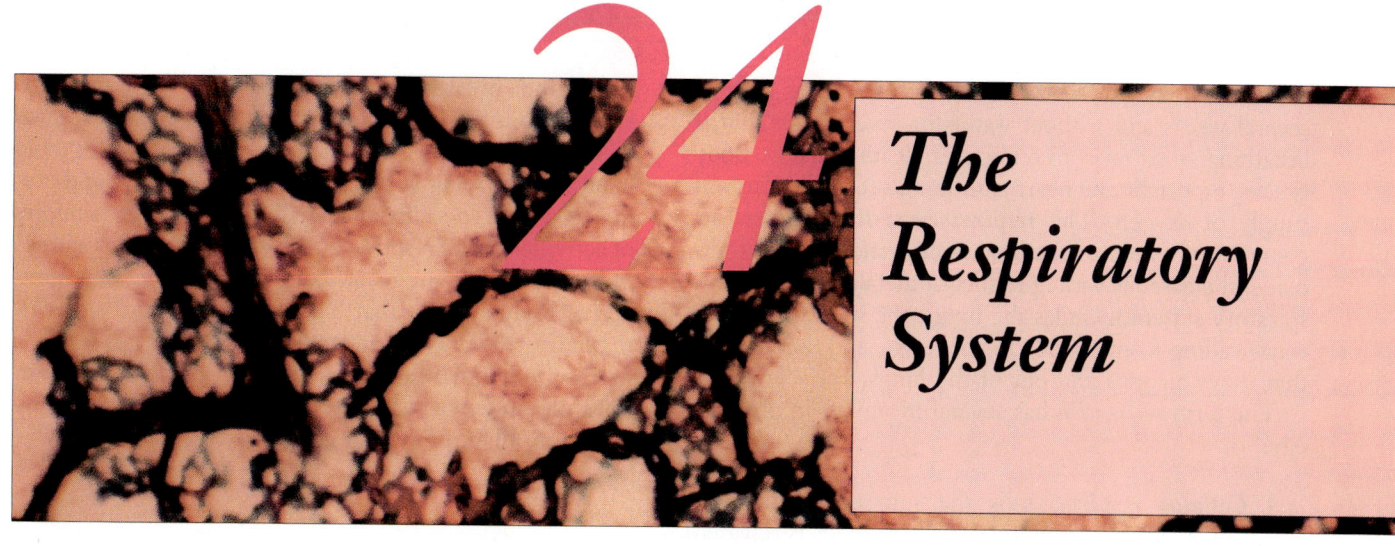

24

The Respiratory System

KEY CONCEPTS

1 Body tissues require a constant supply of oxygen. The respiratory system delivers oxygen to tissues and removes metabolic wastes from tissues via the blood.

2 Muscles used in breathing, especially the diaphragm, assist the lungs in taking air in and forcing it out.

3 Within the lungs are millions of clusters of thin-walled air sacs called alveoli. Gases are exchanged between the interior of the alveoli and the blood in the capillaries surrounding them.

4 Most of the oxygen in blood is combined with hemoglobin.

5 Although breathing can be controlled consciously, it is generally regulated by involuntary nerve impulses.

A human body can survive without food for as long as several weeks and without water for several days, but if breathing stops for 3 to 6 minutes, death is likely. Body tissues, especially the heart and brain, require a constant supply of oxygen. The *respiratory system* delivers air containing oxygen to the blood and removes carbon dioxide, the gaseous waste product of metabolism. The respiratory system includes the lungs, the several passageways leading from outside to the lungs, and the muscles that move air into and out of the lungs.

The term *respiration* has several meanings in physiology:

1 *Cellular respiration* is the sum of biochemical events by which the chemical energy of foods is released to provide energy for life's processes. In cellular respiration, oxygen is utilized as a final electron acceptor in the mitochondria.

2 *External respiration,* which occurs in the lungs, is the exchange of gases between the blood and the lungs; oxygen from the lungs diffuses into the blood, and carbon dioxide diffuses from the blood into the lungs.

3 *Internal respiration* is the exchange of gases in the body tissues; carbon dioxide from the body cells is exchanged for oxygen from the blood.

In contrast to these physiological processes, *ventilation,* or *breathing,* is the mechanical process that moves air into and out of the lungs. It includes two phases: *inspiration* and *expiration.*

GENERAL FUNCTIONS OF THE RESPIRATORY SYSTEM

It is helpful to think of the function of the respiratory system in a series of steps, even though in a living body they all take place at the same time and continuously: (1) Inhaled air, high in oxygen and low in carbon dioxide, travels through the respiratory tract deep into the terminal portions of the lungs; this is *inspiration.* (2) There, oxygen diffuses from the lungs into the blood. (3) From the lungs, oxygenated blood is carried to the heart and then, via the systemic circulatory system, to all parts of the body. During cellular respiration, oxygen moves from the blood into the cells, and carbon dioxide and other wastes are released from the cells into the blood. (4) Finally, deoxygenated venous blood, carrying its load of carbon dioxide, is carried back to the lungs, where carbon dioxide is exhaled during *expiration.*

The *primary* function of the respiratory system is this exchange of oxygen and carbon dioxide, but the system has many *secondary* functions as well. The respiratory tract is the site of sound production. We also use our lungs and breathing muscles as a convenient high-volume, low-pressure pump for dozens of daily jobs, such as inflating balloons, cooling hot coffee, warming cold hands, and playing wind instruments. Laughing and crying, as well as sneezing and coughing, are other examples of activities of the respiratory system. Also, as you will see in later chapters, the respiratory system is one of a number of body systems that help maintain homeostasis by regulating the pH of blood and other body fluids within a narrow functional range. Finally, the muscles of the respiratory system assist in abdominal compression during urination, defecation, and childbirth.

ANATOMY OF THE RESPIRATORY TRACT

Air flows through respiratory passages because of differences in pressure produced by chest and trunk muscles during breathing. Except for the beating of cilia in the respiratory lining, the passageways are simply a series of openings through which air is forced. These passageways and the lungs make up the *respiratory tract* [FIGURE 24.1].

Nose

Air normally enters the respiratory tract through the nose, where (1) the air is warmed, moistened, and filtered; (2) the sense of smell originates; and (3) vocal sounds are produced in the large air chambers. The external nose is supported at the bridge by *nasal bones.* The *septal cartilage* separates the two external openings of the *nostrils,* or *external nares* (NAIR-eez; L. nostrils). Along with the vomer and perpendicular plate of the ethmoid bone, the septal cartilage forms the *nasal septum,* which divides the nasal cavity into bilateral halves.

The *nasal cavity* fills the space between the base of the skull and the roof of the mouth. It is supported above by the *ethmoid bones* and on the sides by the *ethmoid, maxillary,* and *inferior conchae bones* [FIGURE 24.2]. The superior, middle, and inferior conchae bear longitudinal ridges that are covered with highly vascular respiratory mucosa. Between the ridges of the conchae are folds called the *superior, middle,* and *inferior meatuses* (L. passages; sing. *meatus;* mee-AY-tuhss). These meatuses serve as air passageways. The blood vessels in this area, called venous sinuses, bring a large quantity of blood to the mucous membrane that adheres to the underlying periosteum. The blood is a constant source of heat that warms the air inhaled through the nose. Since warm air holds more water than cold air does, the air is also moistened as it moves through the nasal cavity.

The *hard palate* [FIGURE 24.2], strengthened by the

FIGURE 24.1 THE RESPIRATORY SYSTEM ⭐

The interior of the right lung (not in scale) is revealed.

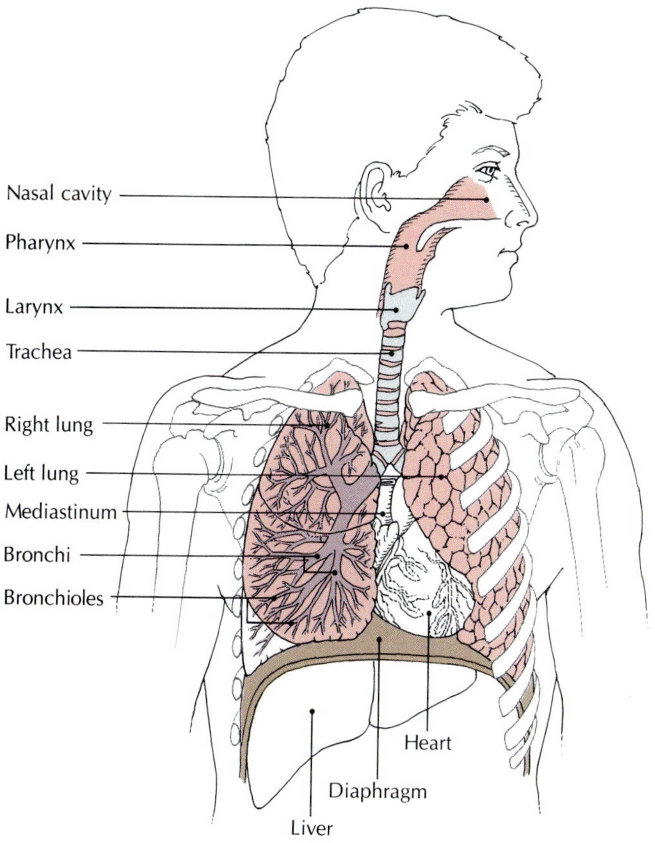

Nasal cavity
Pharynx
Larynx
Trachea
Right lung
Left lung
Mediastinum
Bronchi
Bronchioles
Heart
Diaphragm
Liver

Structure	Major functions
Nasal cavity	Filters, warms, moistens incoming air; passageway to pharynx
Paranasal sinuses	Produce mucus; act as resonators for sound; lighten skull
Pharynx	Passageway for air between nose and larynx and for food from mouth to esophagus
Larynx (voice box)	Passageway for air between pharynx and rest of respiratory tract; produces sound; protects trachea from foreign objects
Trachea	Passageway for air to and from thoracic cavity; traps and expels foreign matter
Diaphragm	Enlarges thoracic cavity to allow for inspiration, returns to original position for expiration
Bronchi	Passageways for air to and from lungs; filter air
Bronchioles	Passageways for air to and from alveoli
Alveoli	Site of gas exchange; functional units of lungs
Lungs	Major respiratory organs
Pleurae	Protect, compartmentalize, and lubricate outer surfaces of lungs; enclose pleural cavities

palatine bones and the palatine process of the maxillary bones, forms the floor of the nasal cavity. If these bones do not grow together normally by the third month of prenatal development, the nasal cavity and mouth are not separated adequately. The condition is known as a *cleft palate*. The immediate problem facing a newborn infant with a cleft palate is that not enough suction can be created to nurse properly. A cleft palate can usually be corrected surgically.

The **soft palate** [FIGURE 24.2] is a flexible muscular sheet that separates the oropharynx from the nasal cavity. It is continuous with the posterior border of the hard palate and consists of several skeletal muscles covered by mucous membrane. Depression of this area during breathing enlarges the air passage into the pharynx. When the soft palate is elevated during swallowing, it prevents food from entering the nasal pharynx.

The *frontal, sphenoidal, maxillary*, and *ethmoid sinuses* are blind sacs; each opens through its own small foramen into the nasal cavity. They are not present in the newborn infant. Together they are called the **paranasal si-**nuses. They give the voice a full, rich tone. A **nasolacrimal** (L. "nose" and "tear") **duct** leads from each eye to the nasal cavity, through which excessive tears from the surface of the eye are drained into the nose; that is why weeping may be accompanied by a watery flow from the nose ("runny nose"). At the back of the nasal cavity, the two *internal nares* open into the pharynx (throat).

The lining of the nasal cavity is specialized. Most of the covering membrane is supplied with mucus-secreting *goblet cells*, which keep the surfaces wet. The mucus catches many of the small dust particles and microorganisms that get past the coarse hairs in the nostrils.

The surface of the nasal cavity is lined with two types of epithelium: (1) *Nasal epithelium* warms and moistens inhaled air. It is a pseudostratified columnar epithelium bearing millions of cilia, which also help capture minute particles and microorganisms [FIGURE 24.3]. (2) *Olfactory epithelium* in the upper part of the nasal cavity contains sensory nerve endings of the neurons whose axons form the olfactory nerve, supporting cells, and basal cells that differentiate into olfactory neurons. Certain chemical

FIGURE 24.2 UPPER RESPIRATORY SYSTEM

[A] Right midsagittal section through the head showing the upper portion of the respiratory system: the nasal cavity, pharynx, larynx, and part of the trachea. [B] A photograph of a sagittal section through the head.

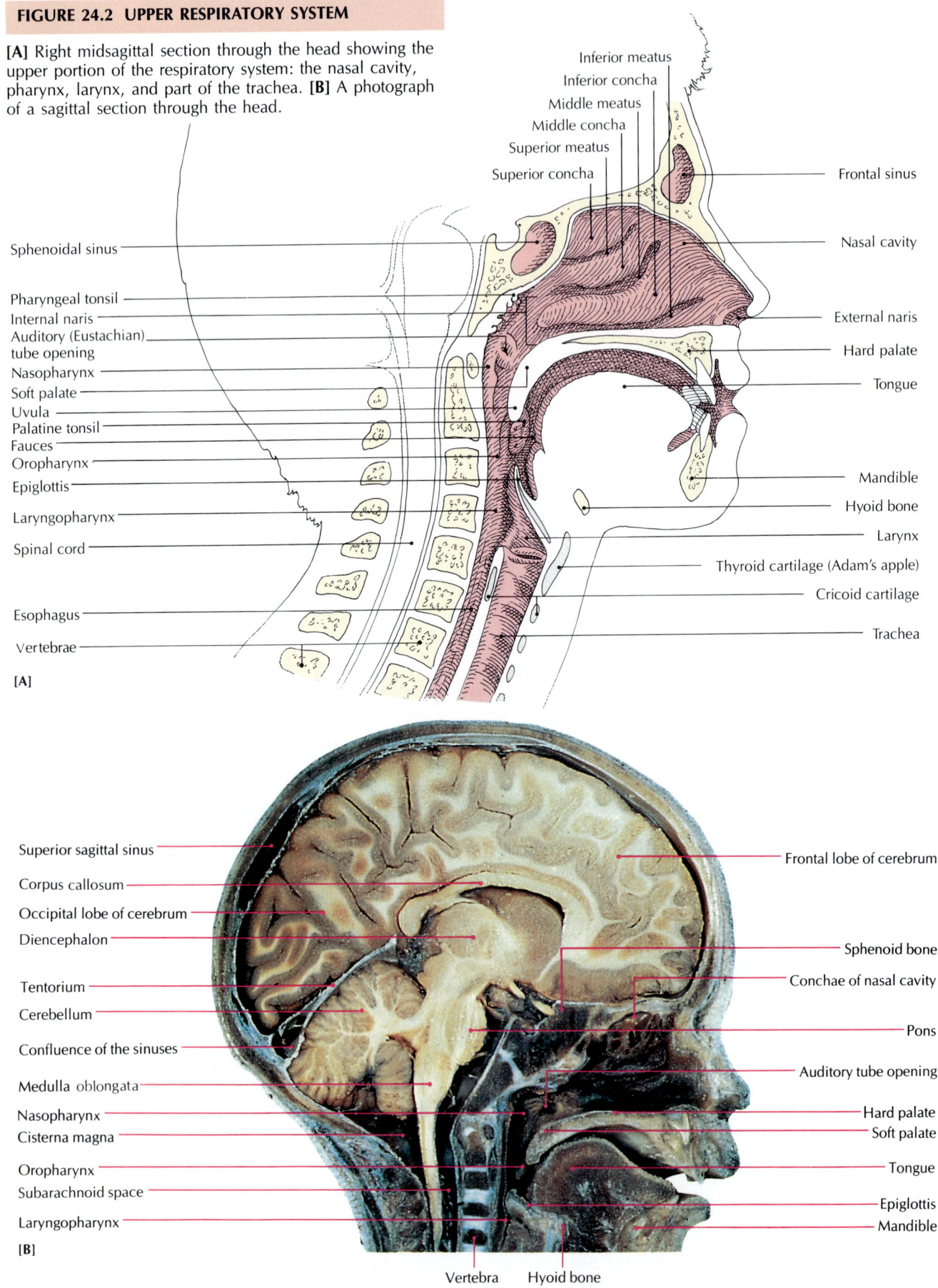

Inferior meatus
Inferior concha
Middle meatus
Middle concha
Superior meatus
Superior concha

Frontal sinus

Nasal cavity

Sphenoidal sinus

Pharyngeal tonsil
Internal naris
Auditory (Eustachian) tube opening
Nasopharynx
Soft palate
Uvula
Palatine tonsil
Fauces
Oropharynx
Epiglottis
Laryngopharynx
Spinal cord
Esophagus
Vertebrae

External naris
Hard palate
Tongue

Mandible
Hyoid bone
Larynx
Thyroid cartilage (Adam's apple)
Cricoid cartilage
Trachea

[A]

Superior sagittal sinus
Corpus callosum
Occipital lobe of cerebrum
Diencephalon
Tentorium
Cerebellum
Confluence of the sinuses
Medulla oblongata
Nasopharynx
Cisterna magna
Oropharynx
Subarachnoid space
Laryngopharynx

Frontal lobe of cerebrum

Sphenoid bone
Conchae of nasal cavity
Pons
Auditory tube opening
Hard palate
Soft palate
Tongue
Epiglottis
Mandible

[B]

Vertebra   Hyoid bone

FIGURE 24.3 NASAL EPITHELIUM

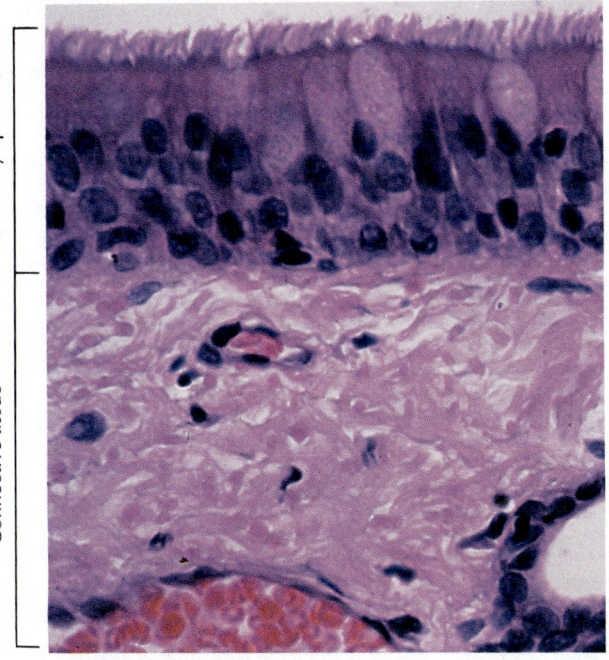

Olfactory epithelium

Connective tissue

[A]

[A] Light micrograph of pseudostratified columnar epithelium in the nasal cavity. ×160. [B] Scanning electron micrograph of the ciliated epithelium that lines the nasal cavity. ×3500.

Cilia

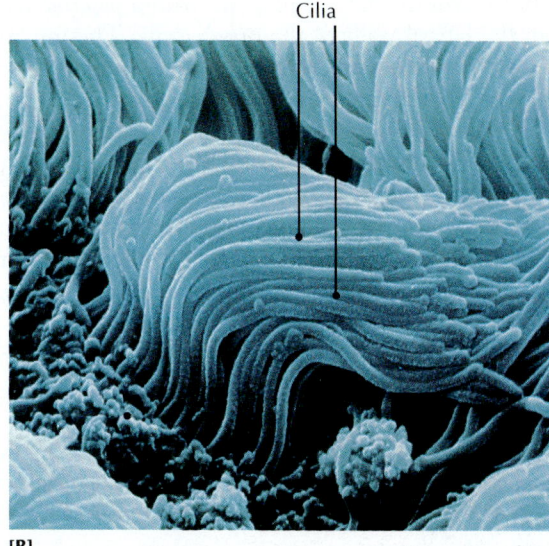

[B]

substances dissolved in the moist coating of the epithelial surface stimulate the nerve endings of the olfactory neurons and cause the sensation of smell.

Pharynx

The *pharynx* (FAR-inks), or *throat*, is a funnel-shaped tube that leads to the respiratory passage and esophagus. Thus the pharynx is a passageway for both air and food, as well as a resonating chamber for sound. The pharynx is connected to the nasal cavity through the internal openings of the nares and is also connected to the mouth, or *oral cavity*. The pharynx is divided anatomically into three portions, superior to inferior: (1) nasopharynx, (2) oropharynx, and (3) laryngopharynx [FIGURE 24.2A]. All three parts of the pharynx are continuous with each other.

Nasopharynx The *nasopharynx* is the upper portion of the pharynx superior to the soft palate into which the two internal nares, as well as the two auditory tubes, open. The two auditory tubes open into the nasopharynx just posterior and lateral to the internal nares. These tubes are air passages that equalize the air pressure on both sides of the eardrums. If these tubes are not open, changes in atmospheric pressure, such as those felt when one goes quickly up or down in an airplane, can cause the ears to "pop." This temporary imbalance can usually be

alleviated by chewing gum or swallowing. Like the nasal cavity, the nasopharynx is lined with ciliated epithelium that aids in cleaning inspired air.

Oropharynx The *oropharynx,* the middle portion of the pharynx, located at the posterior part of the mouth, extends from the soft palate to the epiglottis, where the respiratory and digestive tracts separate into the trachea and esophagus, respectively. Thus this part of the pharynx is a common pathway for both air and food and has both respiratory and digestive functions. The oropharynx is separated from the mouth by a pair of membranous narrow passageways called *fauces* (FAW-seez; sing. *faux*) that spread open during swallowing or panting, when most of the air entering the respiratory tract is drawn in through the mouth rather than the nostrils. Like most tubes that are subject to abrasion, the oropharynx is lined with stratified squamous epithelium, which is constantly being shed and renewed.

Laryngopharynx The lowermost portion of the pharynx is the *laryngopharynx* (luh-RING-go-far-inks). It extends from the level of the hyoid bone posteriorly to the esophagus (food tube) and anteriorly to the larynx (voice box). It is at that point that the respiratory and digestive systems become distinct, with air moving anteriorly into the larynx and food moving posteriorly into the esophagus past the flexible glottis (the opening into the larynx). Like the oropharynx, the laryngopharynx has both respiratory and digestive functions and is also lined with stratified squamous epithelium.

Tonsils *Tonsils* are structures that are part of both the pharynx and lymphatic system. Paired *lingual tonsils* are at the base of the tongue, and paired *palatine tonsils* are in the lateral walls of the oropharynx. The unpaired *pharyngeal tonsil* (adenoid) hangs from the roof of the nasopharynx. These tonsils form a ring of lymphoid tissue (called Waldeyer's ring) around the pharynx. The adenoids hypertrophy and regress by 8 years of age, and the palatine tonsils hypertrophy and regress by puberty. If the adenoid becomes inflamed and swollen, it can interfere with breathing through the nose, may block the opening of the auditory tube, and may even make speech difficult.

Larynx

The *larynx* (voice box; LAR-inks) is a short passageway from the pharynx to the rest of the respiratory tract [FIGURE 24.4] and also produces most of the sound used in speaking and singing.

Structure The larynx is supported by nine cartilages: three unpaired (thyroid, epiglottis, and cricoid) and three paired (arytenoid, corniculate, and cuneiform). The most prominent cartilage is the *thyroid cartilage* (Adam's apple), which consists of two plates that fuse to form an acute angle at the midline. Because the thyroid cartilage is larger in males than in females, it is visible in the front of the neck in men but is less conspicuous in women. The larynx is supported above the thyroid cartilage by ligaments connected to the hyoid bone. Below the thyroid cartilage is the *cricoid cartilage,* a ring of cartilage that connects the thyroid cartilage with the trachea below. The inferior aspect of the cricoid cartilage is the landmark for the insertion of an emergency airway during a *tracheotomy.*

The *epiglottis* (*epi* = above, *glotta* = tongue) is a flap of cartilage that folds down over the opening (called the *glottis*) into the larynx during swallowing and swings back up when the act of swallowing ceases. Since air must pass from the pharynx (which is *posterior* to the mouth) to the larynx (which is *anterior* to the esophagus), there is a crossover between the respiratory and digestive tracts. The epiglottis works well most of the time, mainly because people involuntarily inhale before swallowing, getting air into the lungs, and then exhale after swallowing, thus clearing the air passage. If food does accidentally get into the larynx, it is usually forced out by a strong cough reflex, although it may have to be dislodged by using the Heimlich maneuver (see "The Heimlich Maneuver" on page 768).

Humans produce sound mainly by vibrations of the *vocal folds* (formerly called vocal *cords*) [FIGURE 24.4A–C]. These paired strips of stratified squamous epithelium at the base of the larynx have a front-to-back slit between them, the *glottis.* Above and beside the vocal folds is a pair of *vestibular folds,* usually called *false vocal cords,* that protrude into the vestibule of the larynx, hence the term *vestibular* folds. The true vocal folds ("cords") are held in place and regulated by a pair of **arytenoid** (ar-uh-TEE-noid; Gr. "ladle-shaped") *cartilages,* which in turn are held by a pair of **cuneiform** (KYOO-nee-uh-form; L. "wedge") *cartilages.* Still another pair of cartilages, the **corniculate** (kor-NICK-yoo-lit; L. "little horn") *cartilages,* lies between the arytenoids and the epiglottis.

Sound production Sound production results from a complex coordination of laryngeal skeletal muscles. When these muscles contract, they stretch the vocal folds into the air passageway, so the glottis is narrowed. As air is "exhaled" against the vocal folds, they vibrate and generate sound waves in the air in the pharynx, nose, and mouth. Pitch is regulated mainly by the tension put on the folds, with greater tension producing higher pitch. The actual size of the folds also contributes to pitch. The longer, thicker, less taut folds of most men give lower pitches than the shorter, thinner, more taut folds of most women. The fundamental tone from the folds is only a part of the final quality of the sound, as overtones are added by changes in the positions of lips, tongue, and soft palate.

The variable shapes of the nasal cavity and sinuses give voices their individual qualities. In fact, each person has such a distinct set of vocal overtones that every human voice is as unique as a set of fingerprints. The *intensity,* volume, or "loudness" of vocal sounds is regulated by the amount of air passing over the vocal folds, and that in turn is regulated by the pressure applied to the lungs, mainly by the thoracic and abdominal muscles.

Trachea

The *trachea* (TRAY-kee-uh), or windpipe, is an open tube anterior to the esophagus, extending from the base of the larynx to the top of the lungs, where it divides into two branches, the right and left *bronchi* (BRONG-kee; sing.

Q: *Why are nosebleeds so common?*

A: Bleeding from the nose is common because the nose sticks out far enough to be struck easily, the blood supply is abundant in the nasal membranes, and the mucosal layer is delicate. Bleeding can usually be controlled by packing the external or internal nares or both. Nosebleeds are more common in cold weather because the lower humidity of the inspired air promotes drying, cracking, and bleeding of the nasal epithelium.

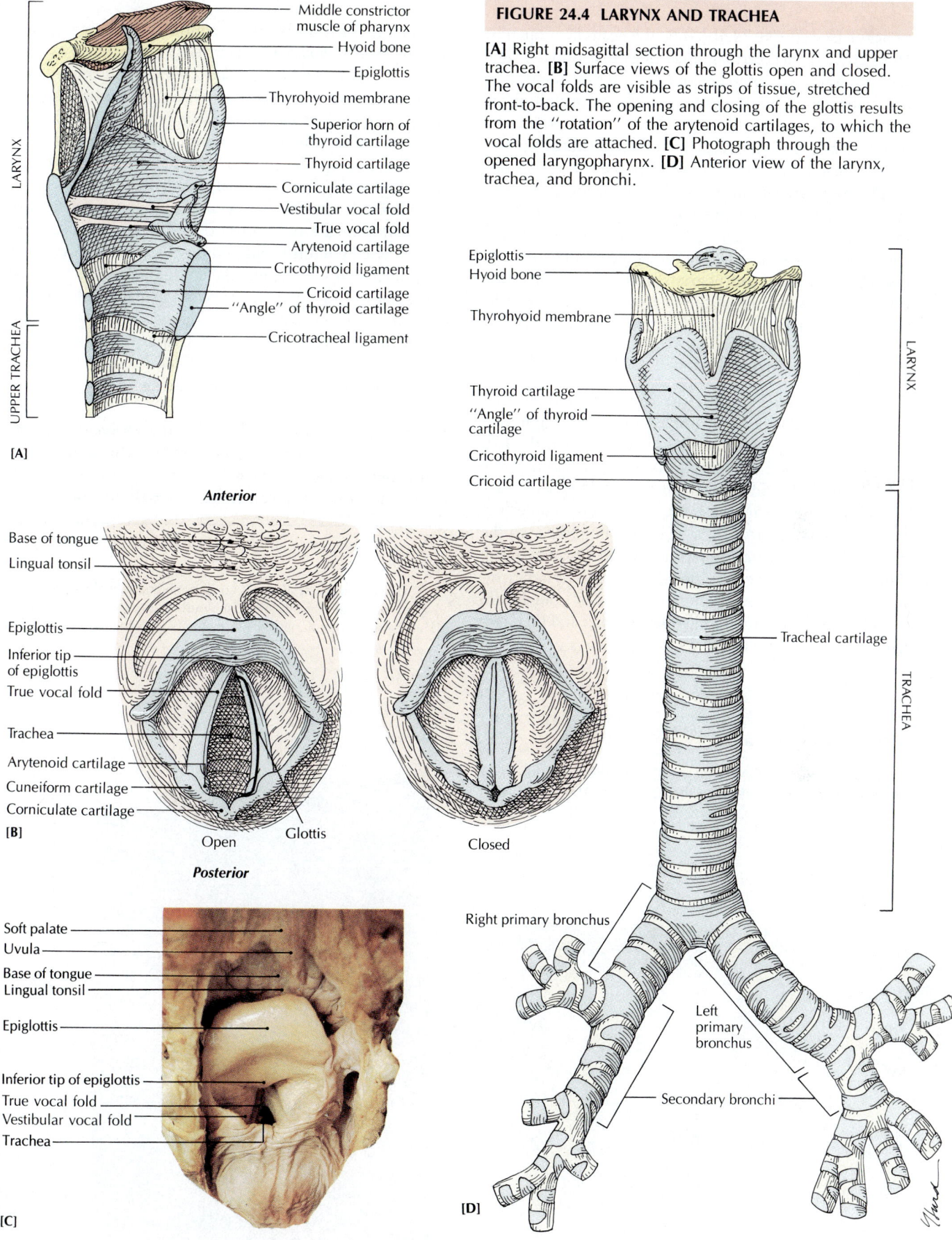

LARYNX

UPPER TRACHEA

Middle constrictor muscle of pharynx
Hyoid bone
Epiglottis
Thyrohyoid membrane
Superior horn of thyroid cartilage
Thyroid cartilage
Corniculate cartilage
Vestibular vocal fold
True vocal fold
Arytenoid cartilage
Cricothyroid ligament
Cricoid cartilage
"Angle" of thyroid cartilage
Cricotracheal ligament

[A]

FIGURE 24.4 LARYNX AND TRACHEA

[A] Right midsagittal section through the larynx and upper trachea. [B] Surface views of the glottis open and closed. The vocal folds are visible as strips of tissue, stretched front-to-back. The opening and closing of the glottis results from the "rotation" of the arytenoid cartilages, to which the vocal folds are attached. [C] Photograph through the opened laryngopharynx. [D] Anterior view of the larynx, trachea, and bronchi.

Epiglottis
Hyoid bone
Thyrohyoid membrane
Thyroid cartilage
"Angle" of thyroid cartilage
Cricothyroid ligament
Cricoid cartilage

LARYNX

Anterior

Base of tongue
Lingual tonsil
Epiglottis
Inferior tip of epiglottis
True vocal fold
Trachea
Arytenoid cartilage
Cuneiform cartilage
Corniculate cartilage

[B]

Open   Glottis   Closed

Posterior

Soft palate
Uvula
Base of tongue
Lingual tonsil
Epiglottis
Inferior tip of epiglottis
True vocal fold
Vestibular vocal fold
Trachea

[C]

Tracheal cartilage

TRACHEA

Right primary bronchus
Left primary bronchus
Secondary bronchi

[D]

bronchus) [see FIGURE 24.4D]. The trachea is kept open by 16 to 20 C-shaped cartilaginous rings that provide rigid support, but are incomplete on the posterior side next to the esophagus. This allows for a slight expansion of the esophagus during swallowing. The rings are connected by fibroelastic connective tissue and longitudinal smooth muscle, making the trachea both flexible and extensible.

The inside surface of the trachea is lined with pseudostratified ciliated columnar epithelium, which produces moist mucus [FIGURE 24.5]. It contains upward-beating cilia. Dust particles and microorganisms that are not caught in the nose and pharynx may be trapped in the trachea and carried up to the pharynx by the cilia to be swallowed or spit out (*expectorated*).

Bronchi and Bronchioles

The trachea branches into the right and left *primary bronchi* that enter the right and left lungs. The right primary bronchus is larger and more vertical than the left, so aspirated objects can become lodged more easily in the right primary bronchus than in the left. Each primary bronchus, on entering the lung, divides into smaller *secondary bronchi*, one to each lobe of the lung (three to the right lung and two to the left lung) [see FIGURE 24.4D]. The secondary bronchi, in turn, divide into *tertiary (segmental) bronchi*. These bronchi continue to branch into smaller and smaller tubes called **bronchioles** and then **terminal bronchioles** [FIGURE 24.6A, B].

The terminal bronchioles continue to branch, ending in *respiratory bronchioles* [FIGURE 24.6B]. They in turn branch into many *alveolar ducts*, which lead into microscopic air sacs called *alveoli*, where gas exchange takes place. (Alveoli are discussed more fully in the following section.) The whole system inside the lungs looks so much like an upside-down tree that it is commonly called the *"respiratory tree"* or *"bronchial tree."*

There are several structural changes in the successive branching of the respiratory tree: (1) The pseudostratified ciliated columnar epithelium in the bronchi changes to nonciliated simple cuboidal epithelium in the terminal bronchioles. (2) The incomplete cartilaginous rings in the primary bronchi become plaques in the secondary bronchi [see FIGURE 24.4D] and disappear in the terminal branches. (3) As the cartilage in the bronchioles disappears, the amount of smooth muscle in the walls increases.

In summary, it can be said that the *air-conducting portion* of the lung is composed of the bronchi and bronchioles. Beyond the terminal bronchioles (the most distal branches of the respiratory tree), the thinner-walled re-

---

Q: *Does bad breath always originate with bacteria in the mouth?*

A: No. Foods such as garlic contain oils that are absorbed from the small intestine into the bloodstream. The odiferous oils are carried to the lungs, where they mix with the exhaled air.

FIGURE 24.5 TRACHEA

[A] Cross section of the trachea. [B] Light micrograph of pseudostratified epithelium in the trachea. ×100.

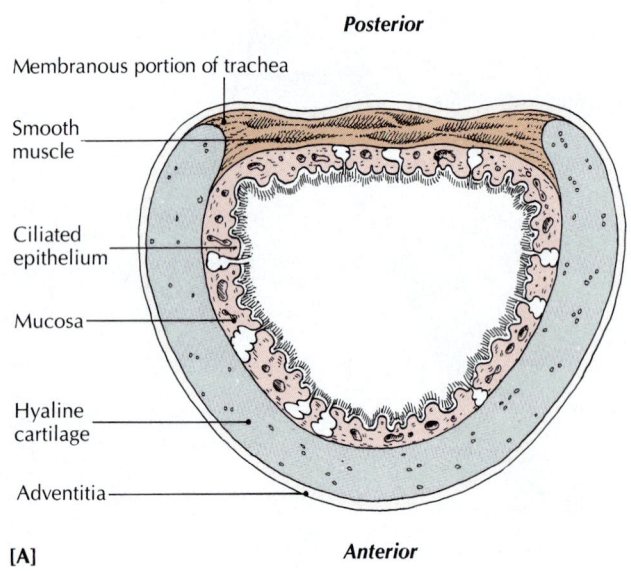

Posterior

Membranous portion of trachea

Smooth muscle

Ciliated epithelium

Mucosa

Hyaline cartilage

Adventitia

[A]   **Anterior**

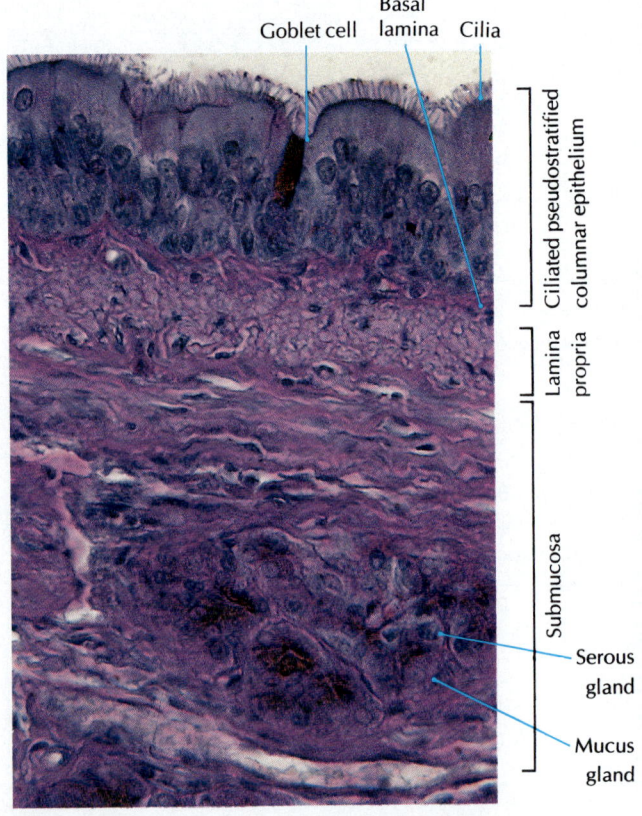

Goblet cell — Basal lamina — Cilia

Ciliated pseudostratified columnar epithelium

Lamina propria

Submucosa

Serous gland

Mucus gland

[B]

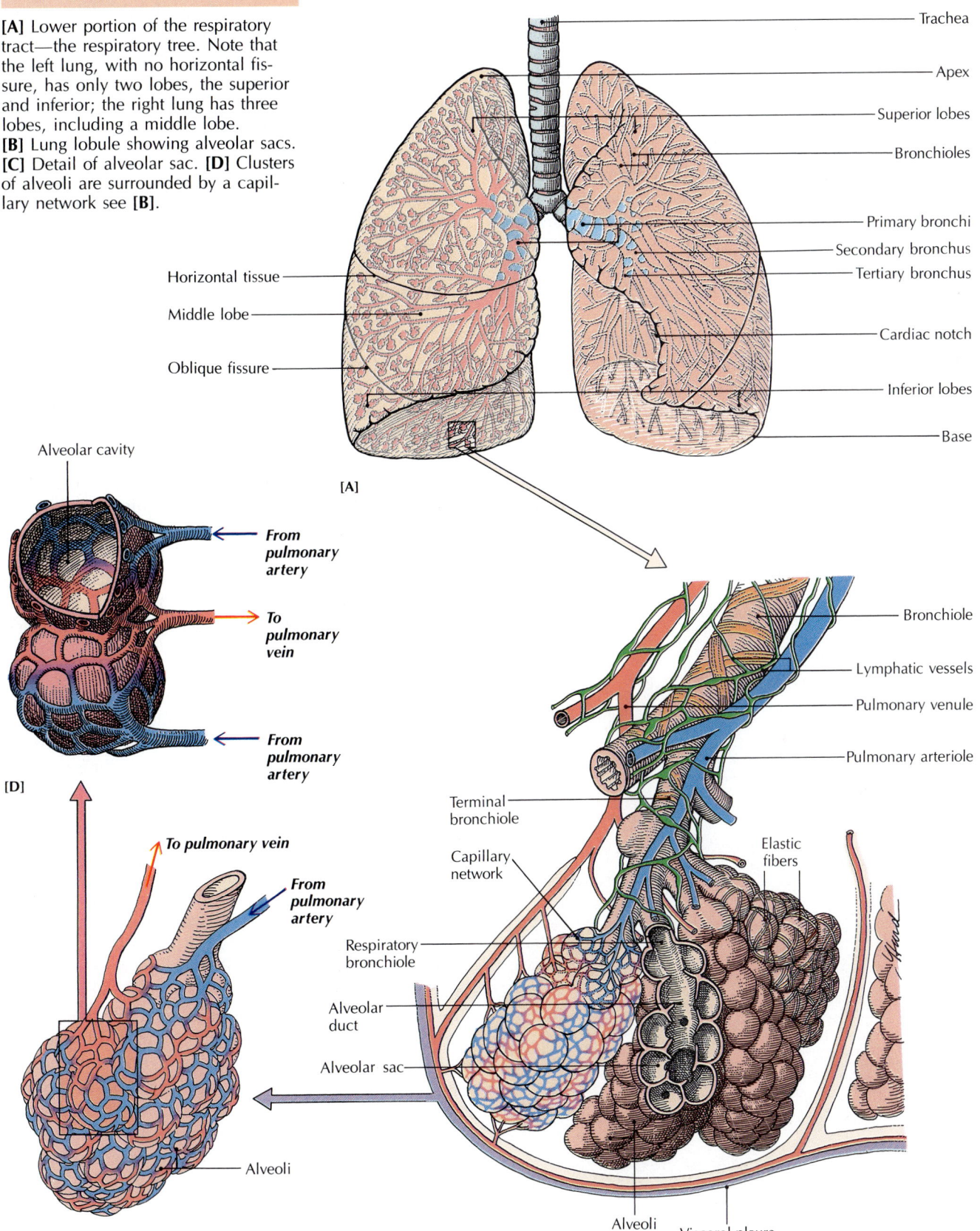

FIGURE 24.6 LUNGS

[A] Lower portion of the respiratory tract—the respiratory tree. Note that the left lung, with no horizontal fissure, has only two lobes, the superior and inferior; the right lung has three lobes, including a middle lobe.
[B] Lung lobule showing alveolar sacs.
[C] Detail of alveolar sac. [D] Clusters of alveoli are surrounded by a capillary network see [B].

Horizontal tissue
Middle lobe
Oblique fissure

Trachea
Apex
Superior lobes
Bronchioles
Primary bronchi
Secondary bronchus
Tertiary bronchus
Cardiac notch
Inferior lobes
Base

[A]

Alveolar cavity

From pulmonary artery
To pulmonary vein
From pulmonary artery

[D]

To pulmonary vein
From pulmonary artery

Terminal bronchiole
Capillary network
Respiratory bronchiole
Alveolar duct
Alveolar sac

Bronchiole
Lymphatic vessels
Pulmonary venule
Pulmonary arteriole

Elastic fibers

Alveoli

Alveoli
Visceral pleura

[C]

[B]

spiratory bronchioles begin the *respiratory portion* of the lung, where the exchange of respiratory gases occurs. The respiratory portion consists of the respiratory bronchioles, alveolar ducts, alveolar sacs, and alveoli.

Alveoli The functional units of the lungs are grapelike clusters called *alveoli* (al-VEE-oh-lye; sing., *alveolus*). A group of two or more alveoli with a common opening into an alveolar duct is called an *alveolar sac.* Each lung contains over 350 million alveoli, each surrounded by many capillaries [FIGURE 24.6B–D], which provide enough surface area to allow for adequate gas exchange. The total surface area of an adult lung is about 60 to 70 m^2, 20 times greater than the surface area of the skin [FIGURE 24.7]. Every time you inhale, you expose to fresh air an area of lung roughly equal to that of a tennis court.

A single alveolus looks like a tiny bubble lined with epithelium; it is supported by a thin, elastic *basement membrane.* The alveolar walls consist of two types of epithelial cells: (1) *Type I alveolar cells* consist mainly of a single layer of squamous cells (squamous pulmonary epithelial cells). (2) There are also *septal cells*, or *type II alve-*

FIGURE 24.7 LUNG TISSUE

Light micrograph of lung tissue. ×40.

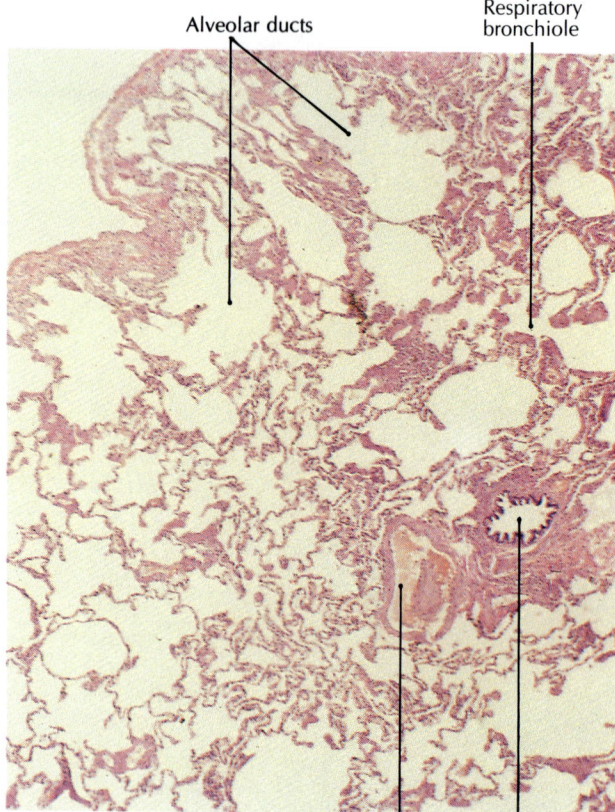

Alveolar ducts

Respiratory bronchiole

Blood vessel   Terminal bronchiole

olar cells, which are smaller, scattered, cuboidal secretory cells [FIGURE 24.8]. Type II cells secrete a detergentlike phospholipid called *lung surfactant,* which helps keep alveoli inflated by reducing surface tension. Alveoli also contain phagocytic *alveolar macrophages* that adhere to the alveolar wall or circulate freely in the lumen of alveoli. These macrophages ingest and destroy microorganisms and other foreign substances that enter the alveoli. The foreign material is moved upward by ciliary action to be expelled by coughing, or it enters lymphatic vessels to be carried to the lymph nodes at the hilum of the lung.

Alveolar-capillary membrane Since the alveoli are the sites of gas exchange with the blood by diffusion, the membranes of both alveolar walls and the capillary walls that line them must be thin enough to allow maximum gas exchange. The layers through which the gases diffuse are known as the *alveolar-capillary (respiratory) membrane.* Its layers include (1) type I and type II alveolar cells, (2) the alveolar epithelial basement membrane, (3) the capillary basement membrane, and (4) the capillary endothelial cells. The alveolar-capillary membrane, despite its several layers, is only about 0.5 μm thick, which is ideal for gas exchange.

Alveoli themselves are about 25 μm in diameter, and their walls are about 0.5 μm thick, which is much thinner than a sheet of paper. The capillaries surrounding each alveolus are also thin-walled for gas exchange. The exchange is facilitated by the small size and the large number of capillaries. Capillaries are so small that red blood cells flow through them in single file, giving each cell maximum exposure to the alveolar walls. There are so many capillaries that at any instant almost a liter of blood is passing through the lungs.

Lungs

The *lungs* fill the thoracic cavity except for a midventral region, called the *mediastinum,* where the heart and major blood vessels lie. Each lung is cone-shaped at the *apex* (the top) and concave at the *base,* where it lies against the diaphragm [FIGURE 24.6A]. Because the heart occupies more space on the left side of the mediastinum than on the right, and because the liver is slightly higher on the right side, the two lungs are not symmetrical. The left one is narrower and longer than the right. On the medial surface of each lung is a region called the *hilus,* through which bronchiole tubes, pulmonary blood vessels, lymphatic vessels, and nerves enter and exit. The hilus, together with the pleura and connective tissue, is called the *root* of the lung. The left lung has a concavity, the *cardiac notch,* into which the heart fits.

Lobes The right lung has three main lobes, the *superior, middle,* and *inferior* [FIGURE 24.9]. Each lobe is fur-

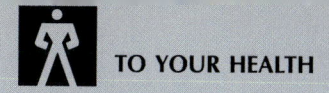

TO YOUR HEALTH

"Smoker's Cough"

The human body has several defenses against polluted air. Hairs in the nose filter out large particles. The mucus that is secreted continuously by the goblet cells in the upper respiratory tract washes out or dissolves smaller irritants and traps particles that enter the respiratory tract. Also, the many cilia in the upper tract sweep away foreign particles and excessive mucus. But air pollutants (especially cigarette smoke, ozone, sulfur dioxide, and nitrogen dioxide) can stiffen, slow, or even destroy the cilia, causing the mucus to clog, a reaction frequently resulting in "smoker's cough." When cilia in the nose are impaired by pollutants or are partially paralyzed by cold air, there is an overproduction of mucus, which is not swept back into the throat as usual. When this happens, the mucus drips for-

ward and causes a "runny nose."

As a result of inefficient cilia, microorganisms and other foreign particles can penetrate the lung tissue and cause respiratory infections and possibly even lung cancer. As lung tissue becomes more irritated by air pollutants, mucus flows more freely to help remove the inhaled pollutants. A normal coughing mechanism expels the contaminated air and mucus. Smoking or air pollution can trigger excessive mucus flow that eventually blocks the air passages, which in turn causes more coughing—a type of positive feedback. Over time, the muscles surrounding the respiratory tree in the lungs can weaken from prolonged, strong coughing so that breathing becomes progressively more difficult. If this positive feedback cycle continues,

chronic bronchitis (a chronic inflammation of the mucous membranes of the respiratory tree) can occur.

According to the Centers for Disease Control, smoking-related deaths are increasing, showing the effects of heavy smoking for 20 or 30 years before. As a result of smoking among women, heart disease is now the leading cause of death among women, replacing breast cancer. Two especially disturbing statistics indicate that about 4000 nonsmokers died in 1991 as a result of inhaling other people's smoke and that almost 3000 babies died as a result of their mothers' smoking habits. (A 1993 study revealed that a cigar has 30 times the carbon monoxide emissions of one cigarette.)

FIGURE 24.8 CELLS OF LUNG TISSUE

This schematic drawing of a section of lung tissue shows an alveolus and surrounding capillaries, as well as various types of cells.

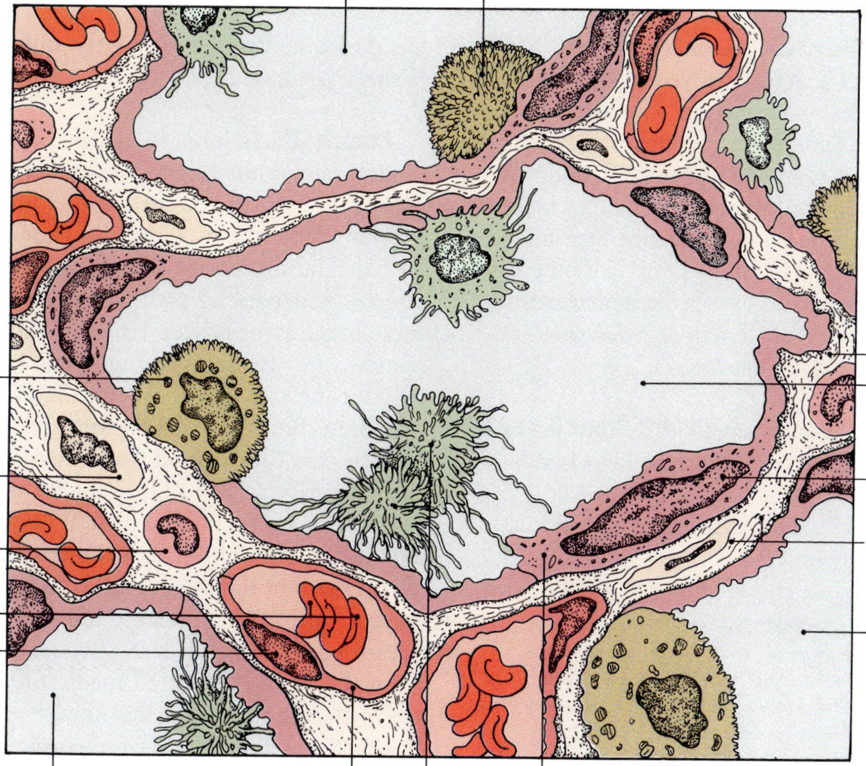

Alveolus · Type II alveolar (septal) cell · Elastic fibers · Alveolar lumen · Type II alveolar (septal) cell · Nucleus of type I alveolar cell · Connective tissue (interstitial) cell · Connective tissue fibers · White blood cell (monocyte) · Red blood cells · Alveolar lumen · Nucleus of capillary endothelium · Alveolus · Capillary endothelium · Type I alveolar (squamous epithelial) cell · Macrophages

TO YOUR HEALTH

The Heimlich Maneuver

Choking on food is such a common accident, and so often causes death from asphyxiation, that many restaurants post notices on how to perform the **Heimlich maneuver.** If a person gets a particle stuck in the glottis or larynx and cannot either exhale or inhale, a second person can help dislodge the particle by using the Heimlich maneuver. The aim is to compress the remaining air in the lungs violently enough to blow the particle free.

If the victim can stand, the rescuer stands behind the victim, places a fist against the abdomen between the ribs and navel, covers the fist with the other hand, and gives a strong, sudden squeeze inward and upward (see drawings). If the victim is unable to stand, the rescuer kneels astride the victim and applies pressure with the heel of one hand between the ribs and navel, pressing down on that lower hand with the other.

Some danger accompanies the Heimlich maneuver because too forceful or badly placed pressure can injure ribs, the diaphragm, or the liver.

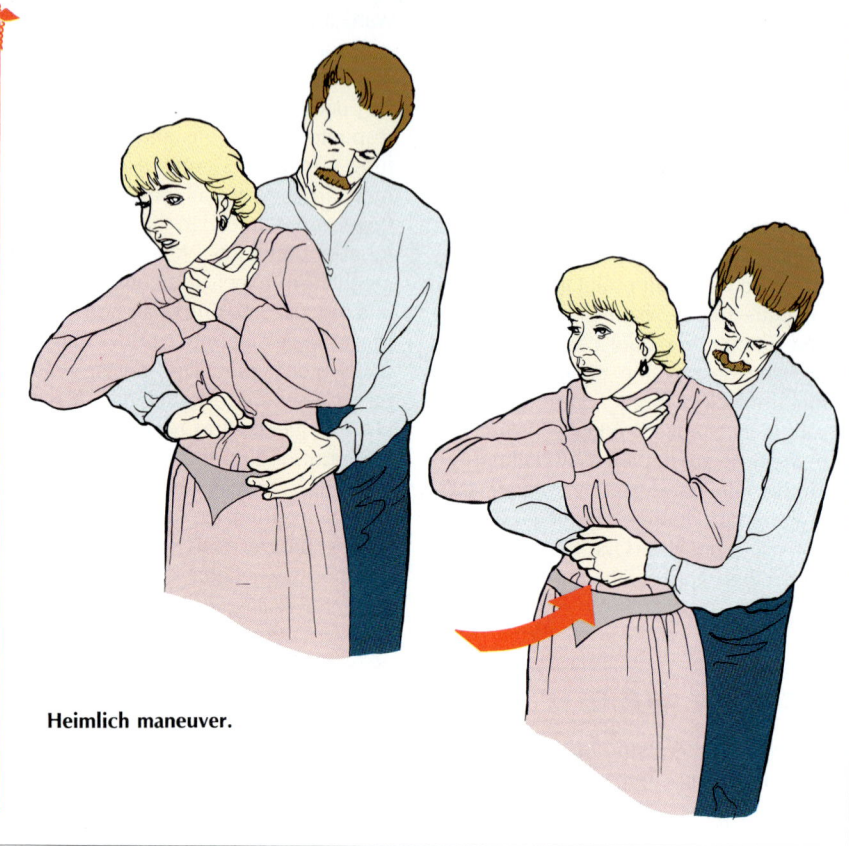

Heimlich maneuver.

ther subdivided into 10 *bronchopulmonary segments* [FIGURE 24.9 C and D], which are served by individual bronchi and bronchioles and function somewhat independently. This partial independence makes it possible to remove one diseased segment without affecting the rest of the lung. Bronchopulmonary segments contain smaller subdivisions, called *lobules*, which are surrounded by elastic connective tissue [FIGURE 24.6B]. A lobule has hundreds of alveoli. It is served by a bronchiole carrying gases, a lymphatic vessel, a small vein (a *pulmonary venule*), and an artery (a *pulmonary arteriole*).

Fissures The left lung is subdivided into its two lobes by the *oblique fissure,* and the right lung is subdi-

vided into its three lobes by the oblique and **horizontal fissures** [FIGURE 24.9C, D].

Pleura Each lung is associated with a continuous pleural membrane that consists of the inner *visceral pleura* directly on the surface of a lung and the *parietal pleura* lining the wall of the thoracic cavity [FIGURE 24.10]. The small fluid-filled potential space between the visceral and parietal pleura is the *pleural cavity*. With each breath, the visceral pleura slides on the parietal pleura, with the necessary lubrication being provided by the thin film of fluid between the two layers.

Three functions are associated with the pleurae: (1) The thin film of fluid from the membranes within the pleural cavity acts as a lubricant for the lungs, which are in constant motion. (The pleural cavity does not contain a large amount of fluid, as is sometimes thought.) (2) The air pressure in the pleural cavity is lower than the atmospheric pressure and thus aids in the mechanics of breathing. (3) The pleurae effectively separate the lungs from the medially located mediastinum, which contains the other thoracic organs. These organs include the heart, the esophagus, the thoracic duct, nerves, and major blood vessels.

Q: *Why does yelling make us hoarse?*

A: Yelling vibrates the vocal folds and slams them together with such strong force that the folds may swell or even bleed. Swollen folds do not close properly, and hoarseness results as air leaks between them. Interestingly, whispering when you have laryngitis may require even more effort than yelling, since the swollen vocal folds have to be held very close together without touching—a difficult muscular maneuver.

FIGURE 24.9 EXTERNAL STRUCTURE OF THE LUNGS

[A] Right lung, anterior view. [B] Left lung, anterior view. [C] Bronchopulmonary segments of the right lung and [D] the left lung. The medial basal segments are hidden in [C] and [D].

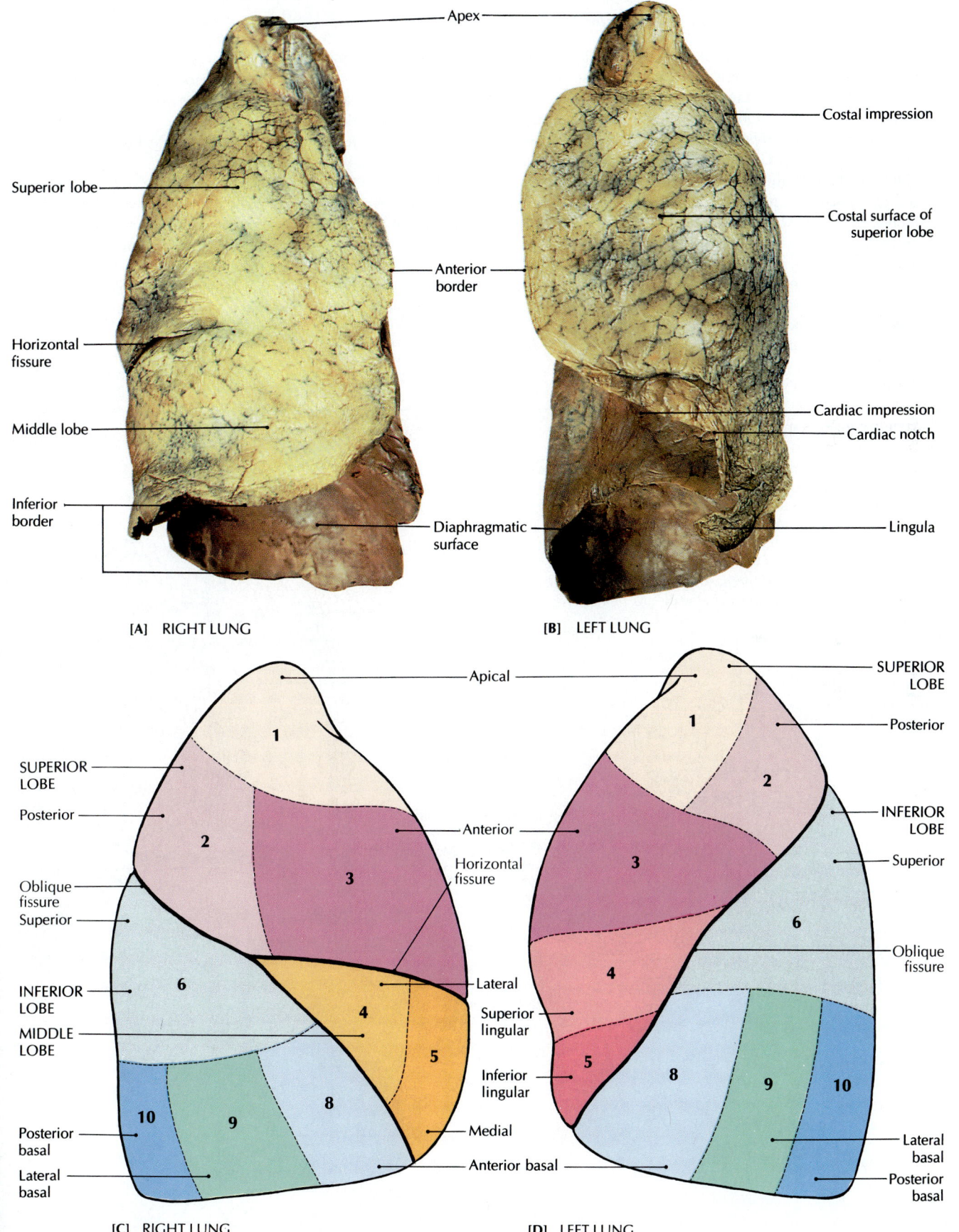

Apex

Costal impression

Superior lobe

Costal surface of superior lobe

Anterior border

Horizontal fissure

Cardiac impression
Cardiac notch

Middle lobe

Inferior border

Diaphragmatic surface

Lingula

[A]   RIGHT LUNG

[B]   LEFT LUNG

Apical

SUPERIOR LOBE

Posterior

SUPERIOR LOBE

Posterior

Anterior

Horizontal fissure

INFERIOR LOBE

Superior

Oblique fissure
Superior

Lateral

Oblique fissure

INFERIOR LOBE

Superior lingular

MIDDLE LOBE

Inferior lingular

Posterior basal

Medial

Lateral basal

Lateral basal

Anterior basal

Posterior basal

[C]   RIGHT LUNG

[D]   LEFT LUNG

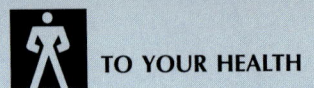

Positive Effects of Physical Activity on the Respiratory System

Anyone who has ever undertaken a sustained program of physical activity knows that some of the most obvious effects are on the respiratory system. Some of the health benefits of exercise are easier and more efficient (and more pleasurable) breathing; increased maximum oxygen uptake; increased vital capacity; increased elasticity of lungs, rib cage, and connective tissue related to breathing; strengthening of breathing muscles; reduction of the negative effects of asthma, emphysema, and other lung disorders; greatest possible oxygenation of brain and body tissues; and a general sense of well-being and good health.

So-called aerobic exercise is any physical activity in which the large skeletal muscles are used for a long enough time without interruption so that you sweat freely and breathe harder than usual. The word *aerobic* means "with oxygen" and relates to physical activities that increase the body's efficiency in distributing oxygen to tissues—more blood, and therefore more oxygen, is carried to the skeletal muscles. Short bursts of energy, such as in sprinting, do not allow the body time to supply the working muscles with the oxygen they have just used up. Such high-intensity bursts of energy are called *anaerobic*, or "without oxygen."

Most physiologists and sports therapists agree that aerobic activity provides the most benefits, but the experts also agree that *any* physical activity is better than none and that the most practical physical activity is probably walking at a brisk pace.

FIGURE 24.10 PLEURAE

[A] Transverse section of the thorax as seen from above, showing the relationship of the lungs and pleurae. [B] Coronal section of the thorax, showing the pleurae and the separation of right and left pleural membranes by the mediastinum. Portions of the parietal pleura are named according to the structures to which the portions fuse: the mediastinal pleura next to the mediastinum, the costal pleura next to the ribs, and the diaphragmatic pleura next to the diaphragm.

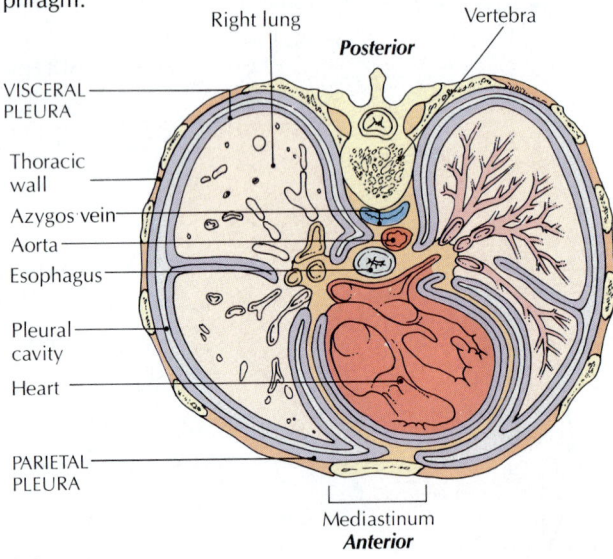

[A]

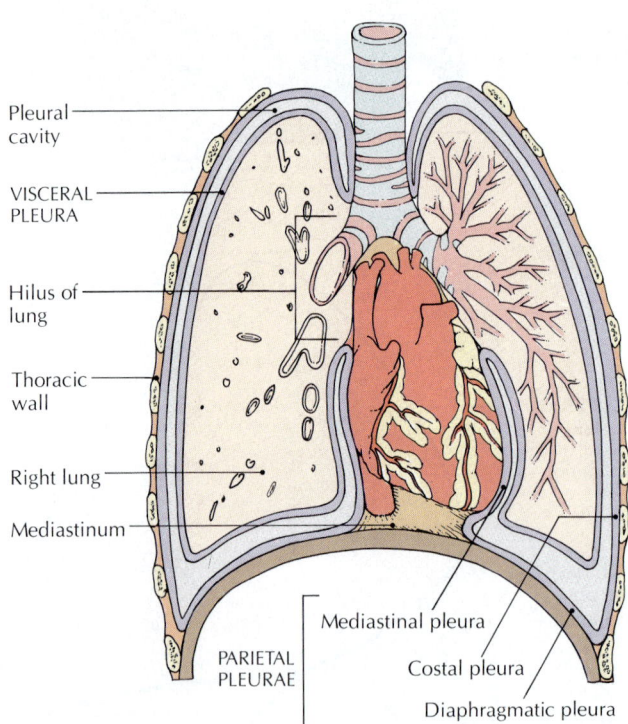

[B]

Innervation The smooth muscle of the tracheo-bronchial tree is innervated by the autonomic nervous system. Branches of the vagus nerve run alongside the pulmonary blood vessels, carrying parasympathetic fibers that constrict the bronchial smooth muscle. Branches from the thoracic sympathetic ganglia carry fibers that dilate the bronchial smooth muscle. Bronchoconstriction may also be caused by airborne agents, acetylcholine, histamine, and some bacterial toxins that act directly on smooth muscle of the bronchioles. In contrast, epinephrine and norepinephrine cause bronchodilation by relaxing the smooth muscle.

Blood supply The respiratory tree is supplied with a dual blood supply by the pulmonary and bronchial circulations. The pulmonary circuit begins in the right ventricle and ends in the left atrium. This circuit includes the pulmonary trunk and arteries, the capillaries of the lungs, and the pulmonary veins. The bronchial circulation, carrying *oxygen-rich* blood, consists of the blood supply to tissues of the lung's air-conducting passageways (terminal bronchioles, supporting tissues, and the outer walls of the pulmonary arteries and veins). The bronchial blood supply comes from the bronchial arteries, which branch off the thoracic aorta.

A S K   Y O U R S E L F

1 What are the basic components of the respiratory system?

2 Why is the lining of the nasal cavity important in respiration?

3 What is the major function of the larynx?

4 What is the function of the epiglottis?

5 What are bronchi, bronchioles, and alveoli?

6 Why are alveoli called the functional units of the lungs?

7 What is the function of lung surfactant?

8 What is the difference between pulmonary and bronchial circulations?

PHYSIOLOGY OF RESPIRATION

The layer of air covering the surface of the earth exerts a pressure of about 760 mmHg, or approximately 1 atmosphere (atm) at sea level. Most of the air that we breathe is inert (chemically unreactive) nitrogen and has no importance in normal respiration. The atmospheric gases of interest in respiration are oxygen and carbon dioxide.

Changes in the size of the thoracic cavity, and thus of the lungs, allow us to inhale and exhale air in a process called *pulmonary ventilation*, or more commonly, *breathing*.

Boyle's Law

Robert Boyle, a seventeenth-century Irish chemist, discovered important relationships between the pressure and volume of a gas. *Boyle's law* predicts that when a gas is compressed to half its *volume*, its *pressure* doubles. The concept that decreasing the volume of a gas increases its pressure, and vice versa, is important for understanding respiratory phenomena. For example, when muscles in the rib cage and abdomen cause lung volume to expand during inspiration, pressure in the lungs falls below that of the atmosphere, and air flows from the area of higher pressure to an area of lower pressure, to enter the respiratory tree. In a similar way, air leaves the lungs when air pressure in the lungs becomes greater than that of the atmosphere by compression of the thoracic wall and abdomen.

Mechanics of Breathing

Inspiration requires continual work by the muscles that increase the volume of the thoracic cavity and expand the lungs. During *inspiration* (inhalation), air is brought into the lungs to equalize a reduction of air pressure caused by the enlarged thoracic cavity. During normal *expiration* (exhalation), these muscles relax (1) to decrease the volume of the thoracic cavity and (2) to move air out of the lungs. These two processes will be described after a brief discussion of respiratory pressures used in breathing.

Pressures involved in breathing The three major pressures in breathing (inspiration and expiration) are (1) atmospheric pressure, (2) intra-alveolar (intrapulmonary) pressure, and (3) intrapleural pressure [FIGURE 24.11].

1 *Atmospheric pressure* is the pressure of air (gases) exerted by the air around us. At sea level this pressure is equal to 760 mmHg.

2 The *intra-alveolar pressure* is the pressure within the alveoli that increases and decreases with each breath, but is always equal to atmospheric pressure at the end of inspiration and expiration.

3 The *intrapleural pressure* is the pressure within the pleural cavity; it fluctuates with each breath. However, the intrapleural pressure is always negative (about 4 mmHg) with respect to atmospheric pressure. This negative pressure is due primarily to factors that cause the visceral and parietal pleura to adhere to each other, causing the tight coupling of the lungs to the thoracic wall in this closed cavity. Loss of this negative pressure

FIGURE 24.11 PRESSURE RELATIONSHIPS

[A] The drawing shows pressure relationships of the atmosphere, lungs, and pleural cavity at the end of an inspiration. At end-inspiration, the pressures in the lungs and in the atmosphere are equal. The elasticity of the lungs maintains a negative pressure in the pleural cavities. **[B]** Changes in intra-alveolar and intrapleural pressures during inspiration and expiration.

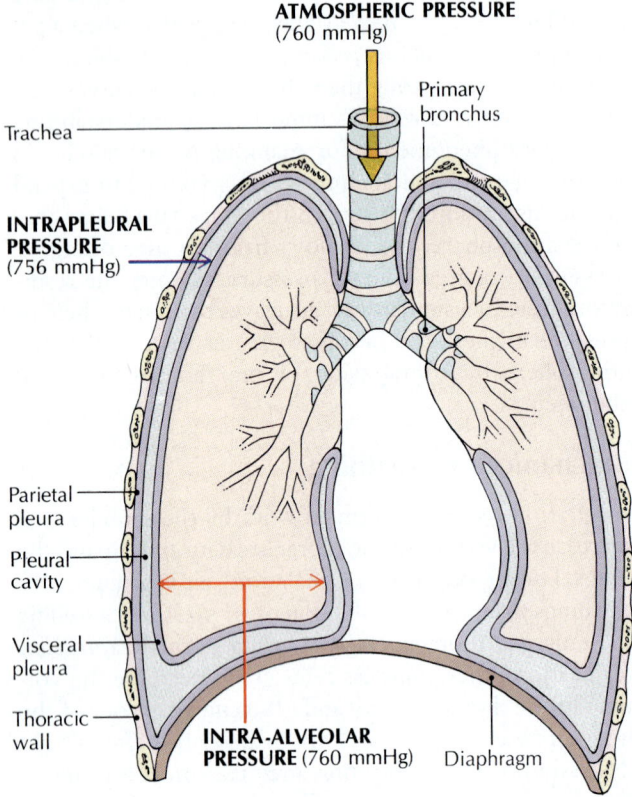

ATMOSPHERIC PRESSURE (760 mmHg)

Primary bronchus

Trachea

INTRAPLEURAL PRESSURE (756 mmHg)

Parietal pleura

Pleural cavity

Visceral pleura

Thoracic wall

INTRA-ALVEOLAR PRESSURE (760 mmHg)   Diaphragm

[A]

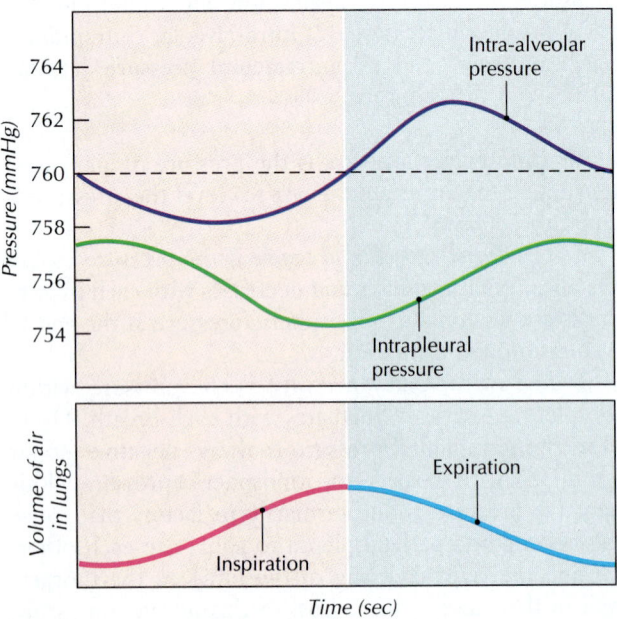

[B]

results in lung collapse or *atelectasis* (AT-uh-leck-tuh-sihss).

Refer to FIGURE 24.11B during the following discussion of inspiration and expiration to see how these pressure changes occur.

Inspiration The major muscles involved in *quiet inspiration* are the diaphragm and the external intercostals [FIGURE 24.12A]. The dome-shaped **diaphragm,** the skeletal muscle that separates the thoracic cavity from the abdominal cavity, is a major muscle of inspiration:

1 As the diaphragm contracts (moves downward), the volume of the thoracic cavity increases and the intra-alveolar pressure decreases from an atmospheric pressure of 760 mmHg to approximately 756 mmHg.

2 A pressure gradient is established and air enters into the lungs. (Remember, according to Boyle's law, as volume increases, pressure decreases.)

3 The **external intercostal muscles** also assist inspiration.* These skeletal muscles extend from rib to rib, and when they contract, (1) they pull the ribs upward and outward, enlarging the transverse dimensions of the thoracic cavity (the upward and outward movement of the ribs has been called the "bucket-handle" inspiratory movement) and (2) they move the lower end of the sternum forward, enlarging the anterior-posterior dimensions of the thoracic cavity. (The forward movement of the lower end of the sternum has been called the "pump-handle" inspiratory movement.)

4 During inspiration, the abdominal muscles are relaxed to allow for movement of the abdominal organs when the diaphragm contracts.†

Several factors, including extreme obesity and advanced pregnancy, may affect the movement of the diaphragm and thus affect inspiration. In these situations, there is a shift from "diaphragmatic" breathing to "intercostal" breathing. Posture also can have a critical effect on inspiration [FIGURE 24.13].

Expiration *Quiet expiration* is primarily a *passive* process, while inspiration is an *active* process [FIGURE 12.12B].

1 Quiet expiration depends on the elastic recoil of the lungs after inspiratory stretching, elastic recoil of the costal cartilages (lowering of rib cage), and the relaxation of the inspiratory muscles.

---

*For the role of the internal interchondral muscles (not discussed in this section) see the tabular material in FIGURE 11.13.

†Many people do not relax their abdominal muscles during inspiration, contracting them instead. Singers are trained to relax their abdominal muscles during inspiration, gradually contracting them during expiration.

FIGURE 24.12 INSPIRATION AND EXPIRATION

The thorax at full inspiration [A] and at the end of expiration [B]. Note the differences in the size of the thoracic cavity, the position of the sternum and ribs, and the shape of the diaphragm and abdominal wall.

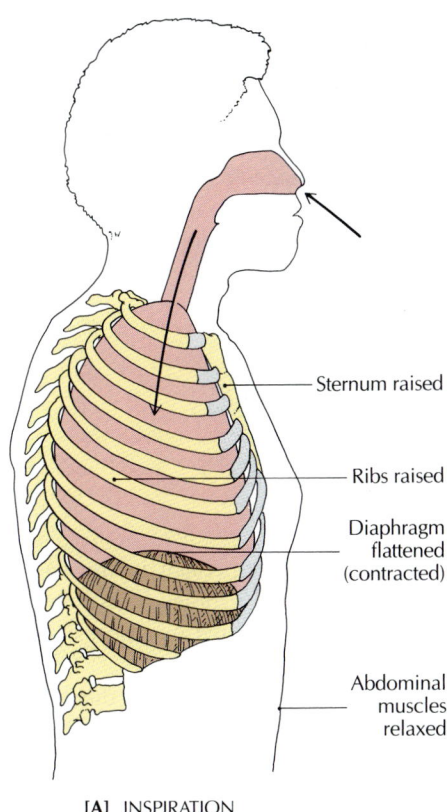

Sternum raised

Ribs raised

Diaphragm flattened (contracted)

Abdominal muscles relaxed

[A] INSPIRATION

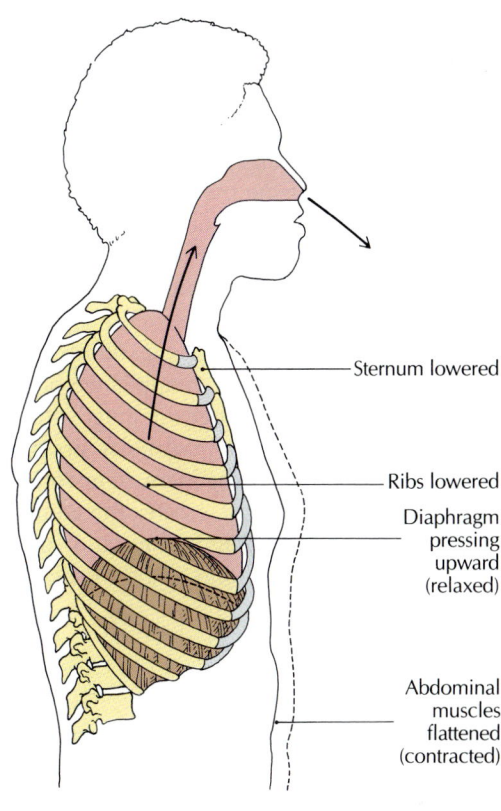

Sternum lowered

Ribs lowered

Diaphragm pressing upward (relaxed)

Abdominal muscles flattened (contracted)

[B] EXPIRATION

FIGURE 24.13 THE EFFECT OF POSTURE ON INSPIRATION

[A] When a person is lying on his back, gravity tends to force the abdominal organs against the diaphragm, making it difficult for the diaphragm to contract (lower). [B] When the person is propped up, the abdominal organs actually assist in lowering the diaphragm, thus assisting inspiration.

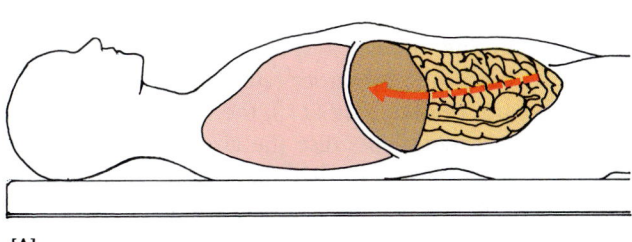

[A]

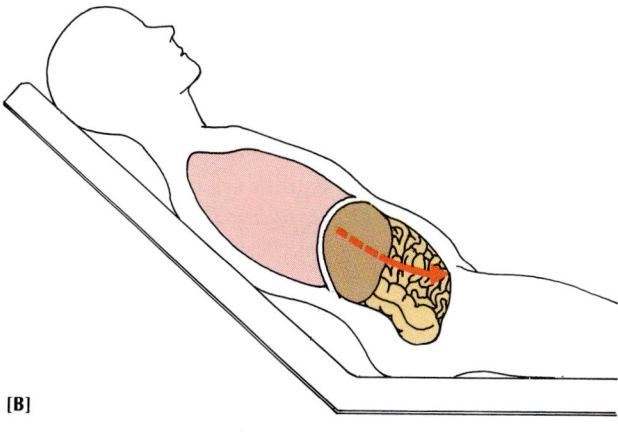

[B]

2 As the inspiratory muscles relax, the volume of the thoracic cage decreases by the elastic recoil of the costal cartilages, and the lungs recoil.

3 The intra-alveolar pressure thus increases above the atmospheric pressure, and air moves out of the lungs. Such normal quiet breathing is called *eupnea* (yoop-NE-uh; *eu* = normal).

In contrast, *forced expiration* is an active process that occurs during strenuous physical activity. The abdominal muscles contract forcibly, pressing the viscera against the passive diaphragm and depressing the rib cage to further reduce lung volume. Also helping to depress the rib cage and decrease volume are the internal intercostals, latissimus dorsi, and quadratus lumborum muscles.

Compliance The expansibility of the thorax, and thus the lungs, is called *compliance.* Compliance of the lung (C_L) is expressed as the change in lung volume (ΔV_L) for every unit change in intra-alveolar pressure (ΔP): $C_L = \Delta V_L/\Delta P$. The greater the increase in volume for a given increase in pressure, the greater the compliance, that is, the lungs and thoracic wall expand easily. Low compliance means that the lungs and thoracic wall are less distensible ("stiff lungs").

Lung Volumes and Capacities

The events of pulmonary ventilation can be described by subdividing the amount of air in the lungs into four volumes and four capacities [FIGURE 24.14]. The apparatus used to measure these amounts is called a *respirometer.* (1) During ordinary (quiet) breathing at rest, both men and women inhale and exhale about 0.5 L with each breath. This is the *tidal volume (TV)*. (2) When the deepest possible breath is taken in, the excess beyond the usual tidal volume is the *inspiratory reserve volume (IRV)*. (3) At the end of an ordinary exhalation, if all possible air is expelled, the quantity beyond the tidal volume is the *expiratory reserve volume (ERV)*. (4) The *residual volume (RV)* is the volume of air remaining in the lungs after the most forceful expiration.

When describing events of air movement, it is desirable at times to consider two or more of the above volumes together. Such combinations are called *pulmonary capacities*. The *inspiratory capacity (IC)* equals the inspiratory reserve volume plus the tidal volume. This is

the amount of air that a person can inhale, beginning at the normal expiratory level and then distending the lungs the maximal amount. The *vital capacity (VC)* equals the inspiratory reserve volume plus the tidal volume plus the expiratory reserve volume. It is the maximum amount of air that a person can expel from the lungs after the lungs have been filled to their maximum extent. The *functional residual capacity (FRC)* equals the expiratory reserve volume plus the residual volume. This is the amount of air that remains in the lungs at the end of normal expiration.

The *total lung capacity (TLC)* equals vital capacity plus residual volume. It is the maximum volume to which the lungs can be expanded with the greatest force.

In summary,

1 $IC = IRV + TV$

2 $VC = IRV + TV + ERV$

3 $FRC = ERV + RV$

4 $TLC = VC + RV$

About 0.15 L of air is present in the nasal passages, trachea, bronchi, and bronchioles. This is known as *anatomical dead space (ADS)* because no gas exchange occurs there.

The total amount of air taken in during 1 minute is called *minute respiratory volume (MRV)*, which is calculated by multiplying the tidal volume (*TV*) per breath by breaths per minute, or frequency (*f*):

$$MRV = TV \times f$$
$$\text{(mL/min)} = \text{(mL/breath)} \times \text{(breaths/min)}$$

Thus the amount of air in the lungs is affected by how deeply and how fast we breathe. During heavy breathing, the tidal volume may increase to as much as 3.1 to 4.8 L. If such maximum breaths are taken once a second, a person may for a short time have a *minute respiratory volume* of 186 to 288 L/min (60 breaths × 3.1, or 4.8 L). During quiet breathing, which normally occurs 12 to 16 times a minute and uses only the tidal volume of 0.5 L, the minute respiratory volume is about 6 to 8 L/min (12 or 16 breaths × 0.5 L).

Although the minute respiratory volume is determined by both the respiratory rate (breaths per minute) and the depth of breathing (*TV*), the two factors are not of equal importance. Because the air in the anatomical dead space is merely pushed back and forth during breathing, the last air expelled from the alveoli does not reach the outside and will be the first air inhaled. If a breath of 0.5 L is taken in and 0.15 L remains in the anatomical dead space, only 0.35 L of fresh air reaches the alveoli. Since the alveoli receive only 0.35 L of the

---

Q: *Why do you sometimes get a "stitch" in your side when you run?*

A: If the diaphragm does not receive enough oxygen or if there is a build-up of lactic acid (as can happen when you are running and breathing heavily), the diaphragm may "tire" and cause you to feel a sharp pain in your side.

FIGURE 24.14 LUNG VOLUMES AND CAPACITIES

Lung volumes and capacities recorded on a spirometer. The pen goes up during the patient's inspiration and down during expiration. Although residual volume and total lung capacity cannot be measured with a spirometer, their values can be calculated easily and are shown here.

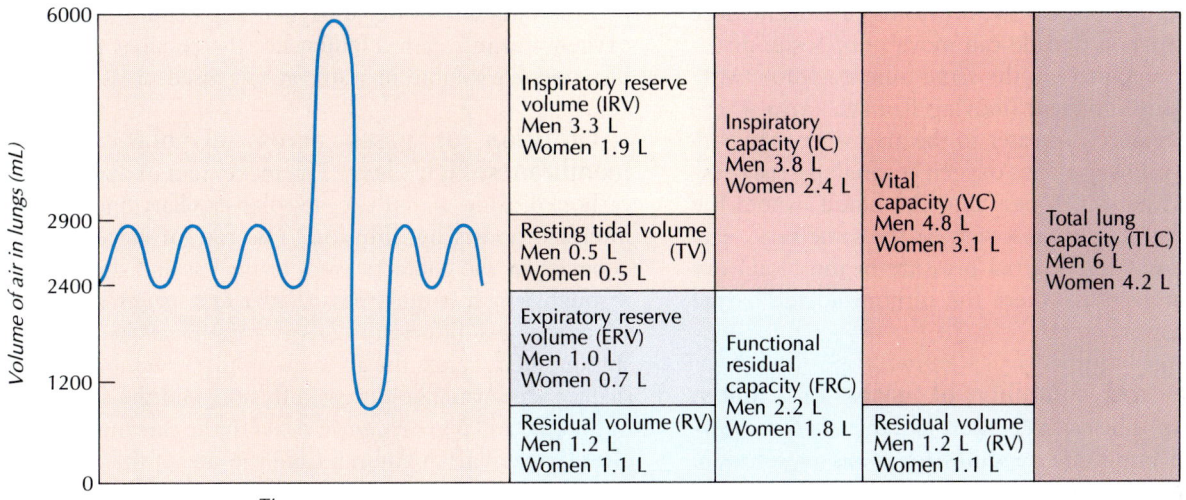

0.5 L *TV*, the actual rate of ***alveolar ventilation (AV)*** is 15 × 0.35, or 5.25 L:

$$AV = (TV - ADS) \times f$$
(mL/min)   (mL/breath)   (breaths/min)

Thus, increasing the depth of breathing can be more effective, in terms of lung volume, than increasing the respiratory rate. The same minute respiratory volume may be achieved by increasing either factor, but deeper breathing is obviously more efficient in gas exchange than breathing faster. When unusually large volumes of air are needed, during strenuous exercise, for example, both respiratory rate and depth of breathing are increased.

A S K   Y O U R S E L F

1 What is Boyle's law?

2 What mechanical events happen during inspiration? Expiration?

3 What is meant by the total lung capacity?

4 What is anatomical dead space?

5 What is vital capacity?

6 How are minute respiratory volume and alveolar ventilation calculated?

Factors Affecting Gas Movement and Solubility

Oxygen and carbon dioxide are *gases* at ordinary temperatures and pressures. In a study of respiration, therefore, we must be aware of the general principles governing the action of gases under different conditions.

Partial pressure of gases in liquids: Dalton's law

Dalton's law, named for the English chemist John Dalton, states that each gas in a mixture of gases exerts its own pressure in proportion to its concentration in the mixture, independent of other gases. Each gas exerts its own *partial pressure*, which is denoted, for example, as P_{O_2} for oxygen and P_{CO_2} for carbon dioxide.

Air is a mixture of gases including oxygen, carbon dioxide, nitrogen, and water vapor, as well as insignificant amounts of several other inert gases such as argon and helium. The sum of the partial pressures of these individual gases equals atmospheric pressure (P_{atm}):

$$P_{atm} = P_{O_2} + P_{CO_2} + P_{N_2} + P_{H_2O}$$

To determine the partial pressures of the individual gases that contribute to atmospheric pressure, multiply the percentage of each gas in the mixture by the atmospheric pressure (760 mmHg). The approximate percentage for each of the atmospheric gases is O_2 = 21 percent, CO_2 = 0.04 percent, N_2 = 78 percent. Therefore,

P_{O_2} = 21% × 760 mmHg = 159.6, or 160 mmHg
P_{CO_2} = 0.04% × 760 mmHg = 0.3 mmHg
P_{N_2} = 78% × 760 mmHg = 593 mmHg

The partial pressures of O_2 and CO_2 in the alveoli and blood differ from those in atmospheric air. There are also different percentages of these gases in inspired and expired air. Inspired air is similar to atmospheric percentages, while expired air has about 16 percent O_2, 4 percent CO_2, and higher water vapor pressures.

Dalton's law applies to the breathing process because oxygen and carbon dioxide can move in opposite directions in the same place at the same time, each following its own pressure gradient (moving from the area where the partial pressure is greater to the area where the partial pressure is lower). Thus oxygen at a high partial pressure in an alveolus can pass into the blood, where the oxygen partial pressure is lower. At the same time, carbon dioxide at a high partial pressure in the blood can pass into an alveolus, where the carbon dioxide partial pressure is lower.

Pressure and solubility of gases in liquids: Henry's law The *partial pressure* and the *solubility coefficient* are the important variables for a gas in solution. According to **Henry's law** (named for the nineteenth-century chemist William Henry), when a mixture of gases is in contact with a liquid, each gas will dissolve in the liquid in proportion to its partial pressure. That is, a liquid can usually dissolve more of a gas at a high partial pressure than it can at a lower partial pressure. Also, the higher the solubility coefficient of the gas, the more gas will stay in solution. The solubility coefficients for carbon dioxide and oxygen are $CO_2 = 0.54$ (high) and $O_2 = 0.024$ (low). This is why champagne shows no bubbles in a tightly corked bottle but fizzes when the bottle is opened. The carbon dioxide (high-solubility coefficient) is under high pressure when corked, but when uncorked it is subject only to ordinary atmospheric pressure, and the carbon dioxide comes out of solution. Under normal physiological conditions, the lungs and blood are not subjected to great variation in pressure. However, when unusual pressure changes do occur, they do affect gas exchange.

The most noticeable effect occurs at high altitudes, where partial pressures of gases decrease in direct proportion to the decrease in atmospheric pressure. At 3800 m (12,500 ft), where the partial pressure of oxygen (P_{O_2}) is the equivalent of 100 mmHg instead of the 160 mmHg at sea level, newcomers to the high altitude become sluggish and feel faint, nauseated, or uncomfortable. In contrast, people who live in high mountains compensate for the reduced oxygen pressure by producing more red blood cells.

Below sea level, higher-than-normal pressures are breathed. Deep-sea divers, for example, may have too much nitrogen dissolved in their blood as a result of the higher atmospheric pressures under water. When they come up to the surface too quickly, the nitrogen comes out of solution in the form of gas bubbles. The bubbles may lodge in joints, creating the painful and incurable malady known as *decompression sickness* or *bends.** The bends can be avoided by decompressing the body so slowly that the nitrogen diffuses out without bubbling, or by breathing a mixture of oxygen and helium instead of oxygen and nitrogen. Helium has the peculiar property of being *less* soluble in water at increased pressure.

Diffusion of gases across alveolar-capillary membranes: Fick's law The movement of oxygen and carbon dioxide across the alveolar-capillary membrane interface occurs by diffusion. The rate of gas diffusion depends on the characteristics of the gas and the barrier through which it must pass. **Fick's law of gas diffusion** states that gas diffusion is *directly proportional* to the pressure gradient across the alveolar-capillary membrane, the surface area available for gas diffusion, and the solubility of the gas, and *inversely proportional* to the thickness of the alveolar-capillary membrane and the size of the gas molecules. Gas diffusion is then faster when small molecules are crossing a thin membrane than it is when large molecules are crossing a thick membrane. Also, diffusion is fast when highly soluble molecules are diffused at a high partial pressure. Fick's law can be written as follows:

$$\text{Rate of gas diffusion} = \frac{\text{difference in partial pressures across alveolar-capillary membrane} \times \frac{\text{surface area for gas diffusion}}{\text{thickness of membrane}} \times \frac{\text{solubility of gas}}{\sqrt{\text{molecular weight of gas}}}}{}$$

or,

$$V_{gas} = (P_1 - P_2) \times A/T \times D$$

where D equals the solubility of the molecule divided by the square root of the molecular weight of the molecule.

In summary, the factors that determine the exchange of oxygen and carbon dioxide across the alveolar-capillary membrane interface by diffusion are (1) the partial pressure on either side of the membrane, (2) the surface area, (3) the thickness of the membrane, and (4) the solubility and size of the gas molecules.

ASK YOURSELF

1 What are Dalton's law and Henry's law?

2 Why is the solubility of oxygen and carbon dioxide important in respiration?

---

*The condition is also called *caisson disease* because it may affect people who work in caissons (pressurized, watertight chambers) during the construction of deep-sea tunnels and other underwater projects.

GAS TRANSPORT

The physiologically important parts of respiration are (1) the exchange of gases between alveoli and capillaries in the lungs and (2) between capillaries and cells in the body tissues. The first exchange is called *external respiration*, and the second is called *internal respiration* [FIGURES 24.15 and 24.16].

External Respiration

Because the lungs do not empty completely with each expiration, the air in the lungs is richer in carbon dioxide and poorer in oxygen than is the outside air [TABLE 24.1]. As a result, the P_{CO_2} in alveoli is equal to about 40 mmHg (instead of the 0.3 mmHg outside), and the P_{O_2} is only about 104 mmHg (instead of the 160 mmHg outside) [FIGURE 24.16]. Blood in the alveolar and bronchiolar capillaries coming from tissues has a high P_{CO_2} of 45 mmHg. The important point is that the carbon dioxide in the blood is at a higher partial pressure and thus is more concentrated than that in the alveoli, while the partial pressure and concentration of oxygen in the blood are lower than those in the alveoli. As each gas flows down its own concentration gradient from a region of high concentration to a region of lower concentration, effective gas exchange occurs.

Blood leaving alveolar capillaries has lost carbon dioxide and gained oxygen; the concentrations of these gases are about the same as in alveolar air; that is, the blood has a P_{CO_2} of 40 mmHg and a P_{O_2} of 104 mmHg. This exchange of oxygen and carbon dioxide between the pulmonary capillaries and alveoli is **external respiration** [FIGURES 24.15B and 24.16], the conversion of deoxygenated blood into oxygenated blood.

Internal Respiration

When oxygenated blood reaches active cells where *cellular respiration* is taking place, **internal respiration** occurs [FIGURE 24.15C]. In an active cell, the P_{O_2} is reduced, and the P_{CO_2} is increased. As blood in the tissue capillaries moves away from the respiring cells, blood has a P_{CO_2} of 45 mmHg and a P_{O_2} of 40 mmHg.

Transport of Oxygen in Blood

Blood carries oxygen in two ways: (1) dissolved in plasma and (2) combined with hemoglobin in red blood cells. Because oxygen is poorly soluble in water, plasma can carry only about 1.5 percent of an adult's oxygen requirement. But **hemoglobin** (Hb), the globular protein in red blood cells, has a high affinity for oxygen, and carries 98.5 percent of the oxygen available in the lungs. In this way, hemoglobin acts as a carrier and makes possible the effective transport of oxygen.

When oxygen is bound to hemoglobin, a new compound is formed: **oxyhemoglobin** (HbO_2). When oxygen is released, the hemoglobin compound is called *reduced hemoglobin*, or **deoxyhemoglobin** (HHb).

Several factors influence the binding and release of oxygen from the hemoglobin molecule: P_{O_2}, P_{CO_2}, pH, temperature, and 2,3-bisphosphoglycerate (2,3-BPG). We will look at each of these factors in turn:

1 P_{O_2}: The most important of these factors is P_{O_2}. As plasma P_{O_2} increases, more hemoglobin becomes saturated with oxygen (partially saturated) until all the hemoglobin becomes saturated (fully saturated). FIGURE 24.17 shows a normal oxygen-hemoglobin dissociation curve (middle curve), in which the relationship between percent saturation of hemoglobin of oxygen and P_{O_2} is shown. As P_{O_2} increases, hemoglobin binds with increasing amounts of oxygen in a nonlinear sigmoid-shaped curve. At a P_{O_2} of 100 mmHg (as in pulmonary capillaries), the hemoglobin is approximately 97 percent saturated with oxygen. At a P_{O_2} of 40 mmHg (cellular level), the hemoglobin is approximately 70 percent saturated. Thus hemoglobin has both the ability to load oxygen (lungs) and unload oxygen (cells) in response to a change in the partial pressure of oxygen.

2 P_{CO_2}: The loading of CO_2 onto the hemoglobin molecule has the ability to cause the unloading of O_2 from hemoglobin. Thus, as P_{CO_2} increases, oxygen is released from the hemoglobin molecule more readily, and the oxygen hemoglobin curve is shifted to the right. As P_{CO_2} decreases, less oxygen is released from hemoglobin, and the curve is shifted to the left [FIGURE 24.17].

3 *pH*: Acid environments (decreased pH, increased P_{CO_2}) lower the affinity of oxygen for hemoglobin (*Bohr*

TABLE 24.1 COMPOSITION OF RESPIRATORY GASES ENTERING AND LEAVING LUNGS (Standard Atmospheric Pressure, Young Adult Male at Rest)			
	Oxygen: %/ partial pressure (mmHg)	Carbon dioxide: %/ partial pressure (mmHg)	Nitrogen: %/ partial pressure (mmHg)
Inspired air	21/160	0.04/0.3	78.0/597
Expired air	16/120	4.0/27	79.2/566
Alveolar air	14/104	5.5/40	79.1/569

Note: Percentages do not add up to 100 because water is also a component of air.

FIGURE 24.15 INTERNAL AND EXTERNAL RESPIRATION

[A] Schematic diagram of internal and external respiration.
[B] External respiration in the lungs. Diagram of gas exchange between an alveolus and a pulmonary capillary.
[C] Internal respiration in body tissues. Diagram of gas exchange between a tissue capillary and respiring cells.

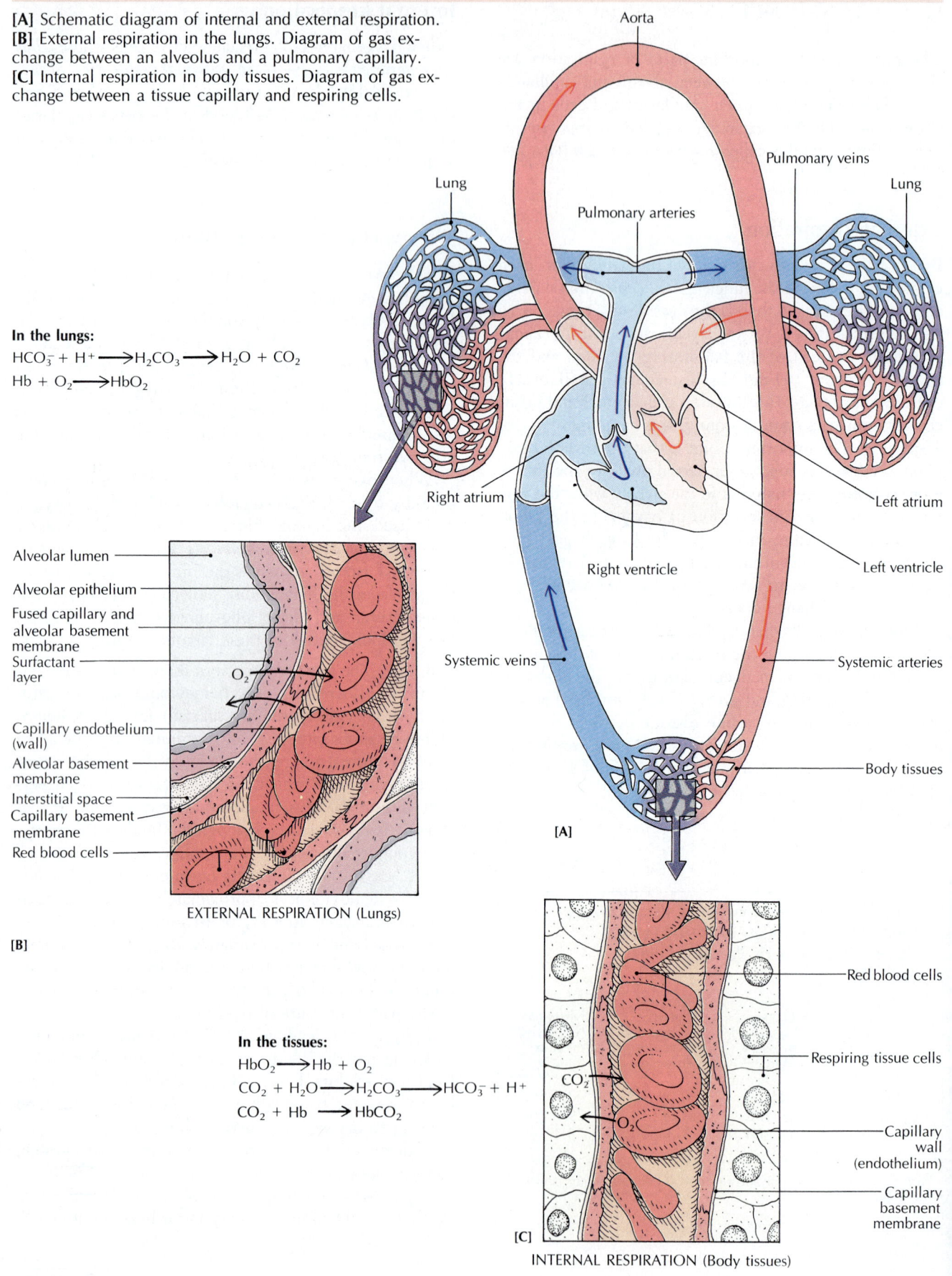

In the lungs:

$$HCO_3^- + H^+ \longrightarrow H_2CO_3 \longrightarrow H_2O + CO_2$$

$$Hb + O_2 \longrightarrow HbO_2$$

Aorta

Lung

Pulmonary veins

Pulmonary arteries

Lung

Right atrium

Left atrium

Right ventricle

Left ventricle

Systemic veins

Systemic arteries

Body tissues

[A]

Alveolar lumen

Alveolar epithelium

Fused capillary and alveolar basement membrane

Surfactant layer

O_2

Capillary endothelium (wall)

Alveolar basement membrane

Interstitial space

Capillary basement membrane

Red blood cells

EXTERNAL RESPIRATION (Lungs)

[B]

In the tissues:

$$HbO_2 \longrightarrow Hb + O_2$$

$$CO_2 + H_2O \longrightarrow H_2CO_3 \longrightarrow HCO_3^- + H^+$$

$$CO_2 + Hb \longrightarrow HbCO_2$$

Red blood cells

Respiring tissue cells

CO_2

O_2

Capillary wall (endothelium)

Capillary basement membrane

[C]

INTERNAL RESPIRATION (Body tissues)

FIGURE 24.16 GAS EXCHANGE

Approximate partial pressures of oxygen and carbon dioxide are given. Red arrows represent oxygen-rich blood; blue arrows represent oxygen-poor blood.

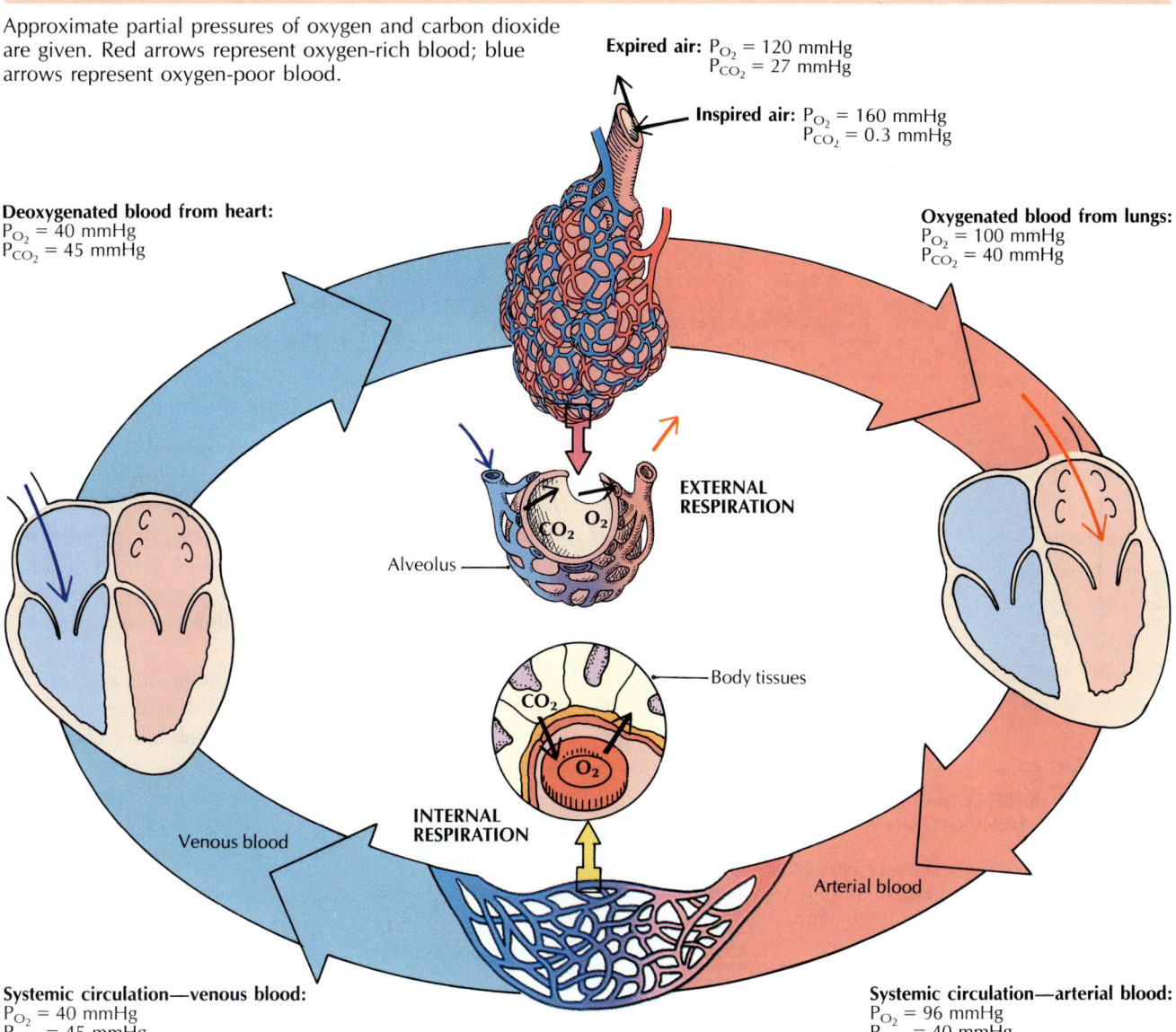

Expired air: P_{O_2} = 120 mmHg
P_{CO_2} = 27 mmHg

Inspired air: P_{O_2} = 160 mmHg
P_{CO_2} = 0.3 mmHg

Deoxygenated blood from heart:
P_{O_2} = 40 mmHg
P_{CO_2} = 45 mmHg

Oxygenated blood from lungs:
P_{O_2} = 100 mmHg
P_{CO_2} = 40 mmHg

EXTERNAL RESPIRATION

CO_2 O_2

Alveolus

Body tissues

CO_2

O_2

INTERNAL RESPIRATION

Venous blood

Arterial blood

Systemic circulation—venous blood:
P_{O_2} = 40 mmHg
P_{CO_2} = 45 mmHg

Systemic circulation—arterial blood:
P_{O_2} = 96 mmHg
P_{CO_2} = 40 mmHg

effect), so more oxygen is released to cells (right curve shift), whereas alkaline environments cause a left curve shift [FIGURE 24.17].

4 *Temperature:* Within limits, as body temperature increases, oxygen release from hemoglobin also increases, resulting in a right curve shift [FIGURE 24.18]. For example, during strenuous exercise, the temperature, P_{CO_2}, and hydrogen-ion concentration all increase, while the affinity of hemoglobin for oxygen at lower P_{O_2} decreases. This means more oxygen is released to the tissues that need oxygen for their increased metabolism. It is important to remember that through all these shifts, the body is attempting to maintain a stable homeostatic environment.

5 *2,3-BPG:* Oxygen binding and release by hemoglobin are further influenced by the presence in red blood cells of 2,3-bisphosphoglycerate (BPG). When BPG binds to hemoglobin, it decreases the affinity of hemoglobin for oxygen, thus increasing the release of oxygen. An increase in temperature, H^+ concentration, or cellular metabolism increases the rate of BPG formation. Certain hormones, such as testosterone, thyroxine, and catecholamines, increase the formation of BPG within red blood cells. The presence of BPG in alveolar capillaries would seem to compete with oxygen for a place on the hemoglobin molecule. However, there is presumably a high enough P_{O_2} in the plasma of such capillaries for the BPG to be replaced by oxygen.

FIGURE 24.17 EFFECTS OF CO₂ ON BINDING OF HEMOGLOBIN AND OXYGEN

These curves show binding of hemoglobin with oxygen at three levels of partial pressure of carbon dioxide. High pressure inhibits the formation of oxyhemoglobin by increasing the hydrogen-ion concentration (lowering pH).

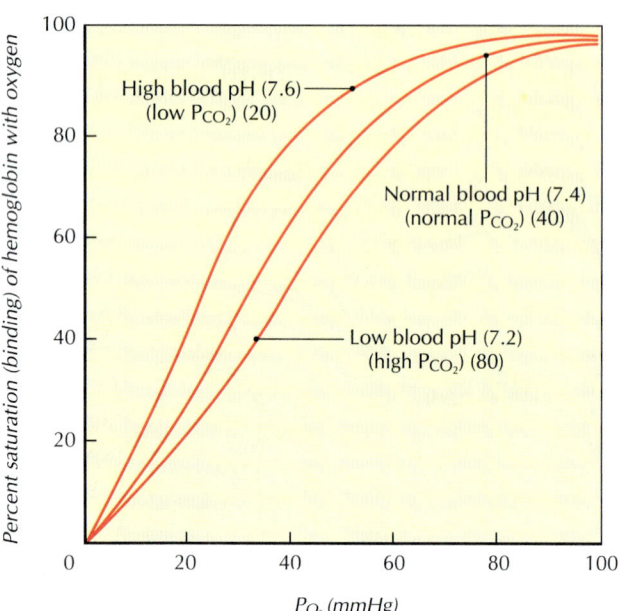

FIGURE 24.18 EFFECTS OF TEMPERATURE ON BINDING OF HEMOGLOBIN AND OXYGEN

These curves show the binding of hemoglobin with oxygen at three temperatures as a function of P_{O_2}. Notice the effects of temperature on percent saturation.

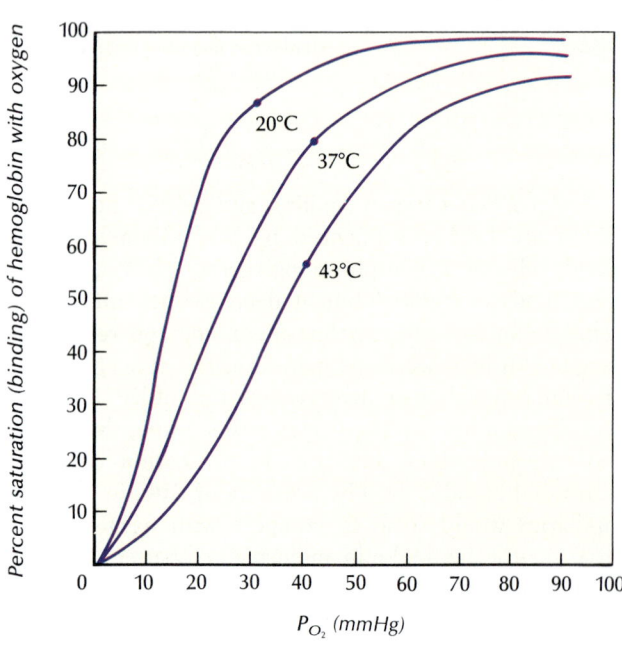

A change in the P_{O_2} of the atmosphere, as when one moves to a high altitude, affects the BPG level of blood. Within a few hours after ascending to a level of about 4500 m (15,000 ft), a person's BPG concentration increases and the affinity for oxygen decreases. This reaction is beneficial at the level of internal respiration. Natives of high altitudes have the reverse response when they descend to low altitudes.

Oxygen uptake in the lungs and its subsequent release in the tissue capillaries are thus regulated by a combination of P_{O_2}, P_{CO_2}, H^+ concentration, temperature, and BPG. Together they enable the blood to carry out effectively one of its main functions: the delivery of oxygen to the cells, where it is needed for cellular respiration.

Transport of Carbon Dioxide in Blood

Carbon dioxide is readily soluble in water. Therefore, when it is generated in respiring cells, it diffuses first into the cytoplasm, then through the plasma membrane into the extracellular fluid, and finally through capillary walls into the blood plasma. There it can be carried in three forms: (1) as dissolved gas in the plasma, (2) as carbaminohemoglobin, or (3) as bicarbonate ions:

1 Only about 7 percent of carbon dioxide is dissolved in plasma. The amount of carbon dioxide carried in this way is related to the P_{CO_2} and the solubility of carbon dioxide.

2 Carbon dioxide combines readily with the amino groups of the amino acids (represented by R below) present in blood proteins, forming *carbamino compounds:*

$$RNH_2 + CO_2 \rightleftharpoons RNHCOO^- + H^+$$

Carbamino compounds are produced by many proteins, but mainly by hemoglobin. The result is *carbaminohemoglobin:*

$$HbNH_2 + CO_2 \rightleftharpoons HbNHCOO^- + H^+$$

Approximately 23 percent of carbon dioxide is carried in the form of carbaminohemoglobin.

While oxygen is being bonded to the heme groups in hemoglobin, carbon dioxide is being bonded to the

---

Q: *What substance do some athletes breathe on the sidelines?*

A: Fatigued football players and other athletes frequently breathe through a plastic mask from a tank of pure oxygen. Current research indicates that athletes gasp for air because of increased carbon dioxide in their bloodstream, not because of a need for extra oxygen, and that pure oxygen is no better than air in helping athletes recover from fatigue. However, oxygen may help reduce the excess lactic acid that builds up during intense physical activity.

amino groups of the same molecule. Although hemoglobin can carry oxygen and carbon dioxide at the same time, the presence of one reduces the bonding power of the other. Oxyhemoglobin is less likely to carry carbon dioxide than deoxygenated hemoglobin is, and deoxygenated hemoglobin carrying carbon dioxide is less likely to bind to oxygen.

3 The greatest proportion of carbon dioxide (approximately 70 percent of the total) is carried in plasma in the form of the bicarbonate ion. When carbon dioxide, activated by the enzyme carbonic anhydrase, reacts with water, it forms carbonic acid (H_2CO_3), which undergoes partial dissociation to yield hydrogen ions (H^+) and bicarbonate ions (HCO_3^-):

$$CO_2 + H_2O \overset{\text{carbonic}}{\underset{\text{anhydrase}}{\rightleftharpoons}} H_2CO_3 \rightleftharpoons H^+ + HCO_3^-$$

This reaction occurs mainly inside the erythrocytes, where carbonic anhydrase is most abundant. (Plasma contains very little carbonic anhydrase.) The resulting bicarbonate ions diffuse from the erythrocytes into the plasma.

Chloride shift The removal of bicarbonate ions from the erythrocytes develops a positive charge inside the cell, and chloride ions (Cl^-) from sodium chloride (NaCl) in the plasma diffuse into the cell in an exchange called the ***chloride shift*** [FIGURE 24.19]. The chloride shift greatly increases the ability of the plasma to carry carbon dioxide. Because of the movement of chloride ions, their concentration in erythrocytes of venous blood is higher than that in arterial erythrocytes. The total osmotic activity of erythrocytes in venous blood is also greater, because although the proteins inside the cell bind many of the hydrogen ions, the reaction of water with carbon dioxide also produces bicarbonate ions, which are exchanged for osmotically active chloride ions. Therefore, in order to maintain osmotic equilibrium within erythrocytes in venous blood, the shift of chloride ions is accomplished by a movement of water that slightly increases the size of the erythrocytes. As a result, the venous hematocrit is slightly greater than the arterial hematocrit. The reverse movement of chloride ions and water occurs in erythrocytes when they enter arterial blood.

Haldane effect The amount of carbon dioxide transported by the blood is also influenced by the P_{O_2} in the blood. When the P_{O_2} decreases, the amount of transported carbon dioxide increases, and when P_{O_2} increases the amount of transported carbon dioxide decreases. This phenomenon is called the ***Haldane effect***. (P_{CO_2} influences the loading and unloading of oxygen by the hemoglobin molecule—the Bohr effect; and P_{O_2} influ-

FIGURE 24.19 THE CHLORIDE SHIFT

[A] In tissue capillaries, chloride ions move from the plasma into red blood cells. **[B]** In alveolar capillaries, chloride ions pass from red blood cells back into the plasma.

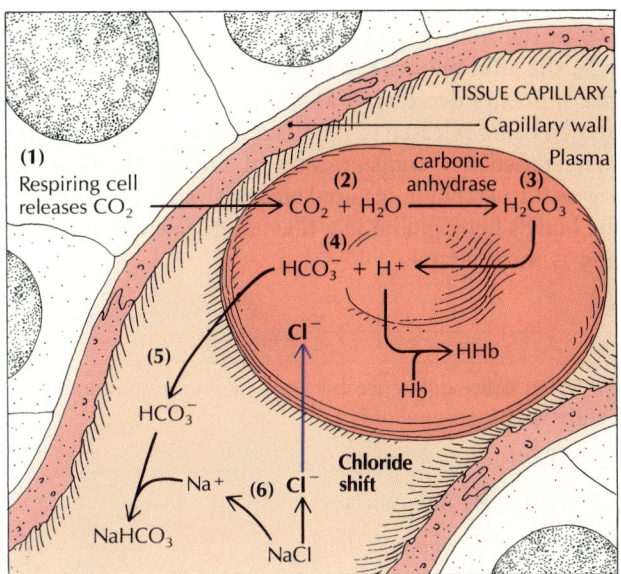

[A]

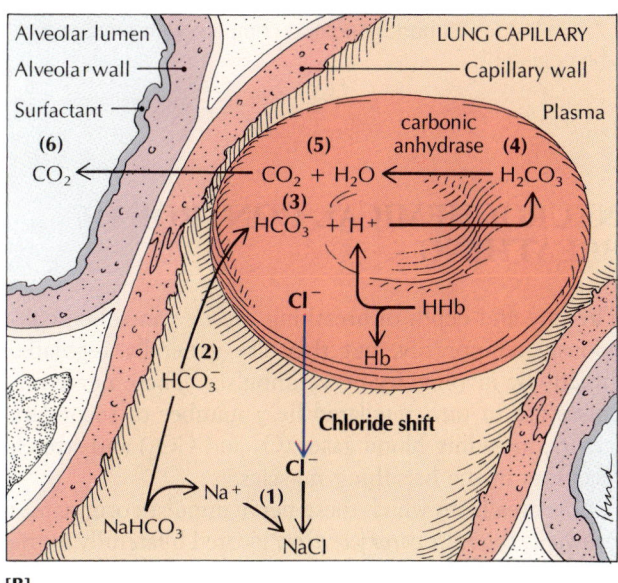

[B]

ences the loading and unloading of carbon dioxide by the hemoglobin molecule—the Haldane effect.)

Effect of Carbon Monoxide on Hemoglobin

Two of the chemical characteristics of carbon monoxide (CO) collectively make it a deadly poison. First, its affinity for hemoglobin is 210 times greater than oxygen's affinity for hemoglobin. As a result, very small amounts of carbon monoxide can compete successfully with oxygen for hemoglobin. Once carbon monoxide binds to hemoglobin as carboxyhemoglobin (COHb), it remains bound indefinitely, preempting binding sites for oxygen. Second, when carbon monoxide combines with one or two of the four heme groups in the hemoglobin molecule, the affinity of the remaining heme groups for oxygen is greatly decreased. This effect results in even less oxygen being released to the tissues. Because of these two characteristics, a concentration of carbon monoxide as low as 0.1 percent in inspired air inactivates 50 percent of the body's hemoglobin in a few hours. Such an inactivation is usually fatal.

A S K Y O U R S E L F

1 What is the difference between internal respiration and external respiration?

2 What is oxyhemoglobin?

3 What is the Bohr effect?

4 What is the chloride shift?

5 How are carbamino compounds formed?

6 What is the Haldane effect?

7 How does carbon monoxide compete with oxygen for hemoglobin?

NEUROCHEMICAL CONTROL OF BREATHING

The rate and depth of breathing can be controlled consciously temporarily, but they are generally regulated directly by involuntary nerve impulses. These nerve impulses are in turn regulated by a number of secondary factors, including blood gases (O_2 and CO_2) and stretch receptors in the breathing muscles.

The area from which these nerve impulses originate is known as the ***respiratory center,*** located bilaterally in the reticular formation of the brainstem [FIGURE 24.20]. The respiratory center contains three subareas or groups of neurons: (1) the *medullary rhythmicity center* in the medulla oblongata, (2) the *pneumotaxic center* in the pons, and (3) the *apneustic center,* also in the pons [FIGURE 24.20B].

Medullary rhythmicity center According to current understanding, which is still incomplete, the ***medullary rhythmicity center*** contains two sets of respiratory neurons in the reticular formation of the medulla oblon-

FIGURE 24.20 RESPIRATORY CENTER IN THE BRAIN

[A] Midsagittal section of brainstem. **[B]** Medullary rhythmicity area. (*Left*) When the inspiratory circuit is active, inspiration occurs and the expiratory circuit is inhibited. (*Right*) When the expiratory circuit is active, expiration occurs and the inspiratory circuit is inhibited. The two circuits act alternately to regulate rhythmic breathing. The small drawing of the brain shows the approximate location of the respiratory center.

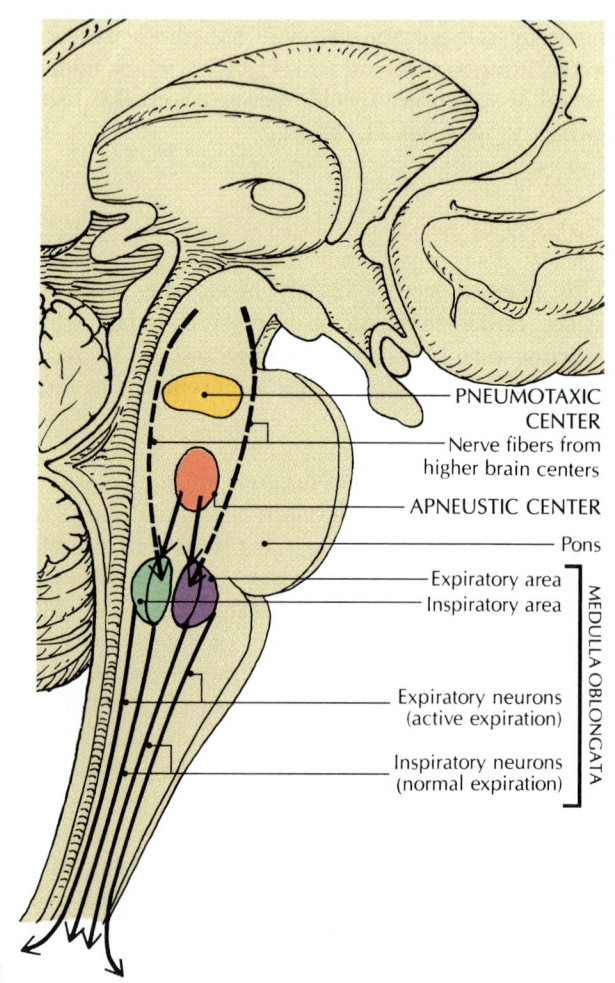

PNEUMOTAXIC CENTER
Nerve fibers from higher brain centers
APNEUSTIC CENTER
Pons
Expiratory area
Inspiratory area
MEDULLA OBLONGATA
Expiratory neurons (active expiration)
Inspiratory neurons (normal expiration)

[A]

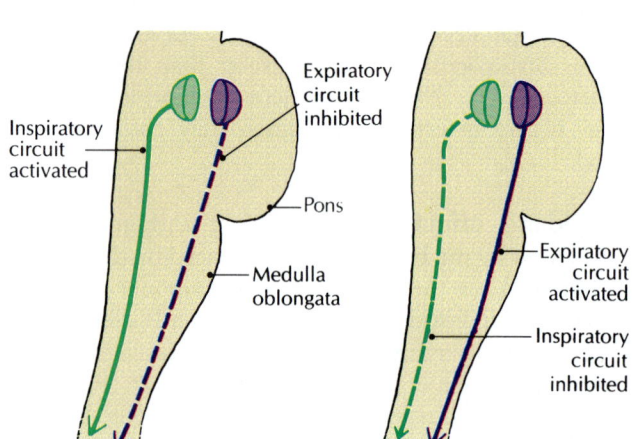

Expiratory circuit inhibited
Inspiratory circuit activated
Pons
Medulla oblongata
Expiratory circuit activated
Inspiratory circuit inhibited

Contraction impulse to muscles of inspiration: chest cavity expands and air enters lungs.

Relaxation impulse to muscles of inspiration: chest cavity contracts and air is forced from lungs.

[B]

gata: (1) the pace-setting, autorhythmic *dorsal respiratory group (DRG)*, consisting mostly of inspiratory neurons, and (2) the *ventral respiratory group (VRG)*, consisting of both inspiratory and expiratory neurons. When the DRG inspiratory neurons fire, they send impulses via the phrenic nerves that cause the diaphragm to contract, and impulses via the intercostal nerves that cause the external intercostal muscles to contract. As a result, inspiration occurs. These inspiratory nerve impulses from the inspiratory center last for about 2 sec, then become inactive for about 3 sec, during which time the inspiratory muscles relax, there is elastic recoil of the lungs and thoracic wall, and expiration occurs. This cyclic on-off firing of the DRG inspiratory neurons leads to a respiratory rate of approximately 12 to 18 breaths per minute. Thus normal, quiet inspiration is an active process, and expiration is a passive process controlled by the rhythmic DRG inspiratory neurons in the medullary respiratory center. The VRG consists of both inspiratory and expiratory neurons, which are inactive during normal quiet breathing. These neurons are involved in active expiration or forced expiratory efforts. They send impulses to the muscles of expiration — the internal intercostals and abdominal muscles — which cause forced expiration by depressing the rib cage and decreasing thoracic volume.

Pneumotaxic and apneustic centers The gross rhythmic pattern of breathing is fine-tuned by two centers in the pons, the pneumotaxic and apneustic centers. The *pneumotaxic center* in the upper pons sends inhibitory impulses to the medullary DRG inspiratory neurons to prevent overinflation of the lungs and to quicken the respiratory rate. In contrast, the *apneustic center* (app-NEW-stick) located in the lower pons, sends stimulatory impulses to the DRG inspiratory center, which cause prolonged inspiratory efforts that result in a deeper, slower pattern of breathing.

Inputs into the Medullary Respiratory Center

The respiratory rhythm is generated by the medullary respiratory center and modified by the pneumotaxic and apneustic centers, but several inputs into the medullary center regulate respiratory functions according to the metabolic demands of the body. These inputs involve higher brain centers, specific reflexes, and chemical factors.

Higher brain centers Situations involving pain, emotion, and excitement (laughing, crying) activate hypothalamic centers and alter ventilation by sending powerful impulses into the medullary respiratory centers. Higher cortical centers can also alter breathing patterns. For example, voluntary breath-holding or altered breathing patterns — as in speaking, playing a wind instrument, or swimming — allow us to hold our breath for a short time. This voluntary control of breathing is limited in time. Carbon dioxide and hydrogen ions build up in the blood, and these are powerful respiratory stimulants that can overpower our strongest "voluntary" attempts to hold our breath.

Irritant reflexes Irritant receptors are located in the bronchioles, and respond to irritating substances in the air such as dust, smoke, and noxious fumes. These receptors, when activated, send afferent input into the medullary respiratory center via branches of the vagus nerve, which results in efferent output that causes bronchiole constriction and/or reflex coughing or sneezing.

Inflation reflex Pulmonary stretch receptors located in the smooth muscle of the walls of the bronchi and bronchioles are activated when stretched by large tidal volumes (greater than 1 L) of air. Afferent fibers in the vagus nerve travel to the inspiratory center to inhibit inspiration. This negative feedback reflex from stretched lungs, known as the *Hering-Breuer reflex*, prevents overinflation of the lungs.

Chemical factors that increase ventilation The three most important chemical factors that increase ventilation are (1) a decrease in arterial P_{O_2}, (2) an increase in arterial P_{CO_2}, and (3) an increase in H^+ concentration. Each of these chemical factors is discussed in detail in the following paragraphs:

1 *Decreased arterial P_{O_2}:* Arterial P_{O_2} is monitored by the *peripheral chemoreceptors* known as the *carotid* and *aortic bodies* located at the bifurcation of the common carotid arteries and in the aortic arch, respectively. These chemoreceptors sense a decrease in arterial P_{O_2} below 60 mmHg. When the P_{O_2} falls to this extent, afferent impulses are sent to the medullary respiratory neurons to reflexly increase the rate and depth of breathing (ventilation). Seldom in real life does the arterial P_{O_2} fall to this level, so this factor does not play a major role in the regulation of minute-to-minute, day-to-day control of

Q: *Why do we yawn?*

A: Everybody yawns, but researchers are not sure exactly what function it serves. Either excitement or boredom can result in insufficient breathing—and we yawn—but the traditional explanation that we yawn to increase our oxygen supply has not been proved to universal satisfaction. In fact, studies have shown that high concentrations of oxygen do not inhibit yawning, and low concentrations of carbon dioxide do not stimulate yawning. The most recent speculation is that an uninhibited yawn (with the mouth wide open) forces blood through blood vessels in the brain, thereby increasing alertness.

FIGURE 24.21 EFFECT OF THREATENINGLY LOW ARTERIAL P$_{O_2}$ ON VENTILATION

The drop in arterial P$_{O_2}$ to less than 60 mmHg inhibits the activity of the central chemoreceptors so they exert no effect on the medullary respiratory center.

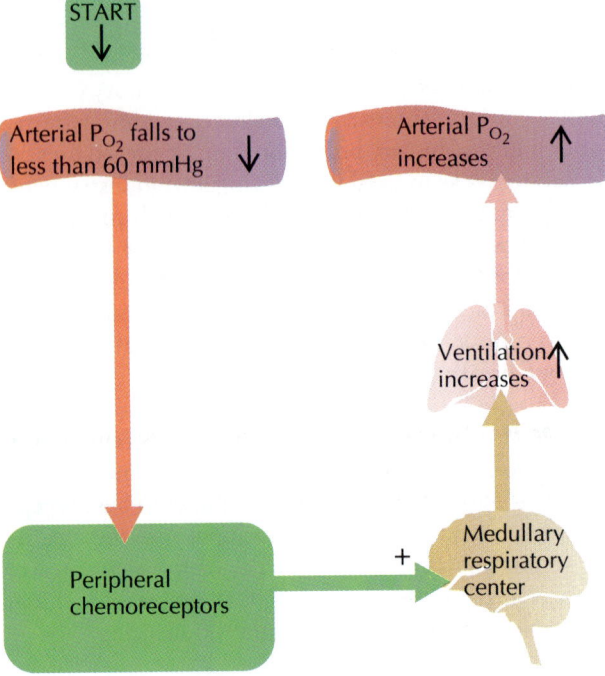

respiration. Also, at a P$_{O_2}$ of 60 mmHg, the hemoglobin is still about 90 percent saturated with oxygen. Thus this reflex operates only in times of dangerously low arterial oxygen levels [FIGURE 24.21].

2 *Increased arterial P$_{CO_2}$:* Small increases in arterial P$_{CO_2}$ induce a significant reflex increase in ventilation to promote the "blowing-off" of excess C$_{O_2}$ into the atmosphere; this action returns the arterial CO$_2$ to approximately 40 mmHg [FIGURE 24.22]. Small decreases in arterial P$_{CO_2}$ reduce ventilation to allow metabolically produced CO$_2$ to increase toward normal. The most sensitive receptors to P$_{CO_2}$ levels are the *central chemoreceptors* located in the medulla oblongata near the respiratory neurons. Carbon dioxide diffuses across the blood-brain barrier into the brain extracellular fluid and combines with water. Thus

$$CO_2 + H_2O \rightleftharpoons H_2CO_3 \rightleftharpoons H^+ + HCO_3^-$$

It is the increased H$^+$ concentration that stimulates the central chemoreceptors to increase ventilation in order to return the P$_{CO_2}$ to normal. (The increased CO$_2$ during breath-holding is the powerful stimulant via the central chemoreceptors that overrides voluntary control.)

3 *Increased arterial H$^+$ concentration:* Increased arterial H$^+$ concentration cannot cross the blood-brain barrier, so it cannot directly stimulate the central chemoreceptors. Hydrogen ions are sensed by the aortic and carotid peripheral chemoreceptors. An increase in arterial H$^+$

FIGURE 24.22 EFFECT OF INCREASED ARTERIAL P$_{CO_2}$ ON VENTILATION

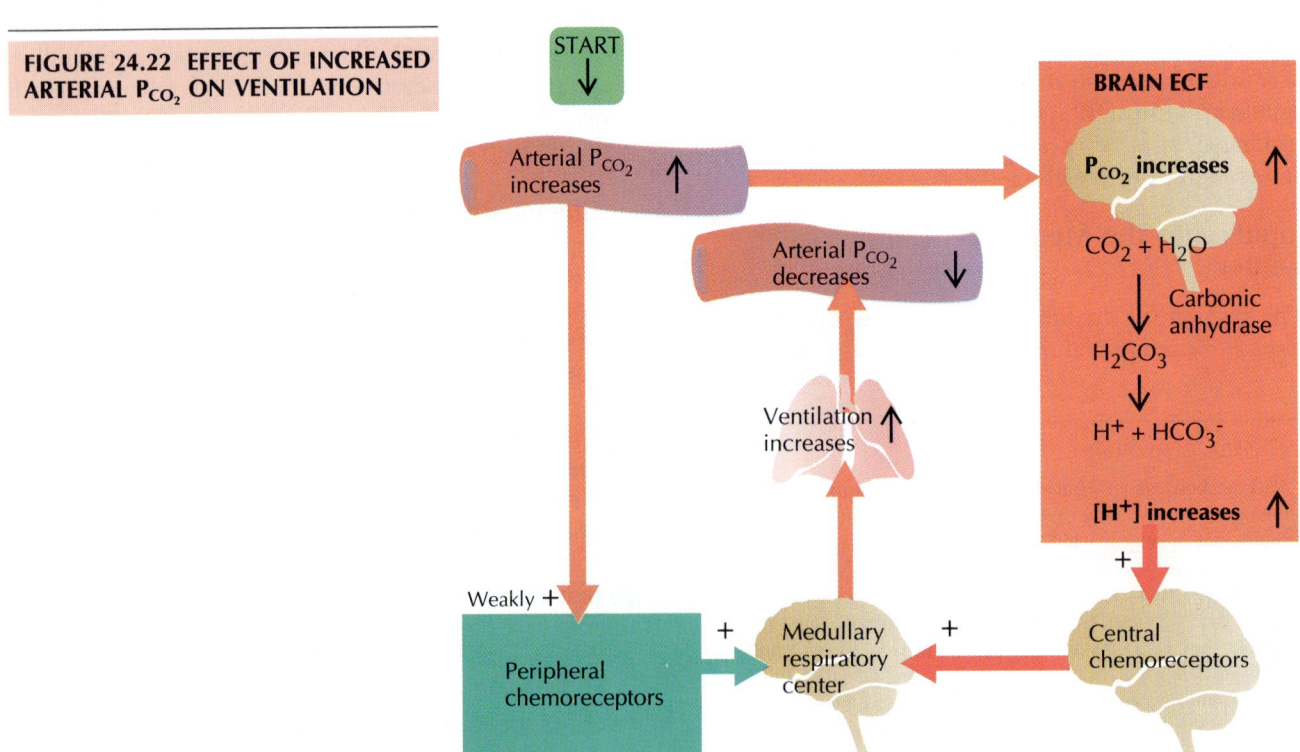

Note that the increased number of H^+ cannot penetrate the blood-brain barrier, and they have no effect on central chemoreceptors.

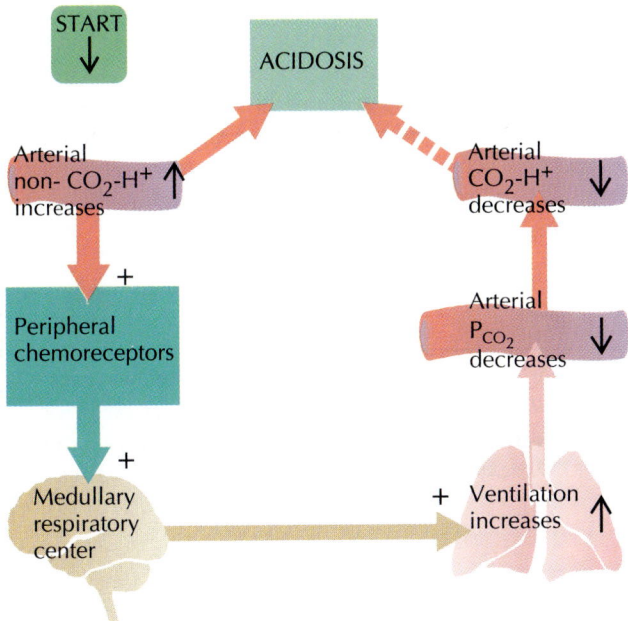

concentration causes a reflex increase in ventilation, and a decrease in arterial H^+ concentration decreases ventilation [FIGURE 24.23].

Other factors that increase ventilation The medullary respiratory center also receives input from muscle and joint receptors (*proprioceptors*) during muscle contraction, which can increase ventilation severalfold. An increase in body temperature and stretching of the anal sphincter muscle also increase ventilation.

OTHER ACTIVITIES OF THE RESPIRATORY SYSTEM

Several other activities involve the respiratory system. A *coughing reflex* involving short, forceful exhalations through the mouth occurs when sensitive parts of the air passages to the lungs are irritated by solids or liquids. The reflex persists even during deep coma.

A *sneeze* is an involuntary explosive expiration, mainly through nasal passages, and is usually caused by irritation of the mucous membranes in the nose. Particles ejected by a robust sneeze may travel up to 165 km/hr (103 mph).

A *hiccup* is a sudden involuntary contraction of the diaphragm that usually results from a disruption in the normal pattern of breathing. When the diaphragm is suddenly contracted, air is "sucked in" so abruptly that the epiglottis snaps shut, producing the sound of a hiccup. Hiccupping usually lasts 2 to 5 min, although some highly unusual cases persist for months or even years. It serves no known function. As a last resort to relieve chronic hiccups, one phrenic nerve in the neck (anterior surface of the anterior scalene muscle) is sometimes crushed. The crushed nerve can regenerate and reinnervate half of the paralyzed diaphragm to restore activity.

Snoring can occur when muscles in the throat relax during sleep, and the loose tissue of the soft palate and uvula (the flap of tissue hanging down from the soft palate) partially obstruct the upper airway.

Sighing, sobbing, crying, yawning, and *laughing* are variants of ordinary breathing and are so closely associated with subjective emotional conditions that they defy exact definition.

THE EFFECTS OF AGING ON THE RESPIRATORY SYSTEM

As people grow older, some alveoli in the lungs are replaced by fibrous tissue, and the exchange of oxygen and carbon dioxide is reduced. Breathing capacity and usable lung capacity are also reduced when muscles in the rib cage are weakened and the lungs lose their full elasticity. (By age 45, lung capacity is usually reduced to about 82 percent of optimum; at age 65, it is about 62 percent; and it is reduced to about 50 percent at age 85.) Aging also causes a decrease in gaseous exchange across pulmonary membranes. Emphysema, lung cancer, chronic bronchitis, and pneumonia also become prevalent with age.

DEVELOPMENTAL ANATOMY OF THE RESPIRATORY TREE

About 4 weeks after fertilization, at about the same time that the laryngotracheal tube and esophagus are separating into distinct structures, an endodermal *bronchial bud* forms at the caudal end of the developing laryngotracheal tube [FIGURE 24.24A]. The single bronchial bud soon divides into two buds, the right one being larger [FIGURE 24.24B]. These bronchial buds develop into the left and right primary bronchi of the future lungs during week 5 [FIGURE 24.24C]. The right bud divides to form three secondary bronchi, and the left bud forms only two. The right bronchus continues to be larger and more branched than the left during embryonic development, and by week 8 the bronchi show their basic mature form, with three secondary bronchi and three lobes in the right lung, but only two secondary bronchi and two lobes in the left lung [FIGURE 24.24D]. The bronchi subsequently branch repeatedly to form more bronchi and bronchioles.

By week 24, the epithelium of the bronchioles thins markedly, and highly vascularized primitive *alveoli* (terminal air sacs) have formed. A baby born prematurely during week 26 has a fairly good chance of surviving without the help of artificial respiratory devices because its lungs have developed an alveolar mechanism for gas exchange. The development of pulmonary blood vessels is as important to the survival of premature infants as is the thinness of the alveolar epithelium, which allows an adequate gas exchange. A newborn infant has only about 15 percent of the adult number of alveoli. The number and size of alveoli continue to increase until a child is about 8 years old. Also at about 24 weeks, type II alveolar epithelial cells begin to secrete *lung surfactant*. After 2 or 3 weeks of secretion, the alveoli are strong enough to retain air and remain open when breathing finally begins at birth.

Q: *Why can a sneeze be stopped by pressing between the upper lip and the base of the nose?*

A: Receptors for sneezing are located there. Firm pressure affects the receptors and suppresses the sneeze.

FIGURE 24.24 EMBRYONIC DEVELOPMENT OF THE BRONCHI

[A] 4 weeks (early), [B] 4 weeks (late), [C] 5 weeks, and [D] 8 weeks. Ventral views.

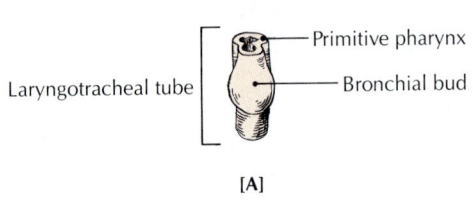

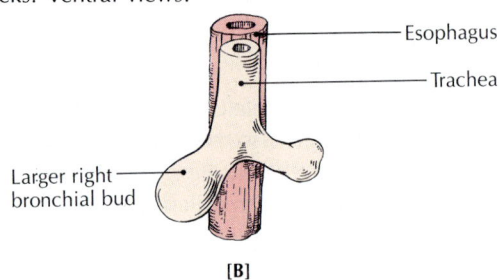

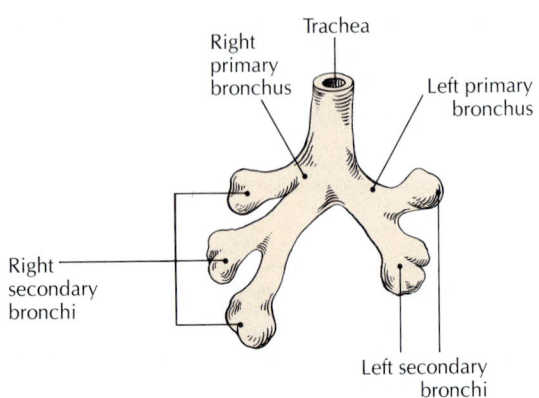

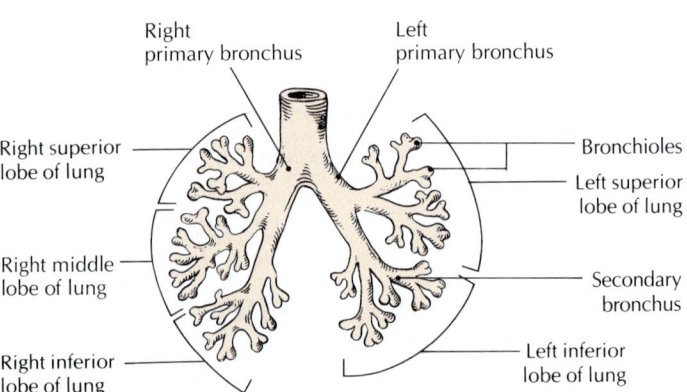

WHEN THINGS GO WRONG

A constant supply of oxygen and the ready removal of waste carbon dioxide are so necessary to continued life that some sort of respiratory failure is the immediate cause of many deaths. Any interference with breathing is serious and requires quick attention.

Rhinitis

Any inflammation of the nose is called **rhinitis** (Gr. *rhin-*, nose). The most common nasal problem is what is known as the "common cold." This virus-induced malady (or several maladies) causes excessive mucus secretion and swelling of the membranes of the nose and pharynx and can spread into the nasolacrimal ducts, sinuses, auditory tubes, and larynx. It can cause fever, general malaise, and breathing difficulty, but is seldom grave except in people weakened by another illness.

Laryngitis

A local infection of the larynx is **laryngitis.** It can totally incapacitate the vocal folds, making normal speech impossible, but it is usually not painful. It may be *idiopathic*, that is, not associated with any other disease. Irritants such as tobacco smoke can cause swelling of the vocal folds to such an extent that chronic hoarseness results.

Asthma

Asthma (Gr. no wind) is a general term for difficulty in breathing. *Bronchial asthma* is a result of the constriction of smooth muscles in the bronchial and bronchiolar walls, accompanied by excess mucus secretion and insufficient recoil of the alveoli. Air is trapped in the lungs, and the victim cannot exhale normally. The condition is usually caused by an allergic reaction but may also be brought on by an emotional upset. Asthma is not always dangerous, though over a long period of time it may result in permanently damaged alveoli. An acute attack, if not treated immediately, can cause death by asphyxiation.

Hay Fever

Hay fever is characterized by abundant tears and a runny nose. Like asthma, it is caused by an allergic reaction. So many people are affected that pollen counts of the atmosphere are regularly reported in the public media when pollen is abundant. The worst plant offenders are wind-pollinated trees, grasses, and ragweed—but, ironically, not hay.

Emphysema

Emphysema (Gr. blown up) is a fairly common disease among the elderly, especially those who are heavy smokers or live in air-polluted cities. It is an abnormal, perma-

nent enlargement of the alveoli, accompanied by the destruction of the alveolar walls. Breathing difficulties, especially during expiration, are caused by the destruction of alveolar septae, the partial collapse of the airway,

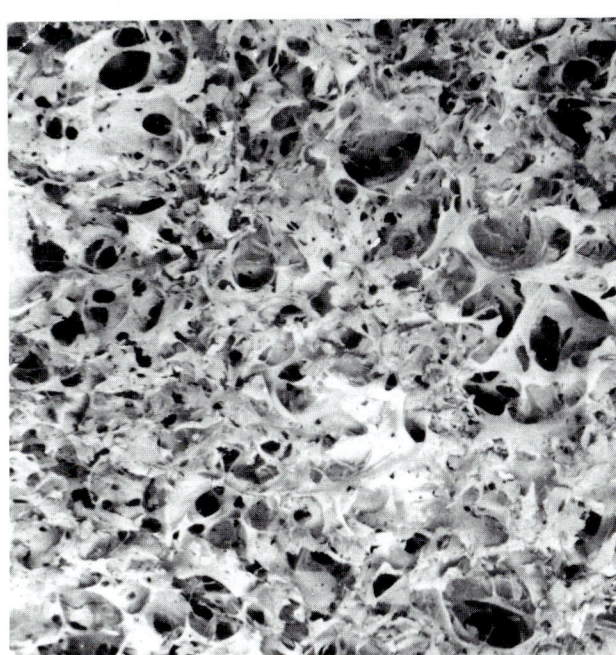

[A]

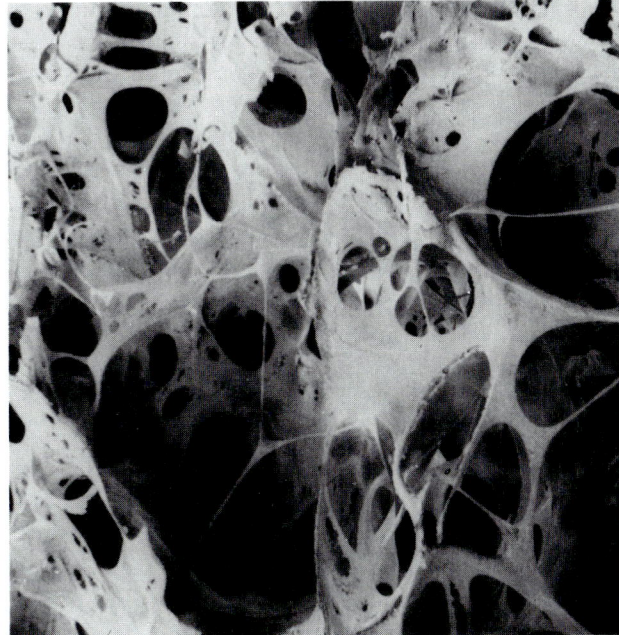

[B]

[A] Normal lung tissue. ×100. [B] Emphysematous lung tissue. ×100.

Tuberculosis: The Return of a Killer

At the start of the twentieth century, the leading killer of people in the United States was tuberculosis (TB). But great improvements were made in housing and sanitation conditions in the first half of the century, and then came the crucial introduction of antibiotics to the general public immediately after World War II.

In the early years of the century, it had been commonplace for TB to claim 100,000 lives a year in the United States alone, but then the number of cases began to decrease rapidly. By 1953, 84,000 cases of TB (not all fatal) were reported, and in 1984 the incidence had dropped to an all-time low: 22,000 cases. Public health officials declared that TB was no longer a serious threat.

Then, in 1985, the number of cases began to rise rapidly; in 1992, 25 percent more cases of TB were reported than had been reported in 1984. To make it worse, some of the new strains were not only much more virulent than their predecessors, they were now also resistant to antibiotics. Although most new cases seemed related to AIDS patients (tuberculosis is a major opportunistic infection) and other people with weakened immune systems—the poor, the homeless, and the drug-addicted—TB experts became alarmed. (In the poorest areas of the United States, TB rates are higher than in the poorest countries in the world.)

Unfortunately, the treatment for TB has not improved significantly since the advent of antibiotics, and new research is probably necessary to keep pace with the rapidly developing resistant strains of bacteria. (A 1992 study in New York City showed that 34 percent of TB patients there were resistant to at least one drug.) Currently, about 37 percent of treated patients die, while about 50 percent of untreated patients die. Drug-resistant TB is, in any event, 50 to 80 percent fatal.

Part of the problem with curing TB is that it usually takes about 6 months of intensive treatment to clear up the infection. Such a rigorous treatment is difficult for most patients to sustain, especially if they are homeless or drug-addicted. (Eighty-seven percent of the patients discharged from New York's Harlem Hospital fail to complete their out-patient treatment.)

Professor Barry Bloom of the Albert Einstein College of Medicine in New York City reflects the discouragement of fellow scientists: "Even if bad things like HIV and homelessness hadn't occurred, TB would have come back. . . . It's a predictable result of abandoning public-health measures that were working."

---

and a loss of elasticity in the lungs. A person suffering from emphysema cannot exhale normal amounts of air even when trying forcibly to do so. As a result, undesirable levels of carbon dioxide may build up. In long-standing cases, the alveoli deteriorate, their walls harden, and alveolar capillaries are damaged (see photos on page 787). Fortunately, because a normal adult has about eight times as many alveoli as needed for routine activities, acute cases of emphysema are not as prevalent as they might be otherwise. As the disease progresses, the respiratory center in the brain loses its effectiveness, the right ventricle becomes overworked in its efforts to force blood through the constricted lung capillaries, and the victim is "short of breath" even after slight physical activity.

The most common cause of emphysema is air pollution, including tobacco smoke, industrial solvents, and agricultural dust. It can develop in elderly people as a result of hardening of alveolar walls, without any apparent environmental stimulus.

Cardiopulmonary Resuscitation (CPR)

In a respiratory emergency, breathing stops or is reduced to the point where the victim's body tissues do not get sufficient oxygen to support life. If breathing is not restored, the heart can stop beating (cardiac arrest). *Cardiopulmonary resuscitation (CPR)* is a technique that combines rescue breathing with external cardiac compression to restore breathing and circulation. *Rescue breathing* supplies oxygen to the victim's lungs, and *cardiac compression* circulates blood by compressing the heart between the sternum and vertebrae.

Pneumonia

Pneumonia is a general term for any condition that results in the filling of the alveoli with fluid. It can be caused by a number of factors—chemicals, bacteria, viruses, protozoa, or fungi—but the usual infective agent is a *Streptococcus* bacterium. Pneumonia most readily attacks people who are already weakened by illness or whose lungs are damaged. Blood infections, chronic alcoholism, inhalation of fluids into the lungs, or even prolonged immobility in bed can predispose a person to microbial infection of the lungs.

Tuberculosis

Tuberculosis, a disease caused by *Mycobacterium tuberculosis,* can occur in almost any part of the body, but because it affects the lungs in 90 percent of the cases, it is usually regarded as a respiratory ailment. The infectious bacterium is usually spread through the air by coughing and sneezing. If the bacterium becomes established in the lungs, they can be damaged in a variety of ways. Most infected people never display any symptoms because a healthy immune system can usually control the bacteria. (See "Tuberculosis: The Return of a Killer" on this page.)

Lung Cancer

Lung cancer, or *pulmonary carcinoma*, usually seems to be caused by environmental factors, cigarette smoking being the one most commonly mentioned. Any inhaled irritant apparently can stimulate some cells to begin abnormal growth. As has been pointed out frequently, early diagnosis is critical. Practically all sufferers from lung cancer cough, most breathe with difficulty, and many have chest pains and spit blood. Any such combination of symptoms calls for chest x-rays, bronchoscopy, and biopsy. Damaged lung tissue shows up in x-rays, sometimes as a discrete region and sometimes in such scattered spots as to suggest a "snowstorm." Lung cancer can spread to other organs, or other organs can send malignant cells to the lungs. The average survival time for a victim of lung cancer after diagnosis is about 9 months. Because it is known that cigarette smokers run about 20 times the risk of non-smokers, cigarette smoking is clearly a risky habit.

Pleurisy

Pleurisy is an infection of the pleurae. It is incapacitating but is not necessarily dangerous in itself. There are several stages and types of the disease. *Fibrinous* or *dry pleurisy* causes intense pain in the parietal pleurae (the visceral pleurae are insensitive) and results in audible crackling or grating when the patient breathes. *Serofibrinous pleurisy* is characterized by deposits of fibrin and by an accumulation of watery fluid (up to 5 L in extreme cases) in the pleural cavity. It can be detected by percussion, but x-rays are more useful. In *purulent pleuritis* or *empyema*, a pus-laden secretion accumulates in the pleural cavity. All types of pleurisy are caused by micro-organisms.

Drowning

One of the most common accidental causes of respiratory failure is *drowning*. A drowning person suffers a laryngeal spasm trying to inhale under water. Little water enters the lungs, but much may be swallowed. Death occurs within minutes unless oxygen is delivered to the tissues promptly.

Cyanosis

Whenever breathing is stopped and the pulse is weak or absent, *cyanosis* (Gr. dark blue) occurs. Cyanosis is the development of a bluish color of the skin, especially the lips, resulting from the build-up of deoxygenated hemoglobin, which is less crimson than oxygenated hemoglobin and appears bluish through the skin.

Hyaline Membrane Disease

Hyaline membrane disease is a failure of newborn infants to produce enough lung surfactant to allow alveoli to fill with air properly. At birth, lungs contain no air, but they must quickly become inflated and stay inflated. In cases of too little lung surfactant, the lungs may be lined with a hyaline (transparent) coating, hence the name of the disease.

Sudden Infant Death Syndrome (SIDS)

Another malady of infants is "crib death," or *sudden infant death syndrome (SIDS)*. It claims almost 10,000 victims under 1 year of age annually in the United States, but its cause remains unknown. (Sleeping face down on a soft pillow may be one of the main causes.) It is included here among respiratory ailments because one likely cause is respiratory failure from spasmodic closure of the air passages or some malfunction of the respiratory center in the brain. The fact that crib death is most common in the autumn seems to indicate that an infectious agent, such as a virus, may be responsible. Another clue offered by a recent study indicates that victims of SIDS show changes in their lungs similar to those accompanying an allergic reaction. Babies who die of SIDS have three times as many immune cells (eosinophils) in their lung tissue as babies who die of other causes. Presumably, toxins released by the eosinophils can cause fever, leakage of fluids into lung tissue, and constriction of alveoli.

Mountain Sickness

Mountain sickness, or *altitude sickness*, is caused mainly by a lack of oxygen — *hypoxia*. As a person reaches higher and higher altitudes above sea level, the atmospheric pressure decreases, as does the concentration of oxygen. The lower levels of oxygen at high altitudes can result in several serious physiological problems, including increased blood flow to the brain, which causes fluid to accumulate; increased levels of blood alkalinity because of overbreathing and the resultant low levels of carbon dioxide; increased production of red blood cells, possibly impeding oxygen flow to tissues and causing blood clots; hemorrhaging in retinal blood vessels, which may overdilate at altitudes above 15,000 ft; fluid accumulation in the alveolar membranes because of hypoxia, which can eventually lead to drowning; and failure of the sodium-potassium pump, producing a depletion of cellular potassium and a resultant edema. Mild forms of mountain sickness usually last only a couple of days.

CHAPTER SUMMARY

General Functions of the Respiratory System (p. 758)

The **respiratory system** delivers oxygen to the body tissues and removes gaseous wastes, mainly carbon dioxide.

Anatomy of the Respiratory Tract (p. 758)

1 The **respiratory tract** includes the *nose, nasal cavities, pharynx, larynx, trachea,* and *bronchi,* which lead by way of *bronchioles* to the *lungs.*

2 Air usually enters the body through the nose. The respiratory mucosa of the **nasal cavity** is specialized to moisten and warm air, and to capture particles such as dust. The olfactory mucosa is specialized to sense odors.

3 The **pharynx** connects the nasal cavity and mouth with the rest of the respiratory tract and the esophagus. The pharynx is divided into the **nasopharynx, oropharynx,** and **laryngopharynx.**

4 The **larynx** contains the **vocal folds,** which are largely responsible for producing sound. At the opening of the larynx, the **glottis** closes during swallowing and opens to allow air to pass.

5 The **trachea,** or windpipe, carries air from the larynx to two **bronchi.** Its mucosal and ciliated lining traps and removes dust particles and microorganisms before they can enter the lungs.

6 The two bronchi divide into smaller and smaller bronchi and then into even smaller **bronchioles,** forming the "respiratory tree." Around the tiniest branches, called respiratory bronchioles, are minute air sacs known as *alveoli.*

7 *Alveoli* are the functional units of the lungs. Exchange of oxygen and carbon dioxide takes place through the walls of the alveoli and the walls of the pulmonary capillaries. The walls of the alveoli are thin enough to permit gases to pass through, but thick enough to remain as open sacs. The tendency of alveolar walls to contract is offset by the action of *lung surfactant* on their inner linings.

8 In the thoracic cavity, the two **lungs** are separated by the *mediastinum.* The right lung has three *lobes,* and the left lung has two. Each lobe is further divided into *bronchopulmonary segments* and then into *lobules.*

9 Each lung is covered by the *visceral pleura,* and the thoracic cavity is lined by the *parietal pleura.* Between these membranes is the **pleural cavity.**

10 The smooth muscle of the tracheobronchial tree is innervated by the autonomic nervous system. The respiratory tree in the lungs is supplied with blood by the pulmonary and bronchial circulations.

Physiology of Respiration (p. 771)

1 *Ventilation* is the mechanical process that moves air into and out of the lungs.

2 During **inspiration,** air is pushed into the lungs by the pressure of the outside air when the size of the thoracic cavity is increased by contraction of the **external intercostal muscles** and the **diaphragm** and the raising of the ribs.

3 During ordinary **expiration,** air is expelled from the lungs when the *respiratory muscles* are relaxed and the volume of the thoracic cavity is reduced. During *forced* expiration, the **abdominal** and **internal intercostal muscles** are contracted. The subsequent reduction in the size of the thorax raises the pressure of the air in the lungs and forces air out.

4 Four *volumes of air movement* are measured in assessing respiratory function. The **residual volume** is the amount of air retained in the lungs at the end of a maximal exhalation. The **tidal volume** is the amount of air inhaled and exhaled during one normal, quiet breath. The **inspiratory reserve volume** is the amount of air greater than the tidal volume that can be taken in during a maximal inhalation. The **expiratory reserve volume** is the amount of air greater than the tidal volume that can be expired during a maximal exhalation.

5 Four *pulmonary capacities* are mea-

sured. The **inspiratory capacity** is inspiratory reserve volume plus tidal volume. The **functional residual capacity** is expiratory reserve volume plus residual volume. The **total lung capacity** is vital capacity plus residual volume. The **vital capacity** is inspiratory reserve volume plus tidal volume plus expiratory volume.

6 The *anatomical dead space* is that part of the respiratory tract in which there is no exchange of respiratory gases: the nasal passages, trachea, bronchi, and bronchioles.

7 The **minute respiratory volume** is the amount of air inspired per minute; it is equal to the *respiratory rate* (the number of breaths per minute) times the volume of air inhaled per breath. **Alveolar ventilation** is the amount of fresh air delivered to the alveoli per breath.

8 According to **Dalton's law,** each gas in a mixture exerts its own pressure, called its *partial pressure,* in proportion to its relative concentration in a mixture of gases.

9 According to **Henry's law,** the quantity of a gas that will dissolve in a liquid is proportional to the partial pressure of the gas. The higher the pressure, the more gas will stay in solution.

10 According to **Fick's law of gas diffusion,** the factors that determine the rate of exchange of oxygen and carbon dioxide across the alveolar-capillary membranes are the partial pressure on either side of the aveolar-capillary membrane, the surface area, the thickness of the membrane, and the solubility and size of the gas molecules.

Gas Transport (p. 777)

1 *External respiration* is the exchange of gases between alveoli and lung capillaries. Because of its higher partial pressure in the alveoli compared with that in the capillaries, oxygen diffuses from the alveoli into the blood. Carbon dioxide moves in the opposite direction.

2 *Internal respiration* is the exchange

of gases in body tissues. There the partial pressures enable the tissues to take up oxygen and release carbon dioxide.

3 About 98 percent of the oxygen carried in blood is bound to the **hemoglobin** of erythrocytes in the form of **oxyhemoglobin.** In tissue capillaries, oxyhemoglobin gives up its oxygen.

4 The binding and release of oxygen are determined by the partial pressure of the gas. A high partial pressure favors binding, and low pressure favors release.

5 Oxygen uptake or release is affected by P_{O_2}, P_{CO_2}, pH, temperature, and presence in erythrocytes of 2,3-bisphosphoglycerate (BPG).

6 Carbon dioxide is carried in blood in three forms: as **dissolved carbon dioxide** (7 percent), as **carbamino compounds,** chiefly **carbaminohemoglobin** (23 percent), and as **bicarbonate ions** (70 percent).

7 When carbonic acid in erythrocytes in tissue capillaries dissociates, the hydrogen ions in the erythrocytes are exchanged for chloride ions from the plasma. In alveolar capillaries, chloride ions are returned to the plasma. This exchange is the **chloride shift.**

Neurochemical Control of Breathing (p. 782)

1 The rate and depth of breathing are controlled by nerve impulses to the breathing muscles from the **respiratory center** in the brain, a system of circuits and controls in the medulla oblongata and pons.

2 The **medullary rhythmicity center** contains two circuits that operate alternately, one stimulating inspiration and the other stimulating expiration.

3 Other breathing controls are in the pons: an **apneustic center** and a **pneumotaxic center,** both of which influence the rhythmicity area.

4 If the concentration of carbon dioxide rises above normal, the concentration of hydrogen ions increases as a result of the dissociation of carbonic acid. This lowering of blood pH directly affects respiratory centers in the brain and is also sensed by carotid and aortic chemoreceptors that send impulses to the medullary rhythmicity center.

5 Respiratory centers are affected by **pulmonary stretch receptors** that send impulses by way of the vagus nerve to the brain to call for a cessation of inhalation and thus prevent overinflation.

Other Activities of the Respiratory System (p. 785)

Other activities of the respiratory system include coughing, sneezing, crying, laughing, yawning, hiccuping, and snoring.

The Effects of Aging on the Respiratory System (p. 785)

1 Aging is usually accompanied by a decrease in gas exchange and a reduction in breathing and lung capacities.

2 Emphysema, lung cancer, chronic bronchitis, and pneumonia may become more prevalent with age.

Developmental Anatomy of the Respiratory Tree (p. 786)

1 Bronchial buds that emerge at about week 4 develop into the left and right primary bronchi of the future lungs. The bronchi show their basic mature form at about week 8, and they branch repeatedly to form more bronchi and bronchioles.

2 Alveoli form by week 24, and lung surfactant begins to be secreted. After 2 or 3 weeks of secretion, the alveoli have expanded enough to be functional.

STUDY AIDS FOR REVIEW

KEY TERMS

alveoli 766
bronchi 762, 764
bronchiole 764
cellular respiration 758
chloride shift 781
diaphragm 772
epiglottis 762
expiration 758, 771, 772
expiratory reserve volume 774
external intercostal muscles 772
external respiration 758, 777
functional residual capacity 774
hemoglobin 777
inspiration 758, 771, 772
inspiratory capacity 774
inspiratory reserve volume 774
internal respiration 758
laryngopharynx 761
larynx 762

lung 766
medullary rhythmicity center 782
minute respiratory volume 774
nasal cavity 758
nasopharynx 761
oropharynx 761
pharynx 761
pleural cavity 768
residual volume 774
respiratory center 782
respiratory tract 758
tidal volume 774
total lung capacity 774
trachea 762
ventilation 758
vital capacity 774

MEDICAL TERMS FOR REFERENCE

ANOXIA Absence of oxygen.

ANTITUSSIVE A drug that controls coughing.

BRONCHITIS Inflammation of the bronchi.

BRONCHODILATOR A drug that dilates the bronchi.

BRONCHOSCOPY Examination of the interior of the bronchi.

DYSPHONIA A pathological condition that affects the vocal folds and causes abnormal sound production.

EMPYEMA A pleural space filled with pus.

EXPECTORANT A drug that encourages the expulsion of mucus from the respiratory tree.

HEMOTHORAX A pleural space filled with blood.

HYPERVENTILATION Excessive movement of air into and out of the lungs.

HYPOXIA Insufficient oxygen in organs and tissues.

LOBECTOMY Removal of one of the lobes of the lungs.

NASAL POLYPS Tumors caused by abnormal growth of the nasal mucous membranes.

PNEUMONECTOMY Removal of one of the lungs.

PULMONARY EDEMA A condition in which there is excessive fluid in the lungs.

SPUTUM (SPYOO-tuhm) Liquid substance containing mucus and saliva.

TRACHEOSCOPY Examination of the interior of the trachea.

TRACHEOTOMY A surgical procedure to make a hole and insert a breathing tube in the trachea.

UNDERSTANDING THE FACTS

1 List the primary and secondary functions of the respiratory system.
2 Contrast external, internal, and cellular respiration.
3 Distinguish between the composition of the conchae, meatuses, and sinuses.
4 Identify the three parts of the pharynx and state the location of each.
5 List the cartilages that compose the larynx.
6 List in order the passages through which inspired air passes between the nasal cavity and the alveoli of the lungs.
7 List the tissue layers in the alveoli through which oxygen and carbon dioxide must diffuse.
8 Distinguish between the visceral and parietal pleurae.
9 What muscles are used during inspiration? During expiration?
10 Contrast tidal volume, residual volume, inspiratory reserve volume, and expiratory reserve volume.
11 What is the difference between vital capacity and total lung capacity?
12 What do we mean by the term partial pressure, as applied to a gas?
13 How is carbon dioxide transported in blood?
14 What is the chloride shift, and what is its effect?
15 What are the components of the respiratory center in the medulla oblongata?
16 What factors can alter the rhythmic rate of quiet breathing?
17 Define pneumonia.
18 Identify the specific location of each of the following:

a nasal cavity
b thyroid cartilage
c vocal folds
d mediastinum
e intercostal muscles
f glottis

UNDERSTANDING THE CONCEPTS

1 How do the structures of the larynx contribute to sound production?
2 How does the structure of the lining of the lumen of the bronchi, bronchioles, and alveoli correlate with their respective functions?
3 What do the pleurae contribute to the functioning of the respiratory system?
4 What is lung surfactant? Which cells secrete it? Why is it important?
5 Why is it important that the capillaries surrounding the alveoli are of small diameter?
6 Why is it important for the respiratory system to have a dual blood supply?

7 What is the force that causes air to enter the lungs during inspiration and to leave the lungs during expiration?

8 Explain how Boyle's law, Dalton's law, Henry's law, and Fick's law pertain to respiration.

9 How do P_{O_2}, P_{CO_2}, and temperature affect the loading and unloading of oxyhemoglobin?

10 Explain why the intake of oxygen can be more effectively increased by an increase in the depth of breathing than by an increase in the rate of respiration.

11 What is the effect of strenuous exercise or an increase in altitude on the loading and unloading of oxyhemoglobin?

12 How does the medullary rhythmicity center regulate the rate of quiet breathing?

13 What effect would an increase of carbon dioxide, a decrease of oxygen, or a decrease in carbon dioxide levels have on the respiratory rate?

SELF-QUIZ
Multiple-Choice Questions

24.1 Which region(s) is (are) lined with stratified squamous epithelium?
 a nasopharynx   d a and b
 b oropharynx   e b and c
 c laryngopharynx

24.2 Which of the following bones is *not* associated with the nasal cavity?
 a ethmoid   d mandible
 b maxillary   e vomer
 c inferior conchae

24.3 Of the following, which is *not* a paranasal sinus?
 a frontal   d maxillary
 b nasolacrimal   e ethmoid
 c sphenoidal

24.4 Lung surfactant is secreted by
 a type I cells   d basement membrane
 b type II cells   e none of the above
 c goblet cells

24.5 Which of the following muscles is/are *not* involved with normal quiet inspiration?
 a internal intercostals   d a and b
 b external intercostals   e b and c
 c diaphragm

24.6 The solubility of a gas is affected by
 a the nature of the gas molecule   d a and b
 b the pressure of the gas   e a, b, and c
 c its partial pressure

24.7 The oxygen-hemoglobin dissociation curve is influenced by all the following *except*
 a increased temperature   d P_{CO_2}
 b increased cellular metabolism   e all of the above
 c pH

24.8 Carbon dioxide is transported in blood primarily
 a in the dissolved state   d as carbaminohemoglobin
 b as HCO_3^-
 c as H_2CO_3 e as BPG

24.9 Carbon monoxide is such a deadly poison because it
 a has a high affinity for hemoglobin
 b preempts binding sites that oxygen might use
 c causes less oxygen to be released to the tissues
 d a and b   e a, b, and c

24.10 As blood passes through the capillaries in the lungs
 a the HCO_3^- concentration increases
 b the amount of oxyhemoglobin decreases
 c blood pH decreases
 d blood P_{O_2} decreases
 e blood P_{CO_2} decreases

24.11 The approximate percentage of O_2 in inhaled air is
 a 10 b 20 c 70 d 80 e 100

24.12 The membrane that covers the lungs is the
 a alveolar membrane   d viscera
 b pleura   e pericardium
 c peritoneum

24.13 Hemoglobin is able to release oxygen in proportion to a tissue's metabolic needs in part because
 a tissue cells are in direct membrane contact with erythrocytes
 b hemoglobin molecules are present in all cells and respond to changes in energy levels
 c the molecule responds to rising CO_2 levels by binding oxygen less tightly
 d the circulation rate of hemoglobin is inversely proportional to the rate of tissue metabolism
 e the capillaries are of larger diameter in regions of high tissue metabolic activity

24.14 A small change in the pH of the blood would affect which of the following?
 a breathing rate
 b the number of oxygen molecules carried by each hemoglobin molecule
 c the excretion rate of CO_2 into the lung
 d the concentration of chloride ions in erythrocytes
 e all of the above

24.15 Oxygen leaves the blood in the capillaries and moves into the tissues because the
 a P_{O_2} in the tissues is lower than the P_{O_2} in the blood
 b P_{CO_2} in the tissues is greater than the P_{CO_2} in the blood
 c P_{CO_2} in the tissues is greater than the P_{O_2} in the blood
 d a and b
 e a, b, and c

Completion Exercises

24.16 Another name for the oral cavity is the _____.

24.17 The anatomical term for the Adam's apple is the _____.

24.18 The _____ is a flap of cartilage that folds down over the opening into the larynx during swallowing.

24.19 The anatomical term for the windpipe is the _____.

24.20 In the lungs, gas exchange occurs within the _____.

24.21 The common term for pulmonary ventilation is _____.

24.22 The exchange of gases between blood and body tissues is termed _____ respiration.

24.23 When oxygen is bound to hemoglobin, _____ is formed.

24.24 The rhythmicity center that controls breathing is located in the _____.

24.25 The vocal folds are held in place by the _____ cartilages.

24.26 The lowermost portion of the pharynx is the _____.

24.27 The respiratory gas that has the greatest effect on the breathing centers is _____.

Matching

  a Boyle's law   **d** Fick's law
  b Dalton's law   **e** compliance
  c Henry's law

24.28 _____ the pressure of a gas is inversely proportional to its volume

24.29 _____ the volume increase for each unit increase in intra-alveolar pressure

24.30 _____ each gas in a mixture can exert its own partial pressure

24.31 _____ a liquid can dissolve more gas at high pressure

24.32 _____ diffusion for small molecules crossing a thin membrane is faster than for large molecules crossing a thick membrane

  a tidal volume   **d** inspiratory capacity
  b residual volume   **e** total lung capacity
  c vital capacity

24.33 _____ vital capacity + residual volume

24.34 _____ normal amount of air taken into lungs during quiet breathing

24.35 _____ air that cannot be expelled from lungs

24.36 _____ inspiratory reserve volume + tidal volume

24.37 _____ inspiratory reserve volume + tidal volume + expiratory reserve volume

A SECOND LOOK

1 In the following drawing, label the hyoid bone, epiglottis, thyroid cartilage, larynx, trachea, and primary and secondary bronchi.

2 In the following drawing, label the capillary network, terminal bronchiole, alveolar duct, and alveolar sac.

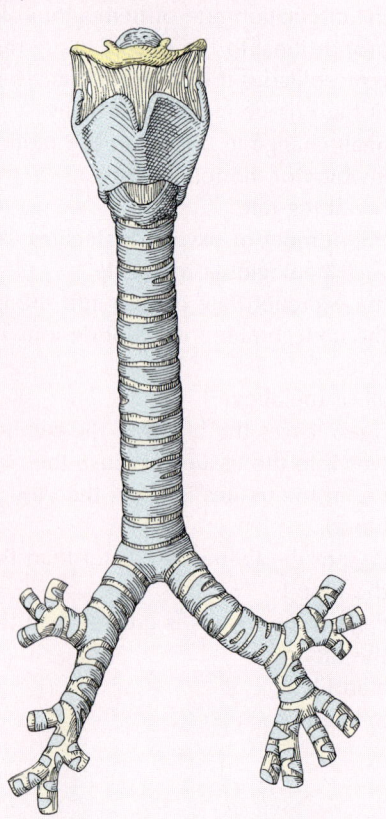

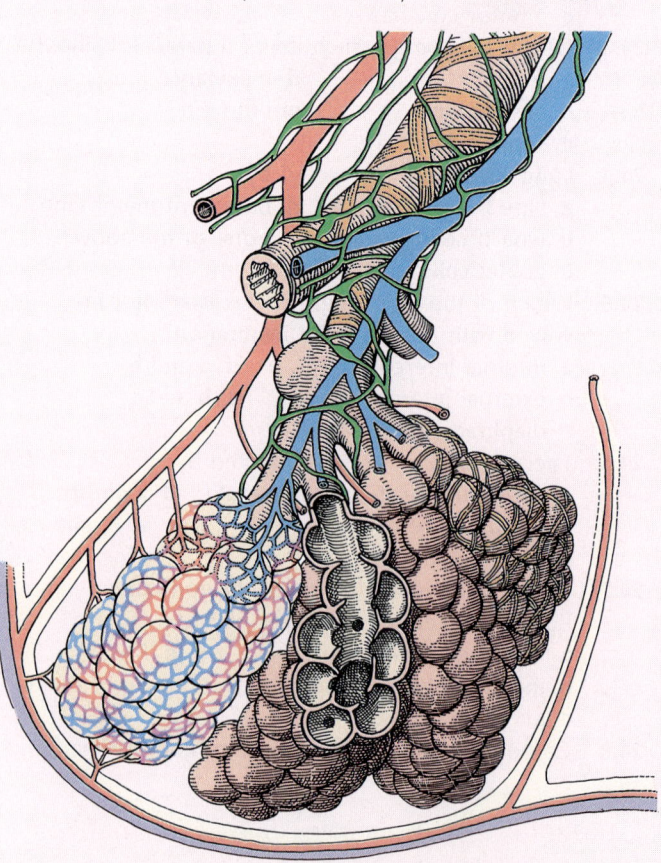

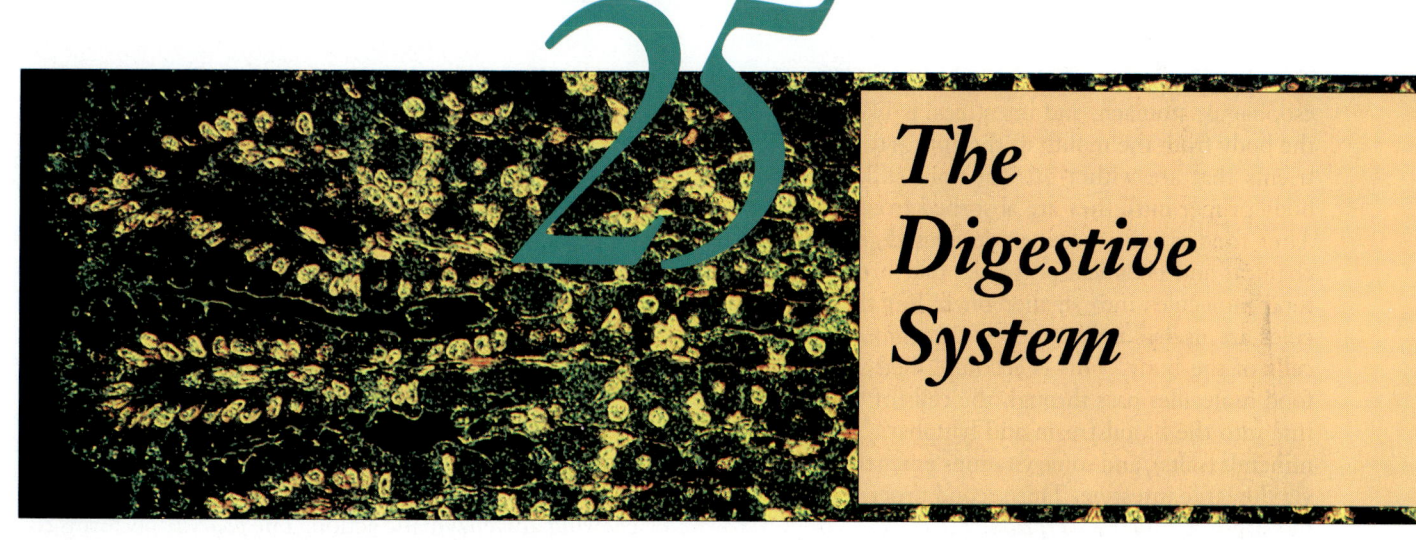

25

The Digestive System

KEY CONCEPTS

1 Digestion involves the breaking up of large food molecules into smaller ones that can be used by cells. Absorption is the process by which these nutrient molecules pass from the small intestine into the bloodstream and lymphatic system.

2 The digestive system is compartmentalized, with each part adapted to a specific function.

3 Various enzymes and hormones are vital to digestion and absorption.

The digestive tract, which includes the mouth, pharynx, esophagus, stomach, and intestines, is like a tube within the body from the mouth to the anus [FIGURE 25.1]. Nutrients that are within this tube are still not inside the body proper until they are absorbed from the intestines. After food is ingested, it undergoes *digestion,* the mechanical and chemical processes that break down large food molecules into smaller ones. But the small molecules are useless unless they can get into the individual cells of the body. This is accomplished when the small food molecules pass through the cells of the small intestine into the bloodstream and lymphatic system. Water, minerals (salts), and some vitamins enter the bloodstream via the large intestine. This second process involving the intestines is called *absorption.*

After digestion and absorption have taken place, the small molecules are ready to be used by the cells of the body. Some of the molecules, such as carbohydrates and fats, are used as a source of energy. Others, such as amino acids, are used by cells to rebuild, repair, and reproduce themselves. Materials that are not digested and absorbed are finally eliminated from the body. Thus the activities of the digestive tract include:

1 *Ingestion,* or eating

2 *Peristalsis,* or the involuntary, sequential smooth muscle contractions that move ingested nutrients along the digestive tract

3 *Digestion,* or the conversion of large nutrient molecules into small molecules

4 *Absorption,* or the passage of usable nutrient molecules from the intestines into the bloodstream and lymphatic system for the final passage into body cells

5 *Defecation,* or the elimination from the body of undigested and unabsorbed material as solid waste

INTRODUCTION TO THE DIGESTIVE SYSTEM

In simple terms, digestion is the process of breaking down large molecules of food that cannot be used by the body into small, soluble molecules that can be absorbed and used by cells. *Chemical digestion* breaks down food particles through a series of metabolic reactions involving enzymes. *Mechanical digestion* involves physical processes such as chewing, peristalsis, and the churning movements of the stomach and small intestine to mix the food with enzymes and digestive juices. The digestive process, assisted by enzymes, takes place in the *digestive tract,* or *alimentary canal.* The part of the digestive tract inferior to the diaphragm (which includes the stomach and intestines) is called the *gastrointestinal (GI) tract.*

The digestive tract (alimentary canal) consists of the mouth, pharynx, esophagus, stomach, small intestine, large intestine, rectum, anal canal, and anus. From mouth to anus this canal is about 9 m (30 ft) long. The *associated structures* of the digestive system include the teeth, lips, tongue, cheeks, salivary glands, pancreas, liver, and gallbladder, all of which will be described later in this chapter.

Basic Functions of the Digestive Tract

The digestive tract is divided into compartments, each one adapted to perform a specific function [FIGURE 25.1A]. In the *mouth,* the breakdown of food begins with chewing and enzymatic action. The *pharynx* performs the act of swallowing, and food passes through the *esophagus* to the stomach. The *stomach* stores food and breaks it down with acid and some enzymes. The *small intestine* is where most of the large molecules are chemically broken down into smaller ones and where most of the nutrients are absorbed into the bloodstream and lymphatic system. The *large intestine* absorbs minerals and some vitamins, carries undigested food, removes additional water from it, and releases solid waste products through the *anus.* The *associated structures* assist in the breakdown and conversion of food particles in a variety of ways, both mechanical and chemical.

Tissue Structure of the Wall of the Digestive Tract

Despite the compartmentalization of the digestive tract, the walls of the various portions of the tract have the same basic organization. The wall of the tube (from the inside out) consists of four main layers of tissue: *mucosa,* *submucosa,* *muscalaris externa,* and *serosa.* They are shown and described in FIGURE 25.2.

MOUTH

By following a mouthful of food through the digestive tract, we can observe the digestive process in detail. Food enters the digestive system through the *mouth* (also called the *oral cavity*) [FIGURE 25.3]. The mouth has two parts: (1) The small, outer *vestibule,* or *buccal* (BUCK-uhl) *cavity,* is bounded by the lips and cheeks on the outside and the teeth and gums on the inside. (2) The *oral cavity proper* extends from behind the teeth and gums to the *fauces* (L. throat), the opening that leads to the *pharynx.* The fauces has sensory nerve endings that trigger the involuntary phase of swallowing. If the sensory region of the fauces is stimulated (with a tongue depressor, for example), the *gag reflex* is produced.

FIGURE 25.1 DIGESTIVE SYSTEM

[A] The drawing shows structures of the digestive tract and associated organs.

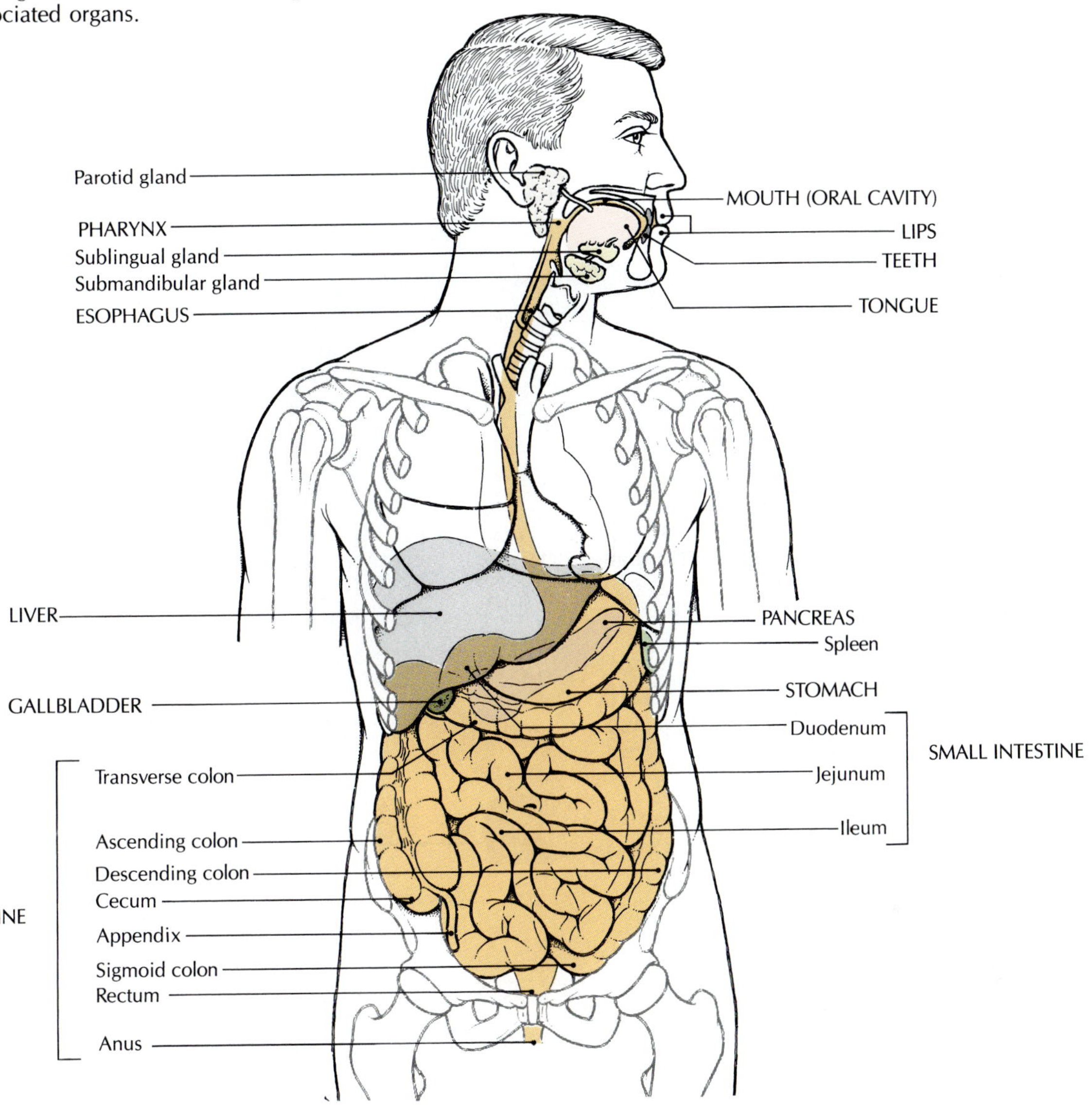

Parotid gland

PHARYNX
Sublingual gland
Submandibular gland
ESOPHAGUS

MOUTH (ORAL CAVITY)
LIPS
TEETH
TONGUE

LIVER

PANCREAS
Spleen

GALLBLADDER

STOMACH
Duodenum

SMALL INTESTINE

Transverse colon

Jejunum

Ascending colon
Descending colon
Cecum

Ileum

LARGE
INTESTINE

Appendix
Sigmoid colon
Rectum

Anus

[A]

Structure	Major digestive secretions	Major digestive functions
Lips and cheeks	None	Hold food in position to be chewed; help identify food textures. Buccal muscles contribute to chewing process.
Teeth	None	Break food into pieces, exposing surface for digestive enzymes.
Tongue	None	Assists chewing action of teeth; helps shape food into bolus, pushes bolus toward pharynx to be swallowed; contains taste buds.
Palate	None	Hard palate helps crush and soften food. Soft palate closes off nasal opening during swallowing.
Salivary glands	Saliva—salivary amylase, mucus, water, various salts	Salivary amylase in saliva begins breakdown of cooked starches into soluble sugars maltose and dextrin. Saliva itself helps form bolus and lubricates it prior to swallowing; dissolves food for tasting; moistens mouth; helps prevent tooth decay.
Pharynx	None	Continues swallowing activity when bolus enters from mouth; propels bolus from mouth into esophagus.

(Figure 25.1 continues on following page)

FIGURE 25.1 DIGESTIVE SYSTEM (Continued)

[B] Photograph of the digestive system, anterior view.

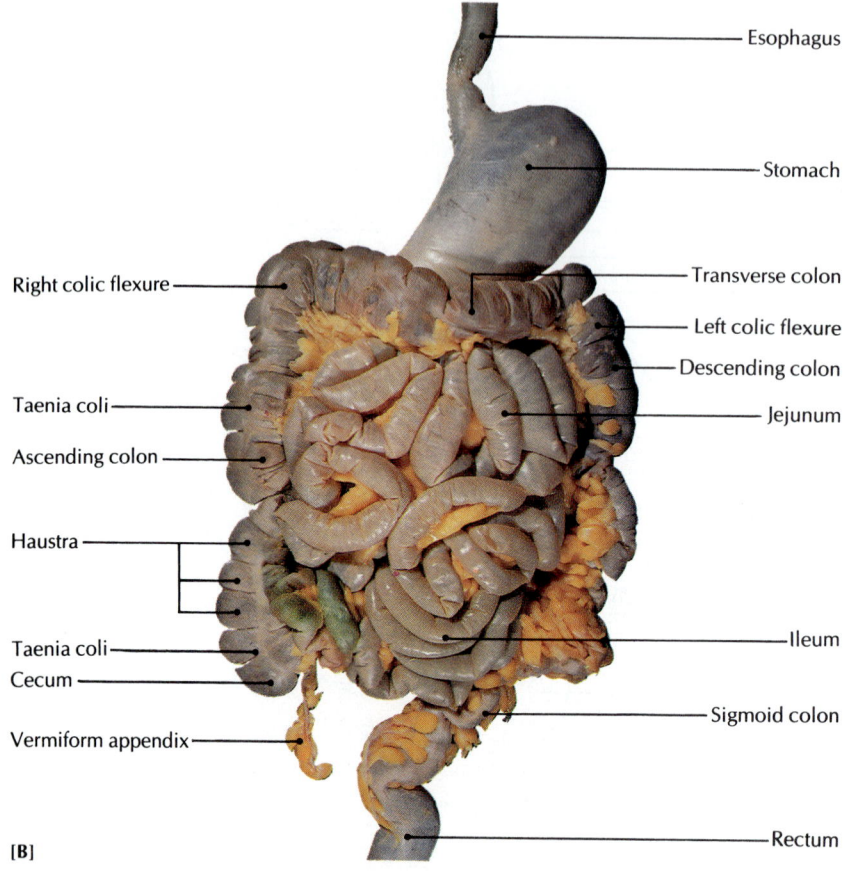

[B]

Structure	Major digestive secretions	Major digestive functions
Esophagus	Mucus	Propels food from pharynx into stomach.
Stomach	Hydrochloric acid, pepsin, mucus, some lipase, gastrin, intrinsic factor	Stores, mixes, digests food, especially protein; regulates even flow of food into small intestine. Acid kills bacteria.
Small intestine	Enzymes: enterokinase, peptidases, maltase, lactase, sucrase, intestinal amylase, intestinal lipase; salts, water, mucus; hormones: cholecystokinin, gastric inhibitory peptide, secretin	Site of most chemical digestion and absorption of most water and nutrients into bloodstream.
Pancreas	Enzymes: trypsin, chymotrypsin, carboxypeptidase, pancreatic amylase, pancreatic lipase, nuclease; bicarbonate	Secretes many digestive enzymes; neutralizes stomach acid with alkaline bicarbonate secretion.
Liver	Bile, bicarbonate	Secretes bile; detoxifies harmful substances; converts nutrients into usable forms and stores them for later use; neutralizes stomach acid with alkaline bicarbonate secretion.
Gallbladder	Mucin	Stores and concentrates bile from liver and releases it when needed.
Large intestine	Mucus	Removes salts and water from undigested food; releases feces through anus; aids synthesis of vitamins B_{12} and K.
Rectum	None	Removes solid wastes by process of defecation.

FIGURE 25.2 THE WALL OF THE DIGESTIVE TRACT

The drawings show the basic organization of the wall of the digestive tract. [A] Cross section. [B] Enlargement of the inner portion (tunica mucosa).

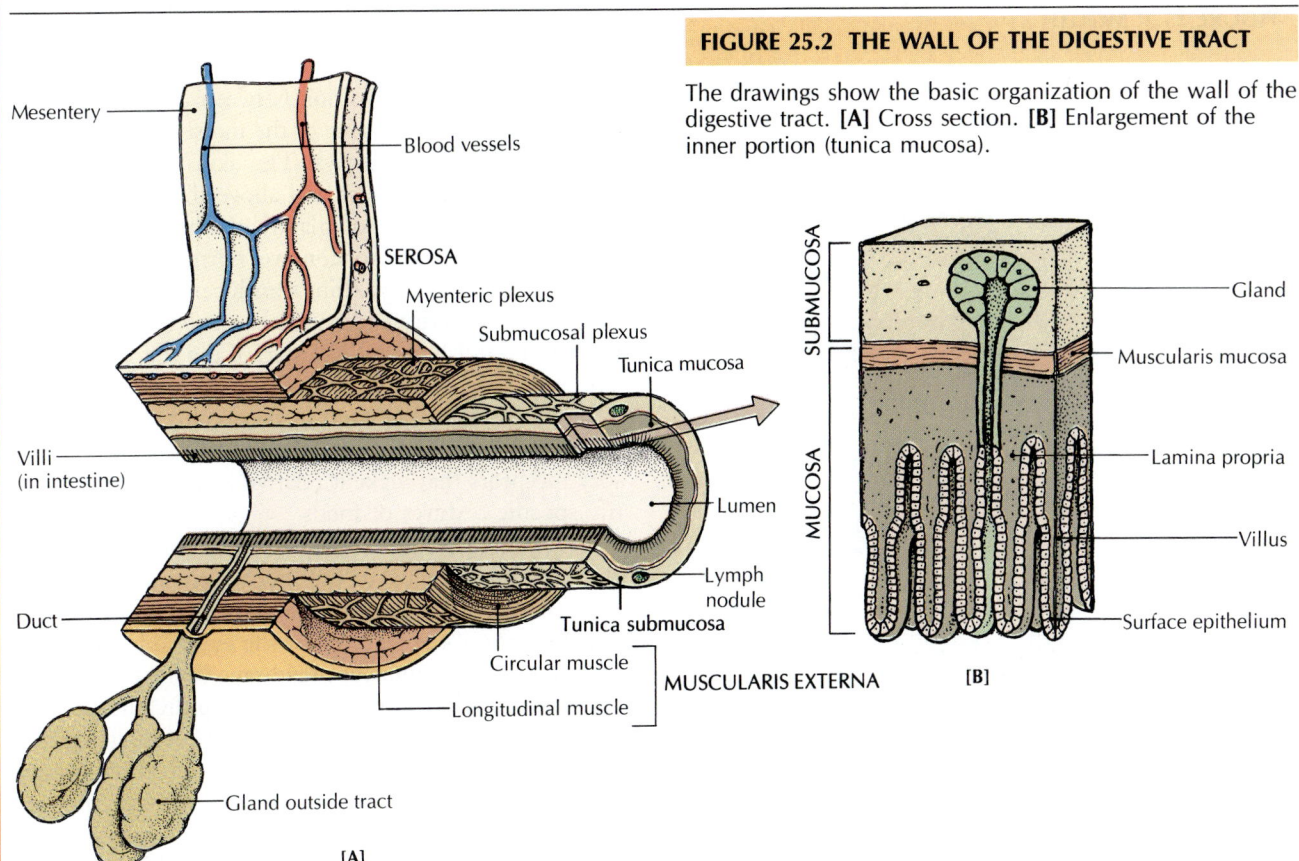

Layer	Description
MUSCOSA (tunica mucosa, mucous membrane)	
Epithelium Lamina propria Muscularis mucosa	Innermost layer of digestive tract. *Epithelium* is site of interaction between ingested food and body. *Lamina propria* is a layer of connective tissue supporting epithelium; contains lymphatic nodules and many lymphocytes. *Muscularis mucosa* consists of two thin muscular layers and contains a nerve plexus. Surface area of epithelial layer is increased by large folds (rugae in the stomach and plicae circulares and villi in the small intestine), indentations (crypts), and glands. Acts as lubricating, secreting, absorbing layer. Lubricates solid contents and facilitates their passage through digestive tract. Contains absorptive cells and secretory cells that produce mucus, enzymes, various ions.
SUBMUCOSA (tunica submucosa)	
	Highly vascular layer of connective tissue between mucosa and muscle layers; contains many nerves and in certain areas, lymphatic nodules. Submucosal glands are present in esophagus and duodenum.
MUSCULARIS EXTERNA (tunica muscularis)	
Circular muscle Longitudinal muscle Oblique muscle	Main muscle layer, consisting in most regions of inner *circular* layer and outer *longitudinal* layer of mostly smooth muscle. Stomach contains additional *oblique* layer internal to other layers. Upper esophagus and sphincters of anus consist of skeletal muscle fibers. Moves food through lumen of digestive tract by waves of muscular contraction called *peristalsis*.
SEROSA (tunica serosa)	
Visceral peritoneum Adventitia	Outermost lamina, consisting of thin connective tissue, and in many places epithelium covering digestive tube and digestive organs. Where epithelium is lacking (as in esophagus), serosa is called *adventitia*. Portion covering viscera is *visceral peritoneum*. Double-layered *mesentery* is portion of serous membrane that connects intestines to dorsal abdominal wall. Contains blood vessels, nerves, and lymphatic vessels.

FIGURE 25.3 MOUTH

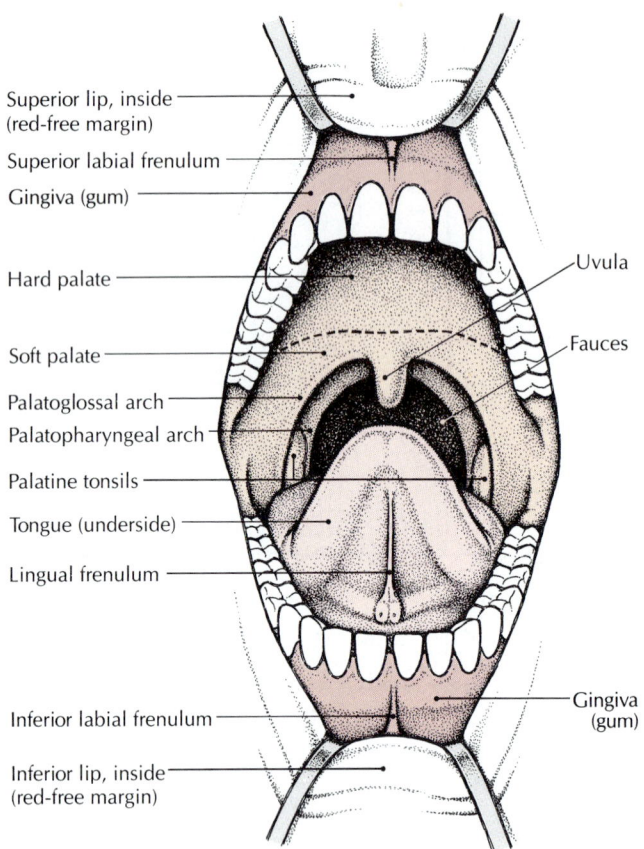

- Superior lip, inside (red-free margin)
- Superior labial frenulum
- Gingiva (gum)
- Hard palate
- Soft palate
- Palatoglossal arch
- Palatopharyngeal arch
- Palatine tonsils
- Tongue (underside)
- Lingual frenulum
- Inferior labial frenulum
- Inferior lip, inside (red-free margin)
- Uvula
- Fauces
- Gingiva (gum)

frenum, bridle) [FIGURE 25.3]. The frenulum under the tongue is the *lingual frenulum*. It limits the backward movement of the tongue.

The lips form a transition between the external skin and the mucous membrane of the moist epithelial lining of the internal passageways. The skin of the outer lip surface (beneath the nose and above the chin) contains the usual sweat glands, oil glands, and hair follicles. The *red free margin*, the portion we normally think of as "lips," is covered by a thin, translucent epidermis that allows the capillaries underneath to show through, giving the lips a reddish color. The skin of the red free margin does not contain sweat glands, oil glands, or hair follicles.

The lips help place food in the mouth and keep it in the proper position for chewing. They also contain sensory receptors that make it relatively easy for us to identify specific textures of foods.

The moist mucous membrane of the inner surface of the lips leads directly to the mucous membrane of the inner surface of the cheeks. The **cheeks** are the fleshy part of either side of the face, below the eye and between the nose and ear. The mucous membrane lining the lips and cheeks is a thick stratified squamous epithelium (nonkeratinized). Such an epithelium is typical of moist epithelial surfaces that are subjected to a great deal of abrasive force. As soon as the surface cells are worn away, they are replaced by the rapidly dividing cells underneath.

The cheeks, like the lips, help hold food in a position where the teeth can chew it conveniently. Also, the muscles of the cheeks contribute to the chewing process.

In the mouth, food is *masticated* (chewed) by the ripping and grinding action of the teeth. As much pressure as 500 kg/sq cm (7000 lb/sq in.) can be exerted by the molars. At the same time, the food is moistened by saliva, which intensifies its taste and eases its passage down the esophagus. Saliva, secreted by the salivary glands, contains an enzyme—**salivary amylase**—that begins the breakdown of large carbohydrate molecules into small sugar molecules. This digestive enzyme is the first of many in the digestive tract.

Lips and Cheeks

The **lips** are the two fleshy, muscular folds that surround the opening of the mouth. They consist mainly of fibroelastic connective tissue and skeletal muscle covered by epithelium. The orbicularis oris muscle makes the lips capable of versatile movement. The lips are also extremely sensitive and are abundantly supplied with blood vessels, lymphatic vessels, and sensory nerve endings from the trigeminal nerve.

Each lip is connected at its midline to the gum by a fold of mucous membrane called a *labial frenulum* (L.

Teeth and Gums

Six months or so after birth, the first **deciduous*** teeth (baby teeth, milk teeth) erupt through the gums [FIGURE 25.4A]. A normal child will eventually have 20 "baby" teeth, each jaw holding 10 teeth: 4 *incisors* (for cutting), 2 *canines* (for tearing), and 4 *premolars* (for grinding). The deciduous teeth are lost when the permanent teeth are ready to emerge. Both sets of teeth are usually present in the gums at birth or shortly afterward, with the permanent teeth lying under the deciduous teeth [FIGURE 25.4B]. By the time a permanent tooth is ready to erupt, the root of the external deciduous tooth above it has been completely resorbed by osteoclasts. The six permanent molars in each jaw have no deciduous predecessors. The shedding of deciduous teeth and the appearance of permanent teeth follow a fairly consistent pattern, as shown in TABLE 25.1.

The 32 permanent teeth (16 in each jaw) are arranged in two arches, one in the upper jaw (maxilla) and the other in the lower jaw (mandible). Each jaw holds four

---

* *Deciduous* (dih-SIHDJ-oo-us) means "to fall off" and is typically used to describe trees that shed their leaves in the autumn.

FIGURE 25.4 TEETH

[A] Deciduous and permanent teeth in the upper and lower jaws. [B] X-ray showing the permanent second premolar (arrow) in place beneath the deciduous tooth of a 10-year-old child. [C] Sagittal section of a tooth.

a   b   c d e

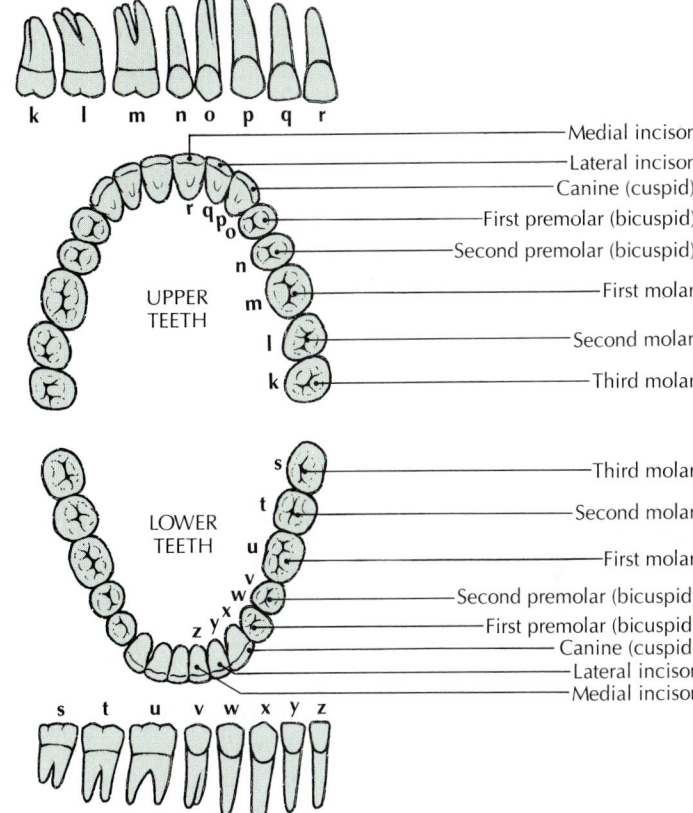

k   l   m   n o p q r

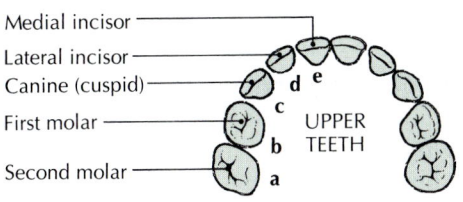

Medial incisor
Lateral incisor
Canine (cuspid)
First molar
Second molar

UPPER TEETH

Medial incisor
Lateral incisor
Canine (cuspid)
First premolar (bicuspid)
Second premolar (bicuspid)
First molar
Second molar
Third molar

UPPER TEETH

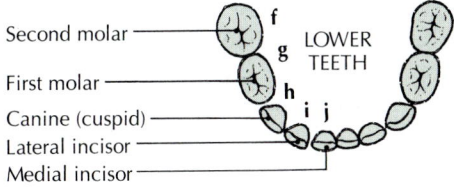

Second molar
First molar
Canine (cuspid)
Lateral incisor
Medial incisor

LOWER TEETH

Third molar
Second molar
First molar
Second premolar (bicuspid)
First premolar (bicuspid)
Canine (cuspid)
Lateral incisor
Medial incisor

LOWER TEETH

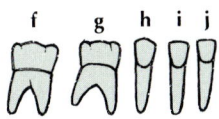

f   g   h i j

s   t   u   v w x y z

[A]   DECIDUOUS "BABY" TEETH

PERMANENT TEETH

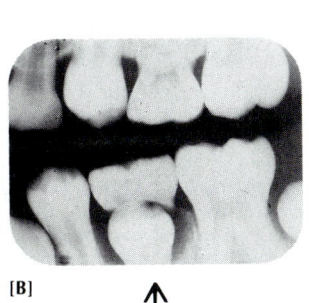

[B]

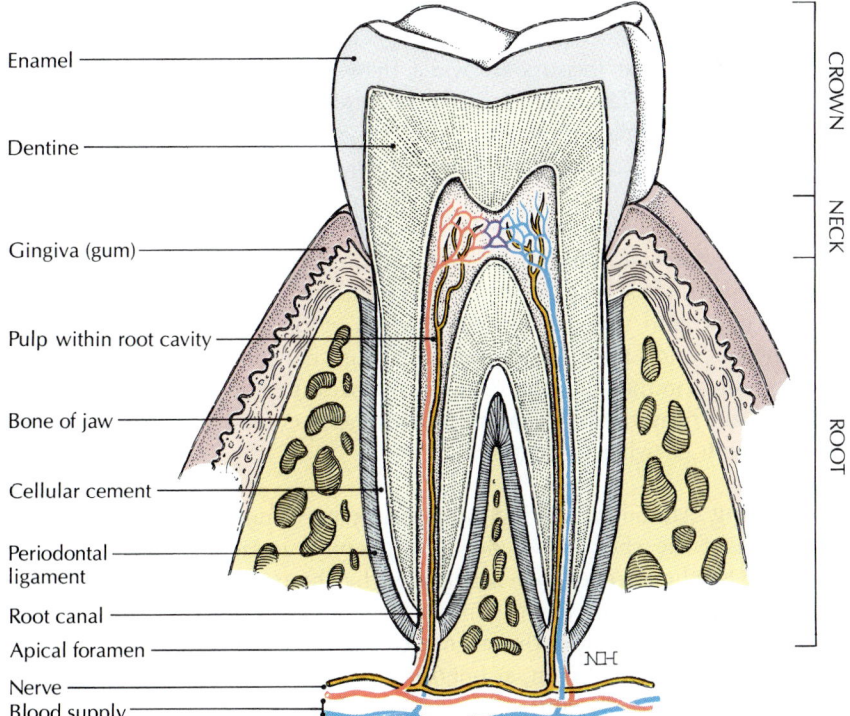

Enamel
Dentine
Gingiva (gum)
Pulp within root cavity
Bone of jaw
Cellular cement
Periodontal ligament
Root canal
Apical foramen
Nerve
Blood supply

CROWN
NECK
ROOT

[C]

TABLE 25.1 ERUPTION AND SHEDDING OF TEETH

| Type of tooth | Deciduous (baby) teeth | | Permanent teeth |
	Approximate time of eruption, months*	Approximate time of shedding, years*	Approximate time of eruption, years*
Medial incisors†	6–12	6–7	6–8
Lateral incisors	9–16	7–8	7–9
Canines	16–23	9–12	9–12
First premolars	14–19	9–11	10–12
Second premolars	23–33	10–12	10–12
First molars	—	—	6–7
Second molars	—	—	11–13
Third molars (wisdom teeth)	—	—	17–21

*According to the American Dental Association.
†Deciduous medial incisors are sometimes present at birth.

incisors (cutting teeth), two *canines* (cuspid, with one point or cusp in its crown), four *premolars* (bicuspids, each with two cusps), and six *molars* (millstone teeth). Third molars, commonly called *wisdom teeth*, usually erupt between the ages of 17 and 21. Sometimes, however, they remain within the alveolar bone and do not erupt. In such cases, the teeth are said to be *impacted* and may have to be removed surgically.

Because the upper incisors are wider than the lower ones, the lower grinding teeth are usually aligned slightly in front of the upper grinders. This arrangement enhances the grinding motion between the upper and lower teeth.

The teeth are held in their sockets by connective tissue called *periodontal ligaments* [FIGURE 25.4C]. The collagenous fibers of each ligament, which extend from the alveolar bone into the cement of the tooth, allow for some normal movement of the teeth during chewing. Nerve endings in the ligaments monitor the pressures of chewing and relay the information to the brain centers involved with chewing movements.

Chewing Although chewing can certainly be voluntary, most of the chewing we do during a meal is an automatic rhythmic reflex that is triggered by the pressure of food against the teeth, gums, tongue, and hard palate. Such pressure causes the jaw muscles to relax and the jaw to drop slightly; then, as opposing muscles contract, the jaw is raised again.

Parts of a tooth All teeth consist of the same three parts: (1) a *root* embedded in a socket (alveolus) in the alveolar process of a jaw bone, (2) a *crown* projecting upward from the gum, and (3) a narrowed *neck* (cervix) between the root and crown, which is surrounded by the gum [FIGURE 25.4C]. The incisors, canines, and premolars have a single root, although the first upper premolar may initially have a double root. The lower molars have two flattened roots, and the upper molars have three conical roots. At the apex of each root is the *apical foramen*, which leads successively into the *root canal* and the *root cavity* (pulp cavity).

Composition of teeth and gums Each tooth is composed of dentine, enamel, cement, and pulp [FIGURE 25.4C]. The *dentine* is the extremely sensitive yellowish portion surrounding the pulp cavity. It forms the bulk of the tooth. The *enamel* is the insensitive white covering of the crown. It is the hardest substance in the body. (In order to cut through enamel, a dentist's drill spins at about half a million revolutions per minute.) The *cement* is the bonelike covering of the neck and root. The *pulp* is the soft core of connective tissue that contains the nerves and blood vessels of the tooth.

Lining the innermost surface of the dentine are odontoblasts of the pulp. Their processes extend through the dentine to the border with the enamel. The teeth's sensitivity to pain is attributed to the odontoblasts, which convey the effects of stimulation to nerve endings in the pulp.

The *gum* (Old Eng. *goma*, palate, jaw), also called the *gingiva* (JIHN-jye-vuh), is the firm connective tissue covered with mucous membrane that surrounds the alveolar processes of the teeth. The stratified squamous epithelium of the gums is slightly keratinized to withstand friction during chewing. The gums are usually attached to the enamel of the tooth somewhere along the crown, but the gum line gradually recedes as we get older. In fact, the gum line may recede so far in elderly people that the gum is attached to the cement instead of the enamel.*

*The expression "long in the tooth," relating to elderly people, refers to this recession of the gums, which exposes more and more of the full length of the tooth.

Q: *Why are canine teeth called "eyeteeth"?*

A: Early anatomical schemes often named body parts according to their relation to other structures or functions. Eyeteeth were so named probably because they lie directly under the eyes. (As another example, people in the Western world wear wedding rings on the fourth finger of the left hand—counting the thumb—because it was believed that this finger is connected directly to the heart.)

Tongue

The *tongue* functions in mechanical digestion, mainly in chewing and helping to move food from the mouth down into the pharynx. The front of the tongue is used to manipulate the food during chewing, and the base of the tongue aids in swallowing. It is also a sensitive tactile organ, plays an important role in speech, and has sensory receptors that help us taste our food.

The tongue is composed mostly of skeletal muscle and is covered by a smooth film of mucous membrane on the underside. The irregular dorsal (top) surface contains papillae, taste buds, and other structures associated with sensing different tastes.

The mucous membrane covering the tongue is ordinarily divided into two sections: the *oral part* on the anterior two-thirds and the *pharyngeal part* on the posterior third. The separate sections are delineated by the V-shaped sulcus terminalis [see FIGURE 17.3]. The oral part, corresponding to the *body* of the tongue, contains the three types of papillae: (1) *Filiform papillae* are located on the anterior two-thirds of the tongue; they appear whitish and contain no taste buds. (2) *Fungiform papillae* are located on the tip of the tongue; they contain taste buds. (3) *Circumvallate papillae* are in the shape of a V on the posterior of the oral part of the tongue; they contain taste buds.

The pharyngeal part, representing the *root* of the tongue, contains the lymphatic nodules of the lingual tonsil.

The interlacing muscles of the tongue are so arranged that it can be moved side to side and in and out. The tongue contains (1) three bilateral pairs of extrinsic muscles (muscles with attachments outside the tongue)—hypoglossus, genioglossus, styloglossus—and (2) four pairs of intrinsic muscles (muscles wholly within the tongue)—longitudinalis superior, longitudinalis inferior, transversus lingual, and verticalis lingual. The extrinsic muscles move food within the mouth to form it into a round mass, or *bolus*, and the intrinsic muscles assist in swallowing.

Palate

The *palate* (PAL-iht), or "roof of the mouth," is one of the many examples of structures that are perfectly suited to their functions. It has two sections [FIGURE 25.3]: (1) The anterior *hard palate,* bordered by the upper teeth, is formed by a portion of the palatine bones and maxillae. Its upper surface forms the floor of the nasal cavity. (2) The posterior *soft palate* is continuous with the posterior border of the hard palate. It extends between the oral and nasal portions of the pharynx, with a small fleshy cone called the *uvula* (YOO-vyoo-luh; L. small grapes) hanging down from the center of its lowermost border [FIGURE 25.3]. The uvula helps keep food

from entering the nasal passages during swallowing. Extending laterally and downward from each side of the base of the soft palate are two curved folds of mucous membranes called the *palatoglossal* and *palatopharyngeal arches.* The palatine tonsils lie between the two arches.

When food is being chewed and moistened with saliva, the tongue is constantly pushing it against the ridged surface of the hard palate, crushing and softening the food before it is swallowed. The hard palate is covered with a firmly anchored mucous membrane and the same tough epithelium of nonkeratinized stratified squamous epithelium as the cheeks.

The soft palate has a very different structure and function. It is composed of interlacing skeletal muscle that allows it to move up and down over the nasopharynx when it is raised. In this way it functions to close off the nasopharynx from the nasal cavity during swallowing, preventing food from being forced into the nasal cavity.

Salivary Glands

Many buccal and minor salivary glands secrete saliva into the oral cavity, but the term *salivary glands* usually refers to the three largest pairs: the parotid, submandibular, and sublingual glands [FIGURE 25.5]. These three pairs of salivary glands secrete more than a liter of saliva daily. *Saliva* is the watery, tasteless mixture of salivary and oral mucous-gland secretions. It lubricates chewed food, moistens the oral walls, contains salts to buffer chemicals in the mouth, and also contains *salivary amylase,* the enzyme that begins the digestion of carbohydrates.

Parotid glands The *parotid glands* (Gr. *parotis,* "near the ear") are the largest of the three main pairs of salivary glands.* They lie in front of the ears, covering the masseter muscle posteriorly [FIGURE 25.5]. The long ducts from the glands (called *parotid* or *Stensen's ducts*) pass forward over the masseter muscle (they can be felt as a ridge by moving the tip of a finger up and down over the muscle) and end in the vestibule alongside the second upper molar tooth. The parotid glands secrete water, salts, and salivary amylase, but unlike the other salivary glands, they do not secrete *mucin* (MYOO-sihn), a protein that forms mucus when dissolved in water. As a result, the saliva from the parotid glands is clear and watery.

Submandibular glands The *submandibular* ("under the mandible") *glands* are located on the medial side of the mandible. They are about the size of a walnut, roughly half the size of the parotid glands. Their ducts,

---

*A specific viral infection of the parotid glands produces *mumps.* The incidence of this highly contagious disease has been considerably reduced since the advent of the combined measles, mumps, and German measles (MMR) vaccine, which is routinely given to children when they are 15 months old.

FIGURE 25.5 MAJOR SALIVARY GLANDS

This right lateral view shows the three major salivary glands: parotid, submandibular, and sublingual.

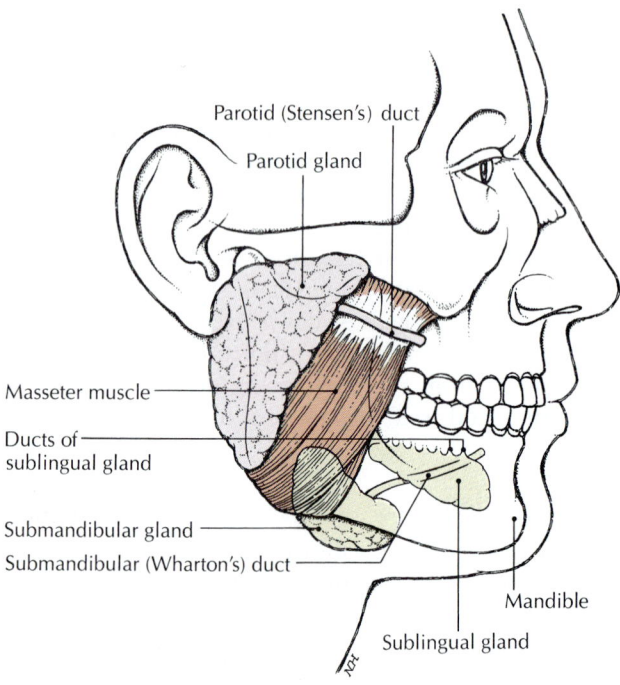

zyme, a bacteriolytic enzyme. The interaction of mucin with the water in saliva produces the highly viscous mucus that moistens and lubricates the food particles so they can slide down the pharynx and esophagus during swallowing. Salivary amylase immediately acts on carbohydrates, breaking them down into smaller carbohydrates. Lysozymes, which are present in small amounts, help destroy oral bacteria.

Saliva also cleanses the mouth and teeth of cellular and food debris, helps keep the soft parts of the mouth supple, and buffers the acidity of the oral cavity [TABLE 25.2]. The bicarbonate concentration of saliva helps reduce **dental caries** (cavities) by neutralizing the acidity of food in the mouth.† Remember also that taste buds cannot be stimulated until the food molecules are dissolved. Saliva provides the necessary solvent. Thus saliva begins the digestion of carbohydrates, but is more important in oral hygiene, taste, and speech.

Control of salivary secretion Because the mouth and throat must be kept moist at all times, there is a continuous, low-level secretion of saliva. The presence of

---

†A substance in saliva called *sialin* seems to neutralize tooth-decaying acids formed by bacteria. People with an abundance of sialin appear to get fewer cavities than do those with only a small amount. The amount of sialin in saliva is determined genetically.

---

called *submandibular (Wharton's) ducts*, open into a papilla on the floor of the mouth beside the lingual frenulum behind the lower incisors. The submandibular glands secrete water, salts, salivary amylase, and mucin. They secrete a thicker saliva than the parotid glands do, with less salivary amylase.

Sublingual glands The *sublingual* ("under the tongue") *glands* are located in the floor of the mouth beneath the tongue.* They are the smallest major salivary glands. Each gland drains through a dozen or so small ducts located on the summit of a fold on either side of the lingual frenulum, just behind the orifice of the submandibular duct. The sublingual glands secrete mostly water, salts, and mucin. Their secretion is the most viscous saliva of the three types of salivary glands. It is high in mucus and low in salivary amylase.

Saliva and its functions Saliva (pH 6.35 to 6.85) contains about 99.5 percent water and 0.5 percent ions (sodium, potassium, chloride, bicarbonate, phosphates) and proteins. The major proteins include *mucin*, which helps form mucus, and the enzymes *salivary amylase*, which initiates the breakdown of carbohydrates, and *lyso-*

---

*In human beings, parts of the submandibular and sublingual glands merge to form a submandibular-sublingual complex.

TABLE 25.2 COMPOSITION AND FUNCTIONS OF SALIVA

Major components	Functions
Water (about 99.5% of total composition of saliva)	Provides solvent for food in order for tasting and digestive reactions to occur; moistens mouth; aids speech.
Bicarbonates	Help maintain pH of saliva at about 6.35 to 6.85.
Chlorides	Activate salivary amylase.
Immunoglobulin A (IgA)	Part of salivary antibacterial system.
Lysozyme	Bacteria-destroying enzyme; prevents dental decay, infection of mucous membrane.
Mucin	A protein that helps form mucus.
Mucus	Lubricates food, helps form bolus; aids swallowing; helps buffer acids and bases.
Phosphates	Help maintain pH of saliva.
Salivary amylase	Catalyzes the breakdown of carbohydrates.
Urea, uric acid	No digestive function; waste products excreted via saliva.

food in the mouth and stomach, the act of chewing (which presses on chemoreceptors and pressoreceptors in the mouth), and even the smell, taste, sight, or thought of food usually stimulate the secretion of saliva [FIGURE 25.6]. The average daily output of saliva is about 1000 to 1500 mL.

Salivary secretion is entirely under nervous control, with no hormonal stimulation. (Other digestive secretions are regulated by both nervous and hormonal control.) The major stimulation of the salivary glands comes from the parasympathetic division of the autonomic nervous system. Afferent (sensory) input from pressoreceptors or chemoreceptors in the mouth and nose and input — associated with the sight, smell, or thought of food — from the cerebral cortex are conveyed to the *superior* and *inferior salivatory nuclei* in the brainstem. Efferent (motor) stimulation via the parasympathetic fibers of the facial (VII) and glossopharyngeal (IX) nerves innervate the salivary glands and increase the secretion of a watery, enzyme-rich saliva [FIGURE 25.6]. In contrast, sympathetic stimulation of the salivary glands decreases their output, resulting in a thick, mucin-rich saliva.

Unpleasant stimuli (such as the smell of rotten food) inhibit the salivary reflex and cause the mouth and pharynx to become dry. The presence of a beautiful, tasty meal increases salivary secretion. Activation of the sympathetic division of the autonomic nervous system, as in fear or rage, decreases salivary secretion, and results in "dry mouth."

FIGURE 25.6 NEURAL PATHWAY FOR THE SALIVARY REFLEX

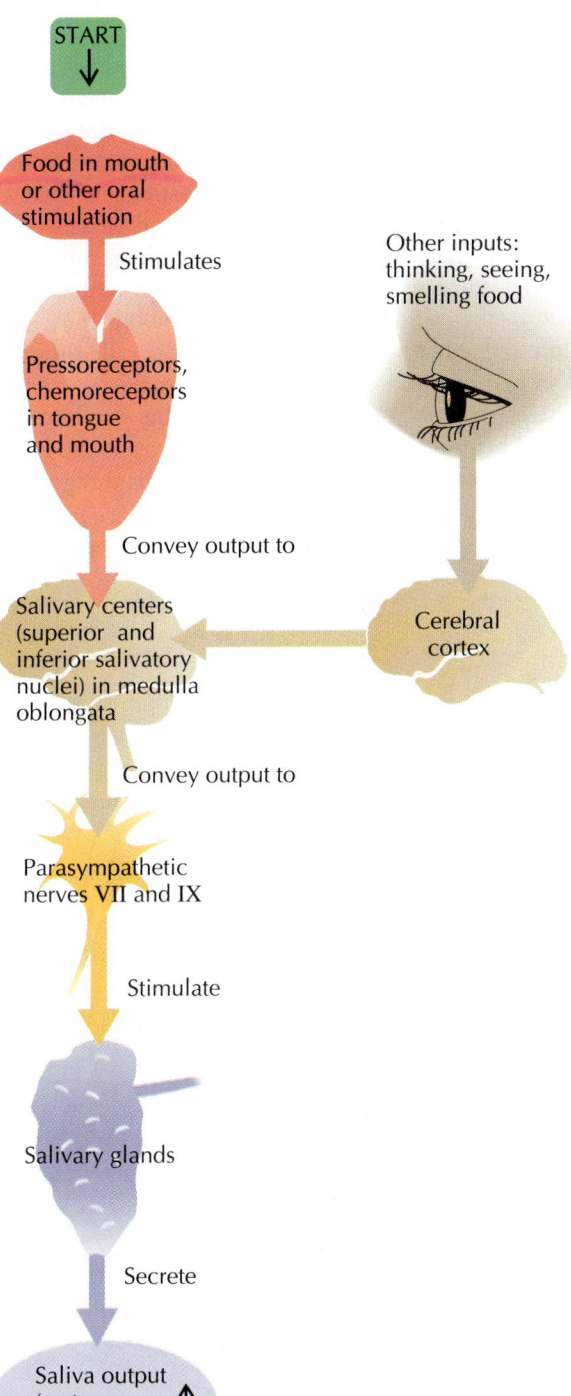

START

Food in mouth or other oral stimulation

Stimulates

Pressoreceptors, chemoreceptors in tongue and mouth

Convey output to

Other inputs: thinking, seeing, smelling food

Salivary centers (superior and inferior salivatory nuclei) in medulla oblongata

Cerebral cortex

Convey output to

Parasympathetic nerves VII and IX

Stimulate

Salivary glands

Secrete

Saliva output (watery, enzyme-rich)

ASK YOURSELF

1 What is the difference between digestion and absorption?

2 What are the parts of the digestive tract?

3 What are the parts of a tooth, and what are they composed of?

4 How do the functions of the hard and soft palates differ?

5 What are the main functions of saliva?

Q: *Why do dental procedures stimulate salivation?*

A: Manipulative activities in the mouth stimulate pressoreceptors that activate salivation by the salivary reflex.

Q: *What causes the dryness of "morning mouth"?*

A: Salivation is reduced during sleep, and the papillae on the tongue trap food and bacteria, producing "morning mouth."

PHARYNX AND ESOPHAGUS

The act of swallowing has a voluntary phase, which occurs in the mouth, and an involuntary phase, which involves the pharynx and esophagus.

Pharynx

Food moves from the mouth into the pharynx. The **pharynx** serves as both an air passage during breathing and a food passage during swallowing. As you saw in Chapter 24, it extends from the base of the skull to the larynx, where it becomes continuous inferiorly with the esophagus and anteriorly with the nasal cavity [FIGURE 25.7; see also FIGURE 25.1]. Thus the pharynx can be divided into three parts: (1) the *nasopharynx*, superior to the soft palate, (2) the *oropharynx*, from the soft palate to the epiglottis, and (3) the *laryngopharynx*, posterior to the epiglottis, which joins the esophagus.

Esophagus

The *esophagus* (ih-SOFF-uh-guss; Gr. gullet) is a muscular, membranous tube, about 25 cm (10 in.) long, through which food passes from the pharynx into the stomach [FIGURE 25.1].

The inner *mucosa* of the esophagus is lined with non-keratinized stratified squamous epithelium arranged in longitudinal folds. Several mucous glands in the mucosa and *submucosa* provide a film of lubricating mucus that facilitates the passage of food to the stomach [FIGURE 25.8]. The submucosa also contains blood vessels. The *muscularis externa* consists entirely of skeletal muscle in the upper third of the esophagus, smooth and skeletal muscle in the middle third, and only smooth muscle in the lower third. The outer fibrous layer is called the *adventitia* because it lacks an epithelial layer.

The esophagus is located in front of the vertebral column and behind the trachea. It passes through the lower neck and thorax before penetrating the diaphragm and joining the stomach.

FIGURE 25.7 PHARYNX

[A] The pharynx extends from behind the nasal cavities to the larynx. It carries both air and food. [B] Muscles of the pharynx. These constrictor muscles propel the bolus via peristalsis. Right lateral views.

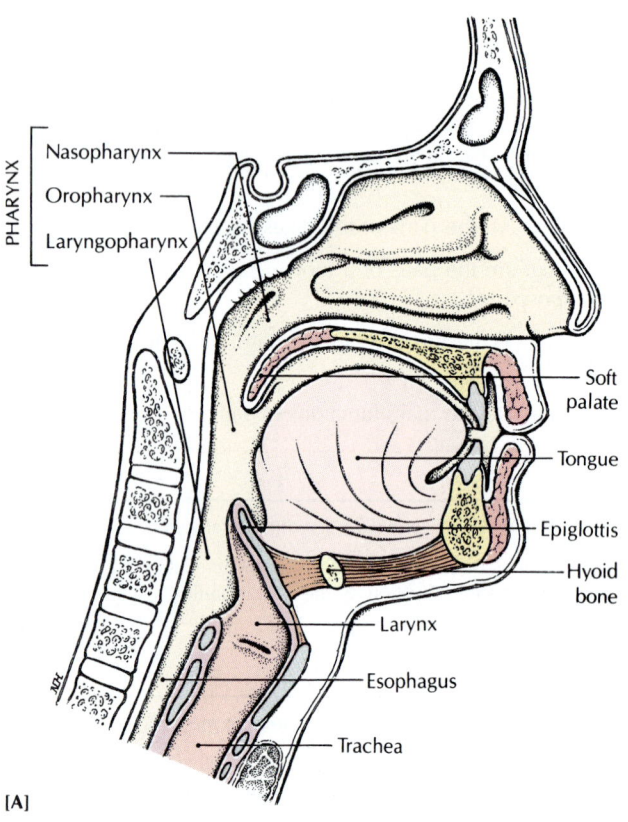

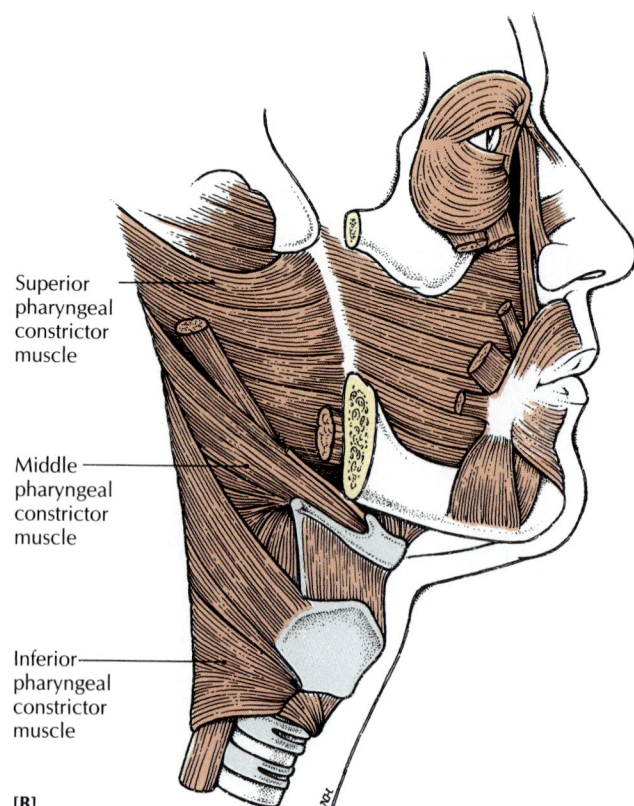

[A]

[B]

FIGURE 25.8 ESOPHAGUS

Scanning electron micrograph of a horizontal section of the esophagus. ×60. (Richard G. Kessel and Randy H. Kardon, *Tissues and Organs: A Text-Atlas of Scanning Electron Microscopy,* San Francisco, W. H. Freeman, 1979.)

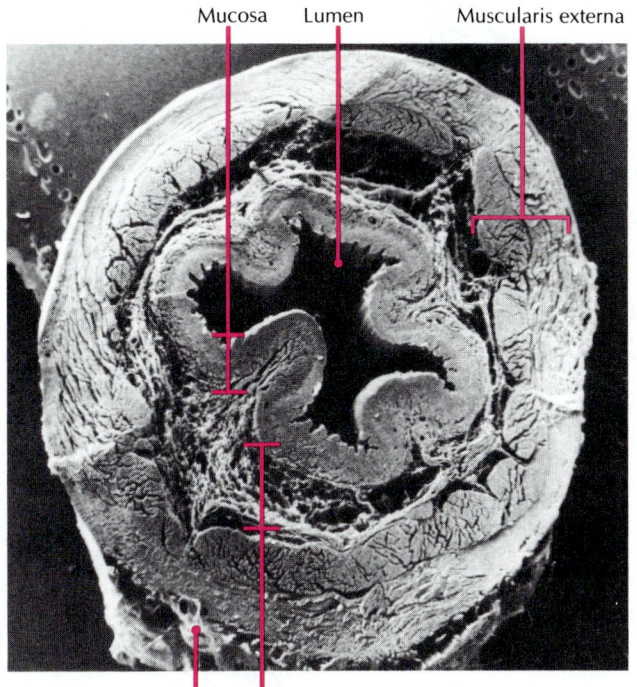

Mucosa   Lumen   Muscularis externa

Adventitia   Submucosa

Each end of the esophagus is closed by a sphincter muscle* when the tube is at rest and collapsed. The upper sphincter is the *superior esophageal sphincter.* Closing of this sphincter is caused not by active muscular contraction but rather by the passive elastic tension in the wall of the esophagus when the esophageal muscles are relaxed. The *lower esophageal sphincter* is a band of smooth muscle that includes the last 4 cm of the esophagus just before it connects to the stomach. The lower sphincter relaxes only long enough to allow food and liquids to pass into the stomach. The rest of the time it is contracted to prevent food and acid from being forced back into the esophagus when pressure increases in the stomach. Such pressure typically increases when the abdominal muscles contract during breathing, during the late stages of pregnancy, and during the normal stomach contractions during digestion. If the lower esophageal sphincter does not close, the acidic contents in the stomach can be forced up into the lower esophagus. The resultant irritation of the lining of the esophagus is commonly known as *heartburn,* so called because the painful sensation is a type of referred pain, and appears to be located near the heart.

*A *sphincter,* which comes from the Greek term for "to bind tight," is usually in a state of contraction, like the tightened drawstrings of a purse or a bag of marbles.

Swallowing (Deglutition)

After food in the mouth is chewed, and moistened and softened by saliva, it is known as a **bolus**; it is then ready to be swallowed. The first stage of swallowing, or **deglutition** (L. *deglutire,* to swallow down), is voluntary. In fact, it is the final voluntary digestive movement until feces are expelled during defecation. The lips, cheeks, and tongue all help form the food into a bolus, which is pushed by the tongue against the hard palate and into the pharynx. The voluntary phase of swallowing ends when the bolus touches the entrance to the pharynx. There it stimulates the glossopharyngeal (IX) nerve to trigger impulses in the "swallowing center" of the medulla oblongata. The medulla oblongata immediately responds by sending a regulated sequence of impulses to the muscles of the pharynx, esophagus, stomach, and breathing muscles.

Swallowing consists of the following three phases [FIGURE 25.9]:

1 *Voluntary oral phase* [FIGURE 25.9A]. Contractions of the mylohyoid and digastric muscles raise the hyoid bone and the tongue toward the hard palate of the mouth. In turn, the intrinsic tongue muscles elevate the tip of the tongue against the upper incisor teeth and maxilla and then squeeze the bolus between the tongue and hard palate to slide toward the pharynx (like toothpaste from a tube).

2 *Involuntary pharyngeal phase* [FIGURE 25.9B]. The pharyngeal phase is triggered by the stimulation of the glossopharyngeal nerve. The soft palate is raised to prevent the bolus from entering the nasopharynx and nasal cavity. (Once the pharyngeal stage of swallowing is initiated, it is impossible to breathe or speak.) The base of the tongue is then thrust backward (retracted), propelling the bolus into the oropharynx. The sequential contractions initiate a muscular wave called **peristalsis** [FIGURE 25.9B; see also FIGURE 25.7B] that propels the bolus along recesses on either side of the larynx on its way to the esophagus. The contractions of the stylohyoid and digastric muscles raise the hyoid bone and draw the larynx under the tongue. The flap of cartilage called the *epiglottis* is pushed from a vertical to a horizontal position, and the larynx closes against the epiglottis. The vocal folds move together [FIGURE 25.9C]. This action can prevent the bolus from entering the trachea. Any food that slips into the trachea causes coughing and choking reflexes until the passage is cleared.

3 *Involuntary esophageal phase* [FIGURE 25.9D and E]. In this phase the inferior pharyngeal muscles contract, initiating a wave of peristalsis that propels the bolus through the esophagus and into the stomach. (Peristalsis continues to assist the passage of food through the rest of the digestive tract.) The relaxation of the palatine, tongue,

FIGURE 25.9 SWALLOWING

Swallowing consists of four simultaneous movements **[A–D]** that prevent food from entering the nasal passages, trachea, or larynx, while allowing its passage into the esophagus. **[E]** Food is conveyed through the esophagus by peristalsis.

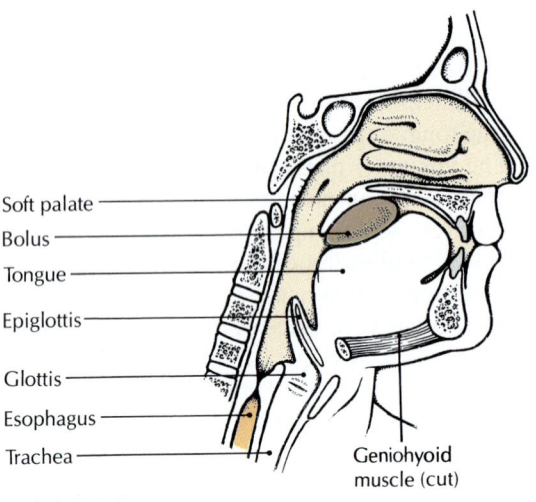

Soft palate
Bolus
Tongue
Epiglottis
Glottis
Esophagus
Trachea

Geniohyoid muscle (cut)

[A]

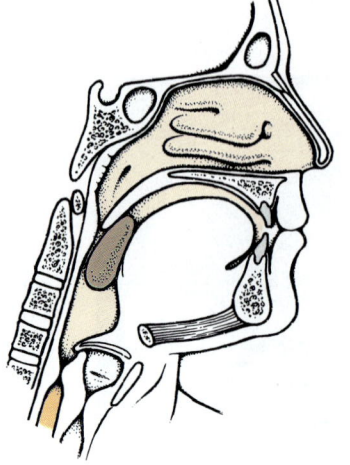

[B]

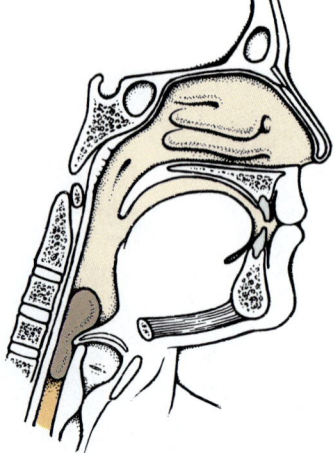

[C]

and pharyngeal muscles results in the opening of the respiratory passages. The larynx is drawn down by the contraction of the infrahyoid muscles, and the contraction of the hyoglossus and genioglossus muscles returns the tongue to the floor of the mouth. The epiglottis returns to its vertical position as the hyoid bone is lowered, and the laryngeal passage is reopened as the vocal folds separate.

Two types of involuntary muscle contractions occur during peristalsis: (1) The bolus is moved forward along the digestive tract by the contraction of the circular smooth muscle layer lying just above and at the top of the bolus. (2) The longitudinal fibers below the bolus contract, shortening this section and increasing the diameter of the esophagus so it can receive the bolus. Peristaltic waves, or contraction-relaxation waves, are repeated down the esophagus into the stomach. The pressure of peristalsis is often higher than arterial blood pressure. It is because of such a strong, one-way force that you can swallow while standing on your head, and an astronaut can swallow with little or no difficulty in a zero-gravity situation.

As much as half a liter of air may be swallowed with a meal, and it is usually released by belching *(eructation)* before it passes farther than the stomach. Air that re-

Q: *How long does it take for food to reach the stomach after being swallowed?*

A: Liquids and very soft foods reach the stomach in about 1 sec. Semisolid food takes about 5 sec, and solid foods take about 10 sec.

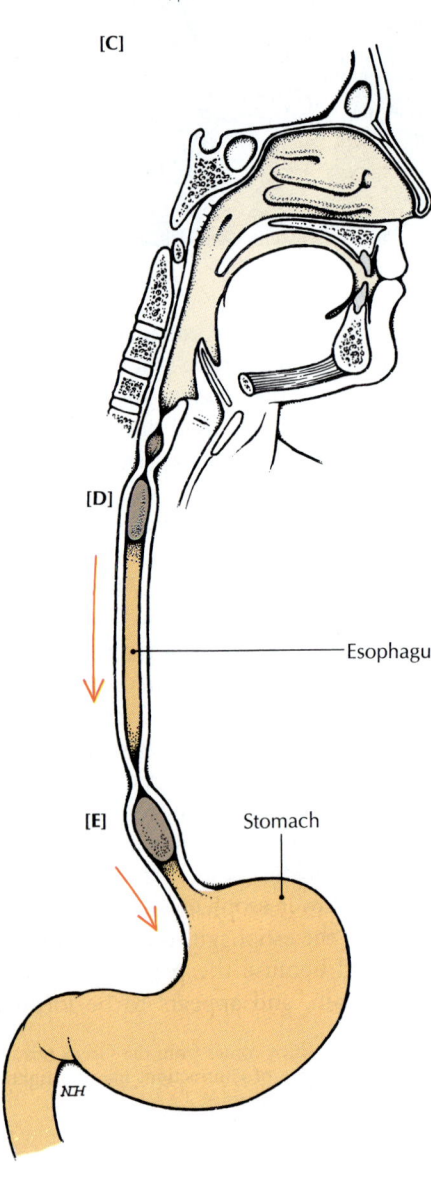

[D]

[E]

Esophagus

Stomach

mains in the stomach or that reaches the small intestine may produce a gurgling sound called *borborygmus* (bore-buh-RIHG-muhss). The sound is caused by the rapid movement of gas and liquid through the intestine. Borborygmus is usually more noticeable when you are hungry because then you salivate more and swallow more air than usual. Excess intestinal gas is relieved by the passing of gas (flatus) from the anus long before the feces are ready to be expelled.

> **A S K   Y O U R S E L F**
>
> **1** What are the functions of the pharynx and esophagus?
>
> **2** What are the different regions of the pharynx?
>
> **3** What is heartburn?
>
> **4** What are the phases of swallowing?
>
> **5** What types of muscular action occur during peristalsis?

ABDOMINAL CAVITY AND PERITONEUM

The ***abdominal cavity*** is the portion of the trunk cavities that lies inferior to the diaphragm [see FIGURE 1.13 on page 24]. If the pelvic cavity is included, it is referred to as the ***abdominopelvic cavity,*** with an imaginary plane separating the abdominal and pelvic cavities. The abdominal viscera include the stomach, liver, gallbladder, pancreas, spleen, kidneys, and small and large intestines. The pelvic viscera include the rectum, urinary bladder, and internal reproductive organs.

Serous membranes line the closed abdominal cavity and cover the viscera contained within the cavity. A serous membrane consists of a smooth sheet of simple squamous epithelium *(mesothelium)* and an adhering layer of loose connective tissue containing capillaries. The serous membrane of the abdominal cavity is the ***peritoneum*** (per-uh-tuh-NEE-uhm). The *parietal peritoneum* lines the abdominal cavity, and the *visceral peritoneum* covers most of the organs in the cavity [FIGURE 25.10].

Between the parietal and visceral peritoneal membranes is a space called the *peritoneal cavity.* It usually contains a small amount of *serous fluid* secreted by the peritoneum, allowing for the nearly frictionless movement of the abdominal organs and their membranes.

Abdominal organs that lie posterior to the peritoneal cavity and are covered but not surrounded by peritoneum are called *retroperitoneal.* Retroperitoneal structures include the pancreas, most of the duodenum, the abdominal aorta, the inferior vena cava, the ascending and descending colons of the large intestine, and the kidneys.

FIGURE 25.10 PERITONEUM

The drawing shows the female peritoneum (blue) and its relationship to some major structures in the abdominopelvic cavity; right sagittal view.

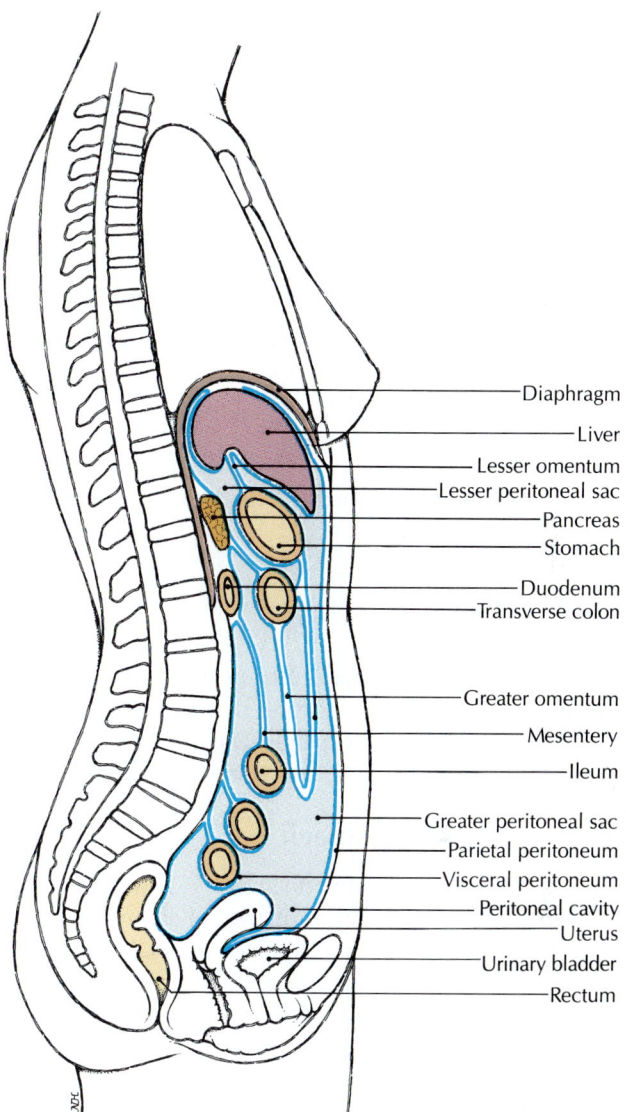

Diaphragm
Liver
Lesser omentum
Lesser peritoneal sac
Pancreas
Stomach
Duodenum
Transverse colon
Greater omentum
Mesentery
Ileum
Greater peritoneal sac
Parietal peritoneum
Visceral peritoneum
Peritoneal cavity
Uterus
Urinary bladder
Rectum

Other organs in the abdominal cavity are suspended from the posterior abdominal wall by two fused layers of serous membrane called ***mesenteries*** [FIGURE 25.11B]. These are the *intraperitoneal organs;* they include the liver, stomach, spleen, most of the small intestine, and the transverse colon of the large intestine. In addition to providing a point of attachment for organs, the mesenteries convey the major arteries, veins, and nerves of the digestive tract, liver, pancreas, and spleen to and from the body wall.

The *greater omentum* (L. "fat skin"; pl. *omenta*) is a large folded membrane that extends from the greater curvature of the stomach to the back wall and down to the pelvic cavity. It contains large amounts of fat and hangs down like a "fatty" apron anterior (ventral) to the abdominal organs, protecting and insulating them [FIGURE 25.11A]. Excess fat in the skin and greater omentum, plus a loss of muscle tone in the abdominal wall, produce the characteristic "potbelly."

The *lesser omentum* extends from the liver to the lesser curvature of the stomach; a small portion extends from the duodenum to the liver [FIGURE 25.11C].

A S K   Y O U R S E L F

1 What constitutes the abdominopelvic cavity?

2 What is the difference between the parietal and visceral peritoneums?

3 What is the function of serous fluid?

4 What is the function of the mesenteries?

5 What is the greater omentum? The lesser omentum?

STOMACH

The bolus of food is propelled from the esophagus into the *stomach,** the saclike portion of the digestive tract. The stomach is the most expandable part of the digestive tract. It stores, mixes, and digests ingested nutrients.

Anatomy of the Stomach

The stomach is usually described as a J-shaped sac. The average adult stomach has a capacity of about 1.5 L but may be distended maximally to about 4 L. There are considerable variations in size and shape. As shown in FIGURE 25.12A, the stomach has a convex lateral surface known as the *greater curvature* and a concave medial surface known as the *lesser curvature.*

The stomach is located much higher than some people think. Rather than lying behind the navel, it is directly under the diaphragm and is protected by the rib cage.

Food from the esophagus enters the stomach through an opening called the *lower esophageal* (cardiac) *orifice,* so called because it is located near the heart. Partially digested nutrients leave the stomach and enter the small intestine through an opening at the other end of the stomach called the *pyloric orifice* (or *pylorus*).† The lower

*The Greek *gaster* for "stomach" gives us the stem *gastero-*, as in *gastric.*

†The ancient anatomists were not totally without a sense of humor when it came to naming the parts of the body. *Pylorus* means "gate-keeper" in Greek.

esophageal and pyloric orifices both contain sphincters. The pyloric sphincter is more powerful than the lower esophageal sphincter because of the greater amount of circular smooth muscle in the pyloric sphincter. The pyloric orifice is usually opened slightly to permit fluids, but not solids, to pass into the duodenum.

The stomach is subdivided into four major regions [FIGURE 25.12A]: (1) The small *cardiac region* is near the lower esophageal (cardiac) orifice. (2) The *fundus* is a small, rounded area above the level of the orifice, which usually contains some swallowed air. (3) The *body* is the large central portion. (4) The *pyloric region,* which includes the *pyloric canal,* is a narrow portion leading to the pyloric orifice.

The *muscularis externa* of the stomach consists of three layers of smooth muscle fibers [FIGURE 25.12B]: (1) The outermost *longitudinal* layer is continuous with the muscles of the esophagus and is most prominent along the curvatures of the stomach. (2) The middle *circular* layer is wrapped around the body of the stomach

---

FIGURE 25.11 MESENTERIES AND GREATER AND LESSER OMENTA

[A] Greater omentum, lifted with an instrument to show the intestines underneath.

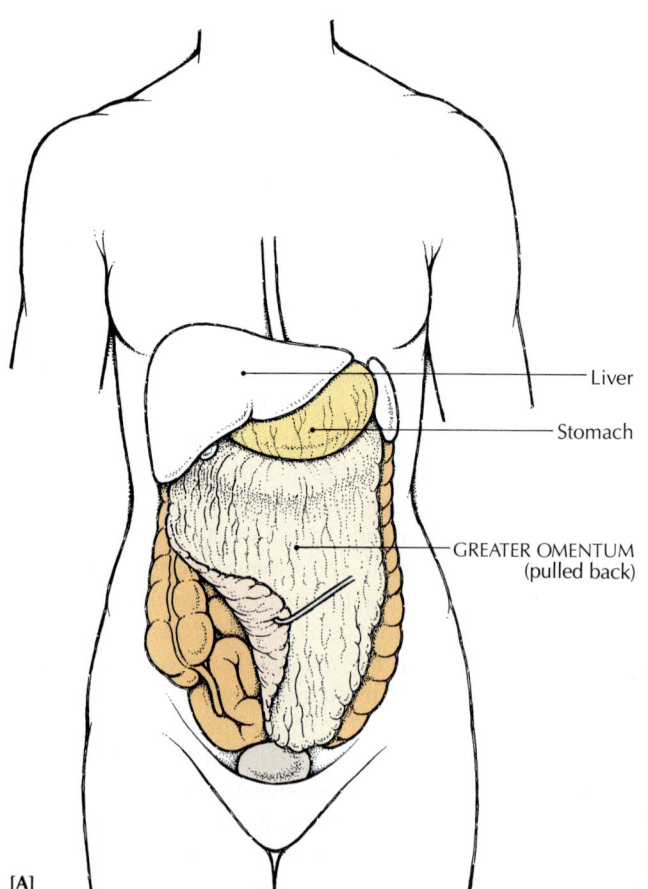

Liver

Stomach

GREATER OMENTUM (pulled back)

[A]

and becomes thickened at the pylorus to form the pyloric sphincter. (3) The innermost *oblique* layer covers the fundus and runs parallel with the lesser curvature of the stomach along the anterior and posterior walls.

When the stomach is empty, its inner mucous membrane contains longitudinal folds called ***rugae*** (ROO-jee; L. folds), which gradually flatten as the stomach becomes filled [FIGURE 25.12A and C]. The stomach is lined with a layer of simple columnar epithelium, which contains millions of narrow channels called *gastric pits*. Extending from each of these pits are three to eight tubular *gastric glands.*

There are three types of gastric glands: (1) the *cardiac* glands near the cardiac orifice, (2) the *pyloric* glands in the pyloric canal, and (3) the *fundic* glands in both the fundus and body of the stomach. Of these three types, the fundic glands are the most numerous.

The six functionally active cell types lining the surface, pits, and glands of the stomach are surface mucous cells, neck mucous cells, parietal cells, zymogenic (chief) cells, enteroendocrine cells, and undifferentiated cells. They are shown in FIGURE 25.13, which also gives a description of each type of cell and its major function.

Functions of the Stomach

The stomach has three main functions:

1 The stomach ***stores*** *ingested nutrients until they can be released into the small intestine* at a rate that is physiologically appropriate for the relatively slow processes of digestion and absorption. The stomach is well suited for storage because its muscles have little tone, and it can expand up to 4 L if necessary.

2 The stomach ***churns*** *ingested nutrients, breaks them up into small particles, and mixes them with gastric juices* to form a soupy liquid mixture called ***chyme*** (KIME).

3 The stomach ***secretes*** *hydrochloric acid and enzymes that initiate the digestion of proteins and kill most of the bacteria that enter the stomach in food.*

[B] Mesentery, with greater omentum deleted for clarity. The intestines are lifted to reveal the mesentery.

[C] Lesser omentum, with greater omentum deleted and liver and gallbladder lifted.

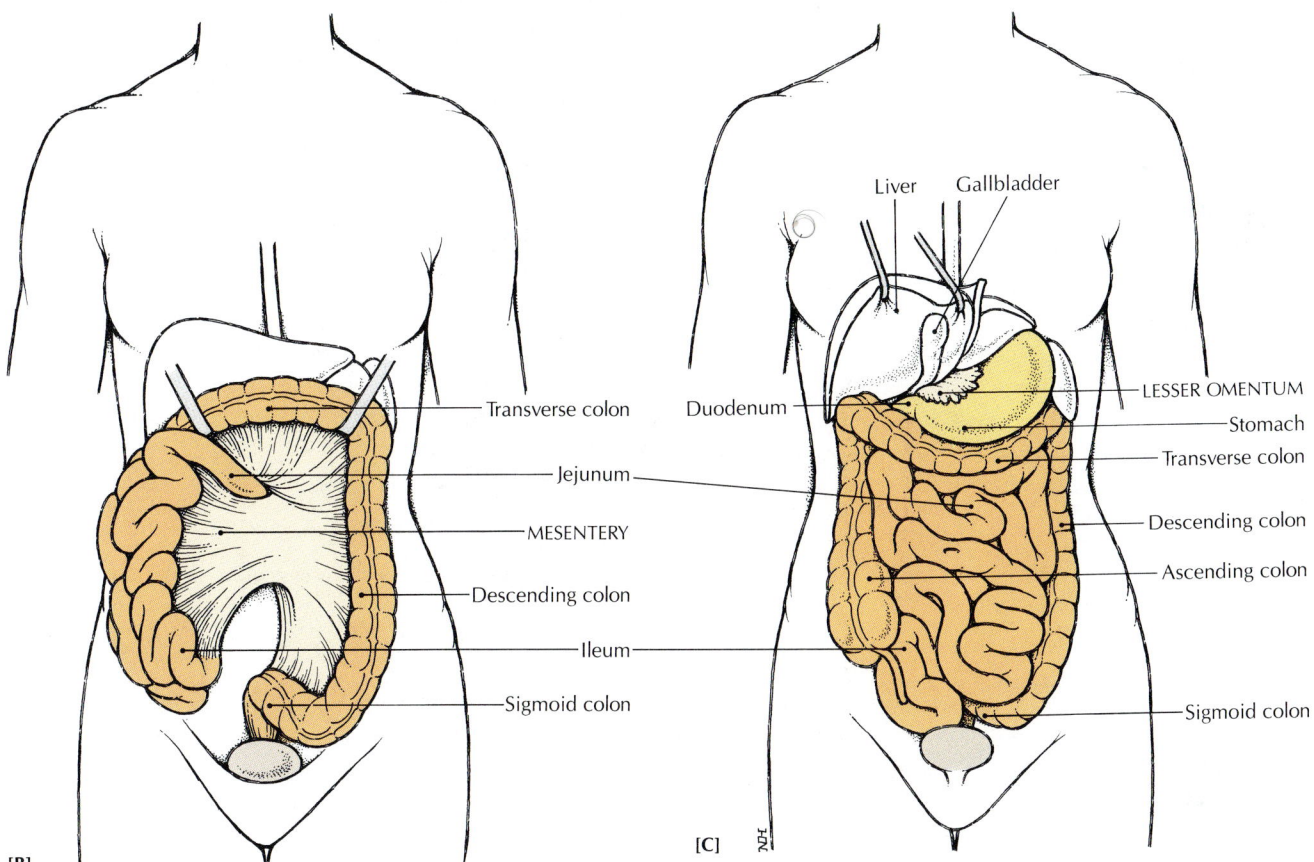

[B]

[C]

FIGURE 25.12 STOMACH

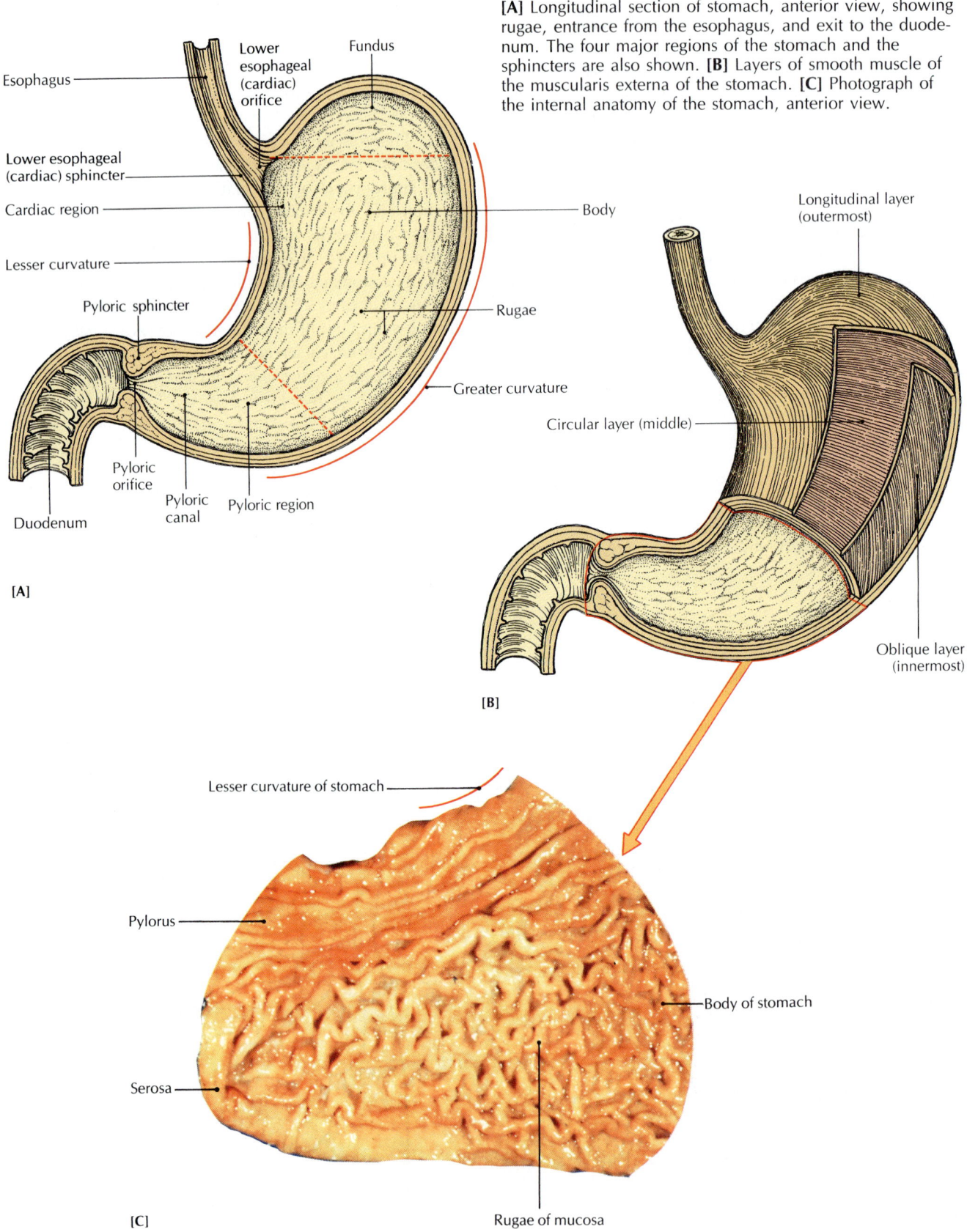

[A] Longitudinal section of stomach, anterior view, showing rugae, entrance from the esophagus, and exit to the duodenum. The four major regions of the stomach and the sphincters are also shown. **[B]** Layers of smooth muscle of the muscularis externa of the stomach. **[C]** Photograph of the internal anatomy of the stomach, anterior view.

Esophagus

Lower esophageal (cardiac) orifice

Fundus

Lower esophageal (cardiac) sphincter

Cardiac region

Lesser curvature

Body

Pyloric sphincter

Rugae

Greater curvature

Pyloric orifice

Pyloric canal

Pyloric region

Duodenum

[A]

Longitudinal layer (outermost)

Circular layer (middle)

Oblique layer (innermost)

[B]

Lesser curvature of stomach

Pylorus

Body of stomach

Serosa

Rugae of mucosa

[C]

FIGURE 25.13 STOMACH WALL

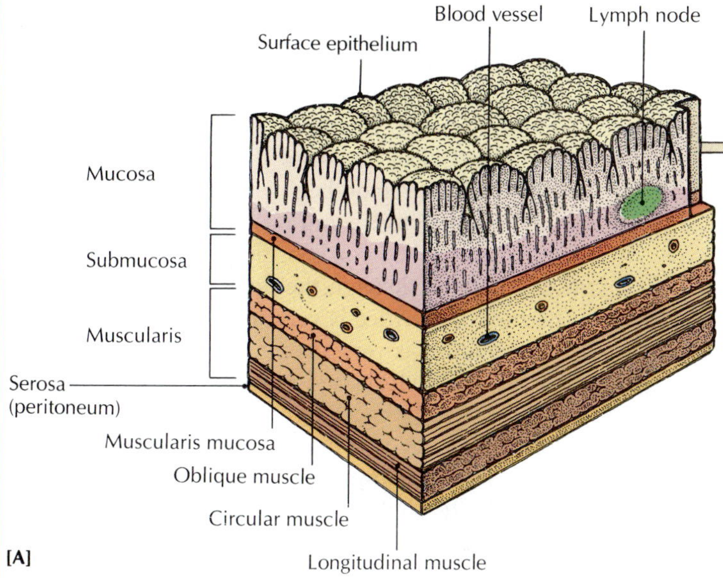

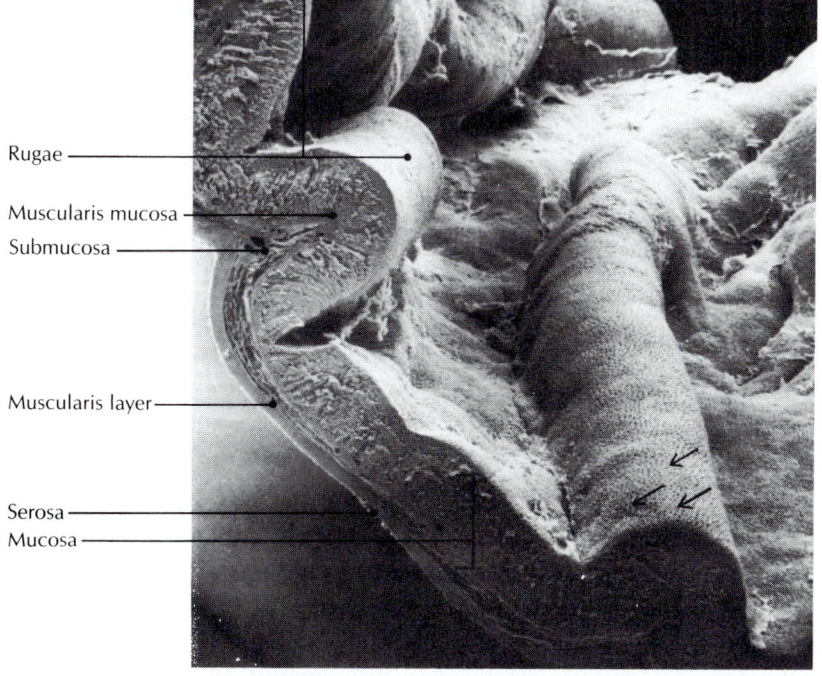

[A]

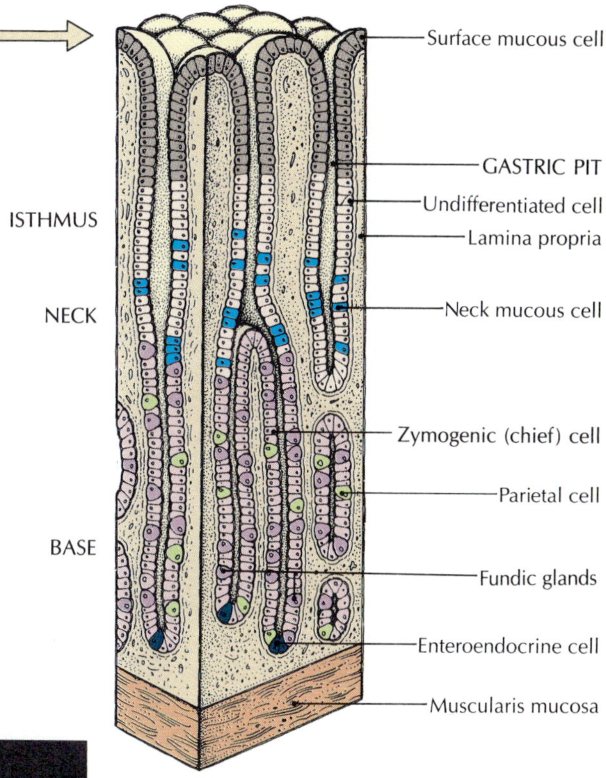

Surface epithelium
Blood vessel
Lymph node

Mucosa

Submucosa

Muscularis

Serosa
(peritoneum)

Muscularis mucosa
Oblique muscle
Circular muscle
Longitudinal muscle

[B]

Surface mucous cell

ISTHMUS

GASTRIC PIT
Undifferentiated cell
Lamina propria

NECK

Neck mucous cell

Zymogenic (chief) cell

BASE

Parietal cell

Fundic glands

Enteroendocrine cell

Muscularis mucosa

Rugae

Muscularis mucosa

Submucosa

Muscularis layer

Serosa
Mucosa

[C]

[A] Transverse section through the stomach wall at the fundus. **[B]** Cells of fundic glands near the body of the stomach at a higher magnification. **[C]** Scanning electron micrograph of the stomach wall. The arrows point to the openings into gastric pits, which are mostly in the mucosa. ×55. (Gene Shih and Richard Kessel, *Living Images: Biological Microstructures Revealed by SEM,* Boston, Science Books International, 1982.)

(Figure 25.13 continues on following page)

FIGURE 25.13 STOMACH WALL *(Continued)*

TYPES OF STOMACH CELLS

Type of cell	Description	Functions
Surface mucous cells	Line the lumen surface of stomach and gastric pits of cardiac, fundic, and pyloric glands; shed into lumen, replaced from below by undifferentiated cells in gastric pits.	Secrete alkaline mucus that protects stomach mucosa from pepsin and high acidity of gastric fluid; prevent ulceration of mucosa. "Parent" cells of all new cells of gastric mucosa.
Neck mucous cells	Line cardiac, pyloric, and fundic glands.	Secrete a more neutral mucus than that of surface mucous cells; replace lost surface cells.
Parietal (oxyntic) cells	Pale, oval cells, found alongside zymogenic cells in lining of fundic glands.	Secrete hydrochloric acid, for conversion of pepsinogen into pepsin; intrinsic factor needed for absorption of vitamin B_{12}.
Zymogenic (chief) cells	Large pyramidal cells that line fundic glands.	Secrete pepsinogen, precursor of protein-digesting enzyme pepsin.
Enteroendocrine cells	Located near base of gastric glands; secretory granules that release secretions into lumen rather than into laminia propria.	Secrete hormones and hormonelike products, including gastrin, histamine, cholecystokin, and somatostatin.
Undifferentiated cells	Cells that are not specialized for any one particular function.	Replace other types of glandular cells when they die.

Digestive movements within the stomach Several types of muscular movements occur within the stomach during digestion, depending on the location and volume of the chyme:

1 A few minutes after food enters the stomach through the lower esophageal orifice, *slow peristaltic mixing waves* start in smooth muscle pacemaker cells in the fundus and body. The pacemaker cells generate action potentials at a rate of three to four per minute that sweep down the stomach toward the pyloric sphincter. This rhythmic rate is known as the *basic electrical rhythm* (BER) of the stomach. These rhythmic contractions produce chyme and push it toward the pyloric region.

2 As the pyloric region begins to fill, *strong peristaltic waves* chop the chyme and propel it through the pyloric canal toward the pyloric orifice. A few milliliters of chyme are pumped through the pyloric orifice with each wave. Because the orifice is very narrow and the pyloric sphincter around it is very strong, most of the chyme is sent back into the pyloric region for further chopping by peristaltic waves. The mechanism that forces chyme past the pyloric sphincter is called the *pyloric pump*.

3 As the stomach empties, peristaltic waves move farther up the body of the stomach, ensuring that all the chyme is pushed into the pyloric region.

Regulation of gastric emptying The rate of *gastric emptying* (the movement of chyme from the stomach into the duodenum) is regulated by both neural and hormonal mechanisms [FIGURE 25.14]. Regulation ensures that the stomach does not become too full and does not empty faster than the small intestine can process the incoming chyme.

Distension of the stomach and the presence of stimulants such as alcohol, caffeine, and partially digested proteins stimulate the secretion of the hormone *gastrin* and increase the activity of the vagus nerve, which stimulates stomach motility and further secretions of gastric juices. The lower esophageal sphincter contracts, the pyloric sphincter relaxes, and chyme passes into the duodenum. Fluid chyme (about the consistency of toothpaste) also increases gastric emptying.

Gastric emptying is *inhibited* primarily by the presence of fat, acid, hypertonicity, or distension in the duodenum. A *neural response* is mediated by the *enterogastric reflex*, which decreases stomach motility and gastric secretion. The *hormonal response* inhibits gastric motility by releasing hormones collectively known as *enterogastrones*. The most important of these hormones are *secretin, cholecystokinin (CCK)*, and *gastric inhibitory peptide (GIP)*. (See TABLE 25.5 on page 827 for a description of digestive hormones.)

Secretion of gastric juices *Gastric juice* is a clear, colorless fluid secreted by the stomach mucosa in response to food. Typically, more than 1.5 L of gastric juice is secreted daily. It is composed of *hydrochloric acid, mucus*, and several enzymes, especially *pepsinogen*, a precursor of the active enzyme *pepsin*. Small amounts of gas-

FIGURE 25.14 GASTRIC EMPTYING

Pathways regulating [A] the stimulation and [B] the inhibition of gastric emptying.

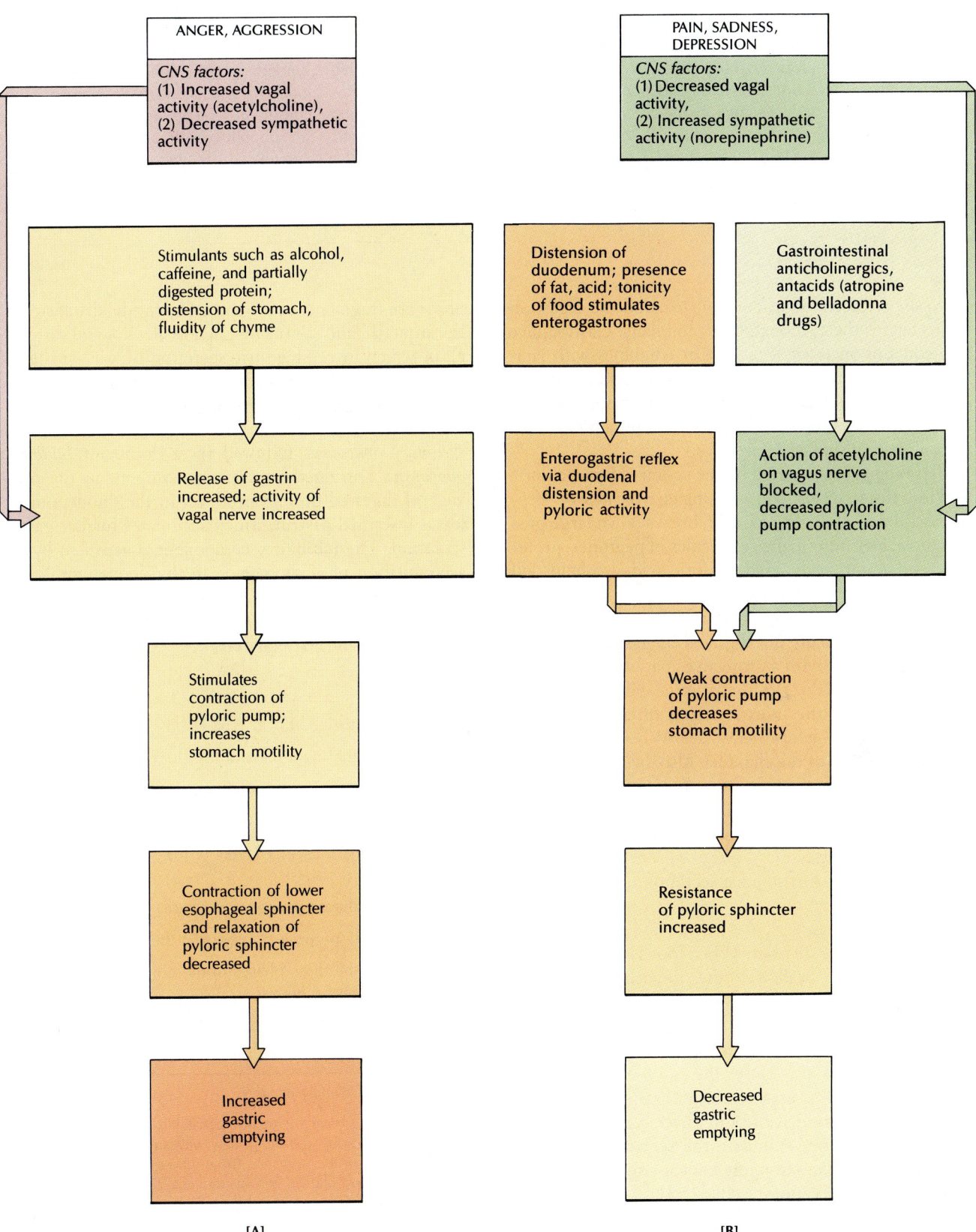

[A]

[B]

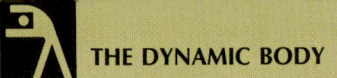
"Pacemakers" in the Digestive Tract

The smooth muscle cells of the gut—like the cardiac muscle cells of the heart—are interconnected by tight junctions and have the same property of spontaneous rhythmic contractions in response to the depolarization of the cell membrane. Two "pacemaker" sites are recognized in the digestive tract: (1) in the fundus of the stomach and (2) in the first part of the duodenum. The frequency of the rhythmic contractions in the stomach is 3 per minute; in the duodenum the rate is 11 per minute. Unlike the heart, the control of contractile activity in the digestive tract is not located at the pacemaker sites but rather in the smooth muscle fibers throughout the gut. The neuronal network of the enteric division contains the components of an intrinsic reflex pathway; the peristaltic reflex is elicited by moderate distension. In the "law of the intestine," distension of the gut causes smooth muscle relaxation below and smooth muscle contraction above the point of distension, thereby moving the intestinal contents caudally. The intrinsic smooth muscle activity of the gut is modulated by the extrinsic excitatory (parasympathetic) and inhibitory (sympathetic) autonomic influences.

---

tric *lipase* are secreted in the stomach, and the digestion of fats begins there, though only minimally. Gastric juice also contains an *intrinsic factor* that combines with vitamin B_{12} (an *extrinsic factor*) from digested food to form an **antianemic factor** necessary for the formation of red blood cells. (See TABLE 25.6 on page 828 for a more complete listing of digestive enzymes.)

Pepsinogen is converted by the action of hydrochloric acid into the active protein-splitting enzyme **pepsin.** Pepsin, assisted by hydrochloric acid, breaks down large protein molecules into smaller molecules of peptones, proteoses, and amino acids. Hydrochloric acid also helps kill bacteria that are swallowed with food.

Although the stomach wall is composed mainly of protein, it is normally not digested by gastric juices because it is covered with a protective coat of mucus that is secreted by the surface epithelial cells and neck mucous cells. A break in the mucous lining often results in a stomach sore, or *gastric ulcer* (see When Things Go Wrong at the end of the chapter). Also, because pepsin is secreted as the inactive pepsinogen, it cannot digest the cells that produce it. The lining of the stomach sheds about 500,000 cells per minute and is completely renewed every 3 days.

The control of gastric juice secretion occurs in three overlapping phases:

1 *Cephalic ("head") phase.* When food is seen, smelled, tasted, chewed, or swallowed, the stomach is stimulated by activity of the vagus nerve, and a small amount of gastric juice is secreted even before the food is swallowed. Vagal stimulation from the cerebral cortex and medulla oblongata also causes the enteroendocrine cells of the pyloric gland area to release gastrin. Gastrin is carried by the blood back to the parietal, chief, and mucous cells to further stimulate the secretion of gastric juice. Gastrin is the strongest known stimulator of hydrochloric acid secretion [TABLE 25.3].

2 *Gastric ("belly") phase.* The stimuli involved in this phase are fragments of protein and peptide, distension of the stomach, and stimulants such as alcohol and caffeine, all of which increase gastric secretion by way of the intrinsic nerve plexuses in the stomach wall, the extrinsic vagal nerve, and the hormone gastrin.

3 *Intestinal phase.* The intestinal phase consists of an *excitatory component* followed by a dominant *inhibitory component.* During the excitatory component, the presence of digested protein products in the duodenum releases intestinal gastrin, which stimulates further gastric secretion. The inhibitory component is activated by the accumulation of acidic chyme in the duodenum, which causes the production of CCK, secretin, and GIP, which are carried by the blood to the stomach, where they turn off the chief and parietal cells.

A S K   Y O U R S E L F

1 What are the main parts of the stomach?

2 What are the six types of stomach cells?

3 What are the functions of the stomach?

4 How do the muscular layers of the stomach contribute to peristalsis?

5 What are the components of gastric juice?

6 What is the purpose of pepsin in the stomach?

7 What are the three phases in the control of the secretion of gastric juices?

Q: *What are "hunger pangs"?*

A: When the stomach is empty, peristaltic waves cease, but after about 10 hours of fasting, new waves may occur in the pyloric region of the stomach. These waves are "hunger pangs" (or hunger pains) elicited by sensory vagal fibers carrying impulses to the brain.

TABLE 25.3 CONTROL OF GASTRIC SECRETION

STIMULATION OF GASTRIC SECRETION

Phase	Stimuli	Excitatory mechanism	Result
Cephalic phase	Stimuli in head: seeing, smelling, tasting, chewing, swallowing food	Vagus nerve is stimulated and, in turn, stimulates intrinsic nerves, which stimulate parietal and chief cells. Vagus nerve also stimulates pyloric region of stomach, which causes increased gastric secretion, which also stimulates parietal and chief cells	Increased gastric secretion
Gastric phase	Stimuli in stomach: protein (peptide fragments), distension, caffeine, alcohol	Same as in cephalic phase plus stimuli in stomach directly stimulate intrinsic nerves and pyloric region	Increased gastric secretion
Excitatory phase	Stimuli in duodenum: digested protein products	Intestinal gastric secretion increases	Increased gastric secretion

INHIBITION OF GASTRIC SECRETION

Region	Stimuli	Inhibitory mechanism	Result
Body, antrum of stomach	Removal of protein; distension as stomach empties	Intrinsic nerves inhibited; vagus nerve inhibited; pyloric region of stomach inhibited, causing decrease in gastrin secretion	Decreased gastric secretion
Antrum of stomach	Accumulation of stomach acid	Pyloric region of stomach inhibited, causing decrease in gastrin secretion	Decreased gastric secretion
Duodenum (inhibitory intestinal phase of gastric secretion)	Fat, acid, hypertonicity, distension	Enterogastric reflex stimulated; Increased enterogastrones (cholecystokinin, secretion, gastric inhibitory peptide)	Decreased gastric secretion

SMALL INTESTINE

After 1 to 3 hours in the stomach, the chyme moves into the *small intestine,* where further contractions continue to mix it. It takes 1 to 6 hours for chyme to move through the 6-m (20-ft) long small intestine [FIGURE 25.15]. It is here that carbohydrate and protein digestion is continued, and fat digestion is initiated and completed. The small intestine absorbs almost all the digested molecules of food into the blood and lymph. One exception is alcohol, which is absorbed in the stomach.

Anatomy of the Small Intestine

The small intestine, like the large intestine, lies within the abdominopelvic cavity.* It is subdivided into three parts: the duodenum, jejunum, and ileum. Most of the remaining digestive processes take place in the duodenum, and most of the absorption of nutrients into the blood and lymph occurs in the duodenum and jejunum.

Duodenum The *duodenum* (doo-oh-DEE-nuhm)† is the C-shaped initial segment of the small intestine [FIGURE 25.15]. It is about 25 cm (10 in.) long and is the shortest of the three parts of the small intestine.

Jejunum and ileum The *jejunum* (jeh-JOO-nuhm),‡ between the duodenum and ileum, is about 2.5 m (8 ft) long. The *ileum* (ILL-ee-uhm) extends from the jejunum to the cecum, the first part of the large intes-

---

*The small intestine and large intestine are named for their relative diameters (about 4 cm and 6 cm, respectively), not their lengths. The small intestine is about 6 m (20 ft) long, and the large intestine is about 1.5 m (5 ft) long.

†*Duodenum* comes from the Greek word meaning "twelve fingers wide," referring to its length, and from the Latin *duodecim,* meaning "twelve."

‡*Jejunum* comes from a Latin word meaning "fasting intestine," so named because it was always found empty when a corpse was dissected.

FIGURE 25.15 SMALL AND LARGE INTESTINES

[A] The drawing shows that the small intestine begins at the pyloric orifice, where it joins the stomach, and ends at the ileocecal valve, where it joins the large intestine. **[B]** Radiograph of the small intestine 3 hours after a barium meal. Anterior views.

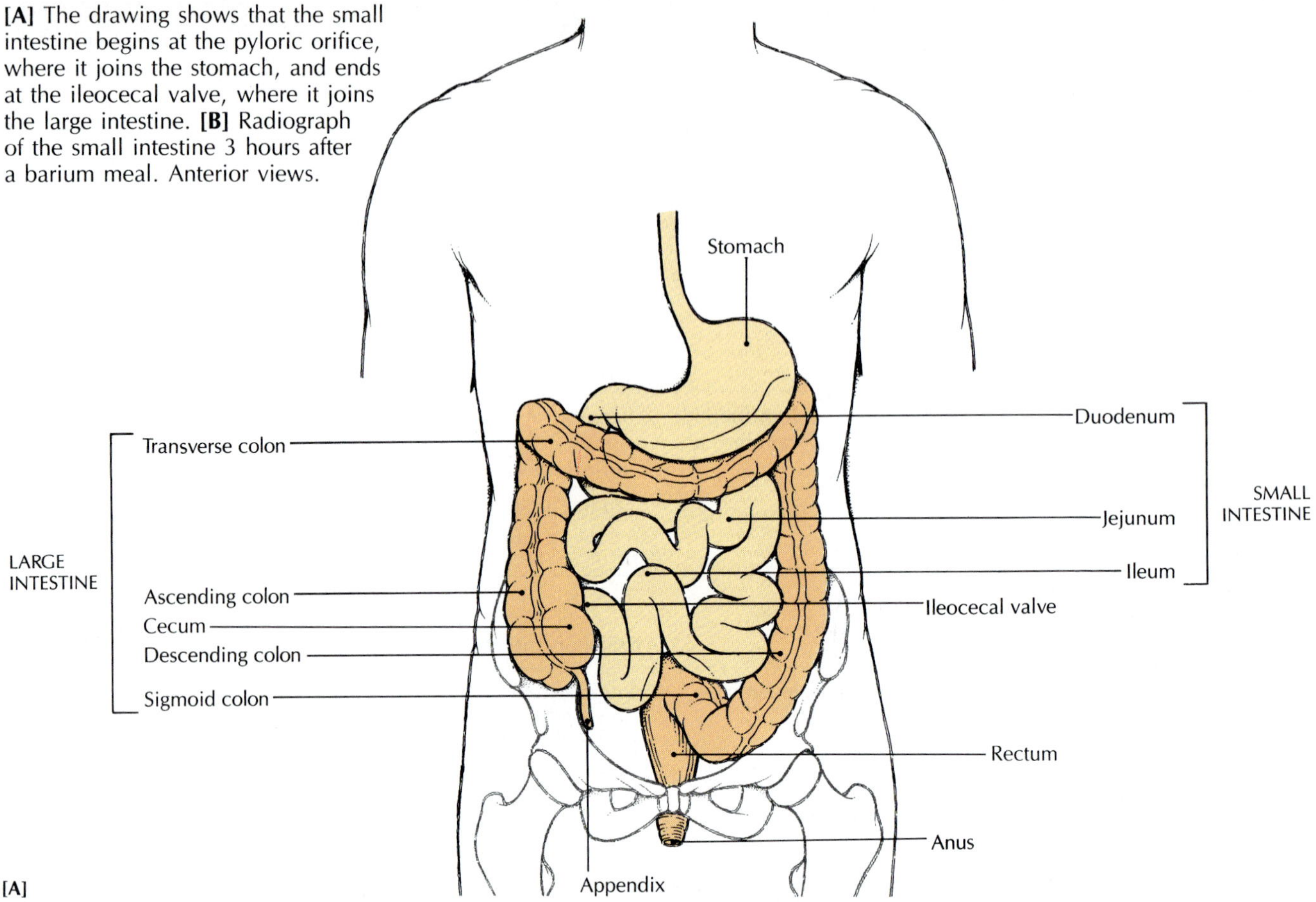

Stomach

Transverse colon

Duodenum

SMALL INTESTINE

Jejunum

Ileum

LARGE INTESTINE

Ascending colon

Ileocecal valve

Cecum

Descending colon

Sigmoid colon

Rectum

Anus

Appendix

[A]

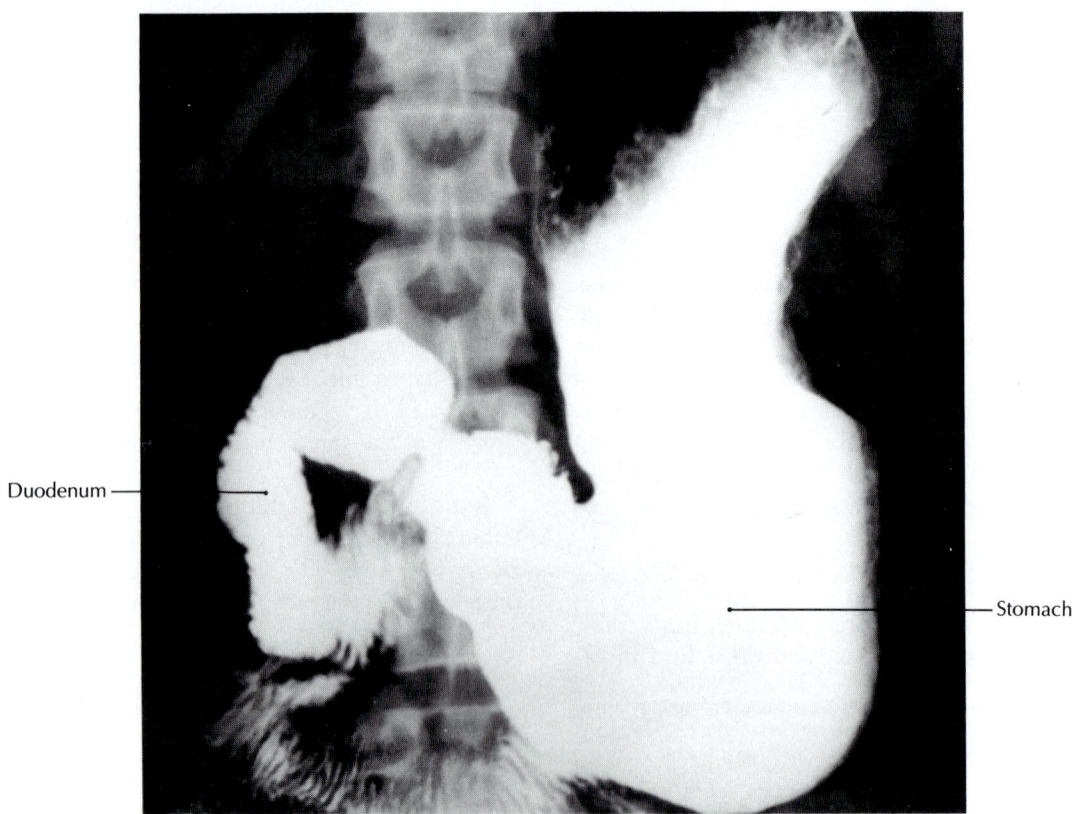

Duodenum

Stomach

[B]

tine [FIGURE 25.15]. It is about 3.5 m (12 ft) long. Both the jejunum and ileum are suspended from the posterior abdominal wall by the mesentery.

The ileum joins the cecum at the *ileocecal valve*, a sphincter that ordinarily remains constricted, regulating the entrance of chyme into the large intestine and preventing the contents of the cecum from flowing back into the ileum.

Adaptations of the mucosa of the small intestine

The wall of the small intestine is composed of the same four layers as the rest of the digestive tract [FIGURES 25.1 and 25.2]. However, the mucosa has three distinctive features that enhance the digestion and absorption processes that take place in the small intestine — projections known as plicae circulares, and villi and glands that secrete intestinal juice:

1 *Plicae circulares* (PLY-see sir-cue-LAR-eez) are circular folds that increase the surface area available for absorption and cause the chyme to spiral rather than move in a straight line; the spiraling motion enhances absorption. Unlike the rugae of the stomach, the plicae circulares are permanent and do not disappear when the mucosa is distended. They are most abundant near the junction of the duodenum and jejunum [FIGURE 25.16A].

2 The digestive and absorptive surface area of the mucosa is greatly increased by millions of long fingerlike protrusions called *villi* (VILL-eye; L. "shaggy hairs"; sing. *villus*), which look like the velvety pile of a rug [FIGURE 25.16A]. The surface area is further increased by the infolding of the epithelium between the bases of the villi, forming tubular *intestinal glands (crypts of Lieberkühn)* [FIGURE 25.16B and C]. Each villus contains blood capillaries and a lymph vessel called a *central lacteal* [FIGURE 25.16C]. Central lacteals are important in the absorption of fats.

In addition to increasing the surface area of the mucosa, the villi aid absorption by adding their constant spiraling wave to the peristaltic movement already occurring throughout the small intestine. The thousands of columnar epithelial cells making up each villus are the units through which absorption from the small intestine takes place. The absorptive surface of each of these epithelial cells contains thousands of *microvilli* (brush border), which further increase the surface area of absorption by about 30-fold [FIGURE 25.16C]. The plicae circulares, villi, and microvilli of the small intestine increase the absorptive surface about 600 times more than a smooth-lined intestine would be, with a total surface area of about 3888 ft^2, compared to the skin's approximately 15 ft^2.

3 Within the *mucosa* are simple, tubular glands that secrete intestinal juice containing several enzymes that aid in the digestion of carbohydrates, proteins, and fats. These mucosal glands are known as intestinal glands (crypts of Lieberkühn) and reach into the lamina propria [FIGURE 25.16B and C]. Glands that reach into the *submucosa* are called *duodenal submucosal glands (Brunner's glands)* and are found only in the duodenum [FIGURE 25.16B]. They secrete viscous, alkaline mucus secretions, which act as a lubricating barrier to protect the mucosa from the acidic chyme.

The lamina propria of the intestinal mucosa in the ileum has regions where lymphoid tissue is concentrated. Large lymphatic nodules form round or oval patches called *aggregated unencapsulated lymph nodules*, which help destroy microorganisms absorbed from the small intestine.

Cell types in the small intestine

The epithelium of the intestinal mucosa is simple columnar epithelium. It contains several types of cells, including (1) columnar absorptive cells, (2) undifferentiated columnar cells, (3) mucous goblet cells, (4) Paneth cells, and (5) enteroendocrine cells.

Columnar absorptive cells are involved in the absorption of sugars, amino acids, and fats from the lumen of the small intestine. They also produce enzymes for the terminal digestion of carbohydrates and proteins.

Undifferentiated columnar cells are found in the depths of the intestinal glands. The entire epithelial surface of the small intestine is replaced about every 5 days. Undifferentiated cells at the base of the villi divide and become differentiated and migrate upward to replace the other cell types as needed. As replaced cells disintegrate into the lumen of the intestine, they discharge their digestive enzymes into the lumen.

The *mucous goblet cells* also appear in the depths of the intestinal glands and migrate upward, secreting and accumulating mucus until they swell into the bulbous shape of a goblet. After these cells release their mucus into the lumen, they die. Mucous goblet cells are especially abundant in the duodenum, where mucus protects the tissue from the high acidity of the chyme entering from the stomach. Because the activity of the mucous goblet cell is inhibited by sympathetic stimulation, the duodenum, especially the superior portion, is particularly susceptible to peptic ulcers caused by nervous stress.

Paneth cells lie deep within the intestinal glands. They help regulate the bacteria in the small intestine by secreting lysozyme, an antibacterial enzyme.

Enteroendocrine cells are found not only in the small intestine, but also in the stomach, large intestine, appendix, the ducts of the pancreas and liver, and even the respiratory passages. They are responsible for the synthesis of more than 20 gastrointestinal hormones.

FIGURE 25.16 WALL AND LINING OF THE SMALL INTESTINE

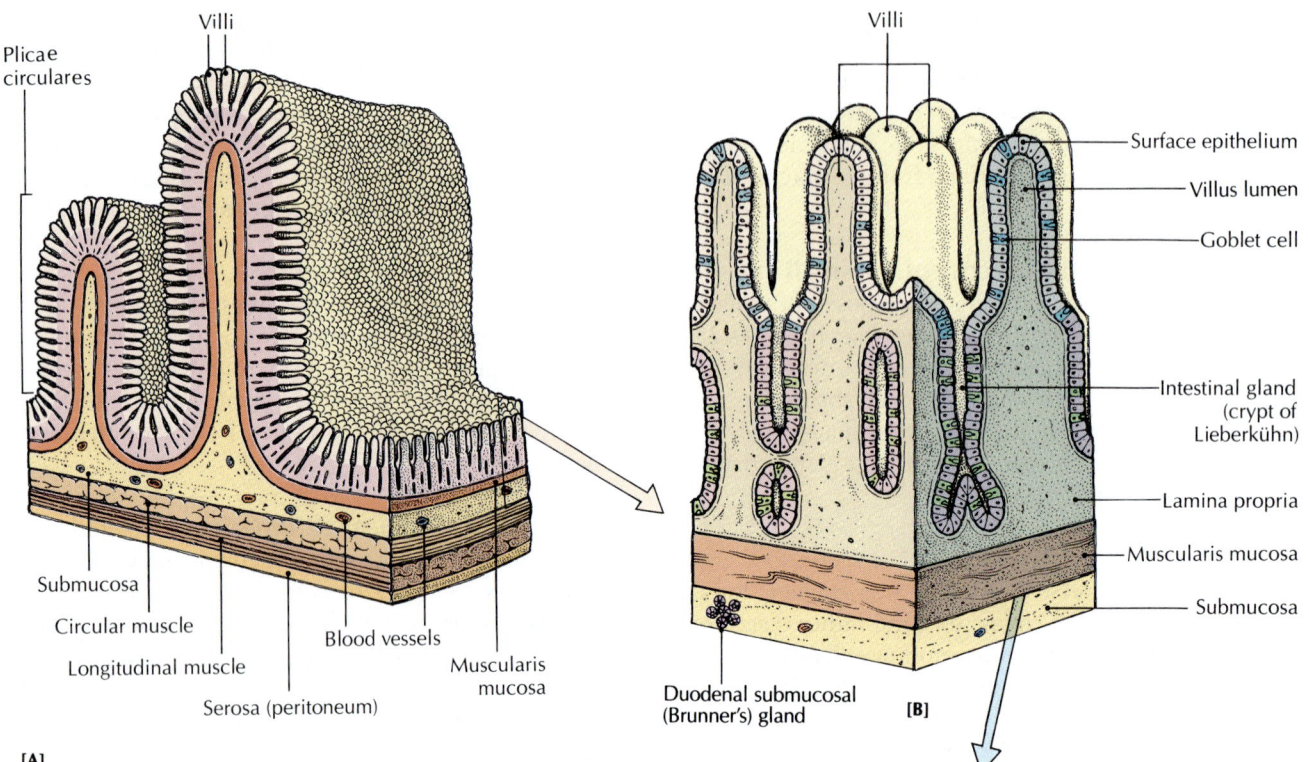

Plicae circulares
Villi
Submucosa
Circular muscle
Longitudinal muscle
Serosa (peritoneum)
Blood vessels
Muscularis mucosa

[A]

Villi
Surface epithelium
Villus lumen
Goblet cell
Intestinal gland (crypt of Lieberkühn)
Lamina propria
Muscularis mucosa
Submucosa
Duodenal submucosal (Brunner's) gland

[B]

Muscularis mucosa   Submucosa   Lumen   Serosa   Villi

[D]

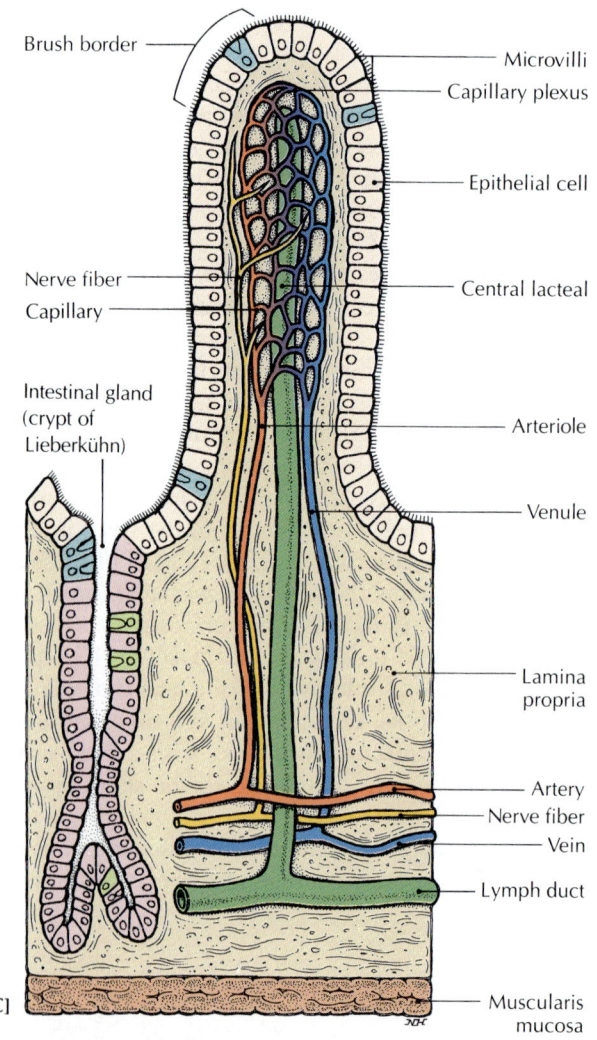

Brush border
Microvilli
Capillary plexus
Epithelial cell
Nerve fiber
Capillary
Central lacteal
Intestinal gland (crypt of Lieberkühn)
Arteriole
Venule
Lamina propria
Artery
Nerve fiber
Vein
Lymph duct
Muscularis mucosa

[C]

[A] Lining of the small intestine. **[B]** Enlarged portion of transverse section of the wall of the duodenum. **[C]** Internal structure of villus. Microvilli extending from the villus form a brush border. **[D]** Scanning electron micrograph of a section of the small intestine. ×30. (Gene Shih and Richard Kessel, *Living Images: Biological Microstructures Revealed by SEM,* Boston, Science Books International, 1982.)

Functions of the Small Intestine

The major functions of the small intestine — digestion and absorption — are made possible by movements of the intestinal smooth muscles and by the biochemical action of enzymes. (Some smooth muscle "pacemaker cells" in the duodenum act as pacemaker sites, with a rhythm of 11 peristaltic waves a minute. See "'Pacemakers' in the Digestive Tract" on page 816.) We will first discuss the digestive movement and enzymes in the small intestine and then take up the process of absorption.

Digestive movements of the small intestine
Rhythmic movements of circular and longitudinal smooth muscle in the small intestine cause the chyme to be mixed and ultimately moved into the large intestine. The mixing movements help reduce large particles within the chyme to smaller ones, exposing as much of the chyme surface to digestive enzymes as possible. The two main types of muscular activity in the small intestine are *segmenting* and *peristaltic contractions.*

Segmenting contractions divide the intestine into segments by sharp contractions of areas of the circular smooth muscle in the intestinal wall [FIGURE 25.17]. The areas between the contracted segments contain the bolus of chyme and are relaxed. These localized contractions mix the chyme with digestive juices and move the food to the mucosa for absorption. The duodenum and ileum start to segment simultaneously when chyme first enters the small intestine. These segmenting contractions are the major movements of the small intestine; they are repeated about 12 to 16 times per minute.

Peristaltic contractions (propulsive contractions) propel chyme through the small intestine and into the large

intestine with weak, repetitive waves that start at the esophagus and stomach and move down the intestine before dying out. Each wave takes 100 to 150 minutes to migrate from its initiation site to the end of the small intestine. Thus chyme may remain in the small intestine for 1 to 6 hours. These weak peristaltic contractions, as well as the stronger segmenting contractions, are controlled by the autonomic nervous system.

Digestive enzymes in the small intestine The exocrine glands in the mucosa of the small intestine secrete about 1.5 L of a dilute aqueous salt and mucous solution into the lumen daily. This solution *contains no enzymes* and is called the *succus entericus,* or **intestinal juice.** Although no enzymes are secreted into the intestinal juice, the small intestine does synthesize digestive enzymes. These enzymes act within the cells or at the *brush borders* of the epithelial cells lining the lumen rather than actually being secreted into the lumen [FIGURE 25.16C].

Within the brush border are three classes of enzymes for the final digestion of carbohydrates and proteins:

1 *Enterokinase* converts the pancreatic enzyme trypsinogen into active trypsin, which, along with several other enzymes, completes the breakdown of peptides into their amino acid components.

2 The *disaccharidases* — sucrase, maltase, and lactase — complete carbohydrate digestion by converting the remaining disaccharides into simple monosaccharides, as follows: (a) Sucrase converts sucrose into glucose and fructose, (b) maltase converts maltose into two molecules of glucose, and (c) lactase converts lactose into glucose and galactose.

3 The *aminopeptidases* aid enterokinase by breaking down peptide fragments into individual amino acids. Aminopeptidases break the peptide bond at the terminal amino acid.

The digestion of carbohydrates and proteins is completed within the brush border, and the products are ready to be absorbed from the small intestine into the blood and lymph. FIGURE 25.18 and TABLE 25.7 (page 836) summarize the digestion of lipids, proteins, and carbohydrates.

Absorption from the Small Intestine

All the products of carbohydrate, lipid, and protein digestion, as well as most of the ingested electrolytes, vitamins, and water, are normally absorbed by the small intestine, with most absorption occurring in the duodenum and jejunum. The ileum is the primary site for the absorption of bile salts and vitamin B_{12}.

The segmenting contractions result from sharp contractions of the circular smooth muscle.

Contracting circular smooth muscle

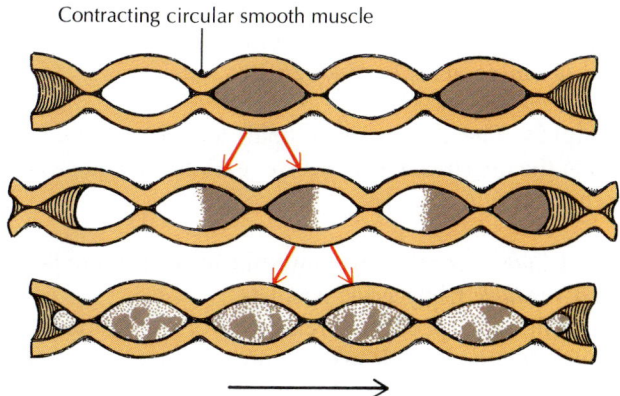

Direction of chyme bolus movement

Lactose Intolerance

Lactose is a disaccharide sugar found in milk, dairy products, and certain foods such as pizza and hot dogs. Some people have no trouble digesting lactose, but many others suffer from **lactose intolerance,** the inability to digest lactose. Lactose intolerance is caused by a deficiency of the lactose-digesting enzyme *lactase.* Lactose intolerance is a relatively common congenital disorder, and it may also appear in later life. Normally, lactase is present at birth, but its concentration declines in most northern Europeans and about 25 percent of Americans when they are between 18 and 36 months old. In addition, a permanent lactase deficiency develops in 80 to 90 percent of blacks and Orientals.

Lactase deficiency may occur more frequently than other deficiencies of digestive enzymes because there is only one enzyme involved with lactose digestion, while several enzymes are involved with the digestion of other dietary sugars. For example, maltose digestion involves three different enzymes.

In complete lactose digestion, lactase splits lactose into glucose and galactose, which are absorbed from the small intestine. Lactase is produced by epithelial cells in the lining of the small intestine. Without sufficient lactase, the ingested lactose remains in the small intestine, where it prevents some water from being absorbed. The water, together with lactose, moves on to the large intestine, where bacteria begin to digest the lactose and metabolize the resultant glucose and galactose. This metabolic activity produces large amounts of gas, causing pain and bloating. Also produced are organic products that inhibit the absorption of ions from the intestine. As a result, the unabsorbed water moves into the lumen of the large intestine, producing diarrhea. The symptoms may range from mild discomfort to severe pain and diarrhea.

Commercially prepared lactase is available without prescription and may be added to milk in a liquid form or taken as pills just prior to milk ingestion. Milk containing a reduced amount of lactose may also be purchased without a prescription.

FIGURE 25.18 COMPARTMENTALIZED DIGESTION

The change from large to small letters P, C, and L indicates the breakdown of nutrients into smaller molecules. Note also how long the bolus remains in each part of the tract.

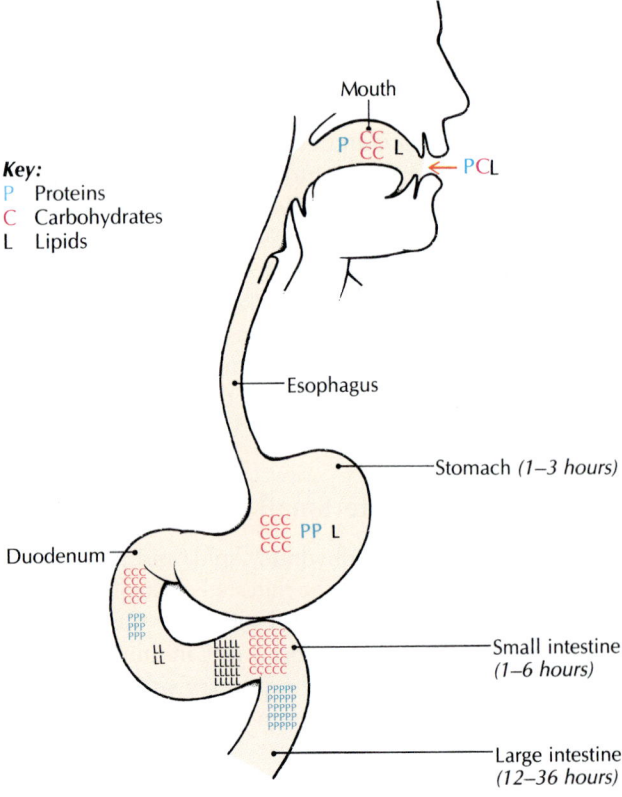

Key:
P Proteins
C Carbohydrates
L Lipids

Carbohydrates Monosaccharides (glucose, fructose, galactose) are readily absorbed through the microvilli on the border of columnar absorptive cells by facilitated diffusion or active transport. Fructose is absorbed by facilitated diffusion. Glucose and galactose are absorbed by **secondary active transport,** in which a transporter carries sodium from the intestinal lumen into the intestinal cell down its concentration gradient. Glucose or galactose is carried along. The glucose or galactose is then concentrated inside the cell and leaves the cell down its concentration gradient into the capillary network in the villus [FIGURE 25.19].

Proteins After proteins have been broken down, their constituent amino acids are absorbed through columnar absorptive cells into capillary networks within the villi. A secondary active transport mechanism, similar to glucose and galactose absorption, is used for the absorption of amino acids [FIGURE 25.20].

Lipids The absorption of lipids into the central lacteal of the intestinal villus involves a complex sequence of events [FIGURE 25.21]:

1 Lipids enter the small intestine in the form of large *water-insoluble* triglyceride droplets.

2 Bile salts speed the breakdown of triglycerides by emulsifying the lipid droplets into smaller ones, exposing additional surface area.

3 Pancreatic lipase starts to break down the triglycer-

FIGURE 25.19 CARBOHYDRATE DIGESTION AND ABSORPTION

DIETARY CARBOHYDRATES

Polysaccharides (starch, glycogen—complex carbohydrates)

Amylase

Disaccharides (sucrose, maltose, lactose)

INTESTINAL LUMEN

Disaccharidases

Na+ cotransport carrier

Microvilli

Monosaccharides (glucose, fructose, galactose—simple sugars)

Monosaccharides

Lacteal

Capillary

Lymph duct

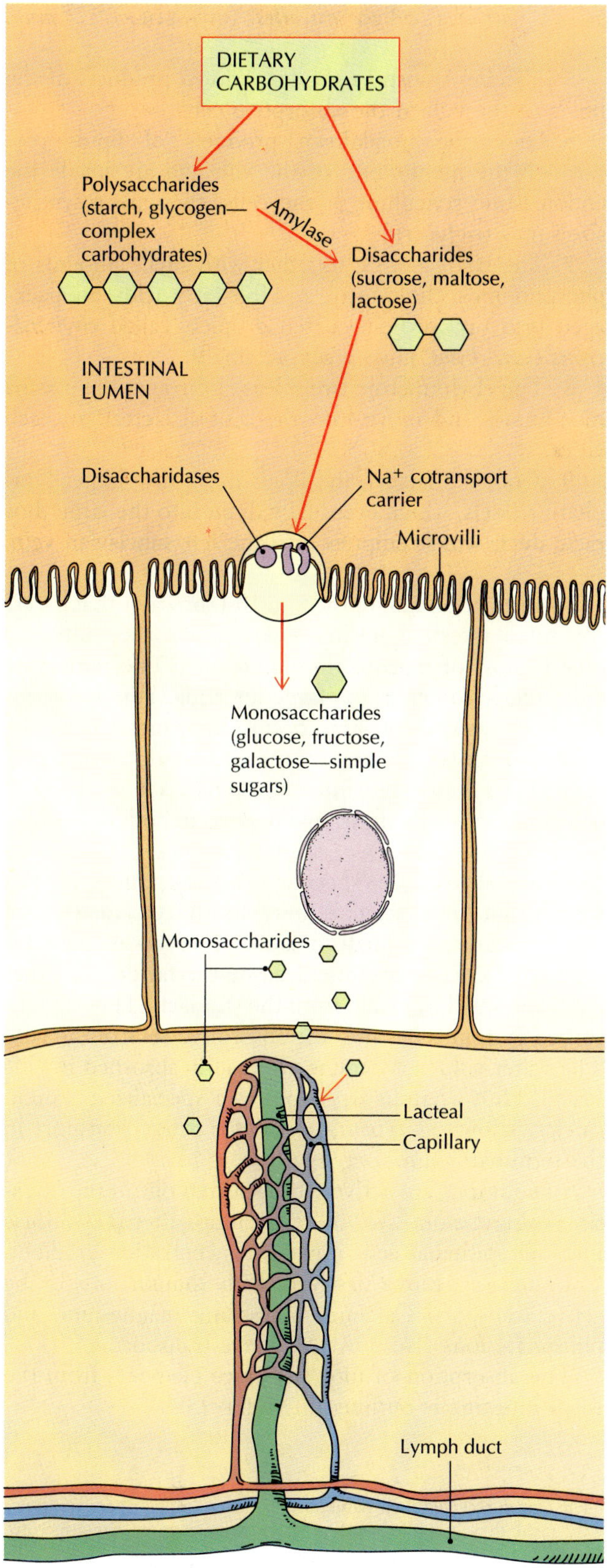

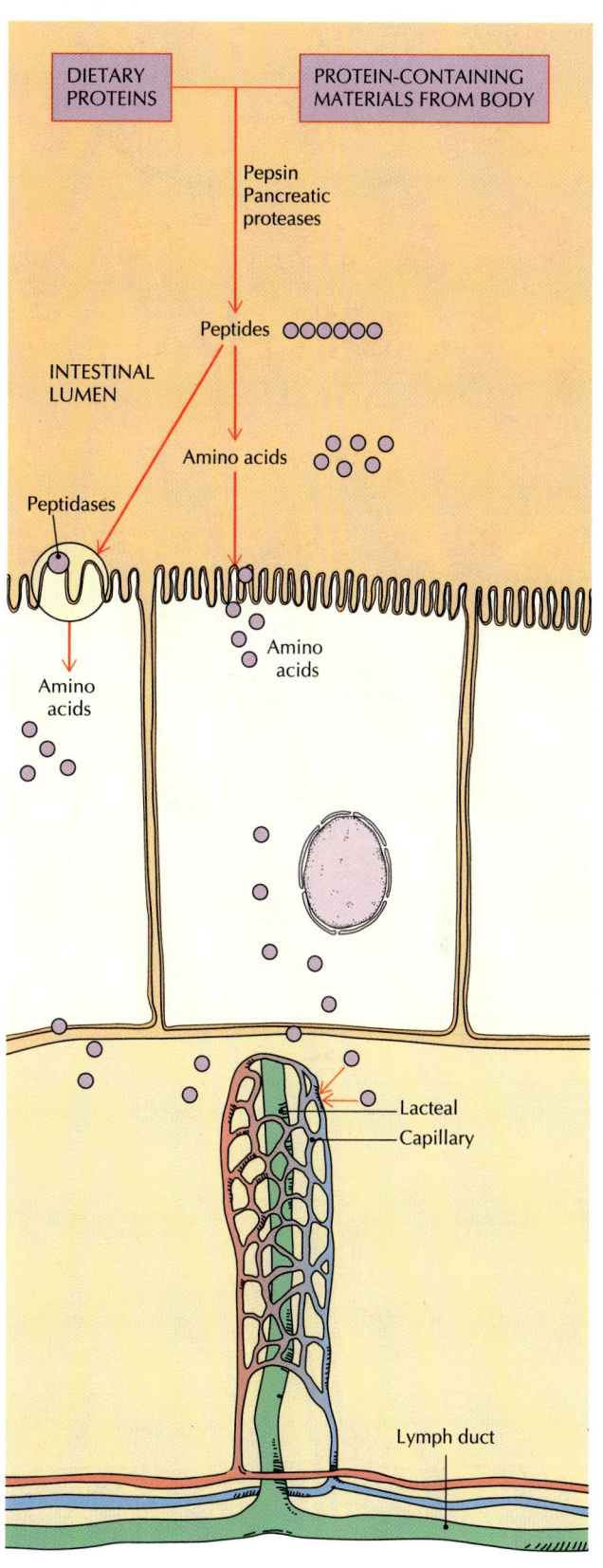

FIGURE 25.20 PROTEIN DIGESTION AND ABSORPTION

DIETARY PROTEINS

PROTEIN-CONTAINING MATERIALS FROM BODY

Pepsin Pancreatic proteases

Peptides

INTESTINAL LUMEN

Amino acids

Peptidases

Amino acids

Amino acids

Lacteal

Capillary

Lymph duct

FIGURE 25.21 LIPID DIGESTION AND ABSORPTION

The drawing summarizes lipid digestion and absorption across the walls of the small intestine. The numerals refer to the text discussion.

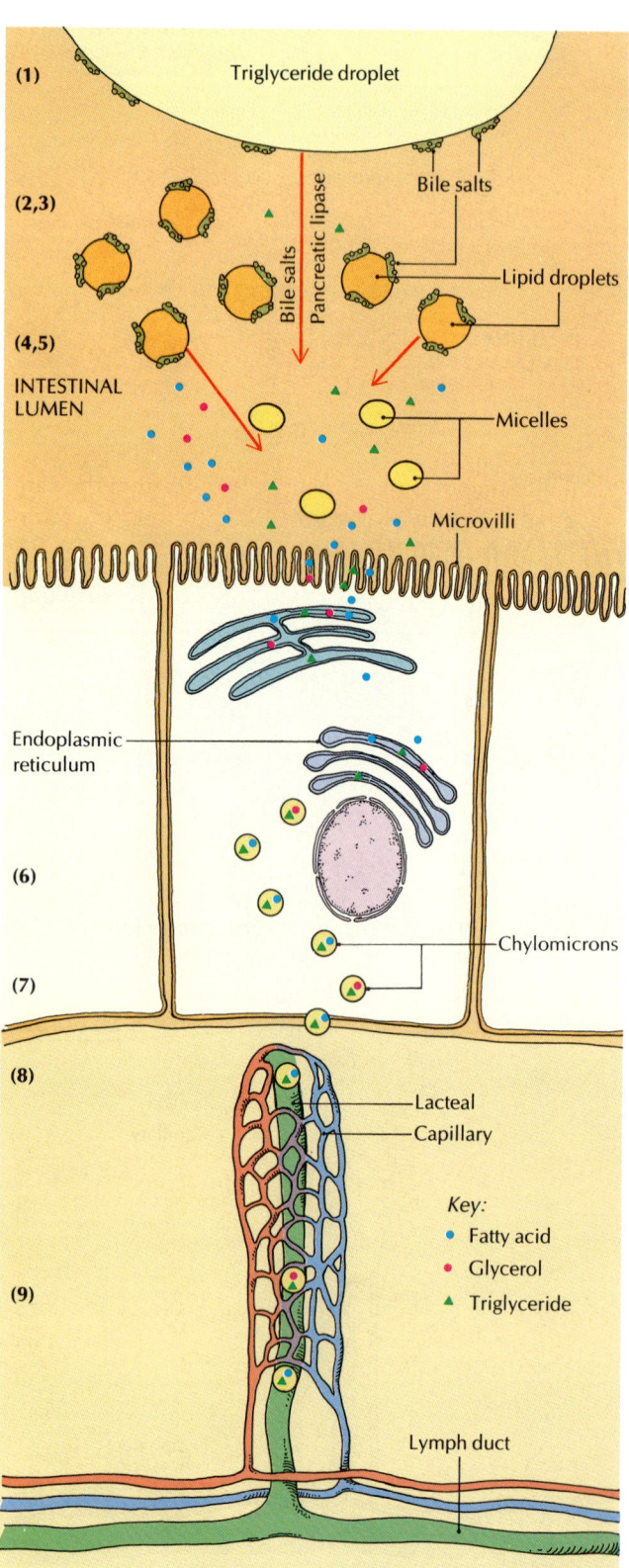

(1) Triglyceride droplet

Bile salts

(2,3) Pancreatic lipase · Bile salts

Lipid droplets

(4,5) INTESTINAL LUMEN

Micelles

Microvilli

Endoplasmic reticulum

(6)

(7) Chylomicrons

(8) Lacteal · Capillary

Key:
- ● Fatty acid
- ● Glycerol
- ▲ Triglyceride

(9)

Lymph duct

---

ide droplets into free fatty acids, glycerol, and monoglycerides.

4 Bile salts also cause fatty acids, phospholipids, and glycerides to combine to form submicroscopic, *water-soluble* particles called **micelles** (my-SEHLZ; L. *mica*, grain).

5 Micelles transport the breakdown products of the lipids to the villi of the absorptive cells.

6 Once the breakdown products of lipids pass through the membranes of the villi and are inside the endoplasmic reticulum of the cells, they are resynthesized into triglycerides.

7 The triglycerides, together with small amounts of phospholipids, cholesterol, and free fatty acids, are packaged into tiny protein-coated droplets called **chylomicrons** (Gr. *chylos*, juice + *micros*, small).

8 The chylomicrons are released through the plasma membranes and move into the central lacteal* of each villus.

9 From the lacteal, the lipids move into larger lymphatic vessels, which eventually drain into the large thoracic duct, which empties into the left subclavian vein.

Water Nearly all of the 5 to 10 L of water that enters the small intestine during a day is absorbed into the blood, most of it from the duodenum. The remaining half liter or so enters the large intestine. Water absorption in the small intestine occurs by osmosis from the osmotic gradient created by the end products of digestion. Water is absorbed into the capillaries in the villi and can move across the intestinal mucosa in both directions.

Other substances Several other substances, including vitamins, nucleic acids, ions, and trace elements, are absorbed from the small intestine. The fat-soluble vitamins (A, D, E, K) are dissolved by the fat droplets that enter the small intestine from the stomach. They are absorbed with micelles into the absorptive cells of the villi. The water-soluble vitamins are readily absorbed by diffusion. Only vitamin B_{12} requires a specialized protein carrier, known as *intrinsic factor*, for active transport in the terminal ileum.

Sodium ions are actively transported out of the intestine, using the Na^+/K^+ ATPase pumps after they diffuse into the epithelial cells down a concentration gradient. Chloride ions move passively with sodium ions or by active transport. Calcium, potassium, magnesium, and phosphate ions also move by active transport.

The absorption of ions and trace elements from the small intestine is outlined in TABLE 25.4.

---

*The *lacteals* received their name, which means "milky," because after a fatty meal they are filled with a milky suspension of chylomicrons.

TABLE 25.4 ABSORPTION OF MAJOR IONS AND TRACE ELEMENTS BY THE SMALL INTESTINE

Substance	Method of absorption
Bicarbonate*	Absorbed by mucosal cells when concentration in lumen is high; secreted into lumen when blood concentration is high. Absorption rate also related to sodium absorption rate in order to maintain electrical neutrality.
Calcium	Active transport; most rapid in duodenum. Rate of absorption varies with dietary intake and plasma concentrations; pregnancy increases rate of absorption, age decreases absorption. Absorption regulated in part by negative feedback with parathyroid hormone. Vitamin D increases calcium absorption.
Chloride*	Passive diffusion when following gradient path of sodium ions from lumen to blood; active transport across plasma membrane coupled with sodium transport.
Copper	Active transport; absorbed mostly by upper portion of small intestine; excess amounts excreted in feces.
Iron	Active transport from lumen into epithelial cells; absorbed in small intestine and stored briefly before being transferred to plasma in response to metabolic requirements. Most of transported iron is stored in the liver as ferritin, released when plasma concentrations of iron drop. Slow rate of absorption increased by vitamin C.
Phosphate	All portions of small intestine absorb phosphate, both actively and passively.
Potassium*	Passive diffusion from lumen to blood; active transport across plasma membrane.
Sodium*	Passive diffusion when gradient is from lumen to blood through tight junctions; active transport through cell, involving two different carrier systems.

*Readily absorbed from the small intestine.

A S K   Y O U R S E L F

1 What are the main parts of the small intestine?

2 In addition to digestion, what is the main purpose of the small intestine?

3 What are the main types of cells in the small intestine?

4 How do segmenting contractions differ from peristaltic contractions?

5 What are the main enzymes in the small intestine?

6 Why is the absorption of fats more complex than the absorption of proteins and carbohydrates?

PANCREAS AS A DIGESTIVE ORGAN

When the chyme is emptied from the stomach into the small intestine, it is mixed with secretions from the pancreas and liver, in addition to juices secreted by the intestinal cells. In the following sections we will discuss the important digestive roles of these accessory organs — the pancreas and the liver, along with the gallbladder — before continuing with a description of the large intestine.

Gross Anatomy

The *pancreas* is a soft, pinkish-gray gland about 12 to 15 cm (5 to 6 in.) long. It lies transversely across the posterior abdominal wall, behind the stomach [FIGURE 25.22A]. It is completely retroperitoneal and is connected, usually by two ducts, to the duodenum.

The pancreas consists of three parts: (1) The broadest part, or *head*, fits into the C-shaped curve of the duodenum. (2) The main part, or *body*, lies between the head and tail of the pancreas, behind the stomach and in front of the second and third lumbar vertebrae. (3) The narrow *tail* extends toward the spleen [FIGURE 25.22A].

Microscopic Anatomy

Pancreatic tissue is composed of both exocrine and endocrine cells [FIGURE 25.22B]. The endocrine function of the pancreas is described in Chapter 18. Approximately 1 percent of the pancreas is made up of endocrine pancreatic islet cells. The remaining 99 percent of the cells are *exocrine* cells, which form groups of cells called *acini* (ASS-ih-nye; L. grapes, so named because they resemble bunches of grapes; sing. *acinus*), which secrete about 1.5 L of fluid and digestive enzymes into the small intestine daily.

FIGURE 25.22 PANCREAS

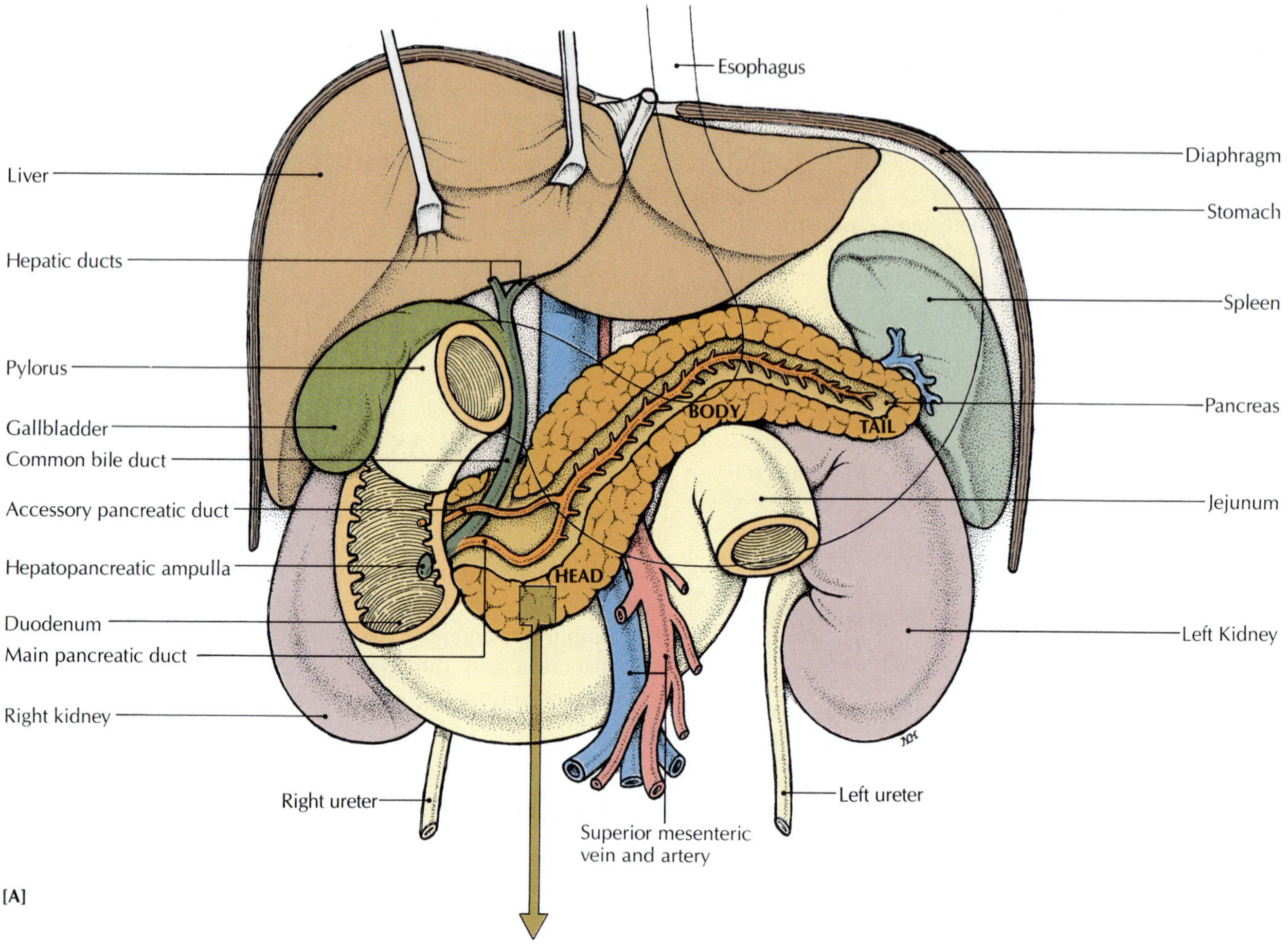

Esophagus

Liver

Diaphragm

Stomach

Hepatic ducts

Spleen

Pylorus

BODY

TAIL

Pancreas

Gallbladder

Common bile duct

Accessory pancreatic duct

Hepatopancreatic ampulla

HEAD

Jejunum

Duodenum

Main pancreatic duct

Left Kidney

Right kidney

Right ureter

Left ureter

Superior mesenteric
vein and artery

[A]

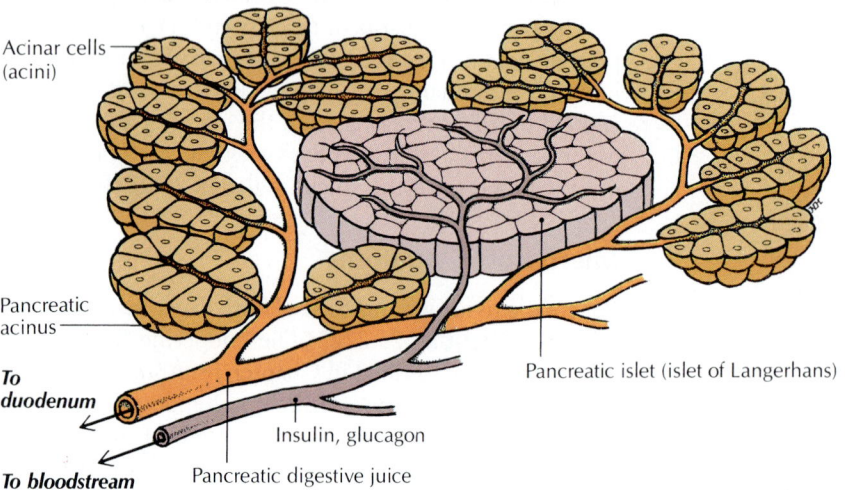

Acinar cells
(acini)

Pancreatic
acinus

*To
duodenum*

Pancreatic islet (islet of Langerhans)

To bloodstream

Insulin, glucagon

Pancreatic digestive juice

[B]

[A] Position of the pancreas, with its head tucked into the curve of the duodenum and its tail extending to the spleen. Note the duct system of the pancreas. The liver, gallbladder, and pancreas all secrete their products into ducts that merge as they enter the duodenum. **[B]** Cells of a pancreatic islet surrounded by acini. Acini secrete digestive fluids that pass into the small intestine.

Pancreatic Secretions

Pancreatic secretions are clear, slightly alkaline (pH 7.1 to 8.2), and contain mostly water, some salts, sodium bicarbonate, and several enzymes or precursors of enzymes. The alkaline secretions buffer the acidic chyme, inhibit the enzyme pepsin, which works best in an acidic environment such as the stomach, and establish the optimum pH for intestinal enzymes. The enzymes produced and secreted by acinar cells for the digestion of lipids, proteins, and carbohydrates are the following:

1 *Pancreatic lipase* splits triglycerides (large-molecule fats) into smaller, absorbable glycerol and free fatty acids.

2 *Pancreatic amylase* converts polysaccharides into monosaccharides and disaccharides, especially into maltose.

3 *Pancreatic proteolytic enzymes* consist of *trypsinogen*, *chymotrypsinogen*, and *procarboxypeptidase*. Each is secreted in an inactive form. After trypsinogen is secreted into the duodenal lumen, it is activated by the enzyme *enterokinase* from the small intestine and becomes *trypsin*. Chymotrypsinogen and procarboxypeptidase are converted into their active forms, *chymotrypsin* and *carboxypeptidase*, in the intestinal lumen by the newly formed trypsin.

See TABLE 25.5 for a description of digestive hormones and TABLE 25.6 for a summary of the major glands, secretions, and enzymes in the digestive system.

Control of Pancreatic Secretions

As with gastric secretions, pancreatic secretions are under both neural and hormonal control. During the cephalic and gastric phases of gastric secretion, efferent output goes to the pancreas (as well as to the stomach) to increase the release of pancreatic enzymes into the small intestine in anticipation of the presence of food.

Two hormones also stimulate the release of pancreatic secretions. The hormone *secretin* is released from the small intestine (enteroendocrine cells) in response to the acidic chyme in the small intestine. Secretin then causes the pancreas to release a bicarbonate-rich pancreatic secretion to neutralize the acid and attain an optimum pH for enzyme activity in the intestinal lumen. The presence of partially digested fats and proteins in the small intestine also causes the release of the hormone *cholecystokinin* (*CCK*; KOHL-eh-SIHSS-toe-kye-nihn) from the intestinal enteroendocrine cells, which causes the release of an enzyme-rich secretion from the pancreas.

A S K   Y O U R S E L F

1 What is the exocrine function of the pancreas?

2 What are the three main groups of pancreatic enzymes?

3 What two hormones stimulate the secretion of pancreatic juices?

LIVER AS A DIGESTIVE ORGAN

The *liver** is a large, compound, tubular gland weighing

---

*The Greek for "liver" is *hepar*, a word used in the combining form *hepato-* in such words as *hepatic* and *hepatitis*.

TABLE 25.5 MAJOR DIGESTIVE HORMONES			
Hormone	**Source**	**Primary stimulus for secretion**	**Major functions**
Cholecystokinin (CCK)	Small intestine (duodenal mucosa)	Proteins, peptides, and fats in duodenum	Stimulates gallbladder to release bile and pancreas to secrete enzymes; inhibits gastrin-stimulated secretion of acid; promotes digestion of all foods.
Gastric inhibitory peptide (GIP)	Small intestine (duodenal mucosa)	Acid or peptides in duodenum	Inhibits secretion of acid by stomach; helps regulate movement of chyme into duodenum by decreasing gastric motility.
Gastrin	Stomach (pyloric region)	Distension of stomach, vagus nerve, protein in stomach	Stimulates secretion of acid by parietal cells, which stimulates secretion of pepsinogen by chief cells; increases motility of pyloric region.
Secretin	Small intestine (duodenal mucosa)	Acid or peptides in duodenum	Stimulates pancreas to secrete juices containing bicarbonate; inhibits gastrin-stimulated secretion of acid; inhibits motility of stomach and duodenum.

TABLE 25.6 MAJOR DIGESTIVE GLANDS, SECRETIONS, AND ENZYMES

Place of digestion	Source	Secretion	Enzyme	pH	Digestive functions of secretion or enzyme
Mouth	Salivary glands	Saliva	Salivary amylase	6–8	Begins carbohydrate digestion; breaks down carbohydrates; inactivated by stomach hydrochloric acid.
	Mucous glands	Mucus		6–7	Lubricates.
Esophagus	Mucous glands	Mucus		6–7	Lubricates.
Stomach	Gastric glands	Gastric juice: contains enzymes that partially digest proteins	Lipase	7–9	Converts triglycerides into fatty acids and glycerol.
			Pepsin	1–3	Converts proteins into polypeptides (proteoses and peptones).
	Gastric mucosa	Hydrochloric acid		0.1	Converts inactive pepsinogen into active pepsin; dissolves minerals; kills microorganisms.
	Mucous glands	Mucus		5–6	Lubricates.
Small intestine	Liver	Bile: emulsifies fats		7–9	Emulsifies fats.
	Pancreas	Pancreatic juice: contains enzymes that digest starch, proteins, fats, nucleic acids	Amylase	6–7	Converts starch into maltose and other disaccharides.
			Chymotrypsin	7–9	Converts proteins into peptides and amino acids.
			Lipase	7–9	Converts triglycerides into fatty acids and glycerol.
			Nuclease	7–9	Converts nucleic acids into mononucleotides.
			Trypsin	7–9	Converts proteins into peptides and amino acids; converts inactive chymotrypsinogen into active chymotrypsin.
			Carboxypeptidase	7–9	Helps digest peptide fragments; frees terminal amino acid from carboxyl end of protein.
	Intestinal glands (crypts of Lieberkühn)	Intestinal juice: contains sugar-digesting enzymes	Enterokinase	7–9	Converts inactive trypsinogen into active trypsin.
			Aminopeptidase	7–9	Cleaves amino acid from peptide fragment.
			Lactase	7–9	Converts lactose into glucose and galactose.
			Maltase	7–9	Converts maltose into glucose.
			Peptidase	7–9	Converts polypeptides into amino acids.
			Sucrase	7–9	Converts sucrose into glucose and fructose.
	Mucous glands	Mucus		7.5–8	Lubricates.
Large intestine	Mucous glands	Mucus		7.5–8	Lubricates.

about 1.5 kg (3 lb) in the average adult.* Though it lies outside the digestive tract, it has many functions that are relevant to digestion and absorption.

Anatomy of the Liver

The liver is reddish, wedge-shaped, and covered by a network of connective tissue *(Glisson's capsule)*. It is located under the diaphragm in the upper right region of the abdominal cavity. The undersurface faces the stomach,

*An infant usually has a pudgy abdomen because of the disproportionately large size of its liver. In most children, the liver occupies about 40 percent of the abdominal cavity and is responsible for approximately 4 percent of the total body weight. In an adult, the liver represents about 2.5 percent of the total body weight.

the first part of the duodenum, and the right side of the large intestine [FIGURE 25.23A; see also FIGURE 25.1].

The liver is divided into two main lobes by the *falciform ligament*, a mesentery attached to the anterior midabdominal wall [FIGURE 25.23B]. The **right lobe,** which is about six times larger than the left lobe, is situated over the right kidney and the right colic (hepatic) flexure of the large intestine. The **left lobe** lies over the stomach. In the free border of the falciform ligament, extending from the liver to the umbilicus, is the *ligamentum teres* (round ligament), a fibrous cord that is a remnant of the left fetal umbilical vein.

The right lobe is further subdivided into a small *quadrate lobe* and a small *caudate lobe* on its ventral (visceral) surface [FIGURE 25.23C]. The quadrate lobe is flanked by

---

FIGURE 25.23 LIVER

[A] Location of the liver. [B] The falciform ligament divides the liver into two lobes. Anterior view. [C] Posterior view.

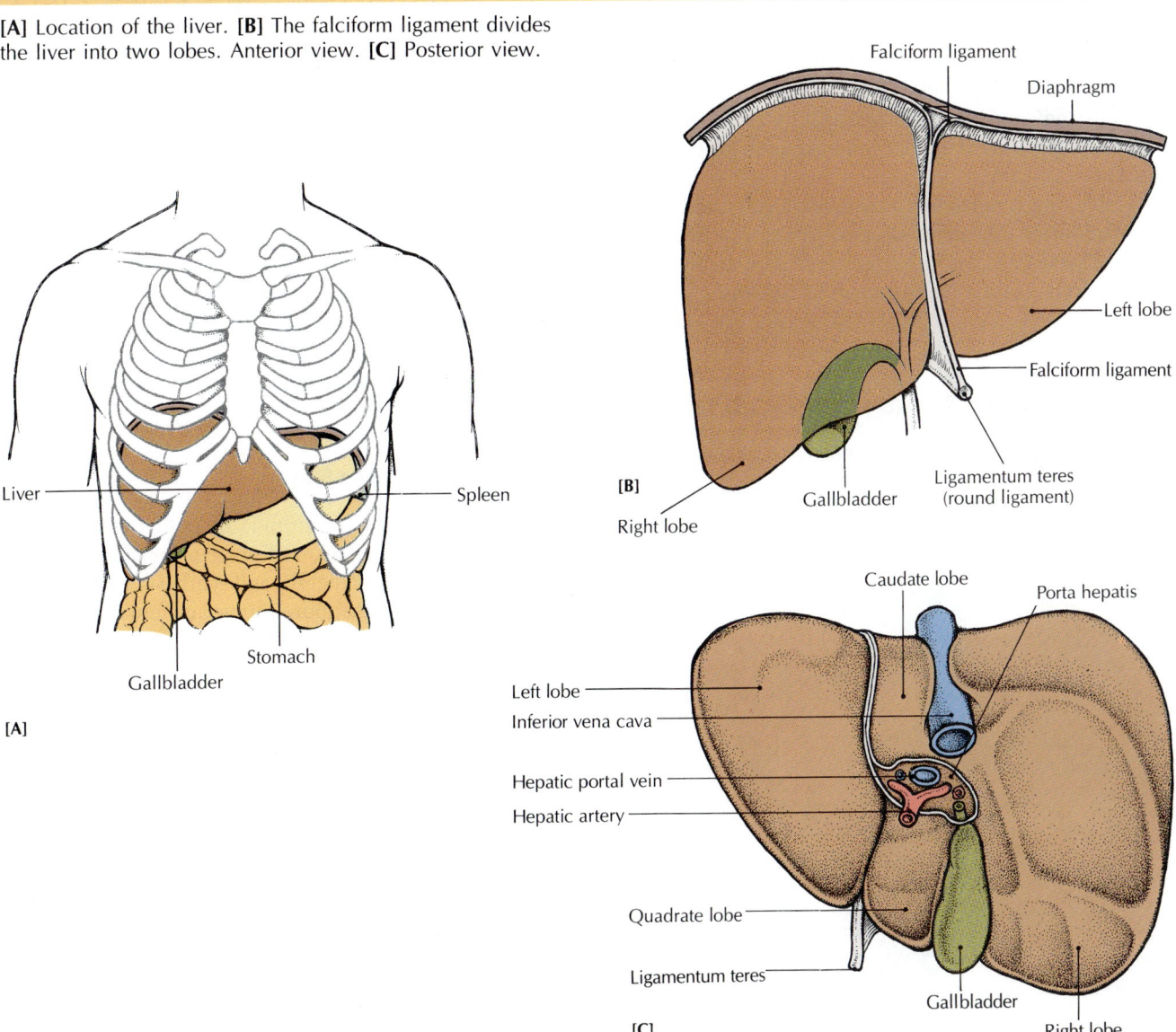

the gallbladder on the right and the ligamentum teres on the left. The quadrate lobe partially envelops and cushions the gallbladder. The caudate lobe is flanked by the inferior vena cava on the right and the *ligamentum venosum* on the left. The ligamentum venosum is a remnant of the fetal ductus venosus, a venous shunt from the umbilical vein to the inferior vena cava.

Vessels of the liver The *porta hepatis* ("liver door") is the area through which the blood vessels, nerves, lymphatic vessels, and ducts enter and leave the liver. It is located between the quadrate and caudate lobes and contains the following vessels and ducts [FIGURE 25.24]:

1 The *hepatic artery* is a branch of the celiac artery of the aorta. It supplies the liver with oxygenated arterial blood, which represents about 20 percent of the blood flow to the liver.

2 The *hepatic portal vein* drains venous blood into the liver from the entire gastrointestinal tract. It supplies the remaining 80 percent of the liver's blood. This blood contains the nutrients absorbed by the small intestine.

3 The *hepatic vein* drains blood from the liver into the inferior vena cava.

4 Bile ducts, called **bile canaliculi,** are formed by the **bile capillaries** that unite after collecting bile from the liver cells [FIGURE 25.25]. Bile is secreted by the liver into the bile canaliculi, which drain into the **right** and **left**

hepatic ducts. The ducts, in turn, converge with the **cystic duct** from the gallbladder to form the **common bile duct.** The common bile duct joins the main pancreatic duct, enlarges into the hepatopancreatic ampulla, and then joins the duodenal papilla, which opens into the duodenum [FIGURE 25.26].

FIGURE 25.24 shows the four main vessels and ducts of the liver, with the hepatic artery and hepatic portal vein entering and the hepatic vein and common bile duct leaving.

Microscopic anatomy The functional units of the liver are five- or six-sided **lobules** that contain the *central vein* running longitudinally through it [FIGURE 25.25]. Most lobules are about 1 mm (0.04 in.) in diameter. Liver cells, known as **hepatic cells** or hepatocytes, within the lobules are arranged in one-cell-thick platelike layers that radiate from the central vein to the edge of the lobule. Each corner of the lobule usually contains a *portal area,* a complex composed of branches of the portal vein, hepatic artery, bile duct, and nerve.

Between the radiating rows of cells are delicate blood channels called **sinusoids,** which transport blood from branches of the portal vein and hepatic artery. Blood flows from the artery and vein in the portal areas into the sinusoids and then to the central vein, which drains it from the lobule into the hepatic vein. The walls of the sinusoids are lined with endothelial cells. Attached to these lining cells are phagocytic stellate reticuloendothelial cells that engulf and digest worn-out red and white blood cells, microorganisms, and other foreign particles passing through the liver.

Functions of the Liver

The liver performs a multitude of functions associated with carbohydrate, lipid, and protein metabolism; storage of vitamins and minerals; and synthesis and excretion of bile. Some of these more important functions will be discussed in the following sections.

Carbohydrate metabolism The liver helps regulate blood glucose by storing glucose in the form of glycogen (glycogenesis) when blood glucose is high. With the help of enzymes, it converts glycogen back into glucose (glycogenolysis) when blood glucose is low.

Lipid metabolism The liver synthesizes lipoproteins, cholesterol, and phospholipids, which are essential components of plasma membranes. Liver cells also use cholesterol in the production of bile. *Beta oxidation* (breakdown of fatty acids into acetyl coenzyme A) and *ketogenesis* (formation of ketone bodies) also occur in the liver.

FIGURE 25.24 VESSELS AND DUCTS OF THE LIVER

Arrows indicate the direction of blood flow, anterior view.

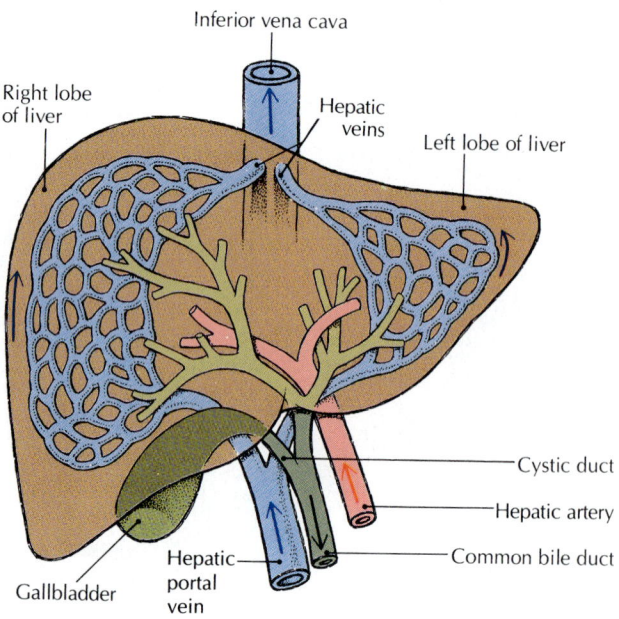

Inferior vena cava

Right lobe of liver

Hepatic veins

Left lobe of liver

Cystic duct

Hepatic artery

Common bile duct

Hepatic portal vein

Gallbladder

FIGURE 25.25 LIVER LOBULES

[A] Cross section showing typical six-sided shape of lobules. [B] Cutaway view of lobule showing channels of flow, including bile canaliculi and sinusoids, which convey their contents in opposite directions. [C] Light micrograph of liver tissue. ×25. [D] Scanning electron micrograph of a liver lobule, showing hepatocytes radiating toward the central vein. ×400.

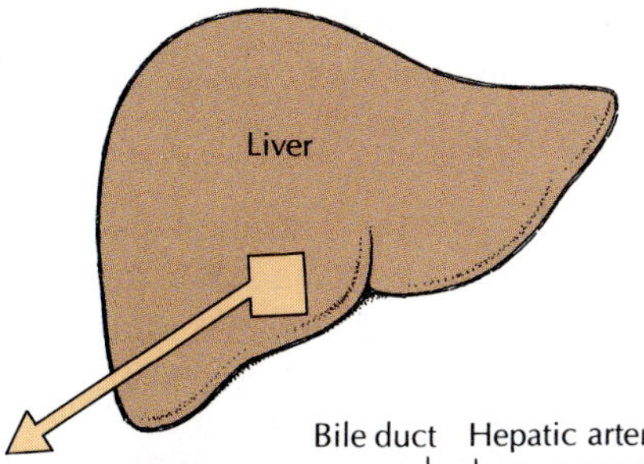

Liver

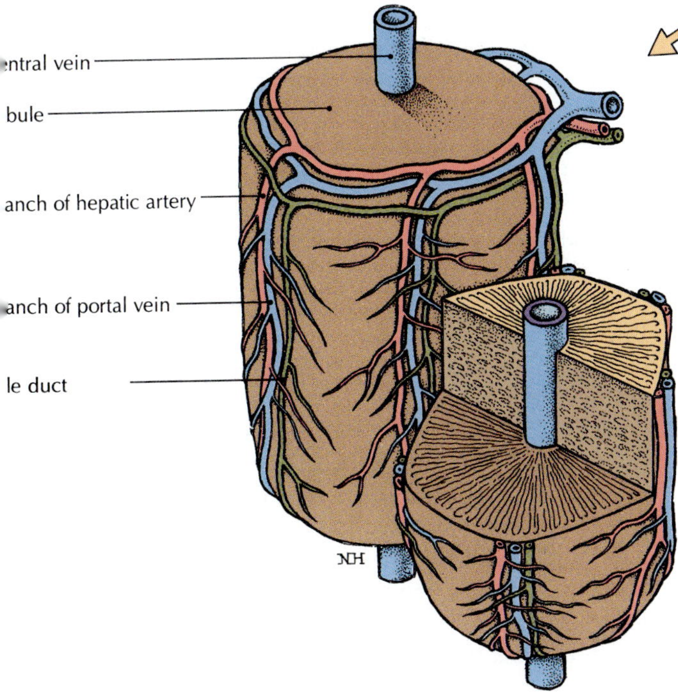

Central vein

bule

Branch of hepatic artery

Branch of portal vein

Bile duct

[A]

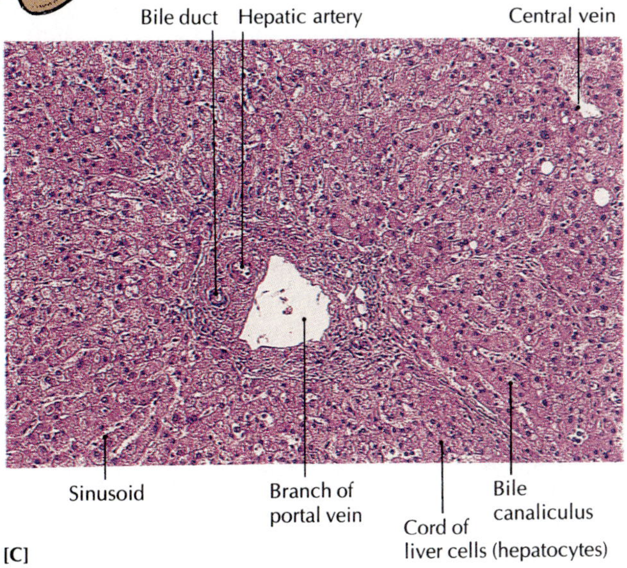

Bile duct   Hepatic artery   Central vein

Sinusoid   Branch of portal vein   Cord of liver cells (hepatocytes)   Bile canaliculus

[C]

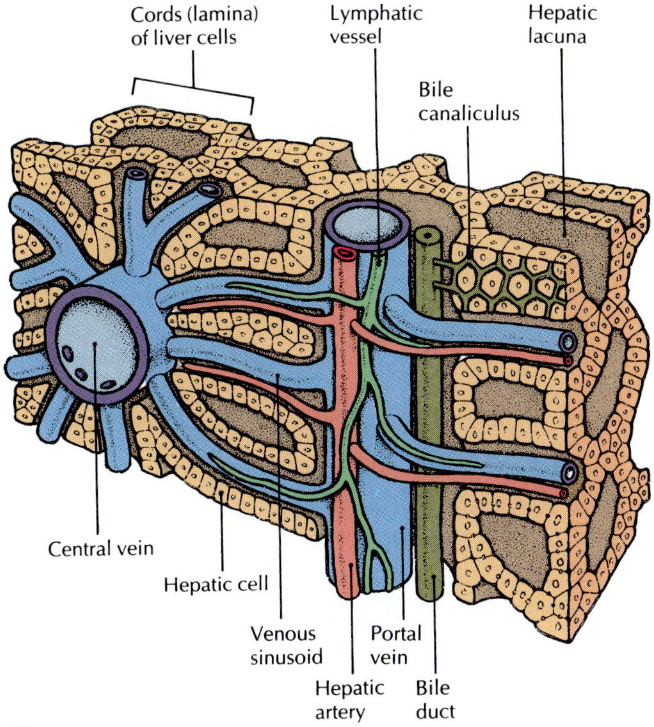

Cords (lamina) of liver cells   Lymphatic vessel   Hepatic lacuna

Bile canaliculus

Central vein

Hepatic cell

Venous sinusoid   Portal vein

Hepatic artery   Bile duct

[B]

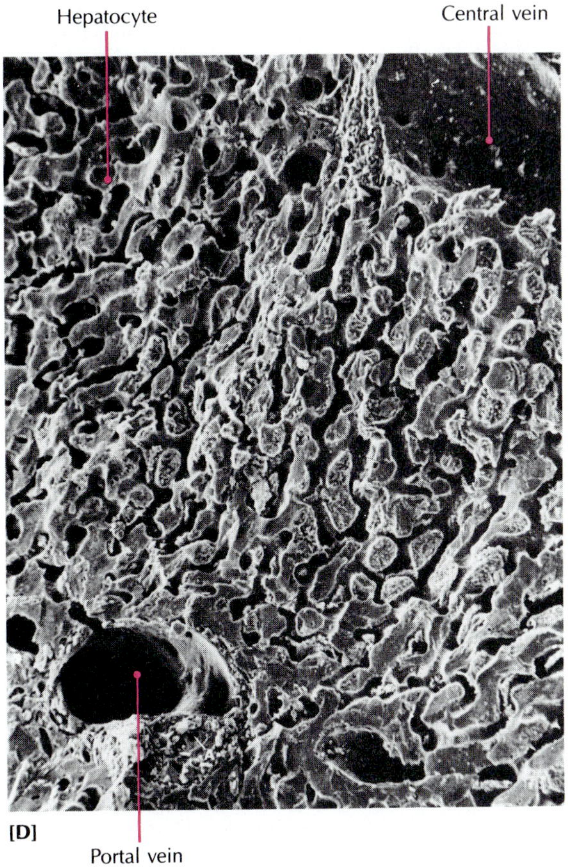

Hepatocyte   Central vein

Portal vein

[D]

Protein metabolism The liver deaminates amino acids (removes the NH_2 group), synthesizes numerous plasma proteins, and converts ammonia into urea for excretion in the urine.

Storage of vitamins and minerals The liver also stores the fat-soluble vitamins (A, D, E, and K), the water-soluble vitamin B_{12} (antianemic factor), and minerals such as iron and copper, all taken from the diet.

Synthesis and secretion of bile One of the liver's main functions as a digestive organ is to secrete *bile*, an alkaline liquid containing water, sodium bicarbonate, bile salts, bile pigments, cholesterol, mucin, lecithin, and bilirubin. As much as 1 L of bile is secreted by the liver every day. It is stored in the gallbladder in a highly concentrated form until it is needed to break down fats, or for any of its many other functions. Bile secretion may be increased by chemical, hormonal (CCK), or neural mechanisms.

Bile salts are derivatives of cholesterol. They are actively secreted into the bile and eventually pass into the duodenum along with other biliary secretions. After these salts participate in fat digestion and absorption, most of them are reabsorbed or taken up into the blood by a special active transport mechanism located in the terminal ileum. From there, bile salts are returned via the hepatic portal system to the liver, which can resecrete them into the bile. This recycling of bile salts—and other biliary secretions—between the small intestine and the liver is called the *enterohepatic circulation* of these substances.

Bile pigments give bile its color. They are derived from the hemoglobin of worn-out red blood cells that are transported to the liver for excretion. The color of feces comes from one of the breakdown products of excreted bile pigments. The condition called *jaundice*, in which the skin has a yellowish color, is caused by excessive amounts of the bile pigment *bilirubin* in extracellular fluids.

A S K   Y O U R S E L F

1 Where is the liver located?

2 What are the vessels of the liver?

3 How does a liver lobule resemble a wheel?

4 What are the components of a portal area?

5 What are some of the major functions of the liver?

6 How does the liver make glucose available as it is required for energy?

7 What are some of the functions of bile?

GALLBLADDER AND BILIARY SYSTEM

The *gallbladder** is a small, pear-shaped saclike organ situated in a depression under the right lobe of the liver [FIGURE 25.26; see also FIGURE 25.22A].

The gallbladder consists of an outer serous peritoneal coat, a middle muscular coat, and an inner mucous membrane that is continuous with the linings of the ducts. The mucous membrane secretes mucin and absorbs water readily. It does not absorb bile salts or bile pigments, but does transport salt out actively, with water following osmotically. As a result, the bile is concentrated. In fact, the organic constituents of bile leaving the gallbladder may be as much as 10 times more concentrated than they were when bile entered from the liver, a factor that may contribute to the formation of gallstones.

The gallbladder has an important storage function. Bile is secreted continuously by the liver. However, more bile is produced than is ordinarily required, and the excess is stored in the gallbladder until needed in the duodenum. The capacity of the gallbladder is between 30 and 60 mL. When the gallbladder is empty, its mucosa is thrown into rugae, which permit the gallbladder to expand to hold the bile.

The so-called *biliary system* consists of (1) the *gallbladder*, (2) the *left* and *right hepatic ducts*, which come together as the *common hepatic duct*, (3) the *cystic duct*, which extends from the gallbladder, and (4) the *common bile duct*, which is formed by the union of the common hepatic duct and the cystic duct [FIGURE 25.26].

The common bile duct and the main pancreatic duct join at an entrance to the duodenum about 10 cm (4 in.) from the pyloric orifice. They fuse there to form the *hepatopancreatic ampulla*. The ampulla travels obliquely through the duodenal wall and opens into the duodenum through the *duodenal papilla*. A sphincter located at the outlet of the common bile duct is called the *sphincter of the common bile duct (sphincter of Boyden)*, and the muscle below it, near the duodenal papilla, is the *sphincter of the hepatopancreatic ampulla (sphincter of Oddi)* [FIGURE 25.26]. The sphincter of the common bile duct appears to be the stronger and more important of the two. About 30 min after a meal, or whenever chyme enters the duodenum, the sphincters relax, the gallbladder contracts, and bile stored in the gallbladder is squirted into the duodenum.

---

* *Gallbladder* is derived from the Latin word *galbinus*, meaning greenish yellow, which is the usual color of *bile. Gall* is the archaic term for bile.

Q: *Is the gallbladder essential for the digestive process?*

A: It is not, and a diseased gallbladder can be removed in a surgical procedure called a cholecystectomy.

The contraction of the gallbladder, the relaxation of the sphincter of the hepatopancreatic ampulla, and the subsequent release of bile take place in the presence of cholecystokinin, a hormone released from the small intestine when fatty acids and amino acids reach the duodenum.

Bile is rich in cholesterol, a rather insoluble fatty substance; concentrated bile in the gallbladder may form crystals of cholesterol that are commonly called *gallstones*. If these crystals grow large enough to block the cystic duct, they can block the flow of bile and produce severe pain. (See When Things Go Wrong at the end of the chapter.)

ASK YOURSELF

1 What is the main function of the gallbladder?

2 What components make up the biliary system?

3 How does cholecystokinin contribute to the function of the gallbladder?

LARGE INTESTINE

The chyme remains in the small intestine for 1 to 6 hours. It then passes through the ileocecal valve into the *cecum*, the first part of the large intestine. By now, diges-

tion is complete and the large intestine functions to remove more water and ions (Na^+, K^+, Cl^-) from the liquid chyme. Removal of water, along with the action of microorganisms, converts liquid wastes into *feces* (FEE-seez; L. *faex*, dregs), a semisolid mixture that is stored in the large intestine until ready to be eliminated through the anus during defecation. The indigestible products of digestion remain in the large intestine from 12 to 36 hours.

Anatomy of the Large Intestine

The *large intestine* is the part of the digestive tract between the ileocecal orifice and the anus. It consists of the cecum; vermiform appendix; the ascending, transverse, descending, and sigmoid colons; and the rectum [FIGURE 25.27A]. The entire large intestine (sometimes called the *colon*) forms a rectangle that frames the tightly packed small intestine.

The ileocecal orifice empties into the *cecum* (SEE-kuhm; L. blind), a cul-de-sac pouch about 6 cm (2.5 in.) long. Opening into the cecum, about 2 cm below the *ileocecal valve*, is the *vermiform appendix* (L. *vermis*, worm, hence "wormlike"), popularly called the *appendix*. It is the narrowest part of the intestines, and can range in length from 5 to 15 cm (2 to 6 in.). Bacteria and indigestible material may become trapped in the appendix, leading to inflammation *(appendicitis)* and one of the most common of all surgical procedures, an *appendectomy*.

FIGURE 25.26 GALLBLADDER AND ITS CONNECTING DUCTS

Bile secreted from the liver reaches the gallbladder via the cystic duct.

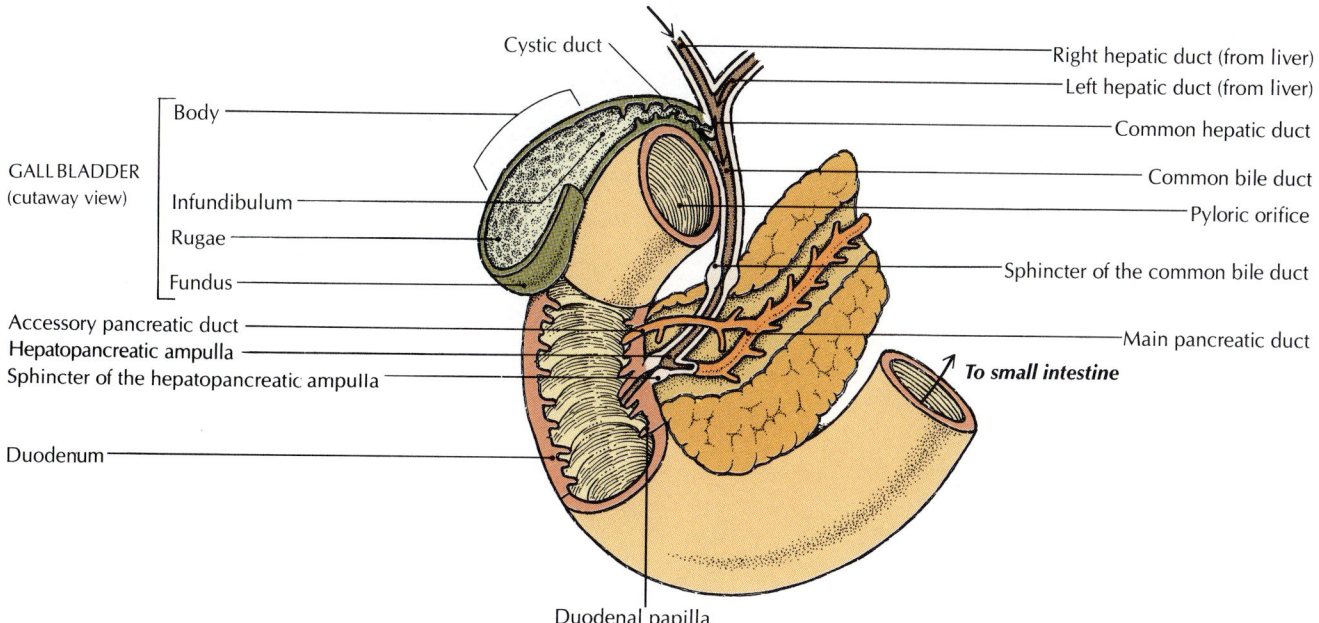

Cystic duct

Right hepatic duct (from liver)
Left hepatic duct (from liver)
Common hepatic duct
Common bile duct
Pyloric orifice
Sphincter of the common bile duct
Main pancreatic duct
To small intestine

GALL BLADDER (cutaway view)

Body
Infundibulum
Rugae
Fundus

Accessory pancreatic duct
Hepatopancreatic ampulla
Sphincter of the hepatopancreatic ampulla

Duodenum

Duodenal papilla

FIGURE 25.27 LARGE INTESTINE

[A] Anterior view with part of the cecum and ascending colon removed to show the ileocecal valve.
[B] Transverse section of large intestine.

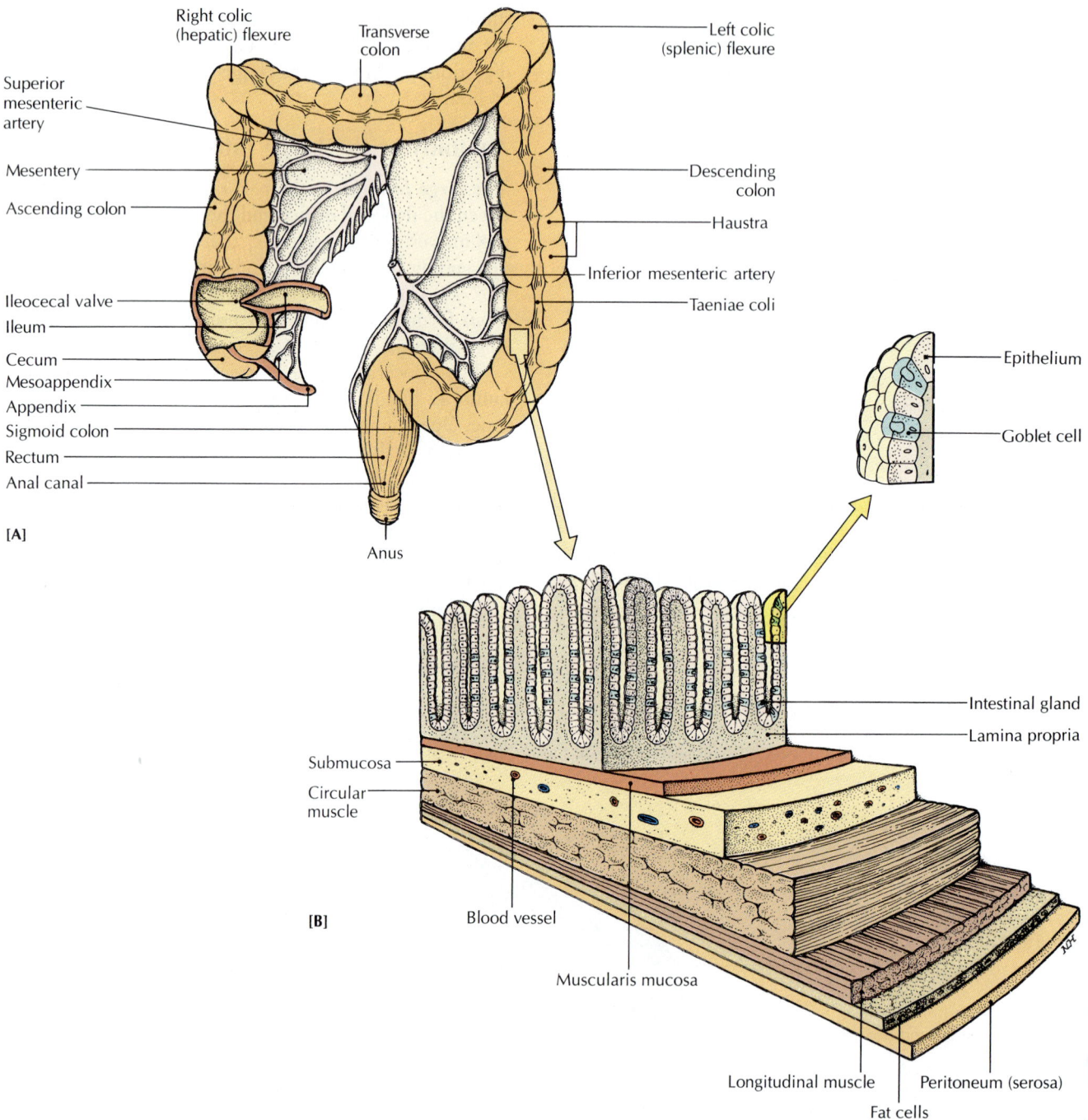

Right colic (hepatic) flexure

Transverse colon

Left colic (splenic) flexure

Superior mesenteric artery

Mesentery

Descending colon

Ascending colon

Haustra

Inferior mesenteric artery

Ileocecal valve

Taeniae coli

Ileum

Cecum

Mesoappendix

Appendix

Sigmoid colon

Rectum

Anal canal

Epithelium

Goblet cell

[A]

Anus

Intestinal gland

Lamina propria

Submucosa

Circular muscle

Blood vessel

[B]

Muscularis mucosa

Longitudinal muscle

Peritoneum (serosa)

Fat cells

The ***ascending colon*** extends upward from the cecum. Under the liver it makes a right-angle bend known as the *right colic* (or *hepatic*) *flexure* [FIGURE 25.27A]. It continues as the ***transverse colon,*** which extends across the abdominal cavity from right to left, where it makes a right-angle downward turn at the spleen, known as the *left colic* (or *splenic*) *flexure.*

The ***descending colon*** extends from the left colic flexure down to the rim of the pelvis, where it becomes the sigmoid colon. The ***sigmoid colon*** (Gr. *sigma*, the letter S) is S-shaped. It travels transversely across the pelvis to the middle of the sacrum, where it continues to the rectum.

External anatomy of the large intestine Besides the differences in diameter and length between the large and small intestines, the large intestine has three distinctive structural differences [FIGURE 25.27]:

1 In the small intestine, an external longitudinal smooth muscle layer completely surrounds the intestine, but in the large intestine, an incomplete layer of longitudinal muscle forms three separate bands of muscle called ***taeniae coli*** (TEE-nee-ee KOHL-eye; L. ribbons + Gr. intestine) along the full length of the intestine.

2 Because the taeniae coli are not as long as the large intestine itself, the wall of the intestine becomes puckered with bulges called ***haustra*** (sing. *haustrum*).

3 Fat-filled pouches called ***epiploic appendages*** are formed at the points where the visceral peritoneum is attached to the taeniae coli in the serous layer.

Microscopic anatomy of the large intestine The microscopic anatomy of the large intestine reflects its primary functions: the reabsorption of any remaining water and some salts and the accumulation and movement (excretion) of undigested substances as feces. Elimination is aided by the secretion of mucus from the numerous goblet cells in the mucosal layer. Mucus acts as a lubricant and protects the mucosa from the semisolid contents.

Because the mucosa of the large intestine does not contain villi or plicae circulares, its smooth absorptive surface is only about 3 percent of the absorptive surface of the small intestine. The lamina propria and submucosa contain lymphatic tissue in the form of many lymphoid nodules. As part of the body's immune system, aggregated lymph nodules, along with epithelial cells, help defend against the ever-changing mixture of antigens, microorganisms, and potentially harmful substances that enter the digestive tract.

Rectum, Anal Canal, and Anus

The terminal segments of the large intestine are the rectum, anal canal, and anus [FIGURE 25.28]. The ***rectum*** (L. rectus, straight) extends about 15 cm (6 in.) from the sigmoid colon to the anus. Despite its name, the rectum is

---

FIGURE 25.28 RECTUM, ANAL CANAL, ANUS

Longitudinal section of rectum, and canal, and anus; anterior view.

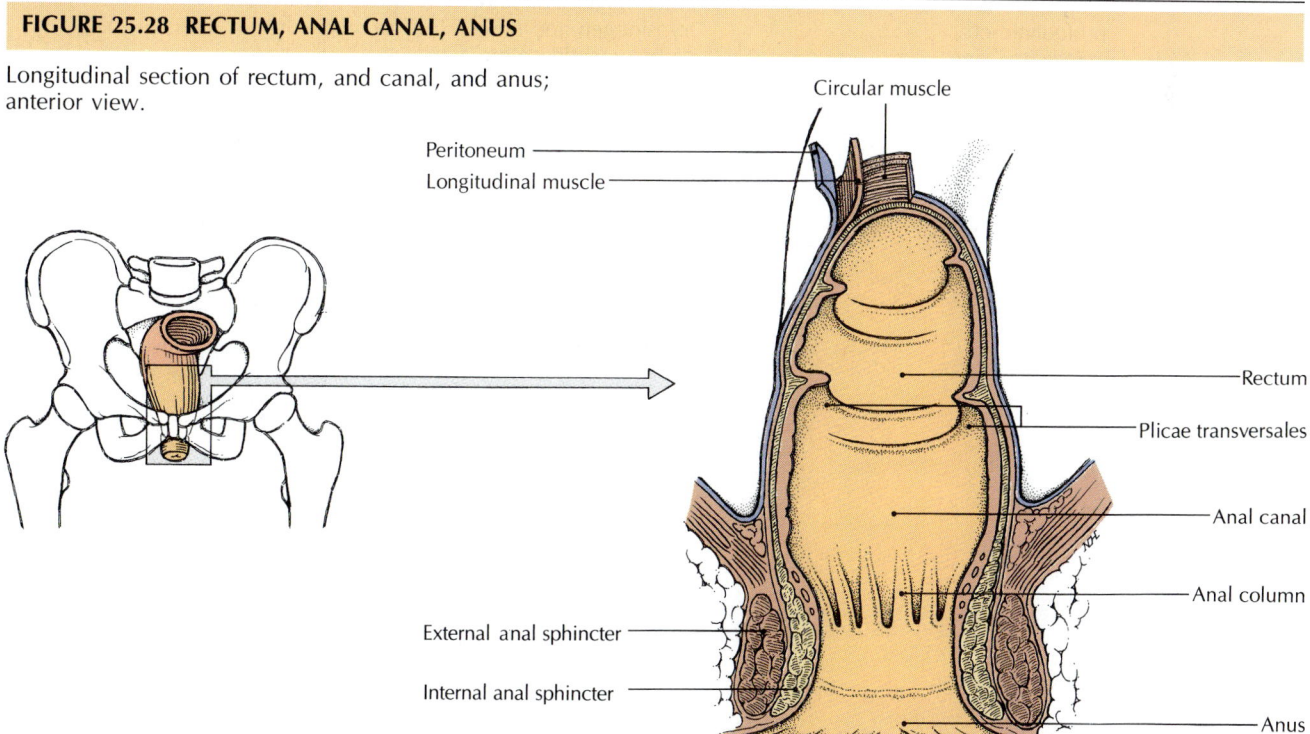

Circular muscle

Peritoneum

Longitudinal muscle

Rectum

Plicae transversales

Anal canal

Anal column

External anal sphincter

Internal anal sphincter

Anus

TABLE 25.7 SUMMARY OF DIGESTION AND ABSORPTION

Region of digestive tract	Secretion	Food substance acted upon	Digestive activity	Absorptive activity
Mouth	*Saliva:* salivary amylase, mucus, lysozyme	Proteins	None	None*
		Carbohydrates	*Salivary amylase* breaks down polysaccharides into maltose and dextrin.	None
		Fats	None	None
Stomach	*Gastric juice:* hydrochloric acid, pepsinogen, lipase, mucus, intrinsic factor	Proteins	*Pepsin,* which results from activation of pepsinogen by *hydrochloric acid,* converts protein into polypeptides (proteoses, peptones, and amino acids). Hydrochloric acid acts to soften (denature) protein.	Little or none†
		Carbohydrates	Some digestion of carbohydrates continues, but salivary amylase is inactivated by stomach acid.	Little or none
		Fats	Gastric lipase hydrolyzes some emulsified fat into fatty acids and glycerol.	Little or none
Small intestine (with assistance of pancreatic and liver secretions)	*Succus entericus* (intestinal juice): Mucus, salt. Microvilli secretions: enterokinase, disaccharidases, aminopeptidases. Pancreatic juice: proteolytic enzymes (trypsinogen, chymotrypsinogen, procarboxypeptidase), pancreatic amylase, pancreatic lipase. Liver secretion: bile	Proteins	*Pancreatic proteolytic enzymes* break down different peptide linkages, producing amino acids and small peptides. (Trypsinogen is activated and converted into trypsin by enterokinase. Chymotrypsinogen and procarboxypeptidase are activated and converted into chymotrypsin and carboxypeptidase by trypsin.) Amino peptidases convert peptides into amino acids.	Absorbed as amino acids into capillary network within villi by active transport.‡
		Carbohydrates	*Pancreatic amylase* converts polysaccharides into monosaccharides and disaccharides, especially into maltose. *Disaccharide maltase* converts maltose into monosaccharide glucose. *Disaccharide lactase* converts lactose into monosaccharide glucose and galactose. *Disaccharide sucrase* converts sucrose into monosaccharide glucose and fructose. *Intestinal amylase* converts complex sugars into disaccharides.	Simple sugars are absorbed by active transport into capillary network within villi. Galactose is the most readily absorbed sugar. Fructose is absorbed by facilitated diffusion.

*Some drugs, such as nitroglycerine and cyanide, are absorbed through the oral mucosa.
†A few highly lipid-soluble substances, such as alcohol, and certain drugs, such as aspirin, are absorbed through the stomach, but generally little absorption occurs there.
‡Electrolytes such as sodium, chloride, potassium, and calcium are absorbed through the wall of the small intestine by active transport.

Region of digestive tract	Secretion	Food substance acted upon	Digestive activity	Absorptive activity
		Fats	*Pancreatic lipase* converts fats into free fatty acids and glycerol. Bile salts convert fats into emulsified fats. *Intestinal lipase* converts fats into fatty acids and glycerol.	Most fats (60 to 70%) are emulsified by bile salts and absorbed as fatty acids and glycerol into lacteals of villi, from which they reach blood circulation through the lymphatic thoracic duct. Remaining fats, which have been broken down by lipases into fatty acids and glycerol, become water-soluble and enter villi on way to liver via hepatic portal system.
Large intestine	*Mucus*	None	None	Little or none§

§Most absorption is completed by the time chyme reaches the large intestine, but some water, electrolytes such as sodium and chloride, and some vitamins are absorbed from the initial segments of the large intestine. Glucose and some drugs can be absorbed when administered through the rectum.

not straight. Three lateral curvatures occur because the longitudinal muscle coat of the rectum is shorter than the other layers of the wall. The rectum is a retroperitoneal structure and has no mesentery, appendixes, epiploic appendages, haustra, or taeniae coli. The rectum continues as the *anal canal*, passes through the muscular pelvic diaphragm, and turns sharply downward and backward. The anal canal is compressed by the anal muscles as it progresses for about 4 cm (1.5 in.) and finally opens to the outside as the *anus* (L. ring).

The anal canal and anus are open only during defecation. At all other times they are held closed by an involuntary *internal anal sphincter* of circular smooth muscle and a complex *external anal sphincter* of skeletal muscle [FIGURE 25.28]. The external sphincter is under voluntary control.

The mucosa and circular muscle of the muscularis of the rectum form shelves within the tract called *plicae transversales*. The plicae must be avoided when diagnostic and therapeutic tubes and instruments are inserted into the rectum. The upper part of the anal canal contains 5 to 10 permanent longitudinal columns of mucous membrane known as *anal* (or *rectal*) *columns*. They are united by folds called *anal valves*. The mucosa above the valves is lined by columnar epithelium, with a scattering of goblet cells and crypts. The mucosa and submucosa of the rectum contain a rich network of veins called the *hemorrhoidal plexus*. When these veins become enlarged, twisted, and blood-filled, the condition is called *hemorrhoids*.

Q: *What is usually meant by "bowels"?*

A: The bowels (L. *botulus*, sausage) are usually considered to be the large intestine and rectum, but they may also refer to the digestive tract below the stomach.

Functions of the Large Intestine

By the time chyme reaches the large intestine, digestion is complete, and only some water, salts, and vitamins remain to be absorbed. [See TABLE 25.7 for a summary of digestion and absorption.] The amount of chyme that enters the large intestine daily varies from less than 100 mL to 500 mL, and only about one-third of it is excreted as feces. The rest (mostly water) is absorbed back into the body in the ascending or transverse colons, and the unabsorbed water becomes part of the feces. Reabsorption of water in the large intestine helps avoid dehydration.

Bacterial activity Intestinal bacteria are harmless as long as they remain in the large intestine; in fact, they are useful in synthesizing vitamins K and B_{12}. Bacteria, both living and dead, make up 25 to 50 percent of the dry weight of feces. Bacterial activity in the large intestine also contributes to the production of intestinal gas, or *flatus* (FLAY-tuss; L. a blowing), which causes *flatulence*. Flatus consists mostly of nitrogen, carbon dioxide, and hydrogen, with small amounts of oxygen, methane, and hydrogen sulfide. The cells in the intestinal glands of the large intestine secrete large amounts of alkaline mucus, which helps neutralize acids produced by intestinal bacteria, and also lubricates the lumen for the easy passage of feces.

Formation of feces Approximately 150 g of feces (about 100 g of water and 50 g of solids) are normally eliminated from the body daily. Besides containing water and bacteria, feces are composed of fat, nitrogen, bile pigments, undigested food such as cellulose, and other waste products from the blood or intestinal wall. The normal brown color of feces is caused by the conversion of products of red blood cells into bile pigments (bilirubin). Excessive fat in the diet causes feces to be a pale color, and blood and other foods containing large amounts of iron will darken feces.

The characteristic odor of feces is caused by *indole* and *skatole*, two substances that result from the decomposition of undigested food residue, unabsorbed amino acids, dead bacteria, and cell debris. A high-protein diet increases the odor of feces because it results in the production of larger quantities of indole and skatole. Hydrogen sulfide and methane also contribute to the odor of feces.

Movements of the large intestine and defecation

Most of the digestive movements of the large intestine are slow and nonpropulsive. The primary motility comes from *haustral contractions* that depend on the autonomous rhythmicity of the smooth muscle cells. These movements are similar to the segmentations of the small intestine but usually take 20 to 30 min between contractions. This slow movement allows for the final absorption of the remaining water and electrolytes.

Three to four times a day, generally following meals, motility increases markedly as large segments of the ascending and tranverse colons contract simultaneously, driving the feces one-half to three-quarters the length of the colon in a few seconds. These "sweeping" peristaltic waves are called *mass movements*. They drive the feces into the descending colon, where waste material is stored until defecation occurs.

The *duodenocolic reflex* moves the contents of the ileum into the large intestine. Mass movements in the colon are produced by the *gastrocolic reflex* when food enters the stomach. The reflex is mediated from the stomach to the colon by the secretion of gastrin and by the external autonomic nerves. The gastrocolic reflex pushes feces into the rectum, triggering the **defecation reflex,** the elimination of waste material from the anus.

The removal of solid wastes in the form of feces is one of the most important functions of the large intestine. **Defecation** normally proceeds as follows:

1 When the rectum is distended by the accumulation of feces, the walls are stretched, pressure in the rectum rises, and receptors in the rectal walls are stimulated.

2 If a conscious decision is made to defecate, the defecation reflex is triggered. The internal anal sphincter relaxes at the same time that peristaltic waves in the descending colon are stimulated, and the rectum and sigmoid colon contract vigorously.

3 Defecation is assisted by the contraction of the abdominal muscles and the diaphragm and a forcible expiration against a closed glottis. These voluntary actions increase the abdominal pressure and create a pushing action that assists in the elimination of feces.

4 The defecation reflex, initiated in the medulla oblongata, passes impulses to the sacral portion of the spinal cord.

5 The external anal sphincter and other perineal muscles relax, and increased peristaltic waves in the sigmoid colon and rectum push the feces past the relaxed internal and external anal sphincters and out of the body through the anus.

6 The initial propulsion of feces from the anus seems to initiate another defecation reflex, which stimulates the expulsion of feces from the sigmoid and descending colons as well as the rectum.

Normal "straining" to assist defecation causes a sharp increase in arterial pressure when the increased pressure in the thoracic cavity is transmitted to the heart. Consequently, the venous return of blood is decreased and cardiac output and arterial pressure drop. Strokes and heart attacks have been known to occur when elderly people strain during defecation.

Defecation is a voluntary act, except in very young children and in people whose spinal cord is severed at the sacral level or above. It can be inhibited by keeping the external anal sphincter contracted. *Voluntary inhibition* of defecation is accomplished in the following way. Once the decision is made *not* to defecate, neural impulses from the central nervous system cause the skeletal muscle fibers of the external anal sphincter to close tightly. Spinal cord impulses also cause the relaxation of the rectum and sigmoid colon, which decreases the tension in the rectal wall and prevents the stretch receptors from being activated. The defecation reflex is avoided, and defecation is delayed either by voluntary control or until the rectum becomes distended to the point where defecation is absolutely necessary. If defecation is postponed voluntarily, water will continue to be absorbed from the feces, and constipation may result. Children younger than 1 or 2 years usually cannot inhibit defecation voluntarily because the appropriate motor neural pathways have not yet reached the proper stage of maturity.

ASK YOURSELF

1 What are the main parts of the large intestine?

2 What are taeniae coli?

3 What are the main functions of the large intestine?

4 How do bacteria in the large intestine serve a useful purpose?

5 What is the usual composition of feces?

THE EFFECTS OF AGING ON THE DIGESTIVE SYSTEM

With age, digestion may be impaired, with such general changes as decreased acid and mucus secretion, decreased neural and hormonal responses, and decreased appetite and taste perception. Specifically, peristalsis slows down, but diseased gums (periodontal disease) and loss of teeth make chewing difficult, and even swallowing becomes more difficult. The small intestine produces fewer digestive enzymes, and intestinal absorption is reduced. Defecation may be less frequent as the muscles of the rectum weaken.

About 50 percent of all people over 65 have *diverticulosis*, the formation of small pockets (diverticula) in the walls of the large intestine. When fecal material becomes lodged in the pockets and the pockets become inflamed, the condition is called *diverticulitis*. *Hemorrhoids* (varicose veins of the lower rectum and anus) frequently occur in people over 50, especially if constipation has been a chronic problem.

Diet and Aging

It is probably safe to say that of all age groups, the elderly are the most poorly nourished. The reasons include poverty, lack of social support, inadequate refrigeration, chewing problems, indigestion due to the decreased secretion of gastric juices or to a hiatal hernia that causes a backflow of gastric juices into the esophagus, the overuse of laxatives that interferes with the absorption of fat-soluble vitamins and other nutrients, and adverse interactions with medications. For example, many elderly people take diuretics to ease hypertension, edema, or congestive heart failure. Unfortunately, diuretics may cause a loss of potassium, calcium, and other necessary minerals.

DEVELOPMENTAL ANATOMY OF THE DIGESTIVE SYSTEM

The embryonic digestive system begins as a hollow tube. The outside of the tube is covered with ectoderm, and the inside is lined with endoderm. The internal cavity develops into the mature digestive system. The mouth forms at about day 22 after fertilization as a depression in the surface ectoderm called the *stomodeum*. The **primitive gut** forms during week 4 as the internal cavity of the embryonic yolk sac is enclosed by the lateral walls of the embryo. It is derived from endoderm and develops into three specific regions of the digestive system: (1) The anterior extension into the head region forms the *foregut*, (2) the central region forms the *midgut*, which opens into

a pouch called the yolk sac, and (3) the portion extending posteriorly forms the *hindgut* [FIGURE 25.29]. At about day 24, the membrane separating the stomodeum from the foregut ruptures, allowing amniotic fluid to enter the embryo's mouth. At about week 5, the yolk sac constricts at the point where it is attached to the midgut and seals off the midgut.

The *foregut* develops into the stomach, the duodenum up to the entrance of the common bile duct, the liver, the gallbladder, and the pancreas. The *midgut* develops into the small intestine, the cecum and appendix, the ascending colon, and most of the proximal portion of the transverse colon. The *hindgut* develops into the remaining distal portion of the transverse colon, the descending colon, the sigmoid colon, the rectum, and the superior portion of the anal canal. The primitive anus develops as an indentation at the caudal end of the digestive tube called the *proctodeum*. The membrane separating the proctodeum from the hindgut ruptures at about week 7, forming a complete tube from the mouth to the anus.

Later, hollow endodermal buds in the foregut move into the mesoderm and develop into the salivary glands, liver, gallbladder, and pancreas. Each of these accessory organs becomes connected to the central digestive tract by way of ducts.

FIGURE 25.29 EARLY DEVELOPMENT OF THE DIGESTIVE SYSTEM

About day 28; sagittal section.

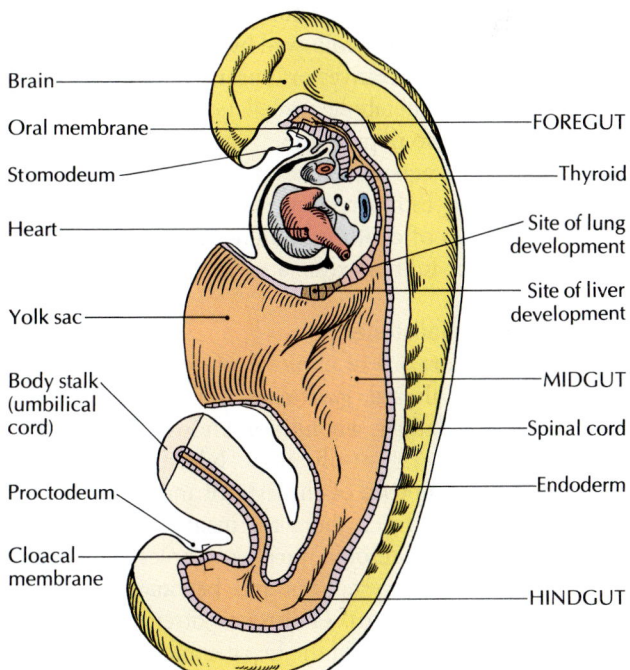

Brain — Oral membrane — Stomodeum — Heart — Yolk sac — Body stalk (umbilical cord) — Proctodeum — Cloacal membrane — FOREGUT — Thyroid — Site of lung development — Site of liver development — MIDGUT — Spinal cord — Endoderm — HINDGUT

WHEN THINGS GO WRONG

Anorexia Nervosa and Bulimia

Anorexia nervosa and bulimia are examples of eating disorders with a psychological basis. *Anorexia nervosa* (Gr. *an*, without + *orexis*, a longing) is characterized by self-imposed starvation and subsequent nutritional deficiency disorders. Victims of this disorder, most often adolescent females from upwardly mobile families, usually are of normal weight but are convinced that they are grossly overweight. Excessive dieting is often accompanied by compulsive physical activity, resulting in a weight loss of 20 percent or more. The exact cause of anorexia nervosa is not known, but it is often brought on by anxiety, fear, or peer or family pressure that encourages thinness. Weight loss and chronic undernourishment are accompanied by atrophy of skeletal muscle, constipation, hypotension, dental caries, irregular or absent menstrual periods, and susceptibility to infection, among other disorders. If the victim does eat, the meal may be followed by self-induced vomiting or an excessive dose of laxatives or diuretics. Such tactics may lead to dehydration or metabolic alkalosis or acidosis. A drop of systolic pressure below 55 mmHg may prove fatal, and electrolyte imbalance may lead to cardiac arrest. Anorexia nervosa is fatal in 5 to 15 percent of cases, and feelings of despondency lead to a higher than normal suicide rate.

Bulimia (Gr. *bous*, ox; *limos*, hunger) also occurs mainly in upwardly mobile young females of normal weight. It is also called *gorge-purge syndrome* and *bulimia nervosa*. It is characterized by an insatiable appetite and gorging on food four or five times a week (ingesting as many as 40,000 calories), followed by self-induced vomiting or an overuse of laxatives or diuretics that may lead to dehydration and metabolic imbalances as in anorexia nervosa. It may also impair kidney and liver function and cause dry skin, frequent infections, muscle spasms, and other disorders. Bulimarectics may suffer from a decreased secretion of cholecystokinin, a hormone that induces a sense of fullness after a meal.

Cholelithiasis (Gallstones)

When the proportions of cholesterol, lecithin, and bile salts in bile are altered, *gallstones* (biliary calculi) may form in the gallbladder, a condition known as *cholelithiasis* (KOH-lee-lih-THIGH-uh-sihss; bile, gall + stone). Gallstones are composed of cholesterol and bilirubin and are usually formed when there are insufficient bile salts to dissolve the cholesterol in micelles. Cholesterol, being insoluble in water, crystallizes and becomes hardened into "gallstones" (photo). Gallstones often pass out of the gallbladder and lodge in the hepatic and common bile ducts, obstructing the flow of bile into the duodenum

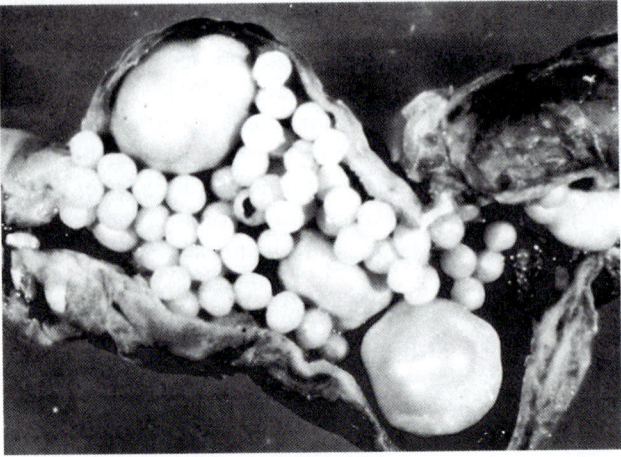

Section of gallbladder showing many gallstones.

and interfering with fat absorption. Jaundice may occur if a stone blocks the common bile duct because bile pigments cannot be excreted, and surgery may be required if stones become impacted in the cystic duct, where pain is usually maximal because the contractions of the gallbladder press on the stones. Other common sites of gallstones are the pancreatic ducts and the hepatopancreatic ampulla.

The production of gallstones is sometimes initiated by pregnancy, diabetes mellitus, celiac disease, cirrhosis of the liver, pancreatitis, or the use of oral contraceptives. Occasionally, gallstones may be dissolved when the naturally occurring bile acid chenodeoxycholic acid (CDCA) is taken orally, but CDCA is expensive and has the undesirable side effect of increasing LDL cholesterol and decreasing HDL cholesterol, increasing the likelihood that the patient will develop atherosclerosis. In most cases, gallstones lodged in ducts are removed surgically, and sometimes the gallbladder itself is removed (cholecystectomy).

A recent nonsurgical technique for removing gallstones employs a device called a *lithotripter* (Gr. "stone crusher"), which uses high-pressure shock waves to pulverize the gallstones after they are located with ultrasound. Endoscopic removal is also becoming common.

Cirrhosis of the Liver

Cirrhosis (sih-ROE-siss; L. "orange-colored disease," from the color of the diseased liver) is a chronic liver disease characterized by the destruction of hepatic cells and their replacement by abnormal fibrous connective tissues. These changes disrupt normal blood and lymph flow and eventually result in blockage of portal circulation, portal hypertension, liver failure, and death. Cirrhosis is more prevalent among chronic alcoholics over age 50, especially if their diets are poor. The mortality rate is high. Among the several types of cirrhosis, the most common is *Laennec's type* (also called portal, nutritional, or alcoholic cirrhosis), which accounts for up to half of all persons suffering from the disease. Among these people, about 90 percent are heavy consumers of alcohol.

Colorectal Cancer

More than half of all cancers of the large intestine (colon) are found in the rectum, and about 75 percent of intestinal cancers are located in the rectum and the intestinal region (sigmoid colon) just above it. Although the number of **colorectal cancer** cases is second only to lung cancer, early diagnosis and treatment make it possible to save 4 of 10 patients. The exact cause of colorectal cancer is unknown, but increasing evidence points to a low-fiber diet containing excessive animal fat (especially beef). A newly discovered mutation of a gene called the *adenomatous polyposis coli (APC)* gene is known to cause an inherited predisposition toward colorectal polyps, a typical forerunner of colorectal cancer. Also, genes responsible for *hereditary nonpolyposis colorectal cancer* (HNPCC), a common inherited disease, have been discovered on chromosomes 2 and 3. Researchers hope to use the new information to develop preventive treatments for HNPCC and other inherited cancers of the digestive tract.

Early symptoms of colorectal cancer are usually vague and unnoticed, but an obstruction may soon develop, producing constipation or diarrhea, rectal bleeding, and dull pain. During later stages, the primary tumor may spread, or metastasize, to other organs in the area, such as the urinary bladder, prostate gland, uterus, and vagina, or even to the liver or lungs.

Colorectal cancer can be diagnosed by several fairly simple methods, including a barium x-ray, digital examination, fecal occult blood test, colonoscopy, and proctoscopy or sigmoidoscopy (examination with tubular instruments that enable the physician to see into the rectum and lower intestine). Unfortunately, 52 percent of all colorectal cancers originate beyond (above) the detection range of flexible sigmoidoscopy, which usually detects cancer of the rectum and descending colon. [See TABLE 25.8 for a summary of the various diagnostic tests for the digestive system.]

Constipation and Diarrhea

Constipation (L. "to press together") is a condition in which feces move through the large intestine too slowly. As a result, too much water is reabsorbed, the feces become hard, and defecation is difficult. Constipation may be caused by spasms that reduce intestinal motility, nervousness, a temporary low-fiber diet, and other factors.

TABLE 25.8 A SUMMARY OF DIAGNOSTIC TESTS OF THE DIGESTIVE SYSTEM	
Test	**Description**
Barium enema (lower GI series)	Barium solution is introduced into patient through rectum; may reveal polyps, cancer, colitis, diverticulosis.
Colonoscopy	Visual inspection of entire colon is made through flexible scope/video monitor; same purpose as barium enema; usually requires some sedation.
Digital rectal examination	Physician inserts gloved finger into patient's rectum; primary method for detecting prostate cancer.
Gallbladder ultrasound	Sound waves are used to reveal gallstones.
Hemoccult	Feces are tested for hidden blood; may reveal bleeding ulcer, polyps, cancer, other gastointestinal diseases.
Sigmoidoscopy	Physician examines patient's rectum and lower colon with flexible scope; may reveal polyps, cancer, source of bleeding.
Upper GI series	Patient swallows barium solution; x-ray shows esophagus, stomach, small intestine; may reveal ulcers, hiatus hernia, gastritis, stomach cancer.

Constipation may produce abdominal discomfort and distension, loss of appetite, and headaches, but it does *not* increase the amount of toxic substances in the body.

Diarrhea (Gr. "a flowing through") is a condition in which watery feces move through the large intestine rapidly and uncontrollably in a reaction sometimes called a *peristaltic rush*. Among the causes are viral, bacterial, or protozoan infections; extreme nervousness; ulcerative colitis; and excessive use of cathartics, all of which increase fluid secretion and intestinal motility. Severe diarrhea is dangerous (especially in infants) because it increases the loss of body fluids and ions.

Crohn's Disease

Crohn's disease (named after Dr. Burrill Crohn, who first diagnosed the disease in the early 1900s) is a chronic intestinal inflammation that affects the small and large intestines. It extends through all layers of the intestinal wall and is generally accompanied by abdominal cramping, fever, and diarrhea. Its onset is often associated with stress. Symptoms may decrease in severity with age. The most common victims are 20- to 40-year-old adults, especially individuals of Jewish descent.

Diverticulosis and Diverticulitis

Diverticula are bulging pouches in the gastrointestinal wall that push the mucosal lining through the muscle in the wall. Diverticula most commonly form in the sigmoid colon, but other likely sites are the duodenum near the hepatopancreatic ampulla and the jejunum. Diverticula appear most often in men over 40, especially if their diet is low in fiber.

In *diverticulosis,* pouches are present but do not present symptoms. In *diverticulitis,* the diverticula are inflamed and may become seriously infected if undigested food and bacteria become lodged in the pouches. The subsequent blockage of blood can lead to infection, peritonitis, and possible hemorrhage.

Food Poisoning

The term *food poisoning* is commonly used to describe gastrointestinal diseases caused by eating food contaminated with either infectious microorganisms or toxic substances produced by the microorganisms. Most food poisonings are associated with the production of toxins.

Hepatitis

Hepatitis, an infection of the liver, may have either a viral or a nonviral cause. Among the symptoms of *viral hepatitis* are loss of appetite, nausea, jaundice, and abdominal pain. Other symptoms, such as headache, dizziness, fever, and rash, may be chemically related. Viral hepatitis has three forms: type A, type B, and type C. *Type A* is infectious, occurs most commonly during the fall and winter in children and young adults, and is transmitted through contaminated food (especially seafood), water, milk, semen, tears, and feces. Onset is sudden, and the incubation period is relatively short (15 to 45 days), but the overall prognosis for a complete cure is good. A person who contracts type A viral hepatitis does not become a carrier, but a patient afflicted with type B does carry the disease for an indefinite period. *Type B* viral hepatitis can affect people of any age at any time of the year; it usually has a slow, prolonged onset (40 to 100 days) and is transmitted through serum, blood and blood products, semen, and feces. Unlike type A, type B becomes worse with age, and permanent immunity does not result. The type B virus not only causes hepatitis B, it also contributes to a tendency to contract liver cancer. *Type C* hepatitis usually results from contaminated commercial blood donations. Because liver cells are capable of regeneration, liver cell destruction resulting from hepatitis is normally overcome.

Nonviral hepatitis is usually caused by exposure to chemicals or drugs such as carbon tetrachloride, poisonous mushrooms, and vinyl chloride. The symptoms usually resemble those of viral hepatitis.

Hernia

A *hiatal hernia* (L. gap + rupture) is a defect in the diaphragm that permits a portion of the stomach to protrude into the thoracic cavity. The common cause of a hernia is connected with a typical weakening of muscle tissue that often comes with aging. The symptoms of hiatal hernia resemble those of a peptic ulcer, with pain in the upper abdomen, and heartburn, especially when the person is lying down. An *inguinal hernia* (L. *inguen,* groin) is the protrusion of the small or large intestine, omentum, or urinary bladder through the inguinal canal in the area of the groin.

Jaundice

Ordinarily, yellow bile pigment (bilirubin) is excreted by the liver in the bile, and not enough of it circulates in the blood to affect the color of the skin. Occasionally, however, pigment levels rise sufficiently to produce a yellowish tint in the skin, mucous membranes, and whites of the eyes. The condition is known as *jaundice* (Old Fr. *jaune,* yellow).

Jaundice may have one of three different causes: (1) When the liver cells remove bilirubin from the blood by hemolysis (destruction) of red blood cells more slowly than it is produced, the pigment builds up in the blood and produces *hemolytic jaundice.* Although the feces usually darken, the urine remains a normal color. (2) In *hepatic jaundice,* the liver's ability to absorb bilirubin, process it metabolically, and secrete it becomes impaired. The urine darkens, and the color of the feces gradually becomes lighter. (3) *Obstructive jaundice* is produced when obstructions in the liver's duct system cause bilirubin to flow backward from the liver cells into the sinusoids. The urine becomes very dark, while the feces are usually the color of light clay.

In each case, the underlying cause is treated rather than the jaundice itself.

Peptic Ulcers

Peptic ulcers (Gr. *peptein,* to digest) are lesions in the mucosa of the digestive tract. They are caused by the digestive action of gastric juice, especially hydrochloric acid. Most peptic ulcers occur in the duodenum, before gastric hydrochloric acid is neutralized by the alkaline secretions of the small intestine.

People who are most susceptible to *duodenal ulcers* secrete high amounts of gastric juices between meals, when there is little or no food in the stomach to buffer the acidity. Men between the ages of 20 and 50 are the most vulnerable. *Gastric ulcers,* in contrast, seem to be most common in men over 50 and among people whose stomach mucosa has a reduced resistance to digestive enzymes rather than an overabundance of gastric secretions. Large amounts of aspirin and alcohol decrease the resistance of the mucosa and thus may lead to gastric ulcers.

Symptoms of ulcers usually include heartburn and indigestion, pain after eating, weight loss, and gastrointestinal bleeding. Treatment may include rest, reduction of strain, antacids, and drug therapy. Hemorrhaging requires emergency treatment to reduce bleeding, and surgery is necessary in severe cases. It is currently believed that some peptic ulcers may be caused by resistant *Helicobacter* bacteria.

Peritonitis

Peritonitis is an inflammation of the peritoneum, usually resulting from a perforation in the gastrointestinal tract that allows bacteria to enter the normally sterile peritoneum. Such a perforation often results from a ruptured appendix, but it can also be caused by diverticulitis, a peptic ulcer, or any other disease or physical trauma that breaks through the wall of the digestive tract. Peritonitis is often limited to a localized abscess rather than a general inflammation because the peritoneum is so resistant to infection. Severe pain, weakness, fever, and decreased intestinal motility are among the more obvious symptoms.

Tooth Decay (Dental Caries)

Bacteria that cause *dental caries,* or tooth decay, produce a gluelike enzyme called *glucosyl transferase (GTF)* that converts ordinary sugar into *dextran,* a sticky substance that clings to the tooth's enamel. GTF also helps the bacteria adhere to the enamel. Dextran is involved in the formation of *plaque,* a destructive film that builds up on teeth. Once plaque is formed, the bacteria on the teeth produce another enzyme that promotes erosion of the enamel.* (Incidentally, apples may keep the doctor away, but not the dentist. They contain more than 10 percent fermentable carbohydrate, a potent producer of tooth decay.)

Vomiting

Vomiting (emesis; Gr. *emein,* to vomit) is the forceful expulsion of part or all of the contents of the stomach and duodenum through the mouth, usually in a series of involuntary spasms. The action is a coordinated reflex controlled by the vomiting center in the medulla oblongata. The vomiting center is activated either by receptors in the stomach or duodenum that respond to foreign constituents in the chyme, or by receptors near the brainstem; receptors near the brainstem respond to chemicals carried by the blood and to other external stimuli, such as motion sickness or extreme pain.

The complex act of vomiting is preceded by increased salivation and sweating, accelerated or irregular heart rate and breathing, discomfort, and nausea, all of which are characteristic of a generalized discharge of the autonomic nervous system. During vomiting, the pyloric region of the stomach goes into spasm, and the usual downward motion of peristalsis is reversed. Meanwhile, the body of the stomach and lower esophageal sphincter are relaxed. The duodenum also goes into spasm, forcing its contents into the stomach.

The final thrust of vomit comes when the abdominal muscles contract, lifting the diaphragm and increasing the pressure in the abdomen. At the same time, slow, deep breathing with the glottis partially closed reduces the pressure in the thorax. As a result, the stomach is squeezed between the diaphragm and compressed abdominal cavity, forcing the contents of the stomach and duodenum past the lower esophageal sphincter into the esophagus and out through the mouth. When this series of events is repeated several times without vomiting, it is called *retching.*

During actual vomiting, the breathing rate decreases, the glottis is closed, and the soft palate is raised, all ensuring that the vomit does not enter the breathing passages and cause suffocation.

Although it seems clear that vomiting may be the body's way of ridding itself of a harmful food substance, it is not fully understood why motion sickness or pain should induce vomiting. Also, it is not certain why some women become nauseated and vomit during early pregnancy ("morning sickness"). It may be caused by hormonal changes, especially the increased secretion of human chorionic gonadotropin, and it may also be related to the changes in carbohydrate metabolism. Both factors are probably involved.

Excessive vomiting, like prolonged diarrhea, can lead to a dangerous depletion of fluids and salts. The loss of acid may lower the overall acidity of the body beyond the point where normal feedback systems can maintain homeostasis.

---

*Our prehistoric ancestors had cavities too, but a study of skeletons that have been dated to before the Iron Age indicates that humans had about 2 to 4 percent decay then, compared with 40 to 70 percent today. However, public health officials predict the virtual end of tooth decay among children and young adults by the end of this century, due to improved dental technology and the widespread use of fluoridation in city water supplies.

Q: *Does eating chocolate cause tooth decay?*

A: There is some evidence that *milk chocolate* may actually help *prevent* cavities. Apparently, the calcium, phosphate, lipids, and protein in milk chocolate decrease the production of cavity-causing mouth acid.

CHAPTER SUMMARY

Food is ingested into the digestive tract and undergoes *digestion,* in which large food molecules are broken down into smaller ones. The small food molecules pass through the plasma membranes of the small intestine into the bloodstream, a process called *absorption.*

Introduction to the Digestive System (p. 796)

1 The digestive process takes place in the *digestive tract,* or *alimentary canal,* which extends from the mouth to the anus.

2 The digestive system is compartmentalized, with each part adapted to a specific function. The digestive tract consists of the mouth, pharynx, esophagus, stomach, small intestine, large intestine, rectum, anal canal, and anus. The associated structures include the teeth, lips, cheeks, salivary glands, pancreas, liver, and gallbladder.

3 The parts of the digestive tract all have the same basic tissue layers. The wall of the tube (from the inside out) is composed of the *mucosa, submucosa, muscularis externa,* and *serosa.*

Mouth (p. 796)

1 The *mouth (oral cavity)* consists of two parts: the small, outer *vestibule (buccal cavity)* and the *oral cavity* proper.

2 The *lips* and *cheeks* aid the digestive process by holding the food in position to be chewed.

3 There are 20 *deciduous* ("baby") *teeth* and 32 *permanent teeth,* with specialized shapes for cutting, tearing, and grinding. Each tooth consists of a *root, crown,* and *neck* and is composed of *dentine, enamel, cement,* and *pulp.* The *gum* is connective tissue that surrounds the alveolar processes of the teeth.

4 The *tongue* helps to form the moistened, chewed food into a *bolus* and pushes it toward the pharynx to be swallowed. It also contains taste buds.

5 The tongue also pushes the food against the *hard palate,* where it is crushed and softened. The movable *soft palate* prevents food from entering the nasal passages by closing over the nasopharynx.

6 The three largest pairs of *salivary glands* are the *parotid, submandibular,* and *sublingual* glands. They secrete saliva, which contains water, salts, proteins, and at least one enzyme, *salivary amylase,* which begins the digestion of carbohydrates. Saliva also moistens and lubricates food so it can be swallowed easily. It allows us to taste food by dissolving food molecules.

Pharynx and Esophagus (p. 806)

1 The *pharynx* leads from the mouth to the esophagus. It is the common pathway for the passage of food and air. The *esophagus* is the tube that carries food from the pharynx to the stomach.

2 Swallowing action, or *deglutition,* in the pharynx initiates a muscular wave called *peristalsis,* which pushes the food along the esophagus into the stomach. Peristalsis continues through the rest of the digestive system.

Abdominal Cavity and Peritoneum (p. 809)

1 The *abdominal cavity* is the portion of the trunk cavities that lies inferior to the diaphragm. When considered as including the pelvic cavity, it is called the *abdominopelvic cavity.*

2 The serous membrane of the abdominal cavity is the *peritoneum.* The *parietal peritoneum* lines the abdominal cavity, and the *visceral peritoneum* covers most of the organs within the cavity.

3 The space between the parietal and visceral membranes is the *peritoneal cavity.* It usually contains a small amount of *peritoneal fluid,* which reduces friction as abdominal organs move.

4 Abdominal organs that lie posterior to the peritoneal cavity and that are covered, but not surrounded, by the peritoneum are called *retroperitoneal.* Those surrounded by the peritoneum are called *intraperitoneal.*

5 Organs in the abdominal cavity are suspended from the cavity wall by the *mesenteries,* or *visceral ligaments.* The mesentery connected to the stomach is the *omentum.*

Stomach (p. 810)

1 The *stomach* stores, mixes, and digests food.

2 Food enters the stomach from the esophagus through the *lower esophageal orifice;* it empties into the small intestine through the *pyloric orifice.*

3 The stomach is divided into the *cardiac region, fundus, body,* and *pyloric region.*

4 The stomach stores large quantities of food, uses a churning action to mix food into a soupy mixture called *chyme,* releases the chyme in regular spurts into the small intestine, and secretes gastric juices that initiate protein digestion.

5 The *muscularis externa* of the stomach has three layers of smooth muscle: the outermost *longitudinal layer,* the middle *circular layer,* and the innermost *oblique layer.*

6 *Gastric juice* is composed of *hydrochloric acid, mucus,* and *enzymes,* including *pepsinogen* (a precursor of the active enzyme *pepsin*). Small amounts of gastric *lipase* are also secreted, initiating fat digestion. The major function of gastric juice is to digest protein.

7 The control of gastric juice secretion occurs in three overlapping phases: *cephalic,* initiated by the sight or taste of food; *gastric,* in response to distension of the stomach; and *intestinal,* with an inhibitory component to prevent oversecretion of gastric juices.

Small Intestine (p. 817)

1 The *small intestine* is subdivided into the *duodenum,* where most of the remaining digestion takes place, and the *jejunum* and *ileum,* where most of the absorption of nutrients and water into the blood and lymph occurs.

2 The absorptive surface of the small intestine is increased by protrusions called *villi,* which contain additional protrusions called *microvilli* and circular folds called *plicae circulares.*

3 *Mucosal* and *submucosal glands* secrete *intestinal juice* and enzymes that aid the digestion of carbohydrates, proteins, and lipids. Large lymphatic nodules combat microorganisms.

4 Muscular activity in the small intestine includes *segmenting contractions* and *peristaltic contractions.*

5 *Intestinal juice* contains *water, salts,* and mucus. Enzymes of the small intestine include *enterokinase, lactase, lipase, maltase, peptidase,* and *sucrase.*

6 The products of carbohydrate, protein, and lipid digestion, as well as electrolytes, vitamins, and water, are absorbed by the small intestine.

7 The absorption of lipids is more complex than that of carbohydrates and proteins. It involves the breakdown of water-insoluble triglyceride droplets into water-soluble particles called *micelles,* which are absorbed by cells. Once inside cells, the breakdown products of lipids are resynthesized into triglycerides and are packaged into tiny droplets called *chylomicrons,* which move from the cells into lymphatic and blood vessels for distribution throughout the body.

Pancreas as a Digestive Organ (p. 825)

1 The *pancreas* consists of a head, body, and tail. It is composed of both exocrine and endocrine secretory cells. The exocrine cells form groups of cells called *acini,* which secrete digestive juices into the small intestine.

2 Secretion of pancreatic juices is stimulated by the detection of food by taste buds and by the secretion of the digestive hormones *secretin* and *cholecystokinin* (CCK) after the chyme passes into the duodenum.

3 The three main types of pancreatic digestive enzymes are *pancreatic lipase,* which acts on fats; *pancreatic amylase,*

which acts on carbohydrates; and *pancreatic proteolytic enzymes* (trypsinogen, chymotrypsinogen, procarboxypeptidase), which break down proteins into amino acids and small peptides.

Liver as a Digestive Organ (p. 827)

1 The liver, the largest glandular organ in the body, is divided into two main lobes by the *falciform ligament,* with the *right lobe* being six times larger than the *left lobe.* The right lobe is further subdivided into a *quadrate* and a *caudate* lobe.

2 The *porta hepatis* is the door through which blood vessels, nerves, and ducts enter and leave the liver. The *hepatic artery* and *hepatic portal vein* enter, and the *hepatic vein* and *bile duct* exit.

3 The functional units of the liver are *lobules* that contain *hepatic cells* arranged in plates that radiate from a *central vein.* Between the rows of cells are blood channels called *sinusoids,* and in the corners of the five- or six-sided lobules are *portal areas* that contain branches of the portal vein, hepatic artery, bile duct, and nerve.

4 The many functions of the liver include removal of amino acids from organic compounds, conversion of excess amino acids into urea, homeostasis of blood, synthesis of certain amino acids, and conversion of carbohydrates and proteins into fat. It also stores glucose in the form of glycogen and secretes *bile,* an alkaline liquid that emulsifies fats.

Gallbladder and Biliary System (p. 832)

1 The *gallbladder* concentrates and stores bile from the liver until it is needed for digestion.

2 The *biliary system* consists of the gallbladder, hepatic ducts, cystic duct, and common bile duct.

3 The contraction of the gallbladder and the subsequent release of bile take place in the presence of cholecystokinin, which is secreted when fatty acids and amino acids reach the duodenum.

Large Intestine (p. 833)

1 When chyme leaves the small intestine, digestion is complete, and the *large intestine* functions to remove water and salts from the liquid chyme. Removal of water converts liquid wastes into *feces.*

2 The large intestine consists of the *cecum* (which contains the *vermiform appendix*), *ascending colon, transverse colon, descending colon, sigmoid colon,* and *rectum.*

3 The large intestine has an incomplete layer of longitudinal smooth muscle that forms three separate muscle bands called *taeniae coli.* The intestinal wall contains bulges called *haustra* and fat-filled pouches called *epiploic appendages.*

4 The terminal segments of the large intestine are the *rectum, anal canal,* and *anus.* The elimination of feces from the anus is called *defecation,* the only voluntary digestive act after the initial stage of swallowing.

The Effects of Aging on the Digestive System (p. 839)

Digestion may become less efficient with age as less hydrochloric acid and fewer enzymes are produced, peristalsis and other muscular actions slow down, and periodontal disease occurs.

Developmental Anatomy of the Digestive System (p. 839)

1 The digestive system begins as a hollow tube lined with endoderm. The *primitive gut* forms during week 4 and develops into the anterior *foregut,* the central *midgut,* and the posterior *hindgut.*

2 The mouth develops from a surface depression called the *stomodeum,* and the anus develops from an indentation at the caudal end of the digestive tract called the *proctodeum.*

3 Accessory organs such as the pancreas remain connected to the central digestive tract via ducts.

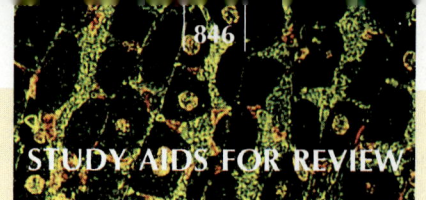

KEY TERMS

absorption 796, 821	large intestine 833
acini 825	liver 827
alimentary canal 796	lower esophageal orifice 810
anal canal 837	mesentery 809
anus 837	mucosal glands 819
bile 832	pancreas 825
biliary system 832	pepsin 816
bolus 807	peristalsis 796, 807
cholecystokinin 827	peritoneum 809
chylomicrons 824	pharynx 806
chyme 811	plicae circulares 819
defecation 796, 838	porta hepatis 830
deglutition 807	pyloric orifice 810
digestion 796	rectum 835
digestive tract 796	salivary amylase 800, 803
duodenum 817	salivary glands 803
esophagus 806	secretin 827
feces 833	small intestine 817
gallbladder 832	soft palate 803
gastric juice 814	stomach 810
hard palate 803	submucosal glands 819
intestinal juice 821	villi 819

MEDICAL TERMS FOR REFERENCE

ACHOLIA (*a*, without + Gr. *khole*, bile) Absence of bile secretion.

ANTIEMETIC (Gr. *anti*, against + *emetos*, vomiting) A substance used to control vomiting.

ANTIFLATULENT (Gr. *anti*, against + L. *flatus*, a breaking wind) A drug that prevents the retention of air in the digestive tract.

ASCITES (uh-SEE-teez; Gr. "bag") An accumulation of serous fluid in the peritoneal cavity.

BARIUM ENEMA EXAMINATION Fluoroscopic and radiographic examination of the large intestine after the administration of a barium sulfate mixture via the rectum. Commonly called *lower GI series*.

CHOLECYSTITIS (Gr. *khole*, bile + *cystis*, bladder + *-itis*, inflammation) Inflammation of the gallbladder.

CHOLESTEROSIS A condition in which an abnormal amount of cholesterol is deposited in tissues.

COLECTOMY (L. *colon*, intestine + *-ectomy*, removal) Surgical removal of the colon. A *hemicolectomy* removes half the colon.

COLITIS Inflammation of the colon.

COLOSTOMY (L. *colon*, intestine + Gr. *stoma*, opening) Surgical creation of an opening in the large intestine through which feces can be eliminated. Also, the opening thus created.

CREPITUS (L. *crepitare*, to crackle) Discharge of flatus from the intestine.

DYSPEPSIA (Gr. *dus*, faulty + *pepsia*, digestion) A condition in which digestion is difficult. Commonly called *indigestion*.

DYSPHAGIA (Gr. *dus*, faulty + *phagein*, to eat) Difficulty in swallowing.

EMETIC (Gr. *emetos*, vomiting) A substance used to induce vomiting.

ENDODONTICS The branch of dentistry that deals with diseases of the dental pulp and associated processes.

ENTERITIS (Gr. *enteron*, intestine + inflammation) Inflammation of the intestines, particularly the small intestine.

GASTRECTOMY (Gr. *gaster*, belly + *-ectomy*, removal) Surgical removal of part or all of the stomach.

GASTROENTEROLOGIST A physician specializing in treating stomach and intestinal (gastrointestinal) disorders.

GASTROSCOPY (Gr. *gaster*, belly + *skopein*, to examine) A procedure in which a viewing device is inserted into the stomach for exploratory purposes.

GASTROSTOMY (Gr. *gaster*, belly + *stoma*, opening) Surgical creation of an opening in the stomach, usually for the purpose of inserting a feeding tube.

GINGIVITIS (L. *gingiva*, gum + inflammation) Inflammation of the gums.

HALITOSIS (L. *halitus*, breath) Bad breath.

HYPEROREXIA (Gr. *hyper*, over + *orexis*, a longing) Abnormal appetite.

HYPOCHLORHYDRIA A deficiency of hydrochloric acid in the stomach.

IRRITABLE BOWEL SYNDROME A generalized gastrointestinal disorder, characterized by a spastic colon, alternating diarrhea and constipation, and excessive mucus in the feces.

LAPAROTOMY (Gr. *lapara*, flank + *tome*, incision) Cutting into the abdomen in order to explore the abdominal cavity.

LAXATIVE (L. *laxus*, loose) A substance used to loosen the contents of the large intestine so that defecation can occur.

MALOCCLUSION (L. *mal*, badly + *occludere*, to close) A condition in which the upper and lower teeth do not fit properly when the jaws are closed.

ORTHODONTIA (Gr. *orthos*, straight + *odous*, tooth) The branch of dentistry that specializes in aligning teeth properly.

PORTAL HYPERTENSION A condition in which there is high blood pressure in the portal circulation of the liver.

PROCTOCELE (Gr. *proktos*, anus + *koilos*, hollow) Hernia of the rectum.

PROCTOSCOPY (Gr. *proktos*, anus + *skopein*, to examine) A procedure in which a viewing device is inserted into the anus and rectum for exploratory purposes.

PROLAPSE OF THE RECTUM A condition where the weakened walls of the rectum fall outward or downward.

PRURITUS ANI (L. *prurire*, to itch + anus) Chronic itching of the anus.

PYORRHEA (L. *pyo*, pus + Gr. *rrhoos*, flowing) A condition in which pus oozes from infected gums.

ROOT CANAL THERAPY The dental procedure that removes infected or damaged pulp and then cleans, sterilizes, fills, and seals the root cavity.

SIALITIS Inflammation of a salivary duct or gland.

SMALL INTESTINE SERIES Serial radiograms of the small intestine during the passage of ingested barium. Commonly called *upper GI series*.

STOMATITIS (Gr. *stoma*, mouth + inflammation) Inflammation of the mouth.

TRACHEOSTOMY (trachea + opening) Surgical creation of a permanent opening in the trachea.

TRENCH MOUTH (VINCENT'S INFECTION) A painful inflammation of the gums, usually including fever, ulcerations, and bleeding.

UNDERSTANDING THE FACTS

1 What is meant by chemical digestion? Mechanical digestion?

2 List the organs of the digestive tract, and state the primary function of each.

3 List the major accessory organs of the digestive tract, and state the primary function of each.

4 List the four main layers in the wall of the digestive tract. Which layers contain contractile tissue?

5 Identify the deciduous and permanent teeth by type, number, and number of roots.

6 Name the three largest pairs of salivary glands.

7 What is a bolus? How is it formed?

8 List in order the sphincters of the digestive tract. How is each stimulated to open?

9 Describe peristalsis. Which digestive organs exhibit peristalsis?

10 What are the components of gastric juice? Which cells secrete each component?

11 Contrast the structure of a gastric gland with that of an intestinal gland.

12 What are the primary cell types in the mucosa of the small intestine? What is the function of each?

13 What is meant by emulsification? Identify the substance that is emulsified during digestion and the emulsifying agent.

14 List the regions of the small intestine and the large intestine, and state the primary functions of each region.

15 What are the digestive enzymes secreted by the pancreas? What is the function of each?

16 Trace the route of bile from its formation to its release within the duodenum.

17 Describe the structure of a liver lobule.

18 List the primary functions of the liver.

19 Define or describe

  a mesentery   e rugae
  b taeniae coli   f greater omentum
  c haustra   g plicae circulares
  d villus   h bile canaliculae

20 Identify the source, site of action, and function of each of the following. Indicate whether each substance is an enzyme or a hormone.

  a gastrin   f CCK
  b trypsin   g lipase
  c pepsin   h secretin
  d enterokinase   i GIP
  e enterogastrone j maltase

UNDERSTANDING THE CONCEPTS

1 Contrast the structure of the mucosa of the esophagus, stomach, and small intestine. How does the structure of each relate to its function?

2 How are each of the following processes regulated?

  a salivation
  b swallowing
  c secretion of gastric juice
  d gastric emptying
  e pancreatic secretion

3 Distinguish the meanings of retroperitoneal and intraperitoneal. Which organs of the digestive system are included in each category?

4 Other than its great length, what structural features contribute to the absorptive surface of the small intestine?

5 Contrast peristalsis with segmenting contractions.

6 Trace the chemical digestion of fats, carbohydrates, and proteins from their entry into the mouth until the time the products of each are absorbed.

7 How are carbohydrates, proteins, fats, and ions absorbed from the lumen of the small intestine?

8 Contrast the role of the exocrine pancreas with that of the liver.

9 Which phases of the digestive process are voluntary and which are involuntary?

10 Which phases of the digestive process are regulated primarily by the nervous system, which are regulated primarily by hormones, and which are regulated by a combination of the two?

11 Distinguish between hepatitis and cirrhosis.

12 Which regions of the digestive tract are the most common sites for ulcers? Explain.

13 What is the major danger posed by severe diarrhea?

SELF-QUIZ
Multiple-Choice Questions

25.1 The walls of the digestive system have the following layers from the inside out
  a serosa, muscularis externa, submucosa, mucosa
  b mucosa, submucosa, muscularis externa, serosa
  c submucosa, mucosa, muscularis externa, serosa
  d serosa, submucosa, mucosa, muscularis externa
  e mucosa, muscularis externa, serosa, submucosa

25.2 Normally, a child has how many baby teeth?
  a 10   d 18
  b 12   e 20
  c 15

25.3 An adult has all of the following *except*
  a four incisors   d two cuspids
  b two canines   e six molars
  c four premolars

25.4 The largest of the three pairs of salivary glands are the
  a parotid   d sublingual
  b submandibular   e subparotid
  c mandibular

25.5 How many sphincters are associated with the esophagus?
  a one   d four
  b two   e five
  c three

25.6 Which of the following is *not* one of the major regions of the stomach?
  a cardiac   d esophageal
  b fundus   e pyloric
  c body

25.7 The stomach contains how many layers of smooth muscle fibers?
  a one   d four
  b two   e five
  c three

25.8 Gastric juice contains all of the following *except*
  a hydrochloric acid   d lysozyme
  b pepsinogen   e intrinisic factor
  c lipase

25.9 Pancreatic juice is secreted in response to intestinal secretion of
  a secretin   d a and b
  b cholecystokinin   e a, b, and c
  c lipase

25.10 Which of the following is *not* part of the biliary system?
  a gallbladder   d pancreatic duct
  b hepatic ducts   e common bile duct
  c cystic ducts

25.11 Someone has just eaten a meal of buttered toast, cream, and eggs. Which of the following would you expect to happen?
  a Gastric motility and secretion of HCl decrease once the food has reached the duodenum.
  b Gastric motility starts to increase while the person is chewing the food (before any swallowing).
  c Fat will be emulsified in the duodenum by the action of bile.
  d Bile will be released from the gallbladder when the fatty meal reaches the duodenum due to stimulation by CCK.
  e All of the above

25.12 Which of the following statements is/are true concerning the control of acid and enzyme secretion by the stomach?
  a In the cephalic phase, gastric secretion increases despite the absence of food in the stomach.
  b Distension of the stomach by increasing the volume of food increases gastric secretion.
  c Protein digestion products, alcohol, and caffeine stimulate the release of the hormone gastrin.
  d a and b
  e a, b, and c

25.13 The rate at which the stomach empties its contents into the small intestines
  a increases with increased fat content of a meal
  b increases with increased secretion of CCK and GIP
  c increases with increasing volume of food in the duodenum
  d increases with increasing volume of food in the stomach
  e increases with increasing levels of acid in the duodenum

25.14 Severe untreated diarrhea may result in death because
  a prolonged water loss can lead to fatal dehydration
  b the absorption of proteins is inhibited
  c diverticulitis and peritonitis can result
  d acute appendicitis frequently develops concomitantly with severe diarrhea
  e a and c

25.15 Before reaching the mucosa of the stomach, a surgeon's scalpel would pass through, in order, the a peritoneum, abdominal cavity, serosa, smooth muscle, and submucosa

b abdominal cavity, serosa, smooth muscle, submucosa, and peritoneum

c peritoneum, abdominal cavity, smooth muscle, serosa, and submucosa

d serosa, peritoneum, smooth muscle, submucosa, and abdominal cavity

Completion Exercises

25.16 _____ is a soupy mixture of churned ingested nutrients and gastric juice.

25.17 A "potbelly" is caused by the deposition of fat in the _____.

25.18 The right and left hepatic ducts converge with the _____ duct to form the common bile duct.

25.19 The delicate blood channels within the liver that transport blood from the branches of the portal vein and the hepatic artery are called _____.

25.20 Sharp contractions of the circular smooth muscle of the small intestine that are important in mixing the chyme with digestive juices are called _____ contractions.

25.21 The part of the digestive tract below the diaphragm is called the _____ tract.

25.22 Teeth are held in their sockets by bundles of connective tissue called _____.

25.23 The portion of the digestive system that serves as both an air passage and a food passage is the _____.

25.24 The serous membrane of the abdominal cavity is the _____.

25.25 Intraperitoneal organs are suspended from the posterior body wall by the _____.

25.26 The absorptive surface area of the mucosa is increased by millions of fingerlike projections called _____.

25.27 The physical breakdown of lipids is called _____.

25.28 The liver is divided into two main lobes by the _____ ligament.

25.29 Bile secretion can be increased by the hormone _____.

25.30 The large intestine contains bulges called _____.

Matching

a dentine
b enamel
c cement
d pulp
e periodontal ligaments

25.31 _____ surrounds the pulp cavity

25.32 _____ connective tissue structures that connect the alveolar bone with the cement of a tooth

25.33 _____ bonelike covering of the neck and root of a tooth

25.34 _____ connective tissue core of teeth, containing blood vessels and nerves

25.35 _____ hard tissue that covers the crowns of teeth

a chief cells
b parietal cells
c enteroendocrine cells
d goblet cells
e acinar cells
f hepatocytes
g surface mucous cells

25.36 _____ liver cells
25.37 _____ secrete cholecystokinin
25.38 _____ secrete pepsinogen
25.39 _____ produce mucus
25.40 _____ secrete hydrochloric acid
25.41 _____ secrete pancreatic enzymes
25.42 _____ stem cells for production of new gastric mucosal cells

A SECOND LOOK

In the following photograph, label the stomach, transverse colon, descending colon, ileum, jejunum, sigmoid colon, and rectum.

Metabolism, Nutrition, and the Regulation of Body Heat

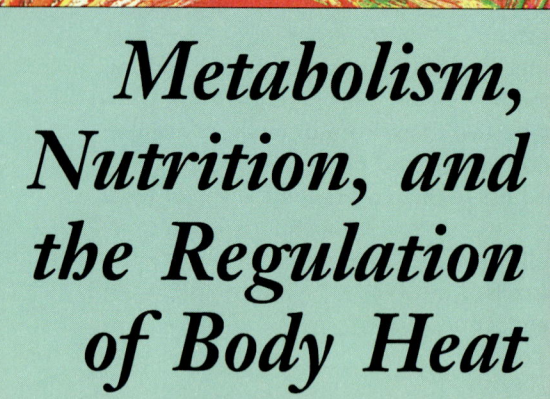

26

KEY CONCEPTS

1 The sum of all cellular activities that maintain our bodies is called metabolism. Energy and raw materials for metabolic processes are obtained from nutrients in food. The breakdown of simple sugars, mainly glucose, is the major source of energy for the body.

2 The initial series of reactions in glucose metabolism is called glycolysis.

3 Following glycolysis, the breakdown of glucose to carbon dioxide and water is completed by the citric acid cycle.

4 Most ATP, which provides energy for metabolism, is formed by the electron transport system.

5 In protein metabolism, the proteins in food are broken down into their constituent amino acids; the amino acids are then used to synthesize proteins needed by the body. Amino acids are modified before they can be used by a cell.

6 Lipid metabolism produces the most concentrated source of food energy.

Living cells are continuously involved in the many chemical and energy transformations that take place during the moment-to-moment, day-to-day activities that keep our bodies alive and healthy. The general term that includes all these cellular activities is **metabolism.** Metabolism maintains life and promotes homeostasis. It includes the conversion of nutrients into the usable energy contained in ATP, the production and replication of nucleic acids, the synthesis of proteins, the physical construction of cells and cell structures, the elimination of cellular wastes, and the production of heat, which helps regulate the temperature of the body.

Everything a human body has or does depends on the **nutrients,** or chemical components, of food, which supply the energy and physical materials a body needs. The food we eat contains six kinds of *essential nutrients* that are vital to our well-being: carbohydrates, proteins, lipids, minerals, vitamins, and water. Nutrients must be supplied by the diet because (1) we break down food to get energy, and (2) many vital substances cannot be synthe-sized by the body. Each type of nutrient provides something different, as you will see in detail later. In general, proteins and some mineral salts are the body-building foods; carbohydrates and lipids are fuels that provide energy; and vitamins and some minerals help regulate cellular activity by enhancing the effects of the other essential nutrients and enzymes.

The energy in food exists in *potential* form. It cannot be used unless it is released by the actions of enzymes in cells. **Cellular respiration*** is the process by which the potential energy in the chemical bonds in food molecules is released. This energy may be used to make biologically useful molecules, especially ATP, or it may be released as heat. The ATP can be used by cells to do the work that keeps the whole body functioning. As shown in FIGURE 26.1, when energy in ATP is released from a cell, ADP is produced. When inorganic phosphate (P_i) and chemical bond energy are added to ADP, ATP is formed.

*When we speak of respiration in the sense of cellular activity, we specifically mean *cellular respiration*, not "breathing."

FIGURE 26.1 ATP AND CELLULAR ACTIVITIES

The ATP molecules produced by catabolism provide energy for anabolism and cellular activities.

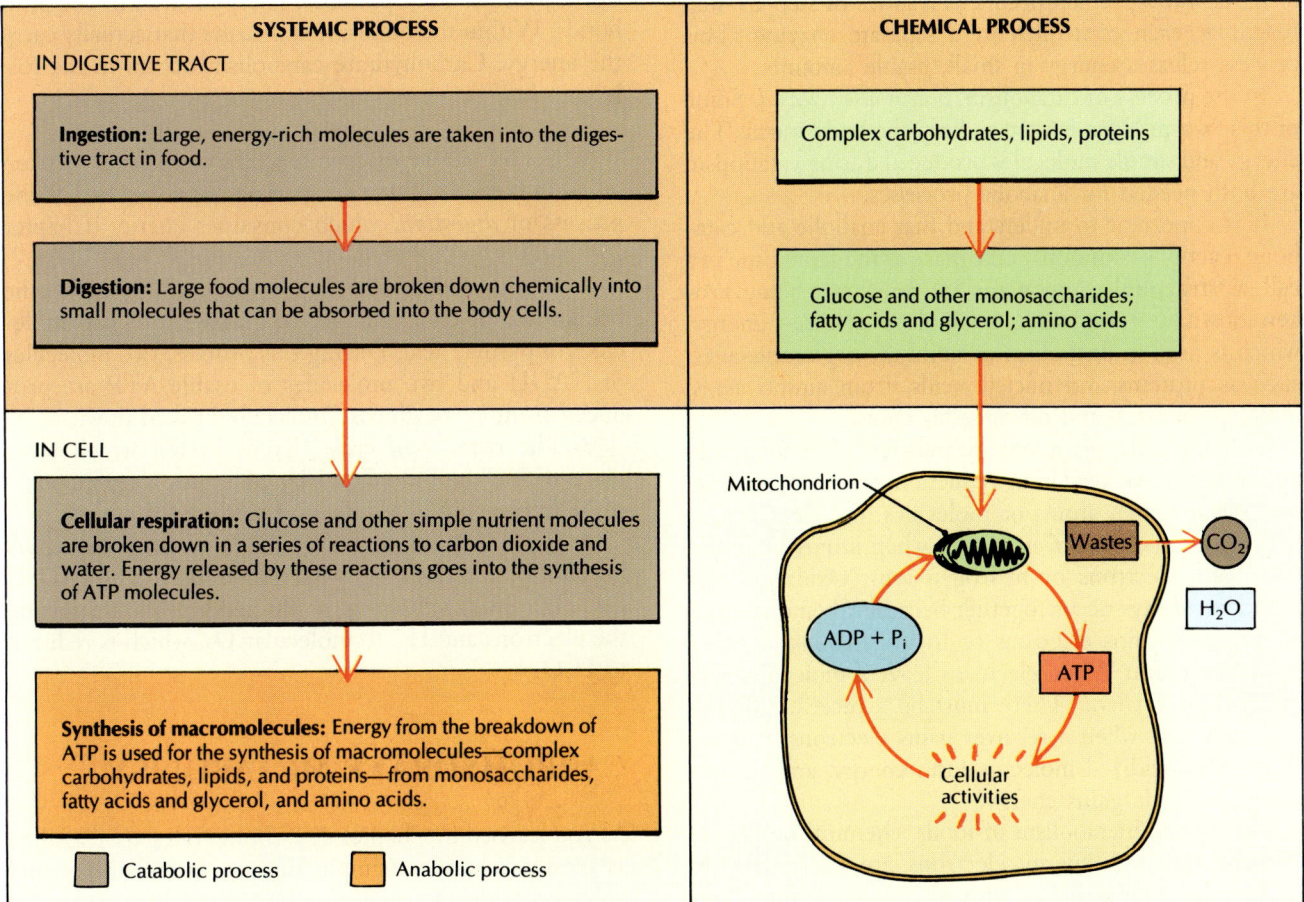

REVIEW OF CHEMICAL REACTIONS

To give you a better understanding of metabolic processes, we begin with a review of several kinds of chemical reactions. Metabolism includes both anabolic and catabolic processes [FIGURE 26.1]. *Anabolism* involves the chemical changes that allow cells to *build up* and repair themselves, or to synthesize macromolecules from smaller precursors, as when amino acids are bonded together and arranged into complex three-dimensional proteins. (Anabolism is also called *biosynthesis*.) Because anabolic reactions synthesize molecules of increased size and complexity, more energy is required because more components are bonded in these macromolecules, and each bond formation *requires energy*. The required energy is provided by the conversion of ATP into ADP and P_i.

Catabolism includes any chemical reactions in which macromolecules are broken down into smaller molecules and energy is released. As you learned in Chapter 25, ingested carbohydrates, proteins, and lipids must be mechanically and chemically broken down into smaller molecules before they can be absorbed by cells. Enzymes are required to complete the catabolism of nutrients during digestion. A more complex form of catabolism is *cellular respiration*, which occurs as a series of step-by-step reactions, each controlled by a separate enzyme. This process releases energy in small, usable amounts.

In the process of catabolism, *heat is also released*. Some of this "waste" heat is used to keep the body warm. The energy and simple molecules produced during catabolism are both needed for anabolic processes.

It is important to understand that anabolic and catabolic reactions commonly take place at the same time in a cell, as structural components are constantly being broken down and replaced. Catabolism releases energy, which is used in anabolism to build up large molecules, such as proteins and nucleic acids, from amino acids, monosaccharides, and nitrogenous bases.

In living cells, many energy transfers of cellular metabolism involve oxidation-reduction reactions. *Oxidation* occurs when atoms or molecules lose electrons or hydrogen ions. *Reduction* occurs when atoms or molecules gain electrons or hydrogen ions. Oxidation and reduction always occur together because when one atom or molecule gains electrons or hydrogen ions, another must lose them. When electrons leave a molecule, that molecule is *oxidized*. There must be a receiver for the electrons, and when a receiver gains electrons, it is *reduced*. The oxidized molecule loses energy, and the reduced molecule gains energy.

During the metabolism of foods, chemical bonds are broken, and high-energy electrons are transferred to nucleotides such as NAD^+ (nicotinamide adenine dinucleotide) and FAD (flavin adenine dinucleotide); that is, these nucleotide molecules actually carry the high-energy electrons. In the process of receiving the electrons, NAD^+ and FAD also accept hydrogen ions (H^+). We say that the nucleotides are *reduced* (receive electrons) and that the foods are *oxidized* (lose electrons). In this way, NAD^+ is converted into $NADH + H^+$, and FAD is converted into $FADH_2$. Each nucleotide receives two electrons and two hydrogen ions. The energy in the reduced nucleotides ($NADH + H^+$ and $FADH_2$) is most often used by mitochondria to make ATP, but may also be used in energy-requiring anabolic reactions.

A S K   Y O U R S E L F

1 Give an example of what anabolic and catabolic reactions accomplish.

2 Is energy mainly required or released in catabolic reactions?

3 What happens during an oxidation-reduction reaction?

INTRODUCTION TO CARBOHYDRATE METABOLISM

Carbohydrates have potential energy stored in chemical bonds. Within the bonds, it is *electrons* that actually carry the energy. Carbohydrate catabolism occurs in the following four stages:

1 Carbohydrate macromolecules in food are broken down into their subunits, especially glucose. This is the process of **digestion**, which consumes energy (Chapter 25).

2 *Glycolysis* occurs in the cytoplasm and results in the breakdown of one molecule of glucose into two molecules of pyruvic acid. During this process, two molecules of NADH and two molecules of usable ATP are produced from every glucose molecule broken down.

3 The *citric acid cycle* (Krebs cycle), which takes place within the mitochondria, generates large amounts of $NADH + H^+$, $FADH_2$, and some ATP.

4 Within the mitochondria, the *electron transport system* converts the energy in $NADH + H^+$ and $FADH_2$ into high-energy chemical bonds in ATP by transporting the electrons and H^+ to molecular O_2, which is reduced to H_2O.

CARBOHYDRATE METABOLISM

As you learned in Chapter 2, carbohydrates are classified as monosaccharides (simple sugars), disaccharides, and polysaccharides. During digestion, the larger molecules are broken down into smaller, more usable molecules, such as glucose. Glucose, a monosaccharide, is an important energy source for the body. The breakdown of glucose in a complex series of metabolic reactions leads to the production of energy. (The breakdown of lipids and proteins, which also contributes to the production of energy, will be described later in the chapter.)

Cellular Respiration

In *cellular respiration,* cells use the potential energy stored in the chemical bonds in food molecules to create ATP from ADP and P_i. The process produces carbon dioxide and water. Although proteins and lipids may enter this process at different points, we begin with the catabolism of glucose. The main steps in glucose catabolism are (1) *glycolysis,* the initial breakdown of glucose and the release of ATP and NADH + H$^+$; (2) the *citric acid cycle,* in which carbon dioxide, ATP, NADH + H$^+$ and FADH$_2$ are released; and (3) the *electron transport system,* in which ATP and water are produced and oxygen is consumed. (See FIGURE 26.10 for a summary of the metabolic processes of all types of nutrients.)

Glycolysis *Glycolysis* is the splitting (*lysis*) of a glucose molecule with 6 carbon atoms into two 3-carbon molecules of pyruvic acid. It occurs in the cytoplasm of a cell in a series of enzymatically controlled steps that do not require oxygen [FIGURE 26.2]. Because glycolysis proceeds without oxygen, it is referred to as *anaerobic respiration.* (*Anaerobic* means "without air-life.") Most usable energy is obtained later from the citric acid cycle and the electron transport system.

Before a molecule of glucose is to be broken down into smaller units, the molecule is *activated* and its energy level is raised. This activation is achieved by the transfer of a phosphate group from an ATP molecule to glucose as ATP becomes ADP. (The transfer of a phosphate group from one compound to another is called *phosphorylation.*) The process of forming ATP during glycolysis and the citric acid cycle is called *substrate-level phosphorylation.* The ATP-forming process that occurs during electron transport is called *oxidative phosphorylation* [see FIGURE 26.5].

FIGURE 26.2 shows the 10 steps of glycolysis:

Steps 1–3 Energy in the form of ATP is used to activate the breakdown of glucose and the rearrangement of the molecules.

Step 4 The 6-carbon molecule of fructose 1,6-diphosphate is split into two 3-carbon molecules: dihydroxyacetone phosphate and glyceraldehyde 3-phosphate.

Step 5 Each of the two 3-carbon molecules created during step 4 may be converted into the other, and some glyceraldehyde 3-phosphate will continue down the pathway.

Step 6 The glyceraldehyde 3-phosphate picks up an inorganic phosphate (P_i) from the cytoplasm and is also

oxidized to form 1,3-bisphosphoglyceric acid. The hydrogen released in the oxidation is picked up by the coenzyme nicotinamide adenine dinucleotide (NAD$^+$) to form NADH + H$^+$, the *reduced* form of NAD$^+$.

Steps 7–10 Four molecules of ATP are generated. Since two molecules of ATP were required to start the activation of the glucose molecule (steps 1 and 3), there is a net gain of only two ATP molecules for each glucose molecule metabolized.

Step 10 The formation of two molecules of pyruvic acid marks the end of glycolysis.

Note that when a glucose molecule has been broken down as far as *pyruvic acid,* the reactions have generated not only a net gain of two ATP molecules but also two molecules of reduced, energy-carrying NADH + H$^+$. The cell has not released any carbon dioxide or used any oxygen. Most of the chemical-bond energy originally present in the glucose is still present in the two molecules of pyruvic acid.

Pyruvic acid can be metabolized by various pathways. A *pathway* is a series of chemical reactions that generates a specific product; glycolysis is a pathway that generates pyruvic acid. Pyruvic acid can then be used in pathways that generate amino acids, oxidize it to carbon dioxide and water, or produce lactic acid.

Citric acid (Krebs) cycle In the major pathway of glucose catabolism, each of the two molecules of pyruvic acid produced by glycolysis combines with a molecule of coenzyme A (CoA) to form a 2-carbon compound called acetyl-coenzyme A (acetyl-CoA). This process releases a molecule of carbon dioxide and two hydrogen atoms. The acetyl-CoA begins the next pathway, a series of reactions called the *citric acid cycle,* or *Krebs cycle,** which occur in the mitochondria of the cell.

FIGURE 26.3 shows the nine steps of the citric acid cycle, an ordered sequence of reactions:

Step 1 The two carbons of acetyl-CoA combine with oxaloacetic acid, a 4-carbon compound already present in mitochondria. The reaction produces a 6-carbon molecule of citric acid.

Steps 2–4 The citric acid is rearranged to form isocitric acid, *losing a pair of hydrogen atoms and a molecule of carbon dioxide,* which is immediately converted into the 5-carbon α-ketoglutaric acid.

Step 5 The α-ketoglutaric acid *loses a pair of hydrogen atoms and a molecule of carbon dioxide,* producing 4-carbon succinyl-coenzyme A.

*The Krebs cycle is named after the British biochemist Sir Hans Adolf Krebs (1900–1981), who during the 1930s first outlined the steps in the complete breakdown of pyruvic acid. Krebs won a Nobel Prize in physiology or medicine in 1953 for his work.

FIGURE 26.2 GLYCOLYSIS

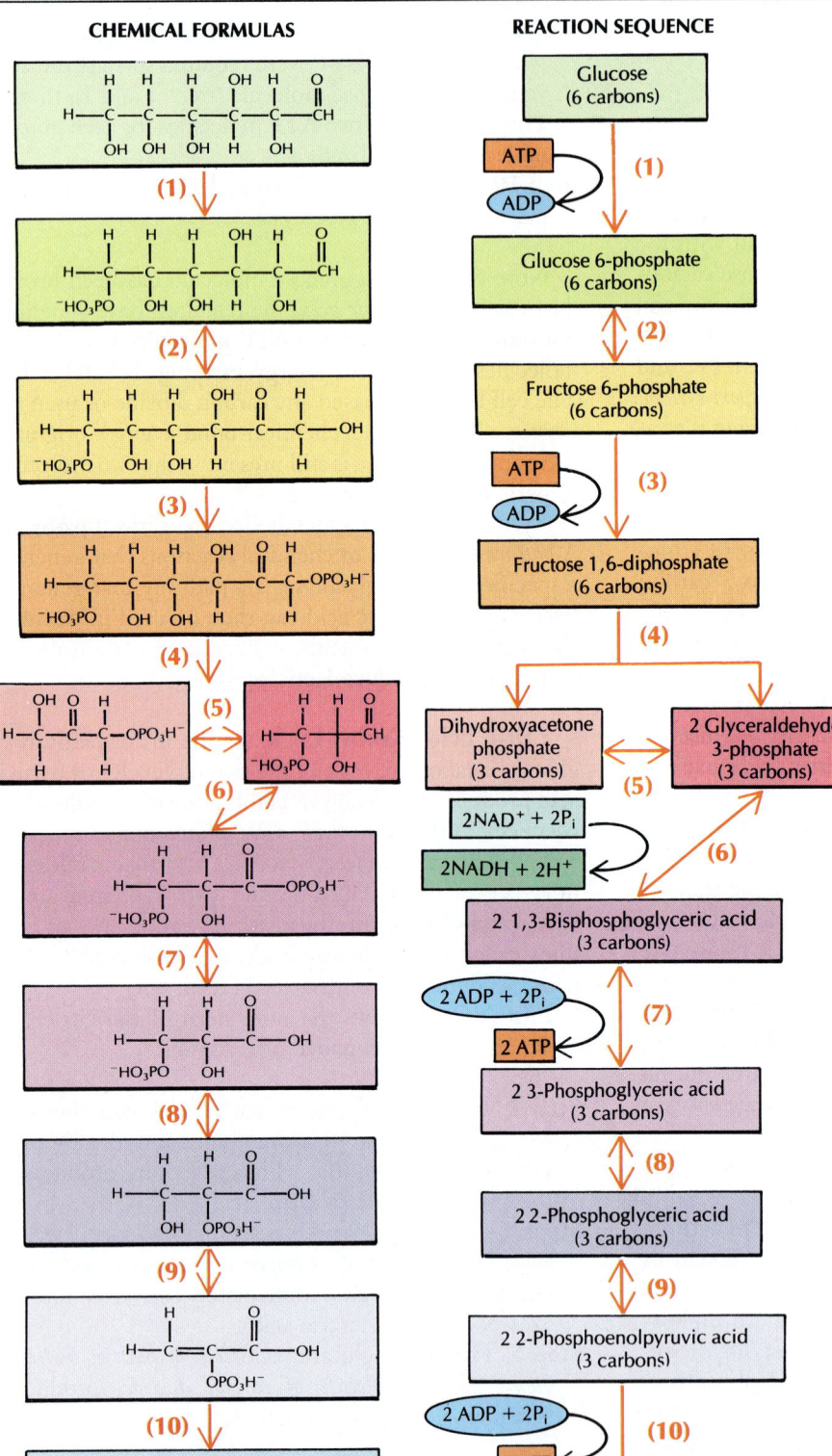

Key to enzymes for each step:
(1) Hexokinase
(2) Phosphoglucose isomerase
(3) Phosphofructokinase
(4) Aldolase
(5) Triose phosphate isomerase
(6) Glyceraldehyde 3-phosphate dehydrogenase
(7) Phosphoglycerate kinase
(8) Phosphoglyceromutase
(9) Enolase
(10) Pyruvate kinase

STEPS

(1) Glucose is phosphorylated (a phosphate group is added) from the breakdown of one ATP to ADP. Phosphorylation provides activation energy for the following reactions and prevents the glucose from diffusing out of the cell.

(2) Glucose 6-phosphate undergoes a molecular rearrangement to form fructose 6-phosphate.

(3) The breakdown of another ATP to ADP adds a second phosphate group to fructose 6-phosphate, forming fructose 1,6-phosphate.

(4) Fructose 1,6-diphosphate is split into two 3-carbon molecules: dihydroxyacetone phosphate and glyceraldehyde 3-phosphate.

(5) Dihydroxyacetone phosphate and glyceraldehyde 3-phosphate are interconvertible.*

(6) The glyceraldehyde 3-phosphate picks up an inorganic phosphate (P_i) from the cytoplasm and is oxidized to form 1,3-bisphosphoglyceric acid. The hydrogen released by the oxidation reaction is picked up by the coenzyme nicotinamide adenine dinucleotide (NAD^+) to form $NADH + H^+$, the reduced form of NAD^+.

(7-10) Four molecules of ATP are generated. Since two molecules of ATP were used at the beginning of the process (steps 1 and 3), there is a net gain of only two ATP molecules for each molecule of glucose metabolized. The formation of two molecules of pyruvic acid marks the end of glycolysis.

*In a closed system, these two compounds would reach an equilibrium where they were present in equal concentrations. However, since glyceraldehyde 3-phosphate undergoes further reaction and disappears from the system, the dihydroxyacetone is converted to glyceraldehyde 3-phosphate.

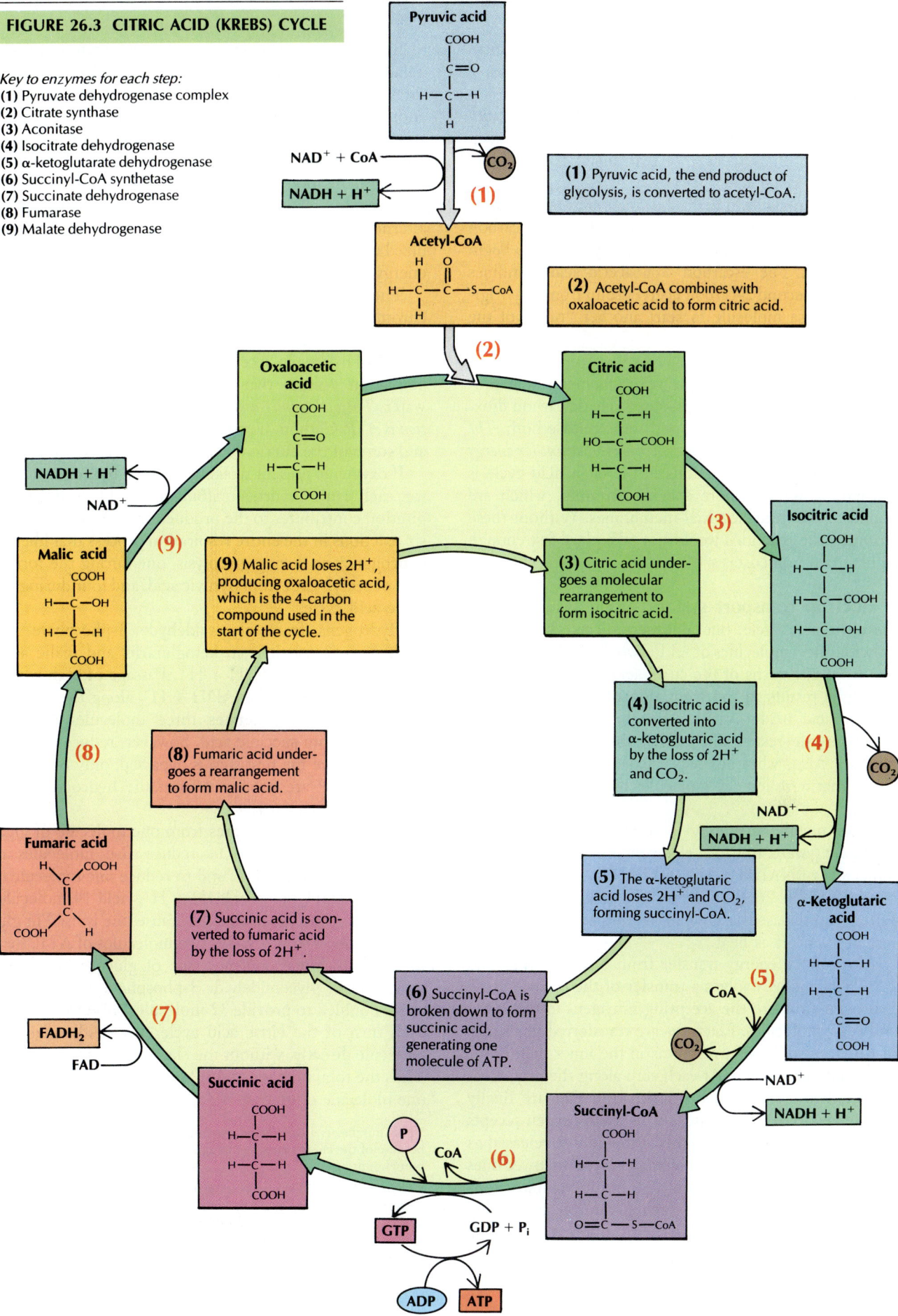

FIGURE 26.3 CITRIC ACID (KREBS) CYCLE

Key to enzymes for each step:
(1) Pyruvate dehydrogenase complex
(2) Citrate synthase
(3) Aconitase
(4) Isocitrate dehydrogenase
(5) α-ketoglutarate dehydrogenase
(6) Succinyl-CoA synthetase
(7) Succinate dehydrogenase
(8) Fumarase
(9) Malate dehydrogenase

Pyruvic acid

$NAD^+ + CoA$

$NADH + H^+$

(1) Pyruvic acid, the end product of glycolysis, is converted to acetyl-CoA.

Acetyl-CoA

(2) Acetyl-CoA combines with oxaloacetic acid to form citric acid.

Oxaloacetic acid

Citric acid

$NADH + H^+$

NAD^+

Isocitric acid

Malic acid

(9) Malic acid loses $2H^+$, producing oxaloacetic acid, which is the 4-carbon compound used in the start of the cycle.

(3) Citric acid undergoes a molecular rearrangement to form isocitric acid.

(4) Isocitric acid is converted into α-ketoglutaric acid by the loss of $2H^+$ and CO_2.

CO_2

NAD^+

$NADH + H^+$

(8) Fumaric acid undergoes a rearrangement to form malic acid.

Fumaric acid

(7) Succinic acid is converted to fumaric acid by the loss of $2H^+$.

(5) The α-ketoglutaric acid loses $2H^+$ and CO_2, forming succinyl-CoA.

α-Ketoglutaric acid

CoA

CO_2

(6) Succinyl-CoA is broken down to form succinic acid, generating one molecule of ATP.

$FADH_2$

FAD

Succinic acid

NAD^+

$NADH + H^+$

P

CoA

Succinyl-CoA

GTP

$GDP + P_i$

ADP

ATP

Step 6 The succinyl-coenzyme A is broken down to form succinic acid. This reaction ultimately generates a molecule of ATP.

Step 7 The succinic acid *loses a pair of hydrogen atoms*, resulting in the formation of 4-carbon fumaric acid.

Step 8 The fumaric acid is rearranged to yield the 4-carbon malic acid.

Step 9 The malic acid *loses a final pair of hydrogen atoms*, producing oxaloacetic acid, the same kind of 4-carbon acid that accepted the 2-carbon acetyl group to begin the cycle. The 4-carbon oxaloacetic acid combines with the second 2-carbon acetyl group from the original glucose molecule to start another "turn" of the cycle.

Some reactions are simple rearrangements that prepare molecules for later steps, some release carbon dioxide, some release hydrogen, and some release both. *The hydrogen-releasing reactions are the most important for energy transfer.* Each of the reactions in the citric acid cycle is assisted by one or more specific enzymes, which are bound to the mitochondrial membranes. Without these enzymes, the reactions would not take place fast enough to maintain homeostasis.

Electron transport system At the completion of the citric acid cycle, one glucose molecule has yielded only four new molecules of ATP: two from glycolysis and two from two turns of the citric acid cycle. (Recall that glycolysis results in the formation of *two* molecules of pyruvic acid made available for the citric acid cycle.) Where is the rest of the energy that is released from glucose? Also, what has happened to all the hydrogen atoms that have left the molecules during the citric acid cycle?

Most metabolic energy is released by the ***electron transport system,*** in which electrons are transferred from molecules of NADH + H$^+$ and FADH$_2$ to molecules of oxygen [FIGURE 26.4]. In the process, some of the chemical energy released is used to synthesize ATP, and the rest is given off as heat.

Most of the energy transfer from NADH + H$^+$ and FADH$_2$ actually involves a transfer of the *electrons* from hydrogen atoms to the accepting nitrogens in the NAD and FAD molecules. Electrons are transferred in a series of oxidation-reduction reactions in the inner membrane of the mitochondrion. At each step along the way, electrons fall to a lower energy state, until they are finally transferred to oxygen. Water forms when oxygen accepts hydrogen and electrons. Some of the energy released as the electrons move from high energy levels to lower ones is ultimately used to generate ATP inside the mitochondrion. The ATP then moves throughout the cell, readily available to power the cell's many different metabolic reactions.

If you look at the electron transport system in greater detail [FIGURE 26.4], you see that when electrons from the hydrogen atoms leave a molecule to be taken on by another molecule, energy goes with them. The electrons are transferred to an orderly arrangement of hydrogen acceptors called *cytochromes* (protein-plus-iron-containing pigment molecules). Some of the cytochromes are accompanied by specific enzymes. Cytochromes can be alternatively reduced (by taking up electrons) and oxidized (by giving up electrons) on the iron group, with an energy loss or gain accompanying any electron transfer.

The transport of electrons from cytochrome to cytochrome is generally coupled to the reactions that bring ADP and inorganic phosphate together to form ATP. The final hydrogen acceptor is oxygen, and when an oxygen atom accepts two hydrogen atoms, the result is water. *The biologically useful final product of cellular respiration is ATP.* Carbon dioxide and water are waste products, and some of the energy from the electrons is lost as heat.

If oxygen is present to act as the final hydrogen acceptor, each pair of hydrogen atoms released by cellular respiration contributes to the production of ATP. There are six reactions in the entire respiration process that liberate hydrogen: one during glycolysis, one during the formation of acetyl-CoA from pyruvic acid, and four during the citric acid cycle.

Hydrogens from glyceraldehyde 3-phosphate and from pyruvic, isocitric, α-ketoglutaric, and malic acids reduce NAD$^+$ to NADH + H$^+$. Passage of each pair of hydrogen atoms from NADH + H$^+$ along the electron transport system generates three molecules of ATP. Hydrogen from succinic acid, however, reduces FAD to FADH$_2$. Since the first step is bypassed, only two molecules of ATP are produced from the hydrogens from FADH$_2$ [FIGURE 26.4].

Enough hydrogen comes from one molecule of glyceraldehyde 3-phosphate to reduce five molecules of NAD$^+$ to NADH + H$^+$, and to reduce one molecule of FAD to FADH$_2$. Five NADH + H$^+$ yield 14 molecules of ATP (not the expected 15, because one is used in glycolysis). One FADH$_2$ yields two molecules of ATP, for a total of 16. Since one molecule of glucose yields two molecules of glyceraldehyde 3-phosphate, the ATP figure is doubled to provide 32 molecules* of ATP. Besides, each turn of the citric acid cycle also yields one ATP molecule directly, without the electron transport system. Thus the total number of ATP molecules produced from one molecule of glucose is 36,† with 2 from glycolysis, 2

---

*This number is a theoretical maximum and is not realized in the cell because of the energy required to shuttle molecules into and out of the mitochondria.

†The number of ATP molecules produced from the oxidation of one glucose molecule may be 38 in some cases, depending on cellular conditions. The additional 2 ATP molecules are produced in the electron transport system from the NADH + H$^+$ produced during glycolysis [FIGURE 26.5].

FIGURE 26.4 ELECTRON TRANSPORT SYSTEM

[A] The transfer of hydrogen atoms and electrons to the compounds of the electron transport system results in the formation of ATP at three places in the system. **[B]** Schematic diagram showing part of the electron transport system in the inner mitochondrial membrane.

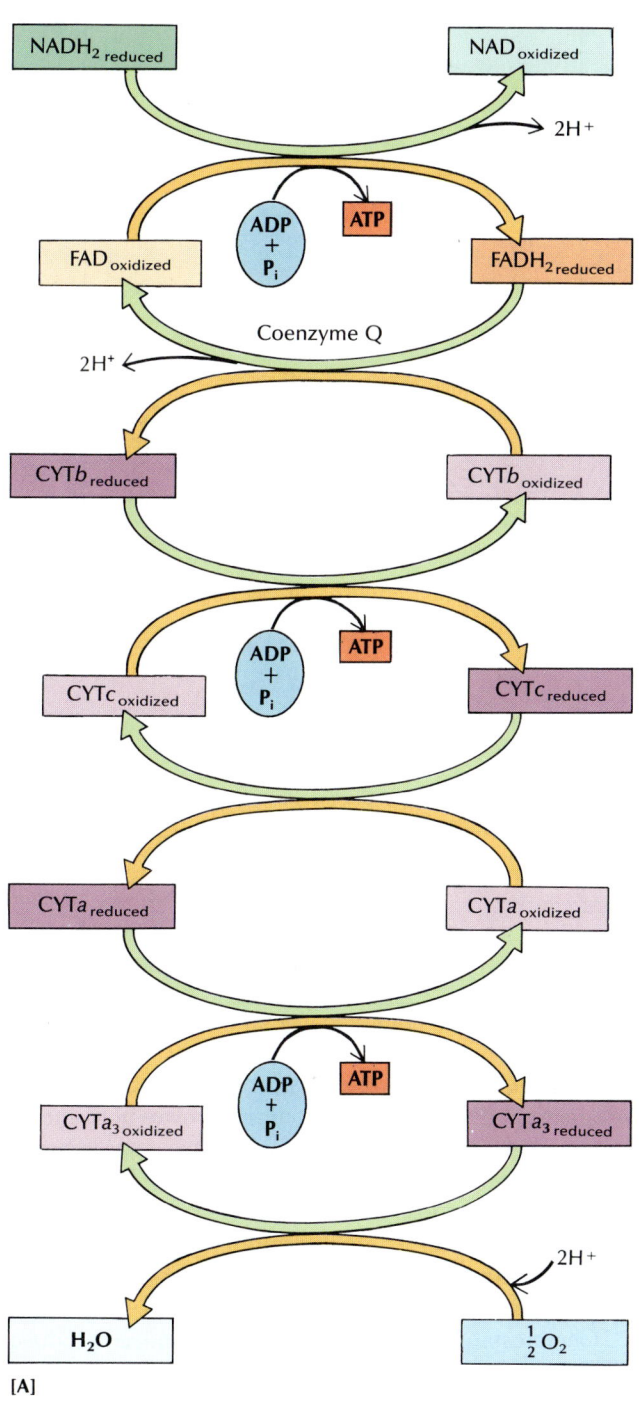

NADH$_2$ reduced → NAD oxidized
$2H^+$
FAD oxidized ← FADH$_2$ reduced
ADP + P$_i$ → ATP
Coenzyme Q
$2H^+$
CYTb reduced ← CYTb oxidized
CYTc oxidized → CYTc reduced
ADP + P$_i$ → ATP
CYTa reduced ← CYTa oxidized
CYTa_3 oxidized ← CYTa_3 reduced
ADP + P$_i$ → ATP
$2H^+$
H$_2$O ← $\frac{1}{2}$ O$_2$

[A]

Dehydrogenase complex
NADH$_2$ → NAD
e^-
H^+
Coenzyme Q
FADH$_2$ → FAD
H^+
Cytochrome c
e^-
Cytochrome b–c$_1$ complex
Cytochrome oxidase complex
$2H^+ + \frac{1}{2}$ O$_2$
H$_2$O
e^-
H^+
INNER COMPARTMENT
H^+
ADP + P$_i$
ATP
ATP synthase
H^+
OUTER COMPARTMENT
Outer membrane
Inner membrane

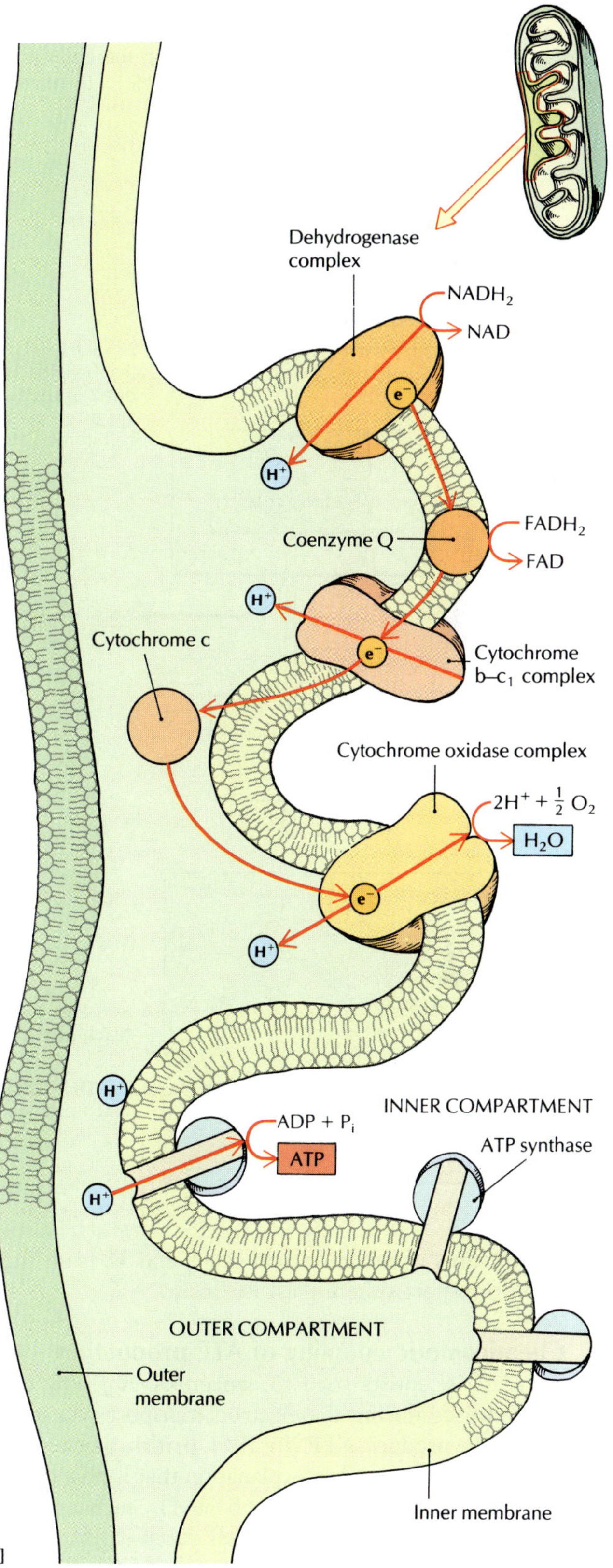

[B]

FIGURE 26.5 YIELD OF ATP FROM OXIDATION OF GLUCOSE

Although a total of 36 ATP molecules is shown as the yield from oxidation of a glucose molecule (2 from glycolysis, 2 from the citric acid cycle, and 32 from the electron transport system), the actual yield may vary, depending on the cellular conditions. For example, the ATP yield from the breakdown of glucose to NADH + H$^+$ may be 6 molecules instead of 4.

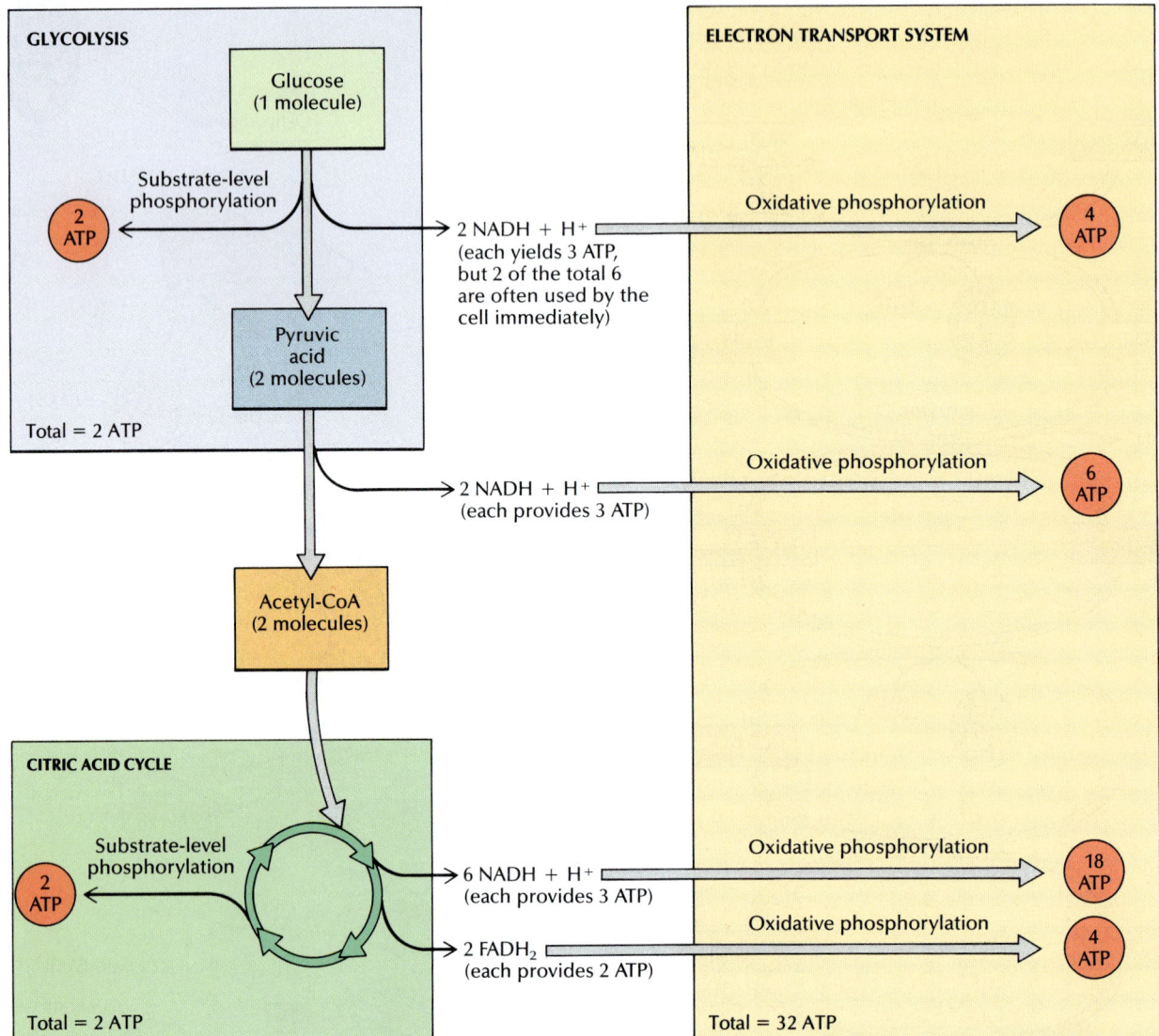

from two turns of the citric acid cycle, and 32 from the electron transport system [FIGURE 26.5].

Chemiosmotic coupling of ATP production For many years, scientists tried to explain exactly how the energy released during the electron transport system is harnessed to produce ATP. In 1961 British biochemist Peter Mitchell proposed an explanation that is now commonly accepted. In fact, it was supported by such a growing mass of evidence that Mitchell received the 1978 Nobel Prize in chemistry for his work. Mitchell's hypothesis, now known as *chemiosmotic coupling of ATP*

production, or *chemiosmosis,* is based on the fact that the production of ATP by mitochondria involves both chemical and transport processes ("chemi" + "osmosis") across a semipermeable mitochondrial membrane [FIGURE 26.6]. Chemiosmosis involves (1) the accumulation of electrochemical energy resulting from an electrochemical proton gradient and (2) the release of that energy to make ATP from ADP and P$_i$.

Chemiosmosis explains how mitochondria produce ATP, and the structure of mitochondria is important in understanding the process. Recall that a mitochondrion has two membranes: an outer one that covers the mito-

FIGURE 26.6 CHEMIOSMOSIS

This is the mechanism that produces ATP as a result of the electron transport system. **[A]** Schematic drawing of a mitochondrion, the site of ATP production. Note how the inner and outer membranes form the matrix and outer compartment. **[B]** Enlarged schematic drawing of the mitochondrial membranes. Energy derived from an electrochemical proton (H^+) gradient is used to drive the enzyme ATP synthase to catalyze the conversion of ADP and P_i into usable ATP. **[C]** Electron micrograph showing F_1 particles on inner mitochondrial membrane. ×320,000.

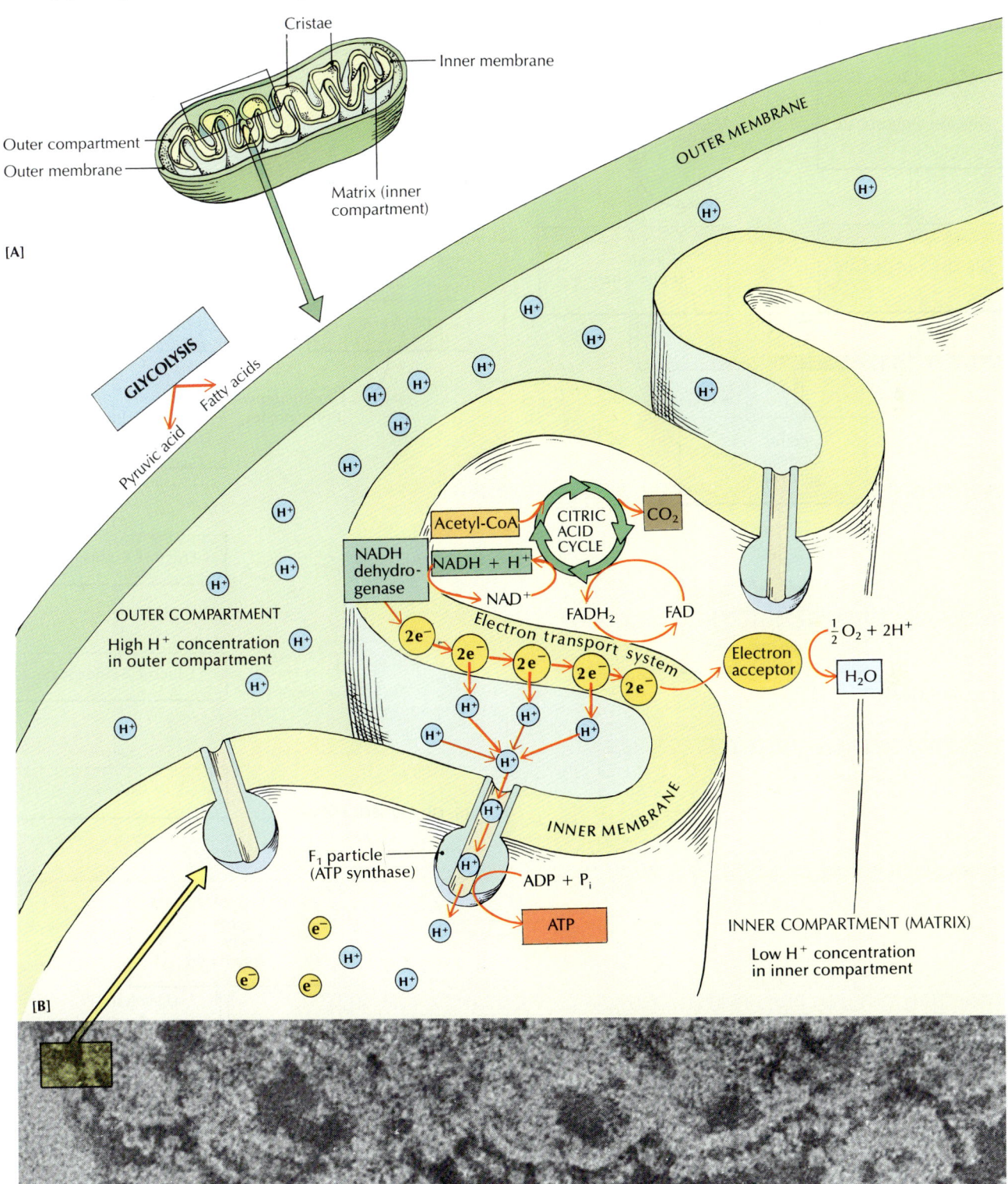

FIGURE 26.7 GLYCOGENESIS, GLYCOGENOLYSIS, GLUCONEOGENESIS

The drawings show a comparison of [A] glycogenesis, [B] glycogenolysis, and [C] gluconeogenesis.

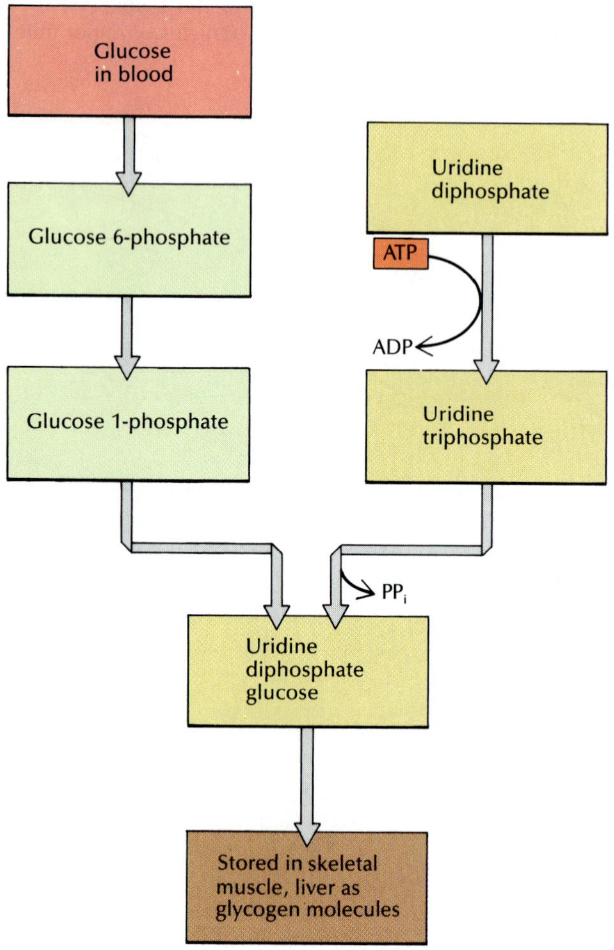

[A] GLYCOGENESIS

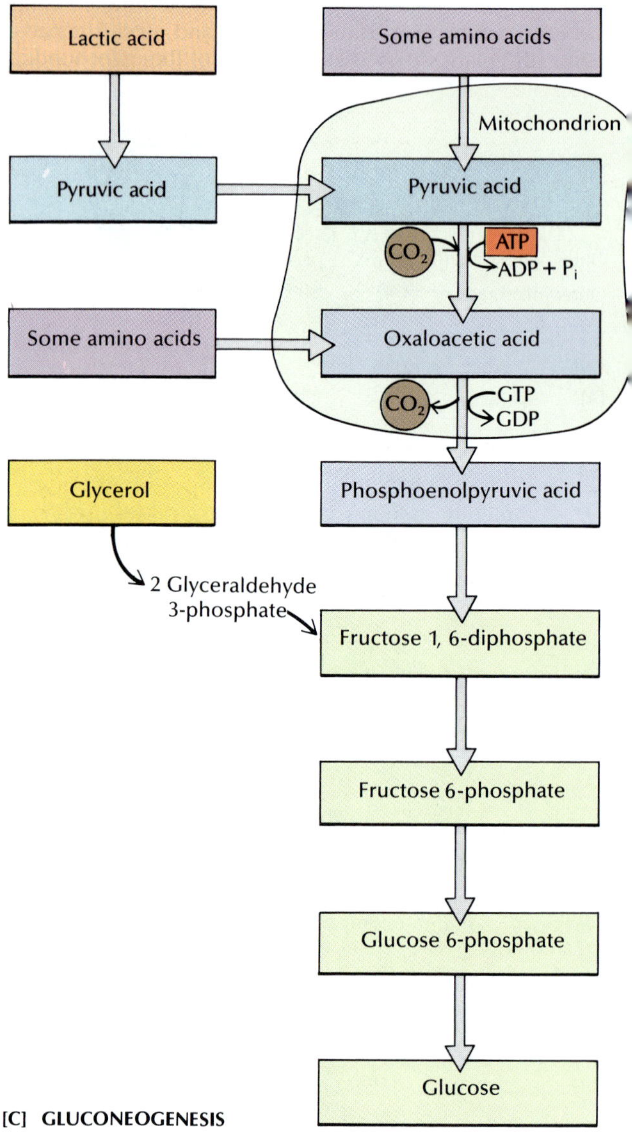

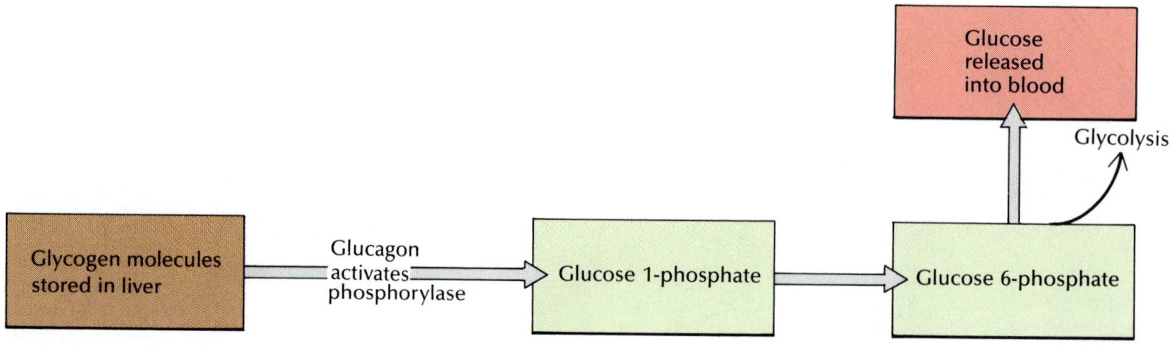

[B] GLYCOGENOLYSIS

[C] GLUCONEOGENESIS

Process	Description
Cellular respiration (glucose catabolism)	Complete oxidation of glucose in cells to produce ATP; 1 molecule of glucose yields 36 molecules of ATP. Includes glycolysis, citric acid cycle, electron transport system.
Glycolysis (anaerobic respiration)	Breakdown of a glucose molecule into two pyruvic acid molecules with a net yield of 2 molecules of ATP; takes place in cytosol; does not require oxygen.
Citric acid cycle (Krebs cycle)	Series of enzymatic reactions that release hydrogen from oxidized acids as a means of energy transfer; produces CO_2, $NADH + H^+$, $FADH_2$, H_2O, and 2 ATP molecules from 2 "turns" of cycle; takes place in the inner membrane of the mitochondria; requires oxygen.
Electron transport system	Series of oxidation-reduction reactions transferring electrons in hydrogen atoms to accepting cytochromes and oxygen; produces most of the ATP of glucose catabolism (32 molecules); occurs in mitochondria.
Glycogenesis	Conversion of glucose to glycogen by the polymerization of glucose molecules.
Glycogenolysis	Splitting of glycogen to yield free glucose.
Gluconeogenesis	Synthesis of glucose from lactic acid, pyruvic acid, or oxaloacetic acid, or from noncarbohydrates.

chondrion and an inner one that forms several folds called *cristae*. The inner membrane is separated from the outer membrane by a space called the *outer compartment*. The space enclosed by the inner membrane is the *inner compartment*, or *matrix*. The inner membrane contains the electron transport system and enzyme complexes called F_1 *particles* (or ATP synthase) that project into the matrix.

In the course of passing through the electron transport system in the inner membrane, electrons fall to lower energy levels, releasing energy. Hydrogen carriers in the membrane transport hydrogen ions (H^+) (protons) and electrons (e^-) from NADH in the matrix to the inner membrane [FIGURE 26.6B]. (NADH generated by glycolysis is also fed into the electron transport system.) Electron carriers in the inner membrane pick up the electrons and return them to the matrix; the protons (hydrogen ions), however, are released into the outer compartment. As a result, there is a higher concentration of protons in the outer compartment than in the matrix. The energy released by the electrons in the electron transport system is harnessed to pump protons across the inner membrane into the outer compartment. The result is an *electrochemical proton gradient* between the inner and outer compartments. For the protons to move back into the inner compartment and reestablish equilibrium, they must pass through channels in the F_1 particles. Within the F_1 particles, energy associated with the proton gradient powers an enzyme, *ATP synthase*, to catalyze the conversion of ADP and P_i into ATP. The production of ATP in the matrix completes the process of **oxidative phosphorylation.**

Conversion of Glucose into Glycogen (Glycogenesis)

Glucose that is not needed immediately by the body cells is removed from the blood and stored in the liver and skeletal muscle cells as **glycogen.** Glycogen is a macromolecule made up of highly branched chains of glucose molecules. The process of converting glucose into glycogen is called **glycogenesis** ("glycogen production") [FIGURE 26.7A]. In glycogenesis, excess blood glucose is converted into glucose-6-phosphate, which is then converted into glucose-1-phosphate. Glucose-1-phosphate combines with uridine triphosphate (UTP) to form uridine diphosphate glucose (UDP glucose), which transfers glucose to preformed glycogen chains. Glycogenesis is stimulated by insulin from the pancreas, and all the reactions are assisted by specific enzymes. Excess glucose that cannot be stored in the liver or muscle cells is usually converted into saturated fat and stored in adipose tissues. Glycogenesis lowers blood glucose.

Conversion of Glycogen into Glucose (Glycogenolysis)

When body cells need extra glucose for energy, the glycogen stored in the liver is reconverted into glucose and released into the blood. This process is called *glycogenolysis* ("splitting of glycogen") [FIGURE 26.7B]. The breakdown of glycogen occurs in liver cells under the endocrine control of glucagon from the pancreas and epinephrine from the adrenal medulla, which causes the activation of the enzyme phosphorylase to convert glycogen into glucose-1-phosphate. The glucose-1-phosphate is then reconverted by another enzyme into glucose-6-phosphate, which can be utilized directly for energy production [FIGURE 26.2] or hydrolyzed into free glucose to permit its distribution via the bloodstream. Glycogenolysis elevates blood glucose.

Production of Glucose from Noncarbohydrates (Gluconeogenesis)

Gluconeogenesis is the "production of new glucose" from noncarbohydrate sources such as lactic acid (lactate), glycerol, and some amino acids [FIGURE 26.7C]. It occurs mainly in the liver but also occurs in bone cells and the cortex of the kidneys. When the level of blood glucose is low and carbohydrates are not readily available, the glucocorticoid hormone cortisol from the adrenal cortex diverts some amino acids from body cells to the liver, where liver cells convert them into glucose. Each amino acid is converted by a slightly different chemical process. Thyroxine may also divert fats from adipose tissue to the liver, where the glycerol portion of the fat molecules is converted into glucose. Only about 60 percent of the amino acids in body proteins can be converted into glucose.

Another very important starting molecule for gluconeogenesis is the lactic acid that originates in skeletal muscle cells as a normal waste product. Once lactic acid arrives in the liver via the bloodstream, it can be converted into glucose. Gluconeogenesis elevates blood glucose.

A S K   Y O U R S E L F

1 What is cellular respiration?

2 What does glycolysis produce?

3 Which phase of glucose catabolism produces the main supply of ATP?

4 What is the maximum number of ATP molecules that can be obtained from one molecule of glucose?

5 What is the main difference between glycogenesis and glycogenolysis?

6 Under what major condition is gluconeogenesis employed?

PROTEIN METABOLISM

The metabolic pathways for carbohydrates, proteins, and lipids are quite similar, differing only in the initial stages of cellular respiration. After proteins are broken down into amino acids, the amino acids undergo certain molecular changes and then, via acetyl-coenzyme A (acetyl-CoA), enter directly into the citric acid cycle, bypassing glycolysis.

After dietary proteins are converted into amino acids by enzymatic action in the digestive tract, the amino acids enter the bloodstream from the small intestine and are carried to the liver via the portal vein. Amino acids are picked up mainly by the liver and may take several different metabolic routes:

1 They can pass into the blood to be used by organs in the synthesis of tissue proteins.

2 They can be used to renew liver cell proteins or be used by the liver to synthesize plasma proteins.

3 They can be degraded and then converted into glucose or glycogen to produce ATP via the citric acid cycle or be converted into lipids and stored.

4 They can be used in the *glucose-alanine cycle* in the liver to maintain the homeostatic balance of blood glucose between meals, being converted into pyruvic acid and then into glucose.

5 They can be used as building blocks to synthesize the nucleotides of nucleic acids, antibodies, hormones, or other nitrogen-containing compounds.

Some reactions of protein metabolism can be carried out by other cells, but not as efficiently or rapidly as by liver cells.

Protein Catabolism

After proteins have been broken down into amino acids, they must be further modified before they can be used for cellular respiration. One type of change in amino acids is called *oxidative deamination:*

$$R-\underset{\underset{\text{amino acid}}{\underset{|}{NH_2}}}{\overset{|}{C}}H-COOH + H_2O \longrightarrow \underset{\alpha\text{-keto acid}}{R-\overset{\overset{O}{\parallel}}{C}-COOH} + \underset{\text{ammonia}}{NH_3} + 2H$$

In the liver, amino acids can undergo oxidative deamination. If the needed coenzyme is present, the carbon chain is oxidized to a *keto acid* (α-ketoglutaric acid), and the amino group is released as free ammonia (NH_3). Then the α-ketoglutaric acid can be converted into pyruvic acid and enter the citric acid cycle [FIGURE 26.8]. The leftover NH_3 group becomes incorporated into *urea* and is excreted. For example, the amino acid alanine can be converted into pyruvic acid, the amino acid aspartic acid

FIGURE 26.8 PROTEIN METABOLISM

Amino acids from ingested proteins are used for building and repairing body tissues. Amino acids may also be diverted and used for energy. Excess amino acids are always used for energy or converted into fat. Certain amino acids can be converted into glucose, an important mechanism that assures sufficient glucose for the brain.

Process	Description
Protein catabolism	Oxidative deamination of amino acids to form keto acids, which can be used to produce glucose or fatty acids.
Protein anabolism	Directed by DNA in the nucleus and carried out by RNA in the cytoplasm. Synthesis by transamination (enzymatic transfer of an amino group from an amino acid to an α-keto acid).

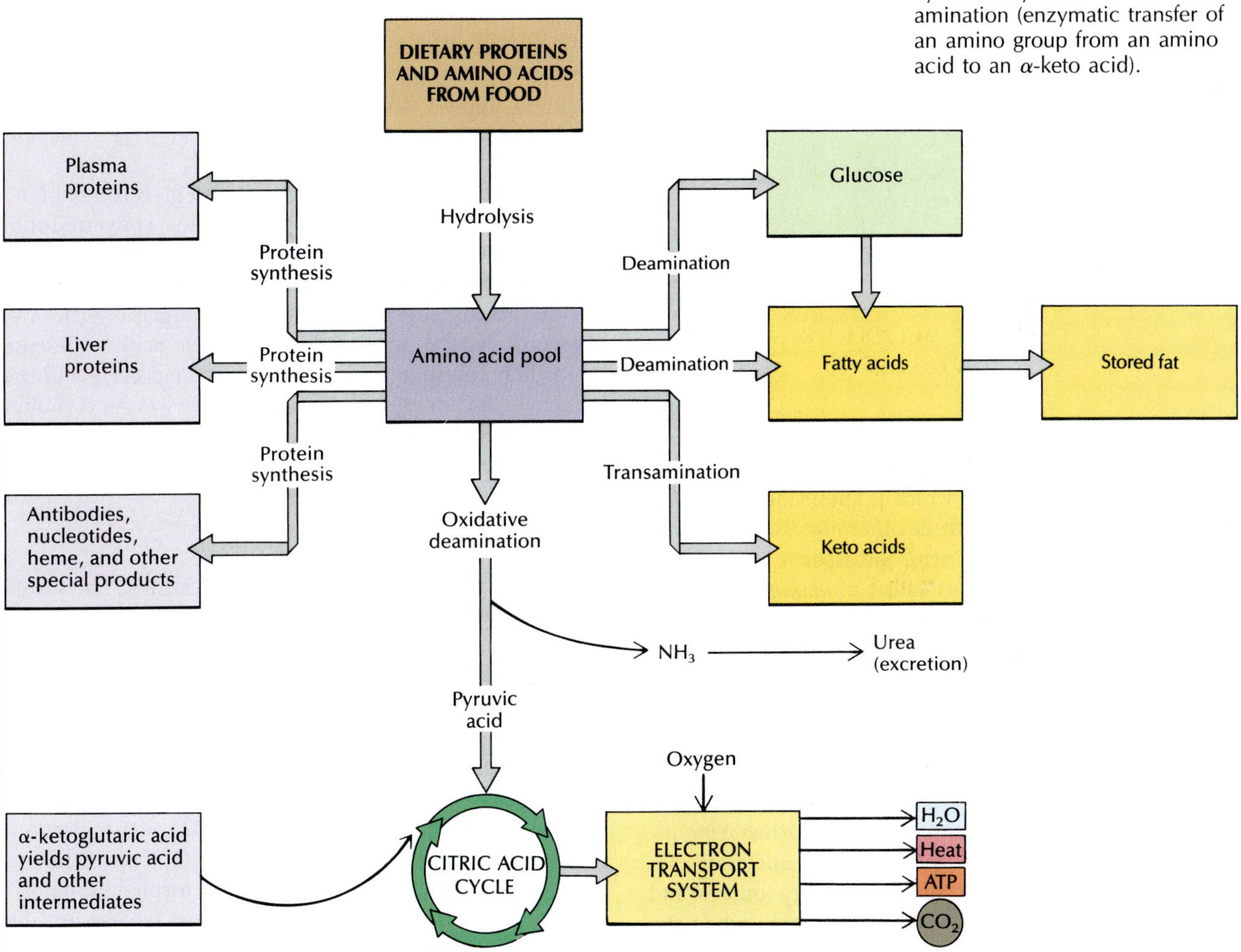

becomes oxaloacetic acid, and glutamine becomes α-ketoglutaric acid. Other, more complex amino acids undergo their own special molecular rearrangements and can also be used in cellular respiration to yield energy in the form of ATP.

Keto acids can be used to produce carbon dioxide and ATP, can be intermediates in the formation of glucose, or can be converted into acetyl-CoA to synthesize fatty acids. Because protein is not stored in cells, such alternative uses of protein are at the expense of its tissue-build-

ing function. When there is a protein deficiency, the prolonged use of protein as a source of glucose, for example, usually causes muscle cells to lose protein mass and the muscles to become smaller (atrophy).

Protein Anabolism

Protein anabolism is carried on by most cells in the body as they synthesize proteins from amino acids. Of the 20 usable amino acids, 11 can be synthesized by cells. These are called *nonessential amino acids.* The other 9, which cannot be synthesized and must be supplied by the diet, are called *essential amino acids.* Nonessential amino acids may also be supplied in part by proteins in the diet. The protein foods that contain all 9 essential amino acids are called *complete proteins.* These include eggs, milk, and meat, which are further classified as *first-class complete proteins* because they contain large quantities of the 9 essential amino acids.

A reaction used to synthesize nonessential amino acids is *transamination,* a process in which the amino group (NH_2) is transferred from an amino acid to a keto acid (a precursor of an amino acid), such as pyruvic acid, producing a different amino acid:

$$
\begin{array}{c}
\overset{\displaystyle H}{\underset{\displaystyle NH_2}{R_1-C-COOH}} + R_2-\overset{\displaystyle O}{C}-COOH \longrightarrow \\[2mm]
\text{amino acid} \qquad\qquad \text{keto acid}
\end{array}
$$

$$
R_1-\overset{\displaystyle O}{C}-COOH + R_2-\overset{\displaystyle H}{\underset{\displaystyle NH_2}{C}}-COOH
$$

new keto acid   new amino acid

Amino groups are usually transferred from the nonessential amino acid glutamine, which is abundant in cells but can also be obtained from three other nonessential amino acids: asparagine, glutamic acid, and aspartic acid. Transamination requires enzymes called *transaminases.*

Protein synthesis is regulated by several hormones: (1) *growth hormone* (GH) increases the rate of synthesis; (2) *insulin* speeds up the transport of amino acids into cells and also increases the available glucose; (3) *glucocorticoids* promote protein catabolism and thereby increase the amount of amino acids in body fluids; they also act on ribosomes to increase translation efficiency, thus improving the rate of protein synthesis; and (4) *thyroxine* increases the rate of protein synthesis when adequate carbohydrates and lipids are available as energy sources, and also degrades proteins to be used for energy when the other nutrients are present in insufficient amounts.

A S K   Y O U R S E L F

1 What is oxidative deamination?

2 What is transamination?

3 What is the difference between an essential and a nonessential amino acid?

4 What are some of the major hormones that regulate protein synthesis?

LIPID METABOLISM

Although we usually use the term *fat* in everyday conversation, *lipid* (Gr. *lipos,* fat) is a more inclusive term for the numerous organic substances that share common solubility properties. The most abundant lipids are the *triglycerides* (also called either *neutral fats* because they are electrochemically uncharged, or *triacylglycerols*), which are composed of three molecules of fatty acids linked to one molecule of glycerol. Triglycerides are synthesized and stored mainly in adipocytes, which are specialized cells in adipose tissue. Other types of lipids include fatty acids; some alcohols; sterols, including cholesterol and ergosterol; hydrocarbons, such as carotenoids; steroid hormones, such as cortisol and aldosterone; and the fat-soluble vitamins (A, D, E, and K). Lipids are generally insoluble in water, but they are soluble in organic solvents such as acetone and alcohol.

Digested fats are broken down into glycerol and fatty acids [FIGURE 26.9]. The glycerol is first converted into an intermediate, dihydroxyacetone phosphate, and then into glyceraldehyde 3-phosphate. Because the latter is on both the glycolytic and the gluconeogenic pathways, glycerol can be converted into pyruvic acid or glucose and then continue through the citric acid cycle and the electron transport system in the same way as carbohydrates. But the fatty acids go directly into the citric acid cycle via acetyl-CoA, bypassing glycolysis.

Lipid Catabolism (Lipolysis)

Adipocytes synthesize and store triglycerides. The stored triglycerides can be hydrolyzed into monoglycerides, fatty acids, and glycerols, which are used in the synthesis of phospholipids and glycolipids, and as fuel molecules. During the catabolism of fatty acids (*beta oxidation*), fatty acids are activated to acetyl-CoA, transported across the inner mitochondrial membrane, and degraded in the mitochondrial matrix by a recurring sequence of four reactions: (1) oxidation linked to FAD, (2) hydration, (3) oxidation linked to NAD^+, and (4) conversion to acetyl-CoA. The $FADH_2$ and NADH formed in the oxidation steps transfer their electrons to oxygen by the electron transport chain, and the acetyl-CoA formed normally enters the citric acid cycle.

After beta oxidation, excess acetyl-CoA not required by the liver is condensed into *acetoacetic acid,* which is then converted into *β-hydroxybutyric acid* and *acetone.* Such substances derived from acetyl-CoA are collectively called *ketone bodies.* The formation of ketone bodies is called *ketogenesis.* Ketone bodies are circulated through the bloodstream to body cells, where they enter the citric acid cycle. During periods when only fats are being metabolized, there may be an excessive accumulation of ketone bodies, a condition called *ketosis* (see page 870).

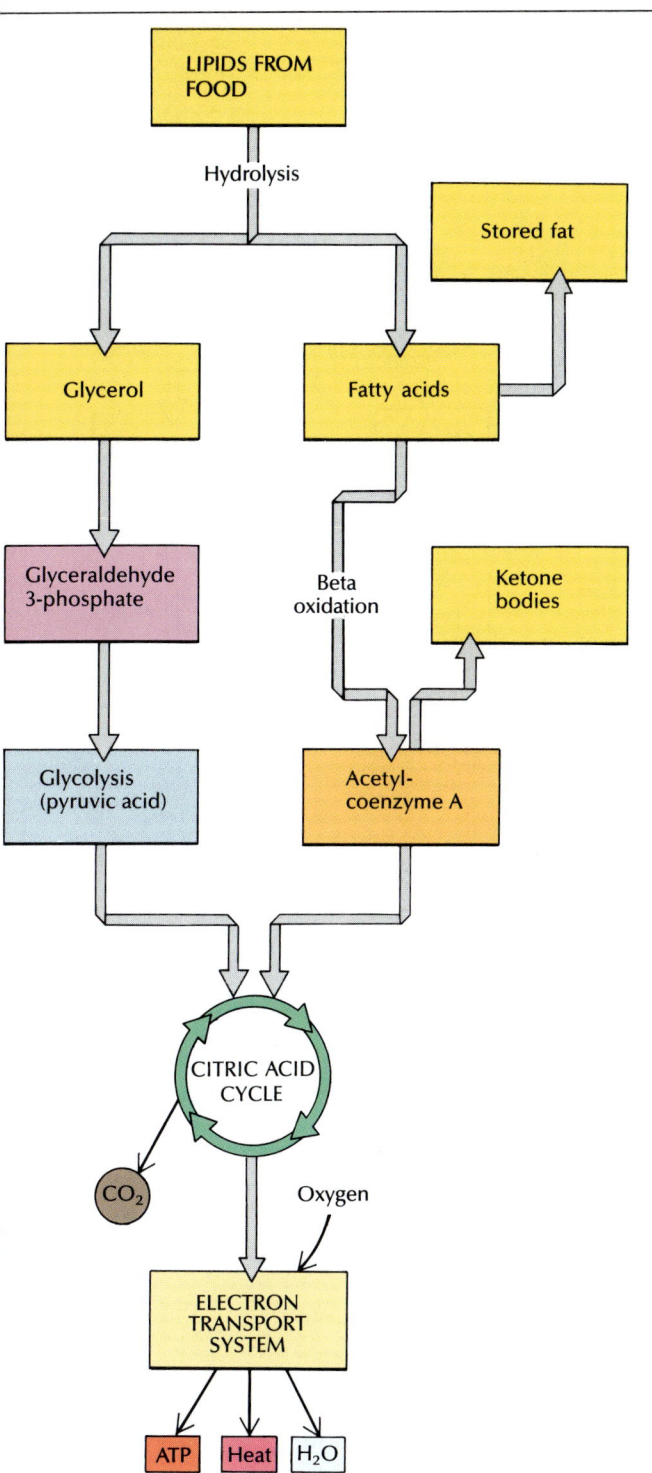

FIGURE 26.9 LIPID METABOLISM

Lipids are hydrolyzed into glycerol and fatty acids, which can be used immediately for energy or stored as fat to be used when energy is required. The pathway with beta oxidation is the most common one.

Process	Description
Lipid catabolism (lipolysis)	Triglycerides hydrolyzed to glycerol, which enters the glycolytic pathway, and fatty acids, which are catabolized by beta oxidation to acetyl-CoA, which enters the citric acid cycle to produce ATP.
Lipid anabolism (lipogenesis)	Synthesis of fats by condensation of acetyl-CoA molecules and reduction to fatty acids and esterification of fatty acids to form triglycerides.

The acetyl-CoA derived from fatty acid catabolism enters the citric acid cycle and is broken down in exactly the same way as the acetyl-CoA from pyruvic acid during the catabolism of glucose. (The catabolism of a typical fatty acid—one containing about 18 carbon atoms—provides the energy for the synthesis of approximately 146 molecules of ATP.) The metabolism of carbohydrates, protein, and lipids is summarized in FIGURE 26.10, and the regulation of metabolism is summarized in TABLE 26.3 on page 880.

Lipid Anabolism (Lipogenesis)

The formation of new fatty acids and their incorporation

FIGURE 26.10 SUMMARY OF METABOLISM

Summary of the metabolism of proteins, carbohydrates, and lipids to produce ATP.
All types of food can be digested to simpler compounds that can be converted
into molecules that can enter the citric acid cycle and be used to produce ATP.

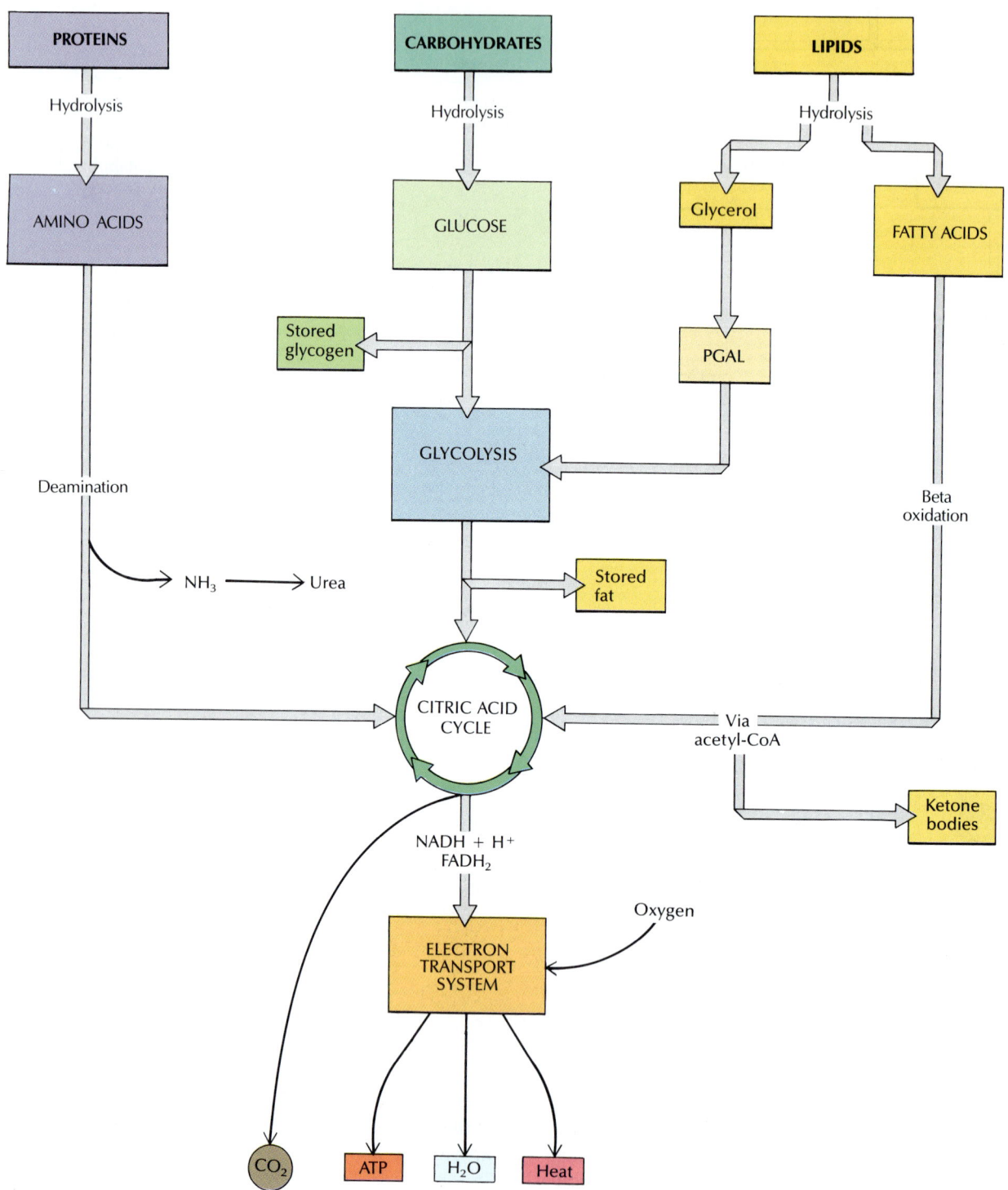

FIGURE 26.11 METABOLIC PATHWAYS OF THE ABSORPTIVE STATE

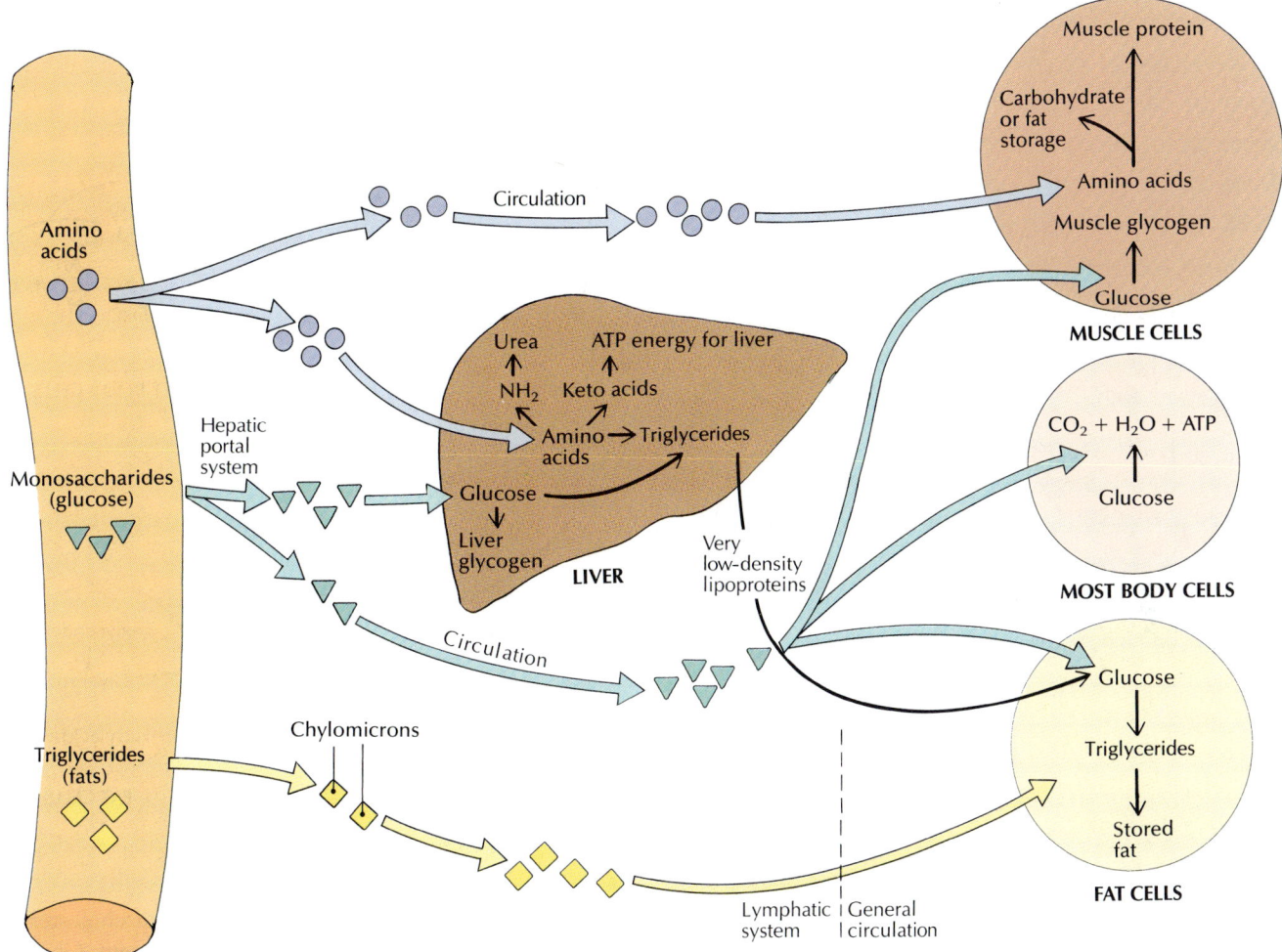

into the fats of adipose tissue utilize enzymes and pathways similar to those found in the liver. Because adipose cells are not supplied as abundantly with cytoplasm or mitochondria as liver cells are, the process of *lipogenesis* in adipose tissue (synthesis of fats) relies primarily on precursor molecules, intermediates, and coenzymes. This is specifically the case in the associated reactions of carbohydrate metabolism leading to the generation of acetyl-CoA, ATP, NADH, and α-glycerophosphate. This is why it is not so much the amount of fat as the amount of carbohydrates in the diet that determines the amount of lipids deposited in the adipocytes.

A S K   Y O U R S E L F

1 What are triglycerides?

2 What are the beginning and end products of beta oxidation? Of fatty acid synthesis?

ABSORPTIVE AND POSTABSORPTIVE STATES

Although we may eat several times a day, with periods of fasting in between, our cells require a continuous supply of ATP. The mechanisms that cope with these alternating periods of feeding and fasting are controlled by hormones, especially insulin and glucagon from the pancreas. The feeding and fasting states are referred to as absorptive and postabsorptive states, as described below.

Absorptive State

The nutrients from food are broken down in the digestive tract. During the ***absorptive state,*** the nutrients are absorbed from the digestive tract into the circulatory and lymphatic systems [FIGURE 26.11]. Following a meal, monosaccharides and amino acids enter the hepatic portal vein and are carried to the liver. But before fats can be

FIGURE 26.12 METABOLIC PATHWAYS OF THE POSTABSORPTIVE STATE

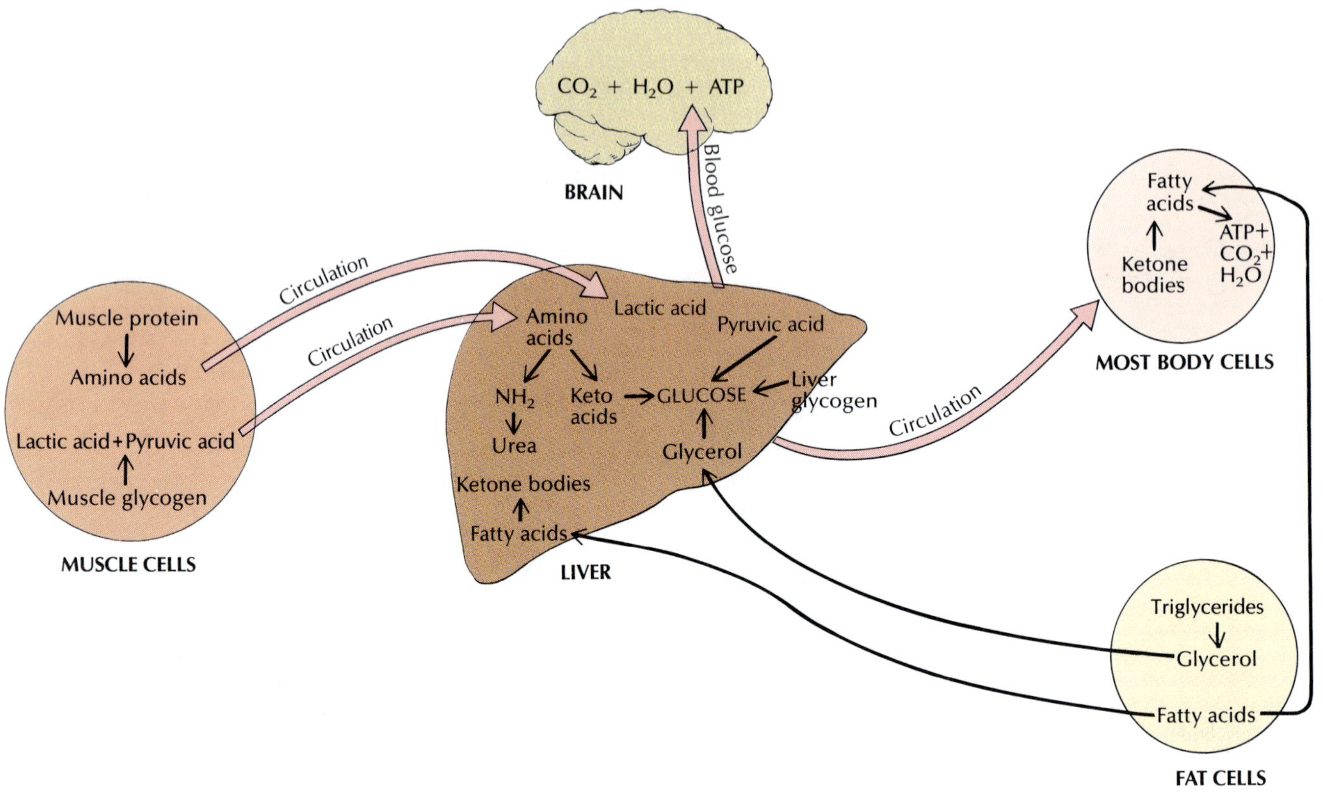

absorbed into the general circulation, they are first absorbed into the lymphatic system.

Monosaccharides After carbohydrates* are absorbed from the digestive tract, they enter liver cells, where they can be converted into glycogen and stored, converted into fat, or used for other purposes. Some fat is stored in the liver, but most is transported by the blood in particles containing protein and other components, called *very low density lipoproteins (VLDL)*. Some of the fat is then transferred to adipose cells. Glucose that does not enter the liver is used almost immediately by most body cells as their main energy source during the absorptive state. Part of the remaining glucose is stored as glycogen in skeletal muscle tissue.

Triglycerides Before ingested lipids are absorbed into the bloodstream, they enter the lymphatic system as *chylomicrons*, which are small lipid droplets composed mainly of fatty acids with small amounts of cholesterol and phospholipids and a thin coating of protein. Chylomicrons are synthesized in intestinal mucosal cells from the products of lipid digestion and are released into lymphatic vessels. When fatty acids are finally absorbed, they are probably taken up by all cells, since they are needed to make membrane phospholipids. During the absorptive state, cells may synthesize triglycerides and other fatty acids from glucose or receive them directly from the VLDL manufactured in the liver.

Amino acids Many of the absorbed amino acids enter the liver cells directly, where they are converted into keto acids by the removal of their NH_2 group. This process can also produce urea, which is excreted in urine. The keto acids enter the citric acid cycle and supply most of the liver's energy during the absorptive state. They may also be converted into triglycerides, which are stored mostly in adipose cells. The amino acids that are not taken up by the liver cells enter other cells to replace the protein that is being broken down continuously. Excess amino acids are stored as carbohydrate or fat.

*Absorbed carbohydrates consist of monosaccharides such as galactose, fructose, and glucose. For the sake of simplicity, we will consider only glucose in this discussion.

Postabsorptive State

In the *postabsorptive* or *fasting state,* the energy requirements of body cells must be satisfied by the nutrients taken in during the absorptive state and by the body's stored nutrients [FIGURE 26.12]. Most importantly, the plasma glucose concentration must be maintained at a homeostatic level of 70 to 110 mg/100 mL. This point is crucial because the brain, kidney medulla, and erythrocytes use glucose as their *only* energy source, but glucose is stored as glycogen in only limited amounts. The body solves this problem in three ways: (1) It breaks down glycogen stored in the liver (glycogenolysis). (2) It forms new glucose from protein and lipids (gluconeogenesis). (3) It shifts the metabolism of most cells so that they produce ATP from fatty acid intermediates rather than from glucose, thus conserving glucose. In fact, most of the body's energy supply during the postabsorptive state comes from the oxidation of stored fat.

The postabsorptive state in human beings usually occurs about 4 hours after the last meal. The complete absorption of a heavy meal takes about 4 hours, so if one eats three meals a day, the absorptive and postabsorptive states each occupy about 12 hours of the day.

Sources of blood glucose The glycogen in the liver is broken down directly to supply glucose to the blood. Triglycerides are broken down into glycerol and fatty acids. The glycerol can be converted into glucose by the liver, but the fatty acids cannot, and they are used directly by other body cells as an energy source.

During long periods of fasting, the body uses protein from skeletal muscles, the liver, the skin, and other parts of the body as an additional source of glucose. Because much of this protein is not essential for the function of muscles and other contributing tissues, it can be broken down into amino acids and converted into glucose. Ordinarily, however, very little structural protein is used for energy production.

Glucose sparing and fat usage With the exception of nerve cells, most body cells shift their metabolism during periods of fasting to use fat instead of glucose, sparing glucose for the nervous system. The triglycerides of adipose tissue are hydrolyzed into glycerol and fatty acids, absorbed into the blood, and taken up by tissue cells to produce ATP during the citric acid cycle. Fatty acids in the liver, however, are converted into ketone bodies, such as acetone, instead of entering the citric acid cycle. The ketone bodies are then released into the blood to be used by tissue cells during the citric acid cycle to produce carbon dioxide, water, and ATP.

As a result of glucose sparing and fat usage, the concentration of blood glucose is reduced by only a few milligrams, and homeostasis is maintained.

ASK YOURSELF

1 What nutrients are the body's major energy sources during the absorptive state?

2 In what form do lipids enter the lymphatic system?

3 Why is it important for the body's glucose level to remain fairly constant?

4 At what time of the day do the postabsorptive states usually occur?

5 How is the brain's glucose supply kept constant?

NUTRITIONAL NEEDS AND BALANCE

The foods we eat contain six classes of *essential nutrients* that are vital to our well-being: carbohydrates, proteins, lipids (fats), vitamins, minerals, and water. Each type of food provides something different, and a well-balanced diet is necessary to ensure that we receive all types of nutrients in sufficient amounts to satisfy our various needs and maintain our health.

Calories and Energy

The energy value of food is measured in terms of calories. A *calorie* (L. *calor,* heat) is the amount of energy required to raise the temperature of 1 g of water 1°C. A calorie, with a small *c,* is also called a *gram calorie.* A *kilocalorie,* also known as a *Calorie* or *kilogram calorie (kcal),* is equal to 1000 calories (*kilo* = thousand). In popular usage, we talk about calories but actually mean Calories (kilocalories), because the larger unit is more useful in measuring the energy values of food. (If an advertisement says that a so-called light beer contains 95 calories per 12 oz, it really means 95,000 calories, or 95 kcal.)

Dietary Carbohydrates

With the exception of lactose, which comes from milk, and glycogen, which is synthesized by the liver, all carbohydrates come from plant sources. Certain fruits supply glucose, fructose, sucrose, and starch.* Vegetables provide starch and some sugars. Syrups and honey supply glucose, fructose, and maltose, and ordinary table sugar is sucrose. Grain products and legumes are good sources of starch. Many plants also supply cellulose, an indigestible polysaccharide. Cellulose is sometimes called dietary

---

*Starch and carbohydrates are often confused. Starch is *one* example of a carbohydrate and is, in fact, the most common carbohydrate in your diet. Some high-starch foods are rice, bread, potatoes, and pasta.

fiber. It assists in the passage of foods through the large intestine and may reduce the risk of cancer of the colon.

Although fats are the most concentrated source of energy in foods, carbohydrates provide the most readily available energy source. Special functions of stored liver glycogen include providing an important source of energy, maintaining the proper blood glucose concentration, and serving as a detoxifying and protective agent. Glucose is the major source of energy for the brain and nervous tissue. Galactosides, which are composed of galactose, a fatty acid, and a nitrogen base, are essential components of brain and nervous tissue. Lactose aids in the absorption of calcium and promotes the growth in the large intestine of desirable bacteria that are involved with some of the B-complex vitamins.

Dietary Proteins

The animal sources of protein include lean meat, poultry, fish and other seafoods, eggs, and milk and milk products, including yogurt and cheese. The plant sources include dried beans and peas, nuts and peanut butter, cereals, bread, and pasta.

Protein is the most abundant of the body's organic compounds. Thousands of different types of proteins exist in cells, with each cell carrying out a specific function regulated by its genes. Proteins form an important part of the blood. They are also important regulators of osmotic pressure and water balance within the body and help maintain the body's pH by acting as chemical buffers.

Amino acids from dietary proteins are used for building and repairing body tissues. Some amino acids are diverted and used for energy in the process of gluconeogenesis. Proteins and carbohydrates each produce about the same amount of energy, yielding about 4 kcal per gram of food metabolized. Excess protein is always used for energy or converted into saturated fat.

Although proteins are a vital part of our diet, we actually need to eat relatively little — about 0.8 g per kilogram of weight for adults. This amount of protein supplies the essential amino acids, plus sufficient nitrogen, to make the nonessential amino acids. The need for protein varies with age, body weight, and amount of physical activity. Ordinarily, men need more protein than women do because men are heavier. Young children need proportionately more protein than adults do because they use a great deal of their protein for the rapid growth of body tissue, while adults use protein mainly to replace the nitrogen lost as body protein is turned over. Pregnant women and nursing mothers require extra protein to provide for the growth of the fetus and the synthesis of milk protein.

Some dietary proteins provide more amino acids than others do because they have an amino acid composition similar to that of body protein. For this reason, less meat, milk, and cheese are required than plant protein such as found in wheat. Dietary protein is most beneficial when it is balanced with many types of food to provide all 20 necessary amino acids.

Dietary Lipids

Neutral fats or triglycerides are contained in cooking fats and oils, butter, margarine, salad dressings and oils, fat in meat and dairy products, nuts, chocolate, and avocados. Fat is visible in foods such as butter, cream, salad dressings, and meats. It is less obvious in foods such as egg yolk, cheese, nuts, and wheat germ.

Lipids are the body's most concentrated source of food energy. They produce about 9 kcal of usable energy per gram, more than twice as much energy as is available from an equal weight of carbohydrate or protein. However, lipids are utilized completely only if they are oxidized along with sugar. Without sufficient sugar, the body is forced to burn stored fat for energy. When only lipids are being metabolized, however, there is an excessive accumulation of the breakdown products (ketone bodies) — acetoacetic acid, β-hydroxybutyric acid, and acetone — producing a condition called **ketosis.** The sweet smell of the acetone that is formed can frequently be detected on the breath when this condition exists. Ketosis occurs frequently if the diet is low in carbohydrates and high in lipids, in starvation, and in uncontrolled diabetes mellitus, when the body switches to the metabolism of fats because glucose is not being metabolized properly.

Some ketone bodies are strong acids and so require large amounts of the body's alkaline supply to buffer them. If these ketone bodies continue to build up in the blood and interstitial fluids, they can easily use up available buffers and cause a drop in the blood pH. In this way, ketosis can lead to **acidosis,** one of the body's most serious acid-base imbalances, which through depression of the nervous system can lead to serious brain damage, coma, and death.

Adipose tissue is a compact storage site consisting chiefly of triglyceride reserves. Large quantities of fat can be stored in the tissues, producing obesity. But not only lipids produce fatty tissue. Excess glucose can also be converted into fat. Obesity results whenever the number of calories in the diet exceeds the body's needs, regardless of whether the extra calories come from fats, carbohydrates, proteins, or alcohol.

Lipids have several functions. They supply a concentrated amount of usable energy in the form of calories, insulate the body and help maintain a constant body temperature, and cushion some internal organs. Some lipids also supply or aid the absorption of the fat-soluble vitamins (A, D, E, K). Lipids add flavor to foods and help

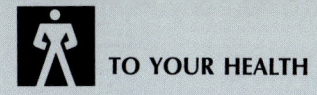

TO YOUR HEALTH

Diet and Cancer

The federal government, the American Heart Association, the American Diabetes Association, and the National Cancer Institute all have issued dietary guidelines in recent years. The guidelines indicate a concern that certain foods may increase the incidence of cancer, especially in the breast, prostate gland, and large intestine and rectum. The dietary recommendations consistently include high fiber, high complex carbohydrates, low fat, low cholesterol, and low calories.

Studies have shown that the incidence of cancer varies according to dietary habits. For example, the incidence of stomach cancer is high in Japan, where salt-cured and smoked foods may produce elevated levels of such cancer-causing substances as hydrocarbons and nitrosamines. However, Japanese people who move to Hawaii often show a statistical increase in breast and colon cancer and a decrease in stomach cancer, apparently because the typical Hawaiian diet is high in fat and low in smoked foods. In Finland, where the diet contains twice as much fiber as the typical American diet, the incidence of colon cancer is one-third the rate in the United States.

Some dietary recommendations for decreasing the risk of cancer include the following. *Eat a low-fat diet,* which reduces the risk of cancer of the large intestine, breast, and prostate gland. *Eat a high-fiber diet,* which speeds the passage of feces through the large intestine and apparently reduces the risk of cancer of the colon. (Also, high-fiber fruits and vegetables contain large amounts of vitamin C and carotene, a precursor of vitamin A, which are believed to be natural inhibitors of the development of cancer.) *Eat fewer salt-cured and smoked foods,* including bacon, smoked fish, and ham. *Drink alcohol only in moderation,* since heavy drinking increases the risk of cancer of the mouth, pharynx, larynx, esophagus, liver, and urinary bladder.

---

promote and maintain healthy skin, normal growth, and reproductive ability. Cholesterol (a precursor of vitamin D) and phospholipids help build cell membranes and intracellular structures. Essential fatty acids are precursors of prostaglandins.

Triglycerides contain three types of fatty acids: saturated (such as palmitic and stearic acids), monounsaturated (such as oleic acid), and polyunsaturated (such as linoleic acid). Saturated fatty acids are solid at room temperature, while unsaturated fatty acids are liquid at room temperature and are generally termed "oils."

The American Heart Association suggests that no more than 30 to 35 percent of our calories should come from fat. Most physicians and dietitians recommend that we consume more polyunsaturated fat than saturated fat and that we restrict our cholesterol intake to 250 mg per day. A low-fat diet may reduce the risk of cancer and atherosclerosis, decrease fluid retention in the body, and promote weight loss.

Vitamins

*Vitamins** are organic compounds required by the body in small amounts for the regulation of metabolism. Many vitamins are converted into coenzymes after they enter the body. They play such an important role in supporting enzymes, especially those involved in providing energy for metabolism, that vitamin deficiencies can cause many different disorders [TABLE 26.1].

Because most vitamins cannot be synthesized by the body, they must be obtained from the diet. However, vitamin K is synthesized by bacteria in the large intestine, and some other vitamins can be synthesized in the body if certain building materials, called *provitamins,* are available in the diet. For example, carotene is a provitamin in the production of vitamin A. Its chemical structure is similar to that of vitamin A, making its conversion into vitamin A by the liver relatively easy.

Vitamins A, D, E, and K are usually classified as **fat-soluble,** and vitamin C and the B vitamins are **water-soluble.** A year's supply of vitamin A can be stored in the liver, and the other fat-soluble vitamins can be stored also,[†] but water-soluble vitamins must be replenished constantly. In fact, water-soluble vitamins are often lost during cooking because they dissolve in the cooking water or are degraded by heat. Water-soluble vitamins are easily absorbed, but the fat-soluble vitamins must be emulsified by bile salts in the gastrointestinal tract before they can be absorbed into the bloodstream. Once vitamins are absorbed, they are usually combined with more complex molecules, frequently in the liver.

---

*The term *vitamine* was suggested in 1912 to describe the life-promoting organic compounds we now call *vitamins.* The original term was a combination of *vita,* meaning life, and *amine,* because all vitamins were thought to be amines. When it was discovered that not all vitamins are amines, the final *e* was discarded.

†The fat-soluble vitamins A, D, and K may accumulate in the body in toxic amounts if they are taken in large, regular doses (*hypervitaminosis*). There is no firm evidence that vitamin E accumulates in toxic levels.

TABLE 26.1 VITAMINS

Vitamin	Major sources	Major functions	Effects of deficiency	Effects of excess
WATER-SOLUBLE VITAMINS				
B_1 (thiamine)*	Organ meats, whole grains, yeast, nuts, molasses, yogurt, oysters	Formation of cocarboxylase needed to convert pyruvic acid into acetyl-CoA in citric acid cycle; helps regulate carbohydrate metabolism; aids production of HCl	Impaired carbohydrate metabolism, beriberi, peripheral nerve changes, edema, heart failure, mental disturbance, paralysis, constipation, anorexia	None reported
B_2 (riboflavin)	Milk, eggs, beef and veal, liver, whole grains and cereals, spinach, nuts, yeast, molasses, beets	Forms two flavin nucleotide enzymes (FAD, FMN) involved in oxidative phosphorylation; aids cellular respiration; releases energy to cells	Sensitivity to light, eye lesions, fissuring of skin, cataracts, vomiting, diarrhea, muscular spasticity	None reported
B_6 (pyridoxine)	Whole grains, liver, milk, molasses, leafy green vegetables, bananas, yeast, tomatoes, corn, yogurt	Coenzyme (pyridoxal phosphate) for amino acid metabolism; transport of amino acids across plasma membranes	Dermatitis, nervous disorders, learning disabilities, anemia, kidney stones, fatty liver	Sensory neuropathy
B_{12} (cyanocobalamin)	Organ meats, milk, cheese, eggs, yeast, fish, oysters; synthesis by intestinal bacteria	Coenzyme needed for RNA synthesis and for erythrocyte formation to prevent anemia	Pernicious anemia, nervous disorders (demyelination), malformed red blood cells, general weakness	None reported
Biotin	Liver, yeast, vegetables, eggs; synthesis by intestinal bacteria	Coenzyme concerned with nucleic acid synthesis, CO_2 fixation, transamination, nitrogen metabolism; aids cell growth, fatty acid production	Scaly dermatitis, muscle pains, weakness, insomnia, depression, fatigue	None reported
Folic acid (folacin, pteroylglutamic acid)	Organ meats, leafy green vegetables, eggs, milk, salmon; synthesis by intestinal bacteria	Needed for transfer of carbon units in nucleic acid metabolism and DNA formation, red blood cell formation; aids growth, reproduction, digestion	Failure of red blood cells to mature, anemia, gastrointestinal disturbances, diarrhea	None reported
Inositol	Fruits, nuts, vegetables, milk, yeast	Aids in metabolism; reduces cholesterol concentration; slows hardening of arteries	Fatty liver, constipation, hair loss, eczema	None reported

*The B group of vitamins includes B_1 through pantothenic acid as listed here.

†The general term *vitamin D* actually refers to a group of steroid vitamins, including vitamin D_3 (cholecalciferol), vitamin D_2 (calciferol or ergocalciferol), and AT_{10} (dihydrotachysterol). Vitamin D_2 and AT_{10} are derived from ergosterol, a steroid found in plants. Although cholecalciferol (vitamin D_3) is the natural animal form of the vitamin, the *active* form of vitamin D is considered to be 1,25-dihydroxycholecalciferol, which is a modified form of cholecalciferol. The active form is 13 times more powerful than cholecalciferol.

Vitamin	Major sources	Major functions	Effects of deficiency	Effects of excess
Niacin (B_3, nicotinic acid)	Whole grains, liver, chicken, yeast, seafood, peanuts, dried beans and legumes	Constituent of NAD^+, NADP; aids carbohydrate metabolism; aids production of sex hormones; reduces cholesterol concentration	Pellagra (skin and gastrointestinal lesions, mental disorders, digestive disturbances, mucous membrane inflammation, hemorrhage)	Flushing, burning, tingling around neck, face, hands
Pantothenic acid (B_5)	Yeast, liver, eggs, nuts, salmon, green vegetables, cereals; synthesis by intestinal bacteria	Forms part of CoA; promotes proper growth, vitamin utilization, energy utilization	Fatigue, sleep disorders, neuromotor disorders, cardiovascular disorders, gastrointestinal distress, eczema	None reported
C (ascorbic acid)	Citrus fruits, butter, tomatoes, green peppers, broccoli, potatoes	Necessary for oxidation reactions, synthesis and maintenance of collagen; aids bone and tooth formation, healing	Scurvy, failure to form normal connective tissue fibers, anemia, low resistance to infection, bruises	Kidney stones
Choline	Leafy vegetables	Part of phospholipids, precursor of acetylcholine	Not known; deficiency is unlikely	None reported

FAT-SOLUBLE VITAMINS				
A (carotene)	Not present in plant foods; provitamin for formation of vitamin A found in egg yolk, carrots, leafy vegetables, fruits, liver, oils, milk products	Needed for synthesis of mucopolysaccharides, growth differential of epithelium, maintenance of normal epithelial structure; formation of visual pigments (major constituent of rhodopsin)	Night blindness, xerophthalmia (keratinization of corneal tissue), skin lesions, allergies, dry hair, fatigue, failure of skeletal growth, reproductive disorders	Vomiting, headache, anorexia, irritability, skin peeling, swelling of long bones, enlarged spleen and liver, patchy hair loss, bone pain and thickening
D†	Fish oils, liver, milk, egg yolk; synthesis of vitamin D_3 (cholecalciferol) results from ultraviolet radiation (sunlight) of 7-dehydrocholesterol in skin	Increases calcium and phosphorus absorption from digestive tract; helps control calcium deposition in bones and teeth; promotes proper heart action	Rickets in children, osteomalacia in adults, diarrhea, insomnia, nervousness	Vomiting, diarrhea, weight loss, kidney damage, calcification of soft tissues
E (tocopherols)	Leafy green vegetables, wheat germ oil, liver, peanuts, eggs	Helps red blood cells resist hemolysis; aids in muscle and nerve maintenance; acts as antioxidant to prevent cell membrane damage of unsaturated fats in organelles	Increased fragility of red blood cells, dry hair, hemolytic anemia in newborns	None reported
K (phylloquinone)	Yogurt, molasses, safflower oil, liver, leafy green vegetables; synthesis by intestinal bacteria	Aids in prothrombin synthesis (clotting factors VII, IX, X) in liver	Failure of blood to clot; severe bleeding, hemorrhages	May cause jaundice, anemia, gastrointestinal problems

Minerals

Minerals are naturally occurring inorganic elements, such as sodium, calcium, and potassium, that are used for the regulation of metabolism [TABLE 26.2]. Minerals make up about 4 to 5 percent of the adult body weight.

Some minerals, such as iodine, copper, zinc, cobalt, iron, and manganese, are needed in the body in such small amounts that they are called *microminerals* or *trace elements*. Although these microminerals are available in the body and in food in only minute amounts, their presence is essential for proper metabolism. Microminerals function primarily as catalysts in enzyme systems. In contrast, calcium, phosphorus, sodium, chlorine, potassium, magnesium, and sulfur are known as *macrominerals* because they are found in significant amounts in body tissue. (Calcium is the most abundant mineral in the body.) At least 100 mg of each macromineral is required in the daily diet. Minerals are frequently bonded chemically with an organic substance, such as iron in hemoglobin and iodine in thyroxine.

Water as a Nutrient

No nutrient is more important than water. About 60 percent of the body is water, so it is a major component. Water participates in many essential biochemical reactions, such as cellular respiration and the hydrolysis of carbohydrates, lipids, and proteins. It transports important chemicals throughout the body. It acts as a solvent and a lubricant. The important role of water in temperature regulation is discussed later in this chapter, and its regulatory role in excretion is covered in Chapters 27 and 28.

Federal Dietary Guidelines

In 1990, the United States government issued a revised set of dietary guidelines, which were published jointly by the Department of Agriculture and the Department of Health and Human Services. The guidelines specify the following:

1 Eat a variety of foods.

2 Maintain a healthy weight.

3 Choose a diet low in fat, saturated fat, and cholesterol.

4 Choose a diet with adequate starch and fiber.

5 Use sugars only in moderation.

6 Use salt and sodium only in moderation.

7 If you drink alcoholic beverages, do so in moderation.

The Federal Dietary Guidelines recommend that carbohydrates be 50 to 60 percent of our total diet, with complex carbohydrates and naturally occurring sugars accounting for 45 to 50 percent, and refined and processed sugars dropping to only 10 percent. Fats should account for 30 percent *or less* of all calories in the diet.

METABOLIC RATE AND TEMPERATURE CONTROL

As you have seen, the energy stored in the chemical bonds of food molecules such as glucose is eventually converted through a series of chemical reactions into energy-producing molecules such as ATP. The rest of the energy is dissipated as heat. Rather than being wasted, however, much of this by-product of heat helps keep our bodies at a constant temperature.

Factors Affecting the Metabolic Rate

One way of measuring energy expenditure in the body as a whole is to determine the *metabolic rate,* which is the amount of energy used by the body during a specific time period. It is measured in terms of the calories used. Among the factors that affect the metabolic rate are type of food eaten, amount of exercise, hormones, and age.

Food The energy used in digesting a meal is known as the *specific dynamic action* of food. Eating a protein-rich meal generally raises the metabolic rate more than eating a meal high in lipids or carbohydrates. The greater increase in metabolic rate after eating protein is probably caused by the relatively rapid deamination of amino acids, among other factors.

Exercise Equally important is the amount of physical activity a person engages in, compared with the energy value of the food consumed, as measured in calories. The number of calories said to be contained in any food is actually the number of calories that the body is able to metabolize and release. If the food you eat contains more calories than you are able to use up by exercising, you will gain weight. If you use up more calories than are provided in your diet, your body will begin to consume stored body fat, and you will probably lose weight.

Hormones Several hormones are involved in a feedback system that helps maintain a steady metabolic rate. For example, when the metabolic rate is too low, the secretion of epinephrine and thyroxine increases. Thyroxine, secreted by the thyroid gland, affects metabolism more than any other hormone does. When the output of thyroid hormone increases, the metabolic rate may be raised as much as 100 percent above normal, and a decreased secretion may cause the metabolic rate to be lowered by as much as 40 percent below normal. The metabolic rate is raised by anxiety and lowered by depression. In situations of fear or anger, for example, the adrenal glands are stimulated to increase epinephrine output, which increases the metabolic rate. The metabolic rate is also increased by growth hormone and sex hormones.

Age and other factors One very important determinant of the metabolic rate is age. (Metabolic rate slows down about 3 percent every 10 years after the age of 20.) The metabolic rate of a child is generally higher than that of an adult because the child requires more energy during its complex growth period. A pregnant or nursing woman also has an increased metabolic rate.

Other factors also affect the metabolic rate. Fever increases the rate. Muscular people tend to have a higher rate than people with a lot of fatty tissue. Race and climate, which were once thought to affect the metabolic rate, are no longer considered important factors.

Basal Metabolic Rate

The *metabolic rate* is the total energy used by the body per unit of time to sustain the minimal normal cellular activity. Because the metabolic rate can be altered by many factors, a person's metabolic rate is usually measured under standard, or *basal*, conditions. The metabolic rate computed under such conditions is called the **basal metabolic rate (BMR)**. It is usually expressed as kilocalories of energy consumed per square meter of body surface area* per hour; BMR = $kcal/m^2/hr$.

*The body surface area may be calculated based on the height and weight of the individual being tested, as follows:

Body surface area = $weight^{0.425} \times height^{0.725} \times 0.007184$

Charts are available to estimate body surface area.

There are several ways to calculate the BMR. The usual technique is to use an instrument called a *respirometer* or *metabolator* to determine the rate of oxygen consumed per unit of time. This method is based on the concept that because 95 to 100 percent of the energy available to the body is derived from cellular reactions involving oxygen, a person's BMR can be estimated accurately from oxygen consumption. It has been determined that for every liter of oxygen consumed, the amount of heat energy released is 4.825 kcal.

The person being tested should have no food for at least 12 hours after a light dinner the night before the test. A good night's sleep, complete rest on the morning of the test, and a moderate environmental temperature (20 to 26.7°C) are important. The person must be in a supine position for at least 20 minutes prior to the test and should be relieved of all mental and physical stimuli that might cause excitement.

A sample calculation of the BMR for a 20-year-old male may proceed as follows. The respirometer shows that 10 L of oxygen is consumed in 1 hour. When 10 is multiplied by the standard figure of 4.825 kcal liberated in 1 hour, the result is 48.25. This figure is then divided by the total body surface area, which has been calculated to be 1.5 m^2, producing a BMR of 32.2. According to standardized charts, the normal value for a 20-year-old male is 25.7, so the BMR of 32.2 is 25.3 percent above normal, and is therefore written as +25.3. If the BMR had been the same percentage below normal, it would have been written as −25.3. Since a BMR of plus or minus 15 percent is judged to be normal, the 20-year-old male being tested is considered to have an abnormally high BMR.

Temperature-Regulating Mechanisms

If the amount of heat usually generated by the resting body of a healthy individual were allowed to accumulate, the temperature of the body would rise about 30°C a day. Such a temperature increase would normally take place in a matter of hours, and during periods of vigorous exercise it would occur in a matter of minutes. Vigorous exercise can increase heat production 20 times above normal in a very short time. Naturally, if we are to maintain homeostasis, such deviations in body temperature must be prevented.

In fact, deviations in body temperature do not usually occur because several control mechanisms are available

---

Q: *Why do you wake up cold when you fall asleep uncovered?*

A: When you are asleep, your body is not as active as when you are awake. As a result, overall metabolism slows down and less body heat is generated. There is also an increase in heat loss because of the dilation of blood vessels in the skin.

TABLE 26.2 MAJOR MINERALS

Mineral	Major sources	Major functions	Effects of deficiency	Effects of excess
MACROMINERALS				
Calcium	Dairy products, eggs, fish, soybeans	Necessary for proper bone structure, normal heart action, blood clotting, muscle contraction, excitability, nerve synapses, mental activity, buffer systems, glycogen metabolism	Tetany of muscles, loss of bone minerals, rickets in children, osteoporosis in adults	Calcium deposits, calcification of soft tissues, heart failure
Chloride	All foods, table salt	Principal anion of extracellular fluid; necessary for acid-base balance, osmotic equilibria	Alkalosis, muscle cramps	Acidosis, edema
Magnesium	Green vegetables, milk, meats, nuts	Necessary for proper bone structure, regulation of nerve and muscle action; catalyst for intracellular enzymatic reactions, especially those related to carbohydrate metabolism	Tetany	None reported
Phosphorus	Dairy products, meats, fish, poultry, beans, grains, eggs	Combines with coenzymes in various metabolic processes; necessary for proper bone structure, intermediary metabolism, buffers, membranes; phosphate bonds essential for energy production (ATP), nucleic acids	Extremely rare; related to rickets, loss of bone mineral	None reported
Potassium	All foods, especially meats, vegetables, milk	Major component of intracellular fluid; necessary for buffering, muscle contraction, nerve impulse transmission	Changes in heart function, alteration in muscle contraction, muscle weakness, alkalosis	Heart block
Sodium	Most foods, table salt	Major component of extracellular fluid; necessary for ionic equilibrium, osmotic gradients, nerve impulse conduction, buffer systems	Dehydration, muscle cramps, kidney failure	Edema, hypertension
Sulfur	All protein-containing foods	Structural, as amino acids are made into proteins	None reported	None reported

Mineral	Major sources	Major functions	Effects of deficiency	Effects of excess
MICROMINERALS (TRACE ELEMENTS)				
Chromium	Meats, vegetables, yeast, beer, unrefined wheat flour, corn oil, shellfish	Necessary for glucose metabolism, formation of insulin for proper blood glucose concentration	Impaired ability to metabolize glucose	Industrial exposure may cause skin and kidney damage
Cobalt	Meats	Necessary for formation of red blood cells	May cause pernicious anemia	Industrial exposure may cause dermatitis, diseases of erythrocytes
Copper	Liver, meats, oysters, margarine, eggs, wheat products	Necessary for hemoglobin formation, maintenance of certain copper-containing enzymes, proper intestinal absorption of iron	Anemia, bone disease (rare), lack of white blood cells	Wilson's disease (rare metabolic condition)
Fluorine	Fluoridated water, toothpastes, milk, tea	Hardens bones and teeth; suppresses bacterial action in mouth	Tendency toward dental caries, osteoporosis	Mottling of teeth
Iodine	Iodized table salt, fish, seaweed	Necessary for synthesis of thyroxine, which is essential for maintenance of normal cellular respiration	Goiter, cretinism in newborns	Inhibits activity of thyroid
Iron	Liver, eggs, red meat, shellfish, beans, nuts, raisins	Component of hemoglobin, myoglobin; necessary for transport of oxygen to tissues, cellular oxidation	Iron-deficiency anemia, fatigue	Cirrhosis of liver, hemosiderin deposits, bloody diarrhea
Manganese	Meats, bananas, bran, beans, leafy vegetables, whole grains, nuts	Necessary for formation of hemoglobin, activation of enzymes	Subnormal tissue respiration, growth retardation, bone and joint abnormalities, nervous system disturbances, reproductive abnormalities	Muscular weakness, nervous system disturbances
Molybdenum	Organ meats, legumes, green leafy vegetables	Component of several enzymes	None reported	Inhibited enzyme activity
Selenium	Most foods, especially liver, other meats, seafood	Enzymes, lipid metabolism, antioxidant (protects plasma membranes from breaking down)	Anemia (rare)	Gastrointestinal disorders, lung irritation
Zinc	Meats, liver, seafood, eggs, legumes, milk, green vegetables	Part of many enzymes	Impaired cell growth and repair, poor wound healing, impaired sense of taste, small reproductive glands	Fever, nausea, vomiting, diarrhea

to regulate body heat. These include the three so-called *gradient mechanisms* — radiation, conduction, and convection — and the evaporation of sweat from the skin (perspiration) and water from surfaces in the upper respiratory tract. Metabolism always produces body heat and evaporation always removes body heat, but gradient mechanisms can either warm the body or cool it. However, as you will see, the gradient mechanisms cannot cool the body when the temperature of the air is higher than the temperature of the skin.

Radiation When your body temperature rises, the temperature control center in the hypothalamus is alerted. Your brain causes the dilation of surface blood vessels, increasing the blood flow to the surface of the skin and carrying body heat along with it. Once the heat is concentrated near the body surface, it can be lost to the atmosphere through the process of *radiation,* the transfer of infrared heat rays from one object to another without physical contact between the objects.

In a normally heated room, 50 to 60 percent of the body's heat loss occurs through radiation. Radiation can also work to warm the body when it is too cool. For example, although you are not in contact with the sun or the fire in your fireplace, you can still be warmed by their heat rays.

Conduction In contrast to radiation, *conduction* is the transfer of heat directly from one object to another. For conduction to occur, the two objects must be at different temperatures and must be in contact with each other. For example, the heat from your body warms the chair you sit in. Also, you warm your cold feet by putting them directly *on* a hot pad, not 6 in. away. Only about 3 percent of the body's heat transfer is accomplished through conduction, an almost negligible amount.

Convection *Convection* is the transfer of heat along a moving gas or liquid. For example, cool air next to the body is heated and then moved away on currents of air. The heated air removes heat from the body and makes way for more cool air, which in turn is heated and moved away. (Because of the importance of convection in aiding heat loss from the body, the *wind-chill index* was devel-

oped. This quantifies the cooling effects of the combination of wind velocity and air temperature.)

Evaporation When the temperature of the air is higher than the temperature of the surface of the body, radiation, conduction, and convection cannot remove heat from the body. The only useful method under that condition is *evaporation,* the conversion of water from a liquid into a gas (such as water vapor). This process requires heat energy.

Evaporation takes place not only on the skin surface but also in respiratory passages, including the nose, mouth, and lungs. For each kilogram of water evaporated from the body, 580 kcal are removed (0.58 kcal for every gram of water). When you are physically active, you perspire, and are aware of being cooled as the sweat evaporates from your skin. Evaporation is effective because heat is required to convert sweat into water vapor. The necessary heat is given up by the body, producing a cooling effect. But you are actually losing body heat through evaporation all the time, either by evaporation from the respiratory tract or through *insensible perspiration*, the undetected evaporation of sweat from your skin.*

Homeostatic Control of Body Temperature

The temperature of the body is regulated in great part by negative feedback mechanisms within the nervous system, especially in the hypothalamus. This system works admirably, with the temperature of the body usually varying no more than about half a degree above or below the accepted "normal" temperature of 37°C (or about 1°F above or below 98.6°F). This negative feedback system has three major components: (1) *temperature receptors* that sense the existing body temperature, (2) *effector organ systems* that control heat production or heat loss, and (3) an *integrator* or *controller* that compares the sensed temperature with a "normal" or "reference" temperature. If the body temperature is too high or too low, the

---

*You lose about 7 kcal an hour through evaporation from the respiratory tract and about 10 kcal an hour through insensible perspiration. During an entire day you lose about 0.7 L of water, enough to liberate approximately 400 kcal.

Q: *Why do you feel warmer on hot, humid days than on hot, dry days?*

A: On hot, humid days the air is already so full of moisture that most of the sweat on your skin cannot evaporate. If the humidity is 90 percent, it means that the air contains 90 percent of the water vapor it can hold. As a result, your evaporation cooling system operates inefficiently. If the humidity reaches 100 percent, your body temperature will actually begin to rise when the outside temperature rises above 32°C (90°F).

Q: *What are some ways in which the "normal" body temperature can vary?*

A: The body temperature of infants is usually higher than 37°C (98.6°F), and that of elderly people lower. Adults wake up with a somewhat lower temperature and go to sleep with a somewhat higher temperature. The normal temperature of a woman is generally 1°F higher about 2 weeks after ovulation. Rectal and armpit temperatures usually are about 1°F higher than oral temperature. In England, the "normal" body temperature is 98.4°F, in contrast to 98.6°F in the United States.

What Is Fever?

Fever is an elevation of body temperature above the normal 37°C (98.6°F). It may be thought of, at least initially, as a nonspecific defense against infection, which is the usual cause of fever. Among the many other causes of fever are drugs, hormonal imbalances, congestive heart failure, anxiety, strenuous physical exertion, disturbances of the immune system or central nervous system, tumors, and injury to tissues. Whatever the cause, metabolic activity increases (about 7 percent per degree Celsius), body heat is retained, and core temperature rises. Any substance capable of producing a fever is called a **pyrogen** (Gr. "fire-producing").

A fever often indicates that the body is fighting a viral or bacterial infection. When harmful bacteria or viruses enter the body, macrophages release the hormone *interleukin-1,* which travels to the temperature-regulating center in the hypothalamus. The hypothalamus then causes heat-sensitive preoptic neurons to release prostaglandins, which reset the body's "hypothalamic thermostat" at a temperature above normal, producing a fever. (Aspirin, which is often used to reduce fever, is a known inhibitor of prostaglandin synthesis.) Apparently, both the fever and the interleukin-1 increase the effectiveness of the immune system in fighting certain infections, sometimes increasing the production of T cells and antibodies as much as 20-fold.

A fever is usually preceded by a body chill and shivering, and cutaneous vasoconstriction causes the skin to remain cool even though the core temperature may already be rising. Shivering increases the production of heat, and at the same time mechanisms for the reduction of heat are inhibited. As a result,

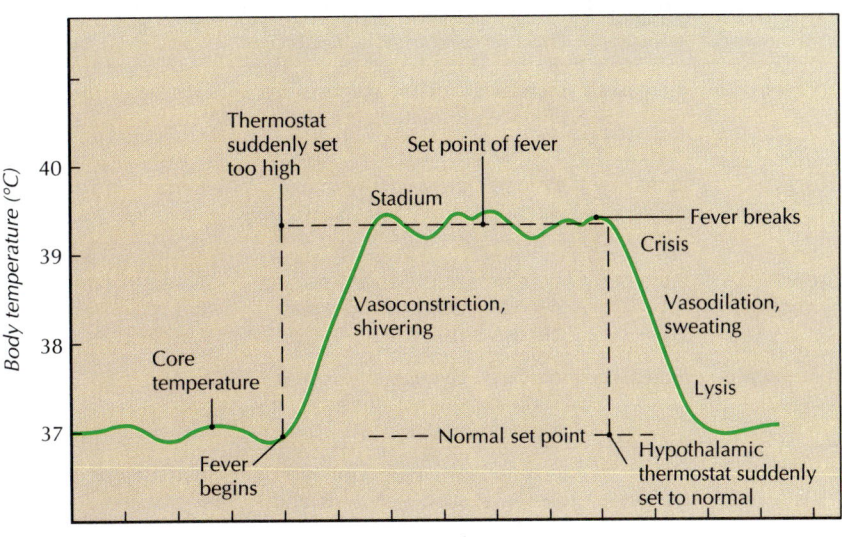

Major events of a fever.

the body temperature is raised and the skin becomes warm.

When the cause of the fever is removed, the original "set point" of the hypothalamus is restored and normal thermoregulation resumes. As the body temperature begins to drop, the skin is flushed and the cardiac output may be increased as excess heat is dissipated. Skin temperature begins to decrease, and the onset of perspiration indicates that the rate of heat loss is greater than the rate of heat production. Normal body temperature is soon restored (see graph.) The period of maximum elevation is called the *stadium* or *fastigium.* It may last from 3 days to 3 weeks. A sudden drop in temperature to normal is called the *crisis,* while a more gradual return to normal is known as *lysis.*

As long as the body temperature is abnormally high, the metabolic rate speeds up and damaged tissues are repaired faster than at normal temperatures. The heat of a fever also helps kill or incapacitate the foreign bacteria or viruses. But when a fever reaches approximately 40°C (104°F), it begins to harm the person as well.

Recent studies indicate that antibiotics are more effective at fever temperatures. High temperature may make the plasma membrane more permeable to antibiotics, or it may weaken the bacteria so that they are less able to resist the antibiotics.

Metabolic activity generally increases in proportion to the extent of the fever, and the body uses increased quantities of carbohydrates, fats, and proteins. A high-calorie, high-protein diet is desirable while the fever persists, especially since some body protein is destroyed during the early stages of a fever.

controller activates the appropriate effector system, returning the body temperature to normal.

It should be noted that normal temperature refers to the *core temperature*, or temperature of the interior of the body, rather than the *surface temperature*, or temperature of the arms, legs, and tissues immediately under the skin. Usually in referring to the core of the body we mean the head and trunk as well as the vital organs housed within those areas. Even though the temperature of the skin and limbs fluctuates, it is important for the temperature of organs such as the heart, brain, and kidneys to remain constant. Interestingly, the temperature of the liver, where many chemical activities take place, is usually higher than that of other organs.

TABLE 26.3 SUMMARY OF HORMONAL REGULATION OF METABOLISM

Hormone	Effect on blood glucose	Effect on carbohydrate metabolism	Effect on protein metabolism	Effect on lipid metabolism
Cortisol	Increases	Increases gluconeogenesis and glycogen formation	Decreases protein synthesis	Increases lipolysis; decreases ketogenesis and lipogenesis
Glucagon	Increases	Increases glycogenolysis and gluconeogenesis; decreases glycogen formation	None	Increases lipolysis and ketogenesis
Growth hormone	Increases	Increases gluconeogenesis and glycogen formation; decreases glucose utilization	Increases protein synthesis and amino acid transport into muscle cells	Increases lipolysis and ketogenesis; decreases lipogenesis
Epinephrine	Increases	Increases glycogenolysis and gluconeogenesis; decreases glycogen formation	None	Increases lipolysis and ketogenesis
Insulin	Decreases	Increases glycogen formation; decreases glycogenolysis and gluconeogenesis	Increases protein synthesis and amino acid transport into muscle cells	Increases lipogenesis; decreases lipolysis and ketogenesis
Testosterone	None	None	Increases protein deposition into muscle cells	None
Thyroxine	None	Increases gluconeogenesis and glucose utilization	Increases protein synthesis	Increases lipolysis

Role of the hypothalamus The hypothalamus plays an important role in preventing the body from overheating. Heat-sensitive neurons in the preoptic area of the hypothalamus speed up their firing rate when the temperature of the body increases. Signals from the preoptic area are combined in the posterior hypothalamus with signals from the rest of the body to bring about a compensating heat loss that restores the normal temperature.

The feedback mechanisms in the hypothalamus cannot operate properly without complementary detectors in the body that are sensitive to temperature changes. The complete heat-controlling mechanism of the hypothalamus and its support systems is known as the *hypothalamic thermostat.*

Some cold-sensitive neurons are found in the hypothalamus, but they are thought to play a minor role in the regulation of body temperature. Instead, cold receptors in the skin, spinal cord, and other areas are sensitive to changes at the cold end of the scale and help prevent the body from becoming too cool.

The hypothalamic thermostat *reduces* body heat in three ways:

1 Blood vessels in the skin are dilated sufficiently so that heat is transferred to the skin from the blood about eight times faster than normal. Such a massive dilation is caused by the inhibition of nerve centers in the posterior hypothalamus that usually cause the constriction of blood vessels.

2 Perspiration is increased, and the resulting increased evaporation of sweat cools the skin. If the body temperature rises 1°C above normal, the rate of perspiration increases enough to remove 10 times the normal body heat.

3 Shivering and other muscular activities that increase body heat are inhibited.

When the body becomes too cold, the hypothalamic thermostat uses three regulatory devices to *increase* body heat:

1 The temperature centers in the posterior hypothalamus cause the constriction of blood vessels, thereby reducing heat loss through the skin.

2 The hypothalamus stimulates shivering and thermogenesis. Shivering occurs when the *primary motor center for shivering* in the posterior hypothalamus stimulates the brainstem, spinal cord, and motor neurons that ultimately cause an increase in muscle tone. Shivering can raise heat production as much as five times above normal.

3 The hypothalamus causes the hairs on the skin to "stand on end." This activity (*piloerection*), along with the accompanying "goose bumps," increases the production of heat slightly, if at all, by increasing the muscle activity of the skin.

Roles of the endocrine system and sympathetic nervous system

When the preoptic area of the hypothalamus is cooled, the secretion of thyrotropin-releasing hormone (TRH) from the hypothalamus is increased. TRH in turn stimulates the secretion of thyroid-stimulating hormone (TSH) from the pituitary gland. TSH then stimulates the thyroid gland to secrete thyroxine, which increases cellular metabolism and body heat [TABLE 26.3].

Increased secretions of epinephrine and norepinephrine from the adrenal glands also help raise the level of cellular metabolism. This effect, known as *chemical thermogenesis*, is caused by the stimulation of the sympathetic nervous system. The effectiveness of chemical (nonshivering) thermogenesis seems to be directly related to the amount of *brown fat* in the body. Because brown fat contains many mitochondria and many small globules of fat instead of one large globule as in white fat, it is capable of a high level of cellular metabolism. An adult body contains little or no brown fat, so it can increase heat production only 10 to 15 percent through chemical thermogenesis. The infant body does contain some brown fat, mostly between and around the shoulder blades, so it can increase heat production 50 to 100 percent through chemical thermogenesis.

ASK YOURSELF

1 What are some important factors that affect the metabolic rate?

2 What is the basal metabolic rate?

3 What are the so-called gradient mechanisms of body-heat regulation?

4 How does evaporation help cool the body?

5 What is the hypothalamic thermostat?

6 What is the function of brown fat?

WHEN THINGS GO WRONG

Obesity

Approximately 50 to 70 million Americans suffer from *obesity* (L. *obesus*, "grown fat by eating"), the condition of being 20 to 25 percent over the recommended body weight. Obesity is usually caused by eating too much and exercising* too little (taking in more calories than are used up) rather than by metabolic disturbances. It is estimated that only about 5 percent of all cases of obesity are caused by hormonal imbalances. It should be noted, however, that the female hormones (estrogen) affect the deposition of fat to establish body contours and may contribute to obesity in some women.

Protein-Calorie Malnutrition

Protein-calorie malnutrition occurs when the body has been deprived of sufficient amino acids and calories for a long period. Two forms of malnutrition are marasmus (protein-calorie deficiency) and kwashiorkor (protein deficiency), both of which are common in underdeveloped countries and in areas where dietary protein is inadequate for proper growth and tissue building. *Marasmus* is caused by a diet that is very low in both calories and protein. It affects infants who are not receiving sufficient breastfeeding or who suffer from chronic diarrhea or other debilitating conditions. It is characterized by tissue wasting and impaired physical growth.

Kwashiorkor occurs when the diet consists primarily of carbohydrates and is very low in protein. It was originally described in young children in Ghana whose diet consisted of maize (corn) gruel. Such a diet lacks the essential amino acids lysine and tryptophan. Kwashiorkor does not affect body growth, but it does cause a decrease in adipose tissue, as fat is metabolized as the main energy source. Abdominal tissue edema gives the patient a puffed-up appearance, although muscle and adipose tissues are actually wasting away. Kwashiorkor is said to afflict as many as 25 percent of African children. Its name comes from a West African word meaning "sickness a child has when another child is born." The shrunken muscles, potbelly, skin sores, bleached and falling hair, and nausea and vomiting frequently begin when a child is weaned from the breast milk of its mother. This condition can be cured, sometimes with dramatic speed, by the addition of sufficient dietary protein containing the proper balance of amino acids.

---

*"Exercise" is meant to include all physical activity—walking, gardening, housecleaning, and any other activity that uses up calories.

Q: *Why do you gain weight when you stop smoking?*

A: Smoking increases the body's metabolic rate. A smoker who quits smoking has a lower metabolic rate and soon gains weight.

Excessive Vitamin Intake

Megadoses of fat-soluble vitamins over a prolonged period may be toxic. Even water-soluble vitamins may be toxic in very large doses. The following conditions have been authenticated. Megadoses of vitamin B_6 can produce peripheral nerve damage. The symptoms of *hypervitaminosis A* include anorexia, headache, irritability enlarged spleen and liver, scaly dermatitis, patchy hair loss, and bone pain and thickening. *Hypervitaminosis D* is characterized by weight loss, calcification of soft tissues, and kidney failure, and *hypervitaminosis K* is associated with anemia, jaundice, and gastrointestinal problems.

Some Common Metabolic Disorders

Cystic fibrosis *Cystic fibrosis* is a metabolic genetic disease typified by a deficiency of the pancreatic enzymes trypsin, amylase, and lipase. It is the most common fatal genetic disease of Caucasian children, and it is estimated that 1 out of every 25 Caucasians in central Europe and North America is a carrier of the disease. People with cystic fibrosis lack functional copies of a protein called the *cystic fibrosis transmembrane conductance regulator (CFTR)*, which acts as a Na^+ channel into and out of cells.

The intestinal tract may be blocked at birth because the neonate does not excrete a viscous intestinal substance called meconium, and older infants have poor intestinal digestion, weight loss, and enlarged abdomens. The young child has poor lipid absorption and develops respiratory problems as a result of an accumulation of thick secretions in the bronchioles and alveoli. Chronic pneumonia, emphysema, or atelectasis (the incomplete expansion of lobules in the alveoli) may eventually cause death.

Phenylketonuria (PKU) *Phenylketonuria (PKU)* is a congenital disease caused by a DNA *mutation*, or genetic error. The mutation prevents the adequate formation of the enzyme phenylalanine hydroxylase, which normally converts the amino acid phenylalanine into the amino acid tyrosine. When phenylalanine is not converted at normal rates, its metabolites begin to accumulate in the blood, generally causing cerebral damage, mental retardation, decreased muscle control, and other problems. Exactly how this disorder leads to retardation is not known. PKU can sometimes be controlled if foods containing phenylalanine are kept out of the diet of a susceptible child.

Tay-Sachs disease When lysosomes are deficient in the enzyme hexosaminidase, they cannot break down fatty substances called gangliosides as they normally would. Gangliosides are products of normal cell growth. As these fatty deposits accumulate around neurons, the neurons become unable to transmit nerve impulses. Neurons that store too much ganglioside become swollen and eventually die. The disorder, which is genetic and hereditary, is called *Tay-Sachs disease* after the physicians who first reported it.

An afflicted baby appears normal at birth, but after 6 months or so the large number of affected neurons begins to cause brain damage. A reddish spot appears on the retina, muscles weaken, vision and hearing are impaired, and mental ability is diminished. Because there is still no effective treatment for the disease, death usually comes before the child is 5 years old.

Tay-Sachs disease occurs most often among Ashkenazi Jews from eastern Europe. About 1 out of every 30 Ashkenazi Jews is a carrier of the disease. A child is affected only if both parents carry the Tay-Sachs gene; carriers do not have the disease themselves. It has been calculated that 1 child out of every 3600 born to Ashkenazi parents will inherit Tay-Sachs disease. Modern techniques make it possible to screen people to find out if they are carriers and if unborn babies have Tay-Sachs disease. Many carriers decide not to have children.

Disorders of Temperature Regulation

Fever *Fever* may not always be a "disorder." In fact, it is often beneficial, as you saw in the essay "What Is Fever?" on page 879.

Frostbite What we call *frostbite* is a localized direct physical injury to plasma membranes. It occurs when water molecules in tissues are taken out of solution and form ice crystals. Blood vessels, especially delicate capillaries, are usually damaged, and the subsequent release of fibrinogen may cause red blood cells to become trapped in a blood clot. The skin in the affected area usually turns white until it thaws, and then it may be purplish-blue or black. The deep color usually indicates the death of tissue, and possibly gangrene. Untreated frostbite will almost certainly lead to gangrene, and amputation may be necessary. Treatment usually consists of gently warming the injured area but never rubbing it, since rubbing can lead to further tissue damage. Frostbite occurs most often in the earlobes, fingers, and toes, where the temperature of the body is at its lowest.

Heat exhaustion/heat prostration *Heat exhaustion,* or *heat prostration,* is caused by a depletion of plasma volume due to sweating, and hypotension in response to reflexively dilated blood vessels in the skin. Decreased blood pressure often causes fainting. Though the heart rate rises and blood pressure falls, the body temperature usually remains normal and the skin is cool. Heat exhaustion is accompanied by thirst, fatigue, giddiness, nausea, cramps, and possibly delirium. Death is rare; it is likely only if water depletion is greater than 20 percent of the total body water (20 percent is about 40 pt). Treatment consists of replacing the lost water and salt.

Heatstroke/sunstroke Unlike heat exhaustion, *heatstroke,* or *sunstroke,* causes core temperatures above 40.5°C (105°F). When the hypothalamus becomes overheated, its heat-regulating system breaks down and sweating stops almost altogether. Excessive loss of water and electrolytes may already have occurred before sweating is halted, and dizziness, nausea, delirium, and coma are likely. If the rising temperature is not controlled quickly, brain damage can result, the cardiovascular system can collapse, and the person can die. Heatstroke is most common in hot, humid weather, especially after a long period of strenuous physical activity.

Heatstroke should be treated as soon as the symptoms become apparent, usually by loosening or removing the person's clothing and applying cold compresses directly to the skin. Cooling is usually stopped when the core temperature drops to 39°C (102°F) to prevent shivering. A drink containing balanced electrolytes may be given if the person is conscious.

Hypothermia *Hypothermia* (*hypo,* under + *thermia,* heat) results when the body mechanisms for producing and conserving heat are exceeded by exposure to severe cold. In contrast to frostbite, which is a localized injury, hypothermia affects the whole body. It is most common among the elderly, the young, heavy drinkers and drug users (because alcohol and barbiturates move blood into capillaries away from body organs), and very thin people who lack a sufficient layer of insulating fat. Body temperature may be reduced to 35 to 36.6°C (95 to 98°F) in mild cases, and to as low as 25 to 28°C (77 to 82.4°F) in severe cases.

Victims exhibit symptoms ranging from shivering in mild cases, to slurred speech in moderate cases, to muscle rigidity and shock in serious cases, to ventricular fibrillations (grossly abnormal heart rhythms) and death in the most severe cases. Hypothermia slows the activity of most organ systems, and if pulse and respiration appear to cease, prolonged cardiopulmonary resuscitation becomes necessary. Gentle warming, starting with the core area to avoid fatal ventricular fibrillations, is recommended, and the condition requires the immediate attention of a physician.

The so-called *dive reflex* is a protective response that explains why children who fall into frozen lakes often survive even though breathing stops. The reflex is triggered by the splashing of cold water on the face, causing the heart rate to drop and increasing the blood flow to the brain and heart. A child's small size also helps. When some icy water is inhaled, it enters the lungs and spreads quickly through the relatively short bloodstream. The blood is chilled, cooling the brain and reducing its need for oxygen. As a result, a child may survive for as long as 40 minutes after breathing stops.

CHAPTER SUMMARY

Metabolism is the overall term for the many chemical and energy transformations that take place within cells to promote homeostasis. *Nutrients,* or chemical components of food, supply the body with energy and materials. *Potential energy* stored in the chemical bonds of nutrient molecules is transformed into usable energy in the form of ATP. ATP is used by cells to carry on metabolic work.

Review of Chemical Reactions (p. 852)

1 *Anabolic reactions* are the chemical changes that allow cells to build and repair tissues through the synthesis of complex molecules from simpler building blocks. These reactions require energy.

2 *Catabolic reactions* break down complex molecules into simpler ones and release energy and heat. Enzymes are usually involved.

3 Anabolic and catabolic reactions take place at the same time within a cell, and structural components are constantly being broken down and replaced.

4 *Oxidation-reduction reactions* are reactions in which an atom or molecule loses hydrogen atoms or electrons (oxidation) or gains hydrogen atoms or electrons (reduction). They always occur together, because when one atom or molecule gains electrons, another loses them.

Carbohydrate Metabolism (p. 852)

1 Carbohydrates have potential energy stored in chemical bonds. Within the bonds, it is *electrons* that carry the energy. During the metabolism of foods, the electrons are transferred to nucleotides such as NAD^+ and FAD.

2 When the nucleotides accept electrons, they also receive hydrogen ions, converting NAD^+ into $NADH + H^+$ and

FAD into $FADH_2$. The reduced nucleotides may be used as direct sources of energy or may be used to make ATP.

3 Glucose is usually the body's preferred nutrient energy source. Energy is released when glucose is catabolized in a series of complex reactions.

4 *Cellular respiration* is the process by which cells are able to release the chemical energy stored in food. The process starts with a food such as glucose; the end products are carbon dioxide, water, and ATP. The main steps in the process are *glycolysis,* the *citric acid cycle,* and the *electron transport system.*

5 *Glycolysis* is the splitting of a 6-carbon molecule of glucose into two 3-carbon molecules of pyruvic acid in a series of anaerobic, enzymatic reactions in the cytosol of a cell. There is a net gain of two molecules of ATP.

6 The *citric acid cycle (Krebs cycle)* is a

series of enzymatic reactions that release hydrogen atoms from oxidized acids as a means of energy transfer. It takes place in the mitochondria, and its end products after two "turns" of the cycle are carbon dioxide, water, two molecules of ATP, six molecules of NADH, and two molecules of FADH$_2$.

7 The **electron transport system** is a series of oxidation-reduction reactions in which electrons are transferred to accepting cytochromes and oxygen. It occurs in mitochondria and produces most of the ATP (32 molecules) from glucose catabolism.

8 In **chemiosmotic coupling of ATP production,** the production of ATP by mitochondria involves the accumulation of electrochemical energy in the form of an electrochemical proton gradient by selectively pumping protons out of the matrix into the inner compartment. The energy in this gradient is then used to drive the activity of the ATP synthase to form ATP from ADP + P$_i$.

9 *Glycogenesis* is the conversion of blood glucose into glycogen, which is stored in the liver and skeletal muscle cells for later use. When the body needs extra glucose, the process of **glycogenolysis** converts glycogen stored in the liver into glucose and releases it into the bloodstream. *Gluconeogenesis* is the production of glucose from noncarbohydrate sources such as lactic acid, glycerol, and amino acids.

Protein Metabolism (p. 862)

1 *Protein catabolism* is the breakdown of proteins into amino acids, which are modified further by **oxidative deamination** (the removal of the amino group and its replacement with oxygen to form a *keto acid*) before they enter the cellular respiratory system. Keto acids can be used to produce carbon dioxide and ATP on the way to forming glucose, or they can be converted into acetyl-CoA to synthesize fatty acids.

2 *Protein anabolism* is the synthesis of protein from amino acids, under the direction of nucleic acids. It takes place on ribosomes and is regulated by several hormones.

3 *Nonessential amino acids* are synthesized by cells in the processes of **transamination. Essential amino acids** cannot be synthesized by cells and therefore must be supplied by dietary protein.

Lipid Metabolism (p. 864)

1 *Lipids* are the body's most concentrated source of food energy. The most abundant lipids are the **triglycerides.** Digested lipids are separated into glycerol and fatty acids.

2 Fatty acids undergo a series of reactions called **beta oxidation** that form acetyl-coenzyme A, which enters the citric acid cycle to produce ATP.

Absorptive and Postabsorptive States (p. 867)

1 During the **absorptive state,** nutrients are absorbed from the gastrointestinal tract into the cardiovascular and lymphatic systems.

2 After absorption, carbohydrates enter liver cells, where they are stored as glycogen or converted into fat. Glucose that does not enter the liver is used almost immediately by most body cells as an energy source.

3 Before absorption, some lipids enter the lymphatic system as *chylomicrons,* which contain triglycerides. Many of the absorbed amino acids enter the liver directly, where they are converted into carbohydrates and used for energy, or converted into fat, which is stored mainly in the liver.

4 In the **postabsorptive** or **fasting state,** the energy requirements of body cells (including the glucose supply of the brain) must be satisfied by nutrients taken in during the absorptive state and from the body's stored nutrients.

5 Body cells shift their metabolism during periods of fasting to use more fat instead of glucose, sparing glucose for the nervous system.

Nutritional Needs and Balance (p. 869)

1 The food we eat contains six kinds of **essential nutrients:** carbohydrates, proteins, lipids, vitamins, minerals, and water.

2 A *calorie* (cal) is the amount of energy required to raise the temperature of 1 g of water 1°C. A *kilocalorie* (kcal) is equal to 1000 calories.

3 The adult body needs relatively little dietary protein to supply the essential amino acids and nitrogen in a form that can be used to synthesize the nonessential amino acids. Dietary protein is most

beneficial when it is balanced with all types of food and provides all 20 of the necessary amino acids. The need for protein varies with age and body weight.

4 *Triglycerides* may contain three types of fatty acids: saturated, monounsaturated, and polyunsaturated.

5 *Vitamins* are organic compounds required by the body in small amounts for the regulation of metabolism. They are important in supporting the work of enzymes. Vitamins are usually classified as *fat-soluble* (A, D, E, K) or *water-soluble* (B group, C).

6 *Minerals* are naturally occurring inorganic elements that are used for the regulation of metabolism. Minerals needed in minute amounts are called *microminerals* or *trace elements*. Those found in significant amounts in body tissue are *macrominerals*.

Metabolic Rate and Temperature Control (p. 874)

1 The *metabolic rate* (number of calories used in a specific time period) is affected by type of food eaten, amount of exercise, hormones, age, and other factors.

2 The amount of energy a person uses daily, under standardized conditions, to maintain only essential bodily functions is the *basal metabolic rate (BMR).*

3 Control mechanisms that regulate body heat are the *gradient mechanisms* (radiation, conduction, and convection) and the evaporation of water from the skin and respiratory tract.

4 The complete heat-controlling mechanism of the hypothalamus and its support system is called the *hypothalamic thermostat.* It reduces heat by dilating skin blood vessels, increasing sweating, and inhibiting shivering and other muscular activities that increase body heat.

5 The secretion of TRH from the hypothalamus starts a series of *hormonal events* that increases body metabolism and heat through the secretion of thyroxine. Metabolism is also increased by the secretion of epinephrine and norepinephrine from the adrenal glands.

6 *Fever* is the condition in which body temperature rises above normal due to pyrogenic substances acting on the hypothalamic thermostat. A moderate fever appears to help the body fight certain viral and bacterial infections.

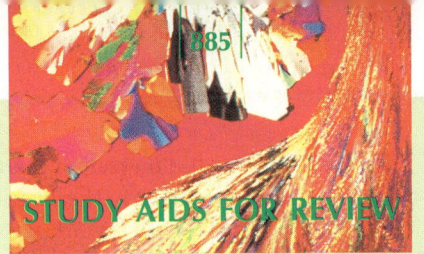

KEY TERMS

absorptive state 867

anabolism 852

basal metabolic rate 875

beta oxidation 864

calorie 869

catabolism 852

cellular respiration 851, 853

chemiosmosis 858

citric acid cycle 852, 853

electron transport
  system 852, 856

essential amino acids 864

gluconeogenesis 862

glycogenesis 861

glycogenolysis 862

glycolysis 852, 853

metabolic rate 874

metabolism 851

nonessential amino acids 864

oxidative deamination 862

postabsorptive state 869

triglycerides 864

MEDICAL TERMS FOR REFERENCE

ACIDOSIS A condition in which the concentration of hydrogen ions in the blood is above normal.

ALKALOSIS A condition in which the concentration of hydrogen ions in the blood is below normal.

HEAT EDEMA A condition in which slight swelling, especially in the ankles, occurs in a person who is unaccustomed to a hot climate; usually caused by salt and water retention.

HYPERCALCEMIA Excessive calcium in the blood.

HYPERKALEMIA Excessive potassium in the blood.

HYPERMAGNESEMIA Excessive magnesium in the blood.

HYPERNATREMIA Excessive sodium in the blood.

HYPERPHOSPHATEMIA Excessive organic phosphate in the blood.

HYPOCALCEMIA A deficiency of calcium in the blood.

HYPOKALEMIA A deficiency of potassium in the blood.

HYPOMAGNESEMIA A deficiency of magnesium in the blood.

HYPONATREMIA A deficiency of sodium in the blood.

HYPOPHOSPHATEMIA A deficiency of organic phosphate in the blood.

UNDERSTANDING THE FACTS

1 Define and contrast metabolism, anabolism, and catabolism.

2 What are the major uses of carbohydrates? Lipids? Proteins?

3 List the stages of carbohydrate catabolism, indicating the energy-rich products of each stage.

4 For each of the following pairs of compounds, indicate which of the pair contains more energy. Explain.

  a ATP and ADP   **c** FAD and $FADH_2$

  b NAD^+ and NADH

5 What stage of catabolism produces the most ATP?

6 Define glycogenesis, glycogenolysis, and gluconeogenesis. When and where does each occur?

7 Why are some amino acids called "essential"?

8 Differentiate between the treatment of monosaccharides, amino acids, and triglycerides during the absorptive state.

9 Distinguish lipogenesis from lipolysis.

10 List the metabolic processes that maintain a steady level of blood glucose during the postabsorptive state.

11 According to the Federal Dietary Guidelines, what are the recommended levels of dietary fats, proteins, and carbohydrates?

12 What is a complete dietary protein?

13 Which vitamins are water-soluble, and which are lipid-soluble?

14 Contrast the metabolic rate with the basal metabolic rate.

15 List the four physical processes by which the body can lose heat, and identify the specific gradient that must exist for each process to occur.

UNDERSTANDING THE CONCEPTS

1 In what ways, if any, does the meaning of the phrase "cellular respiration" differ from that of "carbohydrate catabolism"?

2 How can a cell use pyruvic acid?

3 What is the primary function of each of the following compounds?

  a ADP   **b** ATP   **c** $FADH_2$ **d** NAD^+

4 What are the criteria by which glucose catabolism is divided into four separate stages?

5 What happens to the energy that is released during the electron transport stage of cellular respiration?

6 What processes are coupled during chemiosmosis? What would result from uncoupling the processes?

7 Is all of the energy liberated from the catabolism of a sugar trapped by ATP? Explain.

8 What is accomplished by beta oxidation, oxidative deamination, and transamination?

9 What is the role of the liver during the absorptive and postabsorptive states?

10 What is the physiological significance of glucose sparing?

11 Why are excessive dietary levels of the water-soluble vitamins generally not harmful, whereas excessive dietary levels of the fat-soluble vitamins are?

12 How does the hypothalamus oversee the regulation of body temperature?

SELF-QUIZ

Multiple-Choice Questions

26.1 The nutrient that functions as the body's main energy source is

a protein   c carbohydrate   e oxygen

b fat   d water

26.2 Which process does *not* occur in the liver?

a digestion of starch to glucose

b formation of glycogen from glucose

c formation of glucose from glycogen

d oxidative deamination

e gluconeogenesis

26.3 The conversion of pyruvic acid to lactic acid is accompanied by

a the regeneration of NAD^+ needed in glycolysis

b the initiation of the citric acid cycle reactions

c the formation of a molecule of ATP

d a and b   e a, b, and c

26.4 Which of the following acts as an electron acceptor in the reactions of the citric acid cycle?

a oxygen   c ATP   e FAD

b ADP   d pyruvic acid

26.5 Each of the following substances is both formed and used during cellular respiration *except*

a ATP   c pyruvic acid   e oxygen

b NADH   d oxaloacetic acid

26.6 Protein synthesis is regulated by

a growth hormone   d thyroxine

b insulin   e all of the above

c glucocorticoids

26.7 During the fasting state, the body

a breaks down glycogen stored in the liver

b undergoes gluconeogenesis

c produces ATP from lipids rather than glucose

d a and b

e a, b, and c

26.8 The hypothalamic thermostat reduces body heat by

a dilating blood vessels   d a and b

b increasing perspiration   e a, b, and c

c stimulating shivering

26.9 Chemiosmosis involves an electrochemical gradient produced by the transport through mitochondrial membranes of

a electrons   d sodium ions

b neutrons   e potassium ions

c protons

26.10 Which of the following can apply to a molecule that is oxidized?

a It loses energy.

b It gains energy.

c It loses hydrogen atoms.

d It gains electrons.

e a and c

Completion Exercises

26.11 The liver helps to maintain a stable level of blood glucose by means of the processes of gluconeogenesis, glycogenesis, and _____.

26.12 During the fasting state, all of the body's cells *except* for neurons catabolize _____ as their primary energy source.

26.13 The primary hormone that regulates the metabolic rate is _____.

26.14 Vitamins A and E are examples of _____ vitamins.

26.15 The capacity for chemical thermogenesis is higher when a tissue called _____ is abundant.

26.16 When a molecule gains electrons, it has been _____.

26.17 Cellular _____ is the process in which cells use potential energy stored in the chemical bonds in all kinds of food molecules to produce ATP from ADP and P_i.

26.18 Because glycolysis can proceed without oxygen, it is also referred to as _____ respiration.

26.19 The final electron acceptor in the electron transport chain is _____.

26.20 Specific proteins that catalyze chemical reactions are called _____.

26.21 The small lipid droplets surrounded by a protein coating that result from lipid digestion are called _____.

Matching

a beta oxidation   d glucose sparing

b lipogenesis   e oxidative deamination

c ketogenesis

26.22 _____ the primary method whereby amino acids are converted into metabolites that can enter the citric acid cycle

26.23 _____ the primary method whereby fatty acids are catabolized

26.24 _____ the formation of molecules such as acetone from excess acetyl-CoA

26.25 _____ a process carried on primarily in cells of adipose tissue

26.26 _____ a postabsorptive method that protects the metabolic health of the nervous system

a gluconeogenesis   c glycogenesis

b glycolysis   d glycogenolysis

26.27 _____ the splitting of a glucose molecule into two molecules of pyruvic acid

26.28 _____ the conversion of glucose to a longer-term storage product

26.29 _____ the conversion of glycogen to glucose

26.30 _____ the conversion of lactic acid or amino acids into glucose

27

The Urinary System

KEY CONCEPTS

1 The urinary system maintains the homeostasis of body water and electrolytes and removes waste products in urine.

2 The kidneys filter water and soluble substances from the blood, reabsorb some useful substances back into the blood, and secrete unwanted substances.

3 The kidneys also help regulate the acid-base balance of the blood.

Like other land-dwelling organisms, human beings must deal with the problems of (1) obtaining and conserving water and (2) removing excess salts and metabolic wastes. Our urinary system accomplishes both of these crucial tasks: it removes some waste products through the production of urine, and it carefully regulates the water content of the body.

What we excrete has much more influence on homeostasis than what we eat. *Excretion* (L. *excernere*, to sift out) is the elimination of the waste products of metabolism.

Excretion is performed in several ways by several organs. The lungs excrete carbon dioxide and water when we exhale. The skin releases some salts and organic substances such as urea and ammonia during periods of heavy sweating. The large intestine eliminates a small amount of water, salts, and some microorganisms and undigested food. However, the prime regulator of the proper balance between water and other substances is the urinary system, with the kidneys as its main component.

THE URINARY SYSTEM: COMPONENTS AND FUNCTIONS

The *urinary system,* also called the *renal system* (L. *renes,* kidneys), consists of (1) two *kidneys,* which remove dissolved waste and excess substances from the blood and form urine, (2) two *ureters,* which transport urine from the kidneys to (3) the *urinary bladder,* which acts as a reservoir for urine, and (4) the *urethra,* the duct through which urine from the bladder flows to the outside of the body during urination [FIGURE 27.1].

Every day the kidneys filter about 1700 L of blood. Each kidney contains over a million *nephrons,* the functional units of the kidney. Fortunately, we have many more nephrons than we actually need for these purposes, and if necessary, we could lead a normal life with only one kidney.

What are the major functions of nephrons? First, they *filter* water and soluble components from the blood. Second, they selectively *reabsorb* some of the components back into the blood. In this way, useful substances are conserved and recycled. Finally, they selectively *secrete* inorganic ions (Na^+, K^+, H^+) in order to maintain stable concentrations of these ions in the extracellular fluid. The water and other substances not reabsorbed constitute the urine.

*Only about 47 percent of a person's daily water intake comes from ingested fluids. As much as 39 percent is supplied by so-called solid food. (Meats are 50 to 70 percent water, most vegetables contain more than 90 percent water, and bread is about 35 percent water. The driest food is the baked sunflower seed, which contains about 5 percent water. The wettest food is watermelon, with 97 percent water.)

ANATOMY OF THE KIDNEYS

The average adult takes in about 2.7 L of water every day, most of it from foods and liquids.* Under normal conditions, the same amount of water is lost from the body daily, about a third of it evaporated from the skin and exhaled from the lungs, and the remaining two-thirds excreted as urine [TABLE 27.1]. The organs that maintain this constant water balance are the **kidneys,** a pair of reddish-brown, bean-shaped organs, each one about the size of a large bar of soap.

Location and External Anatomy

Just as the stomach is higher than most people think, so are the kidneys. The paired, bean-shaped kidneys are located in the posterior wall of the abdomen, one on each side of the vertebral column, and extend from vertebrae T12 to L3. The kidneys are protected, at least partially, by the eleventh and twelfth pairs of ribs; and they are capped by the adrenal ("upon the kidney") glands, with the left kidney touching the spleen. The right kidney is in extensive contact with the liver, and because of the liver's large right lobe, the right kidney is slightly lower than the left. The kidneys are retroperitoneal organs [FIGURE 27.2], are well vascularized with blood vessels and lymphatic vessels, and are innervated by nerves.

Each kidney is medially concave and laterally convex. On the medial concave border is the **hilus** (HYE-luhss), a small indentation where an artery, vein, nerves, and the ureter enter and leave the kidney [FIGURE 27.3A].

Covering and supporting the kidneys are three layers of tissue: (1) The innermost layer is a tough, fibrous material called the *renal capsule.* It is continuous with the surface layer of the ureters. (2) The middle layer is the *adipose capsule,* composed of perirenal ("around the kidney") fat, which gives the kidney a protective cushion against impacts and jolts. (3) The outer layer is subserous

TABLE 27.1 WATER INTAKE AND OUTPUT PER DAY*			
Fluid intake, mL		**Water output, mL**	
Ingested liquid	1200–1500	Urine	1200–1700
		Feces	100–250
Ingested food	700–1000	Sweat	100–150
		Epithelial evaporative losses:	
Metabolic oxidation	200–400	Skin	350–400
		Lungs	350–400
Total	2100–2900	Total	2100–2900

*Water intake and output are approximately equal in a healthy person. The figures above are for an average 70-kg adult whose diet provides adequate calories.

FIGURE 27.1 THE URINARY SYSTEM

Anterior view. Differences in the lower portions of the male and female systems are shown in FIGURE 27.18.

Structure	Function
Kidneys	Filter wastes from blood and produce urine
Ureters	Transport urine from kidneys to urinary bladder
Urinary bladder	Stores urine
Urethra	Transports urine from urinary bladder to outside of body

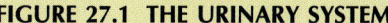

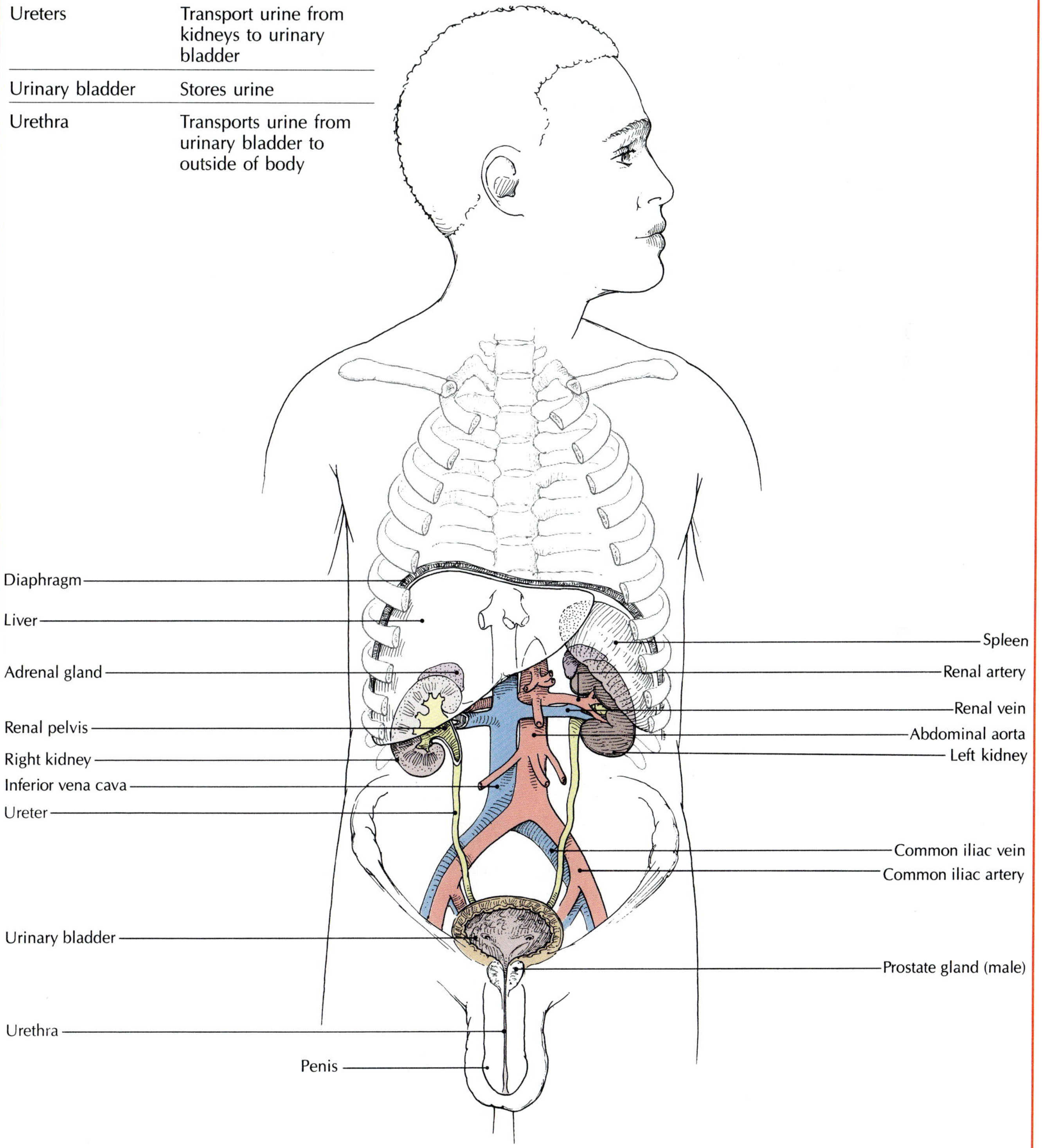

Diaphragm
Liver
Adrenal gland
Renal pelvis
Right kidney
Inferior vena cava
Ureter
Urinary bladder
Urethra
Penis

Spleen
Renal artery
Renal vein
Abdominal aorta
Left kidney
Common iliac vein
Common iliac artery
Prostate gland (male)

FIGURE 27.2 TRANSVERSE SECTION THROUGH UPPER ABDOMEN

This superior view shows the retroperitoneal location of the kidneys. Note the layers of adipose tissue and fasciae surrounding the kidneys.

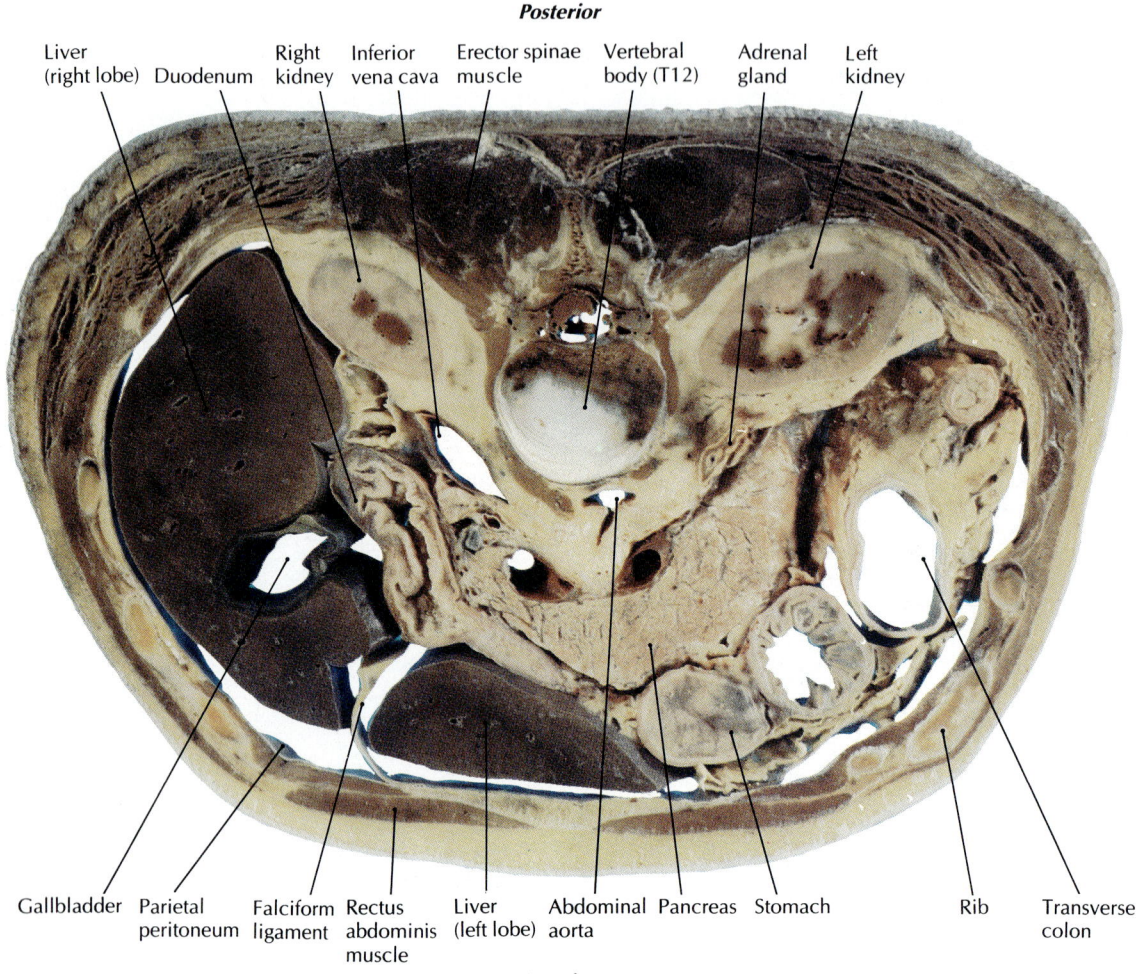

Posterior

Liver (right lobe) — Duodenum — Right kidney — Inferior vena cava — Erector spinae muscle — Vertebral body (T12) — Adrenal gland — Left kidney

Gallbladder — Parietal peritoneum — Falciform ligament — Rectus abdominis muscle — Liver (left lobe) — Abdominal aorta — Pancreas — Stomach — Rib — Transverse colon

Anterior

fascia called the *renal fascia*. Surrounding the renal fascia is another layer of fat, called *pararenal fat*. The renal fascia is composed of connective tissue that surrounds the kidney and pararenal fat; it attaches them firmly to the posterior abdominal wall. Renal fasciae have enough flexibility to permit the kidneys to shift slightly as the diaphragm moves during breathing.*

Internal Anatomy

A sagittal section of a kidney [FIGURE 27.3A] reveals three distinct regions called (from the inside out) the pelvis, medulla, and cortex. The **renal pelvis** is a large urine-collecting space within the kidney, formed from the ex-

panded upper portion of the ureter. The pelvis branches into two smaller cavities: the **major calyces** and **minor calyces** (KAY-luh-seez; sing. *calyx*, KAY-licks; Gr. cup). There are usually 2 to 3 major calyces and 8 to 18 minor calyces in each kidney.

The **renal medulla** is the middle portion of the kidney. It consists of 8 to 18 **renal pyramids,** which are longitudinally striped, cone-shaped areas. The base of each pyramid is adjacent to the outer cortex. The apex of each renal pyramid ends in the *papilla*, which opens into a minor calyx. Renal pyramids consist of tubules and collecting ducts of the nephrons. Urine passes from the collecting ducts in the pyramid to the minor calyces, the major calyces, and the renal pelvis. From there, the urine drains into the ureter and is transported to the urinary bladder.

*At body temperature the pararenal fat is liquid, allowing the kidneys to move up and down slightly in the fluid during breathing.

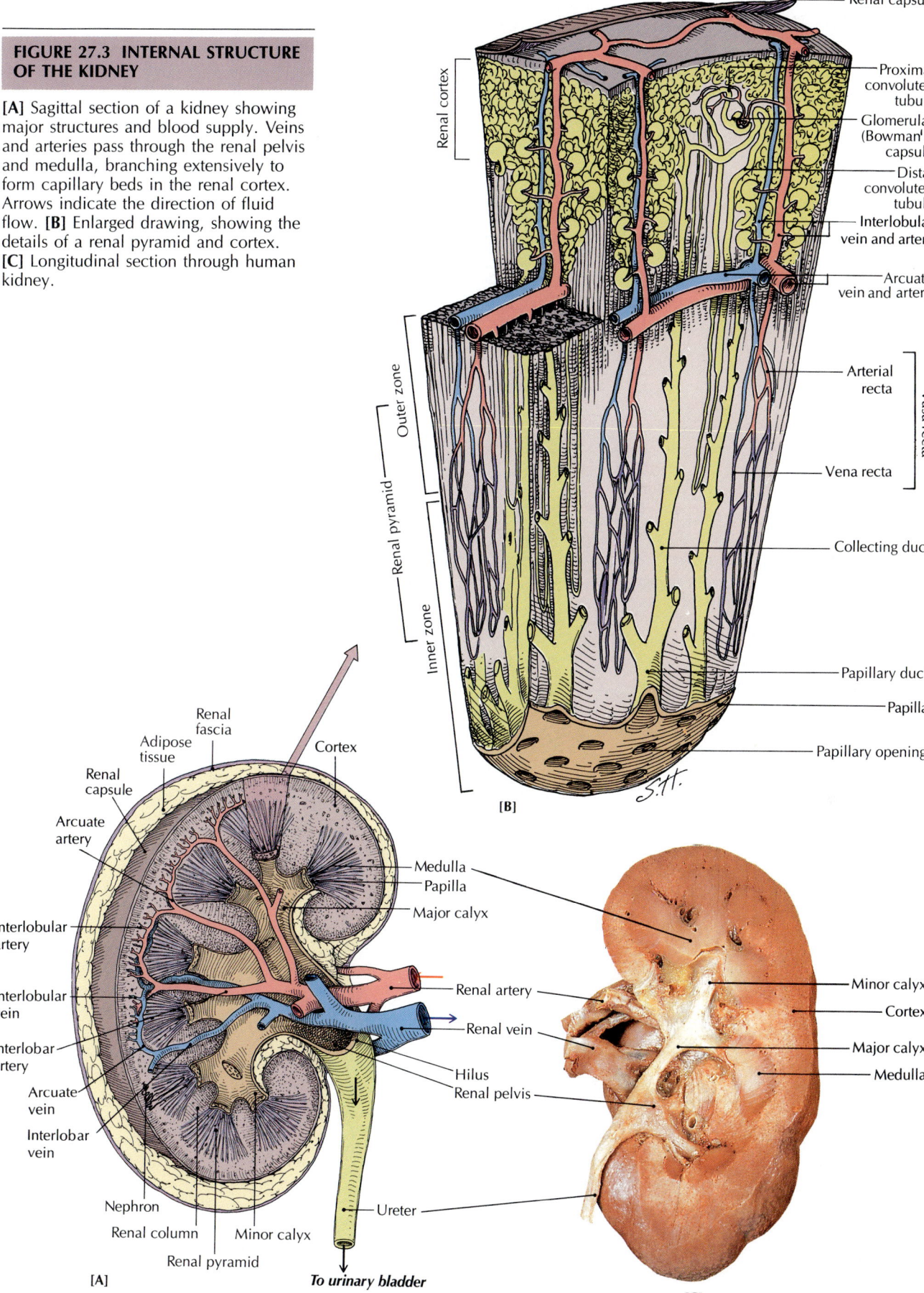

FIGURE 27.3 INTERNAL STRUCTURE OF THE KIDNEY

[A] Sagittal section of a kidney showing major structures and blood supply. Veins and arteries pass through the renal pelvis and medulla, branching extensively to form capillary beds in the renal cortex. Arrows indicate the direction of fluid flow. [B] Enlarged drawing, showing the details of a renal pyramid and cortex. [C] Longitudinal section through human kidney.

Renal capsule

Renal cortex

Proximal convoluted tubule

Glomerular (Bowman's) capsule

Distal convoluted tubule

Interlobular vein and artery

Arcuate vein and artery

Outer zone

Renal pyramid

Arterial recta

Vasa recta

Vena recta

Inner zone

Collecting duct

Papillary duct

Papilla

Papillary opening

[B]

Renal fascia

Adipose tissue

Renal capsule

Cortex

Arcuate artery

Medulla

Papilla

Major calyx

Interlobular artery

Interlobular vein

Renal artery

Interlobar artery

Renal vein

Arcuate vein

Hilus

Renal pelvis

Interlobar vein

Nephron

Renal column

Minor calyx

Renal pyramid

Ureter

To urinary bladder

[A]

Minor calyx

Cortex

Major calyx

Medulla

[C]

FIGURE 27.4 NEPHRON

[A] Entire kidney. [B] Enlarged section
showing locations of juxtamedullary
and cortical nephrons. [C] Structure of
a cortical nephron, including blood
vessels. Arrows indicate the direction
of fluid flow.

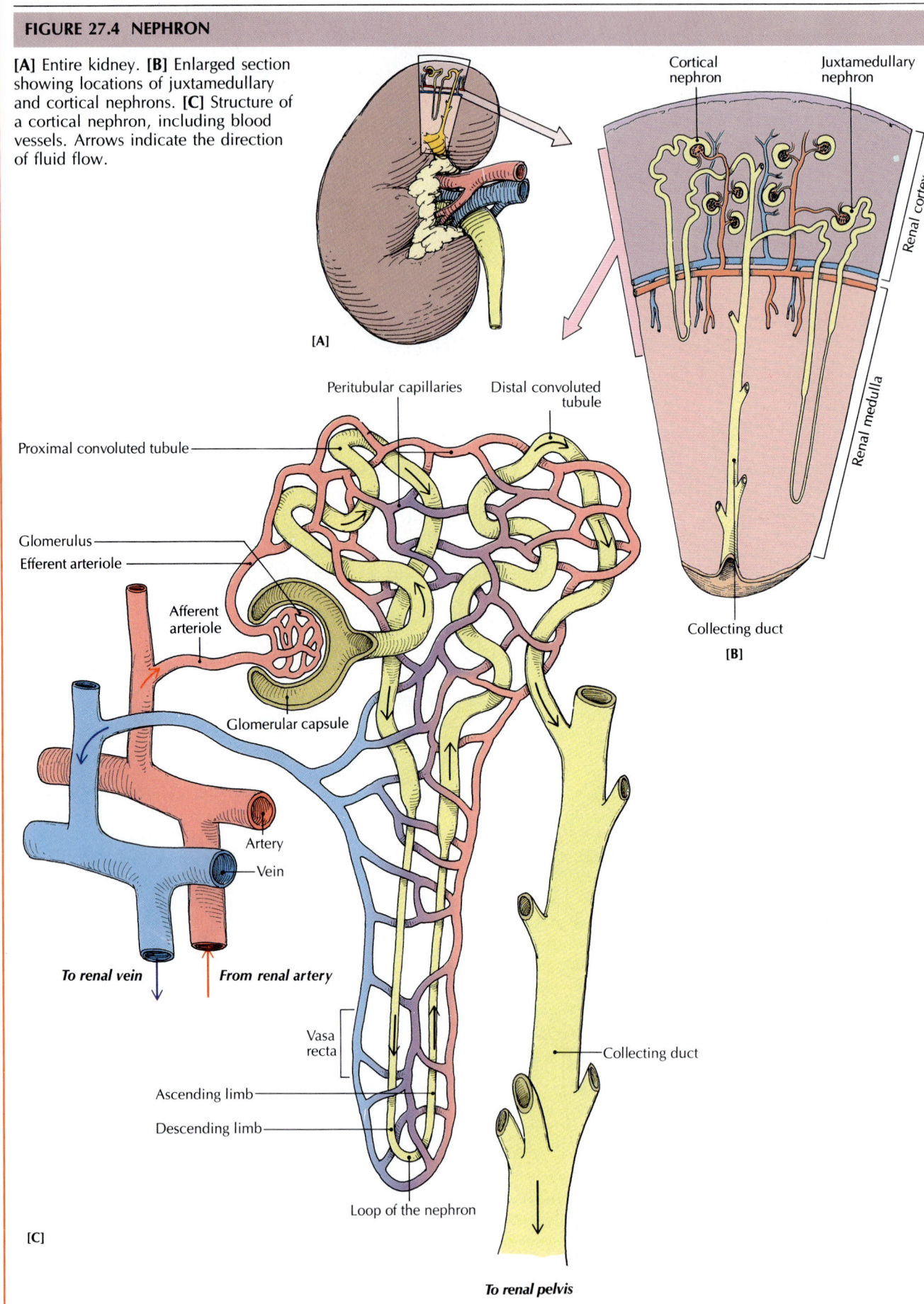

[A]

Cortical
nephron

Juxtamedullary
nephron

Renal cortex

Renal medulla

Collecting duct

[B]

Peritubular capillaries

Distal convoluted
tubule

Proximal convoluted tubule

Glomerulus

Efferent arteriole

Afferent
arteriole

Glomerular capsule

Artery

Vein

To renal vein

From renal artery

Vasa
recta

Collecting duct

Ascending limb

Descending limb

Loop of the nephron

[C]

To renal pelvis

Structure	Major functions
Glomerulus	Vascular (capillary) component of renal corpuscle. Filters (by *hydrostatic pressure*) water, dissolved substances (minus most plasma proteins, blood cells) from blood plasma.
Glomerular capsule	Initial tubular component of nephron. Transports glomerular filtrate to proximal tubule.
Proximal convoluted tubule	Reabsorbs (by *active transport*) Na^+, K^+, Ca^{2+}, amino acids, uric acid, ascorbic acid, ketone bodies, glucose. Reabsorbs (as a result of active H^+ secretion) HPO_4^{2-}, Cl^-, SO_4^{2-}, HCO_3^-. Reabsorbs (by *osmosis*) water. Reabsorbs (by *diffusion*) urea. Actively secretes substances such as penicillin, histamine, organic acids, organic bases.
Descending loop of the nephron	Reabsorbs (by *active transport*) Na^+. Reabsorbs (by *electrochemical gradient*) Cl^-, SO_4^{2-}. Reabsorbs (by *osmosis*) water. Reabsorbs (by *diffusion*) urea.
Ascending loop of the nephron	Reabsorbs (by *active transport*) Na^+, Cl^-. Reabsorbs (by *electrochemical gradient*) HCO_3^-.
Distal convoluted tubule	Reabsorbs (by *active transport*) Na^+. Reabsorbs (by *electrochemical gradient*) PO_4^{3-}, Cl^-, SO_4^{2-}, HCO_3^-. Reabsorbs (by *osmosis*) water. Reabsorbs (by *diffusion*) urea. Actively secretes H^+.
Collecting duct	Reabsorbs (by *active transport*) Na^+. Reabsorbs (by *osmosis*, under control of ADH) water. Actively secrets H^+ and K^+. Actively reabsorbs K^+.

The ***renal cortex*** is the outermost portion of the kidney. It is divided into two regions: the outer *cortical region* and the inner *juxtamedullary* ("next to the medulla") *region*. The cortex has a granular, textured appearance and extends from the outermost *capsule* to the base of the renal pyramids [FIGURE 27.3B]. The granular appearance is caused by spherical bundles of capillaries and associated structures of the nephron that help filter blood. The cortical tissue that penetrates the depth of the renal medulla between the renal pyramids forms ***renal columns***. The columns are composed mainly of interlobar arteries.

Blood and Lymph Vessels

The kidneys receive more blood in proportion to their weight than any other organ in the body. About 20 to 25 percent of the cardiac output goes to the kidneys. Approximately 1.2 L of blood passes through the kidneys every minute, and the body's entire blood volume (4 to 6 L) is filtered through the kidneys 340 times a day. Only a small fraction of this blood is necessary to supply the kidneys' nutritive needs.

Blood comes to each kidney from the abdominal aorta through the ***renal artery*** [FIGURE 27.3A]. Once inside the kidney, the renal artery branches into the ***interlobar arteries***, which pass through the renal columns. When the arteries reach the juncture of the cortex and medulla, they turn and run parallel to the bases of the renal pyramids. At the turning point, the arteries are the ***arcuate arteries*** (L. *arcuare*, to bend like a bow), which make small arcs around the boundary between the cortex and medulla. The arcuate arteries branch further into the ***interlobular arteries***, which ascend into the cortex. Further branching produces numerous small ***afferent arterioles***, which carry blood *to* the sites of filtration (the glomeruli).

Each afferent arteriole branches extensively to form a tightly coiled ball (tuft) of capillaries called a ***glomerulus*** (glow-MARE-yoo-luhss; L. ball) [FIGURE 27.4C]. This capillary bed is where the blood is filtered. Glomerular capillary loops eventually join together to form a single ***efferent arteriole***, which carries blood *away from* the glomerulus. The efferent arteriole is narrower than the afferent arteriole. As a result, the efferent arteriole has a higher resistance to blood flow than the afferent arteriole.

The efferent arteriole eventually branches to form a second capillary bed (the glomerulus was the first one) called the ***peritubular*** ("around the tubules") ***capillaries*** [FIGURE 27.4C]. The peritubular capillaries unite to form the ***interlobular veins***, which carry blood out of the cortex to the ***arcuate veins***. The small arcuate veins join to form the larger ***interlobar veins*** in the renal columns, and the interlobar veins eventually come together to form the single ***renal vein*** that carries blood from each kidney to the inferior vena cava [see FIGURE 27.3A].

The ***arteria recta*** ("straight arteries") are extensions of the efferent arterioles that surround a portion of the nephron called the loop of the nephron (loop of Henle) [FIGURE 27.4C]. They follow the loop of the nephron deep into the medulla and drain into the **vena recta,** which carry blood to the interlobular veins. Only about 1 or 2 percent of the total renal blood flows through these vessels, collectively called the ***vasa recta*** [FIGURE 27.3B]. Nevertheless, the vasa recta play an important role in the concentration of urine, which takes place in the medullary structures of the nephron, as will be seen later.

Like all arterioles, the afferent and efferent arterioles that join the glomerular and peritubular capillary beds contain smooth muscle in their walls (tunica media). These muscles permit the arterioles to constrict or dilate in response to neural or hormonal stimulation.

Lymphatic vessels accompany the larger renal blood vessels and are more prominent around the arteries than the veins. The lymphatic vessels converge in the renal sinus region into several large vessels that leave the kidney at the renal hilus.

Nerve Supply

Sympathetic nerves of the autonomic nervous system enter the kidney at the hilus and follow the arteries to innervate the smooth muscle of the afferent and efferent arterioles. This innervation regulates blood flow through the kidneys by regulating the diameter of blood vessels. These vasomotor nerves help control kidney function by regulating blood pressure in the glomerulus.

In a male, "pain" derived from the kidney and upper ureter can be referred to the testis and skin of the lumbar and inguinal regions, and the upper thigh. Such *referred pain* occurs because the nerves innervating both these organs and skin of these regions terminate in the upper lumbar levels of the spinal cord.

The Nephron

Each ***nephron*** (Gr. *nephros,* kidney) is an independent urine-forming unit, and each kidney contains approximately 1 million nephrons. As the functional unit of the kidney, the nephron accomplishes (1) the initial *filtration* of blood, (2) the selective *reabsorption* back into the blood of filtered substances that are useful to the body, and (3) the *secretion* of unwanted substances from the blood into the filtrate.

Two types of nephrons are recognized: *cortical* and *juxtamedullary* [FIGURE 27.4B]. The tubular structures of the cortical nephron extend only into the base of the renal pyramid of the medulla, while the longer loop of the nephron of the juxtamedullary nephron projects deeper into the renal pyramid. Cortical nephrons are about seven times more numerous than juxtamedullary nephrons.

A nephron consists of (1) a tubular component and (2) an associated vascular component. The *tubular component* starts with the *glomerular (Bowman's) capsule* and includes the excretory tubules, which are the *proximal convoluted tubule; loop of the nephron* (loop of Henle), which consists of a descending and ascending limb; and *distal convoluted tubule* [FIGURE 27.4C]. Most tubules of a nephron are coiled and winding; all the tubules from all the nephrons in the body have a combined length of about 80 km (50 mi). Each of the excretory tubules leads into a large *collecting duct* (which is not part of the nephron) that transports the resulting renal filtrate.

The associated *vascular component* of a nephron is made up of blood vessels. These include the glomerulus and the peritubular capillaries (including the vasa recta), which surround the excretory tubules. The reabsorption and secretion of substances from the excretory tubules into the blood take place in the peritubular capillaries and vasa recta. Each of these parts of a nephron and the associated vascular structure are discussed in detail in the following sections.

Glomerular (Bowman's) capsule The *glomerular (Bowman's) capsule* (originally named for Sir William Bowman, the nineteenth-century English anatomist who first described this structure) is the portion of the nephron that encloses the glomerulus like a hand wrapped around a ball [FIGURE 27.5A]. Together, the capsule and glomerulus form the **renal corpuscle.** The glomerular capsule is always located in the cortex of the kidney and is the first part of a nephron. The outer layer of the glomerular capsule is the *parietal layer* and is composed of simple squamous epithelial cells. The inner *visceral layer* is composed of specialized epithelial cells called **podocytes** (POH-doh-sites; "footlike cells"), which closely surround the glomerular capillaries [FIGURE 27.5B]. The inner and outer walls form a cavity called the *capsular space.*

Filtration of the blood takes place in the renal corpuscle. Vascular fluid must cross three layers before entering the tubular system:

1 The first layer is the *endothelium* of the glomerulus, which contains tiny pores called *fenestrations* ("windows") [FIGURE 27.5B].

2 The middle layer is the *basement membrane* of the glomerulus.

3 The third layer is the *visceral layer* of the glomerular capsule and the podocytes. The podocytes are relatively large cells with a nucleated cell body from which several thousand cell processes spread. These processes branch to form many smaller, fingerlike processes called *foot processes* or **pedicels** (PEHD-ih-sehlz; L. "little feet") [FIGURE 27.5B]. The pedicels cover the basement membrane of the glomerulus. The small regions between pedicels where the underlying basement membrane is exposed are

FIGURE 27.5 GLOMERULAR (BOWMAN'S) CAPSULE STRUCTURE

[A] Drawing of the renal corpuscle. Arrows indicate direction of blood flow. [B] Enlarged drawing of part of a glomerular capillary. [C] Scanning electron micrograph showing podocyte cell processes on glomerular capillaries.

[D] Scanning electron micrograph showing a transected glomerular capillary loop surrounded by podocytes. ×17,200. [E] Electron micrograph of a portion of the wall of a glomerular capillary. ×34,000.

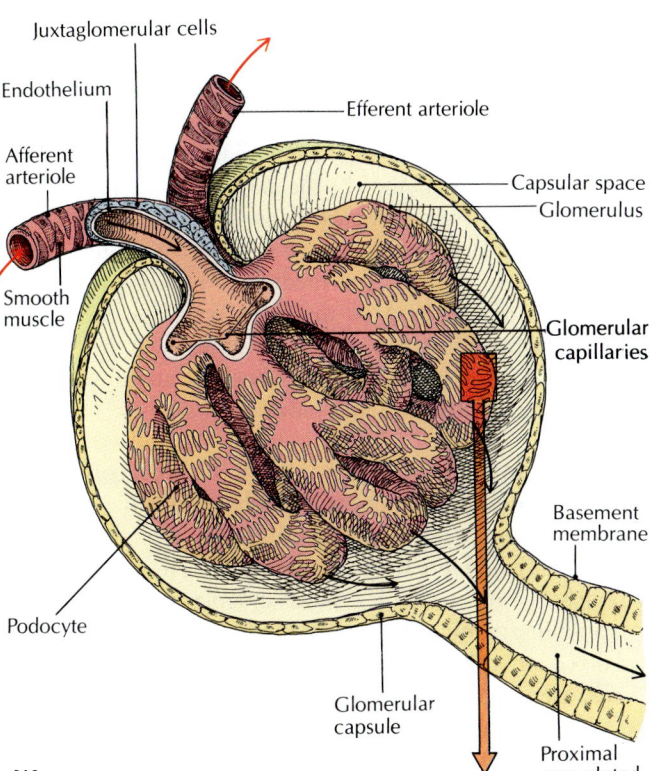

[A]

Juxtaglomerular cells
Endothelium
Afferent arteriole
Smooth muscle
Podocyte
Efferent arteriole
Capsular space
Glomerulus
Glomerular capillaries
Basement membrane
Glomerular capsule
Proximal convoluted tubule

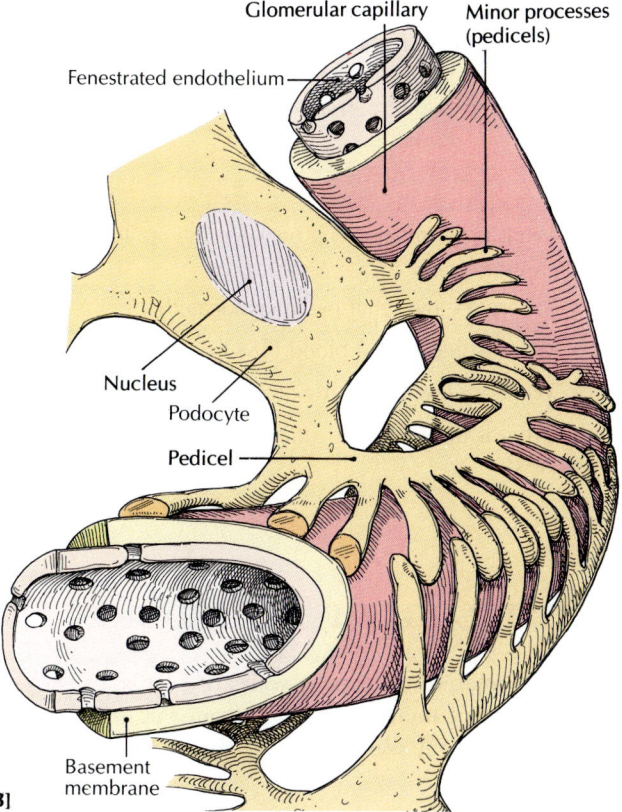

[B]

Glomerular capillary
Minor processes (pedicels)
Fenestrated endothelium
Nucleus
Podocyte
Pedicel
Basement membrane

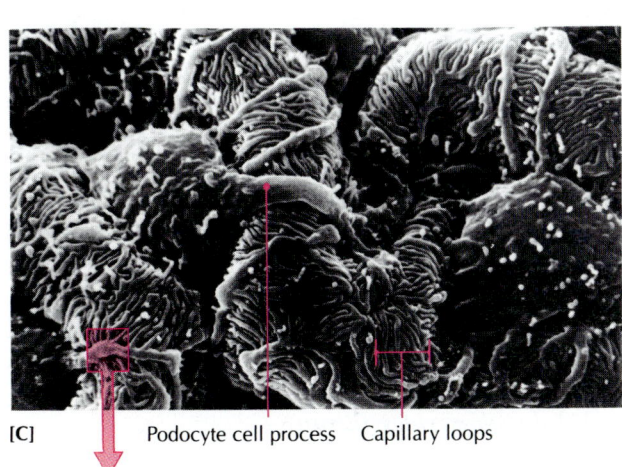

[C]   Podocyte cell process   Capillary loops

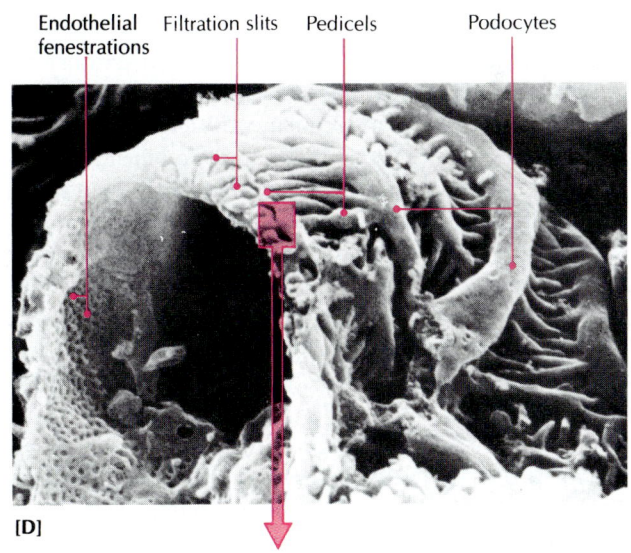

[D]

Endothelial fenestrations   Filtration slits   Pedicels   Podocytes

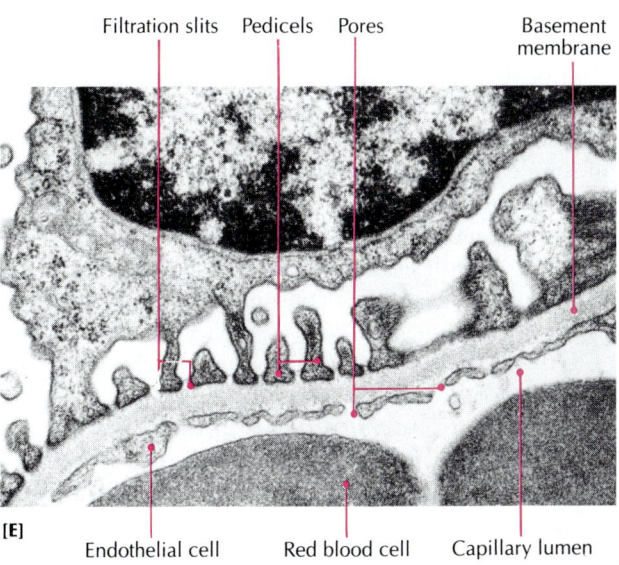

[E]

Filtration slits   Pedicels   Pores   Basement membrane
Endothelial cell   Red blood cell   Capillary lumen

called *filtration slits* [FIGURE 27.5D]. These thin *slit membranes* extend between the foot processes of adjacent cells; the membrane restricts the passage of certain molecules through the filtration slit but allows the passage of others.

This combination of barriers—the fenestrated endothelium, basement membrane, and filtration slits—is called a *filtration barrier.*

The three layers of the renal corpuscle (endothelium, basement membrane, and visceral layer) constitute the *capsular membrane.* The filtration process involves the entire membrane. Although the cellular components of the blood and the large proteins do not normally pass through, water and dissolved solutes (electrolytes, sugars, urea, amino acids, and polypeptides) have no trouble passing from the blood into the capsular space of the glomerular capsule. The fluid filtered from the blood is called the *glomerular filtrate.*

Proximal convoluted tubule   From the glomerular capsule, the glomerular filtrate drains into the *proximal convoluted tubule (PCT)* [FIGURE 27.4C]. The name describes its location (it is the portion of the excretory tubule *proximal* to the glomerular capsule) and its appearance (it is coiled in an irregular, *convoluted* way).

The epithelial cells that make up the tubule are cuboidal epithelia, and the surfaces that face the lumen of the tubule are lined with microvilli, forming a brush border. The microvilli enormously increase the epithelial surface area over which transport can take place. The proximal convoluted tubule is the site for the reabsorption of many substances filtered from the blood, including water, electrolytes, glucose, and some amino acids and small polypeptides.

Loop of the nephron (loop of Henle)   After passing through the proximal convoluted tubule, the remaining glomerular filtrate enters a straight portion of the excretory tubule called the *loop of the nephron (loop of Henle),* originally named for the nineteenth-century German anatomist Friedrich Henle [FIGURE 27.4C]. The loop of the nephron is composed of descending and ascending limbs. The descending limb and the first part of the ascending limb (the *thin ascending limb*) are composed of simple squamous epithelium. The *thick ascending limb* is composed of cuboidal epithelium. Nephrons in the cortex have a thick ascending limb, but no thin ascending limb. As you will see later, the purpose of the loop of the nephron is to generate a concentration gradient.

Distal convoluted tubule   After the glomerular filtrate passes through the loop of the nephron, it moves into the *distal convoluted tubule* in the cortex [FIGURE 27.4C]. As the name suggests, it is an irregularly shaped tubule located far from the glomerular capsule. The cu-

boidal epithelial cells of the distal tubule are similar in size to those of the proximal tubule, but they have very few microvilli. The cells are abundantly supplied with mitochondria near their basal surfaces to provide energy for the active transport of Na^+ and H^+.

Collecting duct   Glomerular filtrate passes from the distal convoluted tubule into the *collecting duct* [FIGURE 27.4B, C]. The collecting duct is lined by two types of epithelial cells: (1) The *light cell* is characterized by apical microvilli and cilia, and (2) the *dark cell* is characterized by branching irregular apical flaps. The distal tubules of many nephrons empty into a single collecting duct, which travels through the medulla, roughly parallel to the limbs of the loop of the nephron. The collecting ducts join to form larger and larger tubes until they reach the minor calyx. From there, the final filtrate (now called *urine*) drains into the renal pelvis.

Juxtaglomerular apparatus   In the renal cortex, the distal convoluted tubule makes contact with the afferent (and sometimes efferent) arterioles of the glomerulus [FIGURE 27.6]. Here the smooth muscle cells of the tunica media of the arterioles have cytoplasm with renin-containing granules instead of myofilaments. These specialized smooth muscle cells, known as *juxtaglomerular cells,* are in contact with a group of epithelial cells of the distal convoluted tubule called the *macula densa* (L. "dense spot"). The cells of the macula densa are longer and narrower than the typical epithelial cells of the distal convoluted tubule.

Together, the juxtaglomerular cells of the afferent arteriole and the macula densa cells of the distal convoluted tubule make up the *juxtaglomerular apparatus* (JGA), which helps regulate systemic blood pressure.

ASK YOURSELF

1 What is the relationship between the renal pelvis and the calyces?

2 Where are the renal pyramids located?

3 What is the path of blood flow through the kidneys?

4 What are the main parts of a nephron?

5 What is the juxtaglomerular apparatus? Why is it so named?

PHYSIOLOGY OF THE KIDNEYS

In the formation of urine, a series of events leads to the elimination of metabolic wastes from the body, regulation of total body water balance, control of the chemical

FIGURE 27.6 JUXTAGLOMERULAR APPARATUS

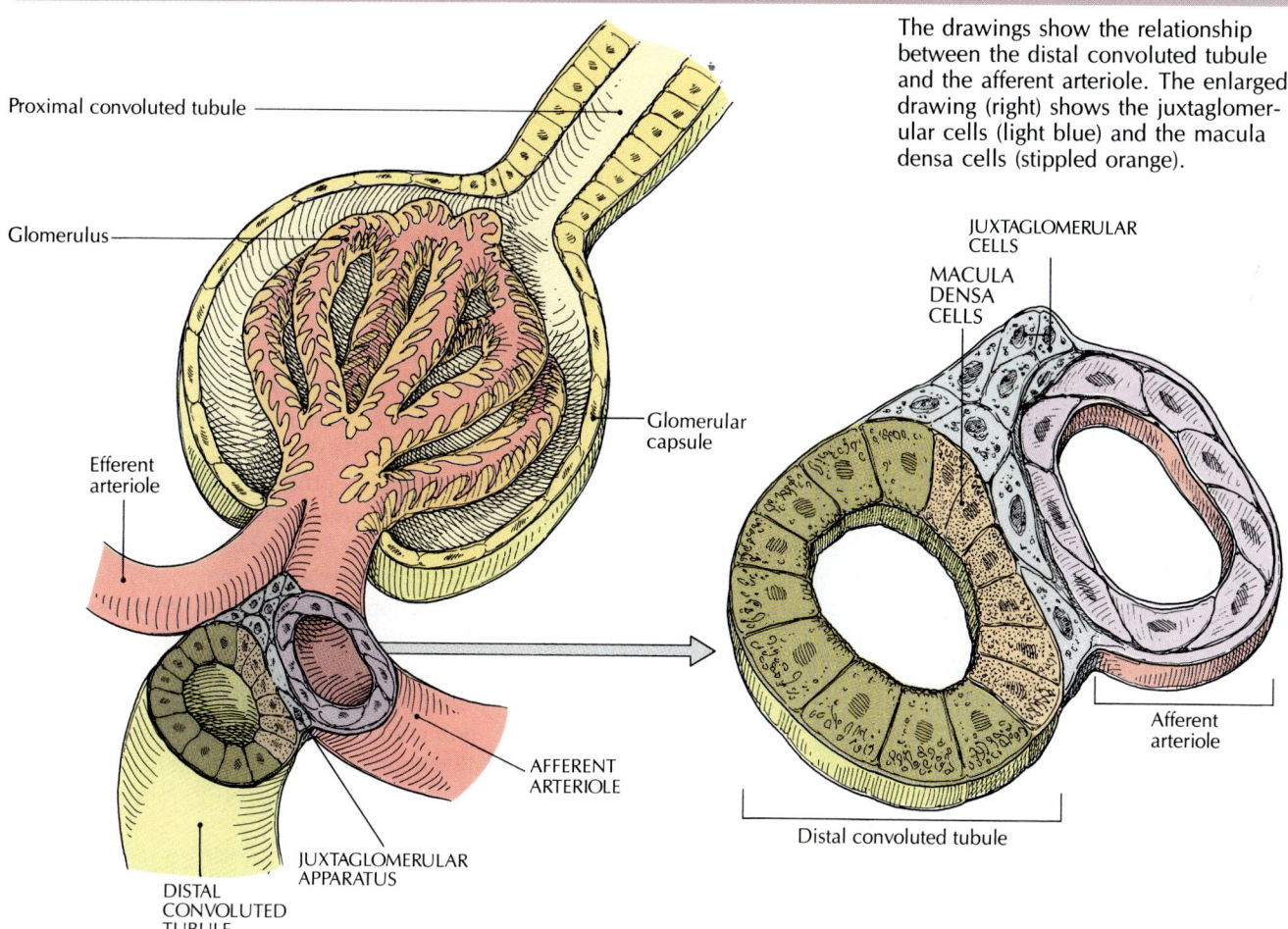

Proximal convoluted tubule

Glomerulus

Efferent arteriole

Glomerular capsule

AFFERENT ARTERIOLE

JUXTAGLOMERULAR APPARATUS

DISTAL CONVOLUTED TUBULE

The drawings show the relationship between the distal convoluted tubule and the afferent arteriole. The enlarged drawing (right) shows the juxtaglomerular cells (light blue) and the macula densa cells (stippled orange).

JUXTAGLOMERULAR CELLS

MACULA DENSA CELLS

Afferent arteriole

Distal convoluted tubule

composition of the blood and other body fluids, regulation of blood pressure, and control of acid-base balance.

By three separate processes, the kidneys produce and modify the glomerular filtrate that is finally excreted from the body as urine [FIGURE 27.7]. These processes are outlined below:

1 *Glomerular filtration:* The kidneys filter blood. When blood flows from the afferent arteriole into the glomerulus, it is under high pressure (about 75 mmHg). This pressure forces *some* of the blood plasma into the glomerular capsule, but the blood cells and large proteins remain within the glomerulus, unable to pass through the endothelial capsular membrane. This process, known as *glomerular filtration,* forms the initial glomerular filtrate.

2 *Tubular reabsorption:* As the glomerular filtrate passes through the length of the nephron tubule, useful substances such as water, sodium ions, glucose, and amino acids that were initially lost from the blood during filtration are returned to the blood by active and passive transport. This process is called *tubular reabsorption.*

3 *Tubular secretion:* Some unwanted ions and substances may be transported (secreted) from the blood in the peritubular capillaries *into* the glomerular filtrate as it passes through the nephron tubules. In this way, products such as potassium ions, hydrogen ions, certain drugs (penicillin, for example), and organic compounds may be excreted. This process is *tubular secretion.* It occurs in a direction opposite to that of tubular reabsorption.

In summary, glomerular filtrate flows through a nephron to the inner medulla of the kidney and then back again to the cortex. In the process, essential substances such as water and glucose are reabsorbed into the blood. Finally, the glomerular filtrate moves to the medulla of the kidney again, where it is now called urine, and passes to the urinary bladder through the ureter. The filtered blood is returned to the body through the renal vein.

FIGURE 27.7 URINE PRODUCTION IN A NEPHRON

In *glomerular filtration* (1) fluid passes from the blood into the glomerular capsule. In *tubular reabsorption* (2) essential substances such as glucose and water pass from the excretory tubule back into the blood of the peritubular capillaries. In *tubular secretion* (3) materials pass from the blood of the peritubular capillaries into the nephron tubule. Note that tubular secretion and reabsorption actually occur simultaneously.

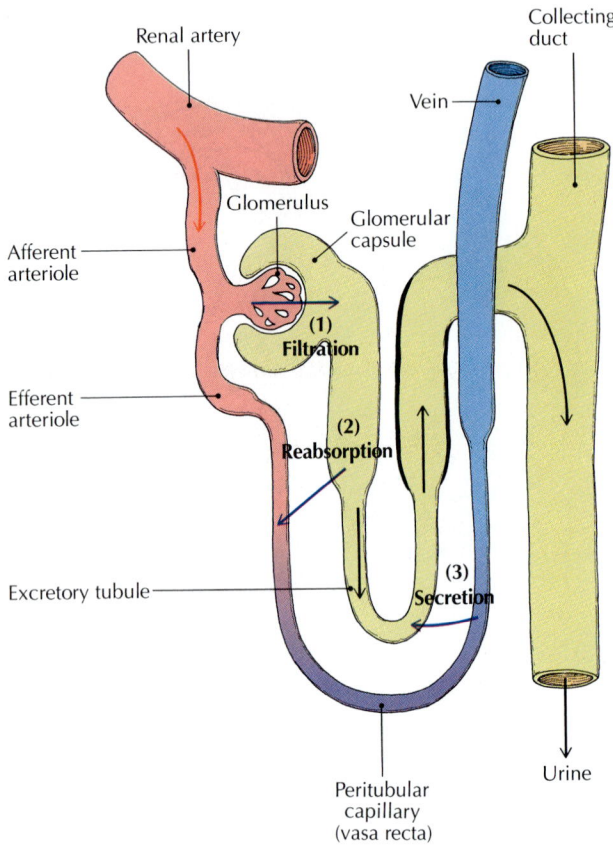

Glomerular Filtration

The filtration that takes place in the renal corpuscle is driven by the high capillary hydrostatic pressure (blood pressure) within the glomerulus. Even though the endothelial capsular membrane of the glomerular capsule is composed of three layers [FIGURE 27.5], it is 100 to 1000 times more permeable to water and ions than an ordinary capillary membrane. Small molecules, ions, and water pass through the fenestrations of the endothelium, the basement membrane, and the filtration slits of the visceral layer of the glomerular capsule. Normally, none of the blood cells or large proteins pass through this barrier.

Mechanics of glomerular filtration *Filtration* is the forcing of a fluid and the substances dissolved in it through a membrane by pressure. The resulting fluid is

called a *filtrate*. Filtration occurs in the glomeruli because blood pressure (glomerular hydrostatic pressure) forces water and some dissolved solutes in the plasma through the filtration barrier. The resulting fluid is called *glomerular filtrate*. The glomeruli send large volumes of glomerular filtrate into the nephron tubules. The blood pressure in the glomerular capillaries is higher than in other capillaries of the body, partly because the afferent arterioles are short, straight branches of the interlobular arteries that can transmit a relatively high pressure, and partly because the efferent arterioles have a smaller diameter.

Net filtration pressure Three factors contribute to the determination of *net filtration pressure (NFP)*, the pressure that promotes glomerular filtration in the kidneys. These factors are (1) glomerular hydrostatic pressure, (2) colloidal osmotic pressure, and (3) capsular hydrostatic pressure. The *glomerular hydrostatic pressure* (blood pressure), at about 55 mmHg, is the major force that moves water and solutes across the glomerular membrane. This force is opposed by the two other forces: the glomerular plasma *colloidal osmotic pressure* of the blood in the glomerular capillaries, which is about 30 mmHg, and the *capsular hydrostatic pressure* (hydrostatic pressure of the glomerular filtrate in the capsular space), which is usually about 15 mmHg [FIGURE 27.8].

Thus the net filtration pressure (NFP) for forming the glomerular filtrate is

NFP = glomerular hydrostatic pressure − (glomerular plasma colloidal osmotic pressure + capsular hydrostatic pressure)

A sample (normal) NFP can be calculated as follows:

NFP = 55 mmHg − (30 mmHg + 15 mmHg)

= 55 mmHg − 45 mmHg

= 10 mmHg

Thus a net filtration pressure of 10 mmHg favors the movement of fluid out of the glomerulus and into the glomerular capsule.

Glomerular filtration rate The amount of filtrate formed in the capsular space each minute is the *glomerular filtration rate (GFR)*. In a normal 70-kg man, the renal blood flow (the rate of blood flow through both kidneys) is about 1200 mL/min, and the renal plasma flow is about 625 mL/min. The percentage of plasma entering the nephrons that actually becomes glomerular filtrate is called the *filtration fraction*. About 20 percent of the plasma is filtered at the glomeruli, giving a glomerular filtration of about 125 mL/min, 7.5 L/hour, or 180 L/day.

FIGURE 27.8 FORCES INVOLVED IN GLOMERULAR FILTRATION

The final net filtration pressure is shown in **[D]**.

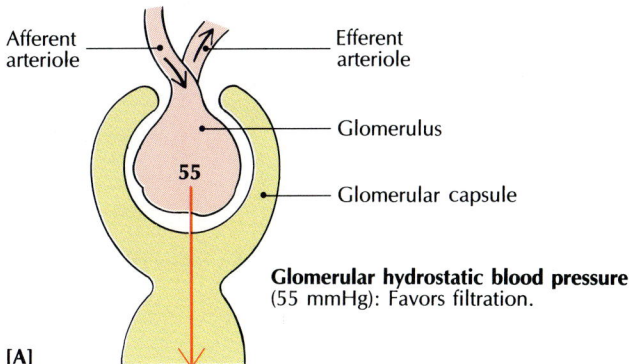

[A]

Afferent arteriole

Efferent arteriole

Glomerulus

Glomerular capsule

55

Glomerular hydrostatic blood pressure
(55 mmHg): Favors filtration.

To proximal convoluted tubule

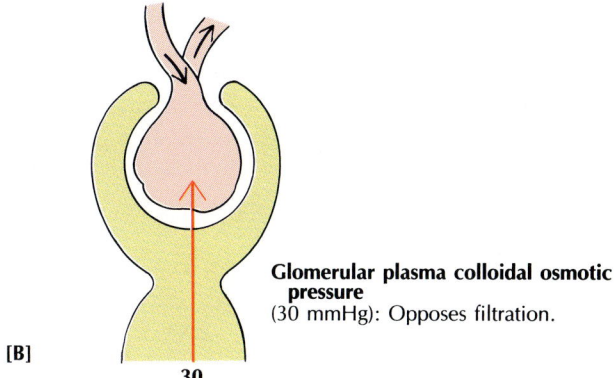

[B]

30

Glomerular plasma colloidal osmotic pressure
(30 mmHg): Opposes filtration.

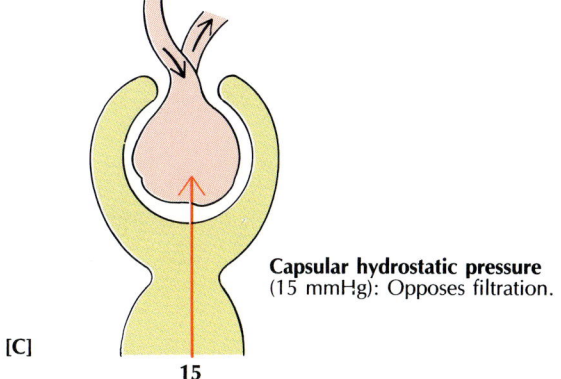

[C]

15

Capsular hydrostatic pressure
(15 mmHg): Opposes filtration.

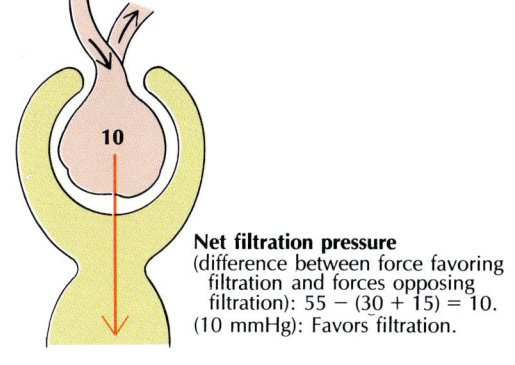

[D]

10

Net filtration pressure
(difference between force favoring filtration and forces opposing filtration): 55 − (30 + 15) = 10. (10 mmHg): Favors filtration.

Factors affecting the glomerular filtration rate

The glomerular filtration rate is dependent on (1) renal autoregulation and (2) sympathetic neural regulation.

1 The kidneys have the ability to maintain a relatively constant glomerular filtration rate in the face of fluctuating blood pressures by an intrinsic (within the kidney) control mechanism known as **renal autoregulation.** This regulation is accomplished by altering the afferent arteriolar diameter. If there is an *increase* in blood pressure, which increases GFR, the net filtration pressure (NFP) and GFR are returned to normal by afferent arteriolar constriction, which decreases blood flow to the glomerulus. If there is a *decrease* in blood pressure and thus a decrease in GFR, glomerular pressure is returned to normal by the dilation of the afferent arteriole. This increases the NFP and returns GFR toward normal.

Renal autoregulation itself may be attributed to two mechanisms: myogenic and tubuloglomerular feedback [FIGURE 27.9]. The **myogenic mechanism** is dependent on the inherent property of vascular smooth muscle to contract in response to increased pressure (stretch) in the vessel. When the afferent arteriole is stretched with increased arterial pressure, it responds by vasoconstricting and reducing blood flow to the glomerulus. In contrast, an unstretched afferent arteriole results in afferent arteriole dilation. The **tubuloglomerular feedback mechanism** operates through negative feedback: When systemic arterial blood pressure drops, there is a decrease in NFP and GFR, which results in a slower rate of flow in the nephrons. The macula densa cells sense this slower flow and inhibit the release of vasoconstrictor agents (such as endothelin) from the juxtaglomerular apparatus. This results in vasodilation of the afferent arterioles, thus increasing blood flow to the glomerular capillaries and also increasing NFP and GFR. If GFR is increased because of an increase in blood pressure, there is an increase in flow in the tubules, which is sensed by the macula densa cells. Vasoconstrictor agents are released to cause afferent arteriolar constriction, thus decreasing GFR toward normal.

These two autoregulatory mechanisms maintain a relatively constant GFR over an arterial pressure range of 80 to 180 mmHg. Systemic blood pressures below or above this range will result in GFR changes proportionate to and in the direction of the blood pressure change.

2 Sympathetic neural control of GFR occurs as sympathetic innervation of the afferent arteriole overrides the autoregulatory mechanisms of control. Increased sympathetic activity results in vasoconstriction of the afferent arteriole, which decreases GFR and urine formation. Such a response occurs in situations like severe hemorrhaging. The drop in blood pressure elicits the baroreceptor reflex, which induces vasoconstriction of

FIGURE 27.9 REGULATION OF THE GLOMERULAR FILTRATION RATE (GFR)

[A] and [B] Autoregulation. [C] Neural (sympathetic) regulation.

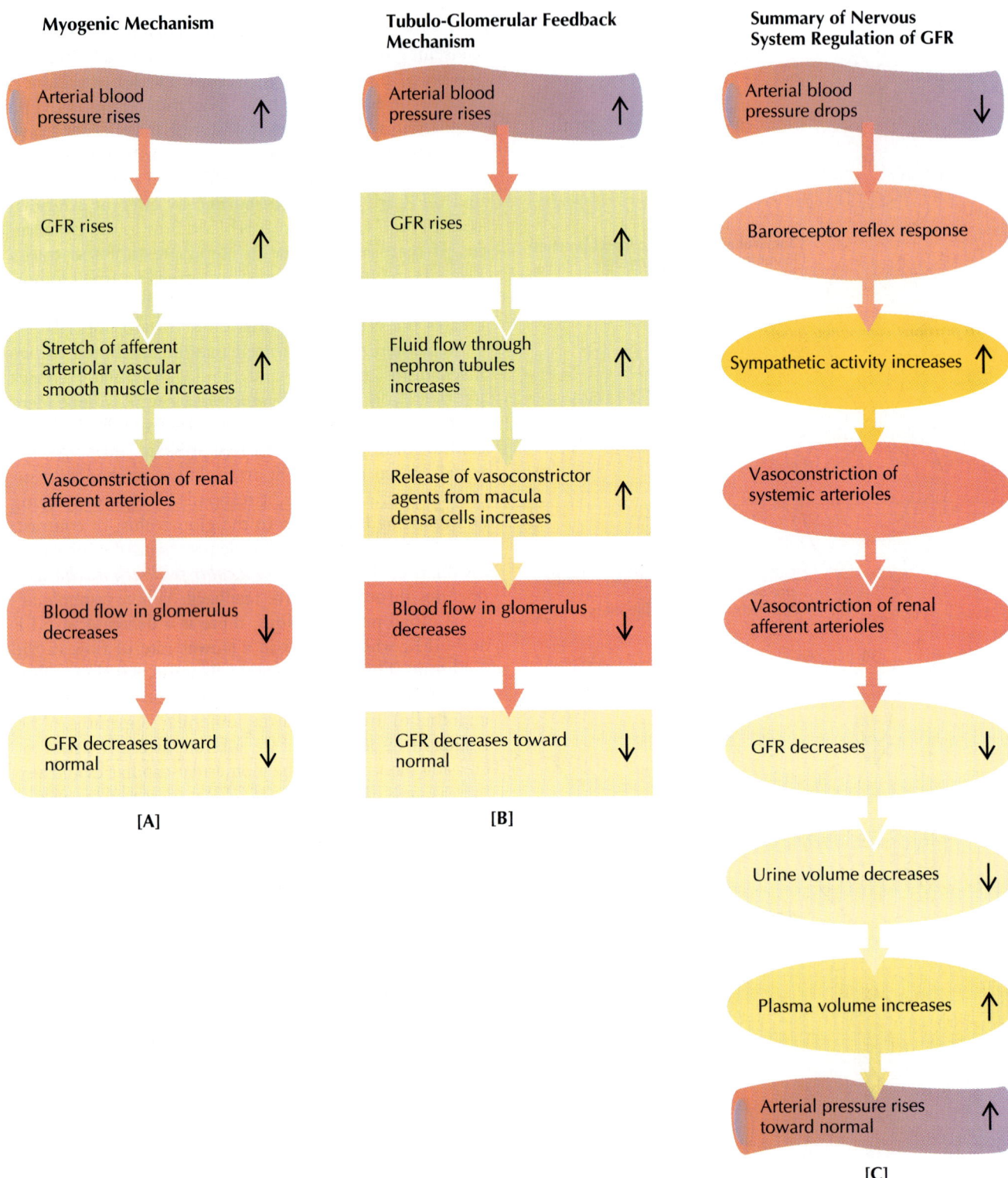

Myogenic Mechanism

Arterial blood pressure rises ↑

GFR rises ↑

Stretch of afferent arteriolar vascular smooth muscle increases ↑

Vasoconstriction of renal afferent arterioles

Blood flow in glomerulus decreases ↓

GFR decreases toward normal ↓

[A]

Tubulo-Glomerular Feedback Mechanism

Arterial blood pressure rises ↑

GFR rises ↑

Fluid flow through nephron tubules increases ↑

Release of vasoconstrictor agents from macula densa cells increases ↑

Blood flow in glomerulus decreases ↓

GFR decreases toward normal ↓

[B]

Summary of Nervous System Regulation of GFR

Arterial blood pressure drops ↓

Baroreceptor reflex response

Sympathetic activity increases ↑

Vasoconstriction of systemic arterioles

Vasocontriction of renal afferent arterioles

GFR decreases ↓

Urine volume decreases ↓

Plasma volume increases ↑

Arterial pressure rises toward normal ↑

[C]

most major arterioles, including the renal afferent arterioles. Thus there is a decrease in GFR and a decrease in urine formation in an attempt to restore plasma volume and blood pressure to normal [FIGURE 27.9c].

Tubular Reabsorption

Of the glomerular filtrate produced each day (180 L), approximately 99 percent is reabsorbed from the nephrons into the peritubular capillaries, with the remaining 1 percent excreted as urine [TABLE 27.2]. Not only is the volume of the filtrate greatly altered as it passes through the nephron, but so is its composition. Present in the glomerular filtrate as it enters the proximal convoluted tubule are water; glucose; amino acids; ions such as Na^+, K^+, Ca^{2+}, Cl^-, HCO_3^-, HPO_4^-; and waste products such as urea, uric acid, and creatinine.

The movement of substances from the nephron tubules back into the peritubular capillaries is called *tubular reabsorption.* The epithelial cells in the tubules of the nephrons carry out this reabsorption process. These tubular epithelial cells have tight junctions at their apical (lumen) borders, which limit the movement of substances between cells [FIGURE 27.10]. Most of the nutritionally important substances such as glucose and amino acids are completely reabsorbed in the proximal convoluted tubule, whereas water and many ions are reabsorbed to varying degrees in various parts of the nephron, depending on hormonal effects.

Reabsorption of substances from the filtrate into the peritubular capillaries occurs by *passive* (nonenergy-requiring) and *active* (energy-requiring) *transport mecha-*

nisms. Substances actively reabsorbed include glucose, amino acids, and most ions. Many of these substances are cotransported with sodium. FIGURE 27.10 demonstrates the active transport of sodium ions in the proximal convoluted tubule. There is a similar transport of sodium in the distal convoluted tubule and collecting duct. There exists a sodium concentration gradient from the tubule lumen (high sodium) to the interior of the epithelial cell (low sodium). There is also an electrical gradient, with the epithelial cell being more negative intracellularly (-70 mV) than is the tubular lumen (-4 mV). Thus sodium moves from the filtrate into the epithelial cell, down its electrochemical gradient. Sodium is also actively pumped across the membranes at the base and sides of the cells (basolateral) into the peritubular capillaries, using the Na^+/K^+ ATPase pump.

At the same time sodium is reabsorbed out of the tubular cells, potassium is moved into the tubular cells from the peritubular capillaries. Sodium reabsorption also causes the movement of water into the peritubular capillary by osmosis. With the osmotic movement of water out of the filtrate into the peritubular capillary, the remaining substances in the filtrate, such as K^+, Cl^-, HCO_3^-, and urea, become more concentrated and then move passively down their concentration gradients.

Almost 100 percent of filtered nutrients such as glucose and amino acids are reabsorbed in the proximal convoluted tubule. These substances are reabsorbed in such a way that the glucose and sodium are both coupled to the same carrier protein (cotransported). Sodium moves down its concentration gradient into the epithelial cells, moving glucose (for example) up its concentration gradi-

TABLE 27.2 COMPARISON OF FILTRATION, REABSORPTION, AND EXCRETION OF VARIOUS PLASMA CONSTITUENTS*

Substance[†]	Amount filtered by kidneys daily	Amount reabsorbed daily	Percentage reabsorbed daily	Amount excreted in urine daily
Glucose (mmol)	800	800	100	0
HCO_3^- (mEq)	4,900	4,900	100	0
Na^+ (mEq)	26,000	25,850	99.4	150
Water (mL)	180,000	179,000	99.4	1000
Cl^- (mEq)	18,000	17,850	99.2	150
Uric acid (mmol)	50	49	98	5
K^+ (mEq)	600	560	93.3	90
Total solute (mOsm)	54,000	53,400	87	700
Urea (mmol)	870	460	53	410
Creatinine (mmol)	12	1	—	12

*For a normal adult on a normal diet.
[†]Substances are arranged in order of percentage reabsorbed. Units used include the milliequivalent (mEq), a unit of measure that expresses the combining activity of an electrolyte (the amount of one electrolyte that will react with a given amount of hydrogen), and the milliosmole (mOsm), a unit of measure that expresses osmotic pressure.

FIGURE 27.10 ACTIVE TRANSPORT OF SODIUM IONS

Active transport of sodium ions across the basal surface of epithelial cells of the proximal convoluted tubule. Subsequent movement of sodium ions may occur by diffusion. Potassium is accumulated within epithelial cells of the distal convoluted tubule by an active transport system in the basal surface membrane. Potassium accumulates partly in exchange for sodium, which is pumped out actively. Potassium diffuses passively across the apical membrane of the tubule cell, entering the lumen of the tubule. Because one step in this process is an active transport, the entire phenomenon of potassium-ion secretion is classified as active. *Solid arrows* indicate active transport; *dashed arrows* indicate passive transport.

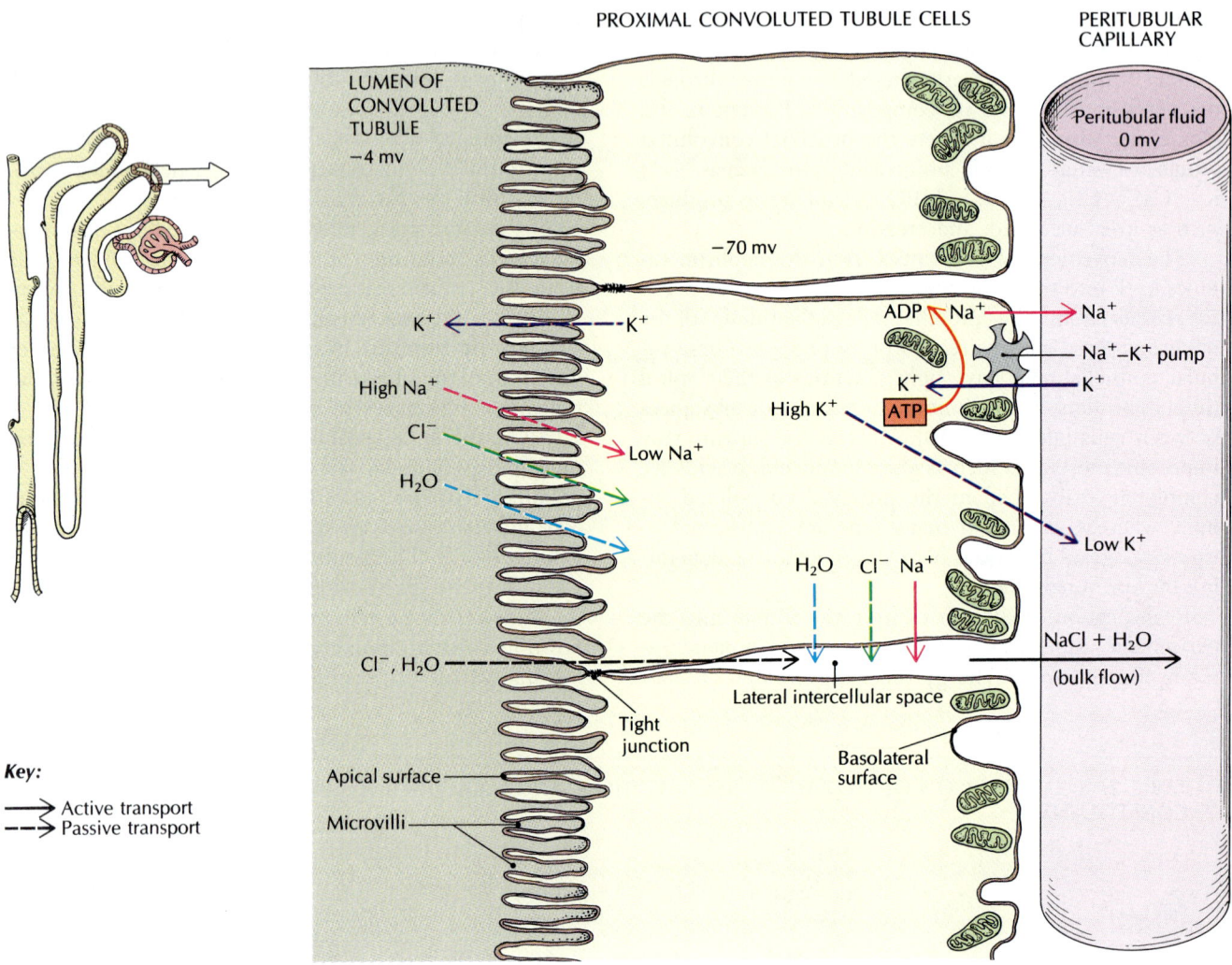

ent. These substances then leave through the basolateral membranes by facilitated diffusion into the peritubular capillaries. The movement of these substances into the peritubular capillaries also pulls water with them by osmosis.

The transport systems for glucose, amino acids, and other filtered nutrients are carrier-specific and transport-limited; that is, there is a *transport maximum (Tm)* for each substance (except sodium). When the carriers for these substances become saturated and the transport system maximum is exceeded, that substance will "spill over" into the urine. If, for example, an individual has an abnormally high blood glucose of 420 mg/100 mL of blood (normal being around 80 to 120 mg/100 mL), the filtered glucose will exceed the calculated Tm for glucose of 375 mg/min, and glucose will appear in the urine. *Renal threshold* (mg/ml) is the plasma level of a substance in excess of its Tm. A substance above its renal threshold concentration will appear in the urine.

Water reabsorption in the descending limb of the nephron occurs by osmosis. The ascending limb is impermeable to water but is very permeable to Na^+ and

FIGURE 27.11 FUNCTIONS OF ADH

Secretion of ADH and the regulation of blood osmolarity.

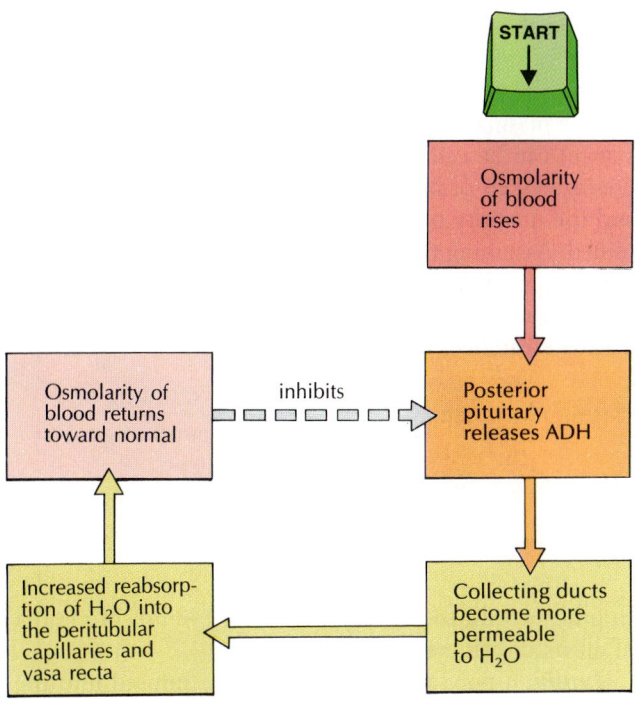

Cl$^-$. These different water permeabilities enable the kidney to fine-tune fluid balance by forming more dilute or more concentrated urines. The distal convoluted tubule and collecting duct are relatively impermeable to water unless antidiuretic hormone (ADH) changes their permeability. When blood osmolarity is high, osmoreceptors in the hypothalamus signal the release of ADH from the posterior pituitary. This hormone increases the water permeability in the distal convoluted tubule and collecting duct, which allows more water to be reabsorbed from the filtrate into the peritubular capillaries, decreasing the blood osmolarity back toward normal [FIGURE 27.11]. If the blood osmolarity is low, ADH secretion is inhibited, water permeability decreases, and a large volume of dilute urine is excreted. The hormone aldosterone, a steroid hormone from the zona glomerulosa of the adrenal cortex, acts on the distal convoluted tubule and collecting duct to increase sodium reabsorption (conserve sodium), with water following passively. Factors that regulate aldosterone release are shown in FIGURE 27.12.

Waste products such as urea and creatinine (nitrogenous by-products of protein and nucleic acid metabolism) are not reabsorbed or are poorly reabsorbed.

Tubular Secretion

The final composition of the urine excreted from the body depends not only on filtration and reabsorption,

but also on the *tubular secretion* of certain substances from the blood *into* the filtrate. Tubular secretion takes place in the proximal and distal convoluted tubules and in the collecting duct, moving substances across epithelial cells using active and passive transport mechanisms similar to those of reabsorption but in the *opposite* direction. Major substances that are secreted are hydrogen and potassium ions.

Hydrogen-ion secretion is very important in the regulation of acid-base balance in the body. Hydrogen ions are secreted when the body fluids are too acidic. Secretion of hydrogen ions is decreased when the body fluids are not acidic.

Potassium ions are actively reabsorbed in the proximal tubule and actively secreted in the distal tubule and collecting duct. The active secretion of potassium is variable and dependent on several factors that ensure a constant plasma potassium level [FIGURE 27.12]: (1) Elevated plasma potassium directly stimulates aldosterone release from the adrenal cortex, which increases potassium secretion. Decreased plasma potassium reduces aldosterone output and conserves potassium. (2) Reduced sodium also stimulates aldosterone release via the renin-angiotensin-aldosterone pathway, which simultaneously promotes sodium reabsorption and potassium secretion. Plasma potassium levels must be closely maintained around normal, since elevated potassium levels

FIGURE 27.12 FACTORS REGULATING POTASSIUM SECRETION

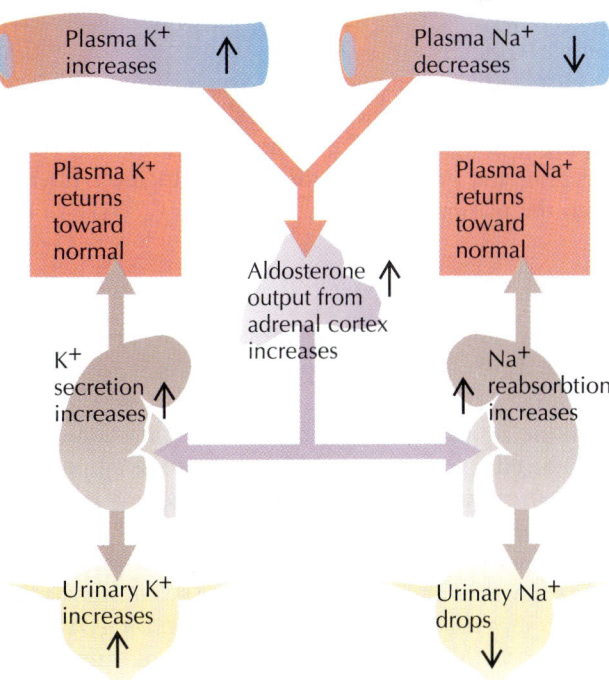

FIGURE 27.13 AVERAGE CONCENTRATIONS OF SUBSTANCES IN DIFFERENT PARTS OF THE NEPHRON

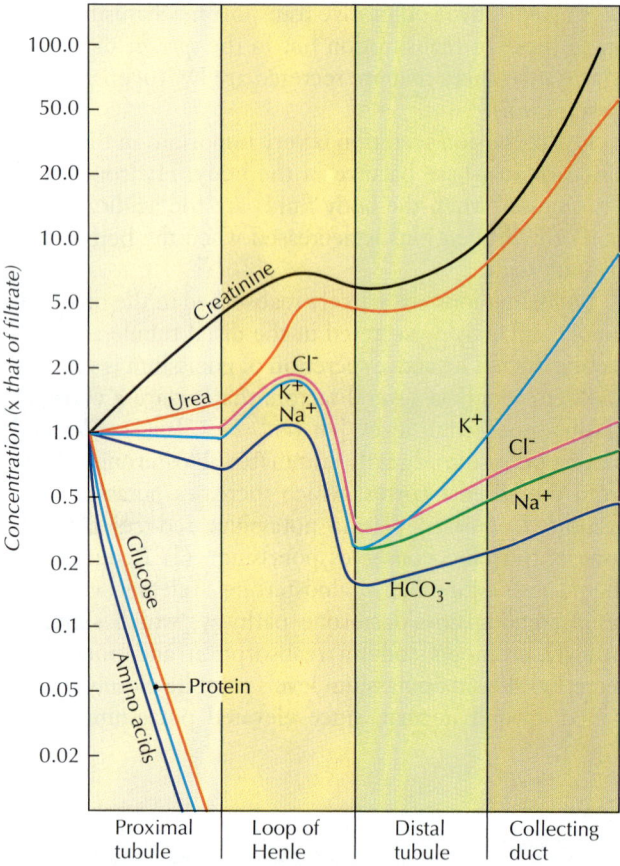

sured in milligrams per minute, is the rate at which the kidneys can excrete the waste compound.)

The excretion of creatinine serves as a good example for plasma clearance, since creatinine can be measured easily in the urine and blood and yields a good estimate of GFR, since it is filtered, not reabsorbed, and only slightly secreted.

The plasma clearance of creatinine (or any other substance) can be calculated as follows: Simultaneous samples are taken of blood and urine excreted each minute, and the quantity of creatinine in each milliliter is determined. Assuming that the kidneys excrete 1.4 mg of creatinine per minute and that the concentration of creatinine in the plasma is 0.01 mg/mL, we can calculate the plasma clearance:

$$\text{Plasma clearance} = \frac{\text{urinary excretion rate (mg/min)}}{\text{plasma concentration (mg/mL)}}$$

$$= \frac{1.4 \text{ mg/min}}{0.01 \text{ mg/mL}}$$

$$= 140 \text{ mL/min}$$

In other words, the kidneys can clear 140 mL of plasma of all its creatinine each minute.

Inulin, a polysaccharide that can be infused into the bloodstream, can be used to determine GFR more accurately, since it is filtered but not reabsorbed or secreted at all. Thus the plasma clearance of inulin is equal to the GFR (125 mL/min):

1 If a substance is filtered, but not reabsorbed or secreted, its plasma clearance equals the GFR.

2 If a substance is filtered and reabsorbed, but not secreted, its plasma clearance is always less than the GFR.

3 If a substance is filtered and secreted, but not reabsorbed, its plasma clearance is always greater than the GFR.

Countercurrent Concentration of Urine

The ***countercurrent mechanism*** is the explanation for the way in which the kidneys form an osmotically concentrated urine. The mechanism depends on (1) the permeability of the different parts of the juxtamedullary nephron tubule (especially the descending and ascending limbs of the loop of the nephron, the vasa recta capillaries, and the collecting duct), (2) the overall structure of the loop of the nephron, (3) the active transport of sodium ions, (4) the concentration gradient in the renal medulla, and (5) the fact that fluid flows first in one direction (*down the descending limb* of the loop of the nephron) and then in the opposite direction (*up the ascending limb*) [FIGURE 27.14]. This is how this mechanism received the name *countercurrent.*

may cause cardiac arrhythmias and cardiac muscle weakness, while low potassium levels cause severe skeletal muscle weakness.

FIGURE 27.13 summarizes the average concentrations of the various substances in the different parts of the nephron, and TABLE 27.2 summarizes filtration, reabsorption, and excretion of the major substances filtered from the plasma.

Plasma Clearance

The processes of glomerular filtration, tubular reabsorption, and tubular secretion produce the excretory product, urine. These processes remove the waste products from the circulating blood that tend to upset homeostasis. The rate per minute at which the kidneys completely clear a particular substance from blood plasma is known as ***plasma clearance.***

In order to quantify the volume of plasma that can be cleared by the kidneys, two values must be known: (1) the *urinary excretion rate* and (2) the *concentration of the substance in the plasma.* (The urinary excretion rate, mea-

FIGURE 27.14 COUNTERCURRENT MULTIPLIER

Operation of the countercurrent multiplier system to produce concentrated urine. Different segments of the loop of the nephron have different permeabilities and transport capacities that affect the countercurrent exchange. Solid arrows indicate active transport, and dashed arrows indicate passive transport. Numerical values are in milliosmoles per liter (mOsm/L).

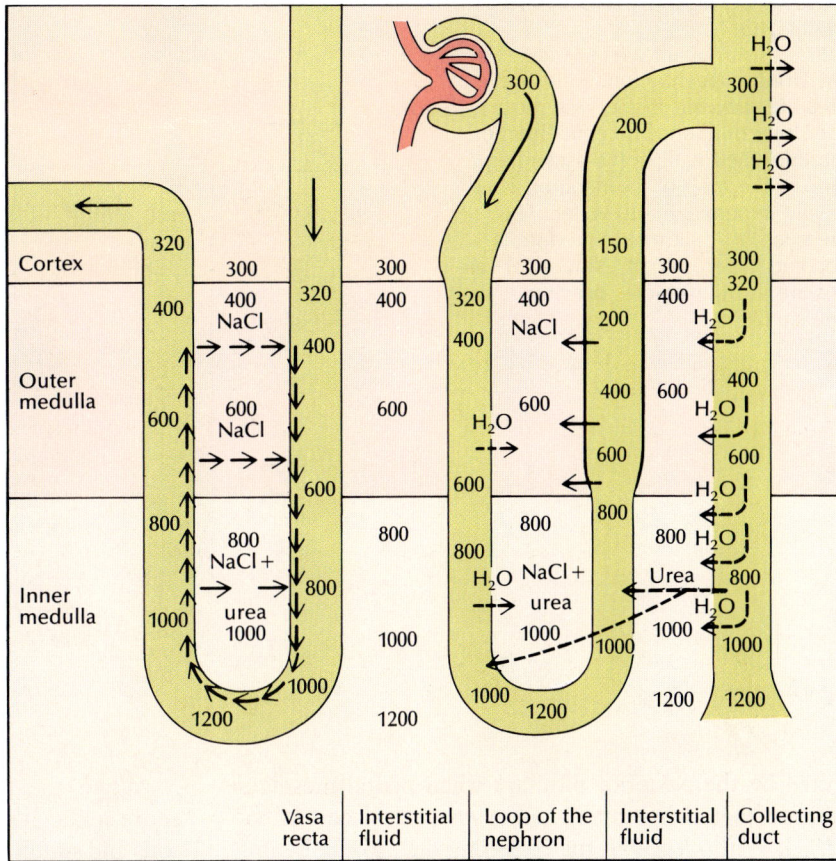

The function of the loop of the nephron is to generate a concentration gradient. The descending limb does not *actively* transport any substance, including sodium ions.* However, sodium ions may enter *passively* from the surrounding tissue [FIGURE 27.15]. The descending limb is also highly permeable to water, which moves *out* by osmosis to the high concentration of solutes in the interstitial fluid. As the glomerular filtrate moves down the descending limb, it loses water passively and becomes more and more concentrated (hyperosmotic). By the time the glomerular filtrate reaches the deep medullary region, it is concentrated to 1200 mOsm/L.

At its most concentrated point, the glomerular filtrate suddenly makes a sharp upward turn into the ascending limb, which runs parallel to the descending limb and close enough to it to allow an exchange of materials. The ascending limb actively cotransports sodium and chloride ions across the basolateral membrane via the sodium-potassium-chloride pump into the surrounding tissue fluid. As the glomerular filtrate moves along the ascending limb, it continues to lose sodium and chloride ions, and the filtrate becomes less and less concentrated (hypoosmotic) as it approaches the distal convoluted tubule. As

*The epithelial cells of the descending limb are small, thin, and flattened. They have none of the microvilli or numerous mitochondria present in epithelial cells that are engaged in active transport.

the glomerular filtrate moves through the nephron, the vasa recta also removes the solutes and water that have been reabsorbed from the loops of the nephron and the collecting duct in order to maintain the cortical-to-medullary concentration gradient of 300-to-1200 mOsm.

A crucial question arises here: If sodium and chloride ions move out of the ascending limb, why doesn't water move out also? The answer is: It would, except that *the walls of the ascending limb are relatively impermeable to water.*

The exchange of sodium and chloride from the ascending limb into the interstitial fluid is called a ***countercurrent multiplier*** [FIGURE 27.14]. This countercurrent multiplier helps maintain the differences in solute concentration from one end of the nephron tubule to the other. When such a system is assisted by active transport, as it is in the loop of the nephron, it actually multiplies the differences in concentration by the countercurrent multiplier system. That is, the more salt the ascending limb transports into the surrounding fluid, the more concentrated will be the fluid that receives it from the descending limb.

In contrast to the ascending limb, the walls of the collecting duct *are* permeable to water. However, this permeability depends on the presence of ADH, which is se-

FIGURE 27.15 MECHANISMS OF URINE CONCENTRATION AND DILUTION

[A] Urine concentration. Salts and urea are recycled in the vasa recta. [B] Urine dilution. The epithelium of the thickened wall of the ascending limb of the loop of the nephron is relatively impermeable to water. The presence of ADH makes the filtrate become hyperosmotic. Numerical values are in milliosmoles per liter (mOsm/L).

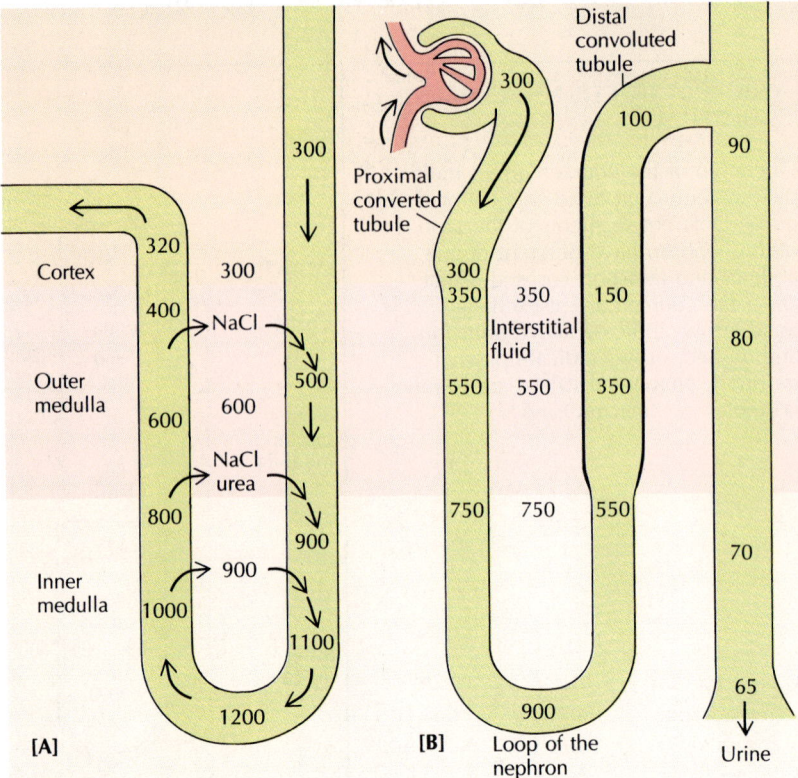

creted by the posterior pituitary when neural messages indicate that dehydration has taken place [see FIGURE 27.11]. Without ADH, the collecting duct would be impermeable, water would remain within the collecting duct, and the urine would be dilute [FIGURE 27.15B]. The presence of ADH enables water to move out of the collecting duct by osmosis, resulting in a final concentrated (hyperosmotic) urine.

In order for this countercurrent multiplier system to remain efficient, only a small number of sodium ions that are transported out of the ascending limb can be carried away by the surrounding vasa recta and peritubular capillaries. A hyperosmotic condition must be maintained in the renal medulla by the vasa recta, which form long loops parallel to the loops of the nephron of the juxtaglomerular nephrons. The sodium (and, to a lesser extent, chloride) that diffuses into the arteria recta subsequently diffuses passively out of the blood vessels in the ascending segments (vena recta) of these vessels. Thus the diffusion of sodium, first into the arteria recta and then out of the vena recta, helps maintain the hyperosmotic condition of the interstitial fluid in the renal medulla.

Acid-Base Regulation

The kidneys help regulate the acid-base balance of the blood primarily by regulating the excretion of hydrogen ions (H^+) and bicarbonate ions (HCO_3^-). This regulatory process helps maintain a homeostatic blood pH of 7.35 to 7.45. The process is summarized here and discussed in detail in the next chapter.

When the blood pH begins to decrease (becomes more acidic), the kidneys secrete the additional hydrogen ions and simultaneously reabsorb the filtered bicarbonate ions. This raises the pH back toward normal. Conversely, when the blood pH begins to increase (becomes more basic), the kidneys decrease their secretion of hydrogen ions and simultaneously decrease reabsorption of the filtered bicarbonate ions. This lowers the pH back toward normal.

ASK YOURSELF

1 What three processes are involved in the overall formation of urine by the kidneys?

2 What is meant by net filtration pressure?

3 What is the glomerular filtration rate, and what factors affect it?

4 What is meant by plasma clearance? How is it calculated?

5 What is the countercurrent multiplier system?

6 How is concentrated urine formed?

ACCESSORY EXCRETORY STRUCTURES

Urine is produced in the kidneys, but accessory structures are required to transfer it, store it, and eventually

FIGURE 27.16 URETER LUMEN AND WALL

The star-shaped lumen and the three tissue layers: mucosa, muscularis, and adventitia. Note that the arrangement of the longitudinal and circular smooth muscle layers is the reverse of that in the intestine. Cross section.

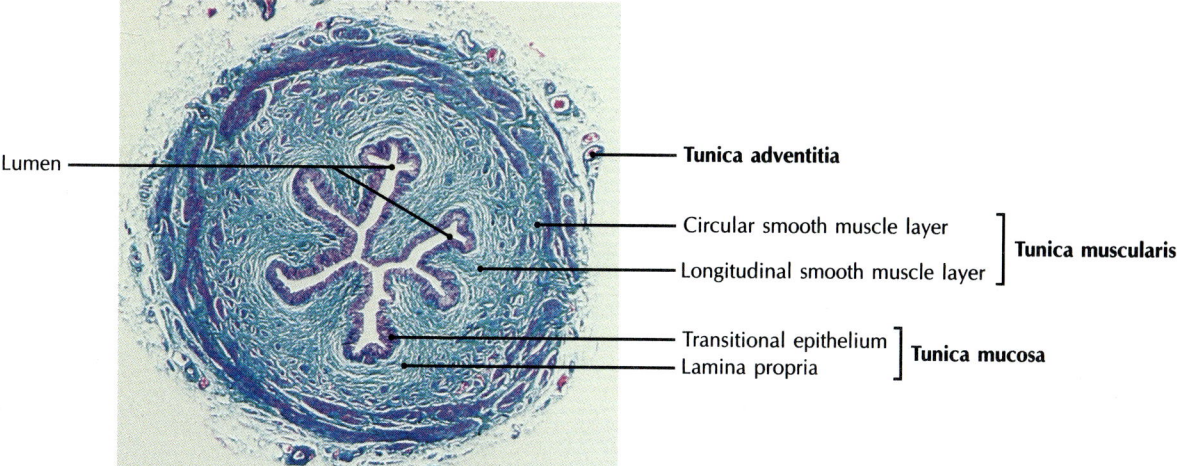

Lumen

Tunica adventitia

Circular smooth muscle layer ⎤
Longitudinal smooth muscle layer ⎦ **Tunica muscularis**

Transitional epithelium ⎤
Lamina propria ⎦ **Tunica mucosa**

eliminate it from the body. These structures are the two ureters, the urinary bladder, and the urethra.

Ureters

Attached to each kidney is a 10- to 12-in. tube called the *ureter,* which transports urine from the renal pelvis to the urinary bladder. The ureters pass from the renal pelvis between the parietal peritoneum and the body wall to the pelvic cavity, where they enter at the base of the urinary bladder on the posterior lateral surfaces. The ureters are narrow near the kidneys and widen near the bladder.

In cross section, the lumen of the ureter has a star shape and three layers of tissue [FIGURE 27.16]:

1 The innermost layer, the *tunica mucosa,* faces the lumen and is made up of transitional epithelium and connective tissue.

2 The middle layer, the *tunica muscularis,* consists of two layers of smooth muscle: an inner longitudinal layer and an outer circular layer. In the lower third of the ureter an additional outer longitudinal layer is present. Periodic peristaltic contractions of this smooth muscle, along with low hydrostatic pressure, help move urine along the ureters into the bladder.

3 The outermost layer is the fibrous *tunica adventitia.* It consists of connective tissue that holds the ureters in place.

Before the ureters end, they run obliquely for a short distance within the bladder wall. This oblique course functions as a check valve. As the bladder fills with urine, it compresses the urethral openings, preventing the backflow of urine from the bladder into the ureters.

Urinary Bladder

The *urinary bladder* is the hollow, muscular organ that collects urine from the ureters and stores it until it is excreted from the body through the urethra. It usually accumulates 300 to 400 mL of urine before being emptied, but it can expand to hold twice that volume. The bladder is located on the floor of the pelvic cavity, and like the kidneys and ureters, it is retroperitoneal. In males, it is anterior to the rectum and above the prostate gland [FIGURE 27.17B]. In females, it is somewhat lower, anterior to the uterus and upper vagina [FIGURE 27.17A].

The urinary bladder is composed of three main layers: (1) It is lined with *transitional epithelium* (tunica mucosa), which allows the bladder to stretch and contract. When the bladder is empty, the tunica mucosa is thrown into folds called *rugae.* (2) The thick middle layer (tunica muscularis) consists of three layers of meshed *smooth muscle:* inner and outer layers of longitudinal fibers and a middle layer of circular fibers. The three layers of muscle are collectively known as the *detrusor muscle.* (3) The outer layer of the bladder is the *adventitia* (tunica serosa). It is derived from the peritoneum and covers only the upper and lateral surfaces of the bladder. It is a moist tissue that lubricates the upper surface, eliminating friction when the full bladder presses against other organs.

The openings of the ureters and urethra into the cavity of the bladder outline a triangular area, the *trigone* [FIGURE 27.18]. Where the urethra leaves the bladder, the smooth muscle in the bladder wall forms spiral, longitudinal, and circular bundles, which contract to prevent the bladder from emptying prematurely. In the middle membranous portion of the urethra, a sphincter of skeletal muscle forms the voluntary *external urethral sphincter* that holds back the urine until urination is convenient.

FIGURE 27.17 URINARY BLADDER AND URETHRA

[A] Sagittal section of the female pelvis showing the urinary bladder and urethra. As the uterus enlarges during pregnancy, it pushes down on the bladder. [B] Sagittal section of the male pelvis. The longer male urethra also carries semen from the testes and accessory reproductive glands through the full length of the penis.

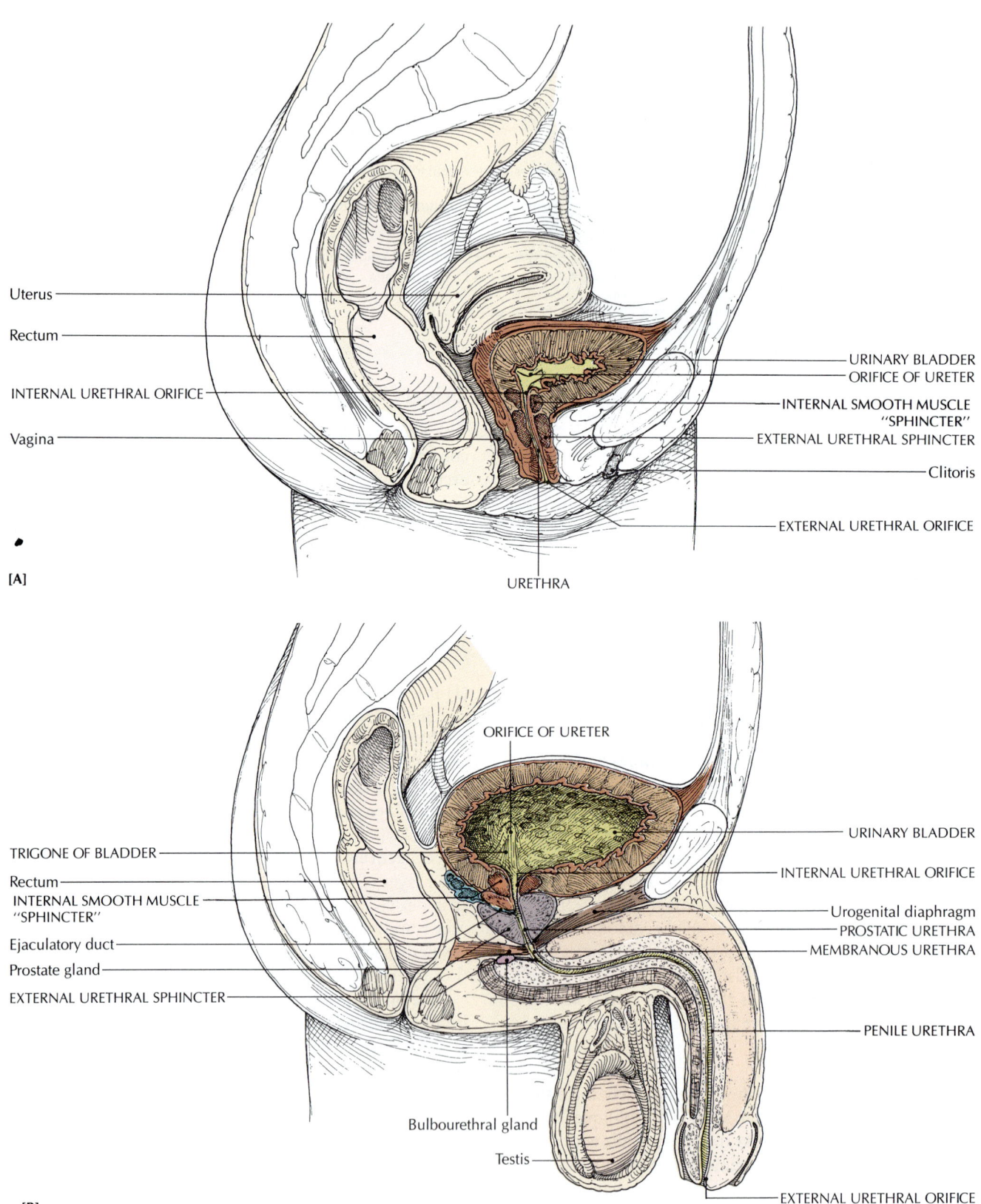

FIGURE 27.18 LONGITUDINAL SECTION THROUGH THE URINARY BLADDER

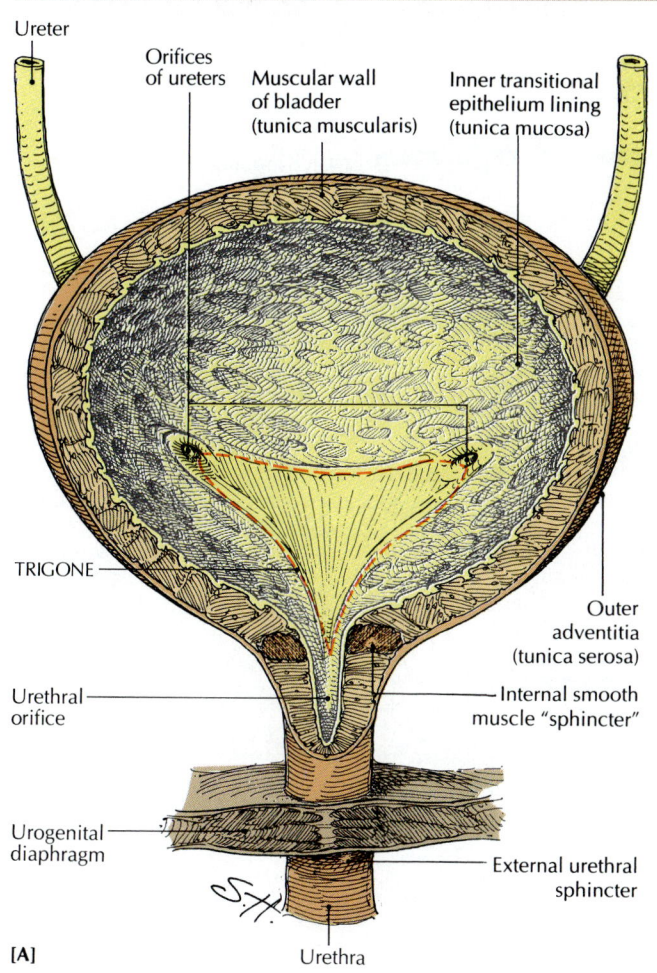

Ureter

Orifices of ureters

Muscular wall of bladder (tunica muscularis)

Inner transitional epithelium lining (tunica mucosa)

TRIGONE

Urethral orifice

Urogenital diaphragm

Outer adventitia (tunica serosa)

Internal smooth muscle "sphincter"

External urethral sphincter

[A]

Urethra

[A] The trigone is a triangular area (dashed lines) inside the urinary bladder formed by the two openings of the ureters and the urethral orifice. When the bladder is distended by urine, the trigone is only slightly distended. **[B]** Section through the bladder wall showing the tissue layers.

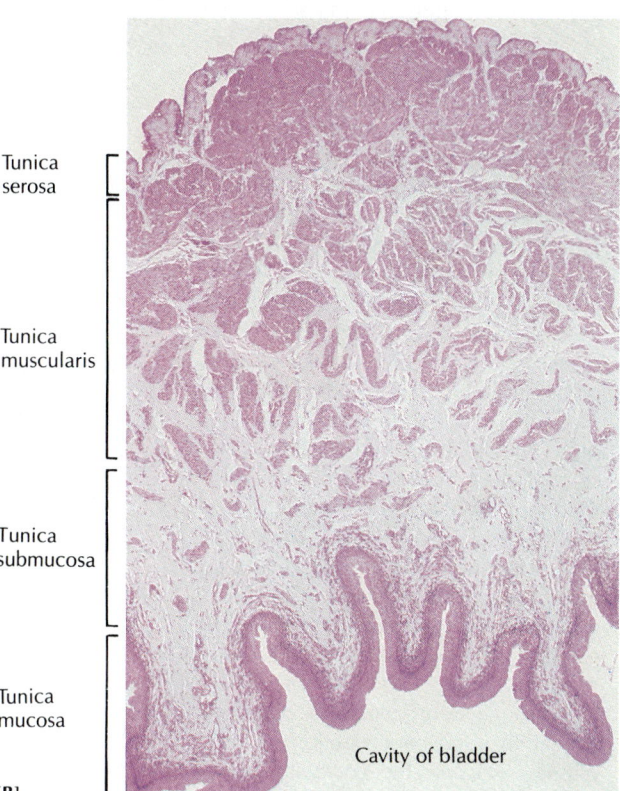

Tunica serosa

Tunica muscularis

Tunica submucosa

Tunica mucosa

Cavity of bladder

[B]

Urethra

The *urethra* is a thin-walled tube of smooth muscle lined internally with mucosa. It joins the bladder at its inferior surface and transports urine outside the body during urination. In the female, the urethra is only about 4 cm (1.5 in.) long, opening to the exterior at an orifice between the clitoris and the vaginal opening [FIGURE 27.17A]. In the male, the urethra is about 20 cm (8 in.) long, extending from the bladder to the external urethral orifice at the tip of the penis.

The male urethra passes through three different regions, and its portions have been named after those regions [FIGURE 27.17B]:

1 The *prostatic portion* passes through the prostate gland and is joined by the ejaculatory duct. Distal to this juncture, the male urethra serves both reproductive and excretory functions.

2 The *membranous portion* is a short segment that passes through the pelvic diaphragm.

3 The longest portion is the *penile,* or *spongy, portion,* which extends the length of the penis from the lower surface of the pelvic diaphragm to the external urethral orifice. The spongy portion is joined by ducts from the *bulbourethral glands* (Cowper's glands) and seminal vesicles, which, together with the prostate gland, secrete fluids into the semen during ejaculation. (Mucus-secreting glands also empty into the urethra along its full length.)

> **ASK YOURSELF**
>
> **1** What is the purpose of the ureters? The urinary bladder?
>
> **2** What is the function of the internal bundles of smooth muscle in the walls of the urinary bladder near the urethra?
>
> **3** What is the urethra? How does it differ in the male and female?

Artificial Kidneys

If the kidneys fail, toxic wastes build up in the body until cells and organs begin to deteriorate and eventually die. Many tens of thousands of Americans suffer from kidney failure, but a large number are leading relatively normal lives because of the successful implantation of donor kidneys or the use of the "artificial kidney"—**hemodialysis therapy** (Gr. *dialuein,* to tear apart).

The principle of diffusion is basic to the working mechanism of the artificial kidney. Very simply, the artificial kidney is a machine that pumps 5 to 6 L of blood from the body through a hollow fiber dialyzer (see drawing). The blood is rinsed by a briny solution, and sodium, potassium, and waste products such as urea, uric acid, excess water, and creatinine diffuse through the dialyzer by osmotic pressure. The cleansed blood is then routed back into the body.

The full procedure is as follows: One tube is permanently implanted into an artery (usually in the arm), and another is implanted into a nearby vein. During dialysis, the tubes are hooked up to the machine. Blood is pumped from the artery through an oxygenated salt solution similar in ionic concentration to body plasma. Because the concentration of wastes is higher than the normal concentration of the plasmalike fluid, the wastes automatically diffuse through the semipermeable membrane of the tubes into this rinsing fluid. The membrane is porous to all blood substances except proteins and red blood cells. The wastes are eliminated from the body, and the purified blood is free to flow back into the body.

Dialysis is sometimes also used to add nutrients to the blood. For instance, large amounts of glucose may be added to the salt solution so that the glucose may be diffused into the blood at the same time that wastes are being removed.

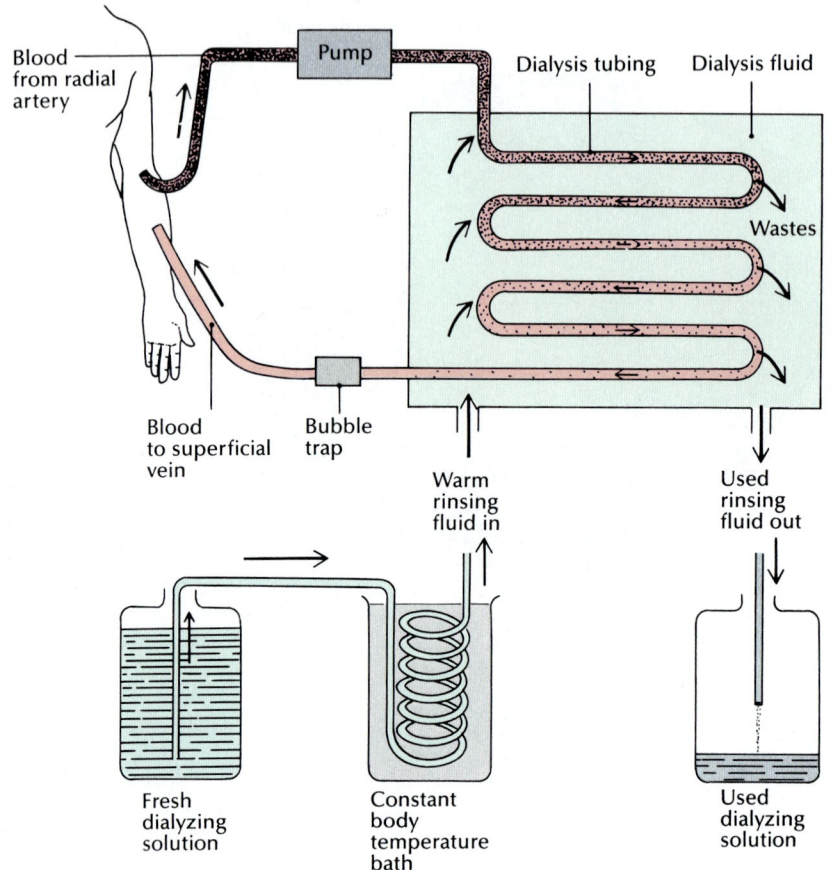

The mechanism of artificial kidneys.

It takes about 6 hours and 20 passes through the bathing fluid to complete a full cycle of dialysis, and most patients receive treatment two or three times a week. Unfortunately, even this highly successful machine, which can remove wastes from the blood 30 times faster than a natural kidney does, provides only partial relief for kidney-failure victims. All patients remain *uremic* (Gr. "urine in the blood") to some degree.

Another method is **peritoneal dialysis,** which uses the patient's peritoneal membrane instead of an artificial membrane for diffusion. The "clean" fluid (dialysis fluid) is injected into the peritoneal cavity each day for 3 or 4 days. Three daytime exchanges remain in the body for 5 hours each before they are removed, and the fourth overnight exchange remains in the body for 8 to 10 hours.

URINE AND URINATION

The laboratory examination of urine (urinalysis) provides an important noninvasive "window" through which a physician can evaluate the condition of many body processes. The composition and physical characteristics of urine are influenced by the diet and nutritional status of the body, the condition of the body's metabolic processes, and the ability of the kidneys to process materials brought to them by the blood.

Composition of Urine

The constituents of urine and their concentrations can vary greatly in a healthy person, depending on diet, exercise, and water consumption, among other factors. In general, however, *urine* is composed mostly of water, urea, chloride, potassium, sodium, creatinine, phosphates, sulfates, and uric acid (see Appendix TABLE D.5).

Normal constituents Water is the major constituent (95 percent) of urine, averaging about 1100 mL per day. The next most abundant constituent is urea, a waste product of amino acid metabolism. Most of the ammonia from excess amino acids is converted by the liver into urea, which is excreted in the urine. However, a small amount of ammonia does appear in the urine.

Abnormal constituents Under normal conditions, there should be no protein in urine. Bleeding in the lower urinary tract and other kidney disorders, however, usually produce large quantities of protein in the urine, a condition called *proteinuria,* or *albuminuria* if the protein is primarily albumin. Kidney diseases often cause an increased permeability of the glomerular membrane. Because more plasma protein is filtered than can be reabsorbed in the proximal tubule, large quantities appear in the urine. In extreme cases, so much protein may be lost that the osmotic pressure of the plasma decreases, causing a net loss of water and solutes from the systemic capillaries. The lost water may accumulate in the tissue spaces, producing a generalized puffiness called *edema.*

Another substance normally absent from urine is glucose, which is totally reabsorbed into the blood from the proximal convoluted tubules. Glucose appears in the urine of diabetics, since glucose uptake by body cells is reduced. As a result, glomerular filtration produces a filtrate containing more glucose than can be reabsorbed by the proximal convoluted tubule. The presence of glucose in the urine is called *glycosuria.*

Other abnormal constituents of the urine are casts and calculi. *Casts* are decomposed blood cells, tubule cells, or fats that form tiny bits of hard material in the nephron tubules. *Calculi* (kidney stones) are formed when the salts in the kidney tubules, ureters, or bladder precipitate to form insoluble masses. They are most common when the diet is consistently high in minerals, is abnormally acidic or alkaline, or when the parathyroid glands are overactive. (See When Things Go Wrong, page 913.)

Physical Properties of Urine

The pH, appearance, specific gravity, and volume of urine also have diagnostic importance when they are considered along with other symptoms. The pH of urine ranges from 5.0 to 8.0 but is usually slightly acidic. Daily variations are closely related to the food eaten. A person who eats meat regularly might have a pH of 5 to 6, and a vegetarian or someone who normally eats a high-fiber, low-protein diet might have a slightly alkaline pH of 7 to 8.

Urine is normally translucent (clear, not cloudy) and yellow or amber in color. The color is caused partly by the presence of bile pigments. However, water intake, diet, and therapeutic drugs (including some antibiotics) can alter the color to dark yellow, brown, or even red. Other factors also influence the color of urine. For example, reddish urine may indicate the presence of hemoglobin from red blood cells (a condition called *hematuria*), and urine with a milky appearance may contain fat droplets or pus *(pyuria).* Any sign of blood or pus in the urine should be taken seriously, since it may indicate the early stages of kidney infection or malfunction.

Specific gravity is the ratio of the weight of a unit volume of a substance to the weight of an equal volume of distilled water. The specific gravity of distilled water is designated as 1.000. The value for urine ranges from 1.001 to 1.035, depending on the total amount of solute in the urine. The more solutes, the higher the specific gravity.

Urine has an aromatic but not unpleasant odor when it is fresh and free of microorganisms. Stale urine that is allowed to stand outside the body, however, is usually contaminated by microorganisms that convert urea into ammonia, and the urine acquires a harsh, stale smell.

Volume of Urine

To maintain the proper osmotic concentrations of the extracellular fluids (including blood), to excrete wastes, and to maintain proper kidney function, the body must excrete *at least* 450 mL of urine per day. A healthy person with normal physical activity and water intake will eliminate from 1000 to 1800 mL of urine every day.

The volume of urine excreted can be influenced by water consumption, diet, external temperature, hormonal and enzymatic actions, blood pressure, diuretics, drugs, and emotional state. When the environmental temperature is high, the body sweats and loses a considerable amount of water. As a result, urine volume is reduced in hot weather.

Therapeutic drugs called *diuretics* (dye-yoo-RET-ihks; Gr. "to urinate through") can increase the volume of urine, primarily by inhibiting sodium reabsorption and thus inhibiting the obligatory water retention that accompanies it. Diuretics are often used in treating hypertension and the edema that occurs with congestive heart failure. Alcohol exerts a diuretic effect by inhibiting the secretion of ADH. (Remember that reabsorption is under the control of ADH.) The caffeine in coffee, tea, and cola drinks causes renal vasodilation, increasing the GFR and urine production.

Urination (Micturition)

Micturition (L. *micturire*, to want to urinate) is the emptying of the urinary bladder. It is commonly called *urination*. Micturition is governed by the micturition reflex but is also under voluntary control. The *micturition reflex* operates in the following way: When stretch receptors in the urinary bladder are activated as urine volume increases (250 to 400 mL of urine), afferent input into the spinal cord stimulates parasympathetic output to the bladder, causing smooth muscle in the bladder to con-

Q: *How often does the average adult urinate each day?*

A: If it were convenient, we would urinate approximately 10 times a day.

Q: *Why does beer stimulate the need to urinate?*

A: Alcohol in the beer inhibits ADH secretion, which increases urine production. Also, another component of beer, lupulin from hops, is a diuretic.

tract. The internal urinary sphincter opens as the bladder contracts, changing the shape of the bladder and pulling open the sphincter. As the bladder fills, *voluntary control* can override the micturition reflex, allowing bladder emptying at the individual's convenience.

ASK YOURSELF

1 What are the main components of urine?

2 What are some of the major physical properties of urine?

3 How does the micturition reflex operate?

THE EFFECTS OF AGING ON THE URINARY SYSTEM

Kidney function usually decreases with age, from 100 percent at 25 years old to about 88 percent at age 45, 78 percent at age 65, and 69 percent at age 85, as arteries to the kidneys become narrowed. Although some kidney tissue may be lost, each kidney needs only about 25 percent of its original tissue to function properly.

Incontinence (lack of control over urination) may result when muscles that control the release of urine from the bladder become weakened. *Frequent urination* in elderly men may be due to enlargement of the prostate gland, which occurs in about 75 percent of men over 55.

Frequent urinary tract infections, kidney stones, and susceptibility to dehydration are other common problems associated with an aging urinary system.

WHEN THINGS GO WRONG

Acute and Chronic Renal Failure

Acute renal failure is the total or nearly total shutdown of kidney function. Little or no urine is produced, and substances that are normally eliminated from the body are retained. It is often caused by a diminished blood supply to the kidneys, which may be brought on by a serious blood loss caused by an injury or hemorrhage, a heart attack, or a thrombosis. Another common cause of acute renal failure is a high level of toxic materials such as mercury, arsenic, carbon tetrachloride, and insecticides that accumulate in the kidneys. Other causes include obstruction (kidney stones, for example) and damage to the

kidneys themselves. After the problem is addressed, recovery usually takes from 7 to 10 days and can be assisted by renal dialysis. Although complete recovery is not uncommon, residual kidney damage may lead to chronic renal failure.

In contrast to acute renal failure, *chronic renal failure* develops slowly and progresses over many years. Its most common causes are bacterial inflammation of the interstitial area and renal pelvis (*pyelonephritis*), kidney inflammation involving the structures around the renal pelvis or glomeruli (*glomerulonephritis*), and renal damage due to high blood pressure or obstructions in the lower urinary tract. The condition is characterized by progressive destruction of nephrons, which may lead to reduced amounts of urine, dilute urine, thirst, severe high blood pressure, poor appetite and vomiting, frequent urination, depletion of bone calcium, coma, and convulsions. A low-protein diet is usually prescribed, and dialysis may be necessary in some cases.

Acute and Chronic Glomerulonephritis

Acute glomerulonephritis (also known as AGN or Bright's disease) is an inflammation of the glomeruli that often follows a streptococcal infection. The body's normal response to this infection produces antibodies that cause an autoimmune inflammation in the glomerulus and damage the endothelial capsular membrane. Symptoms include the presence of blood cells and plasma proteins in the urine (hematuria and proteinuria) and retention of water and salts. Most patients recover fully (especially children), but chronic renal failure is not an uncommon side effect.

Chronic glomerulonephritis (CGN) is a progressive disease that usually leads to renal failure. It is usually irreversible by the time it produces symptoms, which include first hematuria, proteinuria, and hypertension, and later nausea, vomiting, difficulty in breathing, and fatigue. The kidneys atrophy as the disease progresses, and recovery is unlikely.

Pyelonephritis

Acute *pyelonephritis* is one of the most common kidney diseases. It is caused by a bacterial infection, usually from the intestinal bacterium *Escherichia coli*, spreading from the urinary bladder to the ureters and then to the kidneys. The infection begins as an inflammation of the renal tissue between the nephrons and progresses to the glomeruli, tubules, and blood vessels. The disease occurs most often in females, probably because their urethras are short and closer to the rectal and vaginal openings. Symptoms include a fever of 44°C (101°F) or higher, back pain, increased leukocytes in the blood, painful urination (*dysuria*), cloudy urine with an ammonia smell, and the presence of bacteria in the urine. The disease is further complicated for diabetics, whose glucose in the urine (*glycosuria*) provides a ready energy source for bacterial growth.

Renal Calculi

Renal calculi are commonly called *kidney stones*. They may appear anywhere in the urinary tract but are most common in the renal pelvis or calyx. The most usual type of calculus, which accounts for about 80 percent of all cases, forms from the precipitated salts of calcium (calcium oxalate and calcium phosphate). Kidney stones formed from calcium oxalate and trapped in the ureter are usually the ones that cause an intense, stabbing pain

because of their jagged shapes.* Calcium phosphate stones, in contrast, grow quickly and may occupy a large part of the renal pelvis. Kidney stones may also form from the salts of magnesium, uric acid, or cysteine.

The exact cause of the formation of stones is not known, but several conditions are implicated: dehydration, renal infection, obstruction in the urinary tract, hyperparathyroidism, renal tubular acidosis, high levels of uric acid, excessive intake of vitamin D or calcium, and ineffective metabolism of oxalate or cysteine.

Renal calculi are usually revealed in a *pyelogram*, an x-ray of the kidney and ureters after an opaque dye has been introduced into the urinary system. Most renal calculi are small enough to pass out of the urinary system on their own, but in other cases the treatment includes a greatly increased water intake, antibiotics, analgesics, diuretics, a low-calcium diet, surgery, and *extracorporeal* ("outside the body") *shock-wave lithotripsy* that reduces kidney stones to passable particles without harming the body. Lithotripsy is usually effective in the removal of renal calculi less than 2 cm in diameter, which includes most stones. The characteristic extreme pain caused by the passage of stones is called *renal colic* because it resembles intestinal (colic) pain.

Infection of the Urinary Tract

Two forms of a lower urinary tract infection are **cystitis**, inflammation of the bladder, and **urethritis**, inflammation of the urethra. Both are much more prevalent in women than in men; older men are usually affected as they begin to encounter prostate problems. Such inflammations may be trivial, but some urinary infections persist, causing permanent discomfort. The most common cause is an intestinal bacterium such as *Escherichia coli* or *Proteus mirabilis*, but some infections are caused by several different bacteria. Symptoms include frequent urges to urinate, spasms of the bladder, discharge from the penis in males, pain during urination, and excessive urination during the night (*nocturia*).

A form of upper urinary tract infection is *pyelitis*, an inflammation of the renal pelvis and calyces.

Urinary Incontinence

The inability to retain urine in the urinary bladder and control urination is called **urinary incontinence**. Temporary incontinence may be caused by emotional stress. Permanent incontinence usually involves an injury to the nervous system, a bladder infection, or tissue damage to either the bladder or the urethra.

---

*Kidney stones may lodge in one of three sites where the ureters narrow: (1) at the junction with the renal pelvis, (2) where the ureter crosses the pelvic rim, and (3) where the ureters enter the urinary bladder. The ureter distends proximal to the lodged stone, resulting in intense pain.

Kidney and Urinary Bladder Cancer

Kidney cancer occurs most often between the ages of 50 and 60 and is twice as prevalent in men as in women. Early symptoms are blood in the urine *(hematuria)*, pain in one side or the other, and a firm, painless growth of tissue. Kidney cancer often spreads, or *metastasizes*, to other parts of the body. Children under 6 or 7 may develop a variation of adult kidney cancer called *Wilms' tumor*, which causes the abdomen to swell noticeably. It is thought that Wilms' tumor originates in the embryo and then remains dormant for several years.

Urinary bladder cancer appears most often in industrial cities, where such environmental carcinogens as benzidine, nitrates, tobacco smoke, and other chemical inhalants are common. Its symptoms resemble those of kidney cancer, and like kidney cancer, it develops most often in people over 50, especially in men.

Nephroptosis (Floating Kidney)

When a kidney is no longer held in place by the peritoneum, it usually begins to move to the abdominal area above. This condition is called *nephroptosis,* or floating kidney. The moving kidney sometimes twists its ureter, which may lead to blockage of urine. Nephroptosis occurs frequently among truck drivers, horseback riders, and motorcyclists. Apparently, the pain associated with nephroptosis is caused by traction on the renal blood vessels.

Congenital Abnormalities of the Urinary System

Abnormalities of the urinary system occur in approximately 12 percent of all newborns. Some common abnormalities include the absence of a kidney *(renal agenesis)*, location of a kidney in the abdominal region *(renal ectopia)*, and fusion of the kidneys across the midline *(horseshoe kidney)*. When a kidney contains many cysts, it is called a *polycystic kidney.* The most common abnormality of the urethra occurs when the male urethra opens on the ventral surface of the penis instead of the glans *(hypospadias)*. When the urethra fails to close on the dorsal surface of the penis, the condition is called *epispadias.*

Duplication of the ureters (two ureters from each kidney) occurs in about 1 in 200 births.

CHAPTER SUMMARY

Excretion, the elimination of metabolic waste products, is accomplished in part by the lungs, skin, and large intestine, but the prime regulator of water balance and waste elimination is the urinary system.

The Urinary System: Components and Functions (p. 888)

1 The **urinary system** consists of two kidneys, two ureters, the urinary bladder, and the urethra. Urine is formed in the kidneys, carried by the ureters, stored in the urinary bladder, and expelled through the urethra.

2 Each kidney contains over a million **nephrons,** the functional units that filter water and soluble components from the blood, selectively reabsorb some of them back into the blood to maintain a proper balance, and selectively secrete wastes into the urine.

Anatomy of the Kidneys (p. 888)

1 The paired **kidneys** are retroperitoneal, located in the posterior part of the abdomen, lateral to the vertebral column.

2 The medial concave border of each kidney contains a **hilus,** an indented opening where blood vessels, nerves, and ureter join the kidney.

3 The innermost layer of the kidney is the *renal capsule;* the middle layer is the *adipose capsule;* the outer layer is the *renal fascia,* which attaches the kidney to the abdominal wall.

4 The kidney contains three regions: the innermost **renal pelvis,** which branches into the **major** and **minor calyces;** the middle **renal medulla,** consisting of several **renal pyramids,** which open into the calyces; and the outermost **renal cortex.**

5 Blood enters the kidney through the **renal artery,** which branches into **interlobar arteries, arcuate arteries, interlobular arteries,** and then **afferent arterioles,** which carry blood to the filtration site.

6 Each afferent arteriole branches extensively to form a ball of capillaries called a **glomerulus,** where filtration starts. Glomerular capillaries join to form an **efferent arteriole,** which carries blood away from the glomerulus.

7 The efferent arteriole branches to form the **peritubular capillaries,** which unite to form the **interlobular veins, arcuate veins, interlobar veins,** and the **renal vein,** which carries filtered blood from the kidney to the inferior vena cava.

8 *Vasa recta* are extensions of the efferent arterioles that provide the kidney with an emergency system to maintain blood pressure and urine concentration.

9 Nerves from the **renal plexus** help regulate blood pressure in the glomerulus.

10 Each **nephron** is an independent urine-making unit. It consists of a *vascular component* (the glomerulus) and a *tubular component,* including a glomerular capsule, proximal convoluted tubule, loop of the nephron, distal convoluted tubule, and collecting duct. The **renal**

corpuscle consists of the glomerulus and the **glomerular capsule.**

11 Filtration of blood takes place through the three layers of the renal corpuscle (constituting the **endothelial capsular membrane**) from the capillaries of the glomerulus into the glomerular capsule.

12 From the glomerular capsule, the fluid filtered from the blood **(glomerular filtrate)** moves into the **proximal convoluted tubule,** where glucose, proteins, and certain other solutes filtered from the blood are absorbed.

13 The filtrate passes from the proximal convoluted tubule to the **loop of the nephron,** which is responsible for the reabsorption of water and the concentration of urine.

14 From the loop of the nephron, the filtrate moves into the **distal convoluted tubule,** where potassium and hydrogen ions are actively secreted into the filtrate.

15 The filtrate moves from the distal tubule into the **collecting duct,** where the dilute filtrate is concentrated and passed on to the minor calyx. The permeability of the walls of the collecting duct to water is controlled by *antidiuretic hormone* (ADH).

16 The **juxtaglomerular apparatus** is made up of **juxtaglomerular cells** and the **macula densa.** When the composition of the filtrate or the glomerular pressure changes, the juxtaglomerular apparatus secretes the enzyme **renin,** which alters the systemic blood pressure to reinstate normal conditions.

Physiology of the Kidneys (p. 896)

1 The kidneys utilize three processes to produce and modify urine: glomerular filtration, tubular reabsorption, and tubular secretion.

2 **Glomerular filtration** is the process that forces plasma fluid from the glomerulus into the glomerular capsule. In the process, the filtration of blood is begun.

The final hydrostatic pressure in the glomerulus is the **net filtration pressure.** The amount of filtrate formed in the capsular space each minute is the **glomerular filtration rate.**

3 The rate of glomerular filtration depends on the net filtration pressure, stress, total surface area available for filtration, capillary permeability, intrinsic renal autoregulation, and release of renin.

4 **Tubular reabsorption** returns useful substances such as water, some salts, and glucose to the blood by active transport.

5 **Tubular secretion** is the process of transporting wastes from capillaries to tubules in the nephron.

6 The ability of the kidneys to clear wastes from blood plasma is measured by a process called **plasma clearance.**

7 The **countercurrent multiplier system** results from a countercurrent flow in the limbs of the loop of the nephron that helps regulate the solute concentration. It assures that the urine will be more concentrated at the end of the nephron tubule than it was at the beginning.

8 The kidneys help regulate the acid-base balance of the blood, primarily by simultaneously excreting hydrogen ions and reabsorbing bicarbonate ions.

Accessory Excretory Structures (p. 906)

1 The paired **ureters** carry urine from the renal pelvis of the kidney to the urinary bladder. Their tissue layers are the innermost **tunica mucosa,** the middle **tunica muscularis,** and the outermost **tunica adventitia.**

2 The muscular **urinary bladder** is an expandable sac that collects and stores urine until it is excreted. Its tissue layers resemble those of the ureter. At the site where the urethra leaves the bladder, involuntary smooth muscle in the bladder wall contracts to prevent the bladder from emptying prematurely, and a vol-

untary **external urethral sphincter** keeps the urine from leaving the urethra until it is time to urinate.

3 The **urethra** is the tube that transports urine from the bladder to the outside during urination. It is much longer in males than in females. The male urethra contains three **portions,** designated as **prostatic, membranous,** and **penile.**

Urine and Urination (p. 911)

1 **Urine** is composed of water, urea, chloride, potassium, creatinine, phosphates, sulfates, and uric acid. Abnormal constituents include protein, glucose, ketone bodies, casts, and calculi.

2 Urine is usually slightly acidic, with the pH ranging from 5.0 to 8.0. It is normally clear and yellowish, but its color can vary greatly in a healthy person. Its specific gravity ranges from 1.008 to 1.030.

3 A healthy person excretes between 1.0 and 1.8 L of urine daily. In order to maintain homeostasis, an adult must excrete at least 0.45 L of urine daily. The volume and concentration of urine are influenced by diet, diuretics, and other factors and are regulated by ADH, aldosterone, and renin.

4 **Micturition,** or **urination,** is the emptying of the urinary bladder. Urination in an infant is a spinal reflex action initiated by the distension of the bladder. In an adult, the impulses generated by stretch receptors in the bladder are sent to the brainstem and cerebral cortex. Conscious control of micturition must be learned.

The Effects of Aging on the Urinary System (p. 912)

1 Kidney function usually decreases with age, as arteries to the kidneys become narrowed.

2 Elderly men may urinate more frequently as the prostate gland enlarges, and incontinence may occur when the muscles that control urination are weakened.

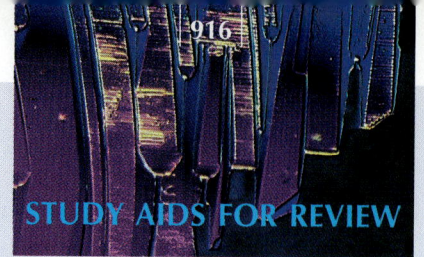
STUDY AIDS FOR REVIEW

KEY TERMS

calyces 890
collecting duct 896
countercurrent multiplier
  system 905
distal convoluted tubule 896
glomerular capsule 894
glomerular filtrate 896, 898
glomerular filtration
  897, 898
glomerular filtration rate 898
glomerulus 893
juxtaglomerular apparatus
  896
loop of the nephron 896
micturition 912
nephron 894
net filtration pressure 898
peritubular capillaries 893

plasma clearance 904
proximal convoluted
  tubule 896
renal artery 893
renal corpuscle 894
renal cortex 893
renal medulla 890
renal pelvis 890
renal vein 893
tubular reabsorption 897, 901
tubular secretion 897, 903
ureter 907
urethra 909
urinary bladder 907
urinary system 888
urination 912
urine 911
vasa recta 894

MEDICAL TERMS FOR REFERENCE

ANURIA The absence of urine, such as when the ureters are obstructed.

CYSTECTOMY Surgical removal of the urinary bladder.

CYSTEINURIA An inborn genetic disorder of amino acid transport that causes excessive excretion of cysteine and other amino acids.

CYSTO- A prefix meaning "bladder."

CYSTOMETRY An examination of the bladder to evaluate its efficiency.

CYSTOSCOPY An examination of the interior of the bladder with a fiber-optic scope.

CYSTOURETHROGRAPHY X-ray examination, after introducing a contrast dye, to determine the size and shape of the bladder and urethra.

DIURETIC Any substance that stimulates urination.

GOUT A metabolic disease characterized by painful urate deposits, usually in the feet and legs.

NEPHRECTOMY Surgical removal of a kidney.

NOCTURNAL ENURESIS Involuntary urination during sleep. Commonly called "bed-wetting."

POLYURIA Frequent urination.

RENAL ANGIOGRAPHY Examination of renal blood vessels, using contrast dye injected into a catheter in the femoral artery or vein.

RENAL HYPERTENSION High blood pressure of the kidney.

RENAL INFARCTION Formation of a clotted area of dead kidney tissue as a result of blockage of renal blood vessels.

RENAL VEIN THROMBOSIS Blood clotting in the renal vein.

RENOVASCULAR HYPERTENSION A rise in systemic blood pressure as a result of blockage of renal arteries.

UREMIA A condition in which waste products that are normally excreted in the urine are found in the blood.

VESICOURETERAL REFLUX A backflow of urine from the bladder into the ureters and renal pelvis.

UNDERSTANDING THE FACTS

1 Describe the components of a nephron, distinguishing the vascular components and the tubular components.

2 List the three primary functions of the nephron, indicating the region of the nephron where each occurs.

3 Identify the location of the renal cortex, renal medulla, renal pelvis, and hilus, indicating which part(s) of the nephron (if any) each contains.

4 Contrast the structure of the renal pyramids with that of renal columns.

5 Trace the route of blood flow through the kidney.

6 How does the composition of glomerular filtrate differ from the composition of plasma and of urine?

7 Define or describe each of the following:
  a renal threshold
  b transport maximum
  c plasma clearance

8 Which substances are reabsorbed? Which are secreted?

9 List the accessory organs of the excretory system, and state the function of each.

10 Contrast the anatomy and the location of the urethra in males and females.

11 What prevents the backflow of urine from the bladder into the ureters?

12 Describe the structure, function, and location of transitional epithelium.

13 State the function and location of each of the following:
  a juxtaglomerular apparatus   **f** trigone
  b vasa recta   **g** internal urethral
  c podocytes   sphincter
  d brush border   **h** external urethral
  e macula densa   sphincter

14 What is urea, where is it synthesized, and what is its function?

15 Contrast acute renal failure with chronic renal failure. What are the causes of each?

16 What is a "floating" kidney?

UNDERSTANDING THE CONCEPTS

1 What are the two major functions of the excretory system?

2 Contrast the diameter of the afferent arteriole with that of the efferent arteriole, and state the significance of the difference.

3 Explain how the structure of the endothelial capsular membrane adapts it for filtration.

4 Explain how the glomerular filtration rate is intrinsically regulated.

5 How and when does the autonomic nervous system regulate kidney filtration?

6 Explain the importance of cotransport in reabsorption.

7 What is meant by the countercurrent multiplier effect?

8 How do the loop of Henle, peritubular capillaries, and vasa recta function in reabsorption?

9 What is the function of the collecting ducts, and how are they regulated?

10 How are the NFP and plasma clearance rate calculated?

11 Distinguish between passive reabsorption and active reabsorption.

12 How does the kidney regulate each of the following?
  a level of potassium ions in the plasma
  b acid-base balance of the body
  c plasma concentration of glucose and amino acids

13 Why is the composition of urine an important diagnostic indicator?

14 How is micturition controlled?

SELF-QUIZ
Multiple-Choice Questions

27.1 The kidneys function in all of the following processes *except*
  a regulation of the hydrogen-ion concentration of blood
  b removal of nitrogenous wastes
  c removal of carbon dioxide
  d maintenance of the osmotic concentration of blood
  e secretion of one or more hormones

27.2 The nephron consists of all of the following *except* the
  a glomerulus   **d** collecting duct
  b Bowman's capsule   **e** peritubular
  c proximal convoluted tubule   capillaries

27.3 The renal corpuscle consists of the
  a glomerular capsule   **d** a and b
  b glomerulus   **e** a, b, and c
  c convoluted tubules

27.4 The apex of each renal pyramid ends in the
  a cortex   **d** capsule
  b papilla   **e** peritubular capillaries
  c juxtaglomerulus

27.5 Specialized epithelial cells in the visceral layer of the glomerular capsule are called
  a slit cells   **d** podocytes
  b fenestrations   **e** macula densa
  c brush border cells

27.6 Renin is a(n) _____ that functions in the regulation of _____ .
  a hormone, filtration
  b enzyme, blood pressure
  c releasing factor, renal reabsorption
  d excretory product, osmotic concentration of plasma
  e nutrient, glomerular filtration rate

27.7 The volume of fluid in the urine represents what percentage of the volume of glomerular filtrate?
  a 1   **b** 10   **c** 20   **d** 50   **e** 99

27.8 The innermost layer of the ureter is the
  a mucosa   **d** trigone
  b muscularis   **e** renal fascia
  c adventitia

27.9 Afferent arterioles of the nephron lead directly into the
  a glomerulus   **d** ureter
  b efferent arterioles   **e** proximal
  c peritubular capillaries   convoluted tubule

27.10 The peritubular capillaries unite to form the
  a glomerulus   **d** interlobular veins
  b afferent arteriole   **e** distal convoluted tubule
  c efferent arteriole

27.11 The vasa recta surround the
  a collecting ducts   **d** proximal convoluted
  b loop of Henle   tubules
  c glomerulus   **e** macula densa

27.12 Tubular secretion is accomplished by all of the following *except* the
  a renal corpuscle   **c** distal convoluted
  b proximal convoluted   tubule
  tubule   **d** collecting duct
       e a and b

27.13 Cells with a brush border are located in the
  a glomerulus
  b proximal convoluted tubule
  c loop of Henle
  d distal convoluted tubule
  e peritubular capillary

27.14 On the average, how many times per day is the entire volume of plasma filtered?
  a 10–20   **d** 1000
  b 50–60   **e** more than 1000
  c 250–300

27.15 Juxtaglomerular cells are part of the wall of the

a proximal convoluted tubule   d macula densa
b distal convoluted tubules   e collecting duct
c afferent arteriole

27.16 The concentration of potassium ion in the plasma is directly related to
a the level of sodium ions in the plasma
b secretion of aldosterone from the adrenal cortex
c the conversion of angiotensin I to angiotensin II
d release of renin by the juxtaglomerular apparatus
e all of the above

Matching

a renal corpuscle
b proximal convoluted tubule
c descending loop of the nephron
d ascending loop of the nephron
e distal convoluted tubule
f collecting duct

27.17 _____ reabsorbs water in presence of ADH
27.18 _____ filters plasma
27.19 _____ site of highest concentration of urea
27.20 _____ actively secretes penicillin and organic acids and bases
27.21 _____ impermeable to water

27.22 _____ site of reabsorption of glucose and most nutritionally important molecules

Completion Exercises

27.23 The amount of filtrate formed in the capsular space each minute is the _____.
27.24 Sodium and chloride ions are recycled in a countercurrent manner in the _____.
27.25 Net filtration pressure is the difference between _____ and the combined effects of glomerular plasma osmotic pressure and capsular hydrostatic pressure.
27.26 Vasoconstriction of the afferent renal arteriole caused by stimulation of the _____ nervous system leads to a _____ in the glomerular filtrate rate.
27.27 Glomerular filtration rate is controlled intrinsically by the myogenic mechanism and by substances released by the _____.
27.28 The transport of two different molecules by one carrier protein is called _____.
27.29 The urinary excretion rate of a molecule divided by its plasma concentration is known as its _____.
27.30 Voluntary control of micturition is regulated by the _____ urethral sphincter.

A SECOND LOOK

1 In the following photograph, label the renal artery and veins, minor and major calyces, cortex, medulla, renal pelvis, and ureter.

2 In the following drawing, label the glomerulus, proximal convoluted tubule, vasa recta, collecting tube, and distal convoluted tubule.

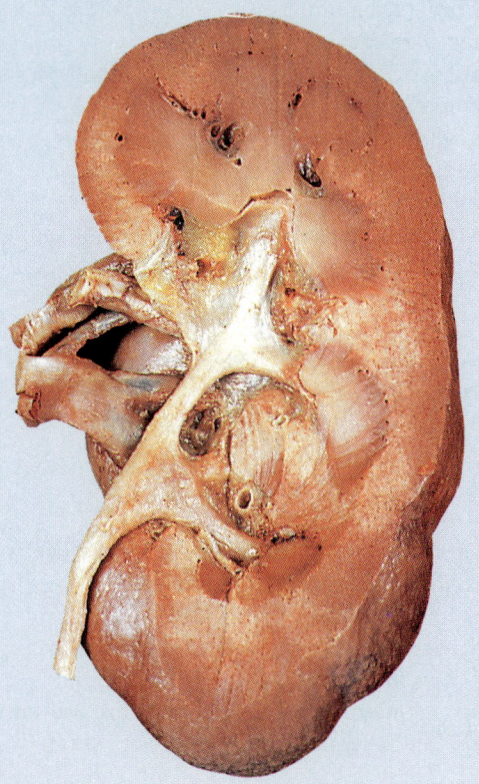

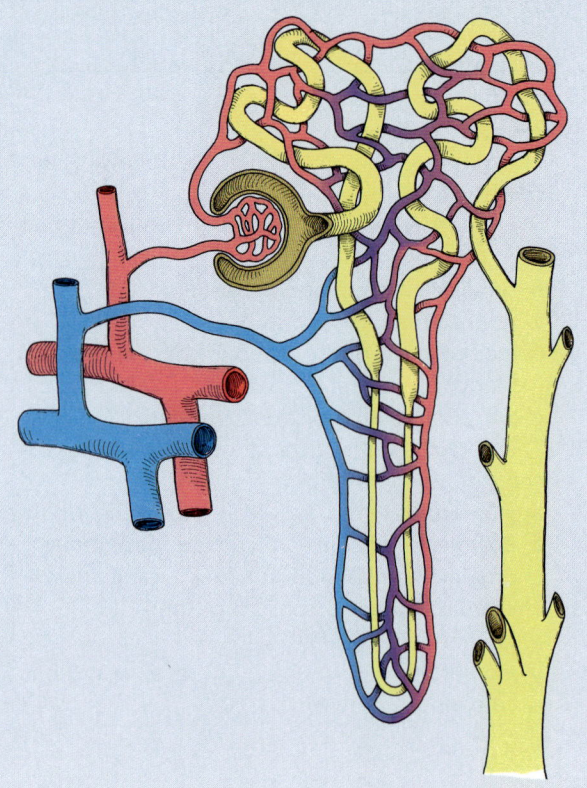

28

Regulation of Body Fluids, Electrolytes, and Acid-Base Balance

KEY CONCEPTS

1 Water is the most abundant constituent of body fluids.

2 Water moves from one body compartment to another in response to pressure changes.

3 The body's acid-base balance is maintained by buffer systems, respiratory regulation, and renal regulation.

4 Chemicals called electrolytes are crucial to the regulation of fluid balance.

The largest single constituent of the body by weight is water. In fact, about 60 percent of your total body weight is water. Your very life depends on maintaining the proper amount of body water, the correct proportion of water and electrolytes in the water, and the proper acid-base balance. Nowhere else in the body is homeostasis so important, and nowhere are imbalances more serious. For example, excessive vomiting or prolonged diarrhea can result in a life-threatening condition if water and electrolytes are not replaced quickly. (For this reason, diarrhea in an infant is more dangerous than constipation.) A loss of less than 10 percent of the total body water usually produces lethargy, fever, and dryness of the mucous membranes. A loss of about 20 percent of the total body water is usually *fatal*. A person with third-degree burns may die, not necessarily from tissue damage, but from the loss of water from the burned areas.

BODY FLUIDS: COMPARTMENTS AND COMPOSITION

All body fluids are either *intracellular* (inside the cell; "intra" = inside) or *extracellular* (outside the cell; "extra" = outside). *Intracellular fluids (ICF)* include all the water and electrolytes enclosed by cell membranes and constitute about 63 percent of the total body water. *Extracellular fluid (ECF)* includes all fluids outside the cells. This fluid can be categorized into three fluid compartments: (1) interstitial ("in the tissue spaces") fluid; (2) blood plasma and lymph; and (3) a specialized compartment of extracellular fluid called *transcellular fluid*, which includes cerebrospinal fluid of the central nervous system; synovial fluid of the joints; aqueous and vitreous humors of the eyes; serous fluids of the pleural, pericardial, and peritoneal cavities; and glandular fluid secretions. These extracellular fluids constitute about 37 percent of the total body water [FIGURE 28.1].

Water is the most abundant constituent of body fluids and is the solvent involved in transporting electrolytes and proteins. FIGURE 28.2 compares the electrolyte and protein concentrations of blood plasma, interstitial fluid, and intracellular fluid. Interstitial fluid and blood plasma

FIGURE 28.1 PROPORTION OF WATER TO BODY WEIGHT

[A] Water normally makes up about 60 percent of the total body weight, with most of it located in the cells. Values are for a 70-kg adult male. [B] Distribution of body water in intracellular and extracellular fluid compartments.

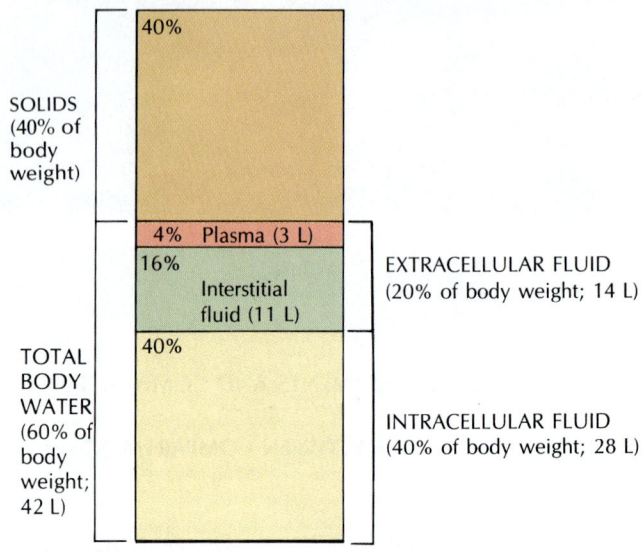

[A]

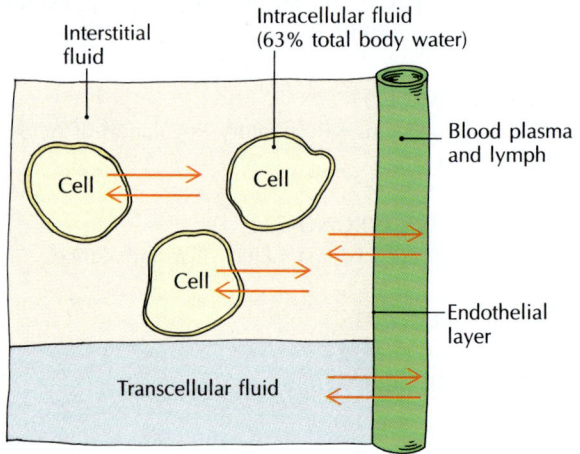

[B]

have relatively high concentrations of sodium, chloride, and bicarbonate ions and lesser concentrations of potassium, calcium, magnesium, phosphate, and sulfate ions.

Q: *What causes a blister?*

A: A blister results when localized damage to capillary walls allows plasma proteins to leak into the interstitial space and into areas where epidermal layers separate because of thermal or mechanical damage. Such a leakage reduces the protein concentration difference between the plasma and interstitial fluid, and water diffuses into the interstitial space, forming a blister.

A S K   Y O U R S E L F

1 What percentage of your body is water?

2 What are the main body fluid compartments?

FIGURE 28.2 ELECTROLYTES AND PROTEIN IN BODY FLUIDS

Comparison of the concentrations of soluble proteins and electrolytes in the three major body fluid compartments. The fluids with lower total solute concentrations have higher water concentrations.

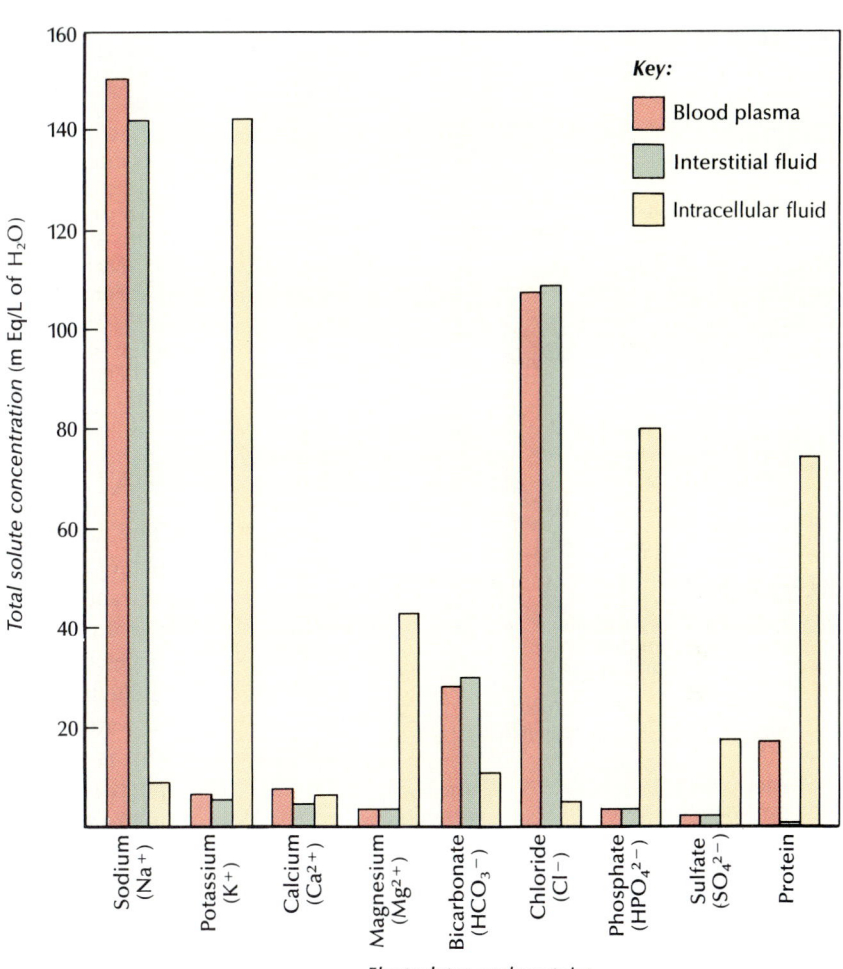

MOVEMENT OF FLUIDS BETWEEN COMPARTMENTS

The movement of water from one body compartment to another is controlled primarily by two forces: (1) hydrostatic pressure and (2) osmotic pressure [FIGURE 28.3]. *Hydrostatic pressure* is the force exerted by a fluid in a system, and *osmotic pressure* is the force required to prevent the movement of pure water across a semipermeable membrane into a solution containing solutes. Because the movement of water is dependent on the total solute concentration, changes in the concentration of major solutes in the plasma or interstitial fluid will have a major effect on water distribution. Conversely, conditions that directly affect water balance (loss or gain in a compartment) will alter the concentrations of the solutes. Thus changes in water and solute concentrations are closely interrelated in their effects on water and solute movements from one body compartment to another.

One example of the movement of fluids and solutes is the movement of materials across capillaries. Fluid leaves the *arterial end* of capillaries and moves into the interstitial spaces because of the net outward force of hydrostatic pressure (blood pressure). Fluid returns to the plasma from the interstitial spaces at the *venous end* of the capillaries because of the net inward force of osmotic pressure. Fluid also leaves the interstitial spaces and moves into the lymphatic capillaries because of the osmotic forces in the lymph vessels. Because of this lymph movement, interstitial fluid can be returned to the plasma [FIGURE 28.3].

Fluid movement between the intracellular and extracellular compartments is also regulated by hydrostatic and osmotic pressures. The hydrostatic pressures within the cells and in the surrounding interstitial fluid are usually equal, and fluid flow between these two compartments is due primarily to changes in osmotic pressure [see FIGURE 28.3].

FIGURE 28.3 MOVEMENT OF FLUID BETWEEN PLASMA AND INTERSTITIAL FLUID IN RESPONSE TO OSMOTIC AND HYDROSTATIC PRESSURE

Arrows indicate the direction of water movement. Ordinarily, there is a state of equilibrium (↔) between the intracellular fluid and interstitial (extracellular) fluid.

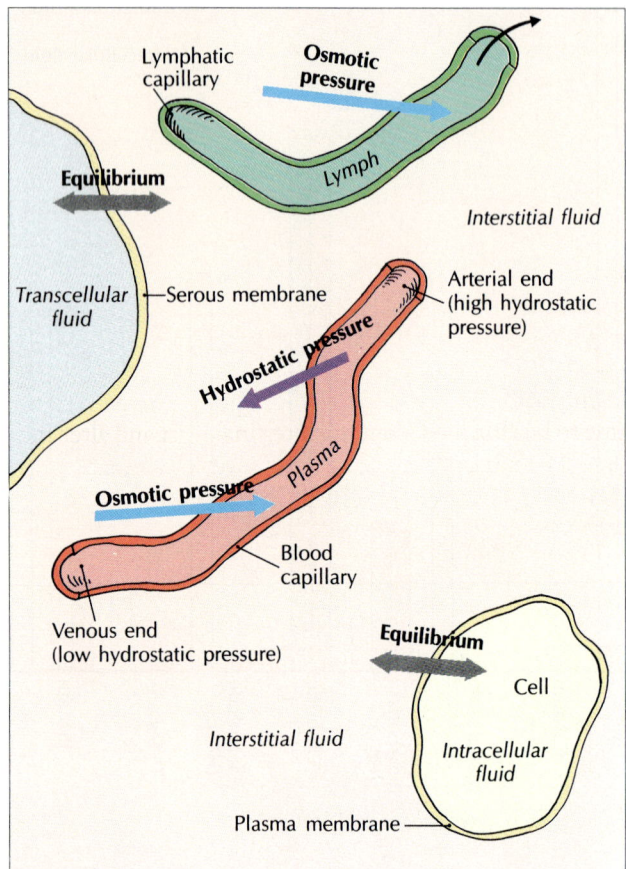

TABLE 28.1 TOTAL BODY WATER AS A PERCENTAGE OF BODY WEIGHT

Age	Total body water, % body weight
Newborn	77
6 months	72
2 years	60
16 years	60
20–39 years	
Male	60
Female	50
40–59 years	
Male	55
Female	47

WATER

The water content of an average adult is about 47 to 60 percent of the total body weight, depending on sex, body fat, and age [TABLE 28.1]. Adult females have less water per unit weight than adult males, primarily because females have more adipose tissue than males, and adipose tissue has a relatively low water content. The percentage of water in the body decreases with age. A newborn infant is nearly 77 percent water. This fact is of great importance in caring for infants. Because so much of the infant's weight is water, any imbalance will have more severe consequences than it would for an adult.

Functions of Water

Water has many important and interrelated functions in the body. The major functions include:

1 Water acts in the *transport of* nutrients to cells and in carrying waste products from cells.

2 It provides a *liquid medium* for intracellular chemical reactions, including overall cellular metabolism.

3 It acts as a *solvent* for electrolytes and other solutes.

4 It helps maintain *body temperature*, aids *digestion*, and promotes proper *excretion*.

Intake of Water

Under normal conditions, water is taken into and excreted from the body in equal amounts. Most water (about 60 percent) is taken into the body through *ingested liquids* (drinking water or other liquids that contain water, such as tea, coffee, and milk). A smaller amount of water (about 30 percent) enters the body as *ingested food* (preformed water), and an even smaller amount (about 10 percent) of *metabolic water* is a by-

Changes in the concentration of major solutes such as sodium in fluids will affect the water balance between cells and the extracellular fluid surrounding them. If extracellular sodium increases outside the cells, there will be a net movement of water from the cells into the interstitial spaces by osmosis, and the cells will shrink. If there is a decrease in plasma sodium, there will be a net movement of water from the plasma into the interstitial fluid and into the cells by osmosis, and the cells will swell.

ASK YOURSELF

1 What are the two main forces that move water between compartments?

2 What forces control the movement of water between the plasma and the interstitial fluid?

3 If the concentration of sodium increases in the plasma, how will the water balance of cells be affected?

FIGURE 28.4 THIRST REGULATION

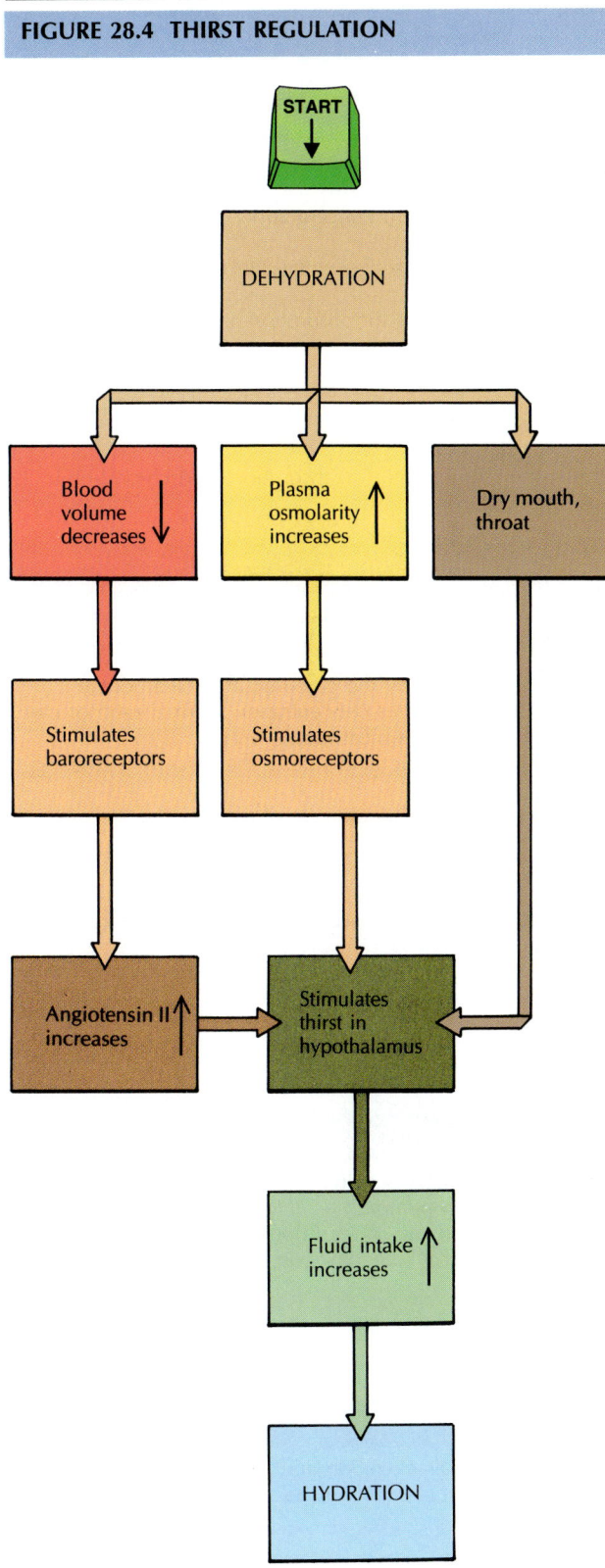

the *thirst center* located in the hypothalamus. Specialized receptors in the thirst center are called *osmoreceptors*. These receptors sense the osmotic pressure of extracellular fluids and control the thirst mechanism in the following way:

1 When there is a loss of water from the body or insufficient intake *(dehydration)*, the plasma osmolarity rises, and the osmoreceptors signal "thirst" to the central nervous system (hypothalamus).

2 The feeling of thirst is typically accompanied by dryness of the mouth and throat, which encourages the ingestion of fluids to reduce plasma osmolarity.

3 A decrease in blood volume, as sensed by volume receptors in key areas of the cardiovascular system, also stimulates the thirst center by activating the release of *angiotensin II*.

4 The stomach distension caused by drinking fluid triggers afferent input into the thirst center to relieve the feeling of thirst, well before the ingested fluid has had time to be absorbed from the intestinal tract and alter the plasma osmolarity. Thus, distension of the stomach prevents us from overingesting fluids, which would alter our fluid balance and cause overhydration.

Output of Water

Under normal conditions, the kidneys are the primary organs responsible for excreting most of the water (about 60 percent) from the body. Smaller but significant volumes of water are lost through the skin as sweat (about 6 percent), through the gastrointestinal tract as feces (about 6 percent), and through the lungs as water vapor in exhaled breath (about 28 percent). These percentages vary with a variety of factors. For example, in extreme heat and during prolonged strenuous physical activity, loss of water through the skin increases. Also, severe diarrhea or vomiting increases the amount of water and certain solutes lost through the gastrointestinal tract and mouth.

Water loss is **regulated** by antidiuretic hormone (ADH), as discussed in the previous chapter. The osmoreceptor-ADH mechanism can decrease the normal urine output of 1500 mL/day to about 500 mL/day during dehydration, and to volumes above 1500 mL/day when excessive amounts of water are ingested (overhydration).

product of the oxidation of food in cells. (Water is produced during electron transport.)

The **thirst mechanism** [FIGURE 28.4], which is incompletely understood, involves a group of neurons called

TABLE 28.2 MAJOR ELECTROLYTES IN THE BODY

Basic functions	Major homeostatic regulators

CALCIUM (Ca^{2+})

Required for building strong and durable bones and teeth.
Essential for blood coagulation.
Decreases neuromuscular irritability.
Promotes normal transmission of nerve impulses.
Establishes thickness and strength of plasma membranes.
Assists in absorption and utilization of vitamin B_{12}.
Activates enzymes that in turn activate chemical reactions within body cells.
Needed for muscle contraction.

Parathyroid hormone raises serum concentration by increasing activity of osteoclasts.
Calcitonin lowers serum concentration by inhibiting osteoclastic action.
Concentration of serum phosphate affects concentration of serum calcium.
Vitamin D is necessary for absorption and utilization of calcium.

CHLORIDE (Cl^-)

Combines with hydrogen in gastric mucosal glands to form HCl.
Diffusion between ECF and ICF helps regulate osmotic pressure differences between compartments.
Assists in transmission of nerve impulses.

Aldosterone regulates Na^+ reabsorption, and chloride follows passively.

HYDROGEN (H^+)

Necessary for healthy cellular function.
Promotes efficient functioning of enzyme systems.
Necessary for the binding of oxygen by hemoglobin.
Determines relative acidity or alkalinity of body fluids.

Buffering, principally by the carbonic acid-bicarbonate buffer system, regulates concentration (normal ratio of carbonic acid to sodium bicarbonate is 1:20).
Lungs regulate carbonic acid side of the ratio by regulating CO_2.
Kidneys regulate bicarbonate side of the ratio.

MAGNESIUM (Mg^{2+})

Activates many enzymes, in particular those associated with vitamin B metabolism and the utilization of potassium, calcium, and protein.
Promotes regulation of serum calcium, phosphorus, and potassium levels.
Essential for integrity of neuromuscular system and function of heart.

Parathyroid hormone increases absorption into blood from intestine.

POTASSIUM (K^+)

Regulates water and electrolyte content of ICF.
Helps promote transmission of nerve impulses, especially within the heart.
Helps promote skeletal muscle function.
Assists in transforming carbohydrates into energy and restructuring amino acids into proteins.
Assists in regulation of acid-base balance by cellular exchange with H^+.

Sodium pump conserves cellular K^+ by actively excluding Na^+ and actively pumping K^+ into cells.
Kidneys (which excrete 80 to 90 percent of K^+) conserve K^+ when cellular K^+ becomes depleted.

SODIUM (Na^+)

Regulates fluid volume within ECF.
Increases plasma membrane permeability.
Maintains blood volume and controls size of vascular space.
Controls body water distribution between ECF and ICF.
Acts as a buffer base (sodium bicarbonate), thereby helping to regulate H^+ concentration.
Stimulates conduction of nerve impulses.
Helps maintain neuromuscular irritability.
Assists in controlling contractility of muscles, in particular, heart muscle.

Aldosterone controls excretion and retention.
Atrial natriuretic peptide stimulates excretion.

FUNCTION AND REGULATION OF SPECIFIC ELECTROLYTES

Electrolytes are compounds that dissociate into ions when in solution, making it possible for the solution to conduct an electric current. The positive ions are called *cations*, and the negative ions are *anions*. A *nonelectrolyte* is a compound that does not form ions in solution. Acids, bases, and salts* are electrolytes [TABLE 28.2]. Although some of the body's electrolytes are found attached to proteins and some are deposited as solids to form bone and teeth, most electrolytes are dissolved in the body fluids. The most physiologically important electrolytes are the cations sodium (Na^+), potassium (K^+), calcium (Ca^{2+}), magnesium (Mg^{2+}), and hydrogen (H^+) and the anions bicarbonate (HCO_3^-), chloride (Cl^-), phosphate (HPO_4^{2-}), and sulfate (SO_4^{2-}).

Electrolytes have four major functions in the body: (1) Many are necessary for normal cell metabolism and contribute to body structures. (2) They facilitate the osmotic movement of water between the body compartments. (3) They help maintain the hydrogen-ion concentration (acid-base balance) required for normal cellular function. (4) They are crucial to the production and maintenance of membrane potentials and action potentials.

Each electrolyte indicated in FIGURE 28.2 is maintained more or less independently at an optimal normal

*When an acid is neutralized by a base, water and a salt are produced:

$$HCl + KOH \rightleftharpoons H_2O + KCl$$

hydrochloric acid (acid)   potassium hydroxide (base)   water   potassium chloride (salt)

Some common salts are sodium chloride (table salt), potassium chloride, sodium bicarbonate (used as an antacid), magnesium sulfate (Epsom salt), and calcium sulfate (plaster of Paris).

concentration in each fluid compartment. The body can excrete in urine electrolytes that become too abundant, or it can restrict excretion to conserve electrolytes that fall below their optimal concentrations. Electrolytes are also lost from the body through the skin in sweat and from the gastrointestinal tract in feces.

Hormones play a key role in the control of the proper concentration of many electrolytes in the plasma. TABLE 28.3 summarizes the effects of antidiuretic hormone (ADH), aldosterone, atrial natriuretic peptide (ANP), and parathyroid hormone (PTH) on water and electrolyte balance.

Sodium

The sodium content of an adult is approximately 142 mEq/L of extracellular fluid. The daily intake of sodium in a typical North American diet greatly exceeds the need. Since sodium is actively removed from cells by the sodium-potassium pump, intracellular fluid has a low sodium concentration (about 5 mEq/L).

Because sodium is a very important ion in osmotic regulation, changes in sodium concentration result in dramatic changes in water distribution between cells and their fluid environment. A low plasma sodium concentration is called *hyponatremia*, and a high concentration is called *hypernatremia*. (See When Things Go Wrong at the end of the chapter.)

The sodium level in the plasma is maintained by three hormones: (1) aldosterone, (2) antidiuretic hormone (ADH), and (3) atrial natriuretic peptide (ANP). Sodium reabsorption is increased by aldosterone, which is secreted by the adrenal cortex. When the extracellular sodium concentration falls below normal, when potassium is elevated, or when blood volume is decreased, ***aldoste-***

TABLE 28.3 HORMONES CONTROLLING THE BALANCE OF WATER AND ELECTROLYTES

Hormone	Source of secretion	Mode and site of action
Antidiuretic hormone (ADH)	Hypothalamus; released by neurohypophysis	Increases water reabsorption in kidney tubules.
Aldosterone	Adrenal glands	Increases sodium reabsorption in kidney tubules.
Atrial natriuretic peptide (ANP)	Heart	Inhibits renin-angiotensin system; inhibits aldosterone secretion from adrenal glands; stimulates sodium excretion in urine.
Parathyroid hormone (PTH)	Parathyroid glands	Increases calcium reabsorption from intestine and in kidney tubules; increases calcium release from bone.

rone is released into the blood and transported to the kidneys. In the kidneys, aldosterone increases sodium reabsorption in the distal convoluted tubules and collecting ducts of the nephrons, thereby reducing the loss of sodium through the urine. If the body's need for sodium is not met in this way, sodium intake must be increased through ingestion. The return of sodium to normal concentrations then inhibits aldosterone secretion. FIGURE 28.5 summarizes the regulation of sodium reabsorption by aldosterone.

A decrease in sodium concentration decreases the release of *antidiuretic hormone (ADH)* from the posterior pituitary, which causes an increased water loss in the urine and returns the sodium concentration to normal.

As you saw in Chapter 18, the peptide hormone *atrial natriuretic peptide (ANP)* is secreted by cardiac muscle cells in the atria of the heart when there is excess plasma volume. An increase in plasma volume causes the atria to stretch. In response, cardiac muscle cells secrete ANP, which inhibits the secretion of renin in the renin-angiotensin system and thus inhibits the secretion of aldosterone from the adrenal glands, which favors sodium loss and reduces the amount of sodium, and subsequently water, in the body.

Potassium

The extracellular (plasma) potassium concentration of an adult is approximately 4.5 mEq/L. Potassium has a much higher intracellular concentration—about 140 mEq/L. Since potassium has a relatively low plasma concentration, it is not as important as sodium in *osmotic regulation.* However, plasma potassium is extremely important in generating and maintaining resting membrane potentials, and also in repolarizing action potentials in cardiac and nervous tissue. While some potassium may be lost in sweat or feces, most loss is due to excretion in the urine. Therefore potassium levels are regulated primarily by the kidneys.

A rise in potassium concentration in the plasma signals the release of aldosterone, causing increased potassium secretion [see FIGURE 28.5]. When plasma potassium is low, aldosterone secretion decreases, and there is a decrease in potassium excretion.

Chloride

Like sodium, chloride has a high concentration in the extracellular fluid. Its normal plasma concentration is

---

Q: *Why do high-sodium diets tend to cause weight gain?*

A: Because the body retains water in an effort to osmotically balance the excess sodium.

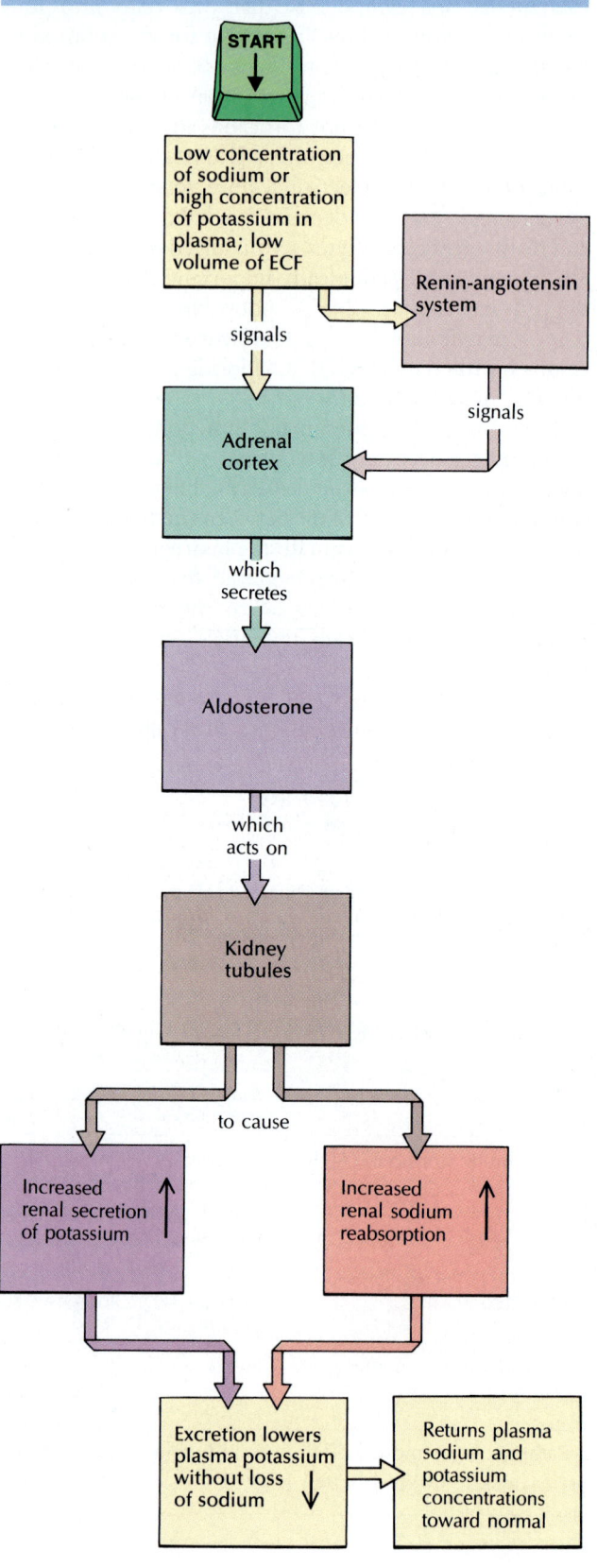

FIGURE 28.5 REGULATION OF SODIUM REABSORPTION AND POTASSIUM SECRETION BY ALDOSTERONE

about 103 mEq/L. Chloride plays a major role in maintaining water balance by balancing the osmotic pressure between fluid compartments. Gastric mucosal cells contain chloride ions, which combine with hydrogen ions to form hydrochloric acid.

Chloride is a negatively charged ion, and so its movement follows the movement of positively charged sodium ions. Under most circumstances, where sodium goes, chloride follows. Since the total positive charges must equal the total negative charges, chloride preserves the ionic equivalence when cations are moved. When sodium is reabsorbed actively, chloride follows passively. When potassium is secreted, chloride accompanies it.

Calcium

The calcium concentration in extracellular fluid is usually between 4.5 and 5.5 mEq/L. Calcium plays important structural and functional roles in the body. It is a major structural component of bones and teeth (about 98 percent) and a crucial factor in blood clotting, muscle contraction, and nerve-impulse transmission.

Plasma calcium concentration is under the control of *parathyroid hormone (PTH* or *parathormone)*, which is produced by the parathyroid glands and secreted into the blood for transport, and *calcitonin (CT)*, produced by special cells in the thyroid gland. Low plasma calcium concentration (*hypocalcemia*) stimulates the production and release of PTH. PTH (1) stimulates the activity of osteoclasts in *bone* to release calcium from bone into the plasma, (2) increases calcium reabsorption from the *gastrointestinal tract* in association with vitamin D, and (3) stimulates the increased absorption of calcium from the *kidneys*.

High plasma levels of calcium (*hypercalcemia*) inhibit the release of PTH and cause the release of calcitonin. Calcitonin lowers plasma calcium by stimulating the activity of osteoblasts, which favors the uptake of calcium into the bones.

Phosphate

The plasma concentration of phosphate is approximately 1.7 to 2.6 mEq/L. Most (about 85 percent) of the body's phosphate is found in bones and teeth in the form of calcium phosphate salts.

The concentration of phosphate ions ($H_2PO_4^-$, HPO_4^{2-}, HPO_4^{3-}) is regulated by PTH and CT. PTH promotes bone absorption and causes the release of large amounts of phosphate ions from bone salts into the extracellular fluid. PTH also causes the kidney tubules to excrete more phosphate ions into the urine. CT lowers plasma phosphate by stimulating the activity of osteoblasts, which removes phosphates from the plasma and deposits them in bone.

Magnesium

The normal plasma concentration of magnesium is about 2 mEq/L. Magnesium is primarily an intracellular electrolyte that plays an important role in the sodium-potassium pump, as well as in the production of ATP energy in mitochondria. Most enzymes that use ATP and ADP in their reactions also need magnesium for optimum enzyme function.

ASK YOURSELF

1 What is an electrolyte?

2 What are two important extracellular electrolytes? Intracellular electrolytes?

3 What is the major organ responsible for controlling electrolyte secretion?

4 How does aldosterone regulate the excretion of sodium and potassium?

5 At what three sites can the plasma concentration of calcium be controlled?

ACID-BASE BALANCE

When one refers to the regulation of the acid-base balance in the body, what is actually meant is the regulation of hydrogen ions in the body fluids, especially the extracellular fluids. In this section we will consider the main mechanisms involved in the regulatory processes that maintain acid-base balance.

Any molecule that dissociates in solution to release a hydrogen ion (H^+) or proton is called an *acid.* Any molecule capable of accepting a hydrogen ion or proton is called a *base.* The hydrogen-ion concentration (measured by the pH scale) can greatly influence every chemical reaction and process in the human body. Enzymes, hormones, and the distribution of ions can all be affected by the concentration of hydrogen ions, so it is not surprising that the concentration is rigorously controlled by the body. The pH of the blood and interstitial fluid is maintained between 7.35 and 7.45. An increase or decrease in the pH value by only a few tenths of a pH unit can be disastrous. (A person can live for only a few hours at the lower or upper limits of 7.0 and 8.0.)

Maintaining an acceptable pH range in the extracellular fluid is accomplished by three mechanisms: (1) specific chemical buffer systems of the body fluids (react very rapidly, in less than a second), (2) respiratory regulation of pH (reacts rapidly, in seconds to minutes), (3) renal regulation of pH (reacts slowly, in minutes to hours).

Strong Acids, Weak Acids, and Equilibrium

A **strong acid** is any compound whose molecules dissociate almost completely in water, yielding hydrogen ions. Hydrochloric acid (HCl) is such an acid. It dissociates almost completely to yield hydrogen and chloride ions according to the equation*

$$HCl \rightleftharpoons H^+ + Cl^-$$

In water, only H^+ and Cl^- exist. There is virtually no intact HCl.

A **weak acid** is any molecule that dissociates only partially in water. Acetic acid is such an acid. In water, only a small fraction of the acetic acid molecules dissociate to yield hydrogen ions and the negatively charged acetate ions:

$$CH_3COOH \rightleftharpoons H^+ + CH_3COO^-$$

acetic acid   hydrogen   acetate
                      ion   ion

In a water solution of acetic acid, all three components of the reaction above

*The preferred direction of the reaction is indicated by the longer arrow.

will be present. There will be a great deal of acetic acid and much smaller amounts of hydrogen ions and acetate ions.

In the dissociation of any weak acid, a stable but dynamic equilibrium is established. The equilibrium represented by the equation above is stable because the processes of dissociation and reassociation are constantly going on at equal rates. The equilibrium is dynamic because if one component is changed, the others will change also. For example, if hydrogen ions (in the form of HCl) are added to the solution of acetic acid, the hydrogen ions will reassociate with the acetate ion, and the equilibrium is said to "shift to the left," forming acetic acid.

In the reassociation of acetate ions and hydrogen ions, the acetate ion acts as a base, accepting a hydrogen ion. In any dissociation of a weak acid that releases hydrogen ions, a weak base is also formed. This is called the **conjugate base** of the weak acid. In the acetic acid solution, if more base is added in the

form of sodium acetate ($NaCH_3COO^-$), the increased concentration of the conjugate base acetate will associate with the hydrogen ions, forcing a shift in the equilibrium to the left, the formation of acetic acid molecules, a reduction of the hydrogen-ion concentration, and a neutralization of the acid (and an increase in pH).

Weak acids are important for two reasons. First, weak acids contribute hydrogen ions to the body and alter the pH if not regulated. Many molecules are weak acids. For example, anaerobic respiration of glucose produces lactic acid, which is a weak acid. Aerobic respiration of glucose produces carbon dioxide, which reacts with water to produce the weak acid carbonic acid. The hydrolysis of fats generates fatty acids, which are weak acids. Second, weak acids, together with their conjugate bases, form acid-base buffer systems that are the body's mechanism for coping with large or abrupt changes in the hydrogen-ion concentration (pH).

Before we begin the discussion of the mechanisms for controlling pH, you may want to review the brief discussion on this page, "Strong Acids, Weak Acids, and Equilibrium."

Acid-Base Buffer Systems

All the **acid-base buffer systems** in the body work by the same principle to *resist changes in pH*. When a weak acid, such as carbonic acid (H_2CO_3), and its conjugate base, sodium bicarbonate ($NaHCO_3$), both exist in body fluids, they constitute a buffer. Addition or depletion of hydrogen ions will result in equilibrium shifts, absorbing added hydrogen ions or releasing hydrogen ions so that the pH remains nearly unchanged. How does this work?

Bicarbonate buffer system The *bicarbonate buffer system,* which is present in the intracellular and extracellular body fluids, helps maintain the appropriate pH of the blood in the following way: Both carbonic acid and bicarbonate are present in higher concentrations than hydrogen ions because they are provided by the body. Carbonic acid dissociates to form hydrogen ions and bicarbonate:

$$H_2CO_3 \rightleftharpoons H^+ + HCO_3^-$$

carbonic acid   bicarbonate

The normal ratio of bicarbonate to carbonic acid in the plasma is 20:1. The equation represents a stable but dynamic equilibrium.

If hydrogen ions are generated by metabolism or by ingestion, they react with the bicarbonate ion (a base) to form more carbonic acid, and the equilibrium shifts toward the formation of the acid:

$$H^+ + HCO_3^- \longrightarrow H_2CO_3$$

The reaction produces slightly more carbonic acid than there was before and slightly less bicarbonate base, but the concentration of free hydrogen ions remains unchanged.

If hydrogen ions are lost (by vomiting, for example), more carbonic acid dissociates:

$$H_2CO_3 \longrightarrow H^+ + HCO_3^-$$

yielding replacement hydrogen ions and bicarbonate ions. The equilibrium shifts, but the concentration of hydrogen ions remains unchanged.

When a strong acid, such as hydrochloric acid, is added to the buffer solution, the following reaction takes place:

$$\underset{\text{strong acid}}{HCl} + \underset{\text{weak base}}{NaHCO_3} \rightleftharpoons \underset{\text{weak acid}}{H_2CO_3} + \underset{\text{salt}}{NaCl}$$

From this equation it can be seen that the strong acid is converted into a weak acid, H_2CO_3. In contrast, if a strong base, such as NaOH, is added to the buffer solution, the following reaction takes place:

$$\underset{\text{strong base}}{NaOH} + \underset{\text{weak acid}}{H_2CO_3} \rightleftharpoons \underset{\text{weak base}}{NaHCO_3} + \underset{\text{water}}{H_2O}$$

The bicarbonate buffer system is important in the body because the concentration of each of its two components (carbonic acid and bicarbonate ions) can be regulated. Carbonic acid is regulated by the respiratory system ($H_2CO_3 \rightarrow CO_2 + H_2O$), and the bicarbonate ion is regulated by the kidneys. (Each of these regulatory systems is discussed later in the chapter.) As a result, the pH of the blood remains relatively constant.

Phosphate buffer system The *phosphate buffer system* is also present in the intracellular and extracellular body fluids, where it regulates the pH within cells (red blood cells, for example) and in the urine, because as seen earlier, the concentration of phosphate in the intracellular fluid is many times that of the extracellular fluid and within the kidney tubules, where phosphate usually becomes greatly concentrated. In the phosphate buffer system, NaH_2PO_4 (sodium dihydrogen phosphate) is the weak acid, while $Na_2HPO_4^-$ (sodium monohydrogen phosphate) is the weak base:

$$H_2PO_4^- \rightleftharpoons H^+ + HPO_4^{2-}$$

When a strong acid, such as HCl, is added to the phosphate buffer system, the following reaction occurs:

$$\underset{\text{strong acid}}{HCl} + \underset{\text{weak base}}{Na_2HPO_4} \rightleftharpoons \underset{\text{weak acid}}{NaH_2PO_4} + \underset{\text{salt}}{NaCl}$$

The net result of this reaction is the exchange of a strong acid for a weak acid, with a relatively small change in pH. In contrast, if a strong base, such as NaOH, is added to the phosphate buffer system, the following reaction occurs:

$$\underset{\text{strong base}}{NaOH} + \underset{\text{weak acid}}{NaH_2PO_4} \rightleftharpoons \underset{\text{weak base}}{Na_2HPO_4} + \underset{\text{water}}{H_2O}$$

In this reaction, sodium hydroxide is buffered to form a weak base and water, allowing the pH to change only slightly.

Protein buffer system The *protein buffer system* is the most abundant and important in body cells and within the plasma because of the high concentration of albumins, hemoglobin, and other proteins. As discussed in Chapter 2, proteins are composed of amino acids bound together by peptide linkages. Some of the amino acid side chains contain a carboxyl group (COOH) that can act as an acid by donating protons as follows:

The hydrogen ion (H^+), or proton, can thus react with any excess hydroxyl ions (OH^-) to form water. (The dotted line signifies a weak bond, and R represents a side chain, or the other component atoms of the amino acid.) The amine group (NH_2) tends to act as a base by accepting protons as follows:

Thus a single amino acid can act as both an acidic and a basic buffer.

It usually takes several hours for the bicarbonate and phosphate buffer systems to be effective. The high concentration and many side chains of amino acids within cells allow the protein buffer system to work instantaneously, making it the most powerful buffer system in the body.

In summary, the acid-base buffer systems accept hydrogen ions, buffering them when body fluids become acidic and releasing them when body fluids become basic, thus allowing only small changes in pH. The stronger acids are converted into weaker acids, and stronger bases are converted into weaker bases.

Respiratory Regulation of Acid-Base Balance

Carbon dioxide is formed continuously within the cells of the body by various metabolic processes. Any increase in the concentration of carbon dioxide in the body fluids as a result of cellular respiration lowers the pH (makes them more acidic). Acidity increases because carbon dioxide reacts with water to form carbonic acid (H_2CO_3), which ionizes into hydrogen (H^+) and bicarbonate (HCO_3^-).

$$\text{Cellular respiration} \longrightarrow \underset{\substack{\text{carbon} \\ \text{dioxide}}}{CO_2} + \underset{\text{water}}{H_2O}$$

$$\underset{\substack{\text{carbon} \\ \text{dioxide}}}{CO_2} + \underset{\text{water}}{H_2O} \rightleftharpoons \underset{\substack{\text{carbonic} \\ \text{acid}}}{H_2CO_3} \rightleftharpoons \underset{\text{hydrogen}}{H^+} + \underset{\text{bicarbonate}}{HCO_3^-}$$

In contrast, a decrease in carbon dioxide concentration raises the pH by decreasing the level of free hydrogen ions.

FIGURE 28.6 RESPIRATORY CONTROL OF pH

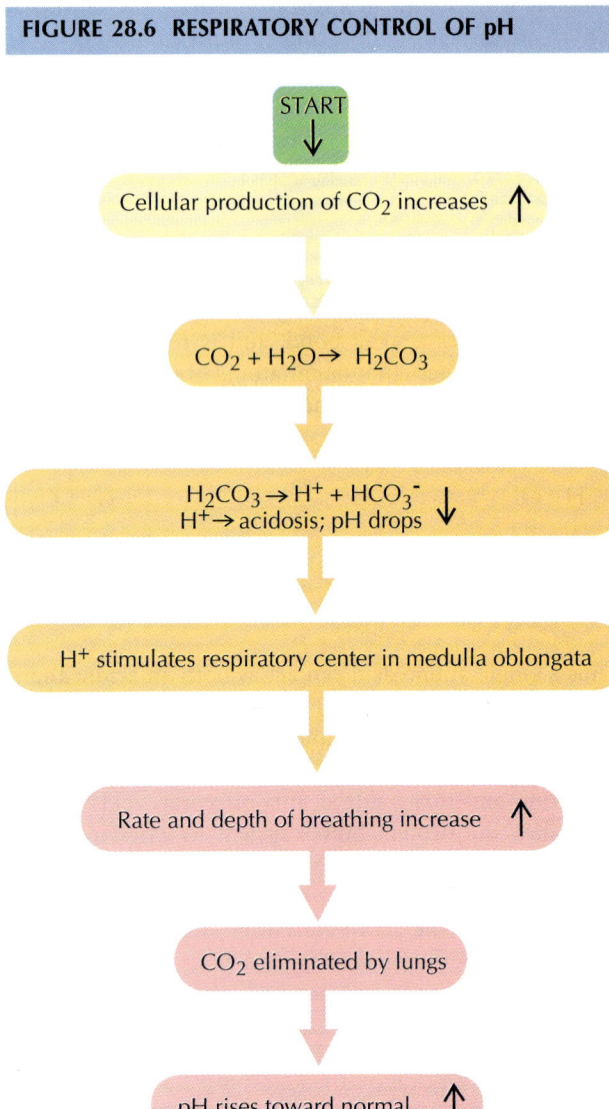

FIGURE 28.7 RENAL CONTROL OF pH

[A] Regulation of acid-base balance by hydrogen-ion secretion in a kidney tubule cell and sodium conservation in exchange for hydrogen ions. Bicarbonate ions for buffering are generated in the body fluids. For each hydrogen ion formed and secreted, a bicarbonate ion is formed and moves into the blood.

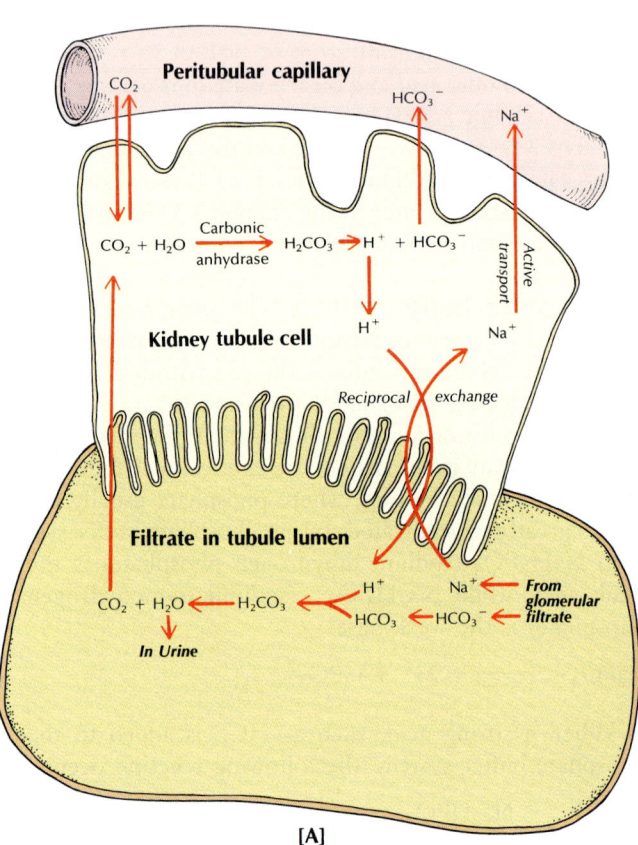

[A]

The respiratory center in the medulla oblongata helps regulate the hydrogen-ion concentration by controlling the rate and depth of respiration. An increase in the rate and/or depth of breathing increases the amount of carbon dioxide exhaled, thereby reducing the carbon dioxide level in the blood [FIGURE 28.6]. This reduces the amount of carbonic acid formed, and the concentration of hydrogen ions is also reduced. Conversely, a decrease in the rate and depth of breathing decreases the carbon dioxide and hydrogen-ion concentration, and increases the pH.

Respiratory adjustments of the pH take 1 to 3 min, much slower than the action of buffer systems. However, the ventilation (rate × depth) can be increased up to eight times the normal rate. This makes the respiratory control of hydrogen-ion concentration a very important short-term regulator of acid-base balance.

Renal Regulation of pH

The body normally consumes more acid-producing foods than base-producing foods, and not only must adjust the pH but also must excrete hydrogen ions. This task is accomplished by renal mechanisms in the kidney tubules, which secrete hydrogen, ammonium, and phosphate ions into the urine. As discussed in the previous chapter, hydrogen ions are secreted into the urine by the proximal and distal convoluted tubules and the collecting ducts.

When a hydrogen ion is secreted into the tubular urine, a sodium ion is simultaneously reabsorbed. This exchange of H⁺ for sodium is important because it

[B] Buffering of excreted hydrogen ions by ammonia. The secreted hydrogen ions combine with ammonia to form ammonium ions, allowing the filtered sodium ions to be exchanged (reabsorbed) reciprocally back into the peritubular capillary (Na^+-H^+ exchange) to combine with the produced HCO_3^-, thus increasing pH.

[C] Transport of excess hydrogen ions into the urine by the phosphate buffer and reabsorption of the sodium ions and generated bicarbonate (HCO_3) into the peritubular capillary to increase pH.

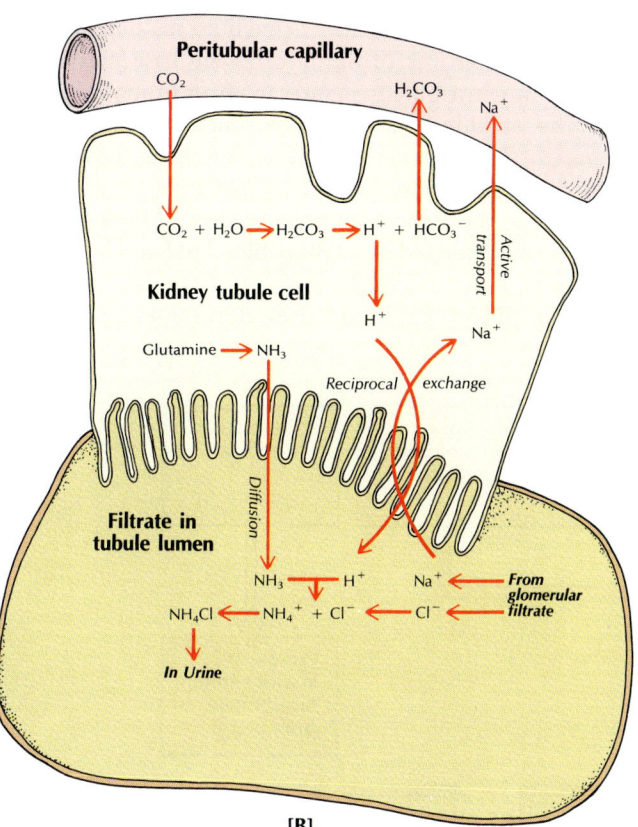

[B]

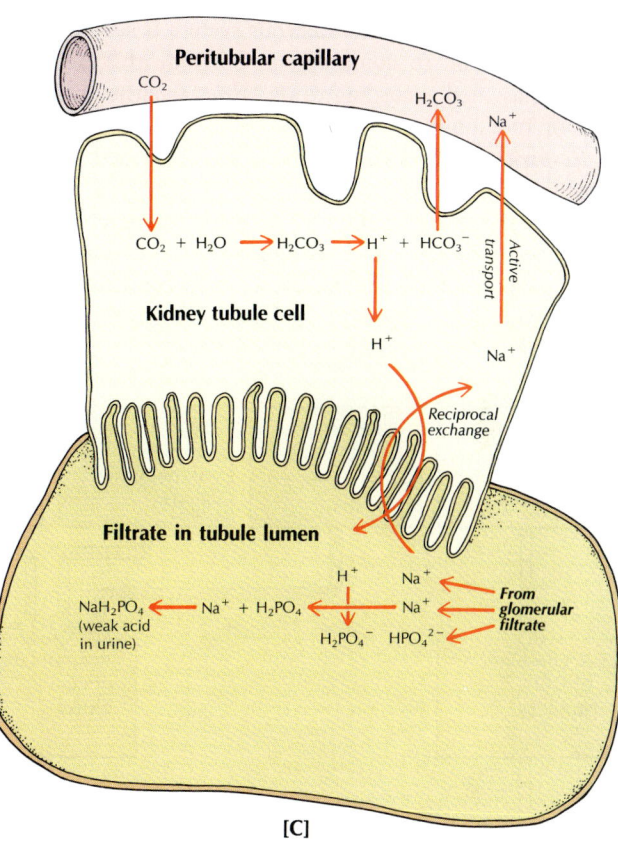

[C]

(1) rids the body of excess H^+ and thus helps maintain pH levels in blood and extracellular fluid, (2) conserves sodium, (3) preserves ionic equivalence, and most importantly, (4) generates sodium bicarbonate for further buffering [FIGURE 28.7A].

Another means for excreting hydrogen ions uses the base ammonia (NH_3) as a vehicle of hydrogen-ion acceptance, forming the ammonium ion (NH_4^+), as described in FIGURE 28.7B. As a result of H^+ and NH_4^+ secretion, the urine usually has a higher hydrogen-ion concentration (lower pH) than the blood. While blood pH is usually about 7.4, the pH of urine may be 6.0 or lower.

A third means by which the kidneys eliminate excess hydrogen ions and help regulate pH is by using the phosphate buffer system. FIGURE 28.7C illustrates how excess hydrogen ions are removed from the tubular fluid by the phosphate buffer, and how this system also functions in the overall process of long-term (1 to 3 days) renal acid-base balance.

A S K   Y O U R S E L F

1 What is a buffer system, and how does it operate?

2 What are the three most important buffer systems in the body?

3 What is the normal pH range for blood?

WHEN THINGS GO WRONG

Acid-Base Imbalances

If the pH of blood deviates from the range of 7.35 to 7.45, serious consequences for the body occur. When the blood pH falls below 7.35, the condition is known as *acidosis* (see drawing A). The pH may fall as low as 7.00 without irreversible damage, but acidosis below 7.00 is usually fatal. (The central nervous system is severely depressed, leading to coma and possibly death.) Acidosis can be classified according to its causes as either respiratory or metabolic.

Respiratory acidosis results from decreased carbon dioxide removal from the lungs, as when a person breathes slowly or shallowly, or when gas exchange is impaired by diseases such as emphysema, chronic bronchitis, or pulmonary edema. Under such conditions, carbon dioxide accumulates in the blood. Respiratory acido-

sis is characterized by a decrease in pH and an increase in P_{CO_2}.

Metabolic acidosis occurs from the metabolic production of acids or the loss of bases. The metabolic production of acids may result from the formation of ketone bodies during increased fat metabolism. Metabolic acidosis may be observed in diabetes mellitus, malnutrition, or starvation. The loss of the bicarbonate base may cause acidosis or contribute to it and can result from prolonged diarrhea or kidney disease. The lungs and kidneys attempt to rid the body of acid as described above. The symptoms and consequences of metabolic acidosis are the same as those for respiratory acidosis. Metabolic acidosis is characterized by a falling blood pH and bicarbonate level.

If the pH of the blood rises above 7.45 because of

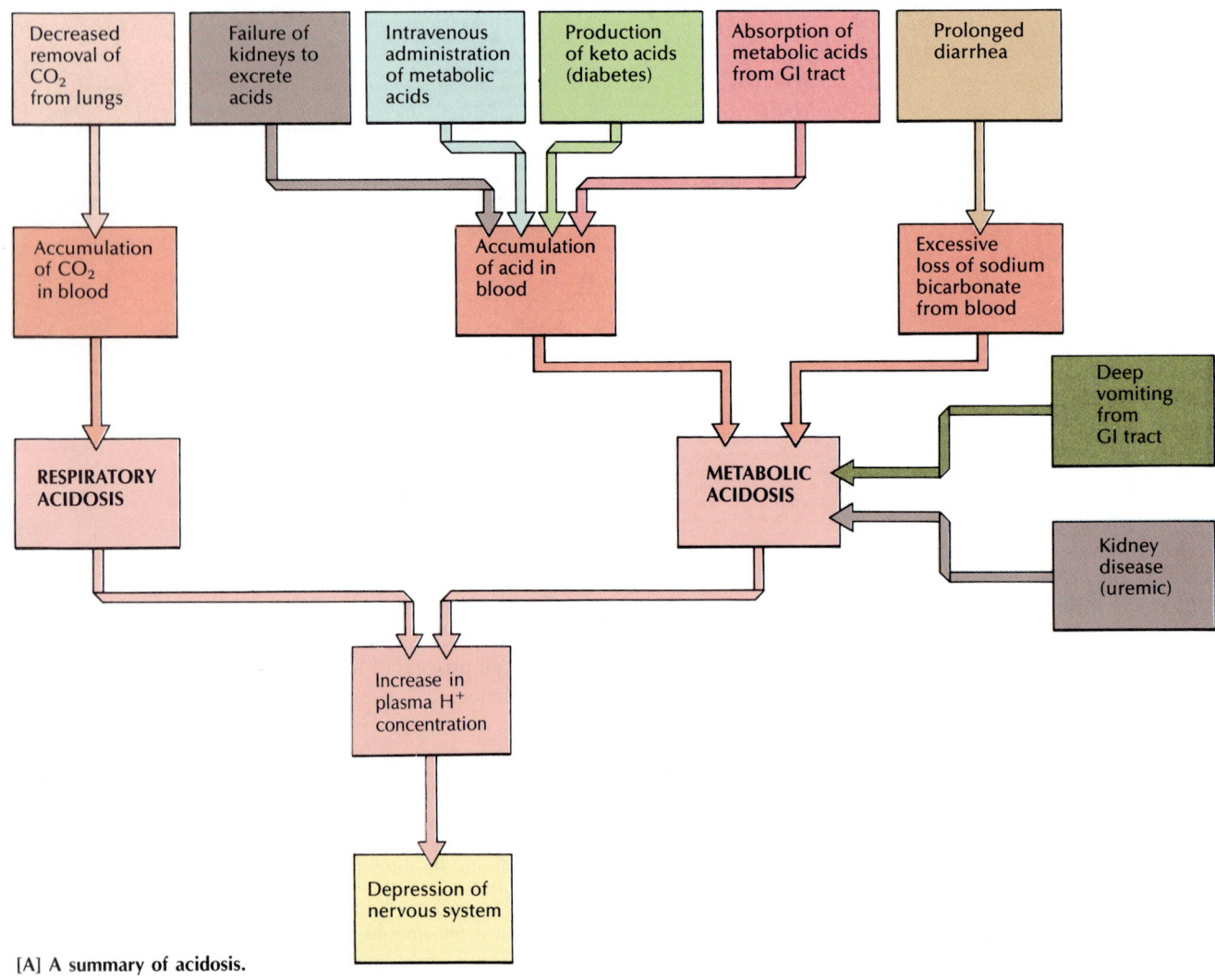

[A] A summary of acidosis.

respiratory or metabolic conditions, *alkalosis* results (see drawing B). ***Respiratory alkalosis*** is caused by an increased loss of carbon dioxide from the lungs at a rate faster than it is produced. This often results from hyperventilation associated with conditions such as emotional disturbances, an overdose of certain drugs, and high altitudes (low P_{O_2}). The kidneys compensate by retaining hydrogen ions and increasing bicarbonate excretion. Decreased carbon dioxide in the lungs will eventually slow the rate of breathing and permit a normal amount of carbon dioxide to be retained in the lungs.

Metabolic alkalosis results from an excessive loss of acid or uptake of alkaline substances. Extensive vomiting of gastric secretions (especially hydrochloric acid) can contribute to this condition, as can excessive ingestion of sodium bicarbonate or other alkaline substances (antacids) that can be absorbed into the blood. In an attempt to lower the pH, the body tries to retain carbon dioxide by decreasing the rate of breathing, and the kidneys increase the retention of hydrogen ions.

The major clinical effect of alkalosis on the body is overexcitability of the central and peripheral nervous systems. Characteristic signs include muscle tetany, convulsions, and sometimes death from respiratory arrest.

Edema

Edema (ih-DEE-muh; Gr. swelling) is an abnormal increase of water within the interstitial fluid compartment. Such an increase in the volume of interstitial fluid produces distension of the tissue, which appears as puffiness on the surface of the body.

Edema may have several causes: (1) Plasma protein may leak across the capillary wall when the capillary lining is damaged. (2) In liver disease, protein synthesis is decreased, plasma protein concentration is reduced, plasma water increases, and net filtration into the interstitial space thus increases. (3) An increase in capillary or venous hydrostatic pressure may occur, which causes increased hydrostatic filtration of water. (4) Lymphatic vessels may become obstructed. (5) In an inflammatory reac-

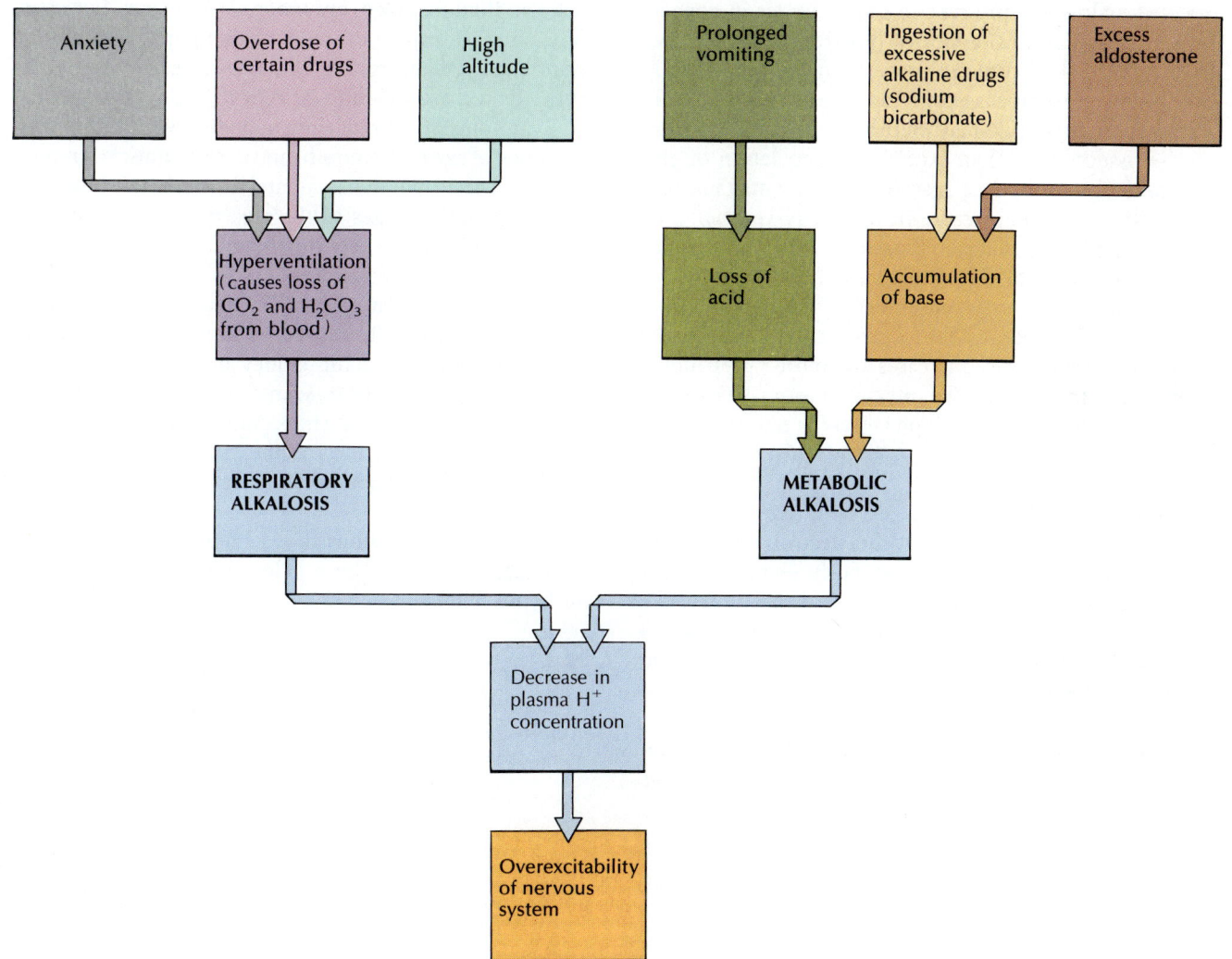

[B] A summary of alkalosis.

tion, in response to infection or tissue damage, capillaries become more permeable to protein.

A decrease in the concentration of plasma protein lowers the osmotic pressure of the plasma and reduces the return of water from the interstitial fluid to the plasma in the venous end of the capillary. The resulting water accumulation in the interstitial space produces the edema.

In varicose veins, the improper closing of venous valves produces edema. When blood accumulates within the veins, there is an increase in the hydrostatic pressure at the venous end of the capillary. This opposes the colloidal osmotic pressure that normally returns water from the interstitial fluid, and the volume of the interstitial water increases.

Obstruction of the lymph vessels that connect the interstitial compartment and drain part of its fluid back into the vascular system also produces edema. Such an obstruction is seen in the parasitic disease called *elephantiasis*. Edema is involved in several other pathological conditions: (1) Obstructed venous flow may cause increased hydrostatic pressure in capillaries (as in congestive heart failure); (2) a lack of blood proteins may decrease the colloid osmotic pressure in the blood (as in starvation); (3) a blockage of lymphatic vessels that drain excess interstitial fluid may increase hydrostatic pressure in the tissues; and (4) an excessive accumulation of glycosaminoglycans in the extracellular matrix may increase the colloid osmotic pressure in tissues *(myxedema)*.

Localized tissue damage (burns) or infection that results in an inflammatory reaction and its attendant release of histamine may cause edema. Histamine makes capillaries more permeable to protein. As a result, the tissue osmotic pressure increases and more water moves from the capillaries into the interstitial area faster than it can be eliminated by the lymph vessels or returned to the venule capillaries.

Electrolyte Imbalances

Imbalances of one or more of the electrolytes, particularly sodium and potassium, are fairly common under certain conditions. Severe imbalances, if left untreated, may cause death. Intravenous infusion of fluid may be necessary to restore depleted electrolytes and prevent serious consequences.

Sodium Low sodium concentration in the extracellular fluids is called *hyponatremia* (under + L. *natrium*,

sodium). It may be due to excessive loss of sodium, reduced intake of sodium, or both. Some causes of hyponatremia include prolonged sweating, vomiting, and diarrhea. Symptoms of hyponatremia include hypotension, tachycardia, muscular weakness, dizziness, muscle cramps, small volume of urine, apprehension, and convulsions. Coma may occur if hyponatremia is severe.

The sudden addition of large amounts of sodium salts causes *hypernatremia*, a higher-than-normal plasma sodium level. Because of the increase in the osmolarity of the plasma, water moves from the cells into the extracellular fluid, resulting in cellular dehydration. The kidneys respond by increasing their sodium excretion in an attempt to maintain homeostasis. Signs of hypernatremia include fatigue, restlessness, extreme thirst, and, if severe, coma.

Plasma concentrations of sodium above 145 mEq/L are associated with hypernatremia, and concentrations below 135 mEq/L are associated with hyponatremia.

Potassium Low potassium concentration in the extracellular fluid is called *hypokalemia* (under + L *kalium*, potassium). It can be caused by a diet poor in potassium, poor intestinal absorption, kidney disease, diarrhea, vomiting, or the use of some diuretics.

Symptoms of hypokalemia include a weakness in contractions of skeletal, smooth, and cardiac muscle; cramps; fatigue; flaccid paralysis; mental confusion; nausea and vomiting; and changes in the electrical conducting system of the heart, as seen in the ECG (lengthened P-R interval and flattened T wave).

High extracellular potassium concentration is called *hyperkalemia*, usually caused by decreased excretion of potassium due to specific kidney diseases, which affect the nephron's capacity to secrete potassium, or diseases of the adrenal cortex, such as Addison's disease. Symptoms include small urine volume, nausea, weakness, irritability, abdominal cramping, diarrhea, and, if severe, cardiac fibrillation.

Plasma concentrations above 5.6 mEq/L are typical of hyperkalemia, while those below 3.5 mEq/L are typical of hypokalemia.

Q: *Why can't we drink saltwater to quench our thirst?*

A: We cannot process the saltwater fast enough to avoid dehydration, and we actually worsen the situation. (We produce about four times as much urine as the saltwater we consume.)

CHAPTER SUMMARY

Body Fluids: Compartments and Composition (p. 920)

1 All body fluids are either *intracellular* (inside the cell) or *extracellular* (outside). The intracellular fluid constitutes about two-thirds of all body fluid.

2 Distinct compartments contain several extracellular fluids, including interstitial fluid, blood plasma and lymph, and transcellular fluid.

3 Plasma and interstitial fluid are more similar to each other in their solute composition than they are to intracellular fluid. However, plasma contains soluble proteins, which are nearly absent from interstitial fluid.

Movement of Fluids between Compartments (p. 921)

1 Movement of water from one body compartment to another depends on *hydrostatic pressure* and *osmotic pressure.* Changes in the electrolyte (solute) concentration can alter the water distribution between compartments.

2 Because fluids are constantly moving into and out of body compartments, the overall concentration of solutes in the plasma and interstitial fluid is maintained.

Water (p. 922)

1 Water makes up between 47 and 77 percent of total body weight, the proportion varying with sex, age, and amount of body fat.

2 Water transports nutrients to cells and removes wastes, provides a medium for intracellular chemical reactions, acts as a solvent, helps maintain body temperature, aids digestion, promotes excretion, and transports enzymes and other substances.

3 Fluid intake and output are approximately equal in a healthy person. Most of the fluid is taken in as food or drink and is excreted as urine.

4 Water intake is regulated by the hypothalamus.

5 The kidneys are the major organs for water excretion. *Antidiuretic hormone (ADH)* controls water excretion by regulating reabsorption of water from the kidney tubules into the plasma.

Function and Regulation of Specific Electrolytes (p. 925)

1 *Electrolytes* are compounds that dissociate into ions when in solution. The most physiologically important electrolytes are sodium, potassium, calcium, magnesium, hydrogen, bicarbonate, chloride, phosphate, and sulfate. Each is normally maintained at an optimal concentration in each body fluid.

2 Electrolytes are required for cell metabolism, contribute to some body structures, facilitate movement of water between compartments, help maintain acid-base balance, and are involved in the production and maintenance of membrane and action potentials.

3 Sodium, potassium, and chloride have the highest concentrations in different body fluids, and therefore have the greatest influence on osmotic relationships.

4 Electrolytes can be lost with water in sweat, feces, or urine. The kidney nephrons control electrolyte excretion through reabsorption and secretion.

5 The nervous and endocrine systems are essential for electrolyte regulation.

6 *Sodium* is low in concentration within cells and high outside cells. Optimal plasma concentration is regulated by aldosterone and atrial natriuretic peptide.

7 *Potassium* is high in concentration within cells and low outside cells. Potassium secretion is regulated by aldosterone.

8 *Chloride* is the most abundant negative ion in the extracellular fluid, and is therefore important osmotically. It moves passively, following the movement of positive ions, especially sodium.

9 The concentration of *calcium* is maintained by absorption from the intestine, kidney excretion or reabsorption, and dissolution from bones, and is regulated by *parathyroid hormone (PTH).*

10 Fluid or water loss may produce imbalances in one or more electrolytes. Prolonged or severe imbalances of sodium or potassium result in an improper fluid balance.

11 The concentration of *phosphate ions* is regulated by PTH, which promotes bone resorption and decreases the transport of phosphate ions.

12 *Magnesium* is primarily an intracellular electrolyte that affects the sodium-potassium pump and the production of ATP energy in mitochondria.

Acid-Base Balance (p. 927)

1 *Acids* release hydrogen ions in water, and *bases* accept hydrogen ions. The hydrogen ion concentration (pH) of body fluids is maintained by buffer systems, respiratory regulation, and renal regulation.

2 *Weak acids* dissociate partially in water solutions, while *strong acids* dissociate completely. Weak acids and their conjugate bases form *acid-base buffer systems,* which resist changes in the pH when hydrogen ions are added or withdrawn.

3 The most important buffers are the *bicarbonate, phosphate,* and *protein systems.*

4 A change in the rate of breathing changes the carbon dioxide level in the blood, and thus affects the hydrogen-ion concentration of body fluids.

5 The kidney nephrons can secrete hydrogen ions into the urine in exchange for reabsorbed sodium ions, can use ammonium ions for hydrogen-ion acceptance, and can use the phosphate buffer system.

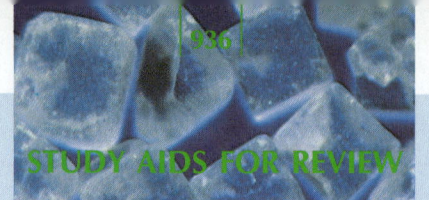

KEY TERMS

acid 927

antidiuretic hormone (ADH) 926

base 927

buffer system 928

electrolytes 925

extracellular fluid 920

hydrostatic pressure 921

intracellular fluid 920

osmotic pressure 921

parathyroid hormone 927

MEDICAL TERMS FOR REFERENCE

ANTACID An alkaline (basic) substance that neutralizes acids.

ANURIA (uh-NYOOR-ee-uh; L. *an,* not + urine) The inability to urinate.

DIURESIS (dye-yoo-REE-sis; Gr. *diourein,* to urinate) An increased excretion of urine.

KETONURIA Excessive ketone bodies (from fat metabolism) in the urine.

KETOSIS Excessive ketone bodies in the body as a result of inadequate metabolism of carbohydrates.

UNDERSTANDING THE FACTS

1 Distinguish between the composition of intracellular and extracellular fluids.

2 Contrast hydrostatic pressure with osmotic pressure.

3 List the major functions of water.

4 List the major functions of electrolytes.

5 List the most important electrolytes in the body and their specific functions.

6 What conditions lead to the loss of extracellular fluid?

7 What is a buffer system?

8 Which buffer system contributes most directly to the maintenance of the pH of blood, and which to the regulation of pH within cells?

9 What is the normal range of pH of blood? What pH values are associated with acidosis and alkalosis?

UNDERSTANDING THE CONCEPTS

1 Explain how hydrostatic pressure and osmotic pressure regulate the movement of fluid in a capillary bed.

2 What is edema? What conditions can contribute to it?

3 Which hormones are the primary regulators of electrolyte balance, and which electrolytes does each hormone control?

4 How does the respiratory system regulate the pH of body fluids?

5 How do the kidneys regulate the pH of body fluid?

6 Explain the respiratory and metabolic causes of acidosis and alkalosis.

SELF-QUIZ

Multiple-Choice Questions

28.1 Which of the following is *not* a major function of water in the body?
 a transports enzymes
 b acts as a solvent
 c is involved in blood coagulation
 d helps maintain body temperature
 e medium for metabolic reactions

28.2 Which of the following are the most similar in composition?
 a plasma and interstitial fluid
 b plasma and intracellular fluid
 c lymph and plasma
 d lymph and intracellular fluid
 e intracellular and extracellular fluids

28.3 Which of the following is present in blood plasma but *not* in interstitial fluid?
 a calcium   d magnesium
 b sodium   e protein
 c potassium

28.4 Movement of water from one body compartment to another is controlled by
 a atmospheric pressure   d a and c
 b hydrostatic pressure   e b and c
 c osmotic pressure

28.5 A solution that has a lower solute concentration than the cells it bathes is
 a hyperosmotic   d hypotonic
 b hypertonic   e isotonic
 c hypoosmotic

28.6 The functions of electrolytes in the body include
 a contributing to body structures
 b facilitating the movement of water between body compartments
 c maintaining acid-base balance
 d a and b
 e a, b, and c

28.7 Parathormone
 a acts on bone cells to cause them to release calcium
 b increases GI tract absorption of calcium
 c increases the reabsorption of calcium from the glomerulofiltrate
 d a and b
 e a, b, and c

28.8 In the body, magnesium
   a is primarily an intracellular electrolyte
   b plays a role in the sodium-potassium pump
   c functions in the production of ATP in mitochondria
   d a and b
   e a, b, and c

28.9 With the bicarbonate buffer system, a strong acid, such as HCl, is converted into
   a a salt   **d** water
   b a weak acid   **e** a strong base
   c a weak base

28.10 The exchange of the H^+ for sodium in the kidney
   a rids the body of excess H^+
   b rids the body of excess sodium
   c generates sodium bicarbonate
   d a and c
   e a, b, and c

28.11 If the hydrogen-ion concentration in the blood plasma increases, the
   a respiratory rate will increase
   b kidney tubes will secrete less NH_4^+
   c protein buffer system slows down
   d a and b
   e a, b, and c

28.12 If a strong base, such as NAOH, is added to a phosphate buffer system, the end product(s) will be
   a NaH_2PO_4 **d** a and b
   b $NaHPO_4$ **e** b and c
   c H_2O

28.13 The addition of HCl to the bicarbonate buffer system produces
   a H_2CO_3 **d** a and b
   b NaCl   **e** b and c
   c $NaHCO_3$

28.14 During severe exertion in a hot environment a person may lose 4 L of hypotonic sweat per hour. This would result in
   a decreased plasma volume
   b decreased plasma osmolarity
   c increased circulation levels of ADH (antidiuretic hormone)
   d return to normal body fluid balance after ingestion of 1000 mL of water
   e a and c

28.15 During severe diarrhea, there is a loss of water, bicarbonate, and sodium from the gastrointestinal tract. Physiological compensation for this would probably include
   a increased alveolar ventilation
   b increased hydrogen-ion secretion by the renal tubules
   c decreased urinary excretion of sodium and water
   d increased renin secretion by the kidneys
   e all of the above

Matching

28.16 _____ NaH_2PO_4 **a** strong acid
28.17 _____ $NaHCO_3$ **b** weak acid
28.18 _____ NaOH   **c** strong base
28.19 _____ HCl   **d** weak base
28.20 _____ H_2CO_3 **e** salt

Completion Exercises

28.21 The loss of about _____ percent of the total body water is usually fatal to an adult.

28.22 _____ is the most abundant single constituent in all body fluids.

28.23 _____ is the abnormal increase of water within the interstitial space.

28.24 Cells of the thirst center of the hypothalamus function as _____ that sense changes in osmolarity.

28.25 In the kidneys, reabsorption of water is under the direct control of the hormone _____.

28.26 _____ are compounds that dissociate in solution.

28.27 Any substance that dissociates in solution to release hydrogen ions (H^+) (protons) is a(n) _____.

28.28 The weak base that is formed in any dissociation of a weak acid is called the _____ base of the weak acid.

28.29 By excreting _____ from the lungs, respiration plays a major role in pH control.

28.30 In the kidneys, when a hydrogen ion is secreted into the tubular urine, a _____ ion is simultaneously exchanged.

iv *case studies*

1

The routine physical for a 2-year-old child was normal. However, during a subsequent conversation with the pediatrician, the mother expressed concern about her child's lack of energy. The little girl always appeared tired and did not play as long or as energetically as her friends. Despite the child's normal blood profile, the doctor ordered an analysis of blood gases. The little girl's arterial oxygen (PaO_2) was low, around 68 mmHg, and it was soon confirmed that she had an atrial-septal defect.

- **What is an atrial-septal defect, and how does this condition alter the route of blood flow?**
- **Why does this defect cause a lowering in the partial pressure of arterial oxygen?**
- **How does this defect arise in the fetus?**

2

While mowing grass, many "seasonal" asthmatics experience tightness in the chest, difficulty in exhalation, and wheezing. Some of these symptoms can be alleviated by using a bronchodilator.

- **What is probably precipitating the asthma attacks in such patients?**
- **Why is exhalation more difficult than inhalation during an asthma attack?**
- **What is the effect of the bronchodilator?**

3

For the sake of convenience, long-distance truck drivers often limit their fluid intake while on cross-country runs. At such times, the urine becomes very dark in color. If fluid intake is increased to about eight glasses of water each day, the urine is much lighter in color.

- **What makes the urine dark?**
- **Why does increased fluid intake cause it to lighten?**
- **Describe the effects of hypotonic (water) intake on antidiuretic hormone (ADH) output.**

4

A middle-aged man was diagnosed with stomach cancer. The treatment of choice was surgery. Two-thirds of his stomach was removed, and the cardiac portion was joined directly to the pyloric portion.

- **What regions of the stomach were removed?**
- **What are the normal enzymatic functions of the stomach?**
- **How was it possible for this patient to survive without the major portion of his stomach?**
- **What changes in lifestyle were required following surgery?**

v

Reproduction and Development

CAREER IN FOCUS: GENETIC COUNSELING

A few years ago, amniocentesis followed by karyotyping could reveal only the presence of such large-scale chromosomal abnormalities as Down's syndrome. Now geneticists seem to expand the list of mapped genes almost monthly. Cystic fibrosis, familial Alzheimer's, Huntington's, ALS (Lou Gehrig's disease), and colon cancer—all are caused by detectable abnormalities in tiny sections of genes that reside in known chromosomal regions. By the turn of the century, the Human Genome Project aims to know the location of every human gene. Of course, knowing the address of a flawed gene on a chromosome does not necessarily mean we can immediately treat, much less cure, its effects. However, mapping is the first step, and geneticists are sure that with time, diagnosis will follow mapping, and treatment will follow diagnosis.

The burst in knowledge of the human genome, coupled with advances in techniques of reproductive biology (in vitro fertilization, genetic analysis of embryos, and reimplantation of embryos into hormonally prepared wombs), has presented society with new choices. Newspapers carry accounts of women who give birth to their own grandchildren by virtue of eggs donated by their daughters. Recently, the press has told us of the possibility of implantation of fetal ovaries into infertile women. Some scientists as well as some laypeople worry that reproduction is becoming more technological, less "natural." But for pro-

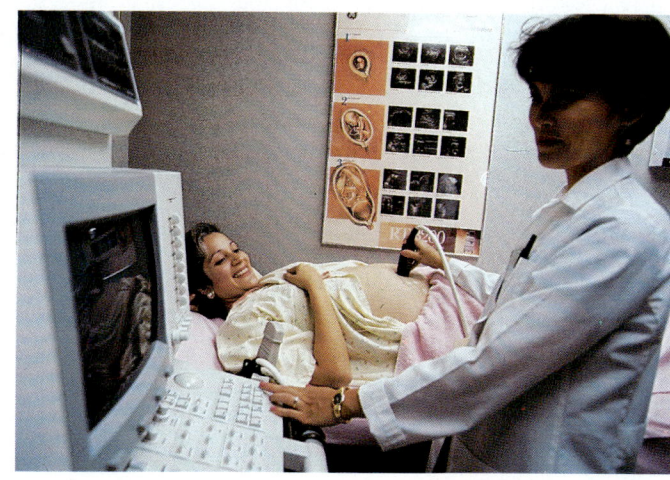

spective parents who believe themselves to be genetic carriers of a debilitating disease, the expanded range of choices made possible by reproductive technology offers hope for a healthy baby.

From our first understanding that some diseases are inherited to the development of new systems of reproductive technology, the genetic counselor has offered impartial, nondirective help to families. Initially using only pedigree analysis to assess parental genotypes, but more recently employing amniocentesis to collect fetal cells for biochemical and genetic analyses, these health professionals assess the genetic risks, explain the medical facts, and present families with the data from which an informed choice can be made.

Board-certified genetic counselors have generally been educated in a 2-year master's degree program. As undergraduates, prospective genetic counselors major in a variety of subjects, including biology, psychology, nursing, and the allied health fields. Either as prerequisites or as a part of the master's program itself, related coursework in-

cludes anatomy and physiology, genetics, developmental biology, reproductive biology, biochemistry of genetic diseases, statistics, and, of course, counseling. In the summer between the 2 academic years, most applicants begin clinical internships that are continued through their second year.

While the majority of genetic counselors work as part of a team of reproductive professionals in hospitals and major medical centers, some are in private practice. Demand for their services has always been strong, but now, with the explosion in knowledge about the human genome, career opportunities for genetic counselors are vastly increased.

Genetic counseling can have a major effect on the well-being of an infant and its parents, sometimes providing a lifesaving service long before the baby is born. The genetic counselor is a health care professional who has a strong background in the basics of the areas of human anatomy and physiology that we look at in this concluding section: reproduction and development.

The
Reproductive
Systems

29

KEY CONCEPTS

1 The reproductive role of the male—to deliver sperm to the female vagina—is simple compared with the complex reproductive role of the female.

2 Under normal conditions, only one sperm enters and fertilizes an egg cell (ovum). The head of the sperm contains the male genetic material.

3 Females usually release one ovum in every menstrual cycle. Complex hormonal feedback systems are involved in regulating the events of the menstrual cycle.

4 The process of meiosis reduces the number of chromosomes in male and female sex cells to half so that the fertilized ovum contains the correct number of chromosomes.

Sexual reproduction produces new human beings and allows hereditary traits to be passed from both parents to their children. Sexual reproduction always involves the union of two parental sex cells: an egg (ovum) from the mother and a sperm from the father. This allows the hereditary material (DNA) from both parents to combine, forming a new individual with a unique combination of genes.

In this chapter we discuss male and female reproductive anatomy and physiology, with special attention to the female's rhythmic menstrual cycle. We also consider the hormonal control of reproductive activities, the formation of sperm and egg cells, conception, and contraception.

MALE REPRODUCTIVE ANATOMY AND PHYSIOLOGY

The reproductive role of the male is to produce sperm and deliver them to the vagina of the female. These functions require four different types of structures [FIGURE 29.1]:

1 The *testes* produce sperm and the primary male sex hormone, testosterone.

2 *Accessory glands* furnish a fluid for carrying the sperm to the penis. This fluid plus sperm is called *semen.*

3 *Accessory ducts* store and carry secretions from the testes and accessory glands to the penis.

4 The *penis* deposits semen into the vagina during sexual intercourse.

Testes

The paired ***testes*** (TESS-teez; sing. *testis*)* are the male reproductive organs (*gonads*), which produce *sperm* [FIGURE 29.2]. Testes are often called *testicles* (from a Latin diminutive form of *testes*).

During fetal development, the testes are formed just below the kidneys inside the abdominopelvic cavity. By the third fetal month each testis has descended from its original site in the abdomen to the *inguinal canal.* During their seventh fetal month the testes pass through the inguinal canal [FIGURE 29.3].

*In Latin, *testis* means "witness." The paired testes were believed to bear witness to a man's virility.

Q: *During a routine physical examination for a male, why does the physician insert a finger in the superficial inguinal ring under the scrotum and ask the patient to cough?*

A: Coughing exerts pressure on the internal abdominal organs, forcing them against the inguinal canal and over the inguinal ligament. A loose inguinal ring, felt by the physician, may indicate a developing inguinal hernia.

The inguinal canal is a passageway leading to the ***scrotum*** (SCROH-tuhm; L. *scrautum,* a leather pouch for arrows), an external sac of skin that hangs between the thighs. The testes complete their descent into the scrotum shortly before or after birth. The inguinal canal is usually sealed off after the testes pass through. If the canal fails to close properly, or if the area is strained or torn, an *inguinal hernia (rupture)* may result.

If the testes remain in the abdominal cavity instead of descending on schedule, surgery is usually performed during early childhood. The condition is called *cryptorchidism* (krip-TOR-kye-dizm; Gr. *kryptos,* hidden + *orchis,* testis, because of the testicular shape of the orchid root). If the condition is not corrected, testosterone will be produced by the testes but no sperm will develop, resulting in sterility. In addition, the risk of testicular cancer increases.

One testis (usually the left) hangs slightly lower than the other, so the testes do not collide uncomfortably during normal activities. Because the testes hang outside the body, their temperature is about 3°F cooler than the body temperature. The lower temperature is necessary for active sperm production and survival. The sperm remain nonviable if they are retained inside the body cavity. For the same reason, a fever can kill hundreds of thousands of sperm, temporarily decreasing a man's fertility. In warm temperatures, the skin of the scrotum hangs loosely, and the testes are held in a low position. In cold temperatures, the *cremaster* muscles under the skin of the scrotum contract and pull the testes closer to the warm body. In this way, the temperature of the testes remains somewhat constant. Sweat glands also help cool the testes.

The interior of the scrotum is divided into two separate compartments by a fibrous *median septum.* One testis lies in each compartment. The line of the median septum is visible on the outside as a ridge of skin, the *perineal raphe* (RAY-fee; Gr. seam), which continues forward to the underside of the penis and backward to the *perineum,* the diamond-shaped area between the legs and back to the anus.

Each testis is oval-shaped and measures about 4.5 cm (1.75 in.) long and 2.5 cm (1 in.) wide in an adult. It is enclosed in a fibrous sac called the ***tunica albuginea*** (al-byoo-JIHN-ee-uh; L. *albus,* white). This sac extends into the testis as *septae,* which divide it into compartments called *lobules* [FIGURE 29.2A, B]. The tunica albuginea is lined by the *tunica vasculosa* and covered by the *tunica vaginalis.* The tunica vasculosa contains a network of blood vessels, and the tunica vaginalis is a continuation of the peritoneal membrane that lines the abdominopelvic cavity.

Each testis contains over 800 tightly coiled ***seminiferous tubules,*** which produce thousands of sperm each *second* in a healthy young man. The combined length of

FIGURE 29.1 MALE REPRODUCTIVE ANATOMY

[A] Sagittal section of the male pelvis. The arrows indicate the path of sperm during ejaculation from the testes through the urethral orifice in the penis; ejaculation takes place via an erect penis (not shown). [B] Anterior view of the male reproductive system with portions of the penis and right testis removed to show the interior structures.

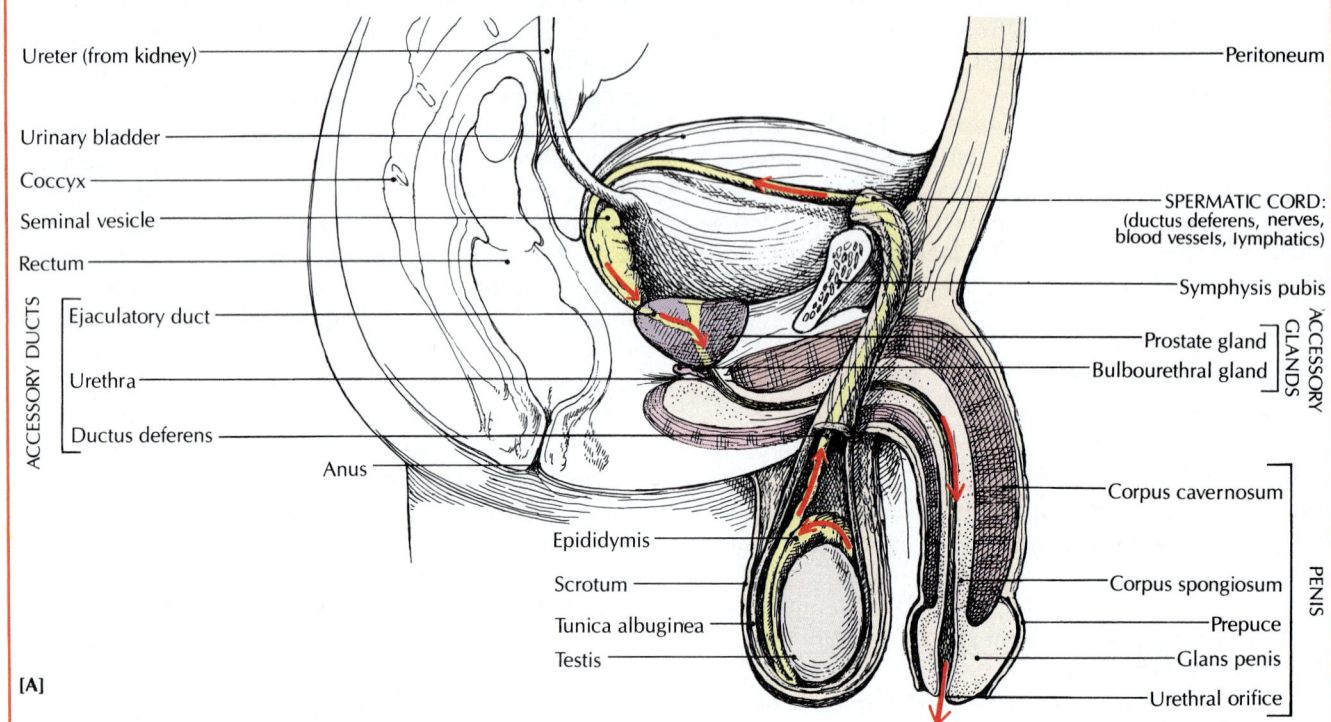

Structure	Function
Testes	Produce sperm and testosterone
Seminal vesicles, prostate gland, bulbourethral gland	Furnish fluid for carrying sperm
Epididymis, ductus deferens, ejaculatory duct, urethra	Store and transport sperm
Penis	Deposits semen into vagina during intercourse

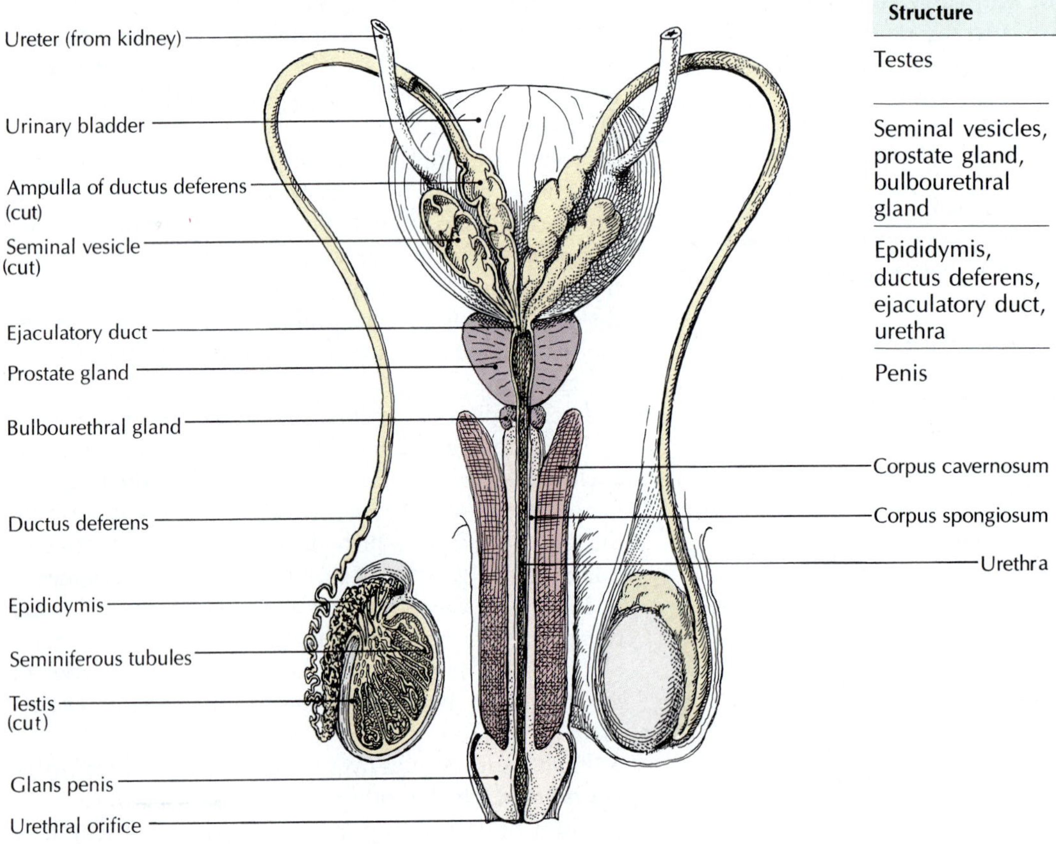

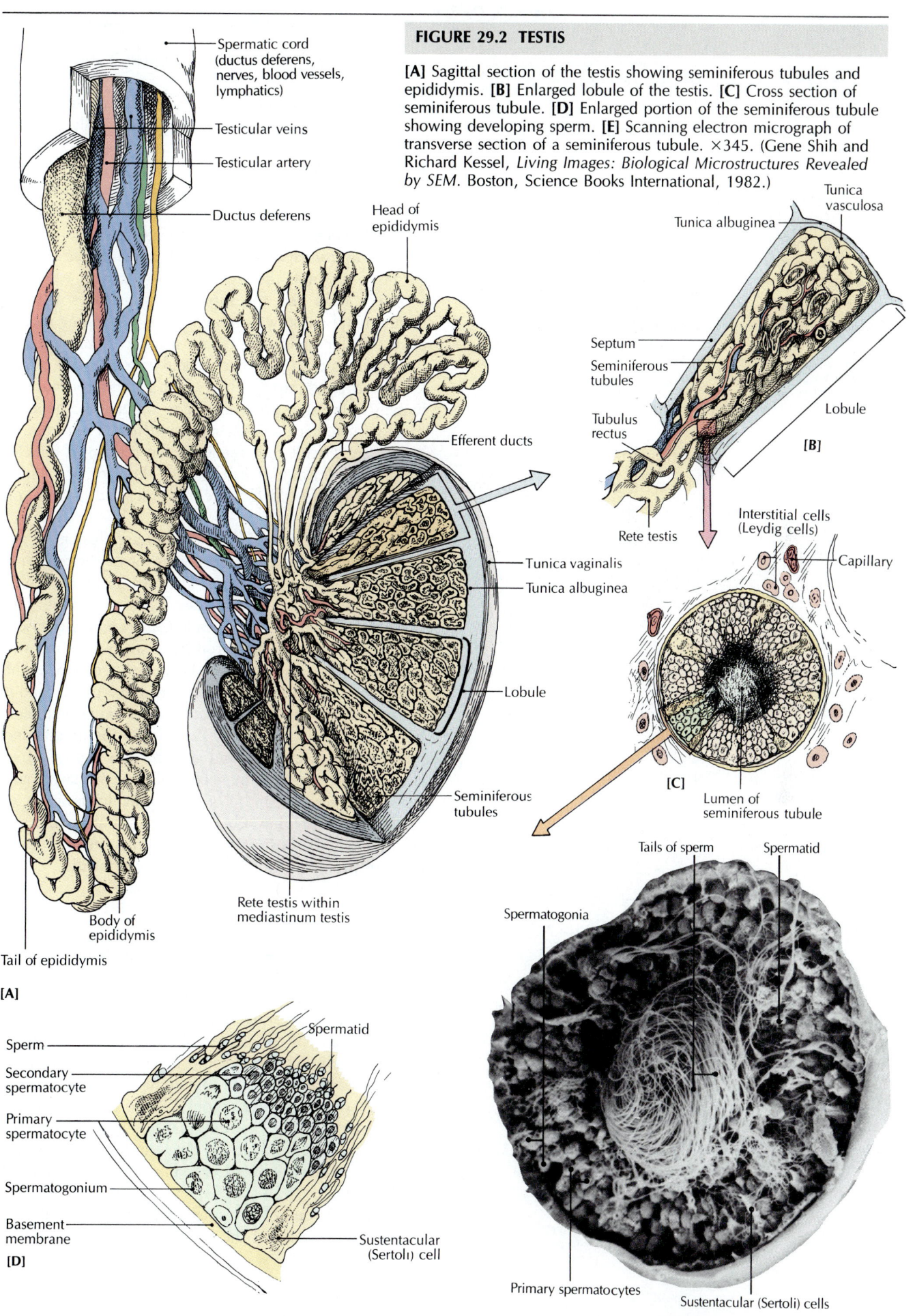

Spermatic cord (ductus deferens, nerves, blood vessels, lymphatics)

Testicular veins

Testicular artery

Ductus deferens

Head of epididymis

Efferent ducts

Body of epididymis

Tail of epididymis

[A]

FIGURE 29.2 TESTIS

[A] Sagittal section of the testis showing seminiferous tubules and epididymis. **[B]** Enlarged lobule of the testis. **[C]** Cross section of seminiferous tubule. **[D]** Enlarged portion of the seminiferous tubule showing developing sperm. **[E]** Scanning electron micrograph of transverse section of a seminiferous tubule. ×345. (Gene Shih and Richard Kessel, *Living Images: Biological Microstructures Revealed by SEM.* Boston, Science Books International, 1982.)

Tunica albuginea

Tunica vasculosa

Septum

Seminiferous tubules

Tubulus rectus

Lobule

Rete testis

[B]

Tunica vaginalis

Tunica albuginea

Lobule

Seminiferous tubules

Rete testis within mediastinum testis

Interstitial cells (Leydig cells)

Capillary

Lumen of seminiferous tubule

[C]

Sperm

Secondary spermatocyte

Primary spermatocyte

Spermatogonium

Basement membrane

Spermatid

Sustentacular (Sertoli) cell

[D]

Tails of sperm

Spermatid

Spermatogonia

Primary spermatocytes

Sustentacular (Sertoli) cells

[E]

FIGURE 29.3 DESCENT OF THE TESTES

The testes descend from the abdominal cavity ino the scrotum during fetal development. **[A]** Six-month fetus. **[B]** Seven-month fetus. **[C]** At birth.

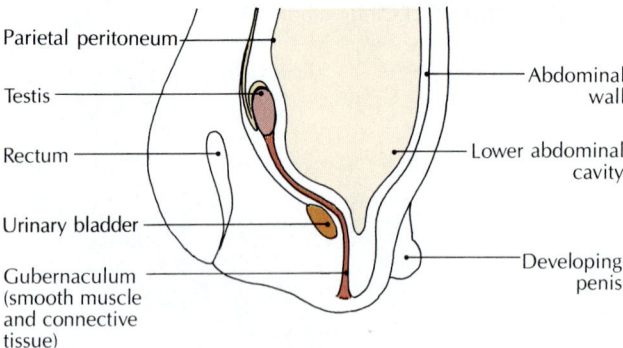

[A] MONTH 6

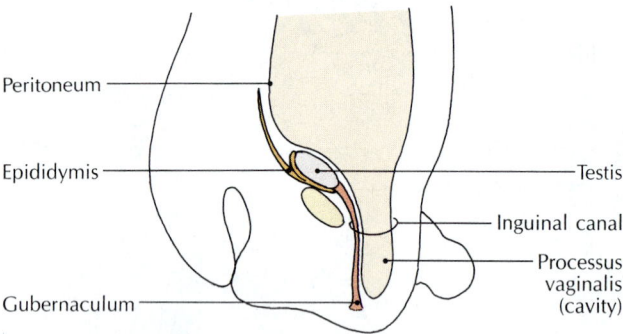

[B] MONTH 7

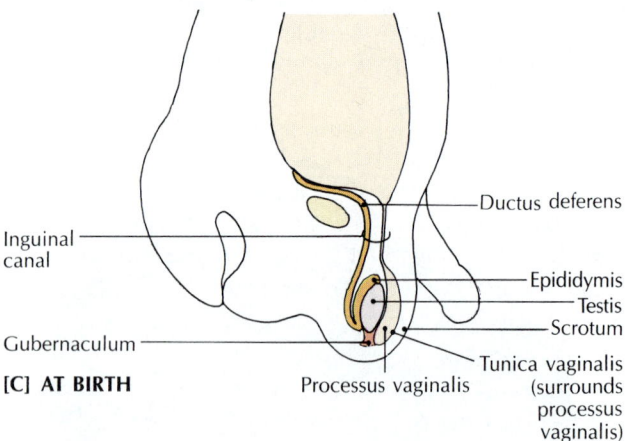

[C] AT BIRTH

the seminiferous tubules in both testes is about 225 m (750 ft).

The walls of the seminiferous tubules are lined with *germinal tissue*, which contains two types of cells: spermatogenic cells and sustentacular (Sertoli) cells. The **spermatogenic cells,** including spermatogonia, spermatocytes, and spermatids, eventually develop into mature sperm [FIGURE 29.2D]. The development of sperm, or **spermatogenesis,** is discussed on page 963.

The **sustentacular cells** provide nourishment for the germinal sperm as they mature. Sustentacular cells also secrete a fluid into the tubules that provides a liquid medium and an outward flow for the developing sperm. Sustentacular cells secrete an *androgen-binding protein* that binds to both testosterone and estrogen, and carries those hormones into the fluid of the seminiferous tubular lumen, where they are available to the maturing sperm. Finally, sustentacular cells secrete *inhibin*, a hormone in a feedback circuit that regulates the release of follicle-stimulating hormone (FSH) from the pituitary gland. (This negative feedback system is described in detail on page 949.)

Between the seminiferous tubules are clusters of endocrine cells called **interstitial cells,** or *Leydig cells* [FIGURE 29.2C]. They secrete the male sex hormones called **androgens,** of which **testosterone** is the most important.

Removal of the testes is called **castration.** Because a castrated male will not produce testosterone, he will eventually lose his sex drive. However, erection of the penis may still be possible after castration. A male who is castrated before puberty will gradually acquire such feminine characteristics as fatty deposits in the breasts and hips, lack of facial hair, and smooth skin texture. If an adult male is castrated, he will usually lose his facial hair, and his bones and muscles will probably diminish in thickness and size, but other feminine characteristics will be minimal. Testosterone therapy may reduce these physical problems, but it cannot restore fertility.

Sperm

Mature **sperm,** or **spermatozoa** (sing. spermatozoan; Gr. *sperma*, seed + *zoon*, animal), have a *head*, a *middle piece*, and a *tail* [FIGURE 29.4]. At the tip of the head is an *acrosome* containing several enzymes that help the sperm penetrate an egg. In the center of the head is a compact *nucleus* containing the chromosomes (and therefore all the genetic material). The middle piece consists mainly of mitochondria, which supply ATP to provide the energy for movement. The beating movement of the undulating tail (flagellum) drives the swimming sperm forward. This ability to swim, called *motility*, is essential for male fertility.

A sperm is one of the smallest cells in the body. Its length from the tip of the head to the tip of the tail is

FIGURE 29.4 SPERM

[A] Schematic drawing of a sperm showing its internal structure and parts, including the head, middle piece, and tail. [B] Cross section of tail.

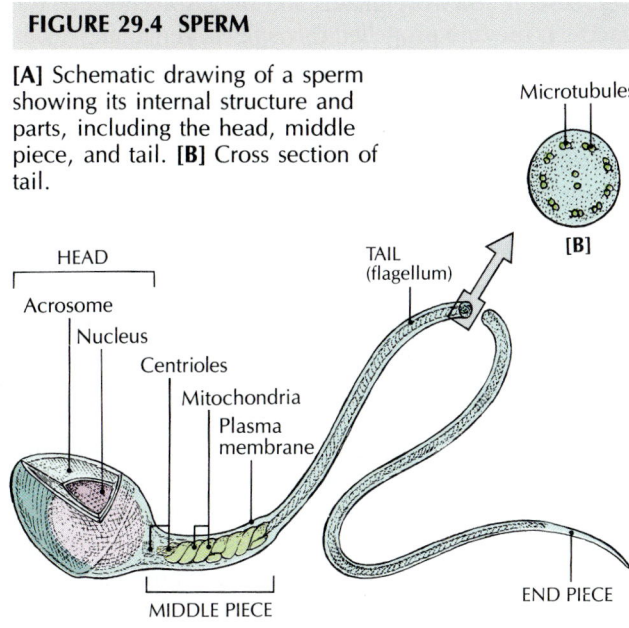

about 0.05 mm. Although it appears to be structurally simple — basically a nucleus with a tail — each sperm requires more than 2 months for its complete development.

Normally, 300 to 500 million sperm are released during an ejaculation. A male who releases less than about 20 to 30 million normal sperm is considered to be *infertile*. Men who produce a subnormal sperm count may still father children, compared with men who are *sterile*, who do not produce viable sperm. Sterility may be the result of sexually transmitted diseases, which may interfere with sperm production, or of diseases such as mumps, which destroy the lining of the seminiferous tubules.

Sperm are continuously being produced in the seminiferous tubules, which always contain sperm in various stages of development [FIGURE 29.2C–E]. The final maturation of a sperm takes place in the epididymis. The epididymis and other accessory ducts and glands are described in the following sections.

Accessory Ducts

The sperm produced in the testes are carried to the point of ejaculation from the penis by a system of ducts. These ducts lead from the testes into the abdominopelvic cavity, where they eventually join the prostatic portion of the urethra, just prior to reaching the penis.

Epididymis   The seminiferous tubules merge in the central posterior portion of the testis [FIGURE 29.2A]. This area is called the *mediastinum testis* (L. "being in the middle of the testis"). The seminiferous tubules straighten to become the ***tubuli recti*** ("straight tubules"),

which open into a network of tiny tubules called the ***rete testis*** (L. *rete*, net). The rete testis, at its upper end, drains into 15 to 20 tubules called the ***efferent ducts.***

The efferent ducts penetrate the tunica albuginea at the upper posterior part of the testis and extend upward into a convoluted mass of tubules that forms a crescent shape as it passes over the top of the testis and down along its side. This coiled tube is the ***epididymis*** (epp-ih-DIHD-ih-miss; pl. *epididymides*; Gr. "upon the twin," the "twin" being the testis) [FIGURES 29.1A and 29.2A]. The tightly coiled epididymis is bunched up along a length of only about 4 cm (1.5 in.), but if it were straightened out, it would extend about 6 m (20 ft). The epididymis has three main functions: (1) It stores sperm until they are mature and ready to be ejaculated. (2) It serves as a duct system for the passage of sperm from the testis to the ejaculatory duct. (3) It contains circular smooth muscle that helps propel mature sperm toward the penis by peristaltic contractions.

Each epididymis has a head, body, and tail. The *head*, which fits over the top of the testis, consists mostly of convoluted efferent ducts. The *body*, extending down the posterolateral border of the testis, consists mainly of the *ductus epididymis*. The *tail* extends almost to the bottom of the testis, where it becomes less and less convoluted until it finally dilates and turns upward as the *ductus deferens*.

Maturing sperm leave the seminiferous tubules and move into the epididymis. The journey through the epididymis may take as little as 10 days or as long as 4 to 5 weeks, until the sperm are mature. While being stored, the sperm are nourished by the lining of the epididymis. When the sperm are mature, they enter the ductus deferens on their way to the ejaculatory duct. Sperm can remain fertile in the epididymis and ductus deferens for about a month. If they are not ejaculated during that time, they degenerate and are resorbed by the body.

Ductus deferens   The ***ductus deferens*** (pl. *ductus deferentia*; L. *deferre*, to carry away), formerly called the *vas deferens* or *sperm duct*, is the dilated continuation of the ductus epididymis. The paired ductus deferentia extend between the epididymis of each testis and the ejaculatory duct [FIGURE 29.1]. Because the ductus deferens can be located easily, it is the site that is cut during a vasectomy. As each one passes from the tail of the epididymis, it is covered by the *spermatic cord*, containing the testicular artery, veins, autonomic nerves, lymphatics, and connective tissue from the anterior abdominal wall. Continuing upward after leaving the scrotum, the ductus deferens enters the lower part of the abdominal wall by way of the inguinal canal. There the ductus deferens becomes free of the spermatic cord and passes behind the urinary bladder, where it travels alongside an accessory gland called the *seminal vesicle* and becomes the *ejaculatory duct* [FIGURE 29.1A].

Just before reaching the seminal vesicle, the ductus deferens widens into an enlarged portion called the *ampulla* [FIGURE 29.1B]. Sperm are stored in the ampulla prior to ejaculation.

The ductus deferens is the main carrier of sperm. It is lined with pseudostratified columnar epithelium and contains three thick layers of smooth muscle. Some of the sympathetic nerves from the pelvic plexus terminate on this smooth muscle. Stimulation of these nerves produces peristaltic contractions that move sperm forward, toward the ejaculatory duct near the base of the penis.

Ejaculatory duct   The ampulla of the ductus deferens is joined by the duct of the seminal vesicle at the *ejaculatory duct* [FIGURES 29.1 and 29.5]. Each of the paired ejaculatory ducts is about 2 cm (1 in.) long. They receive secretions from the seminal vesicles, pass through the prostate gland on its posterior surface, where they receive additional secretions, and finally join the single urethra.

Urethra   The male *urethra* is the final section of the reproductive duct system. It leads from the urinary bladder, through the prostate gland, and into the penis. Its reproductive function is to transport semen outside the body during ejaculation. As you remember, the male urethra also carries urine from the urinary bladder during urination. However, it is physically impossible for a man to urinate and ejaculate at the same time because just prior to ejaculation the internal sphincter closes off the opening of the urinary bladder. The sphincter does not relax until the ejaculation is completed. The closing of this internal sphincter prevents urine from entering the urethra and also prevents the backflow of ejaculatory fluid into the urinary bladder.

The male urethra consists of prostatic, membranous, and penile (spongy) portions. The *prostatic portion* starts at the base of the urinary bladder and proceeds through the prostate gland. Here it receives secretions from the small ducts of the prostate gland and the two ejaculatory ducts. Upon its exit from the prostate gland, the urethra passes through the urogenital diaphragm and is consequently called the *membranous portion*. The external urethral sphincter muscle is located here. The *penile portion* of the urethra extends the full length of the spongy portion of the penis to the external urethral orifice on the glans penis, where either urine or semen leaves the penis. The wall of the urethra has a lining of mucous membrane and a thick outer layer of smooth muscle. Within the wall are *urethral glands*, which secrete mucus into the urethral canal.

Accessory Glands

After the ductus deferens passes around the urinary blad-der, several accessory glands add their secretions to the sperm as they are propelled through the remaining ducts. These accessory glands are the seminal vesicles, prostate gland, and bulbourethral glands. The fluid that results from the combination of sperm and glandular secretions is *semen*. FIGURE 29.1 shows the major complement of accessory ducts and glands.

Seminal vesicles   The paired *seminal vesicles* are secretory sacs that lie next to the ampullae of the ductus deferentia [FIGURE 29.5; see also FIGURE 29.1]. Their alkaline secretions, which provide the bulk of the seminal fluid, contain mostly water, fructose, prostaglandins, and vitamin C. The secretions are produced by the mucous membrane lining of the glands. The seminal vesicles are innervated by sympathetic nerves from the pelvic plexus. Stimulation during sexual excitement and ejaculation causes the seminal fluid to be emptied into the ejaculatory ducts by contractions of the smooth muscle layers. Seminal fluid provides an energy source for sperm and helps neutralize the natural acidity of the vagina.

Prostate gland   The *prostate gland* lies inferior to the urinary bladder and surrounds the first portion of the urethra [FIGURE 29.1]. It is a rounded mass about the size of a chestnut. The smooth muscles of the prostate can contract like a sponge to squeeze the prostatic secretions through tiny openings into the urethra. These secretions help make sperm motile and also help neutralize vaginal acidity.

As the urethra leaves the bladder, it passes through the prostate, where it receives prostatic secretions. Some of these secretions are passed off with the urine, but most of them are released with the semen during ejaculation. The secretions are released continually, as smooth muscle fibers in the wall of the prostate are stimulated by sympathetic nerves from the pelvic plexus. The ejaculatory ducts also pass through the prostate gland and receive its secretions during ejaculation.

The prostate gland is surrounded by a thin but firm capsule of fibrous connective tissue and smooth muscle. Inside, the prostate is made up of many individual glands, which release their secretions into the prostatic urethra through separate ducts [FIGURE 29.6]. There are three types of glands inside the prostate:

1 The inner *mucosal glands* secrete mucus. These small glands sometimes become inflamed and enlarged in older men and make urination difficult by pressing on the urethra.

2 The middle *submucosal glands.*

3 The *main* (external) *prostatic glands* supply the major portion of the prostatic secretions. The secretion contains mainly water, acid phosphatase, cholesterol, buffering salts, and phospholipids. Cancer of the prostate, one

FIGURE 29.5 SEMINAL VESICLE

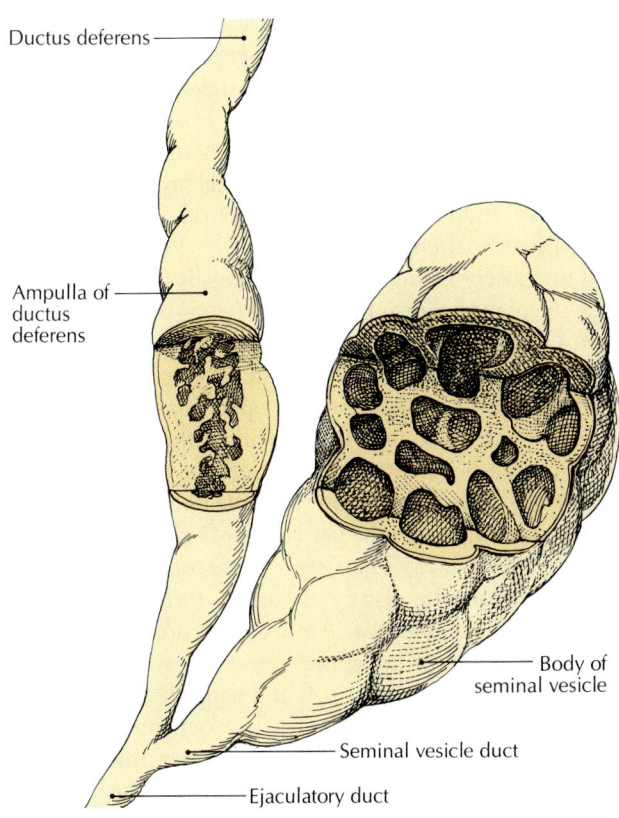

Ductus deferens

Ampulla of ductus deferens

Body of seminal vesicle

Seminal vesicle duct

Ejaculatory duct

FIGURE 29.6 PROSTATE GLAND

This cross section of the prostate gland shows the three types of inner glands and their relationships to the urethra.

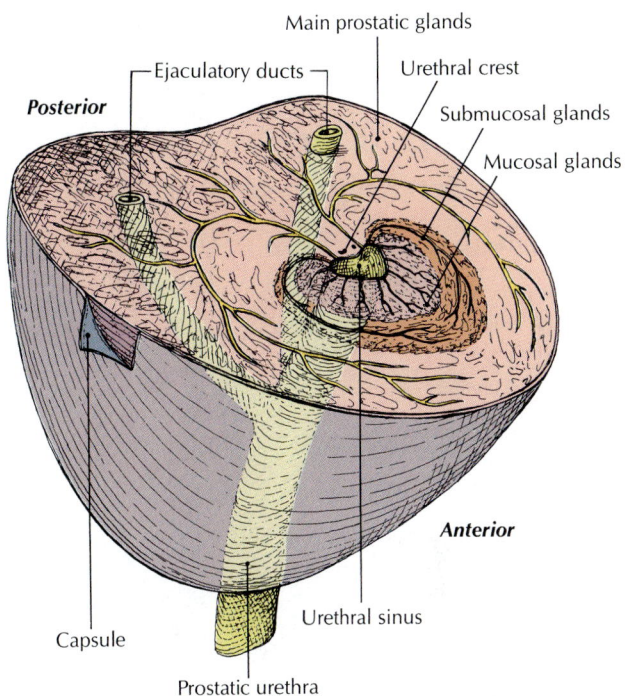

Main prostatic glands

Ejaculatory ducts

Urethral crest

Posterior

Submucosal glands

Mucosal glands

Anterior

Capsule

Urethral sinus

Prostatic urethra

---

of the more common types of cancer, usually occurs only in the main glands.

Bulbourethral glands The paired ***bulbourethral glands*** *(Cowper's glands)* are about the shape and size of a pea. They lie directly below the prostate gland, one on each side of the undersurface of the urethra [FIGURES 29.1 and 29.7]. Each gland has a duct that opens into the penile part of the urethra.

With the onset of sexual excitement, bulbourethral glands secrete clear alkaline fluids into the urethra to neutralize the acidity of any remaining urine. These fluids also act as a lubricant within the urethra to facilitate the ejaculation of semen and to lubricate the tip of the penis prior to sexual intercourse.

Semen

Secretions from the epididymis, seminal vesicles, prostate gland, and bulbourethral glands, together with the sperm, make up the ***semen*** (SEE-muhn; L. seed). Sperm make up only about 1 percent of the semen. The rest is fluid from the accessory glands, which provides fructose to nourish the sperm, an alkaline medium to help neu-

tralize urethral and vaginal acidity that could otherwise inactivate the sperm, and buffering salts and phospholipids that make the sperm motile.

Semen is over 90 percent water but contains many substances, most notably energy-rich fructose. The known vitamins it contains are vitamin C and inositol, and its trace elements include calcium, zinc, magnesium, copper, and sulfur. Semen also contains the highest concentration of prostaglandins in the body. The odor of semen is caused by amines (derivatives of ammonia), which are produced in the testes. The average ejaculation produces about 3 or 4 mL of semen (about a teaspoonful) and contains 300 to 500 million sperm.

Penis

The ***penis*** (L. tail) has two functions: (1) It carries urine through the urethra to the outside during urination, and (2) it transports semen through the urethra during ejaculation. In addition to the urethra, the penis contains three cylindrical strands of *erectile tissue:* two ***corpora cavernosa,*** which run parallel on the dorsal part, and the ***corpus spongiosum,*** which contains the urethra. The corpora cavernosa are surrounded by a dense, relatively

FIGURE 29.7 PENIS

[A] Coronal section of penis. **[B]** Cross section of flaccid penis. **[C]** Cross section of erect penis.

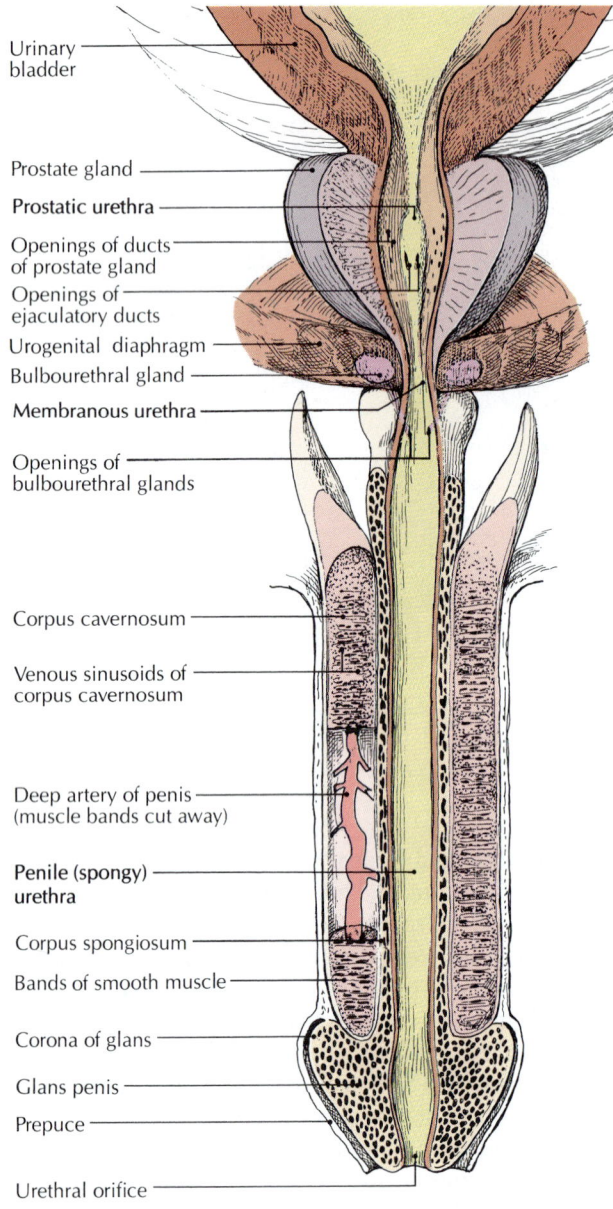

Urinary bladder

Prostate gland

Prostatic urethra

Openings of ducts of prostate gland

Openings of ejaculatory ducts

Urogenital diaphragm

Bulbourethral gland

Membranous urethra

Openings of bulbourethral glands

Corpus cavernosum

Venous sinusoids of corpus cavernosum

Deep artery of penis (muscle bands cut away)

Penile (spongy) urethra

Corpus spongiosum

Bands of smooth muscle

Corona of glans

Glans penis

Prepuce

Urethral orifice

[A]

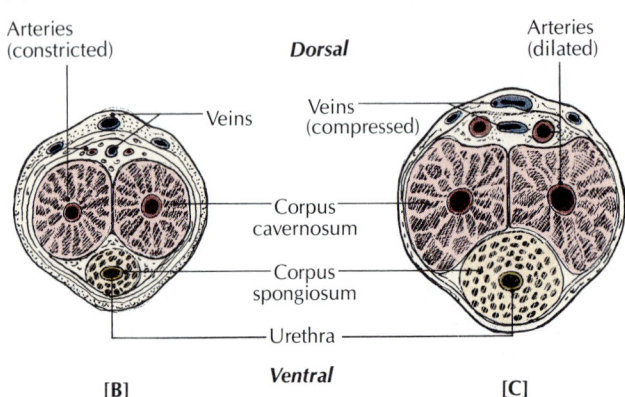

Arteries (constricted)

Dorsal

Arteries (dilated)

Veins

Veins (compressed)

Corpus cavernosum

Corpus spongiosum

Urethra

[B]

Ventral

[C]

inelastic connective tissue called the *tunica albuginea*. The corpus spongiosum does not have such a surrounding layer. The corpora cavernosa contain numerous vascular cavities called *venous sinusoids* [FIGURE 29.7].

The corpus spongiosum extends distally beyond the corpus cavernosa and becomes expanded into the tip of the penis, which is called the **glans penis.** Because the glans penis is a sensitive area containing many nerve endings, it is an important source of sexual arousal. The penile nerve endings are especially rich in the *corona*, the ridged proximal edge of the glans.

The loosely fitting skin of the penis is folded forward over the glans to form the **prepuce** (PREE-pyoos) or **foreskin. Circumcision** is the removal, for religious or health reasons, of the prepuce. Today many circumcisions are performed in the belief that the operation may decrease the occurrence of cancer of the penis, although a controversy currently surrounds this practice.

Just below the corona, on each side of a ridge of tissue called the *frenum*, are the paired *Tyson's glands*, which are modified oil glands. Their secretions, together with old cells shed from the glans and corona, form a cheeselike substance called *smegma*. An accumulation of smegma may be a source of infection.

Ordinarily, the penis is soft and hangs limply. During tactile or mental stimulation, a parasympathetic reflex causes marked vasodilation of the arteries that contain the blood that flows into the sinusoids of the corpora cavernosa, which become engorged with blood under high pressure. The distended sinusoids and the increased pressure compress the veins that usually drain blood away because of the resistance of the enveloping tunica albuginea. This resistance impedes the drainage of blood. In contrast, semen can pass through the urethra during ejaculation because the corpus spongiosum is not surrounded by the inelastic tunica albuginea. This dual action prevents the blood from escaping, and the penis becomes enlarged and firm in an **erection** [FIGURE 29.7B and C].

Two parts of the central nervous system control an erection: (1) the hypothalamus in the brain and (2) the sacral plexus of the spinal cord. Conscious thoughts within the cerebral cortex stimulate the erection center in the hypothalamus, which in turn causes parasympathetic vasodilation of the arterioles. Also, reflex responses in the sacral plexus can cause an erection in an infant or a sleeping adult, especially during the dream state.

Q: *Is it possible for a woman to become pregnant if the man does not ejaculate during sexual intercourse?*

A: Yes, it is, because secretions from the bulbourethral glands and epididymis (released after sexual excitement, but *before* ejaculation) may contain sperm.

FIGURE 29.8 HORMONAL CONTROL OF SPERM PRODUCTION

Gonadotropin-releasing hormone [GnRH] from the hypothalamus stimulates the anterior pituitary to release LH and FSH, which stimulate the secretion of testosterone. Testosterone stimulates sperm production, but also inhibits both the hypothalamus and the anterior pituitary in a negative feedback system (dashed lines).

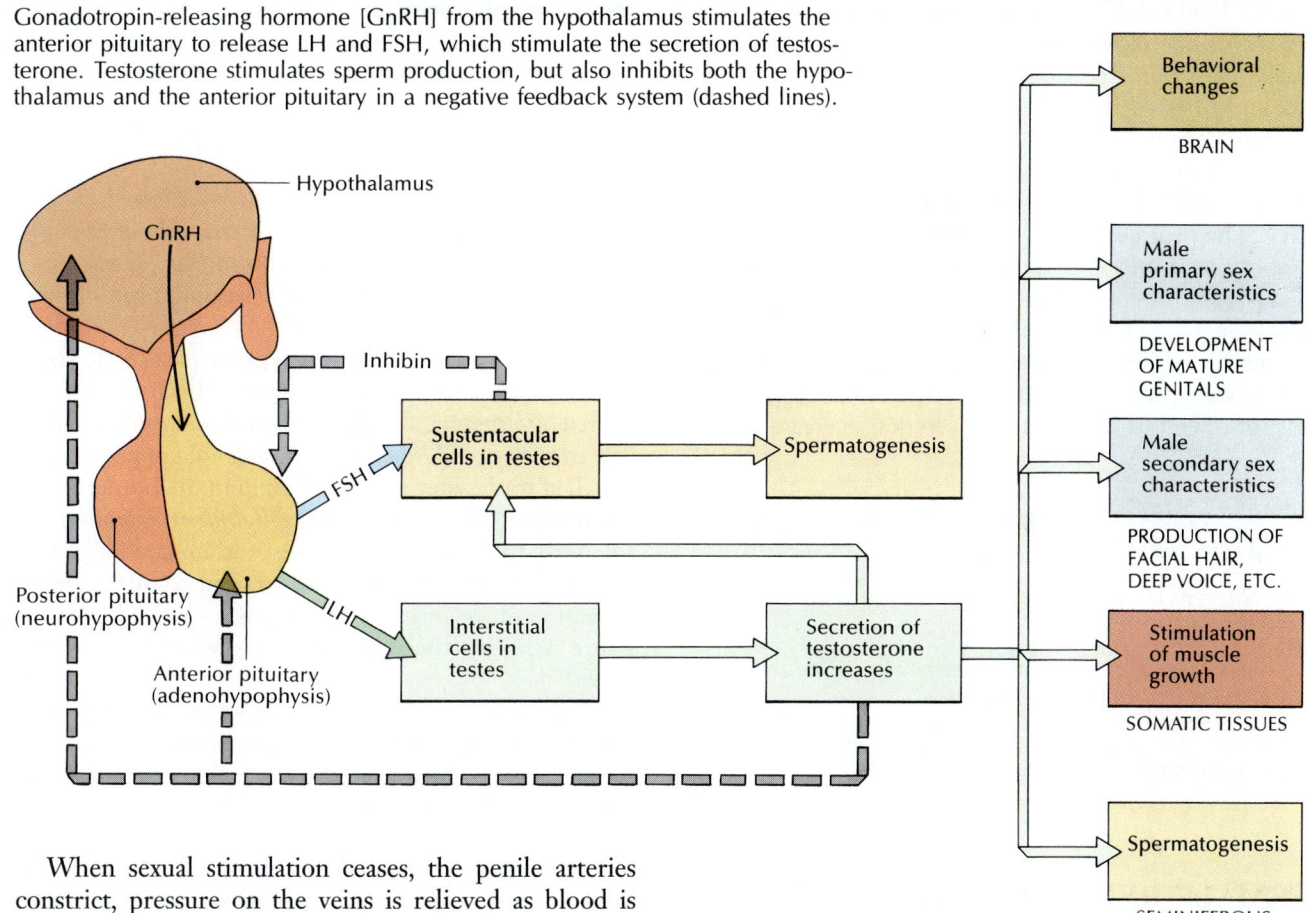

When sexual stimulation ceases, the penile arteries constrict, pressure on the veins is relieved as blood is drained away, and the penis returns to its nonerect, *flaccid* (FLACK-sihd; L. *flaccus*, hanging, flabby) state. This return to the flaccid state is called *detumescence*. The flaccid dimension of a penis bears little relation to the erect dimension, and a small flaccid penis erects to a proportionally larger size than a large flaccid penis. The greatest variation in penis size occurs in the *girth* of the erect penis, not the length.

Hormonal Regulation in the Male

Among the seminiferous tubules in the testes are small masses of ***interstitial cells*** *(Leydig cells)*. These cells secrete testosterone, dihydrotestosterone, and androstenedione. The production and secretion of the main male hormone, ***testosterone,*** are controlled by ***FSH*** (follicle-stimulating hormone) and ***LH*** (luteinizing hormone). The release of FSH and LH is controlled by gonadotropin-releasing hormone *(GnRH)* from the hypothalamus.

Both FSH and LH are produced by the anterior pituitary gland and are chiefly responsible for stimulating spermatogenesis and testosterone secretion. While FSH and LH affect only the testes, testosterone affects not

only spermatogenesis but also the development of the sex organs and the appearance of such secondary male sex characteristics as a deep voice, facial and body hair, skeletal proportions, and distribution of body fat.

Negative feedback in spermatogenesis and testosterone production and secretion FIGURE 29.8 shows the negative feedback system that regulates the process of spermatogenesis and the production and secretion of testosterone. When blood levels of testosterone decrease, the hypothalamus is stimulated to secrete GnRH (gonadotropin-releasing hormone), which travels to the anterior pituitary cells by way of the blood vessels in the hypophyseal portal system. GnRH then stimulates the release of FSH and LH, the two gonadotropic peptide hormones that regulate the function of the gonads. FSH stimulates spermatogenesis in the testes, and FSH plus testosterone stimulate the sustentacular cells in the seminiferous tubules to release *androgen-binding protein (ABP)*, which binds to androgens (testosterone), maintaining a high level of testosterone in the lumen of the seminiferous tubules. This high concentration of testosterone is necessary for continued sperm production. Androgen-binding protein (ABP) keeps testosterone, a highly lipid-soluble steroid, from diffusing out of the lumen of the seminiferous tubules. LH stimulates the interstitial cells to secrete testosterone and a small amount of estrogen.

The negative feedback cycle is completed when testosterone inhibits the release of GnRH from the hypothalamus and the output of LH from the anterior pituitary. The peptide hormone **inhibin** is secreted from the sustentacular cells in the seminiferous tubules along with testosterone. Inhibin senses the state of testicular spermatogenesis. When the sperm count is high, the release of inhibin increases, directly inhibiting the output of FSH. When the sperm count decreases (to 20 million/mL of semen or less), inhibin secretion decreases.

ASK YOURSELF

1 What are the main functions of the testes?

2 What are the main functions of the seminiferous tubules?

3 How do the secretions of the seminal vesicles and prostate gland aid sperm transport?

4 How is the negative feedback loop involved in spermatogenesis and testosterone production?

FEMALE REPRODUCTIVE ANATOMY AND PHYSIOLOGY

The reproductive anatomy and physiology of the female is far more complex than that of the male. Not only do females have a monthly rhythmic menstrual cycle and produce egg cells (ova), but after fertilization they also nourish, carry, and protect the developing embryo. Then their mammary glands produce milk to nurse the infant after it is born.

The female reproductive system consists of a variety of structures with specialized functions [FIGURE 29.9]:

1 Two *ovaries* produce ova and the female sex hormones (estrogen and progesterone).

2 Two *uterine tubes*, also called *Fallopian tubes* or *oviducts*, one from each ovary, carry ova from the ovary to the uterus. Fertilization usually occurs in one of the uterine tubes.

3 The *uterus* houses and nourishes the developing embryo.

4 The *vagina* receives semen from the penis during sexual intercourse, is the exit point for menstrual flow, and is the canal through which the baby passes from the uterus during childbirth.

5 The *external genital organs* (genitalia), collectively called the *vulva*, have protective functions and also play a role in sexual arousal.

6 The *mammary glands*, contained in the paired *breasts*, produce milk for the newborn baby.

Ovaries

The female gonads are the paired **ovaries** (L. *ovum*, egg), which produce ova and female hormones. These elongated, somewhat flattened bodies are about the size and shape of an unshelled almond.

The ovaries are located in the pelvic part of the abdomen, one on each side of the uterus. Each ovary is attached by a mesentery called the **mesovarium** to the back side of each **broad ligament**, a double fold of peritoneum attached to the uterus. A thickening in the border of the mesovarium called the **ovarian ligament** extends from the ovary to the uterus. The mesovarium contains veins, arteries, lymphatics, and nerves to and from the **hilum** (opening) of the ovary. The ovary is suspended from the pelvic wall by the **suspensory ligament** [see FIGURE 29.9B].

The ovaries are covered by a layer of specialized epithelial cells called the **germinal layer**. Beneath it is the **stroma**, a mass of connective tissue which contains ova in various stages of maturity.

Follicles and corpus luteum   A cross section of an ovary reveals a *cortex* and a vascular *medulla*. The cortex contains round epithelial vesicles called **follicles**, the actual centers of ovum production. Each follicle contains an immature ovum called an **oocyte**, which is surrounded by one or more layers of cells. If these cells are one layer thick they are called *follicle cells*, and if they are more than one layer thick, they are called *granulosa cells*.

Follicles are always present in several stages of development [FIGURE 29.10] and are classified according to these different stages:

1 *Primordial follicles* have one layer of squamouslike follicular cells around the oocyte.

2 A *primary follicle* consists of two or more layers of cuboidal or columnar cells around the oocyte.

3 A primary follicle becomes a **secondary follicle** when a fluid-filled space known as the *antrum* develops [FIGURE 29.10].

4 A *vesicular (Graafian) follicle* is the mature follicle with the oocyte located on the "stalk" of granulosa cells. Vesicular follicles bulge onto the surface of the ovary, and one follicle each month releases its oocyte from the ovary in a process called *ovulation*.

5 The follicular structure left behind in the ovary after ovulation is known as the **corpus luteum** ("yellow

FIGURE 29.9 FEMALE REPRODUCTIVE ANATOMY ★

[A] Midsagittal section through the female pelvis. The arrows trace the path of an ovum from the ovary to the uterus and out of the vagina. [B] Anterior view of female reproductive system.

Structure	Function
Ovaries	Produce ova and sex hormones
Uterine tubes	Carry ova from ovary to uterus
Uterus	Houses and nourishes developing embryo
Vagina	Receives sperm during intercourse, exit point for menstrual flow, birth canal
Vulva	Protection, sexual arousal
Mammary glands	Produce milk

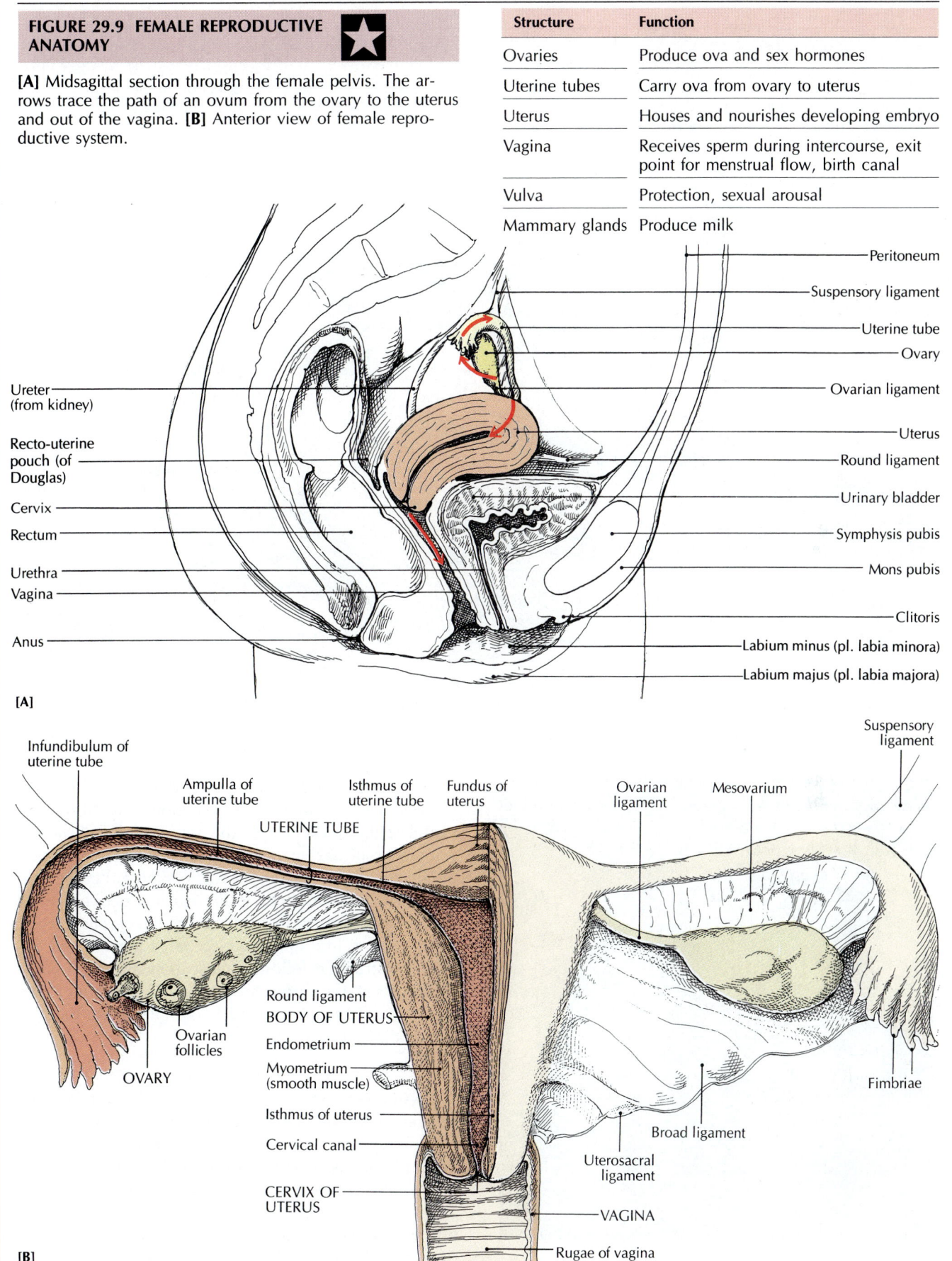

[A]

Ureter (from kidney)
Recto-uterine pouch (of Douglas)
Cervix
Rectum
Urethra
Vagina
Anus

Peritoneum
Suspensory ligament
Uterine tube
Ovary
Ovarian ligament
Uterus
Round ligament
Urinary bladder
Symphysis pubis
Mons pubis
Clitoris
Labium minus (pl. labia minora)
Labium majus (pl. labia majora)

[B]

Infundibulum of uterine tube
Ampulla of uterine tube
Isthmus of uterine tube
Fundus of uterus
UTERINE TUBE
Ovarian ligament
Mesovarium
Suspensory ligament

Round ligament
BODY OF UTERUS
Endometrium
Myometrium (smooth muscle)
Isthmus of uterus
Cervical canal
CERVIX OF UTERUS

Ovarian follicles
OVARY

Broad ligament
Uterosacral ligament
VAGINA
Rugae of vagina

Fimbriae

FIGURE 29.10 OVARY

[A] Schematic diagram of an ovary showing the development and eventual rupture of an ovarian follicle and the subsequent formation and disintegration of the corpus luteum. The sequence of events begins with the primordial follicles and proceeds in the direction of the arrows.
[B] Scanning electron micrograph showing several developing follicles in an ovary.

Egg nests (primordial follicles)

Primary oocytes in progressive stages of primary follicles

Tunica albuginea

Germinal epithelium

Follicle approaching maturity, antrum forming

Mature vesicular (Graafian) follicle

Secondary oocyte

Antrum (cavity) filled with follicular fluid

Hilum containing blood vessels, nerves, lymphatics from mesovarium

Corpus albicans

Mature corpus luteum

Coagulated blood

Developing corpus luteum

Ruptured follicle during ovulation (corpus hemorrhagicum)

Released oocyte

[A]

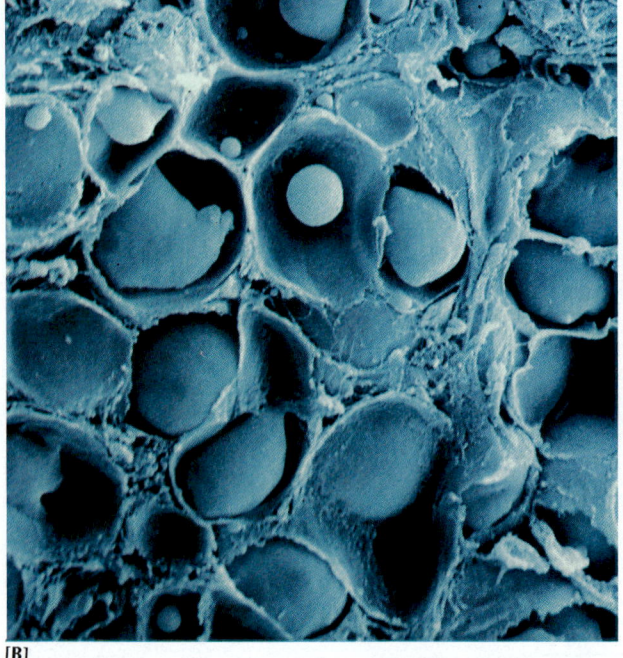

[B]

body"), temporary endocrine tissue that secretes *estrogens* and *progesterone*, which stop additional ovulation and stimulate the thickening of the uterine wall and the development of the mammary glands in anticipation of pregnancy. A high concentration of progesterone also inhibits uterine contractions. If pregnancy does not occur within 14 days after the formation of the corpus luteum, it degenerates into the *corpus albicans*, a form of scar tissue. Menstruation follows almost immediately. If pregnancy occurs, the corpus luteum persists for about 2 to 3 months, and then it eventually degenerates as the placenta takes over its activities.

Ovulation   In response to high concentrations of FSH and LH from the anterior pituitary, the mature follicle ruptures in the process called ***ovulation.*** During an ovulation, an oocyte is released from the surface of one of the ovaries into the serous fluid of the peritoneal cavity.

Beginning at puberty, about 20 ovarian follicles mature each month to the point of being ready for ovulation, but only one (rarely more) actually ruptures and

FIGURE 29.11 UTERINE TUBE

[A] Cross section of the uterine tube. [B] Scanning electron micrograph of the inner lining of a uterine tube showing the many protruding cilia.

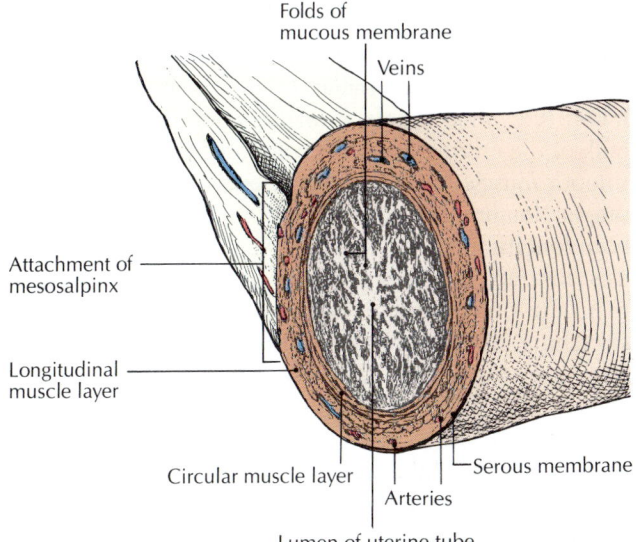

Folds of mucous membrane

Veins

Attachment of mesosalpinx

Longitudinal muscle layer

Circular muscle layer

Arteries

Serous membrane

Lumen of uterine tube

[A]

Cilia

[B]

releases a secondary oocyte. (Apparently there is no particular pattern that selects one ovary over the other each month.) The other follicles degenerate, but only after releasing some progesterone and estrogen. After each oocyte is released from the ovary, the site of the rupture heals. As a result, the surface of the ovary is scarred and pitted, especially in older women.

Although a woman will release no more than 400 to 500 oocytes in her lifetime, she starts out with many more. A 7-month-old female fetus has about 1 million primordial follicles in each ovary, but most of them disintegrate, leaving about 200,000 to 400,000 in each ovary at birth. At puberty (usually between 10 and 14 years of age), each ovary contains about 200,000 oocytes, with the number decreasing as a woman gets older. Most women cease to be fertile during their late forties or early fifties.

Ordinarily, only one oocyte is released each month, but several may be released, especially in women who have already had children, who have taken fertility drugs, or who are genetically predisposed toward the simultaneous release of more than one oocyte. Fraternal twins are the result of the fertilization of two separate oocytes. (See the discussion of multiple births in Chapter 30.)

Uterine Tubes

One of the paired *uterine tubes* receives the secondary oocyte from the ovary and conveys it to the uterus [FIGURE 29.9B]. The tubes are not directly connected to the ovaries. The superior end opens into the abdominal cavity, very close to the ovary, and the inferior end opens into the uterus. Each uterine tube lies in the upper part of a transverse double fold of peritoneum called the *broad ligament*. Three distinct portions of the tube are the funnel-shaped *infundibulum* (L. funnel) near the ovary, the thin-walled middle *ampulla*, and the *isthmus*, which opens into the uterus [FIGURE 29.9B].

The wall of the uterine tube is made up of three layers [FIGURE 29.11A]:

1 The outer *serous membrane* is part of the visceral peritoneum.

2 The middle *muscularis* is composed of an inner layer of spirally oriented smooth muscle fibers and an outer longitudinal layer. Hormonal action produces peristaltic contractions in the muscularis close to the time when a secondary oocyte is released. These contractions help move the oocyte along the uterine tube to the uterus.

3 The inner *mucous membrane* has an epithelium made up of a single layer of columnar cells that alternate irregularly as ciliated and secretory cells. The secretory cells may provide nourishment for the secondary oocyte, and the cilia help propel it toward the uterus.

The infundibulum is fringed with feathery *fimbriae* (FIHM-bree-ee; L. threads), which may actually overlap the ovary. Each month, as a secondary oocyte is released, it is effectively swept across a tiny gap between the tube and the ovary into the infundibulum by the motion of cilia in the fimbriae. Few oocytes are ever lost in the abdominal cavity.

Unlike sperm, the secondary oocyte is unable to move on its own. Instead, it is carried along the uterine tube toward the uterus by the peristaltic contractions of the tube and the waving movements of the cilia in the mucous membrane [FIGURE 29.11B]. Fertilization of the secondary oocyte usually occurs in the ampulla (middle) of the uterine tube. Penetration of the secondary oocyte by the sperm initiates the development of the mature ovum. An unfertilized oocyte will degenerate in the uterine tube. A fertilized oocyte (zygote) continues its journey toward the uterus, where it will become implanted. The zygote's journey takes 4 to 7 days.

Occasionally, a fertilized ovum is delayed as it passes along the uterine tube, and thus adheres to the wall of the tube. This type of implantation is called an *ectopic pregnancy* (Gr. *ektopos*, out of place). This refers to the development of an embryo or fetus *outside* the uterus, usually in the uterine tube, but sometimes in the ovaries, abdominal cavity, or uterine cervix. An ectopic pregnancy is terminated early because of a lack of nourishment and space for the embryo to develop. Unless the implanted embryo is removed, it can rupture the uterine tube, causing further serious complications.

Uterus

The uterine tubes terminate in the *uterus* (L. womb), a hollow, muscular organ located in front of the rectum and behind the urinary bladder [FIGURE 29.9A]. It is shaped like an inverted pear when viewed anteriorly, and it is pear-sized as well. However, it increases three to six times in size during the 9 calendar months of pregnancy.

The uterus is attached to the lateral wall of the pelvis by the *broad ligaments* (double folds of peritoneum), which extend from the uterus to the floor and lateral walls of the pelvic cavity. Within each broad ligament is the *round ligament* of the uterus, which extends as a cord from the junction of the uterine tube and the uterus to the skin of the labium majora of the external genitalia. The round ligament helps keep the uterus tilted forward over the urinary bladder [see FIGURE 29.9A]. Two *uterosacral ligaments* extend from the upper part of the cervix to the sacrum. The *posterior ligament* attaches the uterus to the rectum, and the *anterior ligament* attaches the uterus to the urinary bladder. All these ligaments are involved with the support of the uterus.

The wide upper portion of the uterus is called the *fundus.* The uterine tubes enter the uterus below the fundus [FIGURE 29.9B]. The tapering middle portion of the uterus is the *body,* which terminates in the narrow *cervix* (L. neck), the juncture between the uterus and the vagina. The constricted region between the body and the cervix is the *isthmus.* The *cervical canal,* the interior of the cervix, opens into the vagina. The uterus of a woman who is not pregnant is somewhat flattened, so that the interior of the uterus, or *uterine cavity,* is just a slit [FIGURE 29.12]. In its usual position, the flattened uterus is (1) *anteverted* — the entire uterus leans forward over the

urinary bladder at almost a right angle to the vagina, and (2) *anteflexed* — the uterus is bent back on itself at the isthmus between the cervix and fundus. [FIGURE 29.9A].

The uterus has three layers of tissue [FIGURE 29.12]:

1 The outer *serosal layer* extends to form the two broad ligaments that stretch from the uterus to the lateral walls of the pelvis.

2 The middle, muscular layer, called the *myometrium* (Gr. *myo*, muscle + *metra*, womb), makes up the bulk of the uterine wall. It is composed of three layers of smooth muscle fibers. From the outside in, they are arranged longitudinally, randomly in all directions, and both longitudinally and spirally. The interweaving muscles contract downward during labor with more force than any other muscle. These muscles are capable of stretching during pregnancy to accommodate one or more growing fetuses; they also contract during a woman's orgasm. The myometrium is almost a centimeter thick but becomes even thicker during pregnancy in preparation for childbirth.

3 The innermost layer of the uterus is composed of a specialized mucous membrane called the *endometrium,* which is deep and velvety in texture. It contains an abundant supply of blood vessels and is pitted with simple tubular glands. The endometrium is composed of two layers, the *stratum functionalis* and the *stratum basalis.* Every month, in response to estrogen secretion, the en-

---

FIGURE 29.12 UTERUS

Sagittal view of the uterus. Normally, the uterus is *anteverted* (bent forward on itself) and *anteflexed* (at a slight angle at the junction of its body and the cervix).

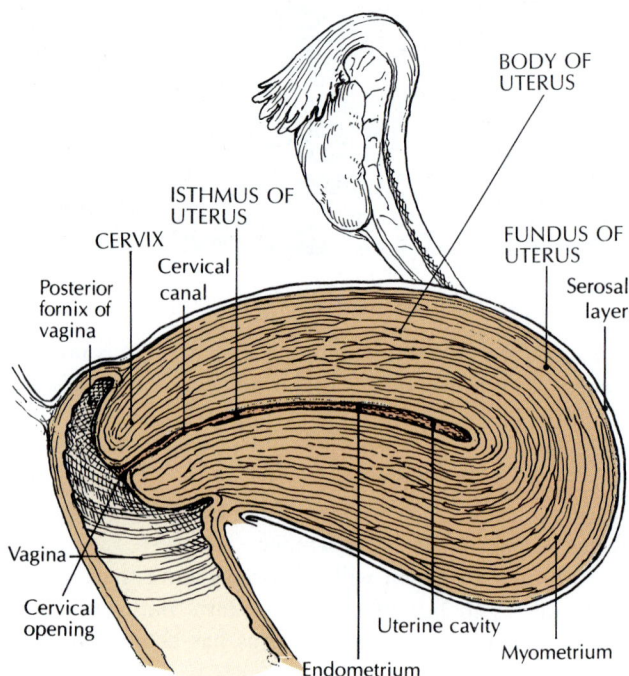

BODY OF UTERUS

ISTHMUS OF UTERUS

CERVIX

Posterior fornix of vagina

Cervical canal

FUNDUS OF UTERUS

Serosal layer

Vagina

Cervical opening

Uterine cavity

Myometrium

Endometrium

FIGURE 29.13 VULVA

Shown are the borders of the diamond-shaped perineum and the urogenital and anal triangles.

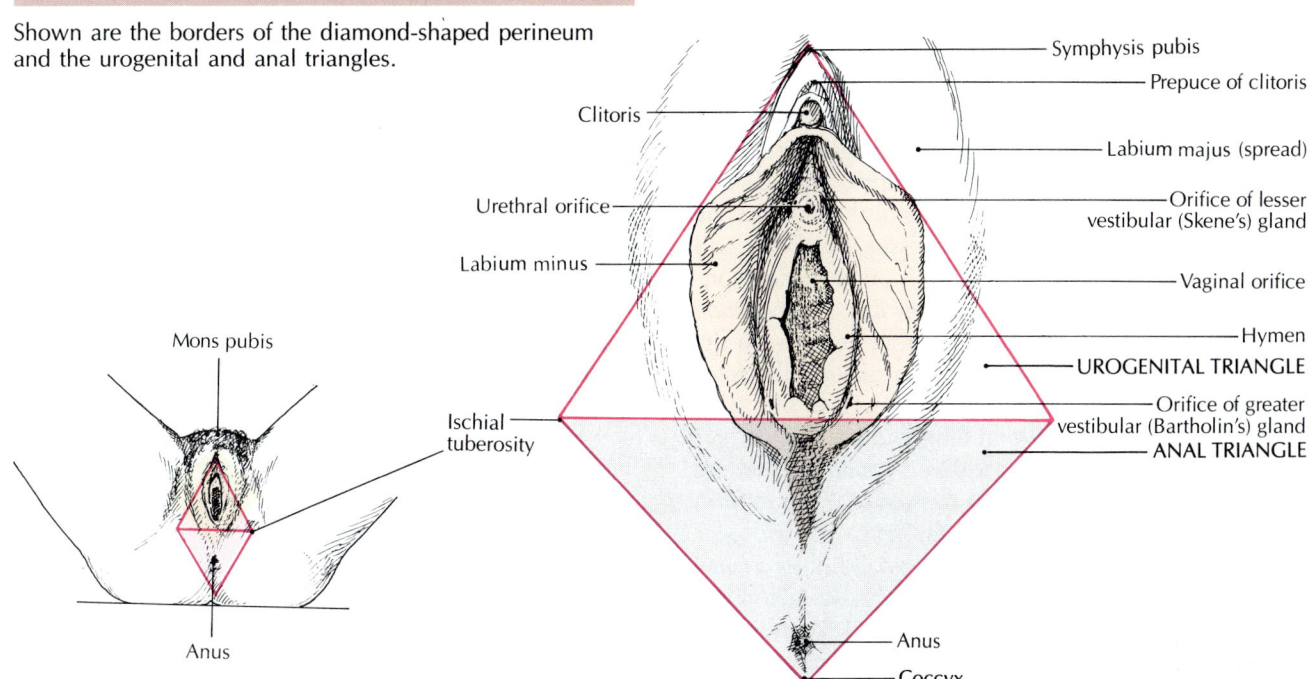

dometrium thickens in preparation for the possible implantation of a fertilized ovum (the beginning of *pregnancy*). Secretions of progesterone help the endometrium develop active glands that make the endometrium rich in nutrients and ready to receive a fertilized ovum. If implantation does not occur, the *stratum functionalis* is shed, together with blood and glandular secretions, through the cervical canal and vagina. This breakdown of the endometrium makes up the **menstrual flow** (L. *mensis*, monthly) in the process of **menstruation.** The *stratum basalis* layer is permanent, and from it a new stratum functionalis regenerates after the 3- to 5-day menstrual period.

If fertilization and implantation do occur, the uterus houses, nourishes, and protects the developing fetus within its muscular walls. As the pregnancy continues, estrogen secretions develop the smooth muscle in the uterine walls in preparation for the expulsive action of childbirth.

Vagina

The uterus leads into the **vagina** (L. sheath), a muscle-lined tube about 8 to 10 cm (3 to 4 in.) long [FIGURE 29.9]. The vagina is the channel for the removal of menstrual flow, the site where semen from the penis is deposited during sexual intercourse, and the birth canal for the baby during childbirth. It lies behind the urinary bladder and urethra, and in front of the rectum and anus, and it angles upward and backward.

The wall of the vagina is composed mainly of smooth muscle and fibroelastic connective tissue. It is lined with mucous membrane containing many rugae. Stratified squamous nonkeratinizing epithelium covers the vagina and also the cervix of the uterus. Ordinarily the wall of the vagina is collapsed, but the vagina can enlarge to accommodate an erect penis during sexual intercourse or the passage of a baby during childbirth.

The mucus that lubricates the vagina comes from glands in the cervix of the uterus; the vagina has no mucous glands. The acidic environment of the vagina, which helps prevent infection, is derived from the fermentation action of bacteria from the vaginal epithelium. The acidic environment is hostile to sperm, but alkaline fluids from male accessory sex glands help neutralize vaginal acidity. Also, the cervical glands of the uterus secrete a more alkaline mucus during the woman's fertile time, around ovulation.

A fold of vaginal mucosa called the *hymen* (Gr. membrane*) partially blocks the vaginal entrance. The hymen is usually ruptured during the female's first sexual intercourse, but it may be broken earlier during other physical activities or by the insertion of a tampon.

External Genital Organs

The **external genitalia** include the mons pubis, labia majora, labia minora, vestibular glands, clitoris, and vestibule of the vagina [FIGURE 29.13]. As a group, these organs are called the **vulva** (L. womb, covering).

*Also, Hymen was the Greek god of marriage.

The *mons pubis* (L. mountain + pubic) is a mound of fatty tissue that covers the symphysis pubis. At puberty, the mons pubis becomes covered with pubic hair. Unlike the pubic hair of a male, which may extend upward in a thin line as far as the navel, the upper limit of female pubic hair lies horizontally across the lower abdomen.

Just below the mons pubis are two longitudinal folds of skin, the *labia majora* ("major lips"; sing. *labium majus*), which form the outer borders of the vulva. They contain fat, smooth muscle, areolar tissue, oil glands, and sensory receptors. After puberty, their outer surface contains hairs. The *labia minora* ("minor lips"; sing. *labium minus*) are two smaller folds of skin that lie between the labia majora. Together with the labia majora, the folds surround the vaginal and urethral openings. The labia minora contain oil glands and blood vessels but no hair or fat. They also contain many nerve endings and are sensitive to the touch. The labia merge at the top to form the *foreskin,* or *prepuce,* of the clitoris.

The *clitoris* (KLIHT-uh-rihss; Gr. small hill) is a small erectile organ at the upper end of the vulva, below the mons pubis, where the labia minora meet. Like the penis, the clitoris contains many nerve endings and has two cor-pora cavernosa that can fill with blood during sexual stimulation, causing the clitoris to become enlarged. It is one of the major sources of sexual arousal. The clitoris is capped by a sensitive *glans,* but does not contain a urethra as the penis does.

The *vestibule* of the vagina is the space between the labia minora. Its floor contains the *greater vestibular glands (Bartholin's glands)* and the openings for the urethra and vagina. Small *lesser vestibular glands (Skene's glands)* open by way of ducts into the anterior part of the vestibule, between the urethral and vaginal orifices. During sexual excitement, the greater and lesser vestibular glands secrete an alkaline mucus solution that provides some lubrication and also offsets some of the natural acidity of the vagina. Most of the lubrication, however, comes from secretions of cervical glands.

The diamond-shaped *perineum* (per-uh-NEE-uhm) is the region bounded anteriorly by the symphysis pubis, posteriorly by the inferior tip of the coccyx, and laterally by the ischial tuberosities. The anterior portion of the perineal diamond is called the *urogenital region (triangle),* and the posterior portion is the *anal region (triangle)* [FIGURE 29.13]. The perineum is present in males also.

FIGURE 29.14 BREAST AND MAMMARY GLANDS

[A] Sagittal section of right female breast showing the mammary gland and areola. [B] Internal anatomy of a female breast showing structures before, during, and after pregnancy.

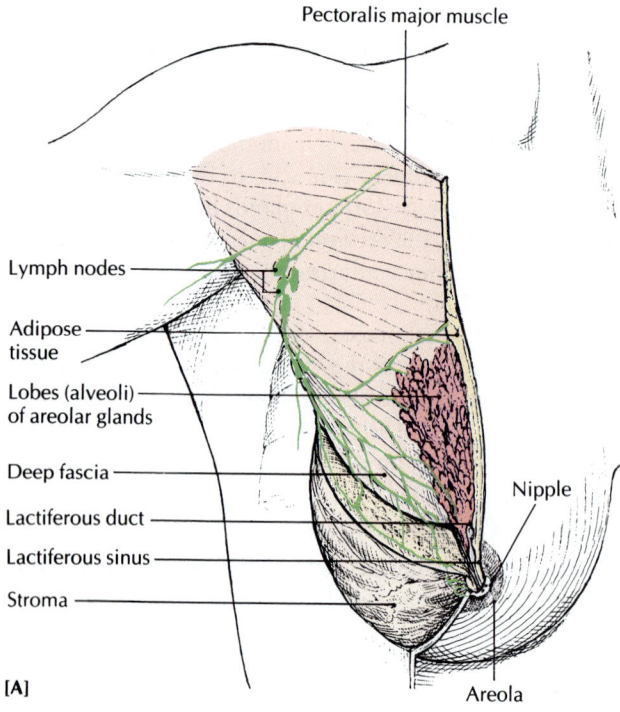

Pectoralis major muscle

Lymph nodes

Adipose tissue

Lobes (alveoli) of areolar glands

Deep fascia

Lactiferous duct

Lactiferous sinus

Stroma

Nipple

Areola

[A]

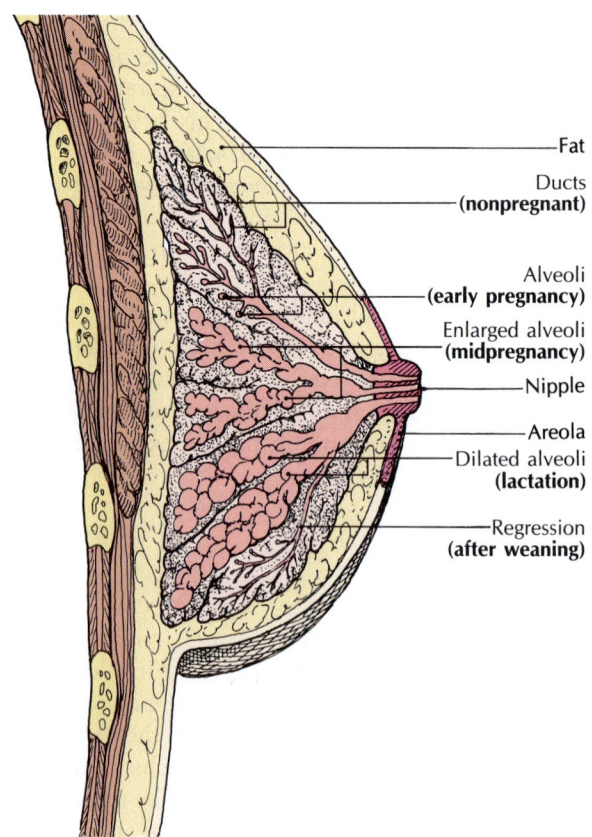

Fat

Ducts **(nonpregnant)**

Alveoli **(early pregnancy)**

Enlarged alveoli **(midpregnancy)**

Nipple

Areola

Dilated alveoli **(lactation)**

Regression **(after weaning)**

[B]

Mammary Glands

The **mammary glands** (L. *mammae*, breasts) within the paired breasts of a woman are modified sweat glands that produce and secrete milk. The breasts rest upon deep fascia covering the pectoralis major and minor muscles, each breast extending from the lateral border of the sternum to the middle of the axilla [FIGURE 29.14]. The breasts are held in place by *suspensory ligaments of the breast (Cooper's ligaments)*. Breasts contain varying amounts of adipose tissue. The amount of adipose tissue determines the size of the breasts, but the amount of mammary tissue does not vary widely from one woman to another.

Each mammary gland is composed of 15 to 20 lobes of compound areolar glands that radiate out from the nipple. These clusters of glands look like bunches of grapes (secretory alveoli), with **lactiferous** ("milk-carrying") **ducts** that convey milk from the glands. The ducts from many glands converge into larger ducts. Just before a lactiferous duct reaches the nipple, it dilates into a **lactiferous sinus** and then constricts as it enters the nipple. Milk is stored in the lactiferous sinuses.

The pigmented area around the nipple is the *areola* (uh-REE-uh-luh; L. dim. of *area*, open place). It enlarges and darkens during pregnancy and retains the darkened color permanently. The nipple consists of dense connective tissue and smooth muscle fibers, as well as many blood vessels and sensitive nerve endings that make it an important area of sexual stimulation.

The breasts contain an extensive drainage system made up of many lymph vessels. Because cells of malignant breast tumors may spread via the lymphatics, frequent self-examination of the breasts for lumps should be a routine practice to locate tumors before they spread.

THE MENSTRUAL CYCLE

The male is continuously fertile from puberty to old age, and throughout that period sex hormones are secreted at a steady rate. The female, however, is fertile only during a few days each month, and the pattern of hormonal secretion is intricately related to the cyclical release of a secondary oocyte from the ovary [TABLE 29.1].

The development of an oocyte in an ovary is timed by a cyclical production of hormones in the hypothalamus [FIGURE 29.15]. **Gonadotropin-releasing hormone (GnRH)** from the hypothalamus acts on the anterior pituitary gland, which then releases two additional hormones—**follicle-stimulating hormone (FSH)** and **luteinizing hormone (LH)**—to bring about the oocyte's maturation and release from the ovary.

One way to understand the hormonal pattern in the normal monthly cycle is to follow the ovarian and uterine

events of the **menstrual cycle** together [FIGURE 29.16]. Toward that end, we have presented the menstrual cycle in two columns alongside FIGURE 29.16, with simultaneous ovarian and uterine events of the menstrual cycle placed side-by-side, in the exact time frame in which they happen. A detailed description of the menstrual cycle follows.

Ovarian Events during the Menstrual Cycle

The average menstrual cycle lasts about 28 days, but it

FIGURE 29.15 HORMONAL REGULATION OF FOLLICLE AND OVUM DEVELOPMENT

GnRH stimulates the release of FSH and LH, which enhances follicle and ovum development in the ovary. They also increase the ovarian output of estrogen. Estrogen stimulates follicle and ovum development and LH secretion; it inhibits release of FSH. Thus two negative feedback systems (dashed lines) and one positive feedback system (one of the few in the body) control the cycle.

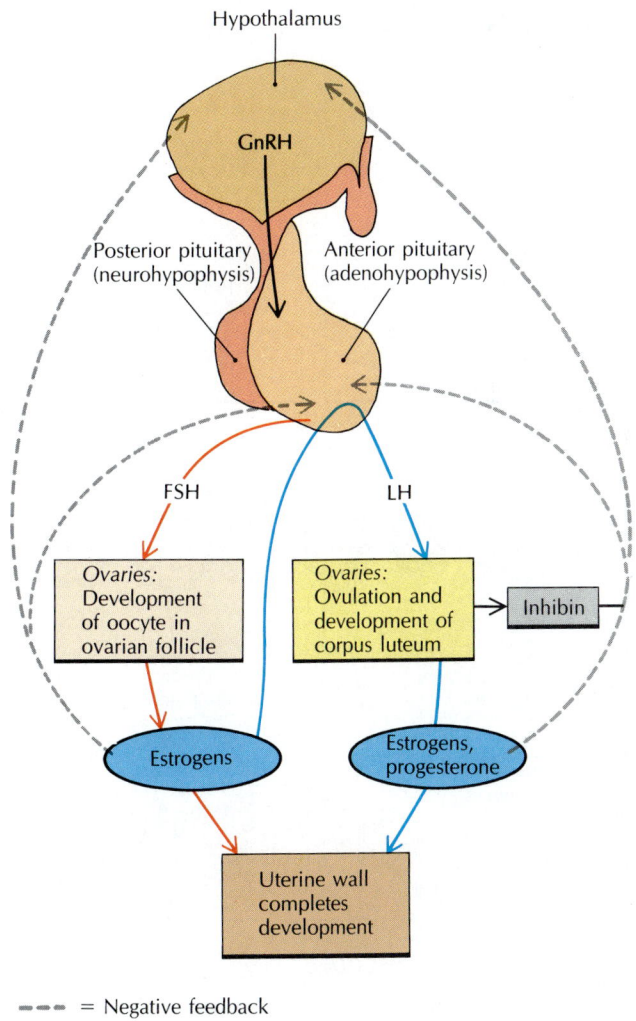

Hypothalamus

GnRH

Posterior pituitary (neurohypophysis)

Anterior pituitary (adenohypophysis)

FSH

LH

Ovaries: Development of oocyte in ovarian follicle

Ovaries: Ovulation and development of corpus luteum

Inhibin

Estrogens

Estrogens, progesterone

Uterine wall completes development

- - - = Negative feedback

—— = Positive feedback

FIGURE 29.16 MENSTRUAL CYCLE

[A] Concentrations of LH and FSH in the blood. [B] Concentrations of estrogen and progesterone in the blood. [C] Relationships between ovarian and uterine changes during the menstrual cycle.

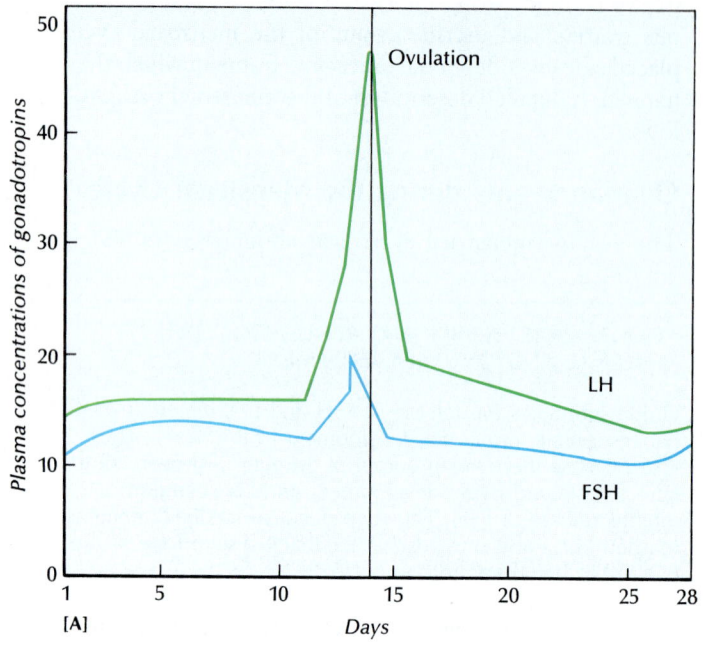

[A]

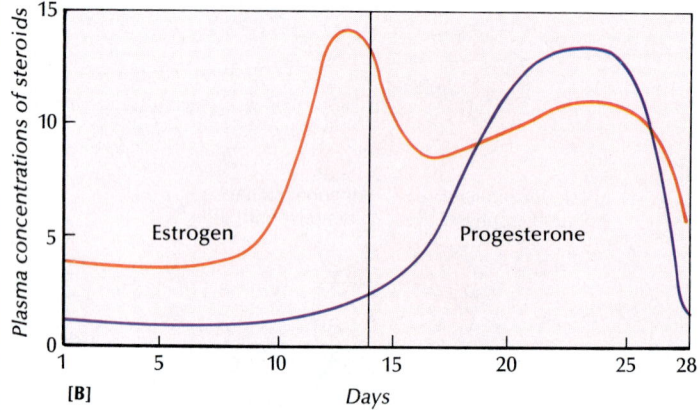

[B]

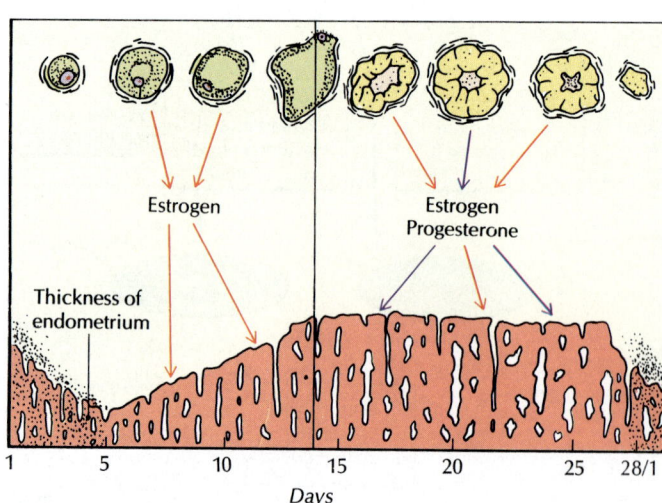

UTERINE PHASES

Menstrual flow	Proliferative phase	Secretory phase

Follicular phase — Luteal phase

OVARIAN PHASES
[C]

Ovulation
(3 to 4 days for events of ovulatory phase)

Ovarian events	Uterine events
Days 1–5: Corpus luteum from previous cycle regresses.	**Days 1–5:** Endometrium sloughs off (*menstruation*).
Estrogen and progesterone concentrations are low.	
GnRH from hypothalamus stimulates the anterior pituitary to secrete previously inhibited FSH and LH. Several ovarian follicles begin to enlarge.	
6: *FSH* promotes development of several follicles within one immature ovarian follicle; one "dominant" follicle is selected.	
7–12: *Estrogen* produced by dominant follicle; further production of FSH inhibited.	**7–12:** Estrogen stimulates thickening of endometrium.
12–13: Elevated estrogen level triggers an *LH* surge from the anterior pituitary.	**12–13:** Endometrium continues to develop.
Oocyte completes the first meiotic cycle. Follicle secretes lytic enzymes and prostaglandin ($F_{2\alpha}$).	
14: LH surge causes the mature follicle to enlarge rapidly and release the oocyte (*ovulation*). LH surge also causes the empty follicle to become an endocrine body, the *corpus luteum*.	**14:** Endometrium continues to develop.
15–25: Corpus luteum secretes *estrogens* and *progesterone*. Secretion of FSH and LH is inhibited by high plasma levels of estrogens and inhibin. As a result, no new follicles develop.	**15–25:** Secretion of estrogens and progesterone from the corpus luteum ensures a glycogen-rich, highly vascular endometrium and maintains it for 10 to 16 days.
25–28: If the released oocyte is not fertilized and implanted in the endometrium, the corpus luteum disintegrates into a *corpus albicans,* and estrogen and progesterone secretion decreases, causing the anterior pituitary to begin active secretion of FSH, which stimulates the development of another follicle. The ovarian events begin again.	**25–28:** Without estrogen and progesterone, the endometrium breaks down and begins to slough off, and *menstruation* occurs. The menstrual flow, which is composed mainly of endometrial tissue and blood, passes out of the body through the vagina. The uterine events begin again.

FOLLICULAR PHASE · OVULATION · LUTEAL PHASE

TABLE 29.1 MAJOR REPRODUCTIVE HORMONES

Hormone	Functions	Source
FEMALE		
Estrogen	Stimulates thickening of uterine wall, stimulates maturation of oocyte, stimulates development of female sex characteristics, inhibits FSH secretion, increases LH secretion	Ovarian follicle, corpus luteum (controlled by FSH)
FSH (follicle-stimulating hormone)	Causes immature oocyte and follicle to develop, increases estrogen secretion, stimulates new gamete formation and development of uterine wall after menstruation	Pituitary gland (controlled by hypothalamus)
GnRH (gonadotropin-releasing hormone	Controls pituitary secretion	Hypothalamus
hCG (human chorionic gonadotropin)	Prevents corpus luteum from disintegrating, stimulates estrogen and progesterone secretion from corpus luteum	Embryonic membranes, placenta
Inhibin	Inhibits secretion of FSH, GnRH	Corpus luteum of ovary
LH (luteinizing hormone)	Stimulates further development of oocyte and follicle, stimulates ovulation, increases progesterone secretion, aids development of corpus luteum	Pituitary gland (controlled by hypothalamus)
Oxytocin	Stimulates uterine contractions during labor, induces mammary glands to eject milk after childbirth	Pituitary gland (controlled by hypothalamus)
Progesterone	Stimulates thickening of uterine wall, stimulates formation of mammary ducts	Corpus luteum (controlled by LH)
Prolactin	Promotes milk production by mammary glands after childbirth	Pituitary gland (controlled by hypothalamus)
Relaxin	Relaxes symphysis pubis, dilates uterine cervix	Placenta
MALE		
FSH	Increases testosterone production, aids sperm maturation	Pituitary gland (controlled by hypothalamus)
GnRH	Controls pituitary secretion	Hypothalamus
Inhibin	Inhibits FSH and GnRH secretion	Sustentacular (Sertoli) cells in testes
LH	Stimulates testosterone secretion	Pituitary gland (controlled by hypothalamus)
Testosterone	Increases sperm production, stimulates development of male primary and secondary sex characteristics, inhibits LH secretion	Interstitial cells (Leydig cells) in testes (controlled by LH)

may be as long as 40 days or as short as 21 days. Since this cycle is really a continuous event, it is convenient to begin with the first day of menstruation as day 1 and divide the ***ovarian events*** of the cycle into three consecutive phases based on changes in the ovaries: The first of the ovarian phases is the (1) *follicular phase*, the time of follicular growth, beginning with the first day of menstruation and ending with ovulation (day 1 through day 10). (2) The second phase is the *ovulation phase*, during which events associated with ovulation occur (day 11 to which events associated with ovulation occur (day 11 to day 14), and (3) the third phase is the *luteal phase*, the time during which the corpus luteum is functional (day 14 through day 28).

Because individual menstrual cycles vary in length, the length of the follicular phase and time of ovulation may also vary, but the length of the luteal phase always remains at 14 days, the time from ovulation to the onset of menstruation. A brief description of what happens during each of these ovarian phases follows; the description is based on a typical 28-day cycle.

Follicular phase The *follicular phase* includes *menstruation,* which occurs from day 1 to day 5 of the average menstrual cycle. The levels of ovarian steroid hormones (estrogens and progesterone) are low, and the ovaries contain primordial follicles. Over time, under pulsatile GnRH release from the hypothalamus, and FSH and LH release from the anterior pituitary, approximately 20 primordial follicles begin to develop, becoming *primary follicles* and then *secondary follicles.* Only one follicle in the ovary reaches maturity and becomes a dominant, fluid-filled *vesicular (Graafian) follicle,* outgrowing the others. The other, less-well-developed follicles undergo degeneration, a process known as *atresia.* As the follicles are growing, the granulosa cells secrete estrogens, which are under the influence of both FSH and LH from the anterior pituitary. LH also causes the release of small amounts of androgens from thecal cells, which diffuse to the granulosa cells where they are converted into estrogens.

As the circulating levels of estrogens rise, there is a *negative feedback* to the anterior pituitary and hypothalamus, decreasing the output of FSH and LH. At the same time, the follicles are becoming more sensitive to FSH and LH and increase their output of estrogens. Thus the initial small increase in circulating estrogens exerts a negative feedback on the hypothalamic-pituitary-ovarian pathway. As the follicles continue to release estrogens, the higher estrogen levels exert a *positive feedback.* The increase in estrogen causes an increase in LH output (and to a lesser extent, FSH) from the anterior pituitary. This *LH surge* begins about 24 hours before ovulation and peaks about 8 hours later, triggering ovulation from the dominant follicle.

Ovulatory phase At about day 14 of the menstrual cycle, *ovulation* (the *ovulatory phase*) occurs, as an oocyte is released from the surface of the ovary into the pelvic cavity and begins its journey toward a uterine tube.

Luteal phase After ovulation, the *luteal phase* occurs. The empty follicle remains on the surface of the ovary and is stimulated by LH to become the corpus luteum, which secretes both estrogens and progesterone. (Remember that the *developing follicles* release only estrogens, whereas the *corpus luteum* releases both estrogens and progesterone.) The rising levels of estrogens and progesterone exert a negative feedback on the secretion of FSH and LH, inhibiting the development of any new follicles that might ovulate later in the cycle. Starting at about day 22, the corpus luteum begins to regress and decreases its output of estrogens and progesterone. Thus new follicles start to develop at the end of one menstrual cycle in preparation for the next cycle. The decreased function of the corpus luteum and the declining levels of estrogens and progesterone cause *menstruation,* the sloughing off of the endometrial lining, which initiates another menstrual phase.

Uterine Events during the Menstrual Cycle

The *uterine events* are closely tied to changes that occur in the endometrium during the menstrual cycle; these changes are simultaneous with changes that are occurring in the ovaries, as described above, and like the ovarian events can be divided into three phases: (1) the *menstrual phase* (days 1 to 5); (2) the *proliferative phase* (days 6 to 14); and (3) the *secretory phase* (days 15 to 28).

Menstrual phase The *menstrual phase* occurs as a result of the drop in ovarian estrogens and progesterone during the ovarian luteal phase. The sloughing off of the endometrial stratum functionalis occurs following vasoconstriction of the spiral arteries. This vasoconstriction may be caused by the release of *prostaglandin* ($F_{2\alpha}$), which accompanies the decrease in progesterone. The detached endometrial lining (*menstrual flow,* composed of tissue and blood) passes from the body through the vaginal canal, usually with a blood loss of 50 to 150 mL.

Proliferative phase The *proliferative phase* occurs while the ovary is in the follicular phase. During the proliferative phase, the stratum functionalis of the endometrium thickens, tubular glands form, and spiral arteries develop. The mucosa thickens and becomes vascular, and estrogen stimulates progesterone receptor proteins in preparation for the next (secretory) phase of the uterine cycle.

Secretory phase Ovulation occurs in the ovary at the end of the proliferative phase. With the beginning of the *secretory phase,* the increased progesterone level stimulates the development of mucous glands; the secretion of estrogens and progesterone ensures a glycogen-rich, vascular-rich uterine lining, which is capable of providing abundant nutrients for the developing embryo if fertilization has occurred. If fertilization has not occurred, the corpus luteum degenerates, resulting in menstruation, and the menstrual phase begins again.

Hormonal Regulation in the Pregnant Female

If, however, fertilization *has* occurred, the corpus luteum is maintained by the production of a hormone called *human chorionic gonadotropin (hCG)* by the trophoblast cells of the blastocyst of the implanted embryo. The hCG stimulates the corpus luteum to continue its production of estrogens and progesterone, which maintains the endometrial lining for the implanted embryo until the placenta is developed well enough to take over the production of estrogens and progesterone. As the placenta develops, the corpus luteum regresses. (Pregnancy tests use antibody or receptor protein tests to detect hCG in the woman's blood or urine as an indication of pregnancy.)

As the embryo develops, other hormones are secreted.

FIGURE 29.17 MEIOSIS

In the course of meiosis, which occurs only in sex cells, there is a single replication of the chromosomes, followed by two cell divisions. The final products of meiosis are four haploid cells.

MEIOSIS I

Interphase: The chromosomes of the diploid cell double (the genetic material replicates), and the centrioles replicate.

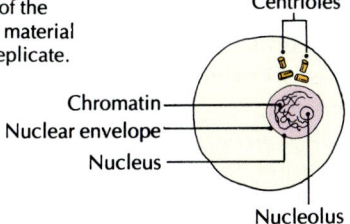

Centrioles

Chromatin
Nuclear envelope
Nucleus

Nucleolus

Early Prophase I: Chromosomes, each consisting of two chromatids, become visible. The nuclear envelope begins to disappear. The nucleolus disappears. Centrioles begin to move toward opposite poles of the cell.

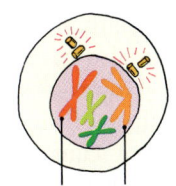

Pair of homologous chromosomes

Late Prophase I: Homologous chromosomes pair up forming tetrads, consisting of four chromatids. During this process, called synapsis, there may be an exchange of parts between paired chromatids.

Spindle   Aster

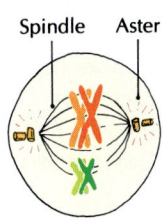

Metaphase I: Homologous chromosome pairs line up across the middle of the cell, on the metaphase plate.

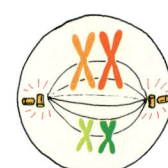

Anaphase I: The pairs of homologous chromosomes separate, moving toward opposite poles of the cell. Each of the doubled chromosomes still consists of two chromatids connected at the centromere.

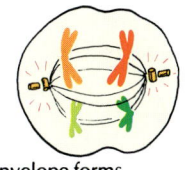

A nuclear envelope forms around each set of chromosomes. Nucleolus reappears.

Telophase I: Cytokinesis occurs as the plasma membrane pinches in, forming two cells. Each new cell contains one chromosome from each homologous pair.

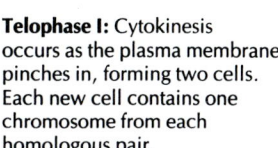
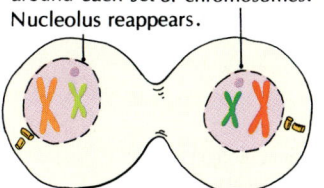

MEIOSIS II

Interphase: Chromosomes may uncoil, but no replication occurs.

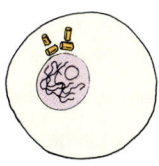
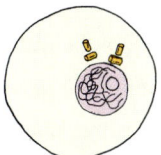

Prophase II: The chromosome content of each cell is the same as it was in telophase I.

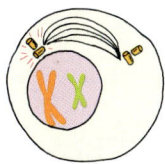

Metaphase II: The chromosomes line up at the metaphase plate.

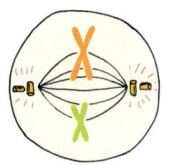
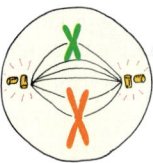

Anaphase II: The centromeres divide, and the attached chromatids separate and move toward opposite poles of the cell.

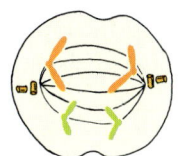
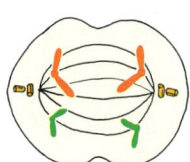

Telophase II: Nuclear envelopes form around the four haploid nuclei.

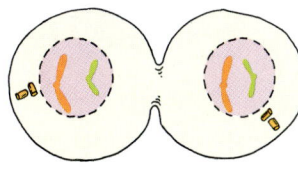
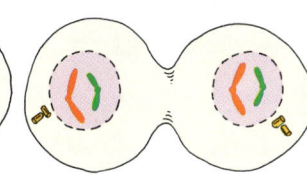

Cytokinesis produces four haploid cells.

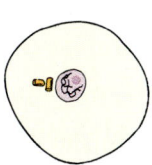

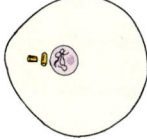
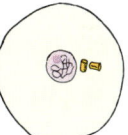

Prolactin and *oxytocin* induce the mammary glands to secrete and eject milk, respectively, after childbirth. Oxytocin also stimulates the uterine contractions that expel the baby from the uterus during childbirth.

A S K   Y O U R S E L F

1 What are the ovarian phases of the menstrual cycle?

2 What are the uterine phases of the menstrual cycle?

3 What is the function of a corpus luteum?

4 Why does menstruation occur?

5 What is the function of hCG?

FORMATION OF SEX CELLS (GAMETOGENESIS)

The formation of *gametes,* or sex cells, is called *gametogenesis.* Male gametogenesis is called *spermatogenesis* and occurs in the testes; female gametogenesis is called *oogenesis* and occurs in the ovaries. Both spermatogenesis and oogenesis involve meiosis, a form of nuclear division. Before discussing gametogenesis we will present a brief description of meiosis and a comparison of mitosis and meiosis.

Meiosis

When body cells are lost, they are replaced by mitosis. Our growth from a single fertilized cell to an adult with over 50 trillion cells also takes place by mitosis. Reproduction, however, involves another kind of cell division called *meiosis* (mye-OH-sihss; Gr. *meioun,* to diminish) [FIGURE 29.17].

As you learned in Chapter 3, *mitosis* guarantees that new daughter cells receive exactly the same genetic information as the parent cell. If the parent cell has 23 *pairs* of chromosomes, or 46 chromosomes, each daughter cell will also have 23 *pairs* of chromosomes.

However, sex cells, or *gametes,* cannot be formed by mitosis. If a sperm contained 46 chromosomes and an ovum contained 46 chromosomes, their combination during fertilization would produce a cell with 92 chromosomes. Such a cell would not be a normal human cell. The problem is solved when *meiosis reduces the number of chromosomes in gametes to half:* Each sperm has 23 chromosomes (not 23 *pairs*), and each ovum has 23 chromosomes (not 23 *pairs*)—so that when ovum and sperm unite, they produce a cell with 46 chromosomes. This cell is called a *zygote.*

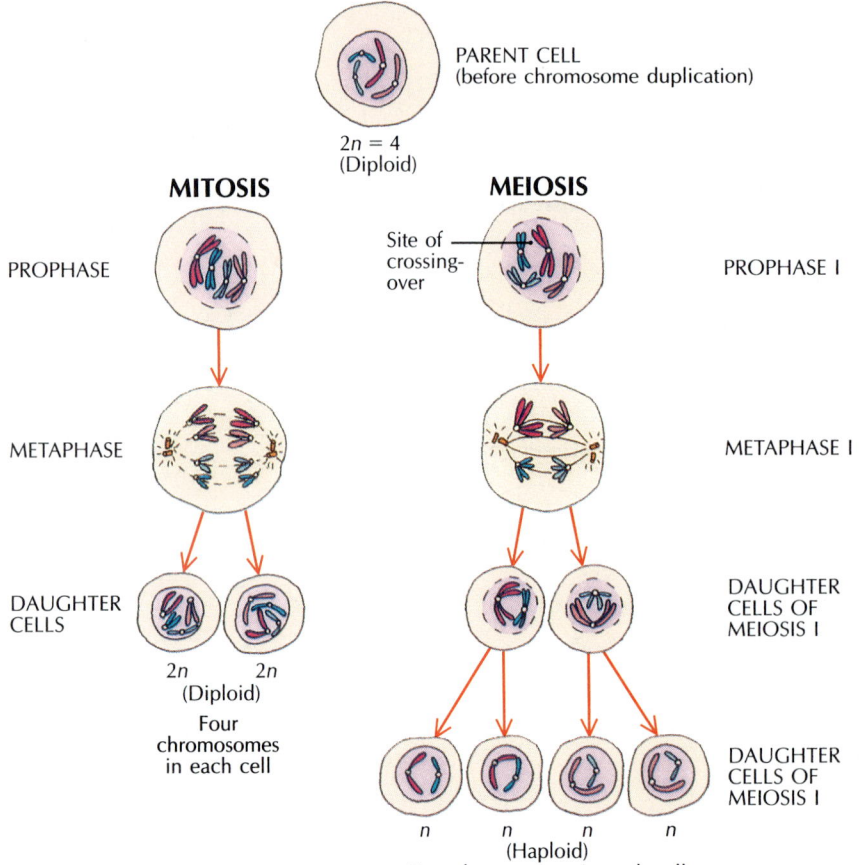

PARENT CELL
(before chromosome duplication)

$2n = 4$
(Diploid)

MITOSIS

PROPHASE

METAPHASE

DAUGHTER CELLS

$2n$ $2n$
(Diploid)
Four chromosomes in each cell

Site of crossing-over

MEIOSIS

PROPHASE I

METAPHASE I

DAUGHTER CELLS OF MEIOSIS I

n n n n
(Haploid)
Two chromosomes in each cell

DAUGHTER CELLS OF MEIOSIS I

FIGURE 29.18 COMPARISON OF MITOSIS AND MEIOSIS

In the simplified example shown here, there are four chromosomes: two long ones (one from the male parent and one from the female parent) and two short ones (one from the male parent and one from the female parent). Mitosis, with one division, results in two cells identical to the original one, with four chromosomes in each cell. Meiosis, with two divisions, results in four cells different from the original one, with two chromosomes in each cell.

Comparison of Mitosis and Meiosis

The stages of meiosis have the same names as the stages of mitosis: *prophase*, *metaphase*, *anaphase*, and *telophase*. However, in mitosis the chromosomes replicate once and the nucleus divides once, while in meiosis a single replication of the chromosomes is followed by *two* nuclear divisions [FIGURE 29.18]. In mitosis, the daughter cells have the same number of chromosomes as the parent cell—the **diploid** (2*n*) number. Body (somatic) cells (all cells except sex cells) are diploid. The diploid number is 46 in humans. In meiosis, daughter cells have only half the parental (diploid) number, or the **haploid** (1*n*) number. Gametes, with only 23 chromosomes, are haploid.

A critical part of meiosis occurs early in the sequence when each double-stranded chromosome from the male parent lines up with a double-stranded chromosome from the female parent. Each male-female set of four chromatids is called a **homologous pair**, or **tetrad** (Gr. *tetra*, four), because it consists of four chromatids [FIGURE 29.19A]. (Remember that each strand is a chromatid.) Homologous pairs are alike in appearance and contain genes that affect the same traits. The chromatids of a homologous pair may become intertwined, and portions of a male chromatid may be exchanged with portions of its female homolog, a process called **crossing-over** [FIGURE 29.19B and C]. Crossing-over results in a rearrangement and redistribution of genes, producing a great variety among human beings.

FIGURE 29.19 CROSSING-OVER

During meiosis 1, homologs connect at crossover points, resulting in an exchange and recombination of genetic material.

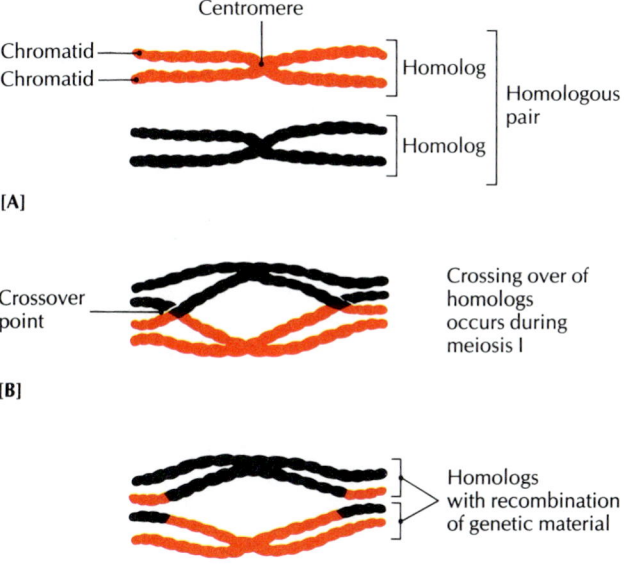

Spermatogenesis

Haploid sperm cells are formed in the testes in a precisely controlled series of events. This process, called **spermatogenesis** ("the birth of seeds"), continuously produces mature sperm in the seminiferous tubules. It proceeds in the following stages [FIGURE 29.20]:

1 With the onset of puberty, when a boy is about 11 to 14 years old, dormant, unspecialized germ cells called **spermatogonia** (sing. *spermatogonium*) are activated by secretions of testosterone.

2 Each spermatogonium divides through *mitosis* to produce two daughter cells, each containing the full complement of 46 chromosomes. (The term "daughter cells" has nothing to do with gender. These cells could just as easily be called "offspring cells.")

3 One of the two daughter cells is a spermatogonium,

FIGURE 29.20 SPERMATOGENESIS

Spermatogenesis in the seminiferous tubules.

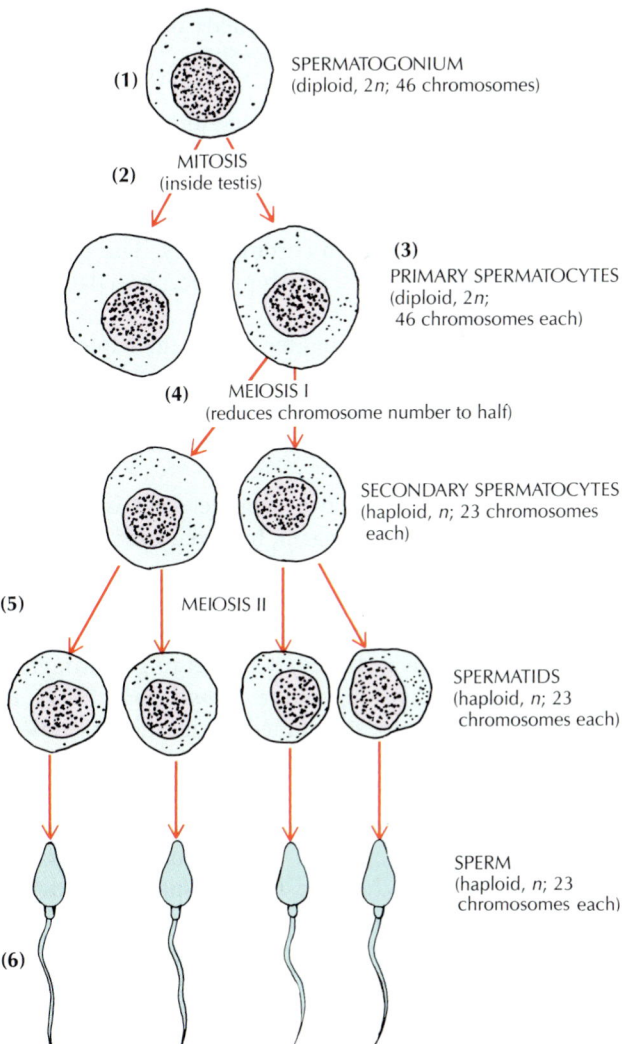

which continues to produce daughter cells. The other daughter cell is a ***primary spermatocyte,*** a large cell that moves toward the lumen of the seminiferous tubule.

4 The primary spermatocyte undergoes *meiosis* to produce two smaller ***secondary spermatocytes,*** each with 23 chromosomes: 22 body chromosomes and 1 X or Y sex chromosome.

5 Both secondary spermatocytes undergo the second meiotic division to form four final primitive germinal cells, the ***spermatids,*** which still have only 23 chromosomes.

6 The spermatids develop into mature ***sperm*** without undergoing any further cell division. Each sperm has 23 chromosomes. The entire process of spermatogenesis takes about 64 days.

Oogenesis

Oogenesis (oh-oh-JEHN-ih-sihss), the maturation of ova in the ovary, differs from the maturation of sperm in several ways. Usually only one ovum at a time matures each month, but billions of sperm may mature during that time. The process of cellular division differs too. Whereas one primary spermatocyte yields four sperm after meiosis, a primary oocyte yields only one ovum. Oogenesis follows the stages in FIGURE 29.21:

1 The ***oogonium*** (oh-oh-GO-nee-uhm), the diploid precursor cell of the ovum, is enclosed in a follicle within the ovary.

2 The oogonium develops into a ***primary oocyte*** (OH-oh-site), which contains 46 chromosomes. The primary oocyte undergoes meiosis, which produces two daughter cells of unequal size.

3 The larger of the daughter cells is the haploid ***secondary oocyte.*** It is perhaps a thousand times as large as the other cell and contains most of the primary oocyte's cytoplasm, which provides nourishment for the developing ovum.

4 The smaller of the two daughter cells is the ***first polar body.*** It may divide again, but eventually it degenerates.

5 The large secondary oocyte leaves the ovarian follicle during ovulation and enters the uterine tube. If the secondary oocyte is fertilized, it begins to go through a second meiotic division, and a ***second polar body*** is "pinched off." It, too, is destined to die. If fertilization does not occur, menstruation follows shortly, and the cycle begins again.

6 During the second meiotic division, the secondary oocyte is completely reduced to the haploid number of 23 chromosomes. When the haploid sperm and ovum nuclei are finally ready to merge, the final stage of nuclear maturity as a ***mature ovum*** is reached.

7 The haploid nuclei of the ovum and sperm unite to form a diploid ***zygote.***

ASK YOURSELF

1 How does meiosis differ from mitosis?

2 What are the main steps in spermatogenesis? In oogenesis? How do they differ?

CONCEPTION

Once every menstrual cycle (28 days or so), a single secondary oocyte (ovum) emerges from the ovary (ovulation) and is carried by a stream of peritoneal fluid generated by the sweeping movements of the fimbriae of the uterine tube to the ampulla of the tube — the usual site of fertilization. From there, the ovum is conveyed slowly by

FIGURE 29.21 OOGENESIS

Oogenesis in the ovary and uterine tube.

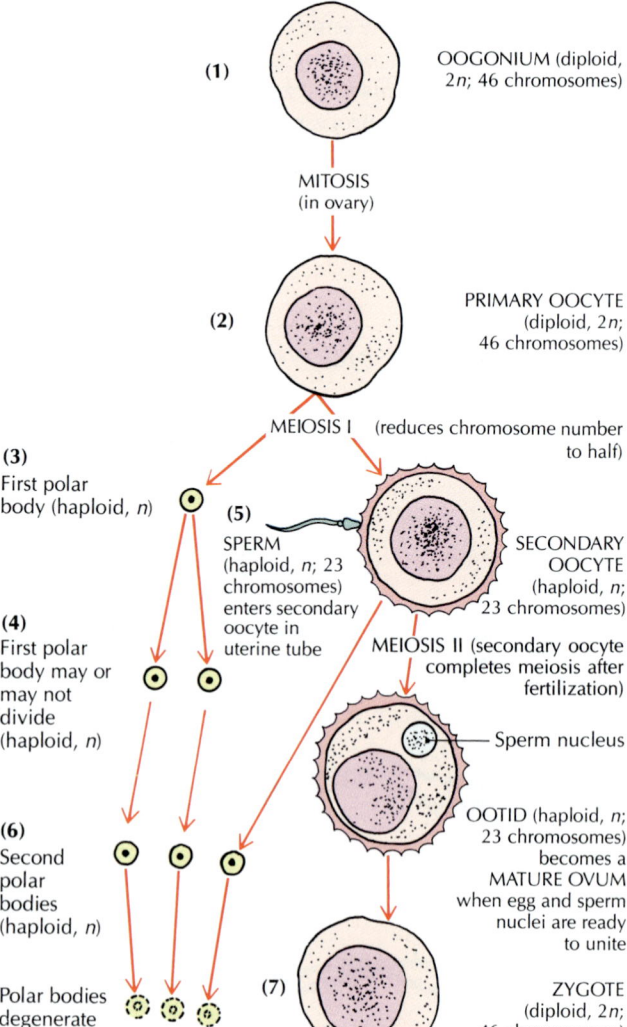

(1) OOGONIUM (diploid, 2n; 46 chromosomes)

MITOSIS (in ovary)

(2) PRIMARY OOCYTE (diploid, 2n; 46 chromosomes)

MEIOSIS I (reduces chromosome number to half)

(3) First polar body (haploid, n)

(5) SPERM (haploid, n; 23 chromosomes) enters secondary oocyte in uterine tube

SECONDARY OOCYTE (haploid, n; 23 chromosomes)

(4) First polar body may or may not divide (haploid, n)

MEIOSIS II (secondary oocyte completes meiosis after fertilization)

Sperm nucleus

(6) Second polar bodies (haploid, n)

OOTID (haploid, n; 23 chromosomes) becomes a MATURE OVUM when egg and sperm nuclei are ready to unite

Polar bodies degenerate

(7) ZYGOTE (diploid, 2n; 46 chromosomes)

In Vitro Fertilization and Surrogate Motherhood

In vitro ("in a test tube") *fertilization* offers women who have difficulty conceiving a possible means of having a successful pregnancy. Essentially, an ovum is removed from the prospective mother's ovaries at the time of ovulation and fertilized in a glass dish or test tube with sperm from a male donor. Two or three days later, after cell division has begun, the embryo is implanted in the woman's uterus through her cervix. This technique has a 10- to 15-percent success rate. It was first done successfully in 1977, resulting in the birth of the first "test-tube baby" in 1978.

Several new techniques attempt to improve on the success rate of *in vitro* fertilization. In *gamete intrafallopian transfer (GIFT),* the ovum and sperm are united in the uterine tube, where fertilization normally takes place. This procedure allows the fertilized ovum more time to develop before it reaches the uterus. Also, the removal of the ovum and its reimplantation take place on the same day, instead of several days apart as in in vitro fertilization. So far, the pregnancy success rate is about 30 percent. In a similar technique—*zygote intrafallopian transfer (ZIFT)*—ova are collected from the prospective mother and combined with sperm in a glass dish, where fertilization occurs. A day later, the resulting zygote is placed in the uterine tube. Another technique, *transvaginal oocyte retrieval,* can be done in a physician's office. The physician obtains ova through the vaginal wall instead of through the abdominal wall. By this method, ova that would otherwise be hidden behind the uterus (a situation that occurs in about 5 to 10 percent of infertile women) can be retrieved.

Surrogate motherhood is an alternative for women who have difficulty either conceiving or carrying a child. For a woman who has a normal uterus but blocked uterine tubes that prevent the ova from being fertilized and traveling to the uterus, a female "surrogate" or donor can be artificially inseminated with sperm from the prospective father. If fertilization occurs, the embryo is transferred to the prospective mother. Also, a fertilized ovum from a woman who cannot produce the hormones necessary to sustain a pregnancy can be transferred to the uterus of another woman, the "surrogate mother," who will carry and give birth to the child.

In a recent development, such a technique has been fairly successful when viable ova from young women are fertilized in vitro and implanted 2 days later into the uterus of postmenopausal women in their early forties.

---

ciliary action and muscular contractions of the uterine tube toward the uterus. The ovum takes 4 to 7 days to travel the 10 cm (4 in.) from the ovary to the uterus.

Fertilization

The process of fertilization and the subsequent establishment of pregnancy are collectively referred to as *conception.* Fertilization must occur no more than 24 hours after ovulation, since the ovum will remain viable for only that period of time.* Sperm usually remain viable in the female tract for up to 72 hours.†

During ejaculation, hundreds of millions of flagellating sperm enter the female's vaginal canal. If sexual intercourse takes place at about the same time as ovulation, some of these sperm travel toward the opposite-moving ovum, but only one sperm may eventually enter and fertilize the ovum [FIGURE 29.22]. If only one sperm may enter, why are millions discharged? More than enough sperm are discharged to increase the chances that one will successfully penetrate the ovum. The sperm must travel from the vagina all the way to the uterine tubes, a trip of 15 to 20 cm that may take as long as a few hours. They must resist the spermicidal acidity of the vagina and overcome opposing fluid currents in the uterus and uter-

*After lying dormant for anywhere from 10 to 50 years, an unfertilized ovum released from an ovary lives only about a day.

†If semen is stored in a laboratory or sperm bank at very cold temperatures, the sperm will remain viable for years.

FIGURE 29.22 MEETING OF SPERM AND OVUM

Although the ovum is enormous compared with the sperm, they both contribute exactly the same amount of genetic material to the zygote. It is believed that only a few hundred sperm manage to reach the ovum. ×1500.

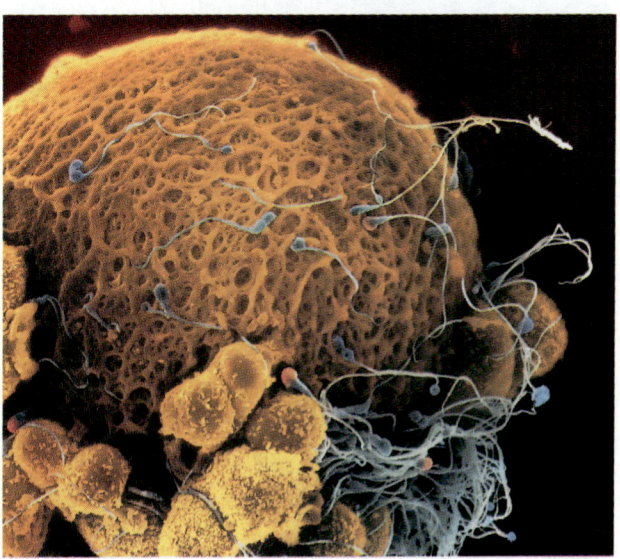

ine tubes. So it is not surprising that the mortality rate of the sperm is enormously high.

Also, the quantity of sperm discharged must be large enough so that several hundred will reach the ovum and

FIGURE 29.23 PENETRATION OF AN OVUM BY A SPERM

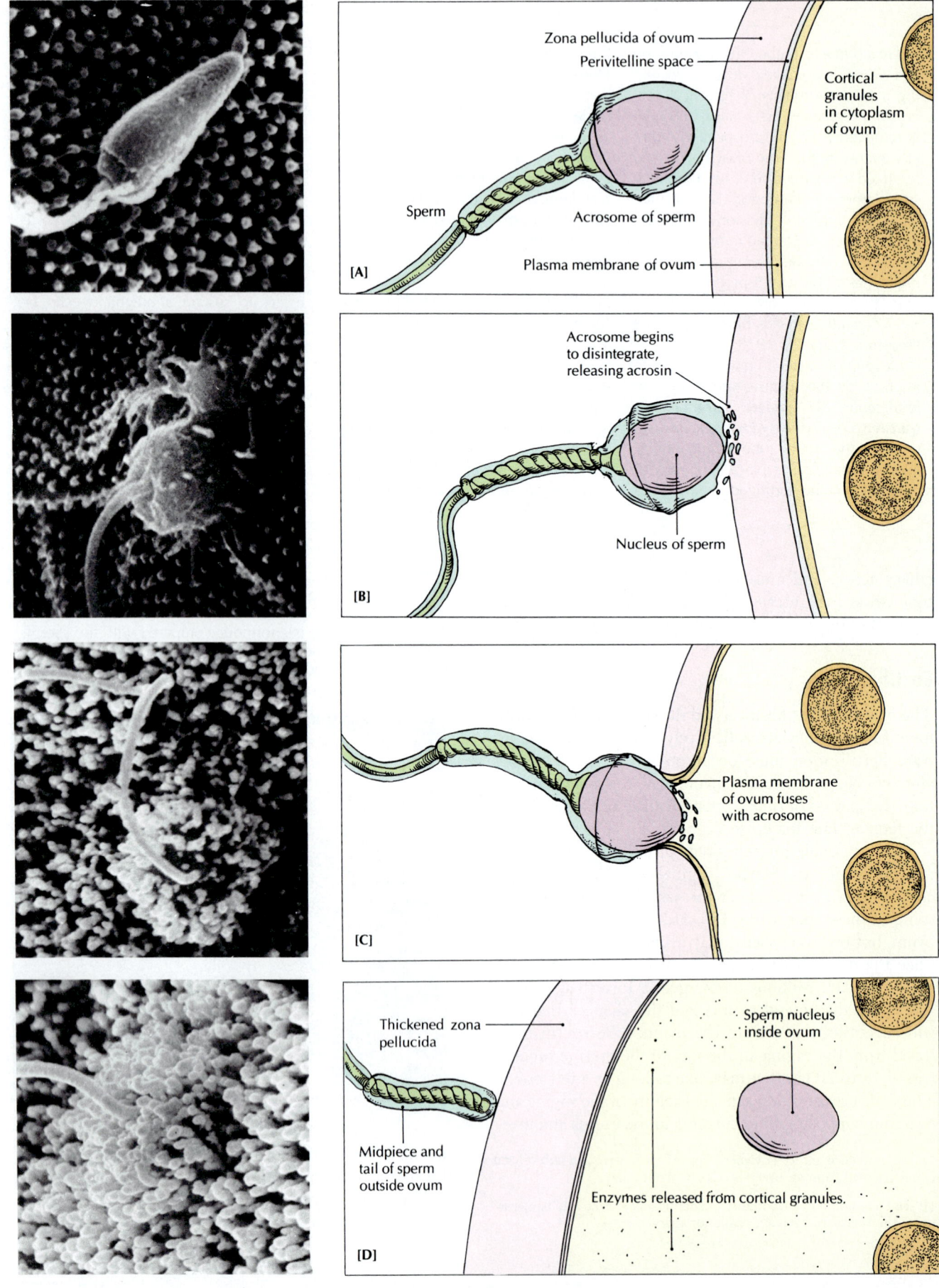

[A]

Zona pellucida of ovum

Perivitelline space

Cortical granules in cytoplasm of ovum

Sperm

Acrosome of sperm

Plasma membrane of ovum

[B]

Acrosome begins to disintegrate, releasing acrosin

Nucleus of sperm

[C]

Plasma membrane of ovum fuses with acrosome

[D]

Thickened zona pellucida

Sperm nucleus inside ovum

Midpiece and tail of sperm outside ovum

Enzymes released from cortical granules.

Scanning electron micrographs and comparative drawings. **[A]** Acrosome of sperm touches the microvilli on the surface of the zona pellucida of an ovum and binds with a sperm-receptor site. (Site not shown in this simplified drawing.) **[B]** The acrosome of the sperm begins to disintegrate, releasing acrosin, which allows the acrosome to move through the zona pellucida. **[C]** The acrosome fuses

with the plasma membrane of the ovum, and the sperm head is engulfed by the plasma membrane. **[D]** The head is pulled into the cytoplasm of the ovum, leaving the midpiece and tail outside. Cortical granules within the ovum release enzymes that make the zona pellucida impenetrable to other sperm. Eventually, the sperm and ovum will fuse to form a zygote.

act together to provide sufficient amounts of *acrosin.* This enzyme, which is found in the head of the sperm, facilitates the entrance of a single sperm through the outer wall of the ovum—the *zona pellucida* (L. "transparent zone")—to allow a single sperm to enter the ovum.

The penetration and fertilization of an ovum by a sperm take place in the following sequence [FIGURE 29.23]:

1 A sperm approaches the ovum and binds with it as specific proteins on the plasma membrane of the sperm form a complementary "fit" with sperm-receptor sites on the surface of the zona pellucida.

2 Immediately after a sperm binds to the zona pellucida, the plasma membrane of the sperm's acrosome begins to break up, releasing vesicles of acrosin. These vesicles digest a path through the zona pellucida and facilitate the entrance of the sperm.

3 The exposed acrosome eventually fuses with the plasma membrane of the ovum. The head passes through the plasma membrane into the cytoplasm of the ovum, leaving the midpiece and tail outside the ovum. Mitochondrial genes are passed on to offspring only from the mother because the mitochondria of the sperm do not enter the ovum at fertilization. Mutations in mitochondrial DNA can cause serious nervous system diseases, including certain forms of central nervous system degeneration and blindness.

4 Cortical granules within the ovum release enzymes that thicken the zona pellucida and make it impenetrable to other sperm. At the same time, the sperm-binding sites on the zona pellucida become inactivated, providing further assurance that only one sperm will enter.

A few hours after the sperm penetrates the ovum, the ovum undergoes a cell division that produces two haploid cells, each with 23 chromosomes. The smaller of the cells disintegrates, and the other cell, now called a *pronucleus,*

fuses with the nucleus of the sperm (now also called a *diploid pronucleus*). Fertilization is completed as a diploid zygote with 46 chromosomes is formed. Now embryonic development begins.

If more than one sperm somehow enters the ovum (a condition called *polyspermy*), the development of the zygote is quickly stopped. This cessation occurs because the extra mitotic spindles cause the abnormal segregation of chromosomes during cleavage.

Human Sex Determination

A new individual begins its development with a full complement of hereditary material. Each parent supplies 23 chromosomes, giving the zygote 23 pairs. In the male and female sets of chromosomes, 22 always match in size and shape and determine the same traits. The 22 matching pairs are called *autosomes.* The other pair are the *sex chromosomes,* and they determine the sex of the new individual. Because the sex chromosomes look alike in females, they are designated *XX.* Because the sex chromosomes do *not* look alike in males, they are designated *XY.*

After meiosis in males, half the sperm contain 22 autosomes and an X chromosome, and the other half contain 22 autosomes and a Y chromosome [FIGURE 29.24]. (After meiosis in the female, *all* ova contain 22 autosomes and an X chromosome.) If an ovum is fertilized by an X-chromosome-bearing sperm, an XX zygote results, and the zygote will develop into a female. Fertilization of an ovum by a Y-bearing sperm produces an XY zygote, which will develop into a male. So the father actually determines the sex of the child, since only his sperm contain the variable, the Y chromosome. Also, it should be noted that the sex of the child is determined only at conception, never before or after.

Q: *Can women be allergic to sperm?*

A: Yes, some women are allergic to their partner's sperm, and in fact, scientists do not completely understand why the immune systems of *all* women do not reject sperm as foreign bodies. Some men are infertile because they produce antibodies against their own sperm, especially following vasectomy.

ASK YOURSELF

1 Why is the mortality rate of sperm high after they enter the female reproductive tract?

2 What is the zona pellucida?

3 What is acrosin?

4 How does the genetic material of the mother and father combine to make a new individual?

5 How is the sex of a new individual determined?

FIGURE 29.24 SEX DETERMINATION

The sex of a child is determined by the type of sperm entering the ovum. An X-bearing sperm fertilizing an ovum (all ova have an X chromosome) produces an XX zygote, which develops into a female. A Y-bearing sperm fertilizing an ovum produces an XY zygote, which develops into a male.

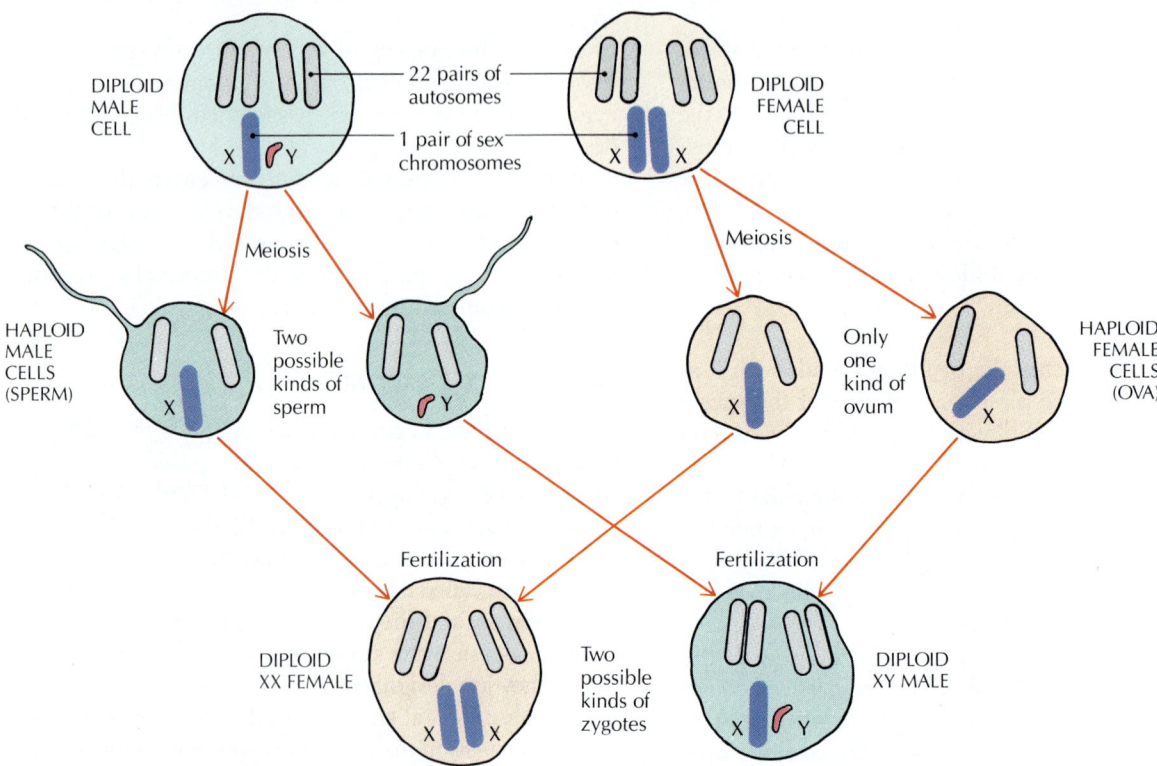

CONTRACEPTION

All *contraceptive* ("against conception") methods aim to prevent pregnancy* by preventing the production of ova or sperm, by keeping ova and sperm from meeting, or by preventing the implantation of an embryo in the uterus. All these methods are fallible, some are dangerous to the female, and some may be considered unacceptable because of psychological factors or religious beliefs. TABLE 29.2 describes the many methods of contraception now available in the United States. Two surgical methods of contraception—*vasectomy* and *tubal ligation*—are illustrated in FIGURE 29.25.

*More than half (58 percent) of the pregnancies in the United States are unplanned.

A S K   Y O U R S E L F

Which contraceptive methods are most effective? Least effective? What are some of the side effects of these methods?

THE EFFECTS OF AGING ON THE REPRODUCTIVE SYSTEMS

As women grow older, their ovaries decrease in weight and begin to atrophy. Estrogen concentration decreases, and symptoms of menopause appear, usually in the late forties or early fifties. The vagina also decreases in length and width, its lining may become dry, and infections and vaginal discharges may result. Fibrous tissue becomes abundant, ovarian cysts are relatively common, and blood vessels harden. The uterus weighs about half as much at age 50 as it did at 30, and its elasticity is lost as elastic tissue is replaced by fibrous tissue. Ligaments that hold the uterus, urinary bladder, and rectum in place may weaken and allow these organs to drop down.

The breasts sag and become flattened as ligaments supporting the breasts become lax, fibrous cells replace milk glands, and fat tissue is lost. The amount of pubic hair gradually decreases, as does the layer of fat under the skin in the pubic region.

Reproductive changes due to age are less evident in men than in woman. Although the testes do not necessarily decrease in size and weight, the specialized cells that

TABLE 29.2 METHODS OF CONTRACEPTION (In Order of Decreasing Effectiveness)

Method	Mode of action	Effectiveness if used correctly	Action required at time of intercourse	Possible side effects or inconveniences
Sterilization Vasectomy Tubal ligation	Prevents release of sperm* Prevents ovum from entering uterus*	Very high Very high	None None	Irreversible† Irreversible†
Foam and male condom	Foam kills sperm, condom prevents sperm from entering vagina	Very high	Must be used before intercourse	None usually; foam may irritate; condom may reduce sensation in male, may interrupt foreplay
Depo-provera (injectable)	Prevents ovulation	Very high	None	May increase risk of developing osteoporosis
Norplant (surgical skin implant)	Prevents follicle maturation and ovulation by secreting progestinlike hormone; may also prevent implantation	Very high	None	Lengthened menstrual period, irregular spotting between periods
Oral contraceptive	Prevents follicle maturation and ovulation by altering hormones present; may also prevent implantation	Very high	None	*Early:* some water retention, breast tenderness, nausea. *Late:* possible clots, hypertension, heart disease, breast cancer
Intrauterine device (coil, loop)	Prevents implantation by stimulating inflammatory response	High	None	Possible expulsion of device, menstrual discomfort, abnormal bleeding, infection, cramps
Foam pad ("Today sponge")	Releases spermicide to kill sperm	High	Must be inserted before intercourse	None as yet
Diaphragm with spermicide	Diaphragm prevents sperm from entering uterus; spermicide kills sperm	High	Must be inserted before intercourse	None, but may cause overlubrication; cannot be fitted to all women
Cervical cap with spermicide	Prevents sperm from entering uterus; spermicide kills sperm	High	Must be inserted before intercourse	May increase rate of abnormal Pap test; may be difficult to insert and remove
Male condom	Prevents sperm from entering vagina	High	Must be put on erect penis before insertion into vagina	Some reduction of sensation in male; may interrupt foreplay
Female condom	Prevents sperm from entering vagina	High	Must be inserted into vagina before intercourse	Some reduction of sensation in female
Temperature rhythm	Prevents sperm from being deposited when ovum is available for fertilization	Medium	None	Requires abstinence during part of cycle; requires good record keeping to determine time of ovulation
Calendar rhythm	Prevents sperm from being deposited when ovum is available for fertilization	Medium to low	None	Requires abstinence during part of cycle; difficult to determine ovulation time
Spermicide (foam, jelly, cream)	Kills sperm	Medium to low	Must be inserted immediately before intercourse	None usually; may irritate
Withdrawal (coitus interruptus)	Prevents sperm from entering vagina	Low	Withdrawal	Frustration in some
Douche	Washes out sperm	Lowest	None	Must be done immediately after intercourse

*Vasectomy and tubal ligation do not affect the production of sex hormones.
†Although new surgical techniques increase the possibility of reversing vasectomies and tubal ligations, both procedures should be considered irreversible at the time they are undertaken.

FIGURE 29.25 SURGICAL STERILIZATION

[A] Male vasectomy removes a small portion of each ductus deferens to block the release of sperm from the testes into the sperm duct. **[B]** Female tubal ligation removes a portion of each uterine tube to prevent the ovum from being fertilized and reaching the uterus.

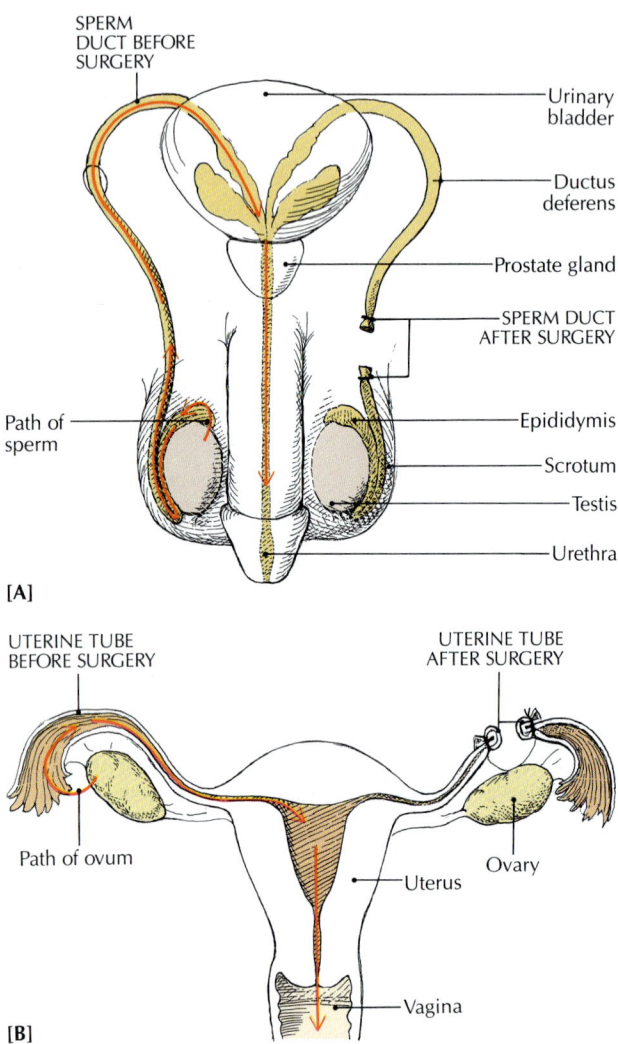

FIGURE 29.26 DEVELOPMENT OF REPRODUCTIVE ORGANS

[A] Development of the external genitalia. **[B]** Development of the internal reproductive systems.

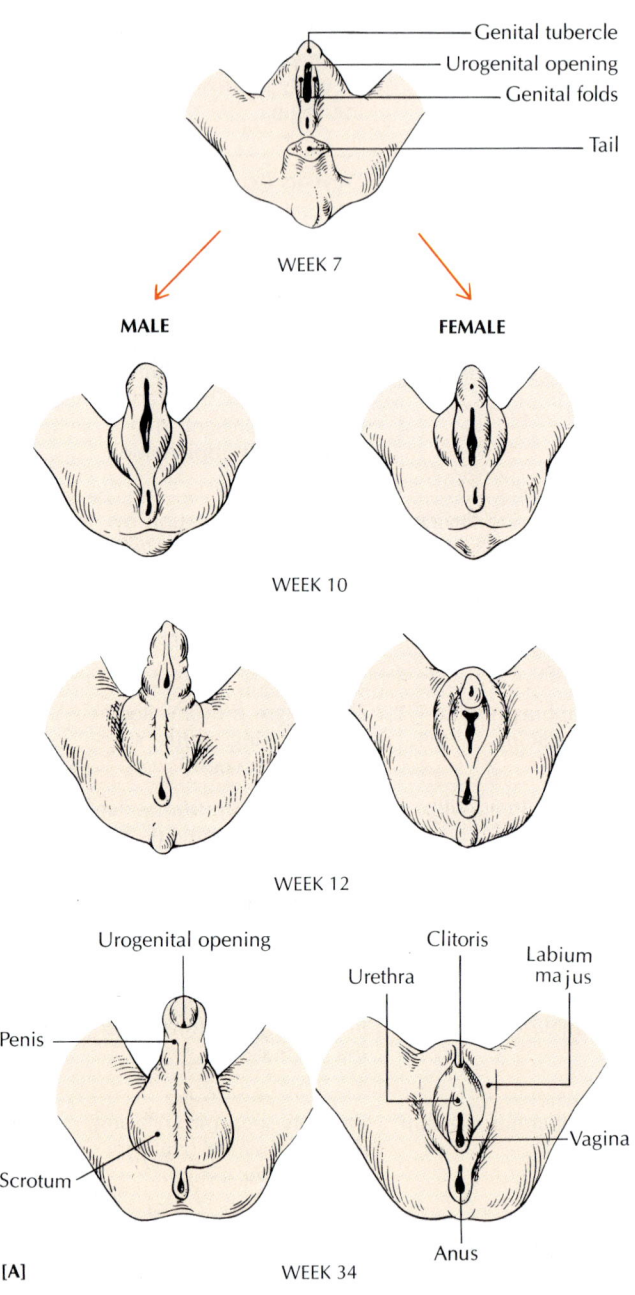

produce and nourish sperm gradually become fewer and less active. Sperm count gradually decreases to about 50 percent, but fertility is frequently retained past the age of 80. Accessory glands and organs such as the seminal vesicles and prostate gland begin to atrophy, and the secretion of testosterone decreases. Sexual desire in healthy men and women usually is not affected by old age.

Certain cancers become more prevalent in elderly men and women. Women over 60 are more likely to get cancer of the breast or uterus than are younger women, and men over 60 are particularly prone to cancer of the prostate gland. A recent study revealed that only 10 percent of women aged 65 to 74 knew that the risk of breast cancer increased with age.

DEVELOPMENTAL ANATOMY OF THE REPRODUCTIVE SYSTEMS

Until about week 10 after fertilization, the external genitalia of the male fetus do not differ greatly from those of the female [FIGURE 29.26A]. By week 12 in the female,

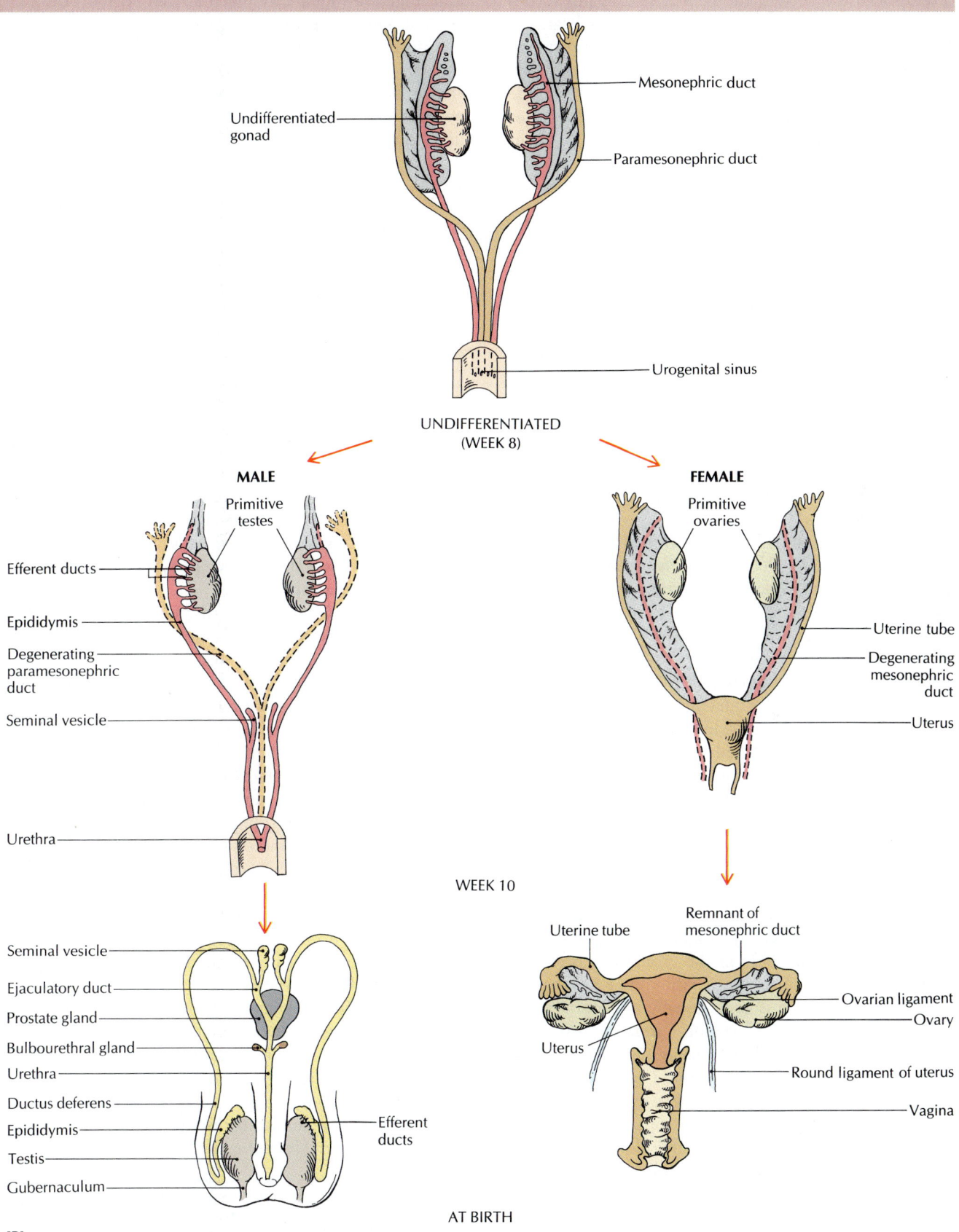

Undifferentiated gonad

Mesonephric duct

Paramesonephric duct

Urogenital sinus

UNDIFFERENTIATED
(WEEK 8)

MALE

Primitive testes

Efferent ducts

Epididymis

Degenerating paramesonephric duct

Seminal vesicle

Urethra

FEMALE

Primitive ovaries

Uterine tube

Degenerating mesonephric duct

Uterus

WEEK 10

Seminal vesicle

Ejaculatory duct

Prostate gland

Bulbourethral gland

Urethra

Ductus deferens

Epididymis

Testis

Gubernaculum

Efferent ducts

Uterine tube

Remnant of mesonephric duct

Ovarian ligament

Ovary

Uterus

Round ligament of uterus

Vagina

AT BIRTH

[B]

the bud that is derived from the genital tubercle begins to develop into the clitoris, the genital fold begins to develop into the labia, and the urogenital opening begins to divide into the separate openings for the urethra and vagina. By week 12 in the male, the bud is clearly developing into the penis, and the genital fold fuses over the urogenital opening to become the scrotum. The testes begin to descend into the scrotum when the fetus is about 7 months old [see FIGURE 29.3]. At week 34, male and female genitals look very much as they will at birth.

Male and female internal reproductive systems are still undifferentiated at week 8 but have undergone distinct changes by week 10 [FIGURE 29.26B] as the undifferentiated gonads develop into testes or ovaries, the mesonephric ducts become the male epididymis or female uterine tube, and male seminal vesicles begin to form.

WHEN THINGS GO WRONG

Sexually Transmitted Diseases

Some microorganisms and agents are transferred from person to person mainly by sexual contact. The diseases they cause are known as **sexually transmitted diseases** (STD) or **venereal** ("of Venus") **diseases** (VD). Contrary to some beliefs, these diseases are rarely contracted by casual, dry contact with persons or objects. The most common sexually transmitted diseases are *nongonococcal urethritis* (NGU), caused by a bacterium; *type II herpes simplex*, caused by a virus; and *gonorrhea* and *syphilis*, both caused by bacteria.*

The most common STD in the United States is **nongonococcal urethritis (NGU)**, or **chlamydia**, named for the bacterium that causes it, *Chlamydia trachomatis*. Although it afflicts between 3 million and 10 million people in the United States every year, NGU is not well known because its symptoms are difficult to diagnose and because the disease is easily confused with gonorrhea.

NGU in men produces infection in the urethra, prostate gland, and epididymis. If left untreated, it can cause sterility. Women are afflicted with a cervical infection and inflammation of the uterine tubes, which if untreated can result in scarring, sterility, or ectopic pregnancy. Women who acquire the disease during pregnancy may transmit it to their babies as they pass through the birth canal. Affected infants may have conjunctivitis or pneumonia, and there is some evidence that the risk of premature or stillborn infants is increased. Both men and women may be affected by conjunctivitis, pneumonia, and enlarged lymph nodes in the groin. The preferred treatment is an extended dose of antibiotics for each partner until the infection is gone, usually in about a week.

Type II herpes (Gr. "a creeping") **simplex** (also called herpes genitalis) ordinarily appears as blisterlike sores on or near the external genitalia about a week after intercourse with an infected partner. Fever, muscle aches, and swollen lymph nodes may also be apparent. When the blisters rupture, they produce painful ulcers and release millions of infectious viruses. The blisters usually heal after a week or two, but the viruses retreat to nerves near the lower spinal cord, where they remain dormant until the next attack. Infected people may harbor the infectious viruses for months, years, or a lifetime.

A 1993 study revealed that about 20 percent of the sexually active population in the United States is infected with a viral sexually transmitted disease. About 31 million people in the United States carry the herpes II viruses; 200,000 to 500,000 new cases are reported each year.

Perhaps the most serious complication of type II herpes simplex is the infection of a baby as it passes through the birth canal. For this reason, if a pregnant woman is known to have type II herpes simplex, the baby is usually delivered by cesarean section.

Gonorrhea (Gr. "flow of seed") is primarily an infection of the urinary and reproductive tracts (especially the urethra and cervix), but any moist part of the body, especially the eyes, may suffer. The causative bacterium, *Neisseria gonorrhoeae*, can be recognized by microscopic examination. Symptoms include urethritis (inflammation of the urethra) and urethral discharge in males and a greenish-yellow cervical discharge in females. Itching, burning, redness, and dysuria (painful urination) are also common symptoms, but some infected people, especially women, are without symptoms. A complication of gonorrhea in men is epididymitis (inflammation of the epididymis), with an accompanying swelling of the testes. The leading cause of arthritis in young adults of both sexes is gonorrhea. If left untreated, gonorrhea becomes difficult to cure, but it can usually be cured by antibiotics, especially if it is diagnosed early.

Before routine treatment of the eyes of newborns with 1 percent silver nitrate was established, thousands of babies were blinded by gonorrheal infection at the time of passage through the vaginal canal.

Syphilis† is a more dangerous disease than gonorrhea. It is caused by a motile, corkscrew-shaped bacterium, *Treponema pallidum*. It begins in the mucous membranes

---

*As you saw in Chapter 23, AIDS can also be transmitted sexually, but it is not exclusively a sexually transmitted disease.

†Syphilis is the title character in a poem by Girolamo Fracastoro, a Veronese physician and poet. In the poem, published in 1530, Syphilis is stricken with a disease that was known at the time as the "great pox." Ever since 1530, however, the disease has been called *syphilis*.

and spreads quickly to lymph nodes and the bloodstream. The early symptom of syphilis is a sore, the hard chancre, at the place where infection occurred. Other symptoms may include fever, general body pain, and skin lesions. Sometimes these symptoms disappear even without treatment, leaving the victim with the false impression that the disease is gone. But later, circulatory or nervous tissue may degenerate so that paralysis, insanity, and death follow.

One of the most unfortunate aspects of sexually transmitted infection is the intrauterine transfer of the microorganisms from a mother to her baby. The syphilis bacterium is able to cross the placenta during pregnancy, whereas the gonorrhea bacterium seems unable to do so. Thus the developing fetus can contract syphilis early in its development and exhibit some of the serious symptoms of the disease at birth. If the baby contracts syphilis during the actual birth process, it is not likely to exhibit any symptoms at the time of birth. However, a baby infected with syphilis will grow poorly, be mentally retarded, and die early.

Like gonorrhea, syphilis can usually be cured by antibiotics if treatment is started early. However, it is well to remember that the occurrence of sexually transmitted diseases in this country has reached epidemic proportions (more than 5 million cases of NGU and gonorrhea are reported annually, and the reported cases of syphilis are increasing—about 120,000 a year), especially among teenagers and young adults. According to public health officials, false feelings of security about not contracting the diseases are prevalent in the population at risk, as is lack of knowledge about the seriousness of the consequences of the diseases. It is true that sexually transmitted diseases can be cured with greater ease than ever before, especially if they are reported and treated early, but the most effective treatment is still intelligent prevention.

Trichomoniasis is a sexually transmitted disease caused by the *Trichomonas vaginalis* protozoan. The infection affects the lower genitourinary tract, especially the vagina and urethra of females and the lower urethra of males. Because the infecting organism is most productive in an alkaline environment, it is aided by excessive douching that disrupts the normal acidity of the vagina, pregnancy, and the use of oral contraceptives. Symptoms include itching, swelling, and frothy vaginal discharge in females, and urethritis and dysuria in males. About 3 million cases are reported annually in the United States.

Pelvic Inflammatory Disease

Pelvic inflammatory disease (PID) is a general term referring to inflammation of the uterus and uterine tubes (*salpingitis*) with or without inflammation of the ovary (*oophoritis*), localized pelvic peritonitis, and abscess formation (*tuboovarian abscess*). It is the most severe complication of the sexually transmitted diseases caused by *N.*

gonorrhoeae and *C. trachomatis*. The spread of the infection can be controlled with antibiotics if treatment is begun early. Severe cases of PID can lead to peritonitis (inflammation of the peritoneal cavity).

Some Common Cancers of the Reproductive Systems

Breast cancer is a common malignant cancer in women and is one of the leading causes of death by cancer among females. (A newborn girl's chance of getting breast cancer during her lifetime is 1 in 8, or 12.5 percent. About 46,000 women in the United States will die of breast cancer this year.) It is especially prevalent between the ages of 55 and 74, and usually does not occur before 35. Breast cancer kills three times as many women as uterine or ovarian cancer, spreading through lymphatics and blood vessels to other parts of the body. (The most common site of breast cancer is the upper lateral quadrant of the breast, close to the lymph nodes.)

The warning signs of breast cancer include a hard lump in the breast, a change in the shape or size of one breast, a change in skin texture, discharge from the nipple, itching or other changes in the nipple, an increase in the skin temperature of the breast, and breast pain. Self-examination of the breasts and regular physical examinations are highly recommended by physicians as a method of early detection of these warning signs.

The recommended treatment for breast cancer depends on the stage at which the cancer is diagnosed. If the diagnosis is early, the cancerous cells can often be removed successfully by a *lumpectomy*, in which the tumor and axillary lymph nodes are removed. In fact, recent studies show that a lumpectomy is often as effective as more extensive surgery, epsecially when combined with radiation. The most drastic surgical treatment is a *radical mastectomy*, in which the entire breast and underlying fascia, the pectoral muscles, and all the axillary lymph nodes are removed. This is performed only as a last resort. Less drastic procedures include a *modified radical mastectomy*, in which the breast and axillary lymph nodes are removed, and a *simple mastectomy*, in which only the breast is removed. Chemotherapy and radiation therapy are frequently used in conjunction with surgery. The current favored treatment for early breast cancer is a lumpectomy and postoperative radiation.

A recent study suggests that mothers who breast-feed their infants may reduce the risk of premenopausal breast cancer.

Cervical cancer (cancer of the cervix) is one of the most common cancers among females. When cancer cells invade the basement membrane and spread to adjacent pelvic areas or to distant sites through lymphatic channels, the cancer is classified as *invasive*. When only the epithelium is affected, it is *preinvasive*. If detected early, preinvasive cancer is curable 75 to 90 percent of the time. While preinvasive cancer produces no apparent

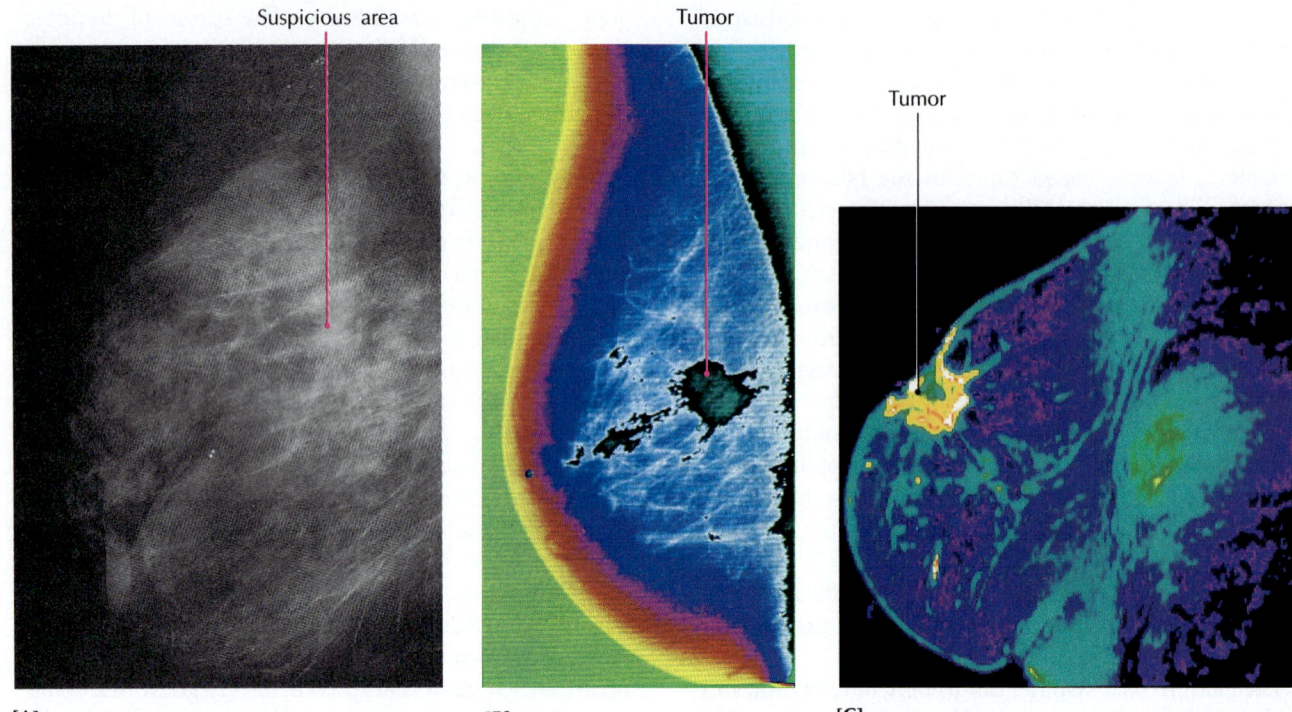

Suspicious area   Tumor   Tumor

[A]   [B]   [C]

[A] X-ray mammograms are the most common technique for detecting early breast cancer. Mammograms usually detect a breast tumor before it can be felt by self-examination.
[B] Digitized mammograms use computers to enhance images, making tumors easier to see.
[C] Magnetic resonance imaging (MRI) uses radio waves and magnetic fields to detect tumors with startling accuracy. Unfortunately, MRI scanning is about 10 times more expensive than conventional x-ray.

symptoms, invasive cancer is characterized by unusual vaginal bleeding or discharge and postcoital pain or bleeding. The most effective method of detection is the Pap test (Papanicolaou stain slide test), a microscopic examination of cells taken from the cervix. Advanced cases of invasive cervical cancer may call for a *hysterectomy*, the surgical removal of the uterus.*

Prostatic cancer is the second most common cancer among males, striking one in eight men (about 200,000) and killing almost 40,000 men in the United States each year. (Lung cancer is still the most common and most frequently fatal cancer among men.) Most prostatic cancers originate in the posterior portion of the gland. When the cancer spreads beyond the prostate itself, it usually travels along the ejaculatory ducts in the spaces between the seminal vesicles. Because prostatic carcinomas rarely produce symptoms until they are well advanced, prostatic cancer is often fatal. Annual or semiannual rectal examinations may detect a small, hard nodule

*A new surgical technique, dramatic in its simplicity, uses laparoscopes fitted with cutting devices to snip the ovaries and the ligaments that hold the uterus in place. The detached ovaries and uterus are drawn out of the abdominal cavity through the vaginal canal. The patient is usually able to resume her normal activities in about 3 weeks.

while it is still localized, and in such cases the recovery rate is high. In the past, many prostate tumors were not detected with rectal examinations until they were inoperable, but now a relatively simple test called *prostate-specific antigen*, or *PSA*, is available. It is 20 to 30 percent more effective than a rectal examination and is more acceptable to patients than the usual rectal examination. PSA is a protein produced by the cells on the surface of the prostate. An enlarged prostate (which can be caused by a tumor) reveals high concentrations of PSA in the blood. Regular examinations and testing are especially important in men over 40.

Current therapies for prostate cancer (all are controversial) are usually dictated by the stage of the disease. (1) *Stage A*, a microscopic cancer confined within the prostate gland, is usually treated with observation, radiation, or a radical prostatectomy (surgical removal of the prostate); (2) *Stage B*, when a lump within the prostate gland can be felt during a rectal exam; necessitates radiation of a radical prostatetectomy; (3) *Stage C*, when a large tumor involves all or most of the prostate gland; calls for radiation, with or without hormal therapy; and (4) Stage D, when the tumor has metastasized; requires hormonal therapy and additional therapy to relive pain and other discomforts.

Testicular cancer, although not considered to be common, is the most common cancer in United States males aged 15 to 35.

Prostate Disorders

The prostate gland may be affected by inflammation

(acute or chronic infections), enlargement, and benign growths. Benign enlargement of the prostate gland is known as *benign prostatic hypertrophy (BPH),* or benign prostatic hyperplasia. It is the most common type of benign tumor, affecting about 75 percent of all men over 50. An enlarged prostate may be caused by a decrease in the secretion of male hormones, inflammation, decreased sexual activity, metabolic and nutritional imbalances, or other factors. When the prostate becomes enlarged, it compresses the urethra and obstructs normal urinary flow. Surgery that removes part or all of the prostate usually relieves the obstruction. A relatively new surgical technique does not sever the nerves near the prostate that influence erections. As a result, sexual potency is retained. Hormonal treatment is under investigation.

Ovarian Cysts

Ovarian cysts generally occur either in the follicles or within the corpus luteum. Although most ovarian cysts are not dangerous, they must be examined thoroughly as possible sites of malignant cancer.

Follicular cysts are usually small, distended bubbles of tissue that are filled with fluid. Ordinarily, small follicular cysts do not produce symptoms unless they rupture, but large or multiple cysts may cause pelvic pain, abnormal uterine bleeding, and irregular ovulation. If follicular cysts are present at menopause, they secrete excessive amounts of estrogen in response to the increased menopausal secretions of LH and FSH.

Granulosa-lutein cysts are produced when an excessive amount of blood accumulates during menstruation. If they appear in early pregnancy, they may cause pain on one side of the pelvis, and if they rupture, there will be massive hemorrhaging within the peritoneum. Granulosa-lutein cysts in nonpregnant women may cause irregular menstrual periods and abnormal bleeding.

Because most ovarian cysts disappear of their own accord, the typical treatment consists of observation to detect early malignancies.

Endometriosis

Endometriosis is the abnormal location of endometrial tissue in sites such as the ovaries, the pelvic peritoneum, and the outer surface of the small intestine. Most cases probably develop as a result of retrograde passage of bits of menstrual endometrium through the opening of the uterine tube into the peritoneal cavity. This condition is usually associated with dysmenorrhea (painful menstruation), pelvic pain, infertility, and dyspareunia (painful coitus). Treatment ranges from symptomatic relief of pain to surgical removal of the endometrial implants, including the use of a laparoscope and lasar beam.

CHAPTER SUMMARY

Male Reproductive Anatomy and Physiology (p. 941)

1 The paired *testes* are held outside the body in the *scrotum,* a sac of skin between the thighs. The *seminiferous tubules* produce sperm, and *interstitial cells* secrete testosterone, the primary male sex hormone.

2 A mature *sperm* consists of a *head* that has an *acrosome,* containing enzymes to aid the sperm in penetrating an ovum, and a *nucleus,* containing DNA; a *middle piece* that contains the mitochondria that provide ATP energy; and a *tail* that drives the sperm forward.

3 The seminiferous tubules merge to form the *tubuli recti,* the *rete testis,* and then the *efferent ducts,* which pass into the tightly coiled epididymis in each testis. The *epididymis* stores sperm as they mature, serves as a duct for the passage of sperm from the testis to the ejaculatory duct, and propels sperm toward the penis with peristaltic contractions.

4 The paired *ductus deferentia* are the dilated continuations of the epididymides. They extend to the ejaculatory ducts and are the main carriers of sperm.

5 The paired *ejaculatory ducts* receive secretions from the seminal vesicles and prostate gland. They carry semen to the urethra during ejaculation.

6 The *urethra* is a tube through the penis, consisting of *prostatic, membranous,* and *penile* portions. It transports sperm out of the body during ejaculation.

7 Secretions of the paired *seminal vesicles* provide nourishment for the sperm and help neutralize vaginal acidity. Secretions of the *prostate gland* make the sperm motile and help neutralize vaginal acidity. Secretions of the paired *bulbourethral glands* neutralize any urine in the urethra and lubricate the urethra.

8 Secretions from the epididymis, seminal vesicles, prostate gland, and bulbourethral glands, together with the sperm, make up the *semen.*

9 The *penis* carries urine to the outside during urination and transports semen during ejaculation. It contains the urethra and three strands of erectile tissue. The tip of the penis is the *glans penis.* During sexual stimulation, the penis becomes enlarged and firm in an *erection.*

10 Testosterone secretion from interstitial cells in the testes is controlled by *FSH* (follicle-stimulating hormone) and *LH* (luteinizing hormone) from the anterior pituitary gland, which is controlled by *GnRH* (gonadotropin-releasing hormone) from the hypothalamus.

Female Reproductive Anatomy and Physiology (p. 950)

1 The female gonads are the paired *ovaries,* which produce ova and the female sex hormones. Ova are found in *follicles,* which are present in various stages of development. *Primordial follicles* are not yet growing. *Vesicular follicles* are almost ready to release an ovum in the monthly process of **ovulation.** After a mature ovum is discharged from a ruptured follicle, the follicle becomes the **corpus luteum,** a temporary endocrine gland that secretes *estrogens and progesterone.*

2 The **uterine tubes** receive mature ova from the ovary and convey them to the uterus. Ova are usually fertilized in the uterine tube, and are then transported to the uterus.

3 The uterine tubes terminate in the **uterus,** a hollow, muscular organ that is the site of implantation of a fertilized ovum. If pregnancy occurs, the uterus houses the developing fetus. The inner lining of the uterus, the **endometrium,** is built up every month for a possible pregnancy. If pregnancy does not occur, the stratum functionalis of the endometrium is shed in the process of **menstruation.**

4 The uterus leads downward to the **vagina,** the muscle-lined tube where sperm from the penis are deposited during sexual intercourse; it is also the exit point for menstrual flow and the birth canal for the baby during childbirth.

5 The **external genital organs,** collectively called the **vulva,** are the mons pubis, labia majora, labia minora, vestibular glands, clitoris, and vestibule of the vagina.

6 *Mammary glands* are modified sweat glands contained in the breasts. They produce and secrete milk for the newborn baby.

The Menstrual Cycle (p. 957)

1 Oocyte development is influenced by **gonadotropin-releasing hormone (GnRH)** from the hypothalamus. *Follicle-stimulating hormone (FSH)* and *luteinizing hormone (LH)* released by the anterior pituitary cause the oocyte's maturation and release from the ovary in the process of **ovulation.**

2 If pregnancy does not take place, the endometrium breaks down and menstruation occurs.

3 If pregnancy occurs, **human chorionic gonadotropin (hCG)** is released from the placenta and the covering membranes of the embryo. The endometrium is maintained, and pregnancy continues.

4 The hormones **prolactin** and **oxytocin** induce the mammary glands to secrete and eject milk after childbirth. Oxytocin also stimulates uterine contractions during childbirth.

Formation of Sex Cells (Gametogenesis) (p. 962)

1 Meiosis reduces the number of chromosomes in sex cells to half, with each sperm having 23 chromosomes and each ovum having 23.

2 When ovum and sperm unite, they produce a cell with 46 chromosomes. A cell with the full number of chromosomes is **diploid,** and a cell with only half is **haploid.**

3 Each male-female set of four chromatids is a **homologous pair,** or **tetrad.**

4 A rearrangement of chromosome parts called **crossing-over** may occur during meiosis.

5 The process of sperm formation is called **spermatogenesis.** A **spermatogonium** divides to produce a primary spermatocyte and another spermatogonium. The **primary spermatocyte** undergoes meiosis to produce two **secondary spermatocytes.** The second meiotic division forms four haploid **spermatids,** which develop into mature **sperm.**

6 The maturation of ova in the ovaries is **oogenesis.** An **oogonium** develops into a diploid **primary oocyte** with 46 chromosomes. The first meiotic division produces a haploid **secondary oocyte** and a **polar body.** The polar body degenerates, and the secondary oocyte leaves the ovary and enters the uterine tube. If the secondary oocyte is fertilized, it undergoes a second meiotic division and produces a second polar body and a haploid **ootid.** When the haploid sperm and ovum nuclei merge, the ootid is considered to be a **mature ovum.**

Conception (p. 964)

1 Hundreds of millions of sperm are ejaculated into the vagina during sexual intercourse, but only one sperm can penetrate the ovum. The release of *acrosin* by many sperm is required to dissolve the outer shell of the ovum to allow penetration.

2 Once one sperm enters the ovum, an impenetrable membrane forms around the ovum so no other sperm can enter.

3 About an hour after fertilization, the haploid (23 chromosomes) nuclei of the ovum and sperm fuse to form a single diploid (46 chromosomes) cell, the **zygote.**

4 From the parents, the child receives 22 pairs of matching chromosomes called **autosomes** and another pair called the **sex chromosomes.** The sex chromosomes look alike in females and are designated **XX.** The sex chromosomes do not look alike in males and are designated **XY.**

5 After meiosis in males, half the sperm contain an X chromosome and half contain a Y chromosome. An ovum fertilized by an X-bearing sperm produces an XX zygote, which develops into a female. Fertilization of an ovum by a Y-bearing sperm produces an XY zygote, which develops into a male.

Contraception (p. 968)

1 The contraceptive aim of preventing pregnancy can be achieved by preventing the production of ova and sperm (as by oral contraceptives), by keeping ova and sperm from meeting (as by vasectomy, tubal ligation, diaphragm, condom, and spermicide), or by preventing the implantation of an embryo in the uterus (by IUD).

2 The most effective contraceptives are sterilization, vasectomy, tubal ligation, and oral and implanted contraceptives.

The Effects of Aging on the Reproductive Systems (p. 968)

1 Major changes in the female reproductive system due to aging include ovarian atrophy, menopause, the prevalence of ovarian cysts and breast or uterine cancer, and the weakening of the ligaments that support the uterus and breasts.

2 Male changes due to aging include a decreased sperm count and decreased testosterone secretion, atrophy of accessory reproductive glands and organs, and an increased susceptibility to cancer of the prostate gland.

Developmental Anatomy of the Reproductive Systems (p. 970)

1 Male and female external genitalia do not differ greatly until about week 10. The genitalia are almost fully formed at week 34.

2 The internal reproductive systems are still undifferentiated at week 8, but have undergone significant anatomical changes by week 10.

STUDY AIDS FOR REVIEW

KEY TERMS

acrosin 967
autosome 967
bulbourethral glands 947
corpora cavernosa 947
corpus luteum 950
corpus spongiosum 947
crossing-over 963
ductus deferens 945
endometrium 954
epididymis 945
follicle 950
FSH 949, 957
GnRH 949, 957
hCG 960
LH 949, 957
mammary glands 957
meiosis 962
menstruation 955
oogenesis 964

ovaries 950
ovulation 952, 960
oxytocin 962
polar body 964
primary oocyte 964
prolactin 962
prostate gland 946
scrotum 941
secondary oocyte 964
semen 947
seminal vesicles 946
seminiferous tubules 941
sex chromosomes 967
sperm 944
spermatogenesis 963
testes 941
uterine tubes 953
uterus 954
vagina 955

MEDICAL TERMS FOR REFERENCE

AMENORRHEA The absence of menstrual flow.

ANOVULATION A condition in which ovulation does not occur.

CYSTOCELE A protrusion of the urinary bladder into the vagina.

DILATION AND CURETTAGE A procedure that dilates the cervix in order to scrape the lining of the uterus. Also called *D and C.*

DYSMENORRHEA Painful or difficult menstruation.

FIBROADENOMA A fibroid breast tumor.

HERMAPHRODITISM A condition in which male and female sex organs are contained in one person.

HYSTERECTOMY The surgical removal of the uterus. A *total hysterectomy* usually removes all the female reproductive organs (including the ovaries and uterine tubes) except the vagina.

IMPOTENCE The inability of a man to have an erection sufficient for sexual intercourse.

LEUKORRHEA A whitish, viscous discharge from the vagina or uterus.

OLIGOSPERMIA A condition in which only a small number of sperm are produced.

OOPHORECTOMY The surgical removal of an ovary.

OOPHORITIS An inflammation of an ovary.

ORCHITIS An inflammation of the testes, produced by a bacillus or staphylococcus infection.

PREMENSTRUAL SYNDROME (PMS) A condition characterized by nervousness, irritability, and abdominal bloating 1 to 2 weeks before menstruation.

PROLAPSED UTERUS A protrusion of the uterus into the vagina.

PROSTATECTOMY The surgical removal of all or part of the prostate gland.

RECTOCELE A protrusion of the rectum that presses on the vagina.

SALPINGECTOMY The surgical removal of a uterine tube.

SPONTANEOUS ABORTION The loss of a fetus through natural causes; usually called a *miscarriage.*

STERILITY The inability to reproduce.

UTERINE LEIOMYOMA A smooth muscle uterine tumor, including myomas, fibromyomas, and fibroids.

VAGINITIS An inflammation of the vagina.

UNDERSTANDING THE FACTS

1 What are the two basic functions of the testes?
2 State the primary function of the
   a spermatogenic cells
   b sustentacular cells
   c interstitial cells
3 Trace the route of a mature sperm from the seminiferous tubule to the penile urethra.
4 List the accessory glands of the male reproductive system, indicating their location and the primary function(s) of each.
5 Describe the structure of a mature sperm, indicating the primary function of each region.
6 List the organs that compose the female reproductive system, indicating the primary function(s) of each.
7 List the stages in the life of an ovarian follicle.
8 A surge in which hormone is the direct cause of ovulation?
9 What are the functions of inhibin, androgen-binding protein, hCG, ocytocin, and prolactin?
10 How many chromosomes are found in human somatic cells? In sperm? In secondary oocytes? In ova?
11 Contrast autosomes with sex chromosomes. How do the sex chromosomes determine the genetic sex of an embryo?

UNDERSTANDING THE CONCEPTS

1 Explain the vascular mechanism by which penile erection occurs.
2 Outline the regulatory feedback pathway by which testosterone levels in the blood and seminiferous tubules are controlled.

3 Contrast the movement of a secondary oocyte through the reproductive ducts of the female with that of a sperm through the ducts of the male.

4 Explain the roles of GnRH, FSH, LH, and testosterone in spermatogenesis.

5 Correlate the levels of FSH, LH, estrogen, and progesterone in the follicular and luteal phases of the menstrual cycle.

6 Correlate the endometrial changes of the menstrual cycle with the cycles of ovarian hormones.

7 Compare and contrast the events of spermatogenesis with those of oogenesis.

8 Contrast a sperm with an ovum with respect to size, motility, and genetic contribution to the zygote.

9 Why is it necessary for so many more sperm to be produced by males than ova by females?

10 Explain why an ovum must be fertilized by only one sperm.

11 List the basic contraceptive methods and compare their effectiveness.

SELF-QUIZ
Multiple-Choice Questions

29.1 The cells that provide nourishment for maturing sperm are called
   a interstitial cells (Leydig cells)
   b spermatogonia
   c sustentacular (Sertoli) cells
   d spermatogenic cells
   e follicle cells

29.2 Which of the following is/are a function of the epididymis?
   a stores sperm
   b serves as a duct system
   c propels sperm toward the ejaculatory duct
   d a and b
   e a, b, and c

29.3 Semen consists of secretions from the testes and
   a epididymis
   b seminal vesicles
   c prostate gland
   d bulbourethral (Cowper's) glands
   e all of the above

29.4 The penis contains how many cylindrical strands of erectile tissue?
   a one   **d** four
   b two   **e** five
   c three

29.5 The male sex hormone testosterone is produced by
   a sustentacular cells
   b tube cells
   c the prostate gland
   d interstitial cells
   e the epididymis

29.6 The production of testosterone is controlled by
   a FSH   **d** a and b
   b LH   **e** a, b, and c
   c ADH

29.7 In the male, the hormone that inhibits the secretion of FSH by the anterior pituitary is
   a LHRH   **d** FSH
   b GnRH   **e** LH
   c inhibin

29.8 The corpus luteum secretes
   a estrogen   **d** a and b
   b progesterone   **e** a, b, and c
   c LH

29.9 In the female, production of FSH is inhibited by
   a estrogen   **d** prolactin
   b LH   **e** a and c
   c progesterone

29.10 Which of the following hormones is used as the basis for pregnancy testing?
   a estrogen   **d** prolactin
   b LH   **e** progesterone
   c hCG

Matching

   a testosterone   **d** FSH
   b androgen-binding protein   **e** LH
   c GnRH

29.11 _____ hormone secreted by the sustentacular cells in the seminiferous tubules

29.12 _____ secreted by the sustentacular cells; maintains a high level of testosterone in the seminiferous tubules

29.13 _____ pituitary hormone that stimulates testosterone secretion in males

29.14 _____ pituitary hormone necessary for sperm maturation

29.15 _____ primary androgenic hormone; produced by cells that lie between the seminiferous tubules

   a GnRH   **d** estrogen
   b LH   **e** progesterone
   c FSH

29.16 _____ ovarian hormone that peaks during the luteal phase of the menstrual cycle

29.17 _____ pituitary hormone that is the primary trigger for ovulation

29.18 _____ primary gonadotropin that triggers the maturation of ovarian primordial follicles

29.19 _____ follicular hormone that has a positive feedback with LH just prior to ovulation

29.20 _____ causes the rise of FSH and LH during the follicular phase of the ovarian cycle

a seminiferous tubules   d ejaculatory ducts
b epididymis   e urethra
c ductus deferentia

29.21 _____ consists of prostatic, membranous, and penile regions

29.22 _____ site of spermatogenesis

29.23 _____ highly coiled sperm duct that lies directly on the testis

29.24 _____ paired ducts that pass through the prostate gland and unite to form the unpaired urethra

29.25 _____ ducts that enter the lower part of the abdominal wall by way of the inguinal canal

a endometrium
b myometrium
c infundibulum
d fundus
e fimbriae

29.26 _____ fringed extensions of the infundibulum that aid the passage of the ovum into the oviduct

29.27 _____ one portion of this layer is shed during menstruation

29.28 _____ funnel-shaped region of the oviduct that is nearest the ovary

29.29 _____ thick, muscular uterine layer

29.30 _____ expanded upper portion of the uterus

Completion Exercises

29.31 Tyson's glands produce _____.

29.32 Circumcision is the removal of the _____.

29.33 The uterus is attached to the lateral wall of the pelvis by two _____.

29.34 The hormone _____ prevents the corpus luteum from degenerating.

29.35 Completion of meiosis by a secondary spermatocyte results in the production of two _____.

29.36 The meiotic division of a primary oocyte results in one _____ and one _____.

29.37 Completion of meiosis by a secondary oocyte results in one _____ and one _____.

29.38 The transparent outer region surrounding an ovum is called the _____.

29.39 Liberated from the head of the sperm, the enzyme _____ creates a path for the sperm to the plasma membrane of the ovum.

29.40 The diploid cell formed by the union of sperm and ovum is called the _____.

A SECOND LOOK

1 In the following drawing, label the ductus deferens, epididymis, prostate gland, bulbourethral gland, spermatic cord, and corpus cavernosum.

2 In the following drawing, label the uterine cervix, vagina, uterine tube, ovary, mons pubis, labium minus, and clitoris.

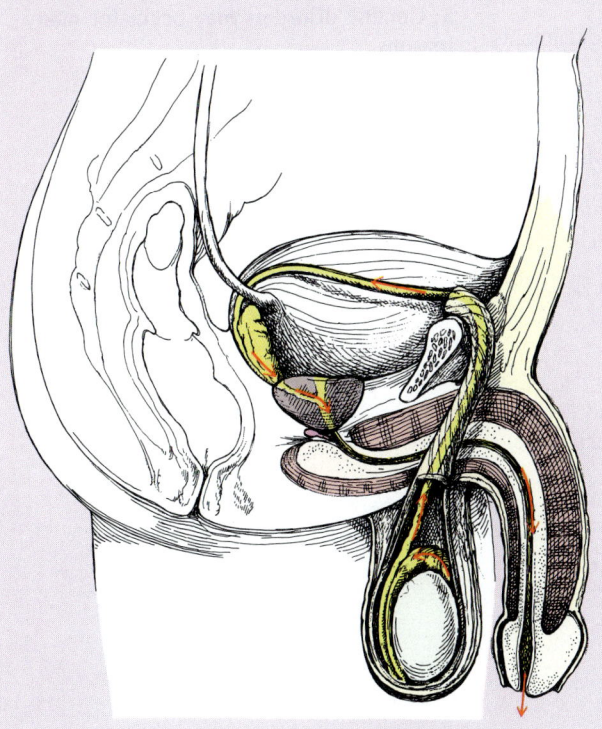

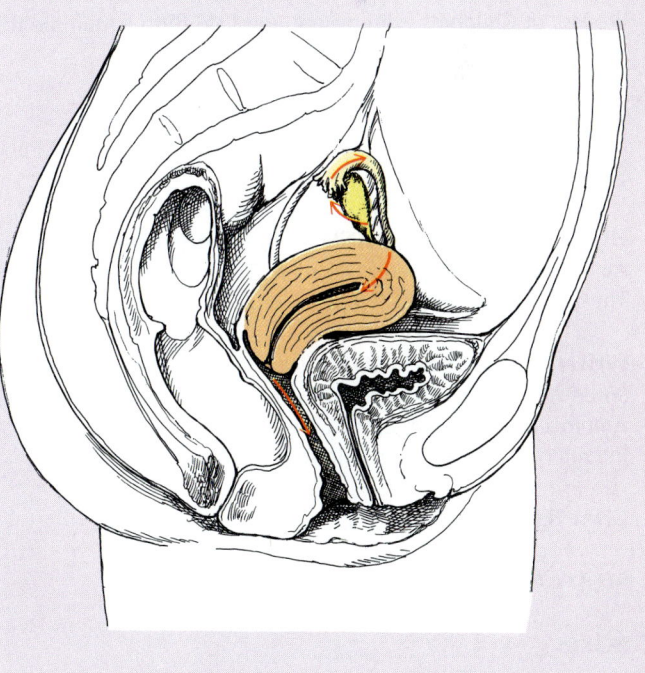

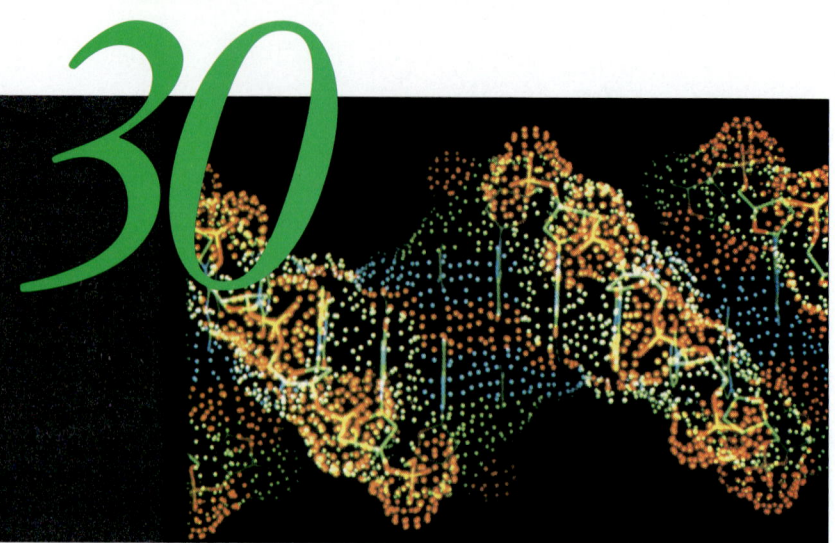

Human Growth and Development

30

KEY CONCEPTS

1 The nuclei of the male and female sex cells unite to form the nucleus of the first cell of a developing individual.

2 The developing individual begins to undergo growth through cell division even before it becomes implanted in the uterus.

3 After the first week, the cells of the developing individual begin to become specialized.

4 The full-term fetus is ready to be born about 266 days after fertilization.

5 Genetic disorders may occur for many reasons.

The adult human body consists of about 50 trillion cells, most of which are specialized in terms of structure and function. In fact, cells are so specialized *structurally* that the liver cells of a human and a horse are more alike than are human liver cells and other types of human cells. Cells and organs are also programmed to carry out specific *functions*. For instance, *all* vertebrate livers do basically the same things. Clearly, structure and function are closely related.

The development of a human being may be divided into **prenatal** ("before birth") and **postnatal** ("after birth") periods. During the prenatal period, the developing individual is called a *zygote* at fertilization, an *embryo* for the next 8 weeks, and a *fetus* from 9 weeks until birth. It becomes an *infant* at birth.

The time of prenatal activity is called the **gestation period** (L. *gestare*, to carry). For as long as the new individual lives, it will never again experience such a dramatic burst of growth and development at it does during this period. During the 9-month gestation period, the weight increases from the fertilized ovum to the newborn baby about 6 *billion* times. In contrast, the body weight increases only about 20 times from birth to age 20. During the first month of prenatal development, the embryo increases in weight about a million percent. From that time on, the rate of growth continues to decrease, though there is a growth spurt during the third month of intrauterine life.

Prenatal development is usually divided into two periods: the **embryonic period,** from fertilization to the end of the eighth week, and the **fetal period,** from the beginning of the ninth week until birth.

EMBRYONIC DEVELOPMENT

When a sperm penetrates a secondary oocyte, it sheds its midpiece and tail and its head is taken into the oocyte's cytoplasm. Meiosis in the oocyte, which had stopped at metaphase of the second division, is resumed after the sperm enters. When meiosis is complete, the second polar body is pinched off; the remaining haploid structure is the *ovum*. The ovum and sperm nuclei then fuse to form the diploid zygote nucleus. This nucleus contains all the genetic material, DNA, which subsequent mitotic cell divisions will distribute equally to all cells of the embryo. A **zygote** is the single cell resulting from the fertilization of an ovum by a sperm.*

---

*The DNA in the nucleus of a zygote is derived from the nuclei of both the sperm and ovum. In contrast, DNA in the zygote's mitochondria is derived solely from the ovum.

Cleavage

After a sperm penetrates an ovum, the ovum immediately develops an impenetrable coat that prevents any other sperm from entering. Fertilization is complete when the haploid nuclei of the sperm and ovum (each called a *pronucleus*) fuse to form a diploid zygote with a complete set of 23 pairs of chromosomes. Then a series of mitotic cell divisions called **cleavage** begins [FIGURE 30.1]. From the beginning of cleavage until the end of the eighth week, the developing individual is called an **embryo.**

Cleavage involves the division of each existing cell in the embryo into two daughter cells. The first cleavage is complete about 36 hours after fertilization, and subsequent cleavages take place about twice a day. The two-celled embryo that results from the first cleavage is approximately 0.1 mm in diameter, still only about the size of the period at the end of this sentence. The daughter cells are called **blastomeres** (Gr. *blastos*, bud + *meros*, a part). The cells do not grow as they divide because they divide so rapidly that there is no growth period between divisions. As each blastomere divides, it forms two cells that are each half the size of the original cell. As a result, the overall size of the embryo stays approximately the same as that of the original zygote.

The first six to eight cleavages occur while the embryo is still enclosed within the zona pellucida. About day 3 after fertilization, a solid ball of about 8 to 50 blastomeres is formed into a mulberry-shaped **morula** (MORE-uh-luh; L. *morum*, mulberry tree) [FIGURE 30.1E]. By this time, the morula is completing its 10-cm (4-in.) journey along the uterine tube and is approaching the entrance to the uterus.

About day 4 to 5 after fertilization, the morula develops into a fluid-filled hollow sphere called a **blastocyst** (Gr. *blasto*, germ + *kystis*, bladder), with an inner cavity called the **blastocoel** (BLASS-toh-seel; Gr. *koilos*, hollow) [FIGURE 30.1F and G]. The distinguishing feature of the blastocyst is that at one pole it has differentiated into an **inner cell mass** from which the embryo will form, and a surrounding epithelial layer called the **trophectoderm** (outer cell mass or trophoblast), which is composed of *trophoblast cells*. The trophectoderm will later develop

Q: *Why doesn't the mother's body reject the implanting blastocyst the way it would reject any other foreign tissue?*

A: The National Institute of Child Health and Human Development has studied this phenomenon in rabbits. The findings indicate that two proteins may be responsible for preventing rejection. One protein, uteroglobin (which has not yet been found in human cells), folds over the foreign antigens on the surface of the embryo and effectively masks them from the mother's immune system. The other protein, transglutaminase (which *is* present in human cells), is a blood-clotting factor that encourages the masking action of uteroglobin. A similar mechanism may exist in human beings.

FIGURE 30.1 CLEAVAGE

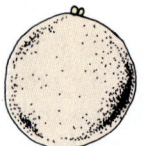 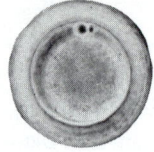

[A] ZYGOTE BEFORE CLEAVAGE

[A] The single-celled zygote still has two polar bodies attached to it. Successive divisions result in two [B], four [C], eight [D] cells, and so on. Continued divisions and rearrangement form a morula [E] and then a blastocyst [F, G]. The cells present at the end of cleavage are much smaller than the zygote, which is a large cell. SEMs ×200.

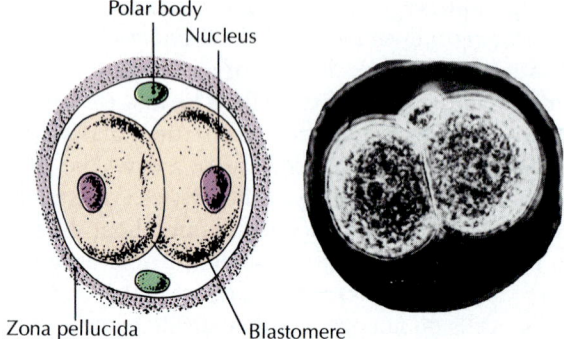

Polar body
Nucleus
Zona pellucida
Blastomere

[B] 36 HOURS AFTER FERTILIZATION
FIRST DIVISION, 2 CELLS

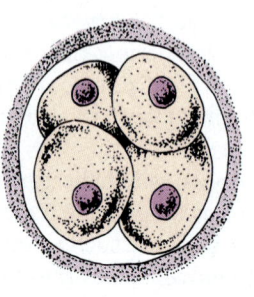 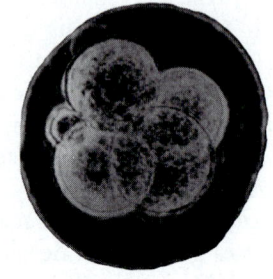

[C] SECOND DIVISION, 4 CELLS

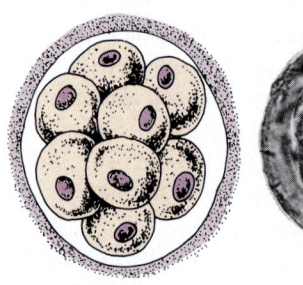 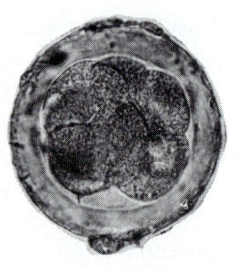

[D] THIRD DIVISION, 8 CELLS

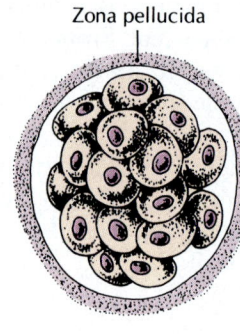 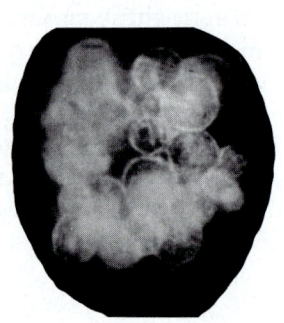

Zona pellucida

[E] 3 TO 5 DAYS AFTER FERTILIZATION
MORULA

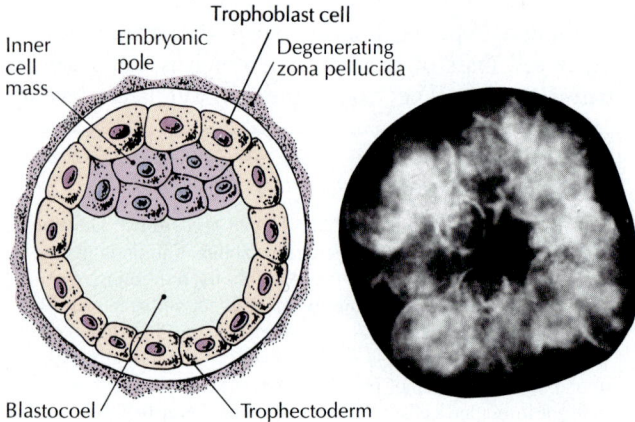

Inner cell mass
Embryonic pole
Trophoblast cell
Degenerating zona pellucida
Blastocoel
Trophectoderm

[F] 5 TO 6 DAYS AFTER FERTILIZATION
BLASTOCYST (EARLY)

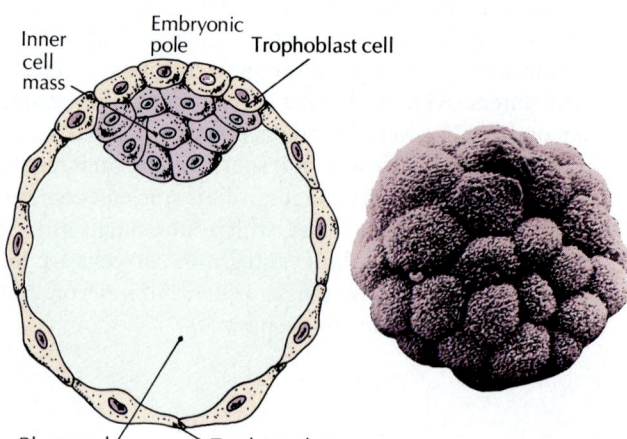

Inner cell mass
Embryonic pole
Trophoblast cell
Blastocoel
Trophectoderm

[G] BLASTOCYST (LATE)

into the *chorion* (a fetal membrane that develops from trophoblast cells) and become part of the membrane system that transports nutrients to the embryo and removes wastes from it.

The blastocyst floats freely in the uterine cavity for 1 or 2 days while it continues to divide and grow. Now the blastocyst is ready to shed its zona pellucida, allowing it to become attached to the maternal uterus. It will soon become embedded in the uterine wall in the process of *implantation.*

Implantation in the Uterus

About 6 or 7 days after fertilization, implantation begins as the blastocyst attaches to the endometrium [FIGURE 30.2A]. The inner cell mass faces toward the epithelium, while the trophectoderm attaches to the epithelium, usually to the superior, posterior wall of the uterus. Within 24 hours after the blastocyst becomes attached to the uterine wall, the trophoblast differentiates into the *cytotrophoblast,* which surrounds the inner cell mass, and the *syncytiotrophoblast,* a large multinucleated extension. The syncytiotrophoblast implants the blastocyst in the uterine wall by invading the nutritious *inner* portion of the endometrium [FIGURE 30.2B]. On about day 9, the blastocyst is completely enclosed by endometrial cells [FIGURE 30.2C], and it can continue to grow and develop. This complete implantation is the point at which **clinical pregnancy** is considered to begin. FIGURE 30.3 shows an overall view of the processes of fertilization, cleavage, and implantation.

Maintenance of the endometrium Before implantation can begin, the trophoblast must secrete the hormone **human chorionic gonadotropin (hCG),** which maintains the corpus luteum at a time when it would otherwise begin to degenerate. The corpus luteum continues its secretion of estrogen and progesterone, preventing menstruation and the breakdown of the endometrium, and allowing implantation to take place.

Eventually, the cytotrophoblast and syncytiotrophoblast will form the embryonic portion of the *placenta* [see FIGURE 30.5], the structure that transfers nutrients from the pregnant woman to the fetus. In the meantime, the embryo receives nutrients from the many broken capillaries in the uterine wall.

As implantation continues during the second week, the inner cell mass changes shape and the blastocyst assumes a flattened disk shape to form the **bilaminar** (two-layered) **embryonic disk.** These two layers are composed of endoderm and ectoderm. As the bilaminar disk is forming, several supporting membranes and other structures also develop. These are the amnion, yolk sac, body stalk, chorion, and allantois, each of which encloses a cavity; they will be described in the next section.

FIGURE 30.2 IMPLANTATION

[A] About day 6 after fertilization, trophoblast cells of the developing blastocyst attach to endometrial cells of the uterine wall. [B] About day 7, implantation proper begins as the syncytiotrophoblast burrows into the endometrial epithelium and the blastocyst begins to move into the uterine wall. Hypoblasts of the inner cell mass will eventually differentiate into fetal tissues. [C] Implantation is almost complete about day 9 as the endometrial epithelium grows over the implanted blastocyst. The amnion, amniotic cavity, primitive yolk sac, bilaminar embryonic disk, and extraembryonic mesoderm are already present.

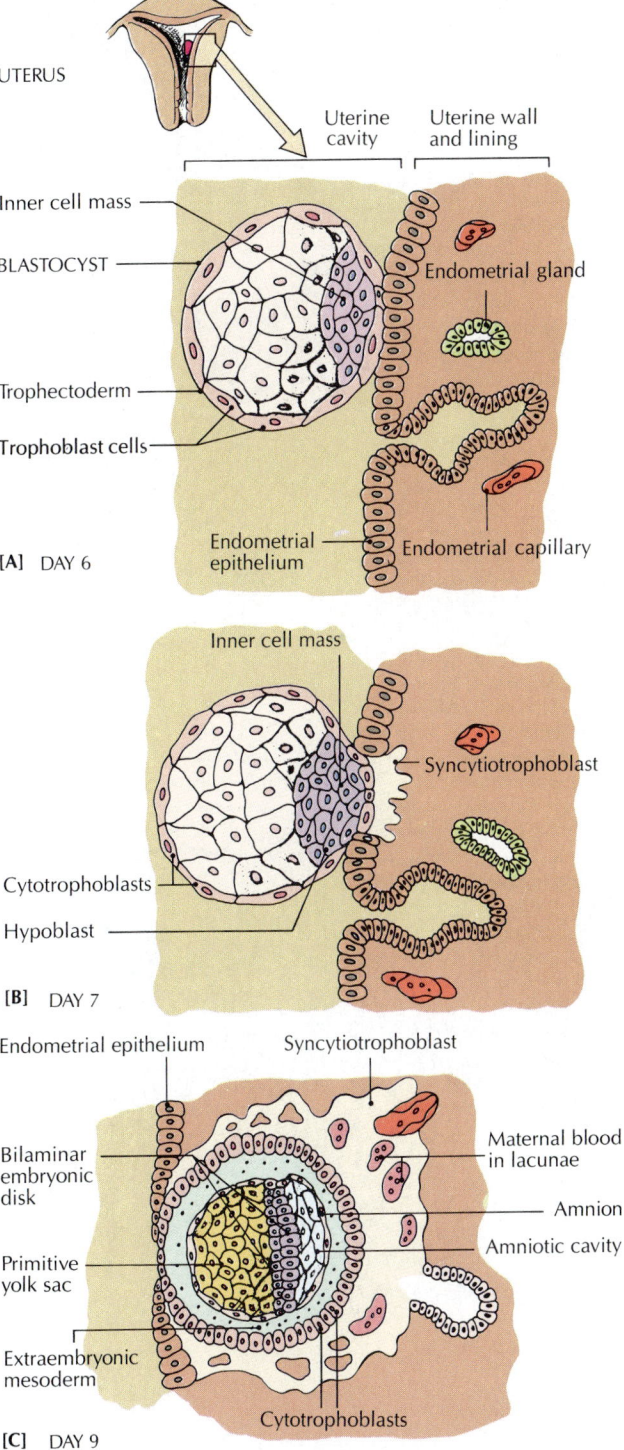

FIGURE 30.3 FERTILIZATION, CLEAVAGE, AND IMPLANTATION

[A] Development from ovulation to implantation: (1) A secondary oocyte is released (ovulation) from the ovary when a mature follicle ruptures. The follicle then becomes a corpus luteum. (2) The secondary oocyte is swept into the uterine tube. (3) Meiosis produces the haploid chromosome number, and the first polar body is formed. The corona radiata consists of follicle cells surrounding the mature oocyte. (4) A sperm penetrates the oocyte. The second meiotic division occurs, and the second polar body is formed. (5) The male and female nuclei fuse, forming the zygote. (6) The

zygote begins cleavage. (7, 8) Cleavage results in the formation of a morula. (9) An early blastocyst is formed. (10) About 4 days after fertilization, a late blastocyst is formed. (11–13) About a week after fertilization, the blastocyst begins the process of implantation into the uterine wall. (14) Once implantation has been accomplished, the uterine wall starts contributing the outer portion of the placenta; the embryo itself contributes the inner part. [B] Photograph of an implanted blastocyst 6 or 8 days after fertilization. ×20.

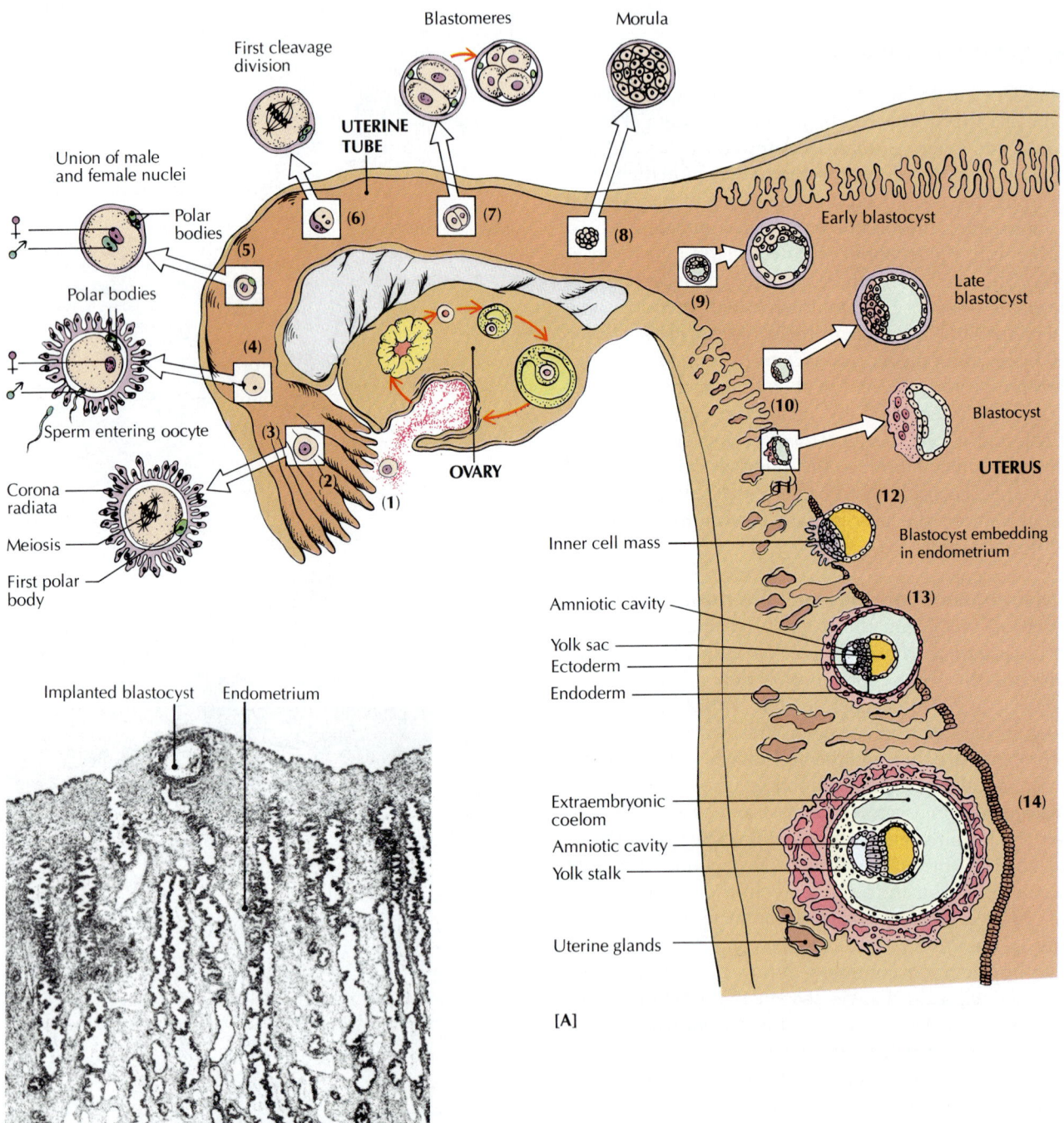

TABLE 30.1 MAJOR STRUCTURES DERIVED FROM PRIMARY GERM LAYERS

Endoderm	Mesoderm	Ectoderm
Epithelium of Pharynx, larynx, trachea, lungs	Most muscle tissue (cardiac, smooth, skeletal), except in iris and sweat glands	All nervous tissue
Tonsils, adenoids	All connective tissue (fibrous, adipose, cartilage, bone, bone marrow)	Epidermis of skin, hair follicles, nails
Part of thyroid; thymus; parathyroids	Synovial and serous membranes	Epithelium and myoepithelial cells of sweat glands, oil glands, mammary glands
Esophagus, stomach, intestines, liver, pancreas, gallbladder, glands of alimentary canal (except salivary)	Lymphoid tissue Tonsils, lymph nodes	Lens of the eye
Urinary bladder, urethra (except terminal male portion), prostate gland, bulbourethral glands	Spleen	Receptor cells of sense organs
Vagina (partial)	Blood cells	Enamel of teeth
Auditory tubes	Reticuloendothelial system	Adrenal medulla
	Dermis of skin	Anterior pituitary
	Teeth (except enamel)	Parafollicular cells of thyroid; melanocytes

	Endothelium of heart, blood vessels, lymphatic vessels	Epithelium of
		  Salivary glands, lips, cheeks, gums, hard palate
		  Nasal cavity, sinuses

	Epithelium of
	  Gonads, reproductive ducts
	  Adrenal cortex
	  Kidneys, ureters
	  Coelom, joint cavities
		Lower third of anal canal

		Terminal portion of male urethra
		Vestibule of vagina and vestibular glands
		Muscle of iris
		Inner ear

Cell division stops briefly after implantation in the uterus. When cell division resumes, *cell differentiation* begins, and the inner cell mass of the blastocyst is rearranged into the three primary germ layers. (See FIGURE 4.1 and TABLE 30.1 for a review of the formation and development of the primary germ layers, first discussed in Chapter 4.)

Extraembryonic Membranes

During the first 2 weeks, the embryo does not have a functional circulatory system. During the implantation process, enzymes released by the trophoblast destroy some maternal capillaries in the endometrium. Blood from these capillaries comes into direct contact with the trophoblasts, providing a temporary source of nutrition. Four membranes begin to form from tissues that were once the trophoblast. Because these membranes are not part of the embryonic body proper, they are called *extraembryonic*, or *fetal, membranes* [FIGURE 30.4]. They include the yolk sac, amnion, allantois, and chorion.

As the trophoblast grows, it branches and extends into the tissue of the uterus. The two kinds of tissue, embryonic and maternal, grow until there is sufficient surface contact to ensure the adequate passage of nourishment and oxygen from the mother, as well as the removal of metabolic wastes, including carbon dioxide, from the embryo. After the embryo is implanted into the endome-trium, small *chorionic villi* grow outward into the maternal tissue from the *chorion,* the protective sac around the embryo. The chorion makes up most of the embryonic side of the placenta, while the endometrium makes up the maternal component. It contains many villi, which allow the exchange of nutrients, gases, and metabolic wastes between the maternal blood and that of the embryo. The fully differentiated villi of the embryo and then the fetus remain to make contact with the maternal endometrium and later become the mature placenta.

The *yolk sac* is a primitive respiratory and digestive structure that is reduced by the sixth week of development to the thin *yolk stalk,* which then becomes incorporated into the umbilical cord via the body stalk. The yolk sac appears to be involved in transporting nutrients to the embryo during the second and third weeks, before the placental transfer is fully developed. It also is the initial site for the formation of blood cells until the liver assumes that responsibility during the fifth week. Finally, *primordial germ cells* in the yolk sac eventually migrate to the developing gonads, where they become spermatogonia or oogonia.

The *amnion* (Gr. sac) is a tough, thin transparent membrane that envelops the embryo. Its interior space, the *amniotic cavity,* becomes filled with a watery *amniotic fluid,* which contains shed fetal epithelial cells, protein, carbohydrates, fats, enzymes, hormones, and pigments. Later, it also contains fetal excretions. The

FIGURE 30.4 DEVELOPMENT OF EXTRAEMBRYONIC MEMBRANES

The extraembryonic membranes are the chorion, yolk sac, and amnion. **[A]** Week 3. **[B]** Week 4. **[C]** Week 10. **[D]** Week 20.

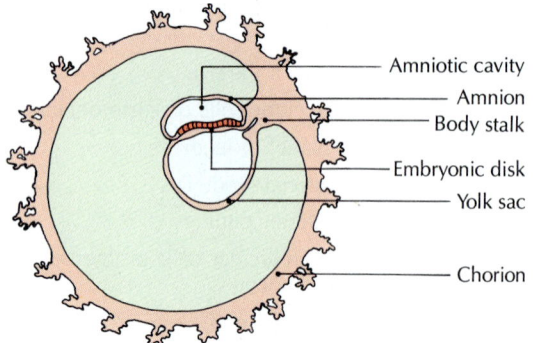

Amniotic cavity
Amnion
Body stalk
Embryonic disk
Yolk sac
Chorion

[A] WEEK 3

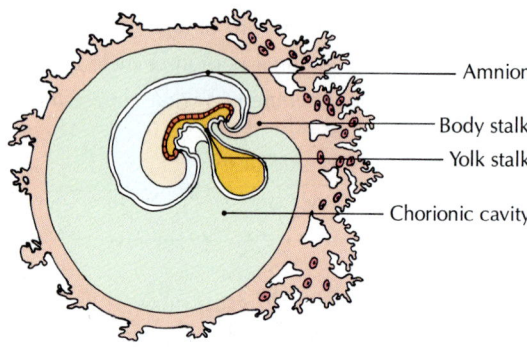

Amnion
Body stalk
Yolk stalk
Chorionic cavity

[B] WEEK 4

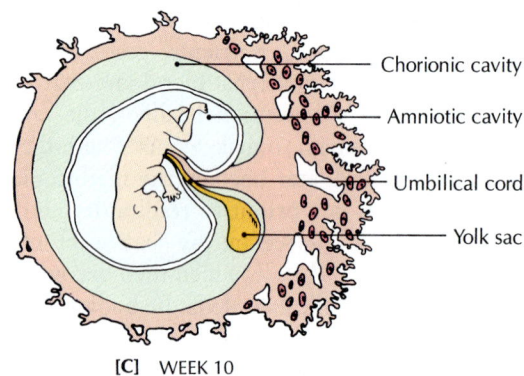

Chorionic cavity
Amniotic cavity
Umbilical cord
Yolk sac

[C] WEEK 10

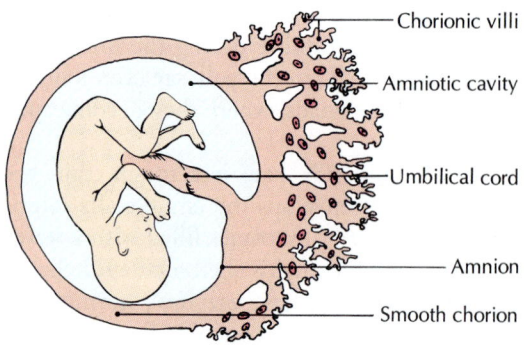

Chorionic villi
Amniotic cavity
Umbilical cord
Amnion
Smooth chorion

[D] WEEK 20

amniotic fluid suspends the embryo in a relatively shock-free environment, prevents the embryo from adhering to the amnion and producing malformations, permits the developing fetus to move freely (which aids the proper development of the muscular and skeletal systems), and provides a relatively stable temperature for the embryo.

The **allantois** (uh-LAN-toh-ihz; Gr. "sausage") usually appears early in the third week as a small fingerlike outpocket of the hind gut. It contributes to the urinary bladder, and its blood vessels become the vein and arteries of the umbilical cord. The allantois is eventually transformed into the median umbilical ligament, which extends from the urinary bladder to the navel.

Placenta and Umbilical Cord

The essential exchange mechanism between mother and embryo is in place by the beginning of the fourth week. At the site of implantation, the chorion joins intimately and intricately with the endometrium to develop into the **placenta** (L. flat cake, because of its disklike shape) [FIGURE 30.5]. The placenta grows rapidly until the fifth month, when it is almost fully developed. A full-term placenta is about 2.5 cm (1 in.) thick and 20.5 cm (8 in.) in diameter. It weighs about 0.45 kg (1 lb dry weight).

The side of the placenta facing the fetus is relatively smooth, with the umbilical cord usually attached somewhere near the center. The side of the placenta that faces the mother has grooves and protuberances. The irregular surface increases the area for the interchange between the fetal and maternal circulations. The surface area of the rough side is about 13 sq m (140 sq ft), more than three times greater than that of the smooth side. The connection between the fetal part of the placenta and the maternal part is close, with thousands of chorionic villi on the fetal part embedded in the maternal part, adding to the contact surface enormously. The fetal capillaries come close to the maternal capillaries to effect the exchange of substances.

The placenta has three main functions:

1 It transports materials between the mother and the embryo and fetus. The transported materials include *gases* (such as oxygen, carbon dioxide), *nutrients* (such as water, vitamins, glucose), *hormones* (especially steroids such as testosterone), *antibodies* (which bestow passive immunity), *wastes* (such as carbon dioxide, urea, uric acid, bilirubin), *drugs* (most drugs pass easily, especially alcohol), and *infectious agents* (such as rubella, measles, encephalitis, poliomyelitis, and the virus that causes AIDS).

2 It synthesizes glycogen and fatty acids, and probably contributes nutrients and energy to the embryo and fetus, especially during the early stages of pregnancy.

3 It produces and secretes hormones, especially the *protein hormones* human chorionic gonadotropin (hCG)

FIGURE 30.5 DEVELOPMENT OF THE PLACENTA

[A] The blastocyst is implanted in the uterine wall. [B] Enlarged view showing the trophoblasts and the penetration of chorionic villi into the endometrium. Arrows indicate developing blood vessels. [C] The trophectoderm differentiates into two layers, and chorionic villi reach the maternal blood supply. [D] The yolk stalk of the developing embryo will eventually become the umbilical cord. [E] At about 5 weeks, the umbilical blood vessels reach the embryo. [F] At about 8 weeks, the chorion and amnion form, and the placenta and umbilical cord are highly developed. [G] Enlargement of the placenta and umbilical cord showing the interrelationships of fetal and maternal tissues.

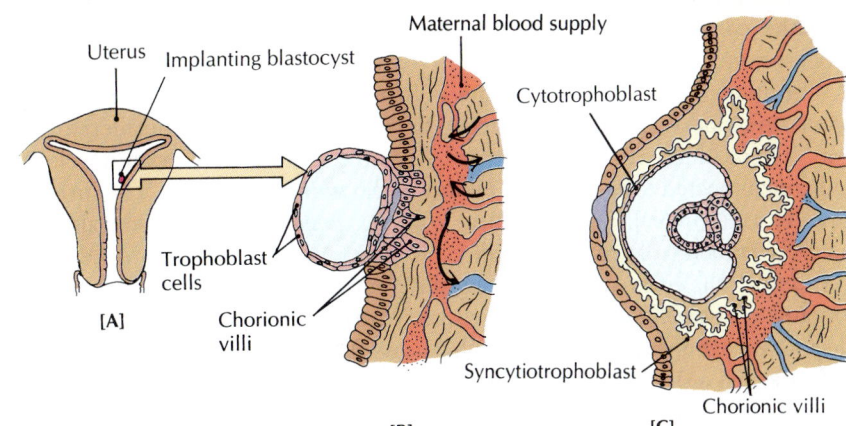

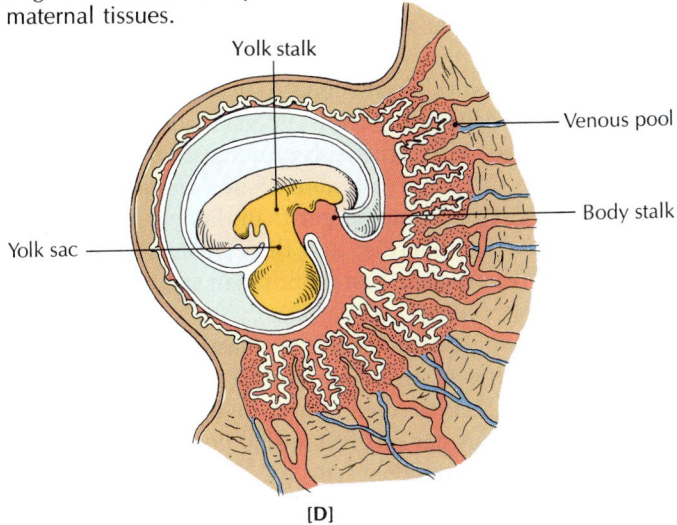

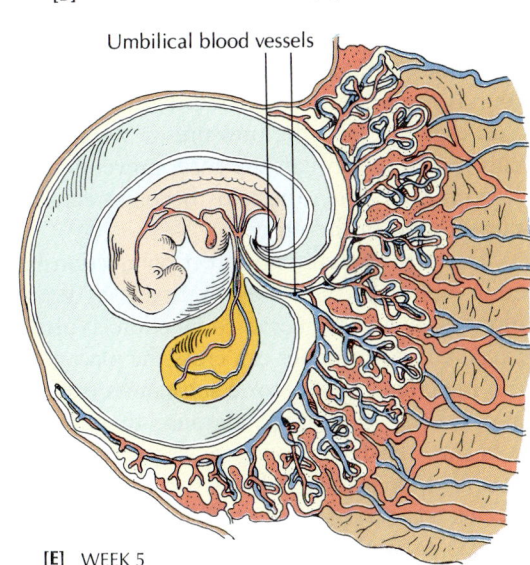

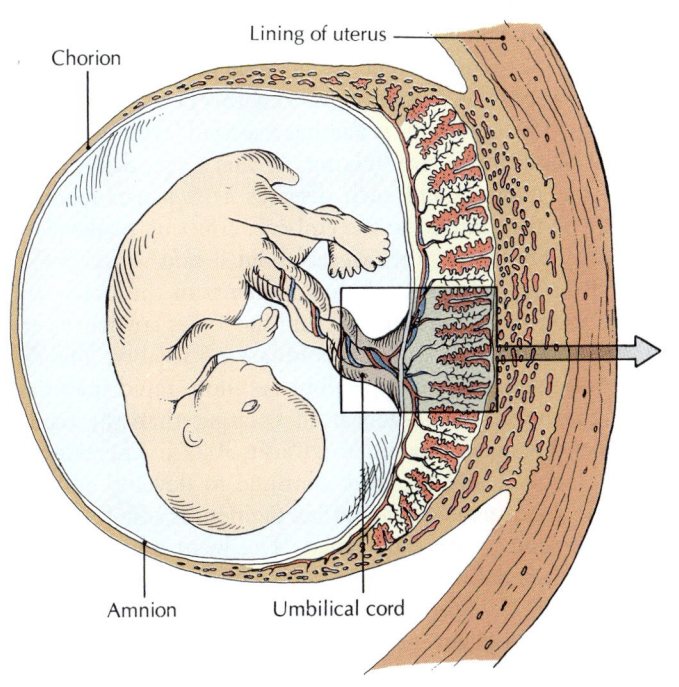

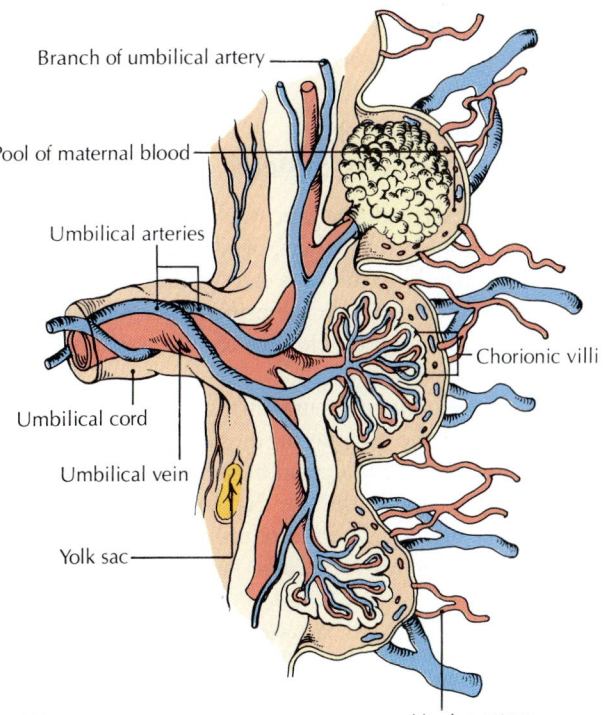

[F] WEEK 8

[G]

and human chorionic somatomammotropin (hCS) and, with the cooperation of the fetus, the *steroid hormones* progesterone and estrogen.

The inner lining of the placenta is made up of the extensive blood vessels and connective tissue of the chorion. These blood vessels are formed from the embryo and are connected to the embryo by way of the **umbilical cord** [FIGURE 30.5G]. The umbilical cord (L. *umbilicus*, navel) is formed from the body stalk, yolk stalk, and chorion during the fifth week. The fully differentiated cord contains two *umbilical arteries*, which carry carbon dioxide and nitrogen wastes from the embryo and fetus to the placenta, and a single *umbilical vein*, which carries oxygen and nutrients from the placenta to the embryo and fetus. (Two embryonic umbilical veins developed; the right vein disappears, while this left umbilical vein persists.) A gelatinous cushion of embryonic connective tissue surrounds the vessels of the umbilical cord. This resilient pad, together with the pressure of blood and other liquids pulsating through the cord, prevents the cord from twisting shut when the fetus becomes active enough to turn around in the uterus.

There is normally no direct connection between the embryonic and the maternal tissue, at least no actual blood flow and no nerve connection. Nutrients, water, oxygen, and hormones can cross the placental barrier, as can infectious agents, toxic substances (such as lead and insecticides), and drugs. Because these substances can pass into the fetal blood, the fetus can be infected, be poisoned, or become addicted to drugs such as cocaine. In fact, a newborn baby can show drug withdrawal symptoms if its mother used cocaine during pregnancy. Nevertheless, the growing embryo is well insulated from most of the possibly harmful influences to which the mother is exposed. The placenta is eventually shed after the baby is born as part of the **afterbirth.***

Weeks 1 through 8: Formation of Body Systems

As already described, the first week of prenatal development is devoted to cleavage, while during the second week of prenatal life the cells start to differentiate structurally and functionally as the primary germ layers begin to form. Within the developing chorionic villi, the first blood vessels are growing.

Rapid development occurs during the third week. Differentiation of the endoderm, mesoderm, and ectoderm is complete, and the body and head can be distinguished. The primitive streak, the site that forms mesoderm, appears about day 15. The mesodermal notochord, a rod-like structure, sends molecular signals to the adjacent

ectoderm; the exact mechanism is not yet known. The resulting action, called *neurulation*, induces the ectoderm to give rise to the entire nervous system [FIGURE 30.6A]. This interaction between the notochord and ectoderm is considered to be one of the most important phenomena in all development.

The primitive body cavities and the cardiovascular system also take form. Villi continue to develop for an improved exchange between the embryo and mother. The primordial heart begins to pulsate, circulating oxygenated blood and other nutrients to the developing embryonic structures. The embryo now measures about 2.5 mm. (All measurements are from crown to rump.)

By the fourth week the embryo is C-shaped. The U-shaped heart has four chambers and pumps blood through a simple system of vessels. Upper limb buds and primitive ears are visible on about day 26. Lower limb buds and primitive eye lenses appear on about day 28. The umbilical cord begins to develop. The intestine is a simple tube, and the liver, gallbladder, pancreas, and lungs begin to form. Also forming are the eyes, nose, and brain. The embryo is about 5 mm from crown to rump, about 10,000 times larger than the fertilized ovum. The relative size increase and the extent of physical change are greater in the first month than at any other time during gestation.

During the fifth week, the head grows disproportionately large as the brain develops rapidly. The forearm is shorter than the hand plate, and finger ridges begin to form about day 33. Primordial nostrils are present, and the tail is prominent [FIGURE 30.6B]. Primordial kidneys, the upper jaw, and the stomach begin to form. The nose continues to develop. The intestine elongates into a loop, and the genital ridge bulges. Primordial blood vessels extend into the head and limbs, and the spleen is visible. Spinal nerves are formed, and cranial nerves are developing. Premuscle masses appear in the head, trunk, and limbs. The epidermis is gaining a second layer. The cerebral hemispheres are bulging. At this stage, drugs taken by the mother, and such diseases as German measles, may be transmitted to the embryo, affecting its development. The embryo measures about 8 mm.

By the sixth week, the components of the upper jaw are prominent but separate, and the lower jaw halves are fused. The limb buds differentiate noticeably, and the development of the upper limbs is more rapid than that of the lower limbs. The head is larger than the trunk, which begins to straighten [FIGURE 30.6C]. The external ears appear, and the eyes continue to develop and become accentuated as the retinal pigment is added. Simple nerve reflexes are established. The heart and lungs acquire their definitive shapes. The embryo is about 12 mm (0.5 in.).

During the seventh and eighth weeks the yolk sac is reduced to the yolk stalk. The face and neck begin to form, and the fingers and toes are differentiated. The

---

*The umbilical cord is usually disposed of after childbirth, but recently umbilical veins have been used successfully as grafts for bypass operations. The veins are removed from the cord, treated with a preservative, and reinforced with a polyester mesh before being used.

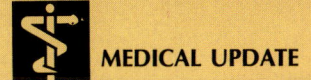

MEDICAL UPDATE

The Gender Gene

Until about the sixth week after fertilization, an embryo does not yet appear male or female. At that point, a genetic signal may occur that triggers the development of male characteristics. Without this signal, the embryo develops female characteristics. It has been known for some time that a Y-bearing chromosome from the male parent produces a male child, and an X-bearing chromosome produces a female child, but until recently the exact mechanism of gender determination was unknown. Now

David C. Page and his coworkers at the Whitehead Institute for Biomedical Research in Cambridge, Massachusetts, have proposed that sex determination depends on a single gene on the Y chromosome. The specific chromosomal segment is called **testis-determining factor (TDF)**.

Apparently TDF regulates the activity of other genes (as yet unidentified) that set in motion a series of biochemical events that lead to the development of male primary sexual characteristics and

even differences in body systems. For example, it is well known that the nervous and immune systems are not identical in males and females.

TDF was discovered when Page and his team studied abnormal men with XX sex chromosomes and abnormal women with XY sex chromosomes. In these rare cases, where a person was genetically infertile but appeared normal, the XX men had a bit of a Y chromosome that contained TDF, and the Y chromosome of the XY women lacked TDF.

---

back straightens, and the upper limbs extend over the chest [FIGURE 30.6D]. The tail is regressing, and the three segments of the upper limbs are evident. The jaws are formed and begin to ossify. The stomach is attaining its final shape, and the brain is becoming large. The muscles are differentiating rapidly throughout the body, assuming their final shapes and relations. The eyelids are beginning to form. At the end of the embryonic period, the embryo is about 17 mm (0.75 in.).

ASK YOURSELF

1 What is cleavage, and when does it occur?

2 What is a morula? A blastocyst?

3 When does implantation take place?

4 What are extraembryonic membranes?

5 What is the difference between the chorion and the amnion?

6 What are the main functions of the placenta?

7 What is the vascular structure of the umbilical cord?

8 Describe the major developmental changes during the first 8 weeks of prenatal life.

FETAL DEVELOPMENT

Typically, the embryonic heart begins to contract by the twenty-second day.* By the end of the eighth week, the major external features (ears, eyes, mouth, upper and lower limbs, fingers, toes) are formed, and the major organ systems have been established. Once this stage is

reached, the developing individual is referred to as a *fetus* (FEE-tuhss). The fetus, although only 3 cm (1.25 in.) long, looks recognizably human after 2 months [FIGURE 30.6E]. After 3 months in the uterus, the fetus is about 5 to 6 cm (2.0 to 2.4 in.) long and contains all the organ systems characteristic of an adult.

The last 6 months of pregnancy are devoted to the increase in size and maturation of the organs formed during the first 3 months. By the time the fetus is 10 cm (4 in.) long, it can move and be felt by the pregnant woman. It is thin, wrinkled, hairy, and moist. As it ages, the fetus loses most of its hair, its bones begin to ossify, it synthesizes fat, and it becomes mature enough to survive on its own. It is said to have come to *term*.

Third Lunar Month[†]

After 8 weeks the embryo is called a *fetus*. By the ninth week, its nose is flat and the eyes are far apart. The tongue muscles are well differentiated, and the earliest taste buds form. The ear canals are distinguishable. The fingers and toes are well formed, and the head is elevating [FIGURE 30.6F]. The growth of the intestines makes the body evenly round. The small intestine is coiling within the umbilical cord, and intestinal villi are developing. The liver is relatively large, and the testes or ovaries are distinguishable as such. The main blood vessels assume their final organization, and there is some bone formation. The muscles of the trunk, limbs, and head are well represented, and the fetus is capable of some movement. All five major subdivisions of the brain are formed, but the brain lacks the convolutions that are characteristic of later stages. The external, middle, and internal ears are assuming their final forms.

---

*The cardiovascular system is one of the first organ systems to differentiate. Although the nervous system begins to differentiate before the heart beats, its growth does not commence until the fourth week, since growth is dependent on the oxygen and nutrients delivered by the slowly circulating blood.

[†]Because the reproductive cycles of women are closer to the lunar (moon) month of 28 days than to the calendar month, fetal development is conventionally described in terms of lunar rather than calendar months.

FIGURE 30.6 STAGES OF EMBRYONIC AND FETAL DEVELOPMENT

The drawings show the real sizes of the embryos and fetuses. [A] Posterior view of three-layered embryo after 3 weeks of development. [B] Embryo and hand after 5 weeks (5–8 mm). [C] Embryo and hand after 6 weeks (10–14 mm). [D] Embryo and hand during seventh week (17–22 mm). [E] Embryo after 8 weeks (25–40 mm). [F] Fetus after 9 weeks (32–50 mm). [G] Fetus at about 10 weeks (43–61 mm). [H] Fetus at about 7 months.

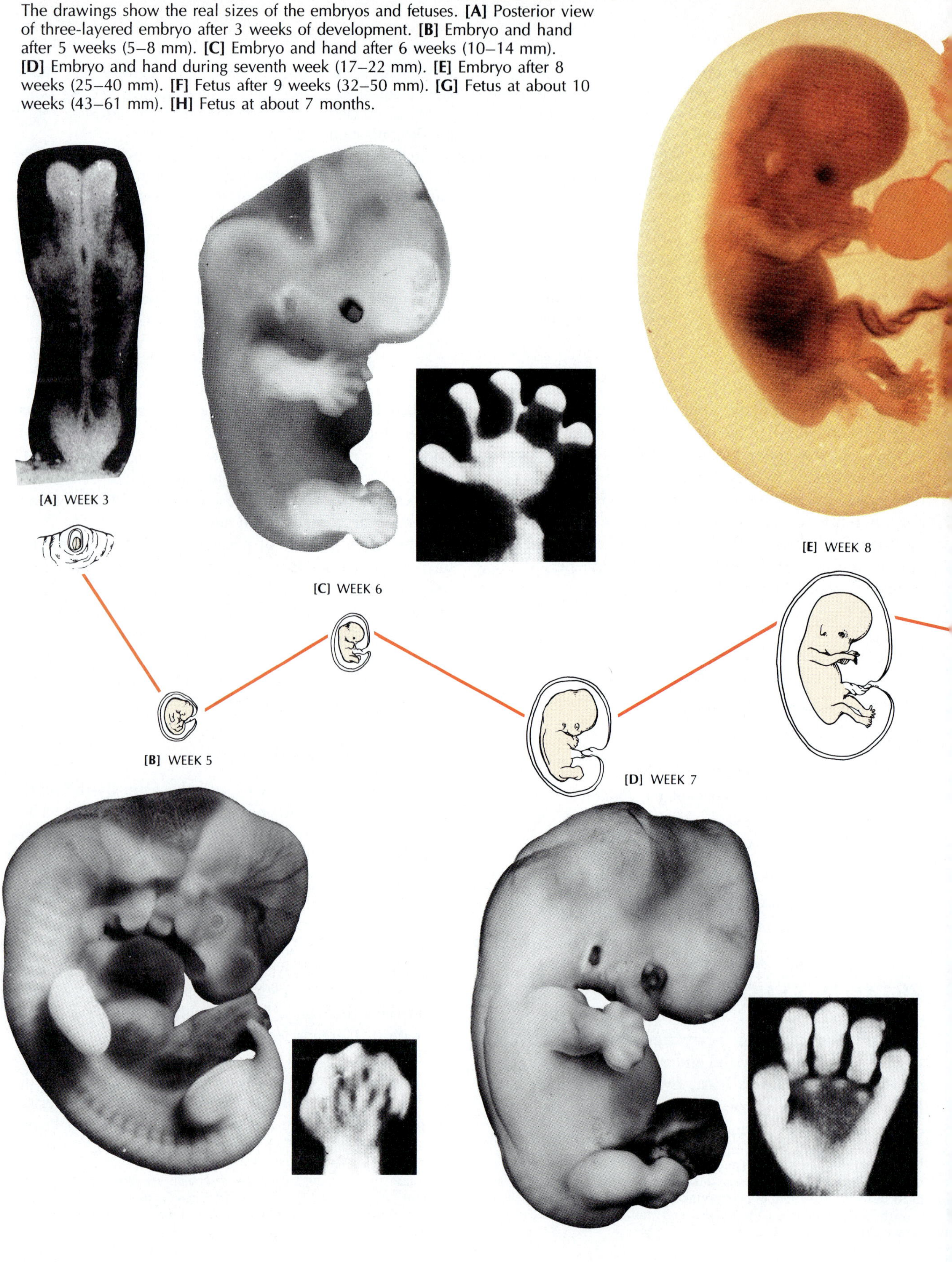

[A] WEEK 3

[C] WEEK 6

[E] WEEK 8

[B] WEEK 5

[D] WEEK 7

[F] WEEK 9

[G] WEEK 10

[H] 7 MONTHS

During early development, the genitals are not yet differentiated, and all embryos are alike. Distinct male characteristics begin to differentiate through the action of male sex hormones (androgens) at about week 6. The differentiation of sexes is completed by the third month of fetal life.

During the tenth to eleventh weeks, the head is held erect. The limbs are nicely molded [FIGURE 30.6G], and the fingernail and toenail folds are indicated. The eyelids are fused, the lips have separated from the jaws, and the nasal passages are partitioned. The intestines withdraw from the umbilical cord, and the anal canal is formed. The kidneys begin to function, the urinary bladder expands as a sac, and urination is now possible. The vaginal sac and rudimentary sex ducts are forming. Early lymph glands appear, and red blood cells predominate in the blood. The earliest hair follicles begin to develop on the face, and tear glands are budding. During this period the spinal cord attains its definitive internal structure. A pulse is now detectable with a stethoscope. The placenta is producing progesterone, which was formerly produced by the corpus luteum. The fetus is about 40 mm (1.5 in.) from crown to rump.

Fourth Lunar Month

The head is still dominant. The nose gains its bridge, tooth buds and bones form, the cheeks are represented, and the nasal glands form. The external genitalia attain their distinctive features. The lungs acquire their final shape, and in the male the prostate gland and seminal vesicles appear. Blood formation begins in the marrow, and some bones are well outlined. The epidermis is triple-layered, and the characteristic organization of the eyes is attained. The brain develops its general structural features. By week 16, all the vital organs are formed. The enlargement of the uterus can be felt by the mother, and the fetus measures about 56 mm (2.25 in.).

Fifth Lunar Month

At the beginning of the fifth lunar month the face looks human, and hair appears on the head. Muscles become active spontaneously, and the body grows faster than the head. The hard and soft palates are differentiating, and the gastric and intestinal glands begin to develop. The kidneys attain their typical shape and plan. In the male, the testes are in position for their later descent into the scrotum. In the female, the uterus and vagina are recognizable as such. Blood formation is active in the spleen, and most bones are distinctly indicated throughout the body. Stretching movements by the fetus are now felt by the mother. The epidermis begins adding other layers to form the skin. Body hair starts developing, and sweat glands appear. The general sense organs begin to differentiate. The crown-to-rump measurement is about 112 mm (4.5 in.).

At the end of the fifth lunar month, the body is covered with downy hair called *lanugo* (luh-NEW-go; L. *lana*, wool). The nasal bones begin to harden. In the female, the vaginal passageway begins to develop. Until birth, blood formation continues to increase in the bone marrow. The heart beats at twice the adult rate. The lungs are formed but do not function. The gripping reflex of the hand begins to develop, and kicking movements and hiccupping may be felt by the mother. The fetus is about 160 mm (6.5 in.).

Sixth Lunar Month

The body is now lean and better proportioned than before. The internal organs occupy their normal positions, and the large intestine becomes recognizable. The nostrils open. The cerebral cortex now gains its typical layers. Thumb sucking may begin.

Seventh Lunar Month

The fetus is lean, wrinkled, and red [FIGURE 30.6H]. The eyelids open, and eyebrows and eyelashes form. Lanugo is prominent. The scrotal sac develops, and the testes begin to descend, concluding their descent in the ninth month. The nervous system is developed enough so that the fetus practices controlled breathing and swallowing movements. The brain enlarges, and cerebral fissures and convolutions are appearing rapidly. The retinal layers of the eyes are completed, so light can be perceived. If delivered at this stage, the baby would have at least a 10-percent chance of survival.

Eighth Lunar Month

The testes settle into the scrotum. Fat is collecting so that wrinkles are smoothing out, and the body is rounding. A sense of taste is present. The weight increase slows down. If delivered at this stage, the baby has about a 70-percent chance of survival.

Ninth Lunar Month

The nails reach the tips of the fingers and toes. Additional fat accumulates.

Tenth Lunar Month

Pulmonary airway branching is only two-thirds complete, and the lungs do not function until birth. Some fetal blood passages are discontinued. The lanugo is shed. The digestive tract is still immature. The lack of space in the mother's uterus causes a decrease in fetal activity. Usually, the fetus turns to a head-down position [FIGURE 30.7]. The maternal blood supplies antibodies.

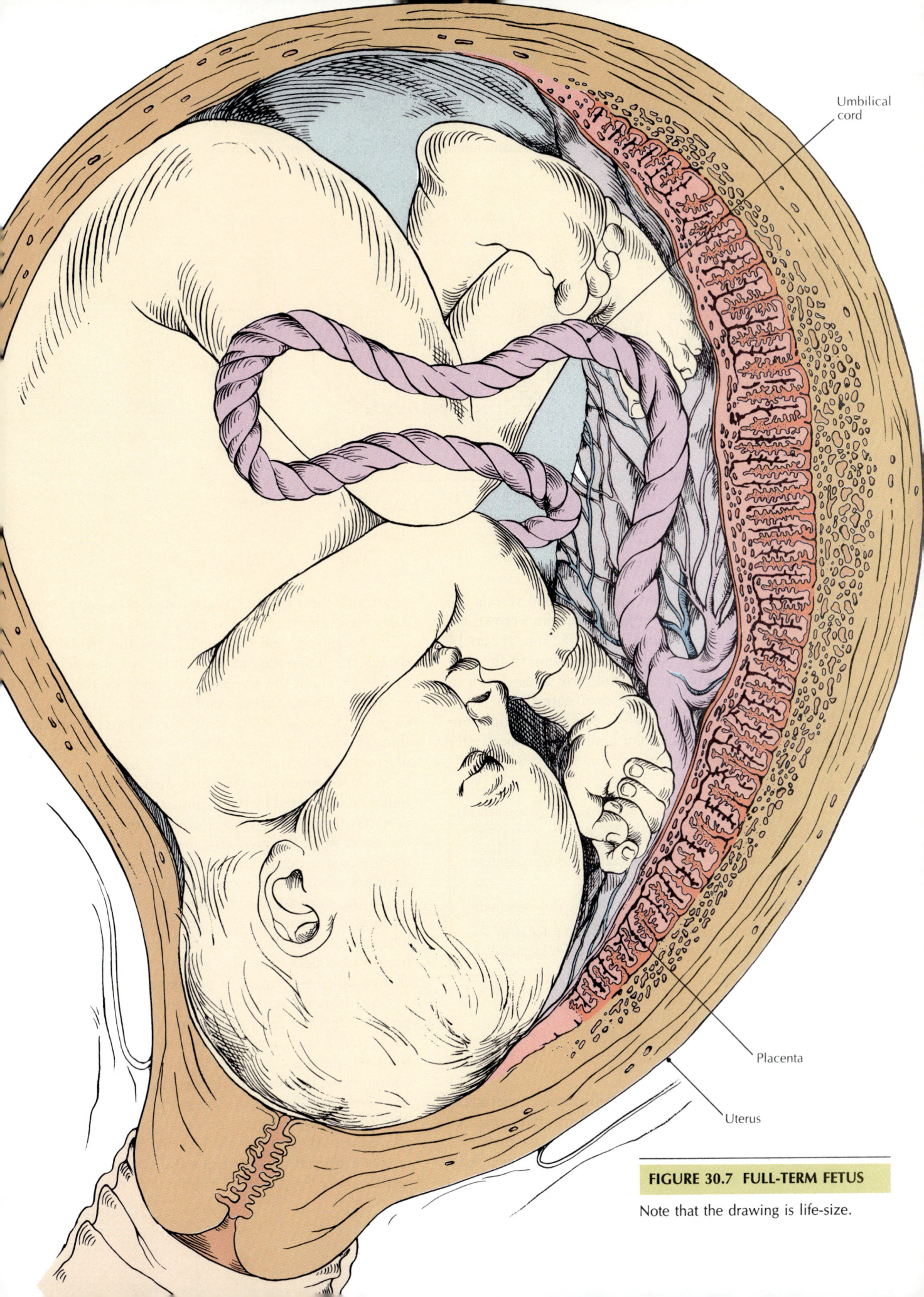

Umbilical cord

Placenta

Uterus

FIGURE 30.7 FULL-TERM FETUS

Note that the drawing is life-size.

FIGURE 30.8 CHANGES DURING PREGNANCY

[A] Changes during pregnancy. [B] The uterus usually returns to its nonpregnant size about 6 weeks after birth.

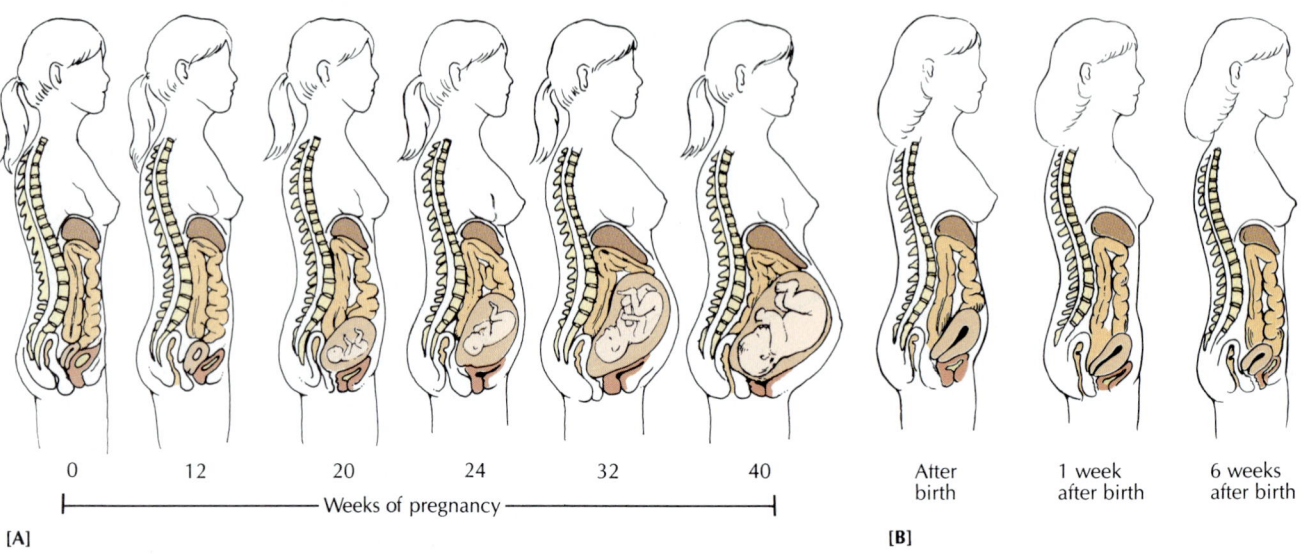

| | | | | | | After birth | 1 week after birth | 6 weeks after birth |

0   12   20   24   32   40

|—————— Weeks of pregnancy ——————|

[A]   [B]

The placenta regresses, and placental blood vessels degenerate. The baby is now at **full term,** with a crown-to-rump measurement of about 300 to 350 mm (12 to 14 in.), and is ready to be born. (Measurements in the final months can vary greatly. The full-term fetus shown in FIGURE 30.7 is slightly smaller than normal but is not premature.) The newborn baby's eyes react to light but will not assume their final color until about a month after birth. In developed countries about 99 percent of full-term infants survive.

A S K   Y O U R S E L F

1 At what point is the developing individual considered to be a fetus?

2 When can the sex of the fetus first be distinguished?

3 At what point is the development of the hand so nearly completed that thumb sucking often begins?

MATERNAL EVENTS OF PREGNANCY

The first sign of pregnancy is usually a missed menstrual period about 3 weeks after sexual intercourse that resulted in conception. Because irregular menstruation may not be a drastic change for some women, the missed period is not a foolproof diagnostic tool.

During the fifth or sixth week of pregnancy, some superficial veins may become prominent, the breasts may begin to enlarge and feel tender, and the areolae may darken. The temporary nausea called "morning sickness" may begin about the seventh week, and "stretch marks" on the breasts may appear as the breasts continue to enlarge. At 8 weeks, a physician can diagnose pregnancy through a physical examination by detecting an enlarged uterus and a softened cervix (Hagar's sign).

At the beginning of the ninth week *colostrum* (a thin, milky fluid secreted by the mammary glands before milk is released) may be squeezed from the breasts, and during the tenth week the physician may be able to feel weak uterine contractions. The nausea that may have started during week 7 usually stops during week 12.

During week 16, the abdomen usually begins to protrude as the uterus and fetus grow larger [FIGURE 30.8]. The genitals may darken and turn bluish, and the forehead and cheeks may also darken in a pigmentation change called the "mask of pregnancy."

Some time between weeks 17 and 20, the mother will probably feel the fetus moving in the uterus. By about week 26 the movements of the fetus may be noticeable from the outside. By week 30 the abdomen protrudes considerably, the breasts are considerably enlarged, and an overall weight gain of almost 9 kg (20 lb) is typical. The cardiac output is now increased, and the heart rate is accelerated. Many other changes are obvious, too. The pregnant woman tends to tire more easily. The need to urinate more frequently is common as the developing fetus puts more and more pressure on the woman's urinary bladder. Varicose veins or hemorrhoids may develop as venous pressure increases. The woman may experience shortness of breath, and her back may ache as the back muscles strain to support the extra abdominal weight.

As the final month of pregnancy approaches, the ribs spread out to accommodate the lungs as they are displaced by the still-expanding uterus. In week 36, 4 weeks before delivery, the uterus has risen to the rib level, and the woman has to lean back to keep her balance. Uterine contractions become more frequent.

About 266 days after fertilization, uterine contractions become regular and strong, the uterine membranes rupture and the amniotic fluid is discharged, the cervix dilates, and after a variable period of labor the baby is born. The placenta is expelled about 30 minutes later, and the pregnancy is over.

Pregnancy Testing

The external signs of pregnancy are significant but not always dependable, and a woman usually does not conclude that she is pregnant until laboratory tests confirm her suspicions. Pregnancy may be detected by some tests as early as 12 days after the first missed menstrual period, but tests are most accurate after day 40 of pregnancy. Most of these tests depend on the fact that *human chorionic gonadotropin* (hCG), a hormone produced by the placenta, is abundant in the blood and urine of a pregnant woman. The level of hCG is highest during the first 3 months of pregnancy and then decreases.

ASK YOURSELF

1 What is usually the first sign of pregnancy?

2 How is hCG related to pregnancy tests?

BIRTH AND LACTATION

Childbirth, or *parturition* (L. *parturire*, to be in labor), usually occurs about 266 days after fertilization, or about 280 days from the first day of the menstrual period preceding fertilization. During the last month of pregnancy, the myometrium of the uterus becomes increasingly sensitive to the hormone oxytocin, which is released by the posterior lobe of the pituitary gland, and a pregnant woman usually begins to experience irregular uterine contractions at that time.

Although the sensitivity to oxytocin may be caused by either an increased concentration of uterine prostaglandins or a decreased concentration of progesterone and estrogens, the exact cause is not known. The signal for the initiation of labor may come from the fetus itself, but no firm evidence supports this idea. One theory for the onset and maintenance of labor is that the fetus finally reaches a point in its development where its head begins to press down on the cervix. This pressure causes the cervix to dilate, which produces uterine contractions and the secretion of oxytocin. Oxytocin stimulates further uterine contractions until the baby is finally pushed down past the cervix and out through the vagina. This is one of the few examples of a *positive* feedback system in the body.

Process of Childbirth

The initial contractions stimulate the release of oxytocin from the posterior pituitary gland, which further stimulates even more powerful uterine contractions. Waves of muscular contractions spread down the walls of the uterus, forcing the fetus toward the cervix [FIGURE 30.9]. By now the cervix is dilated close to its maximum diameter of about 10 cm (4 in.).

The amniotic sac may burst at any time during labor, or it may have to be ruptured by the birth attendant. Either way, the ruptured sac releases the amniotic fluid. If this loss of fluid occurs very early, it may signal the onset of labor. Another possible indication that labor has begun is the release of the cervical plug of mucus from the vagina. Ordinarily, either the bursting of the amniotic sac or the release of the cervical plug (termed "show") will occur early enough to provide ample time to prepare for childbirth, but great variations exist among pregnant women.

During pregnancy, the muscle cells of the uterus grow to as much as 40 times their former size, transforming the uterus into an enormously powerful muscular organ.

Babies are born head first most often, in the so-called *cephalic position,* with the head flexed and the occipital bone facing the mother's spine [FIGURE 30.10]. Other possible birth positions may produce complications that may require the intercession of the birth attendant. Since the skull bones of a newborn are not yet fused, the skull is still pliable. If it were not, the baby's head would not be small enough to pass through the vaginal canal.

After the baby is born, it is still attached to the placenta by the umbilical cord. Immediately after the baby is expelled from the uterus, the umbilical cord is clamped and cut below the clamp. Ordinarily, the umbilical cord is cut immediately after birth, but some physicians believe that the severing of the cord should take place only after the afterbirth has been expelled and all the placental blood has drained into the baby's circulation. Such a procedure gives the baby an additional 80 to 90 mL of blood, about 25 to 30 percent more than it would receive otherwise.

Three stages of *labor* are usually described:

1 The *first stage of labor* starts with regular uterine contractions. Contractions occur every 20 or 30 minutes at first and become more and more frequent until they occur every 2 or 3 minutes. The first stage averages about 14 hours for the birth of a first child and usually gets shorter for subsequent births. Its main function is to dilate the cervix to its maximum.

2 The *second stage of labor* is the actual birth of the child. It lasts anywhere from several hours to several minutes, with the average time being 2 hours. If the fetal membranes have not ruptured during the first stage, they do so now.

3 The *third stage of labor*, which takes about 20 minutes, is the delivery of the placenta, fetal membranes, and any remaining uterine fluid, collectively called the **afterbirth**. It usually follows within 30 minutes of the expulsion of the baby and is accomplished by further uterine contractions. These final contractions also help close off the blood vessels that were ruptured when the placenta was torn away from the uterine wall. About half a liter (1 pt) of maternal blood is usually lost at this time.

Human fetuses are born long before they are able to care for themselves—in a sense, they are always born

---

FIGURE 30.9 BIRTH OF A BABY

The internal events are shown in a series of six models, **[A]**–**[F]**. (Photographs of live births show only external events.)

[A]

[B]

[D]

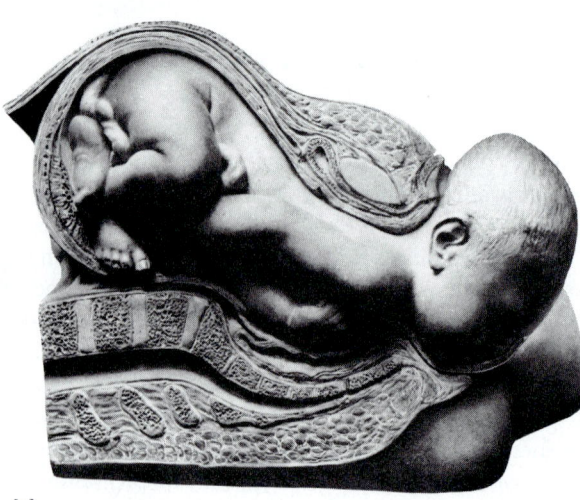

[E]

prematurely. But if gestation continued for longer than 9 or 10 months, the baby's relatively large head (and brain) could not pass through the vaginal canal.

The period after the placenta has been delivered is called the **puerperal period** (pyoo-UR-pur-uhl; L. *puerperus*, bearing young; *puer*, child + *parere*, to give birth to). It is the time when the mother's body reverts back to its nonpregnant state, and it usually lasts at least a month. For example, the uterus and vagina both revert back to their normal sizes about 6 weeks after childbirth [FIGURE 30.8B]. If a woman is not breastfeeding her baby, it will take anywhere from 6 to 24 weeks for the first menstruation.

FIGURE 30.10 THE SECOND STAGE OF LABOR: CHILDBIRTH

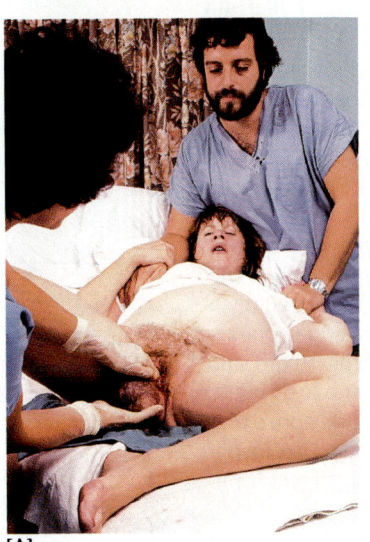

[A]

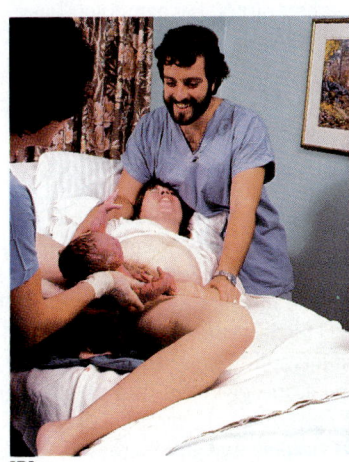

[B]

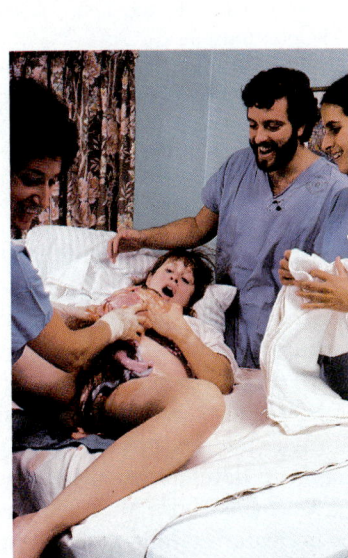

[C]

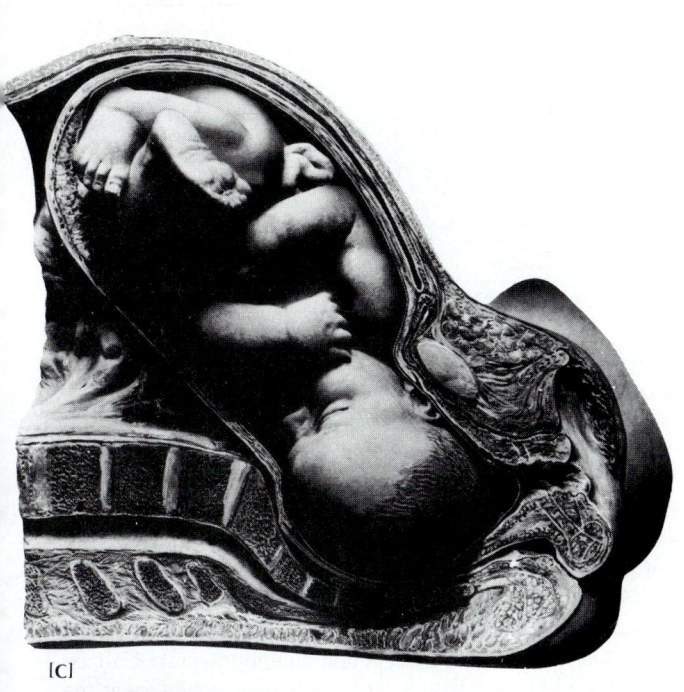

[C]

The Role of the Fetus during Childbirth

Until recently it was thought that the fetus plays no active role in its own birth process. Now there is evidence that the fetus triggers the release of "stress" hormones during childbirth that help it survive the arduous process of childbirth and adjust to life outside its mother's uterus.

In addition to the pressure a fetus feels while passing through the birth canal, it is also deprived of oxygen (hypoxia) periodically when uterine contractions compress the umbilical cord and placenta. During such periods of stress, the fetus produces very high concentrations of epinephrine and norepinephrine, both of which are classified chemically as *catecholamines* (see graph). Catecholamines are generally produced to help an individual react favorably to incidents of extreme stress. For example, the secretion of epinephrine and norepinephrine allows the fetus to counteract hypoxia and other potentially harmful situations throughout most of the gestation period.

The stress situations are actually beneficial, since the presence of unusually high concentrations of catecholamines permits the newborn to adjust to new conditions outside its mother's uterus after delivery. Such postnatal adjustments include the ability to breathe, the breakdown of fat and glycogen into usable fuel for cells, the acceleration of heart rate and output, and the increase of blood flow to the heart, brain, and skeletal muscles.

The surge of catecholamines during childbirth also causes the newborn in-

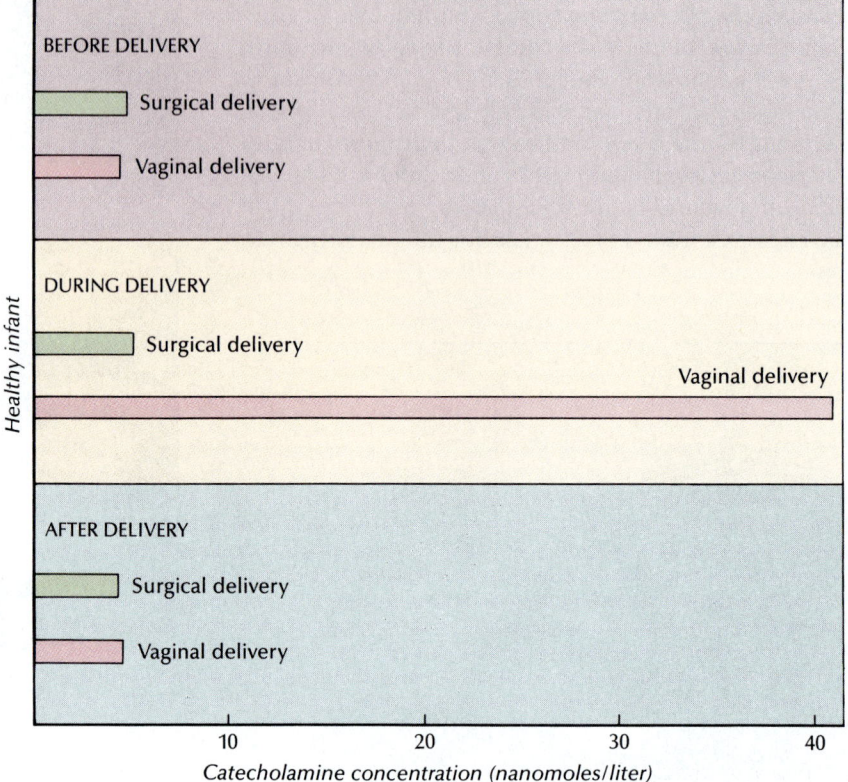

The production of "stress hormones" associated with childbirth is highest during the actual delivery. Normal vaginal delivery elicits a far greater secretion than does abnormal surgical (cesarean) delivery, which places relatively little stress on the fetus.

fant's pupils to become dilated, even when strong light is present. This alertness may help the infant form an early bond with its mother.

The production of fetal catecholamines is a direct result of the adrenal glands' response to stress. In the adult, in contrast, the secretion of these hor-

mones begins with the stimulation of the sympathetic nervous system. The adrenal glands of a fetus are proportionately larger than those in an adult, and fetuses and young children also have extra sources of norepinephrine in specialized tissues (known as paraganglia) near the aorta.

Premature and Late Birth

On the average, pregnancy lasts for about 266 days after fertilization, but many variations occur, and babies have survived after being in the uterus for only 180 to 200 days. These *premature* babies usually grow to be healthy, normal children, but they may require about 3 years to catch up developmentally to their full-term contemporaries. Babies born after 240 days are not considered to be premature as long as they weigh at least 2.4 kg (5.5 lb).

Premature babies are usually poorly proportioned

compared with healthy full-term babies. They usually have a weak cry, poor temperature regulation, wrinkled and dull red skin, closed eyes, and short fingernails and toenails. Because their muscular system is not fully developed, their movements are labored, and breathing may be difficult. Yet a baby born a month early has at least a 70-percent chance of survival. It is interesting that a 1.8-kg (4-lb) baby born after 34 or 35 weeks and nursed by its mother has a better chance of being normal than a full-term baby of the same weight. This is so because when a baby is born prematurely, its mother's milk contains

FIGURE 30.11 MULTIPLE BIRTHS

[A] Fraternal twins result when two ova are fertilized. Each usually has its own placenta and chorion. [B] Identical twins result when the inner cell mass of one fertilized ovum divides in half. Identical twins share the same placenta and chorion.

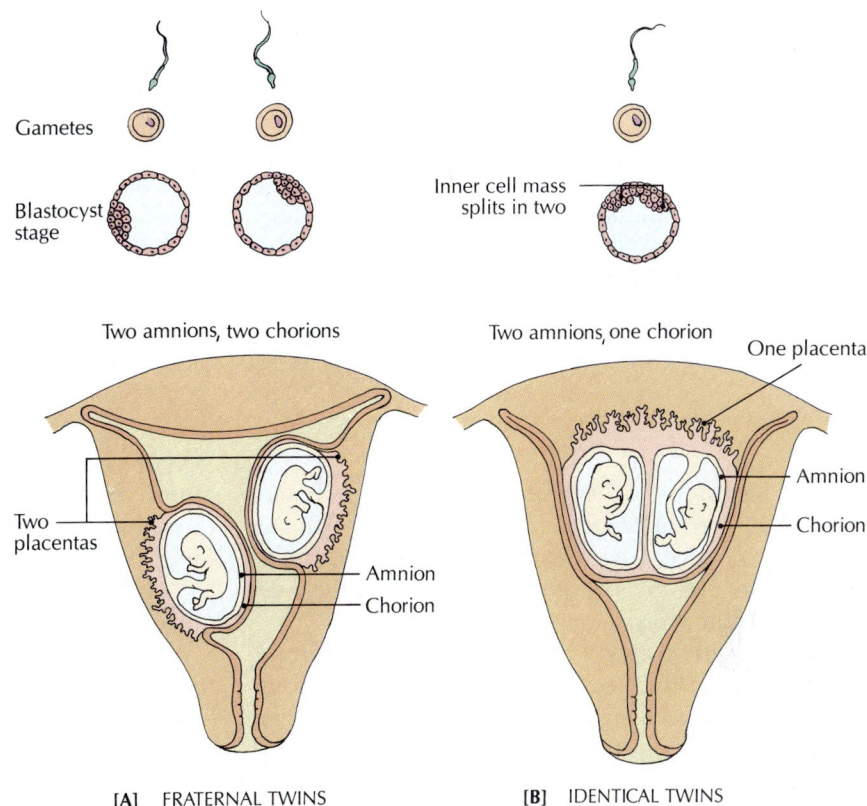

[A] FRATERNAL TWINS   [B] IDENTICAL TWINS

more protein, antibodies, sodium, and chloride than it would have at full term.

When a fetus remains in the uterus for longer than the normal full term, the attending physician may decide to induce labor with intravenous injections of pitocin (synthetic oxytocin) or by other methods, or to perform a cesarean section. Many physicians prefer to let the baby be born whenever it is ready as long as there are no complications. Babies born late who weigh more than 4 or 4.5 kg (9 or 10 lb) usually are not as healthy as normal-sized full-term babies, probably because the mother's body cannot provide adequate nourishment in such circumstances.

Multiple Births

Multiple births are always an exception in human beings. Twins are born once in 86 births, triplets once in 7400 births, and quadruplets once in 635,000 births. The chances of quintuplets being born are about 1 in 55 million. In most cases, one or more of the children born as quintuplets or sextuplets do not survive beyond infancy.

Women who have taken "fertility drugs" (which stimulate the ovaries), are older than 35, and have had children previously may have a higher-than-average rate of multiple births. This increase in multiple births may be caused by irregular ovulatory patterns in older women and the resultant release of more than one ovum at a time. Apparently, the tendency to release more than one ovum at a time is an inherited trait. However, the cell separation that produces identical twins is not due to a hereditary factor, and its cause is not known.

Twins may be fraternal or identical. ***Fraternal (dizygotic) twins*** are formed when more than one ovum is released from the ovary or ovaries at the same time, and two ova are fertilized [FIGURE 30.11A]. Each fraternal twin usually has its own placenta, umbilical cord, chorion, and amnion. Fraternal twins are the most common

Q: *Why are some babies born prematurely?*

A: Babies can be born prematurely for many reasons, including fetal malformations and genetic defects, but the most common cause is the mother's poor health. Maternal factors include failure to consume enough protein or total calories, alcohol or drug abuse, smoking, and hypertension.

Q: *Are both identical twins always either right-handed or left-handed?*

A: No, but about 90 percent of the time they are.

Some Interesting Birth Facts

More babies are born in August, and fewer in April, than any other months.

The most common birth time is between 3 A.M. and 4 A.M., and the least common time is 3 P.M.

Babies born during the first 6 months of the year have a higher rate of infant mortality and congenital defects than those born in the latter half of the year.

For every 100 girls born, 106 boys are born. But childhood death is more common among boys, and by adolescence the numbers have evened out.

On the average, girl babies are carried 1 day longer than boys. The average length of pregnancy for white women is 265 days after fertilization for boys and 265.9 for girls. Black women carry boys for 259.3 days on the average and girls for 260.7 days. (Black newborns, with shorter gestation time, are usually smaller than white ones, but interestingly enough, the newborns in India are even smaller than black newborns yet have a longer gestation time than the larger white children.)

A normal woman, with a normal pregnancy, gains about 12.5 kg (27.5 lb) during pregnancy. The average weight of a newborn child is 3.3 kg (7 lb 4 oz).

On the average, twins are born 19 days earlier than a single child. In the United States, Asians have the highest rate of fraternal twins, with 1 in 40 births; whites have the lowest rate, with 1 in 100 births.

---

multiple births. About 70 percent of all twins are fraternal, and 30 percent are identical. Aside from being the same age, fraternal twins resemble one another no more than any other brothers or sisters do and may be the same sex or different sexes.

FIGURE 30.11B shows the formation of *identical (monozygotic) twins.* Because identical twins result from the same fertilized ovum, they are always the same sex and are genetically identical. Apparently, identical twins are formed when a fertilized ovum divides into two identical cells before implantation (about day 6 or 7), but there is also some evidence that identical twins may also form after implantation. Because human identical twins develop from a single fertilized ovum, they are contained within a common chorionic sac and have a common placenta. However, the umbilical cords are separate and the amnions are also individual, except in rare instances.

Conjoined twins, commonly called *Siamese twins,** are identical twins whose embryonic disks do not separate completely. Sometimes the twins share only skin or tissue, but frequently they share an organ, such as a liver or an anus. Conjoined twins usually die early in life. They are more often females than males and are more common than quintuplets.

Adjustments to Life Outside the Uterus

Most of the body systems of the late fetus are developed and ready to function, but they are not used until the baby is born. Although the sudden shift from total dependency in the uterus to near independence upon birth is drastic, the newborn, or *neonate,* usually adapts to its new environment smoothly and without apparent trauma. The major changes in life outside the uterus affect breathing and circulation.

At birth, the baby must adapt quickly to life in air by making several rapid physiological and anatomical adjustments. It must shift from a fetal circulation, which depends on the placenta, to an essentially adult circulation, which depends on the lungs for gas exchange. Before birth, the fetal heart has an opening, the *foramen ovale,* which conveys blood from the right atrium to the left atrium. Also, the *ductus arteriosus* conveys blood from the pulmonary artery to the thoracic aorta, thus bypassing the lungs [see FIGURE 21.28].

Both the atrial opening and the pulmonary bypass close at birth, allowing the blood to pass through the pulmonary artery to the newly functioning lungs. If either the foramen ovale or the ductus arteriosus does not close off completely, the oxygen content of the blood will be low, and the baby's skin will appear slightly blue. A neonate with this condition is therefore called a "blue baby." Usually a defective ductus arteriosus must be corrected surgically after birth.

The start of actual breathing can be helped if it does not begin immediately at birth. Fluid can be cleaned from the nose and mouth of the neonate, the baby can be held upside down to help drain the breathing passage, and if necessary, it can be given a smack on its bottom to shock it into crying, its first gasp of air. Today it is more common to stimulate breathing with a small whiff of carbon dioxide.

---

*In 1811, Eng and Chang were born as conjoined twins in Bangkok, the capital of Siam (now Thailand). They shared a liver and were joined together at the lower end of the sternum. When they were 18, they were taken to the United States by P. T. Barnum, who made the twins famous in his circus. Eng and Chang (which mean "left" and "right" in Siamese) were not actually Siamese, since their parents where Chinese. The Siamese called them the "Chinese twins."

Q: *Why do nursing babies sometimes refuse milk after the mother has exercised?*

A: Postexercise milk is high in lactic acid, a by-product of exercise. Such lactic acid in breast milk is not harmful, but it has a bitter taste that newborns can detect.

Lactation

Lactation (L. *lactare*, to suckle) includes both production of milk by the mammary glands and release of milk from the breasts. Milk production is controlled by ***prolactin,*** secreted by the anterior pituitary. Milk release is controlled by ***oxytocin,*** secreted by the posterior pituitary. During pregnancy, the breasts enlarge in response to increasing concentrations of estrogen, growth hormone, prolactin, adrenal glucocorticoids, progesterone, and human chorionic somatomammotropin from the placenta. But the most important hormone is prolactin, which stimulates milk production in the mammary glands. Before birth, placental secretions of estrogen and progesterone actually inhibit the production of milk from the mammary glands, but after the placenta has been expelled from the uterus, resulting in a drop of estrogen and progesterone concentrations, the breasts begin to produce copious amounts of milk. The continued production of milk requires most of the mother's other hormones, but most important are prolactin, growth hormone, cortisol, and parathyroid hormone. These are necessary to provide the amino acids, fatty acids, and calcium required for milk production.

The actual release of milk from the mother's breasts does not occur until 1 to 3 days after the baby is born. During those 3 days the suckling baby receives ***colostrum*** (kuh-LAHSS-truhm), a high-protein fluid that is present in the mother's breasts at childbirth. Colostrum may help strengthen the baby's immune system and may also function as a laxative, removing fetal wastes called ***meconium*** (mih-KOH-nee-uhm) (mucus, bile, and epithelial threads) from the intestines. Not only does the mother's milk introduce disease-fighting antibodies into the baby's body, it also contains hormones that aid the proper development of the digestive system, stimulating the growth of cells that line the stomach and intestines.

The concentration of prolactin from the pituitary continues to rise in the mother's blood from the fifth week of pregnancy until birth. Following birth, the concentration returns to normal quickly. However, each time the mother nurses the infant, neural signals from the sucking action on the nipples pass to the mother's hypothalamus, causing a surge of prolactin release that lasts about an hour [FIGURE 30.12]. This prolactin acts on the mammary glands, stimulating them to provide milk for the next nursing period. The infant's suckling stimulates the hypothalamus, which in turn stimulates the posterior pituitary to secrete oxytocin. The oxytocin causes the myoepithelial cells in the breast to contract, which expels the milk within a minute of the baby's initial suckling ("milk letdown"). It is generally believed that almost no milk is secreted from the mammary glands at the actual time of nursing. Instead, milk is secreted and accumulated during the intervals between nursings.

Breasts that are not emptied soon cease to produce milk. It also appears that the nursing of the baby helps the uterus return to its normal size and shape by suppressing gonadotropic and ovarian hormones. Because of this stimulation by the suckling infant, lactation may be prolonged for 2 years or more if nursing continues.

On the average, nursing mothers produce about 1 L of milk each day. Multiple births result in larger secretions, usually about 1 L per day for each suckling baby.

Menstruation usually does not take place for a variable amount of time while the mother is nursing her baby. The baby's suckling action on the nipple stimulates the secretion of prolactin. Prolactin inhibits the secretion of GnRH from the hypothalamus, which suppresses the release of FSH and LH from the pituitary gland. As a result, the concentrations of estrogen and progesterone drop, and ovulation is often inhibited.

If the mother continues to breastfeed, ovulation will occur eventually, and pregnancy can take place during the period of lactation. The milk supply of a pregnant woman who is still lactating will decrease sharply as the pregnancy continues, since the increasing concentrations of estrogen and progesterone she produces will eventually inhibit lactation.

Human milk is usually bluish-white (the blue comes from protein and the white from fat), sweet, and slightly heavier than water. It is composed mostly of water (88 percent) and also contains carbohydrate (6.5 to 8 percent), fat (3 to 5 percent), protein (1 to 2 percent), and salts (0.2 percent).

Q: *Do women with small breasts produce less milk than large-breasted women?*

A: There does not seem to be a relationship between breast size and the volume of milk. Rather, the key factors in the amount of milk are the hormones prolactin (milk production) and oxytocin (milk secretion).

ASK YOURSELF

1 What are the three stages of labor?

2 What are some of the problems that premature babies often face?

3 What are some of the physiological adjustments the neonate has to make?

4 How do fraternal and identical twins differ?

5 What hormones are involved with lactation?

FIGURE 30.12 MILK-EJECTION REFLEX

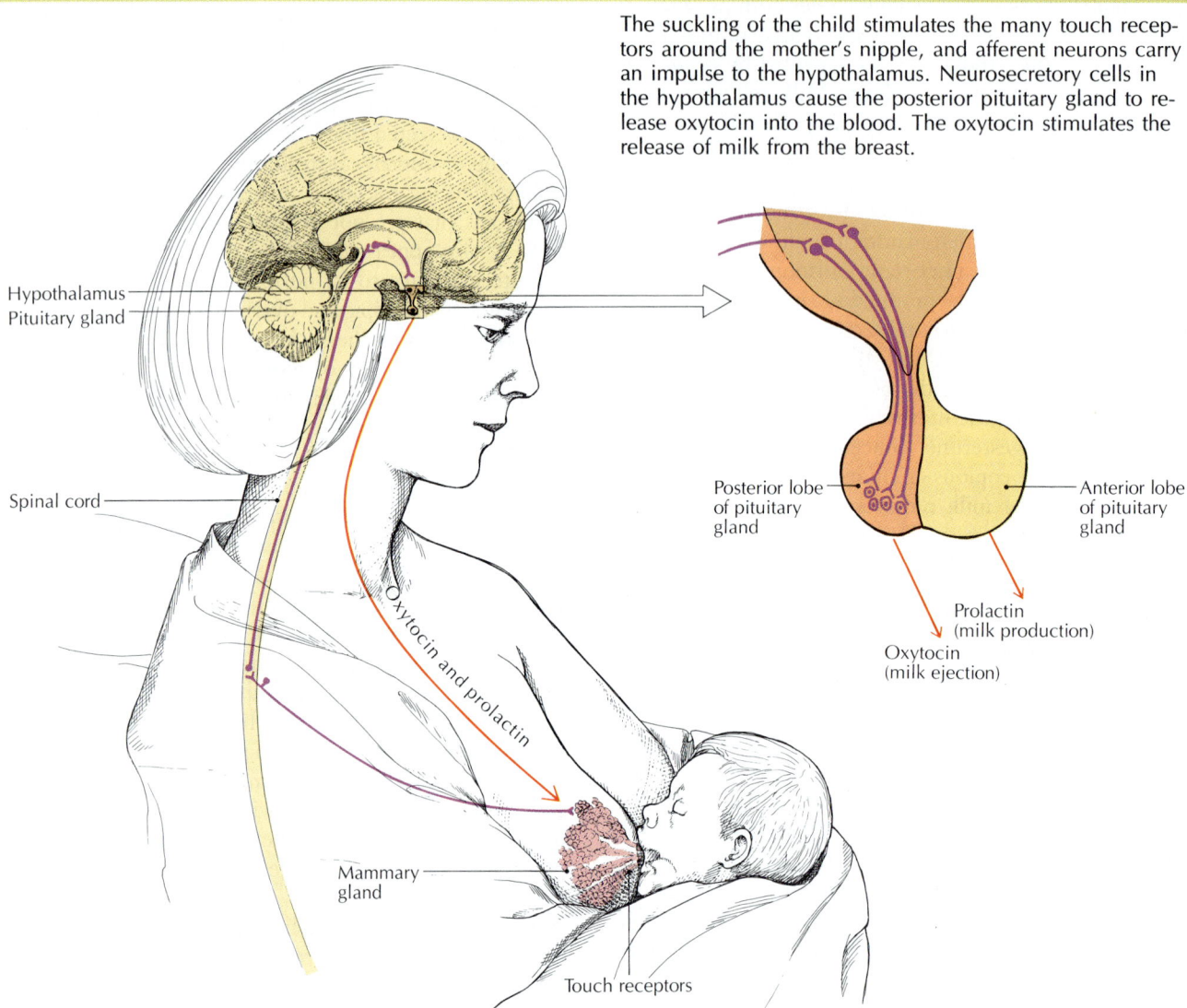

The suckling of the child stimulates the many touch receptors around the mother's nipple, and afferent neurons carry an impulse to the hypothalamus. Neurosecretory cells in the hypothalamus cause the posterior pituitary gland to release oxytocin into the blood. The oxytocin stimulates the release of milk from the breast.

Hypothalamus
Pituitary gland

Spinal cord

Oxytocin and prolactin

Mammary gland

Touch receptors

Posterior lobe of pituitary gland

Anterior lobe of pituitary gland

Prolactin (milk production)

Oxytocin (milk ejection)

POSTNATAL LIFE CYCLE

From the time of fertilization, human development is a highly organized combination of three processes: growth, morphogenesis, and cellular differentiation. *Growth* is an increase in overall size and usually an increase in the number of cells as a result of cell division. These cells must be arranged into specific structures that ultimately form the tissues and organs of the body. The cellular movements that bring about tissue and organ development are called *morphogenesis* (Gr. *morphe*, shape + *gennan*, to produce). After cells have become organized into specific forms, they must become specialized in order to perform the specific functions of tissues. This specialization process is called *cellular differentiation*. *Development* refers to the changes produced by

these processes in successive phases of the life cycle, which begins with fertilization and ends with death. In the earlier sections of this chapter we focused on prenatal development and birth. Here we will outline the stages of *postnatal* ("after birth") development.

Neonatal Stage

The newborn baby is called a *neonate* during its first 4 weeks of life. This period contains many adjustments. In addition to the adjustments in the respiratory and circulatory systems already described, a neonate may face some special functional problems, including the need for an adequate supply of calcium and vitamins C and D to help harden its skeleton and iron to aid the liver in synthesizing red blood cells; unstable temperature regulation and respiration, cardiac output, and metabolic rates; and a delayed functioning of the liver, pancreas, and immune system.

Infancy

The neonatal period is followed by *infancy,* which continues until the infant can sit erectly and walk, usually between 10 and 14 months after birth. The body weight generally triples during the first year.

Gradual physiological changes occur during infancy, especially in the cardiovascular and respiratory systems. Myelinization of the nervous system begins during this period, and motor activities become more and more coordinated. Vision and other functions of the cerebral cortex are refined during the first few postnatal months. The size of the brain increases from one-quarter to one-half of adult size, and the incisor teeth erupt. Because of the immaturity of the liver and kidneys, the infant may be unable to excrete toxic substances, and in general the infant is susceptible to viral and bacterial diseases before its immune system is fully functioning. This is especially true of bottlefed, rather than breastfed, babies.

Childhood

Childhood is the period from the end of infancy to the beginning of adolescence (12 to 13 years). In early childhood, ossification is rapid. Bone growth slows down in late childhood and then increases again during the prepubertal period. Deciduous teeth erupt and are replaced by permanent teeth during late childhood, motor coordination becomes more fully developed, and language and other intellectual skills are refined.

Adolescence and Puberty

The hormone-secreting gonads of both sexes are inactive until the final maturation of the reproductive system. This period of maturation, development, and growth is called *adolescence* (L. *adolescere,* to grow up). It spans the years from the end of childhood to the beginning of adulthood. Adolescence and puberty are not the same. *Puberty* (L. *puber,* adult) is the period when the individual becomes physiologically capable of reproduction. It usually takes place during the early years of adolescence. The average age of puberty is 12 for girls and 14 for boys.

Before puberty, between the ages of 7 and 10, there is usually a slow increase in the secretion of estrogens and androgens. The concentrations of these sex hormones remain inadequate, however, to promote the development of *secondary sex characteristics* such as body hair and enlarged genital organs. At the start of puberty, the hypothalamus begins to release luteinizing hormone-releasing hormone (GnRH), which stimulates the secre-

tion of FSH and LH from the anterior pituitary. It is these pituitary gonadotropic hormones that stimulate the testes and ovaries to secrete androgens and estrogens at full capacity. The present belief is that during the maturation process, the amygdala portion of the brain stimulates the hypothalamus to begin secreting GnRH.

Sequence of body changes The events of puberty usually occur in a definite sequence. The typical pattern for a boy begins about the age of 11, when he may begin to get spontaneous erections with no apparent cause. He may also accumulate deposits of fat prior to the actual changes of puberty. The events of puberty may take as long as 4 years.

The changes that accompany puberty in a girl start about 2 years earlier than those for boys, and they also happen more rapidly and closer together. The whole sequence of changes, starting with the breast buds increasing in size and the nipples beginning to protrude, usually takes 2.5 to 3 years.

Menarche The first menstrual period, the *menarche* (meh-NAR-kee; Gr. "beginning the monthly"), comes during the latter stages of puberty, though ovulation may not take place until a year or so later. The menarche seems to occur when a girl reaches a weight of about 48 kg (106 lb) rather than at a certain age or height. At the start of adolescence, a female has five times as much lean tissue as fat tissue, but the amount of fat more than doubles by the time of menarche. Should there be a shortage of food, the extra fat is enough to nourish a fetus through 9 months of pregnancy and to provide milk for the newborn baby for about a month.

The average age of menarche is between 12 and 13 years. Contrary to some beliefs, the menarche is not hastened by a warm climate, but good nutrition seems to encourage an early menarche and is probably the most important cause of the increasingly lower ages for menarche.

Adulthood

Adulthood spans the years between about 18 or 25 and old age. The maximum secretion of testosterone for males is reached between the ages of 18 and 20, and muscular strength reaches its peak at about 25.

Menopause Women between 45 and 55 usually stop producing and releasing ova, and the monthly menstrual cycle stops. This cessation of menstrual periods is called the *menopause* ("ceasing the monthly"), and it signals the end of reproductive ability. The average age for menopause is currently 52 and is increasing, probably because of improved nutrition. Menopause is not an abrupt change; it usually takes about 2 years of irregular menstrual periods before menstruation and ova production stop permanently.

The decrease of estrogen that accompanies menopause is thought to be the cause of several physical problems, including "hot flashes" of the skin, which result from changes in the vasomotor system that dilate blood vessels and increase blood flow. Estrogen decrease may also cause osteoporosis and other irregularities in bone metabolism, as well as dizziness, fatigue, headaches, chest and neck pains, and insomnia. Before menopause, diseases such as hardening of the arteries are quite rare in women, probably because estrogen lowers the level of cholesterol in the blood. But as estrogen secretions continue to diminish after menopause, the incidence of cardiovascular disease becomes almost equal in men and women. Psychological disturbances, including depression, may accompany the physical problems. Some of the symptoms of menopause may be relieved by small doses of estrogen and progesterone.

Although males usually experience a gradual decrease in testosterone secretion after they reach 40 or 50, they do not experience as drastic hormonal changes as women do at that age. The most likely cause of psychological problems during this period is not hormonal, but rather the fear of impotency and old age. Despite the decrease in testosterone, normal males may retain sexual potency in old age.

Senescence

The indeterminate period when an individual is said to grow old is called *senescence* (L. *senescere*, to grow old). By the age of 70, height is usually a full inch less than it was in the twenties or thirties. Between 70 and 80, body strength decreases to half of what it was at 25, lung capacity decreases to half, and about 65 percent of a person's taste buds become inactive. The nose, ears, and earlobes are longer. Life expectancy is currently 71.4 for males and 78.7 for females. Current evidence suggests that the maximum human life span is about 110 years.

A S K   Y O U R S E L F

1 What are the three main processes of human development?

2 Define neonate, infant, child, and adult.

3 What is the difference between adolescence and puberty?

4 How is the hypothalamus related to the initiation of puberty?

5 What changes occur during puberty in boys and girls?

HUMAN INHERITANCE AND GENETIC ABNORMALITIES

Genetics is the branch of science that studies heredity. It examines the transmission and expression of characteristics from one generation to the next. Much of the study of human genetics has involved inherited diseases or chromosomal abnormalities.

An Example of a Genetic Disorder: Down's Syndrome

Genetic disorders can occur for many reasons. There may be a defective gene that causes an enzyme deficiency (cystic fibrosis), a mutant gene that disrupts normal functioning (hemophilia, color blindness), or an extra chromosome (Down's syndrome), among several other possibilities. As an example of a chromosomal aberration, we will discuss Down's syndrome in the following paragraphs.

As described earlier, chromosomes are arranged in homologous pairs, and each member of the pair is a *homolog*. Occasionally, a pair of homologs fails to separate during meiosis. Thus one of the gametes produced will be missing a chromosome, and one will have an extra chromosome. If the gamete with an extra chromosome unites with a normal gamete that has only one homolog, the zygote will have three copies, a condition called *trisomy*. The most common trisomic condition is **trisomy 21**, also known as **Down's syndrome** [FIGURE 30.13]. This condition is caused by the presence of an extra copy of chromosome 21.

People with Down's syndrome are mentally retarded (it is the most common cause of mental retardation in the United States); are short; have stubby hands and fingers, a rounded face, a lowered resistance to common diseases, some increased risk of visual impairment because of defective lenses, elevated levels of purines that can lead to nervous system disorders, a deficient immune system, and an increased risk of leukemia; and usually suffer respiratory problems and heart defects. In the past, most people with Down's syndrome died early in childhood or rarely lived beyond their twenties. Antibiotics, improved surgical techniques, and special education have allowed many of these afflicted people to live relatively normal lives. Their average life span is now about 30 years, with about 25 percent reaching age 50. However, about one-third still die before they are 10 years old. In addition, about 80 percent of fetuses with Down's syndrome are lost by spontaneous abortion (miscarriage).

The average incidence of Down's syndrome is about 1 in every 650 to 700 live births, but the frequency increases dramatically in older mothers. A 20-year-old mother has about 1 chance in 2000 of having a child with

FIGURE 30.13 A GIRL WITH TRISOMY 21 (DOWN'S SYNDROME)

Down's syndrome. The odds for a 40-year-old mother increase to about 1 in 80, and the odds for a 45-year-old mother go up to about 1 in 45. Although women over 35 account for less than 14 percent of all pregnancies, they give birth to about 45 percent of all children born with Down's syndrome. Many studies have been undertaken in an effort to find out if older fathers are also implicated in the transmission of Down's syndrome, and it appears that the extra chromosome 21 comes from the father in about 5 percent of the cases.

Genetic Counseling

The most common reason for an apparently normal couple to seek out the services of a genetic counselor is the birth of a first child who has a genetic disorder. A genetic counseling service helps prospective parents by estimating the probability that they will produce children with one of the 3000 known genetic disorders.

Genetic counseling relies mainly on three kinds of information. The first is a study of the family *pedigree*, a description of the known patterns of familial traits or diseases. The second is a **karyotype,** or chromosomal analysis of the parents [FIGURE 30.14]. TABLE 30.2 lists some diseases associated with specific chromosomes. The third is a *biochemical analysis* of embryonic blood, urine, or body tissue (including amniocentesis and chorionic villi sampling; see When Things Go Wrong) to detect genetic abnormalities.* Genetic counselors generally look for several types of inheritance patterns in trying to detect abnormalities in a family.

Tay-Sachs disease is an inherited disorder that occurs almost exclusively in Ashkenazi Jews from central or eastern Europe. It causes mental and physical deterioration in a newborn, and the afflicted child usually dies before age 5. Tay-Sachs disease is caused by an enzyme deficiency that impairs the storage of lipids. A biochemical test is available that uses cultured cells to detect altered enzyme levels that indicate whether a person is a carrier of the disease. If both prospective parents are carriers, they can be advised of the possibility of producing a child with Tay-Sachs disease.

Cystic fibrosis is the most common genetically transmitted disease in the United States. About half of those afflicted die before age 20. (It is the most common fatal genetic disease among Caucasian children.) Afflicted individuals produce thick, viscous mucus and other secretions that cause respiratory and digestive problems. It is believed that mucus accumulates because the passage of sodium and chlorine through plasma membranes is somehow affected.

The disease is almost always inherited with one particular DNA variation that can be detected when the affected chromosome is exposed to specific enzymes called *restriction enzymes*. These enzymes cleave the DNA at the points where the defective gene is located, providing a "genetic marker" for that gene. Prospective parents who already have a child with cystic fibrosis, or who suspect the possibility of producing a first child with the disease, can request a test of fetal cells that usually predicts with about 95 percent accuracy whether a defective gene is present in the fetus.

Some other genetic disorders that can be detected before birth are Duchenne's muscular dystrophy, fragile X syndrome, hemophilia, Huntington's disease, polycystic kidney disease, sickle cell anemia, and beta thalassemia. No cures are available for these diseases, but in the future genetic engineering techniques may be able to correct genetic abnormalities right on the DNA strand itself.

It has been determined that many of the same genes produce one defect if they come from the mother and

---

*Genetic defects can now be detected in week-old embryos.

FIGURE 30.14 CHROMOSOMAL ANALYSIS, OR KARYOTYPE

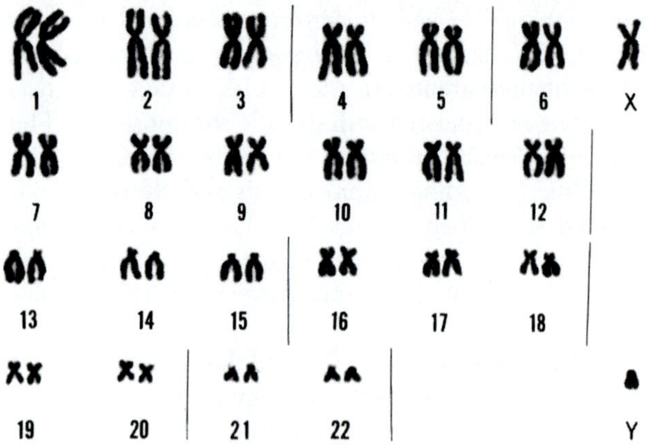

another defect if they come from the father. For example, Prader-Willi syndrome and Angelman syndrome are both caused by the same defective gene on chromosome 15. Children who inherit the mutation from their mothers have Prader-Willi syndrome (characterized by small hands and feet, a small penis in boys, voracious appetite, and mental retardation). Children who inherit the mutation from their fathers have Angelman syndrome, which is characterized by a jerky gait, large mouth, red cheeks, frequent spontaneous laughter, seizures, and profound mental retardation.

The Promise of Gene Therapy

An incurable genetic disorder called severe combined immunodeficiency (SCID) debilitates the immune system and makes it impossible for the body to combat even a relatively simple infection. This rare disorder was brought to the public's attention when "David," a boy with SCID, had to live inside a sterile bubble to avoid being infected. David died in 1984 after he was removed from the protective bubble to be given a bone marrow transplant.

SCID was in the public eye once again in September 1990, when the first federally approved use of gene therapy took place at the Clinical Center of the National Institutes of Health (NIH) in Bethesda, Maryland. The patient was a 4-year-old girl with SCID, which is caused by a deficiency of adenosine deaminase (ADA), an enzyme that breaks down toxic metabolic by-products that otherwise kill B and T cells.

What is gene therapy, and how does it work? Simply put, in **gene therapy** a defective gene is replaced with a healthy one, using the intricate technology of "genetic

TABLE 30.2 DISEASES KNOWN TO BE ASSOCIATED WITH SPECIFIC CHROMOSOMES

Chromosome number	Disease
1	Gaucher's disease: abnormal lipid metabolism caused by enzyme deficiency
2	Hereditary colon cancer
3	Retinitis pigmentosa: progressive degeneration of the retina Hereditary colon cancer
4	Huntington's disease: degenerative nerve disease
5	Familial polyps (growths) of the colon
6	Hemochromatosis: absorption of abnormally high levels of dietary iron Spinocerebellar ataxia: loss of muscle control caused by destruction of brain and spinal nerves
7	Cystic fibrosis: enzyme abnormality that causes mucus to clog ducts of lungs and other organs
8	Multiple exostoses: disorder of bone and cartilage
9	Malignant melanoma: cancer originating in skin
10	Multiple endocrine neoplasia, type 2: tumors of endocrine glands and other tissues
11	Sickle cell anemia: malformation of red blood cells, which clog small blood vessels
12	Phenylketonuria (PKU): abnormal amino acid metabolism caused by enzyme deficiency
13	Retinoblastoma: tumor of the eye
14	Alzheimer's disease: degenerative nerve disease that causes premature senility
15	Tay-Sachs disease: abnormal lipid metabolism caused by enzyme deficiency
16	Polycystic kidney disease: cysts causing enlarged kidneys and renal failure
17	Breast cancer (5–10% of cases)
18	Amyloidosis: accumulation of insoluble protein in tissues
19	Myotonic dystrophy: adult form of muscular dystrophy Familial hypercholesterolemia: extremely high (blood) cholesterol
20	ADA deficiency: susceptibility to infections
21	Amyotrophic lateral sclerosis: degenerative nerve disease Down's syndrome: form of mental deficiency caused by presence of 3 copies of chromosome 21
22	Neurofibromatosis, type 2: tumors of auditory nerves and tissues surrounding the brain
X	Hemophilia: lack of factors needed for blood clotting Duchenne and Becker muscular dystrophy: degenerative disease of muscles
Y	Adrenoleukodystrophy: degenerative nerve disease

Fetal-Tissue Implants

The idea of treating diseases with healthy tissue implants from aborted fetuses is certainly controversial, but it is not new. It started in Italy in 1928, when surgeons used pancreatic tissue from three different fetuses in an unsuccessful attempt to cure a patient with diabetes. Success with fetal-tissue implants first came 40 years later, in 1968, when fetal liver cells were used to treat a rare genetic disorder called DiGeorge syndrome. Now researchers are confident that fetal-tissue implants can be a routine source of permanent cures.

Cells from fetal tissue are particularly useful for transplants because they grow and divide rapidly, making it easier for them to become part of the patient's existing tissue; they are usually not rejected because they have not yet developed surface markers that an immune system cell would interpret as "foreign"; and they may still be undifferentiated, making them capable of becoming more than one type of tissue.

Fetal tissue has been used to treat such serious diseases as juvenile-onset diabetes, Huntington's disease, genetic disorders, and female sterility. Implants of fetal liver and thymus tissue could probably treat leukemia, aplastic anemia, inherited metabolic disorders, radiation injuries, and severe combined immune deficiency (SCID); pancreatic implants could possibly treat diabetes; and neural-tissue implants could possibly treat vision impairment, spinal cord injuries, memory defects, and degenerative syndromes that include Huntington's, Alzheimer's, and Parkinson's diseases.

The treatment of female sterility is another fascinating area of fetal-tissue research. Animal tests have shown that fetal ovaries, which already contain egg cells, may be transplanted to a female who is sterile because of an ovarian problem, enabling her to become pregnant and give birth to healthy, normal offspring.

---

engineering." In the case of the NIH patient, whom we will call Alice, the procedure was as follows:

1 A T cell was removed from Alice's body and exposed to harmless mouse leukemia retroviruses into which healthy human ADA genes had been spliced.

2 The retroviruses were allowed to infect the DNA of Alice's T cells, and the functional genes began to direct the synthesis of ADA.

3 The new T cells were cultured in a laboratory, producing billions of copies, which were introduced into Alice's bloodstream, with each T cell now containing a copy of the specific gene responsible for making ADA.*

It is hoped that the T cells will produce enough ADA to allow Alice to lead a normal life. However, even if the procedure is successful, Alice will have to repeat the treatment every month for at least 2 years.

ADA cannot be injected directly into the bloodstream, since it would break down almost immediately. An alternative, at an annual cost of about $60,000, is the weekly injection of a drug called PEG-ADA, which has a protective coating of *poly*ethylene *g*lycol that keeps the enzyme viable in the bloodstream for a few days. Gene therapy is preferable because it is potentially a one-time, permanent treatment. Many other versions of Alice's treatment are being tested.

The most likely candidates for gene therapy are disorders caused by a single gene, such as hemophilia, cystic fibrosis,† and inherited emphysema. In the future, more complex genetic disorders such as Parkinson's disease, Alzheimer's disease, and cancer may be treatable with gene therapy.

†The single gene that causes cystic fibrosis was discovered in 1989.

*Sex cells are never used in gene therapy because genetic traits introduced into sperm or eggs could be passed on to future generations; this is one of the many ethical issues relating to gene therapy. Instead, body cells are used because they affect only the person being treated.

A S K   Y O U R S E L F

1 What chromosomal abnormality causes Down's syndrome?

2 What is a karyotype?

WHEN THINGS GO WRONG

During the 9-month gestation period *anything* can go wrong, but fortunately, everything of importance usually goes right. Several developmental problems have already been described in this chapter, so we will concentrate on how embryonic and fetal problems are detected and even corrected.

Some Adverse Effects of the Maternal Environment on the Embryo and Fetus

Substances that chemically interfere with development and cause gross deformities without changing the embryonic DNA are called *teratogens* (TERR-uh-toe-jehn; Gr. *teras*, monster). Exposure to teratogens within the first 2 weeks of pregnancy usually causes the death of the embryo through a spontaneous abortion. Death may still result from days 15 to 60, but physical deformities are a more common effect. Later exposure, after the rudiments of all the essential organ systems have been formed, usually produces relatively minor abnormalities.

Two of the most common dangers for the unborn child are *smoking* and *alcoholic consumption* by the pregnant woman. Studies have shown that babies born to women who smoke during pregnancy weigh less than average. Smoking during pregnancy also appears to be related to a higher-than-normal incidence of miscarriage. Interestingly, the babies were lightest when *both* parents smoked, indicating that the smoking mother was also breathing in some of her husband's exhaled smoke and passing it on to the fetus. Smoking also reduces the oxygen supply of the fetus and may hinder mental development.

Women who drink alcohol heavily, especially during the first 3 months (trimester) of pregnancy, may produce children with a set of defects known collectively as *fetal alcohol syndrome,* which includes mental retardation, growth deficiencies, and abnormal facial features. The children may also have abnormal joints and heart disease.

Like most other drugs, alcohol diffuses easily through the mother's placenta and then passes through the fetal circulation. Within the fetus, the alcohol acts to depress the fetal central nervous system. Another factor in fetal alcohol syndrome is that the liver of the fetus is not fully developed and therefore is not an efficient metabolizer of alcohol. As a result, the alcohol remains within the body of the fetus to exert its effects even after it has been eliminated from the mother. In addition, the alcohol may result in intoxication of the unborn infant, and newborn infants may suffer from alcoholic withdrawal. Presumably, the risks are lower for women who consume small quantities of alcohol, but such problems as miscarriage and underweight babies still persist.

Pregnant women are advised to avoid x-rays, since *radiation* can damage the fetal nervous system and eyes and may cause mental retardation. Transplacental infection of the fetus by the rubella (German measles) virus can cause cataracts and deafness and other ear defects.

Most physicians urge their pregnant patients to avoid *drugs* of all kinds during pregnancy. Drugs such as heroin, cocaine, and morphine may actually cause the fetus to be addicted to those drugs at birth, but even such supposedly safe drugs as aspirin, antibiotics, antihistamines, and tranquilizers may cause problems ranging from miscarriage to physical deformation. The antibiotic tetracycline may cause fetal tooth defects if taken by the mother during the final 6 months of pregnancy and may cause cataracts if taken in large quantities.

The maternal environment can even affect the fetus as it is being born. For example, sexually transmitted diseases may be passed on to the baby as it moves through the birth canal. In cases where a sexually transmitted disease is suspected, a *cesarean section* is performed, and the baby is lifted out through the abdominal incision.

Amniocentesis

Amniocentesis (am-nee-oh-sehn-TEE-sihss; Gr. sac + puncture) is the technique of obtaining cells from a fetus. First, the fetus is located by ultrasound (bouncing high-frequency sound waves off it and recording the echoes). This is a relatively safe procedure, unlike the use of potentially dangerous x-rays. Then a hypodermic needle is inserted into the amnion, usually directly through the abdominal wall of the mother (see drawing on page 1009). The amnion surrounding the fetus is filled with fluid in which a number of loose cells float. Because all these cells are derived from the original zygote, they are genetically alike. Some of the amniotic fluid may be grown as a tissue culture. The cells in the fluid may be studied directly to save time, but cultured cells can be made to yield more information.

If the amount of protein that leaks from the neural tube into the amniotic fluid is measured, such diseases as spina bifida and anencephaly (absence of the forebrain) may be detected. Hemolytic diseases (including hemolytic disease of the newborn) may be detected. Inborn errors of metabolism may be identified by studying cell cultures derived from the amniotic fluid.

If chromosomes are examined, the sex of the fetus can be determined, but medically it is more important to discover any chromosomal abnormalities that might indicate possible disorders. The specific chromosome characteristics, or *karyotype,* can be read with fair accuracy.

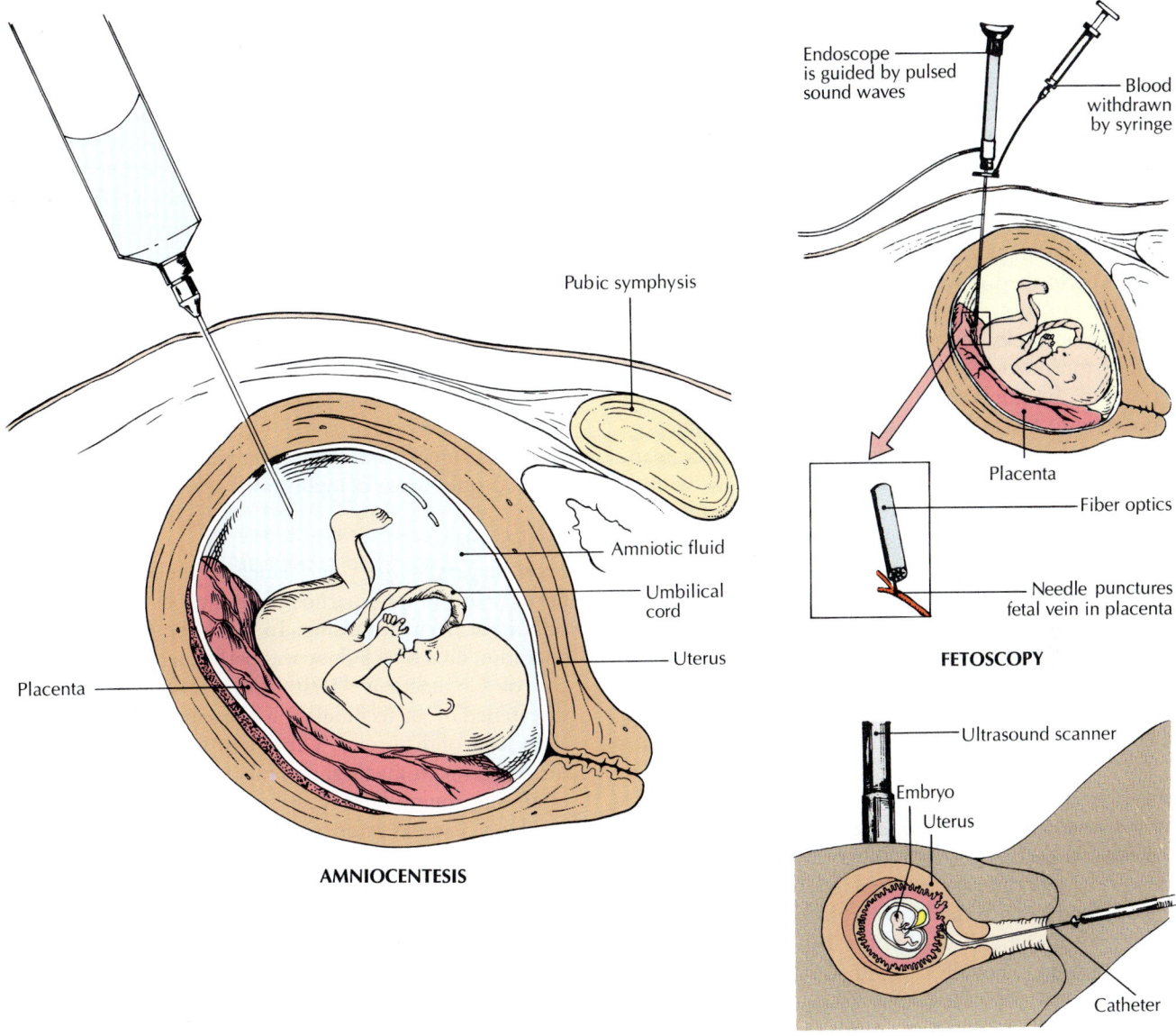

Pubic symphysis

Amniotic fluid

Umbilical cord

Uterus

Placenta

AMNIOCENTESIS

Endoscope is guided by pulsed sound waves

Blood withdrawn by syringe

Placenta

Fiber optics

Needle punctures fetal vein in placenta

FETOSCOPY

Ultrasound scanner

Embryo

Uterus

Catheter

CHORIONIC VILLI SAMPLING

For example, if three chromosomes of chromosome number 13 are present instead of the usual pair, the baby will probably not survive infancy. If there is a history of abnormality in the family, or if the parents are known to be at risk for some disorder, amniocentesis may reassure the parents by showing that the chromosomes are normal.

The procedure of amniocentesis is reasonably safe for both fetus and mother. Amniocentesis is advisable if a mother is known to carry a chromosomal aberration or if she is over 40, since women over 40 have a disproportionately high number of babies with chromosomal defects.

A new technique provides more information than amniocentesis by allowing physicians to locate the fetus by ultrasound and place a needle into the blood vessels of the umbilical cord to take samples of fetal blood, inject drugs, or perform transfusions. This technique can be used as early as week 18 of pregnancy.

Fetoscopy

Fetoscopy ("seeing the fetus") allows the physician to view the fetus directly. The fetus is located with ultrasound waves, and then an *endoscope* (an instrument for examining the interior of hollow organs) is inserted into the uterus through a small abdominal incision (see drawing on this page). Physicians can also insert tiny forceps to take a fetal skin sample from the fetal blood vessels that lie on the surface of the placenta. Like any other intrauterine examination, fetoscopy may damage fetal or maternal tissue if done improperly.

Chorionic Villi Sampling

Chorionic villi sampling is a technique that can be performed in a physician's office as early as the fifth week of pregnancy. A catheter is inserted through the mother's cervix into the uterus, where a sample of chorionic villi tissue is collected (see drawing on page 1009). The tissue is identical to fetal tissue, and its DNA can be examined for such diseases as Down's syndrome and sickle cell anemia. The technique appears to have a low risk factor, and can be performed 6 weeks earlier than amniocentesis.

Alpha-Fetoprotein Test

Alpha-fetoprotein is a substance that is excreted by the fetus into the amniotic fluid and enters the maternal bloodstream. It can be measured by the *alpha-fetoprotein test* on a sample of blood from the pregnant woman. Abnormally high levels of alpha-fetoprotein in the maternal blood indicate that the neural tube of the fetus has not closed completely, allowing large quantities of alpha-fetoprotein to "leak." The test is administered when the fetus is 16 weeks old. It may detect such neurological disorders as *anencephaly*, a condition in which the newborn child has a primitive brain or no brain at all, and *spina bifida*, a condition that exposes a portion of the spine. Neural-tube defects occur in about 1 out of every 1000 births.

Intrauterine Transfusion

An *intrauterine transfusion* is an injection of a concentrate of red blood cells into the peritoneal cavity of the fetus. The concentrate is prepared from Rh-negative whole blood to help the fetus combat hemolytic anemia (such as hemolytic disease of the newborn), a possibly fatal condition. The cells pass into the fetal circulation by way of the diaphragmatic lymphatics.

Ultrasonography

Ultrasonography is a technique that uses high-frequency sound waves to locate and examine the fetus. Ultrasound images are projected onto a viewing monitor. Ultrasonography allows the physician to see the embryonic sac during the embryonic period, the size of the fetus and its placenta, multiple fetuses, and any abnormal fetal positions. Serious abnormalities of the central nervous system, such as hydrocephaly, can also be detected.

Recent refinements in ultrasonography permit physicians to see *inside* the fetal heart. The new technique can detect severe congenital problems, such as abnormal heart chamber and valve formation, and any other condition that would usually produce a "blue baby." This technique allows the attending physician to prepare for treatment soon after birth.

Critics of electronic fetal monitoring, which includes ultrasonography, suggest that ultrasound waves may produce cell damage, and they recommend that sound waves be used only when complications are suspected.

Fetal Surgery

Fetal surgery, surgery performed on the fetus *before* it is born, is still in its experimental stages, but the early results have been promising. In 1982 the first successful operation on a fetal kidney was performed at the University of Connecticut Health Center. The kidney was threatened by an accumulation of excess urine because of a blockage of the ureter. The kidney was drained through the pregnant woman's abdominal wall with the guidance of an ultrasound video scan.

About 3 weeks later, another sonogram showed that fluid was accumulating again, and doctors decided to perform a cesarean section to deliver the 8-month-old fetus so that conventional corrective surgery could be performed. The surgery was successful, and the infant left the hospital a month later with two healthy kidneys.

Fetal surgery has corrected hydrocephalus and other diseases. In an extreme case, a 5-month fetus was removed from the uterus, operated on for a urinary blockage, and then returned to the uterus. It was born healthy 4 months later.

CHAPTER SUMMARY

Embryonic Development (p. 981)

1 The fusion of ovum and sperm nuclei produces the diploid **zygote**. The daughter cells, produced by a process of cell division called **cleavage,** are **blastomeres.** After cleavage begins, the developing organism is called an **embryo** until the end of the eighth week. During the third day, about the time the embryo is entering the uterus, a solid ball of about 16 blastomeres is formed, which is called the **morula.**

2 After day 2 or 3 the morula develops into a fluid-filled sphere called a **blastocyst.** The blastocyst is composed of about 100 cells. Its outer covering of cells (*trophoblasts*) is the **trophectoderm;** the blastocyst cavity is the **blas-** tocoel; a grouping of cells at one pole is the **inner cell mass,** from which the embryo will grow.

3 About a week after fertilization, the blastocyst becomes implanted in the uterine lining. **Implantation** is the actual start of pregnancy.

4 During week 2, the blastocyst develops into the **bilaminar embryonic disk,**

composed of the endoderm and ecto-derm.

5 The *extraembryonic membranes* include the yolk sac, amnion, allantois, and chorion.

6 The *placenta* synthesizes glycogen and other nutrients, transports materials between the mother and embryo, and secretes protein and steroid hormones. The blood vessels in the *umbilical cord* carry nutrients from the maternal placenta to the fetus, and wastes from the fetus to the placenta.

7 The embryonic heart is beating by week 4, and by the end of week 8 the major organ systems have become established. The developing individual is now called a *fetus.*

Fetal Development (p. 989)
The fetal period begins with week 9. Growth is rapid at first but slows down between weeks 17 and 20 as development continues and systems become mature enough to sustain life outside the uterus.

Maternal Events of Pregnancy (p. 994)
1 The first maternal sign of pregnancy is usually a missed menstrual period. Pregnancy tests, which depend on the abundance of hCG in the blood and urine of pregnant women, confirm or deny pregnancy.

2 The pregnant woman's body goes through a series of physical and hormonal changes to accommodate the developing and enlarging fetus. Weak uterine contractions may begin at about week 10, becoming stronger as the pregnancy progresses and frequent during week 36.

3 About day 266, uterine contractions become regular and strong, the uterine membranes rupture and the amniotic fluid is discharged, the cervix dilates, and after a variable period of labor the baby is born. Pregnancy ends about 30 minutes later with the expulsion of the placenta.

Birth and Lactation (p. 995)
1 *Parturition* is the birth of the child. The initial uterine contractions stimulate the pituitary gland to secrete oxytocin, which increases the force of the contractions until the baby is born.

2 The three stages of *labor* are usually described as the start and continuation of regular uterine contractions, the birth of the child, and the delivery of the afterbirth. The period from the delivery of the afterbirth until the mother's body reverts to its nonpregnant state is the *puerperal period.*

3 A baby is considered *premature* if it is born before 240 days and weighs less than 2.4 kg.

4 *Multiple births* are always an exception in human beings. *Fraternal* (dizygotic) *twins* are formed when two ova are released and are fertilized. *Identical* (monozygotic) *twins* result from the same fertilized ovum and are genetically identical and the same sex. *Conjoined twins* are identical twins whose embryonic disks do not separate completely.

5 Among the adjustments the newborn makes to extrauterine life, the major ones are breathing on its own and the shift from a fetal circulation to an adult circulation, which depends on the lungs for gas exchange.

6 *Lactation* includes both milk secretion and milk release from the breasts. The most important hormones for lactation are *prolactin*, which stimulates milk production, and *oxytocin*, which stimulates the release of milk from the breasts.

Postnatal Life Cycle (p. 1002)
1 Human development consists of *growth,* an increase in size and number of cells as a result of mitosis; *morphogenesis,* the cellular movements that bring about tissue and organ development; and *cellular differentiation,* the specialization of cells that allow them to perform the specific functions of tissues.

2 The first 4 weeks after birth are the *neonatal period.* It is followed by *infancy,* which continues until the infant can sit erectly and walk, usually at 10 to 14 months. *Childhood* is the period from the end of infancy to the beginning of adolescence at 12 or 13 years.

3 *Adolescence* is the period of maturation, development, and growth from the end of childhood to the beginning of adulthood. *Puberty* is the period during early adolescence, when an individual becomes physiologically capable of re-production. It occurs about 2 years earlier in females than in males.

4 At the start of puberty, the release of GnRH from the hypothalamus stimulates the secretion of FSH and LH from the pituitary. These pituitary hormones stimulate the testes and ovaries to secrete androgens and estrogens at full capacity.

5 The events of puberty usually occur in a definite sequence. The major body changes are called *secondary sex characteristics.* The first menstrual period, or *menarche,* usually occurs during the latter stages of puberty.

6 *Adulthood* spans the years between about 18 or 25 and old age. The cessation of menstrual periods, *menopause,* usually occurs between the ages of 45 and 55.

7 The indeterminate period when an individual is said to grow old is *senescence.*

Human Inheritance and Genetic Abnormalities (p. 1004)
1 *Genetics* is the science of heredity. It examines the normal transmission and expression of characteristics from one generation to the next and also the *variation* of that normal transmission.

2 Occasionally, chromosomal homologs fail to separate during meiosis. A zygote resulting from the gametes produced may have either only one copy of the chromosome *(monosomy)* or three copies *(trisomy)*. *Down's syndrome* is the most common trisomic condition.

3 *Genetic counseling* provides prospective parents with estimates of the probability that they will produce children with a genetically controlled disease.

4 Genetic counseling relies mainly on information obtained from a family *pedigree* (description of known patterns of familial traits or diseases), a *karyotype* (number, form, and type of chromosomes) for both parents, and a *biochemical analysis* of embryonic blood, urine, or body tissue (including amniocentesis and chorionic villi sampling) to detect genetic abnormalities.

5 *Gene therapy* is a procedure that replaces a defective gene with a healthy one in an effort to correct a genetic disorder.

KEY TERMS

adolescence 1003
adulthood 1003
blastocoel 981
blastocyst 981
blastomere 981
cleavage 981
differentiation 1002
Down's syndrome 1004
embryo 981
extraembryonic
  membranes 985
fetus 989
fraternal twins 999
gene therapy 1006
genetics 1004
identical twins 1000

implantation 983
inner cell mass 981
karyotype 1005
labor 995
lactation 1001
menarche 1003
menopause 1003
morula 981
parturition 995
placenta 986
puberty 1003
secondary sex
  characteristics 1003
senescence 1004
umbilical cord 988
zygote 981

MEDICAL TERMS FOR REFERENCE

ABRUPTIO PLACENTAE The premature separation of the placenta from the uterus.

CESAREAN SECTION An incision through the abdominal wall and the uterine wall to remove a fetus.

CONGENITAL DEFECT A condition existing at birth but not hereditary.

EPISIOTOMY The cutting of the vulva at the lower end of the vaginal orifice at the time of delivery in order to enlarge the birth canal.

FETAL MONITORING The continuous recording of the fetal heart rate, using a transcervical catheter.

FETOLOGY The study of the fetus.

HYSTEROTOMY An incision through the uterine wall.

LAPAROSCOPE (Gr. *lapara*, flank) A long, thin, light-bearing instrument used during a *laparotomy*, which is the surgical incision into any part of the abdominal wall.

LOCHIA (LOE-kee-uh; Gr. *lokhios*, of childbirth) The normal discharge of blood, tissue, and mucus from the vagina after childbirth.

PERINATOLOGY (Gr. *peri*, around + L. *natus*, born) The study and care of the fetus and newborn infant.

PLACENTA PREVIA An abnormal implantation of the placenta at or near the internal opening of the uterine cervix so that the placenta tends to precede the child at birth, usually causing severe maternal hemorrhaging.

PROLAPSE OF THE UMBILICAL CORD Expulsion of the umbilical cord before the fetus is delivered, reducing or cutting off the fetal blood supply.

TOXEMIA OF PREGNANCY A disease occurring in the latter half of pregnancy, characterized by elevated blood pressure, proteinuria, edema, sodium retention, and convulsions; a leading cause of stillbirth.

UNDERSTANDING THE FACTS

1 Define prenatal, postnatal, zygote, blastocyst, embryo, and fetus.

2 Distinguish the trophoblast from the inner cell mass by location in the embryo. What develops from each?

3 When does clinical pregnancy actually begin?

4 When does differentiation begin? What results from the initial differentiation of the cells of the inner cell mass?

5 Where does the blastocyst implant? Be specific.

6 List the extraembryonic membranes, and state the function of each.

7 What are the main functions of the placenta?

8 At what point in gestation does a fetus have a greater than 50 percent chance of survival?

9 List the stages of labor, explaining the major event(s) in each.

10 What role do maternal and placental hormones play in lactation?

11 Which hormone is used as the basis of most pregnancy tests? Where and when is that hormone produced?

12 The decline in which hormone is believed to have the greatest effect in menopause?

13 Define amniocentesis, chorionic villi sampling, and the alpha-fetoprotein test. What kinds of information can be obtained from each procedure?

UNDERSTANDING THE CONCEPTS

1 Explain the role of the syncytiotrophoblast in implantation.

2 What is the significance of differentiation?

3 What role is played by the chorionic villi in the formation of the placenta? Do maternal and fetal blood mix within the placenta? Explain.

4 Describe the blood vessels within the umbilical cord. Which transport nutrient- and oxygen-rich blood, and which transport blood containing fetal waste products?

5 Summarize the major activities of the embryonic period and the major activities of the fetal period.

6 Explain the difference between monozygotic and dizygotic twins.

7 Where are the foramen ovale and ductus arteriosus? What occurs to each at birth?

8 Explain why the resumption of ovulation is delayed in nursing mothers.

9 What are the hormonal changes that trigger puberty?

10 Why is consumption of alcohol by a pregnant woman during the first trimester especially dangerous to her embryo?

11 What kinds of abnormalities can karyotyping diagnose?

SELF-QUIZ

Multiple-Choice Questions

30.1 The embryonic stage that consists of an inner cell mass and a trophoblast is called a

a zygote   d gastrula
b blastomere   e blastocyst
c morula

30.2 The mass of cells that will develop into the chorion is termed the

a trophectoderm   d zona pellucida
b embryoblast   e inner cell mass
c blastocyst

30.3 Ectoderm gives rise to all of the following *except*

a muscle tissue   d spinal cord
b brain   e the anterior
c skin   pituitary gland

30.4 Which of the following is *not* one of the extra-embryonic membranes?

a yolk sac   d allantois
b coelom   e chorion
c amnion

30.5 Which of the following is *not* a function of amniotic fluid?

a suspends the embryo
b prevents the embryo from adhering to the amnion
c permits the developing fetus to change position
d provides nutrients to the fetus
e provides a stable temperature

30.6 Initial uterine contractions in preparation for childbirth are caused by the hormone

a estrogen   d progesterone
b hCG   e a and b
c oxytocin

30.7 Which of the following hormones stimulates milk secretion?

a adrenal glucocorticoids   d pitocin
b progesterone   e estrogen
c prolactin

30.8 The removal of cells from the amniotic cavity for further examination and testing is

a fetoscopy   d ultrasonography
b intrauterine transfusion   e amniocentesis
c endoscopy

30.9 Which of the following is present in the fetal atria but *not* in the neonate?

a foramen ovale   d umbilical vein
b ductus arteriosis   e pulmonary bypass
c umbilical artery

30.10 During pregnancy, estrogen

a is secreted by placental tissues during the last half of the pregnancy
b is secreted by the corpus luteum for the first 4 to 5 months of the pregnancy
c inhibits uterine contractions
d is under control of FSH
e a and b

Matching

a yolk sac   d allantois
b amnion   e all of the above
c chorion

30.11 _____ classified as extraembryonic membranes
30.12 _____ produces villi that grow into the maternal endometrium
30.13 _____ site of formation of the first blood cells
30.14 _____ surrounds the embryo, suspending it in a fluid environment
30.15 _____ produces the vein and arteries of the umbilical cord

a ectoderm   d ectoderm and endoderm
b endoderm   e ectoderm and mesoderm
c mesoderm

30.16 _____ gives rise to most of the lining of the digestive and respiratory systems
30.17 _____ gives rise to the entire lining of the digestive and respiratory systems
30.18 _____ gives rise to the epidermis of the skin and the nervous system
30.19 _____ gives rise to all of the structures of the skin
30.20 _____ gives rise to the entire cardiovascular system

Completion Exercises

30.21 Implantation occurs when the embryo is in the _____ stage.
30.22 The first hormone to be secreted by the placenta is _____.
30.23 Usually monozygotic twins share a common placenta, but they have separate _____ sacs.
30.24 Milk production is controlled by the hormone prolactin, but milk release is regulated by _____.
30.25 A high-protein, antibody-rich fluid that precedes the secretion of milk is called _____.
30.26 Down's syndrome can be detected by a chromosomal analysis called a _____.
30.27 Spina bifida and several other neural-tube defects can often be detected by high levels of _____ in the maternal blood.
30.28 The structure formed by the union of the chorion and the maternal endometrium that functions to support fetal development is called the _____.

1

A 62-year-old math professor noticed that he had to urinate more frequently, but that he voided only small amounts each time he tried to empty his urinary bladder. His sleep was interrupted with the need to void several times during the night. One morning he could barely urinate at all. A dull pain developed in his lower abdomen and intensified with each passing hour; finally, that afternoon he went to see his physician. The physician relieved the professor's pain by inserting a catheter in the bladder and draining a considerable volume of urine. After completing a full physical exam, including a rectal exam and a blood test, the physician assured the professor that there was no obvious sign of cancer.

• **Which organ was obstructing the outflow of urine?**
• **Why did the physician use a rectal exam?**
• **What was the likely diagnosis and treatment for the professor's condition?**

2

After missing a menstrual period, a young woman purchased a commercial pregnancy test. Following a positive reading from the test, she phoned for an appointment with her gynecologist. She was told to bring in a sample of her first morning urine for another test when she came for her first appointment. The second urine test confirmed her pregnancy. After giving her a full physical and pelvic exam, her gynecologist assured her that she was healthy. She was to begin taking supplemental vitamins and minerals and return for her next prenatal visit in one month.

• **Why is a missed menstrual period the first sign of pregnancy?**
• **What is in the urine that indicates pregnancy?**
• **Why is prenatal care of importance so early in the pregnancy of a healthy woman?**

3

A 46-year-old college track coach found a lump in the lateral side of her right breast during a self-examination. Her physician arranged for her to have a mammogram, which was quickly followed by a tissue biopsy. Later, while questioning the coach about her family's medical history, the doctor found that one grandmother and an aunt had had mastectomies years ago. The tissue biopsy contained some malignant cells, and the coach was given information outlining her options. She could have the lump removed (lumpectomy) followed by 6 weeks of daily radiation and chemotherapy, or she could have a modified radical mastectomy followed by chemotherapy. Although the two treatments do not differ in 5-year survival expectancies, the coach chose to have the modified radical mastectomy. After her recovery from surgery and chemotherapy, she returned to her active lifestyle.

• **What is a modified radical mastectomy and why is it followed by chemotherapy?**
• **Was it important to know that the patient had relatives who also had breast cancer?**
• **What were the coach's chances of a cure?**

4

After trying to become pregnant for 2 years, a young couple sought help from a fertility specialist. Testing of both partners showed that the husband had a low sperm count. Although the sperm were motile and well formed, there were fewer than normal in each ejaculate. The fertility specialist instructed the husband to collect several sperm samples at different times. These were then combined and used to artificially inseminate his wife. After several tries using this procedure, the couple achieved a viable pregnancy.

• **How many sperm must be contained within one ejaculate for it to be considered normal?**
• **Why is this number necessary?** • **What prevents an ovum from being fertilized by more than one sperm?**

APPENDIX A

An Overview of Regional Human Anatomy

The following photographs of dissected, preserved human cadavers convey a sense of the actual appearance and relations of grossly observable anatomical structures. The photographs are in contrast to the drawings in textbooks and atlases that, no matter how accurately illustrated, are merely *representations* of the actual body. The photographs provide the student with an opportunity to view an actual dissection of the human body.

These photographs are reminders of the profound role played by the dissection of the human body in the advancement of science. Leonardo da Vinci (1452–1519) produced his legendary (but not always accurate) anatomical sketches during the Renaissance, and in 1542 a scientific breakthrough occurred with the publication of *De corporis humani fabrica* ("On the Structure of the Human Body") by Andreas Vesalius (1514–1564). The anatomically accurate drawings in this monumental work were based on dissections of the human body, and the illustrations of the muscles in particular were so accurate that they are still used by students of anatomy.

FIGURE A.1 ABDOMINAL CAVITY WITH VISCERA EXPOSED

Anterior view.

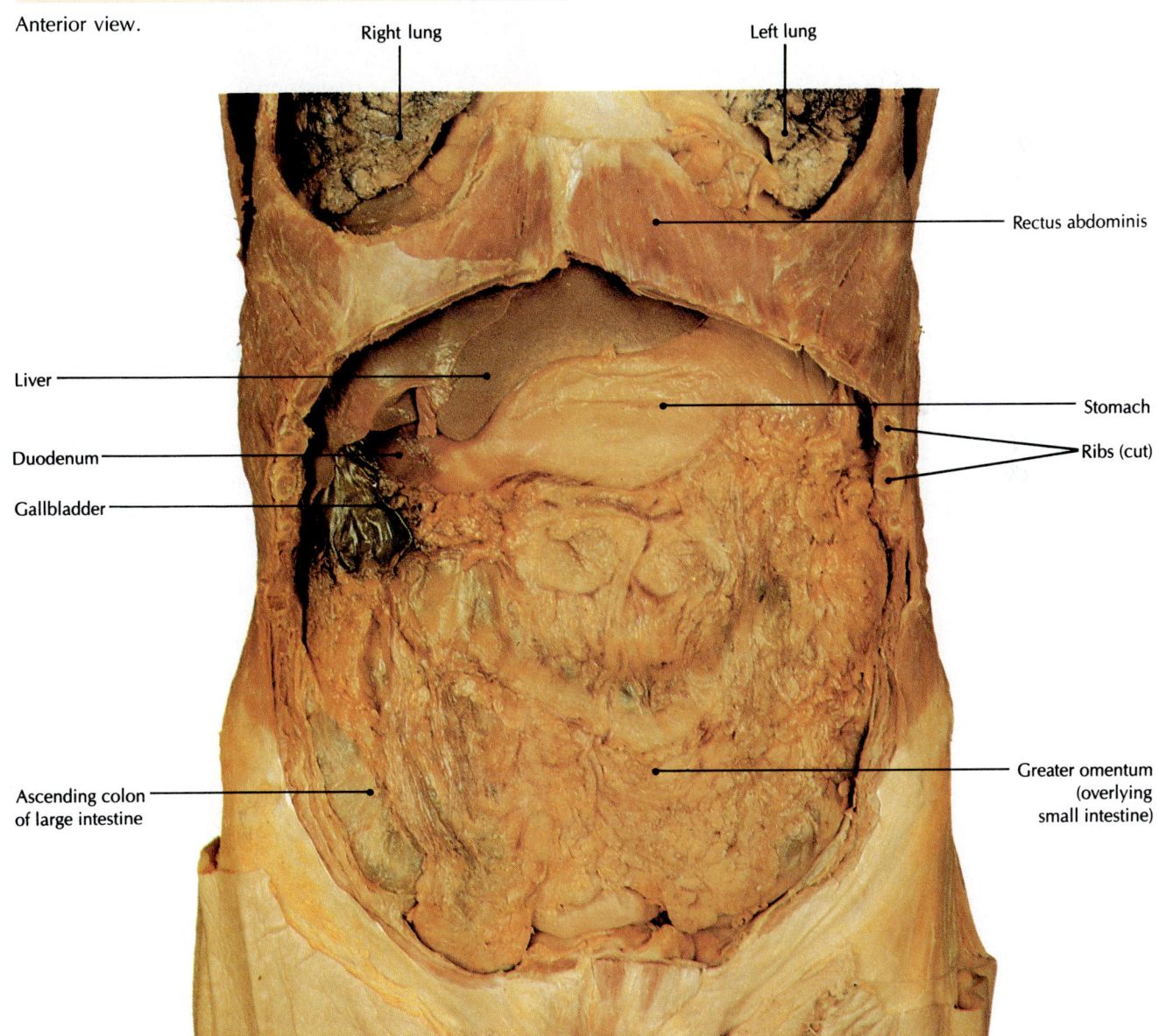

Right lung · Left lung · Rectus abdominis · Liver · Stomach · Duodenum · Ribs (cut) · Gallbladder · Greater omentum (overlying small intestine) · Ascending colon of large intestine

FIGURE A.2 SUPERFICIAL MUSCLE LAYER EXPOSED

Anterior view.

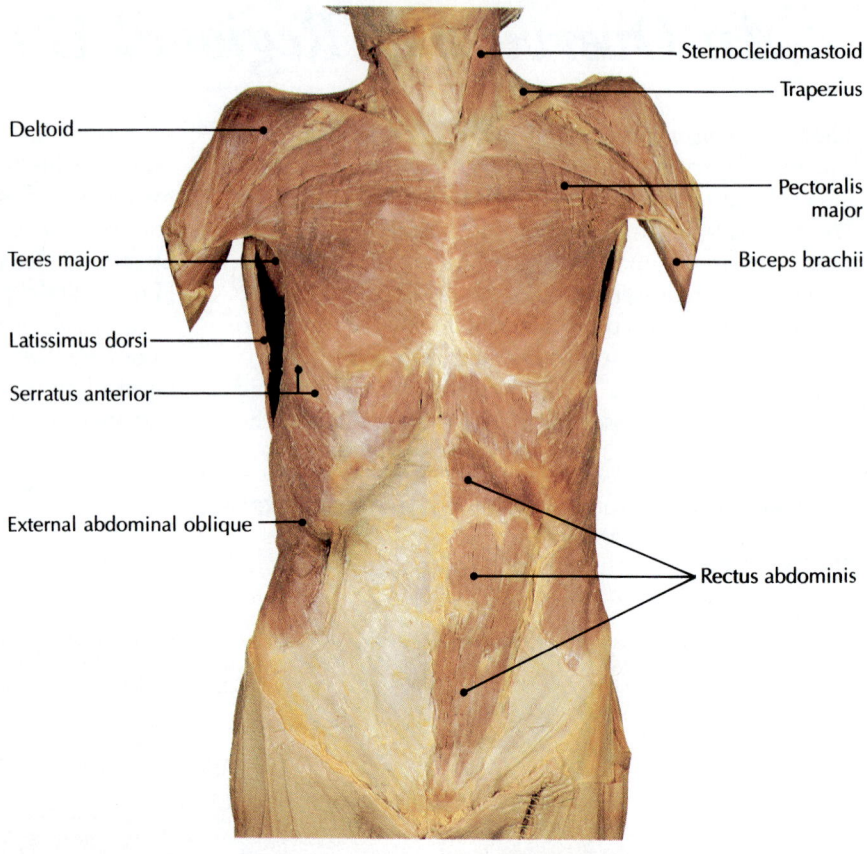

Sternocleidomastoid

Trapezius

Deltoid

Pectoralis major

Teres major

Biceps brachii

Latissimus dorsi

Serratus anterior

External abdominal oblique

Rectus abdominis

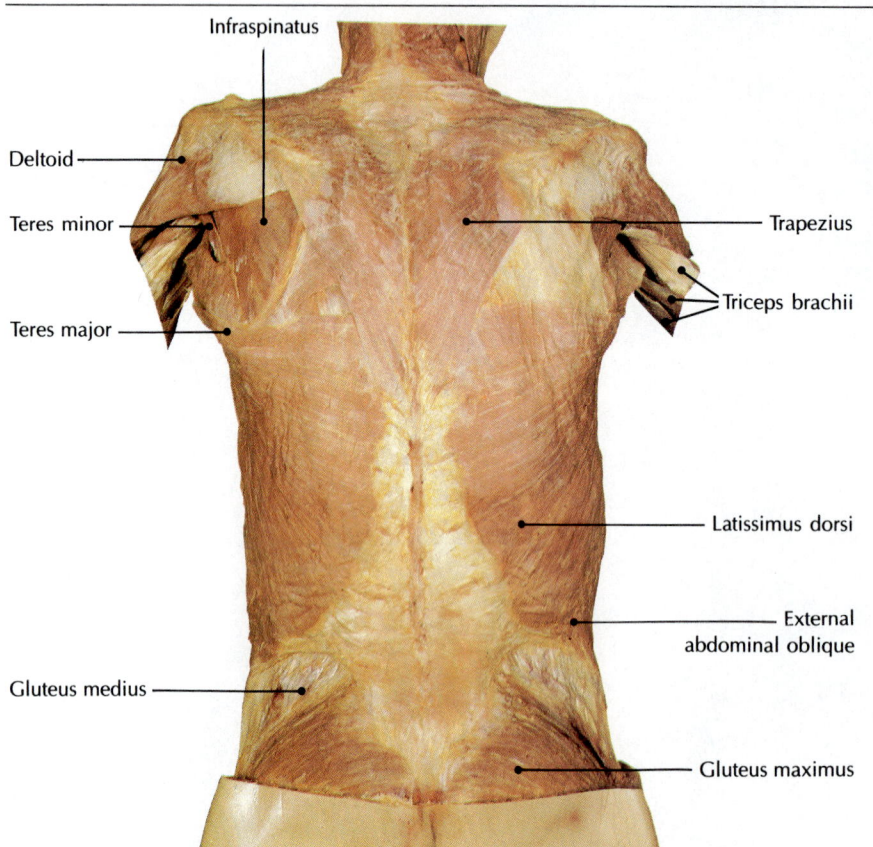

Infraspinatus

Deltoid

Teres minor

Teres major

Trapezius

Triceps brachii

Latissimus dorsi

External abdominal oblique

Gluteus medius

Gluteus maximus

FIGURE A.3 SUPERFICIAL MUSCLE LAYER EXPOSED

Posterior view.

FIGURE A.4 DEEP MUSCLE LAYER EXPOSED

Anterior view.

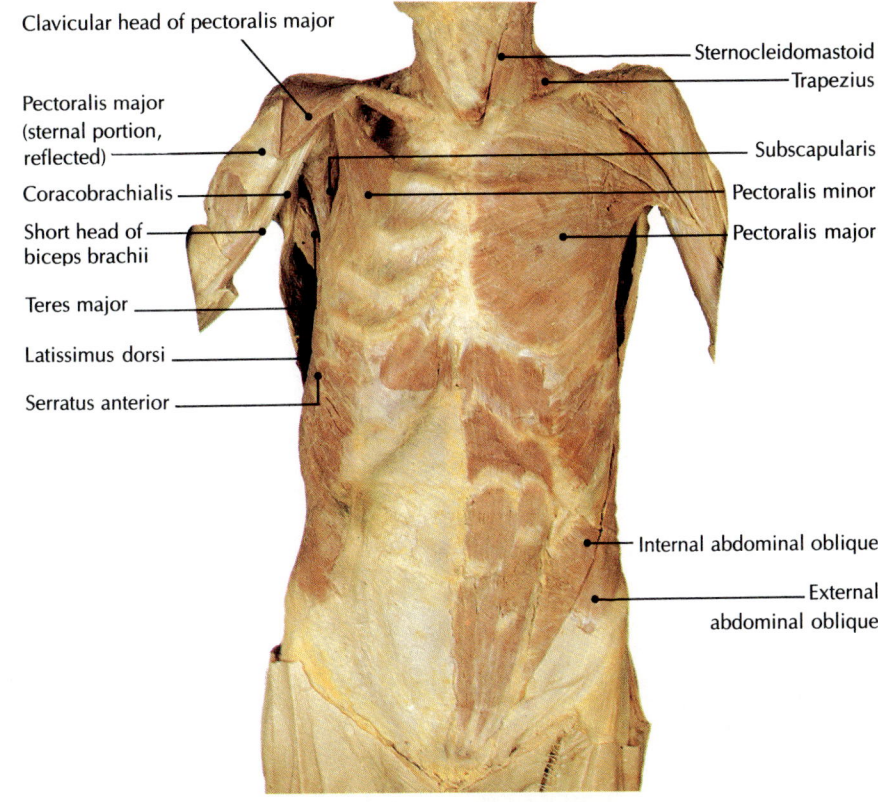

Clavicular head of pectoralis major
Sternocleidomastoid
Trapezius
Pectoralis major (sternal portion, reflected)
Subscapularis
Coracobrachialis
Pectoralis minor
Short head of biceps brachii
Pectoralis major
Teres major
Latissimus dorsi
Serratus anterior
Internal abdominal oblique
External abdominal oblique

FIGURE A.5 DEEP MUSCLE LAYER EXPOSED

Posterior view.

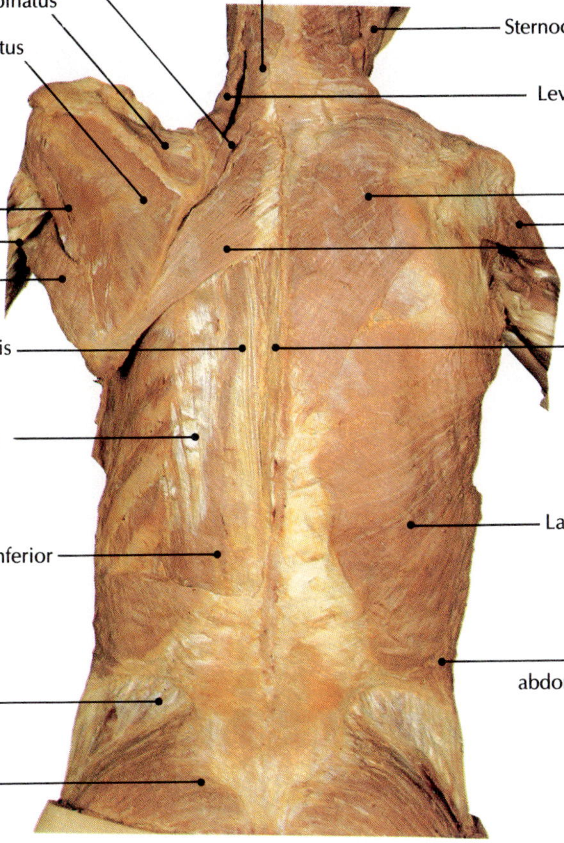

Rhomboid minor
Splenius capitis
Supraspinatus
Infraspinatus
Sternocleidomastoid
Levator scapulae
Teres minor
Triceps brachii
Teres major
Trapezius
Deltoid
Rhomboid major
Longissimus thoracis
Spinalis thoracis
Iliocostalis thoracis
Latissimus dorsi
Serratus posterior inferior
External abdominal oblique
Gluteus medius
Gluteus maximus

FIGURE A.6 EXPOSED THORACIC VISCERA

Anterior view.

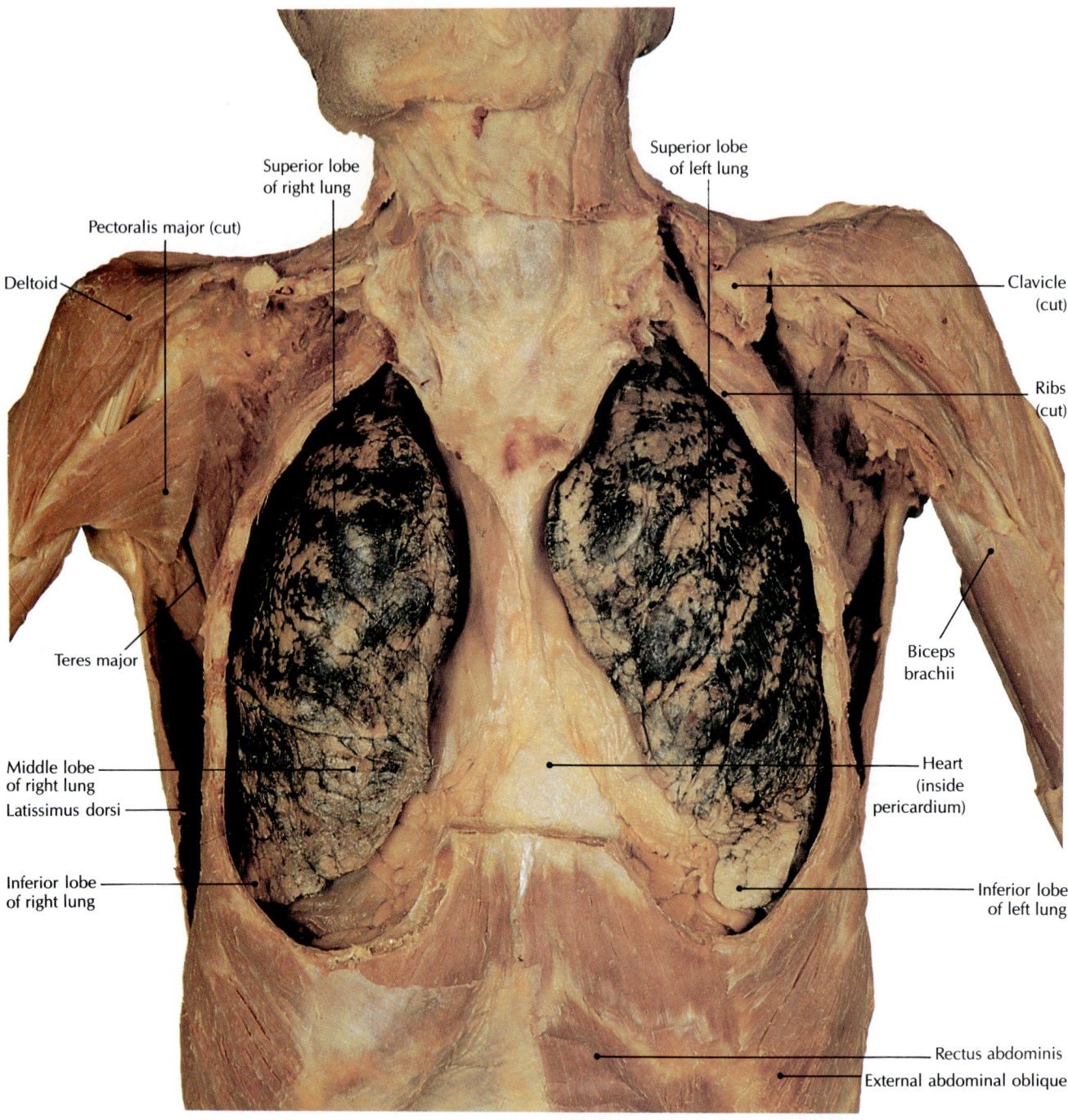

FIGURE A.7 EXPOSED POSTERIOR WALL OF THORAX AND ABDOMEN

Anterior view.

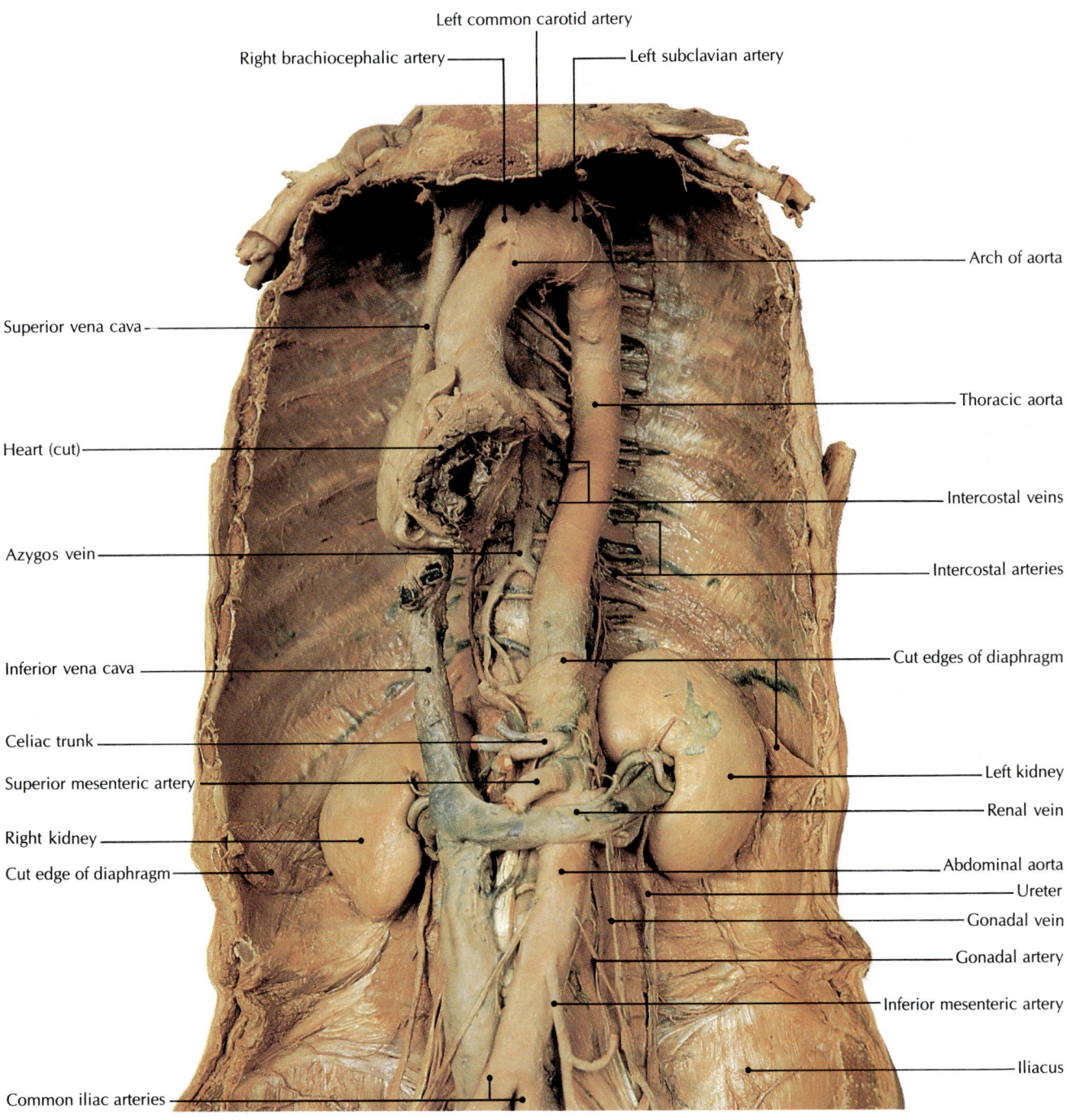

Left common carotid artery

Right brachiocephalic artery

Left subclavian artery

Arch of aorta

Superior vena cava

Thoracic aorta

Heart (cut)

Intercostal veins

Azygos vein

Intercostal arteries

Inferior vena cava

Cut edges of diaphragm

Celiac trunk

Superior mesenteric artery

Left kidney

Renal vein

Right kidney

Abdominal aorta

Cut edge of diaphragm

Ureter

Gonadal vein

Gonadal artery

Inferior mesenteric artery

Iliacus

Common iliac arteries

FIGURE A.8 EXPOSED FEMALE PELVIC CAVITY AND REPRODUCTIVE ORGANS

Right sagittal view.

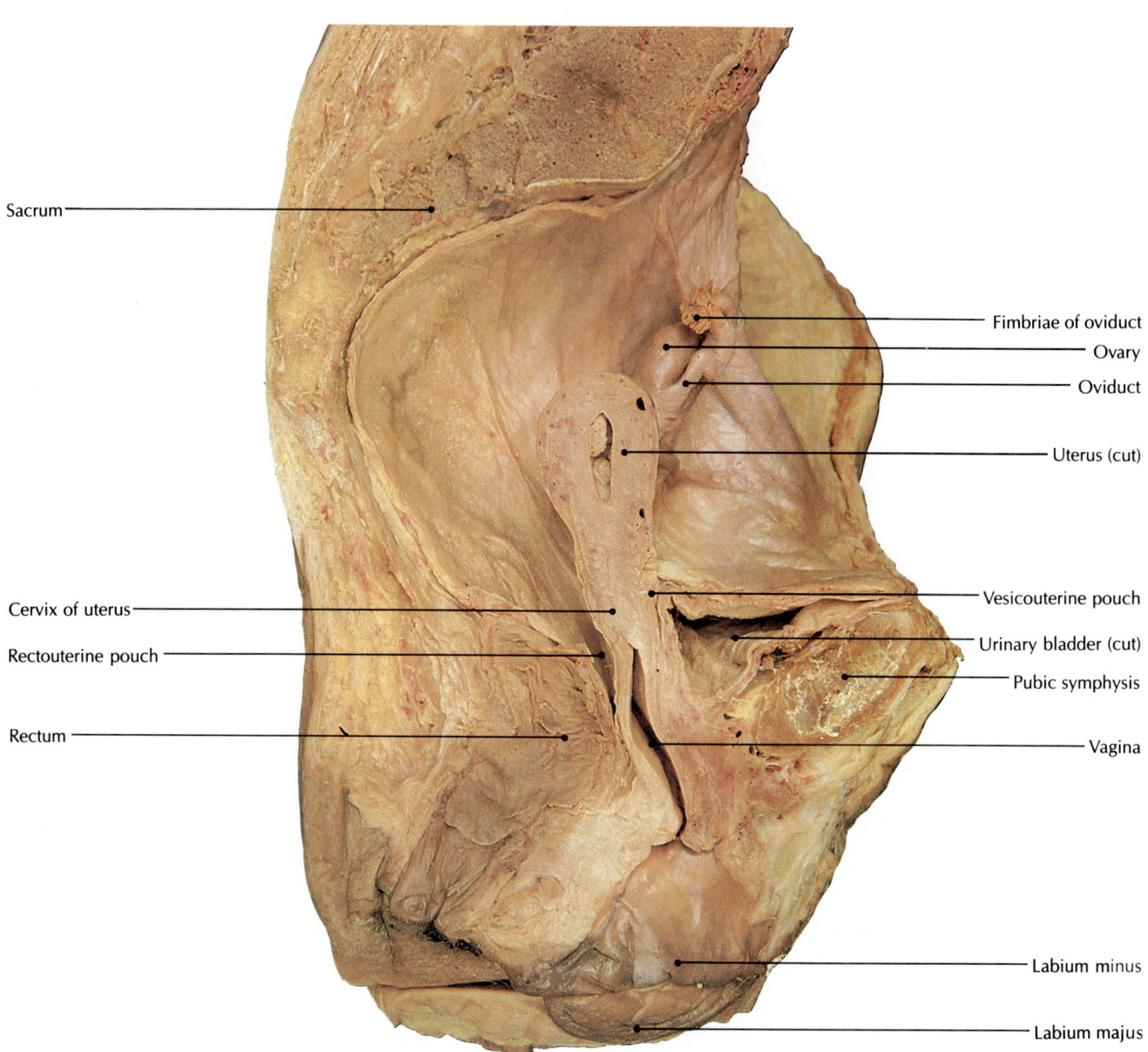

Sacrum

Fimbriae of oviduct

Ovary

Oviduct

Uterus (cut)

Cervix of uterus

Vesicouterine pouch

Rectouterine pouch

Urinary bladder (cut)

Pubic symphysis

Rectum

Vagina

Labium minus

Labium majus

FIGURE A.9 EXPOSED MALE PELVIC CAVITY AND REPRODUCTIVE ORGANS

Right sagittal view.

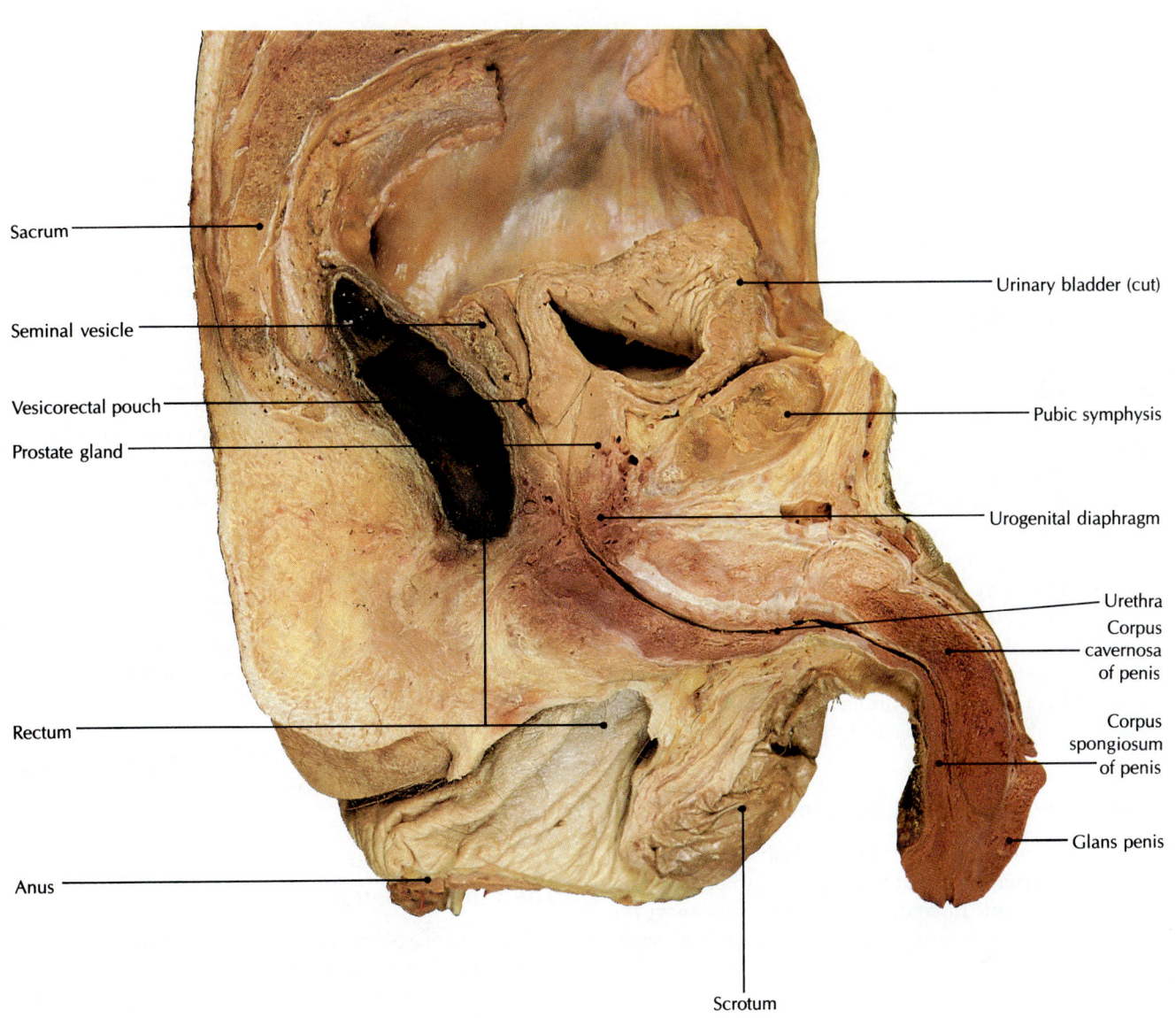

Sacrum

Seminal vesicle

Vesicorectal pouch

Prostate gland

Rectum

Anus

Urinary bladder (cut)

Pubic symphysis

Urogenital diaphragm

Urethra
Corpus cavernosa of penis

Corpus spongiosum of penis

Glans penis

Scrotum

APPENDIX B

Principles of Human Genetics: A Reference Guide

Scattered throughout various chapters of this book there have been references to genes, DNA, chromosomes, inheritance, and other concepts related to human genetics. This appendix will introduce you to some commonly used technical terms that have not been presented in our prior, more general approach to genetics.

Recessive and Dominant Alleles

Genes are located along chromosomes like beads on a string. Each gene has its own place, or *locus*, on a chromosome. Genes for a particular trait come in pairs, one on each chromosome of a homologous pair. Each member of a pair of genes is called an *allele* (ul-LEEL). There are at least two alleles for each trait. Each parent passes on one allele (*A* or *a*) to the child. The allele whose trait is physically expressed most often is considered to be *dominant* (convex nose bridge, for example), and the trait for the allele that is sometimes masked is called *recessive* (concave or straight nose bridge) [TABLE B.1]. The pair of alleles for nose shape may be written as *Aa*, where *A* stands for a convex nose bridge and *a* stands for a concave or straight nose bridge.

Homozygous and Heterozygous

When a person has a pair of identical alleles, such as *AA* or *aa*, for a given trait, that person is said to be *homozygous* for that trait. In contrast, a person is *heterozygous* for a given trait when the pair of alleles does not match (*Aa*).

Genotypes and Phenotypes

The genetic makeup, sometimes hidden, of a person is that person's *genotype* (JEEN-oh-tipe), which may be either homozygous (*AA* or *aa*) or heterozygous (*Aa*) for any one characteristic. The expression of a genetic trait (how the person looks or functions) is that person's *phenotype* (FEE-noh-tipe; Gr. "that which shows") [FIGURE B.1].

TABLE B.1 SOME DOMINANT AND RECESSIVE HUMAN GENETIC TRAITS

Dominant trait	Recessive trait
Normal color vision	Color blindness
Normal night vision	Night blindness
Long eyelashes	Short eyelashes
Dark hair	Light hair
Premature baldness (male)	Normal hair
Normal pigmentation	Albinism
Dimples in cheeks	No dimples in cheeks
Free earlobes	Attached earlobes
Ability to curl tongue	Inability to curl tongue
Convex nose bridge	Concave or straight nose bridge
Achondroplasia (dwarfism)*	Normal height
Polydactylism (extra fingers or toes)*	Normal fingers or toes
Normal arches in feet	Flat feet
Normal sugar metabolism	Diabetes mellitus
Blood group A, B, AB	Blood group O
Normal red blood cells	Sickle-cell trait
Normal blood clotting	Hemophilia
Rh antigen	No Rh antigen
Migraine headaches	Normal

*Although such traits as achondroplasia and polydactylism are dominant, they are also rare. For this reason, most people are of normal height and have 10 fingers and 10 toes.

The Punnett Square

When geneticists know the genotypes of parents for any given trait, they can determine the possible genotypes of the children for that trait by constructing a *Punnett square* [FIGURE B.2], named for the geneticist who first devised the scheme.

Multiple Alleles

Many traits are affected by more than two alleles, or *multiple alleles.* A well-known example of multiple alleles is the *ABO blood-grouping system.* In this system,

FIGURE B.1 GENOTYPES AND PHENOTYPES

Using nose shape as an example, note that a person who is heterozygous for the trait (*Aa*) has a convex nose bridge but carries the recessive allele. The recessive trait could appear in the children of such a person, depending on the genetic makeup of the other parent.

Phenotypes	Convex nose bridge	Convex nose bridge	Concave or straight nose bridge
Genotypes	**AA** (homozygous dominant)	**Aa** (heterozygous)	**aa** (homozygous recessive)
Possible gametes	All **A**	½ **A**, ½ **a**	All **a**

there are three different alleles for blood type: A, B, and O. However, any individual will have only two of these alleles, one on each chromosome of a homologous pair. Different combinations of the three alleles produce four different blood types: A, B, AB, or O. TABLE B.2 shows how the blood types of the parents may be combined to produce the child's blood type.

TABLE B.2 INHERITANCE OF ABO BLOOD TYPES

Blood type of parents	Children's blood type possible	Children's blood type not possible
A + A	A, O	AB, B
A + B	A, B, AB, O	—
A + AB	A, B, AB	O
A + O	A, O	AB, B
B + B	B, O	A, AB
B + AB	A, B, AB	O
B + O	B, O	A, AB
AB + AB	A, B, AB	O
AB + O	A, B	AB, O
O + O	O	A, B, AB

FIGURE B.2 USING A PUNNETT SQUARE

Punnett squares are used to show the possible offspring of parents of known genotypes. To use this scheme, begin by writing the two alleles for one parent above the top pair of boxes. Then write the two alleles for the other parent next to the left pair of boxes. Fill in each of the four boxes with two letters (alleles)—one from above and one from the left. The four boxes in the square show the possible offspring. In [A], one parent is homozygous, and the other parent is heterozygous for a trait. All offspring will show the dominant trait (*A*), but it is possible that half the offspring would be heterozygous and carry the recessive allele too. In [B], the chances are that only one in four children would show the recessive trait, but three out of four would carry the recessive allele.

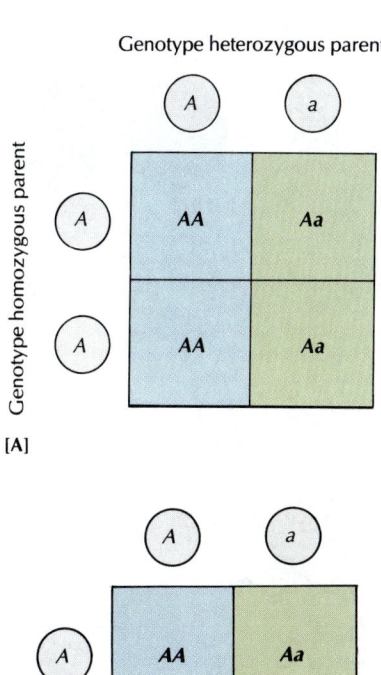

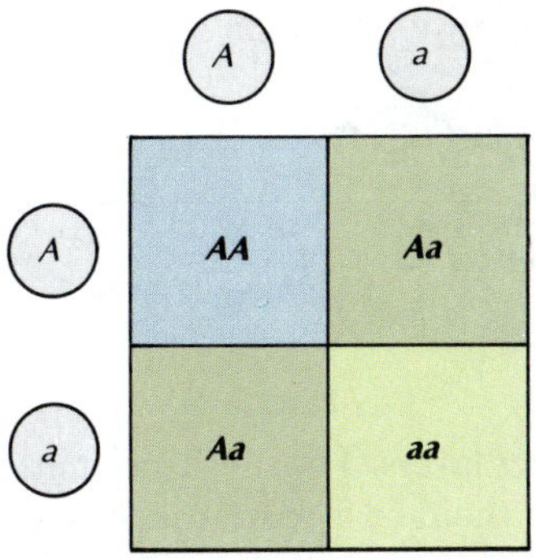

Polygenic Inheritance

Most human traits are affected by *several* pairs of genes, not just one. This type of inheritance is called *polygenic inheritance,* and the genes are called *polygenes.* Polygenes create traits with a wide range, such as skin color [FIGURE B.3A]. Skin color is neither pure black nor pure white. The degree of lightness or darkness depends on the thickness of the skin and the amount of melanin pres-

FIGURE B.3 POLYGENIC INHERITANCE

[A] If skin pigmentation is determined by four pairs of melanin-producing genes, then parents with medium coloration might have four melanin-producing and four neutral genes. As shown in the diagram, their children could have any of nine possible combinations of genes, ranging from eight melanin-producing genes to none. It is thus possible for moderately pigmented parents to produce children who have very dark or very light skin, or any of seven intermediate shades. [B] Height distribution of adult males in the United States. Notice that a typical bell-shaped curve is formed, with the most common heights in the middle of the curve.

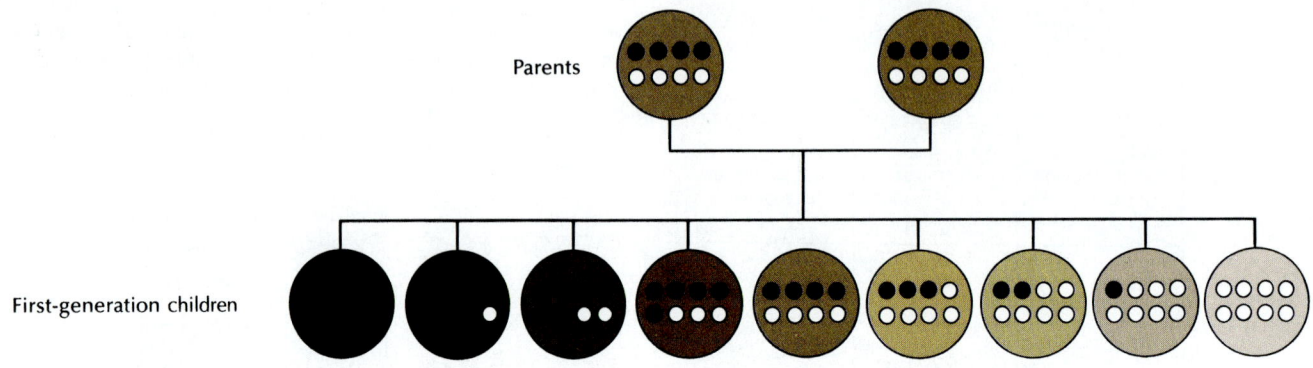

Parents

First-generation children

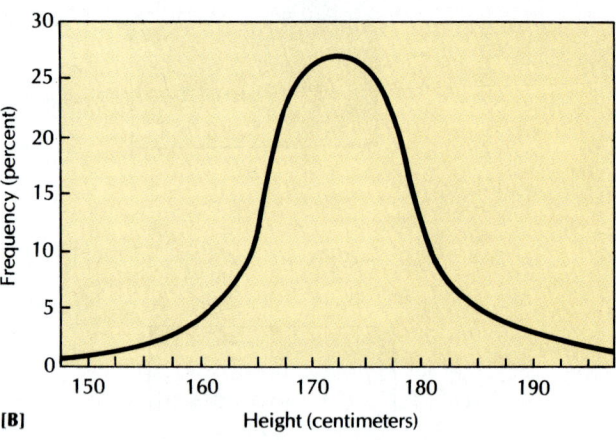

[B]

ent in the outer layers. The more melanin, the darker the skin. Current thought is that at least four genes are involved in the expression of skin color, with the resulting possibility of nine different skin shades [FIGURE B.3A.] Body height is also an example of polygenic inheritance. Consider, for example, that most people are neither giants nor dwarfs. A few are very short, many are of medium height, and a few are very tall [FIGURE B.3B].

Sex-Linked Inheritance

Parental sex chromosomes transmit not only sexual characteristics but also certain other traits in what is known as *sex-linked inheritance.* Most sex-linked genes appear on the X (maternal) chromosome, which is larger than the Y (paternal) chromosome and thus has room for more genes. Y-linked traits are passed from fathers to sons, never to daughters. In contrast, X-linked inheritance is more widespread, with more than 100 X-linked traits known, including hemophilia and color blindness.

APPENDIX C

Review of Chemical Solutions

Some of the most interesting and important substances in nature are solutions. A *solution* is a mixture containing at least two substances, each of which is dispersed evenly throughout. It is a *homogeneous* mixture because the composition of the solution is the same in every part of the whole.

The oceans, which cover approximately three-quarters of the earth's surface and contain a multitude of fascinating life-forms, are actually *liquid solutions*. However, solutions are not limited to liquids. The air we breathe is a *gaseous solution*. Our digestive processes are carried out in solutions, and our blood is a solution. Overall, the homeostasis of all living things is determined by the properties of solutions.

Solvent and Solute

In solutions, the substance responsible for dissolving the other is called the *solvent*. The *solute* is the substance being dissolved. For example, if water and sugar are mixed, water is the solvent and sugar is the solute. At times it is difficult to determine which is the solute and which is the solvent, as, for example, in a mixture of alcohol and water. Usually, however, the solvent is present in larger amounts.

Solubility

The amount of substance dissolved in a given volume of solvent is known as the *concentration* of the solution. The ability of a substance to dissolve in a given volume of solvent is called its *solubility* (L. *solvere*, to loosen). Solubility is expressed in terms of grams of the solute that can be dissolved in 100 mL of solvent — for example, 100 grams in 100 mL of a solvent such as water (100 g/ 100 mL or 1 g/mL). A *dilute solution* is one that contains a small amount of the dissolved substance. A *concentrated solution* is one that contains a large amount of the dissolved substance.

Saturation

As a solute dissolves, the solution becomes increasingly populated with the molecules of the solute. When the solvent holds all the solute that it can take up directly, the solution is said to be *saturated* (L. *saturare*, to fill) at that particular temperature and under these particular conditions. Any solution that contains a smaller amount of solute is said to be *unsaturated*.

Expressing Concentration of Solutions

To determine the amount of solute in solution, the following methods are used:

Ratio	*Example:* 1:1000 *or* 1 part per 1000
Weight per unit volume	*Example:* 10 g/L *or* 10 g in 1000 mL
Percentage	*Example:* 5-percent salt solution *or* 5 g of NaCl diluted in up to 100 mL of water
Volume-volume	*Example:* 5 mL of phenol + 95 mL of water

Molarity of Solutions

Strengths or concentrations of solutions used in physiology are commonly expressed in terms of *molarity*. A *molar solution* is one that contains 1 gram molecular weight (1 mole) of a solute dissolved in 1 liter of solvent. To determine the molecular weight of a compound, let us take the example of sodium chloride (NaCl). In the Periodic Table of the Elements, find the atomic weights of sodium (Na) and chlorine (Cl), round each to the nearest whole number, and add them:

Na = 23   Cl = 35
Molecular weight of NaCl = 58

The molarity of a solution is designated by the letter M. A 1-M NaCl solution contains 1 mole of NaCl dissolved in 1 L (1000 mL) of water. So, 58 g of NaCl dissolved in up to 1000 mL of water is a 1-molar solution. A 0.1-M NaCl solution contains 1/10 of a mole (or 0.1 of the molecular weight) dissolved in 1 L of water. A formula that is helpful in remembering this relationship of solutions is

$$\text{Molar solution} = \frac{\text{moles of solute}}{1000 \text{ mL of solvent}}$$

Molal Solutions

A *molal solution* contains 1 mole of a solute for every 1000 grams of solvent.* This volume, which is *not* a liter, varies with the volume of the formula weight of dissolved solute.

Generally, molal solutions are used whenever the *relative numbers* of solute and solvent molecules are determining factors. That is, molal solutions provide a definite ratio of solute to solvent molecules.

Molal solutions are used in studying various physical properties of solutions, such as vapor pressure, boiling and freezing points, and osmotic pressure.

Normality

Normality (N) is the number of equivalents (eq) or gram equivalent masses of solute dissolved in 1 L of solution. It is expressed by the formula

$$N = \frac{\text{equivalents of solute}}{1 \text{ L of solution}}$$

To prepare 1 L of a normal (1.00 N) solution of a substance, weigh out 1 gram equivalent mass of solute, place the weighed solute in a 1-L flask, and add enough solvent to make the *final volume* of the solution 1 L. The procedure and concept are the same as for molarity, except that 1 gram equivalent mass (1.00 eq) of solution is weighed out instead of 1 gram molecular mass (1.00 mole).

Milliequivalent

A *milliequivalent* (mEq) is the same as a *millimole* (0.001 mol, or 1 formula weight in milligrams) when the *valence*, or the number of charges on the ion, is 1. The milliequivalent is 1/1000 of an equivalent. An *equivalent* (eq) or *equivalent weight* (eq wt) is the atomic weight divided by the valence. The number of milliequivalents of a specific ion in solution (1 L) can also be expressed by the formula

$$\text{mEq/L} = \frac{\begin{array}{c}\text{milligrams of ion}\\ \text{per liter of solution}\end{array}}{\text{atomic weight}} \times \begin{array}{c}\text{number of charges}\\ \text{on the ion (valence)}\end{array}$$

Using the formula above, let us calculate the milliequivalents of calcium in normal blood. In 1 L of plasma

there are 100 mg of calcium, and the Ca^{2+} has two charges (indicated by the $^{2+}$). Using these values in the formula, we have

$$\text{mEq/L} = \frac{100}{40} \times 2 = 5$$

The Mole Concept

Because atoms, molecules, and formula units are so small compared to everyday objects and units, chemists and physiologists prefer to think of them in very large quantities, and they use a special number. This number is called *Avogadro's number* (in honor of the Italian physicist and chemist Amedeo Avogadro, 1776–1856). It is equal to 6.02×10^{23} atoms, molecules, or formula units of a particular chemical substance.

A *mole* (mol) is the amount of substance represented by 6.02×10^{23} atoms, molecules, or formula units of that particular substance. The *gram atomic mass* of an element is the mass in grams of 1.000 mol or 6.02×10^{23} atoms of a naturally occurring mixture of its isotopes. It is numerically the same as the atomic mass in atomic mass units (amu). The *gram molecular mass* in grams of a mainly covalent compound is 1.000×10^{23} molecules. The *gram formula mass* of a mainly ionic compound is the mass in grams of 1.000 mol or 6.02×10^{23} formula units.

Osmolarity

The term *osmolarity* defines particle concentration. An *osmole* is 6.02×10^{23} moleculs of a solution or mixture of solutes of any kind. Thus an osmole of solute dissolved in 1000 mL (1 L) of solution constitutes a *1-osmolar solution*. When determining osmolarity, it is important to consider the effective concentration of all particles. For example, a 0.1-osmolar solution could be a mixture of salts, sugars, amino acids, and proteins. The osmolarity of a solution is always measured in an ideal situation, for example, with the solution separated from pure water by a perfectly semipermeable membrane. Solution A and solution B are *isomolar* if their effective particle concentrations are equal. On the other hand, if solution A has a higher concentration of osmotically active particles than solution B, A is *hyperosmolar* to B and B is *hypoosmolar* to A.

The terms *osmolar* and *osmolarity* refer to osmoles per liter of solution. *Osmolal* and *osmolality* refer to osmoles per kilogram of water. (The distinction is similar to that between the terms *molar* and *molal*.)

---

*A mola*l* solution should not be confused with a mola*r* solution, which contains the gram molecular weight of solute in 1 L of solution.

APPENDIX D

Normal Laboratory Values for Blood and Urine

The following abbreviations are used for normal values:

↑	increased	L	liter	ng	nanogram	μL	microliter
↓	decreased	M	molar	nmol	nanomole	mm	millimeter
>	greater than	mol	mole	pg	picogram	mm^3	cubic millimeter
<	less than	m^2	square meter	sec	second	mmHg	millimeter of mercury
cm^3	cubic centimeter	mCi	millicurie	SI	International	mmol	millimole
mm^3	cubic millimeter	mEq	milliequivalent		System of Units*	mOsm	milliosmole
dL	deciliter	mg	milligram	U	unit	mU	milliunit
g	gram	mg/dL	milligram per deciliter	μm	micrometer	WBCs	white blood cells
IU	international unit	mL	milliliter	μm^3	cubic micrometer		
kg	kilogram	RBCs	red blood cells	μg	microgram		

*SI units are *Système International* units that are used uniformly in the European literature. The benefits consist of scientific standardization in reporting.

TABLE D.1 NORMAL LABORATORY ADULT BLOOD (CHEMISTRY) VALUES

Determination (test)	Normal values	Clinical significance
Acetone (ketone bodies)	Acetone: 0.3–2.0 mg/dL Ketones: 2–4 mg/dL	Values increase in diabetic ketoacidosis (dka), starvation, malnutrition, heat stroke, diarrhea.
Acid phosphatase (ACP)	0.1–2 U/dL (Gutman)	Values increase in prostate cancer, sickle-cell anemia, cirrhosis renal failure, myocardial infarction; values decrease in Down's syndrome.
Alanine amino-transferase (ALT, SGPT)	5–35 U/mL (Frankel)	Values increase in viral hepatitis, cirrhosis, congestive heart failure; values decrease in exercise.
Albumin	3.5–5 g/dL	Values increase in kidney disease, fever, trauma, and myeloma; values decrease in severe burns (proteinuria), starvation, leukemia, chronic liver disease.
Alkaline phosphatase (ALP)	159–400 mg/dL; 1.0–1.6 g/L	Values increase in biliary disease, cancer of liver, cirrhosis, hepatitis, arthritis, ulcers; values decrease in malnutrition, pernicious anemia.
Ammonia	3.2–4.5 g/dL 32–45 g/L (SI units)	Values increase in hepatic failure, hemolytic disease of the newborn, congestive heart failure, emphysema; values decrease in renal failure, hypertension.
Amylase	60–160 U/dL (Somogyi); 111–296 U/L (SI units)	Values increase in pancreatitis, ulcers, cancer, diabetic acidosis, mumps, renal failure, burns, pregnancy; values decrease in necrosis of liver, chronic alcoholism, hepatitis, severe burns.
Aspartate amino-transferase (AST, SGOT)	5–40 U/mL; 4–36 IU/L	Values increase in acute myocardial infarction, hepatitis, liver necrosis, trauma, liver cancer, angina; values decrease in pregnancy, diabetic ketoacidosis, beriberi.
Bilirubin (total and direct)	Total: 0.1–1.2 mg/dL; 1.7–20.5 μmol/L (SI units) Direct: 0.1–0.3 mg/dL; 1.7–5.1 μmol/L (SI units)	Values increase in obstructive biliary disease, liver disease, liver cancer.
Blood urea nitrogen (BUN)	Male: 10–25 mg/dL Female: 8–20 mg/dL	Values increase in dehydration, high protein intake, GI bleeding, prerenal failure, diabetes mellitus, myocardial infarction, kidney disease; values decrease in severe liver damage, low-protein diet, malnutrition.

(Table D.1 continues on following page)

TABLE D.1 NORMAL LABORATORY ADULT BLOOD (CHEMISTRY) VALUES (Continued)

Determination (test)	Normal values	Clinical significance
Calcium (Ca)	4.5–5.5 mEq/L; 9–11 mg/dL; 2.3–2.8 mmol/L (SI units)	Values increase in hypervitaminosis D, bone cancer, lung cancer, multiple fractures, renal calculi, alcoholism; values decrease in diarrhea, laxative abuse, extensive infections, burns, hypoparathyroidism, alcoholism, pancreatitis.
Carbon dioxide combining power (CO_2)	22–30 mEq/L; 22–30 mmol/L (SI units)	Values increase in respiratory diseases, intestinal obstruction, vomiting; values decrease in acidosis, nephritis, diarrhea, starvation.
Chloride (Cl)	95–105 mEq/L; 95–105 mmol/L (SI units)	Values increase in nephritis, Cushing's syndrome, hyperventilation; values decrease in diabetic acidosis, burns.
Cholesterol (total) HDL (α) LDL (β) VLDL (pre-β)	150–250 mg/dL; 3.90–6.50 mmol/L (SI units) May increase with age. 29–77 mg/dL 62–185 mg/dL 0–40 mg/dL	Values increase in acute myocardial infarction, atherosclerosis, hyperthyroidism, biliary obstruction, diabetes mellitus, stress; values decrease in Cushing's disease, starvation, anemia, malabsorption.
Cholinesterase	0.5–1.0 units (RBC); 3–8 units/mL plasma; 6–8 IU/L (RBC); 8–18 IU/L (plasma)	Values increase in nephrotic syndrome; values decrease in insecticide poisoning, liver disorders, acute infection, anemia, carcinomatosis.
Cortisol (hydrocortisone, compound F)	8–10 A.M.: 5–23 μg/dL 4–6 P.M.: 3–13 μg/dL	Values increase in adrenal cancer, stress, pregnancy, obesity, myocardial infarction, diabetic acidosis, hyperthyroidism; values decrease in Addison's disease, respiratory distress syndrome, hypothyroidism.
Creatine phosphokinase (CBK)	Male: 5–35 μg/mL; 15–120 IU/L Female: 5–25 μg/mL; 10–80 IU/L	Values increase in muscular dystrophy, trauma, hypokalemia, myocardial infarction, hemophilia, tetanus, venom toxin; values decrease in pregnancy.
Creatinine (Cr)	0.6–1.2 mg/dL; 53–106 μmol/L (SI units)	Values increase in renal failure, shock, systemic lupus erythematosus, cancer, hypertension, myocardial infarction, diabetic nephropathy, high-protein diet; values decrease in pregnancy, eclampsia.
Gamma-glutamyl transpeptidase (GGTP)	Male: 10–38 IU/L Female: 5–25 IU/L	Values increase in cirrhosis of the liver, alcoholism, viral hepatitis, liver cancer, mononucleosis, diabetes mellitus, myocardial infarction, congestive heart failure, epilepsy.
Glucose (fasting)	70–110 mg/dL	Values increase in diabetes mellitus, liver disease, stress, nephritis, pregnancy, hyperthyroidism; values decrease in hypothyroidism, Addison's disease, pancreatic cancer.
Iron	50–150 μg/dL; 10–27 μmol/L (SI units)	Values increase in hemochromatosis, anemia, liver damage, lead poisoning; values decrease in iron-deficiency anemia, malignancies, arthritis, ulcers, renal failure.
Lactic acid	Arterial blood: 0.5–2.0 mEq/L Venous blood: 0.5–1.5 mEq/L Panic: > 5 mEq/L	Values increase in shock, dehydration, ketoacidosis, severe infections, neoplastic conditions, hepatic failure, renal disease, alcoholism.
Lactic dehydrogenase (LDH/LD)	150–450 U/mL (Wroblewski-LaDue method)	Values increase in myocardial infarction, ketoacidosis, severe infections, neoplastic conditions, hepatic failure, renal disease, alcoholism.
Lipids (fasting) Cholesterol Phospholipids Total fatty acids Total lipids Triglycerides	 120–220 mg/100 mL 9–16 mg/100 mL 190–420 mg/100 mL 450–1000 mg/100 mL 40–150 mg/100 mL	Values increase in hyperlipoproteinemia, myocardial infarction, hypothyroidism, diabetes mellitus, eclampsia; values decrease in chronic obstructive lung disease.
Osmolality	280–300 mOsm/kg H_2O	Values increase in diabetes insipidus, dehydration, hypernatremia, hyperglycemia, uremia; values decrease in excessive fluid intake, bronchus and lung cancer, adrenal cortical hypofunction.
Phosphorus (P_i)	1.7–2.6 mEq/L; 2.5–4.5 mg/dL	Values increase in renal insufficiency, renal failure, hypocalcemia, acromegaly; values decrease in starvation, malabsorption syndrome, hypercalcemia, hyperparathyroidism, diabetic acidosis, myxedema.

TABLE D.1 NORMAL LABORATORY ADULT BLOOD (CHEMISTRY) VALUES (Continued)

Determination (test)	Normal values	Clinical significance
Potassium (K)	3.5–5.0 mEq/L	Values increase in dehydration, starvation, renal failure, diabetic acidosis, Cushing's disease; values decrease in acute renal failure, diarrhea.
Protein		Total protein values increase in severe dehydration, shock; values decrease in hemorrhage and severe malnutrition.
Albumin	3.5–5.0 g/dL	
Globulin	1.5–3.5 g/dL	
A/G ratio	1.5:1 to 2.5:1	
Total	6.0–7.8 g/dL	
Sodium (Na)	135–145 mEq/L; 135–142 mmol/L (SI units)	Values increase in dehydration, severe vomiting and diarrhea, congestive heart failure, Cushing's disease, high-sodium diet; values decrease in vomiting, diarrhea, low-sodium diet, burns, renal disease.
Thyroxine (T_4)	4.5–11.5 μg/dL (by column)	Values increase in hyperthyroidism, viral hepatitis, myasthenia gravis, pregnancy; values decrease in hypothyroidism, protein malnutrition, anterior pituitary hypofunction, exercise.
Uric acid	Male: 3.5–7.8 mg/dL Female: 2.8–6.8 mg/dL	Values increase in gout, alcoholism, leukemia, cancer, diabetes mellitus; values decrease in anemia, burns, pregnancy.

TABLE D.2 LABORATORY ADULT BLOOD (SEROLOGY) VALUES

Determination (test)	Normal values	Clinical significance
Carcinoembryonic antigen (CEA)	< 2.5 mg/mL	Values increase in cancer of the GI system, heart disease, colitis, inflammatory disease.
Heterophile antibody	< 1:28 titers*	Values increase in mononucleosis, serum sickness, viral infections.
Immunoglobulins (Ig)		
Ig	900–2200 mg/dL	Values increase in malnutrition, liver disease, rheumatic fever; values decrease in leukemia, pre-eclampsia, amyloidosis.
IgG	800–1800 mg/dL	Same as Ig.
IgA	100–400 mg/dL	Values increase in autoimmune disorders, rheumatic fever, chronic infections; values decrease in leukemia, malignancies.
IgM	50–150 mg/dL	Values increase in parasitic infections; values decrease in leukemia.
IgD	0.5–3 mg/dL	Values increase in chronic infections and myelomas.
IgE	0.01–0.05 mg/dL	Values increase in allergies.
Rheumatoid factor	< 1:20 titers	Values increase in rheumatoid arthritis, SLE, scleroderma, mononucleosis, tuberculosis, leukemia, hepatitis, syphilis, old age.

*A titer is the concentration of antibody in serum.

TABLE D.3 NORMAL ADULT BLOOD (GAS) VALUES

Arterial O_2 saturation	$\dfrac{O_2 \text{ content}}{O_2 \text{ capacity}} \times 100 = 95\%$	Mixed venous O_2 content	10–16 mL/100 mL of blood
		HCO_3^- content	24–28 mEq/L
Pulmonary arterial O_2 saturation	$\dfrac{O_2 \text{ content}}{O_2 \text{ capacity}} \times 100 = 75\text{–}80\%$	pH arterial plasma	7.35–7.45
		pH venous plasma	7.32–7.43
Whole blood O_2 capacity	17–21 mL of O_2/100 mL of blood	CO_2 combining power (venous plasma)	21–30 mEq/L
Arterial O_2 content	16.5–20.0 mL/100 mL of blood	Arterial CO_2 content (whole blood)	20–25 mEq/L
		Arterial and alveolar CO_2 tension (Pa_{CO_2})	37–41 mmHg
P_{O_2} arterial content	80–95 mmHg	P_{CO_2} arterial tension	35–45 mmHg
P_{O_2} venous content	35–40 mmHg	P_{CO_2} venous tension	38–50 mmHg

TABLE D.4 NORMAL LABORATORY ADULT BLOOD (HEMATOLOGY) VALUES

Determination (test)	Normal values	Clinical significance
Bleeding time	Duke method: 1–5 min Ivy method: 2–7 min	Time increases in liver disease, anemia, leukemia, thrombocytopenia; used to evaluate platelet function.
Capillary fragility	5 or less	Values increase in faulty capillary endothelial integrity, easy bruising, bleeding.
Coagulation time (Lee-White)	5–15 min; average 8 min	Time increases in afibrinogenemia and hyperheparanemia; used to evaluate blood-clotting system.
Erythrocyte sedimentation rate (ESR) (zeta sedimentation ratio; ZSR)	Under 50 years old (Westergren): Male: 0–15 mm/hr Female: 0–20 mm/hr Over 50 years old (Westergren): Male: 0–20 mm/hr Female: 0–30 mm/hr	Values increase in infection, inflammation, tissue necrosis, pregnancy, malignancy; values decrease in sickle-cell anemia, spherocytosis, congestive heart failure.
Hematocrit (HCT) or packed cell volume (PCV)	Male: 40–50%; 0.40–0.54 SI units Female: 36–46%; 0.36–0.46 SI units	Values increase in polycythemia, shock, severe dehydration; values decrease in anemia, blood loss, leukemia, liver disease, hyperthyroidism.
Hemoglobin (Hb or Hgb)	Male: 13.5–18 g/dL Female: 12–16 g/dL	Values increase in obstructive pulmonary disease, congestive heart failure, polycythemia, high altitudes; values decrease in anemia, blood loss, hemolysis, hyperthyroidism, liver disease, pregnancy, excessive fluid intake.
Platelet (thrombocyte) count	150,000–400,000/mm^3 (mean 250,000/mm^3) SI units: 0.15–0.4 × 10^{12} L	Values increase in some anemias, polycythemia, cancer, liver disease, trauma; values decrease in leukemia, aplastic anemia, allergies, cancer chemotherapy.
Prothrombin time (PT)	11–15 sec or 70–100%	Values increase in certain hemorrhagic diseases, liver disease, vitamin K deficiency, various drug use, congenital deficiencies, disseminated intravascular coagulation.
Red blood cell count (RBC)	Male: 4.6–6.0 million/mm^3 Female: 4.0–5.0 million/mm^3	Values increase in dehydration, posthemorrhaging, polycythemia, cor pulmonale, cardiovascular disease; values decrease in anemia, Addison's disease, hemorrhage, chronic infections, leukemia, chronic renal failure, pregnancy.
Reticulocyte count	0.5–1.5% of all RBCs; 25,000–75,000 mm^3 (absolute count)	Values increase in hemolytic and sickle-cell anemia, thalassemia, hemorrhage, iron deficiency, leukemia, hemolytic disease of the newborn, pregnancy; values decrease in pernicious and aplastic anemias, radiation therapy, liver cirrhosis, anterior pituitary hypofunction.
White blood cells (WBCs) or leukocyte total	5000–10,000 mm^3	Values increase in acute infections, myocardial infarction, cirrhosis, burns, cancer, arthritis, gout; values decrease in aplastic and pernicious anemias, viral infections, malaria, alcoholism, uncontrolled diabetes.
White blood cell differential		

Neutrophils
    Segments
    Bands
	50–70% of total WBCs
	50–65%
	0–5%
		Values increase in acute infections, inflammatory disease, lung disease, myocardial infarction.

Eosinophils	0–3%	Values increase in allergies, parasitic diseases, cancer, asthma, kidney disease; values decrease in stress, burns.
Basophils	1–3%	Values increase in inflammation, leukemia; values decrease in stress.
Lymphocytes	25–35%	Values increase in leukemia, viral infections; values decrease in cancer, neurological disorders, renal failure.
Monocytes	2–6%	Values increase in viral diseases, parasitic diseases, leukemia, cancer, arthritis, some anemias; values decrease in lymphocytic leukemia, aplastic anemia.

TABLE D.5 NORMAL LABORATORY ADULT URINE VALUES

Determination (test)	Normal values	Clinical significance
Acetone plus acetoacetate (ketone bodies)	Negative	Values increase in ketonuria, diabetic acidosis, ketoacidosis of alcoholism and diabetes mellitus, fasting, starvation, high-protein diet.
Addis count (12 hr)	Casts: 0–5000 Erythrocytes: 0–500,000 Leukocytes: < 1,000,000	Evaluation of severity and course of glomerulonephritis, during which values increase.
Albumin (quantitative protein)	0–< 20 mg/24 hr	Values increase in renal disease, diabetes mellitus, glomerulonephritis, amyloidosis; values decrease following severe burns.
Ammonia (nitrogen)	20–70 mEq/L	Values increase in diabetes mellitus, liver disease.
Amylase	4–30 IU/2 hr; 24–76 U/mL	Values increase in carcinomas of pancreas, pancreatitis.
Bile (bilirubin)	Negative	Values increase in biliary and liver disease, obstructive disease of biliary tract, cancers.
Calcium	< 300 mg/24 hr	Values increase in hyperparathyroidism; values decrease in hypoparathyroidism.
Concentration test	Specific gravity > 1.026 or 850 mOsm/kg	Used to evaluate renal concentrating ability, a test of tubular function; values decrease in renal disease.
Creatinine	1–2 g/24 hr	Values increase in bacterial infections; values decrease in muscular dystrophy, leukemia, anemia.
Creatinine clearance	150–180 L/day; 100–140 mL/min	Values increase in renal disease.
Fat (lipid)	Negative	Values increase in nephrotic syndrome, renal tubular necrosis, poisoning.
Glucose (sugar)	Negative	Values increase in renal disease; helpful in evaluating glucosuria, renal tubular defects, and in managing diabetes mellitus.
Hemoglobin (free)	Negative	Presence indicates hemoglobinuria as a result of hemolysis.
Nitrite	Negative	Presence indicates bacterial infection, urinary tract infection.
Occult blood	Negative	Presence indicates hematuria, hemoglobinuria.
pH	4.8–7.8	Used to determine subtle presence of distal tubular disease or pyelonephritis, kidney stones. Values increase in vegetarian diet, urinary tract infection, diabetes mellitus, alkalosis; values decrease in acidosis, starvation, emphysema, dehydration.
Phenylalanine screen (phenylpyruvic acid)	Negative	Values increase in phenylketonuria (PKU).
Specific gravity	Range: 1.001–1.035 Normal fluid intake: 1.016–1.022	Values increase (> 1.020) in dehydration, fever, diarrhea, diabetes mellitus, congestive heart failure; values decrease (< 1.009) in diuresis, hypothermia, diabetes insipidus, glomerulonephritis.
Urea	25–35 g/24 hr	Values increase in excessive protein breakdown; values decrease in impaired renal function.
Urea clearance	> 40 mL of blood cleared of urea/minute	Values increase in renal diseases.
Uric acid	0.6–1.0 g/24 hr as urate	Values increase in gout; values decrease in kidney disease.
Urobilinogen (2 hr urine)	Male: 0.3–2.1 mg Female: 0.1–1.1 mg	Values increase in hemolytic anemias, myocardial infarction, liver disease, hyperthyroidism; values decrease in bile duct obstruction, antibiotic overload, hepatitis, starvation.

APPENDIX E

Answers to Case Studies and Self-Quizzes

ANSWERS TO PART-CLOSING CASE STUDIES

Part I (p. 130)

1
- The body cavities include the ventral, thoracic, pleural, and pericardial cavities.
- The bullet passed through the visceral pericardium, parietal pericardium, visceral pleura, and parietal pleura.
- X-rays were probably used to determine where the bullet was lodged.

2
- The cells, like all normal red blood cells, would have looked like biconcave disks.
- The red blood cells in Jake's sample would have looked swollen; some would even have ruptured or hemolyzed. Because Jake diluted his sample with water, the red blood cells in his vial contained a higher concentration of solutes than the suspending solution. Water then entered the cells, causing them to swell and burst.
- Osmosis is the passive movement of water across a semipermeable membrane, from the region in which water is in higher concentration to the region in which it is lower concentration. Because Jake's cells had a higher solute concentration, they had a lower concentration of water, which then flowed down its concentration gradient into the cells.

3
- PET scans analyze gamma-ray emissions from metabolites that have been labeled with radioisotopes, while x-rays are electromagnetic waves that are differentially absorbed by body parts. Gamma radiation is of shorter wavelength and higher energy than x-rays. PET scans can illustrate the level of metabolic activity in active tissue. X-rays are useful for studying structure, especially of the body's hard tissues, but they measure differences in density, not metabolism. Magnetic resonance imaging (MRI) depends on the energy emitted from hydrogen molecules as they change their spin in a magnetic field. MRI is especially useful for visualizing soft tissues, especially those that are encased in a bony protective layer, such as the brain or spinal cord.
- X-rays have been ordered for the elderly woman with the probable bone fracture. The PET scan will be useful in viewing differences in metabolic rate in different areas of the brain in the probable stroke patient, and the MRI will clearly reveal the sclerotic plaques in the patient with multiple sclerosis.
- None of these patients would benefit from a sonogram because the high-frequency sound waves that form the energy source for a sonogram do not penetrate deeply into the body nor do they penetrate bone.

4
- The three major types of skin cancer are (1) basal cell carcinoma, (2) squamous cell carcinoma, and (3) malignant melanoma. Basal cell carcinomas arise from the epidermis. They are the least malignant and most common, accounting for about 75 percent of all skin cancers. Squamous cell carcinomas also arise from the epidermis and often metastasize to the adjacent lymph nodes if not removed. Malignant melanomas arise from melanocytes in the epidermis. They are the least common (about 5 percent of the skin cancers) but the most malignant.
- The term *basal* refers to the epidermal stratum basale, which rests on the basement membrane next to the dermis. The stratum basale consists of a single layer of cuboidal or columnar cells that rapidly divide to provide the skin with cells to replace those that are constantly being shed from the superficial layers.
- The most common treatment for basal cell carcinoma is surgical excision of the affected area.

Part II (p. 350)

1
- A second-degree burn includes redness, blisters, and swelling from fluid that leaves capillaries and enters the tissue fluid in the affected area.
- Both the epidermis and the dermis were damaged.
- Even though the immediate effects of the sunburn disappear within 2 weeks, there is a correlation between severe sunburn and the future development of skin cancer, especially malignant melanoma.

2
- Fingers are easy to x-ray. Each phalanx forms by endochondral ossification and contains a diaphysis and epiphyses. Many of the bones of the skull and the clavicle form by intramembranous ossification and do not have a diaphysis and epiphyses.
- She was looking for a cartilaginous epiphyseal growth plate.
- As long as a region of cartilage remains in an endochondral bone, the bone is capable of growth in length. Once the cartilage has been completely replaced by bone, the bone can grow in girth and can remodel, but it cannot grow longer.

3
- Tennis elbow is an inflammation of the tendon common to the forearm extensors and supinator muscles where they attach to the lateral epicondyle of the humerus.
- Tennis requires more forearm rotation and extension than many other sports. However, the condition may occur with any repetitive motion of these muscles, such as that used by baseball pitchers.
- During a forehand stroke, the forearm flexors contract. They are the antagonists of the extensors and cause stretching of the inflamed extensor tendon.

4
- The clavicle is commonly called the collarbone.
- In his fall, significant force was transmitted up the boy's arm, from the head of the humerus through the acromion of the scapula, where it fits on the distal end of the clavicle. Because this bone is the first to ossify and is brittle in young children, it is frequently broken in falls of this kind.
- The collarbone serves as a brace between the sternum and the clavicle. When the clavicle breaks, the shoulders curl forward and the upper body tends to droop forward.

Part III (p. 582)

1
- The gymnast probably has a herniated or ruptured intervertebral disk. The nucleus pulposus of the injured disk is exerting pressure on the rootlets of the spinal nerves or even on the spinal cord itself.
- The symptoms are those of sciatica, implicating the sciatic nerve. The sciatic nerve arises from the sacral plexus and carries fibers from spinal nerves L4–S3.
- Both the sensory and motor rootlets are being compressed. The numbness and dull, radiating pain are indicative of compression of sensory rootlets; the mild motor loss results from compression of the motor rootlets.
- CT and MRI scans would help confirm the diagnosis of a disk disorder.

2
- The condition is called tic douloureux or trigeminal neuralgia. The nerve involved is the trigeminal nerve (cranial nerve V).
- The trigeminal nerve is composed of the maxillary, mandibular, and ophthalmic branches.
- The excruciating pain of tic douloureux is caused by inflammation of the trigeminal nerve. It can arise from a viral infection or from certain sensory stimuli, such as occur when blasts of air hit the face, putting pressure on the nerve.

3
- The professor is nearsighted (myopic). He is able to see near objects well.
- Accommodation is the reflex adjustment of the lens of the eye to permit rapid changes in focal length.
- For focusing on near objects, the ciliary muscles contract. The ciliary body is pulled forward and inward, reducing the tension on the suspensory ligaments attached to the lens capsule. When the tension on the lens is reduced, the elastic lens becomes thicker and rounder and therefore capable of the greater refraction required for seeing objects that are close. When distant objects are viewed, the ciliary muscles relax. As they relax, the tension on the suspensory ligaments increases, and the lens becomes

thinner (flatter). In that condition, the lens is capable of the lessened refraction that is required for distant objects.
- The professor's problem is age. The lens becomes increasingly less elastic with age, making it difficult to focus on near objects. When he reads from a book or magazine, he is able to move the page far enough away to focus, but notes on a lectern are not as easily moved.

4
- Norepinephrine is released by sympathetic postganglionic cells. Acetylcholine is released by parasympathetic postganglionic neurons. Norepinephrine is an adrenergic agent while acetylcholine is a cholinergic agent.
- Norepinephrine binds to both alpha and beta receptors. Acetylcholine interacts with nicotinic and muscarinic receptors.
- Alpha and beta adrenergic blocking agents, such as phentolamine and propranolol, can block sympathetic activity. Muscarinic cholinergic agents such as atropine can block parasympathetic effects.
- Nicotinic blocking agents, such as curare and hexamethonium, can block both sympathetic and parasympathetic effects at the same time because the preganglionic fibers of both autonomic divisions release acetylcholine as their neurotransmitter. Sympathetic and parasympathetic activity is triggered when acetylcholine binds to nicotinic cholinergic receptors on postganglionic cells.

Part IV (p. 938)

1
- An atrial-septal defect is an abnormal opening in the septum between the two atria. The opening enables a small amount of blood to flow from the right atrium directly into the left atrium, bypassing the right ventricle and thereby missing the oxygenation that takes place in the pulmonary circulation.
- The blood that passes through the atrial-septal opening remains poorly oxygenated. When it mixes with the blood returning to the left atrium from the lungs, it lowers the overall partial pressure of oxygen of the blood that will be pumped from the left ventricle through the systemic circulation.
- An atrial-septal defect results from a weakened atrial septum after the closure of the foramen ovale (the septal opening in the fetal heart between the right and left atria, which normally closes at birth).

2
- Mowing the grass may cause air currents that place the asthmatic in the path of airborne allergens of plant and animal origin. These allergens interact with receptors on bronchial cells, causing contraction of the smooth muscle (bronchospasms and bronchoconstriction), edema of the airways, and increased mucus secretion.
- Normal exhalation is a passive process that results from elastic recoil of the lungs when the diaphragm and intercostal muscles relax. During an asthma attack, the airways are partially constricted, increasing the resistance to the flow of air out of the lungs. Inhalation is an active process in which the diaphragm and intercostal muscles contract. Thus the asthma sufferer has some muscular power to push air against the increased resistance. As a result, it is easier to get air into the lungs than out of the lungs during bronchoconstriction.
- Many bronchodilators contain beta-adrenergic agonists

such as epinephrine, which interact with the beta receptors on the bronchial smooth muscle to cause smooth muscle relaxation (bronchodilation). Albuterol (Proventil) is a more selective beta agonist, selectively interacting with particular beta receptors to initiate bronchodilation.

3

- Concentrated urine is dark. The yellow color is due to pigments produced from bile breakdown (urochromes). The more concentrated the urine, the darker the color.
- Decreased fluid intake produces increased blood osmolarity. Osmoreceptors in the brain sense the increase in osmolarity and trigger an elevated output of ADH, which increases the permeability to water in the distal convoluted tubules and collecting tubules of the kidney. Thus water leaves the urine and is reabsorbed into the blood in an attempt to maintain blood osmolarity within normal limits.
- Ingestion of hypotonic fluids such as water decreases blood osmolarity, which is sensed by osmoreceptors in the hypothalamus. In response, the output of ADH drops and the permeability to water in the distal convoluted tubules and collective tubules decreases. The urine becomes dilute as less water is reabsorbed into the blood.

4

- The fundus and most of the body of the stomach were removed.
- The stomach secretes gastric juice, which is a mixture of mucus, hydrochloric acid, and pepsin. The hydrochloric acid is necessary for conversion of the inactive pepsinogen into the active enzyme pepsin. Pepsin is a protease that hydrolyzes large protein molecules into smaller segments. Small amounts of a lipase are also secreted by the mucosa of the stomach.
- The pancreas secretes a lipase and a set of proteolytic hormones (trypsin, chymotrypsin, and procarboxy-peptidase) into the small intestine. These enzymes duplicate the action of the gastric enzymes and can function without the prior action of gastric enzymes.
- The stomach serves as a mixing and storage organ. Normally, ingested food is mixed and churned with gastric juice, producing a soupy mixture called chyme. The stomach regulates the slow and intermittent passage of chyme into the small intestine, where it undergoes further mixing and chemical digestion. Without the bulk of his stomach, our patient needed to change his eating pattern. Instead of three large meals per day, he had to eat small quantities of food throughout the day. As long as he chewed his food thoroughly, the mixing action and pancreatic enzymes in his small intestine took care of the rest.

Part V (p. 1014)

1

- The prostate gland lies beneath the urinary bladder and completely surrounds a portion of the urethra. The tough outer capsule of the prostate does not stretch, so any enlargement of the organ pushes inward, collapsing the urethra and preventing the outflow of urine.
- The prostate lies immediately anterior to the rectum. The physician can feel the size, shape, and texture of the organ through the rectal wall and can detect any lumps that may be tumors.

- The common enlargement that affects most older men is known as benign prostatic hypertrophy. The pressure on the urethra can be relieved by the surgical procedure of transurethral resection, which removes the prostatic tissue immediately surrounding the urethra, thereby relieving the obstruction. This was likely the treatment in the professor's case.

2

- The menstrual flow begins when the uterine lining is no longer hormonally supported by progesterone and estrogen from the corpus luteum. If a woman is not pregnant, the corpus luteum involutes in about 2 weeks. If a woman is pregnant, the corpus luteum is hormonally supported by human chorionic gonadotropin (hCG) produced by the placenta and the uterine lining does not slough off.
- Pregnancy tests detect hCG, which is produced by the placenta. hCG crosses the uterine veins to be eliminated in the mother's urine. The first urine in the morning is concentrated and would be expected to contain a higher concentration of hCG, giving a more accurate test.
- The early prenatal exams allow the gynecologist to monitor the growth of the uterus and establish baseline levels of blood pressure and blood sugar. Later in the pregnancy these data can be valuable for detecting any complications.

3

- A modified radical mastectomy is the surgical removal of a breast and its lymphatic drainage. The underlying pectoralis major muscle, which is removed in a radical mastectomy, is left intact. The surgery is followed by the administration of a medication intended to kill any tumor cells remaining in the body after surgery.
- There is clearly a genetic component to the development of breast cancer, although most scientists suspect that particular environmental factors also play a role. In some families, women tend to develop breast cancer at a relatively young age. The coach's positive family history alerted her physician to this risk, and as a result, the coach chose the modified radical surgery instead of the lumpectomy.
- If the coach's tumor was removed before any cells had metastacized, her chances for a complete recovery were good. One of the critical indices for predicting complete recovery is remaining disease-free for 5 years. If no metastatic growths appear within 5 years, it is assumed that the surgery was successful.

4

- Although normal fertilization requires one sperm per ovum, a normal ejaculate contains 100 million sperm.
- Many of the sperm in each ejaculate do not reach the ovum. Some migrate up the oviduct toward the nonovulating ovary, others appear to get "lost" in the female reproductive ducts, and still others die in transit. However, many sperm must be in the vicinity of the egg in order to have sufficient concentration of acrosomal enzymes near the ovum. The enzymes function to digest a pathway through the protective layers that surround and protect the ovum.
- The first sperm that penetrates the plasma membrane of an ovum initiates the cortical reaction. During that reaction, cortical granules just under the plasma membrane of the ovum release enzymes that thicken the zona pellucida and strip off its sperm-binding sites, thereby preventing entry by additional sperm.

ANSWERS TO END-OF-CHAPTER SELF-QUIZZES

CHAPTER 1
1.1 e; **1.2** c; **1.3** b; **1.4** c; **1.5** b;
1.6 a; **1.7** c; **1.8** d; **1.9** e; **1.10** c;
1.11 b; **1.12** b; **1.13** d; **1.14** d;
1.15 c; **1.16** g; **1.17** e; **1.18** c;
1.19 i; **1.20** f; **1.21** h; **1.22** d; **1.23** j;
1.24 a; **1.25** b

CHAPTER 2
2.1 d; **2.2** e; **2.3** c; **2.4** b; **2.5** a;
2.6 e; **2.7** c; **2.8** a; **2.9** d; **2.10** b;
2.11 steroid; **2.12** phospholipid;
2.13 monosaccharide;
2.14 polyunsaturated; **2.15** amino acid

CHAPTER 3
3.1 c; **3.2** d; **3.3** a; **3.4** c; **3.5** b;
3.6 e; **3.7** c; **3.8** a; **3.9** e; **3.10** c;
3.11 chromatin; **3.12** mitosis;
3.13 vacuoles; **3.14** glycocalyx;
3.15 selective permeability;
3.16 endoplasmic reticulum;
3.17 lysosomes; **3.18** metastasis;
3.19 carcinogen; **3.20** mutation;
3.21 transcription; **3.22** transfer RNA;
3.23 mRNA, tRNA; **3.24** protein;
3.25 nucleus; **3.26** d; **3.27** b; **3.28** f;
3.29 j; **3.30** h; **3.31** c; **3.32** i; **3.33** g;
3.34 e; **3.35** a; **3.36** k

CHAPTER 4
4.1 d; **4.2** e; **4.3** d; **4.4** e; **4.5** a;
4.6 b; **4.7** c; **4.8** e; **4.9** a; **4.10** d;
4.11 e; **4.12** c; **4.13** d; **4.14** b;
4.15 a; **4.16** d; **4.17** a; **4.18** b;
4.19 c; **4.20** mesothelium;
4.21 secretion; **4.22** pseudostratified;
4.23 elastic; **4.24** fibroblasts;
4.25 fibers; **4.26** mesenchyme;
4.27 chondrocytes; **4.28** hyaline;
4.29 collagen; **4.30** dense, regular;
4.31 loose; **4.32** adipose or fat;
4.33 elastic; **4.34** a

CHAPTER 5
5.1 d; **5.2** b; **5.3** f; **5.4** e; **5.5** a;
5.6 c; **5.7** g; **5.8** a; **5.9** d; **5.10** b;
5.11 c; **5.12** a; **5.13** d;
5.14 epidermis; **5.15** stratum
germinativum; **5.16** stratum corneum;
5.17 third; **5.18** reticular;
5.19 hypodermis; **5.20** melanocytes;
5.21 d

CHAPTER 6
6.1 trabeculae; **6.2** sesamoid;
6.3 medullary cavity; **6.4** central canal;
6.5 endosteum; **6.6** gap junctions;
6.7 nutrient foramen; **6.8** chondrocytes;
6.9 collagen; **6.10** osteoblasts;
6.11 articular; **6.12** periosteum,
endosteum; **6.13** osteon (Haversian
system); **6.14** fat; **6.15** marrow;
6.16 fat storage; **6.17** blood;
6.18 osteoclasts; **6.19** perforating
(Sharpey's); **6.20** osteoclasts;
6.21 intramembranous ossification;
6.22 lamellae; **6.23** periosteum;
6.24 e; **6.25** d; **6.26** a; **6.27** b;
6.28 c; **6.29** h; **6.30** g; **6.31** f; **6.32** f;
6.33 k; **6.34** i

CHAPTER 7
7.1 a; **7.2** b; **7.3** b; **7.4** d; **7.5** d;
7.6 b; **7.7** b; **7.8** c; **7.9** c; **7.10** c;
7.11 e; **7.12** b; **7.13** a; **7.14** c;
7.15 d; **7.16** sutural; **7.17** paranasal
sinuses; **7.18** vomer; **7.19** maxillae;
7.20 mandible; **7.21** malleus, incus,
stapes; **7.22** atlas; **7.23** coccyx;
7.24 12; **7.25** parietal, coccyx; **7.26** b;
7.27 a

CHAPTER 8
8.1 d; **8.2** b; **8.3** b; **8.4** e; **8.5** b;
8.6 e; **8.7** a; **8.8** d; **8.9** g; **8.10** c;
8.11 f; **8.12** b; **8.13** a; **8.14** d;
8.15 e; **8.16** humerus; **8.17** tibia;
8.18 tibia; **8.19** carpal; **8.20** patella;
8.21 interosseous membrane;
8.22 lesser pelvis; **8.23** ilium, ischium,
pubis; **8.24** sacrum, coccyx, hipbones;
8.25 femur

CHAPTER 9
9.1 c; **9.2** d; **9.3** e; **9.4** e; **9.5** e;
9.6 a; **9.7** d; **9.8** b; **9.9** e; **9.10** e;
9.11 b; **9.12** a; **9.13** synchondrosis;
9.14 synovial membrane; **9.15** tendon
sheaths; **9.16** meniscus;
9.17 gomphosis; **9.18** syndesmosis;
9.19 biaxial; **9.20** gliding;
9.21 capsule; **9.22** symphyses;
9.23 plantar-flexing; **9.24** pivot;
9.25 a; **9.26** g; **9.27** e; **9.28** d;
9.29 i; **9.30** c; **9.31** f; **9.32** b; **9.33** h

CHAPTER 10
10.1 e; **10.2** c; **10.3** a; **10.4** e;
10.5 d; **10.6** b; **10.7** f; **10.8** c;
10.9 d; **10.10** g; **10.11** h; **10.12** e;
10.13 i; **10.14** a; **10.15** j; **10.16** b;
10.17 calcium; **10.18** motor unit;
10.19 synaptic cleft; **10.20** single-unit;
10.21 intercalated disks;
10.22 mesoderm; **10.23** myoblasts;
10.24 muscles, bones

CHAPTER 11
11.1 a; **11.2** e; **11.3** b; **11.4** d;
11.5 e; **11.6** b; **11.7** e; **11.8** e;
11.9 e; **11.10** d; **11.11** a; **11.12** e;
11.13 c; **11.14** e; **11.15** d; **11.16** c;
11.17 b; **11.18** a; **11.19** f; **11.20** e;
11.21 belly; **11.22** tendon;
11.23 aponeuroses; **11.24** origin;
11.25 insertion; **11.26** fascicles;
11.27 flexion; **11.28** pronator;
11.29 invertor; **11.30** first; **11.31** j;
11.32 e; **11.33** l; **11.34** f; **11.35** k;
11.36 d; **11.37** g; **11.38** a; **11.39** h;
11.40 c

CHAPTER 12
12.1 d; **12.2** c; **12.3** b; **12.4** c;
12.5 d; **12.6** d; **12.7** b; **12.8** d;
12.9 c; **12.10** a; **12.11** b; **12.12** c;
12.13 calcium; **12.14** synaptic cleft;
12.15 electrical; **12.16** voltage;
12.17 chemically or transmitter;
12.18 all-or-none; **12.19** saltatory
conduction; **12.20** ganglion;
12.21 brain, spinal cord; **12.22** visceral,
somatic; **12.23** d; **12.24** h; **12.25** a;
12.26 b; **12.27** g; **12.28** f; **12.29** e;
12.30 c

CHAPTER 13
13.1 d; **13.2** a; **13.3** e; **13.4** e;
13.5 d; **13.6** e; **13.7** b; **13.8** b;
13.9 a; **13.10** astrocytes;
13.11 arachnoid villi;
13.12 subarachnoid space;
13.13 cerebral cortex; **13.14** corpus
callosum; **13.15** choroid plexus;
13.16 midbrain; **13.17** midbrain;
13.18 cerebral cortex; **13.19** i;
13.20 d; **13.21** c; **13.22** k; **13.23** e;
13.24 a; **13.25** g; **13.26** j; **13.27** h;
13.28 f; **13.29** b; **13.30** l

CHAPTER 14
14.1 d; **14.2** e; **14.3** e; **14.4** e;
14.5 e; **14.6** e; **14.7** e; **14.8** b;
14.9 e; **14.10** d; **14.11** b; **14.12** c;
14.13 c; **14.14** e; **14.15** c; **14.16** b;
14.17 conus terminalis or medullaris;
14.18 filum terminale; **14.19** dorsal
root ganglia or spinal ganglia;
14.20 funiculi; **14.21** decussation;
14.22 reflex; **14.23** denticulate
ligament; **14.24** subarachnoid space

CHAPTER 15
15.1 a; **15.2** c; **15.3** e; **15.4** b;
15.5 a; **15.6** d; **15.7** d; **15.8** e;
15.9 e; **15.10** e; **15.11** motor;
15.12 general somatic afferent;
15.13 general somatic efferent;
15.14 visceral; **15.15** brainstem;
15.16 olfactory; **15.17** optic chiasma;
15.18 III, IV, VI; **15.19** VII, IX;
15.20 superior; **15.21** C7, T1; **15.22** j;
15.23 f; **15.14** a; **15.25** i; **15.26** d;

15.27 b; 15.28 c; 15.29 g; 15.30 e; 15.31 h

CHAPTER 16
16.1 c; 16.2 e; 16.3 e; 16.4 b; 16.5 d; 16.6 b; 16.7 e; 16.8 a; 16.9 c; 16.10 f; 16.11 e; 16.12 g; 16.13 a; 16.14 a; 16.15 b; 16.16 g; 16.17 f; 16.18 sympathetic; 16.19 parasympathetic; 16.20 ganglion; 16.21 catecholamines; 16.22 smooth muscle, cardiac muscle, glands; 16.23 two, one; 16.24 acetylcholine; 16.15 terminal

CHAPTER 17
17.1 b; 17.2 e; 17.3 d; 17.4 e; 17.5 a; 17.6 b; 17.7 c; 17.8 b; 17.9 d; 17.10 e; 17.11 cones; 17.12 visual, auditory; 17.13 receptive field; 17.14 ganglion; 17.15 referred pain; 17.16 labyrinth; 17.17 otoconia 17.18 conjunctiva; 17.19 pigmented epithelium; 17.20 lacrimal gland; 17.21 d; 17.22 a; 17.23 c; 17.24 b; 17.25 f; 17.26 e

CHAPTER 18
18.1 d; 18.2 a; 18.3 c; 18.4 b; 18.5 e; 18.6 a; 18.7 b; 18.8 b; 18.9 c; 18.10 e; 18.11 c; 18.12 b; 18.13 a; 18.14 b; 18.15 glucocorticoids or cortisol; 18.16 melatonin, light; 18.17 testes, ovary; 18.18 cholesterol; 18.19 hypothalamic-hypophyseal portal vessels; 18.20 pancreas, adrenal cortex, insulin; 18.21 neurosecretory; 18.22 hypothalamus; 18.23 thyroid gland; 18.24 epinephrine, norepinephrine

CHAPTER 19
19.1 b; 19.2 e; 19.3 d; 19.4 c; 19.5 a; 19.6 e; 19.7 c; 19.8 d; 19.9 b; 19.10 a; 19.11 e; 19.12 d; 19.13 c; 19.14 c; 19.15 c; 19.16 b; 19.17 a; 19.18 a; 19.19 a; 19.20 b; 19.21 b; 19.22 diapedesis; 19.23 plasma proteins; 19.24 red blood cells; 19.25 erythropoiesis; 19.26 hemolysis; 19.27 leukopoiesis

CHAPTER 20
20.1 b; 20.2 c; 20.3 d; 20.4 c; 20.5 b; 20.6 c; 20.7 e; 20.8 d; 20.9 c; 20.10 d; 20.11 e; 20.12 c; 20.13 c; 20.14 d; 20.15 b; 20.16 a; 20.17 c; 20.18 c; 20.19 b; 20.20 d; 20.21 d; 20.22 a; 20.23 b; 20.24 c; 20.25 SA node; 20.26 calcium; 20.27 atrioventricular bundle; 20.28 AV node; 20.29 heart rate, stroke volume; 20.30 SA node

CHAPTER 21
21.1 d; 21.2 b; 21.3 d; 21.4 b;

21.5 b; 21.6 d; 21.7 c; 21.8 d; 21.9 c; 21.10 e; 21.11 e; 21.12 a; 21.13 d; 21.14 a; 21.15 b; 21.16 c; 21.17 e; 21.18 c; 21.19 e; 21.20 a; 21.21 b; 21.22 e; 21.23 e; 21.24 a; 21.25 b; 21.26 c; 21.27 d; 21.28 vasa vasorum; 21.29 terminal; 21.30 pulse pressure; 21.31 blood pressure or hydrostatic pressure; 21.32 greater; 21.33 pressure; 21.34 blood vessel length, blood viscosity, radius of lumen; 21.35 total peripheral resistance; 21.36 chemoreceptors

CHAPTER 22
22.1 d; 22.2 b; 22.3 c; 22.4 c; 22.5 a; 22.6 b; 22.7 c; 22.8 a; 22.9 e; 22.10 d; 22.11 b; 22.12 a; 22.13 e; 22.14 c; 22.15 d; 22.16 edema; 22.17 lymph nodes; 22.18 hilus; 22.19 palatine, lingual, pharyngeal; 22.20 pharyngeal tonsils; 22.21 spleen; 22.22 interstitial fluid; 22.23 lacteals; 22.24 histiocytes; 22.25 valves; 22.26 B cells, T cells; 22.27 spleen; 22.28 tonsils

CHAPTER 23
23.1 f; 23.2 f; 23.3 d; 23.4 a; 23.5 e; 23.6 c; 23.7 b; 23.8 d; 23.9 b; 23.10 c; 23.11 e; 23.12 a; 23.13 e; 23.14 d; 23.15 a; 23.16 b; 23.17 c; 23.18 e; 23.19 e; 23.20 e; 23.21 e; 23.22 e; 23.23 b; 23.24 c; 23.25 d; 23.26 e; 23.27 c; 23.28 a; 23.29 d

CHAPTER 24
24.1 e; 24.2 d; 24.3 b; 24.4 b; 24.5 a; 24.6 e; 24.7 e; 24.8 a; 24.9 e; 24.10 e; 24.11 b; 24.12 c; 24.13 c; 24.14 e; 24.15 e; 24.16 buccal cavity; 24.17 thyroid cartilage; 24.18 epiglottis; 24.19 trachea; 24.20 alveoli; 24.21 breathing; 24.22 internal; 24.23 oxyhemoglobin; 24.24 medulla oblongata; 24.25 arytenoid; 24.26 laryngopharynx; 24.27 CO_2; 24.28 a; 24.29 e; 24.30 b; 24.31 c; 24.32 d; 24.33 e; 24.34 a; 24.35 b; 24.36 d; 24.37 c

CHAPTER 25
25.1 b; 25.2 e; 25.3 d; 25.4 a; 25.5 b; 25.6 c; 25.7 c; 25.8 d; 25.9 d; 25.10 d; 25.11 e; 25.12 e; 25.13 d; 25.14 a; 25.15 a; 25.16 chyme; 25.17 greater omentum; 25.18 cystic; 25.19 sinusoids; 25.20 segmenting; 25.21 GI; 25.22 periodontal ligaments; 25.23 pharynx; 25.24 peritoneum; 25.25 mesenteries; 25.26 villi; 25.27 emulsification; 25.28 falciform; 25.29 CCK; 25.30 haustra; 25.31 a; 25.32 e; 25.33 c; 25.34 d; 25.35 b;

25.36 f; 25.37 c; 25.38 a; 25.39 d; 25.40 b; 25.41 e; 25.42 g

CHAPTER 26
26.1 c; 26.2 a; 26.3 a; 26.4 e; 26.5 e; 26.6 e; 26.7 e; 26.8 d; 26.9 c; 26.10 e; 26.11 glycogenolysis; 26.12 lipid or fat; 26.13 thyroxine; 26.14 fat-soluble; 26.15 brown fat; 26.16 reduced; 26.17 respiration; 26.18 anaerobic; 26.19 oxygen; 26.20 enzymes; 26.21 chylomicrons; 26.22 e; 26.23 a; 26.24 c; 26.25 b; 26.26 d; 26.27 b; 26.28 c; 26.29 d; 26.30 a

CHAPTER 27
27.1 c; 27.2 d; 27.3 d; 27.4 b; 27.5 d; 27.6 b; 27.7 a; 27.8 a; 27.9 a; 27.10 d 27.11 b; 27.12 a; 27.13 b; 27.14 b; 27.15 c; 27.16 e; 27.17 f; 27.18 a; 27.19 f; 27.20 b; 27.21 d; 27.22 b; 27.23 GFR; 27.24 loop of the nephron (Henle); 27.25 glomerular hydrostatic pressure; 27.26 sympathetic, fall; 27.27 macula densa; 27.28 cotransport; 27.29 plasma clearance; 27.30 external

CHAPTER 28
28.1 c; 28.2 a; 28.3 e; 28.4 e; 28.5 d; 28.6 e; 28.7 e; 28.8 e; 28.9 b; 28.10 d; 28.11 a; 28.12 e; 28.13 d; 28.14 e; 28.15 e; 28.16 b; 28.17 d; 28.18 c; 28.19 a; 28.20 b; 28.21 20; 28.22 Water; 28.23 Edema; 28.24 osmoreceptors; 28.25 ADH; 28.26 Electrolytes; 28.27 acid; 28.28 conjugate; 28.29 CO_2; 28.30 Na^+

CHAPTER 29
29.1 c; 29.2 e; 29.3 e; 29.4 c; 29.5 d; 29.6 d; 29.7 c; 29.8 d; 29.9 e; 29.10 c; 29.11 a; 29.12 b; 29.13 e; 29.14 d; 29.15 c; 29.16 e; 29.17 b; 29.18 c; 29.19 d; 29.20 a; 29.21 e; 29.22 a; 29.23 b; 29.24 d; 29.25 c; 29.26 e; 29.27 a; 29.28 c; 29.29 b; 29.30 d; 29.31 smegma; 29.32 prepuce or foreskin; 29.33 broad ligaments; 29.34 hCG; 29.35 spermatids; 29.36 secondary oocyte, polar body; 29.37 ootid, polar body; 29.38 zona pellucida; 29.39 acrosin; 29.40 zygote

CHAPTER 30
30.1 e; 30.2 a; 30.3 a; 30.4 b; 30.5 d; 30.6 c; 30.7 c; 30.8 e; 30.9 a; 30.10 e; 30.11 e; 30.12 c; 30.13 a; 30.14 b; 30.15 d; 30.16 b; 30.17 d; 30.18 a; 30.19 e; 30.20 b; 30.21 blastocyst; 30.22 hCG; 30.23 amniotic; 30.24 oxytocin; 30.25 colostrum; 30.26 karyotype; 30.27 alpha-fetoprotein; 30.28 placenta

PHOTO CREDITS

PART-OPENING PHOTOGRAPHS
I: Thomas Mayer/Black Star. II: Ed Wheeler/Stock Market. III: Henley and Savage/Stock Market. IV: Barros and Barros/Image Bank. V: Tom McCarthy/Picture Cable.

CHAPTER-OPENING PHOTOGRAPHS
1: Courtesy of Mark S. Ladinsky, University of Colorado, Boulder. 2: M. I. Walker/Photo Researchers. 3: Eric Graves/Science Source/Photo Researchers. 4: E. R. Gegginger/Fran Heyl Associates. 5: CNRI/Phototake. 6–9: Astrid and Hans-Frieder Michler/SPL/Photo Researchers. 10–11: John D. Cunningham/Visuals Unlimited. 12–16: Biophoto Associates/Science Source/Photo Researchers. 17: Ralph Eagle/Photo Researchers. 18: Astrid and Hans-Frieder Michler/Science Photo Library/Photo Researchers. 19–21: Robert Knauff/Biology Media/Photo Researchers. 22–23: R. Becker/Custom Medical Stock. 24: John D. Cunningham/Visuals Unlimited. 25: J. F. Gennaro, L. R. Grillone/Cellular Biology, NYU/Photo Researchers. 26: Photo Researchers. 27: M. I. Walker/Photo Researchers. 28: Charles Falco/Science Source/Photo Researchers. 29: M. I. Walker/Photo Researchers. 30: NIH/Science Source/Photo Researchers.

CHAPTER 1 FIGURE 1.4: Courtesy of Grower Medical Publishing, a division of Lippincott Company, London. From J. A. Gosling et al., *Atlas of Human Anatomy with Integrated Text.* 2d ed., Philadelphia, Lippincott, 1991. FIGURE 1.5 [A] Centre Naztional des Recherches Iconographiques, Paris, France; [B] D. M. Phillips/Taurus Photos; [C] C. Abrahams, M.D./Custom Medical Stock; [D] Lennart Nilsson, *Behold Man*, Boston, Little, Brown, 1974/Bonnier Alba AB, Stockholm, Sweden. FIGURE 1.12: [A] Dan McCoy/Rainbow; [B] Dimensional Medicine, Inc. FIGURE 1.12: [C–E] Dan McCoy/Rainbow; [F] Fischer Imaging Corporation; [G] UCLA School of Medicine; [H] Science Photo Library/Photo Researchers.

CHAPTER 2 FIGURE 2.19: [C] Howard Sochurek/Medichrome/The Stock Shop.

CHAPTER 3 FIGURE 3.3: [B] Courtesy of Dr. Kazuhiko Fujita, Department of Urology, Juntendo University School of Medicine, Tokyo. From *Journal of Urology*, vol. 146, July 1991, no. 1. FIGURE 3.4: [D, 1–3] Dr. Peck-Sun Lin, Department of Therapeutic Radiology, New England Medical Center Hospital. FIGURE 3.5: Courtesy of M. M. Perry, Roslin Institute, Edinburgh, from M. M. Perry and A. B. Gilbert, *J. of Cell Science*, vol. 39, pp. 257–272, 1979. FIGURE 3.6: [C] From Ross and Romrell, *Histology: A Text and Atlas*, 2d ed., 1989, Williams & Wilkins. Reproduced by permission of Williams & Wilkins. FIGURE 3.7: [C] Photo Researchers. FIGURE 3.8: [C] David M. Phillips/Visuals Unlimited. FIGURE 3.9: [C] Don Fawcett, M.D./Photo Researchers. FIGURE 3.10: [B] Porter/Anderson/Photo Research-

ers. FIGURE 3.11: [B] Dr. Joel Rosenbaum, Department of Biology, Yale University. FIGURE 3.12: [C] Don Fawcett, M.D./Photo Researchers. FIGURE 3.13: [C] Don Fawcett, M.D./Penelope Gaddum-Rosse/Photo Researchers. FIGURE 3.14: [C] Keiichi Tanaka, M.D.; [D] Don Fawcett, M.D./Photo Researchers. FIGURE 3.19: Biophoto Associates/Science Source/Photo Researchers. FIGURE 3.22: [A, B] Dr. Leonard Hayflick, University of California, San Francisco Medical School. FIGURE 3.24: [A, B] Dr. Cecil Fox, NCI; [C, D]; Lennart Nilsson/Bonnier Alba AB, Stockholm, Sweden.

CHAPTER 4 FIGURE 4.2: [B] Dennis Kunkel/DNRI/Phototake. FIGURE 4.4: [B] Don Fawcett, M.D./Hirokawa/John Heuser, M.D./Photo Researchers. FIGURE 4.5: (p. 106, left and right) Ed Reschke/Peter Arnold; (p. 107, left) Fred Hossler/Visuals Unlimited; (p. 107, right) Courtesy of Mr. Brett Connors and Dr. Andrew P. Evan, Department of Anatomy, Indiana University. FIGURE 4.6: (p. 108) Ed Reschke/Peter Arnold; (p. 109, left) From Ross & Romrell, *Histology: A Text and Atlas*, 2d ed., 1989, Williams & Wilkins. Reproduced by permission of Williams & Wilkins. (p. 109, right) Michael H. Ross, M.D.; FIGURE 4.7: (left and right) Biophoto Associates/Photo Researchers. FIGURE 4.10: [A] Biophoto Associates/Photo Researchers; [B] Manfred Kage/Peter Arnold. FIGURE 4.11: (p. 117, left) Dr. N. Hoffman/Phototake; (p. 117, right) Biophoto Associates/Photo Researchers; (p. 118, top left) Robert Knauft/Biology Media/Photo Researchers; (p. 118, bottom left) Biophoto Associates/Photo Researchers; (p. 118, right) From Ross and Romrell, *Histology: A Text and Atlas*, 2d ed., 1989, Williams & Wilkins. Reproduced by permission of Wilkins & Wilkins. (p. 119, left) Biophoto Associates/Photo Researchers; (p. 119, right) Ed Reschke/Peter Arnold. FIGURE 4.12: (p. 120, left) M. I. Walker/Photo Researchers; (p. 120, right) Lennart Nilsson, *Behold Man*, Boston, Little, Brown, 1974/Bonnier Alba AB, Stockholm, Sweden; (p. 121) Dr. Mary Notter/Phototake. FIGURE 4.13: [A] Michael Abbey/Science Source/Photo Researchers; [B] M. Abbey/Photo Researchers; [C] Ed Reschke/Peter Arnold. FIGURE 4.14: Biophoto Associates/Photo Researchers.

CHAPTER 5 FIGURE 5.1: [B] Biophoto Associates/Science Source/Photo Researchers. FIGURE 5.4: Dr. Jerome Gross, The Developmental Biology Lab, Lovett Memorial Group, Massachussetts General Hospital. FIGURE 5.8: Biophoto Associates/Photo Researchers. FIGURE 5.11: Lennart Nilsson, *A Child Is Born*, Dell Publishing Company/Bonnier Alba, AB, Stockholm, Sweden. PAGE 148: (all) Lester V. Bergman & Associates. PAGE 151: Lester V. Bergman & Associates. PAGE 152: © Carroll H. Weiss, 1988, all rights reserved.

CHAPTER 6 FIGURE 6.4: [D] Richard G. Kessel and Randy H. Kardon, *Tissues and*

Organs: A Text-Atlas of Scanning Electron Microscopy, San Francisco, Freeman, 1979. FIGURE 6.5: [A] Andreas Feininger, Life Magazine-Time, Inc.; [B] Reprinted with permission from Leon Weiss (ed.), New York, Elsevier, 1983. FIGURE 6.10: Courtesy of David H. Cormack, Ph.D., University of Toronto.

CHAPTER 7 FIGURE 7.3: Custom Medical Stock Photo. FIGURE 7.5 [C] Lester V. Bergman & Associates. FIGURE 7.7: [C] McGraw-Hill photo by Ken Karp. FIGURE 7.9: [C] © Albert Paglialunga, Bergenfield, NJ. FIGURE 7.13: [B] Lester V. Bergman & Associates. FIGURE 7.16: [C] Science Photo Library/Photo Researchers. FIGURE 7.19: [A] Joel Gordon; [B] Custom Medical Stock Photo. PAGE 214: (all) Hospital for Special Surgery.

CHAPTER 8 FIGURE 8.1: Custom Medical Stock Photo: FIGURE 8.5: [C] © Albert Paglialunga, Bergenfield, NJ. FIGURE 8.7: [C] © Albert Paglialunga, Bergenfield, NJ. FIGURE 8.8: Custom Medical Stock Photo. FIGURE 8.10: Custom Medical Stock Photo. FIGURE 8.13: [B] ©Albert Paglialunga, Bergenfield, NJ. FIGURE 8.15: [B] © Albert Paglialunga, Bergenfield, NJ. FIGURE 8.16: [B] Lester V. Bergman & Associates.

CHAPTER 9 FIGURE 9.4: [C] T. Bednarek/Custom Medical Stock Photo. FIGURE 9.6: [B] T. Bednarek/Custom Medical Stock Photo. PAGE 259: Biophoto Associates/Photo Researchers. PAGE 264: (top left) SIU/Peter Arnold. PAGE 264: (bottom left and top right) Lennart Nilsson, *Victorious Body*/Bonnier Alba AB, Stockholm, Sweden. PAGE 264: (bottom left) Carroll Weiss/Camera MD. PAGE 271: (left and center) Photo Researchers. PAGE 271: (right) Peter Arnold.

CHAPTER 10 FIGURE 10.2: [F] Eric V. Gravé/Phototake; [G] C. Franzini-Armstrong, from *J. Cell Biol.* 47(1970):488. FIGURE 10.5: [A] Lennart Nilsson, *Behold Man*, Boston, Little, Brown, 1974/Bonnier Alba AB, Stockholm, Sweden; [E] Courtesy of Dr. E. Heuser, Washington School of Medicine. FIGURE 10.14: [A] Fred Hossler/Visuals Unlimited; [B] Fawcett/Vehara/Photo Researchers. FIGURE 10.16: [A, C] A. W. Ham and D. H. Cormack, *Histology*, 9th ed., Philadelphia, Lippincott, 1987. FIGURE 10.18: [A] W. Bloom/D. W. Fawcett/Photo Researchers.

CHAPTER 11 FIGURE 11.9: [B] Custom Medical Stock Photo. FIGURE 11.14: [B] Custom Medical Stock Photo. FIGURE 11.18: [C] T. Bednarek/Custom Medical Stock Photo. PAGE 326: (left) Focus on Sports; (center) Michell Layton/Duomo; (right) Al Tielemans/Duomo.

CHAPTER 12 FIGURE 12.2: [C] Lennart Nilsson, *Behold Man*, Boston, Little, Brown/Bonnier Alba AB, Stockholm, Sweden. FIGURE 12.3: [B] (top and bottom) Dr. Geoffery McAuliffe, Department of Anatomy, Robert Wood Johnson Medical School. FIGURE 12.4: [B] Richard G. Kessel and Randy H. Kardon, *Tissues and Organs: A*

Text-Atlas of Scanning Electron Microscopy, San Francisco, Freeman, 1979. FIGURE **12.13:** [E] Dr. John E. Heusser, Washington University School of Medicine. PAGE **383:** Dr. John E. Heusser, Washington University School of Medicine.

CHAPTER 13 FIGURE **13.1:** [A, B] Martin M. Rotker, M.D. FIGURE **13.2:** [B] Courtesy of Dr. Delmas J. Allen, Medical College of Ohio. FIGURE **13.7:** Murray L. Barr and John A. Kierman, *The Human Nervous System: An Anatomical Viewpoint,* 5th ed., Philadelphia, Lippincott, 1988. FIGURE **13.13:** British Museum of Natural History.

CHAPTER 14 FIGURE **14.5:** [A] N. Gluhbegovic and T. H. Williams, *The Human Brain: A Photographic Guide,* New York, Harper & Row, 1980.

CHAPTER 15 FIGURE **15.12:** [B] Gene Shih and Richard Kessel, *Living Images: Biological Microstructures Revealed by SEM,* Boston, Science Books International, 1982. FIGURE **15.15:** T. Bednarek/Custom Medical Stock Photo.

CHAPTER 17 FIGURE **17.4:** [C] Gene Shih and Richard Kessel, *Living Images: Biological Microstructures Revealed by SEM,* Boston, Science Books International, 1982. FIGURE **17.5:** [B] Martin Dohrn/Science Photo Library/Photo Researchers. FIGURE **17.14:** [B] Gene Shih and Richard Kessel, *Living Images: Biological Microstructures Revealed by SEM,* Boston, Science Books International, 1982. FIGURE **17.16:** Martin Dohrn/Science Photo Library/Photo Researchers. FIGURE **17.22:** [B] Lester V. Bergman & Associates. FIGURE **17.24:** [F, G] Lennart Nilsson, *A Child Is Born*/Bonnier Alba AB, Stockholm, Sweden.

CHAPTER 18 FIGURE **18.8:** [B] Lester V. Bergman & Associates. FIGURE **18.11:** [B] W. Bloom and D. W. Fawcett, *A Textbook of Histology,* 11th ed., Philadelphia, Saunders, 1986. FIGURE **18.16:** [B] Biophoto Associates/Photo Researchers; [C] Courtesy of K. Kovacs and E. Horvath. PAGE **575:** Michael Serino/Picture Cube. PAGE **576:** (top) Reprinted with permission from "Clinicopathologic Conference," *The American Journal of Medicine,* vol. 20, no. 1, January 1956; (bottom) John Paul Kay/Peter Arnold.

CHAPTER 19 FIGURE **19.1:** Bruce Wetzel and Harry Schaefer, National Cancer Institute. FIGURE **19.5:** (p. 593, top to bottom) Manfred Kage/Peter Arnold; Lester V. Bergman & Associates; Cabisco/Visuals Unlimited; Lester V. Bergman & Associates; John D. Cunningham/Visuals Unlimited; Biophoto Associates/Science Source/Photo Researchers; Lester V. Bergman & Associates. FIGURE **19.6:** [A] Ralph Feder, IBM, Thomas J. Watson Research Center; [B] Hans R. Baumgartner, Hoffman-LaRoche & Company, Basel. FIGURE **19.8:** [A, B] James G. White, M.D.; [B] FIGURE **19.9:** Lennart Nilsson, *Victorious Body*/Bonnier Alba AB, Stockholm, Sweden. PAGE **607:** N. Calder/Science Source/Photo Researchers.

CHAPTER 20 FIGURE **20.4** [B] Martin M. Rotker/Phototake. FIGURE **20.5:** [B] Martin M. Rotker/Phototake. FIGURE **20.6:** [D] Dr. Wallace McAlpine; [F] Lennart Nilsson, *Behold Man,* Boston, Little, Brown/Bonnier Alba AB, Stockholm, Sweden. PAGE **625:**

From H. W. Fischer, *Radiographic Anatomy: A Working Atlas,* New York, McGraw-Hill, 1988. PAGE **631:** Courtesy Hewlett Packard Company.

CHAPTER 21 FIGURE **21.3:** D. W. Fawcett/Photo Researchers. FIGURE **21.6:** [B] Richard G. Kessel and Randy H. Kardon, *Tissues and Organs: A Text-Atlas of Scanning Electron Microscopy,* San Francisco, Freeman, 1979. FIGURE **21.13:** [B] Stanley M. Rogoff, M.D., Department of Radiology, Strong Memorial Hospital, Rochester, NY. FIGURE **21.14:** [B] From H. W. Fischer, *Radiographic Anatomy: A Working Atlas,* New York, McGraw-Hill, 1988. FIGURE **21.15:** [C] Stanley M. Rogoff, M.D., Department of Radiology, Strong Memorial Hospital, Rochester, NY. FIGURE **21.16:** [B] John Watney/Photo Researchers. FIGURE **21.17:** [B] © Albert Paglialunga, Bergenfield, NJ. FIGURE **21.18:** [B] Courtesy of Gower Medical Publishing, a division of Lippincott Company, London. FIGURE **21.19:** [C] From H. W. Fischer, *Radiographic Anatomy: A Working Atlas,* New York, McGraw-Hill, 1988. PAGE **701:** Courtesy of NIH.

CHAPTER 22 FIGURE **22.2:** [B] Courtesy of Gower Medical Publishing, a division of Lippincott Co., London. From J. A. Gosling et al., *Atlas of Human Anatomy with Integrated Text,* 2d ed., Philadelphia, Lippincott, 1991. FIGURE **22.3:** [B] Johannes A. G. Rhodin, M.D., Ph.D., Department of Anatomy, College of Medicine, Tampa, FL. FIGURE **22.4:** [B] Fujita, Tanaka, and Tokunaga, *SEM Atlas of Cells and Tissues,* Tokyo, Igaku-Shoin, 1981. FIGURE **22.5:** [B] Fujita, Tanaka, and Tokunaga, *SEM Atlas of Cells and Tissues,* Tokyo, Igaku-Shoin, 1981. FIGURE **22.8:** [A] Lester V. Bergman & Associates.

CHAPTER 23 FIGURE **23.4:** [B, 1–3] Lennart Nilsson/Bonnier Alba AB, Stockholm, Sweden. FIGURE **23.12:** [B] Andrejs Liepins, Ph.D. FIGURE **23.14:** Lennart Nilsson/Bonnier Alba AB, Stockholm, Sweden. FIGURE **23.15:** [B] David Scharf/Peter Arnold. PAGE **752:** Charles Daguet/Pasteur Institute/Petit Format/Science Source/Photo Researchers.

CHAPTER 24 FIGURE **24.2:** [B] CNRI/Phototake. FIGURE **24.3:** [A] Biophoto Associates/Photo Researchers; [B] Lennart Nilsson, *Behold Man,* Boston, Little, Brown, 1974/Bonnier Alba AB, Stockholm, Sweden. FIGURE **24.4:** [C] Courtesy of Gower Medical Publishing, a division of Lippincott Company, London. From J. A. Gosling et al., *Atlas of Human Anatomy with Integrated Text,* 2d ed. Philadelphia, Lippincott, 1991. FIGURE **24.5:** [B] R. Calentine/Visuals Unlimited. FIGURE **24.7:** Bruce Iverson/Visuals Unlimited. FIGURE **24.9:** [A, B] Courtesy of Gower Medical Publishing, a division of Lippincott Company, London. From J. A. Gosling et al., *Atlas of Human Anatomy with Integrated Text,* 2d ed., Philadelphia, Lippincott, 1991. PAGE **787:** (top and bottom) Kenneth A. Siegesmund, Ph.D., Department of Anatomy and Cellular Biology, Medical College of Wisconsin.

CHAPTER 25 FIGURE **25.1:** [B] Thomas Bednarek/Custom Medical Stock Photo. FIGURE **25.4:** [B] Dr. Leonard B. Zaslow, courtesy of Matthew Carola. FIGURE **25.8:** Rich-

ard G. Kessel and Randy H. Kardon, *Tissues and Organs: A Text-Atlas of Scanning Electron Microscopy,* San Francisco, Freeman, 1979. FIGURE **25.12:** [C] Relam/Custom Medical Stock Photo. FIGURE **25.13:** [C] Gene Shih and Richard Kessel, *Living Images: Biological Microstructures Revealed by SEM,* Boston, Science Books International, 1982. FIGURE **25.15:** [B] Lester V. Bergman & Associates. FIGURE **25.16:** [D] Gene Shih and Richard Kessel, *Living Images: Biological Microstructures Revealed by SEM,* Boston, Science Books International, 1982. FIGURE **25.25:** [C] Lester V. Bergman & Associates; [D] Courtesy of Dr. M. Muto, Department of Anatomy, Niigata University Medical School. PAGE **840:** L. J. Schoenfield/Visuals Unlimited. PAGE **849:** Thomas Bednarek/Custom Medical Stock Photo.

CHAPTER 27 FIGURE **27.2:** Courtesy of Stephen A. Kieffer, M.D., Department of Radiology, SUNY Health Science Center, Syracuse. FIGURE **27.3:** [C] Courtesy of Gower Medical Publishing, a division of Lippincott Company, London. From J. A. Gosling et al., *Atlas of Human Anatomy with Integrated Text,* 2d ed., Philadelphia, Lippincott, 1991. FIGURE **27.5:** [C] David M. Phillips/Visuals Unlimited; [D] Reprinted with permission from Leon Weiss (ed) *Histology: Cell and Tissue Biology,* 5th ed., New York, Elsevier, 1973; [E] Daniel S. Friend, M.D. FIGURE **27.16:** Biophoto Associates/Photo Researchers. FIGURE **27.18:** [B] Michael H. Ross, Ph.D. PAGE **918:** Courtesy of Gower Medical Publishing, a division of Lippincott Company, London. From J. A. Gosling et al., *Atlas of Human Anatomy with Integrated Text,* 2d ed., Philadelphia, Lippincott, 1991.

CHAPTER 29 FIGURE **29.2:** [E] Gene Shih and Richard Kessel, *Living Images: Biological Microstructures Revealed by SEM,* Boston, Science Books International, 1982. FIGURE **29.10:** [B] Manfred Kage/Peter Arnold. FIGURE **29.11:** [B] E. F. Hafez. FIGURE **29.22:** Lennart Nilsson, *A Child Is Born*/Bonnier Alba AB, Stockholm, Sweden. FIGURE **29.23:** [A–D] Mia Tegner, Scripps Institution of Oceanography. PAGE **974:** [A, C] Steven E. Harms, M.D., Director, Magnetic Resonance, Baylor University Medical Center; [B] Peter Sibbald, Toronto, Courtesy of Dr. Martin Yaffe, Sunnybrook Medical Center, Toronto, Canada.

CHAPTER 30 FIGURE **30.1:** [A] Biophoto Associates/Photo Researchers; [B, C, E, F] Landrum B. Shettles, M.D.; [D, G] Petit Format/Nestle/Science Source/Photo Researchers. FIGURE **30.3:** [B] Photo by A. T. Hertig and J. Rock, reprinted with permission from Leon Weiss (ed.), *Histology: Cell and Tissue Biology,* 5th ed., New York, Elsevier, 1973. FIGURE **30.6:** [A–H] (embryos and fetuses) Landrum B. Shettles, M.D.; [B–D] (hands) Carnegie Institute of Washington. FIGURE **30.9:** [A–F] Maternity Center Associations, *The Birth Atlas.* FIGURE **30.10:** [A–C] Jeffery Reed/Medichrome/The Stock Shop. FIGURE **30.13:** Terry Gibbon. FIGURE **30.14:** Dr. J. H. Tijo.

APPENDIX A FIGURES **A.1–A.7:** McGraw-Hill photographs by Custom Medical Stock Photo.

GLOSSARY

A

ABDOMINAL CAVITY The superior portion of the abdominopelvic cavity; contains the liver, gallbladder, stomach, pancreas, spleen, and intestines.

ABDOMINOPELVIC CAVITY The portion of the ventral cavity below the diaphragm; divided by an imaginary line into abdominal and pelvic cavities.

ABDUCTION (L. *ab*, away from) Movement of a limb away from the midline of the body, or of fingers or toes from the medial longitudinal axis of the hand or foot.

ABSORPTION (1) The passage of usable nutrient molecules from the small intestine into the bloodstream and lymphatic system. (2) The taking up of substances by tissues such as those of the skin and mucous membranes.

ABSORPTIVE STATE The metabolic state during which nutrients are broken down in the digestive tract and absorbed into the blood and lymph; see *postabsorptive state*.

ACCOMMODATION A reflex that adjusts the lens of the eye to permit focusing of images from different distances.

ACETABULUM [ass-eh-TAB-yoo-luhm; L. vinegar cup] The socket of the ball-and-socket joint of the femur.

ACETYLCHOLINE [us-SEET-uhl-KOH-leen] A chemical neurotransmitter released by motor neurons at a neuromuscular junction; may be excitatory or inhibitory; plays a major role in muscle contraction.

ACETYLCHOLINESTERASE An enzyme that breaks down acetylcholine into acetic acid (acetate) and choline to halt continuous muscle contractions.

ACID (L. *acidus*, sour) A substance that releases hydrogen ions (H^+) when dissolved in water; any molecule that dissociates in solution to release a hydrogen ion or proton; see *base*.

ACID-BASE BUFFER SYSTEM A regulatory system in the body that helps resist changes in pH.

ACIDOSIS An abnormal metabolic condition in which acidic ketone bodies accumulate in blood and interstitial fluids, requiring a large amount of alkaline buffers, and thus lowering the blood pH below 7.35.

ACINAR GLAND (L. *acinus*, grape, berry) See *alveolar gland*.

ACINI [ASS-ih-nye; sing. **ACINUS**; L. grapes] Smallest lobule of an exocrine gland, as in pancreas.

ACNE VULGARIS (common acne) The most common adolescent skin disorder; it occurs when increased hormonal activity causes sebaceous glands to overproduce sebum and dead keratin cells clog a follicle.

ACQUIRED IMMUNE DEFICIENCY SYNDROME (AIDS) A disease characterized by deficiencies of natural killer T cells that kill virus-infected and tumor cells, and virus-specific killer T cells that kill cells containing a particular virus; victims may die from slight microbial infections.

ACROMEGALY [ak-roh-MEG-uh-lee; Gr. *akros*, extremity + *megas*, big] A form of *giantism* in which growth hormone is oversecreted after the skeletal system is fully developed.

ACROSIN An enzyme found in the head of a sperm that facilitates the entrance of a single sperm through the outer wall of an ovum.

ACROSOME (Gr. *akros*, topmost, extremity + *soma*, body) A region at the tip of the head of a sperm cell, containing enzymes that help the sperm penetrate an ovum.

ACTIN A protein that makes up part of the thin myofilament of muscle fibers; forms light I bands of skeletal muscle.

ACTION POTENTIAL The spread of an impulse along the axon following depolarization of the plasma membrane; also called *nerve impulse*.

ACTIVE TRANSPORT The movement of ions or molecules through a membrane against the concentration gradient; requires energy from the cell.

ACUTE RENAL FAILURE The total or near-total stoppage of kidney function.

ADDISON'S DISEASE A condition caused by an overactive adrenal cortex; characterized by anemia, weakness and fatigue, increased blood potassium levels, and decreased blood sodium levels.

ADDUCTION (L. *ad*, toward) Movement of a limb toward or beyond the midline of the body, or of fingers or toes toward the midline of a body part.

ADENOHYPOPHYSIS The anterior lobe of the pituitary (hypophysis), containing many secretory cells.

ADENOSINE TRIPHOSPHATE (ATP) [uh-DEN-uh-seen try-FOSS-fate] An organic compound containing adenine, ribose, and three phosphate groups; stores energy in chemical bonds and thus serves as an energy source for chemical reactions in cells.

ADIPOSE CELL (L. *adeps*, fat) A fixed connective tissue cell that synthesizes and stores lipids; also called *fat cell*.

ADIPOSE TISSUE Tissue composed almost entirely of clustered adipose cells supported by strands of collagenous and reticular fibers; provides a reserve food supply, cushions organs, and helps prevent loss of body heat.

ADOLESCENCE (L. *adolescere*, to grow up) The period from the end of childhood to the beginning of adulthood.

ADRENAL CORTEX The outer larger portion of the adrenal gland.

ADRENAL GLAND [uh-DREEN-uhl; L. "upon the kidneys"] Paired endocrine glands, one resting upon each kidney.

ADRENAL MEDULLA The inner, smaller portion of the adrenal gland; its main secretion is epinephrine.

ADRENALINE See *epinephrine*.

ADRENERGIC Pertaining to neurons and fibers of the sympathetic division of the autonomic nervous system; so called because the postganglionic neurons and fibers release norepinephrine (noradrenaline); see *cholinergic*.

ADRENOCORTICOTROPIC HORMONE (ACTH) A hormone secreted by the adenohypophysis; it stimulates the production and secretion of adrenal cortex steroids; also called *corticotropin* or *adrenocorticotropin*.

ADRENOCORTICOTROPIN See *adrenocorticotropic hormone*.

ADRENOGENITAL SYNDROME An abnormal condition caused by an adrenocortical tumor that stimulates the excessive production of male sex hormones; characterized by male characteristics in a female and accelerated sexual development in a male.

AFFERENT ARTERIOLES Branchings of the interlobular arteries in the kidney; they carry blood to the glomerulus.

AFFERENT NEURON (L. *ad*, toward + *ferre*, to bring) A nerve cell that conveys nerve impulses from sensory receptors to the central nervous system; also called *sensory neuron*.

AFTERBIRTH The collective name for the placenta, fetal membranes, and any remaining uterine fluid after their postnatal delivery.

AGGLUTININ A protective antibody in blood plasma; also called *isoantibody*.

AGGLUTINOGEN [uh-GLOOT-in-oh-jehn] An A or B antigen on the surface of an erythrocyte; it is the basis for the ABO grouping and Rh system of blood classification; also called *isoantigen.*

AGGREGATED LYMPH NODULES Clusters of unencapsulated lymphoid tissue found in the intestine and appendix; thought to generate plasma cells that secrete antibodies in response to antigens from the intestine; also called *Peyer's patches, gut-associated lymphoid tissue (GALT).*

AGGREGATION The clumping of platelets.

AGONIST (Gr. *agonia,* contest, struggle) A muscle primarily responsible for a movement; also called *prime mover.*

AIDS See *acquired immune deficiency syndrome.*

ALBINISM (L. *albus,* white) An inherited lack of normal skin pigmentation.

ALDOSTERONE The main mineralocorticoid hormone secreted by the adrenal cortex; it acts upon the kidneys to control sodium retention and potassium loss in urine.

ALIMENTARY CANAL The digestive tract, from mouth to anus.

ALKALOSIS A condition in which the blood pH rises above 7.45; see *acidosis.*

ALL-OR-NONE PRINCIPLE The tendency of muscle fibers to contract fully or not at all and of neurons to fire at full power or not at all.

ALLANTOIS [uh-LAN-toh-ihz; Gr. "sausage"] A small fingerlike outpocket of the caudal wall of the embryonic yolk sac; involved with blood cell formation and the development of the urinary bladder.

ALLERGEN An antigen involved in an allergic response.

ALLERGY See *hypersensitivity.*

ALPHA-FETOPROTEIN TEST A measurement of alpha-fetoprotein excreted by the fetus into the amniotic fluid to detect possible neurological disorders.

ALVEOLAR [al-VEE-uh-lur; L. *alveolus,* hollow, cavity] Pertaining to a rounded structure.

ALVEOLAR GLAND An exocrine gland whose secretory portion is rounded; also called *acinar gland.*

ALVEOLI [al-VEE-oh-lie; sing. **ALVEOLUS;** L. a cavity] The functional units of the lungs; the sites of gas exchange.

ALZHEIMER'S DISEASE A neurodegenerative disease characterized by a progressive loss of memory and intellectual function.

AMENORRHEA The absence of menstrual flow.

AMINO ACID [uh-MEE-noh] A chemical compound containing an amino group (—NH$_2$), a carboxyl group (—COOH), and a variable nonacid organic radical;

the structural unit of proteins.

AMNIOCENTESIS [am-nee-oh-sehn-TEE-sihss; Gr. sac + puncture] The technique of obtaining cells of a fetus from its amniotic fluid.

AMNION (Gr. sac) A tough, thin, transparent membrane enveloping the embryo.

AMPHIARTHROSIS (Gr. *amphi,* on both sides + *arthrosis,* articulation) A slightly movable joint.

AMPULLA A small dilation in a canal or duct.

AMYOTROPHIC LATERAL SCLEROSIS The common motor neuron disease causing muscular atrophy; also called *Lou Gehrig's disease.*

ANABOLISM [uh-NAB-uh-lihz-uhm; Gr. *ana,* upward progression] The metabolic process (synthesis) that combines two or more atoms, ions, or molecules into a more complex substance.

ANAL CANAL The continuation of the rectum that pierces the pelvic diaphragm and turns sharply downward and backward, opening into the anus.

ANAL COLUMNS About 5 to 10 permanent longitudinal columns of mucous membrane in the upper part of the anal canal; also called *rectal columns.*

ANALGESIC A pain-relieving drug.

ANAPHASE The third and shortest stage of mitosis; chromosome pairs separate and move toward opposite poles.

ANATOMICAL POSITION The universally accepted position from which locations of body parts can be described. The body is standing erect and facing forward, feet together, arms hanging at sides with palms facing forward.

ANATOMY (Gr. *ana- + temnein,* to cut) The study of the body and its parts.

ANEMIA An abnormal condition in which the number of red blood cells, the concentration of hemoglobin, or the hematocrit is below normal.

ANEURYSM (Gr. "to dilate") A balloonlike, blood-filled dilation of a blood vessel.

ANGINA PECTORIS (L. strangling + chest) An abnormal condition that occurs when insufficient blood reaches the heart because of a damaged or blocked artery, producing chest pain.

ANOREXIA NERVOSA (Gr. *an,* without + *orexis,* a longing) A psychological eating disorder characterized by self-imposed starvation and subsequent nutritional deficiencies.

ANTAGONIST ("against the agonist") A muscle that opposes the movement of a prime mover; see *agonist.*

ANTERIOR (L. *ante,* before) A relative directional term: toward the front of the body; also called *ventral.*

ANTIBODY A protein produced by lymphocytes in response to an antigen; the antibody destroys the antigen by a complex process.

ANTIDIURETIC HORMONE (ADH) A peptide hormone secreted by the neurohypophysis; it helps regulate the body's fluid balance; also called *vasopressin.*

ANTIGEN A substance against which an antibody is produced.

ANUS (L. ring) An opening from the rectum to the outside; the site of feces excretion.

AORTA The major arterial trunk emerging from the left ventricle of the heart; it supplies oxygenated blood to the body.

AORTIC BODY A chemoreceptor in the sinuses of the aorta that responds to a lowered blood pH by increasing the respiratory rate.

APNEUSTIC AREA A breathing-control area in the pons which, when stimulated, causes strong inhalations and weak exhalations; see *respiratory center.*

APONEUROSIS A broad, flat sheet of closely packed connective collagenous fibers; it is the connection between a muscle and its attachment.

APPENDICULAR (L. *appendere,* to hang from) Pertaining to the appendicular part of the body, which includes the upper and lower extremities.

APPENDICULAR SKELETON The bones of the upper and lower extremities; includes the shoulder and pelvic girdles.

APPENDIX See *vermiform appendix.*

AQUEOUS HUMOR A thin, watery fluid in the posterior and anterior chambers of the eye.

ARACHNOID (Gr. cobweblike) The delicate middle layer of the meninges, which covers the brain and spinal cord.

ARBOR VITAE [VYE-tee; L. tree of life] Branched arrangement of white matter within the cerebellum.

AREOLA (L. dim. of *area,* open place) A small space in a tissue; a small, dark-colored area around a center portion; pigmented area around a nipple.

AREOLAR TISSUE [uh-REE-uh-lur; L. *areola,* open place] Most common connective tissue; contains tiny extracellular spaces usually filled with ground substance and tissue fluids.

ARRECTOR PILI The muscle that contracts to pull a follicle and its hair erect, elevating the skin above and producing a "goose bump."

ARRHYTHMIA See *cardiac arrhythmia.*

ARTERIOLE A small artery that branches before reaching capillary networks.

ARTERY A blood vessel that carries blood away from the heart.

ARTHRITIS (Gr. *arthron,* joint + inflammation) Inflammation of a joint; a general term for many specific forms of arthritis.

ARTHROSCOPY A diagnostic and surgical technique in which a small fiberoptic scope is used to look into a joint.

ARTICULAR CAPSULE A fibrous capsule that lines the synovial cavity in the noncartilaginous parts of the joint, permitting considerable movement.

ARTICULAR CARTILAGE The smooth cartilage that caps the bones facing the synovial cavity; also called *hyaline cartilage.*

ARTICULAR DISK A fibrocartilage disk that (1) acts as a shock absorber for a joint, (2) adjusts the uneven articulating surfaces, and (3) allows two kinds of movements to occur simultaneously.

ARTICULATION (L. *articulus,* joint) A joint, the place where bones meet or where cartilages or bones and cartilages meet.

ASCENDING COLON The portion of the large intestine extending upward from the cecum.

ASSOCIATED CELLS Specialized cells, other than neurons, in the CNS and PNS.

ASSOCIATION FIBERS Nerve fibers that link one area of the cerebral cortex to another area of the cortex of the same cerebral hemisphere.

ASTHMA (Gr. no wind) A general term for difficulty in breathing, often relating to bronchial constriction.

ASTIGMATISM (Gr. "without focus") A condition in which the curvature of the lens or cornea is not uniform, producing an image that is partially unfocused.

ASTROCYTE The largest, most numerous glial cell; sustains neurons.

ATHEROSCLEROSIS (Gr. "porridgelike hardening") An abnormal condition characterized by deposits of fat, fibrin, cellular debris, and calcium on the inside of arterial walls.

ATLAS The first cervical vertebra (spinal bone), which supports the head.

ATOM (Gr. *atomos,* indivisible) The basic unit of all matter, consisting of protons, neutrons, and electrons; the smallest unit of an element that retains the chemical characteristics of that element.

ATOMIC NUMBER The number of protons in the nucleus of an atom.

ATOMIC WEIGHT The relative weight of an element compared with that of carbon-12.

ATP See *adenosine triphosphate.*

ATRIAL NATRIURETIC PEPTIDE (ANP) A peptide hormone secreted by secretory granules in cardiac muscle cells; it helps maintain the homeostatic balance of fluids and electrolytes and lowers blood pressure and volume.

ATRIOVENTRICULAR BUNDLE A group of conducting fibers in the interventricular septum that branch into each ventricle, finally branching into hundreds of cardiac conducting myofibers; formerly called *bundle of His.*

ATRIOVENTRICULAR (AV) NODE A mass of specialized heart tissue that delays the electrical activity of the heart a few hundredths of a second before allowing it to pass into the ventricles.

ATRIOVENTRICULAR (AV) VALVE Heart valve that allows blood to flow in one direction from atria to ventricles.

ATRIUM [AY-tree-uhm; L. porch, antechamber] One of two upper heart chambers; see *ventricle.*

AUDITORY CORTEX The portion of the temporal lobe of the cerebrum involved with basic sounds and the feeling of dizziness.

AUDITORY TUBE A tube in the anterior wall of the tympanic cavity leading to the nasopharynx above the soft palate; permits air to pass from the nasal cavity into the middle ear; commonly called the *Eustachian tube.*

AUTOIMMUNITY A condition in which the tolerance to "self" antigens breaks down and the body forms antibodies to its own antigens; results in autoimmune diseases.

AUTONOMIC GANGLIA Clusters of cell bodies and dendrites of the ANS, with synapses that occur outside the CNS.

AUTONOMIC NERVOUS SYSTEM (ANS) The efferent motor division of the visceral nervous system; composed of nerve fibers from the brain and spinal cord that inhibit or excite smooth muscle, cardiac muscle, or glands; consists of sympathetic and parasympathetic nervous systems.

AUTONOMIC PLEXUSES (L. braids) Branched, interlaced networks in the thoracic, abdominal, and pelvic cavities; includes cardiac, pulmonary, celiac, hypogastric, and enteric plexuses.

AUTOSOMES The 22 matching pairs of chromosomes that determine genetic traits, but not gender.

AXIAL (L. *axis,* hub, axis) Pertaining to the axis, or trunk, of the body. The axial part is composed of the head, neck, thorax, abdomen, and pelvis.

AXIAL SKELETON The portion of the skeleton forming the longitudinal axis of the body; includes the skull, vertebral column, sternum, and ribs.

AXIS The second cervical vertebra, on which the atlas rests.

AXON A long, specialized process of a neuron that carries nerve impulses away from the cell body to the next neuron, muscle cell, or gland cell.

B

B CELL A type of lymphocyte that produces specific antibodies; also called *B lymphocyte.*

B LYMPHOCYTE See *B cell.*

BABINSKI'S REFLEX A diagnostic reflex in which the stroking of the lateral part of the sole results in the big toe pointing up and the toes fanning.

BALL-AND-SOCKET JOINT A multiaxial joint in which the globular head of one bone fits into a cuplike cavity of another bone, such as in the hip.

BARORECEPTORS (Gr. "pressure receivers") Clusters of sensory receptor cells in the walls of the aortic arch and carotid artery sinuses; they help maintain blood pressure; also called *pressoreceptors.*

BARTHOLIN'S GLANDS See *greater vestibular glands.*

BASAL Pertaining to, located at, or forming a base.

BASAL GANGLIA (sing. **GANGLION**) Deep, large cores of gray matter beneath the white matter of each cerebral hemisphere; cell bodies of neuron clusters that help coordinate muscle movements. Basal ganglia are nuclei.

BASAL LAMINA The superficial layer of the *basement membrane* (see); it is a homogeneous layer lacking fibers.

BASAL METABOLIC RATE (BMR) The total energy used by the body per unit time to sustain the minimal normal cellular activity.

BASE A chemical substance that releases hydroxyl ions (OH^-) when dissolved in water; any molecule capable of accepting a hydrogen ion or proton; see *acid.*

BASEMENT MEMBRANE Anchors epithelial tissue to underlying connective tissue, provides elastic support, and acts as a partial barrier for diffusion and filtration; consists of a *basal lamina* and a deeper layer containing reticular fibers.

BASOPHIL A type of white blood cell that is involved in allergic reactions and inflammation.

BELLY The bulging part of a muscle between its two ends.

BENIGN NEOPLASM (L. *bene,* well) An abnormal growth of tissue containing cells that appear and act almost normally and are enclosed within a capsule of thick connective tissue.

BICUSPID VALVE Left atrioventricular heart valve; also called the *mitral valve.*

BILAMINAR EMBRYONIC DISK A bilayered structure formed when the blastocyst assumes a flattened shape during the second embryonic week; composed of endoderm and ectoderm.

BILATERAL Pertaining to both sides of the body.

BILE An alkaline liquid secreted by the liver; aids in digestion of lipids.

BILIARY SYSTEM Consists of the gall-bladder, left and right hepatic ducts, cystic duct, and common bile duct.

BIOFEEDBACK The conscious self-regulation of some "involuntary" bodily responses typically under the control of the autonomic nervous system.

BIOPSY (Gr. *bios*, life + *-opsy*, examination) The microscopic examination of living tissue removed from the body.

BLASTOCOEL [BLASS-toh-seel; Gr. bud + *koilos*, hollow] The fluid-filled inner cavity of a blastocyst.

BLASTOCYST (Gr. *blasto*, germ + *kystis*, "bladder") A fluid-filled hollow sphere of cells formed about 4 to 5 days after fertilization.

BLASTOMERE (Gr. *blastos*, bud + *meros*, a part) One of two daughter cells resulting from the first cleavage of a zygote; a cell formed by the repeated division of a fertilized ovum.

BLOOD PRESSURE The force (energy) with which blood is pushed against the walls of blood vessels and circulated throughout the body when the heart contracts.

BLOOD-BRAIN BARRIER A system of tight junctions in the endothelial cells of brain capillaries that forms a semipermeable membrane, allowing only certain substances to enter the brain.

BONE See *osseous tissue*.

BONE MARROW Tissue filling the porous medullary cavity of the diaphysis of bones; also called *myeloid tissue*.

BOWMAN'S CAPSULE See *glomerular capsule*.

BRACHIAL PLEXUS Ventral rami of lower four cervical and first thoracic nerves in lower neck and axilla.

BRAINSTEM The stalk of the brain relays messages between the spinal cord and brain; consists of the medulla oblongata, pons, and midbrain.

BROAD LIGAMENTS Paired double folds of peritoneum that attach the uterus to the lateral wall of the pelvis.

BROCA'S (MOTOR SPEECH) AREA (area 44) A motor area in the frontal lobe of the cerebrum; involved in formulating spoken words; also called *anterior speech area*.

BRONCHI [BRONG-kee; sing. **BRONCHUS;** Gr. throat, windpipe] Branches of the respiratory tree that emerge from the trachea into the lungs.

BRONCHIOLES [BRONG-kee-ohlz] Small tubes emerging from branching bronchi in the lungs; they branch until they end as terminal bronchioles.

BUFFER A substance that regulates acid-base balance; combinations of weak acids or bases and their respective salts in solution that help body fluids resist changes in pH.

BULBOURETHRAL GLANDS Paired male reproductive glands that open into the urethra; secrete alkaline fluids that neutralize the acidity of urine and act as a lubricant at the tip of the penis; also called *Cowper's glands*.

BULBOUS CORPUSCLES (OF KRAUSE) Sensory receptors in the skin believed to be sensors for touch-pressure, position sense of a body part, and movement; probably variants of lamellated (Pacinian) corpuscles; formerly called *Krause's end bulbs*.

BULIMIA (Gr. *bous*, ox + *limos*, hunger) A psychological eating disorder characterized by an insatiable appetite and gorging on food, followed by self-induced vomiting or overuse of laxatives or diuretics; may lead to metabolic imbalances.

BUNDLE OF HIS See *atrioventricular bundle*.

BUNION (Old Fr. *buigne*, bump on the head) A lateral deviation of the great toe toward the second toe, accompanied by the formation of a bursa and callus on the bony prominence of the first metatarsal.

BURSAE [BURR-see; sing. **BURSA,** BURR-sah; Gr. purse] Flattened sacs filled with synovial fluid to help eliminate friction in areas where a muscle or tendon rubs against another muscle, tendon, or bone.

BURSITIS Inflammation of a bursa.

C

CALCITONIN A thyroid hormone that lowers calcium and phosphate levels in blood; also called *thyrocalcitonin*.

CALLUS (L. hard skin) An area of the skin hardened by repeated external pressure or friction.

CALORIE (L. *calor*, heat) The amount of energy required to raise the temperature of 1 g of water 1° Celsius; also called *gram calorie*; see *kilocalorie*.

CALVARIA (L. skull) The roof of the cranium, composed of the brow portion of the frontal bone, the parietal bones, and the occipital bone.

CALYCES [KAY-luh-seez; sing. **CALYX;** KAY-licks; Gr. cups] Two small cavities in the kidney formed from the branching of the renal pelvis.

CANALICULI [KAN-uh-lick-yuh-lie; L. dim. *canalis*, channel] (1) Small channels radiating from lacunae in bone; they transport materials by diffusion. (2) Channels of the liver that drain bile into the biliary system.

CANCER (L. crab) Any of the various malignant neoplasms that spread to new sites.

CAPILLARY (L. *capillus*, hair) Tiny blood vessels that connect the arterial and venous systems.

CAPILLARY FLUID SHIFT A mechanism that regulates arterial pressure by altering blood volume.

CARBOHYDRATE A molecule composed of carbon, hydrogen, and oxygen in a ratio of 1:2:1; the main source of body energy.

CARCINOGEN [kar-SIHN-uh-jehn; Gr. *karkinos*, cancer, crab + *-gen*, producing] A cancer-causing agent.

CARCINOMA [kar-suh-NOH-muh; L. cancerous ulcer + *oma*, tumor] A malignant neoplasm that originates in epithelial tissues and spreads via the lymphatic system.

CARDIAC ARRHYTHMIA A general condition referring to abnormal electrical conduction in heart tissue or changes in heart rate and rhythm.

CARDIAC CONDUCTING MYOFIBERS See *Purkinje fibers*.

CARDIAC CYCLE The sequence of steps commonly referred to as the beating of a heart; it includes contraction (systole) and relaxation (diastole).

CARDIAC MUSCLE TISSUE (Gr. *kardia*, heart) Specialized muscle tissue found only in the heart.

CARDIAC OUTPUT (CO) The quantity of blood pumped by either ventricle, but not both, in 1 minute.

CARDIAC SKELETON Tough connective tissue in the heart that provides attachment sites for the valves and corresponding muscle fibers.

CARDIOPULMONARY RESUSCITATION (CPR) A physical technique used to help reestablish respiration and circulation when a victim's breathing stops; may include mouth-to-mouth resuscitation or external cardiac compression.

CARDIOREGULATORY CENTER ANS nerve centers in the medulla oblongata that regulate the heart rate.

CAROTID BODY A chemoreceptor in the sinuses of the carotid arteries that responds to a lowered blood oxygen by increasing the respiratory rate.

CARPUS (Gr. *karpos*, wrist) The eight short bones connected by ligaments in each wrist.

CARTILAGE A specialized type of connective tissue that provides support and aids movement at joints.

CARTILAGINOUS JOINT A joint in which bones are joined by hyaline cartilage or a fibrocartilaginous disk; allows little or no movement. It includes synchondroses and symphyses.

CASTRATION The surgical removal of the testes.

CATABOLISM [kuh-TAB-uh-lihz-uhm; Gr. *katabole*, a throwing down] The metabolic process that breaks down large molecules into two or more molecules.

CATARACT A condition in which the lens of the eye becomes opaque, possibly producing blindness.

CAUDA EQUINA [KAW-duh ee-KWY-nuh; L. horse's tail] The collection of spinal nerve roots passing caudally below the conus terminalis of the spinal cord.

CECUM (L. blind) A cul-de-sac pouch on the distal side of the ileocecal orifice.

CELL The smallest independent unit of life; the component of tissues.

CELL CYCLE The period from the beginning of one cell division to the beginning of the next; the life span of a cell.

CELL-MEDIATED IMMUNITY Type of immunity that involves direct contact between T cells and antigens.

CELLULAR RESPIRATION The biochemical events within cells by which the chemical energy of foods is released.

CEMENT The bonelike covering of the neck and root of a tooth.

CENTER OF OSSIFICATION The site at which a ring of cells forms around a blood vessel in the initial development of bone tissue.

CENTRAL CANAL A longitudinal channel in the osteon (Haversian system); contains nerves and lymphatic and blood vessels; also called *Haversian canal*.

CENTRAL LOBE See *insula*.

CENTRAL NERVOUS SYSTEM (CNS) The brain and spinal cord.

CENTRIOLES Two small organelles in the *centrosome*, a specialized region of the cytoplasm of a cell; involved with the movement of chromosomes during cell division; form basal body of cilia and flagella.

CEPHALIC [suh-FAL-ihk; Gr. *kephale*, head] Pertaining to the head.

CEREBELLAR CORTEX A surface layer of gray matter covering the cerebellar lobes; composed of a network of billions of neurons.

CEREBELLAR PEDUNCLES [peh-DUNG-kuhlz; L. little feet] Three nerve bundles that attach the cerebellum to the brainstem; composed of the inferior, middle, and superior cerebellar peduncles.

CEREBELLUM (L. little brain) The second-largest part of the brain, composed of the vermis, two small flocculonodular lobes, and two large lateral lobes; located behind the pons in the posterior cranial fossa; refines and coordinates muscular movements.

CEREBRAL CORTEX A surface mantle of gray matter over the cerebrum; it is thin and convoluted, containing about 50 billion neurons and 250 billion glial cells.

CEREBRAL PALSY Impaired muscular power and coordination as a result of brain damage, usually occurring at, before, or shortly after birth.

CEREBRAL PEDUNCLES Two nerve fiber bundles composed of the pyramidal tract and corticobulbar and corticopontine fibers.

CEREBROSPINAL FLUID (CSF) A clear, watery ultrafiltrate solution formed from blood in the capillaries; bathes the ventricles of the brain and cavity of the spinal cord; cushions the brain and spinal cord.

CEREBROVASCULAR ACCIDENT (CVA) A sudden withdrawal of sufficient blood supply to the brain; commonly called a *stroke*.

CEREBRUM [suh-REE-bruhm; L. brain] The largest and most complex structure of the nervous system; consists of two cerebral hemispheres and the diencephalon; each hemisphere is composed of a cortex (gray matter), white matter, and basal ganglia; the cortex is divided into six lobes. All conscious living depends on the cerebrum.

CERVICAL PLEXUS The ventral rami of C1 to C4 nerves in the neck region.

CERVICAL VERTEBRAE The seven small neck bones between the skull and thoracic vertebrae; they support the head and allow it to move.

CERVIX (L. neck) Any neck-shaped anatomical structure; the narrow juncture between the uterus and vagina.

CESAREAN SECTION An incision through the abdominal and uterine walls to remove a fetus.

CHEMICAL SYNAPSE A junction at which two cells communicate by way of a chemical neurotransmitter.

CHEMIOSMOSIS The accumulation of electrical energy built up from an electrochemical proton gradient, and the release of that energy to make ATP from ADP and P_i; occurs in mitochondria.

CHEMORECEPTOR (L. "chemical receiver") A sensory receptor that responds to chemical stimuli.

CHEMOTHERAPY Therapy that uses drugs in treating cancer.

CHLAMYDIAL INFECTION [klah-MIHD-ee-ul] The most common sexually transmitted disease in the United States, caused by the bacterium *Chlamydia trachomatis*.

CHLORIDE SHIFT The condition that results when bicarbonate ions are removed from red blood cells, developing a positive charge inside the cell and allowing chloride ions from sodium chloride in the plasma to diffuse into the cell.

CHOLECYSTOKININ (CCK) [koh-lee-sis-TOE-kine-in] A digestive hormone secreted from the duodenal wall; it stimulates gallbladder contraction and the release of pancreatic enzymes.

CHOLELITHIASIS (bile, gall + stone) A condition in which hard deposits form in the gallbladder, usually caused by insufficient bile salts to dissolve the cholesterol in *micelles*; the cholesterol crystallizes and hardens into "gallstones" (biliary calculi).

CHOLINERGIC Pertaining to neurons of the ANS and motor end plates of motor neurons that release *acetylcholine*.

CHOLINESTERASE An enzyme that causes the breakdown of acetylcholine into choline and acetic acid (acetate).

CHONDROCYTE (Gr. *khondros*, cartilage) A cartilage cell embedded in lacunae within the matrix of cartilage connective tissue.

CHORION The protective sac around the embryo.

CHORIONIC VILLI Fingerlike projections growing outward from the embryonic chorion into maternal tissue; allow the exchange of nutrients and gases between mother and embryo and removal of metabolic wastes from embryo.

CHORIONIC VILLI SAMPLING A technique in which a sample of chorionic villi tissue is collected and examined for genetic diseases.

CHOROID The posterior two-thirds of the vascular layer of the eye, composed of blood vessels and connective tissue between the sclera and retina.

CHROMATOPHILIC SUBSTANCE Rough endoplasmic reticulum and free ribosomes in the cell bodies of neurons; involved in protein synthesis; formerly called *Nissl bodies*.

CHROMOSOME (Gr. "colored body") A threadlike nucleoprotein structure within the nucleus of a cell that contains the DNA.

CHRONIC RENAL FAILURE The progressive destruction of nephrons, possibly leading to stoppage of kidney function.

CHYLOMICRON A tiny, protein-coated droplet containing triglycerides, phospholipids, cholesterol, and free fatty acids.

CHYME [KIME] A soupy mixture formed in the stomach from gastric juices and particles of broken-down food.

CILIA [SIHL-ee-uh; sing. **CILIUM**; L.

eyelid, eyelash] Short processes extending from the surface of some cells; often capable of producing a rhythmic paddling motion.

CILIARY BODY The thickened part of the vascular layer of the eye, connecting the choroid with the iris; contains muscles involved in accommodation.

CIRCULATORY SHOCK A condition in which there is inadequate blood flow throughout the body.

CIRCUMCISION (L. *circumcidere*, to cut around) The surgical removal of the prepuce of the penis.

CIRCUMDUCTION A movement in which the distal end of a bone moves in a circular path while the proximal end remains stable, acting as a pivot.

CIRRHOSIS OF THE LIVER [sih-ROE-siss; L. "orange-colored disease"] A chronic liver disease characterized by the destruction of liver cells and their replacement by abnormal fibrous connective tissue, eventually resulting in liver failure and death.

CITRIC ACID CYCLE A series of intercellular metabolic reactions that begins with pyruvic acid and releases a molecule of carbon dioxide and two hydrogen atoms; also called *Krebs cycle*.

CLEAVAGE A series of mitotic cell divisions occurring in a zygote immediately after fertilization.

CLITORIS [KLIHT-uh-rihss; Gr. small hill] A small erectile organ at the upper end of the vulva, below the mons pubis, where the labia minora meet; a major source of sexual arousal.

CLONAL SELECTION THEORY A theory that an antigen influences the amount of antibody produced but does not affect the combining sites on the antibody's three-dimensional structure. It also proposes that each B cell is genetically programmed to react only to a particular antigen.

COAGULATION (L. *coagulare*, to curdle) The process of blood clotting.

COCCYGEAL PLEXUS The coccygeal nerve plus communications from nerves S4 and S5.

COCCYX The three to five fused vertebrae at the end of the vertebral column, below the sacrum; commonly called *tailbone*.

COCHLEA [KAHK-lee-uh; Gr. *kokhlos*, snail] A spiral bony chamber surrounding the semicircular ducts of the inner ear; divided into the scala vestibuli, scala tympani, and scala media.

COENZYME An organic compound or metal required for enzyme function.

COITUS [KOH-uh-tuhss; L. *coire*, to come together] The act of sexual intercourse; also called *copulation*.

COLLAGEN [KAHL-uh-juhn; Gr. *kolla*,

glue + *genes*, born, produced] The protein found in the fibers of bone, cartilage, and connective tissue proper.

COLLATERAL BRANCHES The side branches of an axon.

COLLECTING DUCT The portion of the renal excretory tubule that receives the glomerular filtrate after it has passed through the distal convoluted tubule.

COLOR BLINDNESS A deficiency in one or another eye pigment necessary to see certain colors.

COLOSTRUM A high-protein fluid present in the mother's breasts before milk production starts about 3 days later.

COMMISSURAL FIBERS The axons that project from a cortical area of one cerebral hemisphere to a corresponding area of the opposite hemisphere.

COMPACT BONE TISSUE The very hard and dense portion of bone.

COMPLEMENT A group of 11 proteins and 8 additional factors found in blood serum; it enhances the body's defensive actions.

COMPOUND (L. *componere*, to put together) Molecules made up of atoms of two or more elements; has different chemical properties from the elements that compose it.

COMPUTERIZED AXIAL TOMOGRAPHY (CT) A scanning procedure that combines x-rays with computer technology to show cross-sectional views of internal body structures; the device is called a *CT scanner*.

CONCENTRATION GRADIENT The difference in solute concentration on either side of the plasma membrane of a cell.

CONCEPTION The process of fertilization and the subsequent establishment of pregnancy.

CONDYLE (L. knuckle) A rounded, knuckle-shaped projection on a bone; may be knoblike or convex.

CONE A color-sensitive photoreceptor cell concentrated in the retina.

CONGENITAL DEFECT A condition existing at birth, but not necessarily hereditary.

CONGESTIVE HEART FAILURE (CHF) A condition that occurs when either ventricle fails to pump blood out of the heart as quickly as it enters the atria, or when the ventricles do not pump equal amounts of blood.

CONJOINED TWINS Identical (monozygotic) twins whose embryonic disks do not separate completely; commonly called *Siamese twins*.

CONJUNCTIVA (pl. **CONJUNCTIVAE;** L. connective) The transparent mucous membrane lining the inner surface of the eyelid and exposed surface of the eyeball.

CONJUNCTIVITIS Inflammation of the

conjunctiva of the eye.

CONNECTIVE TISSUE A supportive and protective tissue consisting of fibers, ground substance, cells, and some extracellular fluid; the most abundant type of tissue.

CONSTIPATION (L. "to press together") A condition in which feces move through the large intestine too slowly, so too much water is reabsorbed, resulting in hard feces and difficult defecation.

CONTRACEPTION ("against conception") The prevention of conception.

CONTRACTILITY The ability of muscle tissue to contract.

CONTRALATERAL Pertaining to opposite sides of the body.

CONUS TERMINALIS The cone-shaped lowermost end of the spinal cord; also called *conus medullaris*.

CONVERGENCE (L. *convergere*, to merge) (1) A condition in which the receptive segment of a postsynaptic neuron is excited or inhibited by the axon terminals of many presynaptic neurons. (2) The coordinated inward turning of the eyes to focus on a nearby object.

CONVOLUTIONS See *gyri*.

COPULATION (L. *copulare*, to fasten together) See *coitus*.

CORNEA (L. *corneus*, horny tissue) The transparent anterior part of the outer layer of the eye, a uniformly thick, convex nonadjustable lens; light enters the eye through the cornea.

CORONAL See *frontal*.

CORONARY ARTERY DISEASE (CAD) An abnormal condition usually brought on by atherosclerosis and characterized by a reduced supply of oxygen and nutrients to the myocardium.

CORONARY CIRCULATION The flow of blood to the heart itself.

CORONARY SINUS A large vein that receives blood from the veins that drain the heart, emptying it into the right atrium.

CORPUS CALLOSUM (L. hard body) The larger of two cerebral commissures connecting the cerebral hemispheres.

CORPUS LUTEUM (L. "yellow body") A temporary ovarian endocrine tissue that secretes female sex hormones; formed as the lining of a follicle grows inward after the discharge of an ovum.

CORPUS STRIATUM (L. furrowed body) The largest mass of gray matter in the cerebral basal ganglia; composed of the caudate and lentiform nuclei, and other nuclei.

CORPUSCLES OF RUFFINI Sensory receptors in skin believed to be sensors for touch-pressure, position sense of a body part, and movement; probably variant of lamellated (Pacinian) corpuscles.

CORTEX (L. bark, shell) The outer layer of an organ or part.

CORTICOSPINAL TRACTS See *pyramidal tracts.*

CORTICOTROPIN See *adrenocorticotropic hormone.*

COUNTERCURRENT MULTIPLIER SYSTEM A positive feedback system that results from a countercurrent flow in the loop of the nephron that helps to regulate the solute concentration.

COVALENT BOND A strong chemical bond formed when atoms share one or more pairs of electrons.

CRANIAL NERVES Twelve pairs of peripheral nerves carrying sensory signals to the brain, and/or motor impulses to various places in the body.

CRANIAL PARASYMPATHETIC OUTFLOW Preganglionic fibers from the cranial portion of the parasympathetic nervous system that leave the brainstem via cranial nerves III, VII, IX, and X.

CRANIOSACRAL DIVISION Visceral efferent nerve fibers that leave the central nervous system via cranial nerves III, VII, IX, and X from the brainstem, or spinal nerves S3 and S4; part of parasympathetic nervous system.

CREATINE PHOSPHATE (CP) A high-energy phosphate similar to ATP; also called *phosphocreatine.*

CRETINISM [CREH-tin-ism] A condition caused by underactivity of the thyroid during fetal development after the twelfth week; characterized by mental retardation and irregular development of bones and muscles.

CRYPTORCHIDISM [krip-TOR-kye-dizm; Gr. *kryptos*, hidden + *orchis*, testis] A condition in which the testes do not descend from their fetal position into the scrotum after birth.

CRYPTS OF LIEBERKÜHN See *intestinal glands.*

CT SCAN See *computerized axial tomography.*

CUSHING'S DISEASE A condition caused by an overactive adrenal cortex; characterized by fattening of the face, chest, and abdomen and a tendency toward diabetes; also called *Cushing's syndrome.*

CYANOSIS (Gr. dark blue) The development of a bluish color of the skin, especially the lips, resulting from the build-up of deoxygenated hemoglobin.

CYCLIC AMP A form of AMP produced by the activities of certain hormones and neurotransmitters; causes target cell to perform specific function.

CYCLIC GUANOSINE MONOPHOSPHATE (cGMP) An intracellular messenger hydrolyzed in rod cells in the retina.

CYST A mass of fluid-filled tissue extending to the dermis or hypodermis; over 1 cm in diameter.

CYSTIC FIBROSIS A genetic disease characterized by thick mucus secretions that impair breathing and digestion.

CYSTITIS Inflammation of the urinary bladder.

CYTOKINESIS (Gr. *cyto*, cell + movement) The separation of the cytoplasm into two parts during this final stage of cell division following mitosis; two genetically identical daughter cells are formed.

CYTOLOGY [sigh-TAHL-uh-jee; Gr. *kytos*, hollow vessel; *cyto*, cell] The microscopic study of cells.

CYTOPLASM The portion of a cell outside the nucleus, where metabolic reactions take place; the fluid portion is cytosol, which contains subcellular organelles.

CYTOPLASMIC INCLUSIONS Solid particles temporarily in cells; usually food or the stored products of the cell's metabolism.

CYTOSKELETON The flexible cellular framework of interconnecting microfilaments, intermediate filaments, and other organelles; involved with support and cell movement; site for the binding of specific enzymes; also called *microtrabecular lattice.*

D

DECUSSATION (L. *decussis*, the number ten, X, indicating a crossing over) The crossing over of the axons of sensory and motor pathways from one side of the spinal cord or brainstem to the other.

DEEP A relative directional term of the body: farther from the surface of the body; opposite of *superficial.*

DEFECATION The discharge of feces from the rectum through the anus.

DEGLUTITION (L. *deglutire*, to swallow down) The act of swallowing.

DELAYED HYPERSENSITIVITY An allergic reaction that takes several days or longer to be expressed.

DENDRITES (Gr. *dendron*, tree) Short, threadlike extensions of the cell body of a neuron; they conduct nerve impulses toward the cell body.

DENTAL CARIES Tooth decay ("cavities") initiated by bacteria that promote the production of an enamel-eroding enzyme.

DENTINE The extremely sensitive yellowish portion of a tooth surrounding the pulp cavity; it forms the bulk of a tooth.

DEOXYRIBONUCLEIC ACID (DNA)

[dee-AHK-see-rye-boh-noo-KLAY-ihk] A double-stranded nucleic acid that is a constituent of chromosomes; contains hereditary information coded in specific sequences of nucleotides.

DEPOLARIZATION A reversal of electrical charges on the plasma membrane of a neuron, giving the inner side of the membrane a positive charge relative to the outer side. (See *action potential.*)

DEPRESSION A movement that lowers a body part, such as opening the mouth.

DERMATOLOGIST A physician specializing in the treatment of skin disorders.

DERMATOME (Gr. *derma*, skin + *tomos*, a cutting) (1) An instrument used in cutting thin slices of the skin, as in skin grafting. (2) A segment of skin with sensory fibers from a single spinal nerve; used in locating injuries to dorsal roots of the spinal cord.

DERMIS (Gr. *derma*, skin) The connective tissue meshwork of collagenous, reticular, and elastic fibers that makes up most of the skin.

DESCENDING COLON The portion of the large intestine extending from the left colic flexure of the transverse colon down to the rim of the pelvis, where it becomes the sigmoid colon.

DESMOSOME (Gr. *desmos*, binding) A junction with no direct contact between adjacent plasma membranes; common in skin; also called *spot desmosome.*

DIABETES INSIPIDUS [in-SIPP-ih-duhss] A condition in which ADH is undersecreted and excessively large amounts of water are excreted in the urine.

DIABETES MELLITUS [MELL-ih-tuhss] A hereditary disease that results when insufficient insulin is produced and glucose accumulates in the blood.

DIALYSIS A passive transport process in which small particles diffuse through a semipermeable membrane, leaving larger particles behind; does not occur in the body naturally.

DIAPHYSIS [die-AHF-uh-siss; Gr. *dia*, between + growth] A tubular shaft of compact bone in most adult long bones; the center of the shaft is filled with marrow, and there is a spongy *epiphysis* at each end of the shaft.

DIARRHEA (Gr. "a flowing through") A condition in which watery feces move through the large intestine rapidly and uncontrollably.

DIARTHROSIS (Gr. *dia*, between + *arthrosis*, articulation) A freely movable joint.

DIASTOLE [die-ASS-toe-lee] The relaxation of the atria and ventricles during the cardiac cycle.

DIASTOLIC PRESSURE The blood pressure measurement during the interval between heartbeats; it is the second number shown in a blood pressure reading; see *systolic pressure*.

DIENCEPHALON (L. "between brain") The deep part of the cerebrum, connecting the midbrain with the cerebral hemispheres; houses the third ventricle and is composed of the thalamus, hypothalamus, epithalamus, and ventral thalamus (subthalamus).

DIFFERENTIATION The process by which cells develop into specialized tissues and organs.

DIFFUSION See *simple diffusion*.

DIGESTION The chemical and mechanical conversion of large nutrient particles into small, absorbable molecules.

DIGITAL SUBTRACTION ANGIOGRAPHY (DSA) (Gr. *angeion*, vessel) A noninvasive exploratory technique that uses a digital computer to produce three-dimensional pictures of blood vessels.

DISACCHARIDE [dye-SACK-uh-ride; Gr. *di*, two + *sakkharon*, sugar] A carbohydrate formed when two monosaccharides are united chemically. Disaccharides have the general formula $C_{12}H_{22}O_{11}$.

DISLOCATION (L. *luxare*, to put out of joint) A displacement of bones in a joint so that two articulating surfaces become separated; also called *luxation*.

DISSOCIATION The tendency for some molecules to break up into ions in water.

DISTAL (from *distant*) A relative directional term: away from the trunk of the body (away from the attached end of a limb).

DISTAL CONVOLUTED TUBULE The portion of the renal excretory tubule that receives the glomerular filtrate after it has passed through the loop of the nephron.

DIURETIC (Gr. "to urinate through") A therapeutic drug that increases the volume of urine by decreasing the water reabsorbed from the renal collecting duct.

DIVERGENCE (L. *divergere*, to bend) A condition in which the transmissive segment of a presynaptic neuron branches to have many synaptic connections with the receptive segments of many other neurons; see *convergence*.

DIVERTICULA (sing. **DIVERTICULUM**) Bulging pouches in the wall of the large intestine that push the mucosal lining through the muscle in the wall; see *diverticulitis*.

DIVERTICULITIS An abnormal condition in which diverticula in the large intestine become inflamed; may become seriously infected if food and bacteria become lodged in the diverticula; in *diverticulosis*, diverticula (pouches) are

present but do not present symptoms.

DIVERTICULOSIS See *diverticulitis*.

DIZYGOTIC TWINS Twins formed when more than one ovum is released from the ovary or ovaries at the same time and two ova are fertilized; also called *fraternal twins*.

DNA See *deoxyribonucleic acid*.

DOMINANT A genetic characteristic that is expressed to the exclusion of a less-powerful recessive characteristic.

DORSAL (L. *dorsalis*, back) A relative directional term: toward the back of the body; the upper surface of the hand or foot; also called *posterior*; opposite of *ventral*.

DORSAL CAVITY The smaller of two main body cavities; contains the cranial and spinal cavities.

DORSIFLEXION Upward flexion of the foot at the ankle joint.

DOWN'S SYNDROME See *trisomy 21*.

DUCTUS ARTERIOSUS A shunt through which fetal blood in the pulmonary artery flows to the aorta, bypassing the nonfunctional fetal lungs.

DUCTUS DEFERENS (pl. **DUCTUS DEFERENTIA**; L. *deferre*, to bring to) The continuation of the ductus epididymis, extending from the epididymis of each testis to the ejaculatory duct; formerly called *vas deferens* or *sperm duct*.

DUCTUS VENOSUS A branch of the umbilical vein (which carries oxygenated blood) that bypasses the fetal liver.

DUODENUM [doo-oh-DEE-numh; Gr. "twelve fingers wide"] The C-shaped initial segment of the small intestine.

DURA MATER [DYOOR-uh MAY-ter; L. hard mother] The tough, fibrous outermost layer of the meninges.

DYNAMIC EQUILIBRIUM See *vestibular apparatus*.

DYNAMIC SPATIAL RECONSTRUCTOR (DSR) A scanning device that produces three-dimensional computer-generated pictures of the active brain. It can be used to view blood flow through the brain.

DYSLEXIA (L. *dys*, faulty + Gr. *lexis*, speech) Extreme difficulty in identifying printed words, usually manifested as a reading and writing disability.

DYSMENORRHEA Painful or difficult menstruation.

DYSURIA Painful urination.

E

ECTODERM (Gr. *ektos*, outside + *derma*, skin) The outermost of the three primary germ layers of an embryo; forms epithelial and nervous tissue.

ECTOPIC PREGNANCY (Gr. *ektopos*, out

of place) An abnormal implantation of a fertilized ovum into the wall of the uterine tube, for example, instead of the uterine wall.

EFFECTOR A muscle or gland that responds to nerve impulses from the central nervous system.

EFFERENT ARTERIOLE (L. *ex*, away from + *ferre*, to bring) A blood vessel in the kidney formed from the joining of glomerular capillary loops; carries blood away from the glomerulus.

EFFERENT NEURON A nerve cell that conveys impulses away from the central nervous system to the effectors; also called *motor neuron*.

EJACULATION (L. *ejaculari*, to throw) The expulsion of semen through the penis via the urethra; the expulsion accompanies orgasm in males.

EJACULATORY DUCT A duct continuing from the ductus deferens, transporting sperm and secretions from the seminal vesicles and prostate gland to the urethra.

ELECTRICAL SYNAPSE A gap junction by which two cells are electrically coupled by tiny intercellular channels.

ELECTROCARDIOGRAM (ECG or EKG) A recording of the electrical waves of the heart, as registered on an electrocardiograph.

ELECTROENCEPHALOGRAM (EEG) ("electric writing in the head") A recording that shows the electrical activity in the brain; produced by an electroencephalograph.

ELECTROENCEPHALOGRAPH (EEG) An instrument used to record the electrical activity of the brain.

ELECTROLYTE Any substance whose solution conducts electricity.

ELECTRON A negatively charged subatomic particle that moves around the nucleus of an atom.

ELECTRON TRANSPORT SYSTEM A group of compounds in mitochondria that transfer electrons from NADH and $FADH_2$ to molecules of oxygen, releasing some chemical energy to synthesize ATP.

ELEMENT A chemical substance that cannot be broken down into simpler substances by ordinary chemical means.

ELEVATION A movement that raises a body part.

EMBOLISM (Gr. *emballein*, to throw in) A traveling blood clot, air bubble, or other blockage that reduces the flow of blood through a blood vessel.

EMBRYONIC PERIOD The prenatal period from fertilization to the end of the eighth week; see *fetal period*.

EMPHYSEMA (Gr. blown up) A condition in which the alveoli in the lungs fail to contract properly, thus expelling insufficient air.

ENAMEL The insensitive white covering of the crown of a tooth.

ENCEPHALITIS A severe inflammation of the brain, usually involving a virus.

ENDOCARDIUM ("inside the heart") The fibrous layer covering the inside cavities of the heart and all associated valves and muscles.

ENDOCHONDRAL OSSIFICATION (Gr. *endon*, within + *khondros*, cartilage) The process by which bone tissue develops by replacing hyaline cartilage.

ENDOCRINE GLAND (Gr. *endon*, within + *krinein*, separate or secrete) An organ with specialized secretory cells but no ducts; the cells release their secretions directly into the bloodstream.

ENDOCYTOSIS (Gr. *endon*, within + *cyto*, cell) The active process that moves large molecules or particles through a plasma membrane when the membrane forms a pocket around the material, enclosing it and drawing it into the cytoplasm within the cell. Includes *pinocytosis*, *receptor-mediated endocytosis*, and *phagocytosis*.

ENDODERM (Gr. *endon*, within + Gr. *derma*, skin) The innermost of the three primary germ layers of an embryo; forms epithelial tissue.

ENDOLYMPH Thin, watery fluid in the membranous labyrinth of the inner ear.

ENDOMETRIOSIS The abnormal location of endometrial tissue in sites such as the ovaries, pelvic peritoneum, and small intestine.

ENDOMETRIUM (inside + Gr. *metra*, womb) The innermost tissue layer of the uterus, composed of specialized mucous membrane.

ENDOMYSIUM (Gr. *endon*, inside, within + muscle) A connective tissue sheath surrounding each muscle fiber.

ENDONEURIUM ("within the nerve") Interstitial connective tissue separating individual nerve fibers.

ENDOPLASMIC RETICULUM (ER) A labyrinth of flattened sheets, sacs, tubules, and double membranes that branch through the cytoplasm, creating channels for intracellular transport.

ENDORPHINS ("endogenous morphine-like substances") Naturally occurring peptides found in several regions of the brain, spinal cord, and pituitary gland; have the pain-killing effects of opiates by suppressing synaptic activity leading to pain sensation.

ENDOSTEUM [end-AHSS-tee-uhm; Gr. *endon*, inside + *osteon*, bone] The membrane lining the internal cavities of bones.

ENKEPHALINS [en-KEFF-uh-lihnz; "in-the-head substances"] Morphinelike

breakdown products of *endorphins*.

ENTEROGASTRIC REFLEX A nervous reflex that decreases stomach motility and gastric secretion.

ENTEROKINASE An enzyme that converts the pancreatic enzyme trypsinogen into active trypsin, which helps complete the breakdown of peptides into amino acids.

ENZYME (Gr. *enzumos*, leavened) A protein catalyst that increases the rate of a chemical reaction.

EOSINOPHIL A type of white blood cell that helps modulate allergic inflammatory reactions and destroys antibody-antigen complexes.

EPENDYMAL CELL A glial cell (see *neuroglia*) that helps form part of the inner membranes of the neural tube during embryonic growth; secretes cerebrospinal fluid.

EPIDERMIS (Gr. *epi*, over + *derma*, skin) The outermost layer of the skin.

EPIDIDYMIS (Gr. "upon the twin") A coiled tube that stores immature sperm until they mature; serves as a duct for the passage of sperm from the testis to the ejaculatory duct.

EPIDURAL SPACE The space between the dura mater of the spinal cord and the periosteum of the vertebrae; contains blood vessels and fat.

EPIGLOTTIS A flap of cartilage that folds down over the opening into the larynx during swallowing and swings back up when swallowing ceases.

EPILEPSY (Gr. *epilambanein*, seize upon) A nervous disorder characterized by recurrrent attacks of motor, sensory, or psychological malfunction, with or without unconsciousness or convulsions.

EPIMYSIUM (*epi*, over, upon + muscle) The connective tissue sheath below the deep fascia surrounding a muscle.

EPINEPHRINE [ep-ih-NEFF-rihn] The main secretion of the adrenal medulla; increases pulse rate, blood pressure, and heart rate; stimulates the contraction of smooth muscle; and increases blood sugar levels; also called *adrenaline*.

EPINEURIUM ("upon the nerve") Connective tissue sheath containing blood and lymphatic vessels that together surround a bundle of nerve fascicles.

EPIPHYSEAL PLATE A thick plate of hyaline cartilage that provides the framework for the construction of spongy bone tissue within the metaphysis; also called *growth plate*.

EPIPHYSIS [ih-PIHF-uh-sis; Gr. "to grow upon"] The roughly spherical end of a long bone; composed of spongy bone tissue.

EPISIOTOMY The cutting of the mother's vulva at the lower end of the vaginal orifice at the time of delivery to enlarge the birth canal.

EPITHELIAL TISSUE Groups of cells that cover or line something; secretes substances that lubricate or take part in chemical reactions; also called *epithelium*.

EPITHELIUM See *epithelial tissue*.

EPONYCHIUM [epp-oh-NICK-ee-uhm; Gr. *epi*, upon + *onyx*, nail] The thin layer of epidermis covering the developing nail; the *cuticle* in the mature nail.

ERECTION The state of the clitoris or penis when its spongy tissue becomes engorged with blood, usually during erotic stimulation.

ERYTHEMA A diffuse redness of the skin.

ERYTHROCYTES [ih-RITH-roh-sites; Gr. *eruthros*, red + cell] Red blood cells, which constitute about half the total volume of blood; contain *hemoglobin*.

ERYTHROPOIESIS [ih-RITH-roh-poy-EE-sis; Gr. *eruthros*, red + to make] The production of erythrocytes in bone marrow.

ERYTHROPOIETIN (Gr. *eruthros*, red + *poiesis*, making) A glycoprotein hormone produced mainly in the kidneys; it is the main controller of the rate of erythrocyte production.

ESOPHAGUS [ih-SOFF-uh-guss; Gr. gullet] A muscular, membranous tube through which food passes from the pharynx into the stomach.

ESSENTIAL AMINO ACIDS The amino acids that cannot be synthesized by the body and must be supplied in the diet.

ESTROGENS A class of ovarian hormones that help regulate the menstrual cycle and the development of the mammary glands and female secondary sex characteristics; include estrin, estrone, and estradiol.

EUSTACHIAN TUBE [yoo-STAY-shun] See *auditory tube*.

EVERSION A movement of the foot in which the great toe is turned downward, and sole of foot faces laterally; opposite of *inversion*.

EXCITABILITY The capacity of a nerve or muscle cell to respond to a stimulus.

EXCITATORY POSTSYNAPTIC POTENTIAL (EPSP) A partial depolarization of a neuron membrane, but not enough to generate an action potential.

EXCRETION (L. *excernere*, to sift out) The elimination of the waste products of metabolism and surplus substances from body tissues.

EXHALATION See *expiration*.

EXOCRINE GLANDS Organs with specialized cells that produce secretions and with ducts that carry the secretions to body surfaces.

EXOCYTOSIS (Gr. *exo*, outside + *cyto*, cell) The active process in which an endosome fuses with the plasma membrane and expels unwanted particles from a cell; opposite of endocytosis.

EXPIRATION The process in which oxygen-poor air from venous blood is exhaled back into the atmosphere; also called *exhalation*.

EXPIRATORY RESERVE VOLUME The volume of air remaining in the lungs after an ordinary exhalation.

EXTENSION A straightening motion that increases the angle of a joint.

EXTERNAL A relative directional term of the body: outside; opposite of *internal*.

EXTERNAL AUDITORY CANAL A slightly curved canal separating the external ear from the middle ear; also called *external auditory meatus*.

EXTERNAL AUDITORY MEATUS See *external auditory canal*.

EXTERNAL EAR Composed of the visible auricle and the external auditory canal.

EXTERNAL NARES [NAIR-eez; L. nostrils] The two external openings of the nostrils.

EXTERNAL RESPIRATION The exchange of respiratory gases in which oxygen from the lungs moves into the blood and carbon dioxide and water move from the blood to the lungs.

EXTEROCEPTOR (L. "received from the outside") A sensory receptor that responds to external stimuli that affect the skin directly.

EXTRACELLULAR FLUID Fluid that surrounds and bathes the body's cells.

EXTRAEMBRYONIC MEMBRANES Four membranes that form outside the embryo during the third week of embryonic development, including the yolk sac, amnion, allantois, and chorion; also called *fetal membranes*.

EXTREMITIES The extremities, or appendages, of the body are the *upper extremities* (shoulders, upper arms, forearms, wrists, hands) and *lower extremities* (thighs, legs, ankles, feet).

EXTRINSIC BLOOD-CLOTTING PATHWAY A rapid blood-clotting system activated when blood vessels are ruptured and tissues are damaged; see *intrinsic blood-clotting pathway*.

F

FACET (Fr. little face) A small flat surface on a bone.

FACILITATED DIFFUSION The passive transport process in which large molecules use carrier proteins to pass through the protein channels of a plasma membrane.

FALLOPIAN TUBE See *uterine tube*.

FASCIA [FASH-ee-uh; pl. **FASCIAE**; L. band] A sheath of fibrous tissue enclosing skeletal muscles, holding them together; may be superficial, deep, or subserous (visceral).

FASCICLE A bundle of nerve or muscle fibers.

FASCICULI [fah-SICK-yoo-lie; L. little bundles] Bundles of fibers divided into tracts within each funiculus in the white matter of the spinal cord.

FAT An energy-rich molecule that is a reserve food or long-term fuel; provides the body with insulation, protection, and cushioning.

FATTY ACID The part of a fat molecule that contains carboxyl groups (COOH); see *glycerol*.

FECES [FEE-seez; L. *faex*, dregs] An indigestible semisolid mixture stored in the large intestine until ready to be eliminated through the anus.

FEEDBACK CONTROL SYSTEM A regulatory system in which changes in the body or environment are fed back through a circular system into a central control unit such as the brain or brainstem, where initial adjustments are made to maintain homeostasis.

FETAL ALCOHOL SYNDROME A set of defects in the newborn caused by an excessive intake of alcohol by the mother during pregnancy.

FETAL MEMBRANES See *extraembryonic membranes*.

FETAL PERIOD The prenatal period from the beginning of the ninth week until birth; see *embryonic period*.

FETOSCOPY ("seeing the fetus") A procedure that allows the physician to view the fetus directly by inserting an endoscope into the mother's uterus.

FETUS [FEE-tuhss] The embryo after 8 weeks of development.

FIBRIN The insoluble, stringy plasma protein whose threads entangle blood cells to form a clot during the blood-clotting process.

FIBRINOGEN A soluble plasma protein converted into insoluble fibrin during blood clotting.

FIBRINOLYSIS ("fibrin breaking") Blood-clot destruction.

FIBROBLASTS (L. *fibro*, fiber + Gr. *blastos*, growth) The most common connective tissue cells and the only cells found in tendons; synthesize matrix materials and are considered to be secretory; assist in wound-healing.

FIBROUS JOINT A joint that lacks a joint cavity and has its bones united by fibrous connective tissue; includes *sutures*, *syndesmoses*, and *gomphoses*.

FILTRATION A passive transport process that forces small molecules through semipermeable membranes with the aid of hydrostatic pressure or some other external force.

FILUM TERMINALE A nonneural fibrous filament extending caudally from the conus terminalis of the spinal cord; attaches to the coccyx.

FIMBRIAE [FIHM-bree-ee; L. threads] Ciliated fringes that help sweep an ovum released by an ovary into the *infundibulum*.

FISSURE (L. *fissio*, split) A groove or cleft, as in bones and the brain.

FIXATOR MUSCLE A muscle that provides a stable base for the action of the prime mover; also called *postural muscle*.

FIXED-MEMBRANE-RECEPTOR MECHANISM A mechanism by which water-soluble hormones (amines and proteins) regulate cellular responses.

FLACCID [FLACK-sihd; L. *flaccus*, hanging, flabby] (1) The loose, nonerect state of the clitoris or penis; lacking muscle tone. (2) The flabby state of skeletal muscle when deprived of innervation (as in poliomyelitis).

FLAGELLA (sing. **FLAGELLUM;** L. "whip") Threadlike appendages of certain cells, usually numbering no more than one or two per cell; used to propel the cell through a fluid environment.

FLEXION A bending motion that decreases an angle at a joint.

FLEXOR REFLEX ARC A withdrawal reflex involving sensory receptors, afferent neurons, interneurons, alpha motor neurons, and voluntary muscles.

FLUID-MOSAIC MODEL The model that proposes that a plasma membrane is a fluid bilayer composed mainly of proteins and phospholipids.

FOLLICLE (of hair) (L. *follicus*, little bag) The tubular structure enclosing the hair *root* and *bulb*.

FOLLICLE-STIMULATING HORMONE (FSH) A hormone secreted by the anterior pituitary; it stimulates spermatogenesis and testosterone secretion in males and estrogen secretion in females.

FOLLICLES The centers of oogenesis in the cortex of the ovaries.

FONTANEL A large membranous area between incompletely ossified bones.

FORAMEN [fuh-RAY-muhn; pl. **FORAMINA;** L. opening] A natural opening into or through a bone.

FORAMEN OVALE ("oval window") An opening in the fetal interatrial septum that closes at birth.

FORESKIN See *prepuce*.

FORMED ELEMENTS The elements that make up the solid part of blood; include erythrocytes, leukocytes, and thrombocytes.

FOSSA (L. trench) A shallow depressed area, as in bones.

FOVEA [FOE-vee-uh; L. small pit] A depressed area in the macula lutea near the center of the retina; contains only cones; image formation and color vision are most acute here.

FRACTURE (L. *fractura,* broken) A broken bone.

FRATERNAL TWINS See *dizygotic twins.*

FREE-RADICAL DAMAGE Part of the aging process, in which highly reactive superoxide radicals and hydrogen peroxide build up due to enzyme deficiencies, and harmful free-radical chain reactions occur within cells.

FRONTAL A plane dividing the body into anterior and posterior sections formed by making a lengthwise cut at right angles to the midsagittal plane; also called *coronal.*

FRONTAL LOBE A cerebral lobe involved with control of voluntary movements (including those associated with speech) and control of a variety of emotional expressions and moral and ethical behavior; also called the *motor lobe.*

FROSTBITE A localized, direct, mechanical injury to plasma membranes that occurs when water molecules in tissues form ice crystals.

FSH See *follicle-stimulating hormone.*

FULCRUM (L. *fulcire,* to support) The point or support on which a lever turns.

FUNCTIONAL RESIDUAL AMOUNT The expiratory reserve volume plus the residual volume; about 1.8 L in women, 2.2 L in men.

FUNDUS (L. bottom) The inner basal surface of an organ farthest from the opening; the wide upper portion of the uterus.

FUNICULI [fyoo-NICK-yoo-lie; sing. **FUNICULUS;** L. little ropes] Three pairs of columns of myelinated fibers that run the length of the white matter of the spinal cord.

"FUNNY BONE" The unprotected part of the ulnar nerve, which extends the entire length of the upper limb into the hand. The nerve is well protected by tissue everywhere but at the elbow, just behind the medial epicondyle of the humerus.

G

GALLSTONES See *cholelithiasis.*

GAMETE (L. *gamos,* marriage) A sex cell; the female gamete is an ovum (egg cell), and the male gamete is a sperm cell.

GAMETE INTRAFALLOPIAN TRANSFER (GIFT) An *in vitro* procedure that unites an ovum and sperm in the uterine tube, where fertilization normally takes place.

GAMETOGENESIS The formation of gametes (sex cells).

GAMMA-AMINOBUTYRIC ACID (GABA) The major inhibitory neurotransmitter of the small local circuit neurons in such structures as the cerebral cortex, cerebellum, and upper brainstem.

GANGLIA (sing. **GANGLION**) Groups of cell bodies located outside the central nervous system.

GAP JUNCTION A junction formed from several links of channel protein connecting two plasma membranes; found in interstitial epithelia.

GASTRIC (Gr. *gaster,* belly, womb) Pertaining to the stomach.

GASTRIN A polypeptide hormone secreted by the stomach mucosa; it produces digestive enzymes and hydrochloric acid in the stomach.

GASTROINTESTINAL (GI) TRACT The part of the digestive tract below the diaphragm.

GENE [JEEN; Gr. *genes,* born, to produce] A segment of DNA that controls a specific cellular function, either by determining which proteins are synthesized or by regulating the action of other genes; a hereditary unit that carries hereditary traits.

GENERAL SENSES The senses of touch-pressure, heat, cold, pain, and body position; also called *somatic senses.*

GENOTYPE [JEEN-oh-tipe] The genetic makeup, sometimes hidden, of a person.

GERIATRICS [Gr. *geras,* old age + *iatrikos,* physician] Study of old age.

GERMINAL EPITHELIUM A specialized layer covering the ovaries and lining the seminiferous tubules of the testes; also called *germinal layer.*

GERONTOLOGY The study of aging.

GESTATION (L. *gestare,* to carry) The period during which developing offspring are carried in the uterus.

GIANTISM A hormonal disorder in which a person grows larger than normal because of an oversecretion of growth hormone.

GINGIVA [jihn-JYE-vuh] See *gum.*

GLANS CLITORIS (L. acorn) The small mass of sensitive tissue at the tip of the clitoris.

GLANS PENIS (L. acorn) The sensitive tip of the penis.

GLAUCOMA [glaw-KOH-muh] A condition that occurs when the aqueous humor of the eye does not drain properly, producing excessive pressure within the eyeball; results in blindness

when retinal neurons are destroyed by the increased pressure.

GLIAL CELLS See *neuroglia.*

GLIDING JOINT A small biaxial joint that usually has only one axis of rotation, permitting side-to-side and back-and-forth movements.

GLOMERULAR CAPSULE The portion of the nephron enclosing the glomerulus; also called *Bowman's capsule.*

GLOMERULAR FILTRATE The fluid filtered from the blood in the kidney.

GLOMERULAR FILTRATION The renal process that forces plasma fluid from the glomerulus into the glomerular capsule.

GLOMERULAR FILTRATION RATE (GFR) The amount of glomerular filtrate formed in the capsular space each minute.

GLOMERULUS [glow-MARE-yoo-luhss; pl. **GLOMERULI,** glow-MARE-you-lie; L. ball] Coiled ball of capillaries in the kidney formed by branching of an afferent arteriole; site of blood filtration.

GLUCAGON [GLOO-kuh-gon] A peptide hormone secreted by alpha cells in the pancreas; raises blood glucose levels by stimulating *glycogenolysis* and *gluconeogenesis.*

GLUCOCORTICOIDS [gloo-koh-KORE-tih-koidz] A group of hormones secreted by the adrenal cortex; include cortisol, cortisone, corticosterone, and 11-deoxycorticosterone; the glucocorticoids affect metabolism, growth, and blood glucose levels.

GLUCONEOGENESIS The process in which glucagon stimulates glucose formation from noncarbohydrate sources.

GLYCEROL The part of a fat molecule containing three hydroxyl (OH^-) groups.

GLYCOCALYX The cell coat; composed of surface carbohydrates, proteins, and lipids.

GLYCOGENESIS ("glycogen production") The conversion of glucose into glycogen.

GLYCOGENOLYSIS The process by which the liver converts glycogen into glucose.

GLYCOLYSIS The anaerobic cellular process that splits a glucose molecule into two molecules of pyruvic acid.

GLYCOSURIA The presence of glucose in the urine.

GOITER See *hypothyroidism.*

GOLGI APPARATUS Flattened stacks of membranes in cytoplasm of most cells; packages glycoproteins for secretion; also called *Golgi complex, Golgi body.*

GOLGI TENDON ORGAN An encapsulated sensory receptor that monitors the tension in muscles and tendons; involved with voluntary muscle reflexes.

GOMPHOSIS (Gr. *gomphos*, bolt) A fibrous joint in which a peg fits into a socket, such as teeth in the maxilla or mandible.

GONADOCORTICOID A steroid hormone secreted by the adrenal cortex; consists mainly of weak androgens and small amounts of estrogens, both of which have a slight effect on the gonads; also called *adrenal sex hormone*.

GONADOTROPIN-RELEASING HORMONE (GnRH) A hormone secreted by the hypothalamus to control pituitary secretion; in the female, it affects the timing of oocyte development by stimulating the secretion of FSH and LH.

GONADS (Gr. *gonos*, offspring) Sex organs: ovaries in females, testes in males.

GONORRHEA (Gr. "flow of seed") A sexually transmitted disease caused by the bacterium *Neisseria gonorrhoeae*; it is primarily an infection of the urinary and reproductive tracts.

GRAAFIAN FOLLICLE See *vesicular ovarian follicle*.

GRADED LOCAL POTENTIAL An electrical potential that spreads its effects passively, fading out a short distance from the site of stimulation; includes postsynaptic potentials.

GRAY COMMISSURE (L. "joining together") The pair of anterior horns that forms the "cross bar" of the H-shaped gray matter in the spinal cord; functions in cross reflexes.

GRAY MATTER The central part of the spinal cord; consists of nerve cell bodies and dendrites of association and efferent neurons, unmyelinated axons of spinal neurons, sensory and motor neurons, and axon terminals.

GRAY RAMI COMMUNICANTES [RAY-mee] Unmyelinated nerve fibers containing postganglionic sympathetic fibers.

GREATER OMENTUM (L. "fat skin") An extensive folded mesentery extending from the greater curvature of the stomach to the back wall and down to the pelvic cavity; filled with fat cells, plasma cells, and other defensive cells; it protects and insulates abdominal organs; see *lesser omentum*.

GREATER VESTIBULAR GLANDS Paired glands located in the floor of the vaginal vestibule; during sexual arousal they secrete an alkaline mucus solution that provides some lubrication and offsets some vaginal acidity; also called *Bartholin's glands*.

GROSS ANATOMY Any branch of anatomy that studies structures that can be examined without a microscope.

GROUND SUBSTANCE A homogeneous, extracellular material of tissues that provides a suitable medium for the passage of nutrients and wastes between cells and the bloodstream.

GROWTH FACTORS Specialized proteins that stimulate cell growth and division.

GROWTH HORMONE (GH) A hormone secreted by the adenohypophysis, affecting growth and metabolism; also called *somatotropin* or *somatotropic hormone (STH)*.

GROWTH PLATE See *epiphyseal plate*.

GUM (Old Eng. *goma*, palate, jaw) The firm connective tissue covered with mucous membrane that surrounds the alveolar processes of the teeth; also called *gingiva*.

GYNECOLOGY (Gr. *gune*, woman + study of) The medical treatment of female diseases, reproductive physiology, and endocrinology.

GYRI [JYR-rye; sing. **GYRUS;** L. circles] Raised ridges of the cerebral cortex; also called *convolutions*.

H

HAIR A specialization of the skin that develops from the epidermis; composed of cornified threads of cells, covering almost the entire body.

HAIR CELLS Specialized proprioceptor cells of the vestibular apparatus in the inner ear.

HAUSTRA Puckered bulges in the wall of the large intestine caused by the uneven pull of the taeniae coli.

HAVERSIAN CANAL See *central canal*.

HAVERSIAN SYSTEM See *osteon*.

HEAD The expanded, rounded surface at the proximal end of a bone; often joined to the shaft by a narrow neck; also called *caput*.

HEAT EXHAUSTION A condition caused by a depletion of plasma volume due to sweating and hypotension due to reflexively dilated blood vessels in the skin; also called *heat prostration*.

HEAT PROSTRATION See *heat exhaustion*.

HEATSTROKE A condition in which the hypothalamus becomes overheated and its heat-regulating system breaks down, halting perspiration; also called *sunstroke*.

HEIMLICH MANEUVER A physical maneuver that dislodges a particle stuck in the glottis or larynx by compressing the remaining air in the lungs violently enough to blow the particle free.

HEMATOCRIT (Gr. "to separate blood") The volume percentage of red blood cells in whole blood.

HEMATOENCEPHALIC BARRIER See *blood-brain barrier*.

HEMATOMA [hee-muh-TOE-muh; L. *hemato*, blood + *oma*, tumor] A localized swelling filled with blood; a blood clot.

HEMATURIA The presence of blood in the urine.

HEME An iron-containing pigment of hemoglobin.

HEMIPLEGIA Unilateral paralysis of upper and lower limbs; usually results from damage to only one side of the spinal cord or serious brain damage on the opposite side.

HEMODIALYSIS THERAPY (Gr. *dialvein*, to tear apart) The use of an artificial device to filter blood or to add nutrients to the blood in the absence of functioning kidneys.

HEMODYNAMICS The study of the principles governing blood flow.

HEMOGLOBIN [HEE-moh-gloh-bihn; Gr. *haima*, blood + L. *globulus*, little globe] A globular iron-containing protein found in erythrocytes; it transports oxygen from the lungs and some carbon dioxide to the lungs.

HEMOLYTIC DISEASE OF THE NEWBORN An abnormal condition in which the fetus's agglutinated erythrocytes break up and release hemoglobin into the blood; formerly called *erythroblastosis fetalis*.

HEMOPHILIA An affliction in which one of several protein clotting factors is absent; also called *bleeder's disease*.

HEMOPOIETIC TISSUE [hee-muh-poy-ET-ihk; Gr. *haima*, blood + *poiein*, to make] Tissues, including red bone marrow and lymphoid tissues, that produce red blood cells, platelets, and certain white blood cells.

HEMORRHAGE (Gr. *haima*, blood + *rhegnunai*, to burst forth) The rapid loss of blood from blood vessels.

HEMOSTASIS The stoppage of bleeding.

HEPARIN An acidic mucopolysaccharide that inhibits blood clotting.

HEPATIC (Gr. *hepatikos*, liver) Pertaining to the liver.

HEPATIC PORTAL SYSTEM (Gr. *hepatikos*, liver) A system of vessels that moves blood from capillary beds of the intestines to sinusoidal beds of the liver.

HEPATITIS A liver infection with either a viral or a nonviral cause.

HERNIA (L. protruded organ) The protrusion of any organ or body part through the muscular wall that usually contains it; also called *rupture*.

HERNIATED DISK A condition in which the pulpy center of an intervertebral disk protrudes through a weakened or torn surrounding outer ring; the pulpy center pushes against a spinal root or even the spinal cord; also called *ruptured* or *slipped disk*.

HIATAL HERNIA See *hernia*.

HIGH BLOOD PRESSURE See *hypertension*.

HILUS [HYE-luhss; L. *hilum*, trifle] A small, indented opening on the medial concave border of the kidney, where arteries, veins, nerves, and the ureter enter and leave the kidney; also called the *hilum*.

HINGE JOINT A joint that resembles hinges on the lid of a box. The convex surface of one bone fits into the concave surface of another bone, permitting only a uniaxial movement around a single axis, such as at the knee joint; also called *ginglymus joint*.

HISTOLOGY [hiss-TAHL-uh-jee; Gr. *histos*, web] The microscopic study of tissues.

HODGKIN'S DISEASE A form of cancer characterized by the presence of large, multinucleate cells in the affected lymphoid tissue.

HOLOCRINE GLAND (Gr. *holos*, whole) An exocrine gland that releases its secretions by the detaching and dying of whole cells, which become the secretion; sebaceous glands are probably the only holocrine glands in the body.

HOMEOSTASIS [ho-mee-oh-STAY-siss; Gr. *homois*, same + *stasis*, standing still] A state of inner balance and stability in the body, which remains relatively constant despite external environmental changes.

HORMONE A chemical "messenger" produced and secreted by endocrine cells or tissues; circulates in the blood to "target" cells, affecting their metabolic activity.

HUMAN CHORIONIC GONADOTROPIN (hCG) A hormone released by the placenta and covering membranes of the embryo; it prevents the corpus luteum from disintegrating during part of the pregnancy, stimulating it to secrete estrogen and progesterone.

HUMORAL IMMUNITY A type of immunity that involves B cells in the production of antibodies; most active against bacteria, viruses, toxins.

HYALINE CARTILAGE [HYE-uh-lihn; Gr. *hyalos*, glassy] The most prevalent type of cartilage; contains collagenous fibers scattered in a network filled in with ground substance.

HYDROCEPHALUS (Gr. *hudor*, water + head) A condition in which there is an excess of cerebrospinal fluid within the ventricles, causing the skull to enlarge and put pressure on the brain; mental retardation is common; commonly called *water on the brain*.

HYDROLYSIS (Gr. *hudor*, water + *lusis*, loosening) The chemical process by which a molecule of water interacts with a reactant, thereby breaking the bonds of the reactant and rearranging it into different molecules.

HYDROSTATIC PRESSURE The force exerted by a fluid against the surface of the compartment containing the fluid.

HYMEN (Gr. membrane; Hymen was the Greek god of marriage) A fold of skin partially blocking the vaginal entrance.

HYPEREXTENSION Excessive extension beyond the straight (anatomical) position.

HYPERKALEMIA (*hyper*, over + L. *kalium*, potassium) A condition in which there is a high potassium concentration in extracellular fluids.

HYPERMAGNESEMIA An abnormally high magnesium level in the blood.

HYPERMETROPIA (Gr. "beyond measure") Farsightedness, occurring when the focus occurs beyond the retina.

HYPERNATREMIA (*hyper*, over + L. *natrium*, sodium) A condition in which there is a high sodium concentration in the extracellular fluids.

HYPERSENSITIVITY An overreaction to an allergen (antigen); also called *allergy*.

HYPERTENSION A condition in which systolic or diastolic pressure is above normal all the time; commonly called *high blood pressure*.

HYPERTHYROIDISM A condition characterized by the overactivity of the thyroid gland.

HYPERTONIC SOLUTION (Gr. *hyper*, above + *tonos*, tension) A solution in which the solute concentration is higher than the solute concentration inside a cell.

HYPERTROPHY (Gr. "overnourished") A condition in which the diameter of muscle fibers is increased as the result of physical activity; opposite of *atrophy*.

HYPODERMIS (Gr. *hypo*, under + *derma*, skin) The layer of loose, fibrous connective tissue lying below the dermis; also called *subcutaneous layer*.

HYPOGLYCEMIA An abnormal condition caused by the excessive secretion of insulin; also called *low blood glucose*.

HYPOKALEMIA (*hypo*, under + L. *kalium*, potassium) A condition in which there is a low potassium concentration in extracellular fluids.

HYPOMAGNESEMIA An abnormally low magnesium level in the blood.

HYPONATREMIA (*hypo*, under + L. *natrium*, sodium) A condition in which there is a low sodium concentration in the extracellular fluids.

HYPOPHYSEAL PORTAL SYSTEM A system of vessels through which the blood flows from the capillary bed of the hypothalamus by way of veins to the capillary bed of the pituitary gland.

HYPOPHYSIS (Gr. undergrowth) See *pituitary gland*.

HYPOTHALAMIC-HYPOPHYSEAL PORTAL SYSTEM An extensive system of blood vessels extending from the hypothalamus to the adenohypophysis.

HYPOTHALAMUS ("under the thalamus") The part of the brain located under the thalamus, forming the floor of the third ventricle; regulates body temperature, some metabolic processes, and other autonomic activities.

HYPOTHERMIA ("under-heat") A non-local condition in which the heat-producing and conserving mechanisms are exceeded by severe cold.

HYPOTHYROIDISM A condition characterized by the underactivity of the thyroid gland; it is usually associated with a goiter caused by an insufficiency of iodine in the diet.

HYPOTONIC SOLUTION (Gr. *hypo*, under) A solution in which the solute concentration is lower than the solute concentration inside a cell.

HYSTERECTOMY The surgical removal of the uterus.

I

IDENTICAL TWINS See *monozygotic twins*.

ILEUM [ILL-ee-uhm] The portion of the small intestine extending from the jejunum to the cecum, the first part of the large intestine.

IMMEDIATE HYPERSENSITIVITY An allergic reaction that occurs within minutes after exposure to an allergen.

IMMUNE SYSTEM The overall defensive system of lymphocytes and lymphatic tissues and organs.

IMMUNITY (L. *immunis*, exempt) An overall protective mechanism that forms antibodies to help protect the body against foreign substances.

IMMUNOGLOBULIN An antibody in the globulin group of proteins; involved with the immune response; five classes are: IgA, IgD, IgE, IgG, and IgM.

IMMUNOSUPPRESSIVE DRUG A drug that helps to suppress an immune reaction.

IMPETIGO (*impetigo contagiosa*) [ihm-puh-TIE-go; L. an attack] A contagious infection of the skin, caused by staphylococcal or streptococcal bacteria and characterized by small red macules that become pus-filled.

IMPLANTATION The process by which the blastocyst becomes embedded in the uterine wall.

IMPOTENCE The inability of a man to have an erection.

IN VITRO ("IN A TEST TUBE") FERTILIZATION A procedure in which an ovum is removed from a prospective mother's ovary at ovulation and fertilized in a glass dish or test tube with sperm from the male donor. The embryo is later implanted in the woman's uterus; a successful procedure results in the birth of a "test-tube baby."

INCONTINENCE See *urinary incontinence*.

INFANCY The period between the first 4 weeks of life and the time the child can sit erect and walk.

INFECTIOUS MONONUCLEOSIS A viral disease caused by the Epstein-Barr virus, a member of the herpes group.

INFERIOR (L. low) A relative directional term: toward the feet; below.

INFERIOR VENA CAVA A large vein that drains blood from the abdomen, pelvis, and lower limbs, emptying it into the right atrium; see *superior vena cava*.

INFLAMMATORY RESPONSE Part of the healing process, including redness, pain, swelling, scavenging by neutrophils and monocytes, and tissue repair by fibroblasts.

INFUNDIBULAR STALK (L. funnel) A stalk of nerve cells and blood vessels that connects the pituitary and the hypothalamus; also called *infundibulum*.

INFUNDIBULUM (L. funnel) The funnel-shaped portion of the uterine tube near the ovary; also see *infundibular stalk*.

INGESTION The taking in of nutrients by eating or drinking.

INGUINAL HERNIA See *hernia*.

INHALATION See *inspiration*.

INHIBIN A hormone secreted by the testes; it inhibits secretion of FSH by the anterior pituitary, maintaining a constant rate of spermatogenesis.

INHIBITING FACTORS Substances secreted by the hypothalamus that inhibit secretions from the adenohypophysis; see *releasing hormones*.

INHIBITORY POSTSYNAPTIC POTENTIAL (IPSP) The effect of a neurotransmitter on a postsynaptic receptor site, increasing the negative charge inside the plasma membrane of a neuron above the resting level.

INNER EAR (internal ear) The portion of the ear that includes the vestibule, the semicircular ducts and canals, and the spiral cochlea; also called *labyrinth*.

INORGANIC COMPOUND A chemical compound composed of relatively small molecules usually bonded ionically; see *organic compound*.

INSERTION The point of attachment of a muscle to the bone it moves.

INSPIRATION The process in which oxygen-rich, carbon dioxide-poor air travels from the atmosphere through the respiratory tract to the terminal portion of the lungs; also called *inhalation*.

INSPIRATORY CAPACITY The inspiratory reserve volume plus the tidal volume; about 2.4 L in women, 3.8 L in men.

INSPIRATORY RESERVE VOLUME The excess air taken into the lungs by the deepest inhalation beyond a normal inspiration.

INSULA (L. island) The cerebral lobe beneath the parietal, frontal, and temporal lobes; it appears to be associated with gastrointestinal and other visceral activities; also called *central lobe*.

INSULIN A peptide hormone secreted by beta cells in the pancreas; it lowers blood glucose concentration by facilitating glucose transport across plasma membranes.

INTERCALATED DISKS (L. *intercalatus*, to insert between) Thickenings of the sarcolemma that separate adjacent cardiac muscle fibers.

INTERFERON Small protein molecules produced by virus-infected cells; prevent viruses from multiplying.

INTERNAL A relative directional term of the body; inside; opposite of *external*.

INTERNAL RESPIRATION The process in which body cells exchange carbon dioxide for oxygen from the blood.

INTEROCEPTOR (L. "received from inside") A specialized sensory receptor responding to stimuli originating in internal organs; also called *visceroceptor*.

INTERPHASE The period between cell divisions during which the activities of growth, cellular respiration, RNA and protein synthesis, and DNA replication take place.

INTERSTITIAL CELLS OF LEYDIG See *interstitial endocrinocytes*.

INTERSTITIAL ENDOCRINOCYTES Clusters of endocrine cells among the seminiferous tubules in the testes that secrete the male sex hormones, including *testosterone*; also called *Leydig cells*.

INTERSTITIAL FLUID The fluid that bathes and nourishes body cells.

INTESTINAL GLANDS Tubular glands in the small intestine, formed by the infolding of the epithelium between the bases of the villi; formerly called *crypts of Lieberkühn*.

INTRACELLULAR FLUIDS (ICF) Fluids inside a cell.

INTRAFUSAL MUSCLE FIBERS Tiny muscles within a neuromuscular spindle.

INTRAMEMBRANOUS OSSIFICATION The process by which bone tissue develops directly from connective tissue.

INTRAUTERINE DEVICE (IUD) A plastic or metal device placed inside the uterus, where it prevents implantation by stimulating an inflammatory response.

INTRAUTERINE TRANSFUSION An injection of a concentrate of red blood cells into the peritoneal cavity of the fetus to help the fetus combat hemolytic anemia.

INTRAVENOUS INFUSION The controlled introduction of relatively large amounts of prescribed fluid into the body of a patient in order to maintain fluid balance, electrolyte concentration, and acid-base balance.

INTRINSIC BLOOD-CLOTTING PATHWAY A blood-clotting mechanism activated when the inner walls of blood vessels become damaged or irregular.

INVERSION A movement of the foot in which the great toe is turned upward and sole of foot faces medially; opposite of *eversion*.

INVOLUNTARY MUSCLE See *smooth muscle tissue*.

ION [EYE-ahn; Gr. "going particle"] An atom that has acquired an electrical charge by gaining or losing electrons. A positive ion is a *cation*, and a negative ion is an *anion*.

IONIC BOND A chemical bond formed when an atom or group of atoms with an electrical charge becomes attracted to an atom or group of atoms with an opposite charge.

IPSILATERAL ("same side") Pertaining to the same side of the body.

IPSILATERAL REFLEX A reflex occurring on the same side of the body and spinal cord as where the stimulus is received.

IRIS (Gr. rainbow) The colored portion of the eye that surrounds the pupil.

ISLETS OF LANGERHANS See *pancreatic islets*.

ISOANTIBODY See *agglutinin*.

ISOANTIGEN See *agglutinogen*.

ISOMETRIC CONTRACTION (Gr. *isos*, equal + *metron*, length) A muscle contraction in which tension increases but muscle length remains the same.

ISOTONIC CONTRACTION (Gr. *isos*, equal + *tonos*, tension) A muscle contraction in which the muscle becomes shorter and thicker, but the tension stays constant.

ISOTONIC SOLUTION (Gr. *isos*, equal) A solution in which the solute concentration is the same inside and outside a cell.

ISOTOPE (Gr. *isos*, equal + *topos*, place) Isotopes of the same element have the same number of protons and electrons, but different numbers of neutrons.

ISTHMUS [ISS-muhss] A narrow passage connecting two larger cavities.

J

JAUNDICE (Fr. "yellow") A syndrome characterized by bile pigment in the skin and mucous membranes, resulting in a yellow appearance.

JEJUNUM [jeh-JOO-nuhm; L. "fasting intestine"] The 2.5-m-long portion of the digestive tract that extends between the duodenum and the ileum.

JUNCTIONAL COMPLEXES Specialized parts that hold cells together, enabling groups of cells to function as a unit; include desmosomes, gap junctions, and tight junctions.

JUXTAGLOMERULAR APPARATUS The portion of the kidney composed of juxtaglomerular cells, the macula densa, afferent and efferent arterioles, and the extraglomerular mesangium.

K

KERATIN (Gr. *keras*, horn) A tough protein forming the outer layer of hair and nails; it is soft in hair and hard in nails.

KETOGENESIS The formation of ketone bodies.

KETONE BODIES Substances such as acetone, acetoacetic acid, and β-hydroxybutyric acid produced from excess acetyl-CoA during lipid catabolism.

KETOSIS An abnormal condition in which only fats are being metabolized, producing an excess of *ketone bodies.*

KIDNEYS A pair of organs located in the posterior wall of the abdominal cavity; maintain water and salt balance in the blood.

KILOCALORIE A unit of measurement equal to 1000 calories; also known as a *Calorie.*

KINESIOLOGY The study of motion.

KINESTHESIA (Gr. *kinema*, motion + sensory ability) The sense of body movement.

KLINEFELTER'S SYNDROME A chromosomal aberration in which an individual has an extra X chromosome, producing an XXY genotype.

KRAUSE'S END BULBS See *bulbous corpuscles (of Krause).*

KREBS CYCLE See *citric acid cycle.*

KYPHOSIS (Gr. "hunchbacked") A condition in which the spine curves backward abnormally, usually at the thoracic level.

L

LABIA MAJORA ("major lips") Two longitudinal folds of skin just below the mons pubis, forming the outer borders of the *vulva.*

LABIA MINORA ("minor lips") Two relatively small folds of skin lying between the larger labia majora.

LABYRINTH (Gr. maze) The intricate interconnecting chambers and passages that make up the inner ear.

LACRIMAL APPARATUS [LACK-ruh-mull; L. tear] Apparatus consisting of lacrimal gland and sac and nasolacrimal duct.

LACRIMAL GLAND The tear gland of the eye, whose secretions keep the eye moist and clean, combat microorganisms, and distribute water and nutrients to the cornea and lens.

LACTATION (L. *lactare*, to suckle) The process that includes both production of milk by mammary glands and release of milk from the breasts.

LACTEALS (L. *lacteus*, of milk) Specialized lymphatic capillaries that extend into the intestinal villi.

LACTIC ACID A toxic waste substance that builds up in muscles as they become fatigued during the anaerobic respiration that accompanies vigorous physical activity.

LACTOGENIC HORMONE See *prolactin.*

LACTOSE INTOLERANCE The inability to digest lactose.

LACUNAE [luh-KYOO-nee; sing. **LA-CUNA**; L. cavities, pods] Small cavities within the connective tissue matrix containing chondrocytes in cartilage and osteocytes in bone.

LAMELLAE (L. "thin plates") Concentric layers of bone that make up cylinders of calcified bone called *osteon.*

LAMELLATED (PACINIAN) CORPUSCLE A sensory receptor involved with vibratory sense and touch-pressure on the skin; formerly called *Pacinian corpuscle.*

LANUGO (L. *lana*, fine wool) Fine, downy hair covering the fetus by the fifth month; shed before birth, except on the eyebrows and scalp, where it becomes thicker.

LARGE INTESTINE The part of the digestive system between the ileocecal orifice of the small intestine and the anus; removes salt and water from undigested food and releases feces through the anus.

LARYNGOPHARYNX The lowest part of the pharynx, extending downward into the larynx.

LARYNX An air passage at the beginning of the respiratory tract where the vocal folds are located; commonly called *voice box.*

LATERAL (L. *lateralis*, side) A relative directional term: away from the midline of the body (toward the *side* of the body); opposite of medial.

LATERAL ROTATION A twisting movement in which the anterior surface of a limb or bone moves away from the body's medial plane; opposite of medial rotation.

LENS An elastic, colorless, transparent body of epithelial cells behind the iris; its shape is adjustable to focus on objects at different distances.

LESSER OMENTUM (L. "fat skin") A fatty mesentery extending from the liver to the lesser curvature of the stomach.

LESSER VESTIBULAR GLANDS Paired glands with ducts that open into the anterior part of the vaginal vestibule; during sexual arousal they secrete an alkaline mucus solution that provides some lubrication and offsets some vaginal acidity; also called *Skene's glands.*

LEUKEMIA (Gr. *leukos*, white and *-emia*, blood) A condition in which a malignant neoplasm originates in blood-forming cells of the bone marrow; characterized by uncontrolled reproduction of white blood cells.

LEUKOCYTE [LOO-koh-site; Gr. *leukos*, clear, white] A white blood cell, usually of the scavenger type, that ingests foreign material in the bloodstream and tissues.

LEYDIG CELLS See *interstitial endocrinocytes.*

LH See *luteinizing hormone.*

LIGAMENT A fairly inelastic fibrous thickening of an articular capsule that joins a bone to its articulating mate, allowing movement at the joint.

LIGHT TOUCH The sense that is perceived when the skin is touched but not deformed.

LIMBIC SYSTEM An assemblage of structures in the cerebrum, diencephalon, and midbrain involved in memory and emotions, and the visceral and behavioral responses associated with them.

LIPID [LIHP-ihd; Gr. *lipos*, fat] Organic compound that is insoluble in water but can be dissolved in organic solvents; includes body fats.

LONG BONE A bone whose length is greater than its width.

LOOP OF HENLE See *loop of the nephron.*

LOOP OF THE NEPHRON Portion of the kidney excretory tubule that receives glomerular filtrate from the proximal convoluted tubule; also called *loop of Henle.*

LORDOSIS (Gr. "bent backward") An exaggerated forward curve of the spine at the lumbar level; also called "swayback."

LOWER EXTREMITIES See *extremities.*

LUMBAR PLEXUS The ventral rami of L1 to L4 nerves in the interior of the posterior abdominal wall.

LUNG One of the paired organs of respiration, on either side of the heart in the thoracic cavity.

LUTEINIZING HORMONE (LH) Hormone secreted by the anterior pituitary; it stimulates ovulation and progesterone secretion in females and testosterone secretion in males.

LUTEOTROPIC HORMONE (LTH) See *prolactin.*

LYMPH (L. "clear water") Interstitial fluid inside lymphatic capillaries.

LYMPH NODES Bodies of lymphoid tissue situated along collecting lymphatic vessels; filter foreign material from lymph.

LYMPHATIC CAPILLARIES Tiny vessels arising in body tissues that collect excess interstitial fluid (lymph).

LYMPHATIC SYSTEM The body system that collects and drains fluid that seeps from the blood and accumulates in the spaces between cells.

LYMPHATICS Collecting vessels that carry lymph from lymphatic capillaries to veins in the neck, where it is returned to the bloodstream.

LYMPHOCYTE Wandering connective tissue cells found under moist epithelial linings of respiratory and intestinal tracts; main producers of antibodies; an activated B lymphocyte becomes a plasma cell.

LYMPHOMA [lihm-FOH-muh] A malignant neoplasm originating in lymph nodes.

LYSOSOME (Gr. "dissolving body") Membrane-bound organelle containing digestive enzymes; protects cell against microorganisms and clears away damaged cells.

M

MACROPHAGE [MACK-roh-fahj; Gr. *makros,* large; *phagein,* to eat] A connective tissue cell that is an active phagocyte; can be fixed or wandering.

MACULA The receptor region of utricles and saccules in the ear; contains hair cells embedded in the gelatinous otolithic membrane.

MACULA DENSA (L. "dense spot") See *juxtaglomerular apparatus.*

MACULA LUTEA (L. "yellow spot") An area in the center of the retina containing only cones.

MAGNETIC RESONANCE IMAGING (MRI) A noninvasive exploratory diagnostic technique that uses a strong magnetic field to detect differences in healthy and unhealthy tissues; also called *nuclear magnetic resonance (NMR).*

MALIGNANT MELANOMA The most serious form of skin cancer, involving the pigment-producing melanocytes.

MALIGNANT NEOPLASM (L. *malus,* bad) An abnormal growth whose uncontrolled cells break out of their connective tissue capsule to invade neighboring tissue and grow rapidly in an uncontrolled pattern.

MAMMARY GLANDS (L. *mammae,* breasts) Paired female modified sweat glands that produce and secrete milk for a newborn child.

MARROW See *bone marrow.*

MAST CELL A wandering connective tissue cell often found near blood vessels; contains secretory granules that produce heparin and histamine.

MASTECTOMY The surgical removal of a breast, usually to prevent a malignant neoplasm from spreading.

MATRIX [MAY-triks; L. womb, mother] (1) The extracellular fibers and ground substance in connective tissues. (2) The thick layer of skin beneath the root of a nail, where new cells are generated.

MEAN ARTERIAL PRESSURE (MAP) The average pressure that drives blood through the systemic circulation.

MEATUS [mee-AY-tuhss; L. passage] A large, tubular channel or opening, not necessarily through a bone; see *external auditory canal.*

MECHANORECEPTOR ("mechanical receiver") A sensory receptor responding to such physical stimuli as touch-pressure, muscle tension, air vibrations, and head movements.

MEDIAL [MEE-dee-uhl; L. *medius,* middle] A relative directional term: toward the midline of the body.

MEDIAL ROTATION A twisting movement in which the anterior surface of a limb or bone moves toward the medial plane of the body.

MEDIASTINUM [mee-dee-as-TIE-nuhm; L. *medius,* middle] The mass of tissues and organs between the lungs. It contains all the contents of the thoracic cavity except the lungs.

MEDULLA [meh-DULL-uh; L. marrow] The inner core of a structure.

MEDULLA OBLONGATA (L. elongated marrow) The lowermost portion of the brainstem, continuous with the spinal cord.

MEDULLARY CAVITY [MED-uh-lehr-ee; L. *medulla,* marrow] The marrow cavity inside the shaft of a long bone.

MEDULLARY RHYTHMICITY AREA A portion of the respiratory center in the brain that regulates inspiration and expiration.

MEIOSIS [mye-OH-sihss; Gr. *meioun,* to diminish] A type of cell division that reduces the number of chromosomes in gametes to half.

MEISSNER'S CORPUSCLES See *tactile (Meissner's) corpuscles.*

MELANIN [MEHL-un-nihn; Gr. *melas,* black] A dark pigment produced by specialized cells called melanocytes; contributes to skin color.

MELANOCYTE-STIMULATING HORMONE (MSH) A chemical substance, probably a precursor of an active hormone.

MELANOMA (Gr. *melas,* black + *-oma,* tumor) A neoplasm composed of cells containing a dark pigment, usually melanin.

MEMBRANES Thin, pliable layers of epithelial and/or connective tissue that line body cavities and cover or separate structures.

MENARCHE [meh-NAR-kee; Gr. "beginning the monthly"] The first menstrual period, usually occurring during the latter stages of puberty.

MENINGES [muh-NIHN-jeez; Gr. pl. of *meninx,* membrane] Three layers of protective membranes (dura mater, arachnoid, pia mater) surrounding the brain and spinal cord.

MENINGITIS Inflammation of the meninges.

MENOPAUSE (Gr. "ceasing the monthly") The cessation of menstrual periods.

MENSTRUAL CYCLE A monthly series of events that prepares the endometrium of the uterus for pregnancy and then discharges the sloughed-off endometrium, mucus, and blood in the menstrual flow if pregnancy does not occur.

MENSTRUATION (L. *mensis,* monthly) The monthly breakdown of the endometrium of a nonpregnant female.

MERKEL'S DISKS See *tactile (Merkel's) corpuscles.*

MEROCRINE GLAND (Gr. *meros,* divide) An exocrine gland that releases its secretions via exocytosis without breaking the plasma membrane.

MESENCHYME [MEHZ-uhn-kime; Gr. *mesos,* middle + L. *enchyma,* cellular tissue] Embryonic mesoderm that develops into connective tissue.

MESENTERY [MEZZ-uhn-ter-ee; Gr. *mes,* middle + *enteron,* intestines] Fused layers of visceral peritoneum that attach abdominopelvic organs to the cavity wall; also called *visceral ligament.*

MESODERM (Gr. *mesos,* middle) The embryonic germ layer between the ectoderm and endoderm, developing into mature epithelial, connective, and muscle tissue.

MESOVARIUM A mesentery that attaches the ovaries to the broad ligament.

METABOLISM [muh-TAB-uh-lihz-uhm; Gr. *metabole,* change] All the chemical processes in the body; may either build up or break down substances (see *anabolism, catabolism*).

METACARPAL BONES (L. behind the wrist) The five miniature long bones constituting the palm of each hand; also called *metacarpus.*

METACARPUS See *metacarpal bones.*

METAPHASE The second stage of mitosis; centromeres double, one going to each chromatid, which is now a single-stranded chromosome.

METAPHYSIS [muh-TAHF-uh-siss; Gr. "to grow beyond"] The area where longitudinal growth continues after birth, between the epiphyseal (growth) plate and the diaphysis.

METASTASIS [muh-TASS-tuh-siss; Gr. *meta,* involving change + *stasis,* state of standing] The spread of malignant cells from the primary site to other parts of the body.

METATARSAL BONES The five miniature long bones in each foot between the ankle (tarsal bones) and toes.

MICELLE [my-SELL; L. *mica,* grain] A submicroscopic water-soluble particle composed of fatty acids, phospholipids, and glycerides.

MICROFILAMENT A solid, rodlike organelle containing actin; provides cellular support and aids movement.

MICROGLIA The smallest glial cell (see *neuroglia*); a macrophage that removes disintegrating products of neurons.

MICROTUBULE A slender organelle that helps support the cell; involved with organelle movement, cellular shape changes, and intracellular transport.

MICROVILLI [my-krow-VILL-eye; Gr. *mikros,* small; L. *villus,* shaggy hair] Microscopic fingerlike projections protruding from plasma membranes of some cells.

MICTURITION (L. *micturire,* to want to urinate) The process of emptying the urinary bladder; also called *urination.*

MIDBRAIN The portion of the brainstem located between the pons and diencephalon; connects the pons and cerebellum with the cerebrum; also called *mesencephalon.*

MIDDLE EAR A small chamber between the tympanic membrane and inner ear; includes the tympanic cavity and contains auditory ossicles.

MIDGET See *pituitary dwarf.*

MIDSAGITTAL The plane that divides the left and right sides of the body lengthwise along the midline into externally symmetrical sections.

MINERAL A naturally occurring, inorganic element used for the regulation of metabolism; microminerals (trace elements) are required in only minute amounts.

MINERALOCORTICOID A steroid hormone secreted by the adrenal cortex.

MINUTE RESPIRATORY VOLUME The total volume of air taken into the lungs, measured in L/min.

MISCARRIAGE See *spontaneous abortion.*

MITOCHONDRION (pl. **MITOCHONDRIA;** Gr. *mitos,* a thread) A double-membraned, saclike organelle; produces most of the energy (in the form of ATP) for cellular metabolism; contains some DNA.

MITOSIS (Gr. *mitos,* a thread) The process of nuclear division; it arranges cellular material for equal distribution to daughter cells and divides the nuclear DNA equally to each new cell.

MITRAL VALVE [MY-truhl] See *bicuspid valve.*

MIXED-TISSUE NEOPLASM A malignant neoplasm derived from tissue capable of differentiating into either epithelial or connective tissue.

MOBILE-RECEPTOR MECHANISM A mechanism of hormone action in which lipid-soluble hormones (steroids) regulate cellular activity through the target cell's protein synthesis.

MODELING The alteration of bone size and shape during the bone's developmental growth.

MOLECULE [MAHL-uh-kyool; L. *moles,* mass, bulk] The chemical combination of two or more atoms; the simplest unit that displays the physical and chemical properties of a compound.

MONOCLONAL ANTIBODY An antibody produced by laboratory-produced clones (identical copies) of B cells fused with cancer cells.

MONOCYTE A type of white blood cell that becomes a phagocytic macrophage that ingests and destroys harmful particles.

MONOSACCHARIDE [mahn-oh-SACK-uh-ride; Gr. *monos,* single + *sakkharon,* sugar] A single-sugar carbohydrate that cannot be decomposed by uniting it with water.

MONOZYGOTIC TWINS Twins that result from the same fertilized ovum that divides into two identical cells before implantation; also called *identical twins.*

MONS PUBIS (L. mountain + pubic) A mound of fatty tissue covering the female symphysis pubis; also called *mons veneris.*

MONS VENERIS (L. mountain + Venus, the Roman goddess of love and beauty) See *mons pubis.*

MORPHOGENESIS (Gr. *morphe,* shape + *gennan,* to produce) The cellular development that brings about tissue and organ development.

MORULA [MORE-uh-luh; L. *morum,* mulberry tree] A solid ball of about 8 to 50 cells produced by cell divisions of a single fertilized ovum.

MOTOR END PLATE The junction (a synapse) between a motor neuron and a muscle fiber.

MOTOR NEURON See *efferent neuron.*

MOTOR UNIT A motor neuron and the muscle fibers it innervates.

MRI See *magnetic resonance imaging.*

MUCOUS MEMBRANE The membrane that lines body passageways that open to the outside of the body.

MUCUS [MYOO-kuhss] The thick, protective liquid secreted by glands in the mucous membranes.

MULTIPLE SCLEROSIS (MS) A progressive demyelination of neurons; interferes with the conduction of nerve impulses and results in impaired sensory perceptions and motor coordination.

MULTIPOLAR NEURON A nerve cell with many dendrites, but with only one axon.

MUSCLE (L. *musculus,* "little mouse") A collection of muscle fibers that can contract and relax to move body parts.

MUSCLE FIBERS Collections of specialized, individual muscle cells that make up skeletal muscle tissue; have a long, cylindrical shape and several nuclei.

MUSCULAR DYSTROPHY The general name for a group of inherited diseases resulting in progressive weakness due to the degeneration of muscles.

MUTATION (L. *mutare,* to change) A change in genetic material that alters the characteristics of a cell, making the cell abnormal.

MYASTHENIA GRAVIS (Gr. *mus,* muscle + *astheneia,* without strength + L. *gravis,* weighty, serious) An autoimmune disease caused by antibodies directed against acetylcholine receptors, producing muscular weakness.

MYELIN [MY-ih-linn; Gr. *myelos,* marrow] A laminated lipid sheath covering an axon.

MYELIN SHEATH A thick pad of insulating myelin surrounding an axon.

MYELINATED FIBER A nerve fiber covered with a myelin sheath.

MYOCARDIAL INFARCTION (L. *infercire,* to stuff) Commonly called a *heart attack;* occurs when blood flow through a coronary artery is reduced and the myocardium is deprived of oxygen, leading to death of the heart tissue.

MYOCARDIUM ("heart muscle") The middle layer of muscle in the heart wall.

MYOFIBRIL Small units or fibers within individual threadlike muscle fibers; suspended in the sarcoplasm along with mitochondria and other multicellular material.

MYOFILAMENT A muscle filament composed of thick and thin threads that make up a myofibril.

MYOGLOBIN A form of hemoglobin found in muscle fibers.

MYOMETRIUM (muscle + Gr. *metra,*

womb) The middle, muscular tissue layer of the uterus.

MYONEURAL JUNCTION See *neuromuscular junction.*

MYOPIA (Gr. "contracting the eyes") Nearsightedness, which occurs when light rays come to a focus before they reach the retina.

MYOSIN A fairly large protein that makes up the thick myofilaments of muscle fibers.

MYXEDEMA [mix-uh-DEE-muh] A condition caused by the underactivity of the thyroid during adulthood.

N

NAIL A modification of the epidermis, composed of hard keratin overlying the tips of fingers and toes.

NASAL CAVITY The cavity that fills the space between the base of the skull and the roof of the mouth.

NASAL SEPTUM A vertical wall dividing the nasal cavity into two bilateral halves.

NASOLACRIMAL DUCT (L. "nose" + "tear") A duct leading from each eye to the nasal cavity; it drains excessive secretions of tears into the nasal cavity.

NASOPHARYNX The part of the pharynx above the soft palate.

NECROBIOSIS (Gr. *nekros,* corpse, death + *biosis,* way of life) The natural degeneration and death of cells and tissues; see *necrosis.*

NECROSIS (Gr. *nekros,* corpse, death) The death of cells or tissues due to disease or injury.

NEGATIVE FEEDBACK SYSTEM A regulatory feedback system that produces a response that changes or reduces the initial stimulus.

NEONATE A newborn child during the first 4 weeks after birth.

NEOPLASM (Gr. *neos,* new + L. *plasma,* form) An abnormal growth of new tissue, which may be benign or malignant.

NEPHRECTOMY The surgical removal of a kidney.

NEPHRON (Gr. *nephros,* kidney) The functional unit of a kidney, each of the approximately 1 million in each kidney operating as an independent urine-making unit.

NEPHROPTOSIS A condition in which a kidney moves because it is no longer held securely in place by the peritoneum; commonly called *floating kidney.*

NERVE A bundle of peripheral nerve fibers enclosed in a sheath.

NERVE IMPULSE See *action potential.*

NEURILEMMA The plasma membrane of a Schwann cell, which enwraps the

axon of an unmyelinated nerve fiber and the myelin layers of a myelinated nerve fiber; also called *Schwann sheath.*

NEUROFILAMENT A semirigid tubular structure in cell bodies and processes of neurons; also called *microfilament.*

NEUROGLIA (Gr. nerve + glue) Nonconducting cells of the central nervous system that protect, nurture, and support the nervous system; also called *glial cells.*

NEUROHYPOPHYSIS The posterior lobe of the pituitary.

NEUROMUSCULAR JUNCTION The junction of a motor-nerve ending with a muscle fiber; also called *myoneural junction.*

NEURON (Gr. nerve) A cell specialized to transmit nerve impulses.

NEUROPEPTIDE A type of chemical neurotransmitter including somatostatin, endorphins, and encephalins.

NEUROTRANSMITTER A chemical substance synthesized by neurons; may produce an excitatory or inhibitory response in a receptor.

NEUROTUBULE A threadlike protein structure in cell bodies and processes of neurons; involved in intracellular transport.

NEUTROPHIL A type of phagocytic white blood cell.

NISSL BODIES See *chromatophilic substance.*

NOCICEPTOR [NO-see; L. "injury receiver"] A sensory receptor responding to stimuli that produce pain.

NODES OF RANVIER Regular gaps in a myelin sheath around a nerve fiber.

NONESSENTIAL AMINO ACIDS Amino acids that can be synthesized by the body; see *essential amino acids.*

NONGONOCOCCAL URETHRITIS (NGU) A sexually transmitted disease characterized by inflammation of the urethra and accompanied by a discharge of pus; also called *nonspecific urethritis (NSU).*

NONSPECIFIC URETHRITIS (NSU) See *nongonococcal urethritis.*

NOREPINEPHRINE [nor-ep-ih-NEFF-rihn] A hormone secreted by the adrenal medulla; it constricts arterioles and increases the metabolic rate; also called *NE, noradrenaline.*

NUCLEAR ENVELOPE The double outer membrane of a cellular nucleus.

NUCLEAR MAGNETIC RESONANCE (NMR) See *magnetic resonance imaging.*

NUCLEIC ACID [noo-KLAY-ihk] Any of two groups of complex compounds composed of nucleotides; the carrier of hereditary material.

NUCLEOLUS [new-KLEE-oh-luhss; pl. **NUCLEOLI;** "little nucleus"] A somewhat spherical mass in the cell nucleus; contains genetic material in the form

of DNA and RNA; a preassembly point for ribosomes.

NUCLEOPLASM The material within the nucleus of a cell.

NUCLEOTIDE [NOO-klay-uh-tide] Structural subunit of nucleic acids; composed of a phosphate group, a sugar, and a nitrogenous base.

NUCLEUS (L. nut, kernel) (1) The central portion of an atom, containing positively charged protons and uncharged neutrons. It is surrounded by negatively charged electrons. (2) The central portion of a cell, containing chromosomes. (3) A collection of nerve cells inside the central nervous system that processes afferent inputs.

NUTRIENT The chemical component of foods, supplying the energy and physical materials a body needs.

O

OBESITY (L. *obesus,* "grown fat by eating") The condition of being 20 to 25 percent over the recommended body weight.

OCCIPITAL LOBE The posterior cerebral lobe; composed of several areas organized for vision and its associated forms of expression.

OIL GLANDS See *sebaceous glands.*

OLFACTORY BULB A stemlike extension of the olfactory region of the brain; receives impulses from nerve fibers stimulated by an odiferous substance.

OLFACTORY TRACT Axons of mitral and tufted cells that carry impulses from the olfactory bulb posteriorly to the olfactory cortex in the brain.

OLIGODENDROCYTE A relatively small glial cell similar to a Schwann cell; produces myelin sheath segments of many nerve fibers, provides a supportive framework, and supplies nutrition for neurons.

OLIGOSPERMIA (Gr. *oligos,* few + sperm) A condition in which only a small amount of sperm is produced.

ONCOGENE (Gr. *onkos,* tumor) A gene that apparently becomes reactivated and causes the growth of cancerous cells.

ONCOLOGY (Gr. *onkos,* tumor + *logy,* the study of) The study of neoplasms; the study of cancer.

ONCOTIC PRESSURE The osmotic pressure of the plasma proteins in a solution.

OOGENESIS The monthly maturation of an ovum in the ovary.

OOPHORECTOMY The surgical removal of an ovary.

OPPOSITION The angular movement of the thumb pad touching and opposing a finger pad; occurs only at the carpometacarpal joint of the thumb; opposite of *reposition*.

OPTIC CHIASMA [kye-AZ-muh; after the X-shaped Greek letter *chi*, KYE] A point in the cranial cavity where half the fibers of each optic nerve of each eye cross over to the other side.

OPTIC TRACT Nerve fibers after they have passed through the *optic chiasma*.

ORAL CAVITY The mouth consists of a smaller outer vestibule (buccal cavity) and a larger inner oral cavity proper.

ORAL CONTRACEPTIVE A hormonal pill taken daily by a woman; prevents follicle maturation and ovulation by altering hormones present; may also prevent implantation.

ORGAN An integrated collection of two or more kinds of tissue that combine to perform a specific function.

ORGAN OF CORTI See *spiral organ (of Corti)*.

ORGANELLE ("little organ") Various subcellular structures with specific structures and functions.

ORGANIC COMPOUND A chemical compound containing carbon and hydrogen, usually bonded covalently. See *inorganic compound*.

ORGASM (Gr. *orgasmos*, to swell with excitement) The climax of sexual excitement.

ORIGIN The end of a muscle attached to the bone that does not move.

OROPHARYNX The part of the pharynx at the back of the mouth, extending from the soft palate to the epiglottis, where the respiratory and alimentary tracts separate.

OSMOSIS (Gr. "pushing") The passive-transport process occurring when water passes through a semipermeable membrane from an area of high concentration to an area of lower concentration.

OSMOTIC PRESSURE The potential pressure developed by a solution separated from another solution by a differentially permeable membrane.

OSSEOUS TISSUE (L. *os*, bone) A tissue composed of cells embedded in a matrix of ground substance, inorganic salts, and collagenous fibers; also called *bone*.

OSSICLES The three small bones of the ear (malleus, incus, stapes).

OSSIFICATION CENTER See *center of ossification*.

OSTEOBLAST (Gr. *osteon*, bone + *blastos*, bud, growth) A bone cell capable of synthesizing and secreting new bone matrix as needed; usually found on growing portions of bones.

OSTEOCLAST (Gr. *osteon*, bone + *klastes*, breaker) A multinuclear bone-destroying cell; usually found where

bone is resorbed during normal growth.

OSTEOCYTE (Gr. *osteon*, bone + cell) A main cell of mature bone tissue; regulates the concentration of calcium in body fluids by helping to release calcium from bone tissue into the blood.

OSTEOGENIC CELL (Gr. *osteon*, bone + *genes*, born) A bone cell capable of being transformed into an *osteoblast*.

OSTEOMALACIA [ahss-teh-oh-muh-LAY-shee-uh; Gr. *osteon*, bone + *malakia*, soft] A skeletal defect caused by a deficiency of vitamin D.

OSTEOMYELITIS (Gr. *osteon*, bone + *myelos*, marrow) An inflammation of bone, and/or bone marrow; frequently caused by bacteria.

OSTEON (Gr. bone) Concentric cylinders of calcified bone that make up compact bone; also called *Haversian system*.

OSTEOPOROSIS (Gr. *osteon*, bone + *poros*, passage) A bone disorder occurring most often in the elderly; the bones grow porous and crumble under ordinary stress.

OTITIS MEDIA Inflammation of the middle ear.

OTOLITHS ("ear stones") Calcium carbonate crystals piled on top of the otolithic membrane; assist in maintaining static equilibrium.

OVARY (L. *ovum*, egg) A paired female gonad that produces ova and female hormones.

OVIDUCT See *uterine tube*.

OVULATION The monthly process in which a mature ovum is ejected from the ovary into the serous fluid of the peritoneal cavity near the uterine tube.

OXIDATION A chemical reaction in which atoms or molecules lose electrons or hydrogen ions.

OXIDATION-REDUCTION REACTION The simultaneous loss of electrons or hydrogen atoms (oxidation) by one substance and the gain of electrons or hydrogen atoms (reduction) by another substance.

OXYHEMOGLOBIN The compound formed when oxygen is bound to hemoglobin.

OXYTOCIN A hormone secreted by the pituitary that stimulates uterine contractions during labor and milk ejection from the mammary glands after childbirth.

P

P WAVE The first activity in an electrocardiogram, caused by the depolarization and contraction of both atria.

PACINIAN CORPUSCLE [pah-SIHN-ee-an]

See *lamellated (Pacinian) corpuscle*.

PAGET'S DISEASE A progressive bone disease in which a pattern of excessive bone destruction followed by bone formation contributes to bone thickening; also called *osteitis deformans*.

PALATE [PAL-iht] The roof of the mouth, divided into the anterior hard palate and the posterior soft palate.

PALMAR A relative directional term of the body: surface of the palm of the hand; also called *volar*.

PALPITATIONS Skipping, pounding, or racing heartbeats.

PANCREAS (Gr. "all flesh") An organ located posterior to the stomach; its endocrine functions include the secretion of insulin and glucagon; it also functions as an exocrine gland.

PANCREATIC ISLETS Clusters of endocrine cells in the pancreas that produce insulin and glucagon; also called *islets of Langerhans*.

PAPANICOLAOU STAIN SLIDE TEST A diagnostic procedure that tests cells scraped from the female genital epithelium to detect malignant and premalignant conditions; also called *Pap test, Pap smear*.

PAPILLAE [puh-PILL-ee, sing. **PAPILLA**; L. nipple; dim. of *papula*, pimple] Projections on the tongue surface, palate, throat, and posterior surface of the epiglottis that contain taste buds.

PARANASAL SINUS An air cavity within a bone, in direct communication with the nasal cavity.

PARAPLEGIA Loss of motor or sensory function in both lower extremities due to damage to the spinal cord.

PARASYMPATHETIC NERVOUS SYSTEM The portion of the autonomic nervous system that directs activities associated with the conservation and restoration of body resources.

PARATHORMONE (PTH) A parathyroid hormone that increases blood levels of calcium and decreases blood levels of phosphate; also called *parathyroid hormone*.

PARATHYROID GLANDS Small endocrine glands embedded in the posterior thyroid; their secretions affect blood levels of calcium and phosphate.

PARATHYROID HORMONE See *parathormone*.

PARIETAL LOBE The cerebral lobe that lies between the frontal and occipital lobes; concerned with the evaluation of the general senses (including an awareness of the body and its relation to the outside world), and of taste.

PARIETAL PLEURA See *pleural cavity*.

PARKINSON'S DISEASE A motor disability characterized by tremors and stiff posture; it results from a deficiency of dopamine; also called *Parkinsonism*.

PAROTID GLANDS (Gr. *parotis*, "near the ear") The largest of the three main pairs of salivary glands; located below the ears.

PARTURITION (L. *parturire*, to be in labor) Childbirth.

PASSIVE TRANSPORT The movement of molecules across cell membranes from areas of high concentration to areas of lower concentration, without the use of cellular energy.

PATELLAR REFLEX (KNEE JERK) A diagnostic reflex in which the tapping of the patellar tendon produces the contraction of the quadriceps femoris muscle, causing the lower leg to jerk upward.

PATENT DUCTUS ARTERIOSUS A condition that occurs when the ductus arteriosus of the fetus does not close off at birth.

PATHOLOGY (Gr. *pathos*, suffering + study) The study of changes in diseased cells and tissues.

PECTORAL GIRDLE The upper limb girdle, consisting of the clavicle and scapula; also called *shoulder girdle*.

PEDICEL (L. little foot) A footlike process in contact with glomerular capillaries.

PELVIC CAVITY The inferior portion of the abdominopelvic cavity. It contains the urinary bladder, rectum, anus, and internal reproductive organs.

PELVIC GIRDLE The paired hip bones (ossa coxae), formed by the ilium, ischium, and pubis; also called *lower limb girdle*.

PELVIC INFLAMMATORY DISEASE (PID) A general term referring to inflammation of the uterus and uterine tubes, with or without inflammation of the ovary, localized pelvic peritonitis, and abscess formation.

PELVIS (L. basin) The bowl-shaped bony structure formed by the sacrum and coccyx posteriorly and the two hip bones anteriorly and laterally.

PENIS [PEE-nihss; L. tail] The male copulatory organ, which transports semen through the urethra during ejaculation, and also carries urine through the urethra to the outside during urination.

PENNATE MUSCLE (L. *penna*, feather) A muscle with many short fascicles set at an angle to a long tendon.

PEPTIC ULCERS (Gr. *peptein*, to digest) Lesions in the mucosa of the digestive tract, caused by the digestive action of gastric juice, especially hydrochloric acid.

PEPTIDE BOND The strong covalent bond formed as a result of the union of amino acids.

PERFORATING CANAL A branch running at a right angle to the central canal, extending the system of nerves and vessels outward to the periosteum, and inward to the endosteum of the bony marrow cavity; also called *nutrient canal, Volkmann's canal*.

PERICARDIAL (Gr. *peri*, around + heart) Pertaining to the membranes enclosing the heart and lining the pericardial cavity.

PERICARDIUM ("around the heart") A protective sac around the heart; also called *pericardial sac*.

PERICHONDRIUM (Gr. *peri*, around + *khondros*, cartilage) A fibrous covering enclosing hyaline and elastic cartilage.

PERILYMPH The thin, watery fluid in the bony labyrinth of the inner ear.

PERIMYSIUM (Gr. *peri*, around + muscle) A connective tissue layer extending inward from the epimysium; it encloses bundles of muscle fibers.

PERINEURIUM ("around the nerve") A thick connective tissue sheath surrounding a primary bundle of nerve fibers.

PERIOSTEUM [pehr-ee-AHSS-tee-uhm; Gr. *peri*, around + *osteon*, bone] A fibrous membrane covering the outer surfaces of bones (except at joints); contains bone-forming cells, nerves, and vessels.

PERIPHERAL A relative directional term used to describe structures other than internal organs that are located or directed away from the central axis of the body.

PERIPHERAL AUTONOMIC NERVOUS SYSTEM Sympathetic and parasympathetic divisions of motor nerves; each division sends efferent nerve fibers to the muscle, gland, or organ it innervates.

PERIPHERAL NERVOUS SYSTEM (PNS) The neurons and their fibers emerging from and going to the brain (cranial nerves) and spinal cord (spinal nerves).

PERIPHERAL RESISTANCE The impediment (friction) to blood flow through a blood vessel.

PERISTALSIS The involuntary, sequential muscular contractions that move food along the digestive tract.

PERITONEUM [per-uh-tuh-NEE-uhm] The serous membrane that lines the abdominal cavity and mesenteries; covers most of the abdominal organs.

PERITONITIS. An inflammation of the peritoneum.

PEROXISOME A membrane-bound organelle containing oxidative enzymes that break down hydrogen peroxide.

PET SCAN See *positron-emission tomography*.

PHAGOCYTOSIS (Gr. *phagein*, to eat + *cyto*, cell) The active process in which large molecules and particles are taken into the cell through the plasma membrane; a form of endocytosis.

PHALANGES [fuh-LAN-jeez; sing. **PHALANX;** FAY-langks, Gr. line of soldiers] The 14 finger bones in each hand; also the 14 toe bones in each foot.

PHARYNX A tube leading from the internal nares (nostrils) and mouth to the larynx and esophagus; serves as an air passage during breathing, and a food passage during swallowing; commonly called the *throat*.

PHENOTYPE [FEE-noh-tipe; Gr. "that which shows"] The expression of a genetic trait (how a person looks or functions).

PHENYLKETONURIA (PKU) A congenital metabolic disease caused by a mutation that prevents the adequate formation of phenylalanine hydroxylase, which normally converts the amino acid phenylalanine into tyrosine.

PHOTORECEPTOR (Gr. "light receiver") A light-sensitive sensory receptor in the retina.

PHYSIOLOGY The study of how the body functions.

PIA MATER (L. tender mother) The thin, highly vascular innermost layer of the meninges, adjacent to the brain and spinal cord.

PINEAL GLAND [PIHN-ee-uhl; L. *pinea*, pine cone] A small gland in the midbrain that converts a signal from the nervous system into an endocrine signal; it produces *melatonin*, but its exact function is uncertain; also called *pineal body, epiphysis cerebri*.

PINOCYTOSIS (Gr. *pinein*, to drink + *cyto*, cell) The nonspecific uptake by a cell of small droplets of extracellular fluid; a form of endocytosis.

PITUITARY DWARF A person whose skeletal development is halted at about the stage of a 6-year-old child because of an undersecretion of growth hormone; also called *midget*.

PITUITARY GLAND An endocrine gland consisting of two lobes, the anterior adenohypophysis, containing many secretory cells, and the posterior neurohypophysis, containing many nerve endings; also called *hypophysis*.

PIVOT JOINT A uniaxial joint that rotates around a central axis.

PLACENTA A thick bed of tissues and blood vessels embedded in the wall of a pregnant woman's uterus; it contains arteries and veins of the mother and fetus and provides an indirect connection between mother and fetus.

PLANTAR A relative directional term of the body: surface of the sole of the foot.

PLANTAR FLEXION The downward

bending of the foot at the ankle.

PLASMA The clear, yellowish liquid part of blood.

PLASMA CELL See *lymphocyte.*

PLASMA MEMBRANE The bilayered membrane surrounding a cell.

PLASMA PROTEIN One of the proteins found dissolved in blood plasma; includes albumins, fibrinogen, and globulins.

PLATELET A type of blood cell that is important in blood clotting; also called *thrombocyte.*

PLATELET AGGREGATION The process in which platelets aggregate, plugging a small hole in a damaged blood vessel; also called *platelet plug.*

PLATELET PLUG See *platelet aggregation.*

PLEURA [PLOOR-uh; Gr. side, rib] The serous membrane lining the pleural cavity and covering the lungs; also lines the thoracic wall.

PLEURAL CAVITY The moisture-filled potential space between the visceral pleura on the lung surface and the parietal pleura lining the thoracic cavity.

PLEURISY An infection of the pleurae covering the lungs and lining the thoracic cavity.

PLEXUS (L. braid) A complex network of interlaced nerves.

PLICAE CIRCULARES [PLY-see sir-cue-LAR-eez] Circular folds in the small intestine that increase the area for absorption.

PMS See *premenstrual syndrome.*

PNEUMONIA A general term for any condition that results in filling the alveoli with fluid.

PNEUMOTAXIC AREA A breathing-control area in the pons which, when stimulated, reduces inhalations and increases exhalations; see *respiratory center.*

PODOCYTE ("footlike cell") A specialized epithelial cell surrounding glomerular capillaries.

POLAR BODY The smaller of two cells resulting from the uneven distribution of cytoplasm during oogenesis.

POLAR MOLECULE A molecule whose two ends have different electrical charges; water is a polar molecule.

POLARIZATION A condition of the plasma membrane of a neuron, when the intracellular fluid is negatively charged relative to the positively charged extracellular fluid.

POLIOMYELITIS A contagious viral infection affecting both the brain and spinal cord, sometimes causing the destruction of neurons; damage to motor neurons in the spinal cord causes paralysis.

POLYPEPTIDE A protein formed by the chemical bonding of many amino

acids.

POLYSACCHARIDE [pah-lee-SACK-uh-ride; gr. *poly,* many + *sakkharon,* sugar] A carbohydrate made up of more than two simple sugars.

POLYURIA An excessive production of urine; frequent urination.

PORTA HEPATIS ("liver door") The area through which blood vessels, nerves, lymphatic vessels, and ducts enter and leave the liver.

POSITIVE FEEDBACK SYSTEM A reaction that reinforces a stimulus rather than inhibiting it; can disrupt homeostasis.

POSITRON EMISSION TOMOGRAPHY (PET) A scanning procedure that produces pictures that reveal the metabolic state of the organ being viewed; the device is called a *PET scanner.*

POSTABSORPTIVE STATE The metabolic state during which the body's energy requirements must be satisfied by the nutrients taken in during the absorptive state and from stored nutrients; also called *fasting state.*

POSTERIOR (L. *post,* behind, after) Toward the back of the body; also called *dorsal;* opposite of anterior.

POSTGANGLIONIC NEURON The second neuron in a two-neuron sequence in the autonomic nervous system; its cell body is in an autonomic ganglion and its unmyelinated axon terminates in a motor ending with smooth or cardiac muscle or a gland.

POSTSYNAPTIC NEURON ("after the synapse") A nerve cell that carries impulses away from a synapse.

POSTSYNAPTIC POTENTIAL (PSP) An electrical potential on the postsynaptic (receptor) membrane of a neuron, muscle fiber, or gland cell.

POTENTIAL DIFFERENCE The difference in electrical charge across the plasma membrane of a neuron.

P-R SEGMENT The recording in an electrocardiogram of the passage of the electrical P wave between the atria and ventricles.

PREGANGLIONIC NEURON ("before the ganglion") The first neuron in a two-neuron sequence in the autonomic nervous system; its cell body is in the brainstem or spinal cord and its myelinated axon terminates in an autonomic ganglion located outside the central nervous system.

PREMENSTRUAL SYNDROME (PMS) A condition characterized by nervousness, irritability, and abdominal bloating 1 or 2 weeks before menstruation.

PREPUCE [PREE-pyoos] The loose-fitting skin folded over the glans of the clitoris and an uncircumcized penis; also called *foreskin.*

PRESBYOPIA (Gr. *presbus,* old man +

eye) A condition in which the lens of the eye becomes less elastic with age, making it difficult to focus on near objects.

PRESSORECEPTORS See *baroreceptors.*

PRESYNAPTIC NEURON ("before the synapse") A nerve cell carrying impulses toward a synapse; it initiates a response in the receptive segment of a postsynaptic neuron.

PREVERTEBRAL GANGLIA Autonomic ganglia lying in front of the vertebrae; also called *collateral ganglia.*

PRIMARY CENTER OF OSSIFICATION The site near the middle of what will become the diaphysis, where bone cell development occurs by the second or third prenatal month.

PRIMARY GERM LAYERS Three layers of embryonic tissue called endoderm, mesoderm, and ectoderm, which form the organs and tissues of the body.

PRIMARY MOTOR CORTEX A motor area in the frontal lobe of the cerebrum controlling specific voluntary muscles or muscle groups; also called *Brodmann area 4.*

PRIMARY SOMESTHETIC AREA The portion of the parietal lobe of the cerebrum that receives information about the general senses from receptors in the skin, joints, muscles, and body organs; also called *general sensory area.*

PRIMARY VISUAL CORTEX (area 17) An area in the occipital lobe of the cerebrum that receives visual images from the retina; conveys visual information to cerebral areas 18 and 19 for further processing and evaluation.

PRIME MOVER See *agonist.*

PRIMORDIAL FOLLICLE An ovarian follicle that is not yet growing.

PROGESTERONE A female hormone secreted by the corpus luteum; stimulates thickening of the uterine wall and the formation of mammary ducts.

PROGESTINS Class of female sex hormones that regulates the menstrual cycle and development of mammary glands; aid in the formation of the placenta during pregnancy.

PROLACTIN A hormone secreted by the adenohypophysis; stimulates the duct system of the mammary glands during pregnancy and milk production after childbirth; also called lactogenic hormone and luteotropic hormone (LTH).

PROLAPSE A condition in which a body organ, especially the uterus, falls or slips out of place.

PRONATION A pivoting movement of the forearm that turns the palm downward or backward, crossing the radius diagonally over the ulna; opposite of supination.

PROPHASE The first stage of mitosis;

doubled chromosomes become visible; centriole pairs move to opposite poles of nucleoplasm, spindle forms, and chromatid pairs move toward center of spindle.

PROPRIOCEPTOR (L. "received from one's self") A sensory receptor responding to stimuli within the body, such as those from muscles and joints.

PROSTAGLANDINS Hormonelike substances made from fatty acids in plasma membranes; increase or decrease effect of cyclic AMP on target cells, raising or lowering blood pressure and regulating digestive secretions.

PROSTATE GLAND (Gr. *prostates,* standing in front of) A male gland whose secretions pass into the semen to make sperm motile and help neutralize vaginal acidity.

PROSTATECTOMY The surgical removal of all or part of the prostate gland.

PROTEIN (Gr. *protos,* first) Any of a group of complex organic compounds that always contain carbon, hydrogen, oxygen, and nitrogen; their basic structural units are amino acids.

PROTEIN-CALORIE MALNUTRITION A condition that occurs when the body has been deprived of sufficient amino acids and calories for a long period; marasmus and kwashiokor are two forms.

PROTHROMBIN An inactive plasma protein that is converted into the active enzyme thrombin during the blood-clotting process.

PROTO-ONCOGENE A gene that helps regulate normal cell division; a mutation can transform it into a cancer-causing *oncogene* (see).

PROTRACTION A forward pushing movement; opposite of retraction.

PROXIMAL (L. *proximus,* nearest) A relative directional term: nearer the trunk of the body (toward the attached end of a limb); used with extremities; opposite of *distal.*

PROXIMAL CONVOLUTED TUBULE Portion of the coiled excretory tubule proximal to the glomerular capsule; receives glomerular filtrate from the glomerular capsule.

PSORIASIS [suh-RYE-uh-siss; Gr. *psorian,* to have the itch] A skin disease occurring when skin cells move from the basal layer to the stratum corneum in 4 days instead of the usual 28, causing red, dry lesions covered with silvery, scaly patches.

PUBERTY (L. *puber,* adult) The developmental period when the person becomes physiologically capable of reproduction.

PULMONARY (L. *pulmo,* lung) Pertaining to the lungs.

PULMONARY ARTERIES Blood vessels carrying oxygen-poor blood from the heart to the lungs.

PULMONARY CIRCULATION The system of blood vessels that carries blood between the heart and the lungs.

PULMONARY TRUNK The major arterial trunk emerging from the right ventricle; it carries blood to the lungs.

PULMONARY VEINS Large veins that drain blood from the lungs into the left atrium.

PULP The soft core of connective tissue that contains the nerves and blood vessels of a tooth.

PULSE The alternating expansion and elastic recoil of the arterial wall that can be felt where an artery lies close to the skin.

PULSE PRESSURE The difference between systolic and diastolic pressure.

PUPIL (L. doll) The opening in the iris that opens (dilates) and closes (constricts) reflexively to adjust to the amount of available light.

PURKINJE FIBERS Modified nerve fibrils in the walls of the ventricles of the heart; help produce a coordinated pumping effort; also called *cardiac conducting myofibers.*

PYRAMID A bilateral elevated ridge in the ventral surface of the medulla oblongata; composed of fibers of motor (pyramidal) tracts from the motor cerebral cortex to the spinal cord.

PYRAMIDAL DECUSSATION (L. *decussare,* from *dec,* ten; the Latin symbol for 10 is X, representing a crossing over) The crossing over of the pyramidal tracts in the lower part of the medulla oblongata to the opposite side of the spinal cord, causing each side of the brain to control the opposite side of the body.

PYRAMIDAL SYSTEM Tracts from the cerebral motor cortex that terminate in the brainstem corticobulbar fibers and in the spinal cord (pyramidal tracts).

PYRAMIDAL TRACTS Descending fibers of motor tracts from the motor cerebral cortex to the spinal cord; also called *corticospinal tracts.*

PYROGEN (Gr. "fire-producing") Substance capable of producing fever.

PYURIA The presence of fat droplets or pus in the urine.

Q

QRS COMPLEX A triple-wave activity recorded during an electrocardiogram as the ventricles are depolarized.

QUADRIPLEGIA Paralysis of all four extremities, as well as any part of the body below the level of injury to the spinal cord; usually results from injury at the C8 to T1 level.

R

RADIATION THERAPY Therapy that uses x-rays or rays from radioactive substances to kill cancer cells.

RADIOGRAPHIC ANATOMY The study of the structures of the body using x-rays.

RAMI COMMUNICANTES [RAY-mee ko-myoo-ni-KAN-teez; sing. **RAMUS COMMUNICANS**] Myelinated or unmyelinated branches of a spinal nerve; composed of sensory and motor nerve fibers associated with the autonomic nervous system.

RAMUS [RAY-muhss; pl. **RAMI,** RAY-mye] A branch of a spinal nerve.

RECEPTOR (L. *recipere,* to receive) The peripheral end of the dendrites of afferent sensory neurons, specialized to receive stimuli and convert them into nerve impulses.

RECEPTOR-MEDIATED ENDOCYTOSIS An active process that involves a specific receptor on a plasma membrane that "recognizes" an extracellular macromolecule and binds to it; a form of endocytosis.

RECESSIVE A genetic trait that is hidden by the presence of a dominant trait.

RECTUM (L. *rectus,* straight) The 15-cm duct in the digestive tract between the sigmoid colon and the anus; removes solid wastes by the process of defecation.

RED BLOOD CELLS See *erythrocytes.*

REDUCTION (1) The gaining of electrons or hydrogen atoms by an atom or molecule. (2) The act of restoring dislocated bones to their normal positions in a joint.

REFERRED PAIN A visceral pain felt subjectively in a somatic area away from the actual source of pain.

REFLEX (L. to bend back) A predictable involuntary response to a stimulus.

REFLEX ARC A sequence of events leading to a reflex response.

REFRACTION The bending of light waves as they pass from one medium to another with a different density.

REFRACTORY PERIOD The brief period after the firing of a nerve impulse when the plasma membrane of a neuron cannot generate another impulse.

REGIONAL ANATOMY The anatomical study of specific regions of the body.

RELAXIN A female hormone secreted by the ovaries; dilates the symphysis pubis and cervix during childbirth and increases sperm motility.

RELEASING HORMONES Substances secreted by the hypothalamus that stimulate secretions from the adenohypophysis.

REMODELING A process of bone replacement or renewal; occurs throughout adult life in response to stress.

RENAL (L. *renes,* kidneys) Pertaining to the kidney.

RENAL CALCULI Kidney stones, usually formed from precipitated calcium salts.

RENAL COLUMN A column in the kidney formed by the tissue of the cortex that penetrates the depth of the medulla; composed mainly of tubules.

RENAL CORPUSCLE The portion of a kidney consisting of the glomerulus and glomerular capsule.

RENAL CORTEX The outermost portion of the kidney, divided into the outer cortical region and the inner juxtamedullary region.

RENAL FASCIA The outer tissue layer of the kidney.

RENAL MEDULLA The middle portion of the kidney, consisting of 8 to 18 renal pyramids.

RENAL PELVIS The large collecting space within the kidney formed from the expanded upper portion of the ureter.

RENAL PYRAMID A group of 8 to 18 longitudinally striped, cone-shaped areas that make up the medulla; consists of tubules and collecting ducts of the nephrons.

RENIN [REE-nihn] The renal enzyme, secreted by the juxtaglomerular apparatus, which alters the systemic blood pressure to maintain homeostasis.

REPOLARIZATION The restoration of a relatively positive charge outside the plasma membrane of a neuron.

RESIDUAL VOLUME The volume of air remaining in the lungs after a forceful expiration.

RESPIRATION The overall exchange of oxygen and carbon dioxide between the atmosphere, blood, lungs, and body cells.

RESPIRATORY CENTER The breathing-control center, including the inspiratory and expiratory circuits in the medulla oblongata and the apneustic and pneumotaxic areas in the pons.

RESTING MEMBRANE POTENTIAL The potential difference across the plasma membrane of a neuron that is not conducting an impulse.

RETICULAR ACTIVATING SYSTEM (RAS) A network of branched nerve cells in the brainstem; involved with the adjustment of many behavioral activities, including the sleep-wake cycle, awareness, levels of sensory perception, emotions, and motivation; also called the *arousal system.*

RETICULAR CELL A flat, star-shaped cell

that forms the cellular framework of bone marrow, lymph nodes, the spleen, and other lymphoid tissues involved in the immune response.

RETICULAR FIBERS Delicately branched networks that make up some connective tissues; similar to collagenous fibers.

RETICULAR FORMATION A network of nerve cells and fibers throughout the brainstem; consists of ascending and descending pathways and cranial nerves; regulates respiratory and cardiovascular centers as well as the brain's awareness level.

RETINA [REH-tin-uh; L. *rete,* net] The innermost layer of the eye, containing a thick layer of photoreceptor cells and other nervous tissue called the *neuro-retina* and a thin layer of pigmented epithelium that prevents reflection.

RETINAL A photosensitive derivative of vitamin A found in rods in the retina.

RETRACTION A backward movement, such as a backward pull of the shoulders; opposite of protraction.

RETROPERITONEAL (L. *retro,* behind + peritoneum) Located behind the abdominopelvic cavity and the peritoneum.

Rh FACTOR A blood factor characterized by an inherited agglutinogen on the surface of erythrocytes.

RHEUMATISM (Gr. *rheumatismos,* to suffer from a flux or stream) Any of several diseases of muscles, tendons, joints, bones, or nerves.

RHODOPSIN A reddish, light-sensitive pigment in rod cells within the retina; contains the light-absorbing organic molecule 11-*cis* retinal and the protein scotopsin; also called *visual purple.*

RIBONUCLEIC ACID (RNA) [rye-boh-noo-KLAY-ihk] A single-stranded nucleic acid containing the sugar ribose; transcribed from DNA; found in the nucleus and cytoplasm.

RIBOSOME A subcellular structure containing RNA and protein; the site of protein synthesis.

RICKETS (variant of Gr. *rhakhitis,* disease of the spine) A childhood disease caused by a deficiency of vitamin D; progresses to skeletal deformity.

ROD A specialized photoreceptor cell in the retina; not sensitive to color, but very sensitive to light.

ROTATION A pivoting movement that twists a body part, arm, or leg on its long axis.

ROUND LIGAMENTS Paired bands of fibrous connective tissue just below the entrance of the uterine tubes into the uterus; help to keep the uterus tilted forward over the urinary bladder.

RUFFINI CORPUSCLES See *corpuscles of*

Ruffini.

RUGAE [ROO-jee; L. folds] Folds or creases of tissue, as in the stomach and vagina.

S

SACRAL PLEXUS The ventral rami of L4, L5, and S1 to S3 nerves in the posterior pelvic wall.

SADDLE JOINT A multiaxial joint in which opposing articular surfaces of both bones are shaped like a saddle.

SAGITTAL (L. *sagitta,* arrow) An off-center longitudinal plane dividing the body into asymmetrical left and right sections.

SALIVA The secretion of salivary glands, composed of about 99 percent water and 1 percent electrolytes and proteins and the enzyme salivary amylase.

SALIVARY GLANDS The three largest pairs of glands that secrete saliva into the oral cavity: parotid, submandibular, and sublingual glands.

SALT The compound (other than water) formed during a neutralization reaction between acids and bases.

SALTATORY CONDUCTION (L. *saltare,* to jump) Conduction along a myelinated nerve fiber, where the impulse jumps from one node of Ranvier to the next.

SARCOLEMMA (Gr. *sarkos,* flesh + *lemma,* husk) A thin membrane enclosing each skeletal muscle fiber.

SARCOMA (Gr. *sarkoun,* to make fleshy) A malignant neoplasm that originates in connective tissue and spreads through the bloodstream.

SARCOMERE (Gr. *sarkos,* flesh + *meros,* part) The fundamental unit of muscle contraction composed of a section of muscle fiber extending from one Z line to the next.

SARCOPLASM (Gr. *sarkos,* flesh) A specialized form of cytoplasm found in skeletal muscle fibers.

SARCOPLASMIC RETICULUM A specialized type of endoplasmic reticulum containing a network of tubes and sacs containing calcium ions.

SCHWANN CELL Cell that surrounds an axon of a peripheral nerve fiber; see *neurilemma.*

SCHWANN SHEATH See *neurilemma.*

SCIATICA [sye-AT-ih-kuh] Nerve inflammation characterized by sharp pains along the sciatic nerve and its branches.

SCLERA (Gr. *skleros,* hard) The outer supporting layer of the eyeball; the "white" of the eye.

SCOLIOSIS (Gr. "crookedness") An abnormal lateral curvature of the spine in the thoracic, lumbar, or thoracolumbar region.

SCOTOPSIN The protein component of the photopigment rhodopsin.

SCROTUM [SCROH-tuhm; L. *scrautum*, a leather pouch for arrows] An external sac of skin that hangs between the male thighs; it contains the testes.

SEBACEOUS GLANDS [sih-BAY-shuhss; L. *sebum*, tallow, fat] Simple, branched alveolar glands in the dermis that secrete sebum; also called *oil glands.*

SEBUM (L. tallow, fat) The oily secretion found at the base of a hair follicle.

SECOND-MESSENGER CONCEPT See *fixed-membrane-receptor mechanism.*

SECONDARY SEX CHARACTERISTICS Sexually distinct characteristics such as body hair and enlarged genitals that develop during puberty.

SECRETIN A polypeptide hormone secreted by the duodenal mucosa; it stimulates the release of pancreatic juice to neutralize stomach acid.

SELLA TURCICA [SEH-luh TUR-sihk-uh; L. *sella*, saddle + Turkish] A deep depression within the sphenoid bone that houses the pituitary gland.

SEMEN [SEE-muhn; L. seed] Male ejaculatory fluid containing sperm and secretions from the epididymis, seminal vesicles, prostate gland, and bulbourethral glands; also called *seminal fluid.*

SEMILUNAR VALVES Heart valves that prevent blood in the pulmonary artery and aorta from flowing back into the ventricles.

SEMINAL FLUID See *semen.*

SEMINAL VESICLES Paired secretory sacs whose secretions provide an energy source for motile sperm and help to neutralize the acidity of the vagina.

SEMINIFEROUS TUBULES Tightly coiled tubules in the testes that produce sperm.

SEMIPERMEABILITY (L. *permeare*, to pass through) The quality of cellular membranes that allows some substances into the cell while keeping others out.

SENESCENCE (L. *senescere*, to grow old) The indeterminate period when an individual is said to grow old; also called *primary aging, biological aging.*

SENILE DEMENTIA (L. *senex*, old + madness) A progressive, abnormally accelerated deterioration of mental faculties in old age; commonly called *senility.*

SENILITY See *senile dementia.*

SENSORY NEURON See *afferent neuron.*

SEPTUM (L. *sepire*, to separate with a hedge) A wall between two cavities, such as in the heart or nose.

SEROUS (L. *serosus*, serum) Pertaining to the secretion of a serumlike fluid.

SEROUS MEMBRANE A double layer of

loose connective tissue covered by a layer of simple squamous epithelium that lines some of the walls of the thoracic and abdominopelvic cavities and covers organs lying within these cavities; includes peritoneum, pericardium, and pleura.

SERTOLI CELL See *sustentacular cell.*

SESAMOID BONE A small bone embedded within a tendon.

SEX CHROMOSOMES A pair of chromosomes that determine the sex of the new individual; an XX pair produces a female and an XY pair a male.

SEXUALLY TRANSMITTED DISEASE (STD) A disease that is caused by the transferral of infectious microorganisms from person to person, mainly by sexual contact; also called *venereal disease.*

SHINGLES Acute inflammation of the dorsal root ganglia along one side of the body; also called *herpes zoster.*

SHORT BONE An irregularly shaped bone about equal in length, width, and thickness.

SIAMESE TWINS See *conjoined twins.*

SICKLE-CELL ANEMIA A hereditary form of anemia characterized by crescent-shaped red blood cells that do not carry or release sufficient oxygen.

SIGMOID COLON (Gr. *sigma*, the letter "S") The S-shaped portion of the large intestine immediately following the descending colon.

SIMPLE DIFFUSION (L. *diffundere*, to spread) A passive transport process in which molecules move randomly from areas of high concentration to areas of lower concentration until they are evenly distributed.

SINOATRIAL (SA) NODE A mass of specialized heart muscle that initiates the heartbeat.

SKELETAL MUSCLE Muscle tissue that can be contracted voluntarily; attached to the skeleton; also known as *striated muscle, voluntary muscle.*

SKENE'S GLANDS See *lesser vestibular glands.*

SKULL Bones of the head, with or without the mandible.

SLIDING-FILAMENT MODEL The model that proposes that in muscle contraction, actin and myosin myofilaments slide past each other, with cross bridges pulling the muscle into a contracted state; the cross-bridge connections are broken when the muscle relaxes.

SMALL INTESTINE The 6-m-long portion of the digestive tract between the stomach and the large intestine; the site of most chemical digestion and absorption.

SMOOTH MUSCLE TISSUE Nonstriated muscle tissue controlled by the auto-

nomic nervous system; forms sheets in the walls of large, hollow organs; also called *involuntary muscle.*

SODIUM-POTASSIUM PUMP Part of a self-regulating transport system within the plasma membrane of a neuron; helps regulate the concentration of sodium and potassium ions inside and outside the membrane.

SOLUTE (L. *solvere*, to loosen) A substance capable of being dissolved in another substance; see *solvent.*

SOLUTION A homogeneous mixture of a solvent and the dissolved solute.

SOLVENT A liquid or gas capable of dissolving another substance; see *solute.*

SOMATIC NERVOUS SYSTEM The portion of the nervous system composed of a motor division that excites skeletal muscles and a sensory division that receives and processes sensory input from the sense organs.

SOMATOTROPHIC HORMONE (STH) See *growth hormone.*

SOMATOTROPIN See *growth hormone.*

SONOGRAPHY See *ultrasound.*

SPECIAL SENSES The senses of sight, hearing, equilibrium, smell, and taste.

SPERM (Gr. *sperma*, seed) Mature male sex cells; also called *spermatozoa.*

SPERMATIC CORD A cord covering the ductus deferens as it passes from the tail of the epididymis; contains the testicular artery, veins, autonomic nerves, lymphatic vessels, and connective tissue.

SPERMATOGENESIS ("the birth of seeds") The continuous process that forms sperm in the testes.

SPERMICIDE A chemical agent that kills sperm.

SPHINCTER (Gr. "that which binds tight") A circular muscle that helps keep an opening or tubular structure closed.

SPHYGMOMANOMETER [sfig-moh-muh-NOM-ih-ter] The instrument used for measuring blood pressure.

SPINA BIFIDA [SPY-nuh BIFF-uh-duh; L. *bifidus*, split into two parts] A congenital malformation in which the two sides of the neural arch of one or more vertebrae do not fuse during embryonic development; commonly called *cleft spine.*

SPINAL CORD The part of the central nervous system extending caudally from the foramen magnum; has 31 pairs of nerves and connects the brain with the rest of the body.

SPINOUS PROCESS A sharp, elongated process of a bone, such as the spine of a vertebra.

SPIRAL ORGAN (OF CORTI) The organ of hearing; formerly called *organ of Corti.*

SPLEEN The largest lymphoid organ, located below the diaphragm on the left side; it filters blood and produces lymphocytes and monocytes.

SPONTANEOUS ABORTION The loss of a fetus through natural causes; also called *miscarriage.*

SPRAIN A tearing of ligaments following the sudden wrenching of a joint.

STATIC EQUILIBRIUM See *vestibular apparatus.*

STEREOGNOSIS [STEHR-ee-oh-NO-siss; Gr. *stereos*, solid, three-dimensional + *gnosis*, knowledge] The ability to identify unseen objects by handling them.

STERILITY A condition in which a mature person is unable to conceive or produce offspring.

STIMULUS (L. a goad) A change in the external or internal environment that affects receptors.

STOMACH The distensible sac that churns food mass into small particles, stores them until they are released into the small intestine; secretes enzymes that initiate protein digestion.

STRATUM BASALE (L. *basis*, base) The layer of the epidermis resting on the basement membrane next to the dermis.

STRESS (Mid. Eng. *stresse*, hardship) Any factor or factors that put pressure on the body to make an adaptive change to maintain homeostasis.

STROKE See *cerebrovascular accident.*

STROKE VOLUME The volume of blood ejected by either ventricle in one *systole.*

SUBARACHNOID SPACE The space between the arachnoid and pia mater layers of the meninges; contains cerebrospinal fluid and blood vessels.

SUBCUTANEOUS LAYER (L. *sub*, under + *cutis*, skin) See *hypodermis.*

SUBDURAL SPACE Potential space between dura mater and arachnoid meninges; contains no cerebrospinal fluid.

SUBLINGUAL GLANDS ("under the tongue") Paired salivary glands located in the floor of the mouth beneath the tongue.

SUBMANDIBULAR GLANDS ("under the mandible") Paired salivary glands located on medial side of the mandible.

SUDORIFEROUS GLANDS (L. *sudor*, sweat) Sweat glands.

SULCUS [pl. **SULCI**, SUHL-kye; L. groove] A deep furrow on the surface of a structure; also called *groove.*

SUMMATION OF TWITCHES The tension achieved when a muscle receives repeated stimuli at a rapid rate so that it cannot relax completely between contractions.

SUPERFICIAL A relative directional term of the body; nearer the surface of the

body; opposite of *deep.*

SUPERIOR (L. *superus*, situated above) A relative directional term: toward the head; above; opposite of *inferior.*

SUPERIOR VENA CAVA The large vein that drains blood from the head, neck, upper limbs, and thorax, emptying it into the right atrium.

SUPINATION A pivoting movement of the forearm that turns the palm forward or upward, making the radius parallel with the ulna; opposite of *pronation.*

SURFACTANT A phospholipid that reduces surface tension in the lung alveoli.

SUSTENTACULAR CELL A type of cell in the seminiferous tubules of testes that provides nourishment for the germinal sperm as they mature; also called *Sertoli cell.*

SUSTENTACULUM A supporting process of a bone.

SUTURAL BONES Separate small bones in the sutures of the calvaria of the skull; also called *Wormian bones.*

SUTURE (L. *sutura*, seam) A seamlike joint that connects skull bones, making them immovable.

SWALLOWING See *deglutition.*

SYMPATHETIC NERVOUS SYSTEM The division of the autonomic nervous system that stimulates activities that are mobilized during emergency and stress situations; also called *thoracolumbar division, adrenergic division.*

SYMPHYSIS (Gr. "growing together") A cartilaginous joint in which two bony surfaces are covered by thin layers of hyaline cartilage and cushioned by fibrocartilaginous disks; also called *secondary synchondrosis.*

SYNAPSE [SIN-apps; Gr. a connection] The electrochemical junction between neurons.

SYNAPTIC BOUTON A tiny swelling on the terminal ends of branches at the distal end of an axon; presynaptic nerve ending at a synapse.

SYNAPTIC CLEFT The narrow space between the terminal ending of a neuron and the receptor site of the postsynaptic cell.

SYNAPTIC DELAY The period during which a neurotransmitter bridges a synaptic cleft.

SYNAPTIC GUTTER The invaginated area of the sarcolemma under and around the axon terminal; also called *synaptic trough.*

SYNARTHROSIS (Gr. *syn*, together + *arthrosis*, articulation) An immovable joint.

SYNCHONDROSIS (Gr. "together with cartilage") A cartilaginous joint that allows growth, not movement; a tem-

porary joint of cartilage that joins the epiphysis and diaphysis of a growing bone; it is eventually replaced by bone; also called *primary cartilaginous joint.*

SYNDESMOSIS (Gr. "to bond together"; *syn + desmos*, bond) A fibrous joint in which bones are held close together, but not touching, by collagenous fibers or interosseous ligaments.

SYNERGISTIC MUSCLE [SIHN-uhr-jist-ihk; Gr. *syn*, together + *ergon*, work] A muscle that complements the action of a prime mover (agonist).

SYNOVIAL CAVITY The space between two articulating bones; also called *joint cavity.*

SYNOVIAL FLUID [sin-OH-vee-uhl; Gr. *syn*, with + L. *ovum*, egg] The thick, lubricating fluid secreted by membranes in joint cavities.

SYNOVIAL JOINT An articulation in which bones move easily on each other; most movable joints of the body are synovial.

SYNOVIAL MEMBRANE A membrane lining the cavities of joints and similar areas where friction needs to be reduced; composed of loose connective and adipose tissues covered by fibrous connective tissue.

SYPHILIS A sexually transmitted disease caused by the bacterium *Treponema pallidum*; may produce degeneration of circulatory or nervous tissue, leading to paralysis, insanity, and death.

SYSTEM A group of organs that work together to perform a major body function.

SYSTEMIC ANATOMY The anatomical study of the systems of the body.

SYSTEMIC CIRCULATION The system of blood vessels that supplies the body with oxygen-rich blood and returns oxygen-poor blood from the body to the heart.

SYSTEMIC LUPUS ERYTHEMATOSUS (SLE) An autoimmune disease that affects the lining of joints and other connective tissue; may be hereditary and caused by a breakdown in the immune system.

SYSTOLE [SISS-toe-lee] The contraction of the atria and ventricles during the cardiac cycle.

SYSTOLIC PRESSURE The portion of blood pressure measurement that represents the highest pressure reached during ventricular ejection; it is the first number shown in a blood pressure reading; see *diastolic pressure.*

T

T CELL A type of lymphocyte that attacks specific foreign cells; able to differentiate into helper T cell, suppressor

T cell, or cytotoxic T cell; also called *T lymphocyte.*

T LYMPHOCYTE See *T cell.*

T WAVE A recorded representation during an electrocardiogram indicating a recovery electrical wave from the ventricles to the atria as the ventricles are repolarized.

TACTILE (MEISSNER'S) CORPUSCLES Sensory receptors in the skin that detect light pressure; formerly called *Meissner's corpuscles.*

TACTILE (MERKEL'S) CORPUSCLES Sensory receptors of light touch, located in the deep epidermal layers of the palms and soles; formerly called *Merkel's disks.*

TAENIAE COLI [TEE-nee-ee KOHL-eye; L. ribbons + Gr. intestine] Three separate bands of longitudinal muscle along the full length of the large intestine.

TARGET CELL A cell with surface receptors that allow it to be affected by a specific hormone.

TARSUS Each of the seven proximally located short bones of each foot.

TAY-SACHS DISEASE A hereditary metabolic disease in which lysosomes are deficient in the enzyme hexosaminidase, preventing the breakdown of fatty substances that accumulate around neurons and make them unable to transmit nerve impulses.

TELERECEPTOR (Gr. "received from a distance") A sensory receptor located in the eyes, ears, and nose that detects distant environmental stimuli.

TELOPHASE The final stage of mitosis; the two sets of chromosomes arrive at the poles and are surrounded by new nuclear envelopes; nucleoli form.

TEMPORAL LOBE The cerebral lobe closest to the ears; critical in hearing, equilibrium, and, to a certain degree, emotion and memory.

TEMPORAL SUMMATION The increased effect of a presynaptic neuron on a postsynaptic neuron from repeated firing.

TENDINITIS Inflammation of a tendon and tendon sheath.

TENDON A strong cord of collagenous fibers that attaches muscle to the periosteum of bone.

TERATOGEN (Gr. *teras,* monster) A substance that chemically interferes with embryonic development and causes gross deformities without changing the embryonic DNA.

TEST-TUBE BABY See *in vitro fertilization.*

TESTES [TESS-teez; sing. **TESTIS,** L. witness] The paired male reproductive organs, which produce sperm.

TESTIS-DETERMINING FACTOR (TDF) A specific chromosomal segment, or gene, on the Y chromosome that apparently determines sex.

TESTOSTERONE The most important of the male sex hormones; stimulates sperm production and development of male sex organs and behavior.

TETANUS A more or less continuous contraction of a muscle; also called *tetanic contraction.*

THALAMUS (Gr. inner chamber) Two masses of gray matter covered by a thin layer of white matter; located directly beneath the cerebrum and above the hypothalamus; processing center for all sensory impulses (except smell) ascending to the cerebral cortex.

THERMORECEPTOR (Gr. "heat receiver") A sensory receptor that responds to temperature changes.

THORACIC (Gr. *thorax,* breastplate) Pertaining to the chest.

THORACIC DUCT The largest lymphatic vessel; drains lymph not drained by the right lymphatic duct into the left subclavian vein; also called the *left lymphatic duct.*

THORACOLUMBAR DIVISION [thuh-RASS-oh-LUM-bar] The portion of the sympathetic nervous system with visceral efferent nerve fibers that leave the central nervous system through thoracic and lumbar spinal nerves.

THORACOLUMBAR OUTFLOW Myelinated nerve fibers emerging from the spinal cord in the ventral nerve roots of the 12 thoracic and first 2 or 3 lumbar spinal nerves.

THORAX (Gr. breastplate) The chest portion of the axial skeleton, including 12 thoracic vertebrae, 12 pairs of ribs, 12 costal cartilages, and the sternum.

THRESHOLD STIMULUS A stimulus strong enough to initiate an impulse in a neuron.

THROMBIN The active enzyme converted from prothrombin during the blood-clotting process; converts the soluble plasma protein fibrinogen into insoluble fibrin.

THROMBOCYTE See *platelet.*

THROMBOPHLEBITIS (Gr. *thrombos,* clot + *phleps,* blood vessel + *-itis,* inflammation) An acute condition characterized by clot formation and inflammation of deep or superficial veins.

THROMBOPLASTIN An enzyme involved in the blood-clotting process.

THROMBUS (Gr. a clotting) A blood clot obstructing a blood vessel or heart cavity; the condition is *thrombosis.*

THYMUS GLAND A ductless mass of flattened lymphoid tissue situated behind the top of the sternum; it forms antibodies in the newborn and is involved in the development of the immune system.

THYROCALCITONIN See *calcitonin.*

THYROID GLAND An endocrine gland involved with the metabolic functions of the body.

THYROID-STIMULATING HORMONE (TSH) A hormone secreted by the adenohypophysis; stimulates the production and secretion of thyroid hormones; also called *thyrotropin, thyrotropic hormone.*

THYROTROPIC HORMONE See *thyroid-stimulating hormone.*

THYROTROPIN See *thyroid-stimulating hormone.*

THYROXINE A thyroid hormone that increases rate of metabolism and sensitivity of the cardiovascular system to the nervous system; also called T_4.

TIDAL VOLUME The volume of air (usually about 0.5 L) inhaled and exhaled into and out of the lungs with each normal breath.

TISSUE An aggregation of many similar cells that perform a specific function; generally classified as epithelial, connective, muscle, or nervous.

TOMOGRAPHY (Gr. *tomos,* a cut or section + *graphein,* to write or draw) A technique for making pictures of a section of a body part, as in computer-assisted tomography; the picture produced is a *tomogram.*

TONSILS Aggregates of lymphatic nodules enclosed in connective tissue (capsule); they destroy microorganisms that enter the digestive and respiratory systems; include pharyngeal, palatine, and lingual tonsils.

TOTAL LUNG CAPACITY The vital capacity plus residual volume; about 4.2 L in women, 6.0 L in men.

TRABECULAE [truh-BECK-yuh-lee; L. dim. *trabs,* beam] (1) Tiny spikes of bone tissue surrounded by calcified bone matrix; prominent in the interior of spongy bone. (2) Bands of connective tissue that extend from the capsule into the interior of an organ.

TRACHEA [TRAY-kee-uh] An open tube extending from the base of the larynx to the top of the lungs, where it forks into two bronchi; commonly called *windpipe.*

TRACT A bundle of nerve fibers and their sheaths within the central nervous system.

TRANSDUCIN A protein activated by the conversion of 11-*cis* retinal into all-*trans* retinal during the absorption of light in the retina; activates the enzyme phosphodiesterase, which hydrolyzes the intracellular messenger *cyclic guanosine monophosphate (cGMP).*

TRANSVAGINAL OOCYTE RETRIEVAL An *in vitro* procedure in which the physician obtains ova through the vaginal wall and unites them with sperm from the male donor; the fertilized ova are then implanted in the uterus.

TRANSVERSE A plane that divides the body horizontally into superior and inferior sections.

TRANSVERSE COLON The portion of the large intestine immediately after the ascending colon, extending across the abdominal cavity from right to left.

TRANSVERSE TUBULES Within a muscle fiber, a series of tubes lined by extensions of the plasma membrane; they cross the sarcoplasmic reticulum at right angles; also called *T tubules*.

TREPPE (L. *trepidus,* alarmed) A type of muscle contraction in which the first few contractions increase in strength when a rested muscle receives repeated stimuli.

TRIAD The combination of a transverse tubule and a terminal cisterna of a sarcoplasmic reticulum of a muscle fiber.

TRICHOMONIASIS A sexually transmitted disease caused by the *Trichomonas vaginalis* protozoan.

TRICUSPID VALVE The right atrioventricular valve.

TRIGONE A triangular area in the urinary bladder outlined by the openings of the ureters and urethra.

TRIIODOTHYRONINE A thyroid hormone that increases the rate of metabolism and the sensitivity of the cardiovascular system; also called *T₃*.

TRISOMY 21 A chromosomal aberration in which a pair of chromosomes fails to separate at meiosis, producing an extra copy of chromosome 21; causes mental retardation and physical abnormalities; also called *Down's syndrome*, previously called *mongolism*.

TROCHANTER Either of the two large, rounded processes below the neck of the femur.

TROPHECTODERM The surrounding epithelial layer of a blastocyst, composed of cells called *trophoblasts*; develops into a fetal membrane system.

TROPHOBLAST See *trophectoderm*.

TUBAL LIGATION A surgical contraceptive procedure that removes a portion of each uterine tube to prevent the ovum from being fertilized.

TUBERCLE (L. small lump) A small, roughly rounded process of a bone.

TUBEROSITY (L. lump) A medium-sized, roughly rounded elevated process of a bone.

TUBULAR REABSORPTION The renal process that returns useful substances to the blood by active transport.

TUBULAR SECRETION The renal process that concentrates waste products in the glomerular filtrate.

TUMOR (L. *tumere,* to swell) An abnormal growth of tissue, in which cells reproduce at a faster rate than normal; also known as *neoplasm*. (Note: *-oma* = tumor.)

TUNICA ADVENTITIA ("outermost covering") The outermost covering of an artery, composed mainly of collagen fibers and elastic tissue.

TUNICA ALBUGINEA [al-byoo-JIHN-ee-uh; L. *albus,* white] The thick, bluish-white fibrous membrane enclosing a testis and forming a thin connective tissue over the cortex of an ovary.

TUNICA INTIMA ("innermost covering") The innermost lining of the lumen of an artery.

TUNICA MEDIA ("middle covering") The middle, and thickest, layer of the arterial wall in large arteries.

TWITCH A momentary spasmodic contraction of a muscle fiber in response to a single stimulus.

TYMPANIC CAVITY (Gr. *tumpanon,* drum) A narrow, irregular, air-filled space in the temporal bone; separated from the external auditory canal by the tympanic membrane, and from the inner ear by the posterior bony wall; also called *middle-ear cavity*.

TYMPANIC MEMBRANE A deflectable membrane between the external and middle ear; vibrates in response to sound waves entering the ear; commonly called the *eardrum*.

U

ULCERS See *peptic ulcers*.

ULTRASONOGRAPHY High-frequency sound waves used to locate and examine a fetus or other internal structures.

ULTRASOUND A noninvasive exploratory technique that sends pulses of ultrahigh-frequency sound waves into designated body cavities; images are formed from echoes of the sound waves; also called *sonography*.

UMBILICAL CORD (L. *umbilicus,* navel) A connecting tube between the embryo and placenta, carrying carbon dioxide and nitrogen wastes from the embryo and oxygen and nutrients to the embryo.

UNIPOLAR NEURON A nerve cell with one process dividing into two branches; one branch extends into the brain or spinal cord, the other extends to a peripheral sensory receptor.

UPPER EXTREMITIES See *extremities*.

UREMIA A condition in which waste products normally excreted in the urine are present in the blood.

URETERS Two tubes that carry urine from the renal pelvis of the kidneys to the urinary bladder.

URETHRA The final section of the male reproductive duct system, leading from the urinary bladder, through the prostate gland, and into the penis, carrying semen outside the body during ejaculation, and urine during urination; in females it conveys urine from the urinary bladder.

URETHRITIS Inflammation of the urethra.

URINARY BLADDER A hollow, muscular sac that collects urine from the ureters and stores it until it is excreted from the body through the urethra.

URINARY INCONTINENCE The inability to retain urine in the urinary bladder and control urination.

URINARY SYSTEM The body system that eliminates the wastes of protein metabolism in the urine and regulates the amount of water and salts in the blood; composed of two kidneys and ureters, the urinary bladder, and the urethra.

URINE (Gr. *ourein,* to urinate) The excretory fluid produced by the kidneys.

UTERINE TUBE One of a pair of tubes that receives the mature ovum from the ovary and conveys it to the uterus; also called *Fallopian tube, oviduct*.

UTERUS (L. womb) A hollow, muscular organ behind the urinary bladder that during pregnancy houses the developing fetus; also called *womb*.

V

VAGINA (L. sheath) A muscle-lined tube from the uterus to the outside; the site where semen is deposited during sexual intercourse, the channel for menstrual flow, and the birth canal for the baby during childbirth.

VALVULAR HEART DISEASE A condition in which one or more cardiac valves operate improperly.

VARICOSE VEINS (L. *varix,* swollen veins) Abnormally dilated and twisted veins, most often the saphenous veins and their branches in the legs.

VAS DEFERENS See *ductus deferens*.

VASA VASORUM ("vessels of the vessels") Small blood vessels that supply the walls of arteries and veins.

VASECTOMY A surgical contraceptive procedure that removes a small portion of each ductus deferens to block the release of sperm from the testes.

VASOCONSTRICTION The process of constricting blood vessels.

VASODILATION The process of dilating blood vessels.

VASOMOTOR CENTER A regulatory center in the lower pons and medulla oblongata that regulates the diameter of blood vessels, especially arterioles.

VASOPRESSIN See *antidiuretic hormone*.

VEIN A vessel that carries blood from the body to the heart.

VENEREAL DISEASE (from *Venus*, Roman goddess of love) See *sexually transmitted disease*.

VENTILATION Breathing.

VENTRAL (L. *venter*, belly) A relative directional term: toward the front of the body; also called *anterior*; opposite of *dorsal*.

VENTRAL CAVITY The larger of two main body cavities; separated into the thoracic and abdominopelvic cavities by the diaphragm.

VENTRICLE (L. "little belly") (1) A cavity in the brain filled with cerebrospinal fluid. (2) The left or right inferior heart chamber.

VENULE [VEHN-yool] A tiny vein into which blood drains from capillaries.

VERMIFORM APPENDIX (L. *vermis*, worm) The short, narrow, wormlike region of the digestive tract opening into the cecum; may be involved with the immune system; commonly called the *appendix*.

VERMIS (L. worm) The midline portion of the cerebellum, separating the hemispheres; involved, together with the flocculonodular lobes, with maintaining muscle tone, equilibrium, and posture.

VERTEBRAE [VER-tuh-bree; L. "something to turn on"] The 26 individual bones in the spinal column.

VERTEBRAL COLUMN Commonly called the spinal column.

VESICULAR OVARIAN FOLLICLE An ovarian follicle that is almost ready to release a mature ovum (ovulation).

VESTIBULAR APPARATUS (L. *vestibulum*, entrance) Specific parts of the inner ear that signal changes in the motion and position of the head with respect to gravity.

VESTIBULAR GLANDS See *greater* and *lesser vestibular glands*.

VESTIBULE (1) Any cavity, chamber, or channel serving as an approach or entrance to another cavity. (2) The space between the labia minora. (3) The central chamber of the labyrinth in the middle ear. (4) The buccal cavity of the mouth.

VILLI [VILL-eye; sing. **VILLUS;** L. "shaggy hairs"] Tiny fingerlike protrusions in the mucosa of the small intestine that increase the absorptive surface area.

VISCERA [VISS-ser-uh; L. body organ] The internal organs of the body.

VISCERAL Pertaining to an internal organ or a body cavity, or describing a membrane covering an internal organ.

VISCERAL NERVOUS SYSTEM The portion of the peripheral nervous system composed of a motor division (autonomic nervous system) and a sensory division.

VISCERAL PLEURA See *pleural cavity*.

VITAL CAPACITY The inspiratory reserve volume plus the tidal volume plus the expiratory reserve volume; about 3.1 L in women, 4.8 L in men.

VITAMIN An organic compound required by the body in small amounts for the regulation of metabolism; classified as *water-soluble* (C and B group) and *fat-soluble* (A, D, E, K).

VITREOUS HUMOR A gelatinous substance within the large vitreous chamber of the eyeball; keeps the eyeball from collapsing as a result of external pressure.

VOCAL FOLDS Paired folds of mucous membrane at the base of the larynx that produce sound when vibrated; commonly called *vocal cords*.

VOICE BOX See *larynx*.

VOLAR See *palmar*.

VOLKMANN'S CANAL See *perforating canal*.

VOLUNTARY MUSCLE See *skeletal muscle*.

VOMITING The forceful expulsion of part or all of the contents of the stomach and duodenum through the mouth, usually in a series of involuntary spasms; also called *emesis*.

VULVA (L. womb) The collective name for the external female genital organs, including the mons pubis, labia majora and minora, vestibular glands, clitoris, and the vestibule of the vagina.

W

WANDERING CELLS Connective tissue cells usually involved with short-term activities such as protection and repair.

WART A benign epithelial tumor caused by various papilloma viruses; also called *verruca*.

WHITE BLOOD CELL See *leukocyte*.

WHITE MATTER The portion of the spinal cord consisting mainly of whitish myelinated nerve fibers.

WHITE RAMI COMMUNICANTES Myelinated preganglionic fibers of the thoracolumbar outflow of the spinal cord that form small nerve bundles.

WINDPIPE See *trachea*.

WOMB See *uterus*.

WORMIAN BONES See *sutural bones*.

Y

YOLK SAC An extraembryonic membrane that is a primitive respiratory and digestive system before the development of the placenta; becomes nonfunctional and incorporated into the umbilical cord by the sixth or seventh week after fertilization.

Z

ZONA PELLUCIDA (L. *perlucere*, to shine through, thus "transparent zone") The outer wall of an ovum.

ZYGOTE (Gr. *zugotos*, joined, yolked) The cell formed by the union of male and female gametes.

INDEX

NOTE: Page numbers in **boldface** type indicate illustrations and usually text as well.

Clinical Applications Index